Compactifications of PEL-Type Shimura Varieties and Kuga Families with Ordinary Loci

Compactifications of PEL-Type Shimura Varieties and Kuga Families with Ordinary Loci

Kai-Wen Lan

University of Minnesota, Twin Cities, USA

World Scientific

NEW JERSEY • LONDON • SINGAPORE • BEIJING • SHANGHAI • HONG KONG • TAIPEI • CHENNAI • TOKYO

Published by

World Scientific Publishing Co. Pte. Ltd.
5 Toh Tuck Link, Singapore 596224
USA office: 27 Warren Street, Suite 401-402, Hackensack, NJ 07601
UK office: 57 Shelton Street, Covent Garden, London WC2H 9HE

British Library Cataloguing-in-Publication Data
A catalogue record for this book is available from the British Library.

COMPACTIFICATIONS OF PEL-TYPE SHIMURA VARIETIES AND KUGA FAMILIES WITH ORDINARY LOCI

ISBN 978-981-3207-32-5

Printed in Singapore

Preface

In the summer of 2011, Michael Harris, Richard Taylor, Jack Thorne, and I found a method for attaching p-adic Galois representations to *all* cohomological automorphic representations of general linear groups over CM or totally real fields, without any polarizability conditions. We were very excited about such a breakthrough. At that time, the arguments were based on my earlier works [57] (which was revised and published as [62]) and [61], which required the prime p to satisfy certain assumptions ensuring that good reduction p-integral models of Shimura varieties and their toroidal compactifications exist at some bottom levels at p. A natural question was whether such assumptions could be removed by improving my earlier works. Fortunately, soon afterwards, I was able to convince myself that the answer is in the affirmative, although a lot still had to be written down to fully justify the ideas. We then decided that we would split up our writing into two parts. One part would remain a four-author collaboration—which ended up being the long article [39]—explaining the most important new arguments under the assumption that certain geometric objects can be constructed in some special cases. The other part would be only by myself—which ended up being this even longer book—detailing the constructions of the geometric objects in the generality they deserve.

The *ordinary loci* in the title of this book are natural generalizations of what are called *Igusa towers* in Hida's works (see, for example, [41]), which are the loci where certain multiplicative-type subgroup schemes of the p-power torsion subgroup schemes of the universal abelian schemes can be rigidified (as in Katz's article [46] on p-adic modular forms). (This is, admittedly, an abuse of language—we do not mean the full loci where the geometric fibers of the universal abelian schemes are ordinary.) Historically, such ordinary loci were almost always built over some *good reduction* p-integral models of Shimura varieties and their toroidal compactifications, which unavoidably imposed some conditions on the residue characteristic p. However, as we shall see in this book, even without such conditions, it is still possible to construct and describe in precise detail some ordinary loci and their partial compactifications, without relying on any good understanding of the whole p-integral models of Shimura varieties and their compactifications. (We shall see some more detailed explanations in the introduction of the main text.) This is good enough for many applications to the construction of p-adic automorphic forms, and justifies the removal of all assumptions on p in the above-mentioned collaboration.

It has been more than five years since the writing started. Although the manuscript already achieved its current form and complexity by the summer of 2013, it has undergone another three years of intermittent reviewing and revising. Despite all the flaws that still remain after all of these, I hope this book will nevertheless provide a clear and reasonably complete picture of a useful theory.

January, 2017, in Minneapolis.

Contents

Preface v

0. Introduction 1

0.1 Background and Aim 1
0.2 Overview 2
0.3 Outline of the Constructions 8
0.4 What is Known, What is New, and What Can Be Studied Next . . 13
0.5 What to Note and to Skip in Special Cases 17
0.6 Notation and Conventions 18
0.7 Acknowledgements 18

1. Theory in Characteristic Zero 21

1.1 PEL-type Moduli Problems and Shimura Varieties 21
1.1.1 Linear Algebraic Data for PEL Structures 21
1.1.2 PEL-type Moduli Problems 23
1.1.3 PEL-type Shimura Varieties 26
1.2 Linear Algebraic Data for Cusps 26
1.2.1 Cusp Labels 26
1.2.2 Cone Decompositions 33
1.2.3 Rational Boundary Components 38
1.2.4 Parameters for Kuga Families 43
1.3 Algebraic Compactifications in Characteristic Zero 57
1.3.1 Toroidal and Minimal Compactifications of PEL-Type Moduli Problems 57
1.3.2 Boundary of PEL-Type Moduli Problems 66
1.3.3 Toroidal Compactifications of PEL-Type Kuga Families and Their Generalizations 95
1.3.4 Justification for the Parameters for Kuga Families 106
1.4 Automorphic Bundles and Canonical Extensions in Characteristic Zero 117

1.4.1 Automorphic Bundles . . . 117
1.4.2 Canonical Extensions . . . 120
1.4.3 Hecke Actions . . . 123
1.5 Comparison with the Analytic Construction . . . 124

2. Flat Integral Models 125

2.1 Auxiliary Choices . . . 125
2.1.1 Auxiliary Choices of Smooth Moduli Problems . . . 125
2.1.2 Auxiliary Choices of Toroidal and Minimal Compactifications . . . 132
2.2 Flat Integral Models as Normalizations and Blow-Ups . . . 141
2.2.1 Flat Integral Models for Minimal Compactifications . . . 141
2.2.2 Flat Integral Models for Projective Toroidal Compactifications . . . 147
2.2.3 Hecke Actions . . . 149
2.2.4 The Case When p is a Good Prime . . . 152

3. Ordinary Loci 155

3.1 Ordinary Semi-Abelian Schemes and Serre's Construction . . . 155
3.1.1 Ordinary Abelian Schemes and Semi-Abelian Schemes . . . 155
3.1.2 Serre's Construction for Ordinary Abelian Schemes . . . 157
3.1.3 Extensibility of Isogenies . . . 160
3.2 Linear Algebraic Data for Ordinary Loci . . . 164
3.2.1 Necessary Data for Ordinary Reductions . . . 164
3.2.2 Maximal Totally Isotropic Submodules at p . . . 168
3.2.3 Compatibility with Cusp Labels . . . 171
3.3 Level Structures . . . 172
3.3.1 Level Structures Away from p . . . 172
3.3.2 Hecke Twists Defined by Level Structures Away from p . . . 175
3.3.3 Ordinary Level Structures at p . . . 177
3.3.4 Hecke Twists Defined by Ordinary Level Structures at p . . . 183
3.3.5 Comparison with Level Structures in Characteristic Zero . . . 192
3.3.6 Valuative Criteria . . . 194
3.4 Ordinary Loci . . . 197
3.4.1 Naive Moduli Problems with Ordinary Level Structures . . . 197
3.4.2 Ordinary Loci as Normalizations . . . 203
3.4.3 Properties of Kodaira–Spencer Morphisms . . . 206
3.4.4 Hecke Actions . . . 208
3.4.5 The Case When p is a Good Prime . . . 212
3.4.6 Quasi-Projectivity of Coarse Moduli . . . 213

4. Degeneration Data and Boundary Charts 215

4.1 Theory of Degeneration Data 215
4.1.1 Degenerating Families of Type $(\mathrm{PE}, \mathcal{O})$ 215
4.1.2 Common Setting for the Theory of Degeneration 216
4.1.3 Degeneration Data for Polarized Abelian Schemes with Endomorphism Structures 216
4.1.4 Degeneration Data for Principal Ordinary Level Structures 217
4.1.5 Degeneration Data for General Ordinary Level Structures . 235
4.1.6 Comparison with Degeneration Data for Level Structures in Characteristic Zero 242
4.2 Boundary Charts of Ordinary Loci 248
4.2.1 Constructions with Level Structures but without Positivity Conditions 248
4.2.2 Toroidal Embeddings, Positivity Conditions, and Mumford Families 267
4.2.3 Extended Kodaira–Spencer Morphisms and Induced Isomorphisms 272

5. Partial Toroidal Compactifications 279

5.1 Approximation and Gluing Along the Ordinary Loci 279
5.1.1 Ordinary Good Formal Models 279
5.1.2 Ordinary Good Algebraic Models 285
5.1.3 Gluing in the Étale Topology 288
5.2 Partial Toroidal Compactifications of Ordinary Loci 294
5.2.1 Main Statements 295
5.2.2 Hecke Actions 300
5.2.3 The Case When p is a Good Prime 305
5.2.4 Boundary of Ordinary Loci 310

6. Partial Minimal Compactifications 331

6.1 Homogeneous Spectra and Their Properties 331
6.1.1 Construction of Quasi-Projective Models 331
6.1.2 Local Structures and Stratifications 338
6.2 Partial Minimal Compactifications of Ordinary Loci 345
6.2.1 Main Statements 345
6.2.2 Hecke Actions 353
6.2.3 Quasi-Projectivity of Partial Toroidal Compactifications . . 359
6.3 Full Ordinary Loci in p-Adic Completions 361
6.3.1 Hasse Invariants 361
6.3.2 Nonordinary and Full Ordinary Loci 365
6.3.3 Nonemptiness of Ordinary Loci 369

7. Ordinary Kuga Families 371

7.1 Partial Toroidal Compactifications . . . 371
7.1.1 Parameters for Ordinary Kuga Families . . . 371
7.1.2 Boundary of Ordinary Loci, Continued . . . 378
7.1.3 Ordinary Kuga Families and Their Generalizations . . . 390
7.1.4 Main Statements . . . 392
7.2 Main Constructions of Compactifications and Morphisms . . . 407
7.2.1 Partial Toroidal Boundary Strata . . . 407
7.2.2 Justification for the Parameters . . . 411
7.2.3 Extensibility of f_κ . . . 421
7.2.4 Properness of f^{tor} . . . 425
7.2.5 Log Smoothness of f^{tor} . . . 430
7.2.6 Equidimensionality of f^{tor} . . . 440
7.2.7 Hecke Actions . . . 443
7.3 Calculation of Formal Cohomology . . . 444
7.3.1 Setting . . . 444
7.3.2 Formal Fibers of f^{tor} . . . 445
7.3.3 Relative Cohomology and Local Freeness . . . 453
7.3.4 Degeneracy of the (Relative) Hodge Spectral Sequence . . 469
7.3.5 Extended Gauss–Manin Connections . . . 469
7.3.6 Identification of $R^b f^{\mathrm{tor}}_*(\mathscr{O}_{\vec{\mathsf{N}}^{\mathrm{ord,tor}}})$ . . . 471

8. Automorphic Bundles and Canonical Extensions 477

8.1 Constructions over the Ordinary Loci . . . 477
8.1.1 Technical Assumptions . . . 477
8.1.2 Automorphic Bundles . . . 480
8.1.3 Canonical Extensions . . . 483
8.1.4 Hecke Actions . . . 488
8.2 Higher Direct Images to the Minimal Compactifications . . . 490
8.2.1 Some Vanishing Theorems . . . 490
8.2.2 Formal Fibers of $\vec{\oint}^{\mathrm{ord}}_{\mathcal{H}}$ . . . 492
8.2.3 Relative Cohomology of Formal Fibers of $\vec{\oint}^{\mathrm{ord}}_{\mathcal{H}}$ . . . 496
8.2.4 Formal Fibers of Canonical Extensions . . . 501
8.2.5 End of the Proof . . . 508
8.3 Constructions over the Total Models . . . 508
8.3.1 Principal Bundles . . . 508
8.3.2 Automorphic Bundles . . . 510
8.3.3 Canonical Extensions . . . 512
8.3.4 Compatibility with the Constructions over the Ordinary Loci . . . 513
8.3.5 Pushforwards to the Total Minimal Compactifications . . . 514

8.3.6 Hecke Actions 518

Bibliography 525

Index 531

Chapter 0

Introduction

0.1 Background and Aim

In [62] (which is a published revision of [57]), based on the theories developed in [82] and [28], we studied the theory of degeneration of abelian varieties with PEL structures, and applied this theory to the construction of toroidal and minimal compactifications of moduli problems defining integral models of PEL-type Shimura varieties, under the assumption that each residue characteristic p is *good* in the sense that it is unramified in all linear algebraic data involved in the definition of the moduli problem of abelian varieties with PEL structures, and under the assumption that the level structures are defined by open compact subgroups of the adelic points of the associated reductive groups that are *hyperspecial* (maximal) at p. In [61], we also constructed toroidal compactifications of PEL-type Kuga families under the same assumptions on the residue characteristics, by realizing such toroidal compactifications in the toroidal boundary of larger PEL-type Shimura varieties. While these have been carried out for all PEL-type Shimura varieties, for practical reasons it is also natural to consider integral models when p is *ramified* in the linear algebraic data and when the level structures are defined by smaller open compact subgroups. Since the theory of degeneration developed in [62] works as long as the generic characteristic is good (and as long as the base of degenerations are noetherian normal), there is, a priori, no reason that we cannot consider compactifications with bad residue characteristics.

However, without the assumption that p is good and that the level structures are defined by an open compact subgroup hyperspecial at p, it is not clear what integral models really mean in general (although reasonably natural definitions can still be made in many special cases). The answer may depend on the applications. For applications involving counting points over finite fields, it seems necessary to have integral models with a specific kind of moduli interpretations, but usually even the flatness of such models are difficult to prove. For studying intersections of cycles, it is desirable to have models that are regular and flat, and we might consider the closures of the generic fibers in moduli problems as a general source of flat models, but the regularity of such models can be beyond reach already at very low levels. In both cases, it is difficult to say much about the integral models of Shimura varieties themselves, let alone their compactifications. On the other hand, for studying modular forms using coherent sheaves, there is already a rich theory using mainly the *ordinary loci*, or more precisely the *ordinary loci where (multiplicative-type) canonical subgroups can be defined*, when p is good in the above sense, and when the levels are certain (analogues of) "$\Gamma_1(p^r)$ levels". The aim of this book is to show that, without insisting on the (perhaps still desirable) moduli interpretations, such a theory can be generalized without the assumption that p is good, after adding sufficiently many p-power roots of unity to the base rings.

For such ordinary loci, we will construct *partial toroidal and minimal compactifications* which admit descriptions analogous to (and compatible with) their analogues in characteristic zero (and in mixed characteristics in the hyperspecial smooth case in [62]). We will also construct partial toroidal compactifications of Kuga families over such partial toroidal and minimal compactifications. Our construction works for all PEL-type Shimura varieties that can possibly admit ordinary loci. We allow p to be ramified (*i.e.*, p is not good in the sense described above), and we allow (analogues of) arbitrarily high "$\Gamma_1(p^r)$ levels". We need the ordinary loci to be defined, but we do not assume that they are nonempty (although the theory is uninteresting otherwise). (In some special cases, we can easily show the nonemptiness of the ordinary loci using the partial toroidal compactifications we construct. See Section 6.3.3.) Unsurprisingly, we started the construction in this work because of some interesting cases in which the nonemptiness of ordinary loci is clear. (See, for example, [39].)

As in characteristic zero and in the good reduction case, we will not need to answer difficult questions about p-divisible groups or p-adic Hodge theory in such a theory. As we shall see below, the difficulty in the constructions lies mainly in the sheer number of objects, morphisms, combinatorial data, and small subtle steps involved. It is not about proving some well-known conjecture that can be readily stated—rather, we want to know as much as possible about the constructions, and (at least for some applications we know) the theorems are useful only when they are detailed enough. It is fair to say that this is just another long exercise like [62]. Assuming that this is still interesting, the marathon begins.

0.2 Overview

Let us briefly describe the various objects to be constructed. (We say *briefly* but it still spans over many pages.)

Algebraic Constructions in Characteristic Zero

Starting with an integral PEL datum $(\mathcal{O}, \star, L, \langle\,\cdot\,,\,\cdot\,\rangle, h_0)$, as in [62], we can construct the following canonical objects (in characteristic zero):

(1) A group scheme G over $\mathrm{Spec}(\mathbb{Z})$, which is smooth and reductive over $\mathrm{Spec}(\mathbb{Z}_{(p)})$ when p is a good prime.

(2) A number field F_0, which is defined as a subfield of $\mathbb{C}$, called the *reflex field*.

(3) A moduli problem $\mathsf{M}_{\mathcal{H}}$ over $\mathsf{S}_0 = \mathrm{Spec}(F_0)$ for each open compact subgroup $\mathcal{H} \subset \mathrm{G}(\hat{\mathbb{Z}})$, parameterizing abelian schemes with PEL structures defined by the integral PEL datum, which is an algebraic stack separated, smooth, and of finite type over $\mathsf{S}_0 := \mathrm{Spec}(F_0)$. When $\mathcal{H}$ is *neat* (see [89, 0.6] or [62, Def. 1.4.1.8]), $\mathsf{M}_{\mathcal{H}}$ is quasi-projective over $\mathrm{Spec}(F_0)$. (In particular, $\mathsf{M}_{\mathcal{H}}$ is a scheme.)

(4) A finite étale surjection $[g] : \mathsf{M}_{\mathcal{H}'} \to \mathsf{M}_{\mathcal{H}}$ over S_0 for all $g \in \mathrm{G}(\mathbb{A}^\infty)$ and open compact subgroups $\mathcal{H}$ and $\mathcal{H}'$ such that $\mathcal{H}' \subset g\mathcal{H}g^{-1}$. This can be interpreted as the Hecke action of $\mathrm{G}(\mathbb{A}^\infty)$ on the collection $\{\mathsf{M}_{\mathcal{H}}\}_{\mathcal{H}}$.

(5) A *toroidal compactification* $\mathsf{M}^{\mathrm{tor}}_{\mathcal{H},\Sigma}$ of $\mathsf{M}_{\mathcal{H}}$ over S_0 for each compatible choice Σ of admissible smooth rational polyhedral cone decomposition data for $\mathsf{M}_{\mathcal{H}}$, which is a collection of combinatorial data which can be defined using only the integral PEL datum. For technical reasons, we assume that Σ is smooth (even in characteristic zero) and satisfies some mild conditions, in which case we can show that $\mathsf{M}^{\mathrm{tor}}_{\mathcal{H},\Sigma}$ is a proper smooth algebraic stack, and that the boundary is a simple normal crossings divisor. If $\mathcal{H}$ is neat, then $\mathsf{M}^{\mathrm{tor}}_{\mathcal{H},\Sigma}$ is an algebraic space. The toroidal compactification admits a stratification (defined in terms of G and Σ), and the structure along its boundary can be described in detail. These are useful for defining and studying modular forms using coherent sheaf cohomology. (See [60] for a survey on this topic.)

(6) A proper log étale surjection $[g]^{\mathrm{tor}} : \mathsf{M}^{\mathrm{tor}}_{\mathcal{H}',\Sigma'} \to \mathsf{M}^{\mathrm{tor}}_{\mathcal{H},\Sigma}$ over S_0 extending $[g]$, for all $g \in \mathrm{G}(\mathbb{A}^\infty)$ and open compact subgroups $\mathcal{H}$ and $\mathcal{H}'$ such that $\mathcal{H}' \subset g\mathcal{H}g^{-1}$, and for each Σ' that is a g-refinement of Σ in a suitable sense. This can be interpreted as the Hecke action of $\mathrm{G}(\mathbb{A}^\infty)$ on the collection $\{\mathsf{M}^{\mathrm{tor}}_{\mathcal{H},\Sigma}\}_{\mathcal{H},\Sigma}$. When $g = 1$ and $\mathcal{H}' = \mathcal{H}$, this means we have proper log étale surjections $[1]^{\mathrm{tor}} : \mathsf{M}^{\mathrm{tor}}_{\mathcal{H},\Sigma'} \to \mathsf{M}^{\mathrm{tor}}_{\mathcal{H},\Sigma}$ when Σ' is a refinement of Σ.

(7) A *minimal compactification* $\mathsf{M}^{\mathrm{min}}_{\mathcal{H}}$ over S_0 of the coarse moduli $[\mathsf{M}_{\mathcal{H}}]$ of $\mathsf{M}_{\mathcal{H}}$ for each open compact subgroup $\mathcal{H} \subset \mathrm{G}(\hat{\mathbb{Z}})$, which is a normal scheme projective over S_0, which admits a canonical (proper and surjective) morphism from the toroidal compactification $\mathsf{M}^{\mathrm{tor}}_{\mathcal{H},\Sigma}$ for each Σ. The stratification of any such toroidal compactification induces a stratification of the minimal compactification, which is independent of the choice of Σ. The strata in such a stratification are called *cusps*.

(8) A finite surjection $[g]^{\mathrm{min}} : \mathsf{M}^{\mathrm{min}}_{\mathcal{H}'} \to \mathsf{M}^{\mathrm{min}}_{\mathcal{H}}$ over S_0 extending the finite surjection $[[g]] : [\mathsf{M}_{\mathcal{H}'}] \to [\mathsf{M}_{\mathcal{H}}]$ between coarse moduli for all $g \in \mathrm{G}(\mathbb{A}^\infty)$ and open compact subgroups $\mathcal{H}$ and $\mathcal{H}'$ such that $\mathcal{H}' \subset g\mathcal{H}g^{-1}$. This can be interpreted as the Hecke action of $\mathrm{G}(\mathbb{A}^\infty)$ on the collection $\{\mathsf{M}^{\mathrm{min}}_{\mathcal{H}}\}_{\mathcal{H}}$.

(9) If $\mathcal{H}$ is neat and if Σ is (smooth and) projective, then we show that $\mathsf{M}^{\mathrm{tor}}_{\mathcal{H},\Sigma}$ is

(smooth and) projective over S_0 by showing that it is the normalization of the blowup of some coherent ideal sheaf $\mathcal{J}_{\mathcal{H},d_0\mathsf{pol}}$ on $\mathsf{M}_{\mathcal{H}}^{\min}$, defined by some integer $d_0 \geq 1$ and some compatible collection pol of polarization functions for Σ. In particular, $\mathsf{M}_{\mathcal{H},\Sigma}^{\mathrm{tor}}$ is a *scheme* in this case.

(10) A collection of *Kuga families* over $\mathsf{M}_{\mathcal{H}}$, which is a collection of abelian schemes including the self-fiber products of the tautological (*i.e.*, universal) abelian scheme as special members, together with toroidal compactifications *projective* over S_0 and satisfying a long list of desirable compatibilities, including in particular the existence (up to refinements of cone decompositions) of compatible proper log smooth morphisms from toroidal compactifications of PEL-type Kuga families to $\mathsf{M}_{\mathcal{H},\Sigma}^{\mathrm{tor}}$. We can enlarge the collection and include objects which are torsors under PEL-type Kuga families over $\mathsf{M}_{\mathcal{H}}$, which we call *generalized Kuga families* over $\mathsf{M}_{\mathcal{H}}$. They share the same nice properties enjoyed by PEL-type Kuga families.

(11) A collection of *automorphic bundles* over $\mathsf{M}_{\mathcal{H}}$, and their *canonical* and *subcanonical extensions* over $\mathsf{M}_{\mathcal{H},\Sigma}^{\mathrm{tor}}$. The (algebraic) construction of such canonical and subcanonical extensions uses the toroidal compactifications of PEL-type Kuga families.

The above constructions use only the theory of degeneration data and standard techniques in algebraic geometry. We call them the *algebraic constructions* over S_0.

Analytic Constructions and Comparison with Them

There is also the *analytic constructions* of analogous objects over S_0, which precedes the algebraic constructions in history. (These are *algebraic objects* constructed using transcendental arguments crucially in their constructions. Such analytic constructions use GAGA [95], but when we compare them to the algebraic constructions, we are not talking about a problem of GAGA anymore.)

In [59] we showed that, for suitable $\mathcal{H}$ and Σ, these analytically constructed objects admit canonical open and closed immersions to the algebraically constructed objects above, respecting all stratifications and descriptions of local structures. (See also [63] for the relation between rational boundary components and cusp labels.)

The Case When p is a Good Prime

As explained in [62], when p is a good prime for the $(\mathcal{O}, \star, L, \langle \cdot , \cdot \rangle, h_0)$, and when $\mathcal{H}$ is of the form $\mathcal{H} = \mathcal{H}^p\mathrm{G}(\mathbb{Z}_p) \subset \mathrm{G}(\hat{\mathbb{Z}}^p) \times \mathrm{G}(\mathbb{Z}_p) = \mathrm{G}(\hat{\mathbb{Z}})$, (note that in this case $\mathrm{G}(\mathbb{Z}_p)$ is a hyperspecial maximal open compact subgroup of $\mathrm{G}(\mathbb{Q}_p)$,) the above algebraically constructed objects admit analogues over $\mathrm{Spec}(\mathcal{O}_{F_0,(p)})$, which we denote by $\mathsf{M}_{\mathcal{H}^p}$, $\mathsf{M}_{\mathcal{H}^p,\Sigma^p}^{\mathrm{tor}}$, $\mathsf{M}_{\mathcal{H}^p}^{\min}$, *etc.* (Such notation makes sense because in the construction of these objects we only use $\mathcal{H}^p$ and an analogue Σ^p of Σ involving only adelic objects away from p.) Then there are canonical morphisms $\mathsf{M}_{\mathcal{H}} \to \mathsf{M}_{\mathcal{H}^p}$, $\mathsf{M}_{\mathcal{H},\Sigma}^{\mathrm{tor}} \to \mathsf{M}_{\mathcal{H}^p,\Sigma^p}^{\mathrm{tor}}$, $\mathsf{M}_{\mathcal{H}}^{\min} \to \mathsf{M}_{\mathcal{H}^p}^{\min}$, *etc*, compatible with each other, and respecting all stratifications and descriptions of local structures. (But we will not assume that p is a good prime for the $(\mathcal{O}, \star, L, \langle \cdot, \cdot \rangle, h_0)$ in what follows.)

Total Models in Mixed Characteristics

Assuming no longer that p is good, we will construct the following objects (in mixed characteristics $(0, p)$):

(1) A normal algebraic stack $\vec{\mathsf{M}}_{\mathcal{H}}$ for each open compact subgroup $\mathcal{H} \subset \mathrm{G}(\hat{\mathbb{Z}})$, flat over $\vec{\mathsf{S}}_0 = \mathrm{Spec}(\mathcal{O}_{F_0,(p)})$ for each open compact subgroup $\mathcal{H}$ as above, which admits a canonical morphism $\mathsf{M}_{\mathcal{H}} \to \vec{\mathsf{M}}_{\mathcal{H}}$. The coarse moduli space $[\vec{\mathsf{M}}_{\mathcal{H}}]$ of $\vec{\mathsf{M}}_{\mathcal{H}}$ is a normal scheme quasi-projective and flat over $\vec{\mathsf{S}}_0 = \mathrm{Spec}(\mathcal{O}_{F_0,(p)})$, which admits a canonical morphism $[\mathsf{M}_{\mathcal{H}}] \to [\vec{\mathsf{M}}_{\mathcal{H}}]$.

(2) A finite surjection $[\vec{g}] : \vec{\mathsf{M}}_{\mathcal{H}'} \to \vec{\mathsf{M}}_{\mathcal{H}}$ over $\vec{\mathsf{S}}_0$ for each $g = (g_0, g_p) \in \mathrm{G}(\mathbb{A}^{\infty,p}) \times \mathrm{G}(\mathbb{Z}_p) \subset \mathrm{G}(\mathbb{A}^\infty)$ and two open compact subgroups $\mathcal{H}$ and $\mathcal{H}'$ such that $\mathcal{H}' \subset g\mathcal{H}g^{-1}$. This can be interpreted as the Hecke action of $\mathrm{G}(\mathbb{A}^{\infty,p}) \times \mathrm{G}(\mathbb{Z}_p)$ on the collection $\{\vec{\mathsf{M}}_{\mathcal{H}}\}_{\mathcal{H}}$.

(3) A normal scheme $\vec{\mathsf{M}}_{\mathcal{H}}^{\min}$ projective and flat over $\vec{\mathsf{S}}_0$ for each open compact subgroup $\mathcal{H} \subset \mathrm{G}(\hat{\mathbb{Z}})$, containing the coarse moduli $[\vec{\mathsf{M}}_{\mathcal{H}}]$ of $\vec{\mathsf{M}}_{\mathcal{H}}$ as an open dense subscheme.

(4) A finite surjection $\vec{[g]}^{\min} : \vec{\mathsf{M}}_{\mathcal{H}'}^{\min} \to \vec{\mathsf{M}}_{\mathcal{H}}^{\min}$ over $\vec{\mathsf{S}}_0$ extending the finite surjection $[[\vec{g}]] : [\vec{\mathsf{M}}_{\mathcal{H}'}] \to [\vec{\mathsf{M}}_{\mathcal{H}}]$ between coarse moduli for each $g \in \mathrm{G}(\mathbb{A}^{\infty,p}) \times \mathrm{G}(\mathbb{Z}_p)$ and two open compact subgroups $\mathcal{H}$ and $\mathcal{H}'$ such that $\mathcal{H}' \subset g\mathcal{H}g^{-1}$. This can be interpreted as the Hecke action of $\mathrm{G}(\mathbb{A}^{\infty,p}) \times \mathrm{G}(\mathbb{Z}_p)$ on the collection $\{\vec{\mathsf{M}}_{\mathcal{H}}^{\min}\}_{\mathcal{H}}$.

(5) A normal scheme $\vec{\mathsf{M}}_{\mathcal{H},d_0\mathsf{pol}}^{\mathrm{tor}}$ projective and flat over $\vec{\mathsf{S}}_0$, which is defined when $\mathcal{H}$ is neat, when Σ is (smooth and) projective with a compatible collection pol of polarization functions, and when $d_0 \geq 1$ is sufficiently large, which is the normalization of the blowup of some coherent ideal sheaf $\vec{\mathcal{J}}_{\mathcal{H},d_0\mathsf{pol}}$ on $\vec{\mathsf{M}}_{\mathcal{H}}^{\min}$, such that the subscheme of $\vec{\mathsf{M}}_{\mathcal{H}}^{\min}$ defined by $\vec{\mathcal{J}}_{\mathcal{H},d_0\mathsf{pol}}$ is the schematic closure of the subscheme of $\mathsf{M}_{\mathcal{H}}^{\min}$ defined by the $\mathcal{J}_{\mathcal{H},d_0\mathsf{pol}}$ above.

(6) A collection of *automorphic bundles* over $\vec{\mathsf{M}}_{\mathcal{H}}$, and their *canonical* and *subcanonical extensions* over $\vec{\mathsf{M}}_{\mathcal{H},d_0\mathsf{pol}}^{\mathrm{tor}}$ when $\vec{\mathsf{M}}_{\mathcal{H},d_0\mathsf{pol}}^{\mathrm{tor}}$ is defined.

(7) For each integer $i \geq 0$, we define $\mathsf{S}_{0,i} := \mathrm{Spec}(F_0[\zeta_{p^i}])$ and $\vec{\mathsf{S}}_{0,i} := \mathrm{Spec}(\mathcal{O}_{F_0,(p)}[\zeta_{p^i}])$, and define $\vec{\mathsf{M}}_{\mathcal{H},i}$ (resp. $\vec{\mathsf{M}}_{\mathcal{H},i}^{\min}$, resp. $\vec{\mathsf{M}}_{\mathcal{H},d_0\mathsf{pol},i}^{\mathrm{tor}}$) to be the normalization of $\vec{\mathsf{M}}_{\mathcal{H}} \underset{\vec{\mathsf{S}}_0}{\times} \vec{\mathsf{S}}_{0,i}$ (resp. $\vec{\mathsf{M}}_{\mathcal{H}}^{\min} \underset{\vec{\mathsf{S}}_0}{\times} \vec{\mathsf{S}}_{0,i}$, resp. $\vec{\mathsf{M}}_{\mathcal{H},d_0\mathsf{pol}}^{\mathrm{tor}} \underset{\vec{\mathsf{S}}_0}{\times} \vec{\mathsf{S}}_{0,i}$).

These constructions require noncanonical auxiliary choices. A priori, it is unclear whether the objects thus constructed are independent of the choices, although it can be proved that they are indeed so. The quasi-projectivity of certain objects that will be canonically constructed below, such as $\vec{\mathsf{M}}_{\mathcal{H}}^{\mathrm{ord},\min}$ over $\vec{\mathsf{S}}_{0,r_{\mathcal{H}}} = \mathrm{Spec}(\mathcal{O}_{F_0,(p)}[\zeta_{p^{r_{\mathcal{H}}}}])$, is proved using the projectivity of such a noncanonically constructed $\vec{\mathsf{M}}_{\mathcal{H}}^{\min}$ over $\vec{\mathsf{S}}_0$. (We do not know any other method for proving such quasi-projectivity.) Such quasi-projectivity over mixed characteristics bases is important for many practical reasons. In particular, it allows us to talk about congruences (between algebro-geometrically defined automorphic forms) using its affine subsets.

The Ordinary Loci in Mixed Characteristics

This is the main theme of this work. With a suitable choice of a maximal totally isotropic filtration D on $L \underset{\mathbb{Z}}{\otimes} \mathbb{Z}_p$, we will construct the following canonical objects (in mixed characteristics $(0,p)$):

(1) A subgroup scheme $\mathrm{P}_{\mathsf{D}}^{\mathrm{ord}}$ of $\mathrm{G} \underset{\mathbb{Z}}{\otimes} \mathbb{Z}_p$ stabilizing the filtration D. (We do not say that $\mathrm{P}_{\mathsf{D}}^{\mathrm{ord}}$ is parabolic because $\mathrm{G} \underset{\mathbb{Z}}{\otimes} \mathbb{Z}_p$ is not smooth in general. But when $\mathrm{G} \underset{\mathbb{Z}}{\otimes} \mathbb{Q}_p$ is connected, which is the case when $\mathcal{O} \underset{\mathbb{Z}}{\otimes} \mathbb{Q}$ involves no factor of type D in the sense of [62, Def. 1.2.1.15], $\mathrm{P}_{\mathsf{D}}^{\mathrm{ord}} \underset{\mathbb{Z}_p}{\otimes} \mathbb{Q}_p$ is indeed a parabolic subgroup scheme of the reductive group scheme $\mathrm{G} \underset{\mathbb{Z}}{\otimes} \mathbb{Q}_p$ in the usual sense.)

(2) A collection of open compact subgroups $\mathcal{H} \subset \mathrm{G}(\hat{\mathbb{Z}})$ of the form $\mathcal{H} = \mathcal{H}^p\mathcal{H}_p \subset \mathrm{G}(\hat{\mathbb{Z}}^p) \times \mathrm{G}(\mathbb{Z}_p) = \mathrm{G}(\hat{\mathbb{Z}})$, such that $\mathcal{H}_p$ satisfies $\mathcal{U}_{p,1}^{\mathrm{bal}}(p^r) \subset \mathcal{H}_p \subset \mathcal{U}_{p,0}(p^r)$ for some integer $r \geq 0$. Here $\mathcal{U}_{p,1}^{\mathrm{bal}}(p^r)$ and $\mathcal{U}_{p,0}(p^r)$ are open compact subgroups of $\mathrm{G}(\mathbb{Z}_p)$ defining the (analogues of) "balanced $\Gamma_1(p^r)$" and "$\Gamma_0(p^r)$" levels at p.

(3) A naive moduli problem $\ddot{\mathsf{M}}_{\mathcal{H}}^{\mathrm{ord}}$ over $\mathrm{Spec}(\mathbb{Z}_{(p)})$, parameterizing abelian schemes with PEL structures away from p, and with certain *ordinary level structures* at p, but without the determinantal condition for Lie algebras in the definition of $\mathsf{M}_{\mathcal{H}}$. (Since p is not assumed to be a good prime, such a condition is not useful.) This $\ddot{\mathsf{M}}_{\mathcal{H}}^{\mathrm{ord}}$ is an algebraic stack separated and of finite over $\mathrm{Spec}(\mathbb{Z}_{(p)})$, with completions of strict local rings the same as those of a group scheme of multiplicative type of finite type over $\mathrm{Spec}(\mathbb{Z}_{(p)})$. (Hence, it is not smooth in general, but the singularity is mild.)

(4) An integer $r_{\mathcal{H}}$ determined by the integral PEL datum $(\mathcal{O}, \star, L, \langle \cdot\,,\,\cdot \rangle, h_0)$, the data D. This integer $r_{\mathcal{H}}$ stays as a constant r_{D} and does not increase with r if $\mathcal{U}_{p,1}(p^r) \subset \mathcal{H}_p$, where $\mathcal{U}_{p,1}(p^r)$ is an open compact subgroup of $\mathrm{G}(\mathbb{Z}_p)$ defining the (analogue of) "$\Gamma_1(p^r)$" levels at p. But it increases with r (and is equal to $\max(r_{\mathsf{D}}, r)$) if, for example, $\mathcal{H}_p = \mathcal{U}_{p,1}^{\mathrm{bal}}(p^r)$.

(5) An algebraic stack $\mathsf{M}_{\mathcal{H}}^{\mathrm{ord}}$ separated, smooth, and of finite type over $\mathsf{S}_{0,r_{\mathcal{H}}} := \mathrm{Spec}(F_0[\zeta_{p^{r_{\mathcal{H}}}}])$ parameterizing certain ordinary level structures in characteristic zero, which is canonically isomorphic to $\mathsf{M}_{\mathcal{H}} \underset{\mathsf{S}_0}{\times} \mathsf{S}_{0,r_{\mathcal{H}}}$, but with the understanding that the usual level structures are turned into the ordinary level structures it parameterizes (with the help of roots of unity in $\mathsf{S}_{0,r_{\mathcal{H}}}$). The universal property of $\ddot{\mathsf{M}}_{\mathcal{H}}^{\mathrm{ord}}$ induces a canonical quasi-finite morphism $\mathsf{M}_{\mathcal{H}}^{\mathrm{ord}} \to \ddot{\mathsf{M}}_{\mathcal{H}}^{\mathrm{ord}}$.

(6) An algebraic stack $\vec{\mathsf{M}}^{\mathrm{ord}}_{\mathcal{H}}$ separated, smooth, and of finite type over $\vec{\mathsf{S}}_{0,r_{\mathcal{H}}} = \operatorname{Spec}(\mathcal{O}_{F_0,(p)}[\zeta_{p^{r_{\mathcal{H}}}}])$, which admits canonical finite morphism to $\dddot{\mathsf{M}}^{\mathrm{ord}}_{\mathcal{H}}$ extending the quasi-finite morphism $\mathsf{M}^{\mathrm{ord}}_{\mathcal{H}} \to \dddot{\mathsf{M}}^{\mathrm{ord}}_{\mathcal{H}}$.

(7) A quasi-finite flat surjection $\vec{[g]}^{\mathrm{ord}} : \vec{\mathsf{M}}^{\mathrm{ord}}_{\mathcal{H}'} \to \vec{\mathsf{M}}^{\mathrm{ord}}_{\mathcal{H}}$ over $\vec{\mathsf{S}}_{0,r_{\mathcal{H}}}$ for each $g = (g_0, g_p) \in \mathrm{G}(\mathbb{A}^{\infty,p}) \times \mathrm{P}^{\mathrm{ord}}_{\mathtt{D}}(\mathbb{Q}_p) \subset \mathrm{G}(\mathbb{A}^\infty)$ and two open compact subgroups $\mathcal{H}$ and $\mathcal{H}'$ such that $\mathcal{H}' \subset g\mathcal{H}g^{-1}$, satisfying some reasonable additional conditions. This can be interpreted as the Hecke action of a semi-subgroup of $\mathrm{G}(\mathbb{A}^{\infty,p}) \times \mathrm{P}^{\mathrm{ord}}_{\mathtt{D}}(\mathbb{Q}_p)$ on the collection $\{\vec{\mathsf{M}}^{\mathrm{ord}}_{\mathcal{H}}\}_{\mathcal{H}}$. (And there are conditions for the morphisms $\vec{[g]}^{\mathrm{ord}}$ to be finite or étale.)

(8) A *partial toroidal compactification* $\vec{\mathsf{M}}^{\mathrm{ord,tor}}_{\mathcal{H},\Sigma^{\mathrm{ord}}}$ of $\vec{\mathsf{M}}^{\mathrm{ord}}_{\mathcal{H}}$ over $\vec{\mathsf{S}}_{0,r_{\mathcal{H}}}$ for each compatible choice Σ^{ord} of admissible smooth rational polyhedral cone decomposition data for $\vec{\mathsf{M}}^{\mathrm{ord}}_{\mathcal{H}}$, which is a collection of combinatorial data defined in a way similar to the case of Σ above. In fact, each Σ as above induces a Σ^{ord}. This $\vec{\mathsf{M}}^{\mathrm{ord,tor}}_{\mathcal{H},\Sigma^{\mathrm{ord}}}$ is an algebraic stack separated, smooth, and of finite type over $\vec{\mathsf{S}}_{0,r_{\mathcal{H}}}$, and the boundary is a simple normal crossings divisor. If $\mathcal{H}^p$ is neat, then $\vec{\mathsf{M}}^{\mathrm{ord,tor}}_{\mathcal{H},\Sigma^{\mathrm{ord}}}$ is an algebraic space. The partial toroidal compactification admits a stratification (defined in terms of G, $\mathtt{D}$, and Σ^{ord}), and the structure along its boundary can be described in detail; both are as in the case of $\mathsf{M}^{\mathrm{tor}}_{\mathcal{H},\Sigma}$.

(9) A surjection $\vec{[g]}^{\mathrm{ord,tor}} : \vec{\mathsf{M}}^{\mathrm{ord,tor}}_{\mathcal{H}',\Sigma^{\mathrm{ord},\prime}} \to \vec{\mathsf{M}}^{\mathrm{ord,tor}}_{\mathcal{H},\Sigma^{\mathrm{ord}}}$ over $\vec{\mathsf{S}}_{0,r_{\mathcal{H}}}$ extending $\vec{[g]}^{\mathrm{ord}}$ for each g as above such that $\vec{[g]}^{\mathrm{ord}}$ is defined, and for each $\Sigma^{\mathrm{ord},\prime}$ that is a g-refinement of Σ^{ord} in a suitable sense. This can be interpreted as the Hecke action of the same semi-subgroup of $\mathrm{G}(\mathbb{A}^{\infty,p}) \times \mathrm{P}^{\mathrm{ord}}_{\mathtt{D}}(\mathbb{Q}_p)$ as above on the collection $\{\vec{\mathsf{M}}^{\mathrm{ord,tor}}_{\mathcal{H},\Sigma^{\mathrm{ord}}}\}_{\mathcal{H},\Sigma^{\mathrm{ord}}}$. (And there are conditions for $\vec{[g]}^{\mathrm{ord,min}}$ to be proper, finite, flat, log étale, or étale.)

(10) A *partial minimal compactification* $\vec{\mathsf{M}}^{\mathrm{ord,min}}_{\mathcal{H}}$ over $\vec{\mathsf{S}}_{0,r_{\mathcal{H}}}$ of the coarse moduli $[\vec{\mathsf{M}}^{\mathrm{ord}}_{\mathcal{H}}]$ of $\vec{\mathsf{M}}^{\mathrm{ord}}_{\mathcal{H}}$, which is a normal scheme quasi-projective and flat over $\vec{\mathsf{S}}_{0,r_{\mathcal{H}}}$, which admits a canonical proper (and surjective) morphism from the partial toroidal compactification $\vec{\mathsf{M}}^{\mathrm{ord,tor}}_{\mathcal{H},\Sigma^{\mathrm{ord}}}$ for each Σ^{ord}. The stratification of any such partial toroidal compactification induces a stratification of the partial minimal compactification, which is independent of the choice of Σ^{ord}. The strata in such a stratification are called *ordinary cusps*.

(11) A quasi-finite surjection $\vec{[g]}^{\mathrm{ord,min}} : \vec{\mathsf{M}}^{\mathrm{ord,min}}_{\mathcal{H}'} \to \vec{\mathsf{M}}^{\mathrm{ord,min}}_{\mathcal{H}}$ over $\vec{\mathsf{S}}_{0,r_{\mathcal{H}}}$ extending the quasi-finite surjection $[\vec{[g]}^{\mathrm{ord}}] : [\vec{\mathsf{M}}^{\mathrm{ord}}_{\mathcal{H}'}] \to [\vec{\mathsf{M}}^{\mathrm{ord}}_{\mathcal{H}}]$ between coarse moduli for each g as above such that $\vec{[g]}^{\mathrm{ord}}$ is defined. This can be interpreted as the Hecke action of the same semi-subgroup of $\mathrm{G}(\mathbb{A}^{\infty,p}) \times \mathrm{P}^{\mathrm{ord}}_{\mathtt{D}}(\mathbb{Q}_p)$ as above on the collection $\{\vec{\mathsf{M}}^{\mathrm{ord,min}}_{\mathcal{H}}\}_{\mathcal{H}}$. (And there are conditions for $\vec{[g]}^{\mathrm{ord,min}}$ to be finite.)

(12) If $\mathcal{H}^p$ is neat and if Σ^{ord} is (smooth and) projective, then we show that

$\vec{\mathsf{M}}^{\mathrm{ord,tor}}_{\mathcal{H},\Sigma^{\mathrm{ord}}}$ is (smooth and) quasi-projective over $\vec{\mathsf{S}}_{0,r_{\mathcal{H}}}$ by showing that it is the normalization of the blowup of some coherent ideal sheaf $\vec{\mathscr{I}}_{\mathcal{H},d_0\mathsf{pol}^{\mathrm{ord}}}$ on $\vec{\mathsf{M}}^{\mathrm{ord,min}}_{\mathcal{H}}$, defined by some integer $d_0 \geq 1$ and some compatible collection $\mathsf{pol}^{\mathrm{ord}}$ of polarization functions for Σ^{ord}. In particular, $\vec{\mathsf{M}}^{\mathrm{ord,tor}}_{\mathcal{H},\Sigma^{\mathrm{ord}}}$ is a quasi-projective *scheme* in this case.

(13) A collection of *ordinary Kuga families* over $\vec{\mathsf{M}}^{\mathrm{ord}}_{\mathcal{H}}$, which is a collection of abelian schemes containing the self-fiber products of the tautological abelian scheme, together with partial toroidal compactifications *quasi-projective* over $\vec{\mathsf{S}}_{0,r_{\mathcal{H}}}$ and satisfying a long list of desirable compatibilities, including in particular the existence (up to refinements of cone decompositions) of compatible proper log smooth morphisms from partial toroidal compactifications of ordinary PEL-type Kuga families to $\vec{\mathsf{M}}^{\mathrm{ord,tor}}_{\mathcal{H},\Sigma^{\mathrm{ord}}}$. We can enlarge the collection and include objects which are torsors under Kuga families over $\vec{\mathsf{M}}^{\mathrm{ord}}_{\mathcal{H}}$, which we call *generalized ordinary Kuga families* over $\vec{\mathsf{M}}^{\mathrm{ord}}_{\mathcal{H}}$. They share the same nice properties enjoyed by ordinary PEL-type Kuga families.

(14) A collection of *automorphic bundles* over $\vec{\mathsf{M}}^{\mathrm{ord}}_{\mathcal{H}}$, and their *canonical* and *subcanonical extensions* over $\vec{\mathsf{M}}^{\mathrm{ord,tor}}_{\mathcal{H},\Sigma^{\mathrm{ord}}}$. The (algebraic) construction of such canonical and subcanonical extensions uses the partial toroidal compactifications of ordinary PEL-type Kuga families (over $\vec{\mathsf{M}}^{\mathrm{ord}}_{\mathcal{H}}$). (The class of automorphic bundles we can construct over $\vec{\mathsf{M}}^{\mathrm{ord}}_{\mathcal{H}}$ is more restrictive than that over $\mathsf{M}_{\mathcal{H}}$.)

These objects are compatible with the algebraically constructed objects in characteristic zero, such as $\mathsf{M}_{\mathcal{H}}$, and with the total models in mixed characteristics, such as $\vec{\mathsf{M}}_{\mathcal{H}}$.

0.3 Outline of the Constructions

The objects above are not constructed in the same order as they are listed. The logical steps we need are as follows:

Algebraic Constructions in Characteristic Zero

We start with all algebraically constructed objects $\mathsf{M}_{\mathcal{H}}$, $\mathsf{M}^{\mathrm{tor}}_{\mathcal{H},\Sigma}$, $\mathsf{M}^{\mathrm{min}}_{\mathcal{H}}$, *etc* over $\mathsf{S}_0 = \mathrm{Spec}(F_0)$. The algebraic construction of $\mathsf{M}^{\mathrm{tor}}_{\mathcal{H},\Sigma}$ by the theory of degeneration endows it with a semi-abelian scheme with PEL structures, which is universal among semi-abelian degenerations of abelian varieties with PEL structures of a particular *degeneration pattern* given by Σ. We call such semi-abelian schemes *degenerating families*.

These are done in [62]. We review them in Chapter 1, because it might be hard to get used to the fact that even the characteristic zero theory can be done so non-canonically with various choices. (We still consider the theory canonical because it hardly favors any particular choices.)

Auxiliary Choices of Good Reduction Models

We make a (noncanonical) auxiliary choice of an integral PEL datum $(\mathcal{O}_{\mathrm{aux}}, \star_{\mathrm{aux}}, L_{\mathrm{aux}}, \langle \cdot, \cdot \rangle_{\mathrm{aux}}, h_{0,\mathrm{aux}})$ for which p is a good prime. This allows us to define a group scheme $\mathrm{G}_{\mathrm{aux}}$ over $\mathrm{Spec}(\mathbb{Z})$ such that $\mathrm{G}_{\mathrm{aux}}(\mathbb{Z}_p)$ is a hyperspecial maximal open compact subgroup of $\mathrm{G}_{\mathrm{aux}}(\mathbb{Q}_p)$, and to construct for each open compact subgroup $\mathcal{H}^p_{\mathrm{aux}} \subset \mathrm{G}_{\mathrm{aux}}(\hat{\mathbb{Z}}^p)$ the objects $\mathsf{M}_{\mathcal{H}^p_{\mathrm{aux}}}$, $\mathsf{M}^{\mathrm{tor}}_{\mathcal{H}^p_{\mathrm{aux}}, \Sigma^p_{\mathrm{aux}}}$, $\mathsf{M}^{\mathrm{min}}_{\mathcal{H}^p_{\mathrm{aux}}}$, *etc* over $\vec{\mathsf{S}}_{0,\mathrm{aux}} = \mathrm{Spec}(\mathcal{O}_{F_{0,\mathrm{aux}},(p)})$. (Here the superscript "p" means "away from p".) The point is that $\mathsf{M}_{\mathcal{H}^p_{\mathrm{aux}}}$ is a moduli problem, $\mathsf{M}^{\mathrm{tor}}_{\mathcal{H}^p_{\mathrm{aux}}, \Sigma^p}$ carries a tautological degenerating family, and $\mathsf{M}^{\mathrm{min}}_{\mathcal{H}^p_{\mathrm{aux}}}$ is projective over $\vec{\mathsf{S}}_{0,\mathrm{aux}}$.

The auxiliary choices are made in Section 2.1. The constructions of the geometric objects are done in [62]. We do not explicitly review them because they are only auxiliary in nature, and because their behaviors are almost identical to those of $\mathsf{M}_{\mathcal{H}}$, $\mathsf{M}^{\mathrm{tor}}_{\mathcal{H},\Sigma}$, $\mathsf{M}^{\mathrm{min}}_{\mathcal{H}}$ above. We will simply cite [62], with the "$\square$" there filled with "p", and with each object there attached with a subscript "aux" and a superscript "p".

The auxiliary objects are chosen so that there is a homomorphism $\mathrm{G} \to \mathrm{G}_{\mathrm{aux}}$ of group schemes over $\mathbb{Z}$, and so that we have morphisms $\mathsf{M}_{\mathcal{H}} \to \mathsf{M}_{\mathcal{H}^p_{\mathrm{aux}}}$, $\mathsf{M}^{\mathrm{min}}_{\mathcal{H}} \to \mathsf{M}^{\mathrm{min}}_{\mathcal{H}^p_{\mathrm{aux}}}$, $\mathsf{M}^{\mathrm{tor}}_{\mathcal{H},\Sigma} \to \mathsf{M}^{\mathrm{tor}}_{\mathcal{H}^p_{\mathrm{aux}}, \Sigma^p}$, *etc* compatible with each other when $\mathcal{H}$ is of the form $\mathcal{H} = \mathcal{H}^p \mathcal{H}_p \subset \mathrm{G}(\hat{\mathbb{Z}}^p) \times \mathrm{G}(\mathbb{Z}_p) = \mathrm{G}(\hat{\mathbb{Z}})$ and is mapped into $\mathcal{H}^p_{\mathrm{aux}} \mathrm{G}(\mathbb{Z}_p) \subset \mathrm{G}_{\mathrm{aux}}(\hat{\mathbb{Z}})$.

Total Models in Mixed Characteristics

We define $\vec{\mathsf{M}}_{\mathcal{H}}$ (resp. $\vec{\mathsf{M}}^{\mathrm{min}}_{\mathcal{H}}$) to be the normalization of $\mathsf{M}_{\mathrm{G}_{\mathrm{aux}}(\hat{\mathbb{Z}}^p)}$ (resp. $\mathsf{M}^{\mathrm{min}}_{\mathrm{G}_{\mathrm{aux}}(\hat{\mathbb{Z}}^p)}$) in $\mathsf{M}_{\mathcal{H}}$ (resp. $\mathsf{M}^{\mathrm{min}}_{\mathcal{H}}$). Then we define $\vec{\mathsf{M}}^{\mathrm{tor}}_{\mathcal{H},d_0\mathsf{pol}}$ as described above as a normalization of a suitable blowup of $\vec{\mathsf{M}}^{\mathrm{min}}_{\mathcal{H}}$ (depending on the choices of pol and d_0). These are algebraic stacks or schemes over $\vec{\mathsf{S}}_0 = \mathrm{Spec}(\mathcal{O}_{F_0,(p)})$, depending on the choice of $(\mathcal{O}_{\mathrm{aux}}, \star_{\mathrm{aux}}, L_{\mathrm{aux}}, \langle \cdot, \cdot \rangle_{\mathrm{aux}}, h_{0,\mathrm{aux}})$. The Hecke actions are induced by the universal property of normalizations.

These are done in Sections 2.2.1, 2.2.2, and 2.2.3. These total models should be considered as auxiliary in nature. Although they can be shown to be canonical by an indirect argument, based on certain techniques developed in [58], their constructions are noncanonical, and we cannot say much about their local structures. We will have to construct the ordinary loci separately, map them to these total models, and then show that suitable normalizations of these total models (after ramified base changes) have smooth open subschemes given by the images of the ordinary loci.

Nevertheless, when p is a good prime, the schemes $\vec{\mathsf{M}}_{\mathcal{H}}$ and $\vec{\mathsf{M}}^{\mathrm{min}}_{\mathcal{H}}$ can be canonically constructed (without the auxiliary objects). This is explained in Section 2.2.4. Such special cases are important because p *is* a good prime for $(\mathcal{O}_{\mathrm{aux}}, \star_{\mathrm{aux}}, L_{\mathrm{aux}}, \langle \cdot, \cdot \rangle_{\mathrm{aux}}, h_{0,\mathrm{aux}})$. In what follows, we can often reduce the proof of important facts to the case of the auxiliary models, and prove them by more direct methods.

Construction of $\vec{\mathsf{M}}_{\mathcal{H}}^{\mathrm{ord}}$

We construct $\vec{\mathsf{M}}_{\mathcal{H}}^{\mathrm{ord}}$ over $\vec{\mathsf{S}}_{0,r_{\mathcal{H}}} = \mathrm{Spec}(\mathcal{O}_{F_0,(p)}[\zeta_{p^{r_{\mathcal{H}}}}])$ as follows:

(1) We investigate, roughly speaking, what happens when an abelian scheme with PEL structures over a scheme over $\mathsf{M}_{\mathcal{H}}$ extends to an *ordinary abelian scheme* over a scheme over $\vec{\mathsf{M}}_{\mathcal{H}}$. We write down the necessary linear algebraic data for this to happen, and turn them into formal definitions. This gives, in particular, a filtration $\mathtt{D}$ on $L \underset{\mathbb{Z}}{\otimes} \mathbb{Z}_p$ satisfying certain properties. These are done in Sections 3.2.1 and 3.2.2.

(2) We develop the notion of *ordinary level structures* at p defined by $\mathtt{D}$, accompanied by usual level structures away from p. (We do not assume that the polarization degree is prime to p.) This is done in Section 3.3.

(3) We define $\dddot{\mathsf{M}}_{\mathcal{H}}^{\mathrm{ord}}$ over $\mathrm{Spec}(\mathbb{Z}_{(p)})$ as a naive moduli problem for abelian schemes with polarizations, endomorphism structures, usual level structures away from p, and with ordinary level structures at p defined by $\mathtt{D}$. The local structures of $\dddot{\mathsf{M}}_{\mathcal{H}}^{\mathrm{ord}}$ can be studied in two ways. At points of characteristic zero, it is the same as in the case of $\mathsf{M}_{\mathcal{H}}$. At points of characteristic p, since all abelian schemes involved are ordinary, we use the *Serre–Tate deformation theory* explained in [47]. This is done in Section 3.4.1.

(4) We define $\mathsf{M}_{\mathcal{H}}^{\mathrm{ord}}$ over $\mathsf{S}_{0,r_{\mathcal{H}}} = \mathrm{Spec}(F_0[\zeta_{p^{r_{\mathcal{H}}}}])$ by turning the level structures at p parameterized by the moduli problem $\mathsf{M}_{\mathcal{H}}$ into ordinary level structures parameterized by $\dddot{\mathsf{M}}_{\mathcal{H}}^{\mathrm{ord}}$. Then we define $\vec{\mathsf{M}}_{\mathcal{H}}^{\mathrm{ord}}$ over $\vec{\mathsf{S}}_{0,r_{\mathcal{H}}} = \mathrm{Spec}(\mathcal{O}_{F_0,(p)}[\zeta_{p^{r_{\mathcal{H}}}}])$ to be the normalization of $\dddot{\mathsf{M}}_{\mathcal{H}}^{\mathrm{ord}}$ in $\mathsf{M}_{\mathcal{H}}^{\mathrm{ord}}$. The point of making the (ramified) base change to $\vec{\mathsf{S}}_{0,r_{\mathcal{H}}}$ is that the normalization $\vec{\mathsf{M}}_{\mathcal{H}}^{\mathrm{ord}}$ is smooth over $\vec{\mathsf{S}}_{0,r_{\mathcal{H}}}$ and regular. These are done in Section 3.4.2.

When p is a good prime, in which case $\mathsf{M}_{\mathcal{H}^p}$ is defined over $\vec{\mathsf{S}}_0 = \mathrm{Spec}(\mathcal{O}_{F_0,(p)})$, we show that $\vec{\mathsf{M}}_{\mathcal{H}}^{\mathrm{ord}}$ can be defined by taking the schematic closure of $\mathsf{M}_{\mathcal{H}}^{\mathrm{ord}}$ (the latter being just a base change of $\mathsf{M}_{\mathcal{H}}$) in a moduli problem schematic and quasi-finite over $\mathsf{M}_{\mathcal{H}^p}$. This is done in Section 3.4.5.

Then one can show the quasi-projectivity of $[\vec{\mathsf{M}}_{\mathcal{H}}^{\mathrm{ord}}]$ over $\vec{\mathsf{S}}_{0,r_{\mathcal{H}}}$ as follows:

(1) Using an auxiliary choice of the filtration $\mathtt{D}_{\mathrm{aux}}$ at p for the $(\mathcal{O}_{\mathrm{aux}}, \star_{\mathrm{aux}}, L_{\mathrm{aux}}, \langle\,\cdot\,,\,\cdot\,\rangle_{\mathrm{aux}}, h_{0,\mathrm{aux}})$ above, we define $\vec{\mathsf{M}}_{\mathcal{H}_{\mathrm{aux}}}^{\mathrm{ord}}$ as above, together with a quasi-finite morphism $\vec{\mathsf{M}}_{\mathcal{H}}^{\mathrm{ord}} \to \vec{\mathsf{M}}_{\mathcal{H}_{\mathrm{aux}}}^{\mathrm{ord}}$.

(2) Since p is a good prime for $(\mathcal{O}_{\mathrm{aux}}, \star_{\mathrm{aux}}, L_{\mathrm{aux}}, \langle\,\cdot\,,\,\cdot\,\rangle_{\mathrm{aux}}, h_{0,\mathrm{aux}})$ by assumption, we obtain a quasi-finite morphism $\vec{\mathsf{M}}_{\mathcal{H}_{\mathrm{aux}}}^{\mathrm{ord}} \to \vec{\mathsf{M}}_{\mathrm{G}(\hat{\mathbb{Z}}^p)}^{\mathrm{ord}}$. (See above.)

(3) Combining the above, we obtain a quasi-finite morphism $\vec{\mathsf{M}}_{\mathcal{H}}^{\mathrm{ord}} \to \vec{\mathsf{M}}_{\mathrm{G}(\hat{\mathbb{Z}}^p)}^{\mathrm{ord}}$,

which induces a quasi-finite morphism $[\vec{\mathsf{M}}^{\mathrm{ord}}_{\mathcal{H}}] \to [\vec{\mathsf{M}}^{\mathrm{ord}}_{\mathrm{G}(\hat{\mathbb{Z}}^p)}]$ between noetherian normal schemes. Then Zariski's Main Theorem implies that we have an open immersion $[\vec{\mathsf{M}}^{\mathrm{ord}}_{\mathcal{H}}] \hookrightarrow [\vec{\mathsf{M}}_{\mathcal{H},r_{\mathcal{H}}}]$, which shows that $[\vec{\mathsf{M}}^{\mathrm{ord}}_{\mathcal{H}}]$ is quasi-projective over $\vec{\mathsf{S}}_{0,r_{\mathcal{H}}}$.

These are done in Section 3.4.6.

Construction of $\vec{\mathsf{M}}^{\mathrm{ord,tor}}_{\mathcal{H},\Sigma^{\mathrm{ord}}}$

We construct $\vec{\mathsf{M}}^{\mathrm{ord,tor}}_{\mathcal{H},\Sigma^{\mathrm{ord}}}$ over $\vec{\mathsf{S}}_{0,r_{\mathcal{H}}}$ as follows:

(1) Following [28] and [62], we develop a theory of degeneration for abelian varieties with ordinary level structures. This will be used in the construction of toroidal boundary charts, and in showing that what we obtained satisfy certain universal property among all degenerations over normal schemes. (This will, in particular, provide us with a valuative criterion over complete discrete valuation rings.) This is done in Section 4.1.

(2) Using the theory of degeneration, we can construct naive toroidal boundary charts over $\vec{\mathsf{S}}_0$ parameterizing the degeneration data for the degeneration of objects parameterized by $\dddot{\mathsf{M}}^{\mathrm{ord}}_{\mathcal{H}}$. These naive toroidal boundary charts are similar to their analogues constructed algebraically over S_0 (as in [62]). We take the normalization of the naive objects in the base changes of the characteristic zero objects to $\mathsf{S}_{0,r_{\mathcal{H}}}$ (as in the construction of $\vec{\mathsf{M}}^{\mathrm{ord}}_{\mathcal{H}}$ above). We can show that these normalizations are smooth over $\vec{\mathsf{S}}_{0,r_{\mathcal{H}}}$ and regular. This is done in Section 4.2.

(3) Then we show that suitable algebraizations of the formal completions of these normalizations can be glued to $\vec{\mathsf{M}}^{\mathrm{ord}}_{\mathcal{H}}$ in the étale topology. This gives us the desired $\vec{\mathsf{M}}^{\mathrm{ord,tor}}_{\mathcal{H},\Sigma^{\mathrm{ord}}}$. This is done in Sections 5.1 and 5.2.

The outline here is simple, but of course these constructions are central to the whole work, without which other technical considerations make no sense. (Otherwise we could have also included the nonordinary loci in our study.) Fortunately, since the theory in [62] is developed in sufficient generality, there is no surprising difficulty in this part of the theory.

Construction of $\vec{\mathsf{M}}^{\mathrm{ord,min}}_{\mathcal{H}}$

We construct $\vec{\mathsf{M}}^{\mathrm{ord,min}}_{\mathcal{H}}$ over $\vec{\mathsf{S}}_{0,r_{\mathcal{H}}}$ as follows:

(1) We start with a partial toroidal compactification $\vec{\mathsf{M}}^{\mathrm{ord,tor}}_{\mathcal{H},\Sigma^{\mathrm{ord}}}$ carrying a semi-abelian scheme G, and we define the so-called *Hodge invertible sheaf*
$$\omega_{\vec{\mathsf{M}}^{\mathrm{ord,tor}}_{\mathcal{H},\Sigma^{\mathrm{ord}}}} := \wedge^{\mathrm{top}}\,\underline{\mathrm{Lie}}^{\vee}_{G/\vec{\mathsf{M}}^{\mathrm{ord,tor}}_{\mathcal{H},\Sigma^{\mathrm{ord}}}} \cong \wedge^{\mathrm{top}}\, e_G^*\Omega^1_{G/\vec{\mathsf{M}}^{\mathrm{ord,tor}}_{\mathcal{H},\Sigma^{\mathrm{ord}}}}$$
as usual. By imitating the construction of $\mathsf{M}^{\mathrm{min}}_{\mathcal{H}}$, we define
$$\vec{\mathsf{M}}^{\mathrm{ord,min}}_{\mathcal{H}} := \mathrm{Proj}\Big(\bigoplus_{k\geq 0}\Gamma(\vec{\mathsf{M}}^{\mathrm{ord,tor}}_{\mathcal{H},\Sigma^{\mathrm{ord}}},\omega^{\otimes k}_{\vec{\mathsf{M}}^{\mathrm{ord,tor}}_{\mathcal{H},\Sigma^{\mathrm{ord}}}})\Big).$$

However, since $\vec{\mathsf{M}}^{\mathrm{ord,tor}}_{\mathcal{H},\Sigma^{\mathrm{ord}}}$ is not proper, we cannot assert that $\vec{\mathsf{M}}^{\mathrm{ord,min}}_{\mathcal{H}}$ is projective over $\vec{\mathsf{S}}_{0,r_{\mathcal{H}}}$. The question is whether we can show that it is *quasi-projective* over $\vec{\mathsf{S}}_{0,r_{\mathcal{H}}}$, and whether we can show that the canonical morphism $\vec{\mathsf{M}}^{\mathrm{ord,tor}}_{\mathcal{H},\Sigma^{\mathrm{ord}}} \to \vec{\mathsf{M}}^{\mathrm{ord,min}}_{\mathcal{H}}$ is *proper*. (We will outline the steps below.)

(2) Once we know this last properness, the familiar arguments for studying the local structures of $\vec{\mathsf{M}}^{\mathrm{ord,min}}_{\mathcal{H}}$ by considering the Stein factorization of $\vec{\mathsf{M}}^{\mathrm{ord,tor}}_{\mathcal{H},\Sigma^{\mathrm{ord}}} \to \vec{\mathsf{M}}^{\mathrm{ord,min}}_{\mathcal{H}}$ (which coincide with itself) work as in [62].

These are done in Sections 6.1 and 6.2.1.

The proof for the properness of $\vec{\mathsf{M}}^{\mathrm{ord,tor}}_{\mathcal{H},\Sigma^{\mathrm{ord}}} \to \vec{\mathsf{M}}^{\mathrm{ord,min}}_{\mathcal{H}}$ and the quasi-projectivity of $\vec{\mathsf{M}}^{\mathrm{ord,min}}_{\mathcal{H}}$ over $\vec{\mathsf{S}}_{0,r_{\mathcal{H}}}$ is somewhat indirect. Therefore, we would like to summarize the steps here too:

(1) We show that the statements can be proved by replacing $\mathcal{H}$ with a higher level that is equally deep at p, and by replacing Σ^{ord} with a refinement.

(2) Take $\mathcal{H}_{\mathrm{aux}}$ to be as deep as $\mathcal{H}$. Using the assumption that p is good for $(\mathcal{O}_{\mathrm{aux}}, \star_{\mathrm{aux}}, L_{\mathrm{aux}}, \langle\,\cdot\,,\,\cdot\,\rangle_{\mathrm{aux}}, h_{0,\mathrm{aux}})$, we explain that, for $\Sigma^{\mathrm{ord}}_{\mathrm{aux}}$ induced by some $\Sigma^{p}_{\mathrm{aux}}$, we can build $\vec{\mathsf{M}}^{\mathrm{ord,tor}}_{\mathcal{H}_{\mathrm{aux}},\Sigma^{\mathrm{ord}}_{\mathrm{aux}}}$ by taking the schematic closure of $\mathsf{M}^{\mathrm{ord}}_{\mathcal{H}_{\mathrm{aux}}}$ in a moduli problem schematic and quasi-finite over $\mathsf{M}^{\mathrm{tor}}_{\mathcal{H}^{p}_{\mathrm{aux}},\Sigma^{p}_{\mathrm{aux}}}$.

(3) Take $\mathcal{H}'_{\mathrm{aux}}$ to be $\mathcal{H}^{p}_{\mathrm{aux}}\mathrm{G}_{\mathrm{aux}}(\mathbb{Z}_p)$, and take the $\Sigma^{\mathrm{ord},\prime}_{\mathrm{aux}}$ also induced by $\Sigma^{p}_{\mathrm{aux}}$. Then, the quasi-finite morphism $\vec{\mathsf{M}}^{\mathrm{ord,tor}}_{\mathcal{H}_{\mathrm{aux}},\Sigma^{\mathrm{ord}}_{\mathrm{aux}}} \to \mathsf{M}^{\mathrm{tor}}_{\mathcal{H}^{p}_{\mathrm{aux}},\Sigma^{p}_{\mathrm{aux}}}$ factors through an open immersion $\vec{\mathsf{M}}^{\mathrm{ord,tor}}_{\mathcal{H}'_{\mathrm{aux}},\Sigma^{\mathrm{ord},\prime}_{\mathrm{aux}}} \hookrightarrow \mathsf{M}^{\mathrm{tor}}_{\mathcal{H}^{p}_{\mathrm{aux}},\Sigma^{p}_{\mathrm{aux}}}$, and there is an induced open immersion $\vec{\mathsf{M}}^{\mathrm{ord,min}}_{\mathcal{H}'_{\mathrm{aux}}} \hookrightarrow \mathsf{M}^{\mathrm{min}}_{\mathcal{H}^{p}_{\mathrm{aux}}}$. This shows that $\vec{\mathsf{M}}^{\mathrm{ord,min}}_{\mathcal{H}'_{\mathrm{aux}}}$ is quasi-projective over $\vec{\mathsf{S}}_{0,r_{\mathcal{H}_{\mathrm{aux}}}}$.

(4) We can use the theory of degeneration to show that $\vec{\mathsf{M}}^{\mathrm{ord,tor}}_{\mathcal{H}'_{\mathrm{aux}},\Sigma^{\mathrm{ord},\prime}_{\mathrm{aux}}}$ is the precise preimage of $\vec{\mathsf{M}}^{\mathrm{ord,min}}_{\mathcal{H}'_{\mathrm{aux}}}$ under the proper morphism $\mathsf{M}^{\mathrm{tor}}_{\mathcal{H}^{p}_{\mathrm{aux}},\Sigma^{p}_{\mathrm{aux}}} \to \mathsf{M}^{\mathrm{min}}_{\mathcal{H}^{p}_{\mathrm{aux}}}$. This shows what $\vec{\mathsf{M}}^{\mathrm{ord,tor}}_{\mathcal{H}'_{\mathrm{aux}},\Sigma^{\mathrm{ord},\prime}_{\mathrm{aux}}} \to \vec{\mathsf{M}}^{\mathrm{ord,min}}_{\mathcal{H}'_{\mathrm{aux}}}$ is proper and surjective.

(5) By studying the fibers of the quasi-finite morphism $\vec{\mathsf{M}}^{\mathrm{ord,tor}}_{\mathcal{H}_{\mathrm{aux}},\Sigma^{\mathrm{ord}}_{\mathrm{aux}}} \to \vec{\mathsf{M}}^{\mathrm{ord,tor}}_{\mathcal{H}'_{\mathrm{aux}},\Sigma^{\mathrm{ord},\prime}_{\mathrm{aux}}}$, we also obtain the properness of the morphism $\vec{\mathsf{M}}^{\mathrm{ord,tor}}_{\mathcal{H}_{\mathrm{aux}},\Sigma^{\mathrm{ord}}_{\mathrm{aux}}} \to \vec{\mathsf{M}}^{\mathrm{ord,min}}_{\mathcal{H}_{\mathrm{aux}}}$. The usual Stein factorization argument then shows that $\vec{\mathsf{M}}^{\mathrm{ord,min}}_{\mathcal{H}_{\mathrm{aux}}}$ is embedded as an open subscheme in $\vec{\mathsf{M}}^{\mathrm{min}}_{\mathcal{H}_{\mathrm{aux}},r_{\mathcal{H}}}$. This shows that $\vec{\mathsf{M}}^{\mathrm{ord,min}}_{\mathcal{H}_{\mathrm{aux}}}$ is quasi-projective over $\vec{\mathsf{S}}_{0,r_{\mathcal{H}}}$.

(6) Under the assumption that $\mathcal{H}$ and $\mathcal{H}_{\mathrm{aux}}$ are equally deep at p, by refining Σ^{ord} if necessary (so that it is compatible with $\Sigma^{\mathrm{ord}}_{\mathrm{aux}}$), we obtain a proper morphism $\vec{\mathsf{M}}^{\mathrm{ord,tor}}_{\mathcal{H},\Sigma^{\mathrm{ord}}} \to \vec{\mathsf{M}}^{\mathrm{ord,tor}}_{\mathcal{H}_{\mathrm{aux}},\Sigma^{\mathrm{ord}}_{\mathrm{aux}}}$. Then we can finish the proof by the usual Stein factorization argument.

These are done in Sections 5.2.3 and 6.1.1. (When we show such quasi-projectivity, we need the noncanonically constructed total models above.)

If $\mathcal{H}^p$ is neat and if Σ^{ord} is (smooth and) projective, then we can construct $\vec{\mathsf{M}}^{\mathrm{ord,tor}}_{\mathcal{H},\Sigma^{\mathrm{ord}}} \to \vec{\mathsf{M}}^{\mathrm{ord,min}}_{\mathcal{H}}$ as the normalization of a blow-up, as explained above, which implies that $\vec{\mathsf{M}}^{\mathrm{ord,tor}}_{\mathcal{H},\Sigma^{\mathrm{ord}}}$ is quasi-projective over $\vec{\mathsf{S}}_{0,r_{\mathcal{H}}}$ in this case. This is done in Section 6.2.3.

Other Constructions

As in [61], the partial toroidal compactifications of *ordinary* Kuga families (and their generalizations) are realized as closures of toroidal boundary strata in the partial toroidal boundary of ordinary loci for a *larger* $\mathsf{M}_{\mathcal{H}}$. This is done in Chapter 7. (See also Sections 1.3.2, 5.2.4, and 7.1.2, where we collect geometric objects appearing along the toroidal boundaries of $\mathsf{M}_{\mathcal{H}}$ and generalized Kuga families, and interpret them as universal spaces for certain degeneration data.)

The constructions of automorphic bundles and their canonical and subcanonical extensions in mixed characteristics are delicate because the group is ramified at p, but some ad hoc constructions are still possible. They are carried out in Chapter 8 (extending the more canonical theory in characteristic zero in Section 1.4).

The constructions of Hecke actions are scattered in Sections 2.2.3, 3.4.4, 5.2.2, 7.2.7, 8.1.4, and 8.3.6, using various universal properties (of moduli problems, normalizations, universal spaces for degenerations, *etc*). They are all based on the same idea of modifying the tautological abelian or semi-abelian schemes by quasi-isogenies (and by forgetting part of the data on the level structures) which we call *Hecke twists*.

On the p-adic completions of the total models of integral models we constructed, we also compare the ordinary loci we use (which are the loci where canonical subgroups can be defined and rigidified by linear algebraic data) and the ordinary loci defined by the subscheme whose geometric points define ordinary abelian varieties on the auxiliary model. (In particular, we provide a simple criterion which guarantees that our theory is not empty in the applications we have in mind.) This is done in Section 6.3.

0.4 What is Known, What is New, and What Can Be Studied Next

Let us relate our techniques of construction to what is known in the literature.

Ordinary Level Structures and "Balanced $\Gamma_1(p^r)$ Levels"

The consideration of the ordinary loci carrying canonical subgroups is influenced by [46] and works of Hida (see, for example, the book [41] and the citations there).

Our use of "balanced $\Gamma_1(p^r)$ levels", and the strategy of studying structures near infinity (*i.e.*, the cusps) after adding sufficiently many roots of unity, are both influenced by Katz and Mazur [49]. We do not know whether "balanced $\Gamma_1(p^r)$ levels" have been seriously considered in general.

In this work we only consider the ordinary loci where the (multiplicative-type) canonical subgroups can be defined. If we also consider maximal totally isotropic subgroups which admits a filtration with graded pieces given by groups of étale and multiplicative (but no other) types, then we can extend the definition of ordinary loci and have a richer theory. The "balanced $\Gamma_1(p^r)$ levels" should be the ideal context for studying such "full ordinary loci". It is also possible to consider ordinary level structures of increasing depth along a flag of subgroup schemes. However, both of these require much heavier notation. We have chosen not to carry this out, because it complicates an already lengthy story.

Theory of Degeneration and Partial Toroidal Compactifications

The theory of degeneration in this work is built on those developed in [82], [28], and [62]. In order to study the ordinary level structures without the assumption that p is good, which means, in particular, that the polarization degree might not be prime to p, we introduced the "balanced Γ_1 levels" and studied the ordinary level structures on the abelian scheme and its dual in a parallel way. This is consistent with the fact that the theory of degeneration data is also "balanced" in the sense that most objects in the theory of degeneration appear in pairs (one for the degenerating abelian scheme, one for its dual).

Our boundary construction for $\vec{\mathsf{M}}^{\mathrm{ord,tor}}_{\mathcal{H},\Sigma^{\mathrm{ord}}}$ is heavily built on [62], which include considerations not readily available in [28]. Since we take normalizations of certain naive models in the models in characteristic zero algebraically constructed in [62], our work requires [62] but does not replace it. (The ordinary level structures in characteristic p is hardly more complicated than the principal levels in characteristic zero. The main reason we need to take normalizations is because of the ramification at p. We first introduced some of the ideas of working with arbitrary ramifications in [58], which we further developed in this work.) Our method of gluing (for the construction of $\vec{\mathsf{M}}^{\mathrm{ord,tor}}_{\mathcal{H},\Sigma^{\mathrm{ord}}}$) is the same as in [28] and [62].

Our construction of the partial toroidal compactifications of Kuga families (and their generalizations) is the same as that of [61], which is different from that of [28]. (It is not clear that [61] and [28, Ch. VI] even construct the same objects.) It is close in spirit to the construction of toroidal compactifications of mixed Shimura varieties in [89], although the construction techniques can hardly be directly compared. (The construction in [89] has arithmetic quotients of symmetric spaces as local charts, which has been developed along the lines of [5] and [4]. On the other hand, the purely algebraic construction in [61] is based on the theory of degeneration in [62]. It was not until [59] that we know these two constructions are compatible.)

The same techniques in this work allow the generalization of the theory of degeneration and the boundary construction to the "full ordinary loci" mentioned above, although one will need to add more roots of unity to the base rings. They also allow the generalization to the case of ordinary level structures of increasing depth along a flag of subgroup schemes. However, as we mentioned above, both will require much heavier notation. We have chosen not to carry them out even though the method is almost identical.

There are still many other cases where one can consider the construction of (total or partial) toroidal compactifications. Our rather simple-minded techniques do not seem to be useful when one seriously considers the nonordinary loci. (See the following discussion on local models.)

Local Models

We learned the Serre–Tate deformation theory of ordinary abelian varieties from [73] and [47]. Together with the deformation theory in the good reduction case in [62] (with no levels at p), these are all that we need for our main constructions.

Although we allow ramification and level structures at p, our consideration is disjoint from the theory of *local models* (involving also nonordinary abelian varieties) in, for example, [91], [85], [86], [87], and [88]. In general, our integral models of Shimura varieties are not even the same. By giving up the moduli interpretation, we obtain normality and flatness for free, but we no longer have enough information about the nonordinary loci.

We note that Stroh's constructions of compactifications of the Siegel moduli with parahoric levels at p (generalizing the "$\Gamma_0(p)$ levels"; see [97], [98], [99]), unlike ours, used the same integral models as in the works mentioned in the previous paragraph, and indeed used results from the theory of local model to deduce the normality he needs. The strength of his work is that he also considered the nonordinary loci. (If the ordinary loci is all one wants, one can just take the normalization of some relatively representable moduli problems of canonical subgroups over the toroidal compactifications with no level at p. In the Siegel case, there is a nice "bottom level" to start with, with no ramification at p at all.) However, since it is unclear to us what "$\Gamma_1(p^r)$ levels" mean at the nonordinary loci (especially) when $r > 1$, we have not generalized his work to the higher levels we want.

Nevertheless, there are special cases where our models at the "bottom level" at p indeed agree with the ones considered in, for example, [85] and [86], in which case we can also describe the local structures of the nonordinary loci of the boundary. For example, we can show that certain toroidal and minimal compactifications are normal and Cohen–Macaulay, and have geometrically normal reductions mod p. (Some modification of our constructions would also allow us to study collections of isogenies defining parahoric levels.) See [65] for more details. (See also [68] and [64] for some more recent improvements.)

After all, over the nonordinary loci, there is still very little we know, and there is ample room for further investigations. We believe that some new ideas might be needed, not just for solving the known difficult problems in the theory of local models, but also for seeing whether one should fundamentally revise the way we construct integral models.

Use of Auxiliary Models

The technical idea of using auxiliary models (such as the Siegel moduli) to study models of Shimura varieties (which are, a priori, analytically defined double coset spaces) has a long history. In characteristic zero, this can be traced back to the work of Shimura and Deligne (see [19] and [21], and their references).

In mixed characteristic, Carayol [13] defined integral models of Shimura curves using integral models of unitary Shimura varieties, including the bad reduction case. For Hodge-type Shimura varieties, one can find such an idea in, for example, [78], [100], [79], [51], and [74].

In the PEL-type cases, with arbitrarily high levels at p, our approach is closer in spirit to the classical work of Deligne and Rapoport [24], in which models with higher levels at p are simply constructed as normalizations. This is the same approach taken in [15] (in which the model with no level at p is constructed by [90] and [23]).

It is fair to say that we are influenced by both. (It is hard not to know the latter because of our upbringing; it is hard not to have heard of the former because of the current fashion trend.) One should keep in mind that we need the auxiliary models mainly as a source of quasi-projectivity—which is otherwise difficult to obtain!

Adelic Language and Mixed Shimura Varieties

The collections of geometric objects we construct do carry Hecke actions, as we have painstakingly gone through their constructions in all relevant sections, but our descriptions of them are somewhat indirect. For many applications, it is also desirable to adopt a language closer to the adelic formulations of double cosets (as in, for example, the theory for mixed Shimura varieties in [89]). We have chosen not to fully carry this out, mainly because in our proofs in mixed characteristics (especially for showing the universal properties of the partial toroidal compactifications, and for showing the quasi-projectivity of the partial minimal compactifications) we need the theory of degeneration, and we want to be able to cite the available results in our previous work [62] without much reformulation or generalization. But we certainly agree that it is helpful to develop a more convenient language after the proofs are done. We leave this as a potential future development. (We believe that, since we have shown that the algebraic picture in mixed characteristics is analogous to the complex analytic picture in characteristic zero, such a task can be done by a person with no knowledge of the theory of degeneration. There is no logical reason that the proofs and the applications have to be in the same language.)

Roots of Unity in the Base Rings

To obtain nice models in mixed characteristics, we added roots of unity to the base rings and performed normalizations whenever needed. We have made the effort to keep track of the precise exponents of roots of unity we need, but in practice, in mixed characteristics $(0, p)$, it might be much easier to add all p-power roots of unity at once. We traded this convenience with some notational complication, partly because in many cases we only need roots of unity of a bounded exponent (and sometimes none at all), and it is still desirable to have a precise formula for such bounds.

0.5 What to Note and to Skip in Special Cases

Readers might naturally wonder whether some of the considerations can be ignored or more easily addressed, or whether the constructions can be shortened or simplified, in some special cases. In what follows we list the sections or subsections that can be skipped in each typical special case, and remark about some convenient special facts. (Certainly, in each of these cases, the work can be further shortened, at least typographically, by simplifying the notation system.)

When p is a Good Prime

The readers can safely skip Sections 2.1.1, 2.1.2, and 6.3.3, and most of Sections 3.1, 3.4.6, 6.1.1, 8.1, and 8.3, because most of the statements or constructions there can be easily achieved using the "bottom level" at p, which is the hyperspecial good reduction case already been explained in [62] and [61]. Whenever the auxiliary models are mentioned, the readers can safely assume that they are in the maximal hyperspecial good reduction case.

Moreover, the reader should focus on the sections titled "The case when p is a good prime," as they provide substantial shortcuts to the various constructions. For example, by Lemma 5.2.3.2, for most levels of practical interest, and for cone decompositions that are admissible also for a level hyperspecial at p defining a good reduction model, the partial toroidal compactifications can be easily constructed over the good reduction model as a relatively representable moduli problem of ordinary level structures.

When the Pairing is Self-Dual at p

In Sections 2.1.1, 2.1.2, 3.4.6, and 6.1.1, the reader can safely assume that L and L_{aux} have exactly the same size, because no Zarhin's trick is really necessary.

The Siegel Cases

For Siegel cases defined for abelian schemes with principal polarization, it is as in the case above when p is a good prime. (There are many other simplifications possible, but it is less clear how we should give the instructions on them.) Moreover, the nonemptiness of the ordinary loci on the characteristic p fibers is trivial.

For Siegel cases defined for abelian schemes without principal polarizations, Zarhin's trick is used in our work, and hence the auxiliary models are still essential. However, one can ignore all treatments concerning the ramification of p in $\mathcal{O}$.

In both cases, it is possible to describe the cone decompositions using a simpler combinatorial language. For example, it is simpler to focus on the point boundary strata (of the partial minimal compactification) given by the rational parabolic subgroup of "maximal rank" (with abelian unipotent radical, and with Levi the product of a general linear group with $\mathbf{G}_{\mathrm{m}}$). Whether trivially true or not, Corollary 6.3.3.2 shows that the ordinary loci on the characteristic p fiber is nonempty.

The "Easier" Unitary Cases

By "easier" unitary groups we mean unitary groups defined by a Hermitian pairing over an imaginary quadratic or CM field F, but not over a noncommutative semisimple algebra. In addition to the above (concerning whether p is good, whether the pairing is self-dual, *etc*), the main simplification possible is that it is also possible to describe the cone decompositions using a simpler combinatorial language. For example, it is also simpler to focus on the point boundary strata (of the partial minimal compactification) given by the rational parabolic subgroup of "maximal rank" (with possibly non-abelian unipotent radical).

All Cases with "Siegel Parabolics"

By a "Siegel parabolic" subgroup we mean a rational parabolic subgroup with an abelian unipotent radical. (These include all kinds of cases involving general semisimple algebras with positive involutions. We do not just consider the Siegel and "easier" unitary cases.) In such cases, the nonemptiness of the ordinary loci on the characteristic p fibers follows from Corollary 6.3.3.2. (Note that many of these are cases with nonquasisplit groups.)

0.6 Notation and Conventions

We shall follow [62, Notation and Conventions] unless otherwise specified. By symplectic isomorphisms between modules with symplectic pairings, we *always* mean isomorphisms between the modules matching the pairings up to an invertible scalar multiple. (These are often called symplectic similitudes, but our understanding is that the codomains of pairings are modules rather than rings, which ought to be matched as well.) Sheaves on schemes, algebraic spaces, or (Deligne–Mumford) algebraic stacks are étale sheaves by default, although for coherent sheaves on schemes it would suffice to work in the Zariski topology.

0.7 Acknowledgements

I would like to thank (in alphabetical order) George Boxer, Brian Conrad, Christopher Skinner, Richard Taylor, Jack Thorne, and Yichao Tian for helpful comments and corrections. I would also like to thank the following people and institutions for the hospitality during my visits: Jian-Shu Li and Shou-Wu Zhang, and the Institute for Advanced Study of the Hong Kong University of Science and Technology (HKUST IAS); Iku Nakamura and Hokkaido University; Richard Taylor and the Institute for Advanced Study (in Princeton); and Chia-Fu Yu, and the National Center for Theoretical Sciences (NCTS) and Academia Sinica (in Taipei).

During the preparation of this work, my research was partially supported by the Learned Society Travel Fund at Princeton University; by a setup funding at the University of Minnesota, Twin Cities; by the National Science Foundation under agreements Nos. DMS-1069154, DMS-1258962, and DMS-1352216; and by an Alfred P. Sloan Research Fellowship. Any opinions, findings, and conclusions or recommendations expressed here are those of mine, and do not necessarily reflect the views of the funding organizations.

Chapter 1

Theory in Characteristic Zero

In this chapter, we review the main definitions and results in [62] and [61] (based on earlier results of others) and specialize them to the case of characteristic zero bases (over the reflex fields). (Also, we take this opportunity to correct or improve certain assertions in [62] and [61].) Readers who are already familiar with these results should feel free to skip this chapter (and return to here only for references).

However, despite the similarity, the theory developed in [19], [21], [5], [4], [38], [89], *etc* (based on arithmetic quotients of Hermitian symmetric domains, for which the compactifications were constructed by gluing using the analytic coordinates) are not directly related to the results reviewed here in this chapter (based on the moduli of polarized abelian varieties, for which the compactifications were constructed by gluing using the theory of degeneration). (It is not completely obvious that the two kinds of theories are compatible along the boundary; see [59].) Some readers might find the definitions and results in this chapter unfamiliar, and might want to at least glance over the notation system and running assumptions.

1.1 PEL-type Moduli Problems and Shimura Varieties

1.1.1 *Linear Algebraic Data for PEL Structures*

Let us begin with the usual (rational) PEL data, which suffice for the definition of complex analytic PEL-type Shimura varieties and their attached moduli problems in characteristic zero or in every *good* characteristic as in [53] (see Definition 1.1.1.6 below).

Let $\mathbb{Z}(1) := \ker(\exp : \mathbb{C} \to \mathbb{C}^\times) = (2\pi\sqrt{-1})\mathbb{Z}$, which is a free $\mathbb{Z}$-module of rank one. Each square-root $\sqrt{-1}$ of -1 in $\mathbb{C}$ determines an isomorphism $(2\pi\sqrt{-1})^{-1}$: $\mathbb{Z}(1) \xrightarrow{\sim} \mathbb{Z}$, but there is no canonical isomorphism between $\mathbb{Z}(1)$ and $\mathbb{Z}$. For each $\mathbb{Z}$-module M, we denote by $M(1)$ the module $M \underset{\mathbb{Z}}{\otimes} \mathbb{Z}(1)$, called the *Tate twist* of M. Note that $M(1)$ and M are noncanonically isomorphic as $\mathbb{Z}$-modules.

For the construction of compactifications using the theory of degeneration in [28] and [62], it is useful to start with a (noncanonical) choice of an *integral PEL datum*:

Definition 1.1.1.1. *An integral PEL datum is a tuple* $(\mathcal{O}, \star, L, \langle\,\cdot\,,\,\cdot\,\rangle, h_0)$ *consisting* *of:*

(1) *An order* $\mathcal{O}$ *in a finite-dimensional semisimple* $\mathbb{Q}$*-algebra with a positive involution* $\star$ *stabilizing* $\mathcal{O}$*. We shall denote the center of* $\mathcal{O} \underset{\mathbb{Z}}{\otimes} \mathbb{Q}$ *by* F*. (Then* F *is a product of number fields.)*

(2) *An* $\mathcal{O}$*-lattice* L*; namely, a finite free* $\mathbb{Z}$*-module* L *with the structure of an* $\mathcal{O}$*-module.*

(3) *An alternating pairing* $\langle \cdot , \cdot \rangle : L \times L \to \mathbb{Z}(1)$ *satisfying* $\langle bx, y\rangle = \langle x, b^\star y\rangle$ *for all* $x, y \in L$ *and* $b \in \mathcal{O}$*, together with an* $\mathbb{R}$*-algebra homomorphism* $h_0 : \mathbb{C} \to \mathrm{End}_{\mathcal{O} \underset{\mathbb{Z}}{\otimes} \mathbb{R}}(L \underset{\mathbb{Z}}{\otimes} \mathbb{R})$*, satisfying:*

   (a) *For all* $z \in \mathbb{C}$ *and* $x, y \in L \underset{\mathbb{Z}}{\otimes} \mathbb{R}$*, we have* $\langle h_0(z)x, y\rangle = \langle x, h_0(z^c)y\rangle$*, where* $z \mapsto z^c$ *is complex conjugation.*

   (b) *The* $\mathbb{R}$*-bilinear pairing* $(2\pi\sqrt{-1})^{-1}\langle \cdot , h_0(\sqrt{-1}) \cdot \rangle$ *on* $L \underset{\mathbb{Z}}{\otimes} \mathbb{R}$ *is (symmetric and) positive definite. (See [62, Def. 1.2.1.3], where* h_0 *was denoted by* h*.)*

Such a tuple $(\mathcal{O}, \star, L, \langle \cdot , \cdot \rangle, h_0)$ is an integral version of the PEL datum $(B, \star, V, \langle \cdot , \cdot \rangle, h_0)$ in [53] and related works.

Definition 1.1.1.2. *The dual lattice* $L^\#$ *of* L *(with respect to the pairing* $\langle \cdot , \cdot \rangle$*) is*

$$L^\# := \{x \in L \underset{\mathbb{Z}}{\otimes} \mathbb{Q} : \langle x, y\rangle \in \mathbb{Z}(1), \forall y \in L\}.$$

One advantage of making the choice of an integral datum is that it fixes the choice of an integral model of the algebraic reductive group in the usual definition of Shimura varieties:

Definition 1.1.1.3. *(See [62, Def. 1.2.1.6].) Let* $\mathcal{O}$ *and* $(L, \langle \cdot , \cdot \rangle)$ *be given as above. For each* $\mathbb{Z}$*-algebra* R*, set*

$$\mathrm{G}(R) := \left\{ \begin{array}{c} (g, r) \in \mathrm{GL}_{\mathcal{O} \underset{\mathbb{Z}}{\otimes} R}(L \underset{\mathbb{Z}}{\otimes} R) \times \mathbf{G}_{\mathrm{m}}(R) : \\ \langle gx, gy\rangle = r\langle x, y\rangle, \ \forall x, y \in L \underset{\mathbb{Z}}{\otimes} R \end{array} \right\}.$$

The assignment is functorial in R *and defines a group functor* G *over* $\mathrm{Spec}(\mathbb{Z})$*. The projection to the second factor* $(g, r) \mapsto r$ *defines a homomorphism* $\nu : \mathrm{G} \to \mathbf{G}_{\mathrm{m}}$*, which we call the* **similitude character***. For simplicity, we shall often denote elements* (g, r) *in* G *by simply* g*, and denote by* $\nu(g)$ *the value of* r *when we need it. (This is an abuse of notation, because the value of* r *is not always determined by* g*.)*

Then we have, for each rational prime number $p > 0$, definitions for $\mathrm{G}(\mathbb{Q})$, $\mathrm{G}(\mathbb{R})$, $\mathrm{G}(\mathbb{A}^\infty)$, $\mathrm{G}(\mathbb{A}^{\infty,p})$, $\mathrm{G}(\mathbb{A})$, $\mathrm{G}(\mathbb{A}^p)$, $\mathrm{G}(\mathbb{Z})$, $\mathrm{G}(\mathbb{Z}/n\mathbb{Z})$, $\mathrm{G}(\hat{\mathbb{Z}})$, $\mathrm{G}(\hat{\mathbb{Z}}^p)$,

$$\mathcal{U}(n) := \ker(\mathrm{G}(\hat{\mathbb{Z}}) \to \mathrm{G}(\hat{\mathbb{Z}}/n\hat{\mathbb{Z}}) = \mathrm{G}(\mathbb{Z}/n\mathbb{Z}))$$

for each integer $n \geq 1$,

$$\mathcal{U}^p(n_0) := \ker(\mathrm{G}(\hat{\mathbb{Z}}^p) \to \mathrm{G}(\hat{\mathbb{Z}}^p/n_0\hat{\mathbb{Z}}^p) = \mathrm{G}(\mathbb{Z}/n_0\mathbb{Z}))$$

for each integer $n_0 \geq 1$ prime to p.

The homomorphism $h_0 : \mathbb{C} \to \mathrm{End}_{\mathcal{O} \underset{\mathbb{Z}}{\otimes} \mathbb{R}}(L \underset{\mathbb{Z}}{\otimes} \mathbb{R})$ defines a Hodge structure of weight -1 on L, with Hodge decomposition

$$L \underset{\mathbb{Z}}{\otimes} \mathbb{C} = V_0 \oplus V_0^c, \tag{1.1.1.4}$$

such that $h_0(z)$ acts by $1 \otimes z$ on V_0, and by $1 \otimes z^c$ on V_0^c. One can check easily that V_0 is (maximal) totally isotropic under the nondegenerate pairing $\langle \cdot, \cdot \rangle$, and hence (1.1.1.4) induces canonically an isomorphism

$$V_0^c \cong V_0^\vee(1) := \mathrm{Hom}_{\mathbb{C}}(V_0, \mathbb{C})(1). \tag{1.1.1.5}$$

Let F_0 be the reflex field of the $\mathcal{O} \underset{\mathbb{Z}}{\otimes} \mathbb{C}$-module V_0. Recall (see [53, p. 389] or [62, Def. 1.2.5.4]) that F_0 is the subfield of $\mathbb{C}$ generated over $\mathbb{Q}$ by the traces $\mathrm{Tr}_{\mathbb{C}}(b|V_0)$ for $b \in \mathcal{O}$.

Definition 1.1.1.6. *We say that a rational prime number $p > 0$ is* **good** *(for the integral PEL datum $(\mathcal{O}, \star, L, \langle \cdot, \cdot \rangle, h_0)$) if it satisfies the following conditions (cf. [53, Sec. 5] or [62, Def. 1.4.1.1]):*

(1) p is unramified in $\mathcal{O}$ (as in [62, Def. 1.1.1.18]).
(2) $p \neq 2$ if $\mathcal{O} \underset{\mathbb{Z}}{\otimes} \mathbb{Q}$ involves simple factors of type D *(as in [62, Def. 1.2.1.15]).*
(3) $p \nmid [L^\# : L]$ (see Definition 1.1.1.2*).*

When p is good, $\mathrm{G} \underset{\mathbb{Z}}{\otimes} \mathbb{Z}_p$ is smooth and unramified (*cf.* [62, Prop. 1.2.3.11 and Cor. 1.2.3.12]).

1.1.2 *PEL-type Moduli Problems*

Let $\mathcal{H}$ be an open compact subgroup of $\mathrm{G}(\hat{\mathbb{Z}})$.

By [62, Def. 1.4.1.4] (with $\Box = \emptyset$ there), the data of $(L, \langle \cdot, \cdot \rangle, h_0)$ and $\mathcal{H}$ define a moduli problem $\mathsf{M}_{\mathcal{H}}$ over $\mathsf{S}_0 = \mathrm{Spec}(F_0)$, parameterizing tuples $(A, \lambda, i, \alpha_{\mathcal{H}})$ over schemes S over S_0 of the following form:

(1) $A \to S$ is an abelian scheme.
(2) $\lambda : A \to A^\vee$ is a polarization.
(3) $i : \mathcal{O} \hookrightarrow \mathrm{End}_S(A)$ is an $\mathcal{O}$-endomorphism structure as in [62, Def. 1.3.3.1].
(4) $\underline{\mathrm{Lie}}_{A/S}$ with its $\mathcal{O} \underset{\mathbb{Z}}{\otimes} \mathbb{Q}$-module structure given naturally by i satisfies the determinantal condition in [62, Def. 1.3.4.1] given by $(L \underset{\mathbb{Z}}{\otimes} \mathbb{R}, \langle \cdot, \cdot \rangle, h_0)$.
(5) $\alpha_{\mathcal{H}}$ is an (integral) level-$\mathcal{H}$ structure of (A, λ, i) of type $(L \underset{\mathbb{Z}}{\otimes} \hat{\mathbb{Z}}, \langle \cdot, \cdot \rangle)$ as in [62, Def. 1.3.7.6].

Remark 1.1.2.1. By [62, Prop. 1.4.3.4], the definition agrees with the one in [53, Sec. 5] over $\mathsf{S}_0 = \mathrm{Spec}(F_0)$. The choice of L in $L \underset{\mathbb{Z}}{\otimes} \mathbb{Q}$ corresponds to the choice of a tautological (or *universal*) abelian scheme A over $\mathsf{M}_{\mathcal{H}}$ within its $\mathbb{Q}^\times$-isogeny class. If we have chosen another PEL-type $\mathcal{O}$-lattice L' in $L \underset{\mathbb{Z}}{\otimes} \mathbb{Q}$ which is also stabilized by $\mathcal{H}$, then we have the corresponding A' (with additional structures) over a moduli problem $\mathsf{M}'_{\mathcal{H}}$ canonically isomorphic to $\mathsf{M}_{\mathcal{H}}$ (see [62, Cor. 1.4.3.8]), together with a $\mathbb{Q}^\times$-isogeny $A \to A'$ (if we identify $\mathsf{M}'_{\mathcal{H}}$ with $\mathsf{M}_{\mathcal{H}}$). In brief, the $\mathbb{Q}^\times$-isogeny class of A is independent of the choice of L in $L \underset{\mathbb{Z}}{\otimes} \mathbb{Q}$. This is useful because every open compact subgroup of $\mathrm{G}(\mathbb{A}^\infty)$ stabilizes some PEL-type $\mathcal{O}$-lattice L', and for every two PEL-type $\mathcal{O}$-lattices L and L' there are common open compact subgroups of $\mathrm{G}(\mathbb{A}^\infty)$ stabilizing both lattices. Hence, we can form a collection $\{\mathsf{M}_{\mathcal{H}}\}_{\mathcal{H}}$, indexed by all open compact subgroups $\mathcal{H}$ of $\mathrm{G}(\mathbb{A}^\infty)$, not just those of $\mathrm{G}(\hat{\mathbb{Z}})$ (with a canonical action of $\mathrm{G}(\mathbb{A}^\infty)$; see [62, Rem. 1.4.3.11]).

By [62, Thm. 1.4.1.11 and Cor. 7.2.3.10], $\mathsf{M}_{\mathcal{H}}$ is an algebraic stack separated, smooth, and of finite type over S_0, which is representable by a scheme quasi-projective (and smooth) over S_0 when $\mathcal{H}$ is *neat*. (See [89, 0.6] or [62, Def. 1.4.1.8] for the definition of neatness.)

Let $(A, \lambda, i, \alpha_{\mathcal{H}}) \to \mathsf{M}_{\mathcal{H}}$ be the tautological tuple over $\mathsf{M}_{\mathcal{H}}$. Consider the *relative de Rham cohomology* $\underline{H}^1_{\mathrm{dR}}(A/\mathsf{M}_{\mathcal{H}})$, with the dual

$$\underline{H}_1^{\mathrm{dR}}(A/\mathsf{M}_{\mathcal{H}}) := \underline{\mathrm{Hom}}_{\mathscr{O}_{\mathsf{M}_{\mathcal{H}}}}(\underline{H}^1_{\mathrm{dR}}(A/\mathsf{M}_{\mathcal{H}}), \mathscr{O}_{\mathsf{M}_{\mathcal{H}}})$$

defined to be the *relative de Rham homology*. Consider the canonical pairing

$$\langle\,\cdot\,,\,\cdot\,\rangle_\lambda : \underline{H}_1^{\mathrm{dR}}(A/\mathsf{M}_{\mathcal{H}}) \times \underline{H}_1^{\mathrm{dR}}(A/\mathsf{M}_{\mathcal{H}}) \to \mathscr{O}_{\mathsf{M}_{\mathcal{H}}}(1) \tag{1.1.2.2}$$

defined by the pullback under $\mathrm{Id} \times \lambda_*$ of the canonical perfect pairing

$$\underline{H}_1^{\mathrm{dR}}(A/\mathsf{M}_{\mathcal{H}}) \times \underline{H}_1^{\mathrm{dR}}(A^\vee/\mathsf{M}_{\mathcal{H}}) \to \mathscr{O}_{\mathsf{M}_{\mathcal{H}}}(1)$$

defined by the first Chern class of the Poincaré invertible sheaf $\mathcal{P}_A$ over $A \underset{\mathsf{M}_{\mathcal{H}}}{\times} A^\vee$. (See, for example, [23, 1.5].) Since $\mathsf{M}_{\mathcal{H}}$ is defined over the characteristic zero base $\mathsf{S}_0 = \mathrm{Spec}(F_0)$, we know that λ is separable, that λ_* is an isomorphism, and hence that the pairing $\langle\,\cdot\,,\,\cdot\,\rangle_\lambda$ above is *perfect*. Let $\langle\,\cdot\,,\,\cdot\,\rangle_\lambda$ also denote the induced pairing on $\underline{H}^1_{\mathrm{dR}}(A/\mathsf{M}_{\mathcal{H}}) \times \underline{H}^1_{\mathrm{dR}}(A/\mathsf{M}_{\mathcal{H}})$ by duality. By [6, Lem. 2.5.3], we have canonical short exact sequences

$$0 \to \underline{\mathrm{Lie}}^\vee_{A^\vee/\mathsf{M}_{\mathcal{H}}}(1) \to \underline{H}_1^{\mathrm{dR}}(A/\mathsf{M}_{\mathcal{H}}) \to \underline{\mathrm{Lie}}_{A/\mathsf{M}_{\mathcal{H}}} \to 0$$

and

$$0 \to \underline{\mathrm{Lie}}^\vee_{A/\mathsf{M}_{\mathcal{H}}} \to \underline{H}^1_{\mathrm{dR}}(A/\mathsf{M}_{\mathcal{H}}) \to \underline{\mathrm{Lie}}_{A^\vee/\mathsf{M}_{\mathcal{H}}}(-1) \to 0.$$

The submodules $\underline{\mathrm{Lie}}^\vee_{A^\vee/\mathsf{M}_{\mathcal{H}}}$ and $\underline{\mathrm{Lie}}^\vee_{A/\mathsf{M}_{\mathcal{H}}}$ are maximal totally isotropic under the pairing $\langle\,\cdot\,,\,\cdot\,\rangle_\lambda$.

Remark 1.1.2.3. The Tate twists in $\underline{\mathrm{Lie}}^\vee_{A^\vee/\mathsf{M}_{\mathcal{H}}}(1)$ and $\underline{\mathrm{Lie}}_{A^\vee/\mathsf{M}_{\mathcal{H}}}(-1)$ are often omitted (and have also been omitted in most of this author's earlier writings).

They signify a sign convention in the (otherwise canonical) identification between $\underline{H}^1(A, \mathscr{O}_A)$ and $\underline{\mathrm{Lie}}_{A^\vee/\mathsf{M}_\mathcal{H}}$, which is nevertheless the same sign convention involved in the definition of the pairing (1.1.2.2). Hence, we shall carry such Tate twists in the notation when the pairing (1.1.2.2) is also involved. However, for the sake of simplicity, we shall still omit them when discussing the Gauss–Manin connections or Kodaira–Spencer morphisms below.

Let $\tilde{\mathsf{M}}_\mathcal{H}^{(m)}$ be the m-th infinitesimal neighborhood of the diagonal image of $\mathsf{M}_\mathcal{H}$ in $\mathsf{M}_\mathcal{H} \underset{\mathsf{S}_0}{\times} \mathsf{M}_\mathcal{H}$, and let $\mathrm{pr}_1, \mathrm{pr}_2 : \tilde{\mathsf{M}}_\mathcal{H}^{(m)} \to \mathsf{M}_\mathcal{H}$ be the two projections. Then we have by definition the canonical morphism $\mathscr{O}_{\mathsf{M}_\mathcal{H}} \to \mathscr{P}^m_{\mathsf{M}_\mathcal{H}/\mathsf{S}_0} := \mathrm{pr}_{1,*} \mathrm{pr}_2^*(\mathscr{O}_{\mathsf{M}_\mathcal{H}})$. The isomorphism $s : \tilde{\mathsf{M}}_\mathcal{H}^{(m)} \to \tilde{\mathsf{M}}_\mathcal{H}^{(m)}$ over $\mathsf{M}_\mathcal{H}$ swapping two components of the fiber product then defines an automorphism s^* of $\mathscr{P}^m_{\mathsf{M}_\mathcal{H}/\mathsf{S}_0}$. When $m = 1$, the kernel of the structural morphism $\mathrm{str}^* : \mathscr{P}^1_{\mathsf{M}_\mathcal{H}/\mathsf{S}_0} \to \mathscr{O}_{\mathsf{M}_\mathcal{H}}$, canonically isomorphic to $\Omega^1_{\mathsf{M}_\mathcal{H}/\mathsf{S}_0}$ by definition, is spanned by the image of $s^* - \mathrm{Id}^*$ (induced by $\mathrm{pr}_1^* - \mathrm{pr}_2^*$).

An important property of the relative de Rham cohomology of a smooth morphism like $A \to \mathsf{M}_\mathcal{H}$ is that, for every two smooth lifts $\tilde{A}_1 \to \tilde{\mathsf{M}}_\mathcal{H}^{(1)}$ and $\tilde{A}_2 \to \tilde{\mathsf{M}}_\mathcal{H}^{(1)}$ of $A \to \mathsf{M}_\mathcal{H}$, there is a canonical isomorphism $\underline{H}^1_{\mathrm{dR}}(\tilde{A}_2/\tilde{\mathsf{M}}_\mathcal{H}^{(1)}) \overset{\sim}{\to} \underline{H}^1_{\mathrm{dR}}(\tilde{A}_1/\tilde{\mathsf{M}}_\mathcal{H}^{(1)})$ lifting the identity morphism on $\underline{H}^1_{\mathrm{dR}}(A/\mathsf{M}_\mathcal{H})$. (See, for example, [62, Prop. 2.1.6.4].) If we consider $\tilde{A}_1 := \mathrm{pr}_1^* A$ and $\tilde{A}_2 := \mathrm{pr}_2^* A$, then we obtain a canonical $\mathrm{pr}_2^* \underline{H}^1_{\mathrm{dR}}(A/\mathsf{M}_\mathcal{H}) \cong \underline{H}^1_{\mathrm{dR}}(\mathrm{pr}_2^* A/\tilde{\mathsf{M}}_\mathcal{H}^{(1)}) \overset{\sim}{\to} \underline{H}^1_{\mathrm{dR}}(\mathrm{pr}_1^* A/\tilde{\mathsf{M}}_\mathcal{H}^{(1)}) \cong \mathrm{pr}_1^* \underline{H}^1_{\mathrm{dR}}(A/\mathsf{M}_\mathcal{H})$, which we denote by Id^* by abuse of notation. On the other hand, pulling back by the swapping automorphism $s : \tilde{\mathsf{M}}_\mathcal{H}^{(1)} \overset{\sim}{\to} \tilde{\mathsf{M}}_\mathcal{H}^{(1)}$ defines another canonical isomorphism $s^* :\ \mathrm{pr}_2^* \underline{H}^1_{\mathrm{dR}}(A/\mathsf{M}_\mathcal{H}) \cong \underline{H}^1_{\mathrm{dR}}(\mathrm{pr}_2^* A/\tilde{\mathsf{M}}_\mathcal{H}^{(1)}) \overset{\sim}{\to} \underline{H}^1_{\mathrm{dR}}(\mathrm{pr}_1^* A/\tilde{\mathsf{M}}_\mathcal{H}^{(1)}) \cong \mathrm{pr}_1^* \underline{H}^1_{\mathrm{dR}}(A/\mathsf{M}_\mathcal{H})$. This allows us to define the Gauss–Manin connection as follows (*cf.* [62, Rem. 2.1.7.4]):

Definition 1.1.2.4. *The* **Gauss–Manin connection**

$$\nabla : \underline{H}^1_{\mathrm{dR}}(A/\mathsf{M}_\mathcal{H}) \to \underline{H}^1_{\mathrm{dR}}(A/\mathsf{M}_\mathcal{H}) \underset{\mathscr{O}_{\mathsf{M}_\mathcal{H}}}{\otimes} \Omega^1_{\mathsf{M}_\mathcal{H}/\mathsf{S}_0} \tag{1.1.2.5}$$

on $\underline{H}^1_{\mathrm{dR}}(A/\mathsf{M}_\mathcal{H})$ *is the composition*

$$\underline{H}^1_{\mathrm{dR}}(A/\mathsf{M}_\mathcal{H}) \overset{\mathrm{pr}_2^*}{\to} \underline{H}^1_{\mathrm{dR}}(\mathrm{pr}_2^* A/\tilde{\mathsf{M}}_\mathcal{H}^{(1)}) \overset{s^* - \mathrm{Id}^*}{\to} \underline{H}^1_{\mathrm{dR}}(A/\mathsf{M}_\mathcal{H}) \underset{\mathscr{O}_{\mathsf{M}_\mathcal{H}}}{\otimes} \Omega^1_{\mathsf{M}_\mathcal{H}/\mathsf{S}_0}.$$

Definition 1.1.2.6. *The composition (ignoring Tate twists; see Remark* 1.1.2.3*)*

$$\begin{aligned} \underline{\mathrm{Lie}}^\vee_{A/\mathsf{M}_\mathcal{H}} &\hookrightarrow \underline{H}^1_{\mathrm{dR}}(A/\mathsf{M}_\mathcal{H}) \\ &\overset{\nabla}{\to} \underline{H}^1_{\mathrm{dR}}(A/\mathsf{M}_\mathcal{H}) \underset{\mathscr{O}_{\mathsf{M}_\mathcal{H}}}{\otimes} \Omega^1_{\mathsf{M}_\mathcal{H}/\mathsf{S}_0} \twoheadrightarrow \underline{\mathrm{Lie}}_{A^\vee/\mathsf{M}_\mathcal{H}} \underset{\mathscr{O}_{\mathsf{M}_\mathcal{H}}}{\otimes} \Omega^1_{\mathsf{M}_\mathcal{H}/\mathsf{S}_0} \end{aligned}$$

defines by duality a morphism

$$\mathrm{KS}_{A/\mathsf{M}_\mathcal{H}/\mathsf{S}_0} : \underline{\mathrm{Lie}}^\vee_{A/\mathsf{M}_\mathcal{H}} \underset{\mathscr{O}_{\mathsf{M}_\mathcal{H}}}{\otimes} \underline{\mathrm{Lie}}^\vee_{A^\vee/\mathsf{M}_\mathcal{H}} \to \Omega^1_{\mathsf{M}_\mathcal{H}/\mathsf{S}_0}, \tag{1.1.2.7}$$

which we call the **Kodaira–Spencer morphism**. *(This definition is compatible with the definition by deformation theory in [62, Def. 2.1.7.9].)*

Definition 1.1.2.8. *(See [62, Def. 2.3.5.1].) The sheaf*

$$\underline{\mathrm{KS}}_{(A,\lambda,i)/\mathsf{M}_{\mathcal{H}}} := \underline{\mathrm{KS}}_{(A,\lambda,i,\alpha_{\mathcal{H}})/\mathsf{M}_{\mathcal{H}}}$$

is the quotient

$$(\underline{\mathrm{Lie}}^{\vee}_{A/\mathsf{M}_{\mathcal{H}}} \underset{\mathscr{O}_{\mathsf{M}_{\mathcal{H}}}}{\otimes} \underline{\mathrm{Lie}}^{\vee}_{A^{\vee}/\mathsf{M}_{\mathcal{H}}})/\begin{pmatrix} \lambda^*(y)\otimes z - \lambda^*(z)\otimes y \\ i(b)^*(x)\otimes y - x\otimes(i(b)^{\vee})^*(y) \end{pmatrix}_{\substack{x\in\underline{\mathrm{Lie}}^{\vee}_{A/\mathsf{M}_{\mathcal{H}}}, \\ y,z\in\underline{\mathrm{Lie}}^{\vee}_{A^{\vee}/\mathsf{M}_{\mathcal{H}}}, \\ b\in\mathcal{O}.}}$$

According to [62, Prop. 2.3.5.2], we have:

Proposition 1.1.2.9. *The Kodaira–Spencer morphism* (1.1.2.7) *factors through the canonical quotient* $\underline{\mathrm{Lie}}^{\vee}_{A/\mathsf{M}_{\mathcal{H}}} \underset{\mathscr{O}_{\mathsf{M}_{\mathcal{H}}}}{\otimes} \underline{\mathrm{Lie}}^{\vee}_{A^{\vee}/\mathsf{M}_{\mathcal{H}}} \twoheadrightarrow \underline{\mathrm{KS}}_{(A,\lambda,i)/\mathsf{M}_{\mathcal{H}}}$ *and induces an isomorphism*

$$\underline{\mathrm{KS}}_{(A,\lambda,i)/\mathsf{M}_{\mathcal{H}}} \xrightarrow{\sim} \Omega^1_{\mathsf{M}_{\mathcal{H}}/\mathsf{S}_0}, \tag{1.1.2.10}$$

which we call the **Kodaira–Spencer isomorphism**, *and denote again (by abuse of notation) by* $\mathrm{KS}_{A/\mathsf{M}_{\mathcal{H}}/\mathsf{S}_0}$.

1.1.3 *PEL-type Shimura Varieties*

Consider the (real analytic) set $\mathsf{X} = \mathrm{G}(\mathbb{R})h_0$ of $\mathrm{G}(\mathbb{R})$-conjugates $h : \mathbb{C} \to \mathrm{End}_{\mathcal{O}\underset{\mathbb{Z}}{\otimes}\mathbb{R}}(L \underset{\mathbb{Z}}{\otimes} \mathbb{R})$ of $h_0 : \mathbb{C} \to \mathrm{End}_{\mathcal{O}\underset{\mathbb{Z}}{\otimes}\mathbb{R}}(L \underset{\mathbb{Z}}{\otimes} \mathbb{R})$. It is well known (see [53, Sec. 8] or [59, Sec. 2]) that there exists a quasi-projective variety $\mathsf{Sh}_{\mathcal{H}}$ over F_0, together with a canonical open and closed immersion

$$\mathsf{Sh}_{\mathcal{H}} \hookrightarrow [\mathsf{M}_{\mathcal{H}}] \tag{1.1.3.1}$$

over $\mathsf{S}_0 = \mathrm{Spec}(F_0)$, where $[\,\cdot\,]$ denotes the coarse moduli space of an algebraic stack (see [62, Sec. A.7.5]), such that the analytification of $\mathsf{Sh}_{\mathcal{H}} \underset{F_0}{\otimes} \mathbb{C}$ (as a complex analytic space) can be canonically identified with the double coset space $\mathrm{G}(\mathbb{Q})\backslash\mathsf{X}\times\mathrm{G}(\mathbb{A}^{\infty})/\mathcal{H}$. (Note that $\mathsf{Sh}_{\mathcal{H}} \hookrightarrow [\mathsf{M}_{\mathcal{H}}]$ is not an isomorphism in general, due to the so-called *failure of Hasse's principle*. See, for example, [53, Sec. 8] and [62, Rem. 1.4.3.12].)

We call both $\mathsf{Sh}_{\mathcal{H}}$ and $\mathsf{Sh}_{\mathcal{H}} \underset{F_0}{\otimes} \mathbb{C}$ the *PEL-type Shimura variety* of level $\mathcal{H}$ associated to the integral PEL datum $(\mathcal{O}, \star, L, \langle\,\cdot\,,\,\cdot\,\rangle, h_0)$. (More precisely, we should call $\mathsf{Sh}_{\mathcal{H}} \underset{F_0}{\otimes} \mathbb{C}$ the complex PEL-type Shimura variety, and call $\mathsf{Sh}_{\mathcal{H}}$ the canonical model.) We will not emphasis their roles in our constructions from now on.

1.2 Linear Algebraic Data for Cusps

1.2.1 *Cusp Labels*

For technical reasons, we shall impose the following technical condition on all PEL-type $\mathcal{O}$-lattices we use:

Condition 1.2.1.1. (*See [62, Cond. 1.4.3.10].*) *The PEL-type $\mathcal{O}$-lattice $(L, \langle \cdot , \cdot \rangle, h_0)$ is chosen such that the action of $\mathcal{O}$ on L extends to an action of some maximal order $\mathcal{O}'$ in $\mathcal{O} \underset{\mathbb{Z}}{\otimes} \mathbb{Q}$ containing $\mathcal{O}$.*

Although there is no *rational boundary components* in the algebraic theory of toroidal and minimal compactifications (constructed by the theory of degeneration, as in [28] and in [62]), we have developed in [62, Sec. 5.4] the notion of *cusp labels* that serves a similar purpose. (While $\mathrm{G}(\mathbb{Q})$ plays an important role in the analytic theory over $\mathbb{C}$, it does not play any obvious role in the algebraic theory of degeneration.)

Unlike in the analytic theory over $\mathbb{C}$, where boundary components are naturally parameterized by group-theoretic objects, the only algebraic machinery we have is the theory of semi-abelian degenerations of abelian varieties with PEL structures. The cusp labels are (by their very design) part of the parameters (which we call the *degeneration data*) for such (semi-abelian) degenerations.

Definition 1.2.1.2. (*See [62, Sec. 1.2.6].*) *Let R be any noetherian $\mathbb{Z}$-algebra. Suppose we have an increasing filtration $\mathtt{F} = \{\mathtt{F}_{-i}\}$ on $L \underset{\mathbb{Z}}{\otimes} R$, indexed by nonpositive integers $-i$, such that $\mathtt{F}_0 = L \underset{\mathbb{Z}}{\otimes} R$.*

(1) We say that $\mathtt{F}$ is **integrable** *if, for every i, $\mathrm{Gr}^{\mathtt{F}}_{-i} := \mathtt{F}_{-i}/\mathtt{F}_{-i-1}$ is integrable in the sense that $\mathrm{Gr}^{\mathtt{F}}_{-i} \cong M_i \underset{\mathbb{Z}}{\otimes} R$ (as R-modules) for some $\mathcal{O}$-lattice M_i.*

(2) We say that $\mathtt{F}$ is **split** *if there exists (noncanonically) some isomorphism $\mathrm{Gr}^{\mathtt{F}} := \underset{i}{\oplus} \mathrm{Gr}^{\mathtt{F}}_{-i} \overset{\sim}{\to} \mathtt{F}_0$ of R-modules.*

(3) We say that $\mathtt{F}$ is **admissible** *if it is both integrable and split.*

(4) Let m be an integer. We say that $\mathtt{F}$ is **m-symplectic with respect to** *$(L, \langle \cdot , \cdot \rangle)$ if, for every i, $\mathtt{F}_{-m+i}$ and $\mathtt{F}_{-i}$ are annihilators of each other under the pairing $\langle \cdot , \cdot \rangle$ on $\mathtt{F}_0$.*

We shall only work with $m = 3$, and we shall suppress m in what follows. The fact that $\hat{\mathbb{Z}}$ (or $\hat{\mathbb{Z}}^p$ for some rational prime number p) almost always involves *bad primes* (*cf.* Definition 1.1.1.6) is the main reason that we may have to allow nonprojective filtrations.

Definition 1.2.1.3. (*See [62, Def. 5.2.7.1].*) *We say that a symplectic admissible filtration $\mathtt{Z}$ on $L \underset{\mathbb{Z}}{\otimes} \hat{\mathbb{Z}}$ is* **fully symplectic with respect to** *$(L, \langle \cdot , \cdot \rangle)$ if there is a symplectic admissible filtration $\mathtt{Z}_{\mathbb{A}} = \{\mathtt{Z}_{-i,\mathbb{A}}\}$ on $L \underset{\mathbb{Z}}{\otimes} \mathbb{A}$ that* **extends** *$\mathtt{Z}$ in the sense that $\mathtt{Z}_{-i,\mathbb{A}} \cap (L \underset{\mathbb{Z}}{\otimes} \hat{\mathbb{Z}}) = \mathtt{Z}_{-i}$ in $L \underset{\mathbb{Z}}{\otimes} \mathbb{A}$ for all i.*

Definition 1.2.1.4. (*See [62, Def. 5.2.7.3].*) *A symplectic-liftable admissible filtration $\mathtt{Z}_n$ on L/nL is called* **fully symplectic-liftable** *with respect to $(L, \langle \cdot , \cdot \rangle)$ if it is the reduction modulo n of some admissible filtration $\mathtt{Z}$ on $L \underset{\mathbb{Z}}{\otimes} \hat{\mathbb{Z}}$ that is fully symplectic with respect to $(L, \langle \cdot , \cdot \rangle)$ as in Definition* 1.2.1.3.

Degenerations into semi-abelian schemes induce filtrations on Tate modules and on Lie algebras of the generic fibers. While the symplectic-liftable admissible filtrations represent (certain orbits of) filtrations on $L \underset{\mathbb{Z}}{\otimes} \hat{\mathbb{Z}}$ induced by filtrations on Tate modules via the level structures, the fully symplectic-liftable ones are equipped with (certain orbits of) filtrations on $L \underset{\mathbb{Z}}{\otimes} \mathbb{R}$ induced by the filtrations on Lie algebras via the Lie algebra condition (see Section 1.1.2). (One may interpret the Lie algebra condition as the "de Rham" (or rather "Hodge") component of a certain "complete level structure", the direct product of whose "ℓ-adic" components being a level structure in the usual sense.) Such (orbits of) filtrations are the crudest invariants of degenerations we consider.

Definition 1.2.1.5. *(See [62, Def. 5.4.1.3].) Given a fully symplectic admissible filtration* Z *on* $L \underset{\mathbb{Z}}{\otimes} \hat{\mathbb{Z}}$ *with respect to* $(L, \langle \cdot, \cdot \rangle)$ *as in Definition* 1.2.1.3, *a* **torus argument** *for* Z *is a tuple*

$$\Phi = (X, Y, \phi, \varphi_{-2}, \varphi_0),$$

where the entries are as follows:

(1) X *and* Y *are* $\mathcal{O}$*-lattices of the same* $\mathcal{O}$*-multi-rank (see [62, Def. 5.2.2.6]), and* $\phi : Y \hookrightarrow X$ *is an* $\mathcal{O}$*-equivariant embedding.*

(2) $\varphi_{-2} : \mathrm{Gr}^{\mathrm{Z}}_{-2} \xrightarrow{\sim} \mathrm{Hom}_{\hat{\mathbb{Z}}}(X \underset{\mathbb{Z}}{\otimes} \hat{\mathbb{Z}}, \hat{\mathbb{Z}}(1))$ *and* $\varphi_0 : \mathrm{Gr}^{\mathrm{Z}}_0 \xrightarrow{\sim} Y \underset{\mathbb{Z}}{\otimes} \hat{\mathbb{Z}}$ *are isomorphisms (of* $\mathcal{O} \underset{\mathbb{Z}}{\otimes} \hat{\mathbb{Z}}$*-modules) such that the pairing* $\langle \cdot, \cdot \rangle_{20} : \mathrm{Gr}^{\mathrm{Z}}_{-2} \times \mathrm{Gr}^{\mathrm{Z}}_0 \to \hat{\mathbb{Z}}(1)$ *defined by* Z *is the pullback of the pairing*

$$\langle \cdot, \cdot \rangle_\phi : \mathrm{Hom}_{\hat{\mathbb{Z}}}(X \underset{\mathbb{Z}}{\otimes} \hat{\mathbb{Z}}, \hat{\mathbb{Z}}(1)) \times (Y \underset{\mathbb{Z}}{\otimes} \hat{\mathbb{Z}}) \to \hat{\mathbb{Z}}(1)$$

defined by the composition

$$\begin{aligned} &\mathrm{Hom}_{\hat{\mathbb{Z}}}(X \underset{\mathbb{Z}}{\otimes} \hat{\mathbb{Z}}, \hat{\mathbb{Z}}(1)) \times (Y \underset{\mathbb{Z}}{\otimes} \hat{\mathbb{Z}}) \\ &\stackrel{\mathrm{Id} \times \phi}{\to} \mathrm{Hom}_{\hat{\mathbb{Z}}}(X \underset{\mathbb{Z}}{\otimes} \hat{\mathbb{Z}}, \hat{\mathbb{Z}}(1)) \times (X \underset{\mathbb{Z}}{\otimes} \hat{\mathbb{Z}}) \to \hat{\mathbb{Z}}(1), \end{aligned}$$

with the sign convention that $\langle \cdot, \cdot \rangle_\phi(x, y) = x(\phi(y)) = (\phi(y))(x)$ *for all* $x \in \mathrm{Hom}_{\hat{\mathbb{Z}}}(X \underset{\mathbb{Z}}{\otimes} \hat{\mathbb{Z}}, \hat{\mathbb{Z}}(1))$ *and* $y \in Y \underset{\mathbb{Z}}{\otimes} \hat{\mathbb{Z}}$.

Definition 1.2.1.6. *(See [62, Def. 5.4.1.4 and 5.4.1.5].) Given a fully symplectic-liftable admissible filtration* Z_n *on* L/nL *with respect to* $(L, \langle \cdot, \cdot \rangle)$ *as in Definition* 1.2.1.4, *a* **torus argument at level** n *for* Z_n *is a tuple*

$$\Phi_n = (X, Y, \phi, \varphi_{-2,n}, \varphi_{0,n}),$$

where:

(1) X *and* Y *are* $\mathcal{O}$*-lattices of the same* $\mathcal{O}$*-multi-rank, and* $\phi : Y \hookrightarrow X$ *is an* $\mathcal{O}$*-equivariant embedding.*

(2) $\varphi_{-2,n} : \mathrm{Gr}^{\mathsf{Z}}_{-2,n} \overset{\sim}{\to} \mathrm{Hom}(X/nX, (\mathbb{Z}/n\mathbb{Z})(1))$ (*resp.* $\varphi_{0,n} : \mathrm{Gr}^{\mathsf{Z}}_{0,n} \overset{\sim}{\to} Y/nY$) *is an isomorphism that is the reduction modulo* n *of some isomorphism* $\varphi_{-2} : \mathrm{Gr}^{\mathsf{Z}}_{-2} \overset{\sim}{\to} \mathrm{Hom}_{\hat{\mathbb{Z}}}(X \underset{\mathbb{Z}}{\otimes} \hat{\mathbb{Z}}, \hat{\mathbb{Z}}(1))$ (*resp.* $\varphi_0 : \mathrm{Gr}^{\mathsf{Z}}_0 \overset{\sim}{\to} (Y \underset{\mathbb{Z}}{\otimes} \hat{\mathbb{Z}})$), *such that* $\Phi = (X, Y, \phi, \varphi_{-2}, \varphi_0)$ *is a torus argument as in Definition* 1.2.1.5.
We say in this case that Φ_n *is the reduction modulo* n *of* Φ.

Two torus arguments $\Phi_n = (X, Y, \phi, \varphi_{-2,n}, \varphi_{0,n})$ *and* $\Phi'_n = (X', Y', \phi', \varphi'_{-2,n}, \varphi'_{0,n})$ *at level* n *are* **equivalent** *if there exists a pair of isomorphisms* $(\gamma_X : X' \overset{\sim}{\to} X, \gamma_Y : Y \overset{\sim}{\to} Y')$ *(of* $\mathcal{O}$*-lattices) such that* $\phi = \gamma_X \phi' \gamma_Y$, $\varphi'_{-2,n} = {}^t\gamma_X \varphi_{-2,n}$, *and* $\varphi'_{0,n} = \gamma_Y \varphi_{0,n}$. *In this case, we say that* Φ_n *and* Φ'_n *are equivalent under the pair of isomorphisms* $\gamma = (\gamma_X, \gamma_Y)$, *which we denote by* $\gamma = (\gamma_X, \gamma_Y) : \Phi_n \overset{\sim}{\to} \Phi'_n$.

The torus arguments record the isomorphism classes of the torus parts of degenerations of abelian schemes with PEL structures. These are the second crudest invariants of degenerations we consider.

Definition 1.2.1.7. (*See [62, Def. 5.4.1.9].*) *A* **(principal) cusp label at level** n *for a PEL-type* $\mathcal{O}$*-lattice* $(L, \langle \cdot, \cdot \rangle, h_0)$, *or a cusp label of the moduli problem* M_n, *is an equivalence class* $[(\mathsf{Z}_n, \Phi_n, \delta_n)]$ *of triples* $(\mathsf{Z}_n, \Phi_n, \delta_n)$, *where:*

(1) Z_n *is an admissible filtration on* L/nL *that is fully symplectic-liftable in the sense of Definition* 1.2.1.4.
(2) Φ_n *is a torus argument at level* n *for* Z_n.
(3) $\delta_n : \mathrm{Gr}^{\mathsf{Z}}_n \overset{\sim}{\to} L/nL$ *is a liftable splitting.*

Two triples $(\mathsf{Z}_n, \Phi_n, \delta_n)$ *and* $(\mathsf{Z}'_n, \Phi'_n, \delta'_n)$ *are equivalent if* Z_n *and* Z'_n *are identical, and if* Φ_n *and* Φ'_n *are equivalent as in Definition* 1.2.1.6.

The liftable splitting δ_n in each triple $(\mathsf{Z}_n, \Phi_n, \delta_n)$ is noncanonical and auxiliary in nature. Such splittings are needed for analyzing the "degeneration of pairings" in general PEL cases (unlike in the special case in Faltings–Chai [28, Ch. IV, Sec. 6]).

To proceed from principal cusp labels at level n to general cusp labels at level $\mathcal{H}$, where $\mathcal{H}$ is an open compact subgroup of $\mathrm{G}(\hat{\mathbb{Z}})$, we form *étale orbits* of the objects we have thus defined. The precise definitions are complicated (see [62, Def. 5.4.2.1, 5.4.2.2, and 5.4.2.4]) but the idea is simple: For each $\mathcal{H}$ as above, consider those $n \geq 1$ sufficiently divisible such that $\mathcal{U}(n) \subset \mathcal{H}$. Then we have a compatible system of finite groups $H_n = \mathcal{H}/\mathcal{U}(n)$, and an object at level $\mathcal{H}$ is simply defined to be a compatible system of étale H_n-orbits of objects at running levels n as above. Then we arrive at the notions of *torus arguments* $\Phi_{\mathcal{H}} = (X, Y, \phi, \varphi_{-2,\mathcal{H}}, \varphi_{0,\mathcal{H}})$ *at level* $\mathcal{H}$, and of *representatives* $(\mathsf{Z}_{\mathcal{H}}, \Phi_{\mathcal{H}}, \delta_{\mathcal{H}})$ *of cusp labels* $[(\mathsf{Z}_{\mathcal{H}}, \Phi_{\mathcal{H}}, \delta_{\mathcal{H}})]$ *at level* $\mathcal{H}$. (The liftability condition is implicit in such a definition, as in the definition of level structures we omitted.) By abuse of language, we call these $\mathcal{H}$-orbits of $\Phi = (X, Y, \phi, \varphi_{-2}, \varphi_0)$, $(\mathsf{Z}, \Phi, \delta)$, and $[(\mathsf{Z}, \Phi, \delta)]$, respectively. (Note that the splitting δ was denoted $\hat{\delta}$ in [62, Sec. 5.2.2].)

For simplicity, we shall often omit $\mathsf{Z}_{\mathcal{H}}$ from the notation.

Lemma 1.2.1.8. *(See [62, Lem. 5.2.7.5].) Let* Z_n *be an admissible filtration on* L/nL *that is fully symplectic-liftable with respect to* $(L, \langle\,\cdot\,,\,\cdot\,\rangle)$. *Let* $(\mathrm{Gr}^{\mathsf{Z}}_{-1}, \langle\,\cdot\,,\,\cdot\,\rangle_{11})$ *be induced by some fully symplectic lifting* Z *of* Z_n, *and let* $(\mathrm{Gr}^{\mathsf{Z}}_{-1,\mathbb{R}}, \langle\,\cdot\,,\,\cdot\,\rangle_{11,\mathbb{R}}, (h_0)_{-1})$ *be determined by [62, Prop. 5.1.2.2] by any extension* $\mathsf{Z}_{\mathbb{A}}$ *in Definition* 1.2.1.3 (*which has the same reflex field* F_0 *as* $(L \underset{\mathbb{Z}}{\otimes} \mathbb{R}, \langle\,\cdot\,,\,\cdot\,\rangle, h_0)$ *does*). *Then there is associated* (*noncanonically*) *a PEL-type* $\mathcal{O}$*-lattice* $(L^{\mathsf{Z}_n}, \langle\,\cdot\,,\,\cdot\,\rangle^{\mathsf{Z}_n}, h_0^{\mathsf{Z}_n})$ *satisfying Condition* 1.2.1.1 *such that we have the following:*

(1) *There exist* (*noncanonical*) $\mathcal{O}$*-equivariant isomorphisms*
$$(\mathrm{Gr}^{\mathsf{Z}}_{-1}, \langle\,\cdot\,,\,\cdot\,\rangle_{11}) \xrightarrow{\sim} (L^{\mathsf{Z}_n} \underset{\mathbb{Z}}{\otimes} \hat{\mathbb{Z}}, \langle\,\cdot\,,\,\cdot\,\rangle^{\mathsf{Z}_n})$$
(*over* $\hat{\mathbb{Z}}$) *and*
$$(\mathrm{Gr}^{\mathsf{Z}}_{-1,\mathbb{R}}, \langle\,\cdot\,,\,\cdot\,\rangle_{11,\mathbb{R}}, (h_0)_{-1}) \xrightarrow{\sim} (L^{\mathsf{Z}_n} \underset{\mathbb{Z}}{\otimes} \mathbb{R}, \langle\,\cdot\,,\,\cdot\,\rangle^{\mathsf{Z}_n}, h_0^{\mathsf{Z}_n})$$
(*over* $\mathbb{R}$).

(2) *The moduli problem* $\mathsf{M}_n^{\mathsf{Z}_n}$ *defined by the noncanonical* $(L^{\mathsf{Z}_n}, \langle\,\cdot\,,\,\cdot\,\rangle^{\mathsf{Z}_n}, h_0^{\mathsf{Z}_n})$ *as in Section* 1.1.2 *is canonical in the sense that it depends* (*up to isomorphism*) *only on* Z_n, *but not on the choice of* $(L^{\mathsf{Z}_n}, \langle\,\cdot\,,\,\cdot\,\rangle^{\mathsf{Z}_n}, h_0^{\mathsf{Z}_n})$.

In fact, Lemma 1.2.1.8 (or rather [62, Lem. 5.2.7.5]) is based on [62, Rem. 5.2.7.2], which asserts the existence of a (noncanonical) boundary lattice $(L^{\mathsf{Z}}, \langle\,\cdot\,,\,\cdot\,\rangle^{\mathsf{Z}}, h_0^{\mathsf{Z}})$ for each fully symplectic admissible filtration Z of $L \underset{\mathbb{Z}}{\otimes} \hat{\mathbb{Z}}$, so that we have, in particular, a canonical isomorphism $\mathrm{Gr}^{\mathsf{Z}}_{-1} \underset{\hat{\mathbb{Z}}}{\otimes} R \cong L \underset{\mathbb{Z}}{\otimes} R$ for each $\hat{\mathbb{Z}}$-algebra R. With any fixed (noncanonical) choice of such $(L^{\mathsf{Z}}, \langle\,\cdot\,,\,\cdot\,\rangle^{\mathsf{Z}}, h_0^{\mathsf{Z}})$, we can make the following:

Definition 1.2.1.9. *We define the group functor* $\mathrm{G}^{\mathsf{Z}} = \mathrm{G}_{(L^{\mathsf{Z}}, \langle\,\cdot\,,\,\cdot\,\rangle^{\mathsf{Z}})}$ *by* $(L^{\mathsf{Z}}, \langle\,\cdot\,,\,\cdot\,\rangle^{\mathsf{Z}})$ *as in Definition* 1.1.1.3, *so that* $\mathrm{G}_{h,\mathsf{Z}} := \mathrm{G}^{\mathsf{Z}} \underset{\mathbb{Z}}{\otimes} \hat{\mathbb{Z}}$ *is well defined and depends only on* Z (*but not on the choice of* $(L^{\mathsf{Z}}, \langle\,\cdot\,,\,\cdot\,\rangle^{\mathsf{Z}}, h_0^{\mathsf{Z}})$).

Definition 1.2.1.10. *For each* $\hat{\mathbb{Z}}$*-algebra* R, *let* $\mathrm{P}_{\mathsf{Z}}(R)$ *denote the subgroup of* $\mathrm{G}(R)$ *consisting of elements* g *such that* $g(\mathsf{Z}_{-2} \underset{\hat{\mathbb{Z}}}{\otimes} R) = \mathsf{Z}_{-2} \underset{\hat{\mathbb{Z}}}{\otimes} R$ *and* $g(\mathsf{Z}_{-1} \underset{\hat{\mathbb{Z}}}{\otimes} R) = \mathsf{Z}_{-1} \underset{\hat{\mathbb{Z}}}{\otimes} R$. *Each element* g *in* $\mathrm{P}_{\mathsf{Z}}(R)$ *defines an isomorphism* $\mathrm{Gr}^{\mathsf{Z}}_{-i}(g) : \mathrm{Gr}^{\mathsf{Z}}_{-i} \underset{\hat{\mathbb{Z}}}{\otimes} R \xrightarrow{\sim} \mathrm{Gr}^{\mathsf{Z}}_{-i} \underset{\hat{\mathbb{Z}}}{\otimes} R$ *for each* $0 \leq i \leq 2$. *Then, under the isomorphism* $\mathrm{Gr}^{\mathsf{Z}}_{-1} \underset{\hat{\mathbb{Z}}}{\otimes} R \cong L \underset{\mathbb{Z}}{\otimes} R$ *above, the isomorphism* $\mathrm{Gr}^{\mathsf{Z}}_{-1}(g)$ *corresponds to an element of* $\mathrm{G}_{h,\mathsf{Z}}(R)$, *and define a group homomorphism* $\mathrm{Gr}^{\mathsf{Z}}_{-1} : \mathrm{P}_{\mathsf{Z}}(R) \to \mathrm{G}_{h,\mathsf{Z}}(R)$.

Definition 1.2.1.11. *For each* $\hat{\mathbb{Z}}$*-algebra* R, *we also define the following quotients of subgroups of* $\mathrm{P}_{\mathsf{Z}}(R)$ (*see Definition* 1.2.1.10):

(1) $\mathrm{Z}_{\mathsf{Z}}(R)$ *is the kernel of the canonical homomorphism* $\mathrm{Gr}^{\mathsf{Z}}_{-1} : \mathrm{P}_{\mathsf{Z}}(R) \to \mathrm{G}_{h,\mathsf{Z}}(R)$. *Then any splitting* δ *as above canonically induces an isomorphism* $\mathrm{P}_{\mathsf{Z}}(R) \cong \mathrm{G}_{h,\mathsf{Z}}(R) \ltimes \mathrm{Z}_{\mathsf{Z}}(R)$.

(2) $\mathrm{U_Z}(R)$ *is the subgroup of* $\mathrm{P_Z}(R)$ *consisting of elements* g *such that* $\mathrm{Gr^Z}(g) = \mathrm{Id}_{\mathrm{Gr^Z} \underset{\hat{\mathbb{Z}}}{\otimes} R}$ *(i.e.,* $\mathrm{Gr^Z_{-i}}(g) = \mathrm{Id}_{\mathrm{Gr^Z_{-i}} \underset{\hat{\mathbb{Z}}}{\otimes} R}$ *for all* i*).*

(3) $\mathrm{U_{2,Z}}(R)$ *is the subgroup of* $\mathrm{P_Z}(R)$ *consisting of elements* g *which induces* $\mathrm{Id}_{\mathsf{Z}_{-1} \underset{\hat{\mathbb{Z}}}{\otimes} R}$ *and* $\mathrm{Id}_{(\mathsf{Z}_0 \underset{\hat{\mathbb{Z}}}{\otimes} R)/(\mathsf{Z}_{-2} \underset{\hat{\mathbb{Z}}}{\otimes} R)}$ *on* $\mathsf{Z}_{-1} \underset{\hat{\mathbb{Z}}}{\otimes} R$ *and* $(\mathsf{Z}_0 \underset{\hat{\mathbb{Z}}}{\otimes} R)/(\mathsf{Z}_{-2} \underset{\hat{\mathbb{Z}}}{\otimes} R)$*, respectively. (Using any splitting* δ *as above, this means* $\delta^{-1} \circ g \circ \delta$ *is of the form* $\begin{pmatrix} 1 & & g_{20} \\ & 1 & \\ & & 1 \end{pmatrix}$ *for some* $g_{20} \in \mathrm{Hom}_{\mathcal{O}}(\mathrm{Gr^Z_0} \underset{\hat{\mathbb{Z}}}{\otimes} R, \mathrm{Gr^Z_{-2}} \underset{\hat{\mathbb{Z}}}{\otimes} R)$*.)*

(4) $\mathrm{U_{1,Z}}(R) := \mathrm{U_Z}(R)/\mathrm{U_{2,Z}}(R)$.

(5) $\mathrm{G}_{l,\mathsf{Z}}(R) := \mathrm{Z_Z}(R)/\mathrm{U_Z}(R)$*, which is canonically isomorphic to the subgroup* $\mathrm{G}'_{l,\mathsf{Z}}(R)$ *of* $\mathrm{GL}_{\mathcal{O}}(\mathrm{Gr^Z_{-2}} \underset{\hat{\mathbb{Z}}}{\otimes} R) \times \mathrm{GL}_{\mathcal{O}}(\mathrm{Gr^Z_0} \underset{\hat{\mathbb{Z}}}{\otimes} R)$ *consisting of elements compatible with the morphism* $\mathrm{Gr^Z_0} \to \mathrm{Hom}_{\hat{\mathbb{Z}}}(\mathrm{Gr^Z_{-2}}, \hat{\mathbb{Z}}(1))$ *induced by* $\langle \cdot, \cdot \rangle$ *(which are therefore the elements compatible with* $\phi : Y \hookrightarrow X$*,* $\varphi_{-2} : \mathrm{Gr^Z_{-2}} \xrightarrow{\sim} \mathrm{Hom}_{\hat{\mathbb{Z}}}(X \underset{\mathbb{Z}}{\otimes} \hat{\mathbb{Z}}, \hat{\mathbb{Z}}(1))$*, and* $\varphi_0 : \mathrm{Gr^Z_0} \xrightarrow{\sim} Y \underset{\mathbb{Z}}{\otimes} \hat{\mathbb{Z}}$ *for any torus argument* Φ *of* Z*; see Definition* 1.2.1.5*).*

(6) $\mathrm{P'_Z}(R)$ *is the kernel of the canonical homomorphism* $(\nu^{-1}\,\mathrm{Gr^Z_{-2}}, \mathrm{Gr^Z_0}) : \mathrm{P_Z}(R) \to \mathrm{G}'_{l,\mathsf{Z}}(R)$*; i.e., the subgroup of* $\mathrm{P_Z}(R)$ *consisting of elements* g *such that* $\mathrm{Gr^Z_{-2}}(g) = \nu(g)\,\mathrm{Id}_{\mathrm{Gr^Z_{-2}} \underset{\hat{\mathbb{Z}}}{\otimes} R}$ *and* $\mathrm{Gr^Z_0}(g) = \mathrm{Id}_{\mathrm{Gr^Z_0} \underset{\hat{\mathbb{Z}}}{\otimes} R}$*. Then any splitting* $\delta : \mathrm{Gr^Z} \xrightarrow{\sim} \mathsf{Z}$ *canonically induces an isomorphism* $\mathrm{P_Z}(R) \cong \mathrm{G}_{l,\mathsf{Z}}(R) \ltimes \mathrm{P'_Z}(R)$*.*

(7) $\mathrm{G_{1,Z}}(R) := \mathrm{P'_Z}(R)/\mathrm{U_{2,Z}}(R)$*, which is (under any splitting* δ *above) isomorphic to* $(\mathrm{G}_{h,\mathsf{Z}} \ltimes \mathrm{U_{1,Z}})(R) := \mathrm{G}_{h,\mathsf{Z}}(R) \ltimes \mathrm{U_{1,Z}}(R)$*.*

(8) $\mathrm{G}'_{h,\mathsf{Z}}(R) := \mathrm{G_{1,Z}}(R)/\mathrm{U_{1,Z}}(R) \cong \mathrm{P'_Z}(R)/\mathrm{U_Z}(R) \cong \mathrm{P_Z}(R)/\mathrm{Z_Z}(R) \cong \mathrm{G}_{h,\mathsf{Z}}(R)$.

Definition 1.2.1.12. *For each open compact subgroup* $\mathcal{H}$ *of* $\mathrm{G}(\hat{\mathbb{Z}})$*, we define:*

(1) $\mathcal{H}_{\mathrm{P_Z}} := \mathcal{H} \cap \mathrm{P_Z}(\hat{\mathbb{Z}})$.
(2) $\mathcal{H}_{\mathrm{Z_Z}} := \mathcal{H} \cap \mathrm{Z_Z}(\hat{\mathbb{Z}})$.
(3) $\mathcal{H}_{\mathrm{G}_{h,\mathsf{Z}}} := \mathcal{H}_{\mathrm{P_Z}}/\mathcal{H}_{\mathrm{Z_Z}}$.
(4) $\mathcal{H}_{\mathrm{U_Z}} := \mathcal{H} \cap \mathrm{U_Z}(\hat{\mathbb{Z}})$.
(5) $\mathcal{H}_{\mathrm{U_{2,Z}}} := \mathcal{H} \cap \mathrm{U_{2,Z}}(\hat{\mathbb{Z}})$.
(6) $\mathcal{H}_{\mathrm{U_{1,Z}}} := \mathcal{H}_{\mathrm{U_Z}}/\mathcal{H}_{\mathrm{U_{2,Z}}}$.
(7) $\mathcal{H}_{\mathrm{G}_{l,\mathsf{Z}}} := \mathcal{H}_{\mathrm{Z_Z}}/\mathcal{H}_{\mathrm{U_Z}}$.
(8) $\mathcal{H}_{\mathrm{P'_Z}} := \mathcal{H} \cap \mathrm{P'_Z}(\hat{\mathbb{Z}})$.
(9) $\mathcal{H}_{\mathrm{G_{1,Z}}} := \mathcal{H}_{\mathrm{P'_Z}}/\mathcal{H}_{\mathrm{U_{2,Z}}}$.
(10) $\mathcal{H}_{\mathrm{G}'_{h,\mathsf{Z}}} := \mathcal{H}_{\mathrm{P'_Z}}/\mathcal{H}_{\mathrm{U_Z}}$.
(11) $\mathcal{H}_{\mathrm{G}'_{l,\mathsf{Z}}} := \mathcal{H}_{\mathrm{P_Z}}/\mathcal{H}_{\mathrm{P'_Z}}$.

Then we have an exact sequence

$$1 \to \mathcal{H}_{\mathrm{U_{1,Z}}} \to \mathcal{H}_{\mathrm{G_{1,Z}}} \to \mathcal{H}_{\mathrm{G}'_{h,\mathsf{Z}}} \to 1 \tag{1.2.1.13}$$

compatible with the canonical exact sequence

$$1 \to \mathrm{U_{1,Z}}(\hat{\mathbb{Z}}) \to \mathrm{G_{1,Z}}(\hat{\mathbb{Z}}) \to \mathrm{G}'_{h,\mathsf{Z}}(\hat{\mathbb{Z}}) \to 1. \tag{1.2.1.14}$$

We shall also extend this definition to the cases of $\hat{\mathbb{Z}}^p$*- and* $\mathbb{Z}_p$*-valued groups above.*

Definition 1.2.1.15. (*See [62, Def. 5.4.2.6 and the errata].*) *The PEL-type* $\mathcal{O}$*-lattice* $(L^{\mathsf{Z}_{\mathcal{H}}}, \langle\,\cdot\,,\,\cdot\,\rangle^{\mathsf{Z}_{\mathcal{H}}}, h_0^{\mathsf{Z}_{\mathcal{H}}})$ *is a fixed (noncanonical) choice of any of the PEL-type* $\mathcal{O}$*-lattice* $(L^{\mathsf{Z}_n}, \langle\,\cdot\,,\,\cdot\,\rangle^{\mathsf{Z}_n}, h_0^{\mathsf{Z}_n})$ *in Lemma* 1.2.1.8 *for any element* Z_n *in any* Z_{H_n} *(in* $\mathsf{Z}_{\mathcal{H}} = \{\mathsf{Z}_{H_n}\}$*, a compatible collection of étale orbits* Z_{H_n} *at various levels* n *such that* $\mathcal{U}(n) \subset \mathcal{H}$*). The elements of* H_n *leaving* Z_n *invariant induce a subgroup of* $\mathrm{G}^{\mathsf{Z}}(\mathbb{Z}/n\mathbb{Z})$*. Let* $\mathcal{H}_h$ *be the preimage of this subgroup under* $\mathrm{G}_{(L^{\mathsf{Z}_n},\langle\,\cdot\,,\,\cdot\,\rangle^{\mathsf{Z}_n})}(\hat{\mathbb{Z}}) \twoheadrightarrow \mathrm{G}_{(L^{\mathsf{Z}_n},\langle\,\cdot\,,\,\cdot\,\rangle^{\mathsf{Z}_n})}(\mathbb{Z}/n\mathbb{Z})$ *(see Definition* 1.2.1.9*). Then we define* $\mathsf{M}_{\mathcal{H}_h}$ *to be the moduli problem defined by* $(L^{\mathsf{Z}_n}, \langle\,\cdot\,,\,\cdot\,\rangle^{\mathsf{Z}_n}, h^{\mathsf{Z}_n})$ *with level-*$\mathcal{H}_h$ *structures as in Lemma* 1.2.1.8*. (The isomorphism class of* $\mathsf{M}_{\mathcal{H}_h}$ *is well defined and independent of the choice of* $(L^{\mathsf{Z}_{\mathcal{H}}}, \langle\,\cdot\,,\,\cdot\,\rangle^{\mathsf{Z}_{\mathcal{H}}}, h^{\mathsf{Z}_{\mathcal{H}}}) = (L^{\mathsf{Z}_n}, \langle\,\cdot\,,\,\cdot\,\rangle^{\mathsf{Z}_n}, h^{\mathsf{Z}_n})$*.) We define* $\mathsf{M}_{\mathcal{H}}^{\Phi_{\mathcal{H}}}$ *to be the quotient of* $\coprod \mathsf{M}_n^{\mathsf{Z}_n}$ *by* H_n*, where the disjoint union is over representatives* $(\mathsf{Z}_n, \Phi_n, \delta_n)$ *(with the same* (X, Y, ϕ)*) in* $(\mathsf{Z}_{\mathcal{H}}, \Phi_{\mathcal{H}}, \delta_{\mathcal{H}})$*, which is finite étale over* $\mathsf{M}_{\mathcal{H}_h}$ *by construction. (The isomorphism class of* $\mathsf{M}_{\mathcal{H}}^{\Phi_{\mathcal{H}}}$ *is independent of the choice of* n *and the representatives* $(\mathsf{Z}_n, \Phi_n, \delta_n)$ *we use.) We then (abusively) define* $\mathsf{M}_{\mathcal{H}}^{\mathsf{Z}_{\mathcal{H}}}$ *to be the quotient of* $\mathsf{M}_{\mathcal{H}}^{\Phi_{\mathcal{H}}}$ *by the subgroup of* Γ_ϕ *stabilizing* $\Phi_{\mathcal{H}}$ *(whose action factors through a finite index subgroup), which depends only on the cusp label* $[(\mathsf{Z}_{\mathcal{H}}, \Phi_{\mathcal{H}}, \delta_{\mathcal{H}})]$*, but not on the choice of the representative* $(\mathsf{Z}_{\mathcal{H}}, \Phi_{\mathcal{H}}, \delta_{\mathcal{H}})$*. By construction, we have finite étale morphisms* $\mathsf{M}_{\mathcal{H}}^{\Phi_{\mathcal{H}}} \to \mathsf{M}_{\mathcal{H}}^{\mathsf{Z}_{\mathcal{H}}} \to \mathsf{M}_{\mathcal{H}_h}$ *(which can be identified with* $\mathsf{M}_{\mathcal{H}'_h} \to \mathsf{M}_{\mathcal{H}''_h} \to \mathsf{M}_{\mathcal{H}_h}$ *for some canonically determined open compact subgroups* $\mathcal{H}'_h \subset \mathcal{H}''_h \subset \mathcal{H}_h$*; see Lemmas* 1.3.2.1 *and* 1.3.2.5 *below).*

Such *boundary moduli problems* $\mathsf{M}_{\mathcal{H}}^{\mathsf{Z}_{\mathcal{H}}}$ are the fundamental building blocks in the construction of toroidal boundary charts for $\mathsf{M}_{\mathcal{H}}$. (They actually appear in the boundary of the minimal compactification of $\mathsf{M}_{\mathcal{H}}$, which we call *cusps*. They are parameterized by the cusp labels of $\mathsf{M}_{\mathcal{H}}$.)

It is important to study the relations among cusp labels of different multi-ranks.

Definition 1.2.1.16. (*See [62, Def. 5.4.1.14].*) *A* **surjection** $(\mathsf{Z}_n, \Phi_n, \delta_n) \twoheadrightarrow (\mathsf{Z}'_n, \Phi'_n, \delta'_n)$ *between representatives of cusp labels at level* n*, where* $\Phi_n = (X, Y, \phi, \varphi_{-2,n}, \varphi_{0,n})$ *and where* $\Phi'_n = (X', Y', \phi', \varphi'_{-2,n}, \varphi'_{0,n})$*, is a pair (of surjections)* $(s_X : X \twoheadrightarrow X', s_Y : Y \twoheadrightarrow Y')$ *(of* $\mathcal{O}$*-lattices) such that we have the following:*

(1) Both s_X *and* s_Y *are admissible surjections (i.e., with kernels defining filtrations that are admissible as in Definition* 1.2.1.2*), and they are compatible with* ϕ *and* ϕ' *in the sense that* $s_X\phi = \phi' s_Y$*.*

(2) $\mathsf{Z}'_{-2,n}$ *is an admissible submodule of* $\mathsf{Z}_{-2,n}$ *(i.e., defining an admissible filtration as in Definition* 1.2.1.2*), and the natural embedding* $\mathrm{Gr}^{\mathsf{Z}'}_{-2,n} \hookrightarrow \mathrm{Gr}^{\mathsf{Z}}_{-2,n}$ *satisfies* $\varphi_{-2,n} \circ (\mathrm{Gr}^{\mathsf{Z}'}_{-2,n} \hookrightarrow \mathrm{Gr}^{\mathsf{Z}}_{-2,n}) = s_X^* \circ \varphi'_{-2,n}$*.*

(3) $\mathsf{Z}_{-1,n}$ *is an admissible submodule of* $\mathsf{Z}'_{-1,n}$*, and the natural surjection* $\mathrm{Gr}^{\mathsf{Z}}_{0,n} \twoheadrightarrow \mathrm{Gr}^{\mathsf{Z}'}_{0,n}$ *satisfies* $s_Y \circ \varphi_{0,n} = \varphi'_{0,n} \circ (\mathrm{Gr}^{\mathsf{Z}}_{0,n} \twoheadrightarrow \mathrm{Gr}^{\mathsf{Z}'}_{0,n})$*.*

In this case, we write $s = (s_X, s_Y) : (\mathsf{Z}_n, \Phi_n, \delta_n) \twoheadrightarrow (\mathsf{Z}'_n, \Phi'_n, \delta'_n)$

By taking orbits as above, there is a corresponding notion for general cusp labels:

Definition 1.2.1.17. *(See [62, Def. 5.4.2.12].) A* **surjection** $(\mathsf{Z}_{\mathcal{H}}, \Phi_{\mathcal{H}}, \delta_{\mathcal{H}}) \twoheadrightarrow (\mathsf{Z}'_{\mathcal{H}}, \Phi'_{\mathcal{H}}, \delta'_{\mathcal{H}})$ *between representatives of cusp labels at level* $\mathcal{H}$*, where* $\Phi_{\mathcal{H}} = (X, Y, \phi, \varphi_{-2,\mathcal{H}}, \varphi_{0,\mathcal{H}})$ *and where* $\Phi'_{\mathcal{H}} = (X', Y', \phi', \varphi'_{-2,\mathcal{H}}, \varphi'_{0,\mathcal{H}})$*, is a pair (of surjections)* $s = (s_X : X \twoheadrightarrow X', s_Y : Y \twoheadrightarrow Y')$ *(of* $\mathcal{O}$*-lattices) such that we have the following:*

(1) Both s_X *and* s_Y *are admissible surjections, and they are compatible with* ϕ *and* ϕ' *in the sense that* $s_X\phi = \phi' s_Y$*.*
(2) $\mathsf{Z}'_{\mathcal{H}}$ *and* $(\varphi'_{-2,\mathcal{H}}, \varphi'_{0,\mathcal{H}})$ *are assigned to* $\mathsf{Z}_{\mathcal{H}}$ *and* $(\varphi_{-2,\mathcal{H}}, \varphi_{0,\mathcal{H}})$ *respectively under* $s = (s_X, s_Y)$ *as in [62, Lem. 5.4.2.11].*

In this case, we write $s = (s_X, s_Y) : (\mathsf{Z}_{\mathcal{H}}, \Phi_{\mathcal{H}}, \delta_{\mathcal{H}}) \twoheadrightarrow (\mathsf{Z}'_{\mathcal{H}}, \Phi'_{\mathcal{H}}, \delta'_{\mathcal{H}})$*.*

Definition 1.2.1.18. *(See [62, Def. 5.4.2.13].) We say that there is a* **surjection** *from a cusp label at level* $\mathcal{H}$ *represented by some* $(\mathsf{Z}_{\mathcal{H}}, \Phi_{\mathcal{H}}, \delta_{\mathcal{H}})$ *to a cusp label at level* $\mathcal{H}$ *represented by some* $(\mathsf{Z}'_{\mathcal{H}}, \Phi'_{\mathcal{H}}, \delta'_{\mathcal{H}})$ *if there is a surjection* (s_X, s_Y) *from* $(\mathsf{Z}_{\mathcal{H}}, \Phi_{\mathcal{H}}, \delta_{\mathcal{H}})$ *to* $(\mathsf{Z}'_{\mathcal{H}}, \Phi'_{\mathcal{H}}, \delta'_{\mathcal{H}})$*.*

This is well defined by [62, Lem. 5.4.1.15].

The surjection among cusp labels can be naturally seen when we have the so-called *two-step degenerations* (see [28, Ch. III, Sec. 10] and [62, Sec. 4.5.6]). This notion will be further developed in Definitions 1.2.2.12, 1.2.2.18, and 1.2.2.19 below.

1.2.2 *Cone Decompositions*

For each torus argument $\Phi_n = (X, Y, \phi, \varphi_{-2,n}, \varphi_{0,n})$ at level n, consider the finitely generated commutative group (*i.e.*, $\mathbb{Z}$-module)

$$\dddot{\mathbf{S}}_{\Phi_n} := ((\tfrac{1}{n}Y) \underset{\mathbb{Z}}{\otimes} X) / \begin{pmatrix} y \otimes \phi(y') - y' \otimes \phi(y) \\ (b\frac{1}{n}y) \otimes \chi - (\frac{1}{n}y) \otimes (b^\star \chi) \end{pmatrix}_{\substack{y,y' \in Y, \\ \chi \in X, b \in \mathcal{O}}} \tag{1.2.2.1}$$

and set $\mathbf{S}_{\Phi_n} := \dddot{\mathbf{S}}_{\Phi_n,\mathrm{free}}$, the free quotient of $\dddot{\mathbf{S}}_{\Phi_n}$. (See [62, (6.2.3.5) and Conv. 6.2.3.20].) Then, for a general torus argument $\Phi_{\mathcal{H}} = (X, Y, \phi, \varphi_{-2,\mathcal{H}}, \varphi_{0,\mathcal{H}})$ at level $\mathcal{H}$, there is a recipe [62, Lem. 6.2.4.4] that gives a corresponding free commutative group $\mathbf{S}_{\Phi_{\mathcal{H}}}$ (which can be identified with a finite index subgroup of some $\mathbf{S}_{\Phi_n}$).

The group $\mathbf{S}_{\Phi_{\mathcal{H}}}$ provides indices for certain "Laurent series expansions" near the boundary strata. In the modular curve case, it is canonically isomorphic to $\mathbb{Z}$, which means there is a canonical parameter q near the boundary—*i.e.*, the *cusps*. The expansion of modular forms with respect to this parameter then gives the familiar q-expansion along the cusps. The compactification of the modular curves can be described locally near each of the cusps by $\mathrm{Spec}(R[q^i]_{i\in\mathbb{Z}}) \hookrightarrow \mathrm{Spec}(R[q^i]_{i\in\mathbb{Z}_{\geq 0}})$ for some suitable base ring R. For $\mathsf{M}_{\mathcal{H}}$, we would like to have an analogous theory in which the torus with the character group $\mathbf{S}_{\Phi_{\mathcal{H}}}$ can be partially compactified by adding normal crossings divisors in a smooth scheme. This is best achieved by the theory of *toroidal embeddings* developed in [50]. Many terminologies in such a theory will naturally show up in our description of the toroidal boundary charts, and we will review them in what follows.

Let $\mathbf{S}_{\Phi_{\mathcal{H}}}^{\vee} := \mathrm{Hom}_{\mathbb{Z}}(\mathbf{S}_{\Phi_{\mathcal{H}}}, \mathbb{Z})$ be the $\mathbb{Z}$-dual of $\mathbf{S}_{\Phi_{\mathcal{H}}}$, and let $(\mathbf{S}_{\Phi_{\mathcal{H}}})_{\mathbb{R}}^{\vee} := \mathbf{S}_{\Phi_{\mathcal{H}}}^{\vee} \underset{\mathbb{Z}}{\otimes} \mathbb{R} = \mathrm{Hom}_{\mathbb{Z}}(\mathbf{S}_{\Phi_{\mathcal{H}}}, \mathbb{R})$. By the construction of $\mathbf{S}_{\Phi_{\mathcal{H}}}$, the $\mathbb{R}$-vector space $(\mathbf{S}_{\Phi_{\mathcal{H}}})_{\mathbb{R}}^{\vee}$ is isomorphic to the space of Hermitian pairings $(\!| \cdot, \cdot |\!) : (Y \underset{\mathbb{Z}}{\otimes} \mathbb{R}) \times (Y \underset{\mathbb{Z}}{\otimes} \mathbb{R}) \to \mathcal{O} \underset{\mathbb{Z}}{\otimes} \mathbb{R}$, by sending a Hermitian pairing $(\!| \cdot, \cdot |\!)$ to the function $y \otimes \phi(y') \mapsto \mathrm{Tr}_{\mathcal{O} \underset{\mathbb{Z}}{\otimes} \mathbb{R}/\mathbb{R}}(\!| y, y' |\!)$ in $\mathrm{Hom}_{\mathbb{R}}((Y \underset{\mathbb{Z}}{\otimes} \mathbb{R}) \times (Y \underset{\mathbb{Z}}{\otimes} \mathbb{R}), \mathbb{R}) \cong (\mathbf{S}_{\Phi_{\mathcal{H}}})_{\mathbb{R}}^{\vee}$. (See [62, Lem. 1.1.4.5].)

Definition 1.2.2.2. *(See [62, Sec. 6.1.1 and Def. 6.1.2.5].)*

(1) A subset of $(\mathbf{S}_{\Phi_{\mathcal{H}}})_{\mathbb{R}}^{\vee}$ is called a **cone** *if it is invariant under the natural multiplication action of $\mathbb{R}_{>0}^{\times}$ on the $\mathbb{R}$-vector space $(\mathbf{S}_{\Phi_{\mathcal{H}}})_{\mathbb{R}}^{\vee}$.*

(2) A cone in $(\mathbf{S}_{\Phi_{\mathcal{H}}})_{\mathbb{R}}^{\vee}$ is **nondegenerate** *if its closure does not contain any nonzero $\mathbb{R}$-vector subspace of $(\mathbf{S}_{\Phi_{\mathcal{H}}})_{\mathbb{R}}^{\vee}$.*

(3) A **rational polyhedral cone** *in $(\mathbf{S}_{\Phi_{\mathcal{H}}})_{\mathbb{R}}^{\vee}$ is a cone in $(\mathbf{S}_{\Phi_{\mathcal{H}}})_{\mathbb{R}}^{\vee}$ of the form $\sigma = \mathbb{R}_{>0}v_1 + \cdots + \mathbb{R}_{>0}v_n$ with $v_1, \ldots, v_n \in (\mathbf{S}_{\Phi_{\mathcal{H}}})_{\mathbb{Q}}^{\vee} = \mathbf{S}_{\Phi_{\mathcal{H}}}^{\vee} \underset{\mathbb{Z}}{\otimes} \mathbb{Q}$.*

(4) A **supporting hyperplane** *of σ is a hyperplane P in $(\mathbf{S}_{\Phi_{\mathcal{H}}})_{\mathbb{R}}^{\vee}$ such that σ does not overlap with both sides of P.*

(5) A **face** *of σ is a rational polyhedral cone τ such that $\overline{\tau} = \overline{\sigma} \cap P$ for some supporting hyperplane P of σ. (Here an overline on a cone means its closure in the ambient space $(\mathbf{S}_{\Phi_{\mathcal{H}}})_{\mathbb{R}}^{\vee}$.)*

(6) The canonical pairing $\langle \cdot, \cdot \rangle : \mathbf{S}_{\Phi_{\mathcal{H}}} \times \mathbf{S}_{\Phi_{\mathcal{H}}}^{\vee} \to \mathbb{Z}$ defines by extension of scalars a canonical pairing $\langle \cdot, \cdot \rangle : \mathbf{S}_{\Phi_{\mathcal{H}}} \times (\mathbf{S}_{\Phi_{\mathcal{H}}})_{\mathbb{R}}^{\vee} \to \mathbb{R}$. Then we define for each rational polyhedral cone σ in $(\mathbf{S}_{\Phi_{\mathcal{H}}})_{\mathbb{R}}^{\vee}$ the following semisubgroups of $\mathbf{S}_{\Phi_{\mathcal{H}}}$:

$$\sigma^{\vee} := \{\ell \in \mathbf{S}_{\Phi_{\mathcal{H}}} : \langle \ell, y \rangle \geq 0 \ \forall y \in \sigma\},$$
$$\sigma_0^{\vee} := \{\ell \in \mathbf{S}_{\Phi_{\mathcal{H}}} : \langle \ell, y \rangle > 0 \ \forall y \in \sigma\},$$
$$\sigma^{\perp} := \{\ell \in \mathbf{S}_{\Phi_{\mathcal{H}}} : \langle \ell, y \rangle = 0 \ \forall y \in \sigma\} \cong \sigma^{\vee}/\sigma_0^{\vee}.$$

Let $\mathbf{P}_{\Phi_{\mathcal{H}}}$ be the subset of $(\mathbf{S}_{\Phi_{\mathcal{H}}})_{\mathbb{R}}^{\vee}$ corresponding to positive semi-definite Hermitian pairings $(\!| \cdot, \cdot |\!) : (Y \underset{\mathbb{Z}}{\otimes} \mathbb{R}) \times (Y \underset{\mathbb{Z}}{\otimes} \mathbb{R}) \to \mathcal{O} \underset{\mathbb{Z}}{\otimes} \mathbb{R}$, with radical (namely the annihilator of the whole space) *admissible* in the sense that it is the $\mathbb{R}$-span of some *admissible submodule* Y' of Y. (Recall that we say that a submodule Y' of Y is admissible if $Y' \subset Y$ defines an admissible filtration on Y; *cf.* Definition 1.2.1.2. In particular, the quotient Y/Y' is also an $\mathcal{O}$-lattice.)

Definition 1.2.2.3. *(See [62, Def. 6.2.4.1 and 5.4.1.6].) The group $\Gamma_{\Phi_{\mathcal{H}}}$ is the subgroup of $\mathrm{GL}_{\mathcal{O}}(X) \times \mathrm{GL}_{\mathcal{O}}(Y)$ consisting of elements $\gamma = (\gamma_X, \gamma_Y)$ satisfying $\phi = \gamma_X \phi \gamma_Y$, $\varphi_{-2,\mathcal{H}} = {}^t\gamma_X \varphi_{-2,\mathcal{H}}$, and $\varphi_{0,\mathcal{H}} = \gamma_Y \varphi_{0,\mathcal{H}}$ (if we view the latter two as collections of orbits).*

The group $\Gamma_{\Phi_{\mathcal{H}}}$ acts on $\mathbf{S}_{\Phi_{\mathcal{H}}}$, and its induced action preserves the subset $\mathbf{P}_{\Phi_{\mathcal{H}}}$ of $(\mathbf{S}_{\Phi_{\mathcal{H}}})_{\mathbb{R}}^{\vee}$. (The group $\Gamma_{\Phi_{\mathcal{H}}}$ is the automorphism group of the torus argument $\Phi_{\mathcal{H}}$.

Such automorphism groups show up naturally because torus arguments are only determined up to isomorphism.)

Definition 1.2.2.4. *(See [62, Def. 6.1.1.10].) A* $\Gamma_{\Phi_{\mathcal{H}}}$**-admissible rational polyhedral cone decomposition** *of* $\mathbf{P}_{\Phi_{\mathcal{H}}}$ *is a collection* $\Sigma_{\Phi_{\mathcal{H}}} = \{\sigma_j\}_{j \in J}$ *with some indexing set* J *such that we have the following:*

(1) Each σ_j *is a nondegenerate rational polyhedral cone.*
(2) $\mathbf{P}_{\Phi_{\mathcal{H}}}$ *is the disjoint union of all the* σ_j*'s in* Σ*. For each* $j \in J$*, the closure of* σ_j *in* $\mathbf{P}_{\Phi_{\mathcal{H}}}$ *is a disjoint union of* σ_k*'s with* $k \in J$*. In other words,* $\mathbf{P}_{\Phi_{\mathcal{H}}} = \coprod_{j \in J} \sigma_j$ *is a stratification of* $\mathbf{P}_{\Phi_{\mathcal{H}}}$*. (Here "*$\coprod$*" only means a set-theoretic disjoint union. The geometric structure of* $\coprod_{j \in J} \sigma_j$ *is still the one inherited from the ambient space* $(\mathbf{S}_{\Phi_{\mathcal{H}}})^\vee_{\mathbb{R}}$ *of* $\mathbf{P}_{\Phi_{\mathcal{H}}}$*.)*
(3) Σ *is invariant under the action of* $\Gamma_{\Phi_{\mathcal{H}}}$ *on* $(\mathbf{S}_{\Phi_{\mathcal{H}}})^\vee_{\mathbb{R}}$*, in the sense that* $\Gamma_{\Phi_{\mathcal{H}}}$ *permutes the cones in* Σ*. Under this action, the set* $\Sigma_{\Phi_{\mathcal{H}}}/\Gamma_{\Phi_{\mathcal{H}}}$ *of* $\Gamma_{\Phi_{\mathcal{H}}}$*-orbits is finite.*

Definition 1.2.2.5. *(See [62, Def. 6.1.1.11].) A rational polyhedral cone* σ *in* $(\mathbf{S}_{\Phi_{\mathcal{H}}})^\vee_{\mathbb{R}}$ *is* **smooth** *with respect to the integral structure given by* $\mathbf{S}^\vee_{\Phi_{\mathcal{H}}}$ *if we have* $\sigma = \mathbb{R}_{>0}v_1 + \cdots + \mathbb{R}_{>0}v_n$ *with* $v_1, \ldots, v_n$ *forming part of a* $\mathbb{Z}$*-basis of* $\mathbf{S}^\vee_{\Phi_{\mathcal{H}}}$*.*

Definition 1.2.2.6. *(See [62, Def. 6.1.1.12].) A* $\Gamma_{\Phi_{\mathcal{H}}}$**-admissible smooth rational polyhedral cone decomposition** *of* $\mathbf{P}_{\Phi_{\mathcal{H}}}$ *is a* $\Gamma_{\Phi_{\mathcal{H}}}$*-admissible rational polyhedral cone decomposition* $\Sigma_{\Phi_{\mathcal{H}}} = \{\sigma_j\}_{j \in J}$ *of* $\mathbf{P}_{\Phi_{\mathcal{H}}}$ *in which every* σ_j *is* **smooth**.

Definition 1.2.2.7. *(See [62, Def. 7.3.1.1].) Let* $\Sigma_{\Phi_{\mathcal{H}}} = \{\sigma_j\}_{j \in J}$ *be any* $\Gamma_{\Phi_{\mathcal{H}}}$*-admissible rational polyhedral cone decomposition of* $\mathbf{P}_{\Phi_{\mathcal{H}}}$*. An* (**invariant**) **polarization function** *on* $\mathbf{P}_{\Phi_{\mathcal{H}}}$ *for the cone decomposition* $\Sigma_{\Phi_{\mathcal{H}}}$ *is a* $\Gamma_{\Phi_{\mathcal{H}}}$*-invariant continuous piecewise linear function* $\mathrm{pol}_{\Phi_{\mathcal{H}}} : \mathbf{P}_{\Phi_{\mathcal{H}}} \to \mathbb{R}_{\geq 0}$ *such that we have the following:*

(1) $\mathrm{pol}_{\Phi_{\mathcal{H}}}$ *is linear (i.e., coincides with a linear function) on each cone* σ_j *in* $\Sigma_{\Phi_{\mathcal{H}}}$*. (In particular,* $\mathrm{pol}_{\Phi_{\mathcal{H}}}(tx) = t\mathrm{pol}_{\Phi_{\mathcal{H}}}(x)$ *for all* $x \in \mathbf{P}_{\Phi_{\mathcal{H}}}$ *and* $t \in \mathbb{R}_{\geq 0}$*.)*
(2) $\mathrm{pol}_{\Phi_{\mathcal{H}}}((\mathbf{P}_{\Phi_{\mathcal{H}}} \cap \mathbf{S}^\vee_{\Phi_{\mathcal{H}}}) - \{0\}) \subset \mathbb{Z}_{>0}$*. (In particular,* $\mathrm{pol}_{\Phi_{\mathcal{H}}}(x) > 0$ *for all nonzero* x *in* $\mathbf{P}_{\Phi_{\mathcal{H}}}$*.)*
(3) $\mathrm{pol}_{\Phi_{\mathcal{H}}}$ *is linear (in the above sense) on a rational polyhedral cone* σ *in* $\mathbf{P}_{\Phi_{\mathcal{H}}}$ *if and only if* σ *is contained in some cone* σ_j *in* $\Sigma_{\Phi_{\mathcal{H}}}$*.*
(4) For all $x, y \in \mathbf{P}_{\Phi_{\mathcal{H}}}$*, we have* $\mathrm{pol}_{\Phi_{\mathcal{H}}}(x + y) \geq \mathrm{pol}_{\Phi_{\mathcal{H}}}(x) + \mathrm{pol}_{\Phi_{\mathcal{H}}}(y)$*. This is called the* **convexity** *of* $\mathrm{pol}_{\Phi_{\mathcal{H}}}$*.*

If such a polarization function exists, then we say that the $\Gamma_{\Phi_{\mathcal{H}}}$*-admissible rational polyhedral cone decomposition* $\Sigma_{\Phi_{\mathcal{H}}}$ *is* **projective**.

Definition 1.2.2.8. *An* **admissible boundary component** *of* $\mathbf{P}_{\Phi_{\mathcal{H}}}$ *is the image of* $\mathbf{P}_{\Phi'_{\mathcal{H}}}$ *under the embedding* $(\mathbf{S}_{\Phi'_{\mathcal{H}}})^\vee_{\mathbb{R}} \hookrightarrow (\mathbf{S}_{\Phi_{\mathcal{H}}})^\vee_{\mathbb{R}}$ *defined by some surjection* $(\Phi_{\mathcal{H}}, \delta_{\mathcal{H}}) \twoheadrightarrow (\Phi'_{\mathcal{H}}, \delta'_{\mathcal{H}})$*. (See Definition* 1.2.1.17.*)*

We shall always assume that the following technical condition is satisfied:

Condition 1.2.2.9. (*See [62, Cond. 6.2.5.25]; cf. [28, Ch. IV, Rem. 5.8(a)].*) *The cone decomposition $\Sigma_{\Phi_{\mathcal{H}}} = \{\sigma_j\}_{j\in J}$ of $\mathbf{P}_{\Phi_{\mathcal{H}}}$ is chosen such that, for each $j \in J$, if $\gamma\overline{\sigma}_j \cap \overline{\sigma}_j \neq \{0\}$ for some $\gamma \in \Gamma_{\Phi_{\mathcal{H}}}$, then γ acts as the identity on the smallest admissible boundary component of $\mathbf{P}_{\Phi_{\mathcal{H}}}$ containing σ_j.*

This condition is used to ensure that there are no self-intersections of toroidal boundary strata when the level $\mathcal{H}$ is neat.

To describe the toroidal boundary of $\mathsf{M}_{\mathcal{H}}$, we will need not only cusp labels but also the cones:

Definition 1.2.2.10. (*See [62, Def. 6.2.6.1].*) *Let $(\Phi_{\mathcal{H}}, \delta_{\mathcal{H}})$ and $(\Phi'_{\mathcal{H}}, \delta'_{\mathcal{H}})$ be two representatives of cusp labels at level $\mathcal{H}$, let $\sigma \subset (\mathbf{S}_{\Phi_{\mathcal{H}}})^\vee_{\mathbb{R}}$, and let $\sigma' \subset (\mathbf{S}_{\Phi'_{\mathcal{H}}})^\vee_{\mathbb{R}}$. We say that the two triples $(\Phi_{\mathcal{H}}, \delta_{\mathcal{H}}, \sigma)$ and $(\Phi'_{\mathcal{H}}, \delta'_{\mathcal{H}}, \sigma')$ are* **equivalent** *if there exists a pair of isomorphisms $\gamma = (\gamma_X : X' \xrightarrow{\sim} X, \gamma_Y : Y \xrightarrow{\sim} Y')$ (of $\mathcal{O}$-lattices) such that we have the following:*

(1) *The two representatives $(\Phi_{\mathcal{H}}, \delta_{\mathcal{H}})$ and $(\Phi'_{\mathcal{H}}, \delta'_{\mathcal{H}})$ are equivalent under γ (as in [62, Def. 5.4.2.4], the general level analogue of Definition* 1.2.1.7).
(2) *The isomorphism $(\mathbf{S}_{\Phi'_{\mathcal{H}}})^\vee_{\mathbb{R}} \xrightarrow{\sim} (\mathbf{S}_{\Phi_{\mathcal{H}}})^\vee_{\mathbb{R}}$ induced by γ sends σ' to σ.*

In this case, we say that the two triples $(\Phi_{\mathcal{H}}, \delta_{\mathcal{H}}, \sigma)$ and $(\Phi'_{\mathcal{H}}, \delta'_{\mathcal{H}}, \sigma')$ are equivalent under the pair of isomorphisms $\gamma = (\gamma_X, \gamma_Y)$.

Definition 1.2.2.11. (*See [62, Def. 6.2.6.2].*) *Let $(\Phi_{\mathcal{H}}, \delta_{\mathcal{H}})$ and $(\Phi'_{\mathcal{H}}, \delta'_{\mathcal{H}})$ be two representatives of cusp labels at level $\mathcal{H}$, and let $\Sigma_{\Phi_{\mathcal{H}}}$ (resp. $\Sigma_{\Phi'_{\mathcal{H}}}$) be a $\Gamma_{\Phi_{\mathcal{H}}}$-admissible (resp. $\Gamma_{\Phi'_{\mathcal{H}}}$-admissible) smooth rational polyhedral cone decomposition of $\mathbf{P}_{\Phi_{\mathcal{H}}}$ (resp. $\mathbf{P}_{\Phi'_{\mathcal{H}}}$). We say that the two triples $(\Phi_{\mathcal{H}}, \delta_{\mathcal{H}}, \Sigma_{\Phi_{\mathcal{H}}})$ and $(\Phi'_{\mathcal{H}}, \delta'_{\mathcal{H}}, \Sigma_{\Phi'_{\mathcal{H}}})$ are* **equivalent** *if $(\Phi_{\mathcal{H}}, \delta_{\mathcal{H}})$ and $(\Phi'_{\mathcal{H}}, \delta'_{\mathcal{H}})$ are equivalent under some pair of isomorphisms $\gamma = (\gamma_X : X' \xrightarrow{\sim} X, \gamma_Y : Y \xrightarrow{\sim} Y')$, and if under one (and hence every) such γ the cone decomposition $\Sigma_{\Phi_{\mathcal{H}}}$ of $\mathbf{P}_{\Phi_{\mathcal{H}}}$ is identified with the cone decomposition $\Sigma_{\Phi'_{\mathcal{H}}}$ of $\mathbf{P}_{\Phi'_{\mathcal{H}}}$. In this case, we say that the two triples $(\Phi_{\mathcal{H}}, \delta_{\mathcal{H}}, \Sigma_{\Phi_{\mathcal{H}}})$ and $(\Phi'_{\mathcal{H}}, \delta'_{\mathcal{H}}, \Sigma_{\Phi'_{\mathcal{H}}})$ are equivalent under the pair of isomorphisms $\gamma = (\gamma_X, \gamma_Y)$.*

The compatibility among cone decompositions over different cusp labels are described as follows:

Definition 1.2.2.12. (*See [62, Def. 6.2.6.4].*) *Let $(\Phi_{\mathcal{H}}, \delta_{\mathcal{H}})$ (resp. $(\Phi'_{\mathcal{H}}, \delta'_{\mathcal{H}})$) be a representative of a cusp label at level $\mathcal{H}$, and let $\Sigma_{\Phi_{\mathcal{H}}}$ (resp. $\Sigma_{\Phi'_{\mathcal{H}}}$) be a $\Gamma_{\Phi_{\mathcal{H}}}$-admissible (resp. $\Gamma_{\Phi'_{\mathcal{H}}}$-admissible) smooth rational polyhedral cone decomposition of $\mathbf{P}_{\Phi_{\mathcal{H}}}$ (resp. $\mathbf{P}_{\Phi'_{\mathcal{H}}}$). A* **surjection** *$(\Phi_{\mathcal{H}}, \delta_{\mathcal{H}}, \Sigma_{\Phi_{\mathcal{H}}}) \twoheadrightarrow (\Phi'_{\mathcal{H}}, \delta'_{\mathcal{H}}, \Sigma_{\Phi'_{\mathcal{H}}})$ is given by a surjection $s = (s_X : X \twoheadrightarrow X', s_Y : Y \twoheadrightarrow Y') : (\Phi_{\mathcal{H}}, \delta_{\mathcal{H}}) \twoheadrightarrow (\Phi'_{\mathcal{H}}, \delta'_{\mathcal{H}})$ (see Definition* 1.2.1.17) *that induces an embedding $\mathbf{P}_{\Phi'_{\mathcal{H}}} \hookrightarrow \mathbf{P}_{\Phi_{\mathcal{H}}}$ such that the restriction $\Sigma_{\Phi_{\mathcal{H}}}|_{\mathbf{P}_{\Phi'_{\mathcal{H}}}}$ of the cone decomposition $\Sigma_{\Phi_{\mathcal{H}}}$ of $\mathbf{P}_{\Phi_{\mathcal{H}}}$ to $\mathbf{P}_{\Phi'_{\mathcal{H}}}$ is the cone decomposition $\Sigma_{\Phi'_{\mathcal{H}}}$ of $\mathbf{P}_{\Phi'_{\mathcal{H}}}$.*

This allows us to define:

Definition 1.2.2.13. *(See [62, Cond. 6.3.3.2 and Def. 6.3.3.4].) A* **compatible choice of admissible smooth rational polyhedral cone decomposition data** *for* $\mathsf{M}_{\mathcal{H}}$ *is a complete set* $\Sigma = \{\Sigma_{\Phi_{\mathcal{H}}}\}_{[(\Phi_{\mathcal{H}},\delta_{\mathcal{H}})]}$ *of* **compatible choices** *of* $\Sigma_{\Phi_{\mathcal{H}}}$ *(satisfying Condition 1.2.2.9) such that, for every surjection* $(\Phi_{\mathcal{H}},\delta_{\mathcal{H}}) \twoheadrightarrow (\Phi'_{\mathcal{H}},\delta'_{\mathcal{H}})$ *of representatives of cusp labels, the cone decompositions* $\Sigma_{\Phi_{\mathcal{H}}}$ *and* $\Sigma_{\Phi'_{\mathcal{H}}}$ *define a surjection* $(\Phi_{\mathcal{H}},\delta_{\mathcal{H}},\Sigma_{\Phi_{\mathcal{H}}}) \twoheadrightarrow (\Phi'_{\mathcal{H}},\delta'_{\mathcal{H}},\Sigma_{\Phi'_{\mathcal{H}}})$ *as in Definition 1.2.2.12.*

Definition 1.2.2.14. *(See [62, Def. 7.3.1.3].) We say that a compatible choice* $\Sigma = \{\Sigma_{\Phi_{\mathcal{H}}}\}_{[(\Phi_{\mathcal{H}},\delta_{\mathcal{H}})]}$ *of admissible smooth rational polyhedral cone decomposition data for* $\mathsf{M}_{\mathcal{H}}$ *(see Definition 1.2.2.13) is* **projective** *if it satisfies the following condition: There is a collection* $\mathsf{pol} = \{\mathsf{pol}_{\Phi_{\mathcal{H}}} : \mathbf{P}_{\Phi_{\mathcal{H}}} \to \mathbb{R}_{\geq 0}\}_{[(\Phi_{\mathcal{H}},\delta_{\mathcal{H}})]}$ *of polarization functions labeled by representatives* $(\Phi_{\mathcal{H}},\delta_{\mathcal{H}})$ *of cusp labels, each* $\mathsf{pol}_{\Phi_{\mathcal{H}}}$ *being a polarization function of the cone decomposition* $\Sigma_{\Phi_{\mathcal{H}}}$ *in* Σ *(see Definition 1.2.2.7), which are* **compatible** *in the following sense: For every surjection* $(\Phi_{\mathcal{H}},\delta_{\mathcal{H}}) \twoheadrightarrow (\Phi'_{\mathcal{H}},\delta'_{\mathcal{H}})$ *of representatives of cusp labels (see Definition 1.2.1.17) inducing an embedding* $\mathbf{P}_{\Phi'_{\mathcal{H}}} \hookrightarrow \mathbf{P}_{\Phi_{\mathcal{H}}}$*, we have* $\mathsf{pol}_{\Phi_{\mathcal{H}}}|_{\mathbf{P}_{\Phi'_{\mathcal{H}}}} = \mathsf{pol}_{\Phi'_{\mathcal{H}}}$.

The most important relations among cone decompositions and among compatible choices of them are the so-called *refinements*:

Definition 1.2.2.15. *(See [62, Def. 6.2.6.3].) Let* $(\Phi_{\mathcal{H}},\delta_{\mathcal{H}})$ *and* $(\Phi'_{\mathcal{H}},\delta'_{\mathcal{H}})$ *be two representatives of cusp labels at level* $\mathcal{H}$*, and let* $\Sigma_{\Phi_{\mathcal{H}}}$ *(resp.* $\Sigma_{\Phi'_{\mathcal{H}}}$*) be a* $\Gamma_{\Phi_{\mathcal{H}}}$*-admissible (resp.* $\Gamma_{\Phi'_{\mathcal{H}}}$*-admissible) smooth rational polyhedral cone decomposition of* $\mathbf{P}_{\Phi_{\mathcal{H}}}$ *(resp.* $\mathbf{P}_{\Phi'_{\mathcal{H}}}$*). We say that the triple* $(\Phi_{\mathcal{H}},\delta_{\mathcal{H}},\Sigma_{\Phi_{\mathcal{H}}})$ *is a* **refinement** *of the triple* $(\Phi'_{\mathcal{H}},\delta'_{\mathcal{H}},\Sigma_{\Phi'_{\mathcal{H}}})$ *if* $(\Phi_{\mathcal{H}},\delta_{\mathcal{H}})$ *and* $(\Phi'_{\mathcal{H}},\delta'_{\mathcal{H}})$ *are equivalent under some pair of isomorphisms* $\gamma = (\gamma_X,\gamma_Y)$*, and if under one (and hence every) such* γ *the cone decomposition* $\Sigma_{\Phi_{\mathcal{H}}}$ *of* $\mathbf{P}_{\Phi_{\mathcal{H}}}$ *is identified with a refinement of the cone decomposition* $\Sigma_{\Phi'_{\mathcal{H}}}$ *of* $\mathbf{P}_{\Phi'_{\mathcal{H}}}$*. In this case, we say that the triple* $(\Phi_{\mathcal{H}},\delta_{\mathcal{H}},\Sigma_{\Phi_{\mathcal{H}}})$ *is a refinement of the triple* $(\Phi'_{\mathcal{H}},\delta'_{\mathcal{H}},\Sigma_{\Phi'_{\mathcal{H}}})$ *under the pair of isomorphisms* $\gamma = (\gamma_X,\gamma_Y)$.

Definition 1.2.2.16. *(See [62, Def. 6.4.2.2].) Let* $\Sigma = \{\Sigma_{\Phi_{\mathcal{H}}}\}_{[(\Phi_{\mathcal{H}},\delta_{\mathcal{H}})]}$ *and* $\Sigma' = \{\Sigma'_{\Phi_{\mathcal{H}}}\}_{[(\Phi_{\mathcal{H}},\delta_{\mathcal{H}})]}$ *be two compatible choices of admissible smooth rational polyhedral cone decomposition data for* $\mathsf{M}_{\mathcal{H}}$*. We say that* Σ *is a* **refinement** *of* Σ' *if the triple* $(\Phi_{\mathcal{H}},\delta_{\mathcal{H}},\Sigma_{\Phi_{\mathcal{H}}})$ *is a refinement of the triple* $(\Phi_{\mathcal{H}},\delta_{\mathcal{H}},\Sigma'_{\Phi_{\mathcal{H}}})$*, as in Definition 1.2.2.15, for* $(\Phi_{\mathcal{H}},\delta_{\mathcal{H}})$ *running through all representatives of cusp labels.*

Proposition 1.2.2.17. *(See [62, Prop. 6.3.3.5 and 7.3.1.4].)*

(1) A compatible choice Σ *of admissible smooth rational polyhedral cone decomposition data for* $\mathsf{M}_{\mathcal{H}}$*, as in Definition 1.2.2.13,* **exists***. Moreover, we may assume that* Σ *is* **projective** *as in Definition 1.2.2.14.*

(2) Given any Σ *and* Σ'*, we can find a common refinement for them, which we may require to be smooth as in Definition 1.2.2.13, or both smooth and*

projective as in Definition 1.2.2.14. *The same is true if we allow varying levels or twists by Hecke actions (see [62, Def. 6.4.2.8 and 6.4.3.2]). We may assume that this common refinement is invariant under any choice of an open compact subgroup* $\mathcal{H}'$ *of* $\mathrm{G}(\mathbb{A}^\infty)$ *normalizing* $\mathcal{H}$.

Proof. The first part has been explained in the proofs of [62, Prop. 6.3.3.5 and 7.3.1.4], by induction on magnitudes of cusp labels (*i.e.*, by starting with cusp labels of smaller multiranks and building cone decompositions and polarization functions along them, which appear as rational boundary components of homogeneous cones attached to cusp labels of larger multiranks). Based on such inductive constructions (which builds the smaller dimensional cones first), the second part can be reduced to questions over each $\mathbf{P}_{\Phi_\mathcal{H}}$ (with prescribed cone decompositions and polarization functions over $\mathbf{P}_{\Phi_\mathcal{H}} - \mathbf{P}^+_{\Phi_\mathcal{H}}$), which is then well known. (See the arguments in [89, 5.21, 5.23, 5.24, 5.25], where the crucial existence of smooth and projective refinements is in turn based on [50, Ch. I, Sec. 2, Thm. 11 on pp. 33–35].) □

Finally, we would like to describe the relations among the equivalence classes $[(\Phi_\mathcal{H}, \delta_\mathcal{H}, \sigma)]$, which will describe the "incidence relations" among (closures of) the toroidal boundary strata.

Definition 1.2.2.18. *(See [62, Def. 6.3.2.13].) Let* $(\Phi_\mathcal{H}, \delta_\mathcal{H})$ *be a representative of a cusp label at level* $\mathcal{H}$, *and let* $\sigma \subset \mathbf{P}^+_{\Phi_\mathcal{H}}$ *be a nondegenerate smooth rational polyhedral cone. We say that a triple* $(\Phi'_\mathcal{H}, \delta'_\mathcal{H}, \sigma')$ *is a* **face** *of* $(\Phi_\mathcal{H}, \delta_\mathcal{H}, \sigma)$, *if:*

(1) $(\Phi'_\mathcal{H}, \delta'_\mathcal{H})$ *is the representative of some cusp label at level* $\mathcal{H}$, *such that there exists a surjection* $s = (s_X, s_Y) : (\Phi_\mathcal{H}, \delta_\mathcal{H}) \twoheadrightarrow (\Phi'_\mathcal{H}, \delta'_\mathcal{H})$ *as in Definition* 1.2.1.17.

(2) $\sigma' \subset \mathbf{P}^+_{\Phi'_\mathcal{H}}$ *is a nondegenerate smooth rational polyhedral cone, such that for one (and hence every) surjection* $s = (s_X, s_Y)$ *as above, the image of* σ' *under the induced embedding* $\mathbf{P}_{\Phi'_\mathcal{H}} \hookrightarrow \mathbf{P}_{\Phi_\mathcal{H}}$ *is contained in the* $\Gamma_{\Phi_\mathcal{H}}$*-orbit of a face of* σ.

Note that this definition is insensitive to the choices of representatives in the classes $[(\Phi_\mathcal{H}, \delta_\mathcal{H}, \sigma)]$ and $[(\Phi'_\mathcal{H}, \delta'_\mathcal{H}, \sigma')]$. This justifies the following:

Definition 1.2.2.19. *(See [62, Def. 6.3.2.14].) We say that the equivalence class* $[(\Phi'_\mathcal{H}, \delta'_\mathcal{H}, \sigma')]$ *of* $(\Phi'_\mathcal{H}, \delta'_\mathcal{H}, \sigma')$ *is a* **face** *of the equivalence class* $[(\Phi_\mathcal{H}, \delta_\mathcal{H}, \sigma)]$ *of* $(\Phi_\mathcal{H}, \delta_\mathcal{H}, \sigma)$ *if some triple equivalent to* $(\Phi'_\mathcal{H}, \delta'_\mathcal{H}, \sigma')$ *is a face of some triple equivalent to* $(\Phi_\mathcal{H}, \delta_\mathcal{H}, \sigma)$.

1.2.3 *Rational Boundary Components*

Now we explain how to associate cusp labels with rational boundary components. This is mainly for readers who are familiar with the notion of rational boundary components of Hermitian symmetric domains. (See, for example, the summaries in [5] or [9].)

Let X_0 be the connected component of X containing h_0, and let $\mathrm{G}(\mathbb{R})_0$ (resp. $\mathrm{G}(\mathbb{Q})_0$) denote its stabilizer in $\mathrm{G}(\mathbb{R})$ (resp. $\mathrm{G}(\mathbb{Q})$). Then $\mathrm{G}(\mathbb{R})_0$ (resp. $\mathrm{G}(\mathbb{Q})_0$) has finite index in $\mathrm{G}(\mathbb{R})$ (resp. $\mathrm{G}(\mathbb{Q})$).

Lemma 1.2.3.1. *Let us fix a choice of an element $g \in \mathrm{G}(\mathbb{A}^\infty)$. Let $L^{(g)}$ denote the $\mathcal{O}$-lattice in $L \underset{\mathbb{Z}}{\otimes} \mathbb{Q}$ such that $L^{(g)} \underset{\mathbb{Z}}{\otimes} \hat{\mathbb{Z}}$ corresponds naturally to the $\mathcal{O} \underset{\mathbb{Z}}{\otimes} \hat{\mathbb{Z}}$-submodule $g(L \underset{\mathbb{Z}}{\otimes} \hat{\mathbb{Z}})$ of $L \underset{\mathbb{Z}}{\otimes} \mathbb{A}^\infty$. Consider the five sets formed respectively by the following five types of data on $(L, \langle \cdot, \cdot \rangle, h_0)$:*

(1) *A rational boundary component of X_0 (as in [5, 1.5]). (For compatibility with formation of products, it is necessary to include X_0 itself as a rational boundary component.)*

(2) *An $\mathcal{O} \underset{\mathbb{Z}}{\otimes} \mathbb{Q}$-submodule $\mathtt{V}_{-2}$ of $L \underset{\mathbb{Z}}{\otimes} \mathbb{Q}$ that is totally isotropic under the pairing $\langle \cdot, \cdot \rangle$.*

(3) *An increasing filtration $\mathtt{V} = \{\mathtt{V}_{-i}\}_{i \in \mathbb{Z}}$ of $L \underset{\mathbb{Z}}{\otimes} \mathbb{Q}$ satisfying the following conditions:*

  (a) *$\mathtt{V}_{-3} = 0$ and $\mathtt{V}_0 = L \underset{\mathbb{Z}}{\otimes} \mathbb{Q}$.*

  (b) *Each graded piece $\mathrm{Gr}^{\mathtt{V}}_{-i} := \mathtt{V}_{-i}/\mathtt{V}_{-i-1}$ is an $\mathcal{O} \underset{\mathbb{Z}}{\otimes} \mathbb{Q}$-module. (In this case, the filtration $\mathtt{V}$ is* **admissible***.)*

  (c) *$\mathtt{V}_{-1}$ and $\mathtt{V}_{-2}$ are annihilators of each other under the pairing $\langle \cdot, \cdot \rangle$. (In this case, the filtration $\mathtt{V}$ is* **symplectic***.)*

(4) *An $\mathcal{O}$-sublattice $\mathtt{F}^{(g)}_{-2}$ of $L^{(g)}$, with $L^{(g)}/\mathtt{F}^{(g)}_{-2}$ torsion-free, that is totally isotropic under the pairing $\langle \cdot, \cdot \rangle$.*

(5) *An increasing filtration $\mathtt{F}^{(g)} = \{\mathtt{F}^{(g)}_{-i}\}_{i \in \mathbb{Z}}$ of $L^{(g)}$ satisfying the following conditions:*

  (a) *$\mathtt{F}^{(g)}_{-3} = 0$ and $\mathtt{F}^{(g)}_0 = L^{(g)}$.*

  (b) *Each graded piece $\mathrm{Gr}^{\mathtt{F}^{(g)}}_{-i} := \mathtt{F}^{(g)}_{-i}/\mathtt{F}^{(g)}_{-i-1}$ is an $\mathcal{O}$-lattice, admitting an splitting $\varepsilon^{(g)} : \mathrm{Gr}^{\mathtt{F}^{(g)}} := \underset{-i \in \mathbb{Z}}{\oplus} \mathrm{Gr}^{\mathtt{F}^{(g)}}_{-i} \xrightarrow{\sim} L^{(g)}$. (In this case, the filtration $\mathtt{F}^{(g)}$ is* **admissible***.)*

  (c) *$\mathtt{F}^{(g)}_{-1}$ and $\mathtt{F}^{(g)}_{-2}$ are annihilators of each other under the pairing $\langle \cdot, \cdot \rangle^{(g)} : L^{(g)} \times L^{(g)} \to \mathbb{Z}(1)$. (In this case, the filtration $\mathtt{F}^{(g)}$ is* **symplectic***.)*

(We allow parabolic subgroups to be the whole group, and we allow totally isotropic submodules to be zero.) Then the five sets are in canonical bijections with each other.

Proof. As explained in [5, 1.5], the rational boundary components of X_0 correspond bijectively to the rational parabolic subgroups of $\mathrm{G} \underset{\mathbb{Z}}{\otimes} \mathbb{Q}$ each of whose images in the $\mathbb{Q}$-simple quotients of $\mathrm{G} \underset{\mathbb{Z}}{\otimes} \mathbb{Q}$ is either a maximal proper parabolic subgroup or the whole group. For simplicity, let us call temporarily such rational parabolic subgroups *maximal*. Given any such rational parabolic subgroup of $\mathrm{G} \underset{\mathbb{Z}}{\otimes} \mathbb{Q}$, the action of the Lie algebra of its unipotent radical defines an isotropic filtration $\mathtt{V}$ of $L \underset{\mathbb{Z}}{\otimes} \mathbb{Q}$. By maximality of the parabolic subgroup, we see that $\mathtt{V}$ is determined by its largest totally isotropic filtered piece. Now the equivalences among the maximal rational parabolic subgroups and the remaining objects in the lemma is elementary. □

For each $g \in \mathrm{G}(\mathbb{A}^\infty)$, let $L^{(g)}$ denote the $\mathcal{O}$-lattice in $L \underset{\mathbb{Z}}{\otimes} \mathbb{Q}$ such that $L^{(g)} \underset{\mathbb{Z}}{\otimes} \hat{\mathbb{Z}}$ corresponds naturally to the $\mathcal{O} \underset{\mathbb{Z}}{\otimes} \hat{\mathbb{Z}}$-submodule $g(L \underset{\mathbb{Z}}{\otimes} \hat{\mathbb{Z}})$ of $L \underset{\mathbb{Z}}{\otimes} \mathbb{A}^\infty$. Then the assignment

$$\begin{aligned} \mathtt{V}_{-2} \mapsto \mathtt{V} &= \{\mathtt{V}_{-i}\}_{i\in\mathbb{Z}} \\ &\mapsto \mathtt{F}^{(g)} := \{\mathtt{F}^{(g)}_{-i} := \mathtt{V}_{-i} \cap L^{(g)}\}_{i\in\mathbb{Z}} \\ &\mapsto \mathtt{Z}^{(g)} := \{\mathtt{Z}^{(g)}_{-i} := g^{-1}(\mathtt{F}^{(g)}_{-i} \underset{\mathbb{Z}}{\otimes} \hat{\mathbb{Z}})\}_{i\in\mathbb{Z}} \\ &\quad = \{(g^{-1}(\mathtt{V}_{-i} \underset{\mathbb{Q}}{\otimes} \mathbb{A}^\infty)) \cap (L \underset{\mathbb{Z}}{\otimes} \hat{\mathbb{Z}})\}_{i\in\mathbb{Z}} \end{aligned} \tag{1.2.3.2}$$

defines an injection from the set of rational boundary components of X_0 to the set of fully symplectic admissible filtrations on $L \underset{\mathbb{Z}}{\otimes} \hat{\mathbb{Z}}$. (See [62, Def. 5.2.7.1].)

The action of $\mathrm{G}(\mathbb{Q})$ on $\mathsf{X} \times \mathrm{G}(\mathbb{A}^\infty)$ induces an action of $\mathrm{G}(\mathbb{Q})$ on $\{\mathtt{V}\} \times \mathrm{G}(\mathbb{A}^\infty)$.

Definition 1.2.3.3. *A rational boundary component of* $\mathsf{X} \times \mathrm{G}(\mathbb{A}^\infty)$ *is a* $\mathrm{G}(\mathbb{Q})$*-orbit of some pair* $(\mathtt{V}, g)$.

By the explicit definition above, pairs in the $\mathrm{G}(\mathbb{Q})$-orbit of $(\mathtt{V}, g)$ define the same fully symplectic admissible filtration on $L \underset{\mathbb{Z}}{\otimes} \hat{\mathbb{Z}}$. This induces a map from the set of rational boundary components of $\mathsf{X} \times \mathrm{G}(\mathbb{A}^\infty)$ to the set of fully symplectic admissible filtrations on $L \underset{\mathbb{Z}}{\otimes} \hat{\mathbb{Z}}$. However, this map is generally far from injective.

For example, if $u \in \mathrm{G}(\hat{\mathbb{Z}})$ is an element preserving $\mathtt{V}_{-2,\mathbb{A}^\infty} := \mathtt{V}_{-2} \underset{\mathbb{Q}}{\otimes} \mathbb{A}^\infty$, then $(\mathtt{V}, g)$ and $(\mathtt{V}, gu)$ define the same filtration $\mathtt{Z}^{(g)} = \mathtt{Z}^{(gu)}$. For the purpose of studying toroidal compactifications, it is important to distinguish between $(\mathtt{V}, g)$ and $(\mathtt{V}, gu)$ by supplying a rigidification on the rational structure of $\mathtt{V}_{-2}$. For each given $(\mathtt{V}, g)$, let us define a torus argument $\Phi^{(g)} = (X^{(g)}, Y^{(g)}, \phi^{(g)}, \varphi^{(g)}_{-2}, \varphi^{(g)}_0)$ for $\mathtt{Z}^{(g)}$ as follows:

(1) $X^{(g)} := \mathrm{Hom}_{\mathbb{Z}}(\mathtt{F}^{(g)}_{-2}, \mathbb{Z}(1)) = \mathrm{Hom}_{\mathbb{Z}}(\mathrm{Gr}^{\mathtt{F}^{(g)}}_{-2}, \mathbb{Z}(1))$.
(2) $Y^{(g)} := \mathrm{Gr}^{\mathtt{F}^{(g)}}_0 = \mathtt{F}^{(g)}_0 / \mathtt{F}^{(g)}_{-1}$.
(3) $\phi^{(g)} : Y^{(g)} \hookrightarrow X^{(g)}$ is equivalent to the nondegenerate pairing

$$\langle\,\cdot\,,\,\cdot\,\rangle^{(g)}_{20} : \mathrm{Gr}^{\mathtt{F}^{(g)}}_{-2} \times \mathrm{Gr}^{\mathtt{F}^{(g)}}_0 \to \mathbb{Z}(1)$$

induced by $\langle\,\cdot\,,\,\cdot\,\rangle^{(g)} : L^{(g)} \times L^{(g)} \to \mathbb{Z}(1)$, with the sign convention $\langle x, y\rangle^{(g)}_{20} = \phi^{(g)}(y)(x)$.

(4) $\varphi_{-2}^{(g)} : \mathrm{Gr}_{-2}^{\mathtt{Z}^{(g)}} \xrightarrow{\sim} \mathrm{Hom}_{\hat{\mathbb{Z}}}(X^{(g)} \underset{\mathbb{Z}}{\otimes} \hat{\mathbb{Z}}, \hat{\mathbb{Z}}(1))$ is the composition

$$\mathrm{Gr}_{-2}^{\mathtt{Z}^{(g)}} \overset{\mathrm{Gr}_{-2}(g)}{\xrightarrow{\sim}} \mathrm{Gr}_{-2}^{\mathtt{F}^{(g)}} \underset{\mathbb{Z}}{\otimes} \hat{\mathbb{Z}} \xrightarrow{\sim} \mathrm{Hom}_{\hat{\mathbb{Z}}}(X^{(g)} \underset{\mathbb{Z}}{\otimes} \hat{\mathbb{Z}}, \hat{\mathbb{Z}}(1)).$$

(5) $\varphi_{0}^{(g)} : \mathrm{Gr}_{0}^{\mathtt{Z}^{(g)}} \xrightarrow{\sim} Y^{(g)} \underset{\mathbb{Z}}{\otimes} \hat{\mathbb{Z}}$ is the composition

$$\mathrm{Gr}_{0}^{\mathtt{Z}^{(g)}} \overset{\mathrm{Gr}_{0}(g)}{\xrightarrow{\sim}} \mathrm{Gr}_{0}^{\mathtt{F}^{(g)}} \underset{\mathbb{Z}}{\otimes} \hat{\mathbb{Z}} \xrightarrow{\sim} Y^{(g)} \underset{\mathbb{Z}}{\otimes} \hat{\mathbb{Z}}.$$

Finally, by Condition 1.2.1.1 and the fact that maximal orders over Dedekind domains are *hereditary* ([93, Thm. 21.4 and Cor. 21.5]), for each $(\mathtt{V}, g)$, the associated filtration $\mathtt{F}^{(g)}$ of $L^{(g)}$ is split by some splitting $\varepsilon^{(g)} : \mathrm{Gr}^{\mathtt{F}^{(g)}} \xrightarrow{\sim} L^{(g)}$. Each splitting $\varepsilon^{(g)}$ defines by base change a splitting $\varepsilon^{(g)} \underset{\mathbb{Z}}{\otimes} \hat{\mathbb{Z}} : \mathrm{Gr}^{\mathtt{F}^{(g)}} \underset{\mathbb{Z}}{\otimes} \hat{\mathbb{Z}} \xrightarrow{\sim} L^{(g)} \underset{\mathbb{Z}}{\otimes} \hat{\mathbb{Z}} = g(L \underset{\mathbb{Z}}{\otimes} \hat{\mathbb{Z}})$, and hence by pre- and post-composition with $\mathrm{Gr}(g)$ and g^{-1} a splitting $\delta^{(g)} : \mathrm{Gr}^{\mathtt{Z}^{(g)}} \xrightarrow{\sim} L \underset{\mathbb{Z}}{\otimes} \hat{\mathbb{Z}}$. This defines an assignment

$$(\mathtt{V}, g, \varepsilon^{(g)}) \mapsto (\mathtt{Z}^{(g)}, \Phi^{(g)}, \delta^{(g)}). \tag{1.2.3.4}$$

Let us define two triples $(\mathtt{V}, g, \varepsilon^{(g)})$ and $(\mathtt{V}', g', (\varepsilon^{(g)})')$ to be equivalent if $\mathtt{V} = \mathtt{V}'$ and $g = g'$, and define two triples $(\mathtt{Z}, \Phi, \delta)$ and $(\mathtt{Z}', \Phi', \delta')$ to be equivalent if $\mathtt{Z} = \mathtt{Z}'$ and if the torus arguments $\Phi = (X, Y, \phi, \varphi_{-2}, \varphi_0)$ and $\Phi' = (X', Y', \phi', \varphi'_{-2}, \varphi'_0)$ are equivalent in the sense that there exists some pair of isomorphisms $(\gamma_X : X' \xrightarrow{\sim} X, \gamma_Y : Y \xrightarrow{\sim} Y')$ matching the remaining data. By definition, the equivalence classes $[(\mathtt{V}, g, \varepsilon^{(g)})]$ of triples $(\mathtt{V}, g, \varepsilon^{(g)})$ correspond exactly to the pairs $(\mathtt{V}, g)$ they define by forgetting the splitting $\varepsilon^{(g)}$. On the other hand, let us denote by $[(\mathtt{Z}^{(g)}, \Phi^{(g)}, \delta^{(g)})]$ the equivalence class defined by $(\mathtt{Z}^{(g)}, \Phi^{(g)}, \delta^{(g)})$, and let us call them the *cusp labels* for $(L, \langle \cdot, \cdot \rangle, h_0)$.

Now we have the assignment $(\mathtt{V}, g) \mapsto [(\mathtt{Z}^{(g)}, \Phi^{(g)}, \delta^{(g)})]$ induced by the assignment $(\mathtt{V}, g, \varepsilon^{(g)}) \mapsto (\mathtt{Z}^{(g)}, \Phi^{(g)}, \delta^{(g)})$. This assignment is still not injective in general, but will suffice for our purpose.

For each $\mathbb{Q}$-algebra R, let us write $\mathtt{V}_{-i,R} := \mathtt{V}_{-i} \underset{\mathbb{Q}}{\otimes} R$ and $\mathrm{Gr}_{-i,R}^{\mathtt{V}} := \mathtt{V}_{-i,R}/\mathtt{V}_{-i-1,R}$. Similarly, for each $\mathbb{Z}$-algebra R, let us write $\mathtt{F}_{-i,R}^{(g)} := \mathtt{F}_{-i}^{(g)} \underset{\mathbb{Z}}{\otimes} R$ and $\mathrm{Gr}_{-i,R}^{\mathtt{F}^{(g)}} := \mathtt{F}_{-i,R}^{(g)}/\mathtt{F}_{-i-1,R}^{(g)}$.

To each boundary component represented by $(\mathtt{V}, g)$, the symplectic filtration $\mathtt{V}$ induces a symplectic lattice $(\mathrm{Gr}_{-1}^{\mathtt{V}}, \langle \cdot, \cdot \rangle_{11})$, and the associated symplectic filtration $\mathtt{F}^{(g)}$ on $L^{(g)}$ induces a symplectic lattice $(\mathrm{Gr}_{-1}^{\mathtt{F}^{(g)}}, \langle \cdot, \cdot \rangle_{11}^{(g)})$. It is clear that $(\mathrm{Gr}_{-1}^{\mathtt{F}^{(g)}} \underset{\mathbb{Z}}{\otimes} \mathbb{Q}, \langle \cdot, \cdot \rangle_{11}^{(g)}) \cong (\mathrm{Gr}_{-1}^{\mathtt{V}}, \langle \cdot, \cdot \rangle_{11})$.

Each $h \in \mathsf{X}$ defines a complex structure $h(\sqrt{-1})$ on $L \underset{\mathbb{Z}}{\otimes} \mathbb{R}$, inducing an isomorphism $L \underset{\mathbb{Z}}{\otimes} \mathbb{R} \xrightarrow{\sim} V_h = (L \underset{\mathbb{Z}}{\otimes} \mathbb{C})/\mathsf{P}_h$ of $\mathbb{C}$-vector spaces. Since $\mathtt{F}_{-2,\mathbb{R}}^{(g)}$ is totally isotropic, and since $-\operatorname{sgn}(h)\sqrt{-1}\,\langle\,\cdot\,,h(\sqrt{-1})\,\cdot\,\rangle$ is positive definite for some $\operatorname{sgn}(h)\in\{\pm1\}$, we have $\mathtt{F}^{(g)}_{-2,\mathbb{R}}\cap h(\sqrt{-1})(\mathtt{F}^{(g)}_{-2,\mathbb{R}})=\{0\}$. Then h defines a $\mathbb{C}$-linear embedding $\mathtt{F}^{(g)}_{-2,\mathbb{C}}\hookrightarrow V_h$, such that the composition $\mathtt{F}^{(g)}_{-2,\mathbb{R}}\overset{h(\sqrt{-1})}{\to}L\underset{\mathbb{Z}}{\otimes}\mathbb{R}\twoheadrightarrow\operatorname{Gr}^{\mathtt{F}^{(g)}}_{0,\mathbb{R}}$ is an isomorphism of $\mathcal{O}\underset{\mathbb{Z}}{\otimes}\mathbb{R}$-modules. By abuse of notation, we shall denote the image of the above embedding $\mathtt{F}^{(g)}_{-2,\mathbb{C}}\hookrightarrow V_h$ by $\mathtt{F}^{(g)}_{-2,h(\mathbb{C})}$. Let

$$(\mathtt{F}^{(g)}_{-2,h(\mathbb{C})})^{\perp}:=\{x\in L\underset{\mathbb{Z}}{\otimes}\mathbb{R}:\langle x,y\rangle=0,\forall y\in\mathtt{F}^{(g)}_{-2,h(\mathbb{C})}\}.$$

Then we obtain an orthogonal direct sum

$$\begin{aligned}&(L\underset{\mathbb{Z}}{\otimes}\mathbb{R},\langle\,\cdot\,,\,\cdot\,\rangle)\\&\cong(\mathtt{F}^{(g)}_{-2,h(\mathbb{C})},\langle\,\cdot\,,\,\cdot\,\rangle|_{\mathtt{F}^{(g)}_{-2,h(\mathbb{C})}})\overset{\perp}{\oplus}((\mathtt{F}^{(g)}_{-2,h(\mathbb{C})})^{\perp},\langle\,\cdot\,,\,\cdot\,\rangle|_{(\mathtt{F}^{(g)}_{-2,h(\mathbb{C})})^{\perp}}),\end{aligned}\tag{1.2.3.5}$$

which induces an isomorphism

$$((\mathtt{F}^{(g)}_{-2,h(\mathbb{C})})^{\perp},\langle\,\cdot\,,\,\cdot\,\rangle|_{(\mathtt{F}^{(g)}_{-2,h(\mathbb{C})})^{\perp}})\overset{\sim}{\to}(\operatorname{Gr}^{\mathtt{F}^{(g)}}_{-1,\mathbb{R}},\langle\,\cdot\,,\,\cdot\,\rangle^{(g)}_{11,\mathbb{R}})\tag{1.2.3.6}$$

of symplectic $\mathcal{O}\underset{\mathbb{Z}}{\otimes}\mathbb{R}$-modules. Since $h(\sqrt{-1})$ preserves $\mathtt{F}^{(g)}_{-2,h(\mathbb{C})}$, the relation

$$\langle h(\sqrt{-1})x,h(\sqrt{-1})y\rangle=\langle x,y\rangle$$

for every $x,y\in L\underset{\mathbb{Z}}{\otimes}\mathbb{R}$ shows that $h(\sqrt{-1})$ also preserves $(\mathtt{F}^{(g)}_{-2,h(\mathbb{C})})^{\perp}$. As a result, the restriction of $h(\sqrt{-1})$ defines a complex structure on $(\mathtt{F}^{(g)}_{-2,h(\mathbb{C})})^{\perp}$, which corresponds via the isomorphism (1.2.3.6) (and [59, Lem. 2.1.2]) to a $\operatorname{sgn}(h)$-polarization h_{-1} of $(\operatorname{Gr}^{\mathtt{F}^{(g)}}_{-1,\mathbb{R}},\langle\,\cdot\,,\,\cdot\,\rangle^{(g)}_{-1,\mathbb{R}})$ (see [59, Def. 2.1.1]), such that

$$\begin{aligned}&((\mathtt{F}^{(g)}_{-2,h(\mathbb{C})})^{\perp},\langle\,\cdot\,,\,\cdot\,\rangle|_{(\mathtt{F}^{(g)}_{-2,h(\mathbb{C})})^{\perp}},h|_{(\mathtt{F}^{(g)}_{-2,h(\mathbb{C})})^{\perp}})\\&\overset{\sim}{\to}(\operatorname{Gr}^{\mathtt{F}^{(g)}}_{-1,\mathbb{R}},\langle\,\cdot\,,\,\cdot\,\rangle^{(g)}_{11,\mathbb{R}},h_{-1})\end{aligned}\tag{1.2.3.7}$$

is an isomorphism of polarized symplectic $\mathcal{O}\underset{\mathbb{Z}}{\otimes}\mathbb{R}$-modules. Hence, the triple $(\operatorname{Gr}^{\mathtt{F}^{(g)}}_{-1},\langle\,\cdot\,,\,\cdot\,\rangle^{(g)},h_{-1})$ is a PEL-type $\mathcal{O}$-lattice. (In particular, this is the case for $h=h_0$.)

Lemma 1.2.3.8. *With notation as in [62, Rem. 5.2.7.2] (with h there replaced with h_0 here), the PEL-type $\mathcal{O}$-lattice $(\operatorname{Gr}^{\mathtt{F}^{(g)}}_{-1},\langle\,\cdot\,,\,\cdot\,\rangle^{(g)},(h_0)_{-1})$ qualifies as a (noncanonical) choice of $(L^{\mathtt{Z}^{(g)}},\langle\,\cdot\,,\,\cdot\,\rangle^{\mathtt{Z}^{(g)}},h_0^{\mathtt{Z}^{(g)}})$, so that $(\operatorname{Gr}^{\mathtt{Z}^{(g)}}_{-1},\langle\,\cdot\,,\,\cdot\,\rangle_{11})\cong(\operatorname{Gr}^{\mathtt{F}^{(g)}}_{-1,\hat{\mathbb{Z}}},\langle\,\cdot\,,\,\cdot\,\rangle^{(g)}_{11})$ and $(\operatorname{Gr}^{\mathtt{Z}^{(g)}}_{-1,\mathbb{R}},\langle\,\cdot\,,\,\cdot\,\rangle^{(g)}_{11,\mathbb{R}},(h_0)_{-1})\overset{\sim}{\to}(\operatorname{Gr}^{\mathtt{F}^{(g)}}_{-1,\mathbb{R}},\langle\,\cdot\,,\,\cdot\,\rangle^{(g)}_{11},(h_0)_{-1})$. (See Remark* 1.2.3.9 *below for the justification of such notation.) In particular, at any neat level $\mathcal{H}$, the scheme $\mathsf{M}^{\mathtt{Z}^{(g)}_{\mathcal{H}}}_{\mathcal{H}}$ can be identified with the moduli problem defined by $(\operatorname{Gr}^{\mathtt{F}^{(g)}}_{-1},\langle\,\cdot\,,\,\cdot\,\rangle^{(g)},(h_0)_{-1})$ at a suitable level (see Lemma* 1.3.2.1 *below).*

Remark 1.2.3.9. The notation $(h_0)_{-1}$ appeared twice in the second isomorphism in Lemma 1.2.3.8. Nevertheless, their constructions are identical because we have to use $\mathtt{F}^{(g)}_{-2,\mathbb{R}}=\operatorname{Hom}_{\mathbb{R}}(X^{(g)}\underset{\mathbb{Z}}{\otimes}\mathbb{R},\mathbb{R}(1))\hookrightarrow L\underset{\mathbb{Z}}{\otimes}\mathbb{R}$ to define $(h_0)_{-1}$ for $(\operatorname{Gr}^{\mathtt{Z}^{(g)}}_{-1,\mathbb{R}},\langle\,\cdot\,,\,\cdot\,\rangle^{(g)}_{11,\mathbb{R}})$ in [62, Prop. 5.1.2.2]. This is why we allow such an identification.

1.2.4 *Parameters for Kuga Families*

For the considerations in Section 1.3.3, we would like to have parameter sets for the toroidal compactifications there.

Let $(\mathcal{O}, \star, L, \langle\,\cdot\,,\,\cdot\,\rangle, h_0)$ be an integral PEL datum as in Definition 1.1.1.1. By our running assumption that $\mathcal{O}$ satisfies 1.2.1.1, the action of $\mathcal{O}$ on L extends to an action of some maximal order $\mathcal{O}'$ in $\mathcal{O} \underset{\mathbb{Z}}{\otimes} \mathbb{Q}$ containing $\mathcal{O}$. Let us fix the choice of such a maximal order $\mathcal{O}'$.

Let Q be an $\mathcal{O}$-lattice. Consider $\mathrm{Diff}^{-1} = \mathrm{Diff}^{-1}_{\mathcal{O}/\mathbb{Z}}$, the inverse different of $\mathcal{O}$ over $\mathbb{Z}$ [62, Def. 1.1.1.8] with its canonical left $\mathcal{O}$-module structure. Since the trace pairing $\mathrm{Diff}^{-1} \times \mathcal{O} \to \mathbb{Z} : (y, x) \mapsto \mathrm{Tr}_{\mathcal{O}/\mathbb{Z}}(yx)$ is perfect by definition, for each $\mathcal{O}$-lattice Q, we may identify $Q^\vee := \mathrm{Hom}_{\mathbb{Z}}(Q, \mathbb{Z})$ with $\mathrm{Hom}_{\mathcal{O}}(Q, \mathrm{Diff}^{-1})$. By composition with the involution $\star : \mathcal{O} \xrightarrow{\sim} \mathcal{O}^{\mathrm{op}}$, the natural right action of $\mathcal{O}$ on Diff^{-1} induced a left action of $\mathcal{O}$ on Diff^{-1}, which commutes with the natural left action of $\mathcal{O}$ on Diff^{-1}. Accordingly, the $\mathbb{Z}$-module $Q^\vee$ is torsion-free and has a canonical left $\mathcal{O}$-structure induced by the right action of $\mathcal{O}^{\mathrm{op}}$ on Diff^{-1} (and $\star : \mathcal{O} \xrightarrow{\sim} \mathcal{O}^{\mathrm{op}}$). In other words, $Q^\vee$ is an $\mathcal{O}$-lattice. Then the trace pairing induces a perfect pairing

$$\langle\,\cdot\,,\,\cdot\,\rangle_Q : Q^\vee \times Q \to \mathbb{Z} : (f, x) \mapsto \mathrm{Tr}_{\mathcal{O}/\mathbb{Z}}(f(x)).$$

For all $b \in \mathcal{O}$, $f \in Q^\vee$, and $x \in Q$, we have

$$\langle bf, x\rangle_Q = \mathrm{Tr}_{\mathcal{O}/\mathbb{Z}}(f(x)b^\star) = \mathrm{Tr}_{\mathcal{O}/\mathbb{Z}}(b^\star f(x)) = \mathrm{Tr}_{\mathcal{O}/\mathbb{Z}}(f(b^\star x)) = \langle f, b^\star x\rangle.$$

Lemma 1.2.4.1. (*See [61, Lem. 2.5].*) *There exists an embedding $j_Q : Q^\vee \hookrightarrow Q$ of $\mathcal{O}$-lattices inducing an isomorphism $j_Q : Q^\vee \underset{\mathbb{Z}}{\otimes} \mathbb{Q} \xrightarrow{\sim} Q \underset{\mathbb{Z}}{\otimes} \mathbb{Q}$ of $\mathcal{O} \underset{\mathbb{Z}}{\otimes} \mathbb{Q}$-modules such that the pairing*

$$\langle j_Q^{-1}(\,\cdot\,), \,\cdot\,\rangle_Q : (Q \underset{\mathbb{Z}}{\otimes} \mathbb{R}) \times (Q \underset{\mathbb{Z}}{\otimes} \mathbb{R}) \to \mathbb{R}$$

is positive definite.

Let $j_Q : Q^\vee \hookrightarrow Q$ be an embedding of $\mathcal{O}$-lattices given by Lemma 1.2.4.1, and let $(\widetilde{L}, \langle\,\cdot\,,\,\cdot\,\rangle^{\sim}, \widetilde{h}_0)$ be the symplectic $\mathcal{O}$-lattice given by the following data:

(1) An $\mathcal{O}$-lattice

$$\widetilde{L} := Q_{-2} \oplus L \oplus Q_0,$$

where

$$Q_{-2} := \mathrm{Hom}_{\mathcal{O}}(Q, \mathrm{Diff}^{-1}_{\mathcal{O}'/\mathbb{Z}}(1)) \subset Q^\vee \underset{\mathbb{Z}}{\otimes} \mathbb{Q}(1)$$

has $\mathcal{O}'$-module structure inherited from the two-sided ideal $\mathrm{Diff}^{-1}_{\mathcal{O}'/\mathbb{Z}}$ of $\mathcal{O}'$, and where

$$Q_0 := \mathcal{O}' \cdot Q \subset Q \underset{\mathbb{Z}}{\otimes} \mathbb{Q}.$$

(Then the perfect pairing $\langle\,\cdot\,,\,\cdot\,\rangle_Q : Q^\vee \times Q \to \mathbb{Z} : (f, x) \mapsto \mathrm{Tr}_{\mathcal{O}/\mathbb{Z}}(f(x))$ induces a perfect pairing $\langle\,\cdot\,,\,\cdot\,\rangle_Q : Q_{-2} \times Q_0 \to \mathbb{Z}(1)$, and $\widetilde{L}$ satisfies Condition 1.2.1.1 by construction.)

(2) A symplectic $\mathcal{O}$-pairing $\langle\,\cdot\,,\,\cdot\,\rangle^{\sim} : \widetilde{L} \times \widetilde{L} \to \mathbb{Z}(1)$ defined (symbolically) by the matrix

$$\langle x, y\rangle^{\sim} := {}^t\begin{pmatrix} x_{-2} \\ x_{-1} \\ x_0 \end{pmatrix} \begin{pmatrix} & & \langle\,\cdot\,,\,\cdot\,\rangle_Q \\ & \langle\,\cdot\,,\,\cdot\,\rangle & \\ -{}^t\langle\,\cdot\,,\,\cdot\,\rangle_Q & & \end{pmatrix} \begin{pmatrix} y_{-2} \\ y_{-1} \\ y_0 \end{pmatrix},$$

namely by

$$\langle x, y\rangle^{\sim} := \langle x_{-2}, y_0\rangle_Q + \langle x_{-1}, y_{-1}\rangle - \langle y_{-2}, x_0\rangle_Q,$$

where $x = \begin{pmatrix} x_{-2} \\ x_{-1} \\ x_0 \end{pmatrix}$ and $y = \begin{pmatrix} y_{-2} \\ y_{-1} \\ y_0 \end{pmatrix}$ are elements of $\widetilde{L} = Q_{-2} \oplus L \oplus Q_0$ expressed (vertically) in terms of components in the direct summands.

(3) An $\mathbb{R}$-algebra homomorphism $\widetilde{h}_0 : \mathbb{C} \to \operatorname{End}_{\mathcal{O} \underset{\mathbb{Z}}{\otimes} \mathbb{R}}(\widetilde{L} \underset{\mathbb{Z}}{\otimes} \mathbb{R})$ defined by

$$z = z_1 + \sqrt{-1}\, z_2 \mapsto \widetilde{h}_0(z) := \begin{pmatrix} z_1 \operatorname{Id}_{Q_{-2} \underset{\mathbb{Z}}{\otimes} \mathbb{R}} & & -z_2((2\pi\sqrt{-1}) \circ j_Q^{-1}) \\ & h_0(z) & \\ z_2(j_Q \circ (2\pi\sqrt{-1})^{-1}) & & z_1 \operatorname{Id}_{Q_0 \underset{\mathbb{Z}}{\otimes} \mathbb{R}} \end{pmatrix},$$

where $2\pi\sqrt{-1} : \mathbb{Z} \overset{\sim}{\to} \mathbb{Z}(1)$ and $(2\pi\sqrt{-1})^{-1} : \mathbb{Z}(1) \overset{\sim}{\to} \mathbb{Z}$ stand for the isomorphisms defined by the choice of $\sqrt{-1}$ in $\mathbb{C}$, and where the matrix acts (symbolically) on elements $x = \begin{pmatrix} x_{-2} \\ x_{-1} \\ x_0 \end{pmatrix}$ of $\widetilde{L} \underset{\mathbb{Z}}{\otimes} \mathbb{R}$ by left multiplication. In other words,

$$\widetilde{h}_0(z) \begin{pmatrix} x_{-2} \\ x_{-1} \\ x_0 \end{pmatrix} = \begin{pmatrix} z_1 x_{-2} - z_2((2\pi\sqrt{-1}) \circ j_Q^{-1})(x_0) \\ h_0(z) x_{-1} \\ z_2(j_Q \circ (2\pi\sqrt{-1})^{-1})(x_{-2}) + z_1 x_0 \end{pmatrix}.$$

Then $\widetilde{h}_0$ is a polarization of $(\widetilde{L}, \langle\,\cdot\,,\,\cdot\,\rangle^{\sim})$ making $(\widetilde{L}, \langle\,\cdot\,,\,\cdot\,\rangle^{\sim}, \widetilde{h}_0)$ a PEL-type $\mathcal{O}$-lattice. Note that the reflex field of $(\widetilde{L} \underset{\mathbb{Z}}{\otimes} \mathbb{R}, \langle\,\cdot\,,\,\cdot\,\rangle^{\sim}, \widetilde{h})$ is also F_0.

Remark 1.2.4.2. If p is a good prime for $(\mathcal{O}, \star, L, \langle\,\cdot\,,\,\cdot\,\rangle, h_0)$ as in Definition 1.1.1.6, then it is also a good prime for $(\mathcal{O}, \star, \widetilde{L}, \langle\,\cdot\,,\,\cdot\,\rangle^{\sim}, \widetilde{h}_0)$.

By the construction of $(\widetilde{L}, \langle\,\cdot\,,\,\cdot\,\rangle^{\sim})$, there is a fully symplectic admissible filtration on $\widetilde{L} \underset{\mathbb{Z}}{\otimes} \hat{\mathbb{Z}}$ induced by

$$0 \subset Q_{-2} \subset Q_{-2} \oplus L \subset Q_{-2} \oplus L \oplus Q_0 = \widetilde{L}.$$

More precisely, we have

$$\widetilde{\mathrm{Z}}_{-3} := 0, \quad \widetilde{\mathrm{Z}}_{-2} := Q_{-2} \underset{\mathbb{Z}}{\otimes} \hat{\mathbb{Z}}, \quad \widetilde{\mathrm{Z}}_{-1} := (Q_{-2} \underset{\mathbb{Z}}{\otimes} \hat{\mathbb{Z}}) \oplus (L \underset{\mathbb{Z}}{\otimes} \hat{\mathbb{Z}}), \quad \text{and}$$
$$\widetilde{\mathrm{Z}}_0 := (Q_{-2} \underset{\mathbb{Z}}{\otimes} \hat{\mathbb{Z}}) \oplus (L \underset{\mathbb{Z}}{\otimes} \hat{\mathbb{Z}}) \oplus (Q_0 \underset{\mathbb{Z}}{\otimes} \hat{\mathbb{Z}}) = \widetilde{L} \underset{\mathbb{Z}}{\otimes} \hat{\mathbb{Z}},$$

so that there are canonical isomorphisms

$$\mathrm{Gr}_{-2}^{\widetilde{\mathrm{Z}}} \cong Q_{-2} \underset{\mathbb{Z}}{\otimes} \hat{\mathbb{Z}}, \quad \mathrm{Gr}_{-1}^{\widetilde{\mathrm{Z}}} \cong L \underset{\mathbb{Z}}{\otimes} \hat{\mathbb{Z}}, \quad \text{and} \quad \mathrm{Gr}_{0}^{\widetilde{\mathrm{Z}}} \cong Q_0 \underset{\mathbb{Z}}{\otimes} \hat{\mathbb{Z}}$$

matching the pairings $\mathrm{Gr}_{-2}^{\widetilde{\mathrm{Z}}} \times \mathrm{Gr}_{0}^{\widetilde{\mathrm{Z}}} \to \hat{\mathbb{Z}}(1)$ and $\mathrm{Gr}_{-1}^{\widetilde{\mathrm{Z}}} \times \mathrm{Gr}_{-1}^{\widetilde{\mathrm{Z}}} \to \hat{\mathbb{Z}}(1)$ induced by $\langle\,\cdot\,,\,\cdot\,\rangle^{\sim}$ with $\langle\,\cdot\,,\,\cdot\,\rangle_Q$ and $\langle\,\cdot\,,\,\cdot\,\rangle$, respectively.

Let $\widetilde{X} := \mathrm{Hom}_{\mathcal{O}}(Q_{-2}, \mathrm{Diff}^{-1}(1))$ and $\widetilde{Y} := Q_0$. The pairing $\langle\,\cdot\,,\,\cdot\,\rangle_Q : Q_{-2} \times Q_0 \to \mathbb{Z}(1)$ induces a canonical embedding $\widetilde{\phi} : \widetilde{Y} \hookrightarrow \widetilde{X}$ and there are canonical isomorphisms $\widetilde{\varphi}_{-2} : \mathrm{Gr}_{-2}^{\widetilde{\mathrm{Z}}} \xrightarrow{\sim} \mathrm{Hom}_{\hat{\mathbb{Z}}}(\widetilde{X} \underset{\mathbb{Z}}{\otimes} \hat{\mathbb{Z}}, \hat{\mathbb{Z}}(1))$ and $\widetilde{\varphi}_0 : \mathrm{Gr}_{0}^{\widetilde{\mathrm{Z}}} \xrightarrow{\sim} \widetilde{Y} \underset{\mathbb{Z}}{\otimes} \hat{\mathbb{Z}}$ (of $\mathcal{O} \underset{\mathbb{Z}}{\otimes} \hat{\mathbb{Z}}$-modules). These data define a torus argument $\widetilde{\Phi} := (\widetilde{X}, \widetilde{Y}, \widetilde{\phi}, \widetilde{\varphi}_{-2}, \widetilde{\varphi}_0)$ for $\widetilde{\mathrm{Z}}$ as in Definition 1.2.1.5.

Let $\widetilde{\delta}$ be the obvious splitting of $\widetilde{\mathrm{Z}}$ induced by the equality $Q_{-2} \oplus L \oplus Q_0 = \widetilde{L}$.

Let $\widetilde{\mathrm{G}}$ be the group functor defined by $(\widetilde{L}, \langle\,\cdot\,,\,\cdot\,\rangle^{\sim})$ as in Definition 1.1.1.3, with the subgroup functor $\widetilde{\mathrm{P}}_{\widetilde{\mathrm{Z}}}$ defined by $\widetilde{\mathrm{Z}}$ as in Definition 1.2.1.10, and quotients $\widetilde{\mathrm{Z}}_{\widetilde{\mathrm{Z}}}(R)$, $\widetilde{\mathrm{U}}_{\widetilde{\mathrm{Z}}}(R)$, $\widetilde{\mathrm{U}}_{2,\widetilde{\mathrm{Z}}}(R)$, $\widetilde{\mathrm{U}}_{1,\widetilde{\mathrm{Z}}}(R)$, $\widetilde{\mathrm{G}}_{l,\widetilde{\mathrm{Z}}}(R)$, $\widetilde{\mathrm{G}}'_{l,\widetilde{\mathrm{Z}}}(R)$, $\widetilde{\mathrm{P}}'_{\widetilde{\mathrm{Z}}}(R)$, $\widetilde{\mathrm{G}}_{1,\widetilde{\mathrm{Z}}}(R)$, and $\widetilde{\mathrm{G}}'_{h,\widetilde{\mathrm{Z}}}(R)$ of subgroups of $\widetilde{\mathrm{P}}_{\widetilde{\mathrm{Z}}}(R)$ defined for each $\hat{\mathbb{Z}}$-algebra R as in Definition 1.2.1.11. Note that we have, by Definition 1.2.1.9, a canonical isomorphism $\mathrm{G}_{h,\widetilde{\mathrm{Z}}}(R) \cong \mathrm{G}'_{h,\widetilde{\mathrm{Z}}}(R) \cong (\mathrm{G} \underset{\mathbb{Z}}{\otimes} \hat{\mathbb{Z}})(R)$. Then we also define:

Definition 1.2.4.3.

(1) $\widehat{\mathrm{U}}(R) := \widetilde{\mathrm{U}}_{1,\widetilde{\mathrm{Z}}}(R) = \widetilde{\mathrm{U}}_{\widetilde{\mathrm{Z}}}(R)/\widetilde{\mathrm{U}}_{2,\widetilde{\mathrm{Z}}}(R)$.

(2) $\widehat{\mathrm{G}}(R) := \widetilde{\mathrm{G}}_{1,\widetilde{\mathrm{Z}}}(R) = \widetilde{\mathrm{P}}'_{\widetilde{\mathrm{Z}}}(R)/\widetilde{\mathrm{U}}_{2,\widetilde{\mathrm{Z}}}(R)$, *which is (under the splitting* $\widetilde{\delta}$ *above) isomorphic to* $(\mathrm{G} \ltimes \widehat{\mathrm{U}})(R) := \mathrm{G}(R) \ltimes \widehat{\mathrm{U}}(R)$.

Definition 1.2.4.4. (*Compare with Definition* 1.2.1.12.) *For each open compact subgroup* $\widetilde{\mathcal{H}}$ *of* $\widetilde{\mathrm{G}}(\hat{\mathbb{Z}})$, *we define* $\widetilde{\mathcal{H}}_{\widetilde{\mathrm{P}}_{\widetilde{\mathrm{Z}}}} := \widetilde{\mathcal{H}} \cap \widetilde{\mathrm{P}}_{\widetilde{\mathrm{Z}}}(\hat{\mathbb{Z}})$, $\widetilde{\mathcal{H}}_{\widetilde{\mathrm{Z}}_{\widetilde{\mathrm{Z}}}} := \widetilde{\mathcal{H}} \cap \widetilde{\mathrm{Z}}_{\widetilde{\mathrm{Z}}}(\hat{\mathbb{Z}})$, $\widetilde{\mathcal{H}}_{\widetilde{\mathrm{G}}_{h,\widetilde{\mathrm{Z}}}} := \widetilde{\mathcal{H}}_{\widetilde{\mathrm{P}}_{\widetilde{\mathrm{Z}}}}/\widetilde{\mathcal{H}}_{\widetilde{\mathrm{Z}}_{\widetilde{\mathrm{Z}}}}$, $\widetilde{\mathcal{H}}_{\widetilde{\mathrm{U}}_{\widetilde{\mathrm{Z}}}} := \widetilde{\mathcal{H}} \cap \widetilde{\mathrm{U}}_{\widetilde{\mathrm{Z}}}(\hat{\mathbb{Z}})$, $\widetilde{\mathcal{H}}_{\widetilde{\mathrm{U}}_{2,\widetilde{\mathrm{Z}}}} := \widetilde{\mathcal{H}} \cap \widetilde{\mathrm{U}}_{2,\widetilde{\mathrm{Z}}}(\hat{\mathbb{Z}})$, $\widetilde{\mathcal{H}}_{\widetilde{\mathrm{U}}_{1,\widetilde{\mathrm{Z}}}} := \widetilde{\mathcal{H}}_{\widetilde{\mathrm{U}}_{\widetilde{\mathrm{Z}}}}/\widetilde{\mathcal{H}}_{\widetilde{\mathrm{U}}_{2,\widetilde{\mathrm{Z}}}}$, $\widetilde{\mathcal{H}}_{\widetilde{\mathrm{G}}_{l,\widetilde{\mathrm{Z}}}} := \widetilde{\mathcal{H}}_{\widetilde{\mathrm{Z}}_{\widetilde{\mathrm{Z}}}}/\widetilde{\mathcal{H}}_{\widetilde{\mathrm{U}}_{\widetilde{\mathrm{Z}}}}$, $\widetilde{\mathcal{H}}_{\widetilde{\mathrm{P}}'_{\widetilde{\mathrm{Z}}}} := \widetilde{\mathcal{H}} \cap \widetilde{\mathrm{P}}'_{\widetilde{\mathrm{Z}}}(\hat{\mathbb{Z}})$, $\widetilde{\mathcal{H}}_{\widetilde{\mathrm{G}}_{1,\widetilde{\mathrm{Z}}}} := \widetilde{\mathcal{H}}_{\widetilde{\mathrm{P}}'_{\widetilde{\mathrm{Z}}}}/\widetilde{\mathcal{H}}_{\widetilde{\mathrm{U}}_{2,\widetilde{\mathrm{Z}}}}$, $\widetilde{\mathcal{H}}_{\widetilde{\mathrm{G}}'_{h,\widetilde{\mathrm{Z}}}} := \widetilde{\mathcal{H}}_{\widetilde{\mathrm{P}}'_{\widetilde{\mathrm{Z}}}}/\widetilde{\mathcal{H}}_{\widetilde{\mathrm{U}}_{\widetilde{\mathrm{Z}}}}$, *and* $\widetilde{\mathcal{H}}_{\widetilde{\mathrm{G}}'_{l,\widetilde{\mathrm{Z}}}} := \widetilde{\mathcal{H}}_{\widetilde{\mathrm{P}}_{\widetilde{\mathrm{Z}}}}/\widetilde{\mathcal{H}}_{\widetilde{\mathrm{P}}'_{\widetilde{\mathrm{Z}}}}$ *as in Definition* 1.2.1.12, *and define*

(1) $\widehat{\mathcal{H}} := \widetilde{\mathcal{H}}_{\widehat{\mathrm{G}}} := \widetilde{\mathcal{H}}_{\widetilde{\mathrm{G}}_{1,\widetilde{\mathrm{Z}}}}$.

(2) $\widehat{\mathcal{H}}_{\widehat{\mathrm{U}}} := \widetilde{\mathcal{H}}_{\widehat{\mathrm{U}}} := \widetilde{\mathcal{H}}_{\widetilde{\mathrm{U}}_{1,\widetilde{\mathrm{Z}}}}$.

(3) $\widehat{\mathcal{H}}_{\mathrm{G}} := \widehat{\mathcal{H}}/\widehat{\mathcal{H}}_{\widehat{\mathrm{U}}} \cong \widetilde{\mathcal{H}}_{\widetilde{\mathrm{G}}'_{h,\widetilde{\mathrm{Z}}}}$.

Then we have an exact sequence

$$1 \to \widehat{\mathcal{H}}_{\widehat{\mathrm{U}}} \to \widehat{\mathcal{H}} \to \widehat{\mathcal{H}}_{\mathrm{G}} \to 1 \tag{1.2.4.5}$$

compatible with the canonical exact sequence

$$1 \to \widehat{\mathrm{U}}(\hat{\mathbb{Z}}) \to \widehat{\mathrm{G}}(\hat{\mathbb{Z}}) \to \mathrm{G}(\hat{\mathbb{Z}}) \to 1. \tag{1.2.4.6}$$

We shall also extend this definition to the cases of $\hat{\mathbb{Z}}^p$- *or* $\mathbb{Z}_p$-*valued groups above.*

Let $\mathcal{H}$ be any open compact subgroup of $\mathrm{G}(\hat{\mathbb{Z}})$, and let $\widetilde{\mathcal{H}}$ be any *neat* open compact subgroup of $\widetilde{\mathrm{G}}(\hat{\mathbb{Z}})$ satisfying the following condition:

Condition 1.2.4.7. $\widehat{\mathcal{H}}_{\mathrm{G}} = \mathrm{Gr}^{\widetilde{\mathsf{Z}}}_{-1}(\widetilde{\mathcal{H}}_{\widetilde{\mathrm{P}}'_{\widetilde{\mathsf{Z}}}}) = \mathrm{Gr}^{\widetilde{\mathsf{Z}}}_{-1}(\widetilde{\mathcal{H}}_{\widetilde{\mathrm{P}}_{\widetilde{\mathsf{Z}}}}) \subset \mathcal{H}$. *(The first equality is just the definition, while the second equality is the essential condition. Then* $\mathrm{Gr}^{\widetilde{\mathsf{Z}}}_{-1}(\widetilde{\mathcal{H}}_{\widetilde{\mathrm{P}}_{\widetilde{\mathsf{Z}}}})$ *is a direct factor of* $\widetilde{\mathcal{H}}_{\widetilde{\mathrm{P}}_{\widetilde{\mathsf{Z}}}}/\widetilde{\mathcal{H}}_{\widetilde{\mathrm{U}}_{\widetilde{\mathsf{Z}}}}$.*)*

For later purposes, we define two more conditions on such $\widetilde{\mathcal{H}}$, or rather on $\widehat{\mathcal{H}}$:

Condition 1.2.4.8. $\widehat{\mathcal{H}}_{\mathrm{G}} = \mathcal{H}$. *(Then Condition* 1.2.4.7 *is redundant, because we always have* $\widehat{\mathcal{H}}_{\mathrm{G}} = \mathrm{Gr}^{\widetilde{\mathsf{Z}}}_{-1}(\widetilde{\mathcal{H}}_{\widetilde{\mathrm{P}}'_{\widetilde{\mathsf{Z}}}}) \subset \mathrm{Gr}^{\widetilde{\mathsf{Z}}}_{-1}(\widetilde{\mathcal{H}}_{\widetilde{\mathrm{P}}_{\widetilde{\mathsf{Z}}}}) \subset \mathcal{H}$.*)*

Condition 1.2.4.9. *The splitting* $\widetilde{\delta}$ *defines a (group-theoretic) splitting of the sequence* (1.2.4.6) *and induces an isomorphism* $\widehat{\mathrm{G}}(\hat{\mathbb{Z}}) \cong \mathrm{G}(\hat{\mathbb{Z}}) \ltimes \widehat{\mathrm{U}}(\hat{\mathbb{Z}})$*, which also defines a (group-theoretic) splitting of the sequence* (1.2.4.5) *and induces an isomorphism* $\widehat{\mathcal{H}} \cong \mathcal{H} \ltimes \widehat{\mathcal{H}}_{\widehat{\mathrm{U}}}$. *(This condition is equivalent to the condition that the splitting* $\mathrm{G}(\hat{\mathbb{Z}}) \to \widehat{\mathrm{G}}(\hat{\mathbb{Z}})$ *defined by* $\widetilde{\delta}$ *maps* $\mathcal{H}$ *to* $\widehat{\mathcal{H}}$.*)*

Remark 1.2.4.10. For each $\mathcal{H}$, there exists $\widetilde{\mathcal{H}}$ satisfying these conditions, because the pairing $\langle\,\cdot\,,\,\cdot\,\rangle^{\sim}$ is the direct sum of the pairings on $Q_{-2} \oplus Q_0$ and on L.

Let $(\widetilde{\mathsf{Z}}, \widetilde{\Phi}, \widetilde{\delta})$ be defined as above, which induces a representative $(\widetilde{\mathsf{Z}}_{\widetilde{\mathcal{H}}}, \widetilde{\Phi}_{\widetilde{\mathcal{H}}} = (\widetilde{X}, \widetilde{Y}, \widetilde{\phi}, \widetilde{\varphi}_{-2,\widetilde{\mathcal{H}}}, \widetilde{\varphi}_{0,\widetilde{\mathcal{H}}}), \widetilde{\delta}_{\widetilde{\mathcal{H}}})$ of a cusp label $[(\widetilde{\mathsf{Z}}_{\widetilde{\mathcal{H}}}, \widetilde{\Phi}_{\widetilde{\mathcal{H}}}, \widetilde{\delta}_{\widetilde{\mathcal{H}}})]$ at level $\widetilde{\mathcal{H}}$. Let $\widetilde{\Sigma}$ be any compatible choice of admissible smooth rational polyhedral cone decomposition data for $\widetilde{\mathsf{M}}_{\widetilde{\mathcal{H}}}$ that is *projective* (see Definitions 1.2.2.13 and 1.2.2.14). Let $\widetilde{\sigma} \subset \mathbf{P}^{+}_{\widetilde{\Phi}_{\widetilde{\mathcal{H}}}}$ be any *top-dimensional* nondegenerate rational polyhedral cone in the cone decomposition $\widetilde{\Sigma}_{\widetilde{\Phi}_{\widetilde{\mathcal{H}}}}$ in $\widetilde{\Sigma}$.

Definition 1.2.4.11.

(1) $\widetilde{\mathbf{K}}^{++}_{Q,\mathcal{H}}$ *is the set of all triples* $\widetilde{\kappa} = (\widetilde{\mathcal{H}}, \widetilde{\Sigma}, \widetilde{\sigma})$ *as above (such that* $\widetilde{\mathcal{H}}$ *satisfies Condition* 1.2.4.7*).*

(2) $\widetilde{\mathbf{K}}^{+}_{Q,\mathcal{H}}$ *is the subset of* $\widetilde{\mathbf{K}}^{++}_{Q,\mathcal{H}}$ *consisting of elements* $\widetilde{\kappa} = (\widetilde{\mathcal{H}}, \widetilde{\Sigma}, \widetilde{\sigma})$ *such that* $\widetilde{\mathcal{H}}$ *satisfies Condition* 1.2.4.8.

(3) $\widetilde{\mathbf{K}}_{Q,\mathcal{H}}$ *is the subset of* $\widetilde{\mathbf{K}}^{+}_{Q,\mathcal{H}}$ *consisting of elements* $\widetilde{\kappa} = (\widetilde{\mathcal{H}}, \widetilde{\Sigma}, \widetilde{\sigma})$ *such that* $\widetilde{\mathcal{H}}$ *also satisfies Condition* 1.2.4.9.

The equivalence classes $[(\breve{\Phi}_{\widetilde{\mathcal{H}}}, \breve{\delta}_{\widetilde{\mathcal{H}}}, \breve{\tau})]$ having $[(\widetilde{\Phi}_{\widetilde{\mathcal{H}}}, \widetilde{\delta}_{\widetilde{\mathcal{H}}}, \widetilde{\sigma})]$ as a face (as in Definition 1.2.2.19) are $\widetilde{\mathcal{H}}$-orbits of data of the following form:

(1) A fully symplectic admissible filtration $\breve{\mathsf{Z}} = \{\breve{\mathsf{Z}}_{-i}\}$ on $\widetilde{L} \underset{\mathbb{Z}}{\otimes} \hat{\mathbb{Z}}$ satisfying

$$\widetilde{\mathsf{Z}}_{-2} \subset \breve{\mathsf{Z}}_{-2} \subset \breve{\mathsf{Z}}_{-1} \subset \widetilde{\mathsf{Z}}_{-1}. \tag{1.2.4.12}$$

Each such filtration $\breve{\mathsf{Z}}$ induces a fully symplectic admissible filtration $\mathsf{Z} = \{\mathsf{Z}_{-i}\}$ on $L \underset{\mathbb{Z}}{\otimes} \hat{\mathbb{Z}}$ by $\mathsf{Z}_{-2} := \breve{\mathsf{Z}}_{-2}/\widetilde{\mathsf{Z}}_{-2}$ and $\mathsf{Z}_{-1} := \breve{\mathsf{Z}}_{-1}/\widetilde{\mathsf{Z}}_{-2}$, so that there is a canonical isomorphism

$$\mathsf{Z}_0/\mathsf{Z}_{-1} \cong \widetilde{\mathsf{Z}}_{-1}/\breve{\mathsf{Z}}_{-1}. \tag{1.2.4.13}$$

Conversely, each fully symplectic admissible filtration Z on $L \underset{\mathbb{Z}}{\otimes} \hat{\mathbb{Z}}$ induces a fully symplectic admissible filtration $\breve{\mathsf{Z}}$ on $\widetilde{L} \underset{\mathbb{Z}}{\otimes} \hat{\mathbb{Z}}$ satisfying (1.2.4.12) and (1.2.4.13).

(2) A torus argument $\breve{\Phi} = (\breve{X}, \breve{Y}, \breve{\phi}, \breve{\varphi}_{-2}, \breve{\varphi}_0)$ for $\breve{\mathsf{Z}}$ (as in Definition 1.2.1.5), together with admissible surjections $s_{\breve{X}} : \breve{X} \twoheadrightarrow \widetilde{X}$ and $s_{\breve{Y}} : \breve{Y} \twoheadrightarrow \widetilde{Y}$ satisfying $s_{\breve{X}}\breve{\phi} = \widetilde{\phi}s_{\breve{Y}}$ and other natural compatibilities with $\breve{\varphi}_{-2}$, $\breve{\varphi}_0$, $\widetilde{\varphi}_{-2}$, and $\widetilde{\varphi}_0$. (See Definitions 1.2.1.16, 1.2.1.17, and 1.2.1.18.)

Any $\breve{\Phi}$, $s_{\breve{X}}$, and $s_{\breve{Y}}$ determine a torus argument $\Phi = (X, Y, \phi, \varphi_{-2}, \varphi_0)$ for Z by $X := \ker(s_{\breve{X}})$, $Y := \ker(s_{\breve{Y}})$, and $\phi := \breve{\phi}|_Y$, so that there is a commutative diagram

$$\begin{array}{ccccccccc} 0 & \longrightarrow & Y & \longrightarrow & \breve{Y} & \xrightarrow{s_{\breve{Y}}} & \widetilde{Y} & \longrightarrow & 0 \\ & & \downarrow{\scriptstyle \phi} & & \downarrow{\scriptstyle \breve{\phi}} & & \downarrow{\scriptstyle \widetilde{\phi}} & & \\ 0 & \longrightarrow & X & \longrightarrow & \breve{X} & \xrightarrow[s_{\breve{X}}]{} & \widetilde{X} & \longrightarrow & 0 \end{array} \tag{1.2.4.14}$$

whose horizontal rows are exact sequences.

(3) The existence of some splitting $\breve{\delta}$ of $\breve{\mathsf{Z}}$, inducing some liftable splitting $\breve{\delta}_{\widetilde{\mathcal{H}}}$ defining the representative $(\breve{\mathsf{Z}}_{\widetilde{\mathcal{H}}}, \breve{\Phi}_{\widetilde{\mathcal{H}}}, \breve{\delta}_{\widetilde{\mathcal{H}}})$ of cusp label $[(\breve{\mathsf{Z}}_{\widetilde{\mathcal{H}}}, \breve{\Phi}_{\widetilde{\mathcal{H}}}, \breve{\delta}_{\widetilde{\mathcal{H}}})]$ at level $\widetilde{\mathcal{H}}$.

Given the splitting $\widetilde{\delta}$, the existence of some splitting $\breve{\delta}$ is equivalent to the existence of some splitting δ of Z. Then, for forming compatible orbits, we have the following

Lemma 1.2.4.15. *There is a canonical assignment from the set of cusp labels $[(\breve{\mathsf{Z}}_{\widetilde{\mathcal{H}}}, \breve{\Phi}_{\widetilde{\mathcal{H}}}, \breve{\delta}_{\widetilde{\mathcal{H}}})]$ at level $\widetilde{\mathcal{H}}$ admitting a surjection to $[(\widetilde{\mathsf{Z}}_{\widetilde{\mathcal{H}}}, \widetilde{\Phi}_{\widetilde{\mathcal{H}}}, \widetilde{\delta}_{\widetilde{\mathcal{H}}})]$, to the set of cusp labels $[(\mathsf{Z}_{\mathcal{H}}, \Phi_{\mathcal{H}}, \delta_{\mathcal{H}})]$ at level $\mathcal{H}$. This assignment is bijective if we assume Condition* 1.2.4.8*, so that, in particular, $\mathrm{Gr}^{\widetilde{\mathsf{Z}}}_{-1}(\widetilde{\mathcal{H}}_{\mathrm{P}_{\widetilde{\mathsf{Z}}}}) = \mathcal{H}$; and is still surjective if we only assume $\mathrm{Gr}^{\widetilde{\mathsf{Z}}}_{-1}(\widetilde{\mathcal{H}}_{\mathrm{P}_{\widetilde{\mathsf{Z}}}}) \subset \mathcal{H}$.*

By definition, we have the following:

Lemma 1.2.4.16. *With the fixed choice of $(\widetilde{\mathsf{Z}}, \widetilde{\Phi}, \widetilde{\delta})$ representing the cusp label $[(\widetilde{\mathsf{Z}}_{\widetilde{\mathcal{H}}}, \widetilde{\Phi}_{\widetilde{\mathcal{H}}}, \widetilde{\delta}_{\widetilde{\mathcal{H}}})]$ at level $\widetilde{\mathcal{H}}$, the choices of representatives $(\breve{\mathsf{Z}}, \breve{\Phi}, \breve{\delta})$ of the cusp label $[(\breve{\mathsf{Z}}_{\widetilde{\mathcal{H}}}, \breve{\Phi}_{\widetilde{\mathcal{H}}}, \breve{\delta}_{\widetilde{\mathcal{H}}})]$ that are compatible with $(\widetilde{\mathsf{Z}}, \widetilde{\Phi}, \widetilde{\delta})$ as above form an $\widehat{\mathcal{H}}$-orbit up to equivalence.*

Therefore, it makes sense to have the following:

Definition 1.2.4.17. *We shall denote any $\widehat{\mathcal{H}}$-orbit as in Lemma* 1.2.4.16 *by* $(\breve{\mathsf{Z}}_{\widehat{\mathcal{H}}}, \breve{\Phi}_{\widehat{\mathcal{H}}} = (\breve{X}, \breve{Y}, \breve{\phi}, \breve{\varphi}_{-2,\widehat{\mathcal{H}}}, \breve{\varphi}_{0,\widehat{\mathcal{H}}}), \breve{\delta}_{\widehat{\mathcal{H}}})$, *and denote its (well-defined) equivalence class by* $[(\breve{\mathsf{Z}}_{\widehat{\mathcal{H}}}, \breve{\Phi}_{\widehat{\mathcal{H}}}, \breve{\delta}_{\widehat{\mathcal{H}}})]$. *We say in this case that* $\breve{\Phi}_{\widehat{\mathcal{H}}}$ *is a torus argument for* $\breve{\mathsf{Z}}_{\widehat{\mathcal{H}}}$ *at level* $\widehat{\mathcal{H}}$. *We can generalize all notions for cusp labels for* $\mathsf{M}_{\mathcal{H}}$ *to the context here, and consider a cusp label* $[(\breve{\mathsf{Z}}_{\widehat{\mathcal{H}}}, \breve{\Phi}_{\widehat{\mathcal{H}}}, \breve{\delta}_{\widehat{\mathcal{H}}})]$ *at level* $\widehat{\mathcal{H}}$ *for* $(\widetilde{L}, \langle\,\cdot\,,\,\cdot\,\rangle^{\sim}, \widetilde{h}_0, \widetilde{\mathsf{Z}})$ *(see Definition* 1.2.1.7 *and [62, Def. 5.4.2.1, 5.4.2.2, and 5.4.2.4]). We shall replace the subscripts "*$\breve{\Phi}_{\widetilde{\mathcal{H}}}$*" with "*$\breve{\Phi}_{\widehat{\mathcal{H}}}$*" in the notation for objects depending only on the $\widehat{\mathcal{H}}$-orbit of $\breve{\Phi}$.*

(4) Let $\Phi_{\mathcal{H}}$ (resp. $\breve{\Phi}_{\widetilde{\mathcal{H}}}$) be the torus argument for $\mathsf{Z}_{\mathcal{H}}$ (resp. $\widetilde{\mathsf{Z}}_{\widetilde{\mathcal{H}}}$) at level $\mathcal{H}$ (resp. $\widetilde{\mathcal{H}}$) induced by Φ (resp. $\breve{\Phi}$). Then (1.2.4.14) induces morphisms

$$\mathbf{S}_{\Phi_{\mathcal{H}}} \hookrightarrow \mathbf{S}_{\breve{\Phi}_{\widetilde{\mathcal{H}}}} \twoheadrightarrow \mathbf{S}_{\widetilde{\Phi}_{\widetilde{\mathcal{H}}}}, \tag{1.2.4.18}$$

where the first morphism is canonical, and where the second morphism is defined by $s_{\breve{X}}$ and $s_{\breve{Y}}$, whose composition is zero. (In general, the morphisms in (1.2.4.18) *do not* form an exact sequence.)
The dual of (1.2.4.18) defines morphisms

$$(\mathbf{S}_{\widetilde{\Phi}_{\widetilde{\mathcal{H}}}})^{\vee}_{\mathbb{R}} \hookrightarrow (\mathbf{S}_{\breve{\Phi}_{\widetilde{\mathcal{H}}}})^{\vee}_{\mathbb{R}} \twoheadrightarrow (\mathbf{S}_{\Phi_{\mathcal{H}}})^{\vee}_{\mathbb{R}}, \tag{1.2.4.19}$$

where the first morphism is defined by $s_{\breve{X}}$ and $s_{\breve{Y}}$, and where the second morphism is canonical, whose composition is zero, inducing morphisms

$$\mathbf{P}^{+}_{\widetilde{\Phi}_{\widetilde{\mathcal{H}}}} \hookrightarrow \mathbf{P}_{\breve{\Phi}_{\widetilde{\mathcal{H}}}} \twoheadrightarrow \mathbf{P}_{\Phi_{\mathcal{H}}}. \tag{1.2.4.20}$$

Then $\breve{\tau} \subset \mathbf{P}^{+}_{\breve{\Phi}_{\widetilde{\mathcal{H}}}}$ is a cone in the cone decomposition $\widetilde{\Sigma}_{\breve{\Phi}_{\widetilde{\mathcal{H}}}}$ having a face $\breve{\sigma}$ that is a $\Gamma_{\breve{\Phi}_{\widetilde{\mathcal{H}}}}$-translation (see Definition 1.2.2.3) of the image of $\widetilde{\sigma} \subset \mathbf{P}^{+}_{\widetilde{\Phi}_{\widetilde{\mathcal{H}}}}$ under the first morphism in (1.2.4.20).

Without loss of generality, let us set $\breve{\sigma}$ to be the image of $\widetilde{\sigma} \subset \mathbf{P}^{+}_{\widetilde{\Phi}_{\widetilde{\mathcal{H}}}}$ under the first morphism in (1.2.4.20), and consider the following:

Definition 1.2.4.21.

(1) $\widetilde{\Sigma}_{\breve{\Phi}_{\widehat{\mathcal{H}}},\breve{\sigma}} = \widetilde{\Sigma}_{\breve{\Phi}_{\widetilde{\mathcal{H}}},\breve{\sigma}}$ *(resp.* $\widetilde{\Sigma}^{+}_{\breve{\Phi}_{\widehat{\mathcal{H}}},\breve{\sigma}} = \widetilde{\Sigma}^{+}_{\breve{\Phi}_{\widetilde{\mathcal{H}}},\breve{\sigma}}$*) is the subset of* $\widetilde{\Sigma}_{\breve{\Phi}_{\widetilde{\mathcal{H}}}}$ *consisting of cones* $\breve{\tau} \subset \mathbf{P}_{\breve{\Phi}_{\widetilde{\mathcal{H}}}}$ *(resp.* $\breve{\tau} \subset \mathbf{P}^{+}_{\breve{\Phi}_{\widetilde{\mathcal{H}}}}$*) having* $\breve{\sigma}$ *(not just a* $\Gamma_{\breve{\Phi}_{\widetilde{\mathcal{H}}}}$*-translation) as a face.*
(2) $\Gamma_{\breve{\Phi}_{\widetilde{\mathcal{H}}},\Phi_{\mathcal{H}}}$ *is the subgroup of* $\Gamma_{\breve{\Phi}_{\widetilde{\mathcal{H}}}}$ *stabilizing (both) X and Y.*
(3) $\Gamma_{\breve{\Phi}_{\widetilde{\mathcal{H}}},\Phi_{\mathcal{H}},\breve{\sigma}}$ *is the subgroup of* $\Gamma_{\breve{\Phi}_{\widetilde{\mathcal{H}}},\Phi_{\mathcal{H}}}$ *stabilizing* $\breve{\sigma}$.
(4) $\Gamma_{\breve{\Phi}_{\widehat{\mathcal{H}}}}$ *is the kernel of the canonical homomorphism* $\Gamma_{\breve{\Phi}_{\widetilde{\mathcal{H}}},\Phi_{\mathcal{H}}} \to \Gamma_{\widetilde{\Phi}_{\widetilde{\mathcal{H}}}}$ *(induced by* $s_{\breve{X}}$ *and* $s_{\breve{Y}}$*).*
(5) $\Gamma_{\breve{\Phi}_{\widehat{\mathcal{H}}},\Phi_{\mathcal{H}}}$ *is the kernel of the canonical homomorphism* $\Gamma_{\breve{\Phi}_{\widetilde{\mathcal{H}}},\Phi_{\mathcal{H}}} \to \Gamma_{\Phi_{\mathcal{H}}} \times \Gamma_{\widetilde{\Phi}_{\widetilde{\mathcal{H}}}}$*, which coincides with the kernel of the canonical homomorphism* $\Gamma_{\breve{\Phi}_{\widehat{\mathcal{H}}}} \to \Gamma_{\Phi_{\mathcal{H}}}$.

By definition, we have the following compatible exact sequences

$$1 \to \Gamma_{\breve{\Phi}_{\widehat{\mathcal{H}}},\Phi_{\mathcal{H}}} \to \Gamma_{\breve{\Phi}_{\widetilde{\mathcal{H}}},\Phi_{\mathcal{H}}} \to \Gamma_{\Phi_{\mathcal{H}}} \times \Gamma_{\widetilde{\Phi}_{\widetilde{\mathcal{H}}}} \tag{1.2.4.22}$$

and

$$1 \to \Gamma_{\breve{\Phi}_{\widehat{\mathcal{H}}},\Phi_{\mathcal{H}}} \to \Gamma_{\breve{\Phi}_{\widehat{\mathcal{H}}}} \to \Gamma_{\Phi_{\mathcal{H}}}. \tag{1.2.4.23}$$

Remark 1.2.4.24. The notation of $\Gamma_{\breve{\Phi}_{\widehat{\mathcal{H}}}}$ and $\Gamma_{\breve{\Phi}_{\widehat{\mathcal{H}}},\Phi_{\mathcal{H}}}$ in Definition 1.2.4.21 is justified because the subgroups $\Gamma_{\breve{\Phi}_{\widehat{\mathcal{H}}}}$ and $\Gamma_{\breve{\Phi}_{\widehat{\mathcal{H}}},\Phi_{\mathcal{H}}}$ of $\Gamma_{\breve{\Phi}_{\widetilde{\mathcal{H}}}}$ depend only on the group $\widehat{\mathcal{H}}$ determined by $\widetilde{\mathcal{H}}$.

Lemma 1.2.4.25. *The subgroups $\Gamma_{\breve{\Phi}_{\widetilde{\mathcal{H}}},\Phi_{\mathcal{H}},\breve{\sigma}}$ and $\Gamma_{\breve{\Phi}_{\widehat{\mathcal{H}}}}$ of $\Gamma_{\breve{\Phi}_{\widetilde{\mathcal{H}}},\Phi_{\mathcal{H}}}$ are identical.*

Proof. This is because the image of $\Gamma_{\breve{\Phi}_{\widetilde{\mathcal{H}}},\Phi_{\mathcal{H}},\breve{\sigma}}$ in $\Gamma_{\widetilde{\Phi}_{\widetilde{\mathcal{H}}}}$ is $\Gamma_{\widetilde{\Phi}_{\widetilde{\mathcal{H}}},\widetilde{\sigma}}$, which is trivial by Conditions 1.2.2.9 and [62, Lem. 6.2.5.27]. □

Corollary 1.2.4.26. *For choosing representatives $(\breve{\Phi}_{\widetilde{\mathcal{H}}}, \breve{\delta}_{\widetilde{\mathcal{H}}}, \breve{\tau})$ of equivalence classes $[(\breve{\Phi}_{\widetilde{\mathcal{H}}}, \breve{\delta}_{\widetilde{\mathcal{H}}}, \breve{\tau})]$ having $[(\widetilde{\Phi}_{\widetilde{\mathcal{H}}}, \widetilde{\delta}_{\widetilde{\mathcal{H}}}, \widetilde{\sigma})]$ as a face (as above), for any given choices of $\breve{\mathrm{Z}}$, $\breve{\Phi}$, $s_{\breve{X}}$, $s_{\breve{Y}}$, and $\breve{\delta}$ (compatible with $\widetilde{\mathrm{Z}}$, $\widetilde{\Phi}$, and $\widetilde{\delta}$), it suffices to take one $\breve{\tau}$ from each $\Gamma_{\breve{\Phi}_{\widehat{\mathcal{H}}}}$-orbit in $\breve{\Sigma}^{+}_{\breve{\Phi}_{\widetilde{\mathcal{H}}},\breve{\sigma}}$.*

Proof. This follows from the above review on equivalence classes $[(\breve{\Phi}_{\widetilde{\mathcal{H}}}, \breve{\delta}_{\widetilde{\mathcal{H}}}, \breve{\tau})]$ having $[(\widetilde{\Phi}_{\widetilde{\mathcal{H}}}, \widetilde{\delta}_{\widetilde{\mathcal{H}}}, \widetilde{\sigma})]$ as a face, from the very definitions of $\Gamma_{\breve{\Phi}_{\widetilde{\mathcal{H}}},\Phi_{\mathcal{H}},\breve{\sigma}}$ and $\breve{\Sigma}^{+}_{\breve{\Phi}_{\widetilde{\mathcal{H}}},\breve{\sigma}}$, and from Lemma 1.2.4.25. □

Lemma 1.2.4.27. *The surjections $s_{\breve{X}} : \breve{X} \twoheadrightarrow \widetilde{X}$ and $s_{\breve{Y}} : \breve{Y} \twoheadrightarrow \widetilde{Y}$ identify $\Gamma_{\breve{\Phi}_{\widetilde{\mathcal{H}}},\Phi_{\mathcal{H}}}$ as a finite index subgroup of $\mathrm{Hom}_{\mathcal{O}}(\widetilde{X}, X)$, whose elements map $\widetilde{\phi}(\widetilde{Y})$ to $\phi(Y)$, by sending each element $(\gamma_{\breve{X}}, \gamma_{\breve{Y}}) \in \Gamma_{\breve{\Phi}_{\widetilde{\mathcal{H}}},\Phi_{\mathcal{H}}}$ to the element in $\mathrm{Hom}_{\mathcal{O}}(\widetilde{X}, X)$ induced by $\gamma_{\breve{X}} - \mathrm{Id}_{\breve{X}} : \breve{X} \to X$ (which contains X in its kernel).*

Proof. The homomorphism from $\Gamma_{\breve{\Phi}_{\widetilde{\mathcal{H}}},\Phi_{\mathcal{H}}}$ to $\mathrm{Hom}_{\mathcal{O}}(\widetilde{X}, X)$ defined in the statement of this lemma is injective because $\gamma_{\breve{X}} - \mathrm{Id}_{\breve{X}} = 0$ exactly when $\gamma_{\breve{X}} = \mathrm{Id}_{\breve{X}}$. The element in $\mathrm{Hom}_{\mathcal{O}}(\widetilde{X}, X)$ induced by $\gamma_{\breve{X}} - \mathrm{Id}_{\breve{X}}$ maps $\widetilde{\phi}(\widetilde{Y})$ to $\phi(Y)$ because its restriction to $\widetilde{\phi}(\widetilde{Y})$ defines the element in $\mathrm{Hom}_{\mathcal{O}}(\widetilde{Y}, Y)$ induced by $\gamma_{\breve{Y}} - \mathrm{Id}_{\breve{Y}}$. Conversely, any element $f_{\breve{X}} \in \mathrm{Hom}_{\mathcal{O}}(\widetilde{X}, X)$ induces an element $\gamma_{\breve{X}} \in \mathrm{GL}_{\mathcal{O}}(\breve{X})$ with image in X by setting $\gamma_{\breve{X}} = \mathrm{Id}_{\breve{X}} + s_{\breve{X}} \circ f_{\breve{X}}$, any element also mapping $\widetilde{\phi}(\widetilde{Y})$ to $\phi(Y)$ induces an element $\gamma_{\breve{Y}} \in \mathrm{GL}_{\mathcal{O}}(\breve{Y})$ with image in $\breve{Y}$ by setting $\gamma_{\breve{Y}} = \mathrm{Id}_{\breve{Y}} + s_{\breve{Y}} \circ f_{\breve{Y}}$, where $f_{\breve{Y}} \in \mathrm{Hom}_{\mathcal{O}}(\widetilde{Y}, Y)$ is induced by $f_{\breve{X}}$, $\widetilde{\phi}$, and ϕ. Since a sufficiently divisible multiple of any element of $\mathrm{Hom}_{\mathcal{O}}(\widetilde{X}, X)$ maps $\widetilde{\phi}(\widetilde{Y})$ to $\phi(Y)$, and since a sufficiently high power of the element $(\gamma_{\breve{X}}, \gamma_{\breve{Y}})$ defined as above has trivial reduction modulo any prescribed integer (which means it can be made to be contained in $\Gamma_{\breve{\Phi}_{\widetilde{\mathcal{H}}},\Phi_{\mathcal{H}}}$), the recipe in the lemma identifies $\Gamma_{\breve{\Phi}_{\widetilde{\mathcal{H}}},\Phi_{\mathcal{H}}}$ as a finite index subgroup of $\mathrm{Hom}_{\mathcal{O}}(\widetilde{X}, X)$, as desired. □

Remark 1.2.4.28. This group $\Gamma_{\breve{\Phi}_{\widehat{\mathcal{H}}},\Phi_{\mathcal{H}}}$ here is the replacement of the group $\Gamma_{\widetilde{\Phi}_{\widetilde{\mathcal{H}}},\Phi_{\mathcal{H}}}$ in [61, Sec. 4A], which was incorrectly defined. (The rest of the arguments in [61] can be fixed with $\Gamma_{\widetilde{\Phi}_{\widetilde{\mathcal{H}}},\Phi_{\mathcal{H}}}$ there replaced with the group $\Gamma_{\breve{\Phi}_{\widehat{\mathcal{H}}},\Phi_{\mathcal{H}}}$ here.)

Definition 1.2.4.29. *We shall denote the kernel of the second morphism in* (1.2.4.18) *by* $\widehat{\mathbf{S}}_{\breve{\Phi}_{\widehat{\mathcal{H}}}}$*, so that the first morphism in* (1.2.4.19) *induces a canonical isomorphism*

$$(\widehat{\mathbf{S}}_{\breve{\Phi}_{\widehat{\mathcal{H}}}})^{\vee}_{\mathbb{R}} \cong (\mathbf{S}_{\breve{\Phi}_{\widehat{\mathcal{H}}}})^{\vee}_{\mathbb{R}}/(\mathbf{S}_{\widetilde{\Phi}_{\widetilde{\mathcal{H}}}})^{\vee}_{\mathbb{R}}. \tag{1.2.4.30}$$

By choosing some (noncanonical) splitting of $s_{\breve{X}} \underset{\mathbb{Z}}{\otimes} \mathbb{Q} : \breve{X} \underset{\mathbb{Z}}{\otimes} \mathbb{Q} \twoheadrightarrow \widetilde{X} \underset{\mathbb{Z}}{\otimes} \mathbb{Q}$ (over $\mathbb{Q}$), we can decompose the real vector space $(\mathbf{S}_{\breve{\Phi}_{\widehat{\mathcal{H}}}})^{\vee}_{\mathbb{R}}$ (noncanonically) as a direct sum

$$(\mathbf{S}_{\widetilde{\Phi}_{\widetilde{\mathcal{H}}}})^{\vee}_{\mathbb{R}} \oplus (\Gamma_{\breve{\Phi}_{\widehat{\mathcal{H}}},\Phi_{\mathcal{H}}})^{\vee}_{\mathbb{R}} \oplus (\mathbf{S}_{\Phi_{\mathcal{H}}})^{\vee}_{\mathbb{R}} \tag{1.2.4.31}$$

(defined over $\mathbb{Q}$), on which the action of $\Gamma_{\breve{\Phi}_{\widehat{\mathcal{H}}},\Phi_{\mathcal{H}}}$ is realized by its canonical translation action on the second factor. In particular, such a (noncanonical) splitting defines a projection

$$\begin{aligned} \mathrm{pr}_{(\widehat{\mathbf{S}}_{\breve{\Phi}_{\widehat{\mathcal{H}}}})^{\vee}_{\mathbb{R}}} : (\mathbf{S}_{\breve{\Phi}_{\widehat{\mathcal{H}}}})^{\vee}_{\mathbb{R}} &\to (\Gamma_{\breve{\Phi}_{\widehat{\mathcal{H}}},\Phi_{\mathcal{H}}})^{\vee}_{\mathbb{R}} \oplus (\mathbf{S}_{\Phi_{\mathcal{H}}})^{\vee}_{\mathbb{R}} \cong (\widehat{\mathbf{S}}_{\breve{\Phi}_{\widehat{\mathcal{H}}}})^{\vee}_{\mathbb{R}} \\ (x,y,z) &\mapsto (y,z) \end{aligned} \tag{1.2.4.32}$$

(the intermediate morphisms are defined over $\mathbb{Q}$, while the whole composition is defined over $\mathbb{Z}$ and independent of the choices of splittings, by Definition 1.2.4.29). Let

$$\widehat{\mathbf{P}}_{\breve{\Phi}_{\widehat{\mathcal{H}}}} := \mathrm{pr}_{(\widehat{\mathbf{S}}_{\breve{\Phi}_{\widehat{\mathcal{H}}}})^{\vee}_{\mathbb{R}}}(\mathbf{P}_{\breve{\Phi}_{\widehat{\mathcal{H}}}}) \tag{1.2.4.33}$$

and

$$\widehat{\mathbf{P}}^{+}_{\breve{\Phi}_{\widehat{\mathcal{H}}}} := \mathrm{pr}_{(\widehat{\mathbf{S}}_{\breve{\Phi}_{\widehat{\mathcal{H}}}})^{\vee}_{\mathbb{R}}}(\mathbf{P}^{+}_{\breve{\Phi}_{\widehat{\mathcal{H}}}}). \tag{1.2.4.34}$$

Lemma 1.2.4.35. *The canonical morphisms* $\mathbf{P}_{\breve{\Phi}_{\widehat{\mathcal{H}}}} \twoheadrightarrow \mathbf{P}_{\Phi_{\mathcal{H}}}$ *and* $\mathbf{P}^{+}_{\breve{\Phi}_{\widehat{\mathcal{H}}}} \twoheadrightarrow \mathbf{P}^{+}_{\Phi_{\mathcal{H}}}$ *(induced by the second morphism in* (1.2.4.20)*) factor through the canonical morphism* $\mathrm{pr}_{(\widehat{\mathbf{S}}_{\breve{\Phi}_{\widehat{\mathcal{H}}}})^{\vee}_{\mathbb{R}}}$ *in* (1.2.4.32) *and induce canonical morphisms*

$$\widehat{\mathbf{P}}_{\breve{\Phi}_{\widehat{\mathcal{H}}}} \twoheadrightarrow \mathbf{P}_{\Phi_{\mathcal{H}}} \tag{1.2.4.36}$$

and

$$\widehat{\mathbf{P}}^{+}_{\breve{\Phi}_{\widehat{\mathcal{H}}}} \twoheadrightarrow \mathbf{P}^{+}_{\Phi_{\mathcal{H}}}, \tag{1.2.4.37}$$

respectively.

Proof. This is because the second morphism in (1.2.4.20) is unchanged under translation by an element of $(\mathbf{S}_{\widetilde{\Phi}_{\widetilde{\mathcal{H}}}})^{\vee}_{\mathbb{R}}$. □

Lemma 1.2.4.38. *Under the projection* $\mathrm{pr}_{(\widehat{\mathbf{S}}_{\breve{\Phi}_{\widetilde{\mathcal{H}}}})^\vee_{\mathbb{R}}}$ *in* (1.2.4.32), *the image* $\mathrm{pr}_{(\widehat{\mathbf{S}}_{\breve{\Phi}_{\widetilde{\mathcal{H}}}})^\vee_{\mathbb{R}}}(\breve{\tau})$ *of each* $\breve{\tau}$ *in* $\widetilde{\Sigma}_{\breve{\Phi}_{\widetilde{\mathcal{H}}},\breve{\sigma}}$ *(resp.* $\widetilde{\Sigma}^+_{\breve{\Phi}_{\widetilde{\mathcal{H}}},\breve{\sigma}}$*) is a nondegenerate rational polyhedral cone in* $\widehat{\mathbf{P}}_{\breve{\Phi}_{\widehat{\mathcal{H}}}}$ *(resp.* $\widehat{\mathbf{P}}^+_{\breve{\Phi}_{\widehat{\mathcal{H}}}}$*).*

Proof. Since $\widetilde{\sigma}$ is a (nondegenerate) top-dimensional smooth rational polyhedral cone in $\mathbf{P}^+_{\widetilde{\Phi}_{\widetilde{\mathcal{H}}}}$, we can find a minimal subset $\{v_1, \ldots, v_r\}$ of $(\mathbf{S}_{\breve{\Phi}_{\widetilde{\mathcal{H}}}})^\vee_{\mathbb{R}}$ such that $\breve{\sigma} = \mathbb{R}_{>0}v_1 + \cdots + \mathbb{R}_{>0}v_r$ and $\mathbb{R}v_1 + \cdots + \mathbb{R}v_r = (\mathbf{S}_{\widetilde{\Phi}_{\widetilde{\mathcal{H}}}})^\vee_{\mathbb{R}}$ (which we view as a subset of $(\mathbf{S}_{\breve{\Phi}_{\widetilde{\mathcal{H}}}})^\vee_{\mathbb{R}}$). Then, for each $\breve{\tau} \in \widetilde{\Sigma}_{\breve{\Phi}_{\widetilde{\mathcal{H}}},\breve{\sigma}}$ (which has $\breve{\sigma}$ as a face), there is a minimal subset $\{v_{r+1}, \ldots, v_{r+s}\}$ of $(\mathbf{S}_{\breve{\Phi}_{\widetilde{\mathcal{H}}}})^\vee_{\mathbb{R}}$ such that

$$\breve{\tau} = \mathbb{R}_{>0}v_1 + \cdots + \mathbb{R}_{>0}v_r + \mathbb{R}_{>0}v_{r+1} + \cdots + \mathbb{R}_{>0}v_{r+s}.$$

Moreover, we can write each x in the closure $\overline{\breve{\tau}}$ of $\breve{\tau}$ as $x = c_1v_1 + \cdots + c_{r+s}v_{r+s}$, where the coordinates $(c_1, \ldots, c_{r+s}) \in \mathbb{R}^{r+s}_{\geq 0}$ are uniquely determined by x.

Suppose there are $x = c_1v_1 + \cdots + c_{r+s}v_{r+s}$ and $y = d_1v_1 + \cdots + d_{r+s}v_{r+s}$ in $\overline{\breve{\tau}}$ such that $\mathrm{pr}_{(\widehat{\mathbf{S}}_{\breve{\Phi}_{\widehat{\mathcal{H}}}})^\vee_{\mathbb{R}}}(x) + \mathrm{pr}_{(\widehat{\mathbf{S}}_{\breve{\Phi}_{\widehat{\mathcal{H}}}})^\vee_{\mathbb{R}}}(y) = 0$. Then $x + y = e_1v_1 + \cdots + e_rv_r$ for some $e_1, \ldots, e_r \in \mathbb{R}$, because $\{v_1, \ldots, v_r\}$ spans the kernel $(\mathbf{S}_{\widetilde{\Phi}_{\widetilde{\mathcal{H}}}})^\vee_{\mathbb{R}}$ of $\mathrm{pr}_{(\widehat{\mathbf{S}}_{\breve{\Phi}_{\widehat{\mathcal{H}}}})^\vee_{\mathbb{R}}}$. By choosing $e \in \mathbb{R}_{\geq 0}$ such that $e + e_i \geq 0$ for all $1 \leq i \leq r$, we obtain an identity of elements $x + y + (ev_1 + \cdots + ev_r) = (e + e_1)v_1 + \cdots + (e + e_r)v_r$ in $\overline{\breve{\tau}}$, and the nonnegative coordinates of both sides must coincide because of the choices of $v_1, \ldots v_{r+s}$. Hence, we must have $c_{r+1} = \cdots = c_{r+s} = 0$ and $d_{r+1} = \cdots = d_{r+s} = 0$ (because they are nonnegative). Thus, the image $\mathrm{pr}_{(\widehat{\mathbf{S}}_{\breve{\Phi}_{\widehat{\mathcal{H}}}})^\vee_{\mathbb{R}}}(\breve{\tau})$ of $\breve{\tau}$ cannot contain any nonzero $\mathbb{R}$-vector subspace; that is, it is a nondegenerate rational polyhedral cone in $\widehat{\mathbf{P}}_{\breve{\Phi}_{\widehat{\mathcal{H}}}}$ (see (1.2.4.33)). If $\breve{\tau} \in \widetilde{\Sigma}^+_{\breve{\Phi}_{\widetilde{\mathcal{H}}},\breve{\sigma}}$, then $(\widehat{\mathbf{S}}_{\breve{\Phi}_{\widehat{\mathcal{H}}}})^\vee_{\mathbb{R}}$ is contained in $\widehat{\mathbf{P}}^+_{\breve{\Phi}_{\widehat{\mathcal{H}}}}$ (see (1.2.4.34)). □

Lemma 1.2.4.39. *There exists a continuous section*

$$\tilde{x}_0 : (\Gamma_{\breve{\Phi}_{\widetilde{\mathcal{H}}},\Phi_{\mathcal{H}}})^\vee_{\mathbb{R}} \oplus (\mathbf{S}_{\Phi_{\mathcal{H}}})^\vee_{\mathbb{R}} \to (\mathbf{S}_{\widetilde{\Phi}_{\widetilde{\mathcal{H}}}})^\vee_{\mathbb{R}} \oplus (\Gamma_{\breve{\Phi}_{\widetilde{\mathcal{H}}},\Phi_{\mathcal{H}}})^\vee_{\mathbb{R}} \oplus (\mathbf{S}_{\Phi_{\mathcal{H}}})^\vee_{\mathbb{R}}$$

$$(y, z) \mapsto (x_0(y, z), y, z)$$

such that $(\tilde{x}_0 \circ \mathrm{pr}_{(\widehat{\mathbf{S}}_{\breve{\Phi}_{\widehat{\mathcal{H}}}})^\vee_{\mathbb{R}}})(\breve{\tau}) \subset \breve{\tau}$ *for all* $\breve{\tau} \in \widetilde{\Sigma}_{\breve{\Phi}_{\widetilde{\mathcal{H}}},\breve{\sigma}}$.

Proof. Let $\{v_1, \ldots, v_r\}$ be as in the proof of Lemma 1.2.4.38. Then we can write the desired function $x_0(y, z)$ as

$$x_0(y, z) = x_{0,1}(y, z)\, v_1 + x_{0,2}(y, z)\, v_2 + \cdots + x_{0,r}(y, z)\, v_r,$$

where each $x_{0,j}(\,\cdot\,,\,\cdot\,)$ is a $\mathbb{R}$-valued continuous function on $(\Gamma_{\breve{\Phi}_{\widetilde{\mathcal{H}}},\Phi_{\mathcal{H}}})^\vee_{\mathbb{R}} \oplus (\mathbf{S}_{\Phi_{\mathcal{H}}})^\vee_{\mathbb{R}}$. For each $\breve{\tau} \in \widetilde{\Sigma}_{\breve{\Phi}_{\widetilde{\mathcal{H}}},\breve{\sigma}}$, let $\{v_{r+1}, \ldots, v_{r+s}\}$ be as in the proof of Lemma 1.2.4.38. For each $i = r + 1, \ldots, r + s$, let $w_i = \mathrm{pr}_{(\widehat{\mathbf{S}}_{\breve{\Phi}_{\widehat{\mathcal{H}}}})^\vee_{\mathbb{R}}}(v_i)$ and write $w'_i := v_i - (0, w_i)$ as a linear combination $w'_i = c_{i1}v_1 + c_{i2}v_2 + \cdots + c_{ir}v_r$, where $c_{i1}, \ldots, c_{ir} \in \mathbb{R}$. By taking $x^{\breve{\tau}}_0(y, z) = x^{\breve{\tau}}_{0,1}(y, z)\, v_1 + x^{\breve{\tau}}_{0,2}(y, z)\, v_2 + \cdots + x^{\breve{\tau}}_{0,r}(y, z)\, v_r$ to be linear on

$\mathbb{R}_{\geq 0}w_{r+1} + \cdots + \mathbb{R}_{\geq 0}w_{r+s}$ and zero elsewhere, and by taking $x_{0,j}^{\check{\tau}}(y,z)$ to satisfy $x_{0,j}^{\check{\tau}}(w_i) > c_{ij}$ for all i and j, we have $(x_0^{\check{\tau}}(w_i), w_i) = (x_0^{\check{\tau}}(w_i) - w_i', 0, 0) + (w_i', w_i) \in \check{\sigma} + v_i$ for all i, and hence $\tilde{x}_0^{\check{\tau}}(y,z) := (x_0^{\check{\tau}}(y,z), y, z)$ satisfies $(\tilde{x}_0^{\check{\tau}} \circ \mathrm{pr}_{(\widehat{\mathbf{S}}_{\check{\Phi}_{\widehat{\mathcal{H}}}})^\vee_{\mathbb{R}}})(\check{\tau}) \subset \check{\tau}$. The same is true if we replace each $x_{0,j}^{\check{\tau}}(y,z)$ with a function with (pointwise) greater value. If $\check{\tau}$ and $\check{\tau}'$ are cones in $\widetilde{\Sigma}_{\check{\Phi}_{\widetilde{\mathcal{H}}},\check{\sigma}}$ meeting some fiber of $\mathrm{pr}_{(\widehat{\mathbf{S}}_{\check{\Phi}_{\widehat{\mathcal{H}}}})^\vee_{\mathbb{R}}}$ above $(y,z) \in (\Gamma_{\check{\Phi}_{\widetilde{\mathcal{H}}},\Phi_{\mathcal{H}}})^\vee_{\mathbb{R}} \oplus (\mathbf{S}_{\Phi_{\mathcal{H}}})^\vee_{\mathbb{R}}$, then the above argument shows that there exists some $v \in \check{\tau}$ such that $v + (0,y,z) \in \check{\tau} \cap \check{\tau}'$, forcing $\check{\tau} = \check{\tau}'$. Hence, there is at most one $\check{\tau}$ in $\widetilde{\Sigma}_{\check{\Phi}_{\widetilde{\mathcal{H}}},\check{\sigma}}$ meeting each fiber of $\mathrm{pr}_{(\widehat{\mathbf{S}}_{\check{\Phi}_{\widehat{\mathcal{H}}}})^\vee_{\mathbb{R}}}$, and so we can take $x_{0,j}(y,z)$ to be any continuous function (pointwise) greater than $x_{0,j}^{\check{\tau}}$ for all $\check{\tau} \in \widetilde{\Sigma}_{\check{\Phi}_{\widetilde{\mathcal{H}}},\check{\sigma}}$. Then we have $(\tilde{x}_0 \circ \mathrm{pr}_{(\widehat{\mathbf{S}}_{\check{\Phi}_{\widehat{\mathcal{H}}}})^\vee_{\mathbb{R}}})(\check{\tau}) \subset \check{\tau}$ for all $\check{\tau} \in \widetilde{\Sigma}_{\check{\Phi}_{\widetilde{\mathcal{H}}},\check{\sigma}}$, as desired. □

Corollary 1.2.4.40. *The set*

$$\widehat{\Sigma}_{\check{\Phi}_{\widehat{\mathcal{H}}}} = \{\mathrm{pr}_{(\widehat{\mathbf{S}}_{\check{\Phi}_{\widehat{\mathcal{H}}}})^\vee_{\mathbb{R}}}(\check{\tau})\}_{\check{\tau} \in \widetilde{\Sigma}_{\check{\Phi}_{\widetilde{\mathcal{H}}},\check{\sigma}}}$$

of rational polyhedral cones defines a $\Gamma_{\check{\Phi}_{\widehat{\mathcal{H}}}}$**-admissible rational polyhedral cone decomposition** (*cf. Definition* 1.2.2.4) *of*

$$\widehat{\mathbf{P}}_{\check{\Phi}_{\widehat{\mathcal{H}}}} = \mathrm{pr}_{(\widehat{\mathbf{S}}_{\check{\Phi}_{\widehat{\mathcal{H}}}})^\vee_{\mathbb{R}}}(\mathbf{P}_{\check{\Phi}_{\widetilde{\mathcal{H}}}}) = \bigcup_{\check{\tau} \in \widetilde{\Sigma}_{\check{\Phi}_{\widetilde{\mathcal{H}}},\check{\sigma}}} \left(\mathrm{pr}_{(\widehat{\mathbf{S}}_{\check{\Phi}_{\widehat{\mathcal{H}}}})^\vee_{\mathbb{R}}}(\check{\tau})\right) \quad (1.2.4.41)$$

in the sense that we have the following:

(1) *Each* $\mathrm{pr}_{(\widehat{\mathbf{S}}_{\check{\Phi}_{\widehat{\mathcal{H}}}})^\vee_{\mathbb{R}}}(\check{\tau})$ *is a nondegenerate rational polyhedral cone.*

(2) *The union* (1.2.4.41) *is disjoint and defines a stratification of* $\widehat{\mathbf{P}}_{\check{\Phi}_{\widehat{\mathcal{H}}}}$.

(3) $\widehat{\Sigma}_{\check{\Phi}_{\widehat{\mathcal{H}}}}$ *is invariant under the action of* $\Gamma_{\check{\Phi}_{\widehat{\mathcal{H}}}}$ *in the sense that* $\Gamma_{\check{\Phi}_{\widehat{\mathcal{H}}}}$ *permutes the cones in it. Under this action, the set of* $\Gamma_{\check{\Phi}_{\widehat{\mathcal{H}}}}$*-orbits is finite.*

Proof. Statement (1) is Lemma 1.2.4.38. As for statement (2), suppose $w \in \mathrm{pr}_{(\widehat{\mathbf{S}}_{\check{\Phi}_{\widehat{\mathcal{H}}}})^\vee_{\mathbb{R}}}(\check{\tau}) \cap \mathrm{pr}_{(\widehat{\mathbf{S}}_{\check{\Phi}_{\widehat{\mathcal{H}}}})^\vee_{\mathbb{R}}}(\check{\tau}') \neq \emptyset$ for some $\check{\tau}, \check{\tau}' \in \widetilde{\Sigma}_{\check{\Phi}_{\widetilde{\mathcal{H}}},\check{\sigma}}$. Then it was shown in the proof of Lemma 1.2.4.39 that $\check{\tau} = \check{\tau}'$. (Alternatively, any continuous section $\tilde{x}_0$ as in the statement of Lemma 1.2.4.39 defines an element $\tilde{x}_0(w) \in \check{\tau} \cap \check{\tau}'$, forcing that $\check{\tau} = \check{\tau}'$.) Hence, the union (1.2.4.41) is disjoint. Consequently, the incidence relations in $\{\mathrm{pr}_{(\widehat{\mathbf{S}}_{\check{\Phi}_{\widehat{\mathcal{H}}}})^\vee_{\mathbb{R}}}(\check{\tau})\}_{\check{\tau} \in \widetilde{\Sigma}_{\check{\Phi}_{\widetilde{\mathcal{H}}},\check{\sigma}}}$ inherits exactly those in $\widetilde{\Sigma}_{\check{\Phi}_{\widetilde{\mathcal{H}}},\check{\sigma}}$, and hence the union (1.2.4.41) defines a stratification (*cf.* (1) of Definition 1). Finally, statement (3) follows from the corresponding statement that $\Gamma_{\check{\Phi}_{\widetilde{\mathcal{H}}}}$ acts on $\widetilde{\Sigma}_{\check{\Phi}_{\widetilde{\mathcal{H}}},\check{\sigma}}$ with a finite number of orbits (*cf.* Corollary 1.2.4.26). □

When only $\widehat{\Sigma}_{\check{\Phi}_{\widehat{\mathcal{H}}}}$ is relevant in the context, we shall denote elements $\mathrm{pr}_{(\widehat{\mathbf{S}}_{\check{\Phi}_{\widehat{\mathcal{H}}}})^\vee_{\mathbb{R}}}(\check{\tau}) \in \widehat{\Sigma}_{\check{\Phi}_{\widehat{\mathcal{H}}}}$ by $\widehat{\tau}$, without reference to the original $\widetilde{\Sigma}_{\check{\Phi}_{\widetilde{\mathcal{H}}},\check{\sigma}}$.

Lemma 1.2.4.42. *The collection* $\widehat{\Sigma} = \{\widehat{\Sigma}_{\check{\Phi}_{\widehat{\mathcal{H}}}}\}_{[(\check{\Phi}_{\widehat{\mathcal{H}}},\check{\delta}_{\widehat{\mathcal{H}}})]}$, *where* $[(\check{\Phi}_{\widehat{\mathcal{H}}},\check{\delta}_{\widehat{\mathcal{H}}})]$ *runs through cusp labels at level* $\widehat{\mathcal{H}}$ *for* $(\widetilde{L}, \langle\,\cdot\,,\,\cdot\,\rangle^{\sim}, \widetilde{h}_0, \widetilde{\mathrm{Z}})$ (*i.e., equivalence classes of*

$\widehat{\mathcal{H}}$-orbits of representatives $(\breve{\Phi}, \breve{\delta})$ compatible with $(\widetilde{\Phi}, \widetilde{\delta})$ as in Definition 1.2.4.17, *with $\breve{\mathsf{Z}}$ and $\widetilde{\mathsf{Z}}$ suppressed in the notation), defines a* **compatible choice of admissible smooth rational polyhedral cone decomposition data** *analogous to the notion for $\mathsf{M}_{\mathcal{H}}$ in Definition* 1.2.2.13. *There is an obvious notion of refinements for such collections, analogous to that in [62, Def. 6.4.2.8].*

Then we can also talk about equivalence classes $[(\breve{\Phi}_{\widehat{\mathcal{H}}}, \breve{\delta}_{\widehat{\mathcal{H}}}, \widehat{\tau})]$ and their facial relations as in Definitions 1.2.2.10 *and* 1.2.2.19, *and their refinements as in [62, Def. 6.4.3.1].*

Proof. This follows from the corresponding facts for $\widetilde{\Sigma} = \{\widetilde{\Sigma}_{\breve{\Phi}_{\widetilde{\mathcal{H}}}}\}_{[(\breve{\Phi}_{\widetilde{\mathcal{H}}}, \breve{\delta}_{\widetilde{\mathcal{H}}})]}$ (with indices running through all cusp labels). □

Remark 1.2.4.43. Here we omit the precise definition of a compatible choice of admissible smooth rational polyhedral cone decomposition data because we can only construct toroidal compactifications of Kuga families for those $\widehat{\Sigma}$ defined by some $\widetilde{\Sigma}$ and $\widetilde{\sigma}$.

Definition 1.2.4.44. *We say that two $\widetilde{\kappa}_1$ and $\widetilde{\kappa}_2$ in $\widetilde{\mathbf{K}}^{++}_{Q,\mathcal{H}}$ (see Definition* 1.2.4.11) *are equivalent if they determine the same $\kappa = (\widehat{\mathcal{H}}, \widehat{\Sigma})$. In this case, we shall abusively write $\kappa = [\widetilde{\kappa}_1] = [\widetilde{\kappa}_2]$. Then we take $\mathbf{K}^{++}_{Q,\mathcal{H}}$ to be the set of all such $\kappa = (\widehat{\mathcal{H}}, \widehat{\Sigma})$, with a partial order $\kappa' = (\widehat{\mathcal{H}}', \widehat{\Sigma}') \succ \kappa = (\widehat{\mathcal{H}}, \widehat{\Sigma})$ when $\widehat{\mathcal{H}}' \subset \widehat{\mathcal{H}}'$ and when $\widehat{\Sigma}'$ is a refinement of $\widehat{\Sigma}$ (see Definition* 1.2.4.42). *We also take the subset $\mathbf{K}^{+}_{Q,\mathcal{H}}$ (resp. $\mathbf{K}_{Q,\mathcal{H}}$) of $\mathbf{K}^{++}_{Q,\mathcal{H}}$ to be the image of the subset $\widetilde{\mathbf{K}}^{+}_{Q,\mathcal{H}}$ (resp. $\widetilde{\mathbf{K}}_{Q,\mathcal{H}}$) of $\widetilde{\mathbf{K}}^{++}_{Q,\mathcal{H}}$ under the canonical surjection $\widetilde{\mathbf{K}}^{++}_{Q,\mathcal{H}} \twoheadrightarrow \mathbf{K}^{++}_{Q,\mathcal{H}}$, with an induced partial order denoted by the same symbol $\succ$.*

Lemma 1.2.4.45. *Every neat open compact subgroup $\widehat{\mathcal{H}}$ of $\mathrm{G}(\hat{\mathbb{Z}})$ is induced by some neat open compact subgroup $\widetilde{\mathcal{H}}$ of $\widetilde{\mathrm{G}}(\hat{\mathbb{Z}})$ as in Definition* 1.2.4.4. *Moreover, we may assume that $\widetilde{\mathcal{H}}$ satisfies Condition* 1.2.4.7.

Proof. Consider any integer $n \geq 3$ such that $\widetilde{\mathcal{U}}(n)_{\widehat{\mathrm{G}}} \subset \widehat{\mathcal{H}}$. Consider the preimage $\widehat{\mathcal{H}}^{+}$ of $\widehat{\mathcal{H}}$ under the canonical homomorphism $\widetilde{\mathrm{P}}'_{\widetilde{\mathsf{Z}}}(\hat{\mathbb{Z}}) \twoheadrightarrow \widetilde{\mathrm{P}}'_{\widetilde{\mathsf{Z}}}(\hat{\mathbb{Z}})/\widetilde{\mathrm{U}}_{2,\widetilde{\mathsf{Z}}}(\hat{\mathbb{Z}})$. Then $\widetilde{\mathcal{H}} := \widehat{\mathcal{H}}^{+}\widetilde{\mathcal{U}}(n)$ induces $\widehat{\mathcal{H}}$ as in Definition 1.2.4.4, and satisfies Condition 1.2.4.7. Since elements of $\widetilde{\mathrm{U}}_{2,\widetilde{\mathsf{Z}}}(\hat{\mathbb{Z}})$ are unipotent, the elements in $\widehat{\mathcal{H}}^{+}$ and $\widehat{\mathcal{H}}$ have the same eigenvalues (up to multiplicity). Since the elements of $\widetilde{\mathcal{H}} = \widehat{\mathcal{H}}^{+}\widetilde{\mathcal{U}}(n)$ are congruent to elements of $\widehat{\mathcal{H}}^{+}$ modulo n by definition, $\widetilde{\mathcal{H}}$ is neat by definition (see [89, 0.6] or [62, Def. 1.4.1.8]), and by Serre's lemma that no nontrivial root of unity can be congruent to 1 modulo n if $n \geq 3$. □

Lemma 1.2.4.46. *For each neat open compact subgroup $\widehat{\mathcal{H}}$ of $\mathrm{G}(\hat{\mathbb{Z}})$, there exists some element $\kappa = (\widehat{\mathcal{H}}, \widehat{\Sigma}) \in \mathbf{K}^{++}_{Q,\mathcal{H}}$, which lies in $\mathbf{K}^{+}_{Q,\mathcal{H}}$ (resp. $\mathbf{K}_{Q,\mathcal{H}}$) if $\widehat{\mathcal{H}}$ satisfies Condition* 1.2.4.8 *(resp. both Conditions* 1.2.4.8 *and* 1.2.4.9).

Proof. By Lemma 1.2.4.45, $\widehat{\mathcal{H}}$ is induced by some neat $\widetilde{\mathcal{H}}$ as in Definition 1.2.4.4, which we assume to also satisfy Condition 1.2.4.7. By Proposition 1.2.2.17, there exists some compatible choice $\widetilde{\Sigma}$ for $\widetilde{\mathsf{M}}_{\widetilde{\mathcal{H}}}$. Let us take $\widehat{\Sigma}$ to be induced by $\widetilde{\Sigma}$ as in Lemma 1.2.4.42, and take $\kappa = (\widehat{\mathcal{H}}, \widehat{\Sigma})$. Then, by definition, we have $\kappa \in \mathbf{K}^{++}_{Q,\mathcal{H}}$. The remaining statements of the lemma also follow by definition. □

Lemma 1.2.4.47. *The partial order $\succ$ among elements in $\mathbf{K}^{++}_{Q,\mathcal{H}}$ (resp. $\mathbf{K}^{+}_{Q,\mathcal{H}}$, resp. $\mathbf{K}_{Q,\mathcal{H}}$) is* **directed***; that is, if we are given two $\kappa = (\widehat{\mathcal{H}}, \widehat{\Sigma})$ and $\kappa' = (\widehat{\mathcal{H}}', \widehat{\Sigma}')$, then there exists some $\kappa'' = (\widehat{\mathcal{H}}'', \widehat{\Sigma}'')$ such that $\kappa'' \succ \kappa$ and $\kappa'' \succ \kappa'$. Moreover, we can take $\widehat{\mathcal{H}}''$ to be any open compact subgroup of $\widehat{\mathcal{H}} \cap \widehat{\mathcal{H}}'$ (which can be $\widehat{\mathcal{H}} \cap \widehat{\mathcal{H}}'$ itself).*

Proof. Let us begin with the set $\mathbf{K}^{++}_{Q,\mathcal{H}}$.

Suppose $\kappa = (\widehat{\mathcal{H}}, \widehat{\Sigma}) = [(\widetilde{\mathcal{H}}, \widetilde{\Sigma}, \widetilde{\sigma})]$ and $\kappa' = (\widehat{\mathcal{H}}', \widehat{\Sigma}') = [(\widetilde{\mathcal{H}}', \widetilde{\Sigma}', \widetilde{\sigma}')]$ are in $\mathbf{K}^{++}_{Q,\mathcal{H}}$, where $(\widetilde{\mathcal{H}}, \widetilde{\Sigma}, \widetilde{\sigma})$ and $(\widetilde{\mathcal{H}}', \widetilde{\Sigma}', \widetilde{\sigma}')$ are in $\widetilde{\mathbf{K}}^{++}_{Q,\mathcal{H}}$, so that $\widehat{\mathcal{H}} = \widetilde{\mathcal{H}}_{\widehat{\mathrm{G}}}$ and $\widehat{\mathcal{H}}' = \widetilde{\mathcal{H}}'_{\widehat{\mathrm{G}}}$. Let $\widehat{\mathcal{H}}''$ be any open compact subgroup of $\widehat{\mathcal{H}} \cap \widehat{\mathcal{H}}'$ (which can be $\widehat{\mathcal{H}} \cap \widehat{\mathcal{H}}'$ itself). Then, as in the proof of Lemma 1.2.4.45, by choosing some integer $n \geq 3$ such that $\widetilde{\mathcal{U}}(n)_{\widehat{\mathrm{G}}} \subset \widehat{\mathcal{H}}''$ and $\widetilde{\mathcal{U}}(n) \subset \widetilde{\mathcal{H}} \cap \widetilde{\mathcal{H}}'$, and by taking $\widehat{\mathcal{H}}''^{,+}$ to be the preimage of $\widehat{\mathcal{H}}''$ under the canonical homomorphism $\widetilde{\mathrm{P}}'_{\widetilde{\mathrm{Z}}}(\hat{\mathbb{Z}}) \twoheadrightarrow \widetilde{\mathrm{P}}'_{\widetilde{\mathrm{Z}}}(\hat{\mathbb{Z}})/\widetilde{\mathrm{U}}_{2,\widetilde{\mathrm{Z}}}(\hat{\mathbb{Z}})$, we obtain a neat open compact subgroup $\widetilde{\mathcal{H}}'' = \widetilde{\mathcal{U}}(n)\widehat{\mathcal{H}}''^{,+}$ of $\widetilde{\mathcal{H}} \cap \widetilde{\mathcal{H}}'$ satisfying Condition 1.2.4.7 (with $\widetilde{\mathcal{H}}''_{\widehat{\mathrm{G}}} = \widehat{\mathcal{H}}''$).

Let $\widehat{\Sigma}^{(\widehat{\mathcal{H}}'')}$ (resp. $\widehat{\Sigma}'^{,(\widehat{\mathcal{H}}'')}$) denote the collection induced by $\widehat{\Sigma}$ (resp. $\widehat{\Sigma}'$) at level $\widehat{\mathcal{H}}''$, as in [62, Constr. 7.3.1.6]. By definition, it is also induced by the collection $\widetilde{\Sigma}^{(\widetilde{\mathcal{H}}'')}$ (resp. $\widetilde{\Sigma}'^{,(\widetilde{\mathcal{H}}'')}$) induced by $\widetilde{\Sigma}$ (resp. $\widetilde{\Sigma}'$) at level $\widetilde{\mathcal{H}}''$. Let $\widehat{\Sigma}''^{,\mathrm{pre}}$ be any common refinement of both $\widehat{\Sigma}^{(\widehat{\mathcal{H}}'')}$ and $\widehat{\Sigma}'^{,(\widehat{\mathcal{H}}'')}$ (which might not be determined by some $(\widetilde{\mathcal{H}}'', \widetilde{\Sigma}'', \widetilde{\sigma}'')$ in $\widetilde{\mathbf{K}}^{++}_{Q,\mathcal{H}}$). The refinement $\widehat{\Sigma}''^{,\mathrm{pre}}$ of $\widehat{\Sigma}^{(\widehat{\mathcal{H}}'')}$ defines (by taking preimages) certain subdivisions of cones in (the cone decompositions in) $\widetilde{\Sigma}^{(\widetilde{\mathcal{H}}'')}$, which can be further subdivided into a projective smooth refinement $\widetilde{\Sigma}''$ of $\widetilde{\Sigma}^{(\widetilde{\mathcal{H}}'')}$. Let $\widetilde{\sigma}''$ be a top-dimensional cone in $\widetilde{\Sigma}''$ such that $\widetilde{\sigma}'' \subset \widetilde{\sigma}$, and let $\kappa'' := (\widehat{\mathcal{H}}'', \widehat{\Sigma}'') := [(\widetilde{\mathcal{H}}'', \widetilde{\Sigma}'', \widetilde{\sigma}'')]$. Then $\widehat{\Sigma}''$ is a refinement of $\widehat{\Sigma}''^{,\mathrm{pre}}$ (and hence of both $\widehat{\Sigma}^{(\widehat{\mathcal{H}}'')}$ and $\widehat{\Sigma}'^{,(\widehat{\mathcal{H}}'')}$). Thus, we have defined an element κ'' in $\mathbf{K}^{++}_{Q,\mathcal{H}}$ satisfying both $\kappa'' \succ \kappa$ and $\kappa'' \succ \kappa'$, as desired.

Then the cases for the sets $\mathbf{K}^{+}_{Q,\mathcal{H}}$ and $\mathbf{K}_{Q,\mathcal{H}}$ also follow, because Condition 1.2.4.8 is clearly compatible with intersections in $\widehat{\mathcal{H}}$; and so is the condition (equivalence to Condition 1.2.4.9) that the splitting $\mathrm{G}(\hat{\mathbb{Z}}) \to \widehat{\mathrm{G}}(\hat{\mathbb{Z}})$ defined by $\widetilde{\delta}$ maps $\mathcal{H}$ to $\widehat{\mathcal{H}}$. □

Now we consider some compatibility conditions between a collection Σ for $\mathsf{M}_{\mathcal{H}}$ and elements of $\widetilde{\mathbf{K}}^{++}_{Q,\mathcal{H}}$ or $\mathbf{K}^{++}_{Q,\mathcal{H}}$.

First consider the following condition on an element $\widetilde{\kappa} = (\widetilde{\mathcal{H}}, \widetilde{\Sigma}, \widetilde{\sigma})$ in $\widetilde{\mathbf{K}}^{++}_{Q,\mathcal{H}}$:

Condition 1.2.4.48. (*Compare with [61, Cond. 3.8].*) *For each $(\breve{\Phi}_{\widetilde{\mathcal{H}}}, \breve{\delta}_{\widetilde{\mathcal{H}}}, \breve{\tau})$ as above, where $\breve{\tau} \subset \mathbf{P}^{+}_{\breve{\Phi}_{\widetilde{\mathcal{H}}}}$ is a cone in the cone decomposition $\widetilde{\Sigma}_{\breve{\Phi}_{\widetilde{\mathcal{H}}}}$ (in $\widetilde{\Sigma}$) having $\breve{\sigma}$ as* *a face, the image of* $\check{\tau}$ *in* $\mathbf{P}_{\Phi_{\mathcal{H}}}$ *under the (canonical) second morphism in* (1.2.4.20) *is contained in some cone* $\tau \subset \mathbf{P}^+_{\Phi_{\mathcal{H}}}$ *in the cone decomposition* $\Sigma_{\Phi_{\mathcal{H}}}$ *(in* Σ*).*

By Lemma 1.2.4.35, if $\kappa = [\widetilde{\kappa}] \in \mathbf{K}^{++}_{Q,\mathcal{H}}$ is the element determined by $\widetilde{\kappa}$, then Condition 1.2.4.48 for $\widetilde{\kappa}$ is equivalent to the following condition for κ:

Condition 1.2.4.49. (*Compare with [28, Ch. VI, Def. 1.3].*) *For each* $\widehat{\tau} \in \widehat{\Sigma}_{\check{\Phi}_{\widehat{\mathcal{H}}}}$ *(where* $\widehat{\tau} = \mathrm{pr}_{(\widehat{\mathbf{S}}_{\check{\Phi}_{\widehat{\mathcal{H}}}})^\vee_{\mathbb{R}}}(\check{\tau})$ *for some* $(\check{\Phi}_{\widetilde{\mathcal{H}}}, \check{\delta}_{\widetilde{\mathcal{H}}}, \check{\tau})$ *is in the cone decomposition* $\widehat{\Sigma}_{\check{\Phi}_{\widehat{\mathcal{H}}}}$ *in* $\widehat{\Sigma}$*), the image of* $\widehat{\tau}$ *in* $\mathbf{P}^+_{\Phi_{\mathcal{H}}}$ *under* (1.2.4.37) *is contained in some cone* $\tau \subset \mathbf{P}^+_{\Phi_{\mathcal{H}}}$ *in the cone decomposition* $\Sigma_{\Phi_{\mathcal{H}}}$ *(in* Σ*).*

Definition 1.2.4.50. *For* $? = ++, +,$ *or* $\emptyset$*, let us take* $\mathbf{K}^?_{Q,\mathcal{H},\Sigma}$ *to be the subset of* $\mathbf{K}^?_{Q,\mathcal{H}}$ *consisting of elements* κ *satisfying Condition* 1.2.4.49.

Since Condition 1.2.4.49 can be achieved by replacing any given $\widehat{\Sigma}$ with a refinement (in the same set), we see that each $\mathbf{K}^?_{Q,\mathcal{H},\Sigma}$ is nonempty and has an induced directed partial order.

Remark 1.2.4.51. (Compare with [61, Rem. 3.10].) Condition 1.2.4.49 is analogous to the condition in [89, 6.25(b)], which is used in, for example, [40, Lem. 1.6.5] and other related works based on [4].

Proposition 1.2.4.52. *Suppose* $\mathcal{H}$ *is any open compact subgroup of* $\mathrm{G}(\hat{\mathbb{Z}})$*. For each* $?_1 = ++, +,$ *or* $\emptyset$*, and for each* $?_2 = \Sigma$ *or* $\emptyset$*, the sets* $\mathbf{K}^{?_1}_{Q,\mathcal{H}}$ *and* $\mathbf{K}^{?_1}_{Q,\mathcal{H},?_2}$ *are nonempty and compatible with each other under the various canonical maps. Common refinements for finite subsets exist in any sets of the form* $\mathbf{K}^{?_1}_{Q,\mathcal{H},?_2}$*. We may allow varying levels or twists by Hecke actions when doing so, and we may vary* $?_1$ *and* $?_2$ *as well (in any order). For any such refinement* $\kappa = (\widehat{\mathcal{H}}, \widehat{\Sigma})$*, we may prescribe* $\widehat{\mathcal{H}}$ *to be any allowed open compact subgroup of* $\widehat{\mathrm{G}}(\hat{\mathbb{Z}})$ *in the context, we may require* $\widehat{\Sigma}$ *to be finer than any cone decomposition* $\widehat{\Sigma}'$*, and we may require* $\widehat{\Sigma}$ *to be invariant under any choice of an open compact subgroup of* $\widehat{\mathrm{G}}(\mathbb{A}^\infty)$ *normalizing* $\widehat{\mathcal{H}}$*.*

Proof. These follow from the corresponding existence and refinement statements in Proposition 1.2.2.17 for collections $\widetilde{\Sigma}$ and $\widetilde{\mathsf{pol}}$ for $\widetilde{\mathsf{M}}_{\widetilde{\mathcal{H}}}$. □

For later references, let us conclude with the following definitions:

Definition 1.2.4.53. (*Compare with Definitions* 1.2.1.11 *and* 1.2.4.3.) *Let* $\check{\mathrm{Z}}$ *be any fully symplectic admissible filtration of* $\widetilde{L} \underset{\mathbb{Z}}{\otimes} \hat{\mathbb{Z}}$ *satisfying* (1.2.4.12)*. For each* $\hat{\mathbb{Z}}$*-algebra* R*, since* $\widetilde{\mathrm{P}}'_{\check{\mathrm{Z}}}(R) \subset \widetilde{\mathrm{P}}'_{\widetilde{\mathrm{Z}}}(R)$ *and* $\widetilde{\mathrm{U}}_{2,\widetilde{\mathrm{Z}}}(R) \subset \widetilde{\mathrm{U}}_{2,\check{\mathrm{Z}}}(R)$*, we define the following quotient of subgroups of* $\widehat{\mathrm{G}}(R) = \widetilde{\mathrm{G}}_{1,\widetilde{\mathrm{Z}}}(R) = \widetilde{\mathrm{P}}'_{\widetilde{\mathrm{Z}}}(R)/\widetilde{\mathrm{U}}_{2,\widetilde{\mathrm{Z}}}(R)$*:*

(1) $\widehat{\mathrm{P}}_{\check{\mathrm{Z}}}(R) := (\widetilde{\mathrm{P}}_{\check{\mathrm{Z}}}(R) \cap \widetilde{\mathrm{P}}'_{\widetilde{\mathrm{Z}}}(R))/\widetilde{\mathrm{U}}_{2,\widetilde{\mathrm{Z}}}(R)$.

(2) $\widehat{\mathrm{Z}}_{\check{\mathrm{Z}}}(R) := (\widetilde{\mathrm{Z}}_{\check{\mathrm{Z}}}(R) \cap \widetilde{\mathrm{P}}'_{\widetilde{\mathrm{Z}}}(R))/\widetilde{\mathrm{U}}_{2,\widetilde{\mathrm{Z}}}(R)$.

(3) $\widehat{\mathrm{G}}_{h,\check{\mathrm{Z}}}(R) := \widehat{\mathrm{P}}_{\check{\mathrm{Z}}}(R)/\widehat{\mathrm{Z}}_{\check{\mathrm{Z}}}(R) \cong \widetilde{\mathrm{G}}_{h,\check{\mathrm{Z}}}(R) \cong \widetilde{\mathrm{P}}_{\check{\mathrm{Z}}}(R)/\widetilde{\mathrm{Z}}_{\check{\mathrm{Z}}}(R)$.
(4) $\widehat{\mathrm{U}}_{\check{\mathrm{Z}}}(R) := \widetilde{\mathrm{U}}_{\check{\mathrm{Z}}}(R)/\widetilde{\mathrm{U}}_{2,\tilde{\mathrm{Z}}}(R)$.
(5) $\widehat{\mathrm{U}}_{2,\check{\mathrm{Z}}}(R) := \widetilde{\mathrm{U}}_{2,\check{\mathrm{Z}}}(R)/\widetilde{\mathrm{U}}_{2,\tilde{\mathrm{Z}}}(R)$.
(6) $\widehat{\mathrm{U}}_{1,\check{\mathrm{Z}}}(R) := \widehat{\mathrm{U}}_{\check{\mathrm{Z}}}(R)/\widehat{\mathrm{U}}_{2,\check{\mathrm{Z}}}(R) \cong \widetilde{\mathrm{U}}_{1,\check{\mathrm{Z}}}(R) = \widetilde{\mathrm{U}}_{\check{\mathrm{Z}}}(R)/\widetilde{\mathrm{U}}_{2,\check{\mathrm{Z}}}(R)$.
(7) $\widehat{\mathrm{G}}_{l,\check{\mathrm{Z}}}(R) := \widehat{\mathrm{Z}}_{\check{\mathrm{Z}}}(R)/\widehat{\mathrm{U}}_{\check{\mathrm{Z}}}(R) \cong (\widetilde{\mathrm{Z}}_{\check{\mathrm{Z}}}(R) \cap \widetilde{\mathrm{P}}'_{\tilde{\mathrm{Z}}}(R))/\widetilde{\mathrm{U}}_{\check{\mathrm{Z}}}(R)$.
(8) $\widehat{\mathrm{P}}'_{\check{\mathrm{Z}}}(R) := \widetilde{\mathrm{P}}'_{\check{\mathrm{Z}}}(R)/\widetilde{\mathrm{U}}_{2,\tilde{\mathrm{Z}}}(R)$.
(9) $\widehat{\mathrm{G}}_{1,\check{\mathrm{Z}}}(R) := \widehat{\mathrm{P}}'_{\check{\mathrm{Z}}}(R)/\widehat{\mathrm{U}}_{2,\check{\mathrm{Z}}}(R) \cong \widetilde{\mathrm{G}}_{1,\check{\mathrm{Z}}}(R) = \widetilde{\mathrm{P}}'_{\check{\mathrm{Z}}}(R)/\widetilde{\mathrm{U}}_{2,\check{\mathrm{Z}}}(R)$.
(10) $\widehat{\mathrm{G}}'_{h,\check{\mathrm{Z}}}(R) := \widehat{\mathrm{P}}'_{\check{\mathrm{Z}}}(R)/\widehat{\mathrm{U}}_{\check{\mathrm{Z}}}(R) \cong \widetilde{\mathrm{G}}'_{h,\check{\mathrm{Z}}}(R) = \widetilde{\mathrm{P}}'_{\check{\mathrm{Z}}}(R)/\widetilde{\mathrm{U}}_{\check{\mathrm{Z}}}(R)$.
(11) $\widehat{\mathrm{G}}'_{l,\check{\mathrm{Z}}}(R) := \widehat{\mathrm{P}}_{\check{\mathrm{Z}}}(R)/\widehat{\mathrm{P}}'_{\check{\mathrm{Z}}}(R) \cong (\widetilde{\mathrm{P}}_{\check{\mathrm{Z}}}(R) \cap \widetilde{\mathrm{P}}'_{\tilde{\mathrm{Z}}}(R))/\widetilde{\mathrm{P}}'_{\check{\mathrm{Z}}}(R)$.

Then the canonical homomorphism $\widehat{\mathrm{G}}(R) \to \mathrm{G}(R)$ induces the following canonical homomorphisms:

(1) $\widehat{\mathrm{P}}_{\check{\mathrm{Z}}}(R) \to \mathrm{P}_{\mathrm{Z}}(R)$.
(2) $\widehat{\mathrm{Z}}_{\check{\mathrm{Z}}}(R) \to \mathrm{Z}_{\mathrm{Z}}(R)$.
(3) $\widehat{\mathrm{G}}_{h,\check{\mathrm{Z}}}(R) \to \mathrm{G}_{h,\mathrm{Z}}(R)$.
(4) $\widehat{\mathrm{U}}_{\check{\mathrm{Z}}}(R) \to \mathrm{U}_{\mathrm{Z}}(R)$.
(5) $\widehat{\mathrm{U}}_{2,\check{\mathrm{Z}}}(R) \to \mathrm{U}_{2,\mathrm{Z}}(R)$.
(6) $\widehat{\mathrm{U}}_{1,\check{\mathrm{Z}}}(R) \to \mathrm{U}_{1,\mathrm{Z}}(R)$.
(7) $\widehat{\mathrm{G}}_{l,\check{\mathrm{Z}}}(R) \to \mathrm{G}_{l,\mathrm{Z}}(R)$.
(8) $\widehat{\mathrm{P}}'_{\check{\mathrm{Z}}}(R) \to \mathrm{P}'_{\mathrm{Z}}(R)$.
(9) $\widehat{\mathrm{G}}_{1,\check{\mathrm{Z}}}(R) \to \mathrm{G}_{1,\mathrm{Z}}(R)$.
(10) $\widehat{\mathrm{G}}'_{h,\check{\mathrm{Z}}}(R) \to \mathrm{G}'_{h,\mathrm{Z}}(R)$.
(11) $\widehat{\mathrm{G}}'_{l,\check{\mathrm{Z}}}(R) \to \mathrm{G}'_{l,\mathrm{Z}}(R)$.

Hence, it makes sense to define $\widehat{\mathcal{H}}_{\widehat{\mathrm{P}}_{\check{\mathrm{Z}}}} := (\widetilde{\mathcal{H}}_{\widetilde{\mathrm{P}}_{\check{\mathrm{Z}}}} \cap \widetilde{\mathcal{H}}_{\widetilde{\mathrm{P}}'_{\tilde{\mathrm{Z}}}})/\widetilde{\mathcal{H}}_{\widetilde{\mathrm{U}}_{2,\tilde{\mathrm{Z}}}}$ etc when $\widehat{\mathcal{H}} = \widetilde{\mathcal{H}}_{\widehat{\mathrm{G}}}$, so that we have $\widehat{\mathcal{H}}_{\widehat{\mathrm{P}}_{\check{\mathrm{Z}}}} \to \mathcal{H}_{\mathrm{P}_{\mathrm{Z}}}$ etc when $\widehat{\mathcal{H}}_{\mathrm{G}} \subset \mathcal{H}$.

Definition 1.2.4.54. *With the setting as in Definition* 1.2.4.53, *consider*

$$\widetilde{\mathrm{P}}_{\check{\mathrm{Z}},\tilde{\mathrm{Z}}}(R) := \widetilde{\mathrm{P}}_{\check{\mathrm{Z}}}(R) \cap \widetilde{\mathrm{P}}_{\tilde{\mathrm{Z}}}(R),$$

and define

$$\widetilde{\mathrm{G}}_{l,\check{\mathrm{Z}},\tilde{\mathrm{Z}}}(R) := \widetilde{\mathrm{P}}_{\check{\mathrm{Z}},\tilde{\mathrm{Z}}}(R)/\widetilde{\mathrm{P}}'_{\check{\mathrm{Z}}}(R),$$

which is the subgroup of $\widetilde{\mathrm{G}}_{l,\check{\mathrm{Z}}}(R)$ consisting of elements preserving the filtrations induced by the admissible surjections $s_{\check{X}} : \check{X} \twoheadrightarrow \widetilde{X}$ and $s_{\check{Y}} : \check{Y} \twoheadrightarrow \widetilde{Y}$. Let

$$\widetilde{\mathcal{H}}_{\widetilde{\mathrm{P}}_{\check{\mathrm{Z}},\tilde{\mathrm{Z}}}} := (\widetilde{\mathcal{H}} \cap \widetilde{\mathrm{P}}_{\check{\mathrm{Z}},\tilde{\mathrm{Z}}}(\hat{\mathbb{Z}}))$$

and

$$\widetilde{\mathcal{H}}_{\widetilde{\mathrm{G}}_{l,\check{\mathrm{Z}},\tilde{\mathrm{Z}}}} := \widetilde{\mathcal{H}}_{\widetilde{\mathrm{P}}_{\check{\mathrm{Z}},\tilde{\mathrm{Z}}}}/\widetilde{\mathcal{H}}_{\widetilde{\mathrm{P}}'_{\check{\mathrm{Z}}}}.$$

Then there are canonical homomorphisms

$$\widetilde{\mathrm{P}}_{\check{\mathrm{Z}},\widetilde{\mathrm{Z}}}(R)/\widetilde{\mathrm{U}}_{2,\check{\mathrm{Z}}}(R) \to \mathrm{P}_{\mathrm{Z}}(R)/\mathrm{U}_{2,\mathrm{Z}}(R)$$

and

$$\widetilde{\mathrm{G}}_{l,\check{\mathrm{Z}},\widetilde{\mathrm{Z}}}(R) \to \mathrm{G}'_{l,\mathrm{Z}}(R) = \mathrm{P}_{\mathrm{Z}}(R)/\mathrm{P}'_{\mathrm{Z}}(R),$$

inducing $\widetilde{\mathcal{H}}_{\widetilde{\mathrm{P}}_{\check{\mathrm{Z}},\widetilde{\mathrm{Z}}}}/\widetilde{\mathcal{H}}_{\widetilde{\mathrm{U}}_{2,\check{\mathrm{Z}}}} \to \mathcal{H}_{\mathrm{P}_{\mathrm{Z}}}/\mathcal{H}_{\mathrm{U}_{2,\mathrm{Z}}}$ *and* $\widetilde{\mathcal{H}}_{\widetilde{\mathrm{G}}_{l,\check{\mathrm{Z}},\widetilde{\mathrm{Z}}}} \to \mathcal{H}_{\mathrm{G}'_{l,\mathrm{Z}}} = \mathcal{H}_{\mathrm{P}_{\mathrm{Z}}}/\mathcal{H}_{\mathrm{P}'_{\mathrm{Z}}}$ *when* $\widetilde{\mathcal{H}}_{\mathrm{G}} \subset \mathcal{H}$.

1.3 Algebraic Compactifications in Characteristic Zero

By algebraic compactifications, we mean compactifications as algebraic varieties, algebraic spaces, or algebraic stacks constructed using the (algebraic) theory of degeneration developed in [82], [28], and [62]. (We do not consider the constructions in [89] and [38] algebraic, because they are based on the analytic construction in [4] and on the theory of canonical models.)

1.3.1 *Toroidal and Minimal Compactifications of PEL-Type Moduli Problems*

Definition 1.3.1.1. (*See [62, Def. 5.3.2.1].*) *Let S be a normal locally noetherian algebraic stack. A tuple $(G, \lambda, i, \alpha_{\mathcal{H}})$ over S is called a* **degenerating family of type** $\mathsf{M}_{\mathcal{H}}$, *or simply a* **degenerating family** *when the context is clear, if there exists a dense subalgebraic stack S_1 of S, such that S_1 is defined over $\mathsf{S}_0 = \mathrm{Spec}(F_0)$, and such that we have the following:*

1. *By viewing group schemes as relative schemes (cf. [37]), G is a semi-abelian scheme over S whose restriction G_{S_1} to S_1 is an abelian scheme. In this case, the dual semi-abelian scheme $G^\vee$ exists (up to unique isomorphism; cf. [80, IV, 7.1] or [62, Thm. 3.4.3.2]), whose restriction $G^\vee_{S_1}$ to S_1 is the dual abelian scheme of G_{S_1}.*
2. *$\lambda : G \to G^\vee$ is a group homomorphism that induces by restriction a polarization λ_{S_1} of G_{S_1}.*
3. *$i : \mathcal{O} \to \mathrm{End}_S(G)$ is a homomorphism that defines by restriction an $\mathcal{O}$-structure $i_{S_1} : \mathcal{O} \to \mathrm{End}_{S_1}(G_{S_1})$ of (G_{S_1}, λ_{S_1}).*
4. *$(G_{S_1}, \lambda_{S_1}, i_{S_1}, \alpha_{\mathcal{H}}) \to S_1$ defines a tuple parameterized by the moduli problem $\mathsf{M}_{\mathcal{H}}$.*

Definition 1.3.1.2. (*See [62, Def. 6.3.1].*) *Let $(G, \lambda, i, \alpha_{\mathcal{H}})$ be a degenerating family of type $\mathsf{M}_{\mathcal{H}}$ over S (as in Definition 1.3.1.1) over $\mathsf{S}_0 = \mathrm{Spec}(F_0)$. Let $\underline{\mathrm{Lie}}^\vee_{G/S} := e^*_G\Omega^1_{G/S}$ be the dual of $\underline{\mathrm{Lie}}_{G/S}$, and let $\underline{\mathrm{Lie}}^\vee_{G^\vee/S} := e^*_G\Omega^1_{G^\vee/S}$ be the dual of $\underline{\mathrm{Lie}}_{G^\vee/S}$. Note that $\lambda : G \to G^\vee$ induces an $\mathcal{O}$-equivariant morphism $\lambda^* : \underline{\mathrm{Lie}}^\vee_{G^\vee/S} \to \underline{\mathrm{Lie}}^\vee_{G/S}$. (Here the $\mathcal{O}$-action on $\underline{\mathrm{Lie}}^\vee_{G/S}$ is a left action after twisted by the involution *.) Then we define the $\mathscr{O}_S$-module $\underline{\mathrm{KS}} = \underline{\mathrm{KS}}_{(G,\lambda,i)/S} = \underline{\mathrm{KS}}_{(G,\lambda,i,\alpha_{\mathcal{H}})/S}$ by setting*

$$\underline{\mathrm{KS}} := (\underline{\mathrm{Lie}}^\vee_{G/S} \underset{\mathscr{O}_S}{\otimes} \underline{\mathrm{Lie}}^\vee_{G^\vee/S}) / \begin{pmatrix} \lambda^*(y) \otimes z - \lambda^*(z) \otimes y \\ (b^\star x) \otimes y - x \otimes (by) \end{pmatrix}_{\substack{x \in \underline{\mathrm{Lie}}^\vee_{G/S}, \\ y,z \in \underline{\mathrm{Lie}}^\vee_{G^\vee/S}, \\ b \in \mathcal{O}}} .$$

Analogues of the $\mathscr{O}_S$-module $\underline{\mathrm{KS}}$ appear naturally in the deformation theory of abelian varieties with PEL structures (without degenerations). The point of Definition 1.3.1.2 is that it extends the conventional definition (for abelian schemes with PEL structures) to the context of (semi-abelian) degenerating families (see Definition 1.3.1.1).

The algebraically constructed toroidal compactifications in characteristic zero can be described as follows:

Theorem 1.3.1.3. *(See [62, Thm. 6.4.1.1].) To each compatible choice* $\Sigma = \{\Sigma_{\Phi_{\mathcal{H}}}\}_{[(\Phi_{\mathcal{H}},\delta_{\mathcal{H}})]}$ *of admissible smooth rational polyhedral cone decomposition data as in Definition* 1.2.2.13*, there is associated an algebraic stack* $\mathsf{M}^{\mathrm{tor}}_{\mathcal{H}} = \mathsf{M}^{\mathrm{tor}}_{\mathcal{H},\Sigma}$ *proper and smooth over* $\mathsf{S}_0 = \mathrm{Spec}(F_0)$*, which is an algebraic space when* $\mathcal{H}$ *is neat (see [89, 0.6] or [62, Def. 1.4.1.8]), containing* $\mathsf{M}_{\mathcal{H}}$ *as an open dense subalgebraic stack, together with a degenerating family* $(G, \lambda, i, \alpha_{\mathcal{H}})$ *over* $\mathsf{M}^{\mathrm{tor}}_{\mathcal{H}}$ *(as in Definition* 1.3.1.1*) such that we have the following:*

1. *(1) The restriction* $(G_{\mathsf{M}_{\mathcal{H}}}, \lambda_{\mathsf{M}_{\mathcal{H}}}, i_{\mathsf{M}_{\mathcal{H}}}, \alpha_{\mathcal{H}})$ *of* $(G, \lambda, i, \alpha_{\mathcal{H}})$ *to* $\mathsf{M}_{\mathcal{H}}$ *is the tautological tuple over* $\mathsf{M}_{\mathcal{H}}$*.*
2. *(2)* $\mathsf{M}^{\mathrm{tor}}_{\mathcal{H}}$ *has a stratification by locally closed subalgebraic stacks*

$$\mathsf{M}^{\mathrm{tor}}_{\mathcal{H}} = \coprod_{[(\Phi_{\mathcal{H}},\delta_{\mathcal{H}},\sigma)]} \mathsf{Z}_{[(\Phi_{\mathcal{H}},\delta_{\mathcal{H}},\sigma)]},$$

with $[(\Phi_{\mathcal{H}}, \delta_{\mathcal{H}}, \sigma)]$ *running through a complete set of equivalence classes of* $(\Phi_{\mathcal{H}}, \delta_{\mathcal{H}}, \sigma)$ *(as in Definition* 1.2.2.10*) with* $\sigma \subset \mathbf{P}^{+}_{\Phi_{\mathcal{H}}}$ *and* $\sigma \in \Sigma_{\Phi_{\mathcal{H}}} \in \Sigma$*. (Here* $\mathsf{Z}_{\mathcal{H}}$ *is suppressed in the notation by our convention. The notation "*$\coprod$*" only means a set-theoretic disjoint union. The algebro-geometric structure is still that of* $\mathsf{M}^{\mathrm{tor}}_{\mathcal{H}}$*.)*
In this stratification, the $[(\Phi'_{\mathcal{H}}, \delta'_{\mathcal{H}}, \sigma')]$*-stratum* $\mathsf{Z}_{[(\Phi'_{\mathcal{H}},\delta'_{\mathcal{H}},\sigma')]}$ *lies in the closure of the* $[(\Phi_{\mathcal{H}}, \delta_{\mathcal{H}}, \sigma)]$*-stratum* $\mathsf{Z}_{[(\Phi_{\mathcal{H}},\delta_{\mathcal{H}},\sigma)]}$ *if and only if* $[(\Phi_{\mathcal{H}}, \delta_{\mathcal{H}}, \sigma)]$ *is a face of* $[(\Phi'_{\mathcal{H}}, \delta'_{\mathcal{H}}, \sigma')]$ *as in Definition* 1.2.2.19 *(see also [62, Rem. 6.3.2.15]).*
The $[(\Phi_{\mathcal{H}}, \delta_{\mathcal{H}}, \sigma)]$*-stratum* $\mathsf{Z}_{[(\Phi_{\mathcal{H}},\delta_{\mathcal{H}},\sigma)]}$ *is smooth over* S_0 *and isomorphic to the support of the formal algebraic stack* $\mathfrak{X}_{\Phi_{\mathcal{H}},\delta_{\mathcal{H}},\sigma}/\Gamma_{\Phi_{\mathcal{H}},\sigma}$ *for every representative* $(\Phi_{\mathcal{H}}, \delta_{\mathcal{H}}, \sigma)$ *of* $[(\Phi_{\mathcal{H}}, \delta_{\mathcal{H}}, \sigma)]$*, where the formal algebraic stack* $\mathfrak{X}_{\Phi_{\mathcal{H}},\delta_{\mathcal{H}},\sigma}$ *(before quotient by* $\Gamma_{\Phi_{\mathcal{H}},\sigma}$*, the subgroup of* $\Gamma_{\Phi_{\mathcal{H}}}$ *formed by elements mapping* σ *to itself; see [62, Def. 6.2.5.23]) admits a canonical structure as the completion of an affine toroidal embedding* $\Xi_{\Phi_{\mathcal{H}},\delta_{\mathcal{H}}}(\sigma)$ *(along its* σ*-stratum* $\Xi_{\Phi_{\mathcal{H}},\delta_{\mathcal{H}},\sigma}$*) of a torus torsor* $\Xi_{\Phi_{\mathcal{H}},\delta_{\mathcal{H}}}$ *over an abelian scheme torsor* $C_{\Phi_{\mathcal{H}},\delta_{\mathcal{H}}}$ *over a finite étale cover* $\mathsf{M}^{\Phi_{\mathcal{H}}}_{\mathcal{H}}$ *of the algebraic stack* $\mathsf{M}^{\mathsf{Z}_{\mathcal{H}}}_{\mathcal{H}}$ *(separated, smooth, and of finite over* S_0*) in Definition* 1.2.1.15*. (Note that* $\mathsf{Z}_{\mathcal{H}}$ *and the isomorphism class of* $\mathsf{M}^{\mathsf{Z}_{\mathcal{H}}}_{\mathcal{H}}$ *depend only on the cusp label* $[(\mathsf{Z}_{\mathcal{H}}, \Phi_{\mathcal{H}}, \delta_{\mathcal{H}})]$*, but not on the choice of the representative* $(\mathsf{Z}_{\mathcal{H}}, \Phi_{\mathcal{H}}, \delta_{\mathcal{H}})$*.)*
In particular, $\mathsf{M}_{\mathcal{H}}$ *is an open dense stratum in this stratification.*
3. *(3) The complement of* $\mathsf{M}_{\mathcal{H}}$ *in* $\mathsf{M}^{\mathrm{tor}}_{\mathcal{H}}$ *(with its reduced structure) is a relative Cartier divisor* $\mathsf{D}_{\infty,\mathcal{H}}$ *with normal crossings, such that each irreducible com-*

ponent of a stratum of $\mathsf{M}_{\mathcal{H}}^{\mathrm{tor}} - \mathsf{M}_{\mathcal{H}}$ *is open dense in an intersection of irreducible components of* $\mathsf{D}_{\infty,\mathcal{H}}$ *(including possible self-intersections). When* $\mathcal{H}$ *is neat, the irreducible components of* $\mathsf{D}_{\infty,\mathcal{H}}$ *have no self-intersections (cf. Condition* 1.2.2.9, *[62, Rem. 6.2.5.26], and [28, Ch. IV, Rem. 5.8(a)]).*

(4) *The extended Kodaira–Spencer morphism [62, Def. 4.6.3.44] for* $G \to \mathsf{M}_{\mathcal{H}}^{\mathrm{tor}}$ *induces an isomorphism*

$$\mathrm{KS}_{G/\mathsf{M}_{\mathcal{H}}^{\mathrm{tor}}/\mathsf{S}_0} : \underline{\mathrm{KS}}_{(G,\lambda,i)/\mathsf{M}_{\mathcal{H}}^{\mathrm{tor}}} \xrightarrow{\sim} \Omega^1_{\mathsf{M}_{\mathcal{H}}^{\mathrm{tor}}/\mathsf{S}_0}[d\log\infty]$$

(see Definition 1.3.1.2*). Here the sheaf* $\Omega^1_{\mathsf{M}_{\mathcal{H}}^{\mathrm{tor}}/\mathsf{S}_0}[d\log\infty]$ *is the sheaf of modules of log* 1*-differentials on* $\mathsf{M}_{\mathcal{H}}^{\mathrm{tor}}$ *over* S_0*, with respect to the relative Cartier divisor* $\mathsf{D}_{\infty,\mathcal{H}}$ *with normal crossings.*

(5) *For every representative* $(\Phi_{\mathcal{H}},\delta_{\mathcal{H}},\sigma)$ *of* $[(\Phi_{\mathcal{H}},\delta_{\mathcal{H}},\sigma)]$*, the formal completion* $(\mathsf{M}_{\mathcal{H}}^{\mathrm{tor}})^{\wedge}_{\mathsf{Z}_{[(\Phi_{\mathcal{H}},\delta_{\mathcal{H}},\sigma)]}}$ *of* $\mathsf{M}_{\mathcal{H}}^{\mathrm{tor}}$ *along the* $[(\Phi_{\mathcal{H}},\delta_{\mathcal{H}},\sigma)]$*-stratum* $\mathsf{Z}_{[(\Phi_{\mathcal{H}},\delta_{\mathcal{H}},\sigma)]}$ *is canonically isomorphic to the formal algebraic stack* $\mathfrak{X}_{\Phi_{\mathcal{H}},\delta_{\mathcal{H}},\sigma}/\Gamma_{\Phi_{\mathcal{H}},\sigma}$*. (To form the formal completion along a given locally closed stratum, we first remove the other strata appearing in the closure of this stratum from the total space, and then form the formal completion of the remaining space along this stratum.)*

This isomorphism respects stratifications in the sense that, given any étale (i.e., formally étale and of finite type; see [35, I, 10.13.3]) morphism $\mathrm{Spf}(R,I) \to \mathfrak{X}_{\Phi_{\mathcal{H}},\delta_{\mathcal{H}},\sigma}/\Gamma_{\Phi_{\mathcal{H}},\sigma}$ *inducing a morphism* $\mathrm{Spec}(R) \to \Xi_{\Phi_{\mathcal{H}},\delta_{\mathcal{H}}}(\sigma)/\Gamma_{\Phi_{\mathcal{H}},\sigma}$*, the stratification of* $\mathrm{Spec}(R)$ *(inherited from* $\Xi_{\Phi_{\mathcal{H}},\delta_{\mathcal{H}}}(\sigma)/\Gamma_{\Phi_{\mathcal{H}},\sigma}$*; see [62, Prop. 6.3.1.6 and Def. 6.3.2.16]) makes the induced morphism* $\mathrm{Spec}(R) \to \mathsf{M}_{\mathcal{H}}^{\mathrm{tor}}$ *a strata-preserving morphism.*

The pullback of the degenerating family $(G,\lambda,i,\alpha_{\mathcal{H}})$ *over* $\mathsf{M}_{\mathcal{H}}^{\mathrm{tor}}$ *to* $(\mathsf{M}_{\mathcal{H}}^{\mathrm{tor}})^{\wedge}_{\mathsf{Z}_{[(\Phi_{\mathcal{H}},\delta_{\mathcal{H}},\sigma)]}}$ *is the Mumford family* $({}^{\heartsuit}G, {}^{\heartsuit}\lambda, {}^{\heartsuit}i, {}^{\heartsuit}\alpha_{\mathcal{H}})$ *over* $\mathfrak{X}_{\Phi_{\mathcal{H}},\delta_{\mathcal{H}},\sigma}/\Gamma_{\Phi_{\mathcal{H}},\sigma}$ *(see [62, Sec. 6.2.5]) after we identify the bases using the isomorphism. (Here both the pullback of* $(G,\lambda,i,\alpha_{\mathcal{H}})$ *and the Mumford family* $({}^{\heartsuit}G, {}^{\heartsuit}\lambda, {}^{\heartsuit}i, {}^{\heartsuit}\alpha_{\mathcal{H}})$ *are considered as relative schemes with additional structures; cf. [37].)*

(6) *Let* S *be an irreducible noetherian normal scheme over* S_0*, and suppose that we have a degenerating family* $(G^\dagger,\lambda^\dagger,i^\dagger,\alpha^\dagger_{\mathcal{H}})$ *of type* $\mathsf{M}_{\mathcal{H}}$ *over* S *as in Definition* 1.3.1.1*. Then* $(G^\dagger,\lambda^\dagger,i^\dagger,\alpha^\dagger_{\mathcal{H}}) \to S$ *is the pullback of* $(G,\lambda,i,\alpha_{\mathcal{H}}) \to \mathsf{M}_{\mathcal{H}}^{\mathrm{tor}}$ *via a (necessarily unique) morphism* $S \to \mathsf{M}_{\mathcal{H}}^{\mathrm{tor}}$ *(over* S_0*) if and only if the following condition is satisfied at each geometric point* $\bar{s}$ *of* S*:*

Consider any dominant morphism $\mathrm{Spec}(V) \to S$ *centered at* $\bar{s}$*, where* V *is a complete discrete valuation ring with fraction field* K*, algebraically closed residue field* k*, and discrete valuation* v*. Let* $(G^\ddagger,\lambda^\ddagger,i^\ddagger,\alpha^\ddagger_{\mathcal{H}}) \to \mathrm{Spec}(V)$ *be the pullback of* $(G^\dagger,\lambda^\dagger,i^\dagger,\alpha^\dagger_{\mathcal{H}}) \to S$*. This pullback family defines an object of* $\mathrm{DEG}_{\mathrm{PEL},\mathsf{M}_{\mathcal{H}}}(V)$*, which corresponds to a tuple* $(B^\ddagger,\lambda_{B^\ddagger},i_{B^\ddagger},\underline{X}^\ddagger,\underline{Y}^\ddagger,\phi^\ddagger,c^\ddagger,c^{\vee,\ddagger},\tau^\ddagger,[\alpha^{\natural,\ddagger}_{\mathcal{H}}])$ *in* $\mathrm{DD}_{\mathrm{PEL},\mathsf{M}_{\mathcal{H}}}(V)$ *under [62,*

Thm. 5.3.1.19]. Then we have a fully symplectic-liftable admissible filtration $\mathsf{Z}^{\ddagger}_{\mathcal{H}}$ *determined by* $[\alpha^{\natural,\ddagger}_{\mathcal{H}}]$. *Moreover, the étale sheaves* $\underline{X}^{\ddagger}$ *and* $\underline{Y}^{\ddagger}$ *are necessarily constant, because the base ring* V *is strict local. Hence, it makes sense to say we also have a uniquely determined torus argument* $\Phi^{\ddagger}_{\mathcal{H}}$ *at level* $\mathcal{H}$ *for* $\mathsf{Z}^{\ddagger}_{\mathcal{H}}$.

On the other hand, we have objects $\underline{\Phi}_{\mathcal{H}}(G^{\ddagger})$, $\underline{\mathbf{S}}_{\underline{\Phi}_{\mathcal{H}}(G^{\ddagger})}$, *and* $\underline{B}(G^{\ddagger})$ *(see [62, Constr. 6.3.1.1]), which define objects* $\Phi^{\ddagger}_{\mathcal{H}}$, $\mathbf{S}_{\Phi^{\ddagger}_{\mathcal{H}}}$, *and in particular* $B^{\ddagger} : \mathbf{S}_{\Phi^{\ddagger}_{\mathcal{H}}} \to \mathrm{Inv}(V)$ *over the special fiber. Then* $\upsilon \circ B^{\ddagger} : \mathbf{S}_{\Phi^{\ddagger}_{\mathcal{H}}} \to \mathbb{Z}$ *defines an element of* $\mathbf{S}^{\vee}_{\Phi^{\ddagger}_{\mathcal{H}}}$, *where* $\upsilon : \mathrm{Inv}(V) \to \mathbb{Z}$ *is the homomorphism induced by the discrete valuation of* V.

Then the condition is that, for each $\mathrm{Spec}(V) \to S$ *as above (centered at* $\bar{s}$*), and for some (and hence every) choice of* $\delta^{\ddagger}_{\mathcal{H}}$, *there is a cone* $\sigma^{\ddagger}$ *in the cone decomposition* $\Sigma_{\Phi^{\ddagger}_{\mathcal{H}}}$ *of* $\mathbf{P}_{\Phi^{\ddagger}_{\mathcal{H}}}$ *(given by the choice of* Σ*; cf. Definition* 1.2.2.13*) such that* $\overline{\sigma}^{\ddagger}$ *contains all* $\upsilon \circ B^{\ddagger}$ *obtained in this way.*

Statement (1) means the tautological tuple over $\mathsf{M}_{\mathcal{H}}$ extends to a degenerating family $(G, \lambda, i, \alpha_{\mathcal{H}})$ over $\mathsf{M}^{\mathrm{tor}}_{\mathcal{H}}$. (Since $\mathsf{M}^{\mathrm{tor}}_{\mathcal{H}}$ is noetherian normal, this extension is unique up to unique isomorphism, by [92, IX, 1.4], [28, Ch. I, Prop. 2.7], or [62, Prop. 3.3.1.5].) Statements (2), (3), (4), and (5) are self-explanatory. Statement (6) can be interpreted as a *universal property* for the degenerating family $(G, \lambda, i, \alpha_{\mathcal{H}}) \to \mathsf{M}^{\mathrm{tor}}_{\mathcal{H}}$ among degenerating families over normal locally noetherian bases, as in Definition 1.3.1.1, satisfying moreover some conditions describing the *degeneration patterns* over pullbacks to complete discrete valuation rings with algebraically closed residue fields. (This *universal property* was crucially used in the proof of Theorem 1.3.3.15 below in [61].)

Remark 1.3.1.4. (Compare with Remark 1.1.2.1.) If we have chosen another PEL-type $\mathcal{O}$-lattice L' in $L \underset{\mathbb{Z}}{\otimes} \mathbb{Q}$ which is also stabilized by $\mathcal{H}$, so that $\mathsf{M}_{\mathcal{H}}$ carries the corresponding abelian scheme A' (with additional structures) as in Remark 1.1.2.1, with a $\mathbb{Q}^{\times}$-isogeny $f : A \to A'$, then a sufficiently divisible multiple Nf of f is an isogeny with finite étale kernel, which we denote by K. Since $A = G_{\mathsf{M}_{\mathcal{H}}}$ (see (1) of Theorem 1.3.1.3), we can take the schematic-closure K^{ext} of K in G, which is quasi-finite étale over $\mathsf{M}^{\mathrm{tor}}_{\mathcal{H},\Sigma}$. Then we can form the quotient $G' := G/K^{\mathrm{ext}}$ by [62, Lem. 3.4.3.1], which is a semi-abelian scheme with an $\mathbb{Q}^{\times}$-isogeny $f^{\mathrm{ext}} : G \to G'$. By [92, IX, 1.4], [28, Ch. I, Prop. 2.7], or [62, Prop. 3.3.1.5], G' is (up to unique isomorphism) independent of the choice of N, and the additional structures λ, i, $\alpha_{\mathcal{H}}$ of G naturally induce the additional structures λ', i', and $\alpha'_{\mathcal{H}}$ of G', which extend those of A' (based on the moduli interpretation of the moduli problem $\mathsf{M}'_{\mathcal{H}}$ defined by L'). Hence, the $\mathbb{Q}^{\times}$-isogeny class of G extends that of A, and carries well-defined additional structures. (It can be verified that $(G', \lambda', i', \alpha'_{\mathcal{H}}) \to \mathsf{M}^{\mathrm{tor}}_{\mathcal{H},\Sigma}$ satisfies the corresponding universal property of the toroidal compactification of $\mathsf{M}'_{\mathcal{H}}$ defined by the corresponding collection of cone decompositions as in (6) of

Theorem 1.3.1.3, so that the theory does not really depend on the choice of L. Then, as in Remark 1.1.2.1, we can define the collection $\{\mathsf{M}^{\mathrm{tor}}_{\mathcal{H},\Sigma}\}_{\mathcal{H},\Sigma}$, indexed by all open compact subgroups $\mathcal{H}$ of $\mathrm{G}(\mathbb{A}^\infty)$ and collections Σ for the corresponding $\mathsf{M}_{\mathcal{H}}$, with a canonical action of $\mathrm{G}(\mathbb{A}^\infty)$; see Proposition 1.3.1.15 below.)

The algebraically constructed minimal compactifications in characteristic zero can be described as follows:

Theorem 1.3.1.5. *(See [62, Thm. 7.2.4.1].) There exists a normal scheme $\mathsf{M}^{\min}_{\mathcal{H}}$ projective and flat over $\mathsf{S}_0 = \mathrm{Spec}(F_0)$, such that we have the following:*

(1) $\mathsf{M}^{\min}_{\mathcal{H}}$ contains the **coarse moduli space** *$[\mathsf{M}_{\mathcal{H}}]$ of $\mathsf{M}_{\mathcal{H}}$ (see [62, Sec. A.7.5]) as an open dense* **subscheme**.

(2) Let $(G_{\mathsf{M}_{\mathcal{H}}}, \lambda_{\mathsf{M}_{\mathcal{H}}}, i_{\mathsf{M}_{\mathcal{H}}}, \alpha_{\mathcal{H}})$ be the tautological tuple over $\mathsf{M}_{\mathcal{H}}$. Let us define the invertible sheaf $\omega_{\mathsf{M}_{\mathcal{H}}} := \wedge^{\mathrm{top}} \underline{\mathrm{Lie}}^\vee_{G_{\mathsf{M}_{\mathcal{H}}}/\mathsf{M}_{\mathcal{H}}} = \wedge^{\mathrm{top}} e^*_{G_{\mathsf{M}_{\mathcal{H}}}} \Omega^1_{G_{\mathsf{M}_{\mathcal{H}}}/\mathsf{M}_{\mathcal{H}}}$ over $\mathsf{M}_{\mathcal{H}}$. Then there is a smallest integer $N_0 \geq 1$ such that $\omega^{\otimes N_0}_{\mathsf{M}_{\mathcal{H}}}$ is the pullback of an ample invertible sheaf $\mathcal{O}(1)$ over $\mathsf{M}^{\min}_{\mathcal{H}}$.

If $\mathcal{H}$ is neat, then $\mathsf{M}_{\mathcal{H}} \to [\mathsf{M}_{\mathcal{H}}]$ is an isomorphism, and induces an embedding of $\mathsf{M}_{\mathcal{H}}$ as an open dense subscheme of $\mathsf{M}^{\min}_{\mathcal{H}}$. Moreover, we have $N_0 = 1$ with a canonical choice of $\mathcal{O}(1)$, and the restriction of $\mathcal{O}(1)$ to $\mathsf{M}_{\mathcal{H}}$ is isomorphic to $\omega_{\mathsf{M}_{\mathcal{H}}}$. We shall denote $\mathcal{O}(1)$ by $\omega_{\mathsf{M}^{\min}_{\mathcal{H}}}$, and interpret it as an extension of $\omega_{\mathsf{M}_{\mathcal{H}}}$ to $\mathsf{M}^{\min}_{\mathcal{H}}$.

By abuse of notation, for each integer k divisible by N_0, we shall denote $\mathcal{O}(1)^{\otimes k/N_0}$ by $\omega^{\otimes k}_{\mathsf{M}^{\min}_{\mathcal{H}}}$, even when $\omega_{\mathsf{M}^{\min}_{\mathcal{H}}}$ itself is not defined.

(3) For each (smooth) arithmetic toroidal compactification $\mathsf{M}^{\mathrm{tor}}_{\mathcal{H}}$ of $\mathsf{M}_{\mathcal{H}}$ as in Theorem 1.3.1.3, with a degenerating family $(G, \lambda, i, \alpha_{\mathcal{H}})$ over $\mathsf{M}^{\mathrm{tor}}_{\mathcal{H}}$ extending the tautological tuple $(G_{\mathsf{M}_{\mathcal{H}}}, \lambda_{\mathsf{M}_{\mathcal{H}}}, i_{\mathsf{M}_{\mathcal{H}}}, \alpha_{\mathcal{H}})$ over $\mathsf{M}_{\mathcal{H}}$, let $\omega_{\mathsf{M}^{\mathrm{tor}}_{\mathcal{H}}} := \wedge^{\mathrm{top}} \underline{\mathrm{Lie}}^\vee_{G/\mathsf{M}^{\mathrm{tor}}_{\mathcal{H}}} = \wedge^{\mathrm{top}} e^*_G \Omega^1_{G/\mathsf{M}^{\mathrm{tor}}_{\mathcal{H}}}$ be the invertible sheaf over $\mathsf{M}^{\mathrm{tor}}_{\mathcal{H}}$ extending $\omega_{\mathsf{M}_{\mathcal{H}}}$ naturally. Then the graded algebra $\underset{k\geq 0}{\oplus} \Gamma(\mathsf{M}^{\mathrm{tor}}_{\mathcal{H}}, \omega^{\otimes k}_{\mathsf{M}^{\mathrm{tor}}_{\mathcal{H}}})$, with its natural algebra structure induced by tensor products, is finitely generated over F_0, and is independent of the choice (of the Σ used in the definition) of $\mathsf{M}^{\mathrm{tor}}_{\mathcal{H}}$. The normal scheme $\mathsf{M}^{\min}_{\mathcal{H}}$ (projective and flat over S_0) is canonically isomorphic to $\mathrm{Proj}\Big(\underset{k\geq 0}{\oplus} \Gamma(\mathsf{M}^{\mathrm{tor}}_{\mathcal{H}}, \omega^{\otimes k}_{\mathsf{M}^{\mathrm{tor}}_{\mathcal{H}}})\Big)$, and there is a canonical morphism $\oint_{\mathcal{H}} : \mathsf{M}^{\mathrm{tor}}_{\mathcal{H}} \to \mathsf{M}^{\min}_{\mathcal{H}}$ determined by $\omega_{\mathsf{M}^{\mathrm{tor}}_{\mathcal{H}}}$ and the universal property of Proj*, such that $\oint^*_{\mathcal{H}} \mathcal{O}(1) \cong \omega^{\otimes N_0}_{\mathsf{M}^{\mathrm{tor}}_{\mathcal{H}}}$ over $\mathsf{M}^{\mathrm{tor}}_{\mathcal{H}}$, and such that the canonical morphism $\mathscr{O}_{\mathsf{M}^{\min}_{\mathcal{H}}} \to \oint_{\mathcal{H},*} \mathscr{O}_{\mathsf{M}^{\mathrm{tor}}_{\mathcal{H}}}$ is an isomorphism. Moreover, when we vary the choices of $\mathsf{M}^{\mathrm{tor}}_{\mathcal{H}}$'s, the morphisms $\oint_{\mathcal{H}}$'s are compatible with the canonical morphisms among the $\mathsf{M}^{\mathrm{tor}}_{\mathcal{H}}$'s as in [62, Prop. 6.4.2.3].*

When $\mathcal{H}$ is neat, we have $\oint^*_{\mathcal{H}} \omega_{\mathsf{M}^{\min}_{\mathcal{H}}} \cong \omega_{\mathsf{M}^{\mathrm{tor}}_{\mathcal{H}}}$ and $\oint_{\mathcal{H},*} \omega_{\mathsf{M}^{\mathrm{tor}}_{\mathcal{H}}} \cong \omega_{\mathsf{M}^{\min}_{\mathcal{H}}}$.

(4) $\mathsf{M}^{\min}_{\mathcal{H}}$ has a natural stratification by locally closed subschemes

$$\mathsf{M}^{\min}_{\mathcal{H}} = \coprod_{[(\Phi_{\mathcal{H}}, \delta_{\mathcal{H}})]} \mathsf{Z}_{[(\Phi_{\mathcal{H}}, \delta_{\mathcal{H}})]},$$

with $[(\Phi_{\mathcal{H}}, \delta_{\mathcal{H}})]$ *running through a complete set of cusp labels (see Definition* 1.2.1.7 *and [62, Def. 5.4.2.4]), such that the* $[(\Phi'_{\mathcal{H}}, \delta'_{\mathcal{H}})]$*-stratum* $\mathsf{Z}_{[(\Phi'_{\mathcal{H}}, \delta'_{\mathcal{H}})]}$ *lies in the closure of the* $[(\Phi_{\mathcal{H}}, \delta_{\mathcal{H}})]$*-stratum* $\mathsf{Z}_{[(\Phi_{\mathcal{H}}, \delta_{\mathcal{H}})]}$ *if and only if there is a surjection from the cusp label* $[(\Phi'_{\mathcal{H}}, \delta'_{\mathcal{H}})]$ *to the cusp label* $[(\Phi_{\mathcal{H}}, \delta_{\mathcal{H}})]$ *as in Definition* 1.2.1.18*. (The notation "*$\coprod$*" only means a set-theoretic disjoint union. The algebro-geometric structure is still that of* $\mathsf{M}_{\mathcal{H}}^{\min}$*.)*

Each $[(\Phi_{\mathcal{H}}, \delta_{\mathcal{H}})]$*-stratum* $\mathsf{Z}_{[(\Phi_{\mathcal{H}}, \delta_{\mathcal{H}})]}$ *is canonically isomorphic to the coarse moduli space* $[\mathsf{M}_{\mathcal{H}}^{\mathsf{Z}_{\mathcal{H}}}]$ *(which is a scheme) of the corresponding algebraic stack* $\mathsf{M}_{\mathcal{H}}^{\mathsf{Z}_{\mathcal{H}}}$ *(separated, smooth, and of finite type over* S_0*) as in Definition* 1.2.1.15*.*

Let us define the $\mathcal{O}$*-multi-rank of a stratum* $\mathsf{Z}_{[(\Phi_{\mathcal{H}}, \delta_{\mathcal{H}})]}$ *to be the* $\mathcal{O}$*-multi-rank of the cusp label represented by* $(\Phi_{\mathcal{H}}, \delta_{\mathcal{H}})$ *(see [62, Def. 5.4.2.7]). The only stratum with* $\mathcal{O}$*-multi-rank zero is the open stratum* $\mathsf{Z}_{[(0,0)]} \cong [\mathsf{M}_{\mathcal{H}}]$*, and those strata* $\mathsf{Z}_{[(\Phi_{\mathcal{H}}, \delta_{\mathcal{H}})]}$ *with nonzero* $\mathcal{O}$*-multi-ranks are called* ***cusps****. (This explains the name of the cusp labels.)*

(5) The restriction of $\oint_{\mathcal{H}}$ *to the stratum* $\mathsf{Z}_{[(\Phi_{\mathcal{H}}, \delta_{\mathcal{H}}, \sigma)]}$ *of* $\mathsf{M}_{\mathcal{H}}^{\mathrm{tor}}$ *is a surjection to the stratum* $\mathsf{Z}_{[(\Phi_{\mathcal{H}}, \delta_{\mathcal{H}})]}$ *of* $\mathsf{M}_{\mathcal{H}}^{\min}$*. This surjection is smooth when* $\mathcal{H}$ *is neat, and is proper if* σ *is top-dimensional in* $\mathbf{P}_{\Phi_{\mathcal{H}}}^{+} \subset (\mathbf{S}_{\Phi_{\mathcal{H}}})_{\mathbb{R}}^{\vee}$*.*

Under the above-mentioned identification $[\mathsf{M}_{\mathcal{H}}^{\mathsf{Z}_{\mathcal{H}}}] \overset{\sim}{\to} \mathsf{Z}_{[(\Phi_{\mathcal{H}}, \delta_{\mathcal{H}})]}$ *on the target, this surjection can be viewed as the quotient by* $\Gamma_{\Phi_{\mathcal{H}}, \sigma}$ *(see [62, Def. 6.2.5.23]) of a torsor under a torus* $E_{\Phi_{\mathcal{H}}, \sigma}$ *over an abelian scheme torsor* $C_{\Phi_{\mathcal{H}}, \delta_{\mathcal{H}}}$ *(see Remark* 1.3.1.6 *below) over the finite étale cover* $\mathsf{M}_{\mathcal{H}}^{\Phi_{\mathcal{H}}}$ *of the algebraic stack* $\mathsf{M}_{\mathcal{H}}^{\mathsf{Z}_{\mathcal{H}}}$ *over the coarse moduli space* $[\mathsf{M}_{\mathcal{H}}^{\mathsf{Z}_{\mathcal{H}}}]$ *(which is a scheme). More precisely, this torus* $E_{\Phi_{\mathcal{H}}, \sigma}$ *is the quotient of the torus* $E_{\Phi_{\mathcal{H}}} := \underline{\mathrm{Hom}}_{\mathbb{Z}}(\mathbf{S}_{\Phi_{\mathcal{H}}}, \mathbf{G}_{\mathrm{m}})$ *corresponding to the subgroup* $\mathbf{S}_{\Phi_{\mathcal{H}}, \sigma} := \{\ell \in \mathbf{S}_{\Phi_{\mathcal{H}}} : \langle \ell, y \rangle = 0, \forall y \in \sigma\}$ *of* $\mathbf{S}_{\Phi_{\mathcal{H}}}$*. (See [62, Lem. 6.2.4.4] for the definition of* $E_{\Phi_{\mathcal{H}}}$*, and see [62, Def. 6.1.2.7] for the definition of* σ*-stratum.)*

Remark 1.3.1.6. In [62, Sec. 6.2.4; see also the errata], we should have considered a subquotient of H_n which is an extension of $H_{n,\mathrm{G}_{h,\mathsf{Z}_n}^{\mathrm{ess}}}$ by $H_{n,\mathrm{U}_{1,\mathsf{Z}_n}^{\mathrm{ess}}}$ (just as $\widehat{\mathcal{H}}$ is an extension of $\widehat{\mathcal{H}}_{\mathrm{G}}$ by $\widehat{\mathcal{H}}_{\widehat{\mathrm{U}}}$ in (1.2.4.5)), which is not necessarily the semi-direct product $H_{n,\mathrm{G}_{h,\mathsf{Z}_n}^{\mathrm{ess}} \ltimes \mathrm{U}_{1,\mathsf{Z}_n}^{\mathrm{ess}}} = H_{n,\mathrm{G}_{h,\mathsf{Z}_n}^{\mathrm{ess}}} \ltimes H_{n,\mathrm{U}_{1,\mathsf{Z}_n}^{\mathrm{ess}}}$ used there. Therefore, it is incorrect to conclude from [62, Lem. 6.2.4.5] that the further quotient $C_{\Phi_{\mathcal{H}}, \delta_{\mathcal{H}}} \to \mathsf{M}_{\mathcal{H}}^{\Phi_{\mathcal{H}}}$ is also an abelian scheme. (The identity section might not descend under the quotient by $H_{n,\mathrm{U}_{1,\mathsf{Z}_n}^{\mathrm{ess}}}$.) Accordingly, in [62, Prop. 6.2.4.7 and later sections], we should only assert that $C_{\Phi_{\mathcal{H}}, \delta_{\mathcal{H}}} \to \mathsf{M}_{\mathcal{H}}^{\Phi_{\mathcal{H}}}$ is an *abelian scheme torsor*. This does not affect the constructions of torus torsors and toroidal embeddings because the existence of identity sections is not logically necessary.

Now suppose $\mathcal{H}$ is *neat*. Let $\Sigma = \{\Sigma_{\Phi_{\mathcal{H}}}\}_{[(\Phi_{\mathcal{H}}, \delta_{\mathcal{H}})]}$ be any *projective* compatible choice of smooth rational polyhedral cone decomposition data, with a compatible collection $\mathsf{pol} = \{\mathsf{pol}_{\Phi_{\mathcal{H}}}\}_{[(\Phi_{\mathcal{H}}, \delta_{\mathcal{H}})]}$ of polarization functions as in Definition 1.2.2.14.

Definition 1.3.1.7. *(See [62, Def. 7.3.3.1].) Let* Σ, pol, *and* $\mathsf{M}_{\mathcal{H}}^{\mathrm{tor}} = \mathsf{M}_{\mathcal{H},\Sigma}^{\mathrm{tor}}$ *be as above. By* (3) *of Theorem* 1.3.1.3, *the complement* $\mathsf{D}_{\infty,\mathcal{H}}$ *of* $\mathsf{M}_{\mathcal{H}}$ *in* $\mathsf{M}_{\mathcal{H}}^{\mathrm{tor}} = \mathsf{M}_{\mathcal{H},\Sigma}^{\mathrm{tor}}$ *(with its reduced structure) is a relative Cartier divisor with normal crossings, each of whose irreducible components is an irreducible component of some* $\overline{\mathsf{Z}}_{[(\Phi_{\mathcal{H}},\delta_{\mathcal{H}},\sigma)]}$ *that is the closure of some strata* $\mathsf{Z}_{[(\Phi_{\mathcal{H}},\delta_{\mathcal{H}},\sigma)]}$ *labeled by the equivalence class* $[(\Phi_{\mathcal{H}},\delta_{\mathcal{H}},\sigma)]$ *of some triple* $(\Phi_{\mathcal{H}},\delta_{\mathcal{H}},\sigma)$ *with* σ *a one-dimensional cone in the cone decomposition* $\Sigma_{\Phi_{\mathcal{H}}}$ *of* $\mathbf{P}_{\Phi_{\mathcal{H}}}$. *Let* $\jmath_{\mathcal{H},\mathsf{pol}}$ *be the invertible sheaf of ideals over* $\mathsf{M}_{\mathcal{H}}^{\mathrm{tor}}$ *supported on* $\mathsf{D}_{\infty,\mathcal{H}}$ *such that the order of* $\jmath_{\mathcal{H},\mathsf{pol}}$ *along each* $\overline{\mathsf{Z}}_{[(\Phi_{\mathcal{H}},\delta_{\mathcal{H}},\sigma)]}$ *is the value of* $\mathsf{pol}_{\Phi_{\mathcal{H}}}$ *at the* $\mathbb{Z}_{>0}$*-generator of* $\sigma \cap \mathbf{S}_{\Phi_{\mathcal{H}}}^{\vee}$ *for some (and hence every) representative* $(\Phi_{\mathcal{H}},\delta_{\mathcal{H}},\sigma)$. *This is well defined because of the compatibility condition for* $\mathsf{pol} = \{\mathsf{pol}_{\Phi_{\mathcal{H}}}\}_{[(\Phi_{\mathcal{H}},\delta_{\mathcal{H}})]}$ *as in Definition* 1.2.2.14.

For each integer $d \geq 1$, *let* $d\mathsf{pol}$ *denote the collection of polarization functions defined by multiplying all polarization functions in the collection* pol *by* d. *Then we have a canonical isomorphism* $\jmath_{\mathcal{H},d\mathsf{pol}} \cong \jmath_{\mathcal{H},\mathsf{pol}}^{\otimes d}$.

Definition 1.3.1.8. *For each integer* $d \geq 1$, *let*

$$\mathcal{J}_{\mathcal{H},d\mathsf{pol}} := \mathcal{J}_{\mathcal{H},\mathsf{pol}}^{(d)} := \oint_{\mathcal{H},*}(\jmath_{\mathcal{H},\mathsf{pol}}^{\otimes d}) \cong \oint_{\mathcal{H},*}(\jmath_{\mathcal{H},d\mathsf{pol}}),$$

where $\oint_{\mathcal{H}} : \mathsf{M}_{\mathcal{H}}^{\mathrm{tor}} \to \mathsf{M}_{\mathcal{H}}^{\min}$ *is the canonical morphism (as in* (3) *of Theorem* 1.3.1.5). *(We introduced the two intermediate objects* $\mathcal{J}_{\mathcal{H},\mathsf{pol}}^{(d)}$ *and* $\oint_{\mathcal{H},*}(\jmath_{\mathcal{H},\mathsf{pol}}^{\otimes d})$ *because this is what was done in [28, Ch. V] and [62, Sec. 7.3]. Later we will mainly use* $\mathcal{J}_{\mathcal{H},d\mathsf{pol}}$ *and* $\jmath_{\mathcal{H},d\mathsf{pol}}$ *in our exposition. Note that* $\mathcal{J}_{\mathcal{H},d\mathsf{pol}}$ *is a coherent* $\mathscr{O}_{\mathsf{M}_{\mathcal{H}}^{\min}}$*-ideal because* $\oint_{\mathcal{H}}$ *is proper and because the canonical morphism* $\mathscr{O}_{\mathsf{M}_{\mathcal{H}}^{\min}} \to \oint_{\mathcal{H},*}\mathscr{O}_{\mathsf{M}_{\mathcal{H}}^{\mathrm{tor}}}$ *is an isomorphism.)*

Let us introduce the following condition for $\Sigma = \{\Sigma_{\Phi_{\mathcal{H}}}\}_{[(\Phi_{\mathcal{H}},\delta_{\mathcal{H}})]}$ and $\mathsf{pol} = \{\mathsf{pol}_{\Phi_{\mathcal{H}}}\}_{[(\Phi_{\mathcal{H}},\delta_{\mathcal{H}})]}$ (*cf.* [62, Lem. 7.3.1.7]):

Condition 1.3.1.9. *(See [62, Cond. 7.3.3.3]; cf. [4, Ch. IV, Sec. 2, p. 329] and [28, Ch. V, Sec. 5, p. 178].) For each representative* $(\Phi_{\mathcal{H}},\delta_{\mathcal{H}})$ *of cusp label and each vertex* ℓ_0 *of* $K_{\mathsf{pol}_{\Phi_{\mathcal{H}}}}^{\vee}$ *corresponding to a top-dimensional cone* σ_0, *we have*

$$\langle \ell_0, x\rangle < \langle \gamma \cdot \ell_0, x\rangle$$

for all $x \in \overline{\sigma}_0 \cap \mathbf{P}_{\Phi_{\mathcal{H}}}^{+}$ *and all* $\gamma \in \Gamma_{\Phi_{\mathcal{H}}}$ *such that* $\gamma \neq 1$.

Theorem 1.3.1.10. *(See [62, Thm. 7.3.3.4]; cf. [4, Ch. IV, Sec. 2.1, Thm.] and [28, Ch. V, Thm. 5.8].) Suppose* $\mathcal{H}$ *is neat, and suppose* Σ *is projective with a compatible collection* pol *of polarization functions as in Definition* 1.2.2.14. *For each integer* $d \geq 1$, *suppose* $\jmath_{\mathcal{H},d\mathsf{pol}}$ *is defined over* $\mathsf{M}_{\mathcal{H}}^{\mathrm{tor}} = \mathsf{M}_{\mathcal{H},\Sigma}^{\mathrm{tor}}$ *as in Definition* 1.3.1.7, *and suppose* $\mathcal{J}_{\mathcal{H},d\mathsf{pol}}$ *is defined over* $\mathsf{M}_{\mathcal{H}}^{\min}$ *as in Definition* 1.3.1.8. *Then there exists an integer* $d_0 \geq 1$ *such that the following are true:*

(1) The canonical morphism $\oint_{\mathcal{H}}^{-1}\mathcal{J}_{\mathcal{H},d_0\mathsf{pol}} \cdot \mathscr{O}_{\mathsf{M}_{\mathcal{H}}^{\mathrm{tor}}} \to \jmath_{\mathcal{H},d_0\mathsf{pol}}$ *of coherent* $\mathscr{O}_{\mathsf{M}_{\mathcal{H}}^{\mathrm{tor}}}$*-ideals is an isomorphism, which induces a canonical morphism*

$$\mathrm{NBl}_{\mathcal{J}_{\mathcal{H},d_0\mathsf{pol}}}(\oint_{\mathcal{H}}) : \mathsf{M}_{\mathcal{H}}^{\mathrm{tor}} \to \mathrm{NBl}_{\mathcal{J}_{\mathcal{H},d_0\mathsf{pol}}}(\mathsf{M}_{\mathcal{H}}^{\min})$$

by the universal property of the normalization of blow-up (see [62, Def. 7.3.2.2]). (Here NBl. $(\cdot)$ *denotes the normalization of the blow-up, or the morphism induced by its universal property.)*

(2) *The canonical morphism* $\mathrm{NBl}_{\mathcal{J}_{\mathcal{H},d_0\mathsf{pol}}}(\oint_{\mathcal{H}})$ *above is an isomorphism.*

In particular, $\mathsf{M}^{\mathrm{tor}}_{\mathcal{H}}$ *is a scheme projective (and smooth) over* S_0. *If Condition* 1.3.1.9 *is satisfied, then the above two statements are true for all* $d_0 \geq 3$.

For technical reasons, we shall enlarge the collection of smooth toroidal compactifications we have in Theorem 1.3.1.3 to the following setup, including certain projective but nonsmooth toroidal compactifications.

Proposition 1.3.1.11. *With assumptions as in Theorem* 1.3.1.10, *suppose* $\mathcal{H}'$ *is an open compact subgroup of* $\mathcal{H}$, *with* Σ' *(resp.* pol'*) at level* $\mathcal{H}'$ *induced by* Σ *(resp.* pol*) as in [62, Constr. 7.3.1.6]. (Note that* Σ' *is not necessarily smooth.) For each integer* $d \geq 1$, *let*

$$\mathcal{J}_{\mathcal{H}',d\mathsf{pol}'} := (\mathsf{M}^{\mathrm{min}}_{\mathcal{H}'} \to \mathsf{M}^{\mathrm{min}}_{\mathcal{H}})^* \mathcal{J}_{\mathcal{H},d\mathsf{pol}}.$$

Suppose $d_0 \geq 1$ *is any integer such that the statements in Theorem* 1.3.1.10 *are true. Then we define*

$$\mathsf{M}^{\mathrm{tor}}_{\mathcal{H}',d_0\mathsf{pol}'} := \mathrm{NBl}_{\mathcal{J}_{\mathcal{H}',d_0\mathsf{pol}'}}(\mathsf{M}^{\mathrm{min}}_{\mathcal{H}'}).$$

With this definition, there is a canonical morphism

$$\mathsf{M}^{\mathrm{tor}}_{\mathcal{H}',d_0\mathsf{pol}'} \to \mathsf{M}^{\mathrm{tor}}_{\mathcal{H},\Sigma} \cong \mathsf{M}^{\mathrm{tor}}_{\mathcal{H},d_0\mathsf{pol}} \tag{1.3.1.12}$$

which is **finite**. *Moreover,* $\mathsf{M}^{\mathrm{tor}}_{\mathcal{H}',d_0\mathsf{pol}'}$ *is canonically isomorphic to the normalization of* $\mathsf{M}^{\mathrm{tor}}_{\mathcal{H},d_0\mathsf{pol}}$ *in* $\mathsf{M}_{\mathcal{H}'}$ *under the composition of canonical morphisms* $\mathsf{M}_{\mathcal{H}'} \to \mathsf{M}_{\mathcal{H}} \hookrightarrow \mathsf{M}^{\mathrm{tor}}_{\mathcal{H},\Sigma} \cong \mathsf{M}^{\mathrm{tor}}_{\mathcal{H},d_0\mathsf{pol}}$, *and is independent of the choices of* pol *and* d_0.

If Σ' *is* **smooth**, *then we have a canonical isomorphism*

$$\mathsf{M}^{\mathrm{tor}}_{\mathcal{H}',\Sigma'} \xrightarrow{\sim} \mathsf{M}^{\mathrm{tor}}_{\mathcal{H}',d_0\mathsf{pol}'}, \tag{1.3.1.13}$$

where $\mathsf{M}^{\mathrm{tor}}_{\mathcal{H}',\Sigma'}$ *is given by Theorem* 1.3.1.3.

Proof. Since $\mathcal{J}_{\mathcal{H}',d_0\mathsf{pol}'}$ is the pullback of $\mathcal{J}_{\mathcal{H},d_0\mathsf{pol}}$ under the finite morphism $\mathsf{M}^{\mathrm{min}}_{\mathcal{H}'} \to \mathsf{M}^{\mathrm{min}}_{\mathcal{H}}$, the canonical morphism (1.3.1.12) exists and is finite by Theorem 1.3.1.10 and by the universal property of the normalization of blow-up. Since $\mathsf{M}^{\mathrm{tor}}_{\mathcal{H}',d_0\mathsf{pol}'}$ is normal, it is canonically isomorphic to the normalization of $\mathsf{M}^{\mathrm{tor}}_{\mathcal{H},d_0\mathsf{pol}}$ in $\mathsf{M}_{\mathcal{H}'}$ by Zariski's main theorem (see [35, III-1, 4.4.3, 4.4.11]).

If Σ' is *smooth*, then we have $\mathsf{M}^{\mathrm{tor}}_{\mathcal{H}',\Sigma'}$ given by Theorem 1.3.1.3. Moreover, $\jmath_{\mathcal{H}',d_0\mathsf{pol}'}$ is defined (as in Definition 1.3.1.7) and is the pullback of $\jmath_{\mathcal{H},d_0\mathsf{pol}}$ under the canonical morphism $\mathsf{M}^{\mathrm{tor}}_{\mathcal{H}',\Sigma'} \to \mathsf{M}^{\mathrm{tor}}_{\mathcal{H},\Sigma}$. Hence, we have a canonical morphism $\mathsf{M}^{\mathrm{tor}}_{\mathcal{H}',\Sigma'} \to \mathsf{M}^{\mathrm{tor}}_{\mathcal{H},d_0\mathsf{pol}}$ by Theorem 1.3.1.10, inducing a canonical morphism $\mathsf{M}^{\mathrm{tor}}_{\mathcal{H}',\Sigma'} \to \mathsf{M}^{\mathrm{tor}}_{\mathcal{H}',d_0\mathsf{pol}'}$, both of which follow from the universal property of normalization of blowup. By Zariski's main theorem again, this last morphism is an isomorphism and gives (1.3.1.13), as desired. □

Then we can describe the so-called *Hecke actions* of $\mathrm{G}(\mathbb{A}^\infty)$ as follows:

Proposition 1.3.1.14. (*See [62, Prop. 7.2.5.1].*) *Suppose we have an element $g \in \mathrm{G}(\mathbb{A}^\infty)$, and suppose we have two open compact subgroups $\mathcal{H}$ and $\mathcal{H}'$ of $\mathrm{G}(\hat{\mathbb{Z}})$ such that $\mathcal{H}' \subset g\mathcal{H}g^{-1}$. Then there is a canonical finite surjection*

$$[g]^{\min} : \mathsf{M}^{\min}_{\mathcal{H}'} \to \mathsf{M}^{\min}_{\mathcal{H}}$$

(over $\mathsf{S}_0 = \mathrm{Spec}(F_0)$) extending the canonical finite surjection $[[g]] : [\mathsf{M}_{\mathcal{H}'}] \to [\mathsf{M}_{\mathcal{H}}]$ induced by the canonical finite surjection

$$[g] : \mathsf{M}_{\mathcal{H}'} \to \mathsf{M}_{\mathcal{H}}$$

defined by the Hecke action of g, such that $\omega^{\otimes k}_{\mathsf{M}^{\min}_{\mathcal{H}}}$ over $\mathsf{M}^{\min}_{\mathcal{H}}$ is pulled back to $\omega^{\otimes k}_{\mathsf{M}^{\min}_{\mathcal{H}'}}$ over $\mathsf{M}^{\min}_{\mathcal{H}'}$ (up to canonical isomorphism) whenever the former is defined.

Moreover, the surjection $[g]^{\min}$ maps the $[(\Phi'_{\mathcal{H}'}, \delta'_{\mathcal{H}'})]$-stratum $\mathsf{Z}_{[(\Phi'_{\mathcal{H}'}, \delta'_{\mathcal{H}'})]}$ of $\mathsf{M}^{\min}_{\mathcal{H}'}$ to the $[(\Phi_{\mathcal{H}}, \delta_{\mathcal{H}})]$-stratum $\mathsf{Z}_{[(\Phi_{\mathcal{H}}, \delta_{\mathcal{H}})]}$ of $\mathsf{M}^{\min}_{\mathcal{H}}$ if and only if there are representatives $(\Phi_{\mathcal{H}}, \delta_{\mathcal{H}})$ and $(\Phi'_{\mathcal{H}'}, \delta'_{\mathcal{H}'})$ of $[(\Phi_{\mathcal{H}}, \delta_{\mathcal{H}})]$ and $[(\Phi'_{\mathcal{H}'}, \delta'_{\mathcal{H}'})]$, respectively, such that $(\Phi_{\mathcal{H}}, \delta_{\mathcal{H}})$ is g-assigned to $(\Phi'_{\mathcal{H}'}, \delta'_{\mathcal{H}'})$ as in [62, Def. 5.4.3.9].

If $g = g_1 g_2$, where g_1 and g_2 are elements of $\mathrm{G}(\mathbb{A}^\infty)$, each having a setup similar to that of g, then we have $[g] = [g_2] \circ [g_1]$, $[[g]] = [[g_2]] \circ [[g_1]]$, and $[g]^{\min} = [g_2]^{\min} \circ [g_1]^{\min}$.

Proposition 1.3.1.15. (*See [62, Prop. 6.4.3.4].*) *With the same setting as in Proposition 1.3.1.14, suppose $\Sigma = \{\Sigma_{\Phi_{\mathcal{H}}}\}_{[(\Phi_{\mathcal{H}}, \delta_{\mathcal{H}})]}$ and $\Sigma' = \{\Sigma'_{\Phi'_{\mathcal{H}'}}\}_{[(\Phi'_{\mathcal{H}'}, \delta'_{\mathcal{H}'})]}$ are two compatible choices of admissible smooth rational polyhedral cone decomposition data for $\mathsf{M}_{\mathcal{H}}$ and $\mathsf{M}_{\mathcal{H}'}$, respectively, such that Σ' is a g-refinement of Σ as in [62, Def. 6.4.3.3]. Then there is a canonical proper surjection*

$$[g]^{\mathrm{tor}} : \mathsf{M}^{\mathrm{tor}}_{\mathcal{H}', \Sigma'} \twoheadrightarrow \mathsf{M}^{\mathrm{tor}}_{\mathcal{H}, \Sigma}$$

(over $\mathsf{S}_0 = \mathrm{Spec}(F_0)$) compatible with the canonical finite surjection

$$[g]^{\min} : \mathsf{M}^{\min}_{\mathcal{H}'} \twoheadrightarrow \mathsf{M}^{\min}_{\mathcal{H}}$$

in Proposition 1.3.1.14 and the canonical proper surjections $\oint_{\mathcal{H}'} : \mathsf{M}^{\mathrm{tor}}_{\mathcal{H}', \Sigma'} \to \mathsf{M}^{\min}_{\mathcal{H}'}$ and $\oint_{\mathcal{H}} : \mathsf{M}^{\mathrm{tor}}_{\mathcal{H}, \Sigma} \to \mathsf{M}^{\min}_{\mathcal{H}}$, such that $\omega_{\mathsf{M}^{\mathrm{tor}}_{\mathcal{H}, \Sigma}}$ over $\mathsf{M}^{\mathrm{tor}}_{\mathcal{H}, \Sigma}$ is pulled back to $\omega_{\mathsf{M}^{\mathrm{tor}}_{\mathcal{H}', \Sigma'}}$ over $\mathsf{M}^{\mathrm{tor}}_{\mathcal{H}', \Sigma'}$ (up to canonical isomorphism).

Moreover, the surjection $[g]^{\mathrm{tor}}$ maps the $[(\Phi'_{\mathcal{H}'}, \delta'_{\mathcal{H}'}, \sigma')]$-stratum $\mathsf{Z}_{[(\Phi'_{\mathcal{H}'}, \delta'_{\mathcal{H}'}, \sigma')]}$ of $\mathsf{M}^{\mathrm{tor}}_{\mathcal{H}', \Sigma'}$ to the $[(\Phi_{\mathcal{H}}, \delta_{\mathcal{H}}, \sigma)]$-stratum $\mathsf{Z}_{[(\Phi_{\mathcal{H}}, \delta_{\mathcal{H}}, \sigma)]}$ of $\mathsf{M}^{\mathrm{tor}}_{\mathcal{H}, \Sigma}$ if and only if there are representatives $(\Phi_{\mathcal{H}}, \delta_{\mathcal{H}}, \sigma)$ and $(\Phi'_{\mathcal{H}'}, \delta'_{\mathcal{H}'}, \sigma')$ of $[(\Phi_{\mathcal{H}}, \delta_{\mathcal{H}}, \sigma)]$ and $[(\Phi'_{\mathcal{H}'}, \delta'_{\mathcal{H}'}, \sigma')]$, respectively, such that $(\Phi'_{\mathcal{H}'}, \delta'_{\mathcal{H}'}, \sigma')$ is a g-refinement of $(\Phi_{\mathcal{H}}, \delta_{\mathcal{H}}, \sigma)$ as in [62, Def. 6.4.3.1].

If $g = g_1 g_2$, where g_1 and g_2 are elements of $\mathrm{G}(\mathbb{A}^\infty)$, each having a setup similar to that of g, then we have $[g]^{\mathrm{tor}} = [g_2]^{\mathrm{tor}} \circ [g_1]^{\mathrm{tor}}$, extending $[g] = [g_2] \circ [g_1]$ and lifting $[g]^{\min} = [g_2]^{\min} \circ [g_1]^{\min}$.

Remark 1.3.1.16. While Proposition 1.3.1.14 is a logical consequence of Proposition 1.3.1.15, they were stated in the reversed order, because the former is easier to describe and understand than the latter. The last statements of Propositions 1.3.1.15 and 1.3.1.14 were not explicitly stated in [62, Prop. 6.4.3.4 and 7.2.5.1], but were implicit in the proofs there.

1.3.2 *Boundary of PEL-Type Moduli Problems*

Let us describe the building blocks of $\mathsf{M}^{\mathrm{tor}}_{\mathcal{H},\Sigma}$ in more detail. In particular, we would like to describe and characterize the algebraic stacks $\mathsf{M}^{\mathsf{Z}_{\mathcal{H}}}_{\mathcal{H}}$, $\mathsf{M}^{\Phi_{\mathcal{H}}}_{\mathcal{H}}$, $C_{\Phi_{\mathcal{H}},\delta_{\mathcal{H}}}$, $\Xi_{\Phi_{\mathcal{H}},\delta_{\mathcal{H}}}$, $\Xi_{\Phi_{\mathcal{H}},\delta_{\mathcal{H}}}(\sigma)$, $\Xi_{\Phi_{\mathcal{H}},\delta_{\mathcal{H}},\sigma}$, $\mathsf{Z}_{[(\Phi_{\mathcal{H}},\delta_{\mathcal{H}},\sigma)]} \cong \Xi_{\Phi_{\mathcal{H}},\delta_{\mathcal{H}},\sigma}/\Gamma_{\Phi_{\mathcal{H}},\sigma}$ and the formal algebraic stacks $\mathfrak{X}_{\Phi_{\mathcal{H}},\delta_{\mathcal{H}},\sigma}$ and $\mathfrak{X}_{\Phi_{\mathcal{H}},\delta_{\mathcal{H}},\sigma}/\Gamma_{\Phi_{\mathcal{H}},\sigma}$ in (2) of Theorem 1.3.1.3 (and (5) of Theorem 1.3.1.5), and to describe canonical Hecke actions on collections of these geometric objects (compatible with those in Proposition 1.3.1.15).

Throughout this subsection, let us fix the choice of a fully symplectic admissible filtration Z of $L \underset{\mathbb{Z}}{\otimes} \hat{\mathbb{Z}}$ as in Definitions 1.2.1.2 and 1.2.1.3. Let us also fix a (noncanonical) choice of $(L^{\mathsf{Z}}, \langle\,\cdot\,,\,\cdot\,\rangle^{\mathsf{Z}}, h_0^{\mathsf{Z}})$, so that G^{Z} can be defined as in Definition 1.2.1.9.

For each open compact subgroup $\mathcal{H}$ of $\mathrm{G}(\hat{\mathbb{Z}})$, we can define the boundary moduli problems $\mathsf{M}_{\mathcal{H}_h}$, $\mathsf{M}^{\Phi_{\mathcal{H}}}_{\mathcal{H}}$, and $\mathsf{M}^{\mathsf{Z}_{\mathcal{H}}}_{\mathcal{H}}$ as in Definition 1.2.1.15. By definition, $\mathsf{M}_{\mathcal{H}_h}$ parameterizes $(B, \lambda_B, i_B, \varphi_{-1,\mathcal{H}})$ appearing as the abelian part in degeneration data. By the construction of $\mathsf{M}^{\Phi_{\mathcal{H}}}_{\mathcal{H}}$ as the quotient of $\coprod \mathsf{M}^{\mathsf{Z}_n}_n$ by $H_n = \mathcal{H}/\mathcal{U}(n)$ (for any integer $n \geq 1$ such that $\mathcal{U}(n) \subset \mathcal{H}$), where the disjoint union is over representatives $(\mathsf{Z}_n, \Phi_n, \delta_n)$ (with the same (X, Y, ϕ)) in $(\mathsf{Z}_{\mathcal{H}}, \Phi_{\mathcal{H}}, \delta_{\mathcal{H}})$, the finite étale cover $\mathsf{M}^{\Phi_{\mathcal{H}}}_{\mathcal{H}} \to \mathsf{M}_{\mathcal{H}_h}$ parameterizes the twisted objects $(\varphi^{\sim}_{-2,\mathcal{H}}, \varphi^{\sim}_{0,\mathcal{H}})$ inducing both $(\varphi_{-2,\mathcal{H}}, \varphi_{0,\mathcal{H}})$ and $\varphi_{-1,\mathcal{H}}$ over $\mathsf{M}_{\mathcal{H}_h}$. Therefore, by the definition of $\mathsf{M}^{\mathsf{Z}_{\mathcal{H}}}_{\mathcal{H}}$ as the quotient of $\mathsf{M}^{\Phi_{\mathcal{H}}}_{\mathcal{H}}$ by $\Gamma_{\Phi_{\mathcal{H}}}$ (see Definition 1.2.2.3), we have the following:

Lemma 1.3.2.1. *Let us fix the choice of a representative* $(\mathsf{Z}, \Phi, \delta)$ *in* $(\mathsf{Z}_{\mathcal{H}}, \Phi_{\mathcal{H}}, \delta_{\mathcal{H}})$. *Let* $\mathcal{H}_{\mathrm{G}_{h,\mathsf{Z}}} = \mathcal{H}_{\mathrm{P}_{\mathsf{Z}}}/\mathcal{H}_{\mathrm{Z}_{\mathsf{Z}}}$ *be the open compact subgroup of* $\mathrm{G}^{\mathsf{Z}}(\hat{\mathbb{Z}}) \cong \mathrm{G}_{h,\mathsf{Z}}(\hat{\mathbb{Z}})$ *as in Definition* 1.2.1.12, *and let* $\mathcal{H}_{\mathrm{G}_{h,\mathsf{Z}},\Phi}$ *denote the image in* $\mathrm{G}_{h,\mathsf{Z}}(\hat{\mathbb{Z}})$ *of the stabilizer* $\mathcal{H}_{\mathrm{P}_{\mathsf{Z}},\Phi}$ *of* $\Phi = (X, Y, \phi, \varphi_{-2}, \varphi_0)$ *in* $\mathcal{H}_{\mathrm{P}_{\mathsf{Z}}}$, *which is an open compact subgroup of* $\mathcal{H}_{\mathrm{G}_{h,\mathsf{Z}}}$ *isomorphic to* $\mathcal{H}_{\mathrm{P}_{\mathsf{Z}},\Phi}/\mathcal{H}_{\mathrm{Z}_{\mathsf{Z}}}$. *Then* $\mathsf{M}_{\mathcal{H}_h} \cong \mathsf{M}_{\mathcal{H}_{\mathrm{G}_{h,\mathsf{Z}}}}$ *and there is a canonical isomorphism*

$$\mathsf{M}^{\mathsf{Z}_{\mathcal{H}}}_{\mathcal{H}} \cong \mathsf{M}_{\mathcal{H}_{\mathrm{G}_{h,\mathsf{Z}},\Phi}}, \tag{1.3.2.2}$$

where $\mathsf{M}_{\mathcal{H}_{\mathrm{G}_{h,\mathsf{Z}},\Phi}}$ *is defined by* $(L^{\mathsf{Z}}, \langle\,\cdot\,,\,\cdot\,\rangle^{\mathsf{Z}}, h_0^{\mathsf{Z}})$ *as in Section* 1.1.2. *If* $\mathcal{H}'$ *is an open compact subgroup of* $\mathcal{H}$, *then the corresponding morphism*

$$\mathsf{M}^{\mathsf{Z}_{\mathcal{H}'}}_{\mathcal{H}'} \to \mathsf{M}^{\mathsf{Z}_{\mathcal{H}}}_{\mathcal{H}} \tag{1.3.2.3}$$

can be canonically identified with the finite étale morphism

$$\mathsf{M}_{\mathcal{H}'_{\mathrm{G}_{h,\mathsf{Z}},\Phi}} \to \mathsf{M}_{\mathcal{H}_{\mathrm{G}_{h,\mathsf{Z}},\Phi}}. \tag{1.3.2.4}$$

The collection $\{\mathsf{M}_{\mathcal{H}_{\mathrm{G}_{h,\mathrm{z}},\Phi}}\}_{\mathcal{H}_{\mathrm{G}_{h,\mathrm{z}},\Phi}}$ *naturally carries a Hecke action by elements* $g_h \in \mathrm{G}^{\mathrm{Z}}(\mathbb{A}^\infty) \cong \mathrm{G}_{h,\mathrm{Z}}(\mathbb{A}^\infty)$, *realized by finite étale surjections pulling tautological objects back to Hecke twists. If moreover* $\mathcal{H}'$ *is a normal subgroup of* $\mathcal{H}$, *then* (1.3.2.4) *is an* $\mathcal{H}_{\mathrm{G}_{h,\mathrm{z}},\Phi}/\mathcal{H}'_{\mathrm{G}_{h,\mathrm{z}},\Phi}$*-torsor.*

Lemma 1.3.2.5. *With the same setting as in Lemma* 1.3.2.1, *let* $\mathcal{H}_{\mathrm{G}'_{h,\mathrm{z}}}$ *be the open compact subgroup of* $\mathrm{G}^{\mathrm{Z}}(\hat{\mathbb{Z}}) \cong \mathrm{G}'_{h,\mathrm{Z}}(\hat{\mathbb{Z}})$ *as in Definition* 1.2.1.12, *which is a normal subgroup of* $\mathcal{H}_{\mathrm{G}_{h,\mathrm{z}},\Phi}$ *(by definition). Then there is a canonical isomorphism*

$$\mathsf{M}_{\mathcal{H}}^{\Phi_{\mathcal{H}}} \cong \mathsf{M}_{\mathcal{H}_{\mathrm{G}'_{h,\mathrm{z}}}}, \tag{1.3.2.6}$$

which is compatible with (1.3.2.2) *and with Hecke actions as in Lemma* 1.3.2.1. *The canonical morphisms* $\mathsf{M}_{\mathcal{H}}^{\Phi_{\mathcal{H}}} \to \mathsf{M}_{\mathcal{H}}^{\mathsf{Z}_{\mathcal{H}}} \to \mathsf{M}_{\mathcal{H}_h}$ *can be identified with the canonical finite étale morphisms* $\mathsf{M}_{\mathcal{H}_{\mathrm{G}'_{h,\mathrm{z}}}} \to \mathsf{M}_{\mathcal{H}_{\mathrm{G}_{h,\mathrm{z}},\Phi}} \to \mathsf{M}_{\mathcal{H}_{\mathrm{G}_{h,\mathrm{z}}}}$, *on which* $\Gamma_{\Phi_{\mathcal{H}}}$ *acts equivariantly (and trivially on the latter two objects) via the canonical homomorphism* $\Gamma_{\Phi_{\mathcal{H}}} \to \mathcal{H}_{\mathrm{G}'_{l,\mathrm{z}}}/\mathcal{H}_{\mathrm{G}_{l,\mathrm{z}}} \cong \mathcal{H}_{\mathrm{G}_{h,\mathrm{z}}}/\mathcal{H}_{\mathrm{G}'_{h,\mathrm{z}}}$ *with image* $\mathcal{H}_{\mathrm{G}_{h,\mathrm{z}},\Phi}/\mathcal{H}_{\mathrm{G}'_{h,\mathrm{z}}}$. *In particular,* $\mathsf{M}_{\mathcal{H}}^{\Phi_{\mathcal{H}}} \to \mathsf{M}_{\mathcal{H}}^{\mathsf{Z}_{\mathcal{H}}} = \mathsf{M}_{\mathcal{H}}^{\Phi_{\mathcal{H}}}/\Gamma_{\Phi_{\mathcal{H}}}$ *is finite étale and an* $\mathcal{H}_{\mathrm{G}_{h,\mathrm{z}},\Phi}/\mathcal{H}_{\mathrm{G}'_{h,\mathrm{z}}}$*-torsor.*

The abelian scheme torsor $C_{\Phi_{\mathcal{H}},\delta_{\mathcal{H}}} \to \mathsf{M}_{\mathcal{H}}^{\Phi_{\mathcal{H}}}$ is, by the construction in [62, Sec. 6.2.3–6.2.4] (see also the correction in Remark 1.3.1.6), the quotient of

$$\coprod C_{\Phi_n,\delta_n} \to \coprod \mathsf{M}_n^{\mathsf{Z}_n}$$

by $H_n = \mathcal{H}/\mathcal{U}(n)$ (for any integer $n \geq 1$ such that $\mathcal{U}(n) \subset \mathcal{H}$), where each $C_{\Phi_n,\delta_n} \to \mathsf{M}_n^{\mathsf{Z}_n}$ has a canonical structure of an abelian scheme which is preserved under the action of $H_{n,\mathrm{U}_{1,\mathrm{Z}_n}^{\mathrm{ess}}} \cong \mathcal{H}_{\mathrm{U}_{1,\mathrm{z}}}/\mathcal{U}(n)_{\mathrm{U}_{1,\mathrm{z}}}$. Therefore, we have the following:

Lemma 1.3.2.7. *The quotient* $C_{\Phi_{\mathcal{H}},\delta_{\mathcal{H}}} \to \mathsf{M}_{\mathcal{H}}^{\Phi_{\mathcal{H}}}$ *depends only on* $\mathcal{H}_{\mathrm{G}_{1,\mathrm{z}}}$, *is an abelian scheme when the splitting of* (1.2.1.14) *defined by any splitting* δ *also splits* (1.2.1.13) *(and induces an isomorphism* $\mathcal{H}_{\mathrm{G}_{1,\mathrm{z}}} \cong \mathcal{H}_{\mathrm{G}'_{h,\mathrm{z}}} \ltimes \mathcal{H}_{\mathrm{U}_{1,\mathrm{z}}}$), *and is a torsor under the abelian scheme* $C^{\mathrm{grp}}_{\Phi_{\mathcal{H}},\delta_{\mathcal{H}}} := C_{\Phi_{\mathcal{H}'},\delta_{\mathcal{H}'}}$ *defined by any* $\mathcal{H}'$ *with* $\mathcal{H}'_{\mathrm{G}_{1,\mathrm{z}}} \cong \mathcal{H}_{\mathrm{G}'_{h,\mathrm{z}}} \ltimes \mathcal{H}_{\mathrm{U}_{1,\mathrm{z}}}$, *which is canonically* $\mathbb{Q}^\times$*-isogenous to* $\underline{\mathrm{Hom}}_{\mathcal{O}}(X, B)^\circ$. *(This clarifies the abelian scheme torsor structure of* $C_{\Phi_{\mathcal{H}},\delta_{\mathcal{H}}} \to \mathsf{M}_{\mathcal{H}}^{\Phi_{\mathcal{H}}}$.) *We deduce from this that there is a canonical isomorphism*

$$\begin{aligned}\Omega^1_{C_{\Phi_{\mathcal{H}},\delta_{\mathcal{H}}}/\mathsf{M}_{\mathcal{H}}^{\mathsf{Z}_{\mathcal{H}}}} &\cong \Omega^1_{C_{\Phi_{\mathcal{H}},\delta_{\mathcal{H}}}/\mathsf{M}_{\mathcal{H}}^{\Phi_{\mathcal{H}}}} \cong \Omega^1_{C^{\mathrm{grp}}_{\Phi_{\mathcal{H}},\delta_{\mathcal{H}}}/\mathsf{M}_{\mathcal{H}}^{\Phi_{\mathcal{H}}}} \\ &\cong (C_{\Phi_{\mathcal{H}},\delta_{\mathcal{H}}} \to \mathsf{M}_{\mathcal{H}}^{\mathsf{Z}_{\mathcal{H}}})^* \underline{\mathrm{Hom}}_{\mathcal{O}}(X, \underline{\mathrm{Lie}}^\vee_{B/\mathsf{M}_{\mathcal{H}}^{\mathsf{Z}_{\mathcal{H}}}}).\end{aligned} \tag{1.3.2.8}$$

If we fix the choice of (Z_n *and)* Φ_n, *then the canonical morphism*

$$C_{\Phi_n,\delta_n} \to C_{\Phi_{\mathcal{H}},\delta_{\mathcal{H}}} \tag{1.3.2.9}$$

is an $\mathcal{H}_{\mathrm{G}_{1,\mathrm{z}}}/\mathcal{U}(n)_{\mathrm{G}_{1,\mathrm{z}}} \cong H_{n,\mathrm{G}^{\mathrm{ess}}_{h,\mathrm{Z}_n} \ltimes \mathrm{U}^{\mathrm{ess}}_{1,\mathrm{Z}_n}}$*-torsor (see [62, Sec. 6.2.4; see also the errata]), where* $\mathcal{H}_{\mathrm{G}_{1,\mathrm{z}}}$ *and* $\mathcal{U}(n)_{\mathrm{G}_{1,\mathrm{z}}}$ *are open compact subgroups of* $\mathrm{G}_{1,\mathrm{Z}}(\hat{\mathbb{Z}})$ *as in Definition* 1.2.1.12, *and induces an isomorphism*

$$C_{\Phi_n,\delta_n}/(\mathcal{H}_{\mathrm{G}_{1,\mathrm{z}}}/\mathcal{U}(n)_{\mathrm{G}_{1,\mathrm{z}}}) \xrightarrow{\sim} C_{\Phi_{\mathcal{H}},\delta_{\mathcal{H}}}. \tag{1.3.2.10}$$

Lemma 1.3.2.11. *The abelian scheme torsor* $S := C_{\Phi_n,\delta_n} \to \mathsf{M}_n^{\Phi_n} := \mathsf{M}_n^{\mathsf{Z}_n}$ *is universal for the additional structures* $(c_n, c_n^\vee)$ *satisfying certain symplectic and liftability conditions, which we review as follows:*

(1) *The homomorphism* $c : X_S \to B^\vee$ *induced by* $c_n : \frac{1}{n}X_S \to B^\vee$ *(by restriction) is equivalent to the data of a semi-abelian scheme* $G^\natural$ *that is an extension of* B *by the split torus* T *with character group* X*, and the lifting* c_n *of* c *is equivalent to the data of a splitting of the canonical short exact sequence*

$$0 \to T[n] \to G^\natural[n] \to B[n] \to 0.$$

(2) *The homomorphism* $c^\vee : Y_S \to B$ *induced by* $c_n^\vee : \frac{1}{n}Y_S \to B$ *(by restriction) is equivalent to the data of a semi-abelian scheme* $G^{\vee,\natural}$ *that is an extension of* $B^\vee$ *by the split torus* $T^\vee$ *with character group* Y*, and the lifting* $c_n^\vee$ *of* $c^\vee$ *is equivalent to the data of a splitting of the canonical short exact sequence*

$$0 \to T^\vee[n] \to G^{\vee,\natural}[n] \to B^\vee[n] \to 0.$$

(3) *The homomorphisms* c, $c^\vee$, $\phi : Y \hookrightarrow X$, *and* $\lambda_B : B \to B^\vee$ *satisfy the compatibility* $\lambda_B c^\vee = c\phi$, *and hence defines a homomorphism* $\lambda^\natural : G^\natural \to G^{\vee,\natural}$ *inducing* $\lambda_T = \phi_S^* : T \to T^\vee$ *and* $\lambda_B : B \to B^\vee$. *All of these are compatible with their* $\mathcal{O}$*-structures.*

(4) *The splittings defined by* c_n *and* $c_n^\vee$ *are not necessarily compatible under the canonical morphism* $G^\natural \to G^{\vee,\natural}$ *induced by* $\lambda_B : B \to B^\vee$ *and* $\phi : Y \hookrightarrow X$. *The failure of such a compatibility can be identified with the nontriviality of the pairing*

$$\mathtt{d}_{10,n} : B[n] \times (\tfrac{1}{n}Y/Y)_S \to \boldsymbol{\mu}_{n,S}$$

(cf. [62, Lem. 5.2.3.12]), which sends $(a, \frac{1}{n}y)$ *to*

$$\mathrm{e}_{B[n]}(a, (\lambda_B c_n^\vee - c_n\phi_n)(\tfrac{1}{n}y)),$$

where $\mathrm{e}_{B[n]} : B[n] \times B^\vee[n] \to \boldsymbol{\mu}_{n,S}$ *is the canonical perfect pairing between* $B[n]$ *and* $B^\vee[n]$, *for any functorial points* a *of* $B[n]$ *and* $\frac{1}{n}y$ *of* $(\frac{1}{n}Y/Y)_S$.

(5) *The symplectic condition for* $(c_n, c_n^\vee)$ *is that, under* $\varphi_{-1,n}$ *and* $\varphi_{0,n}$, *the pairing* $\mathtt{d}_{10,n}$ *above is matched with the pairing*

$$\langle\,\cdot\,,\,\cdot\,\rangle_{10,n} : \mathrm{Gr}_{-1,n}^{\mathsf{Z}} \times \mathrm{Gr}_{0,n}^{\mathsf{Z}} \to ((\mathbb{Z}/n\mathbb{Z})(1))_S$$

induced by $\langle\,\cdot\,,\,\cdot\,\rangle$ *and* δ_n.

(6) *The liftability condition for* $(c_n, c_n^\vee)$ *is that, for each integer* $m \geq 1$ *such that* $n|m$, *and for any lifting* δ_m *of* δ_n, *there exists a finite étale covering of* S *over which there exist* $\varphi_{-1,m}$, $(\varphi_{-2,m}, \varphi_{0,m})$, *and* $(c_m, c_m^\vee)$ *lifting* $\varphi_{-1,n}$, $(\varphi_{-2,n}, \varphi_{0,n})$, *and* $(c_n, c_n^\vee)$, *respectively, and satisfying the symplectic condition defined by* $\langle\,\cdot\,,\,\cdot\,\rangle$ *and* δ_m *as above.*

These can be re-interpreted as follows: $S = C_{\Phi_n,\delta_n} \to \mathsf{M}_n^{\Phi_n} = \mathsf{M}_n^{\mathsf{Z}_n}$ *parameterizes tuples*

$$(G^\natural, \lambda^\natural : G^\natural \to G^{\vee,\natural}, i^\natural, \beta_n^\natural),$$

where:

(a) $G^\natural$ *(resp.* $G^{\vee,\natural}$*) is an extension of* B *(resp.* $B^\vee$*) by* T *(resp.* $T^\vee$*) as above, and* $\lambda^\natural : G^\natural \to G^{\vee,\natural}$ *induces* $\lambda_T = \phi^* : T \to T^\vee$ *and* $\lambda_B : B \to B^\vee$*.*
(b) $i^\natural$ *is a pair of homomorphisms* $\mathcal{O} \to \mathrm{End}_S(G^\natural)$ *and* $\mathcal{O} \to \mathrm{End}_S(G^{\vee,\natural})$ *compatible with each other under* $\lambda^\natural : G^\natural \to G^{\vee,\natural}$*, inducing compatible* $\mathcal{O}$*-structures on* B*,* $B^\vee$*,* T*, and* $T^\vee$*.*
(c) $\beta_n^\natural = (\beta_n^{\natural,0}, \beta_n^{\natural,\#,0}, \nu_n^\natural)$ *is a* **principal level-n structure** *of* $(G^\natural, \lambda^\natural, i^\natural)$ *of type* $(L \underset{\mathbb{Z}}{\otimes} \hat{\mathbb{Z}}, \langle\,\cdot\,,\,\cdot\,\rangle, \mathsf{Z})$*, where* $\beta_n^{\natural,0} : (\mathsf{Z}_{-1,n})_S \overset{\sim}{\to} G^\natural[n]$ *and* $\beta_n^{\natural,\#,0} : (\mathsf{Z}^\#_{-1,n})_S \overset{\sim}{\to} G^{\vee,\natural}[n]$ *are* $\mathcal{O}$*-equivariant isomorphisms preserving filtrations on both sides and inducing on the graded pieces the given data* $\varphi_{-2,n}$*,* $\varphi_{-1,n}$*, and* $\varphi_{0,n}$ *(by duality), respectively; and where* $\nu_n^\natural : ((\mathbb{Z}/n\mathbb{Z})(1))_S \overset{\sim}{\to} \boldsymbol{\mu}_{n,S}$ *is an isomorphism, which are compatible with* $\lambda^\natural$ *and the canonical morphism* $\mathsf{Z}_{-1,n} \to \mathsf{Z}^\#_{-1,n}$ *induced by* $\langle\,\cdot\,,\,\cdot\,\rangle$*. (Here* $\mathsf{Z}^\#$ *is the filtration on* $L^\# \underset{\mathbb{Z}}{\otimes} \hat{\mathbb{Z}}$ *canonically dual to the filtration on* $L \underset{\mathbb{Z}}{\otimes} \hat{\mathbb{Z}}$*, equipped with a canonical morphism* $\mathsf{Z} \to \mathsf{Z}^\#$*, respecting the filtration degrees, induced by* $\langle\,\cdot\,,\,\cdot\,\rangle$*. Then the splitting* δ *corresponds under* $\beta_n^\natural$ *to splittings of* $0 \to T[n] \to G^\natural[n] \to B[n] \to 0$ *and* $0 \to T^\vee[n] \to G^{\vee,\natural}[n] \to B^\vee[n] \to 0$*.) Moreover,* $\beta_n^\natural$ *satisfies the liftability condition that, for each integer* $m \geq 1$ *such that* $n|m$*, there exists a finite étale covering of* S *over which there exists an analogous triple* $\beta_m^\natural$ *lifting the pullback of* $\beta_n^\natural$*.*

Proof. The statements are self-explanatory. □

Proposition 1.3.2.12. *The abelian scheme torsor* $S := C_{\Phi_\mathcal{H},\delta_\mathcal{H}} \to \mathsf{M}_\mathcal{H}^{\Phi_\mathcal{H}}$ *is universal for the additional structures* $(c_\mathcal{H}, c_\mathcal{H}^\vee)$ *satisfying certain symplectic and liftability conditions, which can be interpreted as parameterizing tuples*

$$(G^\natural, \lambda^\natural : G^\natural \to G^{\vee,\natural}, i^\natural, \beta_\mathcal{H}^\natural), \tag{1.3.2.13}$$

where $G^\natural$*,* $G^{\vee,\natural}$*,* $\lambda^\natural$*, and* $i^\natural$ *are as in Lemma* 1.3.2.11*, and where* $\beta_\mathcal{H}^\natural$ *is a* **level-$\mathcal{H}$ structure** *of* $(G^\natural, \lambda^\natural, i^\natural)$ *of type* $(L \underset{\mathbb{Z}}{\otimes} \hat{\mathbb{Z}}, \langle\,\cdot\,,\,\cdot\,\rangle, \mathsf{Z})$*, which is a collection* $\{\beta_{H_n}^\natural\}_n$*, where* $n \geq 1$ *runs over integers such that* $\mathcal{U}(n) \subset \mathcal{H}$*, such that each* $\beta_{H_n}^\natural$ *(where* $H_n := \mathcal{H}/\mathcal{U}(n)$*) is a subscheme of*

$$\coprod \Big(\underline{\mathrm{Isom}}_S((\mathsf{Z}_{-1,n})_S, G^\natural[n]) \underset{S}{\times} \underline{\mathrm{Isom}}_S((\mathsf{Z}^\#_{-1,n})_S, G^{\vee,\natural}[n]) \underset{S}{\times} \underline{\mathrm{Isom}}_S\big(((\mathbb{Z}/n\mathbb{Z})(1))_S, \boldsymbol{\mu}_{n,S}\big)\Big)$$

over S*, where the disjoint union is over representatives* $(\mathsf{Z}_n, \Phi_n, \delta_n)$ *(with the same* (X, Y, ϕ)*) in* $(\mathsf{Z}_\mathcal{H}, \Phi_\mathcal{H}, \delta_\mathcal{H})$*, that becomes the disjoint union of all elements* *in the* H_n*-orbit of some principal level-*n *structure* $\beta_n^\natural$ *of* $(G^\natural, \lambda^\natural, i^\natural)$ *of type* $(L \underset{\mathbb{Z}}{\otimes} \hat{\mathbb{Z}}, \langle \cdot, \cdot \rangle, \mathsf{Z})$*, as in Lemma* 1.3.2.11*, for any* Z *lifting* Z_n*; and where* $\beta_{H_m}^\natural$ *is mapped to* $\beta_{H_n}^\natural$ *(under the canonical morphism, which we omit for simplicity) when* $n|m$*.*

Proof. This follows from the construction of $C_{\Phi_\mathcal{H}, \delta_\mathcal{H}} \to \mathsf{M}_\mathcal{H}^{\Phi_\mathcal{H}}$ as a quotient of $\coprod C_{\Phi_n, \delta_n} \to \coprod \mathsf{M}_n^{\mathsf{Z}_n}$ (over the same index set). □

Proposition 1.3.2.14. (*Compare with Proposition* 1.3.2.12.) *Fix any lifting* $(\mathsf{Z}, \Phi = (X, Y, \phi, \varphi_{-2}, \varphi_0), \delta)$ *of a representative* $(\mathsf{Z}_\mathcal{H}, \Phi_\mathcal{H}, \delta_\mathcal{H})$ *of* $[(\mathsf{Z}_\mathcal{H}, \Phi_\mathcal{H}, \delta_\mathcal{H})]$*. The abelian scheme torsor* $C_{\Phi_\mathcal{H}, \delta_\mathcal{H}} \to \mathsf{M}_\mathcal{H}^{\Phi_\mathcal{H}}$ *is universal for* $\mathbb{Q}^\times$*-isogeny classes of tuples*

$$(G^\natural, \lambda^\natural : G^\natural \to G^{\vee,\natural}, i^\natural, j^\natural, j^{\vee,\natural}, [\hat{\beta}^\natural]_{\mathcal{H}_{\mathrm{G}_{1,\mathsf{z}}}}) \tag{1.3.2.15}$$

over locally noetherian base schemes S*, where:*

(1) $G^\natural$ *(resp.* $G^{\vee,\natural}$*) is a semi-abelian scheme which is the extension of an abelian scheme* B *(resp.* $B^\vee$*) by a split torus* T *(resp.* $T^\vee$*) over* S*, which is equivalent to a homomorphism* $c : \mathbf{X}(T) \to B^\vee$ *(resp.* $c^\vee : \mathbf{X}(T^\vee) \to B$*).*

(2) $\lambda^\natural : G^\natural \to G^{\vee,\natural}$ *is a* $\mathbb{Q}^\times$*-isogeny (i.e., a* $(\mathbb{Q}^\times_{>0})_S$*-multiple of a quasi-finite surjective homomorphism; or cf. [62, Def. 1.3.1.16]) of semi-abelian schemes over* S*, inducing a* $\mathbb{Q}^\times$*-isogeny* $\lambda_T : T \to T^\vee$ *between the torus parts, which is dual to a* $\mathbb{Q}$*-isomorphism* $\lambda_T^* : \mathbf{X}(T^\vee) \underset{\mathbb{Z}}{\otimes} \mathbb{Q} \to \mathbf{X}(T) \underset{\mathbb{Z}}{\otimes} \mathbb{Q}$*, and a* $\mathbb{Q}^\times$*-polarization* $\lambda_B : B \to B^\vee$ *between the abelian parts (cf. [62, Def. 1.3.2.19 and the errata]), so that* $c(N\lambda_T^*) = (N\lambda_B)c^\vee$ *when* N *is any locally constant function over* S *valued in positive integers such that* $(N\lambda_T^*)(\mathbf{X}(T^\vee)) \subset \mathbf{X}(T)$ *and such that* $N\lambda^\natural : G^\natural \to G^{\vee,\natural}$ *is an isogeny.*

(3) $i^\natural : \mathcal{O} \underset{\mathbb{Z}}{\otimes} \mathbb{Q} \to \mathrm{End}_S(G^\natural) \underset{\mathbb{Z}}{\otimes} \mathbb{Q}_S$ *is a homomorphism inducing* $\mathcal{O} \underset{\mathbb{Z}}{\otimes} \mathbb{Q}$*-actions on* $G^{\vee,\natural}$*,* T*,* $T^\vee$*,* B*, and* $B^\vee$ *up to* $\mathbb{Q}^\times$*-isogeny, compatible with each other under the homomorphisms between the objects introduced thus far. In particular, the induced homomorphism* $i_B : \mathcal{O} \underset{\mathbb{Z}}{\otimes} \mathbb{Q} \to \mathrm{End}_S(B) \underset{\mathbb{Z}}{\otimes} \mathbb{Q}_S$ *satisfies the Rosati condition defined by* λ_B *(cf. [62, Def. 1.3.3.1]).*

(4) $j^\natural : X \underset{\mathbb{Z}}{\otimes} \mathbb{Q}_S \xrightarrow{\sim} \mathbf{X}(T) \underset{\mathbb{Z}}{\otimes} \mathbb{Q}_S$ *and* $j^{\vee,\natural} : Y \underset{\mathbb{Z}}{\otimes} \mathbb{Q}_S \xrightarrow{\sim} \mathbf{X}(T^\vee) \underset{\mathbb{Z}}{\otimes} \mathbb{Q}_S$ *are isomorphisms of* $\mathcal{O} \underset{\mathbb{Z}}{\otimes} \mathbb{Q}$*-modules, such that there exists a section* $r(j^\natural, j^{\vee,\natural})$ *of* $(\mathbb{Q}^\times_{>0})_S$ *such that* $j^\natural \circ \phi = r(j^\natural, j^{\vee,\natural})\lambda_T^* \circ j^{\vee,\natural}$*.*

(5) $[\hat{\beta}^\natural]_{\mathcal{H}_{\mathrm{G}_{1,\mathsf{z}}}}$ *is a* **rational level-$\mathcal{H}$ structure** *of* $(G^\natural, \lambda^\natural, i^\natural, j^\natural, j^{\vee,\natural})$ *of type* $(L \underset{\mathbb{Z}}{\otimes} \mathbb{A}^\infty, \langle \cdot, \cdot \rangle, \mathsf{Z} \underset{\hat{\mathbb{Z}}}{\otimes} \mathbb{A}^\infty, \Phi)$*, which is an assignment to each geometric point* $\bar{s}$ *of* S *a rational level-*$\mathcal{H}$ *structure of* $(G^\natural, \lambda^\natural, i^\natural, j^\natural, j^{\vee,\natural})$ *of type* $(L \underset{\mathbb{Z}}{\otimes} \mathbb{A}^\infty, \langle \cdot, \cdot \rangle, \mathsf{Z} \underset{\hat{\mathbb{Z}}}{\otimes} \mathbb{A}^\infty, \Phi)$ **based at** $\bar{s}$ *(cf. [62, Def. 1.3.8.7]), which is a* $\pi_1(S, \bar{s})$*-invariant* $\mathcal{H}_{\mathrm{G}_{1,\mathsf{z}}}$*-orbit* $[\hat{\beta}_{\bar{s}}^\natural]_{\mathcal{H}_{\mathrm{G}_{1,\mathsf{z}}}}$ *of triples* $\hat{\beta}_{\bar{s}}^\natural = (\hat{\beta}_{\bar{s}}^{\natural,0}, \hat{\beta}_{\bar{s}}^{\natural,\#,0}, \hat{\nu}_{\bar{s}}^\natural)$*, such that the assignments at any two geometric points* $\bar{s}$ *and* $\bar{s}'$ *of the same*

connected component of S determine each other (cf. [62, Lem. 1.3.8.6]), where:

(a) $\hat{\beta}^{\natural,0}_{\bar{s}} : \mathsf{Z}_{-1} \underset{\hat{\mathbb{Z}}}{\otimes} \mathbb{A}^{\infty} \xrightarrow{\sim} \mathrm{V}\, G^{\natural}_{\bar{s}}$ *and* $\hat{\beta}^{\natural,\#,0}_{\bar{s}} : \mathsf{Z}^{\#}_{-1} \underset{\hat{\mathbb{Z}}}{\otimes} \mathbb{A}^{\infty} \xrightarrow{\sim} \mathrm{V}\, G^{\vee,\natural}_{\bar{s}}$ *are* $\mathcal{O} \underset{\mathbb{Z}}{\otimes} \mathbb{A}^{\infty}$*-equivariant isomorphisms preserving filtrations on both sides, which are compatible with* $\lambda^{\natural}$ *and the canonical morphism* $\mathsf{Z}_{-1} \underset{\hat{\mathbb{Z}}}{\otimes} \mathbb{A}^{\infty} \to \mathsf{Z}^{\#}_{-1} \underset{\hat{\mathbb{Z}}}{\otimes} \mathbb{A}^{\infty}$ *induced by* $\langle \cdot, \cdot \rangle$.

(b) $\hat{\nu}^{\natural}_{\bar{s}} : \mathbb{A}^{\infty}(1) \xrightarrow{\sim} \mathrm{V}\, \mathbf{G}_{\mathrm{m},\bar{s}}$ *is an isomorphism of* $\mathbb{A}^{\infty}$*-modules such that* $r(j^{\natural}, j^{\vee,\natural})_{\bar{s}}\, \hat{\nu}^{\natural}_{\bar{s}}$ *maps* $\hat{\mathbb{Z}}(1)$ *to* $\mathrm{T}\, \mathbf{G}_{\mathrm{m},\bar{s}}$*, where* $r(j^{\natural}, j^{\vee,\natural})_{\bar{s}}$ *is the value at* $\bar{s}$ *of the above section* $r(j^{\natural}, j^{\vee,\natural})$ *of* $(\mathbb{Q}^{\times}_{>0})_S$ *such that* $j^{\natural} \circ \phi = r(j^{\natural}, j^{\vee,\natural})\, \lambda^{*}_{T} \circ j^{\vee,\natural}$.

(c) *The induced morphisms* $\mathrm{Gr}_{-2}(\hat{\beta}^{\natural,0}_{\bar{s}}) : \mathrm{Gr}^{\mathsf{Z}}_{-2} \underset{\hat{\mathbb{Z}}}{\otimes} \mathbb{A}^{\infty} \xrightarrow{\sim} \mathrm{V}\, T_{\bar{s}}$ *and* $\mathrm{Gr}_{-2}(\hat{\beta}^{\natural,\#,0}_{\bar{s}}) : \mathrm{Gr}^{\mathsf{Z}^{\#}}_{-2} \underset{\hat{\mathbb{Z}}}{\otimes} \mathbb{A}^{\infty} \xrightarrow{\sim} \mathrm{V}\, T^{\vee}_{\bar{s}}$ *coincide with the compositions*

$$\begin{aligned}\mathrm{Gr}^{\mathsf{Z}}_{-2} \underset{\hat{\mathbb{Z}}}{\otimes} \mathbb{A}^{\infty} &\overset{\varphi_{-2} \underset{\hat{\mathbb{Z}}}{\otimes} \mathbb{A}^{\infty}}{\xrightarrow{\sim}} \mathrm{Hom}_{\mathbb{A}^{\infty}}(X \underset{\mathbb{Z}}{\otimes} \mathbb{A}^{\infty}, \mathbb{A}^{\infty}(1)) \\ &\overset{((j^{\natural})^{-1} \underset{\mathbb{Q}}{\otimes} \mathbb{A}^{\infty})^{*}}{\xrightarrow{\sim}} \mathrm{Hom}_{\mathbb{A}^{\infty}}(\mathbf{X}(T) \underset{\mathbb{Z}}{\otimes} \mathbb{A}^{\infty}, \mathbb{A}^{\infty}(1)) \\ &\overset{\hat{\nu}^{\natural}_{\bar{s}}}{\xrightarrow{\sim}} \mathrm{Hom}_{\mathbb{A}^{\infty}}(\mathbf{X}(T) \underset{\mathbb{Z}}{\otimes} \mathbb{A}^{\infty}, \mathrm{V}\, \mathbf{G}_{\mathrm{m},\bar{s}}) \xrightarrow{\sim} \mathrm{V}\, T_{\bar{s}}.\end{aligned}$$

and

$$\begin{aligned}\mathrm{Gr}^{\mathsf{Z}^{\#}}_{-2} \underset{\hat{\mathbb{Z}}}{\otimes} \mathbb{A}^{\infty} &\overset{\varphi^{\#}_{-2} \underset{\hat{\mathbb{Z}}}{\otimes} \mathbb{A}^{\infty}}{\xrightarrow{\sim}} \mathrm{Hom}_{\mathbb{A}^{\infty}}(Y \underset{\mathbb{Z}}{\otimes} \mathbb{A}^{\infty}, \mathbb{A}^{\infty}(1)) \\ &\overset{((j^{\vee,\natural})^{-1} \underset{\mathbb{Q}}{\otimes} \mathbb{A}^{\infty})^{*}}{\xrightarrow{\sim}} \mathrm{Hom}_{\mathbb{A}^{\infty}}(\mathbf{X}(T^{\vee}) \underset{\mathbb{Z}}{\otimes} \mathbb{A}^{\infty}, \mathbb{A}^{\infty}(1)) \\ &\overset{\hat{\nu}^{\natural}_{\bar{s}}}{\xrightarrow{\sim}} \mathrm{Hom}_{\mathbb{A}^{\infty}}(\mathbf{X}(T^{\vee}) \underset{\mathbb{Z}}{\otimes} \mathbb{A}^{\infty}, \mathrm{V}\, \mathbf{G}_{\mathrm{m},\bar{s}}) \xrightarrow{\sim} \mathrm{V}\, T^{\vee}_{\bar{s}},\end{aligned}$$

respectively, where $\varphi^{\#}_{-2} : \mathrm{Gr}^{\mathsf{Z}^{\#}}_{-2} \xrightarrow{\sim} \mathrm{Hom}_{\hat{\mathbb{Z}}}(Y \underset{\mathbb{Z}}{\otimes} \hat{\mathbb{Z}}, \hat{\mathbb{Z}}(1))$ *is induced by* φ_0 *by duality.*

(d) *Together with* $\hat{\nu}_{-1,\bar{s}} := \hat{\nu}^{\natural}_{\bar{s}}$*, the induced morphisms*

$$\hat{\varphi}_{-1,\bar{s}} := \mathrm{Gr}_{-1}(\hat{\beta}^{\natural,0}_{\bar{s}}) : \mathrm{Gr}^{\mathsf{Z}}_{-1} \underset{\hat{\mathbb{Z}}}{\otimes} \mathbb{A}^{\infty} \xrightarrow{\sim} \mathrm{V}\, B_{\bar{s}}$$

and

$$\hat{\varphi}^{\#}_{-1,\bar{s}} := \mathrm{Gr}_{-1}(\hat{\beta}^{\natural,\#,0}_{\bar{s}}) : \mathrm{Gr}^{\mathsf{Z}^{\#}}_{-1} \underset{\hat{\mathbb{Z}}}{\otimes} \mathbb{A}^{\infty} \xrightarrow{\sim} \mathrm{V}\, B^{\vee}_{\bar{s}}$$

determine each other by duality. By varying $\bar{s}$ over geometric points of S, the ($\pi_1(S,\bar{s})$-invariant) $\mathcal{H}_{\mathrm{G}_{h,\mathrm{z}}}$-orbits of

$$(\hat{\varphi}_{-1,\bar{s}}, \hat{\nu}_{-1,\bar{s}})$$

determine a tuple

$$(B, \lambda_B, i_B, \varphi_{-1,\mathcal{H}})$$

whose $\mathbb{Q}^\times$-isogeny class is parameterized by $\mathsf{M}_{\mathcal{H}_h}$ (cf. [62, Prop. 1.4.3.4]), while the ($\pi_1(S,\bar{s})$-invariant) $\mathcal{H}_{\mathrm{G}_{1,\mathrm{z}}}$-orbits of

$$(\hat{\varphi}_{-1,\bar{s}}, \hat{\nu}_{-1,\bar{s}}, \varphi_{-2}, \varphi_0)$$

determine a tuple

$$((B, \lambda_B, i_B, \varphi_{-1,\mathcal{H}}), (\varphi^{\sim}_{-2,\mathcal{H}}, \varphi^{\sim}_{0,\mathcal{H}}))$$

whose $\mathbb{Q}^\times$-isogeny class is parameterized by $\mathsf{M}^{\Phi_\mathcal{H}}_{\mathcal{H}}$.

The $\mathbb{Q}^\times$-isogenies

$$\begin{aligned} &(G^\natural, \lambda^\natural : G^\natural \to G^{\vee,\natural}, i^\natural, j^\natural, j^{\vee,\natural}, [\hat{\beta}^\natural]_{\mathcal{H}_{\mathrm{G}_{1,\mathrm{z}}}}) \\ &\sim_{\mathbb{Q}^\times\text{-isog.}} (G^{\natural,\prime}, \lambda^{\natural,\prime} : G^{\natural,\prime} \to G^{\vee,\natural,\prime}, i^{\natural,\prime}, j^{\natural,\prime}, j^{\vee,\natural,\prime}, [\hat{\beta}^{\natural,\prime}]_{\mathcal{H}_{\mathrm{G}_{1,\mathrm{z}}}}) \end{aligned}$$

between tuples as in (1.3.2.15) *are given by pairs of $\mathbb{Q}^\times$-isogenies*

$$(f^\natural : G^\natural \to G^{\natural,\prime}, f^{\vee,\natural} : G^{\vee,\natural,\prime} \to G^{\vee,\natural})$$

such that we have the following:

(i) *There exists a section $r(f^\natural, f^{\vee,\natural})$ of $(\mathbb{Q}^\times_{>0})_S$ such that $\lambda^\natural = r(f^\natural, f^{\vee,\natural}) f^{\vee,\natural} \circ \lambda^{\natural,\prime} \circ f^\natural$.*
(ii) *$f^\natural$ and $f^{\vee,\natural}$ respect the compatible $\mathcal{O} \underset{\mathbb{Z}}{\otimes} \mathbb{Q}$-actions on $G^\natural$, $G^{\natural,\prime}$, $G^{\vee,\natural}$, and $G^{\vee,\natural,\prime}$ (defined by $i^\natural$ and $i^{\natural,\prime}$).*
(iii) *$j^\natural = (f^\natural)^* \circ j^{\natural,\prime}$ and $j^{\vee,\natural,\prime} = (f^{\vee,\natural})^* \circ j^{\vee,\natural}$.*
(iv) *For each geometric point $\bar{s}$, the morphisms $\mathrm{V}(f^\natural) : \mathrm{V}\, G^\natural_{\bar{s}} \xrightarrow{\sim} \mathrm{V}\, G^{\natural,\prime}_{\bar{s}}$ and $\mathrm{V}(f^{\vee,\natural}) : \mathrm{V}\, G^{\vee,\natural,\prime}_{\bar{s}} \xrightarrow{\sim} \mathrm{V}\, G^{\vee,\natural}_{\bar{s}}$ satisfy the condition that, for any representatives $\hat{\beta}^\natural_{\bar{s}} = (\hat{\beta}^{\natural,0}_{\bar{s}}, \hat{\beta}^{\natural,\#,0}_{\bar{s}}, \hat{\nu}^\natural_{\bar{s}})$ and $\hat{\beta}^{\natural,\prime}_{\bar{s}} = (\hat{\beta}^{\natural,0,\prime}_{\bar{s}}, \hat{\beta}^{\natural,\#,0,\prime}_{\bar{s}}, \hat{\nu}^{\natural,\prime}_{\bar{s}})$ of $[\hat{\beta}^\natural]_{\mathcal{H}_{\mathrm{G}_{1,\mathrm{z}}}}$ and $[\hat{\beta}^{\natural,\prime}]_{\mathcal{H}_{\mathrm{G}_{1,\mathrm{z}}}}$, respectively, the $\mathcal{H}_{\mathrm{G}_{1,\mathrm{z}}}$-orbits of*

$$(\mathrm{V}(f^\natural) \circ \hat{\beta}^{\natural,0}_{\bar{s}}, \mathrm{V}(f^{\vee,\natural})^{-1} \circ \hat{\beta}^{\natural,\#,0}_{\bar{s}}, r(f^\natural, f^{\vee,\natural})^{-1}_{\bar{s}} \hat{\nu}^\natural_{\bar{s}})$$

and

$$(\hat{\beta}^{\natural,0,\prime}_{\bar{s}}, \hat{\beta}^{\natural,\#,0,\prime}_{\bar{s}}, \hat{\nu}^{\natural,\prime}_{\bar{s}})$$

coincide, where $r(f^\natural, f^{\vee,\natural})_{\bar{s}}$ is the value at $\bar{s}$ of the above section $r(f^\natural, f^{\vee,\natural})$ of $(\mathbb{Q}^\times_{>0})_S$ such that $\lambda^\natural = r(f^\natural, f^{\vee,\natural}) f^{\vee,\natural} \circ \lambda^{\natural,\prime} \circ f^\natural$.

Proof. As in [62, Sec. 1.4.3], this can be proved by replacing any tuple as in (1.3.2.15) up to $\mathbb{Q}^\times$-isogeny, as in the statement of this proposition, with a tuple such that $j^\natural : X \underset{\mathbb{Z}}{\otimes} \mathbb{Q} \xrightarrow{\sim} \mathbf{X}(T) \underset{\mathbb{Z}}{\otimes} \mathbb{Q}$ (resp. $j^{\vee,\natural} : Y \underset{\mathbb{Z}}{\otimes} \mathbb{Q} \xrightarrow{\sim} \mathbf{X}(T^\vee) \underset{\mathbb{Z}}{\otimes} \mathbb{Q}$) maps X (resp. Y) to $\mathbf{X}(T)$ (resp. $\mathbf{X}(T^\vee)$), and such that, at each geometric point $\bar{s}$ of S, the assigned $\hat{\beta}^\natural_{\bar{s}} = (\hat{\beta}^{\natural,0}_{\bar{s}}, \hat{\beta}^{\natural,\#,0}_{\bar{s}}, \hat{\nu}^\natural_{\bar{s}})$ satisfies the condition that $\hat{\beta}^{\natural,0}_{\bar{s}}$ (resp. $\hat{\beta}^{\natural,\#,0}_{\bar{s}}$, resp. $\hat{\nu}^\natural_{\bar{s}}$) maps Z_{-1} (resp. $\mathsf{Z}^\#_{-1}$, resp. $\hat{\mathbb{Z}}(1)$) to $\mathrm{T}\, G^\natural_{\bar{s}}$ (resp. $\mathrm{T}\, G^{\vee,\natural}_{\bar{s}}$, resp. $\mathrm{T}\, \mathbf{G}_{\mathrm{m},\bar{s}}$). Then the tuple determines and is determined by a tuple as in (1.3.2.13), as desired. (These can be simultaneously achieved because of the existence of the section $r(j^\natural, j^{\vee,\natural})$ of $(\mathbb{Q}^\times_{>0})_S$. The proof is similar to that of [62, Prop. 1.4.3.4], and hence omitted.) □

Construction 1.3.2.16. Suppose $\mathcal{H}$ is neat. Consider the degenerating family

$$(G, \lambda, i, \alpha_\mathcal{H}) \to \mathsf{M}^{\mathrm{tor}}_{\mathcal{H},\Sigma} \tag{1.3.2.17}$$

of type $\mathsf{M}_\mathcal{H}$ as in Theorem 1.3.1.3. Let $\mathsf{Z} = \mathsf{Z}_{[(\Phi_\mathcal{H},\delta_\mathcal{H},\sigma)]}$ be any stratum of $\mathsf{M}^{\mathrm{tor}}_{\mathcal{H},\Sigma}$ such that $\sigma \subset \mathbf{P}^+_{\Phi_\mathcal{H}}$ is a top-dimensional cone in $\Sigma_{\Phi_\mathcal{H}}$ (in Σ). Let

$$(G^\natural_{\overline{\mathsf{Z}}}, \lambda^\natural_{\overline{\mathsf{Z}}}, i^\natural_{\overline{\mathsf{Z}}}) \to \overline{\mathsf{Z}} \tag{1.3.2.18}$$

denote the pullback of the (G, λ, i) in (1.3.2.17) to $\overline{\mathsf{Z}}$, the closure of Z in $\mathsf{M}^{\mathrm{tor}}_{\mathcal{H},\Sigma}$. Since σ is top-dimensional, the canonical morphism $\mathsf{Z} \to C_{\Phi_\mathcal{H},\delta_\mathcal{H}}$ is an isomorphism. Since $\alpha_\mathcal{H}$ is defined only over $\mathsf{M}_\mathcal{H}$, its pullback to $\overline{\mathsf{Z}}$ is undefined. The goal of this construction is to define a partial pullback, which still retains some information of $\alpha_\mathcal{H}$.

Let $n \geq 1$ be any integer such that $\mathcal{U}(n) \subset \mathcal{H}$, and let us fix any choice of $(\mathsf{Z}_n, \Phi_n, \delta_n)$. Consider any top-dimensional cone σ' contained in σ that is smooth for the integral structure defined by $\mathbf{S}_{\Phi_n}$, we have a canonical morphism $\mathfrak{X}_{\Phi_n,\delta_n,\sigma'} \to \mathfrak{X}_{\Phi_\mathcal{H},\delta_\mathcal{H},\sigma}$ (which might not be finite étale), inducing a morphism from the σ'-stratum $\mathsf{Z}_n = \mathsf{Z}_{[(\Phi_n,\delta_n,\sigma')]}$ of the source to the σ-stratum $\mathsf{Z} = \mathsf{Z}_{[(\Phi_\mathcal{H},\delta_\mathcal{H},\sigma)]}$ of the target (although the scheme-theoretic preimage of latter might not be the former), which can be identified with the canonical morphism (1.3.2.9). Let us denote the pullback of (1.3.2.18) to Z_n by

$$(G^\natural_{\mathsf{Z}_n}, \lambda^\natural_{\mathsf{Z}_n}, i^\natural_{\mathsf{Z}_n}) \to \mathsf{Z}_n. \tag{1.3.2.19}$$

Over each affine open formal subscheme $\mathrm{Spf}(R, I)$ of $\mathfrak{X}_{\Phi_n,\delta_n,\sigma'}$, such that $S_0 = \mathrm{Spec}(R/I)$ is the σ'-stratum of $S = \mathrm{Spec}(R)$, where both R and R/I are regular domains, we have a degenerating family $(G_S, \lambda_S, i_S, \alpha_{n,\eta}) \to S$ of type $\mathsf{M}_n = \mathsf{M}_{\mathcal{U}(n)}$. A priori, the level structure $\alpha_{n,\eta}$ is defined only over the generic point η of S (and it only extends to the largest open subscheme of S over which the pullback of G_S is an abelian scheme). Nevertheless, as explained in [62, Prop. 5.2.2.1], $G_\eta[n]$ (resp. $G^\vee_\eta[n]$) admits a canonical filtration $0 \subset T_\eta[n] \subset G^\natural_\eta[n] \subset G_\eta[n]$ (resp. $0 \subset T^\vee_\eta[n] \subset G^{\vee,\natural}_\eta[n] \subset G^\vee_\eta[n]$), with notation as in Lemma 1.3.2.11, where the subscripts "η" (and similar usages later) mean pullbacks. By the construction of $\mathfrak{X}_{\Phi_n,\delta_n,\sigma'}$, the symplectic isomorphism $\alpha_{n,\eta} : L/nL \xrightarrow{\sim} G_\eta[n]$ sends the filtration Z_n to the above geometric filtration on $G_\eta[n]$, and induces the pair $(\varphi_{-2,n}, \varphi_{0,n})$ in Φ_n

when restricted to the top and bottom filtered pieces. By duality, and by using the isomorphism $\nu_{n,\eta} : ((\mathbb{Z}/n\mathbb{Z})(1))_\eta \xrightarrow{\sim} \boldsymbol{\mu}_{n,\eta}$ (which is part of the data of $\alpha_{n,\eta}$), it also defines a symplectic isomorphism $\alpha^\#_{n,\eta} : L^\#/nL^\# \xrightarrow{\sim} G^\vee_\eta[n]$ which sends the dual filtration $\mathsf{Z}^\#_n$ to the above geometric filtration on $G^\vee_\eta[n]$, which induces (in particular) an object $\varphi^\#_{-2,n}$ dual to $\varphi_{0,n}$ in the obvious sense. These two isomorphisms $\alpha_{n,\eta}$ and $\alpha^\#_{n,\eta}$ are compatible under the canonical morphisms $L \hookrightarrow L^\#$ and $\lambda_\eta : G_\eta \to G^\vee_\eta$.

Since $G^\natural_S[n]$ (resp. $G^{\vee,\natural}_S[n]$, resp. $\boldsymbol{\mu}_{n,S}$) is finite étale over S, the restriction of $\alpha_{n,\eta}$ to Z_{-1} (resp. the restriction of $\alpha^\#_{n,\eta}$ to $\mathsf{Z}^\#_{-1}$, resp. the isomorphism $\nu_{n,\eta}$) over η extends to an isomorphism $\beta^{\natural,0}_{n,S} : (\mathsf{Z}_{-1,n})_S \xrightarrow{\sim} G^\natural_S[n]$ (resp. $\beta^{\natural,\#,0}_{n,S} : (\mathsf{Z}^\#_{-1,n})_S \xrightarrow{\sim} G^{\vee,\natural}_S[n]$, resp. $\nu^\natural_{n,S} : ((\mathbb{Z}/n\mathbb{Z})(1))_S \xrightarrow{\sim} \boldsymbol{\mu}_{n,S}$) over the whole normal scheme S. These two isomorphisms $\beta^{\natural,0}_{n,S}$ and $\beta^{\natural,\#,0}_{n,S}$ are compatible under the canonical morphisms $(\mathsf{Z}_{-1,n})_S \to (\mathsf{Z}^\#_{-1,n})_S$ and $\lambda^\natural_S : G^\natural_S \to G^{\vee,\natural}_S$. Let $\beta^\natural_{n,S} := (\beta^{\natural,0}_{n,S}, \beta^{\natural,\#,0}_{n,S}, \nu^\natural_{n,S})$, and consider its pullback $\beta^\natural_{n,S_0} := (\beta^{\natural,0}_{n,S_0}, \beta^{\natural,\#,0}_{n,S_0}, \nu^\natural_{n,S_0})$ to S_0. By analyzing $\beta^\natural_{n,S}$ as in the case of $\alpha_{n,\eta}$ as in [62, Sec. 5.2.2–5.2.3], we see that $\beta^\natural_{n,S}$ retains almost all information of $\alpha_{n,\eta}$, including the pairing $\mathrm{e}_{10,n}$ to be compared with $\mathsf{d}_{10,n}$, as in [62, Lem. 5.2.3.12 and Thm. 5.2.3.14], except that it loses information about the pairing $\mathrm{e}_{00,n}$ to be compared with $\mathsf{d}_{00,n}$. Hence, if we denote the pullback of (1.3.2.18) to S_0 by $(G^\natural_{S_0}, \lambda^\natural_{S_0}, i^\natural_{S_0}) \to S_0$, then $(G^\natural_{S_0}, \lambda^\natural_{S_0}, i^\natural_{S_0}, \beta^\natural_{n,S_0}) \to S_0$ determines and is determined by (the prescribed $(\mathsf{Z}_n, \Phi_n, \delta_n)$ and) the pullback to S_0 of the tautological object $((B, \lambda_B, i_B, \varphi_{-1,n}), (c_n, c^\vee_n))$ over C_{Φ_n,δ_n} (up to isomorphisms inducing automorphisms of Φ_n; *i.e.*, elements of Γ_{Φ_n}; see Lemma 1.3.2.11). By patching over varying S, we obtain (with $(G^\natural_{\mathsf{Z}_n}, \lambda^\natural_{\mathsf{Z}_n}, i^\natural_{\mathsf{Z}_n})$ already defined as in (1.3.2.19)) a tuple

$$(G^\natural_{\mathsf{Z}_n}, \lambda^\natural_{\mathsf{Z}_n}, i^\natural_{\mathsf{Z}_n}, \beta^\natural_{n,\mathsf{Z}_n}) \to \mathsf{Z}_n \cong C_{\Phi_n,\delta_n} \tag{1.3.2.20}$$

such that the previous sentence is true with S_0 replaced with Z_n.

Since $\mathcal{H}_{\mathrm{G}_{1,\mathrm{z}}}/\mathcal{U}(n)_{\mathrm{G}_{1,\mathrm{z}}}$ acts compatibly on $\beta^\natural_{n,\mathsf{Z}_n}$ and $(\varphi_{-1,n}, c_n, c^\vee_n)$, the latter action being compatible with the $\mathcal{H}_{\mathrm{G}_{1,\mathrm{z}}}/\mathcal{U}(n)_{\mathrm{G}_{1,\mathrm{z}}}$-torsor structure of (1.3.2.9), by forming the $\mathcal{H}_{\mathrm{G}_{1,\mathrm{z}}}/\mathcal{U}(n)_{\mathrm{G}_{1,\mathrm{z}}}$-orbit $\beta^\natural_{\mathcal{H},\mathsf{Z}_n}$ of $\beta^\natural_{n,\mathsf{Z}_n}$, we can descend (1.3.2.20) to a tuple

$$(G^\natural_{\mathsf{Z}}, \lambda^\natural_{\mathsf{Z}}, i^\natural_{\mathsf{Z}}, \beta^\natural_{\mathcal{H},\mathsf{Z}}) \to \mathsf{Z} \cong C_{\Phi_\mathcal{H},\delta_\mathcal{H}}, \tag{1.3.2.21}$$

where the first three entries form the pullback of (1.3.2.18) to Z, which determines and is determined by (the prescribed $(\mathsf{Z}_\mathcal{H}, \Phi_\mathcal{H}, \delta_\mathcal{H})$ and) the tautological object

$$\big((B, \lambda_B, i_B, \varphi_{-1,\mathcal{H}}), (\varphi^\sim_{-2,\mathcal{H}}, \varphi^\sim_{0,\mathcal{H}}), (c_\mathcal{H}, c^\vee_\mathcal{H})\big) \to C_{\Phi_\mathcal{H},\delta_\mathcal{H}} \tag{1.3.2.22}$$

(up to isomorphisms inducing automorphisms of $\Phi_\mathcal{H}$; *i.e.*, elements of $\Gamma_{\Phi_\mathcal{H}}$). Since the tautological object (1.3.2.22) is independent of the choice of n, so is the tuple (1.3.2.21).

By abuse of language, we say that

$$(G^\natural_{\overline{\mathsf{Z}}}, \lambda^\natural_{\overline{\mathsf{Z}}}, i^\natural_{\overline{\mathsf{Z}}}, \beta^\natural_{\mathcal{H},\mathsf{Z}}) \to \overline{\mathsf{Z}} \tag{1.3.2.23}$$

is the *pullback* of the degenerating family (1.3.2.17) to $\overline{\mathsf{Z}}$, with the convention that (as in the case of $(G, \lambda, i, \alpha_\mathcal{H})$ itself) $\beta^\natural_{\mathcal{H},\mathsf{Z}}$ is defined only over Z, while $(G^\natural, \lambda^\natural, i^\natural)$ is defined over all of $\overline{\mathsf{Z}}$ as in (1.3.2.18). (This finishes Construction 1.3.2.16.)

Proposition 1.3.2.24. (*Compare with Proposition* 1.3.1.15.) *By considering compatible* $\mathbb{Q}^\times$*-isogenies* $(f : G^\natural \to G^{\natural,\prime}, f^\vee : G^{\vee,\natural,\prime} \to G^{\vee,\natural})$ *inducing isomorphisms on the torus parts, we can define Hecke twists of the tautological object* $(G^\natural, \lambda^\natural, i^\natural, \beta^\natural_{\mathcal{H}}) \to C_{\Phi_\mathcal{H},\delta_\mathcal{H}}$ *by elements of* $\mathrm{G}_{1,\mathsf{Z}}(\mathbb{A}^\infty)$*, and define the Hecke action of* $\mathrm{G}_{1,\mathsf{Z}}(\mathbb{A}^\infty)$ *on the collection* $\{C_{\Phi_\mathcal{H},\delta_\mathcal{H}}\}_{\mathcal{H}_{\mathrm{G}_{1,\mathsf{Z}}}}$*, realized by finite étale surjections pulling tautological objects back to Hecke twists, which is compatible with the Hecke action of* $\mathrm{G}'_{h,\mathsf{Z}}(\mathbb{A}^\infty)$ *on the collection* $\{\mathsf{M}^{\Phi_\mathcal{H}}_\mathcal{H}\}_{\mathcal{H}_{\mathrm{G}'_{h,\mathsf{Z}}}}$ *under the canonical morphisms* $C_{\Phi_\mathcal{H},\delta_\mathcal{H}} \to \mathsf{M}^{\Phi_\mathcal{H}}_\mathcal{H}$ *(with varying* $\mathcal{H}$*) and the canonical homomorphism* $\mathrm{G}_{1,\mathsf{Z}}(\mathbb{A}^\infty) \to \mathrm{G}'_{h,\mathsf{Z}}(\mathbb{A}^\infty)$*. Over the subcollection indexed by* $\mathcal{H}_{\mathrm{G}_{1,\mathsf{Z}}}$ *with neat* $\mathcal{H}$*, the Hecke action of* $\mathrm{G}_{1,\mathsf{Z}}(\mathbb{A}^\infty)$ *on* $\{C_{\Phi_\mathcal{H},\delta_\mathcal{H}}\}_{\mathcal{H}_{\mathrm{G}_{1,\mathsf{Z}}}}$ *is compatible with the Hecke action of* $\mathrm{P}'_\mathsf{Z}(\mathbb{A}^\infty)$ *on the collection of strata* $\{\mathsf{Z}_{[(\Phi_\mathcal{H},\delta_\mathcal{H},\sigma)]}\}$ *above* $\{C_{\Phi_\mathcal{H},\delta_\mathcal{H}}\}_{\mathcal{H}_{\mathrm{G}_{1,\mathsf{Z}}}}$ *(cf. Proposition* 1.3.1.15*) under the canonical homomorphism* $\mathrm{P}'_\mathsf{Z}(\mathbb{A}^\infty) \to \mathrm{G}_{1,\mathsf{Z}}(\mathbb{A}^\infty) = \mathrm{P}'_\mathsf{Z}(\mathbb{A}^\infty)/\mathrm{U}_{2,\mathsf{Z}}(\mathbb{A}^\infty)$*.*

By also considering $\mathbb{Q}^\times$*-isogenies* $(f : G^\natural \to G^{\natural,\prime}, f^\vee : G^{\vee,\natural,\prime} \to G^{\vee,\natural})$ *inducing* $\mathbb{Q}^\times$*-isogenies on the torus parts, we can also define Hecke twists of the tautological object* $(G^\natural, \lambda^\natural, i^\natural, \beta^\natural_{\mathcal{H}}) \to C_{\Phi_\mathcal{H},\delta_\mathcal{H}}$ *by elements of* $\mathrm{P}_\mathsf{Z}(\mathbb{A}^\infty)/\mathrm{U}_{2,\mathsf{Z}}(\mathbb{A}^\infty)$*, and define the Hecke action of* $\mathrm{P}_\mathsf{Z}(\mathbb{A}^\infty)/\mathrm{U}_{2,\mathsf{Z}}(\mathbb{A}^\infty)$ *on the collection* $\{\coprod C_{\Phi_\mathcal{H},\delta_\mathcal{H}}\}_{\mathcal{H}_{\mathrm{P}_\mathsf{Z}}/\mathcal{H}_{\mathrm{U}_{2,\mathsf{Z}}}}$*, realized by finite étale surjections pulling tautological objects back to Hecke twists, where the disjoint unions are over classes* $[(\mathsf{Z}_\mathcal{H}, \Phi_\mathcal{H}, \delta_\mathcal{H})]$ *sharing the same* $\mathsf{Z}_\mathcal{H}$*, which induces an action of* $\mathrm{G}'_{l,\mathsf{Z}}(\mathbb{A}^\infty) = \mathrm{P}_\mathsf{Z}(\mathbb{A}^\infty)/\mathrm{P}'_\mathsf{Z}(\mathbb{A}^\infty)$ *on the index sets* $\{[(\mathsf{Z}_\mathcal{H}, \Phi_\mathcal{H}, \delta_\mathcal{H})]\}$*, which is compatible with the Hecke action of* $\mathrm{P}_\mathsf{Z}(\mathbb{A}^\infty)/\mathrm{U}_\mathsf{Z}(\mathbb{A}^\infty) \cong \mathrm{G}'_{l,\mathsf{Z}}(\mathbb{A}^\infty) \times \mathrm{G}'_{h,\mathsf{Z}}(\mathbb{A}^\infty)$ *on the collection* $\{\coprod \mathsf{M}^{\Phi_\mathcal{H}}_\mathcal{H}\}_{\mathcal{H}_{\mathrm{G}'_{h,\mathsf{Z}}}}$ *(with the same index sets and the same induced action of* $\mathrm{G}'_{l,\mathsf{Z}}(\mathbb{A}^\infty)$*) under the canonical morphisms* $C_{\Phi_\mathcal{H},\delta_\mathcal{H}} \to \mathsf{M}^{\Phi_\mathcal{H}}_\mathcal{H}$ *(with varying* $\mathcal{H}$*) and the canonical homomorphism* $\mathrm{P}_\mathsf{Z}(\mathbb{A}^\infty)/\mathrm{U}_{2,\mathsf{Z}}(\mathbb{A}^\infty) \to \mathrm{G}'_{l,\mathsf{Z}}(\mathbb{A}^\infty) \times \mathrm{G}'_{h,\mathsf{Z}}(\mathbb{A}^\infty)$*. Over the subcollection indexed by* $\mathcal{H}_{\mathrm{P}_\mathsf{Z}}/\mathcal{H}_{\mathrm{U}_{2,\mathsf{Z}}}$ *with neat* $\mathcal{H}$*, the Hecke action of* $\mathrm{P}_\mathsf{Z}(\mathbb{A}^\infty)/\mathrm{U}_{2,\mathsf{Z}}(\mathbb{A}^\infty)$ *on* $\{\coprod C_{\Phi_\mathcal{H},\delta_\mathcal{H}}\}_{\mathcal{H}_{\mathrm{P}_\mathsf{Z}}/\mathcal{H}_{\mathrm{U}_{2,\mathsf{Z}}}}$ *is compatible with the Hecke action of* $\mathrm{P}_\mathsf{Z}(\mathbb{A}^\infty)(\mathbb{A}^\infty)$ *on the collection of strata* $\{\mathsf{Z}_{[(\Phi_\mathcal{H},\delta_\mathcal{H},\sigma)]}\}$ *above* $\{\coprod C_{\Phi_\mathcal{H},\delta_\mathcal{H}}\}_{\mathcal{H}_{\mathrm{P}_\mathsf{Z}}/\mathcal{H}_{\mathrm{U}_{2,\mathsf{Z}}}}$ *(cf. Proposition* 1.3.1.15*) under the canonical homomorphism* $\mathrm{P}_\mathsf{Z}(\mathbb{A}^\infty)(\mathbb{A}^\infty) \to \mathrm{P}_\mathsf{Z}(\mathbb{A}^\infty)/\mathrm{U}_{2,\mathsf{Z}}(\mathbb{A}^\infty)$*.*

In the $\mathbb{Q}^\times$*-isogeny class language as in Proposition* 1.3.2.14*, the morphism*

$$[g] : C_{\Phi_{\mathcal{H}'},\delta_{\mathcal{H}'}} \to C_{\Phi_\mathcal{H},\delta_\mathcal{H}},$$

for $g \in \mathrm{P}_\mathsf{Z}(\mathbb{A}^\infty)/\mathrm{U}_{2,\mathsf{Z}}(\mathbb{A}^\infty)$ *such that* $\mathcal{H}'_{\mathrm{P}_\mathsf{Z}}/\mathcal{H}'_{\mathrm{U}_{2,\mathsf{Z}}} \subset g(\mathcal{H}_{\mathrm{P}_\mathsf{Z}}/\mathcal{H}_{\mathrm{U}_{2,\mathsf{Z}}})g^{-1}$ *and such that* $[(\Phi_\mathcal{H}, \delta_\mathcal{H})]$ *is* g*-assigned to* $[(\Phi_{\mathcal{H}'}, \delta_{\mathcal{H}'})]$ *with a pair isomorphisms*

$$(f_X : X \underset{\mathbb{Z}}{\otimes} \mathbb{Q} \xrightarrow{\sim} X' \underset{\mathbb{Z}}{\otimes} \mathbb{Q}, f_Y : Y' \underset{\mathbb{Z}}{\otimes} \mathbb{Q} \xrightarrow{\sim} Y \underset{\mathbb{Z}}{\otimes} \mathbb{Q})$$

as in [62, Prop. 5.4.3.8], is characterized by

$$[g]^*(G^\natural, \lambda^\natural : G^\natural \to G^{\vee,\natural}, i^\natural, j^\natural, j^{\vee,\natural}, [\hat{\beta}^\natural]_{\mathcal{H}_{\mathrm{G}_{1,\mathsf{Z}}}})$$
$$\sim_{\mathbb{Q}^\times\text{-isog.}} (G^{\natural,\prime}, \lambda^{\natural,\prime} : G^{\natural,\prime} \to G^{\vee,\natural,\prime}, i^{\natural,\prime}, f_X \circ j^{\natural,\prime}, f_Y^{-1} \circ j^{\vee,\natural,\prime}, [\hat{\beta}^{\natural,\prime} \circ g]_{\mathcal{H}_{\mathrm{G}_{1,\mathsf{Z}}}})$$

over $C_{\Phi_{\mathcal{H}'},\delta_{\mathcal{H}'}}$*, where*

$$(G^\natural, \lambda^\natural : G^\natural \to G^{\vee,\natural}, i^\natural, j^\natural, j^{\vee,\natural}, [\hat{\beta}^\natural]_{\mathcal{H}_{\mathrm{G}_{1,\mathsf{Z}}}})$$

and

$$(G^{\natural,\prime}, \lambda^{\natural,\prime} : G^{\natural,\prime} \to G^{\vee,\natural,\prime}, i^{\natural,\prime}, j^{\natural,\prime}, j^{\vee,\natural,\prime}, [\hat{\beta}^{\natural,\prime}]_{\mathcal{H}'_{\mathrm{G}_{1,\mathsf{Z}}}})$$

are representatives of the universal $\mathbb{Q}^\times$*-isogeny classes over* $C_{\Phi_{\mathcal{H}},\delta_{\mathcal{H}}}$ *and* $C_{\Phi_{\mathcal{H}'},\delta_{\mathcal{H}'}}$*, respectively, and where the rational level-*$\mathcal{H}$ *structure* $[\hat{\beta}^{\natural,\prime} \circ g]_{\mathcal{H}_{\mathrm{G}_{1,\mathsf{Z}}}}$ *of* $(G^{\natural,\prime}, \lambda^{\natural,\prime}, i^{\natural,\prime}, f_X \circ j^{\natural,\prime}, f_Y^{-1} \circ j^{\vee,\natural,\prime})$ *of type* $(L \underset{\mathbb{Z}}{\otimes} \mathbb{A}^\infty, \langle\,\cdot\,,\,\cdot\,\rangle, \mathsf{Z} \underset{\hat{\mathbb{Z}}}{\otimes} \mathbb{A}^\infty, \Phi)$ *is determined at each geometric point* $\bar{s}$ *of* $C_{\Phi_{\mathcal{H}'},\delta_{\mathcal{H}'}}$ *by the* $\mathcal{H}_{\mathrm{G}_{1,\mathsf{Z}}}$*-orbit of* $\hat{\beta}^{\natural,\prime}_{\bar{s}} \circ g$*, where* $\hat{\beta}^{\natural,\prime}_{\bar{s}}$ *is any representative of the rational level-*$\mathcal{H}'$ *structure* $[\hat{\beta}^{\natural,\prime}_{\bar{s}}]_{\mathcal{H}'_{\mathrm{G}_{1,\mathsf{Z}}}}$ *of* $(G^{\natural,\prime}, \lambda^{\natural,\prime}, i^{\natural,\prime}, j^{\natural,\prime}, j^{\vee,\natural,\prime})$ *of type* $(L \underset{\mathbb{Z}}{\otimes} \mathbb{A}^\infty, \langle\,\cdot\,,\,\cdot\,\rangle, \mathsf{Z} \underset{\hat{\mathbb{Z}}}{\otimes} \mathbb{A}^\infty, \Phi')$ *based at* $\bar{s}$ *(assigned to* $\bar{s}$ *by* $[\hat{\beta}^{\natural,\prime}]_{\mathcal{H}'_{\mathrm{G}_{1,\mathsf{Z}}}}$*).*

Proof. The first assertions in both of the first two paragraphs, and the whole third paragraph, can be justified as in the case of $\mathsf{M}_{\mathcal{H}}$, which we omit for simplicity. As for the second assertions in both of the first two paragraphs, it suffices to note that the pullback of the Hecke twist of (1.3.2.17) is the Hecke twist of (1.3.2.21), the latter of which can be identified with the tautological object over $C_{\Phi_{\mathcal{H}},\delta_{\mathcal{H}}}$ under the canonical isomorphism $\mathsf{Z}_{[(\Phi_{\mathcal{H}},\delta_{\mathcal{H}},\sigma)]} \xrightarrow{\sim} C_{\Phi_{\mathcal{H}},\delta_{\mathcal{H}}}$ (for any top-dimensional σ, when $\mathcal{H}$ is neat). □

The torus torsor $\Xi_{\Phi_{\mathcal{H}},\delta_{\mathcal{H}}} \to C_{\Phi_{\mathcal{H}},\delta_{\mathcal{H}}}$ is, by the construction in [62, Sec. 6.2.3–6.2.4; see also the errata], the quotient of

$$\coprod \Xi_{\Phi_n,\delta_n} \to \coprod C_{\Phi_n,\delta_n}$$

by $H_n = \mathcal{H}/\mathcal{U}(n)$ (for any integer $n \geq 1$ such that $\mathcal{U}(n) \subset \mathcal{H}$), where each $\Xi_{\Phi_n,\delta_n} \to C_{\Phi_n,\delta_n}$ has a canonical structure of a torsor under the torus E_{Φ_n} with character group $\mathbf{S}_{\Phi_n}$ (see Section 1.2.2), which is preserved under the action of $H_{n,\mathrm{U}^{\mathrm{ess}}_{2,\mathsf{Z}_n}} \cong \mathcal{H}_{\mathrm{U}_{2,\mathsf{Z}}}/\mathcal{U}(n)_{\mathrm{U}_{2,\mathsf{Z}}}$. Therefore, we have the following:

Lemma 1.3.2.25. *The quotient* $\Xi_{\Phi_{\mathcal{H}},\delta_{\mathcal{H}}}$ *depends only on* $\mathcal{H}_{\mathrm{P}'_{\mathsf{Z}}}$*, and is a torsor under the torus* $E_{\Phi_{\mathcal{H}}}$ *with character group* $\mathbf{S}_{\Phi_{\mathcal{H}}}$*.*

If we fix the choice of $(\mathsf{Z}_n$ *and)* Φ_n*, then the canonical morphism*

$$\Xi_{\Phi_n,\delta_n} \to \Xi_{\Phi_{\mathcal{H}},\delta_{\mathcal{H}}} \tag{1.3.2.26}$$

is an $\mathcal{H}_{\mathrm{P}'_{\mathsf{Z}}}/\mathcal{U}(n)_{\mathrm{P}'_{\mathsf{Z}}} \cong H_{n,\mathrm{G}^{\mathrm{ess}}_{h,\mathsf{Z}_n} \ltimes \mathrm{U}^{\mathrm{ess}}_{\mathsf{Z}_n}}$*-torsor (see [62, Sec. 6.2.4; see also the errata]), where* $\mathcal{H}_{\mathrm{P}'_{\mathsf{Z}}}$ *and* $\mathcal{U}(n)_{\mathrm{P}'_{\mathsf{Z}}}$ *are open compact subgroups of* $\mathrm{P}'_{\mathsf{Z}}(\hat{\mathbb{Z}})$ *as in Definition* 1.2.1.12*, and induces an isomorphism*

$$\Xi_{\Phi_n,\delta_n}/(\mathcal{H}_{\mathrm{P}'_{\mathsf{Z}}}/\mathcal{U}(n)_{\mathrm{P}'_{\mathsf{Z}}}) \xrightarrow{\sim} \Xi_{\Phi_{\mathcal{H}},\delta_{\mathcal{H}}}. \tag{1.3.2.27}$$

Lemma 1.3.2.28. (*Compare with Lemma* 1.3.2.7.) *The torus torsor* $S := \Xi_{\Phi_n,\delta_n} \to C_{\Phi_n,\delta_n}$ *is universal for the additional structure* τ_n *satisfying certain symplectic and liftability conditions, which we review as follows:*

(1) $\tau_n : \mathbf{1}_{\frac{1}{n}Y \times X,S} \xrightarrow{\sim} (c_n^\vee \times c)^* \mathcal{P}_B^{\otimes -1}$ *is an* $\mathcal{O}$*-compatible trivialization of biextensions (as in [62, Lem. 5.2.3.2 and Def. 5.2.7.8]), where* $\mathcal{P}_B$ *is the Poincaré invertible sheaf of* B*, which defines an* $\mathcal{O}$*-equivariant homomorphism* $\iota_n : \frac{1}{n}Y_S \to G^\natural$ *lifting* $c_n^\vee : \frac{1}{n}Y_S \to B$ *(via the canonical homomorphism* $G^\natural \to B$*). Its restriction* $\tau : \mathbf{1}_{Y \times X,S} \xrightarrow{\sim} (c_n^\vee \times c)^* \mathcal{P}_B^{\otimes -1}$ *is a trivialization of biextensions such that* $(\mathrm{Id}_Y \times \phi)^* \tau$ *is symmetric, and such that* $(i_Y(b) \times \mathrm{Id}_X)^* \tau = (\mathrm{Id}_Y \times i_X(b^\star))^* \tau$ *for all* $b \in \mathcal{O}$*; and defines homomorphisms* $\iota : Y_S \to G^\natural$ *and* $\iota^\vee : X_S \to G^{\vee,\natural}$ *compatible with each under* $\phi : Y \hookrightarrow X$ *and* $\lambda^\natural : G^\natural \to G^{\vee,\natural}$*.*

(2) *Let* $[n] : G^\natural \to G^\natural$ *denote the multiplication by* n *morphism on* $G^\natural$*. Then we define*

$$G[n] := [n]^{-1}(\iota(Y))/\iota(Y),$$

where $\iota(Y)$ *is the image of the* $\mathcal{O}$*-equivariant homomorphism* $\iota : Y \to G^\natural$ *induced by* ι_n *by restriction. Note that we defined* $G[n]$ *without actually having a quotient* $G = G^\natural/\iota(Y)$ *over* S *(cf. [20, p. 57]). Then we have an exact sequence*

$$0 \to G^\natural[n] \to G[n] \to (Y/nY)_S \to 0 \qquad (1.3.2.29)$$

of finite flat group schemes over S*. Similarly, we define*

$$G^\vee[n] := [n]^{-1}(\iota^\vee(X))/\iota^\vee(X)$$

without defining $G^\vee$*, together with an exact sequence*

$$0 \to G^{\vee,\natural}[n] \to G^\vee[n] \to (X/nX)_S \to 0 \qquad (1.3.2.30)$$

of finite flat group schemes over S*, which is then equipped with a homomorphism*

$$\lambda : G[n] \to G^\vee[n]$$

without defining $\lambda : G \to G^\vee$*, respecting the filtrations defined by* (1.3.2.29) *and* (1.3.2.30)*. We note that there is a canonical duality between* $G[n]$ *and* $G^\vee[n]$*, just as in the case of usual abelian schemes, but we will not explicitly use this canonical duality for our purpose.*

(3) *The lifting* ι_n *of* ι *defines a splitting of* (1.3.2.29)*. Together with the splittings defined by* $(c_n, c_n^\vee)$ *in Lemma* 1.3.2.11*, we obtain a splitting*

$$\varsigma_n : T[n] \oplus B[n] \oplus (Y/nY)_S \xrightarrow{\sim} G[n].$$

On the other hand, the biextension properties of $\mathcal{P}_B$ *allows* τ_n *to induce a dual trivialization*

$$\tau_n^\vee : \mathbf{1}_{\frac{1}{n}X \times Y,S} \xrightarrow{\sim} (c_n \times c^\vee)^* \mathcal{P}_B^{\otimes -1},$$

by setting

$$\tau_n^\vee(\tfrac{1}{n}\chi, y) : \mathbf{1}_S \xrightarrow{\sim} (c_n(\tfrac{1}{n}\chi, c^\vee(y))^* \mathcal{P}_B^{\otimes -1}$$

to be

$$\mathbf{1}_S \overset{\tau_n(\frac{1}{n}y,\chi)}{\overset{\sim}{\to}} (c_n^\vee(\tfrac{1}{n}y), c_n(\tfrac{1}{n}\chi))^*\mathcal{P}_B^{\otimes -n} \cong (c_n(\tfrac{1}{n}\chi, c^\vee(y))^*\mathcal{P}_B^{\otimes -1}.$$

Then $\tau_n^\vee$ induces a lifting $\iota_n^\vee$ of ι, which defines a splitting of (1.3.2.29). *Together with the splittings defined by $(c_n, c_n^\vee)$ in Lemma* 1.3.2.11*, we obtain a splitting*

$$\varsigma_n^\vee : T^\vee[n] \oplus B^\vee[n] \oplus (X/nX)_S \overset{\sim}{\to} G^\vee[n].$$

However, the two splittings have no reason to be compatible with each other under $\lambda : G[n] \to G^\vee[n]$. While the failure measured by the induced homomorphisms $B[n] \to T^\vee[n]$ and $(Y/nY)_S \to B^\vee[n]$ can be identified (up to a sign convention) with the pairing

$$\mathtt{d}_{10,n} : B[n] \times (\tfrac{1}{n}Y/Y)_S \to \boldsymbol{\mu}_{n,S}$$

defined in Lemma 1.3.2.11*, the failure measured by the induced homomorphism $(Y/nY)_S \to T^\vee[n]$ can be identified (up to a sign convention) with a pairing*

$$\mathtt{d}_{00,n} : (\tfrac{1}{n}Y/Y)_S \times (\tfrac{1}{n}Y/Y)_S \to \boldsymbol{\mu}_{n,S}$$

which sends $(\frac{1}{n}y_1, \frac{1}{n}y_2)$ to

$$\tau_n(\tfrac{1}{n}y_1, \phi(y_2))\tau_n(\tfrac{1}{n}y_2, \phi(y_1))^{-1}$$

for any functorial points y_1 and y_2 of $(\frac{1}{n}Y/Y)_S$ (cf. [62, Lem. 5.2.3.12]).

(4) The symplectic condition for τ_n (or ι_n) is that, under $\varphi_{0,n}$, the pairing $\mathtt{d}_{00,n}$ above is matched with the pairing

$$\langle\,\cdot\,,\,\cdot\,\rangle_{00,n} : \mathrm{Gr}_{0,n}^{\mathrm{Z}} \times \mathrm{Gr}_{0,n}^{\mathrm{Z}} \to ((\mathbb{Z}/n\mathbb{Z})(1))_S$$

induced by $\langle\,\cdot\,,\,\cdot\,\rangle$ and δ_n. Then, together with the symplectic condition for $(c_n, c_n^\vee)$ in Lemma 1.3.2.11*, under $(\varphi_{-2,n}, \varphi_{0,n})$ and $\varphi_{-1,n}$ (equipped with $\nu_{-1,n} : ((\mathbb{Z}/n\mathbb{Z})(1))_S \overset{\sim}{\to} \boldsymbol{\mu}_{n,S}$), we obtain an $\mathcal{O}$-equivariant isomorphisms*

$$\beta_n^0 : (L/nL)_S \overset{\sim}{\to} G[n]$$

and

$$\beta_n^{\#,0} : (L^\#/nL^\#)_S \overset{\sim}{\to} G^\vee[n]$$

respecting filtrations on both sides, together with the isomorphism $\nu_n = \nu_{-1,n}$, which are compatible with the canonical morphisms $L \hookrightarrow L^\#$ and $\lambda : G[n] \to G^\vee[n]$.

(5) The liftability condition for τ_n is that, for each integer m such that $n|m$, and for any lifting δ_m of δ_n, there exists a finite étale covering of S over which there exist $\varphi_{-1,m}$, $(\varphi_{-2,m}, \varphi_{0,m})$, $(c_m, c_m^\vee)$, and τ_m lifting $\varphi_{-1,n}$, $(\varphi_{-2,n}, \varphi_{0,n})$, $(c_n, c_n^\vee)$, τ_n, respectively, and satisfying the symplectic condition defined by $\langle\,\cdot\,,\,\cdot\,\rangle$ and δ_m as above.

(6) *The group of multiplicative type* $\ddot{E}_{\Phi_n}$ *with character group* $\ddot{\mathbf{S}}_{\Phi_n}$ *as in* (1.2.2.1) *define a subgroup of*

$$\underline{\mathrm{Hom}}_{\mathbb{Z}}\big((\tfrac{1}{n}Y)\underset{\mathbb{Z}}{\otimes}X)_S,\mathbf{G}_{\mathrm{m},S}\big)\cong\underline{\mathrm{Hom}}_{\mathbb{Z}}(\tfrac{1}{n}Y_S,T)$$

over S, *which acts on the collection of* $\mathcal{O}$*-compatible* τ_n, *possibly not satisfying the symplectic and liftability conditions above (but preserving the symmetry and* $\mathcal{O}$*-compatibility of the induced* τ), *inducing a translation action on the collection of* $\mathcal{O}$*-equivariant homomorphisms* $\iota_n:\frac{1}{n}Y_S\to G^\natural$. *The subgroup* E_{Φ_n} *of* $\ddot{E}_{\Phi_n}$ *with character group* $\mathbf{S}_{\Phi_n}=\ddot{\mathbf{S}}_{\Phi_n,\mathrm{free}}$, *the free quotient of* $\ddot{\mathbf{S}}_{\Phi_n}$, *preserves in addition the symplectic and liftability conditions satisfied by* τ_n *(see [62, (6.2.3.5) and Conv. 6.2.3.20], and the proofs leading to there), and makes* $S=\Xi_{\Phi_n,\delta_n}\to C_{\Phi_n,\delta_n}$ *a torsor under the torus* E_{Φ_n}, *which is equipped with a homomorphism*

$$\mathbf{S}_{\Phi_n}\to\underline{\mathrm{Pic}}_e(C_{\Phi_n,\delta_n}/\mathsf{M}_n^{\mathsf{Z}_n}):\ell\mapsto\Psi_{\Phi_n,\delta_n}(\ell)$$

(by the torus torsor structure; see [62, Prop. 6.2.3.21 and (6.2.3.22)]), assigning to each $\ell\in\mathbf{S}_{\Phi_n}$ *a rigidified invertible sheaf* $\Psi_{\Phi_n,\delta_n}(\ell)$ *over* C_{Φ_n,δ_n}, *such that*

$$\Xi_{\Phi_n,\delta_n}\cong\underline{\mathrm{Spec}}_{\mathscr{O}_{C_{\Phi_n,\delta_n}}}\bigg(\bigoplus_{\ell\in\mathbf{S}_{\Phi_n}}\Psi_{\Phi_n,\delta_n}(\ell)\bigg).$$

When $\ell=[\frac{1}{n}y\otimes\chi]$ *for some* $y\in Y$ *and* $\chi\in X$, *we have a canonical isomorphism*

$$\Psi_{\Phi_n,\delta_n}(\ell)\cong(c_n^\vee(\tfrac{1}{n}y),c(\chi))^*\mathcal{P}_B.$$

By construction, we have $\mathbf{S}_{\Phi_1}^\vee/\mathbf{S}_{\Phi_n}^\vee\cong\mathrm{U}_{2,\mathsf{Z}}(\hat{\mathbb{Z}})/\mathcal{U}(n)_{\mathrm{U}_{2,\mathsf{Z}}}$.

These can be re-interpreted as follows: $S=\Xi_{\Phi_n,\delta_n}\to C_{\Phi_n,\delta_n}$ *parameterizes tuples*

$$(G^\natural,\lambda^\natural:G^\natural\to G^{\vee,\natural},i^\natural,\tau,\beta_n),$$

where:

(a) $G^\natural$, $G^{\vee,\natural}$, $\lambda^\natural$, *and* $i^\natural$ *are as in Lemma* 1.3.2.11.

(b) $\tau:\mathbf{1}_{Y\times X}\xrightarrow{\sim}(c^\vee\times c)^*\mathcal{P}_B^{\otimes-1}$ *is a trivialization of biextensions such that* $(\mathrm{Id}_Y\times\phi)^*\tau$ *is symmetric, and such that* $(i_Y(b)\times\mathrm{Id}_X)^*\tau=(\mathrm{Id}_Y\times i_X(b^\star))^*\tau$ *for all* $b\in\mathcal{O}$. *Then* τ *induces homomorphisms* $\iota:Y\to G^\natural$ *and* $\iota^\vee:X\to G^{\vee,\natural}$ *compatible with the homomorphisms* $\phi:Y\hookrightarrow X$ *and* $\lambda^\natural:G^\natural\to G^{\vee,\natural}$, *and induces an* $\mathcal{O}$*-equivariant homomorphism* $\lambda:G[n]\to G^\vee[n]$.

(c) $\beta_n=(\beta_n^0,\beta_n^{\#,0},\nu_n)$ *is a* **principal level-n structure** *of* $(G^\natural,\lambda^\natural,i^\natural,\tau)$ *of type* $(L\underset{\mathbb{Z}}{\otimes}\hat{\mathbb{Z}},\langle\cdot,\cdot\rangle,\mathsf{Z})$, *where* $\beta_n^0:(L/nL)_S\xrightarrow{\sim}G[n]$ *and* $\beta_n^{\#,0}:(L^\#/nL^\#)_S\xrightarrow{\sim}G^\vee[n]$ *are* $\mathcal{O}$*-equivariant isomorphisms respecting the canonical filtrations on both sides, and* $\nu_n:((\mathbb{Z}/n\mathbb{Z})(1))_S\xrightarrow{\sim}\boldsymbol{\mu}_{n,S}$ *is an isomorphism, which are compatible with the canonical morphisms* $L\hookrightarrow L^\#$ *and* $\lambda:G[n]\to G^\vee[n]$, *and induce on the graded pieces the given data* $\varphi_{-2,n}$, $\varphi_{-1,n}$, *and* $\varphi_{0,n}$. *Moreover,* β_n *satisfies the liftability condition that, for each integer* $m\geq1$ *such that* $n|m$, *there exists a finite étale covering of* S *over which there exists an analogous triple* β_m *lifting the pullback of* β_n.

Proof. The statements are self-explanatory. □

Proposition 1.3.2.31. (*Compare with Proposition* 1.3.2.12.) *The torus torsor* $S := \Xi_{\Phi_{\mathcal{H}},\delta_{\mathcal{H}}} \to C_{\Phi_{\mathcal{H}},\delta_{\mathcal{H}}}$ *is universal for the additional structure* $\tau_{\mathcal{H}}$ *satisfying certain symplectic and liftability conditions, which can be interpreted as parameterizing tuples*

$$(G^\natural, \lambda^\natural : G^\natural \to G^{\vee,\natural}, i^\natural, \tau, \beta_{\mathcal{H}}),$$

where $G^\natural$, $G^{\vee,\natural}$, $\lambda^\natural$, *and* $i^\natural$ *are as in Lemma* 1.3.2.11, *where* τ *is as in Lemma* 1.3.2.28, *and where* $\beta_{\mathcal{H}}$ *is a* **level-$\mathcal{H}$ structure** *of* $(G^\natural, \lambda^\natural, i^\natural, \tau)$ *of type* $(L \underset{\mathbb{Z}}{\otimes} \hat{\mathbb{Z}}, \langle\,\cdot\,,\,\cdot\,\rangle, \mathsf{Z})$, *which is a collection* $\{\beta_{H_n}\}_n$, *where* $n \geq 1$ *runs over integers such that* $\mathcal{U}(n) \subset \mathcal{H}$, *such that each* β_{H_n} *(where* $H_n := \mathcal{H}/\mathcal{U}(n)$*) is a subscheme of*

$$\coprod \Big(\underline{\mathrm{Isom}}_S((L/nL)_S, G[n]) \underset{S}{\times} \underline{\mathrm{Isom}}_S((L^\#/nL^\#)_S, G^\vee[n]) \underset{S}{\times} \underline{\mathrm{Isom}}_S(((\mathbb{Z}/n\mathbb{Z})(1))_S, \boldsymbol{\mu}_{n,S}) \Big)$$

over S, *where the disjoint union is over representatives* $(\mathsf{Z}_n, \Phi_n, \delta_n)$ *(with the same* (X, Y, ϕ)*) in* $(\mathsf{Z}_{\mathcal{H}}, \Phi_{\mathcal{H}}, \delta_{\mathcal{H}})$, *that becomes the disjoint union of all elements in the* H_n*-orbit of some principal level-*n *structure* β_n *of* $(G^\natural, \lambda^\natural, i^\natural, \tau)$ *of type* $(L \underset{\mathbb{Z}}{\otimes} \hat{\mathbb{Z}}, \langle\,\cdot\,,\,\cdot\,\rangle, \mathsf{Z})$, *as in Lemma* 1.3.2.28, *for any* Z *lifting* Z_n*; and where* β_{H_m} *is mapped to* β_{H_n} *(under the canonical morphism, which we omit for simplicity) when* $n|m$.

Let $\mathbf{S}_{\Phi_{\mathcal{H}}}$ *be the unique lattice in* $\mathbf{S}_{\Phi_1} \underset{\mathbb{Z}}{\otimes} \mathbb{Q}$ *such that* $\mathbf{S}^\vee_{\Phi_1}/\mathbf{S}^\vee_{\Phi_{\mathcal{H}}} \cong \mathrm{U}_{2,\mathsf{Z}}(\hat{\mathbb{Z}})/\mathcal{H}_{\mathrm{U}_{2,\mathsf{Z}}}$. *Then* $S = \Xi_{\Phi_{\mathcal{H}},\delta_{\mathcal{H}}} \to C_{\Phi_{\mathcal{H}},\delta_{\mathcal{H}}}$ *is torsor under the split torus* $E_{\Phi_{\mathcal{H}}}$ *with character group* $\mathbf{S}_{\Phi_{\mathcal{H}}}$, *equipped a homomorphism*

$$\mathbf{S}_{\Phi_{\mathcal{H}}} \to \underline{\mathrm{Pic}}(C_{\Phi_{\mathcal{H}},\delta_{\mathcal{H}}}) : \ell \mapsto \Psi_{\Phi_{\mathcal{H}},\delta_{\mathcal{H}}}(\ell)$$

(by the torus torsor structure; see [62, Prop. 6.2.4.7 and (6.2.4.8); see also the errata]), assigning to each $\ell \in \mathbf{S}_{\Phi_{\mathcal{H}}}$ *an invertible sheaf* $\Psi_{\Phi_{\mathcal{H}},\delta_{\mathcal{H}}}(\ell)$ *over* $C_{\Phi_{\mathcal{H}},\delta_{\mathcal{H}}}$ *(up to isomorphism), together with isomorphisms*

$$\Delta^*_{\Phi_{\mathcal{H}},\delta_{\mathcal{H}},\ell,\ell'} : \Psi_{\Phi_{\mathcal{H}},\delta_{\mathcal{H}}}(\ell) \underset{\mathscr{O}_{C_{\Phi_{\mathcal{H}},\delta_{\mathcal{H}}}}}{\otimes} \Psi_{\Phi_{\mathcal{H}},\delta_{\mathcal{H}}}(\ell') \xrightarrow{\sim} \Psi_{\Phi_{\mathcal{H}},\delta_{\mathcal{H}}}(\ell + \ell')$$

for all $\ell, \ell' \in \mathbf{S}_{\Phi_{\mathcal{H}}}$, *satisfying the necessary compatibilities with each other making* $\underset{\ell \in \mathbf{S}_{\Phi_{\mathcal{H}}}}{\oplus} \Psi_{\Phi_{\mathcal{H}},\delta_{\mathcal{H}}}(\ell)$ *an* $\mathscr{O}_{C_{\Phi_{\mathcal{H}},\delta_{\mathcal{H}}}}$*-algebra, such that*

$$\Xi_{\Phi_{\mathcal{H}},\delta_{\mathcal{H}}} \cong \underline{\mathrm{Spec}}_{\mathscr{O}_{C_{\Phi_{\mathcal{H}},\delta_{\mathcal{H}}}}} \Big(\underset{\ell \in \mathbf{S}_{\Phi_{\mathcal{H}}}}{\oplus} \Psi_{\Phi_{\mathcal{H}},\delta_{\mathcal{H}}}(\ell) \Big).$$

When $\ell = [y \otimes \chi]$ *for some* $y \in Y$ *and* $\chi \in X$, *we have a canonical isomorphism*

$$\Psi_{\Phi_{\mathcal{H}},\delta_{\mathcal{H}}}(\ell) \cong (c^\vee(y), c(\chi))^* \mathcal{P}_B.$$

Proof. This follows from the construction of $\Xi_{\Phi_{\mathcal{H}},\delta_{\mathcal{H}}} \to C_{\Phi_{\mathcal{H}},\delta_{\mathcal{H}}}$ as a quotient of $\coprod \Xi_{\Phi_n,\delta_n} \to \coprod C_{\Phi_n,\delta_n}$ (over the same index set). □

For each rational polyhedral cone $\sigma \subset (\mathbf{S}_{\Phi_{\mathcal{H}}})^{\vee}_{\mathbb{R}}$ as in Definition 1.2.2.2, we have an affine toroidal embedding

$$\Xi_{\Phi_{\mathcal{H}},\delta_{\mathcal{H}}} \hookrightarrow \Xi_{\Phi_{\mathcal{H}},\delta_{\mathcal{H}}}(\sigma) := \underline{\mathrm{Spec}}_{\mathscr{O}_{C_{\Phi_{\mathcal{H}},\delta_{\mathcal{H}}}}}\left(\bigoplus_{\ell\in\sigma^{\vee}} \Psi_{\Phi_{\mathcal{H}},\delta_{\mathcal{H}}}(\ell)\right), \tag{1.3.2.32}$$

both sides being relative affine over $C_{\Phi_{\mathcal{H}},\delta_{\mathcal{H}}}$, where $\Xi_{\Phi_{\mathcal{H}},\delta_{\mathcal{H}}}(\sigma) \to C_{\Phi_{\mathcal{H}},\delta_{\mathcal{H}}}$ is smooth when the cone σ is smooth, with a closed subalgebraic stack defined by

$$\Xi_{\Phi_{\mathcal{H}},\delta_{\mathcal{H}},\sigma} := \underline{\mathrm{Spec}}_{\mathscr{O}_{C_{\Phi_{\mathcal{H}},\delta_{\mathcal{H}}}}}\left(\bigoplus_{\ell\in\sigma^{\perp}} \Psi_{\Phi_{\mathcal{H}},\delta_{\mathcal{H}}}(\ell)\right), \tag{1.3.2.33}$$

which we call the σ-stratum (*cf.* [62, Def. 6.1.2.7]), which is by itself a torsor under the torus $E_{\Phi_{\mathcal{H}},\sigma}$ with character group $\sigma^{\perp}$. For each $\Gamma_{\Phi_{\mathcal{H}}}$-admissible rational polyhedral cone decomposition $\Sigma_{\Phi_{\mathcal{H}}}$ as in Definition 1.2.2.4, we have a toroidal embedding

$$\Xi_{\Phi_{\mathcal{H}},\delta_{\mathcal{H}}} \hookrightarrow \overline{\Xi}_{\Phi_{\mathcal{H}},\delta_{\mathcal{H}}} = \overline{\Xi}_{\Phi_{\mathcal{H}},\delta_{\mathcal{H}},\Sigma_{\Phi_{\mathcal{H}}}}, \tag{1.3.2.34}$$

the right-hand side being only locally of finite type over $C_{\Phi_{\mathcal{H}},\delta_{\mathcal{H}}}$, with an open covering

$$\overline{\Xi}_{\Phi_{\mathcal{H}},\delta_{\mathcal{H}}} = \bigcup_{\sigma\in\Sigma_{\Phi_{\mathcal{H}}}} \Xi_{\Phi_{\mathcal{H}},\delta_{\mathcal{H}}}(\sigma), \tag{1.3.2.35}$$

inducing a stratification

$$\overline{\Xi}_{\Phi_{\mathcal{H}},\delta_{\mathcal{H}}} = \coprod_{\sigma\in\Sigma_{\Phi_{\mathcal{H}}}} \Xi_{\Phi_{\mathcal{H}},\delta_{\mathcal{H}},\sigma}. \tag{1.3.2.36}$$

(The notation "$\coprod$" only means a set-theoretic disjoint union. The algebro-geometric structure is still the one inherited from $\overline{\Xi}_{\Phi_{\mathcal{H}},\delta_{\mathcal{H}}}$.) Concretely, if σ is a face of ρ, then $\rho^{\vee} \subset \sigma^{\vee}$ and $\Xi_{\Phi_{\mathcal{H}},\delta_{\mathcal{H}}}(\sigma) \subset \Xi_{\Phi_{\mathcal{H}},\delta_{\mathcal{H}}}(\rho)$, but $\Xi_{\Phi_{\mathcal{H}},\delta_{\mathcal{H}},\rho}$ is contained in the closure of $\Xi_{\Phi_{\mathcal{H}},\delta_{\mathcal{H}},\sigma}$. The closure of $\Xi_{\Phi_{\mathcal{H}},\delta_{\mathcal{H}},\sigma}$ in $\Xi_{\Phi_{\mathcal{H}},\delta_{\mathcal{H}}}(\rho)$ is

$$\Xi_{\Phi_{\mathcal{H}},\delta_{\mathcal{H}},\sigma}(\rho) := \underline{\mathrm{Spec}}_{\mathscr{O}_{C_{\Phi_{\mathcal{H}},\delta_{\mathcal{H}}}}}\left(\bigoplus_{\ell\in\sigma^{\perp}\cap\rho^{\vee}} \Psi_{\Phi_{\mathcal{H}},\delta_{\mathcal{H}}}(\ell)\right). \tag{1.3.2.37}$$

In this case, the open embedding

$$\Xi_{\Phi_{\mathcal{H}},\delta_{\mathcal{H}},\sigma} \hookrightarrow \Xi_{\Phi_{\mathcal{H}},\delta_{\mathcal{H}},\sigma}(\rho) \tag{1.3.2.38}$$

is an affine toroidal embedding (as in [62, Def. 6.1.2.3]) for the torus torsor $\Xi_{\Phi_{\mathcal{H}},\delta_{\mathcal{H}},\sigma} \to C_{\Phi_{\mathcal{H}},\delta_{\mathcal{H}}}$. Let

$$\mathfrak{X}_{\Phi_{\mathcal{H}},\delta_{\mathcal{H}},\sigma} := (\Xi_{\Phi_{\mathcal{H}},\delta_{\mathcal{H}}}(\sigma))^{\wedge}_{\Xi_{\Phi_{\mathcal{H}},\delta_{\mathcal{H}},\sigma}}, \tag{1.3.2.39}$$

the formal completion of $\Xi_{\Phi_{\mathcal{H}},\delta_{\mathcal{H}}}(\sigma)$ along its σ-stratum $\Xi_{\Phi_{\mathcal{H}},\delta_{\mathcal{H}},\sigma}$. When $\sigma \subset \mathbf{P}^{+}_{\Phi_{\mathcal{H}}}$ appears in $\Sigma_{\Phi_{\mathcal{H}}} \in \Sigma$, the quotient $\mathfrak{X}_{\Phi_{\mathcal{H}},\delta_{\mathcal{H}},\sigma}/\Gamma_{\Phi_{\mathcal{H}},\sigma}$ is isomorphic to the formal completion of $\mathsf{M}^{\mathrm{tor}}_{\mathcal{H},\Sigma}$ along its $[(\Phi_{\mathcal{H}},\delta_{\mathcal{H}},\sigma)]$-stratum $\mathsf{Z}_{[(\Phi_{\mathcal{H}},\delta_{\mathcal{H}},\sigma)]} \cong \Xi_{\Phi_{\mathcal{H}},\delta_{\mathcal{H}},\sigma}/\Gamma_{\Phi_{\mathcal{H}},\sigma}$, as in Theorem 1.3.1.3. If there is a surjection $(\mathsf{Z}'_{\mathcal{H}},\Phi'_{\mathcal{H}},\delta'_{\mathcal{H}}) \twoheadrightarrow (\mathsf{Z}_{\mathcal{H}},\Phi_{\mathcal{H}},\delta_{\mathcal{H}})$ such that σ is mapped to a face of a cone $\rho \subset \mathbf{P}^{+}_{\Phi'_{\mathcal{H}}}$ under the canonical mapping $\mathbf{P}^{+}_{\Phi_{\mathcal{H}}} \to \mathbf{P}_{\Phi'_{\mathcal{H}}}$, and if $\rho \in \Sigma_{\Phi'_{\mathcal{H}}} \in \Sigma$, then $\mathsf{Z}_{[(\Phi'_{\mathcal{H}},\delta'_{\mathcal{H}},\rho)]}$ is contained in the closure $\overline{\mathsf{Z}}_{[(\Phi_{\mathcal{H}},\delta_{\mathcal{H}},\sigma)]}$

of $\mathsf{Z}_{[(\Phi_{\mathcal{H}},\delta_{\mathcal{H}},\sigma)]}$ in $\mathsf{M}^{\mathrm{tor}}_{\mathcal{H},\Sigma}$, and the completion of $\overline{\mathsf{Z}}_{[(\Phi_{\mathcal{H}},\delta_{\mathcal{H}},\sigma)]}$ along $\mathsf{Z}_{[(\Phi'_{\mathcal{H}},\delta'_{\mathcal{H}},\rho)]}$ is canonically isomorphic to

$$\mathfrak{X}_{\Phi_{\mathcal{H}},\delta_{\mathcal{H}},\sigma,\rho} := (\Xi_{\Phi_{\mathcal{H}},\delta_{\mathcal{H}},\sigma}(\rho))^{\wedge}_{\Xi_{\Phi_{\mathcal{H}},\delta_{\mathcal{H}},\rho}}, \tag{1.3.2.40}$$

the formal completion of $\Xi_{\Phi_{\mathcal{H}},\delta_{\mathcal{H}},\sigma}(\rho)$ along its ρ-stratum $\Xi_{\Phi_{\mathcal{H}},\delta_{\mathcal{H}},\rho}$.

Lemma 1.3.2.41. *Let $\mathfrak{X}_{\Phi_{\mathcal{H}},\delta_{\mathcal{H}}} = \mathfrak{X}_{\Phi_{\mathcal{H}},\delta_{\mathcal{H}},\Sigma_{\Phi_{\mathcal{H}}}}$ be the formal completion of $\overline{\Xi}_{\Phi_{\mathcal{H}},\delta_{\mathcal{H}}}$ along the union of the σ-strata $\Xi_{\Phi_{\mathcal{H}},\delta_{\mathcal{H}},\sigma}$ for $\sigma \in \Sigma_{\Phi_{\mathcal{H}}}$ and $\sigma \subset \mathbf{P}^{+}_{\Phi_{\mathcal{H}}}$. Then we have a canonical morphism*

$$\mathfrak{X}_{\Phi_{\mathcal{H}},\delta_{\mathcal{H}}} \to \mathsf{M}^{\mathrm{tor}}_{\mathcal{H},\Sigma} \tag{1.3.2.42}$$

inducing a canonical isomorphism

$$\mathfrak{X}_{\Phi_{\mathcal{H}},\delta_{\mathcal{H}}}/\Gamma_{\Phi_{\mathcal{H}}} \xrightarrow{\sim} (\mathsf{M}^{\mathrm{tor}}_{\mathcal{H},\Sigma})^{\wedge}_{\cup \mathsf{Z}_{[(\Phi_{\mathcal{H}},\delta_{\mathcal{H}},\sigma)]}}, \tag{1.3.2.43}$$

where $\cup \mathsf{Z}_{[(\Phi_{\mathcal{H}},\delta_{\mathcal{H}},\sigma)]}$ is the union of all strata $\mathsf{Z}_{[(\Phi_{\mathcal{H}},\delta_{\mathcal{H}},\sigma)]}$ with $\sigma \in \Sigma_{\Phi_{\mathcal{H}}}$ (and $\sigma \subset \mathbf{P}^{+}_{\Phi_{\mathcal{H}}}$), under which the pullback of $\underline{\mathrm{Lie}}^{\vee}_{G/\mathsf{M}^{\mathrm{tor}}_{\mathcal{H},\Sigma}}$ (resp. $\underline{\mathrm{Lie}}^{\vee}_{G^{\vee}/\mathsf{M}^{\mathrm{tor}}_{\mathcal{H},\Sigma}}$, resp. $\lambda^{*} : \underline{\mathrm{Lie}}^{\vee}_{G^{\vee}/\mathsf{M}^{\mathrm{tor}}_{\mathcal{H},\Sigma}} \to \underline{\mathrm{Lie}}^{\vee}_{G/\mathsf{M}^{\mathrm{tor}}_{\mathcal{H},\Sigma}}$) can be canonically identified with the pullback of $\underline{\mathrm{Lie}}^{\vee}_{G^{\natural}/C_{\Phi_{\mathcal{H}},\delta_{\mathcal{H}}}}$ (resp. $\underline{\mathrm{Lie}}^{\vee}_{G^{\vee,\natural}/C_{\Phi_{\mathcal{H}},\delta_{\mathcal{H}}}}$, resp. $(\lambda^{\natural})^{*} : \underline{\mathrm{Lie}}^{\vee}_{G^{\vee,\natural}/C_{\Phi_{\mathcal{H}},\delta_{\mathcal{H}}}} \to \underline{\mathrm{Lie}}^{\vee}_{G^{\natural}/C_{\Phi_{\mathcal{H}},\delta_{\mathcal{H}}}}$). For each stratum $\mathsf{Z}_{[(\Phi_{\mathcal{H}},\delta_{\mathcal{H}},\sigma)]}$, the isomorphism (1.3.2.43) is compatible with the isomorphism $\mathfrak{X}_{\Phi_{\mathcal{H}},\delta_{\mathcal{H}},\sigma}/\Gamma_{\Phi_{\mathcal{H}},\sigma} \xrightarrow{\sim} (\mathsf{M}^{\mathrm{tor}}_{\mathcal{H},\Sigma})^{\wedge}_{\mathsf{Z}_{[(\Phi_{\mathcal{H}},\delta_{\mathcal{H}},\sigma)]}}$ in (5) of Theorem 1.3.1.3 (under the canonical morphisms $\mathfrak{X}_{\Phi_{\mathcal{H}},\delta_{\mathcal{H}},\sigma}/\Gamma_{\Phi_{\mathcal{H}},\sigma} \to \mathfrak{X}_{\Phi_{\mathcal{H}},\delta_{\mathcal{H}}}/\Gamma_{\Phi_{\mathcal{H}}}$ and $(\mathsf{M}^{\mathrm{tor}}_{\mathcal{H},\Sigma})^{\wedge}_{\mathsf{Z}_{[(\Phi_{\mathcal{H}},\delta_{\mathcal{H}},\sigma)]}} \to (\mathsf{M}^{\mathrm{tor}}_{\mathcal{H},\Sigma})^{\wedge}_{\cup \mathsf{Z}_{[(\Phi_{\mathcal{H}},\delta_{\mathcal{H}},\sigma)]}}$). (Such isomorphisms are induced by strata-preserving isomorphisms from étale neighborhoods of points of $\Xi_{\Phi_{\mathcal{H}},\delta_{\mathcal{H}},\sigma}$ in $\Xi_{\Phi_{\mathcal{H}},\delta_{\mathcal{H}}}(\sigma)$ to étale neighborhoods of points of $\mathsf{Z}_{[(\Phi_{\mathcal{H}},\delta_{\mathcal{H}},\sigma)]}$ in $\mathsf{M}^{\mathrm{tor}}_{\mathcal{H},\Sigma}$.)

Proof. The formal algebraic stack $\mathfrak{X}_{\Phi_{\mathcal{H}},\delta_{\mathcal{H}}}$ admits an open covering by open formal algebraic substacks $\mathfrak{U}_{\sigma}$, where each $\mathfrak{U}_{\sigma}$ is the formal completion of the smooth algebraic stack $\Xi_{\Phi_{\mathcal{H}},\delta_{\mathcal{H}}}(\sigma)$ along its closed sub-algebraic stack formed by the union of $\Xi_{\Phi_{\mathcal{H}},\delta_{\mathcal{H}},\tau}$ such that τ is a face of σ (which can be σ itself), $\tau \in \Sigma_{\Phi_{\mathcal{H}}}$, and $\tau \subset \mathbf{P}^{+}_{\Phi_{\mathcal{H}}}$. Using the Mumford family carried by $\mathfrak{X}_{\Phi_{\mathcal{H}},\delta_{\mathcal{H}}}$ (see [62, Sec. 6.2.5]), by (6) of Theorem 1.3.1.3, three exist morphisms $\mathfrak{U}_{\sigma} \to \mathsf{M}^{\mathrm{tor}}_{\mathcal{H},\Sigma}$ which patch together and form the desired canonical morphism (1.3.2.42), which is unchanged under the canonical action of $\Gamma_{\Phi_{\mathcal{H}}}$ and hence factors through a canonical morphism

$$\mathfrak{X}_{\Phi_{\mathcal{H}},\delta_{\mathcal{H}}}/\Gamma_{\Phi_{\mathcal{H}}} \to (\mathsf{M}^{\mathrm{tor}}_{\mathcal{H},\Sigma})^{\wedge}_{\cup \mathsf{Z}_{[(\Phi_{\mathcal{H}},\delta_{\mathcal{H}},\sigma)]}}. \tag{1.3.2.44}$$

On the other hand, by the construction of $\mathsf{M}^{\mathrm{tor}}_{\mathcal{H},\Sigma}$ by gluing good algebraic models (see [62, Sec. 6.3]) in the étale topology, the pullback of the tautological object $(G, \lambda, i, \alpha_{\mathcal{H}}) \to \mathsf{M}^{\mathrm{tor}}_{\mathcal{H},\Sigma}$ to $(\mathsf{M}^{\mathrm{tor}}_{\mathcal{H},\Sigma})^{\wedge}_{\cup \mathsf{Z}_{[(\Phi_{\mathcal{H}},\delta_{\mathcal{H}},\sigma)]}}$ define degeneration data parameterized by $\mathfrak{X}_{\Phi_{\mathcal{H}},\delta_{\mathcal{H}}}/\Gamma_{\Phi_{\mathcal{H}}}$, and hence there is a canonical morphism giving the inverse of (1.3.2.44) and induces the canonical isomorphism (1.3.2.43). (This explains the last parenthetical remark in the statement of the lemma, because the good algebraic models carry approximations of the degeneration data, which include in particular trivializations of the invertible sheaves $\Psi_{\Phi_{\mathcal{H}},\delta_{\mathcal{H}}}(\ell)$, which determine the stratifications.) Moreover, since the pullbacks of $(G, G^\vee, \lambda) \to \mathsf{M}^{\mathrm{tor}}_{\mathcal{H},\Sigma}$ and $(G^\natural, G^{\vee,\natural}, \lambda^\natural) \to C_{\Phi_{\mathcal{H}},\delta_{\mathcal{H}}}$ induce canonically isomorphic formal completions $(G_{\mathrm{for}}, G^\vee_{\mathrm{for}}, \lambda_{\mathrm{for}}) \to \mathfrak{U}_\sigma$ and $(G^\natural_{\mathrm{for}}, G^{\vee,\natural}_{\mathrm{for}}, \lambda^\natural_{\mathrm{for}}) \to \mathfrak{U}_\sigma$ (by the theory of degeneration) over each $\mathfrak{U}_\sigma$, the pullback of $\underline{\mathrm{Lie}}^\vee_{G/\mathsf{M}^{\mathrm{tor}}_{\mathcal{H},\Sigma}}$ (resp. $\underline{\mathrm{Lie}}^\vee_{G^\vee/\mathsf{M}^{\mathrm{tor}}_{\mathcal{H},\Sigma}}$, resp. $\lambda^* : \underline{\mathrm{Lie}}^\vee_{G^\vee/\mathsf{M}^{\mathrm{tor}}_{\mathcal{H},\Sigma}} \to \underline{\mathrm{Lie}}^\vee_{G/\mathsf{M}^{\mathrm{tor}}_{\mathcal{H},\Sigma}}$) can be canonically identified with the pullback of $\underline{\mathrm{Lie}}^\vee_{G^\natural/C_{\Phi_{\mathcal{H}},\delta_{\mathcal{H}}}}$ (resp. $\underline{\mathrm{Lie}}^\vee_{G^{\vee,\natural}/C_{\Phi_{\mathcal{H}},\delta_{\mathcal{H}}}}$, resp. $(\lambda^\natural)^* : \underline{\mathrm{Lie}}^\vee_{G^{\vee,\natural}/C_{\Phi_{\mathcal{H}},\delta_{\mathcal{H}}}} \to \underline{\mathrm{Lie}}^\vee_{G^\natural/C_{\Phi_{\mathcal{H}},\delta_{\mathcal{H}}}}$) under (1.3.2.43). Since (1.3.2.43) and the canonical isomorphism in (5) of Theorem 1.3.1.3 are both defined by the universal properties given in terms of degeneration data, they are naturally compatible with each other. □

Proposition 1.3.2.45. (*Compare with Propositions* 1.3.1.15 *and* 1.3.2.24.) *By considering compatible $\mathbb{Q}^\times$-isogenies $(f : G^\natural \to G^{\natural,\prime}, f^\vee : G^{\vee,\natural,\prime} \to G^{\vee,\natural})$ compatible with the homomorphisms $(\iota : Y \to G^\natural, \iota^\vee : X \to G^{\vee,\natural})$ inducing isomorphisms on the torus parts T and $T^\vee$ and on the domains of ι and $\iota^\vee$, we can define Hecke twists of the tautological object $(G^\natural, \lambda^\natural, i^\natural, \tau, \beta_{\mathcal{H}}) \to \Xi_{\Phi_{\mathcal{H}},\delta_{\mathcal{H}}}$ by elements of $\mathrm{P}'_{\mathsf{Z}}(\mathbb{A}^\infty)$, and define the Hecke action of $\mathrm{P}'_{\mathsf{Z}}(\mathbb{A}^\infty)$ on the collection $\{\Xi_{\Phi_{\mathcal{H}},\delta_{\mathcal{H}}}\}_{\mathcal{H}_{\mathrm{P}'_{\mathsf{Z}}}}$, realized by finite étale surjections pulling tautological objects back to Hecke twists, which is compatible with the Hecke action of $\mathrm{G}_{1,\mathsf{Z}}(\mathbb{A}^\infty)$ on the collection $\{C_{\Phi_{\mathcal{H}},\delta_{\mathcal{H}}}\}_{\mathcal{H}_{\mathrm{G}_{1,\mathsf{Z}}}}$ under the canonical morphisms $\Xi_{\Phi_{\mathcal{H}},\delta_{\mathcal{H}}} \to C_{\Phi_{\mathcal{H}},\delta_{\mathcal{H}}}$ (with varying $\mathcal{H}$) and the canonical homomorphism $\mathrm{P}'_{\mathsf{Z}}(\mathbb{A}^\infty) \to \mathrm{G}_{1,\mathsf{Z}}(\mathbb{A}^\infty) = \mathrm{P}'_{\mathsf{Z}}(\mathbb{A}^\infty)/\mathrm{U}_{2,\mathsf{Z}}(\mathbb{A}^\infty)$.*

By also considering $\mathbb{Q}^\times$-isogenies $(f : G^\natural \to G^{\natural,\prime}, f^\vee : G^{\vee,\natural,\prime} \to G^{\vee,\natural})$ compatible with the homomorphisms $(\iota : Y \to G^\natural, \iota^\vee : X \to G^{\vee,\natural})$ inducing $\mathbb{Q}^\times$-isogenies on the torus parts T and $T^\vee$ and on the domains of ι and $\iota^\vee$ (possibly varying the isomorphism classes of the $\mathcal{O}$-lattices X and Y), we can also define Hecke twists of the tautological object $(G^\natural, \lambda^\natural, i^\natural, \tau, \beta_{\mathcal{H}}) \to C_{\Phi_{\mathcal{H}},\delta_{\mathcal{H}}}$ by elements of $\mathrm{P}_{\mathsf{Z}}(\mathbb{A}^\infty)$, and define the Hecke action of $\mathrm{P}_{\mathsf{Z}}(\mathbb{A}^\infty)$ on the collection $\{\coprod \Xi_{\Phi_{\mathcal{H}},\delta_{\mathcal{H}}}\}_{\mathcal{H}_{\mathrm{P}_{\mathsf{Z}}}}$, realized by finite étale surjections pulling tautological objects back to Hecke twists, where the disjoint unions are over classes $[(\mathsf{Z}_{\mathcal{H}}, \Phi_{\mathcal{H}}, \delta_{\mathcal{H}})]$ sharing the same $\mathsf{Z}_{\mathcal{H}}$, which induces an action of $\mathrm{G}'_{l,\mathsf{Z}}(\mathbb{A}^\infty) = \mathrm{P}_{\mathsf{Z}}(\mathbb{A}^\infty)/\mathrm{P}'_{\mathsf{Z}}(\mathbb{A}^\infty)$ on the index sets $\{[(\mathsf{Z}_{\mathcal{H}}, \Phi_{\mathcal{H}}, \delta_{\mathcal{H}})]\}$, which is compatible with the Hecke action of $\mathrm{P}_{\mathsf{Z}}(\mathbb{A}^\infty)/\mathrm{U}_{2,\mathsf{Z}}(\mathbb{A}^\infty)$ on the collection $\{\coprod C_{\Phi_{\mathcal{H}},\delta_{\mathcal{H}}}\}_{\mathcal{H}_{\mathrm{P}_{\mathsf{Z}}}/\mathcal{H}_{\mathrm{U}_{2,\mathsf{Z}}}}$ (with the same index sets and the same induced action of $\mathrm{G}'_{l,\mathsf{Z}}(\mathbb{A}^\infty)$) under the canonical morphisms $\Xi_{\Phi_{\mathcal{H}},\delta_{\mathcal{H}}} \to C_{\Phi_{\mathcal{H}},\delta_{\mathcal{H}}}$ (with varying $\mathcal{H}$) and the canonical homomorphism $\mathrm{P}_{\mathsf{Z}}(\mathbb{A}^\infty) \to \mathrm{P}_{\mathsf{Z}}(\mathbb{A}^\infty)/\mathrm{U}_{2,\mathsf{Z}}(\mathbb{A}^\infty)$.

Any such Hecke action

$$[g] : \Xi_{\Phi'_{\mathcal{H}'},\delta'_{\mathcal{H}'}} \to \Xi_{\Phi_{\mathcal{H}},\delta_{\mathcal{H}}}$$

covering $[g] : C_{\Phi'_{\mathcal{H}'},\delta'_{\mathcal{H}'}} \to C_{\Phi_{\mathcal{H}},\delta_{\mathcal{H}}}$ induces a (finite étale) morphism

$$\Xi_{\Phi'_{\mathcal{H}'},\delta'_{\mathcal{H}'}} \to \Xi_{\Phi_{\mathcal{H}},\delta_{\mathcal{H}}} \underset{C_{\Phi_{\mathcal{H}},\delta_{\mathcal{H}}}}{\times} C_{\Phi'_{\mathcal{H}'},\delta'_{\mathcal{H}'}}$$

between torus torsors over $C_{\Phi'_{\mathcal{H}'},\delta'_{\mathcal{H}'}}$, which is equivariant with the morphism $E_{\Phi'_{\mathcal{H}'}} \to E_{\Phi_{\mathcal{H}}}$ dual to the homomorphism $\mathbf{S}_{\Phi_{\mathcal{H}}} \to \mathbf{S}_{\Phi'_{\mathcal{H}'}}$ induced by the pair of mor-

phisms $(f_X : X \underset{\mathbb{Z}}{\otimes} \mathbb{Q} \xrightarrow{\sim} X' \underset{\mathbb{Z}}{\otimes} \mathbb{Q}, f_Y : Y' \underset{\mathbb{Z}}{\otimes} \mathbb{Q} \xrightarrow{\sim} Y \underset{\mathbb{Z}}{\otimes} \mathbb{Q})$ *defining the* g*-assignment* $(\mathsf{Z}'_{\mathcal{H}'}, \Phi'_{\mathcal{H}'}, \delta'_{\mathcal{H}'}) \to_g (\mathsf{Z}_{\mathcal{H}}, \Phi_{\mathcal{H}}, \delta_{\mathcal{H}})$ *of cusp labels (cf. [62, Def. 5.4.3.9]).*

If $g \in \mathrm{P}_{\mathsf{Z}}(\mathbb{A}^\infty)$ *is as above and if* $(\Phi'_{\mathcal{H}'}, \delta'_{\mathcal{H}'}, \rho)$ *is a* g*-refinement of* $(\Phi_{\mathcal{H}}, \delta_{\mathcal{H}}, \sigma)$ *as in [62, Def. 6.4.3.1], then there is a canonical morphism*

$$[g] : \Xi_{\Phi'_{\mathcal{H}'}, \delta'_{\mathcal{H}'}}(\rho) \to \Xi_{\Phi_{\mathcal{H}}, \delta_{\mathcal{H}}}(\sigma) \tag{1.3.2.46}$$

covering $[g] : C_{\Phi'_{\mathcal{H}'}, \delta'_{\mathcal{H}'}} \to C_{\Phi_{\mathcal{H}}, \delta_{\mathcal{H}}}$*, extending* $[g] : \Xi_{\Phi'_{\mathcal{H}'}, \delta'_{\mathcal{H}'}} \to \Xi_{\Phi_{\mathcal{H}}, \delta_{\mathcal{H}}}$*, mapping* $\Xi_{\Phi'_{\mathcal{H}'}, \delta'_{\mathcal{H}'}, \rho}$ *to* $\Xi_{\Phi_{\mathcal{H}}, \delta_{\mathcal{H}}, \sigma}$*, and inducing a canonical morphism*

$$[g] : \mathfrak{X}_{\Phi'_{\mathcal{H}'}, \delta'_{\mathcal{H}'}, \rho} \to \mathfrak{X}_{\Phi_{\mathcal{H}}, \delta_{\mathcal{H}}, \sigma}. \tag{1.3.2.47}$$

If $g \in \mathrm{P}_{\mathsf{Z}}(\mathbb{A}^\infty)$ *is as above and if* $(\Phi'_{\mathcal{H}'}, \delta'_{\mathcal{H}'}, \Sigma'_{\Phi'_{\mathcal{H}'}})$ *is a* g*-refinement of* $(\Phi_{\mathcal{H}}, \delta_{\mathcal{H}}, \Sigma_{\Phi_{\mathcal{H}}})$ *as in [62, Def. 6.4.3.2], then morphisms like* (1.3.2.46) *patch together and define a canonical morphism*

$$[g] : \overline{\Xi}_{\Phi'_{\mathcal{H}'}, \delta'_{\mathcal{H}'}, \Sigma'_{\Phi'_{\mathcal{H}'}}} \to \overline{\Xi}_{\Phi_{\mathcal{H}}, \delta_{\mathcal{H}}, \Sigma_{\Phi_{\mathcal{H}}}} \tag{1.3.2.48}$$

covering $[g] : C_{\Phi'_{\mathcal{H}'}, \delta'_{\mathcal{H}'}} \to C_{\Phi_{\mathcal{H}}, \delta_{\mathcal{H}}}$*, extending* $[g] : \Xi_{\Phi'_{\mathcal{H}'}, \delta'_{\mathcal{H}'}} \to \Xi_{\Phi_{\mathcal{H}}, \delta_{\mathcal{H}}}$*, and inducing a canonical morphism*

$$[g] : \mathfrak{X}_{\Phi'_{\mathcal{H}'}, \delta'_{\mathcal{H}'}, \Sigma'_{\Phi'_{\mathcal{H}'}}} \to \mathfrak{X}_{\Phi_{\mathcal{H}}, \delta_{\mathcal{H}}, \Sigma_{\Phi_{\mathcal{H}}}} \tag{1.3.2.49}$$

compatible with each (1.3.2.47) *as above (under canonical morphisms).*

If $g \in \mathrm{P}_{\mathsf{Z}}(\mathbb{A}^\infty)$ *and if we have a collection* Σ' *for* $\mathsf{M}_{\mathcal{H}'}$ *that is a* g*-refinement of a collection* Σ *for* $\mathsf{M}_{\mathcal{H}}$ *as in [62, Def. 6.4.3.3], then the canonical morphism*

$$[g]^{\mathrm{tor}} : \mathsf{M}^{\mathrm{tor}}_{\mathcal{H}', \Sigma'} \to \mathsf{M}^{\mathrm{tor}}_{\mathcal{H}, \Sigma}$$

as in Proposition 1.3.1.15 *is compatible with* (1.3.2.47) *when* $(\Phi'_{\mathcal{H}'}, \delta'_{\mathcal{H}'}, \rho)$ *is a* g*-refinement of* $(\Phi_{\mathcal{H}}, \delta_{\mathcal{H}}, \sigma)$*, under the canonical isomorphisms as in* (5) *of Theorem* 1.3.1.3*; and is compatible with* (1.3.2.49) *when* $(\Phi'_{\mathcal{H}'}, \delta'_{\mathcal{H}'}, \Sigma'_{\Phi'_{\mathcal{H}'}})$ *is a* g*-refinement of* $(\Phi_{\mathcal{H}}, \delta_{\mathcal{H}}, \Sigma_{\Phi_{\mathcal{H}}})$*, under the canonical isomorphisms as in Lemma* 1.3.2.41*.*

Proof. The assertions in the first two paragraphs can be justified as in the case of $\mathsf{M}_{\mathcal{H}}$. (We omit the details for simplicity.) The third paragraph follows by comparing the torus torsor actions of sufficiently divisible multiples of elements, for which we have explicit descriptions in Lemma 1.3.2.28 and Proposition 1.3.2.31. As for the last paragraph, since the canonical morphisms are defined by universal properties given in terms of degeneration data, their compatibility follows from the fact that (by the theory of degeneration [62, Thm. 5.3.1.19] based on [62, Thm. 5.2.3.14], in particular) the Hecke twist of the tautological tuple over $\mathsf{M}^{\mathrm{tor}}_{\mathcal{H}', \Sigma'}$ by g defined using the level structure $\alpha_{\mathcal{H}'}$ over $\mathsf{M}_{\mathcal{H}'}$ is compatible with the Hecke twist of the tautological tuple over $\Xi_{\Phi'_{\mathcal{H}'}, \delta'_{\mathcal{H}'}}(\rho)$ by g defined using the level structure $\beta_{\mathcal{H}'}$ over $\Xi_{\Phi'_{\mathcal{H}'}, \delta'_{\mathcal{H}'}}$. □

Now let $(\widetilde{L}, \langle \cdot, \cdot \rangle^{\sim}, \widetilde{h}_0)$, $(\widetilde{\mathsf{Z}}, \widetilde{\Phi}, \widetilde{\delta})$, *etc* be chosen as in Section 1.2.4. Let $\widetilde{\kappa} = (\widetilde{\mathcal{H}}, \widetilde{\Sigma}, \widetilde{\sigma})$ be any element in the set $\widetilde{\mathbf{K}}^{++}_{Q,\mathcal{H}}$ as in Definition 1.2.4.11. The data of $\mathcal{O}$, $(\widetilde{L}, \langle \cdot, \cdot \rangle^{\sim}, \widetilde{h}_0)$, and $\widetilde{\mathcal{H}} \subset \widetilde{\mathrm{G}}(\hat{\mathbb{Z}})$ define a moduli problem $\widetilde{\mathsf{M}}_{\widetilde{\mathcal{H}}}$ as in Section 1.1.2. Since $\widetilde{\mathcal{H}}$ is neat and $\widetilde{\Sigma}$ is projective (and smooth), by Theorems 1.3.1.3 and 1.3.1.10, we have a toroidal compactification $\widetilde{\mathsf{M}}^{\mathrm{tor}}_{\widetilde{\mathcal{H}}} = \widetilde{\mathsf{M}}^{\mathrm{tor}}_{\widetilde{\mathcal{H}},\widetilde{\Sigma}}$ of $\widetilde{\mathsf{M}}_{\widetilde{\mathcal{H}}}$ which is projective and smooth over S_0. We are mainly interested in comparing the boundary structures of $\widetilde{\mathsf{M}}^{\mathrm{tor}}_{\widetilde{\mathcal{H}},\widetilde{\Sigma}}$ and $\mathsf{M}^{\mathrm{tor}}_{\mathcal{H},\Sigma}$ under suitable conditions.

In the remainder of this subsection, let us fix the choice of a $\breve{\mathsf{Z}}$ satisfying (1.2.4.12), so that we have the groups and homomorphisms defined in Definitions 1.2.4.53 and 1.2.4.54.

Suppose that the cusp label $[(\mathsf{Z}_{\mathcal{H}}, \Phi_{\mathcal{H}}, \delta_{\mathcal{H}})]$ at level $\mathcal{H}$ is canonically assigned (as in Lemma 1.2.4.15) to a cusp label $[(\breve{\mathsf{Z}}_{\widetilde{\mathcal{H}}}, \breve{\Phi}_{\widetilde{\mathcal{H}}}, \breve{\delta}_{\widetilde{\mathcal{H}}})]$ at level $\widetilde{\mathcal{H}}$ admitting a surjection to $[(\widetilde{\mathsf{Z}}_{\widetilde{\mathcal{H}}}, \widetilde{\Phi}_{\widetilde{\mathcal{H}}}, \widetilde{\delta}_{\widetilde{\mathcal{H}}})]$, so that we have (1.2.4.18), (1.2.4.19), and (1.2.4.20), and the definitions following them.

Lemma 1.3.2.50. (*Compare with [61, Lem. 4.9; see also the errata].*) *By comparing the universal properties, we obtain a canonical morphism*

$$\widetilde{C}_{\breve{\Phi}_{\widetilde{\mathcal{H}}},\breve{\delta}_{\widetilde{\mathcal{H}}}} \to C_{\Phi_{\mathcal{H}},\delta_{\mathcal{H}}}, \tag{1.3.2.51}$$

by sending $(\breve{c}_{\widetilde{\mathcal{H}}}, \breve{c}^{\vee}_{\widetilde{\mathcal{H}}})$, *which is an orbit of étale-locally-defined pairs*

$$(\breve{c}_n : \tfrac{1}{n}\breve{X} \to B^{\vee}, \breve{c}^{\vee}_n : \tfrac{1}{n}\breve{Y} \to B)$$

for some integer $n \geq 1$ *such that* $\widetilde{\mathcal{U}}(n) \subset \widetilde{\mathcal{H}}$, *to the orbit* $(c_{\mathcal{H}}, c^{\vee}_{\mathcal{H}})$ *of étale-locally-defined pairs*

$$(c_n : \tfrac{1}{n}X \to B^{\vee}, c^{\vee}_n : \tfrac{1}{n}Y \to B),$$

with $(c_n, c^{\vee}_n)$ *induced by* $(\breve{c}_n, \breve{c}^{\vee}_n)$ *by restrictions to* $\frac{1}{n}X$ *and* $\frac{1}{n}Y$, *where* X *and* Y *are the kernels of the admissible surjections* $s_{\breve{X}} : \breve{X} \twoheadrightarrow \widetilde{X}$ *and* $s_{\breve{Y}} : \breve{Y} \twoheadrightarrow \widetilde{Y}$, *respectively. (This definition canonically extends to a compatible definition in the* $\mathbb{Q}^{\times}$*-isogeny class language in Proposition* 1.3.2.14, *which we omit for simplicity.)*

The morphisms (1.3.2.51) *and* (1.3.2.52) *are proper and smooth. If* $\widehat{\mathcal{H}}_{\mathrm{G}} = \mathcal{H}$, *then* $\widetilde{\mathsf{M}}^{\widetilde{\Phi}_{\widetilde{\mathcal{H}}}}_{\widetilde{\mathcal{H}}} \cong \mathsf{M}^{\Phi_{\mathcal{H}}}_{\mathcal{H}}$ *and there is a canonical homomorphism*

$$\widetilde{C}^{\mathrm{grp}}_{\breve{\Phi}_{\widetilde{\mathcal{H}}},\breve{\delta}_{\widetilde{\mathcal{H}}}} \to C^{\mathrm{grp}}_{\Phi_{\mathcal{H}},\delta_{\mathcal{H}}} \tag{1.3.2.52}$$

of abelian schemes over $\mathsf{M}^{\Phi_{\mathcal{H}}}_{\mathcal{H}}$, *which can be identified with the canonical homomorphism*

$$\underline{\mathrm{Hom}}_{\mathcal{O}}(\breve{X}, B)^{\circ} \to \underline{\mathrm{Hom}}_{\mathcal{O}}(X, B)^{\circ} \tag{1.3.2.53}$$

up to canonical $\mathbb{Q}^{\times}$*-isogenies over* $\mathsf{M}^{\Phi_{\mathcal{H}}}_{\mathcal{H}}$, *and the* $\widetilde{C}^{\mathrm{grp}}_{\breve{\Phi}_{\widetilde{\mathcal{H}}},\breve{\delta}_{\widetilde{\mathcal{H}}}}$*- and* $C^{\mathrm{grp}}_{\Phi_{\mathcal{H}},\delta_{\mathcal{H}}}$*-torsor structures of* $\widetilde{C}_{\breve{\Phi}_{\widetilde{\mathcal{H}}},\breve{\delta}_{\widetilde{\mathcal{H}}}} \to \mathsf{M}^{\Phi_{\mathcal{H}}}_{\mathcal{H}}$ *and* $C_{\Phi_{\mathcal{H}},\delta_{\mathcal{H}}} \to \mathsf{M}^{\Phi_{\mathcal{H}}}_{\mathcal{H}}$, *respectively, are compatible with each other under* (1.3.2.51) *and* (1.3.2.52). (*See [62, Def. 5.2.3.8 and Prop. 5.2.3.9]* *for the formation of the fiberwise geometric identity components* $\underline{\mathrm{Hom}}_{\mathcal{O}}(\check{X}, B)^\circ$ *and* $\underline{\mathrm{Hom}}_{\mathcal{O}}(X, B)^\circ$. *See also the beginning of Section* 1.3.3 *below.) Moreover, the kernel of* (1.3.2.52) *is an abelian scheme over* $\mathsf{M}_{\mathcal{H}}^{\Phi_{\mathcal{H}}}$, *which is canonically* $\mathbb{Q}^\times$*-isogenous to the kernel* $\underline{\mathrm{Hom}}_{\mathcal{O}}(\widetilde{X}, B)^\circ$ *of* (1.3.2.53), *and* (1.3.2.51) *is a torsor under the pullback to* $C_{\Phi_{\mathcal{H}},\delta_{\mathcal{H}}}$ *of this abelian scheme. We deduce from these that, whether* $\widehat{\mathcal{H}}_{\mathrm{G}} = \mathcal{H}$ *or not, we have*

$$\Omega^1_{\widetilde{C}_{\check{\Phi}_{\widetilde{\mathcal{H}}},\check{\delta}_{\widetilde{\mathcal{H}}}}/C_{\Phi_{\mathcal{H}},\delta_{\mathcal{H}}}} \cong (\widetilde{C}_{\check{\Phi}_{\widetilde{\mathcal{H}}},\check{\delta}_{\widetilde{\mathcal{H}}}} \to \mathsf{M}_{\mathcal{H}}^{\mathsf{Z}_{\mathcal{H}}})^* \underline{\mathrm{Hom}}_{\mathcal{O}}(\widetilde{X}, \underline{\mathrm{Lie}}^\vee_{B/\mathsf{M}_{\mathcal{H}}^{\mathsf{Z}_{\mathcal{H}}}}), \tag{1.3.2.54}$$

and the canonical short exact sequence

$$\begin{aligned} 0 \to (\widetilde{C}_{\check{\Phi}_{\widetilde{\mathcal{H}}},\check{\delta}_{\widetilde{\mathcal{H}}}} \to C_{\Phi_{\mathcal{H}},\delta_{\mathcal{H}}})^* \Omega^1_{C_{\Phi_{\mathcal{H}},\delta_{\mathcal{H}}}/\mathsf{M}_{\mathcal{H}}^{\mathsf{Z}_{\mathcal{H}}}} &\to \Omega^1_{\widetilde{C}_{\check{\Phi}_{\widetilde{\mathcal{H}}},\check{\delta}_{\widetilde{\mathcal{H}}}}/\mathsf{M}_{\mathcal{H}}^{\mathsf{Z}_{\mathcal{H}}}} \\ &\to \Omega^1_{\widetilde{C}_{\check{\Phi}_{\widetilde{\mathcal{H}}},\check{\delta}_{\widetilde{\mathcal{H}}}}/C_{\Phi_{\mathcal{H}},\delta_{\mathcal{H}}}} \to 0 \end{aligned}$$

can be identified with the pullback under $\widetilde{C}_{\check{\Phi}_{\widetilde{\mathcal{H}}},\check{\delta}_{\widetilde{\mathcal{H}}}} \to \mathsf{M}_{\mathcal{H}}^{\mathsf{Z}_{\mathcal{H}}}$ *of the canonical short exact sequence*

$$\begin{aligned} 0 \to \underline{\mathrm{Hom}}_{\mathcal{O}}(X, \underline{\mathrm{Lie}}^\vee_{B/\mathsf{M}_{\mathcal{H}}^{\mathsf{Z}_{\mathcal{H}}}}) &\to \underline{\mathrm{Hom}}_{\mathcal{O}}(\check{X}, \underline{\mathrm{Lie}}^\vee_{B/\mathsf{M}_{\mathcal{H}}^{\mathsf{Z}_{\mathcal{H}}}}) \\ &\to \underline{\mathrm{Hom}}_{\mathcal{O}}(\widetilde{X}, \underline{\mathrm{Lie}}^\vee_{B/\mathsf{M}_{\mathcal{H}}^{\mathsf{Z}_{\mathcal{H}}}}) \to 0 \end{aligned}$$

under canonical morphisms (as in (1.3.2.8) *and* (1.3.2.54)).

The abelian scheme torsor $\widetilde{C}_{\check{\Phi}_{\widetilde{\mathcal{H}}},\check{\delta}_{\widetilde{\mathcal{H}}}} \to \widetilde{\mathsf{M}}_{\widetilde{\mathcal{H}}}^{\check{\Phi}_{\widetilde{\mathcal{H}}}}$ *and the finite étale covering* $\widetilde{\mathsf{M}}_{\widetilde{\mathcal{H}}}^{\check{\Phi}_{\widetilde{\mathcal{H}}}} \to \widetilde{\mathsf{M}}_{\widetilde{\mathcal{H}}}^{\check{\mathsf{Z}}_{\widetilde{\mathcal{H}}}}$ *depend (up to canonical isomorphism) only on* $\widehat{\mathcal{H}} = \widehat{\mathcal{H}}_{\widehat{\mathrm{G}}}$ *and* $(\check{\mathsf{Z}}_{\widehat{\mathcal{H}}}, \check{\Phi}_{\widehat{\mathcal{H}}}, \check{\delta}_{\widehat{\mathcal{H}}})$ *(see Definition* 1.2.4.17). *We shall denote them as* $\widehat{C}_{\check{\Phi}_{\widehat{\mathcal{H}}},\check{\delta}_{\widehat{\mathcal{H}}}} \to \widehat{\mathsf{M}}_{\widehat{\mathcal{H}}}^{\check{\Phi}_{\widehat{\mathcal{H}}}}$ *and* $\widehat{\mathsf{M}}_{\widehat{\mathcal{H}}}^{\check{\Phi}_{\widehat{\mathcal{H}}}} \to \widehat{\mathsf{M}}_{\widehat{\mathcal{H}}}^{\check{\mathsf{Z}}_{\widehat{\mathcal{H}}}}$ *when we want to emphasize this (in)dependence.*

Proof. The first paragraph is self-explanatory. As for the second paragraph, by Lemma 1.3.2.7, it suffices to verify the statements in the case $\mathcal{H} = \mathcal{U}(n)$ and $\widetilde{\mathcal{H}} = \widetilde{\mathcal{U}}(n)$ for some integer $n \geq 1$. (The third paragraph also follows by Lemma 1.3.2.7.) In this case, $\widetilde{C}_{\check{\Phi}_{\widetilde{\mathcal{H}}},\check{\delta}_{\widetilde{\mathcal{H}}}} = \widetilde{C}^{\mathrm{grp}}_{\check{\Phi}_{\widetilde{\mathcal{H}}},\check{\delta}_{\widetilde{\mathcal{H}}}} = \widetilde{C}_{\check{\Phi}_n,\check{\delta}_n}$ and $C_{\Phi_{\mathcal{H}},\delta_{\mathcal{H}}} = C^{\mathrm{grp}}_{\Phi_{\mathcal{H}},\delta_{\mathcal{H}}} = C_{\Phi_n,\delta_n}$ are abelian schemes over $\widetilde{\mathsf{M}}_{\widetilde{\mathcal{H}}}^{\check{\Phi}_{\widetilde{\mathcal{H}}}} = \widetilde{\mathsf{M}}_n^{\check{\mathsf{Z}}_n} \cong \mathsf{M}_{\mathcal{H}}^{\Phi_{\mathcal{H}}} = \mathsf{M}_n^{\mathsf{Z}_n}$. For simplicity, let us denote the kernel of (1.3.2.51) by C, viewed as a scheme over $\mathsf{M}_n^{\mathsf{Z}_n}$.

While the abelian scheme torsor $\widetilde{C}_{\check{\Phi}_n,\check{\delta}_n} \to \mathsf{M}_n^{\mathsf{Z}_n}$ parameterizes liftings (to level n) of pairs of the form $(\check{c} : \check{X} \to B^\vee, \check{c}^\vee : \check{Y} \to B)$ satisfying the compatibility $\check{c}\check{\phi} = \lambda_B \check{c}^\vee$ and the liftability and pairing conditions, and while the abelian scheme torsor $C_{\Phi_n,\delta_n} \to \mathsf{M}_n^{\mathsf{Z}_n}$ parameterizes liftings (to level n) of pairs of the form $(c : X \to B^\vee, c^\vee : Y \to B)$ satisfying the compatibility $c\phi = \lambda_B c^\vee$ and the liftability and pairing conditions, the scheme $C \to \mathsf{M}_n^{\mathsf{Z}_n}$ parameterizes liftings of pairs of the form $(\widetilde{c} : \widetilde{X} \to B^\vee, \widetilde{c}^\vee : \widetilde{Y} \to B)$ satisfying the compatibility $\widetilde{c}\widetilde{\phi} = \lambda_B \widetilde{c}^\vee$ and the liftability and pairing conditions induced by the ones of the pairs over $\widetilde{C}_{\check{\Phi}_n,\check{\delta}_n} \to \mathsf{M}_n^{\mathsf{Z}_n}$. Therefore, the same (component annihilating) argument in [62, Sec. 6.2.3–6.2.4] shows that C is an abelian scheme $\mathbb{Q}^\times$-isogenous to $\underline{\mathrm{Hom}}_{\mathcal{O}}(\widetilde{X}, B)^\circ$.

Consequently, all geometric fibers of the morphism (1.3.2.51) are smooth and have the same dimension (as the relative dimension of $C \to \mathsf{M}_n^{\mathsf{Z}_n}$). Since both $\widetilde{C}_{\breve{\Phi}_n,\breve{\delta}_n}$ and C_{Φ_n,δ_n} are smooth over S_0, the morphism (1.3.2.51) is smooth by [35, IV-3, 15.4.2 e')⇒b), and IV-4, 17.5.1 b)⇒a)]. By [10, Sec. 2.2, Prop. 14], smooth morphisms between schemes have sections étale locally. This shows that (1.3.2.51) is a torsor under the pullback of C to C_{Φ_n,δ_n}. (Regardless of this argument, the morphism (1.3.2.51) is proper because the morphism $C_{\breve{\Phi}_n,\breve{\delta}_n} \to \mathsf{M}_n^{\mathsf{Z}_n}$ is.) □

Proposition 1.3.2.55. *Under the canonical morphisms as in* (1.3.2.51) (*with varying* $\widetilde{\mathcal{H}}$ *and* $\mathcal{H}$), *and under the canonical homomorphisms* $\widehat{\mathrm{G}}_{1,\breve{\mathsf{Z}}}(\mathbb{A}^\infty) \to \mathrm{G}_{1,\mathsf{Z}}(\mathbb{A}^\infty)$ *and* $\widetilde{\mathrm{P}}_{\breve{\mathsf{Z}},\widetilde{\mathsf{Z}}}(\mathbb{A}^\infty)/\widetilde{\mathrm{U}}_{2,\breve{\mathsf{Z}}}(\mathbb{A}^\infty) \to \mathrm{P}_{\mathsf{Z}}(\mathbb{A}^\infty)/\mathrm{U}_{2,\mathsf{Z}}(\mathbb{A}^\infty)$, *the Hecke action of* $\widehat{\mathrm{G}}_{1,\breve{\mathsf{Z}}}(\mathbb{A}^\infty)$ *on the collection* $\{\widehat{C}_{\breve{\Phi}_{\widehat{\mathcal{H}}},\breve{\delta}_{\widehat{\mathcal{H}}}}\}_{\widehat{\mathcal{H}}_{\widehat{\mathrm{G}}_{1,\breve{\mathsf{Z}}}}}$ *is compatible with the Hecke action of* $\mathrm{G}_{1,\mathsf{Z}}(\mathbb{A}^\infty)$ *on the collection* $\{C_{\Phi_{\mathcal{H}},\delta_{\mathcal{H}}}\}_{\mathcal{H}_{\mathrm{G}_{1,\mathsf{Z}}}}$ (*see Proposition* 1.3.2.24); *the Hecke action of* $\widetilde{\mathrm{P}}_{\breve{\mathsf{Z}},\widetilde{\mathsf{Z}}}(\mathbb{A}^\infty)/\widetilde{\mathrm{U}}_{2,\breve{\mathsf{Z}}}(\mathbb{A}^\infty)$ (*see Definition* 1.2.4.54) *on the collection* $\{\coprod \widetilde{C}_{\breve{\Phi}_{\widetilde{\mathcal{H}}},\breve{\delta}_{\widetilde{\mathcal{H}}}}\}_{\widetilde{\mathcal{H}}_{\widetilde{\mathrm{P}}_{\breve{\mathsf{Z}},\widetilde{\mathsf{Z}}}}/\widetilde{\mathcal{H}}_{\widetilde{\mathrm{U}}_{2,\breve{\mathsf{Z}}}}}$ *is compatible with the Hecke action of* $\mathrm{P}_{\mathsf{Z}}(\mathbb{A}^\infty)/\mathrm{U}_{2,\mathsf{Z}}(\mathbb{A}^\infty)$ *on the collection* $\{\coprod C_{\Phi_{\mathcal{H}},\delta_{\mathcal{H}}}\}_{\mathcal{H}_{\mathrm{P}_{\mathsf{Z}}}/\mathcal{H}_{\mathrm{U}_{2,\mathsf{Z}}}}$; *and the induced action of* $\widetilde{\mathrm{G}}_{l,\breve{\mathsf{Z}},\widetilde{\mathsf{Z}}}(\mathbb{A}^\infty)$ *on the index sets* $\{[(\widetilde{\mathsf{Z}}_{\widetilde{\mathcal{H}}},\widetilde{\Phi}_{\widetilde{\mathcal{H}}},\widetilde{\delta}_{\widetilde{\mathcal{H}}})]\}$ *is compatible with the induced action of* $\mathrm{G}'_{l,\mathsf{Z}}(\mathbb{A}^\infty)$ *on the index sets* $\{[(\mathsf{Z}_{\mathcal{H}},\Phi_{\mathcal{H}},\delta_{\mathcal{H}})]\}$ (*again see Proposition* 1.3.2.24) *under the canonical homomorphism* $\widetilde{\mathrm{G}}_{l,\breve{\mathsf{Z}},\widetilde{\mathsf{Z}}}(\mathbb{A}^\infty) \to \mathrm{G}'_{l,\mathsf{Z}}(\mathbb{A}^\infty)$.

These Hecke actions induce a Hecke action of the subgroup $\widehat{\mathrm{P}}_{\breve{\mathsf{Z}}}(\mathbb{A}^\infty)/\widehat{\mathrm{U}}_{2,\breve{\mathsf{Z}}}(\mathbb{A}^\infty)$ *of* $\widetilde{\mathrm{P}}_{\breve{\mathsf{Z}},\widetilde{\mathsf{Z}}}(\mathbb{A}^\infty)/\widetilde{\mathrm{U}}_{2,\breve{\mathsf{Z}}}(\mathbb{A}^\infty)$ *on the collection* $\{\coprod \widehat{C}_{\breve{\Phi}_{\widehat{\mathcal{H}}},\breve{\delta}_{\widehat{\mathcal{H}}}}\}_{\widehat{\mathcal{H}}_{\widehat{\mathrm{P}}_{\breve{\mathsf{Z}}}}/\widehat{\mathcal{H}}_{\widehat{\mathrm{U}}_{2,\breve{\mathsf{Z}}}}}$, *which is compatible with the Hecke action of* $\mathrm{P}_{\mathsf{Z}}(\mathbb{A}^\infty)/\mathrm{U}_{2,\mathsf{Z}}(\mathbb{A}^\infty)$ *on the collection* $\{\coprod C_{\Phi_{\mathcal{H}},\delta_{\mathcal{H}}}\}_{\mathcal{H}_{\mathrm{P}_{\mathsf{Z}}}/\mathcal{H}_{\mathrm{U}_{2,\mathsf{Z}}}}$ *under the canonical morphisms* $\widehat{C}_{\breve{\Phi}_{\widehat{\mathcal{H}}},\breve{\delta}_{\widehat{\mathcal{H}}}} \to C_{\Phi_{\mathcal{H}},\delta_{\mathcal{H}}}$ (*with varying* $\widehat{\mathcal{H}}$ *and* $\mathcal{H}$) *and the canonical homomorphism* $\widehat{\mathrm{P}}_{\breve{\mathsf{Z}}}(\mathbb{A}^\infty)/\widehat{\mathrm{U}}_{2,\breve{\mathsf{Z}}}(\mathbb{A}^\infty) \to \mathrm{P}_{\mathsf{Z}}(\mathbb{A}^\infty)/\mathrm{U}_{2,\mathsf{Z}}(\mathbb{A}^\infty)$; *and the induced action of the subgroup* $\widehat{\mathrm{G}}'_{l,\breve{\mathsf{Z}}}(\mathbb{A}^\infty)$ *of* $\widetilde{\mathrm{G}}_{l,\breve{\mathsf{Z}},\widetilde{\mathsf{Z}}}(\mathbb{A}^\infty)$ *on the index sets* $\{[(\breve{\mathsf{Z}}_{\widehat{\mathcal{H}}},\breve{\Phi}_{\widehat{\mathcal{H}}},\breve{\delta}_{\widehat{\mathcal{H}}})]\}$ *is compatible with the induced action of* $\mathrm{G}'_{l,\mathsf{Z}}(\mathbb{A}^\infty)$ *on the index sets* $\{[(\mathsf{Z}_{\mathcal{H}},\Phi_{\mathcal{H}},\delta_{\mathcal{H}})]\}$ *under the canonical homomorphism* $\widehat{\mathrm{G}}'_{l,\breve{\mathsf{Z}}}(\mathbb{A}^\infty) \to \mathrm{G}'_{l,\mathsf{Z}}(\mathbb{A}^\infty)$.

Proof. The canonical morphisms as in (1.3.2.51) correspond to pushouts of extensions of B (resp. $B^\vee$) by $\breve{T}$ (resp. $\breve{T}^\vee$) under the canonical homomorphism $\breve{T} \to T$ (resp. $\breve{T}^\vee \to T^\vee$) induced by the restriction from $\breve{X}$ (resp. $\breve{Y}$) to X (resp. Y). Hence, the realizations of the Hecke twists are compatible in the desired ways. (We omit the details for simplicity.) □

Suppose $\widetilde{\sigma} \subset \mathbf{P}^+_{\widetilde{\Phi}_{\widetilde{\mathcal{H}}}}$ is a *top-dimensional* nondegenerate rational polyhedral cone in the cone decomposition $\widetilde{\Sigma}_{\widetilde{\Phi}_{\widetilde{\mathcal{H}}}}$ in $\widetilde{\Sigma}$, and suppose $\breve{\sigma}$ is the image of $\widetilde{\sigma} \subset \mathbf{P}^+_{\widetilde{\Phi}_{\widetilde{\mathcal{H}}}}$ under the first morphism in (1.2.4.20). Then we have

$$\breve{\sigma}^\perp = \widehat{\mathbf{S}}_{\breve{\Phi}_{\widehat{\mathcal{H}}}}$$

(see Definition 1.2.4.29) for any such $\breve{\sigma}$, where $\widehat{\mathcal{H}} = \widetilde{\mathcal{H}}_{\widehat{\mathrm{G}}}$, and we have the following:

Proposition 1.3.2.56. (*Compare with Lemma* 1.3.2.28 *and Proposition* 1.3.2.31.) *The scheme*

$$\widetilde{\widetilde{\Xi}}_{\breve{\Phi}_{\widetilde{\mathcal{H}}},\breve{\delta}_{\widetilde{\mathcal{H}}},\breve{\sigma}} \cong \underline{\mathrm{Spec}}_{\mathscr{O}_{\widetilde{C}_{\breve{\Phi}_{\widetilde{\mathcal{H}}},\breve{\delta}_{\widetilde{\mathcal{H}}}}}}\left(\underset{\breve{\ell}\in\breve{\sigma}^{\perp}}{\oplus} \widetilde{\Psi}_{\breve{\Phi}_{\widetilde{\mathcal{H}}},\breve{\delta}_{\widetilde{\mathcal{H}}}}(\breve{\ell})\right)$$

over $\widetilde{C}_{\breve{\Phi}_{\widetilde{\mathcal{H}}},\breve{\delta}_{\widetilde{\mathcal{H}}}}$ *is a torsor under the split torus* $\widetilde{E}_{\breve{\Phi}_{\widetilde{\mathcal{H}}},\breve{\sigma}}$ *with character group* $\breve{\sigma}^{\perp}$*, which is canonically isomorphic to the split torus* $\widehat{E}_{\breve{\Phi}_{\widehat{\mathcal{H}}}}$ *with character group* $\widehat{\mathbf{S}}_{\breve{\Phi}_{\widehat{\mathcal{H}}}}$*, which depends only on* $\widehat{\mathcal{H}}_{\widehat{\mathrm{P}}_{\breve{\mathrm{Z}}}} = (\widetilde{\mathcal{H}}_{\widetilde{\mathrm{P}}_{\breve{\mathrm{Z}}}} \cap \widetilde{\mathcal{H}}_{\widetilde{\mathrm{P}}'_{\breve{\mathrm{Z}}}})/\widetilde{\mathcal{H}}_{\widetilde{\mathrm{U}}_{2,\breve{\mathrm{Z}}}}$ *(see Definition* 1.2.4.53*). We have* $\widehat{\mathbf{S}}^{\vee}_{\breve{\Phi}_1}/\widehat{\mathbf{S}}^{\vee}_{\breve{\Phi}_{\widehat{\mathcal{H}}}} \cong \widehat{\mathrm{U}}_{2,\breve{\mathrm{Z}}}(\widehat{\mathbb{Z}})/\widehat{\mathcal{H}}_{\widehat{\mathrm{U}}_{2,\breve{\mathrm{Z}}}}$*, where* $\widehat{\mathbf{S}}_{\breve{\Phi}_1} := \widehat{\mathbf{S}}_{\breve{\Phi}_{\widehat{\mathrm{G}}(\hat{\mathbb{Z}})}}$ *is the kernel of the canonical homomorphism* $\mathbf{S}_{\breve{\Phi}_1} \twoheadrightarrow \mathbf{S}_{\widetilde{\Phi}_1}$ *(see Definition* 1.2.4.29*).*

The torus torsor $S := \widetilde{\widetilde{\Xi}}_{\breve{\Phi}_{\widetilde{\mathcal{H}}},\breve{\delta}_{\widetilde{\mathcal{H}}},\breve{\sigma}} \to \widetilde{C}_{\breve{\Phi}_{\widetilde{\mathcal{H}}},\breve{\delta}_{\widetilde{\mathcal{H}}}}$ *is universal for the additional structures* $(\widehat{\tau}_{\widehat{\mathcal{H}}}, \widehat{\tau}^{\vee}_{\widehat{\mathcal{H}}})$*, which are* $\widehat{\mathcal{H}}_{\widehat{\mathrm{P}}_{\breve{\mathrm{Z}}}}/\widehat{\mathcal{U}}(n)_{\widehat{\mathrm{P}}_{\breve{\mathrm{Z}}}}$*-orbits of étale-locally-defined pairs* $(\widehat{\tau}_n, \widehat{\tau}^{\vee}_n)$*, where:*

1. $\widehat{\tau}_n : \mathbf{1}_{\frac{1}{n}Y\times\breve{X},S} \xrightarrow{\sim} (\breve{c}^{\vee}_n|_{\frac{1}{n}Y_S} \times \breve{c})^*\mathcal{P}_B^{\otimes -1}$ *is a trivialization of biextensions.*
2. $\widehat{\tau}^{\vee}_n : \mathbf{1}_{\frac{1}{n}\breve{Y}\times X,S} \xrightarrow{\sim} (\breve{c}^{\vee}_n \times \breve{c}|_{X_S})^*\mathcal{P}_B^{\otimes -1}$ *is a trivialization of biextensions.*
3. $\widehat{\tau}_n$ *and* $\widehat{\tau}^{\vee}_n$ *satisfy the analogues of the usual* $\mathcal{O}$*-compatibility condition.*
4. $\widehat{\tau}_n$ *and* $\widehat{\tau}^{\vee}_n$ *satisfy the symmetry condition that* $\widehat{\tau}_n|_{\mathbf{1}_{Y\times\breve{Y},S}}$ *and* $\widehat{\tau}^{\vee}_n|_{\mathbf{1}_{\breve{Y}\times Y,S}}$ *coincide under the canonical isomorphism induced by the swapping isomorphism* $\mathbf{1}_{Y\times\breve{Y},S} \xrightarrow{\sim} \mathbf{1}_{\breve{Y}\times Y,S}$ *and the symmetry automorphism of* $\mathcal{P}_B$*.*
5. $\widehat{\tau}_n|_{\mathbf{1}_{\frac{1}{n}Y\times X,S}} = \widehat{\tau}^{\vee}_n|_{\mathbf{1}_{\frac{1}{n}Y\times X,S}}$*.*

We shall denote $\widetilde{\widetilde{\Xi}}_{\breve{\Phi}_{\widetilde{\mathcal{H}}},\breve{\delta}_{\widetilde{\mathcal{H}}},\breve{\sigma}}$ *by* $\widehat{\Xi}_{\breve{\Phi}_{\widehat{\mathcal{H}}},\breve{\delta}_{\widehat{\mathcal{H}}}}$ *when we want to emphasize that (by Lemma* 1.3.2.25*) it depends only on* $\widehat{\mathcal{H}} = \widehat{\mathcal{H}}_{\widehat{\mathrm{G}}}$ *and* $(\breve{\mathrm{Z}}_{\widehat{\mathcal{H}}}, \breve{\Phi}_{\widehat{\mathcal{H}}}, \breve{\delta}_{\widehat{\mathcal{H}}})$ *(see Definition* 1.2.4.17*), but does not depend on the choice of* $\breve{\sigma}$*.*

The $\widehat{E}_{\breve{\Phi}_{\widehat{\mathcal{H}}}}$*-torsor structure of* $\widehat{\Xi}_{\breve{\Phi}_{\widehat{\mathcal{H}}},\breve{\delta}_{\widehat{\mathcal{H}}}} \to \widehat{C}_{\breve{\Phi}_{\widehat{\mathcal{H}}},\breve{\delta}_{\widehat{\mathcal{H}}}}$ *defines a homomorphism*

$$\widehat{\mathbf{S}}_{\breve{\Phi}_{\widehat{\mathcal{H}}}} \to \underline{\mathrm{Pic}}(\widehat{C}_{\breve{\Phi}_{\widehat{\mathcal{H}}},\breve{\delta}_{\widehat{\mathcal{H}}}}) : \breve{\ell} \mapsto \widehat{\Psi}_{\breve{\Phi}_{\widehat{\mathcal{H}}},\breve{\delta}_{\widehat{\mathcal{H}}}}(\breve{\ell}),$$

assigning to each $\breve{\ell} \in \widehat{\mathbf{S}}_{\breve{\Phi}_{\widehat{\mathcal{H}}}}$ *an invertible sheaf* $\widehat{\Psi}_{\breve{\Phi}_{\widehat{\mathcal{H}}},\breve{\delta}_{\widehat{\mathcal{H}}}}(\breve{\ell})$ *over* $\widehat{C}_{\breve{\Phi}_{\widehat{\mathcal{H}}},\breve{\delta}_{\widehat{\mathcal{H}}}}$ *(up to isomorphism), together with isomorphisms*

$$\widehat{\Delta}^*_{\breve{\Phi}_{\widehat{\mathcal{H}}},\breve{\delta}_{\widehat{\mathcal{H}}},\breve{\ell},\breve{\ell}'} : \widehat{\Psi}_{\breve{\Phi}_{\widehat{\mathcal{H}}},\breve{\delta}_{\widehat{\mathcal{H}}}}(\breve{\ell}) \underset{\mathscr{O}_{\widehat{C}_{\breve{\Phi}_{\widehat{\mathcal{H}}},\breve{\delta}_{\widehat{\mathcal{H}}}}}}{\otimes} \widehat{\Psi}_{\breve{\Phi}_{\widehat{\mathcal{H}}},\breve{\delta}_{\widehat{\mathcal{H}}}}(\breve{\ell}') \xrightarrow{\sim} \widehat{\Psi}_{\breve{\Phi}_{\widehat{\mathcal{H}}},\breve{\delta}_{\widehat{\mathcal{H}}}}(\breve{\ell}+\breve{\ell}')$$

for all $\breve{\ell}, \breve{\ell}' \in \widehat{\mathbf{S}}_{\breve{\Phi}_{\widehat{\mathcal{H}}}}$*, satisfying the necessary compatibilities with each other making* $\underset{\breve{\ell}\in\widehat{\mathbf{S}}_{\breve{\Phi}_{\widehat{\mathcal{H}}}}}{\oplus} \widehat{\Psi}_{\breve{\Phi}_{\widehat{\mathcal{H}}},\breve{\delta}_{\widehat{\mathcal{H}}}}(\breve{\ell})$ *an* $\mathscr{O}_{\widehat{C}_{\breve{\Phi}_{\widehat{\mathcal{H}}},\breve{\delta}_{\widehat{\mathcal{H}}}}}$*-algebra, such that*

$$\widehat{\Xi}_{\breve{\Phi}_{\widehat{\mathcal{H}}},\breve{\delta}_{\widehat{\mathcal{H}}}} \cong \underline{\mathrm{Spec}}_{\mathscr{O}_{\widehat{C}_{\breve{\Phi}_{\widehat{\mathcal{H}}},\breve{\delta}_{\widehat{\mathcal{H}}}}}}\left(\underset{\breve{\ell}\in\widehat{\mathbf{S}}_{\breve{\Phi}_{\widehat{\mathcal{H}}}}}{\oplus} \widehat{\Psi}_{\breve{\Phi}_{\widehat{\mathcal{H}}},\breve{\delta}_{\widehat{\mathcal{H}}}}(\breve{\ell})\right).$$

When $\breve{\ell} = [y \otimes \chi]$ for some $y \in \breve{Y}$ and $\chi \in \breve{X}$ such that either $y \in Y$ or $\chi \in X$, we have a canonical isomorphism

$$\widehat{\Psi}_{\breve{\Phi}_{\widehat{\mathcal{H}}},\breve{\delta}_{\widehat{\mathcal{H}}}}(\breve{\ell}) \cong (\breve{c}^\vee(y), \breve{c}(\chi))^* \mathcal{P}_B.$$

If we fix the choice of ($\breve{\mathrm{Z}}_n$ and) $\breve{\Phi}_n$, then the canonical morphism

$$\widehat{\Xi}_{\breve{\Phi}_n,\breve{\delta}_n} \to \widehat{\Xi}_{\breve{\Phi}_{\widehat{\mathcal{H}}},\breve{\delta}_{\widehat{\mathcal{H}}}} \tag{1.3.2.57}$$

is an $\widehat{\mathcal{H}}_{\widehat{\mathrm{P}}'_{\breve{\mathrm{Z}}}}/\widehat{\mathcal{U}}(n)_{\widehat{\mathrm{P}}'_{\breve{\mathrm{Z}}}}$-torsor, and induces an isomorphism

$$\widehat{\Xi}_{\breve{\Phi}_n,\breve{\delta}_n}/(\widehat{\mathcal{H}}_{\widehat{\mathrm{P}}'_{\breve{\mathrm{Z}}}}/\widehat{\mathcal{U}}(n)_{\widehat{\mathrm{P}}'_{\breve{\mathrm{Z}}}}) \xrightarrow{\sim} \widehat{\Xi}_{\breve{\Phi}_{\widehat{\mathcal{H}}},\breve{\delta}_{\widehat{\mathcal{H}}}} \tag{1.3.2.58}$$

(*cf. Lemma* 1.3.2.25).

Proof. These follow from the corresponding properties of $\widetilde{\Xi}_{\breve{\Phi}_{\widetilde{\mathcal{H}}},\breve{\delta}_{\widetilde{\mathcal{H}}}}$ as in Lemmas 1.3.2.25 and 1.3.2.28, and Proposition 1.3.2.31, because the restriction from $\mathbf{S}_{\breve{\Phi}_{\widetilde{\mathcal{H}}}}$ to the subgroup $\widehat{\mathbf{S}}_{\breve{\Phi}_{\widehat{\mathcal{H}}}}$ (see Definition 1.2.4.29) corresponds to taking orbits of restrictions of $\breve{\tau}_n : \mathbf{1}_{\frac{1}{n}\breve{Y} \times \breve{X},S} \xrightarrow{\sim} (\breve{c}^\vee \times \breve{c})^* \mathcal{P}_B^{\otimes -1}$ to $\mathbf{1}_{\frac{1}{n}Y \times \breve{X},S}$ and $\mathbf{1}_{\frac{1}{n}\breve{Y} \times X,S}$, which form the pairs $(\widehat{\tau}_n, \widehat{\tau}_n^\vee)$ as above. □

For each rational polyhedral cone $\breve{\rho} \subset (\mathbf{S}_{\breve{\Phi}_{\widetilde{\mathcal{H}}}})^\vee_{\mathbb{R}}$ having $\breve{\sigma}$ as a face, we have an affine toroidal embedding

$$\widetilde{\Xi}_{\breve{\Phi}_{\widetilde{\mathcal{H}}},\breve{\delta}_{\widetilde{\mathcal{H}}},\breve{\sigma}} \hookrightarrow \widetilde{\Xi}_{\breve{\Phi}_{\widetilde{\mathcal{H}}},\breve{\delta}_{\widetilde{\mathcal{H}}},\breve{\sigma}}(\breve{\rho}) := \underline{\mathrm{Spec}}_{\mathscr{O}_{\widetilde{C}_{\breve{\Phi}_{\widetilde{\mathcal{H}}},\breve{\delta}_{\widetilde{\mathcal{H}}}}}}\left(\bigoplus_{\breve{\ell} \in \breve{\sigma}^\perp \cap \breve{\rho}^\vee} \widetilde{\Psi}_{\breve{\Phi}_{\widetilde{\mathcal{H}}},\breve{\delta}_{\widetilde{\mathcal{H}}}}(\breve{\ell})\right) \tag{1.3.2.59}$$

as in (1.3.2.32).

In general, for each rational polyhedral cone $\widehat{\rho} \subset (\widehat{\mathbf{S}}_{\breve{\Phi}_{\widehat{\mathcal{H}}}})^\vee_{\mathbb{R}}$, we have an affine toroidal embedding

$$\widehat{\Xi}_{\breve{\Phi}_{\widehat{\mathcal{H}}},\breve{\delta}_{\widehat{\mathcal{H}}}} \hookrightarrow \widehat{\Xi}_{\breve{\Phi}_{\widehat{\mathcal{H}}},\breve{\delta}_{\widehat{\mathcal{H}}}}(\widehat{\rho}) := \underline{\mathrm{Spec}}_{\mathscr{O}_{\widehat{C}_{\breve{\Phi}_{\widehat{\mathcal{H}}},\breve{\delta}_{\widehat{\mathcal{H}}}}}}\left(\bigoplus_{\breve{\ell} \in \widehat{\rho}^\vee} \widehat{\Psi}_{\breve{\Phi}_{\widehat{\mathcal{H}}},\breve{\delta}_{\widehat{\mathcal{H}}}}(\breve{\ell})\right). \tag{1.3.2.60}$$

By Proposition 1.3.2.56, (1.3.2.59) and (1.3.2.60) can be canonically identified when $\widehat{\mathcal{H}} = \widetilde{\mathcal{H}}_{\widehat{\mathrm{G}}}$, when $(\breve{\mathrm{Z}}_{\widehat{\mathcal{H}}}, \breve{\Phi}_{\widehat{\mathcal{H}}}, \breve{\delta}_{\widehat{\mathcal{H}}})$ is determined by $(\breve{\mathrm{Z}}_{\widetilde{\mathcal{H}}}, \breve{\Phi}_{\widetilde{\mathcal{H}}}, \breve{\delta}_{\widetilde{\mathcal{H}}})$ as in Definition 1.2.4.17, and when $\widehat{\rho} = \mathrm{pr}_{(\widehat{\mathbf{S}}_{\breve{\Phi}_{\widehat{\mathcal{H}}}})^\vee_{\mathbb{R}}}(\breve{\rho})$. (Hence, (1.3.2.59) depends only on these induced parameters.)

Both sides of (1.3.2.60) are relative affine over $\widehat{C}_{\breve{\Phi}_{\widehat{\mathcal{H}}},\breve{\delta}_{\widehat{\mathcal{H}}}}$, where $\widehat{\Xi}_{\breve{\Phi}_{\widehat{\mathcal{H}}},\breve{\delta}_{\widehat{\mathcal{H}}}}(\widehat{\rho}) \to \widehat{C}_{\breve{\Phi}_{\widehat{\mathcal{H}}},\breve{\delta}_{\widehat{\mathcal{H}}}}$ is smooth when the cone $\widehat{\rho}$ is smooth. The $\widehat{\rho}$-stratum of $\widehat{\Xi}_{\breve{\Phi}_{\widehat{\mathcal{H}}},\breve{\delta}_{\widehat{\mathcal{H}}}}(\widehat{\rho})$ is

$$\widehat{\Xi}_{\breve{\Phi}_{\widehat{\mathcal{H}}},\breve{\delta}_{\widehat{\mathcal{H}}},\widehat{\rho}} := \underline{\mathrm{Spec}}_{\mathscr{O}_{\widehat{C}_{\breve{\Phi}_{\widehat{\mathcal{H}}},\breve{\delta}_{\widehat{\mathcal{H}}}}}}\left(\bigoplus_{\breve{\ell} \in \widehat{\rho}^\perp} \widehat{\Psi}_{\breve{\Phi}_{\widehat{\mathcal{H}}},\breve{\delta}_{\widehat{\mathcal{H}}}}(\breve{\ell})\right), \tag{1.3.2.61}$$

which is canonically isomorphic to $\widetilde{\Xi}_{\breve{\Phi}_{\widetilde{\mathcal{H}}},\breve{\delta}_{\widetilde{\mathcal{H}}},\breve{\rho}}$ (*cf.* (1.3.2.33)). The affine morphism $\widehat{\Xi}_{\breve{\Phi}_{\widehat{\mathcal{H}}},\breve{\delta}_{\widehat{\mathcal{H}}},\widehat{\rho}} \to \widehat{C}_{\breve{\Phi}_{\widehat{\mathcal{H}}},\breve{\delta}_{\widehat{\mathcal{H}}}}$ is a torsor under the torus $\widehat{E}_{\breve{\Phi}_{\widehat{\mathcal{H}}},\widehat{\rho}} \cong E_{\breve{\Phi}_{\widetilde{\mathcal{H}}},\breve{\rho}}$ with character group $\widehat{\rho}^\perp \cong \breve{\rho}^\perp$. (Note that these two instance of $\perp$ are taken in different ambient spaces.)

For each $\Gamma_{\check{\Phi}_{\widehat{\mathcal{H}}}}$-admissible rational polyhedral cone decomposition $\widehat{\Sigma}_{\check{\Phi}_{\widehat{\mathcal{H}}}}$ of $\widehat{\mathbf{P}}_{\check{\Phi}_{\widehat{\mathcal{H}}}}$ as in Definition 1.2.4.40, we have (as in (1.3.2.34)) a toroidal embedding

$$\widehat{\Xi}_{\check{\Phi}_{\widehat{\mathcal{H}}},\check{\delta}_{\widehat{\mathcal{H}}}} \hookrightarrow \overline{\widehat{\Xi}}_{\check{\Phi}_{\widehat{\mathcal{H}}},\check{\delta}_{\widehat{\mathcal{H}}}} = \overline{\widehat{\Xi}}_{\check{\Phi}_{\widehat{\mathcal{H}}},\check{\delta}_{\widehat{\mathcal{H}}},\widehat{\Sigma}_{\check{\Phi}_{\widehat{\mathcal{H}}}}}, \tag{1.3.2.62}$$

the right-hand side being only locally of finite type over $\widehat{C}_{\check{\Phi}_{\widehat{\mathcal{H}}},\check{\delta}_{\widehat{\mathcal{H}}}}$, with an open covering

$$\overline{\widehat{\Xi}}_{\check{\Phi}_{\widehat{\mathcal{H}}},\check{\delta}_{\widehat{\mathcal{H}}}} = \bigcup_{\widehat{\rho}\in\widehat{\Sigma}_{\check{\Phi}_{\widehat{\mathcal{H}}}}} \widehat{\Xi}_{\check{\Phi}_{\widehat{\mathcal{H}}},\check{\delta}_{\widehat{\mathcal{H}}}}(\widehat{\rho}), \tag{1.3.2.63}$$

(*cf.* (1.3.2.35)) inducing a stratification

$$\overline{\widehat{\Xi}}_{\check{\Phi}_{\widehat{\mathcal{H}}},\check{\delta}_{\widehat{\mathcal{H}}}} = \coprod_{\widehat{\rho}\in\widehat{\Sigma}_{\check{\Phi}_{\widehat{\mathcal{H}}}}} \widehat{\Xi}_{\check{\Phi}_{\widehat{\mathcal{H}}},\check{\delta}_{\widehat{\mathcal{H}}},\widehat{\rho}} \tag{1.3.2.64}$$

(*cf.* (1.3.2.36)). (The notation "$\coprod$" only means a set-theoretic disjoint union. The algebro-geometric structure is still the one inherited from $\overline{\widehat{\Xi}}_{\check{\Phi}_{\widehat{\mathcal{H}}},\check{\delta}_{\widehat{\mathcal{H}}}}$.) Let

$$\widehat{\mathfrak{X}}_{\check{\Phi}_{\widehat{\mathcal{H}}},\check{\delta}_{\widehat{\mathcal{H}}},\widehat{\rho}} := (\widehat{\Xi}_{\check{\Phi}_{\widehat{\mathcal{H}}},\check{\delta}_{\widehat{\mathcal{H}}}}(\widehat{\rho}))^{\wedge}_{\widehat{\Xi}_{\check{\Phi}_{\widehat{\mathcal{H}}},\check{\delta}_{\widehat{\mathcal{H}}},\widehat{\rho}}} \tag{1.3.2.65}$$

(*cf.* (1.3.2.39)), the formal completion of $\widehat{\Xi}_{\check{\Phi}_{\widehat{\mathcal{H}}},\check{\delta}_{\widehat{\mathcal{H}}}}(\widehat{\rho})$ along its $\widehat{\rho}$-stratum $\widehat{\Xi}_{\check{\Phi}_{\widehat{\mathcal{H}}},\check{\delta}_{\widehat{\mathcal{H}}},\widehat{\rho}}$, which is canonically isomorphic to $\widetilde{\mathfrak{X}}_{\check{\Phi}_{\widetilde{\mathcal{H}}},\check{\delta}_{\widetilde{\mathcal{H}}},\check{\sigma},\check{\rho}}$, the formal completion of $\widetilde{\Xi}_{\check{\Phi}_{\widetilde{\mathcal{H}}},\check{\delta}_{\widetilde{\mathcal{H}}},\check{\sigma}}(\check{\rho})$, the closure of $\widetilde{\Xi}_{\check{\Phi}_{\widetilde{\mathcal{H}}},\check{\delta}_{\widetilde{\mathcal{H}}},\check{\sigma}}$ in $\widetilde{\Xi}_{\check{\Phi}_{\widetilde{\mathcal{H}}},\check{\delta}_{\widetilde{\mathcal{H}}}}(\check{\rho})$, along its $\check{\rho}$-stratum $\widetilde{\Xi}_{\check{\Phi}_{\widetilde{\mathcal{H}}},\check{\delta}_{\widetilde{\mathcal{H}}},\check{\rho}}$ (*cf.* (1.3.2.40)). Also, let us define

$$\widehat{\mathfrak{X}}_{\check{\Phi}_{\widehat{\mathcal{H}}},\check{\delta}_{\widehat{\mathcal{H}}}} = \widehat{\mathfrak{X}}_{\check{\Phi}_{\widehat{\mathcal{H}}},\check{\delta}_{\widehat{\mathcal{H}}},\widehat{\Sigma}_{\check{\Phi}_{\widehat{\mathcal{H}}}}} \tag{1.3.2.66}$$

(*cf.* Lemma 1.3.2.41) to be the formal completion of $\overline{\widehat{\Xi}}_{\check{\Phi}_{\widehat{\mathcal{H}}},\check{\delta}_{\widehat{\mathcal{H}}},\widehat{\Sigma}_{\check{\Phi}_{\widehat{\mathcal{H}}}}}$ along the union of the $\widehat{\sigma}$-strata $\widehat{\Xi}_{\check{\Phi}_{\widehat{\mathcal{H}}},\check{\delta}_{\widehat{\mathcal{H}}},\widehat{\rho}}$ for $\widehat{\rho}\in\widehat{\Sigma}_{\check{\Phi}_{\widehat{\mathcal{H}}}}$ and $\widehat{\rho}\subset\widehat{\mathbf{P}}^{+}_{\check{\Phi}_{\widehat{\mathcal{H}}}}$.

Proposition 1.3.2.67. (*Compare with Propositions* 1.3.1.15, 1.3.2.24, 1.3.2.45, *and* 1.3.2.55.) *There is a Hecke action of* $\widehat{\mathrm{P}}'_{\check{\mathrm{Z}}}(\mathbb{A}^\infty)$ *on the collection* $\{\widehat{\Xi}_{\check{\Phi}_{\widehat{\mathcal{H}}},\check{\delta}_{\widehat{\mathcal{H}}}}\}_{\widehat{\mathcal{H}}_{\widehat{\mathrm{P}}'_{\check{\mathrm{Z}}}}}$, *realized by finite étale surjections pulling tautological objects back to Hecke twists, which is compatible with the Hecke action of* $\widehat{\mathrm{G}}_{1,\check{\mathrm{Z}}}(\mathbb{A}^\infty)$ *on the collection* $\{\widehat{C}_{\check{\Phi}_{\widehat{\mathcal{H}}},\check{\delta}_{\widehat{\mathcal{H}}}}\}_{\widehat{\mathcal{H}}_{\widehat{\mathrm{G}}_{1,\check{\mathrm{Z}}}}}$ *under the canonical morphisms* $\widehat{\Xi}_{\check{\Phi}_{\widehat{\mathcal{H}}},\check{\delta}_{\widehat{\mathcal{H}}}}\to\widehat{C}_{\check{\Phi}_{\widehat{\mathcal{H}}},\check{\delta}_{\widehat{\mathcal{H}}}}$ *(with varying* $\widehat{\mathcal{H}}$*) and the canonical homomorphism* $\widehat{\mathrm{P}}'_{\check{\mathrm{Z}}}(\mathbb{A}^\infty)\to\widehat{\mathrm{G}}_{1,\check{\mathrm{Z}}}(\mathbb{A}^\infty)=\widehat{\mathrm{P}}'_{\check{\mathrm{Z}}}(\mathbb{A}^\infty)/\widehat{\mathrm{U}}_{2,\check{\mathrm{Z}}}(\mathbb{A}^\infty)$.

There is also a Hecke action of $\widehat{\mathrm{P}}_{\check{\mathrm{Z}}}(\mathbb{A}^\infty)$ *on the collection* $\{\coprod\widehat{\Xi}_{\check{\Phi}_{\widehat{\mathcal{H}}},\check{\delta}_{\widehat{\mathcal{H}}}}\}_{\widehat{\mathcal{H}}_{\widehat{\mathrm{P}}_{\check{\mathrm{Z}}}}}$, *where the disjoint unions are over classes* $[(\check{\mathrm{Z}}_{\widehat{\mathcal{H}}},\check{\Phi}_{\widehat{\mathcal{H}}},\check{\delta}_{\widehat{\mathcal{H}}})]$ *sharing the same* $\check{\mathrm{Z}}_{\widehat{\mathcal{H}}}$, *realized by finite étale surjections pulling tautological objects back to Hecke twists, which induces an action of* $\widehat{\mathrm{G}}'_{l,\check{\mathrm{Z}}}(\mathbb{A}^\infty)=\widehat{\mathrm{P}}_{\check{\mathrm{Z}}}(\mathbb{A}^\infty)/\widehat{\mathrm{P}}'_{\check{\mathrm{Z}}}(\mathbb{A}^\infty)$ *on the index sets* $\{[(\check{\mathrm{Z}}_{\widehat{\mathcal{H}}},\check{\Phi}_{\widehat{\mathcal{H}}},\check{\delta}_{\widehat{\mathcal{H}}})]\}$, *which is compatible with the Hecke action of* $\widehat{\mathrm{P}}_{\check{\mathrm{Z}}}(\mathbb{A}^\infty)/\widehat{\mathrm{U}}_{2,\check{\mathrm{Z}}}(\mathbb{A}^\infty)$ *on the collection*

$\{\coprod \widehat{C}_{\breve{\Phi}_{\widehat{\mathcal{H}}},\breve{\delta}_{\widehat{\mathcal{H}}}}\}_{\widehat{\mathcal{H}}_{\widehat{\mathrm{P}}_{\breve{\mathrm{Z}}}}/\widehat{\mathcal{H}}_{\widehat{\mathrm{U}}_{2,\breve{\mathrm{Z}}}}}$ *(with the same index sets and the same induced action of* $\widehat{\mathrm{G}}'_{l,\breve{\mathrm{Z}}}(\mathbb{A}^\infty)$*) under the canonical morphisms* $\widehat{\Xi}_{\breve{\Phi}_{\widehat{\mathcal{H}}},\breve{\delta}_{\widehat{\mathcal{H}}}} \to \widehat{C}_{\breve{\Phi}_{\widehat{\mathcal{H}}},\breve{\delta}_{\widehat{\mathcal{H}}}}$ *(with varying* $\widehat{\mathcal{H}}$*) and the canonical homomorphism* $\widehat{\mathrm{P}}_{\breve{\mathrm{Z}}}(\mathbb{A}^\infty) \to \widehat{\mathrm{P}}_{\breve{\mathrm{Z}}}(\mathbb{A}^\infty)/\widehat{\mathrm{U}}_{2,\breve{\mathrm{Z}}}(\mathbb{A}^\infty)$*.*

Any such Hecke action

$$[\widehat{g}] : \widehat{\Xi}_{\breve{\Phi}'_{\widehat{\mathcal{H}}'},\breve{\delta}'_{\widehat{\mathcal{H}}'}} \to \widehat{\Xi}_{\breve{\Phi}_{\widehat{\mathcal{H}}},\breve{\delta}_{\widehat{\mathcal{H}}}}$$

covering $[\widehat{g}] : \widehat{C}_{\breve{\Phi}'_{\widehat{\mathcal{H}}'},\breve{\delta}'_{\widehat{\mathcal{H}}'}} \to \widehat{C}_{\breve{\Phi}_{\widehat{\mathcal{H}}},\breve{\delta}_{\widehat{\mathcal{H}}}}$ *induces a morphism*

$$\widehat{\Xi}_{\breve{\Phi}'_{\widehat{\mathcal{H}}'},\breve{\delta}'_{\widehat{\mathcal{H}}'}} \to \widehat{\Xi}_{\breve{\Phi}_{\widehat{\mathcal{H}}},\breve{\delta}_{\widehat{\mathcal{H}}}} \underset{\widehat{C}_{\breve{\Phi}_{\widehat{\mathcal{H}}},\breve{\delta}_{\widehat{\mathcal{H}}}}}{\times} \widehat{C}_{\breve{\Phi}'_{\widehat{\mathcal{H}}'},\breve{\delta}'_{\widehat{\mathcal{H}}'}}$$

between torus torsors over $\widehat{C}_{\breve{\Phi}'_{\widehat{\mathcal{H}}'},\breve{\delta}'_{\widehat{\mathcal{H}}'}}$*, which is equivariant with the morphism* $\widehat{E}_{\breve{\Phi}'_{\widehat{\mathcal{H}}'}} \to \widehat{E}_{\breve{\Phi}_{\widehat{\mathcal{H}}}}$ *dual to the homomorphism* $\widehat{\mathbf{S}}_{\breve{\Phi}_{\widehat{\mathcal{H}}}} \to \widehat{\mathbf{S}}_{\breve{\Phi}'_{\widehat{\mathcal{H}}'}}$ *induced by the pair of morphisms* $(f_{\breve{X}} : \breve{X} \underset{\mathbb{Z}}{\otimes} \mathbb{Q} \xrightarrow{\sim} \breve{X}' \underset{\mathbb{Z}}{\otimes} \mathbb{Q}, f_{\breve{Y}} : \breve{Y}' \underset{\mathbb{Z}}{\otimes} \mathbb{Q} \xrightarrow{\sim} \breve{Y} \underset{\mathbb{Z}}{\otimes} \mathbb{Q})$ *defining the* $\widehat{g}$*-assignment* $(\breve{\mathrm{Z}}'_{\widehat{\mathcal{H}}'}, \breve{\Phi}'_{\widehat{\mathcal{H}}'}, \breve{\delta}'_{\widehat{\mathcal{H}}'}) \to_{\widehat{g}} (\breve{\mathrm{Z}}_{\widehat{\mathcal{H}}}, \breve{\Phi}_{\widehat{\mathcal{H}}}, \breve{\delta}_{\widehat{\mathcal{H}}})$ *of cusp labels (which is the* $\widetilde{g}$*-assignment for any element* $\widetilde{g} \in \widetilde{\mathrm{P}}_{\breve{\mathrm{Z}}}(\mathbb{A}^\infty) \cap \widetilde{\mathrm{P}}'_{\widetilde{\mathrm{Z}}}(\mathbb{A}^\infty)$ *lifting* $\widehat{g} \in \widehat{\mathrm{P}}_{\breve{\mathrm{Z}}}(\mathbb{A}^\infty) = (\widetilde{\mathrm{P}}_{\breve{\mathrm{Z}}}(\mathbb{A}^\infty) \cap \widetilde{\mathrm{P}}'_{\widetilde{\mathrm{Z}}}(\mathbb{A}^\infty))/\widetilde{\mathrm{U}}_{2,\widetilde{\mathrm{Z}}}(\mathbb{A}^\infty)$*, which is nevertheless independent of the choice of* $\widetilde{g}$*; cf. Lemma* 1.2.4.42 *and [62, Def. 5.4.3.9]).*

If $\widehat{g} \in \widehat{\mathrm{P}}_{\breve{\mathrm{Z}}}(\mathbb{A}^\infty)$ *is as above and if* $(\breve{\Phi}'_{\widehat{\mathcal{H}}'}, \breve{\delta}'_{\widehat{\mathcal{H}}'}, \widehat{\rho}')$ *is a* $\widehat{g}$*-refinement of* $(\breve{\Phi}_{\widehat{\mathcal{H}}}, \breve{\delta}_{\widehat{\mathcal{H}}}, \widehat{\rho})$ *(cf. Lemma* 1.2.4.42 *and [62, Def. 6.4.3.1]), then there is a canonical morphism*

$$[\widehat{g}] : \widehat{\Xi}_{\breve{\Phi}'_{\widehat{\mathcal{H}}'},\breve{\delta}'_{\widehat{\mathcal{H}}'}}(\widehat{\rho}') \to \widehat{\Xi}_{\breve{\Phi}_{\widehat{\mathcal{H}}},\breve{\delta}_{\widehat{\mathcal{H}}}}(\widehat{\rho}) \tag{1.3.2.68}$$

(cf. (1.3.2.46)*) covering* $[\widehat{g}] : \widehat{C}_{\breve{\Phi}'_{\widehat{\mathcal{H}}'},\breve{\delta}'_{\widehat{\mathcal{H}}'}} \to \widehat{C}_{\breve{\Phi}_{\widehat{\mathcal{H}}},\breve{\delta}_{\widehat{\mathcal{H}}}}$*, extending* $[\widehat{g}] : \widehat{\Xi}_{\breve{\Phi}'_{\widehat{\mathcal{H}}'},\breve{\delta}'_{\widehat{\mathcal{H}}'}} \to \widehat{\Xi}_{\breve{\Phi}_{\widehat{\mathcal{H}}},\breve{\delta}_{\widehat{\mathcal{H}}}}$*, mapping* $\widehat{\Xi}_{\breve{\Phi}'_{\widehat{\mathcal{H}}'},\breve{\delta}'_{\widehat{\mathcal{H}}'},\widehat{\rho}'}$ *to* $\widehat{\Xi}_{\breve{\Phi}_{\widehat{\mathcal{H}}},\breve{\delta}_{\widehat{\mathcal{H}}},\widehat{\rho}}$*, and inducing a canonical morphism*

$$[\widehat{g}] : \widehat{\mathfrak{X}}_{\breve{\Phi}'_{\widehat{\mathcal{H}}'},\breve{\delta}'_{\widehat{\mathcal{H}}'},\widehat{\rho}'} \to \widehat{\mathfrak{X}}_{\breve{\Phi}_{\widehat{\mathcal{H}}},\breve{\delta}_{\widehat{\mathcal{H}}},\widehat{\rho}} \tag{1.3.2.69}$$

(cf. (1.3.2.47)*). If* $\widehat{g} \in \widehat{\mathrm{P}}_{\breve{\mathrm{Z}}}(\mathbb{A}^\infty)$ *is as above and if* $(\breve{\Phi}'_{\widehat{\mathcal{H}}'}, \breve{\delta}'_{\widehat{\mathcal{H}}'}, \widehat{\Sigma}'_{\breve{\Phi}'_{\widehat{\mathcal{H}}'}})$ *is a* $\widehat{g}$*-refinement of* $(\breve{\Phi}_{\widehat{\mathcal{H}}}, \breve{\delta}_{\widehat{\mathcal{H}}}, \widehat{\Sigma}_{\breve{\Phi}_{\widehat{\mathcal{H}}}})$ *(cf. Lemma* 1.2.4.42 *and [62, Def. 6.4.3.2]), then morphisms like* (1.3.2.68) *patch together and define a canonical morphism*

$$[\widehat{g}] : \overline{\widehat{\Xi}}_{\breve{\Phi}'_{\widehat{\mathcal{H}}'},\breve{\delta}'_{\widehat{\mathcal{H}}'},\widehat{\Sigma}'_{\breve{\Phi}'_{\widehat{\mathcal{H}}'}}} \to \overline{\widehat{\Xi}}_{\breve{\Phi}_{\widehat{\mathcal{H}}},\breve{\delta}_{\widehat{\mathcal{H}}},\widehat{\Sigma}_{\breve{\Phi}_{\widehat{\mathcal{H}}}}} \tag{1.3.2.70}$$

(cf. (1.3.2.48)*) covering* $[\widehat{g}] : \widehat{C}_{\breve{\Phi}'_{\widehat{\mathcal{H}}'},\breve{\delta}'_{\widehat{\mathcal{H}}'}} \to \widehat{C}_{\breve{\Phi}_{\widehat{\mathcal{H}}},\breve{\delta}_{\widehat{\mathcal{H}}}}$*, extending* $[\widehat{g}] : \widehat{\Xi}_{\breve{\Phi}'_{\widehat{\mathcal{H}}'},\breve{\delta}'_{\widehat{\mathcal{H}}'}} \to \widehat{\Xi}_{\breve{\Phi}_{\widehat{\mathcal{H}}},\breve{\delta}_{\widehat{\mathcal{H}}}}$*, and inducing a canonical morphism*

$$[\widehat{g}] : \widehat{\mathfrak{X}}_{\breve{\Phi}'_{\widehat{\mathcal{H}}'},\breve{\delta}'_{\widehat{\mathcal{H}}'},\widehat{\Sigma}'_{\breve{\Phi}'_{\widehat{\mathcal{H}}'}}} \to \widehat{\mathfrak{X}}_{\breve{\Phi}_{\widehat{\mathcal{H}}},\breve{\delta}_{\widehat{\mathcal{H}}},\widehat{\Sigma}_{\breve{\Phi}_{\widehat{\mathcal{H}}}}} \tag{1.3.2.71}$$

(cf. (1.3.2.49)*) compatible with each* (1.3.2.69) *as above (under canonical morphisms).*

Proof. By Proposition 1.3.2.56 (see in particular (1.3.2.58)), the assertions in the first three paragraphs are reduced to the ones for the principal levels, which then follow from the corresponding assertions for the collection $\{\coprod \widetilde{\Xi}_{\breve{\Phi}_{\widetilde{\mathcal{H}}},\breve{\delta}_{\widetilde{\mathcal{H}}}}\}_{\widetilde{\mathcal{H}}_{\widetilde{\mathrm{P}}'_{\breve{\mathrm{Z}}}}}$ (by restricting the action of $\widetilde{\mathrm{P}}_{\breve{\mathrm{Z}}}(\mathbb{A}^\infty)$ to $\widetilde{\mathrm{P}}_{\breve{\mathrm{Z}}}(\mathbb{A}^\infty)\cap\widetilde{\mathrm{P}}'_{\breve{\mathrm{Z}}}(\mathbb{A}^\infty)$), because the tautological objects over $\widehat{\Xi}_{\breve{\Phi}_n,\breve{\delta}_n} = \widehat{\Xi}_{\breve{\Phi}_{\widehat{\mathcal{U}}(n)},\breve{\delta}_{\widehat{\mathcal{U}}(n)}}$ are canonically induced by those over $\widetilde{\Xi}_{\breve{\Phi}_n,\breve{\delta}_n} = \widetilde{\Xi}_{\breve{\Phi}_{\widetilde{\mathcal{U}}(n)},\breve{\delta}_{\widetilde{\mathcal{U}}(n)}}$. The assertions in the last paragraph then follow from the universal properties of toroidal embeddings (*cf.* [62, Prop. 6.2.5.11]). □

Lemma 1.3.2.72. (*Compare with Lemma* 1.3.2.50.) *By comparing the universal properties, we obtain a canonical morphism*

$$\widetilde{\Xi}_{\breve{\Phi}_{\widetilde{\mathcal{H}}},\breve{\delta}_{\widetilde{\mathcal{H}}}} \to \Xi_{\Phi_{\mathcal{H}},\delta_{\mathcal{H}}} \tag{1.3.2.73}$$

covering (1.3.2.51), *by sending $\breve{\tau}_{\widetilde{\mathcal{H}}}$, which is an orbit of étale-locally-defined trivializations $\breve{\tau}_n : \mathbf{1}_{\frac{1}{n}\breve{Y}\times\breve{X},S} \xrightarrow{\sim} (\breve{c}^\vee\times\breve{c})^*\mathcal{P}_B^{\otimes -1}$ for some integer $n\geq 1$ such that $\widetilde{\mathcal{U}}(n)\subset\widetilde{\mathcal{H}}$, to the orbit $\tau_{\mathcal{H}}$ of étale-locally-defined trivializations $\tau_n = \breve{\tau}_n|_{\mathbf{1}_{\frac{1}{n}Y\times X,S}}$.*

The morphisms (1.3.2.73) *and* (1.3.2.51) *induce a canonical morphism*

$$\widetilde{\Xi}_{\breve{\Phi}_{\widetilde{\mathcal{H}}},\breve{\delta}_{\widetilde{\mathcal{H}}}} \to \Xi_{\Phi_{\mathcal{H}},\delta_{\mathcal{H}}} \underset{C_{\Phi_{\mathcal{H}},\delta_{\mathcal{H}}}}{\times} \widetilde{C}_{\breve{\Phi}_{\widetilde{\mathcal{H}}},\breve{\delta}_{\widetilde{\mathcal{H}}}} \tag{1.3.2.74}$$

between torus torsors over $\widetilde{C}_{\breve{\Phi}_{\widetilde{\mathcal{H}}},\breve{\delta}_{\widetilde{\mathcal{H}}}}$, equivariant with the homomorphism $E_{\breve{\Phi}_{\widetilde{\mathcal{H}}}}\to E_{\Phi_{\mathcal{H}}}$ dual to the canonical homomorphism $\mathbf{S}_{\Phi_{\mathcal{H}}}\to\mathbf{S}_{\breve{\Phi}_{\widetilde{\mathcal{H}}}}$ (see (1.2.4.18)).

Suppose the image of a rational polyhedral cone $\breve{\rho}\subset(\mathbf{S}_{\breve{\Phi}_{\widetilde{\mathcal{H}}}})^\vee_{\mathbb{R}}$ under the (canonical) second morphism in (1.2.4.20) *is contained in some rational polyhedral cone $\rho\subset(\mathbf{S}_{\Phi_{\mathcal{H}}})^\vee_{\mathbb{R}}$. Then there is a canonical morphism*

$$\widetilde{\Xi}_{\breve{\Phi}_{\widetilde{\mathcal{H}}},\breve{\delta}_{\widetilde{\mathcal{H}}}}(\breve{\rho}) \to \Xi_{\Phi_{\mathcal{H}},\delta_{\mathcal{H}}}(\rho) \tag{1.3.2.75}$$

covering (1.3.2.51) *and extending* (1.3.2.73), *mapping $\widetilde{\Xi}_{\breve{\Phi}_{\widetilde{\mathcal{H}}},\breve{\delta}_{\widetilde{\mathcal{H}}},\breve{\rho}}$ to $\Xi_{\Phi_{\mathcal{H}},\delta_{\mathcal{H}},\rho}$ and inducing a canonical morphism*

$$\widetilde{\mathfrak{X}}_{\breve{\Phi}_{\widetilde{\mathcal{H}}},\breve{\delta}_{\widetilde{\mathcal{H}}},\breve{\rho}} \to \mathfrak{X}_{\Phi_{\mathcal{H}},\delta_{\mathcal{H}},\rho}. \tag{1.3.2.76}$$

If $\widetilde{\Sigma}_{\breve{\Phi}_{\widetilde{\mathcal{H}}}}$ and $\Sigma_{\Phi_{\mathcal{H}}}$ are cone decompositions of $\mathbf{P}_{\breve{\Phi}_{\widetilde{\mathcal{H}}}}$ and $\mathbf{P}_{\Phi_{\mathcal{H}}}$, respectively, such that the image of each $\breve{\rho}$ in $\widetilde{\Sigma}_{\breve{\Phi}_{\widetilde{\mathcal{H}}}}$ under the (canonical) second morphism in (1.2.4.20) *is contained in some $\rho\in\Sigma_{\Phi_{\mathcal{H}}}$, then morphisms like* (1.3.2.75) *patch together and define a canonical morphism*

$$\overline{\widetilde{\Xi}}_{\breve{\Phi}_{\widetilde{\mathcal{H}}},\breve{\delta}_{\widetilde{\mathcal{H}}},\widetilde{\Sigma}_{\breve{\Phi}_{\widetilde{\mathcal{H}}}}} \to \overline{\Xi}_{\Phi_{\mathcal{H}},\delta_{\mathcal{H}},\Sigma_{\Phi_{\mathcal{H}}}} \tag{1.3.2.77}$$

covering (1.3.2.51), *extending* (1.3.2.73), *and inducing a canonical morphism*

$$\widetilde{\mathfrak{X}}_{\breve{\Phi}_{\widetilde{\mathcal{H}}},\breve{\delta}_{\widetilde{\mathcal{H}}},\widetilde{\Sigma}_{\breve{\Phi}_{\widetilde{\mathcal{H}}}}} \to \mathfrak{X}_{\Phi_{\mathcal{H}},\delta_{\mathcal{H}},\Sigma_{\Phi_{\mathcal{H}}}} \tag{1.3.2.78}$$

compatible with each (1.3.2.76) *as above (under canonical morphisms).*

Proof. The statements are self-explanatory. □

Lemma 1.3.2.79. (*Compare with Lemmas* 1.3.2.50 *and* 1.3.2.72.) *By comparing the universal properties* (*cf. Proposition* 1.3.2.56), *we obtain a canonical morphism*

$$\widehat{\Xi}_{\breve{\Phi}_{\widehat{\mathcal{H}}},\breve{\delta}_{\widehat{\mathcal{H}}}} \to \Xi_{\Phi_{\mathcal{H}},\delta_{\mathcal{H}}} \tag{1.3.2.80}$$

covering (1.3.2.51), *by sending the pair* $(\widehat{\tau}_{\widehat{\mathcal{H}}}, \widehat{\tau}^\vee_{\widehat{\mathcal{H}}})$, *which is an orbit of étale-locally-defined pairs* $(\widehat{\tau}_n, \widehat{\tau}^\vee_n)$ *for some integer* $n \geq 1$ *such that* $\widetilde{\mathcal{U}}(n) \subset \widetilde{\mathcal{H}}$, *to the orbit* $\tau_{\mathcal{H}}$ *of étale-locally-defined* $\tau_n = \widehat{\tau}_n|_{\mathbf{1}_{\frac{1}{n}Y \times X,S}} = \widehat{\tau}^\vee_n|_{\mathbf{1}_{\frac{1}{n}Y \times X,S}}$, *as in Proposition* 1.3.2.56.

The morphisms (1.3.2.80) *and* (1.3.2.51) *induce a canonical morphism*

$$\widehat{\Xi}_{\breve{\Phi}_{\widehat{\mathcal{H}}},\breve{\delta}_{\widehat{\mathcal{H}}}} \to \Xi_{\Phi_{\mathcal{H}},\delta_{\mathcal{H}}} \underset{C_{\Phi_{\mathcal{H}},\delta_{\mathcal{H}}}}{\times} \widehat{C}_{\breve{\Phi}_{\widehat{\mathcal{H}}},\breve{\delta}_{\widehat{\mathcal{H}}}} \tag{1.3.2.81}$$

between torus torsors over $\widehat{C}_{\breve{\Phi}_{\widehat{\mathcal{H}}},\breve{\delta}_{\widehat{\mathcal{H}}}} = \widetilde{C}_{\breve{\Phi}_{\widetilde{\mathcal{H}}},\breve{\delta}_{\widetilde{\mathcal{H}}}}$, *equivariant with the surjective homomorphism* $\widehat{E}_{\breve{\Phi}_{\widehat{\mathcal{H}}}} \to E_{\Phi_{\mathcal{H}}}$ (*see Proposition* 1.3.2.56) *dual to the canonical injective homomorphism* $\mathbf{S}_{\Phi_{\mathcal{H}}} \hookrightarrow \widehat{\mathbf{S}}_{\breve{\Phi}_{\widehat{\mathcal{H}}}}$ (*see Definition* 1.2.4.29).

Suppose the image of a rational polyhedral cone $\widehat{\rho} \subset (\widehat{\mathbf{S}}_{\breve{\Phi}_{\widetilde{\mathcal{H}}}})^\vee_{\mathbb{R}}$ *under* (1.2.4.37) *is contained in some rational polyhedral cone* $\rho \subset (\mathbf{S}_{\Phi_{\mathcal{H}}})^\vee_{\mathbb{R}}$. *Then there is a canonical morphism*

$$\widehat{\Xi}_{\breve{\Phi}_{\widehat{\mathcal{H}}},\breve{\delta}_{\widehat{\mathcal{H}}}}(\widehat{\rho}) \to \Xi_{\Phi_{\mathcal{H}},\delta_{\mathcal{H}}}(\rho) \tag{1.3.2.82}$$

(*cf.* (1.3.2.75)) *covering* (1.3.2.51) *and extending* (1.3.2.80), *mapping* $\widehat{\Xi}_{\breve{\Phi}_{\widehat{\mathcal{H}}},\breve{\delta}_{\widehat{\mathcal{H}}},\widehat{\rho}}$ *to* $\Xi_{\Phi_{\mathcal{H}},\delta_{\mathcal{H}},\rho}$ *and inducing a canonical morphism*

$$\widehat{\mathfrak{X}}_{\breve{\Phi}_{\widehat{\mathcal{H}}},\breve{\delta}_{\widehat{\mathcal{H}}},\widehat{\rho}} \to \mathfrak{X}_{\Phi_{\mathcal{H}},\delta_{\mathcal{H}},\rho} \tag{1.3.2.83}$$

(*cf.* (1.3.2.76)). *If* $\widehat{\Sigma}_{\breve{\Phi}_{\widehat{\mathcal{H}}}}$ *and* $\Sigma_{\Phi_{\mathcal{H}}}$ *are cone decompositions of* $\widehat{\mathbf{P}}_{\breve{\Phi}_{\widehat{\mathcal{H}}}}$ *and* $\mathbf{P}_{\Phi_{\mathcal{H}}}$, *respectively, such that the image of each* $\widehat{\rho}$ *in* $\widehat{\Sigma}_{\breve{\Phi}_{\widehat{\mathcal{H}}}}$ *under* (1.2.4.37) *is contained in some* $\rho \in \Sigma_{\Phi_{\mathcal{H}}}$, *then morphisms like* (1.3.2.82) *patch together and define a canonical morphism*

$$\overline{\widehat{\Xi}}_{\breve{\Phi}_{\widehat{\mathcal{H}}},\breve{\delta}_{\widehat{\mathcal{H}}},\widehat{\Sigma}_{\breve{\Phi}_{\widehat{\mathcal{H}}}}} \to \overline{\Xi}_{\Phi_{\mathcal{H}},\delta_{\mathcal{H}},\Sigma_{\Phi_{\mathcal{H}}}} \tag{1.3.2.84}$$

(*cf.* (1.3.2.77)) *covering* (1.3.2.51), *extending* (1.3.2.80), *and inducing a canonical morphism*

$$\widehat{\mathfrak{X}}_{\breve{\Phi}_{\widehat{\mathcal{H}}},\breve{\delta}_{\widehat{\mathcal{H}}},\widehat{\Sigma}_{\breve{\Phi}_{\widehat{\mathcal{H}}}}} \to \mathfrak{X}_{\Phi_{\mathcal{H}},\delta_{\mathcal{H}},\Sigma_{\Phi_{\mathcal{H}}}} \tag{1.3.2.85}$$

(*cf.* (1.3.2.78)) *compatible with each* (1.3.2.83) *as above* (*under canonical morphisms*).

By the same argument in [61, Sec. 3C], using the extended Kodaira–Spencer isomorphism as in [62, Prop. 6.2.5.18], the morphism (1.3.2.84) *is log smooth and we have a canonical isomorphism*

$$\begin{aligned}&\overline{\Omega}^1_{\overline{\widehat{\Xi}}_{\breve{\Phi}_{\widehat{\mathcal{H}}},\breve{\delta}_{\widehat{\mathcal{H}}},\widehat{\Sigma}_{\breve{\Phi}_{\widehat{\mathcal{H}}}}}/\overline{\Xi}_{\Phi_{\mathcal{H}},\delta_{\mathcal{H}},\Sigma_{\Phi_{\mathcal{H}}}}} \\ &\cong (\overline{\widehat{\Xi}}_{\breve{\Phi}_{\widehat{\mathcal{H}}},\breve{\delta}_{\widehat{\mathcal{H}}},\widehat{\Sigma}_{\breve{\Phi}_{\widehat{\mathcal{H}}}}} \to C_{\Phi_{\mathcal{H}},\delta_{\mathcal{H}}})^* \underline{\mathrm{Hom}}_{\mathcal{O}}(\widetilde{X}, \underline{\mathrm{Lie}}^\vee_{G^\natural/C_{\Phi_{\mathcal{H}},\delta_{\mathcal{H}}}}),\end{aligned} \tag{1.3.2.86}$$

where

$$\overline{\Omega}^1_{\overline{\widehat{\Xi}}_{\breve{\Phi}_{\widehat{\mathcal{H}}},\breve{\delta}_{\widehat{\mathcal{H}}},\widehat{\Sigma}_{\breve{\Phi}_{\widehat{\mathcal{H}}}}}/\overline{\Xi}_{\Phi_{\mathcal{H}},\delta_{\mathcal{H}},\Sigma_{\Phi_{\mathcal{H}}}}} := (\Omega^1_{\overline{\widehat{\Xi}}_{\breve{\Phi}_{\widehat{\mathcal{H}}},\breve{\delta}_{\widehat{\mathcal{H}}},\widehat{\Sigma}_{\breve{\Phi}_{\widehat{\mathcal{H}}}}}/C_{\Phi_{\mathcal{H}},\delta_{\mathcal{H}}}}[d\log\infty])/$$
$$((\overline{\widehat{\Xi}}_{\breve{\Phi}_{\widehat{\mathcal{H}}},\breve{\delta}_{\widehat{\mathcal{H}}},\widehat{\Sigma}_{\breve{\Phi}_{\widehat{\mathcal{H}}}}} \to \overline{\Xi}_{\Phi_{\mathcal{H}},\delta_{\mathcal{H}},\Sigma_{\Phi_{\mathcal{H}}}})^*\Omega^1_{\overline{\Xi}_{\Phi_{\mathcal{H}},\delta_{\mathcal{H}},\Sigma_{\Phi_{\mathcal{H}}}}/C_{\Phi_{\mathcal{H}},\delta_{\mathcal{H}}}}[d\log\infty])$$

is the sheaf of modules of relative log 1*-differentials, and where* $G^\natural \to C_{\Phi_{\mathcal{H}},\delta_{\mathcal{H}}}$ *is the tautological semi-abelian scheme as in Proposition* 1.3.2.12*. Moreover, the canonical morphism*

$$\overline{\widehat{\Xi}}_{\breve{\Phi}_{\widehat{\mathcal{H}}},\breve{\delta}_{\widehat{\mathcal{H}}},\widehat{\Sigma}_{\breve{\Phi}_{\widehat{\mathcal{H}}}}} \to \overline{\Xi}_{\Phi_{\mathcal{H}},\delta_{\mathcal{H}},\Sigma_{\Phi_{\mathcal{H}}}} \underset{C_{\Phi_{\mathcal{H}},\delta_{\mathcal{H}}}}{\times} \widehat{C}_{\breve{\Phi}_{\widehat{\mathcal{H}}},\breve{\delta}_{\widehat{\mathcal{H}}}} \tag{1.3.2.87}$$

(induced by (1.3.2.51) *and* (1.3.2.84)*) induces a canonical short exact sequence*

$$\begin{aligned} 0 &\to (\overline{\widehat{\Xi}}_{\breve{\Phi}_{\widehat{\mathcal{H}}},\breve{\delta}_{\widehat{\mathcal{H}}},\widehat{\Sigma}_{\breve{\Phi}_{\widehat{\mathcal{H}}}}} \to \widehat{C}_{\breve{\Phi}_{\widehat{\mathcal{H}}},\breve{\delta}_{\widehat{\mathcal{H}}}})^*\Omega^1_{\widehat{C}_{\breve{\Phi}_{\widehat{\mathcal{H}}},\breve{\delta}_{\widehat{\mathcal{H}}}}/C_{\Phi_{\mathcal{H}},\delta_{\mathcal{H}}}} \\ &\to \overline{\Omega}^1_{\overline{\widehat{\Xi}}_{\breve{\Phi}_{\widehat{\mathcal{H}}},\breve{\delta}_{\widehat{\mathcal{H}}},\widehat{\Sigma}_{\breve{\Phi}_{\widehat{\mathcal{H}}}}}/\overline{\Xi}_{\Phi_{\mathcal{H}},\delta_{\mathcal{H}},\Sigma_{\Phi_{\mathcal{H}}}}} \\ &\to \overline{\Omega}^1_{\overline{\widehat{\Xi}}_{\breve{\Phi}_{\widehat{\mathcal{H}}},\breve{\delta}_{\widehat{\mathcal{H}}},\widehat{\Sigma}_{\breve{\Phi}_{\widehat{\mathcal{H}}}}}/(\overline{\Xi}_{\Phi_{\mathcal{H}},\delta_{\mathcal{H}},\Sigma_{\Phi_{\mathcal{H}}}} \underset{C_{\Phi_{\mathcal{H}},\delta_{\mathcal{H}}}}{\times} \widehat{C}_{\breve{\Phi}_{\widehat{\mathcal{H}}},\breve{\delta}_{\widehat{\mathcal{H}}}})} \to 0, \end{aligned} \tag{1.3.2.88}$$

where

$$\overline{\Omega}^1_{\overline{\widehat{\Xi}}_{\breve{\Phi}_{\widehat{\mathcal{H}}},\breve{\delta}_{\widehat{\mathcal{H}}},\widehat{\Sigma}_{\breve{\Phi}_{\widehat{\mathcal{H}}}}}/(\overline{\Xi}_{\Phi_{\mathcal{H}},\delta_{\mathcal{H}},\Sigma_{\Phi_{\mathcal{H}}}} \underset{C_{\Phi_{\mathcal{H}},\delta_{\mathcal{H}}}}{\times} \widehat{C}_{\breve{\Phi}_{\widehat{\mathcal{H}}},\breve{\delta}_{\widehat{\mathcal{H}}}})} := (\Omega^1_{\overline{\widehat{\Xi}}_{\breve{\Phi}_{\widehat{\mathcal{H}}},\breve{\delta}_{\widehat{\mathcal{H}}},\widehat{\Sigma}_{\breve{\Phi}_{\widehat{\mathcal{H}}}}}/\widehat{C}_{\breve{\Phi}_{\widehat{\mathcal{H}}},\breve{\delta}_{\widehat{\mathcal{H}}}}}[d\log\infty])/$$
$$((\overline{\widehat{\Xi}}_{\breve{\Phi}_{\widehat{\mathcal{H}}},\breve{\delta}_{\widehat{\mathcal{H}}},\widehat{\Sigma}_{\breve{\Phi}_{\widehat{\mathcal{H}}}}} \to \overline{\Xi}_{\Phi_{\mathcal{H}},\delta_{\mathcal{H}},\Sigma_{\Phi_{\mathcal{H}}}})^*\Omega^1_{\overline{\Xi}_{\Phi_{\mathcal{H}},\delta_{\mathcal{H}},\Sigma_{\Phi_{\mathcal{H}}}}/C_{\Phi_{\mathcal{H}},\delta_{\mathcal{H}}}}[d\log\infty])$$

is the sheaf of modules of relative log 1*-differentials, which is exact and has locally free terms, and which can be canonically identified with the pullback under* $\overline{\widehat{\Xi}}_{\breve{\Phi}_{\widehat{\mathcal{H}}},\breve{\delta}_{\widehat{\mathcal{H}}},\widehat{\Sigma}_{\breve{\Phi}_{\widehat{\mathcal{H}}}}} \to C_{\Phi_{\mathcal{H}},\delta_{\mathcal{H}}}$ *of the canonical short exact sequence*

$$\begin{aligned} 0 \to \underline{\mathrm{Hom}}_{\mathcal{O}}(\widetilde{X}, \underline{\mathrm{Lie}}^\vee_{B/C_{\Phi_{\mathcal{H}},\delta_{\mathcal{H}}}}) &\to \underline{\mathrm{Hom}}_{\mathcal{O}}(\widetilde{X}, \underline{\mathrm{Lie}}^\vee_{G^\natural/C_{\Phi_{\mathcal{H}},\delta_{\mathcal{H}}}}) \\ &\to \underline{\mathrm{Hom}}_{\mathcal{O}}(\widetilde{X}, \underline{\mathrm{Lie}}^\vee_{T/C_{\Phi_{\mathcal{H}},\delta_{\mathcal{H}}}}) \to 0 \end{aligned} \tag{1.3.2.89}$$

of locally free sheaves (compatible with (1.3.2.54)*). Hence,* (1.3.2.87) *is also log smooth (by [45, 3.12]).*

Proof. The statements are self-explanatory. □

Proposition 1.3.2.90. *(Compare with Proposition* 1.3.2.55.*) Under the canonical morphisms as in* (1.3.2.73) *(with varying* $\widetilde{\mathcal{H}}$ *and* $\mathcal{H}$*), and under the canonical homomorphisms* $\widetilde{\mathrm{P}}'_{\breve{\mathrm{Z}}}(\mathbb{A}^\infty) \to \mathrm{P}'_{\mathrm{Z}}(\mathbb{A}^\infty)$ *and* $\widetilde{\mathrm{P}}_{\breve{\mathrm{Z}},\widetilde{\mathrm{Z}}}(\mathbb{A}^\infty) \to \mathrm{P}_{\mathrm{Z}}(\mathbb{A}^\infty)$*, the Hecke action of* $\widetilde{\mathrm{P}}'_{\breve{\mathrm{Z}}}(\mathbb{A}^\infty)$ *on the collection* $\{\widetilde{\Xi}_{\breve{\Phi}_{\widetilde{\mathcal{H}}},\breve{\delta}_{\widetilde{\mathcal{H}}}}\}_{\widetilde{\mathcal{H}}_{\widetilde{\mathrm{P}}'_{\breve{\mathrm{Z}}}}}$ *is compatible with the Hecke action of* $\mathrm{P}'_{\mathrm{Z}}(\mathbb{A}^\infty)$ *on the collection* $\{\Xi_{\Phi_{\mathcal{H}},\delta_{\mathcal{H}}}\}_{\mathcal{H}_{\mathrm{P}'_{\mathrm{Z}}}}$*; and the Hecke action of* $\widetilde{\mathrm{P}}_{\breve{\mathrm{Z}},\widetilde{\mathrm{Z}}}(\mathbb{A}^\infty)$ *on the collection* $\{\coprod \widetilde{\Xi}_{\breve{\Phi}_{\widetilde{\mathcal{H}}},\breve{\delta}_{\widetilde{\mathcal{H}}}}\}_{\widetilde{\mathcal{H}}_{\widetilde{\mathrm{P}}_{\breve{\mathrm{Z}},\widetilde{\mathrm{Z}}}}}$ *is compatible with the Hecke action of* $\mathrm{P}_{\mathrm{Z}}(\mathbb{A}^\infty)$ *on the collection* $\{\coprod \Xi_{\Phi_{\mathcal{H}},\delta_{\mathcal{H}}}\}_{\mathcal{H}_{\mathrm{P}_{\mathrm{Z}}}}$*, where the index sets are as in Proposition* 1.3.2.45*. These Hecke actions are all compatible with those in Proposition* 1.3.2.55*. They are also compatible with extensions to toroidal embeddings and their formal completions.*

Proof. As in the case of Proposition 1.3.2.55, the canonical morphisms as in (1.3.2.80) correspond to pushouts of extensions of B (resp. $B^\vee$) by $\breve{T}$ (resp. $\breve{T}^\vee$) under the canonical homomorphism $\breve{T} \to T$ (resp. $\breve{T}^\vee \to T^\vee$) induced by the restriction from $\breve{X}$ (resp. $\breve{Y}$) to X (resp. Y). Hence, the realizations of the Hecke twists are compatible in the desired ways. (We omit the details for simplicity.) □

Proposition 1.3.2.91. (*Compare with Propositions* 1.3.2.55 *and* 1.3.2.90.) *Under the canonical morphisms as in* (1.3.2.80) (*with varying* $\widehat{\mathcal{H}}$ *and* $\mathcal{H}$), *and under the canonical homomorphisms* $\widehat{\mathrm{P}}'_{\breve{\mathrm{Z}}}(\mathbb{A}^\infty) \to \mathrm{P}'_{\mathrm{Z}}(\mathbb{A}^\infty)$ *and* $\widehat{\mathrm{P}}_{\breve{\mathrm{Z}}}(\mathbb{A}^\infty) \to \mathrm{P}_{\mathrm{Z}}(\mathbb{A}^\infty)$, *the Hecke action of* $\widehat{\mathrm{P}}'_{\breve{\mathrm{Z}}}(\mathbb{A}^\infty)$ *on the collection* $\{\widehat{\Xi}_{\breve{\Phi}_{\widehat{\mathcal{H}}},\breve{\delta}_{\widehat{\mathcal{H}}}}\}_{\widehat{\mathcal{H}}_{\widehat{\mathrm{P}}'_{\breve{\mathrm{Z}}}}}$ *is compatible with the Hecke action of* $\mathrm{P}'_{\mathrm{Z}}(\mathbb{A}^\infty)$ *on the collection* $\{\Xi_{\Phi_{\mathcal{H}},\delta_{\mathcal{H}}}\}_{\mathcal{H}_{\mathrm{P}'_{\mathrm{Z}}}}$; *and the Hecke action of* $\widehat{\mathrm{P}}_{\breve{\mathrm{Z}}}(\mathbb{A}^\infty)$ *on the collection* $\{\coprod \widehat{\Xi}_{\breve{\Phi}_{\widehat{\mathcal{H}}},\breve{\delta}_{\widehat{\mathcal{H}}}}\}_{\widehat{\mathcal{H}}_{\widehat{\mathrm{P}}_{\breve{\mathrm{Z}}}}}$ *is compatible with the Hecke action of* $\mathrm{P}_{\mathrm{Z}}(\mathbb{A}^\infty)$ *on the collection* $\{\coprod \Xi_{\Phi_{\mathcal{H}},\delta_{\mathcal{H}}}\}_{\mathcal{H}_{\mathrm{P}_{\mathrm{Z}}}}$, *where the index sets are as in Proposition* 1.3.2.45. *These Hecke actions are all compatible with those in Proposition* 1.3.2.55. *They are also compatible with extensions to toroidal embeddings and their formal completions.*

Proof. As in the proof of Proposition 1.3.2.67, the Hecke action of $\widehat{\mathrm{P}}_{\breve{\mathrm{Z}}}(\mathbb{A}^\infty)$ on the collection $\{\coprod \widehat{\Xi}_{\breve{\Phi}_{\widehat{\mathcal{H}}},\breve{\delta}_{\widehat{\mathcal{H}}}}\}_{\widehat{\mathcal{H}}_{\widehat{\mathrm{P}}_{\breve{\mathrm{Z}}}}}$ is induced by the Hecke action of $\widetilde{\mathrm{P}}_{\breve{\mathrm{Z}},\widetilde{\mathrm{Z}}}(\mathbb{A}^\infty)$ on the collection $\{\coprod \widetilde{\Xi}_{\breve{\Phi}_{\widetilde{\mathcal{H}}},\breve{\delta}_{\widetilde{\mathcal{H}}}}\}_{\widetilde{\mathcal{H}}_{\widetilde{\mathrm{P}}'_{\breve{\mathrm{Z}}}}}$. Hence, these statements follow from the corresponding statements of Proposition 1.3.2.90. □

1.3.3 *Toroidal Compactifications of PEL-Type Kuga Families and Their Generalizations*

For simplicity, in the remainder of this subsection, all morphisms between schemes or algebraic stacks over $\mathsf{S}_0 = \mathrm{Spec}(F_0)$ will be defined over S_0, unless otherwise specified.

Let Q be any $\mathcal{O}$-lattice. Consider the abelian scheme $G_{\mathsf{M}_\mathcal{H}}$ over $\mathsf{M}_\mathcal{H}$ in (1) of Theorem 1.3.1.3. By [62, Prop. 5.2.3.9], the group functor $\underline{\mathrm{Hom}}_\mathcal{O}(Q, G_{\mathsf{M}_\mathcal{H}})$ over $\mathsf{M}_\mathcal{H}$ is relatively representable by a proper smooth group scheme which is an extension of a finite étale group scheme, whose rank has no prime factors other than those of $\mathrm{Disc} = \mathrm{Disc}_{\mathcal{O}/\mathbb{Z}}$ [62, Def. 1.1.1.6], by an abelian scheme $\underline{\mathrm{Hom}}_\mathcal{O}(Q, G_{\mathsf{M}_\mathcal{H}})^\circ$, the *fiberwise geometric identity component* of $\underline{\mathrm{Hom}}_\mathcal{O}(Q, G_{\mathsf{M}_\mathcal{H}})$ (see [62, Def. 5.2.3.8]).

Example 1.3.3.1. If $Q \cong \mathcal{O}^{\oplus s}$ for some integer $s \geq 0$, then $\underline{\mathrm{Hom}}_\mathcal{O}(Q, G_{\mathsf{M}_\mathcal{H}})^\circ = \underline{\mathrm{Hom}}_\mathcal{O}(Q, G_{\mathsf{M}_\mathcal{H}}) \cong G_{\mathsf{M}_\mathcal{H}}^{\times s}$ is the s-fold fiber product of $G_{\mathsf{M}_\mathcal{H}}$ over $\mathsf{M}_\mathcal{H}$.

Example 1.3.3.2. If $\mathcal{O} \cong \mathrm{M}_k(\mathcal{O}_F)$ and Q is of finite index in $\mathcal{O}_F^{\oplus k}$ for some integer $k \geq 1$, then the relative dimension of $\underline{\mathrm{Hom}}_\mathcal{O}(Q, G_{\mathsf{M}_\mathcal{H}})^\circ$ over $\mathsf{M}_\mathcal{H}$ is $1/k$ of the relative dimension of $G_{\mathsf{M}_\mathcal{H}}$ over $\mathsf{M}_\mathcal{H}$.

Definition 1.3.3.3. (*See [61, Def. 2.4].*) *A* **Kuga family** *over* $\mathsf{M}_\mathcal{H}$ *is an abelian scheme* $\mathsf{N}^{\mathrm{grp}} \to \mathsf{M}_\mathcal{H}$ *that is* $\mathbb{Q}^\times$*-isogenous to* $\underline{\mathrm{Hom}}_\mathcal{O}(Q, G_{\mathsf{M}_\mathcal{H}})^\circ$ *for some* $\mathcal{O}$*-lattice* Q.

Definition 1.3.3.4. *A* **generalized Kuga family** *over* $\mathsf{M}_{\mathcal{H}}$ *is a torsor* $\mathsf{N} \to \mathsf{M}_{\mathcal{H}}$ *under some Kuga family* $\mathsf{N}^{\mathrm{grp}} \to \mathsf{M}_{\mathcal{H}}$ *as in Definition* 1.3.3.3.

Lemma 1.3.3.5. (*See [61, Lem. 2.6].*) *The abelian scheme* $\underline{\mathrm{Hom}}_{\mathbb{Z}}(Q^\vee, G^\vee_{\mathsf{M}_{\mathcal{H}}})$ *is isomorphic to the dual abelian scheme of* $\underline{\mathrm{Hom}}_{\mathbb{Z}}(Q, G_{\mathsf{M}_{\mathcal{H}}})$.

Lemma 1.3.3.6. (*See [61, Lem. 2.9].*) *Let* $j_Q : Q^\vee \hookrightarrow Q$ *be as in Lemma* 1.2.4.1. *Then the isogeny*

$$\lambda_{\mathsf{M}_{\mathcal{H}}, j_Q, \mathbb{Z}} : \underline{\mathrm{Hom}}_{\mathbb{Z}}(Q, G_{\mathsf{M}_{\mathcal{H}}}) \to \underline{\mathrm{Hom}}_{\mathbb{Z}}(Q^\vee, G^\vee_{\mathsf{M}_{\mathcal{H}}})$$

induced canonically by j_Q *and* $\lambda_{\mathsf{M}_{\mathcal{H}}} : G_{\mathsf{M}_{\mathcal{H}}} \to G^\vee_{\mathsf{M}_{\mathcal{H}}}$ *is a polarization.*

Proposition 1.3.3.7. (*See [61, Prop. 2.10 and Cor. 2.12].*) *The abelian scheme* $\underline{\mathrm{Hom}}_{\mathcal{O}}(Q^\vee, G^\vee_{\mathsf{M}_{\mathcal{H}}})^\circ$ *is* $\mathbb{Q}^\times$*-isogenous to the dual abelian scheme of* $\underline{\mathrm{Hom}}_{\mathcal{O}}(Q, G_{\mathsf{M}_{\mathcal{H}}})^\circ$. *Moreover, given any* $j_Q : Q^\vee \hookrightarrow Q$ *as in Lemma* 1.2.4.1, *the composition*

$$\begin{aligned}\lambda_{\mathsf{M}_{\mathcal{H}}, j_Q} : \underline{\mathrm{Hom}}_{\mathcal{O}}(Q, G_{\mathsf{M}_{\mathcal{H}}})^\circ &\hookrightarrow \underline{\mathrm{Hom}}_{\mathbb{Z}}(Q, G_{\mathsf{M}_{\mathcal{H}}}) \\ &\overset{\lambda_{\mathsf{M}_{\mathcal{H}}, j_Q, \mathbb{Z}}}{\to} \underline{\mathrm{Hom}}_{\mathbb{Z}}(Q^\vee, G^\vee_{\mathsf{M}_{\mathcal{H}}}) \twoheadrightarrow (\underline{\mathrm{Hom}}_{\mathcal{O}}(Q, G_{\mathsf{M}_{\mathcal{H}}})^\circ)^\vee\end{aligned} \tag{1.3.3.8}$$

induced canonically by j_Q *and the polarization* $\lambda_{\mathsf{M}_{\mathcal{H}}} : G_{\mathsf{M}_{\mathcal{H}}} \to G^\vee_{\mathsf{M}_{\mathcal{H}}}$ *is a polarization.*

Definition 1.3.3.9. *Let* $\mathsf{N} \to \mathsf{M}_{\mathcal{H}}$ *be as in Definition* 1.3.3.4. *Then we define the dual* $\mathsf{N}^\vee \to \mathsf{M}_{\mathcal{H}}$ *to be the dual abelian scheme* $\mathsf{N}^{\mathrm{grp},\vee} \to \mathsf{M}_{\mathcal{H}}$ *of* $\mathsf{N}^{\mathrm{grp}} \to \mathsf{M}_{\mathcal{H}}$.

Remark 1.3.3.10. By [92, XIII, Prop. 1.1], $\mathsf{N}^\vee = \mathsf{N}^{\mathrm{grp},\vee} \to \mathsf{M}_{\mathcal{H}}$ is canonically isomorphic to $\underline{\mathrm{Pic}}^0(\mathsf{N}/\mathsf{M}_{\mathcal{H}}) \to \mathsf{M}_{\mathcal{H}}$ (which can be defined as in the case of abelian schemes; *cf.* [62, Def. 1.3.2.1]). Note that this is always a group scheme, with its identity section, even when $\mathsf{N} \to \mathsf{M}_{\mathcal{H}}$ is a nontrivial torsor of $\mathsf{N}^{\mathrm{grp}} \to \mathsf{M}_{\mathcal{H}}$.

Definition 1.3.3.11. *By abuse of notation, we denote by* $\underline{\mathrm{Lie}}_{\mathsf{N}/\mathsf{M}_{\mathcal{H}}}$ (*resp.* $\underline{\mathrm{Lie}}^\vee_{\mathsf{N}/\mathsf{M}_{\mathcal{H}}}$, *resp.* $\underline{\mathrm{Lie}}_{\mathsf{N}^\vee/\mathsf{M}_{\mathcal{H}}}$, *resp.* $\underline{\mathrm{Lie}}^\vee_{\mathsf{N}^\vee/\mathsf{M}_{\mathcal{H}}}$) *the locally free sheaf* $\underline{\mathrm{Lie}}_{\mathsf{N}^{\mathrm{grp}}/\mathsf{M}_{\mathcal{H}}}$ (*resp.* $\underline{\mathrm{Lie}}^\vee_{\mathsf{N}^{\mathrm{grp}}/\mathsf{M}_{\mathcal{H}}}$, *resp.* $\underline{\mathrm{Lie}}_{\mathsf{N}^{\mathrm{grp},\vee}/\mathsf{M}_{\mathcal{H}}}$, *resp.* $\underline{\mathrm{Lie}}^\vee_{\mathsf{N}^{\mathrm{grp},\vee}/\mathsf{M}_{\mathcal{H}}}$) *over* $\mathsf{M}_{\mathcal{H}}$, *although* $\mathsf{N} \to \mathsf{M}_{\mathcal{H}}$ *might have no section.*

Lemma 1.3.3.12. *We have:*

$$\begin{aligned}
\underline{\mathrm{Lie}}^\vee_{\mathsf{N}/\mathsf{M}_{\mathcal{H}}} &\cong \underline{\mathrm{Hom}}_{\mathscr{O}_{\mathsf{M}_{\mathcal{H}}}}(\underline{\mathrm{Lie}}_{\mathsf{N}/\mathsf{M}_{\mathcal{H}}}, \mathscr{O}_{\mathsf{M}_{\mathcal{H}}}),\\
\underline{\mathrm{Lie}}^\vee_{\mathsf{N}^\vee/\mathsf{M}_{\mathcal{H}}} &\cong \underline{\mathrm{Hom}}_{\mathscr{O}_{\mathsf{M}_{\mathcal{H}}}}(\underline{\mathrm{Lie}}_{\mathsf{N}^\vee/\mathsf{M}_{\mathcal{H}}}, \mathscr{O}_{\mathsf{M}_{\mathcal{H}}}),\\
\Omega^1_{\mathsf{N}/\mathsf{M}_{\mathcal{H}}} &\cong (\mathsf{N} \to \mathsf{M}_{\mathcal{H}})^* \underline{\mathrm{Lie}}^\vee_{\mathsf{N}/\mathsf{M}_{\mathcal{H}}},\\
\Omega^1_{\mathsf{N}^\vee/\mathsf{M}_{\mathcal{H}}} &\cong (\mathsf{N}^\vee \to \mathsf{M}_{\mathcal{H}})^* \underline{\mathrm{Lie}}^\vee_{\mathsf{N}^\vee/\mathsf{M}_{\mathcal{H}}},\\
(\mathsf{N} \to \mathsf{M}_{\mathcal{H}})_* \Omega^1_{\mathsf{N}/\mathsf{M}_{\mathcal{H}}} &\cong \underline{\mathrm{Lie}}^\vee_{\mathsf{N}/\mathsf{M}_{\mathcal{H}}},\\
(\mathsf{N}^\vee \to \mathsf{M}_{\mathcal{H}})_* \Omega^1_{\mathsf{N}^\vee/\mathsf{M}_{\mathcal{H}}} &\cong \underline{\mathrm{Lie}}^\vee_{\mathsf{N}^\vee/\mathsf{M}_{\mathcal{H}}},\\
R^1(\mathsf{N} \to \mathsf{M}_{\mathcal{H}})_* \mathscr{O}_{\mathsf{N}} &\cong \underline{\mathrm{Lie}}_{\mathsf{N}^\vee/\mathsf{M}_{\mathcal{H}}},\\
R^1(\mathsf{N}^\vee \to \mathsf{M}_{\mathcal{H}})_* \mathscr{O}_{\mathsf{N}^\vee} &\cong \underline{\mathrm{Lie}}_{\mathsf{N}/\mathsf{M}_{\mathcal{H}}}.
\end{aligned}$$

The relative de Rham cohomology

$$\underline{H}^i_{\mathrm{dR}}(\mathsf{N}/\mathsf{M}_{\mathcal{H}}) := R^i(\mathsf{N} \to \mathsf{M}_{\mathcal{H}})_*(\Omega^\bullet_{\mathsf{N}/\mathsf{M}_{\mathcal{H}}})$$

and its Hodge filtration and Gauss–Manin connection ∇ *are canonically isomorphic to those of* $\underline{H}^i_{\mathrm{dR}}(\mathsf{N}^{\mathrm{grp}}/\mathsf{M}_{\mathcal{H}})$.

Proof. The first two follows from the definition and the corresponding statement for $\underline{\mathrm{Lie}}_{\mathsf{N}^{\mathrm{grp}}/\mathsf{M}_{\mathcal{H}}}$, $\underline{\mathrm{Lie}}^\vee_{\mathsf{N}^{\mathrm{grp}}/\mathsf{M}_{\mathcal{H}}}$, $\underline{\mathrm{Lie}}_{\mathsf{N}^{\mathrm{grp},\vee}/\mathsf{M}_{\mathcal{H}}}$, and $\underline{\mathrm{Lie}}^\vee_{\mathsf{N}^{\mathrm{grp},\vee}/\mathsf{M}_{\mathcal{H}}}$. The remaining ones follow by étale descent from the corresponding ones for $\mathsf{N}^{\mathrm{grp}} \to \mathsf{M}_{\mathcal{H}}$ (*cf.* [62, Cor. 2.1.5.9 and Lem. 2.1.5.11]). □

Corollary 1.3.3.13. (*Compare with [61, Cor. 2.13].*) *If a generalized Kuga family* $\mathsf{N} \to \mathsf{M}_{\mathcal{H}}$ *is a torsor under a Kuga family* $\mathsf{N}^{\mathrm{grp}} \to \mathsf{M}_{\mathcal{H}}$ *which is* $\mathbb{Q}^\times$*-isogenous to* $\underline{\mathrm{Hom}}_{\mathcal{O}}(Q, G_{\mathsf{M}_{\mathcal{H}}})^\circ$ *for some* $\mathcal{O}$*-lattice* Q*, then we have canonical isomorphisms of locally free sheaves over* $\mathsf{M}_{\mathcal{H}}$*:*

$$\begin{aligned}
\underline{\mathrm{Lie}}_{\mathsf{N}/\mathsf{M}_{\mathcal{H}}} &\cong \underline{\mathrm{Hom}}_{\mathcal{O}}(Q, \underline{\mathrm{Lie}}_{G_{\mathsf{M}_{\mathcal{H}}}/\mathsf{M}_{\mathcal{H}}}),\\
\underline{\mathrm{Lie}}_{\mathsf{N}^\vee/\mathsf{M}_{\mathcal{H}}} &\cong \underline{\mathrm{Hom}}_{\mathcal{O}}(Q^\vee, \underline{\mathrm{Lie}}_{G^\vee/\mathsf{M}_{\mathcal{H}}}),\\
\underline{\mathrm{Lie}}^\vee_{\mathsf{N}/\mathsf{M}_{\mathcal{H}}} &\cong \underline{\mathrm{Hom}}_{\mathcal{O}}(Q^\vee, \underline{\mathrm{Lie}}^\vee_{G_{\mathsf{M}_{\mathcal{H}}}/\mathsf{M}_{\mathcal{H}}}),\\
\underline{\mathrm{Lie}}^\vee_{\mathsf{N}^\vee/\mathsf{M}_{\mathcal{H}}} &\cong \underline{\mathrm{Hom}}_{\mathcal{O}}(Q, \underline{\mathrm{Lie}}^\vee_{G^\vee_{\mathsf{M}_{\mathcal{H}}}/\mathsf{M}_{\mathcal{H}}}).
\end{aligned}$$

Remark 1.3.3.14. We do not need to choose a polarization $\mathsf{N}^{\mathrm{grp}} \to \mathsf{N}^{\mathrm{grp},\vee}$ in the isomorphisms in Corollary 1.3.3.13.

The algebraically constructed toroidal compactifications of Kuga families and their generalizations in characteristic zero can be described as follows:

Theorem 1.3.3.15. (*Compare with [61, Thm. 2.15].*) *Let* Q *be any* $\mathcal{O}$*-lattice. Suppose that* $\mathcal{H}$ *is* **neat** (*see [89, 0.6] or [62, Def. 1.4.1.8]*)*, and that* Σ *is as in Definition* 1.2.2.13*, so that the moduli problem* $\mathsf{M}_{\mathcal{H}}$ *is representable by a scheme quasi-projective over* S_0*, and so that (by Theorem* 1.3.1.3*)* $\mathsf{M}^{\mathrm{tor}}_{\mathcal{H}} = \mathsf{M}^{\mathrm{tor}}_{\mathcal{H},\Sigma}$ *is an algebraic space proper and smooth over* S_0*. (By Theorem* 1.3.1.10*, if* Σ *is projective as in Definition* 1.2.2.14*, then* $\mathsf{M}^{\mathrm{tor}}_{\mathcal{H},\Sigma}$ *is projective over* S_0*.) Consider the sets* $\mathbf{K}_{Q,\mathcal{H}} \subset \mathbf{K}^+_{Q,\mathcal{H}} \subset \mathbf{K}^{++}_{Q,\mathcal{H}}$ *and* $\mathbf{K}_{Q,\mathcal{H},\Sigma} \subset \mathbf{K}^+_{Q,\mathcal{H},\Sigma} \subset \mathbf{K}^{++}_{Q,\mathcal{H},\Sigma}$ *as in Definitions* 1.2.4.44 *and* 1.2.4.50*, with compatible directed partial orders. These sets parameterize the following data:*

(1) *For each* $\kappa = (\widehat{\mathcal{H}}, \widehat{\Sigma}) \in \mathbf{K}^{++}_{Q,\mathcal{H}}$*, if* $\mathcal{H}_\kappa := \widehat{\mathcal{H}}_{\mathrm{G}}$ *(which is contained in* $\mathcal{H}$*, so that* $\mathsf{M}_{\mathcal{H}_\kappa}$ *is a finite étale cover of* $\mathsf{M}_{\mathcal{H}}$*; see Definition* 1.2.4.4*), then there is a generalized Kuga family* $\mathsf{N}_\kappa \to \mathsf{M}_{\mathcal{H}_\kappa}$ *(see Definition* 1.3.3.4*), which is a torsor under a Kuga family* $\mathsf{N}^{\mathrm{grp}}_\kappa \to \mathsf{M}_{\mathcal{H}_\kappa}$ *(see Definition* 1.3.3.3*) with a* $\mathbb{Q}^\times$*-isogeny*

$$\kappa^{\mathrm{isog}} : \underline{\mathrm{Hom}}_{\mathcal{O}}(Q, G_{\mathsf{M}_{\mathcal{H}_\kappa}})^\circ \to \mathsf{N}^{\mathrm{grp}}_\kappa$$

of abelian schemes over $\mathsf{M}_{\mathcal{H}_\kappa}$*, together with an open dense immersion*

$$\kappa^{\mathrm{tor}} : \mathsf{N}_\kappa \hookrightarrow \mathsf{N}_\kappa^{\mathrm{tor}}$$

of schemes over S_0*, such that the scheme* $\mathsf{N}_\kappa^{\mathrm{tor}}$ *is* **projective** *and* **smooth** *over* S_0*, and such that the complement of* N_κ *in* $\mathsf{N}_\kappa^{\mathrm{tor}}$ *(with its reduced structure) is a relative Cartier divisor* $\mathsf{E}_{\infty,\kappa}$ *with simple normal crossings. The scheme* $\mathsf{N}_\kappa^{\mathrm{tor}}$ *has a stratification by locally closed subschemes*

$$\mathsf{N}_\kappa^{\mathrm{tor}} = \coprod_{[(\breve{\Phi}_{\widehat{\mathcal{H}}},\breve{\delta}_{\widehat{\mathcal{H}}},\widehat{\tau})]} \widehat{\mathsf{Z}}_{[(\breve{\Phi}_{\widehat{\mathcal{H}}},\breve{\delta}_{\widehat{\mathcal{H}}},\widehat{\tau})]},$$

with $[(\breve{\Phi}_{\widehat{\mathcal{H}}},\breve{\delta}_{\widehat{\mathcal{H}}},\widehat{\tau})]$ *running through a complete set of equivalence classes of* $(\breve{\Phi}_{\widehat{\mathcal{H}}},\breve{\delta}_{\widehat{\mathcal{H}}},\widehat{\tau})$ *(as in Lemma 1.2.4.42) with* $\widehat{\tau} \subset \widehat{\mathbf{P}}^+_{\breve{\Phi}_{\widehat{\mathcal{H}}}}$ *and* $\widehat{\tau} \in \widehat{\Sigma}_{\breve{\Phi}_{\widehat{\mathcal{H}}}} \in \widehat{\Sigma}$*. (Here* $\breve{\mathsf{Z}}_{\widehat{\mathcal{H}}}$ *is suppressed in the notation by our convention. The notation "*$\coprod$*" only means a set-theoretic disjoint union. The algebro-geometric structure is still that of* $\mathsf{N}_\kappa^{\mathrm{tor}}$*.) In this stratification, the* $[(\breve{\Phi}'_{\widehat{\mathcal{H}}},\breve{\delta}'_{\widehat{\mathcal{H}}},\widehat{\tau}')]$*-stratum* $\widehat{\mathsf{Z}}_{[(\breve{\Phi}'_{\widehat{\mathcal{H}}},\breve{\delta}'_{\widehat{\mathcal{H}}},\widehat{\tau}')]}$ *lies in the closure of the* $[(\breve{\Phi}_{\widehat{\mathcal{H}}},\breve{\delta}_{\widehat{\mathcal{H}}},\widehat{\tau})]$*-stratum* $\widehat{\mathsf{Z}}_{[(\breve{\Phi}_{\widehat{\mathcal{H}}},\breve{\delta}_{\widehat{\mathcal{H}}},\widehat{\tau})]}$ *if and only if* $[(\breve{\Phi}_{\widehat{\mathcal{H}}},\breve{\delta}_{\widehat{\mathcal{H}}},\widehat{\tau})]$ *is a face of* $[(\breve{\Phi}'_{\widehat{\mathcal{H}}},\breve{\delta}'_{\widehat{\mathcal{H}}},\widehat{\tau}')]$ *as in Lemma 1.2.4.42. In particular,* $\mathsf{N}_\kappa = \widehat{\mathsf{Z}}_{[(0,0,\{0\})]}$ *is an open dense stratum in this stratification.*

The $[(\breve{\Phi}_{\widehat{\mathcal{H}}},\breve{\delta}_{\widehat{\mathcal{H}}},\widehat{\tau})]$*-stratum* $\widehat{\mathsf{Z}}_{[(\breve{\Phi}_{\widehat{\mathcal{H}}},\breve{\delta}_{\widehat{\mathcal{H}}},\widehat{\tau})]}$ *is smooth over* S_0 *and isomorphic to the support of the formal scheme* $\widehat{\mathfrak{X}}_{\breve{\Phi}_{\widehat{\mathcal{H}}},\breve{\delta}_{\widehat{\mathcal{H}}},\widehat{\tau}}$ *(see (1.3.2.65)) for every representative* $(\breve{\Phi}_{\widehat{\mathcal{H}}},\breve{\delta}_{\widehat{\mathcal{H}}},\widehat{\tau})$ *of* $[(\breve{\Phi}_{\widehat{\mathcal{H}}},\breve{\delta}_{\widehat{\mathcal{H}}},\widehat{\tau})]$*, which is the completion of an affine toroidal embedding* $\widehat{\overline{\Xi}}_{\breve{\Phi}_{\widehat{\mathcal{H}}},\breve{\delta}_{\widehat{\mathcal{H}}}}(\widehat{\tau})$ *(along its* $\widehat{\tau}$*-stratum* $\widehat{\overline{\Xi}}_{\breve{\Phi}_{\widehat{\mathcal{H}}},\breve{\delta}_{\widehat{\mathcal{H}}},\widehat{\tau}}$*) of a torus torsor* $\widehat{\Xi}_{\breve{\Phi}_{\widehat{\mathcal{H}}},\breve{\delta}_{\widehat{\mathcal{H}}}}$ *over an abelian scheme torsor* $\widehat{C}_{\breve{\Phi}_{\widehat{\mathcal{H}}},\breve{\delta}_{\widehat{\mathcal{H}}}}$ *over a finite étale cover* $\widehat{\mathsf{M}}_{\widehat{\mathcal{H}}}^{\breve{\Phi}_{\widehat{\mathcal{H}}}}$ *of the scheme* $\widehat{\mathsf{M}}_{\widehat{\mathcal{H}}}^{\breve{\mathsf{Z}}_{\widehat{\mathcal{H}}}}$ *(quasi-projective over* S_0*) in Lemma 1.3.2.50 and Proposition 1.3.2.56.*

The formal completion $(\mathsf{N}_\kappa^{\mathrm{tor}})^\wedge_{\widehat{\mathsf{Z}}_{[(\breve{\Phi}_{\widehat{\mathcal{H}}},\breve{\delta}_{\widehat{\mathcal{H}}},\widehat{\tau})]}}$ *of* $\mathsf{N}_\kappa^{\mathrm{tor}}$ *along* $\widehat{\mathsf{Z}}_{[(\breve{\Phi}_{\widehat{\mathcal{H}}},\breve{\delta}_{\widehat{\mathcal{H}}},\widehat{\tau})]}$ *is canonically isomorphic to* $\widehat{\mathfrak{X}}_{\breve{\Phi}_{\widehat{\mathcal{H}}},\breve{\delta}_{\widehat{\mathcal{H}}},\widehat{\tau}}$*; and the formal completion* $(\mathsf{N}_\kappa^{\mathrm{tor}})^\wedge_{\cup\widehat{\mathsf{Z}}_{[(\breve{\Phi}_{\widehat{\mathcal{H}}},\breve{\delta}_{\widehat{\mathcal{H}}},\widehat{\tau})]}}$*, where* $\cup\widehat{\mathsf{Z}}_{[(\breve{\Phi}_{\widehat{\mathcal{H}}},\breve{\delta}_{\widehat{\mathcal{H}}},\widehat{\tau})]}$ *is the union of all strata* $\widehat{\mathsf{Z}}_{[(\breve{\Phi}_{\widehat{\mathcal{H}}},\breve{\delta}_{\widehat{\mathcal{H}}},\widehat{\tau})]}$ *with* $\widehat{\tau} \in \widehat{\Sigma}_{\breve{\Phi}_{\widehat{\mathcal{H}}}}$*, is canonically isomorphic to* $\widehat{\mathfrak{X}}_{\breve{\Phi}_{\widehat{\mathcal{H}}},\breve{\delta}_{\widehat{\mathcal{H}}}}/\Gamma_{\breve{\Phi}_{\widehat{\mathcal{H}}}}$ *(cf. (5) of Theorem 1.3.1.3 and Lemma 1.3.2.41). (Such isomorphisms can be induced by strata-preserving isomorphisms between étale neighborhoods of points of* $\widehat{\mathsf{Z}}_{[(\breve{\Phi}_{\widehat{\mathcal{H}}},\breve{\delta}_{\widehat{\mathcal{H}}},\widehat{\tau})]}$ *in* $\mathsf{N}_\kappa^{\mathrm{tor}}$ *and étale neighborhoods of points of* $\widehat{\overline{\Xi}}_{\breve{\Phi}_{\widehat{\mathcal{H}}},\breve{\delta}_{\widehat{\mathcal{H}}},\widehat{\tau}}$ *in* $\widehat{\overline{\Xi}}_{\breve{\Phi}_{\widehat{\mathcal{H}}},\breve{\delta}_{\widehat{\mathcal{H}}}}(\widehat{\tau})$*.)*

Each $\mathsf{N}_\kappa^{\mathrm{tor}}$ *admits a canonical proper surjection* $\mathsf{N}_\kappa^{\mathrm{tor}} \to \mathsf{M}_{\mathcal{H}}^{\mathrm{min}}$ *extending the canonical proper surjection* $\mathsf{N}_\kappa \to \mathsf{M}_{\mathcal{H}}$*, and the latter is the pullback of the former under the canonical morphism* $\mathsf{M}_{\mathcal{H}} \hookrightarrow \mathsf{M}_{\mathcal{H}}^{\mathrm{min}}$ *on the target (see Theorem 1.3.1.5). Such a morphism maps the* $[(\breve{\Phi}_{\widehat{\mathcal{H}}},\breve{\delta}_{\widehat{\mathcal{H}}},\widehat{\tau})]$*-stratum* $\widehat{\mathsf{Z}}_{[(\breve{\Phi}_{\widehat{\mathcal{H}}},\breve{\delta}_{\widehat{\mathcal{H}}},\widehat{\tau})]}$ *of* $\mathsf{N}_\kappa^{\mathrm{tor}}$ *to the* $[(\Phi_{\mathcal{H}},\delta_{\mathcal{H}})]$*-stratum* $\mathsf{Z}_{[(\Phi_{\mathcal{H}},\delta_{\mathcal{H}})]}$ *of* $\mathsf{M}_{\mathcal{H}}^{\mathrm{min}}$ *if and only*

if the cusp label $[(\Phi_{\mathcal{H}}, \delta_{\mathcal{H}})]$ *is assigned to the cusp label* $[(\breve{\Phi}_{\widehat{\mathcal{H}}}, \breve{\delta}_{\widehat{\mathcal{H}}})]$ *as in Lemma* 1.2.4.15.
If $\kappa \in \mathbf{K}^{+}_{Q,\mathcal{H}}$, *then* $\mathcal{H}_\kappa = \mathcal{H}$ *and hence* $\mathsf{M}_{\mathcal{H}_\kappa} = \mathsf{M}_{\mathcal{H}}$. *If* $\kappa \in \mathbf{K}_{Q,\mathcal{H}}$, *then* $\mathsf{N}_\kappa = \mathsf{N}^{\mathrm{grp}}_\kappa \to \mathsf{M}_{\mathcal{H}_\kappa} = \mathsf{M}_{\mathcal{H}}$ *is a Kuga family.*
For each relation $\kappa' \succ \kappa$ *in* $\mathbf{K}^{++}_{Q,\mathcal{H}}$, *there is a proper log étale surjection*

$$f^{\mathrm{tor}}_{\kappa',\kappa} : \mathsf{N}^{\mathrm{tor}}_{\kappa'} \to \mathsf{N}^{\mathrm{tor}}_{\kappa},$$

extending a canonical finite étale surjection

$$f_{\kappa',\kappa} : \mathsf{N}_{\kappa'} \to \mathsf{N}_{\kappa}$$

inducing a canonical finite étale surjection $\mathsf{N}_{\kappa'} \to \mathsf{N}_\kappa \underset{\mathsf{M}_{\mathcal{H}_\kappa}}{\times} \mathsf{M}_{\mathcal{H}_{\kappa'}}$ *equivariant with the canonical* $\mathbb{Q}^\times$*-isogeny*

$$f^{\mathrm{grp}}_{\kappa',\kappa} := \kappa^{\mathrm{isog}} \circ ((\kappa')^{\mathrm{isog}})^{-1} : \mathsf{N}^{\mathrm{grp}}_{\kappa'} \to \mathsf{N}^{\mathrm{grp}}_{\kappa} \underset{\mathsf{M}_{\mathcal{H}_\kappa}}{\times} \mathsf{M}_{\mathcal{H}_{\kappa'}},$$

such that $R^i(f^{\mathrm{tor}}_{\kappa',\kappa})_* \mathscr{O}_{\mathsf{N}^{\mathrm{tor}}_{\kappa'}} = 0$ *for* $i > 0$. *These surjections are compatible with the canonical morphisms to* $\mathsf{M}^{\min}_{\mathcal{H}}$.

(2) *For each* $\kappa \in \mathbf{K}^{++}_{Q,\mathcal{H},\Sigma}$, *the structural morphism* $f_\kappa : \mathsf{N}_\kappa \to \mathsf{M}_{\mathcal{H}}$ *extends (necessarily uniquely) to a surjection* $f^{\mathrm{tor}}_\kappa : \mathsf{N}^{\mathrm{tor}}_\kappa \to \mathsf{M}^{\mathrm{tor}}_{\mathcal{H}} = \mathsf{M}^{\mathrm{tor}}_{\mathcal{H},\Sigma}$, *which is* **proper** *and* **log smooth** *(as in [45, 3.3] and [43, 1.6]) if we equip* $\mathsf{N}^{\mathrm{tor}}_\kappa$ *and* $\mathsf{M}^{\mathrm{tor}}_{\mathcal{H}}$ *with the canonical (fine) log structures given respectively by the relative Cartier divisors with (simple) normal crossings* $\mathsf{E}_{\infty,\kappa}$ *and* $\mathsf{D}_{\infty,\mathcal{H}}$ *(see (1) above and (3) of Theorem* 1.3.1.3). *Then we have the following commutative diagram:*

$$\begin{array}{ccccc}
\mathsf{N}_\kappa & \overset{+\mathrm{NCD}}{\hookrightarrow} & \mathsf{N}^{\mathrm{tor}}_\kappa & & \\
{\scriptstyle \text{proper smooth surjective}}\ \downarrow f_\kappa & & {\scriptstyle \text{proper log smooth surjective}}\ \downarrow f^{\mathrm{tor}}_\kappa & \searrow {\scriptstyle \text{projective smooth}} & \\
\mathsf{M}_{\mathcal{H}} & \underset{+\mathrm{NCD}}{\hookrightarrow} & \mathsf{M}^{\mathrm{tor}}_{\mathcal{H}} & \underset{\text{proper smooth}}{\longrightarrow} & \mathsf{S}_0
\end{array}$$

The morphism f^{tor}_κ *maps the* $[(\breve{\Phi}_{\widehat{\mathcal{H}}}, \breve{\delta}_{\widehat{\mathcal{H}}}, \widehat{\tau})]$*-stratum* $\widehat{\mathsf{Z}}_{[(\breve{\Phi}_{\widehat{\mathcal{H}}}, \breve{\delta}_{\widehat{\mathcal{H}}}, \widehat{\tau})]}$ *of* $\mathsf{N}^{\mathrm{tor}}_\kappa$ *to the* $[(\Phi_{\mathcal{H}}, \delta_{\mathcal{H}}, \tau)]$*-stratum* $\mathsf{Z}_{[(\Phi_{\mathcal{H}}, \delta_{\mathcal{H}}, \tau)]}$ *of* $\mathsf{M}^{\mathrm{tor}}_{\mathcal{H}}$ *if and only if (the cusp label* $[(\Phi_{\mathcal{H}}, \delta_{\mathcal{H}})]$ *is assigned to the cusp label* $[(\breve{\Phi}_{\widehat{\mathcal{H}}}, \breve{\delta}_{\widehat{\mathcal{H}}})]$ *as in Lemma* 1.2.4.15 *and) the image of* $\widehat{\tau} \in \widehat{\Sigma}_{\breve{\Phi}_{\widehat{\mathcal{H}}}}$ *under* (1.2.4.37) *is contained in* $\tau \in \Sigma_{\Phi_{\mathcal{H}}}$ *as in Condition* 1.2.4.49. *In this case, the compatible morphisms* $\widehat{\mathfrak{X}}_{\breve{\Phi}_{\widehat{\mathcal{H}}}, \breve{\delta}_{\widehat{\mathcal{H}}}, \widehat{\tau}} \to \mathfrak{X}_{\Phi_{\mathcal{H}}, \delta_{\mathcal{H}}, \tau}$ *and* $\widehat{\mathfrak{X}}_{\breve{\Phi}_{\widehat{\mathcal{H}}}, \breve{\delta}_{\widehat{\mathcal{H}}}, \widehat{\Sigma}_{\breve{\Phi}_{\widehat{\mathcal{H}}}}} \to \mathfrak{X}_{\Phi_{\mathcal{H}}, \delta_{\mathcal{H}}, \Sigma_{\Phi_{\mathcal{H}}}}$ *induced by* f^{tor}_κ *(and the canonical isomorphisms in* (1) *above and in* (5) *of Theorem* 1.3.1.3) *coincide with the canonical morphisms as in* (1.3.2.83) *and* (1.3.2.85). *(These morphisms can be induced by compatible morphisms between étale neighborhoods of points of the supports of formal schemes in relevant ambient schemes as in* (1) *above, compatible with all stratifications.)*
If $\kappa' \succ \kappa$, *then we have the compatibility* $f^{\mathrm{tor}}_{\kappa'} = f^{\mathrm{tor}}_\kappa \circ f^{\mathrm{tor}}_{\kappa',\kappa}$.

(3) *Suppose* $\kappa \in \mathbf{K}^{+}_{Q,\mathcal{H},\Sigma}$ (*not just in* $\mathbf{K}^{++}_{Q,\mathcal{H},\Sigma}$, *so that* $\mathcal{H}_\kappa = \mathcal{H}$). *For simplicity, let us suppress the subscripts "κ" from the notation.* (*All canonical isomorphisms will be required to be compatible with the canonical isomorphisms defined by pulling back under* $f^{\mathrm{tor}}_{\kappa',\kappa}$ *for each relation* $\kappa' \succ \kappa$ *in* $\mathbf{K}^{+}_{Q,\mathcal{H},\Sigma}$.) *Then the following are true:*

(a) *Let* $\Omega^1_{\mathsf{N}^{\mathrm{tor}}/\mathsf{S}_0}[d\log\infty]$ *and* $\Omega^1_{\mathsf{M}^{\mathrm{tor}}_{\mathcal{H}}/\mathsf{S}_0}[d\log\infty]$ *denote the sheaves of modules of log* 1*-differentials over* S_0 *given by the* (*respective*) *canonical log structures defined in* (2). *Let*

$$\overline{\Omega}^1_{\mathsf{N}^{\mathrm{tor}}/\mathsf{M}^{\mathrm{tor}}_{\mathcal{H}}} := (\Omega^1_{\mathsf{N}^{\mathrm{tor}}/\mathsf{S}_0}[d\log\infty])/((f^{\mathrm{tor}})^*(\Omega^1_{\mathsf{M}^{\mathrm{tor}}_{\mathcal{H}}/\mathsf{S}_0}[d\log\infty])).$$

Then there is a canonical isomorphism

$$(f^{\mathrm{tor}})^*(\underline{\mathrm{Hom}}_{\mathcal{O}}(Q^\vee, \underline{\mathrm{Lie}}^\vee_{G/\mathsf{M}^{\mathrm{tor}}_{\mathcal{H}}})) \cong \overline{\Omega}^1_{\mathsf{N}^{\mathrm{tor}}/\mathsf{M}^{\mathrm{tor}}_{\mathcal{H}}} \tag{1.3.3.16}$$

between locally free sheaves over $\mathsf{N}^{\mathrm{tor}}$, *extending the composition of canonical isomorphisms*

$$f^*(\underline{\mathrm{Hom}}_{\mathcal{O}}(Q^\vee, \underline{\mathrm{Lie}}^\vee_{G_{\mathsf{M}_{\mathcal{H}}}/\mathsf{M}_{\mathcal{H}}})) \cong f^*\underline{\mathrm{Lie}}^\vee_{\mathsf{N}/\mathsf{M}_{\mathcal{H}}} \cong \Omega^1_{\mathsf{N}/\mathsf{M}_{\mathcal{H}}} \tag{1.3.3.17}$$

over N (*see Lemma* 1.3.3.12).

(b) *For each integer* $b \geq 0$, *there exist canonical isomorphisms*

$$\begin{aligned} R^b f^{\mathrm{tor}}_*(\overline{\Omega}^a_{\mathsf{N}^{\mathrm{tor}}/\mathsf{M}^{\mathrm{tor}}_{\mathcal{H}}}) \cong & (\wedge^b(\underline{\mathrm{Hom}}_{\mathcal{O}}(Q^\vee, \underline{\mathrm{Lie}}_{G^\vee/\mathsf{M}^{\mathrm{tor}}_{\mathcal{H}}}))) \\ & \underset{\mathscr{O}_{\mathsf{M}^{\mathrm{tor}}_{\mathcal{H}}}}{\otimes} (\wedge^a(\underline{\mathrm{Hom}}_{\mathcal{O}}(Q^\vee, \underline{\mathrm{Lie}}^\vee_{G/\mathsf{M}^{\mathrm{tor}}_{\mathcal{H}}}))) \end{aligned} \tag{1.3.3.18}$$

and

$$R^b f^{\mathrm{tor}}_*(\overline{\Omega}^a_{\mathsf{N}^{\mathrm{tor}}/\mathsf{M}^{\mathrm{tor}}_{\mathcal{H}}} \underset{\mathscr{O}_{\mathsf{N}^{\mathrm{tor}}}}{\otimes} \mathscr{I}_{\mathsf{E}_\infty}) \cong R^b f^{\mathrm{tor}}_*(\overline{\Omega}^a_{\mathsf{N}^{\mathrm{tor}}/\mathsf{M}^{\mathrm{tor}}_{\mathcal{H}}}) \underset{\mathscr{O}_{\mathsf{M}^{\mathrm{tor}}_{\mathcal{H}}}}{\otimes} \mathscr{I}_{\mathsf{D}_{\infty,\mathcal{H}}} \tag{1.3.3.19}$$

of locally free sheaves over $\mathsf{M}^{\mathrm{tor}}_{\mathcal{H}}$, *where* $\mathscr{I}_{\mathsf{E}_\infty}$ (*resp.* $\mathscr{I}_{\mathsf{D}_{\infty,\mathcal{H}}}$) *is the* $\mathscr{O}_{\mathsf{N}^{\mathrm{tor}}}$*-ideal* (*resp.* $\mathscr{O}_{\mathsf{M}^{\mathrm{tor}}_{\mathcal{H}}}$*-ideal*) *defining the relative Cartier divisor* $\mathsf{E}_\infty = \mathsf{E}_{\infty,\kappa}$ (*resp.* $\mathsf{D}_{\infty,\mathcal{H}}$) (*with its reduced structure*), *compatible with cup products and exterior products, extending the canonical isomorphism over* $\mathsf{M}_{\mathcal{H}}$ *induced by the composition of canonical isomorphisms*

$$R^b f_*(\mathscr{O}_{\mathsf{N}}) \cong \wedge^b \underline{\mathrm{Lie}}_{\mathsf{N}^\vee/\mathsf{M}_{\mathcal{H}}} \cong \wedge^b(\underline{\mathrm{Hom}}_{\mathcal{O}}(Q^\vee, \underline{\mathrm{Lie}}_{G^\vee_{\mathsf{M}_{\mathcal{H}}}/\mathsf{M}_{\mathcal{H}}})). \tag{1.3.3.20}$$

(c) *Let* $\overline{\Omega}^\bullet_{\mathsf{N}^{\mathrm{tor}}/\mathsf{M}^{\mathrm{tor}}_{\mathcal{H}}} := \wedge^\bullet \overline{\Omega}^1_{\mathsf{N}^{\mathrm{tor}}/\mathsf{M}^{\mathrm{tor}}_{\mathcal{H}}}$ *be the log de Rham complex associated with* $f^{\mathrm{tor}} : \mathsf{N}^{\mathrm{tor}} \to \mathsf{M}^{\mathrm{tor}}_{\mathcal{H}}$ (*with differentials inherited from* $\Omega^\bullet_{\mathsf{N}/\mathsf{M}_{\mathcal{H}}}$). *Let the* (*relative*) **log de Rham cohomology** *be defined by*

$$\underline{H}^i_{\text{log-dR}}(\mathsf{N}^{\mathrm{tor}}/\mathsf{M}^{\mathrm{tor}}_{\mathcal{H}}) := R^i f^{\mathrm{tor}}_*(\overline{\Omega}^\bullet_{\mathsf{N}^{\mathrm{tor}}/\mathsf{M}^{\mathrm{tor}}_{\mathcal{H}}}).$$

Then the (*relative*) **Hodge spectral sequence**

$$E_1^{a,b} := R^b f^{\mathrm{tor}}_*(\overline{\Omega}^a_{\mathsf{N}^{\mathrm{tor}}/\mathsf{M}^{\mathrm{tor}}_{\mathcal{H}}}) \Rightarrow \underline{H}^{a+b}_{\text{log-dR}}(\mathsf{N}^{\mathrm{tor}}/\mathsf{M}^{\mathrm{tor}}_{\mathcal{H}}) \tag{1.3.3.21}$$

degenerates at E_1 *terms, and defines a* **Hodge filtration** *on* $\underline{H}^i_{\text{log-dR}}(\mathsf{N}^{\text{tor}}/\mathsf{M}^{\text{tor}}_{\mathcal{H}})$ *with locally free graded pieces given by* $R^b f^{\text{tor}}_*(\overline{\Omega}^a_{\mathsf{N}^{\text{tor}}/\mathsf{M}^{\text{tor}}_{\mathcal{H}}})$ *for integers* $a+b=i$*, extending the canonical Hodge filtration on* $\underline{H}^i_{\text{dR}}(\mathsf{N}/\mathsf{M}_{\mathcal{H}})$.

As a result, for each integer $i \geq 0$*, there is a canonical isomorphism*

$$\wedge^i \underline{H}^1_{\text{log-dR}}(\mathsf{N}^{\text{tor}}/\mathsf{M}^{\text{tor}}_{\mathcal{H}}) \xrightarrow{\sim} \underline{H}^i_{\text{log-dR}}(\mathsf{N}^{\text{tor}}/\mathsf{M}^{\text{tor}}_{\mathcal{H}}),$$

compatible with the Hodge filtrations defined by (1.3.3.21)*, extending the canonical isomorphism* $\wedge^i \underline{H}^1_{\text{dR}}(\mathsf{N}/\mathsf{M}_{\mathcal{H}}) \xrightarrow{\sim} \underline{H}^i_{\text{dR}}(\mathsf{N}/\mathsf{M}_{\mathcal{H}})$ *over* $\mathsf{M}_{\mathcal{H}}$ *(defined by cup product).*

(d) *For each* $j_Q : Q^\vee \hookrightarrow Q$ *as in Lemma* 1.2.4.1*, the* $\mathbb{Q}^\times$*-polarization*

$$\lambda_{\mathsf{M}_{\mathcal{H}},j_Q} : \underline{\mathrm{Hom}}_{\mathcal{O}}(Q, G_{\mathsf{M}_{\mathcal{H}}})^\circ \to (\underline{\mathrm{Hom}}_{\mathcal{O}}(Q, G_{\mathsf{M}_{\mathcal{H}}})^\circ)^\vee$$

in Proposition 1.3.3.7 *induces a* $\mathbb{Q}^\times$*-polarization*

$$\lambda_{\mathsf{N},j_Q} : \mathsf{N}^{\text{grp}} \to \mathsf{N}^{\text{grp},\vee},$$

and hence defines canonically (as in [23, 1.5], by étale descent; see Lemma 1.3.3.12*) a perfect pairing*

$$\langle\,\cdot\,,\,\cdot\,\rangle_{\lambda_{\mathsf{M}_{\mathcal{H}},j_Q}} : \underline{H}^1_{\text{dR}}(\mathsf{N}/\mathsf{M}_{\mathcal{H}}) \times \underline{H}^1_{\text{dR}}(\mathsf{N}/\mathsf{M}_{\mathcal{H}}) \to \mathcal{O}_{\mathsf{M}_{\mathcal{H}}}(1).$$

Then $\underline{H}^1_{\text{log-dR}}(\mathsf{N}^{\text{tor}}/\mathsf{M}^{\text{tor}}_{\mathcal{H}})$ *is (under the restriction morphism) canonically isomorphic to the unique subsheaf of*

$$(\mathsf{M}_{\mathcal{H}} \hookrightarrow \mathsf{M}^{\text{tor}}_{\mathcal{H}})_*(\underline{H}^1_{\text{dR}}(\mathsf{N}/\mathsf{M}_{\mathcal{H}}))$$

satisfying the following conditions:

i. $\underline{H}^1_{\text{log-dR}}(\mathsf{N}^{\text{tor}}/\mathsf{M}^{\text{tor}}_{\mathcal{H}})$ *is locally free of finite rank over* $\mathcal{O}_{\mathsf{M}^{\text{tor}}_{\mathcal{H}}}$.
ii. *The sheaf* $f^{\text{tor}}_*(\overline{\Omega}^1_{\mathsf{N}^{\text{tor}}/\mathsf{M}^{\text{tor}}_{\mathcal{H}}})$ *can be identified with the subsheaf of* $(\mathsf{M}_{\mathcal{H}} \hookrightarrow \mathsf{M}^{\text{tor}}_{\mathcal{H}})_*(f_*(\Omega^1_{\mathsf{N}/\mathsf{M}_{\mathcal{H}}}))$ *formed (locally) by sections that are also sections of* $\underline{H}^1_{\text{log-dR}}(\mathsf{N}^{\text{tor}}/\mathsf{M}^{\text{tor}}_{\mathcal{H}})$*. (Here we view all sheaves canonically as subsheaves of* $(\mathsf{M}_{\mathcal{H}} \hookrightarrow \mathsf{M}^{\text{tor}}_{\mathcal{H}})_*(\underline{H}^1_{\text{dR}}(\mathsf{N}/\mathsf{M}_{\mathcal{H}}))$*.)*
iii. $\underline{H}^1_{\text{log-dR}}(\mathsf{N}^{\text{tor}}/\mathsf{M}^{\text{tor}}_{\mathcal{H}})$ *is self-dual under the push-forward* $(\mathsf{M}_{\mathcal{H}} \hookrightarrow \mathsf{M}^{\text{tor}}_{\mathcal{H}})_*\langle\,\cdot\,,\,\cdot\,\rangle_{\lambda_{\mathsf{M}_{\mathcal{H}},j_Q}}$.

(e) *The Gauss–Manin connection*

$$\nabla : \underline{H}^\bullet_{\text{dR}}(\mathsf{N}/\mathsf{M}_{\mathcal{H}}) \to \underline{H}^\bullet_{\text{dR}}(\mathsf{N}/\mathsf{M}_{\mathcal{H}}) \underset{\mathcal{O}_{\mathsf{M}_{\mathcal{H}}}}{\otimes} \Omega^1_{\mathsf{M}_{\mathcal{H}}/\mathsf{S}_0} \tag{1.3.3.22}$$

extends to an integrable connection

$$\nabla : \underline{H}^\bullet_{\text{log-dR}}(\mathsf{N}^{\text{tor}}/\mathsf{M}^{\text{tor}}_{\mathcal{H}}) \to \underline{H}^\bullet_{\text{log-dR}}(\mathsf{N}^{\text{tor}}/\mathsf{M}^{\text{tor}}_{\mathcal{H}}) \underset{\mathcal{O}_{\mathsf{M}^{\text{tor}}_{\mathcal{H}}}}{\otimes} \overline{\Omega}^1_{\mathsf{M}^{\text{tor}}_{\mathcal{H}}/\mathsf{S}_0} \tag{1.3.3.23}$$

with log poles along $\mathsf{D}_{\infty,\mathcal{H}}$*, called the* **extended Gauss–Manin connection***, satisfying the usual Griffiths transversality with the Hodge filtration defined by* (1.3.3.21).

(4) *(Hecke actions.) Suppose we have an element* $\widehat{g} \in \widehat{\mathrm{G}}(\mathbb{A}^\infty)$ *with image* $g_h \in \mathrm{G}(\mathbb{A}^\infty)$ *under the canonical homomorphism* $\widehat{\mathrm{G}}(\mathbb{A}^\infty) \to \mathrm{G}(\mathbb{A}^\infty)$, *and suppose we have two neat open compact subgroups* $\mathcal{H}$ *and* $\mathcal{H}'$ *of* $\mathrm{G}(\hat{\mathbb{Z}})$ *such that* $\mathcal{H}' \subset g_h \mathcal{H} g_h^{-1}$. *Suppose* $\Sigma' = \{\Sigma'_{\Phi'_{\mathcal{H}'}}\}_{[(\Phi'_{\mathcal{H}'},\delta'_{\mathcal{H}'})]}$ *is a compatible choice of admissible smooth rational polyhedral cone decomposition data for* $\mathsf{M}_{\mathcal{H}'}$, *which is a* g_h*-refinement of* Σ *as in [62, Def. 6.4.3.3]. Consider the sets* $\mathbf{K}_{Q,\mathcal{H}'} \subset \mathbf{K}^+_{Q,\mathcal{H}'} \subset \mathbf{K}^{++}_{Q,\mathcal{H}'}$ *and* $\mathbf{K}_{Q,\mathcal{H}',\Sigma'} \subset \mathbf{K}^+_{Q,\mathcal{H}',\Sigma'} \subset \mathbf{K}^{++}_{Q,\mathcal{H}',\Sigma'}$ *as in Definitions* 1.2.4.44 *and* 1.2.4.50 *(for* $\mathcal{H}'$ *and* Σ'*), with compatible directed partial orders, parameterizing generalized Kuga families and their compactifications with properties as in* (1), (2), *and* (3) *above. The sets* $\mathbf{K}^{++}_{Q,\mathcal{H}}$ *etc and* $\mathbf{K}^{++}_{Q,\mathcal{H}'}$ *etc (and the objects they parameterize) satisfy the compatibility with* $\widehat{g}$ *(and* g_h*) in the sense that the following are true:*

(a) *For each* $\kappa = (\widehat{\mathcal{H}}, \widehat{\Sigma}) \in \mathbf{K}^{++}_{Q,\mathcal{H}}$ *(resp.* $\mathbf{K}^+_{Q,\mathcal{H}}$, *resp.* $\mathbf{K}_{Q,\mathcal{H}}$*), and for each open compact subgroup* $\widehat{\mathcal{H}}' \subset \widehat{\mathrm{G}}(\hat{\mathbb{Z}})$ *such that* $\widehat{\mathcal{H}}' \subset \widehat{g}\widehat{\mathcal{H}}\widehat{g}^{-1}$ *(so that* $\mathcal{H}_\kappa = \widehat{\mathcal{H}}_{\mathrm{G}}$ *and* $\mathcal{H}_{\kappa'} = \widehat{\mathcal{H}}'_{\mathrm{G}}$ *satisfy* $\mathcal{H}_{\kappa'} \subset g_h \mathcal{H}_\kappa g_h^{-1}$*), there exists an element* $\kappa' = (\widehat{\mathcal{H}}', \widehat{\Sigma}') \in \mathbf{K}^{++}_{Q,\mathcal{H}'}$ *(resp.* $\mathbf{K}^+_{Q,\mathcal{H}'}$, *resp.* $\mathbf{K}_{Q,\mathcal{H}'}$*) such that there exists a (necessarily unique) finite étale surjection*

$$[\widehat{g}] : \mathsf{N}_{\kappa'} \to \mathsf{N}_\kappa \tag{1.3.3.24}$$

covering the compatible surjections $[g_h] : \mathsf{M}_{\mathcal{H}'} \to \mathsf{M}_{\mathcal{H}}$ *and* $[g_h] : \mathsf{M}_{\mathcal{H}_{\kappa'}} \to \mathsf{M}_{\mathcal{H}_\kappa}$ *given by [62, Prop. 6.4.3.4] (see Proposition* 1.3.1.15*), inducing a finite étale surjection* $\mathsf{N}_{\kappa'} \to \mathsf{N}_\kappa \underset{\mathsf{M}_{\mathcal{H}_\kappa}}{\times} \mathsf{M}_{\mathcal{H}_{\kappa'}}$ *of abelian scheme torsors equivariant with the isogeny (not just* $\mathbb{Q}^\times$*-isogeny)* $\mathsf{N}^{\mathrm{grp}}_{\kappa'} \to \mathsf{N}^{\mathrm{grp}}_\kappa \underset{\mathsf{M}_{\mathcal{H}_\kappa}}{\times} \mathsf{M}_{\mathcal{H}_{\kappa'}}$ *induced by* $(\kappa')^{\mathrm{isog}}$, κ^{isog}, *and the* $\mathbb{Q}^\times$*-isogeny* $G_{\mathsf{M}_{\mathcal{H}_{\kappa'}}} \to G_{\mathsf{M}_{\mathcal{H}_\kappa}} \underset{\mathsf{M}_{\mathcal{H}_\kappa}}{\times} \mathsf{M}_{\mathcal{H}_{\kappa'}}$ *realizing* $G_{\mathsf{M}_{\mathcal{H}_\kappa}} \underset{\mathsf{M}_{\mathcal{H}_\kappa}}{\times} \mathsf{M}_{\mathcal{H}_{\kappa'}}$ *as a Hecke twist of* $G_{\mathsf{M}_{\mathcal{H}_{\kappa'}}}$ *by* g_h *(which is the pullback of the* $\mathbb{Q}^\times$*-isogeny* $G_{\mathsf{M}_{\mathcal{H}'}} \to G_{\mathsf{M}_{\mathcal{H}}} \underset{\mathsf{M}_{\mathcal{H}}}{\times} \mathsf{M}_{\mathcal{H}'}$ *realizing* $G_{\mathsf{M}_{\mathcal{H}}} \underset{\mathsf{M}_{\mathcal{H}}}{\times} \mathsf{M}_{\mathcal{H}'}$ *as a Hecke twist of* $G_{\mathsf{M}_{\mathcal{H}'}}$ *by* g_h*). (Here all the base changes from* $\mathsf{M}_{\mathcal{H}}$ *to* $\mathsf{M}_{\mathcal{H}'}$ *and from* $\mathsf{M}_{\mathcal{H}_\kappa}$ *to* $\mathsf{M}_{\mathcal{H}_{\kappa'}}$ *use the surjections denoted by* $[g_h]$*.)*

(b) *For each* $\kappa = (\widehat{\mathcal{H}}, \widehat{\Sigma})$ *and* $\widehat{\mathcal{H}}'$ *as in* (4a) *such that* $\kappa \in \mathbf{K}^{++}_{Q,\mathcal{H}}$ *(resp.* $\mathbf{K}^+_{Q,\mathcal{H}}$, *resp.* $\mathbf{K}_{Q,\mathcal{H}}$*), there is an element* $\kappa' = (\widehat{\mathcal{H}}', \widehat{\Sigma}') \in \mathbf{K}^{++}_{Q,\mathcal{H}'}$ *(resp.* $\mathbf{K}^+_{Q,\mathcal{H}'}$, *resp.* $\mathbf{K}_{Q,\mathcal{H}'}$*) such that* $[\widehat{g}]$ *is defined as in* (4a) *(see* (1.3.3.24)*), and such that* $\widehat{\Sigma}'$ *is a* $\widehat{g}$*-refinement of* $\widehat{\Sigma}$ *(cf. Lemma* 1.2.4.42 *and [62, Def. 6.4.3.3]), which extends to a (necessarily unique) proper log étale surjection*

$$[\widehat{g}]^{\mathrm{tor}} : \mathsf{N}^{\mathrm{tor}}_{\kappa'} \to \mathsf{N}^{\mathrm{tor}}_\kappa \tag{1.3.3.25}$$

such that

$$R^i[\widehat{g}]^{\mathrm{tor}}_* \mathscr{O}_{(\mathsf{N}'_{\kappa'})^{\mathrm{tor}}} = 0 \tag{1.3.3.26}$$

for all $i > 0$.

Under (1.3.3.25), *the* $[(\breve{\Phi}'_{\widehat{\mathcal{H}}'}, \breve{\delta}'_{\widehat{\mathcal{H}}'}, \widehat{\tau}')]$*-stratum* $\widehat{\mathsf{Z}}_{[(\breve{\Phi}'_{\widehat{\mathcal{H}}'}, \breve{\delta}'_{\widehat{\mathcal{H}}'}, \widehat{\tau}')]}$ *of* $\mathsf{N}^{\mathrm{tor}}_{\kappa'}$ *is mapped to the* $[(\breve{\Phi}_{\widehat{\mathcal{H}}}, \breve{\delta}_{\widehat{\mathcal{H}}}, \widehat{\tau})]$*-stratum* $\widehat{\mathsf{Z}}_{[(\breve{\Phi}_{\widehat{\mathcal{H}}}, \breve{\delta}_{\widehat{\mathcal{H}}}, \widehat{\tau})]}$ *of* $\mathsf{N}^{\mathrm{tor}}_{\kappa}$ *if and only if there are representatives* $(\breve{\Phi}_{\widehat{\mathcal{H}}}, \breve{\delta}_{\widehat{\mathcal{H}}}, \widehat{\tau})$ *and* $(\breve{\Phi}'_{\widehat{\mathcal{H}}'}, \breve{\delta}'_{\widehat{\mathcal{H}}'}, \widehat{\tau}')$ *of* $[(\breve{\Phi}_{\widehat{\mathcal{H}}}, \breve{\delta}_{\widehat{\mathcal{H}}}, \widehat{\tau})]$ *and* $[(\breve{\Phi}'_{\widehat{\mathcal{H}}'}, \breve{\delta}'_{\widehat{\mathcal{H}}'}, \widehat{\tau}')]$*, respectively, such that* $(\breve{\Phi}'_{\widehat{\mathcal{H}}'}, \breve{\delta}'_{\widehat{\mathcal{H}}'}, \widehat{\tau}')$ *is a* $\widehat{g}$*-refinement of* $(\breve{\Phi}_{\widehat{\mathcal{H}}}, \breve{\delta}_{\widehat{\mathcal{H}}}, \widehat{\tau})$ (*cf. Lemma* 1.2.4.42 *and [62, Def. 6.4.3.1]*). *In this case, the compatible morphisms* $\widehat{\mathfrak{X}}_{\breve{\Phi}'_{\widehat{\mathcal{H}}'}, \breve{\delta}'_{\widehat{\mathcal{H}}'}, \widehat{\tau}'} \to \widehat{\mathfrak{X}}_{\breve{\Phi}_{\widehat{\mathcal{H}}}, \breve{\delta}_{\widehat{\mathcal{H}}}, \widehat{\tau}}$ *and* $\widehat{\mathfrak{X}}_{\breve{\Phi}'_{\widehat{\mathcal{H}}'}, \breve{\delta}'_{\widehat{\mathcal{H}}'}, \widehat{\Sigma}'_{\breve{\Phi}'_{\widehat{\mathcal{H}}'}}} \to \widehat{\mathfrak{X}}_{\breve{\Phi}_{\widehat{\mathcal{H}}}, \breve{\delta}_{\widehat{\mathcal{H}}}, \widehat{\Sigma}_{\breve{\Phi}_{\widehat{\mathcal{H}}}}}$ *induced by* (1.3.3.25) (*and the canonical isomorphisms in* (1) *above*) *coincide with the canonical morphisms as in* (1.3.2.69) *and* (1.3.2.71).

If $\kappa \in \mathbf{K}^{++}_{Q,\mathcal{H},\Sigma}$ (*resp.* $\mathbf{K}^{+}_{Q,\mathcal{H},\Sigma}$, *resp.* $\mathbf{K}_{Q,\mathcal{H},\Sigma}$), *we may assume in the above that* $\kappa' \in \mathbf{K}^{++}_{Q,\mathcal{H}',\Sigma'}$ (*resp.* $\mathbf{K}^{+}_{Q,\mathcal{H}',\Sigma'}$, *resp.* $\mathbf{K}_{Q,\mathcal{H}',\Sigma'}$), *so that* (1.3.3.25) *covers the surjection*

$$[g_h]^{\mathrm{tor}} : \mathsf{M}^{\mathrm{tor}}_{\mathcal{H}',\Sigma'} \to \mathsf{M}^{\mathrm{tor}}_{\mathcal{H},\Sigma}$$

given by [62, Prop. 6.4.3.4],

(c) *Suppose* $[\widehat{g}]^{\mathrm{tor}}$ *is defined as in* (4b) *for some* $\kappa \in \mathbf{K}^{+}_{Q,\mathcal{H},\Sigma}$ *and* $\kappa' \in \mathbf{K}^{+}_{Q,\mathcal{H}',\Sigma'}$ (*not just in* $\mathbf{K}^{++}_{Q,\mathcal{H},\Sigma}$ *and* $\mathbf{K}^{++}_{Q,\mathcal{H}',\Sigma'}$). *Then there is a canonical isomorphism*

$$([\widehat{g}]^{\mathrm{tor}})^* : ([g_h]^{\mathrm{tor}})^* \underline{H}^{a+b}_{\text{log-dR}}(\mathsf{N}^{\mathrm{tor}}_{\kappa}/\mathsf{M}^{\mathrm{tor}}_{\mathcal{H},\Sigma}) \xrightarrow{\sim} \underline{H}^{a+b}_{\text{log-dR}}(\mathsf{N}^{\mathrm{tor}}_{\kappa'}/\mathsf{M}^{\mathrm{tor}}_{\mathcal{H}',\Sigma'})$$

respecting the Hodge filtrations and compatible with the canonical isomorphisms

$$([\widehat{g}]^{\mathrm{tor}})^* : ([\widehat{g}]^{\mathrm{tor}})^* \overline{\Omega}^1_{\mathsf{N}^{\mathrm{tor}}_{\kappa}/\mathsf{M}^{\mathrm{tor}}_{\mathcal{H},\Sigma}} \xrightarrow{\sim} \overline{\Omega}^1_{\mathsf{N}^{\mathrm{tor}}_{\kappa'}/\mathsf{M}^{\mathrm{tor}}_{\mathcal{H}',\Sigma'}},$$

$$([g_h]^{\mathrm{tor}})^* : ([g_h]^{\mathrm{tor}})^* \underline{\mathrm{Lie}}_{G^\vee/\mathsf{M}^{\mathrm{tor}}_{\mathcal{H},\Sigma}} \xrightarrow{\sim} \underline{\mathrm{Lie}}_{G^\vee/\mathsf{M}^{\mathrm{tor}}_{\mathcal{H}',\Sigma'}},$$

$$([g_h]^{\mathrm{tor}})^* : ([g_h]^{\mathrm{tor}})^* \underline{\mathrm{Lie}}^\vee_{G/\mathsf{M}^{\mathrm{tor}}_{\mathcal{H},\Sigma}} \xrightarrow{\sim} \underline{\mathrm{Lie}}^\vee_{G/\mathsf{M}^{\mathrm{tor}}_{\mathcal{H}',\Sigma'}},$$

and the canonical isomorphisms in (3) *for* $\mathsf{N}^{\mathrm{tor}}_{\kappa}$ *and* $\mathsf{N}^{\mathrm{tor}}_{\kappa'}$.

(d) *If we have an element* $\widehat{g}' \in \widehat{\mathrm{G}}(\mathbb{A}^\infty)$ *with image* $g'_h \in \mathrm{G}(\mathbb{A}^\infty)$ *under the canonical homomorphism* $\widehat{\mathrm{G}}(\mathbb{A}^\infty) \to \mathrm{G}(\mathbb{A}^\infty)$, *with a similar setup such that* $[\widehat{g}'] : \mathsf{N}_{\kappa''} \to \mathsf{N}_{\kappa'}$ *and* $[\widehat{g}']^{\mathrm{tor}} : \mathsf{N}^{\mathrm{tor}}_{\kappa''} \to \mathsf{N}^{\mathrm{tor}}_{\kappa'}$ *are compatibly defined for some* $\kappa'' \in \mathbf{K}^{++}_{Q,\mathcal{H}'',\Sigma''}$, *then* $[\widehat{g}'\widehat{g}] : \mathsf{N}_{\kappa''} \to \mathsf{N}_{\kappa}$ *and* $[\widehat{g}'\widehat{g}]^{\mathrm{tor}} : \mathsf{N}^{\mathrm{tor}}_{\kappa''} \to \mathsf{N}^{\mathrm{tor}}_{\kappa}$ *are also compatibly defined and satisfy the identities* $[\widehat{g}'\widehat{g}] = [\widehat{g}] \circ [\widehat{g}']$ *and* $[\widehat{g}'\widehat{g}]^{\mathrm{tor}} = [\widehat{g}]^{\mathrm{tor}} \circ [\widehat{g}']^{\mathrm{tor}}$. *If* $\kappa \in \mathbf{K}^{+}_{Q,\mathcal{H},\Sigma}$, $\kappa' \in \mathbf{K}^{+}_{Q,\mathcal{H}',\Sigma'}$, *and* $\kappa'' \in \mathbf{K}^{+}_{Q,\mathcal{H}'',\Sigma''}$, *we also have* $[\widehat{g}'\widehat{g}]^* = [\widehat{g}']^* \circ [\widehat{g}]^*$ *and* $([\widehat{g}'\widehat{g}]^{\mathrm{tor}})^* = ([\widehat{g}']^{\mathrm{tor}})^* \circ ([\widehat{g}]^{\mathrm{tor}})^*$ (*in both applicable senses above*).

(5) ($\mathbb{Q}^\times$*-isogenies.*) *Let* g_l *be an element of* $\mathrm{GL}_{\mathcal{O} \underset{\mathbb{Z}}{\otimes} \mathbb{A}^\infty}(Q \underset{\mathbb{Z}}{\otimes} \mathbb{A}^\infty)$. *Then the submodule* $g_l(Q \underset{\mathbb{Z}}{\otimes} \hat{\mathbb{Z}})$ *in* $Q \underset{\mathbb{Z}}{\otimes} \mathbb{A}^\infty$ *determines a unique* $\mathcal{O}$*-lattice* Q' (*up to isomorphism*), *together with a unique choice of an isomorphism*

$$[g_l]_Q : Q \underset{\mathbb{Z}}{\otimes} \mathbb{Q} \xrightarrow{\sim} Q' \underset{\mathbb{Z}}{\otimes} \mathbb{Q},$$

inducing an isomorphism $Q \underset{\mathbb{Z}}{\otimes} \mathbb{A}^\infty \xrightarrow{\sim} Q' \underset{\mathbb{Z}}{\otimes} \mathbb{A}^\infty$ *matching* $g_l(Q \underset{\mathbb{Z}}{\otimes} \hat{\mathbb{Z}})$ *with* $Q' \underset{\mathbb{Z}}{\otimes} \hat{\mathbb{Z}}$, *and inducing a canonical* $\mathbb{Q}^\times$*-isogeny*

$$[g_l]_Q^* : \underline{\mathrm{Hom}}_{\mathcal{O}}(Q', G_{\mathsf{M}_{\mathcal{H}}})^\circ \to \underline{\mathrm{Hom}}_{\mathcal{O}}(Q, G_{\mathsf{M}_{\mathcal{H}}})^\circ$$

defined by $[g_l]_Q$. *Consider the sets* $\mathbf{K}_{Q',\mathcal{H}} \subset \mathbf{K}_{Q',\mathcal{H}}^+ \subset \mathbf{K}_{Q',\mathcal{H}}^{++}$ *and* $\mathbf{K}_{Q',\mathcal{H},\Sigma} \subset \mathbf{K}_{Q',\mathcal{H},\Sigma}^+ \subset \mathbf{K}_{Q',\mathcal{H},\Sigma}^{++}$ *as in Definitions* 1.2.4.44 *and* 1.2.4.50 (*with* Q *replaced with* Q'), *with compatible directed partial orders, parameterizing generalized Kuga families and their compactifications with properties as in* (1), (2), *and* (3) *above. The sets* $\mathbf{K}_{Q,\mathcal{H}}^{++}$ *etc and* $\mathbf{K}_{Q',\mathcal{H}}^{++}$ *etc* (*and the objects they parameterize*) *satisfy the compatibility with* g_l *in the sense that the following are true:*

(a) *For each* $\kappa = (\widehat{\mathcal{H}}, \widehat{\Sigma}) \in \mathbf{K}_{Q,\mathcal{H}}^{++}$ (*resp.* $\mathbf{K}_{Q,\mathcal{H}}^+$, *resp.* $\mathbf{K}_{Q,\mathcal{H}}$), *there is an element* $\kappa' = (\widehat{\mathcal{H}}', \widehat{\Sigma}') \in \mathbf{K}_{Q',\mathcal{H},\Sigma}^{++}$ (*resp.* $\mathbf{K}_{Q',\mathcal{H},\Sigma}^+$, *resp.* $\mathbf{K}_{Q',\mathcal{H},\Sigma}$) *such that* $\mathcal{H}_{\kappa'} = \widehat{\mathcal{H}}'_{\mathrm{G}} \subset \mathcal{H}_\kappa = \widehat{\mathcal{H}}_{\mathrm{G}}$, *such that the* $\mathbb{Q}^\times$*-isogeny*

$$[g_l]_{\kappa',\kappa}^{*,\mathrm{grp}} := \kappa^{\mathrm{isog}} \circ [g_l]_Q^* \circ ((\kappa')^{\mathrm{isog}})^{-1} : \mathsf{N}_{\kappa'}^{\mathrm{grp}} \to \mathsf{N}_\kappa^{\mathrm{grp}} \underset{\mathsf{M}_{\mathcal{H}_\kappa}}{\times} \mathsf{M}_{\mathcal{H}_{\kappa'}}$$

is an **isogeny** (*not just a quasi-isogeny*), *and such that there is a* (*necessarily unique*) *finite étale surjection*

$$[g_l]_{\kappa',\kappa}^* : \mathsf{N}_{\kappa'} \to \mathsf{N}_\kappa$$

inducing a finite étale surjection $\mathsf{N}_{\kappa'} \to \mathsf{N}_\kappa \underset{\mathsf{M}_{\mathcal{H}_\kappa}}{\times} \mathsf{M}_{\mathcal{H}_{\kappa'}}$ *of abelian scheme torsors equivariant with the isogeny* $[g_l]_{\kappa',\kappa}^{*,\mathrm{grp}}$.

(b) *For each* $\kappa = (\widehat{\mathcal{H}}, \widehat{\Sigma})$ *as in* (5a), *there is an element* $\kappa' = (\widehat{\mathcal{H}}', \widehat{\Sigma}') \in \mathbf{K}_{Q',\mathcal{H}}^{++}$ (*resp.* $\mathbf{K}_{Q',\mathcal{H}}^+$, *resp.* $\mathbf{K}_{Q',\mathcal{H}}$) *such that* $[g_l]_{\kappa',\kappa}^*$ *is defined as in* (5a) *and extends to a* (*necessarily unique*) *proper log étale surjection*

$$[g_l]_{\kappa',\kappa}^{*,\mathrm{tor}} : \mathsf{N}_{\kappa'}^{\mathrm{tor}} \to \mathsf{N}_\kappa^{\mathrm{tor}}, \tag{1.3.3.27}$$

such that

$$R^i([g_l]_{\kappa',\kappa}^{*,\mathrm{tor}})_* \mathscr{O}_{\mathsf{N}_{\kappa'}^{\mathrm{tor}}} = 0 \tag{1.3.3.28}$$

for all $i > 0$.

If $\kappa \in \mathbf{K}_{Q,\mathcal{H},\Sigma}^{++}$ (*resp.* $\mathbf{K}_{Q,\mathcal{H},\Sigma}^+$, *resp.* $\mathbf{K}_{Q,\mathcal{H},\Sigma}$), *we may assume in the above that* $\kappa' \in \mathbf{K}_{Q',\mathcal{H},\Sigma}^{++}$ (*resp.* $\mathbf{K}_{Q',\mathcal{H},\Sigma}^+$, *resp.* $\mathbf{K}_{Q',\mathcal{H},\Sigma}$). *Then* (1.3.3.27) *is compatible with the canonical morphisms* $f_\kappa^{\mathrm{tor}} : \mathsf{N}_\kappa^{\mathrm{tor}} \to \mathsf{M}_{\mathcal{H},\Sigma}^{\mathrm{tor}}$ *and* $f_{\kappa'}^{\mathrm{tor}} : \mathsf{N}_{\kappa'}^{\mathrm{tor}} \to \mathsf{M}_{\mathcal{H},\Sigma}^{\mathrm{tor}}$.

(c) *Suppose* $[g_l]_{\kappa',\kappa}^{*,\mathrm{tor}}$ *is defined as in* (5b) *for some* $\kappa \in \mathbf{K}_{Q,\mathcal{H},\Sigma}^+$ *and* $\kappa' \in \mathbf{K}_{Q',\mathcal{H},\Sigma}^+$ (*not just in* $\mathbf{K}_{Q,\mathcal{H},\Sigma}^{++}$ *and* $\mathbf{K}_{Q',\mathcal{H},\Sigma}^{++}$). *Then, for each integer* $i \geq 0$, *there is a canonical isomorphism*

$$([g_l]_{\kappa',\kappa}^{*,\mathrm{tor}})^* : \underline{H}_{\log\text{-dR}}^i(\mathsf{N}_\kappa^{\mathrm{tor}}/\mathsf{M}_{\mathcal{H},\Sigma}^{\mathrm{tor}}) \xrightarrow{\sim} \underline{H}_{\log\text{-dR}}^i(\mathsf{N}_{\kappa'}^{\mathrm{tor}}/\mathsf{M}_{\mathcal{H},\Sigma}^{\mathrm{tor}})$$

extending the canonical isomorphism

$$([g_l]^*_{\kappa',\kappa})^* : \underline{H}^i_{\mathrm{dR}}(\mathsf{N}_\kappa/\mathsf{M}_{\mathcal{H}}) \xrightarrow{\sim} \underline{H}^i_{\mathrm{dR}}(\mathsf{N}_{\kappa'}/\mathsf{M}_{\mathcal{H}})$$

induced by $[g_l]_Q$*, respecting the Hodge filtrations and inducing canonical isomorphisms*

$$([g_l]^{*,\mathrm{tor}}_{\kappa',\kappa})^* : R^b f^{\mathrm{tor}}_*(\overline{\Omega}^a_{\mathsf{N}^{\mathrm{tor}}_\kappa/\mathsf{M}^{\mathrm{tor}}_{\mathcal{H}}}) \xrightarrow{\sim} R^b f^{\mathrm{tor}}_*(\overline{\Omega}^a_{\mathsf{N}^{\mathrm{tor}}_{\kappa'}/\mathsf{M}^{\mathrm{tor}}_{\mathcal{H}}})$$

(for integers $a+b=i$*) compatible (under the canonical isomorphisms in* (3) *for* $\mathsf{N}^{\mathrm{tor}}_\kappa$ *and* $\mathsf{N}^{\mathrm{tor}}_{\kappa'}$*) with the canonical isomorphisms*

$$([g_l]^*_Q)^* : \underline{\mathrm{Hom}}_{\mathcal{O}}(Q^\vee, \underline{\mathrm{Lie}}_{G^\vee/\mathsf{M}^{\mathrm{tor}}_{\mathcal{H}}}) \xrightarrow{\sim} \underline{\mathrm{Hom}}_{\mathcal{O}}((Q')^\vee, \underline{\mathrm{Lie}}_{G^\vee/\mathsf{M}^{\mathrm{tor}}_{\mathcal{H}}})$$

and

$$([g_l]^*_Q)^* : \underline{\mathrm{Hom}}_{\mathcal{O}}(Q^\vee, \underline{\mathrm{Lie}}^\vee_{G/\mathsf{M}^{\mathrm{tor}}_{\mathcal{H}}}) \xrightarrow{\sim} \underline{\mathrm{Hom}}_{\mathcal{O}}((Q')^\vee, \underline{\mathrm{Lie}}^\vee_{G/\mathsf{M}^{\mathrm{tor}}_{\mathcal{H}}}).$$

(d) *If we have an element* $g'_l \in \mathrm{GL}_{\mathcal{O}\underset{\mathbb{Z}}{\otimes}\mathbb{A}^\infty}(Q\underset{\mathbb{Z}}{\otimes}\mathbb{A}^\infty)$ *with a similar setup such that* $[g'_l]^*_{\kappa'',\kappa'}$ *and* $[g'_l]^{*,\mathrm{tor}}_{\kappa'',\kappa'}$ *are compatibly defined for some* $\kappa'' \in \mathbf{K}^{++}_{Q'',\mathcal{H},\Sigma}$*, then* $[g_lg'_l]^*_{\kappa'',\kappa}$ *and* $[g_lg'_l]^{*,\mathrm{tor}}_{\kappa'',\kappa}$ *are also compatibly defined and satisfy the identities* $[g_lg'_l]^*_{\kappa'',\kappa} = [g_l]^*_{\kappa',\kappa} \circ [g'_l]^*_{\kappa'',\kappa'}$ *and* $[g_lg'_l]^{*,\mathrm{tor}}_{\kappa'',\kappa} = [g_l]^{*,\mathrm{tor}}_{\kappa',\kappa} \circ [g'_l]^{*,\mathrm{tor}}_{\kappa'',\kappa'}$*. If* $\kappa \in \mathbf{K}^+_{Q,\mathcal{H},\Sigma}$*,* $\kappa' \in \mathbf{K}^+_{Q',\mathcal{H},\Sigma}$*, and* $\kappa'' \in \mathbf{K}^+_{Q'',\mathcal{H},\Sigma}$*, we also have* $([g_lg'_l]^*_{\kappa'',\kappa})^* = ([g'_l]^*_{\kappa'',\kappa'})^* \circ ([g_l]^*_{\kappa',\kappa})^*$ *and* $([g_lg'_l]^{*,\mathrm{tor}}_{\kappa'',\kappa})^* = ([g'_l]^{*,\mathrm{tor}}_{\kappa'',\kappa'})^* \circ ([g_l]^{*,\mathrm{tor}}_{\kappa',\kappa})^*$.

Remark 1.3.3.29. The statements of Theorem 1.3.3.15 are more general than those in [61, Thm. 2.15], because we now consider not just Kuga families, but also generalized Kuga families over a finite étale cover $\mathsf{M}_{\mathcal{H}_\kappa}$ of $\mathsf{M}_{\mathcal{H}}$. Nevertheless, the proof of [61, Thm. 2.15] works almost verbatim for such generalizations. We will explain the necessary modifications in the next section. (We will only need the compactified Kuga families, *i.e.*, those with $\kappa \in \mathbf{K}_{Q,\mathcal{H},\Sigma}$, for the construction of canonical and subcanonical extensions of automorphic bundles in Section 1.4.2, and for all applications we know. We included their generalizations in Theorem 1.3.3.15 only because it seems natural to do so.)

Remark 1.3.3.30. The second, third, and fourth paragraphs of (1) of Theorem 1.3.3.15 follow from the construction of N_κ and $\mathsf{N}^{\mathrm{tor}}_\kappa$ using the toroidal boundary of a larger PEL-type moduli problem $\widetilde{\mathsf{M}}_{\widetilde{\mathcal{H}}}$, and from (2) and (5) of Theorem 1.3.1.3, and from Lemma 1.3.2.41 (and from the justifications provided in Section 1.3.2 and to be provided in Section 1.3.4 below). The second paragraphs of (2) and (4b) of Theorem 1.3.3.15 follow from the construction of f^{tor}_κ and $[\widehat{g}]^{\mathrm{tor}}$ using the universal property of certain suitably chosen $\widetilde{\mathsf{M}}^{\mathrm{tor}}_{\widetilde{\mathcal{H}}} = \widetilde{\mathsf{M}}^{\mathrm{tor}}_{\widetilde{\mathcal{H}},\widetilde{\Sigma}}$ (given by (6) of Theorem 1.3.1.3), which is consistent with the construction of the canonical morphisms in Lemmas 1.3.2.41 and 1.3.2.79 and Proposition 1.3.2.67 using the various universal properties (all given in terms of degeneration data). These statements were not in [61, Thm. 2.15], but are implicit in the theory and can be deduced from other known statements. Similarly, the second last paragraph of statement (1) of Theorem 1.3.3.15 was not in [61, Thm. 2.15], but can be deduced from the other statements (in Theorems 1.3.1.3 and 1.3.1.5, Propositions 1.3.1.15 and 1.3.1.14, and Theorem 1.3.3.15). We omit the proof here because the proof of a subtler statement in mixed characteristics will be given for Theorem 7.1.4.1 (see Proposition 7.2.4.14 below).

Remark 1.3.3.31. The isomorphism (1.3.3.19) (even just in the case of Kuga families) was not in the statement of [61, Thm. 2.15], although it is implicit in the arguments of the (rather lengthy) proof there. We included it here for the sake of completeness. The details of the proof are similar to those for (7.1.4.5) (to be given below in Section 7.3), and hence are omitted here.

Remark 1.3.3.32. Statements (4) and (5) of Theorem 1.3.3.15 were not as explicitly stated in [61, Thm. 2.15], but follow from the same argument of the proof there (based on an analogue of Proposition 1.3.1.15).

Remark 1.3.3.33. (Compare with Remarks 1.1.2.1 and 1.3.1.4.) By varying the choices of L and Q, and hence varying the choices of $\widetilde{L}$, we can (in practice) allow the $\widehat{\mathcal{H}}$ in the parameter $\kappa = (\widehat{\mathcal{H}}, \widehat{\Sigma})$ to be any open compact subgroup of $\widehat{\mathrm{G}}(\mathbb{A}^\infty)$. Nevertheless, this can be achieved by varying the lattice Q alone, and hence is already incorporated in (5) of Theorem 1.3.3.15.

1.3.4 *Justification for the Parameters for Kuga Families*

For later constructions (in Chapter 7), and for some applications, we would like to spell out what $f_\kappa : \mathsf{N}_\kappa \to \mathsf{M}_{\mathcal{H}}$ and $\kappa^{\mathrm{isog}} : \underline{\mathrm{Hom}}_{\mathcal{O}}(Q, G_{\mathsf{M}_{\mathcal{H}_\kappa}})^\circ \to \mathsf{N}_\kappa^{\mathrm{grp}}$ are for each $\kappa \in \mathbf{K}^{++}_{Q,\mathcal{H}}$, and what $f_\kappa^{\mathrm{tor}} : \mathsf{N}_\kappa^{\mathrm{tor}} \to \mathsf{M}_{\mathcal{H}}^{\mathrm{tor}} = \mathsf{M}_{\mathcal{H},\Sigma}^{\mathrm{tor}}$ is for each $\kappa = (\widehat{\mathcal{H}}, \widehat{\Sigma}) \in \mathbf{K}^{++}_{Q,\mathcal{H},\Sigma}$. Let us review some of the materials in [61, Sec. 3]. (We will also take this opportunity to correct a mistake there. See Remark 1.3.4.5 below.)

Let $(\widetilde{L}, \langle\,\cdot\,,\,\cdot\,\rangle^\sim, \widetilde{h}_0)$, $(\widetilde{\mathrm{Z}}, \widetilde{\Phi}, \widetilde{\delta})$, *etc* be chosen as in Section 1.2.4. Let $\widetilde{\kappa} = (\widetilde{\mathcal{H}}, \widetilde{\Sigma}, \widetilde{\sigma})$ be any element in the set $\widetilde{\mathbf{K}}^{++}_{Q,\mathcal{H}}$ as in Definition 1.2.4.11, and let $\kappa = [\widetilde{\kappa}] \in \mathbf{K}^{++}_{Q,\mathcal{H}}$ be as in Definition 1.2.4.44.

The data of $\mathcal{O}$, $(\widetilde{L}, \langle\,\cdot\,,\,\cdot\,\rangle^\sim, \widetilde{h}_0)$, and $\widetilde{\mathcal{H}} \subset \widetilde{\mathrm{G}}(\widehat{\mathbb{Z}})$ define a moduli problem $\widetilde{\mathsf{M}}_{\widetilde{\mathcal{H}}}$ as in Section 1.1.2. Since $\widetilde{\mathcal{H}}$ is neat and $\widetilde{\Sigma}$ is projective (and smooth), by Theorems 1.3.1.3 and 1.3.1.10, we have a toroidal compactification $\widetilde{\mathsf{M}}^{\mathrm{tor}}_{\widetilde{\mathcal{H}}} = \widetilde{\mathsf{M}}^{\mathrm{tor}}_{\widetilde{\mathcal{H}},\widetilde{\Sigma}}$ of $\widetilde{\mathsf{M}}_{\widetilde{\mathcal{H}}}$ which is projective and smooth over S_0, as in the latter half of Section 1.3.2. Since $\widetilde{\mathcal{H}}$ satisfies Condition 1.2.4.7, by Lemmas 1.3.2.1 and 1.3.2.5, we have $\widetilde{\mathsf{M}}^{\widetilde{\Phi}_{\widetilde{\mathcal{H}}}}_{\widetilde{\mathcal{H}}} \cong \widetilde{\mathsf{M}}^{\widetilde{\mathrm{Z}}_{\widetilde{\mathcal{H}}}}_{\widetilde{\mathcal{H}}} \cong \mathsf{M}_{\mathcal{H}_\kappa}$, where $\mathcal{H}_\kappa = \widehat{\mathcal{H}}_{\mathrm{G}} = \mathrm{Gr}^{\widetilde{\mathrm{Z}}}_{-1}(\widetilde{\mathcal{H}}_{\widetilde{\mathrm{P}}'_{\widetilde{\mathrm{Z}}}}) = \mathrm{Gr}^{\widetilde{\mathrm{Z}}}_{-1}(\widetilde{\mathcal{H}}_{\widetilde{\mathrm{P}}_{\widetilde{\mathrm{Z}}}})$. By the construction of $\widetilde{C}_{\widetilde{\Phi}_{\widetilde{\mathcal{H}}},\widetilde{\delta}_{\widetilde{\mathcal{H}}}} \to \widetilde{\mathsf{M}}^{\widetilde{\Phi}_{\widetilde{\mathcal{H}}}}_{\widetilde{\mathcal{H}}}$ in [62, Sec. 6.2.3–6.2.4] (see also the correction in Remark 1.3.1.6), it is a torsor under an abelian scheme $\widetilde{C}^{\mathrm{grp}}_{\widetilde{\Phi}_{\widetilde{\mathcal{H}}},\widetilde{\delta}_{\widetilde{\mathcal{H}}}}$ canonically $\mathbb{Q}^\times$-isogenous to $\underline{\mathrm{Hom}}_{\mathcal{O}}(Q, G_{\mathsf{M}_{\mathcal{H}}})^\circ$. If $\widehat{\mathcal{H}}$ satisfies Condition 1.2.4.8, then we have $\mathcal{H}_\kappa = \mathcal{H}$ and hence $\mathsf{M}_{\mathcal{H}_\kappa} = \mathsf{M}_{\mathcal{H}}$. If $\widehat{\mathcal{H}}$

also satisfies Condition 1.2.4.9, then $\widetilde{C}_{\widetilde{\Phi}_{\widetilde{\mathcal{H}}},\widetilde{\delta}_{\widetilde{\mathcal{H}}}} = \widetilde{C}^{\mathrm{grp}}_{\widetilde{\Phi}_{\widetilde{\mathcal{H}}},\widetilde{\delta}_{\widetilde{\mathcal{H}}}} \to \mathsf{M}_{\mathcal{H}_\kappa} = \mathsf{M}_{\mathcal{H}}$ is an abelian scheme, not just a torsor. (See Remark 1.3.1.6.)

Remark 1.3.4.1. The isomorphism $\widetilde{\mathsf{M}}^{\widetilde{\Phi}_{\widetilde{\mathcal{H}}}}_{\widetilde{\mathcal{H}}} \cong \widetilde{\mathsf{M}}^{\widetilde{\mathsf{Z}}_{\widetilde{\mathcal{H}}}}_{\widetilde{\mathcal{H}}} \cong \mathsf{M}_{\mathcal{H}_\kappa}$ means we do not need to consider nontrivial twisted objects $(\widetilde{\varphi}^{\sim}_{-2,\widetilde{\mathcal{H}}}, \widetilde{\varphi}^{\sim}_{0,\widetilde{\mathcal{H}}})$ above $(\widetilde{\varphi}_{-2,\widetilde{\mathcal{H}}}, \widetilde{\varphi}_{0,\widetilde{\mathcal{H}}})$ and $\widetilde{\varphi}_{-1,\widetilde{\mathcal{H}}} = \alpha_{\mathcal{H}_\kappa}$.

Since $\widetilde{\sigma} \subset \mathbf{P}^+_{\widetilde{\Phi}_{\widetilde{\mathcal{H}}}}$ is a *top-dimensional* nondegenerate rational polyhedral cone in the cone decomposition $\widetilde{\Sigma}_{\widetilde{\Phi}_{\widetilde{\mathcal{H}}}}$ in $\widetilde{\Sigma}$, by (2) of Theorem 1.3.1.3, the locally closed stratum $\widetilde{\mathsf{Z}}_{[(\widetilde{\Phi}_{\widetilde{\mathcal{H}}},\widetilde{\delta}_{\widetilde{\mathcal{H}}},\widetilde{\sigma})]}$ (not its closure) of $\widetilde{\mathsf{M}}^{\mathrm{tor}}_{\widetilde{\mathcal{H}}}$ is a zero-dimensional torus bundle over the abelian scheme torsor $\widetilde{C}_{\widetilde{\Phi}_{\widetilde{\mathcal{H}}},\widetilde{\delta}_{\widetilde{\mathcal{H}}}}$ over $\mathsf{M}_{\mathcal{H}_\kappa}$. In other words, $\widetilde{\mathsf{Z}}_{[(\widetilde{\Phi}_{\widetilde{\mathcal{H}}},\widetilde{\delta}_{\widetilde{\mathcal{H}}},\widetilde{\sigma})]}$ is canonically *isomorphic* to $\widetilde{C}_{\widetilde{\Phi}_{\widetilde{\mathcal{H}}},\widetilde{\delta}_{\widetilde{\mathcal{H}}}}$. Let us define $\mathsf{N}_{\widetilde{\kappa}}$ to be this stratum $\widetilde{\mathsf{Z}}_{[(\widetilde{\Phi}_{\widetilde{\mathcal{H}}},\widetilde{\delta}_{\widetilde{\mathcal{H}}},\widetilde{\sigma})]}$, and denote the canonical morphism $\mathsf{N}_{\widetilde{\kappa}} \to \mathsf{M}_{\mathcal{H}}$ by $f_{\widetilde{\kappa}}$ (which is the composition of the canonical morphisms $\mathsf{N}_{\widetilde{\kappa}} \to \mathsf{M}_{\mathcal{H}_\kappa}$ and $\mathsf{M}_{\mathcal{H}_\kappa} \to \mathsf{M}_{\mathcal{H}}$). Let us denote the canonical $\mathbb{Q}^\times$-isogeny $\underline{\mathrm{Hom}}_{\mathcal{O}}(Q, G_{\mathsf{M}_{\mathcal{H}}})^\circ \to \mathsf{N}^{\mathrm{grp}}_{\widetilde{\kappa}} := \widetilde{C}^{\mathrm{grp}}_{\widetilde{\Phi}_{\widetilde{\mathcal{H}}},\widetilde{\delta}_{\widetilde{\mathcal{H}}}}$ by $\widetilde{\kappa}^{\mathrm{isog}}$. Note that $\mathsf{N}_{\widetilde{\kappa}} = \widetilde{\mathsf{Z}}_{[(\widetilde{\Phi}_{\widetilde{\mathcal{H}}},\widetilde{\delta}_{\widetilde{\mathcal{H}}},\widetilde{\sigma})]}$ is canonically isomorphic to $\widetilde{C}_{\widetilde{\Phi}_{\widetilde{\mathcal{H}}},\widetilde{\delta}_{\widetilde{\mathcal{H}}}}$ for every $\widetilde{\Sigma}$ and every top-dimensional cone $\widetilde{\sigma}$ in $\widetilde{\Sigma}_{\widetilde{\Phi}_{\widetilde{\mathcal{H}}}}$.

Lemma 1.3.4.2. *The abelian scheme torsor* $\widetilde{C}_{\widetilde{\Phi}_{\widetilde{\mathcal{H}}},\widetilde{\delta}_{\widetilde{\mathcal{H}}}}$ *(see* (5) *of Theorem* 1.3.1.5 *and Definition* 1.2.1.15) *and the canonical isogeny* $\underline{\mathrm{Hom}}_{\mathcal{O}}(Q, G_{\mathsf{M}_{\mathcal{H}_\kappa}})^\circ \to \widetilde{C}^{\mathrm{grp}}_{\widetilde{\Phi}_{\widetilde{\mathcal{H}}},\widetilde{\delta}_{\widetilde{\mathcal{H}}}}$ *of abelian schemes over* $\widetilde{\mathsf{M}}^{\widetilde{\Phi}_{\widetilde{\mathcal{H}}}}_{\widetilde{\mathcal{H}}} \cong \mathsf{M}_{\mathcal{H}_\kappa}$ *depend (up to canonical isomorphism) only on the open compact subgroup* $\widehat{\mathcal{H}} = \widetilde{\mathcal{H}}_{\widehat{\mathrm{G}}}$ *of* $\widehat{\mathrm{G}}(\hat{\mathbb{Z}})$ *(see Definitions* 1.2.4.3 *and* 1.2.4.4) *determined by* $\widetilde{\mathcal{H}}$. *Moreover, if* $\widetilde{\mathcal{H}}'$ *is any open compact subgroup of* $\widetilde{\mathrm{G}}(\hat{\mathbb{Z}})$ *still satisfying Condition* 1.2.4.7 *such that* $\widetilde{\mathcal{H}}'_{\widehat{\mathrm{G}}} = \widehat{\mathcal{H}}_{\mathrm{G}} \ltimes \widehat{\mathcal{H}}_{\widehat{\mathrm{U}}}$ *under the isomorphism* $\widehat{\mathrm{G}}(\hat{\mathbb{Z}}) \cong \mathrm{G}(\hat{\mathbb{Z}}) \ltimes \widehat{\mathrm{U}}(\hat{\mathbb{Z}})$ *induced by the splitting* $\widetilde{\delta}$ *(cf. Condition* 1.2.4.9), *then we have* $\widetilde{C}_{\widetilde{\Phi}_{\widetilde{\mathcal{H}}'},\widetilde{\delta}_{\widetilde{\mathcal{H}}'}} = \widetilde{C}^{\mathrm{grp}}_{\widetilde{\Phi}_{\widetilde{\mathcal{H}}'},\widetilde{\delta}_{\widetilde{\mathcal{H}}'}} \cong \widetilde{C}^{\mathrm{grp}}_{\widetilde{\Phi}_{\widetilde{\mathcal{H}}},\widetilde{\delta}_{\widetilde{\mathcal{H}}}}$.

Proof. These follow from the corresponding statements of Lemma 1.3.2.7. □

Consequently, $\mathsf{N}_{\widetilde{\kappa}}$ and $\widetilde{\kappa}^{\mathrm{isog}}$ depend (up to canonical isomorphism) only on the open compact subgroup $\widehat{\mathcal{H}}$ of $\widehat{\mathrm{G}}(\hat{\mathbb{Z}})$ determined by $\widetilde{\mathcal{H}}$ (see Definitions 1.2.4.3 and 1.2.4.4).

Let us take $\mathsf{N}^{\mathrm{tor}}_{\widetilde{\kappa}}$ to be the schematic closure of the locally closed stratum $\widetilde{\mathsf{Z}}_{[(\widetilde{\Phi}_{\widetilde{\mathcal{H}}},\widetilde{\delta}_{\widetilde{\mathcal{H}}},\widetilde{\sigma})]}$ in $\widetilde{\mathsf{M}}^{\mathrm{tor}}_{\widetilde{\mathcal{H}},\widetilde{\Sigma}}$. Then we obtain the canonical open dense immersion $\widetilde{\kappa}^{\mathrm{tor}} : \mathsf{N}_{\widetilde{\kappa}} \hookrightarrow \mathsf{N}^{\mathrm{tor}}_{\widetilde{\kappa}}$. Certainly, $\mathsf{N}^{\mathrm{tor}}_{\widetilde{\kappa}}$ depends not only on $\widehat{\mathcal{H}}$ but also on the choices of $\widetilde{\Sigma}_{\widetilde{\Phi}_{\widetilde{\mathcal{H}}}}$ and $\widetilde{\sigma}$.

Lemma 1.3.4.3. *(See [61, Lem. 3.1].) Under the assumption that* $\widetilde{\mathcal{H}}$ *is neat, the closure of every stratum in* $\widetilde{\mathsf{M}}^{\mathrm{tor}}_{\widetilde{\mathcal{H}},\widetilde{\Sigma}}$ *has no self-intersection.*

Corollary 1.3.4.4. (*Compare with [61, Cor. 3.2].*) *For each* $\widetilde{\kappa} = (\widetilde{\mathcal{H}}, \widetilde{\Sigma}, \widetilde{\sigma}) \in \widetilde{\mathbf{K}}^{++}_{Q,\mathcal{H}}$, *the closure* $\mathsf{N}^{\mathrm{tor}}_{\widetilde{\kappa}}$ *of* $\mathsf{N}_{\widetilde{\kappa}} = \widetilde{\mathsf{Z}}_{[(\widetilde{\Phi}_{\widetilde{\mathcal{H}}}, \widetilde{\delta}_{\widetilde{\mathcal{H}}}, \widetilde{\sigma})]}$ *in* $\widetilde{\mathsf{M}}^{\mathrm{tor}}_{\widetilde{\mathcal{H}}, \widetilde{\Sigma}}$ *is projective and smooth over* S_0, *and the complement of* $\mathsf{N}_{\widetilde{\kappa}}$ *in* $\mathsf{N}^{\mathrm{tor}}_{\widetilde{\kappa}}$ *(with its reduced structure) is a relative Cartier divisor with simple normal crossings.*

Proof. Combine Lemma 1.3.4.3, (3) of Theorem 1.3.1.3, and Theorem 1.3.1.10. □

Remark 1.3.4.5. In [61, Sec. 3], the $\widetilde{\kappa}$ *etc* above were denoted κ *etc* without the tildes. However, the binary relation $\succ$ introduced there is not a directed partial order. We take this opportunity to correct this mistake and (at the same time) provide a formulation better for the applications. The desired parameters should be given by equivalence classes $\kappa = [\widetilde{\kappa}]$ of $\widetilde{\kappa}$, with the natural partial order $\succ$ among them. (See Definitions 1.2.4.44 and 1.3.4.20, and see Proposition 1.3.4.19 below. Before then, we can not yet assert the second half of [61, Cor. 3.2].)

The stratification of $\widetilde{\mathsf{M}}^{\mathrm{tor}}_{\widetilde{\mathcal{H}}}$ induces a stratification of $\mathsf{N}^{\mathrm{tor}}$. By (2) of Theorem 1.3.1.3, the strata of $\mathsf{N}^{\mathrm{tor}}$ are parameterized by equivalence classes $[(\breve{\Phi}_{\widetilde{\mathcal{H}}}, \breve{\delta}_{\widetilde{\mathcal{H}}}, \breve{\tau})]$ having $[(\widetilde{\Phi}_{\widetilde{\mathcal{H}}}, \widetilde{\delta}_{\widetilde{\mathcal{H}}}, \widetilde{\sigma})]$ as a face (as in Definition 1.2.2.19), spelled out in Section 1.2.4. By Lemma 1.2.4.42, they can also be parameterized by the equivalence classes $[(\breve{\Phi}_{\widehat{\mathcal{H}}}, \breve{\delta}_{\widehat{\mathcal{H}}}, \widehat{\tau})]$.

Let $\breve{\sigma}$ be the image of $\widetilde{\sigma} \subset \mathbf{P}^{+}_{\breve{\Phi}_{\widetilde{\mathcal{H}}}}$ under the first morphism in (1.2.4.20). Consider the sets $\widetilde{\Sigma}_{\breve{\Phi}_{\widetilde{\mathcal{H}}}, \breve{\sigma}}$ and $\widetilde{\Sigma}^{+}_{\breve{\Phi}_{\widetilde{\mathcal{H}}}, \breve{\sigma}}$, and the groups $\Gamma_{\breve{\Phi}_{\widetilde{\mathcal{H}}}, \Phi_{\mathcal{H}}}$, $\Gamma_{\breve{\Phi}_{\widetilde{\mathcal{H}}}, \Phi_{\mathcal{H}}, \breve{\sigma}}$, $\Gamma_{\breve{\Phi}_{\widehat{\mathcal{H}}}}$, and $\Gamma_{\breve{\Phi}_{\widehat{\mathcal{H}}}, \Phi_{\mathcal{H}}}$ defined in Definition 1.2.4.21; consider the sets $\widehat{\mathbf{S}}_{\breve{\Phi}_{\widehat{\mathcal{H}}}}$ and $(\widehat{\mathbf{S}}_{\breve{\Phi}_{\widehat{\mathcal{H}}}})^{\vee}_{\mathbb{R}}$ defined in Definition 1.2.4.29; consider the $\Gamma_{\breve{\Phi}_{\widehat{\mathcal{H}}}}$-*admissible rational polyhedral cone decomposition* $\widehat{\Sigma}_{\breve{\Phi}_{\widehat{\mathcal{H}}}}$ of $\widehat{\mathbf{P}}_{\breve{\Phi}_{\widehat{\mathcal{H}}}}$ defined in Corollary 1.2.4.40; consider the collection $\widehat{\Sigma}$ defined in Lemma 1.2.4.42; and consider the set $\mathbf{K}_{Q,\mathcal{H}}$ of equivalence classes $\kappa = (\widehat{\mathcal{H}}, \widehat{\Sigma}) = [\widetilde{\kappa}]$ of elements $\widetilde{\kappa}$ in $\widetilde{\mathbf{K}}_{Q,\mathcal{H}}$, with a directed partial order $\kappa' = (\widehat{\mathcal{H}}', \widehat{\Sigma}') \succ \kappa = (\widehat{\mathcal{H}}, \widehat{\Sigma})$ when $\widehat{\mathcal{H}}' \subset \widehat{\mathcal{H}}'$ and when $\widehat{\Sigma}'$ is a refinement of $\widehat{\Sigma}$, as in Definition 1.2.4.44 and Lemma 1.2.4.47.

Construction 1.3.4.6. For each $\widetilde{\kappa} = (\widetilde{\mathcal{H}}, \widetilde{\Sigma}, \widetilde{\sigma})$ in $\widetilde{\mathbf{K}}^{++}_{Q,\mathcal{H}}$, consider the degenerating family

$$(\widetilde{G}, \widetilde{\lambda}, \widetilde{i}, \widetilde{\alpha}_{\widetilde{\mathcal{H}}}) \to \widetilde{\mathsf{M}}^{\mathrm{tor}}_{\widetilde{\mathcal{H}}, \widetilde{\Sigma}} \tag{1.3.4.7}$$

of type $\widetilde{\mathsf{M}}_{\widetilde{\mathcal{H}}}$ as in Theorem 1.3.1.3. As in (1.3.2.18), let

$$(\widehat{G}, \widehat{\lambda}, \widehat{i}) \to \mathsf{N}^{\mathrm{tor}}_{\widetilde{\kappa}} \tag{1.3.4.8}$$

denote the pullback of (1.3.4.7) to $\mathsf{N}^{\mathrm{tor}}_{\widetilde{\kappa}}$, the closure of $\mathsf{N}_{\widetilde{\kappa}} = \widetilde{\mathsf{Z}}_{[(\widetilde{\Phi}_{\widetilde{\mathcal{H}}}, \widetilde{\delta}_{\widetilde{\mathcal{H}}}, \widetilde{\sigma})]}$ in $\widetilde{\mathsf{M}}^{\mathrm{tor}}_{\widetilde{\mathcal{H}}, \widetilde{\Sigma}}$. Note that $\mathsf{N}_{\widetilde{\kappa}}$ is canonically isomorphic to $\widetilde{C}_{\widetilde{\Phi}_{\widetilde{\mathcal{H}}}, \widetilde{\delta}_{\widetilde{\mathcal{H}}}}$ because $\widetilde{\sigma}$ is top-dimensional. Although $\widetilde{\alpha}_{\widetilde{\mathcal{H}}}$ is defined only over $\widetilde{\mathsf{M}}_{\widetilde{\mathcal{H}}}$, by proceeding as in Construction 1.3.2.16, we can define a (partial) pullback

$$(\widehat{G}, \widehat{\lambda}, \widehat{i}, \widehat{\alpha}_{\widehat{\mathcal{H}}}) := (\widetilde{G}^{\natural}, \widetilde{\lambda}^{\natural}, \widetilde{i}^{\natural}, \widetilde{\beta}^{\natural}_{\widetilde{\mathcal{H}}}) \to \mathsf{N}^{\mathrm{tor}}_{\widetilde{\kappa}} \tag{1.3.4.9}$$

of the degenerating family (1.3.4.7) to $\mathsf{N}^{\mathrm{tor}}_{\widetilde{\kappa}}$, with the convention that (as in the case of $(\widetilde{G}, \widetilde{\lambda}, \widetilde{i}, \widetilde{\alpha}_{\widetilde{\mathcal{H}}})$ itself) $\widehat{\alpha}_{\widehat{\mathcal{H}}}$ is defined only over $\mathsf{N}_{\widetilde{\kappa}}$, while $(\widehat{G}, \widehat{\lambda}, \widehat{i})$ is defined over all of $\mathsf{N}^{\mathrm{tor}}_{\widetilde{\kappa}}$ as in (1.3.4.8). By construction, the pullback

$$(\widehat{G}_{\mathsf{N}_{\widetilde{\kappa}}}, \widehat{\lambda}_{\mathsf{N}_{\widetilde{\kappa}}}, \widehat{i}_{\mathsf{N}_{\widetilde{\kappa}}}, \widehat{\alpha}_{\widehat{\mathcal{H}}}) \to \mathsf{N}_{\widetilde{\kappa}} \cong \widetilde{C}_{\widetilde{\Phi}_{\widetilde{\mathcal{H}}}, \widetilde{\delta}_{\widetilde{\mathcal{H}}}}, \tag{1.3.4.10}$$

of (1.3.4.9) to $\mathsf{N}_{\widetilde{\kappa}}$ determines and is determined by (the prescribed $(\widetilde{\mathsf{Z}}_{\widetilde{\mathcal{H}}}, \widetilde{\Phi}_{\widetilde{\mathcal{H}}}, \widetilde{\delta}_{\widetilde{\mathcal{H}}})$ and) the tautological object

$$((A, \lambda, i, \alpha_{\mathcal{H}_\kappa}), (\widetilde{c}_{\widetilde{\mathcal{H}}}, \widetilde{c}^\vee_{\widetilde{\mathcal{H}}})) \to \widetilde{C}_{\widetilde{\Phi}_{\widetilde{\mathcal{H}}}, \widetilde{\delta}_{\widetilde{\mathcal{H}}}} \tag{1.3.4.11}$$

(up to isomorphisms inducing automorphisms of $\widetilde{\Phi}_{\widetilde{\mathcal{H}}}$; *i.e.*, elements of $\Gamma_{\widetilde{\Phi}_{\widetilde{\mathcal{H}}}}$). Here $(A, \lambda, i, \alpha_{\mathcal{H}_\kappa})$ is the tautological object over $\widetilde{\mathsf{M}}^{\widetilde{\mathsf{Z}}_{\widetilde{\mathcal{H}}}}_{\widetilde{\mathcal{H}}} \cong \mathsf{M}_{\mathcal{H}_\kappa}$. As explained in Remark 1.3.4.1, we do not need to consider nontrivial twisted objects $(\widetilde{\varphi}^{\sim}_{-2,\widetilde{\mathcal{H}}}, \widetilde{\varphi}^{\sim}_{0,\widetilde{\mathcal{H}}})$ above $(\widetilde{\varphi}_{-2,\widetilde{\mathcal{H}}}, \widetilde{\varphi}_{0,\widetilde{\mathcal{H}}})$ and $\widetilde{\varphi}_{-1,\widetilde{\mathcal{H}}} = \alpha_{\mathcal{H}_\kappa}$. With the fixed choice of $(\widetilde{\mathsf{Z}}, \widetilde{\Phi}, \widetilde{\delta})$, the tautological object (1.3.4.11) depends only on $\widehat{\mathcal{H}}$, and hence so is the tuple (1.3.4.10). Thus, the notation $\widehat{\alpha}_{\widehat{\mathcal{H}}}$ is justified.

Construction 1.3.4.12. Let $(\widehat{G}, \widehat{\lambda}, \widehat{i}, \widehat{\alpha}_{\widehat{\mathcal{H}}}) \to \mathsf{N}^{\mathrm{tor}}_{\widetilde{\kappa}}$ be as in (1.3.4.9) in Construction 1.3.4.6. Consider any morphism $\xi : \mathrm{Spec}(V) \to \mathsf{N}^{\mathrm{tor}}_{\widetilde{\kappa}}$ centered at a geometric point $\bar{s}$ of $\mathsf{N}^{\mathrm{tor}}_{\widetilde{\kappa}}$ such that V is a complete discrete valuation ring with fraction field K, and such that $\eta := \mathrm{Spec}(K)$ is mapped to the generic point of the irreducible component containing the image of $\bar{s}$. Suppose the image of $\bar{s}$ lies on the $[(\breve{\Phi}_{\widetilde{\mathcal{H}}}, \breve{\delta}_{\widetilde{\mathcal{H}}}, \breve{\rho})]$-stratum $\widetilde{\mathsf{Z}}_{[(\breve{\Phi}_{\widetilde{\mathcal{H}}}, \breve{\delta}_{\widetilde{\mathcal{H}}}, \breve{\rho})]}$ of $\widetilde{\mathsf{M}}^{\mathrm{tor}}_{\widetilde{\mathcal{H}}, \widetilde{\Sigma}}$, where $[(\breve{\Phi}_{\widetilde{\mathcal{H}}}, \breve{\delta}_{\widetilde{\mathcal{H}}}, \breve{\rho})]$ is represented by some $(\breve{\Phi}_{\widetilde{\mathcal{H}}}, \breve{\delta}_{\widetilde{\mathcal{H}}}, \breve{\rho})$ with $(\breve{\mathsf{Z}}_{\widetilde{\mathcal{H}}}, \breve{\Phi}_{\widetilde{\mathcal{H}}} = (\breve{X}, \breve{Y}, \breve{\phi}, \breve{\varphi}_{-2,\widetilde{\mathcal{H}}}, \breve{\varphi}_{0,\widetilde{\mathcal{H}}}), \breve{\delta}_{\widetilde{\mathcal{H}}})$ representing some cusp label as in Section 1.2.4. (We avoid using the more familiar notation $(\breve{\Phi}_{\widetilde{\mathcal{H}}}, \breve{\delta}_{\widetilde{\mathcal{H}}}, \breve{\tau})$ because the symbol τ will be used for another purpose below.) For simplicity, let us fix compatible choices of representatives $(\widetilde{\mathsf{Z}}, \widetilde{\Phi} = (\widetilde{X}, \widetilde{Y}, \widetilde{\phi}, \widetilde{\varphi}_{-2}, \widetilde{\varphi}_0), \widetilde{\delta})$ and $(\breve{\mathsf{Z}}, \breve{\Phi} = (\breve{X}, \breve{Y}, \breve{\phi}, \breve{\varphi}_{-2}, \breve{\varphi}_0), \breve{\delta})$, as in Section 1.2.4, in their $\widetilde{\mathcal{H}}$-orbits.

Since $\widetilde{\mathfrak{X}}_{\breve{\Phi}_{\widetilde{\mathcal{H}}}, \breve{\delta}_{\widetilde{\mathcal{H}}}, \breve{\rho}}$ is formally smooth over S_0, there exists a complete regular local ring $\widetilde{V}$ and an ideal $\widetilde{I} \subset \widetilde{V}$ such that $\widetilde{V}/\widetilde{I} \cong V$ and such that the morphism $\mathrm{Spec}(V) \to \mathsf{N}^{\mathrm{tor}}_{\widetilde{\kappa}}$ extends to a morphism $\widetilde{\xi} : \mathrm{Spf}(\widetilde{V}, \widetilde{I}) \to \widetilde{\mathfrak{X}}_{\breve{\Phi}_{\widetilde{\mathcal{H}}}, \breve{\delta}_{\widetilde{\mathcal{H}}}, \breve{\rho}}$ which induces a dominant morphism from $\mathrm{Spec}(\widetilde{V})$ to $\mathrm{Spec}(\widetilde{R})$, where $\widetilde{R}$ is the local ring of $\widetilde{\mathfrak{X}}_{\breve{\Phi}_{\widetilde{\mathcal{H}}}, \breve{\delta}_{\widetilde{\mathcal{H}}}, \breve{\rho}}$ at the image of $\bar{s}$. Let

$$(\widetilde{G}^\ddagger, \widetilde{\lambda}^\ddagger, \widetilde{i}^\ddagger, \widetilde{\alpha}^\ddagger_{\widetilde{\mathcal{H}}}) \to \mathrm{Spec}(\widetilde{V}) \tag{1.3.4.13}$$

denote the pullback of (1.3.4.7) under the composition of $\widetilde{\xi}$ with the canonical morphism $\widetilde{\mathfrak{X}}_{\breve{\Phi}_{\widetilde{\mathcal{H}}}, \breve{\delta}_{\widetilde{\mathcal{H}}}, \breve{\rho}} \to \widetilde{\mathsf{M}}^{\mathrm{tor}}_{\widetilde{\mathcal{H}}, \widetilde{\Sigma}}$, and let

$$(\widehat{G}^\ddagger, \widehat{\lambda}^\ddagger, \widehat{i}^\ddagger, \widehat{\alpha}^\ddagger_{\widehat{\mathcal{H}}}) \to \mathrm{Spec}(V) \tag{1.3.4.14}$$

denote the pullback of (1.3.4.9) under ξ.

As in (6) of Theorem 1.3.1.3, (1.3.4.13) defines an object of $\mathrm{DEG}_{\mathrm{PEL},\widetilde{\mathsf{M}}_{\widetilde{\mathcal{H}}}}(\widetilde{V})$, which corresponds to an object

$$(\widetilde{B}^{\ddagger}, \lambda_{\widetilde{B}^{\ddagger}}, i_{\widetilde{B}^{\ddagger}}, \widetilde{X}^{\ddagger}, \widetilde{Y}^{\ddagger}, \widetilde{\phi}^{\ddagger}, \widetilde{c}^{\ddagger}, \widetilde{c}^{\vee,\ddagger}, \widetilde{\tau}^{\ddagger}, [\widetilde{\alpha}^{\natural,\ddagger}_{\widetilde{\mathcal{H}}}])$$

in $\mathrm{DD}_{\mathrm{PEL},\widetilde{\mathsf{M}}_{\widetilde{\mathcal{H}}}}(\widetilde{V})$ under [62, Thm. 5.3.1.19], where $[\widetilde{\alpha}^{\natural,\ddagger}_{\widetilde{\mathcal{H}}}]$ is represented by some

$$\widetilde{\alpha}^{\natural,\ddagger}_{\widetilde{\mathcal{H}}} = (\widetilde{\mathsf{Z}}^{\ddagger}_{\widetilde{\mathcal{H}}}, \widetilde{\varphi}^{\ddagger,\sim}_{-2,\widetilde{\mathcal{H}}}, \widetilde{\varphi}^{\ddagger}_{-1,\widetilde{\mathcal{H}}}, \widetilde{\varphi}^{\ddagger,\sim}_{0,\widetilde{\mathcal{H}}}, \widetilde{\delta}^{\ddagger}_{\widetilde{\mathcal{H}}}, \widetilde{c}^{\ddagger}_{\widetilde{\mathcal{H}}}, \widetilde{c}^{\vee,\ddagger}_{\widetilde{\mathcal{H}}}, \widetilde{\tau}^{\ddagger}_{\widetilde{\mathcal{H}}})$$

as in [62, Def. 5.3.1.14; see also the errata]. Note that $(\widetilde{X}^{\ddagger}, \widetilde{Y}^{\ddagger}, \widetilde{\phi}^{\ddagger}, [\widetilde{\alpha}^{\natural,\ddagger}_{\widetilde{\mathcal{H}}}])$ determines some cusp label $[(\widetilde{\mathsf{Z}}^{\ddagger}_{\widetilde{\mathcal{H}}}, \widetilde{\Phi}^{\ddagger}_{\widetilde{\mathcal{H}}}, \widetilde{\delta}^{\ddagger}_{\widetilde{\mathcal{H}}})]$ equivalent to the cusp label $[(\breve{\mathsf{Z}}_{\widetilde{\mathcal{H}}}, \breve{\Phi}_{\widetilde{\mathcal{H}}}, \breve{\delta}_{\widetilde{\mathcal{H}}})]$ represented by the $\widetilde{\mathcal{H}}$-orbit of the $(\breve{\mathsf{Z}}, \breve{\Phi}, \breve{\delta})$ introduced above (where the $(\widetilde{\varphi}^{\ddagger}_{-2,\widetilde{\mathcal{H}}}, \widetilde{\varphi}^{\ddagger}_{0,\widetilde{\mathcal{H}}})$ in $\breve{\Phi}_{\widetilde{\mathcal{H}}}$ is induced by $(\widetilde{\varphi}^{\ddagger,\sim}_{-2,\widetilde{\mathcal{H}}}, \widetilde{\varphi}^{\ddagger,\sim}_{0,\widetilde{\mathcal{H}}})$ as in the corrected [62, Def. 5.4.2.8] in the errata). For simplicity, we shall use entries in this last representative to replace their isomorphic (or equivalent) objects, and say in this case that $(\widetilde{\varphi}^{\ddagger,\sim}_{-2,\widetilde{\mathcal{H}}}, \widetilde{\varphi}^{\ddagger,\sim}_{0,\widetilde{\mathcal{H}}})$ induces $(\breve{\varphi}_{-2,\widetilde{\mathcal{H}}}, \breve{\varphi}_{0,\widetilde{\mathcal{H}}})$.

By definition, the pullback of $(\widetilde{B}^{\ddagger}, \lambda_{\widetilde{B}^{\ddagger}}, i_{\widetilde{B}^{\ddagger}}, \breve{X}, \breve{Y}, \breve{\phi}, \widetilde{c}^{\ddagger}, \widetilde{c}^{\vee,\ddagger})$ to the subscheme $\mathrm{Spec}(V)$ of $\mathrm{Spec}(\widetilde{V})$ depends only on $(\widehat{G}^{\ddagger}, \widehat{\lambda}^{\ddagger}, \widehat{i}^{\ddagger}) \to \mathrm{Spec}(V)$. Let us denote it by

$$(\widehat{B}^{\ddagger}, \lambda_{\widehat{B}^{\ddagger}}, i_{\widehat{B}^{\ddagger}}, \breve{X}, \breve{Y}, \breve{\phi}, \widehat{c}^{\ddagger}, \widehat{c}^{\vee,\ddagger}).$$

Note that the $\widetilde{\mathcal{H}}$-orbit $(\widetilde{\mathsf{Z}}_{\widetilde{\mathcal{H}}}, \widetilde{\Phi}_{\widetilde{\mathcal{H}}} = (\widetilde{X}, \widetilde{Y}, \widetilde{\phi}, \widetilde{\varphi}_{-2,\widetilde{\mathcal{H}}}, \widetilde{\varphi}_{0,\widetilde{\mathcal{H}}}), \widetilde{\delta}_{\widetilde{\mathcal{H}}})$ is part of the data of $\widetilde{\kappa}$. By Lemma 1.2.4.16, it makes sense to consider $\breve{\mathsf{Z}}_{\widehat{\mathcal{H}}}$, $(\breve{\varphi}_{-2,\widehat{\mathcal{H}}}, \breve{\varphi}_{0,\widehat{\mathcal{H}}})$, and $\breve{\delta}_{\widehat{\mathcal{H}}}$, which are the $\widehat{\mathcal{H}}$-orbits of $\breve{\mathsf{Z}}$, $(\breve{\varphi}_{-2} : \mathrm{Gr}^{\breve{\mathsf{Z}}}_{-2} \xrightarrow{\sim} \mathrm{Hom}_{\hat{\mathbb{Z}}}(\breve{X} \underset{\mathbb{Z}}{\otimes} \hat{\mathbb{Z}}, \hat{\mathbb{Z}}(1)), \breve{\varphi}_0 : \mathrm{Gr}^{\breve{\mathsf{Z}}}_0 \xrightarrow{\sim} \breve{Y} \underset{\mathbb{Z}}{\otimes} \hat{\mathbb{Z}})$, and $\breve{\delta}$, respectively. Moreover, by extending restrictions to subgroups of $\widetilde{L}/n\widetilde{L}$ (with $\widetilde{\mathsf{Z}}_{-1,n}$ replaced with its subgroup $\breve{\mathsf{Z}}_{-1,n}$) as in Construction 1.3.4.6, $\widetilde{\alpha}^{\ddagger}_{\widetilde{\mathcal{H}}}$ induces a level-$\mathcal{H}$ structure $\widetilde{\varphi}^{\ddagger}_{-1,\widetilde{\mathcal{H}}}$ of $(\widehat{B}^{\ddagger}, \lambda_{\widehat{B}^{\ddagger}}, i_{\widehat{B}^{\ddagger}})$ depending only on $\widehat{\alpha}^{\ddagger}_{\widehat{\mathcal{H}}}$, which we denote by $\widehat{\varphi}^{\ddagger}_{-1,\widehat{\mathcal{H}}}$. Then it also makes sense to consider the $\widehat{\mathcal{H}}$-orbit $(\widetilde{\varphi}^{\ddagger,\sim}_{-2,\widehat{\mathcal{H}}}, \widetilde{\varphi}^{\ddagger,\sim}_{0,\widehat{\mathcal{H}}})$, which we denote by $(\widehat{\varphi}^{\ddagger,\sim}_{-2,\widehat{\mathcal{H}}}, \widehat{\varphi}^{\ddagger,\sim}_{0,\widehat{\mathcal{H}}})$, which is a subscheme of $(\breve{\varphi}_{-2,\widehat{\mathcal{H}}}, \breve{\varphi}_{0,\widehat{\mathcal{H}}}) \underset{\breve{\mathsf{Z}}_{\widehat{\mathcal{H}}}}{\times} \widehat{\varphi}^{\ddagger}_{-1,\widehat{\mathcal{H}}}$ which can be identified with a system of $\widehat{\mathcal{H}}/\widehat{\mathcal{U}}(n)$-orbits, where $n \geq 1$ are integers such that $\widehat{\mathcal{U}}(n) := \widetilde{\mathcal{U}}(n)_{\widehat{\mathrm{G}}} \subset \widehat{\mathcal{H}}$, which surjects under the two projections to the orbits defining $(\breve{\varphi}_{-2,\widehat{\mathcal{H}}}, \breve{\varphi}_{0,\widehat{\mathcal{H}}})$ and $\widehat{\varphi}^{\ddagger}_{-1,\widehat{\mathcal{H}}}$. In this case, we say that $(\widehat{\varphi}^{\ddagger,\sim}_{-2,\widehat{\mathcal{H}}}, \widehat{\varphi}^{\ddagger,\sim}_{0,\widehat{\mathcal{H}}})$ induces the $\widehat{\mathcal{H}}$-orbit $(\breve{\varphi}_{-2,\widehat{\mathcal{H}}}, \breve{\varphi}_{0,\widehat{\mathcal{H}}})$.

Let $\widetilde{K} := \mathrm{Frac}(\widetilde{V})$ and $\widetilde{\eta} := \mathrm{Spec}(\widetilde{K})$. By [62, Lem. 4.2.1.7], the trivialization of biextensions $\widetilde{\tau}^{\ddagger} : \mathbf{1}_{\breve{Y} \times \breve{X},\widetilde{\eta}} \xrightarrow{\sim} (\widetilde{c}^{\vee,\ddagger} \times \widetilde{c}^{\ddagger})^{*}\mathcal{P}^{\otimes -1}_{\widetilde{B}^{\ddagger},\widetilde{\eta}}$ determines a homomorphism $\widetilde{\iota}^{\ddagger} : \breve{Y}_{\widetilde{\eta}} \to \widetilde{G}^{\natural,\ddagger}_{\widetilde{\eta}}$ lifting $\widetilde{c}^{\vee,\ddagger} : \breve{Y} \to \widetilde{B}^{\ddagger}$. It also determines a homomorphism $\widetilde{\iota}^{\vee,\ddagger} : \breve{X}_{\widetilde{\eta}} \to \widetilde{G}^{\vee,\natural,\ddagger}_{\widetilde{\eta}}$ lifting $\widetilde{c}^{\ddagger} : \breve{X} \to \widetilde{B}^{\vee,\ddagger}$, which is compatible with $\widetilde{\iota}^{\ddagger}$ under the homomorphisms $\breve{\phi} : \breve{Y} \to \breve{X}$ and $\widetilde{\lambda}_{\widetilde{B}^{\ddagger}} : \widetilde{B}^{\ddagger} \to \widetilde{B}^{\vee,\ddagger}$ (and the homomorphism $\widetilde{\lambda}^{\natural,\ddagger} : \widetilde{G}^{\natural,\ddagger} \to \widetilde{G}^{\vee,\natural,\ddagger}$ determined by them), by symmetry of $\widetilde{\tau}^{\ddagger}$. Let V^{1} be the localization of $\widetilde{V}$ at the (prime) kernel $\widetilde{I}^{1}$ of $\widetilde{V} \to V$. By [62, Prop. 4.5.3.11 and Cor. 4.5.3.12], the restriction

$\widetilde{\iota}^{\ddagger}|_Y$ (resp. $\widetilde{\iota}^{\vee,\ddagger}|_X$) extends to a homomorphism $\widetilde{\iota}^{\ddagger,1} : Y \to \widetilde{G}^{\natural,\ddagger}_{V^1}$ (resp. $\widetilde{\iota}^{\vee,\ddagger,1} : X \to \widetilde{G}^{\vee,\natural,\ddagger}_{V^1}$), whose pullback to the closed point η of $\mathrm{Spec}(V^1)$ is a homomorphism $\widehat{\iota}^{\ddagger} : Y \to \widehat{G}^{\natural,\ddagger}_{\eta}$ (resp. $\widehat{\iota}^{\vee,\ddagger} : X \to \widehat{G}^{\vee,\natural,\ddagger}_{\eta}$). The two homomorphisms $\widetilde{\iota}^{\ddagger,1}$ and $\widetilde{\iota}^{\vee,\ddagger,1}$ (resp. $\widehat{\iota}^{\ddagger}$ and $\widehat{\iota}^{\vee,\ddagger}$) are compatible with each other under the homomorphisms $\phi : Y \to X$ and $\widetilde{\lambda}^{\natural,\ddagger} : \widetilde{G}^{\natural,\ddagger} \to \widetilde{G}^{\vee,\natural,\ddagger}$ (resp. $\widehat{\lambda}^{\natural,\ddagger} : \widehat{G}^{\natural,\ddagger} \to \widehat{G}^{\vee,\natural,\ddagger}$). By the same argument as in the proof of [62, Lem. 4.2.1.7], the pair $(\widehat{\iota}^{\ddagger}, \widehat{\iota}^{\vee,\ddagger})$ determines a pair

$$(\widehat{\tau}^{\ddagger} : \mathbf{1}_{Y \times \check{X},\eta} \xrightarrow{\sim} (\widehat{c}^{\vee,\ddagger}|_Y \times \widehat{c}^{\ddagger})^* \mathcal{P}^{\otimes -1}_{\widehat{B}^{\ddagger},\eta},$$
$$\widehat{\tau}^{\vee,\ddagger} : \mathbf{1}_{\check{Y} \times X,\eta} \xrightarrow{\sim} (\widehat{c}^{\vee,\ddagger} \times \widehat{c}^{\ddagger}|_X)^* \mathcal{P}^{\otimes -1}_{\widehat{B}^{\ddagger},\eta})$$

(satisfying certain familiar compatibility conditions, which we omit for simplicity).

For each integer $n \geq 1$ such that $\widetilde{\mathcal{U}}(n) \subset \widetilde{\mathcal{H}} \subset \widetilde{\mathrm{G}}(\hat{\mathbb{Z}})$, there exists an étale covering $\widetilde{\eta}_n = \mathrm{Spec}(\widetilde{K}_n) \twoheadrightarrow \widetilde{\eta} = \mathrm{Spec}(\widetilde{K})$ and an $\mathcal{H}_{\widetilde{\mathrm{P}}'_{\widetilde{\mathrm{Z}}}}$-orbit of liftings $\widetilde{\tau}^{\ddagger}_n : \mathbf{1}_{\frac{1}{n}\check{Y} \times \check{X},\widetilde{\eta}_n} \xrightarrow{\sim} (c^{\vee,\ddagger}_n \times c^{\ddagger})^* \mathcal{P}^{\otimes -1}_{B^{\ddagger},\widetilde{\eta}_n}$ of $\widetilde{\tau}^{\ddagger}$, which determines orbits of homomorphisms $\widetilde{\iota}^{\ddagger}_n : \frac{1}{n}\check{Y}_{\widetilde{\eta}_n} \to \widetilde{G}^{\natural,\ddagger}_{\widetilde{\eta}_n}$ and $\widetilde{\iota}^{\vee,\ddagger}_n : \frac{1}{n}\check{X}_{\widetilde{\eta}_n} \to \widetilde{G}^{\vee,\natural,\ddagger}_{\widetilde{\eta}_n}$ compatible with liftings $c^{\vee,\ddagger}_n : \frac{1}{n}\check{Y}_{\widetilde{\eta}_n} \to \widetilde{B}^{\ddagger}_{\widetilde{\eta}_n}$ and $c^{\ddagger}_n : \frac{1}{n}\check{X}_{\widetilde{\eta}_n} \to \widetilde{B}^{\vee,\ddagger}_{\widetilde{\eta}_n}$ (and with $\check{\phi}_n : \frac{1}{n}\check{Y} \to \frac{1}{n}\check{X}$, $\widetilde{\lambda}_{\widetilde{B}^{\ddagger}} : \widetilde{B}^{\ddagger} \to \widetilde{B}^{\vee,\ddagger}$, and $\widetilde{\lambda}^{\natural,\ddagger} : \widetilde{G}^{\natural,\ddagger} \to \widetilde{G}^{\vee,\natural,\ddagger}$). Let $\widetilde{V}_n$ be the normalization of $\widetilde{V}$ in $\widetilde{K}_n$, let $\widetilde{I}^1_n := \mathrm{rad}(\widetilde{I}^1 \cdot \widetilde{V}_n)$, let V^1_n be the localization of $\widetilde{V}_n$ at the multiplicative subset complement to $\widetilde{I}^1$, and let K_n be the reduction of V^1_n modulo $V^1_n \cdot \widetilde{I}^1_n$. By the construction of $\widetilde{\mathfrak{X}}_{\widetilde{\Phi}_{\widetilde{\mathcal{H}}},\widetilde{\delta}_{\widetilde{\mathcal{H}}},\widetilde{\sigma}}$ as a completion of the affine toroidal embedding $\widetilde{\Xi}_{\widetilde{\Phi}_{\widetilde{\mathcal{H}}},\widetilde{\delta}_{\widetilde{\mathcal{H}}}}(\widetilde{\sigma})$ along its $\widetilde{\sigma}$-strata, we may choose $\widetilde{V}_n$ such that K_n is a finite étale K-algebra. Let $\eta_n = \mathrm{Spec}(K_n)$. Let V_n be the normalization of V in K_n. By [62, Prop. 4.5.3.11 and Cor. 4.5.3.12], the restriction $\widetilde{\iota}^{\ddagger}_n|_{\frac{1}{n}Y_{\widetilde{\eta}_n}}$ (resp. $\widetilde{\iota}^{\vee,\ddagger}_n|_{\frac{1}{n}X_{\widetilde{\eta}_n}}$) extends to a homomorphism $\widetilde{\iota}^{\ddagger,1}_n : \frac{1}{n}Y_{\widetilde{\eta}_n} \to \widetilde{G}^{\natural,\ddagger}_{V^1_n}$ (resp. $\widetilde{\iota}^{\vee,\ddagger,1}_n : \frac{1}{n}X_{\widetilde{\eta}_n} \to \widetilde{G}^{\vee,\natural,\ddagger}_{V^1_n}$), whose pullback to the dense subscheme $\eta_n = \mathrm{Spec}(K_n)$ of $\mathrm{Spec}(V_n)$ is a homomorphism $\widehat{\iota}^{\ddagger}_n : \frac{1}{n}Y_{\eta_n} \to \widehat{G}^{\natural,\ddagger}_{\eta_n}$ (resp. $\widehat{\iota}^{\vee,\ddagger}_n : \frac{1}{n}X_{\eta_n} \to \widehat{G}^{\vee,\natural,\ddagger}_{\eta_n}$). These two homomorphisms $\widetilde{\iota}^{\ddagger,1}_n$ and $\widetilde{\iota}^{\vee,\ddagger,1}_n$ (resp. $\widehat{\iota}^{\ddagger}_n$ and $\widehat{\iota}^{\vee,\ddagger}_n$) are compatible with each other under the homomorphisms $\phi_n : \frac{1}{n}Y \to \frac{1}{n}X$ and $\widetilde{\lambda}^{\natural,\ddagger} : \widetilde{G}^{\natural,\ddagger} \to \widetilde{G}^{\vee,\natural,\ddagger}$ (resp. $\widehat{\lambda}^{\natural,\ddagger} : \widehat{G}^{\natural,\ddagger} \to \widehat{G}^{\vee,\natural,\ddagger}$), and determine homomorphisms $\widetilde{c}^{\vee,\ddagger,1}_n : \frac{1}{n}Y \to \widetilde{B}^{\ddagger}_{V^1_n}$ and $\widetilde{c}^{\ddagger,1}_n : \frac{1}{n}X \to \widetilde{B}^{\vee,\ddagger}_{V^1_n}$ (resp. $\widehat{c}^{\vee,\ddagger}_n : \frac{1}{n}Y \to \widehat{B}^{\ddagger}_{\eta_n}$ and $\widehat{c}^{\ddagger}_n : \frac{1}{n}X \to \widehat{B}^{\vee,\ddagger}_{\eta_n}$). The $\widehat{\mathcal{H}}$-orbit of $(\widehat{c}^{\ddagger}_n, \widehat{c}^{\vee,\ddagger}_n, \widehat{\iota}^{\ddagger}_n, \widehat{\iota}^{\vee,\ddagger}_n)$ is well defined (*i.e.*, independent of the choice of the representative $(\widetilde{c}^{\ddagger}_n, \widetilde{c}^{\vee,\ddagger}_n, \widetilde{\tau}^{\ddagger}_n)$ in its $\mathcal{H}_{\widetilde{\mathrm{P}}'_{\widetilde{\mathrm{Z}}}}$-orbit) and descends (as compatible subschemes of schemes of homomorphisms) to η. Such a descended object is independent of n, which we shall denote by

$$(\widehat{c}^{\ddagger}_{\widehat{\mathcal{H}}}, \widehat{c}^{\vee,\ddagger}_{\widehat{\mathcal{H}}}, \widehat{\iota}^{\ddagger}_{\widehat{\mathcal{H}}}, \widehat{\iota}^{\vee,\ddagger}_{\widehat{\mathcal{H}}}).$$

Then, as above, the pair $(\widehat{\iota}^{\ddagger}_{\widehat{\mathcal{H}}}, \widehat{\iota}^{\vee,\ddagger}_{\widehat{\mathcal{H}}})$ determines a pair

$$(\widehat{\tau}^{\ddagger}_{\widehat{\mathcal{H}}}, \widehat{\tau}^{\vee,\ddagger}_{\widehat{\mathcal{H}}})$$

(whose detailed definitions we omit for simplicity).

In summary, given the family $(\widehat{G}, \widehat{\lambda}, \widehat{i}, \widehat{\alpha}_{\widehat{\mathcal{H}}}) \to \mathsf{N}^{\mathrm{tor}}_{\widetilde{\kappa}}$ as in Construction 1.3.4.6, each morphism $\xi : \mathrm{Spec}(V) \to \mathsf{N}^{\mathrm{tor}}_{\widetilde{\kappa}}$ as above determines a tuple

$$(\widehat{B}^{\ddagger}, \lambda_{\widehat{B}^{\ddagger}}, i_{\widehat{B}^{\ddagger}}, \breve{X}, \breve{Y}, \breve{\phi}, \widehat{c}^{\ddagger}, \widehat{c}^{\vee,\ddagger}, \widehat{\tau}^{\ddagger}, \widehat{\tau}^{\vee,\ddagger}, [\widehat{\alpha}^{\natural,\ddagger}_{\widehat{\mathcal{H}}}]), \tag{1.3.4.15}$$

where $[\widehat{\alpha}^{\natural,\ddagger}_{\widehat{\mathcal{H}}}]$ is an equivalence class of

$$\widehat{\alpha}^{\natural,\ddagger}_{\widehat{\mathcal{H}}} = (\breve{\mathsf{Z}}_{\widehat{\mathcal{H}}}, \widehat{\varphi}^{\ddagger,\sim}_{-2,\widehat{\mathcal{H}}}, \widehat{\varphi}^{\ddagger}_{-1,\widehat{\mathcal{H}}}, \widehat{\varphi}^{\ddagger,\sim}_{0,\widehat{\mathcal{H}}}, \breve{\delta}_{\widehat{\mathcal{H}}}, \widehat{c}^{\ddagger}_{\widehat{\mathcal{H}}}, \widehat{c}^{\vee,\ddagger}_{\widehat{\mathcal{H}}}, \widehat{\tau}^{\ddagger}_{\widehat{\mathcal{H}}}, \widehat{\tau}^{\vee,\ddagger}_{\widehat{\mathcal{H}}}) \tag{1.3.4.16}$$

(whose precise definitions we omit for simplicity).

Given a tuple as in (1.3.4.15), if we set

$$(B^{\ddagger}, \lambda_{B^{\ddagger}}, i_{B^{\ddagger}}, \varphi_{-1,\mathcal{H}_\kappa}) := (\widehat{B}^{\ddagger}, \lambda_{\widehat{B}^{\ddagger}}, i_{\widehat{B}^{\ddagger}}, \widehat{\varphi}_{-1,\widehat{\mathcal{H}}})$$

and

$$(c^{\ddagger}, c^{\vee,\ddagger}, \tau^{\ddagger}) := (\widehat{c}^{\ddagger}|_X, \widehat{c}^{\vee,\ddagger}|_Y, \widehat{\tau}^{\ddagger}|_{\mathbf{1}_{Y\times X,\eta}}),$$

and define $[\alpha^{\natural,\ddagger}_{\mathcal{H}_\kappa}]$ using similar restrictions, then the tuple

$$(B^{\ddagger}, \lambda_{B^{\ddagger}}, i_{B^{\ddagger}}, X, Y, \phi, c^{\ddagger}, c^{\vee,\ddagger}, \tau^{\ddagger}, [\alpha^{\natural,\ddagger}_{\mathcal{H}_\kappa}]) \tag{1.3.4.17}$$

defines an object of $\mathrm{DD}_{\mathrm{PEL},\mathsf{M}_{\mathcal{H}_\kappa}}(V)$. On the other hand, the pullback $(\widehat{G}^{\ddagger}, \widehat{\lambda}^{\ddagger}, \widehat{i}^{\ddagger}, \widehat{\alpha}^{\ddagger}_{\widehat{\mathcal{H}}}) \to \mathrm{Spec}(V)$ is determined up to isomorphism by its generic fiber $(\widehat{G}^{\ddagger}_\eta, \widehat{\lambda}^{\ddagger}_\eta, \widehat{i}^{\ddagger}_\eta, \widehat{\alpha}^{\ddagger}_{\widehat{\mathcal{H}},\eta}) \to \mathrm{Spec}(K)$, which (up to isomorphism) determines and is determined by a tuple $((G^{\ddagger}_\eta, \lambda^{\ddagger}_\eta, i^{\ddagger}_\eta, \alpha^{\ddagger}_{\mathcal{H}_\kappa,\eta}), (\widetilde{c}^{\ddagger}_{\widetilde{\mathcal{H}},\eta}, \widetilde{c}^{\vee,\ddagger}_{\widetilde{\mathcal{H}},\eta})) \to \mathrm{Spec}(K)$ parameterized by $\widetilde{C}_{\widetilde{\Phi}_{\widetilde{\mathcal{H}}},\widetilde{\delta}_{\widetilde{\mathcal{H}}}}$. The abelian part $(G^{\ddagger}_\eta, \lambda^{\ddagger}_\eta, i^{\ddagger}_\eta, \alpha^{\ddagger}_{\mathcal{H}_\kappa,\eta})$ extends to a degenerating family

$$(G^{\ddagger}, \lambda^{\ddagger}, i^{\ddagger}, \alpha^{\ddagger}_{\mathcal{H}_\kappa}) \tag{1.3.4.18}$$

of type $\mathsf{M}_{\mathcal{H}_\kappa}$ over $\mathrm{Spec}(V)$ (with $\alpha^{\ddagger}_{\mathcal{H}_\kappa}$ still defined only over $\mathrm{Spec}(K)$) which defines an object of $\mathrm{DEG}_{\mathrm{PEL},\mathsf{M}_{\mathcal{H}_\kappa}}(V)$. By the theory of two-step degenerations (see [28, Ch. III, Thm. 10.2] and [62, Sec. 4.5.6]), and by analyzing endomorphism structures and level structures as in [62, Sec. 5.1–5.3], under [62, Thm. 5.3.1.19], this last object (1.3.4.18) corresponds to the above object (1.3.4.17) in $\mathrm{DD}_{\mathrm{PEL},\mathsf{M}_{\mathcal{H}_\kappa}}(V)$.

As for $(\widetilde{c}^{\ddagger}_{\widetilde{\mathcal{H}},\eta}, \widetilde{c}^{\vee,\ddagger}_{\widetilde{\mathcal{H}},\eta})$, they are determined by their values on $\bar{K}$-valued points, where $\bar{K}$ is any fixed algebraic closure of K, which are orbits of compatible homomorphisms $\widetilde{c}^{\ddagger}_n(\bar{K}) : \frac{1}{n}\widetilde{X} \to B^{\vee,\ddagger}(\bar{K})$ and $\widetilde{c}^{\vee,\ddagger}_n(\bar{K}) : \frac{1}{n}\widetilde{Y} \to B^{\ddagger}(\bar{K})$. On the other hand, by the same argument as in the proof of [62, Lem. 4.2.1.7], $(\widehat{\tau}^{\ddagger}_{\widehat{\mathcal{H}}}, \widehat{\tau}^{\vee,\ddagger}_{\widehat{\mathcal{H}}})$ is determined by orbits of compatible homomorphisms $\widehat{\iota}^{\prime,\ddagger}_n(\bar{K}) : \frac{1}{n}\breve{X} \to G^{\vee,\natural,\ddagger}(\bar{K})$ and $\widehat{\iota}^{\vee,\prime,\ddagger}_n(\bar{K}) : \frac{1}{n}\breve{Y} \to G^{\natural,\ddagger}(\bar{K})$, where $G^{\natural,\ddagger}$ and $G^{\vee,\natural,\ddagger}$ are determined by $\widehat{c}^{\ddagger}|_X : X \to \widehat{B}^{\vee,\ddagger} = B^{\vee,\ddagger}$ and $\widehat{c}^{\ddagger}|_Y : Y \to \widehat{B}^{\ddagger} = B^{\ddagger}$, respectively. By definition, $\widehat{\iota}^{\prime,\ddagger}(\bar{K})$ and $\widehat{\iota}^{\vee,\prime,\ddagger}(\bar{K})$ are compatible with the homomorphism $\iota^{\vee,\ddagger}(\bar{K}) : X \to G^{\vee,\natural,\ddagger}(\bar{K})$ and $\iota^{\ddagger}(\bar{K}) : Y \to G^{\natural,\ddagger}(\bar{K})$ defined by $\tau^{\ddagger}$. Given the splitting δ, there exists a subgroup $\breve{X}_n$ (resp. $\breve{Y}_n$) of $\frac{1}{n}\breve{X}$ (resp. $\frac{1}{n}\breve{Y}$) containing X (resp. Y) such that the admissible surjection $s_{\breve{X}} : \breve{X} \twoheadrightarrow \widetilde{X}$ (resp. $s_{\breve{Y}} : \breve{Y} \twoheadrightarrow \widetilde{Y}$) induces an isomorphism $\breve{X}_n/X \cong \frac{1}{n}\widetilde{X}$

(resp. $\breve{Y}_n/Y \cong \frac{1}{n}\widetilde{Y}$). Hence, by [62, Prop. 4.5.5.3], we can form equivariant quotients by the images of $\iota^{\vee,\ddagger}(\bar{K})$ and $\iota^{\ddagger}(\bar{K})$, and obtain (by restrictions) orbits of compatible homomorphisms $\frac{1}{n}\widetilde{X} \cong \breve{X}_n/(\frac{1}{n}X) \to B^{\vee}(\bar{K}) \cong (G^{\vee,\natural,\ddagger}(\bar{K}))/(\iota^{\vee,\ddagger}(\bar{K})(X))$ and $\frac{1}{n}\widetilde{Y} \cong \breve{Y}_n/(\frac{1}{n}Y) \to B(\bar{K}) \cong (G^{\natural,\ddagger}(\bar{K}))/(\iota^{\ddagger}(\bar{K})(Y))$. These coincide with the above orbits of $\widetilde{c}_n^{\ddagger}(\bar{K})$ and $\widetilde{c}_n^{\vee,\ddagger}(\bar{K})$ because of the following reasons: At level one, this follows from the theory of two-step degenerations (see [28, Ch. III, Sec. 10] and [62, Sec. 4.5.6]), because $\widetilde{c}^{\ddagger}$ and $\widetilde{c}^{\vee,\ddagger}$ (which are defined over K) are induced by (reductions of extensions of) compatible homomorphisms $\widetilde{X} \to \widetilde{G}_{\widetilde{\eta}}^{\vee,\natural}$ and $\widetilde{Y} \to \widetilde{G}_{\widetilde{\eta}}^{\natural}$, where $\widetilde{\eta} = \mathrm{Spec}(\widetilde{K}) = \mathrm{Spec}(\mathrm{Frac}\,\widetilde{V})$ is some auxiliary choice as above. At higher levels, this follows from the way we reconstruct level structures from its graded pieces using the various splittings.

In brief, the tuple over $\mathrm{Spec}(V)$ as in (1.3.4.15) determines and is determined by the tuple (1.3.4.14) (up to isomorphism, over $\mathrm{Spec}(K)$).

As in [62, (6.2.5.10)], the pair $(\widehat{\tau}^{\ddagger}, \widehat{\tau}^{\vee,\ddagger})$ defines compatible morphisms $\upsilon_{\widehat{\tau}^{\ddagger}} : Y \times \breve{X} \to \mathbb{Z}$ and $\upsilon_{\widehat{\tau}^{\vee,\ddagger}} : \breve{Y} \times X \to \mathbb{Z}$ (using the discrete valuation $\upsilon : \mathrm{Inv}(V) \to \mathbb{Z}$ of V), which define the same element

$$\upsilon_{\widehat{\tau}^{\ddagger}} = \upsilon_{\widehat{\tau}^{\vee,\ddagger}} \in (\widehat{\mathbf{S}}_{\breve{\Phi}_{\widehat{\mathcal{H}}}})^{\vee}_{\mathbb{R}}$$

(see (1.2.4.29)). On the other hand, as in (6) of Theorem 1.3.1.3, $\widetilde{\tau}^{\ddagger}$ defines a morphism $\upsilon_{\widetilde{\tau}^{\ddagger}} : \breve{Y} \times \breve{X} \to \mathbb{Z}$, which defines an element

$$\upsilon_{\widetilde{\tau}^{\ddagger}} \in \breve{\rho} \subset \mathbf{P}^{+}_{\breve{\Phi}_{\widetilde{\mathcal{H}}}},$$

where $\breve{\rho}$ is as above. Since $(\widehat{\tau}^{\ddagger}, \widehat{\tau}^{\vee,\ddagger})$ is defined by extending restrictions of $\upsilon_{\widetilde{\tau}^{\ddagger}}$, we see that

$$\upsilon_{\widehat{\tau}^{\ddagger}} = \upsilon_{\widehat{\tau}^{\vee,\ddagger}} \in \widehat{\rho} = \mathrm{pr}_{(\widehat{\mathbf{S}}_{\breve{\Phi}_{\widehat{\mathcal{H}}}})^{\vee}_{\mathbb{R}}}(\breve{\rho}) \subset \widehat{\mathbf{P}}_{\breve{\Phi}_{\widehat{\mathcal{H}}}}$$

(see (1.2.4.41)). If $\breve{\rho}$ is replaced with another representative, then $\widehat{\rho}$ is replaced with a translation under the action of $\Gamma_{\breve{\Phi}_{\widehat{\mathcal{H}}}}$. (This finishes Construction 1.3.4.12.)

Proposition 1.3.4.19. *Suppose $\widetilde{\kappa} = (\widetilde{\mathcal{H}}, \widetilde{\Sigma}, \widetilde{\sigma})$ and $\widetilde{\kappa}' = (\widetilde{\mathcal{H}}', \widetilde{\Sigma}', \widetilde{\sigma}')$ are elements in $\widetilde{\mathbf{K}}^{++}_{Q,\mathcal{H}}$ such that $\kappa' = [\widetilde{\kappa}'] \succ \kappa = [\widetilde{\kappa}]$ in $\mathbf{K}^{++}_{Q,\mathcal{H}}$ (see Definition 1.2.4.44). Let $(\widehat{G}, \widehat{\lambda}, \widehat{i}, \widehat{\alpha}_{\widehat{\mathcal{H}}}) \to \mathsf{N}^{\mathrm{tor}}_{\widetilde{\kappa}}$ (resp. $(\widehat{G}', \widehat{\lambda}', \widehat{i}', \widehat{\alpha}'_{\widehat{\mathcal{H}}'}) \to \mathsf{N}^{\mathrm{tor}}_{\widetilde{\kappa}'}$) denote the pullback of the degenerating family $(\widetilde{G}, \widetilde{\lambda}, \widetilde{i}, \widetilde{\alpha}_{\widetilde{\mathcal{H}}}) \to \widetilde{\mathsf{M}}^{\mathrm{tor}}_{\widetilde{\mathcal{H}},\widetilde{\Sigma}}$ (resp. $(\widetilde{G}', \widetilde{\lambda}', \widetilde{i}', \widetilde{\alpha}'_{\widetilde{\mathcal{H}}'}) \to \widetilde{\mathsf{M}}^{\mathrm{tor}}_{\widetilde{\mathcal{H}}',\widetilde{\Sigma}'}$) as in Construction 1.3.4.6. Then there is a canonical morphism $f^{\mathrm{tor}}_{\widetilde{\kappa}',\widetilde{\kappa}} : \mathsf{N}^{\mathrm{tor}}_{\widetilde{\kappa}'} \to \mathsf{N}^{\mathrm{tor}}_{\widetilde{\kappa}}$ such that $(\widehat{G}', \widehat{\lambda}', \widehat{i}', \widehat{\alpha}'_{\widehat{\mathcal{H}}'}) \to \mathsf{N}^{\mathrm{tor}}_{\widetilde{\kappa}'}$ is canonically isomorphic to the pullback of $(\widehat{G}, \widehat{\lambda}, \widehat{i}, \widehat{\alpha}_{\widehat{\mathcal{H}}}) \to \mathsf{N}^{\mathrm{tor}}_{\widetilde{\kappa}}$ under $f^{\mathrm{tor}}_{\widetilde{\kappa}',\widetilde{\kappa}}$.*

In particular, for each $\widetilde{\kappa} = (\widetilde{\mathcal{H}}, \widetilde{\Sigma}, \widetilde{\sigma}) \in \widetilde{\mathbf{K}}^{++}_{Q,\mathcal{H}}$, the closure $\mathsf{N}^{\mathrm{tor}}_{\widetilde{\kappa}}$ of $\mathsf{N}_{\widetilde{\kappa}} = \widetilde{\mathsf{Z}}_{[(\widetilde{\Phi}_{\widetilde{\mathcal{H}}}, \widetilde{\delta}_{\widetilde{\mathcal{H}}}, \widetilde{\sigma})]}$ in $\widetilde{\mathsf{M}}^{\mathrm{tor}}_{\widetilde{\mathcal{H}},\widetilde{\Sigma}}$ and the open embedding $\widetilde{\kappa}^{\mathrm{tor}} : \mathsf{N}_{\widetilde{\kappa}} \hookrightarrow \mathsf{N}^{\mathrm{tor}}_{\widetilde{\kappa}}$ depend (up to canonical isomorphism) only on the pair $\kappa = [\widetilde{\kappa}] = (\widehat{\mathcal{H}}, \widehat{\Sigma})$ in $\mathbf{K}^{++}_{Q,\mathcal{H}}$.

The morphism $f^{\mathrm{tor}}_{\widetilde{\kappa}',\widetilde{\kappa}} : \mathsf{N}^{\mathrm{tor}}_{\widetilde{\kappa}'} \to \mathsf{N}^{\mathrm{tor}}_{\widetilde{\kappa}}$ is étale locally given by equivariant morphisms between toric schemes mapping strata to strata, which is log étale essentially by *definition (see [45, Thm. 3.5]). Moreover, as in [28, Ch. V, Rem. 1.2(b)] and in the proof of [62, Lem. 7.1.1.4], we have $R^i(f^{\mathrm{tor}}_{\widetilde{\kappa}',\widetilde{\kappa}})_*\mathscr{O}_{\mathsf{N}^{\mathrm{tor}}_{\widetilde{\kappa}'}} = 0$ for $i > 0$ by [50, Ch. I, Sec. 3].*

Proof. Suppose $\widehat{\mathcal{H}}$ (resp. $\widehat{\mathcal{H}}'$) is determined by some $\widetilde{\mathcal{H}}$ (resp. $\widetilde{\mathcal{H}}'$) satisfying Condition 1.2.4.7. By Lemma 1.3.4.2, we may replace $\widehat{\mathcal{H}}'$ (resp. $\widetilde{\mathcal{H}}'$) with $\widehat{\mathcal{H}}'\cap\widehat{\mathcal{H}}$ (resp. $\widetilde{\mathcal{H}}'\cap\widetilde{\mathcal{H}}$), in which case we have a canonical (forgetful) morphism $f_{\widetilde{\kappa}',\widetilde{\kappa}} : \mathsf{N}_{\widetilde{\kappa}'} \cong \widetilde{C}_{\widetilde{\Phi}_{\widetilde{\mathcal{H}}'},\widetilde{\delta}_{\widetilde{\mathcal{H}}'}} \to \widetilde{C}_{\widetilde{\Phi}_{\widetilde{\mathcal{H}}},\widetilde{\delta}_{\widetilde{\mathcal{H}}}} \cong \mathsf{N}_{\widetilde{\kappa}}$ (by constructions). Suppose $((G',\lambda',i',\alpha'_{\mathcal{H}_{\kappa'}}),(\widetilde{c}'_{\widetilde{\mathcal{H}}'},\widetilde{c}^{\vee,\prime}_{\widetilde{\mathcal{H}}'}))$ (resp. $((G,\lambda,i,\alpha_{\mathcal{H}_\kappa}),(\widetilde{c}_{\widetilde{\mathcal{H}}},\widetilde{c}^{\vee}_{\widetilde{\mathcal{H}}})))$ is the tautological object over $\widetilde{C}_{\widetilde{\Phi}_{\widetilde{\mathcal{H}}'},\widetilde{\delta}_{\widetilde{\mathcal{H}}'}}$ (resp. $\widetilde{C}_{\widetilde{\Phi}_{\widetilde{\mathcal{H}}},\widetilde{\delta}_{\widetilde{\mathcal{H}}}}$), as in Construction 1.3.4.6, which determines and is determined by $(\widehat{G}'_{\mathsf{N}_{\widetilde{\kappa}'}},\widehat{\lambda}'_{\mathsf{N}_{\widetilde{\kappa}'}},\widehat{i}'_{\mathsf{N}_{\widetilde{\kappa}'}},\widehat{\alpha}'_{\widehat{\mathcal{H}}'}) \to \mathsf{N}_{\widetilde{\kappa}'}$ (resp. $(\widehat{G}_{\mathsf{N}_{\widetilde{\kappa}}},\widehat{\lambda}_{\mathsf{N}_{\widetilde{\kappa}}},\widehat{i}_{\mathsf{N}_{\widetilde{\kappa}}},\widehat{\alpha}_{\widehat{\mathcal{H}}}) \to \mathsf{N}_{\widetilde{\kappa}}$), the pullback of $(\widehat{G}',\widehat{\lambda}',\widehat{i}',\widehat{\alpha}'_{\widehat{\mathcal{H}}'}) \to \mathsf{N}^{\mathrm{tor}}_{\widetilde{\kappa}'}$ (resp. $(\widehat{G},\widehat{\lambda},\widehat{i},\widehat{\alpha}_{\widehat{\mathcal{H}}}) \to \mathsf{N}^{\mathrm{tor}}_{\widetilde{\kappa}}$) to $\mathsf{N}_{\widetilde{\kappa}'}$ (resp. $\mathsf{N}_{\widetilde{\kappa}}$). Then $f_{\widetilde{\kappa}',\widetilde{\kappa}}$ is also the canonical morphism determined by the universal property of $\widetilde{C}_{\widetilde{\Phi}_{\widetilde{\mathcal{H}}},\widetilde{\delta}_{\widetilde{\mathcal{H}}}}$, such that the pullback of $((G,\lambda,i,\alpha_{\mathcal{H}_\kappa}),(\widetilde{c}_{\widetilde{\mathcal{H}}},\widetilde{c}^{\vee}_{\widetilde{\mathcal{H}}}))$ under $f_{\widetilde{\kappa}',\widetilde{\kappa}}$ is canonically isomorphic to the $\widehat{\mathcal{H}}$-orbit $((G',\lambda',i',\alpha'_{\mathcal{H}_\kappa}),(\widetilde{c}'_{\widetilde{\mathcal{H}}},\widetilde{c}^{\vee,\prime}_{\widetilde{\mathcal{H}}}))$ of $((G',\lambda',i',\alpha'_{\mathcal{H}'_0}),(\widetilde{c}'_{\widetilde{\mathcal{H}}'},\widetilde{c}^{\vee,\prime}_{\widetilde{\mathcal{H}}'}))$; or, rather, such that the pullback of $(\widehat{G}_{\mathsf{N}_{\widetilde{\kappa}}},\widehat{\lambda}_{\mathsf{N}_{\widetilde{\kappa}}},\widehat{i}_{\mathsf{N}_{\widetilde{\kappa}}},\widehat{\alpha}_{\widehat{\mathcal{H}}}) \to \mathsf{N}_{\widetilde{\kappa}}$ under $f_{\widetilde{\kappa}',\widetilde{\kappa}}$ is canonically isomorphic to the $\widehat{\mathcal{H}}$-orbit $(\widehat{G}'_{\mathsf{N}_{\widetilde{\kappa}'}},\widehat{\lambda}'_{\mathsf{N}_{\widetilde{\kappa}'}},\widehat{i}'_{\mathsf{N}_{\widetilde{\kappa}'}},\widehat{\alpha}'_{\widehat{\mathcal{H}}}) \to \mathsf{N}_{\widetilde{\kappa}'}$ of $(\widehat{G}'_{\mathsf{N}_{\widetilde{\kappa}'}},\widehat{\lambda}'_{\mathsf{N}_{\widetilde{\kappa}'}},\widehat{i}'_{\mathsf{N}_{\widetilde{\kappa}'}},\widehat{\alpha}'_{\widehat{\mathcal{H}}'}) \to \mathsf{N}_{\widetilde{\kappa}'}$.

Since $\mathsf{N}^{\mathrm{tor}}_{\widetilde{\kappa}'}$ is noetherian normal, by [92, IX, 1.4], [28, Ch. I, Prop. 2.7], or [62, Prop. 3.3.1.5], since $(\widehat{G}'_{\mathsf{N}_{\widetilde{\kappa}'}},\widehat{\lambda}'_{\mathsf{N}_{\widetilde{\kappa}'}},\widehat{i}'_{\mathsf{N}_{\widetilde{\kappa}'}},\widehat{\alpha}'_{\widehat{\mathcal{H}}'}) \to \mathsf{N}_{\widetilde{\kappa}'}$ is canonically isomorphic to the pullback of $(\widehat{G}_{\mathsf{N}_{\widetilde{\kappa}}},\widehat{\lambda}_{\mathsf{N}_{\widetilde{\kappa}}},\widehat{i}_{\mathsf{N}_{\widetilde{\kappa}}},\widehat{\alpha}_{\widehat{\mathcal{H}}}) \to \mathsf{N}_{\widetilde{\kappa}}$ under $f_{\widetilde{\kappa}',\widetilde{\kappa}}$, as soon as $f_{\widetilde{\kappa}',\widetilde{\kappa}} : \mathsf{N}_{\widetilde{\kappa}'} \to \mathsf{N}_{\widetilde{\kappa}}$ extends to a morphism $f^{\mathrm{tor}}_{\widetilde{\kappa}',\widetilde{\kappa}} : \mathsf{N}^{\mathrm{tor}}_{\widetilde{\kappa}'} \to \mathsf{N}^{\mathrm{tor}}_{\widetilde{\kappa}}$, we know that $(\widehat{G}',\widehat{\lambda}',\widehat{i}',\widehat{\alpha}'_{\widehat{\mathcal{H}}'}) \to \mathsf{N}^{\mathrm{tor}}_{\widetilde{\kappa}'}$ is canonically isomorphic to the pullback of $(\widehat{G},\widehat{\lambda},\widehat{i},\widehat{\alpha}_{\widehat{\mathcal{H}}}) \to \mathsf{N}^{\mathrm{tor}}_{\widetilde{\kappa}}$ under $f^{\mathrm{tor}}_{\widetilde{\kappa}',\widetilde{\kappa}}$. Such an extension $f^{\mathrm{tor}}_{\widetilde{\kappa}',\widetilde{\kappa}}$ is necessarily unique, because $\mathsf{N}_{\widetilde{\kappa}}$ (resp. $\mathsf{N}_{\widetilde{\kappa}'}$) is dense in $\mathsf{N}^{\mathrm{tor}}_{\widetilde{\kappa}}$ (resp. $\mathsf{N}^{\mathrm{tor}}_{\widetilde{\kappa}'}$). Hence, it suffices to show that $f_{\widetilde{\kappa}',\widetilde{\kappa}} : \mathsf{N}_{\widetilde{\kappa}'} \to \mathsf{N}_{\widetilde{\kappa}}$ extends locally.

Let $\bar{s}$ be any geometric point of $\mathsf{N}^{\mathrm{tor}}_{\widetilde{\kappa}'}$ on the $\widetilde{\mathsf{Z}}_{[(\breve{\Phi}_{\widetilde{\mathcal{H}}'},\breve{\delta}_{\widetilde{\mathcal{H}}'},\breve{\rho}')]}$-stratum of $\widetilde{\mathsf{M}}^{\mathrm{tor}}_{\widetilde{\mathcal{H}}',\widetilde{\Sigma}'}$, where $[(\breve{\Phi}_{\widetilde{\mathcal{H}}'},\breve{\delta}_{\widetilde{\mathcal{H}}'},\breve{\rho}')]$ is represented by some $(\breve{\Phi}_{\widetilde{\mathcal{H}}'},\breve{\delta}_{\widetilde{\mathcal{H}}'},\breve{\rho}')$ with $(\breve{\mathsf{Z}}_{\widetilde{\mathcal{H}}'},\breve{\Phi}_{\widetilde{\mathcal{H}}'} = (\breve{X},\breve{Y},\breve{\phi},\breve{\varphi}_{-2,\widetilde{\mathcal{H}}'},\breve{\varphi}_{0,\widetilde{\mathcal{H}}'}),\breve{\delta}_{\widetilde{\mathcal{H}}'})$ representing some cusp label as in Section 1.2.4. For simplicity, let us fix compatible choices of representatives $(\widetilde{\mathsf{Z}},\widetilde{\Phi} = (\widetilde{X},\widetilde{Y},\widetilde{\phi},\widetilde{\varphi}_{-2},\widetilde{\varphi}_0),\widetilde{\delta})$ and $(\breve{\mathsf{Z}},\breve{\Phi} = (\breve{X},\breve{Y},\breve{\phi},\breve{\varphi}_{-2},\breve{\varphi}_0),\breve{\delta})$, as in Section 1.2.4, in their $\widetilde{\mathcal{H}}'$-orbits. As in Construction 1.3.4.12, each morphism $\xi' : \mathrm{Spec}(V) \to \mathsf{N}^{\mathrm{tor}}_{\widetilde{\kappa}'}$ centered at a geometric point $\bar{s}$ of $\mathsf{N}^{\mathrm{tor}}_{\widetilde{\kappa}'}$, where V is a complete discrete valuation ring with fraction field K, and where $\eta := \mathrm{Spec}(K)$ is mapped to the generic point of the irreducible component containing the image of $\bar{s}$, determines a tuple

$$(\widehat{B}^{\ddagger},\lambda_{\widehat{B}^{\ddagger}},i_{\widehat{B}^{\ddagger}},\breve{X},\breve{Y},\breve{\phi},\widehat{c}^{\ddagger},\widehat{c}^{\vee,\ddagger},\widehat{\tau}^{\ddagger},\widehat{\tau}^{\vee,\ddagger},[\widehat{\alpha}^{\natural,\ddagger}_{\widehat{\mathcal{H}}'}])$$

as in (1.3.4.15), where $[\widehat{\alpha}^{\natural,\ddagger}_{\widehat{\mathcal{H}}'}]$ is an equivalence class of

$$\widehat{\alpha}^{\natural,\ddagger}_{\widehat{\mathcal{H}}'} = (\breve{\mathsf{Z}}_{\widehat{\mathcal{H}}'},\widehat{\varphi}^{\ddagger,\sim}_{-2,\widehat{\mathcal{H}}'},\widehat{\varphi}^{\ddagger}_{-1,\widehat{\mathcal{H}}'},\widehat{\varphi}^{\ddagger,\sim}_{0,\widehat{\mathcal{H}}'},\breve{\delta}_{\widehat{\mathcal{H}}'},\widehat{c}^{\ddagger}_{\widehat{\mathcal{H}}'},\widehat{c}^{\vee,\ddagger}_{\widehat{\mathcal{H}}'},\widehat{\tau}^{\ddagger}_{\widehat{\mathcal{H}}'},\widehat{\tau}^{\vee,\ddagger}_{\widehat{\mathcal{H}}'})$$

as in (1.3.4.16), and the pair $(\widehat{\tau}^{\ddagger},\widehat{\tau}^{\vee,\ddagger})$ defines an element $\upsilon_{\widehat{\tau}^{\ddagger}} = \upsilon_{\widehat{\tau}^{\vee,\ddagger}}$ in $\widehat{\rho}'$ for some $\widehat{\rho}' \subset \widehat{\mathbf{P}}^{+}_{\breve{\Phi}_{\widehat{\mathcal{H}}'}}$ in $\widehat{\Sigma}_{\breve{\Phi}_{\widehat{\mathcal{H}}'}}$. (We should have denoted all these entries with some extra $\prime$ in their superscripts, because they are determined by the pullback of $(\widehat{G}', \widehat{\lambda}', \widehat{i}', \widehat{\alpha}'_{\widehat{\mathcal{H}}'}) \to \mathsf{N}^{\mathrm{tor}}_{\widetilde{\kappa}'}$. But we omit them for the sake of simplicity.) By forming $\widehat{\mathcal{H}}$-orbits, we obtain a tuple

$$(\widehat{B}^{\ddagger}, \lambda_{\widehat{B}^{\ddagger}}, i_{\widehat{B}^{\ddagger}}, \breve{X}, \breve{Y}, \breve{\phi}, \widehat{c}^{\ddagger}, \widehat{c}^{\vee,\ddagger}, \widehat{\tau}^{\ddagger}, \widehat{\tau}^{\vee,\ddagger}, [\widehat{\alpha}^{\natural,\ddagger}_{\widehat{\mathcal{H}}}]),$$

where $[\widehat{\alpha}^{\natural,\ddagger}_{\widehat{\mathcal{H}}}]$ is an equivalence class of

$$\widehat{\alpha}^{\natural,\ddagger}_{\widehat{\mathcal{H}}} = (\breve{\mathsf{Z}}_{\widehat{\mathcal{H}}}, \widehat{\varphi}^{\ddagger,\sim}_{-2,\widehat{\mathcal{H}}}, \widehat{\varphi}^{\ddagger}_{-1,\widehat{\mathcal{H}}}, \widehat{\varphi}^{\ddagger,\sim}_{0,\widehat{\mathcal{H}}}, \breve{\delta}_{\widehat{\mathcal{H}}}, \widehat{c}^{\ddagger}_{\widehat{\mathcal{H}}}, \widehat{c}^{\vee,\ddagger}_{\widehat{\mathcal{H}}}, \widehat{\tau}^{\ddagger}_{\widehat{\mathcal{H}}}, \widehat{\tau}^{\vee,\ddagger}_{\widehat{\mathcal{H}}}),$$

and the pair $(\widehat{\tau}^{\ddagger}, \widehat{\tau}^{\vee,\ddagger})$ defines the same element $\upsilon_{\widehat{\tau}^{\ddagger}} = \upsilon_{\widehat{\tau}^{\vee,\ddagger}}$ in $\widehat{\rho}'$. By assumption, $\widehat{\Sigma}'$ is a refinement of $\widehat{\Sigma}$. Hence, under the canonical isomorphism $\widehat{\mathbf{P}}_{\breve{\Phi}_{\widehat{\mathcal{H}}'}} \cong \widehat{\mathbf{P}}_{\breve{\Phi}_{\widehat{\mathcal{H}}}}$, we have $\widehat{\rho}' \subset \widehat{\rho}$ for some cone $\widehat{\rho} \subset \widehat{\mathbf{P}}^{+}_{\breve{\Phi}_{\widehat{\mathcal{H}}}}$ in $\widehat{\Sigma}_{\breve{\Phi}_{\widehat{\mathcal{H}}}}$, so that $\upsilon_{\widehat{\tau}^{\ddagger}} = \upsilon_{\widehat{\tau}^{\vee,\ddagger}}$ lies in $\widehat{\rho}$.

By the universal property of $\widetilde{\mathsf{M}}^{\breve{\Phi}_{\widetilde{\mathcal{H}}}}_{\widetilde{\mathcal{H}}}$ (which depends only on $\widehat{\mathcal{H}}$; see Definition 1.2.1.15), the data $(\breve{\mathsf{Z}}_{\widehat{\mathcal{H}}}, \breve{\Phi}_{\widehat{\mathcal{H}}} = (\breve{X}, \breve{Y}, \breve{\phi}, \breve{\varphi}_{-2,\widehat{\mathcal{H}}}, \breve{\varphi}_{0,\widehat{\mathcal{H}}}), \breve{\delta}_{\widehat{\mathcal{H}}})$, $(\widehat{\varphi}^{\ddagger,\sim}_{-2,\widehat{\mathcal{H}}}, \widehat{\varphi}^{\ddagger,\sim}_{0,\widehat{\mathcal{H}}})$, and $(\widehat{B}^{\ddagger}, \lambda_{\widehat{B}^{\ddagger}}, i_{\widehat{B}^{\ddagger}}, \widehat{\varphi}^{\ddagger}_{-1,\widehat{\mathcal{H}}})$ on the torus and abelian parts define a canonical morphism $\xi_1 : \mathrm{Spec}(V) \to \widetilde{\mathsf{M}}^{\breve{\Phi}_{\widetilde{\mathcal{H}}}}_{\widetilde{\mathcal{H}}}$. By the universal property of $\widetilde{C}_{\breve{\Phi}_{\widetilde{\mathcal{H}}}, \breve{\delta}_{\widetilde{\mathcal{H}}}} \to \widetilde{\mathsf{M}}^{\breve{\Phi}_{\widetilde{\mathcal{H}}}}_{\widetilde{\mathcal{H}}}$, the additional data $(\widehat{c}^{\ddagger}_{\widehat{\mathcal{H}}}, \widehat{c}^{\vee,\ddagger}_{\widehat{\mathcal{H}}})$ lifting $(\widehat{c}^{\ddagger}, \widehat{c}^{\vee,\ddagger})$ define a canonical morphism $\xi_0 : \mathrm{Spec}(V) \to \widetilde{C}_{\breve{\Phi}_{\widetilde{\mathcal{H}}}, \breve{\delta}_{\widetilde{\mathcal{H}}}}$ lifting ξ_1. By the construction of

$$\widetilde{\Xi}_{\breve{\Phi}_{\widetilde{\mathcal{H}}}, \breve{\delta}_{\widetilde{\mathcal{H}}}, \breve{\sigma}} \cong \underline{\mathrm{Spec}}_{\mathscr{O}_{\widetilde{C}_{\breve{\Phi}_{\widetilde{\mathcal{H}}}, \breve{\delta}_{\widetilde{\mathcal{H}}}}}} \left(\bigoplus_{\breve{\ell} \in \breve{\sigma}^{\perp}} \widetilde{\Psi}_{\breve{\Phi}_{\widetilde{\mathcal{H}}}, \breve{\delta}_{\widetilde{\mathcal{H}}}}(\breve{\ell}) \right)$$

over $\widetilde{C}_{\breve{\Phi}_{\widetilde{\mathcal{H}}}, \breve{\delta}_{\widetilde{\mathcal{H}}}}$, which we can canonically identify as

$$\widehat{\Xi}_{\breve{\Phi}_{\widetilde{\mathcal{H}}}, \breve{\delta}_{\widetilde{\mathcal{H}}}} \cong \underline{\mathrm{Spec}}_{\mathscr{O}_{\widehat{C}_{\breve{\Phi}_{\widetilde{\mathcal{H}}}, \breve{\delta}_{\widetilde{\mathcal{H}}}}}} \left(\bigoplus_{\breve{\ell} \in \widehat{\mathbf{S}}_{\breve{\Phi}_{\widetilde{\mathcal{H}}}}} \widehat{\Psi}_{\breve{\Phi}_{\widetilde{\mathcal{H}}}, \breve{\delta}_{\widetilde{\mathcal{H}}}}(\breve{\ell}) \right)$$

over $\widehat{C}_{\breve{\Phi}_{\widetilde{\mathcal{H}}}, \breve{\delta}_{\widetilde{\mathcal{H}}}}$ (see Proposition 1.3.2.56), it enjoys the universal property (similar to that of $\widetilde{\Xi}_{\breve{\Phi}_{\widetilde{\mathcal{H}}}, \breve{\delta}_{\widetilde{\mathcal{H}}}} \to \widetilde{C}_{\breve{\Phi}_{\widetilde{\mathcal{H}}}, \breve{\delta}_{\widetilde{\mathcal{H}}}}$) such that the final part of the data $(\widehat{\tau}^{\ddagger}_{\widehat{\mathcal{H}}}, \widehat{\tau}^{\vee,\ddagger}_{\widehat{\mathcal{H}}})$ lifting $(\widehat{\tau}^{\ddagger}, \widehat{\tau}^{\vee,\ddagger})$ determines a canonical morphism $\widetilde{\xi}_K : \mathrm{Spec}(K) \to \widetilde{\Xi}_{\breve{\Phi}_{\widetilde{\mathcal{H}}}, \breve{\delta}_{\widetilde{\mathcal{H}}}, \breve{\sigma}}$ lifting ξ_0 under the canonical morphism $\widetilde{\Xi}_{\breve{\Phi}_{\widetilde{\mathcal{H}}}, \breve{\delta}_{\widetilde{\mathcal{H}}}, \breve{\sigma}} \to \widetilde{C}_{\breve{\Phi}_{\widetilde{\mathcal{H}}}, \breve{\delta}_{\widetilde{\mathcal{H}}}}$. Since the element $\upsilon_{\widehat{\tau}^{\ddagger}} = \upsilon_{\widehat{\tau}^{\vee,\ddagger}}$ defined by $(\widehat{\tau}^{\ddagger}, \widehat{\tau}^{\vee,\ddagger})$ lies in $\widehat{\rho}' \subset \widehat{\rho}$, by the construction of

$$\widetilde{\Xi}_{\breve{\Phi}_{\widetilde{\mathcal{H}}}, \breve{\delta}_{\widetilde{\mathcal{H}}}, \breve{\sigma}}(\breve{\rho}) \cong \underline{\mathrm{Spec}}_{\mathscr{O}_{\widetilde{C}_{\breve{\Phi}_{\widetilde{\mathcal{H}}}, \breve{\delta}_{\widetilde{\mathcal{H}}}}}} \left(\bigoplus_{\breve{\ell} \in \breve{\sigma}^{\perp} \cap \breve{\rho}^{\vee}} \widetilde{\Psi}_{\breve{\Phi}_{\widetilde{\mathcal{H}}}, \breve{\delta}_{\widetilde{\mathcal{H}}}}(\breve{\ell}) \right)$$

(see [61, Sec. 3B]), which we can canonically identify as

$$\widehat{\Xi}_{\breve{\Phi}_{\widetilde{\mathcal{H}}}, \breve{\delta}_{\widetilde{\mathcal{H}}}}(\widehat{\rho}) = \underline{\mathrm{Spec}}_{\mathscr{O}_{\widehat{C}_{\breve{\Phi}_{\widetilde{\mathcal{H}}}, \breve{\delta}_{\widetilde{\mathcal{H}}}}}} \left(\bigoplus_{\breve{\ell} \in \widehat{\rho}^{\vee}} \widehat{\Psi}_{\breve{\Phi}_{\widetilde{\mathcal{H}}}, \breve{\delta}_{\widetilde{\mathcal{H}}}}(\breve{\ell}) \right)$$

(see (1.3.2.60)), which depends only on $\widehat{\mathcal{H}}$ and on $\widehat{\rho}^{\vee} \cong \breve{\sigma}^{\perp} \cap \breve{\rho}^{\vee}$, and by the same argument as in the proof of [62, Prop. 6.2.5.11], the morphism $\widetilde{\xi}_K$ extends to a morphism $\tilde{\xi} : \mathrm{Spec}(V) \to \widetilde{\Xi}_{\breve{\Phi}_{\widetilde{\mathcal{H}}},\breve{\delta}_{\widetilde{\mathcal{H}}},\breve{\sigma}}(\breve{\rho})$ lifting ξ_0 under the canonical morphism $\widetilde{\Xi}_{\breve{\Phi}_{\widetilde{\mathcal{H}}},\breve{\delta}_{\widetilde{\mathcal{H}}},\breve{\sigma}}(\breve{\rho}) \to \widetilde{C}_{\breve{\Phi}_{\widetilde{\mathcal{H}}},\breve{\delta}_{\widetilde{\mathcal{H}}}}$, which maps the special point of $\mathrm{Spec}(V)$ to the $\breve{\rho}$-stratum $\widetilde{\Xi}_{\breve{\Phi}_{\widetilde{\mathcal{H}}},\breve{\delta}_{\widetilde{\mathcal{H}}},\breve{\rho}}$ of $\widetilde{\Xi}_{\breve{\Phi}_{\widetilde{\mathcal{H}}},\breve{\delta}_{\widetilde{\mathcal{H}}},\breve{\sigma}}(\breve{\rho})$. (Alternatively, we can noncanonically lift $\upsilon_{\widehat{\tau}^{\ddagger}} = \upsilon_{\widehat{\tau}^{\vee,\ddagger}}$ to elements of $\breve{\rho} \subset \mathbf{P}^{+}_{\breve{\Phi}_{\widetilde{\mathcal{H}}}}$, work with $\widetilde{\Xi}_{\breve{\Phi}_{\widetilde{\mathcal{H}}},\breve{\delta}_{\widetilde{\mathcal{H}}}}$ and $\widetilde{\Xi}_{\breve{\Phi}_{\widetilde{\mathcal{H}}},\breve{\delta}_{\widetilde{\mathcal{H}}}}(\breve{\rho})$ directly, and invoke the original [62, Prop. 6.2.5.11].) Since V is complete, $\tilde{\xi}$ induces a morphism $\hat{\xi}$ from $\mathrm{Spf}(V)$ to $\widetilde{\mathfrak{X}}_{\breve{\Phi}_{\widetilde{\mathcal{H}}},\breve{\delta}_{\widetilde{\mathcal{H}}},\breve{\sigma},\breve{\rho}}$, the formal completion of $\widetilde{\Xi}_{\breve{\Phi}_{\widetilde{\mathcal{H}}},\breve{\delta}_{\widetilde{\mathcal{H}}},\breve{\sigma}}(\breve{\rho})$ along its $\breve{\rho}$-stratum $\widetilde{\Xi}_{\breve{\Phi}_{\widetilde{\mathcal{H}}},\breve{\delta}_{\widetilde{\mathcal{H}}},\breve{\rho}}$. Then the composition of $\hat{\xi}$ with the canonical morphism $\widetilde{\mathfrak{X}}_{\breve{\Phi}_{\widetilde{\mathcal{H}}},\breve{\delta}_{\widetilde{\mathcal{H}}},\breve{\sigma},\breve{\rho}} \to \mathsf{N}^{\mathrm{tor}}_{\widetilde{\kappa}}$ gives a canonical morphism $\xi : \mathrm{Spec}(V) \to \mathsf{N}^{\mathrm{tor}}_{\widetilde{\kappa}}$.

As explained in Construction 1.3.4.12, $\xi_\eta := \xi|_\eta : \eta = \mathrm{Spec}(K) \to \mathsf{N}_{\widetilde{\kappa}}$ is determined by the pullback $(\widehat{G}'_\eta, \widehat{\lambda}'_\eta, \widehat{i}'_\eta, \widehat{\alpha}'_{\widehat{\mathcal{H}}',\eta}) \to \mathrm{Spec}(K)$ of $(\widehat{G}', \widehat{\lambda}', \widehat{i}', \widehat{\alpha}'_{\widehat{\mathcal{H}}'}) \to \mathsf{N}^{\mathrm{tor}}_{\widetilde{\kappa}'}$ under $\xi'_\eta := \xi'|_\eta : \mathrm{Spec}(K) \to \mathsf{N}^{\mathrm{tor}}_{\widetilde{\kappa}'}$, whose $\widehat{\mathcal{H}}$-orbit $(\widehat{G}'_\eta, \widehat{\lambda}'_\eta, \widehat{i}'_\eta, \widehat{\alpha}'_{\widehat{\mathcal{H}},\eta}) \to \mathrm{Spec}(K)$ is (as explained in the first paragraph of this proof) isomorphic to the pullback of $(\widehat{G}, \widehat{\lambda}, \widehat{i}, \widehat{\alpha}_{\widehat{\mathcal{H}}}) \to \mathsf{N}^{\mathrm{tor}}_{\widetilde{\kappa}}$ under the composition of $\xi'_\eta : \mathrm{Spec}(K) \to \mathsf{N}_{\widetilde{\kappa}'}$ with $f_{\widetilde{\kappa}',\widetilde{\kappa}} : \mathsf{N}^{\mathrm{tor}}_{\widetilde{\kappa}'} \to \mathsf{N}^{\mathrm{tor}}_{\widetilde{\kappa}}$. Hence, $\xi_\eta = f_{\widetilde{\kappa}',\widetilde{\kappa}} \circ \xi'_\eta$ by the universal property of $\mathsf{N}_{\widetilde{\kappa}}$, and $\xi : \mathrm{Spec}(V) \to \mathsf{N}^{\mathrm{tor}}_{\widetilde{\kappa}}$ can be interpreted as a (necessarily unique) extension of $f_{\widetilde{\kappa}',\widetilde{\kappa}} \circ \xi'_\eta : \mathrm{Spec}(K) \to \mathsf{N}_{\widetilde{\kappa}}$.

Since $\xi' : \mathrm{Spec}(V) \to \mathsf{N}^{\mathrm{tor}}_{\widetilde{\kappa}'}$ and $\bar{s}$ (the prescribed center of ξ') are arbitrary, and since $\mathsf{N}^{\mathrm{tor}}_{\widetilde{\kappa}'}$ is noetherian normal, this shows that $f_{\widetilde{\kappa}',\widetilde{\kappa}}$ extends to $f^{\mathrm{tor}}_{\widetilde{\kappa}',\widetilde{\kappa}}$, as desired.

By considering good algebraic models as in the paragraph preceding [61, Lem. 5.10], the morphism $f^{\mathrm{tor}}_{\widetilde{\kappa}',\widetilde{\kappa}} : \mathsf{N}^{\mathrm{tor}}_{\widetilde{\kappa}'} \to \mathsf{N}^{\mathrm{tor}}_{\widetilde{\kappa}}$ is étale locally given by the canonical morphism $\widetilde{\Xi}_{\breve{\Phi}_{\widetilde{\mathcal{H}}'},\breve{\delta}_{\widetilde{\mathcal{H}}'},\breve{\sigma}'}(\breve{\rho}') \to \widetilde{\Xi}_{\breve{\Phi}_{\widetilde{\mathcal{H}}},\breve{\delta}_{\widetilde{\mathcal{H}}},\breve{\sigma}}(\breve{\rho})$, because the tautological data (as in (1.3.4.15)) over $\widetilde{\Xi}_{\breve{\Phi}_{\widetilde{\mathcal{H}}'},\breve{\delta}_{\widetilde{\mathcal{H}}'},\breve{\sigma}'}(\breve{\rho}')$ is the pullback of the one over $\widetilde{\Xi}_{\breve{\Phi}_{\widetilde{\mathcal{H}}},\breve{\delta}_{\widetilde{\mathcal{H}}},\breve{\sigma}}(\breve{\rho})$. By construction, $\widetilde{\Xi}_{\breve{\Phi}_{\widetilde{\mathcal{H}}'},\breve{\delta}_{\widetilde{\mathcal{H}}'},\breve{\sigma}'}(\breve{\rho}') \to \widetilde{\Xi}_{\breve{\Phi}_{\widetilde{\mathcal{H}}},\breve{\delta}_{\widetilde{\mathcal{H}}},\breve{\sigma}}(\breve{\rho})$ is log étale and equivariant with respect to the canonical homomorphism $\widetilde{E}_{\breve{\Phi}_{\widetilde{\mathcal{H}}'},\breve{\sigma}'} \to \widetilde{E}_{\breve{\Phi}_{\widetilde{\mathcal{H}}},\breve{\sigma}}$ between tori, which (by Proposition 1.3.2.56 again) can be canonically identified with the canonical log étale morphism $\widehat{\Xi}_{\breve{\Phi}_{\widehat{\mathcal{H}}'},\breve{\delta}_{\widehat{\mathcal{H}}'}}(\widehat{\rho}') \to \widehat{\Xi}_{\breve{\Phi}_{\widehat{\mathcal{H}}},\breve{\delta}_{\widehat{\mathcal{H}}}}(\widehat{\rho})$, equivariant with respect to the canonical homomorphism $\widehat{E}_{\breve{\Phi}_{\widehat{\mathcal{H}}'}} \to \widehat{E}_{\breve{\Phi}_{\widehat{\mathcal{H}}}}$ between tori (dual to the canonical homomorphism $\widehat{\mathbf{S}}_{\breve{\Phi}_{\widehat{\mathcal{H}}}} \to \widehat{\mathbf{S}}_{\breve{\Phi}_{\widehat{\mathcal{H}}'}}$). The remainder of the proposition then follows. □

Thanks to Lemma 1.3.4.2 and Proposition 1.3.4.19, we can make the following:

Definition 1.3.4.20. *For $\widetilde{\kappa} \in \widetilde{\mathbf{K}}^{++}_{Q,\mathcal{H}}$ which defines $\kappa = [\widetilde{\kappa}] \in \mathbf{K}^{++}_{Q,\mathcal{H}}$ (see Definition 1.2.4.44), we shall denote $\widetilde{\kappa}^{\mathrm{isog}} : \underline{\mathrm{Hom}}_{\mathcal{O}}(Q, G_{\mathsf{M}_{\mathcal{H}_\kappa}})^\circ \to \mathsf{N}^{\mathrm{grp}}_{\widetilde{\kappa}}$ and $\widetilde{\kappa}^{\mathrm{tor}} : \mathsf{N}_{\widetilde{\kappa}} \hookrightarrow \mathsf{N}^{\mathrm{tor}}_{\widetilde{\kappa}}$ by $\kappa^{\mathrm{isog}} : \underline{\mathrm{Hom}}_{\mathcal{O}}(Q, G_{\mathsf{M}_{\mathcal{H}_\kappa}})^\circ \to \mathsf{N}^{\mathrm{grp}}_{\kappa}$ and $\kappa^{\mathrm{tor}} : \mathsf{N}_\kappa \hookrightarrow \mathsf{N}^{\mathrm{tor}}_\kappa$, respectively. For $\widetilde{\kappa}$ and $\widetilde{\kappa}'$ in $\widetilde{\mathbf{K}}^{++}_{Q,\mathcal{H}}$ such that $\kappa' = [\widetilde{\kappa}'] \succ \kappa = [\widetilde{\kappa}]$ in $\mathbf{K}^{++}_{Q,\mathcal{H}}$, we shall denote the canonical morphisms $f_{\widetilde{\kappa}',\widetilde{\kappa}} : \mathsf{N}_{\widetilde{\kappa}'} \to \mathsf{N}_{\widetilde{\kappa}}$, $f^{\mathrm{grp}}_{\widetilde{\kappa}',\widetilde{\kappa}} := \widetilde{\kappa}^{\mathrm{isog}} \circ ((\widetilde{\kappa}')^{\mathrm{isog}})^{-1} : \mathsf{N}^{\mathrm{grp}}_{\widetilde{\kappa}'} \to \mathsf{N}^{\mathrm{grp}}_{\widetilde{\kappa}} \underset{\mathsf{M}_{\mathcal{H}_\kappa}}{\times} \mathsf{M}_{\mathcal{H}_{\kappa'}}$, and $f^{\mathrm{tor}}_{\widetilde{\kappa}',\widetilde{\kappa}} : \mathsf{N}^{\mathrm{tor}}_{\widetilde{\kappa}'} \to \mathsf{N}^{\mathrm{tor}}_{\widetilde{\kappa}}$ by $f_{\kappa',\kappa} : \mathsf{N}_{\kappa'} \to \mathsf{N}_\kappa$, $f^{\mathrm{grp}}_{\kappa',\kappa} := \kappa^{\mathrm{isog}} \circ ((\kappa')^{\mathrm{isog}})^{-1} : \mathsf{N}^{\mathrm{grp}}_{\kappa'} \to \mathsf{N}^{\mathrm{grp}}_{\kappa} \underset{\mathsf{M}_{\mathcal{H}_\kappa}}{\times} \mathsf{M}_{\mathcal{H}_{\kappa'}}$, and $f^{\mathrm{tor}}_{\kappa',\kappa} : \mathsf{N}^{\mathrm{tor}}_{\kappa'} \to \mathsf{N}^{\mathrm{tor}}_\kappa$, respectively. (That is, we drop the tildes in* *all such notations.) We shall denote by* $\widehat{\mathsf{Z}}_{[(\breve{\Phi}_{\widehat{\mathcal{H}}},\breve{\delta}_{\widehat{\mathcal{H}}},\widehat{\tau})]}$ *the* $[(\breve{\Phi}_{\widehat{\mathcal{H}}},\breve{\delta}_{\widehat{\mathcal{H}}},\widehat{\tau})]$*-stratum of* $\mathsf{N}^{\mathrm{tor}}_{\kappa}$*, which is the* $[(\breve{\Phi}_{\widetilde{\mathcal{H}}},\breve{\delta}_{\widetilde{\mathcal{H}}},\breve{\tau})]$*-stratum* $\widetilde{\mathsf{Z}}_{[(\breve{\Phi}_{\widetilde{\mathcal{H}}},\breve{\delta}_{\widetilde{\mathcal{H}}},\breve{\tau})]} \cong \widetilde{\Xi}_{\breve{\Phi}_{\widetilde{\mathcal{H}}},\breve{\delta}_{\widetilde{\mathcal{H}}},\breve{\tau}} \cong \widehat{\Xi}_{\breve{\Phi}_{\widehat{\mathcal{H}}},\breve{\delta}_{\widehat{\mathcal{H}}},\widehat{\tau}}$ *of* $\mathsf{N}^{\mathrm{tor}}_{\widetilde{\kappa}}$ *under the canonical identification between* $\mathsf{N}^{\mathrm{tor}}_{\kappa}$ *and* $\mathsf{N}^{\mathrm{tor}}_{\widetilde{\kappa}}$ *(up to canonical isomorphism) (when* $(\breve{\Phi}_{\widehat{\mathcal{H}}},\breve{\delta}_{\widehat{\mathcal{H}}},\widehat{\tau})$ *is determined by* $(\breve{\Phi}_{\widetilde{\mathcal{H}}},\breve{\delta}_{\widetilde{\mathcal{H}}},\breve{\tau})$ *as in Section* 1.2.4).

Now the question is whether the structural morphism $f_\kappa : \mathsf{N}_\kappa \to \mathsf{M}_{\mathcal{H}}$ extends (necessarily uniquely) to a (proper) morphism $f^{\mathrm{tor}}_\kappa : \mathsf{N}^{\mathrm{tor}}_\kappa \to \mathsf{M}^{\mathrm{tor}}_{\mathcal{H}} = \mathsf{M}^{\mathrm{tor}}_{\mathcal{H},\Sigma}$ between the compactifications.

Let $\mathbf{K}^{++}_{Q,\mathcal{H},\Sigma}$ be the subset of $\mathbf{K}^{++}_{Q,\mathcal{H}}$ defined at the end of Section 1.2.4, which is the subset of $\mathbf{K}^{++}_{Q,\mathcal{H}}$ consisting of elements κ satisfying Condition 1.2.4.49. The main result of [61, Sec. 3B] is the following: For $\kappa = (\widehat{\mathcal{H}},\widehat{\Sigma}) = [\widetilde{\kappa}] = [(\widetilde{\mathcal{H}},\widetilde{\Sigma},\widetilde{\sigma})] \in \mathbf{K}^{++}_{Q,\mathcal{H},\Sigma}$ (which means $\widetilde{\kappa} = (\widetilde{\mathcal{H}},\widetilde{\Sigma},\widetilde{\sigma})$ satisfies Condition 1.2.4.48, for some and hence every representative $\widetilde{\kappa}$ of κ), the structural morphism $f_\kappa : \mathsf{N}_\kappa \to \mathsf{M}_{\mathcal{H}}$ extends to a (unique) morphism $f^{\mathrm{tor}}_\kappa : \mathsf{N}^{\mathrm{tor}}_\kappa \to \mathsf{M}^{\mathrm{tor}}_{\mathcal{H}}$, which is étale locally given by morphisms between toric schemes equivariant under (surjective) morphisms between tori. (The proof does not require $\widehat{\mathcal{H}}$ to satisfy either Conditions 1.2.4.8 or 1.2.4.9.) In the remainder of [61, Sec. 3–5], it was shown that the collection of such extended morphisms satisfy the remaining requirements of Theorem 1.3.3.15. (The proofs of these used a particular representative $\widetilde{\kappa}$ of $\kappa = [\widetilde{\kappa}]$, which nevertheless suffices, by Proposition 1.3.4.19. Also, they assumed that $\kappa \in \mathbf{K}_{Q,\mathcal{H},\Sigma}$, in which case $\mathsf{N}_\kappa \to \mathsf{M}_{\mathcal{H}_\kappa} = \mathsf{M}_{\mathcal{H}}$ is an abelian scheme—but this is not really necessary: For log smoothness, in [61, Sec. 3C], the proof using the extended Kodaira–Spencer isomorphism over $\widetilde{\mathsf{M}}_{\widetilde{\mathcal{H}},\widetilde{\Sigma}}$ is insensitive to whether $\mathsf{N}_\kappa \to \mathsf{M}_{\mathcal{H}}$ is an abelian scheme or not. We note that for the condition on equidimensionality, in [61, Sec. 3D], the proofs there are combinatorial in nature and also insensitive to whether $\mathsf{N}_\kappa \to \mathsf{M}_{\mathcal{H}}$ is an abelian scheme or not. For the statements (4) and (5) of Theorem 1.3.3.15, the proof using the Hecke action of $\widetilde{\mathrm{G}}(\mathbb{A}^\infty)$ on the collection $\{\widetilde{\mathsf{M}}_{\widetilde{\mathcal{H}},\widetilde{\Sigma}}\}_{\widetilde{\mathcal{H}},\widetilde{\Sigma}}$ are also insensitive to whether $\mathsf{N}_\kappa \to \mathsf{M}_{\mathcal{H}}$ is an abelian scheme or not. In [61, Sec. 4–5], the proof for statements in (3) of Theorem 1.3.3.15 can be verified by étale descent, and hence can be proved with the same methods even when we only assume $\kappa \in \mathbf{K}^{+}_{Q,\mathcal{H},\Sigma}$, in which case $\mathsf{N}_\kappa \to \mathsf{M}_{\mathcal{H}_\kappa} = \mathsf{M}_{\mathcal{H}}$ is only an abelian scheme torsor.) Hence, the same methods of the proof of [61, Thm. 2.15; see also the errata] work here for the slightly generalized Theorem 1.3.3.15.

1.4 Automorphic Bundles and Canonical Extensions in Characteristic Zero

1.4.1 *Automorphic Bundles*

Suppose there exists a finite extension F'_0 of F_0 in $\mathbb{C}$ such that there exists an $\mathcal{O} \underset{\mathbb{Z}}{\otimes} F'_0$-module L_0 such that $L_0 \underset{F'_0}{\otimes} \mathbb{C} \cong V_0$, where V_0 is as in (1.1.1.4). Once the choice of F'_0 is fixed, the choice of L_0 is unique up to isomorphism because $\mathcal{O} \underset{\mathbb{Z}}{\otimes} F'_0$-modules are uniquely determined by their multi-ranks. (See [62, Lem. 1.1.3.4

and Def. 1.1.3.5] for the notion of multi-ranks.) Let

$$\langle\,\cdot\,,\,\cdot\,\rangle_{\mathrm{can.}} : (L_0 \oplus L_0^\vee(1)) \times (L_0 \oplus L_0^\vee(1)) \to F_0'(1)$$

be the alternating pairing defined by

$$\langle (x_1, f_1), (x_2, f_2) \rangle_{\mathrm{can.}} := f_2(x_1) - f_1(x_2)$$

(*cf.* [62, Lem. 1.1.4.13]).

Definition 1.4.1.1. (*See [61, Def. 6.2].*) *For each F_0'-algebra R, set*

$$\mathrm{G}_0(R) := \left\{ \begin{array}{c} (g, r) \in \mathrm{GL}_{\mathcal{O} \underset{\mathbb{Z}}{\otimes} R}((L_0 \oplus L_0^\vee(1)) \underset{F_0'}{\otimes} R) \times \mathbf{G}_{\mathrm{m}}(R) : \\ \langle gx, gy \rangle_{\mathrm{can.}} = r \langle x, y \rangle_{\mathrm{can.}}, \forall x, y \in (L_0 \oplus L_0^\vee(1)) \underset{F_0'}{\otimes} R \end{array} \right\},$$

$$\mathrm{P}_0(R) := \{(g, r) \in \mathrm{G}_0(R) : g(L_0^\vee(1) \underset{F_0'}{\otimes} R) = L_0^\vee(1) \underset{F_0'}{\otimes} R\},$$

$$\mathrm{M}_0(R) := \mathrm{GL}_{\mathcal{O} \underset{\mathbb{Z}}{\otimes} R}(L_0^\vee(1) \underset{F_0'}{\otimes} R) \times \mathbf{G}_{\mathrm{m}}(R),$$

where we view $\mathrm{M}_0(R)$ canonically as a quotient of $\mathrm{P}_0(R)$ by

$$\mathrm{P}_0(R) \to \mathrm{M}_0(R) : (g, r) \mapsto (g|_{L_0^\vee(1) \underset{F_0'}{\otimes} R}, r).$$

The assignments are functorial in R and define group functors G_0, P_0, and M_0 over F_0'.

Lemma 1.4.1.2. (*See [61, Lem. 6.3].*) *Suppose R is the algebraic closure of F_0' in $\mathbb{C}$. Then there is an isomorphism*

$$(L \underset{\mathbb{Z}}{\otimes} R, \langle\,\cdot\,,\,\cdot\,\rangle) \cong ((L_0 \oplus L_0^\vee(1)) \underset{F_0'}{\otimes} R, \langle\,\cdot\,,\,\cdot\,\rangle_{\mathrm{can.}}),$$

which induces an isomorphism $\mathrm{G} \underset{\mathbb{Z}}{\otimes} R \cong \mathrm{G}_0 \underset{F_0'}{\otimes} R$ over R. (Consequently, $\mathrm{P}_0(R)$ can be identified with a "parabolic" subgroup of $\mathrm{G}(R)$.)

(In practice, it is not necessary to take R to be algebraically closed. Much smaller rings would suffice for the existence of isomorphisms as in Lemma 1.4.1.2.)

In the remainder of this subsection, by abuse of notation, we shall replace $\mathsf{M}_{\mathcal{H}}$ *etc* with their base changes from $\mathrm{Spec}(F_0)$ to $\mathrm{Spec}(F_0')$, and replace $\mathsf{S}_0 = \mathrm{Spec}(F_0)$ with $\mathrm{Spec}(F_0')$.

Definition 1.4.1.3. *The* **principal** G_0**-bundle** *over $\mathsf{M}_{\mathcal{H}}$ is the relative scheme*

$$\mathcal{E}_{\mathrm{G}_0} := \underline{\mathrm{Isom}}_{\mathcal{O} \underset{\mathbb{Z}}{\otimes} \mathscr{O}_{\mathsf{M}_{\mathcal{H}}}}((\underline{H}_1^{\mathrm{dR}}(G_{\mathsf{M}_{\mathcal{H}}}/\mathsf{M}_{\mathcal{H}}), \langle\,\cdot\,,\,\cdot\,\rangle_\lambda, \mathscr{O}_{\mathsf{M}_{\mathcal{H}}}(1)),$$
$$((L_0 \oplus L_0^\vee(1)) \underset{F_0'}{\otimes} \mathscr{O}_{\mathsf{M}_{\mathcal{H}}}, \langle\,\cdot\,,\,\cdot\,\rangle_{\mathrm{can.}}, \mathscr{O}_{\mathsf{M}_{\mathcal{H}}}(1))),$$

the sheaf of isomorphisms of $\mathscr{O}_{\mathsf{M}_{\mathcal{H}}}$-sheaves of symplectic $\mathcal{O}$-modules, over $\mathsf{M}_{\mathcal{H}}$. (The group G_0 acts as automorphisms on $(L \underset{\mathbb{Z}}{\otimes} \mathscr{O}_{\mathsf{M}_{\mathcal{H}}}, \langle\,\cdot\,,\,\cdot\,\rangle_\lambda, \mathscr{O}_{\mathsf{M}_{\mathcal{H}}}(1))$ by definition. The third entries in the tuples represent the values of the pairings.)

Definition 1.4.1.4. *The* **principal** P_0**-bundle** *over* $\mathsf{M}_{\mathcal{H}}$ *is the relative scheme*

$$\mathcal{E}_{\mathrm{P}_0} := \underline{\mathrm{Isom}}_{\mathcal{O} \underset{\mathbb{Z}}{\otimes} \mathscr{O}_{\mathsf{M}_{\mathcal{H}}}}((\underline{H}_1^{\mathrm{dR}}(G_{\mathsf{M}_{\mathcal{H}}}/\mathsf{M}_{\mathcal{H}}), \langle\,\cdot\,,\,\cdot\,\rangle_\lambda, \mathscr{O}_{\mathsf{M}_{\mathcal{H}}}(1), \underline{\mathrm{Lie}}^\vee_{G^\vee_{\mathsf{M}_{\mathcal{H}}}/\mathsf{M}_{\mathcal{H}}}(1)),$$
$$((L_0 \oplus L_0^\vee(1)) \underset{F_0'}{\otimes} \mathscr{O}_{\mathsf{M}_{\mathcal{H}}}, \langle\,\cdot\,,\,\cdot\,\rangle_{\mathrm{can.}}, \mathscr{O}_{\mathsf{M}_{\mathcal{H}}}(1), L_0^\vee(1) \underset{F_0'}{\otimes} \mathscr{O}_{\mathsf{M}_{\mathcal{H}}})),$$

the sheaf of isomorphisms of $\mathscr{O}_{\mathsf{M}_{\mathcal{H}}}$*-sheaves of symplectic* $\mathcal{O}$*-modules with maximal totally isotropic* $\mathcal{O} \underset{\mathbb{Z}}{\otimes} F_0'$*-submodules, over* $\mathsf{M}_{\mathcal{H}}$. (*The group* P_0 *acts as automorphisms on* $(L \underset{\mathbb{Z}}{\otimes} \mathscr{O}_{\mathsf{M}_{\mathcal{H}}}, \langle\,\cdot\,,\,\cdot\,\rangle_\lambda, \mathscr{O}_{\mathsf{M}_{\mathcal{H}}}(1), L_0^\vee(1) \underset{F_0'}{\otimes} \mathscr{O}_{\mathsf{M}_{\mathcal{H}}})$ *by definition. The third entries in the tuples represent the values of the pairings.*)

Definition 1.4.1.5. *The* **principal** M_0**-bundle** *over* $\mathsf{M}_{\mathcal{H}}$ *is the relative scheme*

$$\mathcal{E}_{\mathrm{M}_0} := \underline{\mathrm{Isom}}_{\mathcal{O} \underset{\mathbb{Z}}{\otimes} \mathscr{O}_{\mathsf{M}_{\mathcal{H}}}}((\underline{\mathrm{Lie}}^\vee_{G^\vee_{\mathsf{M}_{\mathcal{H}}}/\mathsf{M}_{\mathcal{H}}}(1), \mathscr{O}_{\mathsf{M}_{\mathcal{H}}}(1)), (L_0^\vee(1) \underset{F_0'}{\otimes} \mathscr{O}_{\mathsf{M}_{\mathcal{H}}}, \mathscr{O}_{\mathsf{M}_{\mathcal{H}}}(1))),$$

the sheaf of isomorphisms of $\mathscr{O}_{\mathsf{M}_{\mathcal{H}}}$*-sheaves of* $\mathcal{O} \underset{\mathbb{Z}}{\otimes} F_0'$*-modules, over* $\mathsf{M}_{\mathcal{H}}$. (*We view the second entries in the pairs as an additional structure, inherited from the corresponding objects for* P_0. *The group* M_0 *acts as automorphisms on* $(L_0^\vee(1) \underset{F_0'}{\otimes} \mathscr{O}_{\mathsf{M}_{\mathcal{H}}}, \mathscr{O}_{\mathsf{M}_{\mathcal{H}}}(1))$ *by definition.*)

Remark 1.4.1.6. The Tate twists on $\underline{\mathrm{Lie}}^\vee_{G^\vee_{\mathsf{M}_{\mathcal{H}}}/\mathsf{M}_{\mathcal{H}}}(1)$ in Definitions 1.4.1.4 and 1.4.1.5 have been omitted in most of this author's writing so far (in, for example, [61], [59], and [70]), which unfortunately made it unclear whether the duality between $\underline{\mathrm{Lie}}^\vee_{G^\vee_{\mathsf{M}_{\mathcal{H}}}/\mathsf{M}_{\mathcal{H}}}$ and $\underline{\mathrm{Lie}}_{G^\vee_{\mathsf{M}_{\mathcal{H}}}/\mathsf{M}_{\mathcal{H}}}$ involves a Tate twist or not. For the sake of clarity, we have reinstated such Tate twists, as explained in Remark 1.1.2.3.

Lemma 1.4.1.7. *The relative scheme* $\mathcal{E}_{\mathrm{G}_0}$ (*resp.* $\mathcal{E}_{\mathrm{P}_0}$, *resp.* $\mathcal{E}_{\mathrm{M}_0}$) *over* $\mathsf{M}_{\mathcal{H}}$ *is an étale torsor under* (*the pullback of*) *the group scheme* G_0 (*resp.* P_0, *resp.* M_0).

Proof. The existence of sections over geometric points of $\mathsf{M}_{\mathcal{H}}$ is guaranteed by the determinantal condition for $\underline{\mathrm{Lie}}_{A/\mathsf{M}_{\mathcal{H},1}}$. By the infinitesimal deformation theory explained in [62, Ch. 2] (based on well-known ideas due to Grothendieck, Mumford, and others), we have isomorphisms between

$$(\underline{H}_1^{\mathrm{dR}}(A/\mathsf{M}_{\mathcal{H}}), \langle\,\cdot\,,\,\cdot\,\rangle_\lambda, \mathscr{O}_{\mathsf{M}_{\mathcal{H}}}(1), \underline{\mathrm{Lie}}^\vee_{A^\vee/\mathsf{M}_{\mathcal{H}}})$$

and

$$((L_0 \oplus L_0^\vee(1)) \underset{F_0'}{\otimes} \mathscr{O}_{\mathsf{M}_{\mathcal{H}}}, \langle\,\cdot\,,\,\cdot\,\rangle_{\mathrm{can.}}, \mathscr{O}_{\mathsf{M}_{\mathcal{H}}}(1), L_0^\vee(1) \underset{F_0'}{\otimes} \mathscr{O}_{\mathsf{M}_{\mathcal{H}}})$$

over the formal completions of $\mathsf{M}_{\mathcal{H}}$ at points of finite type over S_0. Since the sheaves involved are all coherent, we can algebraize the isomorphisms over formal bases by Grothendieck's formal existence theory [35, III-1, 5.1.2], and obtain sections of these functors over complete local rings (at points of finite type over S_0). Since the base scheme $\mathsf{S}_0 = \mathrm{Spec}(F_0')$ is a point, and since these functors are locally of finite presentation (because they are defined by morphisms between coherent sheaves over the scheme $\mathsf{M}_{\mathcal{H}}$ of finite type over S_0), Artin's approximation theory [3, Thm. 1.10 and Cor. 2.5] implies that they have sections étale locally over $\mathsf{M}_{\mathcal{H}}$, as desired. □

Definition 1.4.1.8. *For each F_0'-algebra R, we denote by $\mathrm{Rep}_R(\mathrm{G}_0)$ (resp. $\mathrm{Rep}_R(\mathrm{P}_0)$, resp. $\mathrm{Rep}_R(\mathrm{M}_0)$) the category of R-modules with algebraic actions of $\mathrm{G}_0 \underset{F_0'}{\otimes} R$ (resp. $\mathrm{P}_0 \underset{F_0'}{\otimes} R$, resp. $\mathrm{M}_0 \underset{F_0'}{\otimes} R$).*

Definition 1.4.1.9. *Let R be any F_0'-algebra. For each $W \in \mathrm{Rep}_R(\mathrm{G}_0)$, we define*

$$\mathcal{E}_{\mathrm{G}_0,R}(W) := (\mathcal{E}_{\mathrm{G}_0} \underset{F_0'}{\otimes} R) \overset{\mathrm{G}_0 \underset{F_0'}{\otimes} R}{\times} W,$$

called the **automorphic sheaf** *over $\mathsf{M}_{\mathcal{H}} \underset{F_0'}{\otimes} R$ associated with W. It is called an* **automorphic bundle** *if W is locally free of finite rank over R. We define similarly $\mathcal{E}_{\mathrm{P}_0,R}(W)$ (resp. $\mathcal{E}_{\mathrm{M}_0,R}(W)$) for $W \in \mathrm{Rep}_R(\mathrm{P}_0)$ (resp. $W \in \mathrm{Rep}_R(\mathrm{M}_0)$) by replacing G_0 with P_0 (resp. M_0) in the above expression.*

Lemma 1.4.1.10. *Let R be any F_0'-algebra.*

(1) *The assignment $\mathcal{E}_{\mathrm{G}_0,R}(\,\cdot\,)$ (resp. $\mathcal{E}_{\mathrm{P}_0,R}(\,\cdot\,)$, resp. $\mathcal{E}_{\mathrm{M}_0,R}(\,\cdot\,)$) defines an* **exact functor** *from $\mathrm{Rep}_R(\mathrm{G}_0)$ (resp. $\mathrm{Rep}_R(\mathrm{P}_0)$, resp. $\mathrm{Rep}_R(\mathrm{M}_0)$) to the category of quasi-coherent sheaves over $\mathsf{M}_{\mathcal{H}}$.*

(2) *If we consider an object $W \in \mathrm{Rep}_R(\mathrm{G}_0)$ as an object of $\mathrm{Rep}_R(\mathrm{P}_0)$ by restriction to P_0, then we have a canonical isomorphism $\mathcal{E}_{\mathrm{G}_0,R}(W) \cong \mathcal{E}_{\mathrm{P}_0,R}(W)$.*

(3) *If we view an object $W \in \mathrm{Rep}_R(\mathrm{M}_0)$ as an object of $\mathrm{Rep}_R(\mathrm{P}_0)$ via the canonical homomorphism $\mathrm{P}_0 \to \mathrm{M}_0$, then we have a canonical isomorphism $\mathcal{E}_{\mathrm{P}_0,R}(W) \cong \mathcal{E}_{\mathrm{M}_0,R}(W)$.*

(4) *Suppose $W \in \mathrm{Rep}_R(\mathrm{P}_0)$ has a decreasing filtration by subobjects $\mathtt{F}^a(W) \subset W$ in $\mathrm{Rep}_R(\mathrm{P}_0)$ such that each graded piece $\mathrm{Gr}^a_{\mathtt{F}}(W) := \mathtt{F}^a(W)/\mathtt{F}^{a+1}(W)$ can be identified with an object of $\mathrm{Rep}_R(\mathrm{M}_0)$. Then $\mathcal{E}_{\mathrm{P}_0,R}(W)$ has a filtration $\mathtt{F}^a(\mathcal{E}_{\mathrm{P}_0,R}(W)) := \mathcal{E}_{\mathrm{P}_0,R}(\mathtt{F}^a(W))$ with graded pieces $\mathcal{E}_{\mathrm{M}_0,R}(\mathrm{Gr}^a_{\mathtt{F}}(W))$.*

The proofs of these statements can be found in [70, Sec. 1.3].

Lemma 1.4.1.11. *For any F_0'-algebra R, the pullback of $\underline{\mathrm{Lie}}_{G_{\mathsf{M}_{\mathcal{H}}}/\mathsf{M}_{\mathcal{H}}}$ (resp. $\underline{\mathrm{Lie}}^{\vee}_{G_{\mathsf{M}_{\mathcal{H}}}/\mathsf{M}_{\mathcal{H}}}$, resp. $\omega_{\mathsf{M}_{\mathcal{H}}} = \wedge^{\mathrm{top}} \underline{\mathrm{Lie}}^{\vee}_{G_{\mathsf{M}_{\mathcal{H}}}/\mathsf{M}_{\mathcal{H}}}$; see (2) of Theorem 1.3.1.5) to $\mathsf{M}_{\mathcal{H}} \underset{F_0'}{\otimes} R$ is canonically isomorphic to $\mathcal{E}_{\mathrm{M}_0,R}(W)$ for $W = L_0 \underset{F_0'}{\otimes} R$ (resp. $L_0^{\vee} \underset{F_0'}{\otimes} R$, resp. $\wedge^{\mathrm{top}} L_0^{\vee} \underset{F_0'}{\otimes} R$).*

Proof. This follows from Definitions 1.4.1.5 and 1.4.1.9, and from Lemma 1.4.1.10. □

1.4.2 *Canonical Extensions*

Now let us explain the construction of canonical extensions using Theorem 1.3.3.15.

By taking $Q = \mathcal{O}$, so that $\underline{\mathrm{Hom}}_{\mathcal{O}}(Q, G_{\mathsf{M}_{\mathcal{H}}})^{\circ} \cong G_{\mathsf{M}_{\mathcal{H}}}$ and so that there exists some $\mathbb{Q}^{\times}$-isogeny $\kappa^{\mathrm{isog}} : G_{\mathsf{M}_{\mathcal{H}}} \to \mathsf{N} = \mathsf{N}^{\mathrm{grp}}_{\kappa} = \mathsf{N}_{\kappa}$ for some $\kappa \in \mathbf{K}_{Q,\mathcal{H},\Sigma}$ as in

Theorem 1.3.3.15, the locally free sheaf $\underline{H}^1_{\mathrm{dR}}(\mathsf{N}/\mathsf{M}_{\mathcal{H}}) \cong \underline{H}^1_{\mathrm{dR}}(G_{\mathsf{M}_{\mathcal{H}}}/\mathsf{M}_{\mathcal{H}})$ extends to the locally free sheaf $\underline{H}^1_{\text{log-dR}}(\mathsf{N}^{\mathrm{tor}}/\mathsf{M}^{\mathrm{tor}}_{\mathcal{H}})$ over $\mathscr{O}_{\mathsf{M}^{\mathrm{tor}}_{\mathcal{H}}}$. Let

$$\underline{H}_1^{\text{log-dR}}(\mathsf{N}^{\mathrm{tor}}/\mathsf{M}^{\mathrm{tor}}_{\mathcal{H}}) := \underline{\mathrm{Hom}}_{\mathscr{O}_{\mathsf{M}^{\mathrm{tor}}_{\mathcal{H}}}}(\underline{H}^1_{\text{log-dR}}(\mathsf{N}^{\mathrm{tor}}/\mathsf{M}^{\mathrm{tor}}_{\mathcal{H}}), \mathscr{O}_{\mathsf{M}^{\mathrm{tor}}_{\mathcal{H}}}).$$

Then this $\underline{H}_1^{\text{log-dR}}(\mathsf{N}^{\mathrm{tor}}/\mathsf{M}^{\mathrm{tor}}_{\mathcal{H}})$ qualifies as the $\underline{H}_1^{\mathrm{dR}}(G_{\mathsf{M}_{\mathcal{H}}}/\mathsf{M}_{\mathcal{H}})^{\mathrm{can}}$ in the following:

Proposition 1.4.2.1. *(See [61, Prop. 6.9].) There exists a* **unique** *locally free sheaf $\underline{H}_1^{\mathrm{dR}}(G_{\mathsf{M}_{\mathcal{H}}}/\mathsf{M}_{\mathcal{H}})^{\mathrm{can}}$ over $\mathscr{O}_{\mathsf{M}^{\mathrm{tor}}_{\mathcal{H}}}$ satisfying the following properties:*

1. *The sheaf $\underline{H}_1^{\mathrm{dR}}(G_{\mathsf{M}_{\mathcal{H}}}/\mathsf{M}_{\mathcal{H}})^{\mathrm{can}}$, canonically identified with a subsheaf of the quasi-coherent sheaf $(\mathsf{M}_{\mathcal{H}} \hookrightarrow \mathsf{M}^{\mathrm{tor}}_{\mathcal{H}})_*(\underline{H}_1^{\mathrm{dR}}(\mathsf{N}/\mathsf{M}_{\mathcal{H}}))$, is self-dual under the pairing $(\mathsf{M}_{\mathcal{H}} \hookrightarrow \mathsf{M}^{\mathrm{tor}}_{\mathcal{H}})_*\langle\cdot,\cdot\rangle_\lambda$. We shall denote the induced pairing by $\langle\cdot,\cdot\rangle^{\mathrm{can}}_\lambda$.*
2. *$\underline{H}_1^{\mathrm{dR}}(G_{\mathsf{M}_{\mathcal{H}}}/\mathsf{M}_{\mathcal{H}})^{\mathrm{can}}$ contains $\underline{\mathrm{Lie}}^\vee_{G^\vee/\mathsf{M}^{\mathrm{tor}}_{\mathcal{H}}}(1)$ as a subsheaf totally isotropic under the pairing $\langle\cdot,\cdot\rangle^{\mathrm{can}}_\lambda$.*
3. *The quotient sheaf $\underline{H}_1^{\mathrm{dR}}(G_{\mathsf{M}_{\mathcal{H}}}/\mathsf{M}_{\mathcal{H}})^{\mathrm{can}}/\underline{\mathrm{Lie}}^\vee_{G^\vee/\mathsf{M}^{\mathrm{tor}}_{\mathcal{H}}}$ can be canonically identified with the subsheaf $\underline{\mathrm{Lie}}_{G/\mathsf{M}^{\mathrm{tor}}_{\mathcal{H}}}$ of $(\mathsf{M}_{\mathcal{H}} \hookrightarrow \mathsf{M}^{\mathrm{tor}}_{\mathcal{H}})_*\underline{\mathrm{Lie}}_{G_{\mathsf{M}_{\mathcal{H}}}/\mathsf{M}_{\mathcal{H}}}$.*
4. *The pairing $\langle\cdot,\cdot\rangle^{\mathrm{can}}_\lambda$ induces an isomorphism $\underline{\mathrm{Lie}}_{G/\mathsf{M}^{\mathrm{tor}}_{\mathcal{H}}} \xrightarrow{\sim} \underline{\mathrm{Lie}}_{G^\vee/\mathsf{M}^{\mathrm{tor}}_{\mathcal{H}}}$ which coincides with $d\lambda$.*
5. *Let $\underline{H}^1_{\mathrm{dR}}(G_{\mathsf{M}_{\mathcal{H}}}/\mathsf{M}_{\mathcal{H}})^{\mathrm{can}} := \underline{\mathrm{Hom}}_{\mathscr{O}_{\mathsf{M}^{\mathrm{tor}}_{\mathcal{H}}}}(\underline{H}_1^{\mathrm{dR}}(G_{\mathsf{M}_{\mathcal{H}}}/\mathsf{M}_{\mathcal{H}})^{\mathrm{can}}, \mathscr{O}_{\mathsf{M}^{\mathrm{tor}}_{\mathcal{H}}})$. Then the Gauss–Manin connection*

   $$\nabla : \underline{H}^1_{\mathrm{dR}}(G_{\mathsf{M}_{\mathcal{H}}}/\mathsf{M}_{\mathcal{H}}) \to \underline{H}^1_{\mathrm{dR}}(G_{\mathsf{M}_{\mathcal{H}}}/\mathsf{M}_{\mathcal{H}}) \underset{\mathscr{O}_{\mathsf{M}_{\mathcal{H}}}}{\otimes} \Omega^1_{\mathsf{M}_{\mathcal{H}}/\mathsf{S}_0}$$

   extends to an integrable connection

   $$\nabla : \underline{H}^1_{\mathrm{dR}}(G_{\mathsf{M}_{\mathcal{H}}}/\mathsf{M}_{\mathcal{H}})^{\mathrm{can}} \to \underline{H}^1_{\mathrm{dR}}(G_{\mathsf{M}_{\mathcal{H}}}/\mathsf{M}_{\mathcal{H}})^{\mathrm{can}} \underset{\mathscr{O}_{\mathsf{M}^{\mathrm{tor}}_{\mathcal{H}}}}{\otimes} \overline{\Omega}^1_{\mathsf{M}^{\mathrm{tor}}_{\mathcal{H}}/\mathsf{S}_0} \tag{1.4.2.2}$$

   with log poles along $\mathsf{D}_{\infty,\mathcal{H}}$, called the **extended Gauss–Manin connection**, *such that the composition (ignoring Tate twists; see Remark* 1.1.2.3)

   $$\begin{aligned} \underline{\mathrm{Lie}}^\vee_{G/\mathsf{M}^{\mathrm{tor}}_{\mathcal{H}}} &\hookrightarrow \underline{H}^1_{\mathrm{dR}}(G_{\mathsf{M}_{\mathcal{H}}}/\mathsf{M}_{\mathcal{H}})^{\mathrm{can}} \\ &\xrightarrow{\nabla} \underline{H}^1_{\mathrm{dR}}(G_{\mathsf{M}_{\mathcal{H}}}/\mathsf{M}_{\mathcal{H}})^{\mathrm{can}} \underset{\mathscr{O}_{\mathsf{M}^{\mathrm{tor}}_{\mathcal{H}}}}{\otimes} \overline{\Omega}^1_{\mathsf{M}^{\mathrm{tor}}_{\mathcal{H}}/\mathsf{S}_0} \\ &\twoheadrightarrow \underline{\mathrm{Lie}}_{G^\vee/\mathsf{M}^{\mathrm{tor}}_{\mathcal{H}}} \underset{\mathscr{O}_{\mathsf{M}^{\mathrm{tor}}_{\mathcal{H}}}}{\otimes} \overline{\Omega}^1_{\mathsf{M}^{\mathrm{tor}}_{\mathcal{H}}/\mathsf{S}_0} \end{aligned} \tag{1.4.2.3}$$

   induces by duality the **extended Kodaira–Spencer morphism**

   $$\underline{\mathrm{Lie}}^\vee_{G/\mathsf{M}^{\mathrm{tor}}_{\mathcal{H}}} \underset{\mathscr{O}_{\mathsf{M}^{\mathrm{tor}}_{\mathcal{H}}}}{\otimes} \underline{\mathrm{Lie}}^\vee_{G^\vee/\mathsf{M}^{\mathrm{tor}}_{\mathcal{H}}} \to \overline{\Omega}^1_{\mathsf{M}^{\mathrm{tor}}_{\mathcal{H}}/\mathsf{S}_0}$$

   as in [62, Def. 4.6.3.44], which factors through $\underline{\mathrm{KS}}$ *(in Definition* 1.3.1.2) *and induces the extended Kodaira–Spencer isomorphism* $\mathrm{KS}_{G/\mathsf{M}^{\mathrm{tor}}_{\mathcal{H}}/\mathsf{S}_0}$ *in* (4) *of Theorem* 1.3.1.3.

With these characterizing properties, we say $(\underline{H}_1^{\mathrm{dR}}(G_{\mathsf{M}_{\mathcal{H}}}/\mathsf{M}_{\mathcal{H}})^{\mathrm{can}}, \nabla)$ *is the* **canonical extension** *of* $(\underline{H}_1^{\mathrm{dR}}(G_{\mathsf{M}_{\mathcal{H}}}/\mathsf{M}_{\mathcal{H}}), \nabla)$.

Remark 1.4.2.4. The notion of *canonical extensions* is closely related to the notion of *regular singularities* of algebraic differential equations. See [61, Rem. 6.12] for a list of references to this notion.

Then the principal bundle $\mathcal{E}_{\mathrm{G}_0}$ extends canonically to a principal bundle $\mathcal{E}_{\mathrm{G}_0}^{\mathrm{can}}$ over $\mathsf{M}_{\mathcal{H}}^{\mathrm{tor}}$ by setting

$$\begin{aligned} \mathcal{E}_{\mathrm{G}_0}^{\mathrm{can}} := \underline{\mathrm{Isom}}_{\mathcal{O} \underset{\mathbb{Z}}{\otimes} \mathscr{O}_{\mathsf{M}_{\mathcal{H}}^{\mathrm{tor}}}} &((\underline{H}_1^{\mathrm{dR}}(G_{\mathsf{M}_{\mathcal{H}}}/\mathsf{M}_{\mathcal{H}})^{\mathrm{can}}, \langle\,\cdot\,,\,\cdot\,\rangle_{\lambda}^{\mathrm{can}}, \mathscr{O}_{\mathsf{M}_{\mathcal{H}}^{\mathrm{tor}}}(1)), \\ &((L_0 \oplus L_0^{\vee}(1)) \underset{F_0'}{\otimes} \mathscr{O}_{\mathsf{M}_{\mathcal{H}}^{\mathrm{tor}}}, \langle\,\cdot\,,\,\cdot\,\rangle_{\mathrm{can.}}, \mathscr{O}_{\mathsf{M}_{\mathcal{H}}^{\mathrm{tor}}}(1))), \end{aligned} \tag{1.4.2.5}$$

the principal bundle $\mathcal{E}_{\mathrm{P}_0}$ extends canonically to a principal bundle $\mathcal{E}_{\mathrm{P}_0}^{\mathrm{can}}$ over $\mathsf{M}_{\mathcal{H}}^{\mathrm{tor}}$ by setting

$$\begin{aligned} \mathcal{E}_{\mathrm{P}_0}^{\mathrm{can}} := \underline{\mathrm{Isom}}_{\mathcal{O} \underset{\mathbb{Z}}{\otimes} \mathscr{O}_{\mathsf{M}_{\mathcal{H}}^{\mathrm{tor}}}} &((\underline{H}_1^{\mathrm{dR}}(G_{\mathsf{M}_{\mathcal{H}}}/\mathsf{M}_{\mathcal{H}})^{\mathrm{can}}, \langle\,\cdot\,,\,\cdot\,\rangle_{\lambda}^{\mathrm{can}}, \mathscr{O}_{\mathsf{M}_{\mathcal{H}}^{\mathrm{tor}}}(1), \underline{\mathrm{Lie}}_{G^{\vee}/\mathsf{M}_{\mathcal{H}}^{\mathrm{tor}}}^{\vee}(1)), \\ &((L_0 \oplus L_0^{\vee}(1)) \underset{F_0'}{\otimes} \mathscr{O}_{\mathsf{M}_{\mathcal{H}}^{\mathrm{tor}}}, \langle\,\cdot\,,\,\cdot\,\rangle_{\mathrm{can.}}, \mathscr{O}_{\mathsf{M}_{\mathcal{H}}^{\mathrm{tor}}}(1), L_0^{\vee}(1) \underset{F_0'}{\otimes} \mathscr{O}_{\mathsf{M}_{\mathcal{H}}^{\mathrm{tor}}})), \end{aligned} \tag{1.4.2.6}$$

and the principal bundle $\mathcal{E}_{\mathrm{M}_0}$ extends canonically to a principal bundle $\mathcal{E}_{\mathrm{M}_0}^{\mathrm{can}}$ over $\mathsf{M}_{\mathcal{H}}^{\mathrm{tor}}$ by setting

$$\begin{aligned} \mathcal{E}_{\mathrm{M}_0}^{\mathrm{can}} := \underline{\mathrm{Isom}}_{\mathcal{O} \underset{\mathbb{Z}}{\otimes} \mathscr{O}_{\mathsf{M}_{\mathcal{H}}^{\mathrm{tor}}}} &((\underline{\mathrm{Lie}}_{G^{\vee}/\mathsf{M}_{\mathcal{H}}^{\mathrm{tor}}}^{\vee}(1), \mathscr{O}_{\mathsf{M}_{\mathcal{H}}^{\mathrm{tor}}}(1)), \\ &(L_0^{\vee}(1) \underset{F_0'}{\otimes} \mathscr{O}_{\mathsf{M}_{\mathcal{H}}^{\mathrm{tor}}}, \mathscr{O}_{\mathsf{M}_{\mathcal{H}}^{\mathrm{tor}}}(1))). \end{aligned} \tag{1.4.2.7}$$

Lemma 1.4.2.8. *The relative scheme* $\mathcal{E}_{\mathrm{G}_0}^{\mathrm{can}}$ *(resp.* $\mathcal{E}_{\mathrm{P}_0}^{\mathrm{can}}$*, resp.* $\mathcal{E}_{\mathrm{M}_0}^{\mathrm{can}}$*) over* $\mathsf{M}_{\mathcal{H}}^{\mathrm{tor}}$ *is an étale torsor under (the pullback of) the group scheme* G_0 *(resp.* P_0*, resp.* M_0*).*

Proof. As in the proof Lemma 1.4.1.7, these define étale torsors by Artin's approximation theory (*cf.* [3, Thm. 1.10 and Cor. 2.5]), because these schemes have sections over the formal completions of $\mathsf{M}_{\mathcal{H}}^{\mathrm{tor}}$ at points of finite type over S_0 (because $\underline{\mathrm{Lie}}_{G^{\vee}/\mathsf{M}_{\mathcal{H}}^{\mathrm{tor}}}^{\vee}$ and $L_0^{\vee}(1) \underset{F_0'}{\otimes} \mathscr{O}_{\mathsf{M}_{\mathcal{H}}^{\mathrm{tor}}}$ can be compared using the Lie algebra condition [62, Def. 1.3.4.1 and Lem. 1.2.5.11], and because the pairings $\langle\,\cdot\,,\,\cdot\,\rangle_{\lambda}^{\mathrm{can}}$ and $\langle\,\cdot\,,\,\cdot\,\rangle_{\mathrm{can.}}$ can be compared using [62, Cor. 1.2.3.10]). □

Definition 1.4.2.9. *Let R be any F_0'-algebra. For each $W \in \mathrm{Rep}_R(\mathrm{G}_0)$, we define*

$$\mathcal{E}_{\mathrm{G}_0,R}^{\mathrm{can}}(W) := (\mathcal{E}_{\mathrm{G}_0}^{\mathrm{can}} \underset{F_0'}{\otimes} R) \overset{\mathrm{G}_0 \underset{F_0'}{\otimes} R}{\times} W,$$

called the **canonical extension** *of* $\mathcal{E}_{\mathrm{G}_0,R}(W)$*, and define*

$$\mathcal{E}^{\mathrm{sub}}_{\mathrm{G}_0,R}(W) := \mathcal{E}^{\mathrm{can}}_{\mathrm{G}_0,R}(W) \underset{\mathscr{O}_{\mathsf{M}^{\mathrm{tor}}_{\mathcal{H}}}}{\otimes} \mathscr{I}_{\mathsf{D}_{\infty,\mathcal{H}}},$$

called the **subcanonical extension** *of* $\mathcal{E}_{\mathrm{G}_0,R}(W)$*, where* $\mathscr{I}_{\mathsf{D}_{\infty,\mathcal{H}}}$ *is the* $\mathscr{O}_{\mathsf{M}^{\mathrm{tor}}_{\mathcal{H}}}$*-ideal defining the relative Cartier divisor* $\mathsf{D}_{\infty,\mathcal{H}}$ *(with its reduced structure) in* (3) *of Theorem* 1.3.1.3*. We define similarly* $\mathcal{E}^{\mathrm{can}}_{\mathrm{P}_0,R}(W)$ *and* $\mathcal{E}^{\mathrm{sub}}_{\mathrm{P}_0,R}(W)$ *(resp.* $\mathcal{E}^{\mathrm{can}}_{\mathrm{M}_0,R}(W)$ *and* $\mathcal{E}^{\mathrm{sub}}_{\mathrm{M}_0,R}(W)$*) with* G_0 *and its principal bundle replaced accordingly with* P_0 *(resp.* M_0*) and its principal bundle.*

Then we have:

Lemma 1.4.2.10. *Lemma* 1.4.1.10 *remains true if we replace the automorphic sheaves with their canonical or subcanonical extensions.*

As remarked in [71, Sec. 4.2], the same proofs for Lemma 1.4.1.10 also work here.

Lemma 1.4.2.11. *(Compare with Lemma* 1.4.1.11.*) For any* F'_0*-algebra* R*, the pullback of* $\underline{\mathrm{Lie}}_{G/\mathsf{M}^{\mathrm{tor}}_{\mathcal{H}}}$ *(resp.* $\underline{\mathrm{Lie}}^{\vee}_{G/\mathsf{M}^{\mathrm{tor}}_{\mathcal{H}}}$*, resp.* $\omega_{\mathsf{M}^{\mathrm{tor}}_{\mathcal{H}}} = \wedge^{\mathrm{top}} \underline{\mathrm{Lie}}^{\vee}_{G/\mathsf{M}^{\mathrm{tor}}_{\mathcal{H}}}$*; see* (3) *of Theorem* 1.3.1.5*) to* $\mathsf{M}^{\mathrm{tor}}_{\mathcal{H}} \underset{F'_0}{\otimes} R$ *is canonically isomorphic to* $\mathcal{E}^{\mathrm{can}}_{\mathrm{M}_0,R}(W)$ *for* $W = L_0 \underset{F'_0}{\otimes} R$ *(resp.* $L_0^{\vee} \underset{F'_0}{\otimes} R$*, resp.* $\wedge^{\mathrm{top}} L_0^{\vee} \underset{F'_0}{\otimes} R$*).*

Proof. This follows from (1.4.2.7), Definition 1.4.2.9, and Lemma 1.4.2.10. □

1.4.3 *Hecke Actions*

Proposition 1.4.3.1. *(Compare with* (4) *of Theorem* 1.3.3.15.*) Let* R *be any* F'_0*-algebra, and consider any* $W \in \mathrm{Rep}_R(\mathrm{G}_0)$*. Suppose we have an element* $g \in \mathrm{G}(\mathbb{A}^{\infty})$*, and suppose we have two open compact subgroups* $\mathcal{H}$ *and* $\mathcal{H}'$ *of* $\mathrm{G}(\hat{\mathbb{Z}})$ *such that* $\mathcal{H}' \subset g\mathcal{H}g^{-1}$*. Then there is (by abuse of notation) a canonical isomorphism*

$$[g]^* : [g]^*\mathcal{E}_{\mathrm{P}_0,R}(W) \xrightarrow{\sim} \mathcal{E}_{\mathrm{P}_0,R}(W) \qquad (1.4.3.2)$$

of coherent sheaves over $\mathsf{M}_{\mathcal{H}'}$*, where the first* $\mathcal{E}_{\mathrm{P}_0,R}(W)$ *is defined over* $\mathsf{M}_{\mathcal{H}}$*, and where the second is defined over* $\mathsf{M}_{\mathcal{H}'}$*.*

Suppose $\Sigma = \{\Sigma_{\Phi_{\mathcal{H}}}\}_{[(\Phi_{\mathcal{H}},\delta_{\mathcal{H}})]}$ *and* $\Sigma' = \{\Sigma'_{\Phi'_{\mathcal{H}'}}\}_{[(\Phi'_{\mathcal{H}'},\delta'_{\mathcal{H}'})]}$ *are compatible choices of admissible smooth rational polyhedral cone decomposition data for* $\mathsf{M}_{\mathcal{H}}$ *and* $\mathsf{M}_{\mathcal{H}'}$*, respectively, such that* Σ' *is a* g*-refinement of* Σ *as in [62, Def. 6.4.3.3], so that* $[g]^{\mathrm{tor}} : \mathsf{M}^{\mathrm{tor}}_{\mathcal{H}',\Sigma'} \to \mathsf{M}^{\mathrm{tor}}_{\mathcal{H},\Sigma}$ *is defined as in Proposition* 1.3.1.15*. Then there is (by abuse of notation) a canonical isomorphism*

$$([g]^{\mathrm{tor}})^* : ([g]^{\mathrm{tor}})^*\mathcal{E}^{\mathrm{can}}_{\mathrm{P}_0,R}(W) \xrightarrow{\sim} \mathcal{E}^{\mathrm{can}}_{\mathrm{P}_0,R}(W) \qquad (1.4.3.3)$$

of coherent sheaves over $\mathsf{M}^{\mathrm{tor}}_{\mathcal{H}',\Sigma'}$*, where the first* $\mathcal{E}^{\mathrm{can}}_{\mathrm{M}_0,R}(W)$ *is defined over* $\mathsf{M}^{\mathrm{tor}}_{\mathcal{H},\Sigma}$*, and where the second is defined over* $\mathsf{M}^{\mathrm{tor}}_{\mathcal{H}',\Sigma'}$*. There is also (by abuse of notation) a canonical morphism*

$$([g]^{\mathrm{tor}})^* : ([g]^{\mathrm{tor}})^*\mathcal{E}^{\mathrm{sub}}_{\mathrm{P}_0,R}(W) \to \mathcal{E}^{\mathrm{sub}}_{\mathrm{P}_0,R}(W) \qquad (1.4.3.4)$$

of coherent sheaves over $\mathsf{M}^{\mathrm{tor}}_{\mathcal{H}',\Sigma'}$ *(which is not an isomorphism in general). The canonical morphisms* (1.4.3.2), (1.4.3.3), *and* (1.4.3.4) *are compatible with each other.*

The same statements are true if we replace P_0 *with* G_0 *or* M_0.

If $g = g_1 g_2$, *where* g_1 *and* g_2 *are elements of* $\mathrm{G}(\mathbb{A}^\infty)$, *each having a setup similar to that of* g, *then we have* $[g]^* = [g_1]^* \circ [g_2]^*$ *and* $([g]^{\mathrm{tor}})^* = ([g_1]^{\mathrm{tor}})^* \circ ([g_2]^{\mathrm{tor}})^*$ *whenever the involved isomorphisms are defined.*

Proof. Thanks to the construction of $\mathcal{E}^{\mathrm{can}}_{\mathrm{P}_0}$ based on the canonical extensions $\underline{H}^{\mathrm{dR}}_1(G_{\mathsf{M}_{\mathcal{H}}}/\mathsf{M}_{\mathcal{H}})^{\mathrm{can}}$ and $\underline{H}^{\mathrm{dR}}_1(G_{\mathsf{M}_{\mathcal{H}'}}/\mathsf{M}_{\mathcal{H}'})^{\mathrm{can}}$ in Proposition 1.4.2.1, which are in turn based on the relative de Rham homology in Theorem 1.3.3.15, we have the isomorphisms (1.4.3.2) and (1.4.3.3) because of (4c) of Theorem 1.3.3.15, the latter of which inducing the morphism (1.4.3.4), and we have the last statement (for P_0) because of (4d) of Theorem 1.3.3.15.

By Lemmas 1.4.1.10 and 1.4.2.10, these statements for P_0 imply the analogous statements for G_0 and M_0. □

1.5 Comparison with the Analytic Construction

All algebraically constructed objects in this chapter (such as $\mathsf{M}^{\mathrm{tor}}_{\mathcal{H},\Sigma}$, $\mathsf{M}^{\mathrm{min}}_{\mathcal{H}}$, $\mathcal{E}_{\mathrm{P}_0,R}$, *etc*) are naturally compatible with their analytically constructed (algebraic) analogues. More precisely, the canonical open and closed immersion (1.1.3.1) extends to strata-preserving open and closed immersions

$$\mathsf{Sh}^{\mathrm{tor}}_{\mathcal{H},\Sigma} \hookrightarrow [\mathsf{M}^{\mathrm{tor}}_{\mathcal{H},\Sigma}] \tag{1.5.1}$$

and

$$\mathsf{Sh}^{\mathrm{min}}_{\mathcal{H}} \hookrightarrow \mathsf{M}^{\mathrm{min}}_{\mathcal{H}} \tag{1.5.2}$$

(over $\mathsf{S}_0 = \mathrm{Spec}(F_0)$), and the same are true for other objects defined on them. (For this to make sense, we can only consider Σ that works both for [62] and for works such as [89].) For more details, see [59].

We will not really need these results in this work. For applications, it suffices to know that one can first compare the analytically constructed (algebraic) objects with the algebraically constructed objects in [62] in characteristic zero, as carried out in [59]. Then one can compare the algebraically constructed objects in characteristic zero with all new objects constructed in this work.

Chapter 2

Flat Integral Models

From now on, let us fix a choice of a rational prime number $p > 0$.

Let $\mathsf{M}_{\mathcal{H}}$ be as in Section 1.1. In this chapter, we explain some general constructions of noetherian normal flat integral models of $\mathsf{M}_{\mathcal{H}}$ and their compactifications reviewed in Section 1.3. Beyond some basic properties due to their constructions, our understanding of their refined local structures is limited. (Nevertheless, in certain special cases, we can deduce the normality of the characteristic p fiber at the *bottom level* at p—*i.e.*, when $\mathcal{H}$ is of the form $\mathcal{H} = \mathcal{H}^p\mathcal{H}_p$ with $\mathcal{H}_p = \mathrm{G}(\mathbb{Z}_p)$—from results in the theory of local models. See, for example, [65, Sec. 14].) Thus, the reader should keep in mind that the schemes constructed in this chapter are only auxiliary in nature. Because of our applications in mind, these integral models will be constructed only over $\mathbb{Z}_{(p)}$, although one can also obtain the models over $\mathbb{Z}$ by essentially the same constructions.

We will cite [62] for the constructions of the various auxiliary models in this subsection. It is tempting to cite only [28] for these constructions, and this is indeed feasible for most constructions in this section. But for the construction of Hecke actions of elements in $\mathrm{G}(\mathbb{A}^\infty)$ in Section 2.2.3, this is no longer logically sufficient, because the construction in [28], by requiring that the cone decompositions are admissible for $\mathrm{GL}_g(\mathbb{Z})$ in the case of Siegel moduli of principally polarized abelian schemes of relative dimension g, only allowed Hecke actions of elements in $\mathrm{G}_{\mathrm{aux}}(\hat{\mathbb{Z}})$.

2.1 Auxiliary Choices

2.1.1 *Auxiliary Choices of Smooth Moduli Problems*

Lemma 2.1.1.1. *For each integer $d \geq 1$, there exist integers $a_1 > 0$ and $a_2 \geq 0$, and a positive definite symmetric bilinear pairing*

$$(\,\cdot\,,\,\cdot\,)_{\mathrm{aux}} : \mathbb{Z}^{\oplus(a_1+a_2)} \times \mathbb{Z}^{\oplus(a_1+a_2)} \to \mathbb{Z} \tag{2.1.1.2}$$

satisfying the following properties:

(1) Suppose that $[L^\# : L] = d^2$. Then, under the canonical embedding

$$L^{\oplus(a_1+a_2)} \hookrightarrow L_{\mathrm{aux}} := L^{\oplus a_1} \oplus (L^\#)^{\oplus a_2} \tag{2.1.1.3}$$

induced by $L \hookrightarrow L^{\#}$*, the alternating pairing* $\langle\,\cdot\,,\,\cdot\,\rangle\otimes(\,\cdot\,,\,\cdot\,)_{\mathrm{aux}}$ *on* $L^{\oplus(a_1+a_2)} \cong L \underset{\mathbb{Z}}{\otimes} \mathbb{Z}^{\oplus(a_1+a_2)}$ *extends to an alternating pairing* $\langle\,\cdot\,,\,\cdot\,\rangle_{\mathrm{aux}}$ *on* L_{aux} *valued in* $\mathbb{Z}(1)$ *that is self-dual at* p *in the sense that* $p \nmid [L^{\#}_{\mathrm{aux}} : L_{\mathrm{aux}}]$.

(2) *Let* A *be a (relative) abelian scheme over an algebraic stack* S*, and let* $\lambda : A \to A^\vee$ *be a polarization such that* $\deg(\lambda) = d^2$*. Let* $A^{\Delta}_{\mathrm{aux}} := A^{\times(a_1+a_2)}$ *and* $A^{\nabla}_{\mathrm{aux}} := A^{\times a_1} \underset{S}{\times} (A^\vee)^{\times a_2}$*, which are fiber products over* S*; and let* $f := \mathrm{Id}_A^{\times a_1} \underset{S}{\times} \lambda^{\times a_2} : A^{\Delta}_{\mathrm{aux}} \to A^{\nabla}_{\mathrm{aux}}$*. Then* $\lambda : A \to A^\vee$ *and the morphism*

$$(\,\cdot\,,\,\cdot\,)^*_{\mathrm{aux}} : \mathbb{Z}^{\oplus(a_1+a_2)} \xrightarrow{\sim} \mathbb{Z}^{\oplus(a_1+a_2)} \tag{2.1.1.4}$$

canonical induced by $(\,\cdot\,,\,\cdot\,)_{\mathrm{aux}}$ *induce a polarization* $\lambda^{\Delta}_{\mathrm{aux}} : A^{\Delta}_{\mathrm{aux}} \to A^{\Delta,\vee}_{\mathrm{aux}}$ *(cf. Lemmas* 1.2.4.1*,* 1.3.3.5*, and* 1.3.3.6*, or rather [61, Lem. 2.5, 2.6, and 2.9, and their proofs]), and* $\lambda^{\nabla}_{\mathrm{aux}} := (f^\vee)^{-1} \circ \lambda^{\Delta}_{\mathrm{aux}} \circ f : A^{\nabla}_{\mathrm{aux}} \to A^{\nabla,\vee}_{\mathrm{aux}}$ *is a polarization (not just a* $\mathbb{Q}^\times$*-polarization) of degree prime to* p*. Moreover,* $\deg(\lambda^{\nabla}_{\mathrm{aux}})$ *depends only on* $\deg(\lambda) = d^2$ *and the choices of* (a_1, a_2) *and* $(\,\cdot\,,\,\cdot\,)_{\mathrm{aux}}$*, but not on* A *and* λ*.*

If $p \nmid d$*, then we can take* $(a_1, a_2) = (1, 0)$ *and take* $(\,\cdot\,,\,\cdot\,)_{\mathrm{aux}} : \mathbb{Z}\times\mathbb{Z} \to \mathbb{Z}$ *to be the pairing sending* $(1, 1)$ *to* 1*. Otherwise, we can take* $(a_1, a_2) = (4, 4)$*, and take* $(\,\cdot\,,\,\cdot\,)_{\mathrm{aux}}$ *to be defined by some* 2×2 *matrix* $\left(\begin{smallmatrix} 1 & x \\ {}^t x & d^2 \end{smallmatrix}\right)$ *over* $\mathrm{M}_4(\mathbb{Z})$ *such that* ${}^t x x = d^2 - 1$*.*

Proof. The statement is obvious when $p \nmid d$. Otherwise, we can arrange that $\langle\,\cdot\,,\,\cdot\,\rangle_{\mathrm{aux}}$ is self-dual (at every prime) by the proof of Zarhin's trick (as in [104, Sec. 2] and [80, IX, 1.1]), by taking $x = \left(\begin{smallmatrix} x_1 & -x_2 & -x_3 & -x_4 \\ x_2 & x_1 & -x_4 & x_3 \\ x_3 & x_4 & x_1 & -x_2 \\ x_4 & -x_3 & x_2 & x_1 \end{smallmatrix}\right)$ for any integers x_1, x_2, x_3, x_4 such that $x_1^2 + x_2^2 + x_3^2 + x_4^2 = d^2 - 1$, which exist by the fact (due to Lagrange) that every nonnegative integer can be written as the sum of four squares of integers. □

Lemma 2.1.1.5. *Let* (Z, λ_Z) *be any polarized abelian scheme over a scheme* S*. Given any integer* $d \geq 1$*, let us fix the choices of* (a_1, a_2) *and* $(\,\cdot\,,\,\cdot\,)_{\mathrm{aux}}$ *as in Lemma* 2.1.1.1*. Then the functor that assigns to each scheme* T *over* S *the set of isomorphism classes of polarized abelian schemes* (A, λ) *over* T *such that* $\deg(\lambda) = d^2$ *and* $(Z, \lambda_Z) \underset{S}{\times} T \cong (A^{\nabla}_{\mathrm{aux}}, \lambda^{\nabla}_{\mathrm{aux}})$ *over* T*, where* $(A^{\nabla}_{\mathrm{aux}}, \lambda^{\nabla}_{\mathrm{aux}})$ *is defined by* (A, λ) *as in* (2) *of Lemma* 2.1.1.1*, is representable by a scheme finite over* S*.*

Proof. By [81, Sec. 16], $\deg(\lambda_Z) = d^2_{\mathrm{aux}}$ for some integer $d_{\mathrm{aux}} \geq 1$. The assertion to prove is trivially true unless the construction in (2) of Lemma 2.1.1.1 assigns to each pair (A, λ) of genus g and polarization degree d^2 a pair $(A^{\nabla}_{\mathrm{aux}}, \lambda^{\nabla}_{\mathrm{aux}})$ of genus $g_{\mathrm{aux}} = (a_1 + a_2)g$ and polarization degree d^2_{aux}. Hence, it suffices to treat the universal case, which we explain as follows.

Consider the Siegel moduli $\mathcal{A}_{g,d}$ (resp. $\mathcal{A}_{g_{\mathrm{aux}},d_{\mathrm{aux}}}$) of genus g (resp. g_{aux}) and polarization degree d^2 (resp. d^2_{aux}), which is an algebraic stack separated and of finite type over $\mathrm{Spec}(\mathbb{Z})$ (see [80, VII, 4.3] or [17, Def. 1.1 and Rem. 1.2]). The assignment of pairs $(A^{\triangledown}_{\mathrm{aux}}, \lambda^{\triangledown}_{\mathrm{aux}})$ to pairs (A, λ) parameterized by $\mathcal{A}_{g,d}$ as in (2) of Lemma 2.1.1.1 is functorial, and defines (by universal property) a morphism

$$\mathcal{A}_{g,d} \to \mathcal{A}_{g_{\mathrm{aux}}, d_{\mathrm{aux}}}. \tag{2.1.1.6}$$

To prove the lemma, it suffices to show that (2.1.1.6) is finite.

Suppose V is the spectrum of a discrete valuation ring V with fraction field K. Suppose (A_K, λ_K) is an object of $\mathcal{A}_{g,d}(\mathrm{Spec}(K))$, and suppose the corresponding object $(A^{\triangledown}_{\mathrm{aux},K}, \lambda^{\triangledown}_{\mathrm{aux},K})$ of $\mathcal{A}_{g_{\mathrm{aux}},d_{\mathrm{aux}}}(\mathrm{Spec}(K))$ extends to an object $(A^{\triangledown}_{\mathrm{aux},V}, \lambda^{\triangledown}_{\mathrm{aux},V})$ of $\mathcal{A}_{g_{\mathrm{aux}},d_{\mathrm{aux}}}(\mathrm{Spec}(V))$. By the semistable reduction theorem (see, for example, [62, Thm. 3.3.2.4]), up to replacing K with a finite extension field and replacing V accordingly, we may assume that A_K extends to a semi-abelian scheme A_V over $\mathrm{Spec}(V)$. By the theory of Néron models (see [10]; *cf.* [92, IX, 1.4], [28, Ch. I, Prop. 2.7], or [62, Prop. 3.3.1.5]), the isogeny $f_K : A^{\triangle}_{\mathrm{aux},K} = A_K^{\times(a_1+a_2)} \to A^{\triangledown}_{\mathrm{aux},K}$ extends to an isogeny $A_V^{\times(a_1+a_2)} \to A^{\triangledown}_{\mathrm{aux},V}$, and (since $a_1 + a_2 > 0$) this is possible only when A_V is an abelian scheme. Also, the polarization λ_K extends to a polarization λ_V of A_V. Consequently, we have an object (A_V, λ_V) of $\mathcal{A}_{g,d}(\mathrm{Spec}(V))$, which must correspond to the unique extension $(A^{\triangledown}_{\mathrm{aux},V}, \lambda^{\triangledown}_{\mathrm{aux},V})$ of $(A^{\triangledown}_{\mathrm{aux},K}, \lambda^{\triangledown}_{\mathrm{aux},K})$ (up to unique isomorphism, by the theory of Néron models again, or by the separateness of $\mathcal{A}_{g_{\mathrm{aux}},d_{\mathrm{aux}}}$). Hence, (2.1.1.6) is proper by the valuative criterion (and the fact that $\mathcal{A}_{g,d}$ and $\mathcal{A}_{g_{\mathrm{aux}},d_{\mathrm{aux}}}$ are separated and of finite type over $\mathrm{Spec}(\mathbb{Z})$).

In order to show that (2.1.1.6) is finite, it suffices to show that the induced proper morphism

$$\mathcal{A}_{g,d} \underset{\mathbb{Z}}{\otimes} \mathbb{Z}[\tfrac{1}{n}] \to \mathcal{A}_{g_{\mathrm{aux}},d_{\mathrm{aux}}} \underset{\mathbb{Z}}{\otimes} \mathbb{Z}[\tfrac{1}{n}] \tag{2.1.1.7}$$

is finite for at least two integers n prime to each other. For each $n \geq 3$, the algebraic stack $\mathcal{A}_{g_{\mathrm{aux}},d_{\mathrm{aux}}} \underset{\mathbb{Z}}{\otimes} \mathbb{Z}[\frac{1}{n}]$ admits a finite étale cover by the quasi-projective scheme $\mathcal{A}_{g_{\mathrm{aux}},d_{\mathrm{aux}},n}$, defined as in [83, Ch. 7], parameterizing isomorphisms $\gamma_{\mathrm{aux},n} : \mathbb{Z}^{\oplus 2g_{\mathrm{aux}}} \xrightarrow{\sim} A_{\mathrm{aux}}[n]$ for each object $(A_{\mathrm{aux}}, \lambda_{\mathrm{aux}})$ of $\mathcal{A}_{g_{\mathrm{aux}},d_{\mathrm{aux}}} \underset{\mathbb{Z}}{\otimes} \mathbb{Z}[\frac{1}{n}]$. (In order to avoid confusion with our later terminologies, we refrain from calling such isomorphisms level structures, because they are not required to respect the pairings on both sides.) Similarly, the algebraic stack $\mathcal{A}_{g,d} \underset{\mathbb{Z}}{\otimes} \mathbb{Z}[\frac{1}{n}]$ admits a finite étale cover by the quasi-projective scheme $\mathcal{A}_{g,d,n,n}$ parameterizing isomorphisms $\gamma_n : \mathbb{Z}^{\oplus 2g} \xrightarrow{\sim} A[n]$ and $\gamma_n^{\vee} : \mathbb{Z}^{\oplus 2g} \xrightarrow{\sim} A^{\vee}[n]$ for each object (A, λ) of $\mathcal{A}_{g,d}$. (This is even more naive—the two isomorphisms γ_n and $\gamma_n^{\vee}$ are not required to be related to each other under λ.) By assigning to each object $(A, \lambda, \gamma_n, \gamma_n^{\vee})$ of $\mathcal{A}_{g,d,n,n}$ the object $(A^{\triangledown}_{\mathrm{aux}}, \lambda^{\triangledown}_{\mathrm{aux}}, \gamma^{\triangledown}_{\mathrm{aux},n} := \gamma_n^{\times a_1} \times (\gamma_n^{\vee})^{\times a_2})$ of $\mathcal{A}_{g_{\mathrm{aux}},d_{\mathrm{aux}},n}$, we obtain a proper morphism

$$\mathcal{A}_{g,d,n,n} \to \mathcal{A}_{g_{\mathrm{aux}},d_{\mathrm{aux}},n} \tag{2.1.1.8}$$

lifting (2.1.1.7). Then it suffices to show that (2.1.1.8) is finite, or rather just quasi-affine, by [35, III-1, 4.4.2].

Let $\omega_{\mathcal{A}_{g,d,n,n}}$ and $\omega_{\mathcal{A}_{g_{\mathrm{aux}},d_{\mathrm{aux}},n}}$ denote the Hodge invertible sheaves over $\mathcal{A}_{g,d,n,n}$ and $\mathcal{A}_{g_{\mathrm{aux}},d_{\mathrm{aux}},n}$, respectively, defined by the top exterior powers of the duals of the relative Lie algebras of the tautological abelian schemes, which are ample by [80, IX, 3.1]. By [80, IX, 2.4] and by the construction of (2.1.1.8), the pullback of a positive power of $\omega_{\mathcal{A}_{g_{\mathrm{aux}},d_{\mathrm{aux}},n}}$ to $\mathcal{A}_{g,d,n,n}$ is isomorphic to a positive power of $\omega_{\mathcal{A}_{g,d,n,n}}$. By [35, II, 5.1.6], these show that (2.1.1.8) is quasi-affine, as desired. □

Consider any integral PEL datum $(\mathcal{O}_{\mathrm{aux}}, \star_{\mathrm{aux}}, L_{\mathrm{aux}}, \langle\,\cdot\,,\,\cdot\,\rangle_{\mathrm{aux}}, h_{0,\mathrm{aux}})$, where $(L_{\mathrm{aux}}, \langle\,\cdot\,,\,\cdot\,\rangle_{\mathrm{aux}})$ is as in Lemma 2.1.1.1, such that $\mathcal{O}_{\mathrm{aux}}$ is a subring of $\mathcal{O}$ stabilized by $\star$, with an induced (positive) involution we denote by $^{\star_{\mathrm{aux}}}$, and such that $h_{0,\mathrm{aux}}$ is canonically induced by h_0 by the isomorphism $L_{\mathrm{aux}} \underset{\mathbb{Z}}{\otimes} \mathbb{R} \cong L^{\oplus(a_1+a_2)} \underset{\mathbb{Z}}{\otimes} \mathbb{R}$ induced by (2.1.1.3). Suppose moreover that p is a *good prime* for the integral PEL datum $(\mathcal{O}_{\mathrm{aux}}, \star_{\mathrm{aux}}, L_{\mathrm{aux}}, \langle\,\cdot\,,\,\cdot\,\rangle_{\mathrm{aux}}, h_{0,\mathrm{aux}})$ (see Definition 1.1.1.6), which is possible because we already know that $p \nmid [L_{\mathrm{aux}}^{\#} : L_{\mathrm{aux}}]$, and that the action of $\mathcal{O}_{\mathrm{aux}}$ on L_{aux} extends to an action of a maximal order $\mathcal{O}'_{\mathrm{aux}}$ in $\mathcal{O}_{\mathrm{aux}} \underset{\mathbb{Z}}{\otimes} \mathbb{Q}$ containing $\mathcal{O}_{\mathrm{aux}}$ (*cf.* Condition 1.2.1.1). These are possible, for example, by taking $\mathcal{O}_{\mathrm{aux}} = \mathbb{Z}$ with trivial involution $^{\star_{\mathrm{aux}}}$. From now on, we shall fix the auxiliary choices of (a_1, a_2), $(\,\cdot\,,\,\cdot\,)_{\mathrm{aux}}$, and $(\mathcal{O}_{\mathrm{aux}}, \star_{\mathrm{aux}}, L_{\mathrm{aux}}, \langle\,\cdot\,,\,\cdot\,\rangle_{\mathrm{aux}}, h_{0,\mathrm{aux}})$.

Lemma 2.1.1.9. *With the assumptions as above, the assignment*

$$(g, r) \mapsto (g^{\times a_1} \times (r\,{}^t g^{-1})^{\times a_2}, r)$$

defines an injective homomorphism

$$\mathrm{G} \to \mathrm{G}_{\mathrm{aux}} \tag{2.1.1.10}$$

of algebraic group functors over $\mathrm{Spec}(\mathbb{Z})$*, where* $\mathrm{G}_{\mathrm{aux}}$ *is the group functor over* $\mathrm{Spec}(\mathbb{Z})$ *defined by the order* $\mathcal{O}_{\mathrm{aux}}$ *(with positive involution* $^{\star_{\mathrm{aux}}}$*), the lattice* L_{aux}*, and the pairing* $\langle\,\cdot\,,\,\cdot\,\rangle_{\mathrm{aux}}$ *as in Definition* 1.1.1.3*, which is compatible with the similitude characters and maps* $\mathrm{G}(\hat{\mathbb{Z}})$ *(which stabilizes* $L \underset{\mathbb{Z}}{\otimes} \hat{\mathbb{Z}}$*) to a subgroup of* $\mathrm{G}_{\mathrm{aux}}(\hat{\mathbb{Z}})$ *(which stabilizes* $L_{\mathrm{aux}} \underset{\mathbb{Z}}{\otimes} \hat{\mathbb{Z}}$*).*

Proof. The assignment is injective because $a_1 > 0$, and defines a homomorphism as asserted because $\mathcal{O}_{\mathrm{aux}}$ is a subring of $\mathcal{O}$, because $^{\star_{\mathrm{aux}}}$ is the restriction of *, and because $\langle x, rg^{-1}y\rangle = \langle gx, y\rangle = \langle x, {}^t g y\rangle$ by the definition of $\nu(g)$. □

Lemma 2.1.1.11. *The reflex field* $F_{0,\mathrm{aux}}$ *defined by the integral PEL datum* $(\mathcal{O}_{\mathrm{aux}}, \star_{\mathrm{aux}}, L_{\mathrm{aux}}, \langle\,\cdot\,,\,\cdot\,\rangle_{\mathrm{aux}}, h_{0,\mathrm{aux}})$ *(see [53, p. 389] or [62, Def. 1.2.5.4]) is contained in* F_0 *(as subfields of* $\mathbb{C}$*).*

Proof. Since $h_{0,\mathrm{aux}}$ is canonically induced by h_0 by the isomorphism $L_{\mathrm{aux}} \underset{\mathbb{Z}}{\otimes} \mathbb{R} \cong (L^{\oplus a_1} \oplus (L^{\#})^{\oplus a_2}) \underset{\mathbb{Z}}{\otimes} \mathbb{R}$ induced by (2.1.1.3), we have a canonical isomorphism $V_{0,\mathrm{aux}} \cong V_0^{\oplus(a_1+a_2)}$ as $\mathcal{O}_{\mathrm{aux}} \underset{\mathbb{Z}}{\otimes} \mathbb{C}$-modules. By [62, Cor. 1.2.5.6], F_0 (resp. $F_{0,\mathrm{aux}}$) is the subfield of $\mathbb{C}$ generated over $\mathbb{Q}$ by the traces $\mathrm{Tr}_{\mathbb{C}}(b|V_0)$ for $b \in \mathcal{O}$ (resp. $\mathrm{Tr}_{\mathbb{C}}(b|V_{0,\mathrm{aux}})$ for $b \in \mathcal{O}_{\mathrm{aux}}$). Hence, $F_{0,\mathrm{aux}}$ is contained in F_0, as desired. □

For each open compact subgroup $\mathcal{H}_{\mathrm{aux}}$ (resp. $\mathcal{H}^p_{\mathrm{aux}}$) of $\mathrm{G}_{\mathrm{aux}}(\hat{\mathbb{Z}})$ (resp. $\mathrm{G}_{\mathrm{aux}}(\hat{\mathbb{Z}}^p)$), let $\mathsf{M}_{\mathcal{H}_{\mathrm{aux}}}$ (resp. $\mathsf{M}_{\mathcal{H}^p_{\mathrm{aux}}}$) denote the moduli problem defined by the integral PEL datum $(\mathcal{O}_{\mathrm{aux}}, \star_{\mathrm{aux}}, L_{\mathrm{aux}}, \langle\,\cdot\,,\,\cdot\,\rangle_{\mathrm{aux}}, h_{0,\mathrm{aux}})$ over $\mathsf{S}_{0,\mathrm{aux}} = \mathrm{Spec}(F_{0,\mathrm{aux}})$ (resp. $\vec{\mathsf{S}}_{0,\mathrm{aux}} = \mathrm{Spec}(\mathcal{O}_{F_{0,\mathrm{aux}},(p)})$), which is an algebraic stack separated, smooth, and of finite type over $\mathsf{S}_{0,\mathrm{aux}}$ (resp. $\vec{\mathsf{S}}_{0,\mathrm{aux}}$) by [62, Thm. 1.4.1.11]. Let $[\mathsf{M}_{\mathcal{H}_{\mathrm{aux}}}]$ (resp. $[\mathsf{M}_{\mathcal{H}^p_{\mathrm{aux}}}]$) denote the coarse moduli space associated with $\mathsf{M}_{\mathcal{H}_{\mathrm{aux}}}$ (resp. $\mathsf{M}_{\mathcal{H}^p_{\mathrm{aux}}}$; see [62, Sec. A.7.5]), which is a scheme quasi-projective over $\mathsf{S}_{0,\mathrm{aux}}$ (resp. $\vec{\mathsf{S}}_{0,\mathrm{aux}}$) by [62, Cor. 7.2.3.10]. Moreover, let $[\mathsf{M}_{\mathcal{H}^p_{\mathrm{aux}}} \underset{\mathbb{Z}}{\otimes} \mathbb{Q}]$ denote the coarse moduli space associated with $\mathsf{M}_{\mathcal{H}^p_{\mathrm{aux}}} \underset{\mathbb{Z}}{\otimes} \mathbb{Q}$, which is canonically isomorphic to $[\mathsf{M}_{\mathcal{H}^p_{\mathrm{aux}}}] \underset{\mathbb{Z}}{\otimes} \mathbb{Q}$, because the association of coarse moduli spaces is compatible with flat base changes.

When $\mathcal{H}^p_{\mathrm{aux}}$ is the image of $\mathcal{H}_{\mathrm{aux}}$ under the canonical homomorphism $\mathrm{G}_{\mathrm{aux}}(\hat{\mathbb{Z}}) \to \mathrm{G}_{\mathrm{aux}}(\hat{\mathbb{Z}}^p)$, we have a canonical finite morphism

$$\mathsf{M}_{\mathcal{H}_{\mathrm{aux}}} \to \mathsf{M}_{\mathcal{H}^p_{\mathrm{aux}}} \underset{\mathbb{Z}}{\otimes} \mathbb{Q}, \tag{2.1.1.12}$$

by forgetting the level structure at p, which factors as a composition of canonical finite morphisms

$$\mathsf{M}_{\mathcal{H}_{\mathrm{aux}}} \to \mathsf{M}_{\mathcal{H}^p_{\mathrm{aux}}\mathrm{G}(\mathbb{Z}_p)} \to \mathsf{M}_{\mathcal{H}^p_{\mathrm{aux}}} \underset{\mathbb{Z}}{\otimes} \mathbb{Q}.$$

Remark 2.1.1.13. There is a subtle difference between $\mathsf{M}_{\mathcal{H}^p_{\mathrm{aux}}} \underset{\mathbb{Z}}{\otimes} \mathbb{Q}$ and the moduli problem $\mathsf{M}_{\mathcal{H}^p_{\mathrm{aux}}\mathrm{G}_{\mathrm{aux}}(\mathbb{Z}_p)}$ over $\mathsf{S}_{0,\mathrm{aux}} = \mathrm{Spec}(F_{0,\mathrm{aux}})$, because the former is not equipped with a level structure at p. Nevertheless, the canonical morphism

$$\mathsf{M}_{\mathcal{H}^p_{\mathrm{aux}}\mathrm{G}_{\mathrm{aux}}(\mathbb{Z}_p)} \to \mathsf{M}_{\mathcal{H}^p_{\mathrm{aux}}} \underset{\mathbb{Z}}{\otimes} \mathbb{Q} \tag{2.1.1.14}$$

is finite étale, which is an isomorphism at least when $\mathcal{O}_{\mathrm{aux}} \underset{\mathbb{Z}}{\otimes} \mathbb{Q}$ is simple, because p is a good prime for $\mathsf{M}_{\mathrm{G}_{\mathrm{aux}}(\hat{\mathbb{Z}}^p)}$ (by [62, Prop. 1.4.4.3] and [53, Sec. 8]). (In what follows, we will not need (2.1.1.14) to be an isomorphism.)

Proposition 2.1.1.15. *With assumptions as above, for any open compact subgroup $\mathcal{H}_{\mathrm{aux}}$ of $\mathrm{G}_{\mathrm{aux}}(\hat{\mathbb{Z}})$ such that $\mathcal{H}$ is mapped into $\mathcal{H}_{\mathrm{aux}}$ under the homomorphism $\mathrm{G}(\hat{\mathbb{Z}}) \to \mathrm{G}_{\mathrm{aux}}(\hat{\mathbb{Z}})$ given by* (2.1.1.10)*, we can define a finite morphism*

$$\mathsf{M}_{\mathcal{H}} \to \mathsf{M}_{\mathcal{H}_{\mathrm{aux}}} \tag{2.1.1.16}$$

over $\mathsf{S}_{0,\mathrm{aux}}$ such that the pullback $(A^{\triangledown}_{\mathrm{aux}}, \lambda^{\triangledown}_{\mathrm{aux}}, i^{\triangledown}_{\mathrm{aux}}, \alpha^{\triangledown}_{\mathcal{H}_{\mathrm{aux}}})$ of the tautological object over $\mathsf{M}_{\mathcal{H}_{\mathrm{aux}}}$ to $\mathsf{M}_{\mathcal{H}}$ satisfies the following properties:

(1) $A^{\triangledown}_{\mathrm{aux}}$ is isomorphic to $A^{\times a_1} \underset{\mathsf{M}_{\mathcal{H}}}{\times} (A^\vee)^{\times a_2}$ for the same integers (a_1, a_2) as in Lemma 2.1.1.1*, which is equipped with an isogeny*

$$f : A^{\vartriangle}_{\mathrm{aux}} := A^{\times(a_1+a_2)} \to A^{\triangledown}_{\mathrm{aux}}$$

induced by $\lambda : A \to A^\vee$.

(2) The polarization $\lambda^{\triangledown}_{\mathrm{aux}} : A^{\triangledown}_{\mathrm{aux}} \to A^{\triangledown,\vee}_{\mathrm{aux}}$ coincides with the composition $(f^\vee)^{-1} \circ \lambda^{\vartriangle}_{\mathrm{aux}} \circ f$ (as $\mathbb{Q}^\times$-isogenies), where $\lambda^{\vartriangle}_{\mathrm{aux}} : A^{\vartriangle}_{\mathrm{aux}} \to A^{\vartriangle,\vee}_{\mathrm{aux}}$ is induced by $\lambda : A \to A^\vee$ and $(\,\cdot\,,\,\cdot\,)_{\mathrm{aux}}$ as in (2) *of Lemma* 2.1.1.1.

(3) The isogeny $f : A^{\triangle}_{\text{aux}} \to A^{\nabla}_{\text{aux}}$ *is compatible with the* $\mathcal{O}_{\text{aux}}$*-actions defined by the* $\mathcal{O}_{\text{aux}}$*-structure* $i^{\triangle}_{\text{aux}} : \mathcal{O}_{\text{aux}} \to \operatorname{End}_{\mathsf{M}_{\mathcal{H}}}(A^{\triangle}_{\text{aux}})$ *induced by the restriction of* $i : \mathcal{O} \to \operatorname{End}_{\mathsf{M}_{\mathcal{H}}}(A)$ *to* $\mathcal{O}_{\text{aux}}$*, and by* $i^{\nabla}_{\text{aux}} : \mathcal{O}_{\text{aux}} \to \operatorname{End}_{\mathsf{M}_{\mathcal{H}}}(A^{\nabla}_{\text{aux}})$.

(4) At each geometric point $\bar{s}$ *of* $\mathsf{M}_{\mathcal{H}}$*, the level structure* $\alpha_{\mathcal{H}}$ *induces an* $\mathcal{H}$*-orbit of isomorphisms* $\hat{\alpha}_{\bar{s}} : L \underset{\mathbb{Z}}{\otimes} \hat{\mathbb{Z}} \xrightarrow{\sim} \mathrm{T}\, A_{\bar{s}}$*, which in turn induces an* $\mathcal{H}_{\text{aux}}$*-orbit of isomorphisms*

$$\hat{\alpha}_{\bar{s}}^{\oplus(a_1+a_2)} \underset{\hat{\mathbb{Z}}}{\otimes} \mathbb{A}^{\infty} : L_{\text{aux}} \underset{\mathbb{Z}}{\otimes} \mathbb{A}^{\infty} \xrightarrow{\sim} \mathrm{V}\, A^{\nabla}_{\text{aux},\bar{s}}$$

(which makes sense because $\mathcal{H}$ *is mapped into* $\mathcal{H}_{\text{aux}}$ *under the homomorphism* $\mathrm{G}(\hat{\mathbb{Z}}) \to \mathrm{G}_{\text{aux}}(\hat{\mathbb{Z}})$ *given by* (2.1.1.10)*). On the other hand, the level structure* $\alpha^{\nabla}_{\mathcal{H}_{\text{aux}}}$ *induces an* $\mathcal{H}_{\text{aux}}$*-orbit of isomorphisms*

$$\hat{\alpha}^{\nabla}_{\bar{s}} \underset{\hat{\mathbb{Z}}}{\otimes} \mathbb{A}^{\infty} : L_{\text{aux}} \underset{\mathbb{Z}}{\otimes} \mathbb{A}^{\infty} \xrightarrow{\sim} \mathrm{V}\, A^{\nabla}_{\text{aux},\bar{s}}.$$

These two $\mathcal{H}_{\text{aux}}$*-orbits of isomorphisms coincide.*

When $\mathcal{H}^p_{\text{aux}}$ *is the image of* $\mathcal{H}_{\text{aux}}$ *under the canonical homomorphism* $\mathrm{G}_{\text{aux}}(\hat{\mathbb{Z}}) \to \mathrm{G}_{\text{aux}}(\hat{\mathbb{Z}}^p)$*, by composition with* (2.1.1.12)*, we obtain a morphism*

$$\mathsf{M}_{\mathcal{H}} \to \mathsf{M}_{\mathcal{H}^p_{\text{aux}}} \underset{\mathbb{Z}}{\otimes} \mathbb{Q} \tag{2.1.1.17}$$

(over $\mathsf{S}_{0,\text{aux}}$*), which induces a finite morphism* $\mathsf{M}_{\mathcal{H}} \to [\mathsf{M}_{\mathcal{H}^p_{\text{aux}}}] \underset{\mathbb{Z}}{\otimes} \mathbb{Q}$*, such that the pullback* $(A^{\nabla}_{\text{aux}}, \lambda^{\nabla}_{\text{aux}}, i^{\nabla}_{\text{aux}}, \alpha^{\nabla}_{\mathcal{H}^p_{\text{aux}}})$ *of the tautological object over* $\mathsf{M}_{\mathcal{H}^p_{\text{aux}}}$ *to* $\mathsf{M}_{\mathcal{H}}$ *satisfies the same properties as above, with* (4) *replaced with the following:*

(4′) *At each geometric point* $\bar{s}$ *of* $\mathsf{M}_{\mathcal{H}}$*, the level structure* $\alpha_{\mathcal{H}}$ *induces an* $\mathcal{H}$*-orbit of isomorphisms* $\hat{\alpha}_{\bar{s}} : L \underset{\mathbb{Z}}{\otimes} \hat{\mathbb{Z}} \xrightarrow{\sim} \mathrm{T}\, A_{\bar{s}}$*, which in turn induces an* $\mathcal{H}^p_{\text{aux}}$*-orbit of isomorphisms*

$$\hat{\alpha}_{\bar{s}}^{\oplus(a_1+a_2)} \underset{\hat{\mathbb{Z}}}{\otimes} \mathbb{A}^{\infty,p} : L_{\text{aux}} \underset{\mathbb{Z}}{\otimes} \mathbb{A}^{\infty,p} \xrightarrow{\sim} \mathrm{V}^p\, A^{\nabla}_{\text{aux},\bar{s}}$$

(which makes sense because $\mathcal{H}$ *is mapped into* $\mathcal{H}^p_{\text{aux}}\mathrm{G}_{\text{aux}}(\mathbb{Z}_p)$ *under the homomorphism* $\mathrm{G}(\hat{\mathbb{Z}}) \to \mathrm{G}_{\text{aux}}(\hat{\mathbb{Z}})$ *given by* (2.1.1.10)*). On the other hand, the level structure* $\alpha^{\nabla}_{\mathcal{H}^p_{\text{aux}}}$ *induces an* $\mathcal{H}^p_{\text{aux}}$*-orbit of isomorphisms*

$$\hat{\alpha}^{\nabla,p}_{\bar{s}} \underset{\hat{\mathbb{Z}}^p}{\otimes} \mathbb{A}^{\infty,p} : L_{\text{aux}} \underset{\mathbb{Z}}{\otimes} \mathbb{A}^{\infty,p} \xrightarrow{\sim} \mathrm{V}^p\, A^{\nabla}_{\text{aux},\bar{s}}.$$

These two $\mathcal{H}^p_{\text{aux}}$*-orbits of isomorphisms coincide.*

Suppose we replace $\mathcal{H}_{\text{aux}}$ *with an open compact subgroup* $\mathcal{H}'_{\text{aux}}$ *such that* $\mathcal{H}'_{\text{aux}}$ *still contains the image of* $\mathcal{H}$ *under the homomorphism* $\mathrm{G}(\hat{\mathbb{Z}}) \to \mathrm{G}_{\text{aux}}(\hat{\mathbb{Z}})$ *given by* (2.1.1.10)*. Then the morphism* $\mathsf{M}_{\mathcal{H}} \to \mathsf{M}_{\mathcal{H}'_{\text{aux}}} \to \mathsf{M}_{\mathcal{H}'^{,p}_{\text{aux}}} \underset{\mathbb{Z}}{\otimes} \mathbb{Q}$ *thus obtained are compatible with* (2.1.1.16) *and* (2.1.1.17)*, and with the (compatible) canonical morphisms* $\mathsf{M}_{\mathcal{H}'_{\text{aux}}} \to \mathsf{M}_{\mathcal{H}_{\text{aux}}}$ *and* $\mathsf{M}_{\mathcal{H}'^{,p}_{\text{aux}}} \underset{\mathbb{Z}}{\otimes} \mathbb{Q} \to \mathsf{M}_{\mathcal{H}^p_{\text{aux}}} \underset{\mathbb{Z}}{\otimes} \mathbb{Q}$.

Proof. Let $A^{\triangle}_{\mathrm{aux}}$, $A^{\nabla}_{\mathrm{aux}}$, $\lambda^{\triangle}_{\mathrm{aux}}$, $\lambda^{\nabla}_{\mathrm{aux}}$, and f be defined as in (2) of Lemma 2.1.1.1 (with $S = \mathsf{M}_{\mathcal{H}}$ there). Since $\mathcal{O}_{\mathrm{aux}} \subset \mathcal{O}$ and since the involution $\star_{\mathrm{aux}}$ is the restriction of $\star$, the $\mathcal{O}$-structure $i : \mathcal{O} \to \mathrm{End}_{\mathsf{M}_{\mathcal{H}}}(A)$ of (A, λ) induces an $\mathcal{O}_{\mathrm{aux}}$-structure $i^{\triangle}_{\mathrm{aux}} : \mathcal{O}_{\mathrm{aux}} \to \mathrm{End}_{\mathsf{M}_{\mathcal{H}}}(A^{\triangle}_{\mathrm{aux}})$ of $(A^{\triangle}_{\mathrm{aux}}, \lambda^{\triangle}_{\mathrm{aux}})$, which in turn induces an $\mathcal{O}_{\mathrm{aux}} \underset{\mathbb{Z}}{\otimes} \mathbb{Q}$-structure $i^{\nabla}_{\mathrm{aux}} : \mathcal{O}_{\mathrm{aux}} \underset{\mathbb{Z}}{\otimes} \mathbb{Q} \to \mathrm{End}_{\mathsf{M}_{\mathcal{H}}}(A^{\nabla}_{\mathrm{aux}}) \underset{\mathbb{Z}}{\otimes} \mathbb{Q}$ of $(A^{\nabla}_{\mathrm{aux}}, \lambda^{\nabla}_{\mathrm{aux}})$ as in [62, Def. 1.3.3.1] by $i^{\nabla}_{\mathrm{aux}}(b) := f \circ i^{\triangle}_{\mathrm{aux}}(b) \circ f^{-1}$ for each $b \in \mathcal{O}_{\mathrm{aux}}$.

At each geometric point $\bar{s}$ of $\mathsf{M}_{\mathcal{H}}$, the level structure $\alpha_{\mathcal{H}}$ lifts to an $\mathcal{O} \underset{\mathbb{Z}}{\otimes} \hat{\mathbb{Z}}$-equivariant isomorphism $\hat{\alpha}_{\bar{s}} : L \underset{\mathbb{Z}}{\otimes} \hat{\mathbb{Z}} \xrightarrow{\sim} \mathrm{T}\, A_{\bar{s}}$, which induces an $\mathcal{O}_{\mathrm{aux}} \underset{\mathbb{Z}}{\otimes} \hat{\mathbb{Z}}$-equivariant isomorphism

$$\hat{\alpha}^{\triangle}_{\bar{s}} := \hat{\alpha}^{\oplus(a_1+a_2)}_{\bar{s}} : (L^{\oplus(a_1+a_2)}) \underset{\mathbb{Z}}{\otimes} \hat{\mathbb{Z}} \xrightarrow{\sim} \mathrm{T}\, A^{\triangle}_{\mathrm{aux},\bar{s}}$$

and an $\mathcal{O}_{\mathrm{aux}} \underset{\mathbb{Z}}{\otimes} \mathbb{A}^{\infty}$-equivariant isomorphism

$$\hat{\alpha}^{\triangle}_{\bar{s}} \underset{\hat{\mathbb{Z}}}{\otimes} \mathbb{A}^{\infty} : (L^{\oplus(a_1+a_2)}) \underset{\mathbb{Z}}{\otimes} \mathbb{A}^{\infty} \xrightarrow{\sim} \mathrm{V}\, A^{\triangle}_{\mathrm{aux},\bar{s}}$$

(all matching similitudes, implicitly). By [62, Lem. 1.3.5.2], under the isomorphism $\hat{\alpha}_{\bar{s}} \underset{\hat{\mathbb{Z}}}{\otimes} \mathbb{A}^{\infty} : L \underset{\mathbb{Z}}{\otimes} \mathbb{A}^{\infty} \xrightarrow{\sim} \mathrm{V}\, A_{\bar{s}}$, the polarization $\lambda_{\bar{s}} : A_{\bar{s}} \to A^{\vee}_{\bar{s}}$ (as an $\mathcal{O}$-equivariant isogeny) corresponds to the open compact subgroup $L^{\#} \underset{\mathbb{Z}}{\otimes} \hat{\mathbb{Z}}$ of $L \underset{\mathbb{Z}}{\otimes} \mathbb{A}^{\infty}$. Hence, the restriction of $\hat{\alpha}^{\triangle}_{\bar{s}} \underset{\hat{\mathbb{Z}}}{\otimes} \mathbb{A}^{\infty}$ induces an $\mathcal{O}_{\mathrm{aux}} \underset{\mathbb{Z}}{\otimes} \hat{\mathbb{Z}}$-equivariant isomorphism

$$\hat{\alpha}^{\nabla}_{\bar{s}} : L_{\mathrm{aux}} \underset{\mathbb{Z}}{\otimes} \hat{\mathbb{Z}} \xrightarrow{\sim} \mathrm{T}\, A^{\nabla}_{\mathrm{aux},\bar{s}}.$$

Since the choices of $\bar{s}$ and $\hat{\alpha}_{\bar{s}}$ are arbitrary, by [62, Lem. 1.3.5.2] again, the $\mathcal{O}_{\mathrm{aux}} \underset{\mathbb{Z}}{\otimes} \mathbb{Q}$-structure $i^{\nabla}_{\mathrm{aux}} : \mathcal{O}_{\mathrm{aux}} \underset{\mathbb{Z}}{\otimes} \mathbb{Q} \to \mathrm{End}_{\mathsf{M}_{\mathcal{H}}}(A^{\nabla}_{\mathrm{aux}}) \underset{\mathbb{Z}}{\otimes} \mathbb{Q}$ induces an $\mathcal{O}_{\mathrm{aux}}$-structure $i^{\nabla}_{\mathrm{aux}} : \mathcal{O}_{\mathrm{aux}} \to \mathrm{End}_{\mathsf{M}_{\mathcal{H}}}(A^{\nabla}_{\mathrm{aux}})$ of $(A^{\nabla}_{\mathrm{aux}}, \lambda^{\nabla}_{\mathrm{aux}})$. Moreover, by forgetting the factor at p, the $\hat{\alpha}^{\nabla}_{\bar{s}}$ above induces an $\mathcal{O}_{\mathrm{aux}} \underset{\mathbb{Z}}{\otimes} \hat{\mathbb{Z}}^p$-equivariant isomorphism

$$\hat{\alpha}^{\nabla,p}_{\bar{s}} : L_{\mathrm{aux}} \underset{\mathbb{Z}}{\otimes} \hat{\mathbb{Z}}^p \xrightarrow{\sim} \mathrm{T}^p\, A^{\nabla}_{\mathrm{aux},\bar{s}}.$$

Since the $\mathcal{H}$-orbit of $\hat{\alpha}_{\bar{s}}$ is $\pi_1(\mathsf{M}_{\mathcal{H}}, \bar{s})$-invariant, and since $\mathcal{H}$ is mapped into $\mathcal{H}_{\mathrm{aux}}$ (resp. $\mathcal{H}^p_{\mathrm{aux}}$) under the homomorphism $\mathrm{G}(\hat{\mathbb{Z}}) \to \mathrm{G}_{\mathrm{aux}}(\hat{\mathbb{Z}})$ (resp. $\mathrm{G}(\hat{\mathbb{Z}}) \to \mathrm{G}_{\mathrm{aux}}(\hat{\mathbb{Z}}^p)$) given by (2.1.1.10), the $\mathcal{H}_{\mathrm{aux}}$-orbit $[\hat{\alpha}^{\nabla}_{\bar{s}}]_{\mathcal{H}_{\mathrm{aux}}}$ of $\hat{\alpha}^{\nabla}_{\bar{s}}$ (resp. $\mathcal{H}^p_{\mathrm{aux}}$-orbit $[\hat{\alpha}^{\nabla,p}_{\bar{s}}]_{\mathcal{H}^p_{\mathrm{aux}}}$ of $\hat{\alpha}^{\nabla,p}_{\bar{s}}$) is $\pi_1(\mathsf{M}_{\mathcal{H}}, \bar{s})$-invariant. By [62, Prop. 1.4.3.4], the tuple $(A^{\nabla}_{\mathrm{aux}}, \lambda^{\nabla}_{\mathrm{aux}}, i^{\nabla}_{\mathrm{aux}}, [\hat{\alpha}^{\nabla}_{\bar{s}}]_{\mathcal{H}_{\mathrm{aux}}})$ (resp. $(A^{\nabla}_{\mathrm{aux}}, \lambda^{\nabla}_{\mathrm{aux}}, i^{\nabla}_{\mathrm{aux}}, [\hat{\alpha}^{\nabla,p}_{\bar{s}}]_{\mathcal{H}^p_{\mathrm{aux}}})$) defines an object $(A^{\nabla}_{\mathrm{aux}}, \lambda^{\nabla}_{\mathrm{aux}}, i^{\nabla}_{\mathrm{aux}}, \alpha^{\nabla}_{\mathcal{H}_{\mathrm{aux}}})$ (resp. $(A^{\nabla}_{\mathrm{aux}}, \lambda^{\nabla}_{\mathrm{aux}}, i^{\nabla}_{\mathrm{aux}}, \alpha^{\nabla}_{\mathcal{H}^p_{\mathrm{aux}}})$) of $\mathsf{M}_{\mathcal{H}_{\mathrm{aux}}}$ (resp. $\mathsf{M}_{\mathcal{H}^p_{\mathrm{aux}}}$) over $\mathsf{M}_{\mathcal{H}}$, which satisfies the properties described in the proposition by its very construction.

We would like to show that $\underline{\mathrm{Lie}}_{A^{\nabla}_{\mathrm{aux}}/\mathsf{M}_{\mathcal{H}}}$ with its $\mathcal{O} \underset{\mathbb{Z}}{\otimes} \mathbb{Q}$-module structure given by $i^{\nabla}_{\mathrm{aux}}$ satisfies the determinantal condition given by $(L_{\mathrm{aux}} \underset{\mathbb{Z}}{\otimes} \mathbb{R}, \langle\,\cdot\,,\,\cdot\,\rangle_{\mathrm{aux}}, h_{0,\mathrm{aux}})$ as in [62, Def. 1.3.4.1]. Since this condition is closed by definition, and is open in characteristic zero by [62, Lem. 1.2.5.11], it suffices to verify it at each $\mathbb{C}$-point t of $\mathsf{M}_{\mathcal{H}}$. Let (A_t, λ_t, i_t) and $(A^{\triangledown}_{\mathrm{aux},t}, \lambda^{\triangledown}_{\mathrm{aux},t}, i^{\triangledown}_{\mathrm{aux},t})$ denote the respective pullbacks of (A, λ, i) and $(A^{\triangledown}_{\mathrm{aux}}, \lambda^{\triangledown}_{\mathrm{aux}}, i^{\triangledown}_{\mathrm{aux}})$ to such a $\mathbb{C}$-point t. By [62, Lem. 1.2.5.11] again, since $\underline{\mathrm{Lie}}_{A/\mathsf{M}_{\mathcal{H}}}$ with its $\mathcal{O} \underset{\mathbb{Z}}{\otimes} \mathbb{Q}$-module structure given by i satisfies the determinantal condition given by $(L \underset{\mathbb{Z}}{\otimes} \mathbb{R}, \langle \cdot\,, \cdot \rangle, h_0)$, we have $\mathrm{Lie}_{A_t} \cong V_0$ as $\mathcal{O} \underset{\mathbb{Z}}{\otimes} \mathbb{C}$-modules, and it suffices to note that $\mathrm{Lie}_{A^{\triangledown}_{\mathrm{aux},t}} \cong \mathrm{Lie}_{A_t}^{\oplus a_1} \oplus \mathrm{Lie}_{A_t^{\vee}}^{\oplus a_2} \cong V_0^{\oplus(a_1+a_2)} \cong V_{0,\mathrm{aux}}$ as $\mathcal{O}_{\mathrm{aux}} \underset{\mathbb{Z}}{\otimes} \mathbb{C}$-modules (*cf.* the proof of Lemma 2.1.1.11).

Thus, we have obtained the desired morphisms (2.1.1.16) and (2.1.1.17) by the moduli interpretations of $\mathsf{M}_{\mathcal{H}_{\mathrm{aux}}}$ and $\mathsf{M}_{\mathcal{H}^p_{\mathrm{aux}}}$, which are compatible with each other under (2.1.1.12) because $\hat{\alpha}^{\triangledown,p}_{\bar{s}}$ is obtained from $\hat{\alpha}^{\triangledown}_{\bar{s}}$ by forgetting the factor at p. The morphisms (2.1.1.16) and (2.1.1.17) between algebraic stacks are schematic and finite by Lemma 2.1.1.5 (for the abelian schemes and polarizations), by [62, Prop. 1.3.3.7] (for the endomorphism structures), and by the fact that the level structures are defined by isomorphisms between finite étale group schemes. □

Lemma 2.1.1.18. *With assumptions as above, suppose the image $\mathcal{H}^p$ of $\mathcal{H}$ under the canonical morphism $\mathrm{G}(\hat{\mathbb{Z}}) \to \mathrm{G}(\hat{\mathbb{Z}}^p)$ is neat (which means, a fortiori, that $\mathcal{H}$ is also neat). Then there exists a neat open compact subgroup $\mathcal{H}^p_{\mathrm{aux}} \subset \mathrm{G}_{\mathrm{aux}}(\hat{\mathbb{Z}}^p)$ such that $\mathcal{H}$ is mapped into $\mathcal{H}_{\mathrm{aux}} = \mathcal{H}^p_{\mathrm{aux}}\mathrm{G}_{\mathrm{aux}}(\mathbb{Z}_p)$ under the injective homomorphism $\mathrm{G}(\hat{\mathbb{Z}}) \to \mathrm{G}_{\mathrm{aux}}(\hat{\mathbb{Z}})$ given by (2.1.1.10). (If we only assume that $\mathcal{H}$ is neat, then we can still find a neat open compact subgroup $\mathcal{H}'_{\mathrm{aux}} \subset \mathrm{G}_{\mathrm{aux}}(\hat{\mathbb{Z}})$ such that $\mathcal{H}$ is mapped into $\mathcal{H}'_{\mathrm{aux}}$ under $\mathrm{G}(\hat{\mathbb{Z}}) \to \mathrm{G}_{\mathrm{aux}}(\hat{\mathbb{Z}})$.)*

Proof. Let $n_0 \geq 3$ be any integer prime to p such that $\mathcal{U}^p(n_0) \subset \mathcal{H}$, and let $\mathcal{H}^p_{\mathrm{aux}}$ be generated by $\mathcal{U}^p_{\mathrm{aux}}(n_0)$ and the image of $\mathcal{H}^p$ under the injective homomorphism $\mathrm{G}(\hat{\mathbb{Z}}^p) \to \mathrm{G}_{\mathrm{aux}}(\hat{\mathbb{Z}}^p)$ given by (2.1.1.10). Then every element of $\mathcal{H}^p_{\mathrm{aux}}$ is congruent modulo n_0 to the image of some element of $\mathcal{H}^p$, which is neat by assumption. Hence, $\mathcal{H}^p_{\mathrm{aux}}$ and $\mathcal{H}_{\mathrm{aux}} = \mathcal{H}^p_{\mathrm{aux}}\mathrm{G}_{\mathrm{aux}}(\mathbb{Z}_p)$ are also neat, by definition (see [89, 0.6] or [62, Def. 1.4.1.8]), and by Serre's lemma that no nontrivial root of unity can be congruent to 1 modulo n if $n \geq 3$. (This is the same argument used in the proof of Lemma 1.2.4.45. The parenthetical remark in the statement of the lemma follows from the method of the proof.) □

2.1.2 *Auxiliary Choices of Toroidal and Minimal Compactifications*

Let us continue with the setting in Section 2.1.1.

Each symplectic admissible filtration $\mathsf{Z} = \{\mathsf{Z}_{-i}\}_i$ of $L \underset{\mathbb{Z}}{\otimes} \hat{\mathbb{Z}}$ (see Definition 1.2.1.2) induces a symplectic admissible filtration $\mathsf{Z}_{\mathrm{aux}} = \{\mathsf{Z}_{\mathrm{aux},-i}\}_i$ of $L_{\mathrm{aux}} \underset{\mathbb{Z}}{\otimes} \hat{\mathbb{Z}}$ by setting

$$\mathsf{Z}_{\mathrm{aux},-i} := \big((\mathsf{Z}_{-i}^{\oplus(a_1+a_2)}) \underset{\hat{\mathbb{Z}}}{\otimes} \mathbb{A}^{\infty}\big) \cap \big(L_{\mathrm{aux}} \underset{\mathbb{Z}}{\otimes} \hat{\mathbb{Z}}\big) \qquad (2.1.2.1)$$

as submodules of $L_{\mathrm{aux}} \underset{\mathbb{Z}}{\otimes} \mathbb{A}^{\infty}$. If Z is fully symplectic (see Definition 1.2.1.3), which means Z extends to a symplectic filtration $\mathsf{Z}_{\mathbb{A}} = \{\mathsf{Z}_{-i,\mathbb{A}}\}_i$ of $L \underset{\mathbb{Z}}{\otimes} \mathbb{A}$, then $\mathsf{Z}_{\mathrm{aux}} = \{\mathsf{Z}_{\mathrm{aux},-i}\}_i$ also extends to a filtration $\mathsf{Z}_{\mathrm{aux},\mathbb{A}} = \{\mathsf{Z}_{\mathrm{aux},-i,\mathbb{A}}\}_i$ on $L_{\mathrm{aux}} \underset{\mathbb{Z}}{\otimes} \mathbb{A}$, by setting

$$\mathsf{Z}_{\mathrm{aux},-i,\mathbb{A}} := \mathsf{Z}_{-i,\mathbb{A}}^{\oplus(a_1+a_2)}.$$

These definitions are compatible with actions of $\mathrm{G}(\mathbb{A})$ and $\mathrm{G}_{\mathrm{aux}}(\mathbb{A})$ (and with the homomorphism $\mathrm{G}(\mathbb{A}) \to \mathrm{G}_{\mathrm{aux}}(\mathbb{A})$ given by (2.1.1.10)), and are compatible with reductions modulo n for any integer $n \geq 1$. Thus, there is a well-defined assignment

$$\mathsf{Z} \mapsto \mathsf{Z}_{\mathrm{aux}}. \tag{2.1.2.2}$$

If $\Phi = (X, Y, \phi, \varphi_{-2}, \varphi_0)$ is a torus argument of Z (see Definition 1.2.1.5), then we define

$$X_{\mathrm{aux}} := X^{\oplus a_1} \oplus Y^{\oplus a_2}$$

and

$$Y_{\mathrm{aux}} := Y^{\oplus a_1} \oplus X^{\oplus a_2}.$$

Lemma 2.1.2.3. *With the setting as above, there exist canonically induced morphisms*

$$\phi_{\mathrm{aux}} : Y_{\mathrm{aux}} \hookrightarrow X_{\mathrm{aux}},$$

$$\varphi_{\mathrm{aux},-2} : \mathrm{Gr}_{-2}^{\mathsf{Z}_{\mathrm{aux}}} \xrightarrow{\sim} \mathrm{Hom}_{\hat{\mathbb{Z}}}(X_{\mathrm{aux}} \underset{\mathbb{Z}}{\otimes} \hat{\mathbb{Z}}, \hat{\mathbb{Z}}(1)),$$

and

$$\varphi_{\mathrm{aux},0} : \mathrm{Gr}_{0}^{\mathsf{Z}_{\mathrm{aux}}} \xrightarrow{\sim} Y_{\mathrm{aux}} \underset{\mathbb{Z}}{\otimes} \hat{\mathbb{Z}}$$

making

$$\Phi_{\mathrm{aux}} := (X_{\mathrm{aux}}, Y_{\mathrm{aux}}, \phi_{\mathrm{aux}}, \varphi_{\mathrm{aux},-2}, \varphi_{\mathrm{aux},0})$$

a torus argument of $\mathsf{Z}_{\mathrm{aux}}$*, and making the diagrams*

$$\begin{array}{ccc} Y^{\oplus(a_1+a_2)} & \xrightarrow[\hookrightarrow]{\mathrm{Id}_Y^{\oplus a_1} \oplus \phi^{\oplus a_2}} & Y_{\mathrm{aux}} \\ {\scriptstyle \phi \otimes (\,\cdot\,,\,\cdot\,)^*_{\mathrm{aux}}} \big\downarrow & & \big\downarrow {\scriptstyle \phi_{\mathrm{aux}}} \\ X^{\oplus(a_1+a_2)} & \xleftarrow[\mathrm{Id}_X^{\oplus a_1} \oplus \phi^{\oplus a_2}]{\hookleftarrow} & X_{\mathrm{aux}} \end{array} \tag{2.1.2.4}$$

$$\begin{array}{ccc} (\mathrm{Gr}_{-2}^{\mathsf{Z}})^{\oplus(a_1+a_2)} & \twoheadrightarrow & \mathrm{Gr}_{-2}^{\mathsf{Z}_{\mathrm{aux}}} \\ {\scriptstyle \varphi_{-2}^{\oplus(a_1+a_2)}} \big\downarrow \wr & & \wr \big\downarrow {\scriptstyle \varphi_{\mathrm{aux},-2}} \\ (\mathrm{Hom}_{\hat{\mathbb{Z}}}(X \underset{\mathbb{Z}}{\otimes} \hat{\mathbb{Z}}, \hat{\mathbb{Z}}(1)))^{\oplus(a_1+a_2)} & \xrightarrow[(\mathrm{Id}_X^{\oplus a_1} \oplus \phi^{\oplus a_2})^*]{} \!\!\!\!\!\rightarrow & \mathrm{Hom}_{\hat{\mathbb{Z}}}(X_{\mathrm{aux}} \underset{\mathbb{Z}}{\otimes} \hat{\mathbb{Z}}, \hat{\mathbb{Z}}(1)) \end{array} \tag{2.1.2.5}$$

and

$$\begin{array}{ccc} (\mathrm{Gr}_0^{\mathsf{Z}})^{\oplus(a_1+a_2)} & \hookrightarrow & \mathrm{Gr}_0^{\mathsf{Z}_{\mathrm{aux}}} \\ {\scriptstyle \varphi_0^{\oplus(a_1+a_2)}}\downarrow \wr & & \wr \downarrow {\scriptstyle \varphi_{\mathrm{aux},0}} \\ (Y \underset{\mathbb{Z}}{\otimes} \hat{\mathbb{Z}})^{\oplus(a_1+a_2)} & \underset{\mathrm{Id}_Y^{\oplus a_1} \oplus \phi^{\oplus a_2}}{\hookrightarrow} & Y_{\mathrm{aux}} \underset{\mathbb{Z}}{\otimes} \hat{\mathbb{Z}} \end{array} \qquad (2.1.2.6)$$

commutative, where $(\,\cdot\,,\,\cdot\,)^*_{\mathrm{aux}}$ *is canonically induced by* $(\,\cdot\,,\,\cdot\,)_{\mathrm{aux}}$ *as in Lemma* 2.1.1.1.

Proof. These follow from Lemma 2.1.1.1 and from the construction of the filtration $\mathsf{Z}_{\mathrm{aux},-i}$ in (2.1.2.1). □

If $\delta : \mathrm{Gr}^{\mathsf{Z}} \xrightarrow{\sim} L \underset{\mathbb{Z}}{\otimes} \hat{\mathbb{Z}}$ is a splitting of the filtration Z of $L \underset{\mathbb{Z}}{\otimes} \hat{\mathbb{Z}}$, then it induces a splitting of the filtration $\mathsf{Z}_{\mathbb{A}^\infty}$ of $L \underset{\mathbb{Z}}{\otimes} \mathbb{A}^\infty$, and hence induces a splitting δ_{aux} of the filtration $\mathsf{Z}_{\mathrm{aux}}$ of $L_{\mathrm{aux}} \underset{\mathbb{Z}}{\otimes} \mathbb{A}^\infty$.

The above assignments are compatible with the formations of orbits. That is, when $\mathcal{H}$ is mapped into $\mathcal{H}_{\mathrm{aux}}$ under the homomorphism $\mathrm{G}(\hat{\mathbb{Z}}) \to \mathrm{G}_{\mathrm{aux}}(\hat{\mathbb{Z}})$ given by (2.1.1.10), we have a well-defined assignment of representatives of cusp labels

$$(\mathsf{Z}_{\mathcal{H}}, \Phi_{\mathcal{H}}, \delta_{\mathcal{H}}) \mapsto (\mathsf{Z}_{\mathcal{H}_{\mathrm{aux}}}, \Phi_{\mathcal{H}_{\mathrm{aux}}}, \delta_{\mathcal{H}_{\mathrm{aux}}}). \qquad (2.1.2.7)$$

This assignment is also compatible with the equivalence relations among representatives of cusp labels, and induces a well-defined assignment of cusp labels

$$[(\mathsf{Z}_{\mathcal{H}}, \Phi_{\mathcal{H}}, \delta_{\mathcal{H}})] \mapsto [(\mathsf{Z}_{\mathcal{H}_{\mathrm{aux}}}, \Phi_{\mathcal{H}_{\mathrm{aux}}}, \delta_{\mathcal{H}_{\mathrm{aux}}})]. \qquad (2.1.2.8)$$

Moreover, by Lemma 2.1.2.3, tensor products with the symmetric bilinear pairing $(\,\cdot\,,\,\cdot\,)_{\mathrm{aux}}$ in Lemma 2.1.1.1 induce an embedding

$$(\mathbf{S}_{\Phi_{\mathcal{H}}})^\vee_{\mathbb{Q}} \hookrightarrow (\mathbf{S}_{\Phi_{\mathcal{H}_{\mathrm{aux}}}})^\vee_{\mathbb{Q}} : y \mapsto y \otimes (\,\cdot\,,\,\cdot\,)_{\mathrm{aux}} \qquad (2.1.2.9)$$

(by forgetting the compatibility of the pairings with $\mathcal{O}$, but retaining only the compatibility of the pairings with $\mathcal{O}_{\mathrm{aux}}$). Since $(\,\cdot\,,\,\cdot\,)_{\mathrm{aux}}$ is positive definite, the embedding

$$(\mathbf{S}_{\Phi_{\mathcal{H}}})^\vee_{\mathbb{R}} \hookrightarrow (\mathbf{S}_{\Phi_{\mathcal{H}_{\mathrm{aux}}}})^\vee_{\mathbb{R}} \qquad (2.1.2.10)$$

induced by (2.1.2.9) maps $\mathbf{P}_{\Phi_{\mathcal{H}}}$ (resp. $\mathbf{P}^+_{\Phi_{\mathcal{H}}}$) to $\mathbf{P}_{\Phi_{\mathcal{H}_{\mathrm{aux}}}}$ (resp. $\mathbf{P}^+_{\Phi_{\mathcal{H}_{\mathrm{aux}}}}$). By construction, the pullback of a nondegenerate rational polyhedral cone σ_{aux} in $\mathbf{P}_{\Phi_{\mathcal{H}_{\mathrm{aux}}}}$ (resp. $\mathbf{P}^+_{\Phi_{\mathcal{H}_{\mathrm{aux}}}}$) under (2.1.2.10) is either empty or a nondegenerate rational polyhedral cone σ in $\mathbf{P}_{\Phi_{\mathcal{H}}}$ (resp. $\mathbf{P}^+_{\Phi_{\mathcal{H}}}$). (However, σ might not be smooth when σ_{aux} is.) The dual of (2.1.2.9) gives a surjection

$$(\mathbf{S}_{\Phi_{\mathcal{H}_{\mathrm{aux}}}})_{\mathbb{Q}} := \mathbf{S}_{\Phi_{\mathcal{H}_{\mathrm{aux}}}} \underset{\mathbb{Z}}{\otimes} \mathbb{Q} \twoheadrightarrow (\mathbf{S}_{\Phi_{\mathcal{H}}})_{\mathbb{Q}} := \mathbf{S}_{\Phi_{\mathcal{H}}} \underset{\mathbb{Z}}{\otimes} \mathbb{Q}, \qquad (2.1.2.11)$$

which induces a homomorphism

$$\mathbf{S}_{\Phi_{\mathcal{H}_{\mathrm{aux}}}} \to \mathbf{S}_{\Phi_{\mathcal{H}}}. \qquad (2.1.2.12)$$

When $\mathcal{H}_{\text{aux}}$ is mapped to $\mathcal{H}^p_{\text{aux}}$ under the homomorphism $\mathrm{G}_{\text{aux}}(\hat{\mathbb{Z}}) \to \mathrm{G}_{\text{aux}}(\hat{\mathbb{Z}}^p)$, by suppressing the factors at p, we obtain compatible assignments

$$(\mathsf{Z}_{\mathcal{H}_{\text{aux}}}, \Phi_{\mathcal{H}_{\text{aux}}}, \delta_{\mathcal{H}_{\text{aux}}}) \mapsto (\mathsf{Z}_{\mathcal{H}^p_{\text{aux}}}, \Phi_{\mathcal{H}^p_{\text{aux}}}, \delta_{\mathcal{H}^p_{\text{aux}}}) \tag{2.1.2.13}$$

and

$$[(\mathsf{Z}_{\mathcal{H}_{\text{aux}}}, \Phi_{\mathcal{H}_{\text{aux}}}, \delta_{\mathcal{H}_{\text{aux}}})] \mapsto [(\mathsf{Z}_{\mathcal{H}^p_{\text{aux}}}, \Phi_{\mathcal{H}^p_{\text{aux}}}, \delta_{\mathcal{H}^p_{\text{aux}}})], \tag{2.1.2.14}$$

together with the canonical homomorphism

$$\mathbf{S}_{\Phi_{\mathcal{H}^p_{\text{aux}}}} \to \mathbf{S}_{\Phi_{\mathcal{H}_{\text{aux}}}}, \tag{2.1.2.15}$$

which induces the canonical isomorphisms

$$(\mathbf{S}_{\Phi_{\mathcal{H}^p_{\text{aux}}}})_{\mathbb{Q}} := \mathbf{S}_{\Phi_{\mathcal{H}^p_{\text{aux}}}} \underset{\mathbb{Z}}{\otimes} \mathbb{Q} \overset{\sim}{\to} (\mathbf{S}_{\Phi_{\mathcal{H}_{\text{aux}}}})_{\mathbb{Q}}, \tag{2.1.2.16}$$

$$(\mathbf{S}_{\Phi_{\mathcal{H}_{\text{aux}}}})^{\vee}_{\mathbb{Q}} \overset{\sim}{\to} (\mathbf{S}_{\Phi_{\mathcal{H}^p_{\text{aux}}}})^{\vee}_{\mathbb{Q}}, \tag{2.1.2.17}$$

and

$$(\mathbf{S}_{\Phi_{\mathcal{H}_{\text{aux}}}})^{\vee}_{\mathbb{R}} \overset{\sim}{\to} (\mathbf{S}_{\Phi_{\mathcal{H}^p_{\text{aux}}}})^{\vee}_{\mathbb{R}}. \tag{2.1.2.18}$$

By composing (2.1.2.7), (2.1.2.8), (2.1.2.15), (2.1.2.16), (2.1.2.9), and (2.1.2.10) with (2.1.2.13), (2.1.2.14), (2.1.2.12), (2.1.2.11), (2.1.2.17), and (2.1.2.18), respectively, we obtain

$$(\mathsf{Z}_{\mathcal{H}}, \Phi_{\mathcal{H}}, \delta_{\mathcal{H}}) \mapsto (\mathsf{Z}_{\mathcal{H}^p_{\text{aux}}}, \Phi_{\mathcal{H}^p_{\text{aux}}}, \delta_{\mathcal{H}^p_{\text{aux}}}), \tag{2.1.2.19}$$

$$[(\mathsf{Z}_{\mathcal{H}}, \Phi_{\mathcal{H}}, \delta_{\mathcal{H}})] \mapsto [(\mathsf{Z}_{\mathcal{H}^p_{\text{aux}}}, \Phi_{\mathcal{H}^p_{\text{aux}}}, \delta_{\mathcal{H}^p_{\text{aux}}})], \tag{2.1.2.20}$$

$$\mathbf{S}_{\Phi_{\mathcal{H}^p_{\text{aux}}}} \to \mathbf{S}_{\Phi_{\mathcal{H}}}, \tag{2.1.2.21}$$

$$(\mathbf{S}_{\Phi_{\mathcal{H}^p_{\text{aux}}}})_{\mathbb{Q}} \twoheadrightarrow (\mathbf{S}_{\Phi_{\mathcal{H}}})_{\mathbb{Q}}, \tag{2.1.2.22}$$

$$(\mathbf{S}_{\Phi_{\mathcal{H}}})^{\vee}_{\mathbb{Q}} \hookrightarrow (\mathbf{S}_{\Phi_{\mathcal{H}^p_{\text{aux}}}})^{\vee}_{\mathbb{Q}}, \tag{2.1.2.23}$$

and

$$(\mathbf{S}_{\Phi_{\mathcal{H}}})^{\vee}_{\mathbb{R}} \hookrightarrow (\mathbf{S}_{\Phi_{\mathcal{H}^p_{\text{aux}}}})^{\vee}_{\mathbb{R}}, \tag{2.1.2.24}$$

respectively.

Definition 2.1.2.25. *Let Σ (resp. Σ_{aux}, resp. Σ^p_{aux}) be a compatible choice of admissible smooth rational polyhedral cone decomposition data for $\mathsf{M}_{\mathcal{H}}$ (resp. $\mathsf{M}_{\mathcal{H}_{\text{aux}}}$, resp. $\mathsf{M}_{\mathcal{H}^p_{\text{aux}}}$). We say that Σ and Σ_{aux} (resp. Σ^p_{aux}) are* **compatible** *with each other if, for each representative $(\mathsf{Z}_{\mathcal{H}}, \Phi_{\mathcal{H}}, \delta_{\mathcal{H}})$ of cusp labels of $\mathsf{M}_{\mathcal{H}}$ with assigned representative $(\mathsf{Z}_{\mathcal{H}_{\text{aux}}}, \Phi_{\mathcal{H}_{\text{aux}}}, \delta_{\mathcal{H}_{\text{aux}}})$ (resp. $(\mathsf{Z}_{\mathcal{H}^p_{\text{aux}}}, \Phi_{\mathcal{H}^p_{\text{aux}}}, \delta_{\mathcal{H}^p_{\text{aux}}})$) of cusp labels of $\mathsf{M}_{\mathcal{H}_{\text{aux}}}$ (resp. $\mathsf{M}_{\mathcal{H}^p_{\text{aux}}}$) as in* (2.1.2.7) *(resp.* (2.1.2.19)*), the image of each $\sigma \in \Sigma_{\Phi_{\mathcal{H}}}$ under the embedding* (2.1.2.10) *(resp.* (2.1.2.24)*) is contained in some cone $\sigma_{\text{aux}} \in \Sigma_{\Phi_{\mathcal{H}_{\text{aux}}}}$ (resp. $\sigma^p_{\text{aux}} \in \Sigma_{\Phi_{\mathcal{H}^p_{\text{aux}}}}$). In this case, we say that $(\Phi_{\mathcal{H}^p_{\text{aux}}}, \delta_{\mathcal{H}^p_{\text{aux}}}, \sigma^p_{\text{aux}})$ is assigned to $(\Phi_{\mathcal{H}}, \delta_{\mathcal{H}}, \sigma)$, and (since this is compatible with the equivalence relations)* *we also say that* $[(\Phi_{\mathcal{H}^p_{\mathrm{aux}}}, \delta_{\mathcal{H}^p_{\mathrm{aux}}}, \sigma^p_{\mathrm{aux}})]$ *is assigned to* $[(\Phi_{\mathcal{H}}, \delta_{\mathcal{H}}, \sigma)]$. *We say that* Σ_{aux} *and* Σ^p_{aux} *are* **compatible** (*resp.* Σ^p_{aux} **induces** Σ_{aux}) *if, for each representative* $(\mathsf{Z}_{\mathcal{H}_{\mathrm{aux}}}, \Phi_{\mathcal{H}_{\mathrm{aux}}}, \delta_{\mathcal{H}_{\mathrm{aux}}})$ *of cusp label of* $\mathsf{M}_{\mathcal{H}_{\mathrm{aux}}}$ *with assigned representative* $(\mathsf{Z}_{\mathcal{H}^p_{\mathrm{aux}}}, \Phi_{\mathcal{H}^p_{\mathrm{aux}}}, \delta_{\mathcal{H}^p_{\mathrm{aux}}})$ *of cusp label of* $\mathsf{M}_{\mathcal{H}^p_{\mathrm{aux}}}$ *as in* (2.1.2.13), *the image of each* $\sigma_{\mathrm{aux}} \in \Sigma_{\Phi_{\mathcal{H}}}$ *under the isomorphism* (2.1.2.18) *is* **contained in** *some cone in* $\Sigma_{\Phi_{\mathcal{H}^p_{\mathrm{aux}}}}$ (*resp. is* **exactly** *some cone* σ^p_{aux} *in* $\Sigma_{\Phi_{\mathcal{H}^p_{\mathrm{aux}}}}$). *In this case, we say that* $(\Phi_{\mathcal{H}^p_{\mathrm{aux}}}, \delta_{\mathcal{H}^p_{\mathrm{aux}}}, \sigma^p_{\mathrm{aux}})$ *is assigned to* $(\Phi_{\mathcal{H}_{\mathrm{aux}}}, \delta_{\mathcal{H}_{\mathrm{aux}}}, \sigma_{\mathrm{aux}})$ (*resp.* $(\Phi_{\mathcal{H}^p_{\mathrm{aux}}}, \delta_{\mathcal{H}^p_{\mathrm{aux}}}, \sigma^p_{\mathrm{aux}})$ *is induced by* $(\Phi_{\mathcal{H}_{\mathrm{aux}}}, \delta_{\mathcal{H}_{\mathrm{aux}}}, \sigma_{\mathrm{aux}})$), *and* (*since this is compatible with the equivalence relations*) *we also say that* $[(\Phi_{\mathcal{H}^p_{\mathrm{aux}}}, \delta_{\mathcal{H}^p_{\mathrm{aux}}}, \sigma^p_{\mathrm{aux}})]$ *is assigned to* $[(\Phi_{\mathcal{H}_{\mathrm{aux}}}, \delta_{\mathcal{H}_{\mathrm{aux}}}, \sigma_{\mathrm{aux}})]$ (*resp.* $[(\Phi_{\mathcal{H}^p_{\mathrm{aux}}}, \delta_{\mathcal{H}^p_{\mathrm{aux}}}, \sigma^p_{\mathrm{aux}})]$ *is induced by* $[(\Phi_{\mathcal{H}_{\mathrm{aux}}}, \delta_{\mathcal{H}_{\mathrm{aux}}}, \sigma_{\mathrm{aux}})]$)

Lemma 2.1.2.26. *If* Σ *and* Σ_{aux} *are compatible, and if* Σ_{aux} *and* Σ^p_{aux} *are compatible, then* Σ *and* Σ^p_{aux} *are also compatible. If* Σ *and* Σ^p_{aux} *are compatible, and if* Σ^p_{aux} *induces* Σ_{aux}, *then* Σ *and* Σ_{aux} *are also compatible.*

Proof. These follow immediately from the definitions. □

Lemma 2.1.2.27. *Suppose* Σ_{aux} *and* Σ^p_{aux} *are* **compatible**. *Then the morphism* (2.1.1.12) *canonically extends to a morphism*

$$\mathsf{M}^{\mathrm{tor}}_{\mathcal{H}_{\mathrm{aux}},\Sigma_{\mathrm{aux}}} \to \mathsf{M}^{\mathrm{tor}}_{\mathcal{H}^p_{\mathrm{aux}},\Sigma^p_{\mathrm{aux}}} \underset{\mathbb{Z}}{\otimes} \mathbb{Q}, \tag{2.1.2.28}$$

where $\mathsf{M}^{\mathrm{tor}}_{\mathcal{H}_{\mathrm{aux}},\Sigma_{\mathrm{aux}}}$ *and* $\mathsf{M}^{\mathrm{tor}}_{\mathcal{H}^p_{\mathrm{aux}},\Sigma^p_{\mathrm{aux}}}$ *are as in Theorem* 1.3.1.3 *and [62, Thm. 6.4.1.1], such that the tautological tuple* $(G_{\mathrm{aux}}, \lambda_{\mathrm{aux}}, i_{\mathrm{aux}}, \alpha_{\mathcal{H}_{\mathrm{aux}}})$ *over* $\mathsf{M}^{\mathrm{tor}}_{\mathcal{H}_{\mathrm{aux}},\Sigma_{\mathrm{aux}}}$ *induces* (*by forgetting the factor at* p *of* $\alpha_{\mathcal{H}_{\mathrm{aux}}}$) *the pullback of the tautological tuple* $(G_{\mathrm{aux}}, \lambda_{\mathrm{aux}}, i_{\mathrm{aux}}, \alpha_{\mathcal{H}^p_{\mathrm{aux}}})$ *over* $\mathsf{M}^{\mathrm{tor}}_{\mathcal{H}_{\mathrm{aux}},\Sigma_{\mathrm{aux}}}$ *over* $\mathsf{M}^{\mathrm{tor}}_{\mathcal{H}^p_{\mathrm{aux}},\Sigma^p_{\mathrm{aux}}}$ (*denoted similarly, by abuse of notation*), *mapping the* $[(\Phi_{\mathcal{H}_{\mathrm{aux}}}, \delta_{\mathcal{H}_{\mathrm{aux}}}, \sigma_{\mathrm{aux}})]$*-stratum* $\mathsf{Z}_{[(\Phi_{\mathcal{H}_{\mathrm{aux}}}, \delta_{\mathcal{H}_{\mathrm{aux}}}, \sigma_{\mathrm{aux}})]}$ *of* $\mathsf{M}^{\mathrm{tor}}_{\mathcal{H}_{\mathrm{aux}},\Sigma_{\mathrm{aux}}}$ *to the* $[(\Phi_{\mathcal{H}^p_{\mathrm{aux}}}, \delta_{\mathcal{H}^p_{\mathrm{aux}}}, \sigma^p_{\mathrm{aux}})]$*-stratum* $\mathsf{Z}_{[(\Phi_{\mathcal{H}^p_{\mathrm{aux}}}, \delta_{\mathcal{H}^p_{\mathrm{aux}}}, \sigma^p_{\mathrm{aux}})]}$ *of* $\mathsf{M}^{\mathrm{tor}}_{\mathcal{H}^p_{\mathrm{aux}},\Sigma^p_{\mathrm{aux}}}$ *when* $[(\Phi_{\mathcal{H}^p_{\mathrm{aux}}}, \delta_{\mathcal{H}^p_{\mathrm{aux}}}, \sigma^p_{\mathrm{aux}})]$ *is assigned to* $[(\Phi_{\mathcal{H}_{\mathrm{aux}}}, \delta_{\mathcal{H}_{\mathrm{aux}}}, \sigma_{\mathrm{aux}})]$.

Proof. This follows by comparing the universal properties of $\mathsf{M}^{\mathrm{tor}}_{\mathcal{H}_{\mathrm{aux}},\Sigma_{\mathrm{aux}}}$ and $\mathsf{M}^{\mathrm{tor}}_{\mathcal{H}^p_{\mathrm{aux}},\Sigma^p_{\mathrm{aux}}}$, as in (6) of Theorem 1.3.1.3 and [62, Thm. 6.4.1.1]. □

Proposition 2.1.2.29. *With assumptions as in Proposition* 2.1.1.15, *there exist compatible choices* Σ, Σ_{aux}, *and* Σ^p_{aux} *of admissible smooth rational polyhedral cone decomposition data for* $\mathsf{M}_{\mathcal{H}}$, $\mathsf{M}_{\mathcal{H}_{\mathrm{aux}}}$, *and* $\mathsf{M}_{\mathcal{H}^p_{\mathrm{aux}}}$, *respectively, such that* Σ, Σ_{aux}, *and* Σ^p_{aux} *are compatible with each other as in Definition* 2.1.2.25, *and such that the morphism* (2.1.1.17) *canonically extends to a morphism*

$$\mathsf{M}^{\mathrm{tor}}_{\mathcal{H},\Sigma} \to \mathsf{M}^{\mathrm{tor}}_{\mathcal{H}_{\mathrm{aux}},\Sigma_{\mathrm{aux}}}, \tag{2.1.2.30}$$

which induces by composition with (2.1.2.28) *a morphism*

$$\mathsf{M}^{\mathrm{tor}}_{\mathcal{H},\Sigma} \to \mathsf{M}^{\mathrm{tor}}_{\mathcal{H}^p_{\mathrm{aux}},\Sigma^p_{\mathrm{aux}}} \underset{\mathbb{Z}}{\otimes} \mathbb{Q}. \tag{2.1.2.31}$$

The morphism (2.1.2.30) (*resp.* (2.1.2.31)) *maps the* $[(\Phi_{\mathcal{H}}, \delta_{\mathcal{H}}, \sigma)]$*-stratum* $\mathsf{Z}_{[(\Phi_{\mathcal{H}}, \delta_{\mathcal{H}}, \sigma)]}$ *of* $\mathsf{M}^{\mathrm{tor}}_{\mathcal{H},\Sigma}$ *to the* $[(\Phi_{\mathcal{H}_{\mathrm{aux}}}, \delta_{\mathcal{H}_{\mathrm{aux}}}, \sigma_{\mathrm{aux}})]$*-stratum* $\mathsf{Z}_{[(\Phi_{\mathcal{H}_{\mathrm{aux}}}, \delta_{\mathcal{H}_{\mathrm{aux}}}, \sigma_{\mathrm{aux}})]}$ *of*

$\mathsf{M}^{\mathrm{tor}}_{\mathcal{H}_{\mathrm{aux}},\Sigma_{\mathrm{aux}}}$ *(resp.* $[(\Phi_{\mathcal{H}^p_{\mathrm{aux}}},\delta_{\mathcal{H}^p_{\mathrm{aux}}},\sigma^p_{\mathrm{aux}})]$*-stratum* $\mathsf{Z}_{[(\Phi_{\mathcal{H}^p_{\mathrm{aux}}},\delta_{\mathcal{H}^p_{\mathrm{aux}}},\sigma^p_{\mathrm{aux}})]}$ *of* $\mathsf{M}^{\mathrm{tor}}_{\mathcal{H}^p_{\mathrm{aux}},\Sigma^p_{\mathrm{aux}}}$*) when* $(\Phi_{\mathcal{H}_{\mathrm{aux}}},\delta_{\mathcal{H}_{\mathrm{aux}}},\sigma_{\mathrm{aux}})$ *(resp.* $(\Phi_{\mathcal{H}^p_{\mathrm{aux}}},\delta_{\mathcal{H}^p_{\mathrm{aux}}},\sigma^p_{\mathrm{aux}})$*) is assigned to* $(\Phi_{\mathcal{H}},\delta_{\mathcal{H}},\sigma)$ *(see Definition* 2.1.2.25*). Let* $(G,\lambda,i,\alpha_{\mathcal{H}})$ *(resp.* $(G_{\mathrm{aux}},\lambda_{\mathrm{aux}},i_{\mathrm{aux}},\alpha_{\mathcal{H}_{\mathrm{aux}}})$*, resp.* $(G_{\mathrm{aux}},\lambda_{\mathrm{aux}},i_{\mathrm{aux}},\alpha_{\mathcal{H}^p_{\mathrm{aux}}})$*) denote the degenerating family of type* $\mathsf{M}_{\mathcal{H}}$ *(resp.* $\mathsf{M}_{\mathcal{H}_{\mathrm{aux}}}$*, resp.* $\mathsf{M}_{\mathcal{H}^p_{\mathrm{aux}}}$*) over* $\mathsf{M}^{\mathrm{tor}}_{\mathcal{H},\Sigma}$ *(resp.* $\mathsf{M}^{\mathrm{tor}}_{\mathcal{H}_{\mathrm{aux}},\Sigma_{\mathrm{aux}}}$*; resp.* $\mathsf{M}^{\mathrm{tor}}_{\mathcal{H}^p_{\mathrm{aux}},\Sigma^p_{\mathrm{aux}}}$*, denoted similarly by abuse of notation) as in Theorem* 1.3.1.3 *(or rather [62, Thm. 6.4.1.1]). Then the pullback of* G_{aux} *(from either* $\mathsf{M}^{\mathrm{tor}}_{\mathcal{H}_{\mathrm{aux}},\Sigma_{\mathrm{aux}}}$ *or* $\mathsf{M}^{\mathrm{tor}}_{\mathcal{H}^p_{\mathrm{aux}},\Sigma^p_{\mathrm{aux}}}$*) to* $\mathsf{M}^{\mathrm{tor}}_{\mathcal{H},\Sigma}$ *is isomorphic to* $G^{\times a_1}\underset{\mathsf{M}^{\mathrm{tor}}_{\mathcal{H},\Sigma}}{\times}(G^\vee)^{\times a_2}$*, and satisfies analogues of the characterizing properties in Proposition* 2.1.1.15*. (In fact, by [92, IX, 1.4], [28, Ch. I, Prop. 2.7], or [62, Prop. 3.3.1.5], the pullbacks of* $(G_{\mathrm{aux}},\lambda_{\mathrm{aux}},i_{\mathrm{aux}},\alpha_{\mathcal{H}_{\mathrm{aux}}})$ *and* $(G_{\mathrm{aux}},\lambda_{\mathrm{aux}},i_{\mathrm{aux}},\alpha_{\mathcal{H}^p_{\mathrm{aux}}})$ *are determined up to unique isomorphisms by their restrictions to* $\mathsf{M}_{\mathcal{H}}$*, which are then characterized by the properties stated in Proposition* 2.1.1.15.*)*

Proof. As in (2) of Lemma 2.1.1.1 and as in the proof of Proposition 2.1.1.15, let $G^{\triangle}_{\mathrm{aux}} := G^{\times(a_1+a_2)}$, $G^{\triangle,\vee}_{\mathrm{aux}} := (G^\vee)^{\times(a_1+a_2)}$, $G^{\nabla}_{\mathrm{aux}} := G^{\times a_1}\underset{\mathsf{M}^{\mathrm{tor}}_{\mathcal{H},\Sigma}}{\times}(G^\vee)^{\times a_2}$, and $G^{\nabla,\vee}_{\mathrm{aux}} := (G^\vee)^{\times a_1}\underset{\mathsf{M}^{\mathrm{tor}}_{\mathcal{H},\Sigma}}{\times}G^{\times a_2}$, which are fiber products over $\mathsf{M}^{\mathrm{tor}}_{\mathcal{H},\Sigma}$, whose pullbacks to $\mathsf{M}_{\mathcal{H}}$ can be canonically identified with $A^{\triangle}_{\mathrm{aux}}$, $A^{\triangle,\vee}_{\mathrm{aux}}$, $A^{\nabla}_{\mathrm{aux}}$, and $A^{\nabla,\vee}_{\mathrm{aux}}$, respectively. Let $f := \mathrm{Id}^{a_1}_G\underset{\mathsf{M}^{\mathrm{tor}}_{\mathcal{H},\Sigma}}{\times}\lambda^{a_2} : G^{\triangle}_{\mathrm{aux}}\to G^{\nabla}_{\mathrm{aux}}$ and $f^\vee := \mathrm{Id}^{a_1}_{G^\vee}\underset{\mathsf{M}^{\mathrm{tor}}_{\mathcal{H},\Sigma}}{\times}\lambda^{a_2} : G^{\nabla,\vee}_{\mathrm{aux}}\to G^{\triangle,\vee}_{\mathrm{aux}}$, whose pullbacks to $\mathsf{M}_{\mathcal{H}}$ are dual isogenies of each other. Let $\lambda^{\triangle}_{\mathrm{aux}}$ be defined by λ and the morphism $(\,\cdot\,,\,\cdot\,)^*_{\mathrm{aux}}$ as in Lemma 2.1.1.1, and let $i^{\triangle}_{\mathrm{aux}} : \mathcal{O}_{\mathrm{aux}}\to\mathrm{End}_{\mathsf{M}^{\mathrm{tor}}_{\mathcal{H},\Sigma}}(G^{\triangle}_{\mathrm{aux}})$ be induced by the restriction of i to $\mathcal{O}_{\mathrm{aux}}$. By (2) of Lemma 2.1.1.1, and by [92, IX, 1.4], [28, Ch. I, Prop. 2.7], or [62, Prop. 3.3.1.5], $\lambda^{\nabla}_{\mathrm{aux}} := (f^\vee)^{-1}\circ\lambda^{\triangle}_{\mathrm{aux}}\circ f : G^{\nabla}_{\mathrm{aux}}\to G^{\nabla,\vee}_{\mathrm{aux}}$ is an isogeny (not just a $\mathbb{Q}^\times$-isogeny) of degree prime to p whose pullback to $\mathsf{M}_{\mathcal{H}}$ is a polarization, and we have an $i^{\nabla}_{\mathrm{aux}} : \mathcal{O}_{\mathrm{aux}}\to\mathrm{End}_{\mathsf{M}^{\mathrm{tor}}_{\mathcal{H},\Sigma}}(G^{\nabla}_{\mathrm{aux}})$ uniquely extending its pullback to $\mathsf{M}_{\mathcal{H}}$. Together with the $\alpha^{\nabla}_{\mathcal{H}_{\mathrm{aux}}}$ (resp. $\alpha^{\nabla}_{\mathcal{H}^p_{\mathrm{aux}}}$) over $\mathsf{M}_{\mathcal{H}}$ constructed in the proof of Proposition 2.1.1.15, we obtain a degenerating family $(G^{\nabla}_{\mathrm{aux}},\lambda^{\nabla}_{\mathrm{aux}},i^{\nabla}_{\mathrm{aux}},\alpha^{\nabla}_{\mathcal{H}_{\mathrm{aux}}})$ (resp. $(G^{\nabla}_{\mathrm{aux}},\lambda^{\nabla}_{\mathrm{aux}},i^{\nabla}_{\mathrm{aux}},\alpha^{\nabla}_{\mathcal{H}^p_{\mathrm{aux}}})$) of type $\mathsf{M}_{\mathcal{H}_{\mathrm{aux}}}$ (resp. $\mathsf{M}_{\mathcal{H}^p_{\mathrm{aux}}}$) over $\mathsf{M}^{\mathrm{tor}}_{\mathcal{H},\Sigma}$.

To show that $(G^{\nabla}_{\mathrm{aux}},\lambda^{\nabla}_{\mathrm{aux}},i^{\nabla}_{\mathrm{aux}},\alpha^{\nabla}_{\mathcal{H}_{\mathrm{aux}}})\to\mathsf{M}^{\mathrm{tor}}_{\mathcal{H},\Sigma}$ is canonically isomorphic to the pullback of $(G_{\mathrm{aux}},\lambda_{\mathrm{aux}},i_{\mathrm{aux}},\alpha_{\mathcal{H}_{\mathrm{aux}}})\to\mathsf{M}^{\mathrm{tor}}_{\mathcal{H}_{\mathrm{aux}},\Sigma_{\mathrm{aux}}}$ under a canonically determined morphism (2.1.2.30), we need to verify the condition as in [62, Thm. 6.4.1.1(6)] (*cf.* (6) of Theorem 1.3.1.3).

In the association of degeneration data, over any $\mathrm{Spec}(V)\to\mathsf{M}^{\mathrm{tor}}_{\mathcal{H},\Sigma}$ such that V is a complete discrete valuation ring with algebraically closed residue field k and valuation $v : \mathrm{Inv}(V)\to\mathbb{Z}$, and such that $\mathrm{Spec}(\mathrm{Frac}(V))$ is mapped to a point s of $\mathsf{M}_{\mathcal{H}}$, and for any lifting $\hat{\alpha}_{\bar{s}} : L\underset{\mathbb{Z}}{\otimes}\hat{\mathbb{Z}}\overset{\sim}{\to}\mathrm{T}\,G_{\bar{s}}$ at a geometric point $\bar{s}$ above s, the (noncanonical) filtration Z is defined to be the pullback of the geometric filtration $0\subset\mathrm{T}\,T_{\bar{s}}\subset\mathrm{T}\,G^{\natural}_{\bar{s}}\subset\mathrm{T}\,G_{\bar{s}}$, whose $\mathcal{H}$-orbit $\mathsf{Z}_{\mathcal{H}}$ is uniquely determined by $\alpha_{\mathcal{H}}$. If we define $\hat{\alpha}^{\nabla}_{\bar{s}} : L_{\mathrm{aux}}\underset{\mathbb{Z}}{\otimes}\hat{\mathbb{Z}}\overset{\sim}{\to}\mathrm{T}\,G^{\nabla}_{\mathrm{aux},\bar{s}}$ by $\hat{\alpha}_{\bar{s}}$ as in the proof of Proposition 2.1.1.15, then the filtration $\mathsf{Z}_{\mathrm{aux}}$ defined by Z as in (2.1.2.1) agrees with the pullback of the geometric filtration $0 \subset \mathrm{T}\, T^{\triangledown}_{\mathrm{aux},\bar{s}} \subset \mathrm{T}\, G^{\triangledown,\natural}_{\mathrm{aux},\bar{s}} \subset \mathrm{T}\, G^{\triangledown}_{\mathrm{aux},\bar{s}}$, because this last filtration on $\mathrm{T}\, G^{\triangledown}_{\mathrm{aux},\bar{s}}$ is induced by the filtration $0 \subset \mathrm{V}\, T^{\triangledown}_{\mathrm{aux},\bar{s}} \subset \mathrm{V}\, G^{\triangledown,\natural}_{\mathrm{aux},\bar{s}} \subset \mathrm{V}\, G^{\triangledown}_{\mathrm{aux},\bar{s}}$ on $\mathrm{V}\, G^{\triangledown}_{\mathrm{aux},\bar{s}}$, whose pullback under the isomorphism $\mathrm{V}(f) : \mathrm{V}\, G^{\triangle}_{\mathrm{aux},\bar{s}} \xrightarrow{\sim} \mathrm{V}\, G^{\triangledown}_{\mathrm{aux},\bar{s}}$ agrees with the filtration $0 \subset \mathrm{V}\, T^{\triangle}_{\mathrm{aux},\bar{s}} \subset \mathrm{V}\, G^{\triangle,\natural}_{\mathrm{aux},\bar{s}} \subset \mathrm{V}\, G^{\triangle}_{\mathrm{aux},\bar{s}}$ on $\mathrm{V}\, G^{\triangle}_{\mathrm{aux},\bar{s}}$ (which naturally agrees with the filtration induced by $0 \subset \mathrm{V}\, T_{\bar{s}} \subset \mathrm{V}\, G^{\natural}_{\bar{s}} \subset \mathrm{V}\, G_{\bar{s}}$ on $\mathrm{V}\, G_{\bar{s}}$).

Suppose, under the equivalence of categories in [62, Thm. 5.3.1.19],

$$(B, \lambda_B, i_B, X, Y, \phi, c, c^\vee, \tau, [\alpha^\natural_{\mathcal{H}}]) \tag{2.1.2.32}$$

is the object of $\mathrm{DD}_{\mathrm{PEL},\mathsf{M}_{\mathcal{H}}}(V)$ associated with the object of $\mathrm{DEG}_{\mathrm{PEL},\mathsf{M}_{\mathcal{H}}}(V)$ defined by the pullback of the degenerating family $(G, \lambda, i, \alpha_{\mathcal{H}}) \to \mathsf{M}^{\mathrm{tor}}_{\mathcal{H},\Sigma}$ under $\mathrm{Spec}(V) \to \mathsf{M}^{\mathrm{tor}}_{\mathcal{H},\Sigma}$, and suppose

$$(B_{\mathrm{aux}}, \lambda_{B_{\mathrm{aux}}}, i_{B_{\mathrm{aux}}}, X_{\mathrm{aux}}, Y_{\mathrm{aux}}, c_{\mathrm{aux}}, c^\vee_{\mathrm{aux}}, \tau_{\mathrm{aux}}, [\alpha^\natural_{\mathcal{H}_{\mathrm{aux}}}]) \tag{2.1.2.33}$$

is the object of $\mathrm{DD}_{\mathrm{PEL},\mathsf{M}_{\mathcal{H}_{\mathrm{aux}}}}(V)$ associated with the object of $\mathrm{DEG}_{\mathrm{PEL},\mathsf{M}_{\mathcal{H}_{\mathrm{aux}}}}(V)$ defined by the pullback of the degenerating family $(G^{\triangledown}_{\mathrm{aux}}, \lambda^{\triangledown}_{\mathrm{aux}}, i^{\triangledown}_{\mathrm{aux}}, \alpha^{\triangledown}_{\mathcal{H}_{\mathrm{aux}}}) \to \mathsf{M}^{\mathrm{tor}}_{\mathcal{H},\Sigma}$ under $\mathrm{Spec}(V) \to \mathsf{M}^{\mathrm{tor}}_{\mathcal{H},\Sigma}$. Then (2.1.2.33) is induced by (2.1.2.32) in a sense that can be made precise, which implies in particular the following: Under the assignment (2.1.2.8), the cusp label $[(\mathsf{Z}_{\mathcal{H}}, \Phi_{\mathcal{H}} = (X, Y, \phi, \varphi_{-2,\mathcal{H}}, \varphi_{0,\mathcal{H}}), \delta_{\mathcal{H}})]$ determined by (2.1.2.32) gives the cusp label $[(\mathsf{Z}_{\mathcal{H}_{\mathrm{aux}}}, \Phi_{\mathcal{H}_{\mathrm{aux}}}, \delta_{\mathcal{H}_{\mathrm{aux}}})]$ determined by (2.1.2.33). If we fix a representative $(\mathsf{Z}_{\mathcal{H}}, \Phi_{\mathcal{H}}, \delta_{\mathcal{H}})$ of $[(\mathsf{Z}_{\mathcal{H}}, \Phi_{\mathcal{H}}, \delta_{\mathcal{H}})]$, then the assignment (2.1.2.7) gives a representative $(\mathsf{Z}_{\mathcal{H}_{\mathrm{aux}}}, \Phi_{\mathcal{H}_{\mathrm{aux}}}, \delta_{\mathcal{H}_{\mathrm{aux}}})$ of $[(\mathsf{Z}_{\mathcal{H}_{\mathrm{aux}}}, \Phi_{\mathcal{H}_{\mathrm{aux}}}, \delta_{\mathcal{H}_{\mathrm{aux}}})]$. With such choices of $(\mathsf{Z}_{\mathcal{H}}, \Phi_{\mathcal{H}}, \delta_{\mathcal{H}})$ and $(\mathsf{Z}_{\mathcal{H}_{\mathrm{aux}}}, \Phi_{\mathcal{H}_{\mathrm{aux}}}, \delta_{\mathcal{H}_{\mathrm{aux}}})$, if $B : \mathbf{S}_{\Phi_{\mathcal{H}}} \to \mathrm{Inv}(V)$ and $B_{\mathrm{aux}} : \mathbf{S}_{\Phi_{\mathcal{H}_{\mathrm{aux}}}} \to \mathrm{Inv}(V)$ are determined by (2.1.2.32) and (2.1.2.33), respectively, then (2.1.2.9) maps $\upsilon \circ B : \mathbf{S}_{\Phi_{\mathcal{H}}} \to \mathbb{Z} \hookrightarrow \mathbb{Q}$ to $\upsilon \circ B_{\mathrm{aux}} : \mathbf{S}_{\Phi_{\mathcal{H}_{\mathrm{aux}}}} \to \mathbb{Z} \hookrightarrow \mathbb{Q}$ because $\lambda^{\triangle}_{\mathrm{aux}}$ is induced by λ and $(\,\cdot\,,\,\cdot\,)_{\mathrm{aux}}$. If $\upsilon \circ B$ defines an element of $\sigma \in \Sigma_{\Phi_{\mathcal{H}}}$, and if the image of σ under (2.1.2.24) is contained in some $\sigma_{\mathrm{aux}} \in \Sigma_{\Phi_{\mathcal{H}_{\mathrm{aux}}}}$, then $\upsilon \circ B_{\mathrm{aux}}$ defines an element of σ_{aux}.

Thus, if Σ and Σ_{aux} are compatible with each other as in Definition 2.1.2.25, by considering all morphisms $\mathrm{Spec}(V) \to \mathsf{M}^{\mathrm{tor}}_{\mathcal{H},\Sigma}$ as above, we see that $(G^{\triangledown}_{\mathrm{aux}}, \lambda^{\triangledown}_{\mathrm{aux}}, i^{\triangledown}_{\mathrm{aux}}, \alpha^{\triangledown}_{\mathcal{H}_{\mathrm{aux}}})$ satisfies the condition as in [62, Thm. 6.4.1.1(6)] (*cf.* (6) of Theorem 1.3.1.3), as desired.

The case for $(G^{\triangledown}_{\mathrm{aux}}, \lambda^{\triangledown}_{\mathrm{aux}}, i^{\triangledown}_{\mathrm{aux}}, \alpha^{\triangledown}_{\mathcal{H}^p_{\mathrm{aux}}}) \to \mathsf{M}^{\mathrm{tor}}_{\mathcal{H},\Sigma}$ and (2.1.2.31) is similar, by suppressing the factors at p in the above argument (and hence the obtained (2.1.2.31) is tautologically compatible with (2.1.2.30)). (Or one may just apply Lemma 2.1.2.27.) □

Consider the invertible sheaves

$$\omega_{\mathsf{M}^{\mathrm{tor}}_{\mathcal{H},\Sigma}} := \wedge^{\mathrm{top}}\, \underline{\mathrm{Lie}}^\vee_{G/\mathsf{M}^{\mathrm{tor}}_{\mathcal{H},\Sigma}} = \wedge^{\mathrm{top}}\, e^*_G \Omega^1_{G/\mathsf{M}^{\mathrm{tor}}_{\mathcal{H},\Sigma}}$$

over $\mathsf{M}^{\mathrm{tor}}_{\mathcal{H},\Sigma}$,

$$\omega_{\mathsf{M}^{\mathrm{tor}}_{\mathcal{H}_{\mathrm{aux}},\Sigma_{\mathrm{aux}}}} := \wedge^{\mathrm{top}}\, \underline{\mathrm{Lie}}^\vee_{G_{\mathrm{aux}}/\mathsf{M}^{\mathrm{tor}}_{\mathcal{H}_{\mathrm{aux}},\Sigma_{\mathrm{aux}}}} = \wedge^{\mathrm{top}}\, e^*_{G_{\mathrm{aux}}} \Omega^1_{G_{\mathrm{aux}}/\mathsf{M}^{\mathrm{tor}}_{\mathcal{H}_{\mathrm{aux}},\Sigma_{\mathrm{aux}}}}$$

over $\mathsf{M}^{\mathrm{tor}}_{\mathcal{H}_{\mathrm{aux}},\Sigma_{\mathrm{aux}}}$, and

$$\omega_{\mathsf{M}^{\mathrm{tor}}_{\mathcal{H}^p_{\mathrm{aux}},\Sigma^p_{\mathrm{aux}}}} := \wedge^{\mathrm{top}}\, \underline{\mathrm{Lie}}^{\vee}_{G_{\mathrm{aux}}/\mathsf{M}^{\mathrm{tor}}_{\mathcal{H}^p_{\mathrm{aux}},\Sigma^p_{\mathrm{aux}}}} = \wedge^{\mathrm{top}}\, e^*_{G_{\mathrm{aux}}}\Omega^1_{G_{\mathrm{aux}}/\mathsf{M}^{\mathrm{tor}}_{\mathcal{H}^p_{\mathrm{aux}},\Sigma^p_{\mathrm{aux}}}}$$

over $\mathsf{M}^{\mathrm{tor}}_{\mathcal{H}^p_{\mathrm{aux}},\Sigma^p_{\mathrm{aux}}}$.

We shall denote the pullback of $\omega_{\mathsf{M}^{\mathrm{tor}}_{\mathcal{H},\Sigma}}$ (resp. $\omega_{\mathsf{M}^{\mathrm{tor}}_{\mathcal{H}_{\mathrm{aux}}}}$, resp. $\omega_{\mathsf{M}^{\mathrm{tor}}_{\mathcal{H}^p_{\mathrm{aux}}}}$) to $\mathsf{M}_{\mathcal{H}}$ (resp. $\mathsf{M}_{\mathcal{H}_{\mathrm{aux}}}$, resp. $\mathsf{M}_{\mathcal{H}^p_{\mathrm{aux}}}$) by $\omega_{\mathsf{M}_{\mathcal{H}}}$ (resp. $\omega_{\mathsf{M}_{\mathcal{H}_{\mathrm{aux}}}}$, resp. $\omega_{\mathsf{M}_{\mathcal{H}^p_{\mathrm{aux}}}}$), which is independent of the choice of Σ (resp. Σ_{aux}, resp. Σ^p_{aux}).

Lemma 2.1.2.34. *The pullback of* $\omega_{\mathsf{M}^{\mathrm{tor}}_{\mathcal{H}^p_{\mathrm{aux}},\Sigma^p_{\mathrm{aux}}}}$ *to* $\mathsf{M}^{\mathrm{tor}}_{\mathcal{H}_{\mathrm{aux}},\Sigma_{\mathrm{aux}}}$ *under* (2.1.2.28) *is canonically isomorphic to* $\omega_{\mathsf{M}^{\mathrm{tor}}_{\mathcal{H}_{\mathrm{aux}},\Sigma_{\mathrm{aux}}}}$.

Proof. This follows from Lemma 2.1.2.27 and the definitions of $\omega_{\mathsf{M}^{\mathrm{tor}}_{\mathcal{H}_{\mathrm{aux}},\Sigma_{\mathrm{aux}}}}$ and $\omega_{\mathsf{M}^{\mathrm{tor}}_{\mathcal{H}^p_{\mathrm{aux}},\Sigma^p_{\mathrm{aux}}}}$. □

Lemma 2.1.2.35. *There exists an integer* $1 \leq a_0 \leq 2$ *such that the pullback of* $\omega^{\otimes a_0}_{\mathsf{M}^{\mathrm{tor}}_{\mathcal{H}_{\mathrm{aux}},\Sigma_{\mathrm{aux}}}}$ *(resp.* $\omega^{\otimes a_0}_{\mathsf{M}^{\mathrm{tor}}_{\mathcal{H}^p_{\mathrm{aux}},\Sigma^p_{\mathrm{aux}}}}$*) to* $\mathsf{M}^{\mathrm{tor}}_{\mathcal{H},\Sigma}$ *under the morphism* (2.1.2.30) *(resp.* (2.1.2.31)*) is isomorphic to* $\omega^{\otimes a}_{\mathsf{M}^{\mathrm{tor}}_{\mathcal{H},\Sigma}}$*, where* $a := a_0(a_1 + a_2)$*. We may take* $a_0 = 1$ *when* a_2 *is even.*

We shall henceforth fix a choice of a_0.

Proof of Lemma 2.1.2.35. Consider also the invertible sheaf

$$\omega'_{\mathsf{M}^{\mathrm{tor}}_{\mathcal{H},\Sigma}} := \wedge^{\mathrm{top}}\, \underline{\mathrm{Lie}}^{\vee}_{G^\vee/\mathsf{M}^{\mathrm{tor}}_{\mathcal{H},\Sigma}} = \wedge^{\mathrm{top}}\, e^*_{G^\vee}\Omega^1_{G^\vee/\mathsf{M}^{\mathrm{tor}}_{\mathcal{H},\Sigma}}.$$

By Proposition 2.1.2.29, the pullback of $\omega_{\mathsf{M}^{\mathrm{tor}}_{\mathcal{H}_{\mathrm{aux}},\Sigma_{\mathrm{aux}}}}$ (resp. $\omega_{\mathsf{M}^{\mathrm{tor}}_{\mathcal{H}^p_{\mathrm{aux}},\Sigma^p_{\mathrm{aux}}}}$) to $\mathsf{M}^{\mathrm{tor}}_{\mathcal{H},\Sigma}$ is canonically isomorphic to

$$\omega^{\otimes a_1}_{\mathsf{M}^{\mathrm{tor}}_{\mathcal{H},\Sigma}} \underset{\mathsf{M}^{\mathrm{tor}}_{\mathcal{H},\Sigma}}{\otimes} (\omega'_{\mathsf{M}^{\mathrm{tor}}_{\mathcal{H},\Sigma}})^{\otimes a_2}.$$

(This is consistent with Lemma 2.1.2.34.) By [80, IX, 2.4, and its proof], there exists an integer $1 \leq a_0 \leq 2$ such that

$$\omega^{\otimes a_0}_{\mathsf{M}^{\mathrm{tor}}_{\mathcal{H},\Sigma}} \cong (\omega'_{\mathsf{M}^{\mathrm{tor}}_{\mathcal{H},\Sigma}})^{\otimes a_0}.$$

Hence, up to replacing a_0 with 1 when a_2 is even, the lemma follows. □

Let $\mathsf{M}^{\min}_{\mathcal{H}_{\mathrm{aux}}}$ (resp. $\mathsf{M}^{\min}_{\mathcal{H}^p_{\mathrm{aux}}}$) denote the minimal compactification of $\mathsf{M}_{\mathcal{H}_{\mathrm{aux}}}$ (resp. $\mathsf{M}_{\mathcal{H}^p_{\mathrm{aux}}}$), which is by construction a projective variety over $\mathsf{S}_{0,\mathrm{aux}} = \mathrm{Spec}(F_{0,\mathrm{aux}})$ (resp. $\vec{\mathsf{S}}_{0,\mathrm{aux}} = \mathrm{Spec}(\mathcal{O}_{F_{0,\mathrm{aux}},(p)})$) containing the coarse moduli space $[\mathsf{M}_{\mathcal{H}_{\mathrm{aux}}}]$ of $\mathsf{M}_{\mathcal{H}_{\mathrm{aux}}}$ (resp. $[\mathsf{M}_{\mathcal{H}^p_{\mathrm{aux}}}]$ of $\mathsf{M}_{\mathcal{H}^p_{\mathrm{aux}}}$) as an open subscheme. By [62, Thm. 7.2.4.1], there exists an integer $N_1 \geq 1$ (depending on $\mathcal{H}^p_{\mathrm{aux}}$, which is 1 when $\mathcal{H}^p_{\mathrm{aux}}$ is neat) such that $\omega^{\otimes N_1}_{\mathsf{M}^{\mathrm{tor}}_{\mathcal{H}^p_{\mathrm{aux}},\Sigma^p_{\mathrm{aux}}}}$ descends to an ample invertible sheaf over $\mathsf{M}^{\min}_{\mathcal{H}^p_{\mathrm{aux}}}$, which we denote by $\omega^{\otimes N_1}_{\mathsf{M}^{\min}_{\mathcal{H}^p_{\mathrm{aux}}}}$ by abuse of notation. In this case, by Lemma 2.1.2.34 and by the universal property of the projective spectra

$$\mathsf{M}^{\min}_{\mathcal{H}_{\mathrm{aux}}} \cong \mathrm{Proj}\Big(\underset{k\geq 0}{\oplus}\, \Gamma(\mathsf{M}^{\mathrm{tor}}_{\mathcal{H}_{\mathrm{aux}},\Sigma_{\mathrm{aux}}}, \omega^{\otimes k}_{\mathsf{M}^{\mathrm{tor}}_{\mathcal{H}_{\mathrm{aux}},\Sigma_{\mathrm{aux}}}})\Big)$$

and

$$\mathsf{M}^{\min}_{\mathcal{H}^p_{\mathrm{aux}}} \cong \operatorname{Proj}\Big(\underset{k\geq 0}{\oplus}\Gamma(\mathsf{M}^{\mathrm{tor}}_{\mathcal{H}^p_{\mathrm{aux}},\Sigma^p_{\mathrm{aux}}},\omega^{\otimes k}_{\mathsf{M}^{\mathrm{tor}}_{\mathcal{H}^p_{\mathrm{aux}},\Sigma^p_{\mathrm{aux}}}})\Big)$$

(see (3) of Theorem 1.3.1.5), $\omega^{\otimes N_1}_{\mathsf{M}^{\mathrm{tor}}_{\mathcal{H}_{\mathrm{aux}},\Sigma_{\mathrm{aux}}}}$ also descends to an ample invertible sheaf over $\mathsf{M}^{\min}_{\mathcal{H}_{\mathrm{aux}}}$, which we denote by $\omega^{\otimes N_1}_{\mathsf{M}^{\min}_{\mathcal{H}_{\mathrm{aux}}}}$ by abuse of notation, and the morphism (2.1.2.28) induces a morphism

$$\mathsf{M}^{\min}_{\mathcal{H}_{\mathrm{aux}}} \to \mathsf{M}^{\min}_{\mathcal{H}^p_{\mathrm{aux}}} \underset{\mathbb{Z}}{\otimes} \mathbb{Q} \tag{2.1.2.36}$$

under which the pullback of $\omega^{\otimes N_1}_{\mathsf{M}^{\min}_{\mathcal{H}^p_{\mathrm{aux}}}}$ is canonically isomorphic to $\omega^{\otimes N_1}_{\mathsf{M}^{\min}_{\mathcal{H}_{\mathrm{aux}}}}$.

On the other hand, since $\mathcal{H}$ is neat, $\omega_{\mathsf{M}^{\mathrm{tor}}_{\mathcal{H},\Sigma}}$ descends to an ample invertible sheaf $\omega_{\mathsf{M}^{\min}_{\mathcal{H}}}$ over $\mathsf{M}^{\min}_{\mathcal{H}}$.

Proposition 2.1.2.37. *With assumptions as in Proposition* 2.1.1.15, *there exists a morphism*

$$\mathsf{M}^{\min}_{\mathcal{H}} \to \mathsf{M}^{\min}_{\mathcal{H}_{\mathrm{aux}}} \tag{2.1.2.38}$$

extending (2.1.1.16) *and compatible with* (2.1.2.30), *which induces by composition with* (2.1.2.36) *a morphism*

$$\mathsf{M}^{\min}_{\mathcal{H}} \to \mathsf{M}^{\min}_{\mathcal{H}^p_{\mathrm{aux}}} \underset{\mathbb{Z}}{\otimes} \mathbb{Q} \tag{2.1.2.39}$$

extending (2.1.1.17) *and compatible with* (2.1.2.31). *The morphism* (2.1.2.38) *(resp.* (2.1.2.39)*) maps the* $[(\Phi_{\mathcal{H}},\delta_{\mathcal{H}})]$*-stratum* $\mathsf{Z}_{[(\Phi_{\mathcal{H}},\delta_{\mathcal{H}})]}$ *of* $\mathsf{M}^{\min}_{\mathcal{H}}$ *to the* $[(\Phi_{\mathcal{H}_{\mathrm{aux}}},\delta_{\mathcal{H}_{\mathrm{aux}}})]$*-stratum* $\mathsf{Z}_{[(\Phi_{\mathcal{H}_{\mathrm{aux}}},\delta_{\mathcal{H}_{\mathrm{aux}}})]}$ *of* $\mathsf{M}^{\min}_{\mathcal{H}_{\mathrm{aux}}}$ *(resp.* $[(\Phi_{\mathcal{H}^p_{\mathrm{aux}}},\delta_{\mathcal{H}^p_{\mathrm{aux}}})]$*-stratum* $\mathsf{Z}_{[(\Phi_{\mathcal{H}^p_{\mathrm{aux}}},\delta_{\mathcal{H}^p_{\mathrm{aux}}})]}$ *of* $\mathsf{M}^{\min}_{\mathcal{H}^p_{\mathrm{aux}}}$*) when* $[(\Phi_{\mathcal{H}_{\mathrm{aux}}},\delta_{\mathcal{H}_{\mathrm{aux}}})]$ *(resp.* $[(\Phi_{\mathcal{H}^p_{\mathrm{aux}}},\delta_{\mathcal{H}^p_{\mathrm{aux}}})]$*) is assigned to* $[(\Phi_{\mathcal{H}},\delta_{\mathcal{H}})]$ *as in* (2.1.2.20) *(with the filtrations* $\mathsf{Z}_{\mathcal{H}}$, $\mathsf{Z}_{\mathcal{H}_{\mathrm{aux}}}$, *and* $\mathsf{Z}_{\mathcal{H}^p_{\mathrm{aux}}}$ *suppressed in the notation). If* $N_1 \geq 1$ *is as above, and if* $a_0 \geq 1$ *and* $a \geq 1$ *are integers as in Lemma* 2.1.2.35, *then the pullback of* $\omega^{\otimes a_0 N_1}_{\mathsf{M}^{\min}_{\mathcal{H}_{\mathrm{aux}}}}$ *(resp.* $\omega^{\otimes a_0 N_1}_{\mathsf{M}^{\min}_{\mathcal{H}^p_{\mathrm{aux}}}}$*) to* $\mathsf{M}^{\min}_{\mathcal{H}}$ *is canonically isomorphic to* $\omega^{\otimes aN_1}_{\mathsf{M}^{\min}_{\mathcal{H}}}$.

Consequently, $\mathsf{M}^{\min}_{\mathcal{H}}$ *is the normalization of* $\mathsf{M}^{\min}_{\mathcal{H}_{\mathrm{aux}}}$ *(resp.* $\mathsf{M}^{\min}_{\mathcal{H}^p_{\mathrm{aux}}} \underset{\mathbb{Z}}{\otimes} \mathbb{Q}$*) in* $\mathsf{M}_{\mathcal{H}}$ *under the morphism* $\mathsf{M}_{\mathcal{H}} \to \mathsf{M}^{\min}_{\mathcal{H}_{\mathrm{aux}}}$ *(resp.* $\mathsf{M}_{\mathcal{H}} \to \mathsf{M}^{\min}_{\mathcal{H}^p_{\mathrm{aux}}} \underset{\mathbb{Z}}{\otimes} \mathbb{Q}$*) induced by* (2.1.1.16) *(resp.* (2.1.1.17)*) and the canonical morphism* $\mathsf{M}_{\mathcal{H}_{\mathrm{aux}}} \to \mathsf{M}^{\min}_{\mathcal{H}_{\mathrm{aux}}}$ *(resp.* $\mathsf{M}_{\mathcal{H}^p_{\mathrm{aux}}} \underset{\mathbb{Z}}{\otimes} \mathbb{Q} \to \mathsf{M}^{\min}_{\mathcal{H}^p_{\mathrm{aux}}} \underset{\mathbb{Z}}{\otimes} \mathbb{Q}$*).*

Proof. The first paragraph follows from Proposition 2.1.2.29, from Lemma 2.1.2.35, and from the universal properties of the projective spectrum

$$\mathsf{M}^{\min}_{\mathcal{H}} \cong \operatorname{Proj}\Big(\underset{k\geq 0}{\oplus}\Gamma(\mathsf{M}^{\mathrm{tor}}_{\mathcal{H},\Sigma},\omega^{\otimes k}_{\mathsf{M}^{\mathrm{tor}}_{\mathcal{H},\Sigma}})\Big)$$

(and the ones for $\mathsf{M}^{\min}_{\mathcal{H}_{\mathrm{aux}}}$ and $\mathsf{M}^{\min}_{\mathcal{H}^p_{\mathrm{aux}}}$ above).

Since $\omega^{\otimes a_0 N_1}_{\mathsf{M}^{\min}_{\mathcal{H}_{\mathrm{aux}}}}$ (resp. $\omega^{\otimes a_0 N_1}_{\mathsf{M}^{\min}_{\mathcal{H}^p_{\mathrm{aux}}}}$) is ample over $\mathsf{M}^{\min}_{\mathcal{H}_{\mathrm{aux}}}$ (resp. $\mathsf{M}^{\min}_{\mathcal{H}^p_{\mathrm{aux}}}$), since $\omega^{\otimes aN_1}_{\mathsf{M}^{\min}_{\mathcal{H}}}$ is ample over $\mathsf{M}^{\min}_{\mathcal{H}}$, and since the pullback of the former is canonically isomorphic to the latter, the canonical morphism from $\mathsf{M}^{\min}_{\mathcal{H}}$ to the normalization of $\mathsf{M}^{\min}_{\mathcal{H}_{\mathrm{aux}}}$ (resp. $\mathsf{M}^{\min}_{\mathcal{H}^p_{\mathrm{aux}}} \underset{\mathbb{Z}}{\otimes} \mathbb{Q}$) in $\mathsf{M}_{\mathcal{H}}$ is *finite* (see [35, II, 5.1.6, and III-1, 4.4.2]). Since both the source and target of this finite morphism are normal, and since they share an open dense subscheme $\mathsf{M}_{\mathcal{H}}$, the second paragraph follows from Zariski's main theorem (see [35, III-1, 4.4.3, 4.4.11]), as desired. □

2.2 Flat Integral Models as Normalizations and Blow-Ups

2.2.1 *Flat Integral Models for Minimal Compactifications*

Proposition 2.2.1.1. *Let $\vec{\mathsf{M}}_{\mathcal{H}}$ denote the normalization of $\mathsf{M}_{\mathrm{G}_{\mathrm{aux}}(\hat{\mathbb{Z}}^p)}$ in $\mathsf{M}_{\mathcal{H}}$ under the morphism $\mathsf{M}_{\mathcal{H}} \to \mathsf{M}_{\mathcal{H}^p_{\mathrm{aux}}}$ induced by* (2.1.1.17) *(with $\mathcal{H}^p_{\mathrm{aux}} = \mathrm{G}_{\mathrm{aux}}(\hat{\mathbb{Z}}^p)$ there). Then $\vec{\mathsf{M}}_{\mathcal{H}}$ is a normal algebraic stack flat over $\vec{\mathsf{S}}_0 := \mathrm{Spec}(\mathcal{O}_{F_0,(p)})$ equipped with a canonical isomorphism $\vec{\mathsf{M}}_{\mathcal{H}} \underset{\vec{\mathsf{S}}_0}{\times} \mathsf{S}_0 \cong \mathsf{M}_{\mathcal{H}}$ over S_0, and with a morphism $\vec{\mathsf{M}}_{\mathcal{H}} \to \mathsf{M}_{\mathcal{H}^p_{\mathrm{aux}}} = \mathsf{M}_{\mathrm{G}_{\mathrm{aux}}(\hat{\mathbb{Z}}^p)}$ extending* (2.1.1.17).

The tautological tuple $(A, \lambda, i, \alpha_{\mathcal{H}})$ over $\mathsf{M}_{\mathcal{H}}$ extends to a degenerating family $(\vec{A}, \vec{\lambda}, \vec{i}, \vec{\alpha}_{\mathcal{H}})$ of type $\mathsf{M}_{\mathcal{H}}$ over $\vec{\mathsf{M}}_{\mathcal{H}}$ (see [62, Def. 5.3.2.1] and Definition 1.3.1.1*), where $(\vec{A}, \vec{\lambda})$ is a polarized abelian scheme with an $\mathcal{O}$-structure $\vec{i}$ such that $\underline{\mathrm{Lie}}_{\vec{A}/\vec{\mathsf{M}}_{\mathcal{H}}}$ with its $\mathcal{O} \underset{\mathbb{Z}}{\otimes} \mathbb{Z}_{(p)}$-module structure given naturally by $\vec{i}$ satisfies the determinantal condition in [62, Def. 1.3.4.1] given by $(L \underset{\mathbb{Z}}{\otimes} \mathbb{R}, \langle \cdot\,, \cdot \rangle, h_0)$, and where $\vec{\alpha}_{\mathcal{H}}$ is defined only over $\mathsf{M}_{\mathcal{H}}$. If we denote by $(\vec{A}_{\mathrm{aux}}, \vec{\lambda}_{\mathrm{aux}}, \vec{i}_{\mathrm{aux}}, \vec{\alpha}_{\mathrm{G}_{\mathrm{aux}}(\hat{\mathbb{Z}}^p)})$ the pullback of the tautological tuple $(A_{\mathrm{aux}}, \lambda_{\mathrm{aux}}, i_{\mathrm{aux}}, \alpha_{\mathrm{G}_{\mathrm{aux}}(\hat{\mathbb{Z}}^p)})$ over $\mathsf{M}_{\mathrm{G}_{\mathrm{aux}}(\hat{\mathbb{Z}}^p)}$ under the morphism $\vec{\mathsf{M}}_{\mathcal{H}} \to \mathsf{M}_{\mathcal{H}^p_{\mathrm{aux}}}$ induced by* (2.1.1.17)*, then $(\vec{A}_{\mathrm{aux}}, \vec{\lambda}_{\mathrm{aux}})$ is isomorphic to the polarized abelian scheme $(\vec{A}'_{\mathrm{aux}}, \vec{\lambda}'_{\mathrm{aux}})$ defined by $(\vec{A}, \vec{\lambda})$ as in* (2) *of Lemma* 2.1.1.1*, $\vec{i}$ is the unique extension of i over the noetherian normal base scheme $\vec{\mathsf{M}}_{\mathcal{H}}$ (by [92, IX, 1.4], [28, Ch. I, Prop. 2.7], or [62, Prop. 3.3.1.5]), and $\vec{\alpha}_{\mathrm{G}_{\mathrm{aux}}(\hat{\mathbb{Z}}^p)}$ is determined by $\alpha_{\mathcal{H}}$ in the sense that its further pullback to $\mathsf{M}_{\mathcal{H}}$ is determined by $\alpha_{\mathcal{H}}$ as in Proposition* 2.1.1.15 *(with $\mathcal{H}^p_{\mathrm{aux}} = \mathrm{G}_{\mathrm{aux}}(\hat{\mathbb{Z}}^p)$ there). Then $\omega_{\mathsf{M}_{\mathcal{H}}}$ extends to the invertible sheaf*

$$\omega_{\vec{\mathsf{M}}_{\mathcal{H}}} := \wedge^{\mathrm{top}}\, \underline{\mathrm{Lie}}^{\vee}_{\vec{A}/\vec{\mathsf{M}}_{\mathcal{H}}} = \wedge^{\mathrm{top}}\, e^*_{\vec{A}} \Omega^1_{\vec{A}/\vec{\mathsf{M}}_{\mathcal{H}}}$$

over $\vec{\mathsf{M}}_{\mathcal{H}}$, which is **ample** *if $\mathcal{H}$ is neat. If $a_0 \geq 1$ and $a \geq 1$ are integers as in Lemma* 2.1.2.35*, then $\omega^{\otimes a}_{\vec{\mathsf{M}}_{\mathcal{H}}}$ is canonically isomorphic to the pullback of $\omega^{\otimes a_0}_{\mathsf{M}_{\mathrm{G}_{\mathrm{aux}}(\hat{\mathbb{Z}}^p)}}$ under the morphism $\vec{\mathsf{M}}_{\mathcal{H}} \to \mathsf{M}_{\mathrm{G}_{\mathrm{aux}}(\hat{\mathbb{Z}}^p)}$ induced by* (2.1.1.17).

The coarse moduli space $[\vec{\mathsf{M}}_{\mathcal{H}}]$ of $\vec{\mathsf{M}}_{\mathcal{H}}$ is canonically isomorphic to the normalization of $[\mathsf{M}_{\mathrm{G}_{\mathrm{aux}}(\hat{\mathbb{Z}}^p)}]$ in $[\mathsf{M}_{\mathcal{H}}]$ under the morphism $[\mathsf{M}_{\mathcal{H}}] \to [\mathsf{M}_{\mathrm{G}_{\mathrm{aux}}(\hat{\mathbb{Z}}^p)}]$ induced by (2.1.1.17)*, which is a normal scheme quasi-projective and flat over $\vec{\mathsf{S}}_0$ equipped with a canonical isomorphism $[\vec{\mathsf{M}}_{\mathcal{H}}] \underset{\vec{\mathsf{S}}_0}{\times} \mathsf{S}_0 \cong [\mathsf{M}_{\mathcal{H}}]$ over S_0. In particular, if $\mathcal{H}$ is neat, then $\vec{\mathsf{M}}_{\mathcal{H}} \cong [\vec{\mathsf{M}}_{\mathcal{H}}]$ is a scheme.*

We obtain the same normalization $\vec{\mathsf{M}}_{\mathcal{H}}$ *(up to canonical isomorphism) satisfying the analogous properties if we replace* $\mathrm{G}_{\mathrm{aux}}(\hat{\mathbb{Z}}^p)$ *with any open compact subgroup* $\mathcal{H}^p_{\mathrm{aux}}$ *of* $\mathrm{G}_{\mathrm{aux}}(\hat{\mathbb{Z}}^p)$ *such that* $\mathcal{H}^p_{\mathrm{aux}}\mathrm{G}_{\mathrm{aux}}(\mathbb{Z}_p)$ *still contains the image of* $\mathcal{H}$ *under the homomorphism* $\mathrm{G}(\hat{\mathbb{Z}}) \to \mathrm{G}_{\mathrm{aux}}(\hat{\mathbb{Z}})$ *given by* (2.1.1.10).

Up to canonical isomorphism, $\vec{\mathsf{M}}_{\mathcal{H}}$ *and hence* $[\vec{\mathsf{M}}_{\mathcal{H}}]$ *depend only on the linear algebraic data defining* $\mathsf{M}_{\mathcal{H}}$*, but not on the auxiliary choices in Section* 2.1 *defining* $\mathsf{M}_{\mathrm{G}_{\mathrm{aux}}(\hat{\mathbb{Z}}^p)}$ *or* $\mathsf{M}_{\mathcal{H}^p_{\mathrm{aux}}}$.

Proof. The first paragraph is self-explanatory. As for the second paragraph, except for the ampleness of $\omega_{\vec{\mathsf{M}}_{\mathcal{H}}}$ when $\mathcal{H}$ is neat, it suffices to show that the tautological (A, λ) over $\mathsf{M}_{\mathcal{H}}$ extends to some polarized abelian scheme $(\vec{A}, \vec{\lambda})$ over $\vec{\mathsf{M}}_{\mathcal{H}}$. (Once this is shown, the remainder of the paragraph will follow from the uniqueness of extensions by [92, IX, 1.4], [28, Ch. I, Prop. 2.7], or [62, Prop. 3.3.1.5].) Since the genus of A and the polarization degree of λ is determined by the level structure $\alpha_{\mathcal{H}}$, the tautological (A, λ) over $\mathsf{M}_{\mathcal{H}}$ defines (by forgetting the additional structures) a morphism from $\mathsf{M}_{\mathcal{H}}$ to the Siegel moduli $\mathcal{A}_{g,d}$ of genus $g = \frac{1}{2}\operatorname{rk}_{\mathbb{Z}}(L)$ and polarization degree $d^2 = [L^\# : L]$, which induces a finite morphism $\mathsf{M}_{\mathcal{H}} \to \mathcal{A}_{g,d} \underset{\mathbb{Z}}{\otimes} \mathbb{Q}$ by [62, Prop. 1.3.3.7, Cor. 2.2.2.8, and Prop. 2.2.2.9]. Similarly, the tautological $(A_{\mathrm{aux}}, \lambda_{\mathrm{aux}})$ defines a morphism from $\mathsf{M}_{\mathrm{G}_{\mathrm{aux}}(\hat{\mathbb{Z}}^p)}$ to the Siegel moduli $\mathcal{A}_{g_{\mathrm{aux}},d_{\mathrm{aux}}}$ of genus $g_{\mathrm{aux}} = \frac{1}{2}\operatorname{rk}_{\mathbb{Z}}(L_{\mathrm{aux}})$ and polarization degree $d^2_{\mathrm{aux}} = [L^\#_{\mathrm{aux}} : L_{\mathrm{aux}}]$, which induces a finite morphism $\mathsf{M}_{\mathrm{G}_{\mathrm{aux}}(\hat{\mathbb{Z}}^p)} \to \mathcal{A}_{g_{\mathrm{aux}},d_{\mathrm{aux}}} \underset{\mathbb{Z}}{\otimes} \mathbb{Z}_{(p)}$. As explained in the proof of Lemma 2.1.1.5, the construction as in (2) of Lemma 2.1.1.1 defines a finite morphism $\mathcal{A}_{g,d} \to \mathcal{A}_{g_{\mathrm{aux}},d_{\mathrm{aux}}}$. By comparing the universal properties, the composition $\mathsf{M}_{\mathcal{H}} \to \mathsf{M}_{\mathrm{G}_{\mathrm{aux}}(\hat{\mathbb{Z}}^p)} \underset{\mathbb{Z}}{\otimes} \mathbb{Q} \to \mathcal{A}_{g_{\mathrm{aux}},d_{\mathrm{aux}}} \underset{\mathbb{Z}}{\otimes} \mathbb{Q}$ of finite morphisms coincides with the composition $\mathsf{M}_{\mathcal{H}} \to \mathcal{A}_{g,d} \underset{\mathbb{Z}}{\otimes} \mathbb{Q} \to \mathcal{A}_{g_{\mathrm{aux}},d_{\mathrm{aux}}} \underset{\mathbb{Z}}{\otimes} \mathbb{Q}$ of finite morphisms. Since $\mathcal{A}_{g,d} \to \mathcal{A}_{g_{\mathrm{aux}},d_{\mathrm{aux}}}$ and $\mathsf{M}_{\mathrm{G}_{\mathrm{aux}}(\hat{\mathbb{Z}}^p)} \to \mathcal{A}_{g_{\mathrm{aux}},d_{\mathrm{aux}}} \underset{\mathbb{Z}}{\otimes} \mathbb{Z}_{(p)}$ are finite, it follows that $\vec{\mathsf{M}}_{\mathcal{H}}$ is canonically isomorphic to the normalization of $\mathcal{A}_{g,d} \underset{\mathbb{Z}}{\otimes} \mathbb{Z}_{(p)}$ under the canonical morphism $\mathsf{M}_{\mathcal{H}} \to \mathcal{A}_{g,d} \underset{\mathbb{Z}}{\otimes} \mathbb{Z}_{(p)}$. In particular, the tautological object (A, λ) over $\mathsf{M}_{\mathcal{H}}$ extends to an object $(\vec{A}, \vec{\lambda})$ parameterized by the canonical morphism $\vec{\mathsf{M}}_{\mathcal{H}} \to \mathcal{A}_{g,d}$. This also shows, as in the last paragraph of the statement of the proposition, that $\vec{\mathsf{M}}_{\mathcal{H}}$ is canonical and independent of the auxiliary choices.

The coarse moduli space $[\vec{\mathsf{M}}_{\mathcal{H}}]$ of $\vec{\mathsf{M}}_{\mathcal{H}}$ is canonically isomorphic to the normalization of $[\mathsf{M}_{\mathrm{G}_{\mathrm{aux}}(\hat{\mathbb{Z}}^p)}]$ in $[\mathsf{M}_{\mathcal{H}}]$ by the universal property of coarse moduli spaces, and by Zariski's main theorem (see [35, III-1, 4.4.3, 4.4.11], or the formulation in [62, Prop. 7.2.3.4] for algebraic spaces). Except for the quasi-projectivity of $[\vec{\mathsf{M}}_{\mathcal{H}}]$ over $\vec{\mathsf{S}}_0$, and for the ampleness of $\omega_{\vec{\mathsf{M}}_{\mathcal{H}}}$ when $\mathcal{H}$ is neat, both of which will follow from Proposition 2.2.1.2 below, the remaining statements of the proposition are self-explanatory. □

Although Proposition 2.2.1.1 is stated without any reference to compactifications, the easiest way to show the quasi-projectivity of $[\vec{\mathsf{M}}_{\mathcal{H}}]$ over $\vec{\mathsf{S}}_0$, and the ampleness of $\omega_{\vec{\mathsf{M}}_{\mathcal{H}}}$ when $\mathcal{H}$ is neat, is to introduce the minimal compactifications. (This is a natural consideration because this is what the minimal compactifications in [5] did over $\mathbb{C}$.)

Proposition 2.2.1.2. *Let* $\vec{\mathsf{M}}_{\mathcal{H}}^{\min}$ *denote the normalization of* $\mathsf{M}_{\mathrm{G}_{\mathrm{aux}}(\hat{\mathbb{Z}}^p)}^{\min}$ *in* $\mathsf{M}_{\mathcal{H}}^{\min}$ *under the morphism* $\mathsf{M}_{\mathcal{H}}^{\min} \to \mathsf{M}_{\mathrm{G}_{\mathrm{aux}}(\hat{\mathbb{Z}}^p)}^{\min}$ *induced by* (2.1.2.39) (*with* $\mathcal{H}_{\mathrm{aux}}^p = \mathrm{G}_{\mathrm{aux}}(\hat{\mathbb{Z}}^p)$ *there*). *Then* $\vec{\mathsf{M}}_{\mathcal{H}}^{\min}$ *is a normal scheme projective and flat over* $\vec{\mathsf{S}}_0 = \mathrm{Spec}(\mathcal{O}_{F_0,(p)})$ *equipped with a canonical isomorphism* $\vec{\mathsf{M}}_{\mathcal{H}}^{\min} \underset{\vec{\mathsf{S}}_0}{\times} \mathsf{S}_0 \cong \mathsf{M}_{\mathcal{H}}^{\min}$ *over* S_0.

By construction, $[\vec{\mathsf{M}}_{\mathcal{H}}]$ *is an open dense subscheme of* $\vec{\mathsf{M}}_{\mathcal{H}}^{\min}$, *because* $[\mathsf{M}_{\mathrm{G}_{\mathrm{aux}}(\hat{\mathbb{Z}}^p)}]$ *is an open dense subscheme of* $\mathsf{M}_{\mathrm{G}_{\mathrm{aux}}(\hat{\mathbb{Z}}^p)}^{\min}$ *(by [62, Thm. 7.2.4.1]).*

If $N_1 \geq 1$ *is as in the paragraph preceding Proposition* 2.1.2.37 (*for* $\mathcal{H}_{\mathrm{aux}}^p = \mathrm{G}_{\mathrm{aux}}(\hat{\mathbb{Z}}^p)$), *and if* $a_0 \geq 1$ *and* $a \geq 1$ *are integers as in Lemma* 2.1.2.35, *then* $\omega_{\mathsf{M}_{\mathcal{H}}}^{\otimes aN_1}$ *and* $\omega_{\mathsf{M}_{\mathcal{H}}^{\min}}^{\otimes aN_1}$ *compatibly extend to an* **ample** *invertible sheaf over* $\vec{\mathsf{M}}_{\mathcal{H}}^{\min}$, *which we denote by* $\omega_{\vec{\mathsf{M}}_{\mathcal{H}}^{\min}}^{\otimes aN_1}$ *by abuse of notation, such that the pullback of* $\omega_{\vec{\mathsf{M}}_{\mathcal{H}}^{\min}}^{\otimes aN_1}$ *to* $\vec{\mathsf{M}}_{\mathcal{H}}$ *is canonically isomorphic to* $\omega_{\vec{\mathsf{M}}_{\mathcal{H}}}^{\otimes aN_1}$, *such that the pullback of* $\omega_{\mathsf{M}_{\mathrm{G}_{\mathrm{aux}}(\hat{\mathbb{Z}}^p)}^{\min}}^{\otimes a_0N_1}$ *to* $\vec{\mathsf{M}}_{\mathcal{H}}^{\min}$ *is canonically isomorphic to* $\omega_{\vec{\mathsf{M}}_{\mathcal{H}}^{\min}}^{\otimes aN_1}$, *and so that there is a canonical isomorphism*

$$\vec{\mathsf{M}}_{\mathcal{H}}^{\min} \cong \mathrm{Proj}\Big(\underset{k\geq 0}{\oplus} \Gamma(\vec{\mathsf{M}}_{\mathcal{H}}^{\min}, \omega_{\vec{\mathsf{M}}_{\mathcal{H}}^{\min}}^{\otimes aN_1k})\Big).$$

We obtain the same normalization $\vec{\mathsf{M}}_{\mathcal{H}}^{\min}$ (*up to canonical isomorphism*) *if we replace* $\mathrm{G}_{\mathrm{aux}}(\hat{\mathbb{Z}}^p)$ *with any open compact subgroup* $\mathcal{H}_{\mathrm{aux}}^p$ *of* $\mathrm{G}_{\mathrm{aux}}(\hat{\mathbb{Z}}^p)$ *such that* $\mathcal{H}_{\mathrm{aux}}^p\mathrm{G}_{\mathrm{aux}}(\mathbb{Z}_p)$ *still contains the image of* $\mathcal{H}$ *under the homomorphism* $\mathrm{G}(\hat{\mathbb{Z}}) \to \mathrm{G}_{\mathrm{aux}}(\hat{\mathbb{Z}})$ *given by* (2.1.1.10), *in which case we might reduce the size of* N_1 *in the above statements. By Lemma* 2.1.1.18, *if the image of* $\mathcal{H}$ *under the canonical homomorphism* $\mathrm{G}(\hat{\mathbb{Z}}) \to \mathrm{G}(\hat{\mathbb{Z}}^p)$ *is neat, then we can choose* $\mathcal{H}_{\mathrm{aux}}^p$ *to be neat, so that* $N_1 = 1$.

We also obtain the same $\vec{\mathsf{M}}_{\mathcal{H}}^{\min}$ (*up to canonical isomorphism*) *if we replace* $\mathsf{M}_{\mathcal{H}}^{\min}$ *with* $[\mathsf{M}_{\mathcal{H}}]$, *and if we replace the morphism* $\mathsf{M}_{\mathcal{H}}^{\min} \to \mathsf{M}_{\mathrm{G}_{\mathrm{aux}}(\hat{\mathbb{Z}}^p)}^{\min}$ *induced by* (2.1.2.39) *with the morphism* $[\mathsf{M}_{\mathcal{H}}] \to \mathsf{M}_{\mathrm{G}_{\mathrm{aux}}(\hat{\mathbb{Z}}^p)}^{\min}$ *induced by* (2.1.1.17) (*cf. the second paragraph of Proposition* 2.1.2.37).

As in the case of $\vec{\mathsf{M}}_{\mathcal{H}}$ in Proposition 2.2.1.1, it is also true that, up to canonical isomorphism, $\vec{\mathsf{M}}_{\mathcal{H}}^{\min}$ depends only on the linear algebraic data defining $\mathsf{M}_{\mathcal{H}}$, but not on the auxiliary choices in Section 2.1 defining $\mathsf{M}_{\mathrm{G}_{\mathrm{aux}}(\hat{\mathbb{Z}}^p)}$ or $\mathsf{M}_{\mathcal{H}_{\mathrm{aux}}^p}$. However, the proof of this is somewhat indirect and will be postponed until Corollary 2.2.1.15 below.

Proof of Proposition 2.2.1.2. By construction as a normalization, we know that $\vec{\mathsf{M}}_{\mathcal{H}}^{\min}$ is normal, and that the morphism $\vec{\mathsf{M}}_{\mathcal{H}}^{\min} \to \mathsf{M}_{\mathrm{G}_{\mathrm{aux}}(\hat{\mathbb{Z}}^p)}^{\min}$ is finite. Since $\omega_{\mathsf{M}_{\mathrm{G}_{\mathrm{aux}}(\hat{\mathbb{Z}}^p)}^{\min}}^{\otimes a_0N_1}$ is ample over $\mathsf{M}_{\mathrm{G}_{\mathrm{aux}}(\hat{\mathbb{Z}}^p)}^{\min}$, its pullback to $\vec{\mathsf{M}}_{\mathcal{H}}^{\min}$ is also ample, which we define as the common extension $\omega^{\otimes aN_1}_{\vec{\mathsf{M}}^{\min}_{\mathcal{H}}}$ of $\omega^{\otimes aN_1}_{\mathsf{M}_{\mathcal{H}}}$ and $\omega^{\otimes aN_1}_{\mathsf{M}^{\min}_{\mathcal{H}}}$. (This is consistent with Lemma 2.1.2.35 and Proposition 2.1.2.37.) This shows in particular that $\vec{\mathsf{M}}^{\min}_{\mathcal{H}}$ is projective over $\vec{\mathsf{S}}_0$. Since the structural sheaf of $\vec{\mathsf{M}}^{\min}_{\mathcal{H}}$ is normal and hence has no p-torsion, it is also flat over $\vec{\mathsf{S}}_0$. Since the pullback of $\omega^{\otimes a_0N_1}_{\mathsf{M}^{\min}_{\mathrm{G_{aux}}(\hat{\mathbb{Z}}^p)}}$ to $\mathsf{M}_{\mathrm{G_{aux}}(\hat{\mathbb{Z}}^p)}$ is canonically isomorphic to $\omega^{\otimes a_0N_1}_{\mathsf{M}_{\mathrm{G_{aux}}(\hat{\mathbb{Z}}^p)}}$, its further pullback to $\vec{\mathsf{M}}_{\mathcal{H}}$, which is canonically isomorphic to the pullback of $\omega^{\otimes aN_1}_{\vec{\mathsf{M}}^{\min}_{\mathcal{H}}}$ by construction, is canonically isomorphic to $\omega^{\otimes aN_1}_{\vec{\mathsf{M}}_{\mathcal{H}}}$ (by the part of Proposition 2.2.1.1 we have proved). The remaining statements of the proposition are self-explanatory. □

Now the proof of Proposition 2.2.1.1 is also complete.

Remark 2.2.1.3. In our constructions (including ones to be given below), taking normalizations will never introduce pathologies, either because we are talking integral closures in (products of) separable field extensions (see [77, Sec. 33, Lem. 1]), or because the schemes in questions are all excellent (being a localization of a scheme of finite type over $\mathbb{Z}$; see [76, Sec. 31–34] for more discussions).

For each stratum $\mathsf{Z}_{[(\Phi_{\mathcal{H}},\delta_{\mathcal{H}})]}$ as in (4) of Theorem 1.3.1.5, consider its closure $\overline{\mathsf{Z}}_{[(\Phi_{\mathcal{H}},\delta_{\mathcal{H}})]}$ in $\mathsf{M}^{\min}_{\mathcal{H}}$ and its closure $\vec{\overline{\mathsf{Z}}}_{[(\Phi_{\mathcal{H}},\delta_{\mathcal{H}})]}$ in $\vec{\mathsf{M}}^{\min}_{\mathcal{H}}$. Then we define a locally closed subscheme

$$\vec{\mathsf{Z}}_{[(\Phi_{\mathcal{H}},\delta_{\mathcal{H}})]} := \vec{\overline{\mathsf{Z}}}_{[(\Phi_{\mathcal{H}},\delta_{\mathcal{H}})]} - \bigcup_{\mathsf{Z}_{[(\Phi_{\mathcal{H}},\delta_{\mathcal{H}})]}\not\subseteq\overline{\mathsf{Z}}_{[(\Phi'_{\mathcal{H}},\delta'_{\mathcal{H}})]}} \vec{\overline{\mathsf{Z}}}_{[(\Phi'_{\mathcal{H}},\delta'_{\mathcal{H}})]} \tag{2.2.1.4}$$

of $\vec{\mathsf{M}}^{\min}_{\mathcal{H}}$. By definition, we have the following:

Lemma 2.2.1.5. *If $\mathsf{Z}_{[(\Phi_{\mathcal{H}},\delta_{\mathcal{H}})]}$ is contained in the closure $\overline{\mathsf{Z}}_{[(\Phi'_{\mathcal{H}},\delta'_{\mathcal{H}})]}$ of $\mathsf{Z}_{[(\Phi'_{\mathcal{H}},\delta'_{\mathcal{H}})]}$, then $\vec{\mathsf{Z}}_{[(\Phi_{\mathcal{H}},\delta_{\mathcal{H}})]}$ is contained in $\vec{\overline{\mathsf{Z}}}_{[(\Phi'_{\mathcal{H}},\delta'_{\mathcal{H}})]}$, and the latter agrees with the closure of $\vec{\mathsf{Z}}_{[(\Phi'_{\mathcal{H}},\delta'_{\mathcal{H}})]}$.*

Remark 2.2.1.6. It is nontrivial that the collection $\{\vec{\mathsf{Z}}_{[(\Phi_{\mathcal{H}},\delta_{\mathcal{H}})]}\}_{[(\Phi_{\mathcal{H}},\delta_{\mathcal{H}})]}$ does define a stratification of $\vec{\mathsf{M}}^{\min}_{\mathcal{H}}$ (see [65, Sec. 12]). Nevertheless, without actually using this fact, we shall still call $\vec{\mathsf{Z}}_{[(\Phi_{\mathcal{H}},\delta_{\mathcal{H}})]}$ the $[(\Phi_{\mathcal{H}},\delta_{\mathcal{H}})]$-*stratum*, by abuse of language.

For showing that $\vec{\mathsf{M}}^{\min}_{\mathcal{H}}$ is canonical and independent of the auxiliary choices, and for many applications, it is desirable to know the following:

Proposition 2.2.1.7. *The image of the canonical morphism*

$$\vec{\mathsf{M}}_{\mathcal{H}} \underset{\mathbb{Z}}{\otimes} \mathbb{F}_p \to \vec{\mathsf{M}}^{\min}_{\mathcal{H}} \underset{\mathbb{Z}}{\otimes} \mathbb{F}_p \tag{2.2.1.8}$$

(see Proposition 2.2.1.2) *is an open and dense subset.*

Proof. Let s be any point of $\vec{\mathsf{M}}^{\min}_{\mathcal{H}}$. Consider any morphism $\xi : \mathrm{Spec}(R) \to \vec{\mathsf{M}}^{\min}_{\mathcal{H}}$, where R is a complete discrete valuation ring with fraction field K of characteristic zero and with algebraically closed residue field k of characteristic p, such that the special point $\mathrm{Spec}(k)$ is mapped to s, and such that the restriction of ξ to the generic point $\mathrm{Spec}(K)$ factors as the composition of a morphism $\xi_K : \mathrm{Spec}(K) \to \mathsf{M}_{\mathcal{H}}$ with the canonical morphism $\mathsf{M}_{\mathcal{H}} \to \vec{\mathsf{M}}_{\mathcal{H}}^{\min}$. (Such morphisms ξ and ξ_K exist because $\mathsf{M}_{\mathcal{H}}$ and $\vec{\mathsf{M}}_{\mathcal{H}}^{\min}$ are of finite type over $\vec{\mathsf{S}}_0$, and because the image $[\mathsf{M}_{\mathcal{H}}]$ of $\mathsf{M}_{\mathcal{H}}$ is open dense in $\vec{\mathsf{M}}_{\mathcal{H}}^{\min}$.) By the *semistable reduction theorem* (see, for example, [28, Ch. I, Thm. 2.6] or [62, Thm. 3.3.2.4]), and by the theory of Néron models (see [10]; *cf.* [92, IX, 1.4], [28, Ch. I, Prop. 2.7], or [62, Prop. 3.3.1.5]), up to replacing K with a finite extension field and replacing R accordingly, the pullback under ξ_K of the tautological tuple $(A, \lambda, i, \alpha_{\mathcal{H}})$ over $\mathsf{M}_{\mathcal{H}}$ extends to a degenerating family $(G^\dagger, \lambda^\dagger, i^\dagger, \alpha_{\mathcal{H}}^\dagger)$ of type $\mathsf{M}_{\mathcal{H}}$ over $\mathrm{Spec}(R)$, where $\alpha_{\mathcal{H}}^\dagger$ is defined only over the generic point $\mathrm{Spec}(K)$.

By applying the construction of *elevators* as in the proof of [58, Thm. 3.1] to $(G^\dagger, \lambda^\dagger, i^\dagger)$, there exists a degenerating family $(\tilde{G}, \tilde{\lambda}, \tilde{i})$ of type $(\mathrm{PE}, \mathcal{O})$ (see [58, Def. 2.1]; see also Definition 4.1.3.2 below) over $S := \mathrm{Spec}(\tilde{R})$, where $\tilde{R}$ is a noetherian integral domain over R which is complete with respect to some ideal $\tilde{I}$ such that $\mathrm{rad}(\tilde{I}) = \tilde{I}$, satisfying the following properties:

(1) There exists a morphism $\mathrm{Spec}(R) \to S$ under which $(G^\dagger, \lambda^\dagger, i^\dagger)$ is isomorphic to the pullback of $(\tilde{G}, \tilde{\lambda}, \tilde{i})$.
(2) There exists an open dense subscheme S_1 of S over which $\tilde{G}$ is an abelian scheme, such that $S_1 \underset{\mathbb{Z}}{\otimes} \mathbb{F}_p$ is nonempty and dense in $S \underset{\mathbb{Z}}{\otimes} \mathbb{F}_p$. (This is because, in the proof of [58, Thm. 3.1] in [58, Sec. 3], the scheme Ξ° is smooth over S and where $\Xi^\circ \underset{\mathbb{Z}}{\otimes} \mathbb{F}_p$ is nonempty and dense in $\Xi^\circ(\sigma) \underset{\mathbb{Z}}{\otimes} \mathbb{F}_p$.)
(3) For any morphism $\mathrm{Spec}(V) \to S$ as in (6) of Theorem 1.3.1.3, centered at the geometric point $\mathrm{Spec}(k) \to S$ induced by the morphism $\mathrm{Spec}(R) \to S$ above, there exist some $(\mathsf{Z}_{\mathcal{H}}^\ddagger, \Phi_{\mathcal{H}}^\ddagger, \delta_{\mathcal{H}}^\ddagger)$ and $\sigma^\ddagger$ such that $\sigma^\ddagger$ is a one-dimensional cone in $\mathbf{P}^+_{\Phi_{\mathcal{H}}^\ddagger}$. (This is because the cone σ in the proof of [58, Thm. 3.1] in [58, Sec. 3] can be taken to be one-dimensional.)

Let $\bar{K}$ be an algebraic closure of K. Then there exists an affine integral scheme $S'_{1,\mathbb{Q}}$ finite étale over $S_{1,\mathbb{Q}}$, together with a morphism $\bar{\eta} := \mathrm{Spec}(\bar{K}) \to S'_{1,\mathbb{Q}}$ lifting the above morphism $\eta \to S_{1,\mathbb{Q}}$, such that $(\tilde{G}_{S'_{1,\mathbb{Q}}}, \tilde{\lambda}_{S'_{1,\mathbb{Q}}}, \tilde{i}_{S'_{1,\mathbb{Q}}})$ satisfies the determinantal condition in [62, Def. 1.3.4.1] given by $(L \underset{\mathbb{Z}}{\otimes} \mathbb{R}, \langle\,\cdot\,,\,\cdot\,\rangle, h_0)$ and is equipped with a level-$\mathcal{H}$ structure $\tilde{\alpha}_{\mathcal{H}, S'_{1,\mathbb{Q}}}$ of type $(L \underset{\mathbb{Z}}{\otimes} \hat{\mathbb{Z}}, \langle\,\cdot\,,\,\cdot\,\rangle)$ as in [62, Def. 1.3.7.6], and such that the pullbacks of $(\tilde{G}_{S'_{1,\mathbb{Q}}}, \tilde{\lambda}_{S'_{1,\mathbb{Q}}}, \tilde{i}_{S'_{1,\mathbb{Q}}}, \tilde{\alpha}_{\mathcal{H}, S'_{1,\mathbb{Q}}}) \to S'_{1,\mathbb{Q}}$ and $(G^\dagger, \lambda^\dagger, i^\dagger, \alpha_{\mathcal{H}}^\dagger) \to \mathrm{Spec}(R)$ to $\bar{\eta}$ are isomorphic to each other. By the universal property of $\mathsf{M}_{\mathcal{H}}$, there is a canonical morphism

$$S'_{1,\mathbb{Q}} \to \mathsf{M}_{\mathcal{H}} \tag{2.2.1.9}$$

under which $(\tilde{G}_{S'_{1,\mathbb{Q}}}, \tilde{\lambda}_{S'_{1,\mathbb{Q}}}, \tilde{i}_{S'_{1,\mathbb{Q}}}, \tilde{\alpha}_{\mathcal{H}, S'_{1,\mathbb{Q}}})$ is isomorphic to the pullback of the tautological tuple $(A, \lambda, i, \alpha_{\mathcal{H}})$ over $\mathsf{M}_{\mathcal{H}}$. By construction, the compositions $\bar{\eta} \to S'_{1,\mathbb{Q}} \to$

$\mathsf{M}_{\mathcal{H}}$ and $\bar{\eta} \to \eta \to \mathsf{M}_{\mathcal{H}}$ coincide with each other (*cf.* the proof of [58, Thm. 4.1]).

Let S' and S'_1 denote the normalizations of S and S_1 under the canonical morphisms $S'_{1,\mathbb{Q}} \to S$ and $S'_{1,\mathbb{Q}} \to S_1$, respectively. Then $(\tilde{G}_{S'_{1,\mathbb{Q}}}, \tilde{\lambda}_{S'_{1,\mathbb{Q}}}, \tilde{i}_{S'_{1,\mathbb{Q}}}, \tilde{\alpha}_{\mathcal{H},S'_{1,\mathbb{Q}}})$ canonically extends to degenerating families $(\tilde{G}_{S'}, \tilde{\lambda}_{S'}, \tilde{i}_{S'}, \tilde{\alpha}_{\mathcal{H},S'}) \to S'$ and $(\tilde{G}_{S'_1}, \tilde{\lambda}_{S'_1}, \tilde{i}_{S'_1}, \tilde{\alpha}_{\mathcal{H},S'_1}) \to S'_1$ of type $\mathsf{M}_{\mathcal{H}}$, where $(\tilde{G}_{S'}, \tilde{\lambda}_{S'}, \tilde{i}_{S'})$ and $(\tilde{G}_{S'_1}, \tilde{\lambda}_{S'_1}, \tilde{i}_{S'_1})$ are just the pullbacks of $(\tilde{G}, \tilde{\lambda}, \tilde{i})$ from S to S' and S'_1, respectively, and where $\tilde{\alpha}_{\mathcal{H},S'}$ and $\tilde{\alpha}_{\mathcal{H},S'_1}$ are defined only over $S'_{1,\mathbb{Q}}$.

By definition, $\vec{\mathsf{M}}_{\mathcal{H}}$ and $\vec{\mathsf{M}}^{\min}_{\mathcal{H}}$ are the normalizations of $\mathsf{M}_{\mathcal{H}^p_{\mathrm{aux}}}$ and $\mathsf{M}^{\min}_{\mathcal{H}^p_{\mathrm{aux}}}$ in $\mathsf{M}_{\mathcal{H}}$, as in Propositions 2.2.1.1 and 2.2.1.2, for some open compact subgroup $\mathcal{H}^p_{\mathrm{aux}}$ of $\mathrm{G}_{\mathrm{aux}}(\hat{\mathbb{Z}}^p)$. By the same argument as in the proof of Proposition 2.1.1.15, by forgetting the factor of $\tilde{\alpha}_{\mathcal{H},S'_1}$ at p, the tuple $(\tilde{G}_{S'_1}, \tilde{\lambda}_{S'_1}, \tilde{i}_{S'_1}, \tilde{\alpha}_{\mathcal{H},S'_1})$ induces a tuple parameterized by $\mathsf{M}_{\mathcal{H}^p_{\mathrm{aux}}}$, and induces a morphism

$$S'_1 \to \mathsf{M}_{\mathcal{H}^p_{\mathrm{aux}}} \tag{2.2.1.10}$$

by the universal property of $\mathsf{M}_{\mathcal{H}^p_{\mathrm{aux}}}$, whose restriction to $S'_{1,\mathbb{Q}}$ coincides with the composition of (2.2.1.9) with the morphism $\mathsf{M}_{\mathcal{H}} \to \mathsf{M}_{\mathcal{H}^p_{\mathrm{aux}}}$ induced by (2.1.1.17) (as in Proposition 2.2.1.1). Consequently, by the definition of $\vec{\mathsf{M}}_{\mathcal{H}}$ as a normalization, (2.2.1.10) induces a morphism

$$S'_1 \to \vec{\mathsf{M}}_{\mathcal{H}} \tag{2.2.1.11}$$

extending (2.2.1.9).

By the same argument as in the proof of Proposition 2.1.2.29, the degenerating family $(\tilde{G}_{S'}, \tilde{\lambda}_{S'}, \tilde{i}_{S'}, \tilde{\alpha}_{\mathcal{H},S'}) \to S'$ of type $\mathsf{M}_{\mathcal{H}}$ induces a degenerating family $(\tilde{G}^{\triangledown}_{\mathrm{aux},S'}, \tilde{\lambda}^{\triangledown}_{\mathrm{aux},S'}, \tilde{i}^{\triangledown}_{\mathrm{aux},S'}, \tilde{\alpha}^{\triangledown}_{\mathcal{H}^p_{\mathrm{aux}},S'}) \to S'$ of type $\mathsf{M}_{\mathcal{H}^p_{\mathrm{aux}}}$, which defines a morphism

$$S' \to \mathsf{M}^{\mathrm{tor}}_{\mathcal{H}^p_{\mathrm{aux}},\Sigma^p_{\mathrm{aux}}} \tag{2.2.1.12}$$

for any compatible choice of Σ^p_{aux} for $\mathsf{M}_{\mathcal{H}^p_{\mathrm{aux}}}$ as in [62, Def. 6.3.3.4] (*cf.* Definition 1.2.2.13), because the property (3) above ensures that the degenerating family $(\tilde{G}^{\triangledown}_{\mathrm{aux},S'}, \tilde{\lambda}^{\triangledown}_{\mathrm{aux},S'}, \tilde{i}^{\triangledown}_{\mathrm{aux},S'}, \tilde{\alpha}^{\triangledown}_{\mathcal{H}^p_{\mathrm{aux}},S'}) \to S'$ satisfies the condition as in [62, Thm. 6.4.1.1(6)] (*cf.* the proof of [62, Prop. 6.3.3.17].) By composition with the canonical morphism $\oint_{\mathcal{H}^p_{\mathrm{aux}}} : \mathsf{M}^{\mathrm{tor}}_{\mathcal{H}^p_{\mathrm{aux}},\Sigma^p_{\mathrm{aux}}} \to \mathsf{M}^{\min}_{\mathcal{H}^p_{\mathrm{aux}}}$ as in [62, Thm. 7.2.4.1(3)] (*cf.* (3) of Theorem 1.3.1.5), (2.2.1.12) induces a morphism

$$S' \to \mathsf{M}^{\min}_{\mathcal{H}^p_{\mathrm{aux}}}, \tag{2.2.1.13}$$

whose restriction to S'_1 is the composition of (2.2.1.11) with the canonical morphism $\vec{\mathsf{M}}_{\mathcal{H}} \to \vec{\mathsf{M}}^{\min}_{\mathcal{H}}$. Consequently, by the definition of $\vec{\mathsf{M}}^{\min}_{\mathcal{H}}$ as a normalization, (2.2.1.13) induces a morphism

$$S' \to \vec{\mathsf{M}}^{\min}_{\mathcal{H}} \tag{2.2.1.14}$$

extending (2.2.1.11).

Since the geometric point $\bar{s} \to S$ lifts to some geometric point $\bar{s} \to S'$ by the finiteness of $S' \to S$, and since $S'_1 \underset{\mathbb{Z}}{\otimes} \mathbb{F}_p$ is nonempty and dense in $S' \underset{\mathbb{Z}}{\otimes} \mathbb{F}_p$ because $S_1 \underset{\mathbb{Z}}{\otimes} \mathbb{F}_p$ is nonempty and dense in $S \underset{\mathbb{Z}}{\otimes} \mathbb{F}_p$ (by the property (2) above), the image s of $\bar{s}$ in $\vec{\mathsf{M}}^{\min} \underset{\mathbb{Z}}{\otimes} \mathbb{F}_p$ is the specialization of some point of $\vec{\mathsf{M}}_{\mathcal{H}} \underset{\mathbb{Z}}{\otimes} \mathbb{F}_p$. Since s is arbitrary, this shows that the open image of (2.2.1.8) is a dense subset, as desired. □

Corollary 2.2.1.15. *Up to canonical isomorphism, the scheme $\vec{\mathsf{M}}_{\mathcal{H}}^{\min}$ constructed in Proposition 2.2.1.2 depends only on the linear algebraic data defining $\mathsf{M}_{\mathcal{H}}$, but not on the auxiliary choices defining $\mathsf{M}_{\mathrm{G}_{\mathrm{aux}}(\hat{\mathbb{Z}}^p)}$ or $\mathsf{M}_{\mathcal{H}^p_{\mathrm{aux}}}$.*

Proof. By Propositions 2.2.1.2 and 2.2.1.7, $\vec{\mathsf{M}}_{\mathcal{H}}^{\min}$ is flat over $\mathbb{Z}_{(p)}$ and is noetherian normal, and the complement of $[\vec{\mathsf{M}}_{\mathcal{H}}] \cup \mathsf{M}_{\mathcal{H}}^{\min}$ in $\vec{\mathsf{M}}_{\mathcal{H}}^{\min}$ is of codimension at least two. Hence, the canonical restriction morphism

$$\Gamma(\vec{\mathsf{M}}_{\mathcal{H}}^{\min}, \omega_{\vec{\mathsf{M}}_{\mathcal{H}}^{\min}}^{\otimes aN_1k}) \to \Gamma([\vec{\mathsf{M}}_{\mathcal{H}}] \cup \mathsf{M}_{\mathcal{H}}^{\min}, (\omega_{\vec{\mathsf{M}}_{\mathcal{H}}^{\min}}^{\otimes aN_1k})|_{[\vec{\mathsf{M}}_{\mathcal{H}}] \cup \mathsf{M}_{\mathcal{H}}^{\min}}) \tag{2.2.1.16}$$

is an isomorphism for each $k \geq 0$. By Propositions 2.2.1.1 and 2.2.1.2, the right-hand side of (2.2.1.16) depends only on the linear algebraic data defining $\mathsf{M}_{\mathcal{H}}$. Since $\vec{\mathsf{M}}_{\mathcal{H}}^{\min} \cong \mathrm{Proj}\Big(\underset{k\geq 0}{\oplus} \Gamma(\vec{\mathsf{M}}_{\mathcal{H}}^{\min}, \omega_{\vec{\mathsf{M}}_{\mathcal{H}}^{\min}}^{\otimes aN_1k})\Big)$, the corollary follows, as desired. □

2.2.2 *Flat Integral Models for Projective Toroidal Compactifications*

Proposition 2.2.2.1. *Let $\mathcal{H}$, Σ, pol, $\jmath_{\mathcal{H},d\mathsf{pol}}$, and $\mathcal{J}_{\mathcal{H},d\mathsf{pol}}$ be as in Theorem 1.3.1 10 (for each integer $d \geq 1$). (In particular, $\mathcal{H}$ is neat and Σ is projective.) For each $d \geq 1$, let $\vec{\mathcal{J}}_{\mathcal{H},d\mathsf{pol}}$ be the coherent $\mathscr{O}_{\vec{\mathsf{M}}_{\mathcal{H}}^{\min}}$-ideal defining the schematic closure in $\vec{\mathsf{M}}_{\mathcal{H}}^{\min}$ of the closed subscheme of $\mathsf{M}_{\mathcal{H}}^{\min}$ defined by the coherent $\mathscr{O}_{\mathsf{M}_{\mathcal{H}}^{\min}}$-ideal $\mathcal{J}_{\mathcal{H},d\mathsf{pol}}$. Suppose $d_0 \geq 1$ is any integer such that the statement in Theorem 1.3.1.10 is true. Let*

$$\vec{\mathsf{M}}_{\mathcal{H},d_0\mathsf{pol}}^{\mathrm{tor}} := \mathrm{NBl}_{\vec{\mathcal{J}}_{\mathcal{H},d_0\mathsf{pol}}}(\vec{\mathsf{M}}_{\mathcal{H}}^{\min}).$$

Then $\vec{\mathsf{M}}_{\mathcal{H},d_0\mathsf{pol}}^{\mathrm{tor}}$ is a normal scheme projective and flat over $\vec{\mathsf{S}}_0 = \mathrm{Spec}(\mathcal{O}_{F_0,(p)})$ equipped with a canonical isomorphism $\vec{\mathsf{M}}_{\mathcal{H},d_0\mathsf{pol}}^{\mathrm{tor}} \underset{\vec{\mathsf{S}}_0}{\times} \mathsf{S}_0 \cong \mathsf{M}_{\mathcal{H},\Sigma}^{\mathrm{tor}}$ over $\mathsf{S}_0 = \mathrm{Spec}(F_0)$. The canonical morphism $\oint_{\mathcal{H}} : \mathsf{M}_{\mathcal{H},\Sigma}^{\mathrm{tor}} \to \mathsf{M}_{\mathcal{H}}^{\min}$ extends to a canonical morphism $\vec{\oint}_{\mathcal{H}} : \vec{\mathsf{M}}_{\mathcal{H},d_0\mathsf{pol}}^{\mathrm{tor}} \to \vec{\mathsf{M}}_{\mathcal{H}}^{\min}$. Moreover, the canonical morphisms

$$\mathscr{O}_{\vec{\mathsf{M}}_{\mathcal{H}}^{\min}} \to \vec{\oint}_{\mathcal{H},*}\mathscr{O}_{\vec{\mathsf{M}}_{\mathcal{H},d_0\mathsf{pol}}^{\mathrm{tor}}} \tag{2.2.2.2}$$

is an isomorphism. Since closed subscheme of $\vec{\mathsf{M}}_{\mathcal{H}}^{\min}$ defined by $\vec{\mathcal{J}}_{\mathcal{H},d\mathsf{pol}}$ necessarily lies in the closed complement of $\vec{\mathsf{M}}_{\mathcal{H}}$ in $\vec{\mathsf{M}}_{\mathcal{H}}^{\min}$, the pullback of $\vec{\oint}_{\mathcal{H}}$ under the canonical morphism $\vec{\mathsf{M}}_{\mathcal{H}} \to \vec{\mathsf{M}}_{\mathcal{H}}^{\min}$ is an isomorphism, which canonically identifies $\vec{\mathsf{M}}_{\mathcal{H}}$ as an open dense subscheme of $\vec{\mathsf{M}}_{\mathcal{H},d_0\mathsf{pol}}^{\mathrm{tor}}$.

By abuse of notation, we denote the pullback of $\omega_{\vec{\mathsf{M}}_{\mathcal{H}}^{\min}}^{\otimes aN_1}$ to $\vec{\mathsf{M}}_{\mathcal{H},d_0\mathsf{pol}}^{\mathrm{tor}}$ by $\omega_{\vec{\mathsf{M}}_{\mathcal{H},d_0\mathsf{pol}}^{\mathrm{tor}}}^{\otimes aN_1}$ (cf. Proposition 2.2.1.2). Then $\omega_{\vec{\mathsf{M}}_{\mathcal{H},d_0\mathsf{pol}}^{\mathrm{tor}}}^{\otimes aN_1k}$ is generated by global sections for sufficiently large $k \geq 1$, and we have a canonical isomorphism

$$\vec{\mathsf{M}}_{\mathcal{H}}^{\min} \cong \mathrm{Proj}\Big(\underset{k\geq 0}{\oplus} \Gamma(\vec{\mathsf{M}}_{\mathcal{H},d_0\mathsf{pol}}^{\mathrm{tor}}, \omega_{\vec{\mathsf{M}}_{\mathcal{H},d_0\mathsf{pol}}^{\mathrm{tor}}}^{\otimes aN_1k})\Big).$$

Proof. The statements in the first paragraph are all self-explanatory. The statement in the second paragraph is true because, for each $k \geq 0$, $\Gamma(\vec{\mathsf{M}}^{\mathrm{tor}}_{\mathcal{H},d_0\mathsf{pol}}, \omega^{\otimes aN_1k}_{\vec{\mathsf{M}}^{\mathrm{tor}}_{\mathcal{H},d_0\mathsf{pol}}}) \cong \Gamma(\vec{\mathsf{M}}^{\min}_{\mathcal{H}}, \omega^{\otimes aN_1k}_{\vec{\mathsf{M}}^{\min}_{\mathcal{H}}})$ by the projection formula [35, 0_{I}, 5.4.10.1], and because $\omega^{\otimes aN_1}_{\vec{\mathsf{M}}^{\min}_{\mathcal{H}}}$ is ample over $\vec{\mathsf{M}}^{\min}_{\mathcal{H}}$ (see Proposition 2.2.1.2). □

Proposition 2.2.2.3. *With the assumptions in Proposition* 2.2.2.1*, suppose* $\mathcal{H}'$*,* Σ'*, and* pol' *are as in Proposition* 1.3.1.11*. Then we define*

$$\vec{\mathscr{J}}_{\mathcal{H}',d_0\mathsf{pol}'} := (\vec{\mathsf{M}}^{\min}_{\mathcal{H}'} \to \vec{\mathsf{M}}^{\min}_{\mathcal{H}})^* \vec{\mathscr{J}}_{\mathcal{H},d_0\mathsf{pol}}$$

and

$$\vec{\mathsf{M}}^{\mathrm{tor}}_{\mathcal{H}',d_0\mathsf{pol}'} := \mathrm{NBl}_{\vec{\mathscr{J}}_{\mathcal{H}',d_0\mathsf{pol}'}}(\vec{\mathsf{M}}^{\min}_{\mathcal{H}'}).$$

Then $\vec{\mathsf{M}}^{\mathrm{tor}}_{\mathcal{H}',d_0\mathsf{pol}'}$ *enjoys analogues of properties of* $\vec{\mathsf{M}}^{\mathrm{tor}}_{\mathcal{H},d_0\mathsf{pol}}$ *in Proposition* 2.2.2.1*, and we have a canonical morphism*

$$\vec{\mathsf{M}}^{\mathrm{tor}}_{\mathcal{H}',d_0\mathsf{pol}'} \to \vec{\mathsf{M}}^{\mathrm{tor}}_{\mathcal{H},d_0\mathsf{pol}} \tag{2.2.2.4}$$

which is **finite**. *Moreover,* $\vec{\mathsf{M}}^{\mathrm{tor}}_{\mathcal{H}',d_0\mathsf{pol}'}$ *is canonically isomorphic to the normalization of* $\vec{\mathsf{M}}^{\mathrm{tor}}_{\mathcal{H},d_0\mathsf{pol}}$ *in* $\mathsf{M}_{\mathcal{H}'}$ *under the composition of canonical morphisms* $\mathsf{M}_{\mathcal{H}'} \to \mathsf{M}_{\mathcal{H}} \hookrightarrow \mathsf{M}^{\mathrm{tor}}_{\mathcal{H},\Sigma} \to \vec{\mathsf{M}}^{\mathrm{tor}}_{\mathcal{H},d_0\mathsf{pol}}$.

If Σ' *is* **smooth**, *then* $\vec{\mathsf{M}}^{\mathrm{tor}}_{\mathcal{H}',d_0\mathsf{pol}'}$ *can also be constructed as in Proposition* 2.2.2.1*, and the schemes we obtain in the two constructions are canonically isomorphic.*

Proof. As in the proof of Proposition 1.3.1.11, since $\vec{\mathscr{J}}_{\mathcal{H}',d_0\mathsf{pol}'}$ is the pullback of $\vec{\mathscr{J}}_{\mathcal{H},d_0\mathsf{pol}}$ under the finite morphism $\vec{\mathsf{M}}^{\min}_{\mathcal{H}'} \to \vec{\mathsf{M}}^{\min}_{\mathcal{H}}$, the morphism (2.2.2.4) exists and is finite, by the universal property of the normalization of blow-up. Since $\vec{\mathsf{M}}^{\mathrm{tor}}_{\mathcal{H}',d_0\mathsf{pol}'}$ is normal, it is canonically isomorphic to the normalization of $\vec{\mathsf{M}}^{\mathrm{tor}}_{\mathcal{H},d_0\mathsf{pol}}$ in $\mathsf{M}_{\mathcal{H}'}$ by Zariski's main theorem (see [35, III-1, 4.4.3, 4.4.11]).

If Σ' is *smooth*, then the $\vec{\mathscr{J}}_{\mathcal{H}',d_0\mathsf{pol}'}$ defined as a pullback is canonically isomorphic to the $\vec{\mathscr{J}}_{\mathcal{H}',d_0\mathsf{pol}'}$ defined on $\vec{\mathsf{M}}^{\min}_{\mathcal{H}'}$ itself, and hence the two constructions of $\vec{\mathsf{M}}^{\mathrm{tor}}_{\mathcal{H}',d_0\mathsf{pol}'}$ by normalizations of blow-ups give canonically isomorphic schemes. □

Remark 2.2.2.5. We introduce the scheme $\vec{\mathsf{M}}^{\mathrm{tor}}_{\mathcal{H},d_0\mathsf{pol}}$ in Proposition 2.2.2.1 mainly for technical reasons, and for the sake of completeness. This is even more so for the scheme $\vec{\mathsf{M}}^{\mathrm{tor}}_{\mathcal{H}',d_0\mathsf{pol}'}$ in Proposition 2.2.2.3. (We will need special cases of them in Sections 5.2.3 and 6.1.1 below.)

For each stratum $\mathsf{Z}_{[(\Phi_{\mathcal{H}},\delta_{\mathcal{H}},\sigma)]}$ as in (2) of Theorem 1.3.1.3, consider its closure $\overline{\mathsf{Z}}_{[(\Phi_{\mathcal{H}},\delta_{\mathcal{H}},\sigma)]}$ in $\mathsf{M}^{\mathrm{tor}}_{\mathcal{H},\Sigma}$ and its closure $\overline{\vec{\mathsf{Z}}}_{[(\Phi_{\mathcal{H}},\delta_{\mathcal{H}},\sigma)],d_0\mathsf{pol}}$ in $\vec{\mathsf{M}}^{\mathrm{tor}}_{\mathcal{H},d_0\mathsf{pol}}$. Then we define a locally closed subscheme

$$\vec{\mathsf{Z}}_{[(\Phi_{\mathcal{H}},\delta_{\mathcal{H}},\sigma)],d_0\mathsf{pol}} := \overline{\vec{\mathsf{Z}}}_{[(\Phi_{\mathcal{H}},\delta_{\mathcal{H}},\sigma)],d_0\mathsf{pol}} - \bigcup_{\mathsf{Z}_{[(\Phi_{\mathcal{H}},\delta_{\mathcal{H}},\sigma)]} \not\subseteq \overline{\mathsf{Z}}_{[(\Phi'_{\mathcal{H}},\delta'_{\mathcal{H}},\tau)]}} \overline{\vec{\mathsf{Z}}}_{[(\Phi'_{\mathcal{H}},\delta'_{\mathcal{H}},\tau)],d_0\mathsf{pol}} \tag{2.2.2.6}$$

of $\vec{\mathsf{M}}^{\mathrm{tor}}_{\mathcal{H},d_0\mathsf{pol}}$. By definition, we have the following:

Lemma 2.2.2.7. *If $\mathsf{Z}_{[(\Phi_{\mathcal{H}},\delta_{\mathcal{H}},\sigma)]}$ is contained in the closure $\overline{\mathsf{Z}}_{[(\Phi'_{\mathcal{H}},\delta'_{\mathcal{H}},\tau)]}$ of $\mathsf{Z}_{[(\Phi'_{\mathcal{H}},\delta'_{\mathcal{H}},\tau)]}$, then $\vec{\mathsf{Z}}_{[(\Phi_{\mathcal{H}},\delta_{\mathcal{H}},\sigma)],d_0\mathsf{pol}}$ is contained in $\overline{\vec{\mathsf{Z}}}_{[(\Phi'_{\mathcal{H}},\delta'_{\mathcal{H}},\tau)],d_0\mathsf{pol}}$, and the latter agrees with the closure of $\vec{\mathsf{Z}}_{[(\Phi'_{\mathcal{H}},\delta'_{\mathcal{H}},\tau)],d_0\mathsf{pol}}$ in $\vec{\mathsf{M}}^{\mathrm{tor}}_{\mathcal{H},d_0\mathsf{pol}}$. Moreover, the canonical morphism $\vec{\oint}_{\mathcal{H}}:\vec{\mathsf{M}}^{\mathrm{tor}}_{\mathcal{H},d_0\mathsf{pol}}\to\vec{\mathsf{M}}^{\mathrm{min}}_{\mathcal{H}}$ maps each $\vec{\mathsf{Z}}_{[(\Phi_{\mathcal{H}},\delta_{\mathcal{H}},\sigma)],d_0\mathsf{pol}}$ to (an open subscheme of) $\vec{\mathsf{Z}}_{[(\Phi_{\mathcal{H}},\delta_{\mathcal{H}})]}$.*

Remark 2.2.2.8. It is not clear whether the collection $\{\vec{\mathsf{Z}}_{[(\Phi_{\mathcal{H}},\delta_{\mathcal{H}},\sigma)],d_0\mathsf{pol}}\}_{[(\Phi_{\mathcal{H}},\delta_{\mathcal{H}},\sigma)]}$ defines a stratification of $\vec{\mathsf{M}}^{\mathrm{tor}}_{\mathcal{H},d_0\mathsf{pol}}$ (*cf.* Remark 2.2.1.6). However, we shall still call $\vec{\mathsf{Z}}_{[(\Phi_{\mathcal{H}},\delta_{\mathcal{H}},\sigma)],d_0\mathsf{pol}}$ the $[(\Phi_{\mathcal{H}},\delta_{\mathcal{H}},\sigma)]$-*stratum*, by abuse of language.

2.2.3 *Hecke Actions*

Proposition 2.2.3.1. (*Compare with Proposition* 1.3.1.14.) *Suppose that $g=(g_0,g_p)\in\mathrm{G}(\mathbb{A}^{\infty,p})\times\mathrm{G}(\mathbb{Z}_p)\subset\mathrm{G}(\mathbb{A}^{\infty})$ and that $\mathcal{H}$ and $\mathcal{H}'$ are two open compact subgroups of $\mathrm{G}(\hat{\mathbb{Z}})$ such that $\mathcal{H}'\subset g\mathcal{H}g^{-1}$. Then there is a canonical finite surjection*

$$\vec{[g]}:\vec{\mathsf{M}}_{\mathcal{H}'}\to\vec{\mathsf{M}}_{\mathcal{H}}$$

(over $\vec{\mathsf{S}}_0=\mathrm{Spec}(\mathcal{O}_{F_0,(p)})$) extending the canonical finite surjection $[g]:\mathsf{M}_{\mathcal{H}'}\to\mathsf{M}_{\mathcal{H}}$ (over $\mathsf{S}_0=\mathrm{Spec}(F_0)$) defined by the Hecke action of g, such that $\omega^{\otimes k}_{\vec{\mathsf{M}}_{\mathcal{H}}}$ over $\vec{\mathsf{M}}_{\mathcal{H}}$ is pulled back to $\omega^{\otimes k}_{\vec{\mathsf{M}}_{\mathcal{H}'}}$ over $\vec{\mathsf{M}}_{\mathcal{H}'}$ (up to canonical isomorphism) whenever k is divisible by the integer a in Lemma 2.1.2.35. *Moreover, there is a canonical finite surjection*

$$\vec{[g]}^{\mathrm{min}}:\vec{\mathsf{M}}^{\mathrm{min}}_{\mathcal{H}'}\to\vec{\mathsf{M}}^{\mathrm{min}}_{\mathcal{H}}$$

(over $\vec{\mathsf{S}}_0$) extending the canonical finite surjection $[[g]]:[\mathsf{M}_{\mathcal{H}'}]\to[\mathsf{M}_{\mathcal{H}}]$ (over S_0) induced by $[g]$, such that $\omega^{\otimes k}_{\vec{\mathsf{M}}^{\mathrm{min}}_{\mathcal{H}}}$ over $\vec{\mathsf{M}}^{\mathrm{min}}_{\mathcal{H}}$ is pulled back to $\omega^{\otimes k}_{\vec{\mathsf{M}}^{\mathrm{min}}_{\mathcal{H}'}}$ over $\vec{\mathsf{M}}^{\mathrm{min}}_{\mathcal{H}'}$ (up to canonical isomorphism, compatible with the previous one) whenever the former is defined. (This canonical morphism is compatible with the canonical isomorphism $([g]^{\mathrm{min}})^*\omega^{\otimes k}_{\mathsf{M}^{\mathrm{min}}_{\mathcal{H}}}\xrightarrow{\sim}\omega^{\otimes k}_{\mathsf{M}^{\mathrm{min}}_{\mathcal{H}'}}$ in Proposition 1.3.1.14.) *By restriction, the surjection $\vec{[g]}^{\mathrm{min}}$ induces the surjection $[\vec{[g]}]:[\vec{\mathsf{M}}_{\mathcal{H}'}]\to[\vec{\mathsf{M}}_{\mathcal{H}}]$ induced by $\vec{[g]}$.*

The surjection $\vec{[g]}^{\mathrm{min}}$ maps the $[(\Phi'_{\mathcal{H}'},\delta'_{\mathcal{H}'})]$-stratum $\vec{\mathsf{Z}}_{[(\Phi'_{\mathcal{H}'},\delta'_{\mathcal{H}'})]}$ of $\vec{\mathsf{M}}^{\mathrm{min}}_{\mathcal{H}'}$ to the $[(\Phi_{\mathcal{H}},\delta_{\mathcal{H}})]$-stratum $\vec{\mathsf{Z}}_{[(\Phi_{\mathcal{H}},\delta_{\mathcal{H}})]}$ of $\vec{\mathsf{M}}^{\mathrm{min}}_{\mathcal{H}}$ if and only if there are representatives $(\Phi_{\mathcal{H}},\delta_{\mathcal{H}})$ and $(\Phi'_{\mathcal{H}'},\delta'_{\mathcal{H}'})$ of $[(\Phi_{\mathcal{H}},\delta_{\mathcal{H}})]$ and $[(\Phi'_{\mathcal{H}'},\delta'_{\mathcal{H}'})]$, respectively, such that $(\Phi_{\mathcal{H}},\delta_{\mathcal{H}})$ is g-assigned to $(\Phi'_{\mathcal{H}'},\delta'_{\mathcal{H}'})$ as in [62, Def. 5.4.3.9].

If $g=g_1g_2$, where $g_1=(g_{1,0},g_{1,p})$ and $g_2=(g_{2,0},g_{2,p})$ are elements of $\mathrm{G}(\mathbb{A}^{\infty,p})\times\mathrm{G}(\mathbb{Z}_p)\subset\mathrm{G}(\mathbb{A}^{\infty})$, each having a setup similar to that of g, then we have $\vec{[g]}=\vec{[g_2]}\circ\vec{[g_1]}$, $[\vec{[g]}]=[\vec{[g_2]}]\circ[\vec{[g_1]}]$, and $\vec{[g]}^{\mathrm{min}}=\vec{[g_2]}^{\mathrm{min}}\circ\vec{[g_1]}^{\mathrm{min}}$.

Proof. Since $\mathcal{H}'\subset g\mathcal{H}g^{-1}$, by considering their images under the canonical homomorphisms $\mathrm{G}(\hat{\mathbb{Z}})\to\mathrm{G}(\hat{\mathbb{Z}}^p)$ and $\mathrm{G}(\hat{\mathbb{Z}})\to\mathrm{G}(\mathbb{Z}_p)$ (and the canonical homomorphisms

$G(\hat{\mathbb{Z}}^p) \to G_{\text{aux}}(\hat{\mathbb{Z}}^p)$ and $G(\mathbb{Z}_p) \to G_{\text{aux}}(\mathbb{Z}_p)$ given by (2.1.1.10)), there exist an open compact subgroup $\mathcal{H}'^{,p}_{\text{aux}}$ of $G_{\text{aux}}(\hat{\mathbb{Z}}^p)$ contained in $g_0 G_{\text{aux}}(\hat{\mathbb{Z}}^p) g_0^{-1}$ such that $\mathcal{H}'^{,p}_{\text{aux}} G_{\text{aux}}(\mathbb{Z}_p)$ contains the image of $\mathcal{H}'$ under the homomorphism $G(\hat{\mathbb{Z}}) \to G_{\text{aux}}(\hat{\mathbb{Z}})$ given by (2.1.1.10).

Since $g_p \in G(\mathbb{Z}_p)$, the Hecke twists of tautological objects over $\mathsf{M}_{\mathcal{H}'}$ and $\mathsf{M}_{\mathcal{H}'^{,p}_{\text{aux}}} \underset{\mathbb{Z}}{\otimes} \mathbb{Q}$ are realized by compatible $\mathbb{Z}_{(p)}^\times$-isogenies, and hence $[g] : \mathsf{M}_{\mathcal{H}'} \to \mathsf{M}_{\mathcal{H}}$ and $[g_0] \underset{\mathbb{Z}}{\otimes} \mathbb{Q} : \mathsf{M}_{\mathcal{H}'^{,p}_{\text{aux}}} \underset{\mathbb{Z}}{\otimes} \mathbb{Q} \to \mathsf{M}_{G_{\text{aux}}(\hat{\mathbb{Z}}^p)} \underset{\mathbb{Z}}{\otimes} \mathbb{Q}$ are compatible (see (2) of Lemma 2.1.1.1 and Proposition 2.1.1.15). Then we have a commutative diagram

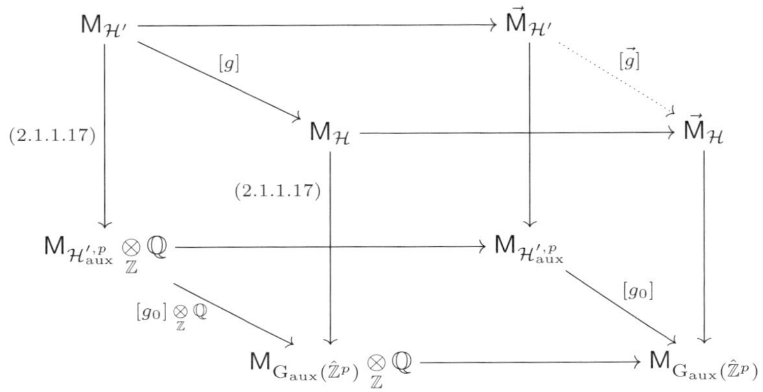

of solid arrows, in which all unnamed morphisms are canonical morphisms, inducing the desired (compatible) dotted arrow.

By taking normalizations and by Zariski's main theorem (see [35, III-1, 4.4.3, 4.4.11]), we obtain a commutative diagram

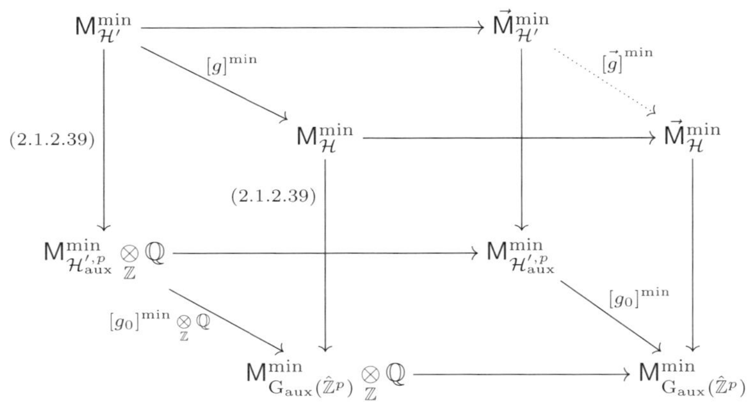

of solid arrows compatible with the previous one, in which all unnamed morphisms are canonical morphisms, inducing the desired dotted arrow (compatible with all the other arrows in both diagrams). The remaining statements in the proposition then follow from the known statements (including those in Proposition 1.3.1.14) and from the various universal properties. □

Corollary 2.2.3.2. (*Compare with [62, Cor. 7.2.5.2].*) *Suppose we have two open compact subgroups $\mathcal{H}$ and $\mathcal{H}'$ of $\mathrm{G}(\hat{\mathbb{Z}})$ such that $\mathcal{H}'$ is a normal subgroup of $\mathcal{H}$. Then the canonical morphisms defined in Proposition* 2.2.3.1 *induce an action of the finite group $\mathcal{H}/\mathcal{H}'$ on $\vec{\mathsf{M}}^{\min}_{\mathcal{H}'}$. The canonical surjection $\vec{[1]}^{\min} : \vec{\mathsf{M}}^{\min}_{\mathcal{H}'} \twoheadrightarrow \vec{\mathsf{M}}^{\min}_{\mathcal{H}}$ defined by Proposition* 2.2.3.1 *can be identified with the quotient of $\vec{\mathsf{M}}^{\min}_{\mathcal{H}'}$ by this action.*

Proof. The existence of such an action is clear. Since $\vec{\mathsf{M}}^{\min}_{\mathcal{H}'}$ is projective over S_0 and normal, the quotient $\vec{\mathsf{M}}^{\min}_{\mathcal{H}'}/(\mathcal{H}/\mathcal{H}')$ exists as a scheme (*cf.* [25, V, 4.1]). Then it follows from Zariski's main theorem (see [35, III-1, 4.4.3, 4.4.11]) that the induced morphism $\vec{\mathsf{M}}^{\min}_{\mathcal{H}'}/(\mathcal{H}/\mathcal{H}') \to \vec{\mathsf{M}}^{\min}_{\mathcal{H}}$ (with noetherian normal target) is an isomorphism, because it is generically so (over $\mathsf{M}^{\min}_{\mathcal{H}}$, by [62, Cor. 7.2.5.2]—in fact, the proof here is part of that of [62, Cor. 7.2.5.2]). ❐

For later references, let us define:

Definition 2.2.3.3. *For each integer $i \geq 0$, we define $\mathsf{S}_{0,i} := \mathrm{Spec}(F_0[\zeta_{p^i}])$ and $\vec{\mathsf{S}}_{0,i} := \mathrm{Spec}(\mathcal{O}_{F_0,(p)}[\zeta_{p^i}])$.*

Definition 2.2.3.4. *For each integer $i \geq 0$, we define $\mathsf{M}_{\mathcal{H},i}$ (resp. $\mathsf{M}^{\min}_{\mathcal{H},i}$, resp. $\mathsf{M}^{\mathrm{tor}}_{\mathcal{H},\Sigma,i}$) to be the base change $\mathsf{M}_{\mathcal{H}} \underset{\mathsf{S}_0}{\times} \mathsf{S}_{0,i}$ (resp. $\mathsf{M}^{\min}_{\mathcal{H}} \underset{\mathsf{S}_0}{\times} \mathsf{S}_{0,i}$, resp. $\mathsf{M}^{\mathrm{tor}}_{\mathcal{H},\Sigma} \underset{\mathsf{S}_0}{\times} \mathsf{S}_{0,i}$) over $\mathsf{S}_{0,i}$. For each locally closed subscheme $\mathsf{Z}_{[(\Phi_{\mathcal{H}},\delta_{\mathcal{H}})]}$ of $\mathsf{M}^{\min}_{\mathcal{H}}$ as in* (4) *of Theorem* 1.3.1.5, *we denote by $\mathsf{Z}_{[(\Phi_{\mathcal{H}},\delta_{\mathcal{H}})],i}$ its pullback under the canonical morphism $\mathsf{M}^{\min}_{\mathcal{H},i} \to \mathsf{M}^{\min}_{\mathcal{H}}$. For each locally closed subalgebraic stack $\mathsf{Z}_{[(\Phi_{\mathcal{H}},\delta_{\mathcal{H}},\sigma)]}$ of $\mathsf{M}^{\mathrm{tor}}_{\mathcal{H},\Sigma}$ as in* (2) *of Theorem* 1.3.1.3, *we denote by $\mathsf{Z}_{[(\Phi_{\mathcal{H}},\delta_{\mathcal{H}},\sigma)],i}$ its pullback under the canonical morphism $\mathsf{M}^{\mathrm{tor}}_{\mathcal{H},\Sigma,i} \to \mathsf{M}^{\mathrm{tor}}_{\mathcal{H},\Sigma}$.*

Definition 2.2.3.5. *For each integer $i \geq 0$, we define $\vec{\mathsf{M}}_{\mathcal{H},i}$ (resp. $\vec{\mathsf{M}}^{\min}_{\mathcal{H},i}$) to be the normalization of $\vec{\mathsf{M}}_{\mathcal{H}} \underset{\vec{\mathsf{S}}_0}{\times} \vec{\mathsf{S}}_{0,i}$ (resp. $\vec{\mathsf{M}}^{\min}_{\mathcal{H}} \underset{\vec{\mathsf{S}}_0}{\times} \vec{\mathsf{S}}_{0,i}$). For each locally closed subscheme $\vec{\mathsf{Z}}_{[(\Phi_{\mathcal{H}},\delta_{\mathcal{H}})]}$ of $\vec{\mathsf{M}}^{\min}_{\mathcal{H}}$ as in* (2.2.1.4), *we denote by $\vec{\mathsf{Z}}_{[(\Phi_{\mathcal{H}},\delta_{\mathcal{H}})],i}$ its pullback under the canonical morphism $\vec{\mathsf{M}}^{\min}_{\mathcal{H},i} \to \vec{\mathsf{M}}^{\min}_{\mathcal{H}}$.*

For each integer $i \geq 0$, and for each $\mathcal{H}$, Σ, pol, and d_0 as in Proposition 2.2.2.1 *such that $\vec{\mathsf{M}}^{\mathrm{tor}}_{\mathcal{H},d_0\mathsf{pol}}$ is defined, we also define $\vec{\mathsf{M}}^{\mathrm{tor}}_{\mathcal{H},d_0\mathsf{pol},i}$ to be the normalization of $\vec{\mathsf{M}}^{\mathrm{tor}}_{\mathcal{H},d_0\mathsf{pol}} \underset{\vec{\mathsf{S}}_0}{\times} \vec{\mathsf{S}}_{0,i}$. For each locally closed subscheme $\vec{\mathsf{Z}}_{[(\Phi_{\mathcal{H}},\delta_{\mathcal{H}},\sigma)],d_0\mathsf{pol}}$ of $\vec{\mathsf{M}}^{\mathrm{tor}}_{\mathcal{H},d_0\mathsf{pol}}$ as in* (2.2.2.6), *we denote by $\vec{\mathsf{Z}}_{[(\Phi_{\mathcal{H}},\delta_{\mathcal{H}},\sigma)],d_0\mathsf{pol},i}$ its pullback under the canonical morphism $\vec{\mathsf{M}}^{\mathrm{tor}}_{\mathcal{H},d_0\mathsf{pol},i} \to \vec{\mathsf{M}}^{\mathrm{tor}}_{\mathcal{H},d_0\mathsf{pol}}$. Then the base change $\vec{\oint}_{\mathcal{H}} \underset{\vec{\mathsf{S}}_0}{\times} \vec{\mathsf{S}}_{0,i} : \vec{\mathsf{M}}^{\mathrm{tor}}_{\mathcal{H},d_0\mathsf{pol}} \underset{\vec{\mathsf{S}}_0}{\times} \vec{\mathsf{S}}_{0,i} \to \vec{\mathsf{M}}^{\min}_{\mathcal{H}} \underset{\vec{\mathsf{S}}_0}{\times} \vec{\mathsf{S}}_{0,i}$ induces a canonical morphism $\vec{\oint}_{\mathcal{H},i} : \vec{\mathsf{M}}^{\mathrm{tor}}_{\mathcal{H},d_0\mathsf{pol},i} \to \vec{\mathsf{M}}^{\min}_{\mathcal{H},i}$, mapping each $\vec{\mathsf{Z}}_{[(\Phi_{\mathcal{H}},\delta_{\mathcal{H}},\sigma)],d_0\mathsf{pol},i}$ as above to $\vec{\mathsf{Z}}_{[(\Phi_{\mathcal{H}},\delta_{\mathcal{H}})],i}$. We naturally extend these definitions to the schemes $\vec{\mathsf{M}}^{\mathrm{tor}}_{\mathcal{H}',d_0\mathsf{pol}'}$ constructed in Proposition* 2.2.2.3.

For all integers $i' \geq i \geq 0$, if $\vec{[g]}^{\min} : \vec{\mathsf{M}}^{\min}_{\mathcal{H}'} \to \vec{\mathsf{M}}^{\min}_{\mathcal{H}}$ is defined as in Proposition 2.2.3.1, *then we denote the canonically induced morphism $\vec{\mathsf{M}}^{\min}_{\mathcal{H}',i'} \to \vec{\mathsf{M}}^{\min}_{\mathcal{H},i}$*

(compatible with $\vec{\mathsf{S}}_{0,i'} \to \vec{\mathsf{S}}_{0,i}$*) by* $[\vec{g}]^{\min}_{i',i}$.

2.2.4 *The Case When p is a Good Prime*

Suppose p is a good prime (for the integral PEL datum $(\mathcal{O}, \star, L, \langle\,\cdot\,,\,\cdot\,\rangle, h_0)$) as in Definition 1.1.1.6. Suppose $\mathcal{H} \subset \mathrm{G}(\hat{\mathbb{Z}})$ is an open compact subgroup. By considering the image of $\mathcal{H}$ under the canonical morphism $\mathrm{G}(\hat{\mathbb{Z}}) \to \mathrm{G}(\hat{\mathbb{Z}}^p)$, we know that there exists some open compact subgroup $\mathcal{H}^p \subset \mathrm{G}(\hat{\mathbb{Z}}^p)$ such that $\mathcal{H} \subset \mathcal{H}' := \mathcal{H}^p\mathrm{G}(\mathbb{Z}_p)$.

By [62, Prop. 1.4.4.3], there is a canonical open and closed immersion

$$\mathsf{M}_{\mathcal{H}'} \hookrightarrow \mathsf{M}_{\mathcal{H}^p} \underset{\vec{\mathsf{S}}_0}{\times} \mathsf{S}_0. \tag{2.2.4.1}$$

(By [53, Sec. 8], this is an isomorphism at least when $\mathcal{O} \underset{\mathbb{Z}}{\otimes} \mathbb{Q}$ is simple, but we do not need to know that.)

Lemma 2.2.4.2. *With assumptions as above, there is a canonical open and closed immersion*

$$\vec{\mathsf{M}}^{\min}_{\mathcal{H}'} \hookrightarrow \mathsf{M}^{\min}_{\mathcal{H}^p} \tag{2.2.4.3}$$

inducing a canonical open and closed immersion

$$[\vec{\mathsf{M}}_{\mathcal{H}'}] \hookrightarrow [\mathsf{M}_{\mathcal{H}^p}], \tag{2.2.4.4}$$

so that $\vec{\mathsf{M}}^{\min}_{\mathcal{H}'}$ *(resp.* $[\vec{\mathsf{M}}_{\mathcal{H}'}]$*) is the scheme closure of* $[\mathsf{M}_{\mathcal{H}'}]$ *in* $\mathsf{M}^{\min}_{\mathcal{H}^p}$ *(resp.* $[\mathsf{M}_{\mathcal{H}^p}]$*) under the canonical morphism induced by* (2.2.4.1).

In particular, the construction of $\vec{\mathsf{M}}^{\min}_{\mathcal{H}'}$ *(resp.* $[\vec{\mathsf{M}}_{\mathcal{H}'}]$*) is independent of the auxiliary choice of* $(\mathcal{O}_{\mathrm{aux}}, \star_{\mathrm{aux}}, L_{\mathrm{aux}}, \langle\,\cdot\,,\,\cdot\,\rangle_{\mathrm{aux}}, h_{0,\mathrm{aux}})$. *The same is true for* $\vec{\mathsf{M}}^{\min}_{\mathcal{H}}$ *(resp.* $[\vec{\mathsf{M}}_{\mathcal{H}}]$*), regardless of the choice of* $\mathcal{H}'$.

Proof. With the setting in Proposition 2.1.1.15, under the additional assumption in this lemma that p is a good prime, we can arrange that the canonical morphism (2.1.1.17) extends to a composition

$$\mathsf{M}_{\mathcal{H}'} \to \mathsf{M}_{\mathcal{H}^p} \to \mathsf{M}_{\mathrm{G}_{\mathrm{aux}}(\hat{\mathbb{Z}}^p)} \tag{2.2.4.5}$$

of canonical morphisms, the latter one being finite by the same argument as in the proof of Proposition 2.1.1.15. (In fact, by Lemma 2.1.1.1, we can take $\mathsf{M}_{\mathrm{G}_{\mathrm{aux}}(\hat{\mathbb{Z}}^p)}$ to be $\mathsf{M}_{\mathrm{G}(\hat{\mathbb{Z}}^p)}$ in this case.) Then (2.2.4.5) induces a composition

$$\mathsf{M}_{\mathcal{H}'} \to \mathsf{M}^{\min}_{\mathcal{H}^p} \to \mathsf{M}^{\min}_{\mathrm{G}_{\mathrm{aux}}(\hat{\mathbb{Z}}^p)} \tag{2.2.4.6}$$

of canonical morphisms, the latter one being finite by the same argument as in the proofs of Proposition 2.1.2.29 and Corollary 2.1.2.37. (Again by Lemma 2.1.1.1, we can take $\mathsf{M}_{\mathrm{G}_{\mathrm{aux}}(\hat{\mathbb{Z}}^p)}$ to be $\mathsf{M}_{\mathrm{G}(\hat{\mathbb{Z}}^p)}$, in which case $\mathsf{M}^{\min}_{\mathrm{G}_{\mathrm{aux}}(\hat{\mathbb{Z}}^p)}$ is $\mathsf{M}^{\min}_{\mathrm{G}(\hat{\mathbb{Z}}^p)}$.) Since $\mathsf{M}^{\min}_{\mathcal{H}^p}$ is normal, by definition of $[\vec{\mathsf{M}}_{\mathcal{H}'}]$ and $\vec{\mathsf{M}}^{\min}_{\mathcal{H}'}$ (see Propositions 2.2.1.1 and 2.2.1.2) and by Zariski's main theorem (see [35, III-1, 4.4.3, 4.4.11]), the open and closed immersion (2.2.4.1) induces the desired open and closed immersions (2.2.4.4) and (2.2.4.3).

The last assertion (concerning $\vec{\mathsf{M}}^{\min}_{\mathcal{H}}$ and $[\vec{\mathsf{M}}_{\mathcal{H}}]$) then follows, because the canonical morphism $[\mathsf{M}_{\mathcal{H}}] \to \mathsf{M}^{\min}_{\mathrm{G}_{\mathrm{aux}}(\hat{\mathbb{Z}}^p)}$ induced by (2.1.1.17) (see Proposition 2.2.1.2) factors as a composition $[\mathsf{M}_{\mathcal{H}}] \to [\mathsf{M}_{\mathcal{H}'}] \to \mathsf{M}^{\min}_{\mathcal{H}^p} \to \mathsf{M}^{\min}_{\mathrm{G}_{\mathrm{aux}}(\hat{\mathbb{Z}}^p)}$ of canonical morphisms (see (2.2.4.6)), and hence $\vec{\mathsf{M}}^{\min}_{\mathcal{H}}$ (resp. $[\vec{\mathsf{M}}_{\mathcal{H}}]$) is the normalization of $\vec{\mathsf{M}}^{\min}_{\mathcal{H}'}$ (resp. $[\vec{\mathsf{M}}_{\mathcal{H}'}]$) under the canonical morphism $[\mathsf{M}_{\mathcal{H}}] \to \vec{\mathsf{M}}^{\min}_{\mathcal{H}'}$ (resp. $[\mathsf{M}_{\mathcal{H}}] \to [\vec{\mathsf{M}}_{\mathcal{H}'}]$). This does not depend on the choice of $\mathcal{H}'$ because replacing $\mathcal{H}'$ with a finite index subgroup only results in finite morphisms between normal schemes, and the construction of $\vec{\mathsf{M}}^{\min}_{\mathcal{H}'}$ (resp. $[\vec{\mathsf{M}}_{\mathcal{H}'}]$) by normalization is insensitive to such morphisms. □

Chapter 3

Ordinary Loci

In this chapter, we introduce the notions of ordinary (semi-)abelian schemes and ordinary level structures, define the moduli problems parameterizing them, and construct the *ordinary loci* by normalizing these moduli problems after suitable base changes. The terminology of ordinary loci is, admittedly, an abuse of language. Nevertheless, these ordinary loci can be embedded into (normalizations of suitable base changes) of the flat integral models constructed in Chapter 2 (which we view as the *total models*). The main point is that, while we cannot describe the local structures of the total models in detail, we can describe the local structures of these ordinary loci rather precisely, because these ordinary loci are constructed as normalizations of moduli problems with explicit and mild singularities.

3.1 Ordinary Semi-Abelian Schemes and Serre's Construction

3.1.1 *Ordinary Abelian Schemes and Semi-Abelian Schemes*

Definition 3.1.1.1. *Let U be a scheme. We say that a quasi-finite flat commutative group scheme H of finite presentation over U is* **of étale-multiplicative type** *if it is étale locally an extension of a (commutative) étale group scheme by a finite flat group scheme of multiplicative type. (For simplicity, we shall often suppress the modifiers such as being commutative or being of finite presentation when we mention group schemes of étale-multiplicative type.)*

Definition 3.1.1.2. *Let U be a scheme. We say that a semi-abelian scheme $Z \to U$ is* **ordinary** *if, for every integer $m \geq 1$, the (commutative) quasi-finite flat group scheme $Z[m]$ (of finite presentation over U) is of étale-multiplicative type. We say an abelian scheme $Z \to U$ is ordinary if it is ordinary as a semi-abelian scheme.*

Remark 3.1.1.3. It suffices to verify this condition over strict local rings of U for rational prime numbers $m > 1$ that are residue characteristics of U.

Remark 3.1.1.4. Suppose U is the spectrum of a strict local ring of residue characteristic $p > 0$. Then it suffices to verify that, at a geometric point above the special point, the connected part of the p-divisible group of the pullback of the semi-abelian scheme is of multiplicative type. This is a condition for the slopes in the Dieudonné–Manin classification of isogeny classes of p-divisible groups over an algebraically closed field of characteristic p (see [75]).

If U is a scheme over $\mathrm{Spec}(\mathbb{Q})$, then every semi-abelian scheme $Z \to U$ is ordinary. If U is the spectrum of an algebraic closed field of characteristic p, then a semi-abelian scheme $Z \to U$, which is an extension of an abelian scheme Z^{ab} by a torus Z^{tor}, is ordinary if and only if Z^{ab} is an ordinary abelian variety.

Lemma 3.1.1.5. *If $Z \to Z'$ is an isogeny between semi-abelian schemes over U, then Z is ordinary if and only if Z' is.*

Proof. This follows from Remarks 3.1.1.3 and 3.1.1.4. □

For later reference, let us define:

Definition 3.1.1.6. *For each scheme S over* $\mathrm{Spec}(\mathbb{Z})$ *and any integer $n \geq 1$, we define a functor on the category of étale sheaves of $\mathbb{Z}/n\mathbb{Z}$-modules over S by setting*

$$\begin{aligned}(\,\cdot\,)^{\mathrm{mult}} &:= \underline{\mathrm{Hom}}_S(\underline{\mathrm{Hom}}_S(\,\cdot\,,((\mathbb{Z}/n\mathbb{Z})(1))_S),\boldsymbol{\mu}_{n,S})\\ &\cong \underline{\mathrm{Hom}}_S(\underline{\mathrm{Hom}}_S(\,\cdot\,,((\mathbb{Z}/n\mathbb{Z})(1))_S),\mathbf{G}_{\mathrm{m},S}).\end{aligned}$$

For constant sheaves of $\mathbb{Z}/n\mathbb{Z}$-modules, we denote $((\,\cdot\,)_S)^{\mathrm{mult}}$ by $(\,\cdot\,)_S^{\mathrm{mult}}$. By functoriality, étale sheaves carrying $\mathcal{O}$-actions are sent to finite flat group schemes of multiplicative type also carrying $\mathcal{O}$-actions.

Definition 3.1.1.7. *When S is a scheme of characteristic p, we extend such a definition to the category of étale sheaves of $\mathbb{Z}_p$-modules over S by setting*

$$(\,\cdot\,)^{\mathrm{mult}} := \varinjlim_r (\,\cdot\,/(p^r\,\cdot\,))^{\mathrm{mult}},$$

which is a p-divisible group of multiplicative type.

Example 3.1.1.8. Let S be a scheme of characteristic p. Let $(\mathbb{Q}_p/\mathbb{Z}_p)_S = \varinjlim_r ((\frac{1}{p^r}\mathbb{Z}_p)/\mathbb{Z}_p)_S$ denote the split rank-one étale p-divisible group over S. (For such constant objects, if S is a geometric point, and if the context is clear, we shall often suppress S from the notation.) Let $\boldsymbol{\mu}_{p^\infty,S} = \varinjlim_r \boldsymbol{\mu}_{p^r,S} = \varinjlim_r \mathbf{G}_{\mathrm{m},S}[p^r]$ denote the split rank-one p-divisible group of multiplicative type over S. Then we have

$$(\mathbb{Z}_p(1))^{\mathrm{mult}} = \varinjlim_r ((\mathbb{Z}/p^r\mathbb{Z})(1))^{\mathrm{mult}} \cong \varinjlim_r \boldsymbol{\mu}_{p^r,S} = \boldsymbol{\mu}_{p^\infty,S},$$

which is the Serre dual of $(\mathbb{Q}_p/\mathbb{Z}_p)_S$.

3.1.2 *Serre's Construction for Ordinary Abelian Schemes*

The following definition has been alluded to when we cited [62, Prop. 5.2.3.9] in Section 1.3.3:

Definition 3.1.2.1. *(See [62, Def. 5.2.3.6].) Let U be a scheme, and let $\underline{N}$ be an étale sheaf of left $\mathcal{O}$-modules that becomes a constant finitely generated $\mathcal{O}$-module N over a finite étale covering of U. Let (Z, λ_Z) be a polarized abelian scheme over U with a left $\mathcal{O}$-action given by some $i_Z : \mathcal{O} \to \mathrm{End}_U(Z)$. Then we denote by $\underline{\mathrm{Hom}}_{\mathcal{O}}(\underline{N}, Z)$ the (commutative) group functor of $\mathcal{O}$-equivariant group homomorphisms from the group functor $\underline{N}$ to the group functor Z.*

The aim of this subsection is to further generalize [62, Prop. 5.2.3.9], to include a treatment of the *fiberwise geometric identity components* and *group schemes of fiberwise geometric connected components* when the base scheme U has residue characteristics ramified in $\mathcal{O}$ and when the abelian scheme $Z \to U$ in question is *ordinary* (see Definition 3.1.1.2).

Lemma 3.1.2.2. *Suppose that W is a commutative proper group scheme of finite presentation over U. Suppose that W_0 is an abelian subscheme of W (i.e. a subgroup scheme that is an abelian scheme), and that there is an integer $m \geq 1$ such that multiplication by m defines a homomorphism $[m] : W \to W$ with schematic image a (closed) subscheme of W_0 and such that $W[m]$, the m-torsion subgroup scheme of W, is finite flat over U. Then, for every geometric point $\bar{s} \to U$, the fiber $(W_0)_{\bar{s}}$ is the reduced subscheme of the connected component of $W_{\bar{s}}$ containing the identity section. Moreover, W is flat and the quotient group functor W/W_0 is representable by a commutative finite flat group scheme E. The group $\pi_0(W_{\bar{s}})$ of connected components can be canonically identified with the $\bar{s}$-valued points of $E_{\bar{s}}$.*

Proof. Since W is commutative and since W_0 is (fppf locally) m-divisible as an abelian scheme, the condition that $[m]$ sends W to W_0 shows that W/W_0 can be identified with the quotient of $W[m]$ by $W_0[m] = W_0 \cap W[m]$. Since W_0 is an abelian scheme, $W_0[m]$ is finite flat over U and the quotient W/W_0 is representable (by [25, V, 4.1]). Since $W[m]$ is finite flat, the quotient $W[m]/W_0[m]$ is also finite flat (by [25, $\mathrm{VI_A}$, 3.2 and 5.4]). (All of these statements are over U.) The statements on the identity components and group of connected components of geometric fibers are obvious. □

Definition 3.1.2.3. *(Compare with [62, Def. 5.2.3.8].) Suppose W_0 is an abelian subscheme of a proper group scheme W of finite presentation over a base scheme U, such that for every geometric point $\bar{s} \to U$, the fiber $(W_0)_{\bar{s}}$ is the* **reduced subscheme** *of the connected component of $W_{\bar{s}}$ containing the identity section. Then we say (for simplicity) that W_0 is the* **fiberwise geometric identity component** *of W (without emphasizing that it is reduced), and denote it by W°. (By [35, IV-2,*

4.5.13], it is also correct to say that W_0 is the fiberwise identity component, without the term **geometric***.)*

Suppose the quotient group functor W/W_0 is representable by a finite group scheme E. Then we say that E is the **group scheme of fiberwise geometric connected components***, and denote it by $\underline{\pi}_0(W/U)$.*

By Lemma 3.1.2.2, the finite group scheme $\underline{\pi}_0(W/U)$ is defined and is finite flat over U if W is commutative and if there is an integer $m \geq 1$ such that multiplication by m defines a homomorphism $[m] : W \to W$ with schematic image a (closed) abelian subscheme of W_0 and such that $W[m]$ is finite flat over U.

Now we can state our (slight) generalization of [62, Prop. 5.2.3.9]:

Proposition 3.1.2.4. *With the setting as in Definition 3.1.2.1, suppose $\underline{N}$ is constant with value some finitely generated $\mathcal{O}$-module N. Then the following are true:*

(1) The group functor $\underline{\mathrm{Hom}}_{\mathcal{O}}(N, Z)$ is representable by a proper subgroup scheme of an abelian scheme over U.
(2) Suppose that N is torsion of order annihilated by some integers $m \geq 1$, and that $Z[m]$ is a finite flat group scheme of étale-multiplicative type over U. Then $\underline{\mathrm{Hom}}_{\mathcal{O}}(N, Z)$ is also finite flat of étale-multiplicative type over U.
(3) If N is projective as an $\mathcal{O}$-module, then $\underline{\mathrm{Hom}}_{\mathcal{O}}(N, Z)$ is representable by an abelian scheme.
(4) Suppose that N is an $\mathcal{O}$-lattice, and that Z is an **ordinary** *abelian scheme over U (see Definition 3.1.1.2). Then $\underline{\mathrm{Hom}}_{\mathcal{O}}(N, Z)$ is representable by a proper flat group scheme which is an extension of a (commutative) finite flat group scheme of étale-multiplicative type, whose rank has no prime factors other than those of the discriminant of* $\mathrm{Disc} = \mathrm{Disc}_{\mathcal{O}/\mathbb{Z}}$ *[62, Def. 1.1.1.6], by an abelian scheme over U.*
Following Definition 3.1.2.3, we shall say that $\underline{\mathrm{Hom}}_{\mathcal{O}}(N, Z)$ is the extension of the finite flat group scheme $\underline{\pi}_0(\underline{\mathrm{Hom}}_{\mathcal{O}}(N, Z)/U)$ of étale-multiplicative type by the abelian scheme $\underline{\mathrm{Hom}}_{\mathcal{O}}(N, Z)^{\circ}$.

We shall still call this *Serre's construction* (as in [62, Prop. 5.2.3.9]).

Proof. (This is essentially the same proof of [62, Prop. 5.2.3.9].) Since $\mathcal{O}$ is (left) noetherian (see, for example, [93, Cor. 2.10]), and since N is finitely generated, there is a *free resolution* $\mathcal{O}^{\oplus r_1} \to \mathcal{O}^{\oplus r_0} \to N \to 0$ for some integers $r_0, r_1 \geq 0$. By taking $\underline{\mathrm{Hom}}_{\mathcal{O}}(\,\cdot\,, Z)$, we obtain an exact sequence

$$0 \to \underline{\mathrm{Hom}}_{\mathcal{O}}(N, Z) \to Z^{r_0} \to Z^{r_1} \qquad (3.1.2.5)$$

(of fppf sheaves) over U, where Z^{r_0} (resp. Z^{r_1}) stands for the fiber products of r_0 (resp. r_1) copies of Z over U, which shows that $\underline{\mathrm{Hom}}_{\mathcal{O}}(N, Z)$ is representable because it is the kernel of the homomorphism $Z^{r_0} \to Z^{r_1}$ between abelian schemes in (3.1.2.5).

To show that $\underline{\mathrm{Hom}}_{\mathcal{O}}(N,Z)$ is proper over U, note that the first homomorphism in (3.1.2.5) is a closed immersion because Z^s is separated over U, and every closed subscheme of Z^r is proper over U. This proves (1) of Proposition 3.1.2.4.

Suppose that N is torsion of order annihilated by some integers $m \geq 1$, and that $Z[m]$ is a finite flat group scheme of étale-multiplicative type over U. Then $\underline{\mathrm{Hom}}_{\mathcal{O}}(N,Z)$ is isomorphic to the closed subscheme $\underline{\mathrm{Hom}}_{\mathcal{O}}(N,Z[m])$ of the finite flat group scheme $\underline{\mathrm{Hom}}_{\mathbb{Z}}(N,Z[m])$ of étale-multiplicative type over U. Over an étale covering of U over which $Z[m]$ admits an $\mathcal{O}$-equivariant filtration by finite flat subgroup schemes whose graded pieces are either constant group schemes or dual to constant group schemes (*i.e.*, split multiplicative), the condition of compatibilities with $\mathcal{O}$-actions on the constant schemes involved is both open and closed. This implies that $\underline{\mathrm{Hom}}_{\mathcal{O}}(N,Z)$ is also finite flat of étale-multiplicative type over U. This proves (2) of Proposition 3.1.2.4.

If N is projective, then it is flat by [93, Cor. 2.16]. This is the same for its dual (right) $\mathcal{O}$-module $N^\vee$. Hence, for every embedding $U \hookrightarrow \tilde{U}$ defined by an ideal $\mathscr{I}$ such that $\mathscr{I}^2 = 0$, the surjectivity of the morphism $Z(\tilde{U}) \to Z(U)$ of $\mathcal{O}$-modules implies the surjectivity of the morphism

$$(N^\vee \underset{\mathcal{O}}{\otimes} Z)(\tilde{U}) \cong N^\vee \underset{\mathcal{O}}{\otimes} Z(\tilde{U}) \to (N^\vee \underset{\mathcal{O}}{\otimes} Z)(U) \cong N^\vee \underset{\mathcal{O}}{\otimes} Z(U).$$

This shows that $\underline{\mathrm{Hom}}_{\mathcal{O}}(N,Z) \to U$ is formally smooth, and hence smooth because it is (locally) of finite presentation (see [35, IV-4, 17.3.1 and 17.5.2]). Moreover, since N is projective, there exists some projective $\mathcal{O}$-module N' such that $N \oplus N' \cong \mathcal{O}^{\oplus r}$ for some $r \geq 0$. Then we have $\underline{\mathrm{Hom}}_{\mathcal{O}}(N,Z) \underset{U}{\times} \underline{\mathrm{Hom}}_{\mathcal{O}}(N',Z) \cong Z^r$, which shows that the geometric fibers of $\underline{\mathrm{Hom}}_{\mathcal{O}}(N,Z) \to U$ are connected. Hence, by definition, $\underline{\mathrm{Hom}}_{\mathcal{O}}(N,Z)$ is an abelian scheme over U. This proves (3) of Proposition 3.1.2.4.

Finally, suppose that N is an $\mathcal{O}$-lattice, and that Z and hence $Z^\vee$ are ordinary abelian schemes over U (see Lemma 3.1.1.5).

Let $\mathcal{O}'$ be any maximal order in $\mathcal{O} \underset{\mathbb{Z}}{\otimes} \mathbb{Q}$ containing $\mathcal{O}$. By [62, Prop. 1.1.1.21], there exists an integer $m \geq 1$, with no prime factors other than those of Disc, such that $m\mathcal{O}' \subset \mathcal{O}$. Consider the intersection K of the kernels of $[b]^\vee : Z^\vee[m] \to Z^\vee[m]$ for all $b \in m\mathcal{O}'$. By working over an étale covering over which $Z[m]$ and $Z^\vee[m]$ admit $\mathcal{O}$-equivariant filtrations by finite flat subgroup schemes whose graded pieces are either constant group schemes or split multiplicative (as in the proof of (2) of Proposition 3.1.2.4 above), we see that K is a finite flat subgroup scheme of $Z^\vee[m]$, and hence so is its orthogonal complement $K^\perp$ in $Z[m]$ (with respect to the canonical pairing $e_{Z[m]} : Z[m] \times Z^\vee[m] \to \boldsymbol{\mu}_{m,U}$). By construction, this $K^\perp$ is the smallest subgroup scheme of $Z[m]$ containing the images of $[b] : Z[m] \to Z[m]$ for all $b \in m\mathcal{O}'$. Therefore, by forming the isogeny $Z \twoheadrightarrow Z' := Z/K^\perp$, the action of $\mathcal{O}$ on Z induces an action of $\mathcal{O}'$ on Z'. In this case, there is also a canonical isogeny $Z' \twoheadrightarrow Z$ whose pre- and post-compositions with the previous isogeny $Z \twoheadrightarrow Z'$ are multiplications by m on Z and Z', respectively.

Let N' be the $\mathcal{O}'$-span of N in $N \underset{\mathbb{Z}}{\otimes} \mathbb{Q}$. Since N' is the $\mathcal{O}'$-space of N, the canonical isogeny $Z \twoheadrightarrow Z'$ induces a canonical homomorphism

$$\underline{\mathrm{Hom}}_{\mathcal{O}}(N, Z) \to \underline{\mathrm{Hom}}_{\mathcal{O}'}(N', Z').$$

On the other hand, the canonical isogeny $Z' \twoheadrightarrow Z$ above induces a canonical homomorphism

$$\underline{\mathrm{Hom}}_{\mathcal{O}'}(N', Z') \to \underline{\mathrm{Hom}}_{\mathcal{O}}(N, Z),$$

whose pre- and post-composition with the previous canonical homomorphism is nothing but the multiplications by m on $\underline{\mathrm{Hom}}_{\mathcal{O}}(N, Z)$ and $\underline{\mathrm{Hom}}_{\mathcal{O}'}(N', Z')$, respectively. As usual, we denote by $[m]$ all such multiplications by m.

Since $\mathcal{O}'$ is maximal, N' is projective as an $\mathcal{O}'$-module by [62, Prop. 1.1.1.23]. By (3) of Proposition 3.1.2.4 proved above, we know that $\underline{\mathrm{Hom}}_{\mathcal{O}'}(N', Z')$ is an abelian scheme. Since $[m] : \underline{\mathrm{Hom}}_{\mathcal{O}}(N, Z) \to \underline{\mathrm{Hom}}_{\mathcal{O}}(N, Z)$ factors as the composition of canonical homomorphisms $\underline{\mathrm{Hom}}_{\mathcal{O}}(N, Z) \to \underline{\mathrm{Hom}}_{\mathcal{O}'}(N', Z') \to \underline{\mathrm{Hom}}_{\mathcal{O}}(N, Z)$, this shows that the schematic image of $[m] : \underline{\mathrm{Hom}}_{\mathcal{O}}(N, Z) \to \underline{\mathrm{Hom}}_{\mathcal{O}}(N, Z)$ is an abelian scheme. On the other hand, by working over an étale covering over which $Z[m]$ admits an $\mathcal{O}$-equivariant filtration by finite flat subgroup schemes whose graded pieces are either constant group schemes or split multiplicative (as above), we see that $\underline{\mathrm{Hom}}_{\mathcal{O}}(N, Z)[m] \cong \underline{\mathrm{Hom}}_{\mathcal{O}}(N, Z[m])$ is finite flat of étale-multiplicative type (of rank dividing a power of m) over U. Hence, by Lemma 3.1.2.2, we see that both $\underline{\pi_0}(\underline{\mathrm{Hom}}_{\mathcal{O}}(N, Z)/U)$ and $\underline{\mathrm{Hom}}_{\mathcal{O}}(N, Z)^{\circ}$ are defined with the desired properties. □

Remark 3.1.2.6. The materials in this subsection generalize naturally to the case when U is an algebraic stack (which is Deligne–Mumford by our convention).

3.1.3 *Extensibility of Isogenies*

Suppose that U is noetherian scheme and that $\overline{U}$ is a noetherian normal scheme containing U as an open dense subscheme. Let $f : Z \to Z'$ be an isogeny of semi-abelian schemes over U (*cf.* Lemma 3.1.1.5), and let $\overline{Z} \to \overline{U}$ be a semi-abelian scheme extending $Z \to U$ (*cf.* Definition 3.1.1.2), in the sense that $\overline{Z}_U = \overline{Z} \underset{\overline{U}}{\times} U$ is isomorphic to U as a group scheme. (Then $\overline{Z}$ is determined up to canonical isomorphism by Z, by noetherian normality of $\overline{U}$ and by [92, IX, 1.4], [28, Ch. I, Prop. 2.7], or [62, Prop. 3.3.1.5].)

Let $K := \ker(f)$, which is quasi-finite flat and of finite presentation over U. Let $\overline{K}$ denote the schematic closure of K in $\overline{Z}$. Let N be an integer such that K is a closed subgroup scheme of $Z[N]$, which exists because U is noetherian. Then $\overline{K}$ is a closed subgroup scheme of $\overline{Z}[N]$ (by the universal property of schematic closures), which is quasi-finite over $\overline{U}$.

Lemma 3.1.3.1. *With assumptions as above, suppose $\overline{U}$ is Dedekind (i.e., noetherian normal and of dimension at most one). Then $\overline{K}$ is a group scheme quasi-finite flat over $\overline{U}$, and there exists a semi-abelian scheme $\overline{Z}' \to \overline{U}$, unique up to unique isomorphism, such that the isogeny $f : Z \to Z'$ over U uniquely extends to an isogeny $\overline{f} : \overline{Z} \to \overline{Z}'$ over $\overline{U}$.*

Proof. Since $\overline{U}$ is one-dimensional, as explained in [10, Sec. 10.1, Prop. 4 and 7; see also the middle of paragraph 2 on p. 310], the (locally of finite type) Néron models of Z and Z' over $\overline{U}$ (uniquely) exist, and the (fiberwise) identity components of them are group schemes $\overline{Z}$ and $\overline{Z}'$ over $\overline{U}$ (the former being up to canonical isomorphism the same $\overline{Z}$ as above) which are commutative, separated, smooth, and of finite type, and have geometrically connected fibers. (Since $\overline{Z}' \to \overline{U}$ admits the identity section, by [35, IV-2, 4.5.13], its connected fibers are also geometrically connected.) Moreover, by the universal property of Néron models and by the definition of identity components, the homomorphism $f : Z \to Z'$ uniquely extends to a homomorphism $\overline{f} : \overline{Z} \to \overline{Z}'$ over $\overline{U}$. Since $\overline{K} = \ker(\overline{f})$ because the latter is the (closed) preimage of the identity section of the separated group scheme $\overline{Z}' \to \overline{U}$, if N is any integer as above such that $K = \ker(f) \subset Z[N]$, then $\overline{K} = \ker(\overline{f}) \subset \overline{Z}[N]$. This forces $\overline{f}$ to be quasi-finite and hence surjective (because it is between schemes with geometrically connected fibers that are separated, smooth, of finite type, and of the same dimension). Therefore, $\overline{K} = \ker(\overline{f})$ is quasi-finite and flat over $\overline{U}$ (see [62, Lem. 1.3.1.11]), and $\overline{Z}' \to \overline{U}$ is also semi-abelian (begin an isogenous quotient of $\overline{Z} \to \overline{U}$, whose fibers are still extensions of abelian varieties by tori), as desired. □

Lemma 3.1.3.2. *With assumptions as above, suppose moreover that $\overline{Z} \to \overline{U}$ is an* **ordinary** *semi-abelian scheme (but suppose no longer that $\overline{U}$ is one-dimensional). Then $\overline{K}$ is a group scheme quasi-finite flat over $\overline{U}$ (of fiber degrees dividing those of K), and there exists a semi-abelian scheme $\overline{Z}' \to \overline{U}$ such that the isogeny $f : Z \to Z'$ over U extends to an isogeny $\overline{f} : \overline{Z} \to \overline{Z}'$ over $\overline{U}$. (Then $\overline{Z}'$ and $\overline{f}$ are determined up to unique isomorphism by Z and f, by noetherian normality of $\overline{U}$ and by [92, IX, 1.4], [28, Ch. I, Prop. 2.7], or [62, Prop. 3.3.1.5].)*

Proof. Since $\overline{U}$ is noetherian normal, by [92, XI, 1.13], $\overline{Z}$ is locally quasi-projective. Hence, by [80, IV, 7.1.2] (see also [62, Lem. 3.4.3.1]), over any open subscheme U' of $\overline{U}$ over which the quasi-finite group scheme $\overline{K}$ is flat, the quotient $(\overline{Z} \underset{\overline{U}}{\times} U')/(\overline{K} \underset{\overline{U}}{\times} U')$ is representable by a semi-abelian scheme over U', which is also ordinary by Lemma 3.1.1.5. Hence, the flatness of $\overline{K}$ and the constructibility of $\overline{Z}'$ and $\overline{f}$ are equivalent conditions over open subschemes of $\overline{U}$.

If $\overline{K}$ were not flat, then it must be so at some point u, and we may take this point to be maximally so. Therefore, to show that $\overline{K}$ is flat, we may replace $\overline{U}$ with its strict localization at an arbitrary point u (see [35, 0_{I}, (6.6.3), and IV-4, 18.8.8(iii)]), and we may enlarge U and assume that U is the full open complement of u in $\overline{U}$. By Lemma 3.1.3.1, we may assume that u has codimension at least two.

Let N be any integer as above such that $K = \ker(f) \subset H := Z[N]$. By the assumption that $\overline{Z} \to \overline{U}$ is ordinary, and by arguing as in [62, Sec. 3.4.1], $\overline{H} := \overline{Z}[N]$ admits a filtration

$$0 \subset \overline{H}^{\mathrm{mult}} \subset \overline{H}^{\mathrm{f}} \subset \overline{H},$$

where $\overline{H}^{\mathrm{mult}}$ is finite flat of multiplicative type, where $\overline{H}^{\mathrm{f}}$ is the maximal finite flat subgroup scheme of $\overline{H}$ over $\overline{U}$ (which is open and closed in $\overline{H}$, because $\overline{U}$ is

Henselian), where the quotient $\overline{H}^{\mathrm{f,\acute{e}t}} := \overline{H}^{\mathrm{f}}/\overline{H}^{\mathrm{mult}}$ is finite étale, and where the quotient $\overline{H}/\overline{H}^{\mathrm{f}}$ is quasi-finite étale (whose special fiber over u consists of only the identity section). Let $H^{\mathrm{mult}} := \overline{H}^{\mathrm{mult}} \underset{\overline{U}}{\times} U$, $H^{\mathrm{f}} := \overline{H}^{\mathrm{f}} \underset{\overline{U}}{\times} U$, $H^{\mathrm{f,\acute{e}t}} := \overline{H}^{\mathrm{f,\acute{e}t}} \underset{\overline{U}}{\times} U$, $K^{\mathrm{mult}} := K \cap H^{\mathrm{mult}}$, and $K^{\mathrm{f}} := K \cap H^{\mathrm{f}}$. Note that K^{f} is (finite and) flat over U because H^{f} is open and closed in H.

Let us denote Cartier duals by superscripts "D". Note that K^{mult} is the kernel of the composition of morphisms $H^{\mathrm{mult}} \hookrightarrow H^{\mathrm{f}} \twoheadrightarrow H^{\mathrm{f}}/K^{\mathrm{f}}$ between finite flat group schemes. (The quotient $H^{\mathrm{f}}/K^{\mathrm{f}}$ is defined because K^{f} is finite flat.) The Cartier dual of this morphism is $(H^{\mathrm{f}}/K^{\mathrm{f}})^{\mathrm{D}} \to (H^{\mathrm{mult}})^{\mathrm{D}}$. Since $(H^{\mathrm{mult}})^{\mathrm{D}}$ is étale over U, its fibers over U are disjoint unions of closed points. Hence, by the fiberwise criterion of flatness as in [35, IV-3, 11.3.10 a)⇒b)], the image of $(H^{\mathrm{f}}/K^{\mathrm{f}})^{\mathrm{D}} \to (H^{\mathrm{mult}})^{\mathrm{D}}$ is finite flat. Then the cokernel of $(H^{\mathrm{f}}/K^{\mathrm{f}})^{\mathrm{D}} \to (H^{\mathrm{mult}})^{\mathrm{D}}$ is defined and also finite flat of finite presentation; and its Cartier dual is K^{mult} and is also finite flat of finite presentation. Thus, we can define $K^{\mathrm{f,\acute{e}t}} := K^{\mathrm{f}}/K^{\mathrm{mult}}$, which is a finite étale subgroup scheme of $H^{\mathrm{f,\acute{e}t}}$.

Let $\overline{K}^{\mathrm{mult}}$ (resp. $\overline{K}^{\mathrm{f,\acute{e}t}}$) be the schematic closure of K^{mult} (resp. $K^{\mathrm{f,\acute{e}t}}$) in H^{mult} (resp. $H^{\mathrm{f,\acute{e}t}}$). Since $\overline{U}$ is strict local and normal, $\overline{H}^{\mathrm{mult}}$ (resp. $\overline{H}^{\mathrm{f,\acute{e}t}}$) is dual to (resp. is) a constant group scheme over $\overline{U}$. Hence, $\overline{K}^{\mathrm{mult}}$ (resp. $\overline{K}^{\mathrm{f,\acute{e}t}}$) is also finite flat of multiplicative (resp. finite étale).

Let $\overline{K}^{+} := (\overline{H}^{\mathrm{f}} \twoheadrightarrow \overline{H}^{\mathrm{f,\acute{e}t}})^{-1}(\overline{K}^{\mathrm{f,\acute{e}t}})$ and let $K^{+} := \overline{K}^{+} \underset{\overline{U}}{\times} U$. Given K^{f} as an extension of $K^{\mathrm{f,\acute{e}t}}$ by K^{mult}, the natural inclusion $H^{\mathrm{mult}} \hookrightarrow K^{+}$ induces an isomorphism $H^{\mathrm{mult}}/K^{\mathrm{mult}} \cong K^{+}/K^{\mathrm{f}}$, and induces a surjection

$$K^{+} \twoheadrightarrow H^{\mathrm{mult}}/K^{\mathrm{mult}}.$$

Note that this surjection determines K^{f} in the sense that its kernel is a finite flat subgroup scheme of H^{f} that is an extension of $K^{\mathrm{f,\acute{e}t}}$ by K^{mult}, which is just K^{f}.

Since $\overline{U}$ is normal and $\overline{K}^{+}$ and $\overline{H}^{\mathrm{mult}}/\overline{K}^{\mathrm{mult}}$ are finite flat, the above surjection extends to a surjection

$$\overline{K}^{+} \twoheadrightarrow \overline{H}^{\mathrm{mult}}/\overline{K}^{\mathrm{mult}},$$

whose kernel defines a finite flat subgroup scheme of $\overline{H}^{\mathrm{f}}$ that is an extension of $\overline{K}^{\mathrm{f,\acute{e}t}}$ by $\overline{K}^{\mathrm{mult}}$, which must coincide with the closure of K^{f} in $\overline{H}^{\mathrm{f}}$, which is nothing but $\overline{K}^{\mathrm{f}} := \overline{K} \cap \overline{H}^{\mathrm{f}}$. Hence, $\overline{K}^{\mathrm{f}}$ is finite flat over $\overline{U}$. Since $\overline{K} - \overline{K}^{\mathrm{f}}$ has an empty fiber over u and coincides with $K - K^{\mathrm{f}}$ over $U = \overline{U} - \{u\}$, we see that $\overline{K}$ is flat over all of $\overline{U}$, as desired. □

Remark 3.1.3.3. Lemma 3.1.3.2 is incorrect if we do not assume that $\overline{Z} \to \overline{U}$ is ordinary. See [18, Sec. 6] for an example even when $\overline{U}$ is regular. (One might as well introduce conditions on $\overline{U}$ as in [101] to ensure that it is *healthy regular*. However, such conditions tend to impose restrictions on the ramification of the universal base ring, which conflicts with our goal.)

Proposition 3.1.3.4. *Let $(Z, \lambda_Z) \to U$ be as in Definition* 3.1.2.1. *Let N be an $\mathcal{O}$-lattice. Suppose Z is ordinary, so that $\underline{\mathrm{Hom}}_{\mathcal{O}}(N, Z)$ is defined and representable by a proper flat subgroup scheme over U, which is an extension of a finite flat group scheme $\underline{\pi}_0(\underline{\mathrm{Hom}}_{\mathcal{O}}(N, Z)/U)$ of étale-multiplicative type by the abelian scheme $\underline{\mathrm{Hom}}_{\mathcal{O}}(N, Z)^\circ$ as in Proposition* 3.1.2.4. *Suppose that $\overline{U}$ is noetherian normal and that $\overline{Z} \to \overline{U}$ is an* **ordinary** *semi-abelian scheme extending $Z \to U$. Then $\underline{\mathrm{Hom}}_{\mathcal{O}}(N, \overline{Z})$ is defined and representable over $\overline{U}$ by an extension of a quasi-finite flat group scheme $\underline{\pi}_0(\underline{\mathrm{Hom}}_{\mathcal{O}}(N, \overline{Z})/\overline{U})$ of étale-multiplicative type by a semi-abelian scheme $\underline{\mathrm{Hom}}_{\mathcal{O}}(N, \overline{Z})^\circ$. The restriction of this extension to U is the extension $\underline{\mathrm{Hom}}_{\mathcal{O}}(N, Z)$ of the finite flat group scheme $\underline{\pi}_0(\underline{\mathrm{Hom}}_{\mathcal{O}}(N, Z)/U)$ of étale-multiplicative type by an abelian scheme $\underline{\mathrm{Hom}}_{\mathcal{O}}(N, Z)^\circ$ in* (4) *of Proposition* 3.1.2.4.

Proof. The same argument of the proof of (1) of Proposition 3.1.2.4 shows that $\underline{\mathrm{Hom}}_{\mathcal{O}}(N, \overline{Z})$ is representable by a (closed) subgroup scheme of a semi-abelian scheme over $\overline{U}$.

The same argument of the proof of (3) of Proposition 3.1.2.4 shows that, when N is projective as an $\mathcal{O}$-module, there exists some projective $\mathcal{O}$-module N' such that $\underline{\mathrm{Hom}}_{\mathcal{O}}(N, \overline{Z}) \underset{\overline{U}}{\times} \underline{\mathrm{Hom}}_{\mathcal{O}}(N', \overline{Z}) \cong \overline{Z}^{\times r}$ for some integer r, which implies that $\underline{\mathrm{Hom}}_{\mathcal{O}}(N, \overline{Z})$ is a semi-abelian scheme over $\overline{U}$, because it is commutative, separated, smooth, and of finite type, and because its geometric fibers are all connected with trivial unipotent radicals (see [10, Sec. 7.3, paragraph following Lem. 1]).

Let $m \geq 1$, $\mathcal{O}'$, $Z \twoheadrightarrow Z'$, $Z' \twoheadrightarrow Z$, and N' be as in the proof of (4) of Proposition 3.1.2.4. Since $\overline{U}$ is noetherian normal, by Lemma 3.1.3.2, there exist a semi-abelian scheme $\overline{Z}'$ and two isogenies $\overline{Z} \twoheadrightarrow \overline{Z}'$ and $\overline{Z}' \twoheadrightarrow \overline{Z}$ over $\overline{U}$ extending the isogenies $Z \twoheadrightarrow Z'$ and $Z' \twoheadrightarrow Z$ over U. By the previous paragraph, we know that $\underline{\mathrm{Hom}}_{\mathcal{O}'}(N', \overline{Z}')$ is representable by a semi-abelian scheme over $\overline{U}$. Moreover, the same argument of the proof of (2) of Proposition 3.1.2.4 shows that $\underline{\mathrm{Hom}}_{\mathcal{O}}(N, \overline{Z})[m]$ is quasi-finite flat of étale-multiplicative type over $\overline{U}$ for every integer $m \geq 1$. Hence, the same argument of the proof of (4) of Proposition 3.1.2.4 shows that the schematic-image of $[m] : \underline{\mathrm{Hom}}_{\mathcal{O}}(N, \overline{Z}) \to \underline{\mathrm{Hom}}_{\mathcal{O}}(N, \overline{Z})$ is an semi-abelian scheme over $\overline{U}$, which is an isogenous quotient of $\underline{\mathrm{Hom}}_{\mathcal{O}'}(N', \overline{Z}')$ (by Lemma 3.1.3.2 again). Let us denote this semi-abelian scheme by $\underline{\mathrm{Hom}}_{\mathcal{O}}(N, \overline{Z})^\circ$.

As in the proof of Lemma 3.1.2.2, since $\underline{\mathrm{Hom}}_{\mathcal{O}}(N, \overline{Z})^\circ$ is (fppf locally) m-divisible (as a semi-abelian scheme) over $\overline{U}$, the quotient $\underline{\mathrm{Hom}}_{\mathcal{O}}(N, \overline{Z})/\underline{\mathrm{Hom}}_{\mathcal{O}}(N, \overline{Z})^\circ$ can be identified (as a fppf sheaf over $\overline{U}$) with $\underline{\mathrm{Hom}}_{\mathcal{O}}(N, \overline{Z})[m]/\underline{\mathrm{Hom}}_{\mathcal{O}}(N, \overline{Z})^\circ[m]$, which is representable by a quasi-finite flat scheme $\underline{\pi}_0(\underline{\mathrm{Hom}}_{\mathcal{O}}(N, \overline{Z})/\overline{U})$ over $\overline{U}$ because $\underline{\mathrm{Hom}}_{\mathcal{O}}(N, \overline{Z})^\circ[m]$ is closed in $\underline{\mathrm{Hom}}_{\mathcal{O}}(N, \overline{Z})[m]$ (by [52, II, 6.16], an algebraic space quasi-finite and separated over a scheme is a scheme).

Thus, $\underline{\mathrm{Hom}}_{\mathcal{O}}(N, \overline{Z})$ is an extension of a quasi-finite flat group scheme $\underline{\pi}_0(\underline{\mathrm{Hom}}_{\mathcal{O}}(N, \overline{Z})/\overline{U})$ of étale-multiplicative type by a semi-abelian scheme $\underline{\mathrm{Hom}}_{\mathcal{O}}(N, \overline{Z})^\circ$ over $\overline{U}$. By its very construction, this extension extends the corresponding extension in (4) of Proposition 3.1.2.4, as desired. □

3.2 Linear Algebraic Data for Ordinary Loci

3.2.1 *Necessary Data for Ordinary Reductions*

Suppose there exists an abelian scheme A over $S = \mathrm{Spec}(R)$, where R is a complete noetherian normal domain with fraction field K of characteristic zero and algebraically closed residue field k of characteristic $p > 0$, and suppose there exist a morphism $\xi : S \to \vec{\mathsf{M}}_{\mathcal{H}}$ (see Proposition 2.2.1.1), which induces a field homomorphism $F_0 \to K$ factoring through $(F_0)_v \to K$ for a place v of F_0 above p. Let $\bar{K}$ be an algebraic closure of K. Let $s := \mathrm{Spec}(k)$, $\eta := \mathrm{Spec}(K)$, and $\bar{\eta} := \mathrm{Spec}(\bar{K})$.

For simplicity, let us assume that $\mathcal{H}$ is neat. Then the restriction $\xi_0 : \eta \to \mathsf{M}_{\mathcal{H}}$ of ξ defines an object $(A_\eta, \lambda_\eta, i_\eta, \alpha_{\mathcal{H},\eta})$ of $\mathsf{M}_{\mathcal{H}}(\eta)$. By [92, IX, 1.4], [28, Ch. I, Prop. 2.7], or [62, Prop. 3.3.1.5], $\lambda_\eta : A_\eta \to A_\eta^\vee$ extends to an isogeny $\lambda : A \to A^\vee$ over S, and $i_\eta : \mathcal{O} \to \mathrm{End}_\eta(A_\eta)$ extends to a homomorphism $i : \mathcal{O} \to \mathrm{End}_S(A)$. By [92, XI, 1.16], the symmetric invertible sheaf $(\mathrm{Id}_A, \lambda)^* \mathcal{P}_A$ is ample because its restriction to η is. Therefore, λ is also a polarization (by definition; *cf.* [23, 1.2, 1.3, 1.4] and [62, Prop. 1.3.2.15 and Def. 1.3.2.16]). Note that the extension i satisfies the Rosati condition defined by λ, because it already does over η. Hence, i is an $\mathcal{O}$-endomorphism structure as in [62, Def. 1.3.3.1]. Moreover, $\underline{\mathrm{Lie}}_{A/S}$ with its $\mathcal{O} \underset{\mathbb{Z}}{\otimes} \mathbb{Z}_{(p)}$-module structure given naturally by i satisfies the determinantal condition in [62, Def. 1.3.4.1] given by $(L \underset{\mathbb{Z}}{\otimes} \mathbb{R}, \langle \cdot\,, \cdot \rangle, h_0)$, because the condition is given by an identity of polynomial functions, and because η is dense in S.

For each p-divisible group $\mathcal{G}$, we denote by $\mathcal{G}^{\mathrm{mult}}$ (resp. $\mathcal{G}^{\mathrm{conn}}$, resp. $\mathcal{G}^{\mathrm{\acute{e}t}}$) the multiplicative-type (resp. connected, resp. étale) part of $\mathcal{G}$, whose formation is functorial and compatible with all automorphisms of $\mathcal{G}$. Let $A_s[p^\infty] = \varinjlim_r A_s[p^r]$ (resp. $A_s^\vee[p^\infty] = \varinjlim_r A_s^\vee[p^r]$) denote the p-divisible group attached to the abelian variety A_s (resp. $A_s^\vee$). The canonical perfect duality

$$\mathrm{e}_{A_s[p^\infty]} : A_s[p^\infty] \times A_s^\vee[p^\infty] \to \boldsymbol{\mu}_{p^\infty,s}$$

induces canonical perfect dualities

$$A_s[p^\infty]^{\mathrm{mult}} \times A_s^\vee[p^\infty]^{\mathrm{\acute{e}t}} \to \boldsymbol{\mu}_{p^\infty,s}$$

and

$$A_s[p^\infty]^{\mathrm{\acute{e}t}} \times A_s^\vee[p^\infty]^{\mathrm{mult}} \to \boldsymbol{\mu}_{p^\infty,s},$$

which induce canonical isomorphisms

$$A_s[p^\infty]^{\mathrm{mult}} \overset{\sim}{\to} \mathrm{Hom}_s(A_s^\vee[p^\infty]^{\mathrm{\acute{e}t}}, \boldsymbol{\mu}_{p^\infty,s})$$

and

$$A_s[p^\infty]^{\mathrm{\acute{e}t}} \overset{\sim}{\to} \mathrm{Hom}_s(A_s^\vee[p^\infty]^{\mathrm{mult}}, \boldsymbol{\mu}_{p^\infty,s}),$$

respectively, compatible with their $\mathcal{O}$-actions induced by i.

Proposition 3.2.1.1. *With the setting as above, suppose moreover that the abelian scheme A over S is* **ordinary** *as in Definition 3.1.1.2. Let us fix the choice of a system of compatible isomorphisms $\{\zeta_{p^r,\bar{\eta}} : (\mathbb{Z}/p^r\mathbb{Z})(1) \xrightarrow{\sim} \boldsymbol{\mu}_{p^r,\bar{\eta}}\}_{r\geq 0}$, which exists because K is of characteristic zero. Then the following are true:*

(1) A_s is an ordinary abelian variety in the usual sense.

(2) The **physical Tate modules** *$\mathrm{T}_p A_s$ and $\mathrm{T}_p A_s^\vee$ are free $\mathbb{Z}_p$-modules of rank $\dim_s A_s$ (when their $\mathcal{O}$-module structures are ignored).*

(3) Let $(\hat{\alpha}, \hat{\nu}) : L \underset{\mathbb{Z}}{\otimes} \hat{\mathbb{Z}} \xrightarrow{\sim} \mathrm{T} A_{\bar{\eta}}$ be any lifting of $\alpha_{\mathcal{H}}$ (whose $\mathcal{H}$ orbit determines $\alpha_{\mathcal{H}}$) (as in [62, Lem. 1.3.6.5]). Then the factor

$$\hat{\alpha}_p : L \underset{\mathbb{Z}}{\otimes} \mathbb{Z}_p \xrightarrow{\sim} \mathrm{T}_p A_{\bar{\eta}}$$

of $\hat{\alpha}$ at p and the canonical $\mathcal{O} \underset{\mathbb{Z}}{\otimes} \mathbb{Z}_p$-equivariant extension

$$0 \to A_s[p^\infty]^{\mathrm{mult}} \to A_s[p^\infty] \to A_s[p^\infty]^{\text{ét}} \to 0$$

define a filtration

$$\mathtt{D}^1 = 0 \subset \mathtt{D}^0 \subset \mathtt{D}^{-1} = L \underset{\mathbb{Z}}{\otimes} \mathbb{Z}_p \tag{3.2.1.2}$$

of $\mathcal{O} \underset{\mathbb{Z}}{\otimes} \mathbb{Z}_p$-modules, whose $\mathcal{H}$-orbit is canonical. Let us set $\mathrm{Gr}_{\mathtt{D}}^{-1} := \mathtt{D}^{-1}/\mathtt{D}^0$ and $\mathrm{Gr}_{\mathtt{D}}^0 := \mathtt{D}^0/\mathtt{D}^1$ as usual. Then there exists canonically induced $\mathcal{O} \underset{\mathbb{Z}}{\otimes} \mathbb{Z}_p$-equivariant isomorphisms

$$(\mathrm{Gr}_{\mathtt{D}}^0)_s^{\mathrm{mult}} \cong A_s[p^\infty]^{\mathrm{mult}}$$

(see Definition 3.1.1.7),

$$\mathrm{Gr}_{\mathtt{D}}^{-1} \cong \mathrm{T}_p A_s,$$

and

$$\mathrm{Gr}_{\mathtt{D}}^{-1} \underset{\mathbb{Z}_p}{\otimes} (\mathbb{Q}_p/\mathbb{Z}_p) \cong A_s[p^\infty]^{\text{ét}}.$$

(4) By duality, we have an analogous filtration

$$\mathtt{D}^{\#,1} = 0 \subset \mathtt{D}^{\#,0} \subset \mathtt{D}^{\#,-1} = L^\# \underset{\mathbb{Z}}{\otimes} \mathbb{Z}_p \tag{3.2.1.3}$$

of $\mathcal{O} \underset{\mathbb{Z}}{\otimes} \mathbb{Z}_p$-modules, together with canonically induced $\mathcal{O} \underset{\mathbb{Z}}{\otimes} \mathbb{Z}_p$-equivariant isomorphisms

$$(\mathrm{Gr}_{\mathtt{D}^\#}^0)_s^{\mathrm{mult}} \cong A_s^\vee[p^\infty]^{\mathrm{mult}},$$

$$\mathrm{Gr}_{\mathtt{D}^\#}^{-1} \cong \mathrm{T}_p A_s^\vee,$$

and

$$\mathrm{Gr}_{\mathtt{D}^\#}^{-1} \underset{\mathbb{Z}_p}{\otimes} (\mathbb{Q}_p/\mathbb{Z}_p) \cong A_s^\vee[p^\infty]^{\text{ét}},$$

where $\mathrm{Gr}_{\mathtt{D}^\#}^{-1} := \mathtt{D}^{\#,-1}/\mathtt{D}^{\#,0}$ and $\mathrm{Gr}_{\mathtt{D}^\#}^0 := \mathtt{D}^{\#,0}/\mathtt{D}^{\#,1}$ as usual.

(5) *The canonical ($\mathcal{O}$-equivariant) embedding $L \hookrightarrow L^\#$ and the polarization $\lambda : A \to A^\vee$ induce canonical $\mathcal{O} \underset{\mathbb{Z}}{\otimes} \mathbb{Z}_p$-equivariant morphisms*

$$(\phi^0_{\mathsf{D}})^{\mathrm{mult}}_s : (\mathrm{Gr}^0_{\mathsf{D}})^{\mathrm{mult}}_s \to (\mathrm{Gr}^0_{\mathsf{D}^\#})^{\mathrm{mult}}_s$$

and

$$\phi^{-1}_{\mathsf{D}} : \mathrm{Gr}^{-1}_{\mathsf{D}} \to \mathrm{Gr}^{-1}_{\mathsf{D}^\#},$$

such that $(\phi^0_{\mathsf{D}})^{\mathrm{mult}}_s$ and $\phi^{-1}_{\mathsf{D}} \underset{\mathbb{Z}_p}{\otimes} (\mathbb{Q}_p/\mathbb{Z}_p)$ are canonically dual to each other, making the diagrams

$$\begin{array}{ccc} (\mathrm{Gr}^0_{\mathsf{D}})^{\mathrm{mult}}_s & \xrightarrow{\sim} & A_s[p^\infty]^{\mathrm{mult}} \\ {\scriptstyle \hat{\nu}^{\mathrm{ord}}(\phi^0_{\mathsf{D}})^{\mathrm{mult}}_s}\downarrow & & \downarrow{\scriptstyle \lambda} \\ (\mathrm{Gr}^0_{\mathsf{D}^\#})^{\mathrm{mult}}_s & \xrightarrow{\sim} & A^\vee_s[p^\infty]^{\mathrm{mult}} \end{array}$$

$$\begin{array}{ccc} \mathrm{Gr}^{-1}_{\mathsf{D}} & \xrightarrow{\sim} & \mathrm{T}_p\, A_s \\ {\scriptstyle \phi^{-1}_{\mathsf{D}}}\downarrow & & \downarrow{\scriptstyle \lambda} \\ \mathrm{Gr}^{-1}_{\mathsf{D}^\#} & \xrightarrow{\sim} & \mathrm{T}_p\, A^\vee_s \end{array}$$

and

$$\begin{array}{ccc} \mathrm{Gr}^{-1}_{\mathsf{D}} \underset{\mathbb{Z}_p}{\otimes} (\mathbb{Q}_p/\mathbb{Z}_p) & \xrightarrow{\sim} & A_s[p^\infty]^{\text{ét}} \\ {\scriptstyle \phi^{-1}_{\mathsf{D}} \underset{\mathbb{Z}_p}{\otimes} (\mathbb{Q}_p/\mathbb{Z}_p)}\downarrow & & \downarrow{\scriptstyle \lambda} \\ \mathrm{Gr}^{-1}_{\mathsf{D}^\#} \underset{\mathbb{Z}_p}{\otimes} (\mathbb{Q}_p/\mathbb{Z}_p) & \xrightarrow{\sim} & A^\vee_s[p^\infty]^{\text{ét}} \end{array}$$

commutative for some canonically determined $\hat{\nu}^{\mathrm{ord}} \in \mathbb{Z}_p^\times$.

(6) *Let us consider $\mathrm{Lie}_{A/S}$ as an $\mathcal{O} \underset{\mathbb{Z}}{\otimes} R$-module using the $\mathcal{O}$-structure i. Then there are canonical isomorphisms*

$$\mathrm{Hom}_{\mathbb{Z}_p}(\mathrm{Gr}^{-1}_{\mathsf{D}^\#}, \mathbb{Z}_p) \underset{\mathbb{Z}_p}{\otimes} R \cong \mathrm{Lie}_{A/S}$$

and

$$\mathrm{Hom}_{\mathbb{Z}_p}(\mathrm{Gr}^{-1}_{\mathsf{D}}, \mathbb{Z}_p) \underset{\mathbb{Z}_p}{\otimes} R \cong \mathrm{Lie}_{A^\vee/S}$$

of $\mathcal{O} \underset{\mathbb{Z}}{\otimes} R$-modules making the diagram

$$\begin{array}{ccc} \mathrm{Hom}_{\mathbb{Z}_p}(\mathrm{Gr}^{-1}_{\mathsf{D}^\#}, \mathbb{Z}_p) \underset{\mathbb{Z}_p}{\otimes} R & \xrightarrow{\sim} & \mathrm{Lie}_{A/S} \\ {\scriptstyle {}^t(\phi^{-1}_{\mathsf{D}})}\downarrow & & \downarrow{\scriptstyle d\lambda} \\ \mathrm{Hom}_{\mathbb{Z}_p}(\mathrm{Gr}^{-1}_{\mathsf{D}}, \mathbb{Z}_p) \underset{\mathbb{Z}_p}{\otimes} R & \xrightarrow{\sim} & \mathrm{Lie}_{A^\vee/S} \end{array}$$

commutative. These are dual to canonical isomorphisms

$$\mathrm{Lie}^{\vee}_{A/S} \xrightarrow{\sim} \mathrm{Gr}^{-1}_{\mathtt{D}^{\#}} \underset{\mathbb{Z}_p}{\otimes} R$$

and

$$\mathrm{Lie}^{\vee}_{A^{\vee}/S} \xrightarrow{\sim} \mathrm{Gr}^{-1}_{\mathtt{D}} \underset{\mathbb{Z}_p}{\otimes} R$$

of $\mathcal{O} \underset{\mathbb{Z}}{\otimes} R$*-modules making the diagram*

$$\begin{array}{ccc} \mathrm{Lie}^{\vee}_{A^{\vee}/S} & \xrightarrow{\sim} & \mathrm{Gr}^{-1}_{\mathtt{D}} \underset{\mathbb{Z}_p}{\otimes} R \\ \Big\downarrow{\scriptstyle \lambda^*} & & \Big\downarrow{\scriptstyle \phi_{\mathtt{D}}^{-1}} \\ \mathrm{Lie}^{\vee}_{A/S} & \xrightarrow{\sim} & \mathrm{Gr}^{-1}_{\mathtt{D}^{\#}} \underset{\mathbb{Z}_p}{\otimes} R \end{array}$$

commutative.

(7) *Let* F_0' *and* L_0 *be as in Section* 1.4*. Then, for each homomorphism* $F_0' \to \bar{K}$ *of extension fields of* F_0*, there is an isomorphism*

$$L_0 \underset{F_0'}{\otimes} \bar{K} \cong \mathrm{Lie}_{A/S} \underset{R}{\otimes} \bar{K}$$

of $\mathcal{O} \underset{\mathbb{Z}}{\otimes} \bar{K}$*-modules. (This follows from the fact that* $(A_\eta, \lambda_\eta, i_\eta, \alpha_{\mathcal{H},\eta})$*, as an object of* $\mathsf{M}_{\mathcal{H}}(\eta)$*, satisfies the Lie algebra condition in [62, Def. 1.3.4.1].) Consequently, we have an isomorphism*

$$L_0 \underset{F_0'}{\otimes} \bar{K} \cong \mathrm{Hom}_{\mathbb{Z}_p}(\mathrm{Gr}^{-1}_{\mathtt{D}^{\#}}, \mathbb{Z}_p) \underset{\mathbb{Z}_p}{\otimes} \bar{K} \tag{3.2.1.4}$$

of $\mathcal{O} \underset{\mathbb{Z}}{\otimes} \bar{K}$*-modules.*

(8) *The canonical homomorphism* $\mathbb{Q}_p \to (F_0)_v$ *is an isomorphism.*

Proof. Statements (1) and (2) are clear form the definitions (see Remark 3.1.1.4). Statements (3), (4), and (5) follow from the rigidity of groups of multiplicative type (see [26, IX, 3.6 and 3.6 bis]), by (uniquely) lifting the torsion subgroup schemes $A_s[p^r]^{\mathrm{mult}}$ and $A_s^{\vee}[p^r]^{\mathrm{mult}}$ to subgroup schemes $A[p^r]^{\mathrm{mult}}$ and $A^{\vee}[p^r]^{\mathrm{mult}}$ of multiplicative type of $A[p^r]$ and $A^{\vee}[p^r]$, respectively, for each $r \geq 0$, so that $\hat{\alpha}$ induces via $\{\zeta_{p^r,\bar{\eta}}\}_{r\geq 0}$ an isomorphism $(\mathrm{Gr}^0_{\mathtt{D}})^{\mathrm{mult}}_{\bar{\eta}} \xrightarrow{\sim} A_{\bar{\eta}}[p^\infty]^{\mathrm{mult}}$, extending to an isomorphism over the normalization $\bar{R}$ of R in $\bar{K}$ and inducing the desired isomorphism $(\mathrm{Gr}^0_{\mathtt{D}})^{\mathrm{mult}}_{s} \xrightarrow{\sim} A_s[p^\infty]^{\mathrm{mult}}$; so that the dual of $\hat{\alpha}$ induces an isomorphism $(\mathrm{Gr}^0_{\mathtt{D}^{\#}})^{\mathrm{mult}}_{\bar{\eta}} \cong A^{\vee}_{\bar{\eta}}[p^\infty]^{\mathrm{mult}}$, compatible with the above isomorphism $(\mathrm{Gr}^0_{\mathtt{D}})^{\mathrm{mult}}_{\bar{\eta}} \xrightarrow{\sim} A_{\bar{\eta}}[p^\infty]^{\mathrm{mult}}$ up to a scalar $\hat{\nu}^{\mathrm{ord}} \in \mathbb{Z}_p^\times$ such that $(\hat{\nu}^{\mathrm{ord}} \bmod p^r)\,\zeta_{p^r,\bar{\eta}} = \hat{\nu} \bmod p^r$ for every $r \geq 0$, extending to an isomorphism over $\bar{R}$ as above and inducing the desired isomorphism $(\mathrm{Gr}^0_{\mathtt{D}})^{\mathrm{mult}}_{s} \xrightarrow{\sim} A_s[p^\infty]^{\mathrm{mult}}$; and so that the rest of the assertions follow by various canonical identifications. Statement (6) follows from [47, 3.4], and statement (7) is self-explanatory. Because of the isomorphism (3.2.1.4), the subfield of $\bar{K}$ generated by traces $\mathrm{Tr}_{\mathbb{C}}(b|L_0 \underset{F_0'}{\otimes} \bar{K})$ for $b \in \mathcal{O}$ is contained in $\mathbb{Q}_p$. Since this is true for every field homomorphism $F_0' \to \bar{K}$, statement (8) follows. □

3.2.2 *Maximal Totally Isotropic Submodules at p*

Now we will turn the observations made in Proposition 3.2.1.1 into formal definitions.

Let $p > 0$ be any rational prime number.

Lemma 3.2.2.1. *Consider the following two kinds of data:*

(1) A filtration

$$\mathtt{D}^1_{\mathbb{Q}_p} = 0 \subset \mathtt{D}^0_{\mathbb{Q}_p} \subset \mathtt{D}^{-1}_{\mathbb{Q}_p} = L \underset{\mathbb{Z}}{\otimes} \mathbb{Q}_p \tag{3.2.2.2}$$

of $\mathcal{O} \underset{\mathbb{Z}}{\otimes} \mathbb{Q}_p$-modules such that $\mathtt{D}^0_{\mathbb{Q}_p}$ is (maximal) totally isotropic under the pairing induced by $\langle \cdot, \cdot \rangle$ and such that $\mathtt{D}^0_{\mathbb{Q}_p}$ is its own annihilator under the pairing. For simplicity, we shall call $\mathtt{D}^0_{\mathbb{Q}_p}$ a **maximal totally isotropic submodule** *of $L \underset{\mathbb{Z}}{\otimes} \mathbb{Q}_p$, without mentioning the $\mathcal{O} \underset{\mathbb{Z}}{\otimes} \mathbb{Q}_p$-submodule structure.*

(2) A filtration

$$\mathtt{D}^1 = 0 \subset \mathtt{D}^0 \subset \mathtt{D}^{-1} = L \underset{\mathbb{Z}}{\otimes} \mathbb{Z}_p \tag{3.2.2.3}$$

of $\mathcal{O} \underset{\mathbb{Z}}{\otimes} \mathbb{Z}_p$-modules such that $\mathtt{D}^0$ is totally isotropic under the pairing induced by $\langle \cdot, \cdot \rangle$, such that the quotient $\mathcal{O} \underset{\mathbb{Z}}{\otimes} \mathbb{Z}_p$-module $\mathrm{Gr}^{-1}_{\mathtt{D}} = \mathtt{D}^{-1}/\mathtt{D}^0$ is torsion-free (as a $\mathbb{Z}_p$-module), and such that $\mathtt{D}^0$ is its own annihilator under the pairing. For simplicity, we shall call $\mathtt{D}^0$ a **maximal totally isotropic submodule** *of $L \underset{\mathbb{Z}}{\otimes} \mathbb{Z}_p$, without mentioning the $\mathcal{O} \underset{\mathbb{Z}}{\otimes} \mathbb{Z}_p$-submodule structure.*

These two kinds of data determine each other in the following way: $\mathtt{D}^0_{\mathbb{Q}_p}$ is the $\mathbb{Q}_p$-span of $\mathtt{D}^0$ in $L \underset{\mathbb{Z}}{\otimes} \mathbb{Q}_p$, while $\mathtt{D}^0$ is the intersection of $\mathtt{D}^0_{\mathbb{Q}_p}$ with $L \underset{\mathbb{Z}}{\otimes} \mathbb{Z}_p$.

Proof. The statements are self-explanatory. □

Lemma 3.2.2.4. *Each choice of a filtration* (3.2.2.2) *(or equivalently a filtration* (3.2.2.3)*; see Lemma* 3.2.2.1*) determines the following list of data:*

(1) A filtration

$$\mathtt{D}^{\#,1} = 0 \subset \mathtt{D}^{\#,0} \subset \mathtt{D}^{\#,-1} = L^{\#} \underset{\mathbb{Z}}{\otimes} \mathbb{Z}_p \tag{3.2.2.5}$$

such that $\mathtt{D}^{\#,0} = \mathtt{D}^{\#,0}_{\mathbb{Q}_p} \cap (L^{\#} \underset{\mathbb{Z}}{\otimes} \mathbb{Z}_p)$ in $L \underset{\mathbb{Z}}{\otimes} \mathbb{Q}_p$, so that $\mathrm{Gr}^{-1}_{\mathtt{D}^{\#}} = \mathtt{D}^{\#,-1}/\mathtt{D}^{\#,0}$ is torsion-free (as a $\mathbb{Z}_p$-module).

(2) A perfect duality

$$\mathrm{Gr}^0_{\mathtt{D}} \times \mathrm{Gr}^{-1}_{\mathtt{D}^{\#}} \to \mathbb{Z}_p(1)$$

induced by $\langle \cdot, \cdot \rangle$.

(3) *A perfect duality*

$$\mathrm{Gr}^0_{\mathtt{D}^\#} \times \mathrm{Gr}^{-1}_{\mathtt{D}} \to \mathbb{Z}_p(1)$$

induced by $\langle\,\cdot\,,\,\cdot\,\rangle$.

(4) *An canonical inclusion*

$$\phi^0_{\mathtt{D}} : \mathrm{Gr}^0_{\mathtt{D}} \hookrightarrow \mathrm{Gr}^0_{\mathtt{D}^\#}$$

(*with finite cokernel*) *dual to a canonical inclusion*

$$\phi^{-1}_{\mathtt{D}} : \mathrm{Gr}^{-1}_{\mathtt{D}} \hookrightarrow \mathrm{Gr}^{-1}_{\mathtt{D}^\#}$$

(*with finite cokernel*).

(5) *For each integer* $r \geq 0$, *we have* $\mathrm{Gr}^0_{\mathtt{D},p^r} := \mathtt{D}^0_{p^r} := \mathtt{D}^0/p^r\mathtt{D}^0$, $\mathrm{Gr}^0_{\mathtt{D}^\#,p^r} := \mathtt{D}^{\#,0}_{p^r} := \mathtt{D}^{\#,0}/p^r\mathtt{D}^{\#,0}$, $\mathrm{Gr}^{-1}_{\mathtt{D},p^r} := \mathrm{Gr}^{-1}_{\mathtt{D}}/p^r\,\mathrm{Gr}^{-1}_{\mathtt{D}}$, *and* $\mathrm{Gr}^{-1}_{\mathtt{D}^\#,p^r} := \mathrm{Gr}^{-1}_{\mathtt{D}^\#}/p^r\,\mathrm{Gr}^{-1}_{\mathtt{D}^\#}$, *together with the morphisms* $\phi^0_{\mathtt{D},p^r} : \mathrm{Gr}^0_{\mathtt{D},p^r} \to \mathrm{Gr}^0_{\mathtt{D}^\#,p^r}$ *and* $\phi^{-1}_{\mathtt{D},p^r} : \mathrm{Gr}^{-1}_{\mathtt{D},p^r} \to \mathrm{Gr}^{-1}_{\mathtt{D}^\#,p^r}$ *induced by* $\phi^0_{\mathtt{D}}$ *and* $\phi^{-1}_{\mathtt{D}}$, *respectively.*

Proof. The statements are self-explanatory. □

Lemma 3.2.2.6. *Under the assumption that* L *satisfies Condition* 1.2.1.1, *any filtration* (3.2.2.3) *as in Lemma* 3.2.2.1 (*noncanonically*) *splits. The filtration* (3.2.2.5) *it determines as in Lemma* 3.2.2.4 *also splits.* (*The splittings might not be compatible with each other under the canonical morphisms induced by* $L \hookrightarrow L^\#$ *and* $\lambda : A \to A^\vee$.)

Proof. By Condition 1.2.1.1, the action of $\mathcal{O}$ on L extends to an action of some maximal order $\mathcal{O}'$ in $\mathcal{O} \underset{\mathbb{Z}}{\otimes} \mathbb{Q}$ containing $\mathcal{O}$. By Lemma 3.2.2.1, $\mathtt{D}^0$ is the intersection of $\mathtt{D}^0_{\mathbb{Q}_p}$ with $L \underset{\mathbb{Z}}{\otimes} \mathbb{Z}_p$. Hence, the action of $\mathcal{O}$ on the submodule $\mathtt{D}^0$ of $L \underset{\mathbb{Z}}{\otimes} \mathbb{Z}_p$ extends to an action of $\mathcal{O}'$ (compatible with those on $\mathtt{D}^0_{\mathbb{Q}_p}$ and $L \underset{\mathbb{Z}}{\otimes} \mathbb{Z}_p$), and the filtration $\mathtt{D}$ of $L \underset{\mathbb{Z}}{\otimes} \mathbb{Z}_p$ is $\mathcal{O}' \underset{\mathbb{Z}}{\otimes} \mathbb{Z}_p$-equivariant. A similar argument shows that the filtration $\mathtt{D}^\#$ on $L^\# \underset{\mathbb{Z}}{\otimes} \mathbb{Z}_p$ is also $\mathcal{O}' \underset{\mathbb{Z}}{\otimes} \mathbb{Z}_p$-equivariant. Since $\mathcal{O}'$ is maximal, $\mathcal{O}' \underset{\mathbb{Z}}{\otimes} \mathbb{Z}_p$ is also maximal, which is *hereditary* in the sense that all $\mathcal{O}' \underset{\mathbb{Z}}{\otimes} \mathbb{Z}_p$-lattices (namely, finitely generated $\mathcal{O}' \underset{\mathbb{Z}}{\otimes} \mathbb{Z}_p$-modules with no p-torsion) are projective $\mathcal{O}' \underset{\mathbb{Z}}{\otimes} \mathbb{Z}_p$-modules (*cf.* [62, Prop. 1.1.1.12 and 1.1.1.23]). Hence, the filtrations (3.2.2.3) and (3.2.2.5) split because $\mathrm{Gr}^{-1}_{\mathtt{D}}$ and $\mathrm{Gr}^{-1}_{\mathtt{D}^\#}$ are torsion-free (as $\mathbb{Z}_p$-modules), as desired. □

Definition 3.2.2.7. *Let* $\mathcal{O}$, $(L, \langle\,\cdot\,,\,\cdot\,\rangle)$ *be given as above. We define for each* $\mathbb{Z}_p$-*algebra* R

$$\mathrm{P}^{\mathrm{ord}}_{\mathtt{D}}(R) := \left\{ \begin{array}{c} (g,r) \in \mathrm{GL}_{\mathcal{O} \underset{\mathbb{Z}}{\otimes} R}(L \underset{\mathbb{Z}}{\otimes} R) \times \mathbf{G}_{\mathrm{m}}(R) : \\ (g,r) \in \mathrm{G}(R),\ g(\mathtt{D} \underset{\mathbb{Z}_p}{\otimes} R) = \mathtt{D} \underset{\mathbb{Z}_p}{\otimes} R \end{array} \right\},$$

$$\mathrm{M}_{\mathsf{D}}^{\mathrm{ord}}(R) := \left\{ \begin{array}{c} (g,r) \in \mathrm{GL}_{\mathcal{O} \underset{\mathbb{Z}}{\otimes} R}(\mathrm{Gr}_{\mathsf{D}} \underset{\mathbb{Z}_p}{\otimes} R) \times \mathbf{G}_{\mathrm{m}}(R) : \\ \langle gx, gy \rangle = r\langle x, y \rangle, \ \forall x, y \in \mathrm{Gr}_{\mathsf{D}} \underset{\mathbb{Z}_p}{\otimes} R \end{array} \right\},$$

$$\mathrm{U}_{\mathsf{D}}^{\mathrm{ord}}(R) := \ker(\mathrm{Gr}_{\mathsf{D}} : \mathrm{P}_{\mathsf{D}}^{\mathrm{ord}}(R) \to \mathrm{M}_{\mathsf{D}}^{\mathrm{ord}}(R)),$$

and

$$\mathrm{U}_{\mathsf{D}}^{\mathrm{ord},i}(R) := \ker(\mathrm{Gr}_{\mathsf{D}}^i : \mathrm{P}_{\mathsf{D}}^{\mathrm{ord}}(R) \to \mathrm{GL}_{\mathcal{O} \underset{\mathbb{Z}}{\otimes} R}(\mathrm{Gr}_{\mathsf{D}}^i \underset{\mathbb{Z}_p}{\otimes} R))$$

for each i. These assignments are functorial in R and define group functors $\mathrm{P}_{\mathsf{D}}^{\mathrm{ord}}$, $\mathrm{M}_{\mathsf{D}}^{\mathrm{ord}}$, $\mathrm{U}_{\mathsf{D}}^{\mathrm{ord}}$, and $\mathrm{U}_{\mathsf{D}}^{\mathrm{ord},i}$ over $\mathrm{Spec}(\mathbb{Z}_p)$. *By definition, $\mathrm{P}_{\mathsf{D}}^{\mathrm{ord}}$ is a subgroup of $\mathrm{G} \underset{\mathbb{Z}}{\otimes} \mathbb{Z}_p$, and (by Lemma* 3.2.2.6*) there is an exact sequence*

$$1 \to \mathrm{U}_{\mathsf{D}}^{\mathrm{ord}} \to \mathrm{P}_{\mathsf{D}}^{\mathrm{ord}} \to \mathrm{M}_{\mathsf{D}}^{\mathrm{ord}} \to 1.$$

As in the case of G, *the projections to the second factor $(g,r) \mapsto r$ define homomorphisms $\nu : \mathrm{P}_{\mathsf{D}}^{\mathrm{ord}} \to \mathbf{G}_{\mathrm{m}} \underset{\mathbb{Z}}{\otimes} \mathbb{Z}_p$, $\nu : \mathrm{M}_{\mathsf{D}}^{\mathrm{ord}} \to \mathbf{G}_{\mathrm{m}} \underset{\mathbb{Z}}{\otimes} \mathbb{Z}_p$, and $\nu : \mathrm{U}_{\mathsf{D}}^{\mathrm{ord},i} \to \mathbf{G}_{\mathrm{m}} \underset{\mathbb{Z}}{\otimes} \mathbb{Z}_p$, which we call the* **similitude characters**. *For simplicity, we shall often denote elements (g,r) by simply g, and denote by $\nu(g)$ the value of r when we need it.*

Definition 3.2.2.8. *For all integers $0 \leq r$ and $0 \leq r_1 \leq r_0$, we set:*

$$\begin{aligned}
\mathcal{U}_p(p^r) &:= \ker(\mathrm{G}(\mathbb{Z}_p) \to \mathrm{G}(\mathbb{Z}/p^r\mathbb{Z})), \\
\mathcal{U}_{p,0}(p^r) &:= (\mathrm{G}(\mathbb{Z}_p) \to \mathrm{G}(\mathbb{Z}/p^r\mathbb{Z}))^{-1}(\mathrm{P}_{\mathsf{D}}^{\mathrm{ord}}(\mathbb{Z}/p^r\mathbb{Z})), \\
\mathcal{U}_{p,1}(p^r) &:= (\mathrm{G}(\mathbb{Z}_p) \to \mathrm{G}(\mathbb{Z}/p^r\mathbb{Z}))^{-1}(\mathrm{U}_{\mathsf{D}}^{\mathrm{ord},-1}(\mathbb{Z}/p^r\mathbb{Z})), \\
\mathcal{U}_{p,1}^{\mathrm{bal}}(p^r) &:= (\mathrm{G}(\mathbb{Z}_p) \to \mathrm{G}(\mathbb{Z}/p^r\mathbb{Z}))^{-1}(\mathrm{U}_{\mathsf{D}}^{\mathrm{ord}}(\mathbb{Z}/p^r\mathbb{Z})), \\
\mathcal{U}_{p,1,0}(p^{r_1}, p^{r_0}) &:= \mathcal{U}_{p,1}(p^{r_1}) \cap \mathcal{U}_{p,0}(p^{r_0}), \\
\mathcal{U}_{p,1,0}^{\mathrm{bal}}(p^{r_1}, p^{r_0}) &:= \mathcal{U}_{p,1}^{\mathrm{bal}}(p^{r_1}) \cap \mathcal{U}_{p,0}(p^{r_0}), \\
\mathcal{U}^{\mathrm{ord}}(p^r) &:= \ker(\mathrm{M}_{\mathsf{D}}^{\mathrm{ord}}(\mathbb{Z}_p) \to \mathrm{M}_{\mathsf{D}}^{\mathrm{ord}}(\mathbb{Z}/p^r\mathbb{Z})).
\end{aligned}$$

Definition 3.2.2.9. *We say that an open compact subgroup $\mathcal{H}_p \subset \mathrm{G}(\mathbb{Z}_p)$ is* **of standard form with respect to** D *if there exists an integer $r \geq 0$ such that*

$$\mathcal{U}_{p,1}^{\mathrm{bal}}(p^r) \subset \mathcal{H}_p \subset \mathcal{U}_{p,0}(p^r).$$

In this case, we say that r is the **depth** *of $\mathcal{H}_p$, and write $r = \mathrm{depth}_{\mathsf{D}}(\mathcal{H}_p)$.*

We say that an open compact subgroup $\mathcal{H} \subset \mathrm{G}(\hat{\mathbb{Z}})$ is **of standard form with respect to** D *if it is of the form $\mathcal{H} = \mathcal{H}^p\mathcal{H}_p$, where $\mathcal{H}^p \subset \mathrm{G}(\hat{\mathbb{Z}}^p)$ and $\mathcal{H}_p \subset \mathrm{G}(\mathbb{Z}_p)$, such that $\mathcal{H}_p$ is of standard form with respect to* D. *In this case, we set* $\mathrm{depth}_{\mathsf{D}}(\mathcal{H}) := \mathrm{depth}_{\mathsf{D}}(\mathcal{H}_p)$.

We say that two open compact subgroups $\mathcal{H}_p$ and $\mathcal{H}'_p$ of $\mathrm{G}(\mathbb{Z}_p)$ (resp. $\mathcal{H}$ and $\mathcal{H}'$ of $\mathrm{G}(\hat{\mathbb{Z}})$) of standard form with respect to D *are* **equally deep** *if* $\mathrm{depth}_{\mathsf{D}}(\mathcal{H}_p) = \mathrm{depth}_{\mathsf{D}}(\mathcal{H}'_p)$ *(resp.* $\mathrm{depth}_{\mathsf{D}}(\mathcal{H}) = \mathrm{depth}_{\mathsf{D}}(\mathcal{H}')$*).*

We shall suppress the term "with respect to D*" when the choice of* D *is clear from the context.*

By Proposition 3.2.1.1, for the filtration $\mathtt{D} = \{\mathtt{D}^i\}_i$ of $\mathcal{O} \underset{\mathbb{Z}}{\otimes} \mathbb{Z}_p$-modules of $L \underset{\mathbb{Z}}{\otimes} \mathbb{Z}_p$ as in Lemma 3.2.2.1 to be useful for our purpose of defining and studying the ordinary loci in mixed characteristics, we need the following:

Assumption 3.2.2.10. *There exists a place v of F_0 above p such that the canonical homomorphism $\mathbb{Q}_p \to (F_0)_v$ is an isomorphism, and there exists an extension field K of $(F_0)_v$ (and F_0), together with a homomorphism $F_0' \to K$ of fields over F_0, such that*

$$L_0 \underset{F_0'}{\otimes} K \cong \mathrm{Hom}_{\mathbb{Z}_p}(\mathrm{Gr}^{-1}_{\mathtt{D}^\#}, \mathbb{Z}_p) \underset{\mathbb{Z}_p}{\otimes} K \tag{3.2.2.11}$$

as $\mathcal{O} \underset{\mathbb{Z}}{\otimes} K$-modules.

This assumption will be made when we define moduli problems for ordinary level structures.

3.2.3 *Compatibility with Cusp Labels*

Let $\mathcal{H} \subset \mathrm{G}(\hat{\mathbb{Z}})$ be of standard form (with respect to $\mathtt{D}$) as in Definition 3.2.2.9, so that $\mathcal{H} = \mathcal{H}^p\mathcal{H}_p$ and $\mathcal{U}^{\mathrm{bal}}_{p,1}(p^r) \subset \mathcal{H}_p \subset \mathcal{U}_{p,0}(p^r)$ for $r = \mathrm{depth}_{\mathtt{D}}(\mathcal{H})$. Let $[(\mathtt{Z}_\mathcal{H}, \Phi_\mathcal{H}, \delta_\mathcal{H})]$ be a cusp label at level $\mathcal{H}$ for the PEL-type $\mathcal{O}$-lattice $(L, \langle \cdot , \cdot \rangle, h_0)$.

By definition, $\mathtt{Z}_\mathcal{H}$ is an $\mathcal{H}$-orbit of strongly symplectic admissible filtrations $\mathtt{Z}$ on $L \underset{\mathbb{Z}}{\otimes} \hat{\mathbb{Z}}$. This includes, in particular, the datum of an $\mathcal{H}_p/\mathcal{U}(p^r)$-orbit of symplectic admissible filtrations $\mathtt{Z} \underset{\hat{\mathbb{Z}}}{\otimes} \mathbb{Z}_p = \{\mathtt{Z}_{-i} \underset{\hat{\mathbb{Z}}}{\otimes} \mathbb{Z}_p\}_i$ on $L \underset{\mathbb{Z}}{\otimes} \mathbb{Z}_p$.

Definition 3.2.3.1. *We say that the cusp label $[(\mathtt{Z}_\mathcal{H}, \Phi_\mathcal{H}, \delta_\mathcal{H})]$ is* **compatible with the filtration** $\mathtt{D}$ *if there exists at least one representative $\mathtt{Z}$ in the $\mathcal{H}$-orbit $\mathtt{Z}_\mathcal{H}$ such that we have*

$$\mathtt{Z}_{-2} \underset{\hat{\mathbb{Z}}}{\otimes} \mathbb{Z}_p \subset \mathtt{D}^0 \subset \mathtt{Z}_{-1} \underset{\hat{\mathbb{Z}}}{\otimes} \mathbb{Z}_p, \tag{3.2.3.2}$$

which induces a filtration $\mathtt{D}_{-1} = \{\mathtt{D}^i_{-1}\}_i$ on $\mathrm{Gr}^{\mathtt{Z}}_{-1} \underset{\hat{\mathbb{Z}}}{\otimes} \mathbb{Z}_p$ given by

$$\mathtt{D}^1_{-1} := 0 \subset \mathtt{D}^0_{-1} := \mathtt{D}^0/(\mathtt{Z}_{-2} \underset{\hat{\mathbb{Z}}}{\otimes} \mathbb{Z}_p) \subset \mathtt{D}^{-1}_{-1} := \mathrm{Gr}^{\mathtt{Z}}_{-1} \underset{\hat{\mathbb{Z}}}{\otimes} \mathbb{Z}_p$$

(serving the same purpose as the filtration $\mathtt{D}$ does for $L \underset{\mathbb{Z}}{\otimes} \mathbb{Z}_p$).

By taking reduction modulo p^r, we have the compatibility

$$\mathtt{Z}_{-2,p^r} \subset \mathtt{D}^0_{p^r} \subset \mathtt{Z}_{-1,p^r}, \tag{3.2.3.3}$$

which induces a filtration $\mathtt{D}_{-1,p^r} = \{\mathtt{D}^i_{-1,p^r}\}_i$ on $\mathrm{Gr}^{\mathtt{Z}}_{-1,p^r}$ given by

$$\mathtt{D}^1_{-1,p^r} := 0 \subset \mathtt{D}^0_{-1,p^r} := \mathtt{D}^0_{p^r}/\mathtt{Z}_{-2,p^r} \subset \mathtt{D}^{-1}_{-1,p^r} := \mathrm{Gr}^{\mathtt{Z}}_{-1,p^r} \tag{3.2.3.4}$$

(serving the same purpose as the filtration $\mathtt{D}_{p^r}$ does for L/p^rL). Similarly, we have the compatibility

$$\mathtt{Z}^\#_{-2,p^r} \subset \mathtt{D}^{\#,0}_{p^r} \subset \mathtt{Z}^\#_{-1,p^r}, \tag{3.2.3.5}$$

which induces a filtration $\mathtt{D}^{\#}_{-1,p^r} = \{\mathtt{D}^{\#,i}_{-1,p^r}\}_i$ on $\mathrm{Gr}^{\mathtt{Z}^{\#}}_{-1,p^r}$ given by

$$\mathtt{D}^{\#,1}_{-1,p^r} := 0 \subset \mathtt{D}^{\#,0}_{-1,p^r} := \mathtt{D}^{\#,0}_{p^r}/\mathtt{Z}^{\#}_{-2,p^r} \subset \mathtt{D}^{\#,-1}_{-1,p^r} := \mathrm{Gr}^{\mathtt{Z}^{\#}}_{-1,p^r}. \tag{3.2.3.6}$$

Remark 3.2.3.7. Since $\mathcal{U}^{\mathrm{bal}}_{p,1}(p^r) \subset \mathcal{H}_p \subset \mathcal{U}_{p,0}(p^r)$, and since the action of $\mathcal{U}_{p,0}(p^r)$ stabilizes $\mathtt{D}^0_{p^r}$ as an $\mathcal{O} \underset{\mathbb{Z}}{\otimes} (\mathbb{Z}/p^r\mathbb{Z})$-submodule of L/p^rL, the compatibilities (3.2.3.2), (3.2.3.3), and (3.2.3.5) are independent of the choice of $\mathtt{Z}$.

Hence, it is justified to have the following:

Definition 3.2.3.8. *We say that a cusp label* $[(\mathtt{Z}_{\mathcal{H}}, \Phi_{\mathcal{H}}, \delta_{\mathcal{H}})]$ *at level* $\mathcal{H}$ *is* **$\mathtt{D}$-ordinary** *if it is compatible with the filtration* $\mathtt{D}$ *as in Definition* 3.2.3.1. *We shall simply say that* $[(\mathtt{Z}_{\mathcal{H}}, \Phi_{\mathcal{H}}, \delta_{\mathcal{H}})]$ *is* **ordinary**, *or an* **ordinary cusp label**, *if the choice of* $\mathtt{D}$ *is clear in the context.*

For later references, let us define:

Definition 3.2.3.9. (*Compare with Definition* 1.2.1.11.) *Suppose* $\mathtt{Z}$ *is compatible with* $\mathtt{D}$ *as in* (3.2.3.2). *For each* $\mathbb{Z}_p$*-algebra* R, *we define the following quotients of subgroups of* $\mathrm{P}_{\mathtt{Z}}(R)$ (*see Definitions* 1.2.1.10 *and* 1.2.1.11)*:*

(1) $\mathrm{P}^{\mathrm{ord}}_{\mathtt{Z},\mathtt{D}}(R) := \mathrm{P}_{\mathtt{Z}}(R) \cap \mathrm{P}^{\mathrm{ord}}_{\mathtt{D}}(R)$.
(2) *Because of the compatibility* (3.2.3.2), $\mathrm{Z}_{\mathtt{Z}}(R) \cap \mathrm{P}^{\mathrm{ord}}_{\mathtt{D}}(R) = \mathrm{Z}_{\mathtt{Z}}(R)$ *does not define a new group. This is similar for* $\mathrm{U}_{\mathtt{Z}}(R)$, $\mathrm{U}_{2,\mathtt{Z}}(R)$, $\mathrm{U}_{1,\mathtt{Z}}(R)$, $\mathrm{G}_{l,\mathtt{Z}}(R)$, *and* $\mathrm{G}'_{l,\mathtt{Z}}(R)$.
(3) $\mathrm{P}^{\mathrm{ord}}_{h,\mathtt{Z},\mathtt{D}}(R) := \mathrm{P}^{\mathrm{ord}}_{\mathtt{Z},\mathtt{D}}(R)/\mathrm{Z}^{\mathrm{ord}}_{\mathtt{Z},\mathtt{D}}(R)$ *is the subgroup of elements of* $\mathrm{G}_{h,\mathtt{Z}}(R)$ *preserving the filtration* $\mathtt{D}_{-1}$ *induced by* $\mathtt{D}$ *on* $\mathrm{Gr}^{\mathtt{Z}}_{-1} \underset{\hat{\mathbb{Z}}}{\otimes} \mathbb{Z}_p$ *as in Definition* 3.2.3.1.
(4) $\mathrm{P}^{\mathrm{ord},\prime}_{\mathtt{Z},\mathtt{D}}(R) := \mathrm{P}'_{\mathtt{Z}}(R) \cap \mathrm{P}^{\mathrm{ord}}_{\mathtt{D}}(R)$ *is the kernel of the canonical homomorphism* $(\nu^{-1}\,\mathrm{Gr}^{\mathtt{Z}}_{-2}, \mathrm{Gr}^{\mathtt{Z}}_{0}) : \mathrm{P}^{\mathrm{ord}}_{\mathtt{Z},\mathtt{D}}(R) \to \mathrm{G}'_{l,\mathtt{Z}}(R)$.
(5) $\mathrm{P}^{\mathrm{ord}}_{1,\mathtt{Z},\mathtt{D}}(R) := \mathrm{P}^{\mathrm{ord},\prime}_{\mathtt{Z},\mathtt{D}}(R)/\mathrm{U}_{2,\mathtt{Z}}(R)$, *which is* (*under any splitting* δ *above*) *isomorphic to* $(\mathrm{P}^{\mathrm{ord}}_{h,\mathtt{Z},\mathtt{D}} \ltimes \mathrm{U}_{1,\mathtt{Z}})(R) := \mathrm{P}^{\mathrm{ord}}_{h,\mathtt{Z},\mathtt{D}}(R) \ltimes \mathrm{U}_{1,\mathtt{Z}}(R)$.
(6) $\mathrm{P}^{\mathrm{ord},\prime}_{h,\mathtt{Z},\mathtt{D}}(R) := \mathrm{P}^{\mathrm{ord}}_{1,\mathtt{Z},\mathtt{D}}(R)/\mathrm{U}_{1,\mathtt{Z}}(R) \cong \mathrm{P}^{\mathrm{ord},\prime}_{\mathtt{Z},\mathtt{D}}(R)/\mathrm{U}_{\mathtt{Z}}(R) \cong \mathrm{P}^{\mathrm{ord}}_{\mathtt{Z},\mathtt{D}}(R)/\mathrm{Z}^{\mathrm{ord}}_{\mathtt{Z},\mathtt{D}}(R) \cong \mathrm{P}^{\mathrm{ord}}_{h,\mathtt{Z},\mathtt{D}}(R)$.

3.3 Level Structures

3.3.1 *Level Structures Away from p*

Suppose $\mathcal{H}^p \subset \mathrm{G}(\hat{\mathbb{Z}}^p)$ is an open compact subgroup. Suppose $n_0 \geq 1$ is an integer prime to p such that $\mathcal{U}^p(n_0) \subset \mathcal{H}^p$. Let $H_{n_0} := \mathcal{H}^p/\mathcal{U}^p(n_0)$.

Definition 3.3.1.1. (*Compare with [62, Def. 1.3.6.1].*) *Let* S *be a scheme over* $\mathrm{Spec}(\mathbb{Z}_{(p)})$. *Let* A *be an abelian scheme over* S, *with a polarization* $\lambda : A \to A^\vee$ *and an* $\mathcal{O}$*-endomorphism structure* $i : \mathcal{O} \hookrightarrow \mathrm{End}_S(A)$ *as in [62, Def. 1.3.3.1]. Let* $\mathcal{H}$, n_0, *and* H_{n_0} *be as above. A* **naive principal level-n_0 structure** *of* (A, λ, i) *of type* $(L/n_0L, \langle\,\cdot\,,\,\cdot\,\rangle)$ *is a pair* $(\alpha_{n_0}, \nu_{n_0})$, *where*

(1) $\alpha_{n_0} : (L/n_0L)_S \xrightarrow{\sim} A[n_0]$ *is an* $\mathcal{O}$*-equivariant isomorphism of (étale) group schemes over* S.
(2) $\nu_{n_0} : ((\mathbb{Z}/n_0\mathbb{Z})(1))_S \xrightarrow{\sim} \boldsymbol{\mu}_{n_0,S}$ *is an isomorphism of group schemes over* S *making the diagram*

$$\begin{array}{ccc} (L/n_0L)_S \underset{S}{\times} (L/n_0L)_S & \xrightarrow{\langle\,\cdot\,,\,\cdot\,\rangle} & ((\mathbb{Z}/n_0\mathbb{Z})(1))_S \\ {\scriptstyle \alpha_{n_0}\times\alpha_{n_0}}\downarrow\wr & & \wr\downarrow{\scriptstyle \nu_{n_0}} \\ A[n_0] \underset{S}{\times} A[n_0] & \xrightarrow[\mathrm{e}^{\lambda}]{} & \boldsymbol{\mu}_{n_0,S} \end{array}$$

commutative, where e^{λ} *is the* λ*-Weil pairing.*

By abuse of notation, we often denote such a symplectic isomorphism by $(\alpha_{n_0}, \nu_{n_0}) : (L/n_0L)_S \xrightarrow{\sim} A[n_0]$, *or simply by* $\alpha_{n_0} : (L/n_0L)_S \xrightarrow{\sim} A[n_0]$, *and denote* ν_{n_0} *by* $\nu(\alpha_{n_0})$ *(although* α_{n_0} *does not always determine* ν_{n_0}*).*

Definition 3.3.1.2. (*Compare with [62, Def. 1.3.6.2].*) *We say a naive principal level-*n_0 *structure* $(\alpha_{n_0}, \nu_{n_0})$ *of* (A, λ, i) *of type* $(L/n_0L, \langle\,\cdot\,,\,\cdot\,\rangle)$ *in Definition* 3.3.1.1 *is a* **principal level-n_0 structure** *of type* $(L \underset{\mathbb{Z}}{\otimes} \hat{\mathbb{Z}}^p, \langle\,\cdot\,,\,\cdot\,\rangle)$ *if it satisfies the following* **symplectic-liftability condition***:*

There exists (noncanonically) a tower $(S_{m_0} \twoheadrightarrow S)_{n_0|m_0, p\nmid m_0}$ *of finite étale coverings such that we have the following:*

(1) $S_{n_0} = S$.
(2) *For each* l_0 *such that* $n_0|l_0$ *and* $l_0|m_0$, *there is a finite étale covering* $S_{m_0} \twoheadrightarrow S_{l_0}$ *whose composition with* $S_{l_0} \to S$ *is the finite étale covering* $S_{m_0} \to S$.
(3) *There is a naive principal level-*m_0 *structure* $(\alpha_{m_0,S_{m_0}}, \nu_{m_0,S_{m_0}})$ *of* $(A, \lambda, i) \underset{S}{\times} S_{m_0}$ *of type* $(L/m_0L, \langle\,\cdot\,,\,\cdot\,\rangle)$ *over each* S_{m_0}.
(4) *For each* l_0 *such that* $n_0|l_0$ *and* $l_0|m_0$, *the pullback of* $(\alpha_{l,S_{l_0}}, \nu_{l_0,S_{l_0}})$ *to* S_{m_0} *is the reduction modulo* l_0 *of* $(\alpha_{m_0,S_{m_0}}, \nu_{m_0,S_{m_0}})$.

Definition 3.3.1.3. (*Compare with [62, Def. 1.3.7.3].*) *Let* S *and* (A, λ, i) *be as in Definition* 3.3.1.1. *Let* $\mathcal{H}^p$, n_0, *and* H_{n_0} *be as above. A* **naive level-H_{n_0} structure** *of* (A, λ, i) *of type* $(L/n_0L, \langle\,\cdot\,,\,\cdot\,\rangle)$ *is an* H_{n_0}*-orbit* $\alpha_{H_{n_0}}$ *of naive principal level-*n_0 *structures* $(L/n_0L)_S \xrightarrow{\sim} A[n_0]$, *namely a (finite étale) subscheme* $\alpha_{H_{n_0}}$ *of the finite étale scheme*

$$\underline{\mathrm{Isom}}_S((L/n_0L)_S, A[n_0]) \underset{S}{\times} \underline{\mathrm{Isom}}_S(((\mathbb{Z}/n_0\mathbb{Z})(1))_S, \boldsymbol{\mu}_{n_0,S})$$

over S *that becomes the disjoint union of elements in some* H_{n_0}*-orbit of naive principal level-*n_0 *structures of type* $(L/n_0L, \langle\,\cdot\,,\,\cdot\,\rangle)$ *after a finite étale surjective base change in* S. *In this case, we denote by* $\nu(\alpha_{H_{n_0}})$ *the projection of* $\alpha_{H_{n_0}}$ *to* $\underline{\mathrm{Isom}}_S(((\mathbb{Z}/n_0\mathbb{Z})(1))_S, \boldsymbol{\mu}_{n_0,S})$, *which is a* $\nu(H_{n_0})$*-orbit of étale-locally-defined isomorphisms with its natural interpretation.*

Definition 3.3.1.4. *(Compare with [62, Def. 1.3.7.6].) Let S and (A, λ, i) be as in Definition 3.3.1.1. Let $\mathcal{H}^p$ be as above. For each integer $n_0 \geq 1$ such that $p \nmid n_0$ and $\mathcal{U}^p(n_0) \subset \mathcal{H}^p$, set $H_{n_0} := \mathcal{H}^p/\mathcal{U}^p(n_0)$ as above. Then a* **level-$\mathcal{H}^p$ structure** *of (A, λ, i) of type $(L \underset{\mathbb{Z}}{\otimes} \hat{\mathbb{Z}}^p, \langle \cdot, \cdot \rangle)$ is a collection $\alpha_{\mathcal{H}^p} = \{\alpha_{H_{n_0}}\}_{n_0}$ labeled by integers $n_0 \geq 1$ such that $p \nmid n_0$ and $\mathcal{U}^p(n_0) \subset \mathcal{H}^p$, with elements $\alpha_{H_{n_0}}$ described as follows:*

(1) For each index n_0, the element $\alpha_{H_{n_0}}$ is a naive level structure of (A, λ, i) of type $(L/n_0L, \langle \cdot, \cdot \rangle)$ and level H_{n_0} as in Definition 3.3.1.3.

(2) For all indices n_0 and m_0 such that $n_0 | m_0$, the H_{n_0}-orbit $\alpha_{H_{n_0}}$ is the schematic image of the H_{m_0}-orbit $\alpha_{H_{m_0}}$ under the canonical (finite étale) morphism

$$\begin{aligned} &\underline{\mathrm{Isom}}_S((L/m_0L)_S, A[m_0]) \underset{S}{\times} \underline{\mathrm{Isom}}_S(((\mathbb{Z}/m_0\mathbb{Z})(1))_S, \boldsymbol{\mu}_{m_0,S}) \\ &\quad \to \underline{\mathrm{Isom}}_S((L/n_0L)_S, A[n_0]) \underset{S}{\times} \underline{\mathrm{Isom}}_S(((\mathbb{Z}/n_0\mathbb{Z})(1))_S, \boldsymbol{\mu}_{n_0,S}), \end{aligned}$$

which is equivalent to the formation of $\mathcal{U}^p(n_0)/\mathcal{U}^p(m_0)$-orbits (see [62, Lem. 1.3.7.5]).

Remark 3.3.1.5. In these definitions, unlike in [62, Sec. 1.3.6], we no longer assume that the polarization λ has degree prime to p. Hence, these level structures away from p do not detect the polarization type of λ.

Lemma 3.3.1.6. *Let S and (A, λ, i) be as in Definition 3.3.1.1. Let $\mathcal{H}^p$ be as above, and let $\alpha_{\mathcal{H}^p}$ be a level-$\mathcal{H}^p$ structure of (A, λ, i) of type $(L \underset{\mathbb{Z}}{\otimes} \hat{\mathbb{Z}}^p, \langle \cdot, \cdot \rangle)$ as Definition 3.3.1.4. Let $\bar{s}$ be any geometric point of S. Then there exists an $\mathcal{O}$-equivariant symplectic isomorphism*

$$(\hat{\alpha}^p_{\bar{s}}, \hat{\nu}^p_{\bar{s}}) : L \underset{\mathbb{Z}}{\otimes} \hat{\mathbb{Z}}^p \xrightarrow{\sim} \mathrm{T}^p A_{\bar{s}} \tag{3.3.1.7}$$

such that, for each integer $n_0 \geq 1$ such that $p \nmid n_0$ and $\mathcal{U}^p(n_0) \subset \mathcal{H}^p$, the H_{n_0}-orbit of the reduction of $(\alpha_{n_0,\bar{s}}, \nu_{n_0,\bar{s}}) : L/n_0L \xrightarrow{\sim} A[n_0]_{\bar{s}}$ of $(\hat{\alpha}^p_{\bar{s}}, \hat{\nu}^p_{\bar{s}})$ modulo n_0 coincides with the pullback of $\alpha_{H_{n_0}}$ from S to $\bar{s}$. We say that this $(\hat{\alpha}^p_{\bar{s}}, \hat{\alpha}^p_{\bar{s}})$, or for simplicity just $\hat{\alpha}^p_{\bar{s}}$, is a **lifting** *of $\alpha_{\mathcal{H}^p}$ at $\bar{s}$. The $\mathcal{H}^p$-orbit $[\hat{\alpha}^p_{\bar{s}}]_{\mathcal{H}^p}$ of $\hat{\alpha}^p_{\bar{s}}$ is unique (i.e., independent of the choice of $\hat{\alpha}^p_{\bar{s}}$).*

If S is locally noetherian, then the $\mathcal{H}^p$-orbit $[\hat{\alpha}^p_{\bar{s}}]_{\mathcal{H}^p}$ is $\pi_1(S, \bar{s})$-invariant. Moreover, we can recover the collection $\alpha_{\mathcal{H}^p} = \{\alpha_{H_{n_0}}\}_{n_0}$ over the connected component of $\bar{s}$ on S from the $\pi_1(S, \bar{s})$-invariant $\mathcal{H}^p$-orbit $[\hat{\alpha}^p_{\bar{s}}]_{\mathcal{H}^p}$.

Proof. The pullback of the compatible collection $\alpha_{H_{n_0}}$ to $\bar{s}$ allows us to choose a compatible collection of $\mathcal{O}$-equivariant isomorphisms $\{(\alpha_{n_0,\bar{s}}, \nu_{n_0,\bar{s}}) : L/n_0L \xrightarrow{\sim} A[n_0]_{\bar{s}}\}_{n_0}$, such that the H_{n_0}-orbit of $(\alpha_{n_0,\bar{s}}, \nu_{n_0,\bar{s}})$ is $\alpha_{H_{n_0}}$ for each n_0, which is equivalent to the desired $(\hat{\alpha}_{\bar{s}}, \hat{\nu}_{\bar{s}})$ in (3.3.1.7). When S is locally noetherian, the $\mathcal{H}^p$-orbit of $(\hat{\alpha}^p_{\bar{s}}, \hat{\nu}^p_{\bar{s}})$ is invariant under the action of $\pi_1(S, \bar{s})$ because the H_{n_0}-orbit of $(\alpha_{n_0,\bar{s}}, \nu_{n_0,\bar{s}})$, or rather the pullback $\alpha_{H_{n_0},\bar{s}}$ of $\alpha_{H_{n_0}}$ to $\bar{s}$, is invariant under $\pi_1(S, \bar{s})$ by definition of $\alpha_{H_{n_0}}$ (see Definition 3.3.1.3) and by definition of $\pi_1(S, \bar{s})$. □

3.3.2 *Hecke Twists Defined by Level Structures Away from p*

Suppose $g_0 \in \mathrm{G}(\mathbb{A}^{\infty,p})$, and suppose we have two open compact subgroups $\mathcal{H}^p$ and $\mathcal{H}'^{,p}$ of $\mathrm{G}(\hat{\mathbb{Z}}^p)$ such that $\mathcal{H}'^{,p} \subset g_0\mathcal{H}^p g_0^{-1}$. Let S and (A, λ, i) be as in Definition 3.3.1.1, and let $\alpha_{\mathcal{H}'^{,p}} = \{\alpha_{H'_{m_0}}\}_{m_0}$ be a level-$\mathcal{H}'^{,p}$ structure of (A, λ, i) of type $(L \underset{\mathbb{Z}}{\otimes} \hat{\mathbb{Z}}^p, \langle \cdot, \cdot \rangle)$ as in Definition 3.3.1.4, indexed by integers $m_0 \geq 1$ such that $p \nmid m_0$ and $\mathcal{U}^p(m_0) \subset \mathcal{H}'^{,p}$, defining $H'_{m_0} := \mathcal{H}'^{,p}/\mathcal{U}^p(m_0)$ for each such m_0.

Proposition 3.3.2.1. *With assumptions as above, there exists a tuple $(A', \lambda', i', \alpha'_{\mathcal{H}^p})$ (over S, unique up to isomorphism), called the* **Hecke twist** *of $(A, \lambda, i, \alpha_{\mathcal{H}'^{,p}})$ by g_0, equipped with a $\mathbb{Z}^\times_{(p)}$-isogeny $[g_0^{-1}] : A \to A'$ of abelian schemes (whose formal inverse we denote by $[g_0] : A' \to A$), satisfying the following characterizing conditions:*

1. *$\lambda' : A' \to A'^{,\vee}$ is a polarization defined by $\lambda' = r_0[g_0]^\vee \circ \lambda \circ [g_0]$ (as positive $\mathbb{Q}^\times$-isogenies), where r_0 is the unique number in $\mathbb{Z}^\times_{(p),>0}$ such that $r_0\nu(g_0)\hat{\mathbb{Z}}^p = \hat{\mathbb{Z}}^p$.*
2. *$i' : \mathcal{O} \to \mathrm{End}_S(A')$ is an $\mathcal{O}$-structure of (A', λ') making $[g_0^{-1}]$ an $\mathcal{O}$-equivariant $\mathbb{Z}^\times_{(p)}$-isogeny.*
3. *$\alpha'_{\mathcal{H}^p}$ is a level-$\mathcal{H}^p$ structure of (A', λ', i') of type $(L \underset{\mathbb{Z}}{\otimes} \hat{\mathbb{Z}}^p, \langle \cdot, \cdot \rangle)$.*
4. *At each geometric point $\bar{s}$, there exist a lifting $(\hat{\alpha}^p_{\bar{s}}, \hat{\nu}^p_{\bar{s}}) : L \underset{\mathbb{Z}}{\otimes} \hat{\mathbb{Z}}^p \overset{\sim}{\to} \mathrm{T}^p A_{\bar{s}}$ (resp. $(\hat{\alpha}^{p,\prime}_{\bar{s}}, \hat{\nu}^{p,\prime}_{\bar{s}}) : L \underset{\mathbb{Z}}{\otimes} \hat{\mathbb{Z}}^p \overset{\sim}{\to} \mathrm{T}^p A'_{\bar{s}}$) of $\alpha_{\mathcal{H}'^{,p}}$ (resp. $\alpha'_{\mathcal{H}^p}$) as in Lemma 3.3.1.6, such that the induced morphisms $\hat{\alpha}^p_{\bar{s},\mathbb{A}^{\infty,p}} : L \underset{\mathbb{Z}}{\otimes} \mathbb{A}^{\infty,p} \overset{\sim}{\to} \mathrm{V}^p A_{\bar{s}}$, $\hat{\alpha}^{p,\prime}_{\bar{s},\mathbb{A}^{\infty,p}} : L \underset{\mathbb{Z}}{\otimes} \mathbb{A}^{\infty,p} \overset{\sim}{\to} \mathrm{V}^p A'_{\bar{s}}$, and $\mathrm{V}^p([g_p]) : \mathrm{V}^p A_{\bar{s}} \overset{\sim}{\to} \mathrm{V}^p A'_{\bar{s}}$ satisfy $\hat{\alpha}^{p,\prime}_{\bar{s},\mathbb{A}^{\infty,p}} = \mathrm{V}^p([g_0^{-1}]) \circ \hat{\alpha}^p_{\bar{s},\mathbb{A}^{\infty,p}} \circ g_0$, and such that $\hat{\nu}^{p,\prime}_{\bar{s}} = \hat{\nu}^p_{\bar{s}} \circ (r_0\nu(g_0))$ as isomorphisms $\hat{\mathbb{Z}}^p(1) \overset{\sim}{\to} \mathrm{T}^p \mathbf{G}_{\mathrm{m},\bar{s}}$, where r_0 as in (1) above. In this case, the $\mathcal{H}'^{,p}$-orbit $[\hat{\alpha}^p_{\bar{s}}]_{\mathcal{H}'^{,p}}$ determines a $(g_0\mathcal{H}^p g_0^{-1})$-orbit $[\hat{\alpha}^p_{\bar{s}}]_{g_0\mathcal{H}^p g_0^{-1}}$ because $\mathcal{H}'^{,p} \subset g_0\mathcal{H}^p g_0^{-1}$, and hence induces an $\mathcal{H}^p$-orbit $[\hat{\alpha}^{p,\prime}_{\bar{s}}]_{\mathcal{H}^p}$.*

We consider such a Hecke twist to be **away from** *p because the two tuples are related by a canonical $\mathbb{Z}^\times_{(p)}$-isogeny $[g_0^{-1}] : A \to A'$ which induces an isomorphism $A[p^r] \to A'[p^r]$ for each integer $r \geq 1$.*

If $g_0 = g_{1,0}g_{2,0}$, where $g_{1,0}$ and $g_{2,0}$ are elements of $\mathrm{G}(\hat{\mathbb{Z}}^p)$, each having a setup analogous to that of g_0, then the Hecke twists by g_0 can be constructed in two steps using Hecke twists by $g_{1,0}$ and $g_{2,0}$, such that $[g_0^{-1}] = [g_{2,0}^{-1}] \circ [g_{1,0}^{-1}]$ (or, equivalently, $[g_0] = [g_{1,0}] \circ [g_{2,0}]$).

Proof. For $[g_0^{-1}]$ to exist, at each geometric point $\bar{s}$ of S and for each lifting $\hat{\alpha}_{\bar{s}}$ of $\alpha_{\mathcal{H}'^{,p}}$, the induced isomorphism $\mathrm{V}^p([g_0^{-1}]_{\bar{s}}) : \mathrm{V}^p A_{\bar{s}} \overset{\sim}{\to} \mathrm{V}^p A'_{\bar{s}}$ must map $\hat{\alpha}_{\bar{s}}(g_0(L \underset{\mathbb{Z}}{\otimes} \hat{\mathbb{Z}}^p))$ to $\mathrm{T}^p A'_{\bar{s}} = \hat{\alpha}'_{\bar{s}}(L \underset{\mathbb{Z}}{\otimes} \hat{\mathbb{Z}}^p)$. (Since $\mathcal{H}^p$ and $\mathcal{H}'^{,p}$ are subgroups of $\mathrm{G}(\hat{\mathbb{Z}}^p)$, this condition is independent of the choice of $\hat{\alpha}_{\bar{s}}$.)

Let us construct $[g_0^{-1}]$ as follows. (When S is locally noetherian, the construction can be much simpler using $\pi_1(S,\bar{s})$-modules, as in [62, Sec. 6.4.3]. But we spell out the details in a more general context, because later we will encounter some analogue of this construction, for which the techniques of $\pi_1(S,\bar{s})$-modules do not work.)

Let n_0, m_0, and N_0 be positive integers prime to p such that $n_0|m_0$, $\mathcal{U}^p(n_0) \subset \mathcal{H}^p$, $\mathcal{U}^p(m_0) \subset \mathcal{H}'^{,p}$, and

$$L \underset{\mathbb{Z}}{\otimes} \hat{\mathbb{Z}}^p \subset N_0^{-1} g_0(L \underset{\mathbb{Z}}{\otimes} \hat{\mathbb{Z}}^p) \subset n_0^{-1} N_0^{-1} g_0(L \underset{\mathbb{Z}}{\otimes} \hat{\mathbb{Z}}^p) \subset m_0^{-1}(L \underset{\mathbb{Z}}{\otimes} \hat{\mathbb{Z}}^p). \tag{3.3.2.2}$$

(Such integers always exist.) Then we have in particular an $\mathcal{O}$-submodule

$$(N_0^{-1} g_0(L \underset{\mathbb{Z}}{\otimes} \hat{\mathbb{Z}}^p))/(L \underset{\mathbb{Z}}{\otimes} \hat{\mathbb{Z}}^p) \subset (m_0^{-1}(L \underset{\mathbb{Z}}{\otimes} \hat{\mathbb{Z}}^p))/(L \underset{\mathbb{Z}}{\otimes} \hat{\mathbb{Z}}^p) \cong L/m_0 L. \tag{3.3.2.3}$$

By (3.3.2.2), we have $\mathcal{U}^p(m_0) \subset g_0 \mathcal{U}^p(n_0) g_0^{-1}$, and the inclusion $\mathcal{H}'^{,p} \hookrightarrow g_0 \mathcal{H}^p g_0^{-1}$ induces a homomorphism $H'_{m_0} = \mathcal{H}'^{,p}/\mathcal{U}^p(m_0) \to H_{n_0} = \mathcal{H}^p/\mathcal{U}^p(n_0)$.

By definition of $\alpha_{\mathcal{H}'^{,p}} = \{\alpha_{H'_{m_0}}\}_{m_0}$, over the scheme $\tilde{S} = \alpha_{H'_{m_0}}$, which is an H'_{m_0}-torsor (finite étale) over S, there is a tautological principal level-m_0 structure $(\alpha_{m_0}, \nu_{m_0})$, where $\alpha_{m_0} : (L/m_0 L)_{\tilde{S}} \xrightarrow{\sim} A_{\tilde{S}}[m_0]$ is an $\mathcal{O}$-equivariant isomorphism of (étale) group schemes over S, and where $\nu_{m_0} : ((\mathbb{Z}/m_0\mathbb{Z})(1))_{\tilde{S}} \xrightarrow{\sim} \boldsymbol{\mu}_{m_0,\tilde{S}}$ is an isomorphism of group schemes over $\tilde{S}$, satisfying the usual symplectic and liftability conditions defining a (principal) level structure.

Let $K_{\tilde{S}}$ be the schematic image of $(N_0^{-1} g_0(L \underset{\mathbb{Z}}{\otimes} \hat{\mathbb{Z}}^p))/(L \underset{\mathbb{Z}}{\otimes} \hat{\mathbb{Z}}^p)$ (see (3.3.2.3)) under α_{m_0}, which is an $\mathcal{O}$-invariant subgroup scheme of $A_{\tilde{S}}[m_0]$, which is finite étale over S_{m_0}. Since the tautological action of H'_{m_0} on $\tilde{S} \to S$ is compatible with the isomorphism α_{m_0}, we can descend $K_{\tilde{S}}$ to a finite étale subgroup scheme K of $A[m_0]$, and define an isogeny

$$A \to A' := A/K. \tag{3.3.2.4}$$

Then we define the $\mathbb{Z}_{(p)}^\times$-isogeny $[g_0^{-1}] : A \to A'$ (see [62, Def. 1.3.1.17]) to be the composition of (3.3.2.4) with $[N_0]^{-1}$, and denote the isogeny (3.3.2.4) as $[N_0 g_0^{-1}]$. (Note that $[N_0 g_0^{-1}] = N_0[g_0^{-1}]$, and $[g_0^{-1}] = [N_0]^{-1}$ if $g_0 = N_0 \operatorname{Id}$.) The $\mathbb{Z}_{(p)}^\times$-isogeny $[g_0^{-1}]$ is independent of the choice of m_0 and N_0. (When S is locally noetherian, we can reformulate the definition of level structures away from p using the language of $\pi_1(S,\bar{s})$-modules, for a geometric point $\bar{s}$ on each connected component of S. Then we can construct $[g_0^{-1}]$ as in [62, Sec. 6.4.3].)

Let $r_0 \in \mathbb{Z}_{(p),>0}^\times$ be such that $r_0 \nu(g_0) \hat{\mathbb{Z}}^p = \hat{\mathbb{Z}}^p$. Note that $N_0^2 r_0 \in \mathbb{Z}_{>0}$ because $L \underset{\mathbb{Z}}{\otimes} \hat{\mathbb{Z}}^p \subset N_0^{-1} g_0(L \underset{\mathbb{Z}}{\otimes} \hat{\mathbb{Z}}^p)$. Then we define a $\mathbb{Q}^\times$-isogeny $\lambda' : A' \to A'^{,\vee}$ by setting $\lambda' := r_0[g_0]^\vee \circ \lambda \circ [g_0] = (N_0^2 r_0)([N_0 g_0^{-1}]^\vee)^{-1} \circ \lambda \circ [N_0 g_0^{-1}]^{-1}$. This $\mathbb{Q}^\times$-isogeny λ' is a $\mathbb{Q}^\times$-polarization by [62, Cor. 1.3.2.18 and 1.3.2.21]. It is an isogeny (and hence a polarization) because we have the inclusions

$$\begin{aligned} L \underset{\mathbb{Z}}{\otimes} \hat{\mathbb{Z}}^p &\subset N_0^{-1} g_0(L \underset{\mathbb{Z}}{\otimes} \hat{\mathbb{Z}}^p) \subset N_0^{-1} g_0(L^\# \underset{\mathbb{Z}}{\otimes} \hat{\mathbb{Z}}^p) \\ &= N_0^{-2} r_0^{-1} (N_0^{-1} g_0(L \underset{\mathbb{Z}}{\otimes} \hat{\mathbb{Z}}^p))^\# \subset N_0^{-2} r_0^{-1}(L^\# \underset{\mathbb{Z}}{\otimes} \hat{\mathbb{Z}}^p), \end{aligned} \tag{3.3.2.5}$$

which by the descent construction of $[N_0 g_0^{-1}]$ corresponds to a factorization of $N_0^2 r_0 \lambda : A \to A^\vee$ as a composition of isogenies

$$A \overset{[N_0 g_0^{-1}]}{\to} A' \overset{\lambda'}{\to} A'^{,\vee} \overset{[N_0 g_0^{-1}]^\vee}{\to} A^\vee,$$

in which $[N_0 g_0^{-1}]$ induces an isomorphism $\ker(N_0^2 r_0 \lambda)[p^r] \overset{\sim}{\to} \ker(\lambda')[p^r]$ for each integer $r \geq 1$.

The above constructions of $[g_0^{-1}] : A \to A'$ and λ' are both compatible with the actions of $\mathcal{O}$. Hence, we obtain an induced $\mathcal{O}$-structure $i' : \mathcal{O} \hookrightarrow \mathrm{End}_S(A')$.

By construction, and by (3.3.2.2), the isomorphism $\alpha_{m_0} \circ g_0$ induces an isomorphism

$$\begin{aligned} \alpha'_{n_0} : L/n_0 L &\cong (n_0^{-1} N_0^{-1} L)/(N_0^{-1} L) \\ &\overset{g_0}{\overset{\sim}{\to}} (n_0^{-1} N_0^{-1} g_0(L \underset{\mathbb{Z}}{\otimes} \hat{\mathbb{Z}}^p))/(N_0^{-1} g_0(L \underset{\mathbb{Z}}{\otimes} \hat{\mathbb{Z}}^p)) \overset{\alpha_{m_0}}{\overset{\sim}{\to}} A'_{\tilde{S}}[n_0], \end{aligned}$$

together with an isomorphism

$$\nu'_{n_0} : ((\mathbb{Z}/n_0\mathbb{Z})(1))_{S_{n_0}} \overset{\sim}{\to} \boldsymbol{\mu}_{n_0, \tilde{S}}$$

induced by restricting $\nu_{m_0} \circ (r_0 \nu(g_0))$. The homomorphism $H'_{m_0} \to H_{n_0}$ induced by $\mathcal{H}'^{,p} \hookrightarrow g_0 \mathcal{H}^p g_0^{-1}$ (*i.e.*, conjugation by g_0) induces a well-defined H_{n_0}-orbit of $(\alpha'_{n_0}, \nu'_{n_0})$ over $\tilde{S}$, which descends to a naive level structure $\alpha'_{H_{n_0}}$ of (A', λ', i') of type $(L/n_0 L, \langle \cdot, \cdot \rangle)$ and level H_{n_0} as in Definition 3.3.1.3. Since we can repeat the above procedure for each integer $n_0 \geq 1$ such that $p \nmid n_0$ and $\mathcal{U}^p(n_0) \subset \mathcal{H}^p$, we obtain a level-$\mathcal{H}^p$ structure $\alpha'_{\mathcal{H}^p} = \{\alpha'_{H_{n_0}}\}_{n_0}$ of (A', λ', i') of type $(L \underset{\mathbb{Z}}{\otimes} \hat{\mathbb{Z}}^p, \langle \cdot, \cdot \rangle)$ as in Definition 3.3.1.4.

This finishes the construction of the Hecke twist $(A', \lambda', i', \alpha'_{\mathcal{H}^p})$.

Since Hecke twists (in this proposition) are constructed using prime-to-p isogenies (and their formal inverses), which are uniquely determined by their behaviors on geometric fibers of torsion subgroup schemes of abelian schemes over S of ranks prime to p (which are finite étale group schemes over S), the last statement of the proposition follows from the characterizing conditions preceding it. □

3.3.3 *Ordinary Level Structures at p*

Let us fix a choice of a filtration $\mathtt{D}$ as in Lemma 3.2.2.1. For each integer $r \geq 0$, the perfect dualities in Lemma 3.2.2.4 induce canonical isomorphisms $(\mathrm{Gr}^0_{\mathtt{D},p^r})^{\mathrm{mult}}_S \cong \mathrm{Hom}_S(\mathrm{Gr}^{-1}_{\mathtt{D}\#,p^r}, \mathbf{G}_{\mathrm{m},S})$ and $(\mathrm{Gr}^0_{\mathtt{D}\#,p^r})^{\mathrm{mult}}_S \cong \mathrm{Hom}_S(\mathrm{Gr}^{-1}_{\mathtt{D},p^r}, \mathbf{G}_{\mathrm{m},S})$ (*cf.* Definition 3.1.1.6), and a morphism $\phi^0_{\mathtt{D},p^r} : \mathrm{Gr}^0_{\mathtt{D},p^r} \to \mathrm{Gr}^0_{\mathtt{D}\#,p^r}$.

Let $\mathcal{H}_p \subset \mathrm{G}(\mathbb{Z}_p)$ be an open compact subgroup of standard form as in Definition 3.2.2.9, so that $\mathcal{U}^{\mathrm{bal}}_{p,1}(p^r) \subset \mathcal{H}_p \subset \mathcal{U}_{p,0}(p^r)$ for $r = \mathrm{depth}_{\mathtt{D}}(\mathcal{H}_p)$, where $\mathcal{U}_{p,0}(p^r)$ and $\mathcal{U}^{\mathrm{bal}}_{p,1}(p^r)$ are as in Definition 3.2.2.8. Let $H^{\mathrm{ord}}_{p^r} := \mathcal{H}_p/\mathcal{U}^{\mathrm{bal}}_{p,1}(p^r)$, which is a subgroup of $\mathrm{M}^{\mathrm{ord}}_{\mathtt{D}}(\mathbb{Z}/p^r\mathbb{Z})$.

Definition 3.3.3.1. *Let S be a scheme over* $\mathrm{Spec}(\mathbb{Z})$. *Let A be an abelian scheme over S, with a polarization $\lambda : A \to A^\vee$ and an $\mathcal{O}$-endomorphism structure $i : \mathcal{O} \hookrightarrow$*

$\mathrm{End}_S(A)$ *as in [62, Def. 1.3.3.1]. Let* $L \underset{\mathbb{Z}}{\otimes} \mathbb{Z}_p$, $\mathtt{D}$, $\mathcal{H}_p$, $H^{\mathrm{ord}}_{p^r}$, *and* $\phi^0_{\mathtt{D},p^r} : \mathrm{Gr}^0_{\mathtt{D},p^r} \to \mathrm{Gr}^0_{\mathtt{D}^\#,p^r}$ *be as above. A* **naive principal ordinary level-p^r structure** *of* (A, λ, i) *of type* $(L/p^rL, \langle\,\cdot\,,\,\cdot\,\rangle, \mathtt{D}_{p^r})$*, or rather of type* $\phi^0_{\mathtt{D},p^r} : \mathrm{Gr}^0_{\mathtt{D},p^r} \to \mathrm{Gr}^0_{\mathtt{D}^\#,p^r}$*, is a triple*

$$\alpha^{\mathrm{ord}}_{p^r} = (\alpha^{\mathrm{ord},0}_{p^r}, \alpha^{\mathrm{ord},\#,0}_{p^r}, \nu^{\mathrm{ord}}_{p^r}),$$

where the first two entries are $\mathcal{O}$-equivariant homomorphisms

$$\alpha^{\mathrm{ord},0}_{p^r} : (\mathrm{Gr}^0_{\mathtt{D},p^r})^{\mathrm{mult}}_S \to A[p^r]$$

and

$$\alpha^{\mathrm{ord},\#,0}_{p^r} : (\mathrm{Gr}^0_{\mathtt{D}^\#,p^r})^{\mathrm{mult}}_S \to A^\vee[p^r]$$

that are closed immersions, and where the third entry $\nu^{\mathrm{ord}}_{p^r}$ *is a section of*

$$\begin{aligned}(\mathbb{Z}/p^r\mathbb{Z})^\times_S &\cong \underline{\mathrm{Isom}}_S(((\mathbb{Z}/p^r\mathbb{Z})(1))_S, ((\mathbb{Z}/p^r\mathbb{Z})(1))_S) \\ &\cong \underline{\mathrm{Isom}}_S(\boldsymbol{\mu}_{p^r,S}, \boldsymbol{\mu}_{p^r,S})\end{aligned}$$

which are symplectic in the sense that the two homomorphisms are compatible with the homomorphisms

$$\nu^{\mathrm{ord}}_{p^r} \circ (\phi^0_{\mathtt{D},p^r})^{\mathrm{mult}}_S : (\mathrm{Gr}^0_{\mathtt{D},p^r})^{\mathrm{mult}}_S \to (\mathrm{Gr}^0_{\mathtt{D}^\#,p^r})^{\mathrm{mult}}_S$$

and

$$\lambda : A[p^r] \to A^\vee[p^r],$$

namely that the following diagram

$$\begin{array}{ccc} (\mathrm{Gr}^0_{\mathtt{D},p^r})^{\mathrm{mult}}_S & \xrightarrow{\alpha^{\mathrm{ord},0}_{p^r}} & A[p^r] \\ {\scriptstyle \nu^{\mathrm{ord}}_{p^r} \circ (\phi^0_{\mathtt{D},p^r})^{\mathrm{mult}}_S} \downarrow & & \downarrow {\scriptstyle \lambda} \\ (\mathrm{Gr}^0_{\mathtt{D}^\#,p^r})^{\mathrm{mult}}_S & \xrightarrow[\alpha^{\mathrm{ord},\#,0}_{p^r}]{} & A^\vee[p^r] \end{array}$$

is commutative, or equivalently that the following diagram

$$\begin{array}{ccc} (\mathrm{Gr}^0_{\mathtt{D},p^r})^{\mathrm{mult}}_S & \xrightarrow{\alpha^{\mathrm{ord},0}_{p^r}} & A[p^r] \\ {\scriptstyle (\phi^0_{\mathtt{D},p^r})^{\mathrm{mult}}_S} \downarrow & & \downarrow {\scriptstyle \lambda} \\ (\mathrm{Gr}^0_{\mathtt{D}^\#,p^r})^{\mathrm{mult}}_S & \xrightarrow[\nu^{\mathrm{ord}}_{p^r} \circ \alpha^{\mathrm{ord},\#,0}_{p^r}]{} & A^\vee[p^r] \end{array}$$

is commutative, and that the schematic images of the two homomorphisms $\alpha^{\mathrm{ord},0}_{p^r}$ *and* $\alpha^{\mathrm{ord},\#,0}_{p^r}$ *are annihilators of each other under the canonical pairing* $\mathrm{e}_{A[p^r]} : A[p^r] \times A^\vee[p^r] \to \boldsymbol{\mu}_{p^r,S}$*. We shall denote* $\nu^{\mathrm{ord}}_{p^r}$ *by* $\nu(\alpha^{\mathrm{ord}}_{p^r})$*. We shall also denote the schematic image of* $\alpha^{\mathrm{ord},0}_{p^r}$ *(resp.* $\alpha^{\mathrm{ord},\#,0}_{p^r}$*), which is a closed subgroup scheme of* A *(resp.* $A^\vee$*), by* $\mathrm{image}(\alpha^{\mathrm{ord},0}_{p^r})$ *(resp.* $\mathrm{image}(\alpha^{\mathrm{ord},\#,0}_{p^r})$*).*

Definition 3.3.3.2. *We say a naive principal ordinary level-p^r structure* $\alpha_{p^r}^{\mathrm{ord}} = (\alpha_{p^r}^{\mathrm{ord},0}, \alpha_{p^r}^{\mathrm{ord},\#,0}, \nu_{p^r}^{\mathrm{ord}})$ *of* (A,λ,i) *of type* $(L/p^rL, \langle\cdot,\cdot\rangle, \mathtt{D}_{p^r})$ *in Definition* 3.3.3.1 *is a* **principal ordinary level-p^r structure** *of type* $(L \underset{\mathbb{Z}}{\otimes} \mathbb{Z}_p, \langle\cdot,\cdot\rangle, \mathtt{D})$ *if it satisfies the following* **symplectic-liftability condition***:*

There exists (noncanonically) a tower $(S_{p^{r'}} \twoheadrightarrow S)_{r'\geq r}$ *of quasi-finite étale coverings such that we have the following:*

1. $S_{p^r} = S$.
2. *For each* r'' *such that* $r' \geq r'' \geq r$, *there is a quasi-finite étale covering* $S_{p^{r'}} \twoheadrightarrow S_{p^{r''}}$ *whose composition with* $S_{p^{r''}} \to S$ *is the quasi-finite étale covering* $S_{p^{r'}} \to S$.
3. *There is a naive principal ordinary level structure* $\alpha_{p^{r'},S_{p^{r'}}}^{\mathrm{ord}}$ *of* $(A,\lambda,i) \underset{S}{\times} S_{p^{r'}}$ *of type* $(L/p^{r'}L, \langle\cdot,\cdot\rangle, \mathtt{D}_{p^{r'}})$ *over each* $S_{p^{r'}}$.
4. *For each* r'' *such that* $r' \geq r'' \geq r$, *the pullback of* $\alpha_{p^{r''},S_{p^{r''}}}^{\mathrm{ord}}$ *to* $S_{p^{r'}}$ *is the reduction modulo* $p^{r''}$ *of* $\alpha_{p^{r'},S_{p^{r'}}}^{\mathrm{ord}}$.

Definition 3.3.3.3. *Let* S *and* (A,λ,i) *be as in Definition* 3.3.3.1, *and let* $H_{p^r}^{\mathrm{ord}}$ *be as above. A* **naive ordinary level-$H_{p^r}^{\mathrm{ord}}$ structure** *of* (A,λ,i) *of type* $(L/p^rL, \langle\cdot,\cdot\rangle, \mathtt{D}_{p^r})$ *is an* $H_{p^r}^{\mathrm{ord}}$*-orbit* $\alpha_{H_{p^r}^{\mathrm{ord}}}^{\mathrm{ord}}$ *of étale-locally-defined (naive) principal ordinary level structures of type* $(L \underset{\mathbb{Z}}{\otimes} \mathbb{Z}_p, \langle\cdot,\cdot\rangle, \mathtt{D})$ *and level* p^r, *namely a (finite étale) subscheme* $\alpha_{H_{p^r}^{\mathrm{ord}}}^{\mathrm{ord}}$ *of the quasi-finite étale scheme*

$$\underline{\mathrm{Hom}}_S\big((\mathrm{Gr}_{\mathtt{D},p^r}^0)_S^{\mathrm{mult}}, A[p^r]\big) \underset{S}{\times} \underline{\mathrm{Hom}}_S\big((\mathrm{Gr}_{\mathtt{D}^\#,p^r}^0)_S^{\mathrm{mult}}, A^\vee[p^r]\big) \underset{S}{\times} (\mathbb{Z}/p^r\mathbb{Z})_S^\times$$

over S *that becomes the (scheme-theoretic)* $H_{p^r}^{\mathrm{ord}}$*-orbit of some naive principal ordinary level structures of type* $(L \underset{\mathbb{Z}}{\otimes} \mathbb{Z}_p, \langle\cdot,\cdot\rangle, \mathtt{D})$ *and level* p^r *(see Definition* 3.3.3.2*) after a finite étale surjective base change in* S.

We shall denote by $\alpha_{H_{p^r}^{\mathrm{ord}}}^{\mathrm{ord},0}$ *(resp.* $\alpha_{H_{p^r}^{\mathrm{ord}}}^{\mathrm{ord},\#,0}$, *resp.* $\nu_{H_{p^r}^{\mathrm{ord}}}^{\mathrm{ord}} = \nu(\alpha_{H_{p^r}^{\mathrm{ord}}}^{\mathrm{ord}})$*) the schematic image of* $\alpha_{H_{p^r}^{\mathrm{ord}}}^{\mathrm{ord}}$ *in* $\underline{\mathrm{Hom}}_S\big((\mathrm{Gr}_{\mathtt{D},p^r}^0)_S^{\mathrm{mult}}, A[p^r]\big)$ *(resp.* $\underline{\mathrm{Hom}}_S\big((\mathrm{Gr}_{\mathtt{D}^\#,p^r}^0)_S^{\mathrm{mult}}, A^\vee[p^r]\big)$, *resp.* $(\mathbb{Z}/p^r\mathbb{Z})_S^\times$*).*

Since the action of $H_{p^r}^{\mathrm{ord}}$ *does not modify the schematic image of* $\alpha_{p^r}^{\mathrm{ord},0}$ *(resp.* $\alpha_{p^r}^{\mathrm{ord},\#,0}$*) in an orbit, it makes sense (by étale descent, and by abuse of language) to consider the common schematic image of* $\alpha_{H_{p^r}^{\mathrm{ord}}}^{\mathrm{ord},0}$ *(resp.* $\alpha_{H_{p^r}^{\mathrm{ord}}}^{\mathrm{ord},\#,0}$*), which is a closed subgroup scheme of* A *(resp.* $A^\vee$*), by* $\mathrm{image}(\alpha_{H_{p^r}^{\mathrm{ord}}}^{\mathrm{ord},0})$ *(resp.* $\mathrm{image}(\alpha_{H_{p^r}^{\mathrm{ord}}}^{\mathrm{ord},\#,0})$*).*

Definition 3.3.3.4. *Let* S *and* (A,λ,i) *be as in Definition* 3.3.3.1, *and let* $H_{p^r}^{\mathrm{ord}}$ *be as above. Consider the open compact subgroup*

$$\mathcal{H}_p^{\mathrm{ord}} := \Big(\mathrm{M}_{\mathtt{D}}^{\mathrm{ord}}(\mathbb{Z}_p) \overset{\mathrm{can.}}{\twoheadrightarrow} \mathrm{M}_{\mathtt{D}}^{\mathrm{ord}}(\mathbb{Z}/p^r\mathbb{Z}) \overset{\mathrm{can.}}{\overset{\sim}{\to}} \big(\mathcal{U}_{p,0}(p^r)/\mathcal{U}_{p,1}^{\mathrm{bal}}(p^r)\big)\Big)^{-1}(H_{p^r}^{\mathrm{ord}}) \quad (3.3.3.5)$$

of $\mathrm{M}_{\mathtt{D}}^{\mathrm{ord}}(\mathbb{Z}_p)$, *so that* $\mathcal{H}_p^{\mathrm{ord}}/\mathcal{U}^{\mathrm{ord}}(p^r) \cong H_{p^r}^{\mathrm{ord}} = \mathcal{H}_p/\mathcal{U}_{p,1}^{\mathrm{bal}}(p^r)$. *Let* S *and* (A,λ,i) *be as in Definition* 3.3.3.1. *For each integer* r' *such that* $r' \geq r$, *set* $H_{p^{r'}}^{\mathrm{ord}} :=$

$\mathcal{H}_p^{\mathrm{ord}}/\mathcal{U}^{\mathrm{ord}}(p^{r'})$, *which is then viewed as a subgroup of* $\mathcal{U}_{p,0}(p^{r'})/\mathcal{U}_{p,1}^{\mathrm{bal}}(p^{r'})$. *Then an* **ordinary level-$\mathcal{H}_p$ structure** $\alpha_{\mathcal{H}_p}^{\mathrm{ord}}$ *of* (A, λ, i) *of type* $(L \underset{\mathbb{Z}}{\otimes} \mathbb{Z}_p, \langle \cdot, \cdot \rangle, \mathtt{D})$ *is a naive ordinary level-*$H_{p^r}^{\mathrm{ord}}$ *structure* $\alpha_{H_{p^r}^{\mathrm{ord}}}^{\mathrm{ord}}$ *of* (A, λ, i) *of type* $(L/p^r L, \langle \cdot, \cdot \rangle, \mathtt{D}_{p^r})$ *that satisfies the following* **symplectic-liftability condition***:*

There exists (noncanonically) a tower $(S_{p^{r'}} \twoheadrightarrow S)_{r' \geq r}$ *of quasi-finite étale coverings such that we have the following:*

1. $S_{p^r} = S$.
2. *For each* r'' *such that* $r' \geq r'' \geq r$*, there is a quasi-finite étale covering* $S_{p^{r'}} \twoheadrightarrow S_{p^{r''}}$ *whose composition with* $S_{p^{r''}} \to S$ *is the quasi-finite étale covering* $S_{p^{r'}} \to S$.
3. *There is a naive ordinary level-*$H_{p^{r'}}^{\mathrm{ord}}$ *structure* $\alpha_{H_{p^{r'}}^{\mathrm{ord}}}^{\mathrm{ord}}$ *of* (A, λ, i) *of type* $(L/p^{r'} L, \langle \cdot, \cdot \rangle, \mathtt{D}_{p^{r'}})$ *over each* $S_{p^{r'}}$.
4. *For each* r'' *such that* $r' \geq r'' \geq r$*, the pullback of* $\alpha_{H_{p^{r''}}^{\mathrm{ord}}, S_{p^{r''}}}^{\mathrm{ord}}$ *to* $S_{p^{r'}}$ *is the reduction modulo* $p^{r''}$ *of* $\alpha_{H_{p^{r'}}^{\mathrm{ord}}, S_{p^{r'}}}^{\mathrm{ord}}$*, in the sense that* $\alpha_{H_{p^{r''}}^{\mathrm{ord}}, S_{p^{r''}}}^{\mathrm{ord}}$ *is the schematic image of* $\alpha_{H_{p^{r'}}^{\mathrm{ord}}, S_{p^{r'}}}^{\mathrm{ord}}$ *under the canonical (quasi-finite étale) morphism*

$$\begin{aligned} &\underline{\mathrm{Hom}}_{S_{p^{r'}}}\big((\mathrm{Gr}^0_{\mathtt{D},p^{r'}})^{\mathrm{mult}}_{S_{p^{r'}}}, A[p^{r'}]\big) \\ &\underset{S_{p^{r'}}}{\times} \underline{\mathrm{Hom}}_{S_{p^{r'}}}\big((\mathrm{Gr}^0_{\mathtt{D}^\#,p^{r'}})^{\mathrm{mult}}_{S_{p^{r'}}}, A^\vee[p^{r'}]\big) \underset{S_{p^{r'}}}{\times} (\mathbb{Z}/p^{r'}\mathbb{Z})^\times_{S_{p^{r'}}} \\ \to &\underline{\mathrm{Hom}}_{S_{p^{r''}}}\big((\mathrm{Gr}^0_{\mathtt{D},p^{r''}})^{\mathrm{mult}}_{S_{p^{r''}}}, A[p^{r''}]\big) \\ &\underset{S_{p^{r''}}}{\times} \underline{\mathrm{Hom}}_{S_{p^{r''}}}\big((\mathrm{Gr}^0_{\mathtt{D}^\#,p^{r''}})^{\mathrm{mult}}_{S_{p^{r''}}}, A^\vee[p^{r''}]\big) \underset{S_{p^{r''}}}{\times} (\mathbb{Z}/p^{r''}\mathbb{Z})^\times_{S_{p^{r''}}} \end{aligned}$$

   defined by restriction to the $p^{r''}$*-torsion in the sources.*

We shall denote $\alpha_{H_{p^r}^{\mathrm{ord}}}^{\mathrm{ord},0}$ *(resp.* $\alpha_{H_{p^r}^{\mathrm{ord}}}^{\mathrm{ord},\#,0}$*, resp.* $\nu_{H_{p^r}^{\mathrm{ord}}}^{\mathrm{ord}} = \nu(\alpha_{H_{p^r}^{\mathrm{ord}}}^{\mathrm{ord}})$*, resp.* $\mathrm{image}(\alpha_{H_{p^r}^{\mathrm{ord}}}^{\mathrm{ord},0})$*, resp.* $\mathrm{image}(\alpha_{H_{p^r}^{\mathrm{ord}}}^{\mathrm{ord},\#,0})$*) by* $\alpha_{\mathcal{H}_p}^{\mathrm{ord},0}$ *(resp.* $\alpha_{\mathcal{H}_p}^{\mathrm{ord},\#,0}$*, resp.* $\nu_{\mathcal{H}_p}^{\mathrm{ord}} = \nu(\alpha_{\mathcal{H}_p}^{\mathrm{ord}})$*, resp.* $\mathrm{image}(\alpha_{\mathcal{H}_p}^{\mathrm{ord},0})$*, resp.* $\mathrm{image}(\alpha_{\mathcal{H}_p}^{\mathrm{ord},\#,0})$*).*

Remark 3.3.3.6. Even when $r = 0$, the existence of a principal ordinary level-p^r structure of (A, λ, i) of type $(L \underset{\mathbb{Z}}{\otimes} \mathbb{Z}_p, \langle \cdot, \cdot \rangle, \mathtt{D})$ in Definition 3.3.3.2 forces A to be *ordinary* (see Definition 3.1.1.2). The same is true for the existence of an ordinary level-$\mathcal{H}_p$ structure of (A, λ, i) of type $(L \underset{\mathbb{Z}}{\otimes} \mathbb{Z}_p, \langle \cdot, \cdot \rangle, \mathtt{D})$ in Definition 3.3.3.4.

Lemma 3.3.3.7. *Let* S *and* (A, λ, i) *be as in Definition* 3.3.3.1. *Suppose moreover that* S *is a scheme over* $\mathrm{Spec}(\mathbb{F}_p)$. *For each integer* $i \geq 0$*, let* $A^{(p^i)} := A \underset{S, \mathrm{F}_S^i}{\times} S$ *(resp.* $A^{\vee,(p^i)} := A^\vee \underset{S, \mathrm{F}_S^i}{\times} S$*) denote the pullback of* A *(resp.* $A^\vee$*) under the* i*-th* *iteration* $\mathrm{F}_S^i : S \to S$ *of the absolute Frobenius morphism* $\mathrm{F}_S : S \to S$*, and let* $\mathrm{F}_{A/S}^{(i)} : A \to A^{(p^i)}$ *(resp.* $\mathrm{F}_{A^\vee/S}^{(i)} : A^\vee \to A^{\vee,(p^i)}$*) denote the relative Frobenius morphism induced by the universal property of* $A^{(p^i)}$ *(resp.* $A^{\vee,(p^i)}$*) as a fiber product. Then* $A^{\vee,(p^i)}$ *is the dual abelian scheme of* $A^{(p^i)}$*, with polarization* $\lambda^{(p^i)} := \lambda \underset{S,\mathrm{F}_S^i}{\times} S$.

Let $\alpha_{\mathcal{H}_p}^{\mathrm{ord}} = (\alpha_{\mathcal{H}_p}^{\mathrm{ord},0}, \alpha_{\mathcal{H}_p}^{\mathrm{ord},\#,0}, \nu_{\mathcal{H}_p}^{\mathrm{ord}})$ *be any ordinary level-*$\mathcal{H}_p$ *structure of* (A, λ, i) *of type* $(L \underset{\mathbb{Z}}{\otimes} \mathbb{Z}_p, \langle \cdot\,, \cdot \rangle, \mathtt{D})$ *as in Definition* 3.3.3.4*. Then we have* $\mathrm{image}(\alpha_{H_{p^r}^{\mathrm{ord}}}^{\mathrm{ord},0})[p^i] = \ker(\mathrm{F}_{A/S}^{(i)})$ *and* $\mathrm{image}(\alpha_{H_{p^r}^{\mathrm{ord}}}^{\mathrm{ord},\#,0})[p^i] = \ker(\mathrm{F}_{A^\vee/S}^{(i)})$*, for each* $0 \leq i \leq r$*. (In particular, we have* $\mathrm{image}(\alpha_{p^r}^{\mathrm{ord},0})[p^i] = \ker(\mathrm{F}_{A/S}^{(i)})$ *and* $\mathrm{image}(\alpha_{p^r}^{\mathrm{ord},\#,0})[p^i] = \ker(\mathrm{F}_{A^\vee/S}^{(i)})$ *for a principal ordinary level-*p^r *structure* $\alpha_{p^r}^{\mathrm{ord}} = (\alpha_{p^r}^{\mathrm{ord},0}, \alpha_{p^r}^{\mathrm{ord},\#,0}, \nu_{p^r}^{\mathrm{ord}})$ *of type* $(L \underset{\mathbb{Z}}{\otimes} \mathbb{Z}_p, \langle \cdot\,, \cdot \rangle, \mathtt{D})$ *as in Definition* 3.3.3.2*.)*

Proof. The first paragraph is nothing but definitions. For the second paragraph, since it is about comparison of finite flat group schemes of finite presentation over S, we may reduce to the case that S is Henselian local, and hence to the case that S is the spectrum of an algebraically closed field of characteristic $p > 0$. Then the assertions follow from the fact that, given their ranks, both $\mathrm{image}(\alpha_{H_{p^r}^{\mathrm{ord}}}^{\mathrm{ord},0})[p^i]$ (resp. $\mathrm{image}(\alpha_{H_{p^r}^{\mathrm{ord}}}^{\mathrm{ord},\#,0})[p^i]$) and $\ker(\mathrm{F}_{A/S}^{(i)})$ (resp. $\ker(\mathrm{F}_{A^\vee/S}^{(i)})$) are the unique maximal subgroup scheme of multiplicative type of the ordinary abelian variety $A[p^i]$ (resp. $A^\vee[p^i]$) over S (see Remark 3.3.3.6). □

Corollary 3.3.3.8. *In Definitions* 3.3.3.2 *and* 3.3.3.4*, if* S *is a scheme over* $\mathrm{Spec}(\mathbb{F}_p)$*, then we may assume that the tower* $(S_{p^{r'}} \twoheadrightarrow S)_{r' \geq r}$ *of quasi-finite étale coverings is* **finite étale***.*

Proof. In this case, for each $r' \geq r$, we know before constructing $\alpha_{H_{p^{r'}}^{\mathrm{ord}}}^{\mathrm{ord}} = (\alpha_{H_{p^{r'}}^{\mathrm{ord}}}^{\mathrm{ord},0}, \alpha_{H_{p^{r'}}^{\mathrm{ord}}}^{\mathrm{ord},\#,0}, \nu_{H_{p^{r'}}^{\mathrm{ord}}}^{\mathrm{ord}})$ that we must have $\mathrm{image}(\alpha_{H_{p^{r'}}^{\mathrm{ord}}}^{\mathrm{ord},0}) = \ker(\mathrm{F}_{A/S}^{(r')})$ and $\mathrm{image}(\alpha_{H_{p^{r'}}^{\mathrm{ord}}}^{\mathrm{ord},\#,0}) = \ker(\mathrm{F}_{A^\vee/S}^{(i)})$, and hence the desired $\alpha_{H_{p^{r'}}^{\mathrm{ord}}}^{\mathrm{ord}}$ (tautologically) exists over some open and closed subscheme $S_{p^{r'}}$ of the finite étale scheme

$$\begin{aligned} &\underline{\mathrm{Isom}}_S\big((\mathrm{Gr}_{\mathtt{D},p^{r'}}^0)_S^{\mathrm{mult}}, \ker(\mathrm{F}_{A/S}^{(r')})\big) \\ &\underset{S}{\times} \underline{\mathrm{Isom}}_S\big((\mathrm{Gr}_{\mathtt{D}^\#,p^{r'}}^0)_S^{\mathrm{mult}}, \ker(\mathrm{F}_{A^\vee/S}^{(r')})\big) \underset{S}{\times} (\mathbb{Z}/p^{r'}\mathbb{Z})_S^\times, \end{aligned}$$

which is finite étale over S (and we can compatibly form a tower of such subschemes such that the necessary compatibility conditions between $S_{p^{r'}}$ and $S_{p^{r''}}$, when $r'' \geq r' \geq r$, are satisfied). □

Lemma 3.3.3.9. *Let* S *and* (A, λ, i) *be as in Definition* 3.3.3.1*. Let* $\mathcal{H}_p$ *and* $H_{p^r}^{\mathrm{ord}}$ *be as above, and let* $\alpha_{\mathcal{H}_p}^{\mathrm{ord}}$ *be an ordinary level-*$\mathcal{H}_p$ *structure of* (A, λ, i) *of type*

$(L \underset{\mathbb{Z}}{\otimes} \mathbb{Z}_p, \langle \cdot, \cdot \rangle, \mathtt{D})$ as Definition 3.3.3.4*. Let $\bar{s}$ be any geometric point of S. Then there exists a triple*

$$\hat{\alpha}_{\bar{s}}^{\mathrm{ord}} = (\hat{\alpha}_{\bar{s}}^{\mathrm{ord},0}, \hat{\alpha}_{\bar{s}}^{\mathrm{ord},\#,0}, \hat{\nu}_{\bar{s}}^{\mathrm{ord}}) \tag{3.3.3.10}$$

where the first two entries are injective $\mathcal{O}$-equivariant homomorphisms

$$\hat{\alpha}_{\bar{s}}^{\mathrm{ord},0} : (\mathrm{Gr}_{\mathtt{D}}^0)_{\bar{s}}^{\mathrm{mult}} \to A_{\bar{s}}[p^\infty]$$

and

$$\hat{\alpha}_{\bar{s}}^{\mathrm{ord},\#,0} : (\mathrm{Gr}_{\mathtt{D}^\#}^0)_{\bar{s}}^{\mathrm{mult}} \to A_{\bar{s}}^\vee[p^\infty],$$

(see Definition 3.1.1.7*) and where the third entry $\hat{\nu}_{\bar{s}}^{\mathrm{ord}}$ is a section of*

$$(\mathbb{Z}_p^\times)_{\bar{s}} \cong \underline{\mathrm{Isom}}_{\bar{s}}(((\mathbb{Z}_p)(1))_{\bar{s}}, ((\mathbb{Z}_p)(1))_{\bar{s}}) \cong \underline{\mathrm{Isom}}_{\bar{s}}(\boldsymbol{\mu}_{p^\infty,\bar{s}}, \boldsymbol{\mu}_{p^\infty,\bar{s}}),$$

satisfying the symplectic condition as in Definition 3.3.3.1 *which we shall spell out below, such that the $H_{p^r}^{\mathrm{ord}}$-orbit of the restriction $\alpha_{p^r,\bar{s}}^{\mathrm{ord}}$ of $\hat{\alpha}_{\bar{s}}^{\mathrm{ord}}$ to the p^r-torsion in the sources coincides with the pullback of $\alpha_{H_{p^r}^{\mathrm{ord}}}^{\mathrm{ord}}$ from S to $\bar{s}$. We say that this $\hat{\alpha}_{\bar{s}}^{\mathrm{ord}}$ is a* **lifting** *of $\alpha_{\mathcal{H}_p}^{\mathrm{ord}}$ at $\bar{s}$. The $\mathcal{H}_p^{\mathrm{ord}}$-orbit $[\hat{\alpha}_{\bar{s}}^{\mathrm{ord}}]_{\mathcal{H}_p^{\mathrm{ord}}}$ of $\hat{\alpha}_{\bar{s}}^{\mathrm{ord}}$ is unique (i.e., independent of the choice of $\hat{\alpha}_{\bar{s}}^{\mathrm{ord}}$).*

The symplectic condition for $\hat{\alpha}_{\bar{s}}^{\mathrm{ord}} = (\hat{\alpha}_{\bar{s}}^{\mathrm{ord},0}, \hat{\alpha}_{\bar{s}}^{\mathrm{ord},\#,0}, \hat{\nu}_{\bar{s}}^{\mathrm{ord}})$ is that the two homomorphisms $\hat{\alpha}_{\bar{s}}^{\mathrm{ord},0}$ and $\hat{\alpha}_{\bar{s}}^{\mathrm{ord},\#,0}$ are compatible with the homomorphisms

$$\hat{\nu}_{\bar{s}}^{\mathrm{ord}} \circ (\phi_{\mathtt{D}}^0)_{\bar{s}}^{\mathrm{mult}} : (\mathrm{Gr}_{\mathtt{D}}^0)_{\bar{s}}^{\mathrm{mult}} \to (\mathrm{Gr}_{\mathtt{D}^\#}^0)_{\bar{s}}^{\mathrm{mult}}$$

and

$$\lambda_{\bar{s}} : A_{\bar{s}}[p^\infty] \to A_{\bar{s}}^\vee[p^\infty],$$

namely that the following diagram

$$\begin{array}{ccc} (\mathrm{Gr}_{\mathtt{D}}^0)_{\bar{s}}^{\mathrm{mult}} & \xrightarrow{\hat{\alpha}_{\bar{s}}^{\mathrm{ord},0}} & A_{\bar{s}}[p^\infty] \\ {\scriptstyle \hat{\nu}_{\bar{s}}^{\mathrm{ord}} \circ (\phi_{\mathtt{D}}^0)_{\bar{s}}^{\mathrm{mult}}} \downarrow & & \downarrow {\scriptstyle \lambda_{\bar{s}}} \\ (\mathrm{Gr}_{\mathtt{D}^\#}^0)_{\bar{s}}^{\mathrm{mult}} & \xrightarrow[\hat{\alpha}_{\bar{s}}^{\mathrm{ord},\#,0}]{} & A_{\bar{s}}^\vee[p^\infty] \end{array}$$

is commutative, or equivalently that the following diagram

$$\begin{array}{ccc} (\mathrm{Gr}_{\mathtt{D}}^0)_{\bar{s}}^{\mathrm{mult}} & \xrightarrow{\hat{\alpha}_{\bar{s}}^{\mathrm{ord},0}} & A_{\bar{s}}[p^\infty] \\ {\scriptstyle (\phi_{\mathtt{D}}^0)_{\bar{s}}^{\mathrm{mult}}} \downarrow & & \downarrow {\scriptstyle \lambda_{\bar{s}}} \\ (\mathrm{Gr}_{\mathtt{D}^\#}^0)_{\bar{s}}^{\mathrm{mult}} & \xrightarrow[\hat{\nu}_{\bar{s}}^{\mathrm{ord}} \circ \hat{\alpha}_{\bar{s}}^{\mathrm{ord},\#,0}]{} & A_{\bar{s}}^\vee[p^\infty] \end{array}$$

is commutative, and that the images of the two homomorphisms $\hat{\alpha}_{\bar{s}}^{\mathrm{ord},0}$ and $\hat{\alpha}_{\bar{s}}^{\mathrm{ord},\#,0}$ are annihilators of each other under the canonical pairing $\mathrm{e}_{A_{\bar{s}}[p^\infty]} : A_{\bar{s}}[p^\infty] \times A_{\bar{s}}^\vee[p^\infty] \to \boldsymbol{\mu}_{p^\infty,\bar{s}}$.

Proof. As in the proof of Lemma 3.3.1.6, the pullback of the compatible tower $(\alpha^{\mathrm{ord}}_{H^{\mathrm{ord}}_{p^{r'}}})_{r'\geq r}$ defined over $(S_{p^{r'}} \twoheadrightarrow S)_{r'\geq r}$ to the geometric point $\bar{s}$ of S (with a compatible choice of liftings to each $S_{p^{r'}} \to S$) allows us to choose a compatible tower $(\alpha^{\mathrm{ord}}_{p^{r'}})_{r'\geq r}$ of principal ordinary level structures, such that the $H^{\mathrm{ord}}_{p^{r'}}$-orbit of each $\alpha^{\mathrm{ord}}_{p^{r'}}$ is $\alpha^{\mathrm{ord}}_{H^{\mathrm{ord}}_{p^{r'}}}$ for each $r' \geq r$, which is equivalent to the desired $\hat{\alpha}^{\mathrm{ord}}_{\bar{s}}$ in (3.3.3.10). The symplectic condition for $\hat{\alpha}^{\mathrm{ord}}_{\bar{s}}$ follows from that of $\alpha^{\mathrm{ord}}_{p^{r'}}$ over each $S_{p^{r'}} \to S$. □

3.3.4 *Hecke Twists Defined by Ordinary Level Structures at p*

For each element g_p of $\mathrm{P}^{\mathrm{ord}}_{\mathsf{D}}(\mathbb{Q}_p) \subset \mathrm{G}(\mathbb{Q}_p)$, we denote by $g^{\mathrm{ord}}_p = (g_{p,0}, g_{p,-1})$ the action of g_p on the graded pieces $\mathrm{Gr}_{\mathsf{D}} \cong \mathrm{Gr}^0_{\mathsf{D}} \oplus \mathrm{Gr}^{-1}_{\mathsf{D}}$, and by $g_{p,\#,0} = {}^t g^{-1}_{p,-1}$, where ${}^t g_{p,-1}$ is the induced action on $\mathrm{Gr}^0_{\mathsf{D}\#}$ by transposition (with respect to the perfect pairing $\mathrm{Gr}^0_{\mathsf{D}\#} \times \mathrm{Gr}^{-1}_{\mathsf{D}} \to \mathbb{Z}_p(1)$; see Lemma 3.2.2.4). Note that $g_{p,\#,0}$ and $\nu(g_p)^{-1} g_{p,0}$ induce the same element in $\mathrm{GL}_{\mathcal{O} \underset{\mathbb{Z}}{\otimes} \mathbb{Q}_p}(\mathsf{D}^0_{\mathbb{Q}_p})$.

Although a more general theory might exist, we shall construct ordinary Hecke twists defined by ordinary level structures at p under a list of conditions. Nevertheless, such ordinary Hecke twists are sufficient for the applications we know.

Suppose we have two open compact subgroups $\mathcal{H}_p$ and $\mathcal{H}'_p$ of $\mathrm{G}(\mathbb{Q}_p)$ of standard form as in Definition 3.2.2.9 such that $r = \mathrm{depth}_{\mathsf{D}}(\mathcal{H}_p) \leq r' = \mathrm{depth}_{\mathsf{D}}(\mathcal{H}'_p)$. Suppose that $g_p \in \mathrm{P}^{\mathrm{ord}}_{\mathsf{D}}(\mathbb{Q}_p)$ satisfies the following conditions:

(1) $\mathcal{H}'_p \subset g_p \mathcal{H}_p g_p^{-1}$.
(2) There exist integers r_0 and $r' \geq r'' \geq r$ such that
$$\mathrm{Gr}^0_{\mathsf{D}} \subset p^{-r_0} g_{p,0}(\mathrm{Gr}^0_{\mathsf{D}}) \subset p^{-r''-r_0} g_{p,0}(\mathrm{Gr}^0_{\mathsf{D}}) \subset p^{-r'}\,\mathrm{Gr}^0_{\mathsf{D}} \tag{3.3.4.1}$$
and
$$\mathrm{Gr}^0_{\mathsf{D}\#} \subset p^{r_0} g_{p,\#,0}(\mathrm{Gr}^0_{\mathsf{D}\#}) \subset p^{-r+r_0} g_{p,\#,0}(\mathrm{Gr}^0_{\mathsf{D}\#}) \subset p^{-r''}\,\mathrm{Gr}^0_{\mathsf{D}\#} \tag{3.3.4.2}$$
Note that (3.3.4.2) is equivalent (by duality) to
$$p^{r''}\,\mathrm{Gr}^{-1}_{\mathsf{D}} \subset p^{r-r_0} g_{p,-1}(\mathrm{Gr}^{-1}_{\mathsf{D}}) \subset p^{-r_0} g_{p,-1}(\mathrm{Gr}^{-1}_{\mathsf{D}}) \subset \mathrm{Gr}^{-1}_{\mathsf{D}}.$$

The relations (3.3.4.1) and (3.3.4.2) define $\mathcal{O}$-submodules
$$(p^{-r_0} g_{p,0}(\mathrm{Gr}^0_{\mathsf{D}}))/\,\mathrm{Gr}^0_{\mathsf{D}} \subset (p^{-r'}\,\mathrm{Gr}^0_{\mathsf{D}})/\,\mathrm{Gr}^0_{\mathsf{D}} \cong \mathrm{Gr}^0_{\mathsf{D},p^{r'}} \tag{3.3.4.3}$$
and
$$(p^{r_0} g_{p,\#,0}(\mathrm{Gr}^0_{\mathsf{D}\#}))/\,\mathrm{Gr}^0_{\mathsf{D}\#} \subset (p^{-r''}\,\mathrm{Gr}^0_{\mathsf{D}\#})/\,\mathrm{Gr}^0_{\mathsf{D}\#} \cong \mathrm{Gr}^0_{\mathsf{D}\#,p^{r''}} \tag{3.3.4.4}$$
respectively.

Since these conditions are complicated, we include some basic examples:

Example 3.3.4.5 (elements in $\mathrm{P}^{\mathrm{ord}}_{\mathsf{D}}(\mathbb{Z}_p)$). Suppose $g_p \in \mathrm{P}^{\mathrm{ord}}_{\mathsf{D}}(\mathbb{Z}_p)$ and $\mathcal{H}'_p \subset g_p \mathcal{H}_p g_p^{-1}$. Then the remaining conditions above are automatic, because $g_{p,0}(\mathrm{Gr}^0_{\mathsf{D}}) = \mathrm{Gr}^0_{\mathsf{D}}$, $g_{p,\#,0}(\mathrm{Gr}^0_{\mathsf{D}\#}) = \mathrm{Gr}^0_{\mathsf{D}\#}$, and $r' \geq r$, and because we can take $r_0 = 0$ and take any $r' \geq r'' \geq r$ for (3.3.4.1) and (3.3.4.4) to hold. (We will continue in Example 3.3.4.18 below.)

Example 3.3.4.6 (multiplication by powers of p). Suppose $g_p \in \mathrm{P}^{\mathrm{ord}}_{\mathtt{D}}(\mathbb{Q}_p)$ acts on $\mathrm{Gr}_{\mathtt{D}_{\mathbb{Q}_p}} \cong \mathrm{Gr}^0_{\mathtt{D}_{\mathbb{Q}_p}} \oplus \mathrm{Gr}^{-1}_{\mathtt{D}_{\mathbb{Q}_p}}$ by $g_p^{\mathrm{ord}} = (g_{p,0}, g_{p,-1}) = (p^{r_0} \mathrm{Id}_{\mathrm{Gr}^0_{\mathtt{D}}}, p^{r_0} \mathrm{Id}_{\mathrm{Gr}^{-1}_{\mathtt{D}}})$ for some integer r_0. Suppose $\mathcal{H}'_p \subset g_p \mathcal{H}_p g_p^{-1} = \mathcal{H}_p$. Then $g_{p,0}(\mathrm{Gr}^0_{\mathtt{D}}) = p^{r_0} \mathrm{Gr}^0_{\mathtt{D}}$, $g_{p,\#,0}(\mathrm{Gr}^0_{\mathtt{D}\#}) = p^{-r_0} \mathrm{Gr}^0_{\mathtt{D}\#}$, $r' \geq r$, and the remaining conditions are automatic, because we can take r_0 as it is and take any $r' \geq r'' \geq r$ for (3.3.4.1) and (3.3.4.4) to hold. (We will continue in Example 3.3.4.19 below.)

Example 3.3.4.7 (U_p operator). Suppose $g_p \in \mathrm{P}^{\mathrm{ord}}_{\mathtt{D}}(\mathbb{Q}_p)$ acts on $\mathrm{Gr}_{\mathtt{D}_{\mathbb{Q}_p}} \cong \mathrm{Gr}^0_{\mathtt{D}_{\mathbb{Q}_p}} \oplus \mathrm{Gr}^{-1}_{\mathtt{D}_{\mathbb{Q}_p}}$ by $g_p^{\mathrm{ord}} = (g_{p,0}, g_{p,-1}) = (p^{-1} \mathrm{Id}_{\mathrm{Gr}^0_{\mathtt{D}}}, \mathrm{Id}_{\mathrm{Gr}^{-1}_{\mathtt{D}}})$. Suppose $\mathcal{H}'_p \subset g_p \mathcal{H}_p g_p^{-1}$. Then $g_{p,0}(\mathrm{Gr}^0_{\mathtt{D}}) = p^{-1} \mathrm{Gr}^0_{\mathtt{D}}$, $g_{p,\#,0}(\mathrm{Gr}^0_{\mathtt{D}\#}) = \mathrm{Gr}^0_{\mathtt{D}\#}$, and $r' > r$. Then the remaining conditions are automatic, because we can take $r_0 = 0$ and take any $r' > r'' \geq r$ for (3.3.4.1) and (3.3.4.4) to hold. (We will continue in Example 3.3.4.20 below.)

Example 3.3.4.8 (generalized U_p operator). Ignoring the $\mathcal{O}$-module structures, suppose $L = \mathbb{Z}^{\oplus 2n}$ for some integer $n \geq 0$, with $\mathtt{D}$ defined by $\mathtt{D}^0 = \mathbb{Z}_p^{\oplus n} \subset L \underset{\mathbb{Z}}{\otimes} \mathbb{Z}_p$. Suppose $g_p \in \mathrm{P}^{\mathrm{ord}}_{\mathtt{D}}(\mathbb{Q}_p)$ acts on $\mathrm{Gr}_{\mathtt{D}_{\mathbb{Q}_p}} \cong \mathrm{Gr}^0_{\mathtt{D}_{\mathbb{Q}_p}} \oplus \mathrm{Gr}^{-1}_{\mathtt{D}_{\mathbb{Q}_p}}$ by $g_p^{\mathrm{ord}} = (g_{p,0}, g_{p,-1}) = (\mathrm{diag}(p^{-r_1}, p^{-r_2}, \ldots, p^{-r_n}), \mathrm{diag}(p^{-r_{n+1}}, p^{-r_{n+2}}, \ldots, p^{-r_{2n}}))$ for some integers $r_1 \geq r_2 \geq \cdots \geq r_n$. Since $r_i + r_{n+i}$ (satisfying $p^{-r_i - r_{n+i}} = \nu(g_p)$) is a constant independent of $1 \leq i \leq n$, this forces $r_{n+1} \leq r_{n+2} \leq \cdots \leq r_{2n}$. Suppose $\mathcal{H}'_p \subset g_p \mathcal{H}_p g_p^{-1}$. Suppose $r_n \geq r_{2n}$ and $r' - r \geq r_1 - r_{n+1}$. Then the remaining conditions hold if we take any r_0 such that $r_n \geq -r_0 \geq r_{2n}$ and take any r'' such that $r' - (r_1 + r_0) \geq r'' \geq r - (r_{n+1} + r_0)$ for (3.3.4.1) and (3.3.4.4) to hold. (This rather elaborate example includes both Examples 3.3.4.6 and 3.3.4.7 as special cases. However, we will not continue this example as in Examples 3.3.4.19 and 3.3.4.20 below.)

Let S and (A, λ, i) be as in Definition 3.3.3.1, and let $\alpha^{\mathrm{ord}}_{\mathcal{H}'_p}$ be an ordinary level-$\mathcal{H}'_p$ structure of (A, λ, i) of type $(L \underset{\mathbb{Z}}{\otimes} \mathbb{Z}_p, \langle \cdot\,, \cdot \rangle, \mathtt{D})$ as in Definition 3.3.3.4, with $\mathcal{H}'^{,\mathrm{ord}}_p$ (resp. $H'^{,\mathrm{ord}}_{p^{r''}}$, for each integer $r'' \geq r'$) defined by $\mathcal{H}'_p$ as $\mathcal{H}^{\mathrm{ord}}_p$ (resp. $H^{\mathrm{ord}}_{p^{r''}}$, for each integer $r'' \geq r$) is defined by $\mathcal{H}_p$.

Proposition 3.3.4.9. *With assumptions as above, there exists a tuple $(A', \lambda', i', \alpha^{\mathrm{ord},\prime}_{\mathcal{H}_p})$ (over S, up to isomorphism), called the* **ordinary Hecke twist** *of $(A, \lambda, i, \alpha^{\mathrm{ord}}_{\mathcal{H}'_p})$ by g_p, equipped with a $\mathbb{Q}^\times$-isogeny $[g_p^{-1}]^{\mathrm{ord}} : A \to A'$ (whose formal inverse we denote by $[g_p]^{\mathrm{ord}} : A' \to A$) satisfying the following characterizing conditions:*

1. *$[g_p^{-1}]^{\mathrm{ord}}$ is the composition $([p^{-r_0} g_{p,-1}]^{\mathrm{ord}})^{-1} \circ [p^{r_0} g_{p,0}^{-1}]^{\mathrm{ord}} \circ [p^{r_0}]^{-1}$, where $[p^{r_0}] : A \to A$ is the multiplication by p^{r_0} on A when $r_0 \geq 0$, or the formal inverse of the multiplication by p^{-r_0} on A when $r_0 < 0$; and where $[p^{r_0} g_{p,0}^{-1}]^{\mathrm{ord}} : A \to A''$ and $[p^{-r_0} g_{p,-1}]^{\mathrm{ord}} : A' \to A''$ are isogenies of p-power*

degrees (whose formal inverses we denote by $[p^{-r_0}g_{p,0}]^{\mathrm{ord}} : A'' \to A$ *and* $[p^{r_0}g_{p,-1}^{-1}]^{\mathrm{ord}} : A'' \to A'$*, respectively).*

(2) $\lambda' : A' \to A'^{,\vee}$ *is a polarization defined by* $\lambda' = r_p([g_p]^{\mathrm{ord}})^{\vee} \circ \lambda \circ [g_p]^{\mathrm{ord}}$ *(as positive* $\mathbb{Q}^{\times}$*-isogenies), where* r_p *is the unique power of* p *such that* $r_p\nu(g_p)\mathbb{Z}_p = \mathbb{Z}_p$.

(3) $i' : \mathcal{O} \to \mathrm{End}_S(A')$ *is an* $\mathcal{O}$*-structures of* (A', λ') *making* $[g_p]^{\mathrm{ord}}$ *an* $\mathcal{O}$*-equivariant* $\mathbb{Q}^{\times}$*-isogeny.*

(4) $\alpha_{\mathcal{H}_p}^{\mathrm{ord},\prime}$ *is an ordinary level-*$\mathcal{H}_p$ *structure of* (A', λ', i') *of type* $(L \underset{\mathbb{Z}}{\otimes} \mathbb{Z}_p, \langle \cdot, \cdot \rangle, \mathtt{D})$.

(5) At each geometric point $\bar{s}$*, there exist a lifting* $\hat{\alpha}_{\bar{s}}^{\mathrm{ord}} = (\hat{\alpha}_{\bar{s}}^{\mathrm{ord},0}, \hat{\alpha}_{\bar{s}}^{\mathrm{ord},\#,0}, \hat{\nu}_{\bar{s}}^{\mathrm{ord}})$ *(resp.* $\hat{\alpha}_{\bar{s}}^{\mathrm{ord},\prime} = (\hat{\alpha}_{\bar{s}}^{\mathrm{ord},0,\prime}, \hat{\alpha}_{\bar{s}}^{\mathrm{ord},\#,0,\prime}, \hat{\nu}_{\bar{s}}^{\mathrm{ord},\prime})$*) of* $\alpha_{\mathcal{H}'_p}^{\mathrm{ord}}$ *(resp.* $\alpha_{\mathcal{H}_p}^{\mathrm{ord},\prime}$*) as in Lemma* 3.3.3.9*, such that* $\hat{\alpha}_{\bar{s}}^{\mathrm{ord},\prime}$ *is compatible with* $\hat{\alpha}_{\bar{s}}^{\mathrm{ord}} \circ g_p^{\mathrm{ord}}$ *under the* $\mathbb{Q}^{\times}$*-isogeny* $[p^{r_0}] : A \to A$ *and the isogenies* $[p^{r_0}g_{p,0}^{-1}]^{\mathrm{ord}} : A \to A''$ *and* $[p^{-r_0}g_{p,-1}]^{\mathrm{ord}} : A' \to A''$*, in the following sense:*

(a) $\ker([p^{r_0}g_{p,0}^{-1}]^{\mathrm{ord}})_{\bar{s}}$ *is the schematic image of the submodule*

$$((p^{-r_0}g_{p,0}(\mathrm{Gr}_{\mathtt{D}}^0))/\,\mathrm{Gr}_{\mathtt{D}}^0)_{\bar{s}}^{\mathrm{mult}}$$

of $(\mathrm{Gr}_{\mathtt{D},p^{r'}}^0)_{\bar{s}}^{\mathrm{mult}} \subset (\mathrm{Gr}_{\mathtt{D}}^0)_{\bar{s}}^{\mathrm{mult}}$ *(see* (3.3.4.3)*) under* $\hat{\alpha}_{\bar{s}}^{\mathrm{ord},0} : (\mathrm{Gr}_{\mathtt{D}}^0)_{\bar{s}}^{\mathrm{mult}} \to A_{\bar{s}}[p^{\infty}]$*. Then*

$$\hat{\alpha}_{\bar{s}}^{\mathrm{ord},-1} : A_{\bar{s}}[p^{\infty}] \to \mathrm{Gr}_{\mathtt{D}}^{-1} \underset{\mathbb{Z}_p}{\otimes} (\mathbb{Q}_p/\mathbb{Z}_p)$$

(which is the Serre dual of $\hat{\alpha}_{\bar{s}}^{\mathrm{ord},\#,0} : (\mathrm{Gr}_{\mathtt{D}^{\#}}^0)_{\bar{s}}^{\mathrm{mult}} \to A_{\bar{s}}^{\vee}[p^{\infty}]$*) satisfies*

$$\ker(\hat{\alpha}_{\bar{s}}^{\mathrm{ord},-1}) = \mathrm{image}(\hat{\alpha}_{\bar{s}}^{\mathrm{ord},0}),$$

and induces an injection

$$\hat{\alpha}_{\bar{s}}^{\mathrm{ord},0,\prime\prime} := [p^{r_0}g_{p,0}^{-1}]_{\bar{s}}^{\mathrm{ord}} \circ \hat{\alpha}_{\bar{s}}^{\mathrm{ord},0} \circ (p^{-r_0}g_{p,0}) : (\mathrm{Gr}_{\mathtt{D}}^0)_{\bar{s}}^{\mathrm{mult}} \to A''_{\bar{s}}[p^{\infty}]$$

and a surjection

$$\hat{\alpha}_{\bar{s}}^{\mathrm{ord},-1,\prime\prime} : A''_{\bar{s}}[p^{\infty}] \twoheadrightarrow \mathrm{Gr}_{\mathtt{D}}^{-1} \underset{\mathbb{Z}_p}{\otimes} (\mathbb{Q}_p/\mathbb{Z}_p)$$

satisfying

$$\ker(\hat{\alpha}_{\bar{s}}^{\mathrm{ord},-1,\prime\prime}) = \mathrm{image}(\hat{\alpha}_{\bar{s}}^{\mathrm{ord},0,\prime\prime}).$$

Consequently,

$$\hat{\alpha}_{\bar{s}}^{\mathrm{ord},\#,0} : (\mathrm{Gr}_{\mathtt{D}^{\#}}^0)_{\bar{s}}^{\mathrm{mult}} \to A_{\bar{s}}^{\vee}[p^{\infty}]$$

is liftable to an injection

$$\hat{\alpha}_{\bar{s}}^{\mathrm{ord},\#,0,\prime\prime} : (\mathrm{Gr}_{\mathtt{D}^{\#}}^0)_{\bar{s}}^{\mathrm{mult}} \to A_{\bar{s}}^{\prime\prime,\vee}[p^{\infty}].$$

(b) *The isogeny* $[p^{-r_0}g_{p,-1}]^{\mathrm{ord}} : A' \to A''$ *is dual to an isogeny* $[p^{-r_0}g_{p,\#,0}^{-1}]^{\mathrm{ord}} : A'',^{\vee} \to A',^{\vee}$, *and* $\ker([p^{-r_0}g_{p,\#,0}^{-1}]^{\mathrm{ord}})_{\bar{s}}$ *is the schematic image of the submodule*

$$((p^{r_0}g_{p,\#,0}(\mathrm{Gr}^0_{\mathtt{D}^\#}))/\,\mathrm{Gr}^0_{\mathtt{D}^\#})^{\mathrm{mult}}_{\bar{s}}$$

of $(\mathrm{Gr}^0_{\mathtt{D}^\#,p^{r''}})^{\mathrm{mult}}_{\bar{s}} \subset (\mathrm{Gr}^0_{\mathtt{D}^\#})^{\mathrm{mult}}_{\bar{s}}$ *(see* (3.3.4.4)) *under* $\hat{\alpha}^{\mathrm{ord},\#,0,\prime\prime}_{\bar{s}}$. *Then*

$$\hat{\alpha}^{\mathrm{ord},\#,0,\prime}_{\bar{s}} : (\mathrm{Gr}^0_{\mathtt{D}^\#})^{\mathrm{mult}}_{\bar{s}} \to A'^{,\vee}_{\bar{s}}[p^\infty]$$

agrees with the composition

$$[p^{-r_0}g_{p,\#,0}^{-1}]_{\bar{s}} \circ \hat{\alpha}^{\mathrm{ord},\#,0,\prime\prime}_{\bar{s}} \circ (p^{r_0}g_{p,\#,0}).$$

(c) *The kernel of the dual surjection*

$$\hat{\alpha}^{\mathrm{ord},\#,-1,\prime\prime}_{\bar{s}} : A''^{,\vee}_{\bar{s}}[p^\infty] \to \mathrm{Gr}^{-1}_{\mathtt{D}^\#} \underset{\mathbb{Z}_p}{\otimes} (\mathbb{Q}_p/\mathbb{Z}_p)$$

of $\hat{\alpha}^{\mathrm{ord},0,\prime\prime}_{\bar{s}}$ *is the schematic image of* $\hat{\alpha}^{\mathrm{ord},\#,0,\prime\prime}_{\bar{s}}$, *which contains* $\ker([p^{r_0}g_{p,\#,0}^{-1}])_{\bar{s}}$. *Hence,* $\hat{\alpha}^{\mathrm{ord},\#,-1,\prime\prime}_{\bar{s}}$ *induces a surjection*

$$A'^{,\vee}_{\bar{s}}[p^\infty] \to \mathrm{Gr}^{-1}_{\mathtt{D}^\#} \underset{\mathbb{Z}_p}{\otimes} (\mathbb{Q}_p/\mathbb{Z}_p),$$

which is dual to an injection

$$(\mathrm{Gr}^0_{\mathtt{D}})^{\mathrm{mult}}_{\bar{s}} \to A'_{\bar{s}}[p^\infty]$$

lifting $\hat{\alpha}^{\mathrm{ord},0,\prime\prime}_{\bar{s}}$. *This injection coincides with* $\hat{\alpha}^{\mathrm{ord},0,\prime}_{\bar{s}}$.

(d) $\hat{\nu}^{\mathrm{ord},\prime}_{\bar{s}} : \boldsymbol{\mu}_{p^\infty,\bar{s}} \xrightarrow{\sim} \boldsymbol{\mu}_{p^\infty,\bar{s}}$ *is induced by* $\hat{\nu}^{\mathrm{ord}}_{\bar{s}} \circ (r_p\nu(g_p))$, *where* r_p *is as in* (2) *above.*

By abuse of language, we say that $\hat{\alpha}^{\mathrm{ord},\prime}_{\bar{s}} = [g_p^{-1}]^{\mathrm{ord}}_{\bar{s}} \circ \hat{\alpha}^{\mathrm{ord}}_{\bar{s}} \circ g_p^{\mathrm{ord}}$. *In this case, the* $\mathcal{H}'^{,\mathrm{ord}}_p$*-orbit* $[\hat{\alpha}^{\mathrm{ord}}_{\bar{s}}]_{\mathcal{H}'^{,\mathrm{ord}}_p}$ *determines a* $(g_p^{\mathrm{ord}}\mathcal{H}^{\mathrm{ord}}_p(g_p^{\mathrm{ord}})^{-1})$*-orbit* $[\hat{\alpha}^{\mathrm{ord}}_{\bar{s}}]_{g_p^{\mathrm{ord}}\mathcal{H}^{\mathrm{ord}}_p(g_p^{\mathrm{ord}})^{-1}}$ *because* $\mathcal{H}'^{,\mathrm{ord}}_p \subset g_p^{\mathrm{ord}}\mathcal{H}^{\mathrm{ord}}_p(g_p^{\mathrm{ord}})^{-1}$, *and hence induces an* $\mathcal{H}^{\mathrm{ord}}_p$*-orbit* $[\hat{\alpha}^{\mathrm{ord},\prime}_{\bar{s}}]_{\mathcal{H}^{\mathrm{ord}}_p}$.

If $g_p = g_{1,p}g_{2,p}$, *where* $g_{1,p}$ *and* $g_{2,p}$ *are elements of* $\mathrm{P}^{\mathrm{ord}}_{\mathtt{D}}(\mathbb{Q}_p)$, *each having a setup similar to that of* g_p, *then the ordinary Hecke twist by* g_p *can be constructed in two steps using ordinary Hecke twists by* $g_{1,p}$ *and* $g_{2,p}$, *such that* $[g_p^{-1}]^{\mathrm{ord}} = [g_{2,p}^{-1}]^{\mathrm{ord}} \circ [g_{1,p}^{-1}]^{\mathrm{ord}}$ *(or, equivalently,* $[g_p]^{\mathrm{ord}} = [g_{1,p}]^{\mathrm{ord}} \circ [g_{2,p}]^{\mathrm{ord}}$).

Proof. Since ordinary Hecke twists (in this proposition) are constructed using p-power isogenies (and their formal inverses), which are uniquely determined by their behaviors on geometric fibers of p-power torsion subgroup schemes of (ordinary) abelian schemes over S (which are finite flat group schemes of étale-multiplicative type over S), the last statement of the proposition follows from the characterizing conditions preceding it. Therefore, we can construct the desired ordinary Hecke twist by g_p in two steps using the ordinary Hecke twists by $p^{-r_0}g_p$ and by $p^{r_0}\,\mathrm{Id}$ (the latter being given by $[p^{-r_0}] = [p^{r_0}]^{-1}$, which is defined for all $r_0 \in \mathbb{Z}$

and does not alter the additional structures λ, i, and $\alpha^{\mathrm{ord}}_{\mathcal{H}'_p}$; see Example 3.3.4.19 below). Hence, our main tasks are to construct $[p^{r_0} g_{p,0}^{-1}]^{\mathrm{ord}}$ and $[p^{-r_0} g_{p,-1}]^{\mathrm{ord}}$, so that $[p^{r_0} g_p^{-1}]^{\mathrm{ord}} = ([p^{-r_0} g_{p,-1}]^{\mathrm{ord}})^{-1} \circ [p^{r_0} g_{p,0}^{-1}]^{\mathrm{ord}}$, and to construct the additional structures λ', i', and $\alpha^{\mathrm{ord},\prime}_{\mathcal{H}_p}$ (the main focus being on $\alpha^{\mathrm{ord},\prime}_{\mathcal{H}_p}$).

By definition of $\alpha^{\mathrm{ord}}_{\mathcal{H}'_p}$, over the scheme $\tilde{S} = \alpha^{\mathrm{ord}}_{H'^{,\mathrm{ord}}_{p^{r'}}}$, which is an $H'^{,\mathrm{ord}}_{p^r}$-torsor (finite étale) over S, there is a tautological principal ordinary level-$p^{r'}$ structure $\alpha^{\mathrm{ord}}_{p^{r'}} = (\alpha^{\mathrm{ord},0}_{p^{r'}}, \alpha^{\mathrm{ord},\#,0}_{p^{r'}}, \nu^{\mathrm{ord}}_{p^{r'}})$ as in Definition 3.3.3.2, where

$$\alpha^{\mathrm{ord},0}_{p^{r'}} : (\mathrm{Gr}^0_{\mathtt{D},p^{r'}})^{\mathrm{mult}}_{\tilde{S}} \to A_{\tilde{S}}[p^{r'}]$$

and

$$\alpha^{\mathrm{ord},\#,0}_{p^{r'}} : (\mathrm{Gr}^0_{\mathtt{D}^\#,p^{r'}})^{\mathrm{mult}}_{\tilde{S}} \to A^\vee_{\tilde{S}}[p^{r'}]$$

are closed immersions, and where the third entry

$$\nu^{\mathrm{ord}}_{p^{r'}} : \boldsymbol{\mu}_{p^{r'},\tilde{S}} \to \boldsymbol{\mu}_{p^{r'},\tilde{S}}$$

is a section of $(\mathbb{Z}/p^{r'}\mathbb{Z})^\times_{\tilde{S}}$, satisfying the usual symplectic and liftability conditions defining a (principal) ordinary level structure.

The schematic image of the submodule

$$((p^{-r_0} g_{p,0}(\mathrm{Gr}^0_{\mathtt{D}}))/\,\mathrm{Gr}^0_{\mathtt{D}})^{\mathrm{mult}}_{\tilde{S}}$$

of $(\mathrm{Gr}^0_{\mathtt{D},p^{r'}})^{\mathrm{mult}}_{\tilde{S}}$ (see (3.3.4.3)) under the closed immersion $\alpha^{\mathrm{ord},0}_{p^{r'}}$ defines a subgroup scheme $K_{0,\tilde{S}}$ of $A_{\tilde{S}}[p^{r'}]$. Since $K_{0,\tilde{S}}$ is isomorphic to $((p^{-r_0} g_{p,0}(\mathrm{Gr}^0_{\mathtt{D}}))/\,\mathrm{Gr}^0_{\mathtt{D}})^{\mathrm{mult}}_{\tilde{S}}$, it is finite flat, of finite presentation, and of multiplicative type over $\tilde{S}$. Since the tautological action of $H'^{,\mathrm{ord}}_{p^r}$ on $\tilde{S} \to S$ is compatible with the homomorphism $\alpha^{\mathrm{ord},0}_{p^{r'}}$, we can descend $K_{0,\tilde{S}}$ to a subgroup scheme K_0 of $A[p^{r'}]$, which is also finite flat, of finite presentation, and of multiplicative type over S. (This is a question of descending the finite étale character group scheme of $K_{0,\tilde{S}}$.) Hence, we can define the isogeny

$$[p^{r_0} g_{p,0}^{-1}]^{\mathrm{ord}} : A \to A'' := A/K_0, \tag{3.3.4.10}$$

inducing an embedding $A[p^{r'}]/K_0 \hookrightarrow A''$. By (3.3.4.1), we know that

$$A''[p^{r''}] \subset A[p^{r'}]/K_0 \tag{3.3.4.11}$$

(as a closed subgroup scheme).

Over $\tilde{S}$, the homomorphism $\alpha^{\mathrm{ord},0}_{p^{r'}}$ induces a homomorphism

$$\alpha^{\mathrm{ord},0,\prime\prime}_{p^{r''}} := \Big([p^{r_0} g_{p,0}^{-1}]^{\mathrm{ord}}_{\tilde{S}} \circ \alpha^{\mathrm{ord},\#,0}_{p^{r'}} \circ (p^{-r_0} g_{p,0})\Big)|_{(\mathrm{Gr}^0_{\mathtt{D},p^{r''}})^{\mathrm{mult}}_{\tilde{S}}} :$$
$$(\mathrm{Gr}^0_{\mathtt{D},p^{r''}})^{\mathrm{mult}}_{\tilde{S}} \to A''_{\tilde{S}}[p^{r''}]$$

that is a closed immersion by (3.3.4.1). Since the kernel $K_{0,\tilde{S}}$ of $[p^{r_0} g_{p,0}^{-1}]^{\mathrm{ord}}_{\tilde{S}} : A_{\tilde{S}} \to A''_{\tilde{S}}$ is contained in the kernel of the surjection

$$\alpha^{\mathrm{ord},-1}_{p^{r'}} : A_{\tilde{S}}[p^{r'}] \to (\mathrm{Gr}^{-1}_{\mathtt{D},p^{r'}})_{\tilde{S}}$$

dual to $\alpha_{p^{r'}}^{\mathrm{ord},\#,0}$, the surjection $\alpha_{p^{r'}}^{\mathrm{ord},-1}$ induces a surjection

$$A_{\tilde{S}}[p^{r'}]/K_{0,\tilde{S}} \to (\mathrm{Gr}_{\mathtt{D},p^{r'}}^{-1})_{\tilde{S}}.$$

By restriction to $A''_{\tilde{S}}[p^{r''}]$ (see (3.3.4.11)), we obtain an induced surjection

$$\alpha_{p^{r''}}^{\mathrm{ord},-1,\prime\prime} : A''_{\tilde{S}}[p^{r''}] \to (\mathrm{Gr}_{\mathtt{D},p^{r''}}^{-1})_{\tilde{S}},$$

which is dual to a homomorphism

$$\alpha_{p^{r''}}^{\mathrm{ord},\#,0,\prime\prime} : (\mathrm{Gr}_{\mathtt{D}\#,p^{r''}}^{0})_{\tilde{S}}^{\mathrm{mult}} \to A_{\tilde{S}}^{\prime\prime,\vee}[p^{r''}]$$

that is a closed immersion, lifting the homomorphism

$$\alpha_{p^{r''}}^{\mathrm{ord},\#,0} : (\mathrm{Gr}_{\mathtt{D}\#,p^{r''}}^{0})_{\tilde{S}}^{\mathrm{mult}} \to A_{\tilde{S}}^{\vee}[p^{r''}]$$

induced by $\alpha_{p^{r'}}^{\mathrm{ord},\#,0}$.

The schematic image of the submodule

$$((p^{r_0}g_{p,\#,0}(\mathrm{Gr}_{\mathtt{D}\#}^{0}))/\operatorname{Gr}_{\mathtt{D}\#}^{0})_{\tilde{S}}^{\mathrm{mult}}$$

of $(\mathrm{Gr}_{\mathtt{D}\#,p^{r''}}^{0})_{\tilde{S}}^{\mathrm{mult}}$ (see (3.3.4.4)) under the closed immersion $\alpha_{p^{r''}}^{\mathrm{ord},\#,0,\prime\prime}$ defines a subgroup scheme $K_{0,\#,\tilde{S}}$ of $A_{\tilde{S}}^{\prime\prime,\vee}[p^{r''}]$. Since $K_{0,\#,\tilde{S}}$ is isomorphic to $((p^{r_0}g_{p,\#,0}(\mathrm{Gr}_{\mathtt{D}\#}^{0}))/\operatorname{Gr}_{\mathtt{D}\#}^{0})_{\tilde{S}}^{\mathrm{mult}}$, it is finite flat, of finite presentation, and of multiplicative type over $\tilde{S}$. Since the tautological action of $H_{p^r}^{\prime,\mathrm{ord}}$ on $\tilde{S} \to S$ is compatible with the homomorphism $\alpha_{p^{r''}}^{\mathrm{ord},\#,0,\prime\prime}$ (and the other homomorphisms involved in its definition), we can descend $K_{0,\#,\tilde{S}}$ to a subgroup scheme $K_{0,\#}$ of $A^{\prime\prime,\vee}[p^{r''}]$, which is also finite flat, of finite presentation, and of multiplicative type over S. (This is a question of descending the finite étale character group scheme of $K_{0,\#,\tilde{S}}$.) Hence, we can define the isogeny

$$[p^{-r_0}g_{p,\#,0}^{-1}]^{\mathrm{ord}} : A^{\prime\prime,\vee} \to A^{\prime,\vee} := A^{\prime\prime,\vee}/K_{0,\#}, \tag{3.3.4.12}$$

and define the isogeny

$$[p^{-r_0}g_{p,-1}]^{\mathrm{ord}} : A' \to A'' \tag{3.3.4.13}$$

to be the dual of $[p^{-r_0}g_{p,\#,0}]^{\mathrm{ord}}$, with kernel isomorphic to the Cartier dual of $K_{0,\#}$. By (3.3.4.2), we know that

$$A^{\prime,\vee}[p^r] \subset A^{\prime\prime,\vee}[p^{r''}]/K_{0,\#} \tag{3.3.4.14}$$

(as a closed subgroup scheme).

Over $\tilde{S}$ again, the homomorphism $\alpha_{p^{r''}}^{\mathrm{ord},\#,0,\prime\prime}$ induces a homomorphism

$$\begin{aligned}\alpha_{p^r}^{\mathrm{ord},\#,0,\prime} := \Big([p^{-r_0}g_{p,\#,0}^{-1}]_{\tilde{S}}^{\mathrm{ord}} \circ \alpha_{p^{r''}}^{\mathrm{ord},\#,0,\prime\prime} \circ (p^{r_0}g_{p,\#,0})\Big)|_{(\mathrm{Gr}_{\mathtt{D}\#,p^r}^{0})_{\tilde{S}}^{\mathrm{mult}}} :\\ (\mathrm{Gr}_{\mathtt{D}\#,p^r}^{0})_{\tilde{S}}^{\mathrm{mult}} \to A_{\tilde{S}}^{\prime,\vee}[p^r]\end{aligned} \tag{3.3.4.15}$$

that is a closed immersion by (3.3.4.2). Since the kernel $K_{0,\#,\tilde{S}}$ of $[p^{-r_0}g_{p,\#,0}^{-1}]_{\tilde{S}}^{\mathrm{ord}} : A_{\tilde{S}}^{\prime\prime,\vee} \to A_{\tilde{S}}^{\prime,\vee}$ is contained in the kernel of the surjection

$$\alpha_{p^{r''}}^{\mathrm{ord},\#,-1,\prime\prime} : A_{\tilde{S}}^{\prime\prime,\vee}[p^{r''}] \to (\mathrm{Gr}_{\mathtt{D}\#,p^{r''}}^{-1})_{\tilde{S}}$$

dual to $\alpha_{p^{r''}}^{\mathrm{ord},0,''}$, the surjection $\alpha_{p^{r''}}^{\mathrm{ord},\#,-1,''}$ induces a surjection

$$A^{'',\vee}[p^{r''}]/K_{0,\#} \to (\mathrm{Gr}_{\mathtt{D}^{\#},p^{r''}}^{-1})_{\tilde{S}}.$$

By restriction to $A_{\tilde{S}}^{',\vee}[p^r]$ (see (3.3.4.14)), we obtain an induced surjection

$$\alpha_{p^r}^{\mathrm{ord},\#,-1,'} : A_{\tilde{S}}^{',\vee}[p^r] \to (\mathrm{Gr}_{\mathtt{D}^{\#},p^r}^{-1})_{\tilde{S}},$$

which is dual to a homomorphism

$$\alpha_{p^r}^{\mathrm{ord},0,'} : (\mathrm{Gr}_{\mathtt{D},p^r}^{0})_{\tilde{S}}^{\mathrm{mult}} \to A'_{\tilde{S}}[p^r] \tag{3.3.4.16}$$

that is a closed immersion, lifting the homomorphism

$$\alpha_{p^r}^{\mathrm{ord},0,''} : (\mathrm{Gr}_{\mathtt{D},p^{r''}}^{0})_{\tilde{S}}^{\mathrm{mult}} \to A''_{\tilde{S}}[p^r]$$

induced by $\alpha_{p^{r''}}^{\mathrm{ord},0,''}$.

Since $r' \geq r$ by assumption, the section $\nu_{p^{r'}}^{\mathrm{ord}} \circ (r_p^{-1}\nu(g_p))$ over $\tilde{S}$, where r_p is as in the statement of the proposition, induces a section

$$\nu_{p^r}^{\mathrm{ord},'} : \boldsymbol{\mu}_{p^r,\tilde{S}} \xrightarrow{\sim} \boldsymbol{\mu}_{p^r,\tilde{S}} \tag{3.3.4.17}$$

of $(\mathbb{Z}/p^r\mathbb{Z})_{\tilde{S}}^{\times}$.

Before moving on, let us first justify the construction of λ' and i' outlined in the statements of the proposition. The only part that is not clear is that the $\mathbb{Q}^{\times}$-polarization λ' is indeed a polarization. This is only a statement of checking whether a $\mathbb{Q}^{\times}$-isogeny is an isogeny, which can be verified after pulled back to geometric points of S, which we can always lift to geometric points of $\tilde{S}$. Hence, it follows from the above construction of $[p^{r_0}g_{p,0}^{-1}]^{\mathrm{ord}}$, $[p^{-r_0}g_{p,\#,0}^{-1}]^{\mathrm{ord}}$, and $[p^{-r_0}g_{p,-1}]^{\mathrm{ord}}$ (see (3.3.4.10), (3.3.4.12), and (3.3.4.13)), and the relations among the various $(\mathcal{O} \underset{\mathbb{Z}}{\otimes} \mathbb{Z}_p)$-submodules of $\mathrm{Gr}_{\mathtt{D}}^0$ and $\mathrm{Gr}_{\mathtt{D}^{\#}}^0$.

Since all the above constructions over $\tilde{S}$ are compatible with the action of the tautological action of $H_{p^r}^{',\mathrm{ord}}$ on $\tilde{S} \to S$, the homomorphism $H_{p^{r'}}^{',\mathrm{ord}} \to H_{p^r}^{\mathrm{ord}}$ induced by $\mathcal{H}_p^{',\mathrm{ord}} \hookrightarrow g_p^{\mathrm{ord}}\mathcal{H}_p^{\mathrm{ord}}(g_p^{\mathrm{ord}})^{-1}$ (*i.e.*, conjugation by g_p^{ord}), or rather by $\mathcal{H}'_p \hookrightarrow g_p\mathcal{H}_pg_p^{-1}$, induces a well-defined $H_{p^r}^{\mathrm{ord}}$-orbit of $\alpha_{p^r}^{\mathrm{ord},'} = (\alpha_{p^r}^{\mathrm{ord},0,'}, \alpha_{p^r}^{\mathrm{ord},\#,0,'}, \nu_{p^r}^{\mathrm{ord},'})$ over $\tilde{S}$ (see (3.3.4.16), (3.3.4.15), and (3.3.4.17)), which descends to a naive ordinary level-$H_{p^r}^{\mathrm{ord}}$ structure $\alpha_{H_{p^r}^{\mathrm{ord}}}^{\mathrm{ord},'}$ of (A', λ', i') of type $(L/p^rL, \langle \cdot\,, \cdot \rangle, \mathtt{D}_{p^r})$ as in Definition 3.3.3.3. Since we can repeat the above procedure when the objects involved are (étale locally) liftable to higher levels, we obtain an ordinary level-$\mathcal{H}_p$ structure $\alpha_{\mathcal{H}_p}^{\mathrm{ord},'}$ of (A', λ', i') of type $(L \underset{\mathbb{Z}}{\otimes} \mathbb{Z}_p, \langle \cdot\,, \cdot \rangle, \mathtt{D})$ as in Definition 3.3.3.4. By construction, at each geometric point $\bar{s}$ of S, there exist liftings of $\alpha_{\mathcal{H}_p}^{\mathrm{ord},'}$ satisfying the characterizations in (5) of the proposition.

This finishes the construction of the ordinary Hecke twist $(A', \lambda', i', \alpha_{\mathcal{H}_p}^{\mathrm{ord},'})$. □

Example 3.3.4.18 (elements in $\mathrm{P}_{\mathsf{D}}^{\mathrm{ord}}(\mathbb{Z}_p)$). (This is a continuation of Example 3.3.4.5.) In this case, the ordinary Hecke twist $(A',\lambda',i',\alpha_{\mathcal{H}_p}^{\mathrm{ord},\prime})$ of $(A,\lambda,i,\alpha_{\mathcal{H}'_p}^{\mathrm{ord}})$ by g_p can be described as follows: The isogeny $[g_p^{-1}] : A \to A'$ is an isomorphism allowing us to identify (A',λ',i') with (A,λ,i). Over the scheme $\tilde{S} = \alpha_{H_{p^{r'}}^{\prime,\mathrm{ord}}}^{\mathrm{ord}}$, where there is a tautological principal ordinary level-$p^{r'}$ structure $\alpha_{p^{r'}}^{\mathrm{ord}} = (\alpha_{p^{r'}}^{\mathrm{ord},0},\alpha_{p^{r'}}^{\mathrm{ord},\#,0},\nu_{p^{r'}}^{\mathrm{ord}})$, we have a twisted triple $\alpha_{p^{r'}}^{\mathrm{ord}} \circ g_p^{\mathrm{ord}} := (\alpha_{p^{r'}}^{\mathrm{ord},0} \circ g_{p,0},\alpha_{p^{r'}}^{\mathrm{ord},\#,0} \circ g_{p,\#,0},\nu_{p^{r'}}^{\mathrm{ord}} \circ \nu(g_p))$, whose reduction modulo p^r defines a triple $\alpha_{p^r}^{\mathrm{ord},\prime}$. The $H_{p^r}^{\mathrm{ord}}$-orbit of $\alpha_{p^r}^{\mathrm{ord},\prime}$ descends to S and agrees with $\alpha_{\mathcal{H}_p}^{\mathrm{ord},\prime}$.

Example 3.3.4.19 (multiplication by powers of p). (This is a continuation of Example 3.3.4.6.) In this case, the ordinary Hecke twist $(A',\lambda',i',\alpha_{\mathcal{H}_p}^{\mathrm{ord},\prime})$ of $(A,\lambda,i,\alpha_{\mathcal{H}'_p}^{\mathrm{ord}})$ by g_p can be described as follows: The triple (A',λ',i') can be identified with (A,λ,i), so that the $\mathbb{Q}^\times$-isogeny $[g_p^{-1}] : A \to A'$ is identified with the $\mathbb{Q}^\times$-isogeny $[p^{-r_0}] = [p^{r_0}]^{-1} : A \to A$, as in (1) of Proposition 3.3.4.9. Over the scheme $\tilde{S} = \alpha_{H_{p^{r'}}^{\prime,\mathrm{ord}}}^{\mathrm{ord}}$, where there is a tautological principal ordinary level-$p^{r'}$ structure $\alpha_{p^{r'}}^{\mathrm{ord}} = (\alpha_{p^{r'}}^{\mathrm{ord},0},\alpha_{p^{r'}}^{\mathrm{ord},\#,0},\nu_{p^{r'}}^{\mathrm{ord}})$, we can take $\alpha_{p^r}^{\mathrm{ord},\prime}$ to be the reduction modulo p^r of the triple $\alpha_{p^{r'}}^{\mathrm{ord}}$. The $H_{p^r}^{\mathrm{ord}}$-orbit of $\alpha_{p^r}^{\mathrm{ord},\prime}$ descends to S and agrees with $\alpha_{\mathcal{H}_p}^{\mathrm{ord},\prime}$.

Example 3.3.4.20 (U_p operator and relative Frobenius). (This is a continuation of Example 3.3.4.7.) In this case, the ordinary Hecke twist $(A',\lambda',i',\alpha_{\mathcal{H}_p}^{\mathrm{ord},\prime})$ of $(A,\lambda,i,\alpha_{\mathcal{H}'_p}^{\mathrm{ord}})$ by g_p can be described as follows: Consider $K_0 := \mathrm{image}(\alpha_{\mathcal{H}'_p}^{\mathrm{ord},0})[p]$, the p-torsion subgroup scheme of $\mathrm{image}(\alpha_{\mathcal{H}'_p}^{\mathrm{ord},0})$. Then $[g_p^{-1}]^{\mathrm{ord}} : A \to A'$ can be identified with the quotient $A \to A/K_0$. If S is a scheme over $\mathrm{Spec}(\mathbb{F}_p)$, we can identify A' with the pullback $A^{(p)}$ of A by the absolute Frobenius morphism $\mathrm{F}_S : S \to S$, and identify $[g_p^{-1}] : A \to A'$ with the relative Frobenius morphism $\mathrm{F}_{A/S} : A \to A^{(p)}$; and, accordingly, we can also identify λ' and i' with the pullbacks $\lambda^{(p)}$ and $i^{(p)}$ by F_S, respectively. Over the scheme $\tilde{S} = \alpha_{H_{p^{r'}}^{\prime,\mathrm{ord}}}^{\mathrm{ord}}$ (but no longer assuming that S is a scheme over $\mathrm{Spec}(\mathbb{F}_p)$), where there is a tautological principal ordinary level-$p^{r'}$ structure $\alpha_{p^{r'}}^{\mathrm{ord}} = (\alpha_{p^{r'}}^{\mathrm{ord},0},\alpha_{p^{r'}}^{\mathrm{ord},\#,0},\nu_{p^{r'}}^{\mathrm{ord}})$, we can take $\alpha_{p^r}^{\mathrm{ord},\prime} = (\alpha_{p^r}^{\mathrm{ord},0,\prime},\alpha_{p^r}^{\mathrm{ord},\#,0,\prime},\nu_{p^{r'}}^{\mathrm{ord},\prime})$ such that $\alpha_{p^r}^{\mathrm{ord},0,\prime}$ is obtained from $\alpha_{p^{r'}}^{\mathrm{ord},0}$ by first taking the quotient of the source and target by the p-torsion subgroup and K_0, respectively, and restrict the induced morphism (which has image in $A'_{\tilde{S}}$) to the p^r-torsion subgroup; such that $\alpha_{p^r}^{\mathrm{ord},\#,0,\prime}$ is the restriction of $\alpha_{p^{r'}}^{\mathrm{ord},\#,0}$ to the p^r-torsion subgroup (whose image in $A_{\tilde{S}}^\vee$ canonically lifts to a subgroup scheme of $A_{\tilde{S}}^{\prime,\vee}$ under the pullback to $\tilde{S}$ of the étale dual morphism $([g_p]^{\mathrm{ord}})^\vee : A^{\prime,\vee} \to A^\vee$ of $[g_p]^{\mathrm{ord}}$); and such that $\nu_{p^{r'}}^{\mathrm{ord},\prime}$ is induced by $\nu_{p^{r'}}^{\mathrm{ord}} \circ (p^{-1}\nu(g_p))$. The $H_{p^r}^{\mathrm{ord}}$-orbit of $\alpha_{p^r}^{\mathrm{ord},\prime}$ descends to S and agrees with $\alpha_{\mathcal{H}_p}^{\mathrm{ord},\prime}$.

Proposition 3.3.4.21. *Suppose that* $g = (g_0, g_p) \in \mathrm{G}(\mathbb{A}^{\infty,p}) \times \mathrm{P}_{\mathsf{D}}^{\mathrm{ord}}(\mathbb{Q}_p) \subset$

$\mathrm{G}(\mathbb{A}^\infty)$ *(see Definition* 3.2.2.7*), and that* $\mathcal{H}$ *and* $\mathcal{H}'$ *are two open compact subgroups of* $\mathrm{G}(\hat{\mathbb{Z}})$ *such that* $\mathcal{H}' \subset g\mathcal{H}g^{-1}$*, and such that* $\mathcal{H}$ *and* $\mathcal{H}'$ *are of standard form as in Definition* 3.2.2.9*. Suppose moreover that* g_p *satisfies the conditions given in Section* 3.3.4*.*

Let S *be a scheme over* $\mathrm{Spec}(\mathbb{Z}_{(p)})$*, let* A *be an abelian scheme over* S*, let* $\lambda : A \to A^\vee$ *be a polarization, let* $i : \mathcal{O} \hookrightarrow \mathrm{End}_S(A)$ *be an* $\mathcal{O}$*-endomorphism structure as in [62, Def. 1.3.3.1], let* $\alpha_{\mathcal{H}'^{,p}} = \{\alpha_{H'_{n_0}}\}_{n_0}$ *be a level-*$\mathcal{H}'^{,p}$ *structure of* (A, λ, i) *of type* $(L \underset{\mathbb{Z}}{\otimes} \hat{\mathbb{Z}}^p, \langle\,\cdot\,,\,\cdot\,\rangle)$ *as in Definition* 3.3.1.4*, and let* $\alpha^{\mathrm{ord}}_{\mathcal{H}'_p}$ *be an ordinary level-*$\mathcal{H}'_p$ *structure of* (A, λ, i) *of type* $(L \underset{\mathbb{Z}}{\otimes} \mathbb{Z}_p, \langle\,\cdot\,,\,\cdot\,\rangle, \mathtt{D})$ *as in Definition* 3.3.3.4*.*

Under these assumptions, the constructions in Propositions 3.3.2.1 *and* 3.3.4.9 *are both applicable and are compatible with each other.*

By Proposition 3.3.2.1*, the tuple* $(A, \lambda, i, \alpha_{\mathcal{H}'^{,p}})$ *admits a Hecke twist by* g_0*, which is a tuple* $(A'', \lambda'', i'', \alpha''_{\mathcal{H}^p})$ *equipped with a* $\mathbb{Z}^\times_{(p)}$*-isogeny* $[g_0^{-1}] : A \to A''$ *compatible with all other structures. Since* $[g_0^{-1}]$ *induces an isomorphism* $A[p^r] \xrightarrow{\sim} A''[p^r]$ *for each* $r \geq 0$*, we have a canonically induced ordinary level structure* $\alpha^{\mathrm{ord},''}_{\mathcal{H}'_p}$ *on* $(A'', \lambda'', i'', \alpha''_{\mathcal{H}^p})$*.*

By Proposition 3.3.4.9*, the tuple* $(A'', \lambda'', i'', \alpha^{\mathrm{ord},''}_{\mathcal{H}'_p})$ *admits an ordinary Hecke twists by* g_p*, which is a tuple* $(A', \lambda', i', \alpha^{\mathrm{ord},'}_{\mathcal{H}_p})$ *equipped with a* $\mathbb{Q}^\times$*-isogeny* $[g_p^{-1}]^{\mathrm{ord}} : A'' \to A'$*, which is the composition* $([p^{-r_0}g_{p,-1}]^{\mathrm{ord}})^{-1} \circ [p^{r_0}g_{p,0}^{-1}]^{\mathrm{ord}} \circ [p^{r_0}]^{-1}$ *of isogenies of* p*-power degrees or their formal inverses. Since* $[g_p^{-1}]^{\mathrm{ord}}$ *induces an isomorphism* $A''[n_0] \xrightarrow{\sim} A'[n_0]$ *for each integer* $n_0 \geq 1$ *such that* $p \nmid n_0$*, we have a canonically induced level structure* $\alpha'_{\mathcal{H}^p}$ *on* (A', λ', i')*.*

Thus, we have obtained a tuple $(A', \lambda', i', \alpha'_{\mathcal{H}^p}, \alpha^{\mathrm{ord},'}_{\mathcal{H}_p})$*, which we call the* **ordinary Hecke twist** *of* $(A, \lambda, i, \alpha_{\mathcal{H}'^{,p}}, \alpha^{\mathrm{ord}}_{\mathcal{H}'_p})$ *by* $g = (g_0, g_p)$*, which is equipped with a* $\mathbb{Q}^\times$*-isogeny* $[g^{-1}]^{\mathrm{ord}} : A \to A'$ *defined by the composition* $[g^{-1}]^{\mathrm{ord}} := [g_p^{-1}]^{\mathrm{ord}} \circ [g_0^{-1}]$*, whose formal inverse we denote by* $[g]^{\mathrm{ord}} : A' \to A$*. By construction, we have* $\lambda' = r([g]^{\mathrm{ord}})^\vee \circ \lambda \circ [g]^{\mathrm{ord}}$ *(as positive* $\mathbb{Q}^\times$*-isogenies), where* r *is the unique number in* $\mathbb{Q}^\times_{>0}$ *such that* $r\nu(g)\hat{\mathbb{Z}} = \hat{\mathbb{Z}}$*.*

If $g = g_1g_2$*, where* $g_1 = (g_{1,0}, g_{1,p})$ *and* $g_2 = (g_{2,0}, g_{2,p})$ *are elements of* $\mathrm{G}(\mathbb{A}^{\infty,p}) \times \mathrm{P}^{\mathrm{ord}}_{\mathtt{D}}(\mathbb{Q}_p)$*, each having a setup similar to that of* g*, then the ordinary Hecke twists by* g *can be constructed in two steps using ordinary Hecke twists by* g_1 *and* g_2*, such that* $[g^{-1}]^{\mathrm{ord}} = [g_2^{-1}]^{\mathrm{ord}} \circ [g_1^{-1}]^{\mathrm{ord}}$ *(or, equivalently,* $[g]^{\mathrm{ord}} = [g_1]^{\mathrm{ord}} \circ [g_2]^{\mathrm{ord}}$*).*

Proof. The statements are self-explanatory. (Since the constructions of isogenies in Propositions 3.3.2.1 and 3.3.4.9 are achieved by quotients by torsion subgroup schemes of prime-to-p and p-power ranks, respectively, and since the quotients by two torsion subgroup schemes of ranks relative prime to each other can be performed in any order, in order to construct the Hecke twists by $g = (g_0, g_p)$, we might as well form the ordinary Hecke twist by g_p first, and form the Hecke twist by g_0 second, so that $[g^{-1}]^{\mathrm{ord}} = [g_p^{-1}]^{\mathrm{ord}} \circ [g_0^{-1}] = [g_0^{-1}] \circ [g_p^{-1}]^{\mathrm{ord}}$, by abuse of notation.

For the same reason, the last statement of the proposition follows from the last statements of Propositions 3.3.2.1 and 3.3.4.9, because $[g^{-1}]^{\mathrm{ord}} = [g_p^{-1}]^{\mathrm{ord}} \circ [g_0^{-1}] = [g_{2,p}^{-1}]^{\mathrm{ord}} \circ [g_{1,p}^{-1}]^{\mathrm{ord}} \circ [g_{2,0}^{-1}] \circ [g_{1,0}^{-1}] = [g_{2,p}^{-1}]^{\mathrm{ord}} \circ [g_{2,0}^{-1}] \circ [g_{1,p}^{-1}]^{\mathrm{ord}} \circ [g_{1,0}^{-1}] = [g_2^{-1}]^{\mathrm{ord}} \circ [g_1^{-1}]^{\mathrm{ord}}$, by a similar abuse of notation.) □

3.3.5 *Comparison with Level Structures in Characteristic Zero*

Let $\mathcal{H} \subset \mathrm{G}(\hat{\mathbb{Z}})$ be of standard form (with respect to D) as in Definition 3.2.2.9, so that $\mathcal{H} = \mathcal{H}^p\mathcal{H}_p$, $\mathcal{U}_{p,1}^{\mathrm{bal}}(p^r) \subset \mathcal{H}_p \subset \mathcal{U}_{p,0}(p^r)$, and $\nu(\mathcal{H}_p) = \ker(\mathbb{Z}_p^\times \to (\mathbb{Z}/p^{r_\nu}\mathbb{Z})^\times)$ for some integer $r_\nu \leq r = \mathrm{depth}_{\mathsf{D}}(\mathcal{H})$. Over each scheme S over $\mathrm{Spec}(\mathbb{Q}[\zeta_{p^{r_\nu}}])$, there exists a canonical isomorphism $\zeta_{p^{r_\nu},S} : ((\mathbb{Z}/p^{r_\nu}\mathbb{Z})(1))_S \xrightarrow{\sim} \boldsymbol{\mu}_{p^{r_\nu},S}$ (which is the pullback of the canonical $\zeta_{p^{r_\nu}}$ over $\mathrm{Spec}(\mathbb{Q}[\zeta_{p^{r_\nu}}])$).

Proposition 3.3.5.1. *Let S be a scheme over $\mathsf{S}_{0,r_\nu} = \mathrm{Spec}(F_0[\zeta_{p^{r_\nu}}])$ (see Definition 2.2.3.3), and let $(A, \lambda, i, \alpha_{\mathcal{H}})$ be an object of $\mathsf{M}_{\mathcal{H}}(S)$ (see Section 1.1.2). Then the level-$\mathcal{H}$ structure $\alpha_{\mathcal{H}} = \{\alpha_{\mathcal{H}_n}\}_n$ (labeled by integers $n \geq 1$ such that $\mathcal{U}(n) \subset \mathcal{H}$) determines the following data:*

(1) *Since $\mathcal{U}_p(p^r) \subset \mathcal{U}_{p,1}^{\mathrm{bal}}(p^r) \subset \mathcal{H}_p$, for each integer $n_0 \geq 1$ such that $p \nmid n_0$ and $\mathcal{U}^p(n_0) \subset \mathcal{H}^p$, we have $\mathcal{U}(n_0p^r) \subset \mathcal{H}$. Let $\alpha_{H_{n_0}}$ be the schematic image of $\alpha_{H_{n_0p^r}}$ under the canonical (reduction modulo n_0) morphism*

$$\begin{aligned}&\underline{\mathrm{Isom}}_S((L/n_0p^rL)_S, A[n_0p^r]) \underset{S}{\times} \underline{\mathrm{Isom}}_S(((\mathbb{Z}/n_0p^r\mathbb{Z})(1))_S, \boldsymbol{\mu}_{n_0p^r,S})\\ &\to \underline{\mathrm{Isom}}_S((L/n_0L)_S, A[n_0]) \underset{S}{\times} \underline{\mathrm{Isom}}_S(((\mathbb{Z}/n_0\mathbb{Z})(1))_S, \boldsymbol{\mu}_{n_0,S}).\end{aligned} \tag{3.3.5.2}$$

Then the collection $\alpha_{\mathcal{H}^p} = \{\alpha_{H_{n_0}}\}_{n_0}$ labeled by integers $n_0 \geq 1$ such that $p \nmid n_0$ and $\mathcal{U}^p(n_0) \subset \mathcal{H}^p$ defines a level-$\mathcal{H}^p$ structure $\alpha_{\mathcal{H}^p}$ of (A, λ, i) of type $(L \underset{\mathbb{Z}}{\otimes} \hat{\mathbb{Z}}^p, \langle \cdot, \cdot \rangle)$ (see Definition 3.3.1.4).

(2) *Let $\alpha_{H_{p^r}}$ be the schematic image of $\alpha_{H_{n_0p^r}}$ under the canonical (reduction modulo p^r) morphism*

$$\begin{aligned}&\underline{\mathrm{Isom}}_S((L/n_0p^rL)_S, A[n_0p^r]) \underset{S}{\times} \underline{\mathrm{Isom}}_S(((\mathbb{Z}/n_0p^r\mathbb{Z})(1))_S, \boldsymbol{\mu}_{n_0p^r,S})\\ &\to \underline{\mathrm{Isom}}_S((L/p^rL)_S, A[p^r]) \underset{S}{\times} \underline{\mathrm{Isom}}_S(((\mathbb{Z}/p^r\mathbb{Z})(1))_S, \boldsymbol{\mu}_{p^r,S}).\end{aligned} \tag{3.3.5.3}$$

Over some finite flat covering $S' \to S$ of finite presentation, suppose that there exists some isomorphism $\zeta_{p^r,S'} : ((\mathbb{Z}/p^r\mathbb{Z})(1))_{S'} \xrightarrow{\sim} \boldsymbol{\mu}_{p^r,S'}$ lifting the pullback $\zeta_{p^{r_\nu},S'}$ of $\zeta_{p^{r_\nu},S} : ((\mathbb{Z}/p^{r_\nu}\mathbb{Z})(1))_S \xrightarrow{\sim} \boldsymbol{\mu}_{p^{r_\nu},S}$ to S'.
Consider the canonical morphism

$$\begin{aligned}&\underline{\mathrm{Isom}}_{S'}((L/p^rL)_{S'}, A_{S'}[p^r]) \underset{S'}{\times} \underline{\mathrm{Isom}}_{S'}(((\mathbb{Z}/p^r\mathbb{Z})(1))_{S'}, \boldsymbol{\mu}_{p^r,S'})\\ &\to \underline{\mathrm{Hom}}_{S'}((\mathrm{Gr}^0_{\mathsf{D},p^r})^{\mathrm{mult}}_{S'}, A_{S'}[p^r]) \underset{S'}{\times} \underline{\mathrm{Hom}}_{S'}((\mathrm{Gr}^0_{\mathsf{D}^\#,p^r})^{\mathrm{mult}}_{S'}, A^\vee_{S'}[p^r])\\ &\qquad \underset{S'}{\times} (\mathbb{Z}/p^r\mathbb{Z})^\times_{S'}\end{aligned} \tag{3.3.5.4}$$

over S' defined by sending

$$(\alpha_{p^r} : (L/p^r L)_{S'} \xrightarrow{\sim} A_{S'}[p^r], \nu_{p^r} : ((\mathbb{Z}/p^r\mathbb{Z})(1))_{S'} \xrightarrow{\sim} \boldsymbol{\mu}_{p^r,S'})$$

in the source to

$$\alpha_{p^r}^{\mathrm{ord}} = (\alpha_{p^r}^{\mathrm{ord},0}, \alpha_{p^r}^{\mathrm{ord},\#,0}, \nu_{p^r}^{\mathrm{ord}})$$

in the target, where:

(a) $\alpha_{p^r}^{\mathrm{ord},0} : (\mathrm{Gr}^0_{\mathtt{D},p^r})^{\mathrm{mult}}_{S'} \to A_{S'}[p^r]$ *is composition of* $\zeta_{p^r,S'}^{-1} : (\mathrm{Gr}^0_{\mathtt{D},p^r})^{\mathrm{mult}}_{S'} \xrightarrow{\sim} (\mathrm{Gr}^0_{\mathtt{D},p^r})_{S'}$ *with the restriction of* α_{p^r} *to* $(\mathrm{Gr}^0_{\mathtt{D},p^r})_{S'}$.

(b) $\alpha_{p^r}^{\mathrm{ord},\#,0} : (\mathrm{Gr}^0_{\mathtt{D}^\#,p^r})^{\mathrm{mult}}_{S'} \to A^\vee_{S'}[p^r]$ *is the restriction of* $\alpha_{p^r}^{\#,\mathrm{mult}}$ *to* $(\mathrm{Gr}^0_{\mathtt{D}^\#,p^r})^{\mathrm{mult}}_{S'}$, *where* $\alpha_{p^r}^{\#,\mathrm{mult}} : (L^\#/p^r L^\#)^{\mathrm{mult}}_{S'} \xrightarrow{\sim} A^\vee_{S'}[p^r]$ *is the inverse of the Cartier dual of* α_{p^r} *over* S'.

(c) $\nu_{p^r}^{\mathrm{ord}}$ *is a section of* $(\mathbb{Z}/p^r\mathbb{Z})^\times_{S'}$ *defined by the composition* $\zeta_{p^r,S'}^{-1} \circ \nu_{p^r} : (\mathbb{Z}/p^r\mathbb{Z})(1))_{S'} \xrightarrow{\sim} ((\mathbb{Z}/p^r\mathbb{Z})(1))_{S'}$.

Then the schematic image of $\alpha_{H_{p^r}} \underset{S}{\times} S'$ *under this canonical morphism is independent of the lifting* $\zeta_{p^r,S'}$ *of* $\zeta_{p^{r_\nu},S'}$ *over* S', *and defines by descent (see [33, VIII, 1.9, 1.11, 5.5]) a subscheme* $\alpha^{\mathrm{ord}}_{H^{\mathrm{ord}}_{p^r}}$ *of*

$$\underline{\mathrm{Hom}}_S((\mathrm{Gr}^0_{\mathtt{D},p^r})^{\mathrm{mult}}_S, A[p^r]) \underset{S}{\times} \underline{\mathrm{Hom}}_S((\mathrm{Gr}^0_{\mathtt{D}^\#,p^r})^{\mathrm{mult}}_S, A^\vee[p^r]) \underset{S}{\times} (\mathbb{Z}/p^r\mathbb{Z})^\times_S$$

that defines an ordinary level-$\mathcal{H}_p$ *structure* $\alpha^{\mathrm{ord}}_{\mathcal{H}_p}$ *of* (A, λ, i) *of type* $(L \underset{\mathbb{Z}}{\otimes} \mathbb{Z}_p, \langle\,\cdot\,,\,\cdot\,\rangle, \mathtt{D})$ *(see Definition* 3.3.3.4*).*

This assignment

$$\alpha_{\mathcal{H}} \mapsto (\alpha_{\mathcal{H}^p}, \alpha^{\mathrm{ord}}_{\mathcal{H}_p}) \tag{3.3.5.5}$$

induces an injection from the set of level-$\mathcal{H}$ *structures of* (A, λ, i) *of type* $(L \underset{\mathbb{Z}}{\otimes} \hat{\mathbb{Z}}, \langle\,\cdot\,,\,\cdot\,\rangle)$ *as in [62, Def. 1.3.7.6] to the set of pairs* $(\alpha_{\mathcal{H}^p}, \alpha^{\mathrm{ord}}_{\mathcal{H}_p})$, *where* $\alpha_{\mathcal{H}^p}$ *is a level-*$\mathcal{H}^p$ *structure* $\alpha_{\mathcal{H}^p}$ *of* (A, λ, i) *of type* $(L \underset{\mathbb{Z}}{\otimes} \hat{\mathbb{Z}}^p, \langle\,\cdot\,,\,\cdot\,\rangle)$ *(see Definition* 3.3.1.4*), and where* $\alpha^{\mathrm{ord}}_{\mathcal{H}_p}$ *is an ordinary level-*$\mathcal{H}_p$ *structure* $\alpha^{\mathrm{ord}}_{\mathcal{H}_p}$ *of* (A, λ, i) *of type* $(L \underset{\mathbb{Z}}{\otimes} \mathbb{Z}_p, \langle\,\cdot\,,\,\cdot\,\rangle, \mathtt{D})$ *(see Definition* 3.3.3.4*).*

Suppose $\hat{\alpha}_{\bar{s}} : L \underset{\mathbb{Z}}{\otimes} \hat{\mathbb{Z}} \xrightarrow{\sim} \mathrm{T}\, A_{\bar{s}}$ *is a lifting of* $\alpha_{\mathcal{H}}$ *as in Lemma* 3.3.1.6 *(or rather in [62, Sec. 1.3.7]). Then* $\hat{\alpha}_{\bar{s}}$ *induces an* $\mathcal{O} \underset{\mathbb{Z}}{\otimes} \mathbb{A}^{\infty,p}$*-equivariant isomorphism* $\hat{\alpha}^p_{\bar{s}} : L \underset{\mathbb{Z}}{\otimes} \hat{\mathbb{Z}}^p \xrightarrow{\sim} \mathrm{T}^p A_{\bar{s}}$, *which is a lifting of* $\alpha_{\mathcal{H}^p}$ *as in Lemma* 3.3.1.6, *because the assignment of* $\alpha_{\mathcal{H}^p}$ *to* $\alpha_{\mathcal{H}}$ *in* (1) *above is defined by the canonical morphisms as in* (3.3.5.2)*. Similarly,* $\hat{\alpha}_{\bar{s}}$ *induces an* $\mathcal{O} \underset{\mathbb{Z}}{\otimes} \mathbb{Z}_p$*-equivariant isomorphism* $\hat{\alpha}_{\bar{s},p} : L \underset{\mathbb{Z}}{\otimes} \mathbb{Z}_p \xrightarrow{\sim} \mathrm{T}_p A_{\bar{s}}$, *which induces by taking graded pieces and by duality a triple* $\hat{\alpha}^{\mathrm{ord}}_{\bar{s}} = (\hat{\alpha}^{\mathrm{ord},0}_{\bar{s}}, \hat{\alpha}^{\mathrm{ord},\#,0}_{\bar{s}}, \hat{\nu}^{\mathrm{ord}}_{\bar{s}})$, *which is a lifting of* $\alpha^{\mathrm{ord}}_{\mathcal{H}_p}$ *as in Lemma* 3.3.3.9, *because the assignment of* $\alpha^{\mathrm{ord}}_{\mathcal{H}_p}$ *to* $\alpha_{\mathcal{H}}$ *in* (2) *above is defined by the canonical morphisms as in* (3.3.5.3) *and* (3.3.5.4)*.*

Proof. It is clear that the assignments define $\alpha_{\mathcal{H}^p}$ and $\alpha^{\mathrm{ord}}_{\mathcal{H}_p}$ as naive level structures. The symplectic-liftability conditions are verified because $\alpha_{\mathcal{H}}$ itself is symplectic-liftable, and its symplectic liftings to higher levels induce the desired symplectic liftings of $\alpha_{\mathcal{H}^p}$ and $\alpha^{\mathrm{ord}}_{\mathcal{H}_p}$.

As for the injectivity of (3.3.5.5), first note that if two geometric points of

$$\underline{\mathrm{Isom}}_S((L/n_0p^rL)_S, A[n_0p^r]) \underset{S}{\times} \underline{\mathrm{Isom}}_S(((\mathbb{Z}/n_0p^r\mathbb{Z})(1))_S, \boldsymbol{\mu}_{n_0p^r,S})$$

have the same images under both the morphisms (3.3.5.2) and (3.3.5.3), then they must be the same. Second note that if two geometric points of

$$\underline{\mathrm{Isom}}_{S'}((L/p^rL)_{S'}, A_{S'}[p^r]) \underset{S'}{\times} \underline{\mathrm{Isom}}_{S'}(((\mathbb{Z}/p^r\mathbb{Z})(1))_{S'}, \boldsymbol{\mu}_{p^r,S'})$$

are mapped to the same point under the morphism (3.3.5.4), then they are in the same $\mathcal{U}_{p,0}(p^r)$-orbit. Since $H^{\mathrm{ord}}_{p^r} = \mathcal{H}_p/\mathcal{U}^{\mathrm{bal}}_{p,1}(p^r)$ by definition (see the beginning of Section 3.3.3), the injectivity of (3.3.5.5) follows.

The last paragraph of the proposition is self-explanatory. □

Proposition 3.3.5.6. *Let g, $\mathcal{H}$, and $\mathcal{H}'$ be as in Proposition* 3.3.4.21. *Suppose $\nu(\mathcal{H}'_p) = \ker(\mathbb{Z}_p^\times \to (\mathbb{Z}/p^{r_{\nu'}}\mathbb{Z})^\times)$. (Then $r_{\nu'} \geq r_\nu$ because $\mathcal{H}'_p \subset \mathcal{H}_p$.) Let S be a scheme over $\mathsf{S}_{0,r_{\nu'}} = \mathrm{Spec}(F_0[\zeta_{p^{r_{\nu'}}}])$, and let $(A, \lambda, i, \alpha_{\mathcal{H}'})$ be an object of $\mathsf{M}_{\mathcal{H}'}(S)$. Let $\alpha_{\mathcal{H}',p}$ and $\alpha^{\mathrm{ord}}_{\mathcal{H}'_p}$ be determined by $\alpha_{\mathcal{H}'}$ as in Proposition* 3.3.5.1. *Let $(A', \lambda', i', \alpha'_{\mathcal{H}})$ be the Hecke twist of $(A, \lambda, i, \alpha_{\mathcal{H}'})$ by g as in [62, Sec. 6.4.3], equipped with a $\mathbb{Q}^\times$-isogeny $[g] : A \to A'$, and let $\alpha'_{\mathcal{H}^p}$ and $\alpha^{\mathrm{ord},\prime}_{\mathcal{H}_p}$ be determined by $\alpha'_{\mathcal{H}}$ as in Proposition* 3.3.5.1. *Let $(A'', \lambda'', i'', \alpha''_{\mathcal{H}^p}, \alpha^{\mathrm{ord},\prime\prime}_{\mathcal{H}_p})$ be the ordinary Hecke twist of $(A, \lambda, i, \alpha_{\mathcal{H}',p}, \alpha^{\mathrm{ord}}_{\mathcal{H}'_p})$ by g as in Proposition* 3.3.4.21. *Then there is a canonical isomorphism between $(A', \lambda', i', \alpha'_{\mathcal{H}^p}, \alpha^{\mathrm{ord},\prime}_{\mathcal{H}_p})$ and $(A'', \lambda'', i'', \alpha''_{\mathcal{H}^p}, \alpha^{\mathrm{ord},\prime\prime}_{\mathcal{H}_p})$ matching $[g^{-1}] : A \to A'$ and $[g^{-1}]^{\mathrm{ord}} : A \to A''$.*

Proof. Since we are in characteristic zero, all kernels of isogenies involved are finite étale group schemes, and all level structures are defined by homomorphisms between étale group schemes. Hence, the statements of this proposition can be verified after pulled back to geometric points of S. At each geometric point $\bar{s}$ of S, the validity of the corresponding statements follows from the last paragraph of Proposition 3.3.5.1, and from the descriptions of the effects of Hecke twists over geometric points in (4) of Proposition 3.3.2.1 and (5) of Proposition 3.3.4.9 (and in the analogue of (4) of Proposition 3.3.2.1 for usual Hecke twists defined by g in characteristic zero). □

3.3.6 *Valuative Criteria*

Definition 3.3.6.1. *We say an element $g_p \in \mathrm{P}^{\mathrm{ord}}_{\mathsf{D}}(\mathbb{Q}_p)$ is* **of** U_p **type** *if it acts on $\mathrm{Gr}_{\mathsf{D}_{\mathbb{Q}_p}} \cong \mathrm{Gr}^0_{\mathsf{D}_{\mathbb{Q}_p}} \oplus \mathrm{Gr}^{-1}_{\mathsf{D}_{\mathbb{Q}_p}}$ by $g^{\mathrm{ord}}_p = (g_{p,0}, g_{p,-1}) = (p^{-a}\,\mathrm{Id}_{\mathrm{Gr}^0_{\mathsf{D}}}, \mathrm{Id}_{\mathrm{Gr}^{-1}_{\mathsf{D}}})$ for some integer $a > 0$, and if for one (and hence every) splitting of the filtration* (3.2.2.3) *(resp.* (3.2.2.5)*), g_p (resp. $g_{p,\#} = {}^t g_p^{-1}$) maps the image of $\mathrm{Gr}^{-1}_{\mathsf{D}}$ (resp. $\mathrm{Gr}^{-1}_{\mathsf{D}^\#}$) to $L \underset{\mathbb{Z}}{\otimes} \mathbb{Z}_p$ (resp. $L^\# \underset{\mathbb{Z}}{\otimes} \mathbb{Z}_p$).*

We say an element $g_p \in \mathrm{P}^{\mathrm{ord}}_{\mathsf{D}}(\mathbb{Q}_p)$ *is* **of twisted** U_p **type** *if it acts on* $\mathrm{Gr}_{\mathsf{D}_{\mathbb{Q}_p}} \cong \mathrm{Gr}^0_{\mathsf{D}_{\mathbb{Q}_p}} \oplus \mathrm{Gr}^{-1}_{\mathsf{D}_{\mathbb{Q}_p}}$ *by* $g_p^{\mathrm{ord}} = (g_{p,0}, g_{p,-1})$ *such that* $g_{p,0}(\mathrm{Gr}^0_{\mathsf{D}}) = p^{-a}\,\mathrm{Gr}^0_{\mathsf{D}}$ *and* $g_{p,-1}(\mathrm{Gr}^{-1}_{\mathsf{D}}) = p^{-b}\,\mathrm{Gr}^{-1}_{\mathsf{D}}$, *so that* $g_{p,\#,0}(\mathrm{Gr}^0_{\mathsf{D}\#}) = {}^t g_{p,-1}^{-1}(\mathrm{Gr}^0_{\mathsf{D}\#}) = p^b\,\mathrm{Gr}^0_{\mathsf{D}\#}$, *for some integers* $a \geq b$. *In this case, we define* $\mathrm{depth}_{\mathsf{D}}(g_p) := a - b$.

Remark 3.3.6.2. Suppose $g_p \in \mathrm{P}^{\mathrm{ord}}_{\mathsf{D}}(\mathbb{Q}_p)$ is of U_p type, then $p^{\mathbb{Z}}\, g_p^{\mathbb{Z}_{\geq 0}}\, \mathrm{P}^{\mathrm{ord}}_{\mathsf{D}}(\mathbb{Z}_p)$ is a subsemigroup of $\mathrm{P}^{\mathrm{ord}}_{\mathsf{D}}(\mathbb{Q}_p)$, whose elements are all of twisted U_p type. (Thus, the elements of twisted U_p type includes Examples 3.3.4.5, 3.3.4.7, and 3.3.4.7 as special cases.)

Remark 3.3.6.3. Suppose that $\mathcal{H}_p$, $\mathcal{H}'_p$, and $g_p \in \mathrm{P}^{\mathrm{ord}}_{\mathsf{D}}(\mathbb{Q}_p)$ satisfies the conditions given in Section 3.3.4, with $r := \mathrm{depth}_{\mathsf{D}}(\mathcal{H}_p)$ and $r' := \mathrm{depth}_{\mathsf{D}}(\mathcal{H}'_p)$, and that g_p is of twisted U_p type as in Definition 3.3.6.1, such that $r' - \mathrm{depth}_{\mathsf{D}}(g_p) = r$. Then, by setting $r_0 := -b$ as in Definition 3.3.6.1 and $r'' := r$, we have

$$\mathrm{Gr}^0_{\mathsf{D}} \subset p^{-r_0} g_{p,0}(\mathrm{Gr}^0_{\mathsf{D}}) \subset p^{-r''-r_0} g_{p,0}(\mathrm{Gr}^0_{\mathsf{D}}) = p^{-r'}\,\mathrm{Gr}^0_{\mathsf{D}} \tag{3.3.6.4}$$

and

$$\mathrm{Gr}^0_{\mathsf{D}\#} = p^{r_0} g_{p,\#,0}(\mathrm{Gr}^0_{\mathsf{D}\#}) \subset p^{-r+r_0} g_{p,\#,0}(\mathrm{Gr}^0_{\mathsf{D}\#}) = p^{-r''}\,\mathrm{Gr}^0_{\mathsf{D}\#} \tag{3.3.6.5}$$

(*cf.* (3.3.4.1) and (3.3.4.2)).

Lemma 3.3.6.6. *Suppose that* $\mathcal{H}_p$, $\mathcal{H}'_p$, $g_p \in \mathrm{P}^{\mathrm{ord}}_{\mathsf{D}}(\mathbb{Q}_p)$, r_0, r, r', r'' *are as in Remark* 3.3.6.3, *and that* $r > 0$. *Suppose* R *is a discrete valuation ring over* $\mathbb{Z}_{(p)}$, *with fraction field* $\mathrm{Frac}(R)$ *and residue field* k *of characteristic* $p > 0$. *Let* $\vec{S} := \mathrm{Spec}(R)$ *and* $S := \mathrm{Spec}(\mathrm{Frac}(R))$. *Suppose that* $(A, \lambda, i, \alpha^{\mathrm{ord}}_{\mathcal{H}'_p}) = (\alpha^{\mathrm{ord},0}_{\mathcal{H}'_p}, \alpha^{\mathrm{ord},\#,0}_{\mathcal{H}'_p}, \nu^{\mathrm{ord}}_{\mathcal{H}'_p})$ *is defined over* S *as in Proposition* 3.3.4.9, *so that the ordinary Hecke twist* $(A', \lambda', i', \alpha^{\mathrm{ord},\prime}_{\mathcal{H}_p} = (\alpha^{\mathrm{ord},0,\prime}_{\mathcal{H}_p}, \alpha^{\mathrm{ord},\#,0,\prime}_{\mathcal{H}_p}, \nu^{\mathrm{ord},\prime}_{\mathcal{H}_p}))$ *of* $(A, \lambda, i, \alpha^{\mathrm{ord}}_{\mathcal{H}'_p})$ *by* g_p *is defined, equipped with a* $\mathbb{Q}^\times$*-isogeny* $[g_p^{-1}]^{\mathrm{ord}} : A \to A'$ *over* S.

Moreover, suppose that A' *extends to a semi-abelian scheme* $\vec{A}'$ *over* $\vec{S}$, *so that* $\vec{A}'^{,\vee}$ *is defined (as in [62, Thm. 3.4.3.2]), and so that* λ' *and* i' *also (uniquely) extend to some* $\vec{\lambda}' : \vec{A}' \to \vec{A}'^{,\vee}$ *and* $\vec{i}' : \mathcal{O} \to \mathrm{End}_{\vec{S}}(\vec{A}')$ *by [92, IX, 1.4], [28, Ch. I, Prop. 2.7], or [62, Prop. 3.3.1.5]; and suppose that* $\mathrm{image}(\alpha^{\mathrm{ord},0,\prime}_{\mathcal{H}_p})$ *(resp.* $\mathrm{image}(\alpha^{\mathrm{ord},\#,0,\prime}_{\mathcal{H}_p})$*) extends to a finite flat subgroup scheme* K' *(resp.* $K^{\#,\prime}$*) of multiplicative type of* $\vec{A}'[p^r]$ *(resp.* $\vec{A}'^{,\vee}[p^r]$*), so that* $\alpha^{\mathrm{ord},\prime}_{\mathcal{H}_p}$ *(uniquely) extends to some* $\vec{\alpha}^{\mathrm{ord},\prime}_{\mathcal{H}_p} = (\vec{\alpha}^{\mathrm{ord},0,\prime}_{\mathcal{H}_p}, \vec{\alpha}^{\mathrm{ord},\#,0,\prime}_{\mathcal{H}_p}, \vec{\nu}^{\mathrm{ord},\prime}_{\mathcal{H}_p})$, *which is étale locally over* S *an* $\mathcal{H}_p/\mathcal{U}^{\mathrm{bal}}_{p,1}(p^r)$*-orbit of some* $\vec{\alpha}^{\mathrm{ord},\prime}_{p^r} = (\vec{\alpha}^{\mathrm{ord},0,\prime}_{p^r}, \vec{\alpha}^{\mathrm{ord},\#,0,\prime}_{p^r}, \vec{\nu}^{\mathrm{ord},\prime}_{p^r})$, *where* $\vec{\alpha}^{\mathrm{ord},0,\prime}_{p^r} : (\mathrm{Gr}^0_{\mathsf{D},p^r})^{\mathrm{mult}}_{\vec{S}} \to \vec{A}'[p^r]$, $\vec{\alpha}^{\mathrm{ord},\#,0,\prime}_{\mathcal{H}_p} : (\mathrm{Gr}^0_{\mathsf{D}\#,p^r})^{\mathrm{mult}}_{\vec{S}} \to \vec{A}'^{,\vee}[p^r]$, *and* $\vec{\nu}^{\mathrm{ord},\prime}_{p^r} \in (\mathbb{Z}/p^r\mathbb{Z})^\times_{\vec{S}}$ *satisfy the same compatibility conditions as in Definitions* 3.3.3.1, 3.3.3.2, 3.3.3.3, *and* 3.3.3.4*; and so that* $\vec{A} \to \vec{S}$ *is ordinary, by Remark* 3.3.3.6.

Then $(A, \lambda, i, \alpha^{\mathrm{ord}}_{\mathcal{H}'_p})$ *also extends to an analogous tuple* $(\vec{A}, \vec{\lambda}, \vec{i}, \vec{\alpha}^{\mathrm{ord}}_{\mathcal{H}'_p})$ *over* $\vec{S}$, *equipped with a* $\mathbb{Q}^\times$*-isogeny* $[g_p^{-1}]^{\mathrm{ord}} : \vec{A} \to \vec{A}'$, *extending* $[g_p^{-1}]^{\mathrm{ord}} : A \to A'$, *between ordinary semi-abelian schemes. If* $\vec{A}'$ *is an abelian scheme, then* $\vec{A}$ *is* *also an abelian scheme;* $\vec{\alpha}^{\mathrm{ord},\prime}_{\mathcal{H}_p}$ *is an ordinary level-*$\mathcal{H}'_p$ *structure of* $(\vec{A},\vec{\lambda},\vec{i})$ *of type* $(L \underset{\mathbb{Z}}{\otimes} \mathbb{Z}_p, \langle\,\cdot\,,\,\cdot\,\rangle, \mathtt{D})$ *as in Definition* 3.3.3.4*; and* $(\vec{A}',\vec{\lambda}',\vec{i}',\vec{\alpha}^{\mathrm{ord},\prime}_{\mathcal{H}_p})$ *is the ordinary Hecke twist of* $(\vec{A},\vec{\lambda},\vec{i},\vec{\alpha}^{\mathrm{ord}}_{\mathcal{H}'_p})$ *by* g_p *as in Proposition* 3.3.4.9.

Proof. By Lemma 3.1.3.1, sufficiently divisible multiples of the formal inverse $[g_p]^{\mathrm{ord}} : A' \to A$ of $[g_p^{-1}]^{\mathrm{ord}}$ extends to sufficiently divisible multiples of the formal inverse $[g_p]^{\mathrm{ord}} : \vec{A}' \to \vec{A}$ of $[g_p^{-1}]^{\mathrm{ord}}$, where $\vec{A}$ is determined by any such extensions by [92, IX, 1.4], [28, Ch. I, Prop. 2.7], or [62, Prop. 3.3.1.5]; and λ and i also (uniquely) extends to some $\vec{\lambda} : \vec{A} \to \vec{A}^\vee$ and $\vec{i} : \mathcal{O} \to \mathrm{End}_{\vec{S}}(\vec{A})$ (by the same references). Moreover, by Lemma 3.1.1.5, $\vec{A} \to \vec{S}$ is ordinary because $\vec{A}' \to \vec{S}$ is.

Hence, the question is whether the schematic closure K (resp. $K^\#$) of $\mathrm{image}(\alpha^{\mathrm{ord},0}_{H^{\mathrm{ord}}_{p^r}})$ (resp. $\mathrm{image}(\alpha^{\mathrm{ord},0}_{H^{\mathrm{ord}}_{p^r}})$) in $\vec{A}$ (resp. $\vec{A}^\vee$), which is quasi-finite flat and of étale-multiplicative type over $\vec{S}$ because $\vec{A} \to \vec{S}$ is ordinary, is of multiplicative type. (The rest of the lemma will be self-explanatory.)

By (3.3.6.4) and (3.3.6.5) in Remark 3.3.6.3, using the crucial condition that $r' - \mathrm{depth}_{\mathtt{D}}(g_p) = r$, we have the following exact sequences of finite flat group schemes

$$0 \to p^r\mathrm{image}(\alpha^{\mathrm{ord},0}_{\mathcal{H}'_p}) \to \mathrm{image}(\alpha^{\mathrm{ord},0}_{\mathcal{H}'_p}) \to \mathrm{image}(\alpha^{\mathrm{ord},0,\prime}_{\mathcal{H}_p}) \to 0$$

and

$$0 \to p^r\mathrm{image}(\alpha^{\mathrm{ord},\#,0}_{\mathcal{H}'_p}) \to \mathrm{image}(\alpha^{\mathrm{ord},\#,0}_{\mathcal{H}'_p}) \to \mathrm{image}(\alpha^{\mathrm{ord},\#,0,\prime}_{\mathcal{H}_p})$$

of multiplicative type over S. By definition of the ordinary level structures, both are étale locally (after forgetting their $\mathcal{O}$-module structures) filtered by subobjects whose graded pieces are isomorphic to

$$0 \to (p^r\mathbb{Z}/p^{r'}\mathbb{Z})^{\mathrm{mult}}_S \to (\mathbb{Z}/p^{r'}\mathbb{Z})^{\mathrm{mult}}_S \to (\mathbb{Z}/p^r\mathbb{Z})^{\mathrm{mult}}_S.$$

Their closures in $\vec{A}$ and $\vec{A}^\vee$, respectively, correspond to the exact sequences

$$0 \to p^rK \to K \to K' \to 0$$

and

$$0 \to p^rK^\# \to K^\# \to K^{\#,\prime} \to 0.$$

By taking normalizations over the étale cover, both are fppf locally (after forgetting their $\mathcal{O}$-module structures) filtered by subobjects whose graded pieces are exact sequences of the form

$$0 \to p^rC \to C \to (\mathbb{Z}/p^r\mathbb{Z})^{\mathrm{mult}}_{\vec{S}} \to 0,$$

where C is cyclic of order $p^{r'}$, and where $r > 0$ by assumption. Hence, C must also be of multiplicative type, by the same argument in the proof of [49, Thm. 6.7.11(2)]; and so are K and $K^\#$, as desired. □

Remark 3.3.6.7. The proof of Lemma 3.3.6.6 shows that, in order to have similar valuative criteria for more general elements $g_p \in \mathrm{P}_{\mathtt{D}}^{\mathrm{ord}}(\mathbb{Q}_p)$, such as those generalized U_p operators in Example 3.3.4.8 (which are, in general, not of twisted U_p type as in Definition 3.3.6.1), we should consider not just the group $\mathcal{U}_{p,0}(p^r)$ stabilizing a maximal totally isotropic subgroup $\mathtt{D}_{p^r}^0$ of $L \underset{\mathbb{Z}}{\otimes} (\mathbb{Z}/p^r)$ (which can be called the *canonical subgroup*), but also a sequence of isotropic subgroups $\mathtt{D}_{p^{r_i}}^i \subset L \underset{\mathbb{Z}}{\otimes} (\mathbb{Z}/p^{r_i})$ such that $r_i \geq r_j$ and $\mathtt{D}_{p^{r_i}}^j = (\mathtt{D}_{p^{r_j}}^j / p^{r_i} \mathtt{D}_{p^{r_j}}^j) \subset \mathtt{D}_{p^{r_i}}^i$ whenever $j \geq i$ (which can be considered a sequence of *partial canonical subgroups of increasing depth*). This is technically possible, but we omit its treatment because it introduces complications that we do not immediate need (in an already lengthy work).

Lemma 3.3.6.8. *In Lemma* 3.3.6.6, *if the characteristic of* $\mathrm{Frac}(R)$ *is* $p > 0$, *then the same conclusion holds without the assumptions that* $r' - \mathrm{depth}_{\mathtt{D}}(g_p) = r$ *in Remark* 3.3.6.3 *and that* $r > 0$ *in Lemma* 3.3.6.6.

Proof. By Lemma 3.3.3.7, since $\vec{S}$ is a scheme over $\mathrm{Spec}(\mathbb{F}_p)$, we have the identities $\mathrm{image}(\alpha_{\mathcal{H}'_p}^{\mathrm{ord},0}) = \ker(\mathrm{F}_{A/S}^{(r')})$ and $\mathrm{image}(\alpha_{\mathcal{H}'_p}^{\mathrm{ord},\#,0}) = \ker(\mathrm{F}_{A^\vee/S}^{(r')})$ over S, which necessarily extends to the identities $K = \ker(\mathrm{F}_{\vec{A}/\vec{S}}^{(r')})$ and $K^\# = \ker(\mathrm{F}_{\vec{A}^\vee/\vec{S}}^{(r')})$ over $\vec{S}$, where K and $K^\#$ are as in the proof of Lemma 3.3.6.6. Hence, K and $K^\#$ are of multiplicative type without the assumption that $r' - \mathrm{depth}_{\mathtt{D}}(g_p) = r > 0$. $\square$

3.4 Ordinary Loci

From now on, we fix the choice of $\mathtt{D}$ as in Lemma 3.2.2.1 that satisfies Assumption 3.2.2.10.

3.4.1 *Naive Moduli Problems with Ordinary Level Structures*

Let $\mathcal{H} \subset \mathrm{G}(\hat{\mathbb{Z}})$ be of standard form as in Definition 3.2.2.9, so that $\mathcal{H} = \mathcal{H}^p \mathcal{H}_p$ and $\mathcal{U}_{p,1}^{\mathrm{bal}}(p^r) \subset \mathcal{H}_p \subset \mathcal{U}_{p,0}(p^r)$ for $r = \mathrm{depth}_{\mathtt{D}}(\mathcal{H})$. Let $H_{p^r}^{\mathrm{ord}} := \mathcal{H}_p / \mathcal{U}_{p,1}^{\mathrm{bal}}(p^r)$ and let $\mathcal{H}_p^{\mathrm{ord}}$ be defined as in (3.3.3.5).

Definition 3.4.1.1. *Let* $\mathcal{H}$, $\mathcal{H}^p$, $\mathcal{H}_p$, r, *and* $\mathcal{H}_p^{\mathrm{ord}}$ *be as above. The moduli problem* $\dddot{\mathsf{M}}_{\mathcal{H}}^{\mathrm{ord}}$ *is defined as the category fibered in groupoids over* $(\mathrm{Sch}/\mathrm{Spec}(\mathbb{Z}_{(p)}))$ *whose fiber over each scheme* S *is the groupoid* $\dddot{\mathsf{M}}_{\mathcal{H}}^{\mathrm{ord}}(S)$ *described as follows: The objects of* $\dddot{\mathsf{M}}_{\mathcal{H}}^{\mathrm{ord}}(S)$ *are tuples* $(A, \lambda, i, \alpha_{\mathcal{H}^p}, \alpha_{\mathcal{H}_p}^{\mathrm{ord}})$, *where:*

(1) A *is an abelian scheme over* S.
(2) $\lambda : A \to A^\vee$ *is a polarization.*
(3) $i : \mathcal{O} \hookrightarrow \mathrm{End}_S(A)$ *is an* $\mathcal{O}$*-endomorphism structure as in [62, Def. 1.3.3.1].*
(4) $\alpha_{\mathcal{H}^p}$ *is a* **level-$\mathcal{H}^p$ structure** *of* (A, λ, i) *of type* $(L \underset{\mathbb{Z}}{\otimes} \hat{\mathbb{Z}}^p, \langle \cdot , \cdot \rangle)$ *(see Definition* 3.3.1.4*).*

(5) $\alpha^{\mathrm{ord}}_{\mathcal{H}_p}$ *is an* **ordinary level-$\mathcal{H}_p$ structure** *of* (A, λ, i) *of type* $(L \underset{\mathbb{Z}}{\otimes} \mathbb{Z}_p, \langle \cdot, \cdot \rangle, \mathtt{D})$ *(see Definition* 3.3.3.4*). (This forces A to be* **ordinary**. *See Remark* 3.3.3.6.)

The morphisms of $\ddot{\mathsf{M}}^{\mathrm{ord}}_{\mathcal{H}}(S)$ *are naive isomorphisms (between the abelian schemes, matching all additional structures).*

If $\mathcal{H}^p = \mathcal{U}^p(n_0)$ *and* $\mathcal{H}_p = \mathcal{U}^{\mathrm{bal}}_{p,1}(p^r)$*, then we denote* $\ddot{\mathsf{M}}^{\mathrm{ord}}_{\mathcal{H}}$ *by* $\ddot{\mathsf{M}}^{\mathrm{ord}}_{n_0 p^r} = \ddot{\mathsf{M}}^{\mathrm{ord}}_{n}$*, and we denote by* α_{n_0} *(resp.* $\alpha^{\mathrm{ord}}_{p^r}$*) the unique principal level-n_0 structure (resp. principal ordinary level-p^r structure) determined by* $\alpha_{\mathcal{H}^p}$ *(resp.* $\alpha^{\mathrm{ord}}_{\mathcal{H}_p}$*).*

As always, the symplectic isomorphisms carry the additional data of isomorphisms between values of pairings, which can be called the similitudes of the symplectic isomorphisms.

If p is a good prime, then an argument similar to that in [62, Ch. 2] shows that the moduli problem $\ddot{\mathsf{M}}^{\mathrm{ord}}_{\mathcal{H}}$ is an algebraic stack separated and of finite type over $\mathrm{Spec}(\mathbb{Z}_{(p)})$. (See the proof of Theorem 3.4.1.9 below.) However, the argument there used the crucial technical result [62, Thm. B.3.11] (due to Artin) to suppress the technical condition [62, Sec. 2.3.4, Cond. 4] in the verification of Artin's criterion, which requires the infinitesimal deformation rings to be noetherian and normal. As we will see below, when p is not a good prime, the infinitesimal deformation rings might not even be geometrically unibranched (and hence not geometrically normal either), a situation to which no variant of [62, Thm. B.3.11] along the lines of [3, Thm. 3.9] seems to apply. To circumvent this difficulty, we shall again introduce some auxiliary moduli problems. (For later references, in this subsection, we will develop more about these auxiliary moduli problems than we need for the proof of representability.)

Construction 3.4.1.2. Let $(\mathcal{O}_{\mathrm{aux}}, \star_{\mathrm{aux}}, L_{\mathrm{aux}}, \langle \cdot, \cdot \rangle_{\mathrm{aux}}, h_{0,\mathrm{aux}})$ be chosen as in the paragraph preceding Lemma 2.1.1.9, where $(L_{\mathrm{aux}} = L^{\oplus a_1} \oplus (L^{\#})^{\oplus a_2}, \langle \cdot, \cdot \rangle_{\mathrm{aux}})$ is as in Lemma 2.1.1.1. Then the filtration $\mathtt{D}$ on $L \underset{\mathbb{Z}}{\otimes} \mathbb{Z}_p$ induces a filtration

$$\mathtt{D}^1_{\mathrm{aux}} = 0 \subset \mathtt{D}^0_{\mathrm{aux}} := (\mathtt{D}^0)^{\oplus a_1} \oplus (\mathtt{D}^{\#,0})^{\oplus a_2} \subset \mathtt{D}^{-1}_{\mathrm{aux}} = L_{\mathrm{aux}} \underset{\mathbb{Z}}{\otimes} \mathbb{Z}_p.$$

Since $\mathtt{D}^0_{\mathrm{aux},\mathbb{Q}_p} := \mathtt{D}^0_{\mathrm{aux}} \underset{\mathbb{Z}_p}{\otimes} \mathbb{Q}_p$ is mapped to $(\mathtt{D}^0)^{\oplus(a_1+a_2)} \underset{\mathbb{Z}_p}{\otimes} \mathbb{Q}_p$ under the canonical isomorphism $L_{\mathrm{aux}} \underset{\mathbb{Z}}{\otimes} \mathbb{Q}_p \cong L^{\oplus(a_1+a_2)} \underset{\mathbb{Z}}{\otimes} \mathbb{Q}_p$, its submodule $\mathtt{D}^0_{\mathrm{aux}} = \mathtt{D}^0_{\mathrm{aux},\mathbb{Q}_p} \cap (L_{\mathrm{aux}} \underset{\mathbb{Z}}{\otimes} \mathbb{Z}_p)$ is maximal totally isotropic under the pairing on $L_{\mathrm{aux}} \underset{\mathbb{Z}}{\otimes} \mathbb{Z}_p$ induced by $\langle \cdot, \cdot \rangle_{\mathrm{aux}}$ (*cf.* Lemma 3.2.2.1). Since $\langle \cdot, \cdot \rangle_{\mathrm{aux}}$ is self-dual at p, the pairing $\langle \cdot, \cdot \rangle_{\mathrm{aux}}$ induces an isomorphism $L_{\mathrm{aux}} \underset{\mathbb{Z}}{\otimes} \mathbb{Z}_p \overset{\sim}{\to} L^{\#}_{\mathrm{aux}} \underset{\mathbb{Z}}{\otimes} \mathbb{Z}_p$ matching $\mathtt{D}_{\mathrm{aux}}$ with $\mathtt{D}^{\#}_{\mathrm{aux}}$ (*cf.* Lemma 3.2.2.4), and induces an isomorphism $\phi^0_{\mathtt{D}_{\mathrm{aux}}} : \mathrm{Gr}^0_{\mathtt{D}_{\mathrm{aux}}} \overset{\sim}{\to} \mathrm{Gr}^0_{\mathtt{D}^{\#}_{\mathrm{aux}}}$. Hence, we have a commutative diagram:

$$\begin{array}{ccc} (\mathrm{Gr}^0_{\mathsf{D}})^{\oplus(a_1+a_2)} & \hookrightarrow & \mathrm{Gr}^0_{\mathsf{D}_{\mathrm{aux}}} \\ {\scriptstyle \phi^0_{\mathsf{D}}\otimes(\cdot,\cdot)^*_{\mathrm{aux}}} \big\downarrow & & \big\downarrow \wr \, {\scriptstyle \phi^0_{\mathsf{D}_{\mathrm{aux}}}} \\ (\mathrm{Gr}^0_{\mathsf{D}^\#})^{\oplus(a_1+a_2)} & \hookleftarrow & \mathrm{Gr}^0_{\mathsf{D}^\#_{\mathrm{aux}}} \end{array} \tag{3.4.1.3}$$

(This finishes Construction 3.4.1.2.)

Suppose $\mathcal{H} = \mathcal{H}^p\mathcal{H}_p$ such that $\mathcal{H}^p \subset \mathrm{G}(\hat{\mathbb{Z}}^p)$ and $\mathcal{H}_p = \mathcal{U}^{\mathrm{bal}}_{p,1}(p^r)$ for some integer $r \geq 0$. Let $\mathcal{H}_{\mathrm{aux}} \subset \mathrm{G}_{\mathrm{aux}}(\hat{\mathbb{Z}})$ be of the form $\mathcal{H}_{\mathrm{aux}} = \mathcal{H}^p_{\mathrm{aux}}\mathcal{H}_{\mathrm{aux},p}$ such that $\mathcal{H}_{p,\mathrm{aux}} = \mathcal{U}^{\mathrm{bal}}_{p,1,\mathrm{aux}}(p^r)$, where $\mathcal{U}_{p,1,\mathrm{aux}}(p^r)$ is defined by the filtration $\mathsf{D}_{\mathrm{aux}}$ on $L_{\mathrm{aux}} \underset{\mathbb{Z}}{\otimes} \mathbb{Z}_p$ as in Definition 3.2.2.8, and such that $\mathcal{H}$ is mapped into $\mathcal{H}_{\mathrm{aux}}$ under the homomorphism $\mathrm{G}(\hat{\mathbb{Z}}) \to \mathrm{G}_{\mathrm{aux}}(\hat{\mathbb{Z}})$ given by (2.1.1.10).

Lemma 3.4.1.4. *Let $\mathcal{H}$ and $\mathcal{H}_{\mathrm{aux}}$ be as above. Then there is a morphism*

$$\dddot{\mathsf{M}}^{\mathrm{ord}}_{\mathcal{H}} \to \dddot{\mathsf{M}}^{\mathrm{ord}}_{\mathcal{H}_{\mathrm{aux}}} \tag{3.4.1.5}$$

compatible with (2.1.1.17).

Proof. The construction of (3.4.1.5) is similar to the construction of (2.1.1.17), but let us still spell out the details in steps where they slightly differ.

Suppose $(A, \lambda, i, \alpha_{\mathcal{H}^p}, \alpha^{\mathrm{ord}}_{\mathcal{H}_p})$ is the tautological tuple over $\dddot{\mathsf{M}}^{\mathrm{ord}}_{\mathcal{H}}$. Then we obtain the abelian schemes $A^{\triangle}_{\mathrm{aux}} := A^{\times(a_1+a_2)}$ and $A^{\triangledown}_{\mathrm{aux}} := A^{\times a_1} \underset{\ddot{\mathsf{M}}^{\mathrm{ord}}_{\mathcal{H}}}{\times} (A^\vee)^{\times a_2}$, the canonical morphism $f : A^{\triangle}_{\mathrm{aux}} \to A^{\triangledown}_{\mathrm{aux}}$, the polarization $\lambda^{\triangle}_{\mathrm{aux}}$ and the $\mathcal{O}_{\mathrm{aux}}$-structure $i^{\triangle}_{\mathrm{aux}}$, and the polarization $\lambda^{\triangledown}_{\mathrm{aux}}$ of degree prime to p and the $\mathcal{O}_{\mathrm{aux}} \underset{\mathbb{Z}}{\otimes} \mathbb{Q}$-structure $i^{\triangledown}_{\mathrm{aux}}$ as in (2) of Lemma 2.1.1.1 and in the proof of Proposition 2.1.1.15.

The kernel K of the canonical isogeny $f : A^{\triangle}_{\mathrm{aux}} \to A^{\triangledown}_{\mathrm{aux}}$ decomposes canonically as a fiber product $K^p \underset{\ddot{\mathsf{M}}^{\mathrm{ord}}_{\mathcal{H}}}{\times} K_p$, where K^p is a finite étale group scheme of rank prime-to-p such that, at each geometric point $\bar{s}$ of $\dddot{\mathsf{M}}^{\mathrm{ord}}_{\mathcal{H}}$, any lifting $\hat{\alpha}^p_{\bar{s}} : L \underset{\mathbb{Z}}{\otimes} \hat{\mathbb{Z}}^p \overset{\sim}{\to} \mathrm{T}^p A_{\bar{s}}$ of $\alpha_{\mathcal{H}^p}$ (as in Lemma 3.3.1.6) defines an isomorphism

$$(L^\#/L)^{\oplus a_2} \underset{\mathbb{Z}}{\otimes} \hat{\mathbb{Z}}^p \overset{\sim}{\to} K^p_{\bar{s}};$$

and where K_p is a finite flat group scheme of p-primary rank such that, at each geometric point $\bar{s}$ of $\dddot{\mathsf{M}}^{\mathrm{ord}}_{\mathcal{H}}$, any lifting $\hat{\alpha}^{\mathrm{ord}}_{\bar{s}} = (\hat{\alpha}^{\mathrm{ord},0}_{\bar{s}}, \hat{\alpha}^{\mathrm{ord},\#,0}_{\bar{s}}, \hat{\nu}^{\mathrm{ord}}_{\bar{s}})$ of $\alpha^{\mathrm{ord}}_{\mathcal{H}_p}$ (as in Lemma 3.3.3.9) defines a short exact sequence

$$0 \to ((\mathrm{Gr}^0_{\mathsf{D}^\#} / \mathrm{Gr}^0_{\mathsf{D}})^{\oplus a_2})^{\mathrm{mult}}_{\bar{s}} \to K_{p,\bar{s}} \to ((\mathrm{Gr}^{-1}_{\mathsf{D}^\#} / \mathrm{Gr}^{-1}_{\mathsf{D}})^{\oplus a_2})_{\bar{s}} \to 0.$$

(Although [62, Lem. 1.3.5.2] is not applicable to isogenies of degree not prime to p, we can use such short exact sequences to study quasi-isogenies formed by isogenies whose kernels are finite flat group schemes of étale-multiplicative type.) Therefore, the $\mathcal{O} \underset{\mathbb{Z}}{\otimes} \mathbb{Q}$-structure $i^{\triangledown}_{\mathrm{aux}} : \mathcal{O}_{\mathrm{aux}} \underset{\mathbb{Z}}{\otimes} \mathbb{Q} \to \mathrm{End}_{\dddot{\mathsf{M}}^{\mathrm{ord}}_{\mathcal{H}}}(A^{\triangledown}_{\mathrm{aux}}) \underset{\mathbb{Z}}{\otimes} \mathbb{Q}$ induces an $\mathcal{O}$-structure $i^{\triangledown}_{\mathrm{aux}} : \mathcal{O}_{\mathrm{aux}} \to \mathrm{End}_{\dddot{\mathsf{M}}^{\mathrm{ord}}_{\mathcal{H}}}(A^{\triangledown}_{\mathrm{aux}})$ of $(A^{\triangledown}_{\mathrm{aux}}, \lambda^{\triangledown}_{\mathrm{aux}})$.

Moreover, away from p, at each geometric point $\bar{s}$ of $\dddot{\mathsf{M}}^{\mathrm{ord}}_{\mathcal{H}}$, the isomorphism $\mathrm{V}^p(f) : \mathrm{V}^p\, A^{\triangle}_{\mathrm{aux},\bar{s}} \overset{\sim}{\to} \mathrm{V}^p\, A^{\triangledown}_{\mathrm{aux},\bar{s}}$ (which can be defined even though $f : A^{\triangle}_{\mathrm{aux}} \to A^{\triangledown}_{\mathrm{aux}}$ is not prime to p) and the lifting $\hat{\alpha}^p_{\bar{s}}$ above induces an $\mathcal{O}_{\mathrm{aux}} \underset{\mathbb{Z}}{\otimes} \hat{\mathbb{Z}}^p$-equivariant isomorphism $\hat{\alpha}^{\triangledown,p}_{\bar{s}} : L_{\mathrm{aux}} \underset{\mathbb{Z}}{\otimes} \hat{\mathbb{Z}}^p \overset{\sim}{\to} \mathrm{T}^p\, A^{\triangledown}_{\mathrm{aux},\bar{s}}$ (matching similitudes, implicitly). Since the $\mathcal{H}^p$-orbit of $\hat{\alpha}^p_{\bar{s}}$ is $\pi_1(\dddot{\mathsf{M}}^{\mathrm{ord}}_{\mathcal{H}}, \bar{s})$-invariant, and since $\mathcal{H}^p$ is mapped into $\mathcal{H}^p_{\mathrm{aux}}$ under the homomorphism $\mathrm{G}(\hat{\mathbb{Z}}^p) \to \mathrm{G}_{\mathrm{aux}}(\hat{\mathbb{Z}}^p)$ given by (2.1.1.10), the $\mathcal{H}^p_{\mathrm{aux}}$-orbit $[\hat{\alpha}^{\triangledown,p}_{\bar{s}}]_{\mathcal{H}^p_{\mathrm{aux}}}$ of $\hat{\alpha}^{\triangledown,p}_{\bar{s}}$ is invariant under $\pi_1(\dddot{\mathsf{M}}^{\mathrm{ord}}_{\mathcal{H}}, \bar{s})$. This allows us to construct a level structure $\alpha^{\triangledown}_{\mathcal{H}^p_{\mathrm{aux}}}$ (away from p) as in the proof of [62, Prop. 1.4.3.4] such that $\hat{\alpha}^{\triangledown,p}_{\bar{s}}$ lifts $\alpha^{\triangledown}_{\mathcal{H}^p_{\mathrm{aux}}}$ (as in Lemma 3.3.1.6).

Finally, at p, since $\mathcal{H}_p = \mathcal{U}^{\mathrm{bal}}_{p,1}(p^r)$ and $\mathcal{H}_{\mathrm{aux},p} = \mathcal{U}^{\mathrm{bal}}_{p,1,\mathrm{aux}}(p^r)$ for some integer $r \geq 0$, we can start with the unique triple

$$\alpha^{\mathrm{ord}}_{p^r} = (\alpha^{\mathrm{ord},0}_{p^r}, \alpha^{\mathrm{ord},\#,0}_{p^r}, \nu^{\mathrm{ord}}_{p^r})$$

determined by $\alpha^{\mathrm{ord}}_{\mathcal{H}_p}$, and define $\alpha^{\mathrm{ord},\triangledown}_{\mathcal{H}_{\mathrm{aux},p}}$ by defining

$$\alpha^{\mathrm{ord},\triangledown}_{p^r} := (\alpha^{\mathrm{ord},0,\triangledown}_{p^r}, \alpha^{\mathrm{ord},\#,0,\triangledown}_{p^r}, \nu^{\mathrm{ord},\triangledown}_{p^r}),$$

where:

(1) $\alpha^{\mathrm{ord},0,\triangledown}_{p^r} := (\alpha^{\mathrm{ord},0,\triangledown}_{p^r})^{\times a_1} \underset{\dddot{\mathsf{M}}^{\mathrm{ord}}_{\mathcal{H}}}{\times} (\alpha^{\mathrm{ord},\#,0,\triangledown}_{p^r})^{\times a_2}$ as homomorphisms from $(\mathrm{Gr}^0_{\mathsf{D}_{\mathrm{aux}},p^r})^{\mathrm{mult}}_{\dddot{\mathsf{M}}^{\mathrm{ord}}_{\mathcal{H}}} \cong ((\mathrm{Gr}^0_{\mathsf{D},p^r})^{\mathrm{mult}}_{\dddot{\mathsf{M}}^{\mathrm{ord}}_{\mathcal{H}}})^{\times a_1} \underset{\dddot{\mathsf{M}}^{\mathrm{ord}}_{\mathcal{H}}}{\times} ((\mathrm{Gr}^0_{\mathsf{D}^\#,p^r})^{\mathrm{mult}}_{\dddot{\mathsf{M}}^{\mathrm{ord}}_{\mathcal{H}}})^{\times a_2}$ to $A^{\triangledown}_{\mathrm{aux}}[p^r]$.

(2) $\alpha^{\mathrm{ord},\#,0,\triangledown}_{p^r} := \lambda^{\triangledown}_{\mathrm{aux}} \circ \alpha^{\mathrm{ord},0,\triangledown}_{p^r} \circ ((\phi^0_{\mathsf{D}_{\mathrm{aux}}})^{\mathrm{mult}}_{\dddot{\mathsf{M}}^{\mathrm{ord}}_{\mathcal{H}}})^{-1}$ (*cf.* (3.4.1.3)) as homomorphisms from $(\mathrm{Gr}^0_{\mathsf{D}^\#_{\mathrm{aux}},p^r})^{\mathrm{mult}}_{\dddot{\mathsf{M}}^{\mathrm{ord}}_{\mathcal{H}}}$ to $A^{\triangledown,\vee}_{\mathrm{aux}}[p^r]$.

(3) $\nu^{\mathrm{ord},\triangledown}_{p^r} := \nu^{\mathrm{ord}}_{p^r}$ as a section of $(\mathbb{Z}/p^r\mathbb{Z})^{\times}_{\dddot{\mathsf{M}}^{\mathrm{ord}}_{\mathcal{H}}}$.

Thus, we assigned to the tautological tuple $(A, \lambda, i, \alpha_{\mathcal{H}^p}, \alpha^{\mathrm{ord}}_{\mathcal{H}_p})$ over $\dddot{\mathsf{M}}^{\mathrm{ord}}_{\mathcal{H}}$ a tuple $(A^{\triangledown}_{\mathrm{aux}}, \lambda^{\triangledown}_{\mathrm{aux}}, i^{\triangledown}_{\mathrm{aux}}, \alpha^{\triangledown}_{\mathcal{H}^p_{\mathrm{aux}}}, \alpha^{\mathrm{ord},\triangledown}_{\mathcal{H}_{\mathrm{aux},p}})$ parameterized by the moduli problem $\dddot{\mathsf{M}}^{\mathrm{ord}}_{\mathcal{H}_{\mathrm{aux}}}$, which then induces the desired morphism (3.4.1.5) by the universal property of $\dddot{\mathsf{M}}^{\mathrm{ord}}_{\mathcal{H}_{\mathrm{aux}}}$. By their very constructions, the pullback to $\dddot{\mathsf{M}}^{\mathrm{ord}}_{\mathcal{H}}$ of the tuple $(A^{\triangledown}_{\mathrm{aux}}, \lambda^{\triangledown}_{\mathrm{aux}}, i^{\triangledown}_{\mathrm{aux}}, \alpha^{\triangledown}_{\mathcal{H}^p_{\mathrm{aux}}})$ constructed in the proof of Proposition 2.1.1.15 gives, via Proposition 3.3.5.1, the pullback to $\dddot{\mathsf{M}}^{\mathrm{ord}}_{\mathcal{H}}$ of the tuple $(A^{\triangledown}_{\mathrm{aux}}, \lambda^{\triangledown}_{\mathrm{aux}}, i^{\triangledown}_{\mathrm{aux}}, \alpha^{\triangledown}_{\mathcal{H}^p_{\mathrm{aux}}}, \alpha^{\mathrm{ord},\triangledown}_{\mathcal{H}_{\mathrm{aux},p}})$ constructed here. (It is nevertheless an abuse of notation when we use the notation $A^{\triangledown}_{\mathrm{aux}}$, $\lambda^{\triangledown}_{\mathrm{aux}}$, and $i^{\triangledown}_{\mathrm{aux}}$ in both of them.) Hence, (3.4.1.5) is compatible with (2.1.1.17), as desired. □

Assuming no longer that $\mathcal{H}_p = \mathcal{U}^{\mathrm{bal}}_{p,1}(p^r)$, we still have:

Lemma 3.4.1.6. *Suppose $\mathcal{H} = \mathcal{H}^p\mathcal{H}_p \subset \mathrm{G}(\hat{\mathbb{Z}})$ and (resp. $\mathcal{H}_{\mathrm{aux}} = \mathcal{H}^p_{\mathrm{aux}}\mathcal{H}_{\mathrm{aux},p} \subset \mathrm{G}_{\mathrm{aux}}(\hat{\mathbb{Z}})$) is an open compact subgroup such that there exists integers $r \geq r'$ such* *that* $\mathcal{U}^{\mathrm{bal}}_{p,1}(p^r) \subset \mathcal{H}_p \subset \mathcal{U}^{\mathrm{bal}}_{p,0}(p^r)$ *and* $\mathcal{U}^{\mathrm{bal}}_{p,1,\mathrm{aux}}(p^{r'}) \subset \mathcal{H}_{\mathrm{aux},p} \subset \mathcal{U}_{p,0,\mathrm{aux}}(p^{r'})$*, and such that* $\mathcal{H}$ *is mapped into* $\mathcal{H}_{\mathrm{aux}}$ *under the homomorphism* $\mathrm{G}(\hat{\mathbb{Z}}) \to \mathrm{G}_{\mathrm{aux}}(\hat{\mathbb{Z}})$ *given by* (2.1.1.10)*. In this case, there is a morphism*

$$\overset{\cdots}{\mathsf{M}}{}^{\mathrm{ord}}_{\mathcal{H}} \to \overset{\cdots}{\mathsf{M}}{}^{\mathrm{ord}}_{\mathcal{H}_{\mathrm{aux}}} \tag{3.4.1.7}$$

compatible with (2.1.1.17) (*and with* (3.4.1.5))*.*

Proof. The operations of taking orbits of level structures on both sides of (3.4.1.5) are compatible with each other. Hence, the morphism (3.4.1.5) at any sufficiently high level induces the morphism (3.4.1.7) by forgetting part of the structures. ❒

Lemma 3.4.1.8. *The morphism* (3.4.1.7) *is schematic, separated, and quasi-finite.*

Proof. As in the proof of Proposition 2.1.1.15, this follows from Lemma 2.1.1.5 (for the abelian schemes and polarizations), from [62, Prop. 1.3.3.7] (for the endomorphism structures), from the fact that the level structures away from p are defined by isomorphisms between finite étale group schemes, and from the fact that the ordinary level structures are defined by morphisms between finite flat group schemes of étale-multiplicative type. □

Theorem 3.4.1.9. *The moduli problem* $\overset{\cdots}{\mathsf{M}}{}^{\mathrm{ord}}_{\mathcal{H}}$ *is an algebraic stack separated and of finite type over* $\mathrm{Spec}(\mathbb{Z}_{(p)})$*. It is representable by an algebraic space if the objects it parameterizes have no nontrivial automorphism, which is, in particular, the case when* $\mathcal{H}^p$ *is neat. Its local structures can be described as follows: Let* $\phi_{\mathsf{D}}^{-1} : \mathrm{Gr}_{\mathsf{D}}^{-1} \to \mathrm{Gr}_{\mathsf{D}^{\#}}^{-1}$ *be as in Lemma 3.2.2.4, and consider the finitely generated* $\mathbb{Z}_p$*-module*

$$\vec{\mathbf{S}}_{\mathsf{D}} := (\mathrm{Gr}_{\mathsf{D}}^{-1} \underset{\mathbb{Z}_p}{\otimes} \mathrm{Gr}_{\mathsf{D}^{\#}}^{-1}) / \begin{pmatrix} d \otimes \phi_{\mathsf{D}}^{-1}(d') - d' \otimes \phi_{\mathsf{D}}^{-1}(d) \\ (bd) \otimes d^{\#} - d \otimes (b^{\star} d^{\#}) \end{pmatrix}_{\substack{d,d' \in \mathrm{Gr}_{\mathsf{D}}^{-1}, \\ d^{\#} \in \mathrm{Gr}_{\mathsf{D}^{\#}}^{-1}, b \in \mathcal{O}}}. \tag{3.4.1.10}$$

Let $\vec{\mathbf{S}}_{\mathsf{D},\mathbb{Z}}$ *be any noncanonical choice of a finitely generated* $\mathbb{Z}$*-module such that* $\vec{\mathbf{S}}_{\mathsf{D},\mathbb{Z}} \underset{\mathbb{Z}}{\otimes} \mathbb{Z}_p \cong \vec{\mathbf{S}}_{\mathsf{D}}$ *and such that the maximal torsion submodule* $\vec{\mathbf{S}}_{\mathsf{D},\mathbb{Z},\mathrm{tor}}$ *of* $\vec{\mathbf{S}}_{\mathsf{D},\mathbb{Z}}$ *is isomorphic to the maximal torsion submodule* $\vec{\mathbf{S}}_{\mathsf{D},\mathrm{tor}}$ *of* $\vec{\mathbf{S}}_{\mathsf{D}}$ (*as* $\mathbb{Z}$*-modules*)*.* (*This latter condition is not necessary for our purpose, but we impose it for simplicity of later exposition.*) *Then the completions of strict local rings of* $\overset{\cdots}{\mathsf{M}}{}^{\mathrm{ord}}_{\mathcal{H}}$ *at geometric points of characteristic* p *are isomorphic to the completions of strict local rings of the group scheme* $E_{\mathsf{D},\mathbb{Z}}$ *of multiplicative type of finite type over* $\mathrm{Spec}(\mathbb{Z}_{(p)})$ *with character group the finitely generated commutative group* $\vec{\mathbf{S}}_{\mathsf{D},\mathbb{Z}}$*.*

Proof. By Lemma 3.4.1.8, $\overset{\cdots}{\mathsf{M}}{}^{\mathrm{ord}}_{\mathcal{H}}$ is schematic, separated, and quasi-finite over $\overset{\cdots}{\mathsf{M}}{}^{\mathrm{ord}}_{\mathcal{H}_{\mathrm{aux}}}$. Since p is a good prime for the moduli problem $\overset{\cdots}{\mathsf{M}}{}^{\mathrm{ord}}_{\mathcal{H}_{\mathrm{aux}}}$, we can show that $\overset{\cdots}{\mathsf{M}}{}^{\mathrm{ord}}_{\mathcal{H}_{\mathrm{aux}}}$ is an algebraic stack separated and of finite type over $\mathrm{Spec}(\mathbb{Z}_{(p)})$ by an argument similar to that in [62, Ch. 2]: The moduli problem $\overset{\cdots}{\mathsf{M}}{}^{\mathrm{ord}}_{\mathcal{H}_{\mathrm{aux}}}$ is an algebraic stack (quasi-separated and) locally of finite type over $\mathrm{Spec}(\mathbb{Z}_{(p)})$ by the theory of infinitesimal deformations and Artin's criterion. Note that, since the infinitesimal deformation rings are smooth (by [62, Prop. 2.2.4.9] and by the fact that $\alpha_{\mathcal{H}^p_{\text{aux}}}$ and $\alpha^{\text{ord}}_{\mathcal{H}_{\text{aux},p}}$ are both parameterized by étale objects), and since the moduli problem $\dddot{\mathsf{M}}^{\text{ord}}_{\mathcal{H}_{\text{aux}}}$ can be extended to a moduli problem over a Dedekind ring having infinitely many residue characteristics, we can suppress the technical condition [62, Sec. 2.3.4, Cond. 4] in Artin's criterion by [62, Thm. B.3.11]. The diagonal 1-morphism $\Delta_{\dddot{\mathsf{M}}^{\text{ord}}_{\mathcal{H}_{\text{aux}}}} : \dddot{\mathsf{M}}^{\text{ord}}_{\mathcal{H}_{\text{aux}}} \to \dddot{\mathsf{M}}^{\text{ord}}_{\mathcal{H}_{\text{aux}}} \underset{\operatorname{Spec}(\mathbb{Z}_{(p)})}{\times} \dddot{\mathsf{M}}^{\text{ord}}_{\mathcal{H}_{\text{aux}}}$ is finite by the theory of Néron models (applied to the abelian schemes and the additional structures—the ordinary level structures, once they exist, have finite automorphism group schemes). This shows that $\dddot{\mathsf{M}}^{\text{ord}}_{\mathcal{H}_{\text{aux}}}$ is separated over $\operatorname{Spec}(\mathbb{Z}_{(p)})$ (by definition). The algebraic stack $\dddot{\mathsf{M}}^{\text{ord}}_{\mathcal{H}_{\text{aux}}}$ is quasi-compact (and hence of finite type over $\operatorname{Spec}(\mathbb{Z}_{(p)})$) because it can be covered by a quasi-compact scheme parameterizing the sections of ample invertible sheaves, the endomorphism structures, and the ordinary level structures. (See [62, Ch. 2] for more details.) Thus, we have shown that $\dddot{\mathsf{M}}^{\text{ord}}_{\mathcal{H}_{\text{aux}}}$ and hence $\dddot{\mathsf{M}}^{\text{ord}}_{\mathcal{H}}$ are algebraic stacks separated and of finite type over $\operatorname{Spec}(\mathbb{Z}_{(p)})$.

If $\mathcal{H}^p$ is neat, then the tuple $(A, \lambda, i, \alpha_{\mathcal{H}^p})$ admits no nontrivial automorphism (regardless of the level structure $\alpha^{\text{ord}}_{\mathcal{H}_p}$ at p), in which case $\dddot{\mathsf{M}}^{\text{ord}}_{\mathcal{H}}$ is representable by an algebraic space.

The existence of ordinary level structures in Definition 3.4.1.1 forces the abelian varieties parameterized by geometric points of characteristic p of $\dddot{\mathsf{M}}^{\text{ord}}_{\mathcal{H}}$ to be ordinary (see Remark 3.3.3.6). Hence, we can describe the completion of strict local rings of $\dddot{\mathsf{M}}^{\text{ord}}_{\mathcal{H}}$ at geometric points of characteristic p using the Serre–Tate deformation theory of ordinary abelian varieties (as in, for example, [47]). By [47, Thm. 2.1, 1) and 2)], the formal moduli of any ordinary abelian variety A over $\bar{s} := \operatorname{Spec}(k)$, where k is an algebraically closed field of characteristic p, is canonically isomorphic to formal torus $\operatorname{Hom}_{\mathbb{Z}_p}(\mathrm{T}_p A \underset{\mathbb{Z}_p}{\otimes} \mathrm{T}_p A^\vee, \hat{\mathbf{G}}_{\text{m}})$, where $\mathrm{T}_p A$ and $\mathrm{T}_p A^\vee$ are the *physical Tate modules* of A and $A^\vee$, which are free $\mathbb{Z}_p$-modules of rank $\dim(A) = \dim(A^\vee)$. If A is part of an object $(A, \lambda, i, \alpha_{\mathcal{H}^p}, \alpha^{\text{ord}}_{\mathcal{H}_p})$ of $\dddot{\mathsf{M}}^{\text{ord}}_{\mathcal{H}}(\bar{s})$, then any lifting $\alpha^{\text{ord}}_{\bar{s}}$ of $\alpha^{\text{ord}}_{\mathcal{H}_p}$ as in Lemma 3.3.3.9 defines isomorphisms $\mathrm{T}_p A \cong \mathrm{Gr}^{-1}_{\mathsf{D}}$ and $\mathrm{T}_p A^\vee \cong \mathrm{Gr}^{-1}_{\mathsf{D}^\#}$ compatible with each other under ϕ^{-1}_{D} and λ (*cf.* Proposition 3.2.1.1). Hence, the formal moduli of A itself is canonically isomorphic to the formal torus $\operatorname{Hom}_{\mathbb{Z}_p}(\mathrm{Gr}^{-1}_{\mathsf{D}} \underset{\mathbb{Z}_p}{\otimes} \mathrm{Gr}^{-1}_{\mathsf{D}^\#}, \hat{\mathbf{G}}_{\text{m}})$. By [47, Thm. 2.1, 3) and 4)], the formal submoduli for liftings of A also carrying liftings of the additional structures λ, i, $\alpha_{\mathcal{H}^p}$, and $\alpha^{\text{ord}}_{\mathcal{H}_p}$ is the formal subgroup scheme of multiplicative type $\operatorname{Hom}_{\mathbb{Z}_p}(\vec{\mathbf{S}}_{\mathsf{D}}, \hat{\mathbf{G}}_{\text{m}})$, where the condition for λ is dual to the relations $d \otimes \phi^{-1}_{\mathsf{D}}(d') - d' \otimes \phi^{-1}_{\mathsf{D}}(d)$ for all $d, d' \in \mathrm{Gr}^{-1}_{\mathsf{D}}$; where the condition for i is dual to the relations $(bd) \otimes d^\# - d \otimes (b^\star d^\#)$ for all $d \in \mathrm{Gr}^{-1}_{\mathsf{D}}$ and $b \in \mathcal{O}$; and where no conditions are needed for the level structures, because they are given by morphisms between finite étale group schemes, or by morphisms between finite flat groups schemes of multiplicative type, which always uniquely lift. Hence, the completion of strict local rings of $\dddot{\mathsf{M}}_{\mathcal{H}}^{\mathrm{ord}}$ can be described as in the statement of the theorem, as desired. (We emphasize that the elegant argument in [47, Sec. 1] requires the prorepresentability as an input. Therefore, it is not true that one can avoid the references as in [62, Ch. 2] to the original theory of deformation of abelian schemes developed by Grothendieck, Mumford, and others.) □

3.4.2 *Ordinary Loci as Normalizations*

Let $\mathcal{H}$ be as at the beginning of Section 3.4.1, so that $\dddot{\mathsf{M}}_{\mathcal{H}}^{\mathrm{ord}}$ is defined as in Definition 3.4.1.1. Suppose that $\nu(\mathcal{H}_p) = \ker(\mathbb{Z}_p^\times \to (\mathbb{Z}/p^{r_\nu}\mathbb{Z})^\times)$ (where $r_\nu \leq r$) (*cf.* also Section 3.3.5).

Definition 3.4.2.1. *Let $\vec{\mathbf{S}}_{\mathsf{D}}$ be as in Theorem 3.4.1.9. Then we define $r_{\mathsf{D}} \geq 0$ to be the smallest nonnegative integer such that $p^{r_{\mathsf{D}}}$ annihilates all (p-primary) torsion elements in $\vec{\mathbf{S}}_{\mathsf{D}}$, and we define $r_{\mathcal{H}} := \max(r_{\mathsf{D}}, r_\nu)$.*

Remark 3.4.2.2. If p is a *good prime* as in Definition 1.1.1.6, then $r_{\mathsf{D}} = 0$ by [62, Prop. 1.2.2.3], and hence $r_{\mathcal{H}} = r_\nu$.

Lemma 3.4.2.3. *The canonical morphism*

$$\mathsf{M}_{\mathcal{H},r_\nu} \to \dddot{\mathsf{M}}_{\mathcal{H}}^{\mathrm{ord}} \underset{\mathbb{Z}}{\otimes} F_0[\zeta_{p^{r_\nu}}] \tag{3.4.2.4}$$

over $\mathsf{S}_{0,r_\nu} = \mathrm{Spec}(F_0[\zeta_{p^{r_\nu}}])$ induced by the assignment (3.3.5.5) is an open and closed immersion.

Proof. The morphism (3.4.2.4) is an open and closed immersion because over F_0 the Lie algebra condition given by $(L \underset{\mathbb{Z}}{\otimes} \mathbb{R}, \langle\,\cdot\,,\,\cdot\,\rangle, h_0)$ (in [62, Def. 1.3.4.1]) is defined and is an open and closed condition by [62, Prop. 2.2.2.9], because all level structures involved are defined by homomorphisms between finite étale group schemes, and because (3.3.5.5) is injective. □

Theorem 3.4.2.5. *Let $\mathsf{M}_{\mathcal{H}}^{\mathrm{ord}}$ denote the open and closed subalgebraic stack of $\dddot{\mathsf{M}}_{\mathcal{H}}^{\mathrm{ord}} \underset{\mathbb{Z}}{\otimes} F_0[\zeta_{p^{r_{\mathcal{H}}}}]$ over $\mathsf{S}_{0,r_{\mathcal{H}}} = \mathrm{Spec}(F_0[\zeta_{p^{r_{\mathcal{H}}}}])$ (see Definition 2.2.3.3) given by the image of the induced canonical open and closed immersion $\mathsf{M}_{\mathcal{H},r_{\mathcal{H}}} \hookrightarrow \dddot{\mathsf{M}}_{\mathcal{H}}^{\mathrm{ord}} \underset{\mathbb{Z}}{\otimes} F_0[\zeta_{p^{r_{\mathcal{H}}}}]$. Then the normalization $\vec{\mathsf{M}}_{\mathcal{H}}^{\mathrm{ord}}$ of $\dddot{\mathsf{M}}_{\mathcal{H}}^{\mathrm{ord}}$ in $\mathsf{M}_{\mathcal{H}}^{\mathrm{ord}}$ under the canonical morphism $\mathsf{M}_{\mathcal{H}}^{\mathrm{ord}} \to \dddot{\mathsf{M}}_{\mathcal{H}}^{\mathrm{ord}}$ is a regular algebraic stack which is separated, smooth, and of finite type over $\vec{\mathsf{S}}_{0,r_{\mathcal{H}}} = \mathrm{Spec}(\mathcal{O}_{F_0,(p)}[\zeta_{p^{r_{\mathcal{H}}}}])$ (see Definition 2.2.3.3). (In particular, the nonsmoothness of the structural morphism $\vec{\mathsf{M}}_{\mathcal{H}}^{\mathrm{ord}} \to \mathrm{Spec}(\mathbb{Z}_{(p)})$ all comes from that of the finite flat morphism $\vec{\mathsf{S}}_{0,r_{\mathcal{H}}} = \mathrm{Spec}(\mathcal{O}_{F_0,(p)}[\zeta_{p^{r_{\mathcal{H}}}}]) \to \mathrm{Spec}(\mathbb{Z}_{(p)})$.) The algebraic stack $\vec{\mathsf{M}}_{\mathcal{H}}^{\mathrm{ord}}$ is representable by an algebraic space when the moduli problem $\dddot{\mathsf{M}}_{\mathcal{H}}^{\mathrm{ord}}$ is (see Theorem 3.4.1.9), which is the case when $\mathcal{H}^p$ is neat.*

Let $\vec{\mathbf{S}}_{\mathsf{D},\mathbb{Z}}$ be noncanonically chosen as in Theorem 3.4.1.9, and let $\vec{\mathbf{S}}_{\mathsf{D},\mathbb{Z},\mathrm{free}}$ be the maximal free quotient of $\vec{\mathbf{S}}_{\mathsf{D},\mathbb{Z}}$. Then the completions of strict local rings of $\vec{\mathsf{M}}_{\mathcal{H}}^{\mathrm{ord}}$ are *isomorphic to the completions of strict local rings of the torus* $E_{\mathtt{D},\mathbb{Z},\mathrm{free}}$ *over* $\vec{\mathsf{S}}_{0,r_{\mathcal{H}}}$ *with character group* $\vec{\mathbf{S}}_{\mathtt{D},\mathbb{Z},\mathrm{free}}$.

If $\mathcal{H}^p = \mathcal{U}^p(n_0)$ *and* $\mathcal{H}_p = \mathcal{U}_{p,1}^{\mathrm{bal}}(p^r)$, *then we denote* $\vec{\mathsf{M}}_{\mathcal{H}}^{\mathrm{ord}}$ *by* $\vec{\mathsf{M}}_{n_0p^r}^{\mathrm{ord}} = \vec{\mathsf{M}}_n^{\mathrm{ord}}$, *where* $n = n_0p^r$, *and we denote by* α_{n_0} *(resp.* $\alpha_{p^r}^{\mathrm{ord}}$*) the unique principal level-*n_0 *structure (resp. principal ordinary level-*p^r *structure) determined by* $\alpha_{\mathcal{H}^p}$ *(resp.* $\alpha_{\mathcal{H}_p}^{\mathrm{ord}}$*).*

Proof. The statements in the first paragraph are self-explanatory. Since all objects involved are excellent (see [35, IV-2, 7.8.3]), the operations of taking formal completions and taking normalizations are interchangeable. Hence, by Theorem 3.4.1.9, it suffices to study the normalization of $E_{\mathtt{D},\mathbb{Z}}$ in $E_{\mathtt{D},\mathbb{Z}} \underset{\mathbb{Z}_{(p)}}{\otimes} F_0[\zeta_{p^{r_{\mathcal{H}}}}]$. With $\vec{\mathbf{S}}_{\mathtt{D},\mathbb{Z}}$, $\vec{\mathbf{S}}_{\mathtt{D},\mathbb{Z},\mathrm{tor}}$, $\vec{\mathbf{S}}_{\mathtt{D},\mathbb{Z},\mathrm{free}}$, $E_{\mathtt{D},\mathbb{Z}}$, and $E_{\mathtt{D},\mathbb{Z},\mathrm{free}}$ as in the statements of Theorem 3.4.1.9 and this theorem, let $E_{\mathtt{D},\mathbb{Z},\mathrm{tor}}$ denote the group scheme of multiplicative type over $\mathrm{Spec}(\mathbb{Z}_{(p)})$ with character group $\vec{\mathbf{S}}_{\mathtt{D},\mathbb{Z},\mathrm{tor}}$. Then we have a short exact sequence

$$0 \to E_{\mathtt{D},\mathbb{Z},\mathrm{free}} \to E_{\mathtt{D},\mathbb{Z}} \to E_{\mathtt{D},\mathbb{Z},\mathrm{tor}} \to 0 \tag{3.4.2.6}$$

of group schemes of multiplicative type over $\mathrm{Spec}(\mathbb{Z}_{(p)})$. Since $r_{\mathcal{H}} \geq r_{\mathtt{D}}$, we have a canonical isomorphism

$$\zeta_{p^{r_{\mathtt{D}}},\vec{\mathsf{S}}_{0,r_{\mathcal{H}}}} : ((\mathbb{Z}/p^{r_{\mathtt{D}}}\mathbb{Z})(1))_{\vec{\mathsf{S}}_{0,r_{\mathcal{H}}}} \xrightarrow{\sim} \boldsymbol{\mu}_{p^{r_{\mathtt{D}}},\vec{\mathsf{S}}_{0,r_{\mathcal{H}}}}$$

over $\mathsf{S}_{0,r_{\mathcal{H}}} = \mathrm{Spec}(F_0[\zeta_{p^{r_{\mathcal{H}}}}])$, inducing a canonical isomorphism

$$\begin{aligned} E_{\mathtt{D},\mathbb{Z},\mathrm{tor}} \underset{\mathbb{Z}_{(p)}}{\otimes} F_0[\zeta_{p^{r_{\mathcal{H}}}}] &\cong \underline{\mathrm{Hom}}_{\mathsf{S}_{0,r_{\mathcal{H}}}}((\mathbf{S}_{\mathtt{D},\mathrm{tor}})_{\mathsf{S}_{0,r_{\mathcal{H}}}}, \boldsymbol{\mu}_{p^{r_{\mathtt{D}}},\mathsf{S}_{0,r_{\mathcal{H}}}}) \\ &\overset{\zeta^{-1}_{p^{r_{\mathtt{D}}},\vec{\mathsf{S}}_{0,r_{\mathcal{H}}}}}{\xrightarrow{\sim}} \underline{\mathrm{Hom}}(\mathbf{S}_{\mathtt{D},\mathrm{tor}}, (\mathbb{Z}/p^{r_{\mathtt{D}}}\mathbb{Z})(1))_{\mathsf{S}_{0,r_{\mathcal{H}}}} \end{aligned}$$

(under the assumption that the torsion submodule $E_{\mathtt{D},\mathbb{Z},\mathrm{tor}}$ is p-primary in Theorem 3.4.1.9), which shows that the underlying scheme of $E_{\mathtt{D},\mathbb{Z},\mathrm{tor}} \underset{\mathbb{Z}_{(p)}}{\otimes} F_0[\zeta_{p^{r_{\mathcal{H}}}}]$ is isomorphic to a disjoint union of duplicates of the base scheme $\mathsf{S}_{0,r_{\mathcal{H}}}$. Therefore, the scheme $E_{\mathtt{D},\mathbb{Z}} \underset{\mathbb{Z}_{(p)}}{\otimes} F_0[\zeta_{p^{r_{\mathcal{H}}}}]$ over $\mathsf{S}_{0,r_{\mathcal{H}}}$ is a disjoint union of duplicates of the scheme $E_{\mathtt{D},\mathbb{Z},\mathrm{free}} \underset{\mathbb{Z}_{(p)}}{\otimes} F_0[\zeta_{p^{r_{\mathcal{H}}}}]$ over $\mathsf{S}_{0,r_{\mathcal{H}}}$, which admit smooth models $E_{\mathtt{D},\mathbb{Z},\mathrm{free}} \underset{\mathbb{Z}_{(p)}}{\otimes} \mathcal{O}_{F_0,(p)}[\zeta_{p^{r_{\mathcal{H}}}}]$ over $\vec{\mathsf{S}}_{0,r_{\mathcal{H}}} = \mathrm{Spec}(\mathcal{O}_{F_0,(p)}[\zeta_{p^{r_{\mathcal{H}}}}])$. This shows that the normalization of $E_{\mathtt{D},\mathbb{Z}}$ in $E_{\mathtt{D},\mathbb{Z}} \underset{\mathbb{Z}_{(p)}}{\otimes} F_0[\zeta_{p^{r_{\mathcal{H}}}}]$ is a disjoint union of duplicates of the smooth scheme $E_{\mathtt{D},\mathbb{Z},\mathrm{free}} \underset{\mathbb{Z}_{(p)}}{\otimes} \mathcal{O}_{F_0,(p)}[\zeta_{p^{r_{\mathcal{H}}}}]$ over $\vec{\mathsf{S}}_{0,r_{\mathcal{H}}} = \mathrm{Spec}(\mathcal{O}_{F_0,(p)}[\zeta_{p^{r_{\mathcal{H}}}}])$, which is regular and satisfies the descriptions of the completions of strict local rings in the proposition, as desired. □

Remark 3.4.2.7. It is not obvious that $\vec{\mathsf{M}}_{\mathcal{H}}^{\mathrm{ord}} \underset{\mathbb{Z}}{\otimes} \mathbb{F}_p$ is nonempty. Nevertheless, in some special cases, we can show the nonemptiness of $\vec{\mathsf{M}}_{\mathcal{H}}^{\mathrm{ord}} \underset{\mathbb{Z}}{\otimes} \mathbb{F}_p$ by constructing its partial toroidal compactifications. See Section 6.3.3 for more discussions.

Remark 3.4.2.8. (Compare with Remark 1.1.2.1.) As in Remark 1.1.2.1, if we have chosen another PEL-type $\mathcal{O}$-lattice L' in $L \underset{\mathbb{Z}}{\otimes} \mathbb{Q}$ which satisfies $L \underset{\mathbb{Z}}{\otimes} \mathbb{Z}_{(p)} = L' \underset{\mathbb{Z}}{\otimes} \mathbb{Z}_{(p)}$, then (by [62, Prop. 1.4.3.4 and Cor. 1.4.3.8]) we have an $\mathbb{Z}_{(p)}^{\times}$-isogeny between the tautological abelian schemes over $\mathsf{M}_{\mathcal{H}}$ (matching their additional structures), and hence also between those over $\mathsf{M}_{\mathcal{H}}^{\mathrm{ord}}$ (in characteristic zero). Since $\vec{\mathsf{M}}_{\mathcal{H}}^{\mathrm{ord}}$ is noetherian normal, and since the tautological abelian scheme $A \to \vec{\mathsf{M}}_{\mathcal{H}}^{\mathrm{ord}}$ is ordinary, by Lemma 3.1.3.2, the $\mathbb{Z}_{(p)}^{\times}$-isogeny between tautological abelian schemes over $\mathsf{M}_{\mathcal{H}}^{\mathrm{ord}}$ extends to a $\mathbb{Z}_{(p)}^{\times}$-isogeny $A \to A'$ relating the corresponding tautological abelian schemes over $\vec{\mathsf{M}}^{\mathrm{ord}}$ (which can be identified with the isomorphic moduli problem defined using L'), and their additional structures are automatically matched by [92, IX, 1.4], [28, Ch. I, Prop. 2.7], or [62, Prop. 3.3.1.5]. Hence, the $\mathbb{Z}_{(p)}^{\times}$-isogenous class of the tautological object depends only on the choice of $L \underset{\mathbb{Z}}{\otimes} \mathbb{Z}_{(p)}$. Then we can define a collection $\{\vec{\mathsf{M}}_{\mathcal{H}}^{\mathrm{ord}}\}_{\mathcal{H}}$ indexed by $\mathcal{H}$ of the form $\mathcal{H}^p\mathcal{H}_p$, with $\mathcal{H}^p$ an arbitrary open compact subgroup of $\mathrm{G}(\mathbb{A}^{\infty,p})$ (not just one of $\mathrm{G}(\hat{\mathbb{Z}}^p)$), and with $\mathcal{H}_p$ satisfying $\mathcal{U}_{p,1}^{\mathrm{bal}}(p^r) \subset \mathcal{H}_p \subset \mathcal{U}_{p,0}(p^r)$ for some integer $r \geq 0$ (carrying a Hecke action as in Proposition 3.4.4.1 below). The choice of $L \underset{\mathbb{Z}}{\otimes} \mathbb{Z}_p$ and its filtration D, however, are more substantial. Modifying the choice of $L \underset{\mathbb{Z}}{\otimes} \mathbb{Z}_p$ will inevitably incur isogenies of degree divisible by p, which can still be done (because Lemma 3.1.3.2 also works) but will make the theory much more complicated.

Convention 3.4.2.9. *To facilitate the language, for S a scheme over $\vec{\mathsf{S}}_{0,r_{\mathcal{H}}} = \mathrm{Spec}(\mathcal{O}_{F_0,(p)}[\zeta_{p^{r_{\mathcal{H}}}}])$, we say that an object $(A, \lambda, i, \alpha_{\mathcal{H}^p}, \alpha_{\mathcal{H}_p}^{\mathrm{ord}})$ of $\dddot{\mathsf{M}}_{\mathcal{H}}^{\mathrm{ord}}(S)$ is parameterized by $\vec{\mathsf{M}}_{\mathcal{H}}^{\mathrm{ord}}$ if the tautological morphism $S \to \dddot{\mathsf{M}}_{\mathcal{H}}^{\mathrm{ord}}$ determined by the universal property factors through $S \to \vec{\mathsf{M}}_{\mathcal{H}}^{\mathrm{ord}}$. Then it also makes sense to consider the pullback of the tautological tuple over $\dddot{\mathsf{M}}_{\mathcal{H}}^{\mathrm{ord}}$ as the tautological tuple over $\vec{\mathsf{M}}_{\mathcal{H}}^{\mathrm{ord}}$.*

Definition 3.4.2.10. (*Compare with [62, Def. 5.3.2.1] and Definition* 1.3.1.1.) *Let S be a normal locally noetherian algebraic stack over $\vec{\mathsf{S}}_{0,r_{\mathcal{H}}}$. A tuple $(G, \lambda, i, \alpha_{\mathcal{H}^p}, \alpha_{\mathcal{H}_p}^{\mathrm{ord}})$ over S is called a* **degenerating family of type** $\vec{\mathsf{M}}_{\mathcal{H}}^{\mathrm{ord}}$, *or simply a* **degenerating family** *when the context is clear, if there exists a dense subalgebraic stack S_1 of S, such that we have the following:*

(1) By viewing group schemes as relative schemes (cf. [37]), G is a semi-abelian scheme over S whose restriction G_{S_1} to S_1 is an abelian scheme. In this case, the dual semi-abelian scheme $G^\vee$ exists (up to unique isomorphism; cf. [80, IV, 7.1] or [62, Thm. 3.4.3.2]), whose restriction $G_{S_1}^\vee$ to S_1 is the dual abelian scheme of G_{S_1}.

(2) $\lambda : G \to G^\vee$ is a group homomorphism that induces by restriction a polarization λ_{S_1} of G_{S_1}.

(3) $i : \mathcal{O} \to \mathrm{End}_S(G)$ is a homomorphism that defines by restriction an $\mathcal{O}$-structure $i_{S_1} : \mathcal{O} \to \mathrm{End}_{S_1}(G_{S_1})$ of (G_{S_1}, λ_{S_1}).

(4) $(G_{S_1}, \lambda_{S_1}, i_{S_1}, \alpha_{\mathcal{H}^p}, \alpha^{\mathrm{ord}}_{\mathcal{H}_p}) \to S_1$ *defines a tuple parameterized by* $\vec{\mathsf{M}}^{\mathrm{ord}}_{\mathcal{H}}$ *(see Convention* 3.4.2.9*).*

(5) *The ordinary level structure* $\alpha^{\mathrm{ord}}_{\mathcal{H}_p}$*, which is an orbit of étale-locally-defined triples* $(\alpha^{\mathrm{ord},0}_{p^r}, \alpha^{\mathrm{ord},\#,0}_{p^r}, \nu^{\mathrm{ord}}_{p^r})$*, where* $\alpha^{\mathrm{ord},0}_{p^r} : (\mathrm{Gr}^0_{\mathtt{D},p^r})^{\mathrm{mult}}_{S_1} \to G_{S_1}[p^r]$ *and* $\alpha^{\mathrm{ord},\#,0}_{p^r} : (\mathrm{Gr}^0_{\mathtt{D}^\#,p^r})^{\mathrm{mult}}_{S_1} \to G^\vee_{S_1}[p^r]$ *are closed immersions and where* $\nu^{\mathrm{ord}}_{p^r}$ *is a section of* $(\mathbb{Z}/p^r\mathbb{Z})^\times_{S_1}$*, extend to an orbit of étale-locally-defined triples* $(\alpha^{\mathrm{ord},0}_{p^r,S}, \alpha^{\mathrm{ord},\#,0}_{p^r,S}, \nu^{\mathrm{ord}}_{p^r,S})$*, where* $\alpha^{\mathrm{ord},0}_{p^r,S} : (\mathrm{Gr}^0_{\mathtt{D},p^r})^{\mathrm{mult}}_S \to G[p^r]$ *and* $\alpha^{\mathrm{ord},\#,0}_{p^r,S} : (\mathrm{Gr}^0_{\mathtt{D}^\#,p^r})^{\mathrm{mult}}_S \to G^\vee[p^r]$ *are closed immersions and where* $\nu^{\mathrm{ord}}_{p^r,S}$ *is a section of* $(\mathbb{Z}/p^r\mathbb{Z})^\times_S$*, that is also symplectic-liftable as in Definition* 3.3.3.4*. By abuse of language, we say in this case that* $\alpha^{\mathrm{ord}}_{\mathcal{H}_p}$ *extends to an ordinary level structure of* (G, λ, i) *over* S*. Since* $\alpha^{\mathrm{ord}}_{\mathcal{H}_p}$ *is determined by its restriction to every dense subalgebraic stack, by abuse of notation, we shall use the same notation for every such restriction.*

Remark 3.4.2.11. The extensibility condition (5) in Definition 3.4.2.10 is nontrivial even when G is an *abelian scheme*, because it forces G to be an ordinary abelian scheme over S. If S_1 is merely (of characteristic zero) defined over S_{0,r_ν}, then G_{S_1} is always ordinary over S_1, but G is not an ordinary abelian scheme over S in general.

Remark 3.4.2.12. Conditions (2), (3), and (4) are closed conditions for structures on *abelian schemes* defined over $\vec{\mathsf{S}}_{0,r_{\mathcal{H}}}$. Hence, the rather weak condition for S_1 in Definition 3.4.2.10 is justified because S_1 can always be replaced with the largest subalgebraic stack of S over $\vec{\mathsf{S}}_{0,r_{\mathcal{H}}}$ (which is open dense in S) such that G_{S_1} is an abelian scheme. (Conditions (2) and (3) are closed by [62, Lem. 4.2.1.6] and by [92, IX, 1.4], [28, Ch. I, Prop. 2.7], or [62, Prop. 3.3.1.5]. Condition (4) is closed thanks to condition (5); see Remark 3.4.2.11.)

3.4.3 *Properties of Kodaira–Spencer Morphisms*

Definition 3.4.3.1. (*Compare with Definitions* 1.1.2.8 *and* 1.3.1.2.) *Let*

$$(G, \lambda, i, \alpha_{\mathcal{H}^p}, \alpha^{\mathrm{ord}}_{\mathcal{H}_p}) \to S$$

be a degenerating family of type $\vec{\mathsf{M}}^{\mathrm{ord}}_{\mathcal{H}}$ *(as in Definition* 3.4.2.10*). Then we define the* $\mathscr{O}_S$*-module*

$$\underline{\mathrm{KS}} = \underline{\mathrm{KS}}_{(G,\lambda,i)/S} = \underline{\mathrm{KS}}_{(G,\lambda,i,\alpha_{\mathcal{H}^p},\alpha^{\mathrm{ord}}_{\mathcal{H}_p})/S}$$

by setting

$$\underline{\mathrm{KS}} := (\underline{\mathrm{Lie}}^\vee_{G/S} \underset{\mathscr{O}_S}{\otimes} \underline{\mathrm{Lie}}^\vee_{G^\vee/S}) \Big/ \begin{pmatrix} \lambda^*(y) \otimes z - \lambda^*(z) \otimes y \\ (b^\star x) \otimes y - x \otimes (by) \end{pmatrix}_{\substack{x \in \underline{\mathrm{Lie}}^\vee_{G/S}, \\ y,z \in \underline{\mathrm{Lie}}^\vee_{G^\vee/S}, \\ b \in \mathcal{O}}} .$$

We also define the $\mathscr{O}_S$*-module*

$$\underline{\mathrm{KS}}_{\mathrm{free}} = \underline{\mathrm{KS}}_{(G,\lambda,i)/S,\mathrm{free}} = \underline{\mathrm{KS}}_{(G,\lambda,i,\alpha_{\mathcal{H}^p},\alpha^{\mathrm{ord}}_{\mathcal{H}_p})/S,\mathrm{free}}$$

to be the quotient of $\underline{\mathrm{KS}}$ *defined as the image of the canonical morphism*

$$\underline{\mathrm{KS}} \to \underline{\mathrm{KS}} \underset{\mathbb{Z}}{\otimes} \mathbb{Q}$$

of $\mathscr{O}_S$*-modules.*

Remark 3.4.3.2. By definition, the sheaf $\underline{\mathrm{KS}}_{(G,\lambda,i)/S,\mathrm{free}}$ contains no p-torsion and hence is flat over $\vec{\mathsf{S}}_{0,r_{\mathcal{H}}} = \mathrm{Spec}(\mathcal{O}_{F_0,(p)}[\zeta_{p^{r_{\mathcal{H}}}}])$ (although it can be pathologically different from $\underline{\mathrm{KS}}_{(G,\lambda,i)/S}$ if S is not flat over $\vec{\mathsf{S}}_{0,r_{\mathcal{H}}}$).

Proposition 3.4.3.3. (*Compare with [62, Prop. 2.3.5.2].*) *Let*

$$(A, \lambda, i, \alpha_{\mathcal{H}^p}, \alpha^{\mathrm{ord}}_{\mathcal{H}_p}) \to S$$

be a tuple parameterized by $\vec{\mathsf{M}}^{\mathrm{ord}}_{\mathcal{H}}$*, with tautological morphism* $f : S \to \vec{\mathsf{M}}^{\mathrm{ord}}_{\mathcal{H}}$*. Suppose that* S *is smooth over* $\vec{\mathsf{S}}_{0,r_{\mathcal{H}}}$*, which implies that* S *is flat over* $\mathrm{Spec}(\mathbb{Z}_{(p)})$*, and that the* $\mathscr{O}_S$*-module* $\Omega^1_{S/\vec{\mathsf{S}}_{0,r_{\mathcal{H}}}}$ *is locally free. Let* $\underline{\mathrm{KS}}_{(A,\lambda,i)/S}$ *and* $\underline{\mathrm{KS}}_{(A,\lambda,i)/S,\mathrm{free}}$ *be defined by* $(A, \lambda, i, \alpha_{\mathcal{H}^p}, \alpha^{\mathrm{ord}}_{\mathcal{H}_p}) \to S$ *as in Definition* 3.4.3.1 (*with* $G = A$ *and* $S_1 = S$*; cf. Definition* 1.1.2.8). *Then the Kodaira–Spencer morphism*

$$\mathrm{KS} = \mathrm{KS}_{A/S/\vec{\mathsf{S}}_{0,r_{\mathcal{H}}}} : \underline{\mathrm{Lie}}^{\vee}_{A/S} \underset{\mathscr{O}_S}{\otimes} \underline{\mathrm{Lie}}^{\vee}_{A^\vee/S} \to \Omega^1_{S/\vec{\mathsf{S}}_{0,r_{\mathcal{H}}}}$$

(*see [62, Def. 2.1.7.9]*) *canonically induces a morphism*

$$\mathrm{KS} : \underline{\mathrm{KS}}_{(A,\lambda,i)/S} \to \Omega^1_{S/\vec{\mathsf{S}}_{0,r_{\mathcal{H}}}}, \tag{3.4.3.4}$$

which factors through

$$\mathrm{KS} : \underline{\mathrm{KS}}_{(A,\lambda,i)/S,\mathrm{free}} \to \Omega^1_{S/\vec{\mathsf{S}}_{0,r_{\mathcal{H}}}}. \tag{3.4.3.5}$$

Moreover, the morphism f *is* **étale** *if and only if it is flat and* (3.4.3.5) *is an* **isomorphism**.

Proof. The canonical morphism (3.4.3.4) exist by the same argument in the proof of [62, Prop. 2.3.5.2], because only the properties of (A, λ, i) are used there. Since $\Omega^1_{S/\vec{\mathsf{S}}_{0,r_{\mathcal{H}}}}$ is locally free over the scheme S flat over $\mathrm{Spec}(\mathbb{Z}_{(p)})$, it is also torsion-free. Hence, (3.4.3.4) factors through (3.4.3.5).

Suppose the morphism f is étale. To show that (3.4.3.5) is an isomorphism over S, it suffice to show it (universally) over $\vec{\mathsf{M}}^{\mathrm{ord}}_{\mathcal{H}}$, or rather over the completions of the strict local rings of $\vec{\mathsf{M}}^{\mathrm{ord}}_{\mathcal{H}}$ at its geometric points. Let us replace S with the spectrum of any such complete local rings, and replace $\Omega^1_{S/\vec{\mathsf{S}}_{0,r_{\mathcal{H}}}}$ with its completion $\widehat{\Omega}^1_{S/\vec{\mathsf{S}}_{0,r_{\mathcal{H}}}}$ (with respect to the topology of the complete local ring). At geometric points of residue characteristic zero, we can conclude the proof by citing [62, Prop. 2.3.5.2]. Hence, we only need to consider geometric points of residue characteristic p. By the construction of $\vec{\mathsf{M}}^{\mathrm{ord}}_{\mathcal{H}}$ (see Theorem 3.4.2.5), the scheme S is the normalization of the spectrum S' of the completion of a strict local ring of $\dddot{\mathsf{M}}^{\mathrm{ord}}_{\mathcal{H}} \underset{\mathbb{Z}_{(p)}}{\otimes} \mathcal{O}_{F_0,(p)}[\zeta_{p^{r_{\mathcal{H}}}}]$, and we may assume that $(A, \lambda, i) \to S$ is the pullback of some $(A', \lambda', i') \to S'$. Let

$\widehat{\Omega}^1_{S'/\vec{\mathsf{S}}_{0,r_{\mathcal{H}}}}$ denote the similar completion of $\Omega^1_{S'/\vec{\mathsf{S}}_{0,r_{\mathcal{H}}}}$. (These are the correct targets of the Kodaira–Spencer morphisms over completed bases which are not necessarily of finite type over $\vec{\mathsf{S}}_{0,r_{\mathcal{H}}}$; *cf.* the proof of [62, Prop. 2.3.5.2].) The canonical morphism

$$(S \to S')^*\widehat{\Omega}^1_{S'/\vec{\mathsf{S}}_{0,r_{\mathcal{H}}}} \to \widehat{\Omega}^1_{S/\vec{\mathsf{S}}_{0,r_{\mathcal{H}}}}$$

can be identified with the pullback (from $\mathrm{Spec}(\mathbb{Z}_{(p)})$ to S) of the canonical morphism

$$\underline{\mathrm{Lie}}^{\vee}_{E_{\mathsf{D},\mathbb{Z}}/\,\mathrm{Spec}(\mathbb{Z}_{(p)})} \to \underline{\mathrm{Lie}}^{\vee}_{E_{\mathsf{D},\mathbb{Z},\mathrm{free}}/\,\mathrm{Spec}(\mathbb{Z}_{(p)})}$$

induced by (3.4.2.6), with kernel the maximal torsion subsheaf of $\underline{\mathrm{Lie}}^{\vee}_{E_{\mathsf{D},\mathbb{Z}}/\,\mathrm{Spec}(\mathbb{Z}_{(p)})}$. This identification is compatible with the Kodaira–Spencer isomorphism

$$\underline{\mathrm{KS}}_{(A',\lambda',i')/S'} \xrightarrow{\sim} \widehat{\Omega}^1_{S'/\vec{\mathsf{S}}_{0,r_{\mathcal{H}}}} \tag{3.4.3.6}$$

induced by the Kodaira–Spencer morphism in the Serre–Tate deformation theory (see [47]) (which is compatible with the usual Kodaira–Spencer morphism defined over smooth schemes), and with the compatible canonical isomorphisms from $\underline{\mathrm{KS}}_{(A',\lambda',i')/S'}$ and $\underline{\mathrm{KS}}_{(A,\lambda,i)/S}$ to pullbacks of $\underline{\mathrm{Lie}}^{\vee}_{E_{\mathsf{D},\mathbb{Z}}/\,\mathrm{Spec}(\mathbb{Z}_{(p)})}$ to S' and S, respectively, induced by (6) of Proposition 3.2.1.1. Thus, we see that the pullback of the isomorphism (3.4.3.6) (from S' to S) induces the desired isomorphism.

Conversely, suppose (3.4.3.5) is an isomorphism. By the previous paragraph, we have an isomorphism $\mathrm{KS} : \underline{\mathrm{KS}}_{(A,\lambda,i)/\vec{\mathsf{M}}^{\mathrm{ord}}_{\mathcal{H}},\mathrm{free}} \xrightarrow{\sim} \Omega^1_{\vec{\mathsf{M}}^{\mathrm{ord}}_{\mathcal{H}}/\vec{\mathsf{S}}_{0,r_{\mathcal{H}}}}$, where by abuse of notation we have also used (A, λ, i) to denote the tautological objects over $\vec{\mathsf{M}}^{\mathrm{ord}}_{\mathcal{H}}$. Since the construction of $\underline{\mathrm{KS}}_{(A,\lambda,i)/S,\mathrm{free}}$ commutes with flat base change (between schemes smooth over $\vec{\mathsf{S}}_{0,r_{\mathcal{H}}}$), and since the association of Kodaira–Spencer morphisms is functorial, the first morphism in the exact sequence

$$f^*\Omega^1_{\vec{\mathsf{M}}_{\mathcal{H},r_{\mathcal{H}}}/\vec{\mathsf{S}}_{0,r_{\mathcal{H}}}} \to \Omega^1_{S/\vec{\mathsf{S}}_{0,r_{\mathcal{H}}}} \to \Omega^1_{S/\vec{\mathsf{M}}_{\mathcal{H},r_{\mathcal{H}}}} \to 0$$

is an isomorphism. This shows that f is unramified, and hence étale because it is flat by assumption. □

3.4.4 *Hecke Actions*

Proposition 3.4.4.1. (*Compare with Propositions* 1.3.1.14 *and* 2.2.3.1.) *Suppose we have an element* $g = (g_0, g_p) \in \mathrm{G}(\mathbb{A}^{\infty,p}) \times \mathrm{P}^{\mathrm{ord}}_{\mathsf{D}}(\mathbb{Q}_p) \subset \mathrm{G}(\mathbb{A}^{\infty})$ (*see Definition* 3.2.2.7), *and suppose we have two open compact subgroups* $\mathcal{H}$ *and* $\mathcal{H}'$ *of* $\mathrm{G}(\hat{\mathbb{Z}})$ *such that* $\mathcal{H}' \subset g\mathcal{H}g^{-1}$, *and such that* $\mathcal{H}$ *and* $\mathcal{H}'$ *are of standard form as in Definition* 3.2.2.9 (*so that* $\dddot{\mathsf{M}}^{\mathrm{ord}}_{\mathcal{H}}$ *and* $\dddot{\mathsf{M}}^{\mathrm{ord}}_{\mathcal{H}'}$ *are defined as in Definition* 3.4.1.1). *Suppose moreover that* g_p *satisfies the conditions given in Section* 3.3.4. *Then the assignment in Proposition* 3.3.4.21 *of ordinary Hecke twists by* g *induces a canonical quasi-finite surjection*

$$\dddot{[g]}^{\mathrm{ord}} : \dddot{\mathsf{M}}^{\mathrm{ord}}_{\mathcal{H}'} \to \dddot{\mathsf{M}}^{\mathrm{ord}}_{\mathcal{H}}$$

(over $\mathrm{Spec}(\mathbb{Z}_{(p)})$*), such that the tautological tuple over* $\dddot{\mathsf{M}}_{\mathcal{H}}^{\mathrm{ord}}$ *is pulled back to the ordinary Hecke twist of the tautological tuple over* $\dddot{\mathsf{M}}_{\mathcal{H}'}^{\mathrm{ord}}$ *by* g*. The surjection* $\dddot{[g]}^{\mathrm{ord}}$ *is finite if the levels* $\mathcal{H}_p$ *and* $\mathcal{H}'_p$ *at* p *are* **equally deep** *as in Definition* 3.2.2.9*, or if* g_p *is of twisted* U_p *type as in Definition* 3.3.6.1 *and* $\mathrm{depth}_{\mathsf{D}}(\mathcal{H}'_p) - \mathrm{depth}_{\mathsf{D}}(g_p) = \mathrm{depth}_{\mathsf{D}}(\mathcal{H}_p) > 0$.

By Proposition 3.3.5.6*, the surjection* $\dddot{[g]}^{\mathrm{ord}} : \dddot{\mathsf{M}}_{\mathcal{H}'}^{\mathrm{ord}} \to \dddot{\mathsf{M}}_{\mathcal{H}}^{\mathrm{ord}}$ *is compatible with the surjection* $[g] : \mathsf{M}_{\mathcal{H}'} \to \mathsf{M}_{\mathcal{H}}$ *over* $\mathsf{M}_{\mathcal{H}'}^{\mathrm{ord}}$*, and induces surjective quasi-finite flat morphisms*

$$[g]^{\mathrm{ord}} : \mathsf{M}_{\mathcal{H}'}^{\mathrm{ord}} \to \mathsf{M}_{\mathcal{H}}^{\mathrm{ord}}$$

(compatible with $\mathsf{S}_{0,r_{\mathcal{H}'}} \to \mathsf{S}_{0,r_{\mathcal{H}}}$*) and*

$$\vec{[g]}^{\mathrm{ord}} : \vec{\mathsf{M}}_{\mathcal{H}'}^{\mathrm{ord}} \to \vec{\mathsf{M}}_{\mathcal{H}}^{\mathrm{ord}}$$

(compatible with $\vec{\mathsf{S}}_{0,r_{\mathcal{H}'}} \to \vec{\mathsf{S}}_{0,r_{\mathcal{H}}}$*) compatible with each other. If* $g_p \in \mathrm{P}_{\mathsf{D}}^{\mathrm{ord}}(\mathbb{Z}_p)$*, then the induced morphisms*

$$[g]_{r_{\mathcal{H}'}}^{\mathrm{ord}} : \mathsf{M}_{\mathcal{H}'}^{\mathrm{ord}} \to \mathsf{M}_{\mathcal{H}}^{\mathrm{ord}} \underset{\vec{\mathsf{S}}_{0,r_{\mathcal{H}}}}{\times} \vec{\mathsf{S}}_{0,r_{\mathcal{H}'}}$$

and

$$\vec{[g]}_{r_{\mathcal{H}'}}^{\mathrm{ord}} : \vec{\mathsf{M}}_{\mathcal{H}'}^{\mathrm{ord}} \to \vec{\mathsf{M}}_{\mathcal{H}}^{\mathrm{ord}} \underset{\vec{\mathsf{S}}_{0,r_{\mathcal{H}}}}{\times} \vec{\mathsf{S}}_{0,r_{\mathcal{H}'}}$$

are quasi-finite étale. These morphisms induced by $\dddot{[g]}^{\mathrm{ord}}$ *are characterized by the property that the pullback of the tautological tuple is the ordinary Hecke twist of the tautological tuple by* g*. They are finite if* $\dddot{[g]}^{\mathrm{ord}}$ *is finite (see the previous paragraph).*

If $g = g_1 g_2$*, where* $g_1 = (g_{1,0}, g_{1,p})$ *and* $g_2 = (g_{2,0}, g_{2,p})$ *are elements of* $\mathrm{G}(\mathbb{A}^{\infty,p}) \times \mathrm{P}_{\mathsf{D}}^{\mathrm{ord}}(\mathbb{Q}_p)$*, each having a setup similar to that of* g*, then we have* $\dddot{[g]}^{\mathrm{ord}} = \dddot{[g_2]}^{\mathrm{ord}} \circ \dddot{[g_1]}^{\mathrm{ord}}$*, inducing* $\vec{[g]}^{\mathrm{ord}} = \vec{[g_2]}^{\mathrm{ord}} \circ \vec{[g_1]}^{\mathrm{ord}}$.

Proof. The morphism $\dddot{[g]}^{\mathrm{ord}} : \dddot{\mathsf{M}}_{\mathcal{H}'}^{\mathrm{ord}} \to \dddot{\mathsf{M}}_{\mathcal{H}}^{\mathrm{ord}}$ (uniquely) exists by the definition of ordinary Hecke twists, and by the definition of $\dddot{\mathsf{M}}_{\mathcal{H}}^{\mathrm{ord}}$ as a moduli problem. Its surjectivity is a consequence of the liftability conditions in the definitions of level structures. Its quasi-finiteness, and its finiteness when the levels $\mathcal{H}_p$ and $\mathcal{H}'_p$ at p are equally deep, follow from the definition of ordinary level structures as orbits of étale-locally-defined principal ordinary level structures (see Definitions 3.3.3.3 and 3.3.3.4). When g_p is of twisted U_p type as in Definition 3.3.6.1 and $\mathrm{depth}_{\mathsf{D}}(\mathcal{H}'_p) - \mathrm{depth}_{\mathsf{D}}(g_p) = \mathrm{depth}_{\mathsf{D}}(\mathcal{H}_p) > 0$, since $\dddot{[g]}^{\mathrm{ord}}$ is (of finite presentation and) quasi-finite, its finiteness follows from Lemma 3.3.6.6 (which verifies its properness by the valuative criterion; *cf.* [35, IV-3, 8.11.1]). Since $\vec{\mathsf{M}}_{\mathcal{H}'}^{\mathrm{ord}}$ and $\vec{\mathsf{M}}_{\mathcal{H}}^{\mathrm{ord}}$ are regular and equidimensional of the same dimension (see Theorem 3.4.2.5), the morphism $[g]^{\mathrm{ord}} : \vec{\mathsf{M}}_{\mathcal{H}'}^{\mathrm{ord}} \to \vec{\mathsf{M}}_{\mathcal{H}}^{\mathrm{ord}}$ is automatically flat (by [35, IV-3, 15.4.2 e′)⇒b)]; *cf.* [62, Lem.

6.3.1.11]). The induced morphism $\vec{[g]}_{r_{\mathcal{H}'}}^{\mathrm{ord}} : \vec{\mathsf{M}}_{\mathcal{H}'}^{\mathrm{ord}} \to \vec{\mathsf{M}}_{\mathcal{H}}^{\mathrm{ord}} \underset{\vec{\mathsf{S}}_{0,r_{\mathcal{H}}}}{\times} \vec{\mathsf{S}}_{0,r_{\mathcal{H}'}}$ is étale because it induces an isomorphism between the completions of strict local rings (again see Theorem 3.4.2.5). The last statement of this proposition follows from the last statement of Proposition 3.3.4.21. The remaining statements of this proposition are self-explanatory. □

Definition 3.4.4.2. *For each algebraic stack or scheme over* $\mathrm{Spec}(\mathbb{Z})$*, such as* $\vec{\mathsf{M}}_{\mathcal{H}}$*, we denote in Fraktur its formal completion along its fiber over* $\mathrm{Spec}(\mathbb{F}_p) \to \mathrm{Spec}(\mathbb{Z})$*, such as* $\vec{\mathfrak{M}}_{\mathcal{H}}$*, and consider it a formal algebraic stack over* $\mathrm{Spf}(\mathbb{Z}_p)$*, with support an algebraic stack over* $\mathrm{Spec}(\mathbb{F}_p)$*. By abuse of notation, we denote the formal completion of* $[\vec{\mathsf{M}}_{\mathcal{H}}]$ *as* $[\vec{\mathfrak{M}}_{\mathcal{H}}]$*.*

Corollary 3.4.4.3. *With the setting as in Proposition* 3.4.4.1*, the morphism*

$$\vec{[g]}^{\mathrm{ord}} : \vec{\mathfrak{M}}_{\mathcal{H}'}^{\mathrm{ord}} \to \vec{\mathfrak{M}}_{\mathcal{H}}^{\mathrm{ord}}$$

induced by $\vec{[g]}^{\mathrm{ord}} : \vec{\mathsf{M}}_{\mathcal{H}'}^{\mathrm{ord}} \to \vec{\mathsf{M}}_{\mathcal{H}}^{\mathrm{ord}}$ *is finite flat. Hence, if* $g_p \in \mathrm{P}_{\mathsf{D}}^{\mathrm{ord}}(\mathbb{Z}_p)$*, then the induced morphism*

$$\vec{[g]}_{r_{\mathcal{H}'}}^{\mathrm{ord}} : \vec{\mathfrak{M}}_{\mathcal{H}'}^{\mathrm{ord}} \to \vec{\mathfrak{M}}_{\mathcal{H}}^{\mathrm{ord}} \underset{\vec{\mathfrak{S}}_{0,r_{\mathcal{H}}}}{\times} \vec{\mathfrak{S}}_{0,r_{\mathcal{H}'}}$$

is finite étale (*because it is quasi-finite étale by Proposition* 3.4.4.1).

Proof. The quasi-finite flat morphisms $\vec{[g]}^{\mathrm{ord}} : \vec{\mathfrak{M}}_{\mathcal{H}'}^{\mathrm{ord}} \to \vec{\mathfrak{M}}_{\mathcal{H}}^{\mathrm{ord}}$ and $\vec{[g]}_{r_{\mathcal{H}'}}^{\mathrm{ord}} : \vec{\mathfrak{M}}_{\mathcal{H}'}^{\mathrm{ord}} \to \vec{\mathfrak{M}}_{\mathcal{H}}^{\mathrm{ord}} \underset{\vec{\mathfrak{S}}_{0,r_{\mathcal{H}}}}{\times} \vec{\mathfrak{S}}_{0,r_{\mathcal{H}'}}$ are finite because the induced morphism

$$\vec{[g]}^{\mathrm{ord}} : \vec{\mathsf{M}}_{\mathcal{H}'}^{\mathrm{ord}} \underset{\mathbb{Z}}{\otimes} \mathbb{F}_p \to \vec{\mathsf{M}}_{\mathcal{H}}^{\mathrm{ord}} \underset{\mathbb{Z}}{\otimes} \mathbb{F}_p$$

is proper by Lemma 3.3.6.8 (*cf.* [35, IV-3, 8.11.5, or IV-4, 8.12.6]). □

Corollary 3.4.4.4. *With the setting as in Proposition* 3.4.4.1*, if* $g = (g_0, g_p) \in \mathrm{G}(\hat{\mathbb{Z}}^p) \times \mathrm{P}_{\mathsf{D}}^{\mathrm{ord}}(\mathbb{Z}_p)$ (*cf. Examples* 3.3.4.5 *and* 3.3.4.18)*, if* $\mathcal{H}'^{,p} = g_0 \mathcal{H}^p g_0^{-1}$ *in* $\mathrm{G}(\hat{\mathbb{Z}}^p)$*, and if* ${\mathcal{H}'_p}^{\mathrm{ord}} = (g_p \mathcal{H}_p g_p^{-1})^{\mathrm{ord}}$ *in* $\mathrm{M}_{\mathsf{D}}^{\mathrm{ord}}(\mathbb{Z}_p)$ (*see* (3.3.3.5))*, then* ($r_{\mathcal{H}'} = r_{\mathcal{H}}$ *and*) *the induced morphism* $\vec{[g]}^{\mathrm{ord}} : \vec{\mathfrak{M}}_{\mathcal{H}'}^{\mathrm{ord}} \to \vec{\mathfrak{M}}_{\mathcal{H}}^{\mathrm{ord}}$ *is an isomorphism.* (*These conditions are true, in particular, when* $g = 1$ *and when* $\mathcal{H} = \mathcal{H}^p \mathcal{H}_p$ *and* $\mathcal{H}' = \mathcal{H}'^{,p} \mathcal{H}'_p$ *satisfy* $\mathcal{H}'^{,p} = \mathcal{H}^p$ *and* ${\mathcal{H}'_p}^{\mathrm{ord}} = \mathcal{H}_p^{\mathrm{ord}}$.)

Proof. By assumption (and by the moduli interpretations), the canonical morphism $\dddot{[g]} : \ddot{\mathsf{M}}_{\mathcal{H}'} \to \ddot{\mathsf{M}}_{\mathcal{H}}$ between the moduli problems induces a bijection $\ddot{\mathsf{M}}_{\mathcal{H}'}(\bar{\mathbb{F}}_p) \to \ddot{\mathsf{M}}_{\mathcal{H}}(\bar{\mathbb{F}}_p)$ between their $\bar{\mathbb{F}}_p$-valued points. Hence, by the description of the local structures of $\ddot{\mathsf{M}}_{\mathcal{H}'}$ and $\ddot{\mathsf{M}}_{\mathcal{H}}$ in Theorem 3.4.2.5 (which asserts that the completions of strict local rings of both of them at their $\bar{\mathbb{F}}_p$-points are isomorphic to the completions of strict local rings of the same group of multiplicative type), and by the definition of

$\vec{\mathsf{M}}^{\mathrm{ord}}_{\mathcal{H}'}$ and $\vec{\mathsf{M}}^{\mathrm{ord}}_{\mathcal{H}}$ as their respective normalizations, we see that $\vec{[g]}^{\mathrm{ord}} : \vec{\mathsf{M}}^{\mathrm{ord}}_{\mathcal{H}'} \to \vec{\mathsf{M}}^{\mathrm{ord}}_{\mathcal{H}}$ also induces a bijection $\vec{\mathsf{M}}^{\mathrm{ord}}_{\mathcal{H}'}(\bar{\mathbb{F}}_p) \to \vec{\mathsf{M}}^{\mathrm{ord}}_{\mathcal{H}}(\bar{\mathbb{F}}_p)$ between their $\bar{\mathbb{F}}_p$-valued points. Hence, by Corollary 3.4.4.3, the induced morphism $\vec{[g]}^{\mathrm{ord}} : \vec{\mathfrak{M}}^{\mathrm{ord}}_{\mathcal{H}'} \to \vec{\mathfrak{M}}^{\mathrm{ord}}_{\mathcal{H}}$ (being finite étale and a bijection on $\bar{\mathbb{F}}_p$-valued points) is an isomorphism. □

Example 3.4.4.5. The condition in Corollary 3.4.4.4 that $\mathcal{H}'^{\,\mathrm{ord}}_p = (g_p \mathcal{H}_p g_p^{-1})^{\mathrm{ord}}$ is also true, for example, when $g_p = 1$, $\mathcal{H}_p = \mathcal{U}_{p,1,0}(p^{r_1}, p^{r_0})$, and $\mathcal{H}'_p = \mathcal{U}_{p,1,0}(p^{r_1}, p^{r'_0})$ for some $r_1 \leq r_0 \leq r'_0$.

Corollary 3.4.4.6 (elements of U_p type). *Suppose in Proposition* 3.4.4.1 *that* $g_0 = 1$ *and* g_p *is* **of U_p type** *as in Definition* 3.3.6.1 (*so that it is of twisted* U_p *type and* $\mathrm{depth}_{\mathsf{D}}(g_p) = 1$). *Then the induced morphism*

$$\vec{[g]}^{\mathrm{ord}} : \vec{\mathsf{M}}^{\mathrm{ord}}_{\mathcal{H}'} \underset{\mathbb{Z}}{\otimes} \mathbb{F}_p \to \vec{\mathsf{M}}^{\mathrm{ord}}_{\mathcal{H}} \underset{\mathbb{Z}}{\otimes} \mathbb{F}_p \tag{3.4.4.7}$$

is finite flat and coincides with the composition of the (*finite flat*) *absolute Frobenius morphism*

$$\mathrm{F}_{\vec{\mathsf{M}}^{\mathrm{ord}}_{\mathcal{H}'} \underset{\mathbb{Z}}{\otimes} \mathbb{F}_p} : \vec{\mathsf{M}}^{\mathrm{ord}}_{\mathcal{H}'} \underset{\mathbb{Z}}{\otimes} \mathbb{F}_p \to \vec{\mathsf{M}}^{\mathrm{ord}}_{\mathcal{H}'} \underset{\mathbb{Z}}{\otimes} \mathbb{F}_p$$

with the canonical finite flat morphism

$$\vec{[1]}^{\mathrm{ord}} : \vec{\mathsf{M}}^{\mathrm{ord}}_{\mathcal{H}'} \underset{\mathbb{Z}}{\otimes} \mathbb{F}_p \to \vec{\mathsf{M}}^{\mathrm{ord}}_{\mathcal{H}} \underset{\mathbb{Z}}{\otimes} \mathbb{F}_p \tag{3.4.4.8}$$

(*see Corollary* 3.4.4.3).

If $\mathcal{H}'^{\,\mathrm{ord}}_p = \mathcal{H}^{\mathrm{ord}}_p$ *as open compact subgroups of* $\mathrm{M}^{\mathrm{ord}}_{\mathsf{D}}(\mathbb{Z}_p)$ (*see* (3.3.3.5)), *then* ($r_{\mathcal{H}'} = r_{\mathcal{H}}$ *and*) *the canonical morphism* (3.4.4.8) *is an isomorphism by Corollary* 3.4.4.4, *and the composition*

$$\vec{\mathsf{M}}^{\mathrm{ord}}_{\mathcal{H}} \underset{\mathbb{Z}}{\otimes} \mathbb{F}_p \overset{(\vec{[1]}^{\mathrm{ord}})^{-1}}{\underset{\sim}{\to}} \vec{\mathsf{M}}^{\mathrm{ord}}_{\mathcal{H}'} \underset{\mathbb{Z}}{\otimes} \mathbb{F}_p \overset{\vec{[g]}^{\mathrm{ord}}}{\to} \vec{\mathsf{M}}^{\mathrm{ord}}_{\mathcal{H}} \underset{\mathbb{Z}}{\otimes} \mathbb{F}_p$$

coincides with the (*finite flat*) *absolute Frobenius morphism*

$$\mathrm{F}_{\vec{\mathsf{M}}^{\mathrm{ord}}_{\mathcal{H}} \underset{\mathbb{Z}}{\otimes} \mathbb{F}_p} : \vec{\mathsf{M}}^{\mathrm{ord}}_{\mathcal{H}} \underset{\mathbb{Z}}{\otimes} \mathbb{F}_p \to \vec{\mathsf{M}}^{\mathrm{ord}}_{\mathcal{H}} \underset{\mathbb{Z}}{\otimes} \mathbb{F}_p.$$

Proof. Since $g_0 = 1$, the ordinary Hecke twist $(A', \lambda', i', \alpha'_{\mathcal{H}^p}, \alpha^{\mathrm{ord}}_{\mathcal{H}_p})$ of the tautological object $(A, \lambda, i, \alpha_{\mathcal{H}'^{,p}}, \alpha^{\mathrm{ord}}_{\mathcal{H}'_p})$ over $\vec{\mathsf{M}}^{\mathrm{ord}}_{\mathcal{H}'} \underset{\mathbb{Z}}{\otimes} \mathbb{F}_p$ by $g = (g_0, g_p)$ is defined essentially only by g_p, equipped with the morphism $[g_p^{-1}] : A \to A'$ which is nothing but the relative Frobenius morphism $\mathrm{F}_{A/\vec{\mathsf{M}}^{\mathrm{ord}}_{\mathcal{H}'} \underset{\mathbb{Z}}{\otimes} \mathbb{F}_p} : A \to A' \cong A^{(p)}$, with the additional structures naturally induced, as explained in Examples 3.3.4.7 and 3.3.4.20. Hence, $(A', \lambda', i', \alpha'_{\mathcal{H}^p}, \alpha^{\mathrm{ord}}_{\mathcal{H}_p})$ coincides with the object naturally induced by the pullback of $(A, \lambda, i, \alpha_{\mathcal{H}'^{,p}}, \alpha^{\mathrm{ord}}_{\mathcal{H}'_p})$ by the absolute Frobenius $\mathrm{F}_{\vec{\mathsf{M}}^{\mathrm{ord}}_{\mathcal{H}'} \underset{\mathbb{Z}}{\otimes} \mathbb{F}_p}$. Hence, the first paragraph of the corollary follows. The second paragraph of the corollary is self-explanatory. □

Remark 3.4.4.9. By Kunz's theorem [54] (*cf.* [76, Sec. 42, Thm. 107]), the absolute Frobenius morphisms $\mathrm{F}_{\vec{\mathsf{M}}^{\mathrm{ord}}_{\mathcal{H}'} \underset{\mathbb{Z}}{\otimes} \mathbb{F}_p}$ and $\mathrm{F}_{\vec{\mathsf{M}}^{\mathrm{ord}}_{\mathcal{H}} \underset{\mathbb{Z}}{\otimes} \mathbb{F}_p}$ in Corollary 3.4.4.6 are flat because $\vec{\mathsf{M}}^{\mathrm{ord}}_{\mathcal{H}'} \underset{\mathbb{Z}}{\otimes} \mathbb{F}_p$ and $\vec{\mathsf{M}}^{\mathrm{ord}}_{\mathcal{H}} \underset{\mathbb{Z}}{\otimes} \mathbb{F}_p$ are *regular* (by smoothness of $\vec{\mathsf{M}}^{\mathrm{ord}}_{\mathcal{H}'}$ and $\vec{\mathsf{M}}^{\mathrm{ord}}_{\mathcal{H}}$ over $\vec{\mathsf{S}}_{0,r_{\mathcal{H}'}}$ and $\vec{\mathsf{S}}_{0,r_{\mathcal{H}}}$, respectively; see Theorem 3.4.2.5).

3.4.5 *The Case When p is a Good Prime*

Let $\mathcal{H}$, $\mathcal{H}^p$, and $\mathcal{H}_p$ be as in Section 3.4.1.

When p is a good prime (for the integral PEL datum $(\mathcal{O}, \star, L, \langle \cdot, \cdot \rangle, h_0)$; see Definition 1.1.1.6), we can define $\vec{\mathsf{M}}^{\mathrm{ord}}_{\mathcal{H}}$ directly without taking the normalization of an object in characteristic zero. In this case, the pairing $\langle \cdot, \cdot \rangle$ is self-dual after base change to $\mathbb{Z}_p$, and hence we can define $\mathsf{M}_{\mathcal{H}^p}$ over $\vec{\mathsf{S}}_0 = \mathrm{Spec}(\mathcal{O}_{F_0,(p)})$ as in [62, Def. 1.4.1.4].

On the other hand, consider $\mathcal{U}_{p,0}(p^0) = \mathrm{G}(\mathbb{Z}_p)$. Then $r_{\mathsf{D}} = 0$ by [62, Prop. 1.2.2.3] and by the assumption that p is a good prime, and $\nu(\mathrm{G}(\mathbb{Z}_p)) = \mathbb{Z}_p^\times$ implies that $r_{\mathcal{H}^p\mathrm{G}(\mathbb{Z}_p)} = r_{\nu(\mathrm{G}(\mathbb{Z}_p))} = 0$ (see Definition 3.4.2.1). Let $\mathcal{H}' := \mathcal{H}^p\mathrm{G}(\mathbb{Z}_p)$. Then we can define $\vec{\mathsf{M}}^{\mathrm{ord}}_{\mathcal{H}'}$ over $\vec{\mathsf{S}}_0$ as in Theorem 3.4.2.5, such that $\vec{\mathsf{M}}^{\mathrm{ord}}_{\mathcal{H}'} \underset{\mathbb{Z}}{\otimes} \mathbb{Q} \cong \mathsf{M}^{\mathrm{ord}}_{\mathcal{H}'} \cong \mathsf{M}_{\mathcal{H}'}$ over S_0.

Lemma 3.4.5.1. *There is a canonical open immersion*

$$\vec{\mathsf{M}}^{\mathrm{ord}}_{\mathcal{H}'} \hookrightarrow \mathsf{M}_{\mathcal{H}^p} \tag{3.4.5.2}$$

such that the pullback of the tautological object $(A, \lambda, i, \alpha_{\mathcal{H}^p})$ *over* $\mathsf{M}_{\mathcal{H}^p}$ *is part of the tautological object over* $\vec{\mathsf{M}}^{\mathrm{ord}}_{\mathcal{H}'}$ *(cf. Convention* 3.4.2.9*).*

Proof. Consider the open immersion (*cf.* [35, IV-4, 17.9.1, and IV-2, 6.15.3])

$$\mathsf{M}^{\mathrm{ord}}_{\mathcal{H}^p} \hookrightarrow \mathsf{M}_{\mathcal{H}^p} \tag{3.4.5.3}$$

representing the ordinary level structure $\alpha^{\mathrm{ord}}_{\mathrm{G}(\mathbb{Z}_p)}$ over $\mathsf{M}_{\mathcal{H}^p}$ (which is unique up to isomorphism if it exists). The two moduli problems $\mathsf{M}^{\mathrm{ord}}_{\mathcal{H}^p}$ and $\dddot{\mathsf{M}}^{\mathrm{ord}}_{\mathcal{H}'}$ (*cf.* Definition 3.4.1.1) are almost identical, except that the former requires the Lie algebra condition in [62, Def. 1.3.4.1] given by $(L \underset{\mathbb{Z}}{\otimes} \mathbb{R}, \langle \cdot, \cdot \rangle, h_0)$ (and hence has to be defined over the finite extension $\mathcal{O}_{F_0,(p)}$ of $\mathbb{Z}_{(p)}$). Therefore, we have a canonical finite morphism

$$\mathsf{M}^{\mathrm{ord}}_{\mathcal{H}^p} \to \dddot{\mathsf{M}}^{\mathrm{ord}}_{\mathcal{H}'}. \tag{3.4.5.4}$$

On the other hand, the tautological tuple over $\vec{\mathsf{M}}^{\mathrm{ord}}_{\mathcal{H}'}$ satisfies the Lie algebra condition, because the condition is given by an identity of polynomials (which is a closed condition), and because it is already satisfied over the generic fiber $\mathsf{M}^{\mathrm{ord}}_{\mathcal{H}'} \cong \mathsf{M}_{\mathcal{H}'}$. Therefore, by Proposition 3.4.3.3 and [62, Prop. 2.3.5.2], and by the valuative criterion of properness, there is a canonical finite étale morphism

$$\vec{\mathsf{M}}^{\mathrm{ord}}_{\mathcal{H}'} \to \mathsf{M}^{\mathrm{ord}}_{\mathcal{H}^p} \tag{3.4.5.5}$$

by the universal property of $\mathsf{M}^{\mathrm{ord}}_{\mathcal{H}^p}$, whose composition with (3.4.5.4) is the canonical finite morphism from $\vec{\mathsf{M}}^{\mathrm{ord}}_{\mathcal{H}'}$ to $\dddot{\mathsf{M}}^{\mathrm{ord}}_{\mathcal{H}'}$. Comparing this with the open and closed immersion (2.2.4.1), we see that (3.4.5.5) is also an open and closed immersion, and that the composition of (3.4.5.5) with (3.4.5.3) gives the desired open immersion (3.4.5.2). □

By the liftability of level structures, we have a canonical finite étale surjection $\mathsf{M}_{\mathcal{H}} \twoheadrightarrow \mathsf{M}_{\mathcal{H}'}$ and a canonical quasi-finite étale surjection $\dddot{\mathsf{M}}^{\mathrm{ord}}_{\mathcal{H}} \twoheadrightarrow \dddot{\mathsf{M}}^{\mathrm{ord}}_{\mathcal{H}'}$, which induces a canonical quasi-finite étale morphism

$$\vec{\mathsf{M}}^{\mathrm{ord}}_{\mathcal{H}} \twoheadrightarrow \vec{\mathsf{M}}^{\mathrm{ord}}_{\mathcal{H}'} \underset{\vec{\mathsf{S}}_0}{\times} \vec{\mathsf{S}}_{0,r_{\mathcal{H}}}. \tag{3.4.5.6}$$

(Here $r_{\mathcal{H}} = r_\nu$ because $r_{\mathtt{D}} = 0$.)

Alternatively, we have:

Proposition 3.4.5.7. *With assumptions as above, we can construct* (3.4.5.6), *or rather the canonical quasi-finite étale morphism*

$$\vec{\mathsf{M}}^{\mathrm{ord}}_{\mathcal{H}} \to \mathsf{M}_{\mathcal{H}^p} \underset{\vec{\mathsf{S}}_0}{\times} \vec{\mathsf{S}}_{0,r_{\mathcal{H}}} \tag{3.4.5.8}$$

(*which is the composition of* (3.4.5.6) *with* (3.4.5.2)), *as a canonical open and closed subalgebraic stack, given by taking the schematic closure of* $\mathsf{M}^{\mathrm{ord}}_{\mathcal{H}}$, *in a relatively representable functor of ordinary level-*$\mathcal{H}_p$ *structures of type* $(L \underset{\mathbb{Z}}{\otimes} \mathbb{Z}_p, \langle\,\cdot\,,\,\cdot\,\rangle, \mathtt{D})$.

Proof. The relative representability, quasi-finiteness, and étaleness are all clear form the definitions. □

3.4.6 *Quasi-Projectivity of Coarse Moduli*

In this subsection, we no longer assume that p is a good prime (for the integral PEL datum $(\mathcal{O}, \star, L, \langle\,\cdot\,,\,\cdot\,\rangle, h_0)$).

Our ultimate source of projectivity or quasi-projectivity is [80, IX, 2.1], or its reformulation in [28, Ch. V, Prop. 2.1] and [62, Prop. 7.2.1.1]. For this purpose, we need a semi-abelian scheme over a proper algebraic stack, and we shall resort to the auxiliary moduli problems (again) and their compactifications (also). (This seems to be our only source of projectivity, since geometric invariant theory as in [83] is not known to be applicable to the construction of minimal compactifications.)

Lemma 3.4.6.1. *Let* $\mathcal{H}$ *and* $\mathcal{H}_{\mathrm{aux}}$ *be as in Lemma* 3.4.1.6. *Let*

$$\omega_{\vec{\mathsf{M}}^{\mathrm{ord}}_{\mathcal{H}}} := \wedge^{\mathrm{top}} \underline{\mathrm{Lie}}^{\vee}_{A/\vec{\mathsf{M}}^{\mathrm{ord}}_{\mathcal{H}}} = \wedge^{\mathrm{top}} e_A^* \Omega^1_{A/\vec{\mathsf{M}}^{\mathrm{ord}}_{\mathcal{H}}},$$

and

$$\omega_{\vec{\mathsf{M}}^{\mathrm{ord}}_{\mathcal{H}_{\mathrm{aux}}}} := \wedge^{\mathrm{top}} \underline{\mathrm{Lie}}^{\vee}_{A_{\mathrm{aux}}/\vec{\mathsf{M}}^{\mathrm{ord}}_{\mathcal{H}_{\mathrm{aux}}}} = \wedge^{\mathrm{top}} e_{A_{\mathrm{aux}}}^* \Omega^1_{A_{\mathrm{aux}}/\vec{\mathsf{M}}^{\mathrm{ord}}_{\mathcal{H}_{\mathrm{aux}}}},$$

where $A \to \vec{\mathsf{M}}^{\mathrm{ord}}$ *(resp.* $A_{\mathrm{aux}} \to \vec{\mathsf{M}}^{\mathrm{ord}}_{\mathcal{H}_{\mathrm{aux}}}$*) is the tautological abelian scheme with identity section* $e_A : \vec{\mathsf{M}}^{\mathrm{ord}} \to A$ *(resp.* $e_{A_{\mathrm{aux}}} : \vec{\mathsf{M}}^{\mathrm{ord}}_{\mathcal{H}_{\mathrm{aux}}} \to A_{\mathrm{aux}}$*). Then the morphism* (3.4.1.7) *canonically induces a quasi-finite morphism*

$$\vec{\mathsf{M}}^{\mathrm{ord}}_{\mathcal{H}} \to \vec{\mathsf{M}}^{\mathrm{ord}}_{\mathcal{H}_{\mathrm{aux}}} \tag{3.4.6.2}$$

compatible with (2.1.1.17)*, such that the pullback of* $\omega^{\otimes a_0}_{\vec{\mathsf{M}}^{\mathrm{ord}}_{\mathcal{H}_{\mathrm{aux}}}}$ *is canonically isomorphic to* $\omega^{\otimes a}_{\vec{\mathsf{M}}^{\mathrm{ord}}_{\mathcal{H}}}$*, where* $a_0 \geq 1$ *and* $a \geq 1$ *are integers as in Lemma* 2.1.2.35.

Proof. These follow from the constructions of the morphisms (3.4.1.5) and (3.4.1.7) (see the proofs of Lemmas 3.4.1.4 and 3.4.1.6) and the two normalizations $\mathsf{M}^{\mathrm{ord}}_{\mathcal{H}}$ and $\mathsf{M}^{\mathrm{ord}}_{\mathcal{H}_{\mathrm{aux}}}$, and from Lemma 3.4.1.8. □

Proposition 3.4.6.3. *With the setting as in Proposition* 3.4.2.5*, there is a canonical morphism*

$$\vec{\mathsf{M}}^{\mathrm{ord}}_{\mathcal{H}} \to \vec{\mathsf{M}}_{\mathcal{H}} \tag{3.4.6.4}$$

(see Proposition 2.2.1.1*) inducing an open immersion*

$$[\vec{\mathsf{M}}^{\mathrm{ord}}_{\mathcal{H}}] \hookrightarrow [\vec{\mathsf{M}}_{\mathcal{H},r_{\mathcal{H}}}] \tag{3.4.6.5}$$

(see Definition 2.2.3.5*). Under* (3.4.6.4)*, the pullback of the tautological* $(\vec{A}, \vec{\lambda}, \vec{i})$ *(see Proposition* 2.2.1.1*) is canonically isomorphic to the tautological* (A, λ, i) *over* $\vec{\mathsf{M}}^{\mathrm{ord}}_{\mathcal{H}}$*. If* $\mathcal{H}^p$ *is neat, then (*$\mathcal{H} = \mathcal{H}^p\mathcal{H}_p$ *is also neat and)* $\vec{\mathsf{M}}_{\mathcal{H},r_{\mathcal{H}}}$ *is a scheme quasi-projective over* $\vec{\mathsf{S}}_{0,r_{\mathcal{H}}}$ *by Proposition* 2.2.1.1*, and the algebraic stack* $\vec{\mathsf{M}}^{\mathrm{ord}}_{\mathcal{H}}$ *(which, a priori, is an algebraic space by Theorem* 3.4.2.5*) is a scheme quasi-projective over* $\vec{\mathsf{S}}_{0,r_{\mathcal{H}}}$ *and is canonically embedded as an open subscheme of* $\vec{\mathsf{M}}_{\mathcal{H},r_{\mathcal{H}}}$*.*

Proof. By Lemma 3.4.6.1, we have a quasi-finite morphism

$$\vec{\mathsf{M}}^{\mathrm{ord}}_{\mathcal{H}} \to \vec{\mathsf{M}}^{\mathrm{ord}}_{\mathrm{G}_{\mathrm{aux}}(\hat{\mathbb{Z}})}$$

(*cf.* (3.4.6.2)). By composition with the canonical open immersion

$$\vec{\mathsf{M}}^{\mathrm{ord}}_{\mathrm{G}_{\mathrm{aux}}(\hat{\mathbb{Z}})} \hookrightarrow \mathsf{M}_{\mathrm{G}_{\mathrm{aux}}(\hat{\mathbb{Z}}^p)}$$

as in (3.4.5.2), we obtain a quasi-finite morphism

$$\vec{\mathsf{M}}^{\mathrm{ord}}_{\mathcal{H}} \to \mathsf{M}_{\mathrm{G}_{\mathrm{aux}}(\hat{\mathbb{Z}}^p)},$$

which induces the morphism (3.4.6.4) by the universal property of $\vec{\mathsf{M}}_{\mathcal{H}}$ as a normalization, and induces the open immersion (3.4.6.5) by Zariski's main theorem (see [35, III-1, 4.4.3, 4.4.11], and the formulation in [62, Prop. 7.2.3.4] for algebraic spaces).

Since the morphism (3.4.6.4) extends the canonical morphism

$$\mathsf{M}^{\mathrm{ord}}_{\mathcal{H}} \cong \mathsf{M}_{\mathcal{H},r_{\mathcal{H}}} \to \mathsf{M}_{\mathcal{H}},$$

the pullback of $(\vec{A}, \vec{\lambda}, \vec{i})$ to $\mathsf{M}^{\mathrm{ord}}_{\mathcal{H}}$ is canonically isomorphic to the pullback of the tautological (A, λ, i) over $\vec{\mathsf{M}}^{\mathrm{ord}}_{\mathcal{H}}$ to $\mathsf{M}^{\mathrm{ord}}_{\mathcal{H}}$. Since $\vec{\mathsf{M}}^{\mathrm{ord}}_{\mathcal{H}}$ is noetherian and normal, and since $\mathsf{M}^{\mathrm{ord}}_{\mathcal{H}}$ is dense in $\vec{\mathsf{M}}^{\mathrm{ord}}_{\mathcal{H}}$, by [92, IX, 1.4], [28, Ch. I, Prop. 2.7], or [62, Prop. 3.3.1.5], it follows that the pullback of $(\vec{A}, \vec{\lambda}, \vec{i})$ under (3.4.6.4) is canonically isomorphic to the tautological (A, λ, i) over $\vec{\mathsf{M}}^{\mathrm{ord}}_{\mathcal{H}}$, as desired. □

Chapter 4

Degeneration Data and Boundary Charts

In this chapter, we explain how to incorporate the considerations of ordinary level structures into the theory of degeneration data and the boundary construction in [62]. (We no longer assume as in Section 3.4.5 that p is a good prime.) This is the technical heart of the whole work. Readers are encouraged to read this chapter only after mastering the earlier results in [82], [28, Ch. II–IV], and [62, Ch. 4–6]. Although the notation is quite heavy, it is designed to be as close as possible to the one in [62, Ch. 4–6], so that readers who are already familiar with the arguments there can easily see what the new considerations here are.

4.1 Theory of Degeneration Data

4.1.1 *Degenerating Families of Type* (PE, $\mathcal{O}$)

Let $\mathcal{O}$ be as above; that is, as in Definition 1.1.1.1, $\mathcal{O}$ is an order in a finite-dimensional semisimple $\mathbb{Q}$-algebra with a positive involution $\star$ stabilizing $\mathcal{O}$.

Definition 4.1.1.1. *Let S be any normal locally noetherian scheme over* $\mathrm{Spec}(\mathbb{Z})$. *A* **degenerating family of type** (PE, $\mathcal{O}$) *is a tuple (G, λ, i) over S such that we have the following:*

(1) G is a semi-abelian scheme over S.

(2) There exists an open dense subscheme S_1 of S such that G_{S_1} is an abelian scheme. In this case, there is a unique semi-abelian scheme $G^\vee$ (up to unique isomorphism), called the dual semi-abelian scheme of G (cf. cf. [80, IV, 7.1] or [62, Thm. 3.4.3.2]), such that $G^\vee_{S_1}$ is the dual abelian scheme of G_{S_1}.

(3) $\lambda : G \to G^\vee$ is a group homomorphism that induces by restriction a polarization λ_{S_1} of G_{S_1}.

(4) $i : \mathcal{O} \to \mathrm{End}_S(G)$ is a map that defines by restriction an $\mathcal{O}$-structure i_{S_1} : $\mathcal{O} \to \mathrm{End}_{S_1}(G_{S_1})$ of (G_{S_1}, λ_{S_1}). (See [62, Def. 1.3.3.1].)

Remark 4.1.1.2. In Definition 4.1.1.1, we allow S to be an arbitrary scheme over $\mathrm{Spec}(\mathbb{Z})$ (without any reference to the reflex field F_0, as opposed to the case in

Definition 1.3.1.1).

4.1.2 *Common Setting for the Theory of Degeneration*

Let R be a noetherian normal domain complete with respect to an ideal I, with $\mathrm{rad}(I) = I$ for convenience. Let $S := \mathrm{Spec}(R)$, $K := \mathrm{Frac}(R)$, $\eta := \mathrm{Spec}(K)$ the generic point of S, and $S_0 := \mathrm{Spec}(R/I)$. We shall denote the pullbacks to η or S_0 with subscripts "η" or "0", respectively.

4.1.3 *Degeneration Data for Polarized Abelian Schemes with Endomorphism Structures*

Definition 4.1.3.1. *(See [62, Def. 5.1.1.2].) With notation and assumptions as in Section* 4.1.2*, the category* $\mathrm{DEG}_{\mathrm{PE},\mathcal{O}}(R, I)$ *has objects consisting of degenerating families* (G, λ, i) *of type* $(\mathrm{PE}, \mathcal{O})$ *(over* $S = \mathrm{Spec}(R)$*) such that* G_0 *is an extension of an abelian scheme* B_0 *by a* **isotrivial** *torus* T_0 *(see [62, Def. 3.1.1.5]) over* S_0*.*

Definition 4.1.3.2. *(See [62, Def. 5.1.1.3].) With notation and assumptions as in Section* 4.1.2*, the category* $\mathrm{DD}_{\mathrm{PE},\mathcal{O}}(R, I)$ *has objects of the form* $(B, \lambda_B, i_B, \underline{X}, \underline{Y}, \phi, c, c^\vee, \tau)$*, with entries described as follows:*

(1) B *is an abelian scheme over* S*,* $\lambda_B : B \to B^\vee$ *is a polarization of* B*, and* $i_B : \mathcal{O} \hookrightarrow \mathrm{End}_S(B)$ *is an* $\mathcal{O}$*-endomorphism structure of* (B, λ_B)*.*

(2) $\underline{X}$ *and* $\underline{Y}$ *are two étale sheaves (of* $\mathcal{O}$*-lattices) canonically dual to two isotrivial tori* T *and* $T^\vee$*, respectively, carrying* $\mathcal{O}$*-actions over* S*, together with an* $\mathcal{O}$*-equivariant embedding* $\phi : \underline{Y} \to \underline{X}$ *with finite cokernel. (We shall denote the actions of an element* $b \in \mathcal{O}$ *on* X *and* Y *by* $i_X(b)$ *and* $i_Y(b)$*, respectively. When the context is clear, we shall simply denote the actions by* b*.)*

(3) $c : \underline{X} \to B^\vee$ *and* $c^\vee : \underline{Y} \to B$ *are two* $\mathcal{O}$*-equivariant morphisms of group schemes over* S*, satisfying the compatibility* $c\phi = \lambda_B c^\vee$*.*

(4) $\tau : \mathbf{1}_{\underline{Y} \times \underline{X},\eta} \xrightarrow{\sim} (c^\vee \times c)^* \mathcal{P}_{B,\eta}^{\otimes -1}$ *is a trivialization of biextensions with symmetric pullback under* $\mathrm{Id}_{\underline{Y}} \times \phi : \underline{Y} \times \underline{Y} \to \underline{Y} \times \underline{X}$*, satisfying the following conditions:*

   (a) (**Compatibility with** $\mathcal{O}$**-actions***:) For each* $b \in \mathcal{O}$*, we have a canonical identification of sections* $(i_Y(b) \times \mathrm{Id}_X)^* \tau = (\mathrm{Id}_Y \times i_X(b^\star))^* \tau$ *under the canonical isomorphism* $(i_B(b) \times \mathrm{Id}_{B^\vee})^* \mathcal{P}_B \cong (\mathrm{Id}_B \times (i_B(b))^\vee)^* \mathcal{P}_B$*.*

   (b) (**Positivity***:) After a finite étale surjective base change in* S *if necessary, let us assume that* $\underline{X}$ *and* $\underline{Y}$ *are constant with values in* X *and* Y*, respectively. For each* $y \in Y$ *and* $\chi \in X$*, the trivialization* $\tau(y, \chi)$ *defines an isomorphism of invertible sheaves from* $(c^\vee(y), c(\chi))^* \mathcal{P}_{B,\eta}$ *to* $\mathbf{1}_\eta$*. Under this isomorphism (which we again denote by* $\tau(y, \chi)$*), the canonical* R*-integral structure* $(c^\vee(y), c(\chi))^* \mathcal{P}_B$ *of* $(c^\vee(y), c(\chi))^* \mathcal{P}_{B,\eta}$

determines an invertible R-submodule $I_{y,\chi}$ of K. Then the positivity condition is that $I_{y,\phi(y)} \subset I$ for all nonzero y in Y. (Clearly, $I_{0,0} = R$.)

By the theory of degeneration data for polarized abelian varieties in [28, Ch. II and III] (explained in [62, Ch. 4]), generalized by functoriality for polarized abelian varieties with endomorphism structures (see [62, Sec. 5.1.1]), the so-called *Mumford's construction* induces an equivalence of categories

$$\begin{aligned} \mathrm{M}_{\mathrm{PE},\mathcal{O}}(R,I) : \ \mathrm{DD}_{\mathrm{PE},\mathcal{O}}(R,I) \to \mathrm{DEG}_{\mathrm{PE},\mathcal{O}}(R,I) : \\ (B,\lambda_B,i_B,\underline{X},\underline{Y},\phi,c,c^\vee,\tau) \mapsto (G,\lambda,i) \end{aligned} \tag{4.1.3.3}$$

realizing (G,λ,i) (up to isomorphism) as the image of an object of $\mathrm{DD}_{\mathrm{PE},\mathcal{O}}(R,I)$.

We say that $(B,\lambda_B,i_B,\underline{X},\underline{Y},\phi,c,c^\vee,\tau)$ is the *degeneration datum* of (G,λ,i). The theory works even when $\underline{X}$ and $\underline{Y}$ are zero. (Then, by [28, Ch. I, 2.8], $G \cong B$ is an abelian scheme over S, and the positivity condition for τ is trivially verified.)

We shall suppress I from the notation when it is clear from the context. (This is the case, for example, when R is a discrete valuation ring.)

4.1.4 *Degeneration Data for Principal Ordinary Level Structures*

Let $S = \mathrm{Spec}(R)$ be as in Section 4.1.2. Let (G,λ,i) be a degenerating family of type $(\mathrm{PE},\mathcal{O})$ over S as in Definition 4.1.3.1, with degeneration datum $(B,\lambda_B,i_B,\underline{X},\underline{Y},\phi,c,c^\vee,\tau)$ given by (4.1.3.3).

Suppose $(G_\eta,\lambda_\eta,i_\eta)$ is equipped with some *naive* principal ordinary level-p^r structure $\alpha_{p^r}^{\mathrm{ord}} = (\alpha_{p^r}^{\mathrm{ord},0},\alpha_{p^r}^{\mathrm{ord},\#,0},\nu_{p^r}^{\mathrm{ord}})$ of type $(L/p^rL,\langle\,\cdot\,,\,\cdot\,\rangle,\mathtt{D}_{p^r})$ (see Definition 3.3.3.1). Explicitly, the first two entries are $\mathcal{O}$-equivariant homomorphisms

$$\alpha_{p^r}^{\mathrm{ord},0} : (\mathrm{Gr}^0_{\mathtt{D}_{p^r}})^{\mathrm{mult}}_\eta \to G[p^r]_\eta$$

and

$$\alpha_{p^r}^{\mathrm{ord},\#,0} : (\mathrm{Gr}^0_{\mathtt{D}^\#_{p^r}})^{\mathrm{mult}}_\eta \to G^\vee[p^r]_\eta$$

that are closed immersions, which, together with the third entry

$$\nu_{p^r}^{\mathrm{ord}} : \boldsymbol{\mu}_{p^r,\eta} \xrightarrow{\sim} \boldsymbol{\mu}_{p^r,\eta},$$

satisfy the symplectic condition

$$\lambda_\eta \circ \alpha_{p^r}^{\mathrm{ord},0} = \nu_{p^r}^{\mathrm{ord}} \circ \alpha_{p^r}^{\mathrm{ord},\#,0} \circ (\phi^0_{\mathtt{D},p^r})^{\mathrm{mult}}_\eta$$

in Definition 3.3.3.2.

Assume in addition that it satisfies the following:

Condition 4.1.4.1. *The schematic image of $\alpha_{p^r}^{\mathrm{ord},0}$ (resp. $\alpha_{p^r}^{\mathrm{ord},\#,0}$) contains the subscheme $T[p^r]_\eta$ (resp. $T^\vee[p^r]_\eta$) of $G[p^r]_\eta$ (resp. $G^\vee[p^r]_\eta$).*

Definition 4.1.4.2. *If the principal ordinary level-p^r structure $\alpha_{p^r}^{\mathrm{ord}}$ satisfies Condition 4.1.4.1, then we say that $\alpha_{p^r}^{\mathrm{ord}}$ is* **compatible with the degeneration**.

One immediate consequence of the assumption that $\alpha^{\mathrm{ord}}_{p^r}$ satisfies Condition 4.1.4.1 (*i.e.*, is compatible with degeneration) is that the pullbacks of the two subgroup schemes of multiplicative type

$$T[p^r]_\eta \subset \mathrm{image}(\alpha^{\mathrm{ord},0}_{p^r}) = \alpha^{\mathrm{ord},0}_{p^r}((\mathrm{Gr}^0_{\mathtt{D}_{p^r}})^{\mathrm{mult}}_\eta)$$

and

$$T^\vee[p^r]_\eta \subset \mathrm{image}(\alpha^{\mathrm{ord},\#,0}_{p^r}) = \alpha^{\mathrm{ord},\#,0}_{p^r}((\mathrm{Gr}^0_{\mathtt{D}^\#_{p^r}})^{\mathrm{mult}}_\eta)$$

under $\alpha^{\mathrm{ord},0}_{p^r}$ and $\alpha^{\mathrm{ord},\#,0}_{p^r}$ determine, respectively, two totally isotropic $\mathcal{O}$-submodules of $\mathrm{Gr}^0_{\mathtt{D},p^r}$ and $\mathrm{Gr}^0_{\mathtt{D}^\#,p^r}$ which are compatible with $\phi^0_{\mathtt{D},p^r} : \mathrm{Gr}^0_{\mathtt{D},p^r} \to \mathrm{Gr}^0_{\mathtt{D}^\#,p^r}$ in the sense that $\phi^0_{\mathtt{D},p^r}$ maps the first submodule to the second submodule. This is equivalent to the determination of a symplectic filtration

$$\mathtt{Z}_{-3,p^r} = 0 \subset \mathtt{Z}_{-2,p^r} \subset \mathtt{Z}_{-1,p^r} \subset \mathtt{Z}_{0,p^r} = L/p^rL \tag{4.1.4.3}$$

of $\mathcal{O}$-submodules, with dual filtration

$$\mathtt{Z}^\#_{-3,p^r} = 0 \subset \mathtt{Z}^\#_{-2,p^r} \subset \mathtt{Z}^\#_{-1,p^r} \subset \mathtt{Z}^\#_{0,p^r} = L^\#/p^rL^\# \tag{4.1.4.4}$$

with respect to the pairing $\langle\,\cdot\,,\,\cdot\,\rangle$ (which implies that canonical morphism $L/p^rL \to L^\#/p^rL^\#$ induces morphisms $\mathtt{Z}_{-i,p^r} \to \mathtt{Z}^\#_{-i,p^r}$ for each i), satisfying the compatibilities

$$\mathtt{Z}_{-2,p^r} \subset \mathtt{D}^0_{p^r} \subset \mathtt{Z}_{-1,p^r} \tag{4.1.4.5}$$

and

$$\mathtt{Z}^\#_{-2,p^r} \subset \mathtt{D}^{\#,0}_{p^r} \subset \mathtt{Z}^\#_{-1,p^r} \tag{4.1.4.6}$$

(*cf.* (3.2.3.3) and (3.2.3.5)) such that the above two submodules of $\mathrm{Gr}^0_{\mathtt{D},p^r}$ and $\mathrm{Gr}^0_{\mathtt{D}^\#,p^r}$ are respectively the two submodules

$$\mathrm{Gr}^{\mathtt{Z}}_{-2,p^r} \subset \mathrm{Gr}^0_{\mathtt{D},p^r} \tag{4.1.4.7}$$

and

$$\mathrm{Gr}^{\mathtt{Z}^\#}_{-2,p^r} \subset \mathrm{Gr}^0_{\mathtt{D}^\#,p^r}, \tag{4.1.4.8}$$

together with two isomorphisms

$$(\varphi_{-2,p^r})^{\mathrm{mult}}_\eta : (\mathrm{Gr}^{\mathtt{Z}}_{-2,p^r})^{\mathrm{mult}}_\eta \xrightarrow{\sim} T[p^r]_\eta \tag{4.1.4.9}$$

and

$$(\varphi^\#_{-2,p^r})^{\mathrm{mult}}_\eta : (\mathrm{Gr}^{\mathtt{Z}^\#}_{-2,p^r})^{\mathrm{mult}}_\eta \xrightarrow{\sim} T^\vee[p^r]_\eta \tag{4.1.4.10}$$

determined by

$$\nu^{\mathrm{ord}}_{p^r} \circ (\varphi_{-2,p^r})^{\mathrm{mult}}_\eta := \alpha^{\mathrm{ord},0}_{p^r}|_{(\mathrm{Gr}^{\mathtt{Z}}_{-2,p^r})^{\mathrm{mult}}_\eta}$$

and

$$(\varphi^\#_{-2,p^r})^{\mathrm{mult}}_\eta := \alpha^{\mathrm{ord},\#,0}_{p^r}|_{(\mathrm{Gr}^{\mathtt{Z}^\#}_{-2,p^r})^{\mathrm{mult}}_\eta}.$$

These two morphisms $(\varphi_{-2,p^r})_\eta^{\text{mult}}$ and $(\varphi^{\#}_{-2,p^r})_\eta^{\text{mult}}$ are equivalent to the two isomorphisms

$$\varphi_{-2,p^r} : \mathrm{Gr}^{\mathtt{Z}}_{-2,p^r} \xrightarrow{\sim} \mathrm{Hom}((\underline{X}/p^r\underline{X})_\eta, (\mathbb{Z}/p^r\mathbb{Z})(1)) \quad (4.1.4.11)$$

and

$$\varphi^{\#}_{-2,p^r} : \mathrm{Gr}^{\mathtt{Z}^{\#}}_{-2,p^r} \xrightarrow{\sim} \mathrm{Hom}((\underline{Y}/p^r\underline{Y})_\eta, (\mathbb{Z}/p^r\mathbb{Z})(1)), \quad (4.1.4.12)$$

respectively. (Thus far, by abuse of notation, we have used notation such as $\mathrm{Gr}^{\mathtt{Z}}_{-2,p^r}$ and $\mathrm{Gr}^{\mathtt{Z}^{\#}}_{-2,p^r}$ for the constant sheaves $(\mathrm{Gr}^{\mathtt{Z}}_{-2,p^r})_\eta$ and $(\mathrm{Gr}^{\mathtt{Z}^{\#}}_{-2,p^r})_\eta$ over the point η. We will adopt a similar abuse of notation in what follows.) By the perfect duality $\langle \cdot, \cdot \rangle_{02} : \mathrm{Gr}^{\mathtt{Z}}_{0,p^r} \times \mathrm{Gr}^{\mathtt{Z}^{\#}}_{-2,p^r} \to (\mathbb{Z}/p^r\mathbb{Z})(1)$ induced by $\langle \cdot, \cdot \rangle$ (by the definition of (3.2.3.3) and (3.2.3.5)), this last isomorphism $\varphi^{\#}_{-2,p^r}$ is canonically equivalent to an isomorphism

$$\varphi_{0,p^r} : \mathrm{Gr}^{\mathtt{Z}}_{0,p^r} \xrightarrow{\sim} (\underline{Y}/p^r\underline{Y})_\eta. \quad (4.1.4.13)$$

The compatibility (4.1.4.5) induces a filtration $\mathtt{D}_{-1,p^r} = \{\mathtt{D}^i_{-1,p^r}\}_i$ on $\mathrm{Gr}^{\mathtt{Z}}_{-1,p^r}$ given by

$$\mathtt{D}^1_{-1,p^r} := 0 \subset \mathtt{D}^0_{-1,p^r} := \mathtt{D}^0_{p^r}/\mathtt{Z}_{-2,p^r} \subset \mathtt{D}^{-1}_{-1,p^r} := \mathrm{Gr}^{\mathtt{Z}}_{-1,p^r} \quad (4.1.4.14)$$

(serving the same purpose as the filtration $\mathtt{D}_{p^r}$ does for L/p^rL). Similarly, the compatibility (4.1.4.6) induces a filtration $\mathtt{D}^{\#}_{-1,p^r} = \{\mathtt{D}^{\#,i}_{-1,p^r}\}_i$ on $\mathrm{Gr}^{\mathtt{Z}^{\#}}_{-1,p^r}$ given by

$$\mathtt{D}^{\#,1}_{-1,p^r} := 0 \subset \mathtt{D}^{\#,0}_{-1,p^r} := \mathtt{D}^{\#,0}_{p^r}/\mathtt{Z}^{\#}_{-2,p^r} \subset \mathtt{D}^{\#,-1}_{-1,p^r} := \mathrm{Gr}^{\mathtt{Z}^{\#}}_{-1,p^r}\,. \quad (4.1.4.15)$$

The filtrations (4.1.4.14) and (4.1.4.15) are dual to each other with respect to the pairing $\langle \cdot, \cdot \rangle_{11} : \mathrm{Gr}^{\mathtt{Z}}_{-1,p^r} \times \mathrm{Gr}^{\mathtt{Z}^{\#}}_{-1,p^r} \to (\mathbb{Z}/p^r\mathbb{Z})(1)$ induced by $\langle \cdot, \cdot \rangle$. Then we have

$$\mathrm{Gr}^0_{\mathtt{D}_{-1,p^r}} = \mathrm{Gr}^0_{\mathtt{D},p^r} \,/\, \mathrm{Gr}^{\mathtt{Z}}_{-2,p^r} \quad (4.1.4.16)$$

and

$$\mathrm{Gr}^0_{\mathtt{D}^{\#}_{-1,p^r}} = \mathrm{Gr}^0_{\mathtt{D}^{\#},p^r} \,/\, \mathrm{Gr}^{\mathtt{Z}^{\#}}_{-2,p^r}, \quad (4.1.4.17)$$

and we have a morphism

$$\phi^0_{\mathtt{D}_{-1,p^r}} : \mathrm{Gr}^0_{\mathtt{D}_{-1,p^r}} \to \mathrm{Gr}^0_{\mathtt{D}^{\#}_{-1,p^r}}\,. \quad (4.1.4.18)$$

Lemma 4.1.4.19. *With the setting as above, if the homomorphisms $\alpha^{\text{ord},0}_{p^r}$ and $\alpha^{\text{ord},\#,0}_{p^r}$ extend to homomorphisms $\alpha^{\text{ord},0}_{p^r,S} : (\mathrm{Gr}^0_{\mathtt{D}_{p^r}})^{\text{mult}}_S \to G[p^r]$ and $\alpha^{\text{ord},\#,0}_{p^r,S} : (\mathrm{Gr}^0_{\mathtt{D}^{\#}_{p^r}})^{\text{mult}}_S \to G^\vee[p^r]$ over S, respectively, then these extensions are closed immersions with schematic images contained in $G^\natural[p^r]$ and $G^{\vee,\natural}[p^r]$, respectively, and $\alpha^{\text{ord}}_{p^r} = (\alpha^{\text{ord},0}_{p^r}, \alpha^{\text{ord},\#,0}_{p^r}, \nu^{\text{ord}}_{p^r})$ satisfies Condition 4.1.4.1.*

Proof. The extension $\alpha^{\mathrm{ord},0}_{p^r,S}$ (resp. $\alpha^{\mathrm{ord},\#,0}_{p^r,S}$) is a closed immersion over S because it is so over η, and the closure of the identity section over η is the identity section over S because G (resp. $G^\vee$) is separated over S. Since $G^\natural[p^r]$ (resp. $G^{\vee,\natural}[p^r]$) is the maximal finite flat closed subgroup scheme of $G[p^r]$ (resp. $G^\vee[p^r]$) (see [62, Sec. 3.4.2], or rather [34, IX, 7.3]), this shows that the schematic image of $\alpha^{\mathrm{ord},0}_{p^r,S}$ (resp. $\alpha^{\mathrm{ord},\#,0}_{p^r,S}$) is contained in $G^\natural[p^r]$ (resp. $G^{\vee,\natural}[p^r]$). Since the schematic images of $\alpha^{\mathrm{ord},0}_{p^r}$ are $\alpha^{\mathrm{ord},\#,0}_{p^r}$ are annihilators of each other under the canonical pairing $e_{G[p^r]_\eta}$: $G[p^r]_\eta \times G^\vee[p^r]_\eta \to \boldsymbol{\mu}_{p^r,\eta}$, and since $T[p^r]_\eta$ (resp. $T^\vee[p^r]_\eta$) is the annihilator of $G^{\vee,\natural}[p^r]_\eta$ (resp. $G^\natural[p^r]_\eta$) under this canonical pairing, we see by duality that $\alpha^{\mathrm{ord}}_{p^r}$ satisfies Condition 4.1.4.1, as desired. □

Lemma 4.1.4.20. *With the setting as above, suppose* $\alpha^{\mathrm{ord}}_{p^r} = (\alpha^{\mathrm{ord},0}_{p^r}, \alpha^{\mathrm{ord},\#,0}_{p^r}, \nu^{\mathrm{ord}}_{p^r})$ *satisfies Condition* 4.1.4.1. *Then there exist homomorphisms*

$$\alpha^{\mathrm{ord},0,\natural}_{p^r} : (\mathrm{Gr}^0_{\mathtt{D}_{p^r}})^{\mathrm{mult}}_\eta \to G^\natural[p^r]_\eta$$

and

$$\alpha^{\mathrm{ord},\#,0,\natural}_{p^r} : (\mathrm{Gr}^0_{\mathtt{D}^\#_{p^r}})^{\mathrm{mult}}_\eta \to G^{\vee,\natural}[p^r]_\eta$$

over η *such that* $\alpha^{\mathrm{ord},0}_{p^r}$ *(resp.* $\alpha^{\mathrm{ord},\#,0}_{p^r}$*) is the composition of* $\alpha^{\mathrm{ord},0,\natural}_{p^r}$ *(resp.* $\alpha^{\mathrm{ord},\#,0,\natural}_{p^r}$*) with the canonical morphism* $G^\natural[p^r]_\eta \hookrightarrow G[p^r]_\eta$ *(resp.* $G^{\vee,\natural}[p^r]_\eta \hookrightarrow G^\vee[p^r]_\eta$*).*

The homomorphism $\alpha^{\mathrm{ord},0}_{p^r}$ *(resp.* $\alpha^{\mathrm{ord},\#,0}_{p^r}$*) extends to a homomorphism* $\alpha^{\mathrm{ord},0}_{p^r,S}$: $(\mathrm{Gr}^0_{\mathtt{D}_{p^r}})^{\mathrm{mult}}_S \to G[p^r]$ *(resp.* $\alpha^{\mathrm{ord},\#,0}_{p^r,S}$: $(\mathrm{Gr}^0_{\mathtt{D}^\#_{p^r}})^{\mathrm{mult}}_S \to G^\vee[p^r]$*) over* S *if and only if the homomorphism* $\alpha^{\mathrm{ord},0,\natural}_{p^r}$ *(resp.* $\alpha^{\mathrm{ord},\#,0,\natural}_{p^r}$*) also extends to a homomorphism* $\alpha^{\mathrm{ord},0,\natural}_{p^r,S} : (\mathrm{Gr}^0_{\mathtt{D}_{p^r}})^{\mathrm{mult}}_S \to G^\natural[p^r]$ *(resp.* $\alpha^{\mathrm{ord},\#,0,\natural}_{p^r,S} : (\mathrm{Gr}^0_{\mathtt{D}^\#_{p^r}})^{\mathrm{mult}}_S \to G^{\vee,\natural}[p^r]$*) over* S.

Proof. Since the schematic image of $\alpha^{\mathrm{ord},0}_{p^r}$ (resp. $\alpha^{\mathrm{ord},\#,0}_{p^r}$) contains the subscheme $T[p^r]_\eta$ (resp. $T^\vee[p^r]_\eta$) of $G[p^r]_\eta$ (resp. $G^\vee[p^r]_\eta$), by duality as in the proof of Lemma 4.1.4.19, the homomorphisms $\alpha^{\mathrm{ord},0}_{p^r}$ and $\alpha^{\mathrm{ord},\#,0}_{p^r}$ have schematic images contained in $G^\natural[p^r]_\eta$ and $G^{\vee,\natural}[p^r]_\eta$, respectively, and induce the two homomorphisms $\alpha^{\mathrm{ord},0,\natural}_{p^r}$ and $\alpha^{\mathrm{ord},\#,0,\natural}_{p^r}$. The assertion about extensions over S is obvious. □

Proposition 4.1.4.21. *Let* $S = \mathrm{Spec}(R)$, $\eta = \mathrm{Spec}(K)$, *and* (G, λ, i) *be as at the beginning of this Section* 4.1.4. *Let* $(B, \lambda_B, i_B, \underline{X}, \underline{Y}, \phi, c, c^\vee, \tau)$ *be the degeneration datum associated with* (G, λ, i) *under the equivalence* (4.1.3.3). *A naive principal ordinary level structure*

$$\alpha^{\mathrm{ord}}_{p^r} = (\alpha^{\mathrm{ord},0}_{p^r}, \alpha^{\mathrm{ord},\#,0}_{p^r}, \nu^{\mathrm{ord}}_{p^r})$$

of type $(L \underset{\mathbb{Z}}{\otimes} \mathbb{Z}_p, \langle\,\cdot\,,\,\cdot\,\rangle, \mathtt{D})$ *and level* p^r *on* $(G_\eta, \lambda_\eta, i_\eta)$ *(see Definition* 3.3.3.1*) satisfying Condition* 4.1.4.1 *determines (up to isomorphism) a tuple*

$$(\mathtt{Z}_{p^r}, (\varphi_{-2,p^r}, \varphi_{0,p^r}), \mathtt{D}_{-1,p^r}, \varphi^{\mathrm{ord}}_{-1,p^r})$$

with entries described as follows:

(1) A symplectic admissible filtration $\mathtt{Z}_{p^r} = \{\mathtt{Z}_{-i,p^r}\}_i$ *of* $\mathcal{O}$*-submodules on* L/p^rL *(see (4.1.4.5) and (4.1.4.6)) satisfying the compatibilities (3.2.3.3) and (3.2.3.5) and determining the* $\mathcal{O}$*-submodules* $\mathrm{Gr}^{\mathtt{Z}}_{-2,p^r} \subset \mathrm{Gr}^0_{\mathtt{D},p^r}$ *and* $\mathrm{Gr}^{\mathtt{Z}^\#}_{-2,p^r} \subset \mathrm{Gr}^0_{\mathtt{D}^\#,p^r}$ *(see (4.1.4.7) and (4.1.4.8)).*

(2) A pair of isomorphisms $(\varphi_{-2,p^r}, \varphi^\#_{-2,p^r})$*, or equivalently a pair of isomorphisms* $(\varphi_{-2,p^r}, \varphi_{0,p^r})$ *(see (4.1.4.11), (4.1.4.12), and (4.1.4.13)), inducing isomorphisms*

$$(\varphi_{-2,p^r})^{\mathrm{mult}}_S : (\mathrm{Gr}^{\mathtt{Z}}_{-2,p^r})^{\mathrm{mult}}_S \xrightarrow{\sim} T[p^r]$$

and

$$(\varphi^\#_{-2,p^r})^{\mathrm{mult}}_S : (\mathrm{Gr}^{\mathtt{Z}^\#}_{-2,p^r})^{\mathrm{mult}}_S \xrightarrow{\sim} T^\vee[p^r]$$

(uniquely extending (4.1.4.9) and (4.1.4.10)), the latter of which is equivalent to an isomorphism

$$\varphi_{0,p^r,S} : (\mathrm{Gr}^{\mathtt{Z}}_{0,p^r})_S \xrightarrow{\sim} (\underline{Y}/p^r\underline{Y})_S$$

(uniquely extending (4.1.4.11)).

(3) A filtration $\mathtt{D}_{-1,p^r} = \{\mathtt{D}^i_{-1,p^r}\}_i$ *on* $\mathrm{Gr}^{\mathtt{Z}}_{-1,p^r}$ *(see (4.1.4.14)) that satisfies the analogous conditions as the filtration* $\mathtt{D}_{p^r}$ *does for* L/p^rL *(see Lemma 3.2.2.1).*

(4) A (naive) principle ordinary level structure

$$\varphi^{\mathrm{ord}}_{-1,p^r} = (\varphi^{\mathrm{ord},0}_{-1,p^r}, \varphi^{\mathrm{ord},\#,0}_{-1,p^r}, \nu^{\mathrm{ord}}_{-1,p^r})$$

of type $\phi^0_{\mathtt{D}_{-1,p^r}} : \mathrm{Gr}^0_{\mathtt{D}_{-1,p^r}} \to \mathrm{Gr}^0_{\mathtt{D}^\#_{-1,p^r}}$ *(see (4.1.4.18)) on* $(B_\eta, \lambda_{B,\eta}, i_{B,\eta})$ *as in Definition 3.3.3.2 such that*

$$\nu^{\mathrm{ord}}_{-1,p^r} : \boldsymbol{\mu}_{p^r,\eta} \xrightarrow{\sim} \boldsymbol{\mu}_{p^r,\eta}$$

is equal to $\nu^{\mathrm{ord}}_{p^r}$*. Explicitly, the first two entries are* $\mathcal{O}$*-equivariant homomorphisms*

$$\varphi^{\mathrm{ord},0}_{-1,p^r} : (\mathrm{Gr}^0_{\mathtt{D}_{-1,p^r}})^{\mathrm{mult}}_\eta \to B[p^r]_\eta$$

and

$$\varphi^{\mathrm{ord},\#,0}_{-1,p^r} : (\mathrm{Gr}^0_{\mathtt{D}^\#_{-1,p^r}})^{\mathrm{mult}}_\eta \to B^\vee[p^r]_\eta$$

that are closed immersions, satisfying the symplectic condition

$$\lambda_{B,\eta} \circ \varphi^{\mathrm{ord},0}_{-1,p^r} = \nu^{\mathrm{ord}}_{-1,p^r} \circ \varphi^{\mathrm{ord},\#,0}_{-1,p^r} \circ (\phi^0_{\mathtt{D}_{-1,p^r}})^{\mathrm{mult}}_\eta$$

as in Definition 3.3.3.2.

The homomorphism $\alpha^{\mathrm{ord},0}_{p^r}$ *(resp.* $\alpha^{\mathrm{ord},\#,0}_{p^r}$*, resp.* $\nu^{\mathrm{ord}}_{p^r}$*) extends to a homomorphism* $\alpha^{\mathrm{ord},0}_{p^r,S} : (\mathrm{Gr}^0_{\mathtt{D}_{p^r}})^{\mathrm{mult}}_S \to G[p^r]$ *(resp.* $\alpha^{\mathrm{ord},\#,0}_{p^r,S} : (\mathrm{Gr}^0_{\mathtt{D}^\#_{p^r}})^{\mathrm{mult}}_S \to G^\vee[p^r]$*, resp.* $\nu^{\mathrm{ord}}_{p^r,S} : \boldsymbol{\mu}_{p^r,S} \xrightarrow{\sim} \boldsymbol{\mu}_{p^r,S}$*) over* S *if and only if the homomorphism* $\varphi^{\mathrm{ord},0}_{-1,p^r}$ *(resp.* $\varphi^{\mathrm{ord},\#,0}_{-1,p^r}$*,* *resp.* $\nu^{\mathrm{ord}}_{-1,p^r}$) *also extends to a homomorphism* $\varphi^{\mathrm{ord},0}_{-1,p^r,S} : (\mathrm{Gr}^0_{\mathtt{D}_{-1,p^r}})^{\mathrm{mult}}_S \to B[p^r]$ (*resp.* $\varphi^{\mathrm{ord},\#,0}_{-1,p^r,S} : (\mathrm{Gr}^0_{\mathtt{D}^\#_{-1,p^r}})^{\mathrm{mult}}_S \to B^\vee[p^r]$, *resp.* $\nu^{\mathrm{ord}}_{-1,p^r,S} : \boldsymbol{\mu}_{p^r,S} \xrightarrow{\sim} \boldsymbol{\mu}_{p^r,S}$) *over* S.

If $\alpha^{\mathrm{ord}}_{p^r}$ *is a principal ordinary level-*p^r *structure of type* $(L \underset{\mathbb{Z}}{\otimes} \mathbb{Z}_p, \langle\,\cdot\,,\,\cdot\,\rangle, \mathtt{D})$ *such that, for each integer* $r' \geq r$, *there exists some lifting to level* $p^{r'}$ (*over some extension of* η) *compatible with degeneration* (*i.e., satisfying the analogue of Condition* 4.1.4.1 *at level* $p^{r'}$), *then all of the above data determined by* $\alpha^{\mathrm{ord}}_{p^r}$ *are compatibly liftable to their analogues over* $\mathbb{Z}_p$.

Proof. Only statement (4) of the proposition requires some explanation: The two homomorphisms $\varphi^{\mathrm{ord},0}_{-1,p^r}$ and $\varphi^{\mathrm{ord},\#,0}_{-1,p^r}$ are induced, respectively, by the two homomorphisms $\alpha^{\mathrm{ord},0,\natural}_{p^r}$ and $\alpha^{\mathrm{ord},\#,0,\natural}_{p^r}$ in Lemma 4.1.4.20. (The statements on extensions over S also follow from there. The statements on liftability are obvious.) □

However, not all tuples $(\mathtt{Z}_{p^r}, (\varphi_{-2,p^r}, \varphi_{0,p^r}), \mathtt{D}_{-1,p^r}, \varphi^{\mathrm{ord}}_{-1,p^r})$ as in Proposition 4.1.4.21 come from (naive) principal ordinary level structures. To formulate the additional condition needed, we introduce splittings both for the constant side, namely the filtrations $0 \subset \mathrm{Gr}^{\mathtt{Z}}_{-2,p^r} \subset \mathrm{Gr}^0_{\mathtt{D},p^r}$ and $0 \subset \mathrm{Gr}^{\mathtt{Z}^\#}_{-2,p^r} \subset \mathrm{Gr}^0_{\mathtt{D}^\#,p^r}$ (see (4.1.4.7) and (4.1.4.8)), and for the geometric side, namely the filtrations $0 \subset T[p^r]_\eta \subset \alpha^{\mathrm{ord},0}_{p^r}((\mathrm{Gr}^0_{\mathtt{D}_{p^r}})^{\mathrm{mult}}_\eta)$ and $0 \subset T^\vee[p^r]_\eta \subset \alpha^{\mathrm{ord},\#,0}_{p^r}((\mathrm{Gr}^0_{\mathtt{D}^\#_{p^r}})^{\mathrm{mult}}_\eta)$ of group schemes of multiplicative type over η.

Lemma 4.1.4.22. *Let* $\mathrm{Gr}^{\mathtt{Z}}_{-2,p^r} \subset \mathrm{Gr}^0_{\mathtt{D},p^r}$, $\mathrm{Gr}^{\mathtt{Z}^\#}_{-2,p^r} \subset \mathrm{Gr}^0_{\mathtt{D}^\#,p^r}$, $\mathrm{Gr}^0_{\mathtt{D}_{-1,p^r}}$, *and* $\mathrm{Gr}^0_{\mathtt{D}^\#_{-1,p^r}}$ *be determined by* $\alpha^{\mathrm{ord}}_{p^r} = (\alpha^{\mathrm{ord},0}_{p^r}, \alpha^{\mathrm{ord},\#,0}_{p^r}, \nu^{\mathrm{ord}}_{p^r})$ *as above* (*see* (4.1.4.7), (4.1.4.8), (4.1.4.16), *and* (4.1.4.17)). *Suppose that Condition* 1.2.1.1 *holds, and that, for each integer* $r' \geq r$, *there exists some lifting to level* $p^{r'}$ (*over some extension of* η) *satisfying the analogue of Condition* 4.1.4.1. *Then there are splittings*

$$\delta^{\mathrm{ord},0}_{p^r} : \mathrm{Gr}^{\mathtt{Z}}_{-2,p^r} \oplus \mathrm{Gr}^0_{\mathtt{D}_{-1,p^r}} \xrightarrow{\sim} \mathrm{Gr}^0_{\mathtt{D},p^r} \tag{4.1.4.23}$$

and

$$\delta^{\mathrm{ord},\#,0}_{p^r} : \mathrm{Gr}^{\mathtt{Z}^\#}_{-2,p^r} \oplus \mathrm{Gr}^0_{\mathtt{D}^\#_{-1,p^r}} \xrightarrow{\sim} \mathrm{Gr}^0_{\mathtt{D}^\#,p^r}, \tag{4.1.4.24}$$

of $\mathcal{O}$*-modules, which are compatible with* $\phi^0_{\mathtt{D},p^r} : \mathrm{Gr}^0_{\mathtt{D},p^r} \to \mathrm{Gr}^0_{\mathtt{D}^\#,p^r}$ *only in the sense that* $\phi^0_{\mathtt{D},p^r}(\mathrm{Gr}^{\mathtt{Z}}_{-2,p^r}) \subset \mathrm{Gr}^{\mathtt{Z}^\#}_{-2,p^r}$ (*inducing the expected morphisms* $\phi^* : \mathrm{Gr}^{\mathtt{Z}}_{-2,p^r} \to \mathrm{Gr}^{\mathtt{Z}^\#}_{-2,p^r}$ *and* $\phi^0_{\mathtt{D}_{-1,p^r}} : \mathrm{Gr}^0_{\mathtt{D}_{-1,p^r}} \to \mathrm{Gr}^0_{\mathtt{D}^\#_{-1,p^r}}$ *between the subquotients*), *but not that* $\phi^0_{\mathtt{D},p^r}(\delta^{\mathrm{ord},0}_{p^r}(\mathrm{Gr}^0_{\mathtt{D}_{-1,p^r}})) \subset \delta^{\mathrm{ord},\#,0}_{p^r}(\mathrm{Gr}^0_{\mathtt{D}^\#_{-1,p^r}})$. *These splittings are liftable to splittings over* $\mathbb{Z}_p$ *with analogous compatibility properties.*

Proof. By Condition 1.2.1.1, the action of $\mathcal{O}$ on L extends to an action of some maximal order $\mathcal{O}'$ in $\mathcal{O} \underset{\mathbb{Z}}{\otimes} \mathbb{Q}$ containing $\mathcal{O}$, and Lemma 3.2.2.6 and its proof show that the filtrations $\mathtt{D}$ and $\mathtt{Z} \underset{\hat{\mathbb{Z}}}{\otimes} \mathbb{Z}_p$ on $L \underset{\mathbb{Z}}{\otimes} \mathbb{Z}_p$ and the filtrations $\mathtt{D}^\#$ and

$\mathsf{Z}^{\#} \underset{\hat{\mathbb{Z}}}{\otimes} \mathbb{Z}_p$ on $L^{\#} \underset{\mathbb{Z}}{\otimes} \mathbb{Z}_p$ are all $\mathcal{O}' \underset{\mathbb{Z}}{\otimes} \mathbb{Z}_p$-equivariant, whose graded pieces are projective $\mathcal{O}' \underset{\mathbb{Z}}{\otimes} \mathbb{Z}_p$-modules. Hence, there exist (noncanonical) splittings of the short exact sequences $0 \to \mathrm{Gr}^{\mathsf{Z}}_{-2} \to \mathrm{Gr}^0_{\mathsf{D}} \to \mathrm{Gr}^0_{\mathsf{D}_{-1}} \to 0$ and $0 \to \mathrm{Gr}^{\mathsf{Z}^{\#}}_{-2} \to \mathrm{Gr}^0_{\mathsf{D}^{\#}} \to \mathrm{Gr}^0_{\mathsf{D}^{\#}_{-1}} \to 0$ of $\mathcal{O}' \underset{\mathbb{Z}}{\otimes} \mathbb{Z}_p$-lattices, which induce the desired splittings (4.1.4.23) and (4.1.4.24) by reduction modulo p^r. □

Lemma 4.1.4.25. *Choices of the splittings $\delta^{\mathrm{ord},0}_{p^r}$ and $\delta^{\mathrm{ord},\#,0}_{p^r}$ as in* (4.1.4.23) *and* (4.1.4.24) *define a morphism*

$$\mathrm{Gr}^0_{\mathsf{D}_{-1,p^r}} \to \mathrm{Gr}^{\mathsf{Z}^{\#}}_{-2,p^r}$$

of $\mathcal{O}$-modules by sending $d \in \mathrm{Gr}^0_{\mathsf{D}_{-1,p^r}}$ to

$$(\mathrm{pr}_{\mathrm{Gr}^{\mathsf{Z}^{\#}}_{-2,p^r}} \circ (\delta^{\mathrm{ord},\#,0}_{p^r})^{-1} \circ \phi^0_{\mathsf{D},p^r} \circ \delta^{\mathrm{ord},0}_{p^r})(0,d) \in \mathrm{Gr}^{\mathsf{Z}^{\#}}_{-2,p^r},$$

which (by the perfect duality $\langle \cdot\,,\,\cdot \rangle_{02} : \mathrm{Gr}^{\mathsf{Z}}_{0,p^r} \times \mathrm{Gr}^{\mathsf{Z}^{\#}}_{-2,p^r} \to (\mathbb{Z}/p^r\mathbb{Z})(1)$ induced by $\langle \cdot\,,\,\cdot \rangle$) is equivalent to a pairing

$$\langle \cdot\,,\,\cdot \rangle^{\mathrm{ord}}_{10,p^r} : \mathrm{Gr}^0_{\mathsf{D}_{-1,p^r}} \times \mathrm{Gr}^{\mathsf{Z}}_{0,p^r} \to (\mathbb{Z}/p^r\mathbb{Z})(1) \tag{4.1.4.26}$$

satisfying $\langle b\,\cdot\,,\,\cdot \rangle^{\mathrm{ord}}_{10,p^r} = \langle \cdot\,,\,b^\star\,\cdot \rangle^{\mathrm{ord}}_{10,p^r}$ for every $b \in \mathcal{O}$.

In other words, the pairing $\langle \cdot\,,\,\cdot \rangle^{\mathrm{ord}}_{10,p^r}$ measures the failure of the condition that

$$\phi^0_{\mathsf{D},p^r}(\delta^{\mathrm{ord},0}_{p^r}(\mathrm{Gr}^0_{\mathsf{D}_{-1,p^r}})) \subset \delta^{\mathrm{ord},\#,0}_{p^r}(\mathrm{Gr}^0_{\mathsf{D}^{\#}_{-1,p^r}}).$$

On the other hand, we would like to have splittings of the filtrations

$$0 \subset T[p^r]_\eta \subset \alpha^{\mathrm{ord},0}_{p^r}((\mathrm{Gr}^0_{\mathsf{D}_{p^r}})^{\mathrm{mult}}_\eta)$$

and

$$0 \subset T^\vee[p^r]_\eta \subset \alpha^{\mathrm{ord},\#,0}_{p^r}((\mathrm{Gr}^0_{\mathsf{D}^{\#}_{p^r}})^{\mathrm{mult}}_\eta)$$

of group schemes of multiplicative type, which are $\mathcal{O}$-equivariant isomorphisms

$$\varsigma^{\mathrm{ord},0}_{p^r} : T[p^r]_\eta \oplus \varphi^{\mathrm{ord},0}_{-1,p^r}((\mathrm{Gr}^0_{\mathsf{D}_{-1,p^r}})^{\mathrm{mult}}_\eta) \xrightarrow{\sim} \alpha^{\mathrm{ord},0}_{p^r}((\mathrm{Gr}^0_{\mathsf{D}_{p^r}})^{\mathrm{mult}}_\eta) \tag{4.1.4.27}$$

and

$$\varsigma^{\mathrm{ord},\#,0}_{p^r} : T^\vee[p^r]_\eta \oplus \varphi^{\mathrm{ord},\#,0}_{-1,p^r}((\mathrm{Gr}^0_{\mathsf{D}^{\#}_{-1,p^r}})^{\mathrm{mult}}_\eta) \xrightarrow{\sim} \alpha^{\mathrm{ord},\#,0}_{p^r}((\mathrm{Gr}^0_{\mathsf{D}^{\#}_{p^r}})^{\mathrm{mult}}_\eta) \tag{4.1.4.28}$$

respecting the subgroup schemes $T[p^r]_\eta$ and $T^\vee[p^r]_\eta$, respectively. In particular, these splittings correspond, respectively, to $\mathcal{O}$-equivariant homomorphisms

$$\tilde{\varphi}^{\mathrm{ord},0}_{-1,p^r} : (\mathrm{Gr}^0_{\mathsf{D}_{-1,p^r}})^{\mathrm{mult}}_\eta \hookrightarrow G^\natural[p^r]_\eta \hookrightarrow G^\natural_\eta \tag{4.1.4.29}$$

and

$$\tilde{\varphi}^{\mathrm{ord},\#,0}_{-1,p^r} : (\mathrm{Gr}^0_{\mathsf{D}^{\#}_{-1,p^r}})^{\mathrm{mult}}_\eta \hookrightarrow G^{\vee,\natural}[p^r]_\eta \hookrightarrow G^{\vee,\natural}_\eta \tag{4.1.4.30}$$

that are closed immersions lifting $\varphi^{\mathrm{ord},0}_{-1,p^r}$ and $\varphi^{\mathrm{ord},\#,0}_{-1,p^r}$, respectively. (In this last step we are not asserting any compatibilities between (4.1.4.29) and (4.1.4.30) other than those between $\varphi^{\mathrm{ord},0}_{-1,p^r}$ and $\varphi^{\mathrm{ord},\#,0}_{-1,p^r}$.)

By abuse of language, let us define the canonical isogenies

$$B_\eta \twoheadrightarrow B^{\mathrm{ord}}_{\eta,p^r} := B_\eta/\mathrm{image}(\varphi^{\mathrm{ord},0}_{-1,p^r}) = B_\eta/\varphi^{\mathrm{ord},0}_{-1,p^r}((\mathrm{Gr}^0_{\mathtt{D}_{-1,p^r}})^{\mathrm{mult}}_\eta) \tag{4.1.4.31}$$

and

$$B^\vee_\eta \twoheadrightarrow B^{\vee,\mathrm{ord}}_{\eta,p^r} := B^\vee_\eta/\mathrm{image}(\varphi^{\mathrm{ord},\#,0}_{-1,p^r}) = B^\vee_\eta/\varphi^{\mathrm{ord},\#,0}_{-1,p^r}((\mathrm{Gr}^0_{\mathtt{D}^\#_{-1,p^r}})^{\mathrm{mult}}_\eta). \tag{4.1.4.32}$$

(These definitions depend on the principal ordinary level structure $\varphi^{\mathrm{ord}}_{-1,p^r}$.)

The close immersion $\varphi^{\mathrm{ord},0}_{-1,p^r} : (\mathrm{Gr}^0_{\mathtt{D}_{-1,p^r}})^{\mathrm{mult}}_\eta \to B[p^r]_\eta$ of finite flat group schemes is dual to the surjection $B^\vee[p^r]_\eta \to (\mathrm{Gr}^{-1}_{\mathtt{D}^\#_{-1,p^r}})_\eta$ with kernel the schematic image of $\varphi^{\mathrm{ord},\#,0}_{-1,p^r} : (\mathrm{Gr}^0_{\mathtt{D}^\#_{-1,p^r}})^{\mathrm{mult}}_\eta \to B^\vee[p^r]_\eta$. Therefore, $(\mathrm{Gr}^{-1}_{\mathtt{D}^\#_{-1,p^r}})_\eta$ is embedded as an étale subgroup scheme of $B^{\vee,\mathrm{ord}}_{\eta,p^r}$, and the canonical isogeny (4.1.4.31) is dual to the (separable) isogeny

$$B^{\vee,\mathrm{ord}}_{\eta,p^r} \twoheadrightarrow B^{\vee,\mathrm{ord}}_{\eta,p^r}/(\mathrm{Gr}^{-1}_{\mathtt{D}^\#_{-1,p^r}})_\eta \cong B^\vee_\eta/B^\vee[p^r]_\eta \cong B^\vee_\eta. \tag{4.1.4.33}$$

Similarly, the canonical isogeny (4.1.4.32) is dual to the (separable) isogeny

$$B^{\mathrm{ord}}_{\eta,p^r} \twoheadrightarrow B^{\mathrm{ord}}_{\eta,p^r}/(\mathrm{Gr}^{-1}_{\mathtt{D}_{-1,p^r}})_\eta \cong B_\eta/B[p^r]_\eta \cong B_\eta. \tag{4.1.4.34}$$

Let us record our observations as follows:

Lemma 4.1.4.35. *The abelian schemes $B^{\mathrm{ord}}_{\eta,p^r}$ and $B^{\vee,\mathrm{ord}}_{\eta,p^r}$ over η are canonically dual to each other. The canonical isogenies* (4.1.4.31) *and* (4.1.4.33) (*resp.* (4.1.4.32) *and* (4.1.4.34)) *are dual to each other.*

Lemma 4.1.4.36.

(1) The embeddings $\tilde{\varphi}^{\mathrm{ord},0}_{-1,p^r}$ as in (4.1.4.29) *correspond to liftings of $c_\eta : \underline{X}_\eta \to B^\vee_\eta$ to*

$$c^{\mathrm{ord}}_{p^r} : \tfrac{1}{p^r}\underline{X}_\eta \to B^{\vee,\mathrm{ord}}_{\eta,p^r},$$

where the morphisms $\underline{X}_\eta \hookrightarrow \frac{1}{p^r}\underline{X}_\eta$ and $B^\vee_\eta \twoheadrightarrow B^{\vee,\mathrm{ord}}_{\eta,p^r}$ are the canonical ones (*see* (4.1.4.32)).

(2) The embeddings $\tilde{\varphi}^{\mathrm{ord},\#,0}_{-1,p^r}$ as in (4.1.4.30) *correspond to liftings of $c^\vee_\eta : \underline{Y}_\eta \to B_\eta$ to*

$$c^{\vee,\mathrm{ord}}_{p^r} : \tfrac{1}{p^r}\underline{Y}_\eta \to B^{\mathrm{ord}}_{\eta,p^r},$$

where the morphisms $\underline{Y}_\eta \hookrightarrow \frac{1}{p^r}\underline{Y}_\eta$ and $B_\eta \twoheadrightarrow B^{\mathrm{ord}}_{\eta,p^r}$ are the canonical ones (*see* (4.1.4.31)).

Proof. The proof is similar to that of [62, Lem. 5.2.3.1]. Let us explain only the proof of the first part, because the proof for the second part is essentially the same.

An embedding $\tilde{\varphi}^{\mathrm{ord},0}_{-1,p^r}$ as in (4.1.4.29) defines in particular an isogeny

$$G^\natural_\eta \to G^{\natural\,\prime}_\eta := G^\natural_\eta / \tilde{\varphi}^{\mathrm{ord},0}_{-1,p^r}((\mathrm{Gr}^0_{\mathrm{D}_{-1,p^r}})^{\mathrm{mult}}_\eta).$$

The subgroup scheme T_η of $G^\natural_\eta$ embeds into a subgroup scheme T'_η of $G^\natural_\eta$ because the pullback to $T_\eta = \ker(G^\natural_\eta \to B_\eta)$ of the schematic image $\varphi^{\mathrm{ord},0}_{-1,p^r}((\mathrm{Gr}^0_{\mathrm{D}_{-1,p^r}})^{\mathrm{mult}}_\eta)$ in $G^\natural_\eta$ is trivial. Hence, we have a commutative diagram:

$$\begin{array}{ccccccccc}
0 & \longrightarrow & T_\eta & \longrightarrow & G^\natural_\eta & \longrightarrow & B_\eta & \longrightarrow & 0 \\
 & & \downarrow \wr & & \downarrow & & \downarrow {\scriptstyle (4.1.4.31)\ \mathrm{mod}\ \varphi^{\mathrm{ord},0}_{-1,p^r}((\mathrm{Gr}^0_{\mathrm{D}_{-1,p^r}})^{\mathrm{mult}}_\eta)} & & \\
0 & \longrightarrow & T'_\eta & \longrightarrow & G^{\natural\,\prime}_\eta & \longrightarrow & B^{\mathrm{ord}}_{\eta,p^r} & \longrightarrow & 0
\end{array}$$

We can complete this into a diagram

$$\begin{array}{ccccccccc}
0 & \longrightarrow & T_\eta & \longrightarrow & G^\natural_\eta & \longrightarrow & B_\eta & \longrightarrow & 0 \\
 & & \downarrow \wr & & \downarrow & & \downarrow {\scriptstyle (4.1.4.31)\ \mathrm{mod}\ \varphi^{\mathrm{ord},0}_{-1,p^r}((\mathrm{Gr}^0_{\mathrm{D}_{-1,p^r}})^{\mathrm{mult}}_\eta)} & & \\
0 & \longrightarrow & T'_\eta & \longrightarrow & G^{\natural\,\prime}_\eta & \longrightarrow & B^{\mathrm{ord}}_{\eta,p^r} & \longrightarrow & 0 \\
 & & {\scriptstyle \mathrm{mod}\ T'_\eta[p^r]} \downarrow & & \downarrow & & \downarrow {\scriptstyle (4.1.4.34)\ \mathrm{mod}\ (\mathrm{Gr}^{-1}_{\mathrm{D}_{-1,p^r}})_\eta} & & \\
0 & \longrightarrow & T_\eta & \longrightarrow & G^\natural_\eta & \longrightarrow & B_\eta & \longrightarrow & 0
\end{array}$$

in which every composition of two vertical arrows is the multiplication by p^r. Therefore, finding an embedding of the form (4.1.4.29) is equivalent to finding an isogeny $G^{\natural\,\prime}_\eta \twoheadrightarrow G^\natural_\eta$ of the form:

$$\begin{array}{ccccccccc}
0 & \longrightarrow & T'_\eta & \longrightarrow & G^{\natural\,\prime}_\eta & \longrightarrow & B^{\mathrm{ord}}_{\eta,p^r} & \longrightarrow & 0 \\
 & & {\scriptstyle \mathrm{mod}\ T'_\eta[p^r]} \downarrow & & \downarrow & & \downarrow {\scriptstyle (4.1.4.34)\ \mathrm{mod}(\mathrm{Gr}^{-1}_{\mathrm{D}_{-1,p^r}})_S} & & \\
0 & \longrightarrow & T_\eta & \longrightarrow & G^\natural_\eta & \longrightarrow & B_\eta & \longrightarrow & 0
\end{array}$$

Since the surjection $T'_\eta \twoheadrightarrow T_\eta$ is the dual of the inclusion $\underline{X}_\eta \hookrightarrow \frac{1}{p^r}\underline{X}_\eta$, and since the isogeny (4.1.4.34) is dual to the isogeny (4.1.4.32) by Lemma 4.1.4.35, by [62, Prop. 3.1.5.1], isogenies $G^{\natural\,\prime}_\eta \to G^\natural_\eta$ of the above form are equivalent to liftings $c^{\mathrm{ord}}_{p^r} : \frac{1}{p^r}\underline{X}_\eta \to B^{\vee,\mathrm{ord}}_{p^r}$ over η of the homomorphism $c : \underline{X} \to B^\vee$ defining the extension structure of $0 \to T \to G^\natural \to B \to 0$. Since all the homomorphisms we consider above are $\mathcal{O}$-equivariant, the lifting $c^{\mathrm{ord}}_{p^r}$ is also $\mathcal{O}$-equivariant by functoriality of [62, Prop. 3.1.5.1]. □

Lemma 4.1.4.37. *Choices of the embeddings* $\tilde{\varphi}^{\mathrm{ord},0}_{-1,p^r}$ *and* $\tilde{\varphi}^{\mathrm{ord},\#,0}_{-1,p^r}$ *as in* (4.1.4.29) *and* (4.1.4.30) *define splittings* $\varsigma^{\mathrm{ord},0}_{p^r}$ *and* $\varsigma^{\mathrm{ord},\#,0}_{p^r}$ *as in* (4.1.4.27) *and* (4.1.4.28), *respectively, and define an* $\mathcal{O}$*-equivariant homomorphism*

$$\varphi^{\mathrm{ord},0}_{-1,p^r}((\mathrm{Gr}^0_{\mathrm{D}_{-1,p^r}})^{\mathrm{mult}}_\eta) \to T^\vee[p^r]_\eta$$

of group schemes of multiplicative type by sending $a \in \varphi^{\mathrm{ord},0}_{-1,p^r}((\mathrm{Gr}^0_{\mathtt{D}_{-1,p^r}})^{\mathrm{mult}}_\eta)$ *to*

$$(\mathrm{pr}_{T^\vee[p^r]_\eta} \circ (\varsigma^{\mathrm{ord},\#,0}_{p^r})^{-1} \circ \lambda_\eta \circ \tilde{\varphi}^{\mathrm{ord},0}_{-1,p^r} \circ (\varphi^{\mathrm{ord},0}_{-1,p^r})^{-1})(a) \in T^\vee[p^r]_\eta,$$

which (by the perfect duality $(\underline{Y}/p^r\underline{Y})_\eta \times T^\vee[p^r]_\eta \to \boldsymbol{\mu}_{p^r,\eta}$*) is equivalent to a pairing*

$$\mathrm{e}^{\mathrm{ord}}_{10,p^r} : \varphi^{\mathrm{ord},0}_{-1,p^r}((\mathrm{Gr}^0_{\mathtt{D}_{-1,p^r}})^{\mathrm{mult}}_\eta) \times (\underline{Y}/p^r\underline{Y})_\eta \to \boldsymbol{\mu}_{p^r,\eta} \tag{4.1.4.38}$$

satisfying $\mathrm{e}^{\mathrm{ord}}_{10,p^r}(b\,\cdot\,,\,\cdot\,) = \mathrm{e}^{\mathrm{ord}}_{10,p^r}(\,\cdot\,,b^\star\,\cdot\,)$ *for every* $b \in \mathcal{O}$.

In other words, the pairing $\mathrm{e}^{\mathrm{ord}}_{10,p^r}$ measures the failure of the condition that

$$\lambda(\varsigma^{\mathrm{ord},0}_{p^r}(\varphi^{\mathrm{ord},0}_{-1,p^r}((\mathrm{Gr}^0_{\mathtt{D}_{-1,p^r}})^{\mathrm{mult}}_\eta))) \subset \varsigma^{\mathrm{ord},\#,0}_{p^r}(\varphi^{\mathrm{ord},\#,0}_{-1,p^r}((\mathrm{Gr}^0_{\mathtt{D}^\#_{-1,p^r}})^{\mathrm{mult}}_\eta)).$$

There is (a priori) a second pairing analogous to (4.1.4.38): Since $\lambda_{B,\eta} : B_\eta \to B^\vee_\eta$ maps $\varphi^{\mathrm{ord},0}_{-1,p^r}((\mathrm{Gr}^0_{\mathtt{D}_{-1,p^r}})^{\mathrm{mult}}_\eta)$ to $\varphi^{\mathrm{ord},\#,0}_{-1,p^r}((\mathrm{Gr}^0_{\mathtt{D}^\#_{-1,p^r}})^{\mathrm{mult}}_\eta)$, it induces a polarization

$$\lambda^{\mathrm{ord}}_{B,\eta,p^r} : B^{\mathrm{ord}}_{\eta,p^r} \to B^{\vee,\mathrm{ord}}_{\eta,p^r} \tag{4.1.4.39}$$

compatible with the two isogenies (4.1.4.31) and (4.1.4.32).

Let us extend $\phi : \underline{Y} \to \underline{X}$ naturally to $\phi_{p^r} : \frac{1}{p^r}\underline{Y} \to \frac{1}{p^r}\underline{X}$. Since $\lambda_B c^\vee = c\phi$, for every section y of $\underline{Y}_\eta$, we see that

$$(\lambda^{\mathrm{ord}}_{B,\eta,p^r} c^{\vee,\mathrm{ord}}_{p^r} - c^{\mathrm{ord}}_{p^r}\phi_{p^r,\eta})(\tfrac{1}{p^r}y)$$

defines a section of

$$\ker(B^{\vee,\mathrm{ord}}_{\eta,p^r} \twoheadrightarrow B^\vee_\eta) \cong (\mathrm{Gr}^{-1}_{\mathtt{D}^\#_{-1,p^r}})_\eta$$

(see (4.1.4.33) and Lemma 4.1.4.35). Therefore:

Lemma 4.1.4.40. *Choices of the embeddings* $\tilde{\varphi}^{\mathrm{ord},0}_{-1,p^r}$ *and* $\tilde{\varphi}^{\mathrm{ord},\#,0}_{-1,p^r}$ *as in* (4.1.4.29) *and* (4.1.4.30) *define by Lemma* 4.1.4.36 *an* $\mathcal{O}$*-equivariant homomorphism*

$$\lambda^{\mathrm{ord}}_{B,\eta,p^r} c^{\vee,\mathrm{ord}}_{p^r} - c^{\mathrm{ord}}_{p^r}\phi_{p^r,\eta} : (\underline{Y}/p^r\underline{Y})_\eta \to (\mathrm{Gr}^{-1}_{\mathtt{D}^\#_{-1,p^r}})_\eta,$$

which (by the perfect duality $(\mathrm{Gr}^0_{\mathtt{D}_{-1,p^r}})^{\mathrm{mult}}_\eta \times (\mathrm{Gr}^{-1}_{\mathtt{D}^\#_{-1,p^r}})_\eta \to \boldsymbol{\mu}_{p^r,\eta}$*; cf. Lemma* 3.2.2.4*) is equivalent to a pairing*

$$\mathtt{d}^{\mathrm{ord}}_{10,p^r} : \varphi^{\mathrm{ord},0}_{-1,p^r}((\mathrm{Gr}^0_{\mathtt{D}_{-1,p^r}})^{\mathrm{mult}}_\eta) \times (\underline{Y}/p^r\underline{Y})_\eta \to \boldsymbol{\mu}_{p^r,\eta} \tag{4.1.4.41}$$

satisfying $\mathtt{d}^{\mathrm{ord}}_{10,p^r}(b\,\cdot\,,\,\cdot\,) = \mathtt{d}^{\mathrm{ord}}_{10,p^r}(\,\cdot\,,b^\star\,\cdot\,)$ *for every* $b \in \mathcal{O}$.

The two pairings in (4.1.4.38) and (4.1.4.41) are, without surprise, related:

Proposition 4.1.4.42. *With the same setting as in Lemmas* 4.1.4.37 *and* 4.1.4.40*, we have* $\mathtt{d}^{\mathrm{ord}}_{10,p^r} = \mathrm{e}^{\mathrm{ord}}_{10,p^r}$.

Proof. Since $\mathtt{d}^{\mathrm{ord}}_{10,p^r} = \mathrm{e}^{\mathrm{ord}}_{10,p^r}$ is a statement about equalities, we may perform an injective continuous base change from R to a noetherian complete local ring, so that (by [76, 31.C, Cor. 2]) the integral closure of R in any finite extension of $K = \mathrm{Frac}(R)$ is finite over R. (This base change is unnecessary when $\mathrm{char}(K) = 0$ or when R is excellent.) By replacing R with a finite étale extension, we may assume that $\underline{X}$ and $\underline{Y}$ are constant with values in X and Y, respectively.

Consider the canonical morphisms

$$\underline{\mathrm{Hom}}_{\mathbb{Z}}(\tfrac{1}{p^r}X, B^\vee_\eta) \twoheadrightarrow \underline{\mathrm{Hom}}_{\mathbb{Z}}(\tfrac{1}{p^r}X, B^{\vee,\mathrm{ord}}_{\eta,p^r}) \tag{4.1.4.43}$$

and

$$\underline{\mathrm{Hom}}_{\mathbb{Z}}(\tfrac{1}{p^r}Y, B_\eta) \twoheadrightarrow \underline{\mathrm{Hom}}_{\mathbb{Z}}(\tfrac{1}{p^r}Y, B^{\mathrm{ord}}_{\eta,p^r}) \tag{4.1.4.44}$$

induced by (4.1.4.32) and (4.1.4.31), respectively. These are isogenies between abelian schemes, because X and Y are $\mathbb{Z}$-lattices. Consider also the canonical homomorphism

$$\underline{\mathrm{Hom}}_{\mathbb{Z}}(\tfrac{1}{p^r}Y, G^\natural_\eta) \to \underline{\mathrm{Hom}}_{\mathbb{Z}}(Y, G^\natural_\eta), \tag{4.1.4.45}$$

which factors through

$$\underline{\mathrm{Hom}}_{\mathbb{Z}}(\tfrac{1}{p^r}Y, G^\natural_\eta) \to \underline{\mathrm{Hom}}_{\mathbb{Z}}(Y, G^\natural_\eta) \underset{\underline{\mathrm{Hom}}_{\mathbb{Z}}(Y, B_\eta)}{\times} \underline{\mathrm{Hom}}_{\mathbb{Z}}(\tfrac{1}{p^r}Y, B^{\mathrm{ord}}_{\eta,p^r}) \tag{4.1.4.46}$$

qy(where the canonical morphism $\underline{\mathrm{Hom}}_{\mathbb{Z}}(\frac{1}{p^r}Y, G^\natural_\eta) \to \underline{\mathrm{Hom}}_{\mathbb{Z}}(\frac{1}{p^r}Y, B^{\mathrm{ord}}_{\eta,p^r})$ is the composition of the canonical morphism $\underline{\mathrm{Hom}}_{\mathbb{Z}}(\frac{1}{p^r}Y, G^\natural_\eta) \to \underline{\mathrm{Hom}}_{\mathbb{Z}}(\frac{1}{p^r}Y, B_\eta)$ and (4.1.4.44)). These are isogenies with finite kernels between semi-abelian schemes. (The kernel of (4.1.4.45) is canonically isomorphic to $\underline{\mathrm{Hom}}_{\mathbb{Z}}(\frac{1}{p^r}Y/Y, G^\natural[p^r]_\eta)$. The kernel of (4.1.4.46) is canonically isomorphic to $\underline{\mathrm{Hom}}_{\mathbb{Z}}(\frac{1}{p^r}Y/Y, \alpha^{\mathrm{ord},0,\natural}_{p^r}((\mathrm{Gr}^0_{\mathtt{D}_{p^r}})^{\mathrm{mult}}_\eta))$; see Lemma 4.1.4.20.)

The two homomorphisms $c^{\mathrm{ord}}_{p^r} : \frac{1}{p^r}X_\eta \to B^{\vee,\mathrm{ord}}_{p^r}$ and $c^{\vee,\mathrm{ord}}_{p^r} : \frac{1}{p^r}Y \to B^{\mathrm{ord}}_{\eta,p^r}$ (with their $\mathcal{O}$-equivariance ignored) define η-valued points of $\underline{\mathrm{Hom}}_{\mathbb{Z}}(X, B^{\vee,\mathrm{ord}}_{\eta,p^r})$ and $\underline{\mathrm{Hom}}_{\mathbb{Z}}(Y, B^{\mathrm{ord}}_{\eta,p^r})$, respectively. Moreover, the pair $(\iota, c^{\mathrm{ord}}_{p^r})$, where $\iota : Y \to G^\natural_\eta$ is the homomorphism determined by $\tau : \mathbf{1}_{Y \times X,\eta} \overset{\sim}{\to} (c^\vee \times c)^* \mathcal{P}^{\otimes -1}_{B,\eta}$ (which is compatible with $c^\vee$ by definition), determines a η-valued point of the target of (4.1.4.46). Since (4.1.4.32) and (4.1.4.46) are isogenies with finite kernels between semi-abelian varieties over η, there exists a finite extension $\tilde{K}$ of K such that, with $\tilde{\eta} := \mathrm{Spec}(\tilde{K})$, the η-valued points of the targets of (4.1.4.43) and (4.1.4.46) lift to $\tilde{\eta}$-valued points of the sources. Since the integral closure $\tilde{R}$ of R in the finite extension $\tilde{K}$ of $K = \mathrm{Frac}(R)$ is finite over R, we may replace R (resp. K) with $\tilde{R}$ (resp. $\tilde{K}$) and assume that $c^{\mathrm{ord}}_{p^r}$ and $c^{\vee,\mathrm{ord}}_{p^r}$ lift to homomorphisms

$$c_{p^r} : \tfrac{1}{p^r}X \to B^\vee_\eta \tag{4.1.4.47}$$

and

$$c^\vee_{p^r} : \tfrac{1}{p^r}Y \to B_\eta, \tag{4.1.4.48}$$

respectively, and that ι lifts to a homomorphism

$$\iota_{p^r} : \tfrac{1}{p^r}Y \to G_\eta^\natural \tag{4.1.4.49}$$

compatible with $c_{p^r}^\vee$ (by composition with the canonical morphism $G_\eta^\natural \to B_\eta$), which is equivalent to a lifting $\tau_{p^r} : \mathbf{1}_{\frac{1}{p^r}Y \times X,\eta} \xrightarrow{\sim} (c_{p^r}^\vee, c_\eta)^* \mathcal{P}_{B,\eta}^{\otimes -1}$ of τ.

By [62, Prop. 5.2.3.3], and by comparing its proof with that of Lemma 4.1.4.36, the lifting $(c_{p^r}, c_{p^r}^\vee, \iota_{p^r})$ of $(c, c^\vee, \iota)$ as in (4.1.4.47), (4.1.4.48), and (4.1.4.49) defines a splitting

$$\varsigma_{p^r} : T[p^r]_\eta \oplus B[p^r]_\eta \oplus (\tfrac{1}{p^r}Y/Y)_\eta \xrightarrow{\sim} G[p^r]_\eta$$

which is not necessarily $\mathcal{O}$-equivariant, which nevertheless induces the two $\mathcal{O}$-equivariant splittings (4.1.4.27) and (4.1.4.28) by taking graded pieces with respect to the filtration defined by $\mathrm{image}(\alpha_{p^r}^{\mathrm{ord},0}) = \alpha_{p^r}^{\mathrm{ord},0}((\mathrm{Gr}_{\mathtt{D}_{p^r}}^0)_\eta^{\mathrm{mult}})$, and by duality.

By [62, Thm. 5.2.3.14], we have

$$\begin{aligned} \mathtt{d}_{10,p^r}(a, \tfrac{1}{p^r}y) &= \mathrm{e}_{B[p^r]_\eta}(a, (\lambda_{B,\eta} c_{p^r}^\vee - c_{p^r}\phi_{p^r})(\tfrac{1}{p^r}y)) \\ &= \mathrm{e}_{10,p^r}(a, \tfrac{1}{p^r}y) = \mathrm{e}^{\lambda_\eta}(\varsigma_{p^r}(0,a,0), \varsigma_{p^r}(0,0,\tfrac{1}{p^r}y)) \end{aligned} \tag{4.1.4.50}$$

for all $a \in B[p^r]_\eta$ and $\frac{1}{p^r}y \in (\frac{1}{p^r}Y/Y)_\eta$. If we consider only $a \in \mathrm{image}(\varphi_{-1,p^r}^{\mathrm{ord},0}) = \varphi_{-1,p^r}^{\mathrm{ord},0}((\mathrm{Gr}_{\mathtt{D}_{-1,p^r}}^0)_\eta^{\mathrm{mult}})$, then the canonical pairings

$$\mathrm{e}_{B[p^r]_\eta} : B[p^r]_\eta \times B^\vee[p^r]_\eta \to \boldsymbol{\mu}_{p^r,\eta}$$

and

$$\mathrm{e}^{\lambda_\eta} : G[p^r]_\eta \times G[p^r]_\eta \to \boldsymbol{\mu}_{p^r,\eta}$$

induce the canonical pairings

$$\mathrm{e}_{\mathrm{image}(\varphi_{-1,p^r}^{\mathrm{ord},0})} : \varphi_{-1,p^r}^{\mathrm{ord},0}((\mathrm{Gr}_{\mathtt{D}_{-1,p^r}}^0)_\eta^{\mathrm{mult}}) \times (\mathrm{Gr}_{\mathtt{D}_{-1,p^r}^\#}^{-1})_\eta \to \boldsymbol{\mu}_{p^r,\eta},$$

and

$$\mathrm{e}_{\mathrm{image}(\alpha_{p^r}^{\mathrm{ord},0})}^{\lambda_\eta} : \alpha_{p^r}^{\mathrm{ord},0}((\mathrm{Gr}_{\mathtt{D}_{p^r}}^0)_\eta^{\mathrm{mult}}) \times (\mathrm{Gr}_{\mathtt{D}_{p^r}}^{-1})_\eta \to \boldsymbol{\mu}_{p^r,\eta},$$

respectively, and the relation (4.1.4.50) implies that

$$\begin{aligned} \mathtt{d}_{10,p^r}^{\mathrm{ord}}(a, \tfrac{1}{p^r}y) &= \mathrm{e}_{\mathrm{image}(\varphi_{-1,p^r}^{\mathrm{ord},0})}(a, (\lambda_{B,\eta}^{\mathrm{ord}} c_{p^r}^{\vee,\mathrm{ord}} - c_{p^r}^{\mathrm{ord}}\phi_{p^r})(\tfrac{1}{p^r}y)) \\ &= \mathrm{e}_{\mathrm{image}(\alpha_{p^r}^{\mathrm{ord},0})}^{\lambda_\eta}(\varsigma_{p^r}^{\mathrm{ord},0}(0,a), \varsigma_{p^r}^{\mathrm{ord},-1}(\tfrac{1}{p^r}y, 0)) \end{aligned} \tag{4.1.4.51}$$

for all $a \in \mathrm{image}(\varphi_{-1,p^r}^{\mathrm{ord},0})$ and $\frac{1}{p^r}y \in (\frac{1}{p^r}Y/Y)_\eta$, where $\varsigma_{p^r}^{\mathrm{ord},0}$ is as in (4.1.4.27), and where

$$\varsigma_{p^r}^{\mathrm{ord},-1} : (\tfrac{1}{p^r}Y/Y)_\eta \oplus (\mathrm{Gr}_{\mathtt{D}_{-1,p^r}}^{-1})_\eta \xrightarrow{\sim} (\mathrm{Gr}_{\mathtt{D}_{p^r}}^{-1})_\eta$$

is canonically dual to the inverse of the $\varsigma_{p^r}^{\mathrm{ord},\#,0}$ as in (4.1.4.28). By comparison with the definition in Lemma 4.1.4.37, we have

$$\mathrm{e}_{10,p^r}^{\mathrm{ord}}(a, \tfrac{1}{p^r}y) = \mathrm{e}_{\mathrm{image}(\alpha_{p^r}^{\mathrm{ord},0})}^{\lambda_\eta}(\varsigma_{p^r}^{\mathrm{ord},0}(0,a), \varsigma_{p^r}^{\mathrm{ord},-1}(\tfrac{1}{p^r}y, 0)).$$

Thus, the proposition follows from (4.1.4.51), as desired. □

Now it is natural to compare the pairing (4.1.4.26) with the pairing (4.1.4.41).

Proposition 4.1.4.52. *(This is a continuation of Proposition* 4.1.4.21.*) With the same setting as in Proposition* 4.1.4.21*, the tuple* $(\mathtt{Z}_{p^r}, (\varphi_{-2,p^r}, \varphi_{0,p^r}), \mathtt{D}_{-1,p^r}, \varphi^{\mathrm{ord}}_{-1,p^r})$ *there satisfies the following condition:*

For each pair of splittings $\delta^{\mathrm{ord}}_{p^r} := (\delta^{\mathrm{ord},0}_{p^r}, \delta^{\mathrm{ord},\#,0}_{p^r})$ *as in Lemma* 4.1.4.22*, which determines a pairing*

$$\langle\,\cdot\,,\,\cdot\,\rangle^{\mathrm{ord}}_{10,p^r} : \mathrm{Gr}^0_{\mathtt{D}_{-1,p^r}} \times \mathrm{Gr}^{\mathtt{Z}}_{0,p^r} \to (\mathbb{Z}/p^r\mathbb{Z})(1)$$

as in Lemma 4.1.4.25*, and accordingly a pairing*

$$(\langle\,\cdot\,,\,\cdot\,\rangle^{\mathrm{ord}}_{10,p^r})^{\mathrm{mult}}_{\eta} : (\mathrm{Gr}^0_{\mathtt{D}_{-1,p^r}})^{\mathrm{mult}}_{\eta} \times (\mathrm{Gr}^{\mathtt{Z}}_{0,p^r})_{\eta} \to \boldsymbol{\mu}_{p^r,\eta},$$

there exists a (necessarily unique) pair of liftings

$$(c^{\mathrm{ord}}_{p^r} : \tfrac{1}{p^r}\underline{X}_{\eta} \to B^{\vee,\mathrm{ord}}_{\eta,p^r}, c^{\vee,\mathrm{ord}}_{p^r} : \tfrac{1}{p^r}\underline{Y}_{\eta} \to B^{\mathrm{ord}}_{\eta,p^r})$$

of $(c : \underline{X} \to B^{\vee}, c^{\vee} : \underline{Y} \to B)$*, which by Lemma* 4.1.4.36 *is equivalent to a pair of embeddings* $(\tilde{\varphi}^{\mathrm{ord},0}_{-1,p^r}, \tilde{\varphi}^{\mathrm{ord},\#,0}_{-1,p^r})$ *as in* (4.1.4.29) *and* (4.1.4.30)*, which determines a pairing*

$$\mathtt{d}^{\mathrm{ord}}_{10,p^r} : \varphi^{\mathrm{ord},0}_{-1,p^r}((\mathrm{Gr}^0_{\mathtt{D}_{-1,p^r}})^{\mathrm{mult}}_{\eta}) \times (\underline{Y}/p^r\underline{Y})_{\eta} \to \boldsymbol{\mu}_{p^r,\eta}$$

as in Lemma 4.1.4.40*, such that*

$$(\varphi^{\mathrm{ord},0}_{-1,p^r}, \varphi_{0,p^r})^*(\mathtt{d}^{\mathrm{ord}}_{10,p^r}) = \nu^{\mathrm{ord}}_{-1,p^r} \circ (\langle\,\cdot\,,\,\cdot\,\rangle^{\mathrm{ord}}_{10,p^r})^{\mathrm{mult}}_{\eta}.$$

This condition is independent of the choice of $(\delta^{\mathrm{ord},0}_{p^r}, \delta^{\mathrm{ord},\#,0}_{p^r})$.

Conversely, each tuple $(\mathtt{Z}_{p^r}, (\varphi_{-2,p^r}, \varphi_{0,p^r}), \mathtt{D}_{-1,p^r}, \varphi^{\mathrm{ord}}_{-1,p^r})$ *that satisfies this condition comes from some naive principal ordinary level-p^r structure* $\alpha^{\mathrm{ord}}_{p^r} = (\alpha^{\mathrm{ord},0}_{p^r}, \alpha^{\mathrm{ord},\#,0}_{p^r}, \nu^{\mathrm{ord}}_{p^r})$ *of type* $(L/p^rL, \langle\,\cdot\,,\,\cdot\,\rangle, \mathtt{D}_{p^r})$ *on* $(G_{\eta}, \lambda_{\eta}, i_{\eta})$ *(see Definition* 3.3.3.2*) as in Proposition* 4.1.4.21*. If the tuple is liftable (satisfying the analogous conditions for the liftings), then it comes from some principal ordinary level-p^r structure* $\alpha^{\mathrm{ord}}_{p^r}$ *of type* $(L \underset{\mathbb{Z}}{\otimes} \mathbb{Z}_p, \langle\,\cdot\,,\,\cdot\,\rangle, \mathtt{D})$.

Explicitly, we can recover the triple $\alpha^{\mathrm{ord}}_{p^r} = (\alpha^{\mathrm{ord},0}_{p^r}, \alpha^{\mathrm{ord},\#,0}_{p^r}, \nu^{\mathrm{ord}}_{p^r})$ *from the tuple* $(\mathtt{Z}_{p^r}, (\varphi_{-2,p^r}, \varphi_{0,p^r}), \mathtt{D}_{-1,p^r}, \varphi^{\mathrm{ord}}_{-1,p^r})$ *as follows:*

Suppose $(\delta^{\mathrm{ord},0}_{p^r}, \delta^{\mathrm{ord},\#,0}_{p^r})$ *and* $(\tilde{\varphi}^{\mathrm{ord},0}_{-1,p^r}, \tilde{\varphi}^{\mathrm{ord},\#,0}_{-1,p^r})$ *have been chosen such that the above condition is satisfied, the latter of which determines a pair of splittings* $(\varsigma^{\mathrm{ord},0}_{p^r}, \varsigma^{\mathrm{ord},\#,0}_{p^r})$ *as in* (4.1.4.27) *and* (4.1.4.28)*. Then we have the following defining relations:*

(1) The homomorphism $\alpha^{\mathrm{ord},0}_{p^r} : (\mathrm{Gr}^0_{\mathtt{D},p^r})^{\mathrm{mult}}_{\eta} \to G[p^r]_{\eta}$ *is defined as follows: The sum of the composition*

$$\nu^{\mathrm{ord}}_{-1,p^r} \circ (\varphi_{-2,p^r})^{\mathrm{mult}}_{\eta} : (\mathrm{Gr}^{\mathtt{Z}}_{-2,p^r})^{\mathrm{mult}}_{\eta} \xrightarrow{\sim} T[p^r]_{\eta} \overset{\mathrm{can.}}{\hookrightarrow} G^{\natural}[p^r]_{\eta}$$

and

$$\tilde{\varphi}^{\mathrm{ord},0}_{-1,p^r} : (\mathrm{Gr}^0_{\mathtt{D}_{-1,p^r}})^{\mathrm{mult}}_{\eta} \to G^{\natural}[p^r]_{\eta}$$

defines a homomorphism

$$(\mathrm{Gr}^{\mathtt{Z}}_{-2,p^r})^{\mathrm{mult}}_{\eta} \oplus (\mathrm{Gr}^0_{\mathtt{D}_{-1,p^r}})^{\mathrm{mult}}_{\eta} \to G^{\natural}[p^r]_{\eta}$$

that is a closed embedding, and the pre-composition of this homomorphism with

$$((\delta^{\mathrm{ord},0}_{p^r})^{\mathrm{mult}}_{\eta})^{-1} : (\mathrm{Gr}^0_{\mathtt{D},p^r})^{\mathrm{mult}}_{\eta} \xrightarrow{\sim} (\mathrm{Gr}^{\mathtt{Z}}_{-2,p^r})^{\mathrm{mult}}_{\eta} \oplus (\mathrm{Gr}^0_{\mathtt{D}_{-1,p^r}})^{\mathrm{mult}}_{\eta}$$

gives the homomorphism

$$\alpha^{\mathrm{ord},0,\natural}_{p^r} : (\mathrm{Gr}^0_{\mathtt{D},p^r})^{\mathrm{mult}}_{\eta} \to G^{\natural}[p^r]_{\eta},$$

and $\alpha^{\mathrm{ord},0}_{p^r}$ *is the composition of* $\alpha^{\mathrm{ord},0,\natural}_{p^r}$ *with the canonical embedding* $G^{\natural}[p^r]_{\eta} \hookrightarrow G[p^r]_{\eta}$.

(2) *The homomorphism* $\alpha^{\mathrm{ord},\#,0}_{p^r} : (\mathrm{Gr}^0_{\mathtt{D}^{\#},p^r})^{\mathrm{mult}}_{\eta} \to G^{\vee}[p^r]_{\eta}$ *is defined as follows: The sum of the composition*

$$(\varphi^{\#}_{-2,p^r})^{\mathrm{mult}}_{\eta} : (\mathrm{Gr}^{\mathtt{Z}^{\#}}_{-2,p^r})^{\mathrm{mult}}_{\eta} \xrightarrow{\sim} T^{\vee}[p^r]_{\eta} \overset{\mathrm{can.}}{\hookrightarrow} G^{\vee,\natural}[p^r]_{\eta}$$

(note that we do not compose with $\nu^{\mathrm{ord}}_{-1,p^r}$ *here) and*

$$\tilde{\varphi}^{\mathrm{ord},\#,0}_{-1,p^r} : (\mathrm{Gr}^0_{\mathtt{D}^{\#}_{-1,p^r}})^{\mathrm{mult}}_{\eta} \to G^{\vee,\natural}[p^r]_{\eta}$$

defines a homomorphism

$$(\mathrm{Gr}^{\mathtt{Z}^{\#}}_{-2,p^r})^{\mathrm{mult}}_{\eta} \oplus (\mathrm{Gr}^0_{\mathtt{D}^{\#}_{-1,p^r}})^{\mathrm{mult}}_{\eta} \to G^{\natural}[p^r]_{\eta}$$

that is a closed embedding, and the pre-composition of this homomorphism with

$$((\delta^{\mathrm{ord},\#,0}_{p^r})^{\mathrm{mult}}_{\eta})^{-1} : (\mathrm{Gr}^0_{\mathtt{D}^{\#},p^r})^{\mathrm{mult}}_{\eta} \xrightarrow{\sim} (\mathrm{Gr}^{\mathtt{Z}^{\#}}_{-2,p^r})^{\mathrm{mult}}_{\eta} \oplus (\mathrm{Gr}^0_{\mathtt{D}^{\#}_{-1,p^r}})^{\mathrm{mult}}_{\eta}$$

gives the homomorphism

$$\alpha^{\mathrm{ord},\#,0,\natural}_{p^r} : (\mathrm{Gr}^0_{\mathtt{D}^{\#},p^r})^{\mathrm{mult}}_{\eta} \to G^{\vee,\natural}[p^r]_{\eta},$$

and $\alpha^{\mathrm{ord},\#,0}_{p^r}$ *is the composition of* $\alpha^{\mathrm{ord},\#,0,\natural}_{p^r}$ *with the canonical embedding* $G^{\vee,\natural}[p^r]_{\eta} \hookrightarrow G^{\vee}[p^r]_{\eta}$.

(3) *The isomorphism* $\nu^{\mathrm{ord}}_{p^r} : \boldsymbol{\mu}_{p^r,\eta} \xrightarrow{\sim} \boldsymbol{\mu}_{p^r,\eta}$ *is equal to* $\nu^{\mathrm{ord}}_{-1,p^r}$, *where* $\nu^{\mathrm{ord}}_{-1,p^r} : \boldsymbol{\mu}_{p^r,\eta} \xrightarrow{\sim} \boldsymbol{\mu}_{p^r,\eta}$ *is part of the data of* $\varphi^{\mathrm{ord}}_{-1,p^r}$.

Proof. The statements are self-explanatory. □

Remark 4.1.4.53. In the theory of degeneration for naive principal ordinary level structures, there is no need to consider some lifting $\tau^{\mathrm{ord}}_{p^r}$ of τ as in the theory for principal level structures (away from p) as in [62, Ch. 5].

Definition 4.1.4.54. (*See [62, Prop. 5.2.2.1].*) *Let* $\phi : \underline{Y} \to \underline{X}$ *be an* $\mathcal{O}$*-equivariant embedding with finite cokernel between étale sheaves of* $\mathcal{O}$*-lattices, which is dual to an* $\mathcal{O}$*-equivariant isogeny* $T \twoheadrightarrow T^\vee$ *of tori with* $\mathcal{O}$*-actions. Then we define for each integer* m *the pairing*

$$\mathrm{e}^{\phi}_m : T[m]_\eta \times (\underline{Y}/m\underline{Y})_\eta \to \boldsymbol{\mu}_{m,\eta}$$

to be the canonical pairing

$$\begin{aligned} T[m]_\eta \times (\underline{Y}/m\underline{Y})_\eta &\overset{\text{can.}}{\overset{\sim}{\to}} (\underline{X}/m\underline{X})^{\mathrm{mult}}_\eta \times (\underline{Y}/m\underline{Y})_\eta \\ &\overset{\mathrm{Id}\times\phi}{\to} (\underline{X}/m\underline{X})^{\mathrm{mult}}_\eta \times (\underline{X}/m\underline{X})_\eta \overset{\text{can.}}{\to} \boldsymbol{\mu}_{m,\eta} \end{aligned}$$

with the sign convention that $\mathrm{e}^{\phi}_m(t,y) = t(\phi(y)) = (\phi(y))(t)$ *for every section* t *of* $T[m]_\eta$ *and* y *of* $(\underline{Y}/m\underline{Y})_\eta$.

Definition 4.1.4.55. (*Compare with [62, Def. 5.2.7.8].*) *With the setting as in Section* 4.1.2, *suppose we are given a tuple* $(B, \lambda_B, i_B, \underline{X}, \underline{Y}, \phi, c, c^\vee, \tau)$ *in* $\mathrm{DD}_{\mathrm{PE},\mathcal{O}}(R, I)$. *A* **naive ordinary pre-level-**n **structure datum of type** $(L/nL, \langle\,\cdot\,,\,\cdot\,\rangle, \mathsf{D}_{p^r})$ *over* η *is a tuple*

$$\alpha^{\natural,\mathrm{ord}}_n := (\mathsf{Z}_n, \varphi_{-2,n}, \varphi^{\mathrm{ord}}_{-1,n}, \varphi_{0,n}, \delta^{\mathrm{ord}}_n, c^{\mathrm{ord}}_n, c^{\vee,\mathrm{ord}}_n, \tau^{\mathrm{ord}}_n)$$

consists of the following data:

(1) *A symplectic admissible filtration* Z_n *on* L/nL. *This determines a dual symplectic admissible filtration* $\mathsf{Z}^\#_n$ *on* $L^\#/nL^\#$, *with induced perfect pairings*

$$\langle\,\cdot\,,\,\cdot\,\rangle_{ij,n} : \mathrm{Gr}^{\mathsf{Z}^\#}_{-i,n} \times \mathrm{Gr}^{\mathsf{Z}}_{j,n} \to (\mathbb{Z}/n\mathbb{Z})(1),$$

for $i + j = 2$, *inducing (possibly nonperfect) pairings*

$$\langle\,\cdot\,,\,\cdot\,\rangle_{ij,n} : \mathrm{Gr}^{\mathsf{Z}}_{-i,n} \times \mathrm{Gr}^{\mathsf{Z}}_{-j,n} \to (\mathbb{Z}/n\mathbb{Z})(1)$$

denoted by the same symbols (by abuse of notation).

By reduction modulo n_0 *(resp.* p^r*), we also obtain an admissible symplectic filtration and pairings between the graded pieces, with* n *replaced with* n_0 *(resp.* p^r*) in the above notation.*

The filtration Z_{p^r} *is compatible with* D_{p^r} *in the sense that*

$$\mathsf{Z}_{-2,p^r} \subset \mathsf{D}_{p^r} \subset \mathsf{Z}_{-1,p^r},$$

and hence induces a filtration $\mathsf{D}_{-1,p^r} = \{\mathsf{D}^i_{-1,p^r}\}_i$ *on* $\mathrm{Gr}^{\mathsf{Z}}_{-1,p^r}$ *(see* (4.1.4.14)*) that satisfies the analogous conditions as the filtration* D_{p^r} *does for* L/p^rL *(see Lemma* 3.2.2.1*).*

(2) *A pair* $\varphi^{\mathrm{ord}}_{-1,n} = (\varphi_{-1,n_0}, \varphi^{\mathrm{ord}}_{-1,p^r})$ *consisting of:*

(a) *A (naive) principal level-*n_0 *structure*

$$\varphi_{-1,n_0} : \mathrm{Gr}^{\mathsf{Z}}_{-1,n_0} \overset{\sim}{\to} B[n_0]_\eta$$

of $(B_\eta, \lambda_{B,\eta}, i_{B,\eta})$ *of type* $(\mathrm{Gr}^{\mathsf{Z}}_{-1,n_0}, \langle\,\cdot\,,\,\cdot\,\rangle_{11,n_0})$ *over* η, *equipped with an isomorphism*

$$\nu_{-1,n_0} : (\mathbb{Z}/n_0\mathbb{Z})(1) \overset{\sim}{\to} \boldsymbol{\mu}_{n_0,\eta},$$

as in Definition 3.3.1.2.

(b) *A (naive) principle ordinary level structure*

$$\varphi^{\mathrm{ord}}_{-1,p^r} = (\varphi^{\mathrm{ord},0}_{-1,p^r}, \varphi^{\mathrm{ord},\#,0}_{-1,p^r}, \nu^{\mathrm{ord}}_{-1,p^r})$$

of $(B_\eta, \lambda_{B,\eta}, i_{B,\eta})$ *of type* $\phi^0_{\mathsf{D}_{-1,p^r}} : \mathrm{Gr}^0_{\mathsf{D}_{-1,p^r}} \to \mathrm{Gr}^0_{\mathsf{D}^\#_{-1,p^r}}$ *(see* (4.1.4.18)*)* *as in Definition* 3.3.3.2 *(see also* (4) *of Proposition* 4.1.4.21*).*

(3) *A pair of $\mathcal{O}$-equivariant isomorphisms*

$$\varphi_{-2,n} : \mathrm{Gr}^{\mathsf{Z}}_{-2,n} \xrightarrow{\sim} \underline{\mathrm{Hom}}_\eta((\underline{X}/n\underline{X})_\eta, (\mathbb{Z}/n\mathbb{Z})(1))$$

and

$$\varphi_{0,n} : \mathrm{Gr}^{\mathsf{Z}}_{0,n} \xrightarrow{\sim} (\underline{Y}/n\underline{Y})_\eta$$

satisfying

$$((\varphi_{-2,n})^{\mathrm{mult}}_\eta \times \varphi_{0,n})^* \mathrm{e}^\phi_n = ((\langle\cdot,\cdot\rangle_{20,n})^{\mathrm{mult}}_\eta,$$

where

$$(\varphi_{-2,n})^{\mathrm{mult}}_\eta : (\mathrm{Gr}^{\mathsf{Z}}_{-2,n})^{\mathrm{mult}}_\eta \xrightarrow{\sim} T[n]_\eta$$

and

$$((\langle\cdot,\cdot\rangle_{20,n})^{\mathrm{mult}}_\eta : (\mathrm{Gr}^{\mathsf{Z}}_{-2,n})^{\mathrm{mult}}_\eta \times (\mathrm{Gr}^{\mathsf{Z}}_{0,n})_\eta \to \boldsymbol{\mu}_{n,\eta}$$

are canonically induced by $\varphi_{-2,n}$ and $\langle\cdot,\cdot\rangle_{20,n}$.
By reduction modulo n_0, we obtain isomorphisms and pairings with n replaced with n_0 in the notation. By reduction modulo p^r, we obtain isomorphisms over η that are equivalent to the pullbacks to η of the isomorphisms

$$(\varphi_{-2,p^r})^{\mathrm{mult}}_S : (\mathrm{Gr}^{\mathsf{Z}}_{-2,p^r})^{\mathrm{mult}}_S \xrightarrow{\sim} T[p^r]$$

and

$$(\varphi^\#_{-2,p^r})^{\mathrm{mult}}_S : (\mathrm{Gr}^{\mathsf{Z}^\#}_{-2,p^r})^{\mathrm{mult}}_S \xrightarrow{\sim} T^\vee[p^r]$$

over S.

(4) *A pair $\delta^{\mathrm{ord}}_n = (\delta_{n_0}, \delta^{\mathrm{ord}}_{p^r})$ consisting of:*

(a) *A splitting $\delta_{n_0} : \mathrm{Gr}^{\mathsf{Z}}_{n_0} \xrightarrow{\sim} L/n_0L$ of $\mathcal{O}$-modules, which determines the pairings*

$$\langle\cdot,\cdot\rangle_{10,n_0} : \mathrm{Gr}^{\mathsf{Z}}_{-1,n_0} \times \mathrm{Gr}^{\mathsf{Z}}_{0,n_0} \to (\mathbb{Z}/n_0\mathbb{Z})(1)$$

and

$$\langle\cdot,\cdot\rangle_{00,n_0} : \mathrm{Gr}^{\mathsf{Z}}_{0,n_0} \times \mathrm{Gr}^{\mathsf{Z}}_{0,n_0} \to (\mathbb{Z}/n_0\mathbb{Z})(1).$$

(b) *A pair $\delta^{\mathrm{ord}}_{p^r} = (\delta^{\mathrm{ord},0}_{p^r}, \delta^{\mathrm{ord},\#,0}_{p^r})$ of splittings*

$$\delta^{\mathrm{ord},0}_{p^r} : \mathrm{Gr}^{\mathsf{Z}}_{-2,p^r} \oplus \mathrm{Gr}^0_{\mathsf{D}_{-1,p^r}} \xrightarrow{\sim} \mathrm{Gr}^0_{\mathsf{D},p^r}$$

and

$$\delta^{\mathrm{ord},\#,0}_{p^r} : \mathrm{Gr}^{\mathsf{Z}^\#}_{-2,p^r} \oplus \mathrm{Gr}^0_{\mathsf{D}^\#_{-1,p^r}} \xrightarrow{\sim} \mathrm{Gr}^0_{\mathsf{D}^\#,p^r}$$

of $\mathcal{O}$-modules as in Lemma 4.1.4.22*, which determines a pairing*

$$\langle\cdot,\cdot\rangle^{\mathrm{ord}}_{10,p^r} : \mathrm{Gr}^0_{\mathsf{D}_{-1,p^r}} \times \mathrm{Gr}^{\mathsf{Z}}_{0,p^r} \to (\mathbb{Z}/p^r\mathbb{Z})(1)$$

as in Lemma 4.1.4.25*, and accordingly a pairing*

$$(\langle\cdot,\cdot\rangle^{\mathrm{ord}}_{10,p^r})^{\mathrm{mult}}_\eta : (\mathrm{Gr}^0_{\mathsf{D}_{-1,p^r}})^{\mathrm{mult}}_\eta \times (\mathrm{Gr}^{\mathsf{Z}}_{0,p^r})_\eta \to \boldsymbol{\mu}_{p^r,\eta}.$$

(5) Liftings

$$c_n^{\mathrm{ord}} : \tfrac{1}{n}\underline{X}_\eta \to B_{\eta,p^r}^{\vee,\mathrm{ord}} := B_\eta^\vee/\varphi_{-1,p^r}^{\mathrm{ord},\#,0}((\mathrm{Gr}^0_{\mathtt{D}^\#_{-1,p^r}})^{\mathrm{mult}}_\eta),$$

$$c_n^{\vee,\mathrm{ord}} : \tfrac{1}{n}\underline{Y}_\eta \to B_{\eta,p^r}^{\mathrm{ord}} := B_\eta/\varphi_{-1,p^r}^{\mathrm{ord},0}((\mathrm{Gr}^0_{\mathtt{D}_{-1,p^r}})^{\mathrm{mult}}_\eta),$$

and

$$\tau_n^{\mathrm{ord}} := \tau_{n_0} : \mathbf{1}_{\frac{1}{n_0}\underline{Y}\underset{S}{\times}\underline{X},\eta} \xrightarrow{\sim} (c_{n_0}^\vee, c_\eta)^*\mathcal{P}_{B,\eta}^{\otimes -1}$$

of $c : \underline{X} \to B^\vee$, $c^\vee : \underline{Y} \to B$ *and* $\tau : \mathbf{1}_{\underline{Y}\underset{S}{\times}\underline{X},\eta} \xrightarrow{\sim} (c^\vee, c_\eta)^*\mathcal{P}_{B,\eta}^{\otimes -1}$ *over* η, *respectively.*

The liftings c_n^{ord} *and* $c_n^{\vee,\mathrm{ord}}$ *determine and are determined by liftings* $c_{n_0} : \frac{1}{n_0}\underline{X} \to B^\vee$, $c_{n_0}^\vee : \frac{1}{n_0}\underline{Y} \to B$, $c_{p^r}^{\mathrm{ord}} : \frac{1}{p^r}\underline{X}_\eta \to B_{\eta,p^r}^{\vee,\mathrm{ord}}$, *and* $c_{p^r}^{\vee,\mathrm{ord}} : \frac{1}{p^r}\underline{Y}_\eta \to B_{\eta,p^r}^{\mathrm{ord}}$ *of* $c : \underline{X} \to B^\vee$, $c^\vee : \underline{Y} \to B$, $c : \underline{X} \to B^\vee$, *and* $c^\vee : \underline{Y} \to B$ *over* η, *respectively.*

By [62, Lem. 5.2.3.12], the liftings c_{n_0}, $c_{n_0}^\vee$, *and* τ_{n_0} *defines two pairings*

$$\mathtt{d}_{10,n_0} : B[n_0]_\eta \times (\underline{Y}/n_0\underline{Y})_\eta \to \boldsymbol{\mu}_{n_0,\eta}$$

and

$$\mathtt{d}_{00,n_0} : (\underline{Y}/n_0\underline{Y})_\eta \times (\underline{Y}/n_0\underline{Y})_\eta \to \boldsymbol{\mu}_{n_0,\eta}$$

by setting

$$\mathtt{d}_{10,n_0}(a, \tfrac{1}{n_0}y) := \mathrm{e}_{B[n_0]}(a, (\lambda_B c_{n_0}^\vee - c_{n_0}\phi_{n_0})(\tfrac{1}{n_0}y) \in \boldsymbol{\mu}_{n_0}(\eta)$$

for sections a *of* $B[n_0]_\eta$ *and* $\frac{1}{n_0}y$ *of* $\frac{1}{n_0}\underline{Y}$, *and by setting*

$$\mathtt{d}_{00,n_0}(\tfrac{1}{n_0}y, \tfrac{1}{n_0}y') := \tau_{n_0}(\tfrac{1}{n_0}y, \phi(y'))\tau_{n_0}(\tfrac{1}{n_0}y', \phi(y))^{-1} \in \boldsymbol{\mu}_m(\eta)$$

for sections $\frac{1}{n_0}y$ *and* $\frac{1}{n_0}y'$ *of* $\frac{1}{n_0}\underline{Y}$.

By Lemmas 4.1.4.36 *and* 4.1.4.40, *the liftings* $c_{p^r}^{\mathrm{ord}}$ *and* $c_{p^r}^{\vee,\mathrm{ord}}$ *define a pairing*

$$\mathtt{d}_{10,p^r}^{\mathrm{ord}} : \varphi_{-1,p^r}^{\mathrm{ord},0}((\mathrm{Gr}^0_{\mathtt{D}_{-1,p^r}})^{\mathrm{mult}}_\eta) \times (\underline{Y}/p^r\underline{Y})_\eta \to \boldsymbol{\mu}_{p^r,\eta}.$$

We say that the naive ordinary pre-level-n structure datum $\alpha_n^{\natural,\mathrm{ord}}$ *is* **symplectic**, *and call it a* **naive ordinary level-n structure datum of type** $(L/nL, \langle\,\cdot\,,\,\cdot\,\rangle, \mathtt{D}_{p^r})$ *over* η, *if the following conditions are satisfied:*

$$(\varphi_{-1,n_0} \times \varphi_{0,n_0})^*(\mathtt{d}_{10,n_0}) = \nu_{-1,n_0} \circ \langle\,\cdot\,,\,\cdot\,\rangle_{10,n_0},$$

$$(\varphi_{0,n_0} \times \varphi_{0,n_0})^*(\mathtt{d}_{00,n_0}) = \nu_{-1,n_0} \circ \langle\,\cdot\,,\,\cdot\,\rangle_{00,n_0},$$

and

$$(\varphi_{-1,p^r}^{\mathrm{ord},0}, \varphi_{0,p^r})^*(\mathtt{d}_{10,p^r}^{\mathrm{ord}}) = \nu_{-1,p^r}^{\mathrm{ord}} \circ (\langle\,\cdot\,,\,\cdot\,\rangle_{10,p^r}^{\mathrm{ord}})^{\mathrm{mult}}_\eta.$$

*We remove "***naive***" from the above terminologies, and call them* **ordinary pre-level-n structure datum of type** $(L \underset{\mathbb{Z}}{\otimes} \hat{\mathbb{Z}}, \langle\,\cdot\,,\,\cdot\,\rangle, \mathtt{D})$ *and* **ordinary level-n structure datum of type** $(L \underset{\mathbb{Z}}{\otimes} \hat{\mathbb{Z}}, \langle\,\cdot\,,\,\cdot\,\rangle, \mathtt{D})$, *when the data are compatibly liftable to data (at all higher levels) satisfying the analogous conditions.*

Proposition 4.1.4.56. *With the setting in Propositions* 4.1.4.21 *and* 4.1.4.52, *suppose moreover that* η *is a scheme over* $\mathrm{Spec}(\mathbb{Z}_{(p)})$. *Then each ordinary level-*n *structure datum* $\alpha_n^{\natural,\mathrm{ord}}$ *of type* $(L \underset{\mathbb{Z}}{\otimes} \hat{\mathbb{Z}}, \langle\,\cdot\,,\,\cdot\,\rangle, \mathtt{D})$ (*in Definition* 4.1.4.55) *defines an extension of the triple* $(G_\eta, \lambda_\eta, i_\eta)$ *to an object* $(G_\eta, \lambda_\eta, i_\eta, \alpha_{n_0}, \alpha_{p^r}^{\mathrm{ord}})$ *of* $\ddot{\mathsf{M}}_n^{\mathrm{ord}}(\eta)$ (*see Definition* 3.4.1.1). *Moreover, each such pair* $(\alpha_{n_0}, \alpha_{p^r}^{\mathrm{ord}})$ *of level structures comes from some* $\alpha_n^{\natural,\mathrm{ord}}$ *in this way.*

Proof. The only thing not explained yet is the statements concerning the principal level-n_0 structure α_{n_0}. In [62, Sec. 5.2], it was assumed that the polarization degree is prime to the residue characteristics, and that the generic point is defined over the rings of integers of the reflex field, but these assumptions were not really used in a substantial way. (They were only used to ensure that one obtains an object over η parameterized by the PEL-type moduli problem.) Therefore, the arguments there (with the liftability condition away from p) still work here, and allows us to construct α_{n_0}, as desired. □

However, the assignment of $(\alpha_{n_0}, \alpha_{p^r}^{\mathrm{ord}})$ to $\alpha_n^{\natural,\mathrm{ord}}$ is not one to one. (The freedom comes from the choice of various splittings.) We can define as in [62, Def. 5.2.7.11] a notion of equivalence classes $[\alpha_n^{\natural,\mathrm{ord}}]$ of objects like $\alpha_n^{\natural,\mathrm{ord}}$. (The definition can be made precise, but we will not record the details here.)

Definition 4.1.4.57. *With the setting as in Section* 4.1.2, *suppose moreover that* η *is a scheme over* $\mathrm{Spec}(\mathbb{Z}_{(p)})$. *The category* $\mathrm{DEG}_{\mathrm{PEL},\ddot{\mathsf{M}}_n^{\mathrm{ord}}}(R,I)$ *has objects of the form* $(G, \lambda, i, \alpha_{n_0}, \alpha_{p^r}^{\mathrm{ord}})$ (*over* S), *where:*

1. (G, λ, i) *defines an object of* $\mathrm{DEG}_{\mathrm{PE},\mathcal{O}}(R,I)$ (*see Definition* 4.1.3.1).
2. $(G_\eta, \lambda_\eta, i_\eta, \alpha_{n_0}, \alpha_{p^r}^{\mathrm{ord}})$ *defines an object of* $\ddot{\mathsf{M}}_n^{\mathrm{ord}}(\eta)$ (*see Definition* 3.4.1.1).
3. $\alpha_{p^r}^{\mathrm{ord}}$ *is a principal ordinary level-*p^r *structure of type* $(L \underset{\mathbb{Z}}{\otimes} \mathbb{Z}_p, \langle\,\cdot\,,\,\cdot\,\rangle, \mathtt{D})$ *such that, for each integer* $r' \geq r$, *there exists some lifting to level* $p^{r'}$ (*over some étale extension of* η) *compatible with degeneration* (*i.e., satisfying the analogue of Condition* 4.1.4.1 *at level* $p^{r'}$).

Definition 4.1.4.58. *With the setting as in Section* 4.1.2, *suppose moreover that* η *is a scheme over* $\mathrm{Spec}(\mathbb{Z}_{(p)})$. *The category* $\mathrm{DD}_{\mathrm{PEL},\ddot{\mathsf{M}}_n^{\mathrm{ord}}}(R,I)$ *has objects of the form*

$$(B, \lambda_B, i_B, \underline{X}, \underline{Y}, \phi, c, c^\vee, \tau, [\alpha_n^{\natural,\mathrm{ord}}]),$$

where:

1. $(B, \lambda_B, i_B, \underline{X}, \underline{Y}, \phi, c, c^\vee, \tau)$ *defines an object of* $\mathrm{DD}_{\mathrm{PE},\mathcal{O}}(R,I)$ (*see Definition* 4.1.3.2).
2. $[\alpha_n^{\natural,\mathrm{ord}}]$ *is an equivalence class of ordinary level-*n *structure data* $\alpha_n^{\natural,\mathrm{ord}}$ *of type* $(L \underset{\mathbb{Z}}{\otimes} \hat{\mathbb{Z}}, \langle\,\cdot\,,\,\cdot\,\rangle, \mathtt{D})$ *defined over* η (*see Definition* 4.1.4.55).

Now it follows from Propositions 4.1.4.21, 4.1.4.52, and 4.1.4.56 that we have the following:

Theorem 4.1.4.59. *There is an equivalence of categories*

$$\mathrm{M}_{\mathrm{PEL},\ddot{\mathsf{M}}_n^{\mathrm{ord}}}(R,I):\ \mathrm{DD}_{\mathrm{PEL},\ddot{\mathsf{M}}_n^{\mathrm{ord}}}(R,I)\to\mathrm{DEG}_{\mathrm{PEL},\ddot{\mathsf{M}}_n^{\mathrm{ord}}}(R,I):$$
$$(B,\lambda_B,i_B,\underline{X},\underline{Y},\phi,c,c^\vee,\tau,[\alpha_n^{\natural,\mathrm{ord}}])\mapsto(G,\lambda,i,\alpha_{n_0},\alpha_{p^r}^{\mathrm{ord}}).$$

4.1.5 *Degeneration Data for General Ordinary Level Structures*

Definition 4.1.5.1. *(See [62, Def. 1.3.7.1].) For each $\hat{\mathbb{Z}}$-algebra R, set*

$$\mathrm{G}^{\mathrm{ess}}(R):=\mathrm{image}(\mathrm{G}(\hat{\mathbb{Z}})\to\mathrm{G}(R)).$$

Definition 4.1.5.2. *(See [62, Def. 5.3.1.4].) Let Z_n be a symplectic filtration on L/nL. Then we define the following subgroups or quotients of subgroups of $\mathrm{G}^{\mathrm{ess}}(\mathbb{Z}/n\mathbb{Z})$:*

$$\mathrm{P}^{\mathrm{ess}}_{\mathsf{Z}_n}:=\{g_n\in\mathrm{G}^{\mathrm{ess}}(\mathbb{Z}/n\mathbb{Z}):g_n^{-1}(\mathsf{Z}_n)=\mathsf{Z}_n\},$$
$$\mathrm{Z}^{\mathrm{ess}}_{\mathsf{Z}_n}:=\{g_n\in\mathrm{P}^{\mathrm{ess}}_{\mathsf{Z}_n}:\mathrm{Gr}^{\mathsf{Z}_n}_{-1}(g_n)=\mathrm{Id}_{\mathrm{Gr}^{\mathsf{Z}_n}_{-1}}\text{ and }\nu(g_n)=1\},$$
$$\mathrm{U}^{\mathrm{ess}}_{\mathsf{Z}_n}:=\{g_n\in\mathrm{P}^{\mathrm{ess}}_{\mathsf{Z}_n}:\mathrm{Gr}^{\mathsf{Z}_n}(g_n)=\mathrm{Id}_{\mathrm{Gr}^{\mathsf{Z}_n}}\text{ and }\nu(g_n)=1\},$$
$$\mathrm{G}^{\mathrm{ess}}_{h,\mathsf{Z}_n}:=\left\{\begin{matrix}(g_{-1,n},r_n)\in\mathrm{GL}_{\mathcal{O}}(\mathrm{Gr}^{\mathsf{Z}_n}_{-1})\times\mathbf{G}_{\mathrm{m}}(\mathbb{Z}/n\mathbb{Z}):\\ \exists g_n\in\mathrm{P}^{\mathrm{ess}}_{\mathsf{Z}_n}\text{ s.t. }\mathrm{Gr}^{\mathsf{Z}_n}_{-1}(g_n)=g_{-1,n}\text{ and }\nu(g_n)=r_n\end{matrix}\right\},$$
$$\mathrm{G}^{\mathrm{ess}}_{l,\mathsf{Z}_n}:=\left\{\begin{matrix}(g_{-2,n},g_{0,n})\in\mathrm{GL}_{\mathcal{O}}(\mathrm{Gr}^{\mathsf{Z}_n}_{-2})\times\mathrm{GL}_{\mathcal{O}}(\mathrm{Gr}^{\mathsf{Z}_n}_{0}):\\ \exists g_n\in\mathrm{Z}^{\mathrm{ess}}_{\mathsf{Z}_n}\text{ s.t. }\mathrm{Gr}^{\mathsf{Z}_n}_{-2}(g_n)=g_{-2,n}\text{ and }\mathrm{Gr}^{\mathsf{Z}_n}_{0}(g_n)=g_{0,n}\end{matrix}\right\},$$
$$\mathrm{U}^{\mathrm{ess}}_{2,\mathsf{Z}_n}:=\left\{\begin{matrix}g_{20,n}\in\mathrm{Hom}_{\mathcal{O}}(\mathrm{Gr}^{\mathsf{Z}_n}_{0},\mathrm{Gr}^{\mathsf{Z}_n}_{-2}):\\ \exists g_n\in\mathrm{U}^{\mathrm{ess}}_{\mathsf{Z}_n}\text{ s.t. }\delta_n^{-1}\circ g_n\circ\delta_n=\begin{pmatrix}1&&g_{20,n}\\&1&\\&&1\end{pmatrix}\end{matrix}\right\},$$
$$\mathrm{U}^{\mathrm{ess}}_{1,\mathsf{Z}_n}:=\left\{\begin{matrix}(g_{21,n},g_{10,n})\in\mathrm{Hom}_{\mathcal{O}}(\mathrm{Gr}^{\mathsf{Z}_n}_{-1},\mathrm{Gr}^{\mathsf{Z}_n}_{-2})\times\mathrm{Hom}_{\mathcal{O}}(\mathrm{Gr}^{\mathsf{Z}_n}_{0},\mathrm{Gr}^{\mathsf{Z}_n}_{-1}):\\ \exists g_n\in\mathrm{U}^{\mathrm{ess}}_{\mathsf{Z}_n}\text{ s.t. }\delta_n^{-1}\circ g_n\circ\delta_n=\begin{pmatrix}1&g_{21,n}&g_{20,n}\\&1&g_{10,n}\\&&1\end{pmatrix},\ \textit{for some }g_{20,n}\end{matrix}\right\}.$$

We define similar subgroups and quotients of subgroups with n replaced with either n_0 or p^r.

Remark 4.1.5.3. Since $\nu(\mathrm{Gr}^{\mathsf{Z}_n}_{-1}(g_n))=\nu(g_n)$ by definition, the condition $\nu(g_n)=1$ in the definition of $\mathrm{Z}^{\mathrm{ess}}_{\mathsf{Z}_n}$ is redundant if we interpret $\mathrm{Gr}^{\mathsf{Z}_n}_{-1}(g_n)=\mathrm{Id}_{\mathrm{Gr}^{\mathsf{Z}_n}_{-1}}$ as an identity of symplectic isomorphisms (which are required to preserve the similitude isomorphisms; see [62, Def. 1.1.4.8]).

Lemma 4.1.5.4. *(See [62, Lem. 5.3.1.6].) By definition, there are natural inclusions*

$$\mathrm{U}^{\mathrm{ess}}_{2,\mathsf{Z}_n}\subset\mathrm{U}^{\mathrm{ess}}_{\mathsf{Z}_n}\subset\mathrm{Z}^{\mathrm{ess}}_{\mathsf{Z}_n}\subset\mathrm{P}^{\mathrm{ess}}_{\mathsf{Z}_n}\subset\mathrm{G}^{\mathrm{ess}}_{\mathsf{Z}_n},\tag{4.1.5.5}$$

and natural exact sequences:

$$1 \to \mathrm{Z}^{\mathrm{ess}}_{\mathsf{Z}_n} \to \mathrm{P}^{\mathrm{ess}}_{\mathsf{Z}_n} \to \mathrm{G}^{\mathrm{ess}}_{h,\mathsf{Z}_n} \to 1, \tag{4.1.5.6}$$

$$1 \to \mathrm{U}^{\mathrm{ess}}_{\mathsf{Z}_n} \to \mathrm{Z}^{\mathrm{ess}}_{\mathsf{Z}_n} \to \mathrm{G}^{\mathrm{ess}}_{l,\mathsf{Z}_n} \to 1, \tag{4.1.5.7}$$

$$1 \to \mathrm{U}^{\mathrm{ess}}_{2,\mathsf{Z}_n} \to \mathrm{U}^{\mathrm{ess}}_{\mathsf{Z}_n} \to \mathrm{U}^{\mathrm{ess}}_{1,\mathsf{Z}_n} \to 1. \tag{4.1.5.8}$$

We have similar statements with n replaced with either n_0 or p^r.

Definition 4.1.5.9. *(See [62, Def. 5.3.1.11].) Let H_n be a subgroup of $\mathrm{G}^{\mathrm{ess}}(\mathbb{Z}/n\mathbb{Z})$. For each of the subgroups $\circledast$ in (4.1.5.5), we define $H_{n,\circledast} := H_n \cap \circledast$. For each of the quotients of two groups $\circledast = \circledast_1/\circledast_2$ in (4.1.5.5), (4.1.5.6), (4.1.5.7), or (4.1.5.8), we define $H_{n,\circledast} := H_{n,\circledast_1}/H_{n,\circledast_2}$. Thus, we have defined the groups $H_{n,\mathrm{P}^{\mathrm{ess}}_{\mathsf{Z}_n}}$, $H_{n,\mathrm{Z}^{\mathrm{ess}}_{\mathsf{Z}_n}}$, $H_{n,\mathrm{U}^{\mathrm{ess}}_{\mathsf{Z}_n}}$, $H_{n,\mathrm{G}^{\mathrm{ess}}_{h,\mathsf{Z}_n}}$, $H_{n,\mathrm{G}^{\mathrm{ess}}_{l,\mathsf{Z}_n}}$, $H_{n,\mathrm{U}^{\mathrm{ess}}_{2,\mathsf{Z}_n}}$, and $H_{n,\mathrm{U}^{\mathrm{ess}}_{1,\mathsf{Z}_n}}$, so that we have the natural inclusions*

$$H_{n,\mathrm{U}^{\mathrm{ess}}_{2,\mathsf{Z}_n}} \subset H_{n,\mathrm{U}^{\mathrm{ess}}_{\mathsf{Z}_n}} \subset H_{n,\mathrm{Z}^{\mathrm{ess}}_{\mathsf{Z}_n}} \subset H_{n,\mathrm{P}^{\mathrm{ess}}_{\mathsf{Z}_n}} \subset H_n$$

and natural exact sequences

$$1 \to H_{n,\mathrm{Z}^{\mathrm{ess}}_{\mathsf{Z}_n}} \to H_{n,\mathrm{P}^{\mathrm{ess}}_{\mathsf{Z}_n}} \to H_{n,\mathrm{G}^{\mathrm{ess}}_{h,\mathsf{Z}_n}} \to 1,$$

$$1 \to H_{n,\mathrm{U}^{\mathrm{ess}}_{\mathsf{Z}_n}} \to H_{n,\mathrm{Z}^{\mathrm{ess}}_{\mathsf{Z}_n}} \to H_{n,\mathrm{G}^{\mathrm{ess}}_{l,\mathsf{Z}_n}} \to 1,$$

$$1 \to H_{n,\mathrm{U}^{\mathrm{ess}}_{2,\mathsf{Z}_n}} \to H_{n,\mathrm{U}^{\mathrm{ess}}_{\mathsf{Z}_n}} \to H_{n,\mathrm{U}^{\mathrm{ess}}_{1,\mathsf{Z}_n}} \to 1.$$

We define similar subgroups with n replaced with either n_0 or p^r.

Proposition 4.1.5.10. *Let $\mathcal{H}^p \subset \mathrm{G}(\hat{\mathbb{Z}}^p)$ be an open compact subgroup such that $\mathcal{U}^p(n_0) \subset \mathcal{H}^p$ for some integer $n_0 \geq 1$ is an integer prime to p. Let $H_{n_0} := \mathcal{H}^p/\mathcal{U}^p(n_0)$. Let $S = \mathrm{Spec}(R)$ be as in Section 4.1.2. Let (G, λ, i) be a degenerating family of type $(\mathrm{PE}, \mathcal{O})$ over S as in Definition 4.1.3.1, with degeneration datum $(B, \lambda_B, i_B, \underline{X}, \underline{Y}, \phi, c, c^\vee, \tau)$ given by (4.1.3.3). Suppose moreover that η is a scheme over $\mathrm{Spec}(\mathbb{Z}_{(p)})$. Let Z_n be a symplectic-liftable filtration on L/nL.*

Consider compositions of finite étale morphisms of schemes

$$\begin{aligned} \tau_{H_{n_0}} \twoheadrightarrow (c_{H_{n_0}}, c^\vee_{H_{n_0}}) \twoheadrightarrow (\varphi^\sim_{-2,H_{n_0}}, \varphi^\sim_{0,H_{n_0}}) \\ \twoheadrightarrow \varphi_{-1,H_{n_0}} \twoheadrightarrow \delta_{H_{n_0}} \xrightarrow{\sim} \mathsf{Z}_{H_{n_0}} \twoheadrightarrow \eta \end{aligned} \tag{4.1.5.11}$$

such that we have the following:

(1) $\mathsf{Z}_{H_{n_0}} \twoheadrightarrow \eta$ is an H_{n_0}-orbit of étale-locally-defined filtrations Z_{n_0}, which is isomorphic to (the pullback of) the constant scheme $H_{n_0,\mathrm{P}^{\mathrm{ess}}_{\mathsf{Z}_{n_0}}} \backslash H_{n_0}$ over some finite étale extension of η.

(2) $\delta_{H_{n_0}} \xrightarrow{\sim} \mathsf{Z}_{H_{n_0}}$ is an isomorphism giving choices of splittings of filtrations parameterized by $\mathsf{Z}_{H_{n_0}}$.

(3) $\varphi_{-1,H_{n_0}} \twoheadrightarrow \delta_{H_{n_0}}$ is an $H_{n_0,\mathrm{G}^{\mathrm{ess}}_{h,\mathsf{Z}_{n_0}}}$-torsor realized as a finite étale subscheme of the pullback to $\delta_{H_{n_0}}$ of the finite étale scheme

$$\underline{\mathrm{Isom}}_\eta((L/n_0L)_\eta, B[n_0]_\eta) \underset{\eta}{\times} \underline{\mathrm{Isom}}_\eta(((\mathbb{Z}/n_0\mathbb{Z})(1))_\eta, \boldsymbol{\mu}_{n_0,\eta})$$

over η, which is an $H_{n_0,\mathrm{G}^{\mathrm{ess}}_{h,\mathsf{Z}_{n_0}}}$-torsor giving an $H_{n_0,\mathrm{G}^{\mathrm{ess}}_{h,\mathsf{Z}_{n_0}}}$-orbit of étale-locally-defined φ_{-1,n_0}. (This defines some naive level structure on the pullback of $(B_\eta, \lambda_{B,\eta}, i_{B,\eta})$ to $\delta_{H_{n_0}}$; cf. Definition 3.3.1.3.)

(4) $(\varphi^{\sim}_{-2,H_{n_0}}, \varphi^{\sim}_{0,H_{n_0}}) \twoheadrightarrow \varphi_{-1,H_{n_0}}$ *is an* $H_{n_0,\mathrm{G}^{\mathrm{ess}}_{l,\mathsf{Z}_{n_0}}}$*-torsor giving an* $H_{n_0,\mathrm{G}^{\mathrm{ess}}_{l,\mathsf{Z}_{n_0}}}$*-orbit of étale-locally-defined pairs* $(\varphi_{-2,n_0}, \varphi_{0,n_0})$.

(5) $(c_{H_{n_0}}, c^{\vee}_{H_{n_0}}) \twoheadrightarrow (\varphi^{\sim}_{-2,H_{n_0}}, \varphi^{\sim}_{0,H_{n_0}})$ *is an* $H_{n_0,\mathrm{U}^{\mathrm{ess}}_{1,\mathsf{Z}_{n_0}}}$*-torsor giving an* $H_{n_0,\mathrm{U}^{\mathrm{ess}}_{1,\mathsf{Z}_{n_0}}}$*-orbit of étale-locally-defined pairs* $(c_{n_0}, c^{\vee}_{n_0})$.

(6) $\tau_{H_{n_0}} \twoheadrightarrow (c_{H_{n_0}}, c^{\vee}_{H_{n_0}})$ *is an* $H_{n_0,\mathrm{U}^{\mathrm{ess}}_{2,\mathsf{Z}_{n_0}}}$*-torsor giving an* $H_{n_0,\mathrm{U}^{\mathrm{ess}}_{2,\mathsf{Z}_{n_0}}}$*-orbit of étale-locally-defined* τ_{n_0}.

(Each of the datum or pairs of data is built on top of the earlier ones.) Then each such scheme $\tau_{H_{n_0}} \twoheadrightarrow \eta$ *determines a naive level-*H_{n_0} *structure* $\alpha_{H_{n_0}}$ *of* $(G_\eta, \lambda_\eta, i_\eta)$ *of type* $(L/n_0L, \langle\cdot,\cdot\rangle)$ *(see Definition* 3.3.1.3*). If these schemes are orbits of liftable objects, then they determine a level-*$\mathcal{H}^p$ *structure* $\alpha_{\mathcal{H}^p}$ *of* $(G_\eta, \lambda_\eta, i_\eta)$ *of type* $(L \underset{\mathbb{Z}}{\otimes} \hat{\mathbb{Z}}^p, \langle\cdot,\cdot\rangle)$ *(see Definition* 3.3.1.4*).*

Conversely, each level-$\mathcal{H}^p$ *structure* $\alpha_{\mathcal{H}^p}$ *of* $(G_\eta, \lambda_\eta, i_\eta)$ *of type* $(L \underset{\mathbb{Z}}{\otimes} \hat{\mathbb{Z}}^p, \langle\cdot,\cdot\rangle)$ *arises this way for some (noncanonical choice of) symplectic-liftable* Z_{n_0}*. (The* H_{n_0}*-orbit of* Z_{n_0}*, or rather the* $\mathcal{H}^p$*-orbit of the lifting* Z^p*, is nevertheless canonically determined by* $\alpha_{\mathcal{H}^p}$*.)*

Proof. This follows from the same arguments as in [62, Sec. 5.3]. □

Definition 4.1.5.12. *With the setting in Definition* 4.1.5.2*, let* $\mathsf{Z}_{p^r} := \mathsf{Z}_n/p^r\mathsf{Z}_n$*, and assume moreover that* Z_{p^r} *satisfies the compatibility* $\mathsf{Z}_{-2,p^r} \subset \mathsf{D}^0_{p^r} \subset \mathsf{Z}_{-1,p^r}$ *(see* (3.2.3.3)*). Then we define the following subgroups or quotients of subgroups of* $\mathrm{G}^{\mathrm{ess}}(\mathbb{Z}/p^r\mathbb{Z})$*:*

$$\mathrm{P}^{\mathrm{ess}}_{\mathsf{D}_{p^r}} := \{g_{p^r} \in \mathrm{G}^{\mathrm{ess}}(\mathbb{Z}/p^r\mathbb{Z}) : g_{p^r}^{-1}(\mathsf{D}^0_{p^r}) = \mathsf{D}^0_{p^r}\} \cong \mathcal{U}_{p,0}(p^r)/\mathcal{U}_p(p^r),$$

$$\mathrm{P}^{\mathrm{ess}}_{\mathsf{Z}_{p^r},\mathsf{D}_{p^r}} := \{g_{p^r} \in \mathrm{P}^{\mathrm{ess}}_{\mathsf{D}_{p^r}} : g_{p^r}^{-1}(\mathsf{Z}_{p^r}) = \mathsf{Z}_{p^r}\},$$

$$\mathrm{Z}^{\mathrm{ess}}_{\mathsf{Z}_{p^r},\mathsf{D}_{p^r}} := \{g_{p^r} \in \mathrm{P}^{\mathrm{ess}}_{\mathsf{Z}_{p^r},\mathsf{D}_{p^r}} : \mathrm{Gr}^{\mathsf{Z}_{p^r}}_{-1}(g_{p^r}) = \mathrm{Id}_{\mathrm{Gr}^{\mathsf{Z}_{p^r}}_{-1}} \text{ and } \nu(g_{p^r}) = 1\},$$

$$\mathrm{U}^{\mathrm{ess}}_{\mathsf{Z}_{p^r},\mathsf{D}_{p^r}} := \{g_{p^r} \in \mathrm{P}^{\mathrm{ess}}_{\mathsf{Z}_{p^r},\mathsf{D}_{p^r}} : \mathrm{Gr}^{\mathsf{Z}_{p^r}}(g_{p^r}) = \mathrm{Id}_{\mathrm{Gr}^{\mathsf{Z}_{p^r}}} \text{ and } \nu(g_{p^r}) = 1\},$$

$$\mathrm{G}^{\mathrm{ess}}_{l,\mathsf{Z}_{p^r},\mathsf{D}_{p^r}} := \left\{ \begin{array}{c} (g_{-2,p^r}, g_{0,p^r}) \in \mathrm{GL}_{\mathcal{O}}(\mathrm{Gr}^{\mathsf{Z}_{p^r}}_{-2}) \times \mathrm{GL}_{\mathcal{O}}(\mathrm{Gr}^{\mathsf{Z}_{p^r}}_{0}) : \\ \exists g_{p^r} \in \mathrm{Z}^{\mathrm{ess}}_{\mathsf{Z}_{p^r},\mathsf{D}_{p^r}} \text{ s.t.} \\ \mathrm{Gr}^{\mathsf{Z}_{p^r}}_{-2}(g_{p^r}) = g_{-2,p^r} \text{ and } \mathrm{Gr}^{\mathsf{Z}_{p^r}}_{0}(g_{p^r}) = g_{0,p^r} \end{array} \right\},$$

$$\mathrm{U}^{\mathrm{ess}}_{2,\mathsf{Z}_{p^r},\mathsf{D}_{p^r}} := \left\{ \begin{array}{c} g_{20,p^r} \in \mathrm{Hom}_{\mathcal{O}}(\mathrm{Gr}^{\mathsf{Z}_{p^r}}_{0}, \mathrm{Gr}^{\mathsf{Z}_{p^r}}_{-2}) : \\ \exists g_{p^r} \in \mathrm{U}^{\mathrm{ess}}_{\mathsf{Z}_{p^r},\mathsf{D}_{p^r}} \text{ s.t. } \delta_{p^r}^{-1} \circ g_{p^r} \circ \delta_{p^r} = \begin{pmatrix} 1 & & g_{20,p^r} \\ & 1 & \\ & & 1 \end{pmatrix} \end{array} \right\},$$

$$\mathrm{U}^{\mathrm{ess}}_{1,\mathsf{Z}_{p^r},\mathsf{D}_{p^r}} := \left\{ \begin{array}{c} (g_{21,p^r}, g_{10,p^r}) \in \begin{array}{c} \mathrm{Hom}_{\mathcal{O}}(\mathrm{Gr}^{\mathsf{Z}_{p^r}}_{-1}, \mathrm{Gr}^{\mathsf{Z}_{p^r}}_{-2}) \\ \times \mathrm{Hom}_{\mathcal{O}}(\mathrm{Gr}^{\mathsf{Z}_{p^r}}_{0}, \mathrm{Gr}^{\mathsf{Z}_{p^r}}_{-1}) \end{array} : \\ \exists g_{p^r} \in \mathrm{U}^{\mathrm{ess}}_{\mathsf{Z}_{p^r},\mathsf{D}_{p^r}} \text{ s.t. } \delta_{p^r}^{-1} \circ g_{p^r} \circ \delta_{p^r} = \begin{pmatrix} 1 & g_{21,p^r} & g_{20,p^r} \\ & 1 & g_{10,p^r} \\ & & 1 \end{pmatrix}, \\ \text{for some } g_{20,p^r} \end{array} \right\}.$$

Note that $\mathrm{Z}^{\mathrm{ess}}_{\mathtt{Z}_{p^r},\mathtt{D}_{p^r}} = \mathrm{Z}^{\mathrm{ess}}_{\mathtt{Z}_{p^r}}$, $\mathrm{U}^{\mathrm{ess}}_{\mathtt{Z}_{p^r},\mathtt{D}_{p^r}} = \mathrm{U}^{\mathrm{ess}}_{\mathtt{Z}_{p^r}}$, *and* $\mathrm{G}^{\mathrm{ess}}_{l,\mathtt{Z}_{p^r},\mathtt{D}_{p^r}} = \mathrm{G}^{\mathrm{ess}}_{l,\mathtt{Z}_{p^r}}$.

Let us consider

$$\overline{\mathcal{U}}^{\mathrm{bal}}_{p,1}(p^r) := \mathcal{U}^{\mathrm{bal}}_{p,1}(p^r)/\mathcal{U}_p(p^r)$$

as a subgroup of $\mathrm{G}^{\mathrm{ess}}_{\mathtt{D}_{p^r}}$. *Then we also define the following subgroups or quotients of subgroups of* $\mathrm{G}^{\mathrm{ess}}(\mathbb{Z}/p^r\mathbb{Z})$:

$$\begin{aligned}
\mathrm{M}^{\mathrm{ess,ord}}_{\mathtt{Z}_{p^r},\mathtt{D}_{p^r}} &:= \mathrm{P}^{\mathrm{ess}}_{\mathtt{D}_{p^r}}/\overline{\mathcal{U}}^{\mathrm{bal}}_{p,1}(p^r) \cong \mathcal{U}_{p,0}(p^r)/\mathcal{U}^{\mathrm{bal}}_{p,1}(p^r),\\
\mathrm{P}^{\mathrm{ess,ord}}_{\mathtt{Z}_{p^r},\mathtt{D}_{p^r}} &:= \text{image of } \mathrm{P}^{\mathrm{ess}}_{\mathtt{Z}_{p^r},\mathtt{D}_{p^r}} \text{ under } \mathrm{P}^{\mathrm{ess}}_{\mathtt{D}_{p^r}} \overset{\text{can.}}{\to} \mathrm{M}^{\mathrm{ess,ord}}_{\mathtt{Z}_{p^r},\mathtt{D}_{p^r}},\\
\mathrm{Z}^{\mathrm{ess,ord}}_{\mathtt{Z}_{p^r},\mathtt{D}_{p^r}} &:= \text{image of } \mathrm{Z}^{\mathrm{ess}}_{\mathtt{Z}_{p^r},\mathtt{D}_{p^r}} = \mathrm{Z}^{\mathrm{ess}}_{\mathtt{Z}_{p^r}} \text{ under } \mathrm{P}^{\mathrm{ess}}_{\mathtt{D}_{p^r}} \overset{\text{can.}}{\to} \mathrm{M}^{\mathrm{ess,ord}}_{\mathtt{Z}_{p^r},\mathtt{D}_{p^r}},\\
\mathrm{U}^{\mathrm{ess,ord}}_{\mathtt{Z}_{p^r},\mathtt{D}_{p^r}} &:= \text{image of } \mathrm{U}^{\mathrm{ess}}_{\mathtt{Z}_{p^r},\mathtt{D}_{p^r}} = \mathrm{U}^{\mathrm{ess}}_{\mathtt{Z}_{p^r}} \text{ under } \mathrm{P}^{\mathrm{ess}}_{\mathtt{D}_{p^r}} \overset{\text{can.}}{\to} \mathrm{M}^{\mathrm{ess,ord}}_{\mathtt{Z}_{p^r},\mathtt{D}_{p^r}},\\
\mathrm{G}^{\mathrm{ess,ord}}_{h,\mathtt{Z}_{p^r},\mathtt{D}_{p^r}} &:= \mathrm{P}^{\mathrm{ess,ord}}_{\mathtt{Z}_{p^r},\mathtt{D}_{p^r}}/\mathrm{Z}^{\mathrm{ess,ord}}_{\mathtt{Z}_{p^r},\mathtt{D}_{p^r}},\\
\mathrm{G}^{\mathrm{ess,ord}}_{l,\mathtt{Z}_{p^r},\mathtt{D}_{p^r}} &:= \mathrm{Z}^{\mathrm{ess,ord}}_{\mathtt{Z}_{p^r},\mathtt{D}_{p^r}}/\mathrm{U}^{\mathrm{ess,ord}}_{\mathtt{Z}_{p^r},\mathtt{D}_{p^r}} \cong \mathrm{Z}^{\mathrm{ess}}_{\mathtt{Z}_{p^r}}/\mathrm{U}^{\mathrm{ess}}_{\mathtt{Z}_{p^r}} = \mathrm{G}^{\mathrm{ess}}_{l,\mathtt{Z}_{p^r}},\\
\mathrm{U}^{\mathrm{ess,ord}}_{2,\mathtt{Z}_{p^r},\mathtt{D}_{p^r}} &:= \{\mathrm{Id}\},\\
\mathrm{U}^{\mathrm{ess,ord}}_{1,\mathtt{Z}_{p^r},\mathtt{D}_{p^r}} &:= \mathrm{U}^{\mathrm{ess,ord}}_{\mathtt{Z}_{p^r},\mathtt{D}_{p^r}}.
\end{aligned}$$

Lemma 4.1.5.13. *By definition, there are natural inclusions*

$$\mathrm{U}^{\mathrm{ess,ord}}_{2,\mathtt{Z}_{p^r},\mathtt{D}_{p^r}} \subset \mathrm{U}^{\mathrm{ess,ord}}_{\mathtt{Z}_{p^r},\mathtt{D}_{p^r}} \subset \mathrm{Z}^{\mathrm{ess,ord}}_{\mathtt{Z}_{p^r},\mathtt{D}_{p^r}} \subset \mathrm{P}^{\mathrm{ess,ord}}_{\mathtt{Z}_{p^r},\mathtt{D}_{p^r}} \subset \mathrm{M}^{\mathrm{ess,ord}}_{\mathtt{Z}_{p^r},\mathtt{D}_{p^r}}, \tag{4.1.5.14}$$

and natural exact sequences:

$$1 \to \mathrm{Z}^{\mathrm{ess,ord}}_{\mathtt{Z}_{p^r},\mathtt{D}_{p^r}} \to \mathrm{P}^{\mathrm{ess,ord}}_{\mathtt{Z}_{p^r},\mathtt{D}_{p^r}} \to \mathrm{G}^{\mathrm{ess,ord}}_{h,\mathtt{Z}_{p^r},\mathtt{D}_{p^r}} \to 1, \tag{4.1.5.15}$$

$$1 \to \mathrm{U}^{\mathrm{ess,ord}}_{\mathtt{Z}_{p^r},\mathtt{D}_{p^r}} \to \mathrm{Z}^{\mathrm{ess,ord}}_{\mathtt{Z}_{p^r},\mathtt{D}_{p^r}} \to \mathrm{G}^{\mathrm{ess,ord}}_{l,\mathtt{Z}_{p^r},\mathtt{D}_{p^r}} \to 1, \tag{4.1.5.16}$$

$$1 \to \mathrm{U}^{\mathrm{ess,ord}}_{2,\mathtt{Z}_{p^r},\mathtt{D}_{p^r}} \to \mathrm{U}^{\mathrm{ess,ord}}_{\mathtt{Z}_{p^r},\mathtt{D}_{p^r}} \to \mathrm{U}^{\mathrm{ess,ord}}_{1,\mathtt{Z}_{p^r},\mathtt{D}_{p^r}} \to 1. \tag{4.1.5.17}$$

Definition 4.1.5.18. *Let* $\mathcal{H}_p$ *be an open compact subgroup of* $\mathrm{G}(\mathbb{Z}_p)$ *such that* $\mathcal{U}^{\mathrm{bal}}_{p,1}(p^r) \subset \mathcal{H}_p \subset \mathcal{U}_{p,0}(p^r)$, *which defines a subgroup* $H_{p^r} := \mathcal{H}_p/\mathcal{U}_p(p^r)$ *of* $\mathrm{P}^{\mathrm{ess}}_{\mathtt{D}_{p^r}}$, *and defines a subgroup* $H^{\mathrm{ord}}_{p^r} := \mathcal{H}_p/\mathcal{U}^{\mathrm{bal}}_{p,1}(p^r)$ *of* $\mathrm{M}^{\mathrm{ess,ord}}_{\mathtt{Z}_{p^r},\mathtt{D}_{p^r}}$. *For each of the subgroups* $\circledast$ *in* (4.1.5.14), *we define* $H^{\mathrm{ord}}_{p^r,\circledast} := H^{\mathrm{ord}}_{p^r} \cap \circledast$. *For each of the quotients of two groups* $\circledast = \circledast_1/\circledast_2$ *in* (4.1.5.14), (4.1.5.15), (4.1.5.16), *or* (4.1.5.17), *we define* $H^{\mathrm{ord}}_{p^r,\circledast} := H^{\mathrm{ord}}_{p^r,\circledast_1}/H^{\mathrm{ord}}_{p^r,\circledast_2}$. *Thus, we have defined the groups* $H^{\mathrm{ord}}_{p^r,\mathrm{P}^{\mathrm{ess,ord}}_{\mathtt{Z}_{p^r},\mathtt{D}_{p^r}}}$, $H^{\mathrm{ord}}_{p^r,\mathrm{Z}^{\mathrm{ess,ord}}_{\mathtt{Z}_{p^r},\mathtt{D}_{p^r}}}$, $H^{\mathrm{ord}}_{p^r,\mathrm{U}^{\mathrm{ess,ord}}_{\mathtt{Z}_{p^r},\mathtt{D}_{p^r}}}$, $H^{\mathrm{ord}}_{p^r,\mathrm{G}^{\mathrm{ess,ord}}_{h,\mathtt{Z}_{p^r},\mathtt{D}_{p^r}}}$, $H^{\mathrm{ord}}_{p^r,\mathrm{G}^{\mathrm{ess,ord}}_{l,\mathtt{Z}_{p^r},\mathtt{D}_{p^r}}}$, $H^{\mathrm{ord}}_{p^r,\mathrm{U}^{\mathrm{ess,ord}}_{2,\mathtt{Z}_{p^r},\mathtt{D}_{p^r}}}$, *and* $H^{\mathrm{ord}}_{p^r,\mathrm{U}^{\mathrm{ess,ord}}_{1,\mathtt{Z}_{p^r},\mathtt{D}_{p^r}}}$, *so that we have the natural inclusions*

$$H^{\mathrm{ord}}_{p^r,\mathrm{U}^{\mathrm{ess,ord}}_{2,\mathtt{Z}_{p^r},\mathtt{D}_{p^r}}} \subset H^{\mathrm{ord}}_{p^r,\mathrm{U}^{\mathrm{ess,ord}}_{\mathtt{Z}_{p^r},\mathtt{D}_{p^r}}} \subset H^{\mathrm{ord}}_{p^r,\mathrm{Z}^{\mathrm{ess,ord}}_{\mathtt{Z}_{p^r},\mathtt{D}_{p^r}}} \subset H^{\mathrm{ord}}_{p^r,\mathrm{P}^{\mathrm{ess,ord}}_{\mathtt{Z}_{p^r},\mathtt{D}_{p^r}}} \subset H^{\mathrm{ord}}_{p^r}$$

and natural exact sequences

$$1 \to H^{\mathrm{ord}}_{p^r,\mathrm{Z}^{\mathrm{ess,ord}}_{\mathtt{Z}_{p^r},\mathtt{D}_{p^r}}} \to H^{\mathrm{ord}}_{p^r,\mathrm{P}^{\mathrm{ess,ord}}_{\mathtt{Z}_{p^r},\mathtt{D}_{p^r}}} \to H^{\mathrm{ord}}_{p^r,\mathrm{G}^{\mathrm{ess,ord}}_{h,\mathtt{Z}_{p^r},\mathtt{D}_{p^r}}} \to 1,$$

$$1 \to H^{\mathrm{ord}}_{p^r,\mathrm{U}^{\mathrm{ess,ord}}_{\mathtt{Z}_{p^r},\mathtt{D}_{p^r}}} \to H^{\mathrm{ord}}_{p^r,\mathrm{Z}^{\mathrm{ess,ord}}_{\mathtt{Z}_{p^r},\mathtt{D}_{p^r}}} \to H^{\mathrm{ord}}_{p^r,\mathrm{G}^{\mathrm{ess,ord}}_{l,\mathtt{Z}_{p^r},\mathtt{D}_{p^r}}} \to 1,$$

$$1 \to H^{\mathrm{ord}}_{p^r,\mathrm{U}^{\mathrm{ess,ord}}_{2,\mathtt{Z}_{p^r},\mathtt{D}_{p^r}}} \to H^{\mathrm{ord}}_{p^r,\mathrm{U}^{\mathrm{ess,ord}}_{\mathtt{Z}_{p^r},\mathtt{D}_{p^r}}} \to H^{\mathrm{ord}}_{p^r,\mathrm{U}^{\mathrm{ess,ord}}_{1,\mathtt{Z}_{p^r},\mathtt{D}_{p^r}}} \to 1.$$

Remark 4.1.5.19. By definition, we have $H^{\mathrm{ord}}_{p^r,\mathrm{G}^{\mathrm{ess,ord}}_{l,\mathrm{Z}_{p^r},\mathrm{D}_{p^r}}} \cong H_{p^r,\mathrm{G}^{\mathrm{ess}}_{l,\mathrm{Z}_{p^r}}}$ under the canonical isomorphism $\mathrm{G}^{\mathrm{ess,ord}}_{l,\mathrm{Z}_{p^r},\mathrm{D}_{p^r}} \cong \mathrm{G}^{\mathrm{ess}}_{l,\mathrm{Z}_{p^r}}$.

Proposition 4.1.5.20. *Let $H^{\mathrm{ord}}_{p^r}$ be as in Definition* 4.1.5.18. *Let $S = \mathrm{Spec}(R)$ be as in Section* 4.1.2. *Let (G,λ,i) be a degenerating family of type* (PE, $\mathcal{O}$) *over S as in Definition* 4.1.3.1, *with degeneration datum $(B,\lambda_B,i_B,\underline{X},\underline{Y},\phi,c,c^\vee,\tau)$ given by* (4.1.3.3). *Let Z_{p^r} be a symplectic filtration on L/p^rL, and assume moreover that Z_{p^r} satisfies the compatibility $\mathrm{Z}_{-2,p^r} \subset \mathrm{D}^0_{p^r} \subset \mathrm{Z}_{-1,p^r}$ (see* (3.2.3.3)).

Consider compositions of finite étale morphisms of schemes

$$\begin{aligned}\tau^{\mathrm{ord}}_{H^{\mathrm{ord}}_{p^r}} &\overset{\sim}{\to} (c^{\mathrm{ord}}_{H^{\mathrm{ord}}_{p^r}}, c^{\vee,\mathrm{ord}}_{H^{\mathrm{ord}}_{p^r}}) \twoheadrightarrow (\varphi^{\mathrm{ord}}_{-2,H^{\mathrm{ord}}_{p^r}}, \varphi^{\mathrm{ord}}_{0,H^{\mathrm{ord}}_{p^r}}) \\ &\twoheadrightarrow \varphi^{\mathrm{ord}}_{-1,H^{\mathrm{ord}}_{p^r}} \twoheadrightarrow \delta^{\mathrm{ord}}_{H^{\mathrm{ord}}_{p^r}} \overset{\sim}{\to} \mathrm{Z}_{H^{\mathrm{ord}}_{p^r}} \twoheadrightarrow \eta,\end{aligned} \tag{4.1.5.21}$$

such that we have the following:

(1) *$\mathrm{Z}_{H^{\mathrm{ord}}_{p^r}} \twoheadrightarrow \eta$ is an $H^{\mathrm{ord}}_{p^r}$-orbit (or equivalently an H_{p^r}-orbit) of étale-locally-defined filtrations Z_{p^r} that are compatible with D_{p^r} in the sense that $\mathrm{Z}_{-2,p^r} \subset \mathrm{D}_{p^r} \subset \mathrm{Z}_{-1,p^r}$, which is isomorphic to (the pullback of) the constant scheme $H^{\mathrm{ord}}_{p^r,\mathrm{P}^{\mathrm{ess,ord}}_{\mathrm{Z}_{p^r},\mathrm{D}_{p^r}}} \backslash H^{\mathrm{ord}}_{p^r}$ over some finite étale extension of η. This also determines an $H^{\mathrm{ord}}_{p^r}$-orbit of étale-locally-defined filtrations conjugate to $\mathrm{D}_{-1,p^r} = \{\mathrm{D}^i_{-1,p^r}\}_i$ on $\mathrm{Gr}^{\mathrm{Z}}_{-1,p^r}$.*

(2) *$\delta^{\mathrm{ord}}_{H^{\mathrm{ord}}_{p^r}} = (\delta^{\mathrm{ord},0}_{H^{\mathrm{ord}}_{p^r}}, \delta^{\mathrm{ord},\#,0}_{H^{\mathrm{ord}}_{p^r}}) \overset{\sim}{\to} \mathrm{Z}_{H^{\mathrm{ord}}_{p^r}}$ gives choices of pairs of splittings $\delta^{\mathrm{ord}}_{p^r} = (\delta^{\mathrm{ord},0}_{p^r}, \delta^{\mathrm{ord},\#,0}_{p^r})$.*

(3) *$\varphi^{\mathrm{ord}}_{-1,H^{\mathrm{ord}}_{p^r}} \twoheadrightarrow \delta^{\mathrm{ord}}_{H^{\mathrm{ord}}_{p^r}}$ is an $H^{\mathrm{ord}}_{p^r,\mathrm{G}^{\mathrm{ess,ord}}_{h,\mathrm{Z}_{p^r},\mathrm{D}_{p^r}}}$-torsor realized as a (finite étale) subscheme of the quasi-finite étale scheme*

$$\begin{aligned}&\underline{\mathrm{Hom}}_{\delta^{\mathrm{ord}}_{H^{\mathrm{ord}}_{p^r}}}\big((\mathrm{Gr}^0_{\mathrm{D},H^{\mathrm{ord}}_{p^r}})^{\mathrm{mult}}, B[p^r]_{\delta^{\mathrm{ord}}_{H^{\mathrm{ord}}_{p^r}}}\big) \\ &\underset{\delta^{\mathrm{ord}}_{H^{\mathrm{ord}}_{p^r}}}{\times} \underline{\mathrm{Hom}}_{\delta^{\mathrm{ord}}_{H^{\mathrm{ord}}_{p^r}}}\big((\mathrm{Gr}^0_{\mathrm{D}^\#,H^{\mathrm{ord}}_{p^r}})^{\mathrm{mult}}, B^\vee[p^r]_{\delta^{\mathrm{ord}}_{H^{\mathrm{ord}}_{p^r}}}\big) \underset{\delta^{\mathrm{ord}}_{H^{\mathrm{ord}}_{p^r}}}{\times} (\mathbb{Z}/p^r\mathbb{Z})^\times_{\delta^{\mathrm{ord}}_{H^{\mathrm{ord}}_{p^r}}}\end{aligned}$$

over $\delta^{\mathrm{ord}}_{H^{\mathrm{ord}}_{p^r}}$, which is an $H^{\mathrm{ord}}_{p^r,\mathrm{G}^{\mathrm{ess,ord}}_{h,\mathrm{Z}_{p^r},\mathrm{D}_{p^r}}}$-orbit of étale-locally-defined triples $\varphi^{\mathrm{ord}}_{-1,p^r} = (\varphi^{\mathrm{ord},0}_{-1,p^r}, \varphi^{\mathrm{ord},\#,0}_{-1,p^r}, \nu^{\mathrm{ord}}_{-1,p^r})$.

(4) *$(\varphi^{\mathrm{ord}}_{-2,H^{\mathrm{ord}}_{p^r}}, \varphi^{\mathrm{ord}}_{0,H^{\mathrm{ord}}_{p^r}}) \twoheadrightarrow \varphi_{-1,H^{\mathrm{ord}}_{p^r}}$ is an $H^{\mathrm{ord}}_{p^r,\mathrm{G}^{\mathrm{ess,ord}}_{l,\mathrm{Z}_{p^r},\mathrm{D}_{p^r}}}$-torsor giving an $H^{\mathrm{ord}}_{p^r,\mathrm{G}^{\mathrm{ess,ord}}_{l,\mathrm{Z}_{p^r},\mathrm{D}_{p^r}}}$-orbit of étale-locally-defined pairs $(\varphi_{-2,p^r}, \varphi_{0,p^r})$.*

(5) *$(c^{\mathrm{ord}}_{H^{\mathrm{ord}}_{p^r}}, c^{\vee,\mathrm{ord}}_{H^{\mathrm{ord}}_{p^r}}) \twoheadrightarrow (\varphi^{\mathrm{ord}}_{-2,H^{\mathrm{ord}}_{p^r}}, \varphi^{\mathrm{ord}}_{0,H^{\mathrm{ord}}_{p^r}})$ is an $H^{\mathrm{ord}}_{p^r,\mathrm{U}^{\mathrm{ess,ord}}_{1,\mathrm{Z}_{p^r},\mathrm{D}_{p^r}}}$-torsor giving an $H^{\mathrm{ord}}_{p^r,\mathrm{U}^{\mathrm{ess,ord}}_{1,\mathrm{Z}_{p^r},\mathrm{D}_{p^r}}}$-orbit of étale-locally-defined pairs $(c^{\mathrm{ord}}_{p^r}, c^{\vee,\mathrm{ord}}_{p^r})$.*

(6) *$\tau^{\mathrm{ord}}_{H^{\mathrm{ord}}_{p^r}} \overset{\sim}{\to} (c^{\mathrm{ord}}_{H^{\mathrm{ord}}_{p^r}}, c^{\vee,\mathrm{ord}}_{H^{\mathrm{ord}}_{p^r}})$ is an $H^{\mathrm{ord}}_{p^r,\mathrm{U}^{\mathrm{ess,ord}}_{2,\mathrm{Z}_{p^r},\mathrm{D}_{p^r}}}$-torsor; i.e., it is an isomorphism giving no new structure.*

(Each of the datum or pairs of data is built on top of the earlier ones.) Then each such scheme $\tau^{\mathrm{ord}}_{H^{\mathrm{ord}}_{p^r}} \twoheadrightarrow \eta$ *determines a naive ordinary level-*$H^{\mathrm{ord}}_{p^r}$ *structure* $\alpha^{\mathrm{ord}}_{H^{\mathrm{ord}}_{p^r}}$ *of* $(G_\eta, \lambda_\eta, i_\eta)$ *of type* $(L/p^rL, \langle\,\cdot\,,\,\cdot\,\rangle, \mathtt{D}_{p^r})$ *(see Definition* 3.3.3.3*). If these schemes are orbits of liftable objects, then they determine an ordinary level-*$\mathcal{H}_p$ *structure* $\alpha^{\mathrm{ord}}_{\mathcal{H}_p}$ *of* $(G_\eta, \lambda_\eta, i_\eta)$ *of type* $(L \underset{\mathbb{Z}}{\otimes} \mathbb{Z}_p, \langle\,\cdot\,,\,\cdot\,\rangle, \mathtt{D})$ *(see Definition* 3.3.3.4*). (This forces* G *to be* **ordinary**. *See Remark* 3.3.3.6.*)*

Conversely, each ordinary level-$\mathcal{H}_p$ *structure* $\alpha^{\mathrm{ord}}_{\mathcal{H}_p}$ *of* $(G_\eta, \lambda_\eta, i_\eta)$ *of type* $(L \underset{\mathbb{Z}}{\otimes} \mathbb{Z}_p, \langle\,\cdot\,,\,\cdot\,\rangle, \mathtt{D})$ *arises this way for some (noncanonical choice of) symplectic-liftable* $\mathtt{Z}_{p^r}$. *(The* $(\mathcal{H}_p/\mathcal{U}_p(p^r))$*-orbit of* $\mathtt{Z}_{p^r}$*, or rather the* $\mathcal{H}_p$*-orbit of the lifting to a symplectic filtration on* $L \underset{\mathbb{Z}}{\otimes} \mathbb{Z}_p$*, is nevertheless well defined.)*

Proof. This follows from arguments similar to those in [62, Sec. 5.3]. □

Definition 4.1.5.22. *(Compare with [62, Def. 5.3.1.12].) With the setting as in Section* 4.1.2*, suppose we are given a tuple* $(B, \lambda_B, i_B, \underline{X}, \underline{Y}, \phi, c, c^\vee, \tau)$ *in* $\mathrm{DD}_{\mathrm{PE},\mathcal{O}}(R, I)$*. Let* $\mathcal{H}$, $\mathcal{H}^p$, $\mathcal{H}_p$*, and* r *be as in Definition* 3.4.1.1*. Let* n_0 *and* $H_{n_0} := \mathcal{H}^p/\mathcal{U}^p(n_0)$ *be as in Proposition* 4.1.5.10*, and let* $H^{\mathrm{ord}}_{p^r}$ *be as in Definition* 4.1.5.18 *(and hence in Proposition* 4.1.5.20*). Let* $n = n_0p^r$*, and let* $H_n := \mathcal{H}/\mathcal{U}(n)$*. By an* H_n**-orbit of étale-locally-defined naive ordinary level-**n **structure data of type** $(L/nL, \langle\,\cdot\,,\,\cdot\,\rangle, \mathtt{D}_{p^r})$*, we mean a scheme*

$$\alpha^{\natural,\mathrm{ord}}_{H_n} = (\mathtt{Z}_{H_n}, \varphi_{-2,H_n}, \varphi^{\mathrm{ord}}_{-1,H_n}, \varphi_{0,H_n}, \delta^{\mathrm{ord}}_{H_n}, c^{\mathrm{ord}}_{H_n}, c^{\vee,\mathrm{ord}}_{H_n}, \tau^{\mathrm{ord}}_{H_n})$$

(or rather just $\tau^{\mathrm{ord}}_{H_n}$*) finite étale over* η*, which is a composition of schemes*

$$\tau^{\mathrm{ord}}_{H_n} \twoheadrightarrow (c^{\mathrm{ord}}_{H_n}, c^{\vee,\mathrm{ord}}_{H_n}) \twoheadrightarrow (\varphi^{\mathrm{ord}}_{-2,H_n}, \varphi^{\mathrm{ord}}_{0,H_n}) \twoheadrightarrow \varphi^{\mathrm{ord}}_{-1,H_n} \twoheadrightarrow \delta^{\mathrm{ord}}_{H_n} \overset{\sim}{\to} \mathtt{Z}_{H_n} \twoheadrightarrow \eta$$

where we have

(1) $\tau^{\mathrm{ord}}_{H_n} \cong \tau_{H_{n_0}} \underset{\eta}{\times} \tau^{\mathrm{ord}}_{H^{\mathrm{ord}}_{p^r}}$,

(2) $(c^{\mathrm{ord}}_{H_n}, c^{\vee,\mathrm{ord}}_{H_n}) \cong (c_{H_{n_0}}, c^\vee_{H_{n_0}}) \underset{\eta}{\times} (c^{\mathrm{ord}}_{H^{\mathrm{ord}}_{p^r}}, c^{\vee,\mathrm{ord}}_{H^{\mathrm{ord}}_{p^r}})$,

(3) $(\varphi^{\mathrm{ord}}_{-2,H_n}, \varphi^{\mathrm{ord}}_{0,H_n}) \cong (\varphi^{\sim}_{-2,H_{n_0}}, \varphi^{\sim}_{0,H_{n_0}}) \underset{\eta}{\times} (\varphi^{\mathrm{ord}}_{-2,H^{\mathrm{ord}}_{p^r}}, \varphi^{\mathrm{ord}}_{0,H^{\mathrm{ord}}_{p^r}})$,

(4) $\varphi^{\mathrm{ord}}_{-1,H_n} \cong \varphi_{-1,H_{n_0}} \underset{\eta}{\times} \varphi^{\mathrm{ord}}_{-1,H^{\mathrm{ord}}_{p^r}}$,

(5) $\delta^{\mathrm{ord}}_{H_n} \cong \delta_{H_{n_0}} \underset{\eta}{\times} \delta^{\mathrm{ord}}_{H^{\mathrm{ord}}_{p^r}}$, *and*

(6) $\mathtt{Z}_{H_n} \cong \mathtt{Z}_{H_{n_0}} \underset{\eta}{\times} \mathtt{Z}_{H^{\mathrm{ord}}_{p^r}}$,

with objects at the right-hand sides in some composition (4.1.5.11) *in Proposition* 4.1.5.10 *and some composition* (4.1.5.21) *in Proposition* 4.1.5.20*. We use the same terminology* H_n**-orbit of étale-locally-defined** *for each of the entries in* $\alpha^{\natural,\mathrm{ord}}_{H_n}$.

*We remove "***naive***" from the terminology, and call it an* H_n**-orbit of étale-locally-defined ordinary level-**n **structure data of type** $(L \underset{\mathbb{Z}}{\otimes} \hat{\mathbb{Z}}, \langle\,\cdot\,,\,\cdot\,\rangle, \mathtt{D})$*, when the data are compatibly liftable to data (at all higher levels) satisfying the analogous conditions.*

As in [62, Def. 5.3.1.13], the equivalence relations among ordinary level-n structure data of type $(L \underset{\mathbb{Z}}{\otimes} \hat{\mathbb{Z}}, \langle \cdot, \cdot \rangle, \mathtt{D})$ over η then induce equivalence relations among their H_n-orbits.

Definition 4.1.5.23. (*Compare with Definition [62, Def. 5.3.1.14].*) *With the setting as in Section* 4.1.2, *suppose we are given a tuple* $(B, \lambda_B, i_B, \underline{X}, \underline{Y}, \phi, c, c^\vee, \tau)$ *in* $\mathrm{DD}_{\mathrm{PE},\mathcal{O}}(R, I)$. *Let* $\mathcal{H}$, $\mathcal{H}^p$, $\mathcal{H}_p$, *and* r *be as in Definition* 3.4.1.1. *For each integer* $n_0 \geq 1$ *such that* $p \nmid n$ *and* $\mathcal{U}^p(n_0) \subset \mathcal{H}^p$, *set* $H_{n_0} := \mathcal{H}^p/\mathcal{U}^p(n)$ *and* $H_{n_0 p^r} := \mathcal{H}/\mathcal{U}(n_0 p^r)$ *as usual. Then an* **ordinary level-$\mathcal{H}$ structure datum of type** $(L \underset{\mathbb{Z}}{\otimes} \hat{\mathbb{Z}}, \langle \cdot, \cdot \rangle, \mathtt{D})$ *over* η *is a collection* $\alpha_{\mathcal{H}}^{\natural,\mathrm{ord}} = \{\alpha_{H_{n_0 p^r}}^{\natural,\mathrm{ord}}\}_{n_0}$ *indexed by integers* $n_0 \geq 1$ *such that* $p \nmid n_0$ *and* $\mathcal{U}^p(n_0) \subset \mathcal{H}^p$, *with elements* $\alpha_{H_{n_0 p^r}}^{\natural,\mathrm{ord}}$ *described as follows:*

(1) *For each index* n_0, *the element* $\alpha_{H_{n_0 p^r}}^{\natural,\mathrm{ord}}$ *is an* $H_{n_0 p^r}$*-orbit of étale-locally-defined ordinary level-n structure data of type* $(L \underset{\mathbb{Z}}{\otimes} \hat{\mathbb{Z}}, \langle \cdot, \cdot \rangle, \mathtt{D})$ *as in Definition* 4.1.5.22.

(2) *For all indices* n_0 *and* m_0 *such that* $n_0 | m_0$, *the* $H_{n_0 p^r}$*-orbit* $\alpha_{H_{n_0 p^r}}^{\natural,\mathrm{ord}}$ *is determined by the* $H_{m_0 p^r}$*-orbit* $\alpha_{H_{m_0 p^r}}^{\natural,\mathrm{ord}}$ *by reduction modulo* $n_0 p^r$.

It is customary to denote $\alpha_{\mathcal{H}}^{\natural,\mathrm{ord}}$ *by a tuple*

$$\alpha_{\mathcal{H}}^{\natural,\mathrm{ord}} = (\mathtt{Z}_{\mathcal{H}}, \varphi_{-2,\mathcal{H}}, \varphi_{-1,\mathcal{H}}^{\mathrm{ord}}, \varphi_{0,\mathcal{H}}, \delta_{\mathcal{H}}^{\mathrm{ord}}, c_{\mathcal{H}}^{\mathrm{ord}}, c_{\mathcal{H}}^{\vee,\mathrm{ord}}, \tau_{\mathcal{H}}^{\mathrm{ord}}),$$

each subtuple or entry being a collection indexed by n_0 *as* $\alpha_{\mathcal{H}}^{\natural,\mathrm{ord}}$ *is, and to denote by* $\iota_{\mathcal{H}}^{\mathrm{ord}}$ *the collection corresponding to* $\tau_{\mathcal{H}}^{\mathrm{ord}}$. *For convenience, we also write* $\tau_{\mathcal{H}}^{\mathrm{ord}} \cong \tau_{\mathcal{H}^p} \underset{\eta}{\times} \tau_{\mathcal{H}_p}^{\mathrm{ord}}$ *etc as in Definition* 4.1.5.22.

As in [62, Def. 5.3.1.13], the equivalence relations among naive ordinary level-n structure data of type $(L \underset{\mathbb{Z}}{\otimes} \hat{\mathbb{Z}}, \langle \cdot, \cdot \rangle, \mathtt{D})$ over η then induce equivalence relations among their H_n-orbits.

Convention 4.1.5.24. (*Compare with [62, Conv. 5.3.1.15].*) *To facilitate the language, we shall call* $\alpha_{\mathcal{H}}^{\natural,\mathrm{ord}}$ *an* $\mathcal{H}$*-orbit, with similar usages applied to other objects with subscripts "$\mathcal{H}$". If we have two open compact subgroups* $\mathcal{H}' \subset \mathcal{H}$ *for which ordinary level structures at those levels make sense, and if we have an object* $\alpha_{\mathcal{H}'}^{\natural,\mathrm{ord}}$ *at level* $\mathcal{H}'$, *then there is a natural meaning of the object* $\alpha_{\mathcal{H}}^{\natural,\mathrm{ord}}$ *at level* $\mathcal{H}$ *determined by* $\alpha_{\mathcal{H}'}^{\natural,\mathrm{ord}}$. *We say in this case that* $\alpha_{\mathcal{H}}^{\natural,\mathrm{ord}}$ *is the* $\mathcal{H}$*-orbit of* $\alpha_{\mathcal{H}'}^{\natural,\mathrm{ord}}$.

As in [62, Def. 5.3.1.16], and as above, the equivalence relations among naive ordinary level-n structure data then also induce equivalent relations among ordinary level-$\mathcal{H}$ structure data.

Definition 4.1.5.25. (*Compare with Definition* 4.1.4.57.) *With the setting as in Section* 4.1.2, *suppose moreover that* η *is a scheme over* $\mathrm{Spec}(\mathbb{Z}_{(p)})$. *The category* $\mathrm{DEG}_{\mathrm{PEL},\ddot{\mathsf{M}}_{\mathcal{H}}^{\mathrm{ord}}}(R, I)$ *has objects of the form* $(G, \lambda, i, \alpha_{\mathcal{H}^p}, \alpha_{\mathcal{H}_p}^{\mathrm{ord}})$ *(over* S*), where:*

(1) (G, λ, i) *defines an object of* $\mathrm{DEG}_{\mathrm{PE},\mathcal{O}}(R, I)$ (*see Definition* 4.1.3.1).

(2) $(G_\eta, \lambda_\eta, i_\eta, \alpha_{\mathcal{H}^p}, \alpha^{\mathrm{ord}}_{\mathcal{H}_p})$ *defines an object of* $\ddot{\mathsf{M}}^{\mathrm{ord}}_{\mathcal{H}}(\eta)$ (*see Definition* 3.4.1.1).

(3) $\alpha^{\mathrm{ord}}_{\mathcal{H}_p}$ *is defined by a (scheme-theoretic)* $H^{\mathrm{ord}}_{p^r}$*-orbit of some principal ordinary level-*p^r *structure of type* $(L \underset{\mathbb{Z}}{\otimes} \mathbb{Z}_p, \langle \cdot\,, \cdot \rangle, \mathtt{D})$ *such that, for each integer* $r' \geq r$, *there exists some lifting to level* $p^{r'}$ (*over some étale extension of* η) *compatible with degeneration (i.e., satisfying the analogue of Condition* 4.1.4.1 *at level* $p^{r'}$).

Definition 4.1.5.26. (*Compare with Definition* 4.1.4.58.) *With the setting as in Section* 4.1.2, *suppose moreover that* η *is a scheme over* $\mathrm{Spec}(\mathbb{Z}_{(p)})$. *The category* $\mathrm{DD}_{\mathrm{PEL},\ddot{\mathsf{M}}^{\mathrm{ord}}_{\mathcal{H}}}(R, I)$ *has objects of the form*

$$(B, \lambda_B, i_B, \underline{X}, \underline{Y}, \phi, c, c^\vee, \tau, [\alpha^{\natural,\mathrm{ord}}_{\mathcal{H}}]),$$

where:

(1) $(B, \lambda_B, i_B, \underline{X}, \underline{Y}, \phi, c, c^\vee, \tau)$ *defines an object of* $\mathrm{DD}_{\mathrm{PE},\mathcal{O}}(R, I)$ (*see Definition* 4.1.3.2).

(2) $[\alpha^{\natural,\mathrm{ord}}_{\mathcal{H}}]$ *is an equivalence class of ordinary level-n structure data* $\alpha^{\natural,\mathrm{ord}}_{\mathcal{H}}$ *of type* $(L \underset{\mathbb{Z}}{\otimes} \hat{\mathbb{Z}}, \langle \cdot\,, \cdot \rangle, \mathtt{D})$ *defined over* η. (*See Definition* 4.1.5.23.)

Now it follows from Propositions 4.1.4.21, 4.1.4.52, and 4.1.4.56 that we have the following:

Theorem 4.1.5.27. *There is an equivalence of categories*

$$\mathrm{M}_{\mathrm{PEL},\ddot{\mathsf{M}}^{\mathrm{ord}}_{\mathcal{H}}}(R, I) : \ \mathrm{DD}_{\mathrm{PEL},\ddot{\mathsf{M}}^{\mathrm{ord}}_{\mathcal{H}}}(R, I) \to \mathrm{DEG}_{\mathrm{PEL},\ddot{\mathsf{M}}^{\mathrm{ord}}_{\mathcal{H}}}(R, I) :$$
$$(B, \lambda_B, i_B, \underline{X}, \underline{Y}, \phi, c, c^\vee, \tau, [\alpha^{\natural,\mathrm{ord}}_{\mathcal{H}}]) \mapsto (G, \lambda, i, \alpha_{\mathcal{H}^p}, \alpha^{\mathrm{ord}}_{\mathcal{H}_p}).$$

4.1.6 *Comparison with Degeneration Data for Level Structures in Characteristic Zero*

Let $\mathcal{H}$, $\mathcal{H}^p$, $\mathcal{H}_p$, r, and r_ν be as in beginning of Section 3.3.5. Let n_0 be some integer prime to p such that $\mathcal{U}^p(n_0) \subset \mathcal{H}^p$. Let $H_{n_0} := \mathcal{H}^p/\mathcal{U}^p(n_0)$ and $H_{p^r} := \mathcal{H}_p/\mathcal{U}_p(p^r)$, and let $H^{\mathrm{ord}}_{p^r} := \mathcal{H}_p/\mathcal{U}^{\mathrm{bal}}_{p,1}(p^r)$ as in Section 3.3.3. Let $n := n_0p^r$. Then $\mathcal{U}(n) \subset \mathcal{H}$ and we set $H_n := \mathcal{H}/\mathcal{U}(n)$.

Let $S = \mathrm{Spec}(R)$ be as in Section 4.1.2. Assume moreover that the generic point η of S is a point over $\mathsf{S}_{0,p^{r_\nu}} = \mathrm{Spec}(F_0[\zeta_{p^{r_\nu}}])$ (see Proposition 3.3.5.1). By normality of S, this forces S to be a scheme over $\mathrm{Spec}(\mathbb{Z}[\zeta_{p^{r_\nu}}])$, so that there exists a canonical isomorphism $\zeta_{p^{r_\nu},S'} : ((\mathbb{Z}/p^{r_\nu}\mathbb{Z})(1))_{S'} \overset{\sim}{\to} \boldsymbol{\mu}_{p^{r_\nu},S'}$ for each scheme $S' \to S$ (which is the pullback of the canonical $\zeta_{p^{r_\nu}}$ over $\mathrm{Spec}(\mathbb{Z}[\zeta_{p^{r_\nu}}])$).

Let (G, λ, i) be a degenerating family of type $(\mathrm{PE}, \mathcal{O})$ over S as in Definition 4.1.3.1, with degeneration datum $(B, \lambda_B, i_B, \underline{X}, \underline{Y}, \phi, c, c^\vee, \tau)$ given by (4.1.3.3). For simplicity, let us assume (until we finish the proof of Proposition 4.1.6.1) that $\underline{X}$ and $\underline{Y}$ are constant with values X and Y, respectively.

Let $[(\mathsf{Z}_{\mathcal{H}}, \Phi_{\mathcal{H}}, \delta_{\mathcal{H}})]$ be an ordinary cusp label at level $\mathcal{H}$ for the PEL-type $\mathcal{O}$-lattice $(L, \langle\,\cdot\,,\,\cdot\,\rangle, h_0)$ (see Definition 3.2.3.8). Let

$$\alpha^{\natural}_{\mathcal{H}} = (\mathsf{Z}_{\mathcal{H}}, \varphi^{\sim}_{-2,\mathcal{H}}, \varphi_{-1,\mathcal{H}}, \varphi^{\sim}_{0,\mathcal{H}}, \delta_{\mathcal{H}}, c_{\mathcal{H}}, c^{\vee}_{\mathcal{H}}, \tau_{\mathcal{H}})$$

be a *level-$\mathcal{H}$ structure datum of type* $(L \underset{\mathbb{Z}}{\otimes} \hat{\mathbb{Z}}, \langle\,\cdot\,,\,\cdot\,\rangle)$, as in [62, Def. 5.3.1.14], with $(\varphi^{\sim}_{-2,\mathcal{H}}, \varphi^{\sim}_{0,\mathcal{H}})$ inducing the $(\varphi_{-2,\mathcal{H}}, \varphi_{0,\mathcal{H}})$ in a representative $(\mathsf{Z}_{\mathcal{H}}, \Phi_{\mathcal{H}} = (X, Y, \phi, \varphi_{-2,\mathcal{H}}, \varphi_{0,\mathcal{H}}), \delta_{\mathcal{H}})$ of $[(\mathsf{Z}_{\mathcal{H}}, \Phi_{\mathcal{H}}, \delta_{\mathcal{H}})]$, as in the corrected [62, Def. 5.4.2.8] in the errata. Then

$$(\mathsf{Z}_{\mathcal{H}}, (X, Y, \phi, \varphi^{\sim}_{-2,\mathcal{H}}, \varphi^{\sim}_{0,\mathcal{H}}), (B, \lambda_B, i_B, \varphi_{-1,\mathcal{H}}), \delta_{\mathcal{H}}, (c_{\mathcal{H}}, c^{\vee}_{\mathcal{H}}, \tau_{\mathcal{H}}))$$

is an object of $\mathrm{DD}^{\text{fil.-spl.}}_{\mathrm{PEL},\mathsf{M}_{\mathcal{H}}}(R, I)$, as in [62, Lem. 5.4.2.10; see also the errata]. This means we have a composition of finite étale morphisms

$$\tau_{H_n} \twoheadrightarrow (c_{H_n}, c^{\vee}_{H_n}) \twoheadrightarrow (\varphi^{\sim}_{-2,H_n}, \varphi^{\sim}_{0,H_n}) \twoheadrightarrow \varphi_{-1,H_n} \twoheadrightarrow \delta_{H_n} \overset{\sim}{\to} \mathsf{Z}_{H_n} \twoheadrightarrow \eta$$

as in [62, Sec. 5.3]. By taking reduction modulo n_0 or restrictions of various objects, we obtain an induced composition as in (4.1.5.11) in Proposition 4.1.5.10. We claim that we can also obtain an induced composition as in (4.1.5.21) in Proposition 4.1.5.20.

By definition, $\mathsf{Z}_{\mathcal{H}}$ is an $\mathcal{H}$-orbit of strongly symplectic admissible filtrations Z on $L \underset{\mathbb{Z}}{\otimes} \hat{\mathbb{Z}}$. This includes, in particular, the datum of an $\mathcal{H}_p$-orbit of a symplectic admissible filtration $\mathsf{Z} \underset{\hat{\mathbb{Z}}}{\otimes} \mathbb{Z}_p = \{\mathsf{Z}_{-i} \underset{\hat{\mathbb{Z}}}{\otimes} \mathbb{Z}_p\}_i$ on $L \underset{\mathbb{Z}}{\otimes} \mathbb{Z}_p$. By Definition 3.2.3.1, and by replacing Z with another representative in the $\mathcal{H}$-orbit $\mathsf{Z}_{\mathcal{H}}$ if necessary, we shall assume that

$$\mathsf{Z}_{-2} \underset{\hat{\mathbb{Z}}}{\otimes} \mathbb{Z}_p \subset \mathsf{D}^0 \subset \mathsf{Z}_{-1} \underset{\hat{\mathbb{Z}}}{\otimes} \mathbb{Z}_p$$

(see (3.2.3.2)), which induces a filtration $\mathsf{D}_{-1} = \{\mathsf{D}^i_{-1}\}_i$ on $\mathrm{Gr}^{\mathsf{Z}}_{-1} \underset{\hat{\mathbb{Z}}}{\otimes} \mathbb{Z}_p$. By taking reduction modulo p^r, we have the compatibility (3.2.3.3) (resp. (3.2.3.5)), which induces a filtration $\mathsf{D}_{-1,p^r} = \{\mathsf{D}^i_{-1,p^r}\}_i$ on $\mathrm{Gr}^{\mathsf{Z}}_{-1,p^r}$ given by (3.2.3.4) (resp. (3.2.3.6)). Note that (3.2.3.3) (resp. (3.2.3.4), resp. (3.2.3.5), resp. (3.2.3.6)) is a special case of (4.1.4.5) (resp. (4.1.4.14), resp. (4.1.4.6), resp. (4.1.4.15)) in the sense that we did not assume that the latter comes from some symplectic-liftable admissible filtration.

Since $\mathcal{U}^{\mathrm{bal}}_{p,1}(p^r) \subset \mathcal{H}_p \subset \mathcal{U}_{p,0}(p^r)$, and since the action of $\mathcal{U}_{p,0}(p^r)$ stabilizes $\mathsf{D}^0_{p^r}$ as an $\mathcal{O} \underset{\mathbb{Z}}{\otimes} (\mathbb{Z}/p^r\mathbb{Z})$-submodule of L/p^rL, the compatibility (3.2.3.3) is independent of the choice of Z, once Z exists (*cf.* Remark 3.2.3.7). Moreover, the $H^{\mathrm{ord}}_{p^r}$-orbit $\mathsf{Z}_{H^{\mathrm{ord}}_{p^r}}$ (which is equivalently an H_{p^r}-orbit) of $\mathsf{Z}_{p^r} = \{\mathsf{Z}_{-i,p^r}\}_i$ determines the $H^{\mathrm{ord}}_{p^r}$-orbits of the two $\mathcal{O}$-submodules $\mathrm{Gr}^{\mathsf{Z}}_{-2,p^r} \subset \mathrm{Gr}^0_{\mathsf{D},p^r}$ and $\mathrm{Gr}^{\mathsf{Z}^{\#}}_{-2,p^r} \subset \mathrm{Gr}^0_{\mathsf{D}^{\#},p^r}$ as in (4.1.4.7) and (4.1.4.8), which in turn define $\mathrm{Gr}^0_{\mathsf{D}_{-1,p^r}} = \mathrm{Gr}^0_{\mathsf{D},p^r} / \mathrm{Gr}^{\mathsf{Z}}_{-2,p^r}$ and $\mathrm{Gr}^0_{\mathsf{D}^{\#}_{-1,p^r}} = \mathrm{Gr}^0_{\mathsf{D}^{\#},p^r} / \mathrm{Gr}^{\mathsf{Z}^{\#}}_{-2,p^r}$ as in (4.1.4.16) and (4.1.4.17). These are all independent of the choice of Z.

Thus, we have obtained a well-defined assignment of $\mathsf{Z}_{H^{\mathrm{ord}}_{p^r}}$ to $\mathsf{Z}_{\mathcal{H}}$.

The datum $\delta_{\mathcal{H}}$ in the cusp label $[(\mathsf{Z}_{\mathcal{H}}, \Phi_{\mathcal{H}}, \delta_{\mathcal{H}})]$ is by definition the $\mathcal{H}$-orbit of some splitting δ, which includes in particular the datum of the H_{p^r}-orbit of some splitting

$$\delta_{p^r} : \mathrm{Gr}^{\mathsf{Z}}_{-2,p^r} \oplus \mathrm{Gr}^{\mathsf{Z}}_{-1,p^r} \oplus \mathrm{Gr}^{\mathsf{Z}}_{0,p^r} \xrightarrow{\sim} L/p^r L.$$

Since we have the compatibility (3.2.3.3), we have splittings

$$\delta^{\mathrm{ord},0}_{p^r} : \mathrm{Gr}^{\mathsf{Z}}_{-2,p^r} \oplus \mathrm{Gr}^{0}_{\mathsf{D}_{-1,p^r}} \xrightarrow{\sim} \mathrm{Gr}^{0}_{\mathsf{D},p^r}$$

and

$$\delta^{\mathrm{ord},\#,0}_{p^r} : \mathrm{Gr}^{\mathsf{Z}^\#}_{-2,p^r} \oplus \mathrm{Gr}^{0}_{\mathsf{D}^\#_{-1,p^r}} \xrightarrow{\sim} \mathrm{Gr}^{0}_{\mathsf{D}^\#,p^r}$$

as in (4.1.4.23) and (4.1.4.24). Note that the H_{p^r}-orbit of δ_{p^r} determines and is determined by the $H^{\mathrm{ord}}_{p^r}$-orbit of the pair

$$\delta^{\mathrm{ord}}_{p^r} := (\delta^{\mathrm{ord},0}_{p^r}, \delta^{\mathrm{ord},\#,0}_{p^r}).$$

(Since the splitting δ_{p^r} does not respect pairings, the two splittings $\delta^{\mathrm{ord},0}_{p^r}$ and $\delta^{\mathrm{ord},\#,0}_{p^r}$ are compatible with $\phi^0_{\mathsf{D},p^r} : \mathrm{Gr}^0_{\mathsf{D},p^r} \to \mathrm{Gr}^0_{\mathsf{D}^\#,p^r}$ only in the sense that $\phi^0_{\mathsf{D},p^r}(\mathrm{Gr}^{\mathsf{Z}}_{-2,p^r}) \subset \mathrm{Gr}^{\mathsf{Z}^\#}_{-2,p^r}$; *cf.* Lemma 4.1.4.22.) By abuse of notation, let us denote the $H^{\mathrm{ord}}_{p^r}$-orbit of the pair $\delta^{\mathrm{ord}}_{p^r} = (\delta^{\mathrm{ord},0}_{p^r}, \delta^{\mathrm{ord},\#,0}_{p^r})$ by $\delta^{\mathrm{ord}}_{H^{\mathrm{ord}}_{p^r}} = (\delta^{\mathrm{ord},0}_{H^{\mathrm{ord}}_{p^r}}, \delta^{\mathrm{ord},\#,0}_{H^{\mathrm{ord}}_{p^r}})$. This is the same $\delta^{\mathrm{ord}}_{H^{\mathrm{ord}}_{p^r}}$ as in Proposition 4.1.5.20.

Thus, we have obtained a well-defined assignment of $\delta^{\mathrm{ord}}_{H^{\mathrm{ord}}_{p^r}}$ to $\delta_{\mathcal{H}}$.

The assignment of $\varphi^{\mathrm{ord}}_{-1,H^{\mathrm{ord}}_{p^r}}$ as in Proposition 4.1.5.20 to $\varphi_{-1,\mathcal{H}}$ or rather some $\varphi_{-1,H_{p^r}}$ is as in Proposition 3.3.5.1 for $\mathsf{M}_{\mathcal{H}_h}$ and an obvious analogue $\dddot{\mathsf{M}}^{\mathrm{ord}}_{\mathcal{H}_h}$ of $\dddot{\mathsf{M}}^{\mathrm{ord}}_{\mathcal{H}}$ (*cf.* Definitions 1.2.1.15 and 3.4.1.9). This is where we need η to be defined over $\mathsf{S}_{0,p^{r_\nu}} = \mathrm{Spec}(F_0[\zeta_{p^{r_\nu}}])$.

The datum $\Phi_{\mathcal{H}}$ in the cusp label $[(\mathsf{Z}_{\mathcal{H}}, \Phi_{\mathcal{H}}, \delta_{\mathcal{H}})]$ is by definition the $\mathcal{H}$-orbit of some tuple $(X, Y, \phi, \varphi_{-2}, \varphi_0)$ (where the action of $\mathcal{H} \subset \mathrm{G}(\hat{\mathbb{Z}})$ does not modify the first three entries), which includes in particular the datum of an H_{p^r}-orbit, or equivalently an $H^{\mathrm{ord}}_{p^r}$-orbit, of pairs of isomorphisms $(\varphi_{-2,p^r}, \varphi_{0,p^r})$, or equivalently pairs of isomorphisms $(\varphi_{-2,p^r}, \varphi^\#_{-2,p^r})$, as in (4.1.4.11), (4.1.4.12), and (4.1.4.13). The two isomorphisms φ_{-2,p^r} and $\varphi^\#_{-2,p^r}$ induce respectively the two isomorphisms

$$(\varphi_{-2,p^r})^{\mathrm{mult}}_S : (\mathrm{Gr}^{\mathsf{Z}}_{-2,p^r})^{\mathrm{mult}}_S \xrightarrow{\sim} T[p^r]$$

and

$$(\varphi^\#_{-2,p^r})^{\mathrm{mult}}_S : (\mathrm{Gr}^{\mathsf{Z}^\#}_{-2,p^r})^{\mathrm{mult}}_S \xrightarrow{\sim} T^\vee[p^r]$$

as in (2) of Proposition 4.1.4.21. By abuse of notation, let us denote the (scheme-theoretic) $H^{\mathrm{ord}}_{p^r}$-orbit of the pair $((\varphi_{-2,p^r})^{\mathrm{mult}}_S, (\varphi^\#_{-2,p^r})^{\mathrm{mult}}_S)$ by $((\varphi_{-2,H^{\mathrm{ord}}_{p^r}})^{\mathrm{mult}}_S, (\varphi^\#_{-2,H^{\mathrm{ord}}_{p^r}})^{\mathrm{mult}}_S)$, which determines and is determined by a scheme $(\varphi_{-2,H^{\mathrm{ord}}_{p^r}}, \varphi_{0,H^{\mathrm{ord}}_{p^r}})$ over η.

If we also include $\varphi_{-1,H_{p^r}}$ and $\varphi^{\mathrm{ord}}_{-1,H^{\mathrm{ord}}_{p^r}}$ into consideration, then we have a subscheme $(\varphi^{\sim}_{-2,H_{p^r}}, \varphi^{\sim}_{0,H_{p^r}})$ of $(\varphi_{-2,H_{p^r}}, \varphi_{0,H_{p^r}}) \underset{\mathsf{Z}_{H^{\mathrm{ord}}_{p^r}}}{\times} \varphi_{-1,H_{p^r}}$ which is an H_{p^r}-orbit of étale-locally-defined $((\varphi_{-2,p^r}, \varphi_{0,p^r}), \varphi_{-1,p^r})$ and surjects under the two projections to $(\varphi_{-2,H_{p^r}}, \varphi_{0,H_{p^r}})$ and $\varphi_{-1,H_{p^r}}$. Moreover, it induces a subscheme $((\varphi^{\mathrm{ord}}_{-2,H^{\mathrm{ord}}_{p^r}})^{\mathrm{mult}}_S, (\varphi^{\mathrm{ord},\#}_{-2,H^{\mathrm{ord}}_{p^r}})^{\mathrm{mult}}_S)$ of $((\varphi_{-2,H^{\mathrm{ord}}_{p^r}})^{\mathrm{mult}}_S, (\varphi^{\#}_{-2,H^{\mathrm{ord}}_{p^r}})^{\mathrm{mult}}_S) \underset{\mathsf{Z}_{H^{\mathrm{ord}}_{p^r}}}{\times} \varphi_{-1,H^{\mathrm{ord}}_{p^r}}$, which determines and is determined by a subscheme $(\varphi^{\mathrm{ord}}_{-2,H^{\mathrm{ord}}_{p^r}}, \varphi^{\mathrm{ord}}_{0,H^{\mathrm{ord}}_{p^r}})$ of $(\varphi_{-2,H^{\mathrm{ord}}_{p^r}}, \varphi_{0,H^{\mathrm{ord}}_{p^r}}) \underset{\mathsf{Z}_{H^{\mathrm{ord}}_{p^r}}}{\times} \varphi_{-1,H^{\mathrm{ord}}_{p^r}}$, both subschemes being $H^{\mathrm{ord}}_{p^r}$-orbits of étale-locally-defined objects inducing surjections under the two projections.

Thus, we have obtained well-defined assignments of $(\varphi_{-2,H^{\mathrm{ord}}_{p^r}}, \varphi_{0,H^{\mathrm{ord}}_{p^r}})$ to $\Phi_{\mathcal{H}}$, or rather just to $(\varphi_{-2,\mathcal{H}}, \varphi_{0,\mathcal{H}})$; and of $((\varphi^{\mathrm{ord}}_{-2,H^{\mathrm{ord}}_{p^r}})^{\mathrm{mult}}_S, (\varphi^{\mathrm{ord},\#}_{-2,H^{\mathrm{ord}}_{p^r}})^{\mathrm{mult}}_S)$ and $(\varphi^{\mathrm{ord}}_{-2,H^{\mathrm{ord}}_{p^r}}, \varphi^{\mathrm{ord}}_{0,H^{\mathrm{ord}}_{p^r}})$ to $(\varphi^{\sim}_{-2,\mathcal{H}}, \varphi^{\sim}_{0,\mathcal{H}})$, which are compatible with each other.

The pair $(c_{H_n}, c^{\vee}_{H_n})$ as a scheme over $(\varphi^{\sim}_{-2,H_n}, \varphi^{\sim}_{0,H_n})$ is an $H_{n,\mathrm{U}^{\mathrm{ess}}_{1,\mathsf{Z}_n}}$-torsor giving an $H_{n,\mathrm{U}^{\mathrm{ess}}_{1,\mathsf{Z}_n}}$-orbit of étale-locally-defined pairs $(c_n, c^{\vee}_n)$. Each $c_n : \frac{1}{n}X \to B^{\vee}_{\eta}$ determines by restriction to $\frac{1}{p^r}X$ and by composition with (4.1.4.32) a homomorphism

$$c^{\mathrm{ord}}_{p^r} : \tfrac{1}{n}X \to B^{\vee}_{\eta} \to B^{\vee,\mathrm{ord}}_{\eta,p^r} = B^{\vee}_{\eta}/\varphi^{\mathrm{ord},\#,0}_{-1,p^r}((\mathrm{Gr}^0_{\mathrm{D}^{\#}_{-1,p^r}})^{\mathrm{mult}}_{\eta})$$

lifting $c : X \to B^{\vee}$, while each $c^{\vee}_n : \frac{1}{n}Y \to B_{\eta}$ determines by restriction to $\frac{1}{p^r}Y$ and by composition with (4.1.4.31) a homomorphism

$$c^{\vee,\mathrm{ord}}_{p^r} : \tfrac{1}{p^r}Y \to B^{\mathrm{ord}}_{\eta,p^r} = B_{\eta}/\varphi^{\mathrm{ord},0}_{-1,p^r}((\mathrm{Gr}^0_{\mathrm{D}_{-1,p^r}})^{\mathrm{mult}}_{\eta})$$

lifting $c^{\vee} : Y \to B$. Since the actions of $H_{n,\mathrm{U}^{\mathrm{ess}}_{1,\mathsf{Z}_n}}$ and $H^{\mathrm{ord}}_{p^r,\mathrm{U}^{\mathrm{ess},\mathrm{ord}}_{1,\mathsf{Z}_{p^r},\mathrm{D}_{p^r}}}$ respect pairings and are compatible, the $H^{\mathrm{ord}}_{p^r,\mathrm{U}^{\mathrm{ess},\mathrm{ord}}_{1,\mathsf{Z}_{p^r},\mathrm{D}_{p^r}}}$-orbit of $(c^{\mathrm{ord}}_{p^r}, c^{\vee,\mathrm{ord}}_{p^r})$ is independent of choices, and defines a scheme $(c^{\mathrm{ord}}_{H^{\mathrm{ord}}_{p^r}}, c^{\vee,\mathrm{ord}}_{H^{\mathrm{ord}}_{p^r}})$ over $(\varphi^{\mathrm{ord}}_{-2,H^{\mathrm{ord}}_{p^r}}, \varphi^{\mathrm{ord}}_{0,H^{\mathrm{ord}}_{p^r}})$.

Thus, we have a well-defined assignment of $(c^{\mathrm{ord}}_{H^{\mathrm{ord}}_{p^r}}, c^{\vee,\mathrm{ord}}_{H^{\mathrm{ord}}_{p^r}})$ to $(c_{\mathcal{H}}, c^{\vee}_{\mathcal{H}})$.

Finally, the scheme $\tau_{H_n} \twoheadrightarrow (c_{H_n}, c^{\vee}_{H_n})$ is an $H_{n,\mathrm{U}^{\mathrm{ess}}_{2,\mathsf{Z}_n}}$-torsor giving an $H_{n_0,\mathrm{U}^{\mathrm{ess}}_{2,\mathsf{Z}_n}}$-orbit of étale-locally-defined τ_n, which induces the (trivial) $H^{\mathrm{ord}}_{p^r,\mathrm{U}^{\mathrm{ess},\mathrm{ord}}_{2,\mathsf{Z}_{p^r},\mathrm{D}_{p^r}}}$-torsor $\tau^{\mathrm{ord}}_{H^{\mathrm{ord}}_{p^r}} \xrightarrow{\sim} (c^{\mathrm{ord}}_{H^{\mathrm{ord}}_{p^r}}, c^{\vee,\mathrm{ord}}_{H^{\mathrm{ord}}_{p^r}})$. (Note that the group $H^{\mathrm{ord}}_{p^r,\mathrm{U}^{\mathrm{ess},\mathrm{ord}}_{2,\mathsf{Z}_{p^r},\mathrm{D}_{p^r}}}$ is trivial.) This is consistent with the convention that $\tau^{\mathrm{ord}}_n := \tau_{n_0}$ does not see any information at p.

Thus, we have verified our claim that we can also obtain an induced composition as in (4.1.5.21) in Proposition 4.1.5.20.

Proposition 4.1.6.1. *With the assumptions as above, there is a commutative dia-*

gram of finite étale morphisms:

$$
\begin{array}{ccc}
\tau_{H_n} & \longrightarrow & \tau^{\mathrm{ord}}_{H_n} \cong \tau_{H_{n_0}} \underset{\eta}{\times} \tau^{\mathrm{ord}}_{H^{\mathrm{ord}}_{p^r}} \\
\Big\downarrow {\scriptstyle \mathrm{mod}\ H_{n,\mathrm{U}^{\mathrm{ess}}_{2,\mathsf{Z}_n}}} & & \Big\downarrow {\scriptstyle \mathrm{mod}\ H_{n_0,\mathrm{U}^{\mathrm{ess}}_{2,\mathsf{Z}_{n_0}}} \times H^{\mathrm{ord}}_{p^r,\mathrm{U}^{\mathrm{ess,ord}}_{2,\mathsf{Z}_{p^r},\mathsf{D}_{p^r}}}} \\
(c_{H_n}, c^{\vee}_{H_n}) & \longrightarrow & (c^{\mathrm{ord}}_{H_n}, c^{\vee,\mathrm{ord}}_{H_n}) \cong (c_{H_{n_0}}, c^{\vee}_{H_{n_0}}) \underset{\eta}{\times} (c^{\mathrm{ord}}_{H^{\mathrm{ord}}_{p^r}}, c^{\vee,\mathrm{ord}}_{H^{\mathrm{ord}}_{p^r}}) \\
\Big\downarrow {\scriptstyle \mathrm{mod}\ H_{n,\mathrm{U}^{\mathrm{ess}}_{1,\mathsf{Z}_n}}} & & \Big\downarrow {\scriptstyle \mathrm{mod}\ H_{n_0,\mathrm{U}^{\mathrm{ess}}_{1,\mathsf{Z}_{n_0}}} \times H^{\mathrm{ord}}_{p^r,\mathrm{U}^{\mathrm{ess,ord}}_{1,\mathsf{Z}_{p^r},\mathsf{D}_{p^r}}}} \\
(\varphi^{\sim}_{-2,H_n}, \varphi^{\sim}_{0,H_n}) & \rightarrow & (\varphi^{\mathrm{ord}}_{-2,H_n}, \varphi^{\mathrm{ord}}_{0,H_n}) \cong (\varphi^{\sim}_{-2,H_{n_0}}, \varphi^{\sim}_{0,H_{n_0}}) \underset{\eta}{\times} (\varphi^{\mathrm{ord}}_{-2,H^{\mathrm{ord}}_{p^r}}, \varphi^{\mathrm{ord}}_{0,H^{\mathrm{ord}}_{p^r}}) \\
\Big\downarrow {\scriptstyle \mathrm{mod}\ H_{n_0,\mathrm{G}^{\mathrm{ess}}_{l,\mathsf{Z}_{n_0}}}} & & \Big\downarrow {\scriptstyle \mathrm{mod}\ H_{n_0,\mathrm{G}^{\mathrm{ess}}_{l,\mathsf{Z}_{n_0}}} \times H^{\mathrm{ord}}_{p^r,\mathrm{G}^{\mathrm{ess,ord}}_{l,\mathsf{Z}_{p^r},\mathsf{D}_{p^r}}}} \\
\varphi_{-1,H_n} & \longrightarrow & \varphi^{\mathrm{ord}}_{-1,H_n} \cong \varphi_{-1,H_{n_0}} \underset{\eta}{\times} \varphi^{\mathrm{ord}}_{-1,H^{\mathrm{ord}}_{p^r}} \\
\Big\downarrow {\scriptstyle \mathrm{mod}\ H_{n,\mathrm{G}^{\mathrm{ess}}_{h,\mathsf{Z}_n}}} & & \Big\downarrow {\scriptstyle \mathrm{mod}\ H_{n_0,\mathrm{G}^{\mathrm{ess}}_{h,\mathsf{Z}_{n_0}}} \times H^{\mathrm{ord}}_{p^r,\mathrm{G}^{\mathrm{ess,ord}}_{h,\mathsf{Z}_{p^r},\mathsf{D}_{p^r}}}} \\
\delta_{H_n} & \underset{\sim}{\longrightarrow} & \delta^{\mathrm{ord}}_{H_n} \cong \delta_{H_{n_0}} \underset{\eta}{\times} \delta^{\mathrm{ord}}_{H^{\mathrm{ord}}_{p^r}} \\
\Big\downarrow \wr & & \Big\downarrow \wr \\
\mathsf{Z}_{H_n} & = & \mathsf{Z}_{H_n} \cong \mathsf{Z}_{H_{n_0}} \underset{\eta}{\times} \mathsf{Z}_{H^{\mathrm{ord}}_{p^r}} \\
\Big\downarrow & & \Big\downarrow \\
\eta & = & \eta
\end{array}
$$

In this diagram, the objects at the right-hand sides form an H_n-orbit of étale-locally-defined naive ordinary level-n structure data (see Definition 4.1.5.22), and all horizontal morphisms over Z_{H_n} are torsors of the expected constant finite groups:

(1) The induced morphism

$$\tau_{H_n} \to (c_{H_n}, c^{\vee}_{H_n}) \underset{(c^{\mathrm{ord}}_{H_n}, c^{\vee,\mathrm{ord}}_{H_n})}{\times} \tau^{\mathrm{ord}}_{H_n}$$

is a torsor under

$$\begin{aligned} \mathrm{U}^{\mathrm{ess}}_{2,\mathsf{Z}_{p^r},\mathsf{D}_{p^r}} &\cong \ker(\mathrm{U}^{\mathrm{ess}}_{2,\mathsf{Z}_{p^r},\mathsf{D}_{p^r}} \twoheadrightarrow \mathrm{U}^{\mathrm{ess,ord}}_{2,\mathsf{Z}_{p^r},\mathsf{D}_{p^r}}) \\ &\cong \ker(H_{p^r,\mathrm{U}^{\mathrm{ess}}_{2,\mathsf{Z}_{p^r}}} \twoheadrightarrow H^{\mathrm{ord}}_{p^r,\mathrm{U}^{\mathrm{ess,ord}}_{2,\mathsf{Z}_{p^r},\mathsf{D}_{p^r}}}). \end{aligned}$$

(2) The induced morphism

$$(c_{H_n}, c^{\vee}_{H_n}) \to (\varphi^{\sim}_{-2,H_n}, \varphi^{\sim}_{0,H_n}) \underset{(\varphi^{\mathrm{ord}}_{-2,H_n}, \varphi^{\mathrm{ord}}_{0,H_n})}{\times} (c^{\mathrm{ord}}_{H_n}, c^{\vee,\mathrm{ord}}_{H_n})$$

is a torsor under

$$\ker(\mathrm{U}^{\mathrm{ess}}_{1,\mathsf{Z}_{p^r},\mathsf{D}_{p^r}} \twoheadrightarrow \mathrm{U}^{\mathrm{ess,ord}}_{1,\mathsf{Z}_{p^r},\mathsf{D}_{p^r}}) \cong \ker(H_{p^r,\mathrm{U}^{\mathrm{ess}}_{1,\mathsf{Z}_{p^r}}} \twoheadrightarrow H^{\mathrm{ord}}_{p^r,\mathrm{U}^{\mathrm{ess,ord}}_{1,\mathsf{Z}_{p^r},\mathsf{D}_{p^r}}}).$$

(3) The induced morphism

$$(\varphi^{\sim}_{-2,H_n}, \varphi^{\sim}_{0,H_n}) \to \varphi_{-1,H_n} \underset{\varphi^{\mathrm{ord}}_{-1,H_n}}{\times} (\varphi^{\mathrm{ord}}_{-2,H_n}, \varphi^{\mathrm{ord}}_{0,H_n})$$

is an isomorphism, because $H_{p^r,\mathrm{G}^{\mathrm{ess}}_{l,\mathsf{Z}_{p^r}}} \cong H^{\mathrm{ord}}_{p^r,\mathrm{G}^{\mathrm{ess,ord}}_{l,\mathsf{Z}_{p^r},\mathsf{D}_{p^r}}}$ *under the canonical isomorphism* $\mathrm{G}^{\mathrm{ess}}_{l,\mathsf{Z}_{p^r},\mathsf{D}_{p^r}} \cong \mathrm{G}^{\mathrm{ess,ord}}_{l,\mathsf{Z}_{p^r},\mathsf{D}_{p^r}}$.

(4) The morphism

$$\varphi_{-1,H_n} \to \varphi^{\mathrm{ord}}_{-1,H_n}$$

is a torsor under

$$\ker(\mathrm{P}^{\mathrm{ess}}_{h,\mathsf{Z}_{p^r},\mathsf{D}_{p^r}} \twoheadrightarrow \mathrm{G}^{\mathrm{ess,ord}}_{h,\mathsf{Z}_{p^r},\mathsf{D}_{p^r}}) \cong \ker(H_{p^r,\mathrm{G}^{\mathrm{ess}}_{h,\mathsf{Z}_{p^r}}} \twoheadrightarrow H^{\mathrm{ord}}_{p^r,\mathrm{G}^{\mathrm{ess,ord}}_{h,\mathsf{Z}_{p^r},\mathsf{D}_{p^r}}}).$$

In particular, all horizontal morphisms in the above commutative diagram and all the induced morphisms are étale and surjective.

Proof. Since η is a point over $\mathsf{S}_{0,p^{r_\nu}} = \mathrm{Spec}(F_0[\zeta_{p^{r_\nu}}])$, the assignment to the data on the right-hand sides determines the data on the left-hand side. (However, this does not imply that all data as in (4.1.5.21) in Proposition 4.1.5.20 comes from such an assignment.) The statements on the induced morphisms follow from their definitions as forgetful functors. The isomorphisms between various kernels follow from the very definitions of the groups (see Definition 4.1.5.12). □

Theorem 4.1.6.2. *With notation and assumptions as in the first two paragraphs of this subsection, let* $\mathrm{DD}_{\mathrm{PEL},\mathsf{M}^{\mathrm{ord}}_{\mathcal{H}}}(R,I)$ *be the full subcategory of* $\mathrm{DD}_{\mathrm{PEL},\mathsf{M}_{\mathcal{H}}}(R,I)$ *formed by objects each of whose underlying* $\mathsf{Z}_{\mathcal{H}}$ *is* **ordinary**. *(That is, each such* $\mathsf{Z}_{\mathcal{H}}$ *is compatible with the filtration* D *in the sense that, over an étale extension of* η *over which* $\mathsf{Z}_{\mathcal{H}}$ *becomes split and becomes part of a representative of a cusp label, it is compatible with* D *as in Definition* 3.2.3.1.*) Let* $\mathrm{DEG}_{\mathrm{PEL},\mathsf{M}^{\mathrm{ord}}_{\mathcal{H}}}(R,I)$ *be the essential image of* $\mathrm{DD}_{\mathrm{PEL},\mathsf{M}^{\mathrm{ord}}_{\mathcal{H}}}(R,I)$ *under the equivalence of categories*

$$\mathrm{M}_{\mathrm{PEL},\mathsf{M}_{\mathcal{H}}(R,I)} : \mathrm{DD}_{\mathrm{PEL},\mathsf{M}_{\mathcal{H}}}(R,I) \to \mathrm{DEG}_{\mathrm{PEL},\mathsf{M}_{\mathcal{H}}}(R,I) \tag{4.1.6.3}$$

in [62, Thm. 5.3.1.19], which induces an equivalence of categories

$$\mathrm{M}_{\mathrm{PEL},\mathsf{M}^{\mathrm{ord}}_{\mathcal{H}}(R,I)} : \mathrm{DD}_{\mathrm{PEL},\mathsf{M}^{\mathrm{ord}}_{\mathcal{H}}}(R,I) \to \mathrm{DEG}_{\mathrm{PEL},\mathsf{M}^{\mathrm{ord}}_{\mathcal{H}}}(R,I). \tag{4.1.6.4}$$

Then there is a commutative diagram

$$\begin{array}{ccccc}
\mathrm{DD}_{\mathrm{PEL},\mathsf{M}_{\mathcal{H}}}(R,I) & \longleftarrow & \mathrm{DD}_{\mathrm{PEL},\mathsf{M}^{\mathrm{ord}}_{\mathcal{H}}}(R,I) & \longrightarrow & \mathrm{DD}_{\mathrm{PEL},\ddot{\mathsf{M}}^{\mathrm{ord}}_{\mathcal{H}}}(R,I) \\
{\scriptstyle \mathrm{M}_{\mathrm{PEL},\mathsf{M}_{\mathcal{H}}(R,I)}}\downarrow\wr & & {\scriptstyle \mathrm{M}_{\mathrm{PEL},\mathsf{M}^{\mathrm{ord}}_{\mathcal{H}}(R,I)}}\downarrow\wr & & \wr\downarrow{\scriptstyle \mathrm{M}_{\mathrm{PEL},\ddot{\mathsf{M}}^{\mathrm{ord}}_{\mathcal{H}}}(R,I)} \\
\mathrm{DEG}_{\mathrm{PEL},\mathsf{M}_{\mathcal{H}}}(R,I) & \leftarrow & \mathrm{DEG}_{\mathrm{PEL},\mathsf{M}^{\mathrm{ord}}_{\mathcal{H}}}(R,I) & \rightarrow & \mathrm{DEG}_{\mathrm{PEL},\ddot{\mathsf{M}}^{\mathrm{ord}}_{\mathcal{H}}}(R,I)
\end{array}$$

where $\mathrm{M}_{\mathrm{PEL},\ddot{\mathsf{M}}^{\mathrm{ord}}_{\mathcal{H}}}(R,I)$ *is the equivalence of categories as in Theorem* 4.1.5.27. *All horizontal morphisms in this commutative diagram are fully faithful.*

Proof. This follows from Proposition 4.1.6.1 by étale descent (which allows us to reduce to the case where $\underline{X}$ and $\underline{Y}$ are constant), and from the proof of Proposition 4.1.4.42 (which shows that the pairing conditions for $\mathrm{DD}_{\mathrm{PEL},\mathsf{M}_{\mathcal{H}}}(R,I)$ and for $\mathrm{DD}_{\mathrm{PEL},\ddot{\mathsf{M}}^{\mathrm{ord}}_{\mathcal{H}}}(R,I)$ are compatible with each other at principal levels). □

4.2 Boundary Charts of Ordinary Loci

In this section, let us continue with the settings in Section 4.1.6. Let $r_{\mathcal{H}}$ be as in Definition 3.4.2.1. Let us fix a representative $(\mathsf{Z}_{\mathcal{H}}, \Phi_{\mathcal{H}}, \delta_{\mathcal{H}})$ of an *ordinary* cusp label at level $\mathcal{H}$ for the PEL-type $\mathcal{O}$-lattice $(L, \langle \cdot, \cdot \rangle, h_0)$ (see Definition 3.2.3.8). Let us fix the choice of a representative $(\mathsf{Z}_n, \Phi_n, \delta_n)$ in the H_n-orbit $(\mathsf{Z}_{H_n}, \Phi_{H_n}, \delta_{H_n})$.

4.2.1 *Constructions with Level Structures but without Positivity Conditions*

The moduli problem $\dddot{\mathsf{M}}_n^{\mathrm{ord}}$ has a boundary version $\dddot{\mathsf{M}}_n^{\mathrm{ord},\mathsf{Z}_n}$, with tautological object

$$(B, \lambda_B, i_B, \varphi_{-1,n_0}, \varphi_{-1,p^r}^{\mathrm{ord}})$$

giving the abelian parts of degenerations. Then we have isogenies

$$B \twoheadrightarrow B_{p^r}^{\mathrm{ord}} := B/\mathrm{image}(\varphi_{-1,p^r}^{\mathrm{ord},0}) = B/\varphi_{-1,p^r}^{\mathrm{ord},0}((\mathrm{Gr}^0_{\mathsf{D}_{-1,p^r}})^{\mathrm{mult}}_{\dddot{\mathsf{M}}_n^{\mathrm{ord},\mathsf{Z}_n}}) \qquad (4.2.1.1)$$

and

$$B^\vee \twoheadrightarrow B_{p^r}^{\vee,\mathrm{ord}} := B^\vee/\mathrm{image}(\varphi_{-1,p^r}^{\mathrm{ord},\#,0}) = B^\vee/\varphi_{-1,p^r}^{\mathrm{ord},\#,0}((\mathrm{Gr}^0_{\mathsf{D}^\#_{-1,p^r}})^{\mathrm{mult}}_{\dddot{\mathsf{M}}_n^{\mathrm{ord},\mathsf{Z}_n}}) \qquad (4.2.1.2)$$

as in (4.1.4.31) and (4.1.4.32), respectively, together with isogenies

$$B_{p^r}^{\mathrm{ord}} \twoheadrightarrow B_{p^r}^{\mathrm{ord}}/(\mathrm{Gr}^{-1}_{\mathsf{D}_{-1,p^r}})_{\dddot{\mathsf{M}}_n^{\mathrm{ord},\mathsf{Z}_n}} \cong B/B[p^r] \cong B \qquad (4.2.1.3)$$

and

$$B_{p^r}^{\vee,\mathrm{ord}} \twoheadrightarrow B_{p^r}^{\vee,\mathrm{ord}}/(\mathrm{Gr}^{-1}_{\mathsf{D}^\#_{-1,p^r}})_{\dddot{\mathsf{M}}_n^{\mathrm{ord},\mathsf{Z}_n}} \cong B^\vee/B^\vee[p^r] \cong B^\vee, \qquad (4.2.1.4)$$

as in (4.1.4.34) and (4.1.4.33), respectively.

Consider the canonical homomorphism

$$\underline{\mathrm{Hom}}_{\mathcal{O}}(\tfrac{1}{n}X, B_{p^r}^{\vee,\mathrm{ord}}) \to \underline{\mathrm{Hom}}_{\mathcal{O}}(X, B^\vee) \qquad (4.2.1.5)$$

defined by pre-composition with $X \overset{\sim}{\to} \frac{1}{n}X$ and by post-composition with

$$B_{p^r}^{\vee,\mathrm{ord}} \overset{(4.2.1.4)}{\twoheadrightarrow} B^\vee \overset{[n_0]}{\to} B^\vee. \qquad (4.2.1.6)$$

(We can compare these with the following: The canonical morphism $\underline{\mathrm{Hom}}_{\mathcal{O}}(\frac{1}{n}X, B^\vee) \to \underline{\mathrm{Hom}}_{\mathcal{O}}(X, B^\vee)$ induced by restriction to X can be defined alternatively by pre-composition with $X \overset{\sim}{\to} \frac{1}{n}X$ and by post-composition with $[n] : B^\vee \twoheadrightarrow B^\vee$.) Similarly, consider the canonical homomorphism

$$\underline{\mathrm{Hom}}_{\mathcal{O}}(\tfrac{1}{n}Y, B_{p^r}^{\mathrm{ord}}) \to \underline{\mathrm{Hom}}_{\mathcal{O}}(Y, B) \qquad (4.2.1.7)$$

defined by pre-composition with $Y \overset{\sim}{\to} \frac{1}{n}Y$ and by post-composition with

$$B_{p^r}^{\mathrm{ord}} \overset{(4.2.1.3)}{\twoheadrightarrow} B \overset{[n_0]}{\to} B. \qquad (4.2.1.8)$$

Consider also the canonical homomorphisms $\underline{\mathrm{Hom}}_{\mathcal{O}}(X, B^\vee) \to \underline{\mathrm{Hom}}_{\mathcal{O}}(Y, B^\vee)$ (resp. $\underline{\mathrm{Hom}}_{\mathcal{O}}(Y, B) \to \underline{\mathrm{Hom}}_{\mathcal{O}}(Y, B^\vee)$) defined by pre-composition with the morphism $\phi : Y \to X$ (resp. by post-composition with $\lambda_B : B \to B^\vee$). All these group functors defined by $\underline{\mathrm{Hom}}_{\mathcal{O}}(\,\cdot\,,\,\cdot\,)$ are relatively representable by Proposition 3.1.2.4, because the abelian schemes involved are all ordinary (see Definition 3.1.1.2).

Lemma 4.2.1.9. *The kernels of* (4.2.1.5) *and* (4.2.1.7) *are finite étale group schemes over* $\ddot{\mathsf{M}}_n^{\mathrm{ord},\mathsf{Z}_n}$. *The kernels of* $\underline{\mathrm{Hom}}_{\mathcal{O}}(X, B^\vee) \to \underline{\mathrm{Hom}}_{\mathcal{O}}(Y, B^\vee)$ *and* $\underline{\mathrm{Hom}}_{\mathcal{O}}(Y, B) \to \underline{\mathrm{Hom}}_{\mathcal{O}}(Y, B^\vee)$ *are finite flat group schemes of étale-multiplicative type over* $\ddot{\mathsf{M}}_n^{\mathrm{ord},\mathsf{Z}_n}$.

Proof. Since $\ker((4.2.1.4)) \cong (\mathrm{Gr}^{-1}_{\mathsf{D}^{\#}_{-1,p^r}})_{\ddot{\mathsf{M}}_n^{\mathrm{ord},\mathsf{Z}_n}}$ and $\ker([n_0] : B^\vee \to B^\vee) \cong B^\vee[n_0]$ are finite étale, $\ker((4.2.1.6))$ is also finite étale. Hence, the kernel of $\underline{\mathrm{Hom}}_{\mathcal{O}}(\frac{1}{n}X, B^{\vee,\mathrm{ord}}_{p^r}) \to \underline{\mathrm{Hom}}_{\mathcal{O}}(X, B^\vee)$ is finite étale because it is a finite flat subgroup scheme of $\underline{\mathrm{Hom}}_{\mathbb{Z}}(\frac{1}{n}X/X, \ker((4.2.1.6)))$. Similarly, since $\ker((4.2.1.3)) \cong (\mathrm{Gr}^{-1}_{\mathsf{D}_{-1,p^r}})_{\ddot{\mathsf{M}}_n^{\mathrm{ord},\mathsf{Z}_n}}$ and $\ker([n_0] : B \to B) \cong B[n_0]$ are finite étale, the kernel of $\underline{\mathrm{Hom}}_{\mathcal{O}}(\frac{1}{n}Y, B^{\mathrm{ord}}_{p^r}) \to \underline{\mathrm{Hom}}_{\mathcal{O}}(Y, B)$ is also finite étale.

Since B (resp. $B^\vee$) is ordinary, its (commutative) finite flat subgroup schemes are all of étale-multiplicative type. Hence, the kernel of $\underline{\mathrm{Hom}}_{\mathcal{O}}(X, B^\vee) \to \underline{\mathrm{Hom}}_{\mathcal{O}}(Y, B^\vee)$ (resp. $\underline{\mathrm{Hom}}_{\mathcal{O}}(Y, B) \to \underline{\mathrm{Hom}}_{\mathcal{O}}(Y, B^\vee)$) is a finite flat group scheme of étale-multiplicative type over $\ddot{\mathsf{M}}_n^{\mathrm{ord},\mathsf{Z}_n}$ because it is a finite flat subgroup scheme of $\underline{\mathrm{Hom}}_{\mathbb{Z}}(X/Y, B^\vee)$ (resp. $\underline{\mathrm{Hom}}_{\mathbb{Z}}(Y, \ker(\lambda_B))$). □

Let $\ddot{C}^{\mathrm{ord}}_{\Phi_n}$ be the (relative) proper flat group scheme over $\ddot{\mathsf{M}}_n^{\mathrm{ord},\mathsf{Z}_n}$ representing the fiber product

$$\underline{\mathrm{Hom}}_{\mathcal{O}}(\tfrac{1}{n}X, B^{\vee,\mathrm{ord}}_{p^r}) \underset{\underline{\mathrm{Hom}}_{\mathcal{O}}(Y,B^\vee)}{\times} \underline{\mathrm{Hom}}_{\mathcal{O}}(\tfrac{1}{n}Y, B^{\mathrm{ord}}_{p^r}). \qquad (4.2.1.10)$$

Then $\ddot{C}^{\mathrm{ord}}_{\Phi_n}$ carries a tautological pair

$$(c_n^{\mathrm{ord}} : \tfrac{1}{n}X \to B^{\vee,\mathrm{ord}}_{p^r}, c_n^{\vee,\mathrm{ord}} : \tfrac{1}{n}Y \to B^{\mathrm{ord}}_{p^r})$$

of liftings of $(c : X \to B^\vee, c^\vee : Y \to B)$ satisfying the compatibility $\lambda_B c^\vee = c\phi$, which is equivalent to two tautological pairs

$$(c_{n_0} : \tfrac{1}{n_0}X \to B^\vee, c^\vee_{n_0} : \tfrac{1}{n_0}Y \to B)$$

and

$$(c_{p^r}^{\mathrm{ord}} : \tfrac{1}{p^r}X \to B^{\vee,\mathrm{ord}}_{p^r}, c_{p^r}^{\vee,\mathrm{ord}} : \tfrac{1}{p^r}Y \to B^{\mathrm{ord}}_{p^r})$$

of liftings of $(c : X \to B^\vee, c^\vee : Y \to B)$.

Let us extend $\phi : Y \to X$ naturally to $\phi_n : \frac{1}{n}Y \to \frac{1}{n}X$.

Proposition 4.2.1.11.

(1) *Let* $\underline{\mathrm{Hom}}_{\mathcal{O}}(\frac{1}{n}X, B^{\mathrm{ord}}_{p^r})^\circ$ *be the (reduced) fiberwise geometric identity component of* $\underline{\mathrm{Hom}}_{\mathcal{O}}(\frac{1}{n}X, B^{\mathrm{ord}}_{p^r})$ *(see (4) of Proposition 3.1.2.4). Then the canonical homomorphism*

$$\underline{\mathrm{Hom}}_{\mathcal{O}}(\tfrac{1}{n}X, B^{\mathrm{ord}}_{p^r})^\circ \to \underline{\mathrm{Hom}}_{\mathcal{O}}(\tfrac{1}{n}X, B^{\vee,\mathrm{ord}}_{p^r}) \underset{\underline{\mathrm{Hom}}_{\mathcal{O}}(Y,B^\vee)}{\times} \underline{\mathrm{Hom}}_{\mathcal{O}}(\tfrac{1}{n}Y, B^{\mathrm{ord}}_{p^r})$$

over $\ddot{\mathsf{M}}_n^{\mathrm{ord}}$ *has kernel the finite flat group scheme*

$$\underline{\mathrm{Hom}}_{\mathcal{O}}(\tfrac{1}{n}X/\phi_n(\tfrac{1}{n}Y), \ker(\lambda_{B,p^r}^{\mathrm{ord}})) \cap \underline{\mathrm{Hom}}_{\mathcal{O}}(\tfrac{1}{n}X, B_{p^r}^{\mathrm{ord}})^\circ$$

(*see* (4.1.4.39)) *of étale-multiplicative type* (*see Definition* 3.1.1.1) *and schematic image an abelian subscheme* $\ddot{C}_{\Phi_n}^{\mathrm{ord},\circ}$ *of* $\ddot{C}_{\Phi_n}^{\mathrm{ord}}$. (*See Lemma* 3.1.2.2 *and Definition* 3.1.2.3.)

(2) There exists an integer $m \geq 1$ *such that multiplication by* m *maps* $\ddot{C}_{\Phi_n}^{\mathrm{ord}}$ *scheme-theoretically to a subscheme of* $\ddot{C}_{\Phi_n}^{\mathrm{ord},\circ}$, *so that the group scheme* $\underline{\pi}_0(\ddot{C}_{\Phi_n}^{\mathrm{ord}}/\ddot{\mathsf{M}}_n^{\mathrm{ord}})$ *of fiberwise connected components of* $\ddot{C}_{\Phi_n}^{\mathrm{ord}}$ *over* $\ddot{\mathsf{M}}_n^{\mathrm{ord}}$ *is defined and is of étale-multiplicative type.* (*See Lemma* 3.1.2.2 *and Definition* 3.1.2.3.) *Moreover, the rank of* $\underline{\pi}_0(\ddot{C}_{\Phi_n}^{\mathrm{ord}}/\ddot{\mathsf{M}}_n^{\mathrm{ord}})$ *has no prime factors other than those of* Disc, n, $[X : \phi(Y)]$, *and the rank of* $\ker(\lambda_B)$ (*or rather* $\ker(\lambda_{B,p^r}^{\mathrm{ord}})$). (*This implies that the rank of* $\underline{\pi}_0(\ddot{C}_{\Phi_n}^{\mathrm{ord}}/\ddot{\mathsf{M}}_n^{\mathrm{ord}})$ *does not contain prime factors other than those of* Disc, n *and* $[L^\# : L]$.)

Proof. The first claim of the lemma is clear, because the finite flat group scheme

$$\begin{aligned}&\underline{\mathrm{Hom}}_{\mathcal{O}}(\tfrac{1}{n}X/\phi_n(\tfrac{1}{n}Y), \ker(\lambda_{B,p^r}^{\mathrm{ord}}))\\ &= \underline{\mathrm{Hom}}_{\mathcal{O}}(\tfrac{1}{n}X, \ker(\lambda_{B,p^r}^{\mathrm{ord}})) \cap \underline{\mathrm{Hom}}_{\mathcal{O}}(\tfrac{1}{n}X/\phi_n(\tfrac{1}{n}Y), B_{p^r}^{\mathrm{ord}})\end{aligned}$$

is the kernel of

$$\underline{\mathrm{Hom}}_{\mathcal{O}}(\tfrac{1}{n}X, B_{p^r}^{\mathrm{ord}}) \to \underline{\mathrm{Hom}}_{\mathcal{O}}(\tfrac{1}{n}X, B_{p^r}^{\vee,\mathrm{ord}}) \underset{\underline{\mathrm{Hom}}_{\mathcal{O}}(Y,B^\vee)}{\times} \underline{\mathrm{Hom}}_{\mathcal{O}}(\tfrac{1}{n}Y, B_{p^r}^{\mathrm{ord}}),$$

and because B and hence $B_{p^r}^{\mathrm{ord}}$ are ordinary (by Lemma 3.1.1.5).

For the second claim, let $\underline{\mathrm{Hom}}_{\mathcal{O}}(\tfrac{1}{n}X, B_{p^r}^{\vee,\mathrm{ord}})^\circ$, $\underline{\mathrm{Hom}}_{\mathcal{O}}(\tfrac{1}{n}Y, B_{p^r}^{\mathrm{ord}})^\circ$, and $\underline{\mathrm{Hom}}_{\mathcal{O}}(Y, B^\vee)^\circ$ denote respectively the fiberwise geometric identity components of $\underline{\mathrm{Hom}}_{\mathcal{O}}(\tfrac{1}{n}X, B_{p^r}^{\vee,\mathrm{ord}})$, $\underline{\mathrm{Hom}}_{\mathcal{O}}(\tfrac{1}{n}Y, B_{p^r}^{\mathrm{ord}})$, and $\underline{\mathrm{Hom}}_{\mathcal{O}}(Y, B^\vee)^\circ$, and let $\ddot{C}_{\Phi_n}^{\mathrm{ord},\circ\circ\circ}$ denote the proper smooth group scheme representing the fiber product

$$\underline{\mathrm{Hom}}_{\mathcal{O}}(\tfrac{1}{n}X, B_{p^r}^{\vee,\mathrm{ord}})^\circ \underset{\underline{\mathrm{Hom}}_{\mathcal{O}}(Y,B^\vee)^\circ}{\times} \underline{\mathrm{Hom}}_{\mathcal{O}}(\tfrac{1}{n}Y, B_{p^r}^{\mathrm{ord}})^\circ.$$

By (4) of Proposition 3.1.2.4, the group schemes $\underline{\pi}_0(\ddot{C}_{\Phi_n}^{\mathrm{ord}}/\ddot{\mathsf{M}}_n^{\mathrm{ord}})$ and $\underline{\pi}_0(\ddot{C}_{\Phi_n}^{\mathrm{ord},\circ\circ\circ}/\ddot{\mathsf{M}}_n^{\mathrm{ord}})$ are defined, and their ranks differ up to multiplication by numbers having only prime factors of those of Disc. Therefore, it suffices to show that the rank of $\underline{\pi}_0(\ddot{C}_{\Phi_n}^{\mathrm{ord},\circ\circ\circ}/\ddot{\mathsf{M}}_n^{\mathrm{ord}})$ has no prime factors other than those of n, $[X : \phi(Y)]$, and the rank of $\ker(\lambda_B)$ (or rather $\ker(\lambda_{B,p^r}^{\mathrm{ord}})$).

The kernel K_n of the canonical homomorphism $\ddot{C}_{\Phi_n}^{\mathrm{ord},\circ\circ\circ} \to \underline{\mathrm{Hom}}_{\mathcal{O}}(Y, B^\vee)^\circ$ is given by a fiber product $K_{n,1} \underset{\mathsf{M}_n^{\mathsf{Z}_n}}{\times} K_{n,2}$, where

$$K_{n,1} := \underline{\mathrm{Hom}}_{\mathcal{O}}(\tfrac{1}{n}X/\phi(Y), B_{p^r}^{\vee,\mathrm{ord}}) \cap \underline{\mathrm{Hom}}_{\mathcal{O}}(\tfrac{1}{n}X, B_{p^r}^{\vee,\mathrm{ord}})^\circ$$

and

$$K_{n,2} := \underline{\mathrm{Hom}}_{\mathcal{O}}(\tfrac{1}{n}Y, \ker(\lambda_{B,p^r}^{\mathrm{ord}})) \cap \underline{\mathrm{Hom}}_{\mathcal{O}}(\tfrac{1}{n}Y, B_{p^r}^{\mathrm{ord}})^\circ.$$

Since $\underline{\mathrm{Hom}}_{\mathcal{O}}(Y, B^\vee)^\circ$ is an abelian scheme, the group $\underline{\pi}_0(\dddot{C}^{\mathrm{ord},\circ\circ\circ}_{\Phi_n}/\dddot{\mathsf{M}}^{\mathrm{ord}}_n)$ can be identified with a quotient of K_n. Since the rank of K_n is the product of the ranks of $K_{n,1}$ and of $K_{n,2}$, it has no prime factors other than those of n, $X/\phi(Y)$, and the rank of $\ker(\lambda^{\mathrm{ord}}_{B,p^r})$, as desired. □

Let us consider the finitely generated commutative group (*cf.* [62, (6.2.3.5)])

$$\dddot{\mathbf{S}}^{\mathrm{ord}}_{\Phi_n} := \dddot{\mathbf{S}}_{\Phi_{n_0}} := ((\tfrac{1}{n_0}Y)\underset{\mathbb{Z}}{\otimes} X)/\begin{pmatrix} y\otimes\phi(y') - y'\otimes\phi(y) \\ (b\tfrac{1}{n_0}y)\otimes\chi - (\tfrac{1}{n_0}y)\otimes(b^\star\chi)\end{pmatrix}_{\substack{y,y'\in Y,\\ \chi\in X, b\in\mathcal{O}}}. \tag{4.2.1.12}$$

As in [62, Sec. 6.2.2–6.2.3], the formal properties of the pullbacks of the Poincaré biextension (as in [62, Lem. 6.2.2.5]) allow us to assign to each

$$\ell = \sum_{1\leq i\leq k} [(\tfrac{1}{n_0}y_i)\otimes\chi_i] \in \dddot{\mathbf{S}}^{\mathrm{ord}}_{\Phi_n}$$

a well-defined rigidified invertible sheaf

$$\Psi^{\mathrm{ord}}_n(\ell) := \underset{\mathscr{O}_{\dddot{C}^{\mathrm{ord}}_{\Phi_n}},1\leq i\leq k}{\otimes} (c^\vee_{n_0}(\tfrac{1}{n_0}y_i), c(\chi_i))^*\mathcal{P}_B \tag{4.2.1.13}$$

over $\dddot{C}^{\mathrm{ord}}_{\Phi_n}$, together with canonical isomorphisms

$$\Delta^{\mathrm{ord},*}_{n,\ell,\ell'} : \Psi^{\mathrm{ord}}_n(\ell) \underset{\mathscr{O}_{\dddot{C}^{\mathrm{ord}}_{\Phi_n}}}{\otimes} \Psi^{\mathrm{ord}}_n(\ell') \overset{\sim}{\to} \Psi^{\mathrm{ord}}_n(\ell+\ell')$$

for all $\ell, \ell' \in \dddot{\mathbf{S}}^{\mathrm{ord}}_{\Phi_n}$, satisfying the necessary compatibilities with each other making $\underset{\ell\in\dddot{\mathbf{S}}^{\mathrm{ord}}_{\Phi_n}}{\oplus} \Psi^{\mathrm{ord}}_n(\ell)$ an $\mathscr{O}_{\dddot{C}^{\mathrm{ord}}_{\Phi_n}}$-algebra, so that we can define

$$\dddot{\Xi}^{\mathrm{ord}}_{\Phi_n} := \underline{\mathrm{Spec}}_{\mathscr{O}_{\dddot{C}^{\mathrm{ord}}_{\Phi_n}}}\Big(\underset{\ell\in\dddot{\mathbf{S}}^{\mathrm{ord}}_{\Phi_n}}{\oplus} \Psi^{\mathrm{ord}}_n(\ell)\Big).$$

If we denote by $\dddot{E}^{\mathrm{ord}}_{\Phi_n} = \underline{\mathrm{Hom}}_{\mathbb{Z}}(\dddot{\mathbf{S}}^{\mathrm{ord}}_{\Phi_n}, \mathbf{G}_{\mathrm{m}})$ the group of multiplicative type of finite type with character group $\dddot{\mathbf{S}}^{\mathrm{ord}}_{\Phi_n}$ over $\mathrm{Spec}(\mathbb{Z})$, then $\dddot{\Xi}^{\mathrm{ord}}_{\Phi_n}$ is an $\dddot{E}^{\mathrm{ord}}_{\Phi_n}$-torsor, and we have tautological trivializations

$$\tau^{\mathrm{ord}}_n = \tau_{n_0} : \mathbf{1}_{(\frac{1}{n_0}Y)\times X} \overset{\sim}{\to} (c^\vee_{n_0}\times c)^*\mathcal{P}^{\otimes\,-1}_B$$

over $\dddot{\Xi}^{\mathrm{ord}}_{\Phi_n}$, which corresponds to a tautological homomorphism

$$\iota^{\mathrm{ord}}_n = \iota_{n_0} : \tfrac{1}{n_0}Y \to G^\natural.$$

Let $\tau : \mathbf{1}_{Y\times X} \overset{\sim}{\to} (c^\vee\times c)^*\mathcal{P}^{\otimes\,-1}_B$ be the restriction of τ^{ord}_n to $\mathbf{1}_{Y\times X}$, which corresponds to a tautological homomorphism $\iota : Y \to G^\natural$.

Let $\dddot{\mathbf{S}}^{\mathrm{ord}}_{\Phi_n,\mathrm{tor}}$ denote the torsion subgroup of $\dddot{\mathbf{S}}^{\mathrm{ord}}_{\Phi_n}$, and let $\dddot{\mathbf{S}}^{\mathrm{ord}}_{\Phi_n,\mathrm{free}}$ denote the quotient of $\dddot{\mathbf{S}}^{\mathrm{ord}}_{\Phi_n}$ by $\dddot{\mathbf{S}}^{\mathrm{ord}}_{\Phi_n,\mathrm{tor}}$, namely the free commutative quotient group of $\dddot{\mathbf{S}}^{\mathrm{ord}}_{\Phi_n}$. Let $\dddot{E}^{\mathrm{ord}}_{\Phi_n,\mathrm{tor}} := \underline{\mathrm{Hom}}_{\mathbb{Z}}(\dddot{\mathbf{S}}^{\mathrm{ord}}_{\Phi_n,\mathrm{tor}}, \mathbf{G}_{\mathrm{m}})$ (resp. $\dddot{E}^{\mathrm{ord}}_{\Phi_n,\mathrm{free}} := \underline{\mathrm{Hom}}_{\mathbb{Z}}(\dddot{\mathbf{S}}^{\mathrm{ord}}_{\Phi_n,\mathrm{free}}, \mathbf{G}_{\mathrm{m}})$) be the group of multiplicative type of finite type with character group $\ddot{\mathbf{S}}^{\mathrm{ord}}_{\Phi_n,\mathrm{tor}}$ (resp. $\ddot{\mathbf{S}}^{\mathrm{ord}}_{\Phi_n,\mathrm{free}}$) over $\mathrm{Spec}(\mathbb{Z})$. Then the exact sequence

$$0 \to \ddot{\mathbf{S}}^{\mathrm{ord}}_{\Phi_n,\mathrm{tor}} \to \ddot{\mathbf{S}}^{\mathrm{ord}}_{\Phi_n} \to \ddot{\mathbf{S}}^{\mathrm{ord}}_{\Phi_n,\mathrm{free}} \to 0$$

induces an exact sequence

$$0 \to \ddot{E}^{\mathrm{ord}}_{\Phi_n,\mathrm{free}} \to \ddot{E}^{\mathrm{ord}}_{\Phi_n} \to \ddot{E}^{\mathrm{ord}}_{\Phi_n,\mathrm{tor}} \to 0$$

in the reversed direction. Since $\ddot{\mathbf{S}}^{\mathrm{ord}}_{\Phi_n,\mathrm{free}}$ is a finitely generated free commutative group, $\ddot{E}^{\mathrm{ord}}_{\Phi_n,\mathrm{free}}$ is by definition a torus (*cf.* [62, Def. 3.1.1.5]).

As in Definition 4.1.5.9, the choice of Z_n in the H_n-orbit Z_{H_n} determines the groups

$$H_{n,\mathrm{U}^{\mathrm{ess}}_{2,\mathsf{Z}_n}} \subset H_{n,\mathrm{U}^{\mathrm{ess}}_{\mathsf{Z}_n}} \subset H_{n,\mathrm{Z}^{\mathrm{ess}}_{\mathsf{Z}_n}} \subset H_{n,\mathrm{P}^{\mathrm{ess}}_{\mathsf{Z}_n}} \subset H_n,$$

with short exact sequences

$$1 \to H_{n,\mathrm{Z}^{\mathrm{ess}}_{\mathsf{Z}_n}} \to H_{n,\mathrm{P}^{\mathrm{ess}}_{\mathsf{Z}_n}} \to H_{n,\mathrm{G}^{\mathrm{ess}}_{h,\mathsf{Z}_n}} \to 1,$$
$$1 \to H_{n,\mathrm{U}^{\mathrm{ess}}_{\mathsf{Z}_n}} \to H_{n,\mathrm{Z}^{\mathrm{ess}}_{\mathsf{Z}_n}} \to H_{n,\mathrm{G}^{\mathrm{ess}}_{l,\mathsf{Z}_n}} \to 1,$$
$$1 \to H_{n,\mathrm{U}^{\mathrm{ess}}_{2,\mathsf{Z}_n}} \to H_{n,\mathrm{U}^{\mathrm{ess}}_{\mathsf{Z}_n}} \to H_{n,\mathrm{U}^{\mathrm{ess}}_{1,\mathsf{Z}_n}} \to 1,$$

together with similar subgroups or quotients of subgroups when n is replaced with n_0 or p^r. Note that the quotient $H_{n,\mathrm{P}^{\mathrm{ess}}_{\mathsf{Z}_n}}\backslash H_n$ describes elements in the orbit Z_{H_n}. The fiber of $\Phi_{H_n} \to \mathsf{Z}_{H_n}$ at Z_n is naturally an orbit under the image $H'_{n,\mathrm{G}^{\mathrm{ess}}_{l,\mathsf{Z}_n}}$ of $H_{n,\mathrm{P}^{\mathrm{ess}}_{\mathsf{Z}_n}}$ in $\mathrm{G}^{\mathrm{ess}}_{l,\mathsf{Z}_n}$. By viewing the semidirect product $\mathrm{G}^{\mathrm{ess}}_{h,\mathsf{Z}_n} \ltimes \mathrm{U}^{\mathrm{ess}}_{\mathsf{Z}_n}$ as a subgroup of $\mathrm{G}^{\mathrm{ess}}(\mathbb{Z}/n\mathbb{Z})$ using the splitting δ_n, and by viewing $\mathrm{G}^{\mathrm{ess}}_{h,\mathsf{Z}_n} \ltimes \mathrm{U}^{\mathrm{ess}}_{1,\mathsf{Z}_n}$ as its quotient by $\mathrm{U}^{\mathrm{ess}}_{2,\mathsf{Z}_n}$, we can define as in Definition 4.1.5.9 the groups $H_{n,\mathrm{G}^{\mathrm{ess}}_{h,\mathsf{Z}_n}\ltimes\mathrm{U}^{\mathrm{ess}}_{\mathsf{Z}_n}}$ and $H_{n,\mathrm{G}^{\mathrm{ess}}_{h,\mathsf{Z}_n}\ltimes\mathrm{U}^{\mathrm{ess}}_{1,\mathsf{Z}_n}}$, fitting into short exact sequences

$$1 \to H_{n,\mathrm{G}^{\mathrm{ess}}_{h,\mathsf{Z}_n}\ltimes\mathrm{U}^{\mathrm{ess}}_{\mathsf{Z}_n}} \to H_{n,\mathrm{P}^{\mathrm{ess}}_{\mathsf{Z}_n}} \to H'_{n,\mathrm{G}^{\mathrm{ess}}_{l,\mathsf{Z}_n}} \to 1$$

and

$$1 \to H_{n,\mathrm{U}^{\mathrm{ess}}_{2,\mathsf{Z}_n}} \to H_{n,\mathrm{G}^{\mathrm{ess}}_{h,\mathsf{Z}_n}\ltimes\mathrm{U}^{\mathrm{ess}}_{\mathsf{Z}_n}} \to H_{n,\mathrm{G}^{\mathrm{ess}}_{h,\mathsf{Z}_n}\ltimes\mathrm{U}^{\mathrm{ess}}_{1,\mathsf{Z}_n}} \to 1.$$

Let $H'_{n,\mathrm{G}^{\mathrm{ess}}_{h,\mathsf{Z}_n}}$ denote the canonical image of $H_{n,\mathrm{G}^{\mathrm{ess}}_{h,\mathsf{Z}_n}\ltimes\mathrm{U}^{\mathrm{ess}}_{1,\mathsf{Z}_n}}$ in $\mathrm{G}^{\mathrm{ess}}_{h,\mathsf{Z}_n}$, so that we have an exact sequence

$$1 \to H_{n,\mathrm{U}^{\mathrm{ess}}_{1,\mathsf{Z}_n}} \to H_{n,\mathrm{G}^{\mathrm{ess}}_{h,\mathsf{Z}_n}\ltimes\mathrm{U}^{\mathrm{ess}}_{1,\mathsf{Z}_n}} \to H'_{n,\mathrm{G}^{\mathrm{ess}}_{h,\mathsf{Z}_n}} \to 1.$$

Let us also define similar subgroups or quotients of subgroups when n is replaced with n_0 or p^r.

As in Definition 4.1.5.18, the compatibility of Z_n with D_{p^r} allows us to define subgroups

$$H^{\mathrm{ord}}_{p^r,\mathrm{U}^{\mathrm{ess,ord}}_{2,\mathsf{Z}_{p^r},\mathsf{D}_{p^r}}} \subset H^{\mathrm{ord}}_{p^r,\mathrm{U}^{\mathrm{ess,ord}}_{\mathsf{Z}_{p^r},\mathsf{D}_{p^r}}} \subset H^{\mathrm{ord}}_{p^r,\mathrm{Z}^{\mathrm{ess,ord}}_{\mathsf{Z}_{p^r},\mathsf{D}_{p^r}}} \subset H^{\mathrm{ord}}_{p^r,\mathrm{P}^{\mathrm{ess,ord}}_{\mathsf{Z}_{p^r},\mathsf{D}_{p^r}}} \subset H^{\mathrm{ord}}_{p^r},$$

with short exact sequences

$$1 \to H^{\mathrm{ord}}_{p^r,\mathrm{Z}^{\mathrm{ess,ord}}_{\mathsf{Z}_{p^r},\mathsf{D}_{p^r}}} \to H^{\mathrm{ord}}_{p^r,\mathrm{P}^{\mathrm{ess,ord}}_{\mathsf{Z}_{p^r},\mathsf{D}_{p^r}}} \to H^{\mathrm{ord}}_{p^r,\mathrm{G}^{\mathrm{ess,ord}}_{h,\mathsf{Z}_{p^r},\mathsf{D}_{p^r}}} \to 1,$$

$$1 \to H^{\mathrm{ord}}_{p^r,\mathrm{U}^{\mathrm{ess,ord}}_{\mathsf{Z}_{p^r},\mathsf{D}_{p^r}}} \to H^{\mathrm{ord}}_{p^r,\mathrm{Z}^{\mathrm{ess,ord}}_{\mathsf{Z}_{p^r},\mathsf{D}_{p^r}}} \to H^{\mathrm{ord}}_{p^r,\mathrm{G}^{\mathrm{ess,ord}}_{l,\mathsf{Z}_{p^r},\mathsf{D}_{p^r}}} \to 1,$$

$$1 \to H^{\mathrm{ord}}_{p^r,\mathrm{U}^{\mathrm{ess,ord}}_{2,\mathsf{Z}_{p^r},\mathsf{D}_{p^r}}} \to H^{\mathrm{ord}}_{p^r,\mathrm{U}^{\mathrm{ess,ord}}_{\mathsf{Z}_{p^r},\mathsf{D}_{p^r}}} \to H^{\mathrm{ord}}_{p^r,\mathrm{U}^{\mathrm{ess,ord}}_{1,\mathsf{Z}_{p^r},\mathsf{D}_{p^r}}} \to 1.$$

As in Remark 4.1.5.19, note that

$$H^{\mathrm{ord}}_{p^r,\mathrm{G}^{\mathrm{ess,ord}}_{l,\mathsf{Z}_{p^r},\mathsf{D}_{p^r}}} \cong H_{p^r,\mathrm{G}^{\mathrm{ess}}_{l,\mathsf{Z}_{p^r}}}.$$

Let $H^{\mathrm{ord},\prime}_{p^r,\mathrm{G}^{\mathrm{ess,ord}}_{l,\mathsf{Z}_{p^r},\mathsf{D}_{p^r}}}$ denote the image of $H^{\mathrm{ord}}_{p^r,\mathrm{P}^{\mathrm{ess,ord}}_{\mathsf{Z}_{p^r},\mathsf{D}_{p^r}}}$ under the canonical homomorphism $\mathrm{P}^{\mathrm{ess,ord}}_{\mathsf{Z}_{p^r},\mathsf{D}_{p^r}} \to \mathrm{G}^{\mathrm{ess,ord}}_{l,\mathsf{Z}_{p^r},\mathsf{D}_{p^r}}$. By definition, we have

$$H^{\mathrm{ord},\prime}_{p^r,\mathrm{G}^{\mathrm{ess,ord}}_{l,\mathsf{Z}_{p^r},\mathsf{D}_{p^r}}} \cong H'_{p^r,\mathrm{G}^{\mathrm{ess}}_{l,\mathsf{Z}_{p^r}}}$$

under the canonical isomorphism $\mathrm{G}^{\mathrm{ess,ord}}_{l,\mathsf{Z}_{p^r},\mathsf{D}_{p^r}} \cong \mathrm{G}^{\mathrm{ess}}_{l,\mathsf{Z}_{p^r}}$.

Let $\delta^{\mathrm{ord}}_{p^r}$ be induced by δ_{p^r} as in Section 4.1.6. Then we can view the semidirect product $\mathrm{G}^{\mathrm{ess,ord}}_{h,\mathsf{Z}_{p^r},\mathsf{D}_{p^r}} \ltimes \mathrm{U}^{\mathrm{ess,ord}}_{\mathsf{Z}_{p^r},\mathsf{D}_{p^r}}$ as a subgroup of $\mathrm{M}^{\mathrm{ess,ord}}_{\mathsf{Z}_{p^r},\mathsf{D}_{p^r}}$ using $\delta^{\mathrm{ord}}_{p^r}$, which coincides with the image of $\mathrm{P}^{\mathrm{ess}}_{\mathsf{Z}_{p^r},\mathsf{D}_{p^r}} \cap (\mathrm{G}^{\mathrm{ess}}_{h,\mathsf{Z}_n} \ltimes \mathrm{U}^{\mathrm{ess}}_{\mathsf{Z}_n})$ under the canonical homomorphism $\mathrm{P}^{\mathrm{ess}}_{\mathsf{D}_{p^r}} \to \mathrm{M}^{\mathrm{ess,ord}}_{\mathsf{Z}_{p^r},\mathsf{D}_{p^r}}$. Note that $\mathrm{G}^{\mathrm{ess,ord}}_{h,\mathsf{Z}_{p^r},\mathsf{D}_{p^r}} \ltimes \mathrm{U}^{\mathrm{ess,ord}}_{1,\mathsf{Z}_{p^r},\mathsf{D}_{p^r}}$ is the (isomorphic) quotient of $\mathrm{G}^{\mathrm{ess,ord}}_{h,\mathsf{Z}_{p^r},\mathsf{D}_{p^r}} \ltimes \mathrm{U}^{\mathrm{ess,ord}}_{\mathsf{Z}_{p^r},\mathsf{D}_{p^r}}$ by $\mathrm{U}^{\mathrm{ess,ord}}_{2,\mathsf{Z}_{p^r},\mathsf{D}_{p^r}} = \{\mathrm{Id}\}$. Hence, we can define as in Definition 4.1.5.18 the groups $H^{\mathrm{ord}}_{p^r,\mathrm{G}^{\mathrm{ess,ord}}_{h,\mathsf{Z}_{p^r},\mathsf{D}_{p^r}} \ltimes \mathrm{U}^{\mathrm{ess,ord}}_{\mathsf{Z}_{p^r},\mathsf{D}_{p^r}}}$ and $H^{\mathrm{ord}}_{p^r,\mathrm{G}^{\mathrm{ess,ord}}_{h,\mathsf{Z}_{p^r},\mathsf{D}_{p^r}} \ltimes \mathrm{U}^{1,\mathrm{ess,ord}}_{\mathsf{Z}_{p^r},\mathsf{D}_{p^r}}}$, fitting into short exact sequences

$$1 \to H^{\mathrm{ord}}_{p^r,\mathrm{G}^{\mathrm{ess,ord}}_{h,\mathsf{Z}_{p^r},\mathsf{D}_{p^r}} \ltimes \mathrm{U}^{\mathrm{ess,ord}}_{\mathsf{Z}_{p^r},\mathsf{D}_{p^r}}} \to H^{\mathrm{ord}}_{p^r,\mathrm{P}^{\mathrm{ess,ord}}_{\mathsf{Z}_{p^r},\mathsf{D}_{p^r}}} \to H^{\mathrm{ord},\prime}_{p^r,\mathrm{G}^{\mathrm{ess,ord}}_{l,\mathsf{Z}_{p^r},\mathsf{D}_{p^r}}} \to 1$$

and

$$1 \to H^{\mathrm{ord}}_{p^r,\mathrm{U}^{\mathrm{ess,ord}}_{2,\mathsf{Z}_{p^r},\mathsf{D}_{p^r}}} \to H^{\mathrm{ord}}_{p^r,\mathrm{G}^{\mathrm{ess,ord}}_{h,\mathsf{Z}_{p^r},\mathsf{D}_{p^r}} \ltimes \mathrm{U}^{\mathrm{ess,ord}}_{\mathsf{Z}_{p^r},\mathsf{D}_{p^r}}} \to H^{\mathrm{ord}}_{p^r,\mathrm{G}^{\mathrm{ess,ord}}_{h,\mathsf{Z}_{p^r},\mathsf{D}_{p^r}} \ltimes \mathrm{U}^{\mathrm{ess,ord}}_{1,\mathsf{Z}_{p^r},\mathsf{D}_{p^r}}} \to 1.$$

(Certainly, $H^{\mathrm{ord}}_{p^r,\mathrm{U}^{\mathrm{ess,ord}}_{2,\mathsf{Z}_{p^r},\mathsf{D}_{p^r}}} = \{\mathrm{Id}\}$, so that $H^{\mathrm{ord}}_{p^r,\mathrm{G}^{\mathrm{ess,ord}}_{h,\mathsf{Z}_{p^r},\mathsf{D}_{p^r}} \ltimes \mathrm{U}^{\mathrm{ess,ord}}_{\mathsf{Z}_{p^r},\mathsf{D}_{p^r}}}$ is isomorphic to $H^{\mathrm{ord}}_{p^r,\mathrm{G}^{\mathrm{ess,ord}}_{h,\mathsf{Z}_{p^r},\mathsf{D}_{p^r}} \ltimes \mathrm{U}^{\mathrm{ess,ord}}_{1,\mathsf{Z}_{p^r},\mathsf{D}_{p^r}}}$ via the last short exact sequence above.) Let $H^{\mathrm{ord},\prime}_{p^r,\mathrm{G}^{\mathrm{ess,ord}}_{h,\mathsf{Z}_{p^r},\mathsf{D}_{p^r}}}$ denote the canonical image of $H^{\mathrm{ord}}_{p^r,\mathrm{G}^{\mathrm{ess,ord}}_{h,\mathsf{Z}_{p^r},\mathsf{D}_{p^r}} \ltimes \mathrm{U}^{\mathrm{ess,ord}}_{1,\mathsf{Z}_{p^r},\mathsf{D}_{p^r}}}$ in $\mathrm{G}^{\mathrm{ess,ord}}_{h,\mathsf{Z}_{p^r},\mathsf{D}_{p^r}}$, so that we have an exact sequence

$$1 \to H^{\mathrm{ord}}_{p^r,\mathrm{U}^{\mathrm{ess,ord}}_{1,\mathsf{Z}_{p^r},\mathsf{D}_{p^r}}} \to H^{\mathrm{ord}}_{p^r,\mathrm{G}^{\mathrm{ess,ord}}_{h,\mathsf{Z}_{p^r},\mathsf{D}_{p^r}} \ltimes \mathrm{U}^{\mathrm{ess,ord}}_{1,\mathsf{Z}_{p^r},\mathsf{D}_{p^r}}} \to H^{\mathrm{ord},\prime}_{p^r,\mathrm{G}^{\mathrm{ess,ord}}_{h,\mathsf{Z}_{p^r},\mathsf{D}_{p^r}}} \to 1.$$

Then we define the following groups:

$$\begin{aligned}
H_n^{\mathrm{ord}} &:= H_{n_0} \times H_{p^r}^{\mathrm{ord}}, \\
H_{n,\mathrm{P}_{\mathsf{Z}_n}^{\mathrm{ess}}}^{\mathrm{ord}} &:= H_{n_0,\mathrm{P}_{\mathsf{Z}_{n_0}}^{\mathrm{ess}}} \times H_{p^r,\mathrm{P}_{\mathsf{Z}_{p^r},\mathsf{D}_{p^r}}^{\mathrm{ess,ord}}}^{\mathrm{ord}}, \\
H_{n,\mathrm{U}_{\mathsf{Z}_n}^{\mathrm{ess}}}^{\mathrm{ord}} &:= H_{n,\mathrm{U}_{\mathsf{Z}_n}^{\mathrm{ess}}} \times H_{p^r,\mathrm{U}_{\mathsf{Z}_{p^r},\mathsf{D}_{p^r}}^{\mathrm{ess,ord}}}^{\mathrm{ord}}, \\
H_{n,\mathrm{Z}_{\mathsf{Z}_n}^{\mathrm{ess}}}^{\mathrm{ord}} &:= H_{n_0,\mathrm{Z}_{\mathsf{Z}_{n_0}}^{\mathrm{ess}}} \times H_{p^r,\mathrm{Z}_{\mathsf{Z}_{p^r},\mathsf{D}_{p^r}}^{\mathrm{ess,ord}}}^{\mathrm{ord}}, \\
H_{n,\mathrm{G}_{h,\mathsf{Z}_n}^{\mathrm{ess}}}^{\mathrm{ord}} &:= H_{n_0,\mathrm{G}_{h,\mathsf{Z}_{n_0}}^{\mathrm{ess}}} \times H_{p^r,\mathrm{G}_{h,\mathsf{Z}_{p^r},\mathsf{D}_{p^r}}^{\mathrm{ess,ord}}}^{\mathrm{ord}}, \\
H_{n,\mathrm{G}_{h,\mathsf{Z}_n}^{\mathrm{ess}}}^{\mathrm{ord},\prime} &:= H'_{n_0,\mathrm{G}_{h,\mathsf{Z}_{n_0}}^{\mathrm{ess}}} \times H_{p^r,\mathrm{G}_{h,\mathsf{Z}_{p^r},\mathsf{D}_{p^r}}^{\mathrm{ess,ord}}}^{\mathrm{ord},\prime}, \\
H_{n,\mathrm{G}_{l,\mathsf{Z}_n}^{\mathrm{ess}}}^{\mathrm{ord}} &:= H_{n_0,\mathrm{G}_{l,\mathsf{Z}_{n_0}}^{\mathrm{ess}}} \times H_{p^r,\mathrm{G}_{l,\mathsf{Z}_{p^r},\mathsf{D}_{p^r}}^{\mathrm{ess,ord}}}^{\mathrm{ord}}, \\
H_{n,\mathrm{G}_{l,\mathsf{Z}_n}^{\mathrm{ess}}}^{\mathrm{ord},\prime} &:= H'_{n_0,\mathrm{G}_{l,\mathsf{Z}_{n_0}}^{\mathrm{ess}}} \times H_{p^r,\mathrm{G}_{l,\mathsf{Z}_{p^r},\mathsf{D}_{p^r}}^{\mathrm{ess,ord}}}^{\mathrm{ord},\prime}, \\
H_{n,\mathrm{U}_{1,\mathsf{Z}_n}^{\mathrm{ess}}}^{\mathrm{ord}} &:= H_{n_0,\mathrm{U}_{1,\mathsf{Z}_{n_0}}^{\mathrm{ess}}} \times H_{p^r,\mathrm{U}_{1,\mathsf{Z}_{p^r},\mathsf{D}_{p^r}}^{\mathrm{ess,ord}}}^{\mathrm{ord}}, \\
H_{n,\mathrm{U}_{2,\mathsf{Z}_n}^{\mathrm{ess}}}^{\mathrm{ord}} &:= H_{n_0,\mathrm{U}_{2,\mathsf{Z}_{n_0}}^{\mathrm{ess}}} \times H_{p^r,\mathrm{U}_{2,\mathsf{Z}_{p^r},\mathsf{D}_{p^r}}^{\mathrm{ess,ord}}}^{\mathrm{ord}}, \\
H_{n,\mathrm{G}_{h,\mathsf{Z}_n}^{\mathrm{ess}} \ltimes \mathrm{U}_{\mathsf{Z}_n}^{\mathrm{ess}}}^{\mathrm{ord}} &:= H_{n_0,\mathrm{G}_{h,\mathsf{Z}_{n_0}}^{\mathrm{ess}} \ltimes \mathrm{U}_{\mathsf{Z}_n}^{\mathrm{ess}}} \times H_{p^r,\mathrm{G}_{h,\mathsf{Z}_{p^r},\mathsf{D}_{p^r}}^{\mathrm{ess,ord}} \ltimes \mathrm{U}_{\mathsf{Z}_{p^r},\mathsf{D}_{p^r}}^{\mathrm{ess,ord}}}^{\mathrm{ord}}, \\
H_{n,\mathrm{G}_{h,\mathsf{Z}_n}^{\mathrm{ess}} \ltimes \mathrm{U}_{1,\mathsf{Z}_n}^{\mathrm{ess}}}^{\mathrm{ord}} &:= H_{n_0,\mathrm{G}_{h,\mathsf{Z}_{n_0}}^{\mathrm{ess}} \ltimes \mathrm{U}_{1,\mathsf{Z}_n}^{\mathrm{ess}}} \times H_{p^r,\mathrm{G}_{h,\mathsf{Z}_{p^r},\mathsf{D}_{p^r}}^{\mathrm{ess,ord}} \ltimes \mathrm{U}_{1,\mathsf{Z}_{p^r},\mathsf{D}_{p^r}}^{\mathrm{ess,ord}}}^{\mathrm{ord}}.
\end{aligned}$$

Note that some of these are not new:

$$\begin{aligned}
H_{n,\mathrm{G}_{l,\mathsf{Z}_n}^{\mathrm{ess}}}^{\mathrm{ord}} &\cong H_{n_0,\mathrm{G}_{l,\mathsf{Z}_{n_0}}^{\mathrm{ess}}} \times H_{p^r,\mathrm{G}_{l,\mathsf{Z}_{p^r}}^{\mathrm{ess}}} \cong H_{n,\mathrm{G}_{l,\mathsf{Z}_n}^{\mathrm{ess}}}, \\
H_{n,\mathrm{G}_{l,\mathsf{Z}_n}^{\mathrm{ess}}}^{\mathrm{ord},\prime} &\cong H'_{n_0,\mathrm{G}_{l,\mathsf{Z}_{n_0}}^{\mathrm{ess}}} \times H'_{p^r,\mathrm{G}_{l,\mathsf{Z}_{p^r}}^{\mathrm{ess}}} \cong H'_{n,\mathrm{G}_{l,\mathsf{Z}_n}^{\mathrm{ess}}}, \\
H_{n,\mathrm{U}_{2,\mathsf{Z}_n}^{\mathrm{ess}}}^{\mathrm{ord}} &\cong H_{n_0,\mathrm{U}_{2,\mathsf{Z}_{n_0}}^{\mathrm{ess}}}.
\end{aligned}$$

Consider the following commutative diagram, in which every square is Cartesian:

$$\begin{array}{ccccccc}
\ddot{\Xi}_{\Phi_n}^{\mathrm{ord}} & \twoheadrightarrow & \ddot{\Xi}_{\Phi_n}^{\mathrm{ord}}/H_{n,\mathrm{U}_{2,\mathsf{Z}_n}^{\mathrm{ess}}}^{\mathrm{ord}} & \twoheadrightarrow & \ddot{\Xi}_{\Phi_n}^{\mathrm{ord}}/H_{n,\mathrm{U}_{\mathsf{Z}_n}^{\mathrm{ess}}}^{\mathrm{ord}} & \twoheadrightarrow & \ddot{\Xi}_{\Phi_n}^{\mathrm{ord}}/H_{n,\mathrm{G}_{h,\mathsf{Z}_n}^{\mathrm{ess}} \ltimes \mathrm{U}_{\mathsf{Z}_n}^{\mathrm{ess}}}^{\mathrm{ord}} \\
& \searrow & \downarrow & & \downarrow & & \downarrow \\
& & \dddot{C}_{\Phi_n}^{\mathrm{ord}} & \twoheadrightarrow & \dddot{C}_{\Phi_n}^{\mathrm{ord}}/H_{n,\mathrm{U}_{1,\mathsf{Z}_n}^{\mathrm{ess}}}^{\mathrm{ord}} & \twoheadrightarrow & \dddot{C}_{\Phi_n}^{\mathrm{ord}}/H_{n,\mathrm{G}_{h,\mathsf{Z}_n}^{\mathrm{ess}} \ltimes \mathrm{U}_{1,\mathsf{Z}_n}^{\mathrm{ess}}}^{\mathrm{ord}} \\
& & & \searrow & \downarrow & & \downarrow \\
& & & & \dddot{\mathsf{M}}_n^{\mathrm{ord},\mathsf{Z}_n} & \twoheadrightarrow & \dddot{\mathsf{M}}_n^{\mathrm{ord},\mathsf{Z}_n}/H_{n,\mathrm{G}_{h,\mathsf{Z}_n}^{\mathrm{ess}}}^{\mathrm{ord},\prime}
\end{array} \tag{4.2.1.14}$$

Now we consider the equivariant quotient of

$$\begin{aligned}
\coprod(\ddot{\Xi}_{\Phi_n}^{\mathrm{ord}}/H_{n,\mathrm{G}_{h,\mathsf{Z}_n}^{\mathrm{ess}} \ltimes \mathrm{U}_{\mathsf{Z}_n}^{\mathrm{ess}}}^{\mathrm{ord}}) &\to \coprod(\dddot{C}_{\Phi_n}^{\mathrm{ord}}/H_{n,\mathrm{G}_{h,\mathsf{Z}_n}^{\mathrm{ess}} \ltimes \mathrm{U}_{1,\mathsf{Z}_n}^{\mathrm{ess}}}^{\mathrm{ord}}) \\
&\to \coprod(\dddot{\mathsf{M}}_n^{\mathrm{ord},\mathsf{Z}_n}/H_{n,\mathrm{G}_{h,\mathsf{Z}_n}^{\mathrm{ess}}}^{\mathrm{ord},\prime})
\end{aligned}$$

by $H^{\mathrm{ord},\prime}_{n,\mathrm{G}^{\mathrm{ess}}_{l,\mathsf{Z}_n}} \cong H'_{n,\mathrm{G}^{\mathrm{ess}}_{l,\mathsf{Z}_n}}$, where the disjoint unions are over elements Φ_n (with the same (X, Y, ϕ)) in the fiber of $\Phi_{H_n} \to \mathsf{Z}_{H_n}$ above Z_n, which is a torsor under $H^{\mathrm{ord},\prime}_{n,\mathrm{G}^{\mathrm{ess}}_{l,\mathsf{Z}_n}} \cong H'_{n,\mathrm{G}^{\mathrm{ess}}_{l,\mathsf{Z}_n}}$. This is (up to canonical isomorphisms) the same as the equivariant quotient of

$$\coprod \ddot{\Xi}^{\mathrm{ord}}_{\Phi_n} \to \coprod \ddot{C}^{\mathrm{ord}}_{\Phi_n} \to \coprod \ddot{\mathsf{M}}_n^{\mathrm{ord},\mathsf{Z}_n}$$

by H_n (or rather by H_n^{ord}, since the kernel $\mathcal{U}^{\mathrm{bal}}_{p,1}(p^r)$ of $H_{p^r} \to H^{\mathrm{ord}}_{p^r}$ acts trivially), where the disjoint unions are over representatives $(\mathsf{Z}_n, \Phi_n, \delta_n)$ (with the same (X, Y, ϕ)) in the H_n-orbit $(\mathsf{Z}_{\mathcal{H}}, \Phi_{\mathcal{H}}, \delta_{\mathcal{H}})$. Let us denote this equivariant quotient (up to canonical isomorphisms) by

$$\ddot{\Xi}^{\mathrm{ord}}_{\Phi_{\mathcal{H}},n} \to \ddot{C}^{\mathrm{ord}}_{\Phi_{\mathcal{H}},n} \to \ddot{\mathsf{M}}_{\mathcal{H}}^{\mathrm{ord},\Phi_{\mathcal{H}}}, \tag{4.2.1.15}$$

whose terms carry compatible actions of $\Gamma_{\Phi_{\mathcal{H}}}$ (see Definition 1.2.2.3). (We keep the subscripts "n" in the notation because $\ddot{\Xi}^{\mathrm{ord}}_{\Phi_{\mathcal{H}},n}$ and $\ddot{C}^{\mathrm{ord}}_{\Phi_{\mathcal{H}},n}$ depend on the choice of n.)

By construction, $\ddot{\Xi}^{\mathrm{ord}}_{\Phi_{\mathcal{H}},n}$ is universal for tuples

$$\begin{aligned}(\mathsf{Z}_{\mathcal{H}}, (X, Y, \phi, \varphi^{\mathrm{ord}}_{-2,\mathcal{H}}, \varphi^{\mathrm{ord}}_{0,\mathcal{H}}),\\ (B, \lambda_B, i_B, \varphi_{-1,\mathcal{H}^p}, \varphi^{\mathrm{ord}}_{-1,\mathcal{H}_p}), \delta^{\mathrm{ord}}_{\mathcal{H}}, (c^{\mathrm{ord}}_{\mathcal{H}}, c^{\vee,\mathrm{ord}}_{\mathcal{H}}, \tau^{\mathrm{ord}}_{\mathcal{H}})),\end{aligned}$$

up to automorphism by $\Gamma_{\Phi_{\mathcal{H}}}$, describing degeneration data without positivity condition, where $(\varphi^{\mathrm{ord}}_{-2,\mathcal{H}}, \varphi^{\mathrm{ord}}_{0,\mathcal{H}})$ (resp. $\delta^{\mathrm{ord}}_{\mathcal{H}}$) induces the same $(\varphi_{-2,\mathcal{H}}, \varphi_{0,\mathcal{H}})$ (resp. $\delta_{\mathcal{H}}$) in the representative $(\mathsf{Z}_{\mathcal{H}}, \Phi_{\mathcal{H}} = (X, Y, \phi, \varphi_{-2,\mathcal{H}}, \varphi_{0,\mathcal{H}}), \delta_{\mathcal{H}})$ we have fixed. The canonical morphisms $\ddot{\Xi}^{\mathrm{ord}}_{\Phi_{\mathcal{H}},n} \to \ddot{C}^{\mathrm{ord}}_{\Phi_{\mathcal{H}},n}$ and $\ddot{C}^{\mathrm{ord}}_{\Phi_{\mathcal{H}},n} \to \ddot{\mathsf{M}}_{\mathcal{H}}^{\mathrm{ord},\Phi_{\mathcal{H}}}$ forget the data $\tau^{\mathrm{ord}}_{\mathcal{H}}$ and $(c^{\mathrm{ord}}_{\mathcal{H}}, c^{\vee,\mathrm{ord}}_{\mathcal{H}})$, respectively.

Lemma 4.2.1.16. *The canonical morphism*

$$\ddot{\mathsf{M}}_n^{\mathrm{ord},\mathsf{Z}_n}/H^{\mathrm{ord},\prime}_{n,\mathrm{G}^{\mathrm{ess}}_{h,\mathsf{Z}_n}} \to \ddot{\mathsf{M}}_{\mathcal{H}}^{\mathrm{ord},\Phi_{\mathcal{H}}} \tag{4.2.1.17}$$

is an isomorphism, and hence the canonical morphisms

$$\ddot{\mathsf{M}}_{\mathcal{H}}^{\mathrm{ord},\Phi_{\mathcal{H}}} \to \ddot{\mathsf{M}}_{\mathcal{H}}^{\mathrm{ord},\mathsf{Z}_{\mathcal{H}}} := \ddot{\mathsf{M}}_{\mathcal{H}}^{\mathrm{ord},\Phi_{\mathcal{H}}}/\Gamma_{\Phi_{\mathcal{H}}} \to \ddot{\mathsf{M}}^{\mathrm{ord}}_{\mathcal{H}_h}, \tag{4.2.1.18}$$

(cf. Definition 1.2.1.15 *and Lemmas* 1.3.2.1 *and* 1.3.2.5*) are finite étale. If, for some (and hence every) choice of a representative* $(\mathsf{Z}_n, \Phi_n, \delta_n)$ *in* $(\mathsf{Z}_{\mathcal{H}}, \Phi_{\mathcal{H}}, \delta_{\mathcal{H}})$*, the image* $H_{n,\mathrm{P}^{\mathrm{ess}}_{\mathsf{Z}_n}}$ *in* $\mathrm{G}^{\mathrm{ess}}_{h,\mathsf{Z}_n} \times \mathrm{G}^{\mathrm{ess}}_{l,\mathsf{Z}_n}$ *is the direct product* $H_{n,\mathrm{G}^{\mathrm{ess}}_{h,\mathsf{Z}_n}} \times H_{n,\mathrm{G}^{\mathrm{ess}}_{l,\mathsf{Z}_n}}$*, then we have* $H^{\mathrm{ord},\prime}_{n,\mathrm{G}^{\mathrm{ess}}_{h,\mathsf{Z}_n}} = H^{\mathrm{ord}}_{n,\mathrm{G}^{\mathrm{ess}}_{h,\mathsf{Z}_n}}$ *and* $H^{\mathrm{ord},\prime}_{n,\mathrm{G}^{\mathrm{ess}}_{l,\mathsf{Z}_n}} = H^{\mathrm{ord}}_{n,\mathrm{G}^{\mathrm{ess}}_{l,\mathsf{Z}_n}}$*, and the canonical morphisms in* (4.2.1.18) *are isomorphisms.*

Proof. The first statement is true because $\ddot{\mathsf{M}}_n^{\mathrm{ord},\mathsf{Z}_n}/H^{\mathrm{ord},\prime}_{n,\mathrm{G}^{\mathrm{ess}}_{h,\mathsf{Z}_n}}$ is finite étale over $\ddot{\mathsf{M}}^{\mathrm{ord}}_{\mathcal{H}_h} \cong \ddot{\mathsf{M}}_n^{\mathrm{ord},\mathsf{Z}_n}/H^{\mathrm{ord}}_{n,\mathrm{G}^{\mathrm{ess}}_{h,\mathsf{Z}_n}}$ (since $H^{\mathrm{ord}}_{n,\mathrm{G}^{\mathrm{ess}}_{h,\mathsf{Z}_n}}$ and $H^{\mathrm{ord},\prime}_{n,\mathrm{G}^{\mathrm{ess}}_{h,\mathsf{Z}_n}}$ act by forming orbits of ordinary level structures), because the index set of the disjoint union

$\coprod(\ddot{\mathsf{M}}_n^{\mathrm{ord},\mathsf{Z}_n}/H^{\mathrm{ord},\prime}_{n,\mathrm{G}^{\mathrm{ess}}_{h,\mathsf{Z}_n}})$ is a torsor under $H^{\mathrm{ord},\prime}_{n,\mathrm{G}^{\mathrm{ess}}_{l,\mathsf{Z}_n}} \cong H'_{n,\mathrm{G}^{\mathrm{ess}}_{l,\mathsf{Z}_n}}$, and because $\Gamma_{\Phi_{\mathcal{H}}}$ acts on $\ddot{\mathsf{M}}_{\mathcal{H}}^{\mathrm{ord},\Phi_{\mathcal{H}}}$ via the canonical homomorphisms

$$\Gamma_{\Phi_{\mathcal{H}}} \to H^{\mathrm{ord},\prime}_{n,\mathrm{G}^{\mathrm{ess}}_{l,\mathsf{Z}_n}}/H^{\mathrm{ord}}_{n,\mathrm{G}^{\mathrm{ess}}_{l,\mathsf{Z}_n}} \cong H^{\mathrm{ord}}_{n,\mathrm{G}^{\mathrm{ess}}_{h,\mathsf{Z}_n}}/H^{\mathrm{ord},\prime}_{n,\mathrm{G}^{\mathrm{ess}}_{h,\mathsf{Z}_n}}.$$

The second statement follows from the definitions of $H^{\mathrm{ord},\prime}_{n,\mathrm{G}^{\mathrm{ess}}_{h,\mathsf{Z}_n}}$ and $H^{\mathrm{ord}}_{n,\mathrm{G}^{\mathrm{ess}}_{h,\mathsf{Z}_n}}$. □

Lemma 4.2.1.19.

(1) The canonical morphism

$$\ddot{C}^{\mathrm{ord}}_{\Phi_n}/H^{\mathrm{ord}}_{n,\mathrm{G}^{\mathrm{ess}}_{h,\mathsf{Z}_n}\ltimes\mathrm{U}^{\mathrm{ess}}_{1,\mathsf{Z}_n}} \to \ddot{C}^{\mathrm{ord}}_{\Phi_{\mathcal{H}},n} \tag{4.2.1.20}$$

is an isomorphism, compatible with (4.2.1.17).

(2) Suppose that, for some (and hence every) choice of a representative $(\mathsf{Z}_n,\Phi_n,\delta_n)$ in $(\mathsf{Z}_{\mathcal{H}},\Phi_{\mathcal{H}},\delta_{\mathcal{H}})$, the splitting of the canonical homomorphism $\mathrm{G}^{\mathrm{ess}}_{h,\mathsf{Z}_n}\ltimes\mathrm{U}^{\mathrm{ess}}_{1,\mathsf{Z}_n} \twoheadrightarrow \mathrm{G}^{\mathrm{ess}}_{h,\mathsf{Z}_n}$ defined by δ_n induces a splitting of the canonical homomorphism $H_{n,\mathrm{G}^{\mathrm{ess}}_{h,\mathsf{Z}_n}\ltimes\mathrm{U}^{\mathrm{ess}}_{1,\mathsf{Z}_n}} \twoheadrightarrow H'_{n,\mathrm{G}^{\mathrm{ess}}_{h,\mathsf{Z}_n}}$, and hence an isomorphism $H_{n,\mathrm{G}^{\mathrm{ess}}_{h,\mathsf{Z}_n}\ltimes\mathrm{U}^{\mathrm{ess}}_{1,\mathsf{Z}_n}} \cong H'_{n,\mathrm{G}^{\mathrm{ess}}_{h,\mathsf{Z}_n}} \ltimes H_{n,\mathrm{U}^{\mathrm{ess}}_{1,\mathsf{Z}_n}}$. In this case, the splitting of the canonical homomorphism

$$(\mathrm{G}^{\mathrm{ess}}_{h,\mathsf{Z}_{n_0}}\times\mathrm{G}^{\mathrm{ess,ord}}_{h,\mathsf{Z}_{p^r},\mathsf{D}_{p^r}})\ltimes(\mathrm{U}^{\mathrm{ess}}_{1,\mathsf{Z}_{n_0}}\times\mathrm{U}^{\mathrm{ess,ord}}_{1,\mathsf{Z}_{p^r},\mathsf{D}_{p^r}}) \twoheadrightarrow (\mathrm{G}^{\mathrm{ess}}_{h,\mathsf{Z}_{n_0}}\times\mathrm{G}^{\mathrm{ess,ord}}_{h,\mathsf{Z}_{p^r},\mathsf{D}_{p^r}}) \tag{4.2.1.21}$$

defined by $\delta^{\mathrm{ord}}_n = (\delta_{n_0},\delta^{\mathrm{ord}}_{p^r})$ induces a splitting of the canonical homomorphism $H^{\mathrm{ord}}_{n,\mathrm{G}^{\mathrm{ess}}_{h,\mathsf{Z}_n}\ltimes\mathrm{U}^{\mathrm{ess}}_{1,\mathsf{Z}_n}} \twoheadrightarrow H^{\mathrm{ord},\prime}_{n,\mathrm{G}^{\mathrm{ess}}_{h,\mathsf{Z}_n}}$, and hence an isomorphism $H^{\mathrm{ord}}_{n,\mathrm{G}^{\mathrm{ess}}_{h,\mathsf{Z}_n}\ltimes\mathrm{U}^{\mathrm{ess}}_{1,\mathsf{Z}_n}} \cong H^{\mathrm{ord},\prime}_{n,\mathrm{G}^{\mathrm{ess}}_{h,\mathsf{Z}_n}} \ltimes H^{\mathrm{ord}}_{n,\mathrm{U}^{\mathrm{ess}}_{1,\mathsf{Z}_n}}$. Under this assumption, $\ddot{C}^{\mathrm{ord}}_{\Phi_{\mathcal{H}},n}$ is a proper flat group scheme over $\ddot{\mathsf{M}}_{\mathcal{H}}^{\mathrm{ord},\Phi_{\mathcal{H}}}$ such that there exists an integer $m \geq 1$ such that multiplication by m maps $\ddot{C}^{\mathrm{ord}}_{\Phi_{\mathcal{H}},n}$ scheme-theoretically to a subgroup scheme $\ddot{C}^{\mathrm{ord},\circ}_{\Phi_{\mathcal{H}},n}$ that is an abelian scheme over $\ddot{\mathsf{M}}_{\mathcal{H}}^{\mathrm{ord},\Phi_{\mathcal{H}}}$, so that the group scheme $\underline{\pi}_0(\ddot{C}^{\mathrm{ord}}_{\Phi_{\mathcal{H}},n}/\ddot{\mathsf{M}}_{\mathcal{H}}^{\mathrm{ord},\Phi_{\mathcal{H}}})$ of fiberwise connected components of $\ddot{C}^{\mathrm{ord}}_{\Phi_{\mathcal{H}},n}$ over $\ddot{\mathsf{M}}_{\mathcal{H}}^{\mathrm{ord},\Phi_{\mathcal{H}}}$ is defined and is of étale-multiplicative type. (See Lemma 3.1.2.2 *and Definition* 3.1.2.3*; cf. Proposition* 4.2.1.11.) *The rank of $\underline{\pi}_0(\ddot{C}^{\mathrm{ord}}_{\Phi_{\mathcal{H}},n}/\ddot{\mathsf{M}}_{\mathcal{H}}^{\mathrm{ord},\Phi_{\mathcal{H}}})$ has no prime factors other than those of* Disc, *n, $[X:\phi(Y)]$, and the rank of* $\ker(\lambda_B)$. *(This implies that the rank of $\underline{\pi}_0(\ddot{C}^{\mathrm{ord}}_{\Phi_{\mathcal{H}},n}/\ddot{\mathsf{M}}_{\mathcal{H}}^{\mathrm{ord},\Phi_{\mathcal{H}}})$ does not contain prime factors other than those of* Disc, *n and $[L^{\#}:L]$.)*

(3) In general (no longer making the assumption on splittings as in statement (2)*), the morphism $\ddot{C}^{\mathrm{ord}}_{\Phi_{\mathcal{H}},n} \to \ddot{\mathsf{M}}_{\mathcal{H}}^{\mathrm{ord},\Phi_{\mathcal{H}}}$ is a torsor under a proper flat group scheme $\ddot{C}^{\mathrm{ord,grp}}_{\Phi_{\mathcal{H}},n} \to \ddot{\mathsf{M}}_{\mathcal{H}}^{\mathrm{ord},\Phi_{\mathcal{H}}}$ satisfying the properties as in statement* (2), *for which $\underline{\pi}_0(\ddot{C}^{\mathrm{ord,grp}}_{\Phi_{\mathcal{H}},n}/\ddot{\mathsf{M}}_{\mathcal{H}}^{\mathrm{ord},\Phi_{\mathcal{H}}})$ is defined. Then, the Stein factorization (see [35, III-1, 4.3.3]) of $\ddot{C}^{\mathrm{ord}}_{\Phi_{\mathcal{H}},n} \to \ddot{\mathsf{M}}_{\mathcal{H}}^{\mathrm{ord},\Phi_{\mathcal{H}}}$, which we denote abusively as $\ddot{C}^{\mathrm{ord}}_{\Phi_{\mathcal{H}},n} \to \underline{\pi}_0(\ddot{C}^{\mathrm{ord}}_{\Phi_{\mathcal{H}},n}/\ddot{\mathsf{M}}_{\mathcal{H}}^{\mathrm{ord},\Phi_{\mathcal{H}}}) \to \ddot{\mathsf{M}}_{\mathcal{H}}^{\mathrm{ord},\Phi_{\mathcal{H}}}$, is the*

composition of an abelian scheme torsor $\ddot{C}^{\mathrm{ord}}_{\Phi_{\mathcal{H}},n} \to \underline{\pi}_0(\ddot{C}^{\mathrm{ord}}_{\Phi_{\mathcal{H}},n}/\ddot{\mathsf{M}}^{\mathrm{ord},\Phi_{\mathcal{H}}}_{\mathcal{H}})$ *with a torsor* $\underline{\pi}_0(\ddot{C}^{\mathrm{ord}}_{\Phi_{\mathcal{H}},n}/\ddot{\mathsf{M}}^{\mathrm{ord},\Phi_{\mathcal{H}}}_{\mathcal{H}}) \to \ddot{\mathsf{M}}^{\mathrm{ord},\Phi_{\mathcal{H}}}_{\mathcal{H}}$ *under the group scheme* $\underline{\pi}_0(\ddot{C}^{\mathrm{ord},\mathrm{grp}}_{\Phi_{\mathcal{H}},n}/\ddot{\mathsf{M}}^{\mathrm{ord},\Phi_{\mathcal{H}}}_{\mathcal{H}}) \to \ddot{\mathsf{M}}^{\mathrm{ord},\Phi_{\mathcal{H}}}_{\mathcal{H}}$.

Proof. Statement (1) is true because the common index set of the disjoint unions $\coprod(\ddot{C}^{\mathrm{ord}}_{\Phi_n}/H^{\mathrm{ord}}_{n,\mathrm{G}^{\mathrm{ess}}_{h,\mathsf{Z}_n}\ltimes\mathrm{U}^{\mathrm{ess}}_{1,\mathsf{Z}_n}})$ and $\coprod(\ddot{\mathsf{M}}^{\mathrm{ord},\mathsf{Z}_n}_n/H^{\mathrm{ord},\prime}_{n,\mathrm{G}^{\mathrm{ess}}_{h,\mathsf{Z}_n}})$ is a torsor under $H^{\mathrm{ord},\prime}_{n,\mathrm{G}^{\mathrm{ess}}_{l,\mathsf{Z}_n}} \cong H'_{n,\mathrm{G}^{\mathrm{ess}}_{l,\mathsf{Z}_n}}$. Then the morphism $\ddot{C}^{\mathrm{ord}}_{\Phi_{\mathcal{H}},n} \to \ddot{\mathsf{M}}^{\mathrm{ord},\Phi_{\mathcal{H}}}_{\mathcal{H}}$ can be canonically identified with the equivariant quotient $\ddot{C}^{\mathrm{ord}}_{\Phi_n}/H^{\mathrm{ord}}_{n,\mathrm{G}^{\mathrm{ess}}_{h,\mathsf{Z}_n}\ltimes\mathrm{U}^{\mathrm{ess}}_{1,\mathsf{Z}_n}} \to \ddot{\mathsf{M}}^{\mathrm{ord},\mathsf{Z}_n}_n/H^{\mathrm{ord},\prime}_{n,\mathrm{G}^{\mathrm{ess}}_{h,\mathsf{Z}_n}}$ of $\ddot{C}^{\mathrm{ord}}_{\Phi_n} \to \ddot{\mathsf{M}}^{\mathrm{ord},\mathsf{Z}_n}_n$.

Via the splitting of (4.2.1.21) defined by $\delta^{\mathrm{ord}}_n = (\delta_{n_0}, \delta^{\mathrm{ord}}_{p^r})$, the equivariant action of $\mathrm{G}^{\mathrm{ess}}_{h,\mathsf{Z}_{n_0}} \times \mathrm{G}^{\mathrm{ess},\mathrm{ord}}_{h,\mathsf{Z}_{p^r},\mathsf{D}_{p^r}}$ is compatible with the group scheme structure of $\ddot{C}^{\mathrm{ord}}_{\Phi_n} \to \ddot{\mathsf{M}}^{\mathrm{ord},\mathsf{Z}_n}_n$. Since the action of $H^{\mathrm{ord}}_{n,\mathrm{U}^{\mathrm{ess}}_{1,\mathsf{Z}_n}}$ on $\ddot{C}^{\mathrm{ord}}_{\Phi_n}$ (modifying the tautological object $(c^{\mathrm{ord}}_n : \frac{1}{n}X \to B^{\vee,\mathrm{ord}}_{p^r}, c^{\vee,\mathrm{ord}}_n : \frac{1}{n}Y \to B^{\mathrm{ord}}_{p^r})$) is realized by the translation action of a (commutative) finite flat subgroup scheme of $\ddot{C}^{\mathrm{ord}}_{\Phi_n} \to \ddot{\mathsf{M}}^{\mathrm{ord},\mathsf{Z}_n}_n$, the quotient $\ddot{C}^{\mathrm{ord}}_{\Phi_n}/H^{\mathrm{ord}}_{n,\mathrm{U}^{\mathrm{ess}}_{1,\mathsf{Z}_n}} \to \ddot{\mathsf{M}}^{\mathrm{ord},\mathsf{Z}_n}_n$ is a proper flat group scheme. Under the assumptions on the splittings as in statement (2), the equivariant action of $H^{\mathrm{ord}}_{n,\mathrm{G}^{\mathrm{ess}}_{h,\mathsf{Z}_n}}$ is compatible with the group scheme structure of $\ddot{C}^{\mathrm{ord}}_{\Phi_n}/H^{\mathrm{ord}}_{n,\mathrm{U}^{\mathrm{ess}}_{1,\mathsf{Z}_n}} \to \ddot{\mathsf{M}}^{\mathrm{ord},\mathsf{Z}_n}_n$, and hence the equivariant quotient $\ddot{C}^{\mathrm{ord}}_{\Phi_n}/H^{\mathrm{ord}}_{n,\mathrm{G}^{\mathrm{ess}}_{h,\mathsf{Z}_n}\ltimes\mathrm{U}^{\mathrm{ess}}_{1,\mathsf{Z}_n}} \to \ddot{\mathsf{M}}^{\mathrm{ord},\mathsf{Z}_n}_n/H^{\mathrm{ord},\prime}_{n,\mathrm{G}^{\mathrm{ess}}_{h,\mathsf{Z}_n}}$ (see the diagram (4.2.1.14), in which every square is Cartesian) is again a proper flat group scheme, which inherits from $\ddot{C}^{\mathrm{ord}}_{\Phi_n} \to \ddot{\mathsf{M}}^{\mathrm{ord},\mathsf{Z}_n}_n$ (see Proposition 4.2.1.11) the properties described as in statement (2).

In general, without the assumptions on the splittings as in statement (2), $H^{\mathrm{ord}}_{n,\mathrm{G}^{\mathrm{ess}}_{h,\mathsf{Z}_n}\ltimes\mathrm{U}^{\mathrm{ess}}_{1,\mathsf{Z}_n}}$ and $H^{\mathrm{ord},\prime}_{n,\mathrm{G}^{\mathrm{ess}}_{h,\mathsf{Z}_n}} \ltimes H^{\mathrm{ord}}_{n,\mathrm{U}^{\mathrm{ess}}_{1,\mathsf{Z}_n}}$ are two different subgroups of $(\mathrm{G}^{\mathrm{ess}}_{h,\mathsf{Z}_{n_0}} \times \mathrm{G}^{\mathrm{ess},\mathrm{ord}}_{h,\mathsf{Z}_{p^r},\mathsf{D}_{p^r}}) \ltimes (\mathrm{U}^{\mathrm{ess}}_{1,\mathsf{Z}_{n_0}} \times \mathrm{U}^{\mathrm{ess},\mathrm{ord}}_{1,\mathsf{Z}_{p^r},\mathsf{D}_{p^r}})$. By the same reasoning as in the previous paragraph, the quotient $\ddot{C}^{\mathrm{ord}}_{\Phi_n}/(H^{\mathrm{ord},\prime}_{n,\mathrm{G}^{\mathrm{ess}}_{h,\mathsf{Z}_n}} \ltimes H^{\mathrm{ord}}_{n,\mathrm{U}^{\mathrm{ess}}_{1,\mathsf{Z}_n}}) \to \ddot{\mathsf{M}}^{\mathrm{ord},\mathsf{Z}_n}_n/H^{\mathrm{ord},\prime}_{n,\mathrm{G}^{\mathrm{ess}}_{h,\mathsf{Z}_n}}$ is a proper flat group scheme, with properties described as in statement (2). The group scheme structure of $\ddot{C}^{\mathrm{ord}}_{\Phi_n}/H^{\mathrm{ord}}_{n,\mathrm{U}^{\mathrm{ess}}_{1,\mathsf{Z}_n}} \to \ddot{\mathsf{M}}^{\mathrm{ord},\mathsf{Z}_n}_n$ might not descend to $\ddot{C}^{\mathrm{ord}}_{\Phi_n}/H^{\mathrm{ord}}_{n,\mathrm{G}^{\mathrm{ess}}_{h,\mathsf{Z}_n}\ltimes\mathrm{U}^{\mathrm{ess}}_{1,\mathsf{Z}_n}} \to \ddot{\mathsf{M}}^{\mathrm{ord},\mathsf{Z}_n}_n/H^{\mathrm{ord},\prime}_{n,\mathrm{G}^{\mathrm{ess}}_{h,\mathsf{Z}_n}}$, but nevertheless makes the latter a torsor under $\ddot{C}^{\mathrm{ord}}_{\Phi_n}/(H^{\mathrm{ord},\prime}_{n,\mathrm{G}^{\mathrm{ess}}_{h,\mathsf{Z}_n}} \ltimes H^{\mathrm{ord}}_{n,\mathrm{U}^{\mathrm{ess}}_{1,\mathsf{Z}_n}}) \to \ddot{\mathsf{M}}^{\mathrm{ord},\mathsf{Z}_n}_n/H^{\mathrm{ord},\prime}_{n,\mathrm{G}^{\mathrm{ess}}_{h,\mathsf{Z}_n}}$. Hence, statement (3) follows if we define $\ddot{C}^{\mathrm{ord},\mathrm{grp}}_{\Phi_{\mathcal{H}},n}$ to be the quotient of $\coprod(\ddot{C}^{\mathrm{ord}}_{\Phi_n}/(H^{\mathrm{ord},\prime}_{n,\mathrm{G}^{\mathrm{ess}}_{h,\mathsf{Z}_n}} \ltimes H^{\mathrm{ord}}_{n,\mathrm{U}^{\mathrm{ess}}_{1,\mathsf{Z}_n}}))$ by $H^{\mathrm{ord},\prime}_{n,\mathrm{G}^{\mathrm{ess}}_{l,\mathsf{Z}_n}} \cong H'_{n,\mathrm{G}^{\mathrm{ess}}_{l,\mathsf{Z}_n}}$, where the disjoint unions are over elements Φ_n in the fiber of $\Phi_{H_n} \to \mathsf{Z}_{H_n}$ above Z_n, as in the definition of $\ddot{C}^{\mathrm{ord}}_{\Phi_{\mathcal{H}},n} \to \ddot{\mathsf{M}}^{\mathrm{ord},\Phi_{\mathcal{H}}}_{\mathcal{H}}$. □

Lemma 4.2.1.22. *The quotient by the action of* $H^{\mathrm{ord}}_{n,\mathrm{U}^{\mathrm{ess}}_{2,\mathsf{Z}_n}}$ *on* $\ddot{E}^{\mathrm{ord}}_{\Phi_n}$ *is realized by the translation action of a (commutative) finite flat subgroup scheme, so that the* *quotient* $\dddot{E}^{\mathrm{ord}}_{\Phi_{\mathcal{H}},n} := \dddot{E}^{\mathrm{ord}}_{\Phi_n}/H^{\mathrm{ord}}_{n,\mathrm{U}^{\mathrm{ess}}_{2,\mathrm{Z}_n}}$ *is a group scheme of multiplicative type of finite type with character group* $\dddot{\mathbf{S}}^{\mathrm{ord}}_{\Phi_{\mathcal{H}},n}$ *a subgroup of* $\dddot{\mathbf{S}}^{\mathrm{ord}}_{\Phi_n}$. *Let* $\dddot{\mathbf{S}}^{\mathrm{ord}}_{\Phi_{\mathcal{H}},n,\mathrm{tor}}$ *be the torsion subgroup of* $\dddot{\mathbf{S}}^{\mathrm{ord}}_{\Phi_{\mathcal{H}},n}$, *and let* $\dddot{\mathbf{S}}^{\mathrm{ord}}_{\Phi_{\mathcal{H}},n,\mathrm{free}}$ *be the free quotient of* $\dddot{\mathbf{S}}^{\mathrm{ord}}_{\Phi_{\mathcal{H}},n}$. *Then the canonical short exact sequence*

$$0 \to \dddot{\mathbf{S}}^{\mathrm{ord}}_{\Phi_{\mathcal{H}},n,\mathrm{tor}} \to \dddot{\mathbf{S}}^{\mathrm{ord}}_{\Phi_{\mathcal{H}},n} \to \dddot{\mathbf{S}}^{\mathrm{ord}}_{\Phi_{\mathcal{H}},n,\mathrm{free}} \to 0$$

induces a short exact sequence

$$0 \to \dddot{E}^{\mathrm{ord}}_{\Phi_{\mathcal{H}},n,\mathrm{free}} \to \dddot{E}^{\mathrm{ord}}_{\Phi_{\mathcal{H}},n} \to \dddot{E}^{\mathrm{ord}}_{\Phi_{\mathcal{H}},n,\mathrm{tor}} \to 0.$$

Then we have canonical isomorphisms $\dddot{\mathbf{S}}^{\mathrm{ord}}_{\Phi_{\mathcal{H}},n,\mathrm{free}} \cong \mathbf{S}_{\Phi_{\mathcal{H}}}$ *and* $\dddot{E}^{\mathrm{ord}}_{\Phi_{\mathcal{H}},n,\mathrm{free}} \cong E_{\Phi_{\mathcal{H}}} \cong \underline{\mathrm{Hom}}_{\mathbb{Z}}(\mathbf{S}_{\Phi_{\mathcal{H}}}, \mathbf{G}_{\mathrm{m}})$, *where* $\mathbf{S}_{\Phi_{\mathcal{H}}}$ *and* $E_{\Phi_{\mathcal{H}}}$ *are defined as in [62, Lem. 6.2.4.4].*

Proof. These statements are about group schemes of multiplicative type of finite type over $\mathrm{Spec}(\mathbb{Z})$ with constant character groups, and hence they can be verified after base change to $\mathrm{Spec}(\mathbb{Q})$ (or any base ring of residue characteristics prime to the order of $\dddot{\mathbf{S}}^{\mathrm{ord}}_{\Phi_{\mathcal{H}},n,\mathrm{tor}}$). Then they all follow from [62, Lem. 6.2.4.4]. □

Lemma 4.2.1.23.

(1) The canonical morphism

$$\ddddot{\Xi}^{\mathrm{ord}}_{\Phi_n}/H^{\mathrm{ord}}_{n,\mathrm{G}^{\mathrm{ess}}_{h,\mathrm{Z}_n}\ltimes \mathrm{U}^{\mathrm{ess}}_{\mathrm{Z}_n}} \to \ddddot{\Xi}^{\mathrm{ord}}_{\Phi_{\mathcal{H}},n} \tag{4.2.1.24}$$

is an isomorphism, compatible with (4.2.1.17) *and* (4.2.1.20).

(2) The morphism $\ddddot{\Xi}^{\mathrm{ord}}_{\Phi_{\mathcal{H}},n} \to \dddot{C}^{\mathrm{ord}}_{\Phi_{\mathcal{H}},n}$ *is a torsor under the pullback of* $\dddot{E}^{\mathrm{ord}}_{\Phi_{\mathcal{H}},n}$, *which factors as a composition* $\ddddot{\Xi}^{\mathrm{ord}}_{\Phi_{\mathcal{H}},n} \to \ddddot{\Xi}^{\mathrm{ord}}_{\Phi_{\mathcal{H}},n,\mathrm{tor}} \to \dddot{C}^{\mathrm{ord}}_{\Phi_{\mathcal{H}},n}$, *in which the morphism* $\ddddot{\Xi}^{\mathrm{ord}}_{\Phi_{\mathcal{H}},n} \to \ddddot{\Xi}^{\mathrm{ord}}_{\Phi_{\mathcal{H}},n,\mathrm{tor}}$ *is a torsor under the pullback of the torus* $\dddot{E}^{\mathrm{ord}}_{\Phi_{\mathcal{H}},n,\mathrm{free}} \cong E_{\Phi_{\mathcal{H}}} \cong \underline{\mathrm{Hom}}_{\mathbb{Z}}(\mathbf{S}_{\Phi_{\mathcal{H}}}, \mathbf{G}_{\mathrm{m}})$, *and in which the morphism* $\ddddot{\Xi}^{\mathrm{ord}}_{\Phi_{\mathcal{H}},n,\mathrm{tor}} \to \dddot{C}^{\mathrm{ord}}_{\Phi_{\mathcal{H}},n}$ *is a torsor under the pullback of the finite flat group scheme* $\dddot{E}^{\mathrm{ord}}_{\Phi_{\mathcal{H}},n,\mathrm{tor}}$ *of multiplicative type.*

Proof. Statement (1) is true because the common index set of the disjoint unions $\coprod(\ddddot{\Xi}^{\mathrm{ord}}_{\Phi_n}/H^{\mathrm{ord}}_{n,\mathrm{G}^{\mathrm{ess}}_{h,\mathrm{Z}_n}\ltimes \mathrm{U}^{\mathrm{ess}}_{\mathrm{Z}_n}})$, $\coprod(\dddot{C}^{\mathrm{ord}}_{\Phi_n}/H^{\mathrm{ord}}_{n,\mathrm{G}^{\mathrm{ess}}_{h,\mathrm{Z}_n}\ltimes \mathrm{U}^{\mathrm{ess}}_{1,\mathrm{Z}_n}})$, and $\coprod(\dddot{\mathrm{M}}^{\mathrm{ord},\mathrm{Z}_n}_n/H^{\mathrm{ord},\prime}_{n,\mathrm{G}^{\mathrm{ess}}_{h,\mathrm{Z}_n}})$ is a torsor under $H^{\mathrm{ord},\prime}_{n,\mathrm{G}^{\mathrm{ess}}_{l,\mathrm{Z}_n}} \cong H'_{n,\mathrm{G}^{\mathrm{ess}}_{l,\mathrm{Z}_n}}$. Then the morphism $\ddddot{\Xi}^{\mathrm{ord}}_{\Phi_{\mathcal{H}},n} \to \dddot{C}^{\mathrm{ord}}_{\Phi_{\mathcal{H}},n}$ can be canonically identified with the equivariant quotient $\ddddot{\Xi}^{\mathrm{ord}}_{\Phi_n}/H^{\mathrm{ord}}_{n,\mathrm{G}^{\mathrm{ess}}_{h,\mathrm{Z}_n}\ltimes \mathrm{U}^{\mathrm{ess}}_{\mathrm{Z}_n}} \to \dddot{C}^{\mathrm{ord}}_{\Phi_n}/H^{\mathrm{ord}}_{n,\mathrm{G}^{\mathrm{ess}}_{h,\mathrm{Z}_n}\ltimes \mathrm{U}^{\mathrm{ess}}_{1,\mathrm{Z}_n}}$ of $\ddddot{\Xi}^{\mathrm{ord}}_{\Phi_n} \to \dddot{C}^{\mathrm{ord}}_{\Phi_n}$. Since the action of $H^{\mathrm{ord}}_{n,\mathrm{U}^{\mathrm{ess}}_{2,\mathrm{Z}_n}}$ (modifying the tautological object $\tau^{\mathrm{ord}}_n = \tau_{n_0} : \mathbf{1}_{(\frac{1}{n_0}Y)\times X} \xrightarrow{\sim} (c^\vee_{n_0} \times c)^*\mathcal{P}_B^{\otimes -1})$ is realized as the same action of a finite flat subgroup scheme of $\dddot{E}^{\mathrm{ord}}_{\Phi_n}$ as in Lemma 4.2.1.22, we see that $\ddddot{\Xi}^{\mathrm{ord}}_{\Phi_n}/H^{\mathrm{ord}}_{n,\mathrm{U}^{\mathrm{ess}}_{2,\mathrm{Z}_n}} \to \dddot{C}^{\mathrm{ord}}_{\Phi_n}$ is a torsor under the pullback of $\dddot{E}^{\mathrm{ord}}_{\Phi_{\mathcal{H}},n}$, and hence so is $\ddddot{\Xi}^{\mathrm{ord}}_{\Phi_n}/H^{\mathrm{ord}}_{n,\mathrm{G}^{\mathrm{ess}}_{h,\mathrm{Z}_n}\ltimes \mathrm{U}^{\mathrm{ess}}_{\mathrm{Z}_n}} \to \dddot{C}^{\mathrm{ord}}_{\Phi_n}/H^{\mathrm{ord}}_{n,\mathrm{G}^{\mathrm{ess}}_{h,\mathrm{Z}_n}\ltimes \mathrm{U}^{\mathrm{ess}}_{1,\mathrm{Z}_n}}$ after equivariant quotient by $H^{\mathrm{ord}}_{n,\mathrm{G}^{\mathrm{ess}}_{h,\mathrm{Z}_n}\ltimes \mathrm{U}^{\mathrm{ess}}_{1,\mathrm{Z}_n}}$. (See the diagram (4.2.1.14), in which every square is Cartesian.) Then the factorization of $\ddddot{\Xi}^{\mathrm{ord}}_{\Phi_{\mathcal{H}},n} \to \dddot{C}^{\mathrm{ord}}_{\Phi_{\mathcal{H}},n}$ follows. □

By [62, Prop. 6.2.4.7; see also the errata], there is an algebraic stack $\Xi_{\Phi_{\mathcal{H}},\delta_{\mathcal{H}}}$ separated, smooth, and schematic over $\mathsf{M}_{\mathcal{H}}^{\mathsf{Z}_{\mathcal{H}}}$ over $\mathsf{S}_0 = \mathrm{Spec}(F_0)$, whose quotient by $\Gamma_{\Phi_{\mathcal{H}}}$ is universal for tuples

$$(\mathsf{Z}_{\mathcal{H}}, (X, Y, \phi, \varphi_{-2,\mathcal{H}}^{\sim}, \varphi_{0,\mathcal{H}}^{\sim}), (B, \lambda_B, i_B, \varphi_{-1,\mathcal{H}}), \delta_{\mathcal{H}}, (c_{\mathcal{H}}, c_{\mathcal{H}}^{\vee}, \tau_{\mathcal{H}}))$$

up to automorphism by $\Gamma_{\Phi_{\mathcal{H}}}$, describing degeneration data without positivity condition, such that $(\mathsf{Z}_{\mathcal{H}}, \Phi_{\mathcal{H}}^{\sim} = (X, Y, \phi, \varphi_{-2,\mathcal{H}}^{\sim}, \varphi_{0,\mathcal{H}}^{\sim}), \delta_{\mathcal{H}})$ induces the same representative $(\mathsf{Z}_{\mathcal{H}}, \Phi_{\mathcal{H}} = (X, Y, \phi, \varphi_{-2,\mathcal{H}}, \varphi_{0,\mathcal{H}}), \delta_{\mathcal{H}})$ we have fixed, as in [62, Lem. 5.4.2.10; see also the errata]. The structural morphism $\Xi_{\Phi_{\mathcal{H}},\delta_{\mathcal{H}}} \to \mathsf{M}_{\mathcal{H}}^{\mathsf{Z}_{\mathcal{H}}}$ factorizes canonically as the composition $\Xi_{\Phi_{\mathcal{H}},\delta_{\mathcal{H}}} \to C_{\Phi_{\mathcal{H}},\delta_{\mathcal{H}}} \to \mathsf{M}_{\mathcal{H}}^{\Phi_{\mathcal{H}}} \to \mathsf{M}_{\mathcal{H}}^{\mathsf{Z}_{\mathcal{H}}}$, where $\Xi_{\Phi_{\mathcal{H}},\delta_{\mathcal{H}}} \to C_{\Phi_{\mathcal{H}},\delta_{\mathcal{H}}}$ is a torsor under the pullback of the torus $E_{\Phi_{\mathcal{H}}} \cong \underline{\mathrm{Hom}}_{\mathbb{Z}}(\mathbf{S}_{\Phi_{\mathcal{H}}}, \mathbf{G}_{\mathrm{m}})$, where $C_{\Phi_{\mathcal{H}},\delta_{\mathcal{H}}} \to \mathsf{M}_{\mathcal{H}}^{\Phi_{\mathcal{H}}}$ is an abelian scheme torsor, and where $\mathsf{M}_{\mathcal{H}}^{\Phi_{\mathcal{H}}} \to \mathsf{M}_{\mathcal{H}}^{\mathsf{Z}_{\mathcal{H}}}$ is finite étale. The morphisms $\Xi_{\Phi_{\mathcal{H}},\delta_{\mathcal{H}}} \to C_{\Phi_{\mathcal{H}},\delta_{\mathcal{H}}}$ and $C_{\Phi_{\mathcal{H}},\delta_{\mathcal{H}}} \to \mathsf{M}_{\mathcal{H}}^{\Phi_{\mathcal{H}}}$ forget the data $\tau_{\mathcal{H}}$ and $(c_{\mathcal{H}}, c_{\mathcal{H}}^{\vee})$, respectively. By the construction in [62, Sec. 6.2.4; see also the errata], $\Xi_{\Phi_{\mathcal{H}},\delta_{\mathcal{H}}}$, $C_{\Phi_{\mathcal{H}},\delta_{\mathcal{H}}}$, $\mathsf{M}_{\mathcal{H}}^{\Phi_{\mathcal{H}}}$, and $\mathsf{M}_{\mathcal{H}}^{\mathsf{Z}_{\mathcal{H}}}$, are, respectively, quotients of objects at principal level n by suitable subgroups of $\mathcal{H}/\mathcal{U}(n)$, which are subgroups of $(\mathrm{G}(\hat{\mathbb{Z}}^p) \times \mathcal{U}_{p,0}(p^r))/(\mathcal{U}^p(n_0) \times \mathcal{U}_p(p^r))$. (See also the descriptions and characterizations of these objects in Section 1.3.2.)

Let $\mathsf{M}_{\mathcal{H}}^{\mathrm{ord},\mathsf{Z}_{\mathcal{H}}}$ be the open subalgebraic stack of $\dddot{\mathsf{M}}_{\mathcal{H}}^{\mathrm{ord},\mathsf{Z}_{\mathcal{H}}} \underset{\mathbb{Z}}{\otimes} F_0[\zeta_{p^{r_{\mathcal{H}}}}]$ given by the image of the canonical open immersion

$$\mathsf{M}_{\mathcal{H},r_{\mathcal{H}}}^{\mathsf{Z}_{\mathcal{H}}} := \mathsf{M}_{\mathcal{H}}^{\mathsf{Z}_{\mathcal{H}}} \underset{\mathsf{S}_0}{\times} \mathsf{S}_{0,r_{\mathcal{H}}} = \mathsf{M}_{\mathcal{H}}^{\mathsf{Z}_{\mathcal{H}}} \underset{F_0}{\otimes} F_0[p^{r_{\mathcal{H}}}] \hookrightarrow \dddot{\mathsf{M}}_{\mathcal{H}}^{\mathrm{ord},\mathsf{Z}_{\mathcal{H}}} \underset{\mathbb{Z}}{\otimes} F_0[\zeta_{p^{r_{\mathcal{H}}}}]$$

(*cf.* Theorem 3.4.2.5), and let $\vec{\mathsf{M}}_{\mathcal{H}}^{\mathrm{ord},\mathsf{Z}_{\mathcal{H}}}$ be the normalization of $\dddot{\mathsf{M}}_{\mathcal{H}}^{\mathrm{ord},\mathsf{Z}_{\mathcal{H}}}$ in $\mathsf{M}_{\mathcal{H}}^{\mathrm{ord},\mathsf{Z}_{\mathcal{H}}}$ under the canonical morphism

$$\mathsf{M}_{\mathcal{H}}^{\mathrm{ord},\mathsf{Z}_{\mathcal{H}}} \to \dddot{\mathsf{M}}_{\mathcal{H}}^{\mathrm{ord},\mathsf{Z}_{\mathcal{H}}}$$

(with properties analogous to those of $\vec{\mathsf{M}}_{\mathcal{H}}^{\mathrm{ord}}$ in Theorem 3.4.2.5).

Let

$$\Xi_{\Phi_{\mathcal{H}},\delta_{\mathcal{H}}}^{\mathrm{ord}} := \Xi_{\Phi_{\mathcal{H}},\delta_{\mathcal{H}}} \underset{\mathsf{M}_{\mathcal{H}}^{\mathsf{Z}_{\mathcal{H}}}}{\times} \mathsf{M}_{\mathcal{H}}^{\mathrm{ord},\mathsf{Z}_{\mathcal{H}}}, \tag{4.2.1.25}$$

$$C_{\Phi_{\mathcal{H}},\delta_{\mathcal{H}}}^{\mathrm{ord}} := C_{\Phi_{\mathcal{H}},\delta_{\mathcal{H}}} \underset{\mathsf{M}_{\mathcal{H}}^{\mathsf{Z}_{\mathcal{H}}}}{\times} \mathsf{M}_{\mathcal{H}}^{\mathrm{ord},\mathsf{Z}_{\mathcal{H}}}, \tag{4.2.1.26}$$

and

$$\mathsf{M}_{\mathcal{H}}^{\mathrm{ord},\Phi_{\mathcal{H}}} := \mathsf{M}_{\mathcal{H}}^{\Phi_{\mathcal{H}}} \underset{\mathsf{M}_{\mathcal{H}}^{\mathsf{Z}_{\mathcal{H}}}}{\times} \mathsf{M}_{\mathcal{H}}^{\mathrm{ord},\mathsf{Z}_{\mathcal{H}}}. \tag{4.2.1.27}$$

By Theorem 4.1.6.2, they carry tuples parameterized by $\dddot{\Xi}_{\Phi_{\mathcal{H}},n}^{\mathrm{ord}}$, $\dddot{C}_{\Phi_{\mathcal{H}},n}^{\mathrm{ord}}$, and $\dddot{\mathsf{M}}_{\mathcal{H}}^{\mathrm{ord},\Phi_{\mathcal{H}}}$, respectively, and induce (by universal properties) canonical morphisms

$$\Xi_{\Phi_{\mathcal{H}},\delta_{\mathcal{H}}}^{\mathrm{ord}} \to \dddot{\Xi}_{\Phi_{\mathcal{H}},n}^{\mathrm{ord}},$$

$$C_{\Phi_{\mathcal{H}},\delta_{\mathcal{H}}}^{\mathrm{ord}} \to \dddot{C}_{\Phi_{\mathcal{H}},n}^{\mathrm{ord}},$$

and

$$\mathsf{M}_{\mathcal{H}}^{\mathrm{ord},\Phi_{\mathcal{H}}} \to \dddot{\mathsf{M}}_{\mathcal{H}}^{\mathrm{ord},\Phi_{\mathcal{H}}}$$

compatible with each other, and with their canonical morphisms to $\mathsf{M}_{\mathcal{H}}^{\mathrm{ord},\mathsf{Z}_{\mathcal{H}}}$ and $\dddot{\mathsf{M}}_{\mathcal{H}}^{\mathrm{ord},\mathsf{Z}_{\mathcal{H}}}$. Let $\vec{\Xi}^{\mathrm{ord}}_{\Phi_{\mathcal{H}},\delta_{\mathcal{H}}}$ (resp. $\vec{C}^{\mathrm{ord}}_{\Phi_{\mathcal{H}},\delta_{\mathcal{H}}}$, resp. $\vec{\mathsf{M}}_{\mathcal{H}}^{\mathrm{ord},\Phi_{\mathcal{H}}}$) be the normalization of $\dddot{\Xi}^{\mathrm{ord}}_{\Phi_{\mathcal{H}},n}$ (resp. $\dddot{C}^{\mathrm{ord}}_{\Phi_{\mathcal{H}},n}$, resp. $\dddot{\mathsf{M}}_{\mathcal{H}}^{\mathrm{ord},\Phi_{\mathcal{H}}}$) in $\Xi^{\mathrm{ord}}_{\Phi_{\mathcal{H}},\delta_{\mathcal{H}}}$ (resp. $C^{\mathrm{ord}}_{\Phi_{\mathcal{H}},\delta_{\mathcal{H}}}$, resp. $\mathsf{M}_{\mathcal{H}}^{\mathrm{ord},\Phi_{\mathcal{H}}}$) under the canonical morphism. Then we obtain the following commutative diagram

$$\begin{array}{ccccc}
\Xi^{\mathrm{ord}}_{\Phi_{\mathcal{H}},\delta_{\mathcal{H}}} & \longrightarrow & \vec{\Xi}^{\mathrm{ord}}_{\Phi_{\mathcal{H}},\delta_{\mathcal{H}}} & \longrightarrow & \dddot{\Xi}^{\mathrm{ord}}_{\Phi_{\mathcal{H}},n} \\
\downarrow & & \downarrow & & \downarrow \\
C^{\mathrm{ord}}_{\Phi_{\mathcal{H}},\delta_{\mathcal{H}}} & \longrightarrow & \vec{C}^{\mathrm{ord}}_{\Phi_{\mathcal{H}},\delta_{\mathcal{H}}} & \longrightarrow & \dddot{C}^{\mathrm{ord}}_{\Phi_{\mathcal{H}},n} \\
\downarrow & & \downarrow & & \downarrow \\
\mathsf{M}_{\mathcal{H}}^{\mathrm{ord},\Phi_{\mathcal{H}}} & \longrightarrow & \vec{\mathsf{M}}_{\mathcal{H}}^{\mathrm{ord},\Phi_{\mathcal{H}}} & \longrightarrow & \dddot{\mathsf{M}}_{\mathcal{H}}^{\mathrm{ord},\Phi_{\mathcal{H}}} \\
\downarrow & & \downarrow & & \downarrow \\
\mathsf{M}_{\mathcal{H}}^{\mathrm{ord},\mathsf{Z}_{\mathcal{H}}} & \longrightarrow & \vec{\mathsf{M}}_{\mathcal{H}}^{\mathrm{ord},\mathsf{Z}_{\mathcal{H}}} & \longrightarrow & \dddot{\mathsf{M}}_{\mathcal{H}}^{\mathrm{ord},\mathsf{Z}_{\mathcal{H}}} \\
\downarrow & & \downarrow & & \downarrow \\
\mathsf{S}_{0,r_{\mathcal{H}}} = \mathrm{Spec}(F_0[\zeta_{p^{r_{\mathcal{H}}}}]) & \to & \vec{\mathsf{S}}_{0,r_{\mathcal{H}}} = \mathrm{Spec}(\mathcal{O}_{F_0,(p)}[\zeta_{p^{r_{\mathcal{H}}}}]) & \to & \mathrm{Spec}(\mathbb{Z}_{(p)})
\end{array} \quad (4.2.1.28)$$

of canonically induced morphisms (which is generally not Cartesian). The objects in this diagram all carry compatible canonical actions of $\Gamma_{\Phi_{\mathcal{H}}}$. (The actions on those in the bottom two rows are all trivial.)

Proposition 4.2.1.29. *In* (4.2.1.28), *the canonical morphism* $\vec{\mathsf{M}}_{\mathcal{H}}^{\mathrm{ord},\Phi_{\mathcal{H}}} \to \vec{\mathsf{M}}_{\mathcal{H}}^{\mathrm{ord},\mathsf{Z}_{\mathcal{H}}}$ *is finite étale, which induces an isomorphism* $\vec{\mathsf{M}}_{\mathcal{H}}^{\mathrm{ord},\Phi_{\mathcal{H}}}/\Gamma_{\Phi_{\mathcal{H}}} \xrightarrow{\sim} \vec{\mathsf{M}}_{\mathcal{H}}^{\mathrm{ord},\mathsf{Z}_{\mathcal{H}}}$. *Moreover, the canonical morphism* $\vec{\mathsf{M}}_{\mathcal{H}}^{\mathrm{ord},\Phi_{\mathcal{H}}} \to \dddot{\mathsf{M}}_{\mathcal{H}}^{\mathrm{ord},\Phi_{\mathcal{H}}} \underset{\dddot{\mathsf{M}}_{\mathcal{H}}^{\mathrm{ord},\mathsf{Z}_{\mathcal{H}}}}{\times} \vec{\mathsf{M}}_{\mathcal{H}}^{\mathrm{ord},\mathsf{Z}_{\mathcal{H}}}$ *is an isomorphism. If the condition in the second statement of Lemma* 4.2.1.16 *is satisfied, then the canonical morphism* $\vec{\mathsf{M}}_{\mathcal{H}}^{\mathrm{ord},\Phi_{\mathcal{H}}} \to \vec{\mathsf{M}}_{\mathcal{H}}^{\mathrm{ord},\mathsf{Z}_{\mathcal{H}}}$ *is an isomorphism.*

Proof. The morphism $\mathsf{M}_{\mathcal{H}}^{\mathrm{ord},\Phi_{\mathcal{H}}} \to \dddot{\mathsf{M}}_{\mathcal{H}}^{\mathrm{ord},\Phi_{\mathcal{H}}} \underset{\dddot{\mathsf{M}}_{\mathcal{H}}^{\mathrm{ord},\mathsf{Z}_{\mathcal{H}}}}{\times} \mathsf{M}_{\mathcal{H}}^{\mathrm{ord},\mathsf{Z}_{\mathcal{H}}}$ is an isomorphism essentially by construction. Since the canonical morphism $\dddot{\mathsf{M}}_{\mathcal{H}}^{\mathrm{ord},\Phi_{\mathcal{H}}} \to \dddot{\mathsf{M}}_{\mathcal{H}}^{\mathrm{ord},\mathsf{Z}_{\mathcal{H}}}$ is finite étale by Lemma 4.2.1.16, and since $\vec{\mathsf{M}}_{\mathcal{H}}^{\mathrm{ord},\Phi_{\mathcal{H}}}$ is defined by normalization, the remainder of the first two statements follows. The last statement also follows from Lemma 4.2.1.16. □

Proposition 4.2.1.30. *In* (4.2.1.28), *the morphism* $\vec{C}^{\mathrm{ord}}_{\Phi_{\mathcal{H}},\delta_{\mathcal{H}}} \to \vec{\mathsf{M}}_{\mathcal{H}}^{\mathrm{ord},\Phi_{\mathcal{H}}}$ *is an abelian scheme torsor. Moreover, the canonical morphism* $\vec{C}^{\mathrm{ord}}_{\Phi_{\mathcal{H}},\delta_{\mathcal{H}}} \to$

$\dddot{C}^{\mathrm{ord}}_{\Phi_{\mathcal{H}},n} \underset{\dddot{\mathsf{M}}^{\mathrm{ord},\Phi_{\mathcal{H}}}_{\mathcal{H}}}{\times} \vec{\mathsf{M}}^{\mathrm{ord},\Phi_{\mathcal{H}}}_{\mathcal{H}}$ *is a closed immersion. If the condition in* (2) *of Lemma* 4.2.1.19 *is satisfied, then the abelian scheme torsor* $\vec{C}^{\mathrm{ord}}_{\Phi_{\mathcal{H}},\delta_{\mathcal{H}}} \to \vec{\mathsf{M}}^{\mathrm{ord},\Phi_{\mathcal{H}}}_{\mathcal{H}}$ *is an abelian scheme.*

Proof. By construction, we have the following commutative diagram, in which the vertical columns in the diagram are Stein factorizations, by Lemma 4.2.1.19:

$$\begin{array}{ccccc}
C^{\mathrm{ord}}_{\Phi_{\mathcal{H}},\delta_{\mathcal{H}}} & \longrightarrow & \dddot{C}^{\mathrm{ord}}_{\Phi_{\mathcal{H}},n} \underset{\dddot{\mathsf{M}}^{\mathrm{ord},\Phi_{\mathcal{H}}}_{\mathcal{H}}}{\times} \mathsf{M}^{\mathrm{ord},\Phi_{\mathcal{H}}}_{\mathcal{H}} & \longrightarrow & \dddot{C}^{\mathrm{ord}}_{\Phi_{\mathcal{H}},n} \underset{\dddot{\mathsf{M}}^{\mathrm{ord},\Phi_{\mathcal{H}}}_{\mathcal{H}}}{\times} \vec{\mathsf{M}}^{\mathrm{ord},\Phi_{\mathcal{H}}}_{\mathcal{H}} \\
 & & \downarrow & & \downarrow \\
 & & \underline{\pi}_0(\dddot{C}^{\mathrm{ord}}_{\Phi_{\mathcal{H}},n}/\dddot{\mathsf{M}}^{\mathrm{ord},\Phi_{\mathcal{H}}}_{\mathcal{H}}) \underset{\dddot{\mathsf{M}}^{\mathrm{ord},\Phi_{\mathcal{H}}}_{\mathcal{H}}}{\times} \mathsf{M}^{\mathrm{ord},\Phi_{\mathcal{H}}}_{\mathcal{H}} & \to & \underline{\pi}_0(\dddot{C}^{\mathrm{ord}}_{\Phi_{\mathcal{H}},n}/\dddot{\mathsf{M}}^{\mathrm{ord},\Phi_{\mathcal{H}}}_{\mathcal{H}}) \underset{\dddot{\mathsf{M}}^{\mathrm{ord},\Phi_{\mathcal{H}}}_{\mathcal{H}}}{\times} \vec{\mathsf{M}}^{\mathrm{ord},\Phi_{\mathcal{H}}}_{\mathcal{H}} \\
 & & \downarrow & & \downarrow \\
 & \searrow & \mathsf{M}^{\mathrm{ord},\Phi_{\mathcal{H}}}_{\mathcal{H}} & \longrightarrow & \vec{\mathsf{M}}^{\mathrm{ord},\Phi_{\mathcal{H}}}_{\mathcal{H}}
\end{array}$$

Since $C^{\mathrm{ord}}_{\Phi_{\mathcal{H}},\delta_{\mathcal{H}}} \to \mathsf{M}^{\mathrm{ord},\Phi_{\mathcal{H}}}_{\mathcal{H}}$ is an abelian scheme torsor, the canonical morphism $C^{\mathrm{ord}}_{\Phi_{\mathcal{H}},\delta_{\mathcal{H}}} \to \dddot{C}^{\mathrm{ord}}_{\Phi_{\mathcal{H}},n} \underset{\dddot{\mathsf{M}}^{\mathrm{ord},\Phi_{\mathcal{H}}}_{\mathcal{H}}}{\times} \mathsf{M}^{\mathrm{ord},\Phi_{\mathcal{H}}}_{\mathcal{H}}$ induces (by considering their Stein factorizations over $\mathsf{M}^{\mathrm{ord},\Phi_{\mathcal{H}}}_{\mathcal{H}}$) a section

$$\mathsf{M}^{\mathrm{ord},\Phi_{\mathcal{H}}}_{\mathcal{H}} \to \underline{\pi}_0(\dddot{C}^{\mathrm{ord}}_{\Phi_{\mathcal{H}},n}/\dddot{\mathsf{M}}^{\mathrm{ord},\Phi_{\mathcal{H}}}_{\mathcal{H}}) \underset{\dddot{\mathsf{M}}^{\mathrm{ord},\Phi_{\mathcal{H}}}_{\mathcal{H}}}{\times} \mathsf{M}^{\mathrm{ord},\Phi_{\mathcal{H}}}_{\mathcal{H}}. \tag{4.2.1.31}$$

Since $\underline{\pi}_0(\dddot{C}^{\mathrm{ord}}_{\Phi_{\mathcal{H}},n}/\dddot{\mathsf{M}}^{\mathrm{ord},\Phi_{\mathcal{H}}}_{\mathcal{H}}) \to \dddot{\mathsf{M}}^{\mathrm{ord},\Phi_{\mathcal{H}}}_{\mathcal{H}}$ is a torsor under the group scheme $\underline{\pi}_0(\dddot{C}^{\mathrm{ord,grp}}_{\Phi_{\mathcal{H}},n}/\dddot{\mathsf{M}}^{\mathrm{ord},\Phi_{\mathcal{H}}}_{\mathcal{H}}) \to \dddot{\mathsf{M}}^{\mathrm{ord},\Phi_{\mathcal{H}}}_{\mathcal{H}}$ in Lemma 4.2.1.19, which is (finite flat, of finite presentation, and) of étale-multiplicative type, and since $\vec{\mathsf{M}}^{\mathrm{ord},\Phi_{\mathcal{H}}}_{\mathcal{H}}$ is normal (by definition), the schematic closure of the image of (4.2.1.31) defines a section

$$\vec{\mathsf{M}}^{\mathrm{ord},\Phi_{\mathcal{H}}}_{\mathcal{H}} \to \underline{\pi}_0(\dddot{C}^{\mathrm{ord}}_{\Phi_{\mathcal{H}},n}/\dddot{\mathsf{M}}^{\mathrm{ord},\Phi_{\mathcal{H}}}_{\mathcal{H}}) \underset{\dddot{\mathsf{M}}^{\mathrm{ord},\Phi_{\mathcal{H}}}_{\mathcal{H}}}{\times} \vec{\mathsf{M}}^{\mathrm{ord},\Phi_{\mathcal{H}}}_{\mathcal{H}}. \tag{4.2.1.32}$$

Since $\dddot{C}^{\mathrm{ord}}_{\Phi_{\mathcal{H}},n} \to \dddot{\mathsf{M}}^{\mathrm{ord},\Phi_{\mathcal{H}}}_{\mathcal{H}}$ is a torsor under the proper flat group scheme $\dddot{C}^{\mathrm{ord,grp}}_{\Phi_{\mathcal{H}},n} \to \dddot{\mathsf{M}}^{\mathrm{ord},\Phi_{\mathcal{H}}}_{\mathcal{H}}$ in Lemma 4.2.1.19, which is the extension of $\underline{\pi}_0(\dddot{C}^{\mathrm{ord,grp}}_{\Phi_{\mathcal{H}},n}/\dddot{\mathsf{M}}^{\mathrm{ord},\Phi_{\mathcal{H}}}_{\mathcal{H}})$ by an abelian scheme $\dddot{C}^{\mathrm{ord,grp},\circ}_{\Phi_{\mathcal{H}},n}$, the pullback of the morphism

$$\dddot{C}^{\mathrm{ord}}_{\Phi_{\mathcal{H}},n} \underset{\dddot{\mathsf{M}}^{\mathrm{ord},\Phi_{\mathcal{H}}}_{\mathcal{H}}}{\times} \vec{\mathsf{M}}^{\mathrm{ord},\Phi_{\mathcal{H}}}_{\mathcal{H}} \to \underline{\pi}_0(\dddot{C}^{\mathrm{ord}}_{\Phi_{\mathcal{H}},n}/\dddot{\mathsf{M}}^{\mathrm{ord},\Phi_{\mathcal{H}}}_{\mathcal{H}}) \underset{\dddot{\mathsf{M}}^{\mathrm{ord},\Phi_{\mathcal{H}}}_{\mathcal{H}}}{\times} \vec{\mathsf{M}}^{\mathrm{ord},\Phi_{\mathcal{H}}}_{\mathcal{H}} \tag{4.2.1.33}$$

under the section (4.2.1.32) is a torsor $C \to \vec{\mathsf{M}}^{\mathrm{ord},\Phi_{\mathcal{H}}}_{\mathcal{H}}$ under the abelian scheme $\dddot{C}^{\mathrm{ord,grp},\circ}_{\Phi_{\mathcal{H}},n} \underset{\dddot{\mathsf{M}}^{\mathrm{ord},\Phi_{\mathcal{H}}}_{\mathcal{H}}}{\times} \vec{\mathsf{M}}^{\mathrm{ord},\Phi_{\mathcal{H}}}_{\mathcal{H}}$, which is in particular separated, smooth, and of finite type. By the construction of $\vec{C}^{\mathrm{ord}}_{\Phi_{\mathcal{H}},\delta_{\mathcal{H}}}$ and C, the canonical morphism $\vec{C}^{\mathrm{ord}}_{\Phi_{\mathcal{H}},\delta_{\mathcal{H}}} \to \dddot{C}^{\mathrm{ord}}_{\Phi_{\mathcal{H}},n} \underset{\dddot{\mathsf{M}}^{\mathrm{ord},\Phi_{\mathcal{H}}}_{\mathcal{H}}}{\times} \vec{\mathsf{M}}^{\mathrm{ord},\Phi_{\mathcal{H}}}_{\mathcal{H}}$ factors through a canonical morphism $\vec{C}^{\mathrm{ord}}_{\Phi_{\mathcal{H}},\delta_{\mathcal{H}}} \to C$, which must be an isomorphism because $\vec{C}^{\mathrm{ord}}_{\Phi_{\mathcal{H}},\delta_{\mathcal{H}}}$ is defined by normalization. This also shows that $\vec{C}^{\mathrm{ord}}_{\Phi_{\mathcal{H}},\delta_{\mathcal{H}}} \to \dddot{C}^{\mathrm{ord}}_{\Phi_{\mathcal{H}},n} \underset{\dddot{\mathsf{M}}^{\mathrm{ord},\Phi_{\mathcal{H}}}_{\mathcal{H}}}{\times} \vec{\mathsf{M}}^{\mathrm{ord},\Phi_{\mathcal{H}}}_{\mathcal{H}}$ is a closed immersion (because the section (4.2.1.32) is), and that $\vec{C}^{\mathrm{ord}}_{\Phi_{\mathcal{H}},\delta_{\mathcal{H}}} \to \vec{\mathsf{M}}^{\mathrm{ord},\Phi_{\mathcal{H}}}_{\mathcal{H}}$ is an abelian scheme torsor.

If the condition in (2) of Lemma 4.2.1.19 is satisfied, then $C_{\Phi_{\mathcal{H}},\delta_{\mathcal{H}}} \to \mathsf{M}^{\Phi_{\mathcal{H}}}_{\mathcal{H}}$ is an abelian scheme (see [62, Prop. 6.2.4.7; see also the errata]), and so is its pullback $C^{\mathrm{ord}}_{\Phi_{\mathcal{H}},\delta_{\mathcal{H}}} \to \mathsf{M}^{\mathrm{ord},\Phi_{\mathcal{H}}}_{\mathcal{H}}$. By taking the closure of the identity section $\mathsf{M}^{\mathrm{ord},\Phi_{\mathcal{H}}}_{\mathcal{H}} \to C^{\mathrm{ord}}_{\Phi_{\mathcal{H}},\delta_{\mathcal{H}}}$, we obtain a section $\vec{\mathsf{M}}^{\mathrm{ord},\Phi_{\mathcal{H}}}_{\mathcal{H}} \to \vec{C}^{\mathrm{ord}}_{\Phi_{\mathcal{H}},\delta_{\mathcal{H}}}$, which shows that the abelian scheme torsor $\vec{C}^{\mathrm{ord}}_{\Phi_{\mathcal{H}},\delta_{\mathcal{H}}} \to \vec{\mathsf{M}}^{\mathrm{ord},\Phi_{\mathcal{H}}}_{\mathcal{H}}$ is also an abelian scheme, as desired. □

Proposition 4.2.1.34. *Let* $\vec{C}^{\mathrm{ord}}_{\Phi_1,\delta_1}$, $\vec{\mathsf{M}}^{\mathrm{ord},\Phi_1}_1$, *and* $\vec{\mathsf{M}}^{\mathrm{ord},\mathsf{Z}_1}_1$ *denote the analogues of* $\vec{C}^{\mathrm{ord}}_{\Phi_{\mathcal{H}},\delta_{\mathcal{H}}}$, $\vec{\mathsf{M}}^{\mathrm{ord},\Phi_{\mathcal{H}}}_{\mathcal{H}}$, *and* $\vec{\mathsf{M}}^{\mathrm{ord},\mathsf{Z}_{\mathcal{H}}}_{\mathcal{H}}$ *at principal level* 1 *(i.e., with* $\mathcal{H}$ *replaced with* $\mathcal{U}(1) = \mathcal{U}^p(1)\mathcal{U}^{\mathrm{bal}}_{p,1}(p^0) = \mathrm{G}(\hat{\mathbb{Z}})$*). Then* $\vec{\mathsf{M}}^{\mathrm{ord},\Phi_1}_1 = \vec{\mathsf{M}}^{\mathrm{ord},\mathsf{Z}_1}_1$ *(by definition), and the canonical morphism* $\vec{C}^{\mathrm{ord}}_{\Phi_{\mathcal{H}},\delta_{\mathcal{H}}} \to \vec{C}^{\mathrm{ord}}_{\Phi_1,\delta_1} \underset{\vec{\mathsf{M}}^{\mathrm{ord},\Phi_1}_1}{\times} \vec{\mathsf{M}}^{\mathrm{ord},\Phi_{\mathcal{H}}}_{\mathcal{H}}$ *is finite étale. (Since* $\vec{\mathsf{M}}^{\mathrm{ord},\Phi_{\mathcal{H}}}_{\mathcal{H}} \to \vec{\mathsf{M}}^{\mathrm{ord},\mathsf{Z}_{\mathcal{H}}}_{\mathcal{H}}$ *is finite étale by Proposition* 4.2.1.29*, the canonical morphism* $\vec{C}^{\mathrm{ord}}_{\Phi_{\mathcal{H}},\delta_{\mathcal{H}}} \to \vec{C}^{\mathrm{ord}}_{\Phi_1,\delta_1} \underset{\vec{\mathsf{M}}^{\mathrm{ord},\mathsf{Z}_1}_1}{\times} \vec{\mathsf{M}}^{\mathrm{ord},\mathsf{Z}_{\mathcal{H}}}_{\mathcal{H}}$ *is also finite étale.)*

Proof. By Lemma 4.2.1.9, for any $n = n_0p^r$ such that $\mathcal{U}^p(n_0) \subset \mathcal{H}^p$ and $\mathcal{U}^{\mathrm{bal}}_{p,1}(p^r) \subset \mathcal{H}_p \subset \mathcal{U}_{p,0}(p^r)$, the canonical morphism $\dddot{C}^{\mathrm{ord}}_{\Phi_n} \to \dddot{C}^{\mathrm{ord}}_{\Phi_1} \underset{\dddot{\mathsf{M}}^{\mathrm{ord},\mathsf{Z}_1}_1}{\times} \dddot{\mathsf{M}}^{\mathrm{ord},\mathsf{Z}_n}_n$ is unramified, because it is a homomorphism with a finite étale kernel. Since this morphism factorizes as a composition of canonical morphisms

$$\dddot{C}^{\mathrm{ord}}_{\Phi_n} \to \dddot{C}^{\mathrm{ord}}_{\Phi_{\mathcal{H}},n} \underset{\dddot{\mathsf{M}}^{\mathrm{ord},\Phi_{\mathcal{H}}}_{\mathcal{H}}}{\times} \dddot{\mathsf{M}}^{\mathrm{ord},\mathsf{Z}_n}_n \to \dddot{C}^{\mathrm{ord}}_{\Phi_1} \underset{\dddot{\mathsf{M}}^{\mathrm{ord},\mathsf{Z}_1}_1}{\times} \dddot{\mathsf{M}}^{\mathrm{ord},\mathsf{Z}_n}_n, \tag{4.2.1.35}$$

in which the first one is surjective by definition (see (4.2.1.15) and (1) of Lemma 4.2.1.19), we see (by étale descent) that the canonical morphism $\dddot{C}^{\mathrm{ord}}_{\Phi_{\mathcal{H}},n} \to \dddot{C}^{\mathrm{ord}}_{\Phi_1} \underset{\dddot{\mathsf{M}}^{\mathrm{ord},\mathsf{Z}_1}_1}{\times} \dddot{\mathsf{M}}^{\mathrm{ord},\Phi_{\mathcal{H}}}_{\mathcal{H}}$, which is the equivariant finite étale quotient by $H^{\mathrm{ord},\prime}_{n,\mathrm{G}^{\mathrm{ess}}_{h,\mathsf{Z}_n}}$ of the second morphism in (4.2.1.35), is also unramified. Consequently, the pullback $\dddot{C}^{\mathrm{ord}}_{\Phi_{\mathcal{H}},n} \underset{\dddot{\mathsf{M}}^{\mathrm{ord},\Phi_{\mathcal{H}}}_{\mathcal{H}}}{\times} \vec{\mathsf{M}}^{\mathrm{ord},\Phi_{\mathcal{H}}}_{\mathcal{H}} \to \dddot{C}^{\mathrm{ord}}_{\Phi_1} \underset{\dddot{\mathsf{M}}^{\mathrm{ord},\mathsf{Z}_1}_1}{\times} \vec{\mathsf{M}}^{\mathrm{ord},\Phi_{\mathcal{H}}}_{\mathcal{H}}$ is unramified. By Proposition 4.2.1.30, the canonical morphisms $\vec{C}^{\mathrm{ord}}_{\Phi_{\mathcal{H}},\delta_{\mathcal{H}}} \to \dddot{C}^{\mathrm{ord}}_{\Phi_{\mathcal{H}},n} \underset{\dddot{\mathsf{M}}^{\mathrm{ord},\Phi_{\mathcal{H}}}_{\mathcal{H}}}{\times} \vec{\mathsf{M}}^{\mathrm{ord},\Phi_{\mathcal{H}}}_{\mathcal{H}}$ and $\vec{C}^{\mathrm{ord}}_{\Phi_1,\delta_1} \underset{\vec{\mathsf{M}}^{\mathrm{ord},\Phi_1}_1}{\times} \vec{\mathsf{M}}^{\mathrm{ord},\Phi_{\mathcal{H}}}_{\mathcal{H}} \to \dddot{C}^{\mathrm{ord}}_{\Phi_1} \underset{\dddot{\mathsf{M}}^{\mathrm{ord},\mathsf{Z}_1}_1}{\times} \vec{\mathsf{M}}^{\mathrm{ord},\Phi_{\mathcal{H}}}_{\mathcal{H}}$ are closed immersions. Hence, the morphism $\vec{C}^{\mathrm{ord}}_{\Phi_{\mathcal{H}},\delta_{\mathcal{H}}} \to \vec{C}^{\mathrm{ord}}_{\Phi_1,\delta_1} \underset{\vec{\mathsf{M}}^{\mathrm{ord},\Phi_1}_1}{\times} \vec{\mathsf{M}}^{\mathrm{ord},\Phi_{\mathcal{H}}}_{\mathcal{H}}$ (compatible with the above two closed immersions) is also unramified. Since this is a morphism between abelian scheme torsors equivariant with a homomorphism of abelian schemes of the same relative dimension (which is automatically surjective), by [35, IV-3, 11.3.10 a)⇒b) and 15.4.2 e′)⇒b)] (*cf.* the proof of [62, Lem. 1.3.1.11]), it is automatically flat, and hence finite étale, as desired. □

Corollary 4.2.1.36. *Suppose $\mathcal{H}' \subset \mathcal{H}$ and suppose (with similar assumptions at level $\mathcal{H}'$) $\vec{C}^{\mathrm{ord}}_{\Phi_{\mathcal{H}'},\delta_{\mathcal{H}'}} \to \vec{\mathsf{M}}^{\mathrm{ord},\Phi_{\mathcal{H}'}}_{\mathcal{H}'}$ is also defined. Then the canonical morphism $\vec{C}^{\mathrm{ord}}_{\Phi_{\mathcal{H}'},\delta_{\mathcal{H}'}} \to \vec{C}^{\mathrm{ord}}_{\Phi_{\mathcal{H}},\delta_{\mathcal{H}}} \underset{\vec{\mathsf{M}}^{\mathrm{ord},\Phi_{\mathcal{H}}}_{\mathcal{H}}}{\times} \vec{\mathsf{M}}^{\mathrm{ord},\Phi_{\mathcal{H}'}}_{\mathcal{H}'}$ is finite étale.*

Proof. By Proposition 4.2.1.34, the composition of canonical morphisms $\vec{C}^{\mathrm{ord}}_{\Phi_{\mathcal{H}'},\delta_{\mathcal{H}'}} \to \vec{C}^{\mathrm{ord}}_{\Phi_{\mathcal{H}},\delta_{\mathcal{H}}} \underset{\vec{\mathsf{M}}^{\mathrm{ord},\Phi_{\mathcal{H}}}_{\mathcal{H}}}{\times} \vec{\mathsf{M}}^{\mathrm{ord},\Phi_{\mathcal{H}'}}_{\mathcal{H}'} \to \vec{C}^{\mathrm{ord}}_{\Phi_1,\delta_1} \underset{\vec{\mathsf{M}}^{\mathrm{ord},\Phi_1}_{1}}{\times} \vec{\mathsf{M}}^{\mathrm{ord},\Phi_{\mathcal{H}'}}_{\mathcal{H}'}$ is finite étale. Since these are morphisms between abelian scheme torsors equivariant under homomorphisms of abelian schemes of the same relative dimension, they are both surjective. Hence, they are both unramified, and (as in the proof of Proposition 4.2.1.34) finite étale. □

Proposition 4.2.1.37. *In (4.2.1.28), the morphism $\vec{\Xi}^{\mathrm{ord}}_{\Phi_{\mathcal{H}},\delta_{\mathcal{H}}} \to \vec{C}^{\mathrm{ord}}_{\Phi_{\mathcal{H}},\delta_{\mathcal{H}}}$ is a torsor under the pullback to $\vec{C}^{\mathrm{ord}}_{\Phi_{\mathcal{H}},\delta_{\mathcal{H}}}$ of the torus $E_{\Phi_{\mathcal{H}}} \cong \underline{\mathrm{Hom}}_{\mathbb{Z}}(\mathbf{S}_{\Phi_{\mathcal{H}}}, \mathbf{G}_{\mathrm{m}})$ (see Lemma 4.2.1.22). Moreover, the canonical morphism $\vec{\Xi}^{\mathrm{ord}}_{\Phi_{\mathcal{H}},\delta_{\mathcal{H}}} \to \dddot{\Xi}^{\mathrm{ord}}_{\Phi_{\mathcal{H}},n} \underset{\dddot{C}^{\mathrm{ord}}_{\Phi_{\mathcal{H}},n}}{\times} \vec{C}^{\mathrm{ord}}_{\Phi_{\mathcal{H}},\delta_{\mathcal{H}}}$ is a closed immersion.*

Proof. By construction, we have the following commutative diagram:

$$\begin{array}{ccccc}
\Xi^{\mathrm{ord}}_{\Phi_{\mathcal{H}},\delta_{\mathcal{H}}} & \longrightarrow & \dddot{\Xi}^{\mathrm{ord}}_{\Phi_{\mathcal{H}},n} \underset{\dddot{C}^{\mathrm{ord}}_{\Phi_{\mathcal{H}},n}}{\times} C^{\mathrm{ord}}_{\Phi_{\mathcal{H}},\delta_{\mathcal{H}}} & \longrightarrow & \dddot{\Xi}^{\mathrm{ord}}_{\Phi_{\mathcal{H}},n} \underset{\dddot{C}^{\mathrm{ord}}_{\Phi_{\mathcal{H}},n}}{\times} \vec{C}^{\mathrm{ord}}_{\Phi_{\mathcal{H}},\delta_{\mathcal{H}}} \\
 & & \downarrow & & \downarrow \\
 & & \dddot{\Xi}^{\mathrm{ord}}_{\Phi_{\mathcal{H}},n,\mathrm{tor}} \underset{\dddot{C}^{\mathrm{ord}}_{\Phi_{\mathcal{H}},n}}{\times} C^{\mathrm{ord}}_{\Phi_{\mathcal{H}},\delta_{\mathcal{H}}} & \longrightarrow & \dddot{\Xi}^{\mathrm{ord}}_{\Phi_{\mathcal{H}},n,\mathrm{tor}} \underset{\dddot{C}^{\mathrm{ord}}_{\Phi_{\mathcal{H}},n}}{\times} \vec{C}^{\mathrm{ord}}_{\Phi_{\mathcal{H}},\delta_{\mathcal{H}}} \\
 & & \downarrow & & \downarrow \\
 & \searrow & C^{\mathrm{ord}}_{\Phi_{\mathcal{H}},\delta_{\mathcal{H}}} & \longrightarrow & \vec{C}^{\mathrm{ord}}_{\Phi_{\mathcal{H}},\delta_{\mathcal{H}}}
\end{array}$$

Since $\Xi^{\mathrm{ord}}_{\Phi_{\mathcal{H}},\delta_{\mathcal{H}}} \to C^{\mathrm{ord}}_{\Phi_{\mathcal{H}},\delta_{\mathcal{H}}}$ is a torsor under the pullback to $C^{\mathrm{ord}}_{\Phi_{\mathcal{H}},\delta_{\mathcal{H}}}$ of the torus $E_{\Phi_{\mathcal{H}}} \cong \underline{\mathrm{Hom}}_{\mathbb{Z}}(\mathbf{S}_{\Phi_{\mathcal{H}}}, \mathbf{G}_{\mathrm{m}})$, the canonical morphism $\Xi^{\mathrm{ord}}_{\Phi_{\mathcal{H}},\delta_{\mathcal{H}}} \to \dddot{\Xi}^{\mathrm{ord}}_{\Phi_{\mathcal{H}},n,\mathrm{tor}} \underset{\dddot{C}^{\mathrm{ord}}_{\Phi_{\mathcal{H}},n}}{\times} C^{\mathrm{ord}}_{\Phi_{\mathcal{H}},\delta_{\mathcal{H}}}$ induces a section

$$\Xi^{\mathrm{ord}}_{\Phi_{\mathcal{H}},\delta_{\mathcal{H}}} \to \dddot{\Xi}^{\mathrm{ord}}_{\Phi_{\mathcal{H}},n,\mathrm{tor}} \underset{\dddot{C}^{\mathrm{ord}}_{\Phi_{\mathcal{H}},n}}{\times} C^{\mathrm{ord}}_{\Phi_{\mathcal{H}},\delta_{\mathcal{H}}}. \tag{4.2.1.38}$$

(It suffices to show that the schematic image of $\Xi^{\mathrm{ord}}_{\Phi_{\mathcal{H}},\delta_{\mathcal{H}}} \to \dddot{\Xi}^{\mathrm{ord}}_{\Phi_{\mathcal{H}},n,\mathrm{tor}} \underset{\dddot{C}^{\mathrm{ord}}_{\Phi_{\mathcal{H}},n}}{\times} C^{\mathrm{ord}}_{\Phi_{\mathcal{H}},\delta_{\mathcal{H}}}$ is isomorphic to $C^{\mathrm{ord}}_{\Phi_{\mathcal{H}},\delta_{\mathcal{H}}}$ under the structural morphism $\dddot{\Xi}^{\mathrm{ord}}_{\Phi_{\mathcal{H}},n,\mathrm{tor}} \underset{\dddot{C}^{\mathrm{ord}}_{\Phi_{\mathcal{H}},n}}{\times} C^{\mathrm{ord}}_{\Phi_{\mathcal{H}},\delta_{\mathcal{H}}} \to C^{\mathrm{ord}}_{\Phi_{\mathcal{H}},\delta_{\mathcal{H}}}$, which can be verified after making an étale localization that trivializes the torsors.) Since $\dddot{\Xi}^{\mathrm{ord}}_{\Phi_{\mathcal{H}},n,\mathrm{tor}} \underset{\dddot{C}^{\mathrm{ord}}_{\Phi_{\mathcal{H}},n}}{\times} \vec{C}^{\mathrm{ord}}_{\Phi_{\mathcal{H}},\delta_{\mathcal{H}}} \to \vec{C}^{\mathrm{ord}}_{\Phi_{\mathcal{H}},\delta_{\mathcal{H}}}$ is a torsor under $\dddot{E}^{\mathrm{ord}}_{\Phi_{\mathcal{H}},n,\mathrm{tor}}$, which is finite flat and of multiplicative type, and since $\vec{C}^{\mathrm{ord}}_{\Phi_{\mathcal{H}},\delta_{\mathcal{H}}}$ is normal (by definition), the schematic closure of the image of (4.2.1.38) defines (as above) a section

$$\vec{C}^{\mathrm{ord}}_{\Phi_{\mathcal{H}},\delta_{\mathcal{H}}} \to \dddot{\Xi}^{\mathrm{ord}}_{\Phi_{\mathcal{H}},n,\mathrm{tor}} \underset{\dddot{C}^{\mathrm{ord}}_{\Phi_{\mathcal{H}},n}}{\times} \vec{C}^{\mathrm{ord}}_{\Phi_{\mathcal{H}},\delta_{\mathcal{H}}}. \tag{4.2.1.39}$$

Since the morphism $\dddot{\Xi}^{\mathrm{ord}}_{\Phi_{\mathcal{H}},n} \to \dddot{\Xi}^{\mathrm{ord}}_{\Phi_{\mathcal{H}},n,\mathrm{tor}}$ is a torsor under the pullback to $\dddot{\Xi}^{\mathrm{ord}}_{\Phi_{\mathcal{H}},n,\mathrm{tor}}$ of the torus $E_{\Phi_{\mathcal{H}}}$ (see Lemma 4.2.1.23), the pullback of the morphism

$$\dddot{\Xi}^{\mathrm{ord}}_{\Phi_{\mathcal{H}},n} \underset{\dddot{C}^{\mathrm{ord}}_{\Phi_{\mathcal{H}},n}}{\times} \vec{C}^{\mathrm{ord}}_{\Phi_{\mathcal{H}},\delta_{\mathcal{H}}} \to \dddot{\Xi}^{\mathrm{ord}}_{\Phi_{\mathcal{H}},n,\mathrm{tor}} \underset{\dddot{C}^{\mathrm{ord}}_{\Phi_{\mathcal{H}},n}}{\times} \vec{C}^{\mathrm{ord}}_{\Phi_{\mathcal{H}},\delta_{\mathcal{H}}} \tag{4.2.1.40}$$

under the section (4.2.1.39) is a torsor $\Xi \to \vec{C}^{\mathrm{ord}}_{\Phi_{\mathcal{H}},\delta_{\mathcal{H}}}$ under the pullback of $E_{\Phi_{\mathcal{H}}}$ to $\vec{C}^{\mathrm{ord}}_{\Phi_{\mathcal{H}},\delta_{\mathcal{H}}}$, which is in particular separated, smooth, and of finite type. By the construction of $\vec{\Xi}^{\mathrm{ord}}_{\Phi_{\mathcal{H}},\delta_{\mathcal{H}}}$ and Ξ, the canonical morphism $\vec{\Xi}^{\mathrm{ord}}_{\Phi_{\mathcal{H}},\delta_{\mathcal{H}}} \to \dddot{\Xi}^{\mathrm{ord}}_{\Phi_{\mathcal{H}},n} \underset{\dddot{C}^{\mathrm{ord}}_{\Phi_{\mathcal{H}},n}}{\times} \vec{C}^{\mathrm{ord}}_{\Phi_{\mathcal{H}},\delta_{\mathcal{H}}}$ factors through a canonical morphism $\vec{\Xi}^{\mathrm{ord}}_{\Phi_{\mathcal{H}},\delta_{\mathcal{H}}} \to \Xi$, which must be an isomorphism because $\vec{\Xi}^{\mathrm{ord}}_{\Phi_{\mathcal{H}},\delta_{\mathcal{H}}}$ is defined by normalization. This shows that $\vec{\Xi}^{\mathrm{ord}}_{\Phi_{\mathcal{H}},\delta_{\mathcal{H}}} \to \vec{C}^{\mathrm{ord}}_{\Phi_{\mathcal{H}},\delta_{\mathcal{H}}}$ is a torsor under the pullback of the torus $E_{\Phi_{\mathcal{H}}}$ to $\vec{C}^{\mathrm{ord}}_{\Phi_{\mathcal{H}},\delta_{\mathcal{H}}}$. This also shows that $\vec{\Xi}^{\mathrm{ord}}_{\Phi_{\mathcal{H}},\delta_{\mathcal{H}}} \to \dddot{\Xi}^{\mathrm{ord}}_{\Phi_{\mathcal{H}},n} \underset{\dddot{C}^{\mathrm{ord}}_{\Phi_{\mathcal{H}},n}}{\times} \vec{C}^{\mathrm{ord}}_{\Phi_{\mathcal{H}},\delta_{\mathcal{H}}}$ is a closed immersion (because the section (4.2.1.39) is), as desired. □

Proposition 4.2.1.41. *Let $\vec{\Xi}^{\mathrm{ord}}_{\Phi_1,\delta_1}$ denote the analogue of $\vec{\Xi}^{\mathrm{ord}}_{\Phi_{\mathcal{H}},\delta_{\mathcal{H}}}$ at principal level 1 (i.e., with $\mathcal{H}$ replaced with $\mathcal{U}(1) = \mathcal{U}^p(1)\mathcal{U}^{\mathrm{bal}}_{p,1}(p^0) = \mathrm{G}(\hat{\mathbb{Z}})$). Then the canonical morphism $\vec{\Xi}^{\mathrm{ord}}_{\Phi_{\mathcal{H}},\delta_{\mathcal{H}}} \to \vec{\Xi}^{\mathrm{ord}}_{\Phi_1,\delta_1} \underset{\vec{C}^{\mathrm{ord}}_{\Phi_1,\delta_1}}{\times} \vec{C}^{\mathrm{ord}}_{\Phi_{\mathcal{H}},\delta_{\mathcal{H}}}$ is finite étale.*

Proof. By definition (see (4.2.1.12) and (4.2.1.15)), for any $n = n_0p^r$ such that $\mathcal{U}^p(n_0) \subset \mathcal{H}^p$ and $\mathcal{U}^{\mathrm{bal}}_{p,1}(p^r) \subset \mathcal{H}_p \subset \mathcal{U}_{p,0}(p^r)$, the canonical morphism $\dddot{\Xi}^{\mathrm{ord}}_{\Phi_n} \to \dddot{\Xi}^{\mathrm{ord}}_{\Phi_1} \underset{\dddot{C}^{\mathrm{ord}}_{\Phi_1}}{\times} \dddot{C}^{\mathrm{ord}}_{\Phi_n}$ is unramified, because (as a morphism between torus torsors) it is étale locally the canonical homomorphism $\ddot{E}^{\mathrm{ord}}_{\Phi_n} = \underline{\mathrm{Hom}}_{\mathbb{Z}}(\ddot{\mathbf{S}}^{\mathrm{ord}}_{\Phi_n}, \mathbf{G}_{\mathrm{m}}) = \ddot{E}_{\Phi_{n_0}} = \underline{\mathrm{Hom}}_{\mathbb{Z}}(\ddot{\mathbf{S}}_{\Phi_{n_0}}, \mathbf{G}_{\mathrm{m}}) \to \ddot{E}^{\mathrm{ord}}_{\Phi_1} = \underline{\mathrm{Hom}}_{\mathbb{Z}}(\ddot{\mathbf{S}}^{\mathrm{ord}}_{\Phi_1}, \mathbf{G}_{\mathrm{m}}) = \ddot{E}_{\Phi_1} = \underline{\mathrm{Hom}}_{\mathbb{Z}}(\ddot{\mathbf{S}}_{\Phi_1}, \mathbf{G}_{\mathrm{m}})$ with a finite étale kernel. Since this morphism factorizes as a composition of canonical morphisms

$$\dddot{\Xi}^{\mathrm{ord}}_{\Phi_n} \to \dddot{\Xi}^{\mathrm{ord}}_{\Phi_{\mathcal{H}},n} \underset{\dddot{C}^{\mathrm{ord}}_{\Phi_{\mathcal{H}},n}}{\times} \dddot{C}^{\mathrm{ord}}_{\Phi_n} \to \dddot{\Xi}^{\mathrm{ord}}_{\Phi_1} \underset{\dddot{C}^{\mathrm{ord}}_{\Phi_1}}{\times} \dddot{C}^{\mathrm{ord}}_{\Phi_n}, \tag{4.2.1.42}$$

in which the first one is surjective by definition (see (4.2.1.15) and (1) of Lemma 4.2.1.23), we see (by étale descent) that the canonical morphism $\dddot{\Xi}^{\mathrm{ord}}_{\Phi_{\mathcal{H}},n} \to \dddot{\Xi}^{\mathrm{ord}}_{\Phi_1} \underset{\dddot{C}^{\mathrm{ord}}_{\Phi_1}}{\times} \dddot{C}^{\mathrm{ord}}_{\Phi_{\mathcal{H}},n}$, which is the equivariant finite étale quotient by $H^{\mathrm{ord}}_{n,\mathrm{G}^{\mathrm{ess}}_{h,\mathbb{Z}_n} \ltimes \mathrm{U}^{\mathrm{ess}}_{1,\mathbb{Z}_n}}$ of the second morphism in (4.2.1.42), is also unramified. Consequently, the pullback $\dddot{\Xi}^{\mathrm{ord}}_{\Phi_{\mathcal{H}},n} \underset{\dddot{C}^{\mathrm{ord}}_{\Phi_{\mathcal{H}},n}}{\times} \vec{C}^{\mathrm{ord}}_{\Phi_{\mathcal{H}},\delta_{\mathcal{H}}} \to \dddot{\Xi}^{\mathrm{ord}}_{\Phi_1} \underset{\dddot{C}^{\mathrm{ord}}_{\Phi_1}}{\times} \vec{C}^{\mathrm{ord}}_{\Phi_{\mathcal{H}},\delta_{\mathcal{H}}}$ is unramified. By Proposition 4.2.1.37, the canonical morphisms $\vec{\Xi}^{\mathrm{ord}}_{\Phi_{\mathcal{H}},\delta_{\mathcal{H}}} \to \dddot{\Xi}^{\mathrm{ord}}_{\Phi_{\mathcal{H}},n} \underset{\dddot{C}^{\mathrm{ord}}_{\Phi_{\mathcal{H}},n}}{\times} \vec{C}^{\mathrm{ord}}_{\Phi_{\mathcal{H}},\delta_{\mathcal{H}}}$ and

$\vec{\Xi}^{\mathrm{ord}}_{\Phi_1,\delta_1} \underset{\vec{C}^{\mathrm{ord}}_{\Phi_1,\delta_1}}{\times} \vec{C}^{\mathrm{ord}}_{\Phi_{\mathcal{H}},\delta_{\mathcal{H}}} \to \ddot{\Xi}^{\mathrm{ord}}_{\Phi_1} \underset{\ddot{C}^{\mathrm{ord}}_{\Phi_1}}{\times} \vec{C}^{\mathrm{ord}}_{\Phi_{\mathcal{H}},\delta_{\mathcal{H}}}$ are closed immersions. Hence, the morphism $\vec{\Xi}^{\mathrm{ord}}_{\Phi_{\mathcal{H}},\delta_{\mathcal{H}}} \to \vec{\Xi}^{\mathrm{ord}}_{\Phi_1,\delta_1} \underset{\vec{C}^{\mathrm{ord}}_{\Phi_1,\delta_1}}{\times} \vec{C}^{\mathrm{ord}}_{\Phi_{\mathcal{H}},\delta_{\mathcal{H}}}$ (compatible with the above two closed immersions) is also unramified. Since this is a morphism between torus torsors equivariant with a homomorphism of tori of the same relative dimension (which is automatically surjective), by [35, IV-3, 11.3.10 a)⇒b) and 15.4.2 e′)⇒b)] (*cf.* the proof of [62, Lem. 1.3.1.11]), it is automatically flat, and hence finite étale, as desired. □

Corollary 4.2.1.43. *Suppose $\mathcal{H}' \subset \mathcal{H}$ and suppose (with similar assumptions at level $\mathcal{H}'$) $\vec{\Xi}^{\mathrm{ord}}_{\Phi_{\mathcal{H}'},\delta_{\mathcal{H}'}} \to \vec{C}^{\mathrm{ord}}_{\Phi_{\mathcal{H}'},\delta_{\mathcal{H}'}}$ are also defined. Then the canonical morphism $\vec{\Xi}^{\mathrm{ord}}_{\Phi_{\mathcal{H}'},\delta_{\mathcal{H}'}} \to \vec{\Xi}^{\mathrm{ord}}_{\Phi_{\mathcal{H}},\delta_{\mathcal{H}}} \underset{\vec{C}^{\mathrm{ord}}_{\Phi_{\mathcal{H}},\delta_{\mathcal{H}}}}{\times} \vec{C}^{\mathrm{ord}}_{\Phi_{\mathcal{H}'},\delta_{\mathcal{H}'}}$ is finite étale.*

Proof. By Proposition 4.2.1.41, the composition of canonical morphisms $\vec{\Xi}^{\mathrm{ord}}_{\Phi_{\mathcal{H}'},\delta_{\mathcal{H}'}} \to \vec{\Xi}^{\mathrm{ord}}_{\Phi_{\mathcal{H}},\delta_{\mathcal{H}}} \underset{\vec{C}^{\mathrm{ord}}_{\Phi_{\mathcal{H}},\delta_{\mathcal{H}}}}{\times} \vec{C}^{\mathrm{ord}}_{\Phi_{\mathcal{H}'},\delta_{\mathcal{H}'}} \to \vec{\Xi}^{\mathrm{ord}}_{\Phi_1,\delta_1} \underset{\vec{C}^{\mathrm{ord}}_{\Phi_1,\delta_1}}{\times} \vec{C}^{\mathrm{ord}}_{\Phi_{\mathcal{H}'},\delta_{\mathcal{H}'}}$ is finite étale. Since these are homomorphisms between torus torsors (equivariant with homomorphisms between tori) of the same relative dimension, they are both surjective, and hence finite étale, as desired. □

Corollary 4.2.1.44. *The algebraic stacks $\vec{\Xi}^{\mathrm{ord}}_{\Phi_{\mathcal{H}},\delta_{\mathcal{H}}}$, $\vec{C}^{\mathrm{ord}}_{\Phi_{\mathcal{H}},\delta_{\mathcal{H}}}$, $\vec{\mathsf{M}}^{\mathrm{ord},\Phi_{\mathcal{H}}}_{\mathcal{H}}$, and $\vec{\mathsf{M}}^{\mathrm{ord},\mathsf{Z}_{\mathcal{H}}}_{\mathcal{H}}$ are all separated, smooth, and of finite type over $\vec{\mathsf{S}}_{0,r_{\mathcal{H}}}$. If $\mathcal{H}^p$ is neat, they are all quasi-projective over $\vec{\mathsf{S}}_{0,r_{\mathcal{H}}}$.*

Proof. These follow from Propositions 3.4.6.3, 4.2.1.30, and 4.2.1.37. □

Convention 4.2.1.45. (*Compare with Convention* 3.4.2.9.) *We say that an object over a scheme S over $\mathsf{S}_{0,r_{\mathcal{H}}}$ parameterized by $\ddot{\Xi}^{\mathrm{ord}}_{\Phi_{\mathcal{H}},n}$ is parameterized by $\vec{\Xi}^{\mathrm{ord}}_{\Phi_{\mathcal{H}},\delta_{\mathcal{H}}}$ if the tautological morphism $S \to \ddot{\Xi}^{\mathrm{ord}}_{\Phi_{\mathcal{H}},n}$ determined by the universal property factors through $S \to \vec{\Xi}^{\mathrm{ord}}_{\Phi_{\mathcal{H}},\delta_{\mathcal{H}}}$. Then it also makes sense to consider the tautological tuple over $\vec{\Xi}^{\mathrm{ord}}_{\Phi_{\mathcal{H}},\delta_{\mathcal{H}}}$. We shall adopt the same convention for $\vec{C}^{\mathrm{ord}}_{\Phi_{\mathcal{H}},\delta_{\mathcal{H}}}$, $\vec{\mathsf{M}}^{\mathrm{ord},\Phi_{\mathcal{H}}}_{\mathcal{H}}$, and $\vec{\mathsf{M}}^{\mathrm{ord},\mathsf{Z}_{\mathcal{H}}}_{\mathcal{H}}$.*

It follows from the constructions above that we have the following proposition, in which $\vec{\Xi}^{\mathrm{ord}}_{\Phi_{\mathcal{H}},\delta_{\mathcal{H}}}$ *etc* are explicitly realized as normalizations in the paragraph preceding (4.2.1.28) (and hence they are, up to canonical isomorphisms, independent of the auxiliary choice of n):

Proposition 4.2.1.46. (*Compare with [62, Prop. 6.2.4.7; see also the errata].*) *Let $\mathcal{H}$, $\mathcal{H}^p$, and $\mathcal{H}_p$ be as in beginning of Section* 3.3.5, *and let $r_{\mathcal{H}}$ be as in Definition* 3.4.2.1. *Let us fix a representative $(\mathsf{Z}_{\mathcal{H}}, \Phi_{\mathcal{H}}, \delta_{\mathcal{H}})$ of an ordinary cusp label at level $\mathcal{H}$ (see Definition* 3.2.3.8), *where $\Phi_{\mathcal{H}} = (X, Y, \phi, \varphi_{-2,\mathcal{H}}, \varphi_{0,\mathcal{H}})$, which defines a finite étale cover $\vec{\mathsf{M}}^{\mathrm{ord},\Phi_{\mathcal{H}}}_{\mathcal{H}}$ of an algebraic stack $\vec{\mathsf{M}}^{\mathrm{ord},\mathsf{Z}_{\mathcal{H}}}_{\mathcal{H}}$ over $\vec{\mathsf{S}}_{0,r_{\mathcal{H}}} = \mathrm{Spec}(\mathcal{O}_{F_0,(p)}[\zeta_{p^{r_{\mathcal{H}}}}])$.*

(Then we can talk about the tautological tuples over $\vec{\mathsf{M}}_{\mathcal{H}}^{\mathrm{ord},\Phi_{\mathcal{H}}}$ *and* $\vec{\mathsf{M}}_{\mathcal{H}}^{\mathrm{ord},\mathsf{Z}_{\mathcal{H}}}$ *as in Convention* 4.2.1.45.*)*

Let us consider the category fibered in groupoids over the category of locally noetherian normal schemes over $\vec{\mathsf{M}}_{\mathcal{H}}^{\mathrm{ord},\mathsf{Z}_{\mathcal{H}}}$ *that are flat over* $\vec{\mathsf{S}}_{0,r_{\mathcal{H}}}$ *whose fiber over each scheme* S *(with the conditions just described) has objects the tuples*

$$(\mathsf{Z}_{\mathcal{H}}, (X,Y,\phi,\varphi_{-2,\mathcal{H}}^{\mathrm{ord}},\varphi_{0,\mathcal{H}}^{\mathrm{ord}}), (B,\lambda_B,i_B,\varphi_{-1,\mathcal{H}^p},\varphi_{-1,\mathcal{H}_p}^{\mathrm{ord}}), \delta_{\mathcal{H}}^{\mathrm{ord}}, (c_{\mathcal{H}}^{\mathrm{ord}}, c_{\mathcal{H}}^{\vee,\mathrm{ord}}, \tau_{\mathcal{H}}^{\mathrm{ord}})) \quad (4.2.1.47)$$

describing degeneration data without positivity condition over S *such that* $(B,\lambda_B,i_B,\varphi_{-1,\mathcal{H}^p},\varphi_{-1,\mathcal{H}_p}^{\mathrm{ord}})$ *is the pullback of the tautological tuple over* $\vec{\mathsf{M}}_{\mathcal{H}}^{\mathrm{ord},\mathsf{Z}_{\mathcal{H}}}$*, such that* $(\varphi_{-2,\mathcal{H}}^{\mathrm{ord}},\varphi_{0,\mathcal{H}}^{\mathrm{ord}})$ *(resp.* $\delta_{\mathcal{H}}^{\mathrm{ord}}$*) induces the* $(\varphi_{-2,\mathcal{H}},\varphi_{0,\mathcal{H}})$ *in* $\Phi_{\mathcal{H}}$ *(resp.* $\delta_{\mathcal{H}}$*), and such that the pullbacks of each tuple as in* (4.2.1.47) *to maximal points (see [36, 0, 2.1.2]) of* S *are induced as in Section* 4.1.6 *by the pullbacks of the corresponding tuple*

$$(\mathsf{Z}_{\mathcal{H}}, (X,Y,\phi,\varphi_{-2,\mathcal{H}}^{\sim},\varphi_{0,\mathcal{H}}^{\sim}), (B,\lambda_B,i_B,\varphi_{-1,\mathcal{H}}), \delta_{\mathcal{H}}, (c_{\mathcal{H}}, c_{\mathcal{H}}^{\vee}, \tau_{\mathcal{H}})) \quad (4.2.1.48)$$

parameterized by $\Xi_{\Phi_{\mathcal{H}},\delta_{\mathcal{H}}}$ *(see [62, Prop. 6.2.4.7; see also the errata]).*

Then there is an algebraic stack $\vec{\Xi}_{\Phi_{\mathcal{H}},\delta_{\mathcal{H}}}^{\mathrm{ord}}$ *separated, smooth, and schematic over* $\vec{\mathsf{M}}_{\mathcal{H}}^{\mathrm{ord},\mathsf{Z}_{\mathcal{H}}}$*, together with a tautological tuple and a natural action of* $\Gamma_{\Phi_{\mathcal{H}}}$ *on* $\vec{\Xi}_{\Phi_{\mathcal{H}},\delta_{\mathcal{H}}}^{\mathrm{ord}}$*, such that the quotient* $\vec{\Xi}_{\Phi_{\mathcal{H}},\delta_{\mathcal{H}}}^{\mathrm{ord}}/\Gamma_{\Phi_{\mathcal{H}}}$ *is isomorphic to the category described above (as categories fibered in groupoids over* $\vec{\mathsf{M}}_{\mathcal{H}}^{\mathrm{ord},\mathsf{Z}_{\mathcal{H}}}$*). Equivalently, for each tuple as in* (4.2.1.47) *over a scheme* S *over* $\vec{\mathsf{M}}_{\mathcal{H}}^{\mathrm{ord},\mathsf{Z}_{\mathcal{H}}}$ *(with properties described above), there is a morphism* $S \to \vec{\Xi}_{\Phi_{\mathcal{H}},\delta_{\mathcal{H}}}^{\mathrm{ord}}$ *(over* $\vec{\mathsf{M}}_{\mathcal{H}}^{\mathrm{ord},\mathsf{Z}_{\mathcal{H}}}$*), which is unique after we fix an isomorphism* $(f_Y : Y \xrightarrow{\sim} Y, f_X : X \xrightarrow{\sim} X)$ *in* $\Gamma_{\Phi_{\mathcal{H}}}$*, such that the tuple over* S *is the pullback of the tautological tuple over* $\vec{\Xi}_{\Phi_{\mathcal{H}},\delta_{\mathcal{H}}}^{\mathrm{ord}}$ *if we identify* X *by* f_X *and* Y *by* f_Y*.*

The structural morphism $\vec{\Xi}_{\Phi_{\mathcal{H}},\delta_{\mathcal{H}}}^{\mathrm{ord}} \to \vec{\mathsf{M}}_{\mathcal{H}}^{\mathrm{ord},\mathsf{Z}_{\mathcal{H}}}$ *factorizes as the composition* $\vec{\Xi}_{\Phi_{\mathcal{H}},\delta_{\mathcal{H}}}^{\mathrm{ord}} \to \vec{C}_{\Phi_{\mathcal{H}},\delta_{\mathcal{H}}}^{\mathrm{ord}} \to \vec{\mathsf{M}}_{\mathcal{H}}^{\mathrm{ord},\Phi_{\mathcal{H}}} \to \vec{\mathsf{M}}_{\mathcal{H}}^{\mathrm{ord},\mathsf{Z}_{\mathcal{H}}}$ *compatible with the natural actions of* $\Gamma_{\Phi_{\mathcal{H}}}$ *(trivial on* $\vec{\mathsf{M}}_{\mathcal{H}}^{\mathrm{ord},\mathsf{Z}_{\mathcal{H}}}$*), where* $\vec{\Xi}_{\Phi_{\mathcal{H}},\delta_{\mathcal{H}}}^{\mathrm{ord}} \to \vec{C}_{\Phi_{\mathcal{H}},\delta_{\mathcal{H}}}^{\mathrm{ord}}$ *is a torsor under the torus* $E_{\Phi_{\mathcal{H}}} \cong \underline{\mathrm{Hom}}_{\mathbb{Z}}(\mathbf{S}_{\Phi_{\mathcal{H}}}, \mathbf{G}_{\mathrm{m}})$*; where* $\vec{C}_{\Phi_{\mathcal{H}},\delta_{\mathcal{H}}}^{\mathrm{ord}} \to \vec{\mathsf{M}}_{\mathcal{H}}^{\mathrm{ord},\Phi_{\mathcal{H}}}$ *is an abelian scheme torsor, which is an abelian scheme when the condition in* (2) *of Lemma* 4.2.1.19 *is satisfied; and where* $\vec{\mathsf{M}}_{\mathcal{H}}^{\mathrm{ord},\Phi_{\mathcal{H}}} \to \vec{\mathsf{M}}_{\mathcal{H}}^{\mathrm{ord},\mathsf{Z}_{\mathcal{H}}}$ *is as above (and is finite étale), which is an isomorphism when the condition in the second statement of Lemma* 4.2.1.16 *is satisfied, inducing an isomorphism* $\vec{\mathsf{M}}_{\mathcal{H}}^{\mathrm{ord},\Phi_{\mathcal{H}}}/\Gamma_{\Phi_{\mathcal{H}}} \xrightarrow{\sim} \vec{\mathsf{M}}_{\mathcal{H}}^{\mathrm{ord},\mathsf{Z}_{\mathcal{H}}}$*. The* $E_{\Phi_{\mathcal{H}}}$*-torsor structure of* $\vec{\Xi}_{\Phi_{\mathcal{H}},\delta_{\mathcal{H}}}^{\mathrm{ord}}$ *defines a canonical homomorphism*

$$\mathbf{S}_{\Phi_{\mathcal{H}}} \to \underline{\mathrm{Pic}}(\vec{C}_{\Phi_{\mathcal{H}},\delta_{\mathcal{H}}}^{\mathrm{ord}}) : \ell \mapsto \vec{\Psi}_{\Phi_{\mathcal{H}},\delta_{\mathcal{H}}}^{\mathrm{ord}}(\ell), \quad (4.2.1.49)$$

assigning to each $\ell \in \mathbf{S}_{\Phi_{\mathcal{H}}}$ *an invertible sheaf* $\vec{\Psi}_{\Phi_{\mathcal{H}},\delta_{\mathcal{H}}}^{\mathrm{ord}}(\ell)$ *over* $\vec{C}_{\Phi_{\mathcal{H}},\delta_{\mathcal{H}}}^{\mathrm{ord}}$ *(up to isomorphism), together with isomorphisms*

$$\vec{\Delta}_{\Phi_{\mathcal{H}},\delta_{\mathcal{H}},\ell,\ell'}^{\mathrm{ord},*} : \vec{\Psi}_{\Phi_{\mathcal{H}},\delta_{\mathcal{H}}}^{\mathrm{ord}}(\ell) \underset{\mathscr{O}_{\vec{C}_{\Phi_{\mathcal{H}},\delta_{\mathcal{H}}}^{\mathrm{ord}}}}{\otimes} \vec{\Psi}_{\Phi_{\mathcal{H}},\delta_{\mathcal{H}}}^{\mathrm{ord}}(\ell') \xrightarrow{\sim} \vec{\Psi}_{\Phi_{\mathcal{H}},\delta_{\mathcal{H}}}^{\mathrm{ord}}(\ell+\ell')$$

for all $\ell, \ell' \in \mathbf{S}_{\Phi_{\mathcal{H}}}$*, satisfying the necessary compatibilities with each other making* $\underset{\ell \in \mathbf{S}_{\Phi_{\mathcal{H}}}}{\oplus} \vec{\Psi}^{\mathrm{ord}}_{\Phi_{\mathcal{H}},\delta_{\mathcal{H}}}(\ell)$ *an* $\mathscr{O}_{\vec{C}^{\mathrm{ord}}_{\Phi_{\mathcal{H}},\delta_{\mathcal{H}}}}$*-algebra, such that*

$$\vec{\Xi}^{\mathrm{ord}}_{\Phi_{\mathcal{H}},\delta_{\mathcal{H}}} \cong \underline{\mathrm{Spec}}_{\mathscr{O}_{\vec{C}^{\mathrm{ord}}_{\Phi_{\mathcal{H}},\delta_{\mathcal{H}}}}} \Big(\underset{\ell \in \mathbf{S}_{\Phi_{\mathcal{H}}}}{\oplus} \vec{\Psi}^{\mathrm{ord}}_{\Phi_{\mathcal{H}},\delta_{\mathcal{H}}}(\ell) \Big).$$

Remark 4.2.1.50. The condition that $(B, \lambda_B, i_B, \varphi^{\mathrm{ord}}_{-1,\mathcal{H}})$ is the pullback of the tautological tuple over $\vec{\mathsf{M}}^{\mathrm{ord},\mathsf{Z}_{\mathcal{H}}}_{\mathcal{H}}$ means in particular that $\varphi^{\mathrm{ord}}_{-1,\mathcal{H}}$ extends over all of S, not just at a maximal point (see [36, 0, 2.1.2]) over $\mathsf{S}_{0,p^{r_{\mathcal{H}}}}$ (*cf.* condition (5) in Definition 3.4.2.10). In general, this is a nontrivial condition even when the cusp label $[(\mathsf{Z}_{\mathcal{H}}, \Phi_{\mathcal{H}}, \delta_{\mathcal{H}})]$ has $\mathcal{O}$-multi-rank zero [62, Def. 5.4.2.7], which implies that B is an ordinary abelian scheme over S (*cf.* Remark 3.4.2.11). (The condition of being ordinary is irrelevant over the maximal points over $\mathsf{S}_{0,p^{r_{\mathcal{H}}}}$ because they are of characteristic zero.)

4.2.2 *Toroidal Embeddings, Positivity Conditions, and Mumford Families*

As in [62, Sec. 6.2.5], let $\mathbf{S}^{\vee}_{\Phi_{\mathcal{H}}} := \mathrm{Hom}_{\mathbb{Z}}(\mathbf{S}_{\Phi_{\mathcal{H}}}, \mathbb{Z})$ be the $\mathbb{Z}$-dual of $\mathbf{S}_{\Phi_{\mathcal{H}}}$, and let $(\mathbf{S}_{\Phi_{\mathcal{H}}})^{\vee}_{\mathbb{R}} := \mathbf{S}^{\vee}_{\Phi_{\mathcal{H}}} \underset{\mathbb{Z}}{\otimes} \mathbb{R} \cong \mathrm{Hom}_{\mathbb{Z}}(\mathbf{S}_{\Phi_{\mathcal{H}}}, \mathbb{R})$. By definition of $\mathbf{S}_{\Phi_{\mathcal{H}}}$ (in [62, Lem. 6.2.4.4]), the $\mathbb{R}$-vector space $(\mathbf{S}_{\Phi_{\mathcal{H}}})^{\vee}_{\mathbb{R}}$ is isomorphic to the $\mathbb{R}$-vector space of Hermitian pairings $(\!| \cdot, \cdot |\!) : (Y \underset{\mathbb{Z}}{\otimes} \mathbb{R}) \times (Y \underset{\mathbb{Z}}{\otimes} \mathbb{R}) \to \mathcal{O} \underset{\mathbb{Z}}{\otimes} \mathbb{R}$, by sending a Hermitian pairing $(\!| \cdot, \cdot |\!)$ to the element in $(\mathbf{S}_{\Phi_{\mathcal{H}}})^{\vee}_{\mathbb{R}}$ induced by the assignment $y \otimes \phi(y') \mapsto \mathrm{Tr}_{(\mathcal{O} \underset{\mathbb{Z}}{\otimes} \mathbb{R})/\mathbb{R}}(\!| y, y' |\!)$. (See [62, Lem. 1.1.4.5].) Let $\mathbf{P}_{\Phi_{\mathcal{H}}}$ (resp. $\mathbf{P}^{+}_{\Phi_{\mathcal{H}}}$) be the subset of $(\mathbf{S}_{\Phi_{\mathcal{H}}})^{\vee}_{\mathbb{R}}$ corresponding to positive semi-definite (resp. positive definite) Hermitian pairings with admissible radicals (see [62, Def. 6.2.5.6]). Then both $\mathbf{P}_{\Phi_{\mathcal{H}}}$ and $\mathbf{P}^{+}_{\Phi_{\mathcal{H}}}$ are cones in $(\mathbf{S}_{\Phi_{\mathcal{H}}})^{\vee}_{\mathbb{R}}$.

Let $\Sigma_{\Phi_{\mathcal{H}}}$ be a $\Gamma_{\Phi_{\mathcal{H}}}$-admissible smooth rational polyhedral cone decomposition of $\mathbf{P}_{\Phi_{\mathcal{H}}}$ with respect to the integral structure given by $\mathbf{S}^{\vee}_{\Phi_{\mathcal{H}}}$ in $(\mathbf{S}_{\Phi_{\mathcal{H}}})^{\vee}_{\mathbb{R}}$ (see Definition 1.2.2.6). For each $\sigma \in \Sigma_{\Phi_{\mathcal{H}}}$, consider the affine toroidal embedding

$$\vec{\Xi}^{\mathrm{ord}}_{\Phi_{\mathcal{H}},\delta_{\mathcal{H}}}(\sigma) := \underline{\mathrm{Spec}}_{\mathscr{O}_{\vec{C}^{\mathrm{ord}}_{\Phi_{\mathcal{H}},\delta_{\mathcal{H}}}}} \Big(\underset{\ell \in \sigma^{\vee} \subset \mathbf{S}_{\Phi_{\mathcal{H}}}}{\oplus} \vec{\Psi}^{\mathrm{ord}}_{\Phi_{\mathcal{H}},\delta_{\mathcal{H}}}(\ell) \Big), \tag{4.2.2.1}$$

with the σ-stratum defined by

$$\vec{\Xi}^{\mathrm{ord}}_{\Phi_{\mathcal{H}},\delta_{\mathcal{H}},\sigma} := \underline{\mathrm{Spec}}_{\mathscr{O}_{\vec{C}^{\mathrm{ord}}_{\Phi_{\mathcal{H}},\delta_{\mathcal{H}}}}} \Big(\underset{\ell \in \sigma^{\perp} \subset \mathbf{S}_{\Phi_{\mathcal{H}}}}{\oplus} \vec{\Psi}^{\mathrm{ord}}_{\Phi_{\mathcal{H}},\delta_{\mathcal{H}}}(\ell) \Big). \tag{4.2.2.2}$$

Then we have canonical morphisms

$$\vec{\Xi}^{\mathrm{ord}}_{\Phi_{\mathcal{H}},\delta_{\mathcal{H}}}(\tau) \to \vec{\Xi}^{\mathrm{ord}}_{\Phi_{\mathcal{H}},\delta_{\mathcal{H}}}(\sigma) \tag{4.2.2.3}$$

when $\overline{\tau} \subset \overline{\sigma}$, which is an open immersion when τ is a face of σ. By gluing $\vec{\Xi}^{\mathrm{ord}}_{\Phi_{\mathcal{H}},\delta_{\mathcal{H}}}(\sigma)$ over cones σ in $\Sigma_{\Phi_{\mathcal{H}}}$ using such open immersions along the faces, we obtain the toroidal embedding

$$\vec{\Xi}^{\mathrm{ord}}_{\Phi_{\mathcal{H}},\delta_{\mathcal{H}}} \hookrightarrow \overline{\vec{\Xi}}^{\mathrm{ord}}_{\Phi_{\mathcal{H}},\delta_{\mathcal{H}}} = \overline{\vec{\Xi}}^{\mathrm{ord}}_{\Phi_{\mathcal{H}},\delta_{\mathcal{H}},\Sigma_{\Phi_{\mathcal{H}}}} \tag{4.2.2.4}$$

defined by $\Sigma_{\Phi_{\mathcal{H}}}$ (as in [62, Def. 6.1.2.3]).

Proposition 4.2.2.5. (*Compare with [28, Ch. IV, p. 102] and [62, Prop. 6.2.5.8].*) *By construction, $\overline{\vec{\Xi}}^{\mathrm{ord}}_{\Phi_{\mathcal{H}},\delta_{\mathcal{H}}}$ has the properties described in [62, Thm. 6.1.2.8], with the following additional ones:*

(1) *There are constructible $\Gamma_{\Phi_{\mathcal{H}}}$-equivariant étale constructible sheaves (of $\mathcal{O}$-lattices) $\underline{X}$ and $\underline{Y}$ on $\overline{\vec{\Xi}}^{\mathrm{ord}}_{\Phi_{\mathcal{H}},\delta_{\mathcal{H}}}$, together with an ($\mathcal{O}$-equivariant) embedding $\underline{\phi} : \underline{Y} \hookrightarrow \underline{X}$, which are defined as follows:*
Each admissible surjection $X \twoheadrightarrow X'$ of $\mathcal{O}$-lattices (see Definition 1.2.1.2 and [62, Def. 1.2.6.7]) determines a surjection from $(\mathsf{Z}_{\mathcal{H}}, \Phi_{\mathcal{H}}, \delta_{\mathcal{H}})$ to some representative $(\mathsf{Z}'_{\mathcal{H}}, \Phi'_{\mathcal{H}}, \delta'_{\mathcal{H}})$ of a cusp label at level $\mathcal{H}$ by [62, Lem. 5.4.2.11], which is also compatible with the filtration D as in Definition 3.2.3.1 and hence also ordinary as in Definition 3.2.3.8, where $\mathsf{Z}'_{\mathcal{H}}$ and $\Phi'_{\mathcal{H}} = (X', Y', \phi', \varphi'_{-2,\mathcal{H}}, \varphi'_{0,\mathcal{H}})$ are uniquely determined by the construction. Consequently, it makes sense to define $\mathbf{P}_{\Phi'_{\mathcal{H}}}$ and an embedding $\mathbf{P}_{\Phi'_{\mathcal{H}}} \hookrightarrow \mathbf{P}_{\Phi_{\mathcal{H}}}$ for each admissible surjection $X \twoheadrightarrow X'$.
Over the locally closed stratum $\vec{\Xi}^{\mathrm{ord}}_{\Phi_{\mathcal{H}},\delta_{\mathcal{H}},\sigma_j}$, the sheaf $\underline{X}$ is the constant quotient sheaf X_{σ_j} of X, with the quotient $X \twoheadrightarrow X_{\sigma_j}$ an admissible surjection defining a pair $(\mathsf{Z}_{\mathcal{H},\sigma_j}, \Phi_{\mathcal{H},\sigma_j} = (X_{\sigma_j}, Y_{\sigma_j}, \phi_{\sigma_j}, \varphi_{-2,\mathcal{H},\sigma_j}, \varphi_{0,\mathcal{H},\sigma_j}))$ such that σ_j is contained in the image of the embedding $\mathbf{P}^+_{\phi_{\sigma_j}} \hookrightarrow \mathbf{P}_{\Phi_{\mathcal{H}}}$. We shall interpret this as having a sheaf version of $\Phi_{\mathcal{H}}$, written as $\underline{\Phi}_{\mathcal{H}} = (\underline{X}, \underline{Y}, \underline{\phi}, \underline{\varphi}_{-2,\mathcal{H}}, \underline{\varphi}_{0,\mathcal{H}})$.

(2) *The formation of $\mathbf{S}_{\Phi_{\mathcal{H}}}$ from $\Phi_{\mathcal{H}}$ applies to $\underline{\Phi}_{\mathcal{H}}$ and defines a sheaf $\underline{\mathbf{S}}_{\underline{\Phi}_{\mathcal{H}}}$.*

(3) *There is a tautological homomorphism $\underline{B} : \underline{\mathbf{S}}_{\underline{\Phi}_{\mathcal{H}}} \to \underline{\mathrm{Inv}}(\overline{\vec{\Xi}}^{\mathrm{ord}}_{\Phi_{\mathcal{H}},\delta_{\mathcal{H}}})$ of constructible sheaves of groups (see [62, Def. 4.2.4.1]) which sends the class of $\ell \in \mathbf{S}_{\Phi_{\mathcal{H},\sigma_j}}$ to the sheaf of ideals $\mathscr{O}_{\vec{\Xi}^{\mathrm{ord}}_{\Phi_{\mathcal{H}},\delta_{\mathcal{H}}}(\sigma_j)} \underset{\mathscr{O}_{\vec{C}^{\mathrm{ord}}_{\Phi_{\mathcal{H}},\delta_{\mathcal{H}}}}}{\otimes} \vec{\Psi}^{\mathrm{ord}}_{\Phi_{\mathcal{H}},\delta_{\mathcal{H}}}(\ell)$ on $\vec{\Xi}^{\mathrm{ord}}_{\Phi_{\mathcal{H}},\delta_{\mathcal{H}}}(\sigma_j)$, such that we have the following:*
   (a) *This homomorphism $\underline{B}$ is $\Gamma_{\Phi_{\mathcal{H}}}$-equivariant (because it is compatible with twists of the identification of $\Phi_{\mathcal{H}}$) and $E_{\Phi_{\mathcal{H}}}$-invariant (because $\vec{\Psi}^{\mathrm{ord}}_{\Phi_{\mathcal{H}},\delta_{\mathcal{H}}}(\ell)$ corresponds to a weight subsheaf of the $\mathscr{O}_{\vec{C}^{\mathrm{ord}}_{\Phi_{\mathcal{H}},\delta_{\mathcal{H}}}}$-algebra $\mathscr{O}_{\vec{\Xi}^{\mathrm{ord}}_{\Phi_{\mathcal{H}},\delta_{\mathcal{H}}}}$ under the action of $E_{\Phi_{\mathcal{H}}}$), and is trivial on the open subscheme $\vec{\Xi}^{\mathrm{ord}}_{\Phi_{\mathcal{H}},\delta_{\mathcal{H}}}$ of $\overline{\vec{\Xi}}^{\mathrm{ord}}_{\Phi_{\mathcal{H}},\delta_{\mathcal{H}}}$.*
   (b) *For each local section y of $\underline{Y}$, the support of $\underline{B}(y \otimes \phi(y))$ is effective, and is the same as the support of y. This is because $\sigma(y, \phi(y)) \geq 0$ for all $\sigma \subset \mathbf{P}_{\Phi_{\mathcal{H}}}$ and $y \in Y$, and $\sigma(y, \phi(y)) > 0$ when $\sigma \subset \mathbf{P}^+_{\Phi_{\mathcal{H},\sigma}}$ and $0 \neq y \in Y_\sigma$.*

Moreover, each open subalgebraic stack $\vec{\Xi}^{\mathrm{ord}}_{\Phi_{\mathcal{H}},\delta_{\mathcal{H}}}(\sigma)$ of $\overline{\vec{\Xi}}^{\mathrm{ord}}_{\Phi_{\mathcal{H}},\delta_{\mathcal{H}}}$ enjoys the following universal property as in [62, Prop. 6.2.5.11]:

Let R be a noetherian normal domain with fraction field K, and suppose we have a morphism $t_R : \mathrm{Spec}(R) \to \vec{C}^{\mathrm{ord}}_{\Phi_{\mathcal{H}},\delta_{\mathcal{H}}}$ that is liftable over $\mathrm{Spec}(K)$ to a morphism $\tilde{t}_K : \mathrm{Spec}(K) \to \vec{\Xi}^{\mathrm{ord}}_{\Phi_{\mathcal{H}},\delta_{\mathcal{H}}}$. By abuse of notation, let us denote by $\vec{\Psi}^{\mathrm{ord}}_{\Phi_{\mathcal{H}},\delta_{\mathcal{H}}}(\ell)_R$ the R-invertible module defined by the pullback under t_R of the invertible sheaf $\vec{\Psi}^{\mathrm{ord}}_{\Phi_{\mathcal{H}},\delta_{\mathcal{H}}}(\ell)$ over $\vec{C}^{\mathrm{ord}}_{\Phi_{\mathcal{H}},\delta_{\mathcal{H}}}$, and denote $\vec{\Psi}^{\mathrm{ord}}_{\Phi_{\mathcal{H}},\delta_{\mathcal{H}}}(\ell)_R \underset{R}{\otimes} K$ by $\vec{\Psi}^{\mathrm{ord}}_{\Phi_{\mathcal{H}},\delta_{\mathcal{H}}}(\ell)_K$. Since $\vec{\Xi}^{\mathrm{ord}}_{\Phi_{\mathcal{H}},\delta_{\mathcal{H}}} \cong \underline{\mathrm{Spec}}_{\mathscr{O}_{\vec{C}^{\mathrm{ord}}_{\Phi_{\mathcal{H}},\delta_{\mathcal{H}}}}}\Big(\underset{\ell\in\mathbf{S}_{\Phi_{\mathcal{H}}}}{\oplus} \vec{\Psi}^{\mathrm{ord}}_{\Phi_{\mathcal{H}},\delta_{\mathcal{H}}}(\ell)\Big)$, the morphism $\tilde{t}_K$ defines isomorphisms $\vec{\Psi}^{\mathrm{ord}}_{\Phi_{\mathcal{H}},\delta_{\mathcal{H}}}(\ell)_K \overset{\sim}{\to} K$, which defines an embedding of $\vec{\Psi}^{\mathrm{ord}}_{\Phi_{\mathcal{H}},\delta_{\mathcal{H}}}(\ell)_R$ as an R-invertible submodule I_ℓ of K. Therefore, the pullback of the homomorphism (4.2.1.49) in Proposition 4.2.1.46 determines a homomorphism

$$B : \mathbf{S}_{\Phi_{\mathcal{H}}} \to \mathrm{Inv}(R) : \ell \mapsto I_\ell \tag{4.2.2.6}$$

(see [62, Def. 4.2.4.1]). If $\ell = [y \otimes \chi]$ for some $y \in Y$ and $\chi \in X$, then $\vec{\Psi}^{\mathrm{ord}}_{\Phi_{\mathcal{H}},\delta_{\mathcal{H}}}(\ell) \cong (c^\vee(y), c(\chi))^* \mathcal{P}_B$ by construction, and hence $I_\ell = I_{y,\chi}$ as R-invertible submodules of K (see [62, Def. 4.2.4.6]; *cf.* (4b) of Definition 4.1.3.2). For each discrete valuation $\upsilon : K^\times \to \mathbb{Z}$ of K, since I_ℓ is locally principal for every ℓ, it makes sense to consider the composition

$$\upsilon \circ B : \mathbf{S}_{\Phi_{\mathcal{H}}} \to \mathbb{Z} : \ell \mapsto \upsilon(I_\ell), \tag{4.2.2.7}$$

which is an element in $\mathbf{S}^\vee_{\Phi_{\mathcal{H}}}$ (*cf.* [62, (6.2.5.9) and (6.2.5.10)]).

Proposition 4.2.2.8. (*Compare with [62, Prop. 6.2.5.11].*) *With assumptions and notation as above, the universal property of* $\vec{\Xi}^{\mathrm{ord}}_{\Phi_{\mathcal{H}},\delta_{\mathcal{H}}}(\sigma)$ *is as follows: The morphism* $\tilde{t}_K : \mathrm{Spec}(K) \to \vec{\Xi}^{\mathrm{ord}}_{\Phi_{\mathcal{H}},\delta_{\mathcal{H}}}$ *extends to a morphism* $\tilde{t}_R : \mathrm{Spec}(R) \to \vec{\Xi}^{\mathrm{ord}}_{\Phi_{\mathcal{H}},\delta_{\mathcal{H}}}(\sigma)$ *if and only if, for every discrete valuation* $\upsilon : K^\times \to \mathbb{Z}$ *of* K *such that* $\upsilon(R) \geq 0$, *the corresponding homomorphism* $\upsilon \circ B : \mathbf{S}_{\Phi_{\mathcal{H}}} \to \mathbb{Z}$ *as in* (4.2.2.7) (*or rather its composition with* $\mathbb{Z} \hookrightarrow \mathbb{R}$) *lies in the closure* $\overline{\sigma}$ *of* σ *in* $(\mathbf{S}_{\Phi_{\mathcal{H}}})^\vee_{\mathbb{R}}$.

Proof. Since $\vec{\Xi}^{\mathrm{ord}}_{\Phi_{\mathcal{H}},\delta_{\mathcal{H}}}(\sigma) \cong \underline{\mathrm{Spec}}_{\mathscr{O}_{\vec{C}^{\mathrm{ord}}_{\Phi_{\mathcal{H}},\delta_{\mathcal{H}}}}}\Big(\underset{\ell\in\sigma^\vee}{\oplus} \vec{\Psi}^{\mathrm{ord}}_{\Phi_{\mathcal{H}},\delta_{\mathcal{H}}}(\ell)\Big)$ is relatively affine over $\vec{C}^{\mathrm{ord}}_{\Phi_{\mathcal{H}},\delta_{\mathcal{H}}}$, the morphism $\tilde{t}_K$ extends to a morphism $\tilde{t}_R : \mathrm{Spec}(R) \to \vec{\Xi}^{\mathrm{ord}}_{\Phi_{\mathcal{H}},\delta_{\mathcal{H}}}(\sigma)$ if $I_\ell \subset R$ for every $\ell \in \sigma^\vee$. Since R is noetherian and normal, this is true if $(\upsilon \circ B)(\ell) \geq 0$ for every discrete valuation υ of K such that $\upsilon(R) \geq 0$ and for every $\ell \in \sigma^\vee$, or equivalently if $\upsilon \circ B$ pairs nonnegatively with $\sigma^\vee$ under the canonical pairing between $\mathbf{S}^\vee_{\Phi_{\mathcal{H}}}$ and $(\mathbf{S}_{\Phi_{\mathcal{H}}})^\vee_{\mathbb{R}}$, or equivalently if $\upsilon \circ B$ lies in $\overline{\sigma}$, as desired. $\square$

Remark 4.2.2.9. If $\tilde{t}_K$ extends to $\tilde{t}_R$, then the homomorphism $B : \mathbf{S}_{\Phi_{\mathcal{H}}} \to \mathrm{Inv}(R)$ agrees with the pullback of the homomorphism $\underline{B} : \underline{\mathbf{S}}_{\Phi_{\mathcal{H}}} \to \underline{\mathrm{Inv}}(\vec{\Xi}^{\mathrm{ord}}_{\Phi_{\mathcal{H}},\delta_{\mathcal{H}}})$ under $\tilde{t}_R$. (Thus, the notation is consistent when B and $\underline{B}$ can be compared over R.)

Remark 4.2.2.10. Recall that the σ-stratum $\vec{\Xi}^{\mathrm{ord}}_{\Phi_{\mathcal{H}},\delta_{\mathcal{H}},\sigma}$ of $\vec{\Xi}^{\mathrm{ord}}_{\Phi_{\mathcal{H}},\delta_{\mathcal{H}}}(\sigma)$ is defined (see [62, Def. 6.1.2.7]) by the sheaf of ideals $\mathscr{I}^{\mathrm{ord}}_{\Phi_{\mathcal{H}},\delta_{\mathcal{H}},\sigma} \cong \underset{\ell\in\sigma_0^\vee}{\oplus} \vec{\Psi}^{\mathrm{ord}}_{\Phi_{\mathcal{H}},\delta_{\mathcal{H}}}(\ell)$ in $\mathscr{O}_{\vec{\Xi}^{\mathrm{ord}}_{\Phi_{\mathcal{H}},\delta_{\mathcal{H}}}(\sigma)} \cong$

$\underset{\ell\in\sigma^\vee}{\oplus}\vec{\Psi}^{\mathrm{ord}}_{\Phi_{\mathcal{H}},\delta_{\mathcal{H}}}(\ell)$ (*cf.* [62, Conv. 6.2.3.20]). Since $\sigma \subset \mathbf{P}_{\Phi_{\mathcal{H}}}$ is positive semidefinite, we have $\sigma(\ell) \geq 0$ for every ℓ of the form $[y \otimes \phi(y)]$. As a result, the trivialization

$$\tau(y,\phi(y)) : \mathscr{O}_{\vec{\Xi}^{\mathrm{ord}}_{\Phi_{\mathcal{H}},\delta_{\mathcal{H}}}} \underset{\mathscr{O}_{\vec{C}^{\mathrm{ord}}_{\Phi_{\mathcal{H}},\delta_{\mathcal{H}}}}}{\otimes} \vec{\Psi}^{\mathrm{ord}}_{\Phi_{\mathcal{H}},\delta_{\mathcal{H}}}(y\otimes\phi(y)) \xrightarrow{\sim} \mathscr{O}_{\vec{\Xi}^{\mathrm{ord}}_{\Phi_{\mathcal{H}},\delta_{\mathcal{H}}}}$$

over $\vec{\Xi}^{\mathrm{ord}}_{\Phi_{\mathcal{H}},\delta_{\mathcal{H}}}$ extends to a section

$$\tau(y,\phi(y)) : \mathscr{O}_{\vec{\Xi}^{\mathrm{ord}}_{\Phi_{\mathcal{H}},\delta_{\mathcal{H}}}(\sigma)} \underset{\mathscr{O}_{\vec{C}^{\mathrm{ord}}_{\Phi_{\mathcal{H}},\delta_{\mathcal{H}}}}}{\otimes} \vec{\Psi}^{\mathrm{ord}}_{\Phi_{\mathcal{H}},\delta_{\mathcal{H}}}(y\otimes\phi(y)) \to \mathscr{O}_{\vec{\Xi}^{\mathrm{ord}}_{\Phi_{\mathcal{H}},\delta_{\mathcal{H}}}(\sigma)} \tag{4.2.2.11}$$

over $\vec{\Xi}^{\mathrm{ord}}_{\Phi_{\mathcal{H}},\delta_{\mathcal{H}}}(\sigma)$. If $\sigma \subset \mathbf{P}^+_{\Phi_{\mathcal{H}}}$, then by [62, Lem. 6.2.5.7], we have $\sigma(y\otimes\phi(y)) > 0$ for every $y \neq 0$. In this case, the section $\tau(y,\phi(y))$ over $\vec{\Xi}^{\mathrm{ord}}_{\Phi_{\mathcal{H}},\delta_{\mathcal{H}}}(\sigma)$ as in (4.2.2.11) has image contained in $\mathscr{I}^{\mathrm{ord}}_{\Phi_{\mathcal{H}},\delta_{\mathcal{H}},\sigma}$. This is almost the positivity condition, except that the base scheme is not completed along $\mathscr{I}^{\mathrm{ord}}_{\Phi_{\mathcal{H}},\delta_{\mathcal{H}},\sigma}$.

Let $\vec{\mathfrak{X}}^{\mathrm{ord}}_{\Phi_{\mathcal{H}},\delta_{\mathcal{H}}} = \vec{\mathfrak{X}}^{\mathrm{ord}}_{\Phi_{\mathcal{H}},\delta_{\mathcal{H}},\Sigma_{\Phi_{\mathcal{H}}}}$ be the formal completion of $\overline{\vec{\Xi}}^{\mathrm{ord}}_{\Phi_{\mathcal{H}},\delta_{\mathcal{H}}}$ along the union of the σ-strata $\vec{\Xi}^{\mathrm{ord}}_{\Phi_{\mathcal{H}},\delta_{\mathcal{H}},\sigma}$ for $\sigma \subset \mathbf{P}^+_{\Phi_{\mathcal{H}}}$. For each $\sigma \subset \mathbf{P}^+_{\Phi_{\mathcal{H}}}$, let $\vec{\mathfrak{X}}^{\mathrm{ord}}_{\Phi_{\mathcal{H}},\delta_{\mathcal{H}},\sigma}$ be the formal completion of $\vec{\Xi}^{\mathrm{ord}}_{\Phi_{\mathcal{H}},\delta_{\mathcal{H}}}(\sigma)$ along the σ-stratum $\vec{\Xi}^{\mathrm{ord}}_{\Phi_{\mathcal{H}},\delta_{\mathcal{H}},\sigma}$. Then, using the language of relative schemes over formal algebraic stacks (see [37]), there are tautological tuples of the form

$$\begin{aligned}&(\mathsf{Z}_{\mathcal{H}},(X,Y,\phi,\varphi^{\mathrm{ord}}_{-2,\mathcal{H}},\varphi^{\mathrm{ord}}_{0,\mathcal{H}}),\\ &\qquad(B,\lambda_B,i_B,\varphi_{-1,\mathcal{H}^p},\varphi^{\mathrm{ord}}_{-1,\mathcal{H}_p}),\delta^{\mathrm{ord}}_{\mathcal{H}},(c^{\mathrm{ord}}_{\mathcal{H}},c^{\vee,\mathrm{ord}}_{\mathcal{H}},\tau^{\mathrm{ord}}_{\mathcal{H}})).\end{aligned} \tag{4.2.2.12}$$

over both the formal algebraic stacks $\vec{\mathfrak{X}}^{\mathrm{ord}}_{\Phi_{\mathcal{H}},\delta_{\mathcal{H}}}$ and $\vec{\mathfrak{X}}^{\mathrm{ord}}_{\Phi_{\mathcal{H}},\delta_{\mathcal{H}},\sigma}$, the one on the latter being the pullback of the one on the former under the canonical morphism $\vec{\mathfrak{X}}^{\mathrm{ord}}_{\Phi_{\mathcal{H}},\delta_{\mathcal{H}},\sigma} \to \vec{\mathfrak{X}}^{\mathrm{ord}}_{\Phi_{\mathcal{H}},\delta_{\mathcal{H}}}$. Moreover, this tautological tuple over $\vec{\mathfrak{X}}^{\mathrm{ord}}_{\Phi_{\mathcal{H}},\delta_{\mathcal{H}}}$ satisfies the *positivity condition* in the following sense: We have a functorial assignment that, to each connected affine formal scheme $\mathfrak{U}$ with an étale (*i.e.*, formally étale and of finite type; see [35, I, 10.13.3]) morphism $\mathfrak{U} \to \vec{\mathfrak{X}}^{\mathrm{ord}}_{\Phi_{\mathcal{H}},\delta_{\mathcal{H}}}$, assigns a tuple of the form (4.2.2.12) (with positivity condition) over the (normal) scheme $\mathrm{Spec}(\Gamma(\mathfrak{U},\mathscr{O}_{\mathfrak{U}}))$ (smooth over $\vec{\mathsf{S}}_{0,r_{\mathcal{H}}}$). Let $R = \Gamma(\mathfrak{U},\mathscr{O}_{\mathfrak{U}})$, and let I be (the radical of) its ideal of definition. Then R and I satisfies the requirement in Section 4.1.2, and we obtain a tuple defining an object of $\mathrm{DD}_{\mathrm{PEL},\mathsf{M}^{\mathrm{ord}}_{\mathcal{H}}}(R,I)$.

By Theorems 4.1.5.27 and 4.1.6.2, Mumford's construction defines an object

$$({}^\heartsuit G,\ {}^\heartsuit\lambda,\ {}^\heartsuit i,\ {}^\heartsuit\alpha_{\mathcal{H}^p},\ {}^\heartsuit\alpha^{\mathrm{ord}}_{\mathcal{H}_p}) \to \mathrm{Spec}(R) \tag{4.2.2.13}$$

in $\mathrm{DEG}_{\mathrm{PEL},\mathsf{M}^{\mathrm{ord}}_{\mathcal{H}}}(R,I)$, which comes from an object

$$({}^\heartsuit G,\ {}^\heartsuit\lambda,\ {}^\heartsuit i,\ {}^\heartsuit\alpha_{\mathcal{H}}) \to \mathrm{Spec}(R) \tag{4.2.2.14}$$

in $\mathrm{DEG}_{\mathrm{PEL},\mathsf{M}_{\mathcal{H}}}(R,I)$ in the sense that $({}^\heartsuit\alpha_{\mathcal{H}^p},\ {}^\heartsuit\alpha^{\mathrm{ord}}_{\mathcal{H}_p})$ is assigned to ${}^\heartsuit\alpha_{\mathcal{H}}$ under (3.3.5.5) over the generic point $\mathrm{Spec}(\mathrm{Frac}(R))$ of $\mathrm{Spec}(R)$. Since the tuple (4.2.2.14) is an object of $\mathrm{DEG}_{\mathrm{PEL},\mathsf{M}_{\mathcal{H}}}(R,I)$, it is a degenerating family of type $\mathsf{M}_{\mathcal{H}}$ as in Definition 1.3.1.1. Since $\varphi^{\mathrm{ord}}_{-1,\mathcal{H}}$ is defined over all of R, the tuple (4.2.2.13) is a degenerating family of type $\vec{\mathsf{M}}^{\mathrm{ord}}_{\mathcal{H}}$ as in Definition 3.4.2.10. Moreover, the torus part of each fiber of ${}^{\heartsuit}G$ over the support of $\mathfrak{U}$ is split with character group X. If we have an étale (*i.e.*, formally étale and of finite type; see [35, I, 10.13.3]) morphism $\mathrm{Spf}(R_1) \to \mathrm{Spf}(R_2)$ and if the degeneration datum over $\mathrm{Spec}(R_2)$ pulls back to the degeneration datum over $\mathrm{Spec}(R_1)$, then the family constructed by Mumford's construction over $\mathrm{Spec}(R_1)$ pulls back to a family over $\mathrm{Spec}(R_2)$ with the same degeneration datum as the datum over $\mathrm{Spec}(R_2)$. The functoriality in [62, Thm. 4.4.16] over $\mathrm{Spec}(R_2)$ then assures that this pullback family agrees with the family constructed from the datum over $\mathrm{Spec}(R_2)$. In particular, we see that the assignment of $({}^{\heartsuit}G, {}^{\heartsuit}\lambda, {}^{\heartsuit}i, {}^{\heartsuit}\alpha_{\mathcal{H}^p}, {}^{\heartsuit}\alpha^{\mathrm{ord}}_{\mathcal{H}_p}) \to \mathrm{Spec}(\Gamma(\mathfrak{U}, \mathscr{O}_{\mathfrak{U}}))$ to $\mathfrak{U}$ is functorial. Hence, the assignment defines a (relative) degenerating family

$$({}^{\heartsuit}G, {}^{\heartsuit}\lambda, {}^{\heartsuit}i, {}^{\heartsuit}\alpha_{\mathcal{H}^p}, {}^{\heartsuit}\alpha^{\mathrm{ord}}_{\mathcal{H}_p}) \to \vec{\mathfrak{X}}^{\mathrm{ord}}_{\Phi_{\mathcal{H}},\delta_{\mathcal{H}}}, \tag{4.2.2.15}$$

which comes from a degenerating family

$$({}^{\heartsuit}G, {}^{\heartsuit}\lambda, {}^{\heartsuit}i, {}^{\heartsuit}\alpha_{\mathcal{H}}) \to \vec{\mathfrak{X}}^{\mathrm{ord}}_{\Phi_{\mathcal{H}},\delta_{\mathcal{H}}} \tag{4.2.2.16}$$

in the sense that it is so over each affine formal scheme $\mathfrak{U}$ as above. Since the cone decomposition $\Sigma_{\Phi_{\mathcal{H}}}$ is $\Gamma_{\Phi_{\mathcal{H}}}$-admissible, the group $\Gamma_{\Phi_{\mathcal{H}}}$ acts naturally on all the objects involved in the degeneration data, and hence by functoriality on the degenerating family $({}^{\heartsuit}G, {}^{\heartsuit}\lambda, {}^{\heartsuit}i, {}^{\heartsuit}\alpha_{\mathcal{H}^p}, {}^{\heartsuit}\alpha^{\mathrm{ord}}_{\mathcal{H}_p}) \to \vec{\mathfrak{X}}^{\mathrm{ord}}_{\Phi_{\mathcal{H}},\delta_{\mathcal{H}}}$.

For each $\sigma \subset \mathbf{P}_{\Phi_{\mathcal{H}}}$, let $\Gamma_{\Phi_{\mathcal{H}},\sigma}$ be defined as in [62, Def. 6.2.5.23], namely the subgroup of $\Gamma_{\Phi_{\mathcal{H}}}$ consisting of elements that maps σ to itself under the natural action of $\Gamma_{\Phi_{\mathcal{H}}}$ on $\mathbf{P}_{\Phi_{\mathcal{H}}}$. Then we have similarly the degenerating family

$$({}^{\heartsuit}G, {}^{\heartsuit}\lambda, {}^{\heartsuit}i, {}^{\heartsuit}\alpha_{\mathcal{H}^p}, {}^{\heartsuit}\alpha^{\mathrm{ord}}_{\mathcal{H}_p}) \to \vec{\mathfrak{X}}^{\mathrm{ord}}_{\Phi_{\mathcal{H}},\delta_{\mathcal{H}},\sigma}, \tag{4.2.2.17}$$

together with an equivariant action of $\Gamma_{\Phi_{\mathcal{H}},\sigma}$, which comes from a degenerating family

$$({}^{\heartsuit}G, {}^{\heartsuit}\lambda, {}^{\heartsuit}i, {}^{\heartsuit}\alpha_{\mathcal{H}}) \to \vec{\mathfrak{X}}^{\mathrm{ord}}_{\Phi_{\mathcal{H}},\delta_{\mathcal{H}},\sigma} \tag{4.2.2.18}$$

(in a sense analogous to that of (4.2.2.15) and (4.2.2.16)), together with a compatible equivariant action of $\Gamma_{\Phi_{\mathcal{H}},\sigma}$.

By [62, Rem. 6.2.5.26 and Lem. 6.2.5.27], if the cone decomposition $\Sigma_{\Phi_{\mathcal{H}}}$ is chosen such that [62, Cond. 6.2.5.25] is satisfied, then each $\Gamma_{\Phi_{\mathcal{H}},\sigma}$ is finite and $\vec{\mathfrak{X}}^{\mathrm{ord}}_{\Phi_{\mathcal{H}},\delta_{\mathcal{H}},\sigma}/\Gamma_{\Phi_{\mathcal{H}},\sigma}$ is a formal (Deligne–Mumford) algebraic stack. Moreover, if $\mathcal{H}$ is neat (which is the case, for example, when $\mathcal{H}^p$ is neat), then $\Gamma_{\Phi_{\mathcal{H}},\sigma}$ is trivial and $\vec{\mathfrak{X}}^{\mathrm{ord}}_{\Phi_{\mathcal{H}},\delta_{\mathcal{H}},\sigma}/\Gamma_{\Phi_{\mathcal{H}},\sigma} = \vec{\mathfrak{X}}^{\mathrm{ord}}_{\Phi_{\mathcal{H}},\delta_{\mathcal{H}},\sigma}$ is a formal algebraic space.

From now on, as always, let us assume that the cone decomposition $\Sigma_{\Phi_{\mathcal{H}}}$ is chosen such that [62, Cond. 6.2.5.25] is satisfied. This is possible by refining any given cone decomposition $\Sigma_{\Phi_{\mathcal{H}}}$. Then the compatible equivariant action of $\Gamma_{\Phi_{\mathcal{H}},\sigma}$ on $({}^{\heartsuit}G, {}^{\heartsuit}\lambda, {}^{\heartsuit}i, {}^{\heartsuit}\alpha_{\mathcal{H}^p}, {}^{\heartsuit}\alpha^{\mathrm{ord}}_{\mathcal{H}_p}) \to \vec{\mathfrak{X}}^{\mathrm{ord}}_{\Phi_{\mathcal{H}},\delta_{\mathcal{H}},\sigma}$ and $({}^{\heartsuit}G, {}^{\heartsuit}\lambda, {}^{\heartsuit}i, {}^{\heartsuit}\alpha_{\mathcal{H}}) \to \vec{\mathfrak{X}}^{\mathrm{ord}}_{\Phi_{\mathcal{H}},\delta_{\mathcal{H}},\sigma}$ imply that we have a descended family

$$({}^{\heartsuit}G, {}^{\heartsuit}\lambda, {}^{\heartsuit}i, {}^{\heartsuit}\alpha_{\mathcal{H}^p}, {}^{\heartsuit}\alpha^{\mathrm{ord}}_{\mathcal{H}_p}) \to \vec{\mathfrak{X}}^{\mathrm{ord}}_{\Phi_{\mathcal{H}},\delta_{\mathcal{H}},\sigma}/\Gamma_{\Phi_{\mathcal{H}},\sigma}, \tag{4.2.2.19}$$

which comes from a descended family

$$({}^{\heartsuit}G,\ {}^{\heartsuit}\lambda,\ {}^{\heartsuit}i,\ {}^{\heartsuit}\alpha_{\mathcal{H}}) \to \vec{\mathfrak{X}}^{\mathrm{ord}}_{\Phi_{\mathcal{H}},\delta_{\mathcal{H}},\sigma}/\Gamma_{\Phi_{\mathcal{H}},\sigma} \tag{4.2.2.20}$$

(in a sense analogous to that of (4.2.2.15) and (4.2.2.16)).

Definition 4.2.2.21. (*Compare with [62, Def. 6.2.5.28].*) *All the degenerating families* (4.2.2.15), (4.2.2.16), (4.2.2.17), (4.2.2.18), (4.2.2.19), *and* (4.2.2.20) *constructed above are called* **Mumford families**.

Remark 4.2.2.22. (Compare with [62, Rem. 6.2.5.29].) By abuse of notation, we will use the same notation $({}^{\heartsuit}G,\ {}^{\heartsuit}\lambda,\ {}^{\heartsuit}i,\ {}^{\heartsuit}\alpha_{\mathcal{H}^p},\ {}^{\heartsuit}\alpha^{\mathrm{ord}}_{\mathcal{H}_p})$ and $({}^{\heartsuit}G,\ {}^{\heartsuit}\lambda,\ {}^{\heartsuit}i,\ {}^{\heartsuit}\alpha_{\mathcal{H}})$ for the Mumford families over various bases.

Remark 4.2.2.23. The analogues of [62, Rem. 6.2.5.30 and 6.2.5.31] are also true in this context.

4.2.3 *Extended Kodaira–Spencer Morphisms and Induced Isomorphisms*

Since we have the tautological presence of $G^{\natural}$ and ι (defined by the tautological tuple $(B, \underline{X}, \underline{Y}, c, c^{\vee}, \tau)$) over the algebraic stack $\vec{\Xi}^{\mathrm{ord}}_{\Phi_{\mathcal{H}},\delta_{\mathcal{H}}}$ separated, smooth, and locally of finite type over $\vec{\mathsf{S}}_{0,r_{\mathcal{H}}}$, we can define (by étale descent if necessary) as in [62, Sec. 4.6.2] the extended Kodaira–Spencer morphism

$$\begin{aligned}&\mathrm{KS}_{(G^{\natural},\iota)/\vec{\Xi}^{\mathrm{ord}}_{\Phi_{\mathcal{H}},\delta_{\mathcal{H}}}/\vec{\mathsf{S}}_{0,r_{\mathcal{H}}}} :\\ &\underline{\mathrm{Lie}}^{\vee}_{G^{\natural}/\vec{\Xi}^{\mathrm{ord}}_{\Phi_{\mathcal{H}},\delta_{\mathcal{H}}}} \underset{\mathscr{O}_{\vec{\Xi}^{\mathrm{ord}}_{\Phi_{\mathcal{H}},\delta_{\mathcal{H}}}}}{\otimes} \underline{\mathrm{Lie}}^{\vee}_{G^{\vee,\natural}/\vec{\Xi}^{\mathrm{ord}}_{\Phi_{\mathcal{H}},\delta_{\mathcal{H}}}} \to \Omega^{1}_{\vec{\Xi}^{\mathrm{ord}}_{\Phi_{\mathcal{H}},\delta_{\mathcal{H}}}/\vec{\mathsf{S}}_{0,r_{\mathcal{H}}}}[d\log\infty].\end{aligned} \tag{4.2.3.1}$$

Let $\lambda^{\natural} : G^{\natural} \to G^{\vee,\natural}$ be the homomorphism defined by the tautological data $\lambda_B : B \to B^{\vee}$ and $\phi : Y \to X$. Then $\lambda^{\natural}$ induces an $\mathcal{O}$-equivariant morphism $(\lambda^{\natural})^* : \underline{\mathrm{Lie}}^{\vee}_{G^{\vee,\natural}/\vec{\Xi}^{\mathrm{ord}}_{\Phi_{\mathcal{H}},\delta_{\mathcal{H}}}} \to \underline{\mathrm{Lie}}^{\vee}_{G^{\natural}/\vec{\Xi}^{\mathrm{ord}}_{\Phi_{\mathcal{H}},\delta_{\mathcal{H}}}}$. Let $i^{\natural} : \mathcal{O} \to \mathrm{End}_{\vec{\Xi}^{\mathrm{ord}}_{\Phi_{\mathcal{H}},\delta_{\mathcal{H}}}}(G^{\natural})$ denote the tautological $\mathcal{O}$-action morphism on $G^{\natural}$.

Definition 4.2.3.2. (*Compare with [62, Def. 2.3.5.1] and Definitions* 1.1.2.8, 1.3.1.2, *and* 3.4.3.1.) *The* $\mathscr{O}_{\vec{\Xi}^{\mathrm{ord}}_{\Phi_{\mathcal{H}},\delta_{\mathcal{H}}}}$*-module*

$$\underline{\mathrm{KS}} = \underline{\mathrm{KS}}_{(G^{\natural},\lambda^{\natural},i^{\natural})/\vec{\Xi}^{\mathrm{ord}}_{\Phi_{\mathcal{H}},\delta_{\mathcal{H}}}}$$

is the quotient of

$$\underline{\mathrm{Lie}}^{\vee}_{G^{\natural}/\vec{\Xi}^{\mathrm{ord}}_{\Phi_{\mathcal{H}},\delta_{\mathcal{H}}}} \underset{\mathscr{O}_{\vec{\Xi}^{\mathrm{ord}}_{\Phi_{\mathcal{H}},\delta_{\mathcal{H}}}}}{\otimes} \underline{\mathrm{Lie}}^{\vee}_{G^{\vee,\natural}/\vec{\Xi}^{\mathrm{ord}}_{\Phi_{\mathcal{H}},\delta_{\mathcal{H}}}}$$

by the $\mathscr{O}_{\vec{\Xi}^{\mathrm{ord}}_{\Phi_{\mathcal{H}},\delta_{\mathcal{H}}}}$*-submodule spanned by*

$$(\lambda^{\natural})^*(y) \otimes z - (\lambda^{\natural})^*(z) \otimes y$$

and

$$(i^\natural(b))^*(x) \otimes y - x \otimes (i^\natural(b)^\vee)^*(y),$$

for $x \in \underline{\mathrm{Lie}}^\vee_{G^\natural/\vec{\Xi}^{\mathrm{ord}}_{\Phi_\mathcal{H},\delta_\mathcal{H}}}$, $y, z \in \underline{\mathrm{Lie}}^\vee_{G^{\vee,\natural}/\vec{\Xi}^{\mathrm{ord}}_{\Phi_\mathcal{H},\delta_\mathcal{H}}}$, *and* $b \in \mathcal{O}$.

Remark 4.2.3.3. Unlike in the good reduction case (see [62, Rem. 6.2.5.17]), the formation of $\underline{\mathrm{KS}}$ here may produce torsion elements.

Therefore, we also introduce the following:

Definition 4.2.3.4. (*Compare with [62, Def. 6.2.5.16] and Definition* 3.4.3.1.) *The* $\mathscr{O}_{\vec{\Xi}^{\mathrm{ord}}_{\Phi_\mathcal{H},\delta_\mathcal{H}}}$*-module*

$$\underline{\mathrm{KS}}_{\mathrm{free}} = \underline{\mathrm{KS}}_{(G^\natural,\lambda^\natural,i^\natural)/\vec{\Xi}^{\mathrm{ord}}_{\Phi_\mathcal{H},\delta_\mathcal{H}},\mathrm{free}}$$

is the quotient of $\underline{\mathrm{KS}}_{(G^\natural,\lambda^\natural,i^\natural)/\vec{\Xi}^{\mathrm{ord}}_{\Phi_\mathcal{H},\delta_\mathcal{H}}}$ *defined as the image of the canonical morphism*

$$\underline{\mathrm{KS}}_{(G^\natural,\lambda^\natural,i^\natural)/\vec{\Xi}^{\mathrm{ord}}_{\Phi_\mathcal{H},\delta_\mathcal{H}}} \to \underline{\mathrm{KS}}_{(G^\natural,\lambda^\natural,i^\natural)/\vec{\Xi}^{\mathrm{ord}}_{\Phi_\mathcal{H},\delta_\mathcal{H}}} \underset{\mathbb{Z}}{\otimes} \mathbb{Q}$$

of $\mathscr{O}_{\vec{\Xi}^{\mathrm{ord}}_{\Phi_\mathcal{H},\delta_\mathcal{H},\mathrm{free}}}$*-modules.*

By definition, the sheaf $\underline{\mathrm{KS}}_{\mathrm{free}} = \underline{\mathrm{KS}}_{(G^\natural,\lambda^\natural,i^\natural)/\vec{\Xi}^{\mathrm{ord}}_{\Phi_\mathcal{H},\delta_\mathcal{H}},\mathrm{free}}$ contains no p-torsion and hence is flat over $\vec{\mathsf{S}}_{0,r_\mathcal{H}} = \mathrm{Spec}(\mathcal{O}_{F_0,(p)}[\zeta_{p^{r_\mathcal{H}}}])$.

Proposition 4.2.3.5. (*Compare with [62, Prop. 6.2.5.18].*) *The Kodaira–Spencer morphism* (4.2.3.1) *factors through the sheaf* $\underline{\mathrm{KS}}_{\mathrm{free}}$ *defined in Definition* 4.2.3.4, *and induces an isomorphism*

$$\underline{\mathrm{KS}}_{\mathrm{free}} \xrightarrow{\sim} \Omega^1_{\vec{\Xi}^{\mathrm{ord}}_{\Phi_\mathcal{H},\delta_\mathcal{H}}/\vec{\mathsf{S}}_{0,r_\mathcal{H}}}[d\log\infty]. \tag{4.2.3.6}$$

In particular, the $\mathscr{O}_{\vec{\Xi}^{\mathrm{ord}}_{\Phi_\mathcal{H},\delta_\mathcal{H}}}$*-module* $\underline{\mathrm{KS}}_{\mathrm{free}}$ *is locally free of finite rank.*

Proof. Let us analyze the structural morphism $\vec{\Xi}^{\mathrm{ord}}_{\Phi_\mathcal{H},\delta_\mathcal{H}} \to \vec{\mathsf{S}}_{0,r_\mathcal{H}}$ as a composition of *smooth* morphisms:

$$\vec{\Xi}^{\mathrm{ord}}_{\Phi_\mathcal{H},\delta_\mathcal{H}} \xrightarrow{\pi_0} \vec{C}^{\mathrm{ord}}_{\Phi_\mathcal{H},\delta_\mathcal{H}} \xrightarrow{\pi_1} \vec{\mathsf{M}}^{\mathrm{ord},\Phi_\mathcal{H}}_\mathcal{H} \xrightarrow{\pi_2} \vec{\mathsf{S}}_{0,r_\mathcal{H}}.$$

For simplicity, let us denote the composition $\pi_1 \circ \pi_0$ by π_{10}. Then $\Omega^1_{\vec{\Xi}^{\mathrm{ord}}_{\Phi_\mathcal{H},\delta_\mathcal{H}}/\vec{\mathsf{S}}_{0,r_\mathcal{H}}}[d\log\infty]$ has an *increasing filtration*

$$0 \subset \pi_{01}^*\,\Omega^1_{\vec{\mathsf{M}}^{\mathrm{ord},\Phi_\mathcal{H}}_\mathcal{H}/\vec{\mathsf{S}}_{0,r_\mathcal{H}}} \subset \pi_0^*\,\Omega^1_{\vec{C}^{\mathrm{ord}}_{\Phi_\mathcal{H},\delta_\mathcal{H}}/\vec{\mathsf{S}}_{0,r_\mathcal{H}}} \subset \Omega^1_{\vec{\Xi}^{\mathrm{ord}}_{\Phi_\mathcal{H},\delta_\mathcal{H}}/\vec{\mathsf{S}}_{0,r_\mathcal{H}}}[d\log\infty],$$

with graded pieces given by $\pi_{01}^*\ \Omega^1_{\vec{\mathsf{M}}^{\mathrm{ord},\Phi_\mathcal{H}}_\mathcal{H}/\vec{\mathsf{S}}_{0,r_\mathcal{H}}}$, $\pi_0^*\ \Omega^1_{\vec{C}^{\mathrm{ord}}_{\Phi_\mathcal{H},\delta_\mathcal{H}}/\vec{\mathsf{M}}^{\mathrm{ord},\Phi_\mathcal{H}}_\mathcal{H}}$, and $\Omega^1_{\vec{\Xi}^{\mathrm{ord}}_{\Phi_\mathcal{H},\delta_\mathcal{H}}/\vec{C}^{\mathrm{ord}}_{\Phi_\mathcal{H},\delta_\mathcal{H}}}[d\log\infty]$, all of which are locally free of finite rank.

On the other hand, the sheaf $\underline{\mathrm{KS}}_{\mathrm{free}} = \underline{\mathrm{KS}}_{(G^\natural,\lambda^\natural,i^\natural)/\overline{\Xi}^{\mathrm{ord}}_{\Phi_{\mathcal{H}},\delta_{\mathcal{H}}},\mathrm{free}}$ has an increasing filtration given by $\pi_{01}^* \underline{\mathrm{KS}}_{(B,\lambda_B,i_B)/\vec{\mathsf{M}}^{\mathrm{ord},\Phi_{\mathcal{H}}}_{\mathcal{H}},\mathrm{free}}$, the pullback (under π_0) of the free quotient $\underline{\mathrm{KS}}_{(B,c,c^\vee,\lambda^\natural,i^\natural)/\vec{C}^{\mathrm{ord}}_{\Phi_{\mathcal{H}},\delta_{\mathcal{H}}},\mathrm{free}}$ of the quotient $\underline{\mathrm{KS}}_{(B,c,c^\vee,\lambda^\natural,i^\natural)/\vec{C}^{\mathrm{ord}}_{\Phi_{\mathcal{H}},\delta_{\mathcal{H}}}}$ of

$$\begin{aligned}&(\underline{\mathrm{Lie}}^\vee_{G^\natural/\vec{C}^{\mathrm{ord}}_{\Phi_{\mathcal{H}},\delta_{\mathcal{H}}}} \underset{\mathscr{O}_{\vec{C}^{\mathrm{ord}}_{\Phi_{\mathcal{H}},\delta_{\mathcal{H}}}}}{\otimes} \underline{\mathrm{Lie}}^\vee_{B^\vee/\vec{C}^{\mathrm{ord}}_{\Phi_{\mathcal{H}},\delta_{\mathcal{H}}}})\\ &+ (\underline{\mathrm{Lie}}^\vee_{B/\vec{C}^{\mathrm{ord}}_{\Phi_{\mathcal{H}},\delta_{\mathcal{H}}}} \underset{\mathscr{O}_{\vec{C}^{\mathrm{ord}}_{\Phi_{\mathcal{H}},\delta_{\mathcal{H}}}}}{\otimes} \underline{\mathrm{Lie}}^\vee_{G^{\vee,\natural}/\vec{C}^{\mathrm{ord}}_{\Phi_{\mathcal{H}},\delta_{\mathcal{H}}}})\end{aligned}$$

(as an $\mathscr{O}_{\vec{C}^{\mathrm{ord}}_{\Phi_{\mathcal{H}},\delta_{\mathcal{H}}}}$-submodule of $\underline{\mathrm{Lie}}^\vee_{G^\natural/\vec{C}^{\mathrm{ord}}_{\Phi_{\mathcal{H}},\delta_{\mathcal{H}}}} \underset{\mathscr{O}_{\vec{C}^{\mathrm{ord}}_{\Phi_{\mathcal{H}},\delta_{\mathcal{H}}}}}{\otimes} \underline{\mathrm{Lie}}^\vee_{G^{\vee,\natural}/\vec{C}^{\mathrm{ord}}_{\Phi_{\mathcal{H}},\delta_{\mathcal{H}}}}$) by relations as in Definition 4.2.3.4, and the whole sheaf $\underline{\mathrm{KS}}_{\mathrm{free}}$. Hence, it suffices to show that the morphism (4.2.3.1) respects the filtrations and matches isomorphically the (free quotients of) the graded pieces.

By Proposition 3.4.3.3, since $\vec{\mathsf{M}}^{\mathrm{ord},\Phi_{\mathcal{H}}}_{\mathcal{H}}$ is étale over the base change $\vec{\mathsf{M}}^{\mathrm{ord}}_{\mathcal{H}_h,r_{\mathcal{H}}}$ of $\vec{\mathsf{M}}^{\mathrm{ord}}_{\mathcal{H}_h}$ (defined by $\mathsf{M}_{\mathcal{H}_h}$ and $\dddot{\mathsf{M}}^{\mathrm{ord}}_{\mathcal{H}_h}$ as in Theorem 3.4.2.5) to $\vec{\mathsf{S}}_{0,r_{\mathcal{H}}}$, the Kodaira–Spencer morphism $\mathrm{KS}_{B/\vec{\mathsf{M}}^{\mathrm{ord},\Phi_{\mathcal{H}}}_{\mathcal{H}}/\vec{\mathsf{S}}_{0,r_{\mathcal{H}}}}$ for B over $\vec{\mathsf{M}}^{\mathrm{ord},\Phi_{\mathcal{H}}}_{\mathcal{H}}$ induces an isomorphism

$$\underline{\mathrm{KS}}_{(B,\lambda_B,i_B)/\vec{\mathsf{M}}^{\mathrm{ord},\Phi_{\mathcal{H}}}_{\mathcal{H}},\mathrm{free}} \xrightarrow{\sim} \Omega^1_{\vec{\mathsf{M}}^{\mathrm{ord},\Phi_{\mathcal{H}}}_{\mathcal{H}}/\vec{\mathsf{S}}_{0,r_{\mathcal{H}}}}, \tag{4.2.3.7}$$

and hence the same remains true after pulled back by π_{01}. Since the Kodaira–Spencer morphism $\mathrm{KS}_{B/\overline{\Xi}^{\mathrm{ord}}_{\Phi_{\mathcal{H}},\delta_{\mathcal{H}}}/\vec{\mathsf{S}}_{0,r_{\mathcal{H}}}} = \pi_{01}^* \mathrm{KS}_{B/\vec{\mathsf{M}}^{\mathrm{ord},\Phi_{\mathcal{H}}}_{\mathcal{H}}/\vec{\mathsf{S}}_{0,r_{\mathcal{H}}}}$ for B over $\overline{\Xi}^{\mathrm{ord}}_{\Phi_{\mathcal{H}},\delta_{\mathcal{H}}}$ is the restriction of the Kodaira–Spencer morphism KS in (4.2.3.1) (*cf.* the compatibility statements in [62, Sec. 4.6.2]), we see that the first filtered pieces are respected.

By the deformation-theoretic interpretation of the Kodaira–Spencer morphisms $\mathrm{KS}_{(B,c)/\vec{C}^{\mathrm{ord}}_{\Phi_{\mathcal{H}},\delta_{\mathcal{H}}}/\vec{\mathsf{S}}_{0,r_{\mathcal{H}}}}$ and $\mathrm{KS}_{(B^\vee,c^\vee)/\vec{C}^{\mathrm{ord}}_{\Phi_{\mathcal{H}},\delta_{\mathcal{H}}}/\vec{\mathsf{S}}_{0,r_{\mathcal{H}}}}$ in [62, Sec. 4.6.1] (see in particular [62, Def. 4.6.1.2]), we see that the restrictions of both of them to $\underline{\mathrm{Lie}}^\vee_{B/\vec{C}^{\mathrm{ord}}_{\Phi_{\mathcal{H}},\delta_{\mathcal{H}}}} \underset{\mathscr{O}_{\vec{C}^{\mathrm{ord}}_{\Phi_{\mathcal{H}},\delta_{\mathcal{H}}}}}{\otimes} \underline{\mathrm{Lie}}^\vee_{B^\vee/\vec{C}^{\mathrm{ord}}_{\Phi_{\mathcal{H}},\delta_{\mathcal{H}}}}$ agree with $\mathrm{KS}_{B/\vec{C}^{\mathrm{ord}}_{\Phi_{\mathcal{H}},\delta_{\mathcal{H}}}/\vec{\mathsf{S}}_{0,r_{\mathcal{H}}}}$, which induces a surjection onto $\pi_1^* \Omega^1_{\vec{\mathsf{M}}^{\mathrm{ord},\Phi_{\mathcal{H}}}_{\mathcal{H}}/\vec{\mathsf{S}}_{0,r_{\mathcal{H}}}}$. Hence, they define a morphism

$$\begin{aligned}&(\underline{\mathrm{Lie}}^\vee_{G^\natural/\vec{C}^{\mathrm{ord}}_{\Phi_{\mathcal{H}},\delta_{\mathcal{H}}}} \underset{\mathscr{O}_{\vec{C}^{\mathrm{ord}}_{\Phi_{\mathcal{H}},\delta_{\mathcal{H}}}}}{\otimes} \underline{\mathrm{Lie}}^\vee_{B^\vee/\vec{C}^{\mathrm{ord}}_{\Phi_{\mathcal{H}},\delta_{\mathcal{H}}}})\\ &+ (\underline{\mathrm{Lie}}^\vee_{B/\vec{C}^{\mathrm{ord}}_{\Phi_{\mathcal{H}},\delta_{\mathcal{H}}}} \underset{\mathscr{O}_{\vec{C}^{\mathrm{ord}}_{\Phi_{\mathcal{H}},\delta_{\mathcal{H}}}}}{\otimes} \underline{\mathrm{Lie}}^\vee_{G^{\vee,\natural}/\vec{C}^{\mathrm{ord}}_{\Phi_{\mathcal{H}},\delta_{\mathcal{H}}}}) \to \Omega^1_{\vec{C}^{\mathrm{ord}}_{\Phi_{\mathcal{H}},\delta_{\mathcal{H}}}/\vec{\mathsf{S}}_{0,r_{\mathcal{H}}}}\end{aligned} \tag{4.2.3.8}$$

compatible with the pullback of $\mathrm{KS}_{B/\vec{\mathsf{M}}^{\mathrm{ord},\Phi_{\mathcal{H}}}_{\mathcal{H}}/\vec{\mathsf{S}}_{0,r_{\mathcal{H}}}}$, which induces (after replacing the source of (4.2.3.8) with its quotient by $\underline{\mathrm{Lie}}^\vee_{B/\vec{C}^{\mathrm{ord}}_{\Phi_{\mathcal{H}},\delta_{\mathcal{H}}}} \underset{\mathscr{O}_{\vec{C}^{\mathrm{ord}}_{\Phi_{\mathcal{H}},\delta_{\mathcal{H}}}}}{\otimes} \underline{\mathrm{Lie}}^\vee_{B^\vee/\vec{C}^{\mathrm{ord}}_{\Phi_{\mathcal{H}},\delta_{\mathcal{H}}}}$) a morphism

$$\begin{aligned}&(\underline{\mathrm{Lie}}^\vee_{T/\vec{C}^{\mathrm{ord}}_{\Phi_{\mathcal{H}},\delta_{\mathcal{H}}}} \underset{\mathscr{O}_{\vec{C}^{\mathrm{ord}}_{\Phi_{\mathcal{H}},\delta_{\mathcal{H}}}}}{\otimes} \underline{\mathrm{Lie}}^\vee_{B^\vee/\vec{C}^{\mathrm{ord}}_{\Phi_{\mathcal{H}},\delta_{\mathcal{H}}}})\\ &+ (\underline{\mathrm{Lie}}^\vee_{B/\vec{C}^{\mathrm{ord}}_{\Phi_{\mathcal{H}},\delta_{\mathcal{H}}}} \underset{\mathscr{O}_{\vec{C}^{\mathrm{ord}}_{\Phi_{\mathcal{H}},\delta_{\mathcal{H}}}}}{\otimes} \underline{\mathrm{Lie}}^\vee_{T^\vee/\vec{C}^{\mathrm{ord}}_{\Phi_{\mathcal{H}},\delta_{\mathcal{H}}}}) \to \Omega^1_{\vec{C}^{\mathrm{ord}}_{\Phi_{\mathcal{H}},\delta_{\mathcal{H}}}/\vec{\mathsf{M}}^{\mathrm{ord},\Phi_{\mathcal{H}}}_{\mathcal{H}}}.\end{aligned} \tag{4.2.3.9}$$

Since c and $c^\vee$ satisfies the compatibility $c\phi = \lambda_B c^\vee$, the morphism (4.2.3.8) is compatible with quotients by relations as in Definition 4.2.3.4, and induces a morphism

$$\underline{\mathrm{KS}}_{(B,c,c^\vee,\lambda^\natural,i^\natural)/\vec{C}^{\mathrm{ord}}_{\Phi_{\mathcal{H}},\delta_{\mathcal{H}}}} \to \Omega^1_{\vec{C}^{\mathrm{ord}}_{\Phi_{\mathcal{H}},\delta_{\mathcal{H}}}/\vec{\mathsf{S}}_{0,r_{\mathcal{H}}}} \tag{4.2.3.10}$$

compatible with the pullback of the isomorphism (4.2.3.7). Then the morphism (4.2.3.9) induces a morphism

$$(\underline{\mathrm{KS}}_{(B,c,c^\vee,\lambda^\natural,i^\natural)/\vec{C}^{\mathrm{ord}}_{\Phi_{\mathcal{H}},\delta_{\mathcal{H}}}}/\underline{\mathrm{KS}}_{(B,\lambda_B,i_B)/\vec{C}^{\mathrm{ord}}_{\Phi_{\mathcal{H}},\delta_{\mathcal{H}}}})_{\mathrm{free}} \to \Omega^1_{\vec{C}^{\mathrm{ord}}_{\Phi_{\mathcal{H}},\delta_{\mathcal{H}}}/\vec{\mathsf{M}}^{\mathrm{ord},\Phi_{\mathcal{H}}}_{\mathcal{H}}}, \tag{4.2.3.11}$$

where $\underline{\mathrm{KS}}_{(B,\lambda_B,i_B)/\vec{C}^{\mathrm{ord}}_{\Phi_{\mathcal{H}},\delta_{\mathcal{H}}}} \cong \pi_1^* \underline{\mathrm{KS}}_{(B,\lambda_B,i_B)/\vec{\mathsf{M}}^{\mathrm{ord},\Phi_{\mathcal{H}}}_{\mathcal{H}}}$. By Propositions 4.2.1.29 and 4.2.1.34, the morphism (4.2.3.11) is the pullback of the corresponding morphism

$$(\underline{\mathrm{KS}}_{(B,c,c^\vee,\lambda^\natural,i^\natural)/\vec{C}^{\mathrm{ord}}_{\Phi_1,\delta_1}}/\underline{\mathrm{KS}}_{(B,\lambda_B,i_B)/\vec{C}^{\mathrm{ord}}_{\Phi_1,\delta_1}})_{\mathrm{free}} \to \Omega^1_{\vec{C}^{\mathrm{ord}}_{\Phi_1,\delta_1}/\vec{\mathsf{M}}^{\mathrm{ord},\mathsf{Z}_1}_1} \tag{4.2.3.12}$$

at principal level 1. Since $\ddot{C}^{\mathrm{ord}}_{\Phi_1}$ is the universal space for $(c,c^\vee)$ over $\ddot{\mathsf{M}}^{\mathrm{ord}}_1$ (satisfying the compatibility $c\phi = \lambda_B c^\vee$), the source of (4.2.3.12) can be canonically identified with the pullback of the free quotient $\Omega^1_{(\ddot{C}^{\mathrm{ord}}_{\Phi_1} \underset{\ddot{\mathsf{M}}^{\mathrm{ord},\mathsf{Z}_1}_1}{\times} \vec{\mathsf{M}}^{\mathrm{ord},\mathsf{Z}_1}_1)/\vec{\mathsf{M}}^{\mathrm{ord},\mathsf{Z}_1}_1,\mathrm{free}}$ of $\Omega^1_{(\ddot{C}^{\mathrm{ord}}_{\Phi_1} \underset{\ddot{\mathsf{M}}^{\mathrm{ord},\mathsf{Z}_1}_1}{\times} \vec{\mathsf{M}}^{\mathrm{ord},\mathsf{Z}_1}_1)/\vec{\mathsf{M}}^{\mathrm{ord},\mathsf{Z}_1}_1}$ under $\vec{C}^{\mathrm{ord}}_{\Phi_1,\delta_1} \to \ddot{C}^{\mathrm{ord}}_{\Phi_1} \underset{\ddot{\mathsf{M}}^{\mathrm{ord},\mathsf{Z}_1}_1}{\times} \vec{\mathsf{M}}^{\mathrm{ord},\mathsf{Z}_1}_1$. By the construction of $\vec{C}^{\mathrm{ord}}_{\Phi_1,\delta_1} \to \vec{\mathsf{M}}^{\mathrm{ord},\mathsf{Z}_1}_1$ as the pullback of (4.2.1.33) under the section (4.2.1.32) in the proof of Proposition 4.2.1.30, the canonical morphism

$$\begin{aligned}(\vec{C}^{\mathrm{ord}}_{\Phi_1,\delta_1} \to \ddot{C}^{\mathrm{ord}}_{\Phi_1} \underset{\ddot{\mathsf{M}}^{\mathrm{ord},\mathsf{Z}_1}_1}{\times} \vec{\mathsf{M}}^{\mathrm{ord},\mathsf{Z}_1}_1)^*\Omega^1_{(\ddot{C}^{\mathrm{ord}}_{\Phi_1} \underset{\ddot{\mathsf{M}}^{\mathrm{ord},\mathsf{Z}_1}_1}{\times} \vec{\mathsf{M}}^{\mathrm{ord},\mathsf{Z}_1}_1)/\vec{\mathsf{M}}^{\mathrm{ord},\mathsf{Z}_1}_1,\mathrm{free}}& \\ \to \Omega^1_{\vec{C}^{\mathrm{ord}}_{\Phi_1,\delta_1}/\vec{\mathsf{M}}^{\mathrm{ord},\mathsf{Z}_1}_1}&\end{aligned}$$

is an isomorphism. Thus, we see that the morphism (4.2.3.12) is an isomorphism, and hence that the morphism (4.2.3.11) is an isomorphism and induces (by a simple diagram chasing) the composition of canonical isomorphisms

$$\begin{aligned}&\underline{\mathrm{KS}}_{(B,c,c^\vee,\lambda^\natural,i^\natural)/\vec{C}^{\mathrm{ord}}_{\Phi_{\mathcal{H}},\delta_{\mathcal{H}}},\mathrm{free}}/\underline{\mathrm{KS}}_{(B,\lambda_B,i_B)/\vec{C}^{\mathrm{ord}}_{\Phi_{\mathcal{H}},\delta_{\mathcal{H}}},\mathrm{free}} \\ &\overset{\sim}{\to} (\underline{\mathrm{KS}}_{(B,c,c^\vee,\lambda^\natural,i^\natural)/\vec{C}^{\mathrm{ord}}_{\Phi_{\mathcal{H}},\delta_{\mathcal{H}}}}/\underline{\mathrm{KS}}_{(B,\lambda_B,i_B)/\vec{C}^{\mathrm{ord}}_{\Phi_{\mathcal{H}},\delta_{\mathcal{H}}}})_{\mathrm{free}} \\ &\overset{\sim}{\to} \Omega^1_{\vec{C}^{\mathrm{ord}}_{\Phi_{\mathcal{H}},\delta_{\mathcal{H}}}/\vec{\mathsf{M}}^{\mathrm{ord},\Phi_{\mathcal{H}}}_{\mathcal{H}}}.\end{aligned} \tag{4.2.3.13}$$

This shows that the morphism (4.2.3.10) also induces an isomorphism

$$\underline{\mathrm{KS}}_{(B,c,c^\vee,\lambda^\natural,i^\natural)/\vec{C}^{\mathrm{ord}}_{\Phi_{\mathcal{H}},\delta_{\mathcal{H}}},\mathrm{free}} \overset{\sim}{\to} \Omega^1_{\vec{C}^{\mathrm{ord}}_{\Phi_{\mathcal{H}},\delta_{\mathcal{H}}}/\vec{\mathsf{S}}_{0,r_{\mathcal{H}}}}. \tag{4.2.3.14}$$

Since the pullback of this isomorphism (4.2.3.14) (under π_0) to $\vec{\Xi}^{\mathrm{ord}}_{\Phi_{\mathcal{H}},\delta_{\mathcal{H}}}$ is induced by the restriction of the extended Kodaira–Spencer morphism KS in (4.2.3.1) (*cf.* [62, Rem. 4.6.2.7]), we see that the second filtered pieces are also respected, with an induced isomorphism between the second graded pieces.

Finally, we arrive at the top filtered pieces, and the question is about the induced morphism

$$\underline{\mathrm{Lie}}^{\vee}_{T/\overline{\Xi}^{\mathrm{ord}}_{\Phi_{\mathcal{H}},\delta_{\mathcal{H}}}} \underset{\mathscr{O}_{\overline{\Xi}^{\mathrm{ord}}_{\Phi_{\mathcal{H}},\delta_{\mathcal{H}}}}}{\otimes} \underline{\mathrm{Lie}}^{\vee}_{T^{\vee}/\overline{\Xi}^{\mathrm{ord}}_{\Phi_{\mathcal{H}},\delta_{\mathcal{H}}}} \to \Omega^{1}_{\overline{\Xi}^{\mathrm{ord}}_{\Phi_{\mathcal{H}},\delta_{\mathcal{H}}}/\vec{C}^{\mathrm{ord}}_{\Phi_{\mathcal{H}},\delta_{\mathcal{H}}}}[d\log\infty] \tag{4.2.3.15}$$

between the top graded pieces. Let us denote by $\underline{\mathrm{KS}}_{(T,\lambda_T,i_T)/\vec{\mathsf{S}}_{0,r_{\mathcal{H}}},\mathrm{free}}$ the free quotient of $\underline{\mathrm{Lie}}^{\vee}_{T/\vec{\mathsf{S}}_{0,r_{\mathcal{H}}}} \underset{\mathscr{O}_{\vec{\mathsf{S}}_{0,r_{\mathcal{H}}}}}{\otimes} \underline{\mathrm{Lie}}^{\vee}_{T^{\vee}/\vec{\mathsf{S}}_{0,r_{\mathcal{H}}}}$ by relations as in Definition 4.2.3.2, and by $\underline{\mathrm{KS}}_{(T,\lambda_T,i_T)/\overline{\Xi}^{\mathrm{ord}}_{\Phi_{\mathcal{H}},\delta_{\mathcal{H}}},\mathrm{free}}$ and $\underline{\mathrm{KS}}_{(T,\lambda_T,i_T)/\vec{\Xi}^{\mathrm{ord}}_{\Phi_{\mathcal{H}},\delta_{\mathcal{H}}},\mathrm{free}}$ their pullbacks to $\overline{\vec{\Xi}}^{\mathrm{ord}}_{\Phi_{\mathcal{H}},\delta_{\mathcal{H}}}$ and $\vec{\Xi}^{\mathrm{ord}}_{\Phi_{\mathcal{H}},\delta_{\mathcal{H}}}$, respectively.

Let us first consider the restriction

$$\underline{\mathrm{Lie}}^{\vee}_{T/\vec{\Xi}^{\mathrm{ord}}_{\Phi_{\mathcal{H}},\delta_{\mathcal{H}}}} \underset{\mathscr{O}_{\vec{\Xi}^{\mathrm{ord}}_{\Phi_{\mathcal{H}},\delta_{\mathcal{H}}}}}{\otimes} \underline{\mathrm{Lie}}^{\vee}_{T^{\vee}/\vec{\Xi}^{\mathrm{ord}}_{\Phi_{\mathcal{H}},\delta_{\mathcal{H}}}} \to \Omega^{1}_{\vec{\Xi}^{\mathrm{ord}}_{\Phi_{\mathcal{H}},\delta_{\mathcal{H}}}/\vec{C}^{\mathrm{ord}}_{\Phi_{\mathcal{H}},\delta_{\mathcal{H}}}} \tag{4.2.3.16}$$

of (4.2.3.15) to $\vec{\Xi}^{\mathrm{ord}}_{\Phi_{\mathcal{H}},\delta_{\mathcal{H}}}$, which is induced by the Kodaira–Spencer morphism $\mathrm{KS}_{(G^{\natural},\iota)/\vec{\Xi}^{\mathrm{ord}}_{\Phi_{\mathcal{H}},\delta_{\mathcal{H}}}/\vec{\mathsf{S}}_{0,r_{\mathcal{H}}}}$ defined deformation-theoretically as in [62, Def. 4.6.2.6]. Since τ is compatible with the polarizations and $\mathcal{O}$-endomorphism structures, the morphism (4.2.3.16) induces a morphism

$$\underline{\mathrm{KS}}_{(T,\lambda_T,i_T)/\vec{\Xi}^{\mathrm{ord}}_{\Phi_{\mathcal{H}},\delta_{\mathcal{H}}},\mathrm{free}} \to \Omega^{1}_{\vec{\Xi}^{\mathrm{ord}}_{\Phi_{\mathcal{H}},\delta_{\mathcal{H}}}/\vec{C}^{\mathrm{ord}}_{\Phi_{\mathcal{H}},\delta_{\mathcal{H}}}}. \tag{4.2.3.17}$$

By Proposition 4.2.1.41, the morphism (4.2.3.17) is the pullback of the corresponding morphism

$$\underline{\mathrm{KS}}_{(T,\lambda_T,i_T)/\vec{\Xi}^{\mathrm{ord}}_{\Phi_1,\delta_1},\mathrm{free}} \to \Omega^{1}_{\vec{\Xi}^{\mathrm{ord}}_{\Phi_1,\delta_1}/\vec{C}^{\mathrm{ord}}_{\Phi_1,\delta_1}}. \tag{4.2.3.18}$$

Since $\dddot{\Xi}^{\mathrm{ord}}_{\Phi_1}$ is the universal space for ι over $\dddot{C}^{\mathrm{ord}}_{\Phi_1}$ (with symmetry condition, but without positivity condition), the source of (4.2.3.18) can be canonically identified with the pullback of the free quotient $\Omega^{1}_{(\dddot{\Xi}^{\mathrm{ord}}_{\Phi_1}\underset{\dddot{C}^{\mathrm{ord}}_{\Phi_1}}{\times}\vec{C}^{\mathrm{ord}}_{\Phi_1,\delta_1})/\vec{C}^{\mathrm{ord}}_{\Phi_1,\delta_1},\mathrm{free}}$ of $\Omega^{1}_{(\dddot{\Xi}^{\mathrm{ord}}_{\Phi_1}\underset{\dddot{C}^{\mathrm{ord}}_{\Phi_1}}{\times}\vec{C}^{\mathrm{ord}}_{\Phi_1,\delta_1})/\vec{C}^{\mathrm{ord}}_{\Phi_1,\delta_1}}$ under $\vec{\Xi}^{\mathrm{ord}}_{\Phi_1,\delta_1} \to \dddot{\Xi}^{\mathrm{ord}}_{\Phi_1}\underset{\dddot{C}^{\mathrm{ord}}_{\Phi_1}}{\times}\vec{C}^{\mathrm{ord}}_{\Phi_1,\delta_1}$. By the construction of $\vec{\Xi}^{\mathrm{ord}}_{\Phi_1,\delta_1} \to \vec{C}^{\mathrm{ord}}_{\Phi_1,\delta_1}$ as the pullback of (4.2.1.40) under the section (4.2.1.39) in the proof of Proposition 4.2.1.37, the canonical morphism

$$(\vec{\Xi}^{\mathrm{ord}}_{\Phi_1,\delta_1} \to \dddot{\Xi}^{\mathrm{ord}}_{\Phi_1}\underset{\dddot{C}^{\mathrm{ord}}_{\Phi_1}}{\times}\vec{C}^{\mathrm{ord}}_{\Phi_1,\delta_1})^{*}\Omega^{1}_{(\dddot{\Xi}^{\mathrm{ord}}_{\Phi_1}\underset{\dddot{C}^{\mathrm{ord}}_{\Phi_1}}{\times}\vec{C}^{\mathrm{ord}}_{\Phi_1,\delta_1})/\vec{C}^{\mathrm{ord}}_{\Phi_1,\delta_1},\mathrm{free}} \to \Omega^{1}_{\vec{\Xi}^{\mathrm{ord}}_{\Phi_{\mathcal{H}},\delta_{\mathcal{H}}}/\vec{C}^{\mathrm{ord}}_{\Phi_{\mathcal{H}},\delta_{\mathcal{H}}}}$$

is an isomorphism. Thus, we see that the morphism (4.2.3.18) is an isomorphism, and hence that the morphism (4.2.3.17) is an isomorphism

$$\underline{\mathrm{KS}}_{(T,\lambda_T,i_T)/\vec{\Xi}^{\mathrm{ord}}_{\Phi_{\mathcal{H}},\delta_{\mathcal{H}}},\mathrm{free}} \xrightarrow{\sim} \Omega^{1}_{\vec{\Xi}^{\mathrm{ord}}_{\Phi_{\mathcal{H}},\delta_{\mathcal{H}}}/\vec{C}^{\mathrm{ord}}_{\Phi_{\mathcal{H}},\delta_{\mathcal{H}}}}. \tag{4.2.3.19}$$

If we work over $\vec{\overline{\Xi}}^{\mathrm{ord}}_{\Phi_{\mathcal{H}},\delta_{\mathcal{H}}}$, then the morphism (4.2.3.15) is induced by the extended Kodaira–Spencer morphism $\mathrm{KS}_{(G^\natural,\iota)/\vec{\overline{\Xi}}^{\mathrm{ord}}_{\Phi_{\mathcal{H}},\delta_{\mathcal{H}}}/\vec{\mathsf{S}}_{0,r_{\mathcal{H}}}}$ defined as in [62, Def. 4.6.2.12]. Since its image in $\Omega^1_{\vec{\overline{\Xi}}^{\mathrm{ord}}_{\Phi_{\mathcal{H}},\delta_{\mathcal{H}}}/\vec{C}^{\mathrm{ord}}_{\Phi_{\mathcal{H}},\delta_{\mathcal{H}}}}[d\log\infty]$ contains $d\log(\vec{\Psi}^{\mathrm{ord}}_{\Phi_{\mathcal{H}},\delta_{\mathcal{H}}}(\ell))$ for all $\ell \in \mathbf{S}_{\Phi_{\mathcal{H}}}$, which are exactly the generators, we see that (4.2.3.15) induces an isomorphism

$$\underline{\mathrm{KS}}_{(T,\lambda_T,i_T)/\vec{\overline{\Xi}}^{\mathrm{ord}}_{\Phi_{\mathcal{H}},\delta_{\mathcal{H}}},\mathrm{free}} \xrightarrow{\sim} \Omega^1_{\vec{\overline{\Xi}}^{\mathrm{ord}}_{\Phi_{\mathcal{H}},\delta_{\mathcal{H}}}/\vec{C}^{\mathrm{ord}}_{\Phi_{\mathcal{H}},\delta_{\mathcal{H}}}}[d\log\infty]$$

and (by a simple diagram chasing) the composition of canonical isomorphisms

$$\begin{aligned}
&\underline{\mathrm{KS}}_{\mathrm{free}}/\pi_0^*\,\underline{\mathrm{KS}}_{(B,c,c^\vee,\lambda^\natural,i^\natural)/\vec{C}^{\mathrm{ord}}_{\Phi_{\mathcal{H}},\delta_{\mathcal{H}}},\mathrm{free}}\\
&\xrightarrow{\sim} \big(\underline{\mathrm{KS}}/\pi_0^*\,\underline{\mathrm{KS}}_{(B,c,c^\vee,\lambda^\natural,i^\natural)/\vec{C}^{\mathrm{ord}}_{\Phi_{\mathcal{H}},\delta_{\mathcal{H}}}}\big)_{\mathrm{free}} \cong \underline{\mathrm{KS}}_{(T,\lambda_T,i_T)/\vec{\overline{\Xi}}^{\mathrm{ord}}_{\Phi_{\mathcal{H}},\delta_{\mathcal{H}}},\mathrm{free}}\\
&\xrightarrow{\sim} \Omega^1_{\vec{\overline{\Xi}}^{\mathrm{ord}}_{\Phi_{\mathcal{H}},\delta_{\mathcal{H}}}/\vec{C}^{\mathrm{ord}}_{\Phi_{\mathcal{H}},\delta_{\mathcal{H}}}}[d\log\infty]
\end{aligned}$$

between the top graded pieces. Hence, (4.2.3.6) is an isomorphism, as desired. □

Chapter 5

Partial Toroidal Compactifications

The goal of this chapter is to construct the partial toroidal compactifications for the ordinary loci defined in Chapter 3, based on the theory of degeneration and the construction of boundary charts given in Chapter 4.

5.1 Approximation and Gluing Along the Ordinary Loci

In this section, let us continue with the setting in Sections 4.1.6 and 4.2.1, including the choices of a level $\mathcal{H} = \mathcal{H}^p\mathcal{H}_p$ and an integer $r_\mathcal{H}$ determined by $\mathcal{H}$ as in Definition 3.4.2.1. The materials in this section follows those of [62, Sec. 6.3] very closely. However, we spell out the precise statements to make sure that the definitions and constructions can indeed be generalized after some subtle modifications.

5.1.1 *Ordinary Good Formal Models*

Construction 5.1.1.1. Let S be an excellent normal algebraic stack that is *flat* over $\vec{\mathsf{S}}_{0,r_\mathcal{H}} = \mathrm{Spec}(\mathcal{O}_{F_0,(p)}[\zeta_{p^{r_\mathcal{H}}}])$, and let $(G, \lambda, i, \alpha_{\mathcal{H}^p}, \alpha^{\mathrm{ord}}_{\mathcal{H}_p})$ be a degenerating family of type $\vec{\mathsf{M}}^{\mathrm{ord}}_\mathcal{H}$ over S (see Definition 3.4.2.10). (As remarked in [62, Constr. 6.3.1.1], the excellence assumption on S might be removed by direct limit arguments, but we do not need this generality for our purpose.) By definition (see Condition 4 of Definition 3.4.2.10), $(G, \lambda, i, \alpha_{\mathcal{H}^p}, \alpha^{\mathrm{ord}}_{\mathcal{H}_p}) \to S$ defines a tuple parameterized by $\vec{\mathsf{M}}^{\mathrm{ord}}_\mathcal{H}$ (see Convention 3.4.2.9). By the construction of $\vec{\mathsf{M}}^{\mathrm{ord}}_\mathcal{H}$ (see Theorem 3.4.2.5), this implies that there is a level-$\mathcal{H}$ structure $\alpha_\mathcal{H}$ on $(G, \lambda, i) \underset{\mathbb{Z}}{\otimes} \mathbb{Q} \to S \underset{\mathbb{Z}}{\otimes} \mathbb{Q}$ such that $(\alpha_{\mathcal{H}^p}, \alpha^{\mathrm{ord}}_{\mathcal{H}_p}) \underset{\mathbb{Z}}{\otimes} \mathbb{Q}$ is assigned to $\alpha_\mathcal{H}$ under (3.3.5.5) over $S \underset{\mathbb{Z}}{\otimes} \mathbb{Q}$. Then $(G, \lambda, i, \alpha_\mathcal{H}) \to S$ qualifies as a degenerating family of type $\mathsf{M}_\mathcal{H}$ over S (by choosing S_1 to be an open dense subscheme of $S \underset{\mathbb{Z}}{\otimes} \mathbb{Q}$ in Definition 1.3.1.1), and [62, Constr. 6.3.1.1] applies and defines the sheaf objects $\underline{\Phi}_\mathcal{H} = (\underline{X}(G), \underline{Y}(G), \underline{\phi}(G), \underline{\varphi}_{-2,\mathcal{H}}(G), \underline{\varphi}_{0,\mathcal{H}}(G))$, $\mathbf{S}_{\underline{\Phi}_\mathcal{H}(G)}$, and

$$\underline{B}(G) : \underline{\mathbf{S}}_{\underline{\Phi}_\mathcal{H}(G)} \to \underline{\mathrm{Inv}}(S). \tag{5.1.1.2}$$

(This finishes Construction 5.1.1.1.)

Let $(\mathsf{Z}_{\mathcal{H}}, \Phi_{\mathcal{H}}, \delta_{\mathcal{H}})$ be a representative of an ordinary cusp label at level $\mathcal{H}$ (see Definition 3.2.3.8). Then we have the Mumford families $({}^{\heartsuit}G, {}^{\heartsuit}\lambda, {}^{\heartsuit}i, {}^{\heartsuit}\alpha_{\mathcal{H}^p}, {}^{\heartsuit}\alpha^{\mathrm{ord}}_{\mathcal{H}_p}) \to \vec{\mathfrak{X}}^{\mathrm{ord}}_{\Phi_{\mathcal{H}},\delta_{\mathcal{H}},\sigma}/\Gamma_{\Phi_{\mathcal{H}},\sigma}$ and $({}^{\heartsuit}G, {}^{\heartsuit}\lambda, {}^{\heartsuit}i, {}^{\heartsuit}\alpha_{\mathcal{H}}) \to \vec{\mathfrak{X}}^{\mathrm{ord}}_{\Phi_{\mathcal{H}},\delta_{\mathcal{H}},\sigma}/\Gamma_{\Phi_{\mathcal{H}},\sigma}$ as in (4.2.2.19) and (4.2.2.20). Let us summarize their properties as follows:

Proposition 5.1.1.3. *(Compare with [62, Prop. 6.3.1.6].) Let $S_{\mathrm{for}} = \mathrm{Spf}(R, I)$ be an affine formal scheme, with an étale (i.e., formally étale and of finite type; see [35, I, 10.13.3]) morphism $\hat{f} : S_{\mathrm{for}} \to \vec{\mathfrak{X}}^{\mathrm{ord}}_{\Phi_{\mathcal{H}},\delta_{\mathcal{H}},\sigma}/\Gamma_{\Phi_{\mathcal{H}},\sigma}$ inducing a morphism $f : S = \mathrm{Spec}(R) \to \vec{\Xi}^{\mathrm{ord}}_{\Phi_{\mathcal{H}},\delta_{\mathcal{H}}}(\sigma)/\Gamma_{\Phi_{\mathcal{H}},\sigma}$ mapping the support $\mathrm{Spec}(R/I)$ of S_{for} to the σ-stratum $\vec{\Xi}^{\mathrm{ord}}_{\Phi_{\mathcal{H}},\delta_{\mathcal{H}},\sigma}/\Gamma_{\Phi_{\mathcal{H}},\sigma}$ of $\vec{\Xi}^{\mathrm{ord}}_{\Phi_{\mathcal{H}},\delta_{\mathcal{H}}}(\sigma)/\Gamma_{\Phi_{\mathcal{H}},\sigma}$. (In this case, the subscheme $\mathrm{Spec}(R/I)$ of S is the scheme-theoretic preimage of its image under f.) Let $({}^{\diamondsuit}G, {}^{\diamondsuit}\lambda, {}^{\diamondsuit}i, {}^{\diamondsuit}\alpha_{\mathcal{H}^p}, {}^{\diamondsuit}\alpha^{\mathrm{ord}}_{\mathcal{H}_p}) \to S = \mathrm{Spec}(R)$ (resp. $({}^{\diamondsuit}G, {}^{\diamondsuit}\lambda, {}^{\diamondsuit}i, {}^{\diamondsuit}\alpha_{\mathcal{H}}) \to S$) be the pullback of $({}^{\heartsuit}G, {}^{\heartsuit}\lambda, {}^{\heartsuit}i, {}^{\heartsuit}\alpha_{\mathcal{H}^p}, {}^{\heartsuit}\alpha^{\mathrm{ord}}_{\mathcal{H}_p}) \to \vec{\mathfrak{X}}^{\mathrm{ord}}_{\Phi_{\mathcal{H}},\delta_{\mathcal{H}},\sigma}/\Gamma_{\Phi_{\mathcal{H}},\sigma}$ (resp. $({}^{\heartsuit}G, {}^{\heartsuit}\lambda, {}^{\heartsuit}i, {}^{\heartsuit}\alpha_{\mathcal{H}}) \to \vec{\mathfrak{X}}^{\mathrm{ord}}_{\Phi_{\mathcal{H}},\delta_{\mathcal{H}},\sigma}/\Gamma_{\Phi_{\mathcal{H}},\sigma}$) to S_{for} under $\hat{f}$ (by abuse of language). In this case, $({}^{\diamondsuit}\alpha_{\mathcal{H}^p}, {}^{\diamondsuit}\alpha^{\mathrm{ord}}_{\mathcal{H}_p})$ is assigned to ${}^{\diamondsuit}\alpha_{\mathcal{H}}$ under (3.3.5.5) over the generic point $\eta = \mathrm{Spec}(K)$ of $\mathrm{Spec}(R)$, where K is the fraction field of R. Then R is an I-adically complete excellent ring, which is formally smooth over the abelian scheme $\vec{C}^{\mathrm{ord}}_{\Phi_{\mathcal{H}},\delta_{\mathcal{H}}}$, and hence also formally smooth over $\vec{\mathsf{S}}_{0,r_{\mathcal{H}}} = \mathrm{Spec}(\mathcal{O}_{F_0,(p)}[\zeta_{p^{r_{\mathcal{H}}}}])$.*

1. *The stratification of $\vec{\Xi}^{\mathrm{ord}}_{\Phi_{\mathcal{H}},\delta_{\mathcal{H}}}(\sigma)/\Gamma_{\Phi_{\mathcal{H}},\sigma}$ determines a stratification of $S = \mathrm{Spec}(R)$ parameterized by $\{\text{faces } \tau \text{ of } \sigma\}/\Gamma_{\Phi_{\mathcal{H}},\sigma}$ such that each stratum of S (with its reduced structure, namely, its structure as an open subscheme in a closed subscheme with reduced structure) is the scheme-theoretic preimage of the corresponding stratum of $\vec{\Xi}^{\mathrm{ord}}_{\Phi_{\mathcal{H}},\delta_{\mathcal{H}}}(\sigma)/\Gamma_{\Phi_{\mathcal{H}},\sigma}$ under f.*
2. *The formal completion of ${}^{\diamondsuit}G$ along the preimage of $\mathrm{Spec}(R/I)$ is canonically isomorphic to the pullback of $G^{\natural}$ under $\hat{f}$ (as a formal algebraic stack, rather than a relative scheme).*
3. *The étale sheaf $\underline{\mathbf{X}}({}^{\diamondsuit}G)$ (see [62, Thm. 3.3.1.9]) is the quotient sheaf of the constant sheaf X such that, over the $(\tau \mod \Gamma_{\Phi_{\mathcal{H}},\sigma})$-stratum, the sheaf $\underline{\mathbf{X}}({}^{\diamondsuit}G)$ is a constant quotient $X_{(\tau \mod \Gamma_{\Phi_{\mathcal{H}},\sigma})}$ of X, with an admissible surjection $X \twoheadrightarrow X_{(\tau \mod \Gamma_{\Phi_{\mathcal{H}},\sigma})}$ inducing a torus argument $\Phi_{\mathcal{H},(\tau \mod \Gamma_{\Phi_{\mathcal{H}},\sigma})}$ from $\Phi_{\mathcal{H}}$ as in [62, Lem. 5.4.2.11], such that τ is contained in the $\Gamma_{\Phi_{\mathcal{H}}}$-orbit of the image of the induced embedding $\mathbf{P}^+_{\Phi_{\mathcal{H},(\tau \mod \Gamma_{\Phi_{\mathcal{H}},\sigma})}} \hookrightarrow \mathbf{P}_{\Phi_{\mathcal{H}}}$. (We know the surjection is admissible because of the existence of level-$\mathcal{H}$ structures; see [62, Lem. 5.2.2.2 and 5.2.2.4].) This produces a sheaf version $\underline{\Phi}_{\mathcal{H}}({}^{\diamondsuit}G)$ of $\Phi_{\mathcal{H}}$ over S.*
   The formation of $\mathbf{S}_{\Phi_{\mathcal{H}}}$ from $\Phi_{\mathcal{H}}$ applies to $\underline{\Phi}_{\mathcal{H}}({}^{\diamondsuit}G)$ and defines a sheaf $\underline{\mathbf{S}}_{\underline{\Phi}_{\mathcal{H}}({}^{\diamondsuit}G)}$ (cf. Proposition 4.2.2.5).
   Then $\underline{\Phi}_{\mathcal{H}}({}^{\diamondsuit}G)$ is equivalent (see Definition 1.2.1.6 and [62, Def. 5.4.2.2]) to the pullback of the tautological $\underline{\Phi}_{\mathcal{H}}$ on $\vec{\Xi}^{\mathrm{ord}}_{\Phi_{\mathcal{H}},\delta_{\mathcal{H}}}(\sigma)/\Gamma_{\Phi_{\mathcal{H}},\sigma}$ (see Proposition 4.2.2.5) under f.
4. *Under the equivalence between $\underline{\Phi}_{\mathcal{H}}({}^{\diamondsuit}G)$ and the pullback of $\underline{\Phi}_{\mathcal{H}}$ above, the*

pullback $f^*(\underline{B}) : \mathbf{S}_{\underline{\Phi}_{\mathcal{H}}(\,^\diamond G)} \to \underline{\mathrm{Inv}}(S)$ *of the tautological homomorphism* $\underline{B}$ *over* $\vec{\Xi}^{\mathrm{ord}}_{\Phi_{\mathcal{H}},\delta_{\mathcal{H}}}(\sigma)/\Gamma_{\Phi_{\mathcal{H}},\sigma}$ *(see Proposition* 4.2.2.5*) under* f *agrees with the homomorphism* $\underline{B}(\,^\diamond G)$ *defined by* $(\,^\diamond G,\,^\diamond\lambda,\,^\diamond i,\,^\diamond\alpha_{\mathcal{H}^p},\,^\diamond\alpha^{\mathrm{ord}}_{\mathcal{H}_p}) \to S$ *(or rather by* $(\,^\diamond G,\,^\diamond\lambda,\,^\diamond i,\,^\diamond\alpha_{\mathcal{H}}) \to S$*) as in Construction* 5.1.1.1 *(or rather as in [62, Constr. 6.3.1.1]).*

(5) *Let* $\underline{\mathrm{KS}}_{(\,^\diamond G,\,^\diamond\lambda,\,^\diamond i)/S,\mathrm{free}}$ *be the* $\mathscr{O}_S$*-module (flat over* $\vec{\mathsf{S}}_{0,r_{\mathcal{H}}}$*) defined by* $(\,^\diamond G,\,^\diamond\lambda,\,^\diamond i,\,^\diamond\alpha_{\mathcal{H}^p},\,^\diamond\alpha^{\mathrm{ord}}_{\mathcal{H}_p}) \to S$ *as in Definition* 3.4.3.1. *As in [62, Sec. 4.6.3], let* $\widehat{\Omega}^1_{S/\vec{\mathsf{S}}_{0,r_{\mathcal{H}}}}$ *denote the completion of* $\Omega^1_{S/\vec{\mathsf{S}}_{0,r_{\mathcal{H}}}}$ *with respect to the topology of* R *defined by* I*, which is locally free of finite rank over* $\mathscr{O}_S$ *(cf. [35,* 0_{IV}*, 20.4.9]), and let* $\widehat{\Omega}^1_{S/\vec{\mathsf{S}}_{0,r_{\mathcal{H}}}}[d\log\infty]$ *be the subsheaf of* $(\eta \hookrightarrow S)_*(\eta \hookrightarrow S)^*\widehat{\Omega}^1_{S/\vec{\mathsf{S}}_{0,r_{\mathcal{H}}}}$ *generated locally by* $\widehat{\Omega}^1_{S/\vec{\mathsf{S}}_{0,r_{\mathcal{H}}}}$ *and those* $d\log q$ *where* q *is a local generator of an irreducible component of the normal crossings divisor of* $\mathrm{Spec}(R)$ *induced by the corresponding normal crossings divisor of* $\vec{\Xi}^{\mathrm{ord}}_{\Phi_{\mathcal{H}},\delta_{\mathcal{H}}}(\sigma)$ *(cf. [62, Thm. 6.1.2.8(5)]). Then the extended Kodaira–Spencer morphism (see [62, Def. 4.6.3.44]) defines an isomorphism*

$$\mathrm{KS}_{\,^\diamond G/S/\vec{\mathsf{S}}_{0,r_{\mathcal{H}}}} : \underline{\mathrm{KS}}_{(\,^\diamond G,\,^\diamond\lambda,\,^\diamond i)/S,\mathrm{free}} \xrightarrow{\sim} \widehat{\Omega}^1_{S/\vec{\mathsf{S}}_{0,r_{\mathcal{H}}}}[d\log\infty].$$

(6) *The morphism* $\hat{f} : S_{\mathrm{for}} = \mathrm{Spf}(R,I) \to \vec{\mathfrak{X}}^{\mathrm{ord}}_{\Phi_{\mathcal{H}},\delta_{\mathcal{H}},\sigma}/\Gamma_{\Phi_{\mathcal{H}},\sigma}$*, or rather the morphism* $f : S = \mathrm{Spec}(R) \to \vec{\Xi}^{\mathrm{ord}}_{\Phi_{\mathcal{H}},\delta_{\mathcal{H}}}(\sigma)/\Gamma_{\Phi_{\mathcal{H}},\sigma}$*, is tautological with respect to the universal property of* $\vec{\Xi}^{\mathrm{ord}}_{\Phi_{\mathcal{H}},\delta_{\mathcal{H}}}(\sigma)/\Gamma_{\Phi_{\mathcal{H}},\sigma}$*, in the following sense:*
The setting is as follows: The base ring R *and the ideal* I *satisfy the setting of Section* 4.1.6. *Let* $(G,\lambda,i,\alpha_{\mathcal{H}^p},\alpha^{\mathrm{ord}}_{\mathcal{H}_p})$ *be a degenerating family of type* $\vec{\mathsf{M}}^{\mathrm{ord}}_{\mathcal{H}}$ *over* S *that defines an object in the essential image of the canonical morphism* $\mathrm{DEG}_{\mathrm{PEL},\mathsf{M}^{\mathrm{ord}}_{\mathcal{H}}}(R,I) \to \mathrm{DEG}_{\mathrm{PEL},\ddot{\mathsf{M}}^{\mathrm{ord}}_{\mathcal{H}}}(R,I)$ *in Theorem* 4.1.6.2. *By Theorem* 4.1.6.2*, the family determines an object in the essential image of* $\mathrm{DD}_{\mathrm{PEL},\mathsf{M}^{\mathrm{ord}}_{\mathcal{H}}}(R,I) \to \mathrm{DD}_{\mathrm{PEL},\ddot{\mathsf{M}}^{\mathrm{ord}}_{\mathcal{H}}}(R,I)$*, which determines an object of* $\mathrm{DD}_{\mathrm{PEL},\mathsf{M}^{\mathrm{ord}}_{\mathcal{H}}}(R,I)$ *up to isomorphism, which is by definition an object of* $\mathrm{DD}_{\mathrm{PEL},\mathsf{M}_{\mathcal{H}}}(R,I)$*, the latter determining, in particular, an ordinary cusp label. Suppose* $(\mathsf{Z}_{\mathcal{H}},\Phi_{\mathcal{H}},\delta_{\mathcal{H}})$ *is a representative of this ordinary cusp label. By [62, Lem. 5.4.2.10; see also the errata], there exists a tuple*

$$(B,\lambda_B,i_B,X,Y,\phi,c,c^\vee,\tau,[\alpha^\natural_{\mathcal{H}}])$$

defining the above object of $\mathrm{DD}_{\mathrm{PEL},\mathsf{M}_{\mathcal{H}}}(R,I)$*, together with a representative*

$$\alpha^\natural_{\mathcal{H}} = (\mathsf{Z}_{\mathcal{H}},\varphi^{\sim}_{-2,\mathcal{H}},\varphi_{-1,\mathcal{H}},\varphi^{\sim}_{0,\mathcal{H}},\delta_{\mathcal{H}},c_{\mathcal{H}},c^\vee_{\mathcal{H}},\tau_{\mathcal{H}})$$

of $[\alpha^\natural_{\mathcal{H}}]$*, such that* $(\varphi^{\sim}_{-2,\mathcal{H}},\varphi^{\sim}_{0,\mathcal{H}})$ *induces the* $(\varphi_{-2,\mathcal{H}},\varphi_{0,\mathcal{H}})$ *in* $\Phi_{\mathcal{H}}$*, as in the corrected [62, Def. 5.4.2.8] in the errata. By Theorem* 4.1.6.2*, this tuple is an object of* $\mathrm{DD}_{\mathrm{PEL},\mathsf{M}^{\mathrm{ord}}_{\mathcal{H}}}(R,I)$*, and determines an object*

$$(B,\lambda_B,i_B,\underline{X},\underline{Y},\phi,c,c^\vee,\tau,[\alpha^{\natural,\mathrm{ord}}_{\mathcal{H}}])$$

in $\mathrm{DD}_{\mathrm{PEL},\ddot{\mathsf{M}}^{\mathrm{ord}}_{\mathcal{H}}}(R,I)$*, with* $[\alpha^{\natural,\mathrm{ord}}_{\mathcal{H}}]$ *represented by the tuple*

$$\alpha^{\natural,\mathrm{ord}}_{\mathcal{H}} = (\mathsf{Z}_{\mathcal{H}},\varphi^{\mathrm{ord}}_{-2,\mathcal{H}},\varphi^{\mathrm{ord}}_{-1,\mathcal{H}} = (\varphi_{-1,\mathcal{H}^p},\varphi^{\mathrm{ord}}_{-1,\mathcal{H}_p}),\varphi^{\mathrm{ord}}_{0,\mathcal{H}},\delta^{\mathrm{ord}}_{\mathcal{H}},c^{\mathrm{ord}}_{\mathcal{H}},c^{\vee,\mathrm{ord}}_{\mathcal{H}},\tau^{\mathrm{ord}}_{\mathcal{H}})$$

determined by $\alpha^\natural_{\mathcal{H}}$ *as in Section* 4.1.6, *so that* $(\varphi^{\mathrm{ord}}_{-2,\mathcal{H}}, \varphi^{\mathrm{ord}}_{0,\mathcal{H}})$ *(resp.* $\delta^{\mathrm{ord}}_{\mathcal{H}}$*) induces the same* $(\varphi_{-2,\mathcal{H}}, \varphi_{0,\mathcal{H}})$ *in* $\Phi_{\mathcal{H}}$ *(resp.* $\delta_{\mathcal{H}}$*). By Proposition* 4.2.1.46, *this tuple without its positivity condition defines a morphism* $\mathrm{Spec}(K) \to \vec{\Xi}^{\mathrm{ord}}_{\Phi_{\mathcal{H}},\delta_{\mathcal{H}}}$ *that is unique up to an action of* $\Gamma_{\Phi_{\mathcal{H}}}$ *on the identification of* $\Phi_{\mathcal{H}}$*, whose composition with* $\vec{\Xi}^{\mathrm{ord}}_{\Phi_{\mathcal{H}},\delta_{\mathcal{H}}} \to \vec{C}^{\mathrm{ord}}_{\Phi_{\mathcal{H}},\delta_{\mathcal{H}}}$ *extends to a morphism* $\mathrm{Spec}(R) \to \vec{C}^{\mathrm{ord}}_{\Phi_{\mathcal{H}},\delta_{\mathcal{H}}}$. *Let* $\underline{B}(G) : \mathbf{S}_{\underline{\Phi}_{\mathcal{H}}(G)} \to \underline{\mathrm{Inv}}(S)$ *be the homomorphism defined as in Construction* 5.1.1.1.

Then the universal property is as follows: Suppose there exists an identification of $\Phi_{\mathcal{H}}$ *such that, for each discrete valuation* $\upsilon : \underline{\mathrm{Inv}}(S) \to \mathbb{Z}$ *defined by a height-one prime of* R*, the composition* $\upsilon \circ \underline{B}(G) : \mathbf{S}_{\underline{\Phi}_{\mathcal{H}}(G)} \to \mathbb{Z}$ *defines an element in the closure* $\overline{\sigma}$ *of* σ *in* $(\mathbf{S}_{\Phi_{\mathcal{H}}})^\vee_{\mathbb{R}}$. *Such an identification of* $\Phi_{\mathcal{H}}$ *is unique up to an element in* $\Gamma_{\Phi_{\mathcal{H}},\sigma}$*, and all morphisms* $\mathrm{Spec}(K) \to \vec{\Xi}^{\mathrm{ord}}_{\Phi_{\mathcal{H}},\delta_{\mathcal{H}}}$ *as above induce the same morphism* $\mathrm{Spec}(K) \to \vec{\Xi}^{\mathrm{ord}}_{\Phi_{\mathcal{H}},\delta_{\mathcal{H}}}/\Gamma_{\Phi_{\mathcal{H}},\sigma}$ *if they respect such identifications of* $\Phi_{\mathcal{H}}$. *Then this morphism extends to a (necessarily unique) morphism* $f : S = \mathrm{Spec}(R) \to \vec{\Xi}^{\mathrm{ord}}_{\Phi_{\mathcal{H}},\delta_{\mathcal{H}}}(\sigma)/\Gamma_{\Phi_{\mathcal{H}},\sigma}$*, sending the subscheme* $\mathrm{Spec}(R/I)$ *to the* σ*-stratum of* $\vec{\Xi}^{\mathrm{ord}}_{\Phi_{\mathcal{H}},\delta_{\mathcal{H}}}(\sigma)/\Gamma_{\Phi_{\mathcal{H}},\sigma}$ *and hence inducing a morphism* $\hat{f} : S_{\mathrm{for}} = \mathrm{Spf}(R, I) \to \vec{\mathfrak{X}}^{\mathrm{ord}}_{\Phi_{\mathcal{H}},\delta_{\mathcal{H}},\sigma}/\Gamma_{\Phi_{\mathcal{H}},\sigma}$ *between formal algebraic stacks, such that* $(G, \lambda, i, \alpha_{\mathcal{H}^p}, \alpha^{\mathrm{ord}}_{\mathcal{H}_p}) \to S$ *is isomorphic to the pullback* $({}^\diamondsuit G, {}^\diamondsuit\lambda, {}^\diamondsuit i, {}^\diamondsuit\alpha_{\mathcal{H}^p}, {}^\diamondsuit\alpha^{\mathrm{ord}}_{\mathcal{H}_p}) \to S$ *of* $({}^\heartsuit G, {}^\heartsuit\lambda, {}^\heartsuit i, {}^\heartsuit\alpha_{\mathcal{H}^p}, {}^\heartsuit\alpha^{\mathrm{ord}}_{\mathcal{H}_p}) \to \vec{\mathfrak{X}}^{\mathrm{ord}}_{\Phi_{\mathcal{H}},\delta_{\mathcal{H}},\sigma}/\Gamma_{\Phi_{\mathcal{H}},\sigma}$ *under* $\hat{f}$ *(and so that* $(G, \lambda, i, \alpha_{\mathcal{H}}) \to S$ *is isomorphic to the pullback* $({}^\diamondsuit G, {}^\diamondsuit\lambda, {}^\diamondsuit i, {}^\diamondsuit\alpha_{\mathcal{H}}) \to S$ *of* $({}^\heartsuit G, {}^\heartsuit\lambda, {}^\heartsuit i, {}^\heartsuit\alpha_{\mathcal{H}}) \to \vec{\mathfrak{X}}^{\mathrm{ord}}_{\Phi_{\mathcal{H}},\delta_{\mathcal{H}},\sigma}/\Gamma_{\Phi_{\mathcal{H}},\sigma}$ *under* $\hat{f}$*, by the injectivity of the assignment* (3.3.5.5)*)*.

Proof. The proof of [62, Prop. 6.3.1.6] based on [62, Thm. 4.6.3.16] works here if we replace the reference to [62, Prop. 6.2.5.18] there with an analogous reference to Proposition 4.2.3.5. □

As a byproduct of our usage of Proposition 4.2.3.5 in the proof:

Corollary 5.1.1.4. *(Compare with [62, Cor. 6.3.1.8].) Suppose* $\hat{f} : S_{\mathrm{for}} = \mathrm{Spf}(R, I) \to \vec{\mathfrak{X}}^{\mathrm{ord}}_{\Phi_{\mathcal{H}},\delta_{\mathcal{H}},\sigma}/\Gamma_{\Phi_{\mathcal{H}},\sigma}$ *is a morphism between noetherian formal schemes formally smooth over* $\vec{\mathsf{S}}_{0,r_{\mathcal{H}}}$*, with induced morphism* $f : S = \mathrm{Spec}(R) \to \vec{\Xi}^{\mathrm{ord}}_{\Phi_{\mathcal{H}},\delta_{\mathcal{H}}}(\sigma)/\Gamma_{\Phi_{\mathcal{H}},\sigma}$ *such that the support* $\mathrm{Spec}(R/I)$ *of* S_{for} *is the scheme-theoretic preimage under* f *of some subalgebraic stack* Z *of the* σ*-stratum* $\vec{\Xi}^{\mathrm{ord}}_{\Phi_{\mathcal{H}},\delta_{\mathcal{H}},\sigma}/\Gamma_{\Phi_{\mathcal{H}},\sigma}$ *of* $\vec{\Xi}^{\mathrm{ord}}_{\Phi_{\mathcal{H}},\delta_{\mathcal{H}}}(\sigma)/\Gamma_{\Phi_{\mathcal{H}},\sigma}$. *Suppose moreover that the pullback of the stratification of* $\vec{\Xi}^{\mathrm{ord}}_{\Phi_{\mathcal{H}},\delta_{\mathcal{H}}}(\sigma)/\Gamma_{\Phi_{\mathcal{H}},\sigma}$ *induces a stratification of* $S = \mathrm{Spec}(R)$ *such that each stratum of* $S = \mathrm{Spec}(R)$ *(with its reduced structure, as in* (1) *of Proposition* 5.1.1.3*) is the scheme-theoretic preimage of a stratum of* $\vec{\Xi}^{\mathrm{ord}}_{\Phi_{\mathcal{H}},\delta_{\mathcal{H}}}(\sigma)/\Gamma_{\Phi_{\mathcal{H}},\sigma}$. *Under these assumptions, we have an induced morphism* $f_0 : \mathrm{Spec}(R/I) \to Z$*, and we can define (as in [62, Thm. 4.6.3.16] and* (5) *of Proposition* 5.1.1.3*) the extended Kodaira–Spencer morphism*

$$\mathrm{KS}_{{}^\diamondsuit G/S/\vec{\mathsf{S}}_{0,r_{\mathcal{H}}}} : \underline{\mathrm{KS}}_{({}^\diamondsuit G, {}^\diamondsuit\lambda, {}^\diamondsuit i)/S,\mathrm{free}} \to \widehat{\Omega}^1_{S/\vec{\mathsf{S}}_{0,r_{\mathcal{H}}}}[d\log\infty], \tag{5.1.1.5}$$

where $\underline{\mathrm{KS}}_{({}^\diamond G,\,{}^\diamond\lambda,\,{}^\diamond i)/S,\mathrm{free}}$ *is the sheaf defined as in Definition* 3.4.3.1 *by the pullback* $({}^\diamond G, {}^\diamond\lambda, {}^\diamond i, {}^\diamond\alpha_{\mathcal{H}^p}, {}^\diamond\alpha^{\mathrm{ord}}_{\mathcal{H}_p}) \to S$ *of* $({}^\heartsuit G, {}^\heartsuit\lambda, {}^\heartsuit i, {}^\heartsuit\alpha_{\mathcal{H}^p}, {}^\heartsuit\alpha^{\mathrm{ord}}_{\mathcal{H}_p}) \to \vec{\mathfrak{X}}^{\mathrm{ord}}_{\Phi_{\mathcal{H}},\delta_{\mathcal{H}},\sigma}/\Gamma_{\Phi_{\mathcal{H}},\sigma}$ *in the sense of relative schemes. Then the morphism* $\hat{f}$ *is* **formally étale** *if and only if it satisfies the conditions that* f *is flat, that* f_0 *is formally smooth, and that the morphism* $\mathrm{KS}_{{}^\diamond G/S/\vec{\mathsf{S}}_{0,r_{\mathcal{H}}}}$ *in* (5.1.1.5) *is surjective.*

Proof. The proof of [62, Cor. 6.3.1.8] works here if we replace the reference to [62, Prop. 6.2.5.18] there with an analogous reference to Proposition 4.2.3.5. □

Corollary 5.1.1.6. (*Compare with [62, Cor. 6.3.1.14].*) *In the context of Corollary* 5.1.1.4, *suppose that* R *is a strict local ring with* (*separably closed*) *residue field* k, *so that the morphism* f *induces a morphism* $\tilde{f} : \mathrm{Spec}(R) \to \mathrm{Spec}(\tilde{R})$ *mapping* $\mathrm{Spec}(k)$ *to* $\mathrm{Spec}(\tilde{k})$, *where* $\tilde{R}$ *is a strict local ring of* $\vec{\Xi}^{\mathrm{ord}}_{\Phi_{\mathcal{H}},\delta_{\mathcal{H}}}(\sigma)/\Gamma_{\Phi_{\mathcal{H}},\sigma}$ *with* (*separably closed*) *residue field* $\tilde{k}$, *and suppose that* k *is of finite type over* $\tilde{k}$. *Then* $\tilde{f}$ *is formally étale if and only if* R *and* f *satisfy the conditions that* R *is equidimensional and has the same dimension as* $\vec{\Xi}^{\mathrm{ord}}_{\Phi_{\mathcal{H}},\delta_{\mathcal{H}}}(\sigma)/\Gamma_{\Phi_{\mathcal{H}},\sigma}$, *and that the induced canonical morphism* (5.1.1.5) *is surjective.* (*This last condition forces the induced homomorphism* $\tilde{k} \to k$ *to be an isomorphism.*)

Proof. The proof of [62, Cor. 6.3.1.14] also works here. □

Definition 5.1.1.7. (*Compare with [28, Ch. IV, Sec. 3] and [62, Def. 6.3.1.15].*) *Let* $(\Phi_{\mathcal{H}}, \delta_{\mathcal{H}})$ *be a representative of an ordinary cusp label at level* $\mathcal{H}$ (*see Definition* 3.2.3.8), *and let* $\sigma \subset \mathbf{P}^+_{\Phi_{\mathcal{H}}}$ *be a nondegenerate smooth rational polyhedral cone. An* **ordinary good formal** $(\Phi_{\mathcal{H}}, \delta_{\mathcal{H}}, \sigma)$**-model** *is a degenerating family* $({}^\diamond G, {}^\diamond\lambda, {}^\diamond i, {}^\diamond\alpha_{\mathcal{H}^p}, {}^\diamond\alpha^{\mathrm{ord}}_{\mathcal{H}_p})$ *of type* $\vec{\mathsf{M}}^{\mathrm{ord}}_{\mathcal{H}}$ *over* $\mathrm{Spec}(R)$ (*see Definition* 3.4.2.10) *where we have the following:*

(1) *R is a strict local ring that is complete with respect to an ideal* $I = \mathrm{rad}(I)$, *together with a stratification of* $\mathrm{Spec}(R)$ *with strata parameterized by* $\Gamma_{\Phi_{\mathcal{H}},\sigma}$*-orbits of faces of* σ.
(2) *There exists a morphism* $f : \mathrm{Spec}(R) \to \vec{\Xi}^{\mathrm{ord}}_{\Phi_{\mathcal{H}},\delta_{\mathcal{H}}}(\sigma)/\Gamma_{\Phi_{\mathcal{H}},\sigma}$ *such that* $\mathrm{Spec}(R/I)$ *is the scheme-theoretic preimage of the* σ*-stratum under* f, *satisfying the following properties:*
   (a) *The morphism* f *makes* R *isomorphic to the completion of a strict local ring of* $\vec{\Xi}^{\mathrm{ord}}_{\Phi_{\mathcal{H}},\delta_{\mathcal{H}}}(\sigma)/\Gamma_{\Phi_{\mathcal{H}},\sigma}$ *with respect to the ideal defining the* σ*-stratum.*
   (b) *The stratification of* $\mathrm{Spec}(R)$ *is strictly compatible with that of* $\vec{\Xi}^{\mathrm{ord}}_{\Phi_{\mathcal{H}},\delta_{\mathcal{H}}}(\sigma)/\Gamma_{\Phi_{\mathcal{H}},\sigma}$ *in the sense that each stratum of* $\mathrm{Spec}(R)$ (*with its reduced structure, as in* (1) *of Proposition* 5.1.1.3) *is the scheme-theoretic preimage of the corresponding stratum of* $\vec{\Xi}^{\mathrm{ord}}_{\Phi_{\mathcal{H}},\delta_{\mathcal{H}}}(\sigma)/\Gamma_{\Phi_{\mathcal{H}},\sigma}$.
   (c) *The degenerating family* $({}^\diamond G, {}^\diamond\lambda, {}^\diamond i, {}^\diamond\alpha_{\mathcal{H}^p}, {}^\diamond\alpha^{\mathrm{ord}}_{\mathcal{H}_p})$ *defines an object of* $\mathrm{DEG}_{\mathrm{PEL},\ddot{\mathsf{M}}^{\mathrm{ord}}_{\mathcal{H}}}(R, I)$ *in the essential image of the canonical morphism* $\mathrm{DEG}_{\mathrm{PEL},\mathsf{M}^{\mathrm{ord}}_{\mathcal{H}}}(R, I) \to \mathrm{DEG}_{\mathrm{PEL},\ddot{\mathsf{M}}^{\mathrm{ord}}_{\mathcal{H}}}(R, I)$ *in Theorem* 4.1.6.2,

and (by abuse of language) $({}^\diamondsuit G, {}^\diamondsuit\lambda, {}^\diamondsuit i, {}^\diamondsuit\alpha_{\mathcal{H}^p}, {}^\diamondsuit\alpha^{\mathrm{ord}}_{\mathcal{H}_p}) \to \mathrm{Spec}(R)$ *is the pullback of the Mumford family* $({}^\heartsuit G, {}^\heartsuit\lambda, {}^\heartsuit i, {}^\heartsuit\alpha_{\mathcal{H}^p}, {}^\heartsuit\alpha^{\mathrm{ord}}_{\mathcal{H}_p}) \to \vec{\mathfrak{X}}^{\mathrm{ord}}_{\Phi_\mathcal{H},\delta_\mathcal{H},\sigma}/\Gamma_{\Phi_\mathcal{H},\sigma}$ *under the morphism* $\hat{f} : \mathrm{Spf}(R, I) \to \vec{\mathfrak{X}}^{\mathrm{ord}}_{\Phi_\mathcal{H},\delta_\mathcal{H},\sigma}/\Gamma_{\Phi_\mathcal{H},\sigma}$ *induced by* f.

Remark 5.1.1.8. (Compare with [62, Rem. 6.3.1.16].) As in Proposition 5.1.1.3, the morphism $\hat{f} : \mathrm{Spf}(R, I) \to \vec{\mathfrak{X}}^{\mathrm{ord}}_{\Phi_\mathcal{H},\delta_\mathcal{H},\sigma}/\Gamma_{\Phi_\mathcal{H},\sigma}$ in Definition 5.1.1.7 (making $({}^\diamondsuit G, {}^\diamondsuit\lambda, {}^\diamondsuit i, {}^\diamondsuit\alpha_{\mathcal{H}^p}, {}^\diamondsuit\alpha^{\mathrm{ord}}_{\mathcal{H}_p}) \to \mathrm{Spec}(R)$ the pullback of the Mumford family $({}^\heartsuit G, {}^\heartsuit\lambda, {}^\heartsuit i, {}^\heartsuit\alpha_{\mathcal{H}^p}, {}^\heartsuit\alpha^{\mathrm{ord}}_{\mathcal{H}_p}) \to \vec{\mathfrak{X}}^{\mathrm{ord}}_{\Phi_\mathcal{H},\delta_\mathcal{H},\sigma}/\Gamma_{\Phi_\mathcal{H},\sigma}$) is necessarily unique.

Remark 5.1.1.9. (Compare with [62, Rem. 6.3.1.17].) By the universal property of $\vec{\Xi}^{\mathrm{ord}}_{\Phi_\mathcal{H},\delta_\mathcal{H}}(\sigma)/\Gamma_{\Phi_\mathcal{H},\sigma}$ (see Proposition 4.2.2.8 and (6) of Proposition 5.1.1.3), the morphism $f : \mathrm{Spec}(R) \to \vec{\Xi}^{\mathrm{ord}}_{\Phi_\mathcal{H},\delta_\mathcal{H}}(\sigma)/\Gamma_{\Phi_\mathcal{H},\sigma}$ in Definition 5.1.1.7 (with the desired properties) is tautological for the induced morphism $\mathrm{Spec}(R) \to \vec{C}^{\mathrm{ord}}_{\Phi_\mathcal{H},\delta_\mathcal{H}}(\sigma)/\Gamma_{\Phi_\mathcal{H},\sigma}$ and the homomorphism $\underline{B}({}^\diamondsuit G) : \mathbf{S}_{\underline{\Phi}_\mathcal{H}({}^\diamondsuit G)} \to \underline{\mathrm{Inv}}(\mathrm{Spec}(R))$.

Corollary 5.1.1.10. (*Compare with [62, Cor. 6.3.1.18].*) *Suppose R is a regular strict local ring complete with respect to an ideal* $I = \mathrm{rad}(I)$, *together with a morphism* $\hat{f} : S_{\mathrm{for}} := \mathrm{Spf}(R, I) \to \vec{\mathfrak{X}}^{\mathrm{ord}}_{\Phi_\mathcal{H},\delta_\mathcal{H},\sigma}/\Gamma_{\Phi_\mathcal{H},\sigma}$ *inducing a morphism* $f : S := \mathrm{Spec}(R) \to \vec{\Xi}^{\mathrm{ord}}_{\Phi_\mathcal{H},\delta_\mathcal{H}}(\sigma)/\Gamma_{\Phi_\mathcal{H},\sigma}$ *such that* $\mathrm{Spec}(R/I)$ *is the scheme-theoretic preimage of the σ-stratum under f, and inducing an isomorphism between separable closures of residue fields. Then we can verify the statement that f makes R isomorphic to the completion of a strict local ring of* $\vec{\Xi}^{\mathrm{ord}}_{\Phi_\mathcal{H},\delta_\mathcal{H}}(\sigma)/\Gamma_{\Phi_\mathcal{H},\sigma}$ *with respect to the ideal defining the σ-stratum by verifying the following conditions:*

(1) *The scheme S has the same dimension as* $\vec{\Xi}^{\mathrm{ord}}_{\Phi_\mathcal{H},\delta_\mathcal{H}}(\sigma)/\Gamma_{\Phi_\mathcal{H},\sigma}$.
(2) *The stratification of* $\vec{\Xi}^{\mathrm{ord}}_{\Phi_\mathcal{H},\delta_\mathcal{H}}(\sigma)/\Gamma_{\Phi_\mathcal{H},\sigma}$ *induces a stratification of* $\mathrm{Spec}(R)$ *that is strictly compatible with that of* $\vec{\Xi}^{\mathrm{ord}}_{\Phi_\mathcal{H},\delta_\mathcal{H}}(\sigma)/\Gamma_{\Phi_\mathcal{H},\sigma}$ *in the sense that each stratum of S (with its reduced structure, as in* (1) *of Proposition* 5.1.1.3) *is the scheme-theoretic preimage of the corresponding stratum of* $\vec{\Xi}^{\mathrm{ord}}_{\Phi_\mathcal{H},\delta_\mathcal{H}}(\sigma)/\Gamma_{\Phi_\mathcal{H},\sigma}$.
(3) *The extended Kodaira–Spencer morphism (see [62, Def. 4.6.3.44]) induces an isomorphism*

$$\mathrm{KS}_{{}^\diamondsuit G/S/\vec{\mathsf{S}}_{0,r_\mathcal{H}}} : \underline{\mathrm{KS}}_{({}^\diamondsuit G, {}^\diamondsuit\lambda, {}^\diamondsuit i)/S,\mathrm{free}} \xrightarrow{\sim} \widehat{\Omega}^1_{S/\vec{\mathsf{S}}_{0,r_\mathcal{H}}}[d\log\infty],$$

where $({}^\diamondsuit G, {}^\diamondsuit\lambda, {}^\diamondsuit i, {}^\diamondsuit\alpha_{\mathcal{H}^p}, {}^\diamondsuit\alpha^{\mathrm{ord}}_{\mathcal{H}_p}) \to S$ *is the pullback of the Mumford family* $({}^\heartsuit G, {}^\heartsuit\lambda, {}^\heartsuit i, {}^\heartsuit\alpha_{\mathcal{H}^p}, {}^\heartsuit\alpha^{\mathrm{ord}}_{\mathcal{H}_p}) \to \vec{\mathfrak{X}}^{\mathrm{ord}}_{\Phi_\mathcal{H},\delta_\mathcal{H},\sigma}/\Gamma_{\Phi_\mathcal{H},\sigma}$ *under $\hat{f}$ (by abuse of language), and where* $\underline{\mathrm{KS}}_{({}^\diamondsuit G, {}^\diamondsuit\lambda, {}^\diamondsuit i)/S,\mathrm{free}}$ *is defined as in Definition* 3.4.3.1.

Remark 5.1.1.11. (Compare with [62, Rem. 6.3.1.19].) The various morphisms from $\mathrm{Spec}(R/I)$ to the support of the formal algebraic stack $\vec{\mathfrak{X}}^{\mathrm{ord}}_{\Phi_\mathcal{H},\delta_\mathcal{H},\sigma}/\Gamma_{\Phi_\mathcal{H}\sigma}$, for the various ordinary good formal $(\Phi_\mathcal{H}, \delta_\mathcal{H}, \sigma)$-models, *cover* the whole σ-stratum.

Remark 5.1.1.12. (Compare with [62, Rem. 6.3.1.20].) An ordinary good formal $(\Phi_{\mathcal{H}}, \delta_{\mathcal{H}}, \sigma)$-model is an ordinary good formal $(\Phi'_{\mathcal{H}}, \delta'_{\mathcal{H}}, \sigma')$-model if and only if $(\Phi_{\mathcal{H}}, \delta_{\mathcal{H}}, \sigma)$ is equivalent to $(\Phi'_{\mathcal{H}}, \delta'_{\mathcal{H}}, \sigma')$ (see Definition 1.2.2.10).

Remark 5.1.1.13. (Compare with [62, Rem. 6.3.1.21].) For two smooth rational polyhedral cones $\sigma, \sigma' \in \mathbf{P}^+_{\Phi_{\mathcal{H}}}$ such that $\sigma \subset \sigma'$, an ordinary good formal $(\Phi_{\mathcal{H}}, \delta_{\mathcal{H}}, \sigma)$-model is not necessarily an ordinary good formal $(\Phi_{\mathcal{H}}, \delta_{\mathcal{H}}, \sigma')$-model (*cf.* [62, Rem. 6.2.5.31] and Remark 4.2.2.23).

5.1.2 *Ordinary Good Algebraic Models*

Proposition 5.1.2.1. *(Compare with [62, Prop. 6.3.2.1].) Let $(\Phi_{\mathcal{H}}, \delta_{\mathcal{H}})$ be a representative of an ordinary cusp label at level $\mathcal{H}$ (see Definition* 3.2.3.8*), and let $\sigma \subset \mathbf{P}^+_{\Phi_{\mathcal{H}}}$ be a nondegenerate smooth rational polyhedral cone. Let R be the strict local ring of a geometric point $\bar{x}$ of the σ-stratum of $\vec{\Xi}^{\mathrm{ord}}_{\Phi_{\mathcal{H}},\delta_{\mathcal{H}}}(\sigma)/\Gamma_{\Phi_{\mathcal{H}},\sigma}$ for some $\sigma \subset \mathbf{P}^+_{\Phi_{\mathcal{H}}}$, let $R^\wedge$ be the completion of R with respect to the ideal I defining the σ-stratum, and let $I^\wedge := I \cdot R^\wedge \subset R^\wedge$. Suppose $({}^\diamond G, {}^\diamond\lambda, {}^\diamond i, {}^\diamond\alpha_{\mathcal{H}^p}, {}^\diamond\alpha^{\mathrm{ord}}_{\mathcal{H}_p}) \to \mathrm{Spec}(R^\wedge)$ defines an ordinary good formal $(\Phi_{\mathcal{H}}, \delta_{\mathcal{H}}, \sigma)$-model over $S^\wedge := \mathrm{Spec}(R^\wedge)$. Then we can find (noncanonically) a degenerating family $(G, \lambda, i, \alpha_{\mathcal{H}^p}, \alpha^{\mathrm{ord}}_{\mathcal{H}_p})$ of type $\vec{\mathsf{M}}^{\mathrm{ord}}_{\mathcal{H}}$ over $S :=$ $\mathrm{Spec}(R)$ as in Definition* 3.4.2.10*, which approximates $({}^\diamond G, {}^\diamond\lambda, {}^\diamond i, {}^\diamond\alpha_{\mathcal{H}^p}, {}^\diamond\alpha^{\mathrm{ord}}_{\mathcal{H}_p})$ in the following sense:*

1. *Over $\mathrm{Spec}(R/I)$, we have $({}^\diamond G, {}^\diamond\lambda, {}^\diamond i) \underset{R}{\otimes} (R/I) \cong (G, \lambda, i) \underset{R}{\otimes} (R/I)$. (We do not compare $({}^\diamond\alpha_{\mathcal{H}^p}, {}^\diamond\alpha_{\mathcal{H}_p})$ and $(\alpha_{\mathcal{H}^p}, \alpha^{\mathrm{ord}}_{\mathcal{H}_p})$ here, because they are not defined over $\mathrm{Spec}(R/I)$.)*
2. *Under the canonical homomorphism $R \hookrightarrow R^\wedge$, the pullbacks of the objects $\underline{\Phi}_{\mathcal{H}}(G)$, $\mathbf{S}_{\underline{\Phi}(G)}$, and $\underline{B}(G)$ defined as in Construction* 5.1.1.1 *are isomorphic to the objects $\Phi_{\mathcal{H}}({}^\diamond G)$, $\mathbf{S}_{\underline{\Phi}({}^\diamond G)}$, and $\underline{B}({}^\diamond G)$ defined as in Proposition* 5.1.1.3*, respectively.*
3. *The pullback of $(G, \lambda, i, \alpha_{\mathcal{H}^p}, \alpha^{\mathrm{ord}}_{\mathcal{H}_p}) \to S$ under the canonical homomorphism $R \hookrightarrow R^\wedge$ defines an ordinary good formal $(\Phi_{\mathcal{H}}, \delta_{\mathcal{H}}, \sigma)$-model $(G, \lambda, i, \alpha_{\mathcal{H}^p}, \alpha^{\mathrm{ord}}_{\mathcal{H}_p}) \underset{R}{\otimes} R^\wedge \to S^\wedge$, and can be realized as the pullback of the Mumford family $({}^\heartsuit G, {}^\heartsuit\lambda, {}^\heartsuit i, {}^\heartsuit\alpha_{\mathcal{H}^p}, {}^\heartsuit\alpha^{\mathrm{ord}}_{\mathcal{H}_p}) \to \vec{\mathfrak{X}}^{\mathrm{ord}}_{\Phi_{\mathcal{H}},\delta_{\mathcal{H}},\sigma}/\Gamma_{\Phi_{\mathcal{H}},\sigma}$ via a canonically defined morphism $\mathrm{Spf}(R^\wedge, I^\wedge) \to \vec{\mathfrak{X}}^{\mathrm{ord}}_{\Phi_{\mathcal{H}},\delta_{\mathcal{H}},\sigma}/\Gamma_{\Phi_{\mathcal{H}},\sigma}$. Comparing this isomorphism with the original morphism $\mathrm{Spf}(R^\wedge, I^\wedge) \to \vec{\mathfrak{X}}^{\mathrm{ord}}_{\Phi_{\mathcal{H}},\delta_{\mathcal{H}},\sigma}/\Gamma_{\Phi_{\mathcal{H}},\sigma}$ making the ordinary good formal $(\Phi_{\mathcal{H}}, \delta_{\mathcal{H}}, \sigma)$-model $({}^\diamond G, {}^\diamond\lambda, {}^\diamond i, {}^\diamond\alpha_{\mathcal{H}^p}, {}^\diamond\alpha^{\mathrm{ord}}_{\mathcal{H}_p}) \to S^\wedge$ a pullback of the Mumford family, we see that they are approximate in the sense that the induced morphisms from $\mathrm{Spec}(R/I)$ to the σ-stratum of $\vec{\mathfrak{X}}^{\mathrm{ord}}_{\Phi_{\mathcal{H}},\delta_{\mathcal{H}},\sigma}/\Gamma_{\Phi_{\mathcal{H}},\sigma}$ (between the supports of the formal schemes)* **coincide**.
4. *The extended Kodaira–Spencer morphism (see [62, Def. 4.6.3.44]) for the above pullback $(G, \lambda, i, \alpha_{\mathcal{H}^p}, \alpha^{\mathrm{ord}}_{\mathcal{H}_p}) \underset{R}{\otimes} R^\wedge \to S^\wedge$ induces (cf. the proof of [62,*

Thm. 4.6.3.43]) an isomorphism

$$\mathrm{KS}_{G/S/\vec{\mathsf{S}}_{0,r_{\mathcal{H}}}} : \underline{\mathrm{KS}}_{(G,\lambda,i)/S,\mathrm{free}} \xrightarrow{\sim} \widetilde{\Omega}^1_{S/\vec{\mathsf{S}}_{0,r_{\mathcal{H}}}}[d\log\infty],$$

where $\underline{\mathrm{KS}}_{(G,\lambda,i)/S,\mathrm{free}}$ *is defined as in Definition* 3.4.3.1, *and where* $\widetilde{\Omega}^1_{S/\vec{\mathsf{S}}_{0,r_{\mathcal{H}}}}[d\log\infty]$ *is defined by* $\widetilde{\Omega}^1_{S/\vec{\mathsf{S}}_{0,r_{\mathcal{H}}}}$, *the coherent sheaf associated with the module of universal finite differentials* $\widetilde{\Omega}^1_{R/\mathcal{O}_{F_0,(p)}[\zeta_{p^{r_{\mathcal{H}}}}]}$ *(see [55, Sec. 11–12]), and by the normal crossings divisor of* $S = \mathrm{Spec}(R)$ *induced by the one of* $\vec{\Xi}^{\mathrm{ord}}_{\Phi_{\mathcal{H}},\delta_{\mathcal{H}}}(\sigma)/\Gamma_{\Phi_{\mathcal{H}},\sigma}$ *(as in* (5) *of Proposition* 5.1.1.3, *with* $\widehat{\Omega}^1_{S/\vec{\mathsf{S}}_{0,r_{\mathcal{H}}}}$ *there replaced with* $\widetilde{\Omega}^1_{S/\vec{\mathsf{S}}_{0,r_{\mathcal{H}}}}$ *here).*

Proof. The proof of [62, Prop. 6.3.2.1] using Artin's approximation theory (*cf.* [2, Thm. 1.10] and [62, Prop. 6.3.2.2]) and [28, Ch. IV, Lem. 4.2] also works here. □

Definition 5.1.2.2. *(Compare with [62, Def. 6.3.2.5].) Let* $(\Phi_{\mathcal{H}}, \delta_{\mathcal{H}})$ *be a representative of an ordinary cusp label at level* $\mathcal{H}$ *(see Definition* 3.2.3.8*), and let* $\sigma \subset \mathbf{P}^+_{\Phi_{\mathcal{H}}}$ *be a nondegenerate smooth rational polyhedral cone. An* **ordinary good algebraic** $(\Phi_{\mathcal{H}}, \delta_{\mathcal{H}}, \sigma)$**-model** *consists of the following data:*

(1) An affine scheme $S = \mathrm{Spec}(R_{\mathrm{alg}})$, *together with a stratification of* S *with strata parameterized by* $\Gamma_{\Phi_{\mathcal{H}},\sigma}$*-orbits of faces of* σ.

(2) A strata-preserving morphism $S \to \vec{\Xi}^{\mathrm{ord}}_{\Phi_{\mathcal{H}},\delta_{\mathcal{H}}}(\sigma)/\Gamma_{\Phi_{\mathcal{H}},\sigma}$ *making* S *an étale neighborhood of some geometric point* $\bar{x}$ *of* $\vec{\Xi}^{\mathrm{ord}}_{\Phi_{\mathcal{H}},\delta_{\mathcal{H}}}(\sigma)/\Gamma_{\Phi_{\mathcal{H}},\sigma}$ *at the* σ*-stratum.*
Let $R^\wedge$ *be the completion of the strict local ring of* $\vec{\Xi}^{\mathrm{ord}}_{\Phi_{\mathcal{H}},\delta_{\mathcal{H}}}(\sigma)/\Gamma_{\Phi_{\mathcal{H}},\sigma}$ *at* $\bar{x}$ *with respect to the ideal defining the* σ*-stratum. Then there is a "natural inclusion"* $\imath^{\mathrm{nat}} : R_{\mathrm{alg}} \hookrightarrow R^\wedge$.

(3) A degenerating family $(G, \lambda, i, \alpha_{\mathcal{H}^p}, \alpha^{\mathrm{ord}}_{\mathcal{H}_p})$ *of type* $\vec{\mathsf{M}}^{\mathrm{ord}}_{\mathcal{H}}$ *over* S *as in Definition* 3.4.2.10, *together with an embedding* $\imath^{\mathrm{alg}} : R_{\mathrm{alg}} \hookrightarrow R^\wedge$, *such that we have the following:*

   (a) There are isomorphisms between the objects $\underline{\Phi}_{\mathcal{H}}(G)$, $\underline{\mathbf{S}}_{\Phi(G)}$, *and* $\underline{B}(G)$ *(see Construction* 5.1.1.1*), and the pullbacks of the tautological objects* $\underline{\Phi}_{\mathcal{H}}$, $\underline{\mathbf{S}}$, *and* $\underline{B}$ *over* $\vec{\Xi}^{\mathrm{ord}}_{\Phi_{\mathcal{H}},\delta_{\mathcal{H}}}(\sigma)/\Gamma_{\Phi_{\mathcal{H}},\sigma}$ *(see Proposition* 4.2.2.5*) under* $S \to \vec{\Xi}^{\mathrm{ord}}_{\Phi_{\mathcal{H}},\delta_{\mathcal{H}}}(\sigma)/\Gamma_{\Phi_{\mathcal{H}},\sigma}$.

   (b) The embedding $\imath^{\mathrm{alg}} : R_{\mathrm{alg}} \hookrightarrow R^\wedge$ *is* **close to the natural inclusion** $\imath^{\mathrm{nat}}$ *in the sense that the following two morphisms* $\mathrm{Spf}(R^\wedge, I) \to \vec{\mathfrak{X}}^{\mathrm{ord}}_{\Phi_{\mathcal{H}},\delta_{\mathcal{H}},\sigma}/\Gamma_{\Phi_{\mathcal{H}},\sigma}$ **coincide** *over the* σ*-stratum:*

      i. The pullback $(G, \lambda, i, \alpha_{\mathcal{H}^p}, \alpha^{\mathrm{ord}}_{\mathcal{H}_p}) \underset{R_{\mathrm{alg}},\imath^{\mathrm{nat}}}{\otimes} R^\wedge \to S^\wedge := \mathrm{Spec}(R^\wedge)$ *defines an ordinary good formal* $(\Phi_{\mathcal{H}}, \delta_{\mathcal{H}})$*-model by the isomorphisms in* (3a) *above, and hence defines a canonical morphism* $\mathrm{Spf}(R^\wedge, I) \to \vec{\mathfrak{X}}^{\mathrm{ord}}_{\Phi_{\mathcal{H}},\delta_{\mathcal{H}},\sigma}/\Gamma_{\Phi_{\mathcal{H}},\sigma}$.

ii. *The embedding* $\imath^{\mathrm{alg}}: R_{\mathrm{alg}} \hookrightarrow R^{\wedge}$ *defines a composition*

$$S^{\wedge} \overset{\mathrm{Spec}(\imath^{\mathrm{alg}})}{\to} S \to \vec{\Xi}^{\mathrm{ord}}_{\Phi_{\mathcal{H}},\delta_{\mathcal{H}}}(\sigma)/\Gamma_{\Phi_{\mathcal{H}},\sigma},$$

inducing a morphism $\mathrm{Spf}(R^{\wedge}, I) \to \vec{\mathfrak{X}}^{\mathrm{ord}}_{\Phi_{\mathcal{H}},\delta_{\mathcal{H}},\sigma}/\Gamma_{\Phi_{\mathcal{H}},\sigma}$.

(c) *The extended Kodaira–Spencer morphism (see [62, Def. 4.6.3.44]) induces an isomorphism*

$$\mathrm{KS}_{G/S/\vec{\mathsf{S}}_{0,r_{\mathcal{H}}}} : \underline{\mathrm{KS}}_{(G,\lambda,i)/S,\mathrm{free}} \overset{\sim}{\to} \Omega^1_{S/\vec{\mathsf{S}}_{0,r_{\mathcal{H}}}}[d\log\infty],$$

where $\underline{\mathrm{KS}}_{(G,\lambda,i)/S,\mathrm{free}}$ *is defined as in Definition* 3.4.3.1.

Proposition 5.1.2.3. *(Compare with [62, Prop. 6.3.2.6].) There exist ordinary good algebraic* $(\Phi_{\mathcal{H}}, \delta_{\mathcal{H}}, \sigma)$*-models such that the morphisms from them to* $\vec{\Xi}^{\mathrm{ord}}_{\Phi_{\mathcal{H}},\delta_{\mathcal{H}}}(\sigma)/\Gamma_{\Phi_{\mathcal{H}},\sigma}$ **cover** *the* σ*-stratum* $\vec{\Xi}^{\mathrm{ord}}_{\Phi_{\mathcal{H}},\delta_{\mathcal{H}},\sigma}/\Gamma_{\Phi_{\mathcal{H}},\sigma}$ *of* $\vec{\Xi}^{\mathrm{ord}}_{\Phi_{\mathcal{H}},\delta_{\mathcal{H}}}(\sigma)/\Gamma_{\Phi_{\mathcal{H}},\sigma}$.

Proof. The proof of [62, Prop. 6.3.2.6] works verbatim here. □

Remark 5.1.2.4. (Compare with [62, Rem. 6.3.2.7].) What is implicit behind Proposition 5.1.2.3 is that, although we need to approximate the (possibly infinitely many) good formal models at all geometric points of the σ-stratum, we only need *finitely many* good algebraic models to cover it, by quasi-compactness of $\vec{\Xi}^{\mathrm{ord}}_{\Phi_{\mathcal{H}},\delta_{\mathcal{H}}}(\sigma)$.

Remark 5.1.2.5. (Compare with [62, Rem. 6.3.2.8].) An ordinary good algebraic $(\Phi_{\mathcal{H}}, \delta_{\mathcal{H}}, \sigma)$-model is an ordinary good algebraic $(\Phi'_{\mathcal{H}}, \delta'_{\mathcal{H}}, \sigma')$-model if and only if $(\Phi_{\mathcal{H}}, \delta_{\mathcal{H}}, \sigma)$ is equivalent to $(\Phi'_{\mathcal{H}}, \delta'_{\mathcal{H}}, \sigma')$ (see Remark 5.1.1.12).

Remark 5.1.2.6. (Compare with [62, Rem. 6.3.2.9].) For two smooth rational polyhedral cones $\sigma, \sigma' \in \mathbf{P}^+_{\Phi_{\mathcal{H}}}$ such that $\sigma \subset \sigma'$, an ordinary good algebraic $(\Phi_{\mathcal{H}}, \delta_{\mathcal{H}}, \sigma)$-model is not necessarily an ordinary good algebraic $(\Phi_{\mathcal{H}}, \delta_{\mathcal{H}}, \sigma')$-model (see [62, Rem. 6.2.5.31] and Remark 5.1.1.13).

Proposition 5.1.2.7. *(Compare with [62, Prop. 6.3.2.10].) Suppose* $\bar{x}$ *is any geometric point in the* $(\tau \bmod \Gamma_{\Phi_{\mathcal{H}},\mathcal{H},\sigma})$*-stratum of an ordinary good algebraic* $(\Phi_{\mathcal{H}}, \delta_{\mathcal{H}}, \sigma)$*-model* $(G, \lambda, i, \alpha_{\mathcal{H}^p}, \alpha^{\mathrm{ord}}_{\mathcal{H}_p}) \to \mathrm{Spec}(R_{\mathrm{alg}}) \to \vec{\Xi}^{\mathrm{ord}}_{\Phi_{\mathcal{H}},\delta_{\mathcal{H}}}(\sigma)/\Gamma_{\Phi_{\mathcal{H}},\sigma}$, *where* τ *is a face of* σ. *By pulling back to the completion* $R^{\wedge}_{\bar{x}}$ *of the strict local ring of* R_{alg} *at* $\bar{x}$ *with respect to the ideal defining the* $(\tau \bmod \Gamma_{\Phi_{\mathcal{H}},\mathcal{H},\sigma})$*-stratum, we obtain a good formal* $(\Phi'_{\mathcal{H}}, \delta'_{\mathcal{H}}, \tau')$*-model, where:*

(1) $\Phi'_{\mathcal{H}} = (X', Y', \phi', \varphi'_{-2,\mathcal{H}}, \varphi'_{0,\mathcal{H}})$ *is the pullback of* $\underline{\Phi}_{\mathcal{H}}$ *to* $\bar{x}$, *which comes equipped with a surjection* $(s_X : X \twoheadrightarrow X', s_Y : Y \twoheadrightarrow Y')$ *(as in Definition* 1.2.1.17*) by definition of* $\underline{\Phi}_{\mathcal{H}}$.

(2) $\delta'_{\mathcal{H}}$ *is any splitting that makes* $(\Phi'_{\mathcal{H}}, \delta'_{\mathcal{H}})$ *a representative of a cusp label. Then there is a surjection* $(\Phi_{\mathcal{H}}, \delta_{\mathcal{H}}) \twoheadrightarrow (\Phi'_{\mathcal{H}}, \delta'_{\mathcal{H}})$ *(the actual choice of* $\delta'_{\mathcal{H}}$ *does not matter).*

(3) $\tau' \subset \mathbf{P}^+_{\Phi'_{\mathcal{H}}}$ *is any nondegenerate smooth rational polyhedral cone whose image under the embedding* $\mathbf{P}_{\Phi'_{\mathcal{H}}} \hookrightarrow \mathbf{P}_{\Phi_{\mathcal{H}}}$ *induced by the surjection* (s_X, s_Y) *is the translation of* τ *by an element of* $\Gamma_{\Phi_{\mathcal{H}}}$.

(*This is the so-called* **openness of versality**.)

Proof. The proof of [62, Prop. 6.3.2.10] also works here (with the ingredients there replaced with their analogues above). □

Remark 5.1.2.8. (Compare with [62, Rem. 6.3.2.15].) Suppose $(\Phi'_{\mathcal{H}}, \delta'_{\mathcal{H}}, \sigma')$ is a *face* of $(\Phi_{\mathcal{H}}, \delta_{\mathcal{H}}, \sigma)$, so that σ' is identified with some face τ of σ under some surjection $(s_X : X \twoheadrightarrow X', s_Y : Y \twoheadrightarrow Y') : (\Phi_{\mathcal{H}}, \delta_{\mathcal{H}}) \twoheadrightarrow (\Phi'_{\mathcal{H}}, \delta'_{\mathcal{H}})$. Then there always exists some ordinary good algebraic $(\Phi_{\mathcal{H}}, \delta_{\mathcal{H}}, \sigma)$-model that has a nonempty $(\tau \mod \Gamma_{\Phi_{\mathcal{H}},\sigma})$-stratum on the base scheme.

For later reference, we shall also make the following definition:

Definition 5.1.2.9. (*Compare with [62, Def. 6.3.2.16].*) *Let* $(\Phi_{\mathcal{H}}, \delta_{\mathcal{H}})$ *be a representative of an ordinary cusp label at level* $\mathcal{H}$ *(see Definition* 3.2.3.8*), and let* $\sigma \subset \mathbf{P}^{+}_{\Phi_{\mathcal{H}}}$ *be a nondegenerate smooth rational polyhedral cone. Suppose* $(\Phi'_{\mathcal{H}}, \delta'_{\mathcal{H}}, \sigma')$ *is a face of* $(\Phi_{\mathcal{H}}, \delta_{\mathcal{H}}, \sigma)$ *such that the image of* σ' *under the embedding* $\mathbf{P}_{\Phi'_{\mathcal{H}}} \hookrightarrow \mathbf{P}_{\Phi_{\mathcal{H}}}$ *induced by some surjection* $(s_X : X \twoheadrightarrow X', s_Y : Y \twoheadrightarrow Y') : (\Phi_{\mathcal{H}}, \delta_{\mathcal{H}}) \twoheadrightarrow (\Phi'_{\mathcal{H}}, \delta'_{\mathcal{H}})$ *is a* $\Gamma_{\Phi_{\mathcal{H}}}$*-translation of a face* τ *of* σ *(which can be* σ *itself). Then we shall call the* $(\tau \mod \Gamma_{\Phi_{\mathcal{H}},\sigma})$*-stratum of* $\Xi^{\mathrm{ord}}_{\Phi_{\mathcal{H}},\delta_{\mathcal{H}}}(\sigma)/\Gamma_{\Phi_{\mathcal{H}},\sigma}$ *the* $[(\Phi'_{\mathcal{H}}, \delta'_{\mathcal{H}}, \tau')]$**-stratum**. *(In this case,* $[(\Phi'_{\mathcal{H}}, \delta'_{\mathcal{H}}, \tau')]$ *is a* **face** *of* $[(\Phi_{\mathcal{H}}, \delta_{\mathcal{H}}, \sigma)]$*; see Definition* 1.2.2.19.*) We shall also call the induced* $(\tau \mod \Gamma_{\Phi_{\mathcal{H}},\sigma})$*-strata of ordinary good formal* $(\Phi_{\mathcal{H}}, \delta_{\mathcal{H}}, \sigma)$*-models and ordinary good algebraic good* $(\Phi_{\mathcal{H}}, \delta_{\mathcal{H}}, \sigma)$*-models (see Definitions* 5.1.1.7 *and* 5.1.2.2*) their* $[(\Phi'_{\mathcal{H}}, \delta'_{\mathcal{H}}, \tau')]$*-strata.*

5.1.3 *Gluing in the Étale Topology*

Definition 5.1.3.1. (*Compare with Definition* 1.2.2.13.) *A* **compatible choice of admissible smooth rational polyhedral cone decomposition data** *for* $\vec{\mathsf{M}}^{\mathrm{ord}}_{\mathcal{H}}$ *is a complete set* $\Sigma^{\mathrm{ord}} = \{\Sigma_{\Phi_{\mathcal{H}}}\}_{[(\Phi_{\mathcal{H}},\delta_{\mathcal{H}})]}$ *of* **compatible choices** *of* $\Sigma_{\Phi_{\mathcal{H}}}$ *as in Definition* 1.2.2.13*, but with* $\Sigma_{\Phi_{\mathcal{H}}}$ *defined only for representatives* $(\mathsf{Z}_{\mathcal{H}}, \Phi_{\mathcal{H}}, \delta_{\mathcal{H}})$ *of ordinary cusp labels (see Definition* 3.2.3.8*).*

Proposition 5.1.3.2. (*Compare with [62, Prop. 6.3.3.5] and Proposition* 1.2.2.17.) *A compatible choice* Σ^{ord} *of admissible smooth rational polyhedral cone decomposition data for* $\vec{\mathsf{M}}^{\mathrm{ord}}_{\mathcal{H}}$ **exists**. *Moreover, each* Σ^{ord} *for* $\vec{\mathsf{M}}^{\mathrm{ord}}_{\mathcal{H}}$ *extends to some* Σ *for* $\mathsf{M}_{\mathcal{H}}$ *(as in Definition* 1.2.2.13*), and we may assume that* Σ *is a refinement of any given collection* Σ' *also inducing* Σ^{ord}*. The same is true if we allow varying levels or twists by Hecke actions (see [62, Def. 6.4.2.8 and 6.4.3.2]). We may also assume that* Σ^{ord} *or* Σ *is invariant under any choice of an open compact subgroup* $\mathcal{H}'$ *of* $\mathrm{G}(\mathbb{A}^{\infty})$ *normalizing* $\mathcal{H}$*. Conversely, each* Σ *for* $\mathsf{M}_{\mathcal{H}}$ *induces (by restriction to ordinary cusp labels) a valid* Σ^{ord} *for* $\vec{\mathsf{M}}^{\mathrm{ord}}_{\mathcal{H}}$.

Proof. By considering only ordinary cusp labels, which makes sense because ordinary cusp labels only surject to ordinary cusp labels (by definition; see Definition

3.2.3.8), the same argument of the proof of [62, Prop. 6.3.3.5] (by induction on the magnitude of cusp labels) works here and shows that some Σ^{ord} exists. That is, assertions in Proposition 1.2.2.17 are true when we only consider the ordinary cusps. The same argument also shows that, by starting with cone decompositions in each given Σ^{ord} for $\vec{\mathsf{M}}_{\mathcal{H}}^{\mathrm{ord}}$ and by extending them to cone decompositions at other cusp labels (which may surject to either ordinary or nonordinary cusp labels), we can extend Σ^{ord} to some Σ for $\mathsf{M}_{\mathcal{H}}$, which can be a refinement of any given collection Σ'. The last statement follows immediately from the definitions. □

Definition 5.1.3.3. (*Compare with Definition* 1.2.2.14.) *A compatible choice* $\Sigma^{\mathrm{ord}} = \{\Sigma_{\Phi_{\mathcal{H}}}\}_{[(\Phi_{\mathcal{H}},\delta_{\mathcal{H}})]}$ *of admissible smooth rational polyhedral cone decomposition data for* $\vec{\mathsf{M}}_{\mathcal{H}}^{\mathrm{ord}}$ (*see Definition* 5.1.3.1) *is* **projective** *if there is a collection* $\mathsf{pol}^{\mathrm{ord}} = \{\mathsf{pol}_{\Phi_{\mathcal{H}}}\}_{[(\Phi_{\mathcal{H}},\delta_{\mathcal{H}})]}$ *of polarization functions labeled by representatives* $(\Phi_{\mathcal{H}}, \delta_{\mathcal{H}})$ *as in Definition* 1.2.2.14, *but with* ($\Sigma_{\Phi_{\mathcal{H}}}$ *and*) $\mathsf{pol}_{\Phi_{\mathcal{H}}}$ *defined only for representatives* $(\mathsf{Z}_{\mathcal{H}}, \Phi_{\mathcal{H}}, \delta_{\mathcal{H}})$ *of ordinary cusp labels* (*see Definition* 3.2.3.8).

Proposition 5.1.3.4. (*Compare with [62, Prop. 7.3.1.4] and Propositions* 1.2.2.17 *and* 5.1.3.2.) *There exists a compatible choice* $\Sigma^{\mathrm{ord}} = \{\Sigma_{\Phi_{\mathcal{H}}}\}_{[(\Phi_{\mathcal{H}},\delta_{\mathcal{H}})]}$ *of admissible smooth rational polyhedral cone decomposition data for* $\vec{\mathsf{M}}_{\mathcal{H}}^{\mathrm{ord}}$ (*see Definition* 5.1.3.1) *that is* **projective**, *carrying a compatible collection of polarization functions* $\mathsf{pol}^{\mathrm{ord}}$ *as in Definition* 5.1.3.3. *Moreover, each such* $(\Sigma^{\mathrm{ord}}, \mathsf{pol}^{\mathrm{ord}})$ *extends to some* (Σ, pol) *for* $\mathsf{M}_{\mathcal{H}}$ (*as in Definition* 1.2.2.14), *and we may assume that* Σ *is a refinement of any given collection* Σ' *also inducing* Σ^{ord}. *The same is true if we allow varying levels or twists by Hecke actions* (*see [62, Def. 6.4.2.8 and 6.4.3.2]*). *We may also assume that* Σ^{ord} *and* $\mathsf{pol}^{\mathrm{ord}}$, *or* Σ *and* pol, *are invariant under any choice of an open compact subgroup* $\mathcal{H}'$ *of* $\mathrm{G}(\mathbb{A}^\infty)$ *normalizing* $\mathcal{H}$. *Conversely, each* (Σ, pol) *for* $\mathsf{M}_{\mathcal{H}}$ *induces* (*by restriction to ordinary cusp labels*) *a valid* $(\Sigma^{\mathrm{ord}}, \mathsf{pol}^{\mathrm{ord}})$ *for* $\vec{\mathsf{M}}_{\mathcal{H}}^{\mathrm{ord}}$.

Proof. As in the proof of Proposition 5.1.3.2, by considering only ordinary cusp labels, the same argument of the proofs of [62, Prop. 6.3.3.5 and 7.3.1.4] (by induction on the magnitude of cusp labels) works here and shows that some Σ^{ord} and $\mathsf{pol}^{\mathrm{ord}}$ exist. The same argument also shows that, by starting with cone decompositions and polarization functions in each given $(\Sigma^{\mathrm{ord}}, \mathsf{pol}^{\mathrm{ord}})$ for $\vec{\mathsf{M}}_{\mathcal{H}}^{\mathrm{ord}}$ and by extending them to ones at other cusp labels (which may surject to either ordinary or nonordinary cusp labels), we can extend $(\Sigma^{\mathrm{ord}}, \mathsf{pol}^{\mathrm{ord}})$ to some (Σ, pol) for $\mathsf{M}_{\mathcal{H}}$, where Σ can be a refinement of any given collection Σ'. And we may assume that these satisfy the additional requirements as in the proposition. The last statement follows immediately from the definitions. □

Let $\Sigma^{\mathrm{ord}} = \{\Sigma_{\Phi_{\mathcal{H}}}\}_{[(\Phi_{\mathcal{H}},\delta_{\mathcal{H}})]}$ be any compatible choice of admissible smooth rational polyhedral cone decomposition data for $\vec{\mathsf{M}}_{\mathcal{H}}^{\mathrm{ord}}$. To construct the desired $\vec{\mathsf{M}}_{\mathcal{H},\Sigma^{\mathrm{ord}}}^{\mathrm{ord,tor}}$ as an algebraic stack, it suffices to give an étale presentation $\vec{\mathsf{U}}_{\mathcal{H}}^{\mathrm{ord}} \twoheadrightarrow \vec{\mathsf{M}}_{\mathcal{H}}^{\mathrm{ord,tor}}$ such that $\vec{\mathsf{R}}_{\mathcal{H}}^{\mathrm{ord}} := \vec{\mathsf{U}}_{\mathcal{H}}^{\mathrm{ord}} \underset{\vec{\mathsf{M}}_{\mathcal{H},\Sigma^{\mathrm{ord}}}^{\mathrm{ord,tor}}}{\times} \vec{\mathsf{U}}_{\mathcal{H}}^{\mathrm{ord}}$ is étale over $\vec{\mathsf{U}}_{\mathcal{H}}^{\mathrm{ord}}$ via the two projections (see [62,

Prop. A.7.1.1 and Def. A.7.1.3]). Equivalently, it suffices to construct the $\vec{\mathsf{U}}^{\mathrm{ord}}_{\mathcal{H}}$ and $\vec{\mathsf{R}}^{\mathrm{ord}}_{\mathcal{H}}$ that satisfy the required groupoid relations, which then realizes $\vec{\mathsf{M}}^{\mathrm{ord,tor}}_{\mathcal{H},\Sigma^{\mathrm{ord}}}$ as the quotient of $\vec{\mathsf{U}}^{\mathrm{ord}}_{\mathcal{H}}$ by $\vec{\mathsf{R}}^{\mathrm{ord}}_{\mathcal{H}}$. Let us first explain our choices of $\vec{\mathsf{U}}^{\mathrm{ord}}_{\mathcal{H}}$ and $\vec{\mathsf{R}}^{\mathrm{ord}}_{\mathcal{H}}$, then show that they have the desired properties.

Construction 5.1.3.5. (Compare with [62, Constr. 6.3.3.1 and 6.3.3.9].) We shall construct $\vec{\mathsf{U}}^{\mathrm{ord}}_{\mathcal{H}}$ and a stratification on it as follows:

(1) Choose a complete set of (mutually inequivalent) representatives $(\Phi_{\mathcal{H}}, \delta_{\mathcal{H}})$ of ordinary cusp labels at level $\mathcal{H}$.
This is a finite set because there is already a finite set of representatives for all cusp labels (including nonordinary ones, when we constructed toroidal compactifications for $\mathsf{M}_{\mathcal{H}}$; see the explanation in [62, Constr. 6.3.3.1]).

(2) For each $(\Phi_{\mathcal{H}}, \delta_{\mathcal{H}})$ chosen above, choose a complete set of (mutually inequivalent) representatives σ in $\Sigma_{\Phi_{\mathcal{H}}}/\Gamma_{\Phi_{\mathcal{H}}}$, where $\Sigma_{\Phi_{\mathcal{H}}}$ is the $\Gamma_{\Phi_{\mathcal{H}}}$-admissible smooth rational polyhedral cone decomposition chosen in Σ^{ord}. This gives a complete set of representatives $(\Phi_{\mathcal{H}}, \delta_{\mathcal{H}}, \sigma)$ of equivalence classes $[(\Phi_{\mathcal{H}}, \delta_{\mathcal{H}}, \sigma)]$ defined in Definition 1.2.2.10 such that the cusp label $[(\Phi_{\mathcal{H}}, \delta_{\mathcal{H}})]$ is ordinary. This is a finite set by the $\Gamma_{\Phi_{\mathcal{H}}}$-admissibility (see Definition 1.2.2.4) of each $\Sigma_{\Phi_{\mathcal{H}}}$.

(3) For each representative $(\Phi_{\mathcal{H}}, \delta_{\mathcal{H}}, \sigma)$ above that satisfies moreover $\sigma \subset \mathbf{P}^{+}_{\Phi_{\mathcal{H}}}$, choose *finitely many* ordinary good algebraic $(\Phi_{\mathcal{H}}, \delta_{\mathcal{H}}, \sigma)$-models $\mathrm{Spec}(R_{\mathrm{alg}})$ (see Definition 5.1.2.2) such that the corresponding étale morphisms from the various $\mathrm{Spec}(R_{\mathrm{alg}}/I)$'s to the σ-stratum $\vec{\Xi}^{\mathrm{ord}}_{\Phi_{\mathcal{H}},\delta_{\mathcal{H}},\sigma}/\Gamma_{\Phi_{\mathcal{H}},\sigma}$ of $\vec{\Xi}^{\mathrm{ord}}_{\Phi_{\mathcal{H}},\delta_{\mathcal{H}}}(\sigma)/\Gamma_{\Phi_{\mathcal{H}},\sigma}$, where I denotes the ideal of R_{alg} defining the σ-stratum of $\mathrm{Spec}(R_{\mathrm{alg}})$, *cover* the whole σ-stratum (see Proposition 5.1.2.3 and Remark 5.1.2.4).
This is possible by the quasi-compactness of $\vec{\Xi}^{\mathrm{ord}}_{\Phi_{\mathcal{H}},\delta_{\mathcal{H}},\sigma}/\Gamma_{\Phi_{\mathcal{H}},\sigma}$, because $\Gamma_{\Phi_{\mathcal{H}},\sigma}$ is finite (by [62, Rem. 6.2.5.26 and Lem. 6.2.5.27], because the cone decomposition $\Sigma_{\Phi_{\mathcal{H}}}$ is chosen such that [62, Cond. 6.2.5.25] is satisfied), and because $\vec{\Xi}^{\mathrm{ord}}_{\Phi_{\mathcal{H}},\delta_{\mathcal{H}},\sigma}$ is a torus torsor over an abelian scheme torsor over a finite étale cover of the algebraic stack $\vec{\mathsf{M}}^{\mathrm{ord},\mathrm{Z}_{\mathcal{H}}}_{\mathcal{H}}$ separated and of finite type over $\vec{\mathsf{S}}_{0,r_{\mathcal{H}}} = \mathrm{Spec}(\mathcal{O}_{F_0,(p)}[\zeta_{p^{r_{\mathcal{H}}}}])$ (see Section 4.2 and Theorem 3.4.2.5).

(4) Let us form the scheme

$$\vec{\mathsf{U}}^{\mathrm{ord}}_{\mathcal{H}} = \begin{pmatrix} \text{disjoint union of the (finitely many)} \\ \text{ordinary good algebraic } (\Phi_{\mathcal{H}}, \delta_{\mathcal{H}}, \sigma)\text{-models} \\ \mathrm{Spec}(R_{\mathrm{alg}}) \text{ chosen above} \end{pmatrix},$$

(smooth over $\vec{\mathsf{S}}_{0,r_{\mathcal{H}}} = \mathrm{Spec}(\mathcal{O}_{F_0,(p)}[\zeta_{p^{r_{\mathcal{H}}}}])$) which comes equipped with a natural stratification labeled as follows:
On an ordinary good algebraic $(\Phi_{\mathcal{H}}, \delta_{\mathcal{H}}, \sigma)$-model $\mathrm{Spec}(R_{\mathrm{alg}})$ used in the construction of $\vec{\mathsf{U}}^{\mathrm{ord}}_{\mathcal{H}}$ above, its stratification inherited from $\vec{\Xi}^{\mathrm{ord}}_{\Phi_{\mathcal{H}},\delta_{\mathcal{H}}}(\sigma)/\Gamma_{\Phi_{\mathcal{H}},\sigma}$ can be relabeled using equivalence classes $[(\Phi'_{\mathcal{H}}, \delta'_{\mathcal{H}}, \tau')]$

(see Definition 1.2.2.10) following the recipe in Definition 5.1.2.9, which are *faces* of $[(\Phi_{\mathcal{H}}, \delta_{\mathcal{H}}, \sigma)]$ (see Definition 1.2.2.19). Then we define the stratification on the disjoint union $\vec{\mathsf{U}}^{\mathrm{ord}}_{\mathcal{H}}$ to be induced by those on the ordinary good algebraic models.

By the compatibility of the choice of Σ^{ord} (see Definitions 1.2.2.13 and 5.1.3.1), we know that in each representative $(\Phi'_{\mathcal{H}}, \delta'_{\mathcal{H}}, \tau')$ of each face $[(\Phi'_{\mathcal{H}}, \delta'_{\mathcal{H}}, \tau')]$ of $[(\Phi_{\mathcal{H}}, \delta_{\mathcal{H}}, \sigma)]$, the cone τ' is in the cone decomposition $\Sigma_{\Phi'_{\mathcal{H}}}$ we have in Σ^{ord}. Hence, we may label all the strata by the equivalence classes of triples we have taken in the construction of $\vec{\mathsf{U}}^{\mathrm{ord}}_{\mathcal{H}}$. For simplicity, we call the $[(0, 0, \{0\})]$-stratum the $[0]$-stratum of $\vec{\mathsf{U}}^{\mathrm{ord}}_{\mathcal{H}}$, which we denote by $\vec{\mathsf{U}}^{\mathrm{ord},[0]}_{\mathcal{H}}$.

(This finishes Construction 5.1.3.5.)

The ordinary good algebraic models $(G, \lambda, i, \alpha_{\mathcal{H}^p}, \alpha^{\mathrm{ord}}_{\mathcal{H}_p})$ over the various $\mathrm{Spec}(R_{\mathrm{alg}})$'s define (by taking union) a degenerating family $(G, \lambda, i, \alpha_{\mathcal{H}^p}, \alpha^{\mathrm{ord}}_{\mathcal{H}_p})$ of type $\vec{\mathsf{M}}^{\mathrm{ord}}_{\mathcal{H}}$ over $\vec{\mathsf{U}}^{\mathrm{ord}}_{\mathcal{H}}$ as in Definition 3.4.2.10, whose restriction to the $[0]$-stratum $\vec{\mathsf{U}}^{\mathrm{ord},[0]}_{\mathcal{H}}$ is a tuple $(G^{[0]}, \lambda^{[0]}, i^{[0]}, \alpha_{\mathcal{H}^p}, \alpha^{\mathrm{ord}}_{\mathcal{H}_p})$ parameterized by $\vec{\mathsf{M}}^{\mathrm{ord}}_{\mathcal{H}}$ (see Convention 3.4.2.9). This determines a canonical morphism $\vec{\mathsf{U}}^{\mathrm{ord},[0]}_{\mathcal{H}} \to \vec{\mathsf{M}}^{\mathrm{ord}}_{\mathcal{H}}$. This morphism is étale because $\vec{\mathsf{U}}^{\mathrm{ord},[0]}_{\mathcal{H}}$ is locally of finite presentation, and the morphism $\vec{\mathsf{U}}^{\mathrm{ord},[0]}_{\mathcal{H}} \to \vec{\mathsf{M}}^{\mathrm{ord}}_{\mathcal{H}}$ is formally étale at every geometric point of $\vec{\mathsf{U}}^{\mathrm{ord},[0]}_{\mathcal{H}}$ by the calculation of Kodaira–Spencer morphisms (using (3c) of Definition 5.1.2.2, [62, Thm. 4.6.3.16], and Proposition 3.4.3.3). As a result, the morphism $\vec{\mathsf{U}}^{\mathrm{ord},[0]}_{\mathcal{H}} \to \vec{\mathsf{M}}^{\mathrm{ord}}_{\mathcal{H}}$ (surjective by definition) defines an étale presentation of $\vec{\mathsf{M}}^{\mathrm{ord}}_{\mathcal{H}}$. This identifies $\vec{\mathsf{M}}^{\mathrm{ord}}_{\mathcal{H}}$ with the quotient of $\vec{\mathsf{U}}^{\mathrm{ord},[0]}_{\mathcal{H}}$ by the étale groupoid $\vec{\mathsf{R}}^{\mathrm{ord},[0]}_{\mathcal{H}}$ over $\vec{\mathsf{U}}^{\mathrm{ord},[0]}_{\mathcal{H}}$ defined by the representable functor

$$\begin{aligned}\vec{\mathsf{R}}^{\mathrm{ord},[0]}_{\mathcal{H}} := \underline{\mathrm{Isom}}_{\vec{\mathsf{U}}^{\mathrm{ord},[0]}_{\mathcal{H}} \underset{\vec{\mathsf{S}}_{0,r_{\mathcal{H}}}}{\times} \vec{\mathsf{U}}^{\mathrm{ord},[0]}_{\mathcal{H}}}\big(&\mathrm{pr}_1^*(G^{[0]}, \lambda^{[0]}, i^{[0]}, \alpha_{\mathcal{H}^p}, \alpha^{\mathrm{ord}}_{\mathcal{H}_p}),\\ &\mathrm{pr}_2^*(G^{[0]}, \lambda^{[0]}, i^{[0]}, \alpha_{\mathcal{H}^p}, \alpha^{\mathrm{ord}}_{\mathcal{H}_p})\big),\end{aligned} \tag{5.1.3.6}$$

where $\mathrm{pr}_1, \mathrm{pr}_2 : \vec{\mathsf{U}}^{\mathrm{ord},[0]}_{\mathcal{H}} \underset{\vec{\mathsf{S}}_{0,r_{\mathcal{H}}}}{\times} \vec{\mathsf{U}}^{\mathrm{ord},[0]}_{\mathcal{H}} \to \vec{\mathsf{U}}^{\mathrm{ord},[0]}_{\mathcal{H}}$ denote, respectively, the two projections.

Proposition 5.1.3.7. (*Compare with [62, Prop. 6.3.3.11].*) *Suppose R is a noetherian normal complete local domain with fraction field K and algebraically closed residue field k. Assume that* $\mathrm{Spec}(R)$ *is flat over* $\vec{\mathsf{S}}_{0,r_{\mathcal{H}}}$, *and that we have a degenerating family* $(G^\ddagger, \lambda^\ddagger, i^\ddagger, \alpha^\ddagger_{\mathcal{H}^p}, \alpha^{\mathrm{ord},\ddagger}_{\mathcal{H}_p})$ *of type* $\vec{\mathsf{M}}^{\mathrm{ord}}_{\mathcal{H}}$ *over* $\mathrm{Spec}(R)$ *as in Definition* 3.4.2.10. *Then the following conditions are equivalent:*

(1) *There exists a morphism* $\mathrm{Spec}(R) \to \vec{\mathsf{U}}^{\mathrm{ord}}_{\mathcal{H}}$ *sending the generic point* $\mathrm{Spec}(K)$ *to the* $[0]$*-stratum such that* $(G^\ddagger, \lambda^\ddagger, i^\ddagger, \alpha^\ddagger_{\mathcal{H}^p}, \alpha^{\mathrm{ord},\ddagger}_{\mathcal{H}_p}) \to \mathrm{Spec}(R)$ *is the pullback of* $(G, \lambda, i, \alpha_{\mathcal{H}^p}, \alpha^{\mathrm{ord}}_{\mathcal{H}_p}) \to \vec{\mathsf{U}}^{\mathrm{ord}}_{\mathcal{H}}$.

(2) *The degenerating family* $(G^\ddagger, \lambda^\ddagger, i^\ddagger, \alpha^\ddagger_{\mathcal{H}^p}, \alpha^{\mathrm{ord},\ddagger}_{\mathcal{H}_p}) \to \mathrm{Spec}(R)$ *is the pullback of the Mumford family* $({}^\heartsuit G, {}^\heartsuit\lambda, {}^\heartsuit i, {}^\heartsuit\alpha_{\mathcal{H}^p}, {}^\heartsuit\alpha^{\mathrm{ord}}_{\mathcal{H}_p}) \to \vec{\mathfrak{X}}^{\mathrm{ord}}_{\Phi_\mathcal{H},\delta_\mathcal{H},\sigma}/\Gamma_{\Phi_\mathcal{H},\sigma}$ *via a morphism* $\mathrm{Spf}(R) \to \vec{\mathfrak{X}}^{\mathrm{ord}}_{\Phi_\mathcal{H},\delta_\mathcal{H},\sigma}/\Gamma_{\Phi_\mathcal{H},\sigma}$*, or equivalently a morphism* $\mathrm{Spec}(R) \to \vec{\Xi}^{\mathrm{ord}}_{\Phi_\mathcal{H},\delta_\mathcal{H}}(\sigma)/\Gamma_{\Phi_\mathcal{H},\sigma}$*, for some* $(\Phi_\mathcal{H}, \delta_\mathcal{H}, \sigma)$ *(which can be assumed to be a triple used in the construction of* $\vec{\mathsf{U}}^{\mathrm{ord}}_\mathcal{H}$*).*

(3) *The degenerating family* $(G^\ddagger, \lambda^\ddagger, i^\ddagger, \alpha^\ddagger_{\mathcal{H}^p}, \alpha^{\mathrm{ord},\ddagger}_{\mathcal{H}_p})$ *over* $\mathrm{Spec}(R)$ *defines an object of the essential image of* $\mathrm{DEG}_{\mathrm{PEL},\mathsf{M}^{\mathrm{ord}}_\mathcal{H}}(R) \to \mathrm{DEG}_{\mathrm{PEL},\ddot{\mathsf{M}}^{\mathrm{ord}}_\mathcal{H}}(R)$*, which corresponds to a tuple*

$$(B^\ddagger, \lambda_{B^\ddagger}, i_{B^\ddagger}, \underline{X}^\ddagger, \underline{Y}^\ddagger, \phi^\ddagger, c^\ddagger, c^{\vee,\ddagger}, \tau^\ddagger, [\alpha^{\natural,\mathrm{ord},\ddagger}_\mathcal{H}])$$

in the essential image of $\mathrm{DD}_{\mathrm{PEL},\mathsf{M}^{\mathrm{ord}}_\mathcal{H}}(R) \to \mathrm{DD}_{\mathrm{PEL},\ddot{\mathsf{M}}^{\mathrm{ord}}_\mathcal{H}}(R)$ *under* (4.1.6.4) *in Theorem* 4.1.6.2*. Then we have a fully symplectic-liftable admissible filtration* $\mathsf{Z}^\ddagger_\mathcal{H}$ *determined by* $[\alpha^{\natural,\mathrm{ord},\ddagger}_\mathcal{H}]$*, which is compatible with the filtration* D *as in Definition* 3.2.3.1*. Moreover, the étale sheaves* $\underline{X}^\ddagger$ *and* $\underline{Y}^\ddagger$ *are necessarily constant, because the base scheme* R *is strict local. Hence, it makes sense to say that we also have a uniquely determined torus argument* $\Phi^\ddagger_\mathcal{H}$ *at level* $\mathcal{H}$ *for* $\mathsf{Z}^\ddagger_\mathcal{H}$*.*

On the other hand, we have objects $\underline{\Phi}_\mathcal{H}(G^\ddagger)$*,* $\underline{\mathbf{S}}_{\underline{\Phi}_\mathcal{H}(G^\ddagger)}$*, and* $\underline{B}(G^\ddagger)$*, which define objects* $\Phi^\ddagger_\mathcal{H}$*,* $\mathbf{S}_{\Phi^\ddagger_\mathcal{H}}$*, and in particular,* $B^\ddagger : \mathbf{S}_{\Phi^\ddagger_\mathcal{H}} \to \mathrm{Inv}(R)$ *over the special fiber.*

If $v : K^\times \to \mathbb{Z}$ *is any discrete valuation defined by a height-one prime of* R*, then* $v \circ B^\ddagger : \mathbf{S}_{\Phi^\ddagger_\mathcal{H}} \to \mathbb{Z}$ *makes sense and defines an element of* $\mathbf{S}^\vee_{\Phi^\ddagger_\mathcal{H}}$*.*

Then the condition is that, for some (and hence every) choice of $\delta^\ddagger_\mathcal{H}$ *making* $(\mathsf{Z}^\ddagger_\mathcal{H}, \Phi^\ddagger_\mathcal{H}, \delta^\ddagger_\mathcal{H})$ *a representative of an ordinary cusp label (see Definition* 3.2.3.8*), there is a cone* $\sigma^\ddagger$ *in the cone decomposition* $\Sigma_{\Phi^\ddagger_\mathcal{H}}$ *of* $\mathbf{P}_{\Phi^\ddagger_\mathcal{H}}$ *(given by the choice of* Σ^{ord}*; cf. Definition* 5.1.3.1*) such that the closure* $\overline{\sigma}^\ddagger$ *of* $\sigma^\ddagger$ *in* $(\mathbf{S}_{\Phi^\ddagger_\mathcal{H}})^\vee_\mathbb{R}$ *contains all* $v \circ B^\ddagger$ *obtained in this way.*

Proof. The implication from (1) to (2) is clear, as the morphism from $\mathrm{Spec}(R)$ to $\vec{\mathsf{U}}^{\mathrm{ord}}_\mathcal{H}$ necessarily factors through the completion of some strict local ring of $\vec{\mathsf{U}}^{\mathrm{ord}}_\mathcal{H}$.

The implication from (2) to (3) is analogous to Proposition 5.1.1.3.

For the implication from (3) to (1), suppose there exists a cone $\sigma^\ddagger$ in the cone decomposition $\Sigma_{\Phi^\ddagger_\mathcal{H}}$ of $\mathbf{P}_{\Phi^\ddagger_\mathcal{H}}$ such that $\overline{\sigma}^\ddagger$ contains all the $v \circ B^\ddagger$'s. Up to replacing $\sigma^\ddagger$ with another cone in $\Sigma_{\Phi^\ddagger_\mathcal{H}}$, let us assume that $\sigma^\ddagger$ is a minimal one. Then some linear combination with positive coefficients of the $v \circ B^\ddagger$'s lie in $\sigma^\ddagger$, the interior of $\overline{\sigma}^\ddagger$. On the other hand, by the positivity condition of $\tau^\ddagger$, such a linear combination with positive coefficients must be positive definite on $Y^\ddagger$. Hence, $\sigma^\ddagger \subset \mathbf{P}^+_{\Phi^\ddagger_\mathcal{H}}$. Then there exists a unique triple $(\Phi_\mathcal{H}, \delta_\mathcal{H}, \sigma)$ chosen in the construction $\vec{\mathsf{U}}^{\mathrm{ord}}_\mathcal{H}$ (see Construction 5.1.3.5) such that $(\Phi^\ddagger_\mathcal{H}, \delta^\ddagger_\mathcal{H}, \sigma^\ddagger)$ and $(\Phi_\mathcal{H}, \delta_\mathcal{H}, \sigma)$ are equivalent (see Definition 1.2.2.10).

Since $\vec{\mathsf{M}}^{\mathrm{ord},\mathsf{Z}_\mathcal{H}}_\mathcal{H}$ is finite étale over the base change $\vec{\mathsf{M}}^{\mathrm{ord}}_{\mathcal{H}_h,r_\mathcal{H}}$ of $\vec{\mathsf{M}}^{\mathrm{ord}}_{\mathcal{H}_h}$ (defined by $\mathsf{M}_{\mathcal{H}_h}$ and $\dddot{\mathsf{M}}^{\mathrm{ord}}_{\mathcal{H}_h}$ as in Theorem 3.4.2.5) to $\vec{\mathsf{S}}_{0,r_\mathcal{H}}$, since $B^\ddagger$ is defined over R, and since R is noetherian, normal, and flat over $\vec{\mathsf{S}}_{0,r_\mathcal{H}}$, as pointed out in Remark 3.4.2.12, the tuple $(B^\ddagger, \lambda_{B^\ddagger}, i_{B^\ddagger}, \varphi^\ddagger_{-1,\mathcal{H}^p}, \varphi^{\mathrm{ord},\ddagger}_{-1,\mathcal{H}_p})$ and the $\Gamma_{\Phi^\ddagger_\mathcal{H}}$-orbit of $(\varphi^{\mathrm{ord},\ddagger}_{-2,\mathcal{H}}, \varphi^{\mathrm{ord},\ddagger}_{0,\mathcal{H}})$ define a morphism $\mathrm{Spec}(R) \to \vec{\mathsf{M}}^{\mathrm{ord},\mathsf{Z}_\mathcal{H}}_\mathcal{H}$ as soon as its restriction to $\mathrm{Spec}(K)$ defines a morphism $\mathrm{Spec}(K) \to \vec{\mathsf{M}}^{\mathrm{ord},\mathsf{Z}_\mathcal{H}}_\mathcal{H}$. By Proposition 4.2.1.46, the degeneration datum (without the positivity condition) associated with $(G^\ddagger, \lambda^\ddagger, i^\ddagger, \alpha^\ddagger_{\mathcal{H}^p}, \alpha^{\mathrm{ord},\ddagger}_{\mathcal{H}_p}) \to \mathrm{Spec}(R)$ determines (by the universal properties of $\vec{\Xi}^{\mathrm{ord}}_{\Phi_\mathcal{H},\delta_\mathcal{H}}$ and $\vec{C}^{\mathrm{ord}}_{\Phi_\mathcal{H},\delta_\mathcal{H}}$) a morphism $\mathrm{Spec}(K) \to \vec{\Xi}^{\mathrm{ord}}_{\Phi_\mathcal{H},\delta_\mathcal{H}}$, whose composition with the (relatively affine) structural morphism $\vec{\Xi}^{\mathrm{ord}}_{\Phi_\mathcal{H},\delta_\mathcal{H}} \to \vec{C}^{\mathrm{ord}}_{\Phi_\mathcal{H},\delta_\mathcal{H}}$ extends to a morphism $\mathrm{Spec}(R) \to \vec{C}^{\mathrm{ord}}_{\Phi_\mathcal{H},\delta_\mathcal{H}}$. By Proposition 4.2.2.8 and the assumption on the $\upsilon \circ B^\ddagger$'s, the morphism $\mathrm{Spec}(K) \to \vec{\Xi}^{\mathrm{ord}}_{\Phi_\mathcal{H},\delta_\mathcal{H}}$ extends to a morphism $\mathrm{Spec}(R) \to \vec{\Xi}^{\mathrm{ord}}_{\Phi_\mathcal{H},\delta_\mathcal{H}}(\sigma)$, which identifies $\underline{B}(G^\ddagger)$ with the pullback of $\underline{B}$ under an identification of $\underline{\Phi}_\mathcal{H}(G^\ddagger)$ with the pullback of $\underline{\Phi}_\mathcal{H}$. The ambiguity of the identifications can be removed (or rather intrinsically incorporated) if we form the quotient $\vec{\Xi}^{\mathrm{ord}}_{\Phi_\mathcal{H},\delta_\mathcal{H}}(\sigma)/\Gamma_{\Phi_\mathcal{H},\sigma}$. Hence, we have a uniquely determined strata-preserving morphism $\mathrm{Spec}(R) \to \vec{\Xi}^{\mathrm{ord}}_{\Phi_\mathcal{H},\delta_\mathcal{H}}(\sigma)/\Gamma_{\Phi_\mathcal{H},\sigma}$, which is independent of the identification of $\Phi^\ddagger_\mathcal{H}$ with $\Phi_\mathcal{H}$ we have chosen. This determines a morphism $\mathrm{Spf}(R) \to \vec{\mathfrak{X}}^{\mathrm{ord}}_{\Phi_\mathcal{H},\delta_\mathcal{H},\sigma}/\Gamma_{\Phi_\mathcal{H},\sigma}$ as in (2).

Let us denote the image of the closed point of $\mathrm{Spec}(R)$ by x, which necessarily lies in the σ-stratum (thanks to the minimality of the choice of $\sigma^\ddagger$). By construction, there is some ordinary good algebraic $(\Phi_\mathcal{H}, \delta_\mathcal{H}, \sigma)$-model $(G, \lambda, i, \alpha_{\mathcal{H}^p}, \alpha_{\mathcal{H}_p}) \to \mathrm{Spec}(R_{\mathrm{alg}})$ used in the construction of $\vec{\mathsf{U}}^{\mathrm{ord}}_\mathcal{H}$ such that the image of the structural morphism $\mathrm{Spec}(R_{\mathrm{alg}}) \to \vec{\Xi}^{\mathrm{ord}}_{\Phi_\mathcal{H},\delta_\mathcal{H}}(\sigma)/\Gamma_{\Phi_\mathcal{H},\sigma}$ contains x. Let $R_{\mathrm{alg}}{}^\wedge$ be the completion of R_{alg} with respect to the ideal defining the σ-stratum of $\mathrm{Spec}(R_{\mathrm{alg}})$, and let $I^\wedge$ be the induced ideal of definition. Then the étale morphism $\mathrm{Spec}(R_{\mathrm{alg}}) \to \vec{\Xi}^{\mathrm{ord}}_{\Phi_\mathcal{H},\delta_\mathcal{H}}(\sigma)/\Gamma_{\Phi_\mathcal{H},\sigma}$ induces a formally étale morphism $\mathrm{Spf}(R_{\mathrm{alg}}{}^\wedge, I^\wedge) \to \vec{\mathfrak{X}}^{\mathrm{ord}}_{\Phi_\mathcal{H},\delta_\mathcal{H},\sigma}/\Gamma_{\Phi_\mathcal{H},\sigma}$. By formal étaleness, the morphism $\mathrm{Spf}(R) \to \vec{\mathfrak{X}}^{\mathrm{ord}}_{\Phi_\mathcal{H},\delta_\mathcal{H},\sigma}/\Gamma_{\Phi_\mathcal{H},\sigma}$ can be uniquely lifted to a morphism $\mathrm{Spf}(R) \to \mathrm{Spf}(R_{\mathrm{alg}}{}^\wedge, I^\wedge)$. The underlying morphism $\mathrm{Spec}(R) \to \mathrm{Spec}(R_{\mathrm{alg}}{}^\wedge)$ identifies the degeneration datum associated with $(G^\ddagger, \lambda^\ddagger, i^\ddagger, \alpha^\ddagger_{\mathcal{H}^p}, \alpha^{\mathrm{ord},\ddagger}_{\mathcal{H}_p}) \to \mathrm{Spec}(R)$ with the degeneration datum associated with the pullback of $(G, \lambda, i, \alpha_{\mathcal{H}^p}, \alpha^{\mathrm{ord}}_{\mathcal{H}_p}) \underset{R_{\mathrm{alg}}}{\otimes} R_{\mathrm{alg}}{}^\wedge \to \mathrm{Spec}(R_{\mathrm{alg}}{}^\wedge)$. Hence, the morphism $\mathrm{Spec}(R) \to \mathrm{Spec}(R_{\mathrm{alg}}{}^\wedge) \to \mathrm{Spec}(R_{\mathrm{alg}})$ identifies $(G^\ddagger, \lambda^\ddagger, i^\ddagger, \alpha^\ddagger_{\mathcal{H}^p}, \alpha^{\mathrm{ord},\ddagger}_{\mathcal{H}_p}) \to \mathrm{Spec}(R)$ with the pullback of $(G, \lambda, i, \alpha_{\mathcal{H}^p}, \alpha^{\mathrm{ord}}_{\mathcal{H}_p}) \to \mathrm{Spec}(R_{\mathrm{alg}})$, as desired. □

The key to the gluing process is the following:

Proposition 5.1.3.8. (*Compare with [62, Prop. 6.3.3.13].*) *The two projections from* $\vec{\mathsf{R}}^{\mathrm{ord}}_\mathcal{H} := \vec{\mathsf{U}}^{\mathrm{ord}}_\mathcal{H} \underset{\vec{\mathsf{M}}^{\mathrm{ord},\mathrm{tor}}_{\mathcal{H},\Sigma^{\mathrm{ord}}}}{\times} \vec{\mathsf{U}}^{\mathrm{ord}}_\mathcal{H}$ *to* $\vec{\mathsf{U}}^{\mathrm{ord}}_\mathcal{H}$ *are* **étale**.

Proof. The same argument of the proof of [62, Prop. 6.3.3.13] works here, with good formal and algebraic models replaced with ordinary good formal and algebraic models, with algebraic stacks such as $\Xi_{\Phi_{\mathcal{H}},\delta_{\mathcal{H}}}(\sigma)/\Gamma_{\Phi_{\mathcal{H}},\sigma}$ replaced with $\vec{\Xi}^{\mathrm{ord}}_{\Phi_{\mathcal{H}},\delta_{\mathcal{H}}}(\sigma)/\Gamma_{\Phi_{\mathcal{H}},\sigma}$, with the *openness of versality* provided by Proposition 5.1.2.7 (instead of [62, Prop. 6.3.2.10]), and with the theory of degeneration provided by Theorems 4.1.5.27 and 4.1.6.2 (instead of [62, Thm. 5.3.1.19]). □

Corollary 5.1.3.9. (*Compare with [62, Cor. 6.3.3.14].*) *The scheme* $\vec{\mathsf{R}}^{\mathrm{ord}}_{\mathcal{H}}$ *over* $\vec{\mathsf{U}}^{\mathrm{ord}}_{\mathcal{H}}$ *defines an étale groupoid space (see [62, Def. A.5.1.2]), which extends the étale groupoid space* $\vec{\mathsf{R}}^{\mathrm{ord},[0]}_{\mathcal{H}}$ *over* $\vec{\mathsf{U}}^{\mathrm{ord},[0]}_{\mathcal{H}}$*. The scheme* $\vec{\mathsf{R}}^{\mathrm{ord}}_{\mathcal{H}}$ *is* **finite** *over* $\vec{\mathsf{U}}^{\mathrm{ord}}_{\mathcal{H}} \underset{\vec{\mathsf{S}}_{0,r_{\mathcal{H}}}}{\times} \vec{\mathsf{U}}^{\mathrm{ord}}_{\mathcal{H}}$*, and hence (by [62, Lem. A.7.2.9])* $\vec{\mathsf{U}}^{\mathrm{ord}}_{\mathcal{H}}/\vec{\mathsf{R}}^{\mathrm{ord}}_{\mathcal{H}}$ *defines an algebraic stack* **separated** *over* $\vec{\mathsf{S}}_{0,r_{\mathcal{H}}} = \mathrm{Spec}(\mathcal{O}_{F_0,(p)}[\zeta_{p^{r_{\mathcal{H}}}}])$.

Proof. The same argument of the proof of [62, Cor. 6.3.3.14] works here, except that for the finiteness of $\vec{\mathsf{R}}^{\mathrm{ord},[0]}_{\mathcal{H}}$ over $\vec{\mathsf{U}}^{\mathrm{ord}}_{\mathcal{H}} \underset{\vec{\mathsf{S}}_{0,r_{\mathcal{H}}}}{\times} \vec{\mathsf{U}}^{\mathrm{ord}}_{\mathcal{H}}$ we need not only the property of the Isom functor of abelian schemes as in [62, Sec. 2.3.4, Cond. 2], but also the fact that the abelian schemes are ordinary (so that the ordinary level structures are defined by isomorphisms between finite étale group schemes, or rather their duals). □

Definition 5.1.3.10. (*Compare with [62, Def. 6.3.3.15].*) *The separated algebraic stack* $\vec{\mathsf{U}}^{\mathrm{ord}}_{\mathcal{H}}/\vec{\mathsf{R}}^{\mathrm{ord}}_{\mathcal{H}}$ *(see [62, Prop. A.7.1.1 and Def. A.7.1.3]) will be denoted by* $\vec{\mathsf{M}}^{\mathrm{ord},\mathrm{tor}}_{\mathcal{H}}$ *(or* $\vec{\mathsf{M}}^{\mathrm{ord},\mathrm{tor}}_{\mathcal{H},\Sigma^{\mathrm{ord}}}$*, to emphasize its dependence on the compatible choice* $\Sigma^{\mathrm{ord}} = \{\Sigma_{\Phi_{\mathcal{H}}}\}_{[(\Phi_{\mathcal{H}},\delta_{\mathcal{H}})]}$ *of cone decompositions).*

Corollary 5.1.3.11. (*Compare with [62, Cor. 6.3.3.16].*) *Both the degenerating family* $(G,\lambda,i,\alpha_{\mathcal{H}^p},\alpha^{\mathrm{ord}}_{\mathcal{H}_p}) \to \vec{\mathsf{U}}^{\mathrm{ord}}_{\mathcal{H}}$ *and the stratification over* $\vec{\mathsf{U}}^{\mathrm{ord}}_{\mathcal{H}}$ *(see Construction 5.1.3.5) descend to* $\vec{\mathsf{M}}^{\mathrm{ord},\mathrm{tor}}_{\mathcal{H},\Sigma^{\mathrm{ord}}}$*, which we again denote by the same notation. This realizes* $\vec{\mathsf{M}}^{\mathrm{ord}}_{\mathcal{H}}$ *as the* $[0]$*-stratum in the stratification, and identifies the restriction of* $(G,\lambda,i,\alpha_{\mathcal{H}^p},\alpha^{\mathrm{ord}}_{\mathcal{H}_p})$ *to* $\vec{\mathsf{M}}^{\mathrm{ord}}_{\mathcal{H}}$ *with the tautological tuple over* $\vec{\mathsf{M}}^{\mathrm{ord}}_{\mathcal{H}}$.

Proof. The same argument of the proof of [62, Cor. 6.3.3.16] works here. □

Remark 5.1.3.12. (Compare with [62, Prop. 6.3.3.17].) Since $\vec{\mathsf{U}}^{\mathrm{ord}}_{\mathcal{H}}$ is smooth and of finite type over $\vec{\mathsf{S}}_{0,r_{\mathcal{H}}} = \mathrm{Spec}(\mathcal{O}_{F_0,(p)}[\zeta_{p^{r_{\mathcal{H}}}}])$, by Proposition 5.1.3.8 and Corollary 5.1.3.9, the algebraic stack $\vec{\mathsf{M}}^{\mathrm{ord},\mathrm{tor}}_{\mathcal{H},\Sigma^{\mathrm{ord}}}$ is separated, smooth, and of finite type over $\vec{\mathsf{S}}_{0,r_{\mathcal{H}}}$. But $\vec{\mathsf{M}}^{\mathrm{ord},\mathrm{tor}}_{\mathcal{H},\Sigma^{\mathrm{ord}}}$ is almost never proper over $\vec{\mathsf{S}}_{0,r_{\mathcal{H}}}$. (See Proposition 6.3.2.2 below.)

5.2 Partial Toroidal Compactifications of Ordinary Loci

In this section, let $\mathcal{H}$, $\mathcal{H}^p$, $\mathcal{H}_p$, and r be as in beginning of Section 3.3.5, and let $r_{\mathcal{H}}$ be as in Definition 3.4.2.1.

5.2.1 *Main Statements*

The partial toroidal compactifications of $\vec{\mathsf{M}}^{\mathrm{ord}}_{\mathcal{H}}$ can be described as follows:

Theorem 5.2.1.1. (*Compare with [62, Thm. 6.4.1.1] and Theorem* 1.3.1.3.) *With settings as above, to each compatible choice $\Sigma^{\mathrm{ord}} = \{\Sigma_{\Phi_{\mathcal{H}}}\}_{[(\Phi_{\mathcal{H}},\delta_{\mathcal{H}})]}$ of admissible smooth rational polyhedral cone decomposition data as in Definition* 5.1.3.1, *there is associated an algebraic stack $\vec{\mathsf{M}}^{\mathrm{ord,tor}}_{\mathcal{H}} = \vec{\mathsf{M}}^{\mathrm{ord,tor}}_{\mathcal{H},\Sigma^{\mathrm{ord}}}$ separated, smooth, and of finite type over $\vec{\mathsf{S}}_{0,r_{\mathcal{H}}} = \mathrm{Spec}(\mathcal{O}_{F_0,(p)}[\zeta_{p^{r_{\mathcal{H}}}}])$ (see Definition* 2.2.3.3), *which is an algebraic space when $\mathcal{H}^p$ is neat (see [89, 0.6] or [62, Def. 1.4.1.8]), containing $\vec{\mathsf{M}}^{\mathrm{ord}}_{\mathcal{H}}$ as an open fiberwise dense subalgebraic stack, together with a degenerating family $(G,\lambda,i,\alpha_{\mathcal{H}^p},\alpha^{\mathrm{ord}}_{\mathcal{H}_p})$ of type $\vec{\mathsf{M}}^{\mathrm{ord}}_{\mathcal{H}}$ over $\vec{\mathsf{M}}^{\mathrm{ord,tor}}_{\mathcal{H}}$ (as in Definition* 3.4.2.10) *such that we have the following:*

(1) *The restriction $(G_{\vec{\mathsf{M}}^{\mathrm{ord}}_{\mathcal{H}}},\lambda_{\vec{\mathsf{M}}^{\mathrm{ord}}_{\mathcal{H}}},i_{\vec{\mathsf{M}}^{\mathrm{ord}}_{\mathcal{H}}},\alpha_{\mathcal{H}^p},\alpha^{\mathrm{ord}}_{\mathcal{H}_p})$ of the degenerating family $(G,\lambda,i,\alpha_{\mathcal{H}^p},\alpha^{\mathrm{ord}}_{\mathcal{H}_p})$ to $\vec{\mathsf{M}}^{\mathrm{ord}}_{\mathcal{H}}$ is the tautological tuple over $\vec{\mathsf{M}}^{\mathrm{ord}}_{\mathcal{H}}$ (see Convention* 3.4.2.9).

(2) *$\vec{\mathsf{M}}^{\mathrm{ord,tor}}_{\mathcal{H}}$ has a stratification by locally closed subalgebraic stacks*

$$\vec{\mathsf{M}}^{\mathrm{ord,tor}}_{\mathcal{H}} = \coprod_{[(\Phi_{\mathcal{H}},\delta_{\mathcal{H}},\sigma)]} \vec{\mathsf{Z}}^{\mathrm{ord}}_{[(\Phi_{\mathcal{H}},\delta_{\mathcal{H}},\sigma)]},$$

with $[(\Phi_{\mathcal{H}},\delta_{\mathcal{H}},\sigma)]$ running through a complete set of equivalence classes of $(\Phi_{\mathcal{H}},\delta_{\mathcal{H}},\sigma)$ (as in Definition 1.2.2.10) *with $[(\Phi_{\mathcal{H}},\delta_{\mathcal{H}})]$ an ordinary cusp label (as in Definition* 3.2.3.8) *and with $\sigma \subset \mathbf{P}^+_{\Phi_{\mathcal{H}}}$ and $\sigma \in \Sigma_{\Phi_{\mathcal{H}}} \in \Sigma^{\mathrm{ord}}$. (Here $\mathsf{Z}_{\mathcal{H}}$ is suppressed in the notation by our convention. The notation "$\coprod$" only means a set-theoretic disjoint union. The algebro-geometric structure is still that of $\vec{\mathsf{M}}^{\mathrm{ord,tor}}_{\mathcal{H}}$.)*

In this stratification, the $[(\Phi'_{\mathcal{H}},\delta'_{\mathcal{H}},\sigma')]$-stratum $\vec{\mathsf{Z}}^{\mathrm{ord}}_{[(\Phi'_{\mathcal{H}},\delta'_{\mathcal{H}},\sigma')]}$ lies in the closure of the $[(\Phi_{\mathcal{H}},\delta_{\mathcal{H}},\sigma)]$-stratum $\vec{\mathsf{Z}}^{\mathrm{ord}}_{[(\Phi_{\mathcal{H}},\delta_{\mathcal{H}},\sigma)]}$ if and only if $[(\Phi_{\mathcal{H}},\delta_{\mathcal{H}},\sigma)]$ is a face of $[(\Phi'_{\mathcal{H}},\delta'_{\mathcal{H}},\sigma')]$ as in Definition 1.2.2.19 (*see also Remark* 5.1.2.8). *The analogous assertion holds after pulled back to fibers over $\vec{\mathsf{S}}_{0,r_{\mathcal{H}}}$.*

The $[(\Phi_{\mathcal{H}},\delta_{\mathcal{H}},\sigma)]$-stratum $\vec{\mathsf{Z}}^{\mathrm{ord}}_{[(\Phi_{\mathcal{H}},\delta_{\mathcal{H}},\sigma)]}$ is smooth over $\vec{\mathsf{S}}_{0,r_{\mathcal{H}}}$ and isomorphic to the support of the formal algebraic stack $\vec{\mathfrak{X}}^{\mathrm{ord}}_{\Phi_{\mathcal{H}},\delta_{\mathcal{H}},\sigma}/\Gamma_{\Phi_{\mathcal{H}},\sigma}$ for every representative $(\Phi_{\mathcal{H}},\delta_{\mathcal{H}},\sigma)$ of $[(\Phi_{\mathcal{H}},\delta_{\mathcal{H}},\sigma)]$, where the formal algebraic stack $\vec{\mathfrak{X}}^{\mathrm{ord}}_{\Phi_{\mathcal{H}},\delta_{\mathcal{H}},\sigma}$ (before quotient by $\Gamma_{\Phi_{\mathcal{H}},\sigma}$, the subgroup of $\Gamma_{\Phi_{\mathcal{H}}}$ formed by elements mapping σ to itself; see [62, Def. 6.2.5.23]) admits a canonical structure as the completion of an affine toroidal embedding $\vec{\Xi}^{\mathrm{ord}}_{\Phi_{\mathcal{H}},\delta_{\mathcal{H}}}(\sigma)$ (along its σ-stratum $\vec{\Xi}^{\mathrm{ord}}_{\Phi_{\mathcal{H}},\delta_{\mathcal{H}},\sigma}$) of a torus torsor $\vec{\Xi}^{\mathrm{ord}}_{\Phi_{\mathcal{H}},\delta_{\mathcal{H}}}$ over an abelian scheme torsor $\vec{C}^{\mathrm{ord}}_{\Phi_{\mathcal{H}},\delta_{\mathcal{H}}}$ over a finite étale cover $\vec{\mathsf{M}}^{\mathrm{ord},\Phi_{\mathcal{H}}}_{\mathcal{H}}$ of the regular algebraic stack $\vec{\mathsf{M}}^{\mathrm{ord},\mathsf{Z}_{\mathcal{H}}}_{\mathcal{H}}$ separated, smooth, and of finite type over $\vec{\mathsf{S}}_{0,r_{\mathcal{H}}}$ (as in Propositions 4.2.1.29, 4.2.1.30, *and* 4.2.1.37). (*Note that $\mathsf{Z}_{\mathcal{H}}$ and the isomorphism class of $\vec{\mathsf{M}}^{\mathrm{ord},\mathsf{Z}_{\mathcal{H}}}_{\mathcal{H}}$ depend only on the class $[(\Phi_{\mathcal{H}},\delta_{\mathcal{H}},\sigma)]$, but not on the choice of the representative $(\Phi_{\mathcal{H}},\delta_{\mathcal{H}},\sigma)$.)*

In particular, $\vec{\mathsf{M}}^{\mathrm{ord}}_{\mathcal{H}}$ *is an open fiberwise dense stratum in this stratification.*

(3) *The complement of* $\vec{\mathsf{M}}^{\mathrm{ord}}_{\mathcal{H}}$ *in* $\vec{\mathsf{M}}^{\mathrm{ord,tor}}_{\mathcal{H}}$ *(with its reduced structure) is a relative Cartier divisor* $\vec{\mathsf{D}}^{\mathrm{ord}}_{\infty,\mathcal{H}}$ *with normal crossings, such that each irreducible component of a stratum of* $\vec{\mathsf{M}}^{\mathrm{ord,tor}}_{\mathcal{H}} - \vec{\mathsf{M}}^{\mathrm{ord}}_{\mathcal{H}}$ *is open dense in an intersection of irreducible components of* $\vec{\mathsf{D}}^{\mathrm{ord}}_{\infty,\mathcal{H}}$ *(including possible self-intersections). When* $\mathcal{H}^p$ *is neat, the irreducible components of* $\vec{\mathsf{D}}^{\mathrm{ord}}_{\infty,\mathcal{H}}$ *have no self-intersections (cf. Condition* 1.2.2.9*, [62, Rem. 6.2.5.26], and [28, Ch. IV, Rem. 5.8(a)]).*

(4) *The extended Kodaira–Spencer morphism [62, Def. 4.6.3.44] for* $G \to \vec{\mathsf{M}}^{\mathrm{ord,tor}}_{\mathcal{H}}$ *induces an isomorphism*

$$\mathrm{KS}_{G/\vec{\mathsf{M}}^{\mathrm{ord,tor}}_{\mathcal{H}}/\vec{\mathsf{S}}_{0,r_{\mathcal{H}}}} : \underline{\mathrm{KS}}_{(G,\lambda,i)/\vec{\mathsf{M}}^{\mathrm{ord,tor}}_{\mathcal{H}},\mathrm{free}} \xrightarrow{\sim} \Omega^1_{\vec{\mathsf{M}}^{\mathrm{ord,tor}}_{\mathcal{H}}/\vec{\mathsf{S}}_{0,r_{\mathcal{H}}}}[d\log\infty],$$

(see Definition 3.4.3.1*). Here the sheaf* $\Omega^1_{\vec{\mathsf{M}}^{\mathrm{ord,tor}}_{\mathcal{H}}/\vec{\mathsf{S}}_{0,r_{\mathcal{H}}}}[d\log\infty]$ *is the sheaf of modules of log* 1*-differentials on* $\vec{\mathsf{M}}^{\mathrm{ord,tor}}_{\mathcal{H}}$ *over* $\vec{\mathsf{S}}_{0,r_{\mathcal{H}}}$*, with respect to the relative Cartier divisor* $\vec{\mathsf{D}}^{\mathrm{ord}}_{\infty,\mathcal{H}}$ *with normal crossings.*

(5) *For every representative* $(\Phi_{\mathcal{H}}, \delta_{\mathcal{H}}, \sigma)$ *of* $[(\Phi_{\mathcal{H}}, \delta_{\mathcal{H}}, \sigma)]$*, the formal completion* $(\vec{\mathsf{M}}^{\mathrm{ord,tor}}_{\mathcal{H}})^{\wedge}_{\vec{\mathsf{Z}}^{\mathrm{ord}}_{[(\Phi_{\mathcal{H}},\delta_{\mathcal{H}},\sigma)]}}$ *of* $\vec{\mathsf{M}}^{\mathrm{ord,tor}}_{\mathcal{H}}$ *along the* $[(\Phi_{\mathcal{H}}, \delta_{\mathcal{H}}, \sigma)]$*-stratum* $\vec{\mathsf{Z}}^{\mathrm{ord}}_{[(\Phi_{\mathcal{H}},\delta_{\mathcal{H}},\sigma)]}$ *is canonically isomorphic to the formal algebraic stack* $\vec{\mathfrak{X}}^{\mathrm{ord}}_{\Phi_{\mathcal{H}},\delta_{\mathcal{H}},\sigma}/\Gamma_{\Phi_{\mathcal{H}},\sigma}$*. This isomorphism respects stratifications in the sense that, given any étale (i.e., formally étale and of finite type; see [35, I, 10.13.3]) morphism* $\mathrm{Spf}(R, I) \to \vec{\mathfrak{X}}^{\mathrm{ord}}_{\Phi_{\mathcal{H}},\delta_{\mathcal{H}},\sigma}/\Gamma_{\Phi_{\mathcal{H}},\sigma}$ *inducing a morphism* $\mathrm{Spec}(R) \to \vec{\Xi}^{\mathrm{ord}}_{\Phi_{\mathcal{H}},\delta_{\mathcal{H}}}(\sigma)/\Gamma_{\Phi_{\mathcal{H}},\sigma}$*, the stratification of* $\mathrm{Spec}(R)$ *inherited from* $\vec{\Xi}^{\mathrm{ord}}_{\Phi_{\mathcal{H}},\delta_{\mathcal{H}}}(\sigma)/\Gamma_{\Phi_{\mathcal{H}},\sigma}$ *(see Proposition* 5.1.1.3 *and Definition* 5.1.2.9*) makes the induced morphism* $\mathrm{Spec}(R) \to \vec{\mathsf{M}}^{\mathrm{ord,tor}}_{\mathcal{H}}$ *a strata-preserving morphism.*
The pullback of the degenerating family $(G, \lambda, i, \alpha_{\mathcal{H}^p}, \alpha^{\mathrm{ord}}_{\mathcal{H}_p})$ *over* $\vec{\mathsf{M}}^{\mathrm{ord,tor}}_{\mathcal{H}}$ *to* $(\vec{\mathsf{M}}^{\mathrm{ord,tor}}_{\mathcal{H}})^{\wedge}_{\vec{\mathsf{Z}}^{\mathrm{ord}}_{[(\Phi_{\mathcal{H}},\delta_{\mathcal{H}},\sigma)]}}$ *is the Mumford family* $({}^{\heartsuit}G, {}^{\heartsuit}\lambda, {}^{\heartsuit}i, {}^{\heartsuit}\alpha_{\mathcal{H}^p}, {}^{\heartsuit}\alpha^{\mathrm{ord}}_{\mathcal{H}_p})$ *over* $\vec{\mathfrak{X}}^{\mathrm{ord}}_{\Phi_{\mathcal{H}},\delta_{\mathcal{H}},\sigma}/\Gamma_{\Phi_{\mathcal{H}},\sigma}$ *(see Definition* 4.2.2.21*) after we identify the bases using the isomorphism. (Here both the pullback of* $(G, \lambda, i, \alpha_{\mathcal{H}^p}, \alpha^{\mathrm{ord}}_{\mathcal{H}_p})$ *and the Mumford family* $({}^{\heartsuit}G, {}^{\heartsuit}\lambda, {}^{\heartsuit}i, {}^{\heartsuit}\alpha_{\mathcal{H}^p}, {}^{\heartsuit}\alpha^{\mathrm{ord}}_{\mathcal{H}_p})$ *are considered as relative schemes with additional structures; cf. [37].)*

(6) *Let* S *be an irreducible noetherian normal scheme* **flat** *over* $\vec{\mathsf{S}}_{0,r_{\mathcal{H}}}$*, and suppose that we have a degenerating family* $(G^{\dagger}, \lambda^{\dagger}, i^{\dagger}, \alpha^{\dagger}_{\mathcal{H}^p}, \alpha^{\mathrm{ord},\dagger}_{\mathcal{H}_p})$ *of type* $\vec{\mathsf{M}}^{\mathrm{ord}}_{\mathcal{H}}$ *over* S *as in Definition* 3.4.2.10*. Then* $(G^{\dagger}, \lambda^{\dagger}, i^{\dagger}, \alpha^{\dagger}_{\mathcal{H}^p}, \alpha^{\mathrm{ord},\dagger}_{\mathcal{H}_p}) \to S$ *is the pullback of* $(G, \lambda, i, \alpha_{\mathcal{H}^p}, \alpha^{\mathrm{ord}}_{\mathcal{H}_p}) \to \vec{\mathsf{M}}^{\mathrm{ord,tor}}_{\mathcal{H}}$ *via a (necessarily unique) morphism* $S \to \vec{\mathsf{M}}^{\mathrm{ord,tor}}_{\mathcal{H}}$ *(over* $\vec{\mathsf{S}}^{\mathrm{ord}}_0$*) if and only if the following condition is satisfied at each geometric point* $\bar{s}$ *of* S*:*
Consider any dominant morphism $\mathrm{Spec}(V) \to S$ *centered at* $\bar{s}$*, where* V *is a complete discrete valuation ring with fraction field* K*, algebraically closed residue field* k*, and discrete valuation* υ*. Let* $(G^{\ddagger}, \lambda^{\ddagger}, i^{\ddagger}, \alpha^{\ddagger}_{\mathcal{H}^p}, \alpha^{\mathrm{ord},\ddagger}_{\mathcal{H}_p}) \to \mathrm{Spec}(V)$ *be the pullback of* $(G^{\dagger}, \lambda^{\dagger}, i^{\dagger}, \alpha^{\dagger}_{\mathcal{H}^p}, \alpha^{\mathrm{ord},\dagger}_{\mathcal{H}_p}) \to S$*. This pullback*

family defines an object in the essential image of $\mathrm{DEG}_{\mathrm{PEL},\mathsf{M}_{\mathcal{H}}^{\mathrm{ord}}}(V) \to \mathrm{DEG}_{\mathrm{PEL},\ddot{\mathsf{M}}_{\mathcal{H}}^{\mathrm{ord}}}(V)$*, which corresponds to a tuple*

$$(B^{\ddagger}, \lambda_{B^{\ddagger}}, i_{B^{\ddagger}}, \underline{X}^{\ddagger}, \underline{Y}^{\ddagger}, \phi^{\ddagger}, c^{\ddagger}, c^{\vee,\ddagger}, \tau^{\ddagger}, [\alpha_{\mathcal{H}}^{\natural,\mathrm{ord},\ddagger}])$$

in the essential image of $\mathrm{DD}_{\mathrm{PEL},\mathsf{M}_{\mathcal{H}}^{\mathrm{ord}}}(V) \to \mathrm{DD}_{\mathrm{PEL},\ddot{\mathsf{M}}_{\mathcal{H}}^{\mathrm{ord}}}(V)$ *under* (4.1.6.4) *in Theorem* 4.1.6.2. *Then we have a fully symplectic-liftable admissible filtration* $\mathsf{Z}_{\mathcal{H}}^{\ddagger}$ *determined by* $[\alpha_{\mathcal{H}}^{\natural,\mathrm{ord},\ddagger}]$. *Moreover, the étale sheaves* $\underline{X}^{\ddagger}$ *and* $\underline{Y}^{\ddagger}$ *are necessarily constant, because the base ring* R *is strict local. Hence, it makes sense to say we also have a uniquely determined torus argument* $\Phi_{\mathcal{H}}^{\ddagger}$ *at level* $\mathcal{H}$ *for* $\mathsf{Z}_{\mathcal{H}}^{\ddagger}$.

On the other hand, we have objects $\underline{\Phi}_{\mathcal{H}}(G^{\ddagger})$, $\underline{\mathbf{S}}_{\underline{\Phi}_{\mathcal{H}}(G^{\ddagger})}$, *and* $\underline{B}(G^{\ddagger})$ *(see [62, Constr. 6.3.1.1]), which define objects* $\Phi_{\mathcal{H}}^{\ddagger}$, $\mathbf{S}_{\Phi_{\mathcal{H}}^{\ddagger}}$, *and in particular* $B^{\ddagger} : \mathbf{S}_{\Phi_{\mathcal{H}}^{\ddagger}} \to \mathrm{Inv}(V)$ *over the special fiber. Then* $\upsilon \circ B^{\ddagger} : \mathbf{S}_{\Phi_{\mathcal{H}}^{\ddagger}} \to \mathbb{Z}$ *defines an element of* $\mathbf{S}_{\Phi_{\mathcal{H}}^{\ddagger}}^{\vee}$, *where* $\upsilon : \mathrm{Inv}(V) \to \mathbb{Z}$ *is the homomorphism induced by the discrete valuation of* V.

Then the condition is that, for each $\mathrm{Spec}(V) \to S$ *as above (centered at* $\bar{s}$*), and for some (and hence every) choice of* $\delta_{\mathcal{H}}^{\ddagger}$, *there is a cone* $\sigma^{\ddagger}$ *in the cone decomposition* $\Sigma_{\Phi_{\mathcal{H}}^{\ddagger}}$ *of* $\mathbf{P}_{\Phi_{\mathcal{H}}^{\ddagger}}$ *(given by the choice of* Σ^{ord}*; cf. Definition* 5.1.3.1*) such that* $\overline{\sigma}^{\ddagger}$ *contains all* $\upsilon \circ B^{\ddagger}$ *obtained in this way.*

(7) If Σ^{ord} *extends to a compatible choice* Σ *of admissible smooth rational polyhedral cone decomposition data for* $\mathsf{M}_{\mathcal{H}}$ *(cf. Proposition* 5.1.3.2*), then there is a canonical open immersion*

$$\vec{\mathsf{M}}_{\mathcal{H},\Sigma^{\mathrm{ord}}}^{\mathrm{ord,tor}} \underset{\mathbb{Z}}{\otimes} \mathbb{Q} \hookrightarrow \mathsf{M}_{\mathcal{H},\Sigma,r_{\mathcal{H}}}^{\mathrm{tor}} \tag{5.2.1.2}$$

(see Definition 2.2.3.4*) over* $\mathsf{S}_{0,r_{\mathcal{H}}}$ *extending the canonical isomorphism* $\mathsf{M}_{\mathcal{H}}^{\mathrm{ord}} \cong \mathsf{M}_{\mathcal{H},r_{\mathcal{H}}}$ *over* $\mathsf{S}_{0,r_{\mathcal{H}}}$ *(see the definition of* $\mathsf{M}_{\mathcal{H}}^{\mathrm{ord}}$ *in Theorem* 3.4.2.5*), such that the pullback of* $(G, \lambda, i, \alpha_{\mathcal{H}^p}, \alpha_{\mathcal{H}_p}^{\mathrm{ord}}) \to \vec{\mathsf{M}}_{\mathcal{H}}^{\mathrm{ord,tor}}$ *is canonically determined by the pullback of* $(G, \lambda, i, \alpha_{\mathcal{H}}) \to \mathsf{M}_{\mathcal{H},\Sigma}^{\mathrm{tor}}$ *(see Theorem* 1.3.1.3*) in the sense that the triples* (G, λ, i) *are isomorphic over* $\mathsf{S}_{0,r_{\mathcal{H}}}$, *and in the sense that the pullback of* $\alpha_{\mathcal{H}}$ *determines the pullback of* $(\alpha_{\mathcal{H}^p}, \alpha_{\mathcal{H}_p}^{\mathrm{ord}}) \underset{\mathbb{Z}}{\otimes} \mathbb{Q}$ *as in Proposition* 3.3.5.1. *The open immersion* (5.2.1.2) *induces isomorphisms*

$$\vec{\mathsf{Z}}_{[(\Phi_{\mathcal{H}},\delta_{\mathcal{H}},\sigma)]}^{\mathrm{ord}} \underset{\mathbb{Z}}{\otimes} \mathbb{Q} \xrightarrow{\sim} \mathsf{Z}_{[(\Phi_{\mathcal{H}},\delta_{\mathcal{H}},\sigma)],r_{\mathcal{H}}} \tag{5.2.1.3}$$

(see Definition 2.2.3.4*) when the cusp label* $[(\Phi_{\mathcal{H}}, \delta_{\mathcal{H}})]$ *is ordinary; otherwise, the pullback of* $\mathsf{Z}_{[(\Phi_{\mathcal{H}},\delta_{\mathcal{H}},\sigma)],r_{\mathcal{H}}}$ *under* (5.2.1.2) *is empty.*

Remark 5.2.1.4. Although statement (6) resembles a valuative criterion, it does not imply that $\vec{\mathsf{M}}_{\mathcal{H}}^{\mathrm{ord,tor}}$ is proper as in the proof of [62, Prop. 6.3.3.17], because the condition of being a degenerating family of type $\vec{\mathsf{M}}_{\mathcal{H}}^{\mathrm{ord}}$ requires the condition (5) in Definition 3.4.2.10, which does not hold in general (*cf.* Remark 3.4.2.11).

Proof of Theorem 5.2.1.1. The proof is almost identical to that of [62, Thm. 6.4.1.1]. However, since this is one of the most important theorems in this work, we repeat the arguments here for the sake of certainty.

Let $\vec{\mathsf{U}}^{\mathrm{ord}}_{\mathcal{H}}$ and $\vec{\mathsf{R}}^{\mathrm{ord}}_{\mathcal{H}}$ be constructed (noncanonically) as in Section 5.1.3, and let us take the separated algebraic stack $\vec{\mathsf{M}}^{\mathrm{ord,tor}}_{\mathcal{H}} = \vec{\mathsf{M}}^{\mathrm{ord,tor}}_{\mathcal{H},\Sigma^{\mathrm{ord}}}$ to be the groupoid quotient $\vec{\mathsf{U}}^{\mathrm{ord}}_{\mathcal{H}}/\vec{\mathsf{R}}^{\mathrm{ord}}_{\mathcal{H}}$ (see [62, Prop. A.7.1.1 and Def. A.7.1.3]) as in Definition 5.1.3.10.

Statements (1) and (2) follow from Corollaries 5.1.3.9 and 5.1.3.11. Statements (3) and (4) are étale local in nature, and hence are inherited from the étale presentation $\vec{\mathsf{U}}^{\mathrm{ord}}_{\mathcal{H}}$ of $\vec{\mathsf{M}}^{\mathrm{ord,tor}}_{\mathcal{H}}$ (with descent data over $\vec{\mathsf{R}}^{\mathrm{ord}}_{\mathcal{H}}$) by construction.

Let us prove statement (6) by explaining why it is essentially a restatement of Proposition 5.1.3.7. Suppose we have a degenerating family $(G^\dagger, \lambda^\dagger, i^\dagger, \alpha^\dagger_{\mathcal{H}^p}, \alpha^{\mathrm{ord},\dagger}_{\mathcal{H}_p}) \to S$ as in the statement. Then there is an open dense subscheme S_1 of S such that the restriction of the family defines an object parameterized by $\vec{\mathsf{M}}^{\mathrm{ord}}_{\mathcal{H}}$ (*cf.* Convention 3.4.2.9), together with a morphism $S_1 \to \vec{\mathsf{M}}^{\mathrm{ord}}_{\mathcal{H}}$. The question is whether this morphism extends to a morphism $S \to \vec{\mathsf{M}}^{\mathrm{ord,tor}}_{\mathcal{H}}$. By [92, IX, 1.4], [28, Ch. I, Prop. 2.7], or [62, Prop. 3.3.1.5], if this is the case, then $(G^\dagger, \lambda^\dagger, i^\dagger, \alpha^\dagger_{\mathcal{H}^p}, \alpha^{\mathrm{ord},\dagger}_{\mathcal{H}_p}) \to S$ is isomorphic to the pullback of the tautological tuple $(G, \lambda, i, \alpha_{\mathcal{H}^p}, \alpha^{\mathrm{ord}}_{\mathcal{H}_p}) \to \vec{\mathsf{M}}^{\mathrm{ord,tor}}_{\mathcal{H}}$ under this morphism, and the condition in the statement certainly holds. Conversely, assume that the condition holds. Since all objects involved are locally of finite presentation, we can apply [62, Thm. 1.3.1.3] and assume that S is excellent. Since extendability is a local question (because $\vec{\mathsf{M}}^{\mathrm{ord,tor}}_{\mathcal{H}}$ is separated over $\vec{\mathsf{S}}_{0,r_{\mathcal{H}}}$), we can work with $\vec{\mathsf{U}}^{\mathrm{ord}}_{\mathcal{H}}$ and apply Proposition 5.1.3.7 (to pullbacks of $(G^\dagger, \lambda^\dagger, i^\dagger, \alpha^\dagger_{\mathcal{H}^p}, \alpha^{\mathrm{ord},\dagger}_{\mathcal{H}_p}) \to S$ to the completions of local rings of S).

Next, let us prove statement (5). By statement (6) we have just proved, we know that there is a unique morphism from $\vec{\mathfrak{X}}^{\mathrm{ord}}_{\Phi_{\mathcal{H}},\delta_{\mathcal{H}},\sigma}/\Gamma_{\Phi_{\mathcal{H}},\sigma}$ to $\vec{\mathsf{M}}^{\mathrm{ord,tor}}_{\mathcal{H}}$. (More precisely, we apply statement (6) to an étale covering of $\vec{\mathfrak{X}}^{\mathrm{ord}}_{\Phi_{\mathcal{H}},\delta_{\mathcal{H}},\sigma}/\Gamma_{\Phi_{\mathcal{H}},\sigma}$ by affine formal schemes with descent data.) This induces a canonical morphism $\vec{\mathfrak{X}}^{\mathrm{ord}}_{\Phi_{\mathcal{H}},\delta_{\mathcal{H}},\sigma}/\Gamma_{\Phi_{\mathcal{H}},\sigma} \to (\vec{\mathsf{M}}^{\mathrm{ord,tor}}_{\mathcal{H}})^\wedge_{\vec{\mathsf{Z}}^{\mathrm{ord}}_{[(\Phi_{\mathcal{H}},\delta_{\mathcal{H}},\sigma)]}}$. For an inverse morphism, note that by construction there is a canonical morphism from the formal completion of $\vec{\mathsf{U}}^{\mathrm{ord}}_{\mathcal{H}}$ along its $[(\Phi_{\mathcal{H}}, \delta_{\mathcal{H}}, \sigma)]$-stratum to $\vec{\mathfrak{X}}^{\mathrm{ord}}_{\Phi_{\mathcal{H}},\delta_{\mathcal{H}},\sigma}/\Gamma_{\Phi_{\mathcal{H}},\sigma}$. Since this canonical morphism is determined by the degeneration data associated with the pullback of the tautological tuple $(G, \lambda, i, \alpha_{\mathcal{H}^p}, \alpha^{\mathrm{ord}}_{\mathcal{H}_p})$ to the completion, and since the two pullbacks of the tautological tuple to $\vec{\mathsf{R}}^{\mathrm{ord}}_{\mathcal{H}}$ are tautologically isomorphic by definition of $\vec{\mathsf{R}}^{\mathrm{ord}}_{\mathcal{H}}$, we see that the morphism from the completion of $\vec{\mathsf{U}}^{\mathrm{ord}}_{\mathcal{H}}$ along its $[(\Phi_{\mathcal{H}}, \delta_{\mathcal{H}}, \sigma)]$-stratum to $\vec{\mathfrak{X}}^{\mathrm{ord}}_{\Phi_{\mathcal{H}},\delta_{\mathcal{H}},\sigma}/\Gamma_{\Phi_{\mathcal{H}},\sigma}$ descends to a morphism $(\vec{\mathsf{M}}^{\mathrm{ord,tor}}_{\mathcal{H}})^\wedge_{\vec{\mathsf{Z}}^{\mathrm{ord}}_{[(\Phi_{\mathcal{H}},\delta_{\mathcal{H}},\sigma)]}} \to \vec{\mathfrak{X}}^{\mathrm{ord}}_{\Phi_{\mathcal{H}},\delta_{\mathcal{H}},\sigma}/\Gamma_{\Phi_{\mathcal{H}},\sigma}$. Then it follows from the constructions that these two canonical morphisms are inverses of each other.

Now, let us prove statement (7). In every step of our construction of $\vec{\mathsf{M}}^{\mathrm{ord,tor}}_{\mathcal{H}}$, the characteristic zero fiber of the boundary charts we have used are the pullback from $\mathsf{S}_0 = \mathrm{Spec}(F_0)$ to $\mathsf{S}_{0,r_{\mathcal{H}}} = \mathrm{Spec}(F_0[\zeta_{p^{r_{\mathcal{H}}}}])$ of the corresponding boundary charts of

$\mathsf{M}^{\mathrm{tor}}_{\mathcal{H},\Sigma}$, and the gluing process for $\vec{\mathsf{M}}^{\mathrm{ord,tor}}_{\mathcal{H}}$ are also compatible with the corresponding gluing process for $\mathsf{M}^{\mathrm{tor}}_{\mathcal{H},\Sigma}$. (See Sections 4.2 and 5.1.) The only difference is that we only consider ordinary cusp labels in the construction for $\vec{\mathsf{M}}^{\mathrm{ord,tor}}_{\mathcal{H}}$, while we need to consider all cusp labels in the construction for $\mathsf{M}^{\mathrm{tor}}_{\mathcal{H},\Sigma}$. Hence, we have a canonical open immersion $\vec{\mathsf{M}}^{\mathrm{ord,tor}}_{\mathcal{H},\Sigma^{\mathrm{ord}}} \underset{\mathbb{Z}}{\otimes} \mathbb{Q} \hookrightarrow \mathsf{M}^{\mathrm{tor}}_{\mathcal{H},\Sigma,r_{\mathcal{H}}}$ respecting their natural stratifications. (Without matching the stratifications explicitly, it still follows from (6) of Theorem 1.3.1.3 and statement (6) of this theorem that $\vec{\mathsf{M}}^{\mathrm{ord,tor}}_{\mathcal{H},\Sigma^{\mathrm{ord}}} \underset{\mathbb{Z}}{\otimes} \mathbb{Q}$ is canonically isomorphic to the open subalgebraic stack of $\mathsf{M}^{\mathrm{tor}}_{\mathcal{H},\Sigma,r_{\mathcal{H}}}$ formed by its strata associated with ordinary cusp labels (with cones), because they enjoy the same universal properties. However, the proofs of these universal properties are also based on the corresponding boundary chart constructions and gluing processes, from which we can directly deduce the full statement we need.)

Finally, suppose that $\mathcal{H}^p$ is neat. Then $\mathcal{H} = \mathcal{H}^p\mathcal{H}_p$ is neat, and $\vec{\mathfrak{X}}^{\mathrm{ord}}_{\Phi_{\mathcal{H}},\delta_{\mathcal{H}},\sigma}$ is a formal algebraic space by [62, Lem. 6.2.5.27] because we have assumed in Definition 5.1.3.1 that each cone decomposition $\Sigma_{\Phi_{\mathcal{H}}}$ in Σ^{ord} satisfies Condition 1.2.2.9. By statements (2) and (5), it follows that points of $\vec{\mathsf{M}}^{\mathrm{ord,tor}}_{\mathcal{H}}$ have no nontrivial automorphisms. Since the diagonal 1-morphism $\Delta_{\vec{\mathsf{M}}^{\mathrm{ord,tor}}_{\mathcal{H}}} : \vec{\mathsf{M}}^{\mathrm{ord,tor}}_{\mathcal{H}} \to \vec{\mathsf{M}}^{\mathrm{ord,tor}}_{\mathcal{H}} \underset{\vec{\mathsf{S}}_{0,r_{\mathcal{H}}}}{\times} \vec{\mathsf{M}}^{\mathrm{ord,tor}}_{\mathcal{H}}$ is finite (by Corollary 5.1.3.9), it must be a closed immersion. Hence, $\vec{\mathsf{M}}^{\mathrm{ord,tor}}_{\mathcal{H}}$ is an algebraic space when $\mathcal{H}^p$ is neat, as desired. □

Remark 5.2.1.5. (Compare with Remarks 1.1.2.1, 1.3.1.4, and 3.4.2.8.) Suppose we have chosen another lattice L' in $L \underset{\mathbb{Z}}{\otimes} \mathbb{Q}$ which nevertheless satisfies $L \underset{\mathbb{Z}}{\otimes} \mathbb{Z}_{(p)} = L' \underset{\mathbb{Z}}{\otimes} \mathbb{Z}_{(p)}$, so that $\vec{\mathsf{M}}^{\mathrm{ord}}_{\mathcal{H}}$ carries the corresponding abelian scheme A' (with additional structures) as in Remark 3.4.2.8, with a $\mathbb{Z}^{\times}_{(p)}$-isogeny $f : A \to A'$. Since $\vec{\mathsf{M}}^{\mathrm{ord,tor}}_{\mathcal{H},\Sigma^{\mathrm{ord}}}$ is noetherian normal, since the tautological semi-abelian scheme $G \to \vec{\mathsf{M}}^{\mathrm{ord,tor}}_{\mathcal{H},\Sigma^{\mathrm{ord}}}$ is ordinary, and since $A = G_{\vec{\mathsf{M}}^{\mathrm{ord}}_{\mathcal{H}}}$, by Lemma 3.1.3.2 and by [92, IX, 1.4], [28, Ch. I, Prop. 2.7], or [62, Prop. 3.3.1.5], f extends to a $\mathbb{Z}^{\times}_{(p)}$-isogeny $f^{\mathrm{ext}} : G \to G'$, and the additional structures λ, i, $\alpha_{\mathcal{H}^p}$, and $\alpha^{\mathrm{ord}}_{\mathcal{H}^p}$ of G naturally induce the additional structures λ', i', $\alpha'_{\mathcal{H}^p}$, and $\alpha^{\mathrm{ord},\prime}_{\mathcal{H}_p}$ of G', which extend those of A'. Hence, the $\mathbb{Z}^{\times}_{(p)}$-isogeny class of G over $\vec{\mathsf{M}}^{\mathrm{ord,tor}}_{\mathcal{H},\Sigma^{\mathrm{ord}}}$ extends that of A over $\vec{\mathsf{M}}^{\mathrm{ord}}_{\mathcal{H}}$, and carries well-defined additional structures. (It can be verified that $(G', \lambda', i', \alpha'_{\mathcal{H}^p}, \alpha^{\mathrm{ord},\prime}_{\mathcal{H}_p}) \to \vec{\mathsf{M}}^{\mathrm{ord,tor}}_{\mathcal{H},\Sigma^{\mathrm{ord}}}$ satisfies the corresponding universal property defined by L' and the corresponding collection of cone decompositions as in (6) of Theorem 5.2.1.1, so that the theory does not really depend on the choice of L within $L \underset{\mathbb{Z}}{\otimes} \mathbb{Z}_{(p)}$. Then we can define a collection $\{\vec{\mathsf{M}}^{\mathrm{ord,tor}}_{\mathcal{H},\Sigma^{\mathrm{ord}}}\}_{\mathcal{H}}$ indexed by $\mathcal{H}$ as in Remark 3.4.2.8 and collections Σ^{ord} for the corresponding $\vec{\mathsf{M}}^{\mathrm{ord}}_{\mathcal{H}}$, carrying a Hecke action as in Proposition 5.2.2.2 below). However, as mentioned in Remark 3.4.2.8, modifying the choice of $L \underset{\mathbb{Z}}{\otimes} \mathbb{Z}_p$ and its filtration $\mathtt{D}$ will make the theory much more complicated.

5.2.2 *Hecke Actions*

Definition 5.2.2.1. (*Compare with [62, Def. 6.4.3.3].*) *Suppose we have an element* $g = (g_0, g_p) \in \mathrm{G}(\mathbb{A}^{\infty,p}) \times \mathrm{P}^{\mathrm{ord}}_{\mathsf{D}}(\mathbb{Q}_p) \subset \mathrm{G}(\mathbb{A}^\infty)$ *(see Definition* 3.2.2.7*), and suppose we have two open compact subgroups* $\mathcal{H}$ *and* $\mathcal{H}'$ *of* $\mathrm{G}(\hat{\mathbb{Z}})$ *such that* $\mathcal{H}' \subset g\mathcal{H}g^{-1}$*, and such that* $\mathcal{H}$ *and* $\mathcal{H}'$ *are of standard form as in Definition* 3.2.2.9*. Suppose moreover that* g_p *satisfies the conditions given in Section* 3.3.4*, so that* $\vec{[g]}^{\mathrm{ord}} : \vec{\mathsf{M}}^{\mathrm{ord}}_{\mathcal{H}'} \to \vec{\mathsf{M}}^{\mathrm{ord}}_{\mathcal{H}}$ *is defined (see Proposition* 3.4.4.1*). Let* $\Sigma^{\mathrm{ord}} = \{\Sigma_{\Phi_{\mathcal{H}}}\}_{[(\Phi_{\mathcal{H}},\delta_{\mathcal{H}})]}$ *and* $\Sigma^{\mathrm{ord},\prime} = \{\Sigma'_{\Phi'_{\mathcal{H}'}}\}_{[(\Phi'_{\mathcal{H}'},\delta'_{\mathcal{H}'})]}$ *be compatible choices of admissible smooth rational polyhedral cone decomposition data for* $\vec{\mathsf{M}}^{\mathrm{ord}}_{\mathcal{H}}$ *and* $\vec{\mathsf{M}}^{\mathrm{ord}}_{\mathcal{H}'}$*, respectively. We say that* $\Sigma^{\mathrm{ord},\prime}$ *is a* g**-refinement** *of* Σ^{ord} *if, for each* g*-assignment* $(f_X, f_Y) : (\Phi_{\mathcal{H}'}, \delta_{\mathcal{H}'}) \to_g (\Phi'_{\mathcal{H}}, \delta'_{\mathcal{H}})$ *of a representative* $(\Phi'_{\mathcal{H}}, \delta'_{\mathcal{H}})$ *of cusp label at level* $\mathcal{H}$ *to a representative* $(\Phi_{\mathcal{H}'}, \delta_{\mathcal{H}'})$ *of cusp label at level* $\mathcal{H}'$ *as in [62, Def. 5.4.3.9], the first cusp label (and hence both of them) being ordinary as in Definition* 3.2.3.8*, the triple* $(\Phi'_{\mathcal{H}'}, \delta'_{\mathcal{H}'}, \Sigma'_{\Phi'_{\mathcal{H}'}})$ *is a* g*-refinement of* $(\Phi_{\mathcal{H}}, \delta_{\mathcal{H}}, \Sigma_{\Phi_{\mathcal{H}}})$ *(under the pair of isomorphisms* (f_X, f_Y)*) as in [62, Def. 6.4.3.2]; namely, the cone decomposition* $\Sigma'_{\Phi'_{\mathcal{H}'}}$ *of* $\mathbf{P}_{\Phi'_{\mathcal{H}'}}$ *is a refinement of the cone decomposition* $\Sigma_{\Phi_{\mathcal{H}}}$ *of* $\mathbf{P}_{\Phi_{\mathcal{H}}}$ *under the identification between* $\mathbf{P}_{\Phi'_{\mathcal{H}'}}$ *and* $\mathbf{P}_{\Phi_{\mathcal{H}}}$ *defined by* (f_X, f_Y)*. We say that* $\Sigma^{\mathrm{ord},\prime}$ *is* g**-induced** *by* Σ^{ord} *if each* $\Sigma'_{\Phi'_{\mathcal{H}'}}$ *above is induced by* $\Sigma_{\Phi_{\mathcal{H}}}$ *under the identification between* $\mathbf{P}_{\Phi'_{\mathcal{H}'}}$ *and* $\mathbf{P}_{\Phi_{\mathcal{H}}}$ *defined by* (f_X, f_Y)*. (This might not be possible because the running assumptions on cone decompositions, such as smoothness, might be incompatibly defined.)*

Proposition 5.2.2.2. (*Compare with Propositions* 1.3.1.15 *and* 3.4.4.1.) *Let* $g = (g_0, g_p)$*,* $\mathcal{H}$*,* $\mathcal{H}'$*,* $\Sigma^{\mathrm{ord}} = \{\Sigma_{\Phi_{\mathcal{H}}}\}_{[(\Phi_{\mathcal{H}},\delta_{\mathcal{H}})]}$*, and* $\Sigma^{\mathrm{ord},\prime} = \{\Sigma'_{\Phi'_{\mathcal{H}'}}\}_{[(\Phi'_{\mathcal{H}'},\delta'_{\mathcal{H}'})]}$ *be as in Definition* 5.2.2.1*, such that* $\Sigma^{\mathrm{ord},\prime}$ *is a* g**-refinement** *of* Σ^{ord}*. Then the ordinary Hecke twist of the family* $(G, \lambda, i, \alpha_{\mathcal{H}'^{,p}}, \alpha^{\mathrm{ord}}_{\mathcal{H}'_p}) \to \vec{\mathsf{M}}^{\mathrm{ord,tor}}_{\mathcal{H}',\Sigma^{\mathrm{ord},\prime}}$ *by* g *(defined by Proposition* 3.3.4.21 *and Lemma* 3.1.3.2*) is the pullback of* $(G, \lambda, i, \alpha_{\mathcal{H}^p}, \alpha^{\mathrm{ord}}_{\mathcal{H}_p}) \to \vec{\mathsf{M}}^{\mathrm{ord,tor}}_{\mathcal{H},\Sigma^{\mathrm{ord}}}$ *via a (unique) surjection*

$$\vec{[g]}^{\mathrm{ord,tor}} : \vec{\mathsf{M}}^{\mathrm{ord,tor}}_{\mathcal{H}',\Sigma^{\mathrm{ord},\prime}} \to \vec{\mathsf{M}}^{\mathrm{ord,tor}}_{\mathcal{H},\Sigma^{\mathrm{ord}}}.$$

The pullback of $\vec{[g]}^{\mathrm{ord,tor}}$ *to* $\vec{\mathsf{M}}^{\mathrm{ord}}_{\mathcal{H}}$ *(on the target) coincides with the surjection*

$$\vec{[g]}^{\mathrm{ord}} : \vec{\mathsf{M}}^{\mathrm{ord}}_{\mathcal{H}'} \to \vec{\mathsf{M}}^{\mathrm{ord}}_{\mathcal{H}}$$

defined in Proposition 3.4.4.1*. The morphism* $\vec{[g]}^{\mathrm{ord,tor}}$ *is quasi-finite flat if* $\Sigma^{\mathrm{ord},\prime}$ *is* g*-induced by* Σ^{ord} *as in Definition* 5.2.2.1*. (Here the flatness follows automatically from the quasi-finiteness, by [35, IV-3, 15.4.2 e′)⇒b)]; cf. [62, Lem. 6.3.1.11].)*

The surjection $\vec{[g]}^{\mathrm{ord,tor}}$ *is proper if the levels* $\mathcal{H}_p$ *and* $\mathcal{H}'_p$ *at* p *are* **equally deep** *as in Definition* 3.2.2.9*, or if* g_p *is of twisted* U_p *type as in Definition* 3.3.6.1 *and* $\mathrm{depth}_{\mathsf{D}}(\mathcal{H}'_p) - \mathrm{depth}_{\mathsf{D}}(g_p) = \mathrm{depth}_{\mathsf{D}}(\mathcal{H}_p) > 0$*. (Note that* $\vec{[g]}^{\mathrm{ord,tor}}$ *is finite if it is both* *quasi-finite and proper, by [35, IV-3, 8.11.1].) If* $g_p \in \mathrm{P}_{\mathsf{D}}^{\mathrm{ord}}(\mathbb{Z}_p)$*, then the induced morphism*

$$\vec{[g]}^{\mathrm{ord,tor}}_{r_{\mathcal{H}'}} : \vec{\mathsf{M}}^{\mathrm{ord,tor}}_{\mathcal{H}',\Sigma^{\mathrm{ord},\prime}} \to \vec{\mathsf{M}}^{\mathrm{ord,tor}}_{\mathcal{H},\Sigma^{\mathrm{ord}}} \underset{\vec{\mathsf{S}}_{0,r_{\mathcal{H}}}}{\times} \vec{\mathsf{S}}_{0,r_{\mathcal{H}'}}$$

is log étale, which is (quasi-finite) étale if $\Sigma^{\mathrm{ord},\prime}$ *is* g*-induced by* Σ^{ord} *as in Definition* 5.2.2.1*. In particular, when* $g = 1$ *and* $\mathcal{H}' = \mathcal{H}$*, we have proper log étale surjections*

$$\vec{[1]}^{\mathrm{ord,tor}} : \vec{\mathsf{M}}^{\mathrm{ord,tor}}_{\mathcal{H},\Sigma^{\mathrm{ord},\prime}} \to \vec{\mathsf{M}}^{\mathrm{ord,tor}}_{\mathcal{H},\Sigma^{\mathrm{ord}}}$$

when $\Sigma^{\mathrm{ord},\prime}$ *is a refinement of* Σ^{ord}*.*

Moreover, the surjection $\vec{[g]}^{\mathrm{ord,tor}}$ *maps the* $[(\Phi'_{\mathcal{H}'}, \delta'_{\mathcal{H}'}, \sigma')]$*-stratum* $\vec{\mathsf{Z}}^{\mathrm{ord}}_{[(\Phi'_{\mathcal{H}'}, \delta'_{\mathcal{H}'}, \sigma')]}$ *of* $\vec{\mathsf{M}}^{\mathrm{ord,tor}}_{\mathcal{H}',\Sigma^{\mathrm{ord},\prime}}$ *to the* $[(\Phi_{\mathcal{H}}, \delta_{\mathcal{H}}, \sigma)]$*-stratum* $\vec{\mathsf{Z}}^{\mathrm{ord}}_{[(\Phi_{\mathcal{H}}, \delta_{\mathcal{H}}, \sigma)]}$ *of* $\vec{\mathsf{M}}^{\mathrm{ord,tor}}_{\mathcal{H},\Sigma^{\mathrm{ord}}}$ *if and only if there are representatives* $(\Phi_{\mathcal{H}}, \delta_{\mathcal{H}}, \sigma)$ *and* $(\Phi'_{\mathcal{H}'}, \delta'_{\mathcal{H}'}, \sigma')$ *of* $[(\Phi_{\mathcal{H}}, \delta_{\mathcal{H}}, \sigma)]$ *and* $[(\Phi'_{\mathcal{H}'}, \delta'_{\mathcal{H}'}, \sigma')]$*, respectively, such that* $(\Phi'_{\mathcal{H}'}, \delta'_{\mathcal{H}'}, \sigma')$ *is a* g*-refinement of* $(\Phi_{\mathcal{H}}, \delta_{\mathcal{H}}, \sigma)$ *as in [62, Def. 6.4.3.1].*

If $g = g_1 g_2$*, where* $g_1 = (g_{1,0}, g_{1,p})$ *and* $g_2 = (g_{2,0}, g_{2,p})$ *are elements of* $\mathrm{G}(\mathbb{A}^{\infty,p}) \times \mathrm{P}_{\mathsf{D}}^{\mathrm{ord}}(\mathbb{Q}_p)$*, each having a setup similar to that of* g*, then we have* $\vec{[g]}^{\mathrm{ord,tor}} = \vec{[g_2]}^{\mathrm{ord,tor}} \circ \vec{[g_1]}^{\mathrm{ord,tor}}$*, extending the identity* $\vec{[g]}^{\mathrm{ord}} = \vec{[g_2]}^{\mathrm{ord}} \circ \vec{[g_1]}^{\mathrm{ord}}$ *in Proposition* 3.4.4.1*.*

Finally, there exists Σ *and* Σ' *extending* Σ^{ord} *and* $\Sigma^{\mathrm{ord},\prime}$*, respectively, as in Proposition* 5.1.3.2*, such that* Σ' *is a* g*-refinement of* Σ*, and so that we have the canonical surjection* $[g]^{\mathrm{tor}} : \mathsf{M}^{\mathrm{tor}}_{\mathcal{H}',\Sigma'} \to \mathsf{M}^{\mathrm{tor}}_{\mathcal{H},\Sigma}$ *as in Proposition* 1.3.1.15*. Let* $[g]^{\mathrm{tor}}_{r_{\mathcal{H}'},r_{\mathcal{H}}} : \mathsf{M}^{\mathrm{tor}}_{\mathcal{H}',\Sigma',r_{\mathcal{H}'}} \to \mathsf{M}^{\mathrm{tor}}_{\mathcal{H},\Sigma,r_{\mathcal{H}}}$ *denote the canonically induced morphism. Then* $\vec{[g]}^{\mathrm{tor}} \underset{\mathbb{Z}}{\otimes} \mathbb{Q}$ *can be identified with the pullback of* $[g]^{\mathrm{tor}}_{r_{\mathcal{H}'},r_{\mathcal{H}}}$ *to* $\vec{\mathsf{M}}^{\mathrm{ord,tor}}_{\mathcal{H},\Sigma^{\mathrm{ord}}} \underset{\mathbb{Z}}{\otimes} \mathbb{Q}$ *(on the target) under* (5.2.1.2) *in* (7) *of Theorem* 5.2.1.1*. In particular,* $\vec{[g]}^{\mathrm{tor}} \underset{\mathbb{Z}}{\otimes} \mathbb{Q}$ *is proper log étale.*

Proof. The existence of $\vec{[g]}^{\mathrm{ord,tor}}$ follows from a combination of [62, Prop. 5.4.3.8] and (6) of Theorem 5.2.1.1:

Since $\vec{\mathsf{M}}^{\mathrm{ord,tor}}_{\mathcal{H}',\Sigma^{\mathrm{ord},\prime}}$ is noetherian normal, and since the extensibility condition (5) in Definition 3.4.2.10 implies that G is an ordinary semi-abelian scheme as in Definition 3.1.1.2, by Lemma 3.1.3.2, any isogenous quotient of $G_{\vec{\mathsf{M}}^{\mathrm{ord}}_{\mathcal{H}'}}$ extends uniquely (up to isomorphism) to an isogenous quotient of G. Hence, the usual ordinary Hecke twist defined by Proposition 3.3.4.21 over $\vec{\mathsf{M}}^{\mathrm{ord}}_{\mathcal{H}'}$ extends to the *ordinary Hecke twist* $(G', \lambda', i', \alpha'_{\mathcal{H}^p}, \alpha^{\mathrm{ord},\prime}_{\mathcal{H}_p}) \to \vec{\mathsf{M}}^{\mathrm{ord,tor}}_{\mathcal{H}',\Sigma^{\mathrm{ord},\prime}}$ of $(G, \lambda, i, \alpha_{\mathcal{H}',p}, \alpha^{\mathrm{ord}}_{\mathcal{H}'_p}) \to \vec{\mathsf{M}}^{\mathrm{ord,tor}}_{\mathcal{H}',\Sigma^{\mathrm{ord},\prime}}$ by $g = (g_0, g_p)$.

By construction, the restriction of $(G', \lambda', i', \alpha'_{\mathcal{H}^p}, \alpha^{\mathrm{ord},\prime}_{\mathcal{H}_p}) \to \mathsf{M}^{\mathrm{tor}}_{\mathcal{H}',\Sigma^{\mathrm{ord},\prime}}$ to $\vec{\mathsf{M}}^{\mathrm{ord}}_{\mathcal{H}'}$ determines the canonical surjection $\vec{[g]}^{\mathrm{ord}} : \vec{\mathsf{M}}^{\mathrm{ord}}_{\mathcal{H}'} \twoheadrightarrow \vec{\mathsf{M}}^{\mathrm{ord}}_{\mathcal{H}}$ defined in Proposition 3.4.4.1. By [62, Prop. 5.4.3.8] and (6) of Theorem 5.2.1.1, the restriction of $(G', \lambda', i', \alpha'_{\mathcal{H}^p}, \alpha^{\mathrm{ord},\prime}_{\mathcal{H}_p}) \to \vec{\mathsf{M}}^{\mathrm{ord,tor}}_{\mathcal{H}',\Sigma^{\mathrm{ord},\prime}}$ to étale local charts of $\vec{\mathsf{M}}^{\mathrm{ord,tor}}_{\mathcal{H}',\Sigma^{\mathrm{ord},\prime}}$ admit unique morphisms to $(G,\lambda,i,\alpha_{\mathcal{H}^p},\alpha^{\mathrm{ord}}_{\mathcal{H}_p}) \to \vec{\mathsf{M}}^{\mathrm{ord,tor}}_{\mathcal{H},\Sigma^{\mathrm{ord}}}$, by our assumption that $\Sigma^{\mathrm{ord},\prime}$ is a g-refinement of Σ^{ord}. (More precisely, the cones containing pairings of the form $\upsilon \circ B' : Y' \times X' \to \mathbb{Z}$ are carried to cones containing pairings of the form $\upsilon \circ B : Y \times X \to \mathbb{Z}$ under the identification between $\mathbf{P}_{\Phi'_{\mathcal{H}'}}$ and $\mathbf{P}_{\Phi_{\mathcal{H}}}$ defined by $(f_X : X' \underset{\mathbb{Z}}{\otimes} \mathbb{Z}_{(p)} \overset{\sim}{\to} X \underset{\mathbb{Z}}{\otimes} \mathbb{Z}_{(p)}, f_Y : Y \underset{\mathbb{Z}}{\otimes} \mathbb{Z}_{(p)} \overset{\sim}{\to} Y' \underset{\mathbb{Z}}{\otimes} \mathbb{Z}_{(p)})$, when we have the objects as in the context of Definition 5.2.2.1.) These morphisms patch uniquely, and hence descend to $\vec{\mathsf{M}}^{\mathrm{ord,tor}}_{\mathcal{H}',\Sigma^{\mathrm{ord},\prime}}$. Therefore, there exists a unique morphism $\vec{[g]}^{\mathrm{ord,tor}}$: $\vec{\mathsf{M}}^{\mathrm{ord,tor}}_{\mathcal{H}',\Sigma^{\mathrm{ord},\prime}} \to \vec{\mathsf{M}}^{\mathrm{ord,tor}}_{\mathcal{H},\Sigma^{\mathrm{ord}}}$ extending $\vec{[g]}^{\mathrm{ord}}$, which pulls $(G,\lambda,i,\alpha_{\mathcal{H}^p},\alpha^{\mathrm{ord}}_{\mathcal{H}_p}) \to \vec{\mathsf{M}}^{\mathrm{ord,tor}}_{\mathcal{H},\Sigma^{\mathrm{ord}}}$ back to $(G',\lambda',i',\alpha'_{\mathcal{H}^p},\alpha^{\mathrm{ord},\prime}_{\mathcal{H}_p}) \to \vec{\mathsf{M}}^{\mathrm{ord,tor}}_{\mathcal{H}',\Sigma^{\mathrm{ord},\prime}}$. Since $\vec{\mathsf{M}}^{\mathrm{ord}}_{\mathcal{H}}$ is dense in $\vec{\mathsf{M}}^{\mathrm{ord,tor}}_{\mathcal{H},\Sigma^{\mathrm{ord}}}$, and since the condition in (6) of Theorem 5.2.1.1 does not involve level structures, starting with a degeneration over a complete discrete valuation ring V centered at an arbitrary geometric point $\bar{s}$ of $\vec{\mathsf{M}}^{\mathrm{ord,tor}}_{\mathcal{H},\Sigma^{\mathrm{ord}}}$, we can construct degenerations over a complete discrete valuation ring V' finite flat over V centered at a geometric point of $\vec{\mathsf{M}}^{\mathrm{ord,tor}}_{\mathcal{H}',\Sigma^{\mathrm{ord},\prime}}$ as soon as we can lift the level structures on the generic points. Therefore, since $\vec{[g]}^{\mathrm{ord}}$ is surjective (by Proposition 3.4.4.1—in fact, this surjectivity is essentially a consequence of the liftability conditions in the definitions of level structures), the morphism $\vec{[g]}^{\mathrm{ord,tor}}$ is also surjective.

Moreover, if the levels $\mathcal{H}_p$ and $\mathcal{H}'_p$ at p are *equally deep*, in which case $\vec{[g]}^{\mathrm{ord}}$ is finite by Proposition 3.4.4.1, then the (separated) morphism $\vec{[g]}^{\mathrm{ord,tor}}$ is proper, because in this case the above argument also verifies the valuative criterion. On the other hand, if g_p is of twisted U_p type as in Definition 3.3.6.1 and $\mathrm{depth}_{\mathsf{D}}(\mathcal{H}'_p) - \mathrm{depth}_{\mathsf{D}}(g_p) = \mathrm{depth}_{\mathsf{D}}(\mathcal{H}_p) > 0$, then the desired valuative criterion follows from Lemma 3.3.6.6.

If $g = g_1 g_2$ as in the last statement of this proposition, then we can also construct the ordinary Hecke twist $(G',\lambda',i',\alpha'_{\mathcal{H}^p},\alpha^{\mathrm{ord},\prime}_{\mathcal{H}_p})$ of $(G,\lambda,i,\alpha_{\mathcal{H}',p},\alpha^{\mathrm{ord}}_{\mathcal{H}'_p})$ in two steps, as in the last statement of Proposition 3.3.4.21. Hence, the induced morphisms between partial toroidal compactifications satisfy the desired identity $\vec{[g]}^{\mathrm{ord,tor}} = \vec{[g_2]}^{\mathrm{ord,tor}} \circ \vec{[g_1]}^{\mathrm{ord,tor}}$, extending the identity $\vec{[g]}^{\mathrm{ord}} = \vec{[g_2]}^{\mathrm{ord}} \circ \vec{[g_1]}^{\mathrm{ord}}$ in Proposition 3.4.4.1.

The assertions on the local structures can be verified by comparing the étale local structures, which then follows from the gluing construction of the partial toroidal compactifications (*cf.* Proposition 5.1.3.7).

As for the last paragraph, the existence of Σ and Σ' follows from Proposition 5.1.3.2. Since $[g]^{\mathrm{tor}}$ is constructed in [62, Prop. 6.4.3.4] using Hecke twists of the tautological object, and since the tautological objects over $\mathsf{M}^{\mathrm{tor}}_{\mathcal{H},\Sigma,r_{\mathcal{H}}}$ and $\mathsf{M}^{\mathrm{tor}}_{\mathcal{H}',\Sigma',r_{\mathcal{H}'}}$ induce the tautological objects over $\vec{\mathsf{M}}^{\mathrm{ord,tor}}_{\mathcal{H},\Sigma^{\mathrm{ord}}} \underset{\mathbb{Z}}{\otimes} \mathbb{Q}$ and $\vec{\mathsf{M}}^{\mathrm{ord,tor}}_{\mathcal{H}',\Sigma^{\mathrm{ord},\prime}} \underset{\mathbb{Z}}{\otimes} \mathbb{Q}$, respectively, we see that $\vec{[g]}^{\mathrm{tor}} \underset{\mathbb{Z}}{\otimes} \mathbb{Q}$ can be identified with the restriction of $[g]^{\mathrm{tor}}_{r_{\mathcal{H}'},r_{\mathcal{H}}}$ to $\vec{\mathsf{M}}^{\mathrm{ord,tor}}_{\mathcal{H}',\Sigma^{\mathrm{ord},\prime}} \underset{\mathbb{Z}}{\otimes} \mathbb{Q}$.

This coincides with the pullback of $[g]^{\mathrm{tor}}_{r_{\mathcal{H}'},r_{\mathcal{H}}}$ to $\vec{\mathsf{M}}^{\mathrm{ord,tor}}_{\mathcal{H},\Sigma^{\mathrm{ord}}} \underset{\mathbb{Z}}{\otimes} \mathbb{Q}$ (on the target) under (5.2.1.2) because $\vec{\mathsf{M}}^{\mathrm{ord,tor}}_{\mathcal{H}',\Sigma^{\mathrm{ord},\prime}} \underset{\mathbb{Z}}{\otimes} \mathbb{Q}$ is the preimage of $\vec{\mathsf{M}}^{\mathrm{ord,tor}}_{\mathcal{H},\Sigma^{\mathrm{ord}}} \underset{\mathbb{Z}}{\otimes} \mathbb{Q}$, by the statements concerning the strata in (7) of Theorem 5.2.1.1 and in Proposition 1.3.1.15.

The remaining assertions of the proposition are self-explanatory. □

Corollary 5.2.2.3. (*Compare with Corollary* 3.4.4.3.) *With the setting as in Proposition* 5.2.2.2, *the morphism*

$$\vec{[g]}^{\mathrm{ord,tor}} : \vec{\mathfrak{M}}^{\mathrm{ord,tor}}_{\mathcal{H}',\Sigma^{\mathrm{ord},\prime}} \to \vec{\mathfrak{M}}^{\mathrm{ord,tor}}_{\mathcal{H},\Sigma^{\mathrm{ord}}}$$

(*cf. Definition* 3.4.4.2) *induced by* $\vec{[g]}^{\mathrm{ord,tor}} : \vec{\mathsf{M}}^{\mathrm{ord,tor}}_{\mathcal{H}',\Sigma^{\mathrm{ord},\prime}} \to \vec{\mathsf{M}}^{\mathrm{ord,tor}}_{\mathcal{H},\Sigma^{\mathrm{ord}}}$ *is proper, which is finite flat if* $\Sigma^{\mathrm{ord},\prime}$ *is g-induced by* Σ^{ord} *as in Definition* 5.2.2.1. *If* $g_p \in \mathrm{P}^{\mathrm{ord}}_{\mathsf{D}}(\mathbb{Z}_p)$, *then the induced morphism*

$$\vec{[g]}^{\mathrm{ord,tor}}_{r_{\mathcal{H}'}} : \vec{\mathfrak{M}}^{\mathrm{ord,tor}}_{\mathcal{H}',\Sigma^{\mathrm{ord},\prime}} \to \vec{\mathfrak{M}}^{\mathrm{ord,tor}}_{\mathcal{H},\Sigma^{\mathrm{ord}}} \underset{\vec{\mathfrak{S}}_{0,r_{\mathcal{H}}}}{\times} \vec{\mathfrak{S}}_{0,r_{\mathcal{H}'}}$$

is proper log étale (*because it is log étale by Proposition* 5.2.2.2). *If moreover* $\Sigma^{\mathrm{ord},\prime}$ *is g-induced by* Σ^{ord} *as in Definition* 5.2.2.1, *then* $\vec{[g]}^{\mathrm{ord,tor}}_{r_{\mathcal{H}'}} : \vec{\mathfrak{M}}^{\mathrm{ord,tor}}_{\mathcal{H}',\Sigma^{\mathrm{ord},\prime}} \to \vec{\mathfrak{M}}^{\mathrm{ord,tor}}_{\mathcal{H},\Sigma^{\mathrm{ord}}} \underset{\vec{\mathfrak{S}}_{0,r_{\mathcal{H}}}}{\times} \vec{\mathfrak{S}}_{0,r_{\mathcal{H}'}}$ *is finite étale* (*because it is quasi-finite étale by Proposition* 5.2.2.2; *cf. [35, IV-3, 8.11.5, or IV-4, 8.12.6]*).

Proof. These follow from the fact that the induced morphism

$$\vec{[g]}^{\mathrm{ord,tor}} : \vec{\mathsf{M}}^{\mathrm{ord,tor}}_{\mathcal{H}',\Sigma^{\mathrm{ord},\prime}} \underset{\mathbb{Z}}{\otimes} \mathbb{F}_p \to \vec{\mathsf{M}}^{\mathrm{ord,tor}}_{\mathcal{H},\Sigma^{\mathrm{ord}}} \underset{\mathbb{Z}}{\otimes} \mathbb{F}_p$$

over $\mathrm{Spec}(\mathbb{F}_p)$ is proper by Lemma 3.3.6.8. □

Corollary 5.2.2.4. (*Compare with Corollary* 3.4.4.4 *and Example* 3.4.4.5.) *With the setting as in Proposition* 5.2.2.2, *if* $g = (g_0, g_p) \in \mathrm{G}(\hat{\mathbb{Z}}^p) \times \mathrm{P}^{\mathrm{ord}}_{\mathsf{D}}(\mathbb{Z}_p)$, *if* $\mathcal{H}'^{,p} = g_0\mathcal{H}^p g_0^{-1}$ *in* $\mathrm{G}(\hat{\mathbb{Z}}^p)$, *if* $\mathcal{H}'^{\mathrm{ord}}_p = (g_p\mathcal{H}_p g_p^{-1})^{\mathrm{ord}}$ *in* $\mathrm{M}^{\mathrm{ord}}_{\mathsf{D}}(\mathbb{Z}_p)$ (*see* (3.3.3.5)), *and if* $\Sigma^{\mathrm{ord},\prime}$ *is g-induced by* Σ^{ord} *as in Definition* 5.2.2.1, *then* ($r_{\mathcal{H}'} = r_{\mathcal{H}}$ *and*) *the induced morphism* $\vec{[g]}^{\mathrm{ord,tor}} : \vec{\mathfrak{M}}^{\mathrm{ord,tor}}_{\mathcal{H}',\Sigma^{\mathrm{ord},\prime}} \to \vec{\mathfrak{M}}^{\mathrm{ord,tor}}_{\mathcal{H},\Sigma^{\mathrm{ord}}}$ *is an isomorphism.* (*See the remark at the end of Corollary* 3.4.4.4.)

Proof. By Corollary 5.2.2.3, the induced morphism $\vec{[g]}^{\mathrm{ord,tor}} : \vec{\mathfrak{M}}^{\mathrm{ord,tor}}_{\mathcal{H}',\Sigma^{\mathrm{ord},\prime}} \to \vec{\mathfrak{M}}^{\mathrm{ord,tor}}_{\mathcal{H},\Sigma^{\mathrm{ord}}}$ is finite étale. By Corollary 3.4.4.4, the restriction of $\vec{[g]}^{\mathrm{ord,tor}}$ to $\vec{\mathfrak{M}}^{\mathrm{ord}}_{\mathcal{H}'}$ is an isomorphism. Since $\vec{\mathsf{M}}^{\mathrm{ord,tor}}_{\mathcal{H}',\Sigma^{\mathrm{ord},\prime}} \underset{\mathbb{Z}}{\otimes} \mathbb{F}_p$ and $\vec{\mathsf{M}}^{\mathrm{ord,tor}}_{\mathcal{H},\Sigma^{\mathrm{ord}}} \underset{\mathbb{Z}}{\otimes} \mathbb{F}_p$ are regular (by smoothness of $\vec{\mathsf{M}}^{\mathrm{ord,tor}}_{\mathcal{H}',\Sigma^{\mathrm{ord},\prime}}$ and $\vec{\mathsf{M}}^{\mathrm{ord,tor}}_{\mathcal{H},\Sigma^{\mathrm{ord}}}$ over $\vec{\mathsf{S}}_{0,r_{\mathcal{H}'}}$ and $\vec{\mathsf{S}}_{0,r_{\mathcal{H}}}$, respectively; see Theorem 5.2.1.1), by Zariski's main theorem (see [35, III-1, 4.4.3, 4.4.11]), the induced finite morphism $\vec{[g]}^{\mathrm{ord,tor}} : \vec{\mathsf{M}}^{\mathrm{ord,tor}}_{\mathcal{H}',\Sigma^{\mathrm{ord},\prime}} \underset{\mathbb{Z}}{\otimes} \mathbb{F}_p \to \vec{\mathsf{M}}^{\mathrm{ord,tor}}_{\mathcal{H},\Sigma^{\mathrm{ord}}} \underset{\mathbb{Z}}{\otimes} \mathbb{F}_p$ is necessarily an isomorphism. Hence, $\vec{[g]}^{\mathrm{ord,tor}} : \vec{\mathfrak{M}}^{\mathrm{ord,tor}}_{\mathcal{H}',\Sigma^{\mathrm{ord},\prime}} \to \vec{\mathfrak{M}}^{\mathrm{ord,tor}}_{\mathcal{H},\Sigma^{\mathrm{ord}}}$ (being finite étale and an isomorphism between the fibers over $\mathrm{Spec}(\mathbb{F}_p)$) is an isomorphism. □

Corollary 5.2.2.5 (elements of U_p type). (*Compare with Corollary* 3.4.4.6.) *Suppose in Proposition* 5.2.2.2 *that* $g_0 = 1$ *and* g_p *is* **of** U_p **type** *as in Definition* 3.3.6.1 (*so that it is of twisted* U_p *type and* $\mathrm{depth}_{\mathsf{D}}(g_p) = 1$). *Then* $\Sigma^{\mathrm{ord},\prime}$ *is also a* 1*-refinement of* Σ^{ord}, *and the induced morphism*

$$\vec{[g]}^{\mathrm{ord,tor}} : \vec{\mathsf{M}}^{\mathrm{ord,tor}}_{\mathcal{H}',\Sigma^{\mathrm{ord},\prime}} \underset{\mathbb{Z}}{\otimes} \mathbb{F}_p \to \vec{\mathsf{M}}^{\mathrm{ord,tor}}_{\mathcal{H},\Sigma^{\mathrm{ord}}} \underset{\mathbb{Z}}{\otimes} \mathbb{F}_p \tag{5.2.2.6}$$

is proper and coincides with the composition of the absolute Frobenius morphism

$$\mathrm{F}_{\vec{\mathsf{M}}^{\mathrm{ord,tor}}_{\mathcal{H}',\Sigma^{\mathrm{ord},\prime}} \underset{\mathbb{Z}}{\otimes} \mathbb{F}_p} : \vec{\mathsf{M}}^{\mathrm{ord,tor}}_{\mathcal{H}',\Sigma^{\mathrm{ord},\prime}} \underset{\mathbb{Z}}{\otimes} \mathbb{F}_p \to \vec{\mathsf{M}}^{\mathrm{ord,tor}}_{\mathcal{H}',\Sigma^{\mathrm{ord},\prime}} \underset{\mathbb{Z}}{\otimes} \mathbb{F}_p$$

with the canonical proper morphism

$$\vec{[1]}^{\mathrm{ord,tor}} : \vec{\mathsf{M}}^{\mathrm{ord,tor}}_{\mathcal{H}',\Sigma^{\mathrm{ord},\prime}} \underset{\mathbb{Z}}{\otimes} \mathbb{F}_p \to \vec{\mathsf{M}}^{\mathrm{ord,tor}}_{\mathcal{H},\Sigma^{\mathrm{ord}}} \underset{\mathbb{Z}}{\otimes} \mathbb{F}_p. \tag{5.2.2.7}$$

Suppose moreover that $\Sigma^{\mathrm{ord},\prime}$ *is* g*-induced by* Σ^{ord} *as in Definition* 5.2.2.1. *Then* $\Sigma^{\mathrm{ord},\prime}$ *is also* 1*-induced by* Σ^{ord}, *and the above morphisms* (5.2.2.6) *and* (5.2.2.7) *are both finite flat.*

If $\mathcal{H}'^{\,\mathrm{ord}}_p = \mathcal{H}^{\mathrm{ord}}_p$ *as open compact subgroups of* $\mathrm{M}^{\mathrm{ord}}_{\mathsf{D}}(\mathbb{Z}_p)$ (*see* (3.3.3.5)), *then we can take* $\Sigma^{\mathrm{ord},\prime}$ *to be* g*-induced by* Σ^{ord} *as in Definition* 5.2.2.1, *so that* ($r_{\mathcal{H}'} = r_{\mathcal{H}}$ *and*) *the canonical morphism* (5.2.2.7) *is an isomorphism by Corollary* 5.2.2.4, *and so that the composition*

$$\vec{\mathsf{M}}^{\mathrm{ord,tor}}_{\mathcal{H},\Sigma^{\mathrm{ord}}} \underset{\mathbb{Z}}{\otimes} \mathbb{F}_p \overset{(\vec{[1]}^{\mathrm{ord,tor}})^{-1}}{\underset{\sim}{\to}} \vec{\mathsf{M}}^{\mathrm{ord,tor}}_{\mathcal{H}',\Sigma^{\mathrm{ord},\prime}} \underset{\mathbb{Z}}{\otimes} \mathbb{F}_p \overset{\vec{[g]}^{\mathrm{ord,tor}}}{\to} \vec{\mathsf{M}}^{\mathrm{ord,tor}}_{\mathcal{H},\Sigma^{\mathrm{ord}}} \underset{\mathbb{Z}}{\otimes} \mathbb{F}_p$$

coincides with the (*finite flat*) *absolute Frobenius morphism*

$$\mathrm{F}_{\vec{\mathsf{M}}^{\mathrm{ord,tor}}_{\mathcal{H},\Sigma^{\mathrm{ord}}} \underset{\mathbb{Z}}{\otimes} \mathbb{F}_p} : \vec{\mathsf{M}}^{\mathrm{ord,tor}}_{\mathcal{H},\Sigma^{\mathrm{ord}}} \underset{\mathbb{Z}}{\otimes} \mathbb{F}_p \to \vec{\mathsf{M}}^{\mathrm{ord,tor}}_{\mathcal{H},\Sigma^{\mathrm{ord}}} \underset{\mathbb{Z}}{\otimes} \mathbb{F}_p.$$

Proof. First note that g_p acts by scalars on $\mathrm{Gr}^0_{\mathsf{D}_{\mathbb{Q}_p}}$ and $\mathrm{Gr}^{-1}_{\mathsf{D}_{\mathbb{Q}_p}}$, and hence $g = (1, g_p)$ preserves any filtration $\mathsf{Z} \underset{\hat{\mathbb{Z}}}{\otimes} \mathbb{A}^\infty$ of $L \underset{\hat{\mathbb{Z}}}{\otimes} \mathbb{A}^\infty$ satisfying $\mathsf{Z}_{-2} \underset{\hat{\mathbb{Z}}}{\otimes} \mathbb{Q}_p \subset \mathsf{D}_{\mathbb{Q}_p} \subset \mathsf{Z}_0 \underset{\hat{\mathbb{Z}}}{\otimes} \mathbb{Q}_p$. Then any identification between $\mathbf{P}_{\Phi'_{\mathcal{H}'}}$ and $\mathbf{P}_{\Phi_{\mathcal{H}}}$ as in Definition 5.2.2.1 is just some (positive) scalar multiplication, and hence being a g-refinement and a 1-refinement (or, being g-induced and 1-induced) are exactly the same notion. This explains all relevant statements in this corollary. Since the ordinary Hecke twist in Proposition 5.2.2.2 is realized as the relative Frobenius morphism (with naturally induced additional structures) over the dense subscheme $\vec{\mathsf{M}}^{\mathrm{ord}}_{\mathcal{H}'} \underset{\mathbb{Z}}{\otimes} \mathbb{F}_p$ as in Corollary 3.4.4.6 (or rather its proof), it must be so over the whole $\vec{\mathsf{M}}^{\mathrm{ord,tor}}_{\mathcal{H}',\Sigma^{\mathrm{ord},\prime}} \underset{\mathbb{Z}}{\otimes} \mathbb{F}_p$. Hence, the first paragraph of the corollary follows. The second paragraph of the corollary follows from the first paragraph and from Corollary 5.2.2.3. The third paragraph of the corollary follows from the second paragraph and from Corollary 5.2.2.4. □

Remark 5.2.2.8. (Compare with Remark 3.4.4.9.) By Kunz's theorem [54] (*cf.* [76, Sec. 42, Thm. 107]), the absolute Frobenius morphisms $F_{\vec{\mathsf{M}}^{\mathrm{ord,tor}}_{\mathcal{H}',\Sigma^{\mathrm{ord},\prime}} \underset{\mathbb{Z}}{\otimes} \mathbb{F}_p}$ and $F_{\vec{\mathsf{M}}^{\mathrm{ord,tor}}_{\mathcal{H},\Sigma^{\mathrm{ord}}} \underset{\mathbb{Z}}{\otimes} \mathbb{F}_p}$ in Corollary 5.2.2.5 are flat because $\vec{\mathsf{M}}^{\mathrm{ord,tor}}_{\mathcal{H}',\Sigma^{\mathrm{ord},\prime}} \underset{\mathbb{Z}}{\otimes} \mathbb{F}_p$ and $\vec{\mathsf{M}}^{\mathrm{ord,tor}}_{\mathcal{H},\Sigma^{\mathrm{ord}}} \underset{\mathbb{Z}}{\otimes} \mathbb{F}_p$ are regular (by smoothness of $\vec{\mathsf{M}}^{\mathrm{ord,tor}}_{\mathcal{H}',\Sigma^{\mathrm{ord},\prime}}$ and $\vec{\mathsf{M}}^{\mathrm{ord,tor}}_{\mathcal{H},\Sigma^{\mathrm{ord}}}$ over $\vec{\mathsf{S}}_{0,r_{\mathcal{H}'}}$ and $\vec{\mathsf{S}}_{0,r_{\mathcal{H}}}$, respectively; see Theorem 5.2.1.1).

5.2.3 *The Case When p is a Good Prime*

As in Section 3.4.5, suppose that p is a good prime (see Definition 1.1.1.6). Then we can also construct $\vec{\mathsf{M}}^{\mathrm{ord,tor}}_{\mathcal{H},\Sigma^{\mathrm{ord}}}$ using the toroidal compactifications already constructed in [62, Thm. 6.4.1.1] in mixed characteristics, *provided that* $\Sigma^{\mathrm{ord}} = \{\Sigma_{\Phi_{\mathcal{H}}}\}_{[(\Phi_{\mathcal{H}},\delta_{\mathcal{H}})]}$ is induced (in a natural sense similar to that in Definition 2.1.2.25) by a compatible choice Σ' of admissible smooth rational polyhedral cone decomposition data for $\mathsf{M}_{\mathcal{H}'}$ as in Definition 1.2.2.13, where $\mathcal{H}' = \mathcal{H}^p\mathrm{G}(\mathbb{Z}_p)$, as in [62, Constr. 7.3.1.6]. If Σ' is projective with a collection pol' of polarization functions as in Definitions 1.2.2.7 and 1.2.2.14, then Σ^{ord} is also projective with an induced collection $\mathsf{pol}^{\mathrm{ord}}$ of polarization functions as in Definition 5.1.3.3.

However, since we need Σ^{ord} to be smooth, we shall make the following:

Assumption 5.2.3.1 (for this subsection). *The smoothness conditions defined by $\mathcal{H}$ on Σ^{ord} and by $\mathcal{H}'$ on Σ' are compatible with each other.*

Lemma 5.2.3.2. *Assumption* 5.2.3.1 *is satisfied, for example, when the group $\mathcal{H}_p$ is of the form $\mathcal{U}_{p,0}(p^r)$, $\mathcal{U}_{p,1}(p^r)$, $\mathcal{U}^{\mathrm{bal}}_{p,1}(p^r)$, $\mathcal{U}_{p,1,0}(p^{r_1},p^{r_0})$, or $\mathcal{U}^{\mathrm{bal}}_{p,1,0}(p^{r_1},p^{r_0})$, for some integers $0 \leq r$ and $0 \leq r_1 \leq r_0$ as in Definition* 3.2.2.8.

Proof. In all such cases, we can canonically identify $\mathbf{S}_{\Phi_{\mathcal{H}}}$ with $\frac{1}{N}\mathbf{S}_{\Phi_{\mathcal{H}'}}$ in $\mathbf{S}_{\Phi_{\mathcal{H}'}} \underset{\mathbb{Z}}{\otimes} \mathbb{Q}$ for some integer $N \geq 1$, where $\Phi_{\mathcal{H}}$ and $\Phi_{\mathcal{H}'}$ are torus arguments for admissible filtrations compatible with the filtration D as in Definition 3.2.3.1. □

Then it is indeed possible that Σ^{ord} is induced by a smooth Σ' for $\mathsf{M}_{\mathcal{H}'}$. By construction, we also obtain an extension of the compatible choice Σ^{ord} (resp. $\mathsf{pol}^{\mathrm{ord}}$, if defined) for $\vec{\mathsf{M}}^{\mathrm{ord}}_{\mathcal{H}}$ to a compatible choice Σ (resp. pol) for $\mathsf{M}_{\mathcal{H}}$, although we cannot assert that Σ is smooth in this case. (Thus, we will need the constructions in Propositions 1.3.1.11 and 2.2.2.3.) We also have an induced smooth $\Sigma^{\mathrm{ord},\prime}$, together with $\mathsf{pol}^{\mathrm{ord},\prime}$ if pol' is also defined. Suppose that there is a smooth Σ^p (resp. pol^p) for $\mathsf{M}_{\mathcal{H}^p}$ such that Σ' (resp. pol', if defined) is induced by Σ^p (resp. pol^p) in the natural sense (similar to that in Definition 2.1.2.25). (These assumptions on Σ' and $\Sigma^{\mathrm{ord},\prime}$ can be met by refining any given cone decomposition Σ^{ord} for $\vec{\mathsf{M}}^{\mathrm{ord}}_{\mathcal{H}}$.) We shall assume that such choices have been made in the remainder of this subsection.

Proposition 5.2.3.3. (*Compare with Proposition* 3.4.5.7.) *With assumptions as above, let $\mathsf{M}_{\mathcal{H}^p}$, $\mathsf{M}^{\mathrm{tor}}_{\mathcal{H}^p,\Sigma^p}$, and $\mathsf{M}^{\mathrm{min}}_{\mathcal{H}^p}$ be constructed over $\vec{\mathsf{S}}_0 = \mathrm{Spec}(\mathcal{O}_{F_0,(p)})$ as in [62].*

(This is possible because p is a good prime as in Definition 1.1.1.6*, in which case* $r_{\mathtt{D}} = 0$ *and* $r_{\mathcal{H}} = r_\nu$*; see Definition* 3.4.2.1.*)*

Then we obtain a canonical open immersion

$$\vec{\mathsf{M}}^{\mathrm{ord,tor}}_{\mathcal{H}',\Sigma^{\mathrm{ord},\prime}} \hookrightarrow \mathsf{M}^{\mathrm{tor}}_{\mathcal{H}^p,\Sigma^p} \tag{5.2.3.4}$$

extending the canonical open immersion $\vec{\mathsf{M}}^{\mathrm{ord}}_{\mathcal{H}'} \hookrightarrow \mathsf{M}_{\mathcal{H}^p}$ *(as in* (3.4.5.2)*), a canonical quasi-finite étale surjective morphism*

$$\vec{\mathsf{M}}^{\mathrm{ord,tor}}_{\mathcal{H},\Sigma^{\mathrm{ord}}} \twoheadrightarrow \vec{\mathsf{M}}^{\mathrm{ord,tor}}_{\mathcal{H}',\Sigma^{\mathrm{ord},\prime}} \underset{\vec{\mathsf{S}}_0}{\times} \vec{\mathsf{S}}_{0,r_{\mathcal{H}}} \tag{5.2.3.5}$$

extending the canonical quasi-finite étale surjective morphism $\vec{\mathsf{M}}^{\mathrm{ord}}_{\mathcal{H}} \twoheadrightarrow \vec{\mathsf{M}}^{\mathrm{ord}}_{\mathcal{H}'} \underset{\vec{\mathsf{S}}_0}{\times} \vec{\mathsf{S}}_{0,r_{\mathcal{H}}}$ *(as in* (3.4.5.6)*), and by composing* (5.2.3.5) *with* (5.2.3.4) *a canonical quasi-finite étale morphism*

$$\vec{\mathsf{M}}^{\mathrm{ord,tor}}_{\mathcal{H},\Sigma^{\mathrm{ord}}} \to \mathsf{M}^{\mathrm{tor}}_{\mathcal{H}^p,\Sigma^p} \underset{\vec{\mathsf{S}}_0}{\times} \vec{\mathsf{S}}_{0,r_{\mathcal{H}}} \tag{5.2.3.6}$$

extending the canonical quasi-finite étale morphism $\vec{\mathsf{M}}^{\mathrm{ord}}_{\mathcal{H}} \to \mathsf{M}_{\mathcal{H}^p} \underset{\vec{\mathsf{S}}_0}{\times} \vec{\mathsf{S}}_{0,r_{\mathcal{H}}}$ *(as in* (3.4.5.8)*).*

Alternatively, we can construct the morphisms (5.2.3.4)*,* (5.2.3.5)*, and* (5.2.3.6) *as a canonical open and closed subalgebraic stack, given by taking the schematic closure of* $\mathsf{M}^{\mathrm{ord}}_{\mathcal{H}}$*, in a relatively representable functor of ordinary level-*$\mathcal{H}_p$ *structures of type* $(L \underset{\mathbb{Z}}{\otimes} \mathbb{Z}_p, \langle\,\cdot\,,\,\cdot\,\rangle, \mathtt{D})$ *and their (unique if existent) extensions over degenerations.*

Under the open immersion (5.2.3.4)*, the pullback of the degenerating family* $(G, \lambda, i, \alpha_{\mathcal{H}^p})$ *of type* $\mathsf{M}_{\mathcal{H}^p}$ *over* $\mathsf{M}^{\mathrm{tor}}_{\mathcal{H}^p,\Sigma^p}$ *admits (up to isomorphism) a unique extension to the degenerating family* $(G, \lambda, i, \alpha_{\mathcal{H}',p}, \alpha^{\mathrm{ord}}_{\mathcal{H}'_p})$ *of type* $\vec{\mathsf{M}}^{\mathrm{ord}}_{\mathcal{H}'}$ *over* $\vec{\mathsf{M}}^{\mathrm{ord,tor}}_{\mathcal{H}',\Sigma^{\mathrm{ord},\prime}}$*. (Note that* $\mathcal{H}'^{,p} = \mathcal{H}^p$*.) Under the quasi-finite morphism* (5.2.3.5)*, the pullback of the degenerating family* $(G, \lambda, i, \alpha_{\mathcal{H}',p}, \alpha^{\mathrm{ord}}_{\mathcal{H}'_p})$ *of type* $\vec{\mathsf{M}}^{\mathrm{ord}}_{\mathcal{H}'}$ *over* $\vec{\mathsf{M}}^{\mathrm{ord,tor}}_{\mathcal{H}',\Sigma^{\mathrm{ord},\prime}}$ *is canonically isomorphic to the degenerating family* $(G, \lambda, i, \alpha_{\mathcal{H}^p}, \alpha^{\mathrm{ord}}_{\mathcal{H}_p})$ *of type* $\vec{\mathsf{M}}^{\mathrm{ord}}_{\mathcal{H}}$ *over* $\vec{\mathsf{M}}^{\mathrm{ord,tor}}_{\mathcal{H},\Sigma^{\mathrm{ord}}}$*. Consequently, the pullback of*

$$\omega_{\mathsf{M}^{\mathrm{tor}}_{\mathcal{H}^p,\Sigma^p}} := \wedge^{\mathrm{top}}\, \underline{\mathrm{Lie}}^\vee_{G/\mathsf{M}^{\mathrm{tor}}_{\mathcal{H}^p,\Sigma^p}} \cong \wedge^{\mathrm{top}}\, e_G^*\Omega^1_{G/\mathsf{M}^{\mathrm{tor}}_{\mathcal{H}^p,\Sigma^p}}$$

under (5.2.3.4) *is canonically isomorphism to*

$$\omega_{\vec{\mathsf{M}}^{\mathrm{ord,tor}}_{\mathcal{H}',\Sigma^{\mathrm{ord},\prime}}} := \wedge^{\mathrm{top}}\, \underline{\mathrm{Lie}}^\vee_{G/\vec{\mathsf{M}}^{\mathrm{ord,tor}}_{\mathcal{H}',\Sigma^{\mathrm{ord},\prime}}} \cong \wedge^{\mathrm{top}}\, e_G^*\Omega^1_{G/\vec{\mathsf{M}}^{\mathrm{ord,tor}}_{\mathcal{H}',\Sigma^{\mathrm{ord},\prime}}},$$

and the pullback of $\omega_{\vec{\mathsf{M}}^{\mathrm{ord,tor}}_{\mathcal{H}',\Sigma^{\mathrm{ord},\prime}}}$ *under* (5.2.3.5) *is canonically isomorphic to*

$$\omega_{\vec{\mathsf{M}}^{\mathrm{ord,tor}}_{\mathcal{H},\Sigma^{\mathrm{ord}}}} := \wedge^{\mathrm{top}}\, \underline{\mathrm{Lie}}^\vee_{G/\vec{\mathsf{M}}^{\mathrm{ord,tor}}_{\mathcal{H},\Sigma^{\mathrm{ord}}}} \cong \wedge^{\mathrm{top}}\, e_G^*\Omega^1_{G/\vec{\mathsf{M}}^{\mathrm{ord,tor}}_{\mathcal{H},\Sigma^{\mathrm{ord}}}}.$$

Proof. Consider the open immersion

$$\mathsf{M}^{\mathrm{ord,tor}}_{\mathcal{H}^p,\Sigma^p} \hookrightarrow \mathsf{M}^{\mathrm{tor}}_{\mathcal{H}^p,\Sigma^p} \tag{5.2.3.7}$$

representing the extensions of the ordinary level structure $\alpha^{\mathrm{ord}}_{\mathrm{G}(\mathbb{Z}_p)}$ over $\mathsf{M}^{\mathrm{ord}}_{\mathcal{H}^p}$ (extending (3.4.5.3)) as in condition (5) of Definition 3.4.2.10 (which is unique up to isomorphism if it exists). By comparing the universal property of $\mathsf{M}^{\mathrm{ord,tor}}_{\mathcal{H}^p,\Sigma^p}$ given by (6) of Theorem 5.2.1.1 with the universal property of $\mathsf{M}^{\mathrm{tor}}_{\mathcal{H}^p,\Sigma^p}$ given by [62, Thm. 6.4.1.1(6)], we obtain the desired open and closed immersion (5.2.3.4) extending the open and closed immersion (3.4.5.5).

The canonical morphism (5.2.3.5) exists by comparing the universal properties of $\vec{\mathsf{M}}^{\mathrm{ord,tor}}_{\mathcal{H},\Sigma^{\mathrm{ord}}}$ and $\vec{\mathsf{M}}^{\mathrm{ord,tor}}_{\mathcal{H}',\Sigma^{\mathrm{ord},\prime}}$ given by (6) of Theorem 5.2.1.1, which is quasi-finite and étale by comparing the sheaves of log 1-differentials using (4) of Theorem 5.2.1.1 (which is in turn based on Proposition 4.2.3.5) and by the assumption that both Σ^{ord} and $\Sigma^{\mathrm{ord},\prime}$ are induced by a smooth Σ' for $\mathsf{M}_{\mathcal{H}'}$. Then we obtain (5.2.3.6) by composing (5.2.3.5) with (5.2.3.4) as in the proposition.

Alternatively, we can construct (5.2.3.5) as a relatively representable functor parameterizing liftings of ordinary level structures and their extensions over degenerations (*cf.* condition (5) of Definition 3.4.2.10). If we construct the similar relative representable functor over the whole of $\mathsf{M}^{\mathrm{tor}}_{\mathcal{H}^p,\Sigma^p}$, then we can also construct (5.2.3.6) as an open and closed subalgebraic stack.

The remaining statements about the pullbacks of $(G, \lambda, i, \alpha_{\mathcal{H}^p})$ are clear because the morphisms (5.2.3.7) and (5.2.3.5) are defined by comparing the degenerating families over the various partial toroidal compactifications, or by their interpretations as open and closed subalgebraic stacks in relative representable functors over $\mathsf{M}^{\mathrm{tor}}_{\mathcal{H}^p,\Sigma^p}$. □

Lemma 5.2.3.8. *With assumptions as in Proposition* 5.2.3.3, *the composition of* (5.2.3.4) *with the canonical morphism* $\oint_{\mathcal{H}^p} : \mathsf{M}^{\mathrm{tor}}_{\mathcal{H}^p,\Sigma^p} \to \mathsf{M}^{\mathrm{min}}_{\mathcal{H}^p}$ *induces a morphism* $\vec{\mathsf{M}}^{\mathrm{ord,tor}}_{\mathcal{H}',\Sigma^{\mathrm{ord},\prime}} \to \mathsf{M}^{\mathrm{min}}_{\mathcal{H}^p}$ *with* **open image**. *Let* $\vec{\mathsf{M}}^{\mathrm{ord,min}}_{\mathcal{H}'}$ *denote this image (with its canonical open subscheme structure). Then* $\vec{\mathsf{M}}^{\mathrm{ord,min}}_{\mathcal{H}'}$ *is quasi-projective over* $\vec{\mathsf{S}}_{0,r_{\mathcal{H}'}}$ *and* $\oint^{-1}_{\mathcal{H}^p}(\vec{\mathsf{M}}^{\mathrm{ord,min}}_{\mathcal{H}'})$ *is (set-theoretically) the image of* (5.2.3.4) *in* $\mathsf{M}^{\mathrm{tor}}_{\mathcal{H}^p,\Sigma^p}$, *and hence the induced morphism* $\vec{\oint}^{\mathrm{ord}}_{\mathcal{H}'} : \vec{\mathsf{M}}^{\mathrm{ord,tor}}_{\mathcal{H}',\Sigma^{\mathrm{ord},\prime}} \to \vec{\mathsf{M}}^{\mathrm{ord,min}}_{\mathcal{H}'}$ *over* $\vec{\mathsf{S}}_{0,r_{\mathcal{H}'}}$ *is proper and surjective.*

Consequently, since $\omega_{\vec{\mathsf{M}}^{\mathrm{ord,tor}}_{\mathcal{H}',\Sigma^{\mathrm{ord},\prime}}}$ *is isomorphic to the pullback of* $\omega_{\mathsf{M}^{\mathrm{tor}}_{\mathcal{H}^p,\Sigma^p}}$ *(see Proposition* 5.2.3.3), *we have a canonical isomorphism*

$$\vec{\mathsf{M}}^{\mathrm{ord,min}}_{\mathcal{H}'} \cong \mathrm{Proj}\Big(\bigoplus_{k\geq 0} \Gamma(\vec{\mathsf{M}}^{\mathrm{ord,tor}}_{\mathcal{H}',\Sigma^{\mathrm{ord},\prime}}, \omega^{\otimes k}_{\vec{\mathsf{M}}^{\mathrm{ord,tor}}_{\mathcal{H}',\Sigma^{\mathrm{ord},\prime}}})\Big),$$

and the proper morphism $\vec{\oint}^{\mathrm{ord}}_{\mathcal{H}'} : \vec{\mathsf{M}}^{\mathrm{ord,tor}}_{\mathcal{H}',\Sigma^{\mathrm{ord},\prime}} \to \vec{\mathsf{M}}^{\mathrm{ord,min}}_{\mathcal{H}'}$ *is the Stein factorization (see [35, III-1, 4.3.3]) of itself (and hence has nonempty connected geometric fibers, by [35, III-1, 4.3.1, 4.3.3, 4.3.4] and its natural generalization to the context of algebraic stacks).*

Proof. Suppose $\bar{x}$ is a geometric point of the $[(\Phi_{\mathcal{H}^p}, \delta_{\mathcal{H}^p}, \sigma)]$-stratum $\mathsf{Z}_{[(\Phi_{\mathcal{H}^p},\delta_{\mathcal{H}^p},\sigma)]}$ of $\mathsf{M}^{\mathrm{tor}}_{\mathcal{H}^p,\Sigma^p}$. By comparing (6) of Theorem 5.2.1.1 with [62, Thm. 6.4.1.1(6)], or rather by the theory of degeneration on which the constructions are based (see Theorem 4.1.6.2), the condition of being in the image of (5.2.3.4) is a condition for the induced cusp label $[(\Phi_{\mathcal{H}^p}, \delta_{\mathcal{H}^p})]$ and on the image of $\bar{x}$ under the canonical morphism $\mathsf{Z}_{[(\Phi_{\mathcal{H}^p}, \delta_{\mathcal{H}^p}, \sigma)]} \to [\mathsf{M}_{\mathcal{H}^p}^{\mathsf{Z}_{\mathcal{H}^p}}] \to \mathsf{M}_{\mathcal{H}^p}^{\min}$, which is nothing but $\oint_{\mathcal{H}^p}(\bar{x})$. Therefore, once $\oint_{\mathcal{H}^p}(\bar{x}) \in \vec{\mathsf{M}}_{\mathcal{H}'}^{\mathrm{ord},\min}$, all other points in $\oint_{\mathcal{H}^p}^{-1}(\oint_{\mathcal{H}^p}(\bar{x}))$ is in the image of (5.2.3.4). This shows that $\oint_{\mathcal{H}^p}^{-1}(\vec{\mathsf{M}}_{\mathcal{H}'}^{\mathrm{ord},\min})$ is (set-theoretically) the image of (5.2.3.4) in $\mathsf{M}_{\mathcal{H}^p,\Sigma^p}^{\mathrm{tor}}$. Since the complement of the (open) image of (5.2.3.4) in $\mathsf{M}_{\mathcal{H}^p,\Sigma^p}^{\mathrm{tor}}$ is closed, and since $\oint_{\mathcal{H}^p} : \mathsf{M}_{\mathcal{H}^p,\Sigma^p}^{\mathrm{tor}} \twoheadrightarrow \mathsf{M}_{\mathcal{H}^p}^{\min}$ is proper, we see that $\vec{\mathsf{M}}_{\mathcal{H}'}^{\mathrm{ord},\min}$ is open in $\mathsf{M}_{\mathcal{H}^p}^{\min}$, and that the induced morphism $\vec{\oint}_{\mathcal{H}'}^{\mathrm{ord}} : \vec{\mathsf{M}}_{\mathcal{H}',\Sigma^{\mathrm{ord},\prime}}^{\mathrm{ord},\mathrm{tor}} \to \vec{\mathsf{M}}_{\mathcal{H}'}^{\mathrm{ord},\min}$ is also proper.

Since $\oint_{\mathcal{H}^p} : \mathsf{M}_{\mathcal{H}^p,\Sigma^p}^{\mathrm{tor}} \to \mathsf{M}_{\mathcal{H}^p}^{\min} \cong \mathrm{Proj}\Big(\underset{k\geq 0}{\oplus} \Gamma(\mathsf{M}_{\mathcal{H}^p,\Sigma^p}^{\mathrm{tor}}, \omega_{\mathsf{M}_{\mathcal{H}^p,\Sigma^p}^{\mathrm{tor}}}^{\otimes k})\Big)$ is defined (as in [62, Sec. 7.2.3]) as the Stein factorization of the morphism from $\mathsf{M}_{\mathcal{H}^p,\Sigma^p}^{\mathrm{tor}}$ to a projective space defined by global sections of a sufficiently divisible power of $\omega_{\mathsf{M}_{\mathcal{H}^p,\Sigma^p}^{\mathrm{tor}}}$, its restriction to an open subscheme (of the source) containing all fibers of $\oint_{\mathcal{H}^p}$ with which it overlaps (or, equivalently, its pullback to an open subscheme of the target) is a proper morphism over its image, which can also be defined by a Stein factorization. Hence, we can identify $\vec{\oint}_{\mathcal{H}'}^{\mathrm{ord}}$ with the canonical morphism

$$\vec{\mathsf{M}}_{\mathcal{H}',\Sigma^{\mathrm{ord},\prime}}^{\mathrm{ord},\mathrm{tor}} \to \vec{\mathsf{M}}_{\mathcal{H}'}^{\mathrm{ord},\min} \cong \mathrm{Proj}\Big(\underset{k\geq 0}{\oplus} \Gamma(\vec{\mathsf{M}}_{\mathcal{H}',\Sigma^{\mathrm{ord},\prime}}^{\mathrm{ord},\mathrm{tor}}, \omega_{\vec{\mathsf{M}}_{\mathcal{H}',\Sigma^{\mathrm{ord},\prime}}^{\mathrm{ord},\mathrm{tor}}}^{\otimes k})\Big),$$

which is the Stein factorization of itself, as desired. □

Lemma 5.2.3.9. *With assumptions as in Proposition* 5.2.3.3, *suppose moreover that $\mathcal{H}^p$ is neat; that Σ^p is projective, smooth, and equipped with a polarization function pol^p, inducing a polarization function pol' for Σ'; and that $\vec{\mathsf{M}}_{\mathcal{H}',d_0\mathsf{pol}'}^{\mathrm{tor}}$ is defined as in Proposition* 2.2.2.1 (*for some integer $d_0 \geq 1$*). *Then there is a canonical open and closed immersion*

$$\vec{\mathsf{M}}_{\mathcal{H}',d_0\mathsf{pol}'}^{\mathrm{tor}} \hookrightarrow \mathsf{M}_{\mathcal{H}^p,\Sigma^p}^{\mathrm{tor}} \tag{5.2.3.10}$$

extending the canonical open and closed immersion $\vec{\mathsf{M}}_{\mathcal{H}'} \hookrightarrow \mathsf{M}_{\mathcal{H}^p}$ (as in (2.2.4.4), *now that $\mathcal{H}^p$ is neat), which is compatible with* (2.2.4.3) *and with the canonical morphisms $\vec{\oint}_{\mathcal{H}'} : \vec{\mathsf{M}}_{\mathcal{H}',d_0\mathsf{pol}'}^{\mathrm{tor}} \to \vec{\mathsf{M}}_{\mathcal{H}'}^{\min}$ and $\oint_{\mathcal{H}^p} : \mathsf{M}_{\mathcal{H}^p,\Sigma^p}^{\mathrm{tor}} \to \mathsf{M}_{\mathcal{H}^p}^{\min}$. Moreover, the canonical finite morphism*

$$\vec{\mathsf{M}}_{\mathcal{H},d_0\mathsf{pol}}^{\mathrm{tor}} \to \vec{\mathsf{M}}_{\mathcal{H}',d_0\mathsf{pol}'}^{\mathrm{tor}} \tag{5.2.3.11}$$

(*cf.* (2.2.2.4), *with $\vec{\mathsf{M}}_{\mathcal{H},d_0\mathsf{pol}}^{\mathrm{tor}}$ constructed as in Proposition* 2.2.2.3) *induces by composition with* (5.2.3.10) *a finite morphism*

$$\vec{\mathsf{M}}_{\mathcal{H},d_0\mathsf{pol}}^{\mathrm{tor}} \to \mathsf{M}_{\mathcal{H}^p,\Sigma^p}^{\mathrm{tor}}. \tag{5.2.3.12}$$

Consequently, $\vec{\mathsf{M}}_{\mathcal{H}',d_0\mathsf{pol}'}^{\mathrm{tor}}$ is independent of the choices of pol' and d_0, and $\vec{\mathsf{M}}_{\mathcal{H},d_0\mathsf{pol}}^{\mathrm{tor}}$ is also independent of the choices of pol and d_0, because it is canonically isomorphic to the normalization of $\mathsf{M}_{\mathcal{H}^p,\Sigma^p}^{\mathrm{tor}}$ in $\mathsf{M}_{\mathcal{H}}$ under the composition of canonical morphisms $\mathsf{M}_{\mathcal{H}} \to \mathsf{M}_{\mathcal{H}'} \hookrightarrow \mathsf{M}_{\mathcal{H}',\Sigma'}^{\mathrm{tor}} \to \mathsf{M}_{\mathcal{H}^p,\Sigma^p}^{\mathrm{tor}}$ (cf. Proposition 2.2.2.3).

Proof. By comparing (6) of Theorem 1.3.1.3 with [62, Thm. 6.4.1.1(6)], there is a canonical open and closed immersion

$$\mathsf{M}^{\mathrm{tor}}_{\mathcal{H}',\Sigma'} \hookrightarrow \mathsf{M}^{\mathrm{tor}}_{\mathcal{H}^p,\Sigma^p} \underset{\mathbb{Z}}{\otimes} \mathbb{Q} \tag{5.2.3.13}$$

(over S_0) compatible with the canonical open and closed immersion

$$\mathsf{M}^{\min}_{\mathcal{H}'} \hookrightarrow \mathsf{M}^{\min}_{\mathcal{H}^p} \underset{\mathbb{Z}}{\otimes} \mathbb{Q}, \tag{5.2.3.14}$$

which can be identified with the pullback of (2.2.4.3) to S_0, and with the canonical morphisms $\oint_{\mathcal{H}'} : \mathsf{M}^{\mathrm{tor}}_{\mathcal{H}',\Sigma'} \to \mathsf{M}^{\min}_{\mathcal{H}'}$ and $\oint_{\mathcal{H}^p} \underset{\mathbb{Z}}{\otimes} \mathbb{Q} : \mathsf{M}^{\mathrm{tor}}_{\mathcal{H}^p,\Sigma^p} \underset{\mathbb{Z}}{\otimes} \mathbb{Q} \to \mathsf{M}^{\min}_{\mathcal{H}^p} \underset{\mathbb{Z}}{\otimes} \mathbb{Q}$ over S_0. For each integer $d \geq 0$, we can compatibly construct $\jmath_{\mathcal{H}',d\mathsf{pol}'}$, $\jmath_{\mathcal{H}^p,d\mathsf{pol}^p}$, $\mathcal{J}_{\mathcal{H}',d\mathsf{pol}'}$, and $\mathcal{J}_{\mathcal{H}^p,d\mathsf{pol}^p}$ as in Definition 1.3.1.7 and [62, Def. 7.3.3.1], so that $\jmath_{\mathcal{H}',d\mathsf{pol}'}$ (resp. $\mathcal{J}_{\mathcal{H}',d\mathsf{pol}'}$) is isomorphic to the pullback of $\jmath_{\mathcal{H}^p,d\mathsf{pol}^p}$ (resp. $\mathcal{J}_{\mathcal{H}^p,d\mathsf{pol}^p}$) under (5.2.3.13) (resp. (5.2.3.14)).

By [62, Thm. 7.3.3.4], there is a canonical isomorphism

$$\mathrm{NBl}_{\mathcal{J}_{\mathcal{H}^p,d_0\mathsf{pol}^p}}(\oint_{\mathcal{H}^p}) : \mathsf{M}^{\mathrm{tor}}_{\mathcal{H}^p,\Sigma^p} \xrightarrow{\sim} \mathrm{NBl}_{\mathcal{J}_{\mathcal{H}^p,d_0\mathsf{pol}^p}}(\mathsf{M}^{\min}_{\mathcal{H}^p}). \tag{5.2.3.15}$$

By its very construction, this canonical isomorphism extends the canonical isomorphism

$$\mathrm{NBl}_{\mathcal{J}_{\mathcal{H}',d_0\mathsf{pol}'}}(\oint_{\mathcal{H}'}) : \mathsf{M}^{\mathrm{tor}}_{\mathcal{H}',\Sigma'} \xrightarrow{\sim} \mathrm{NBl}_{\mathcal{J}_{\mathcal{H}',d_0\mathsf{pol}'}}(\mathsf{M}^{\min}_{\mathcal{H}'}) \tag{5.2.3.16}$$

(as in Theorem 1.3.1.10, for the same d_0).

Since $\jmath_{\mathcal{H}',d_0\mathsf{pol}'}$ and $\jmath_{\mathcal{H}^p,d_0\mathsf{pol}^p}$ are defined by the vanishing orders on boundary divisors in a compatible way, we see that $\vec{\mathcal{J}}_{\mathcal{H}',d_0\mathsf{pol}'}$ (see Proposition 2.2.2.1) is isomorphic to the pullback of $\mathcal{J}_{\mathcal{H}^p,d_0\mathsf{pol}^p}$ under the canonical isomorphism (2.2.4.3). Thus, we have the desired open and closed immersion (5.2.3.10) pulling the isomorphism (5.2.3.15) back to

$$\mathrm{NBl}_{\vec{\mathcal{J}}_{\mathcal{H}',d_0\mathsf{pol}'}}(\vec{\oint}_{\mathcal{H}'}) : \vec{\mathsf{M}}^{\mathrm{tor}}_{\mathcal{H}',d_0\mathsf{pol}'} \xrightarrow{\sim} \mathrm{NBl}_{\vec{\mathcal{J}}_{\mathcal{H}',d_0\mathsf{pol}'}}(\vec{\mathsf{M}}^{\min}_{\mathcal{H}'}) \tag{5.2.3.17}$$

in a way compatible with (2.2.4.3) and with the canonical morphisms $\vec{\oint}_{\mathcal{H}'}$ and $\oint_{\mathcal{H}^p}$.

The statements concerning $\vec{\mathsf{M}}^{\mathrm{tor}}_{\mathcal{H},d_0\mathsf{pol}}$ are self-explanatory. □

Proposition 5.2.3.18. *With the assumptions as in Lemma 5.2.3.9, there exists a canonical open immersion*

$$\vec{\mathsf{M}}^{\mathrm{ord,tor}}_{\mathcal{H},\Sigma^{\mathrm{ord}}} \hookrightarrow \vec{\mathsf{M}}^{\mathrm{tor}}_{\mathcal{H},d_0\mathsf{pol},r_{\mathcal{H}}} \tag{5.2.3.19}$$

(*see Definition* 2.2.3.5) *compatible with* (5.2.3.12) *and* (5.2.3.6) (*and with the canonical morphism* $\vec{\mathsf{M}}^{\mathrm{tor}}_{\mathcal{H},d_0\mathsf{pol},r_{\mathcal{H}}} \to \vec{\mathsf{M}}^{\mathrm{tor}}_{\mathcal{H},d_0\mathsf{pol}}$), *whose composition with the canonical morphism* $\vec{\oint}_{\mathcal{H},r_{\mathcal{H}}} : \vec{\mathsf{M}}^{\mathrm{tor}}_{\mathcal{H},d_0\mathsf{pol},r_{\mathcal{H}}} \to \vec{\mathsf{M}}^{\min}_{\mathcal{H},r_{\mathcal{H}}}$ *induces a morphism* $\vec{\mathsf{M}}^{\mathrm{ord,tor}}_{\mathcal{H},\Sigma^{\mathrm{ord}}} \to \vec{\mathsf{M}}^{\min}_{\mathcal{H},r_{\mathcal{H}}}$ *with* **open image**. *Let* $\vec{\mathsf{M}}^{\mathrm{ord,min}}_{\mathcal{H}}$ *denote this image* (*with its canonical open subscheme structure*). *Then* $\vec{\mathsf{M}}^{\mathrm{ord,min}}_{\mathcal{H}}$ *is quasi-projective and* $\vec{\oint}^{-1}_{\mathcal{H},r_{\mathcal{H}}}(\vec{\mathsf{M}}^{\mathrm{ord,min}}_{\mathcal{H}})$ *is* (*set-theoretically*) *the image of* (5.2.3.19) *in* $\vec{\mathsf{M}}^{\mathrm{tor}}_{\mathcal{H},d_0\mathsf{pol},r_{\mathcal{H}}}$, *and hence the induced morphism* $\vec{\oint}^{\mathrm{ord}}_{\mathcal{H}} : \vec{\mathsf{M}}^{\mathrm{ord,tor}}_{\mathcal{H},\Sigma^{\mathrm{ord}}} \to \vec{\mathsf{M}}^{\mathrm{ord,min}}_{\mathcal{H}}$ *over* $\vec{\mathsf{S}}_{0,r_{\mathcal{H}}}$ *is* **proper** *and surjective.*

Consequently, since $\omega_{\vec{\mathsf{M}}^{\mathrm{ord,tor}}_{\mathcal{H},\Sigma^{\mathrm{ord}}}}$ *is canonically isomorphic to the pullback of* $\omega_{\mathsf{M}^{\mathrm{tor}}_{\mathcal{H}^p,\Sigma^p}}$ *under* (5.2.3.6) (*see Proposition* 5.2.3.3), *we have a canonical isomorphism*

$$\vec{\mathsf{M}}^{\mathrm{ord,min}}_{\mathcal{H}} \cong \mathrm{Proj}\Big(\underset{k\geq 0}{\oplus}\Gamma(\vec{\mathsf{M}}^{\mathrm{ord,tor}}_{\mathcal{H},\Sigma^{\mathrm{ord}}}, \omega^{\otimes k}_{\vec{\mathsf{M}}^{\mathrm{ord,tor}}_{\mathcal{H},\Sigma^{\mathrm{ord}}}})\Big),$$

and the proper morphism $\vec{\oint}^{\mathrm{ord}}_{\mathcal{H}'} : \vec{\mathsf{M}}^{\mathrm{ord,tor}}_{\mathcal{H}',\Sigma^{\mathrm{ord},\prime}} \to \vec{\mathsf{M}}^{\mathrm{ord,min}}_{\mathcal{H}'}$ *is the Stein factorization of itself* (*and hence has nonempty connected geometric fibers, by [35, III-1, 4.3.1, 4.3.3, 4.3.4]*).

Proof. The existence of the canonical open immersion (5.2.3.19) follows from Proposition 5.2.3.3, from the fact that $\vec{\mathsf{M}}^{\mathrm{tor}}_{\mathcal{H},d_0\mathsf{pol},r_{\mathcal{H}}}$ is a normalization (see Lemma 5.2.3.9), and from Zariski's main theorem (see [35, III-1, 4.4.3, 4.4.11]).

To prove the remaining statements, as in the proof of Lemma 5.2.3.8, the most crucial step is to show that $\vec{\oint}^{-1}_{\mathcal{H},r_{\mathcal{H}}}(\vec{\mathsf{M}}^{\mathrm{ord,min}}_{\mathcal{H}})$ is set-theoretically the image of (5.2.3.19) in $\vec{\mathsf{M}}^{\mathrm{tor}}_{\mathcal{H},d_0\mathsf{pol},r_{\mathcal{H}}}$. It suffices to verify this statement after pulled back to $\mathsf{S}_{0,r_{\mathcal{H}}}$ and $\vec{\mathsf{S}}_{0,r_{\mathcal{H}}} \underset{\mathbb{Z}_p}{\otimes} \mathbb{F}_p$ (in two cases). In the former case, we have $\vec{\mathsf{M}}^{\mathrm{tor}}_{\mathcal{H},d_0\mathsf{pol},r_{\mathcal{H}}} \underset{\mathbb{Z}}{\otimes} \mathbb{Q} \cong \mathsf{M}^{\mathrm{tor}}_{\mathcal{H},\Sigma,r_{\mathcal{H}}}$ (see Definition 2.2.3.4), and the proof is similar to the proof of Lemma 5.2.3.8 using the theory of degeneration, by comparing (6) of Theorem 5.2.1.1 with (6) of Theorem 1.3.1.3. In the latter case, the proof follows from Lemma 5.2.3.8; from the density of $\vec{\mathsf{M}}^{\mathrm{ord}}_{\mathcal{H}} \underset{\mathbb{Z}_p}{\otimes} \mathbb{F}_p$ in $\vec{\mathsf{M}}^{\mathrm{ord,tor}}_{\mathcal{H},\Sigma^{\mathrm{ord}}} \underset{\mathbb{Z}_p}{\otimes} \mathbb{F}_p$ (see Theorem 5.2.1.1); and from the fact that the canonical quasi-finite morphism $\vec{\mathsf{M}}^{\mathrm{ord,tor}}_{\mathcal{H},\Sigma^{\mathrm{ord}}} \underset{\mathbb{Z}_p}{\otimes} \mathbb{F}_p \to \vec{\mathsf{M}}^{\mathrm{ord,tor}}_{\mathcal{H}',\Sigma^{\mathrm{ord},\prime}} \underset{\mathbb{Z}_p}{\otimes} \mathbb{F}_p$ induced by (5.2.3.5) is actually *finite*, because it is proper by Lemma 3.3.6.8 (with $g_p = 1$ there; *cf.* the proof of Corollary 5.2.2.3, with $g = 1$ there).

Once we know that $\vec{\oint}^{-1}_{\mathcal{H},r_{\mathcal{H}}}(\vec{\mathsf{M}}^{\mathrm{ord,min}}_{\mathcal{H}})$ is set-theoretically the (open) image of (5.2.3.19) in $\vec{\mathsf{M}}^{\mathrm{tor}}_{\mathcal{H},d_0\mathsf{pol},r_{\mathcal{H}}}$, we know that the induced morphism $\vec{\oint}^{\mathrm{ord}}_{\mathcal{H}} : \vec{\mathsf{M}}^{\mathrm{ord,tor}}_{\mathcal{H},\Sigma^{\mathrm{ord}}} \to \vec{\mathsf{M}}^{\mathrm{ord,min}}_{\mathcal{H}}$ over $\vec{\mathsf{S}}_{0,r_{\mathcal{H}}}$ is proper and surjective. By the same Stein factorization argument as in the proof of Lemma 5.2.3.8, the last statement of the proposition follows from the assertions on $\omega_{\vec{\mathsf{M}}^{\mathrm{ord,tor}}_{\mathcal{H},\Sigma^{\mathrm{ord}}}}$, $\omega_{\vec{\mathsf{M}}^{\mathrm{ord,tor}}_{\mathcal{H}',\Sigma^{\mathrm{ord},\prime}}}$, and $\omega_{\mathsf{M}^{\mathrm{tor}}_{\mathcal{H}^p,\Sigma^p}}$ in Proposition 5.2.3.3, and from the finiteness of (5.2.3.12) (or rather (5.2.3.11)) in Lemma 5.2.3.9. □

5.2.4 *Boundary of Ordinary Loci*

As in Section 1.3.2, let us describe the building blocks of $\mathsf{M}^{\mathrm{ord,tor}}_{\mathcal{H},\Sigma^{\mathrm{ord}}}$ in more detail. In particular, we would like to describe and characterize the algebraic stacks $\vec{\mathsf{M}}^{\mathrm{ord,Z}_{\mathcal{H}}}_{\mathcal{H}}$, $\vec{\mathsf{M}}^{\mathrm{ord,Z}_{\mathcal{H}}}_{\mathcal{H}}$, $\vec{C}^{\mathrm{ord}}_{\Phi_{\mathcal{H}},\delta_{\mathcal{H}}}$, $\vec{\Xi}^{\mathrm{ord}}_{\Phi_{\mathcal{H}},\delta_{\mathcal{H}}}$, $\vec{\Xi}^{\mathrm{ord}}_{\Phi_{\mathcal{H}},\delta_{\mathcal{H}}}(\sigma)$, $\vec{\Xi}^{\mathrm{ord}}_{\Phi_{\mathcal{H}},\delta_{\mathcal{H}},\sigma}$, $\vec{\mathsf{Z}}^{\mathrm{ord}}_{[(\Phi_{\mathcal{H}},\delta_{\mathcal{H}},\sigma)]} \cong \vec{\Xi}^{\mathrm{ord}}_{\Phi_{\mathcal{H}},\delta_{\mathcal{H}},\sigma}/\Gamma_{\Phi_{\mathcal{H}},\sigma}$ and the formal algebraic stacks $\vec{\mathfrak{X}}^{\mathrm{ord}}_{\Phi_{\mathcal{H}},\delta_{\mathcal{H}},\sigma}$ and $\vec{\mathfrak{X}}^{\mathrm{ord}}_{\Phi_{\mathcal{H}},\delta_{\mathcal{H}},\sigma}/\Gamma_{\Phi_{\mathcal{H}},\sigma}$ in (2) of Theorem 5.2.1.1, and to describe canonical Hecke actions on collections of these geometric objects (compatible with those in Proposition 5.2.2.2).

Throughout this subsection, let us fix the choice of a fully symplectic admissible filtration Z of $L \underset{\mathbb{Z}}{\otimes} \hat{\mathbb{Z}}$ as in Definitions 1.2.1.2 and 1.2.1.3, which we assume to be compatible with $\mathtt{D}$ as in Definition 3.2.3.1. Let us also fix a (noncanonical) choice of $(L^{\mathsf{Z}}, \langle\,\cdot\,,\,\cdot\,\rangle^{\mathsf{Z}}, h_0^{\mathsf{Z}})$ so that G^{Z} can be defined as in Definition 1.2.1.9. Then we also have a boundary filtration $\mathtt{D}^{\mathsf{Z}} \cong \mathtt{D}_{-1}$ for $L^{\mathsf{Z}} \underset{\mathbb{Z}}{\otimes} \mathbb{Z}_p \cong \mathrm{Gr}^{\mathsf{Z}}_{-1} \underset{\hat{\mathbb{Z}}}{\otimes} \mathbb{Z}_p$ determined by $\mathtt{D}^{\mathsf{Z},0} \cong \mathtt{D}^0_{-1} = \mathtt{D}^0/(\mathsf{Z}_{-2} \underset{\hat{\mathbb{Z}}}{\otimes} \mathbb{Z}_p)$, which defines subgroups $\mathrm{P}^{\mathrm{ord}}_{\mathtt{D}^{\mathsf{Z}}}$ *etc* of $\mathrm{G}^{\mathsf{Z}} \underset{\mathbb{Z}}{\otimes} \mathbb{Z}_p$ as in Definition 3.2.2.7, and defines quotients of subgroups of $\mathrm{P}^{\mathrm{ord}}_{\mathsf{Z},\mathtt{D}}(R) = \mathrm{P}_{\mathsf{Z}}(R) \cap \mathrm{P}^{\mathrm{ord}}_{\mathtt{D}}(R)$ for each $\mathbb{Z}$-algebra R as in Definition 3.2.3.9, so that $\mathrm{P}^{\mathrm{ord}}_{\mathtt{D}^{\mathsf{Z}}}(R) \cong \mathrm{P}^{\mathrm{ord}}_{h,\mathsf{Z},\mathtt{D}}(R)$.

For each open compact subgroup $\mathcal{H}$ of $\mathrm{G}(\hat{\mathbb{Z}})$ of standard form with respect to $\mathtt{D}$ as in Definition 3.2.2.9, we can define $\vec{\mathsf{M}}^{\mathrm{ord},\Phi_{\mathcal{H}}}_{\mathcal{H}}$ and $\vec{\mathsf{M}}^{\mathrm{ord},\mathsf{Z}_{\mathcal{H}}}_{\mathcal{H}}$ as in the paragraphs containing and preceding (4.2.1.28), which are finite étale over the base change $\vec{\mathsf{M}}^{\mathrm{ord}}_{\mathcal{H}_h,r_{\mathcal{H}}}$ of $\vec{\mathsf{M}}^{\mathrm{ord}}_{\mathcal{H}_h}$ (defined by $\mathsf{M}_{\mathcal{H}_h}$ and $\ddot{\mathsf{M}}^{\mathrm{ord}}_{\mathcal{H}_h}$ as in Theorem 3.4.2.5) to $\vec{\mathsf{S}}_{0,r_{\mathcal{H}}}$. By Proposition 4.2.1.29, the canonical morphism $\vec{\mathsf{M}}^{\mathrm{ord},\Phi_{\mathcal{H}}}_{\mathcal{H}} \to \vec{\mathsf{M}}^{\mathrm{ord},\mathsf{Z}_{\mathcal{H}}}_{\mathcal{H}}$ is finite étale and induces a canonical isomorphism $\vec{\mathsf{M}}^{\mathrm{ord},\Phi_{\mathcal{H}}}_{\mathcal{H}}/\Gamma_{\Phi_{\mathcal{H}}} \xrightarrow{\sim} \vec{\mathsf{M}}^{\mathrm{ord},\mathsf{Z}_{\mathcal{H}}}_{\mathcal{H}}$. If $\mathcal{H} = \mathcal{U}^{\mathrm{bal}}_1(n) := \mathcal{U}^p(n_0)\mathcal{U}^{\mathrm{bal}}_1(p^r)$ for some integer $n_0 \geq 1$ prime to p and some integer $r \geq 0$, then $\vec{\mathsf{M}}^{\mathrm{ord},\Phi_{\mathcal{H}}}_{\mathcal{H}} \cong \vec{\mathsf{M}}^{\mathrm{ord},\mathsf{Z}_{\mathcal{H}}}_{\mathcal{H}}$ because there is a unique $(\varphi^{\mathrm{ord}}_{-2,\mathcal{H}}, \varphi^{\mathrm{ord}}_{0,\mathcal{H}})$ inducing the prescribed $(\varphi_{-2,\mathcal{H}}, \varphi_{0,\mathcal{H}})$. We shall set $\vec{\mathsf{M}}^{\mathrm{ord},\Phi_n}_n := \vec{\mathsf{M}}^{\mathrm{ord},\mathsf{Z}_n}_n := \vec{\mathsf{M}}^{\mathrm{ord},\mathsf{Z}_n}_{\mathcal{U}^{\mathrm{bal}}_1(n)}$. Moreover, for general $\mathcal{H} \subset \mathrm{G}(\hat{\mathbb{Z}})$ (of standard form with respect to $\mathtt{D}$ as in Definition 3.2.2.9), by construction and by Proposition 3.4.4.1, we have the following two lemmas:

Lemma 5.2.4.1. (*Compare with Lemma* 1.3.2.1.) *Let* $(\mathsf{Z}, \Phi, \delta)$ *and* $\mathcal{H}_{\mathrm{G}_{h,\mathsf{Z}},\Phi}$ *be as in Lemma* 1.3.2.1. *Then* $\mathcal{H}_{\mathrm{G}_{h,\mathsf{Z}},\Phi}$ *is of standard form with respect to* $\mathtt{D}^{\mathsf{Z}}$, *which is a normal subgroup of* $\mathcal{H}_{\mathrm{G}_{h,\mathsf{Z}}}$ (*as in Definition* 1.2.1.12) *of equal depth* (*by definition— note that* $\mathcal{U}^{\mathrm{bal}}_1(n)_{\mathrm{G}_{h,\mathsf{Z}}} = \mathcal{U}^{\mathrm{bal}}_1(n)_{\mathrm{G}'_{h,\mathsf{Z}}}$ *for all integers* $n \geq 1$), *and there is a canonical isomorphism*

$$\vec{\mathsf{M}}^{\mathrm{ord},\mathsf{Z}_{\mathcal{H}}}_{\mathcal{H}} \cong \vec{\mathsf{M}}^{\mathrm{ord}}_{\mathcal{H}_{\mathrm{G}_{h,\mathsf{Z}},\Phi},r_{\mathcal{H}}}, \tag{5.2.4.2}$$

where $\vec{\mathsf{M}}^{\mathrm{ord}}_{\mathcal{H}_{\mathrm{G}_{h,\mathsf{Z}},\Phi}}$ *is defined by* $(L^{\mathsf{Z}}, \langle\,\cdot\,,\,\cdot\,\rangle^{\mathsf{Z}}, h_0^{\mathsf{Z}})$ *as in Theorem* 3.4.2.5, *and where the subscript "*$r_{\mathcal{H}}$*" means base change to* $\vec{\mathsf{S}}_{0,r_{\mathcal{H}}}$. *If* $\mathcal{H}'$ *is an open compact subgroup of* $\mathcal{H}$ *of standard form* (*with respect to* $\mathtt{D}$), *then the corresponding morphism*

$$\vec{\mathsf{M}}^{\mathrm{ord},\mathsf{Z}_{\mathcal{H}'}}_{\mathcal{H}'} \to \vec{\mathsf{M}}^{\mathrm{ord},\mathsf{Z}_{\mathcal{H}}}_{\mathcal{H}} \tag{5.2.4.3}$$

can be canonically identified with the quasi-finite flat morphism

$$\vec{\mathsf{M}}^{\mathrm{ord}}_{\mathcal{H}'_{\mathrm{G}_{h,\mathsf{Z}},\Phi},r_{\mathcal{H}'}} \to \vec{\mathsf{M}}^{\mathrm{ord}}_{\mathcal{H}_{\mathrm{G}_{h,\mathsf{Z}},\Phi},r_{\mathcal{H}}}. \tag{5.2.4.4}$$

The collection $\{\vec{\mathsf{M}}^{\mathrm{ord}}_{\mathcal{H}_{\mathrm{G}_{h,\mathsf{Z}},\Phi}}\}_{\mathcal{H}_{\mathrm{G}_{h,\mathsf{Z}},\Phi}}$ (*with* $\mathcal{H}$ *of standard form as above*) *naturally carries a Hecke action by elements* $g_h = (g_{h,0}, g_{h,p}) \in \mathrm{G}^{\mathsf{Z}}(\mathbb{A}^{\infty,p}) \times \mathrm{P}^{\mathrm{ord}}_{\mathtt{D}^{\mathsf{Z}}}(\mathbb{Q}_p) \cong \mathrm{G}_{h,\mathsf{Z}}(\mathbb{A}^{\infty,p}) \times \mathrm{P}^{\mathrm{ord}}_{h,\mathsf{Z},\mathtt{D}}(\mathbb{Q}_p) \subset \mathrm{G}_{h,\mathsf{Z}}(\mathbb{A}^{\infty})$ *with* $g_{h,p}$ *satisfying the conditions defined by* $\mathtt{D}^{\mathsf{Z}} \cong \mathtt{D}_{-1}$ *on* $L^{\mathsf{Z}} \underset{\mathbb{Z}}{\otimes} \mathbb{Z}_p \cong \mathrm{Gr}^{\mathsf{Z}}_{-1} \underset{\hat{\mathbb{Z}}}{\otimes} \mathbb{Z}_p$ *as in Section* 3.3.4, *realized by quasi-finite flat surjections pulling tautological objects back to ordinary Hecke twists. Such a Hecke action enjoys the properties* (*under various conditions*) *concerning étaleness, finiteness, being isomorphisms between formal completions along fibers over* $\mathrm{Spec}(\mathbb{F}_p)$, *and inducing absolute Frobenius morphisms on fibers over* $\mathrm{Spec}(\mathbb{F}_p)$ *for elements of* U_p *type as in Proposition* 5.2.2.2 *and Corollaries* 5.2.2.3, 5.2.2.4, *and* 5.2.2.5. *(We omit the details for simplicity.) If moreover* $\mathcal{H}'$ *is a normal subgroup of* $\mathcal{H}$ *of standard form and equal depth as in Definition* 3.2.2.9, *then the canonical morphism* $\vec{\mathsf{M}}^{\mathrm{ord}}_{\mathcal{H}'_{\mathrm{G}_{h,\mathrm{z}},\Phi},r_{\mathcal{H}'}} \to \vec{\mathsf{M}}^{\mathrm{ord}}_{\mathcal{H}_{\mathrm{G}_{h,\mathrm{z}},\Phi},r_{\mathcal{H}'}}$ *induced by* (5.2.4.4) *(is finite étale and) is an* $\mathcal{H}_{\mathrm{G}_{h,\mathrm{z}},\Phi}/\mathcal{H}'_{\mathrm{G}_{h,\mathrm{z}},\Phi}$*-torsor.*

Lemma 5.2.4.5. *(Compare with Lemma* 1.3.2.5.*) Let* $\mathcal{H}_{\mathrm{G}'_{h,\mathrm{z}}}$ *be as in Definition* 1.2.1.12, *which is a normal subgroup of* $\mathcal{H}_{\mathrm{G}_{h,\mathrm{z}}}$ *(and hence of* $\mathcal{H}_{\mathrm{G}_{h,\mathrm{z}},\Phi}$*) of equal depth (by definition—again, note that* $\mathcal{U}_1^{\mathrm{bal}}(n)_{\mathrm{G}_{h,\mathrm{z}}} = \mathcal{U}_1^{\mathrm{bal}}(n)_{\mathrm{G}'_{h,\mathrm{z}}}$ *for all integers* $n \geq 1$*). Then there is a canonical isomorphism*

$$\vec{\mathsf{M}}^{\mathrm{ord},\Phi_{\mathcal{H}}}_{\mathcal{H}} \cong \vec{\mathsf{M}}^{\mathrm{ord}}_{\mathcal{H}_{\mathrm{G}'_{h,\mathrm{z}}},r_{\mathcal{H}}}, \tag{5.2.4.6}$$

which is compatible with (5.2.4.2) *and with Hecke actions as in Lemma* 5.2.4.1. *The canonical morphisms* $\vec{\mathsf{M}}^{\mathrm{ord},\Phi_{\mathcal{H}}}_{\mathcal{H}} \to \vec{\mathsf{M}}^{\mathrm{ord},\mathrm{Z}_{\mathcal{H}}}_{\mathcal{H}} \to \vec{\mathsf{M}}^{\mathrm{ord}}_{\mathcal{H}_h,r_{\mathcal{H}}}$ *can be identified with the canonical finite étale morphisms* $\vec{\mathsf{M}}^{\mathrm{ord}}_{\mathcal{H}_{\mathrm{G}'_{h,\mathrm{z}}},r_{\mathcal{H}}} \to \vec{\mathsf{M}}^{\mathrm{ord}}_{\mathcal{H}_{\mathrm{G}_{h,\mathrm{z}},\Phi},r_{\mathcal{H}}} \to \vec{\mathsf{M}}^{\mathrm{ord}}_{\mathcal{H}_{\mathrm{G}_{h,\mathrm{z}}},r_{\mathcal{H}}}$, *on which* $\Gamma_{\Phi_{\mathcal{H}}}$ *acts equivariantly (and trivially on the latter two objects) via the canonical homomorphism* $\Gamma_{\Phi_{\mathcal{H}}} \to \mathcal{H}_{\mathrm{G}'_{l,\mathrm{z}}}/\mathcal{H}_{\mathrm{G}_{l,\mathrm{z}}} \cong \mathcal{H}_{\mathrm{G}_{h,\mathrm{z}}}/\mathcal{H}_{\mathrm{G}'_{h,\mathrm{z}}}$ *with image* $\mathcal{H}_{\mathrm{G}_{h,\mathrm{z}},\Phi}/\mathcal{H}_{\mathrm{G}'_{h,\mathrm{z}}}$. *In particular, the finite étale morphism* $\vec{\mathsf{M}}^{\mathrm{ord},\Phi_{\mathcal{H}}}_{\mathcal{H}} \to \vec{\mathsf{M}}^{\mathrm{ord},\mathrm{Z}_{\mathcal{H}}}_{\mathcal{H}}$ *is an* $\mathcal{H}_{\mathrm{G}_{h,\mathrm{z}},\Phi}/\mathcal{H}_{\mathrm{G}'_{h,\mathrm{z}}}$*-torsor.*

By Proposition 4.2.1.30 and its proof, we have the following:

Lemma 5.2.4.7. *(Compare with Lemma* 1.3.2.7.*) The morphism* $\vec{C}^{\mathrm{ord}}_{\Phi_{\mathcal{H}},\delta_{\mathcal{H}}} \to \vec{\mathsf{M}}^{\mathrm{ord},\Phi_{\mathcal{H}}}_{\mathcal{H}}$ *depends only on* $\mathcal{H}_{\mathrm{G}_{1,\mathrm{z}}}$, *is an abelian scheme when the splitting of* (1.2.1.14) *defined by any splitting* δ *also splits* (1.2.1.13) *(and induces an isomorphism* $\mathcal{H}_{\mathrm{G}_{1,\mathrm{z}}} \cong \mathcal{H}_{\mathrm{G}'_{h,\mathrm{z}}} \ltimes \mathcal{H}_{\mathrm{U}_{1,\mathrm{z}}}$; *cf. the condition in* (2) *of Lemma* 4.2.1.19*), and is a torsor under the abelian scheme* $\vec{C}^{\mathrm{ord,grp}}_{\Phi_{\mathcal{H}},\delta_{\mathcal{H}}} := \vec{C}^{\mathrm{ord}}_{\Phi_{\mathcal{H}'},\delta_{\mathcal{H}'}}$ *defined by any* $\mathcal{H}'$ *of standard form with* $\mathcal{H}'_{\mathrm{G}_{1,\mathrm{z}}} \cong \mathcal{H}_{\mathrm{G}'_{h,\mathrm{z}}} \ltimes \mathcal{H}_{\mathrm{U}_{1,\mathrm{z}}}$, *which is* $\mathbb{Q}^\times$*-isogenous to* $\underline{\mathrm{Hom}}_{\mathcal{O}}(X,B)^\circ$. *(This clarifies the abelian scheme torsor structure of* $\vec{C}^{\mathrm{ord}}_{\Phi_{\mathcal{H}},\delta_{\mathcal{H}}} \to \vec{\mathsf{M}}^{\mathrm{ord},\Phi_{\mathcal{H}}}_{\mathcal{H}}$.*) If* $p \nmid [L^\# : L]$, *so that the index of* $\phi : Y \hookrightarrow X$ *and the degree of* $\lambda_B : B \to B^\vee$ *are both prime to* p, *then there is a canonical* $\mathbb{Z}_{(p)}^\times$*-isogeny*

$$\underline{\mathrm{Hom}}_{\mathcal{O}}(X, B^{\mathrm{ord}}_{p^r})^\circ / H^{\mathrm{ord}}_{p^r,\mathrm{U}^{\mathrm{ess,ord}}_{1,\mathrm{Z}_{p^r},\mathrm{D}_{p^r}}} \to \vec{C}^{\mathrm{ord,grp}}_{\Phi_{\mathcal{H}},\delta_{\mathcal{H}}}$$

(cf. (4.1.4.31)*), where* $r := \mathrm{depth}_{\mathtt{D}}(\mathcal{H})$ *(see Definition* 3.2.2.9*) and* $H^{\mathrm{ord}}_{p^r,\mathrm{U}^{\mathrm{ess,ord}}_{1,\mathrm{Z}_{p^r},\mathrm{D}_{p^r}}}$ *(see Definition* 4.1.5.18*) can be canonically embedded in the kernel of the canonical separable isogeny* $\underline{\mathrm{Hom}}_{\mathcal{O}}(X, B^{\mathrm{ord}}_{p^r})^\circ \to \underline{\mathrm{Hom}}_{\mathcal{O}}(X,B)^\circ$ *(induced by* (4.1.4.34)*).*

If $r = \mathrm{depth}_{\mathtt{D}}(\mathcal{H})$ *and if* $n_0 \geq 1$ *is any integer prime to* p *such that* $\mathcal{U}^p(n_0) \subset \mathcal{H}^p$, *so that* $\mathcal{U}_1^{\mathrm{bal}}(n) = \mathcal{U}^p(n_0)\mathcal{U}_{p,1}^{\mathrm{bal}}(p^r) \subset \mathcal{H}$ *are both of standard form and equally deep, and if we fix the choice of* $(\mathsf{Z}_n$ *and)* Φ_n, *then the canonical morphism*

$$\vec{C}^{\mathrm{ord}}_{\Phi_n,\delta_n} := \vec{C}^{\mathrm{ord}}_{\Phi_{\mathcal{U}_1^{\mathrm{bal}}(n)},\delta_{\mathcal{U}_1^{\mathrm{bal}}(n)}} \to \vec{C}^{\mathrm{ord}}_{\Phi_{\mathcal{H}},\delta_{\mathcal{H}},r_n} = \vec{C}^{\mathrm{ord}}_{\Phi_{\mathcal{H}},\delta_{\mathcal{H}}} \underset{\vec{\mathsf{S}}_{0,r_{\mathcal{H}}}}{\times} \vec{\mathsf{S}}_{0,r_n}, \tag{5.2.4.8}$$

where $r_n := r_{\mathcal{U}_1^{\mathrm{bal}}(n)} = \max(r_{\mathsf{D}}, r)$, *(is finite étale and) is an* $\mathcal{H}_{\mathrm{G}_{1,\mathrm{z}}}/\mathcal{U}_1^{\mathrm{bal}}(n)_{\mathrm{G}_{1,\mathrm{z}}} \cong H^{\mathrm{ord}}_{n,\mathrm{G}^{\mathrm{ess}}_{h,\mathrm{Z}_n} \ltimes \mathrm{U}^{\mathrm{ess}}_{1,\mathrm{Z}_n}}$*-torsor (see Definition* 4.1.5.18 *and Section* 4.2.1 *for the definition of* $H^{\mathrm{ord}}_{n,\mathrm{G}^{\mathrm{ess}}_{h,\mathrm{Z}_n} \ltimes \mathrm{U}^{\mathrm{ess}}_{1,\mathrm{Z}_n}}$*), where* $\mathcal{H}_{\mathrm{G}_{1,\mathrm{z}}}$ *and* $\mathcal{U}_1^{\mathrm{bal}}(n)_{\mathrm{G}_{1,\mathrm{z}}}$ *are as in Definition* 1.2.1.12, *and induces an isomorphism*

$$\vec{C}^{\mathrm{ord}}_{\Phi_n,\delta_n}/(\mathcal{H}_{\mathrm{G}_{1,\mathrm{z}}}/\mathcal{U}_1^{\mathrm{bal}}(n)_{\mathrm{G}_{1,\mathrm{z}}}) \xrightarrow{\sim} \vec{C}^{\mathrm{ord}}_{\Phi_{\mathcal{H}},\delta_{\mathcal{H}},r_n}. \tag{5.2.4.9}$$

Proof. The statements are self-explanatory. □

Lemma 5.2.4.10. *(Compare with Lemma* 1.3.2.11.*) Suppose* $n = n_0 p^r$ *for some integer* $n_0 \geq 1$ *prime to* p *and some integer* $r \geq 0$*. Suppose* $(B, \lambda_B, i_B, \varphi^{\mathrm{ord}}_{-1,n})$ *and* $(\mathsf{Z}_n, \Phi_n = (X, Y, \phi : Y \hookrightarrow X, \varphi_{-2,n}, \varphi_{0,n}), \delta_n)$ *are the tautological objects over* $\vec{\mathsf{M}}^{\mathrm{ord},\Phi_n}_n = \vec{\mathsf{M}}^{\mathrm{ord},\mathsf{Z}_n}_n$*. The abelian scheme torsor* $S := \vec{C}^{\mathrm{ord}}_{\Phi_n,\delta_n} \to \vec{\mathsf{M}}^{\mathrm{ord},\Phi_n}_n$ *is universal for the additional structures* $(c_n^{\mathrm{ord}}, c_n^{\vee,\mathrm{ord}})$ *over noetherian normal schemes over* $\vec{\mathsf{M}}^{\mathrm{ord},\Phi_n}_n$ *(inducing dominant morphisms over irreducible components) satisfying certain symplectic and liftability conditions, which can be re-interpreted as follows:* $S = \vec{C}^{\mathrm{ord}}_{\Phi_n,\delta_n} \to \vec{\mathsf{M}}^{\mathrm{ord},\Phi_n}_n = \vec{\mathsf{M}}^{\mathrm{ord},\mathsf{Z}_n}_n$ *parameterizes tuples*

$$(G^\natural, \lambda^\natural : G^\natural \to G^{\vee,\natural}, i^\natural, \beta^\natural_{n_0}, \beta^{\natural,\mathrm{ord}}_{p^r})$$

over noetherian normal schemes flat over $\vec{\mathsf{S}}_{0,r_{\mathcal{H}}}$*, where:*

(a) $G^\natural$ *(resp.* $G^{\vee,\natural}$*) is an extension of* B *(resp.* $B^\vee$*) by the split torus* T *(resp.* $T^\vee$*) with character group* X *(resp.* Y*), and* $\lambda^\natural : G^\natural \to G^{\vee,\natural}$ *induces* $\lambda_T = \phi^* : T \to T^\vee$ *and* $\lambda_B : B \to B^\vee$*.*

(b) $i^\natural = (i^{\natural,0}, i^{\natural,\#,0})$ *is a pair of homomorphisms* $i^{\natural,0} : \mathcal{O} \to \mathrm{End}_S(G^\natural)$ *and* $i^{\natural,\#,0} : \mathcal{O} \to \mathrm{End}_S(G^{\vee,\natural})$ *compatible with each other under* $\lambda^\natural : G^\natural \to G^{\vee,\natural}$*, inducing compatible* $\mathcal{O}$*-structures on* B*,* $B^\vee$*,* T*, and* $T^\vee$*.*

(c) $\beta^\natural_{n_0} = (\beta^{\natural,0}_{n_0}, \beta^{\natural,\#,0}_{n_0}, \nu^\natural_{n_0})$ *is a* **principal level-n_0 structure** *of* $(G^\natural, \lambda^\natural, i^\natural)$ *of type* $(L \underset{\mathbb{Z}}{\otimes} \hat{\mathbb{Z}}^p, \langle\,\cdot\,,\,\cdot\,\rangle, \mathsf{Z} \underset{\hat{\mathbb{Z}}}{\otimes} \hat{\mathbb{Z}}^p)$*, where* $\beta^{\natural,0}_{n_0} : (\mathsf{Z}_{-1,n_0})_S \xrightarrow{\sim} G^\natural[n_0]$ *and* $\beta^{\natural,\#,0}_{n_0} : (\mathsf{Z}^\#_{-1,n_0})_S \xrightarrow{\sim} G^{\vee,\natural}[n_0]$ *are* $\mathcal{O}$*-equivariant isomorphisms preserving filtrations on both sides and inducing on the graded pieces the* φ_{-2,n_0}*,* φ_{-1,n_0}*, and* φ_{0,n_0} *(by duality), respectively, induced by the given* $\varphi_{-2,n}$*,* $\varphi^{\mathrm{ord}}_{-1,n}$*, and* $\varphi_{0,n}$*; and where* $\nu^\natural_{n_0} : ((\mathbb{Z}/n_0\mathbb{Z})(1))_S \xrightarrow{\sim} \boldsymbol{\mu}_{n_0,S}$ *is an isomorphism, which are compatible with* $\lambda^\natural$ *and the canonical morphism* $\mathsf{Z}_{-1,n_0} \to \mathsf{Z}^\#_{-1,n_0}$ *induced by* $\langle\,\cdot\,,\,\cdot\,\rangle$*. (Here* $\mathsf{Z}^\#$ *is the filtration on* $L^\# \underset{\mathbb{Z}}{\otimes} \hat{\mathbb{Z}}$ *canonically dual to the filtration on* $L \underset{\mathbb{Z}}{\otimes} \hat{\mathbb{Z}}$*, equipped with a canonical morphism* $\mathsf{Z} \to \mathsf{Z}^\#$*, respecting the filtration degrees, induced by* $\langle\,\cdot\,,\,\cdot\,\rangle$*. Then the splitting* δ_{n_0} *induced by* δ_n *corresponds under* $\beta^\natural_{n_0}$ *to splittings of* $0 \to T[n_0] \to G^\natural[n_0] \to B[n_0] \to 0$ *and* $0 \to T^\vee[n_0] \to G^{\vee,\natural}[n_0] \to B^\vee[n_0] \to 0$*.) Moreover,* $\beta^\natural_{n_0}$ *satisfies the liftability condition that, for each integer* $m_0 \geq 1$ *such that* $n_0 | m_0$*, there exists a finite étale covering of* S *over which there exists an analogous triple* $\beta^\natural_{m_0}$ *lifting the pullback of* $\beta^\natural_{n_0}$*.*

(d) $\beta_{p^r}^{\natural,\mathrm{ord}} = (\beta_{p^r}^{\natural,\mathrm{ord},0}, \beta_{p^r}^{\natural,\mathrm{ord},\#,0}, \nu_{p^r}^{\natural,\mathrm{ord}})$ *is a* **principle ordinary level-p^r structure** *of* $(G^\natural, \lambda^\natural, i^\natural)$ *of type* $(L \underset{\mathbb{Z}}{\otimes} \mathbb{Z}_p, \langle\,\cdot\,,\,\cdot\,\rangle, \mathtt{Z} \underset{\hat{\mathbb{Z}}}{\otimes} \mathbb{Z}_p, \mathtt{D})$*, where* $\beta_{p^r}^{\natural,\mathrm{ord},0} : (\mathrm{Gr}^0_{\mathtt{D},p^r})^{\mathrm{mult}}_S \to G^\natural[p^r])$ *and* $\beta_{p^r}^{\natural,\mathrm{ord},\#,0} : (\mathrm{Gr}^0_{\mathtt{D}^\#,p^r})^{\mathrm{mult}}_S \to G^{\vee,\natural}[p^r])$ *are* $\mathcal{O}$*-equivariant homomorphisms inducing closed immersions, preserving filtrations on both sides, and inducing on the graded pieces the* φ_{-2,p^r}*,* $\varphi^{\mathrm{ord}}_{-1,p^r}$*, and* φ_{0,p^r} *(by duality), respectively, induced by the given* $\varphi_{-2,n}$*,* $\varphi^{\mathrm{ord}}_{-1,n}$*, and* $\varphi_{0,n}$*; and where* $\nu_{p^r}^{\natural,\mathrm{ord}} \in (\mathbb{Z}/p^r\mathbb{Z})^\times_S$ *is a section, which are compatible with* $\lambda^\natural$ *and the canonical morphism* $\phi^0_{\mathtt{D}} : \mathrm{Gr}^0_{\mathtt{D}} \to \mathrm{Gr}^0_{\mathtt{D}^\#}$ *induced by* $\langle\,\cdot\,,\,\cdot\,\rangle$*. (Then the splitting* $\delta^{\mathrm{ord}}_{p^r}$ *induced by* δ_n *corresponds under* $\beta_{p^r}^{\natural,\mathrm{ord}}$ *to splittings of* $0 \to T[n] \to \mathrm{image}(\beta_{p^r}^{\natural,\mathrm{ord},0}) \to \mathrm{image}(\varphi^{\mathrm{ord},0}_{-1,p^r}) \to 0$ *and* $0 \to \mathrm{image}(\beta_{p^r}^{\natural,\mathrm{ord},\#,0}) \to \mathrm{image}(\varphi^{\mathrm{ord},\#,0}_{-1,p^r}) \to 0$*.) Moreover,* $\beta_{p^r}^{\natural,\mathrm{ord}}$ *satisfies the liftability condition that, for each integer* $r' \geq r$*, there exists a quasi-finite étale covering of* S *over which there exists an analogous triple* $\beta_{p^{r'}}^{\natural,\mathrm{ord}}$ *lifting the pullback of* $\beta_{p^r}^{\natural,\mathrm{ord}}$*.*

Proof. The statements are self-explanatory. □

Proposition 5.2.4.11. (*Compare with Proposition* 1.3.2.12.) *The abelian scheme torsor* $S := \vec{C}^{\mathrm{ord}}_{\Phi_\mathcal{H},\delta_\mathcal{H}} \to \vec{\mathsf{M}}^{\mathrm{ord},\Phi_\mathcal{H}}_\mathcal{H}$ *is universal for the additional structures* $(c^{\mathrm{ord}}_\mathcal{H}, c^{\vee,\mathrm{ord}}_\mathcal{H})$ *over noetherian normal schemes over* $\vec{\mathsf{M}}^{\mathrm{ord},\Phi_\mathcal{H}}_\mathcal{H}$ *(inducing dominant morphisms over irreducible components) satisfying certain symplectic and liftability conditions, which can be interpreted as parameterizing tuples*

$$(G^\natural, \lambda^\natural : G^\natural \to G^{\vee,\natural}, i^\natural, \beta^\natural_{\mathcal{H}^p}, \beta^{\natural,\mathrm{ord}}_{\mathcal{H}_p}), \tag{5.2.4.12}$$

where:

(1) $G^\natural$, $G^{\vee,\natural}$, $\lambda^\natural$, *and* $i^\natural$ *are as in Lemma* 5.2.4.10.

(2) $\beta^\natural_{\mathcal{H}^p}$ *is a* **level-$\mathcal{H}^p$ structure** *of* $(G^\natural, \lambda^\natural, i^\natural)$ *of type* $(L \underset{\mathbb{Z}}{\otimes} \hat{\mathbb{Z}}^p, \langle\,\cdot\,,\,\cdot\,\rangle, \mathtt{Z} \underset{\hat{\mathbb{Z}}}{\otimes} \hat{\mathbb{Z}}^p)$*, which is a collection* $\{\beta^\natural_{H_{n_0}}\}_{n_0}$*, where* $n_0 \geq 1$ *runs over integers prime to* p *such that* $\mathcal{U}^p(n_0) \subset \mathcal{H}^p$*, such that each* $\beta^\natural_{H_{n_0}}$ *(where* $H_{n_0} := \mathcal{H}^p/\mathcal{U}^p(n_0)$*) is a subscheme of*

$$\coprod \Big(\underline{\mathrm{Isom}}_S(\mathtt{Z}_{-1,n_0,S}, G^\natural[n_0]) \underset{S}{\times} \underline{\mathrm{Isom}}_S(\mathtt{Z}^\#_{-1,n_0,S}, G^{\vee,\natural}[n_0]) \underset{S}{\times} \underline{\mathrm{Isom}}_S\big(((\mathbb{Z}/n_0\mathbb{Z})(1))_S, \boldsymbol{\mu}_{n_0,S}\big)\Big)$$

over S*, where the disjoint union is over representatives* $(\mathtt{Z}_{n_0}, \Phi_{n_0}, \delta_{n_0})$ *(with the same* (X, Y, ϕ)*) in* $(\mathtt{Z}_{\mathcal{H}^p}, \Phi_{\mathcal{H}^p}, \delta_{\mathcal{H}^p})$ *induced by* $(\mathtt{Z}_\mathcal{H}, \Phi_\mathcal{H}, \delta_\mathcal{H})$*, that becomes the disjoint union of all elements in the* H_{n_0}*-orbit of some principal level-*n_0 *structure* $\beta^\natural_{n_0}$ *of* $(G^\natural, \lambda^\natural, i^\natural)$ *of type* $(L \underset{\mathbb{Z}}{\otimes} \hat{\mathbb{Z}}^p, \langle\,\cdot\,,\,\cdot\,\rangle, \mathtt{Z} \underset{\hat{\mathbb{Z}}}{\otimes} \hat{\mathbb{Z}}^p)$*, as in Lemma* 5.2.4.10*, for any* $\mathtt{Z}$ *lifting* $\mathtt{Z}_n$*; and where* $\beta^\natural_{H_{m_0}}$ *is mapped to* $\beta^\natural_{H_{n_0}}$ *(under the canonical morphism, which we omit for simplicity) when* $p \nmid m_0$ *and* $n_0|m_0$*. The* H_{n_0}*-orbit of the* φ_{-1,n_0} *determined by such* $\beta^\natural_{n_0}$

as in Lemma 5.2.4.10 *then defines the level structure* $\varphi_{-1,\mathcal{H}^p}$ *of* (B,λ_B,i_B) *(where* $(B,\lambda_B,i_B,\varphi_{-1,\mathcal{H}^p},\varphi^{\mathrm{ord}}_{-1,\mathcal{H}_p})$ *is the pullback of the tautological tuple over* $\vec{\mathsf{M}}^{\mathrm{ord},\mathsf{Z}_{\mathcal{H}}}_{\mathcal{H}}$*).*

(3) $\beta^{\natural,\mathrm{ord}}_{\mathcal{H}_p}$ *is an* **ordinary level-$\mathcal{H}_p$ structure** *of* $(G^\natural,\lambda^\natural,i^\natural)$ *of type* $(L\underset{\mathbb{Z}}{\otimes}\mathbb{Z}_p,\langle\cdot,\cdot\rangle,\mathsf{Z}\underset{\hat{\mathbb{Z}}}{\otimes}\mathbb{Z}_p,\mathsf{D})$*, which is a subscheme of*

$$\coprod\Big(\underline{\mathrm{Hom}}_S\big((\mathrm{Gr}^0_{\mathsf{D},p^r})^{\mathrm{mult}}_S,G^\natural[p^r]\big)\underset{S}{\times}\underline{\mathrm{Hom}}_S\big((\mathrm{Gr}^0_{\mathsf{D}^\#,p^r})^{\mathrm{mult}}_S,G^{\vee,\natural}[p^r]\big)\underset{S}{\times}(\mathbb{Z}/p^r\mathbb{Z})^\times_S\Big)$$

over S*, where the disjoint union is over representatives* $(\mathsf{Z}_{p^r},\Phi_{p^r},\delta_{p^r})$ *(with the same* (X,Y,ϕ)*) in the* $(\mathsf{Z}_{\mathcal{H}_p},\Phi_{\mathcal{H}_p},\delta_{\mathcal{H}_p})$ *induced by* $(\mathsf{Z}_{\mathcal{H}},\Phi_{\mathcal{H}},\delta_{\mathcal{H}})$*, that becomes the disjoint union of all elements in the* $H^{\mathrm{ord}}_{p^r}$*-orbit (where* $H^{\mathrm{ord}}_{p^r}:=\mathcal{H}_p/\mathcal{U}^{\mathrm{bal}}_{p,1}(p^r)$*) of some principle ordinary level-*p^r *structure* $\beta^{\natural,\mathrm{ord}}_{p^r}$ *of* $(G^\natural,\lambda^\natural,i^\natural)$ *of type* $(L\underset{\mathbb{Z}}{\otimes}\mathbb{Z}_p,\langle\cdot,\cdot\rangle,\mathsf{D})$*, as in Lemma* 5.2.4.10*. The* $H^{\mathrm{ord}}_{p^r}$*-orbit of the* $\varphi^{\mathrm{ord}}_{-1,p^r}$ *determined by such* $\beta^{\natural,\mathrm{ord}}_{p^r}$ *as in Lemma* 5.2.4.10 *then defines the ordinary level structure* $\varphi^{\mathrm{ord}}_{-1,\mathcal{H}_p}$ *of* (B,λ_B,i_B)*.*

Proof. This follows from Lemmas 5.2.4.7 and 5.2.4.10, by realizing $\vec{C}^{\mathrm{ord}}_{\Phi_{\mathcal{H}},\delta_{\mathcal{H}},r_n}$ as a quotient of some $\vec{C}^{\mathrm{ord}}_{\Phi_n,\delta_n}$, and by finite flat descent. □

Proposition 5.2.4.13. (*Compare with Propositions* 1.3.2.14 *and* 5.2.4.11.) *Fix any lifting* $(\mathsf{Z},\Phi=(X,Y,\phi,\varphi_{-2},\varphi_0),\delta)$ *of a representative* $(\mathsf{Z}_{\mathcal{H}},\Phi_{\mathcal{H}},\delta_{\mathcal{H}})$ *of* $[(\mathsf{Z}_{\mathcal{H}},\Phi_{\mathcal{H}},\delta_{\mathcal{H}})]$*. The abelian scheme torsor* $\vec{C}^{\mathrm{ord}}_{\Phi_{\mathcal{H}},\delta_{\mathcal{H}}}\to\vec{\mathsf{M}}^{\mathrm{ord},\Phi_{\mathcal{H}}}_{\mathcal{H}}$ *is universal for* $\mathbb{Z}^\times_{(p)}$*-isogeny classes of tuples*

$$(G^\natural,\lambda^\natural:G^\natural\to G^{\vee,\natural},i^\natural,j^\natural,j^{\vee,\natural},[\hat{\beta}^{p,\natural}]_{\mathcal{H}^p_{\mathrm{G}_{1,\mathsf{Z}}}},\beta^{\natural,\mathrm{ord}}_{\mathcal{H}_p}) \tag{5.2.4.14}$$

over noetherian normal base schemes S *over* $\vec{\mathsf{M}}^{\mathrm{ord},\Phi_{\mathcal{H}}}_{\mathcal{H}}$ *(inducing dominant morphisms over irreducible components), where:*

(1) $G^\natural$ *(resp.* $G^{\vee,\natural}$*) is a semi-abelian scheme which is the extension of an abelian scheme* B *(resp.* $B^\vee$*) by a split torus* T *(resp.* $T^\vee$*) over* S*, which is equivalent to a homomorphism* $c:\mathbf{X}(T)\to B^\vee$ *(resp.* $c^\vee:\mathbf{X}(T^\vee)\to B$*).*

(2) $\lambda^\natural:G^\natural\to G^{\vee,\natural}$ *is* $\mathbb{Q}^\times$*-isogeny which is up to* $\mathbb{Z}^\times_{(p)}$*-isogenies an isogeny of semi-abelian schemes over* S*, inducing a* $\mathbb{Q}^\times$*-isogeny* $\lambda_T:T\to T^\vee$ *between the torus parts, which is dual to a* $\mathbb{Q}$*-isomorphism* $\lambda_T^*:\mathbf{X}(T^\vee)\underset{\mathbb{Z}}{\otimes}\mathbb{Q}\to\mathbf{X}(T)\underset{\mathbb{Z}}{\otimes}\mathbb{Q}$*, and a* $\mathbb{Q}^\times$*-polarization* $\lambda_B:B\to B^\vee$ *between the abelian parts (cf. [62, Def. 1.3.2.19 and the errata]) which is up to* $\mathbb{Z}^\times_{(p)}$*-isogenies a polarization, so that* $c(N\lambda_T^*)=(N\lambda_B)c^\vee$ *when* N *is any locally constant function over* S *valued in positive integers such that* $(N\lambda_T^*)(\mathbf{X}(T^\vee))\subset\mathbf{X}(T)$ *and such that* $N\lambda^\natural:G^\natural\to G^{\vee,\natural}$ *is an isogeny.*

(3) $i^\natural : \mathcal{O} \underset{\mathbb{Z}}{\otimes} \mathbb{Z}_{(p)} \to \mathrm{End}_S(G^\natural) \underset{\mathbb{Z}}{\otimes} \mathbb{Z}_{(p)}{}_S$ *is a homomorphism inducing* $\mathcal{O} \underset{\mathbb{Z}}{\otimes} \mathbb{Z}_{(p)}$*-actions on* $G^{\vee,\natural}$, T, $T^\vee$, B, *and* $B^\vee$ *up to* $\mathbb{Z}_{(p)}{}^\times$*-isogeny, compatible with each other under the homomorphisms between these objects introduced thus far. In particular, the induced homomorphism* $i_B : \mathcal{O} \underset{\mathbb{Z}}{\otimes} \mathbb{Z}_{(p)} \to \mathrm{End}_S(B) \underset{\mathbb{Z}}{\otimes} \mathbb{Z}_{(p)}{}_S$ *satisfies the Rosati condition defined by* λ_B *(cf. [62, Def. 1.3.3.1]).*

(4) $j^\natural : X \underset{\mathbb{Z}}{\otimes} \mathbb{Z}_{(p)}{}_S \overset{\sim}{\to} \mathbf{X}(T) \underset{\mathbb{Z}}{\otimes} \mathbb{Z}_{(p)}{}_S$ *and* $j^{\vee,\natural} : Y \underset{\mathbb{Z}}{\otimes} \mathbb{Z}_{(p)}{}_S \overset{\sim}{\to} \mathbf{X}(T^\vee) \underset{\mathbb{Z}}{\otimes} \mathbb{Z}_{(p)}{}_S$ *are isomorphisms of* $\mathcal{O} \underset{\mathbb{Z}}{\otimes} \mathbb{Z}_{(p)}$*-modules, such that there exists a section* $r(j^\natural, j^{\vee,\natural})$ *of* $(\mathbb{Z}^\times_{(p),>0})_S$ *such that* $j^\natural \circ \phi = r(j^\natural, j^{\vee,\natural})\lambda_T^* \circ j^{\vee,\natural}$.

(5) $\beta^{\natural,\mathrm{ord}}_{\mathcal{H}_p}$ *is an* **ordinary level-$\mathcal{H}_p$ structure** *of* $(G^\natural, \lambda^\natural, i^\natural)$ *of type* $(L \underset{\mathbb{Z}}{\otimes} \mathbb{Z}_p, \langle\,\cdot\,,\,\cdot\,\rangle, \mathtt{Z} \underset{\hat{\mathbb{Z}}}{\otimes} \mathbb{Z}_p, \mathtt{D})$ *as in Proposition* 5.2.4.11. *(Note that the definition of* $\beta^{\natural,\mathrm{ord}}_{\mathcal{H}_p}$ *is insensitive to* $\mathbb{Z}^\times_{(p)}$*-isogenies.)*

(6) $[\hat{\beta}^{p,\natural}]_{\mathcal{H}^p_{\mathrm{G}_{1,\mathrm{Z}}}}$ *is a* **rational level-$\mathcal{H}^p$ structure** *of* $(G^\natural, \lambda^\natural, i^\natural, j^\natural, j^{\vee,\natural})$ *of type* $(L \underset{\mathbb{Z}}{\otimes} \mathbb{A}^{\infty,p}, \langle\,\cdot\,,\,\cdot\,\rangle, \mathtt{Z} \underset{\hat{\mathbb{Z}}}{\otimes} \mathbb{A}^{\infty,p}, \Phi)$*, which is an assignment to each geometric point* $\bar{s}$ *of* S *a rational level-*$\mathcal{H}^p$ *structure of* $(G^\natural, \lambda^\natural, i^\natural, j^\natural, j^{\vee,\natural})$ *of type* $(L \underset{\mathbb{Z}}{\otimes} \mathbb{A}^{\infty,p}, \langle\,\cdot\,,\,\cdot\,\rangle, \mathtt{Z} \underset{\hat{\mathbb{Z}}}{\otimes} \mathbb{A}^{\infty,p}, \Phi)$ **based at** $\bar{s}$ *(cf. [62, Def. 1.3.8.7]), which is a* $\pi_1(S, \bar{s})$*-invariant* $\mathcal{H}^p_{\mathrm{G}_{1,\mathrm{Z}}}$*-orbit* $[\hat{\beta}^{p,\natural}_{\bar{s}}]_{\mathcal{H}^p_{\mathrm{G}_{1,\mathrm{Z}}}}$ *of triples*

$$\hat{\beta}^\natural_{\bar{s}} = (\hat{\beta}^{p,\natural,0}_{\bar{s}}, \hat{\beta}^{p,\natural,\#,0}_{\bar{s}}, \hat{\nu}^{p,\natural}_{\bar{s}}),$$

such that the assignments at any two geometric points $\bar{s}$ *and* $\bar{s}'$ *of the same connected component of* S *determine each other (cf. [62, Lem. 1.3.8.6]), where:*

(a) $\hat{\beta}^{p,\natural,0}_{\bar{s}} : \mathtt{Z}_{-1} \underset{\hat{\mathbb{Z}}}{\otimes} \mathbb{A}^{\infty,p} \overset{\sim}{\to} \mathrm{V}^p\, G^\natural_{\bar{s}}$ *and* $\hat{\beta}^{p,\natural,\#,0}_{\bar{s}} : \mathtt{Z}^\#_{-1} \underset{\hat{\mathbb{Z}}}{\otimes} \mathbb{A}^{\infty,p} \overset{\sim}{\to} \mathrm{V}^p\, G^{\vee,\natural}_{\bar{s}}$ *are* $\mathcal{O} \underset{\mathbb{Z}}{\otimes} \mathbb{A}^{\infty,p}$*-equivariant isomorphisms preserving filtrations on both sides, which are compatible with* $\lambda^\natural$ *and the canonical morphism* $\mathtt{Z}_{-1} \underset{\hat{\mathbb{Z}}}{\otimes} \mathbb{A}^\infty \to \mathtt{Z}^\#_{-1} \underset{\hat{\mathbb{Z}}}{\otimes} \mathbb{A}^\infty$ *induced by* $\langle\,\cdot\,,\,\cdot\,\rangle$.

(b) $\hat{\nu}^{p,\natural}_{\bar{s}} : \mathbb{A}^{\infty,p}(1) \overset{\sim}{\to} \mathrm{V}^p\, \mathbf{G}_{\mathrm{m},\bar{s}}$ *is an isomorphism of* $\mathbb{A}^{\infty,p}$*-modules such that* $r(j^\natural, j^{\vee,\natural})_{\bar{s}}\, \hat{\nu}^\natural_{\bar{s}}$ *maps* $\hat{\mathbb{Z}}^p(1)$ *to* $\mathrm{T}^p\, \mathbf{G}_{\mathrm{m},\bar{s}}$*, where* $r(j^\natural, j^{\vee,\natural})_{\bar{s}}$ *is the value at* $\bar{s}$ *of the above section* $r(j^\natural, j^{\vee,\natural})$ *of* $(\mathbb{Z}^\times_{(p),>0})_S$ *such that* $j^\natural \circ \phi = r(j^\natural, j^{\vee,\natural})\lambda_T^* \circ j^{\vee,\natural}$.

(c) *The induced morphisms* $\mathrm{Gr}_{-2}(\hat{\beta}^{p,\natural,0}_{\bar{s}}) : \mathrm{Gr}^{\mathtt{Z}}_{-2} \underset{\hat{\mathbb{Z}}}{\otimes} \mathbb{A}^{\infty,p} \overset{\sim}{\to} \mathrm{V}^p\, T_{\bar{s}}$ *and*

$\mathrm{Gr}_{-2}(\hat{\beta}^{p,\natural,\#,0}_{\bar{s}}) : \mathrm{Gr}^{\mathrm{Z}^{\#}}_{-2} \underset{\hat{\mathbb{Z}}}{\otimes} \mathbb{A}^{\infty,p} \xrightarrow{\sim} \mathrm{V}^p\, T^{\vee}_{\bar{s}}$ *coincide with the compositions*

$$\mathrm{Gr}^{\mathrm{Z}}_{-2} \underset{\hat{\mathbb{Z}}}{\otimes} \mathbb{A}^{\infty,p} \overset{\varphi_{-2} \underset{\hat{\mathbb{Z}}}{\otimes} \mathbb{A}^{\infty,p}}{\xrightarrow{\sim}} \mathrm{Hom}_{\mathbb{A}^{\infty,p}}(X \underset{\mathbb{Z}}{\otimes} \mathbb{A}^{\infty,p}, \mathbb{A}^{\infty,p}(1))$$
$$\overset{((j^{\natural})^{-1} \underset{\mathbb{Z}_{(p)}}{\otimes} \mathbb{A}^{\infty,p})^*}{\xrightarrow{\sim}} \mathrm{Hom}_{\mathbb{A}^{\infty,p}}(\mathbf{X}(T) \underset{\mathbb{Z}}{\otimes} \mathbb{A}^{\infty,p}, \mathbb{A}^{\infty,p}(1))$$
$$\overset{\hat{\nu}^{p,\natural}_{\bar{s}}}{\xrightarrow{\sim}} \mathrm{Hom}_{\mathbb{A}^{\infty,p}}(\mathbf{X}(T) \underset{\mathbb{Z}}{\otimes} \mathbb{A}^{\infty,p}, \mathrm{V}^p\, \mathbf{G}_{\mathrm{m},\bar{s}}) \xrightarrow{\sim} \mathrm{V}^p\, T_{\bar{s}}$$

and

$$\mathrm{Gr}^{\mathrm{Z}^{\#}}_{-2} \underset{\hat{\mathbb{Z}}}{\otimes} \mathbb{A}^{\infty,p} \overset{\varphi^{\#}_{-2} \underset{\hat{\mathbb{Z}}}{\otimes} \mathbb{A}^{\infty,p}}{\xrightarrow{\sim}} \mathrm{Hom}_{\mathbb{A}^{\infty,p}}(Y \underset{\mathbb{Z}}{\otimes} \mathbb{A}^{\infty,p}, \mathbb{A}^{\infty,p}(1))$$
$$\overset{((j^{\vee,\natural})^{-1} \underset{\mathbb{Z}_{(p)}}{\otimes} \mathbb{A}^{\infty,p})^*}{\xrightarrow{\sim}} \mathrm{Hom}_{\mathbb{A}^{\infty}}(\mathbf{X}(T^{\vee}) \underset{\mathbb{Z}}{\otimes} \mathbb{A}^{\infty,p}, \mathbb{A}^{\infty,p}(1))$$
$$\overset{\hat{\nu}^{p,\natural}_{\bar{s}}}{\xrightarrow{\sim}} \mathrm{Hom}_{\mathbb{A}^{\infty}}(\mathbf{X}(T^{\vee}) \underset{\mathbb{Z}}{\otimes} \mathbb{A}^{\infty,p}, \mathrm{V}^p\, \mathbf{G}_{\mathrm{m},\bar{s}}) \xrightarrow{\sim} \mathrm{V}^p\, T^{\vee}_{\bar{s}},$$

respectively, where $\varphi^{\#}_{-2} : \mathrm{Gr}^{\mathrm{Z}^{\#}}_{-2} \xrightarrow{\sim} \mathrm{Hom}_{\hat{\mathbb{Z}}}(Y \underset{\mathbb{Z}}{\otimes} \hat{\mathbb{Z}}, \hat{\mathbb{Z}}(1))$ *is induced by* φ_0 *by duality.*

(d) *Together with* $\hat{\nu}^{p}_{-1,\bar{s}} := \hat{\nu}^{p,\natural}_{\bar{s}}$*, the induced morphisms*

$$\hat{\varphi}^{p}_{-1,\bar{s}} := \mathrm{Gr}_{-1}(\hat{\beta}^{p,\natural,0}_{\bar{s}}) : \mathrm{Gr}^{\mathrm{Z}}_{-1} \underset{\hat{\mathbb{Z}}}{\otimes} \mathbb{A}^{\infty,p} \xrightarrow{\sim} \mathrm{V}^p\, B_{\bar{s}}$$

and

$$\hat{\varphi}^{p,\#}_{-1,\bar{s}} := \mathrm{Gr}_{-1}(\hat{\beta}^{p,\natural,\#,0}_{\bar{s}}) : \mathrm{Gr}^{\mathrm{Z}^{\#}}_{-1} \underset{\hat{\mathbb{Z}}}{\otimes} \mathbb{A}^{\infty,p} \xrightarrow{\sim} \mathrm{V}^p\, B^{\vee}_{\bar{s}}$$

determine each other by duality. By varying $\bar{s}$ *over geometric points of* S*, the* ($\pi_1(S,\bar{s})$*-invariant*) $\mathcal{H}^{p,\prime}_{\mathrm{G}_{h,\mathrm{Z}}}$*-orbits of*

$$(\hat{\varphi}^{p}_{-1,\bar{s}}, \hat{\nu}^{p}_{-1,\bar{s}}, \varphi^{\mathrm{ord}}_{-1,\mathcal{H}_p}, \varphi_{-2}, \varphi_0)$$

(*where* $\varphi^{\mathrm{ord}}_{-1,\mathcal{H}_p}$ *is determined by* $\beta^{\natural,\mathrm{ord}}_{\mathcal{H}_p}$ *as in Proposition* 5.2.4.11) *determine a tuple*

$$((B, \lambda_B, i_B, \varphi_{-1,\mathcal{H}^p}, \varphi^{\mathrm{ord}}_{-1,\mathcal{H}_p}), (\varphi^{\mathrm{ord}}_{-2,\mathcal{H}}, \varphi^{\mathrm{ord}}_{0,\mathcal{H}}))$$

whose $\mathbb{Z}^{\times}_{(p)}$*-isogeny class is parameterized by* $\vec{\mathsf{M}}^{\mathrm{ord},\Phi_{\mathcal{H}}}_{\mathcal{H}}$ *(cf. [62, Prop. 1.4.3.4]).*

The $\mathbb{Z}_{(p)}^{\times}$-isogenies

$$(G^{\natural}, \lambda^{\natural} : G^{\natural} \to G^{\vee,\natural}, i^{\natural}, j^{\natural}, j^{\vee,\natural}, [\hat{\beta}^{p,\natural}]_{\mathcal{H}^p_{\mathrm{G}_{1,\mathrm{z}}}}, \beta^{\natural,\mathrm{ord}}_{\mathcal{H}_p})$$
$$\sim_{\mathbb{Z}_{(p)}^{\times}\text{-isog.}} (G^{\natural,\prime}, \lambda^{\natural,\prime} : G^{\natural,\prime} \to G^{\vee,\natural,\prime}, i^{\natural,\prime}, j^{\natural,\prime}, j^{\vee,\natural,\prime}, [\hat{\beta}^{p,\natural,\prime}]_{\mathcal{H}^p_{\mathrm{G}_{1,\mathrm{z}}}}, \beta^{\natural,\mathrm{ord},\prime}_{\mathcal{H}_p})$$

between tuples as in (5.2.4.14) *are given by pairs of $\mathbb{Z}_{(p)}^{\times}$-isogenies*

$$(f^{\natural} : G^{\natural} \to G^{\natural,\prime}, f^{\vee,\natural} : G^{\vee,\natural,\prime} \to G^{\vee,\natural})$$

such that we have the following:

(i) *There exists a section $r(f^{\natural}, f^{\vee,\natural})$ of $(\mathbb{Z}_{(p),>0}^{\times})_S$ such that $\lambda^{\natural} = r(f^{\natural}, f^{\vee,\natural}) f^{\vee,\natural} \circ \lambda^{\natural,\prime} \circ f^{\natural}$.*

(ii) *$f^{\natural}$ and $f^{\vee,\natural}$ respect the compatible $\mathcal{O} \underset{\mathbb{Z}}{\otimes} \mathbb{Z}_{(p)}$-actions on $G^{\natural}$, $G^{\natural,\prime}$, $G^{\vee,\natural}$, and $G^{\vee,\natural,\prime}$ (defined by $i^{\natural}$ and $i^{\natural,\prime}$).*

(iii) *$j^{\natural} = (f^{\natural})^* \circ j^{\natural,\prime}$ and $j^{\vee,\natural,\prime} = (f^{\vee,\natural})^* \circ j^{\vee,\natural}$.*

(iv) *$\beta^{\natural,\mathrm{ord}}_{\mathcal{H}_p}$ and $\beta^{\natural,\mathrm{ord},\prime}_{\mathcal{H}_p}$ are canonically identified under the $\mathbb{Z}_{(p)}^{\times}$-isogenies $f^{\natural}$ and $f^{\vee,\natural}$.*

(v) *For each geometric point $\bar{s}$, the morphisms $\mathrm{V}^p(f^{\natural}) : \mathrm{V}^p\, G^{\natural}_{\bar{s}} \xrightarrow{\sim} \mathrm{V}^p\, G^{\natural,\prime}_{\bar{s}}$ and $\mathrm{V}^p(f^{\vee,\natural}) : \mathrm{V}^p\, G^{\vee,\natural,\prime}_{\bar{s}} \xrightarrow{\sim} \mathrm{V}^p\, G^{\vee,\natural}_{\bar{s}}$ satisfy the condition that, for any representatives $\hat{\beta}^{p,\natural}_{\bar{s}} = (\hat{\beta}^{p,\natural,0}_{\bar{s}}, \hat{\beta}^{p,\natural,\#,0}_{\bar{s}}, \hat{\nu}^{p,\natural}_{\bar{s}})$ and $\hat{\beta}^{p,\natural,\prime}_{\bar{s}} = (\hat{\beta}^{p,\natural,0,\prime}_{\bar{s}}, \hat{\beta}^{p,\natural,\#,0,\prime}_{\bar{s}}, \hat{\nu}^{p,\natural,\prime}_{\bar{s}})$ of $[\hat{\beta}^{p,\natural}]_{\mathcal{H}^p_{\mathrm{G}_{1,\mathrm{z}}}}$ and $[\hat{\beta}^{p,\natural,\prime}]_{\mathcal{H}^p_{\mathrm{G}_{1,\mathrm{z}}}}$, respectively, the $\mathcal{H}^p_{\mathrm{G}_{1,\mathrm{z}}}$-orbits of*

$$(\mathrm{V}^p(f^{\natural}) \circ \hat{\beta}^{p,\natural,0}_{\bar{s}}, \mathrm{V}^p(f^{\vee,\natural})^{-1} \circ \hat{\beta}^{p,\natural,\#,0}_{\bar{s}}, r(f^{\natural}, f^{\vee,\natural})^{-1}_{\bar{s}} \hat{\nu}^{p,\natural}_{\bar{s}})$$

and

$$(\hat{\beta}^{p,\natural,0,\prime}_{\bar{s}}, \hat{\beta}^{p,\natural,\#,0,\prime}_{\bar{s}}, \hat{\nu}^{p,\natural,\prime}_{\bar{s}})$$

coincide, where $r(f^{\natural}, f^{\vee,\natural})_{\bar{s}}$ is the value at $\bar{s}$ of the above section $r(f^{\natural}, f^{\vee,\natural})$ of $(\mathbb{Z}_{(p),>0}^{\times})_S$ such that $\lambda^{\natural} = r(f^{\natural}, f^{\vee,\natural}) f^{\vee,\natural} \circ \lambda^{\natural,\prime} \circ f^{\natural}$.

Proof. As in [62, Sec. 1.4.3] and in the proof of Proposition 1.3.2.14, this can be proved by replacing any tuple as in (5.2.4.14) up to $\mathbb{Z}_{(p)}^{\times}$-isogeny, as in the statement of this proposition, with a tuple such that $j^{\natural} : X \underset{\mathbb{Z}}{\otimes} \mathbb{Z}_{(p)} \xrightarrow{\sim} \mathbf{X}(T) \underset{\mathbb{Z}}{\otimes} \mathbb{Z}_{(p)}$ (resp. $j^{\vee,\natural} : Y \underset{\mathbb{Z}}{\otimes} \mathbb{Z}_{(p)} \xrightarrow{\sim} \mathbf{X}(T^{\vee}) \underset{\mathbb{Z}}{\otimes} \mathbb{Z}_{(p)}$) maps X (resp. Y) to $\mathbf{X}(T)$ (resp. $\mathbf{X}(T^{\vee})$), and such that, at each geometric point $\bar{s}$ of S, the assigned $\hat{\beta}^{p,\natural}_{\bar{s}} = (\hat{\beta}^{p,\natural,0}_{\bar{s}}, \hat{\beta}^{p,\natural,\#,0}_{\bar{s}}, \hat{\nu}^{p,\natural}_{\bar{s}})$ satisfies the condition that $\hat{\beta}^{p,\natural,0}_{\bar{s}}$ (resp. $\hat{\beta}^{p,\natural,\#,0}_{\bar{s}}$, resp. $\hat{\nu}^{p,\natural}_{\bar{s}}$) maps $\mathsf{Z}_{-1} \underset{\hat{\mathbb{Z}}}{\otimes} \hat{\mathbb{Z}}^p$ (resp. $\mathsf{Z}^{\#}_{-1} \underset{\hat{\mathbb{Z}}}{\otimes} \hat{\mathbb{Z}}^p$, resp. $\hat{\mathbb{Z}}^p(1)$) to $\mathrm{T}^p\, G^{\natural}_{\bar{s}}$ (resp. $\mathrm{T}^p\, G^{\vee,\natural}_{\bar{s}}$, resp. $\mathrm{T}^p\, \mathbf{G}_{\mathrm{m},\bar{s}}$). Then the tuple determines and is determined by a tuple as in (5.2.4.12), as desired. (These can be simultaneously achieved because of the existence of the section $r(j^{\natural}, j^{\vee,\natural})$ of $(\mathbb{Z}_{(p),>0}^{\times})_S$. The proof is similar to that of [62, Prop. 1.4.3.4], and hence omitted.) □

Construction 5.2.4.15. (Compare with Construction 1.3.2.16.) Suppose that $\mathcal{H} = \mathcal{H}^p\mathcal{H}_p$ is of standard form as in Definition 3.2.2.9, and that $\mathcal{H}^p$ is neat. Consider the degenerating family

$$(G, \lambda, i, \alpha_{\mathcal{H}^p}, \alpha^{\mathrm{ord}}_{\mathcal{H}_p}) \to \vec{\mathsf{M}}^{\mathrm{ord,tor}}_{\mathcal{H},\Sigma^{\mathrm{ord}}} \tag{5.2.4.16}$$

of type $\vec{\mathsf{M}}^{\mathrm{ord}}_{\mathcal{H}}$ as in Theorem 5.2.1.1. Let $\vec{\mathsf{Z}}^{\mathrm{ord}} = \vec{\mathsf{Z}}^{\mathrm{ord}}_{[(\Phi_{\mathcal{H}},\delta_{\mathcal{H}},\sigma)]}$ be any stratum of $\vec{\mathsf{M}}^{\mathrm{ord,tor}}_{\mathcal{H},\Sigma^{\mathrm{ord}}}$ such that $\sigma \subset \mathbf{P}^+_{\Phi_{\mathcal{H}}}$ is a top-dimensional cone in $\Sigma_{\Phi_{\mathcal{H}}}$ (in Σ^{ord}). Let

$$(G^{\natural}_{\overline{\vec{\mathsf{Z}}}^{\mathrm{ord}}}, \lambda^{\natural}_{\overline{\vec{\mathsf{Z}}}^{\mathrm{ord}}}, i^{\natural}_{\overline{\vec{\mathsf{Z}}}^{\mathrm{ord}}}) \to \overline{\vec{\mathsf{Z}}}^{\mathrm{ord}} \tag{5.2.4.17}$$

denote the pullback of the (G, λ, i) in (5.2.4.16) to $\overline{\vec{\mathsf{Z}}}^{\mathrm{ord}}$, the closure of $\vec{\mathsf{Z}}^{\mathrm{ord}}$ in $\vec{\mathsf{M}}^{\mathrm{ord,tor}}_{\mathcal{H},\Sigma^{\mathrm{ord}}}$. Since σ is top-dimensional, the canonical morphism $\vec{\mathsf{Z}}^{\mathrm{ord}} \to \vec{C}^{\mathrm{ord}}_{\Phi_{\mathcal{H}},\delta_{\mathcal{H}}}$ is an isomorphism. Since $\alpha_{\mathcal{H}^p}$ is defined only over $\vec{\mathsf{M}}^{\mathrm{ord}}_{\mathcal{H}}$, its pullback to $\overline{\vec{\mathsf{Z}}}^{\mathrm{ord}}$ is undefined. On the other hand, by condition (5) of Definition 3.4.2.10, $\alpha^{\mathrm{ord}}_{\mathcal{H}_p}$ does extend (necessarily uniquely) to the whole $\vec{\mathsf{M}}^{\mathrm{ord,tor}}_{\mathcal{H},\Sigma^{\mathrm{ord}}}$. The goal of this construction is to define a partial pullback of $\alpha_{\mathcal{H}^p}$ to $\vec{\mathsf{Z}}^{\mathrm{ord}}$, which still retains some information of $\alpha_{\mathcal{H}^p}$, and to give a more precise description of the pullback of (the unique extension of) $\alpha^{\mathrm{ord}}_{\mathcal{H}_p}$ to $\overline{\vec{\mathsf{Z}}}^{\mathrm{ord}}$. (The argument will be very similar to that in Construction 1.3.2.16, but we will spell out the details for the sake of certainty.)

Let $n \geq 1$ be any integer such that $\mathcal{U}(n) \subset \mathcal{H}$, such that $n = n_0 p^r$, where $n_0 \geq 1$ is an integer such that $\mathcal{U}(n_0) \subset \mathcal{H}^p$, and where $r = \mathrm{depth}_{\mathsf{D}}(\mathcal{H}_p) \geq 0$ (*cf.* Definition 3.2.2.9) is the integer such that $\mathcal{U}^{\mathrm{bal}}_{p,1}(p^r) \subset \mathcal{H}_p \subset \mathcal{U}_{p,0}(p^r)$, and let us fix any choice of $(\mathsf{Z}_n, \Phi_n, \delta_n)$. Consider any top-dimensional cone σ' contained in σ that is smooth for the integral structure defined by $\mathbf{S}_{\Phi_n}$, we have a canonical morphism $\vec{\mathfrak{X}}^{\mathrm{ord}}_{\Phi_n,\delta_n,\sigma'} \to \vec{\mathfrak{X}}^{\mathrm{ord}}_{\Phi_{\mathcal{H}},\delta_{\mathcal{H}},\sigma}$ (which might not be finite étale), inducing a morphism from the σ'-stratum $\vec{\mathsf{Z}}^{\mathrm{ord}}_n = \vec{\mathsf{Z}}^{\mathrm{ord}}_{[(\Phi_n,\delta_n,\sigma')]}$ of the source to the σ-stratum $\vec{\mathsf{Z}}^{\mathrm{ord}} = \vec{\mathsf{Z}}^{\mathrm{ord}}_{[(\Phi_{\mathcal{H}},\delta_{\mathcal{H}},\sigma)]}$ of the target (although the scheme-theoretic preimage of latter might not be the former), which can be identified with the canonical morphism (5.2.4.8). Let us denote the pullback of (5.2.4.17) to $\vec{\mathsf{Z}}^{\mathrm{ord}}_n$ by

$$(G^{\natural}_{\mathsf{Z}_n}, \lambda^{\natural}_{\mathsf{Z}_n}, i^{\natural}_{\mathsf{Z}_n}) \to \vec{\mathsf{Z}}^{\mathrm{ord}}_n. \tag{5.2.4.18}$$

Over each affine open formal subscheme $\mathrm{Spf}(R, I)$ of $\vec{\mathfrak{X}}^{\mathrm{ord}}_{\Phi_n,\delta_n,\tau}$, such that $S_0 = \mathrm{Spec}(R/I)$ is the τ-stratum of $S = \mathrm{Spec}(R)$, where both R and R/I are regular domains, we have a degenerating family $(G_S, \lambda_S, i_S, \alpha_{n_0,\eta}, \alpha^{\mathrm{ord}}_{p^r,\eta}) \to S$ of type $\vec{\mathsf{M}}^{\mathrm{ord}}_n = \vec{\mathsf{M}}^{\mathrm{ord}}_{\mathcal{U}^{\mathrm{bal}}_1(n)}$. A priori, the level structure $\alpha_{n_0,\eta}$ is defined only over the generic point η of S (and it only extends to the largest open subscheme of S over which the pullback of G_S is an abelian scheme). But since n_0 is prime to the residue characteristics of S, by the same argument as in Construction 1.3.2.16, it induces a triple $\beta^{\natural}_{n_0,S} := (\beta^{\natural,0}_{n_0,S}, \beta^{\natural,\#,0}_{n_0,S}, \nu^{\natural}_{n_0,S})$ over S, with pullback $\beta^{\natural}_{n_0,S_0} := (\beta^{\natural,0}_{n_0,S_0}, \beta^{\natural,\#,0}_{n_0,S_0}, \nu^{\natural}_{n_0,S_0})$

to S_0. On the other hand, since (5.2.4.16) satisfies condition (5) of Definition 3.4.2.10, by Lemmas 4.1.4.19 and 4.1.4.20,

$$\alpha^{\natural,\mathrm{ord}}_{p^r,\eta} = (\alpha^{\mathrm{ord},0}_{p^r,\eta} : (\mathrm{Gr}^0_{\mathrm{D}_{p^r}})^{\mathrm{mult}}_{\eta} \to G_\eta[p^r],$$
$$\alpha^{\mathrm{ord},\#,0}_{p^r,\eta} : (\mathrm{Gr}^0_{\mathrm{D}^{\#}_{p^r}})^{\mathrm{mult}}_{\eta} \to G^\vee_\eta[p^r], \nu^{\mathrm{ord}}_{p^r,\eta} \in (\mathbb{Z}/p^r\mathbb{Z})^\times_\eta)$$

satisfies Condition 4.1.4.1 and extends to a triple

$$\beta^{\mathrm{ord}}_{p^r,S} = (\beta^{\natural,\mathrm{ord},0}_{p^r,S} : (\mathrm{Gr}^0_{\mathrm{D}_{p^r}})^{\mathrm{mult}}_{S} \to G^\natural_S[p^r],$$
$$\beta^{\natural,\mathrm{ord},\#,0}_{p^r,S} : (\mathrm{Gr}^0_{\mathrm{D}^{\#}_{p^r}})^{\mathrm{mult}}_{S} \to G^{\vee,\natural}_S[p^r], \nu^{\natural,\mathrm{ord}}_{p^r,S} \in (\mathbb{Z}/p^r\mathbb{Z})^\times_S),$$

with pullback to S_0 denoted by $\beta^{\mathrm{ord}}_{p^r,S_0} = (\beta^{\natural,\mathrm{ord},0}_{p^r,S_0}, \beta^{\natural,\mathrm{ord},\#,0}_{p^r,S_0}, \nu^{\natural,\mathrm{ord}}_{p^r,S_0})$. By the construction of $\vec{\mathfrak{X}}^{\mathrm{ord}}_{\Phi_n,\delta_n,\tau}$, the triple $\beta^{\mathrm{ord}}_{p^r,S}$ induces, as in Proposition 4.1.4.21, the prescribed pair $(\varphi_{-2,p^r}, \varphi_{0,p^r})$ in Φ_{p^r} (induced by Φ_n).

As in Construction 1.3.2.16, by analyzing $\beta_{n_0,S}$ as in the case of $\alpha_{n_0,\eta}$ as in [62, Sec. 5.2.2–5.2.3], we see that $\beta_{n_0,S}$ retains almost all information of $\alpha_{n_0,\eta}$, including the pairing e_{10,n_0} to be compared with $\mathtt{d}_{10,n_0}$, as in [62, Lem. 5.2.3.12 and Thm. 5.2.3.14], except that it loses information about the pairing e_{00,n_0} to be compared with $\mathtt{d}_{00,n_0}$. On the other hand, by construction, $\beta^{\mathrm{ord}}_{p^r,S}$ retains all information of $\alpha^{\mathrm{ord}}_{p^r,\eta}$. Hence, if we denote the pullback of (5.2.4.17) to S_0 by $(G^\natural_{S_0}, \lambda^\natural_{S_0}, i^\natural_{S_0}) \to S_0$, then $(G^\natural_{S_0}, \lambda^\natural_{S_0}, i^\natural_{S_0}, \alpha^\natural_{n_0,S_0}, \alpha^{\natural,\mathrm{ord}}_{p^r,S_0}) \to S_0$ determines and is determined by (the prescribed $(\mathsf{Z}_n, \Phi_n, \delta_n)$ and) the pullback to S_0 of the tautological object $((B, \lambda_B, i_B, \varphi^{\mathrm{ord}}_{-1,n}), (c^{\mathrm{ord}}_n, c^{\vee,\mathrm{ord}}_n))$ over $\vec{C}^{\mathrm{ord}}_{\Phi_n,\delta_n}$ (up to isomorphisms inducing automorphisms of Φ_n; *i.e.*, elements of Γ_{Φ_n}; see Lemma 1.3.2.11). By patching over varying S, we obtain (with $(G^\natural_{\vec{\mathsf{Z}}^{\mathrm{ord}}_n}, \lambda^\natural_{\vec{\mathsf{Z}}^{\mathrm{ord}}_n}, i^\natural_{\vec{\mathsf{Z}}^{\mathrm{ord}}_n})$ already defined as in (5.2.4.18)) a tuple

$$(G^\natural_{\vec{\mathsf{Z}}^{\mathrm{ord}}_n}, \lambda^\natural_{\vec{\mathsf{Z}}^{\mathrm{ord}}_n}, i^\natural_{\vec{\mathsf{Z}}^{\mathrm{ord}}_n}, \beta^\natural_{n_0,\vec{\mathsf{Z}}^{\mathrm{ord}}_n}, \beta^{\natural,\mathrm{ord}}_{p^r,\vec{\mathsf{Z}}^{\mathrm{ord}}_n}) \to \vec{\mathsf{Z}}^{\mathrm{ord}}_n \cong \vec{C}^{\mathrm{ord}}_{\Phi_n,\delta_n} \tag{5.2.4.19}$$

such that the previous sentence is true with S_0 replaced with $\vec{\mathsf{Z}}^{\mathrm{ord}}_n$. Since $\mathcal{H}_{\mathrm{G}_{1,\mathrm{z}}}/\mathcal{U}^{\mathrm{bal}}_1(n)_{\mathrm{G}_{1,\mathrm{z}}}$ acts compatibly on $(\beta^\natural_{n_0,\vec{\mathsf{Z}}^{\mathrm{ord}}_n}, \beta^{\natural,\mathrm{ord}}_{p^r,\vec{\mathsf{Z}}^{\mathrm{ord}}_n})$ and $(\varphi^{\mathrm{ord}}_{-1,n}, c^{\mathrm{ord}}_n, c^{\vee,\mathrm{ord}}_n)$, the latter action being compatible with the $\mathcal{H}_{\mathrm{G}_{1,\mathrm{z}}}/\mathcal{U}^{\mathrm{bal}}_1(n)_{\mathrm{G}_{1,\mathrm{z}}}$-torsor structure of (5.2.4.8), by forming the $\mathcal{H}_{\mathrm{G}_{1,\mathrm{z}}}/\mathcal{U}^{\mathrm{bal}}_1(n)_{\mathrm{G}_{1,\mathrm{z}}}$-orbit $(\beta^\natural_{\mathcal{H}^p,\vec{\mathsf{Z}}^{\mathrm{ord}}_n}, \beta^{\natural,\mathrm{ord}}_{\mathcal{H}_p,\vec{\mathsf{Z}}^{\mathrm{ord}}_n})$ of $(\beta^\natural_{n_0,\vec{\mathsf{Z}}^{\mathrm{ord}}_n}, \beta^{\natural,\mathrm{ord}}_{p^r,\vec{\mathsf{Z}}^{\mathrm{ord}}_n})$, we can descend (5.2.4.19) to a tuple

$$(G^\natural_{\vec{\mathsf{Z}}^{\mathrm{ord}}}, \lambda^\natural_{\vec{\mathsf{Z}}^{\mathrm{ord}}}, i^\natural_{\vec{\mathsf{Z}}^{\mathrm{ord}}}, \beta^\natural_{\mathcal{H}^p,\vec{\mathsf{Z}}^{\mathrm{ord}}}, \beta^{\natural,\mathrm{ord}}_{\mathcal{H}_p,\vec{\mathsf{Z}}^{\mathrm{ord}}}) \to \vec{\mathsf{Z}}^{\mathrm{ord}} \cong \vec{C}^{\mathrm{ord}}_{\Phi_{\mathcal{H}},\delta_{\mathcal{H}}}, \tag{5.2.4.20}$$

where the first three entries form the pullback of (5.2.4.17) to $\vec{\mathsf{Z}}^{\mathrm{ord}}$, which determines and is determined by (the prescribed $(\mathsf{Z}_{\mathcal{H}}, \Phi_{\mathcal{H}}, \delta_{\mathcal{H}})$ and) the tautological object

$$((B, \lambda_B, i_B, \varphi^{\mathrm{ord}}_{-1,\mathcal{H}}), (\varphi^{\mathrm{ord}}_{-2,\mathcal{H}}, \varphi^{\mathrm{ord}}_{0,\mathcal{H}}), (c^{\mathrm{ord}}_{\mathcal{H}}, c^{\vee,\mathrm{ord}}_{\mathcal{H}})) \to \vec{C}^{\mathrm{ord}}_{\Phi_{\mathcal{H}},\delta_{\mathcal{H}}} \tag{5.2.4.21}$$

(up to isomorphisms inducing automorphisms of $\Phi_{\mathcal{H}}$; *i.e.*, elements of $\Gamma_{\Phi_{\mathcal{H}}}$). Since the tautological object (5.2.4.21) is independent of the choice of n_0, so is the tuple (5.2.4.20).

By abuse of language, we say that

$$(G^{\natural}_{\vec{\overline{\mathsf{Z}}}^{\mathrm{ord}}}, \lambda^{\natural}_{\vec{\overline{\mathsf{Z}}}^{\mathrm{ord}}}, i^{\natural}_{\vec{\overline{\mathsf{Z}}}^{\mathrm{ord}}}, \beta^{\natural}_{\mathcal{H}^p, \vec{\mathsf{Z}}^{\mathrm{ord}}}, \beta^{\natural,\mathrm{ord}}_{\mathcal{H}_p, \vec{\mathsf{Z}}^{\mathrm{ord}}}) \to \vec{\overline{\mathsf{Z}}}^{\mathrm{ord}} \tag{5.2.4.22}$$

is the *pullback* of the degenerating family (5.2.4.16) to $\vec{\overline{\mathsf{Z}}}^{\mathrm{ord}}$, with the convention that (as in the case of $(G, \lambda, i, \alpha_{\mathcal{H}^p}, \alpha^{\mathrm{ord}}_{\mathcal{H}_p})$ itself) $\beta^{\natural}_{\mathcal{H}^p, \vec{\mathsf{Z}}^{\mathrm{ord}}}$ is defined only over $\vec{\mathsf{Z}}^{\mathrm{ord}}$, while $(G^{\natural}, \lambda^{\natural}, i^{\natural})$ (resp. $\beta^{\natural,\mathrm{ord}}_{\mathcal{H}_p, \vec{\mathsf{Z}}^{\mathrm{ord}}}$) is defined (resp. extends) over all of $\vec{\overline{\mathsf{Z}}}^{\mathrm{ord}}$ as in (5.2.4.17).

As in the proof of (7) of Theorem 5.2.1.1, in every step of our construction of $\vec{\mathsf{M}}^{\mathrm{ord,tor}}_{\mathcal{H},\Sigma^{\mathrm{ord}}}$, the characteristic zero fiber of the degenerating families over the boundary charts we have used are the pullback from $\mathsf{S}_0 = \mathrm{Spec}(F_0)$ to $\mathsf{S}_{0,r_{\mathcal{H}}} = \mathrm{Spec}(F_0[\zeta_{p^{r_{\mathcal{H}}}}])$ of the corresponding ones over toroidal boundary charts for $\mathsf{M}_{\mathcal{H}}$, the only deference being that we have only considered ordinary cusp labels in the construction for $\vec{\mathsf{M}}^{\mathrm{ord,tor}}_{\mathcal{H},\Sigma^{\mathrm{ord}}}$. Therefore, there is a degenerating family

$$(G, \lambda, i, \alpha_{\mathcal{H}}) \to \vec{\mathsf{M}}^{\mathrm{ord,tor}}_{\mathcal{H},\Sigma^{\mathrm{ord}}} \tag{5.2.4.23}$$

of type $\mathsf{M}_{\mathcal{H}}$, with the same (G, λ, i) as in (5.2.4.16), where $\alpha_{\mathcal{H}}$ is defined only over $\vec{\mathsf{M}}^{\mathrm{ord}}_{\mathcal{H}} \underset{\mathbb{Z}}{\otimes} \mathbb{Q}$, such that the pair $(\alpha_{\mathcal{H}^p}, \alpha^{\mathrm{ord}}_{\mathcal{H}_p}) \underset{\mathbb{Z}}{\otimes} \mathbb{Q}$ is induced by $\alpha_{\mathcal{H}}$ as in Proposition 3.3.5.1. Therefore, by repeating the argument as in Construction 1.3.2.16 for $\alpha_{\mathcal{H}}$, we obtain a pullback

$$(G^{\natural}_{\vec{\overline{\mathsf{Z}}}^{\mathrm{ord}}}, \lambda^{\natural}_{\vec{\overline{\mathsf{Z}}}^{\mathrm{ord}}}, i^{\natural}_{\vec{\overline{\mathsf{Z}}}^{\mathrm{ord}}}, \beta^{\natural}_{\mathcal{H}, \vec{\mathsf{Z}}^{\mathrm{ord}} \underset{\mathbb{Z}}{\otimes} \mathbb{Q}}) \to \vec{\overline{\mathsf{Z}}}^{\mathrm{ord}} \tag{5.2.4.24}$$

as in (5.2.4.22), where $\beta^{\natural}_{\mathcal{H}, \vec{\mathsf{Z}}^{\mathrm{ord}} \underset{\mathbb{Z}}{\otimes} \mathbb{Q}}$ is defined only over $\vec{\mathsf{Z}}^{\mathrm{ord}} \underset{\mathbb{Z}}{\otimes} \mathbb{Q}$, such that the pair $(\beta^{\natural}_{\mathcal{H}^p, \vec{\mathsf{Z}}^{\mathrm{ord}}}, \beta^{\natural,\mathrm{ord}}_{\mathcal{H}_p, \vec{\mathsf{Z}}^{\mathrm{ord}}}) \underset{\mathbb{Z}}{\otimes} \mathbb{Q}$ is induced by $\beta^{\natural}_{\mathcal{H}, \vec{\mathsf{Z}}^{\mathrm{ord}} \underset{\mathbb{Z}}{\otimes} \mathbb{Q}}$ by an obvious analogue of Proposition 3.3.5.1. (This finishes Construction 5.2.4.15.)

Proposition 5.2.4.25. (*Compare with Proposition* 1.3.2.24.) *By considering compatible* $\mathbb{Q}^{\times}$*-isogenies* $(f : G^{\natural} \to G^{\natural,\prime}, f^{\vee} : G^{\vee,\natural,\prime} \to G^{\vee,\natural})$ *inducing isomorphisms on the torus parts, we can define ordinary Hecke twists of the tautological object* $(G^{\natural}, \lambda^{\natural}, i^{\natural}, \beta^{\natural}_{\mathcal{H}^p}, \beta^{\natural,\mathrm{ord}}_{\mathcal{H}_p}) \to \vec{C}^{\mathrm{ord}}_{\Phi_{\mathcal{H}},\delta_{\mathcal{H}}}$ *by elements* $g = (g_0, g_p) \in \mathrm{G}_{1,\mathsf{Z}}(\mathbb{A}^{\infty,p}) \times \mathrm{P}^{\mathrm{ord}}_{1,\mathsf{Z},\mathsf{D}}(\mathbb{Q}_p) \subset \mathrm{G}_{1,\mathsf{Z}}(\mathbb{A}^{\infty})$ *such that the image of* g_p *under the canonical homomorphism* $\mathrm{P}^{\mathrm{ord}}_{1,\mathsf{Z},\mathsf{D}}(\mathbb{Q}_p) \to \mathrm{P}^{\mathrm{ord}}_{h,\mathsf{D}}(\mathbb{Q}_p)$ *satisfies the condition defined by the filtration* D_{-1} *on* $\mathrm{Gr}^{\mathsf{Z}}_{-1} \underset{\hat{\mathbb{Z}}}{\otimes} \mathbb{Z}_p$ *as in Section* 3.3.4*, and define the Hecke action of (such elements of)* $\mathrm{G}_{1,\mathsf{Z}}(\mathbb{A}^{\infty,p}) \times \mathrm{P}^{\mathrm{ord}}_{1,\mathsf{Z},\mathsf{D}}(\mathbb{Q}_p)$ *on the collection* $\{\vec{C}^{\mathrm{ord}}_{\Phi_{\mathcal{H}},\delta_{\mathcal{H}}}\}_{\mathcal{H}_{\mathrm{G}_{1,\mathsf{Z}}}}$ *with* $\mathcal{H}$ *of standard form, realized by quasi-finite flat surjections pulling tautological objects back to ordinary Hecke twists, which is compatible with the Hecke action of (suitable elements of)* $\mathrm{G}_{h,\mathsf{Z}}(\mathbb{A}^{\infty,p}) \times \mathrm{P}^{\mathrm{ord}}_{h,\mathsf{Z},\mathsf{D}}(\mathbb{Q}_p)$ *on the collection* $\{\vec{\mathsf{M}}^{\mathrm{ord},\Phi_{\mathcal{H}}}_{\mathcal{H}}\}_{\mathcal{H}_{\mathrm{G}'_{h,\mathsf{Z}}}}$ *(with* $\mathcal{H}$ *of standard form) under the canonical morphisms* $\vec{C}^{\mathrm{ord}}_{\Phi_{\mathcal{H}},\delta_{\mathcal{H}}} \to \vec{\mathsf{M}}^{\mathrm{ord},\Phi_{\mathcal{H}}}_{\mathcal{H}}$ *(with varying* $\mathcal{H}$*) and the canonical homomorphism* $\mathrm{G}_{1,\mathsf{Z}}(\mathbb{A}^{\infty,p}) \times \mathrm{P}^{\mathrm{ord}}_{1,\mathsf{Z},\mathsf{D}}(\mathbb{Q}_p) \to \mathrm{G}'_{h,\mathsf{Z}}(\mathbb{A}^{\infty,p}) \times \mathrm{P}^{\mathrm{ord},\prime}_{h,\mathsf{Z},\mathsf{D}}(\mathbb{Q}_p)$.

Such a Hecke action enjoys the properties (under various conditions) concerning étaleness, finiteness, being isomorphisms between formal completions along fibers over $\mathrm{Spec}(\mathbb{F}_p)$, *and inducing absolute Frobenius morphisms on fibers over* $\mathrm{Spec}(\mathbb{F}_p)$ *for elements of* U_p *type as in Proposition* 5.2.2.2 *and Corollaries* 5.2.2.3, 5.2.2.4, *and* 5.2.2.5. *(We omit the details for simplicity.) Over the subcollection indexed by* $\mathcal{H}_{\mathrm{G}_{1,\mathsf{Z}}}$ *with neat* $\mathcal{H}^p$ *(for* $\mathcal{H} = \mathcal{H}^p\mathcal{H}_p$ *of standard form), the Hecke action of (suitable elements of)* $\mathrm{G}_{1,\mathsf{Z}}(\mathbb{A}^{\infty,p}) \times \mathrm{P}^{\mathrm{ord}}_{1,\mathsf{Z},\mathsf{D}}(\mathbb{Q}_p)$ *on* $\{\vec{C}^{\mathrm{ord}}_{\Phi_{\mathcal{H}},\delta_{\mathcal{H}}}\}_{\mathcal{H}_{\mathrm{G}_{1,\mathsf{Z}}}}$ *is compatible with the Hecke action of (suitable elements of)* $\mathrm{P}'_{\mathsf{Z}}(\mathbb{A}^{\infty,p}) \times \mathrm{P}^{\mathrm{ord},\prime}_{\mathsf{Z},\mathsf{D}}(\mathbb{Q}_p)$ *on the collection of strata* $\{\vec{\mathsf{Z}}^{\mathrm{ord}}_{[(\Phi_{\mathcal{H}},\delta_{\mathcal{H}},\sigma)]}\}$ *above* $\{\vec{C}^{\mathrm{ord}}_{\Phi_{\mathcal{H}},\delta_{\mathcal{H}}}\}_{\mathcal{H}_{\mathrm{G}_{1,\mathsf{Z}}}}$ *(cf. Proposition* 5.2.2.2*) under the canonical homomorphism* $\mathrm{P}'_{\mathsf{Z}}(\mathbb{A}^{\infty,p}) \times \mathrm{P}^{\mathrm{ord},\prime}_{\mathsf{Z},\mathsf{D}}(\mathbb{Q}_p) \to \mathrm{G}_{1,\mathsf{Z}}(\mathbb{A}^{\infty,p}) \times \mathrm{P}^{\mathrm{ord}}_{1,\mathsf{Z},\mathsf{D}}(\mathbb{Q}_p) \cong (\mathrm{P}'_{\mathsf{Z}}(\mathbb{A}^{\infty,p}) \times \mathrm{P}^{\mathrm{ord},\prime}_{\mathsf{Z},\mathsf{D}}(\mathbb{Q}_p))/\mathrm{U}_{2,\mathsf{Z}}(\mathbb{A}^{\infty})$.

By also considering $\mathbb{Q}^\times$*-isogenies* $(f : G^\natural \to G^{\natural,\prime}, f^\vee : G^{\vee,\natural,\prime} \to G^{\vee,\natural})$ *inducing* $\mathbb{Q}^\times$*-isogenies on the torus parts, we can also define ordinary Hecke twists of the tautological object* $(G^\natural, \lambda^\natural, i^\natural, \beta^\natural_{\mathcal{H}^p}, \beta^{\natural,\mathrm{ord}}_{\mathcal{H}_p}) \to \vec{C}^{\mathrm{ord}}_{\Phi_{\mathcal{H}},\delta_{\mathcal{H}}}$ *by elements* $g = (g_0, g_p) \in (\mathrm{P}_{\mathsf{Z}}(\mathbb{A}^{\infty,p}) \times \mathrm{P}^{\mathrm{ord}}_{\mathsf{Z},\mathsf{D}}(\mathbb{Q}_p))/\mathrm{U}_{2,\mathsf{Z}}(\mathbb{A}^{\infty})$ *such that the image of* g_p *under the canonical homomorphism* $\mathrm{P}^{\mathrm{ord}}_{\mathsf{Z},\mathsf{D}}(\mathbb{Q}_p)/\mathrm{U}_{2,\mathsf{Z}}(\mathbb{Q}_p) \to \mathrm{P}^{\mathrm{ord}}_{h,\mathsf{D}}(\mathbb{Q}_p)$ *satisfies the condition defined by the filtration* D_{-1} *on* $\mathrm{Gr}^{\mathsf{Z}}_{-1} \underset{\hat{\mathbb{Z}}}{\otimes} \mathbb{Z}_p$ *as in Section* 3.3.4, *and define the Hecke action of (such elements of)* $(\mathrm{P}_{\mathsf{Z}}(\mathbb{A}^{\infty,p}) \times \mathrm{P}^{\mathrm{ord}}_{\mathsf{Z},\mathsf{D}}(\mathbb{Q}_p))/\mathrm{U}_{2,\mathsf{Z}}(\mathbb{A}^{\infty})$ *on the collection* $\{\coprod \vec{C}^{\mathrm{ord}}_{\Phi_{\mathcal{H}},\delta_{\mathcal{H}}}\}_{\mathcal{H}_{\mathrm{P}_{\mathsf{Z}}}/\mathcal{H}_{\mathrm{U}_{2,\mathsf{Z}}}}$ *(with* $\mathcal{H}$ *of standard form), where the disjoint unions are over classes* $[(\mathsf{Z}_{\mathcal{H}}, \Phi_{\mathcal{H}}, \delta_{\mathcal{H}})]$ *sharing the same* $\mathsf{Z}_{\mathcal{H}}$ *compatible with* D, *realized by quasi-finite flat surjections pulling tautological objects back to ordinary Hecke twists, which induces an action of* $\mathrm{G}'_{l,\mathsf{Z}}(\mathbb{A}^{\infty}) = \mathrm{P}_{\mathsf{Z}}(\mathbb{A}^{\infty})/\mathrm{P}'_{\mathsf{Z}}(\mathbb{A}^{\infty}) \cong (\mathrm{P}_{\mathsf{Z}}(\mathbb{A}^{\infty,p}) \times \mathrm{P}^{\mathrm{ord}}_{\mathsf{Z},\mathsf{D}}(\mathbb{Q}_p))/(\mathrm{P}'_{\mathsf{Z}}(\mathbb{A}^{\infty,p}) \times \mathrm{P}^{\mathrm{ord},\prime}_{\mathsf{Z},\mathsf{D}}(\mathbb{Q}_p))$ *on the index sets* $\{[(\mathsf{Z}_{\mathcal{H}}, \Phi_{\mathcal{H}}, \delta_{\mathcal{H}})]\}$, *which is compatible with the Hecke action of (suitable elements of)* $(\mathrm{P}_{\mathsf{Z}}(\mathbb{A}^{\infty,p}) \times \mathrm{P}^{\mathrm{ord}}_{\mathsf{Z},\mathsf{D}}(\mathbb{Q}_p))/\mathrm{U}_{\mathsf{Z}}(\mathbb{A}^{\infty}) \cong \mathrm{G}'_{l,\mathsf{Z}}(\mathbb{A}^{\infty}) \times (\mathrm{G}'_{h,\mathsf{Z}}(\mathbb{A}^{\infty,p}) \times \mathrm{P}^{\mathrm{ord},\prime}_{h,\mathsf{Z},\mathsf{D}}(\mathbb{Q}_p))$ *on the collection* $\{\coprod \vec{\mathsf{M}}^{\mathrm{ord},\Phi_{\mathcal{H}}}_{\mathcal{H}}\}_{\mathcal{H}_{\mathrm{G}'_{h,\mathsf{Z}}}}$ *(with* $\mathcal{H}$ *of standard form, with the same index sets and the same induced action of* $\mathrm{G}'_{l,\mathsf{Z}}(\mathbb{A}^{\infty})$*) under the canonical morphisms* $\vec{C}^{\mathrm{ord}}_{\Phi_{\mathcal{H}},\delta_{\mathcal{H}}} \to \vec{\mathsf{M}}^{\mathrm{ord},\Phi_{\mathcal{H}}}_{\mathcal{H}}$ *(with varying* $\mathcal{H}$*) and the canonical homomorphism* $(\mathrm{P}_{\mathsf{Z}}(\mathbb{A}^{\infty,p}) \times \mathrm{P}^{\mathrm{ord}}_{\mathsf{Z},\mathsf{D}}(\mathbb{Q}_p))/\mathrm{U}_{2,\mathsf{Z}}(\mathbb{A}^{\infty}) \to \mathrm{G}'_{l,\mathsf{Z}}(\mathbb{A}^{\infty}) \times (\mathrm{G}'_{h,\mathsf{Z}}(\mathbb{A}^{\infty,p}) \times \mathrm{P}^{\mathrm{ord},\prime}_{h,\mathsf{Z},\mathsf{D}}(\mathbb{Q}_p))$. *Over the subcollection indexed by* $\mathcal{H}_{\mathrm{P}_{\mathsf{Z}}}/\mathcal{H}_{\mathrm{U}_{2,\mathsf{Z}}}$ *with neat* $\mathcal{H}^p$ *(for* $\mathcal{H} = \mathcal{H}^p\mathcal{H}_p$ *of standard form), the Hecke action of (suitable elements of)* $(\mathrm{P}_{\mathsf{Z}}(\mathbb{A}^{\infty,p}) \times \mathrm{P}^{\mathrm{ord}}_{\mathsf{Z},\mathsf{D}}(\mathbb{Q}_p))/\mathrm{U}_{2,\mathsf{Z}}(\mathbb{A}^{\infty})$ *on* $\{\coprod \vec{C}^{\mathrm{ord}}_{\Phi_{\mathcal{H}},\delta_{\mathcal{H}}}\}_{\mathcal{H}_{\mathrm{P}_{\mathsf{Z}}}/\mathcal{H}_{\mathrm{U}_{2,\mathsf{Z}}}}$ *is compatible with the Hecke action of (suitable elements of)* $\mathrm{P}_{\mathsf{Z}}(\mathbb{A}^{\infty,p}) \times \mathrm{P}^{\mathrm{ord}}_{\mathsf{Z},\mathsf{D}}(\mathbb{Q}_p)$ *on the collection of strata* $\{\vec{\mathsf{Z}}^{\mathrm{ord}}_{[(\Phi_{\mathcal{H}},\delta_{\mathcal{H}},\sigma)]}\}$ *above* $\{\coprod \vec{C}^{\mathrm{ord}}_{\Phi_{\mathcal{H}},\delta_{\mathcal{H}}}\}_{\mathcal{H}_{\mathrm{P}_{\mathsf{Z}}}/\mathcal{H}_{\mathrm{U}_{2,\mathsf{Z}}}}$ *(cf. Proposition* 5.2.2.2*) under the canonical homomorphism* $\mathrm{P}_{\mathsf{Z}}(\mathbb{A}^{\infty,p}) \times \mathrm{P}^{\mathrm{ord}}_{\mathsf{Z},\mathsf{D}}(\mathbb{Q}_p) \to (\mathrm{P}_{\mathsf{Z}}(\mathbb{A}^{\infty,p}) \times \mathrm{P}^{\mathrm{ord}}_{\mathsf{Z},\mathsf{D}}(\mathbb{Q}_p))/\mathrm{U}_{2,\mathsf{Z}}(\mathbb{A}^{\infty})$.

In the $\mathbb{Z}^\times_{(p)}$*-isogeny class language as in Proposition* 5.2.4.13, *the morphism*

$$[g] : \vec{C}^{\mathrm{ord}}_{\Phi_{\mathcal{H}'},\delta_{\mathcal{H}'}} \to \vec{C}^{\mathrm{ord}}_{\Phi_{\mathcal{H}},\delta_{\mathcal{H}}},$$

for $g = (g_0, g_p) \in (\mathrm{P}_{\mathsf{Z}}(\mathbb{A}^{\infty,p}) \times \mathrm{P}^{\mathrm{ord}}_{\mathsf{Z},\mathsf{D}}(\mathbb{Q}_p))/\mathrm{U}_{2,\mathsf{Z}}(\mathbb{A}^{\infty})$ *as above such that* $\mathcal{H}'_{\mathrm{P}_{\mathsf{Z}}}/\mathcal{H}'_{\mathrm{U}_{2,\mathsf{Z}}} \subset g(\mathcal{H}_{\mathrm{P}_{\mathsf{Z}}}/\mathcal{H}_{\mathrm{U}_{2,\mathsf{Z}}})g^{-1}$ *and such that* $[(\Phi_{\mathcal{H}}, \delta_{\mathcal{H}})]$ *is* g*-assigned to* $[(\Phi_{\mathcal{H}'}, \delta_{\mathcal{H}'})]$ *with a pair* *isomorphisms*

$$(f_X : X \underset{\mathbb{Z}}{\otimes} \mathbb{Q} \xrightarrow{\sim} X' \underset{\mathbb{Z}}{\otimes} \mathbb{Q}, f_Y : Y' \underset{\mathbb{Z}}{\otimes} \mathbb{Q} \xrightarrow{\sim} Y \underset{\mathbb{Z}}{\otimes} \mathbb{Q})$$

as in [62, Prop. 5.4.3.8], is characterized by

$$\begin{aligned}
&[g]^*(G^\natural, \lambda^\natural : G^\natural \to G^{\vee,\natural}, i^\natural, j^\natural, j^{\vee,\natural}, [\hat{\beta}^{p,\natural}]_{\mathcal{H}^p_{\mathrm{G}_{1,\mathrm{Z}}}}, \beta^{\natural,\mathrm{ord}}_{\mathcal{H}_p}) \\
&\sim_{\mathbb{Z}^\times_{(p)}\text{-isog.}} (G^{\natural,\prime\prime}, \lambda^{\natural,\prime\prime} : G^{\natural,\prime\prime} \to G^{\vee,\natural,\prime\prime}, i^{\natural,\prime\prime}, \\
&\qquad\qquad f_X \circ j^{\natural,\prime\prime}, f_Y^{-1} \circ j^{\vee,\natural,\prime\prime}, [\hat{\beta}^{p,\natural,\prime\prime} \circ g_0]_{\mathcal{H}^p_{\mathrm{G}_{1,\mathrm{Z}}}}, \beta^{\natural,\mathrm{ord},\prime\prime}_{\mathcal{H}_p})
\end{aligned}$$

over $C_{\Phi_{\mathcal{H}'},\delta_{\mathcal{H}'}}$, *where*

$$(G^\natural, \lambda^\natural : G^\natural \to G^{\vee,\natural}, i^\natural, j^\natural, j^{\vee,\natural}, [\hat{\beta}^{p,\natural}]_{\mathcal{H}^p_{\mathrm{G}_{1,\mathrm{Z}}}}, \beta^{\natural,\mathrm{ord}}_{\mathcal{H}_p})$$

and

$$(G^{\natural,\prime}, \lambda^{\natural,\prime} : G^{\natural,\prime} \to G^{\vee,\natural,\prime}, i^{\natural,\prime}, j^{\natural,\prime}, j^{\vee,\natural,\prime}, [\hat{\beta}^{p,\natural,\prime}]_{\mathcal{H}'^{,p}_{\mathrm{G}_{1,\mathrm{Z}}}}, \beta^{\natural,\mathrm{ord}}_{\mathcal{H}'_p})$$

are representatives of the tautological $\mathbb{Z}^\times_{(p)}$*-isogeny classes over* $\vec{C}^{\mathrm{ord}}_{\Phi_{\mathcal{H}},\delta_{\mathcal{H}}}$ *and* $\vec{C}^{\mathrm{ord}}_{\Phi_{\mathcal{H}'},\delta_{\mathcal{H}'}}$, *respectively, where*

$$(G^{\natural,\prime\prime}, \lambda^{\natural,\prime\prime} : G^{\natural,\prime\prime} \to G^{\vee,\natural,\prime\prime}, i^{\natural,\prime\prime}, j^{\natural,\prime\prime}, j^{\vee,\natural,\prime\prime}, [\hat{\beta}^{p,\natural,\prime\prime}]_{\mathcal{H}'^{,p}_{\mathrm{G}_{1,\mathrm{Z}}}}, \beta^{\natural,\mathrm{ord},\prime\prime}_{\mathcal{H}_p})$$

is the ordinary Hecke twists of the latter by g_p *(defined as in Proposition* 3.3.4.9, *with details omitted for simplicity) realized by some pair*

$$(G^{\natural,\prime} \to G^{\natural,\prime\prime}, G^{\vee,\natural,\prime\prime} \to G^{\vee,\natural,\prime})$$

of isogenies of p*-power degrees (with canonically induced additional structures), and where the rational level-*$\mathcal{H}^p$ *structure*

$$[\hat{\beta}^{p,\natural,\prime\prime} \circ g_0]_{\mathcal{H}^p_{\mathrm{G}_{1,\mathrm{Z}}}}$$

of $(G^{\natural,\prime\prime}, \lambda^{\natural,\prime\prime}, i^{\natural,\prime\prime}, f_X \circ j^{\natural,\prime\prime}, f_Y^{-1} \circ j^{\vee,\natural,\prime\prime})$ *of type* $(L \underset{\mathbb{Z}}{\otimes} \mathbb{A}^{\infty,p}, \langle\cdot,\cdot\rangle, \mathrm{Z} \underset{\hat{\mathbb{Z}}}{\otimes} \mathbb{A}^{\infty,p}, \Phi)$ *is determined at each geometric point* $\bar{s}$ *of* $\vec{C}^{\mathrm{ord}}_{\Phi_{\mathcal{H}'},\delta_{\mathcal{H}'}}$ *by the* $\mathcal{H}^p_{\mathrm{G}_{1,\mathrm{Z}}}$*-orbit of* $\hat{\beta}^{p,\natural,\prime\prime}_{\bar{s}} \circ g_0$, *where* $\hat{\beta}^{p,\natural,\prime\prime}_{\bar{s}}$ *is any representative of the rational level-*$\mathcal{H}'^{,p}$ *structure* $[\hat{\beta}^{p,\natural,\prime\prime}_{\bar{s}}]_{\mathcal{H}'^{,p}_{\mathrm{G}_{1,\mathrm{Z}}}}$ *of* $(G^{\natural,\prime\prime}, \lambda^{\natural,\prime\prime}, i^{\natural,\prime\prime}, j^{\natural,\prime\prime}, j^{\vee,\natural,\prime\prime})$ *of type* $(L \underset{\mathbb{Z}}{\otimes} \mathbb{A}^{\infty,p}, \langle\cdot,\cdot\rangle, \mathrm{Z} \underset{\hat{\mathbb{Z}}}{\otimes} \mathbb{A}^{\infty,p}, \Phi')$ *based at* $\bar{s}$ *(assigned to* $\bar{s}$ *by* $[\hat{\beta}^{\natural,\prime\prime}]_{\mathcal{H}'^{,p}_{\mathrm{G}_{1,\mathrm{Z}}}}$*).*

Proof. As in the proof of Proposition 1.3.2.24, the first assertions of both of the first two paragraphs, and the whole third paragraph, can be justified as in the case of $\vec{\mathsf{M}}^{\mathrm{ord}}_{\mathcal{H}}$. (We omit the details for simplicity.) As for the second assertions in both of the first two paragraphs, it suffices to note that the pullback of the ordinary Hecke twist of (5.2.4.16) is the ordinary Hecke twist of (5.2.4.20), the latter of which can be identified with the tautological object over $\vec{C}^{\mathrm{ord}}_{\Phi_{\mathcal{H}},\delta_{\mathcal{H}}}$ under the canonical isomorphism $\vec{\mathsf{Z}}^{\mathrm{ord}}_{[(\Phi_{\mathcal{H}},\delta_{\mathcal{H}},\sigma)]} \xrightarrow{\sim} \vec{C}^{\mathrm{ord}}_{\Phi_{\mathcal{H}},\delta_{\mathcal{H}}}$ (for any top-dimensional σ, when $\mathcal{H}^p$ and hence $\mathcal{H}$ are neat). □

By Proposition 4.2.1.37 and its proof, we have the following:

Lemma 5.2.4.26. (*Compare with Lemma* 1.3.2.25.) *The quotient* $\vec{\Xi}^{\mathrm{ord}}_{\Phi_{\mathcal{H}},\delta_{\mathcal{H}}}$ *depends only on* $\mathcal{H}_{\mathrm{P}'_{\mathsf{Z}}}$, *and is a torsor under the torus* $E_{\Phi_{\mathcal{H}}}$ *with character group* $\mathbf{S}_{\Phi_{\mathcal{H}}}$.

If $r = \mathrm{depth}_{\mathsf{D}}(\mathcal{H})$ *and if* $n_0 \geq 1$ *is any integer prime to* p *such that* $\mathcal{U}^p(n_0) \subset \mathcal{H}^p$, *so that* $\mathcal{U}_1^{\mathrm{bal}}(n) = \mathcal{U}^p(n_0)\mathcal{U}_{p,1}^{\mathrm{bal}}(p^r) \subset \mathcal{H}$ *are both of standard form and equally deep, and if we fix the choice of* (Z_n *and*) Φ_n, *then the canonical morphism*

$$\vec{\Xi}^{\mathrm{ord}}_{\Phi_n,\delta_n} \to \vec{\Xi}^{\mathrm{ord}}_{\Phi_{\mathcal{H}},\delta_{\mathcal{H}},r_n} = \vec{\Xi}^{\mathrm{ord}}_{\Phi_{\mathcal{H}},\delta_{\mathcal{H}}} \underset{\vec{\mathsf{S}}_{0,r_{\mathcal{H}}}}{\times} \vec{\mathsf{S}}_{0,r_n}, \tag{5.2.4.27}$$

where $r_n = r_{\mathcal{U}_1^{\mathrm{bal}}(n)} = \max(r_{\mathsf{D}}, r)$, (*is finite étale and*) *is an* $\mathcal{H}_{\mathrm{P}'_{\mathsf{Z}}}/\mathcal{U}_1^{\mathrm{bal}}(n)_{\mathrm{P}'_{\mathsf{Z}}} \cong H^{\mathrm{ord}}_{n,\mathrm{G}^{\mathrm{ess}}_{h,\mathsf{Z}_n} \ltimes \mathrm{U}^{\mathrm{ess}}_{\mathsf{Z}_n}}$-*torsor* (*see page* 254 *for the definition of* $H^{\mathrm{ord}}_{n,\mathrm{G}^{\mathrm{ess}}_{h,\mathsf{Z}_n} \ltimes \mathrm{U}^{\mathrm{ess}}_{\mathsf{Z}_n}}$), *where* $\mathcal{H}_{\mathrm{P}'_{\mathsf{Z}}}$ *and* $\mathcal{U}_1^{\mathrm{bal}}(n)_{\mathrm{P}'_{\mathsf{Z}}}$ *are as in Definition* 1.2.1.12, *and induces an isomorphism*

$$\vec{\Xi}^{\mathrm{ord}}_{\Phi_n,\delta_n}/(\mathcal{H}_{\mathrm{P}'_{\mathsf{Z}}}/\mathcal{U}_1^{\mathrm{bal}}(n)_{\mathrm{P}'_{\mathsf{Z}}}) \xrightarrow{\sim} \vec{\Xi}^{\mathrm{ord}}_{\Phi_{\mathcal{H}},\delta_{\mathcal{H}},r_n}. \tag{5.2.4.28}$$

Proof. The statements are self-explanatory. □

Lemma 5.2.4.29. (*Compare with Lemmas* 1.3.2.28 *and* 5.2.4.7.) *Suppose* $n = n_0p^r$ *for some integer* $n_0 \geq 1$ *prime to* p *and some integer* $r \geq 0$. *Suppose* $(B, \lambda_B, i_B, \varphi^{\mathrm{ord}}_{-1,n})$, $(\mathsf{Z}_n, \Phi_n = (X, Y, \phi : Y \hookrightarrow X, \varphi_{-2,n}, \varphi_{0,n}), \delta_n)$, *and* $(c_n^{\mathrm{ord}}, c_n^{\vee,\mathrm{ord}})$ *are the tautological objects over* $\vec{C}^{\mathrm{ord}}_{\Phi_n,\delta_n}$. *The torus torsor* $S := \vec{\Xi}^{\mathrm{ord}}_{\Phi_n,\delta_n} \to \vec{C}^{\mathrm{ord}}_{\Phi_n,\delta_n}$ *is universal for the additional structure* τ_n^{ord} *over noetherian normal schemes over* $\vec{C}^{\mathrm{ord}}_{\Phi_n,\delta_n}$ (*inducing dominant morphisms over irreducible components*) *satisfying certain symplectic and liftability conditions, which can be re-interpreted as follows:* $S = \vec{\Xi}^{\mathrm{ord}}_{\Phi_n,\delta_n} \to \vec{C}^{\mathrm{ord}}_{\Phi_n,\delta_n}$ *parameterizes tuples*

$$(G^\natural, \lambda^\natural : G^\natural \to G^{\vee,\natural}, i^\natural, \tau, \beta_{n_0}, \beta^{\mathrm{ord}}_{p^r}),$$

where:

(a) $G^\natural$, $G^{\vee,\natural}$, $\lambda^\natural$, *and* $i^\natural$ *are as in Lemma* 5.2.4.10.

(b) $\tau : \mathbf{1}_{Y\times X} \xrightarrow{\sim} (c^\vee \times c)^*\mathcal{P}_B^{\otimes -1}$ *is a trivialization of biextensions such that* $(\mathrm{Id}_Y \times \phi)^*\tau$ *is symmetric, and such that* $(i_Y(b) \times \mathrm{Id}_X)^*\tau = (\mathrm{Id}_Y \times i_X(b^\star))^*\tau$ *for all* $b \in \mathcal{O}$. *Then* τ *induces homomorphisms* $\iota : Y \to G^\natural$ *and* $\iota^\vee : X \to G^{\vee,\natural}$ *compatible with the homomorphisms* $\phi : Y \hookrightarrow X$ *and* $\lambda^\natural : G^\natural \to G^{\vee,\natural}$, *and induces an* $\mathcal{O}$-*equivariant homomorphism* $\lambda : G[n] \to G^\vee[n]$.

(c) $\beta_{n_0} = (\beta^0_{n_0}, \beta^{\#,0}_{n_0}, \nu_{n_0})$ *is a* **principal level-n_0 structure** *of* $(G^\natural, \lambda^\natural, i^\natural, \tau)$ *of type* $(L \underset{\mathbb{Z}}{\otimes} \hat{\mathbb{Z}}^p, \langle\cdot,\cdot\rangle, \mathsf{Z} \underset{\hat{\mathbb{Z}}}{\otimes} \hat{\mathbb{Z}}^p)$, *where* $\beta^0_{n_0} : (L/n_0L)_S \xrightarrow{\sim} G[n_0]$ *and* $\beta^{\#,0}_{n_0} : (L^\#/n_0L^\#)_S \xrightarrow{\sim} G^\vee[n_0]$ *are* $\mathcal{O}$-*equivariant isomorphism respecting the canonical filtrations on both sides, and* $\nu_{n_0} : ((\mathbb{Z}/n_0\mathbb{Z})(1))_S \xrightarrow{\sim} \boldsymbol{\mu}_{n_0,S}$ *is an isomorphism, inducing on the graded pieces the* φ_{-2,n_0}, φ_{-1,n_0}, *and* φ_{0,n_0}, *respectively, induced by the given* $\varphi_{-2,n}$, $\varphi_{-1,n}$, *and* $\varphi_{0,n}$, *which are compatible with the canonical morphisms* $L \hookrightarrow L^\#$ *and* $\lambda : G[n_0] \to G^\vee[n_0]$. *Moreover,* β_{n_0} *satisfies the liftability condition that, for each integer* $m_0 \geq 1$ *such that* $n_0|m_0$, *there exists a finite étale covering of* S *over which there exists an analogous triple* β_{m_0} *lifting the pullback of* β_{n_0}.

(d) $\beta^{\mathrm{ord}}_{p^r} = \beta^{\natural,\mathrm{ord}}_{p^r}$ *is, as in Lemma* 5.2.4.10, *an ordinary level-*$\mathcal{H}_p$ *structure of type* $(L \underset{\mathbb{Z}}{\otimes} \mathbb{Z}_p, \langle\,\cdot\,,\,\cdot\,\rangle, \mathsf{Z} \underset{\hat{\mathbb{Z}}}{\otimes} \mathbb{Z}_p, \mathsf{D})$*, whose definition does not require* τ *(and the group schemes* $G[p^r]$ *and* $G^\vee[p^r]$*).*

Proof. The statements are self-explanatory. □

Proposition 5.2.4.30. (*Compare with Propositions* 1.3.2.31 *and* 5.2.4.11.) *The torus torsor* $S := \vec{\Xi}^{\mathrm{ord}}_{\Phi_\mathcal{H},\delta_\mathcal{H}} \to \vec{C}^{\mathrm{ord}}_{\Phi_\mathcal{H},\delta_\mathcal{H}}$ *is universal for the additional structure* $\tau_\mathcal{H}$ *over noetherian normal schemes over* $\vec{C}^{\mathrm{ord}}_{\Phi_\mathcal{H},\delta_\mathcal{H}}$ *(inducing dominant morphisms over irreducible components) satisfying certain symplectic and liftability conditions, which can be interpreted as parameterizing tuples*

$$(G^\natural, \lambda^\natural : G^\natural \to G^{\vee,\natural}, i^\natural, \tau, \beta_{\mathcal{H}^p}, \beta^{\mathrm{ord}}_{\mathcal{H}_p}),$$

where $G^\natural$, $G^{\vee,\natural}$, $\lambda^\natural$, *and* $i^\natural$ *are as in Lemma* 5.2.4.10, *where* τ *is as in Lemma* 5.2.4.29, *where* $\beta^{\mathrm{ord}}_{\mathcal{H}_p} = \beta^{\natural,\mathrm{ord}}_{\mathcal{H}_p}$ *is an* **ordinary level-**$\mathcal{H}_p$ **structure** *of* $(G^\natural, \lambda^\natural, i^\natural)$ *of type* $(L \underset{\mathbb{Z}}{\otimes} \mathbb{Z}_p, \langle\,\cdot\,,\,\cdot\,\rangle, \mathsf{Z} \underset{\hat{\mathbb{Z}}}{\otimes} \mathbb{Z}_p, \mathsf{D})$ *as in Lemma* 5.2.4.11, *for any* $\mathsf{Z} \underset{\hat{\mathbb{Z}}}{\otimes} \mathbb{Z}_p$ *lifting* Z_{p^r} *and compatible with* D*; and where* $\beta_{\mathcal{H}^p}$ *is a* **level-**$\mathcal{H}^p$ **structure** *of* $(G^\natural, \lambda^\natural, i^\natural, \tau)$ *of type* $(L \underset{\mathbb{Z}}{\otimes} \hat{\mathbb{Z}}^p, \langle\,\cdot\,,\,\cdot\,\rangle, \mathsf{Z} \underset{\hat{\mathbb{Z}}}{\otimes} \hat{\mathbb{Z}}^p)$*, which is a collection* $\{\beta_{H_{n_0}}\}_{n_0}$*, where* $n_0 \geq 1$ *runs over integers prime to* p *such that* $\mathcal{U}^p(n_0) \subset \mathcal{H}^p$*, such that each* $\beta_{H_{n_0}}$ *(where* $H_{n_0} := \mathcal{H}^p/\mathcal{U}^p(n_0)$*) is a subscheme of*

$$\coprod \Big(\underline{\mathrm{Isom}}_S((L/n_0L)_S, G[n_0]) \underset{S}{\times} \underline{\mathrm{Isom}}_S((L^\#/n_0L^\#)_S, G^\vee[n_0]) \underset{S}{\times} \underline{\mathrm{Isom}}_S(((\mathbb{Z}/n_0\mathbb{Z})(1))_S, \boldsymbol{\mu}_{n_0,S}) \Big)$$

over S*, where the disjoint union is over representatives* $(\mathsf{Z}_{n_0}, \Phi_{n_0}, \delta_{n_0})$ *(with the same* (X, Y, ϕ)*) in the* $(\mathsf{Z}_{\mathcal{H}^p}, \Phi_{\mathcal{H}^p}, \delta_{\mathcal{H}^p})$ *induced by* $(\mathsf{Z}_\mathcal{H}, \Phi_\mathcal{H}, \delta_\mathcal{H})$*, that becomes the disjoint union of all elements in the* H_{n_0}*-orbit of some principal level-*n_0 *structure* β_{n_0} *of* $(G^\natural, \lambda^\natural, i^\natural, \tau)$ *of type* $(L \underset{\mathbb{Z}}{\otimes} \hat{\mathbb{Z}}^p, \langle\,\cdot\,,\,\cdot\,\rangle, \mathsf{Z} \underset{\hat{\mathbb{Z}}}{\otimes} \hat{\mathbb{Z}}^p)$*, as in Lemma* 5.2.4.29, *for any* $\mathsf{Z} \underset{\hat{\mathbb{Z}}}{\otimes} \hat{\mathbb{Z}}^p$ *lifting* Z_{n_0}*; and where* $\beta_{H_{m_0}}$ *is mapped to* $\beta_{H_{n_0}}$ *(under the canonical morphism, which we omit for simplicity) when* $p \nmid m_0$ *and* $n_0 | m_0$*.*

Let $\mathbf{S}_{\Phi_\mathcal{H}}$ *be the unique lattice in* $\mathbf{S}_{\Phi_1} \underset{\mathbb{Z}}{\otimes} \mathbb{Q}$ *such that* $\mathbf{S}^\vee_{\Phi_1}/\mathbf{S}^\vee_{\Phi_\mathcal{H}} \cong \mathrm{U}_{2,\mathsf{Z}}(\hat{\mathbb{Z}})/\mathcal{H}_{\mathrm{U}_{2,\mathsf{Z}}}$. *Then* $S = \vec{\Xi}^{\mathrm{ord}}_{\Phi_\mathcal{H},\delta_\mathcal{H}} \to \vec{C}^{\mathrm{ord}}_{\Phi_\mathcal{H},\delta_\mathcal{H}}$ *is torsor under the split torus* $E_{\Phi_\mathcal{H}}$ *with character group* $\mathbf{S}_{\Phi_\mathcal{H}}$*, equipped a homomorphism*

$$\mathbf{S}_{\Phi_\mathcal{H}} \to \underline{\mathrm{Pic}}(\vec{C}^{\mathrm{ord}}_{\Phi_\mathcal{H},\delta_\mathcal{H}}) : \ell \mapsto \vec{\Psi}^{\mathrm{ord}}_{\Phi_\mathcal{H},\delta_\mathcal{H}}(\ell)$$

(by the torus torsor structure; see Proposition 4.2.1.46 *and* (4.2.1.49)*), assigning to each* $\ell \in \mathbf{S}_{\Phi_\mathcal{H}}$ *an invertible sheaf* $\vec{\Psi}^{\mathrm{ord}}_{\Phi_\mathcal{H},\delta_\mathcal{H}}(\ell)$ *over* $\vec{C}^{\mathrm{ord}}_{\Phi_\mathcal{H},\delta_\mathcal{H}}$ *(up to isomorphism), together with isomorphisms*

$$\vec{\Delta}^{\mathrm{ord},*}_{\Phi_\mathcal{H},\delta_\mathcal{H},\ell,\ell'} : \vec{\Psi}^{\mathrm{ord}}_{\Phi_\mathcal{H},\delta_\mathcal{H}}(\ell) \underset{\mathscr{O}_{\vec{C}^{\mathrm{ord}}_{\Phi_\mathcal{H},\delta_\mathcal{H}}}}{\otimes} \vec{\Psi}^{\mathrm{ord}}_{\Phi_\mathcal{H},\delta_\mathcal{H}}(\ell') \xrightarrow{\sim} \vec{\Psi}^{\mathrm{ord}}_{\Phi_\mathcal{H},\delta_\mathcal{H}}(\ell + \ell')$$

for all $\ell, \ell' \in \mathbf{S}_{\Phi_{\mathcal{H}}}$, *satisfying the necessary compatibilities with each other making* $\underset{\ell \in \mathbf{S}_{\Phi_{\mathcal{H}}}}{\oplus} \vec{\Psi}^{\mathrm{ord}}_{\Phi_{\mathcal{H}},\delta_{\mathcal{H}}}(\ell)$ *an* $\mathscr{O}_{\vec{C}^{\mathrm{ord}}_{\Phi_{\mathcal{H}},\delta_{\mathcal{H}}}}$*-algebra, such that*

$$\vec{\Xi}^{\mathrm{ord}}_{\Phi_{\mathcal{H}},\delta_{\mathcal{H}}} \cong \underline{\mathrm{Spec}}_{\mathscr{O}_{\vec{C}^{\mathrm{ord}}_{\Phi_{\mathcal{H}},\delta_{\mathcal{H}}}}}\left(\underset{\ell \in \mathbf{S}_{\Phi_{\mathcal{H}}}}{\oplus} \vec{\Psi}^{\mathrm{ord}}_{\Phi_{\mathcal{H}},\delta_{\mathcal{H}}}(\ell)\right).$$

When $\ell = [y \otimes \chi]$ *for some* $y \in Y$ *and* $\chi \in X$, *we have a canonical isomorphism*

$$\vec{\Psi}^{\mathrm{ord}}_{\Phi_{\mathcal{H}},\delta_{\mathcal{H}}}(\ell) \cong (c^{\vee}(y), c(\chi))^{*}\mathcal{P}_{B}.$$

Proof. This follows from Lemmas 5.2.4.26 and 5.2.4.29, by realizing $\vec{\Xi}^{\mathrm{ord}}_{\Phi_{\mathcal{H}},\delta_{\mathcal{H}},r_n}$ as a quotient of some $\vec{\Xi}^{\mathrm{ord}}_{\Phi_n,\delta_n}$, and by finite flat descent. □

For each rational polyhedral cone $\sigma \subset (\mathbf{S}_{\Phi_{\mathcal{H}}})^{\vee}_{\mathbb{R}}$ as in Definition 1.2.2.2, we have an affine toroidal embedding

$$\vec{\Xi}^{\mathrm{ord}}_{\Phi_{\mathcal{H}},\delta_{\mathcal{H}}} \hookrightarrow \vec{\Xi}^{\mathrm{ord}}_{\Phi_{\mathcal{H}},\delta_{\mathcal{H}}}(\sigma) \cong \underline{\mathrm{Spec}}_{\mathscr{O}_{\vec{C}^{\mathrm{ord}}_{\Phi_{\mathcal{H}},\delta_{\mathcal{H}}}}}\left(\underset{\ell \in \sigma^{\vee}}{\oplus} \vec{\Psi}^{\mathrm{ord}}_{\Phi_{\mathcal{H}},\delta_{\mathcal{H}}}(\ell)\right) \tag{5.2.4.31}$$

(*cf.* (1.3.2.32) and (4.2.2.1)), both sides being relative affine over $\vec{C}^{\mathrm{ord}}_{\Phi_{\mathcal{H}},\delta_{\mathcal{H}}}$, where $\vec{\Xi}^{\mathrm{ord}}_{\Phi_{\mathcal{H}},\delta_{\mathcal{H}}}(\sigma) \to \vec{C}^{\mathrm{ord}}_{\Phi_{\mathcal{H}},\delta_{\mathcal{H}}}$ is smooth when the cone σ is smooth, with its σ-stratum $\vec{\Xi}^{\mathrm{ord}}_{\Phi_{\mathcal{H}},\delta_{\mathcal{H}},\sigma} = \underline{\mathrm{Spec}}_{\mathscr{O}_{\vec{C}^{\mathrm{ord}}_{\Phi_{\mathcal{H}},\delta_{\mathcal{H}}}}}\left(\underset{\ell \in \sigma^{\perp}}{\oplus} \vec{\Psi}^{\mathrm{ord}}_{\Phi_{\mathcal{H}},\delta_{\mathcal{H}}}(\ell)\right)$ as in (4.2.2.2) (*cf.* (1.3.2.33)), which is by itself a torsor under the torus $E_{\Phi_{\mathcal{H}},\sigma}$ with character group $\sigma^{\perp}$. For each $\Gamma_{\Phi_{\mathcal{H}}}$-admissible rational polyhedral cone decomposition $\Sigma_{\Phi_{\mathcal{H}}}$ as in Definition 1.2.2.4, we have the toroidal embedding

$$\vec{\Xi}^{\mathrm{ord}}_{\Phi_{\mathcal{H}},\delta_{\mathcal{H}}} \hookrightarrow \overline{\vec{\Xi}}^{\mathrm{ord}}_{\Phi_{\mathcal{H}},\delta_{\mathcal{H}}} = \overline{\vec{\Xi}}^{\mathrm{ord}}_{\Phi_{\mathcal{H}},\delta_{\mathcal{H}},\Sigma_{\Phi_{\mathcal{H}}}}$$

as in (4.2.2.4) (*cf.* (1.3.2.34)), the right-hand side being only locally of finite type over $\vec{C}^{\mathrm{ord}}_{\Phi_{\mathcal{H}},\delta_{\mathcal{H}}}$, with an open covering

$$\overline{\vec{\Xi}}^{\mathrm{ord}}_{\Phi_{\mathcal{H}},\delta_{\mathcal{H}}} = \underset{\sigma \in \Sigma_{\Phi_{\mathcal{H}}}}{\cup} \vec{\Xi}^{\mathrm{ord}}_{\Phi_{\mathcal{H}},\delta_{\mathcal{H}}}(\sigma), \tag{5.2.4.32}$$

(*cf.* (1.3.2.35)) inducing a stratification

$$\overline{\vec{\Xi}}^{\mathrm{ord}}_{\Phi_{\mathcal{H}},\delta_{\mathcal{H}}} = \coprod_{\sigma \in \Sigma_{\Phi_{\mathcal{H}}}} \vec{\Xi}^{\mathrm{ord}}_{\Phi_{\mathcal{H}},\delta_{\mathcal{H}},\sigma} \tag{5.2.4.33}$$

(*cf.* (1.3.2.36)). (The notation "$\coprod$" only means a set-theoretic disjoint union. The algebro-geometric structure is still the one inherited from $\overline{\vec{\Xi}}^{\mathrm{ord}}_{\Phi_{\mathcal{H}},\delta_{\mathcal{H}}}$.) Concretely, if σ is a face of ρ, then $\rho^{\vee} \subset \sigma^{\vee}$ and $\vec{\Xi}^{\mathrm{ord}}_{\Phi_{\mathcal{H}},\delta_{\mathcal{H}}}(\sigma) \subset \vec{\Xi}^{\mathrm{ord}}_{\Phi_{\mathcal{H}},\delta_{\mathcal{H}}}(\rho)$, but $\vec{\Xi}^{\mathrm{ord}}_{\Phi_{\mathcal{H}},\delta_{\mathcal{H}},\rho}$ is contained in the closure of $\vec{\Xi}^{\mathrm{ord}}_{\Phi_{\mathcal{H}},\delta_{\mathcal{H}},\sigma}$. The closure of $\vec{\Xi}^{\mathrm{ord}}_{\Phi_{\mathcal{H}},\delta_{\mathcal{H}},\sigma}$ in $\vec{\Xi}^{\mathrm{ord}}_{\Phi_{\mathcal{H}},\delta_{\mathcal{H}}}(\rho)$ is

$$\vec{\Xi}^{\mathrm{ord}}_{\Phi_{\mathcal{H}},\delta_{\mathcal{H}},\sigma}(\rho) := \underline{\mathrm{Spec}}_{\mathscr{O}_{\vec{C}^{\mathrm{ord}}_{\Phi_{\mathcal{H}},\delta_{\mathcal{H}}}}}\left(\underset{\ell \in \sigma^{\perp} \cap \rho^{\vee}}{\oplus} \vec{\Psi}^{\mathrm{ord}}_{\Phi_{\mathcal{H}},\delta_{\mathcal{H}}}(\ell)\right) \tag{5.2.4.34}$$

(*cf.* (1.3.2.37)). In this case, the open embedding

$$\vec{\Xi}^{\mathrm{ord}}_{\Phi_{\mathcal{H}},\delta_{\mathcal{H}},\sigma} \hookrightarrow \vec{\Xi}^{\mathrm{ord}}_{\Phi_{\mathcal{H}},\delta_{\mathcal{H}},\sigma}(\rho) \tag{5.2.4.35}$$

(*cf.* (1.3.2.38)) is an affine toroidal embedding (as in [62, Def. 6.1.2.3]) for the torus torsor $\vec{\Xi}^{\mathrm{ord}}_{\Phi_{\mathcal{H}},\delta_{\mathcal{H}},\sigma} \to \vec{C}^{\mathrm{ord}}_{\Phi_{\mathcal{H}},\delta_{\mathcal{H}}}$.

In Section 4.2.2, we have also defined

$$\vec{\mathfrak{X}}^{\mathrm{ord}}_{\Phi_{\mathcal{H}},\delta_{\mathcal{H}},\sigma} = (\vec{\Xi}^{\mathrm{ord}}_{\Phi_{\mathcal{H}},\delta_{\mathcal{H}}}(\sigma))^{\wedge}_{\vec{\Xi}^{\mathrm{ord}}_{\Phi_{\mathcal{H}},\delta_{\mathcal{H}},\sigma}}, \tag{5.2.4.36}$$

(*cf.* (1.3.2.39)) the formal completion of $\vec{\Xi}^{\mathrm{ord}}_{\Phi_{\mathcal{H}},\delta_{\mathcal{H}}}(\sigma)$ along its σ-stratum $\vec{\Xi}^{\mathrm{ord}}_{\Phi_{\mathcal{H}},\delta_{\mathcal{H}},\sigma}$. When $\sigma \subset \mathbf{P}^+_{\Phi_{\mathcal{H}}}$ appears in $\Sigma_{\Phi_{\mathcal{H}}} \in \Sigma^{\mathrm{ord}}$, the quotient $\vec{\mathfrak{X}}^{\mathrm{ord}}_{\Phi_{\mathcal{H}},\delta_{\mathcal{H}},\sigma}/\Gamma_{\Phi_{\mathcal{H}},\sigma}$ is isomorphic to the formal completion of $\vec{\mathsf{M}}^{\mathrm{ord,tor}}_{\mathcal{H},\Sigma^{\mathrm{ord}}}$ along its $[(\Phi_{\mathcal{H}},\delta_{\mathcal{H}},\sigma)]$-stratum $\vec{\mathsf{Z}}^{\mathrm{ord}}_{[(\Phi_{\mathcal{H}},\delta_{\mathcal{H}},\sigma)]} \cong \vec{\Xi}^{\mathrm{ord}}_{\Phi_{\mathcal{H}},\delta_{\mathcal{H}},\sigma}/\Gamma_{\Phi_{\mathcal{H}},\sigma}$, as in Theorem 5.2.1.1. If there is a surjection $(\mathsf{Z}'_{\mathcal{H}},\Phi'_{\mathcal{H}},\delta'_{\mathcal{H}}) \twoheadrightarrow (\mathsf{Z}_{\mathcal{H}},\Phi_{\mathcal{H}},\delta_{\mathcal{H}})$ such that σ is mapped to a face of a cone $\rho \subset \mathbf{P}^+_{\Phi'_{\mathcal{H}}}$ under the canonical mapping $\mathbf{P}^+_{\Phi_{\mathcal{H}}} \to \mathbf{P}_{\Phi'_{\mathcal{H}}}$, and if $\rho \in \Sigma_{\Phi'_{\mathcal{H}}} \in \Sigma^{\mathrm{ord}}$, then $\vec{\mathsf{Z}}^{\mathrm{ord}}_{[(\Phi'_{\mathcal{H}},\delta'_{\mathcal{H}},\rho)]}$ is contained in the closure $\overline{\vec{\mathsf{Z}}}^{\mathrm{ord}}_{[(\Phi_{\mathcal{H}},\delta_{\mathcal{H}},\sigma)]}$ of $\vec{\mathsf{Z}}^{\mathrm{ord}}_{[(\Phi_{\mathcal{H}},\delta_{\mathcal{H}},\sigma)]}$ in $\vec{\mathsf{M}}^{\mathrm{ord,tor}}_{\mathcal{H},\Sigma^{\mathrm{ord}}}$, and the completion of $\overline{\vec{\mathsf{Z}}}^{\mathrm{ord}}_{[(\Phi_{\mathcal{H}},\delta_{\mathcal{H}},\sigma)]}$ along $\vec{\mathsf{Z}}^{\mathrm{ord}}_{[(\Phi'_{\mathcal{H}},\delta'_{\mathcal{H}},\rho)]}$ is canonically isomorphic to

$$\vec{\mathfrak{X}}^{\mathrm{ord}}_{\Phi_{\mathcal{H}},\delta_{\mathcal{H}},\sigma,\rho} := (\vec{\Xi}^{\mathrm{ord}}_{\Phi_{\mathcal{H}},\delta_{\mathcal{H}},\sigma}(\rho))^{\wedge}_{\vec{\Xi}^{\mathrm{ord}}_{\Phi_{\mathcal{H}},\delta_{\mathcal{H}},\rho}} \tag{5.2.4.37}$$

(*cf.* (1.3.2.40)), the formal completion of $\vec{\Xi}^{\mathrm{ord}}_{\Phi_{\mathcal{H}},\delta_{\mathcal{H}},\sigma}(\rho)$ along its ρ-stratum $\vec{\Xi}^{\mathrm{ord}}_{\Phi_{\mathcal{H}},\delta_{\mathcal{H}},\rho}$.

Lemma 5.2.4.38. (*Compare with Lemma* 1.3.2.41.) *Consider* $\vec{\mathfrak{X}}^{\mathrm{ord}}_{\Phi_{\mathcal{H}},\delta_{\mathcal{H}}} = \vec{\mathfrak{X}}^{\mathrm{ord}}_{\Phi_{\mathcal{H}},\delta_{\mathcal{H}},\Sigma_{\Phi_{\mathcal{H}}}}$, *the formal completion of* $\overline{\vec{\Xi}}^{\mathrm{ord}}_{\Phi_{\mathcal{H}},\delta_{\mathcal{H}}}$ *along the union of the* σ*-strata* $\vec{\Xi}^{\mathrm{ord}}_{\Phi_{\mathcal{H}},\delta_{\mathcal{H}},\sigma}$ *for* $\sigma \in \Sigma_{\Phi_{\mathcal{H}}}$ *and* $\sigma \subset \mathbf{P}^+_{\Phi_{\mathcal{H}}}$. *Then we have a canonical morphism*

$$\vec{\mathfrak{X}}^{\mathrm{ord}}_{\Phi_{\mathcal{H}},\delta_{\mathcal{H}}} \to \vec{\mathsf{M}}^{\mathrm{ord,tor}}_{\mathcal{H},\Sigma^{\mathrm{ord}}} \tag{5.2.4.39}$$

(*cf.* (1.3.2.42)) *inducing a canonical isomorphism*

$$\vec{\mathfrak{X}}^{\mathrm{ord}}_{\Phi_{\mathcal{H}},\delta_{\mathcal{H}}}/\Gamma_{\Phi_{\mathcal{H}}} \xrightarrow{\sim} (\vec{\mathsf{M}}^{\mathrm{ord,tor}}_{\mathcal{H},\Sigma^{\mathrm{ord}}})^{\wedge}_{\cup\,\vec{\mathsf{Z}}^{\mathrm{ord}}_{[(\Phi_{\mathcal{H}},\delta_{\mathcal{H}},\sigma)]}} \tag{5.2.4.40}$$

(*cf.* (1.3.2.43)), *where* $\cup\,\vec{\mathsf{Z}}^{\mathrm{ord}}_{[(\Phi_{\mathcal{H}},\delta_{\mathcal{H}},\sigma)]}$ *is the union of all strata* $\vec{\mathsf{Z}}^{\mathrm{ord}}_{[(\Phi_{\mathcal{H}},\delta_{\mathcal{H}},\sigma)]}$ *with* $\sigma \in \Sigma_{\Phi_{\mathcal{H}}}$ (*and* $\sigma \subset \mathbf{P}^+_{\Phi_{\mathcal{H}}}$), *under which the pullback of* $\underline{\mathrm{Lie}}^{\vee}_{G/\vec{\mathsf{M}}^{\mathrm{ord,tor}}_{\mathcal{H},\Sigma^{\mathrm{ord}}}}$ (*resp.* $\underline{\mathrm{Lie}}^{\vee}_{G^{\vee}/\vec{\mathsf{M}}^{\mathrm{ord,tor}}_{\mathcal{H},\Sigma^{\mathrm{ord}}}}$, *resp.* $\lambda^* : \underline{\mathrm{Lie}}^{\vee}_{G^{\vee}/\vec{\mathsf{M}}^{\mathrm{ord,tor}}_{\mathcal{H},\Sigma^{\mathrm{ord}}}} \to \underline{\mathrm{Lie}}^{\vee}_{G/\vec{\mathsf{M}}^{\mathrm{ord,tor}}_{\mathcal{H},\Sigma^{\mathrm{ord}}}}$) *can be canonically identified with the pullback of* $\underline{\mathrm{Lie}}^{\vee}_{G^{\natural}/\vec{C}^{\mathrm{ord}}_{\Phi_{\mathcal{H}},\delta_{\mathcal{H}}}}$ (*resp.* $\underline{\mathrm{Lie}}^{\vee}_{G^{\vee,\natural}/\vec{C}^{\mathrm{ord}}_{\Phi_{\mathcal{H}},\delta_{\mathcal{H}}}}$, *resp.* $(\lambda^{\natural})^* : \underline{\mathrm{Lie}}^{\vee}_{G^{\vee,\natural}/\vec{C}^{\mathrm{ord}}_{\Phi_{\mathcal{H}},\delta_{\mathcal{H}}}} \to \underline{\mathrm{Lie}}^{\vee}_{G^{\natural}/\vec{C}^{\mathrm{ord}}_{\Phi_{\mathcal{H}},\delta_{\mathcal{H}}}}$). *For each stratum* $\vec{\mathsf{Z}}^{\mathrm{ord}}_{[(\Phi_{\mathcal{H}},\delta_{\mathcal{H}},\sigma)]}$, *the isomorphism* (5.2.4.40) *is compatible with the isomorphism* $\vec{\mathfrak{X}}^{\mathrm{ord}}_{\Phi_{\mathcal{H}},\delta_{\mathcal{H}},\sigma}/\Gamma_{\Phi_{\mathcal{H}},\sigma} \xrightarrow{\sim} (\vec{\mathsf{M}}^{\mathrm{ord,tor}}_{\mathcal{H},\Sigma})^{\wedge}_{\vec{\mathsf{Z}}^{\mathrm{ord}}_{[(\Phi_{\mathcal{H}},\delta_{\mathcal{H}},\sigma)]}}$ *in* (5) *of Theorem* 5.2.1.1 (*under the canonical morphisms* $\vec{\mathfrak{X}}^{\mathrm{ord}}_{\Phi_{\mathcal{H}},\delta_{\mathcal{H}},\sigma}/\Gamma_{\Phi_{\mathcal{H}},\sigma} \to \vec{\mathfrak{X}}^{\mathrm{ord}}_{\Phi_{\mathcal{H}},\delta_{\mathcal{H}}}/\Gamma_{\Phi_{\mathcal{H}}}$ *and* $(\vec{\mathsf{M}}^{\mathrm{ord,tor}}_{\mathcal{H},\Sigma^{\mathrm{ord}}})^{\wedge}_{\vec{\mathsf{Z}}^{\mathrm{ord}}_{[(\Phi_{\mathcal{H}},\delta_{\mathcal{H}},\sigma)]}} \to (\vec{\mathsf{M}}^{\mathrm{ord,tor}}_{\mathcal{H},\Sigma^{\mathrm{ord}}})^{\wedge}_{\cup\,\vec{\mathsf{Z}}^{\mathrm{ord}}_{[(\Phi_{\mathcal{H}},\delta_{\mathcal{H}},\sigma)]}}$). (*Such isomorphisms are induced by strata-preserving isomorphisms from étale neighborhoods of points of* $\vec{\Xi}^{\mathrm{ord}}_{\Phi_{\mathcal{H}},\delta_{\mathcal{H}},\sigma}$ *in* $\vec{\Xi}^{\mathrm{ord}}_{\Phi_{\mathcal{H}},\delta_{\mathcal{H}}}(\sigma)$ *to étale neighborhoods of points of* $\vec{\mathsf{Z}}^{\mathrm{ord}}_{[(\Phi_{\mathcal{H}},\delta_{\mathcal{H}},\sigma)]}$ *in* $\vec{\mathsf{M}}^{\mathrm{ord,tor}}_{\mathcal{H},\Sigma^{\mathrm{ord}}}$.)

Proof. By using the various universal properties, the same argument in the proof of Lemma 1.3.2.41 also works here. □

Proposition 5.2.4.41. (*Compare with Propositions* 1.3.1.15, 1.3.2.24, 1.3.2.45, 5.2.2.2, 5.2.4.25.) *By considering compatible* $\mathbb{Q}^\times$*-isogenies* $(f : G^\natural \to G^{\natural,\prime}, f^\vee : G^{\vee,\natural,\prime} \to G^{\vee,\natural})$ *compatible with the homomorphisms* $(\iota : Y \to G^\natural, \iota^\vee : X \to G^{\vee,\natural})$ *inducing isomorphisms on the torus parts* T *and* $T^\vee$ *and on the domains of* ι *and* $\iota^\vee$*, we can define ordinary Hecke twists of the tautological object* $(G^\natural, \lambda^\natural, i^\natural, \tau, \beta_{\mathcal{H}^p}, \beta^{\mathrm{ord}}_{\mathcal{H}_p}) \to \vec{\Xi}^{\mathrm{ord}}_{\Phi_{\mathcal{H}},\delta_{\mathcal{H}}}$ *by elements* $g = (g_0, g_p) \in \mathrm{P}'_{\mathsf{Z}}(\mathbb{A}^{\infty,p}) \times \mathrm{P}^{\mathrm{ord},\prime}_{\mathsf{Z},\mathsf{D}}(\mathbb{Q}_p)$ *such that the image of* g_p *under the canonical homomorphism* $\mathrm{P}^{\mathrm{ord},\prime}_{\mathsf{Z},\mathsf{D}}(\mathbb{Q}_p) \to \mathrm{P}^{\mathrm{ord}}_{h,\mathsf{D}}(\mathbb{Q}_p)$ *satisfies the condition defined by the filtration* D_{-1} *on* $\mathrm{Gr}^{\mathsf{Z}}_{-1} \underset{\hat{\mathbb{Z}}}{\otimes} \mathbb{Z}_p$ *as in Section* 3.3.4*, and define the Hecke action of (such elements of)* $\mathrm{P}'_{\mathsf{Z}}(\mathbb{A}^{\infty,p}) \times \mathrm{P}^{\mathrm{ord},\prime}_{\mathsf{Z},\mathsf{D}}(\mathbb{Q}_p)$ *on the collection* $\{\vec{\Xi}^{\mathrm{ord}}_{\Phi_{\mathcal{H}},\delta_{\mathcal{H}}}\}_{\mathcal{H}_{\mathrm{P}'_{\mathsf{Z}}}}$ *with* $\mathcal{H}$ *of standard form, realized by quasi-finite flat surjections pulling tautological objects back to ordinary Hecke twists, which is compatible with the Hecke action of (suitable elements of)* $\mathrm{P}'_{\mathsf{Z}}(\mathbb{A}^{\infty,p}) \times \mathrm{P}^{\mathrm{ord},\prime}_{\mathsf{Z},\mathsf{D}}(\mathbb{Q}_p)$ *on the collection* $\{\vec{C}^{\mathrm{ord}}_{\Phi_{\mathcal{H}},\delta_{\mathcal{H}}}\}_{\mathcal{H}_{\mathrm{G}_{1,\mathsf{Z}}}}$ *(with* $\mathcal{H}$ *of standard form) under the canonical morphisms* $\vec{\Xi}^{\mathrm{ord}}_{\Phi_{\mathcal{H}},\delta_{\mathcal{H}}} \to \vec{C}^{\mathrm{ord}}_{\Phi_{\mathcal{H}},\delta_{\mathcal{H}}}$ *(with varying* $\mathcal{H}$*) and the canonical homomorphism* $\mathrm{P}'_{\mathsf{Z}}(\mathbb{A}^{\infty,p}) \times \mathrm{P}^{\mathrm{ord},\prime}_{\mathsf{Z},\mathsf{D}}(\mathbb{Q}_p) \to \mathrm{G}_{1,\mathsf{Z}}(\mathbb{A}^{\infty,p}) \times \mathrm{P}^{\mathrm{ord}}_{1,\mathsf{Z},\mathsf{D}}(\mathbb{Q}_p) \cong (\mathrm{P}'_{\mathsf{Z}}(\mathbb{A}^{\infty,p}) \times \mathrm{P}^{\mathrm{ord},\prime}_{\mathsf{Z},\mathsf{D}}(\mathbb{Q}_p))/\mathrm{U}_{2,\mathsf{Z}}(\mathbb{A}^\infty)$*. Such a Hecke action enjoys the properties (under various conditions) concerning étaleness, finiteness, being isomorphisms between formal completions along fibers over* $\mathrm{Spec}(\mathbb{F}_p)$*, and inducing absolute Frobenius morphisms on fibers over* $\mathrm{Spec}(\mathbb{F}_p)$ *for elements of* U_p *type as in Proposition* 5.2.2.2 *and Corollaries* 5.2.2.3, 5.2.2.4, *and* 5.2.2.5. *(We omit the details for simplicity.)*

By also considering $\mathbb{Q}^\times$*-isogenies* $(f : G^\natural \to G^{\natural,\prime}, f^\vee : G^{\vee,\natural,\prime} \to G^{\vee,\natural})$ *compatible with the homomorphisms* $(\iota : Y \to G^\natural, \iota^\vee : X \to G^{\vee,\natural})$ *inducing* $\mathbb{Q}^\times$ *on the torus parts* T *and* $T^\vee$ *and on the domains of* ι *and* $\iota^\vee$ *(possibly varying the isomorphism classes of the* $\mathcal{O}$*-lattices* X *and* Y*), we can also define ordinary Hecke twists of the tautological object* $(G^\natural, \lambda^\natural, i^\natural, \tau, \beta_{\mathcal{H}^p}, \beta^{\mathrm{ord}}_{\mathcal{H}_p}) \to \vec{C}^{\mathrm{ord}}_{\Phi_{\mathcal{H}},\delta_{\mathcal{H}}}$ *by elements* $g = (g_0, g_p) \in \mathrm{P}_{\mathsf{Z}}(\mathbb{A}^{\infty,p}) \times \mathrm{P}^{\mathrm{ord}}_{\mathsf{Z},\mathsf{D}}(\mathbb{Q}_p)$ *such that the image of* g_p *under the canonical homomorphism* $\mathrm{P}^{\mathrm{ord}}_{\mathsf{Z},\mathsf{D}}(\mathbb{Q}_p) \to \mathrm{P}^{\mathrm{ord}}_{h,\mathsf{D}}(\mathbb{Q}_p)$ *satisfies the condition defined by the filtration* D_{-1} *on* $\mathrm{Gr}^{\mathsf{Z}}_{-1} \underset{\hat{\mathbb{Z}}}{\otimes} \mathbb{Z}_p$ *as in Section* 3.3.4*, and define the Hecke action of (such elements of)* $\mathrm{P}_{\mathsf{Z}}(\mathbb{A}^{\infty,p}) \times \mathrm{P}^{\mathrm{ord}}_{\mathsf{Z},\mathsf{D}}(\mathbb{Q}_p)$ *on the collection* $\{\coprod \vec{\Xi}^{\mathrm{ord}}_{\Phi_{\mathcal{H}},\delta_{\mathcal{H}}}\}_{\mathcal{H}_{\mathrm{P}_{\mathsf{Z}}}}$ *(with* $\mathcal{H}$ *of standard form), where the disjoint unions are over classes* $[(\mathsf{Z}_{\mathcal{H}}, \Phi_{\mathcal{H}}, \delta_{\mathcal{H}})]$ *sharing the same* $\mathsf{Z}_{\mathcal{H}}$ *compatible with* D*, realized by quasi-finite flat surjections pulling tautological objects back to ordinary Hecke twists, which induces an action of* $\mathrm{G}'_{l,\mathsf{Z}}(\mathbb{A}^\infty) = \mathrm{P}_{\mathsf{Z}}(\mathbb{A}^\infty)/\mathrm{P}'_{\mathsf{Z}}(\mathbb{A}^\infty) \cong (\mathrm{P}_{\mathsf{Z}}(\mathbb{A}^{\infty,p}) \times \mathrm{P}^{\mathrm{ord}}_{\mathsf{Z},\mathsf{D}}(\mathbb{Q}_p))/(\mathrm{P}'_{\mathsf{Z}}(\mathbb{A}^{\infty,p}) \times \mathrm{P}^{\mathrm{ord},\prime}_{\mathsf{Z},\mathsf{D}}(\mathbb{Q}_p))$ *on the index sets* $\{[(\mathsf{Z}_{\mathcal{H}}, \Phi_{\mathcal{H}}, \delta_{\mathcal{H}})]\}$*, which is compatible with the Hecke action of (suitable elements of)* $(\mathrm{P}_{\mathsf{Z}}(\mathbb{A}^{\infty,p}) \times \mathrm{P}^{\mathrm{ord}}_{\mathsf{Z},\mathsf{D}}(\mathbb{Q}_p))/\mathrm{U}_{2,\mathsf{Z}}(\mathbb{A}^\infty)$ *on the collection* $\{\coprod \vec{C}^{\mathrm{ord}}_{\Phi_{\mathcal{H}},\delta_{\mathcal{H}}}\}_{\mathcal{H}_{\mathrm{P}_{\mathsf{Z}}}/\mathcal{H}_{\mathrm{U}_{2,\mathsf{Z}}}}$ *(with* $\mathcal{H}$ *of standard form, with the same index sets and the same induced action of* $\mathrm{G}'_{l,\mathsf{Z}}(\mathbb{A}^\infty)$*) under the canonical morphisms* $\vec{\Xi}^{\mathrm{ord}}_{\Phi_{\mathcal{H}},\delta_{\mathcal{H}}} \to \vec{C}^{\mathrm{ord}}_{\Phi_{\mathcal{H}},\delta_{\mathcal{H}}}$ *(with varying* $\mathcal{H}$*) and the canonical homomorphism* $\mathrm{P}_{\mathsf{Z}}(\mathbb{A}^{\infty,p}) \times \mathrm{P}^{\mathrm{ord}}_{\mathsf{Z},\mathsf{D}}(\mathbb{Q}_p) \to$

$(\mathrm{P}_{\mathsf{Z}}(\mathbb{A}^{\infty,p}) \times \mathrm{P}^{\mathrm{ord}}_{\mathsf{Z},\mathsf{D}}(\mathbb{Q}_p))/\mathrm{U}_{2,\mathsf{Z}}(\mathbb{A}^\infty)$.

Any such Hecke action

$$[\vec{g}]^{\mathrm{ord}} : \vec{\Xi}^{\mathrm{ord}}_{\Phi'_{\mathcal{H}'},\delta'_{\mathcal{H}'}} \to \vec{\Xi}^{\mathrm{ord}}_{\Phi_{\mathcal{H}},\delta_{\mathcal{H}}}$$

covering $[\vec{g}]^{\mathrm{ord}} : \vec{C}^{\mathrm{ord}}_{\Phi'_{\mathcal{H}'},\delta'_{\mathcal{H}'}} \to \vec{C}^{\mathrm{ord}}_{\Phi_{\mathcal{H}},\delta_{\mathcal{H}}}$ *induces a (finite flat) morphism*

$$\vec{\Xi}^{\mathrm{ord}}_{\Phi'_{\mathcal{H}'},\delta'_{\mathcal{H}'}} \to \vec{\Xi}^{\mathrm{ord}}_{\Phi_{\mathcal{H}},\delta_{\mathcal{H}}} \underset{\vec{C}^{\mathrm{ord}}_{\Phi_{\mathcal{H}},\delta_{\mathcal{H}}}}{\times} \vec{C}^{\mathrm{ord}}_{\Phi'_{\mathcal{H}'},\delta'_{\mathcal{H}'}}$$

between torus torsors over $\vec{C}^{\mathrm{ord}}_{\Phi'_{\mathcal{H}'},\delta'_{\mathcal{H}'}}$, *which is equivariant with the morphism* $E_{\Phi'_{\mathcal{H}'}} \to E_{\Phi_{\mathcal{H}}}$ *dual to the homomorphism* $\mathbf{S}_{\Phi_{\mathcal{H}}} \to \mathbf{S}_{\Phi'_{\mathcal{H}'}}$ *induced by the pair of morphisms* $(f_X : X \underset{\mathbb{Z}}{\otimes} \mathbb{Q} \overset{\sim}{\to} X' \underset{\mathbb{Z}}{\otimes} \mathbb{Q}, f_Y : Y' \underset{\mathbb{Z}}{\otimes} \mathbb{Q} \overset{\sim}{\to} Y \underset{\mathbb{Z}}{\otimes} \mathbb{Q})$ *defining the g-assignment* $(\mathsf{Z}'_{\mathcal{H}'}, \Phi'_{\mathcal{H}'}, \delta'_{\mathcal{H}'}) \to_g (\mathsf{Z}_{\mathcal{H}}, \Phi_{\mathcal{H}}, \delta_{\mathcal{H}})$ *of (ordinary) cusp labels (cf. [62, Def. 5.4.3.9]).*

If $g \in \mathrm{P}_{\mathsf{Z}}(\mathbb{A}^{\infty,p}) \times \mathrm{P}^{\mathrm{ord}}_{\mathsf{Z},\mathsf{D}}(\mathbb{Q}_p)$ *is as above and if* $(\Phi'_{\mathcal{H}'}, \delta'_{\mathcal{H}'}, \rho)$ *is a g-refinement of* $(\Phi_{\mathcal{H}}, \delta_{\mathcal{H}}, \sigma)$ *as in [62, Def. 6.4.3.1], then there is a canonical morphism*

$$[\vec{g}]^{\mathrm{ord}} : \vec{\Xi}^{\mathrm{ord}}_{\Phi'_{\mathcal{H}'},\delta'_{\mathcal{H}'}}(\rho) \to \vec{\Xi}^{\mathrm{ord}}_{\Phi_{\mathcal{H}},\delta_{\mathcal{H}}}(\sigma) \tag{5.2.4.42}$$

(cf. (1.3.2.46)) *covering* $[\vec{g}]^{\mathrm{ord}} : \vec{C}^{\mathrm{ord}}_{\Phi'_{\mathcal{H}'},\delta'_{\mathcal{H}'}} \to \vec{C}^{\mathrm{ord}}_{\Phi_{\mathcal{H}},\delta_{\mathcal{H}}}$, *extending* $[\vec{g}]^{\mathrm{ord}} : \vec{\Xi}^{\mathrm{ord}}_{\Phi'_{\mathcal{H}'},\delta'_{\mathcal{H}'}} \to \vec{\Xi}^{\mathrm{ord}}_{\Phi_{\mathcal{H}},\delta_{\mathcal{H}}}$, *mapping* $\vec{\Xi}^{\mathrm{ord}}_{\Phi'_{\mathcal{H}'},\delta'_{\mathcal{H}'},\rho}$ *to* $\vec{\Xi}^{\mathrm{ord}}_{\Phi_{\mathcal{H}},\delta_{\mathcal{H}},\sigma}$, *and inducing a canonical morphism*

$$[\vec{g}]^{\mathrm{ord}} : \vec{\mathfrak{X}}^{\mathrm{ord}}_{\Phi'_{\mathcal{H}'},\delta'_{\mathcal{H}'},\rho} \to \vec{\mathfrak{X}}^{\mathrm{ord}}_{\Phi_{\mathcal{H}},\delta_{\mathcal{H}},\sigma} \tag{5.2.4.43}$$

(cf. (1.3.2.47)). *If* $g \in \mathrm{P}_{\mathsf{Z}}(\mathbb{A}^{\infty,p}) \times \mathrm{P}^{\mathrm{ord}}_{\mathsf{Z},\mathsf{D}}(\mathbb{Q}_p)$ *is as above and if* $(\Phi'_{\mathcal{H}'}, \delta'_{\mathcal{H}'}, \Sigma'_{\Phi'_{\mathcal{H}'}})$ *is a g-refinement of* $(\Phi_{\mathcal{H}}, \delta_{\mathcal{H}}, \Sigma_{\Phi_{\mathcal{H}}})$ *as in [62, Def. 6.4.3.2], then morphisms like* (5.2.4.42) *patch together and define a canonical morphism*

$$[\vec{g}]^{\mathrm{ord}} : \overline{\Xi}_{\Phi'_{\mathcal{H}'},\delta'_{\mathcal{H}'},\Sigma'_{\Phi'_{\mathcal{H}'}}} \to \overline{\Xi}_{\Phi_{\mathcal{H}},\delta_{\mathcal{H}},\Sigma_{\Phi_{\mathcal{H}}}} \tag{5.2.4.44}$$

(cf. (1.3.2.48)) *covering* $[\vec{g}]^{\mathrm{ord}} : \vec{C}^{\mathrm{ord}}_{\Phi'_{\mathcal{H}'},\delta'_{\mathcal{H}'}} \to \vec{C}^{\mathrm{ord}}_{\Phi_{\mathcal{H}},\delta_{\mathcal{H}}}$, *extending* $[g] : \vec{\Xi}^{\mathrm{ord}}_{\Phi'_{\mathcal{H}'},\delta'_{\mathcal{H}'}} \to \vec{\Xi}^{\mathrm{ord}}_{\Phi_{\mathcal{H}},\delta_{\mathcal{H}}}$, *and inducing a canonical morphism*

$$[\vec{g}]^{\mathrm{ord}} : \vec{\mathfrak{X}}^{\mathrm{ord}}_{\Phi'_{\mathcal{H}'},\delta'_{\mathcal{H}'},\Sigma'_{\Phi'_{\mathcal{H}'}}} \to \vec{\mathfrak{X}}^{\mathrm{ord}}_{\Phi_{\mathcal{H}},\delta_{\mathcal{H}},\Sigma_{\Phi_{\mathcal{H}}}} \tag{5.2.4.45}$$

(cf. (1.3.2.49)) *compatible with each* (1.3.2.47) *as above (under canonical morphisms).*

If $g \in \mathrm{P}_{\mathsf{Z}}(\mathbb{A}^{\infty,p}) \times \mathrm{P}_{\mathsf{Z}}(\mathbb{Q}_p)$ *is as above and if we have a collection* $\Sigma^{\mathrm{ord},\prime}$ *for* $\vec{\mathsf{M}}^{\mathrm{ord}}_{\mathcal{H}'}$ *that is a g-refinement of a collection* Σ^{ord} *for* $\vec{\mathsf{M}}^{\mathrm{ord}}_{\mathcal{H}}$ *as in Definition* 5.2.2.1, *then the canonical morphism*

$$[\vec{g}]^{\mathrm{ord,tor}} : \vec{\mathsf{M}}^{\mathrm{ord,tor}}_{\mathcal{H}',\Sigma^{\mathrm{ord},\prime}} \to \vec{\mathsf{M}}^{\mathrm{ord,tor}}_{\mathcal{H},\Sigma^{\mathrm{ord}}}$$

as in Proposition 5.2.2.2 *is compatible with* (5.2.4.43) *when* $(\Phi'_{\mathcal{H}'}, \delta'_{\mathcal{H}'}, \rho)$ *is a g-refinement of* $(\Phi_{\mathcal{H}}, \delta_{\mathcal{H}}, \sigma)$, *under the canonical isomorphisms as in* (5) *of Theorem* 5.2.1.1; *and is compatible with* (5.2.4.45) *when* $(\Phi'_{\mathcal{H}'}, \delta'_{\mathcal{H}'}, \Sigma'_{\Phi'_{\mathcal{H}'}})$ *is a g-refinement of* $(\Phi_{\mathcal{H}}, \delta_{\mathcal{H}}, \Sigma_{\Phi_{\mathcal{H}}})$, *under the canonical isomorphisms as in Lemma* 5.2.4.38.

Proof. The assertions in the first two paragraphs can be justified as in the case of $\vec{\mathsf{M}}^{\mathrm{ord}}_{\mathcal{H}}$. (We omit the details for simplicity.) The third paragraph follows by comparing the torus torsor actions of sufficiently divisible multiples of elements, for which we have explicit descriptions in Lemma 5.2.4.29 and Proposition 5.2.4.30. As for the last paragraph, since the canonical morphisms are defined by universal properties given in terms of degeneration data, their compatibility follows from the fact that (by the theory of degeneration as in Theorem 4.1.6.2, based on [62, Thm. 5.2.3.14] and (4.1.4.50), in particular) the ordinary Hecke twist of the tautological tuple over $\vec{\mathsf{M}}^{\mathrm{ord,tor}}_{\mathcal{H}',\Sigma^{\mathrm{ord},\prime}}$ by g defined using the ordinary level structure $(\alpha_{\mathcal{H}'^{,p}}, \alpha^{\mathrm{ord}}_{\mathcal{H}'_p})$ over $\vec{\mathsf{M}}^{\mathrm{ord}}_{\mathcal{H}'}$ is compatible with the ordinary Hecke twist of the tautological tuple over $\vec{\Xi}^{\mathrm{ord}}_{\Phi'_{\mathcal{H}'},\delta'_{\mathcal{H}'}}(\rho)$ by g defined using the ordinary level structure $(\beta_{\mathcal{H}'^{,p}}, \beta^{\mathrm{ord}}_{\mathcal{H}'_p})$ over $\vec{\Xi}^{\mathrm{ord}}_{\Phi'_{\mathcal{H}'},\delta'_{\mathcal{H}'}}$. □

We will continue the generalization of Section 1.3.2 in Section 7.1.2 below.

Remark 5.2.4.46. Since all objects and morphisms in this subsection are defined by normalizations and by the various universal properties extending their analogues in characteristic zero, they are canonically compatible with the corresponding objects and morphisms in Section 1.3.2.

Chapter 6

Partial Minimal Compactifications

The first goal of this chapter is to construct the partial minimal compactifications for the ordinary loci defined in Chapter 3, based on the partial toroidal compactifications constructed in Chapter 5 and on the total minimal compactifications constructed in Chapter 2 (which is in turn based on the projective minimal compactifications constructed in [62] in the good reduction case, for the auxiliary models). The second goal is to show that the partial toroidal compactifications are *quasi-projective* when the levels are neat away from p and when the compatible choices of smooth admissible cone decompositions are *projective*. The third goal is to show that the reductions of the partial minimal compactifications modulo powers of p are affine. These are all indispensable for the application of our work to the construction of p-adic modular forms as in, for example, [39].

6.1 Homogeneous Spectra and Their Properties

In this section, we continue to assume the same settings as in Section 5.2. (We do not assume as in Section 5.2.3 that p is a good prime for the integral PEL datum $(\mathcal{O}, \star, L, \langle\,\cdot\,,\,\cdot\,\rangle, h_0)$ as in Definition 1.1.1.6.)

6.1.1 *Construction of Quasi-Projective Models*

Let $\vec{\mathsf{M}}^{\mathrm{ord,tor}}_{\mathcal{H},\Sigma^{\mathrm{ord}}}$ be as in Theorem 5.2.1.1, and let

$$\omega_{\vec{\mathsf{M}}^{\mathrm{ord,tor}}_{\mathcal{H},\Sigma^{\mathrm{ord}}}} := \wedge^{\mathrm{top}}\,\underline{\mathrm{Lie}}^{\vee}_{G/\vec{\mathsf{M}}^{\mathrm{ord,tor}}_{\mathcal{H},\Sigma^{\mathrm{ord}}}} \cong \wedge^{\mathrm{top}}\, e_G^*\Omega^1_{G/\vec{\mathsf{M}}^{\mathrm{ord,tor}}_{\mathcal{H},\Sigma^{\mathrm{ord}}}} \tag{6.1.1.1}$$

be the *Hodge invertible sheaf* as usual. By [80, IX, 2.1] (*cf.* [28, Ch. V, Prop. 2.1] and [62, Prop. 7.2.1.1]), there exists an integer $N_0 \geq 1$ such that $\omega^{\otimes N_0}_{\vec{\mathsf{M}}^{\mathrm{ord,tor}}_{\mathcal{H},\Sigma^{\mathrm{ord}}}}$ is generated by its global sections. Then the global sections of $\omega^{\otimes N_0 k}_{\vec{\mathsf{M}}^{\mathrm{ord,tor}}_{\mathcal{H},\Sigma^{\mathrm{ord}}}}$, for $k \geq 0$, define a morphism

$$\vec{\mathsf{M}}^{\mathrm{ord,tor}}_{\mathcal{H},\Sigma^{\mathrm{ord}}} \to \mathrm{Proj}\Big(\underset{k\geq 0}{\oplus}\,\Gamma(\vec{\mathsf{M}}^{\mathrm{ord,tor}}_{\mathcal{H},\Sigma^{\mathrm{ord}}}, \omega^{\otimes N_0 k}_{\vec{\mathsf{M}}^{\mathrm{ord,tor}}_{\mathcal{H},\Sigma^{\mathrm{ord}}}})\Big) \tag{6.1.1.2}$$

over $\vec{\mathsf{S}}_{0,r_{\mathcal{H}}}$ (see [35, II, 3.7.4]). Let

$$\vec{\mathsf{M}}_{\mathcal{H}}^{\mathrm{ord,min}} := \mathrm{Proj}\Big(\underset{k\geq 0}{\oplus}\,\Gamma(\vec{\mathsf{M}}_{\mathcal{H},\Sigma^{\mathrm{ord}}}^{\mathrm{ord,tor}},\omega_{\vec{\mathsf{M}}_{\mathcal{H},\Sigma^{\mathrm{ord}}}^{\mathrm{ord,tor}}}^{\otimes k})\Big) \tag{6.1.1.3}$$

(as in [35, II, Sec. 2]), which is a scheme over $\vec{\mathsf{S}}_{0,r_{\mathcal{H}}}$. By [35, II, 2.4.7], we have a canonical isomorphism

$$\vec{\mathsf{M}}_{\mathcal{H}}^{\mathrm{ord,min}} \cong \mathrm{Proj}\Big(\underset{k\geq 0}{\oplus}\,\Gamma(\vec{\mathsf{M}}_{\mathcal{H},\Sigma^{\mathrm{ord}}}^{\mathrm{ord,tor}},\omega_{\vec{\mathsf{M}}_{\mathcal{H},\Sigma^{\mathrm{ord}}}^{\mathrm{ord,tor}}}^{\otimes Nk})\Big)$$

for each integer $N \geq 1$. Hence, the right-hand side of (6.1.1.2) is independent of the integer $N_0 \geq 1$ above, and (6.1.1.2) induces a canonical morphism

$$\vec{\oint}_{\mathcal{H}}^{\mathrm{ord}} : \vec{\mathsf{M}}_{\mathcal{H},\Sigma^{\mathrm{ord}}}^{\mathrm{ord,tor}} \to \vec{\mathsf{M}}_{\mathcal{H}}^{\mathrm{ord,min}} \tag{6.1.1.4}$$

over $\vec{\mathsf{S}}_{0,r_{\mathcal{H}}}$. (We have seen special cases of this in Propositions 5.2.3.8 and 5.2.3.18.) Since we do not know the finite generation of the graded algebra $\underset{k\geq 0}{\oplus}\,\Gamma(\vec{\mathsf{M}}_{\mathcal{H},\Sigma^{\mathrm{ord}}}^{\mathrm{ord,tor}},\omega_{\vec{\mathsf{M}}_{\mathcal{H},\Sigma^{\mathrm{ord}}}^{\mathrm{ord,tor}}}^{\otimes k})$ over $\mathcal{O}_{F_0,(p)}[\zeta_{p^{r_{\mathcal{H}}}}]$, we cannot assert that $\vec{\mathsf{M}}_{\mathcal{H}}^{\mathrm{ord,min}}$ is projective over $\vec{\mathsf{S}}_{0,r_{\mathcal{H}}}$. (We cannot even assert the quasi-projectivity of $\vec{\mathsf{M}}_{\mathcal{H}}^{\mathrm{ord,min}}$ over $\vec{\mathsf{S}}_{0,r_{\mathcal{H}}}$ at this moment. As we will see soon, $\vec{\mathsf{M}}_{\mathcal{H}}^{\mathrm{ord,min}}$ is indeed quasi-projective over $\vec{\mathsf{S}}_{0,r_{\mathcal{H}}}$, but almost never projective over $\vec{\mathsf{S}}_{0,r_{\mathcal{H}}}$.)

To justify the absence of Σ^{ord} in the notation of $\vec{\mathsf{M}}_{\mathcal{H}}^{\mathrm{ord,min}}$:

Lemma 6.1.1.5. *The definition of* $\vec{\mathsf{M}}_{\mathcal{H}}^{\mathrm{ord,min}}$ *in* (6.1.1.3) *is independent of the choice of* Σ^{ord}.

Proof. Suppose $\Sigma^{\mathrm{ord},\prime}$ is a refinement of Σ^{ord}, and suppose $\vec{[1]}^{\mathrm{ord,tor}} : \vec{\mathsf{M}}_{\mathcal{H},\Sigma^{\mathrm{ord},\prime}}^{\mathrm{ord,tor}} \to \vec{\mathsf{M}}_{\mathcal{H},\Sigma^{\mathrm{ord}}}^{\mathrm{ord,tor}}$ is the proper log étale surjection as in Proposition 5.2.2.2 such that the family $(G,\lambda,i,\alpha_{\mathcal{H}^p},\alpha_{\mathcal{H}_p}^{\mathrm{ord}}) \to \vec{\mathsf{M}}_{\mathcal{H},\Sigma^{\mathrm{ord},\prime}}^{\mathrm{ord,tor}}$ is the pullback of $(G,\lambda,i,\alpha_{\mathcal{H}^p},\alpha_{\mathcal{H}_p}^{\mathrm{ord}}) \to \vec{\mathsf{M}}_{\mathcal{H},\Sigma^{\mathrm{ord}}}^{\mathrm{ord,tor}}$. Then we have $\omega_{\vec{\mathsf{M}}_{\mathcal{H},\Sigma^{\mathrm{ord},\prime}}^{\mathrm{ord,tor}}} \cong (\vec{[1]}^{\mathrm{ord,tor}})^*\omega_{\vec{\mathsf{M}}_{\mathcal{H},\Sigma^{\mathrm{ord}}}^{\mathrm{ord,tor}}}$, by definition. Moreover, as in [28, Ch. V, Rem. 1.2(b)] and in the proof of [62, Lem. 7.1.1.4], we have $R^i\vec{[1]}_*^{\mathrm{ord,tor}}\mathscr{O}_{\mathsf{M}_{\mathcal{H},\Sigma'}^{\mathrm{tor}}} = 0$ for all $i > 0$ and $\vec{[1]}_*^{\mathrm{ord,tor}}\mathscr{O}_{\mathsf{M}_{\mathcal{H},\Sigma'}^{\mathrm{tor}}} = \mathscr{O}_{\mathsf{M}_{\mathcal{H},\Sigma}^{\mathrm{tor}}}$ by [50, Ch. I, Sec. 3], which implies by the projection formula [35, 0_{I}, 5.4.10.1] that the canonical morphism $\Gamma(\vec{\mathsf{M}}_{\mathcal{H},\Sigma^{\mathrm{ord}}}^{\mathrm{ord,tor}},\omega_{\vec{\mathsf{M}}_{\mathcal{H},\Sigma^{\mathrm{ord}}}^{\mathrm{ord,tor}}}^{\otimes k}) \to \Gamma(\vec{\mathsf{M}}_{\mathcal{H},\Sigma^{\mathrm{ord},\prime}}^{\mathrm{ord,tor}},\omega_{\vec{\mathsf{M}}_{\mathcal{H},\Sigma^{\mathrm{ord},\prime}}^{\mathrm{ord,tor}}}^{\otimes k})$ is an isomorphism for each $k \geq 0$. Hence, the canonical morphism

$$\mathrm{Proj}\Big(\underset{k\geq 0}{\oplus}\,\Gamma(\vec{\mathsf{M}}_{\mathcal{H},\Sigma^{\mathrm{ord},\prime}}^{\mathrm{ord,tor}},\omega_{\vec{\mathsf{M}}_{\mathcal{H},\Sigma^{\mathrm{ord},\prime}}^{\mathrm{ord,tor}}}^{\otimes k})\Big) \to \mathrm{Proj}\Big(\underset{k\geq 0}{\oplus}\,\Gamma(\vec{\mathsf{M}}_{\mathcal{H},\Sigma^{\mathrm{ord}}}^{\mathrm{ord,tor}},\omega_{\vec{\mathsf{M}}_{\mathcal{H},\Sigma^{\mathrm{ord}}}^{\mathrm{ord,tor}}}^{\otimes k})\Big)$$

is also an isomorphism, as desired. □

To construct minimal compactifications as in [62, Sec. 7.2.3] using the technique of Stein factorizations, it is desirable to start with a proper morphism with target a scheme quasi-projective over $\vec{\mathsf{S}}_{0,r_{\mathcal{H}}}$. However, this is not straightforward for $\vec{\mathsf{M}}_{\mathcal{H}}^{\mathrm{ord,tor}}$, because it is not proper over $\vec{\mathsf{S}}_{0,r_{\mathcal{H}}}$ in general. We need the help of the auxiliary moduli problems and their compactifications as in Section 2.2 (see also Section 3.4.6).

Proposition 6.1.1.6. *Suppose* $\mathcal{H} = \mathcal{H}^p\mathcal{H}_p$ *such that* $\mathcal{H}^p \subset \mathrm{G}(\hat{\mathbb{Z}}^p)$ *and* $\mathcal{H}_p = \mathcal{U}^{\mathrm{bal}}_{p,1}(p^r)$ *for some integer* $r \geq 0$. *Up to replacing* Σ^{ord} *with a refinement, we may assume that it is smooth and projective, and that there exists a cone decomposition* $\Sigma^{\mathrm{ord}}_{\mathrm{aux}}$ *and an analogous partial toroidal compactification* $\vec{\mathsf{M}}^{\mathrm{ord,tor}}_{\mathcal{H}_{\mathrm{aux}},\Sigma^{\mathrm{ord}}_{\mathrm{aux}}}$ *over* $\vec{\mathsf{S}}_{0,\mathrm{aux},r} := \mathrm{Spec}(\mathcal{O}_{F_{0,\mathrm{aux}},(p)}[\zeta_{p^r}])$ *such that there is a (necessarily unique)* **proper** *morphism*

$$\vec{\mathsf{M}}^{\mathrm{ord,tor}}_{\mathcal{H},\Sigma^{\mathrm{ord}}} \to \vec{\mathsf{M}}^{\mathrm{ord,tor}}_{\mathcal{H}_{\mathrm{aux}},\Sigma^{\mathrm{ord}}_{\mathrm{aux}}} \tag{6.1.1.7}$$

(over $\vec{\mathsf{S}}_{0,\mathrm{aux},r}$*) extending* (3.4.6.2)*, mapping the* $[(\Phi_{\mathcal{H}},\delta_{\mathcal{H}},\sigma)]$*-strata* $\vec{\mathsf{Z}}^{\mathrm{ord}}_{[(\Phi_{\mathcal{H}},\delta_{\mathcal{H}},\sigma)]}$ *of* $\vec{\mathsf{M}}^{\mathrm{ord,tor}}_{\mathcal{H},\Sigma^{\mathrm{ord}}}$ *to the* $[(\Phi_{\mathcal{H}_{\mathrm{aux}}},\delta_{\mathcal{H}_{\mathrm{aux}}},\sigma_{\mathrm{aux}})]$*-stratum* $\vec{\mathsf{Z}}^{\mathrm{ord}}_{[(\Phi_{\mathcal{H}_{\mathrm{aux}}},\delta_{\mathcal{H}_{\mathrm{aux}}},\sigma_{\mathrm{aux}})]}$ *of* $\vec{\mathsf{M}}^{\mathrm{ord,tor}}_{\mathcal{H}_{\mathrm{aux}},\Sigma^{\mathrm{ord}}_{\mathrm{aux}}}$ *when* $(\Phi_{\mathcal{H}_{\mathrm{aux}}},\delta_{\mathcal{H}_{\mathrm{aux}}},\sigma_{\mathrm{aux}})$ *is assigned to* $(\Phi_{\mathcal{H}},\delta_{\mathcal{H}},\sigma)$ *(see Definition* 2.1.2.25*). Under* (6.1.1.7)*, the pullback of* $\omega^{\otimes a_0}_{\vec{\mathsf{M}}^{\mathrm{ord,tor}}_{\mathcal{H}_{\mathrm{aux}},\Sigma^{\mathrm{ord}}_{\mathrm{aux}}}}$ *is canonically isomorphic to* $\omega^{\otimes a}_{\vec{\mathsf{M}}^{\mathrm{ord,tor}}_{\mathcal{H},\Sigma^{\mathrm{ord}}}}$ *for the integers* $a_0 \geq 1$ *and* $a \geq 1$ *as in Lemma* 2.1.2.35.

Proof. Starting with the degenerating family $(G,\lambda,i,\alpha_{\mathcal{H}^p},\alpha^{\mathrm{ord}}_{\mathcal{H}_p})$ of type $\vec{\mathsf{M}}^{\mathrm{ord}}_{\mathcal{H}}$ over $\vec{\mathsf{M}}^{\mathrm{ord,tor}}_{\mathcal{H}}$, we would like to construct a degenerating family $(G^{\triangledown}_{\mathrm{aux}},\lambda^{\triangledown}_{\mathrm{aux}},i^{\triangledown}_{\mathrm{aux}},\alpha^{\triangledown}_{\mathcal{H}^p_{\mathrm{aux}}},\alpha^{\mathrm{ord},\triangledown}_{\mathcal{H}_{\mathrm{aux},p}})$ of type $\vec{\mathsf{M}}^{\mathrm{ord}}_{\mathcal{H}_{\mathrm{aux}}}$ over $\vec{\mathsf{M}}^{\mathrm{ord,tor}}_{\mathcal{H},\Sigma^{\mathrm{ord}}}$, and show that it is the pullback of the tautological degenerating family $(G_{\mathrm{aux}},\lambda_{\mathrm{aux}},i_{\mathrm{aux}},\alpha_{\mathcal{H}^p_{\mathrm{aux}}},\alpha^{\mathrm{ord}}_{\mathcal{H}_{\mathrm{aux},p}})$ of type $\vec{\mathsf{M}}^{\mathrm{ord}}_{\mathcal{H}_{\mathrm{aux}}}$ over $\vec{\mathsf{M}}^{\mathrm{ord,tor}}_{\mathcal{H},\Sigma^{\mathrm{ord}}}$ under some (necessarily unique) morphism (6.1.1.7) extending (3.4.6.2). For this purpose, we first construct $G^{\triangledown}_{\mathrm{aux}}$, $\lambda^{\triangledown}_{\mathrm{aux}}$, and $i^{\triangledown}_{\mathrm{aux}}$ as in the proof of Proposition 2.1.2.29. Then we can construct $\alpha^{\triangledown}_{\mathcal{H}^p_{\mathrm{aux}}}$ and $\alpha^{\mathrm{ord},\triangledown}_{\mathcal{H}_{\mathrm{aux},p}}$ as in the proof of Lemma 3.4.1.4, verify as in the proof of Proposition 2.1.2.29 the universal property of $\vec{\mathsf{M}}^{\mathrm{ord,tor}}_{\mathcal{H}_{\mathrm{aux}},\Sigma^{\mathrm{ord}}_{\mathrm{aux}}}$ stated as in (6) of Theorem 5.2.1.1, and obtain the desired morphism (6.1.1.7) mapping $\vec{\mathsf{Z}}^{\mathrm{ord}}_{[(\Phi_{\mathcal{H}},\delta_{\mathcal{H}},\sigma)]}$ to $\vec{\mathsf{Z}}^{\mathrm{ord}}_{[(\Phi_{\mathcal{H}_{\mathrm{aux}}},\delta_{\mathcal{H}_{\mathrm{aux}}},\sigma_{\mathrm{aux}})]}$ of $\vec{\mathsf{M}}^{\mathrm{ord,tor}}_{\mathcal{H}_{\mathrm{aux}},\Sigma^{\mathrm{ord}}_{\mathrm{aux}}}$ when $(\Phi_{\mathcal{H}_{\mathrm{aux}}},\delta_{\mathcal{H}_{\mathrm{aux}}},\sigma_{\mathrm{aux}})$ is assigned to $(\Phi_{\mathcal{H}},\delta_{\mathcal{H}},\sigma)$ as in Definition 2.1.2.25.

To show that the morphism (6.1.1.7) is proper, we apply the valuative criterion over the spectrum of a complete discrete valuation ring with algebraically closed residue field, with generic point mapped to the open dense subscheme $\vec{\mathsf{M}}^{\mathrm{ord}}_{\mathcal{H}}$ of $\vec{\mathsf{M}}^{\mathrm{ord,tor}}_{\mathcal{H},\Sigma^{\mathrm{ord}}}$, and verify the universal property of $\vec{\mathsf{M}}^{\mathrm{ord,tor}}_{\mathcal{H},\Sigma^{\mathrm{ord}}}$ stated as in (6) of Theorem 5.2.1.1. As in the proof of [62, Prop. 6.3.3.17], since the base ring is a complete discrete valuation ring, the cone decomposition Σ^{ord} does not impose any condition in the verification of this universal property. Thus, the only condition to verify is the extensibility condition (5) in the definition of degenerating families, which is satisfied in this case because the construction of the degenerating family $(G^{\triangledown}_{\mathrm{aux}},\lambda^{\triangledown}_{\mathrm{aux}},i^{\triangledown}_{\mathrm{aux}},\alpha^{\triangledown}_{\mathcal{H}^p_{\mathrm{aux}}},\alpha^{\mathrm{ord},\triangledown}_{\mathcal{H}_{\mathrm{aux},p}})$ above is compatible with (and implicitly used) this extensibility condition. □

Proposition 6.1.1.8. *In Proposition* 6.1.1.6*, we can choose* Σ^{ord}, $\Sigma^{\mathrm{ord}}_{\mathrm{aux}}$, *and*

$\vec{\mathsf{M}}^{\mathrm{ord,tor}}_{\mathcal{H}_{\mathrm{aux}},\Sigma^{\mathrm{ord}}_{\mathrm{aux}}}$ *such that the canonical morphism*

$$\vec{\oint}^{\mathrm{ord}}_{\mathcal{H}_{\mathrm{aux}}} : \vec{\mathsf{M}}^{\mathrm{ord,tor}}_{\mathcal{H}_{\mathrm{aux}},\Sigma^{\mathrm{ord}}_{\mathrm{aux}}} \to \vec{\mathsf{M}}^{\mathrm{ord,min}}_{\mathcal{H}_{\mathrm{aux}}} := \mathrm{Proj}\Big(\mathop{\oplus}_{k\geq 0} \Gamma(\vec{\mathsf{M}}^{\mathrm{ord,tor}}_{\mathcal{H}_{\mathrm{aux}},\Sigma^{\mathrm{ord}}_{\mathrm{aux}}}, \omega^{\otimes k}_{\vec{\mathsf{M}}^{\mathrm{ord,tor}}_{\mathcal{H}_{\mathrm{aux}},\Sigma^{\mathrm{ord}}_{\mathrm{aux}}}})\Big)$$

is proper (and surjective), and is the Stein factorization (see [35, III-1, 4.3.3]) of itself (and hence has nonempty connected geometric fibers, by [35, III-1, 4.3.1, 4.3.3, 4.3.4] and its natural generalization to the context of algebraic stacks).

Concretely, up to replacing $\mathcal{H}^p$ with an open compact subgroup, we may assume moreover that $\mathcal{H}^p_{\mathrm{aux}}$ is neat, that $\Sigma^{\mathrm{ord}}_{\mathrm{aux}}$ is projective with a collection $\mathsf{pol}^{\mathrm{ord}}_{\mathrm{aux}}$ of polarization functions, and that $\mathcal{H}_{\mathrm{aux}}$, $\Sigma^{\mathrm{ord}}_{\mathrm{aux}}$, and $\mathsf{pol}^{\mathrm{ord}}_{\mathrm{aux}}$ fit into the setup of the beginning of Section 5.2.3 *(with Assumption* 5.2.3.1 *automatically satisfied by Lemma* 5.2.3.2*), so that $\Sigma^{\mathrm{ord}}_{\mathrm{aux}}$ (resp. $\mathsf{pol}^{\mathrm{ord}}_{\mathrm{aux}}$) extends to some projective (but possibly nonsmooth) Σ_{aux} (resp. $\mathsf{pol}_{\mathrm{aux}}$) such that $\vec{\mathsf{M}}^{\mathrm{tor}}_{\mathcal{H}_{\mathrm{aux}},d_{0,\mathrm{aux}}\mathsf{pol}_{\mathrm{aux}}}$ is defined (for some integer $d_{0,\mathrm{aux}} \geq 1$; see Proposition* 2.2.2.3*), and so that there is a canonical open immersion*

$$\vec{\mathsf{M}}^{\mathrm{ord,tor}}_{\mathcal{H}_{\mathrm{aux}},\Sigma^{\mathrm{ord}}_{\mathrm{aux}}} \hookrightarrow \vec{\mathsf{M}}^{\mathrm{tor}}_{\mathcal{H}_{\mathrm{aux}},d_{0,\mathrm{aux}}\mathsf{pol}_{\mathrm{aux}},r}$$

(cf. (5.2.3.19)*—here $r_{\mathcal{H}_{\mathrm{aux}}} = r_{\nu(\mathcal{H}_{\mathrm{aux}})} = r$ by definition, as explained in Remark* 3.4.2.2*) inducing by composition with the structural morphism $\vec{\mathsf{M}}^{\mathrm{tor}}_{\mathcal{H}_{\mathrm{aux}},d_{0,\mathrm{aux}}\mathsf{pol}_{\mathrm{aux}},r} \to \vec{\mathsf{M}}^{\mathrm{min}}_{\mathcal{H}_{\mathrm{aux}},r}$ a canonical morphism*

$$\vec{\mathsf{M}}^{\mathrm{ord,tor}}_{\mathcal{H}_{\mathrm{aux}},\Sigma^{\mathrm{ord}}_{\mathrm{aux}}} \to \vec{\mathsf{M}}^{\mathrm{min}}_{\mathcal{H}_{\mathrm{aux}},r},$$

inducing a commutative diagram

$$\begin{array}{ccccc} \vec{\mathsf{M}}^{\mathrm{ord,tor}}_{\mathcal{H},\Sigma^{\mathrm{ord}}} & \longrightarrow & \vec{\mathsf{M}}^{\mathrm{ord,tor}}_{\mathcal{H}_{\mathrm{aux}},\Sigma^{\mathrm{ord}}_{\mathrm{aux}}} & \longrightarrow & \vec{\mathsf{M}}^{\mathrm{tor}}_{\mathcal{H}_{\mathrm{aux}},d_{0,\mathrm{aux}}\mathsf{pol}_{\mathrm{aux}},r} \\ \downarrow & & \downarrow & & \downarrow \\ \vec{\mathsf{M}}^{\mathrm{ord,min}}_{\mathcal{H}} & \longrightarrow & \vec{\mathsf{M}}^{\mathrm{ord,min}}_{\mathcal{H}_{\mathrm{aux}}} & \longrightarrow & \vec{\mathsf{M}}^{\mathrm{min}}_{\mathcal{H}_{\mathrm{aux}},r} \end{array}$$

in which the vertical and the left top horizontal arrows are all proper, in which the left bottom horizontal arrow is finite, and in which the right top and bottom arrows are open immersions making the right-hand square Cartesian, such that the canonical morphisms

$$\mathscr{O}_{\vec{\mathsf{M}}^{\mathrm{ord,min}}_{\mathcal{H}}} \to \vec{\oint}^{\mathrm{ord}}_{\mathcal{H},*}\mathscr{O}_{\vec{\mathsf{M}}^{\mathrm{ord,tor}}_{\mathcal{H},\Sigma^{\mathrm{ord}}}}$$

and

$$\mathscr{O}_{\vec{\mathsf{M}}^{\mathrm{ord,min}}_{\mathcal{H}_{\mathrm{aux}}}} \to \vec{\oint}^{\mathrm{ord}}_{\mathcal{H}_{\mathrm{aux}},*}\mathscr{O}_{\vec{\mathsf{M}}^{\mathrm{ord,tor}}_{\mathcal{H}_{\mathrm{aux}},\Sigma^{\mathrm{ord}}_{\mathrm{aux}}}}$$

are isomorphisms. Consequently, $\vec{\mathsf{M}}^{\mathrm{ord,min}}_{\mathcal{H}}$ is quasi-projective over $\vec{\mathsf{S}}_{0,r_{\mathcal{H}}}$.

Proof. Suppose $\mathcal{H}'^{,p}$ is an open compact subgroup of $\mathcal{H}^p$, which defines an open compact subgroup $\mathcal{H}' = \mathcal{H}'^{,p}\mathcal{H}_p$ of $\mathcal{H} = \mathcal{H}^p\mathcal{H}_p$. Suppose $\Sigma^{\mathrm{ord},\prime}$ is a 1-refinement of Σ^{ord} as in Definition 5.2.2.1. Then the canonical surjection $\vec{[1]}^{\mathrm{ord,tor}} : \vec{\mathsf{M}}^{\mathrm{ord,tor}}_{\mathcal{H}',\Sigma^{\mathrm{ord}}} \to$

$\vec{\mathsf{M}}^{\mathrm{ord,tor}}_{\mathcal{H},\Sigma^{\mathrm{ord}}}$ is *proper* by Proposition 5.2.2.2. Thus, to prove the first paragraph in the proposition, we may compatibly replace $\mathcal{H}^p$ and $\mathcal{H}^p_{\mathrm{aux}}$ with sufficiently small subgroups as above, and assume that $\mathcal{H}^p_{\mathrm{aux}}$ is also neat. We are also allowed to replace Σ^{ord} with suitable refinements.

By Proposition 5.2.3.18, by suitably choosing $\Sigma^{\mathrm{ord}}_{\mathrm{aux}}$ (possibly at the expense of replacing Σ^{ord} with a refinement), we can construct $\vec{\mathsf{M}}^{\mathrm{ord,tor}}_{\mathcal{H}_{\mathrm{aux}},\Sigma^{\mathrm{ord}}_{\mathrm{aux}}} \to \vec{\mathsf{M}}^{\mathrm{ord,min}}_{\mathcal{H}_{\mathrm{aux}}}$ such that it is the (proper and surjective) pullback of $\vec{\mathsf{M}}^{\mathrm{tor}}_{\mathcal{H}_{\mathrm{aux}},d_{0,\mathrm{aux}}\mathsf{pol}_{\mathrm{aux}},r} \to \vec{\mathsf{M}}^{\mathrm{min}}_{\mathcal{H}_{\mathrm{aux}},r}$ under $\vec{\mathsf{M}}^{\mathrm{ord,min}}_{\mathcal{H}_{\mathrm{aux}}} \to \vec{\mathsf{M}}^{\mathrm{min}}_{\mathcal{H}_{\mathrm{aux}},r}$ (on the target), and such that it satisfies the other statements (concerning $\vec{\mathsf{M}}^{\mathrm{ord,tor}}_{\mathcal{H}_{\mathrm{aux}},\Sigma^{\mathrm{ord}}_{\mathrm{aux}}}$, $\vec{\mathsf{M}}^{\mathrm{ord,min}}_{\mathcal{H}_{\mathrm{aux}}}$, $\vec{\mathsf{M}}^{\mathrm{tor}}_{\mathcal{H}_{\mathrm{aux}},d_{0,\mathrm{aux}}\mathsf{pol}_{\mathrm{aux}},r}$, and $\vec{\mathsf{M}}^{\mathrm{min}}_{\mathcal{H}_{\mathrm{aux}},r}$) in this proposition. Then the remaining statements (concerning $\vec{\mathsf{M}}^{\mathrm{ord,tor}}_{\mathcal{H},\Sigma^{\mathrm{ord}}}$ and $\vec{\mathsf{M}}^{\mathrm{ord,min}}_{\mathcal{H}}$) follow from Proposition 6.1.1.6 as formal consequences. □

Lemma 6.1.1.9. (*Compare with Lemma* 3.4.6.1.) *With the setting as at the beginning of this section (but no longer assuming that $\mathcal{H}_p = \mathcal{U}^{\mathrm{bal}}_{p,1}(p^r)$), suppose $\mathcal{H}$ and $\mathcal{H}_{\mathrm{aux}}$ are as in Lemma* 3.4.1.6. *Then there exist compatible choices of smooth and projective Σ^{ord} and $\Sigma^{\mathrm{ord}}_{\mathrm{aux}}$, which can be achieved by compatibly replacing any given choices with refinements, such that $\vec{\mathsf{M}}^{\mathrm{ord,tor}}_{\mathcal{H},\Sigma^{\mathrm{ord}}}$ and $\vec{\mathsf{M}}^{\mathrm{ord,tor}}_{\mathcal{H}_{\mathrm{aux}},\Sigma^{\mathrm{ord}}_{\mathrm{aux}}}$ are defined as in Theorem* 5.2.1.1, *and such that there is a (necessarily unique, but possibly nonproper) morphism*

$$\vec{\mathsf{M}}^{\mathrm{ord,tor}}_{\mathcal{H},\Sigma^{\mathrm{ord}}} \to \vec{\mathsf{M}}^{\mathrm{ord,tor}}_{\mathcal{H}_{\mathrm{aux}},\Sigma^{\mathrm{ord}}_{\mathrm{aux}}} \tag{6.1.1.10}$$

extending (3.4.6.2), *mapping the $[(\Phi_{\mathcal{H}},\delta_{\mathcal{H}},\sigma)]$-strata $\vec{\mathsf{Z}}^{\mathrm{ord}}_{[(\Phi_{\mathcal{H}},\delta_{\mathcal{H}},\sigma)]}$ of $\vec{\mathsf{M}}^{\mathrm{ord,tor}}_{\mathcal{H},\Sigma^{\mathrm{ord}}}$ to the $[(\Phi_{\mathcal{H}_{\mathrm{aux}}},\delta_{\mathcal{H}_{\mathrm{aux}}},\sigma_{\mathrm{aux}})]$-stratum $\vec{\mathsf{Z}}^{\mathrm{ord}}_{[(\Phi_{\mathcal{H}_{\mathrm{aux}}},\delta_{\mathcal{H}_{\mathrm{aux}}},\sigma_{\mathrm{aux}})]}$ of $\vec{\mathsf{M}}^{\mathrm{ord,tor}}_{\mathcal{H}_{\mathrm{aux}},\Sigma^{\mathrm{ord}}_{\mathrm{aux}}}$ when $(\Phi_{\mathcal{H}_{\mathrm{aux}}},\delta_{\mathcal{H}_{\mathrm{aux}}},\sigma_{\mathrm{aux}})$ is assigned to $(\Phi_{\mathcal{H}},\delta_{\mathcal{H}},\sigma)$ (see Definition* 2.1.2.25). *Under* (6.1.1.10), *the pullback of $\omega^{\otimes a_0}_{\vec{\mathsf{M}}^{\mathrm{ord,tor}}_{\mathcal{H}_{\mathrm{aux}},\Sigma^{\mathrm{ord}}_{\mathrm{aux}}}}$ is canonically isomorphic to $\omega^{\otimes a}_{\vec{\mathsf{M}}^{\mathrm{ord,tor}}_{\mathcal{H},\Sigma^{\mathrm{ord}}}}$ for the integers $a_0 \geq 1$ and $a \geq 1$ as in Lemma* 2.1.2.35.

If $\mathcal{H}_{\mathrm{aux}} = \mathrm{G}_{\mathrm{aux}}(\hat{\mathbb{Z}})$, we may assume that $\Sigma^{\mathrm{ord}}_{\mathrm{aux}}$ is induced by some smooth Σ^p_{aux} for $\mathsf{M}_{\mathrm{G}_{\mathrm{aux}}(\hat{\mathbb{Z}}^p)}$, so that Proposition 5.2.3.3 *and Lemma* 5.2.3.8 *apply (with* (5.2.3.4) *there being $\vec{\mathsf{M}}^{\mathrm{ord,tor}}_{\mathrm{G}_{\mathrm{aux}}(\hat{\mathbb{Z}}),\Sigma^{\mathrm{ord}}_{\mathrm{aux}}} \hookrightarrow \mathsf{M}^{\mathrm{tor}}_{\mathrm{G}_{\mathrm{aux}}(\hat{\mathbb{Z}}^p),\Sigma^p_{\mathrm{aux}}}$ in our notation here).*

Proof. As in the proof of Lemma 3.4.1.6, this is because the morphism (6.1.1.7) at sufficiently higher levels and with sufficiently refined cone decompositions induces the morphism (6.1.1.10). □

Corollary 6.1.1.11. *With the assumptions as in Lemma* 6.1.1.9, *there exists a*

commutative diagram

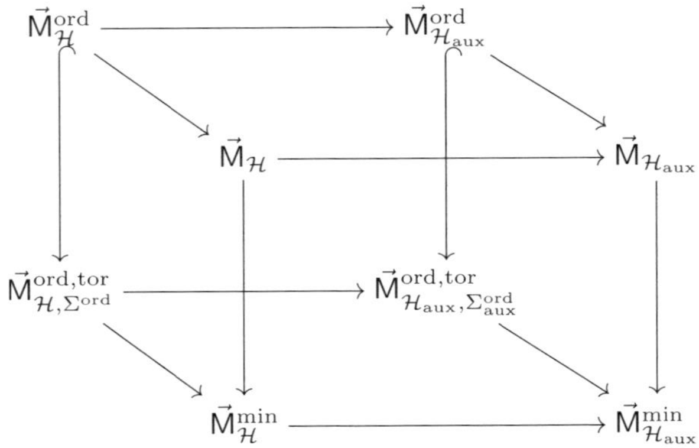

extending the morphisms (3.4.6.2) *and* (6.1.1.10), *which is compatible with other canonical morphisms.*

For $N_1 \geq 1$ *as in Proposition* 2.2.1.2 *and for integers* $a_0 \geq 1$ *and* $a \geq 1$ *as in Lemma* 2.1.2.35, *the pullback of* $\omega^{\otimes aN_1}_{\vec{\mathsf{M}}^{\mathrm{min}}_{\mathcal{H}}}$ *to* $\vec{\mathsf{M}}^{\mathrm{ord,tor}}_{\mathcal{H},\Sigma^{\mathrm{ord}}}$ *is canonically isomorphic to* $\omega^{\otimes aN_1}_{\vec{\mathsf{M}}^{\mathrm{ord,tor}}_{\mathcal{H},\Sigma^{\mathrm{ord}}}}$.

Proof. The upper-left square and its commutativity follow from Lemmas 3.4.6.1 and 6.1.1.9. The canonical morphisms $\vec{\mathsf{M}}_{\mathcal{H}} \to \vec{\mathsf{M}}^{\mathrm{min}}_{\mathcal{H}}$ and $\vec{\mathsf{M}}_{\mathcal{H}_{\mathrm{aux}}} \to \vec{\mathsf{M}}^{\mathrm{min}}_{\mathcal{H}_{\mathrm{aux}}}$ are as in Proposition 2.2.1.2. The canonical morphisms $\vec{\mathsf{M}}^{\mathrm{ord}}_{\mathcal{H}} \to \vec{\mathsf{M}}_{\mathcal{H}}$ and $\vec{\mathsf{M}}^{\mathrm{ord}}_{\mathcal{H}_{\mathrm{aux}}} \to \vec{\mathsf{M}}_{\mathcal{H}_{\mathrm{aux}}}$ are as in Proposition 3.4.6.3. The remaining morphisms are compatibly induced by the universal properties of $\vec{\mathsf{M}}^{\mathrm{ord}}_{\mathcal{H}_{\mathrm{aux}}}$, $\vec{\mathsf{M}}^{\mathrm{min}}_{\mathcal{H}}$, $\vec{\mathsf{M}}^{\mathrm{min}}_{\mathcal{H}}$, and $\vec{\mathsf{M}}^{\mathrm{min}}_{\mathcal{H}_{\mathrm{aux}}}$ as normalizations.

For the second paragraph, we may assume that $\mathcal{H}_{\mathrm{aux}} = \mathrm{G}_{\mathrm{aux}}(\hat{\mathbb{Z}})$, and that $\Sigma^{\mathrm{ord}}_{\mathrm{aux}}$ is induced by some smooth $\Sigma^{p}_{\mathrm{aux}}$ for $\mathsf{M}_{\mathrm{G}_{\mathrm{aux}}(\hat{\mathbb{Z}}^p)}$, as in the last paragraph of Lemma 6.1.1.9. By Proposition 2.2.1.2, $\omega^{\otimes aN_1}_{\vec{\mathsf{M}}^{\mathrm{min}}_{\mathcal{H}}}$ is canonically isomorphic to the pullback of $\omega^{\otimes a_0N_1}_{\mathsf{M}^{\mathrm{min}}_{\mathrm{G}_{\mathrm{aux}}(\hat{\mathbb{Z}}^p)}}$. On the other hand, by [62, Thm. 7.2.4.1], the pullback of $\omega^{\otimes a_0N_1}_{\mathsf{M}^{\mathrm{min}}_{\mathrm{G}_{\mathrm{aux}}(\hat{\mathbb{Z}}^p)}}$ to $\mathsf{M}^{\mathrm{tor}}_{\mathrm{G}_{\mathrm{aux}}(\hat{\mathbb{Z}}^p),\Sigma^p_{\mathrm{aux}}}$ is canonically isomorphic to $\omega^{\otimes a_0N_1}_{\mathsf{M}^{\mathrm{tor}}_{\mathrm{G}_{\mathrm{aux}}(\hat{\mathbb{Z}}^p),\Sigma^p_{\mathrm{aux}}}}$; by Lemma 5.2.3.8, the pullback of $\omega^{\otimes a_0N_1}_{\mathsf{M}^{\mathrm{tor}}_{\mathrm{G}_{\mathrm{aux}}(\hat{\mathbb{Z}}^p),\Sigma^p_{\mathrm{aux}}}}$ to $\vec{\mathsf{M}}^{\mathrm{ord,tor}}_{\mathcal{H}_{\mathrm{aux}},\Sigma^{\mathrm{ord}}_{\mathrm{aux}}} = \vec{\mathsf{M}}^{\mathrm{ord,tor}}_{\mathrm{G}_{\mathrm{aux}}(\hat{\mathbb{Z}}),\Sigma^{\mathrm{ord}}_{\mathrm{aux}}}$ is canonically isomorphic to $\omega^{\otimes a_0N_1}_{\vec{\mathsf{M}}^{\mathrm{ord,tor}}_{\mathcal{H}_{\mathrm{aux}},\Sigma^{\mathrm{ord}}_{\mathrm{aux}}}}$; and, by Lemma 6.1.1.9, the pullback of $\omega^{\otimes a_0N_1}_{\vec{\mathsf{M}}^{\mathrm{ord,tor}}_{\mathcal{H}_{\mathrm{aux}},\Sigma^{\mathrm{ord}}_{\mathrm{aux}}}}$ to $\vec{\mathsf{M}}^{\mathrm{ord,tor}}_{\mathcal{H},\Sigma^{\mathrm{ord}}}$ is canonically isomorphic to $\omega^{\otimes aN_1}_{\vec{\mathsf{M}}^{\mathrm{ord,tor}}_{\mathcal{H},\Sigma^{\mathrm{ord}}}}$. Since $\vec{\mathsf{M}}^{\mathrm{min}}_{\mathcal{H}}$ and $\vec{\mathsf{M}}^{\mathrm{min}}_{\mathcal{H}_{\mathrm{aux}}}$ are defined as normalizations of $\mathsf{M}^{\mathrm{min}}_{\mathrm{G}_{\mathrm{aux}}(\hat{\mathbb{Z}}^p)}$, we see that the pullback of $\omega^{\otimes aN_1}_{\vec{\mathsf{M}}^{\mathrm{min}}_{\mathcal{H}}}$ to $\vec{\mathsf{M}}^{\mathrm{ord,tor}}_{\mathcal{H},\Sigma^{\mathrm{ord}}}$ is canonically isomorphic to $\omega^{\otimes aN_1}_{\vec{\mathsf{M}}^{\mathrm{ord,tor}}_{\mathcal{H},\Sigma^{\mathrm{ord}}}}$, as desired. □

Theorem 6.1.1.12. *With the setting as at the beginning of this section* (*but no longer assuming that* $\mathcal{H}_p = \mathcal{U}^{\mathrm{bal}}_{p,1}(p^r)$), *the canonical morphism* (6.1.1.4)

$$\vec{\oint}^{\mathrm{ord}}_{\mathcal{H}} : \vec{\mathsf{M}}^{\mathrm{ord,tor}}_{\mathcal{H},\Sigma^{\mathrm{ord}}} \to \vec{\mathsf{M}}^{\mathrm{ord,min}}_{\mathcal{H}} = \mathrm{Proj}\Big(\bigoplus_{k\geq 0} \Gamma(\vec{\mathsf{M}}^{\mathrm{ord,tor}}_{\mathcal{H},\Sigma^{\mathrm{ord}}}, \omega^{\otimes k}_{\vec{\mathsf{M}}^{\mathrm{ord,tor}}_{\mathcal{H},\Sigma^{\mathrm{ord}}}})\Big)$$

is **proper** *and is the Stein factorization of itself. Consequently,* $\vec{\oint}^{\mathrm{ord}}_{\mathcal{H}}$ *(is surjective and) has nonempty connected geometric fibers (by [35, III-1, 4.3.1, 4.3.3, 4.3.4] and its natural generalization to the context of algebraic stacks), and the canonically induced morphism*

$$\mathscr{O}_{\vec{\mathsf{M}}^{\mathrm{ord,min}}_{\mathcal{H}}} \to \vec{\oint}^{\mathrm{ord}}_{\mathcal{H},*}\mathscr{O}_{\vec{\mathsf{M}}^{\mathrm{ord,tor}}_{\mathcal{H},\Sigma^{\mathrm{ord}}}} \tag{6.1.1.13}$$

is an isomorphism. Moreover, the invertible sheaf $\omega^{\otimes N_0}_{\vec{\mathsf{M}}^{\mathrm{ord,tor}}_{\mathcal{H},\Sigma^{\mathrm{ord}}}}$ *over* $\vec{\mathsf{M}}^{\mathrm{ord,tor}}_{\mathcal{H},\Sigma^{\mathrm{ord}}}$ *descends to an ample invertible sheaf* $\mathcal{O}(1)$ *over* $\vec{\mathsf{M}}^{\mathrm{ord,min}}_{\mathcal{H}}$. *By abuse of notation, we shall denote* $\mathcal{O}(1)$ *by* $\omega^{\otimes N_0}_{\vec{\mathsf{M}}^{\mathrm{ord,min}}_{\mathcal{H}}}$, *even when* $\omega_{\vec{\mathsf{M}}^{\mathrm{ord,min}}_{\mathcal{H}}}$ *itself is not defined.*

The canonical morphism $\vec{\mathsf{M}}^{\mathrm{ord,tor}}_{\mathcal{H},\Sigma^{\mathrm{ord}}} \to \vec{\mathsf{M}}^{\mathrm{min}}_{\mathcal{H}}$ *in Corollary* 6.1.1.11 *induces a canonical morphism*

$$\vec{\mathsf{M}}^{\mathrm{ord,tor}}_{\mathcal{H},\Sigma^{\mathrm{ord}}} \to \vec{\mathsf{M}}^{\mathrm{min}}_{\mathcal{H},r_{\mathcal{H}}}, \tag{6.1.1.14}$$

which maps the $[(\Phi_{\mathcal{H}},\delta_{\mathcal{H}},\sigma)]$*-stratum* $\vec{\mathsf{Z}}^{\mathrm{ord}}_{[(\Phi_{\mathcal{H}},\delta_{\mathcal{H}},\sigma)]}$ *(see* (2) *of Theorem* 5.2.1.1) *to the* $[(\Phi_{\mathcal{H}},\delta_{\mathcal{H}})]$*-stratum* $\vec{\mathsf{Z}}_{[(\Phi_{\mathcal{H}},\delta_{\mathcal{H}})],r_{\mathcal{H}}}$ *(see Definition* 2.2.3.5), *and which factors canonically as a composition*

$$\vec{\mathsf{M}}^{\mathrm{ord,tor}}_{\mathcal{H},\Sigma^{\mathrm{ord}}} \to \vec{\mathsf{M}}^{\mathrm{ord,min}}_{\mathcal{H}} \to \vec{\mathsf{M}}^{\mathrm{min}}_{\mathcal{H},r_{\mathcal{H}}} \tag{6.1.1.15}$$

inducing a canonical open immersion

$$\vec{\mathsf{M}}^{\mathrm{ord,min}}_{\mathcal{H}} \hookrightarrow \vec{\mathsf{M}}^{\mathrm{min}}_{\mathcal{H},r_{\mathcal{H}}}, \tag{6.1.1.16}$$

under which the pullback of $\omega^{\otimes k}_{\vec{\mathsf{M}}^{\mathrm{min}}_{\mathcal{H}}}$ *to* $\vec{\mathsf{M}}^{\mathrm{ord,min}}_{\mathcal{H}}$ *is canonically isomorphic to* $\omega^{\otimes k}_{\vec{\mathsf{M}}^{\mathrm{ord,min}}_{\mathcal{H}}}$, *when both are defined for some integer* k *(divisible by both* aN_1 *and* N_0). *In particular,* $\vec{\mathsf{M}}^{\mathrm{ord,min}}_{\mathcal{H}}$ *is quasi-projective over* $\vec{\mathsf{S}}_{0,r_{\mathcal{H}}}$.

Proof. Suppose $\mathcal{U}^{\mathrm{bal}}_{p,1}(p^r) \subset \mathcal{H}_p \subset \mathcal{U}_{p,0}(p^r)$. Let $\mathcal{H}' = \mathcal{H}^p\mathcal{H}_p = \mathcal{U}^{\mathrm{bal}}_{p,1}(p^r)$, and let $\Sigma^{\mathrm{ord},\prime}$ be a compatible collection for $\vec{\mathsf{M}}^{\mathrm{ord}}_{\mathcal{H}'}$ such that $\Sigma^{\mathrm{ord},\prime}$ is a 1-refinement of Σ^{ord} as in Definition 5.2.2.1. Then the canonical surjection $\vec{[1]}^{\mathrm{ord,tor}} : \vec{\mathsf{M}}^{\mathrm{ord,tor}}_{\mathcal{H}',\Sigma^{\mathrm{ord}}} \to \vec{\mathsf{M}}^{\mathrm{ord,tor}}_{\mathcal{H},\Sigma^{\mathrm{ord}}}$ is *proper* by Proposition 5.2.2.2. Thus, to show that (6.1.1.4) is proper, it suffices to show that it is so with $\mathcal{H}$ (resp. Σ^{ord}) replaced with $\mathcal{H}'$ (resp. $\Sigma^{\mathrm{ord},\prime}$), which follows from the first paragraph of Proposition 6.1.1.8.

Once the properness of (6.1.1.4) is known, since the canonical morphism (6.1.1.4) (for the original $\mathcal{H}$ and Σ^{ord}) is defined by global sections of $\omega^{\otimes N_0 k}_{\vec{\mathsf{M}}^{\mathrm{ord,tor}}_{\mathcal{H},\Sigma^{\mathrm{ord}}}}$, for $k \geq 0$, it follows that (6.1.1.4) is the Stein factorization of itself, and that, for each $k \geq 0$, the invertible sheaf $\omega^{\otimes N_0 k}_{\vec{\mathsf{M}}^{\mathrm{ord,tor}}_{\mathcal{H},\Sigma^{\mathrm{ord}}}}$ over $\vec{\mathsf{M}}^{\mathrm{ord,tor}}_{\mathcal{H},\Sigma^{\mathrm{ord}}}$ descends to an ample invertible sheaf $\mathcal{O}(1)$ over $\vec{\mathsf{M}}^{\mathrm{ord,min}}_{\mathcal{H}}$ (by definition of $\vec{\mathsf{M}}^{\mathrm{ord,min}}_{\mathcal{H}}$; see (6.1.1.3)).

The induced morphism (6.1.1.14) maps $\vec{\mathsf{Z}}^{\mathrm{ord}}_{[(\Phi_{\mathcal{H}},\delta_{\mathcal{H}},\sigma)]}$ to $\vec{\mathsf{Z}}_{[(\Phi_{\mathcal{H}},\delta_{\mathcal{H}})],r_{\mathcal{H}}}$ by the definition of the latter by taking closures and exclusions. The canonical factorization (6.1.1.15) exists by the last paragraph of Corollary 6.1.1.11. The induced morphism (6.1.1.16) is an open immersion by Zariski's main theorem (see [35, III-1,

4.4.3, 4.4.11]), by the last paragraph of Corollary 6.1.1.11, and by the fact that it is an open immersion over $[\vec{\mathsf{M}}_{\mathcal{H}}^{\mathrm{ord}}]$ (*cf.* (3.4.6.5), Proposition 2.2.1.2, and Corollary 6.1.1.11). When k is divisible by both aN_1 and N_0, the pullback of $\omega_{\vec{\mathsf{M}}_{\mathcal{H}}^{\mathrm{min}}}^{\otimes k}$ to $\vec{\mathsf{M}}_{\mathcal{H}}^{\mathrm{ord,min}}$ under (6.1.1.16) is canonically isomorphic to $\omega_{\vec{\mathsf{M}}_{\mathcal{H}}^{\mathrm{ord,min}}}^{\otimes k}$, because their pullbacks to $\vec{\mathsf{M}}_{\mathcal{H},\Sigma^{\mathrm{ord}}}^{\mathrm{ord,tor}}$ are isomorphic, and because (6.1.1.13) is an isomorphism. (See the argument at the end of the proof of [62, Thm. 7.2.4.1], which is based on the projection formula (see [35, 0_{I}, 5.4.10.1]), used in [62, Lem. 7.2.2.1].) □

6.1.2 *Local Structures and Stratifications*

Proposition 6.1.2.1. (*Compare with [62, Prop. 7.2.3.3].*) $\vec{\mathsf{M}}_{\mathcal{H}}^{\mathrm{ord,min}}$ *is normal.*

Proof. Since $\vec{\mathsf{M}}_{\mathcal{H},\Sigma^{\mathrm{ord}}}^{\mathrm{ord,tor}}$ is normal because it is smooth over the normal base scheme $\vec{\mathsf{S}}_{0,r_{\mathcal{H}}} = \mathrm{Spec}(\mathcal{O}_{F_0,(p)}[\zeta_{p^{r_{\mathcal{H}}}}])$, and since the canonical morphism (6.1.1.13) is an isomorphism by Theorem 6.1.1.12, the proposition follows from [62, Lem. 7.2.3.1]; or, alternatively, from the second half of the proof of [10, Sec. 6.7, Lem. 2] (ignoring the statement about finite generation). □

Corollary 6.1.2.2. (*Compare with [62, the paragraph following Prop. 7.2.3.3].*) $\vec{\mathsf{M}}_{\mathcal{H}}^{\mathrm{ord,min}}$ *is flat over* $\vec{\mathsf{S}}_{0,r_{\mathcal{H}}} = \mathrm{Spec}(\mathcal{O}_{F_0,(p)}[\zeta_{p^{r_{\mathcal{H}}}}])$.

Proof. Since $\mathcal{O}_{F_0,(p)}[\zeta_{p^{r_{\mathcal{H}}}}]$ is a localization of the ring of integers of a number field, $\vec{\mathsf{M}}_{\mathcal{H}}^{\mathrm{ord,min}} \to \vec{\mathsf{S}}_{0,r_{\mathcal{H}}}$ is flat because $\vec{\mathsf{M}}_{\mathcal{H}}^{\mathrm{ord,min}}$ is normal and all its maximal points (see [36, 0, 2.1.2]) are of characteristic zero (as those of $\vec{\mathsf{M}}_{\mathcal{H},\Sigma^{\mathrm{ord}}}^{\mathrm{ord,tor}}$ are). □

By Theorem 6.1.1.12, the canonical morphism (6.1.1.4) has nonempty connected geometric fibers, and the pullback of $\omega_{\vec{\mathsf{M}}_{\mathcal{H},\Sigma^{\mathrm{ord}}}^{\mathrm{ord,tor}}}^{\otimes N_0}$ to each such connected geometric fiber is trivial. By [28, Ch. V, Prop. 2.2] or [62, Prop. 7.2.1.2], this shows that the isomorphism class of the abelian part of G is constant on each of such fibers. In particular, if a geometric fiber of $\vec{\oint}_{\mathcal{H}}^{\mathrm{ord}}$ meets $\vec{\mathsf{M}}_{\mathcal{H}}^{\mathrm{ord}}$, then it has only one closed point.

Since $\vec{\mathsf{M}}_{\mathcal{H}}^{\mathrm{ord}}$ is open in $\vec{\mathsf{M}}_{\mathcal{H},\Sigma^{\mathrm{ord}}}^{\mathrm{ord,tor}}$, and since the formation of coarse moduli spaces commutes with flat base change, we see that $[\vec{\mathsf{M}}_{\mathcal{H}}^{\mathrm{ord}}]$ is an open subalgebraic space of $[\vec{\mathsf{M}}_{\mathcal{H},\Sigma^{\mathrm{ord}}}^{\mathrm{ord,tor}}]$. The morphism $\vec{\oint}_{\mathcal{H}}^{\mathrm{ord}} : \vec{\mathsf{M}}_{\mathcal{H},\Sigma^{\mathrm{ord}}}^{\mathrm{ord,tor}} \to \vec{\mathsf{M}}_{\mathcal{H}}^{\mathrm{ord,min}}$ factors as

$$\vec{\mathsf{M}}_{\mathcal{H},\Sigma^{\mathrm{ord}}}^{\mathrm{ord,tor}} \to [\vec{\mathsf{M}}_{\mathcal{H},\Sigma^{\mathrm{ord}}}^{\mathrm{ord,tor}}] \xrightarrow{[\vec{\oint}_{\mathcal{H}}^{\mathrm{ord}}]} \vec{\mathsf{M}}_{\mathcal{H}}^{\mathrm{ord,min}},$$

whose restriction to $\vec{\mathsf{M}}_{\mathcal{H}}^{\mathrm{ord}}$ is the factorization

$$\vec{\mathsf{M}}_{\mathcal{H}}^{\mathrm{ord}} \to [\vec{\mathsf{M}}_{\mathcal{H}}^{\mathrm{ord}}] \xrightarrow{[\vec{\oint}_{\mathcal{H}}^{\mathrm{ord}}|_{\vec{\mathsf{M}}_{\mathcal{H}}^{\mathrm{ord}}}]} \vec{\mathsf{M}}_{\mathcal{H}}^{\mathrm{ord,min}}.$$

Applying Zariski's main theorem (see [35, III-1, 4.4.3, 4.4.11], and the formulation in [62, Prop. 7.2.3.4] for algebraic spaces) to $[\vec{\oint}_{\mathcal{H}}^{\mathrm{ord}}]$, and taking into account the fact that $\vec{\mathsf{M}}^{\mathrm{ord,min}}_{\mathcal{H}}$ is normal (see Proposition 6.1.2.1), we see that $[\vec{\oint}^{\mathrm{ord}}_{\mathcal{H}}]$ is an isomorphism over an open subscheme of $\vec{\mathsf{M}}^{\mathrm{ord,min}}_{\mathcal{H}}$ containing the image of $[\vec{\mathsf{M}}^{\mathrm{ord}}_{\mathcal{H}}]$. (We will see below that the image of $[\vec{\mathsf{M}}^{\mathrm{ord}}_{\mathcal{H}}]$ is actually open, with complements given by closed subschemes, and hence $[\vec{\oint}^{\mathrm{ord}}_{\mathcal{H}}|_{\vec{\mathsf{M}}^{\mathrm{ord}}_{\mathcal{H}}}]$ is an open immersion.)

More generally, suppose that a fiber of $\vec{\oint}^{\mathrm{ord}}_{\mathcal{H}}$ meets the $[(\Phi_{\mathcal{H}}, \delta_{\mathcal{H}}, \sigma)]$-stratum $\vec{\mathsf{Z}}^{\mathrm{ord}}_{[(\Phi_{\mathcal{H}}, \delta_{\mathcal{H}}, \sigma)]}$. Let $(\Phi_{\mathcal{H}}, \delta_{\mathcal{H}}, \sigma)$ be any representative of the class $[(\Phi_{\mathcal{H}}, \delta_{\mathcal{H}}, \sigma)]$. By (5) of Theorem 5.2.1.1, the formal completion $(\vec{\mathsf{M}}^{\mathrm{ord,tor}}_{\mathcal{H}})^{\wedge}_{\vec{\mathsf{Z}}^{\mathrm{ord}}_{[(\Phi_{\mathcal{H}}, \delta_{\mathcal{H}}, \sigma)]}}$ of $\vec{\mathsf{M}}^{\mathrm{ord,tor}}_{\mathcal{H}}$ along $\vec{\mathsf{Z}}^{\mathrm{ord}}_{[(\Phi_{\mathcal{H}}, \delta_{\mathcal{H}}, \sigma)]}$ is canonically isomorphic to $\vec{\mathfrak{X}}^{\mathrm{ord}}_{\Phi_{\mathcal{H}}, \delta_{\mathcal{H}}, \sigma}/\Gamma_{\Phi_{\mathcal{H}}, \sigma}$. Let ${}^{\heartsuit}\omega$ denote the pullback of $\omega_{\vec{\mathsf{M}}^{\mathrm{ord,tor}}_{\mathcal{H}, \Sigma^{\mathrm{ord}}}}$ to $\vec{\mathfrak{X}}^{\mathrm{ord}}_{\Phi_{\mathcal{H}}, \delta_{\mathcal{H}}, \sigma}/\Gamma_{\Phi_{\mathcal{H}}, \sigma}$. Let $\omega_{\vec{B}} := \omega_{\vec{B}/\vec{\mathsf{M}}^{\mathrm{ord}, \mathsf{Z}_{\mathcal{H}}}_{\mathcal{H}}} := \wedge^{\mathrm{top}} \underline{\mathrm{Lie}}^{\vee}_{\vec{B}/\vec{\mathsf{M}}^{\mathrm{ord}, \mathsf{Z}_{\mathcal{H}}}_{\mathcal{H}}}$, where $\vec{B}$ is the tautological abelian scheme over $\vec{\mathsf{M}}^{\mathrm{ord}, \mathsf{Z}_{\mathcal{H}}}_{\mathcal{H}}$. By abuse of notation, we shall also denote the pullback of $\omega_{\vec{B}}$ by the same notation.

Lemma 6.1.2.3. (*Compare with [62, Lem. 7.1.2.1].*) *There is a canonical isomorphism* ${}^{\heartsuit}\omega \cong (\wedge^{\mathrm{top}}_{\mathbb{Z}} X) \underset{\mathbb{Z}}{\otimes} \omega_{\vec{B}}$ *over the formal algebraic stack* $\vec{\mathfrak{X}}^{\mathrm{ord}}_{\Phi_{\mathcal{H}}, \delta_{\mathcal{H}}, \sigma}/\Gamma_{\Phi_{\mathcal{H}}, \sigma}$.

Proof. The proof of [62, Lem. 7.1.2.1] works verbatim here. □

Let us denote the structural morphism $\vec{C}^{\mathrm{ord}}_{\Phi_{\mathcal{H}}, \delta_{\mathcal{H}}} \to \vec{\mathsf{M}}^{\mathrm{ord}, \mathsf{Z}_{\mathcal{H}}}_{\mathcal{H}}$ by $\vec{\mathrm{p}}^{\mathrm{ord}}_{\Phi_{\mathcal{H}}, \delta_{\mathcal{H}}}$, which is proper and smooth because it is an abelian scheme torsor over the finite étale cover $\vec{\mathsf{M}}^{\mathrm{ord}, \Phi_{\mathcal{H}}}_{\mathcal{H}}$ of $\vec{\mathsf{M}}^{\mathrm{ord}, \mathsf{Z}_{\mathcal{H}}}_{\mathcal{H}}$. For simplicity of notation, as in [62, Def. 7.1.2.2], for each $\ell \in \mathbf{S}_{\Phi_{\mathcal{H}}}$, let

$$\underline{\mathrm{FJ}}^{\mathrm{ord}, (\ell)}_{\Phi_{\mathcal{H}}, \delta_{\mathcal{H}}} := (\vec{\mathrm{p}}^{\mathrm{ord}}_{\Phi_{\mathcal{H}}, \delta_{\mathcal{H}}})_*(\vec{\Psi}^{\mathrm{ord}}_{\Phi_{\mathcal{H}}, \delta_{\mathcal{H}}}(\ell)). \tag{6.1.2.4}$$

Consider the following composition of canonical morphisms (*cf.* [62, (7.1.2.3)]):

$$\begin{aligned}
&\Gamma(\vec{\mathsf{M}}^{\mathrm{ord,tor}}_{\mathcal{H}}, \omega^{\otimes k}_{\vec{\mathsf{M}}^{\mathrm{ord,tor}}_{\mathcal{H}, \Sigma^{\mathrm{ord}}}}) \\
&\to \Gamma((\vec{\mathsf{M}}^{\mathrm{ord,tor}}_{\mathcal{H}})^{\wedge}_{\vec{\mathsf{Z}}^{\mathrm{ord}}_{[(\Phi_{\mathcal{H}}, \delta_{\mathcal{H}}, \sigma)]}}, \omega^{\otimes k}_{\vec{\mathsf{M}}^{\mathrm{ord,tor}}_{\mathcal{H}, \Sigma^{\mathrm{ord}}}}) \cong \Gamma(\vec{\mathfrak{X}}^{\mathrm{ord}}_{\Phi_{\mathcal{H}}, \delta_{\mathcal{H}}, \sigma}/\Gamma_{\Phi_{\mathcal{H}}, \sigma}, {}^{\heartsuit}\omega^{\otimes k}) \\
&\to \Big[\prod_{\ell \in \sigma^{\vee}} \Gamma(\vec{C}^{\mathrm{ord}}_{\Phi_{\mathcal{H}}, \delta_{\mathcal{H}}}, \vec{\Psi}^{\mathrm{ord}}_{\Phi_{\mathcal{H}}, \delta_{\mathcal{H}}}(\ell) \underset{\mathscr{O}_{\vec{C}^{\mathrm{ord}}_{\Phi_{\mathcal{H}}, \delta_{\mathcal{H}}}}}{\otimes} ((\wedge^{\mathrm{top}}_{\mathbb{Z}} X) \underset{\mathbb{Z}}{\otimes} \omega_{\vec{B}})^{\otimes k})\Big]^{\Gamma_{\Phi_{\mathcal{H}}, \sigma}} \\
&\cong \Big[\prod_{\ell \in \sigma^{\vee}} \Gamma(\vec{\mathsf{M}}^{\mathrm{ord}, \mathsf{Z}_{\mathcal{H}}}_{\mathcal{H}}, \underline{\mathrm{FJ}}^{\mathrm{ord}, (\ell)}_{\Phi_{\mathcal{H}}, \delta_{\mathcal{H}}} \underset{\mathscr{O}_{\vec{\mathsf{M}}^{\mathrm{ord}, \mathsf{Z}_{\mathcal{H}}}_{\mathcal{H}}}}{\otimes} ((\wedge^{\mathrm{top}}_{\mathbb{Z}} X) \underset{\mathbb{Z}}{\otimes} \omega_{\vec{B}})^{\otimes k})\Big]^{\Gamma_{\Phi_{\mathcal{H}}, \sigma}}.
\end{aligned} \tag{6.1.2.5}$$

Definition 6.1.2.6. (*Compare with [62, Def. 7.1.2.4].*) *The above composition* (6.1.2.5) *is called the* **Fourier–Jacobi morphism** *along* $(\Phi_{\mathcal{H}}, \delta_{\mathcal{H}}, \sigma)$, *which we denote by* $\mathrm{FJ}^{\mathrm{ord}}_{\Phi_{\mathcal{H}}, \delta_{\mathcal{H}}, \sigma}$. *The image of an element* $f \in \Gamma(\vec{\mathsf{M}}^{\mathrm{ord,tor}}_{\mathcal{H}}, \omega^{\otimes k}_{\vec{\mathsf{M}}^{\mathrm{ord,tor}}_{\mathcal{H}, \Sigma^{\mathrm{ord}}}})$ *has a natural expansion*

$$\mathrm{FJ}^{\mathrm{ord}}_{\Phi_{\mathcal{H}}, \delta_{\mathcal{H}}, \sigma}(f) = \sum_{\ell \in \sigma^{\vee}} \mathrm{FJ}^{\mathrm{ord}, (\ell)}_{\Phi_{\mathcal{H}}, \delta_{\mathcal{H}}, \sigma}(f)$$

where the sum can be infinite and where each $\mathrm{FJ}^{\mathrm{ord},(\ell)}_{\Phi_{\mathcal{H}},\delta_{\mathcal{H}},\sigma}(f)$ *lies in*

$$\mathrm{FJC}^{\mathrm{ord},(\ell)}_{\Phi_{\mathcal{H}},\delta_{\mathcal{H}}}(k) := \Gamma(\vec{\mathsf{M}}^{\mathrm{ord},\mathrm{Z}_{\mathcal{H}}}_{\mathcal{H}}, \underline{\mathrm{FJ}}^{\mathrm{ord},(\ell)}_{\Phi_{\mathcal{H}},\delta_{\mathcal{H}}} \underset{\mathscr{O}_{\vec{\mathsf{M}}^{\mathrm{ord},\mathrm{Z}_{\mathcal{H}}}_{\mathcal{H}}}}{\otimes} ((\wedge^{\mathrm{top}}_{\mathbb{Z}} X) \underset{\mathbb{Z}}{\otimes} \omega_{\vec{B}})^{\otimes k}).$$

The expansion $\mathrm{FJ}^{\mathrm{ord}}_{\Phi_{\mathcal{H}},\delta_{\mathcal{H}},\sigma}(f)$ *is called the* **Fourier–Jacobi expansion** *of* f *along* $(\Phi_{\mathcal{H}},\delta_{\mathcal{H}},\sigma)$, *with* **Fourier–Jacobi coefficients** $\mathrm{FJ}^{\mathrm{ord},(\ell)}_{\Phi_{\mathcal{H}},\delta_{\mathcal{H}},\sigma}(f)$ *of each degree* $\ell \in \sigma^{\vee}$.

By the same argument as in [62, Sec. 7.1.2], we do not really need the Fourier–Jacobi coefficients of degrees outside $\mathbf{P}^{\vee}_{\Phi_{\mathcal{H}}} = \underset{\sigma\in\Sigma_{\Phi_{\mathcal{H}}}}{\cap} \sigma^{\vee}$, and the Fourier–Jacobi expansions are naturally invariant under the action of $\Gamma_{\Phi_{\mathcal{H}}}$. We have an induced morphism

$$\mathrm{FJ}^{\mathrm{ord}}_{\Phi_{\mathcal{H}},\delta_{\mathcal{H}}} : \Gamma(\vec{\mathsf{M}}^{\mathrm{ord},\mathrm{tor}}_{\mathcal{H}}, \omega^{\otimes k}_{\vec{\mathsf{M}}^{\mathrm{ord},\mathrm{tor}}_{\mathcal{H},\Sigma^{\mathrm{ord}}}}) \to \Big[\prod_{\ell\in\mathbf{P}^{\vee}_{\Phi_{\mathcal{H}}}} \mathrm{FJC}^{\mathrm{ord},(\ell)}_{\Phi_{\mathcal{H}},\delta_{\mathcal{H}}}(k)\Big]^{\Gamma_{\Phi_{\mathcal{H}}}} \tag{6.1.2.7}$$

(*cf.* [62, (7.1.2.6)]).

Definition 6.1.2.8. (*Compare with [62, Def. 7.1.2.7].*) *The above morphism* (6.1.2.7) *is called the* **Fourier–Jacobi morphism** *along* $(\Phi_{\mathcal{H}},\delta_{\mathcal{H}})$, *which we denote by* $\mathrm{FJ}^{\mathrm{ord}}_{\Phi_{\mathcal{H}},\delta_{\mathcal{H}}}$ *as above. The image of an element* $f \in \Gamma(\vec{\mathsf{M}}^{\mathrm{ord},\mathrm{tor}}_{\mathcal{H}}, \omega^{\otimes k}_{\vec{\mathsf{M}}^{\mathrm{ord},\mathrm{tor}}_{\mathcal{H},\Sigma^{\mathrm{ord}}}})$ *has a natural expansion*

$$\mathrm{FJ}^{\mathrm{ord}}_{\Phi_{\mathcal{H}},\delta_{\mathcal{H}}}(f) = \sum_{\ell\in\mathbf{P}^{\vee}_{\Phi_{\mathcal{H}}}} \mathrm{FJ}^{\mathrm{ord},(\ell)}_{\Phi_{\mathcal{H}},\delta_{\mathcal{H}}}(f),$$

where each $\mathrm{FJ}^{\mathrm{ord},(\ell)}_{\Phi_{\mathcal{H}},\delta_{\mathcal{H}}}(f)$ *lies in* $\mathrm{FJC}^{\mathrm{ord},(\ell)}_{\Phi_{\mathcal{H}},\delta_{\mathcal{H}}}(k)$. *The expansion* $\mathrm{FJ}^{\mathrm{ord}}_{\Phi_{\mathcal{H}},\delta_{\mathcal{H}}}(f)$ *is called the* **Fourier–Jacobi expansion** *of* f *along* $(\Phi_{\mathcal{H}},\delta_{\mathcal{H}})$, *with* **Fourier–Jacobi coefficients** $\mathrm{FJ}^{\mathrm{ord},(\ell)}_{\Phi_{\mathcal{H}},\delta_{\mathcal{H}}}(f)$ *of each degree* $\ell \in \mathbf{P}^{\vee}_{\Phi_{\mathcal{H}}}$.

Definition 6.1.2.9. (*Compare with [62, Def. 7.1.2.10].*) *The* **constant term** *of a Fourier–Jacobi expansion* $\mathrm{FJ}^{\mathrm{ord}}_{\Phi_{\mathcal{H}},\delta_{\mathcal{H}}}(f)$ *of an element* $f \in \Gamma(\vec{\mathsf{M}}^{\mathrm{ord},\mathrm{tor}}_{\mathcal{H}}, \omega^{\otimes k}_{\vec{\mathsf{M}}^{\mathrm{ord},\mathrm{tor}}_{\mathcal{H},\Sigma^{\mathrm{ord}}}})$ *is the Fourier–Jacobi coefficient* $\mathrm{FJ}^{\mathrm{ord},(0)}_{\Phi_{\mathcal{H}},\delta_{\mathcal{H}}}(f) \in \mathrm{FJC}^{\mathrm{ord},(0)}_{\Phi_{\mathcal{H}},\delta_{\mathcal{H}}}(k)$ *in degree zero.*

The same arguments as in [62, Sec. 7.1.2] gives the following:

Proposition 6.1.2.10. (*Compare with [62, Prop. 7.1.2.8, 7.1.2.9, and 7.1.2.13; see also the errata].*) *The Fourier–Jacobi morphism* $\mathrm{FJ}_{\Phi_{\mathcal{H}},\delta_{\mathcal{H}}}$ *satisfies the following properties:*

(1) $\mathrm{FJ}_{\Phi_{\mathcal{H}},\delta_{\mathcal{H}}}$ *can be computed by any* $\mathrm{FJ}_{\Phi_{\mathcal{H}},\delta_{\mathcal{H}},\sigma}$ *as in Definition* 6.1.2.6. *The definition is independent of the* σ *we use.*
(2) $\mathrm{FJ}_{\Phi_{\mathcal{H}},\delta_{\mathcal{H}}}$ *is independent of the* $\Gamma_{\Phi_{\mathcal{H}}}$*-admissible smooth rational polyhedral cone decomposition* $\Sigma_{\Phi_{\mathcal{H}}}$ *of* $\mathbf{P}_{\Phi_{\mathcal{H}}}$ *we use.*

(3) The value of each element $f \in \Gamma(\vec{\mathsf{M}}^{\mathrm{ord,tor}}_{\mathcal{H}}, \omega^{\otimes k}_{\vec{\mathsf{M}}^{\mathrm{ord,tor}}_{\mathcal{H},\Sigma^{\mathrm{ord}}}})$ *along the* $[(\Phi_{\mathcal{H}}, \delta_{\mathcal{H}}, \sigma)]$*-stratum* $\vec{\mathsf{Z}}^{\mathrm{ord}}_{[(\Phi_{\mathcal{H}},\delta_{\mathcal{H}},\sigma)]}$ *of* $\vec{\mathsf{M}}^{\mathrm{ord,tor}}_{\mathcal{H},\Sigma^{\mathrm{ord}}}$ *is determined by its constant term* $\mathrm{FJ}^{\mathrm{ord},(0)}_{\Phi_{\mathcal{H}},\delta_{\mathcal{H}}}(f)$*, which is a* $\Gamma_{\Phi_{\mathcal{H}}}$*-invariant element in* $\mathrm{FJC}^{\mathrm{ord},(0)}_{\Phi_{\mathcal{H}},\delta_{\mathcal{H}}}(k)$*. In particular, since* $\underline{\mathrm{FJ}}^{\mathrm{ord},(0)}_{\Phi_{\mathcal{H}},\delta_{\mathcal{H}}} \cong (\vec{\mathsf{M}}^{\mathrm{ord},\Phi_{\mathcal{H}}}_{\mathcal{H}} \to \vec{\mathsf{M}}^{\mathrm{ord},\mathsf{Z}_{\mathcal{H}}}_{\mathcal{H}})_* \mathscr{O}_{\vec{\mathsf{M}}^{\mathrm{ord},\Phi_{\mathcal{H}}}_{\mathcal{H}}}$ *and* $\vec{\mathsf{M}}^{\mathrm{ord},\Phi_{\mathcal{H}}}_{\mathcal{H}}/\Gamma_{\Phi_{\mathcal{H}}} \cong \vec{\mathsf{M}}^{\mathrm{ord},\mathsf{Z}_{\mathcal{H}}}_{\mathcal{H}}$ *(see Proposition 4.2.1.29), the value of* f *is* **constant along the fibers** *of the structural morphism* $\vec{\mathsf{Z}}^{\mathrm{ord}}_{[(\Phi_{\mathcal{H}},\delta_{\mathcal{H}},\sigma)]} \to \vec{\mathsf{M}}^{\mathrm{ord},\mathsf{Z}_{\mathcal{H}}}_{\mathcal{H}}$*. We say in this case that it depends only on the* **abelian part** *of* $(G, \lambda, i, \alpha_{\mathcal{H}^p}, \alpha^{\mathrm{ord}}_{\mathcal{H}_p})$ *over* $\vec{\mathsf{Z}}^{\mathrm{ord}}_{[(\Phi_{\mathcal{H}},\delta_{\mathcal{H}},\sigma)]}$.

Proof. The same arguments as in the proofs of [62, Prop. 7.1.2.8, 7.1.2.9, and 7.1.2.13] work verbatim here. (The error in the statement of [62, Prop. 7.1.2.13] is due to changes necessitated by errors in other parts of the book, which does not invalidate the argument of the proof there.) □

By (3) of Proposition 6.1.2.10, applied to those $k \geq 0$ divisible by N_0, we see that $\vec{\oint}^{\mathrm{ord}}_{\mathcal{H}}|_{\vec{\mathsf{Z}}^{\mathrm{ord}}_{[(\Phi_{\mathcal{H}},\delta_{\mathcal{H}},\sigma)]}} : \vec{\mathsf{Z}}^{\mathrm{ord}}_{[(\Phi_{\mathcal{H}},\delta_{\mathcal{H}},\sigma)]} \to \vec{\mathsf{M}}^{\mathrm{ord,min}}_{\mathcal{H}}$ factors through $\vec{\mathsf{Z}}^{\mathrm{ord}}_{[(\Phi_{\mathcal{H}},\delta_{\mathcal{H}},\sigma)]} \to \vec{\mathsf{M}}^{\mathrm{ord},\mathsf{Z}_{\mathcal{H}}}_{\mathcal{H}}$. This induces a morphism

$$\vec{\mathsf{M}}^{\mathrm{ord},\mathsf{Z}_{\mathcal{H}}}_{\mathcal{H}} \to \vec{\mathsf{M}}^{\mathrm{ord,min}}_{\mathcal{H}} \tag{6.1.2.11}$$

from an algebraic stack to a scheme, each of whose geometric fibers has only one single point.

The argument used in proving (1) of Proposition 6.1.2.10 (or rather [62, Prop. 7.1.2.8]) shows the following:

Lemma 6.1.2.12. (*Compare with [62, Lem. 7.2.3.6].*) *Two restrictions* $\vec{\oint}^{\mathrm{ord}}_{\mathcal{H}}|_{\vec{\mathsf{Z}}^{\mathrm{ord}}_{[(\Phi_{\mathcal{H}},\delta_{\mathcal{H}},\sigma)]}} : \vec{\mathsf{Z}}^{\mathrm{ord}}_{[(\Phi_{\mathcal{H}},\delta_{\mathcal{H}},\sigma)]} \to \vec{\mathsf{M}}^{\mathrm{ord,min}}_{\mathcal{H}}$ *and* $\vec{\oint}^{\mathrm{ord}}_{\mathcal{H}}|_{\vec{\mathsf{Z}}^{\mathrm{ord}}_{[(\Phi'_{\mathcal{H}},\delta'_{\mathcal{H}},\sigma')]}} : \vec{\mathsf{Z}}^{\mathrm{ord}}_{[(\Phi'_{\mathcal{H}},\delta'_{\mathcal{H}},\sigma')]} \to \vec{\mathsf{M}}^{\mathrm{ord,min}}_{\mathcal{H}}$ *have the same image and induce the same morphism as in* (6.1.2.11) *(up to the canonical identification between the sources) when there exist representatives* $(\Phi_{\mathcal{H}}, \delta_{\mathcal{H}}, \sigma)$ *and* $(\Phi'_{\mathcal{H}}, \delta'_{\mathcal{H}}, \sigma')$ *of* $[(\Phi_{\mathcal{H}}, \delta_{\mathcal{H}}, \sigma)]$ *and* $[(\Phi'_{\mathcal{H}}, \delta'_{\mathcal{H}}, \sigma')]$*, respectively, such that* $(\Phi_{\mathcal{H}}, \delta_{\mathcal{H}})$ *and* $(\Phi'_{\mathcal{H}}, \delta'_{\mathcal{H}})$ *are equivalent and represent the same cusp label* $[(\Phi_{\mathcal{H}}, \delta_{\mathcal{H}})] = [(\Phi'_{\mathcal{H}}, \delta'_{\mathcal{H}})]$.

Let us denote this common image by $\vec{\mathsf{Z}}^{\mathrm{ord}}_{[(\Phi_{\mathcal{H}},\delta_{\mathcal{H}})]} = \vec{\mathsf{Z}}^{\mathrm{ord}}_{[(\Phi'_{\mathcal{H}},\delta'_{\mathcal{H}})]}$. By Theorem 6.1.1.12, $\vec{\mathsf{Z}}^{\mathrm{ord}}_{[(\Phi_{\mathcal{H}},\delta_{\mathcal{H}})]}$ is an open subscheme of $\vec{\mathsf{Z}}_{[(\Phi_{\mathcal{H}},\delta_{\mathcal{H}})],r_{\mathcal{H}}}$ (see Definition 2.2.3.5).

We claim that the converse is also true:

Proposition 6.1.2.13. (*Compare with [62, Prop. 7.2.3.7].*) *If the intersection of* $\vec{\mathsf{Z}}^{\mathrm{ord}}_{[(\Phi_{\mathcal{H}},\delta_{\mathcal{H}})]} := \mathrm{image}(\vec{\oint}^{\mathrm{ord}}_{\mathcal{H}}|_{\vec{\mathsf{Z}}^{\mathrm{ord}}_{[(\Phi_{\mathcal{H}},\delta_{\mathcal{H}},\sigma)]}})$ *and* $\vec{\mathsf{Z}}^{\mathrm{ord}}_{[(\Phi'_{\mathcal{H}},\delta'_{\mathcal{H}})]} := \mathrm{image}(\vec{\oint}^{\mathrm{ord}}_{\mathcal{H}}|_{\vec{\mathsf{Z}}^{\mathrm{ord}}_{[(\Phi'_{\mathcal{H}},\delta'_{\mathcal{H}},\sigma')]}})$ *is nonempty, then the two cusp labels* $[(\Phi_{\mathcal{H}}, \delta_{\mathcal{H}})]$ *and* $[(\Phi'_{\mathcal{H}}, \delta'_{\mathcal{H}})]$ *are the same. (In this case, we saw above that* $\vec{\mathsf{Z}}^{\mathrm{ord}}_{[(\Phi_{\mathcal{H}},\delta_{\mathcal{H}})]} = \vec{\mathsf{Z}}^{\mathrm{ord}}_{[(\Phi'_{\mathcal{H}},\delta'_{\mathcal{H}})]}$.)

Proof. The proof of [62, Prop. 7.2.3.7] works almost verbatim here. But let us spell out the details for the sake of certainty.

Suppose there exists a geometric point $\bar{x}$ in the intersection of $\vec{\mathsf{Z}}^{\mathrm{ord}}_{[(\Phi_{\mathcal{H}},\delta_{\mathcal{H}})]}$ and $\vec{\mathsf{Z}}^{\mathrm{ord}}_{[(\Phi'_{\mathcal{H}},\delta'_{\mathcal{H}})]}$. Let C be any proper irreducible curve in the fiber of $\vec{\oint}^{\mathrm{ord}}_{\mathcal{H}} : \vec{\mathsf{M}}^{\mathrm{ord,tor}}_{\mathcal{H},\Sigma^{\mathrm{ord}}} \to \vec{\mathsf{M}}^{\mathrm{ord,min}}_{\mathcal{H}}$ over $\bar{x}$. By [28, Ch. V, Prop. 2.2] or [62, Prop. 7.2.1.2] as before, the pullback of $G \to \vec{\mathsf{M}}^{\mathrm{ord,tor}}_{\mathcal{H},\Sigma^{\mathrm{ord}}}$ to C is globally an extension of an isotrivial abelian scheme by a torus. If we take any geometric point $\bar{z}$ of C, and take the pullback of $(G, \lambda, i, \alpha_{\mathcal{H}^p}, \alpha^{\mathrm{ord}}_{\mathcal{H}_p}) \to \vec{\mathsf{M}}^{\mathrm{ord,tor}}_{\mathcal{H},\Sigma^{\mathrm{ord}}}$ to the strict local ring of $\vec{\mathsf{M}}^{\mathrm{ord,tor}}_{\mathcal{H},\Sigma^{\mathrm{ord}}}$ at $\bar{z}$ completed along the curve C, then we obtain a degenerating family of type $\vec{\mathsf{M}}^{\mathrm{ord}}_{\mathcal{H}}$ over a base ring $R_{\bar{z}}$ that fits into the setting of Section 4.1.6. Then, by Theorem 4.1.6.2, this pullback defines an object in the essential image of $\mathrm{DEG}_{\mathrm{PEL},\vec{\mathsf{M}}^{\mathrm{ord}}_{\mathcal{H}}}(R_{\bar{z}}) \to \mathrm{DEG}_{\mathrm{PEL},\ddot{\mathsf{M}}^{\mathrm{ord}}_{\mathcal{H}}}(R_{\bar{z}})$ and hence, in particular, a cusp label $(\Phi_{\mathcal{H}}, \delta_{\mathcal{H}})$. (The key point here is that the pullback of $G \to \vec{\mathsf{M}}^{\mathrm{ord,tor}}_{\mathcal{H},\Sigma^{\mathrm{ord}}}$ to C is globally an extension of an abelian scheme by a *split* torus.) Thus, there is a locally constant association of a cusp label $[(\Phi_{\mathcal{H}}, \delta_{\mathcal{H}})]$ over each such proper irreducible curve C. Since the fiber of $\vec{\oint}^{\mathrm{ord}}_{\mathcal{H}}$ over $\bar{x}$ is connected, we see that the associated cusp label $[(\Phi_{\mathcal{H}}, \delta_{\mathcal{H}})]$ must be globally constant over the whole fiber. This forces $[(\Phi_{\mathcal{H}}, \delta_{\mathcal{H}})] = [(\Phi'_{\mathcal{H}}, \delta'_{\mathcal{H}})]$, as desired. □

Corollary 6.1.2.14. *(Compare with [62, Cor. 7.2.3.8].) The subschemes $\vec{\mathsf{Z}}^{\mathrm{ord}}_{[(\Phi_{\mathcal{H}},\delta_{\mathcal{H}})]}$ form a stratification*

$$\vec{\mathsf{M}}^{\mathrm{ord,min}}_{\mathcal{H}} = \coprod_{[(\Phi_{\mathcal{H}},\delta_{\mathcal{H}})]} \vec{\mathsf{Z}}^{\mathrm{ord}}_{[(\Phi_{\mathcal{H}},\delta_{\mathcal{H}})]} \tag{6.1.2.15}$$

of $\vec{\mathsf{M}}^{\mathrm{ord,min}}_{\mathcal{H}}$ by locally closed subscheme, with $[(\Phi_{\mathcal{H}}, \delta_{\mathcal{H}})]$ running through a complete set of ordinary cusp labels (see Definition 3.2.3.8*), such that the $[(\Phi'_{\mathcal{H}}, \delta'_{\mathcal{H}})]$-stratum $\vec{\mathsf{Z}}^{\mathrm{ord}}_{[(\Phi'_{\mathcal{H}},\delta'_{\mathcal{H}})]}$ lies in the closure of the $[(\Phi_{\mathcal{H}}, \delta_{\mathcal{H}})]$-stratum $\vec{\mathsf{Z}}^{\mathrm{ord}}_{[(\Phi_{\mathcal{H}},\delta_{\mathcal{H}})]}$ if and only if there is a surjection from the cusp label $[(\Phi'_{\mathcal{H}}, \delta'_{\mathcal{H}})]$ to the cusp label $[(\Phi_{\mathcal{H}}, \delta_{\mathcal{H}})]$ as in Definition* 1.2.1.18*. (The notation "$\coprod$" only means a set-theoretic disjoint union. The algebro-geometric structure is still that of $\vec{\mathsf{M}}^{\mathrm{ord,min}}_{\mathcal{H}}$.)*

Proof. According to (2) of Theorem 5.2.1.1, the closure of the $[(\Phi_{\mathcal{H}}, \delta_{\mathcal{H}}, \sigma)]$-stratum $\vec{\mathsf{Z}}^{\mathrm{ord}}_{[(\Phi_{\mathcal{H}},\delta_{\mathcal{H}},\sigma)]}$ in $\vec{\mathsf{M}}^{\mathrm{ord,tor}}_{\mathcal{H},\Sigma^{\mathrm{ord}}}$ is the union of the $[(\Phi'_{\mathcal{H}}, \delta'_{\mathcal{H}}, \sigma')]$-strata $\vec{\mathsf{Z}}^{\mathrm{ord}}_{[(\Phi'_{\mathcal{H}},\delta'_{\mathcal{H}},\sigma')]}$ such that $[(\Phi'_{\mathcal{H}}, \delta'_{\mathcal{H}}, \sigma')]$ is a face of $[(\Phi_{\mathcal{H}}, \delta_{\mathcal{H}}, \sigma)]$ as in Definition 1.2.2.19. Since the morphism $\vec{\oint}^{\mathrm{ord}}_{\mathcal{H}} : \vec{\mathsf{M}}^{\mathrm{ord,tor}}_{\mathcal{H},\Sigma^{\mathrm{ord}}} \to \vec{\mathsf{M}}^{\mathrm{ord,min}}_{\mathcal{H}}$ is proper, we see that the closure of $\vec{\mathsf{Z}}^{\mathrm{ord}}_{[(\Phi_{\mathcal{H}},\delta_{\mathcal{H}},\sigma)]}$ in $\vec{\mathsf{M}}^{\mathrm{ord,tor}}_{\mathcal{H},\Sigma^{\mathrm{ord}}}$ is mapped to the closure of $\vec{\mathsf{Z}}^{\mathrm{ord}}_{[(\Phi_{\mathcal{H}},\delta_{\mathcal{H}})]}$ in $\vec{\mathsf{M}}^{\mathrm{ord,min}}_{\mathcal{H}}$, which is by definition the union of those $\vec{\mathsf{Z}}^{\mathrm{ord}}_{[(\Phi'_{\mathcal{H}},\delta'_{\mathcal{H}})]}$ such that there is a surjection from $[(\Phi_{\mathcal{H}}, \delta_{\mathcal{H}})]$ to $[(\Phi'_{\mathcal{H}}, \delta'_{\mathcal{H}})]$. By Proposition 6.1.2.13, this union is disjoint. Hence, we may conclude (by induction on the incidence relations in the stratification of $\vec{\mathsf{M}}^{\mathrm{ord,tor}}_{\mathcal{H},\Sigma^{\mathrm{ord}}}$) that (6.1.2.15) is indeed a stratification of $\vec{\mathsf{M}}^{\mathrm{ord,min}}_{\mathcal{H}}$. □

As a byproduct:

Corollary 6.1.2.16. (*Compare with [62, Cor. 7.2.3.11].*) *If σ is top-dimensional in $\mathbf{P}^+_{\Phi_\mathcal{H}} \subset (\mathbf{S}_{\Phi_\mathcal{H}})^\vee_\mathbb{R}$, then the morphism $\vec{\oint}^{\mathrm{ord}}_\mathcal{H}|_{\vec{\mathsf{Z}}^{\mathrm{ord}}_{[(\Phi_\mathcal{H},\delta_\mathcal{H},\sigma)]}} : \vec{\mathsf{Z}}^{\mathrm{ord}}_{[(\Phi_\mathcal{H},\delta_\mathcal{H},\sigma)]} \to \vec{\mathsf{Z}}^{\mathrm{ord}}_{[(\Phi_\mathcal{H},\delta_\mathcal{H})]}$ (induced by restriction) is proper.*

Proof. Since σ is a top-dimensional cone, $[(\Phi_\mathcal{H},\delta_\mathcal{H},\sigma)]$ can be a face of another $[(\Phi'_\mathcal{H},\delta'_\mathcal{H},\sigma')]$ (see Definition 1.2.2.19) only when $[(\Phi_\mathcal{H},\delta_\mathcal{H})] \neq [(\Phi'_\mathcal{H},\delta'_\mathcal{H})]$. Then (2) of Theorem 5.2.1.1 and Proposition 6.1.2.13 imply that $\vec{\mathsf{Z}}^{\mathrm{ord}}_{[(\Phi_\mathcal{H},\delta_\mathcal{H},\sigma)]}$ is a closed subalgebraic stack of the preimage $(\vec{\oint}^{\mathrm{ord}}_\mathcal{H})^{-1}(\vec{\mathsf{Z}}^{\mathrm{ord}}_{[(\Phi_\mathcal{H},\delta_\mathcal{H})]})$. Since $\vec{\mathsf{M}}^{\mathrm{ord,tor}}_{\mathcal{H},\Sigma^{\mathrm{ord}}}$ is proper over $\vec{\mathsf{S}}_{0,r_\mathcal{H}}$, the induced morphism $\vec{\oint}^{\mathrm{ord}}_\mathcal{H}|_{\vec{\mathsf{Z}}^{\mathrm{ord}}_{[(\Phi_\mathcal{H},\delta_\mathcal{H},\sigma)]}} : \vec{\mathsf{Z}}^{\mathrm{ord}}_{[(\Phi_\mathcal{H},\delta_\mathcal{H},\sigma)]} \to \vec{\mathsf{Z}}^{\mathrm{ord}}_{[(\Phi_\mathcal{H},\delta_\mathcal{H})]}$ is also proper, as desired. □

Combining Corollary 6.1.2.16 with Lemma 6.1.2.12 and with Zariski's main theorem (see [35, III-1, 4.4.3, 4.4.11], and the formulation in [62, Prop. 7.2.3.4] for algebraic spaces), we obtain the following:

Corollary 6.1.2.17. (*Compare with [62, Cor. 7.2.3.12; see also the errata].*) *The morphism $[\vec{\mathsf{M}}^{\mathrm{ord},\mathsf{Z}_\mathcal{H}}_\mathcal{H}] \to \vec{\mathsf{Z}}^{\mathrm{ord}}_{[(\Phi_\mathcal{H},\delta_\mathcal{H})]}$ induced by (6.1.2.11) is finite and induces a bijection on geometric points.*

Proposition 6.1.2.18. (*Compare with [62, Prop. 7.2.3.13].*) *Let $\vec{\mathsf{M}}^{\mathrm{ord},1}_\mathcal{H}$ be the open subscheme of $\vec{\mathsf{M}}^{\mathrm{ord,min}}_\mathcal{H}$ formed by the union of the strata in (6.1.2.15) of codimension at most one. Then the pullback to $\vec{\mathsf{M}}^{\mathrm{ord},1}_\mathcal{H}$ of the canonical surjection $[\vec{\oint}^{\mathrm{ord}}_\mathcal{H}] : [\vec{\mathsf{M}}^{\mathrm{ord,tor}}_{\mathcal{H},\Sigma^{\mathrm{ord}}}] \twoheadrightarrow \vec{\mathsf{M}}^{\mathrm{ord,min}}_\mathcal{H}$ induced by $\vec{\oint}^{\mathrm{ord}}_\mathcal{H}$ is an isomorphism (regardless of the choice of Σ^{ord} in the construction of $\vec{\mathsf{M}}^{\mathrm{ord,min}}_\mathcal{H}$).*

Proof. The proof of [62, Prop. 7.2.3.13] also works here. □

Proposition 6.1.2.19. (*Compare with [62, Prop. 7.2.3.16; see also the errata].*) *Let $[(\Phi_\mathcal{H},\delta_\mathcal{H})]$ be an ordinary cusp label, and let $(\Phi_\mathcal{H},\delta_\mathcal{H})$ be a representative of $[(\Phi_\mathcal{H},\delta_\mathcal{H})]$. Let $\bar{x}$ be a geometric point of $\vec{\mathsf{M}}^{\mathrm{ord,min}}_\mathcal{H}$ over the $[(\Phi_\mathcal{H},\delta_\mathcal{H})]$-stratum $\vec{\mathsf{Z}}^{\mathrm{ord}}_{[(\Phi_\mathcal{H},\delta_\mathcal{H})]}$, which by abuse of notation we also identify as a geometric point of $[\vec{\mathsf{M}}^{\mathrm{ord},\mathsf{Z}_\mathcal{H}}_\mathcal{H}]$ by Corollary 6.1.2.17. Let $\mathrm{Aut}(\bar{x})$ be the group of automorphisms of $\bar{x} \to \vec{\mathsf{M}}^{\mathrm{ord},\mathsf{Z}_\mathcal{H}}_\mathcal{H}$ (cf. [62, Sec. A.7.5]). Let $(\vec{\mathsf{M}}^{\mathrm{ord,min}}_\mathcal{H})^\wedge_{\bar{x}}$ denote the completion of the strict localization of $\vec{\mathsf{M}}^{\mathrm{ord,min}}_\mathcal{H}$ at $\bar{x}$. Let $([\vec{\mathsf{M}}^{\mathrm{ord},\mathsf{Z}_\mathcal{H}}_\mathcal{H}])^\wedge_{\bar{x}}$ denote the completion of the strict localization of $[\vec{\mathsf{M}}^{\mathrm{ord},\mathsf{Z}_\mathcal{H}}_\mathcal{H}]$ at $\bar{x}$ (as a geometric point of $\vec{\mathsf{M}}^{\mathrm{ord},\mathsf{Z}_\mathcal{H}}_\mathcal{H}$), and let $(\underline{\mathrm{FJ}}^{\mathrm{ord},(\ell)}_{\Phi_\mathcal{H},\delta_\mathcal{H}})^\wedge_{\bar{x}}$ denote the pullback of $\underline{\mathrm{FJ}}^{\mathrm{ord},(\ell)}_{\Phi_\mathcal{H},\delta_\mathcal{H}}$ under the canonical morphism $(\vec{\mathsf{M}}^{\mathrm{ord},\mathsf{Z}_\mathcal{H}}_\mathcal{H})^\wedge_{\bar{x}} \to \vec{\mathsf{M}}^{\mathrm{ord},\mathsf{Z}_\mathcal{H}}_\mathcal{H}$. For convenience, let us also use the notation of the various sheaves supported on $\bar{x}$ to denote their underlying rings or modules. Then we have a canonical isomorphism*

$$\mathscr{O}_{(\vec{\mathsf{M}}^{\mathrm{ord,min}}_\mathcal{H})^\wedge_{\bar{x}}} \cong \Big[\prod_{\ell \in \mathbf{P}^\vee_{\Phi_\mathcal{H}}} (\underline{\mathrm{FJ}}^{\mathrm{ord},(\ell)}_{\Phi_\mathcal{H},\delta_\mathcal{H}})^\wedge_{\bar{x}} \Big]^{\mathrm{Aut}(\bar{x}) \times \Gamma_{\Phi_\mathcal{H}}} \tag{6.1.2.20}$$

of rings, which is adic if we interpret the product on the right-hand side as the completion of the elements that are finite sums with respect to the ideal generated by the elements without constant terms (i.e., with trivial projection to $(\underline{\mathrm{FJ}}^{\mathrm{ord},(0)}_{\Phi_{\mathcal{H}},\delta_{\mathcal{H}}})^{\wedge}_{\bar{x}}$*). Let us denote by* $(\vec{\mathsf{Z}}^{\mathrm{ord}}_{[(\Phi_{\mathcal{H}},\delta_{\mathcal{H}})]})^{\wedge}_{\bar{x}}$ *the completion of the strict localization of* $\mathsf{Z}_{[(\Phi_{\mathcal{H}},\delta_{\mathcal{H}})]}$ *at* $\bar{x}$*. Then* (6.1.2.20) *induces a* **structural morphism** *from* $(\vec{\mathsf{M}}^{\mathrm{ord,min}}_{\mathcal{H}})^{\wedge}_{\bar{x}}$ *to* $([\vec{\mathsf{M}}^{\mathrm{ord},\mathsf{Z}_{\mathcal{H}}}_{\mathcal{H}}])^{\wedge}_{\bar{x}}$*, whose pre-composition with the canonical morphism* $(\vec{\mathsf{Z}}^{\mathrm{ord}}_{[(\Phi_{\mathcal{H}},\delta_{\mathcal{H}})]})^{\wedge}_{\bar{x}} \to (\vec{\mathsf{M}}^{\mathrm{ord,min}}_{\mathcal{H}})^{\wedge}_{\bar{x}}$ *defines a canonical isomorphism* $(\vec{\mathsf{Z}}^{\mathrm{ord}}_{[(\Phi_{\mathcal{H}},\delta_{\mathcal{H}})]})^{\wedge}_{\bar{x}} \overset{\sim}{\to} ([\vec{\mathsf{M}}^{\mathrm{ord},\mathsf{Z}_{\mathcal{H}}}_{\mathcal{H}}])^{\wedge}_{\bar{x}}$*.*

Proof. The proof of [62, Prop. 7.2.3.16] works almost verbatim here. However, given the importance of this proposition, we shall spell out the details.

By [35, III-1, 4.1.5 and 4.3.3], with natural generalizations to the context of algebraic stacks, the ring $\mathscr{O}_{(\vec{\mathsf{M}}^{\mathrm{ord,min}}_{\mathcal{H}})^{\wedge}_{\bar{x}}}$ is isomorphic to the $\mathrm{Aut}(\bar{x})$-invariants in the ring of regular functions over the completion of $\vec{\mathsf{M}}^{\mathrm{ord,tor}}_{\mathcal{H},\Sigma^{\mathrm{ord}}}$ along the fiber of $\vec{\oint}^{\mathrm{ord}}_{\mathcal{H}} : \vec{\mathsf{M}}^{\mathrm{ord,tor}}_{\mathcal{H},\Sigma^{\mathrm{ord}}} \to \vec{\mathsf{M}}^{\mathrm{ord,min}}_{\mathcal{H}}$ at $\bar{x}$. By Proposition 6.1.2.13, the preimage $\tilde{\vec{\mathsf{Z}}}^{\mathrm{ord}}_{[(\Phi_{\mathcal{H}},\delta_{\mathcal{H}})]} := (\vec{\oint}^{\mathrm{ord}}_{\mathcal{H}})^{-1}(\vec{\mathsf{Z}}^{\mathrm{ord}}_{[(\Phi_{\mathcal{H}},\delta_{\mathcal{H}})]})$ of $\vec{\mathsf{Z}}^{\mathrm{ord}}_{[(\Phi_{\mathcal{H}},\delta_{\mathcal{H}})]}$ under $\vec{\oint}^{\mathrm{ord}}_{\mathcal{H}}$ is the union

$$\tilde{\vec{\mathsf{Z}}}^{\mathrm{ord}}_{[(\Phi_{\mathcal{H}},\delta_{\mathcal{H}})]} = \bigcup_{[(\Phi_{\mathcal{H}},\delta_{\mathcal{H}},\sigma)]} \vec{\mathsf{Z}}^{\mathrm{ord}}_{[(\Phi_{\mathcal{H}},\delta_{\mathcal{H}},\sigma)]}$$

of those strata $\vec{\mathsf{Z}}^{\mathrm{ord}}_{[(\Phi_{\mathcal{H}},\delta_{\mathcal{H}},\sigma)]}$ over $\vec{\mathsf{Z}}^{\mathrm{ord}}_{[(\Phi_{\mathcal{H}},\delta_{\mathcal{H}})]}$. According to (5) of Theorem 5.2.1.1 and [62, Lem. 6.2.5.27], there is a canonical isomorphism $(\vec{\mathsf{M}}^{\mathrm{ord,tor}}_{\mathcal{H},\Sigma^{\mathrm{ord}}})^{\wedge}_{\vec{\mathsf{Z}}^{\mathrm{ord}}_{[(\Phi_{\mathcal{H}},\delta_{\mathcal{H}},\sigma)]}} \cong \vec{\mathfrak{X}}^{\mathrm{ord}}_{\Phi_{\mathcal{H}},\delta_{\mathcal{H}},\sigma}/\Gamma_{\Phi_{\mathcal{H}},\sigma}$ for each representative $(\Phi_{\mathcal{H}},\delta_{\mathcal{H}},\sigma)$ of $[(\Phi_{\mathcal{H}},\delta_{\mathcal{H}},\sigma)]$. Therefore, the ring of regular functions over the completion of $\vec{\mathsf{M}}^{\mathrm{ord,tor}}_{\mathcal{H},\Sigma^{\mathrm{ord}}}$ along the fiber of $\vec{\oint}^{\mathrm{ord}}_{\mathcal{H}}$ at $\bar{x}$ is isomorphic to the *common intersection* of the rings of regular functions over the various completions of $\vec{\mathfrak{X}}^{\mathrm{ord}}_{\Phi_{\mathcal{H}},\delta_{\mathcal{H}},\sigma}/\Gamma_{\Phi_{\mathcal{H}},\sigma}$ along the fibers of the structural morphisms $\vec{\mathfrak{X}}^{\mathrm{ord}}_{\Phi_{\mathcal{H}},\delta_{\mathcal{H}},\sigma}/\Gamma_{\Phi_{\mathcal{H}},\sigma} \to \vec{\mathsf{M}}^{\mathrm{ord},\mathsf{Z}_{\mathcal{H}}}_{\mathcal{H}}$. In other words, it is isomorphic to the common intersection of the $\Gamma_{\Phi_{\mathcal{H}},\sigma}$-invariants in the completions of $\hat{\bigoplus}_{\ell\in\sigma^{\vee}} \underline{\mathrm{FJ}}^{\mathrm{ord},(\ell)}_{\Phi_{\mathcal{H}},\delta_{\mathcal{H}}}$ along $\bar{x}$. Note that the identifications $\vec{\mathfrak{X}}^{\mathrm{ord}}_{\Phi_{\mathcal{H}},\delta_{\mathcal{H}},\sigma} \cong \vec{\mathfrak{X}}^{\mathrm{ord}}_{\Phi'_{\mathcal{H}},\delta'_{\mathcal{H}},\sigma'}$ for equivalent triples $(\Phi_{\mathcal{H}},\delta_{\mathcal{H}},\sigma)$ and $(\Phi'_{\mathcal{H}},\delta'_{\mathcal{H}},\sigma')$ involve the canonical actions of $\Gamma_{\Phi_{\mathcal{H}}}$ on the structural sheaves. Hence, the process of taking a common intersection also involves the process of taking $\Gamma_{\Phi_{\mathcal{H}}}$-invariants. This shows the existence of (6.1.2.20).

The claim that (6.1.2.20) is adic and that the composition $(\vec{\mathsf{Z}}^{\mathrm{ord}}_{[(\Phi_{\mathcal{H}},\delta_{\mathcal{H}})]})^{\wedge}_{\bar{x}} \to (\vec{\mathsf{M}}^{\mathrm{ord,min}}_{\mathcal{H}})^{\wedge}_{\bar{x}} \to ([\vec{\mathsf{M}}^{\mathrm{ord},\mathsf{Z}_{\mathcal{H}}}_{\mathcal{H}}])^{\wedge}_{\bar{x}}$ is an isomorphism follows from the fact that the support $\vec{\mathsf{Z}}^{\mathrm{ord}}_{[(\Phi_{\mathcal{H}},\delta_{\mathcal{H}},\sigma)]}$ of each formal completion $(\vec{\mathsf{M}}^{\mathrm{ord,tor}}_{\mathcal{H},\Sigma^{\mathrm{ord}}})^{\wedge}_{\vec{\mathsf{Z}}^{\mathrm{ord}}_{[(\Phi_{\mathcal{H}},\delta_{\mathcal{H}},\sigma)]}} \cong \vec{\mathfrak{X}}^{\mathrm{ord}}_{\Phi_{\mathcal{H}},\delta_{\mathcal{H}},\sigma}/\Gamma_{\Phi_{\mathcal{H}},\sigma}$ is defined by the vanishing of the ideal $\hat{\bigoplus}_{\ell\in\sigma_0^{\vee}} \vec{\Psi}^{\mathrm{ord}}_{\Phi_{\mathcal{H}},\delta_{\mathcal{H}}}(\ell)$ of $\hat{\bigoplus}_{\ell\in\sigma^{\vee}} \vec{\Psi}^{\mathrm{ord}}_{\Phi_{\mathcal{H}},\delta_{\mathcal{H}}}(\ell)$, and that $\mathbf{P}^{\vee}_{\Phi_{\mathcal{H}}} - \{0\} = \bigcap_{\sigma\in\Sigma_{\Phi_{\mathcal{H}}},\sigma\subset\mathbf{P}^{+}_{\Phi_{\mathcal{H}}}} \sigma_0^{\vee}$ (because $\mathbf{P}^{\vee}_{\Phi_{\mathcal{H}}} - \{0\} \subset \sigma_0^{\vee}$ for every $\sigma \subset \mathbf{P}^{+}_{\Phi_{\mathcal{H}}}$ and because $\mathbf{P}^{\vee}_{\Phi_{\mathcal{H}}} = \bigcap_{\sigma\in\Sigma_{\Phi_{\mathcal{H}}}} \sigma^{\vee}$ as explained in [62, Sec. 7.1.2]). Then we can conclude the proof by taking $\mathrm{Aut}(\bar{x}) \times \Gamma_{\Phi_{\mathcal{H}}}$-invariants and by noting that $((\underline{\mathrm{FJ}}^{\mathrm{ord},(0)}_{\Phi_{\mathcal{H}},\delta_{\mathcal{H}}})^{\wedge}_{\bar{x}})^{\mathrm{Aut}(\bar{x}) \times \Gamma_{\Phi_{\mathcal{H}}}} \cong (\mathscr{O}_{(\vec{\mathsf{M}}^{\mathrm{ord},\mathsf{Z}_{\mathcal{H}}}_{\mathcal{H}})^{\wedge}_{\bar{x}}})^{\mathrm{Aut}(\bar{x})} \cong \mathscr{O}_{[(\vec{\mathsf{M}}^{\mathrm{ord},\mathsf{Z}_{\mathcal{H}}}_{\mathcal{H}})^{\wedge}_{\bar{x}}]} \cong \mathscr{O}_{([\vec{\mathsf{M}}^{\mathrm{ord},\mathsf{Z}_{\mathcal{H}}}_{\mathcal{H}}])^{\wedge}_{\bar{x}}}$. □

Corollary 6.1.2.21. (*Compare with [62, Cor. 7.2.3.18].*) *The canonical finite surjection* $[\vec{\mathsf{M}}^{\mathrm{ord},\mathsf{Z}_{\mathcal{H}}}_{\mathcal{H}}] \twoheadrightarrow \vec{\mathsf{Z}}^{\mathrm{ord}}_{[(\Phi_{\mathcal{H}},\delta_{\mathcal{H}})]}$ *defined by* $\vec{\oint}^{\mathrm{ord}}_{\mathcal{H}}$ *is an isomorphism.*

Proof. The proof of Proposition 6.1.2.19 shows that the composition of the completion $([\vec{\mathsf{M}}^{\mathrm{ord},\mathsf{Z}_{\mathcal{H}}}_{\mathcal{H}}])^{\wedge}_{\bar{x}} \to (\vec{\mathsf{Z}}^{\mathrm{ord}}_{[(\Phi_{\mathcal{H}},\delta_{\mathcal{H}})]})^{\wedge}_{\bar{x}}$ of the finite surjection $[\vec{\mathsf{M}}^{\mathrm{ord},\mathsf{Z}_{\mathcal{H}}}_{\mathcal{H}}] \twoheadrightarrow \vec{\mathsf{Z}}^{\mathrm{ord}}_{[(\Phi_{\mathcal{H}},\delta_{\mathcal{H}})]}$ defined by $\vec{\oint}^{\mathrm{ord}}_{\mathcal{H}}$ (described in Corollary 6.1.2.17) with the canonical *structural isomorphism* $(\vec{\mathsf{Z}}^{\mathrm{ord}}_{[(\Phi_{\mathcal{H}},\delta_{\mathcal{H}})]})^{\wedge}_{\bar{x}} \xrightarrow{\sim} ([\vec{\mathsf{M}}^{\mathrm{ord},\mathsf{Z}_{\mathcal{H}}}_{\mathcal{H}}])^{\wedge}_{\bar{x}}$ is the identity isomorphism. This forces $[\vec{\mathsf{M}}^{\mathrm{ord},\mathsf{Z}_{\mathcal{H}}}_{\mathcal{H}}] \twoheadrightarrow \vec{\mathsf{Z}}^{\mathrm{ord}}_{[(\Phi_{\mathcal{H}},\delta_{\mathcal{H}})]}$ to be an isomorphism as the property of being an isomorphism can be verified over the formal completions of the target. □

6.2 Partial Minimal Compactifications of Ordinary Loci

In this section, we continue to assume the same settings as in Section 5.2.

6.2.1 *Main Statements*

The partial minimal compactifications of $\vec{\mathsf{M}}^{\mathrm{ord}}_{\mathcal{H}}$ can be described as follows:

Theorem 6.2.1.1. (*Compare with [62, Thm. 7.2.4.1] and Theorem* 1.3.1.5.) *There exists a normal scheme* $\vec{\mathsf{M}}^{\mathrm{ord},\mathrm{min}}_{\mathcal{H}}$ *quasi-projective and flat over* $\vec{\mathsf{S}}_{0,r_{\mathcal{H}}} = \mathrm{Spec}(\mathcal{O}_{F_{0,(p)}}[\zeta_{p^{r_{\mathcal{H}}}}])$ (*see Definition* 2.2.3.3), *such that we have the following:*

(1) $\vec{\mathsf{M}}^{\mathrm{ord},\mathrm{min}}_{\mathcal{H}}$ *contains the* **coarse moduli space** $[\vec{\mathsf{M}}^{\mathrm{ord}}_{\mathcal{H}}]$ *of* $\vec{\mathsf{M}}^{\mathrm{ord}}_{\mathcal{H}}$ *as an open fiberwise dense* **subscheme**.

(2) *Let* $(G_{\vec{\mathsf{M}}^{\mathrm{ord}}_{\mathcal{H}}}, \lambda_{\vec{\mathsf{M}}^{\mathrm{ord}}_{\mathcal{H}}}, i_{\vec{\mathsf{M}}^{\mathrm{ord}}_{\mathcal{H}}}, \alpha_{\mathcal{H}^p}, \alpha^{\mathrm{ord}}_{\mathcal{H}_p})$ *be the tautological tuple over* $\vec{\mathsf{M}}^{\mathrm{ord}}_{\mathcal{H}}$. *Let us define the invertible sheaf*

$$\omega_{\vec{\mathsf{M}}^{\mathrm{ord}}_{\mathcal{H}}} := \wedge^{\mathrm{top}} \underline{\mathrm{Lie}}^{\vee}_{G_{\vec{\mathsf{M}}^{\mathrm{ord}}_{\mathcal{H}}}/\vec{\mathsf{M}}^{\mathrm{ord}}_{\mathcal{H}}} = \wedge^{\mathrm{top}} e^{*}_{G_{\vec{\mathsf{M}}^{\mathrm{ord}}_{\mathcal{H}}}} \Omega^{1}_{G_{\vec{\mathsf{M}}^{\mathrm{ord}}_{\mathcal{H}}}/\vec{\mathsf{M}}^{\mathrm{ord}}_{\mathcal{H}}}$$

over $\vec{\mathsf{M}}^{\mathrm{ord}}_{\mathcal{H}}$. *Then there is a smallest integer* $N_0 \geq 1$ *such that* $\omega^{\otimes N_0}_{\vec{\mathsf{M}}^{\mathrm{ord}}_{\mathcal{H}}}$ *is the pullback of an ample invertible sheaf* $\mathcal{O}(1)$ *over* $\vec{\mathsf{M}}^{\mathrm{ord},\mathrm{min}}_{\mathcal{H}}$.

If $\mathcal{H}^p$ *is neat, then* $\vec{\mathsf{M}}^{\mathrm{ord}}_{\mathcal{H}} \to [\vec{\mathsf{M}}^{\mathrm{ord}}_{\mathcal{H}}]$ *is an isomorphism, and induces an embedding of* $\vec{\mathsf{M}}^{\mathrm{ord}}_{\mathcal{H}}$ *as an open fiberwise dense subscheme of* $\vec{\mathsf{M}}^{\mathrm{ord},\mathrm{min}}_{\mathcal{H}}$. *Moreover, we have* $N_0 = 1$ *with a canonical choice of* $\mathcal{O}(1)$, *and the restriction of* $\mathcal{O}(1)$ *to* $\vec{\mathsf{M}}^{\mathrm{ord}}_{\mathcal{H}}$ *is isomorphic to* $\omega_{\vec{\mathsf{M}}^{\mathrm{ord}}_{\mathcal{H}}}$. *We shall denote* $\mathcal{O}(1)$ *by* $\omega_{\vec{\mathsf{M}}^{\mathrm{ord},\mathrm{min}}_{\mathcal{H}}}$, *and interpret it as an extension of* $\omega_{\vec{\mathsf{M}}^{\mathrm{ord}}_{\mathcal{H}}}$ *to* $\vec{\mathsf{M}}^{\mathrm{ord},\mathrm{min}}_{\mathcal{H}}$.

By abuse of notation, for each integer k *divisible by* N_0, *we shall denote* $\mathcal{O}(1)^{\otimes k/N_0}$ *by* $\omega^{\otimes k}_{\vec{\mathsf{M}}^{\mathrm{ord},\mathrm{min}}_{\mathcal{H}}}$ *even when* $\omega_{\vec{\mathsf{M}}^{\mathrm{ord},\mathrm{min}}_{\mathcal{H}}}$ *itself is not defined.*

(3) *For each (smooth) partial toroidal compactification* $\vec{\mathsf{M}}^{\mathrm{ord},\mathrm{tor}}_{\mathcal{H}}$ *of* $\vec{\mathsf{M}}^{\mathrm{ord}}_{\mathcal{H}}$ *as in Theorem* 5.2.1.1, *with a degenerating family* $(G, \lambda, i, \alpha_{\mathcal{H}^p}, \alpha^{\mathrm{ord}}_{\mathcal{H}_p})$ *over*

$\vec{\mathsf{M}}_{\mathcal{H}}^{\mathrm{ord,tor}}$ *extending the tautological tuple* $(G_{\vec{\mathsf{M}}_{\mathcal{H}}^{\mathrm{ord}}}, \lambda_{\vec{\mathsf{M}}_{\mathcal{H}}^{\mathrm{ord}}}, i_{\vec{\mathsf{M}}_{\mathcal{H}}^{\mathrm{ord}}}, \alpha_{\mathcal{H}^p}, \alpha_{\mathcal{H}_p}^{\mathrm{ord}})$ *over* $\vec{\mathsf{M}}_{\mathcal{H}}^{\mathrm{ord}}$*, let*

$$\omega_{\vec{\mathsf{M}}_{\mathcal{H}}^{\mathrm{ord,tor}}} := \wedge^{\mathrm{top}} \underline{\mathrm{Lie}}^{\vee}_{G/\vec{\mathsf{M}}_{\mathcal{H}}^{\mathrm{ord,tor}}} = \wedge^{\mathrm{top}} e_G^* \Omega^1_{G/\vec{\mathsf{M}}_{\mathcal{H}}^{\mathrm{ord,tor}}}$$

be the invertible sheaf over $\vec{\mathsf{M}}_{\mathcal{H}}^{\mathrm{ord,tor}}$ *extending* $\omega_{\vec{\mathsf{M}}_{\mathcal{H}}^{\mathrm{ord}}}$ *naturally. Then the graded algebra* $\underset{k\geq 0}{\oplus} \Gamma(\vec{\mathsf{M}}_{\mathcal{H}}^{\mathrm{ord,tor}}, \omega^{\otimes k}_{\vec{\mathsf{M}}_{\mathcal{H}}^{\mathrm{ord,tor}}})$*, with its natural algebra structure induced by tensor products, is independent of the choice (of the* Σ^{ord} *used in the definition) of* $\vec{\mathsf{M}}_{\mathcal{H}}^{\mathrm{ord,tor}}$*.*

The normal scheme $\vec{\mathsf{M}}_{\mathcal{H}}^{\mathrm{ord,min}}$ *(quasi-projective and flat over* $\vec{\mathsf{S}}_{0,r_{\mathcal{H}}}$*) is canonically isomorphic to* $\mathrm{Proj}\Big(\underset{k\geq 0}{\oplus} \Gamma(\vec{\mathsf{M}}_{\mathcal{H}}^{\mathrm{ord,tor}}, \omega^{\otimes k}_{\vec{\mathsf{M}}_{\mathcal{H}}^{\mathrm{ord,tor}}})\Big)$*, and there is a canonical* **proper** *morphism* $\vec{\oint}_{\mathcal{H}}^{\mathrm{ord}} : \vec{\mathsf{M}}_{\mathcal{H}}^{\mathrm{ord,tor}} \to \vec{\mathsf{M}}_{\mathcal{H}}^{\mathrm{ord,min}}$ *determined by* $\omega_{\vec{\mathsf{M}}_{\mathcal{H}}^{\mathrm{ord,tor}}}$ *and the universal property of* Proj*, such that* $\vec{\oint}_{\mathcal{H}}^{\mathrm{ord},*} \mathcal{O}(1) \cong \omega^{\otimes N_0}_{\vec{\mathsf{M}}_{\mathcal{H}}^{\mathrm{ord,tor}}}$ *over* $\vec{\mathsf{M}}_{\mathcal{H}}^{\mathrm{ord,tor}}$*, and such that the canonical morphism* $\mathscr{O}_{\vec{\mathsf{M}}_{\mathcal{H}}^{\mathrm{ord,min}}} \to \vec{\oint}_{\mathcal{H},*}^{\mathrm{ord}} \mathscr{O}_{\vec{\mathsf{M}}_{\mathcal{H}}^{\mathrm{ord,tor}}}$ *is an isomorphism. Moreover, when we vary the choices of* $\vec{\mathsf{M}}_{\mathcal{H}}^{\mathrm{ord,tor}}$*'s, the morphisms* $\vec{\oint}_{\mathcal{H}}^{\mathrm{ord}}$*'s are compatible with the canonical morphisms among the* $\vec{\mathsf{M}}_{\mathcal{H}}^{\mathrm{ord,tor}}$*'s as in Proposition* 5.2.2.2*.*

When $\mathcal{H}^p$ *is neat, we have* $\vec{\oint}_{\mathcal{H}}^{\mathrm{ord},*} \omega^{\otimes k}_{\vec{\mathsf{M}}_{\mathcal{H}}^{\mathrm{ord,min}}} \cong \omega_{\vec{\mathsf{M}}_{\mathcal{H}}^{\mathrm{ord,tor}}}$ *and* $\vec{\oint}_{\mathcal{H},*}^{\mathrm{ord}} \omega_{\vec{\mathsf{M}}_{\mathcal{H}}^{\mathrm{ord,tor}}} \cong \omega^{\otimes k}_{\vec{\mathsf{M}}_{\mathcal{H}}^{\mathrm{ord,min}}}$*.*

(4) $\vec{\mathsf{M}}_{\mathcal{H}}^{\mathrm{ord,min}}$ *has a natural stratification by locally closed subschemes*

$$\vec{\mathsf{M}}_{\mathcal{H}}^{\mathrm{ord,min}} = \coprod_{[(\Phi_{\mathcal{H}},\delta_{\mathcal{H}})]} \vec{\mathsf{Z}}^{\mathrm{ord}}_{[(\Phi_{\mathcal{H}},\delta_{\mathcal{H}})]},$$

with $[(\Phi_{\mathcal{H}}, \delta_{\mathcal{H}})]$ *running through a complete set of ordinary cusp labels (see Definition* 1.2.1.7*, [62, Def. 5.4.2.4], and Definition* 3.2.3.8*), such that the* $[(\Phi'_{\mathcal{H}}, \delta'_{\mathcal{H}})]$*-stratum* $\vec{\mathsf{Z}}^{\mathrm{ord}}_{[(\Phi'_{\mathcal{H}},\delta'_{\mathcal{H}})]}$ *lies in the closure of the* $[(\Phi_{\mathcal{H}}, \delta_{\mathcal{H}})]$*-stratum* $\vec{\mathsf{Z}}^{\mathrm{ord}}_{[(\Phi_{\mathcal{H}},\delta_{\mathcal{H}})]}$ *if and only if there is a surjection from the cusp label* $[(\Phi'_{\mathcal{H}}, \delta'_{\mathcal{H}})]$ *to the cusp label* $[(\Phi_{\mathcal{H}}, \delta_{\mathcal{H}})]$ *as in Definition* 1.2.1.18*. (The notation "*$\coprod$*" only means a set-theoretic disjoint union. The algebro-geometric structure is still that of* $\vec{\mathsf{M}}_{\mathcal{H}}^{\mathrm{ord,min}}$*.) The analogous assertion holds after pulled back to fibers over* $\vec{\mathsf{S}}_{0,r_{\mathcal{H}}}$*.*

Each $[(\Phi_{\mathcal{H}}, \delta_{\mathcal{H}})]$*-stratum* $\vec{\mathsf{Z}}^{\mathrm{ord}}_{[(\Phi_{\mathcal{H}},\delta_{\mathcal{H}})]}$ *is canonically isomorphic to the coarse moduli space* $[\vec{\mathsf{M}}_{\mathcal{H}}^{\mathrm{ord},\mathsf{Z}_{\mathcal{H}}}]$ *(which is a scheme) of the corresponding algebraic stack* $\vec{\mathsf{M}}_{\mathcal{H}}^{\mathrm{ord},\mathsf{Z}_{\mathcal{H}}}$ *(separated, smooth, and of finite type over* $\vec{\mathsf{S}}_{0,r_{\mathcal{H}}}$*) associated with the cusp label* $[(\Phi_{\mathcal{H}}, \delta_{\mathcal{H}})]$ *as in* (4.2.1.28)*.*

Let us define the $\mathcal{O}$*-multi-rank of a stratum* $\vec{\mathsf{Z}}^{\mathrm{ord}}_{[(\Phi_{\mathcal{H}},\delta_{\mathcal{H}})]}$ *to be the* $\mathcal{O}$*-multi-rank of the cusp label represented by* $(\Phi_{\mathcal{H}}, \delta_{\mathcal{H}})$ *(see [62, Def. 5.4.2.7]). The only stratum with* $\mathcal{O}$*-multi-rank zero is the open stratum* $\vec{\mathsf{Z}}^{\mathrm{ord}}_{[(0,0)]} \cong [\vec{\mathsf{M}}_{\mathcal{H}}^{\mathrm{ord}}]$*,* *and those strata* $\vec{\mathsf{Z}}^{\mathrm{ord}}_{[(\Phi_{\mathcal{H}},\delta_{\mathcal{H}})]}$ *with nonzero* $\mathcal{O}$*-multi-ranks are called* **ordinary cusps**.

(5) *The restriction of* $\vec{\oint}^{\mathrm{ord}}_{\mathcal{H}}$ *to the stratum* $\vec{\mathsf{Z}}^{\mathrm{ord}}_{[(\Phi_{\mathcal{H}},\delta_{\mathcal{H}},\sigma)]}$ *of* $\vec{\mathsf{M}}^{\mathrm{ord,tor}}_{\mathcal{H}}$ *is a surjection to the stratum* $\vec{\mathsf{Z}}^{\mathrm{ord}}_{[(\Phi_{\mathcal{H}},\delta_{\mathcal{H}})]}$ *of* $\mathsf{M}^{\mathrm{ord,min}}_{\mathcal{H}}$. *This surjection is smooth when* $\mathcal{H}^p$ *is neat, and is proper if* σ *is top-dimensional in* $\mathbf{P}^+_{\Phi_{\mathcal{H}}} \subset (\mathbf{S}_{\Phi_{\mathcal{H}}})^\vee_{\mathbb{R}}$.
Under the above-mentioned identification $[\vec{\mathsf{M}}^{\mathrm{ord},\mathsf{Z}_{\mathcal{H}}}_{\mathcal{H}}] \xrightarrow{\sim} \vec{\mathsf{Z}}^{\mathrm{ord}}_{[(\Phi_{\mathcal{H}},\delta_{\mathcal{H}})]}$ *on the target, this surjection can be viewed as the quotient by* $\Gamma_{\Phi_{\mathcal{H}},\sigma}$ *(see [62, Def. 6.2.5.23]) of a torsor under a torus* $E_{\Phi_{\mathcal{H}},\sigma}$ *over an abelian scheme torsor* $\vec{C}^{\mathrm{ord}}_{\Phi_{\mathcal{H}},\delta_{\mathcal{H}}}$ *over the finite étale cover* $\vec{\mathsf{M}}^{\mathrm{ord},\Phi_{\mathcal{H}}}_{\mathcal{H}}$ *of the algebraic stack* $\vec{\mathsf{M}}^{\mathrm{ord},\mathsf{Z}_{\mathcal{H}}}_{\mathcal{H}}$ *over the coarse moduli space* $[\vec{\mathsf{M}}^{\mathrm{ord},\mathsf{Z}_{\mathcal{H}}}_{\mathcal{H}}]$ *(which is a scheme), where the torus* $E_{\Phi_{\mathcal{H}},\sigma}$ *is as in* (5) *of Theorem* 1.3.1.5.

(6) *There is a canonical open immersion*

$$\vec{\mathsf{M}}^{\mathrm{ord,min}}_{\mathcal{H}} \underset{\mathbb{Z}}{\otimes} \mathbb{Q} \hookrightarrow \mathsf{M}^{\mathrm{min}}_{\mathcal{H},r_{\mathcal{H}}} \tag{6.2.1.2}$$

(see Definition 2.2.3.4*) over* $\mathsf{S}_{0,r_{\mathcal{H}}}$ *extending the canonical isomorphism* $[\mathsf{M}^{\mathrm{ord}}_{\mathcal{H}}] \cong [\mathsf{M}_{\mathcal{H},r_{\mathcal{H}}}] \cong [\mathsf{M}_{\mathcal{H}}] \underset{\mathsf{S}_0}{\times} \mathsf{S}_{0,r_{\mathcal{H}}}$ *over* $\mathsf{S}_{0,r_{\mathcal{H}}}$ *(see the definition of* $\mathsf{M}^{\mathrm{ord}}_{\mathcal{H}}$ *in Theorem* 3.4.2.5*), which is compatible with any canonical open immersion* (5.2.1.2) *in* (7) *of Theorem* 5.2.1.1 *and with the canonical morphisms* $\vec{\oint}^{\mathrm{ord}}_{\mathcal{H}} : \vec{\mathsf{M}}^{\mathrm{ord,tor}}_{\mathcal{H},\Sigma^{\mathrm{ord}}} \to \vec{\mathsf{M}}^{\mathrm{ord,min}}_{\mathcal{H}}$ *and* $\oint_{\mathcal{H}} : \mathsf{M}^{\mathrm{tor}}_{\mathcal{H},\Sigma} \to \mathsf{M}^{\mathrm{min}}_{\mathcal{H}}$. *Under* (6.2.1.2)*, the pullback of* $\omega^{\otimes k}_{\mathsf{M}^{\mathrm{min}}_{\mathcal{H},r_{\mathcal{H}}}}$ *is canonically isomorphic to the pullback of* $\omega^{\otimes k}_{\vec{\mathsf{M}}^{\mathrm{ord,min}}_{\mathcal{H}}}$, *when both are defined for some integer* k. *The open immersion* (6.2.1.2) *induces isomorphisms*

$$\vec{\mathsf{Z}}^{\mathrm{ord}}_{[(\Phi_{\mathcal{H}},\delta_{\mathcal{H}})]} \underset{\mathbb{Z}}{\otimes} \mathbb{Q} \xrightarrow{\sim} \mathsf{Z}_{[(\Phi_{\mathcal{H}},\delta_{\mathcal{H}})],r_{\mathcal{H}}} \tag{6.2.1.3}$$

(see Definition 2.2.3.4*), compatible with* (5.2.1.3)*, when the cusp label* $[(\Phi_{\mathcal{H}},\delta_{\mathcal{H}})]$ *is ordinary; otherwise, the pullback of* $\mathsf{Z}_{[(\Phi_{\mathcal{H}},\delta_{\mathcal{H}})],r_{\mathcal{H}}}$ *under* (6.2.1.2) *is empty.*
The canonical open immersion (6.2.1.2) *extends to a canonical open immersion*

$$\vec{\mathsf{M}}^{\mathrm{ord,min}}_{\mathcal{H}} \hookrightarrow \vec{\mathsf{M}}^{\mathrm{min}}_{\mathcal{H},r_{\mathcal{H}}} \tag{6.2.1.4}$$

(see Definition 2.2.3.5*) over* $\vec{\mathsf{S}}_{0,r_{\mathcal{H}}}$. *Under* (6.2.1.4)*, the pullback of* $\omega^{\otimes k}_{\vec{\mathsf{M}}^{\mathrm{min}}_{\mathcal{H},r_{\mathcal{H}}}}$ *is canonically isomorphic to* $\omega^{\otimes k}_{\vec{\mathsf{M}}^{\mathrm{ord,min}}_{\mathcal{H}}}$, *when both are defined for some integer* k; *and the* $[(\Phi_{\mathcal{H}},\delta_{\mathcal{H}})]$*-stratum* $\vec{\mathsf{Z}}^{\mathrm{ord}}_{[(\Phi_{\mathcal{H}},\delta_{\mathcal{H}})]}$ *of* $\vec{\mathsf{M}}^{\mathrm{ord,min}}_{\mathcal{H}}$ *is the pullback of the* $[(\Phi_{\mathcal{H}},\delta_{\mathcal{H}})]$*-stratum* $\vec{\mathsf{Z}}_{[(\Phi_{\mathcal{H}},\delta_{\mathcal{H}})],r_{\mathcal{H}}}$ *of* $\vec{\mathsf{M}}^{\mathrm{min}}_{\mathcal{H},r_{\mathcal{H}}}$.

Proof. With the ingredients we have provided, the proof is almost identical to that of [62, Thm. 7.2.4.1]. However, since the construction of $\vec{\mathsf{M}}^{\mathrm{ord,min}}_{\mathcal{H}}$ is rather indirect, we shall spell out the details for the sake of certainty.

Let us take $\vec{\mathsf{M}}_{\mathcal{H}}^{\mathrm{ord,min}}$ to be the normal scheme (quasi-projective and flat over $\vec{\mathsf{S}}_{0,r_{\mathcal{H}}}$) constructed in Section 6.1.1. The first concern is whether its properties as described in the theorem depend on the choice of Σ^{ord} for the toroidal compactification $\vec{\mathsf{M}}_{\mathcal{H}}^{\mathrm{ord,tor}} = \vec{\mathsf{M}}_{\mathcal{H},\Sigma^{\mathrm{ord}}}^{\mathrm{ord,tor}}$ involved in the construction. It is clear that statements (1), (4), and (5) are satisfied regardless of the choice of Σ^{ord}. Let us verify that this is also the case for statements (2) and (3).

Suppose $\Sigma^{\mathrm{ord},\prime}$ is a refinement of Σ^{ord} as in Definition 1.2.2.16, suppose the morphism $[\vec{1}]^{\mathrm{ord,tor}} : \vec{\mathsf{M}}_{\mathcal{H},\Sigma^{\mathrm{ord},\prime}}^{\mathrm{ord,tor}} \to \vec{\mathsf{M}}_{\mathcal{H},\Sigma^{\mathrm{ord}}}^{\mathrm{ord,tor}}$ is the proper log étale surjection as in Proposition 5.2.2.2, and suppose the invertible sheaves $\omega_{\vec{\mathsf{M}}_{\mathcal{H},\Sigma^{\mathrm{ord}}}^{\mathrm{ord,tor}}}$ and $\omega_{\vec{\mathsf{M}}_{\mathcal{H},\Sigma^{\mathrm{ord},\prime}}^{\mathrm{ord,tor}}}$ are defined as in (6.1.1.1). Let $\vec{\oint}_{\mathcal{H},\Sigma}^{\mathrm{ord}} : \vec{\mathsf{M}}_{\mathcal{H},\Sigma^{\mathrm{ord}}}^{\mathrm{ord,tor}} \to \vec{\mathsf{M}}_{\mathcal{H}}^{\mathrm{ord,min}}$ and $\vec{\oint}_{\mathcal{H},\Sigma'}^{\mathrm{ord}} : \vec{\mathsf{M}}_{\mathcal{H},\Sigma^{\mathrm{ord},\prime}}^{\mathrm{ord,tor}} \to \vec{\mathsf{M}}_{\mathcal{H}}^{\mathrm{ord,min}}$ be the two canonical morphisms. Then $\vec{\oint}_{\mathcal{H},\Sigma^{\mathrm{ord},\prime}}^{\mathrm{ord}} = \vec{\oint}_{\mathcal{H},\Sigma^{\mathrm{ord}}}^{\mathrm{ord}} \circ [\vec{1}]^{\mathrm{ord,tor}}$ and $[\vec{1}]_*^{\mathrm{ord,tor}} \mathscr{O}_{\vec{\mathsf{M}}_{\mathcal{H},\Sigma^{\mathrm{ord},\prime}}^{\mathrm{ord,tor}}} \cong \mathscr{O}_{\vec{\mathsf{M}}_{\mathcal{H},\Sigma^{\mathrm{ord}}}^{\mathrm{ord,tor}}}$ implies that $(\vec{\oint}_{\mathcal{H},\Sigma}^{\mathrm{ord}})^*\mathcal{O}(1) \cong \omega_{\vec{\mathsf{M}}_{\mathcal{H},\Sigma^{\mathrm{ord}}}^{\mathrm{ord,tor}}}^{\otimes N_0}$ if and only if $(\vec{\oint}_{\mathcal{H},\Sigma^{\mathrm{ord},\prime}}^{\mathrm{ord}})^*\mathcal{O}(1) \cong \omega_{\vec{\mathsf{M}}_{\mathcal{H},\Sigma^{\mathrm{ord},\prime}}^{\mathrm{ord,tor}}}^{\otimes N_0}$ (for the same $\mathcal{O}(1)$ and N_0). In other words, we can move freely between different choices of Σ by taking pullbacks or push-forwards, and there is a choice of $\mathcal{O}(1)$ with the smallest value of $N_0 \geq 1$ that works for all Σ.

From now on, let us fix a choice of Σ^{ord} and suppress it from the notation. We would like to show that $\omega_{\vec{\mathsf{M}}_{\mathcal{H}}^{\mathrm{ord}}}$ extends to an ample invertible sheaf over $\vec{\mathsf{M}}_{\mathcal{H}}^{\mathrm{ord,min}}$ when $\mathcal{H}^p$ is *neat*.

By Proposition 6.1.2.18, the pullback of $\vec{\oint}_{\mathcal{H}}^{\mathrm{ord}} : \vec{\mathsf{M}}_{\mathcal{H}}^{\mathrm{ord,tor}} \to \vec{\mathsf{M}}_{\mathcal{H}}^{\mathrm{ord,min}}$ to $\vec{\mathsf{M}}_{\mathcal{H}}^{\mathrm{ord},1}$ is an *isomorphism* because the canonical morphism $\vec{\mathsf{M}}_{\mathcal{H}}^{\mathrm{ord,tor}} \to [\vec{\mathsf{M}}_{\mathcal{H}}^{\mathrm{ord,tor}}]$ is an isomorphism when $\mathcal{H}^p$ is neat. Therefore, we can view $\vec{\mathsf{M}}_{\mathcal{H}}^{\mathrm{ord},1}$ as an open subspace of $\vec{\mathsf{M}}_{\mathcal{H}}^{\mathrm{ord,tor}}$ and consider the restriction $\omega_{\vec{\mathsf{M}}_{\mathcal{H}}^{\mathrm{ord,tor}}}|_{\mathsf{M}_{\mathcal{H}}^1}$, where $\omega_{\vec{\mathsf{M}}_{\mathcal{H}}^{\mathrm{ord,tor}}}$ is as in statement (3). Since the complement of $\vec{\mathsf{M}}_{\mathcal{H}}^{\mathrm{ord},1}$ in $\vec{\mathsf{M}}_{\mathcal{H}}^{\mathrm{ord,min}}$ has codimension at least two (by definition of $\vec{\mathsf{M}}_{\mathcal{H}}^{\mathrm{ord},1}$) and since $\vec{\mathsf{M}}_{\mathcal{H}}^{\mathrm{ord,min}}$ is noetherian and normal, it suffices to show that the coherent sheaf (see [32, VIII, Prop. 3.2])

$$\omega_{\vec{\mathsf{M}}_{\mathcal{H}}^{\mathrm{ord,min}}} := (\vec{\mathsf{M}}_{\mathcal{H}}^{\mathrm{ord},1} \hookrightarrow \vec{\mathsf{M}}_{\mathcal{H}}^{\mathrm{ord,min}})_*(\omega_{\vec{\mathsf{M}}_{\mathcal{H}}^{\mathrm{ord,tor}}}|_{\vec{\mathsf{M}}_{\mathcal{H}}^{\mathrm{ord},1}})$$

is an invertible sheaf. By fpqc descent (see [33, VIII, 1.11]), it suffices to verify this statement over the completions of strict localizations of $\vec{\mathsf{M}}_{\mathcal{H}}^{\mathrm{ord,min}}$.

Let $\bar{x}$ be a geometric point over some $[(\Phi_{\mathcal{H}}, \delta_{\mathcal{H}})]$-stratum $\vec{\mathsf{Z}}_{[(\Phi_{\mathcal{H}},\delta_{\mathcal{H}})]}^{\mathrm{ord}}$ in $\vec{\mathsf{M}}_{\mathcal{H}}^{\mathrm{ord,min}}$, and consider any $[(\Phi_{\mathcal{H}}, \delta_{\mathcal{H}}, \sigma)]$-stratum $\vec{\mathsf{Z}}_{[(\Phi_{\mathcal{H}},\delta_{\mathcal{H}},\sigma)]}^{\mathrm{ord}}$ in $\vec{\mathsf{M}}_{\mathcal{H}}^{\mathrm{ord,tor}}$ that maps surjectively to $\vec{\mathsf{Z}}_{[(\Phi_{\mathcal{H}},\delta_{\mathcal{H}})]}^{\mathrm{ord}}$. Let $(\Phi_{\mathcal{H}}, \delta_{\mathcal{H}}, \sigma)$ be any representative of $[(\Phi_{\mathcal{H}}, \delta_{\mathcal{H}}, \sigma)]$. Since $\mathcal{H}^p$ is neat, $\mathcal{H} = \mathcal{H}^p\mathcal{H}_p$ is also neat, and our choice of Σ^{ord} (see Definitions 1.2.2.13 and 5.1.3.1) forces $\Gamma_{\Phi_{\mathcal{H}},\sigma}$ to act trivially on $\vec{\mathfrak{X}}_{\Phi_{\mathcal{H}},\delta_{\mathcal{H}},\sigma}^{\mathrm{ord}}$ (by [62, Lem. 6.2.5.27]). Therefore, we have $(\vec{\mathsf{M}}_{\mathcal{H}}^{\mathrm{ord,tor}})^{\wedge}_{\vec{\mathsf{Z}}_{[(\Phi_{\mathcal{H}},\delta_{\mathcal{H}},\sigma)]}^{\mathrm{ord}}} \cong \vec{\mathfrak{X}}_{\Phi_{\mathcal{H}},\delta_{\mathcal{H}},\sigma}^{\mathrm{ord}}$ (by (5) of Theorem 5.2.1.1). Let $(\vec{\mathsf{M}}_{\mathcal{H}}^{\mathrm{ord,min}})^{\wedge}_{\vec{\mathsf{Z}}_{[(\Phi_{\mathcal{H}},\delta_{\mathcal{H}})]}^{\mathrm{ord}}}$ denote the formal completion of $\vec{\mathsf{M}}_{\mathcal{H}}^{\mathrm{ord,min}}$ along the $[(\Phi_{\mathcal{H}}, \delta_{\mathcal{H}})]$-stratum $\vec{\mathsf{Z}}_{[(\Phi_{\mathcal{H}},\delta_{\mathcal{H}})]}^{\mathrm{ord}}$. Then we have a composition of canonical morphisms $\vec{\mathfrak{X}}^{\mathrm{ord}}_{\Phi_{\mathcal{H}},\delta_{\mathcal{H}},\sigma} \cong (\vec{\mathsf{M}}^{\mathrm{ord,tor}}_{\mathcal{H}})^{\wedge}_{\vec{\mathsf{Z}}^{\mathrm{ord}}_{[(\Phi_{\mathcal{H}},\delta_{\mathcal{H}},\sigma)]}} \to (\vec{\mathsf{M}}^{\mathrm{ord,min}}_{\mathcal{H}})^{\wedge}_{\vec{\mathsf{Z}}^{\mathrm{ord}}_{[(\Phi_{\mathcal{H}},\delta_{\mathcal{H}})]}}$. By abuse of notation, let us denote the pullback of this composition from $(\vec{\mathsf{M}}^{\mathrm{ord,min}}_{\mathcal{H}})^{\wedge}_{\vec{\mathsf{Z}}^{\mathrm{ord}}_{[(\Phi_{\mathcal{H}},\delta_{\mathcal{H}})]}}$ to the completion $(\vec{\mathsf{M}}^{\mathrm{ord,min}}_{\mathcal{H}})^{\wedge}_{\bar{x}}$ of the strict localization of $\vec{\mathsf{M}}^{\mathrm{ord,min}}_{\mathcal{H}}$ at $\bar{x}$ by $(\vec{\mathfrak{X}}^{\mathrm{ord}}_{\Phi_{\mathcal{H}},\delta_{\mathcal{H}},\sigma})^{\wedge}_{\bar{x}} \cong (\vec{\mathsf{M}}^{\mathrm{ord,tor}}_{\mathcal{H}})^{\wedge}_{\bar{x}} \to (\vec{\mathsf{M}}^{\mathrm{ord,min}}_{\mathcal{H}})^{\wedge}_{\bar{x}}$. According to Proposition 6.1.2.19, there is a *structural morphism* $(\vec{\mathsf{M}}^{\mathrm{ord,min}}_{\mathcal{H}})^{\wedge}_{\bar{x}} \to (\vec{\mathsf{M}}^{\mathrm{ord},\mathsf{Z}_{\mathcal{H}}}_{\mathcal{H}})^{\wedge}_{\bar{x}}$ such that the further composition $(\vec{\mathfrak{X}}^{\mathrm{ord}}_{\Phi_{\mathcal{H}},\delta_{\mathcal{H}},\sigma})^{\wedge}_{\bar{x}} \cong (\vec{\mathsf{M}}^{\mathrm{ord,tor}}_{\mathcal{H}})^{\wedge}_{\bar{x}} \to (\vec{\mathsf{M}}^{\mathrm{ord,min}}_{\mathcal{H}})^{\wedge}_{\bar{x}} \to (\vec{\mathsf{M}}^{\mathrm{ord},\mathsf{Z}_{\mathcal{H}}}_{\mathcal{H}})^{\wedge}_{\bar{x}}$ agrees with the morphism $(\vec{\mathfrak{X}}^{\mathrm{ord}}_{\Phi_{\mathcal{H}},\delta_{\mathcal{H}},\sigma})^{\wedge}_{\bar{x}} \to (\vec{\mathsf{M}}^{\mathrm{ord},\mathsf{Z}_{\mathcal{H}}}_{\mathcal{H}})^{\wedge}_{\bar{x}}$ induced by the structural morphism $\vec{\mathfrak{X}}^{\mathrm{ord}}_{\Phi_{\mathcal{H}},\delta_{\mathcal{H}},\sigma} \to \vec{\mathsf{M}}^{\mathrm{ord},\mathsf{Z}_{\mathcal{H}}}_{\mathcal{H}}$ of $\vec{\mathfrak{X}}^{\mathrm{ord}}_{\Phi_{\mathcal{H}},\delta_{\mathcal{H}},\sigma}$. Over $(\vec{\mathfrak{X}}^{\mathrm{ord}}_{\Phi_{\mathcal{H}},\delta_{\mathcal{H}},\sigma})^{\wedge}_{\bar{x}}$, the pullback ${}^{\heartsuit}\omega$ of $\omega_{\vec{\mathsf{M}}^{\mathrm{ord,tor}}_{\mathcal{H}}}$ from $\vec{\mathsf{M}}^{\mathrm{ord,tor}}_{\mathcal{H}}$ is isomorphic to $(\wedge^{\mathrm{top}}_{\mathbb{Z}} X) \underset{\mathbb{Z}}{\otimes} \omega_{\vec{B}}$ by Lemma 6.1.2.3, which does descend to $(\vec{\mathsf{M}}^{\mathrm{ord,min}}_{\mathcal{H}})^{\wedge}_{\bar{x}}$, because the pullback of $\omega_{\vec{B}}$ from $\vec{\mathsf{M}}^{\mathrm{ord},\mathsf{Z}_{\mathcal{H}}}_{\mathcal{H}}$ also makes sense there. Since the complement of $\vec{\mathsf{M}}^{\mathrm{ord},1}_{\mathcal{H}}$ in the normal scheme $\vec{\mathsf{M}}^{\mathrm{ord,min}}_{\mathcal{H}}$ has codimension at least two, the pullback of $\omega_{\vec{\mathsf{M}}^{\mathrm{ord,min}}_{\mathcal{H}}}$ (from $\vec{\mathsf{M}}^{\mathrm{ord,min}}_{\mathcal{H}}$ to $(\vec{\mathsf{M}}^{\mathrm{ord,min}}_{\mathcal{H}})^{\wedge}_{\bar{x}}$) has to agree with the pullback of $(\wedge^{\mathrm{top}}_{\mathbb{Z}} X) \underset{\mathbb{Z}}{\otimes} \omega_{\vec{B}}$ to $(\vec{\mathsf{M}}^{\mathrm{ord,min}}_{\mathcal{H}})^{\wedge}_{\bar{x}}$. In particular, it is invertible, as desired.

Since $\vec{\oint}^{\mathrm{ord}}_{\mathcal{H}} : \vec{\mathsf{M}}^{\mathrm{ord,tor}}_{\mathcal{H}} \to \vec{\mathsf{M}}^{\mathrm{ord,min}}_{\mathcal{H}}$ satisfies $\mathscr{O}_{\vec{\mathsf{M}}^{\mathrm{ord,min}}_{\mathcal{H}}} \overset{\sim}{\to} \vec{\oint}^{\mathrm{ord}}_{\mathcal{H},*}\mathscr{O}_{\vec{\mathsf{M}}^{\mathrm{ord,tor}}_{\mathcal{H}}}$ by Theorem 6.1.1.12, we see that two locally free sheaves $\mathcal{E}$ and $\mathcal{F}$ of finite rank over $\vec{\mathsf{M}}^{\mathrm{ord,min}}_{\mathcal{H}}$ are isomorphic if and only if $(\vec{\oint}^{\mathrm{ord}}_{\mathcal{H}})^{*}\mathcal{E} \cong (\vec{\oint}^{\mathrm{ord}}_{\mathcal{H}})^{*}\mathcal{F}$. Indeed, for the nontrivial implication we just need $\mathcal{E} \cong \vec{\oint}^{\mathrm{ord}}_{\mathcal{H},*}(\vec{\oint}^{\mathrm{ord}}_{\mathcal{H}})^{*}\mathcal{E} \cong \vec{\oint}^{\mathrm{ord}}_{\mathcal{H},*}(\vec{\oint}^{\mathrm{ord}}_{\mathcal{H}})^{*}\mathcal{F} \cong \mathcal{F}$ (by the projection formula (see [35, 0_{I}, 5.4.10.1])). Since $(\vec{\oint}^{\mathrm{ord}}_{\mathcal{H}})^{*}\omega_{\vec{\mathsf{M}}^{\mathrm{ord,min}}_{\mathcal{H}}} \cong \omega_{\vec{\mathsf{M}}^{\mathrm{ord,tor}}_{\mathcal{H}}}$, we have $\vec{\oint}^{\mathrm{ord}}_{\mathcal{H},*}\omega_{\vec{\mathsf{M}}^{\mathrm{ord,tor}}_{\mathcal{H}}} \cong \omega_{\vec{\mathsf{M}}^{\mathrm{ord,min}}_{\mathcal{H}}}$, and the $\mathcal{O}(1)$ above such that $(\vec{\oint}^{\mathrm{ord}}_{\mathcal{H}})^{*}\mathcal{O}(1) \cong \omega^{\otimes N_0}_{\vec{\mathsf{M}}^{\mathrm{ord,tor}}_{\mathcal{H}}}$ has to satisfy $\mathcal{O}(1) \cong \omega^{\otimes N_0}_{\vec{\mathsf{M}}^{\mathrm{ord,min}}_{\mathcal{H}}}$. This shows that $\omega_{\vec{\mathsf{M}}^{\mathrm{ord,min}}_{\mathcal{H}}}$ is ample and finishes the verification of statements (2) and (3).

Finally, statement (6) is a consequence of Proposition 2.2.1.2, (7) of Theorem 5.2.1.1, Theorem 6.1.1.12, Lemma 6.1.2.12, Proposition 6.1.2.13, and Corollary 6.1.2.14. □

Remark 6.2.1.5. When p is a good prime, $\vec{\mathsf{M}}^{\mathrm{ord,min}}_{\mathcal{H}}$ can be constructed as in Section 5.2.3, without any logical dependence on the construction by normalization in Section 2.2.

Proposition 6.2.1.6 (base change properties). (*Compare with [62, Prop. 7.2.4.3].*) *We can repeat the construction of* $\vec{\mathsf{M}}^{\mathrm{ord,min}}_{\mathcal{H}}$ *with* $\vec{\mathsf{S}}_{0,r_{\mathcal{H}}} = \mathrm{Spec}(\mathcal{O}_{F_0,(p)}[\zeta_{p^{r_{\mathcal{H}}}}])$ *replaced with each (quasi-separated) locally noetherian* **normal** *scheme* S *over* $\vec{\mathsf{S}}_{0,r_{\mathcal{H}}}$, *and obtain a normal scheme* $\vec{\mathsf{M}}^{\mathrm{ord,min}}_{\mathcal{H},S}$ *quasi-projective and flat over* S, *with analogous characterizing properties described as in Theorem* 6.2.1.1 *(with* $\mathrm{Proj}(\,\cdot\,)$ *replaced with* $\underline{\mathrm{Proj}}_S(\,\cdot\,)$, *and with* $\Gamma(\,\cdot\,)$ *replaced with direct images* *over S), together with a canonical* **finite** *morphism*

$$\vec{\mathsf{M}}^{\mathrm{ord,min}}_{\mathcal{H},S} \to \vec{\mathsf{M}}^{\mathrm{ord,min}}_{\mathcal{H}} \underset{\vec{\mathsf{S}}_{0,r_{\mathcal{H}}}}{\times} S. \tag{6.2.1.7}$$

If $S' \to S$ is a morphism between locally noetherian normal schemes, then we also have a canonical finite morphism

$$\vec{\mathsf{M}}^{\mathrm{ord,min}}_{\mathcal{H},S'} \to \vec{\mathsf{M}}^{\mathrm{ord,min}}_{\mathcal{H},S} \underset{S}{\times} S'. \tag{6.2.1.8}$$

Moreover, these finite morphisms satisfy the following properties:

1. *If $S \to \vec{\mathsf{S}}_{0,r_{\mathcal{H}}}$ (resp. $S' \to S$) is flat, then* (6.2.1.7) (*resp.* (6.2.1.8)) *is an isomorphism.*
2. *If $\vec{\mathsf{M}}^{\mathrm{ord,min}}_{\mathcal{H}} \underset{\vec{\mathsf{S}}_{0,r_{\mathcal{H}}}}{\times} S$ (resp. $\vec{\mathsf{M}}^{\mathrm{ord,min}}_{\mathcal{H}} \underset{\vec{\mathsf{S}}_{0,r_{\mathcal{H}}}}{\times} S'$) is noetherian and normal, then* (6.2.1.7) (*resp.* (6.2.1.8)) *is an isomorphism* (*by Zariski's main theorem; see [35, III-1, 4.4.3, 4.4.11]*).
3. *Suppose $\bar{s}$ is a geometric point of S. Then* (6.2.1.8) (*with S' replaced with $\bar{s}$) is an isomorphism if the following condition is satisfied:*

   $$\forall \text{ geometric points } \bar{x} \text{ of } \vec{\mathsf{M}}^{\mathrm{ord,min}}_{\mathcal{H}} \underset{\vec{\mathsf{S}}_{0,r_{\mathcal{H}}}}{\times} \bar{s},\ \mathrm{char}(\bar{s}) \nmid \#\,\mathrm{Aut}(\bar{x}). \tag{6.2.1.9}$$

   (*As in Proposition* 6.1.2.19, $\mathrm{Aut}(\bar{x})$ *is the group of automorphisms of $\bar{x} \to \vec{\mathsf{M}}^{\mathrm{ord},\mathsf{Z}_{\mathcal{H}}}_{\mathcal{H}} \underset{\vec{\mathsf{S}}_{0,r_{\mathcal{H}}}}{\times} \bar{s}$, or equivalently that of $\bar{x} \to \vec{\mathsf{M}}^{\mathrm{ord},\mathsf{Z}_{\mathcal{H}}}_{\mathcal{H}} \underset{\vec{\mathsf{S}}_{0,r_{\mathcal{H}}}}{\times} S$ or $\bar{x} \to \vec{\mathsf{M}}^{\mathrm{ord},\mathsf{Z}_{\mathcal{H}}}_{\mathcal{H}}$, if $\bar{x}$ is over the $[(\Phi_{\mathcal{H}}, \delta_{\mathcal{H}})]$-stratum $\vec{\mathsf{Z}}^{\mathrm{ord}}_{[(\Phi_{\mathcal{H}},\delta_{\mathcal{H}})]}$ of $\vec{\mathsf{M}}^{\mathrm{ord,min}}_{\mathcal{H}}$.) In this case, the geometric fiber $\vec{\mathsf{M}}^{\mathrm{ord,min}}_{\mathcal{H},S} \underset{S}{\times} \bar{s}$ is normal because $\vec{\mathsf{M}}^{\mathrm{ord,min}}_{\mathcal{H},\bar{s}}$ is.*
4. *Suppose* (6.2.1.9) *is satisfied by all geometric points $\bar{s}$ of S.* (*This is the case, for example, if $\mathcal{H}^p$ is neat. In general, there is a nonzero constant c depending only on the linear algebraic data defining $\vec{\mathsf{M}}^{\mathrm{ord}}_{\mathcal{H}}$ such that $\#\,\mathrm{Aut}(\bar{x})|c$ for all geometric points $\bar{x}$ of $\vec{\mathsf{M}}^{\mathrm{ord,min}}_{\mathcal{H}}$*). *Then the scheme $\vec{\mathsf{M}}^{\mathrm{ord,min}}_{\mathcal{H}} \underset{\vec{\mathsf{S}}_{0,r_{\mathcal{H}}}}{\times} S$ is normal,* (6.2.1.7) *is an isomorphism* (*by property* (2) *above*), *and the morphism $\vec{\mathsf{M}}^{\mathrm{ord,min}}_{\mathcal{H},S} \to S$ is* **normal** (*i.e., flat with geometrically normal fibers; see [35, IV-2, 6.8.1 and 6.7.8]*). *Moreover, for every locally noetherian normal scheme S' over S, the scheme $\vec{\mathsf{M}}^{\mathrm{ord,min}}_{\mathcal{H},S} \underset{S}{\times} S'$ is normal,* (6.2.1.8) *is an isomorphism* (*again, by property* (2) *above*), *and the morphism $\vec{\mathsf{M}}^{\mathrm{ord,min}}_{\mathcal{H},S'} \to S'$ is normal.*

Proof. We may assume that S and S' are affine, noetherian normal, and connected, because property (1) (and the convention that all schemes are quasi-separated) allows us to patch the construction of $\vec{\mathsf{M}}^{\mathrm{ord,min}}_{\mathcal{H},S}$ along intersections of affine open subschemes of S.

Let us take any $\vec{\mathsf{M}}^{\mathrm{ord,tor}}_{\mathcal{H}} = \vec{\mathsf{M}}^{\mathrm{ord,tor}}_{\mathcal{H},\Sigma^{\mathrm{ord}}}$ as in the construction of $\vec{\mathsf{M}}^{\mathrm{ord,min}}_{\mathcal{H}}$, so that we have the canonical surjection

$$\vec{\oint}^{\mathrm{ord}}_{\mathcal{H}} : \vec{\mathsf{M}}^{\mathrm{ord,tor}}_{\mathcal{H}} \twoheadrightarrow \vec{\mathsf{M}}^{\mathrm{ord,min}}_{\mathcal{H}} = \mathrm{Proj}\Big(\underset{k\geq 0}{\oplus} \Gamma(\vec{\mathsf{M}}^{\mathrm{ord,tor}}_{\mathcal{H}}, \omega^{\otimes k}_{\vec{\mathsf{M}}^{\mathrm{ord,tor}}_{\mathcal{H}}})\Big)$$

which is proper and is the Stein factorization of itself, by Theorem 6.1.1.12. If we repeat the construction of $\vec{\mathsf{M}}_{\mathcal{H}}^{\mathrm{ord,min}}$ over S, then we obtain a canonical morphism

$$\vec{\oint}^{\mathrm{ord}}_{\mathcal{H},S} : \vec{\mathsf{M}}_{\mathcal{H}}^{\mathrm{ord,tor}} \underset{\vec{\mathsf{S}}_{0,r_{\mathcal{H}}}}{\times} S \twoheadrightarrow \vec{\mathsf{M}}_{\mathcal{H},S}^{\mathrm{ord,min}} := \mathrm{Proj}\Big(\underset{k\geq 0}{\oplus} \Gamma(\vec{\mathsf{M}}_{\mathcal{H}}^{\mathrm{ord,tor}} \underset{\vec{\mathsf{S}}_{0,r_{\mathcal{H}}}}{\times} S, (\omega_{\vec{\mathsf{M}}_{\mathcal{H}}^{\mathrm{ord,tor}}} \underset{\mathscr{O}_{\vec{\mathsf{S}}_{0,r_{\mathcal{H}}}}}{\otimes} \mathscr{O}_S)^{\otimes k})\Big).$$

Since the base change morphism $\vec{\oint}^{\mathrm{ord}}_{\mathcal{H}} \underset{\vec{\mathsf{S}}_{0,r_{\mathcal{H}}}}{\times} S : \vec{\mathsf{M}}_{\mathcal{H}}^{\mathrm{ord,tor}} \underset{\vec{\mathsf{S}}_{0,r_{\mathcal{H}}}}{\times} S \to \vec{\mathsf{M}}_{\mathcal{H}}^{\mathrm{ord,min}} \underset{\vec{\mathsf{S}}_{0,r_{\mathcal{H}}}}{\times} S$ is proper, and since $\omega^{\otimes N_0}_{\vec{\mathsf{M}}_{\mathcal{H}}^{\mathrm{ord,tor}}} \underset{\mathscr{O}_{\vec{\mathsf{S}}_{0,r_{\mathcal{H}}}}}{\otimes} \mathscr{O}_S$ is the pullback of the ample invertible sheaf $\omega^{\otimes N_0}_{\vec{\mathsf{M}}_{\mathcal{H}}^{\mathrm{ord,min}}} \underset{\mathscr{O}_{\vec{\mathsf{S}}_{0,r_{\mathcal{H}}}}}{\otimes} \mathscr{O}_S$ for the same integer $N_0 \geq 1$ as in (2) of Theorem 6.2.1.1, the Stein factorization of $\vec{\oint}^{\mathrm{ord}}_{\mathcal{H}} \underset{\vec{\mathsf{S}}_{0,r_{\mathcal{H}}}}{\times} S$ can be identified with the composition of $\vec{\oint}^{\mathrm{ord}}_{\mathcal{H},S}$ with a canonical finite morphism $\vec{\mathsf{M}}_{\mathcal{H},S}^{\mathrm{ord,min}} \to \vec{\mathsf{M}}_{\mathcal{H}}^{\mathrm{ord,min}} \underset{\vec{\mathsf{S}}_{0,r_{\mathcal{H}}}}{\times} S$, which is the desired morphism (6.2.1.7). Consequently, $\vec{\mathsf{M}}_{\mathcal{H},S}^{\mathrm{ord,min}}$ is a normal scheme quasi-projective over S, and $\vec{\oint}^{\mathrm{ord}}_{\mathcal{H},S}$ is also proper and is the Stein factorization of itself.

The morphism $S \to \vec{\mathsf{S}}_{0,r_{\mathcal{H}}}$ either is flat or factors through a closed point s of $\vec{\mathsf{S}}_{0,r_{\mathcal{H}}}$. In the former case, the morphism $\vec{\mathsf{M}}_{\mathcal{H},S}^{\mathrm{ord,min}} \to S$ is the pullback of $\vec{\mathsf{M}}_{\mathcal{H}}^{\mathrm{ord,min}} \to \vec{\mathsf{S}}_{0,r_{\mathcal{H}}}$, which is flat by Theorem 6.2.1.1. In the latter case, the morphism $\vec{\mathsf{M}}_{\mathcal{H},S}^{\mathrm{ord,min}} \to S$ is the pullback of $\vec{\mathsf{M}}_{\mathcal{H},s}^{\mathrm{ord,min}} \to s$, which is automatically flat. Thus, $\vec{\mathsf{M}}_{\mathcal{H},S}^{\mathrm{ord,min}} \to S$ is always flat.

The case of (6.2.1.8) is similar, with $\vec{\mathsf{S}}_{0,r_{\mathcal{H}}}$ (resp. S) replaced with S (resp. S').

Now, property (1) has already been explained. Property (2) is self-explanatory, because $\vec{\mathsf{M}}_{\mathcal{H},S}^{\mathrm{ord,min}}$ and $\vec{\mathsf{M}}_{\mathcal{H},S'}^{\mathrm{ord,min}}$ are noetherian normal by construction.

Let us prove property (3). Suppose that the condition (6.2.1.9) is satisfied. Since (6.2.1.7) is an isomorphism if it is so over the completions of strict local rings at geometric points of the target, and since the formation of $\mathrm{Aut}(\bar{x})$-invariants commutes with the base change from S to $\bar{s}$ because $\mathrm{char}(\bar{s}) \nmid \# \mathrm{Aut}(\bar{x})$, by Proposition 6.1.2.19 (see, in particular, (6.1.2.20)), it suffices to show that, for each $\ell_0 \in \mathbf{P}^{\vee}_{\Phi_{\mathcal{H}}}$ with stabilizer $\Gamma_{\Phi_{\mathcal{H}},\ell_0}$ in $\Gamma_{\Phi_{\mathcal{H}}}$, the formation of $\Gamma_{\Phi_{\mathcal{H}},\ell_0}$-invariants in $(\underline{\mathrm{FJ}}^{\mathrm{ord},(\ell_0)}_{\Phi_{\mathcal{H}},\delta_{\mathcal{H}}})^{\wedge}_{\bar{x}} \cong \Gamma((\vec{C}^{\mathrm{ord}}_{\Phi_{\mathcal{H}},\delta_{\mathcal{H}}})^{\wedge}_{\bar{x}}, (\vec{\Psi}^{\mathrm{ord}}_{\Phi_{\mathcal{H}},\delta_{\mathcal{H}}}(\ell_0))^{\wedge}_{\bar{x}})$ also commutes with the base change from S to $\bar{s}$. By Proposition 4.2.1.29, there exists a finite index normal subgroup $\Gamma'_{\Phi_{\mathcal{H}}}$ of $\Gamma_{\Phi_{\mathcal{H}}}$ such that $\Gamma'_{\Phi_{\mathcal{H}}}$ acts trivially on $\vec{\mathsf{M}}_{\mathcal{H}}^{\mathrm{ord},\Phi_{\mathcal{H}}}$, and such that the induced action of $\Gamma_{\Phi_{\mathcal{H}}}/\Gamma'_{\Phi_{\mathcal{H}}}$ on $\vec{\mathsf{M}}_{\mathcal{H}}^{\mathrm{ord},\Phi_{\mathcal{H}}}$ makes $\vec{\mathsf{M}}_{\mathcal{H}}^{\mathrm{ord},\Phi_{\mathcal{H}}} \to \vec{\mathsf{M}}_{\mathcal{H}}^{\mathrm{ord},\mathsf{Z}_{\mathcal{H}}}$ an étale $(\Gamma_{\Phi_{\mathcal{H}}}/\Gamma'_{\Phi_{\mathcal{H}}})$-torsor. Hence, it suffices to show that, for each geometric point $\bar{y} \to \vec{\mathsf{M}}_{\mathcal{H}}^{\mathrm{ord},\Phi_{\mathcal{H}}}$ lifting $\bar{x} \to \vec{\mathsf{M}}_{\mathcal{H}}^{\mathrm{ord},\mathsf{Z}_{\mathcal{H}}}$, the formation of invariants of $\Gamma'_{\Phi_{\mathcal{H}},\ell_0} = \Gamma_{\Phi_{\mathcal{H}},\ell_0} \cap \Gamma'_{\Phi_{\mathcal{H}}}$ in $\Gamma((\vec{C}^{\mathrm{ord}}_{\Phi_{\mathcal{H}},\delta_{\mathcal{H}}})^{\wedge}_{\bar{y}}, (\vec{\Psi}^{\mathrm{ord}}_{\Phi_{\mathcal{H}},\delta_{\mathcal{H}}}(\ell_0))^{\wedge}_{\bar{y}})$, where $(\,\cdot\,)^{\wedge}_{\bar{y}}$ denote the pullback to the completion of the strict localization of $\vec{\mathsf{M}}_{\mathcal{H}}^{\mathrm{ord},\Phi_{\mathcal{H}}}$ at $\bar{y}$, commutes with the base change from S to $\bar{s}$, for each $\ell_0 \in \mathbf{P}^{\vee}_{\Phi_{\mathcal{H}}}$.

Let X' and Y' be admissible sub-$\mathcal{O}$-lattices of X and Y, respectively, such that $\phi(Y') \subset X'$, such that ℓ' lies in the subgroup $\mathbf{S}'_{\Phi_{\mathcal{H}}}$ of $\mathbf{S}_{\Phi_{\mathcal{H}}}$ defined by the same construction of $\mathbf{S}_{\Phi_{\mathcal{H}}}$ using the embedding $\phi' : Y' \to X'$ induced by ϕ, and such that ℓ_0 is positive in $\Phi'_{\mathcal{H}}$ in the sense that, up to choosing a $\mathbb{Z}$-basis $y_1, \ldots, y_r$ of Y', and by completion of squares for quadratic forms, there exists some integer $N \geq 1$ such that $N \cdot \ell_0$ can be represented as a positive definite matrix of the form $ue\,{}^t u$, where e and u are matrices with integer coefficients, and where $e = \operatorname{diag}(e_1, \ldots, e_r)$ is diagonal with positive entries. In this case, $\Gamma'_{\Phi_{\mathcal{H}},\ell_0}$ acts on $\Phi'_{\mathcal{H}}$ via a discrete subgroup $\overline{\Gamma}'_{\Phi_{\mathcal{H}},\ell_0}$ of the compact orthogonal subgroup of $\mathrm{GL}_{\mathbb{R}}(Y' \underset{\mathbb{Z}}{\otimes} \mathbb{R})$ preserving the above-mentioned positive definite matrix by conjugation, which is necessarily finite. Consider the abelian scheme torsor $\vec{C}^{\mathrm{ord},\prime}_{\Phi_{\mathcal{H}},\delta_{\mathcal{H}}} \to \vec{\mathsf{M}}^{\mathrm{ord},\Phi_{\mathcal{H}}}_{\mathcal{H}}$ defined by the same construction of $\vec{C}^{\mathrm{ord}}_{\Phi_{\mathcal{H}},\delta_{\mathcal{H}}}$, using the embedding $\phi' : Y' \to X'$ instead of $\phi : Y \to X$, with a canonical morphism $\vec{C}^{\mathrm{ord}}_{\Phi_{\mathcal{H}},\delta_{\mathcal{H}}} \to \vec{C}^{\mathrm{ord},\prime}_{\Phi_{\mathcal{H}},\delta_{\mathcal{H}}}$ over $\vec{\mathsf{M}}^{\mathrm{ord},\Phi_{\mathcal{H}}}_{\mathcal{H}}$ which is also an abelian scheme torsor, under which the $\vec{\Psi}^{\mathrm{ord}}_{\Phi_{\mathcal{H}},\delta_{\mathcal{H}}}(\ell_0)$ descends to an invertible sheaf $\vec{\Psi}^{\mathrm{ord},\prime}_{\Phi_{\mathcal{H}},\delta_{\mathcal{H}}}(\ell_0)$, which is relatively ample over $\vec{\mathsf{M}}^{\mathrm{ord},\Phi_{\mathcal{H}}}_{\mathcal{H}}$ because some positive tensor power of $\vec{\Psi}^{\mathrm{ord},\prime}_{\Phi_{\mathcal{H}},\delta_{\mathcal{H}}}(\ell_0)$ is isomorphic to the pullback of the line bundle $\underset{1\leq i\leq r}{\otimes} (\mathrm{pr}_i^*(\mathrm{Id}_B, \lambda_B)^*\mathcal{P}_B)^{\otimes e_i}$ over B under the finite morphism given by the composition $\vec{C}^{\mathrm{ord},\prime}_{\Phi_{\mathcal{H}},\delta_{\mathcal{H}}} \xrightarrow{\mathrm{can.}} \underline{\mathrm{Hom}}_{\mathbb{Z}}(Y, B) \xrightarrow{u^*} \underline{\mathrm{Hom}}_{\mathbb{Z}}(Y, B)$ over $\vec{\mathsf{M}}^{\mathrm{ord},\Phi_{\mathcal{H}}}_{\mathcal{H}}$, because λ_B is a polarization (*cf.* [62, Def. 1.3.2.16]), and because all the e_i's are positive. Then $\Gamma'_{\Phi_{\mathcal{H}},\ell_0}$ acts via the finite quotient $\overline{\Gamma}'_{\Phi_{\mathcal{H}},\ell_0}$ introduced above on $\Gamma((\vec{C}^{\mathrm{ord}}_{\Phi_{\mathcal{H}},\delta_{\mathcal{H}}})^\wedge_{\bar{y}}, (\vec{\Psi}^{\mathrm{ord}}_{\Phi_{\mathcal{H}},\delta_{\mathcal{H}}}(\ell_0))^\wedge_{\bar{y}}) \cong \Gamma((\vec{C}^{\mathrm{ord},\prime}_{\Phi_{\mathcal{H}},\delta_{\mathcal{H}}})^\wedge_{\bar{y}}, (\vec{\Psi}^{\mathrm{ord},\prime}_{\Phi_{\mathcal{H}},\delta_{\mathcal{H}}}(\ell_0))^\wedge_{\bar{y}})$.

If $\mathcal{H}$ is neat, then $\overline{\Gamma}'_{\Phi_{\mathcal{H}},\ell_0}$ is also neat and must be trivial. More generally, since $\vec{\Psi}^{\mathrm{ord},\prime}_{\Phi_{\mathcal{H}},\delta_{\mathcal{H}}}(\ell_0)$ is relatively ample over $\vec{\mathsf{M}}^{\mathrm{ord},\Phi_{\mathcal{H}}}_{\mathcal{H}}$, the action of $\overline{\Gamma}'_{\Phi_{\mathcal{H}},\ell_0}$ on $\underset{N\geq 0}{\oplus} \Gamma((\vec{C}^{\mathrm{ord},\prime}_{\Phi_{\mathcal{H}},\delta_{\mathcal{H}}})^\wedge_{\bar{y}}, (\vec{\Psi}^{\mathrm{ord},\prime}_{\Phi_{\mathcal{H}},\delta_{\mathcal{H}}}(N \cdot \ell_0))^\wedge_{\bar{y}})$ induces a faithful action of $\overline{\Gamma}'_{\Phi_{\mathcal{H}},\ell_0}$ on $(\vec{C}^{\mathrm{ord},\prime}_{\Phi_{\mathcal{H}},\delta_{\mathcal{H}}})^\wedge_{\bar{y}}$ (*cf.* [81, Sec. 21, Thm. 5]). By construction, $(\vec{C}^{\mathrm{ord},\prime}_{\Phi_{\mathcal{H}},\delta_{\mathcal{H}}})^\wedge_{\bar{y}}$ appears (up to some identification) in the partial toroidal boundary construction of $\vec{\mathsf{M}}^{\mathrm{ord},\mathsf{Z}''_{\mathcal{H}}}_{\mathcal{H}}$, where $\vec{\mathsf{M}}^{\mathrm{ord},\mathsf{Z}''_{\mathcal{H}}}_{\mathcal{H}}$ is isomorphic to the stratum of $\vec{\mathsf{M}}^{\mathrm{ord},\min}_{\mathcal{H}}$ labeled by the cusp label $[(\mathsf{Z}''_{\mathcal{H}}, \Phi''_{\mathcal{H}}, \delta''_{\mathcal{H}})]$ induced by $[(\mathsf{Z}_{\mathcal{H}}, \Phi_{\mathcal{H}}, \delta_{\mathcal{H}})]$ by the admissible surjections $X \to X'' := X/X'$ and $Y \to Y'' := Y/Y'$ (see [62, Lem. 5.3.1.14 and 5.4.2.11]). Therefore, there exists a degeneration (over a complete discrete valuation ring with fraction field $k(z)$) of an object parameterized by some functorial point $z \to \vec{\mathsf{M}}^{\mathrm{ord},\mathsf{Z}''_{\mathcal{H}}}_{\mathcal{H}}$ such that $\overline{\Gamma}'_{\Phi_{\mathcal{H}},\ell_0}$ is a subquotient of $\mathrm{Aut}(\bar{z})$ for any geometric point $\bar{z} \to \vec{\mathsf{M}}^{\mathrm{ord},\mathsf{Z}''_{\mathcal{H}}}_{\mathcal{H}}$ above z. Since $\mathrm{char}(\bar{s}) \nmid \#\mathrm{Aut}(\bar{z})$ by the assumption that the condition (6.2.1.9) is satisfied, it follows that $\mathrm{char}(\bar{s}) \nmid \#\overline{\Gamma}'_{\Phi_{\mathcal{H}},\ell_0}$. Therefore, the formation of $\overline{\Gamma}'_{\Phi_{\mathcal{H}},\ell_0}$-invariants in $\Gamma((\vec{C}^{\mathrm{ord},\prime}_{\Phi_{\mathcal{H}},\delta_{\mathcal{H}}})^\wedge_{\bar{y}}, (\vec{\Psi}^{\mathrm{ord},\prime}_{\Phi_{\mathcal{H}},\delta_{\mathcal{H}}}(\ell_0))^\wedge_{\bar{y}})$ commutes with the base change from S to $\bar{s}$, for each $\ell_0 \in \mathbf{P}^\vee_{\Phi_{\mathcal{H}}}$, and property (3) follows.

It remains to prove property (4). Note that the assertions involving S' (in the last sentence) follow from the assertions involving only S, by [35, IV-2, 6.8.2 and 6.14.1] and property (2). To prove the assertions involving only S, we may replace S with its localizations, and assume that it is local. Let S_1 be the localization of $\vec{\mathsf{S}}_{0,r_{\mathcal{H}}}$ at the image under $S \to \vec{\mathsf{S}}_{0,r_{\mathcal{H}}}$ of the closed point of S. Since S_1 is a localization of $\vec{\mathsf{S}}_{0,r_{\mathcal{H}}}$, we know by property (1) that the canonical morphism $\vec{\mathsf{M}}^{\mathrm{ord,min}}_{\mathcal{H},\mathsf{S}_1} \to \vec{\mathsf{M}}^{\mathrm{ord,min}}_{\mathcal{H}} \underset{\vec{\mathsf{S}}_{0,r_{\mathcal{H}}}}{\times} \mathsf{S}_1$ is an isomorphism, so that $\vec{\mathsf{M}}^{\mathrm{ord,min}}_{\mathcal{H},\mathsf{S}_1} \to \mathsf{S}_1$ is the pullback of $\vec{\mathsf{M}}^{\mathrm{ord,min}}_{\mathcal{H}} \to \vec{\mathsf{S}}_{0,r_{\mathcal{H}}}$. Since the geometric points of S_1 are either of characteristic zero or dominated by those of S, by [35, IV-2, 6.7.7], the normality of fibers of $\vec{\mathsf{M}}^{\mathrm{ord,min}}_{\mathcal{H},\mathsf{S}_1} \to \mathsf{S}_1$ follows from the normality of the geometric fibers of $\vec{\mathsf{M}}^{\mathrm{ord,min}}_{\mathcal{H},\mathsf{S}_1} \underset{\mathsf{S}_1}{\times} S \cong \vec{\mathsf{M}}^{\mathrm{ord,min}}_{\mathcal{H}} \underset{\vec{\mathsf{S}}_{0,r_{\mathcal{H}}}}{\times} S \to S$, the latter of which follows from property (3) (and from the assumption that (6.2.1.9) is satisfied by all geometric points $\bar{s}$ of S). Thus, the (flat) morphism $\vec{\mathsf{M}}^{\mathrm{ord,min}}_{\mathcal{H},\mathsf{S}_1} \to \mathsf{S}_1$ is normal. By [35, IV-2, 6.8.2], the pullback $\vec{\mathsf{M}}^{\mathrm{ord,min}}_{\mathcal{H},\mathsf{S}_1} \underset{\mathsf{S}_1}{\times} S \to S$ is also a normal morphism. By [35, IV-2, 6.14.1], the scheme $\vec{\mathsf{M}}^{\mathrm{ord,min}}_{\mathcal{H},\mathsf{S}_1} \underset{\mathsf{S}_1}{\times} S \cong \vec{\mathsf{M}}^{\mathrm{ord,min}}_{\mathcal{H}} \underset{\vec{\mathsf{S}}_{0,r_{\mathcal{H}}}}{\times} S$ is normal. By property (2), this implies that (6.2.1.7) is an isomorphism, and hence that the morphism $\vec{\mathsf{M}}^{\mathrm{ord,min}}_{\mathcal{H},S} \to S$ is normal, as desired. □

6.2.2 *Hecke Actions*

Let us state the following analogue of Proposition 5.2.2.2 for partial minimal compactifications.

Proposition 6.2.2.1. (*Compare with [62, Prop. 7.2.5.1] and Propositions* 1.3.1.14, 2.2.3.1, 3.4.4.1 *and* 5.2.2.2.) *Suppose we have an element* $g = (g_0, g_p) \in \mathrm{G}(\mathbb{A}^{\infty,p}) \times \mathrm{P}^{\mathrm{ord}}_{\mathsf{D}}(\mathbb{Q}_p) \subset \mathrm{G}(\mathbb{A}^{\infty})$ *(see Definition* 3.2.2.7*), and suppose we have two open compact subgroups* $\mathcal{H}$ *and* $\mathcal{H}'$ *of* $\mathrm{G}(\hat{\mathbb{Z}})$ *such that* $\mathcal{H}' \subset g\mathcal{H}g^{-1}$*, and such that* $\mathcal{H}$ *and* $\mathcal{H}'$ *are of standard form as in Definition* 3.2.2.9*. Let* $r_{\mathcal{H}}$ *(resp.* $r_{\mathcal{H}'}$*) be defined by* $\mathcal{H}$ *(resp.* $\mathcal{H}'$*) as in Definition* 3.4.2.1*. Suppose moreover that* g_p *satisfies the conditions given in Section* 3.3.4*, so that* $\vec{[g]}^{\mathrm{ord}} : \vec{\mathsf{M}}^{\mathrm{ord}}_{\mathcal{H}'} \to \vec{\mathsf{M}}^{\mathrm{ord}}_{\mathcal{H}}$ *is defined (see Proposition* 3.4.4.1*). Then there is a canonical quasi-finite surjection*

$$\vec{[g]}^{\mathrm{ord,min}} : \vec{\mathsf{M}}^{\mathrm{ord,min}}_{\mathcal{H}'} \to \vec{\mathsf{M}}^{\mathrm{ord,min}}_{\mathcal{H}}$$

extending the canonical quasi-finite surjection $[\vec{[g]}^{\mathrm{ord}}] : [\vec{\mathsf{M}}^{\mathrm{ord}}_{\mathcal{H}'}] \to [\vec{\mathsf{M}}^{\mathrm{ord}}_{\mathcal{H}}]$ *induced by the canonical quasi-finite surjection* $\vec{[g]}^{\mathrm{ord}} : \vec{\mathsf{M}}^{\mathrm{ord}}_{\mathcal{H}'} \to \vec{\mathsf{M}}^{\mathrm{ord}}_{\mathcal{H}}$ *defined in Proposition* 3.4.4.1*. If the levels* $\mathcal{H}_p$ *and* $\mathcal{H}'_p$ *at* p *are* **equally deep** *as in Definition* 3.2.2.9*, or if* g_p *is of twisted* U_p *type as in Definition* 3.3.6.1 *and* $\mathrm{depth}_{\mathsf{D}}(\mathcal{H}'_p) - \mathrm{depth}_{\mathsf{D}}(g_p) = \mathrm{depth}_{\mathsf{D}}(\mathcal{H}_p) > 0$*, then the surjection* $\vec{[g]}^{\mathrm{ord,min}}$ *is finite.*

If $L \underset{\mathbb{Z}}{\otimes} \mathbb{Z}_p \subset g_p(L \underset{\mathbb{Z}}{\otimes} \mathbb{Z}_p)$*, then there is a canonical morphism*

$$(\vec{[g]}^{\mathrm{ord,min}})^* : (\vec{[g]}^{\mathrm{ord,min}})^* \omega^{\otimes k}_{\vec{\mathsf{M}}^{\mathrm{ord,min}}_{\mathcal{H}}} \to \omega^{\otimes k}_{\vec{\mathsf{M}}^{\mathrm{ord,min}}_{\mathcal{H}'}} \tag{6.2.2.2}$$

whenever $\omega^{\otimes k}_{\vec{\mathsf{M}}^{\mathrm{ord,min}}_{\mathcal{H}}}$ *is defined (which is compatible with the canonical isomorphism between the pullback of* $\omega^{\otimes k}_{\mathsf{M}_{\mathcal{H}}}$ *and* $\omega^{\otimes k}_{\mathsf{M}_{\mathcal{H}'}}$ *over* $\mathsf{M}^{\mathrm{ord}}_{\mathcal{H}'} \cong \mathsf{M}_{\mathcal{H}',r_{\mathcal{H}'}}$*). If* $g_p \in \mathrm{P}^{\mathrm{ord}}_{\mathsf{D}}(\mathbb{Z}_p)$*,* *then the following diagram*

$$\begin{array}{ccc} \vec{\mathsf{M}}^{\mathrm{ord,min}}_{\mathcal{H}'} & \hookrightarrow & \vec{\mathsf{M}}^{\mathrm{min}}_{\mathcal{H}',r_{\mathcal{H}'}} \\ {\scriptstyle \overrightarrow{[g]}^{\mathrm{ord,min}}}\big\downarrow & & \big\downarrow{\scriptstyle \overrightarrow{[g]}^{\mathrm{min}}_{r_{\mathcal{H}'},r_{\mathcal{H}}}} \\ \vec{\mathsf{M}}^{\mathrm{ord,min}}_{\mathcal{H}} & \hookrightarrow & \vec{\mathsf{M}}^{\mathrm{min}}_{\mathcal{H},r_{\mathcal{H}}} \end{array} \tag{6.2.2.3}$$

(see Definition 2.2.3.5*) is (commutative and) Cartesian, and* (6.2.2.2) *is an isomorphism compatible with the corresponding one in Proposition* 2.2.3.1.

Moreover, the surjection $\overrightarrow{[g]}^{\mathrm{ord,min}}$ *maps the* $[(\Phi'_{\mathcal{H}'},\delta'_{\mathcal{H}'})]$*-stratum* $\vec{\mathsf{Z}}^{\mathrm{ord}}_{[(\Phi'_{\mathcal{H}'},\delta'_{\mathcal{H}'})]}$ *of* $\vec{\mathsf{M}}^{\mathrm{ord,min}}_{\mathcal{H}'}$ *to the* $[(\Phi_{\mathcal{H}},\delta_{\mathcal{H}})]$*-stratum* $\vec{\mathsf{Z}}^{\mathrm{ord}}_{[(\Phi_{\mathcal{H}},\delta_{\mathcal{H}})]}$ *of* $\vec{\mathsf{M}}^{\mathrm{ord,min}}_{\mathcal{H}}$ *if and only if there are representatives* $(\Phi_{\mathcal{H}},\delta_{\mathcal{H}})$ *and* $(\Phi'_{\mathcal{H}'},\delta'_{\mathcal{H}'})$ *of* $[(\Phi_{\mathcal{H}},\delta_{\mathcal{H}})]$ *and* $[(\Phi'_{\mathcal{H}'},\delta'_{\mathcal{H}'})]$*, respectively, such that* $(\Phi_{\mathcal{H}},\delta_{\mathcal{H}})$ *is* g*-assigned to* $(\Phi'_{\mathcal{H}'},\delta'_{\mathcal{H}'})$ *as in [62, Def. 5.4.3.9].*

If $\Sigma^{\mathrm{ord}} = \{\Sigma_{\Phi_{\mathcal{H}}}\}_{[(\Phi_{\mathcal{H}},\delta_{\mathcal{H}})]}$ *and* $\Sigma^{\mathrm{ord},\prime} = \{\Sigma'_{\Phi'_{\mathcal{H}'}}\}_{[(\Phi'_{\mathcal{H}'},\delta'_{\mathcal{H}'})]}$ *are two compatible choices of admissible smooth rational polyhedral cone decomposition data for* $\vec{\mathsf{M}}^{\mathrm{ord}}_{\mathcal{H}}$ *and* $\vec{\mathsf{M}}^{\mathrm{ord}}_{\mathcal{H}'}$*, respectively, such that* $\Sigma^{\mathrm{ord},\prime}$ *is a* g*-refinement of* Σ^{ord} *as in Definition* 5.2.2.1*, then the canonical surjection* $\overrightarrow{[g]}^{\mathrm{ord,min}} : \vec{\mathsf{M}}^{\mathrm{ord,min}}_{\mathcal{H}'} \twoheadrightarrow \vec{\mathsf{M}}^{\mathrm{ord,min}}_{\mathcal{H}}$ *is compatible with the surjection* $\overrightarrow{[g]}^{\mathrm{ord,tor}} : \vec{\mathsf{M}}^{\mathrm{ord,tor}}_{\mathcal{H}',\Sigma^{\mathrm{ord},\prime}} \twoheadrightarrow \vec{\mathsf{M}}^{\mathrm{ord,tor}}_{\mathcal{H},\Sigma^{\mathrm{ord}}}$ *given by Proposition* 5.2.2.2.

If $g = g_1g_2$*, where* $g_1 = (g_{1,0},g_{1,p})$ *and* $g_2 = (g_{2,0},g_{2,p})$ *are elements of* $\mathrm{G}(\mathbb{A}^{\infty,p}) \times \mathrm{P}^{\mathrm{ord}}_{\mathsf{D}}(\mathbb{Q}_p)$*, each having a setup similar to that of* g*, then we have* $\overrightarrow{[g]}^{\mathrm{ord,min}} = \overrightarrow{[g_2]}^{\mathrm{ord,min}} \circ \overrightarrow{[g_1]}^{\mathrm{ord,min}}$*, extending the identity* $\overrightarrow{[g]}^{\mathrm{ord}} = \overrightarrow{[g_2]}^{\mathrm{ord}} \circ \overrightarrow{[g_1]}^{\mathrm{ord}}$ *in Proposition* 3.4.4.1.

Finally, the finite surjection $[g]^{\mathrm{min}} : \mathsf{M}^{\mathrm{min}}_{\mathcal{H}'} \to \mathsf{M}^{\mathrm{min}}_{\mathcal{H}}$ *in Proposition* 1.3.1.14 *canonically induces a finite surjection* $[g]^{\mathrm{min}}_{r_{\mathcal{H}'},r_{\mathcal{H}}} : \mathsf{M}^{\mathrm{min}}_{\mathcal{H}',r_{\mathcal{H}'}} \to \mathsf{M}^{\mathrm{min}}_{\mathcal{H},r_{\mathcal{H}}}$*. Then* $\overrightarrow{[g]}^{\mathrm{min}} \underset{\mathbb{Z}}{\otimes} \mathbb{Q}$ *can be identified with the pullback of* $[g]^{\mathrm{min}}_{r_{\mathcal{H}'},r_{\mathcal{H}}}$ *to* $\vec{\mathsf{M}}^{\mathrm{ord,min}}_{\mathcal{H}} \underset{\mathbb{Z}}{\otimes} \mathbb{Q}$ *(on the target) under* (6.2.1.2) *in* (6) *of Theorem* 6.2.1.1*. In particular,* $\overrightarrow{[g]}^{\mathrm{min}} \underset{\mathbb{Z}}{\otimes} \mathbb{Q}$ *is finite.*

Proof. Let $\Sigma^{\mathrm{ord}} = \{\Sigma_{\Phi_{\mathcal{H}}}\}_{[(\Phi_{\mathcal{H}},\delta_{\mathcal{H}})]}$ and $\Sigma^{\mathrm{ord},\prime} = \{\Sigma'_{\Phi'_{\mathcal{H}'}}\}_{[(\Phi'_{\mathcal{H}'},\delta'_{\mathcal{H}'})]}$ be any two compatible choices of admissible smooth rational polyhedral cone decomposition data for $\vec{\mathsf{M}}^{\mathrm{ord}}_{\mathcal{H}}$ and $\vec{\mathsf{M}}^{\mathrm{ord}}_{\mathcal{H}'}$, respectively, such that $\Sigma^{\mathrm{ord},\prime}$ is a g-refinement of Σ^{ord} as in Definition 5.2.2.1. (Such compatible choices of cone decompositions always exist after refinements, by Proposition 5.1.3.2.) Let $\vec{\oint}^{\mathrm{ord}}_{\mathcal{H}} : \vec{\mathsf{M}}^{\mathrm{ord,tor}}_{\mathcal{H},\Sigma^{\mathrm{ord}}} \twoheadrightarrow \vec{\mathsf{M}}^{\mathrm{ord,min}}_{\mathcal{H}}$ and $\vec{\oint}^{\mathrm{ord}}_{\mathcal{H}'} : \vec{\mathsf{M}}^{\mathrm{ord,tor}}_{\mathcal{H}',\Sigma^{\prime,\mathrm{ord}}} \twoheadrightarrow \vec{\mathsf{M}}^{\mathrm{ord,min}}_{\mathcal{H}'}$ be the surjections given by (3) of Theorem 6.2.1.1. Let $\overrightarrow{[g]}^{\mathrm{ord,tor}} : \vec{\mathsf{M}}^{\mathrm{ord,tor}}_{\mathcal{H}',\Sigma^{\mathrm{ord},\prime}} \twoheadrightarrow \vec{\mathsf{M}}^{\mathrm{ord,tor}}_{\mathcal{H},\Sigma^{\mathrm{ord}}}$ be the canonical surjection given by Proposition 5.2.2.2 extending the canonical morphism $\overrightarrow{[g]}^{\mathrm{ord}} : \vec{\mathsf{M}}^{\mathrm{ord}}_{\mathcal{H}'} \to \vec{\mathsf{M}}^{\mathrm{ord}}_{\mathcal{H}}$ defined by the Hecke action of g, under which the ordinary Hecke twist of the tautological family $(G,\lambda,i,\alpha_{\mathcal{H}'^{,p}},\alpha^{\mathrm{ord}}_{\mathcal{H}'_p}) \to \vec{\mathsf{M}}^{\mathrm{ord,tor}}_{\mathcal{H}',\Sigma^{\mathrm{ord},\prime}}$ by g (defined by Proposition 3.3.4.21 and Lemma 3.1.3.2) is the pullback $(G',\lambda',i',\alpha'_{\mathcal{H}^p},\alpha^{\mathrm{ord},\prime}_{\mathcal{H}_p}) \to \vec{\mathsf{M}}^{\mathrm{ord,tor}}_{\mathcal{H}',\Sigma^{\mathrm{ord},\prime}}$ of the tautological family $(G, \lambda, i, \alpha_{\mathcal{H}^p}, \alpha^{\mathrm{ord}}_{\mathcal{H}_p}) \to \vec{\mathsf{M}}^{\mathrm{ord,tor}}_{\mathcal{H},\Sigma^{\mathrm{ord}}}$, equipped with a $\mathbb{Q}^\times$-isogeny $[g^{-1}]^{\mathrm{ord}} : G \to G'$, which is up to $\mathbb{Z}^\times_{(p)}$-isogeny an isogeny (not just a $\mathbb{Q}^\times$-isogeny) when $L \underset{\mathbb{Z}}{\otimes} \mathbb{Z}_p \subset g_p(L \underset{\mathbb{Z}}{\otimes} \mathbb{Z}_p)$.

Consider the invertible sheaf

$$\omega'_{\vec{\mathsf{M}}^{\mathrm{ord,tor}}_{\mathcal{H}',\Sigma^{\mathrm{ord},\prime}}} := \wedge^{\mathrm{top}} \underline{\mathrm{Lie}}^\vee_{G'/\vec{\mathsf{M}}^{\mathrm{ord,tor}}_{\mathcal{H}',\Sigma^{\mathrm{ord},\prime}}} = \wedge^{\mathrm{top}} e^*_G \Omega^1_{G'/\vec{\mathsf{M}}^{\mathrm{ord,tor}}_{\mathcal{H}',\Sigma^{\mathrm{ord},\prime}}}$$

over $\vec{\mathsf{M}}^{\mathrm{ord,tor}}_{\mathcal{H}',\Sigma^{\mathrm{ord},\prime}}$, which is the pullback of $\omega_{\vec{\mathsf{M}}^{\mathrm{ord,tor}}_{\mathcal{H},\Sigma^{\mathrm{ord}}}}$ under $\vec{[g]}^{\mathrm{ord,tor}}$. We claim that there exists an integer $N_0' > 0$ such that, for each k divisible by N_0', the k-th tensor power $(\omega'_{\vec{\mathsf{M}}^{\mathrm{ord,tor}}_{\mathcal{H}',\Sigma^{\mathrm{ord},\prime}}})^{\otimes k}$ of $\omega'_{\vec{\mathsf{M}}^{\mathrm{ord,tor}}_{\mathcal{H}',\Sigma^{\mathrm{ord},\prime}}}$ descends to a (necessarily unique) invertible sheaf over $\vec{\mathsf{M}}^{\mathrm{ord,min}}_{\mathcal{H}'}$ (see [62, Lem. 7.2.2.1]), which we abusively denote by $(\omega'_{\vec{\mathsf{M}}^{\mathrm{ord,min}}_{\mathcal{H}'}})^{\otimes k}$. By the same argument as in the proof of Theorem 6.2.1.1, it suffices to show that there exists an integer $N_0' > 0$ such that, for each stratum $\vec{\mathsf{Z}}^{\mathrm{ord}}_{[(\Phi_{\mathcal{H}'},\delta_{\mathcal{H}'})]}$ in $\vec{\mathsf{M}}^{\mathrm{ord,min}}_{\mathcal{H}'}$, for each stratum $\vec{\mathsf{Z}}^{\mathrm{ord}}_{[(\Phi_{\mathcal{H}'},\delta_{\mathcal{H}'},\sigma')]}$ in $\vec{\mathsf{M}}^{\mathrm{ord,tor}}_{\mathcal{H}',\Sigma^{\mathrm{ord},\prime}}$ that maps surjectively to $\vec{\mathsf{Z}}^{\mathrm{ord}}_{[(\Phi_{\mathcal{H}'},\delta_{\mathcal{H}'})]}$, and for each representative $(\Phi_{\mathcal{H}'}, \delta_{\mathcal{H}'}, \sigma')$ of $[(\Phi_{\mathcal{H}'}, \delta_{\mathcal{H}'}, \sigma')]$, the N_0'-th tensor power $({}^\heartsuit\omega')^{\otimes N_0'}$ of the pullback ${}^\heartsuit\omega'$ of $\omega'_{\vec{\mathsf{M}}^{\mathrm{ord,tor}}_{\mathcal{H}',\Sigma^{\mathrm{ord},\prime}}}$ under the canonical morphism $\vec{\mathfrak{X}}^{\mathrm{ord}}_{\Phi_{\mathcal{H}'},\delta_{\mathcal{H}'},\sigma'} \to \vec{\mathsf{M}}^{\mathrm{ord,tor}}_{\mathcal{H}',\Sigma^{\mathrm{ord},\prime}}$ (see (5) of Theorem 5.2.1.1) descends to an invertible sheaf under the canonical morphism $\vec{\mathfrak{X}}^{\mathrm{ord}}_{\Phi_{\mathcal{H}'},\delta_{\mathcal{H}'},\sigma'} \to [\vec{\mathsf{M}}^{\mathrm{ord},\mathsf{Z}_{\mathcal{H}'}}_{\mathcal{H}}]$. By considering the pullback of the $\mathbb{Q}^\times$-isogeny $[g^{-1}]^{\mathrm{ord}} : G \to G'$ to $\vec{\mathfrak{X}}^{\mathrm{ord}}_{\Phi_{\mathcal{H}'},\delta_{\mathcal{H}'},\sigma'}$, there exists an $\mathcal{O}'$-lattice X' and an abelian scheme B' over $\vec{\mathsf{M}}^{\mathrm{ord},\mathsf{Z}_{\mathcal{H}'}}_{\mathcal{H}}$ such that ${}^\heartsuit\omega' \cong (\wedge^{\mathrm{top}}_{\mathbb{Z}} X') \underset{\mathbb{Z}}{\otimes} \omega_{B'}$, where $\omega_{B'} := \wedge^{\mathrm{top}} \underline{\mathrm{Lie}}^\vee_{B'/\vec{\mathsf{M}}^{\mathrm{ord},\mathsf{Z}_{\mathcal{H}'}}_{\mathcal{H}}}$ (see Lemma 6.1.2.3, or rather the proof of [62, Lem. 7.1.2.1]). Hence, it suffices to take any integer $N_0' > 0$ such that $\Gamma_{\mathcal{H}'}$ acts trivially on $(\wedge^{\mathrm{top}}_{\mathbb{Z}} X')^{\otimes N_0'}$, and such that $\omega_{B'}^{\otimes N_0'}$ descends to $[\vec{\mathsf{M}}^{\mathrm{ord},\mathsf{Z}_{\mathcal{H}'}}_{\mathcal{H}}]$, for any X' and B' as above, which exists by applying the analogue of (2) of Theorem 6.2.1.1 to the finitely many strata of $\vec{\mathsf{M}}^{\mathrm{ord,min}}_{\mathcal{H}'}$.

On the other hand, by (2) of Theorem 6.2.1.1, there exists an integer $N_0 > 0$ such that $\omega^{\otimes k}_{\vec{\mathsf{M}}^{\mathrm{ord,min}}_{\mathcal{H}}}$ is defined for each k divisible by N_0, in which case the k-th tensor power $(\omega'_{\vec{\mathsf{M}}^{\mathrm{ord,tor}}_{\mathcal{H}',\Sigma^{\mathrm{ord},\prime}}})^{\otimes k}$ of $\omega'_{\vec{\mathsf{M}}^{\mathrm{ord,tor}}_{\mathcal{H}',\Sigma^{\mathrm{ord},\prime}}}$ is the pullback of $\omega^{\otimes k}_{\vec{\mathsf{M}}^{\mathrm{ord,min}}_{\mathcal{H}}}$ under the composition $\vec{\oint}^{\mathrm{ord}}_{\mathcal{H}} \circ \vec{[g]}^{\mathrm{ord,tor}} : \vec{\mathsf{M}}^{\mathrm{ord,tor}}_{\mathcal{H}',\Sigma^{\mathrm{ord},\prime}} \to \vec{\mathsf{M}}^{\mathrm{ord,min}}_{\mathcal{H}}$, and hence $\vec{\oint}^{\mathrm{ord}}_{\mathcal{H}} \circ \vec{[g]}^{\mathrm{ord,tor}}$ can be identified with the composition

$$\begin{aligned} \vec{\mathsf{M}}^{\mathrm{ord,tor}}_{\mathcal{H}',\Sigma^{\mathrm{ord},\prime}} &\to \mathrm{Proj}\Big(\bigoplus_{k \geq 0,\, N_0 N_0' | k} \Gamma(\vec{\mathsf{M}}^{\mathrm{ord,tor}}_{\mathcal{H}',\Sigma^{\mathrm{ord},\prime}}, (\omega'_{\vec{\mathsf{M}}^{\mathrm{ord,tor}}_{\mathcal{H}',\Sigma^{\mathrm{ord},\prime}}})^{\otimes k}) \Big) \\ &\to \vec{\mathsf{M}}^{\mathrm{ord,min}}_{\mathcal{H}} \cong \mathrm{Proj}\Big(\bigoplus_{k \geq 0,\, N_0 | k} \Gamma(\vec{\mathsf{M}}^{\mathrm{ord,min}}_{\mathcal{H}}, \omega^{\otimes k}_{\vec{\mathsf{M}}^{\mathrm{ord,min}}_{\mathcal{H}}}) \Big) \end{aligned} \tag{6.2.2.4}$$

of canonical morphisms. Since $(\omega'_{\vec{\mathsf{M}}^{\mathrm{ord,tor}}_{\mathcal{H}',\Sigma^{\mathrm{ord},\prime}}})^{\otimes k}$ descends to an invertible sheaf $(\omega'_{\vec{\mathsf{M}}^{\mathrm{ord,min}}_{\mathcal{H}'}})^{\otimes k}$ over $\vec{\mathsf{M}}^{\mathrm{ord,min}}_{\mathcal{H}'}$ when $N_0'|k$, the above morphism (6.2.2.4) factors through $\vec{\oint}^{\,\mathrm{ord}}_{\mathcal{H}'} : \vec{\mathsf{M}}^{\mathrm{ord,tor}}_{\mathcal{H}',\Sigma^{\mathrm{ord},\prime}} \to \vec{\mathsf{M}}^{\mathrm{ord,min}}_{\mathcal{H}'}$ and induces a (necessarily surjective) morphism $\vec{[g]}^{\mathrm{ord,min}} : \vec{\mathsf{M}}^{\mathrm{ord,min}}_{\mathcal{H}'} \to \vec{\mathsf{M}}^{\mathrm{ord,min}}_{\mathcal{H}}$, under which $(\omega'_{\vec{\mathsf{M}}^{\mathrm{ord,min}}_{\mathcal{H}'}})^{\otimes k}$ is the pullback of $\omega^{\otimes k}_{\vec{\mathsf{M}}^{\mathrm{ord,min}}_{\mathcal{H}}}$ when $N_0 N_0' | k$.

But then $(\omega'_{\vec{\mathsf{M}}^{\mathrm{ord,tor}}_{\mathcal{H}',\Sigma^{\mathrm{ord},\prime}}})^{\otimes k}$ descends to a (necessarily unique) invertible sheaf $(\omega'_{\vec{\mathsf{M}}^{\mathrm{ord,min}}_{\mathcal{H}'}})^{\otimes k}$ over $\vec{\mathsf{M}}^{\mathrm{ord,min}}_{\mathcal{H}'}$, which is just the pullback of $\omega^{\otimes k}_{\vec{\mathsf{M}}^{\mathrm{ord,min}}_{\mathcal{H}'}}$ under $\vec{[g]}^{\mathrm{ord,min}}$, whenever $N_0|k$. Since $\omega'_{\vec{\mathsf{M}}^{\mathrm{ord,tor}}_{\mathcal{H}',\Sigma^{\mathrm{ord},\prime}}} \cong (\vec{[g]}^{\mathrm{ord,tor}})^* \omega_{\vec{\mathsf{M}}^{\mathrm{ord,tor}}_{\mathcal{H},\Sigma^{\mathrm{ord}}}}$, any sufficiently large p-power multiple of $[g^{-1}]^{\mathrm{ord}}$, which can be $[g^{-1}]^{\mathrm{ord}}$ itself when $L \underset{\mathbb{Z}}{\otimes} \mathbb{Z}_p \subset g_p(L \underset{\mathbb{Z}}{\otimes} \mathbb{Z}_p)$, induces a morphism

$$(\vec{[g]}^{\mathrm{ord,tor}})^* : (\vec{[g]}^{\mathrm{ord,tor}})^* \omega_{\vec{\mathsf{M}}^{\mathrm{ord,tor}}_{\mathcal{H},\Sigma^{\mathrm{ord}}}} \to \omega_{\vec{\mathsf{M}}^{\mathrm{ord,tor}}_{\mathcal{H}',\Sigma^{\mathrm{ord},\prime}}}, \tag{6.2.2.5}$$

which in turn induces the desired morphism (6.2.2.2) whenever $N_0|k$.

If the levels $\mathcal{H}_p$ and $\mathcal{H}'_p$ at p are *equally deep*, or if g_p is of twisted U_p type as in Definition 3.3.6.1 and $\mathrm{depth}_{\mathsf{D}}(\mathcal{H}'_p) - \mathrm{depth}_{\mathsf{D}}(g_p) = \mathrm{depth}_{\mathsf{D}}(\mathcal{H}_p) > 0$, then the surjection $\vec{[g]}^{\mathrm{ord,tor}}$ is proper by Proposition 5.2.2.2, and hence the induced surjection $\vec{[g]}^{\mathrm{ord,min}}$ is finite.

Suppose $g_p \in \mathrm{P}^{\mathrm{ord}}_{\mathsf{D}}(\mathbb{Z}_p)$ (in this paragraph). Then the $\mathbb{Q}^\times$-isogeny $[g^{-1}]^{\mathrm{ord}} : G \to G'$ is a $\mathbb{Z}^\times_{(p)}$-isogeny, and hence the morphisms (6.2.2.5) and (6.2.2.2) induced by $[g^{-1}]^{\mathrm{ord}}$ itself (not by nontrivial p-power multiples) are isomorphisms. Moreover, we know that $\vec{[g]}^{\mathrm{ord,min}}$ is finite by the previous paragraph. By the construction of $\vec{[g]}^{\mathrm{min}}_{r_{\mathcal{H}'},r_{\mathcal{H}}}$ by various universal properties (see Definition 2.2.3.4 and the proof of Proposition 2.2.3.1), we obtain the commutative diagram (6.2.2.3), and the canonical isomorphism (6.2.2.2) is compatible with the corresponding one in Proposition 2.2.3.1. By the fact that the restriction of $\vec{\oint}^{\,\mathrm{ord}}_{\mathcal{H}'}$ to $\vec{\mathsf{M}}^{\mathrm{ord}}_{\mathcal{H}'}$ is the canonical morphism $\vec{\mathsf{M}}^{\mathrm{ord}}_{\mathcal{H}'} \to [\vec{\mathsf{M}}^{\mathrm{ord}}_{\mathcal{H}'}]$, we see that the restriction of $\vec{[g]}^{\mathrm{ord,min}}$ to $[\vec{\mathsf{M}}^{\mathrm{ord}}_{\mathcal{H}'}]$ is the canonical surjection $[\vec{[g]}^{\mathrm{ord}}] : [\vec{\mathsf{M}}^{\mathrm{ord}}_{\mathcal{H}'}] \to [\vec{\mathsf{M}}^{\mathrm{ord}}_{\mathcal{H}}]$ induced by the canonical surjection $\vec{[g]}^{\mathrm{ord}} : \vec{\mathsf{M}}^{\mathrm{ord}}_{\mathcal{H}'} \twoheadrightarrow \vec{\mathsf{M}}^{\mathrm{ord}}_{\mathcal{H}}$ defined by the Hecke action of g. Consequently, by noetherian normality of $\vec{\mathsf{M}}^{\mathrm{ord,min}}_{\mathcal{H}}$ and $\vec{\mathsf{M}}^{\mathrm{ord,min}}_{\mathcal{H}'}$ (and by Zariski's main theorem; see [35, III-1, 4.4.3, 4.4.11]), $\vec{\mathsf{M}}^{\mathrm{ord,min}}_{\mathcal{H}'}$ coincides with the normalization of $\vec{\mathsf{M}}^{\mathrm{ord,min}}_{\mathcal{H}}$ in $\vec{\mathsf{M}}^{\mathrm{ord}}_{\mathcal{H}'}$ under the composition of canonical morphisms $\vec{\mathsf{M}}^{\mathrm{ord}}_{\mathcal{H}'} \overset{\vec{[g]}^{\mathrm{ord}}}{\to} \vec{\mathsf{M}}^{\mathrm{ord}}_{\mathcal{H}} \hookrightarrow \vec{\mathsf{M}}^{\mathrm{ord,min}}_{\mathcal{H}}$, or equivalently (by the construction of $\vec{\mathsf{M}}^{\mathrm{ord}}_{\mathcal{H}'}$ and $\vec{\mathsf{M}}^{\mathrm{ord}}_{\mathcal{H}}$ in Theorem 3.4.2.5, and by the construction of $[g]^{\mathrm{ord}}$ and $\vec{[g]}^{\mathrm{ord}}$ in Proposition 3.4.4.1) the normalization of $\vec{\mathsf{M}}^{\mathrm{ord,min}}_{\mathcal{H}}$ in $\mathsf{M}^{\mathrm{ord}}_{\mathcal{H}'} \cong \mathsf{M}_{\mathcal{H}',r_{\mathcal{H}'}}$ under the composition of canonical morphisms $\mathsf{M}^{\mathrm{ord}}_{\mathcal{H}'} \overset{[g]^{\mathrm{ord}}}{\to} \mathsf{M}^{\mathrm{ord}}_{\mathcal{H}} \to \vec{\mathsf{M}}^{\mathrm{ord,min}}_{\mathcal{H}}$. Hence, the commutative diagram (6.2.2.3) is Cartesian, because $\vec{[g]}^{\mathrm{min}}_{r_{\mathcal{H}'},r_{\mathcal{H}}}$ is the normalization of $\vec{\mathsf{M}}^{\mathrm{min}}_{\mathcal{H},r_{\mathcal{H}}}$ in $\mathsf{M}^{\mathrm{ord}}_{\mathcal{H}'}$ under the canonical morphism $\mathsf{M}^{\mathrm{ord}}_{\mathcal{H}'} \overset{[g]^{\mathrm{ord}}}{\to} \mathsf{M}^{\mathrm{ord}}_{\mathcal{H}} \to \vec{\mathsf{M}}^{\mathrm{ord,min}}_{\mathcal{H}} \hookrightarrow \vec{\mathsf{M}}^{\mathrm{min}}_{\mathcal{H},r_{\mathcal{H}}}$ (by the construction in Definition

2.2.3.5 and the proof of Proposition 2.2.3.1, and by Proposition 3.4.4.1 again for the comparison between $[g]^{\mathrm{ord}}$ and $[g]$).

The statements about the images of the strata of $\vec{\mathsf{M}}_{\mathcal{H}'}^{\mathrm{ord,min}}$ under $\vec{[g]}^{\mathrm{ord,min}}$ follow from the corresponding statements about the images of the strata of $\vec{\mathsf{M}}_{\mathcal{H}',\Sigma^{\mathrm{ord}}}^{\mathrm{ord,tor}}$ under $\vec{[g]}^{\mathrm{ord,tor}}$. The last two paragraphs of this proposition follows from the last two paragraphs of Proposition 5.2.2.2 (by choosing compatible choices of cone decompositions, which is always possible after refinements, by Proposition 5.1.3.2, such that $\vec{[g]}^{\mathrm{ord,tor}}$, $\vec{[g_1]}^{\mathrm{ord,tor}}$, and $\vec{[g_2]}^{\mathrm{ord,tor}}$ are all defined). □

Corollary 6.2.2.6. (*Compare with [62, Cor. 7.2.5.2] and Corollary* 2.2.3.2.) *Suppose we have two open compact subgroups $\mathcal{H}$ and $\mathcal{H}'$ of $\mathrm{G}(\hat{\mathbb{Z}})$ such that $\mathcal{H}$ and $\mathcal{H}'$ are of standard form as in Definition* 3.2.2.9, *such that $\mathcal{H}'$ is a normal subgroup of $\mathcal{H}$, and such that $\mathrm{depth}_{\mathsf{D}}(\mathcal{H}) = \mathrm{depth}_{\mathsf{D}}(\mathcal{H}')$. Then the canonical morphisms defined in Proposition* 6.2.2.1 *induces a canonical action of the finite group $\mathcal{H}/\mathcal{H}'$ on $\vec{\mathsf{M}}_{\mathcal{H}'}^{\mathrm{ord,min}}$, and the canonical surjection $\vec{[1]}^{\mathrm{ord,min}} : \vec{\mathsf{M}}_{\mathcal{H}'}^{\mathrm{ord,min}} \twoheadrightarrow \vec{\mathsf{M}}_{\mathcal{H},r_{\mathcal{H}'}}^{\mathrm{ord,min}}$, where $\vec{\mathsf{M}}_{\mathcal{H},r_{\mathcal{H}'}}^{\mathrm{ord,min}}$ is the normalization of $\vec{\mathsf{M}}_{\mathcal{H}}^{\mathrm{ord,min}} \underset{\vec{\mathsf{S}}_{0,r_{\mathcal{H}}}}{\times} \vec{\mathsf{S}}_{0,r_{\mathcal{H}'}}$, can be identified with the quotient of $\vec{\mathsf{M}}_{\mathcal{H}'}^{\mathrm{ord,min}}$ by this action.*

Proof. The existence of such an action is clear. Since $\vec{\mathsf{M}}_{\mathcal{H}'}^{\mathrm{ord,min}}$ is quasi-projective over $\vec{\mathsf{S}}_{0,r_{\mathcal{H}'}}$ and normal, the quotient $\vec{\mathsf{M}}_{\mathcal{H}'}^{\mathrm{ord,min}}/(\mathcal{H}/\mathcal{H}')$ exists as a scheme over $\vec{\mathsf{S}}_{0,r_{\mathcal{H}'}}$ (*cf.* [25, V, 4.1]). Then it follows from Zariski's main theorem (see [35, III-1, 4.4.3, 4.4.11]) that the induced morphism $\vec{\mathsf{M}}_{\mathcal{H}'}^{\mathrm{ord,min}}/(\mathcal{H}/\mathcal{H}') \to \vec{\mathsf{M}}_{\mathcal{H},r_{\mathcal{H}'}}^{\mathrm{ord,min}}$ over $\vec{\mathsf{S}}_{0,r_{\mathcal{H}'}}$ (with noetherian normal target) is an isomorphism, because it is finite by Proposition 6.2.2.1, and because it is generically an isomorphism (over $\mathsf{M}_{\mathcal{H}}^{\mathrm{ord}}$, by the moduli interpretations of $\mathsf{M}_{\mathcal{H}'}^{\mathrm{ord}}$ and $\mathsf{M}_{\mathcal{H}}^{\mathrm{ord}}$ as in Theorem 3.4.2.5, and by the characterization of coarse moduli spaces as geometric and uniform categorical quotients in the category of algebraic spaces; see [62, Sec. A.7.5]). □

Corollary 6.2.2.7. (*Compare with Corollaries* 3.4.4.3 *and* 5.2.2.3.) *With the setting as in Proposition* 6.2.2.1, *the morphism*

$$\vec{[g]}^{\mathrm{ord,min}} : \mathfrak{M}_{\mathcal{H}'}^{\mathrm{ord,min}} \to \mathfrak{M}_{\mathcal{H}}^{\mathrm{ord,min}}$$

(*cf. Definition* 3.4.4.2) *induced by $\vec{[g]}^{\mathrm{ord,min}} : \vec{\mathsf{M}}_{\mathcal{H}'}^{\mathrm{ord,min}} \to \vec{\mathsf{M}}_{\mathcal{H}}^{\mathrm{ord,min}}$ is finite.*

Proof. As in the proof of Proposition 6.2.2.1, we may assume that $\vec{[g]}^{\mathrm{ord,min}}$: $\vec{\mathsf{M}}_{\mathcal{H}'}^{\mathrm{ord,min}} \to \vec{\mathsf{M}}_{\mathcal{H}}^{\mathrm{ord,min}}$ is induced by some $\vec{[g]}^{\mathrm{ord,tor}} : \vec{\mathsf{M}}_{\mathcal{H}',\Sigma^{\mathrm{ord},\prime}}^{\mathrm{ord,tor}} \to \vec{\mathsf{M}}_{\mathcal{H},\Sigma^{\mathrm{ord}}}^{\mathrm{ord,tor}}$. Since the canonical morphisms $\vec{\oint}_{\mathcal{H}'}^{\mathrm{ord}} : \vec{\mathsf{M}}_{\mathcal{H}',\Sigma^{\mathrm{ord},\prime}}^{\mathrm{ord,tor}} \to \vec{\mathsf{M}}_{\mathcal{H}'}^{\mathrm{ord,min}}$ and $\vec{\oint}_{\mathcal{H}}^{\mathrm{ord}} : \vec{\mathsf{M}}_{\mathcal{H},\Sigma^{\mathrm{ord}}}^{\mathrm{ord,tor}} \to \vec{\mathsf{M}}_{\mathcal{H}}^{\mathrm{ord,min}}$ are proper and surjective, and since the canonical morphism $\vec{[g]}^{\mathrm{ord,tor}}$:

$\vec{\mathsf{M}}^{\mathrm{ord,tor}}_{\mathcal{H}',\Sigma^{\mathrm{ord},\prime}} \underset{\mathbb{Z}}{\otimes} \mathbb{F}_p \to \vec{\mathsf{M}}^{\mathrm{ord,tor}}_{\mathcal{H},\Sigma^{\mathrm{ord}}} \underset{\mathbb{Z}}{\otimes} \mathbb{F}_p$ is proper by Lemma 3.3.6.8 (see the proof of Corollary 5.2.2.3), it follows that the quasi-finite morphism

$$\vec{[g]}^{\mathrm{ord,min}} : \vec{\mathsf{M}}^{\mathrm{ord,min}}_{\mathcal{H}'} \underset{\mathbb{Z}}{\otimes} \mathbb{F}_p \to \vec{\mathsf{M}}^{\mathrm{ord,min}}_{\mathcal{H}} \underset{\mathbb{Z}}{\otimes} \mathbb{F}_p$$

is also proper, which is then finite (*cf.* [35, IV-3, 8.11.1]). Hence, the morphism $\vec{[g]}^{\mathrm{ord,min}} : \vec{\mathfrak{M}}^{\mathrm{ord,min}}_{\mathcal{H}'} \to \vec{\mathfrak{M}}^{\mathrm{ord,min}}_{\mathcal{H}}$ is also finite, as desired. □

Corollary 6.2.2.8. (*Compare with Corollaries* 3.4.4.4 *and* 5.2.2.4, *and Example* 3.4.4.5.) *With the setting as in Proposition* 6.2.2.1, *if* $g = (g_0, g_p) \in \mathrm{G}(\hat{\mathbb{Z}}^p) \times \mathrm{P}^{\mathrm{ord}}_{\mathtt{D}}(\mathbb{Z}_p)$, *if* $\mathcal{H}'^{,p} = g_0 \mathcal{H}^p g_0^{-1}$ *in* $\mathrm{G}(\hat{\mathbb{Z}}^p)$, *if* $\mathcal{H}'^{\mathrm{ord}}_p = (g_p \mathcal{H}_p g_p^{-1})^{\mathrm{ord}}$ *in* $\mathrm{M}^{\mathrm{ord}}_{\mathtt{D}}(\mathbb{Z}_p)$ (*see* (3.3.3.5)), *and if* $\mathcal{H}^p$ *and hence* $\mathcal{H}'^{,p}$ *are neat, then* ($r_{\mathcal{H}'} = r_{\mathcal{H}}$ *and*) *the induced morphism* $\vec{[g]}^{\mathrm{ord,min}} : \vec{\mathfrak{M}}^{\mathrm{ord,min}}_{\mathcal{H}'} \to \vec{\mathfrak{M}}^{\mathrm{ord,min}}_{\mathcal{H}}$ *is an isomorphism.* (*See the remark at the end of Corollary* 3.4.4.4.) *Consequently, the morphism* $\vec{[g]}^{\mathrm{ord,min}} : \vec{\mathsf{M}}^{\mathrm{ord,min}}_{\mathcal{H}'} \to \vec{\mathsf{M}}^{\mathrm{ord,min}}_{\mathcal{H}}$ *is étale in a* (*Zariski*) *neighborhood of* $\vec{\mathsf{M}}^{\mathrm{ord,min}}_{\mathcal{H}'} \underset{\mathbb{Z}}{\otimes} \mathbb{F}_p$. (*Despite the last statement of Proposition* 6.2.2.1, *the morphism* $[g]^{\mathrm{min}} : \mathsf{M}^{\mathrm{min}}_{\mathcal{H}'} \to \mathsf{M}^{\mathrm{min}}_{\mathcal{H}}$ *in Proposition* 1.3.1.14 *might not be étale along cusps* $\mathsf{Z}_{[(\Phi_{\mathcal{H}'}, \delta_{\mathcal{H}'})]}$ *labeled by cusp labels* $[(\Phi_{\mathcal{H}'}, \delta_{\mathcal{H}'})]$ *not compatible with* $\mathtt{D}$, *which do not meet* $\vec{\mathsf{M}}^{\mathrm{ord,min}}_{\mathcal{H}'} \underset{\mathbb{Z}}{\otimes} \mathbb{Q}$.)

Proof. By Corollary 6.2.2.7, the induced morphism $\vec{[g]}^{\mathrm{ord,min}} : \vec{\mathfrak{M}}^{\mathrm{ord,min}}_{\mathcal{H}'} \to \vec{\mathfrak{M}}^{\mathrm{ord,min}}_{\mathcal{H}}$ is finite. By Corollary 3.4.4.4, the restriction of $\vec{[g]}^{\mathrm{ord,min}}$ to $\vec{\mathfrak{M}}^{\mathrm{ord}}_{\mathcal{H}'}$ is an isomorphism. Since $\mathcal{H}^p$ and $\mathcal{H}'^{,p}$ are neat, by (4) of Proposition 6.2.1.6, $\vec{\mathsf{M}}^{\mathrm{ord,min}}_{\mathcal{H}'} \underset{\mathbb{Z}}{\otimes} \mathbb{F}_p$ and $\vec{\mathsf{M}}^{\mathrm{ord,min}}_{\mathcal{H}} \underset{\mathbb{Z}}{\otimes} \mathbb{F}_p$ are normal. Hence, by Zariski's main theorem (see [35, III-1, 4.4.3, 4.4.11]), the induced finite morphism $\vec{[g]}^{\mathrm{ord,min}} : \vec{\mathsf{M}}^{\mathrm{ord,tor}}_{\mathcal{H}'} \underset{\mathbb{Z}}{\otimes} \mathbb{F}_p \to \vec{\mathsf{M}}^{\mathrm{ord,min}}_{\mathcal{H}} \underset{\mathbb{Z}}{\otimes} \mathbb{F}_p$ is necessarily an isomorphism. Hence, $\vec{[g]}^{\mathrm{ord,min}} : \vec{\mathfrak{M}}^{\mathrm{ord,min}}_{\mathcal{H}'} \to \vec{\mathfrak{M}}^{\mathrm{ord,min}}_{\mathcal{H}}$ (being a finite morphism between formal schemes flat over $\mathbb{Z}_p$, which is an isomorphism between the fibers over $\mathrm{Spec}(\mathbb{F}_p)$) is an isomorphism. As for the last statement, $\vec{[g]}^{\mathrm{ord,min}} : \vec{\mathsf{M}}^{\mathrm{ord,min}}_{\mathcal{H}'} \to \vec{\mathsf{M}}^{\mathrm{ord,min}}_{\mathcal{H}}$ is étale at the points of $\vec{\mathsf{M}}^{\mathrm{ord,tor}}_{\mathcal{H}'} \underset{\mathbb{Z}}{\otimes} \mathbb{F}_p$ because it induces isomorphisms between the formal completions (see [35, IV-4, 17.6.3]), and hence it is étale at a neighborhood of $\vec{\mathsf{M}}^{\mathrm{ord,min}}_{\mathcal{H}'} \underset{\mathbb{Z}}{\otimes} \mathbb{F}_p$ because étaleness is an open condition on the source of a morphism (see [33, I, 4.5]). □

Corollary 6.2.2.9 (elements of U_p type). (*Compare with Corollaries* 3.4.4.6 *and* 5.2.2.5.) *Suppose in Proposition* 6.2.2.1 *that* $g_0 = 1$ *and* g_p *is* **of** U_p **type** *as in Definition* 3.3.6.1 (*so that it is of twisted* U_p *type and* $\mathrm{depth}_{\mathtt{D}}(g_p) = 1$). *Then the induced morphism*

$$\vec{[g]}^{\mathrm{ord,min}} : \vec{\mathsf{M}}^{\mathrm{ord,min}}_{\mathcal{H}'} \underset{\mathbb{Z}}{\otimes} \mathbb{F}_p \to \vec{\mathsf{M}}^{\mathrm{ord,min}}_{\mathcal{H}} \underset{\mathbb{Z}}{\otimes} \mathbb{F}_p \qquad (6.2.2.10)$$

is finite and coincides with the composition of the absolute Frobenius morphism

$$\mathrm{F}_{\vec{\mathsf{M}}^{\mathrm{ord,min}}_{\mathcal{H}'} \underset{\mathbb{Z}}{\otimes} \mathbb{F}_p} : \vec{\mathsf{M}}^{\mathrm{ord,min}}_{\mathcal{H}'} \underset{\mathbb{Z}}{\otimes} \mathbb{F}_p \to \vec{\mathsf{M}}^{\mathrm{ord,min}}_{\mathcal{H}'} \underset{\mathbb{Z}}{\otimes} \mathbb{F}_p$$

with the canonical finite morphism

$$[\vec{1}]^{\mathrm{ord,min}} : \vec{\mathsf{M}}^{\mathrm{ord,min}}_{\mathcal{H}'} \underset{\mathbb{Z}}{\otimes} \mathbb{F}_p \to \vec{\mathsf{M}}^{\mathrm{ord,min}}_{\mathcal{H}} \underset{\mathbb{Z}}{\otimes} \mathbb{F}_p. \tag{6.2.2.11}$$

If ${\mathcal{H}'_p}^{\mathrm{ord}} = \mathcal{H}^{\mathrm{ord}}_p$ *as open compact subgroups of* $\mathrm{M}^{\mathrm{ord}}_{\mathsf{D}}(\mathbb{Z}_p)$ *(see* (3.3.3.5)*), then* ($r_{\mathcal{H}} = r_{\mathcal{H}'}$ *and) the canonical morphism* (6.2.2.11) *is an isomorphism by Corollary* 6.2.2.8*, and the composition*

$$\vec{\mathsf{M}}^{\mathrm{ord,min}}_{\mathcal{H}} \underset{\mathbb{Z}}{\otimes} \mathbb{F}_p \overset{([\vec{1}]^{\mathrm{ord,min}})^{-1}}{\overset{\sim}{\to}} \vec{\mathsf{M}}^{\mathrm{ord,min}}_{\mathcal{H}'} \underset{\mathbb{Z}}{\otimes} \mathbb{F}_p \overset{[\vec{g}]^{\mathrm{ord,min}}}{\to} \vec{\mathsf{M}}^{\mathrm{ord,min}}_{\mathcal{H}} \underset{\mathbb{Z}}{\otimes} \mathbb{F}_p$$

coincides with the absolute Frobenius morphism

$$\mathrm{F}_{\vec{\mathsf{M}}^{\mathrm{ord,min}}_{\mathcal{H}} \underset{\mathbb{Z}}{\otimes} \mathbb{F}_p} : \vec{\mathsf{M}}^{\mathrm{ord,min}}_{\mathcal{H}} \underset{\mathbb{Z}}{\otimes} \mathbb{F}_p \to \vec{\mathsf{M}}^{\mathrm{ord,min}}_{\mathcal{H}} \underset{\mathbb{Z}}{\otimes} \mathbb{F}_p.$$

Proof. The first paragraph of the corollary follows from the corresponding first paragraph of Corollary 5.2.2.5, because $[\vec{g}]^{\mathrm{ord,min}} : \vec{\mathsf{M}}^{\mathrm{ord,min}}_{\mathcal{H}'} \to \vec{\mathsf{M}}^{\mathrm{ord,min}}_{\mathcal{H}}$ is induced by some $[\vec{g}]^{\mathrm{ord,tor}} : \vec{\mathsf{M}}^{\mathrm{ord,tor}}_{\mathcal{H}',\Sigma^{\mathrm{ord},\prime}} \to \vec{\mathsf{M}}^{\mathrm{ord,tor}}_{\mathcal{H},\Sigma^{\mathrm{ord}}}$ as in the proof of Proposition 6.2.2.1. The second paragraph of the corollary follows from the first paragraph and from Corollary 6.2.2.8. □

Remark 6.2.2.12. (Compare with Remarks 3.4.4.9 and 5.2.2.8.) By Kunz's theorem [54] (*cf.* [76, Sec. 42, Thm. 107]), the absolute Frobenius morphisms $\mathrm{F}_{\vec{\mathsf{M}}^{\mathrm{ord,min}}_{\mathcal{H}'} \underset{\mathbb{Z}}{\otimes} \mathbb{F}_p}$ and $\mathrm{F}_{\vec{\mathsf{M}}^{\mathrm{ord,min}}_{\mathcal{H}} \underset{\mathbb{Z}}{\otimes} \mathbb{F}_p}$ in Corollary 6.2.2.9 are *not flat* in general, because $\vec{\mathsf{M}}^{\mathrm{ord,min}}_{\mathcal{H}'} \underset{\mathbb{Z}}{\otimes} \mathbb{F}_p$ and $\vec{\mathsf{M}}^{\mathrm{ord,min}}_{\mathcal{H}} \underset{\mathbb{Z}}{\otimes} \mathbb{F}_p$ are not regular in general (except in very special cases). Therefore, while familiar facts in the modular curve case might remain true in general, some proofs might have to be modified due to the failure of flatness of such absolute Frobenius morphisms.

6.2.3 *Quasi-Projectivity of Partial Toroidal Compactifications*

Theorem 6.2.3.1. (*Compare with [62, Thm. 7.3.3.4] and Theorem* 1.3.1.10.) *Suppose* $\mathcal{H}^p$ *is neat, and suppose* Σ^{ord} *is projective with a compatible collection* $\mathsf{pol}^{\mathrm{ord}}$ *of polarization functions as in Definition* 5.1.3.3. (*Such* Σ^{ord} *and* $\mathsf{pol}^{\mathrm{ord}}$ *exist by Proposition* 5.1.3.4.) *For each integer* $d \geq 1$*, suppose* $\vec{\jmath}_{\mathcal{H},d\mathsf{pol}^{\mathrm{ord}}}$ *is defined over* $\vec{\mathsf{M}}^{\mathrm{ord,tor}}_{\mathcal{H}} = \vec{\mathsf{M}}^{\mathrm{ord,tor}}_{\mathcal{H},\Sigma^{\mathrm{ord}}}$ *as in Definition* 1.3.1.7*, and suppose* $\vec{\mathcal{J}}_{\mathcal{H},d\mathsf{pol}^{\mathrm{ord}}}$ *is defined over* $\vec{\mathsf{M}}^{\mathrm{ord,min}}_{\mathcal{H}}$ *as in Definition* 1.3.1.8*. Then there exists an integer* $d_0 \geq 1$ *such that the following are true:*

(1) The canonical morphism $(\vec{\oint}^{\mathrm{ord}}_{\mathcal{H}})^{-1}\vec{\mathscr{J}}_{\mathcal{H},d_0\mathsf{pol}^{\mathrm{ord}}}\cdot\mathscr{O}_{\vec{\mathsf{M}}^{\mathrm{ord,tor}}_{\mathcal{H}}}\to\vec{\mathscr{J}}_{\mathcal{H},d_0\mathsf{pol}^{\mathrm{ord}}}$ *of coherent* $\mathscr{O}_{\vec{\mathsf{M}}^{\mathrm{ord,tor}}_{\mathcal{H}}}$*-ideals is an isomorphism, which induces a canonical morphism*

$$\mathrm{NBl}_{\vec{\mathscr{J}}_{\mathcal{H},d_0\mathsf{pol}^{\mathrm{ord}}}}(\vec{\oint}^{\mathrm{ord}}_{\mathcal{H}}):\vec{\mathsf{M}}^{\mathrm{ord,tor}}_{\mathcal{H}}\to\mathrm{NBl}_{\vec{\mathscr{J}}_{\mathcal{H},d_0\mathsf{pol}^{\mathrm{ord}}}}(\vec{\mathsf{M}}^{\mathrm{ord,min}}_{\mathcal{H}})$$

by the universal property of the normalization of blow-up (see [62, Def. 7.3.2.2]).

(2) The canonical morphism $\mathrm{NBl}_{\vec{\mathscr{J}}_{\mathcal{H},d_0\mathsf{pol}^{\mathrm{ord}}}}(\vec{\oint}^{\mathrm{ord}}_{\mathcal{H}})$ *above is an isomorphism.*

In particular, the morphism $\vec{\oint}^{\mathrm{ord}}_{\mathcal{H}}:\vec{\mathsf{M}}^{\mathrm{ord,tor}}_{\mathcal{H}}\to\vec{\mathsf{M}}^{\mathrm{ord,min}}_{\mathcal{H}}$ *is projective, and hence* $\vec{\mathsf{M}}^{\mathrm{ord,tor}}_{\mathcal{H}}$ *is a scheme quasi-projective (and smooth) over* $\vec{\mathsf{S}}_{0,r_{\mathcal{H}}}$*. If Condition* 1.3.1.9 *is satisfied, then the above two statements* (1) *and* (2) *are true for all* $d_0\geq 3$.

Proof. Let us begin with the reduction to the case Condition 1.3.1.9 (*cf.* [4, Ch. IV, Sec. 2, p. 329], [28, Ch. V, Sec. 5, p. 178], and [62, Cond. 7.3.3.3]) is satisfied.

By [62, Lem. 7.3.1.7], or rather by the original [28, Ch. V, Lem. 5.3], there exists a normal open compact subgroup $\mathcal{H}'^{,p}$ of $\mathcal{H}^p$ such that, for $\mathcal{H}'=\mathcal{H}'^{,p}\mathcal{H}_p$ (while $\mathcal{H}=\mathcal{H}^p\mathcal{H}_p$, with the same $\mathcal{H}_p$), Condition 1.3.1.9 is satisfied by the $\Sigma^{\mathrm{ord},(\mathcal{H}')}=\{\Sigma_{\Phi_{\mathcal{H}'}}\}_{[(\Phi_{\mathcal{H}'},\delta_{\mathcal{H}'})]}$ and $\mathsf{pol}^{\mathrm{ord},(\mathcal{H}')}=\{\mathsf{pol}_{\Phi_{\mathcal{H}'}}\}_{[(\Phi_{\mathcal{H}'},\delta_{\mathcal{H}'})]}$ induced by Σ^{ord} and $\mathsf{pol}^{\mathrm{ord}}$ as in [62, Constr. 7.3.1.6], and such that $\Sigma^{\mathrm{ord},(\mathcal{H}')}$ is smooth.

Suppose that Theorem 6.2.3.1 is true for $\vec{\mathsf{M}}^{\mathrm{ord,tor}}_{\mathcal{H}'}=\vec{\mathsf{M}}^{\mathrm{ord,tor}}_{\mathcal{H}',\Sigma^{\mathrm{ord},(\mathcal{H}')}}$, $\vec{\mathscr{J}}^{\mathrm{ord}}_{\mathcal{H}',d_0'\mathsf{pol}^{(\mathcal{H}')}}$, and $\vec{\mathscr{J}}^{\mathrm{ord}}_{\mathcal{H}',d_0'\mathsf{pol}^{(\mathcal{H}')}}$ for some integer $d_0'\geq 1$. In particular, $\vec{\mathsf{M}}^{\mathrm{ord,tor}}_{\mathcal{H}'}$ is quasi-projective and smooth over $\vec{\mathsf{S}}_{0,r_{\mathcal{H}}}$.

By construction, the surjections $\vec{\Xi}^{\mathrm{ord}}_{\Phi_{\mathcal{H}'},\delta_{\mathcal{H}'}}(\sigma)\twoheadrightarrow\vec{\Xi}^{\mathrm{ord}}_{\Phi_{\mathcal{H}},\delta_{\mathcal{H}}}(\sigma)$ are finite flat (with possible ramification along the boundary strata) whenever $(\Phi_{\mathcal{H}},\delta_{\mathcal{H}})$ is induced by $(\Phi_{\mathcal{H}'},\delta_{\mathcal{H}'})$ and σ is a cone in the cone decomposition $\Sigma_{\Phi_{\mathcal{H}'}}=\Sigma_{\Phi_{\mathcal{H}}}$ of $\mathbf{P}_{\Phi_{\mathcal{H}'}}\cong\mathbf{P}_{\Phi_{\mathcal{H}}}$. Therefore, the canonical surjection $\vec{[1]}^{\mathrm{ord,tor}}:\vec{\mathsf{M}}^{\mathrm{ord,tor}}_{\mathcal{H}',\Sigma^{\mathrm{ord},(\mathcal{H}')}}\twoheadrightarrow\vec{\mathsf{M}}^{\mathrm{ord,tor}}_{\mathcal{H},\Sigma^{\mathrm{ord}}}$ (given by Proposition 5.2.2.2) is finite flat. It is the unique finite flat extension of the canonical (finite étale) surjection $\vec{\mathsf{M}}^{\mathrm{ord}}_{\mathcal{H}'}\twoheadrightarrow\vec{\mathsf{M}}^{\mathrm{ord}}_{\mathcal{H}}$ (since $\mathcal{H}'=\mathcal{H}'^{,p}\mathcal{H}_p$ and $\mathcal{H}=\mathcal{H}^p\mathcal{H}_p$ only differ away from p). Since $\vec{\mathsf{M}}^{\mathrm{ord,tor}}_{\mathcal{H}',\Sigma^{\mathrm{ord},(\mathcal{H}')}}$ is quasi-projective and smooth over $\vec{\mathsf{S}}_{0,r_{\mathcal{H}}}$, the quotient by $\mathcal{H}/\mathcal{H}'$ is also quasi-projective and isomorphic to $\vec{\mathsf{M}}^{\mathrm{ord,tor}}_{\mathcal{H},\Sigma^{\mathrm{ord}}}$ over $\vec{\mathsf{S}}_{0,r_{\mathcal{H}}}$ (by [25, V, 4.1] and by Zariski's main theorem [35, III-1, 4.4.3, 4.4.11] as in the proof of [62, Cor. 7.2.5.2] and Corollary 6.2.2.6). Moreover, we know that $\vec{\mathscr{J}}^{\mathrm{ord}}_{\mathcal{H}',\mathsf{pol}^{\mathrm{ord},(\mathcal{H}')}}\cong(\vec{[1]}^{\mathrm{ord,tor}})^*\vec{\mathscr{J}}^{\mathrm{ord}}_{\mathcal{H},\mathsf{pol}^{\mathrm{ord}}}$ by construction. Hence, we have verified all the assumptions of [62, Prop. 7.3.2.3], whose application completes the reduction.

Since the description of $\vec{\oint}^{\mathrm{ord}}_{\mathcal{H}}:\vec{\mathsf{M}}^{\mathrm{ord,tor}}_{\mathcal{H}}\to\vec{\mathsf{M}}^{\mathrm{ord,min}}_{\mathcal{H}}$ parallels that of $\oint_{\mathcal{H}}:\mathsf{M}^{\mathrm{tor}}_{\mathcal{H}}\to\mathsf{M}^{\mathrm{min}}_{\mathcal{H}}$ (see Proposition 6.1.2.19 and [62, Prop. 7.2.3.16]), the remainder of the proof, namely the verification of the theorem under Condition 1.3.1.9, is analogous to that of [62, Thm. 7.3.3.4; see also the errata]. □

Corollary 6.2.3.2. *Under the assumptions in Theorem* 6.1.1.12, *suppose moreover that* $\mathcal{H}^p$ *is neat, that* Σ^{ord} *is projective smooth and extends to a projective (but possibly nonsmooth)* Σ *for* $\mathsf{M}_{\mathcal{H}}$, *with a collection* pol *of polarization functions, and that* $\vec{\mathsf{M}}^{\mathrm{tor}}_{\mathcal{H},d_0\mathsf{pol}}$ *is still defined as in Proposition* 2.2.2.3, *for some integer* $d_0 \geq 1$. *(If we extend* Σ^{ord} *to a projective smooth* Σ *for* $\mathsf{M}_{\mathcal{H}}$, *which is possible by Proposition* 5.1.3.4, *then we only need Proposition* 2.2.2.1.) *Then we have a commutative diagram*

$$\begin{array}{ccc} \vec{\mathsf{M}}^{\mathrm{ord,tor}}_{\mathcal{H},\Sigma^{\mathrm{ord}}} & \hookrightarrow & \vec{\mathsf{M}}^{\mathrm{tor}}_{\mathcal{H},d_0\mathsf{pol},r_{\mathcal{H}}} \\ \big\downarrow & & \big\downarrow \\ \vec{\mathsf{M}}^{\mathrm{ord,min}}_{\mathcal{H}} & \hookrightarrow & \vec{\mathsf{M}}^{\mathrm{min}}_{\mathcal{H},r_{\mathcal{H}}} \end{array} \tag{6.2.3.3}$$

of canonical morphisms over $\vec{\mathsf{S}}_{0,r_{\mathcal{H}}}$, *in which the top horizontal arrow is also an open immersion (over* $\vec{\mathsf{S}}_{0,r_{\mathcal{H}}}$) *extending the open immersion* $\vec{\mathsf{M}}^{\mathrm{ord,tor}}_{\mathcal{H},\Sigma^{\mathrm{ord}}} \underset{\vec{\mathsf{S}}_{0,r_{\mathcal{H}}}}{\times} \mathsf{S}_{0,r_{\mathcal{H}}} \hookrightarrow \mathsf{M}^{\mathrm{tor}}_{\mathcal{H},\Sigma,r_{\mathcal{H}}}$ *(over* $\mathsf{S}_{0,r_{\mathcal{H}}}$) *in* (7) *of Theorem* 5.2.1.1, *and a fortiori the induced canonical morphism*

$$\vec{\mathsf{M}}^{\mathrm{ord,tor}}_{\mathcal{H},\Sigma^{\mathrm{ord}}} \to \vec{\mathsf{M}}^{\mathrm{ord,min}}_{\mathcal{H}} \underset{\vec{\mathsf{M}}^{\mathrm{min}}_{\mathcal{H},r_{\mathcal{H}}}}{\times} \vec{\mathsf{M}}^{\mathrm{tor}}_{\mathcal{H},d_0\mathsf{pol},r_{\mathcal{H}}} \tag{6.2.3.4}$$

is an isomorphism. (That is, the diagram (6.2.3.3) *is Cartesian.)*

Proof. The existence of the commutative diagram (6.2.3.3) in which the top horizontal arrow is an open immersion is implied by the existence of the isomorphism (6.2.3.4). By assumption, $\mathsf{pol}^{\mathrm{ord}}$ is the restriction of pol, so that

$$\vec{\mathscr{J}}_{\mathcal{H},d_0\mathsf{pol}^{\mathrm{ord}}} \cong (\vec{\mathsf{M}}^{\mathrm{ord,min}}_{\mathcal{H}} \to \vec{\mathsf{M}}^{\mathrm{min}}_{\mathcal{H}})^* \vec{\mathscr{J}}_{\mathcal{H},d_0\mathsf{pol}}$$

because they both define the schematic closure of the closed subscheme of $\vec{\mathsf{M}}^{\mathrm{ord,min}}_{\mathcal{H}} \underset{\mathbb{Z}}{\otimes} \mathbb{Q}$ defined by $(\vec{\mathsf{M}}^{\mathrm{ord,min}}_{\mathcal{H}} \underset{\mathbb{Z}}{\otimes} \mathbb{Q} \to \mathsf{M}^{\mathrm{min}}_{\mathcal{H}})^* \mathscr{J}_{\mathcal{H},d_0\mathsf{pol}}$. Hence, we obtain a canonical isomorphism

$$\mathrm{NBl}_{\vec{\mathscr{J}}_{\mathcal{H},d_0\mathsf{pol}^{\mathrm{ord}}}}(\vec{\mathsf{M}}^{\mathrm{ord,min}}_{\mathcal{H}}) \to \mathrm{NBl}_{(\vec{\mathsf{M}}^{\mathrm{ord,min}}_{\mathcal{H}} \to \vec{\mathsf{M}}^{\mathrm{min}}_{\mathcal{H}})^* \vec{\mathscr{J}}_{\mathcal{H},d_0\mathsf{pol}}}(\vec{\mathsf{M}}^{\mathrm{ord,min}}_{\mathcal{H}}),$$

which can be identified with the desired isomorphism (6.2.3.4) by Theorem 6.2.3.1 and by the construction in Proposition 2.2.2.3 (or Proposition 2.2.2.1, if Σ is projective smooth). □

6.3 Full Ordinary Loci in p-Adic Completions

6.3.1 *Hasse Invariants*

Consider any toroidal compactification $\mathsf{M}^{\mathrm{tor}}_{\mathrm{G}_{\mathrm{aux}}(\hat{\mathbb{Z}}^p)}$ (over $\vec{\mathsf{S}}_{0,\mathrm{aux}} = \mathrm{Spec}(\mathcal{O}_{F_{0,\mathrm{aux}},(p)})$) carrying a tautological semi-abelian scheme G_{aux} as in [62, Thm. 6.4.1.1]. (In this subsection, we shall suppress the notation for cone decompositions for simplicity.)

Let $\mathrm{F}_{\mathrm{abs}} : \mathsf{M}^{\mathrm{tor}}_{\mathrm{G}_{\mathrm{aux}}(\hat{\mathbb{Z}}^p)} \underset{\mathbb{Z}}{\otimes} \mathbb{F}_p \to \mathsf{M}^{\mathrm{tor}}_{\mathrm{G}_{\mathrm{aux}}(\hat{\mathbb{Z}}^p)} \underset{\mathbb{Z}}{\otimes} \mathbb{F}_p$ denote the absolute Frobenius morphism, and let

$$\mathrm{F}_{\mathrm{aux}} : G_{\mathrm{aux}} \underset{\mathbb{Z}}{\otimes} \mathbb{F}_p \to (G_{\mathrm{aux}} \underset{\mathbb{Z}}{\otimes} \mathbb{F}_p)^{(p)} := \mathrm{F}^*_{\mathrm{abs}}(G_{\mathrm{aux}} \underset{\mathbb{Z}}{\otimes} \mathbb{F}_p)$$

be the relative Frobenius morphism of $G_{\mathrm{aux}} \underset{\mathbb{Z}}{\otimes} \mathbb{F}_p \to \mathsf{M}^{\mathrm{tor}}_{\mathrm{G}_{\mathrm{aux}}(\hat{\mathbb{Z}}^p)} \underset{\mathbb{Z}}{\otimes} \mathbb{F}_p$, which is an isogeny of semi-abelian schemes over $\mathsf{M}^{\mathrm{tor}}_{\mathrm{G}_{\mathrm{aux}}(\hat{\mathbb{Z}}^p)} \underset{\mathbb{Z}}{\otimes} \mathbb{F}_p$. Since $G_{\mathrm{aux}} \underset{\mathbb{Z}}{\otimes} \mathbb{F}_p \to \mathsf{M}^{\mathrm{tor}}_{\mathrm{G}_{\mathrm{aux}}(\hat{\mathbb{Z}}^p)} \underset{\mathbb{Z}}{\otimes} \mathbb{F}_p$ is flat, by [25, VII$_{\mathrm{A}}$, 4.3], there is a canonical morphism

$$\mathrm{V}_{\mathrm{aux}} : (G_{\mathrm{aux}} \underset{\mathbb{Z}}{\otimes} \mathbb{F}_p)^{(p)} \to G_{\mathrm{aux}} \underset{\mathbb{Z}}{\otimes} \mathbb{F}_p,$$

called the (relative) *Verschiebung* morphism of $G_{\mathrm{aux}} \underset{\mathbb{Z}}{\otimes} \mathbb{F}_p \to \mathsf{M}^{\mathrm{tor}}_{\mathrm{G}_{\mathrm{aux}}(\hat{\mathbb{Z}}^p)} \underset{\mathbb{Z}}{\otimes} \mathbb{F}_p$, such that $\mathrm{V}_{\mathrm{aux}} \circ \mathrm{F}_{\mathrm{aux}} = [p]$, the multiplication by p on $G_{\mathrm{aux}} \underset{\mathbb{Z}}{\otimes} \mathbb{F}_p$. Then $\mathrm{V}_{\mathrm{aux}}$ is also an isogeny, and induces a morphism

$$\begin{aligned}&\mathrm{Lie}^\vee(\mathrm{V}_{\mathrm{aux}}) : \underline{\mathrm{Lie}}^\vee_{(G_{\mathrm{aux}} \underset{\mathbb{Z}}{\otimes} \mathbb{F}_p)/(\mathsf{M}^{\mathrm{tor}}_{\mathrm{G}_{\mathrm{aux}}(\hat{\mathbb{Z}}^p)} \underset{\mathbb{Z}}{\otimes} \mathbb{F}_p)} \\ &\to \underline{\mathrm{Lie}}^\vee_{(G_{\mathrm{aux}} \underset{\mathbb{Z}}{\otimes} \mathbb{F}_p)^{(p)}/(\mathsf{M}^{\mathrm{tor}}_{\mathrm{G}_{\mathrm{aux}}(\hat{\mathbb{Z}}^p)} \underset{\mathbb{Z}}{\otimes} \mathbb{F}_p)} \cong \mathrm{F}^*_{\mathrm{abs}}(\underline{\mathrm{Lie}}^\vee_{(G_{\mathrm{aux}} \underset{\mathbb{Z}}{\otimes} \mathbb{F}_p)/(\mathsf{M}^{\mathrm{tor}}_{\mathrm{G}_{\mathrm{aux}}(\hat{\mathbb{Z}}^p)} \underset{\mathbb{Z}}{\otimes} \mathbb{F}_p)}).\end{aligned}$$

By taking top exterior powers (of the same degree), we obtain a morphism

$$\begin{aligned}&\wedge^{\mathrm{top}} \mathrm{Lie}^\vee(\mathrm{V}_{\mathrm{aux}}) : \omega_{\mathsf{M}^{\mathrm{tor}}_{\mathrm{G}_{\mathrm{aux}}(\hat{\mathbb{Z}}^p)}} \underset{\mathbb{Z}}{\otimes} \mathbb{F}_p \\ &\to \mathrm{F}^*_{\mathrm{abs}}(\omega_{\mathsf{M}^{\mathrm{tor}}_{\mathrm{G}_{\mathrm{aux}}(\hat{\mathbb{Z}}^p)}} \underset{\mathbb{Z}}{\otimes} \mathbb{F}_p) \cong \omega^{\otimes p}_{\mathsf{M}^{\mathrm{tor}}_{\mathrm{G}_{\mathrm{aux}}(\hat{\mathbb{Z}}^p)}} \underset{\mathbb{Z}}{\otimes} \mathbb{F}_p,\end{aligned}$$

or equivalently a section

$$\mathrm{Hasse}_{\mathrm{aux}} \in \Gamma(\mathsf{M}^{\mathrm{tor}}_{\mathrm{G}_{\mathrm{aux}}(\hat{\mathbb{Z}}^p)} \underset{\mathbb{Z}}{\otimes} \mathbb{F}_p, \omega^{\otimes(p-1)}_{\mathsf{M}^{\mathrm{tor}}_{\mathrm{G}_{\mathrm{aux}}(\hat{\mathbb{Z}}^p)}} \underset{\mathbb{Z}}{\otimes} \mathbb{F}_p). \tag{6.3.1.1}$$

The value of $\mathrm{Hasse}_{\mathrm{aux}}$ at each geometric point $\bar{s}$ of $\mathsf{M}^{\mathrm{tor}}_{\mathrm{G}_{\mathrm{aux}}(\hat{\mathbb{Z}}^p)} \underset{\mathbb{Z}}{\otimes} \mathbb{F}_p$ is the so-called *Hasse invariant* of the pullback $(G_{\mathrm{aux}})_{\bar{s}}$ of the semi-abelian scheme G_{aux} to $\bar{s}$, which is nonzero exactly when $(G_{\mathrm{aux}})_{\bar{s}}$ is an ordinary semi-abelian variety, or equivalently when the *abelian part* of $(G_{\mathrm{aux}})_{\bar{s}}$ is an ordinary abelian variety (because the nonvanishing of the Hasse invariant is equivalent to the separability of the Verschiebung morphism, which is in turn equivalent to the triviality of the *local-local part*, as in [81, p. 147], of the p-torsion subgroup scheme of the abelian part of $(G_{\mathrm{aux}})_{\bar{s}}$).

Remark 6.3.1.2. By definition, the formation of Hasse invariants is compatible with separable isogenies, because they induce isomorphisms between sheaves of differentials.

Let us also consider any smooth partial toroidal compactification $\vec{\mathsf{M}}^{\mathrm{ord,tor}}_{\mathcal{H}}$ over $\vec{\mathsf{S}}_{0,r_{\mathcal{H}}}$ carrying a tautological semi-abelian scheme G as in Theorem 5.2.1.1. Then, similar to the case of $\mathrm{Hasse}_{\mathrm{aux}}$ above, we can define a section

$$\mathrm{Hasse}^{\mathrm{ord}}_{\mathcal{H}} \in \Gamma(\vec{\mathsf{M}}^{\mathrm{ord,tor}}_{\mathcal{H}} \underset{\mathbb{Z}}{\otimes} \mathbb{F}_p, \omega^{\otimes(p-1)}_{\vec{\mathsf{M}}^{\mathrm{ord,tor}}_{\mathcal{H}}} \underset{\mathbb{Z}}{\otimes} \mathbb{F}_p). \tag{6.3.1.3}$$

The value of $\mathrm{Hasse}^{\mathrm{ord}}_{\mathcal{H}}$ at each geometric point $\bar{s}$ of $\vec{\mathsf{M}}^{\mathrm{ord,tor}}_{\mathcal{H}} \underset{\mathbb{Z}}{\otimes} \mathbb{F}_p$ is the Hasse invariant of the pullback $G_{\bar{s}}$ of the semi-abelian scheme G to $\bar{s}$, which is always nonzero because the abelian part of $G_{\bar{s}}$ is always an ordinary abelian variety. This is consistent with the following:

Lemma 6.3.1.4. *With the setting as in Lemma* 6.1.1.9*, suppose (with suitable choices of cone decompositions, up to refinement if necessary) there is a morphism* $\vec{\mathsf{M}}^{\mathrm{ord,tor}}_{\mathcal{H}} \to \mathsf{M}^{\mathrm{tor}}_{\mathrm{G_{aux}}(\hat{\mathbb{Z}}^p)}$ *extending the morphism* $\vec{\mathsf{M}}^{\mathrm{ord}}_{\mathcal{H}} \to [\mathsf{M}_{\mathrm{G_{aux}}(\hat{\mathbb{Z}}^p)}]$ *given by the composition* $\vec{\mathsf{M}}^{\mathrm{ord}}_{\mathcal{H}} \overset{\mathrm{can.}}{\to} \vec{\mathsf{M}}_{\mathcal{H}} \to \mathsf{M}_{\mathrm{G_{aux}}(\hat{\mathbb{Z}}^p)} \overset{\mathrm{can.}}{\to} [\mathsf{M}_{\mathrm{G_{aux}}(\hat{\mathbb{Z}}^p)}]$ *(see Propositions* 2.2.1.1 *and* 3.4.6.3*). Then the pullback of* $\mathrm{Hasse}_{\mathrm{aux}}$ *to* $\vec{\mathsf{M}}^{\mathrm{ord,tor}}_{\mathcal{H}} \underset{\mathbb{Z}}{\otimes} \mathbb{F}_p$ *under the canonical morphism* $\vec{\mathsf{M}}^{\mathrm{ord,tor}}_{\mathcal{H}} \underset{\mathbb{Z}}{\otimes} \mathbb{F}_p \to (\vec{\mathsf{M}}^{\mathrm{tor}}_{\mathcal{H},d_0\mathsf{pol}} \underset{\vec{\mathsf{S}}_0}{\times} \vec{\mathsf{S}}_{0,r_{\mathcal{H}}}) \underset{\mathbb{Z}}{\otimes} \mathbb{F}_p$ *is nowhere zero.*

If $a_0 \geq 1$ *and* $a \geq 1$ *are integers as in Lemma* 2.1.2.35*, then the pullback of* $\mathrm{Hasse}^{a_0}_{\mathrm{aux}}$ *to* $\vec{\mathsf{M}}^{\mathrm{ord,tor}}_{\mathcal{H}} \underset{\mathbb{Z}}{\otimes} \mathbb{F}_p$ *is the multiple of* $(\mathrm{Hasse}^{\mathrm{ord}}_{\mathcal{H}})^a$ *by a global unit.*

Proof. The pullback of G_{aux} (resp. $G^{\vee}_{\mathrm{aux}}$) under $\vec{\mathsf{M}}^{\mathrm{ord,tor}}_{\mathcal{H}} \to \mathsf{M}^{\mathrm{tor}}_{\mathrm{G_{aux}}(\hat{\mathbb{Z}}^p)}$ must be isomorphic to $G^{\times a_1} \underset{\vec{\mathsf{M}}^{\mathrm{ord,tor}}_{\mathcal{H}}}{\times} (G^{\vee})^{\times a_2}$ (resp. $G^{\times a_2} \underset{\vec{\mathsf{M}}^{\mathrm{ord,tor}}_{\mathcal{H}}}{\times} (G^{\vee})^{\times a_1}$), because it is already so over the open dense subscheme $\mathsf{M}^{\mathrm{ord}}_{\mathcal{H}}$ by Proposition 2.1.1.15 (and by [92, IX, 1.4], [28, Ch. I, Prop. 2.7], or [62, Prop. 3.3.1.5]). Then the lemma follows from the fact that an abelian variety is ordinary if and only if its dual is (*cf.* Lemma 3.1.1.5). □

Remark 6.3.1.5. The definition of $\mathrm{Hasse}^{\mathrm{ord}}$ and Lemma 6.3.1.4 may seem redundant, but we will indeed need them in Proposition 6.3.2.2 below.

Proposition 6.3.1.6. *Suppose that* $\mathcal{H}^p_{\mathrm{aux}} \subset \mathrm{G_{aux}}(\hat{\mathbb{Z}}^p)$ *is a* **neat** *open compact subgroup. Then we can replace* $\mathsf{M}_{\mathrm{G_{aux}}(\hat{\mathbb{Z}}^p)}$ *(resp.* $\mathsf{M}^{\mathrm{min}}_{\mathrm{G_{aux}}(\hat{\mathbb{Z}}^p)}$*, resp.* $\mathsf{M}^{\mathrm{tor}}_{\mathrm{G_{aux}}(\hat{\mathbb{Z}}^p)}$*) with the scheme* $\mathsf{M}_{\mathcal{H}^p_{\mathrm{aux}}}$ *(resp.* $\mathsf{M}^{\mathrm{min}}_{\mathcal{H}^p_{\mathrm{aux}}}$*, resp.* $\mathsf{M}^{\mathrm{tor}}_{\mathcal{H}^p_{\mathrm{aux}}}$ *with any choice of cone decompositions) over* $\vec{\mathsf{S}}_{0,\mathrm{aux}}$ *in the constructions above. The invertible sheaf* $\omega_{\mathsf{M}^{\mathrm{tor}}_{\mathcal{H}^p_{\mathrm{aux}}}}$ *descends to an ample invertible sheaf* $\omega_{\mathsf{M}^{\mathrm{min}}_{\mathcal{H}^p_{\mathrm{aux}}}}$ *on* $\mathsf{M}^{\mathrm{min}}_{\mathcal{H}^p_{\mathrm{aux}}}$ *(by [62, Thm. 7.2.4.1]), and for each integer* $k \geq 0$ *the canonical morphism*

$$\Gamma(\mathsf{M}^{\mathrm{min}}_{\mathcal{H}^p_{\mathrm{aux}}} \underset{\mathbb{Z}}{\otimes} \mathbb{F}_p, \omega^{\otimes k}_{\mathsf{M}^{\mathrm{min}}_{\mathcal{H}^p_{\mathrm{aux}}}} \underset{\mathbb{Z}}{\otimes} \mathbb{F}_p) \to \Gamma(\mathsf{M}^{\mathrm{tor}}_{\mathcal{H}^p_{\mathrm{aux}}} \underset{\mathbb{Z}}{\otimes} \mathbb{F}_p, \omega^{\otimes k}_{\mathsf{M}^{\mathrm{tor}}_{\mathcal{H}^p_{\mathrm{aux}}}} \underset{\mathbb{Z}}{\otimes} \mathbb{F}_p)$$

is an isomorphism, under which we can pullback $\mathrm{Hasse}_{\mathrm{aux}}$ *and its powers (which are a priori defined on* $\mathsf{M}^{\mathrm{tor}}_{\mathrm{G_{aux}}(\hat{\mathbb{Z}}^p)} \underset{\mathbb{Z}}{\otimes} \mathbb{F}_p$*) to* $\mathsf{M}^{\mathrm{min}}_{\mathcal{H}^p_{\mathrm{aux}}} \underset{\mathbb{Z}}{\otimes} \mathbb{F}_p$*.*

Proof. This follows from [62, Prop. 7.2.4.3 and Cor. 7.2.4.8]. □

Corollary 6.3.1.7. *Let* $a_0 \geq 1$ *and* $a \geq 1$ *be integers as in Lemma* 2.1.2.35*, and let* N_1 *be as in Proposition* 2.2.1.2 *(for some choice of* $\mathcal{H}^p_{\mathrm{aux}}$*). Then we can* *pullback* $\mathrm{Hasse}_{\mathrm{aux}}^{a_0 N_1}$ *to a section of* $\Gamma(\vec{\mathsf{M}}_{\mathcal{H}}^{\min} \underset{\mathbb{Z}}{\otimes} \mathbb{F}_p, \omega_{\vec{\mathsf{M}}_{\mathcal{H}}^{\min}}^{\otimes\, aN_1(p-1)} \underset{\mathbb{Z}}{\otimes} \mathbb{F}_p)$*, which we denote by* $\mathrm{Hasse}_{\mathcal{H}}^{aN_1}$ *by abuse of notation.*

Suppose moreover that the image of $\mathcal{H}$ *under the canonical homomorphism* $\mathrm{G}(\hat{\mathbb{Z}}) \to \mathrm{G}(\hat{\mathbb{Z}}^p)$ *is neat (which means, a fortiori, that* $\mathcal{H}$ *is also neat), so that (by Lemma* 2.1.1.18*) there exists some neat open compact subgroup* $\mathcal{H}_{\mathrm{aux}}^p \subset \mathrm{G}_{\mathrm{aux}}(\hat{\mathbb{Z}}^p)$ *such that* $\mathcal{H}$ *is mapped into* $\mathcal{H}_{\mathrm{aux}} = \mathcal{H}_{\mathrm{aux}}^p \mathrm{G}_{\mathrm{aux}}(\mathbb{Z}_p)$ *under the homomorphism* $\mathrm{G}(\hat{\mathbb{Z}}) \to \mathrm{G}_{\mathrm{aux}}(\hat{\mathbb{Z}})$ *given by* (2.1.1.10)*. (This condition is satisfied when* $\mathcal{H} = \mathcal{H}^p\mathcal{H}_p$ *is of standard form as in Definition* 3.2.2.9 *and* $\mathcal{H}^p$ *is neat.) Then (with this choice of* $\mathcal{H}_{\mathrm{aux}}^p$*) we can take* $N_1 = 1$ *in the above paragraph.*

Proof. This follows from Proposition 6.3.1.6, because the canonical morphism $\vec{\mathsf{M}}_{\mathcal{H}}^{\min} \to \mathsf{M}_{\mathrm{G}_{\mathrm{aux}}(\hat{\mathbb{Z}}^p)}^{\min}$ factors through $\vec{\mathsf{M}}_{\mathcal{H}}^{\min} \to \mathsf{M}_{\mathcal{H}_{\mathrm{aux}}^p}^{\min}$ in this case. (The second paragraph is self-explanatory.) □

Corollary 6.3.1.8. *Suppose we have an element* $g = (g_0, g_p) \in \mathrm{G}(\mathbb{A}^{\infty,p}) \times \mathrm{G}(\mathbb{Z}_p) \subset \mathrm{G}(\mathbb{A}^\infty)$*, and suppose we have two open compact subgroups* $\mathcal{H}$ *and* $\mathcal{H}'$ *of* $\mathrm{G}(\hat{\mathbb{Z}})$ *such that* $\mathcal{H}' \subset g\mathcal{H}g^{-1}$*, so that we have a canonical finite surjection* $\vec{[g]}^{\min} : \vec{\mathsf{M}}_{\mathcal{H}'}^{\min} \to \vec{\mathsf{M}}_{\mathcal{H}}^{\min}$ *(over* $\vec{\mathsf{S}}_0$*) and a canonical isomorphism*

$$(\vec{[g]}^{\min})^* \omega_{\vec{\mathsf{M}}_{\mathcal{H}}^{\min}}^{\otimes\, k(p-1)} \xrightarrow{\sim} \omega_{\vec{\mathsf{M}}_{\mathcal{H}'}^{\min}}^{\otimes\, k(p-1)} \tag{6.3.1.9}$$

over $\vec{\mathsf{M}}_{\mathcal{H}'}^{\min}$ *whenever the right-hand side is defined for some* $k \geq 1$*, as in Proposition* 2.2.3.1*. Up to replacing* k *with a more divisible integer, suppose that both* $\mathrm{Hasse}_{\mathcal{H}}^k$ *and* $\mathrm{Hasse}_{\mathcal{H}'}^k$ *are defined as (some powers of the ones defined) in Corollary* 6.3.1.7*. Then the canonical morphism*

$$\begin{aligned} (\vec{[g]}^{\min})^* : \Gamma(\vec{\mathsf{M}}_{\mathcal{H}}^{\min} \underset{\mathbb{Z}}{\otimes} \mathbb{F}_p, \omega_{\vec{\mathsf{M}}_{\mathcal{H}}^{\min}}^{\otimes\, k(p-1)} \underset{\mathbb{Z}}{\otimes} \mathbb{F}_p) \\ \to \Gamma(\vec{\mathsf{M}}_{\mathcal{H}'}^{\min} \underset{\mathbb{Z}}{\otimes} \mathbb{F}_p, \omega_{\vec{\mathsf{M}}_{\mathcal{H}'}^{\min}}^{\otimes\, k(p-1)} \underset{\mathbb{Z}}{\otimes} \mathbb{F}_p) \end{aligned} \tag{6.3.1.10}$$

induced by (6.3.1.9) *sends* $\mathrm{Hasse}_{\mathcal{H}}^k$ *to* $\mathrm{Hasse}_{\mathcal{H}'}^k$*.*

Proof. By the construction of $\vec{[g]}^{\min}$ (see Proposition 2.2.3.1), it is induced by some $[g_0]^{\min} : \mathsf{M}_{\mathcal{H}_{\mathrm{aux}}'^{,p}}^{\min} \to \mathsf{M}_{\mathrm{G}_{\mathrm{aux}}(\hat{\mathbb{Z}}^p)}^{\min}$, which is in turn induced by some $[g_0]^{\mathrm{tor}} : \mathsf{M}_{\mathcal{H}_{\mathrm{aux}}'^{,p}}^{\mathrm{tor}} \to \mathsf{M}_{\mathrm{G}_{\mathrm{aux}}(\hat{\mathbb{Z}}^p)}^{\mathrm{tor}}$ (with some suitable cone decompositions). By definition of $\mathrm{Hasse}_{\mathcal{H}}^k$ and $\mathrm{Hasse}_{\mathcal{H}'}^k$ in Corollary 6.3.1.8, by Proposition 6.3.1.6, and by the density of $\mathsf{M}_{\mathcal{H}_{\mathrm{aux}}'^{,p}} \underset{\mathbb{Z}}{\otimes} \mathbb{F}_p$ (resp. $\mathsf{M}_{\mathrm{G}_{\mathrm{aux}}(\hat{\mathbb{Z}}^p)} \underset{\mathbb{Z}}{\otimes} \mathbb{F}_p$) in $\mathsf{M}_{\mathcal{H}_{\mathrm{aux}}'^{,p}}^{\mathrm{tor}} \underset{\mathbb{Z}}{\otimes} \mathbb{F}_p$ (resp. $\mathsf{M}_{\mathrm{G}_{\mathrm{aux}}(\hat{\mathbb{Z}}^p)}^{\mathrm{tor}} \underset{\mathbb{Z}}{\otimes} \mathbb{F}_p$), the morphism (6.3.1.10) is induced by the corresponding morphism

$$\begin{aligned} [g_0]^* : \Gamma(\mathsf{M}_{\mathrm{G}_{\mathrm{aux}}(\hat{\mathbb{Z}}^p)} \underset{\mathbb{Z}}{\otimes} \mathbb{F}_p, \omega_{\mathsf{M}_{\mathrm{G}_{\mathrm{aux}}(\hat{\mathbb{Z}}^p)}}^{\otimes\, k_{\mathrm{aux}}(p-1)} \underset{\mathbb{Z}}{\otimes} \mathbb{F}_p) \\ \to \Gamma(\mathsf{M}_{\mathcal{H}_{\mathrm{aux}}'^{,p}} \underset{\mathbb{Z}}{\otimes} \mathbb{F}_p, \omega_{\mathsf{M}_{\mathcal{H}_{\mathrm{aux}}'^{,p}}}^{\otimes\, k_{\mathrm{aux}}(p-1)} \underset{\mathbb{Z}}{\otimes} \mathbb{F}_p) \end{aligned}$$

(for some integer $k_{\mathrm{aux}} \geq 1$ related to k), and it suffices to show that this morphism sends (the restriction of) $\mathrm{Hasse}_{\mathrm{aux}}$ to its pullback under the canonical morphism

$\mathsf{M}_{\mathcal{H}'^{,p}_{\mathrm{aux}}} \underset{\mathbb{Z}}{\otimes} \mathbb{F}_p \to \mathsf{M}_{\mathrm{G}_{\mathrm{aux}}(\hat{\mathbb{Z}}^p)} \underset{\mathbb{Z}}{\otimes} \mathbb{F}_p$. Since the morphism $[g_0]^{\mathrm{tor}}$ is defined by a Hecke twist of the universal object on $\mathsf{M}_{\mathcal{H}'^{,p}_{\mathrm{aux}}}$ (see [62, Sec. 6.4.3]) using its level structure away from p (see Proposition 3.3.2.1 for how it is realized as an $\mathbb{Z}^\times_{(p)}$-isogeny), this follows from the fact that the formation of Hasse invariants is compatible with separable isogenies (see Remark 6.3.1.2). $\square$

6.3.2 *Nonordinary and Full Ordinary Loci*

Definition 6.3.2.1. *Let* $\mathrm{Hasse}_{\mathrm{aux}}$ *be defined as in* (6.3.1.1).

(1) The locus where $\mathrm{Hasse}_{\mathrm{aux}}$ *vanishes on* $\mathsf{M}^{\mathrm{tor}}_{\mathrm{G}_{\mathrm{aux}}(\hat{\mathbb{Z}}^p)} \underset{\mathbb{Z}}{\otimes} \mathbb{F}_p$ *defines a closed subalgebraic stack with reduced structures, which we denote by* $(\mathsf{M}^{\mathrm{tor}}_{\mathrm{G}_{\mathrm{aux}}(\hat{\mathbb{Z}}^p)} \underset{\mathbb{Z}}{\otimes} \mathbb{F}_p)^{\text{non-ord}}$.

(2) The set-theoretic image of $(\mathsf{M}^{\mathrm{tor}}_{\mathrm{G}_{\mathrm{aux}}(\hat{\mathbb{Z}}^p)} \underset{\mathbb{Z}}{\otimes} \mathbb{F}_p)^{\text{non-ord}}$ *under the proper canonical morphism* $\mathsf{M}^{\mathrm{tor}}_{\mathrm{G}_{\mathrm{aux}}(\hat{\mathbb{Z}}^p)} \underset{\mathbb{Z}}{\otimes} \mathbb{F}_p \to \mathsf{M}^{\mathrm{min}}_{\mathrm{G}_{\mathrm{aux}}(\hat{\mathbb{Z}}^p)} \underset{\mathbb{Z}}{\otimes} \mathbb{F}_p$ *defines a closed subscheme with reduced structures, which we denote by* $(\mathsf{M}^{\mathrm{min}}_{\mathrm{G}_{\mathrm{aux}}(\hat{\mathbb{Z}}^p)} \underset{\mathbb{Z}}{\otimes} \mathbb{F}_p)^{\text{non-ord}}$.

(3) The set-theoretic preimage of $(\mathsf{M}^{\mathrm{min}}_{\mathrm{G}_{\mathrm{aux}}(\hat{\mathbb{Z}}^p)} \underset{\mathbb{Z}}{\otimes} \mathbb{F}_p)^{\text{non-ord}}$ *in* $\vec{\mathsf{M}}^{\mathrm{min}}_{\mathcal{H}} \underset{\mathbb{Z}}{\otimes} \mathbb{F}_p$ *under the canonical morphism in Proposition* 2.2.1.2 *defines a closed subscheme with reduced structures, which we denote by* $(\vec{\mathsf{M}}^{\mathrm{min}}_{\mathcal{H}} \underset{\mathbb{Z}}{\otimes} \mathbb{F}_p)^{\text{non-ord}}$. *We define in the same way* $(\vec{\mathsf{M}}_{\mathcal{H}} \underset{\mathbb{Z}}{\otimes} \mathbb{F}_p)^{\text{non-ord}}$, $([\vec{\mathsf{M}}_{\mathcal{H}} \underset{\mathbb{Z}}{\otimes} \mathbb{F}_p])^{\text{non-ord}}$, *and* $(\vec{\mathsf{M}}^{\mathrm{tor}}_{\mathcal{H},d_0\mathsf{pol}} \underset{\mathbb{Z}}{\otimes} \mathbb{F}_p)^{\text{non-ord}}$ *using other canonical morphisms in Propositions* 2.2.1.1, 2.2.1.2, *and* 2.2.2.1. *We use similar notation for their base changes to other rings.*

(4) Consider the notation of formal completions in Fraktur as in Definition 3.4.4.2, *we shall also denote the schemes defined in* (3) *by* $\vec{\mathfrak{M}}^{\text{min,non-ord}}_{\mathcal{H}}$, $\vec{\mathfrak{M}}^{\text{non-ord}}_{\mathcal{H}}$, $[\vec{\mathfrak{M}}_{\mathcal{H}}]^{\text{non-ord}}$, *and* $\vec{\mathfrak{M}}^{\text{tor,non-ord}}_{\mathcal{H},d_0\mathsf{pol}}$, *which we now view as closed subschemes with reduced structures in their obvious ambient formal schemes or formal algebraic stacks, respectively.*

We call these closed subschemes the **nonordinary loci** *of the various schemes. We call the open complements of these closed subschemes the* **full ordinary loci** *of the various schemes of formal schemes, and denote them with the superscripts* "full-ord" *(replacing* "non-ord"*).*

Proposition 6.3.2.2. *The open immersion* (6.1.1.16) *induces an open immersion*

$$\vec{\mathfrak{M}}^{\text{ord,min}}_{\mathcal{H}} \hookrightarrow \vec{\mathfrak{M}}^{\mathrm{min}}_{\mathcal{H},r_{\mathcal{H}}},$$

whose image factors through an open immersion

$$\vec{\mathfrak{M}}^{\text{ord,min}}_{\mathcal{H}} \hookrightarrow \vec{\mathfrak{M}}^{\text{min,full-ord}}_{\mathcal{H},r_{\mathcal{H}}} \tag{6.3.2.3}$$

that is also **closed**.

Suppose that $\mathcal{H}^p$ is neat, that Σ^{ord} extends to a (projective smooth) cone decomposition Σ for $\mathsf{M}_{\mathcal{H}}$, with a collection pol *of polarization functions (which is possible by Proposition 5.1.3.4), so that $\vec{\mathsf{M}}^{\mathrm{tor}}_{\mathcal{H},d_0\mathsf{pol}}$ is defined, for some integer $d_0 \geq 1$, as in Proposition 2.2.2.1. Then the commutative diagram (6.2.3.3) induces a commutative diagram*

$$\begin{array}{ccccc}
\vec{\mathfrak{M}}^{\mathrm{ord,tor}}_{\mathcal{H},\Sigma^{\mathrm{ord}}} & \hookrightarrow & \vec{\mathfrak{M}}^{\mathrm{tor,full\text{-}ord}}_{\mathcal{H},d_0\mathsf{pol},r_{\mathcal{H}}} & \hookrightarrow & \vec{\mathfrak{M}}^{\mathrm{tor}}_{\mathcal{H},d_0\mathsf{pol},r_{\mathcal{H}}} \\
\downarrow & & \downarrow & & \downarrow \\
\vec{\mathfrak{M}}^{\mathrm{ord,min}}_{\mathcal{H}} & \hookrightarrow & \vec{\mathfrak{M}}^{\mathrm{min,full\text{-}ord}}_{\mathcal{H},r_{\mathcal{H}}} & \hookrightarrow & \vec{\mathfrak{M}}^{\mathrm{min}}_{\mathcal{H},r_{\mathcal{H}}}
\end{array}$$

of formal schemes, in which the vertical arrows are all proper and surjective, in which all horizontal arrows are open immersions, and in which the two horizontal arrows at the left-hand side are also closed immersions.

Proof. There are two kinds of statements to be proved. The first kind is to show that the open image of the canonical morphism $\vec{\mathfrak{M}}^{\mathrm{ord,min}}_{\mathcal{H}} \hookrightarrow \vec{\mathfrak{M}}^{\mathrm{min}}_{\mathcal{H},r_{\mathcal{H}}}$ does not meet $\vec{\mathfrak{M}}^{\mathrm{min,non\text{-}ord}}_{\mathcal{H},r_{\mathcal{H}}}$, and that in the setup of the second paragraph the open image of the canonical morphism $\vec{\mathfrak{M}}^{\mathrm{ord,tor}}_{\mathcal{H},\Sigma^{\mathrm{ord}}} \hookrightarrow \vec{\mathfrak{M}}^{\mathrm{tor}}_{\mathcal{H},d_0\mathsf{pol},r_{\mathcal{H}}}$ does not meet $\vec{\mathfrak{M}}^{\mathrm{tor,non\text{-}ord}}_{\mathcal{H},d_0\mathsf{pol},r_{\mathcal{H}}}$. (The latter statement implies the former statement.) The second kind is to show that the induced open immersion $\vec{\mathfrak{M}}^{\mathrm{ord,min}}_{\mathcal{H}} \hookrightarrow \vec{\mathfrak{M}}^{\mathrm{min,full\text{-}ord}}_{\mathcal{H},r_{\mathcal{H}}}$ is closed, and that in the setup of the second paragraph the induced open immersion $\vec{\mathfrak{M}}^{\mathrm{ord,tor}}_{\mathcal{H},\Sigma^{\mathrm{ord}}} \hookrightarrow \vec{\mathfrak{M}}^{\mathrm{tor,full\text{-}ord}}_{\mathcal{H},d_0\mathsf{pol},r_{\mathcal{H}}}$ is also closed. (Again, the latter statement implies the former statement.)

For statements of both kinds, we are allowed to replace $\mathcal{H} = \mathcal{H}^p\mathcal{H}_p$ with some $\mathcal{H}' = \mathcal{H}'^{,p}\mathcal{H}'_p$, where $\mathcal{H}'^{,p}$ is a neat open compact subgroup of $\mathcal{H}^p$, and where $\mathcal{H}'_p = \mathcal{U}^{\mathrm{bal}}_{p,1}(p^r)$ if $\mathcal{U}^{\mathrm{bal}}_{p,1}(p^r) \subset \mathcal{H}_p \subset \mathcal{U}_{p,0}(p^r)$ for some integer $r \geq 0$ (and replace Σ^{ord} and Σ with suitable cone decompositions), because then the morphisms from the new setup to the current setup will consist of proper morphisms compatible with each other (*cf.* Propositions 5.2.2.2 and 6.2.2.1 for the case of $\vec{\mathfrak{M}}^{\mathrm{ord,tor}}_{\mathcal{H},\Sigma^{\mathrm{ord}}}$ and $\vec{\mathfrak{M}}^{\mathrm{ord,min}}_{\mathcal{H}}$—the case of $\vec{\mathfrak{M}}^{\mathrm{tor}}_{\mathcal{H},d_0\mathsf{pol},r_{\mathcal{H}}}$ and $\vec{\mathfrak{M}}^{\mathrm{min}}_{\mathcal{H},r_{\mathcal{H}}}$ are obvious because they are proper by themselves), such that the nonordinary loci in the new setup is the precise pullback from the current setup. Then we may assume that we are in the setup of the second paragraph, so that both Lemma 6.3.1.4 and Corollary 6.2.3.2 are applicable.

Hence, the statements of the first kind follow immediately from Lemma 6.3.1.4, because the pullback of every positive power of $\mathrm{Hasse}_{\mathrm{aux}}$ is nowhere zero on $\vec{\mathfrak{M}}^{\mathrm{ord,tor}}_{\mathcal{H},\Sigma^{\mathrm{ord}}}$.

As for statements of the second kind, since the vertical arrows are proper, it suffices to show that the canonical open immersion $\vec{\mathsf{M}}^{\mathrm{ord,tor}}_{\mathcal{H},\Sigma^{\mathrm{ord}}} \underset{\mathbb{Z}}{\otimes} \mathbb{F}_p \hookrightarrow \vec{\mathsf{M}}^{\mathrm{tor,full\text{-}ord}}_{\mathcal{H},d_0\mathsf{pol},r_{\mathcal{H}}} \underset{\mathbb{Z}}{\otimes} \mathbb{F}_p$ is closed. Note that this statement can be verified by replacing d_0 with a sufficiently large multiple $d_{0,\mathrm{aux}}$. By Proposition 6.1.1.6, it suffices to show that the canonical open immersion $\vec{\mathsf{M}}^{\mathrm{ord,tor}}_{\mathcal{H}_{\mathrm{aux}},\Sigma^{\mathrm{ord}}_{\mathrm{aux}}} \underset{\mathbb{Z}}{\otimes} \mathbb{F}_p \hookrightarrow \vec{\mathsf{M}}^{\mathrm{tor,full\text{-}ord}}_{\mathcal{H}_{\mathrm{aux}},d_{0,\mathrm{aux}}\mathsf{pol}_{\mathrm{aux}},r} \underset{\mathbb{Z}}{\otimes} \mathbb{F}_p$ is closed (for sufficiently large $d_{0,\mathrm{aux}} \geq 1$ such that $\vec{\mathsf{M}}^{\mathrm{tor,full\text{-}ord}}_{\mathcal{H}_{\mathrm{aux}},d_{0,\mathrm{aux}}\mathsf{pol}_{\mathrm{aux}},r}$ can be compatibly defined). Then we may assume that there is a semi-abelian scheme A_{aux} with additional structures λ_{aux} and i_{aux} over $\vec{\mathsf{M}}^{\mathrm{tor,full\text{-}ord}}_{\mathcal{H},d_0\mathsf{pol},r_{\mathcal{H}}} \underset{\mathbb{Z}}{\otimes} \mathbb{F}_p$. After pulled back to any strict local base scheme S, the quasi-finite flat group scheme $G_{\mathrm{aux},S}[p^r]$ (resp. $G^\vee_{\mathrm{aux},S}[p^r]$) admits a canonical subgroup scheme $G_{\mathrm{aux},S}[p^r]^{\mathrm{mult}}$ (resp. $G^\vee_{\mathrm{aux},S}[p^r]^{\mathrm{mult}}$) of multiplicative type, which is finite flat and of the same rank as $(\mathrm{Gr}^0_{\mathsf{D}_{\mathrm{aux}},p^r})^{\mathrm{mult}}_S$ (resp. $(\mathrm{Gr}^0_{\mathsf{D}^\#_{\mathrm{aux}},p^r})^{\mathrm{mult}}_S$), such that the quotient $G_{\mathrm{aux},S}[p^r]/G_{\mathrm{aux},S}[p^r]^{\mathrm{mult}}$ (resp. $G^\vee_{\mathrm{aux},S}[p^r]/G^\vee_{\mathrm{aux},S}[p^r]^{\mathrm{mult}}$) is a quasi-finite étale group scheme. (This is possible because S is in characteristic $p > 0$, and because the abelian part of every fiber of G_{aux} is ordinary. Then we can construct such a subgroup scheme by putting together the torus part and the multiplicative part of the abelian part of the torsion points; see [62, Sec. 3.4.1] for a review of the definition of the torus part and the abelian parts.) Thus, the existence of principal ordinary level-p^r structures is an open and closed condition, as desired. □

Proposition 6.3.2.4. *For any given integers $i \geq 0$ and $j \geq 1$, the scheme $\vec{\mathfrak{M}}^{\mathrm{min,full\text{-}ord}}_{\mathcal{H},i} \underset{\mathbb{Z}}{\otimes} (\mathbb{Z}/p^j\mathbb{Z}) \cong (\vec{\mathsf{M}}^{\mathrm{min}}_{\mathcal{H},i} \underset{\mathbb{Z}}{\otimes} (\mathbb{Z}/p^j\mathbb{Z}))^{\mathrm{full\text{-}ord}}$ is affine.*

Proof. By Corollary 6.3.1.7 and Definition 6.3.2.1, with N_1 as in Proposition 2.2.1.2 (for some choice of $\mathcal{H}^p_{\mathrm{aux}}$), the section $\mathrm{Hasse}^{aN_1} \in \Gamma(\vec{\mathsf{M}}^{\mathrm{min}}_{\mathcal{H}} \underset{\mathbb{Z}}{\otimes} \mathbb{F}_p, \omega^{\otimes\, aN_1(p-1)}_{\vec{\mathsf{M}}^{\mathrm{min}}_{\mathcal{H}}} \underset{\mathbb{Z}}{\otimes} \mathbb{F}_p)$ defines the subscheme $(\vec{\mathsf{M}}^{\mathrm{min}}_{\mathcal{H}} \underset{\mathbb{Z}}{\otimes} \mathbb{F}_p)^{\mathrm{non\text{-}ord}}$ of $\vec{\mathsf{M}}^{\mathrm{min}}_{\mathcal{H}} \underset{\mathbb{Z}}{\otimes} \mathbb{F}_p$ (as its vanishing locus), and its pullback to $\vec{\mathsf{M}}^{\mathrm{min}}_{\mathcal{H},i} \underset{\mathbb{Z}}{\otimes} \mathbb{F}_p$ defines the subscheme $(\vec{\mathsf{M}}^{\mathrm{min}}_{\mathcal{H},i} \underset{\mathbb{Z}}{\otimes} \mathbb{F}_p)^{\mathrm{non\text{-}ord}}$ of $\vec{\mathsf{M}}^{\mathrm{min}}_{\mathcal{H},i} \underset{\mathbb{Z}}{\otimes} \mathbb{F}_p$. Since $\omega^{\otimes\, aN_1}_{\vec{\mathsf{M}}^{\mathrm{min}}_{\mathcal{H}}}$ is ample over $\vec{\mathsf{M}}^{\mathrm{min}}_{\mathcal{H}}$ (see Proposition 2.2.1.2), and since the canonical morphism $\vec{\mathsf{M}}^{\mathrm{min}}_{\mathcal{H},i} \to \vec{\mathsf{M}}^{\mathrm{min}}_{\mathcal{H}}$ is finite (for each $i \geq 0$), this shows that $(\vec{\mathsf{M}}^{\mathrm{min}}_{\mathcal{H},i} \underset{\mathbb{Z}}{\otimes} \mathbb{F}_p)^{\mathrm{full\text{-}ord}}$ is affine. Since $x \equiv y \pmod{p}$ implies $x^{p^{j-1}} \equiv y^{p^{j-1}} \pmod{p^j}$ (in any ring), by patching over affine open subschemes of $\vec{\mathsf{M}}^{\mathrm{min}}_{\mathcal{H}} \underset{\mathbb{Z}}{\otimes} (\mathbb{Z}/p^j\mathbb{Z})$, we can uniquely lift $\mathrm{Hasse}^{p^{j-1}aN_1}$ to $\Gamma(\vec{\mathsf{M}}^{\mathrm{min}}_{\mathcal{H}} \underset{\mathbb{Z}}{\otimes} (\mathbb{Z}/p^j\mathbb{Z}), \omega^{\otimes\, p^{j-1}aN_1(p-1)}_{\vec{\mathsf{M}}^{\mathrm{min}}_{\mathcal{H}}} \underset{\mathbb{Z}}{\otimes} (\mathbb{Z}/p^j\mathbb{Z}))$. Then the same argument as above shows that $(\vec{\mathsf{M}}^{\mathrm{min}}_{\mathcal{H},i} \underset{\mathbb{Z}}{\otimes} (\mathbb{Z}/p^j\mathbb{Z}))^{\mathrm{full\text{-}ord}}$ is affine (for each $i \geq 0$), as desired. □

Corollary 6.3.2.5. *Suppose that $\mathcal{H}^p$ is neat (so that $\mathcal{H} = \mathcal{H}^p\mathcal{H}_p$ satisfies the condition in Corollary 6.3.1.7). Then, for each integer $j \geq 1$, both $\vec{\mathfrak{M}}^{\mathrm{ord,min}}_{\mathcal{H}} \underset{\mathbb{Z}}{\otimes} (\mathbb{Z}/p^j\mathbb{Z}) \cong \vec{\mathsf{M}}^{\mathrm{ord,min}}_{\mathcal{H}} \underset{\mathbb{Z}}{\otimes} (\mathbb{Z}/p^j\mathbb{Z})$ and $\vec{\mathfrak{M}}^{\mathrm{min,full\text{-}ord}}_{\mathcal{H},r_{\mathcal{H}}} \underset{\mathbb{Z}}{\otimes} (\mathbb{Z}/p^j\mathbb{Z}) \cong (\vec{\mathsf{M}}^{\mathrm{min}}_{\mathcal{H},r_{\mathcal{H}}} \underset{\mathbb{Z}}{\otimes} (\mathbb{Z}/p^j\mathbb{Z}))^{\mathrm{full\text{-}ord}}$ are affine.*

Proof. This follows from Propositions 6.3.2.2 and 6.3.2.4. □

For the sake of completeness, let us include a condition which ensures that the open and closed immersion (6.3.2.3) is actually an isomorphism when $\mathcal{H}_p = \mathrm{G}(\mathbb{Z}_p)$.

Condition 6.3.2.6. *The group $\mathrm{G}(\mathbb{Z}_p)$ acts transitively on the set of maximal totally isotropic $\mathcal{O} \underset{\mathbb{Z}}{\otimes} \mathbb{Z}_p$-modules D' of $L \underset{\mathbb{Z}}{\otimes} \mathbb{Z}_p$ satisfying the same conditions as D does as* *in Lemma* 3.2.2.1 *and Assumption* 3.2.2.10.

Lemma 6.3.2.7. *Suppose that Condition* 6.3.2.6 *holds, and that* $\mathcal{H} = \mathcal{H}^p\mathcal{H}_p$ *with* $\mathcal{H}_p = \mathrm{G}(\mathbb{Z}_p)$*. Then the canonical morphism*

$$\vec{\mathsf{M}}^{\mathrm{ord}}_{\mathcal{H}} \underset{\mathbb{Z}}{\otimes} \mathbb{F}_p \to (\vec{\mathsf{M}}_{\mathcal{H},r_{\mathcal{H}}} \underset{\mathbb{Z}}{\otimes} \mathbb{F}_p)^{\text{full-ord}} \tag{6.3.2.8}$$

(*induced by* (3.4.6.4)) *is a bijection on geometric points, and hence* (*by Zariski density of the image*) *the open and closed immersion* (6.3.2.3) *is an isomorphism. When* $\mathcal{H}^p$ *is neat, by Zariski's main theorem* (*see [35, III-1, 4.4.3, 4.4.11]*), *the canonical morphism* $\vec{\mathsf{M}}^{\mathrm{ord}}_{\mathcal{H}} \to \vec{\mathsf{M}}_{\mathcal{H},r_{\mathcal{H}}} - (\vec{\mathsf{M}}_{\mathcal{H},r_{\mathcal{H}}} \underset{\mathbb{Z}}{\otimes} \mathbb{F}_p)^{\text{non-ord}}$ *between noetherian normal schemes over* $\mathrm{Spec}(\mathbb{Z}_{(p)})$*, lifting the morphism* (6.3.2.8) *over* $\mathrm{Spec}(\mathbb{F}_p)$*, is also an isomorphism.*

Proof. Given a geometric point $s = \mathrm{Spec}(k) \to (\vec{\mathsf{M}}_{\mathcal{H},r_{\mathcal{H}}} \underset{\mathbb{Z}}{\otimes} \mathbb{F}_p)^{\text{full-ord}}$, by the construction of $\vec{\mathsf{M}}_{\mathcal{H},r_{\mathcal{H}}}$ as a normalization, we may assume that there exists an abelian scheme A over $S = \mathrm{Spec}(R)$, where R is a complete discrete valuation ring with fraction field K of characteristic zero and algebraically closed residue field k of characteristic $p > 0$, and assume that there exist a morphism $\xi : S \to \vec{\mathsf{M}}_{\mathcal{H},r_{\mathcal{H}}}$ (see Proposition 2.2.1.1) lifting $\mathrm{Spec}(k) \to (\vec{\mathsf{M}}_{\mathcal{H},r_{\mathcal{H}}} \underset{\mathbb{Z}}{\otimes} \mathbb{F}_p)^{\text{non-ord}}$, as in Section 3.2.1. Then, by extending the pullback of the tautological object over $\mathsf{M}_{\mathcal{H}}$ to the noetherian normal S (by [92, IX, 1.4], [28, Ch. I, Prop. 2.7], or [62, Prop. 3.3.1.5], and by extending isomorphisms between finite étale group schemes), A also carries a polarization λ and an $\mathcal{O}$-endomorphism structure i, and a level $\mathcal{H}^p$-structure $\alpha_{\mathcal{H}^p}$ (see [62, Def. 1.3.7.6] and Definition 3.3.1.4). Moreover, since $\mathcal{H}_p = \mathrm{G}(\mathbb{Z}_p)$, by Proposition 3.2.1.1, (A, λ, i) also carries an ordinary level-$\mathcal{H}_p$ structure $\alpha^{\mathrm{ord},\prime}_{\mathcal{H}_p}$ of (A, λ, i) of type $(L \underset{\mathbb{Z}}{\otimes} \mathbb{Z}_p, \langle\,\cdot\,,\,\cdot\,\rangle, \mathtt{D}')$ (as in Definition 3.3.3.4), for some filtration $\mathtt{D}'$ of $L \underset{\mathbb{Z}}{\otimes} \mathbb{Z}_p$ satisfying the same conditions as $\mathtt{D}$ does as in Lemma 3.2.2.1 and Assumption 3.2.2.10. In general, $\mathtt{D}'$ can be different from $\mathtt{D}$.

Since $\mathcal{H}_p = \mathrm{G}(\mathbb{Z}_p)$ acts transitively on the set of such $\mathtt{D}'$, there exists $g_p \in \mathrm{G}(\mathbb{Z}_p)$ which defines an isomorphism matching $(L \underset{\mathbb{Z}}{\otimes} \mathbb{Z}_p, \langle\,\cdot\,,\,\cdot\,\rangle, \mathtt{D}, \mathtt{D}^{\#}, \phi_{\mathtt{D}})$ with $(L \underset{\mathbb{Z}}{\otimes} \mathbb{Z}_p, \langle\,\cdot\,,\,\cdot\,\rangle, \mathtt{D}', \mathtt{D}'^{,\#}, \phi'_{\mathtt{D}})$, and hence (without modifying $(A, \lambda, i, \alpha_{\mathcal{H}^p})$) the tuple $(A, \lambda, i, \alpha_{\mathcal{H}^p}, \alpha^{\mathrm{ord},\prime}_{\mathcal{H}_p}) \to S$ (with $\alpha^{\mathrm{ord},\prime}_{\mathcal{H}_p}$ defined by $\mathtt{D}'$) can be canonically identified with a tuple $(A, \lambda, i, \alpha_{\mathcal{H}^p}, \alpha^{\mathrm{ord}}_{\mathcal{H}_p}) \to S$ parameterized by $\vec{\mathsf{M}}^{\mathrm{ord}}_{\mathcal{H}}$ (with $\alpha^{\mathrm{ord}}_{\mathcal{H}_p}$ defined by $\mathtt{D}$), with $(\alpha_{\mathcal{H}^p}, \alpha^{\mathrm{ord}}_{\mathcal{H}_p}) \underset{\mathbb{Z}}{\otimes} \mathbb{Q}$ induced by $\alpha_{\mathcal{H}}$ as in Proposition 3.3.5.1. Thus, $\xi : S \to \vec{\mathsf{M}}_{\mathcal{H}}$ must factor through $S \to \vec{\mathsf{M}}^{\mathrm{ord}}_{\mathcal{H}} \to \vec{\mathsf{M}}_{\mathcal{H},r_{\mathcal{H}}}$ (see (3.4.6.4)). Since the geometric point $s = \mathrm{Spec}(k) \to (\vec{\mathsf{M}}_{\mathcal{H},r_{\mathcal{H}}} \underset{\mathbb{Z}}{\otimes} \mathbb{F}_p)^{\text{non-ord}}$ is arbitrary, the lemma follows, as desired. □

Lemma 6.3.2.9. *Condition* 6.3.2.6 *is true when either* p *is a good prime as in Definition* 1.1.1.6, *or when the Iwasawa decomposition* $\mathrm{G}(\mathbb{Q}_p) = \mathrm{P}^{\mathrm{ord}}_{\mathtt{D}}(\mathbb{Q}_p)\mathrm{G}(\mathbb{Z}_p)$ *holds. By [12, Prop. 4.4.3]* (*see also [14, (18) on p. 392] for a more explicit statement*), *the latter is true when, for example,* $\mathrm{G} \underset{\mathbb{Z}}{\otimes} \mathbb{Q}_p$ *is connected (which is the case when* $\mathcal{O} \underset{\mathbb{Z}}{\otimes} \mathbb{Q}$ *involves no simple factor of type* D *as in [62, Def. 1.2.1.15]; cf. [53, Sec. 7, p. 393]) and when* $\mathrm{G}(\mathbb{Z}_p)$ *is maximal open compact in* $\mathrm{G}(\mathbb{Q}_p)$.

Proof. When p is a good prime, this follows from the Gram–Schmidt process as in [62, Prop. 1.2.4.5] (with $R = \mathbb{Z}_p$ and $k = \mathbb{F}_p$). Otherwise, by the same [62, Prop. 1.2.4.5] (but with $R = k = \mathbb{Q}_p$ there), we know that there exists some element $g_p \in \mathrm{G}(\mathbb{Q}_p)$ such that $\mathsf{D}'^{,0} \underset{\mathbb{Z}_p}{\otimes} \mathbb{Q}_p = g_p(\mathsf{D}^0 \underset{\mathbb{Z}_p}{\otimes} \mathbb{Q}_p)$. By the Iwasawa decomposition $\mathrm{G}(\mathbb{Q}_p) = \mathrm{P}^{\mathrm{ord}}_{\mathsf{D}}(\mathbb{Q}_p)\mathrm{G}(\mathbb{Z}_p)$, since $\mathrm{P}^{\mathrm{ord}}_{\mathsf{D}}(\mathbb{Q}_p)$ stabilizes $\mathsf{D} \underset{\mathbb{Z}_p}{\otimes} \mathbb{Q}_p$, we may assume that $g_p \in \mathrm{G}(\mathbb{Z}_p)$, as desired. (We note that the underlying $\mathcal{O} \underset{\mathbb{Z}}{\otimes} \mathbb{Z}_p$-modules Gr_{D} and $\mathrm{Gr}_{\mathsf{D}'}$ are isomorphic for more basic reasons: As in the proof of Lemma 3.2.2.6, we may assume that they are $\mathcal{O}' \underset{\mathbb{Z}}{\otimes} \mathbb{Z}_p$-modules, for some maximal $\mathcal{O}'$ as in Condition 1.2.1.1. Then it follows from [93, Thm. 18.10] that they are isomorphic because their $\mathbb{Q}_p$-spans are.) □

Remark 6.3.2.10. Although Lemma 6.3.2.9 can be improved, we omit the further discussions for the sake of simplicity.

6.3.3 *Nonemptiness of Ordinary Loci*

So far we have not touched upon the question of whether $\vec{\mathsf{M}}^{\mathrm{ord}}_{\mathcal{H}} \underset{\mathbb{Z}}{\otimes} \mathbb{F}_p$ is even *nonempty*. When p is good (as in Definition 1.1.1.6), one can show that it follows from [102] (and the surjectivity of (3.4.5.6)) that $\vec{\mathsf{M}}^{\mathrm{ord}}_{\mathcal{H}} \underset{\mathbb{Z}}{\otimes} \mathbb{F}_p$ is nonempty if and only if Assumption 3.2.2.10 holds. (It suffices to verify this when $\mathcal{H} = \mathcal{H}^p\mathrm{G}(\mathbb{Z}_p)$. By [102], the full ordinary locus of $\mathsf{M}_{\mathcal{H}^p} \underset{\mathbb{Z}}{\otimes} \mathbb{F}_p$ is open and dense when the first half of Assumption 3.2.2.10 holds. Such a full ordinary locus contains a nonempty ordinary locus when the second half of Assumption 3.2.2.10 also holds—note that, by Lemmas 6.3.2.7 and 6.3.2.9, this is automatic except when $\mathcal{O} \underset{\mathbb{Z}}{\otimes} \mathbb{Q}$ involves some factor of type D, or when $\mathrm{G}(\mathbb{Z}_p)$ fails to be a maximal open compact in $\mathrm{G}(\mathbb{Q}_p)$—by [12, Cor. 3.3.2], up to modifying the choice of the integral PEL datum, the latter can always be avoided.) This simple criterion, however, does not necessarily apply when p is not good. We shall record in this subsection some simple-minded (but rather restrictive) implication of the construction of partial toroidal compactifications.

Proposition 6.3.3.1. *In the construction of* $\vec{\mathsf{M}}^{\mathrm{ord,tor}}_{\mathcal{H},\Sigma^{\mathrm{ord}}}$ *in Theorem* 5.2.1.1, *if* $\vec{\mathsf{M}}^{\mathrm{ord},\mathsf{Z}_{\mathcal{H}}}_{\mathcal{H}} \underset{\mathbb{Z}}{\otimes} \mathbb{F}_p$ *is nonempty for some* $\mathsf{Z}_{\mathcal{H}}$ *(forming part of the representative of some cusp label* $[(\mathsf{Z}_{\mathcal{H}}, \Phi_{\mathcal{H}}, \delta_{\mathcal{H}})]$*), then* $\vec{\mathsf{M}}^{\mathrm{ord}}_{\mathcal{H}} \underset{\mathbb{Z}}{\otimes} \mathbb{F}_p$ *and* $\vec{\mathsf{M}}^{\mathrm{ord,tor}}_{\mathcal{H},\Sigma^{\mathrm{ord}}} \underset{\mathbb{Z}}{\otimes} \mathbb{F}_p$ *are nonempty.*

Proof. By (5) of Theorem 5.2.1.1, the formal completion $(\vec{\mathsf{M}}^{\mathrm{ord,tor}}_{\mathcal{H}})^{\wedge}_{\vec{\mathsf{Z}}^{\mathrm{ord}}_{[(\Phi_{\mathcal{H}},\delta_{\mathcal{H}},\sigma)]}}$ of $\vec{\mathsf{M}}^{\mathrm{ord,tor}}_{\mathcal{H}}$ along the $[(\Phi_{\mathcal{H}}, \delta_{\mathcal{H}}, \sigma)]$-stratum $\vec{\mathsf{Z}}^{\mathrm{ord}}_{[(\Phi_{\mathcal{H}},\delta_{\mathcal{H}},\sigma)]}$ is canonically isomorphic to the formal algebraic stack $\vec{\mathfrak{X}}^{\mathrm{ord}}_{\Phi_{\mathcal{H}},\delta_{\mathcal{H}},\sigma}/\Gamma_{\Phi_{\mathcal{H}},\sigma}$. Since $\vec{\Xi}^{\mathrm{ord}}_{\Phi_{\mathcal{H}},\delta_{\mathcal{H}}}$ and $\vec{\Xi}^{\mathrm{ord}}_{\Phi_{\mathcal{H}},\delta_{\mathcal{H}},\sigma}$ are both torus torsors over the abelian scheme torsor $\vec{C}^{\mathrm{ord}}_{\Phi_{\mathcal{H}},\delta_{\mathcal{H}}}$ over the nontrivial finite étale cover $\vec{\mathsf{M}}^{\mathrm{ord},\Phi_{\mathcal{H}}}_{\mathcal{H}}$ of $\vec{\mathsf{M}}^{\mathrm{ord},\mathsf{Z}_{\mathcal{H}}}_{\mathcal{H}}$, and since $\vec{\Xi}^{\mathrm{ord}}_{\Phi_{\mathcal{H}},\delta_{\mathcal{H}}} \hookrightarrow \vec{\Xi}^{\mathrm{ord}}_{\Phi_{\mathcal{H}},\delta_{\mathcal{H}}}(\sigma)$ is an affine toroidal embedding over $\vec{C}^{\mathrm{ord}}_{\Phi_{\mathcal{H}},\delta_{\mathcal{H}}}$, if $\vec{\mathsf{M}}^{\mathrm{ord},\mathsf{Z}_{\mathcal{H}}}_{\mathcal{H}} \underset{\mathbb{Z}}{\otimes} \mathbb{F}_p$ is nonempty, then $\vec{\Xi}^{\mathrm{ord}}_{\Phi_{\mathcal{H}},\delta_{\mathcal{H}},\sigma} \underset{\mathbb{Z}}{\otimes} \mathbb{F}_p$ is nonempty, and $\vec{\Xi}^{\mathrm{ord}}_{\Phi_{\mathcal{H}},\delta_{\mathcal{H}}} \underset{\mathbb{Z}}{\otimes} \mathbb{F}_p$ is nonempty and open dense in $\vec{\Xi}^{\mathrm{ord}}_{\Phi_{\mathcal{H}},\delta_{\mathcal{H}}}(\sigma) \underset{\mathbb{Z}}{\otimes} \mathbb{F}_p$. Since $\vec{\mathfrak{X}}^{\mathrm{ord}}_{\Phi_{\mathcal{H}},\delta_{\mathcal{H}},\sigma}$ is the formal completion of $\vec{\Xi}^{\mathrm{ord}}_{\Phi_{\mathcal{H}},\delta_{\mathcal{H}}}(\sigma)$ (which contains $\vec{\Xi}^{\mathrm{ord}}_{\Phi_{\mathcal{H}},\delta_{\mathcal{H}}}$ as an open subalgebraic stack) along $\vec{\Xi}^{\mathrm{ord}}_{\Phi_{\mathcal{H}},\delta_{\mathcal{H}},\sigma}$, this shows that $\vec{\mathfrak{X}}^{\mathrm{ord}}_{\Phi_{\mathcal{H}},\delta_{\mathcal{H}},\sigma} \underset{\mathbb{Z}}{\otimes} \mathbb{F}_p$ is nonempty, which implies that $\vec{\mathsf{M}}^{\mathrm{ord,tor}}_{\mathcal{H},\Sigma^{\mathrm{ord}}} \underset{\mathbb{Z}}{\otimes} \mathbb{F}_p$ is nonempty. Moreover, the nonemptiness and open density of $\vec{\Xi}^{\mathrm{ord}}_{\Phi_{\mathcal{H}},\delta_{\mathcal{H}}} \underset{\mathbb{Z}}{\otimes} \mathbb{F}_p$ in $\vec{\Xi}^{\mathrm{ord}}_{\Phi_{\mathcal{H}},\delta_{\mathcal{H}}}(\sigma) \underset{\mathbb{Z}}{\otimes} \mathbb{F}_p$ implies that $\vec{\mathsf{M}}^{\mathrm{ord}}_{\mathcal{H}} \underset{\mathbb{Z}}{\otimes} \mathbb{F}_p$ is also nonempty. □

Corollary 6.3.3.2. *If there exists a fully symplectic admissible filtration* Z *on* $L \underset{\mathbb{Z}}{\otimes} \hat{\mathbb{Z}}$ *with respect to* $(L, \langle \cdot, \cdot \rangle)$ *as in Definition* 1.2.1.3 *that is compatible with* D *as in Definition* 3.2.3.1 *such that* $\mathrm{Gr}^{\mathsf{Z}}_{-1} = \mathsf{Z}_{-1}/\mathsf{Z}_{-2} = \{0\}$, *then* $\vec{\mathsf{M}}^{\mathrm{ord}}_{\mathcal{H}} \underset{\mathbb{Z}}{\otimes} \mathbb{F}_p$ *and* $\vec{\mathsf{M}}^{\mathrm{ord,tor}}_{\mathcal{H},\Sigma^{\mathrm{ord}}} \underset{\mathbb{Z}}{\otimes} \mathbb{F}_p$ *are nonempty.*

Proof. This follows from Proposition 6.3.3.1 because, in this case, we can extend the $\mathcal{H}$-orbit $\mathsf{Z}_{\mathcal{H}}$ of Z to a representative $(\mathsf{Z}_{\mathcal{H}}, \Phi_{\mathcal{H}}, \delta_{\mathcal{H}})$ of an ordinary cusp label (as in Definition 3.2.3.8), and the zero-dimensional $\vec{\mathsf{M}}^{\mathrm{ord},\mathsf{Z}_{\mathcal{H}}}_{\mathcal{H}} \underset{\mathbb{Z}}{\otimes} \mathbb{F}_p$ is trivially nonempty. □

Remark 6.3.3.3. In practice, in the setup of Corollary 6.3.3.2, we may choose D after finding a Z such that $\mathrm{Gr}^{\mathsf{Z}}_{-1}$ is trivial. Although this seems very restrictive, it is applicable whenever G admits a rational parabolic subgroup with abelian unipotent radical. This is thus applicable, for example, to the construction of Galois representations for cohomological automorphic representations of general linear groups over CM or totally real fields (without any polarizability condition) in [39].

Chapter 7

Ordinary Kuga Families

In this chapter, we continue to assume the same settings as in Section 5.2. Our main goal is to generalize the results in Section 1.3.3 to the context of ordinary loci. We will first introduce some analogues of the definitions and results in Section 1.3.3, and explain how the proofs in [61] can be translated into this context.

7.1 Partial Toroidal Compactifications

7.1.1 *Parameters for Ordinary Kuga Families*

Let Q, $\mathcal{O}'$, Q_{-2}, Q_0, $(\widetilde{L}, \langle\,\cdot\,,\,\cdot\,\rangle^{\sim}, \widetilde{h}_0)$, $(\widetilde{\mathsf{Z}}, \widetilde{\Phi}, \widetilde{\delta})$, *etc* be chosen as in Section 1.2.4. (To make sure Theorems 1.3.3.15 and 7.1.4.1 below are compatible, we need to make identical choices.) The filtration $\mathtt{D}$ of $L \underset{\mathbb{Z}}{\otimes} \mathbb{Z}_p$ defines a filtration $\widetilde{\mathtt{D}}$ of $\widetilde{L} \underset{\mathbb{Z}}{\otimes} \mathbb{Z}_p$ by setting

$$\widetilde{\mathtt{D}}^1 = 0 \subset \widetilde{\mathtt{D}}^0 = (\widetilde{\mathsf{Z}}_{-2} \underset{\hat{\mathbb{Z}}}{\otimes} \mathbb{Z}_p) \oplus \mathtt{D}^0 \subset \widetilde{\mathtt{D}}^{-1} = \widetilde{L} \underset{\mathbb{Z}}{\otimes} \mathbb{Z}_p. \tag{7.1.1.1}$$

Then $(\widetilde{\mathsf{Z}}, \widetilde{\Phi}, \widetilde{\delta})$ is compatible with $\widetilde{\mathtt{D}}$ (as in Definition 3.2.3.1).

Consider any open compact subgroup $\widetilde{\mathcal{H}} \subset \widetilde{\mathrm{G}}(\hat{\mathbb{Z}})$ of standard form with respect to $\widetilde{\mathtt{D}}$ as in Definition 3.2.2.9, so that $\widetilde{\mathcal{H}} = \widetilde{\mathcal{H}}^p \widetilde{\mathcal{H}}_p$ and $\widetilde{\mathcal{U}}^{\mathrm{bal}}_{p,1}(p^r) \subset \widetilde{\mathcal{H}}_p \subset \widetilde{\mathcal{U}}_{p,0}(p^r)$ for $r = \mathrm{depth}_{\widetilde{\mathtt{D}}}(\widetilde{\mathcal{H}})$; and $\widetilde{\nu}(\widetilde{\mathcal{H}}_p) = \ker(\mathbb{Z}_p^\times \to (\mathbb{Z}/p^{r_{\widetilde{\nu}}}\mathbb{Z})^\times)$ (where $r_{\widetilde{\nu}} \leq r$). (Here $\widetilde{\mathcal{U}}^{\mathrm{bal}}_{p,1}(p^r)$ and $\widetilde{\mathcal{U}}_{p,0}(p^r)$ are subgroups of $\widetilde{\mathrm{G}}(\mathbb{Z}_p)$ as in Definition 3.2.2.8.) Let $r_{\widetilde{\mathcal{H}}}$ be as in Definition 3.4.2.1.

Note that $\widehat{\mathcal{H}} = \widetilde{\mathcal{H}}_{\widehat{\mathrm{G}}}$ (see Definition 1.2.4.4) is *of standard form* of depth r in the following sense:

Definition 7.1.1.2. (*Compare with Definition* 3.2.2.9.) *For any integer $r \geq 0$, let us write $\widehat{\mathcal{U}}^{\mathrm{bal}}_{p,1}(p^r) := \widetilde{\mathcal{U}}^{\mathrm{bal}}_{p,1}(p^r)_{\widehat{\mathrm{G}}}$ and $\widehat{\mathcal{U}}_{p,0}(p^r) := \widetilde{\mathcal{U}}_{p,0}(p^r)_{\widehat{\mathrm{G}}}$ (cf. Definition* 1.2.4.4).

We say that an open compact subgroup $\widehat{\mathcal{H}}_p \subset \widehat{\mathrm{G}}(\mathbb{Z}_p)$ is **of standard form with respect to** $\mathtt{D}$ *if there exists an integer $r \geq 0$ such that*

$$\widehat{\mathcal{U}}^{\mathrm{bal}}_{p,1}(p^r) \subset \widehat{\mathcal{H}}_p \subset \widehat{\mathcal{U}}_{p,0}(p^r).$$

In this case, we say that r is the **depth** *of $\widehat{\mathcal{H}}_p$, and write $r = \mathrm{depth}_{\mathtt{D}}(\widehat{\mathcal{H}}_p)$. (The notation makes sense because $\widetilde{\mathtt{D}}$ is uniquely determined by $\mathtt{D}$.)*

We say that an open compact subgroup $\widehat{\mathcal{H}} \subset \widehat{\mathrm{G}}(\hat{\mathbb{Z}})$ *is* **of standard form with respect to** D *if it is of the form* $\widehat{\mathcal{H}} = \widehat{\mathcal{H}}^p\widehat{\mathcal{H}}_p$*, where* $\widehat{\mathcal{H}}^p \subset \widehat{\mathrm{G}}(\hat{\mathbb{Z}}^p)$ *and* $\widehat{\mathcal{H}}_p \subset \widehat{\mathrm{G}}(\mathbb{Z}_p)$*, such that* $\widehat{\mathcal{H}}_p$ *is of standard form with respect to* D*. In this case, we set* $\mathrm{depth}_{\mathsf{D}}(\widehat{\mathcal{H}}) := \mathrm{depth}_{\mathsf{D}}(\widehat{\mathcal{H}}_p)$.

We say that two open compact subgroups $\widehat{\mathcal{H}}_p$ *and* $\widehat{\mathcal{H}}'_p$ *of* $\widehat{\mathrm{G}}(\mathbb{Z}_p)$ *(resp.* $\widehat{\mathcal{H}}$ *and* $\widehat{\mathcal{H}}'$ *of* $\widehat{\mathrm{G}}(\hat{\mathbb{Z}})$*) of standard form with respect to* D *are* **equally deep** *if* $\mathrm{depth}_{\mathsf{D}}(\widehat{\mathcal{H}}_p) = \mathrm{depth}_{\mathsf{D}}(\widehat{\mathcal{H}}'_p)$ *(resp.* $\mathrm{depth}_{\mathsf{D}}(\widehat{\mathcal{H}}) = \mathrm{depth}_{\mathsf{D}}(\widehat{\mathcal{H}}')$*).*

We shall suppress the term "with respect to D*" when the choice of* D *is clear from the context.*

Remark 7.1.1.3. By definition, if $\widehat{\mathcal{H}}_p$ (resp. $\widehat{\mathcal{H}}$) is of standard form, then so is $(\widehat{\mathcal{H}}_p)_{\mathrm{G}}$ (resp. $\widehat{\mathcal{H}}_{\mathrm{G}}$; see Definition 1.2.4.4), and $\mathrm{depth}_{\mathsf{D}}(\widehat{\mathcal{H}}_p) = \mathrm{depth}_{\mathsf{D}}((\widehat{\mathcal{H}}_p)_{\mathrm{G}})$ (resp. $\mathrm{depth}_{\mathsf{D}}(\widehat{\mathcal{H}}) = \mathrm{depth}_{\mathsf{D}}(\widehat{\mathcal{H}}_{\mathrm{G}})$).

Let $\mathcal{H} \subset \mathrm{G}(\hat{\mathbb{Z}})$ be of standard form (with respect to D) as in Definition 3.2.2.9, and let $r_{\mathcal{H}}$ be as in Definition 3.4.2.1 (so that Theorem 5.2.1.1 and its consequences hold). Let $\widetilde{\mathcal{H}}$, $\widetilde{\mathcal{H}}^p$, and $r_{\widetilde{\mathcal{H}}}$ be as in the previous paragraph, such that $\widetilde{\mathcal{H}}^p$ is *neat* and such that $\widetilde{\mathcal{H}}$ satisfies Condition 1.2.4.7 (which involves $\mathcal{H}$). Then $\widehat{\mathcal{H}} = \widetilde{\mathcal{U}}_{\widehat{\mathrm{G}}}$ (as we have seen above) satisfies the following:

Condition 7.1.1.4. $\widehat{\mathcal{H}}$ *is of standard form with respect to* D *as in Definition* 7.1.1.2*,* $\widehat{\mathcal{H}}^p$ *is neat, and* $\widehat{\mathcal{H}}_{\mathrm{G}}$ *(see Remark* 7.1.1.3*) is also of standard form with respect to* D *as in Definition* 3.2.2.9*, so that* $r_{\widehat{\mathcal{H}}_{\mathrm{G}}}$ *is defined as in Definition* 3.4.2.1*.*

(This is a condition when we consider $\widehat{\mathcal{H}}$ alone, without referring to $\widetilde{\mathcal{H}}$.) We shall assume that $\widetilde{\mathcal{H}}$ (or $\widehat{\mathcal{H}}$) satisfies moreover the following:

Condition 7.1.1.5. $r_{\widetilde{\mathcal{H}}} = r_{\widehat{\mathcal{H}}_{\mathrm{G}}}$ $(\geq r_{\mathcal{H}})$.

Remark 7.1.1.6. (Compare with Remark 1.2.4.10.) For each $\mathcal{H}$ as above, there exists $\widetilde{\mathcal{H}}$ satisfying these conditions, which we may also require to satisfy Conditions 1.2.4.8 and 1.2.4.9, because the pairing $\langle\,\cdot\,,\,\cdot\,\rangle^{\sim}$ is the direct sum of the pairings on $Q_{-2} \oplus Q_0$ and on L.

Since $(\widetilde{\mathsf{Z}}, \widetilde{\Phi}, \widetilde{\delta})$ is compatible with $\widetilde{\mathsf{D}}$ (as in Definition 3.2.3.1), it induces a representative $(\widetilde{\mathsf{Z}}_{\widetilde{\mathcal{H}}}, \widetilde{\Phi}_{\widetilde{\mathcal{H}}} = (\widetilde{X}, \widetilde{Y}, \widetilde{\phi}, \widetilde{\varphi}_{-2,\widetilde{\mathcal{H}}}, \widetilde{\varphi}_{0,\widetilde{\mathcal{H}}}), \widetilde{\delta}_{\widetilde{\mathcal{H}}})$ of an *ordinary* cusp label $[(\widetilde{\mathsf{Z}}_{\widetilde{\mathcal{H}}}, \widetilde{\Phi}_{\widetilde{\mathcal{H}}}, \widetilde{\delta}_{\widetilde{\mathcal{H}}})]$ at level $\widetilde{\mathcal{H}}$ (as in Definition 3.2.3.8). Let $\widetilde{\Sigma}^{\mathrm{ord}}$ be any compatible choice of admissible smooth rational polyhedral cone decomposition data for $\vec{\widetilde{\mathsf{M}}}^{\mathrm{ord}}_{\widetilde{\mathcal{H}}}$ that is *projective* (see Definitions 5.1.3.1 and 5.1.3.3). Let $\widetilde{\sigma} \subset \mathbf{P}^+_{\widetilde{\Phi}_{\widetilde{\mathcal{H}}}}$ be any *top-dimensional* nondegenerate rational polyhedral cone in the cone decomposition $\widetilde{\Sigma}_{\widetilde{\Phi}_{\widetilde{\mathcal{H}}}}$ in $\widetilde{\Sigma}^{\mathrm{ord}}$.

Definition 7.1.1.7. (*Compare with Definition* 1.2.4.11.)

(1) $\widetilde{\mathbf{K}}^{\mathrm{ord},++}_{Q,\mathcal{H}}$ *is the set of all triples* $\widetilde{\kappa} = (\widetilde{\mathcal{H}}, \widetilde{\Sigma}^{\mathrm{ord}}, \widetilde{\sigma})$ *as above (such that* $\widetilde{\mathcal{H}}$ *satisfies Conditions* 1.2.4.7 *and* 7.1.1.5*).*

(2) $\widetilde{\mathbf{K}}^{\mathrm{ord},+}_{Q,\mathcal{H}}$ *is the subset of* $\widetilde{\mathbf{K}}^{\mathrm{ord},++}_{Q,\mathcal{H}}$ *consisting of elements* $\widetilde{\kappa} = (\widetilde{\mathcal{H}}, \widetilde{\Sigma}^{\mathrm{ord}}, \widetilde{\sigma})$ *such that* $\widetilde{\mathcal{H}}$ *satisfies Condition* 1.2.4.8.

(3) $\widetilde{\mathbf{K}}^{\mathrm{ord}}_{Q,\mathcal{H}}$ *is the subset of* $\widetilde{\mathbf{K}}^{\mathrm{ord},+}_{Q,\mathcal{H}}$ *consisting of elements* $\widetilde{\kappa} = (\widetilde{\mathcal{H}}, \widetilde{\Sigma}^{\mathrm{ord}}, \widetilde{\sigma})$ *such that* $\widetilde{\mathcal{H}}$ *also satisfies Condition* 1.2.4.9.

The equivalence classes $[(\breve{\Phi}_{\widetilde{\mathcal{H}}}, \breve{\delta}_{\widetilde{\mathcal{H}}}, \breve{\tau})]$ having $[(\widetilde{\Phi}_{\widetilde{\mathcal{H}}}, \widetilde{\delta}_{\widetilde{\mathcal{H}}}, \widetilde{\sigma})]$ as a face has been described in Section 1.2.4 (following Definition 1.2.4.11), and the *ordinary cusp labels* $[(\breve{\mathsf{Z}}_{\widetilde{\mathcal{H}}}, \breve{\Phi}_{\widetilde{\mathcal{H}}}, \breve{\delta}_{\widetilde{\mathcal{H}}})]$ (see Definition 3.2.3.8) having $[(\widetilde{\mathsf{Z}}_{\widetilde{\mathcal{H}}}, \widetilde{\Phi}_{\widetilde{\mathcal{H}}}, \widetilde{\delta}_{\widetilde{\mathcal{H}}})]$ as a face have representatives $(\breve{\mathsf{Z}}_{\widetilde{\mathcal{H}}}, \breve{\Phi}_{\widetilde{\mathcal{H}}}, \breve{\delta}_{\widetilde{\mathcal{H}}})$ which are $\widetilde{\mathcal{H}}$-orbits of triples $(\breve{\mathsf{Z}}, \breve{\Phi}, \breve{\delta})$ such that $\breve{\mathsf{Z}}$ (as in (1.2.4.12)) satisfies the additional property that it is compatible with $\widetilde{\mathsf{D}}$ (as in Definition 3.2.3.1). In this case, the filtration Z induced by $\breve{\mathsf{Z}}$ as in Section 1.2.4 (following Definition 1.2.4.11) is compatible with D (*cf.* (1.2.4.13)). Hence, we have:

Lemma 7.1.1.8. *Under the canonical surjective assignment (given the splitting* $\widetilde{\delta}$*; see* (3) *following Definition* 1.2.4.11*) from the set of cusp labels* $[(\breve{\mathsf{Z}}_{\widetilde{\mathcal{H}}}, \breve{\Phi}_{\widetilde{\mathcal{H}}}, \breve{\delta}_{\widetilde{\mathcal{H}}})]$ *at level* $\widetilde{\mathcal{H}}$ *admitting a surjection to* $[(\widetilde{\mathsf{Z}}_{\widetilde{\mathcal{H}}}, \widetilde{\Phi}_{\widetilde{\mathcal{H}}}, \widetilde{\delta}_{\widetilde{\mathcal{H}}})]$*, to the set of cusp labels* $[(\mathsf{Z}_{\mathcal{H}}, \Phi_{\mathcal{H}}, \delta_{\mathcal{H}})]$ *at level* $\mathcal{H}$*, the* **ordinary ones** *are mapped to ordinary ones, and the preimage of the set of ordinary ones is the set of ordinary ones. The induced assignment from ordinary ones to ordinary ones is bijective if we assume Condition* 1.2.4.8*, and is still surjective if we only assume* $\mathrm{Gr}^{\widetilde{\mathsf{Z}}}_{-1}(\widetilde{\mathcal{H}}_{\mathrm{P}_{\widetilde{\mathsf{Z}}}}) \subset \mathcal{H}$.

Then we have the diagram (1.2.4.14) and the morphisms (1.2.4.18), (1.2.4.19), and (1.2.4.20); we define $\breve{\sigma}$ to be the image of $\widetilde{\sigma} \subset \mathbf{P}^{+}_{\widetilde{\Phi}_{\widetilde{\mathcal{H}}}}$ under the first morphism in (1.2.4.20); we consider as in Definition 1.2.4.21 the subsets $\widetilde{\Sigma}_{\breve{\Phi}_{\widetilde{\mathcal{H}}},\breve{\sigma}}$ and $\widetilde{\Sigma}^{+}_{\breve{\Phi}_{\widetilde{\mathcal{H}}},\breve{\sigma}}$ of $\widetilde{\Sigma}_{\breve{\Phi}_{\widetilde{\mathcal{H}}}}$ and the subgroups $\Gamma_{\breve{\Phi}_{\widetilde{\mathcal{H}}},\Phi_{\mathcal{H}}}$, $\Gamma_{\breve{\Phi}_{\widetilde{\mathcal{H}}},\Phi_{\mathcal{H}},\breve{\sigma}}$, $\Gamma_{\breve{\Phi}_{\widehat{\mathcal{H}}}}$, and $\Gamma_{\breve{\Phi}_{\widehat{\mathcal{H}}},\Phi_{\mathcal{H}}}$ of $\Gamma_{\breve{\Phi}_{\widetilde{\mathcal{H}}}}$; and, most importantly, we define $\widehat{\mathbf{S}}_{\breve{\Phi}_{\widehat{\mathcal{H}}}}$, $(\widehat{\mathbf{S}}_{\breve{\Phi}_{\widehat{\mathcal{H}}}})^{\vee}_{\mathbb{R}}$, $\mathrm{pr}_{(\widehat{\mathbf{S}}_{\breve{\Phi}_{\widehat{\mathcal{H}}}})^{\vee}_{\mathbb{R}}} : (\mathbf{S}_{\breve{\Phi}_{\widetilde{\mathcal{H}}}})^{\vee}_{\mathbb{R}} \to (\widehat{\mathbf{S}}_{\breve{\Phi}_{\widehat{\mathcal{H}}}})^{\vee}_{\mathbb{R}}$, $\widehat{\mathbf{P}}_{\breve{\Phi}_{\widehat{\mathcal{H}}}}$, and $\widehat{\mathbf{P}}^{+}_{\breve{\Phi}_{\widehat{\mathcal{H}}}}$ as in Definition 1.2.4.29, (1.2.4.32), (1.2.4.33), and (1.2.4.34), and define the $\Gamma_{\breve{\Phi}_{\widehat{\mathcal{H}}}}$-*admissible rational polyhedral cone decomposition* $\widehat{\Sigma}_{\breve{\Phi}_{\widehat{\mathcal{H}}}} = \{\mathrm{pr}_{(\widehat{\mathbf{S}}_{\breve{\Phi}_{\widehat{\mathcal{H}}}})^{\vee}_{\mathbb{R}}}(\breve{\tau})\}_{\breve{\tau}\in\widetilde{\Sigma}_{\breve{\Phi}_{\widetilde{\mathcal{H}}},\breve{\sigma}}}$ of $\widehat{\mathbf{P}}_{\breve{\Phi}_{\widehat{\mathcal{H}}}}$ as in Corollary 1.2.4.40. (All of these are verbatim as in Section 1.2.4.) Then we have:

Lemma 7.1.1.9. (*Compare with Lemma* 1.2.4.42.) *The collection* $\widehat{\Sigma}^{\mathrm{ord}} = \{\widehat{\Sigma}_{\breve{\Phi}_{\widehat{\mathcal{H}}}}\}_{[(\breve{\Phi}_{\widehat{\mathcal{H}}},\breve{\delta}_{\widehat{\mathcal{H}}})]}$*, where* $[(\breve{\Phi}_{\widehat{\mathcal{H}}}, \breve{\delta}_{\widehat{\mathcal{H}}})]$ *runs through equivalence classes of* $\widehat{\mathcal{H}}$*-orbits of representatives* $(\breve{\Phi}, \breve{\delta})$ *compatible with* $(\widetilde{\Phi}, \widetilde{\delta})$ *as in Definition* 1.2.4.17 *(with* $\breve{\mathsf{Z}}$ *and* $\widetilde{\mathsf{Z}}$ *suppressed in the notation), such that* $\breve{\mathsf{Z}}$ *is compatible with* $\widetilde{\mathsf{D}}$ *(as in Definition* 3.2.3.1*) (so that the cusp label* $[(\breve{\mathsf{Z}}_{\widetilde{\mathcal{H}}}, \breve{\Phi}_{\widetilde{\mathcal{H}}}, \breve{\delta}_{\widetilde{\mathcal{H}}})]$ *is ordinary), defines a* **compatible choice of admissible smooth rational polyhedral cone decomposition data** *analogous to the notion for* $\mathsf{M}_{\mathcal{H}}$ *in Definition* 1.2.2.13*. There is an obvious notion of refinements for such collections, analogous to that in [62, Def. 6.4.2.8].*

Moreover, each such $\widehat{\Sigma}^{\mathrm{ord}}$ *extends to some* $\widehat{\Sigma}$ *as in Lemma* 1.2.4.42. *Conversely, each* $\widehat{\Sigma}$ *as in Lemma* 1.2.4.42 *induces (by restriction to ordinary cusp labels) a valid* $\widehat{\Sigma}^{\mathrm{ord}}$ *for this lemma.*

Proof. This follows from the corresponding facts for $\widetilde{\Sigma}^{\mathrm{ord}} = \{\widetilde{\Sigma}_{\breve{\Phi}_{\widetilde{\mathcal{H}}}}\}_{[(\breve{\Phi}_{\widetilde{\mathcal{H}}},\breve{\delta}_{\widetilde{\mathcal{H}}})]}$ (with indices running through all ordinary cusp labels). (The statements concerning extensibility and restrictions follow from the corresponding ones in Propositions 5.1.3.2 and 5.1.3.4.) □

Remark 7.1.1.10. (Compare with Remark 1.2.4.43.) Here we omit the precise definition of a compatible choice of admissible smooth rational polyhedral cone decomposition data because we can only construct partial toroidal compactifications of Kuga families for those $\widehat{\Sigma}^{\mathrm{ord}}$ defined by some $\widetilde{\Sigma}^{\mathrm{ord}}$ and $\widetilde{\sigma}$.

Definition 7.1.1.11. (*Compare with Definition* 1.2.4.44.) *We say that two* $\widetilde{\kappa}_1$ *and* $\widetilde{\kappa}_2$ *in* $\widetilde{\mathbf{K}}_{Q,\mathcal{H}}^{\mathrm{ord},++}$ (*see Definition* 7.1.1.7) *are equivalent if they determine the same* $\kappa = (\widehat{\mathcal{H}}, \widehat{\Sigma}^{\mathrm{ord}})$. *In this case, we shall abusively write* $\kappa = [\widetilde{\kappa}_1] = [\widetilde{\kappa}_2]$. *Then we take* $\mathbf{K}_{Q,\mathcal{H}}^{\mathrm{ord},++}$ *to be the set of all such* $\kappa = (\widehat{\mathcal{H}}, \widehat{\Sigma}^{\mathrm{ord}})$, *with a partial order* $\kappa' = (\widehat{\mathcal{H}}', \widehat{\Sigma}^{\mathrm{ord},\prime}) \succ \kappa = (\widehat{\mathcal{H}}, \widehat{\Sigma}^{\mathrm{ord}})$ *when* $\widehat{\mathcal{H}}' \subset \widehat{\mathcal{H}}'$ *and when* $\widehat{\Sigma}^{\mathrm{ord},\prime}$ *is a refinement of* $\widehat{\Sigma}^{\mathrm{ord}}$ (*see Definition* 7.1.1.9). *We also take the subset* $\mathbf{K}_{Q,\mathcal{H}}^{\mathrm{ord},+}$ (*resp.* $\mathbf{K}_{Q,\mathcal{H}}^{\mathrm{ord}}$) *of* $\mathbf{K}_{Q,\mathcal{H}}^{\mathrm{ord},++}$ *to be the image of the subset* $\widetilde{\mathbf{K}}_{Q,\mathcal{H}}^{\mathrm{ord},+}$ (*resp.* $\widetilde{\mathbf{K}}_{Q,\mathcal{H}}^{\mathrm{ord}}$) *of* $\widetilde{\mathbf{K}}_{Q,\mathcal{H}}^{\mathrm{ord},++}$ *under the canonical surjection* $\widetilde{\mathbf{K}}_{Q,\mathcal{H}}^{\mathrm{ord},++} \twoheadrightarrow \mathbf{K}_{Q,\mathcal{H}}^{\mathrm{ord},++}$, *with an induced partial order denoted by the same symbol* $\succ$.

Lemma 7.1.1.12. (*Compare with Lemma* 1.2.4.45.) *In Lemma* 1.2.4.45, *if* $\widehat{\mathcal{H}}_{\mathrm{G}} \subset \mathrm{G}(\hat{\mathbb{Z}})$ *satisfies Condition* 7.1.1.4, *then we may assume that* $\widetilde{\mathcal{H}} = \widetilde{\mathcal{H}}^p\widetilde{\mathcal{H}}_p$ *is of standard form with neat* $\widetilde{\mathcal{H}}^p$, *and satisfies both Conditions* 1.2.4.7 *and* 7.1.1.5.

Proof. This follows from the proof of Lemma 1.2.4.45, because the restriction of $\widetilde{\nu} : \widetilde{\mathrm{G}} \to \mathbf{G}_{\mathrm{m}}$ to $\widetilde{\mathrm{P}}'_{\widetilde{\mathrm{Z}}}$ factors as $\widetilde{\mathrm{P}}'_{\widetilde{\mathrm{Z}}} \to \mathrm{G} \overset{\nu}{\to} \mathbf{G}_{\mathrm{m}}$. □

Lemma 7.1.1.13. (*Compare with Lemma* 1.2.4.46.) *For each neat open compact subgroup* $\widehat{\mathcal{H}}$ *of* $\mathrm{G}(\hat{\mathbb{Z}})$ *satisfying Condition* 7.1.1.4, *there exists some element* $\kappa = (\widehat{\mathcal{H}}, \widehat{\Sigma}^{\mathrm{ord}}) \in \mathbf{K}_{Q,\mathcal{H}}^{\mathrm{ord},++}$, *which lies in* $\mathbf{K}_{Q,\mathcal{H}}^{\mathrm{ord},+}$ (*resp.* $\mathbf{K}_{Q,\mathcal{H}}^{\mathrm{ord}}$) *if* $\widehat{\mathcal{H}}$ *also satisfies Condition* 1.2.4.8 (*resp. both Conditions* 1.2.4.8 *and* 1.2.4.9).

Proof. By Lemmas 1.2.4.45 and 7.1.1.12, $\widehat{\mathcal{H}}$ is induced by some $\widetilde{\mathcal{H}} = \widetilde{\mathcal{H}}^p\widetilde{\mathcal{H}}_p$ with neat $\widetilde{\mathcal{H}}^p$ as in Definition 1.2.4.4, which we assume to also satisfy Conditions 1.2.4.7 and 7.1.1.5. By Propositions 5.1.3.2 and 5.1.3.4, there exists some $\widetilde{\Sigma}^{\mathrm{ord}}$ for $\vec{\mathsf{M}}_{\widetilde{\mathcal{H}}}^{\mathrm{ord}}$. Let us take $\widehat{\Sigma}^{\mathrm{ord}}$ to be induced by $\widetilde{\Sigma}^{\mathrm{ord}}$ as in Lemma 7.1.1.9, and take $\kappa = (\widehat{\mathcal{H}}, \widehat{\Sigma}^{\mathrm{ord}})$. Then, by definition, we have $\kappa \in \mathbf{K}_{Q,\mathcal{H}}^{\mathrm{ord},++}$. The remaining statements of the lemma also follow by definition. □

Lemma 7.1.1.14. (*Compare with Lemma* 1.2.4.47.) *The partial order* $\succ$ *among elements in* $\mathbf{K}_{Q,\mathcal{H}}^{\mathrm{ord},++}$ (*resp.* $\mathbf{K}_{Q,\mathcal{H}}^{\mathrm{ord},+}$, *resp.* $\mathbf{K}_{Q,\mathcal{H}}^{\mathrm{ord}}$) *is* **directed**; *that is, if we are given* *two* $\kappa = (\widehat{\mathcal{H}}, \widehat{\Sigma}^{\mathrm{ord}})$ *and* $\kappa' = (\widehat{\mathcal{H}}', \widehat{\Sigma}^{\mathrm{ord},\prime})$*, then there exists some* $\kappa'' = (\widehat{\mathcal{H}}'', \widehat{\Sigma}^{\mathrm{ord},\prime\prime})$ *such that* $\kappa'' \succ \kappa$ *and* $\kappa'' \succ \kappa'$*. Moreover, we can take* $\widehat{\mathcal{H}}''$ *to be any open compact subgroup of* $\widehat{\mathcal{H}} \cap \widehat{\mathcal{H}}'$ *(which can be* $\widehat{\mathcal{H}} \cap \widehat{\mathcal{H}}'$ *itself) satisfying Condition* 7.1.1.4 *(with* $\widehat{\mathcal{H}}$ *there replaced with* $\widehat{\mathcal{H}}''$ *here).*

Proof. The same argument of the proof of Lemma 1.2.4.47 works here. □

Now we consider some compatibility conditions between a collection Σ^{ord} for $\vec{\mathsf{M}}^{\mathrm{ord}}_{\mathcal{H}}$ and elements of $\widetilde{\mathbf{K}}^{\mathrm{ord},++}_{Q,\mathcal{H}}$ or $\mathbf{K}^{\mathrm{ord},++}_{Q,\mathcal{H}}$.

First consider the following condition on an element $\widetilde{\kappa} = (\widetilde{\mathcal{H}}, \widetilde{\Sigma}^{\mathrm{ord}}, \widetilde{\sigma})$ in $\widetilde{\mathbf{K}}^{\mathrm{ord},++}_{Q,\mathcal{H}}$:

Condition 7.1.1.15. (*Compare with [61, Cond. 3.8] and Condition* 1.2.4.48.) *For each* $(\breve{\Phi}_{\widetilde{\mathcal{H}}}, \breve{\delta}_{\widetilde{\mathcal{H}}}, \breve{\tau})$ *(with* $\breve{\mathsf{Z}}_{\widetilde{\mathcal{H}}}$ *suppressed in the notation) such that* $(\breve{\mathsf{Z}}_{\widetilde{\mathcal{H}}}, \breve{\Phi}_{\widetilde{\mathcal{H}}}, \breve{\delta}_{\widetilde{\mathcal{H}}})$ *is a representative of an ordinary cusp label which admits a surjection* $(s_{\breve{X}} : \breve{X} \twoheadrightarrow \widetilde{X}, s_{\breve{Y}} : \breve{Y} \twoheadrightarrow \widetilde{Y})$ *to* $(\widetilde{\mathsf{Z}}_{\widetilde{\mathcal{H}}}, \widetilde{\Phi}_{\widetilde{\mathcal{H}}}, \widetilde{\delta}_{\widetilde{\mathcal{H}}})$*, and such that* $\breve{\tau} \subset \mathbf{P}^+_{\breve{\Phi}_{\widetilde{\mathcal{H}}}}$ *is a cone in the cone decomposition* $\widetilde{\Sigma}_{\breve{\Phi}_{\widetilde{\mathcal{H}}}}$ *(in* $\widetilde{\Sigma}^{\mathrm{ord}}$*) having a face* $\breve{\sigma}$ *that is a* $\Gamma_{\breve{\Phi}_{\widetilde{\mathcal{H}}}}$*-translation (see Definition* 1.2.2.3*) of the image of* $\widetilde{\sigma} \subset \mathbf{P}^+_{\widetilde{\Phi}_{\widetilde{\mathcal{H}}}}$ *under the first morphism in* (1.2.4.20) *(induced by* $(s_{\breve{X}}, s_{\breve{Y}})$*), the image of* $\breve{\tau}$ *in* $\mathbf{P}_{\Phi_{\mathcal{H}}}$ *under the (canonical) second morphism in* (1.2.4.20) *is contained in some cone* $\tau \subset \mathbf{P}^+_{\Phi_{\mathcal{H}}}$ *in the cone decomposition* $\Sigma_{\Phi_{\mathcal{H}}}$ *(in* Σ^{ord}*).*

Remark 7.1.1.16. The only difference between Conditions 1.2.4.48 and 7.1.1.15 is that we only consider ordinary cusp labels in the latter.

By Lemma 1.2.4.35 (and Remark 7.1.1.16), if $\kappa = [\widetilde{\kappa}] \in \mathbf{K}^{\mathrm{ord},++}_{Q,\mathcal{H}}$ is the element determined by $\widetilde{\kappa}$, then Condition 7.1.1.15 for $\widetilde{\kappa}$ is equivalent to the following condition for κ:

Condition 7.1.1.17. (*Compare with [28, Ch. VI, Def. 1.3] and Condition* 1.2.4.49.) *For each* $\widehat{\tau} \in \widehat{\Sigma}_{\breve{\Phi}_{\widehat{\mathcal{H}}}}$ *(where* $\widehat{\tau} = \mathrm{pr}_{(\widehat{\mathbf{S}}_{\breve{\Phi}_{\widehat{\mathcal{H}}}})^{\vee}_{\mathbb{R}}}(\breve{\tau})$ *for some* $(\breve{\Phi}_{\widetilde{\mathcal{H}}}, \breve{\delta}_{\widetilde{\mathcal{H}}}, \breve{\tau})$ *is in the cone decomposition* $\widehat{\Sigma}_{\breve{\Phi}_{\widehat{\mathcal{H}}}}$ *in* $\widehat{\Sigma}^{\mathrm{ord}}$*), the image of* $\widehat{\tau}$ *in* $\mathbf{P}^+_{\Phi_{\mathcal{H}}}$ *under* (1.2.4.37) *is contained in some cone* $\tau \subset \mathbf{P}^+_{\Phi_{\mathcal{H}}}$ *in the cone decomposition* $\Sigma_{\Phi_{\mathcal{H}}}$ *(in* Σ^{ord}*).*

Remark 7.1.1.18. The only difference between Conditions 1.2.4.49 and 7.1.1.17 is that we only consider ordinary cusp labels in the latter.

Definition 7.1.1.19. (*Compare with Definition* 1.2.4.50.) *For* $? = ++$, $+$, *or* $\emptyset$, *let us take* $\mathbf{K}^{\mathrm{ord},?}_{Q,\mathcal{H},\Sigma^{\mathrm{ord}}}$ *to be the subset of* $\mathbf{K}^{\mathrm{ord},?}_{Q,\mathcal{H}}$ *consisting of elements* κ *satisfying Condition* 7.1.1.17.

Since Condition 7.1.1.17 can be achieved by replacing any given $\widehat{\Sigma}^{\mathrm{ord}}$ with a refinement (in the same set), we see that each $\mathbf{K}^{\mathrm{ord},?}_{Q,\mathcal{H},\Sigma}$ is nonempty and has an induced directed partial order.

Thus, we have defined analogues of all the sets $\widetilde{\mathbf{K}}_{Q,\mathcal{H}} \subset \widetilde{\mathbf{K}}^+_{Q,\mathcal{H}} \subset \widetilde{\mathbf{K}}^{++}_{Q,\mathcal{H}}$, $\mathbf{K}_{Q,\mathcal{H}} \subset \mathbf{K}^+_{Q,\mathcal{H}} \subset \mathbf{K}^{++}_{Q,\mathcal{H}}$, and $\mathbf{K}_{Q,\mathcal{H},\Sigma} \subset \mathbf{K}^+_{Q,\mathcal{H},\Sigma} \subset \mathbf{K}^{++}_{Q,\mathcal{H},\Sigma}$ in Definitions 1.2.4.11, 1.2.4.44, and 1.2.4.50. We would like to analyze the relation between the sets we defined there and here.

Definition 7.1.1.20. (*Compare with Definitions* 1.2.4.11, 1.2.4.44, *and* 1.2.4.50.) *For each* $?_1 = ++$, $+$, *or* $\emptyset$, *and for each* $?_2 = \Sigma$ *or* $\emptyset$:

(1) $\widetilde{\mathbf{K}}^{\mathrm{std},?_1}_{Q,\mathcal{H}}$ *is the (nonempty) set of all* $\widetilde{\kappa} = (\widetilde{\mathcal{H}}, \widetilde{\Sigma}, \widetilde{\sigma}) \in \widetilde{\mathbf{K}}^{?_1}_{Q,\mathcal{H}}$ *such that* $\widetilde{\mathcal{H}}$ *is of standard from with respect to* $\widetilde{\mathtt{D}}$ *as in Definition* 3.2.2.9.

(2) $\mathbf{K}^{\mathrm{std},?_1}_{Q,\mathcal{H},?_2}$ *is the (nonempty) set of all* $\kappa = (\widehat{\mathcal{H}}, \widehat{\Sigma}) \in \mathbf{K}^{?_1}_{Q,\mathcal{H},?_2}$ *such that* $\widehat{\mathcal{H}}$ *is of standard from with respect to* $\mathtt{D}$ *as in Definition* 7.1.1.2.

Proposition 7.1.1.21. (*Compare with Proposition* 1.2.4.52.) *Suppose* $\mathcal{H}$ *is of standard form with respect to* $\mathtt{D}$ *as above. For each* $?_1 = ++$, $+$, *or* $\emptyset$, *and for each* $?_2 = \Sigma$ *or* $\emptyset$, *the sets* $\widetilde{\mathbf{K}}^{\mathrm{ord},?_1}_{Q,\mathcal{H}}$ *and* $\mathbf{K}^{\mathrm{ord},?_1}_{Q,\mathcal{H},?_2}$ *are nonempty, and the natural canonical morphisms*

$$\mathrm{ord} : \widetilde{\mathbf{K}}^{\mathrm{std},?_1}_{Q,\mathcal{H}} \to \widetilde{\mathbf{K}}^{\mathrm{ord},?_1}_{Q,\mathcal{H}}$$

and

$$\mathrm{ord} : \mathbf{K}^{\mathrm{std},?_1}_{Q,\mathcal{H},?_2} \to \mathbf{K}^{\mathrm{ord},?_1}_{Q,\mathcal{H},?_2}$$

are **surjective** *and compatible with each other under the various canonical maps. Common refinements for finite subsets exist in any sets of the form* $\mathbf{K}^{\mathrm{std},?_1}_{Q,\mathcal{H},?_2}$ *or* $\mathbf{K}^{\mathrm{ord},?_1}_{Q,\mathcal{H},?_2}$. *When doing so, we may allow varying levels or twists by Hecke actions, and we may vary* $?_1$ *and* $?_2$ *as well (in any order). For any such refinement* $\kappa = (\widehat{\mathcal{H}}, \widehat{\Sigma})$ *or* $(\widehat{\mathcal{H}}, \widehat{\Sigma}^{\mathrm{ord}})$, *we may prescribe* $\widehat{\mathcal{H}}$ *to be any allowed open compact subgroup of* $\widehat{\mathrm{G}}(\hat{\mathbb{Z}})$ *in the context, we may require* $\widehat{\Sigma}$ *to be finer than any cone decomposition* $\widehat{\Sigma}'$, *and we may require* $\widehat{\Sigma}$ *or* $\widehat{\Sigma}^{\mathrm{ord}}$ *to be invariant under any choice of an open compact subgroup of* $\widehat{\mathrm{G}}(\mathbb{A}^\infty)$ *normalizing* $\widehat{\mathcal{H}}$. *If* $\kappa = (\widehat{\mathcal{H}}, \widehat{\Sigma}^{\mathrm{ord}}) \in \mathbf{K}^{\mathrm{ord},?_1}_{Q,\mathcal{H},?_2}$, *if* $\widehat{\mathcal{H}}' \subset \widehat{\mathcal{H}}$ *is an open compact subgroup such that the integral structures on* $(\widehat{\mathbf{S}}_{\breve{\Phi}_{\widehat{\mathcal{H}}'}})^\vee_{\mathbb{R}}$ *defined by* $\widehat{\mathbf{S}}_{\breve{\Phi}_{\widehat{\mathcal{H}}'}}$ *and* $\widehat{\mathbf{S}}_{\breve{\Phi}_{\widehat{\mathcal{H}}}}$ *are identical for each ordinary cusp label* $[(\breve{\Phi}_{\widehat{\mathcal{H}}'}, \breve{\delta}_{\widehat{\mathcal{H}}'})]$ *at level* $\widehat{\mathcal{H}}'$ *inducing an ordinary cusp label* $[(\breve{\Phi}_{\widehat{\mathcal{H}}}, \breve{\delta}_{\widehat{\mathcal{H}}})]$ *at level* $\widehat{\mathcal{H}}$, *and if* $\widehat{\Sigma}^{\mathrm{ord},\prime}$ *is the collection induced by* $\widehat{\Sigma}^{\mathrm{ord}}$ *at level* $\widehat{\mathcal{H}}'$, *then* $\kappa' := (\widehat{\mathcal{H}}', \widehat{\Sigma}^{\mathrm{ord},\prime})$ *belongs to the same set* $\mathbf{K}^{\mathrm{ord},?_1}_{Q,\mathcal{H},?_2}$ *(without the need to refine* $\widehat{\Sigma}^{\mathrm{ord},\prime}$*). If* $\kappa = (\widehat{\mathcal{H}}, \widehat{\Sigma}^{\mathrm{ord}}) \in \mathbf{K}^{\mathrm{ord},?_1}_{Q,\mathcal{H},?_2}$ *and* $\kappa' = (\widehat{\mathcal{H}}', \widehat{\Sigma}') \in \mathbf{K}^{\mathrm{std},?_1}_{Q,\mathcal{H},?_2}$ *such that* $\kappa \succ \mathrm{ord}(\kappa')$, *then there exists an element* $\kappa'' = (\widehat{\mathcal{H}}'', \widehat{\Sigma}'') \in \mathbf{K}^{\mathrm{std},?_1}_{Q,\mathcal{H},?_2}$ *such that* $\mathrm{ord}(\kappa'') = \kappa$ *and* $\kappa'' \succ \kappa'$, *and we may assume that* $\kappa'' \succ \kappa'''$ *for any* $\kappa''' \in \mathbf{K}^{\mathrm{std},?_1}_{Q,\mathcal{H},?_2}$ *such that* $\mathrm{ord}(\kappa''') = \kappa$.

Proof. These follow from the corresponding existence, refinement, and extensibility statements in Propositions 1.2.2.17, 5.1.3.2, and 5.1.3.4 for collections $\widetilde{\Sigma}$ and $\widetilde{\mathsf{pol}}$ for $\widetilde{\mathsf{M}}_{\widetilde{\mathcal{H}}}$ and for collections $\widetilde{\Sigma}^{\mathrm{ord}}$ and $\widetilde{\mathsf{pol}}^{\mathrm{ord}}$ for $\vec{\widetilde{\mathsf{M}}}^{\mathrm{ord}}_{\widetilde{\mathcal{H}}}$. The second last statement is obvious. As for the last statement, note that ordinary cusp labels (by their very definition, see Definition 3.2.3.8) can only admit surjections to ordinary cusp labels, and hence refinements of cone decompositions over non-ordinary cusp labels do not necessitate further refinements over the ordinary cusp labels. □

For later references, let us also define the following:

Definition 7.1.1.22. *For each $\mathbb{Z}_p$-algebra R, we also define the following quotients of subgroups of* $\widetilde{\mathrm{P}}_{\widetilde{\mathrm{Z}}}(R)$ *and* $\widetilde{\mathrm{P}}^{\mathrm{ord}}_{\widetilde{\mathrm{D}}}(R)$ *(see Definitions* 1.2.1.10, 1.2.1.11, 1.2.4.3, 3.2.2.7, *and* 3.2.3.9*):*

(1) $\widetilde{\mathrm{P}}^{\mathrm{ord}}_{\widetilde{\mathrm{Z}},\widetilde{\mathrm{D}}}(R) := \widetilde{\mathrm{P}}_{\widetilde{\mathrm{Z}}}(R) \cap \widetilde{\mathrm{P}}^{\mathrm{ord}}_{\widetilde{\mathrm{D}}}(R)$.

(2) $\widetilde{\mathrm{P}}^{\mathrm{ord},\prime}_{\widetilde{\mathrm{Z}},\widetilde{\mathrm{D}}}(R) := \widetilde{\mathrm{P}}'_{\widetilde{\mathrm{Z}}}(R) \cap \widetilde{\mathrm{P}}^{\mathrm{ord}}_{\widetilde{\mathrm{D}}}(R)$.

(3) $\widehat{\mathrm{P}}^{\mathrm{ord}}_{\mathrm{D}}(R) := \widetilde{\mathrm{P}}^{\mathrm{ord}}_{1,\widetilde{\mathrm{Z}},\widetilde{\mathrm{D}}}(R) = \widetilde{\mathrm{P}}^{\mathrm{ord},\prime}_{\widetilde{\mathrm{Z}},\widetilde{\mathrm{D}}}(R)/\widetilde{\mathrm{U}}_{2,\widetilde{\mathrm{Z}}}(R)$.

(4) $\widehat{\mathrm{U}}^{\mathrm{ord}}_{\mathrm{D}}(R) := \widetilde{\mathrm{U}}_{\widetilde{\mathrm{Z}}}(R)/(\widetilde{\mathrm{U}}_{\widetilde{\mathrm{Z}}}(R) \cap \widetilde{\mathrm{U}}^{\mathrm{ord}}_{\widetilde{\mathrm{D}}}(R))$, *which can be canonically identified with the image of* $\widetilde{\mathrm{U}}_{\widetilde{\mathrm{Z}}}(R)$ *under the canonical homomorphism* $\widetilde{\mathrm{P}}^{\mathrm{ord}}_{\widetilde{\mathrm{D}}}(R) \to \widetilde{\mathrm{M}}^{\mathrm{ord}}_{\widetilde{\mathrm{D}}}(R)$.

(5) $\widehat{\mathrm{M}}^{\mathrm{ord}}_{\mathrm{D}}(R) := \widetilde{\mathrm{P}}^{\mathrm{ord},\prime}_{\widetilde{\mathrm{Z}},\widetilde{\mathrm{D}}}(R)/\widetilde{\mathrm{U}}^{\mathrm{ord}}_{\widetilde{\mathrm{D}}}(R)$, *which can be canonically identified with the image of* $\widetilde{\mathrm{P}}^{\mathrm{ord},\prime}_{\widetilde{\mathrm{Z}},\widetilde{\mathrm{D}}}(R)$ *under the canonical homomorphism* $\widetilde{\mathrm{P}}^{\mathrm{ord}}_{\widetilde{\mathrm{D}}}(R) \to \widetilde{\mathrm{M}}^{\mathrm{ord}}_{\widetilde{\mathrm{D}}}(R)$, *which is (under the splitting* $\widetilde{\delta}$ *above) isomorphic to* $(\mathrm{M}^{\mathrm{ord}}_{\mathrm{D}} \ltimes \widehat{\mathrm{U}}^{\mathrm{ord}}_{\mathrm{D}})(R) := \mathrm{M}^{\mathrm{ord}}_{\mathrm{D}}(R) \ltimes \widehat{\mathrm{U}}^{\mathrm{ord}}_{\mathrm{D}}(R)$.

Definition 7.1.1.23. *For each open compact subgroup* $\widetilde{\mathcal{H}}_p$ *of* $\widetilde{\mathrm{G}}(\mathbb{Z}_p)$ *of standard form as in Definition* 3.2.2.9, *with* $r = \mathrm{depth}_{\widetilde{\mathrm{D}}}(\widetilde{\mathcal{H}}_p)$, *we define* $\widehat{\mathcal{H}}_p = (\widetilde{\mathcal{H}}_p)_{\widehat{\mathrm{G}}}$ *as in Definition* 1.2.4.4, *so that (by definition)* $r = \mathrm{depth}_{\mathrm{D}}(\widehat{\mathcal{H}}_p) = \mathrm{depth}_{\widetilde{\mathrm{D}}}(\widetilde{\mathcal{H}}_p)$. *Then*

$$\widehat{H}_{p^r} := \widehat{\mathcal{H}}_p/\widehat{\mathcal{U}}^{\mathrm{bal}}_{p,1}(p^r)$$

is a subgroup of

$$\widehat{\mathcal{U}}_{p,0}(p^r)/\widehat{\mathcal{U}}^{\mathrm{bal}}_{p,1}(p^r) \cong \widehat{\mathrm{M}}^{\mathrm{ord}}_{\mathrm{D}}(\mathbb{Z}/p^r\mathbb{Z}),$$

and we set

$$\widehat{\mathcal{H}}^{\mathrm{ord}}_p := \left(\widehat{\mathrm{M}}^{\mathrm{ord}}_{\mathrm{D}}(\mathbb{Z}_p) \xrightarrow{\mathrm{can.}} \widehat{\mathrm{M}}^{\mathrm{ord}}_{\mathrm{D}}(\mathbb{Z}/p^r\mathbb{Z}) \overset{\mathrm{can.}}{\xrightarrow{\sim}} \left(\widehat{\mathcal{U}}_{p,0}(p^r)/\widehat{\mathcal{U}}^{\mathrm{bal}}_{p,1}(p^r)\right)\right)^{-1}(\widehat{H}_{p^r}) \quad (7.1.1.24)$$

(cf. (3.3.3.5)*),*

$$\widehat{\mathcal{H}}^{\mathrm{ord}}_{p,\widehat{\mathrm{U}}^{\mathrm{ord}}_{\mathrm{D}}} := \widehat{\mathcal{H}}^{\mathrm{ord}}_p \cap \widehat{\mathrm{U}}^{\mathrm{ord}}_{\mathrm{D}}(\mathbb{Z}_p),$$

and

$$\widehat{\mathcal{H}}^{\mathrm{ord}}_{p,\mathrm{M}^{\mathrm{ord}}_{\mathrm{D}}} := \widehat{\mathcal{H}}^{\mathrm{ord}}_p/\widehat{\mathcal{H}}^{\mathrm{ord}}_{p,\widehat{\mathrm{U}}^{\mathrm{ord}}_{\mathrm{D}}}.$$

Then (by definition)

$$\widehat{\mathcal{H}}^{\mathrm{ord}}_{p,\mathrm{M}^{\mathrm{ord}}_{\mathrm{D}}} = ((\widehat{\mathcal{H}}_p)_{\mathrm{G}})^{\mathrm{ord}}$$

(where the right-hand side is defined using Definition 1.2.4.4 *and* (3.3.3.5)*) as subgroups of* $\mathrm{M}^{\mathrm{ord}}_{\mathrm{D}}(\mathbb{Z}_p)$, *and we have an exact sequence*

$$1 \to \widehat{\mathcal{H}}^{\mathrm{ord}}_{p,\widehat{\mathrm{U}}^{\mathrm{ord}}_{\mathrm{D}}} \to \widehat{\mathcal{H}}^{\mathrm{ord}}_p \to \widehat{\mathcal{H}}^{\mathrm{ord}}_{p,\mathrm{M}^{\mathrm{ord}}_{\mathrm{D}}} \to 1 \quad (7.1.1.25)$$

compatible with the canonical exact sequence

$$1 \to \widehat{\mathrm{U}}^{\mathrm{ord}}_{\mathrm{D}}(\mathbb{Z}_p) \to \widehat{\mathrm{M}}^{\mathrm{ord}}_{\mathrm{D}}(\mathbb{Z}_p) \to \mathrm{M}^{\mathrm{ord}}_{\mathrm{D}}(\mathbb{Z}_p) \to 1. \quad (7.1.1.26)$$

Definition 7.1.1.27. (*Compare with Definitions* 1.2.4.53, 1.2.4.54, *and* 3.2.3.9.) *Suppose* $\breve{\mathsf{Z}}$ *is compatible with* $\widetilde{\mathsf{D}}$ *as in* (3.2.3.2). *For each* $\mathbb{Z}_p$*-algebra* R, *we define the following quotients of subgroups of* $\widehat{\mathrm{P}}_{\breve{\mathsf{Z}}}(R)$ (*see Definition* 1.2.4.53)*:*

(1) $\widehat{\mathrm{P}}^{\mathrm{ord}}_{\breve{\mathsf{Z}},\mathsf{D}}(R) := (\widetilde{\mathrm{P}}^{\mathrm{ord}}_{\breve{\mathsf{Z}},\widetilde{\mathsf{D}}}(R) \cap \widetilde{\mathrm{P}}'_{\widetilde{\mathsf{Z}}}(R))/\widetilde{\mathrm{U}}_{2,\widetilde{\mathsf{Z}}}(R)$. (*Because of the compatibility between* $\breve{\mathsf{Z}}$ *and* $\widetilde{\mathsf{D}}$, *we do not define new groups for* $\widehat{\mathrm{Z}}_{\breve{\mathsf{Z}}}(R)$, $\widehat{\mathrm{U}}_{\breve{\mathsf{Z}}}(R)$, $\widehat{\mathrm{U}}_{2,\breve{\mathsf{Z}}}(R)$, $\widehat{\mathrm{U}}_{1,\breve{\mathsf{Z}}}(R)$, $\widehat{\mathrm{G}}_{l,\breve{\mathsf{Z}}}(R)$, *and* $\widehat{\mathrm{G}}'_{l,\breve{\mathsf{Z}}}(R)$ *here.*)

(2) $\widehat{\mathrm{P}}^{\mathrm{ord}}_{h,\breve{\mathsf{Z}},\mathsf{D}}(R) := \widetilde{\mathrm{P}}^{\mathrm{ord}}_{\breve{\mathsf{Z}},\widetilde{\mathsf{D}}}(R)/\widetilde{\mathrm{Z}}_{\breve{\mathsf{Z}}}(R) \cong \widehat{\mathrm{P}}^{\mathrm{ord}}_{\breve{\mathsf{Z}},\widetilde{\mathsf{D}}}(R)/\widehat{\mathrm{Z}}_{\breve{\mathsf{Z}}}(R)$ *is the subgroup of elements of* $\widehat{\mathrm{G}}_{h,\breve{\mathsf{Z}}}(R) \cong \mathrm{G}_{h,\mathsf{Z}}(R)$ *preserving the filtration* D_{-1} *induced by* $\widetilde{\mathsf{D}}$ *on* $\mathrm{Gr}^{\breve{\mathsf{Z}}}_{-1} \underset{\hat{\mathbb{Z}}}{\otimes} \mathbb{Z}_p \cong \mathrm{Gr}^{\mathsf{Z}}_{-1} \underset{\hat{\mathbb{Z}}}{\otimes} \mathbb{Z}_p$ *as in Definition* 3.2.3.1.

(3) $\widehat{\mathrm{P}}^{\mathrm{ord},\prime}_{\breve{\mathsf{Z}},\mathsf{D}}(R) := \widetilde{\mathrm{P}}^{\mathrm{ord},\prime}_{\breve{\mathsf{Z}},\widetilde{\mathsf{D}}}(R)/\widetilde{\mathrm{U}}_{2,\widetilde{\mathsf{Z}}}(R)$ *is the kernel of the canonical homomorphism* $(\widetilde{\nu}^{-1}\,\mathrm{Gr}^{\breve{\mathsf{Z}}}_{-2}, \mathrm{Gr}^{\breve{\mathsf{Z}}}_{0}) : \widehat{\mathrm{P}}^{\mathrm{ord}}_{\breve{\mathsf{Z}},\mathsf{D}}(R) \to \widehat{\mathrm{G}}'_{l,\breve{\mathsf{Z}}}(R)$.

(4) $\widehat{\mathrm{P}}^{\mathrm{ord}}_{1,\breve{\mathsf{Z}},\mathsf{D}}(R) := \widetilde{\mathrm{P}}^{\mathrm{ord},\prime}_{\breve{\mathsf{Z}},\widetilde{\mathsf{D}}}(R) \cong \widehat{\mathrm{P}}^{\mathrm{ord},\prime}_{\breve{\mathsf{Z}},\mathsf{D}}(R)/\widehat{\mathrm{U}}_{2,\breve{\mathsf{Z}}}(R)$.

(5) $\widehat{\mathrm{P}}^{\mathrm{ord},\prime}_{h,\breve{\mathsf{Z}},\mathsf{D}}(R) := \widetilde{\mathrm{P}}^{\mathrm{ord}}_{1,\breve{\mathsf{Z}},\widetilde{\mathsf{D}}}(R)/\widetilde{\mathrm{U}}_{1,\breve{\mathsf{Z}}}(R) \cong \widehat{\mathrm{P}}^{\mathrm{ord}}_{1,\breve{\mathsf{Z}},\mathsf{D}}(R)/\widehat{\mathrm{U}}_{1,\breve{\mathsf{Z}}}(R) \cong \widetilde{\mathrm{P}}^{\mathrm{ord},\prime}_{\breve{\mathsf{Z}},\widetilde{\mathsf{D}}}(R)/\widetilde{\mathrm{U}}_{\breve{\mathsf{Z}}}(R) \cong \widehat{\mathrm{P}}^{\mathrm{ord},\prime}_{1,\breve{\mathsf{Z}},\mathsf{D}}(R)/\widehat{\mathrm{U}}_{\breve{\mathsf{Z}}}(R) \cong \widetilde{\mathrm{P}}^{\mathrm{ord}}_{\breve{\mathsf{Z}},\widetilde{\mathsf{D}}}(R)/\widetilde{\mathrm{Z}}^{\mathrm{ord}}_{\breve{\mathsf{Z}},\widetilde{\mathsf{D}}}(R) \cong \widehat{\mathrm{P}}^{\mathrm{ord}}_{\breve{\mathsf{Z}},\mathsf{D}}(R)/\widehat{\mathrm{Z}}_{\breve{\mathsf{Z}}}(R) \cong \widehat{\mathrm{P}}^{\mathrm{ord}}_{h,\breve{\mathsf{Z}},\mathsf{D}}(R)$.

Then the canonical homomorphism $\widehat{\mathrm{G}}(R) \to \mathrm{G}(R)$ *induces the following canonical homomorphisms:*

(1) $\widehat{\mathrm{P}}^{\mathrm{ord}}_{\breve{\mathsf{Z}},\mathsf{D}}(R) \to \mathrm{P}^{\mathrm{ord}}_{\mathsf{Z},\mathsf{D}}(R)$.

(2) $\widehat{\mathrm{P}}^{\mathrm{ord}}_{h,\breve{\mathsf{Z}},\mathsf{D}}(R) \to \mathrm{P}^{\mathrm{ord}}_{h,\mathsf{Z},\mathsf{D}}(R)$.

(3) $\widehat{\mathrm{P}}^{\mathrm{ord},\prime}_{\breve{\mathsf{Z}},\mathsf{D}}(R) \to \mathrm{P}^{\mathrm{ord},\prime}_{\mathsf{Z},\mathsf{D}}(R)$.

(4) $\widehat{\mathrm{P}}^{\mathrm{ord}}_{1,\breve{\mathsf{Z}},\mathsf{D}}(R) \to \mathrm{P}^{\mathrm{ord}}_{1,\mathsf{Z},\mathsf{D}}(R)$.

(5) $\widehat{\mathrm{P}}^{\mathrm{ord},\prime}_{h,\breve{\mathsf{Z}},\mathsf{D}}(R) \to \mathrm{P}^{\mathrm{ord},\prime}_{h,\mathsf{Z},\mathsf{D}}(R)$.

We also consider (*see Definition* 1.2.4.54)

$$\widetilde{\mathrm{P}}^{\mathrm{ord}}_{\breve{\mathsf{Z}},\widetilde{\mathsf{Z}},\widetilde{\mathsf{D}}}(R) := \widetilde{\mathrm{P}}_{\breve{\mathsf{Z}}}(R) \cap \widetilde{\mathrm{P}}_{\widetilde{\mathsf{Z}}}(R) \cap \widetilde{\mathrm{P}}^{\mathrm{ord}}_{\widetilde{\mathsf{D}}}(R).$$

Then we have the canonical isomorphism

$$\widetilde{\mathrm{P}}^{\mathrm{ord}}_{\breve{\mathsf{Z}},\widetilde{\mathsf{Z}},\widetilde{\mathsf{D}}}(R)/\widetilde{\mathrm{P}}^{\mathrm{ord},\prime}_{\breve{\mathsf{Z}},\widetilde{\mathsf{D}}}(R) \xrightarrow{\sim} \widetilde{\mathrm{P}}_{\breve{\mathsf{Z}},\widetilde{\mathsf{Z}}}(R)/\widetilde{\mathrm{P}}'_{\breve{\mathsf{Z}}}(R) = \widetilde{\mathrm{G}}_{l,\breve{\mathsf{Z}},\widetilde{\mathsf{Z}}}(R),$$

and the canonical homomorphism

$$\widetilde{\mathrm{P}}^{\mathrm{ord}}_{\breve{\mathsf{Z}},\widetilde{\mathsf{Z}},\widetilde{\mathsf{D}}}(R)/\widetilde{\mathrm{U}}_{2,\breve{\mathsf{Z}}}(R) \to \mathrm{P}^{\mathrm{ord}}_{\mathsf{Z},\mathsf{D}}(R)/\mathrm{U}_{2,\mathsf{Z}}(R).$$

7.1.2 *Boundary of Ordinary Loci, Continued*

Let us continue the study in Section 5.2.4, which generalized part of Section 1.3.2.

Let us continue with the setup of Section 7.1.1, with the same choices of $(\widetilde{L}, \langle\,\cdot\,,\,\cdot\,\rangle^{\sim}, \widetilde{h}_0)$, $(\widetilde{\mathsf{Z}}, \widetilde{\Phi}, \widetilde{\delta})$, *etc* as in Section 1.2.4. Let $\widetilde{\kappa} = (\widetilde{\mathcal{H}}, \widetilde{\Sigma}^{\mathrm{ord}}, \widetilde{\sigma})$ be any element in the set $\widetilde{\mathbf{K}}^{\mathrm{ord},++}_{Q,\mathcal{H}}$ as in Definition 7.1.1.7. The data of $\mathcal{O}$, $(\widetilde{L}, \langle\,\cdot\,,\,\cdot\,\rangle^{\sim}, \widetilde{h}_0)$, $\widetilde{\mathsf{D}}$, and

$\widetilde{\mathcal{H}} \subset \widetilde{\mathrm{G}}(\hat{\mathbb{Z}})$ (of standard form with respect to $\widetilde{\mathsf{D}}$) then define $\vec{\widetilde{\mathsf{M}}}^{\mathrm{ord}}_{\widetilde{\mathcal{H}}}$ as in Theorem 3.4.2.5. Since $\widetilde{\mathcal{H}}^p$ is neat and $\widetilde{\Sigma}^{\mathrm{ord}}$ is projective (and smooth), by Theorems 5.2.1.1 and 6.2.3.1, we have a partial toroidal compactification $\vec{\widetilde{\mathsf{M}}}^{\mathrm{ord,tor}}_{\widetilde{\mathcal{H}}} = \vec{\widetilde{\mathsf{M}}}^{\mathrm{ord,tor}}_{\widetilde{\mathcal{H}},\widetilde{\Sigma}^{\mathrm{ord}}}$ of $\vec{\widetilde{\mathsf{M}}}^{\mathrm{ord}}_{\widetilde{\mathcal{H}}}$ which is quasi-projective and smooth over $\vec{\mathsf{S}}_{0,r_{\widetilde{\mathcal{H}}}}$. We are mainly interested in comparing the boundary structures of $\vec{\widetilde{\mathsf{M}}}^{\mathrm{ord,tor}}_{\widetilde{\mathcal{H}},\widetilde{\Sigma}^{\mathrm{ord}}}$ and $\vec{\mathsf{M}}^{\mathrm{ord,tor}}_{\mathcal{H},\Sigma^{\mathrm{ord}}}$ under suitable conditions.

In the remainder of this subsection, let us fix the choice of a $\breve{\mathsf{Z}}$ satisfying (1.2.4.12), so that we have the groups and homomorphisms defined in Definitions 1.2.4.53, 1.2.4.54, and 7.1.1.27.

Suppose that an *ordinary* cusp label $[(\mathsf{Z}_{\mathcal{H}}, \Phi_{\mathcal{H}}, \delta_{\mathcal{H}})]$ at level $\mathcal{H}$ is canonically assigned (as in Lemma 1.2.4.15) to a cusp label $[(\breve{\mathsf{Z}}_{\widetilde{\mathcal{H}}}, \breve{\Phi}_{\widetilde{\mathcal{H}}}, \breve{\delta}_{\widetilde{\mathcal{H}}})]$ at level $\widetilde{\mathcal{H}}$ (necessarily also ordinary) admitting a surjection to $[(\widetilde{\mathsf{Z}}_{\widetilde{\mathcal{H}}}, \widetilde{\Phi}_{\widetilde{\mathcal{H}}}, \widetilde{\delta}_{\widetilde{\mathcal{H}}})]$, so that we have (1.2.4.18), (1.2.4.19), and (1.2.4.20), and the definitions following them.

Lemma 7.1.2.1. (*Compare with Lemma* 1.3.2.50.) *By comparing the universal properties, we obtain a canonical morphism*

$$\vec{\widetilde{C}}^{\mathrm{ord}}_{\breve{\Phi}_{\widetilde{\mathcal{H}}},\breve{\delta}_{\widetilde{\mathcal{H}}}} \to \vec{C}^{\mathrm{ord}}_{\Phi_{\mathcal{H}},\delta_{\mathcal{H}}}, \tag{7.1.2.2}$$

by sending $(\breve{c}^{\mathrm{ord}}_{\widetilde{\mathcal{H}}}, \breve{c}^{\vee,\mathrm{ord}}_{\widetilde{\mathcal{H}}})$, *which is an orbit of étale-locally-defined pairs*

$$(\breve{c}^{\mathrm{ord}}_n : \tfrac{1}{n}\breve{X} \to B^{\vee,\mathrm{ord}}_{p^r}, \breve{c}^{\vee,\mathrm{ord}}_n : \tfrac{1}{n}\breve{Y} \to B^{\mathrm{ord}}_{p^r})$$

for some integer $n = n_0 p^r$ *where* $n_0 \geq 1$ *is an integer prime to* p *such that* $\widetilde{\mathcal{U}}^p(n_0) \subset \widetilde{\mathcal{H}}^p$ *and where* $r = \mathrm{depth}_{\widetilde{\mathsf{D}}}(\widetilde{\mathcal{H}}_p)$, *to the orbit* $(c^{\mathrm{ord}}_{\mathcal{H}}, c^{\vee,\mathrm{ord}}_{\mathcal{H}})$ *of étale-locally-defined pairs*

$$(c^{\mathrm{ord}}_n : \tfrac{1}{n}X \to B^{\vee,\mathrm{ord}}_{p^r}, c^{\vee,\mathrm{ord}}_n : \tfrac{1}{n}Y \to B^{\mathrm{ord}}_{p^r}),$$

with $(c^{\mathrm{ord}}_n, c^{\vee,\mathrm{ord}}_n)$ *induced by* $(\breve{c}^{\mathrm{ord}}_n, \breve{c}^{\vee,\mathrm{ord}}_n)$ *by restrictions to* $\frac{1}{n}X$ *and* $\frac{1}{n}Y$, *where* X *and* Y *are the kernels of the admissible surjections* $s_{\breve{X}} : \breve{X} \twoheadrightarrow \widetilde{X}$ *and* $s_{\breve{Y}} : \breve{Y} \twoheadrightarrow \widetilde{Y}$, *respectively.* (*This definition canonically extends to a compatible definition in the* $\mathbb{Z}^\times_{(p)}$*-isogeny class language in Proposition* 5.2.4.13, *which we omit for simplicity.*)

If $\widehat{\mathcal{H}}_{\mathrm{G}} = \mathcal{H}$, *then* $\vec{\widetilde{\mathsf{M}}}^{\mathrm{ord},\widetilde{\Phi}_{\widetilde{\mathcal{H}}}}_{\widetilde{\mathcal{H}}} \cong \vec{\mathsf{M}}^{\mathrm{ord},\Phi_{\mathcal{H}}}_{\mathcal{H}}$ *and there is a canonical homomorphism*

$$\vec{\widetilde{C}}^{\mathrm{ord,grp}}_{\breve{\Phi}_{\widetilde{\mathcal{H}}},\breve{\delta}_{\widetilde{\mathcal{H}}}} \to \vec{C}^{\mathrm{ord,grp}}_{\Phi_{\mathcal{H}},\delta_{\mathcal{H}}} \tag{7.1.2.3}$$

(*cf.* (1.3.2.52)) *of abelian schemes over* $\vec{\mathsf{M}}^{\mathrm{ord},\Phi_{\mathcal{H}}}_{\mathcal{H}}$, *which can be identified with the canonical homomorphism*

$$\underline{\mathrm{Hom}}_{\mathcal{O}}(\breve{X}, B)^\circ \to \underline{\mathrm{Hom}}_{\mathcal{O}}(X, B)^\circ \tag{7.1.2.4}$$

(*cf.* (1.3.2.53)) *up to canonical* $\mathbb{Q}^\times$*-isogenies over* $\vec{\mathsf{M}}^{\mathrm{ord},\Phi_{\mathcal{H}}}_{\mathcal{H}}$, *and the* $\vec{\widetilde{C}}^{\mathrm{ord,grp}}_{\breve{\Phi}_{\widetilde{\mathcal{H}}},\breve{\delta}_{\widetilde{\mathcal{H}}}}$*- and* $\vec{C}^{\mathrm{ord,grp}}_{\Phi_{\mathcal{H}},\delta_{\mathcal{H}}}$*-torsor structures of* $\vec{\widetilde{C}}^{\mathrm{ord}}_{\breve{\Phi}_{\widetilde{\mathcal{H}}},\breve{\delta}_{\widetilde{\mathcal{H}}}} \to \vec{\mathsf{M}}^{\mathrm{ord},\Phi_{\mathcal{H}}}_{\mathcal{H}}$ *and* $\vec{C}^{\mathrm{ord}}_{\Phi_{\mathcal{H}},\delta_{\mathcal{H}}} \to \vec{\mathsf{M}}^{\mathrm{ord},\Phi_{\mathcal{H}}}_{\mathcal{H}}$, *respectively, are compatible with each other under* (7.1.2.2) *and* (7.1.2.3). *Moreover,* *the kernel of* (7.1.2.3) *is an abelian scheme over* $\vec{\mathsf{M}}_{\mathcal{H}}^{\mathrm{ord},\Phi_{\mathcal{H}}}$, *which is canonically* $\mathbb{Q}^\times$*-isogenous to the kernel* $\underline{\mathrm{Hom}}_{\mathcal{O}}(\widetilde{X},B)^\circ$ *of* (7.1.2.4), *and* (7.1.2.2) *is a torsor under the pullback to* $\vec{C}_{\Phi_{\mathcal{H}},\delta_{\mathcal{H}}}^{\mathrm{ord}}$ *of this abelian scheme.*

The abelian scheme torsor $\vec{\widetilde{C}}_{\check{\Phi}_{\widetilde{\mathcal{H}}},\check{\delta}_{\widetilde{\mathcal{H}}}}^{\mathrm{ord}} \to \vec{\widetilde{\mathsf{M}}}_{\widetilde{\mathcal{H}}}^{\mathrm{ord},\check{\Phi}_{\widetilde{\mathcal{H}}}}$ *and the finite étale covering* $\vec{\widetilde{\mathsf{M}}}_{\widetilde{\mathcal{H}}}^{\mathrm{ord},\check{\Phi}_{\widetilde{\mathcal{H}}}} \to \vec{\widetilde{\mathsf{M}}}_{\widetilde{\mathcal{H}}}^{\mathrm{ord},\check{\mathsf{Z}}_{\widetilde{\mathcal{H}}}}$ *depend (up to canonical isomorphism) only on* $\widehat{\mathcal{H}} = \widetilde{\mathcal{H}}_{\widehat{\mathrm{G}}}$ *and* $(\check{\mathsf{Z}}_{\widehat{\mathcal{H}}},\check{\Phi}_{\widehat{\mathcal{H}}},\check{\delta}_{\widehat{\mathcal{H}}})$ *(see Definition* 1.2.4.17*). We shall denote them as* $\vec{\widehat{C}}_{\check{\Phi}_{\widehat{\mathcal{H}}},\check{\delta}_{\widehat{\mathcal{H}}}}^{\mathrm{ord}} \to \vec{\widehat{\mathsf{M}}}_{\widehat{\mathcal{H}}}^{\mathrm{ord},\check{\Phi}_{\widehat{\mathcal{H}}}}$ *and* $\vec{\widehat{\mathsf{M}}}_{\widehat{\mathcal{H}}}^{\mathrm{ord},\check{\Phi}_{\widehat{\mathcal{H}}}} \to \vec{\widehat{\mathsf{M}}}_{\widehat{\mathcal{H}}}^{\mathrm{ord},\check{\mathsf{Z}}_{\widehat{\mathcal{H}}}}$ *when we want to emphasize this (in)dependence.*

Proof. The first paragraph is self-explanatory. As for the second paragraph, by Lemma 5.2.4.7, it suffices to verify the statements in the case $\mathcal{H} = \mathcal{U}_1^{\mathrm{bal}}(n)$ and $\widetilde{\mathcal{H}} = \widetilde{\mathcal{U}}_1^{\mathrm{bal}}(n)$ for some integer $n = n_0p^r$. (Then the third paragraph also follows by Lemma 5.2.4.7.) In this case, $\vec{\widetilde{C}}_{\check{\Phi}_{\widetilde{\mathcal{H}}},\check{\delta}_{\widetilde{\mathcal{H}}}}^{\mathrm{ord}} = \vec{\widetilde{C}}_{\check{\Phi}_n,\check{\delta}_n}^{\mathrm{ord}}$ and $\vec{C}_{\Phi_{\mathcal{H}},\delta_{\mathcal{H}}}^{\mathrm{ord}} = \vec{C}_{\Phi_n,\delta_n}^{\mathrm{ord}}$ are abelian schemes over $\vec{\widetilde{\mathsf{M}}}_{\widetilde{\mathcal{H}}}^{\mathrm{ord},\check{\Phi}_{\widetilde{\mathcal{H}}}} = \vec{\widetilde{\mathsf{M}}}_n^{\mathrm{ord},\check{\mathsf{Z}}_n} \cong \vec{\mathsf{M}}_{\mathcal{H}}^{\mathrm{ord},\Phi_{\mathcal{H}}} = \vec{\mathsf{M}}_n^{\mathrm{ord},\mathsf{Z}_n}$. For simplicity, let us denote the kernel of (7.1.2.2) by C, viewed as a scheme over $\vec{\mathsf{M}}_n^{\mathrm{ord},\mathsf{Z}_n}$.

By Proposition 4.2.1.30 and its proof, C is necessarily the extension of a finite flat group scheme $\underline{\pi}_0(C/\vec{\mathsf{M}}_n^{\mathrm{ord},\mathsf{Z}_n})$ of étale-multiplicative type by an abelian scheme over $\vec{\mathsf{M}}_n^{\mathrm{ord},\mathsf{Z}_n}$. By Lemma 7.1.2.1, we know that $C \underset{\mathbb{Z}}{\otimes} \mathbb{Q}$ is an abelian scheme $\mathbb{Q}^\times$-isogenous to the pullback of the ordinary abelian scheme $\underline{\mathrm{Hom}}_{\mathcal{O}}(\widetilde{X},B)^\circ \to \vec{\mathsf{M}}_{\mathcal{H}}^{\mathrm{ord},\Phi_{\mathcal{H}}}$. Hence, $\underline{\pi}_0(C/\vec{\mathsf{M}}_n^{\mathrm{ord},\mathsf{Z}_n})$ must be trivial, and C is necessarily an abelian scheme, which is $\mathbb{Q}^\times$-isogenous to $\underline{\mathrm{Hom}}_{\mathcal{O}}(\widetilde{X},B)^\circ \to \vec{\mathsf{M}}_{\mathcal{H}}^{\mathrm{ord},\Phi_{\mathcal{H}}}$ over all of $\vec{\mathsf{M}}_n^{\mathrm{ord},\mathsf{Z}_n}$ (by noetherian normality of $\vec{\mathsf{M}}_n^{\mathrm{ord},\mathsf{Z}_n}$ and by [92, IX, 1.4], [28, Ch. I, Prop. 2.7], or [62, Prop. 3.3.1.5]). Hence, h is a torsor under the pullback of C to $\vec{C}_{\Phi_n,\delta_n}^{\mathrm{ord}}$, as explained at the end of the proof of Lemma 1.3.2.50. □

Proposition 7.1.2.5. (*Compare with Proposition* 1.3.2.55.) *Under the canonical morphisms as in* (7.1.2.2), *and under the canonical homomorphisms*

$$\widehat{\mathrm{G}}_{1,\check{\mathsf{Z}}}(\mathbb{A}^{\infty,p}) \times \widehat{\mathrm{P}}_{1,\check{\mathsf{Z}},\mathsf{D}}^{\mathrm{ord}}(\mathbb{Q}_p) \to \mathrm{G}_{1,\mathsf{Z}}(\mathbb{A}^{\infty,p}) \times \mathrm{P}_{1,\mathsf{Z},\mathsf{D}}^{\mathrm{ord}}(\mathbb{Q}_p)$$

and

$$(\widetilde{\mathrm{P}}_{\check{\mathsf{Z}},\widetilde{\mathsf{Z}}}(\mathbb{A}^{\infty,p}) \times \widetilde{\mathrm{P}}_{\check{\mathsf{Z}},\widetilde{\mathsf{Z}},\widetilde{\mathsf{D}}}^{\mathrm{ord}}(\mathbb{Q}_p))/\widetilde{\mathrm{U}}_{2,\check{\mathsf{Z}}}(\mathbb{A}^\infty) \to (\mathrm{P}_{\mathsf{Z}}(\mathbb{A}^{\infty,p}) \times \mathrm{P}_{\mathsf{Z},\mathsf{D}}^{\mathrm{ord}}(\mathbb{Q}_p))/\mathrm{U}_{2,\mathsf{Z}}(\mathbb{A}^\infty),$$

the Hecke action of (suitable elements of) $\widehat{\mathrm{G}}_{1,\check{\mathsf{Z}}}(\mathbb{A}^{\infty,p}) \times \widehat{\mathrm{P}}_{1,\check{\mathsf{Z}},\mathsf{D}}^{\mathrm{ord}}(\mathbb{Q}_p)$ *on the collection* $\{\vec{\widehat{C}}_{\check{\Phi}_{\widehat{\mathcal{H}}},\check{\delta}_{\widehat{\mathcal{H}}}}^{\mathrm{ord}}\}_{\widehat{\mathcal{H}}_{\widehat{\mathrm{G}}_{1,\check{\mathsf{Z}}}}}$ *(with* $\widehat{\mathcal{H}}$ *of standard form) is compatible with the Hecke action of (suitable elements of)* $\mathrm{G}_{1,\mathsf{Z}}(\mathbb{A}^{\infty,p}) \times \mathrm{P}_{1,\mathsf{Z},\mathsf{D}}^{\mathrm{ord}}(\mathbb{Q}_p)$ *on the collection* $\{\vec{C}_{\Phi_{\mathcal{H}},\delta_{\mathcal{H}}}^{\mathrm{ord}}\}_{\mathcal{H}_{\mathrm{G}_{1,\mathsf{Z}}}}$ *(with* $\mathcal{H}$ *of standard form; see Proposition* 5.2.4.25*); the Hecke action of (suitable elements of)* $(\widetilde{\mathrm{P}}_{\check{\mathsf{Z}},\widetilde{\mathsf{Z}}}(\mathbb{A}^{\infty,p}) \times \widetilde{\mathrm{P}}_{\check{\mathsf{Z}},\widetilde{\mathsf{Z}},\widetilde{\mathsf{D}}}^{\mathrm{ord}}(\mathbb{Q}_p))/\widetilde{\mathrm{U}}_{2,\check{\mathsf{Z}}}(\mathbb{A}^\infty)$ *on the collection* $\{\coprod \vec{\widetilde{C}}_{\check{\Phi}_{\widetilde{\mathcal{H}}},\check{\delta}_{\widetilde{\mathcal{H}}}}^{\mathrm{ord}}\}_{\widetilde{\mathcal{H}}_{\widetilde{\mathrm{P}}_{\check{\mathsf{Z}},\widetilde{\mathsf{Z}}}}/\widetilde{\mathcal{H}}_{\widetilde{\mathrm{U}}_{2,\check{\mathsf{Z}}}}}$

(with $\widetilde{\mathcal{H}}$ of standard form) is compatible with the Hecke action of (suitable elements of) $(\mathrm{P_Z}(\mathbb{A}^{\infty,p}) \times \mathrm{P}^{\mathrm{ord}}_{\mathrm{Z,D}}(\mathbb{Q}_p))/\mathrm{U_{2,Z}}(\mathbb{A}^\infty)$ *on the collection* $\{\coprod \vec{C}^{\mathrm{ord}}_{\Phi_\mathcal{H},\delta_\mathcal{H}}\}_{\mathcal{H}_{\mathrm{P_Z}}/\mathcal{H}_{\mathrm{U_{2,Z}}}}$ *(with $\mathcal{H}$ of standard form); and the induced action of* $\widetilde{\mathrm{G}}_{l,\breve{\mathrm{Z}},\widetilde{\mathrm{Z}}}(\mathbb{A}^\infty)$ *on the index sets* $\{[(\widetilde{\mathrm{Z}}_{\widetilde{\mathcal{H}}}, \widetilde{\Phi}_{\widetilde{\mathcal{H}}}, \widetilde{\delta}_{\widetilde{\mathcal{H}}})]\}$ *is compatible with the induced action of* $\mathrm{G}'_{l,\mathrm{Z}}(\mathbb{A}^\infty)$ *on the index sets* $\{[(\mathrm{Z}_\mathcal{H}, \Phi_\mathcal{H}, \delta_\mathcal{H})]\}$ *(again see Proposition 5.2.4.25) under the canonical homomorphism* $\widetilde{\mathrm{G}}_{l,\breve{\mathrm{Z}},\widetilde{\mathrm{Z}}}(\mathbb{A}^\infty) \to \mathrm{G}'_{l,\mathrm{Z}}(\mathbb{A}^\infty)$.

These Hecke actions induce a Hecke action of (suitable elements of) the subgroup $(\widehat{\mathrm{P}}_{\breve{\mathrm{Z}}}(\mathbb{A}^{\infty,p}) \times \widehat{\mathrm{P}}^{\mathrm{ord}}_{\breve{\mathrm{Z}},\mathrm{D}}(\mathbb{Q}_p))/\widehat{\mathrm{U}}_{2,\breve{\mathrm{Z}}}(\mathbb{A}^\infty)$ *of* $(\widetilde{\mathrm{P}}_{\breve{\mathrm{Z}},\widetilde{\mathrm{Z}}}(\mathbb{A}^{\infty,p}) \times \widetilde{\mathrm{P}}^{\mathrm{ord}}_{\breve{\mathrm{Z}},\widetilde{\mathrm{Z}},\widetilde{\mathrm{D}}}(\mathbb{Q}_p))/\widetilde{\mathrm{U}}_{2,\breve{\mathrm{Z}}}(\mathbb{A}^\infty)$ *on the collection* $\{\coprod \vec{\widetilde{C}}^{\mathrm{ord}}_{\breve{\Phi}_{\widehat{\mathcal{H}}},\breve{\delta}_{\widehat{\mathcal{H}}}}\}_{\widehat{\mathcal{H}}_{\widehat{\mathrm{P}}_{\breve{\mathrm{Z}}}}/\widehat{\mathcal{H}}_{\widehat{\mathrm{U}}_{2,\breve{\mathrm{Z}}}}}$ *(with $\widehat{\mathcal{H}}$ of standard form), which is compatible with the Hecke action of (suitable elements of)* $(\mathrm{P_Z}(\mathbb{A}^{\infty,p}) \times \mathrm{P}^{\mathrm{ord}}_{\mathrm{Z,D}}(\mathbb{Q}_p))/\mathrm{U_{2,Z}}(\mathbb{A}^\infty)$ *on the collection* $\{\coprod \vec{C}^{\mathrm{ord}}_{\Phi_\mathcal{H},\delta_\mathcal{H}}\}_{\mathcal{H}_{\mathrm{P_Z}}/\mathcal{H}_{\mathrm{U_{2,Z}}}}$ *(with $\mathcal{H}$ of standard form) under the canonical morphisms* $\vec{\widetilde{C}}^{\mathrm{ord}}_{\breve{\Phi}_{\widehat{\mathcal{H}}},\breve{\delta}_{\widehat{\mathcal{H}}}} \to \vec{C}^{\mathrm{ord}}_{\Phi_\mathcal{H},\delta_\mathcal{H}}$ *(with varying $\widehat{\mathcal{H}}$ and $\mathcal{H}$) and the canonical homomorphism* $(\widehat{\mathrm{P}}_{\breve{\mathrm{Z}}}(\mathbb{A}^{\infty,p}) \times \widehat{\mathrm{P}}^{\mathrm{ord}}_{\breve{\mathrm{Z}},\mathrm{D}}(\mathbb{Q}_p))/\widehat{\mathrm{U}}_{2,\breve{\mathrm{Z}}}(\mathbb{A}^\infty) \to (\mathrm{P_Z}(\mathbb{A}^{\infty,p}) \times \mathrm{P}^{\mathrm{ord}}_{\mathrm{Z,D}}(\mathbb{Q}_p))/\mathrm{U_{2,Z}}(\mathbb{A}^\infty)$*; and the induced action of the subgroup* $\widehat{\mathrm{G}}'_{l,\breve{\mathrm{Z}}}(\mathbb{A}^\infty)$ *of* $\widetilde{\mathrm{G}}_{l,\breve{\mathrm{Z}},\widetilde{\mathrm{Z}}}(\mathbb{A}^\infty)$ *on the index sets* $\{[(\breve{\mathrm{Z}}_{\widehat{\mathcal{H}}}, \breve{\Phi}_{\widehat{\mathcal{H}}}, \breve{\delta}_{\widehat{\mathcal{H}}})]\}$ *is compatible with the induced action of* $\mathrm{G}'_{l,\mathrm{Z}}(\mathbb{A}^\infty)$ *on the index sets* $\{[(\mathrm{Z}_\mathcal{H}, \Phi_\mathcal{H}, \delta_\mathcal{H})]\}$ *under the canonical homomorphism* $\widehat{\mathrm{G}}'_{l,\breve{\mathrm{Z}}}(\mathbb{A}^\infty) \to \mathrm{G}'_{l,\mathrm{Z}}(\mathbb{A}^\infty)$.

Proof. As in the case of (1.3.2.51), the canonical morphisms as in (7.1.2.2) correspond to pushouts of extensions of B (resp. $B^\vee$) by $\breve{T}$ (resp. $\breve{T}^\vee$) under the canonical homomorphism $\breve{T} \to T$ (resp. $\breve{T}^\vee \to T^\vee$) induced by the restriction from $\breve{X}$ (resp. $\breve{Y}$) to X (resp. Y). Hence, the realizations of the Hecke twists are compatible in the desired ways. (We omit the details for simplicity.) □

Suppose $\widetilde{\sigma} \subset \mathbf{P}^+_{\widetilde{\Phi}_{\widetilde{\mathcal{H}}}}$ is a *top-dimensional* nondegenerate rational polyhedral cone in the cone decomposition $\widetilde{\Sigma}_{\widetilde{\Phi}_{\widetilde{\mathcal{H}}}}$ in $\widetilde{\Sigma}^{\mathrm{ord}}$, and suppose $\breve{\sigma}$ is the image of $\widetilde{\sigma} \subset \mathbf{P}^+_{\widetilde{\Phi}_{\widetilde{\mathcal{H}}}}$ under the first morphism in (1.2.4.20). Then we have $\breve{\sigma}^\perp = \widehat{\mathbf{S}}_{\breve{\Phi}_{\widehat{\mathcal{H}}}}$ (see Definition 1.2.4.29) for any such $\breve{\sigma}^\vee$, where $\widehat{\mathcal{H}} = \widetilde{\mathcal{H}}_{\widehat{\mathrm{G}}}$, and we have the following:

Proposition 7.1.2.6. *(Compare with Proposition 1.3.2.56, Lemma 5.2.4.29 and Proposition 5.2.4.30.) The scheme*

$$\vec{\widetilde{\Xi}}^{\mathrm{ord}}_{\breve{\Phi}_{\widetilde{\mathcal{H}}},\breve{\delta}_{\widetilde{\mathcal{H}}},\breve{\sigma}} \cong \underline{\mathrm{Spec}}_{\mathscr{O}_{\vec{\widetilde{C}}^{\mathrm{ord}}_{\breve{\Phi}_{\widetilde{\mathcal{H}}},\breve{\delta}_{\widetilde{\mathcal{H}}}}}} \left(\bigoplus_{\breve{\ell} \in \breve{\sigma}^\perp} \vec{\widetilde{\Psi}}^{\mathrm{ord}}_{\breve{\Phi}_{\widetilde{\mathcal{H}}},\breve{\delta}_{\widetilde{\mathcal{H}}}}(\breve{\ell}) \right)$$

over $\vec{\widetilde{C}}^{\mathrm{ord}}_{\breve{\Phi}_{\widetilde{\mathcal{H}}},\breve{\delta}_{\widetilde{\mathcal{H}}}}$ *is a torsor under the split torus* $\widetilde{E}_{\breve{\Phi}_{\widetilde{\mathcal{H}}},\breve{\sigma}}$ *with character group* $\breve{\sigma}^\perp$*, which is canonically isomorphic to the split torus* $\widehat{E}_{\breve{\Phi}_{\widehat{\mathcal{H}}}}$ *with character group* $\widehat{\mathbf{S}}_{\breve{\Phi}_{\widehat{\mathcal{H}}}}$*, which depends only on* $\widehat{\mathcal{H}}_{\widehat{\mathrm{P}}_{\breve{\mathrm{Z}}'}} = \widetilde{\mathcal{H}}_{\widetilde{\mathrm{P}}'_{\breve{\mathrm{Z}}}}/\widetilde{\mathcal{H}}_{\widetilde{\mathrm{U}}_{2,\breve{\mathrm{Z}}}}$ *(see Definition 1.2.4.53). We have* $\widehat{\mathbf{S}}^\vee_{\breve{\Phi}_1}/\widehat{\mathbf{S}}^\vee_{\breve{\Phi}_{\widehat{\mathcal{H}}}} \cong \widehat{\mathrm{U}}_{2,\breve{\mathrm{Z}}}(\hat{\mathbb{Z}})/\widehat{\mathcal{H}}_{\widehat{\mathrm{U}}_{2,\breve{\mathrm{Z}}}}$*, where* $\widehat{\mathbf{S}}_{\breve{\Phi}_1} := \widehat{\mathbf{S}}_{\breve{\Phi}_{\widehat{\mathrm{G}}(\hat{\mathbb{Z}})}}$ *is the kernel of the canonical homomorphism* $\mathbf{S}_{\breve{\Phi}_1} \twoheadrightarrow \mathbf{S}_{\widetilde{\Phi}_1}$ *(see Definition 1.2.4.29).*

The torus torsor $S := \vec{\widetilde{\Xi}}^{\mathrm{ord}}_{\check{\Phi}_{\widetilde{\mathcal{H}}},\check{\delta}_{\widetilde{\mathcal{H}}},\check{\sigma}} \to \vec{\widetilde{C}}^{\mathrm{ord}}_{\check{\Phi}_{\widetilde{\mathcal{H}}},\check{\delta}_{\widetilde{\mathcal{H}}}}$ *is universal for the additional structures* $(\widehat{\tau}^{\mathrm{ord}}_{\widehat{\mathcal{H}}}, \widehat{\tau}^{\vee,\mathrm{ord}}_{\widehat{\mathcal{H}}})$, *which are* $\widehat{\mathcal{H}}_{\widehat{\mathrm{P}}_{\check{\mathrm{Z}}}}/\widehat{\mathcal{U}}^{\mathrm{bal}}_1(n)_{\widehat{\mathrm{P}}_{\check{\mathrm{Z}}}}$*-orbits of étale-locally-defined pairs* $(\widehat{\tau}_n, \widehat{\tau}^{\vee}_n)$ *(for some integer* $n = n_0 p^r$ *where* $n_0 \geq 1$ *is an integer prime to* p *such that* $\widehat{\mathcal{U}}^p(n_0) \subset \widehat{\mathcal{H}}^p$ *and where* $r = \mathrm{depth}_{\mathsf{D}}(\widehat{\mathcal{H}}_p)$*, in which case we set* $\widehat{\mathcal{U}}^{\mathrm{bal}}_1(n) := \widehat{\mathcal{U}}^p(n_0)\widehat{\mathcal{U}}^{\mathrm{bal}}_{p,1}(p^r)$*), where:*

(1) $\widehat{\tau}^{\mathrm{ord}}_n = \widehat{\tau}_{n_0} : \mathbf{1}_{\frac{1}{n_0}Y \times \check{X},S} \xrightarrow{\sim} (\check{c}^{\vee}_{n_0}|_{\frac{1}{n}Y_S} \times \check{c})^* \mathcal{P}^{\otimes -1}_B$ *is a trivialization of biextensions.*
(2) $\widehat{\tau}^{\vee,\mathrm{ord}}_n = \widehat{\tau}^{\vee}_{n_0} : \mathbf{1}_{\frac{1}{n_0}\check{Y} \times X,S} \xrightarrow{\sim} (\check{c}^{\vee}_{n_0} \times \check{c}|_{X_S})^* \mathcal{P}^{\otimes -1}_B$ *is a trivialization of biextensions.*
(3) $\widehat{\tau}_{n_0}$ *and* $\widehat{\tau}^{\vee}_{n_0}$ *satisfy the analogues of the usual* $\mathcal{O}$*-compatibility condition.*
(4) $\widehat{\tau}_{n_0}$ *and* $\widehat{\tau}^{\vee}_{n_0}$ *satisfy the symmetry condition that* $\widehat{\tau}_{n_0}|_{\mathbf{1}_{Y \times \check{Y},S}}$ *and* $\widehat{\tau}^{\vee}_{n_0}|_{\mathbf{1}_{\check{Y} \times Y,S}}$ *coincide under the canonical isomorphism induced by the swapping isomorphism* $\mathbf{1}_{Y \times \check{Y},S} \xrightarrow{\sim} \mathbf{1}_{\check{Y} \times Y,S}$ *and the symmetry automorphism of* $\mathcal{P}_B$.
(5) $\widehat{\tau}_{n_0}|_{\mathbf{1}_{\frac{1}{n_0}Y \times X,S}} = \widehat{\tau}^{\vee}_{n_0}|_{\mathbf{1}_{\frac{1}{n_0}Y \times X,S}}$.

We shall denote $\vec{\widetilde{\Xi}}^{\mathrm{ord}}_{\check{\Phi}_{\widetilde{\mathcal{H}}},\check{\delta}_{\widetilde{\mathcal{H}}},\check{\sigma}}$ *by* $\vec{\widehat{\Xi}}^{\mathrm{ord}}_{\check{\Phi}_{\widehat{\mathcal{H}}},\check{\delta}_{\widehat{\mathcal{H}}}}$ *when we want to emphasize that (by Lemma* 5.2.4.26*) it depends only on* $\widehat{\mathcal{H}} = \widehat{\mathcal{H}}_{\widehat{\mathrm{G}}}$ *and* $(\check{\mathrm{Z}}_{\widehat{\mathcal{H}}}, \check{\Phi}_{\widehat{\mathcal{H}}}, \check{\delta}_{\widehat{\mathcal{H}}})$ *(see Definition* 1.2.4.17*), but does not depend on the choice of* $\check{\sigma}$.

The $\widehat{E}_{\check{\Phi}_{\widehat{\mathcal{H}}}}$*-torsor structure of* $\vec{\widehat{\Xi}}^{\mathrm{ord}}_{\check{\Phi}_{\widehat{\mathcal{H}}},\check{\delta}_{\widehat{\mathcal{H}}}} \to \vec{\widehat{C}}^{\mathrm{ord}}_{\check{\Phi}_{\widehat{\mathcal{H}}},\check{\delta}_{\widehat{\mathcal{H}}}}$ *defines a homomorphism*

$$\widehat{\mathbf{S}}_{\check{\Phi}_{\widehat{\mathcal{H}}}} \to \underline{\mathrm{Pic}}(\vec{\widehat{C}}^{\mathrm{ord}}_{\check{\Phi}_{\widehat{\mathcal{H}}},\check{\delta}_{\widehat{\mathcal{H}}}}) : \check{\ell} \mapsto \vec{\widehat{\Psi}}^{\mathrm{ord}}_{\check{\Phi}_{\widehat{\mathcal{H}}},\check{\delta}_{\widehat{\mathcal{H}}}}(\check{\ell}),$$

assigning to each $\check{\ell} \in \widehat{\mathbf{S}}_{\check{\Phi}_{\widehat{\mathcal{H}}}}$ *an invertible sheaf* $\vec{\widehat{\Psi}}^{\mathrm{ord}}_{\check{\Phi}_{\widehat{\mathcal{H}}},\check{\delta}_{\widehat{\mathcal{H}}}}(\check{\ell})$ *over* $\vec{\widehat{C}}^{\mathrm{ord}}_{\check{\Phi}_{\widehat{\mathcal{H}}},\check{\delta}_{\widehat{\mathcal{H}}}}$ *(up to isomorphism), together with isomorphisms*

$$\vec{\widehat{\Delta}}^{\mathrm{ord},*}_{\check{\Phi}_{\widehat{\mathcal{H}}},\check{\delta}_{\widehat{\mathcal{H}}},\check{\ell},\check{\ell}'} : \vec{\widehat{\Psi}}^{\mathrm{ord}}_{\check{\Phi}_{\widehat{\mathcal{H}}},\check{\delta}_{\widehat{\mathcal{H}}}}(\check{\ell}) \underset{\mathscr{O}_{\vec{\widehat{C}}^{\mathrm{ord}}_{\check{\Phi}_{\widehat{\mathcal{H}}},\check{\delta}_{\widehat{\mathcal{H}}}}}}{\otimes} \vec{\widehat{\Psi}}^{\mathrm{ord}}_{\check{\Phi}_{\widehat{\mathcal{H}}},\check{\delta}_{\widehat{\mathcal{H}}}}(\check{\ell}') \xrightarrow{\sim} \vec{\widehat{\Psi}}^{\mathrm{ord}}_{\check{\Phi}_{\widehat{\mathcal{H}}},\check{\delta}_{\widehat{\mathcal{H}}}}(\check{\ell} + \check{\ell}')$$

for all $\check{\ell}, \check{\ell}' \in \widehat{\mathbf{S}}_{\check{\Phi}_{\widehat{\mathcal{H}}}}$*, satisfying the necessary compatibilities with each other making* $\underset{\check{\ell} \in \widehat{\mathbf{S}}_{\check{\Phi}_{\widehat{\mathcal{H}}}}}{\oplus} \vec{\widehat{\Psi}}^{\mathrm{ord}}_{\check{\Phi}_{\widehat{\mathcal{H}}},\check{\delta}_{\widehat{\mathcal{H}}}}(\check{\ell})$ *an* $\mathscr{O}_{\vec{\widehat{C}}^{\mathrm{ord}}_{\check{\Phi}_{\widehat{\mathcal{H}}},\check{\delta}_{\widehat{\mathcal{H}}}}}$*-algebra, such that*

$$\vec{\widehat{\Xi}}^{\mathrm{ord}}_{\check{\Phi}_{\widehat{\mathcal{H}}},\check{\delta}_{\widehat{\mathcal{H}}}} \cong \underline{\mathrm{Spec}}_{\mathscr{O}_{\vec{\widehat{C}}^{\mathrm{ord}}_{\check{\Phi}_{\widehat{\mathcal{H}}},\check{\delta}_{\widehat{\mathcal{H}}}}}} \left(\underset{\check{\ell} \in \widehat{\mathbf{S}}_{\check{\Phi}_{\widehat{\mathcal{H}}}}}{\oplus} \vec{\widehat{\Psi}}^{\mathrm{ord}}_{\check{\Phi}_{\widehat{\mathcal{H}}},\check{\delta}_{\widehat{\mathcal{H}}}}(\check{\ell}) \right).$$

When $\check{\ell} = [y \otimes \chi]$ *for some* $y \in \check{Y}$ *and* $\chi \in \check{X}$ *such that either* $y \in Y$ *or* $\chi \in X$*, we have a canonical isomorphism*

$$\vec{\widehat{\Psi}}^{\mathrm{ord}}_{\check{\Phi}_{\widehat{\mathcal{H}}},\check{\delta}_{\widehat{\mathcal{H}}}}(\check{\ell}) \cong (\check{c}^{\vee}(y), \check{c}(\chi))^* \mathcal{P}_B.$$

If we fix the choice of ($\check{\mathsf{Z}}_n$ *and*) $\check{\Phi}_n$, *then the canonical morphism*

$$\vec{\widehat{\Xi}}^{\mathrm{ord}}_{\check{\Phi}_n,\check{\delta}_n} \to \vec{\widehat{\Xi}}^{\mathrm{ord}}_{\check{\Phi}_{\widehat{\mathcal{H}}},\check{\delta}_{\widehat{\mathcal{H}}},r_n} = \vec{\widehat{\Xi}}^{\mathrm{ord}}_{\check{\Phi}_{\widehat{\mathcal{H}}},\check{\delta}_{\widehat{\mathcal{H}}}} \underset{\vec{\mathsf{S}}_{0,r_{\widehat{\mathcal{H}}}}}{\times} \vec{\mathsf{S}}_{0,r_n} \tag{7.1.2.7}$$

(*cf.* (1.3.2.57)), *where* $r_n = r_{\widehat{\mathcal{U}}_1^{\mathrm{bal}}(n)} = r_{\mathcal{U}_1^{\mathrm{bal}}(n)} = \max(r_{\mathsf{D}}, r)$, *is an* $\widehat{\mathcal{H}}_{\widehat{\mathsf{P}}'_{\check{\mathsf{Z}}}}/\widehat{\mathcal{U}}_1^{\mathrm{bal}}(n)_{\widehat{\mathsf{P}}'_{\check{\mathsf{Z}}}}$*-torsor, and induces an isomorphism*

$$\vec{\widehat{\Xi}}^{\mathrm{ord}}_{\check{\Phi}_n,\check{\delta}_n}/(\widehat{\mathcal{H}}_{\widehat{\mathsf{P}}'_{\check{\mathsf{Z}}}}/\widehat{\mathcal{U}}_1^{\mathrm{bal}}(n)_{\widehat{\mathsf{P}}'_{\check{\mathsf{Z}}}}) \xrightarrow{\sim} \vec{\widehat{\Xi}}^{\mathrm{ord}}_{\check{\Phi}_{\widehat{\mathcal{H}}},\check{\delta}_{\widehat{\mathcal{H}}},r_n} \tag{7.1.2.8}$$

(*cf. Lemma* 5.2.4.26*; cf. also* (1.3.2.58)).

Proof. These follow from the corresponding properties of $\vec{\widetilde{\Xi}}^{\mathrm{ord}}_{\check{\Phi}_{\widetilde{\mathcal{H}}},\check{\delta}_{\widetilde{\mathcal{H}}}}$ as in Lemmas 5.2.4.26 and 5.2.4.29, and Proposition 5.2.4.30, because the restriction from $\mathbf{S}_{\check{\Phi}_{\widetilde{\mathcal{H}}}}$ to the subgroup $\widehat{\mathbf{S}}_{\check{\Phi}_{\widehat{\mathcal{H}}}}$ (see Definition 1.2.4.29) corresponds to taking orbits of restrictions of $\check{\tau}_{n_0} : \mathbf{1}_{\frac{1}{n_0}\check{Y}\times\check{X},S} \xrightarrow{\sim} (\check{c}^\vee\times\check{c})^*\mathcal{P}_B^{\otimes -1}$ to $\mathbf{1}_{\frac{1}{n_0}Y\times\check{X},S}$ and $\mathbf{1}_{\frac{1}{n_0}\check{Y}\times X,S}$, which form the pairs $(\widehat{\tau}_n^{\mathrm{ord}}, \widehat{\tau}_n^{\vee,\mathrm{ord}}) = (\widehat{\tau}_{n_0}, \widehat{\tau}_{n_0}^\vee)$ as above. □

For each rational polyhedral cone $\check{\rho} \subset (\mathbf{S}_{\check{\Phi}_{\widetilde{\mathcal{H}}}})^\vee_{\mathbb{R}}$ having $\check{\sigma}$ as a face, we have an affine toroidal embedding

$$\vec{\widetilde{\Xi}}^{\mathrm{ord}}_{\check{\Phi}_{\widetilde{\mathcal{H}}},\check{\delta}_{\widetilde{\mathcal{H}}},\check{\sigma}} \hookrightarrow \vec{\widetilde{\Xi}}^{\mathrm{ord}}_{\check{\Phi}_{\widetilde{\mathcal{H}}},\check{\delta}_{\widetilde{\mathcal{H}}},\check{\sigma}}(\check{\rho}) := \underline{\mathrm{Spec}}_{\mathscr{O}_{\widetilde{C}_{\check{\Phi}_{\widetilde{\mathcal{H}}},\check{\delta}_{\widetilde{\mathcal{H}}}}}}\left(\bigoplus_{\check{\ell}\in\check{\sigma}^\perp\cap\check{\rho}^\vee} \vec{\widetilde{\Psi}}^{\mathrm{ord}}_{\check{\Phi}_{\widetilde{\mathcal{H}}},\check{\delta}_{\widetilde{\mathcal{H}}}}(\check{\ell})\right) \tag{7.1.2.9}$$

(*cf.* (1.3.2.59)) as in (5.2.4.31).

In general, for each rational polyhedral cone $\widehat{\rho} \subset (\widehat{\mathbf{S}}_{\check{\Phi}_{\widehat{\mathcal{H}}}})^\vee_{\mathbb{R}}$, we have an affine toroidal embedding

$$\vec{\widehat{\Xi}}^{\mathrm{ord}}_{\check{\Phi}_{\widehat{\mathcal{H}}},\check{\delta}_{\widehat{\mathcal{H}}}} \hookrightarrow \vec{\widehat{\Xi}}^{\mathrm{ord}}_{\check{\Phi}_{\widehat{\mathcal{H}}},\check{\delta}_{\widehat{\mathcal{H}}}}(\widehat{\rho}) := \underline{\mathrm{Spec}}_{\mathscr{O}_{\vec{\widehat{C}}^{\mathrm{ord}}_{\check{\Phi}_{\widehat{\mathcal{H}}},\check{\delta}_{\widehat{\mathcal{H}}}}}}\left(\bigoplus_{\check{\ell}\in\widehat{\rho}^\vee} \vec{\widehat{\Psi}}^{\mathrm{ord}}_{\check{\Phi}_{\widehat{\mathcal{H}}},\check{\delta}_{\widehat{\mathcal{H}}}}(\check{\ell})\right) \tag{7.1.2.10}$$

(*cf.* (1.3.2.60)).

By Proposition 7.1.2.6, (7.1.2.9) and (7.1.2.10) can be canonically identified when $\widehat{\mathcal{H}} = \widetilde{\mathcal{H}}_{\widehat{\mathsf{G}}}$, when $(\check{\mathsf{Z}}_{\widehat{\mathcal{H}}}, \check{\Phi}_{\widehat{\mathcal{H}}}, \check{\delta}_{\widehat{\mathcal{H}}})$ is determined by $(\check{\mathsf{Z}}_{\widetilde{\mathcal{H}}}, \check{\Phi}_{\widetilde{\mathcal{H}}}, \check{\delta}_{\widetilde{\mathcal{H}}})$ as in Definition 1.2.4.17, and when $\widehat{\rho} = \mathrm{pr}_{(\widehat{\mathbf{S}}_{\check{\Phi}_{\widehat{\mathcal{H}}}})^\vee_{\mathbb{R}}}(\check{\rho})$. (Hence, (7.1.2.9) depends only on these induced parameters.)

Both sides of (7.1.2.10) are relative affine over $\vec{\widehat{C}}^{\mathrm{ord}}_{\check{\Phi}_{\widehat{\mathcal{H}}},\check{\delta}_{\widehat{\mathcal{H}}}}$, where $\vec{\widehat{\Xi}}^{\mathrm{ord}}_{\check{\Phi}_{\widehat{\mathcal{H}}},\check{\delta}_{\widehat{\mathcal{H}}}}(\widehat{\rho}) \to \vec{\widehat{C}}^{\mathrm{ord}}_{\check{\Phi}_{\widehat{\mathcal{H}}},\check{\delta}_{\widehat{\mathcal{H}}}}$ is smooth when the cone $\widehat{\rho}$ is smooth. The $\widehat{\rho}$-stratum of $\vec{\widehat{\Xi}}^{\mathrm{ord}}_{\check{\Phi}_{\widehat{\mathcal{H}}},\check{\delta}_{\widehat{\mathcal{H}}}}(\widehat{\rho})$ is

$$\vec{\widehat{\Xi}}^{\mathrm{ord}}_{\check{\Phi}_{\widehat{\mathcal{H}}},\check{\delta}_{\widehat{\mathcal{H}}},\widehat{\rho}} := \underline{\mathrm{Spec}}_{\mathscr{O}_{\vec{\widehat{C}}^{\mathrm{ord}}_{\check{\Phi}_{\widehat{\mathcal{H}}},\check{\delta}_{\widehat{\mathcal{H}}}}}}\left(\bigoplus_{\check{\ell}\in\widehat{\rho}^\perp} \vec{\widehat{\Psi}}^{\mathrm{ord}}_{\check{\Phi}_{\widehat{\mathcal{H}}},\check{\delta}_{\widehat{\mathcal{H}}}}(\check{\ell})\right) \tag{7.1.2.11}$$

(*cf.* (1.3.2.61)), which is canonically isomorphic to $\vec{\widetilde{\Xi}}^{\mathrm{ord}}_{\check{\Phi}_{\widetilde{\mathcal{H}}},\check{\delta}_{\widetilde{\mathcal{H}}},\check{\rho}}$. The affine morphism $\vec{\widehat{\Xi}}^{\mathrm{ord}}_{\check{\Phi}_{\widehat{\mathcal{H}}},\check{\delta}_{\widehat{\mathcal{H}}},\widehat{\rho}} \to \vec{\widehat{C}}^{\mathrm{ord}}_{\check{\Phi}_{\widehat{\mathcal{H}}},\check{\delta}_{\widehat{\mathcal{H}}}}$ is a torsor under the torus $\widehat{E}_{\check{\Phi}_{\widehat{\mathcal{H}}},\widehat{\rho}} \cong E_{\check{\Phi}_{\widetilde{\mathcal{H}}},\check{\rho}}$ with character group $\widehat{\rho}^{\perp} \cong \breve{\rho}^{\perp}$. (Note that these two instance of $\perp$ are taken in different ambient spaces.) For each $\Gamma_{\breve{\Phi}_{\widehat{\mathcal{H}}}}$-admissible rational polyhedral cone decomposition $\widehat{\Sigma}_{\breve{\Phi}_{\widehat{\mathcal{H}}}}$ of $\widehat{\mathbf{P}}_{\breve{\Phi}_{\widehat{\mathcal{H}}}}$ as in Definition 1.2.4.40, we have (as in (4.2.2.4)) a toroidal embedding

$$\vec{\widehat{\Xi}}^{\mathrm{ord}}_{\breve{\Phi}_{\widehat{\mathcal{H}}}, \breve{\delta}_{\widehat{\mathcal{H}}}} \hookrightarrow \vec{\overline{\widehat{\Xi}}}^{\mathrm{ord}}_{\breve{\Phi}_{\widehat{\mathcal{H}}}, \breve{\delta}_{\widehat{\mathcal{H}}}} = \vec{\overline{\widehat{\Xi}}}^{\mathrm{ord}}_{\breve{\Phi}_{\widehat{\mathcal{H}}}, \breve{\delta}_{\widehat{\mathcal{H}}}, \widehat{\Sigma}_{\breve{\Phi}_{\widehat{\mathcal{H}}}}} \tag{7.1.2.12}$$

(*cf.* (1.3.2.62)), the right-hand side being only locally of finite type over $\vec{\widehat{C}}^{\mathrm{ord}}_{\breve{\Phi}_{\widehat{\mathcal{H}}}, \breve{\delta}_{\widehat{\mathcal{H}}}}$, with an open covering

$$\vec{\overline{\widehat{\Xi}}}^{\mathrm{ord}}_{\breve{\Phi}_{\widehat{\mathcal{H}}}, \breve{\delta}_{\widehat{\mathcal{H}}}} = \bigcup_{\widehat{\rho} \in \widehat{\Sigma}_{\breve{\Phi}_{\widehat{\mathcal{H}}}}} \vec{\widehat{\Xi}}^{\mathrm{ord}}_{\breve{\Phi}_{\widehat{\mathcal{H}}}, \breve{\delta}_{\widehat{\mathcal{H}}}}(\widehat{\rho}) \tag{7.1.2.13}$$

(*cf.* (1.3.2.63)) inducing a stratification

$$\vec{\overline{\widehat{\Xi}}}^{\mathrm{ord}}_{\breve{\Phi}_{\widehat{\mathcal{H}}}, \breve{\delta}_{\widehat{\mathcal{H}}}} = \coprod_{\widehat{\rho} \in \widehat{\Sigma}_{\breve{\Phi}_{\widehat{\mathcal{H}}}}} \vec{\widehat{\Xi}}^{\mathrm{ord}}_{\breve{\Phi}_{\widehat{\mathcal{H}}}, \breve{\delta}_{\widehat{\mathcal{H}}}, \widehat{\rho}} \tag{7.1.2.14}$$

(*cf.* (1.3.2.64)). (The notation "$\coprod$" only means a set-theoretic disjoint union. The algebro-geometric structure is still the one inherited from $\vec{\overline{\widehat{\Xi}}}^{\mathrm{ord}}_{\breve{\Phi}_{\widehat{\mathcal{H}}}, \breve{\delta}_{\widehat{\mathcal{H}}}}$.) Let

$$\vec{\widehat{\mathfrak{X}}}^{\mathrm{ord}}_{\breve{\Phi}_{\widehat{\mathcal{H}}}, \breve{\delta}_{\widehat{\mathcal{H}}}, \widehat{\rho}} := (\vec{\widehat{\Xi}}^{\mathrm{ord}}_{\breve{\Phi}_{\widehat{\mathcal{H}}}, \breve{\delta}_{\widehat{\mathcal{H}}}}(\widehat{\rho}))^{\wedge}_{\vec{\widehat{\Xi}}^{\mathrm{ord}}_{\breve{\Phi}_{\widehat{\mathcal{H}}}, \breve{\delta}_{\widehat{\mathcal{H}}}, \widehat{\rho}}} \tag{7.1.2.15}$$

(*cf.* (1.3.2.65)), the formal completion of $\vec{\widehat{\Xi}}^{\mathrm{ord}}_{\breve{\Phi}_{\widehat{\mathcal{H}}}, \breve{\delta}_{\widehat{\mathcal{H}}}}(\widehat{\rho})$ along its $\widehat{\rho}$-stratum $\vec{\widehat{\Xi}}^{\mathrm{ord}}_{\breve{\Phi}_{\widehat{\mathcal{H}}}, \breve{\delta}_{\widehat{\mathcal{H}}}, \widehat{\rho}}$, which is canonically isomorphic to $\vec{\widetilde{\mathfrak{X}}}^{\mathrm{ord}}_{\breve{\Phi}_{\widehat{\mathcal{H}}}, \breve{\delta}_{\widehat{\mathcal{H}}}, \breve{\sigma}, \breve{\rho}}$, the formal completion of $\vec{\widetilde{\Xi}}^{\mathrm{ord}}_{\breve{\Phi}_{\widehat{\mathcal{H}}}, \breve{\delta}_{\widehat{\mathcal{H}}}, \breve{\sigma}}(\breve{\rho})$, the closure of $\vec{\widetilde{\Xi}}^{\mathrm{ord}}_{\breve{\Phi}_{\widehat{\mathcal{H}}}, \breve{\delta}_{\widehat{\mathcal{H}}}, \breve{\sigma}}$ in $\vec{\widetilde{\Xi}}^{\mathrm{ord}}_{\breve{\Phi}_{\widehat{\mathcal{H}}}, \breve{\delta}_{\widehat{\mathcal{H}}}}(\breve{\rho})$, along its $\breve{\rho}$-stratum $\vec{\widetilde{\Xi}}^{\mathrm{ord}}_{\breve{\Phi}_{\widehat{\mathcal{H}}}, \breve{\delta}_{\widehat{\mathcal{H}}}, \breve{\rho}}$ (*cf.* (1.3.2.40)). Also, let us define

$$\vec{\widehat{\mathfrak{X}}}^{\mathrm{ord}}_{\breve{\Phi}_{\widehat{\mathcal{H}}}, \breve{\delta}_{\widehat{\mathcal{H}}}} = \vec{\widehat{\mathfrak{X}}}^{\mathrm{ord}}_{\breve{\Phi}_{\widehat{\mathcal{H}}}, \breve{\delta}_{\widehat{\mathcal{H}}}, \widehat{\Sigma}_{\breve{\Phi}_{\widehat{\mathcal{H}}}}} \tag{7.1.2.16}$$

(*cf.* (1.3.2.66) and Lemma 5.2.4.38) to be the formal completion of $\vec{\overline{\widehat{\Xi}}}^{\mathrm{ord}}_{\breve{\Phi}_{\widehat{\mathcal{H}}}, \breve{\delta}_{\widehat{\mathcal{H}}}, \widehat{\Sigma}_{\breve{\Phi}_{\widehat{\mathcal{H}}}}}$ along the union of the $\widehat{\sigma}$-strata $\vec{\widehat{\Xi}}^{\mathrm{ord}}_{\breve{\Phi}_{\widehat{\mathcal{H}}}, \breve{\delta}_{\widehat{\mathcal{H}}}, \widehat{\rho}}$ for $\widehat{\rho} \in \widehat{\Sigma}_{\breve{\Phi}_{\widehat{\mathcal{H}}}}$ and $\widehat{\rho} \subset \widehat{\mathbf{P}}^{+}_{\breve{\Phi}_{\widehat{\mathcal{H}}}}$.

Proposition 7.1.2.17. (*Compare with Propositions* 1.3.2.67, 5.2.2.2, 5.2.4.25, 5.2.4.41, *and* 7.1.2.5.) *There is a Hecke action of (suitable elements of)* $\widehat{\mathrm{P}}'_{\breve{\mathrm{Z}}}(\mathbb{A}^{\infty,p}) \times \widehat{\mathrm{P}}^{\mathrm{ord},\prime}_{\breve{\mathrm{Z}},\mathrm{D}}(\mathbb{Q}_p)$ *on the collection* $\{\vec{\overline{\widehat{\Xi}}}^{\mathrm{ord}}_{\breve{\Phi}_{\widehat{\mathcal{H}}}, \breve{\delta}_{\widehat{\mathcal{H}}}}\}_{\widehat{\mathcal{H}}_{\widehat{\mathrm{P}}'_{\breve{\mathrm{Z}}}}}$ *(with* $\widehat{\mathcal{H}}$ *of standard form), realized by quasi-finite flat surjections pulling tautological objects back to ordinary Hecke twists, which is compatible with the Hecke action of (suitable elements of)* $\widehat{\mathrm{G}}_{1,\breve{\mathrm{Z}}}(\mathbb{A}^{\infty,p}) \times \widehat{\mathrm{P}}^{\mathrm{ord}}_{1,\breve{\mathrm{Z}},\mathrm{D}}(\mathbb{Q}_p)$ *on the collection* $\{\vec{\widehat{C}}^{\mathrm{ord}}_{\breve{\Phi}_{\widehat{\mathcal{H}}}, \breve{\delta}_{\widehat{\mathcal{H}}}}\}_{\widehat{\mathcal{H}}_{\widehat{\mathrm{G}}_{1,\breve{\mathrm{Z}}}}}$ *(with* $\widehat{\mathcal{H}}$ *of standard* *form) under the canonical morphisms* $\vec{\widehat{\Xi}}^{\mathrm{ord}}_{\breve{\Phi}_{\widehat{\mathcal{H}}},\breve{\delta}_{\widehat{\mathcal{H}}}} \to \vec{\widehat{C}}^{\mathrm{ord}}_{\breve{\Phi}_{\widehat{\mathcal{H}}},\breve{\delta}_{\widehat{\mathcal{H}}}}$ *(with varying* $\widehat{\mathcal{H}}$*) and the canonical homomorphism* $\widehat{\mathrm{P}}'_{\breve{\mathrm{Z}}}(\mathbb{A}^{\infty,p}) \times \widehat{\mathrm{P}}^{\mathrm{ord},\prime}_{\breve{\mathrm{Z}},\mathrm{D}}(\mathbb{Q}_p) \to \widehat{\mathrm{G}}_{1,\breve{\mathrm{Z}}}(\mathbb{A}^{\infty,p}) \times \widehat{\mathrm{G}}^{\mathrm{ord}}_{1,\breve{\mathrm{Z}}}(\mathbb{Q}_p) = (\widehat{\mathrm{P}}'_{\breve{\mathrm{Z}}}(\mathbb{A}^{\infty}) \times \widehat{\mathrm{P}}^{\mathrm{ord},\prime}_{\breve{\mathrm{Z}},\mathrm{D}})/\widehat{\mathrm{U}}_{2,\breve{\mathrm{Z}}}(\mathbb{A}^{\infty})$*. Such a Hecke action enjoys the properties (under various conditions) concerning étaleness, finiteness, being isomorphisms between formal completions along fibers over* $\mathrm{Spec}(\mathbb{F}_p)$*, and inducing absolute Frobenius morphisms on fibers over* $\mathrm{Spec}(\mathbb{F}_p)$ *for elements of* U_p *type as in Proposition* 5.2.2.2 *and Corollaries* 5.2.2.3, 5.2.2.4, *and* 5.2.2.5. *(We omit the details for simplicity.)*

There is also a Hecke action of (suitable elements of) $\widehat{\mathrm{P}}_{\breve{\mathrm{Z}}}(\mathbb{A}^{\infty,p}) \times \widehat{\mathrm{P}}^{\mathrm{ord}}_{\breve{\mathrm{Z}},\mathrm{D}}(\mathbb{Q}_p)$ *on the collection* $\{\coprod \vec{\widehat{\Xi}}^{\mathrm{ord}}_{\breve{\Phi}_{\widehat{\mathcal{H}}},\breve{\delta}_{\widehat{\mathcal{H}}}}\}_{\widehat{\mathcal{H}}_{\widehat{\mathrm{P}}_{\breve{\mathrm{Z}}}}}$ *(with* $\widehat{\mathcal{H}}$ *of standard form), where the disjoint unions are over classes* $[(\breve{\mathrm{Z}}_{\widehat{\mathcal{H}}}, \breve{\Phi}_{\widehat{\mathcal{H}}}, \breve{\delta}_{\widehat{\mathcal{H}}})]$ *sharing the same* $\breve{\mathrm{Z}}_{\widehat{\mathcal{H}}}$ *compatible with* D*, realized by quasi-finite flat surjections pulling tautological objects back to ordinary Hecke twists, which induces an action of* $\widehat{\mathrm{G}}'_{l,\breve{\mathrm{Z}}}(\mathbb{A}^{\infty}) = \widehat{\mathrm{P}}_{\breve{\mathrm{Z}}}(\mathbb{A}^{\infty})/\widehat{\mathrm{P}}'_{\breve{\mathrm{Z}}}(\mathbb{A}^{\infty})$ *on the index sets* $\{[(\breve{\mathrm{Z}}_{\widehat{\mathcal{H}}}, \breve{\Phi}_{\widehat{\mathcal{H}}}, \breve{\delta}_{\widehat{\mathcal{H}}})]\}$*, which is compatible with the Hecke action of (suitable elements of)* $(\widehat{\mathrm{P}}_{\breve{\mathrm{Z}}}(\mathbb{A}^{\infty,p}) \times \widehat{\mathrm{P}}^{\mathrm{ord}}_{\breve{\mathrm{Z}},\mathrm{D}}(\mathbb{Q}_p))/\widehat{\mathrm{U}}_{2,\breve{\mathrm{Z}}}(\mathbb{A}^{\infty})$ *on the collection* $\{\coprod \vec{\widehat{C}}^{\mathrm{ord}}_{\breve{\Phi}_{\widehat{\mathcal{H}}},\breve{\delta}_{\widehat{\mathcal{H}}}}\}_{\widehat{\mathcal{H}}_{\widehat{\mathrm{P}}_{\breve{\mathrm{Z}}}}/\widehat{\mathcal{H}}_{\widehat{\mathrm{U}}_{2,\breve{\mathrm{Z}}}}}$ *(with* $\widehat{\mathcal{H}}$ *of standard form, with the same index sets and the same induced action of* $\widehat{\mathrm{G}}'_{l,\breve{\mathrm{Z}}}(\mathbb{A}^{\infty})$*) under the canonical morphisms* $\widehat{\Xi}_{\breve{\Phi}_{\widehat{\mathcal{H}}},\breve{\delta}_{\widehat{\mathcal{H}}}} \to \widehat{C}_{\breve{\Phi}_{\widehat{\mathcal{H}}},\breve{\delta}_{\widehat{\mathcal{H}}}}$ *(with varying* $\widehat{\mathcal{H}}$*) and the canonical homomorphism* $\widehat{\mathrm{P}}_{\breve{\mathrm{Z}}}(\mathbb{A}^{\infty,p}) \times \widehat{\mathrm{P}}_{\breve{\mathrm{Z}}}(\mathbb{Q}_p) \to (\widehat{\mathrm{P}}_{\breve{\mathrm{Z}}}(\mathbb{A}^{\infty,p}) \times \widehat{\mathrm{P}}^{\mathrm{ord}}_{\breve{\mathrm{Z}},\mathrm{D}}(\mathbb{Q}_p))/\widehat{\mathrm{U}}_{2,\breve{\mathrm{Z}}}(\mathbb{A}^{\infty})$.

Any such Hecke action

$$[\vec{\widehat{g}}]^{\mathrm{ord}} : \vec{\widehat{\Xi}}^{\mathrm{ord}}_{\breve{\Phi}'_{\widehat{\mathcal{H}}'},\breve{\delta}'_{\widehat{\mathcal{H}}'}} \to \vec{\widehat{\Xi}}^{\mathrm{ord}}_{\breve{\Phi}_{\widehat{\mathcal{H}}},\breve{\delta}_{\widehat{\mathcal{H}}}}$$

covering $[\vec{\widehat{g}}]^{\mathrm{ord}} : \vec{\widehat{C}}^{\mathrm{ord}}_{\breve{\Phi}'_{\widehat{\mathcal{H}}'},\breve{\delta}'_{\widehat{\mathcal{H}}'}} \to \vec{\widehat{C}}^{\mathrm{ord}}_{\breve{\Phi}_{\widehat{\mathcal{H}}},\breve{\delta}_{\widehat{\mathcal{H}}}}$ *induces a morphism*

$$\vec{\widehat{\Xi}}^{\mathrm{ord}}_{\breve{\Phi}'_{\widehat{\mathcal{H}}'},\breve{\delta}'_{\widehat{\mathcal{H}}'}} \to \vec{\widehat{\Xi}}^{\mathrm{ord}}_{\breve{\Phi}_{\widehat{\mathcal{H}}},\breve{\delta}_{\widehat{\mathcal{H}}}} \underset{\vec{\widehat{C}}^{\mathrm{ord}}_{\breve{\Phi}_{\widehat{\mathcal{H}}},\breve{\delta}_{\widehat{\mathcal{H}}}}}{\times} \vec{\widehat{C}}^{\mathrm{ord}}_{\breve{\Phi}'_{\widehat{\mathcal{H}}'},\breve{\delta}'_{\widehat{\mathcal{H}}'}}$$

between torus torsors over $\vec{\widehat{C}}^{\mathrm{ord}}_{\breve{\Phi}'_{\widehat{\mathcal{H}}'},\breve{\delta}'_{\widehat{\mathcal{H}}'}}$*, which is equivariant with the morphism* $\widehat{E}_{\breve{\Phi}'_{\widehat{\mathcal{H}}'}} \to \widehat{E}_{\breve{\Phi}_{\widehat{\mathcal{H}}}}$ *dual to the homomorphism* $\widehat{\mathbf{S}}_{\breve{\Phi}_{\widehat{\mathcal{H}}}} \to \widehat{\mathbf{S}}_{\breve{\Phi}'_{\widehat{\mathcal{H}}'}}$ *induced by the pair of morphisms* $(f_{\breve{X}} : \breve{X} \underset{\mathbb{Z}}{\otimes} \mathbb{Q} \xrightarrow{\sim} \breve{X}' \underset{\mathbb{Z}}{\otimes} \mathbb{Q}, f_{\breve{Y}} : \breve{Y}' \underset{\mathbb{Z}}{\otimes} \mathbb{Q} \xrightarrow{\sim} \breve{Y} \underset{\mathbb{Z}}{\otimes} \mathbb{Q})$ *defining the* $\widehat{g}$*-assignment* $(\breve{\mathrm{Z}}'_{\widehat{\mathcal{H}}'}, \breve{\Phi}'_{\widehat{\mathcal{H}}'}, \breve{\delta}'_{\widehat{\mathcal{H}}'}) \to_{\widehat{g}} (\breve{\mathrm{Z}}_{\widehat{\mathcal{H}}}, \breve{\Phi}_{\widehat{\mathcal{H}}}, \breve{\delta}_{\widehat{\mathcal{H}}})$ *of cusp labels (which is the* $\widetilde{g}$*-assignment for any element* $\widetilde{g} \in \widetilde{\mathrm{P}}_{\breve{\mathrm{Z}}}(\mathbb{A}^{\infty}) \cap \widetilde{\mathrm{P}}'_{\breve{\mathrm{Z}}}(\mathbb{A}^{\infty})$ *lifting* $\widehat{g} \in \widehat{\mathrm{P}}_{\breve{\mathrm{Z}}}(\mathbb{A}^{\infty}) = (\widetilde{\mathrm{P}}_{\breve{\mathrm{Z}}}(\mathbb{A}^{\infty}) \cap \widetilde{\mathrm{P}}'_{\breve{\mathrm{Z}}}(\mathbb{A}^{\infty}))/\widetilde{\mathrm{U}}_{2,\breve{\mathrm{Z}}}(\mathbb{A}^{\infty})$*, which is nevertheless independent of the choice of* $\widetilde{g}$*; cf. Lemma* 1.2.4.42 *and [62, Def. 5.4.3.9]).*

If $\widehat{g} \in \widehat{\mathrm{P}}_{\breve{\mathrm{Z}}}(\mathbb{A}^{\infty,p}) \times \widehat{\mathrm{P}}^{\mathrm{ord}}_{\breve{\mathrm{Z}},\mathrm{D}}(\mathbb{Q}_p)$ *is as above and if* $(\breve{\Phi}'_{\widehat{\mathcal{H}}'}, \breve{\delta}'_{\widehat{\mathcal{H}}'}, \widehat{\rho}')$ *is a* $\widehat{g}$*-refinement of* $(\breve{\Phi}_{\widehat{\mathcal{H}}}, \breve{\delta}_{\widehat{\mathcal{H}}}, \widehat{\rho})$ *(cf. Lemma* 1.2.4.42 *and [62, Def. 6.4.3.1]), then there is a canonical morphism*

$$[\vec{\widehat{g}}]^{\mathrm{ord}} : \vec{\widehat{\Xi}}^{\mathrm{ord}}_{\breve{\Phi}'_{\widehat{\mathcal{H}}'},\breve{\delta}'_{\widehat{\mathcal{H}}'}}(\widehat{\rho}') \to \vec{\widehat{\Xi}}^{\mathrm{ord}}_{\breve{\Phi}_{\widehat{\mathcal{H}}},\breve{\delta}_{\widehat{\mathcal{H}}}}(\widehat{\rho}) \tag{7.1.2.18}$$

(cf. (1.3.2.68) *and* (5.2.4.42)) *covering* $[\vec{\widehat{g}}]^{\mathrm{ord}} : \vec{\widehat{C}}^{\mathrm{ord}}_{\breve{\Phi}'_{\widehat{\mathcal{H}}'},\breve{\delta}'_{\widehat{\mathcal{H}}'}} \to \vec{\widehat{C}}^{\mathrm{ord}}_{\breve{\Phi}_{\widehat{\mathcal{H}}},\breve{\delta}_{\widehat{\mathcal{H}}}}$, *extending* $[\vec{\widehat{g}}]^{\mathrm{ord}} : \vec{\widehat{\Xi}}^{\mathrm{ord}}_{\breve{\Phi}'_{\widehat{\mathcal{H}}'},\breve{\delta}'_{\widehat{\mathcal{H}}'}} \to \vec{\widehat{\Xi}}^{\mathrm{ord}}_{\breve{\Phi}_{\widehat{\mathcal{H}}},\breve{\delta}_{\widehat{\mathcal{H}}}}$, *mapping* $\vec{\widehat{\Xi}}^{\mathrm{ord}}_{\breve{\Phi}'_{\widehat{\mathcal{H}}'},\breve{\delta}'_{\widehat{\mathcal{H}}'},\widehat{\rho}'}$ *to* $\vec{\widehat{\Xi}}^{\mathrm{ord}}_{\breve{\Phi}_{\widehat{\mathcal{H}}},\breve{\delta}_{\widehat{\mathcal{H}}},\widehat{\rho}}$, *and inducing a canonical morphism*

$$[\vec{\widehat{g}}]^{\mathrm{ord}} : \vec{\widehat{\mathfrak{X}}}^{\mathrm{ord}}_{\breve{\Phi}'_{\widehat{\mathcal{H}}'},\breve{\delta}'_{\widehat{\mathcal{H}}'},\widehat{\rho}'} \to \vec{\widehat{\mathfrak{X}}}^{\mathrm{ord}}_{\breve{\Phi}_{\widehat{\mathcal{H}}},\breve{\delta}_{\widehat{\mathcal{H}}},\widehat{\rho}} \tag{7.1.2.19}$$

(cf. (1.3.2.69) *and* (5.2.4.43)). *If* $\widehat{g} \in \widehat{\mathrm{P}}_{\breve{\mathrm{Z}}}(\mathbb{A}^{\infty,p}) \times \widehat{\mathrm{P}}^{\mathrm{ord}}_{\breve{\mathrm{Z}}}(\mathbb{Q}_p)$ *is as above and if* $(\breve{\Phi}'_{\widehat{\mathcal{H}}'}, \breve{\delta}'_{\widehat{\mathcal{H}}'}, \widehat{\Sigma}'_{\breve{\Phi}'_{\widehat{\mathcal{H}}'}})$ *is a* $\widehat{g}$*-refinement of* $(\breve{\Phi}_{\widehat{\mathcal{H}}}, \breve{\delta}_{\widehat{\mathcal{H}}}, \widehat{\Sigma}_{\breve{\Phi}_{\widehat{\mathcal{H}}}})$ *(cf. Lemma* 1.2.4.42 *and [62, Def. 6.4.3.2]), then morphisms like* (7.1.2.18) *patch together and define a canonical morphism*

$$[\vec{\widehat{g}}]^{\mathrm{ord}} : \vec{\widehat{\overline{\Xi}}}^{\mathrm{ord}}_{\breve{\Phi}'_{\widehat{\mathcal{H}}'},\breve{\delta}'_{\widehat{\mathcal{H}}'},\widehat{\Sigma}'_{\breve{\Phi}'_{\widehat{\mathcal{H}}'}}} \to \vec{\widehat{\overline{\Xi}}}^{\mathrm{ord}}_{\breve{\Phi}_{\widehat{\mathcal{H}}},\breve{\delta}_{\widehat{\mathcal{H}}},\widehat{\Sigma}_{\breve{\Phi}_{\widehat{\mathcal{H}}}}} \tag{7.1.2.20}$$

(cf. (1.3.2.70) *and* (5.2.4.44)) *covering* $[\vec{\widehat{g}}]^{\mathrm{ord}} : \vec{\widehat{C}}^{\mathrm{ord}}_{\breve{\Phi}'_{\widehat{\mathcal{H}}'},\breve{\delta}'_{\widehat{\mathcal{H}}'}} \to \vec{\widehat{C}}^{\mathrm{ord}}_{\breve{\Phi}_{\widehat{\mathcal{H}}},\breve{\delta}_{\widehat{\mathcal{H}}}}$, *extending* $[\vec{\widehat{g}}]^{\mathrm{ord}} : \vec{\widehat{\Xi}}^{\mathrm{ord}}_{\breve{\Phi}'_{\widehat{\mathcal{H}}'},\breve{\delta}'_{\widehat{\mathcal{H}}'}} \to \vec{\widehat{\Xi}}^{\mathrm{ord}}_{\breve{\Phi}_{\widehat{\mathcal{H}}},\breve{\delta}_{\widehat{\mathcal{H}}}}$, *and inducing a canonical morphism*

$$[\vec{\widehat{g}}]^{\mathrm{ord}} : \vec{\widehat{\mathfrak{X}}}^{\mathrm{ord}}_{\breve{\Phi}'_{\widehat{\mathcal{H}}'},\breve{\delta}'_{\widehat{\mathcal{H}}'},\widehat{\Sigma}'_{\breve{\Phi}'_{\widehat{\mathcal{H}}'}}} \to \vec{\widehat{\mathfrak{X}}}^{\mathrm{ord}}_{\breve{\Phi}_{\widehat{\mathcal{H}}},\breve{\delta}_{\widehat{\mathcal{H}}},\widehat{\Sigma}_{\breve{\Phi}_{\widehat{\mathcal{H}}}}} \tag{7.1.2.21}$$

(cf. (1.3.2.71) *and* (5.2.4.45)) *compatible with each* (7.1.2.19) *as above (under canonical morphisms).*

Proof. By Proposition 7.1.2.6 (see in particular (7.1.2.8)), and by finite flat descent, the assertions in the first three paragraphs are reduced to the ones for the principal levels, which then follow from the corresponding assertions for the collection $\{\coprod \vec{\widetilde{\overline{\Xi}}}^{\mathrm{ord}}_{\breve{\Phi}_{\widetilde{\mathcal{H}}},\breve{\delta}_{\widetilde{\mathcal{H}}}}\}_{\widetilde{\mathcal{H}}_{\widetilde{\mathrm{P}}'_{\breve{\mathrm{Z}}}}}$ (by restricting the action of suitable elements of $\widetilde{\mathrm{P}}_{\breve{\mathrm{Z}}}(\mathbb{A}^{\infty,p}) \times \widetilde{\mathrm{P}}^{\mathrm{ord}}_{\breve{\mathrm{Z}},\widetilde{\mathrm{D}}}(\mathbb{Q}_p)$ to suitable elements of $(\widetilde{\mathrm{P}}_{\breve{\mathrm{Z}}}(\mathbb{A}^{\infty,p}) \cap \widetilde{\mathrm{P}}'_{\widetilde{\mathrm{Z}}}(\mathbb{A}^{\infty,p})) \times (\widetilde{\mathrm{P}}_{\breve{\mathrm{Z}}}(\mathbb{Q}_p) \cap \widetilde{\mathrm{P}}'_{\widetilde{\mathrm{Z}}}(\mathbb{Q}_p) \cap \widetilde{\mathrm{P}}^{\mathrm{ord}}_{\widetilde{\mathrm{D}}}(\mathbb{Q}_p)))$, because the tautological objects over $\vec{\widehat{\Xi}}^{\mathrm{ord}}_{\breve{\Phi}_n,\breve{\delta}_n} = \vec{\widehat{\Xi}}^{\mathrm{ord}}_{\breve{\Phi}_{\widehat{\mathcal{U}}^{\mathrm{bal}}_1(n)},\breve{\delta}_{\widehat{\mathcal{U}}^{\mathrm{bal}}_1(n)}}$ are canonically induced by those over $\vec{\widetilde{\Xi}}^{\mathrm{ord}}_{\breve{\Phi}_n,\breve{\delta}_n} = \vec{\widetilde{\Xi}}^{\mathrm{ord}}_{\breve{\Phi}_{\widetilde{\mathcal{U}}^{\mathrm{bal}}_1(n)},\breve{\delta}_{\widetilde{\mathcal{U}}^{\mathrm{bal}}_1(n)}}$. The assertions in the last paragraph then follow from the universal properties of toroidal embeddings (*cf.* [62, Prop. 6.2.5.11]). □

Lemma 7.1.2.22. *(Compare with Lemmas* 1.3.2.72 *and* 7.1.2.1.) *By comparing the universal properties, we obtain a canonical morphism*

$$\vec{\widetilde{\Xi}}^{\mathrm{ord}}_{\breve{\Phi}_{\widetilde{\mathcal{H}}},\breve{\delta}_{\widetilde{\mathcal{H}}}} \to \vec{\Xi}^{\mathrm{ord}}_{\Phi_{\mathcal{H}},\delta_{\mathcal{H}}} \tag{7.1.2.23}$$

covering (7.1.2.2), *by sending* $\breve{\tau}^{\mathrm{ord}}_{\widetilde{\mathcal{H}}}$, *which is an orbit of étale-locally-defined trivializations* $\breve{\tau}^{\mathrm{ord}}_n = \breve{\tau}_{n_0} : \mathbf{1}_{\frac{1}{n_0}\breve{Y} \times \breve{X},S} \xrightarrow{\sim} (\breve{c}^{\vee}_{n_0} \times \breve{c})^* \mathcal{P}^{\otimes -1}_B$ *(for some integer* $n = n_0 p^r$ *where*

$n_0 \geq 1$ is an integer prime to p such that $\widetilde{\mathcal{U}}^p(n_0) \subset \widetilde{\mathcal{H}}^p$ and where $r = \mathrm{depth}_{\widetilde{\mathsf{D}}}(\widetilde{\mathcal{H}}_p)$), to the orbit $\tau^{\mathrm{ord}}_{\mathcal{H}}$ of étale-locally-defined trivializations $\tau^{\mathrm{ord}}_n = \tau_{n_0} = \breve{\tau}_{n_0}|_{\mathbf{1}_{\frac{1}{n_0}Y \times X, S}}$.

The morphisms (7.1.2.23) *and* (7.1.2.2) *induce a canonical morphism*

$$\vec{\widetilde{\Xi}}^{\mathrm{ord}}_{\breve{\Phi}_{\widetilde{\mathcal{H}}},\breve{\delta}_{\widetilde{\mathcal{H}}}} \to \vec{\Xi}^{\mathrm{ord}}_{\Phi_{\mathcal{H}},\delta_{\mathcal{H}}} \underset{\vec{C}^{\mathrm{ord}}_{\Phi_{\mathcal{H}},\delta_{\mathcal{H}}}}{\times} \vec{\widetilde{C}}^{\mathrm{ord}}_{\breve{\Phi}_{\widetilde{\mathcal{H}}},\breve{\delta}_{\widetilde{\mathcal{H}}}} \tag{7.1.2.24}$$

between torus torsors over $\vec{\widetilde{C}}^{\mathrm{ord}}_{\breve{\Phi}_{\widetilde{\mathcal{H}}},\breve{\delta}_{\widetilde{\mathcal{H}}}}$, equivariant with the homomorphism $E_{\breve{\Phi}_{\widetilde{\mathcal{H}}}} \to E_{\Phi_{\mathcal{H}}}$ dual to the canonical homomorphism $\mathbf{S}_{\Phi_{\mathcal{H}}} \to \mathbf{S}_{\breve{\Phi}_{\widetilde{\mathcal{H}}}}$ (see (1.2.4.18)).

Suppose the image of a rational polyhedral cone $\breve{\rho} \subset (\mathbf{S}_{\breve{\Phi}_{\widetilde{\mathcal{H}}}})^\vee_{\mathbb{R}}$ under the (canonical) second morphism in (1.2.4.20) *is contained in some rational polyhedral cone $\rho \subset (\mathbf{S}_{\Phi_{\mathcal{H}}})^\vee_{\mathbb{R}}$. Then there is a canonical morphism*

$$\vec{\widetilde{\Xi}}^{\mathrm{ord}}_{\breve{\Phi}_{\widetilde{\mathcal{H}}},\breve{\delta}_{\widetilde{\mathcal{H}}}}(\breve{\rho}) \to \vec{\Xi}^{\mathrm{ord}}_{\Phi_{\mathcal{H}},\delta_{\mathcal{H}}}(\rho) \tag{7.1.2.25}$$

(cf. (1.3.2.75)) *covering* (7.1.2.2) *and extending* (7.1.2.23), *mapping $\vec{\widetilde{\Xi}}^{\mathrm{ord}}_{\breve{\Phi}_{\widetilde{\mathcal{H}}},\breve{\delta}_{\widetilde{\mathcal{H}}},\breve{\rho}}$ to $\vec{\Xi}^{\mathrm{ord}}_{\Phi_{\mathcal{H}},\delta_{\mathcal{H}},\rho}$ and inducing a canonical morphism*

$$\vec{\widetilde{\mathfrak{X}}}^{\mathrm{ord}}_{\breve{\Phi}_{\widetilde{\mathcal{H}}},\breve{\delta}_{\widetilde{\mathcal{H}}},\breve{\rho}} \to \vec{\mathfrak{X}}^{\mathrm{ord}}_{\Phi_{\mathcal{H}},\delta_{\mathcal{H}},\rho} \tag{7.1.2.26}$$

(cf. (1.3.2.76)). *If $\widetilde{\Sigma}_{\breve{\Phi}_{\widetilde{\mathcal{H}}}}$ and $\Sigma_{\Phi_{\mathcal{H}}}$ are cone decompositions of $\mathbf{P}_{\breve{\Phi}_{\widetilde{\mathcal{H}}}}$ and $\mathbf{P}_{\Phi_{\mathcal{H}}}$, respectively, such that the image of each $\breve{\rho}$ in $\widetilde{\Sigma}_{\breve{\Phi}_{\widetilde{\mathcal{H}}}}$ under the (canonical) second morphism in* (1.2.4.20) *is contained in some $\rho \in \Sigma_{\Phi_{\mathcal{H}}}$, then morphisms like* (7.1.2.25) *patch together and define a canonical morphism*

$$\overline{\vec{\widetilde{\Xi}}}^{\mathrm{ord}}_{\breve{\Phi}_{\widetilde{\mathcal{H}}},\breve{\delta}_{\widetilde{\mathcal{H}}},\widetilde{\Sigma}_{\breve{\Phi}_{\widetilde{\mathcal{H}}}}} \to \overline{\vec{\Xi}}^{\mathrm{ord}}_{\Phi_{\mathcal{H}},\delta_{\mathcal{H}},\Sigma_{\Phi_{\mathcal{H}}}} \tag{7.1.2.27}$$

(cf. (1.3.2.77)) *covering* (7.1.2.2), *extending* (7.1.2.23), *and inducing a canonical morphism*

$$\vec{\widetilde{\mathfrak{X}}}^{\mathrm{ord}}_{\breve{\Phi}_{\widetilde{\mathcal{H}}},\breve{\delta}_{\widetilde{\mathcal{H}}},\widetilde{\Sigma}_{\breve{\Phi}_{\widetilde{\mathcal{H}}}}} \to \vec{\mathfrak{X}}^{\mathrm{ord}}_{\Phi_{\mathcal{H}},\delta_{\mathcal{H}},\Sigma_{\Phi_{\mathcal{H}}}} \tag{7.1.2.28}$$

(cf. (1.3.2.78)) *compatible with each* (7.1.2.26) *as above (under canonical morphisms).*

Proof. The statements are self-explanatory. □

Lemma 7.1.2.29. (*Compare with Lemmas* 1.3.2.79, 7.1.2.1, *and* 7.1.2.22.) *By comparing the universal properties (cf. Proposition* 7.1.2.6), *we obtain a canonical morphism*

$$\vec{\widehat{\Xi}}^{\mathrm{ord}}_{\breve{\Phi}_{\widehat{\mathcal{H}}},\breve{\delta}_{\widehat{\mathcal{H}}}} \to \vec{\Xi}^{\mathrm{ord}}_{\Phi_{\mathcal{H}},\delta_{\mathcal{H}}} \tag{7.1.2.30}$$

covering (7.1.2.2), *by sending the pair* $(\widehat{\tau}^{\mathrm{ord}}_{\widehat{\mathcal{H}}}, \widehat{\tau}^{\vee,\mathrm{ord}}_{\widehat{\mathcal{H}}})$, *which is an orbit of étale-locally-defined pairs* $(\widehat{\tau}^{\mathrm{ord}}_n, \widehat{\tau}^{\vee,\mathrm{ord}}_n) = (\widehat{\tau}_{n_0}, \widehat{\tau}^{\vee}_{n_0})$ *(for some integer* $n = n_0p^r$ *where* $n_0 \geq 1$ *is an integer prime to* p *such that* $\widehat{\mathcal{U}}^p(n_0) \subset \widehat{\mathcal{H}}^p$ *and where* $r = \mathrm{depth}_{\mathsf{D}}(\widehat{\mathcal{H}}_p)$), *to the orbit* $\tau^{\mathrm{ord}}_{\mathcal{H}}$ *of étale-locally-defined* $\tau^{\mathrm{ord}}_n = \tau_{n_0} = \widehat{\tau}_{n_0}|_{\mathbf{1}_{\frac{1}{n_0}Y \times X, S}} = \widehat{\tau}^{\vee}_{n_0}|_{\mathbf{1}_{\frac{1}{n_0}Y \times X, S}}$, *as in Proposition* 7.1.2.6.

The morphisms (7.1.2.30) *and* (7.1.2.2) *induce a canonical morphism*

$$\vec{\widehat{\Xi}}^{\mathrm{ord}}_{\breve{\Phi}_{\widehat{\mathcal{H}}}, \breve{\delta}_{\widehat{\mathcal{H}}}} \to \vec{\Xi}^{\mathrm{ord}}_{\Phi_{\mathcal{H}}, \delta_{\mathcal{H}}} \underset{\vec{C}^{\mathrm{ord}}_{\Phi_{\mathcal{H}}, \delta_{\mathcal{H}}}}{\times} \vec{\widehat{C}}^{\mathrm{ord}}_{\breve{\Phi}_{\widehat{\mathcal{H}}}, \breve{\delta}_{\widehat{\mathcal{H}}}} \tag{7.1.2.31}$$

(*cf.* (1.3.2.81)) *between torus torsors over* $\vec{\widehat{C}}^{\mathrm{ord}}_{\breve{\Phi}_{\widehat{\mathcal{H}}}, \breve{\delta}_{\widehat{\mathcal{H}}}}$, *equivariant with the surjective homomorphism* $\widehat{E}_{\breve{\Phi}_{\widehat{\mathcal{H}}}} \to E_{\Phi_{\mathcal{H}}}$ (*see Proposition* 7.1.2.6) *dual to the canonical injective homomorphism* $\mathbf{S}_{\Phi_{\mathcal{H}}} \hookrightarrow \widehat{\mathbf{S}}_{\breve{\Phi}_{\widehat{\mathcal{H}}}}$ (*see Definition* 1.2.4.29).

Suppose the image of a rational polyhedral cone $\widehat{\rho} \subset (\widehat{\mathbf{S}}_{\breve{\Phi}_{\widehat{\mathcal{H}}}})^{\vee}_{\mathbb{R}}$ *under* (1.2.4.37) *is contained in some rational polyhedral cone* $\rho \subset (\mathbf{S}_{\Phi_{\mathcal{H}}})^{\vee}_{\mathbb{R}}$. *Then there is a canonical morphism*

$$\vec{\widehat{\Xi}}^{\mathrm{ord}}_{\breve{\Phi}_{\widehat{\mathcal{H}}}, \breve{\delta}_{\widehat{\mathcal{H}}}}(\widehat{\rho}) \to \vec{\Xi}^{\mathrm{ord}}_{\Phi_{\mathcal{H}}, \delta_{\mathcal{H}}}(\rho) \tag{7.1.2.32}$$

(*cf.* (1.3.2.82) *and* (7.1.2.25)) *covering* (7.1.2.2) *and extending* (7.1.2.30), *mapping* $\vec{\widehat{\Xi}}^{\mathrm{ord}}_{\breve{\Phi}_{\widehat{\mathcal{H}}}, \breve{\delta}_{\widehat{\mathcal{H}}}, \widehat{\rho}}$ *to* $\vec{\Xi}^{\mathrm{ord}}_{\Phi_{\mathcal{H}}, \delta_{\mathcal{H}}, \rho}$ *and inducing a canonical morphism*

$$\vec{\widehat{\mathfrak{X}}}^{\mathrm{ord}}_{\breve{\Phi}_{\widehat{\mathcal{H}}}, \breve{\delta}_{\widehat{\mathcal{H}}}, \widehat{\rho}} \to \vec{\mathfrak{X}}^{\mathrm{ord}}_{\Phi_{\mathcal{H}}, \delta_{\mathcal{H}}, \rho} \tag{7.1.2.33}$$

(*cf.* (1.3.2.83) *and* (7.1.2.26)). *If* $\widehat{\Sigma}_{\breve{\Phi}_{\widehat{\mathcal{H}}}}$ *and* $\Sigma_{\Phi_{\mathcal{H}}}$ *are cone decompositions of* $\widehat{\mathbf{P}}_{\breve{\Phi}_{\widehat{\mathcal{H}}}}$ *and* $\mathbf{P}_{\Phi_{\mathcal{H}}}$, *respectively, such that the image of each* $\widehat{\rho}$ *in* $\widehat{\Sigma}_{\breve{\Phi}_{\widehat{\mathcal{H}}}}$ *under* (1.2.4.37) *is contained in some* $\rho \in \Sigma_{\Phi_{\mathcal{H}}}$, *then morphisms like* (7.1.2.32) *patch together and define a canonical morphism*

$$\vec{\widehat{\Xi}}^{\mathrm{ord}}_{\breve{\Phi}_{\widehat{\mathcal{H}}}, \breve{\delta}_{\widehat{\mathcal{H}}}, \widehat{\Sigma}_{\breve{\Phi}_{\widehat{\mathcal{H}}}}} \to \vec{\overline{\Xi}}^{\mathrm{ord}}_{\Phi_{\mathcal{H}}, \delta_{\mathcal{H}}, \Sigma_{\Phi_{\mathcal{H}}}} \tag{7.1.2.34}$$

(*cf.* (1.3.2.84) *and* (7.1.2.27)) *covering* (7.1.2.2), *extending* (7.1.2.30), *and inducing a canonical morphism*

$$\vec{\widehat{\mathfrak{X}}}^{\mathrm{ord}}_{\breve{\Phi}_{\widehat{\mathcal{H}}}, \breve{\delta}_{\widehat{\mathcal{H}}}, \widehat{\Sigma}_{\breve{\Phi}_{\widehat{\mathcal{H}}}}} \to \vec{\mathfrak{X}}^{\mathrm{ord}}_{\Phi_{\mathcal{H}}, \delta_{\mathcal{H}}, \Sigma_{\Phi_{\mathcal{H}}}} \tag{7.1.2.35}$$

(*cf.* (1.3.2.85) *and* (7.1.2.28)) *compatible with each* (7.1.2.33) *as above* (*under canonical morphisms*). (*See Lemma* 7.2.5.34 *below for statements concerning the relative log* 1*-differentials of* (7.1.2.34).)

Proof. The statements are self-explanatory. □

Proposition 7.1.2.36. (*Compare with Propositions* 1.3.2.90 *and* 7.1.2.5.) *Under the canonical morphisms as in* (7.1.2.23) (*with varying* $\widetilde{\mathcal{H}}$ *and* $\mathcal{H}$), *and under the canonical homomorphisms* $\widetilde{\mathrm{P}}'_{\breve{\mathsf{Z}}}(\mathbb{A}^{\infty,p}) \times \widetilde{\mathrm{P}}^{\mathrm{ord},\prime}_{\breve{\mathsf{Z}},\widetilde{\mathsf{D}}}(\mathbb{Q}_p) \to \mathrm{P}'_{\mathsf{Z}}(\mathbb{A}^{\infty,p}) \times \mathrm{P}^{\mathrm{ord},\prime}_{\mathsf{Z},\mathsf{D}}(\mathbb{Q}_p)$ *and* $\widetilde{\mathrm{P}}_{\breve{\mathsf{Z}},\widetilde{\mathsf{Z}}}(\mathbb{A}^{\infty,p}) \times \widetilde{\mathrm{P}}^{\mathrm{ord}}_{\breve{\mathsf{Z}},\widetilde{\mathsf{Z}},\widetilde{\mathsf{D}}}(\mathbb{Q}_p) \to \mathrm{P}_{\mathsf{Z}}(\mathbb{A}^{\infty,p}) \times \mathrm{P}^{\mathrm{ord}}_{\mathsf{Z},\mathsf{D}}(\mathbb{Q}_p)$, *the Hecke action of* (*suitable elements of*) $\widetilde{\mathrm{P}}'_{\breve{\mathsf{Z}}}(\mathbb{A}^{\infty,p}) \times \widetilde{\mathrm{P}}^{\mathrm{ord},\prime}_{\breve{\mathsf{Z}},\widetilde{\mathsf{D}}}(\mathbb{Q}_p)$ *on the collection* $\{\vec{\widetilde{\Xi}}^{\mathrm{ord}}_{\breve{\Phi}_{\widetilde{\mathcal{H}}},\breve{\delta}_{\widetilde{\mathcal{H}}}}\}_{\widetilde{\mathcal{H}}_{\widetilde{\mathrm{P}}'_{\breve{\mathsf{Z}}}}}$ (*with* $\widetilde{\mathcal{H}}$ *of standard form*) *is compatible with the Hecke action of* (*suitable elements of*) $\mathrm{P}'_{\mathsf{Z}}(\mathbb{A}^{\infty,p}) \times \mathrm{P}^{\mathrm{ord},\prime}_{\mathsf{Z},\mathsf{D}}(\mathbb{Q}_p)$ *on the collection* $\{\vec{\Xi}^{\mathrm{ord}}_{\Phi_{\mathcal{H}},\delta_{\mathcal{H}}}\}_{\mathcal{H}_{\mathrm{P}'_{\mathsf{Z}}}}$ (*with* $\mathcal{H}$ *of standard form*); *and the Hecke action of* (*suitable elements of*) $\widetilde{\mathrm{P}}_{\breve{\mathsf{Z}},\widetilde{\mathsf{Z}}}(\mathbb{A}^{\infty,p}) \times \widetilde{\mathrm{P}}^{\mathrm{ord}}_{\breve{\mathsf{Z}},\widetilde{\mathsf{Z}},\widetilde{\mathsf{D}}}(\mathbb{Q}_p)$ *on the collection* $\{\coprod \vec{\widetilde{\Xi}}^{\mathrm{ord}}_{\breve{\Phi}_{\widetilde{\mathcal{H}}},\breve{\delta}_{\widetilde{\mathcal{H}}}}\}_{\widetilde{\mathcal{H}}_{\widetilde{\mathrm{P}}_{\breve{\mathsf{Z}},\widetilde{\mathsf{Z}}}}}$ (*with* $\widetilde{\mathcal{H}}$ *of standard form*) *is compatible with the Hecke action of* (*suitable elements of*) $\mathrm{P}_{\mathsf{Z}}(\mathbb{A}^{\infty,p}) \times \mathrm{P}^{\mathrm{ord}}_{\mathsf{Z},\mathsf{D}}(\mathbb{Q}_p)$ *on the collection* $\{\coprod \vec{\Xi}^{\mathrm{ord}}_{\Phi_{\mathcal{H}},\delta_{\mathcal{H}}}\}_{\mathcal{H}_{\mathrm{P}_{\mathsf{Z}}}}$ (*with* $\mathcal{H}$ *of standard form*), *where the index sets are as in Proposition* 5.2.4.41. *These Hecke actions are all compatible with those in Proposition* 7.1.2.5. *They are also compatible with extensions to toroidal embeddings and their formal completions.*

Proof. As in the case of Proposition 7.1.2.5, the canonical morphisms as in (7.1.2.30) correspond to pushouts of extensions of B (resp. $B^\vee$) by $\breve{T}$ (resp. $\breve{T}^\vee$) under the canonical homomorphism $\breve{T} \to T$ (resp. $\breve{T}^\vee \to T^\vee$) induced by the restriction from $\breve{X}$ (resp. $\breve{Y}$) to X (resp. Y). Hence, the realizations of the Hecke twists are compatible in the desired ways. (We omit the details for simplicity.) □

Proposition 7.1.2.37. (*Compare with Propositions* 1.3.2.91, 7.1.2.5 *and* 7.1.2.36.) *Under the canonical morphisms as in* (7.1.2.30) (*with varying* $\widehat{\mathcal{H}}$ *and* $\mathcal{H}$), *and under the canonical homomorphisms* $\widehat{\mathrm{P}}'_{\breve{\mathsf{Z}}}(\mathbb{A}^{\infty,p}) \times \widehat{\mathrm{P}}^{\mathrm{ord},\prime}_{\breve{\mathsf{Z}},\mathsf{D}}(\mathbb{Q}_p) \to \mathrm{P}'_{\mathsf{Z}}(\mathbb{A}^{\infty,p}) \times \mathrm{P}^{\mathrm{ord},\prime}_{\mathsf{Z},\mathsf{D}}(\mathbb{Q}_p)$ *and* $\widehat{\mathrm{P}}_{\breve{\mathsf{Z}}}(\mathbb{A}^{\infty,p}) \times \widehat{\mathrm{P}}^{\mathrm{ord}}_{\breve{\mathsf{Z}},\mathsf{D}}(\mathbb{Q}_p) \to \mathrm{P}_{\mathsf{Z}}(\mathbb{A}^{\infty,p}) \times \mathrm{P}^{\mathrm{ord}}_{\mathsf{Z},\mathsf{D}}(\mathbb{Q}_p)$, *the Hecke action of* (*suitable elements of*) $\widehat{\mathrm{P}}'_{\breve{\mathsf{Z}}}(\mathbb{A}^{\infty,p}) \times \widehat{\mathrm{P}}^{\mathrm{ord},\prime}_{\breve{\mathsf{Z}},\mathsf{D}}(\mathbb{Q}_p)$ *on the collection* $\{\vec{\widehat{\Xi}}^{\mathrm{ord}}_{\breve{\Phi}_{\widehat{\mathcal{H}}},\breve{\delta}_{\widehat{\mathcal{H}}}}\}_{\widehat{\mathcal{H}}_{\widehat{\mathrm{P}}'_{\breve{\mathsf{Z}}}}}$ (*with* $\widehat{\mathcal{H}}$ *of standard form*) *is compatible with the Hecke action of* (*suitable elements of*) $\mathrm{P}'_{\mathsf{Z}}(\mathbb{A}^{\infty,p}) \times \mathrm{P}^{\mathrm{ord},\prime}_{\mathsf{Z},\mathsf{D}}(\mathbb{Q}_p)$ *on the collection* $\{\vec{\Xi}^{\mathrm{ord}}_{\Phi_{\mathcal{H}},\delta_{\mathcal{H}}}\}_{\mathcal{H}_{\mathrm{P}'_{\mathsf{Z}}}}$ (*with* $\mathcal{H}$ *of standard form*); *and the Hecke action of* (*suitable elements of*) $\widehat{\mathrm{P}}_{\breve{\mathsf{Z}}}(\mathbb{A}^{\infty,p}) \times \widehat{\mathrm{P}}^{\mathrm{ord}}_{\breve{\mathsf{Z}},\mathsf{D}}(\mathbb{Q}_p)$ *on the collection* $\{\coprod \vec{\widehat{\Xi}}^{\mathrm{ord}}_{\breve{\Phi}_{\widehat{\mathcal{H}}},\breve{\delta}_{\widehat{\mathcal{H}}}}\}_{\widehat{\mathcal{H}}_{\widehat{\mathrm{P}}_{\breve{\mathsf{Z}}}}}$ (*with* $\widehat{\mathcal{H}}$ *of standard form*) *is compatible with the Hecke action of* (*suitable elements of*) $\mathrm{P}_{\mathsf{Z}}(\mathbb{A}^{\infty,p}) \times \mathrm{P}^{\mathrm{ord}}_{\mathsf{Z},\mathsf{D}}(\mathbb{Q}_p)$ *on the collection* $\{\coprod \vec{\Xi}^{\mathrm{ord}}_{\Phi_{\mathcal{H}},\delta_{\mathcal{H}}}\}_{\mathcal{H}_{\mathrm{P}_{\mathsf{Z}}}}$ (*with* $\mathcal{H}$ *of standard form*), *where the index sets are as in Proposition* 5.2.4.41. *These Hecke actions are all compatible with those in Proposition* 7.1.2.5. *They are also compatible with extensions to toroidal embeddings and their formal completions.*

Proof. As in the proofs of Propositions 1.3.2.91 and 7.1.2.17, the Hecke action of (suitable elements of) $\widehat{\mathrm{P}}_{\breve{\mathsf{Z}}}(\mathbb{A}^{\infty,p}) \times \widehat{\mathrm{P}}^{\mathrm{ord}}_{\breve{\mathsf{Z}},\mathsf{D}}(\mathbb{Q}_p)$ on the collection $\{\coprod \vec{\widehat{\Xi}}^{\mathrm{ord}}_{\breve{\Phi}_{\widehat{\mathcal{H}}},\breve{\delta}_{\widehat{\mathcal{H}}}}\}_{\widehat{\mathcal{H}}_{\widehat{\mathrm{P}}_{\breve{\mathsf{Z}}}}}$ (with $\widehat{\mathcal{H}}$ of standard form) is induced by the Hecke action of (suitable elements of)

$\widetilde{\mathsf{P}}'_{\breve{\mathsf{Z}}}(\mathbb{A}^{\infty,p}) \times \widetilde{\mathsf{P}}^{\mathrm{ord},\prime}_{\breve{\mathsf{Z}},\widetilde{\mathsf{D}}}(\mathbb{Q}_p)$ on the collection $\{\coprod \vec{\widetilde{\Xi}}^{\mathrm{ord}}_{\breve{\Phi}_{\widetilde{\mathcal{H}}},\breve{\delta}_{\widetilde{\mathcal{H}}}}\}_{\widetilde{\mathcal{H}}_{\widetilde{\mathsf{P}}'_{\breve{\mathsf{Z}}}}}$ (with $\widetilde{\mathcal{H}}$ of standard form). Hence, these statements follow from the corresponding statements of Proposition 7.1.2.36. □

Remark 7.1.2.38. (Compare with Remark 5.2.4.46.) As in Remark 5.2.4.46, since all objects and morphisms in this subsection are defined by normalizations and by the various universal properties extending their analogues in characteristic zero, they are canonically compatible with the corresponding objects and morphisms in Section 1.3.2.

7.1.3 *Ordinary Kuga Families and Their Generalizations*

Consider the abelian scheme $G_{\vec{\mathsf{M}}^{\mathrm{ord}}_{\mathcal{H}}}$ over $\vec{\mathsf{M}}^{\mathrm{ord}}_{\mathcal{H}}$ and its semi-abelian extension G over $\vec{\mathsf{M}}^{\mathrm{ord,tor}}_{\mathcal{H}} = \vec{\mathsf{M}}^{\mathrm{ord,tor}}_{\mathcal{H},\Sigma^{\mathrm{ord}}}$ as in Theorem 5.2.1.1. Let Q be any $\mathcal{O}$-lattice. By (4) of Proposition 3.1.2.4, the abelian scheme $\underline{\mathrm{Hom}}_{\mathcal{O}}(Q, G_{\vec{\mathsf{M}}^{\mathrm{ord}}_{\mathcal{H}}})^{\circ} \to \vec{\mathsf{M}}^{\mathrm{ord}}_{\mathcal{H}}$ is defined and is ordinary.

Definition 7.1.3.1. (*Compare with Definition* 1.3.3.3.) *An* **ordinary Kuga family** *over* $\vec{\mathsf{M}}^{\mathrm{ord}}_{\mathcal{H}}$ *is an (ordinary) abelian scheme* $\mathsf{N}^{\mathrm{grp}} \to \vec{\mathsf{M}}^{\mathrm{ord}}_{\mathcal{H}}$ *that is* $\mathbb{Q}^{\times}$*-isogenous to* $\underline{\mathrm{Hom}}_{\mathcal{O}}(Q, G_{\vec{\mathsf{M}}^{\mathrm{ord}}_{\mathcal{H}}})^{\circ}$ *for some* $\mathcal{O}$*-lattice* Q.

Definition 7.1.3.2. (*Compare with Definition* 1.3.3.4.) *An* **generalized ordinary Kuga family** *over* $\vec{\mathsf{M}}^{\mathrm{ord}}_{\mathcal{H}}$ *is a torsor* $\mathsf{N} \to \vec{\mathsf{M}}^{\mathrm{ord}}_{\mathcal{H}}$ *under some ordinary Kuga family* $\mathsf{N}^{\mathrm{grp}} \to \vec{\mathsf{M}}^{\mathrm{ord}}_{\mathcal{H}}$ *as in Definition* 7.1.3.1.

Then the following four lemmas and proposition can be proved by the same arguments as before:

Lemma 7.1.3.3. (*Compare with [61, Lem. 2.6] and Lemma* 1.3.3.5.) *The abelian scheme* $\underline{\mathrm{Hom}}_{\mathbb{Z}}(Q^{\vee}, G^{\vee}_{\vec{\mathsf{M}}^{\mathrm{ord}}_{\mathcal{H}}})$ *is isomorphic to the dual abelian scheme of* $\underline{\mathrm{Hom}}_{\mathbb{Z}}(Q, G_{\vec{\mathsf{M}}^{\mathrm{ord}}_{\mathcal{H}}})$.

Lemma 7.1.3.4. (*Compare with [61, Lem. 2.9] and Lemma* 1.3.3.6.) *Let* $j_Q : Q^{\vee} \hookrightarrow Q$ *be as in Lemma* 1.2.4.1. *Then the isogeny*

$$\lambda_{\vec{\mathsf{M}}^{\mathrm{ord}}_{\mathcal{H}},j_Q,\mathbb{Z}} : \underline{\mathrm{Hom}}_{\mathbb{Z}}(Q, G_{\vec{\mathsf{M}}^{\mathrm{ord}}_{\mathcal{H}}}) \to \underline{\mathrm{Hom}}_{\mathbb{Z}}(Q^{\vee}, G^{\vee}_{\vec{\mathsf{M}}^{\mathrm{ord}}_{\mathcal{H}}})$$

induced canonically by j_Q *and* $\lambda_{\vec{\mathsf{M}}^{\mathrm{ord}}_{\mathcal{H}}} : G_{\vec{\mathsf{M}}^{\mathrm{ord}}_{\mathcal{H}}} \to G^{\vee}_{\vec{\mathsf{M}}^{\mathrm{ord}}_{\mathcal{H}}}$ *is a polarization.*

Proposition 7.1.3.5. (*Compare with [61, Prop. 2.10 and Cor. 2.12] and Proposition* 1.3.3.7.) *The abelian scheme* $\underline{\mathrm{Hom}}_{\mathcal{O}}(Q^{\vee}, G^{\vee}_{\vec{\mathsf{M}}^{\mathrm{ord}}_{\mathcal{H}}})^{\circ}$ *is* $\mathbb{Q}^{\times}$*-isogenous to the dual abelian scheme of* $\underline{\mathrm{Hom}}_{\mathcal{O}}(Q, G_{\vec{\mathsf{M}}^{\mathrm{ord}}_{\mathcal{H}}})^{\circ}$. *Moreover, given any* $j_Q : Q^{\vee} \hookrightarrow Q$ *as in Lemma* 1.2.4.1, *the composition*

$$\begin{aligned}\lambda_{\vec{\mathsf{M}}^{\mathrm{ord}}_{\mathcal{H}},j_Q} : \underline{\mathrm{Hom}}_{\mathcal{O}}(Q, G_{\vec{\mathsf{M}}^{\mathrm{ord}}_{\mathcal{H}}})^{\circ} &\hookrightarrow \underline{\mathrm{Hom}}_{\mathbb{Z}}(Q, G_{\vec{\mathsf{M}}^{\mathrm{ord}}_{\mathcal{H}}}) \\ &\xrightarrow{\lambda_{\vec{\mathsf{M}}^{\mathrm{ord}}_{\mathcal{H}},j_Q,\mathbb{Z}}} \underline{\mathrm{Hom}}_{\mathbb{Z}}(Q^{\vee}, G^{\vee}_{\vec{\mathsf{M}}^{\mathrm{ord}}_{\mathcal{H}}}) \twoheadrightarrow (\underline{\mathrm{Hom}}_{\mathcal{O}}(Q, G_{\vec{\mathsf{M}}^{\mathrm{ord}}_{\mathcal{H}}})^{\circ})^{\vee}\end{aligned} \tag{7.1.3.6}$$

induced canonically by j_Q *and the polarization* $\lambda_{\vec{\mathsf{M}}^{\mathrm{ord}}_{\mathcal{H}}} : G_{\vec{\mathsf{M}}^{\mathrm{ord}}_{\mathcal{H}}} \to G^\vee_{\vec{\mathsf{M}}^{\mathrm{ord}}_{\mathcal{H}}}$ *is a polarization.*

Definition 7.1.3.7. (*Compare with Definition* 1.3.3.9.) *Let* $\mathsf{N} \to \vec{\mathsf{M}}^{\mathrm{ord}}_{\mathcal{H}}$ *be as in Definition* 7.1.3.2. *Then we define the dual* $\mathsf{N}^\vee \to \vec{\mathsf{M}}^{\mathrm{ord}}_{\mathcal{H}}$ *to be the dual abelian scheme* $\mathsf{N}^{\mathrm{grp},\vee} \to \vec{\mathsf{M}}^{\mathrm{ord}}_{\mathcal{H}}$ *of* $\mathsf{N}^{\mathrm{grp}} \to \vec{\mathsf{M}}^{\mathrm{ord}}_{\mathcal{H}}$.

Remark 7.1.3.8. (Compare with Remark 1.3.3.10.) By [92, XIII, Prop. 1.1], $\mathsf{N}^\vee = \mathsf{N}^{\mathrm{grp},\vee} \to \vec{\mathsf{M}}^{\mathrm{ord}}_{\mathcal{H}}$ is canonically isomorphic to $\underline{\mathrm{Pic}}^0(\mathsf{N}/\vec{\mathsf{M}}^{\mathrm{ord}}_{\mathcal{H}}) \to \vec{\mathsf{M}}^{\mathrm{ord}}_{\mathcal{H}}$ (which can be defined as in the case of abelian schemes; *cf.* [62, Def. 1.3.2.1]). Note that this is always a group scheme, with its identity section, even when $\mathsf{N} \to \vec{\mathsf{M}}^{\mathrm{ord}}_{\mathcal{H}}$ is a nontrivial torsor of $\mathsf{N}^{\mathrm{grp}} \to \vec{\mathsf{M}}^{\mathrm{ord}}_{\mathcal{H}}$.

Definition 7.1.3.9. (*Compare with Definition* 1.3.3.11.) *By abuse of notation, we denote by* $\underline{\mathrm{Lie}}_{\mathsf{N}/\vec{\mathsf{M}}^{\mathrm{ord}}_{\mathcal{H}}}$ (*resp.* $\underline{\mathrm{Lie}}^\vee_{\mathsf{N}/\vec{\mathsf{M}}^{\mathrm{ord}}_{\mathcal{H}}}$, *resp.* $\underline{\mathrm{Lie}}_{\mathsf{N}^\vee/\vec{\mathsf{M}}^{\mathrm{ord}}_{\mathcal{H}}}$, *resp.* $\underline{\mathrm{Lie}}^\vee_{\mathsf{N}^\vee/\vec{\mathsf{M}}^{\mathrm{ord}}_{\mathcal{H}}}$) *the locally free sheaf* $\underline{\mathrm{Lie}}_{\mathsf{N}^{\mathrm{grp}}/\vec{\mathsf{M}}^{\mathrm{ord}}_{\mathcal{H}}}$ (*resp.* $\underline{\mathrm{Lie}}^\vee_{\mathsf{N}^{\mathrm{grp}}/\vec{\mathsf{M}}^{\mathrm{ord}}_{\mathcal{H}}}$, *resp.* $\underline{\mathrm{Lie}}_{\mathsf{N}^{\mathrm{grp},\vee}/\vec{\mathsf{M}}^{\mathrm{ord}}_{\mathcal{H}}}$, *resp.* $\underline{\mathrm{Lie}}^\vee_{\mathsf{N}^{\mathrm{grp},\vee}/\vec{\mathsf{M}}^{\mathrm{ord}}_{\mathcal{H}}}$) *over* $\vec{\mathsf{M}}^{\mathrm{ord}}_{\mathcal{H}}$, *although* $\mathsf{N} \to \vec{\mathsf{M}}^{\mathrm{ord}}_{\mathcal{H}}$ *might have no section.*

Lemma 7.1.3.10. (*Compare with Lemma* 1.3.3.12.) *We have:*

$$\begin{aligned}
\underline{\mathrm{Lie}}^\vee_{\mathsf{N}/\vec{\mathsf{M}}^{\mathrm{ord}}_{\mathcal{H}}} &\cong \underline{\mathrm{Hom}}_{\mathscr{O}_{\vec{\mathsf{M}}^{\mathrm{ord}}_{\mathcal{H}}}}(\underline{\mathrm{Lie}}_{\mathsf{N}/\vec{\mathsf{M}}^{\mathrm{ord}}_{\mathcal{H}}}, \mathscr{O}_{\vec{\mathsf{M}}^{\mathrm{ord}}_{\mathcal{H}}}),\\
\underline{\mathrm{Lie}}^\vee_{\mathsf{N}^\vee/\vec{\mathsf{M}}^{\mathrm{ord}}_{\mathcal{H}}} &\cong \underline{\mathrm{Hom}}_{\mathscr{O}_{\vec{\mathsf{M}}^{\mathrm{ord}}_{\mathcal{H}}}}(\underline{\mathrm{Lie}}_{\mathsf{N}^\vee/\vec{\mathsf{M}}^{\mathrm{ord}}_{\mathcal{H}}}, \mathscr{O}_{\vec{\mathsf{M}}^{\mathrm{ord}}_{\mathcal{H}}}),\\
\Omega^1_{\mathsf{N}/\vec{\mathsf{M}}^{\mathrm{ord}}_{\mathcal{H}}} &\cong (\mathsf{N} \to \vec{\mathsf{M}}^{\mathrm{ord}}_{\mathcal{H}})^* \underline{\mathrm{Lie}}^\vee_{\mathsf{N}/\vec{\mathsf{M}}^{\mathrm{ord}}_{\mathcal{H}}},\\
\Omega^1_{\mathsf{N}^\vee/\vec{\mathsf{M}}^{\mathrm{ord}}_{\mathcal{H}}} &\cong (\mathsf{N}^\vee \to \vec{\mathsf{M}}^{\mathrm{ord}}_{\mathcal{H}})^* \underline{\mathrm{Lie}}^\vee_{\mathsf{N}^\vee/\vec{\mathsf{M}}^{\mathrm{ord}}_{\mathcal{H}}},\\
(\mathsf{N} \to \vec{\mathsf{M}}^{\mathrm{ord}}_{\mathcal{H}})_* \Omega^1_{\mathsf{N}/\vec{\mathsf{M}}^{\mathrm{ord}}_{\mathcal{H}}} &\cong \underline{\mathrm{Lie}}^\vee_{\mathsf{N}/\vec{\mathsf{M}}^{\mathrm{ord}}_{\mathcal{H}}},\\
(\mathsf{N}^\vee \to \vec{\mathsf{M}}^{\mathrm{ord}}_{\mathcal{H}})_* \Omega^1_{\mathsf{N}^\vee/\vec{\mathsf{M}}^{\mathrm{ord}}_{\mathcal{H}}} &\cong \underline{\mathrm{Lie}}^\vee_{\mathsf{N}^\vee/\vec{\mathsf{M}}^{\mathrm{ord}}_{\mathcal{H}}},\\
R^1(\mathsf{N} \to \vec{\mathsf{M}}^{\mathrm{ord}}_{\mathcal{H}})_* \mathscr{O}_{\mathsf{N}} &\cong \underline{\mathrm{Lie}}_{\mathsf{N}^\vee/\vec{\mathsf{M}}^{\mathrm{ord}}_{\mathcal{H}}},\\
R^1(\mathsf{N}^\vee \to \vec{\mathsf{M}}^{\mathrm{ord}}_{\mathcal{H}})_* \mathscr{O}_{\mathsf{N}^\vee} &\cong \underline{\mathrm{Lie}}_{\mathsf{N}/\vec{\mathsf{M}}^{\mathrm{ord}}_{\mathcal{H}}}.
\end{aligned}$$

The relative de Rham cohomology

$$\underline{H}^i_{\mathrm{dR}}(\mathsf{N}/\vec{\mathsf{M}}^{\mathrm{ord}}_{\mathcal{H}}) := R^i(\mathsf{N} \to \vec{\mathsf{M}}^{\mathrm{ord}}_{\mathcal{H}})_*(\Omega^\bullet_{\mathsf{N}/\vec{\mathsf{M}}^{\mathrm{ord}}_{\mathcal{H}}})$$

and its Hodge filtration and Gauss–Manin connection ∇ *are canonically isomorphic to those of* $\underline{H}^i_{\mathrm{dR}}(\mathsf{N}^{\mathrm{grp}}/\vec{\mathsf{M}}^{\mathrm{ord}}_{\mathcal{H}})$.

In Theorem 1.3.3.15, we used the isomorphisms in Corollary 1.3.3.13 and denoted, for example, the extension of $\underline{\mathrm{Hom}}_{\mathcal{O}}(Q, \underline{\mathrm{Lie}}_{G_{\mathsf{M}_{\mathcal{H}}}/\mathsf{M}_{\mathcal{H}}})$ to $\mathsf{M}^{\mathrm{tor}}_{\mathcal{H}}$ as $\underline{\mathrm{Hom}}_{\mathcal{O}}(Q, \underline{\mathrm{Lie}}_{G/\mathsf{M}^{\mathrm{tor}}_{\mathcal{H}}})$. However, such isomorphisms involve $\mathbb{Q}^\times$-isogenies which might not induce isomorphisms between Lie algebras (or their duals) in mixed characteristics, and this is indeed a concern because we allow the residue characteristics to ramify in the integral PEL datum. Therefore, we need to be more precise in the sheaves of modules we use. Indeed, it is now better to have not only the semi-abelian extension $G \to \vec{\mathsf{M}}_{\mathcal{H}}^{\mathrm{ord,tor}}$ of $G_{\vec{\mathsf{M}}_{\mathcal{H}}^{\mathrm{ord}}} \to \vec{\mathsf{M}}_{\mathcal{H}}^{\mathrm{ord}}$, but also the semi-abelian extensions of the Kuga families $\mathsf{N}^{\mathrm{grp}} \to \vec{\mathsf{M}}_{\mathcal{H}}^{\mathrm{ord}}$ as in Definition 7.1.3.1, their torsors $\mathsf{N} \to \vec{\mathsf{M}}_{\mathcal{H}}^{\mathrm{ord}}$ as in Definition 7.1.3.2, and the duals of all these, to $\vec{\mathsf{M}}_{\mathcal{H}}^{\mathrm{ord,tor}}$.

By Proposition 3.1.3.4, the abelian subscheme $\underline{\mathrm{Hom}}_{\mathcal{O}}(Q, G_{\vec{\mathsf{M}}_{\mathcal{H}}^{\mathrm{ord}}})^\circ$ of the abelian scheme $\underline{\mathrm{Hom}}_{\mathbb{Z}}(Q, G_{\vec{\mathsf{M}}_{\mathcal{H}}^{\mathrm{ord}}}) \cong G_{\vec{\mathsf{M}}_{\mathcal{H}}^{\mathrm{ord}}}^{\times \mathrm{rk}_{\mathbb{Z}}(Q)}$ over $\vec{\mathsf{M}}_{\mathcal{H}}^{\mathrm{ord}}$ extends to the semi-abelian subscheme $\underline{\mathrm{Hom}}_{\mathcal{O}}(Q, G)^\circ$ of the semi-abelian scheme $\underline{\mathrm{Hom}}_{\mathbb{Z}}(Q, G) \cong G^{\times \mathrm{rk}_{\mathbb{Z}}(Q)}$ over $\vec{\mathsf{M}}_{\mathcal{H}}^{\mathrm{ord,tor}}$.

Let $\mathsf{N} \to \vec{\mathsf{M}}_{\mathcal{H}}^{\mathrm{ord}}$ be a generalized ordinary Kuga family as in Definition 7.1.3.2, which is a torsor under some ordinary Kuga family $\mathsf{N}^{\mathrm{grp}} \to \vec{\mathsf{M}}_{\mathcal{H}}^{\mathrm{ord}}$ as in Definition 7.1.3.1, together with a $\mathbb{Q}^\times$-isogeny $h : Z := \underline{\mathrm{Hom}}_{\mathcal{O}}(Q, G_{\vec{\mathsf{M}}_{\mathcal{H}}^{\mathrm{ord}}})^\circ \to \mathsf{N}^{\mathrm{grp}}$ over $\vec{\mathsf{M}}_{\mathcal{H}}^{\mathrm{ord}}$. Let $Z^{\mathrm{tor}} := \underline{\mathrm{Hom}}_{\mathcal{O}}(Q, G)^\circ$ be as in Proposition 3.1.3.4. By definition, there exist an integer $N \geq 1$ such that Nh is an isogeny. Since $\vec{\mathsf{M}}_{\mathcal{H}}^{\mathrm{ord}}$ is noetherian normal, by Lemma 3.1.3.2, Nh extends to an isogeny $Z^{\mathrm{tor}} \to \mathsf{N}^{\mathrm{ext}}$ over $\vec{\mathsf{M}}_{\mathcal{H}}^{\mathrm{ord,tor}}$, and we formally define the $\mathbb{Q}^\times$-isogeny h^{tor} to be N^{-1} times this extended isogeny. (Since $\vec{\mathsf{M}}_{\mathcal{H}}^{\mathrm{ord}}$ is noetherian normal, this is well defined by [92, IX, 1.4], [28, Ch. I, Prop. 2.7], or [62, Prop. 3.3.1.5].)

Definition 7.1.3.11. *We can say that the semi-abelian scheme* $\mathsf{N}^{\mathrm{ext}} \to \vec{\mathsf{M}}_{\mathcal{H}}^{\mathrm{ord,tor}}$ *is the* **extended ordinary Kuga family** *over* $\vec{\mathsf{M}}_{\mathcal{H}}^{\mathrm{ord,tor}}$*. It is determined (up to isomorphism) by its restriction* $\mathsf{N}^{\mathrm{grp}} \to \vec{\mathsf{M}}_{\mathcal{H}}^{\mathrm{ord}}$ *to* $\vec{\mathsf{M}}_{\mathcal{H}}^{\mathrm{ord}}$*. (It does not depend on the structure of* $\mathsf{N} \to \vec{\mathsf{M}}_{\mathcal{H}}^{\mathrm{ord}}$ *as a torsor of* $\mathsf{N}^{\mathrm{grp}} \to \vec{\mathsf{M}}_{\mathcal{H}}^{\mathrm{ord}}$*.)*

By [80, IV, 7.1] (see also [62, Thm. 3.4.3.2]), there is also a dual semi-abelian scheme $\mathsf{N}^{\mathrm{ext},\vee} \to \vec{\mathsf{M}}_{\mathcal{H}}^{\mathrm{ord,tor}}$ extending the dual abelian scheme $\mathsf{N}^{\mathrm{grp},\vee} \to \vec{\mathsf{M}}_{\mathcal{H}}^{\mathrm{ord}}$.

7.1.4 *Main Statements*

The partial toroidal compactifications of Kuga families and their generalizations can be described as follows:

Theorem 7.1.4.1. (*Compare with [61, Thm. 2.15] and Theorem* 1.3.3.15.) *Let* Q *be as in Theorem* 1.3.3.15. *Let* $\mathcal{H} = \mathcal{H}^p\mathcal{H}_p$ *be as at the beginning of Section* 3.3.5, *and let* $r_{\mathcal{H}}$ *be as in Definition* 3.4.2.1. *Suppose that* $\mathcal{H}^p$ *is* **neat**, *and that* Σ^{ord} *is as in Definition* 5.1.3.1, *so that (by Theorem* 5.2.1.1) $\vec{\mathsf{M}}_{\mathcal{H}}^{\mathrm{ord,tor}} = \vec{\mathsf{M}}_{\mathcal{H},\Sigma^{\mathrm{ord}}}^{\mathrm{ord,tor}}$ *is an algebraic space separated, smooth, and of finite type over* $\vec{\mathsf{S}}_{0,r_{\mathcal{H}}}$. *(By Theorem* 6.2.3.1, *if* Σ^{ord} *is projective as in Definition* 5.1.3.3, *then* $\vec{\mathsf{M}}_{\mathcal{H},\Sigma^{\mathrm{ord}}}^{\mathrm{ord,tor}}$ *is quasi-projective over* $\vec{\mathsf{S}}_{0,r_{\mathcal{H}}}$.) *Consider the abelian scheme* $G_{\vec{\mathsf{M}}_{\mathcal{H}}^{\mathrm{ord}}}$ *over* $\vec{\mathsf{M}}_{\mathcal{H}}^{\mathrm{ord}}$ *in* (1) *of Theorem* 5.2.1.1. *Consider the sets* $\mathbf{K}_{Q,\mathcal{H}}^{\mathrm{ord}} \subset \mathbf{K}_{Q,\mathcal{H}}^{\mathrm{ord},+} \subset \mathbf{K}_{Q,\mathcal{H}}^{\mathrm{ord},++}$ *and* $\mathbf{K}_{Q,\mathcal{H},\Sigma^{\mathrm{ord}}}^{\mathrm{ord}} \subset \mathbf{K}_{Q,\mathcal{H},\Sigma^{\mathrm{ord}}}^{\mathrm{ord},+} \subset \mathbf{K}_{Q,\mathcal{H},\Sigma^{\mathrm{ord}}}^{\mathrm{ord},++}$ *as in Definitions* 7.1.1.11 *and* 7.1.1.19, *with compatible directed partial orders. These sets parameterize the following data:*

(1) *For each* $\kappa = (\widehat{\mathcal{H}}, \widehat{\Sigma}^{\mathrm{ord}}) \in \mathbf{K}^{\mathrm{ord},++}_{Q,\mathcal{H}}$, *let* $\mathcal{H}_\kappa := \widehat{\mathcal{H}}_{\mathrm{G}}$ *(which satisfies Condition* 7.1.1.4 *and is contained in* $\mathcal{H}$*; see Definition* 1.2.4.4) *and* $r_\kappa := r_{\mathcal{H}_\kappa}$, *so that* $\vec{\mathsf{M}}^{\mathrm{ord}}_{\mathcal{H}_\kappa}$ *is a quasi-finite étale cover of* $\vec{\mathsf{M}}^{\mathrm{ord}}_{\mathcal{H},r_\kappa} := \vec{\mathsf{M}}^{\mathrm{ord}}_{\mathcal{H}} \underset{\vec{\mathsf{S}}_{0,r_{\mathcal{H}}}}{\times} \vec{\mathsf{S}}_{0,r_\kappa}$, *inducing a quasi-finite flat morphism* $\vec{\mathsf{M}}^{\mathrm{ord}}_{\mathcal{H}_\kappa} \to \vec{\mathsf{M}}^{\mathrm{ord}}_{\mathcal{H}}$, *which is finite when* $\mathcal{H}_\kappa$ *and* $\mathcal{H}$ *are* **equally deep** *as in Definition* 3.2.2.9. *(This is the case, for example, when* $\kappa = (\widehat{\mathcal{H}}, \widehat{\Sigma}^{\mathrm{ord}}) \in \mathbf{K}^{\mathrm{ord},+}_{Q,\mathcal{H}}$ *and hence* $\mathcal{H}_\kappa = \mathcal{H}$.)

Then there is an generalized ordinary Kuga family $\vec{\mathsf{N}}^{\mathrm{ord}}_\kappa \to \vec{\mathsf{M}}^{\mathrm{ord}}_{\mathcal{H}_\kappa}$ *(see Definition* 7.1.3.2), *which is a torsor under an ordinary Kuga family* $\vec{\mathsf{N}}^{\mathrm{ord,grp}}_\kappa \to \vec{\mathsf{M}}^{\mathrm{ord}}_{\mathcal{H}_\kappa}$ *(see Definition* 7.1.3.1) *with a* $\mathbb{Q}^\times$*-isogeny*

$$\kappa^{\mathrm{isog}} : \underline{\mathrm{Hom}}_{\mathcal{O}}(Q, G_{\vec{\mathsf{M}}^{\mathrm{ord}}_{\mathcal{H}_\kappa}})^\circ \to \vec{\mathsf{N}}^{\mathrm{ord,grp}}_\kappa$$

of abelian schemes over $\vec{\mathsf{M}}^{\mathrm{ord}}_{\mathcal{H}_\kappa}$, *together with an open fiberwise dense immersion*

$$\kappa^{\mathrm{tor}} : \vec{\mathsf{N}}^{\mathrm{ord}}_\kappa \hookrightarrow \vec{\mathsf{N}}^{\mathrm{ord,tor}}_\kappa$$

of schemes over $\vec{\mathsf{S}}_{0,r_\kappa}$, *such that the scheme* $\vec{\mathsf{N}}^{\mathrm{ord,tor}}_\kappa$ *is* **quasi-projective** *and* **smooth** *over* $\vec{\mathsf{S}}_{0,r_\kappa}$, *and such that the complement of* $\vec{\mathsf{N}}^{\mathrm{ord}}_\kappa$ *in* $\vec{\mathsf{N}}^{\mathrm{ord,tor}}_\kappa$ *(with its reduced structure) is a relative Cartier divisor* $\vec{\mathsf{E}}^{\mathrm{ord}}_{\infty,\kappa}$ *with simple normal crossings.*

The scheme $\vec{\mathsf{N}}^{\mathrm{ord,tor}}_\kappa$ *has a stratification by locally closed subschemes*

$$\vec{\mathsf{N}}^{\mathrm{ord,tor}}_\kappa = \coprod_{[(\breve{\Phi}_{\widehat{\mathcal{H}}}, \breve{\delta}_{\widehat{\mathcal{H}}}, \widehat{\tau})]} \vec{\widehat{\mathsf{Z}}}^{\mathrm{ord}}_{[(\breve{\Phi}_{\widehat{\mathcal{H}}}, \breve{\delta}_{\widehat{\mathcal{H}}}, \widehat{\tau})]},$$

with $[(\breve{\Phi}_{\widehat{\mathcal{H}}}, \breve{\delta}_{\widehat{\mathcal{H}}}, \widehat{\tau})]$ *running through a complete set of equivalence classes of* $(\breve{\Phi}_{\widehat{\mathcal{H}}}, \breve{\delta}_{\widehat{\mathcal{H}}}, \widehat{\tau})$ *(as in Lemma* 1.2.4.42) *with the underlying* $\breve{\mathsf{Z}}_{\widehat{\mathcal{H}}}$ *(suppressed in the notation by our convention) compatible with* $\mathtt{D}$, *and with* $\widehat{\tau} \subset \widehat{\mathbf{P}}^+_{\breve{\Phi}_{\widehat{\mathcal{H}}}}$ *and* $\widehat{\tau} \in \widehat{\Sigma}_{\breve{\Phi}_{\widehat{\mathcal{H}}}} \in \widehat{\Sigma}$. *(The notation "*$\coprod$*" only means a set-theoretic disjoint union. The algebro-geometric structure is still that of* $\vec{\mathsf{N}}^{\mathrm{ord,tor}}_\kappa$.) *In this stratification, the* $[(\breve{\Phi}'_{\widehat{\mathcal{H}}}, \breve{\delta}'_{\widehat{\mathcal{H}}}, \widehat{\tau}')]$*-stratum* $\vec{\widehat{\mathsf{Z}}}^{\mathrm{ord}}_{[(\breve{\Phi}'_{\widehat{\mathcal{H}}}, \breve{\delta}'_{\widehat{\mathcal{H}}}, \widehat{\tau}')]}$ *lies in the closure of the* $[(\breve{\Phi}_{\widehat{\mathcal{H}}}, \breve{\delta}_{\widehat{\mathcal{H}}}, \widehat{\tau})]$*-stratum* $\vec{\widehat{\mathsf{Z}}}^{\mathrm{ord}}_{[(\breve{\Phi}_{\widehat{\mathcal{H}}}, \breve{\delta}_{\widehat{\mathcal{H}}}, \widehat{\tau})]}$ *if and only if* $[(\breve{\Phi}_{\widehat{\mathcal{H}}}, \breve{\delta}_{\widehat{\mathcal{H}}}, \widehat{\tau})]$ *is a face of* $[(\breve{\Phi}'_{\widehat{\mathcal{H}}}, \breve{\delta}'_{\widehat{\mathcal{H}}}, \widehat{\tau}')]$ *as in Lemma* 1.2.4.42. *The analogous assertion holds after pulled back to fibers over* $\vec{\mathsf{S}}_{0,r_\kappa}$. *In particular,* $\vec{\mathsf{N}}^{\mathrm{ord}}_\kappa = \vec{\widehat{\mathsf{Z}}}^{\mathrm{ord}}_{[(0,0,\{0\})]}$ *is an open fiberwise dense stratum in this stratification.*

The $[(\breve{\Phi}_{\widehat{\mathcal{H}}}, \breve{\delta}_{\widehat{\mathcal{H}}}, \widehat{\tau})]$*-stratum* $\vec{\widehat{\mathsf{Z}}}^{\mathrm{ord}}_{[(\breve{\Phi}_{\widehat{\mathcal{H}}}, \breve{\delta}_{\widehat{\mathcal{H}}}, \widehat{\tau})]}$ *is smooth over* $\vec{\mathsf{S}}_{0,r_\kappa}$ *and isomorphic to the support of the formal scheme* $\vec{\widehat{\mathfrak{X}}}^{\mathrm{ord}}_{\breve{\Phi}_{\widehat{\mathcal{H}}}, \breve{\delta}_{\widehat{\mathcal{H}}}, \widehat{\tau}}$ *(see* (7.1.2.15)) *for every representative* $(\breve{\Phi}_{\widehat{\mathcal{H}}}, \breve{\delta}_{\widehat{\mathcal{H}}}, \widehat{\tau})$ *of* $[(\breve{\Phi}_{\widehat{\mathcal{H}}}, \breve{\delta}_{\widehat{\mathcal{H}}}, \widehat{\tau})]$, *which is the completion of an affine toroidal embedding* $\vec{\widehat{\Xi}}^{\mathrm{ord}}_{\breve{\Phi}_{\widehat{\mathcal{H}}}, \breve{\delta}_{\widehat{\mathcal{H}}}}(\widehat{\tau})$ *(along its* $\widehat{\tau}$*-stratum* $\vec{\widehat{\Xi}}^{\mathrm{ord}}_{\breve{\Phi}_{\widehat{\mathcal{H}}}, \breve{\delta}_{\widehat{\mathcal{H}}}, \widehat{\tau}}$) *of a* *torus torsor* $\vec{\widehat{\Xi}}^{\mathrm{ord}}_{\breve{\Phi}_{\widehat{\mathcal{H}}},\breve{\delta}_{\widehat{\mathcal{H}}}}$ *over an abelian scheme torsor* $\vec{\widehat{C}}^{\mathrm{ord}}_{\breve{\Phi}_{\widehat{\mathcal{H}}},\breve{\delta}_{\widehat{\mathcal{H}}}}$ *over a finite étale cover* $\vec{\widehat{\mathsf{M}}}^{\mathrm{ord},\breve{\Phi}_{\widehat{\mathcal{H}}}}_{\widehat{\mathcal{H}}}$ *of the scheme* $\vec{\widehat{\mathsf{M}}}^{\mathrm{ord},\breve{\mathsf{Z}}_{\widehat{\mathcal{H}}}}_{\widehat{\mathcal{H}}}$ *(quasi-projective over* $\vec{\mathsf{S}}_{0,r_\kappa}$*) in Lemma* 7.1.2.1 *and Proposition* 7.1.2.6.

The formal completion $(\vec{\mathsf{N}}^{\mathrm{ord,tor}}_{\kappa})^{\wedge}_{\vec{\widehat{\mathsf{Z}}}^{\mathrm{ord}}_{[(\breve{\Phi}_{\widehat{\mathcal{H}}},\breve{\delta}_{\widehat{\mathcal{H}}},\widehat{\tau})]}}$ *of* $\vec{\mathsf{N}}^{\mathrm{ord,tor}}_{\kappa}$ *along* $\vec{\widehat{\mathsf{Z}}}^{\mathrm{ord}}_{[(\breve{\Phi}_{\widehat{\mathcal{H}}},\breve{\delta}_{\widehat{\mathcal{H}}},\widehat{\tau})]}$ *is canonically isomorphic to* $\vec{\widehat{\mathfrak{X}}}^{\mathrm{ord}}_{\breve{\Phi}_{\widehat{\mathcal{H}}},\breve{\delta}_{\widehat{\mathcal{H}}},\widehat{\tau}}$*; and the formal completion* $(\vec{\mathsf{N}}^{\mathrm{ord,tor}}_{\kappa})^{\wedge}_{\cup\vec{\widehat{\mathsf{Z}}}^{\mathrm{ord}}_{[(\breve{\Phi}_{\widehat{\mathcal{H}}},\breve{\delta}_{\widehat{\mathcal{H}}},\widehat{\tau})]}}$*, where* $\cup\vec{\widehat{\mathsf{Z}}}^{\mathrm{ord}}_{[(\breve{\Phi}_{\widehat{\mathcal{H}}},\breve{\delta}_{\widehat{\mathcal{H}}},\widehat{\tau})]}$ *is the union of all strata* $\vec{\widehat{\mathsf{Z}}}^{\mathrm{ord}}_{[(\breve{\Phi}_{\widehat{\mathcal{H}}},\breve{\delta}_{\widehat{\mathcal{H}}},\widehat{\tau})]}$ *with* $\widehat{\tau}\in\widehat{\Sigma}_{\breve{\Phi}_{\widehat{\mathcal{H}}}}$*, is canonically isomorphic to* $\vec{\widehat{\mathfrak{X}}}^{\mathrm{ord}}_{\breve{\Phi}_{\widehat{\mathcal{H}}},\breve{\delta}_{\widehat{\mathcal{H}}}}/\Gamma_{\breve{\Phi}_{\widehat{\mathcal{H}}}}$ *(cf.* (5) *of Theorem* 5.2.1.1 *and Lemma* 5.2.4.38*). (Such isomorphisms can be induced by strata-preserving isomorphisms between étale neighborhoods of points of* $\vec{\widehat{\mathsf{Z}}}^{\mathrm{ord}}_{[(\breve{\Phi}_{\widehat{\mathcal{H}}},\breve{\delta}_{\widehat{\mathcal{H}}},\widehat{\tau})]}$ *in* $\vec{\mathsf{N}}^{\mathrm{ord,tor}}_{\kappa}$ *and étale neighborhoods of points of* $\vec{\widehat{\Xi}}^{\mathrm{ord}}_{\breve{\Phi}_{\widehat{\mathcal{H}}},\breve{\delta}_{\widehat{\mathcal{H}}},\widehat{\tau}}$ *in* $\vec{\widehat{\Xi}}^{\mathrm{ord}}_{\breve{\Phi}_{\widehat{\mathcal{H}}},\breve{\delta}_{\widehat{\mathcal{H}}}}(\widehat{\tau})$*.)*

Each $\vec{\mathsf{N}}^{\mathrm{ord,tor}}_{\kappa}$ *admits a canonical surjection* $\vec{\mathsf{N}}^{\mathrm{ord,tor}}_{\kappa}\to\vec{\mathsf{M}}^{\mathrm{ord,min}}_{\mathcal{H}}$ *extending the canonical surjection* $\vec{\mathsf{N}}^{\mathrm{ord}}_{\kappa}\to\vec{\mathsf{M}}^{\mathrm{ord}}_{\mathcal{H}}$*, and the latter is the pullback of the former under the canonical morphism* $\vec{\mathsf{M}}^{\mathrm{ord}}_{\mathcal{H}}\hookrightarrow\vec{\mathsf{M}}^{\mathrm{ord,min}}_{\mathcal{H}}$ *on the target (see Theorem* 6.2.1.1*). Both* $\vec{\mathsf{N}}^{\mathrm{ord,tor}}_{\kappa}\to\vec{\mathsf{M}}^{\mathrm{ord,min}}_{\mathcal{H}}$ *and* $\vec{\mathsf{N}}^{\mathrm{ord}}_{\kappa}\to\vec{\mathsf{M}}^{\mathrm{ord}}_{\mathcal{H}}$ *are proper when* $\mathcal{H}_\kappa$ *and* $\mathcal{H}_{\kappa'}$ *are* **equally deep** *as in Definition* 3.2.2.9*. Such a morphism maps the* $[(\breve{\Phi}_{\widehat{\mathcal{H}}},\breve{\delta}_{\widehat{\mathcal{H}}},\widehat{\tau})]$*-stratum* $\vec{\widehat{\mathsf{Z}}}^{\mathrm{ord}}_{[(\breve{\Phi}_{\widehat{\mathcal{H}}},\breve{\delta}_{\widehat{\mathcal{H}}},\widehat{\tau})]}$ *of* $\vec{\mathsf{N}}^{\mathrm{ord,tor}}_{\kappa}$ *to the* $[(\Phi_{\mathcal{H}},\delta_{\mathcal{H}})]$*-stratum* $\vec{\mathsf{Z}}^{\mathrm{ord}}_{[(\Phi_{\mathcal{H}},\delta_{\mathcal{H}})]}$ *of* $\vec{\mathsf{M}}^{\mathrm{ord,min}}_{\mathcal{H}}$ *if and only if the cusp label* $[(\Phi_{\mathcal{H}},\delta_{\mathcal{H}})]$ *is assigned to the cusp label* $[(\breve{\Phi}_{\widehat{\mathcal{H}}},\breve{\delta}_{\widehat{\mathcal{H}}})]$ *as in Lemma* 1.2.4.15*.*

If $\kappa\in\mathbf{K}^{\mathrm{ord},+}_{Q,\mathcal{H}}$*, then* $\mathcal{H}_\kappa=\mathcal{H}$*, and hence* $r_\kappa=r_{\mathcal{H}}$ *and* $\vec{\mathsf{M}}^{\mathrm{ord}}_{\mathcal{H}_\kappa}=\vec{\mathsf{M}}^{\mathrm{ord}}_{\mathcal{H}}$*. If* $\kappa\in\mathbf{K}^{\mathrm{ord}}_{Q,\mathcal{H}}$*, then* $\vec{\mathsf{N}}^{\mathrm{ord}}_{\kappa}=\vec{\mathsf{N}}^{\mathrm{ord,grp}}_{\kappa}\to\vec{\mathsf{M}}^{\mathrm{ord}}_{\mathcal{H}_\kappa}=\vec{\mathsf{M}}^{\mathrm{ord}}_{\mathcal{H}}$ *is an ordinary Kuga family.*

For each relation $\kappa'=(\widehat{\mathcal{H}}',\widehat{\Sigma}^{\mathrm{ord},\prime})\succ\kappa=(\widehat{\mathcal{H}},\widehat{\Sigma})$ *in* $\mathbf{K}^{\mathrm{ord},++}_{Q,\mathcal{H}}$*, we have* $\mathcal{H}_{\kappa'}=\widehat{\mathcal{H}}'_{\mathrm{G}}\subset\mathcal{H}_\kappa=\widehat{\mathcal{H}}_{\mathrm{G}}$ *and hence* $r_{\kappa'}\geq r_\kappa$*; and there is a surjection*

$$f^{\mathrm{tor}}_{\kappa',\kappa}:\vec{\mathsf{N}}^{\mathrm{ord,tor}}_{\kappa'}\to\vec{\mathsf{N}}^{\mathrm{ord,tor}}_{\kappa}$$

extending a canonical quasi-finite flat surjection

$$f_{\kappa',\kappa}:\vec{\mathsf{N}}^{\mathrm{ord}}_{\kappa'}\to\vec{\mathsf{N}}^{\mathrm{ord}}_{\kappa},$$

inducing a canonical log étale surjection $\vec{\mathsf{N}}^{\mathrm{ord,tor}}_{\kappa'}\to\vec{\mathsf{N}}^{\mathrm{ord,tor}}_{\kappa}\underset{\vec{\mathsf{S}}_{0,r_\kappa}}{\times}\vec{\mathsf{S}}_{0,r_{\kappa'}}$ *extending a canonical finite étale surjection* $\vec{\mathsf{N}}^{\mathrm{ord}}_{\kappa'}\to\vec{\mathsf{N}}^{\mathrm{ord}}_{\kappa}\underset{\vec{\mathsf{M}}^{\mathrm{ord}}_{\mathcal{H}_\kappa}}{\times}\vec{\mathsf{M}}^{\mathrm{ord}}_{\mathcal{H}_{\kappa'}}$*, equivariant with the canonical* $\mathbb{Q}^\times$*-isogeny*

$$f^{\mathrm{grp}}_{\kappa',\kappa}:=\kappa^{\mathrm{isog}}\circ((\kappa')^{\mathrm{isog}})^{-1}:\vec{\mathsf{N}}^{\mathrm{ord,grp}}_{\kappa'}\to\vec{\mathsf{N}}^{\mathrm{ord,grp}}_{\kappa}\underset{\vec{\mathsf{M}}^{\mathrm{ord}}_{\mathcal{H}_\kappa}}{\times}\vec{\mathsf{M}}^{\mathrm{ord}}_{\mathcal{H}_{\kappa'}},$$

such that $R^i(f^{\mathrm{tor}}_{\kappa',\kappa})_*\mathscr{O}_{\vec{\mathsf{N}}^{\mathrm{tor}}_{\kappa'}} = 0$ *for* $i > 0$. *These surjections are compatible with the canonical morphisms to* $\vec{\mathsf{M}}^{\mathrm{ord,min}}_{\mathcal{H}}$. *The morphism* $f^{\mathrm{tor}}_{\kappa',\kappa}$ *is proper log étale, and the morphism* $f_{\kappa',\kappa} : \vec{\mathsf{N}}^{\mathrm{ord}}_{\kappa'} \to \vec{\mathsf{N}}^{\mathrm{ord}}_{\kappa}$ *is finite étale, if* $\mathcal{H}_\kappa$ *and* $\mathcal{H}_{\kappa'}$ *are* **equally deep** *as in Definition* 3.2.2.9.

(2) *For each* $\kappa \in \mathbf{K}^{\mathrm{ord},++}_{Q,\mathcal{H},\Sigma^{\mathrm{ord}}}$, *the structural morphism*

$$f_\kappa : \vec{\mathsf{N}}^{\mathrm{ord}}_{\kappa} \to \vec{\mathsf{M}}^{\mathrm{ord}}_{\mathcal{H},r_\kappa} = \vec{\mathsf{M}}^{\mathrm{ord}}_{\mathcal{H}} \underset{\vec{\mathsf{S}}_{0,r_{\mathcal{H}}}}{\times} \vec{\mathsf{S}}_{0,r_\kappa}$$

(*the composition of the structural morphism* $\vec{\mathsf{N}}^{\mathrm{ord}}_{\kappa} \to \vec{\mathsf{M}}^{\mathrm{ord}}_{\mathcal{H}_\kappa}$ *in* (1) *with the canonical morphism* $\vec{\mathsf{M}}^{\mathrm{ord}}_{\mathcal{H}_\kappa} \to \vec{\mathsf{M}}^{\mathrm{ord}}_{\mathcal{H},r_\kappa}$) *extends* (*necessarily uniquely*) *to a surjection*

$$f^{\mathrm{tor}}_\kappa : \vec{\mathsf{N}}^{\mathrm{ord,tor}}_{\kappa} \to \vec{\mathsf{M}}^{\mathrm{ord,tor}}_{\mathcal{H},r_\kappa} = \vec{\mathsf{M}}^{\mathrm{ord,tor}}_{\mathcal{H},\Sigma^{\mathrm{ord}},r_\kappa} = \vec{\mathsf{M}}^{\mathrm{ord,tor}}_{\mathcal{H},\Sigma^{\mathrm{ord}}} \underset{\vec{\mathsf{S}}_{0,r_{\mathcal{H}}}}{\times} \vec{\mathsf{S}}_{0,r_\kappa},$$

which is **log smooth** (*as in [45, 3.3] and [43, 1.6]*) *if we equip* $\vec{\mathsf{N}}^{\mathrm{ord,tor}}_{\kappa}$ *and* $\mathsf{M}^{\mathrm{tor}}_{\mathcal{H},r_\kappa}$ *with the canonical* (*fine*) *log structures given respectively by the relative Cartier divisors with* (*simple*) *normal crossings* $\vec{\mathsf{E}}^{\mathrm{ord}}_{\infty,\kappa}$ *and* $\vec{\mathsf{D}}^{\mathrm{ord}}_{\infty,\mathcal{H},r_\kappa} := \vec{\mathsf{D}}^{\mathrm{ord}}_{\infty,\mathcal{H}} \underset{\vec{\mathsf{S}}_{0,r_{\mathcal{H}}}}{\times} \vec{\mathsf{S}}_{0,r_\kappa}$ (*see* (1) *above and* (3) *of Theorem* 5.2.1.1). *Then we have the following commutative diagram:*

$$\begin{array}{ccccc}
\vec{\mathsf{N}}^{\mathrm{ord}}_{\kappa} & \xhookrightarrow{\text{+NCD}} & \vec{\mathsf{N}}^{\mathrm{ord,tor}}_{\kappa} & & \\
{\scriptstyle \text{smooth surjective}}\ \downarrow f_\kappa & & {\scriptstyle \text{log smooth surjective}}\ \downarrow f^{\mathrm{tor}}_\kappa & \searrow^{\text{quasi-projective smooth}} & \\
\vec{\mathsf{M}}^{\mathrm{ord}}_{\mathcal{H},r_\kappa} & \xhookrightarrow[\text{+NCD}]{} & \vec{\mathsf{M}}^{\mathrm{ord,tor}}_{\mathcal{H},r_\kappa} & \xrightarrow[\text{separated smooth of finite type}]{} & \vec{\mathsf{S}}_{0,r_\kappa}
\end{array}$$

The morphism f^{tor}_κ *maps the* $[(\breve{\Phi}_{\widehat{\mathcal{H}}},\breve{\delta}_{\widehat{\mathcal{H}}},\widehat{\tau})]$*-stratum* $\widehat{\vec{\mathsf{Z}}}^{\mathrm{ord}}_{[(\breve{\Phi}_{\widehat{\mathcal{H}}},\breve{\delta}_{\widehat{\mathcal{H}}},\widehat{\tau})]}$ *of* $\vec{\mathsf{N}}^{\mathrm{ord,tor}}_{\kappa}$ *to the* $[(\Phi_{\mathcal{H}},\delta_{\mathcal{H}},\tau)]$*-stratum* $\vec{\mathsf{Z}}^{\mathrm{ord}}_{[(\Phi_{\mathcal{H}},\delta_{\mathcal{H}},\tau)]}$ *of* $\vec{\mathsf{M}}^{\mathrm{ord,tor}}_{\mathcal{H}}$ *if and only if* (*the cusp label* $[(\Phi_{\mathcal{H}},\delta_{\mathcal{H}})]$ *is assigned to the cusp label* $[(\breve{\Phi}_{\widehat{\mathcal{H}}},\breve{\delta}_{\widehat{\mathcal{H}}})]$ *as in Lemma* 1.2.4.15 *and*) *the image of* $\widehat{\tau} \in \widehat{\Sigma}_{\breve{\Phi}_{\widehat{\mathcal{H}}}}$ *under* (1.2.4.37) *is contained in* $\tau \in \Sigma_{\Phi_{\mathcal{H}}}$ *as in Condition* 1.2.4.49. *In this case, the compatible morphisms* $\widehat{\vec{\mathfrak{X}}}^{\mathrm{ord}}_{\breve{\Phi}_{\widehat{\mathcal{H}}},\breve{\delta}_{\widehat{\mathcal{H}}},\widehat{\tau}} \to \vec{\mathfrak{X}}^{\mathrm{ord}}_{\Phi_{\mathcal{H}},\delta_{\mathcal{H}},\tau}$ *and* $\widehat{\vec{\mathfrak{X}}}^{\mathrm{ord}}_{\breve{\Phi}_{\widehat{\mathcal{H}}},\breve{\delta}_{\widehat{\mathcal{H}}},\widehat{\Sigma}_{\breve{\Phi}_{\widehat{\mathcal{H}}}}} \to \vec{\mathfrak{X}}^{\mathrm{ord}}_{\Phi_{\mathcal{H}},\delta_{\mathcal{H}},\Sigma_{\Phi_{\mathcal{H}}}}$ *induced by* f^{tor}_κ (*and the canonical isomorphisms in* (1) *above and in* (5) *of Theorem* 5.2.1.1) *coincide with the canonical morphisms as in* (7.1.2.33) *and* (7.1.2.35). (*These morphisms can be induced by compatible morphisms between étale neighborhoods of points of the supports of formal schemes in relevant ambient schemes as in* (1) *above, compatible with all stratifications.*)

The morphisms $f_\kappa : \vec{\mathsf{N}}^{\mathrm{ord}}_{\kappa} \to \vec{\mathsf{M}}^{\mathrm{ord}}_{\mathcal{H}_\kappa}$ *and* $f^{\mathrm{tor}}_\kappa : \vec{\mathsf{N}}^{\mathrm{ord,tor}}_{\kappa} \to \vec{\mathsf{M}}^{\mathrm{ord,tor}}_{\mathcal{H},r_\kappa}$ *are proper when* $\mathcal{H}_\kappa$ *and* $\mathcal{H}$ *are* **equally deep** *as in Definition* 3.2.2.9 (*e.g., when* $\kappa \in \mathbf{K}^{\mathrm{ord},+}_{Q,\mathcal{H},\Sigma^{\mathrm{ord}}}$; *cf.* (1) *above*).

If $\kappa' \succ \kappa$, *then we have the compatibility* $f^{\mathrm{tor}}_{\kappa'} = f^{\mathrm{tor}}_{\kappa,r_{\kappa'}} \circ f^{\mathrm{tor}}_{\kappa',\kappa}$, *where* $f^{\mathrm{tor}}_{\kappa,r_{\kappa'}} := f^{\mathrm{tor}}_{\kappa} \underset{\vec{\mathsf{S}}_{0,r_\kappa}}{\times} \vec{\mathsf{S}}_{0,r_{\kappa'}}$.

(3) *Suppose* $\kappa \in \mathbf{K}^{\mathrm{ord},+}_{Q,\mathcal{H},\Sigma^{\mathrm{ord}}}$ (*not just in* $\mathbf{K}^{\mathrm{ord},++}_{Q,\mathcal{H},\Sigma^{\mathrm{ord}}}$, *so that* $\mathcal{H}_\kappa = \mathcal{H}$ *and* $r_\kappa = r_{\mathcal{H}}$, *and so that the base change from* $\vec{\mathsf{S}}_{0,r_{\mathcal{H}}}$ *to* $\vec{\mathsf{S}}_{0,r_\kappa}$ *in* (2) *is unnecessary*). *In this case, we also consider as in Definition* 7.1.3.11 *the extended ordinary Kuga family* $\vec{\mathsf{N}}^{\mathrm{ord,ext}}_{\kappa} \to \vec{\mathsf{M}}^{\mathrm{ord,tor}}_{\mathcal{H}}$, *a semi-abelian scheme extending the ordinary Kuga family* $\vec{\mathsf{N}}^{\mathrm{ord,grp}}_{\kappa} \to \vec{\mathsf{M}}^{\mathrm{ord}}_{\mathcal{H}}$, *together with the semi-abelian scheme* $\vec{\mathsf{N}}^{\mathrm{ord,ext},\vee}_{\kappa} \to \vec{\mathsf{M}}^{\mathrm{ord,tor}}_{\mathcal{H}}$ *extending the dual abelian scheme* $\vec{\mathsf{N}}^{\mathrm{ord,grp},\vee}_{\kappa} \to \vec{\mathsf{M}}^{\mathrm{ord}}_{\mathcal{H}}$, *so that the locally free sheaves* $\underline{\mathrm{Lie}}_{\vec{\mathsf{N}}^{\mathrm{ord,ext}}_{\kappa}/\vec{\mathsf{M}}^{\mathrm{ord,tor}}_{\mathcal{H}}}$, $\underline{\mathrm{Lie}}^{\vee}_{\vec{\mathsf{N}}^{\mathrm{ord,ext}}_{\kappa}/\vec{\mathsf{M}}^{\mathrm{ord,tor}}_{\mathcal{H}}}$, $\underline{\mathrm{Lie}}_{\vec{\mathsf{N}}^{\mathrm{ord,ext},\vee}_{\kappa}/\vec{\mathsf{M}}^{\mathrm{ord,tor}}_{\mathcal{H}}}$, *and* $\underline{\mathrm{Lie}}^{\vee}_{\vec{\mathsf{N}}^{\mathrm{ord,ext},\vee}_{\kappa}/\vec{\mathsf{M}}^{\mathrm{ord,tor}}_{\mathcal{H}}}$ *over* $\vec{\mathsf{M}}^{\mathrm{ord,tor}}_{\mathcal{H}}$ *extend the locally free sheaves* $\underline{\mathrm{Lie}}_{\vec{\mathsf{N}}^{\mathrm{ord}}_{\kappa}/\vec{\mathsf{M}}^{\mathrm{ord}}_{\mathcal{H}}}$, $\underline{\mathrm{Lie}}^{\vee}_{\vec{\mathsf{N}}^{\mathrm{ord}}_{\kappa}/\vec{\mathsf{M}}^{\mathrm{ord}}_{\mathcal{H}}}$, $\underline{\mathrm{Lie}}_{\vec{\mathsf{N}}^{\mathrm{ord},\vee}_{\kappa}/\vec{\mathsf{M}}^{\mathrm{ord}}_{\mathcal{H}}}$, *and* $\underline{\mathrm{Lie}}^{\vee}_{\vec{\mathsf{N}}^{\mathrm{ord},\vee}_{\kappa}/\vec{\mathsf{M}}^{\mathrm{ord}}_{\mathcal{H}}}$ *over* $\vec{\mathsf{M}}^{\mathrm{ord}}_{\mathcal{H}}$, *respectively.* (*See Definition* 7.1.3.9.)

For simplicity, let us suppress the subscripts "κ*" from the notation.* (*All canonical isomorphisms will be required to be compatible with the canonical isomorphisms defined by pulling back under* $f^{\mathrm{tor}}_{\kappa',\kappa}$ *for each relation* $\kappa' \succ \kappa$ *in* $\mathbf{K}^{\mathrm{ord},+}_{Q,\mathcal{H},\Sigma}$.)

Then the following are true:

(a) *Let* $\Omega^1_{\vec{\mathsf{N}}^{\mathrm{ord,tor}}/\vec{\mathsf{S}}_{0,r_{\mathcal{H}}}}[d\log\infty]$ *and* $\Omega^1_{\vec{\mathsf{M}}^{\mathrm{ord,tor}}_{\mathcal{H}}/\vec{\mathsf{S}}_{0,r_{\mathcal{H}}}}[d\log\infty]$ *denote the sheaves of modules of log* 1*-differentials over* $\vec{\mathsf{S}}_{0,r_{\mathcal{H}}}$ *given by the* (*respective*) *canonical log structures defined in* (2). *Let*

$$\begin{aligned}&\overline{\Omega}^1_{\vec{\mathsf{N}}^{\mathrm{ord,tor}}/\vec{\mathsf{M}}^{\mathrm{ord,tor}}_{\mathcal{H}}}\\&:= (\Omega^1_{\vec{\mathsf{N}}^{\mathrm{ord,tor}}/\vec{\mathsf{S}}_{0,r_{\mathcal{H}}}}[d\log\infty])/((f^{\mathrm{tor}})^*(\Omega^1_{\vec{\mathsf{M}}^{\mathrm{ord,tor}}_{\mathcal{H}}/\vec{\mathsf{S}}_{0,r_{\mathcal{H}}}}[d\log\infty])).\end{aligned}$$

Then there is a canonical isomorphism

$$(f^{\mathrm{tor}})^*\underline{\mathrm{Lie}}^{\vee}_{\vec{\mathsf{N}}^{\mathrm{ord,ext}}/\vec{\mathsf{M}}^{\mathrm{ord,tor}}_{\mathcal{H}}} \cong \overline{\Omega}^1_{\vec{\mathsf{N}}^{\mathrm{ord,tor}}/\vec{\mathsf{M}}^{\mathrm{ord,tor}}_{\mathcal{H}}} \tag{7.1.4.2}$$

between locally free sheaves over $\vec{\mathsf{N}}^{\mathrm{ord,tor}}$, *extending the canonical isomorphism*

$$f^*\underline{\mathrm{Lie}}^{\vee}_{\vec{\mathsf{N}}^{\mathrm{ord}}/\vec{\mathsf{M}}^{\mathrm{ord}}_{\mathcal{H}}} \cong \Omega^1_{\vec{\mathsf{N}}^{\mathrm{ord}}/\vec{\mathsf{M}}^{\mathrm{ord}}_{\mathcal{H}}} \tag{7.1.4.3}$$

over $\vec{\mathsf{N}}^{\mathrm{ord}}$ (*see Lemma* 7.1.3.10).

(b) *For each integer* $b \geq 0$, *there exist canonical isomorphisms*

$$\begin{aligned}&R^b f^{\mathrm{tor}}_*(\overline{\Omega}^a_{\vec{\mathsf{N}}^{\mathrm{ord,tor}}/\vec{\mathsf{M}}^{\mathrm{ord,tor}}_{\mathcal{H}}})\\&\cong (\wedge^b\underline{\mathrm{Lie}}_{\vec{\mathsf{N}}^{\mathrm{ord,ext},\vee}/\vec{\mathsf{M}}^{\mathrm{ord,tor}}_{\mathcal{H}}}) \underset{\mathscr{O}_{\vec{\mathsf{M}}^{\mathrm{ord,tor}}_{\mathcal{H}}}}{\otimes} (\wedge^a\underline{\mathrm{Lie}}^{\vee}_{\vec{\mathsf{N}}^{\mathrm{ord,ext}}/\vec{\mathsf{M}}^{\mathrm{ord,tor}}_{\mathcal{H}}}).\end{aligned} \tag{7.1.4.4}$$

and

$$\begin{aligned}&R^b f^{\mathrm{tor}}_*(\overline{\Omega}^a_{\vec{\mathsf{N}}^{\mathrm{ord,tor}}/\vec{\mathsf{M}}^{\mathrm{ord,tor}}_{\mathcal{H}}} \underset{\mathscr{O}_{\vec{\mathsf{N}}^{\mathrm{ord,tor}}}}{\otimes} \mathscr{I}_{\vec{\mathsf{E}}^{\mathrm{ord}}_{\infty}})\\&\cong R^b f^{\mathrm{tor}}_*(\overline{\Omega}^a_{\vec{\mathsf{N}}^{\mathrm{ord,tor}}/\vec{\mathsf{M}}^{\mathrm{ord,tor}}_{\mathcal{H}}}) \underset{\mathscr{O}_{\vec{\mathsf{M}}^{\mathrm{ord,tor}}_{\mathcal{H}}}}{\otimes} \mathscr{I}_{\vec{\mathsf{D}}^{\mathrm{ord}}_{\infty,\mathcal{H}}}\end{aligned} \tag{7.1.4.5}$$

of locally free sheaves over $\vec{\mathsf{M}}_{\mathcal{H}}^{\mathrm{ord,tor}}$, *where* $\mathscr{I}_{\vec{\mathsf{E}}_\infty^{\mathrm{ord}}}$ *(resp.* $\mathscr{I}_{\vec{\mathsf{D}}_{\infty,\mathcal{H}}^{\mathrm{ord}}}$*) is the* $\mathscr{O}_{\vec{\mathsf{N}}^{\mathrm{ord,tor}}}$*-ideal (resp.* $\mathscr{O}_{\vec{\mathsf{M}}_{\mathcal{H}}^{\mathrm{ord,tor}}}$*-ideal) defining the relative Cartier divisor* $\vec{\mathsf{E}}_\infty^{\mathrm{ord}} = \vec{\mathsf{E}}_{\infty,\kappa}^{\mathrm{ord}}$ *(resp.* $\vec{\mathsf{D}}_{\infty,\mathcal{H}}^{\mathrm{ord}}$*) (with its reduced structure), compatible with cup products and exterior products, extending the canonical isomorphism over* $\vec{\mathsf{M}}_{\mathcal{H}}^{\mathrm{ord}}$ *induced by the canonical isomorphism*

$$R^b f_*(\mathscr{O}_{\vec{\mathsf{N}}^{\mathrm{ord}}}) \cong \wedge^b \underline{\mathrm{Lie}}_{\vec{\mathsf{N}}^{\mathrm{ord},\vee}/\vec{\mathsf{M}}_{\mathcal{H}}^{\mathrm{ord}}}. \tag{7.1.4.6}$$

(c) *Let* $\overline{\Omega}^{\bullet}_{\vec{\mathsf{N}}^{\mathrm{ord,tor}}/\vec{\mathsf{M}}_{\mathcal{H}}^{\mathrm{ord,tor}}} := \wedge^{\bullet}\overline{\Omega}^{1}_{\vec{\mathsf{N}}^{\mathrm{ord,tor}}/\vec{\mathsf{M}}_{\mathcal{H}}^{\mathrm{ord,tor}}}$ *be the log de Rham complex associated with* $f^{\mathrm{tor}} : \vec{\mathsf{N}}^{\mathrm{ord,tor}} \to \vec{\mathsf{M}}_{\mathcal{H}}^{\mathrm{ord,tor}}$ *(with differentials inherited from* $\Omega^{\bullet}_{\vec{\mathsf{N}}^{\mathrm{ord}}/\vec{\mathsf{M}}_{\mathcal{H}}^{\mathrm{ord}}}$*). Let the (relative)* **log de Rham cohomology** *be defined by*

$$\underline{H}^i_{\text{log-dR}}(\vec{\mathsf{N}}^{\mathrm{ord,tor}}/\vec{\mathsf{M}}_{\mathcal{H}}^{\mathrm{ord,tor}}) := R^i f^{\mathrm{tor}}_*(\overline{\Omega}^{\bullet}_{\vec{\mathsf{N}}^{\mathrm{ord,tor}}/\vec{\mathsf{M}}_{\mathcal{H}}^{\mathrm{ord,tor}}}).$$

Then the (relative) **Hodge spectral sequence**

$$E_1^{a,b} := R^b f^{\mathrm{tor}}_*(\overline{\Omega}^{a}_{\vec{\mathsf{N}}^{\mathrm{ord,tor}}/\vec{\mathsf{M}}_{\mathcal{H}}^{\mathrm{ord,tor}}}) \Rightarrow \underline{H}^{a+b}_{\text{log-dR}}(\vec{\mathsf{N}}^{\mathrm{ord,tor}}/\vec{\mathsf{M}}_{\mathcal{H}}^{\mathrm{ord,tor}}) \tag{7.1.4.7}$$

degenerates at E_1 *terms, and defines a* **Hodge filtration** *on* $\underline{H}^i_{\text{log-dR}}(\vec{\mathsf{N}}^{\mathrm{ord,tor}}/\vec{\mathsf{M}}_{\mathcal{H}}^{\mathrm{ord,tor}})$ *with locally free graded pieces given by* $R^b f^{\mathrm{tor}}_*(\overline{\Omega}^{a}_{\vec{\mathsf{N}}^{\mathrm{ord,tor}}/\vec{\mathsf{M}}_{\mathcal{H}}^{\mathrm{ord,tor}}})$ *for integers* $a+b=i$*, extending the canonical Hodge filtration on* $\underline{H}^i_{\mathrm{dR}}(\vec{\mathsf{N}}^{\mathrm{ord}}/\vec{\mathsf{M}}_{\mathcal{H}}^{\mathrm{ord}})$.

As a result, for each integer $i \geq 0$*, there is a canonical isomorphism*

$$\wedge^i \underline{H}^1_{\text{log-dR}}(\vec{\mathsf{N}}^{\mathrm{ord,tor}}/\vec{\mathsf{M}}_{\mathcal{H}}^{\mathrm{ord,tor}}) \xrightarrow{\sim} \underline{H}^i_{\text{log-dR}}(\vec{\mathsf{N}}^{\mathrm{ord,tor}}/\vec{\mathsf{M}}_{\mathcal{H}}^{\mathrm{ord,tor}}),$$

compatible with the Hodge filtrations defined by (7.1.4.7)*, extending the canonical isomorphism*

$$\wedge^i \underline{H}^1_{\mathrm{dR}}(\vec{\mathsf{N}}^{\mathrm{ord}}/\vec{\mathsf{M}}_{\mathcal{H}}^{\mathrm{ord}}) \xrightarrow{\sim} \underline{H}^i_{\mathrm{dR}}(\vec{\mathsf{N}}^{\mathrm{ord}}/\vec{\mathsf{M}}_{\mathcal{H}}^{\mathrm{ord}})$$

over $\vec{\mathsf{M}}_{\mathcal{H}}^{\mathrm{ord}}$ *(defined by cup product).*

(d) *For each* $j_Q : Q^\vee \hookrightarrow Q$ *as in Lemma 1.2.4.1, the* $\mathbb{Q}^\times$*-polarization* $\lambda_{\mathsf{M}_{\mathcal{H}},j_Q}$ *in Proposition 1.3.3.7 extends canonically to a* $\mathbb{Q}^\times$*-polarization*

$$\lambda_{\vec{\mathsf{M}}_{\mathcal{H}}^{\mathrm{ord}},j_Q} : \underline{\mathrm{Hom}}_{\mathcal{O}}(Q, G_{\vec{\mathsf{M}}_{\mathcal{H}}^{\mathrm{ord}}})^\circ \to (\underline{\mathrm{Hom}}_{\mathcal{O}}(Q, G_{\vec{\mathsf{M}}_{\mathcal{H}}^{\mathrm{ord}}})^\circ)^\vee,$$

which induces a $\mathbb{Q}^\times$*-polarization*

$$\lambda_{\vec{\mathsf{N}}^{\mathrm{ord}},j_Q} : \vec{\mathsf{N}}^{\mathrm{ord,grp}} \to \vec{\mathsf{N}}^{\mathrm{ord,grp},\vee}$$

and defines canonically (as in [23, 1.5]) a perfect pairing

$$\langle \cdot, \cdot \rangle_{\lambda_{\vec{\mathsf{M}}_{\mathcal{H}}^{\mathrm{ord}},j_Q}} \underset{\mathbb{Z}}{\otimes} \mathbb{Q} : (\underline{H}^1_{\mathrm{dR}}(\vec{\mathsf{N}}^{\mathrm{ord}}/\vec{\mathsf{M}}_{\mathcal{H}}^{\mathrm{ord}}) \underset{\mathbb{Z}}{\otimes} \mathbb{Q}) \times (\underline{H}^1_{\mathrm{dR}}(\vec{\mathsf{N}}^{\mathrm{ord}}/\vec{\mathsf{M}}_{\mathcal{H}}^{\mathrm{ord}}) \underset{\mathbb{Z}}{\otimes} \mathbb{Q}) \to \mathscr{O}_{\vec{\mathsf{M}}_{\mathcal{H}}^{\mathrm{ord}}}(1) \underset{\mathbb{Z}}{\otimes} \mathbb{Q}.$$

(This is an abuse of notation because $\langle \cdot, \cdot \rangle_{\lambda_{\vec{\mathsf{M}}_{\mathcal{H}}^{\mathrm{ord}},j_Q}}$ *is not yet defined.)*

Then $\underline{H}^1_{\text{log-dR}}(\vec{\mathsf{N}}^{\mathrm{ord,tor}}/\vec{\mathsf{M}}_{\mathcal{H}}^{\mathrm{ord,tor}})$ *is (under the restriction morphism) canonically isomorphic to the unique subsheaf of*

$$(\vec{\mathsf{M}}_{\mathcal{H}}^{\mathrm{ord}} \hookrightarrow \vec{\mathsf{M}}_{\mathcal{H}}^{\mathrm{ord,tor}})_*(\underline{H}^1_{\mathrm{dR}}(\vec{\mathsf{N}}^{\mathrm{ord}}/\vec{\mathsf{M}}_{\mathcal{H}}^{\mathrm{ord}}))$$

satisfying the following conditions:

i. $\underline{H}^1_{\text{log-dR}}(\vec{\mathsf{N}}^{\text{ord,tor}}/\vec{\mathsf{M}}^{\text{ord,tor}}_{\mathcal{H}})$ *is locally free of finite rank over* $\mathscr{O}_{\vec{\mathsf{M}}^{\text{ord,tor}}_{\mathcal{H}}}$.

ii. The sheaf $f^{\text{tor}}_*(\overline{\Omega}^1_{\vec{\mathsf{N}}^{\text{ord,tor}}/\vec{\mathsf{M}}^{\text{ord,tor}}_{\mathcal{H}}})$ *can be identified with the subsheaf of* $(\vec{\mathsf{M}}^{\text{ord}}_{\mathcal{H}} \hookrightarrow \vec{\mathsf{M}}^{\text{ord,tor}}_{\mathcal{H}})_*(f_*(\Omega^1_{\vec{\mathsf{N}}^{\text{ord}}/\vec{\mathsf{M}}^{\text{ord}}_{\mathcal{H}}}))$ *formed (locally) by sections that are also sections of* $\underline{H}^1_{\text{log-dR}}(\vec{\mathsf{N}}^{\text{ord,tor}}/\vec{\mathsf{M}}^{\text{ord,tor}}_{\mathcal{H}})$. *(Here we view all sheaves canonically as subsheaves of* $(\vec{\mathsf{M}}^{\text{ord}}_{\mathcal{H}} \hookrightarrow \vec{\mathsf{M}}^{\text{ord,tor}}_{\mathcal{H}})_*(\underline{H}^1_{\text{dR}}(\vec{\mathsf{N}}^{\text{ord}}/\vec{\mathsf{M}}^{\text{ord}}_{\mathcal{H}}))$.*)*

iii. $\underline{H}^1_{\text{log-dR}}(\vec{\mathsf{N}}^{\text{ord,tor}}/\vec{\mathsf{M}}^{\text{ord,tor}}_{\mathcal{H}}) \underset{\mathbb{Z}}{\otimes} \mathbb{Q}$ *is self-dual under the push-forward* $(\vec{\mathsf{M}}^{\text{ord}}_{\mathcal{H}} \hookrightarrow \vec{\mathsf{M}}^{\text{ord,tor}}_{\mathcal{H}})_*\langle\,\cdot\,,\,\cdot\,\rangle_{\lambda_{\vec{\mathsf{M}}^{\text{ord}}_{\mathcal{H}}},j_Q} \underset{\mathbb{Z}}{\otimes} \mathbb{Q}$.

(e) The Gauss–Manin connection

$$\nabla : \underline{H}^\bullet_{\text{dR}}(\vec{\mathsf{N}}^{\text{ord}}/\vec{\mathsf{M}}^{\text{ord}}_{\mathcal{H}}) \to \underline{H}^\bullet_{\text{dR}}(\vec{\mathsf{N}}^{\text{ord}}/\vec{\mathsf{M}}^{\text{ord}}_{\mathcal{H}}) \underset{\mathscr{O}_{\vec{\mathsf{M}}^{\text{ord}}_{\mathcal{H}}}}{\otimes} \Omega^1_{\vec{\mathsf{M}}^{\text{ord}}_{\mathcal{H}}/\vec{\mathsf{S}}_{0,r_{\mathcal{H}}}} \tag{7.1.4.8}$$

extends to an integrable connection

$$\begin{aligned}\nabla : {} & \underline{H}^\bullet_{\text{log-dR}}(\vec{\mathsf{N}}^{\text{ord,tor}}/\vec{\mathsf{M}}^{\text{ord,tor}}_{\mathcal{H}}) \\ & \to \underline{H}^\bullet_{\text{log-dR}}(\vec{\mathsf{N}}^{\text{ord,tor}}/\vec{\mathsf{M}}^{\text{ord,tor}}_{\mathcal{H}}) \underset{\mathscr{O}_{\vec{\mathsf{M}}^{\text{ord,tor}}_{\mathcal{H}}}}{\otimes} \overline{\Omega}^1_{\vec{\mathsf{M}}^{\text{ord,tor}}_{\mathcal{H}}/\vec{\mathsf{S}}_{0,r_{\mathcal{H}}}}\end{aligned} \tag{7.1.4.9}$$

with log poles along $\vec{\mathsf{D}}^{\text{ord}}_{\infty,\mathcal{H}}$, *called the* **extended Gauss–Manin connection**, *satisfying the usual Griffiths transversality with the Hodge filtration defined by* (7.1.4.7).

(4) (Hecke actions; cf. Propositions 3.4.4.1 *and* 5.2.2.2*) Suppose we have an element* $\widehat{g} = (\widehat{g}_0, \widehat{g}_p) \in \widehat{\mathrm{G}}(\mathbb{A}^{\infty,p}) = \times\, \widehat{\mathrm{P}}^{\text{ord}}_{\mathsf{D}}(\mathbb{Q}_p)$ *with image* $g_h = (g_{h,0}, g_{h,p}) \in \mathrm{G}(\mathbb{A}^{\infty,p}) \times \mathrm{P}^{\text{ord}}_{\mathsf{D}}(\mathbb{Q}_p)$ *(see Definition* 3.2.2.7*) under the canonical homomorphism* $\widehat{\mathrm{G}}(\mathbb{A}^{\infty,p}) \times \widehat{\mathrm{P}}^{\text{ord}}_{\mathsf{D}}(\mathbb{Q}_p) \to \mathrm{G}(\mathbb{A}^{\infty,p}) \times \mathrm{P}^{\text{ord}}_{\mathsf{D}}(\mathbb{Q}_p)$, *and suppose we have two open compact subgroups* $\mathcal{H}$ *and* $\mathcal{H}'$ *of* $\mathrm{G}(\hat{\mathbb{Z}})$ *such that* $\mathcal{H}' \subset g_h \mathcal{H} g_h^{-1}$, *and such that* $\mathcal{H}$ *and* $\mathcal{H}'$ *are of standard form as in Definition* 3.2.2.9. *Suppose moreover that* $g_{h,p}$ *satisfies the conditions given in Section* 3.3.4 *(with* g_p *there replaced with* $g_{h,p}$ *here), so that* $[\vec{g_h}]^{\text{ord}} : \vec{\mathsf{M}}^{\text{ord}}_{\mathcal{H}'} \to \vec{\mathsf{M}}^{\text{ord}}_{\mathcal{H}}$ *is defined (see Proposition* 3.4.4.1*). Suppose* $\Sigma^{\text{ord},\prime} = \{\Sigma'_{\Phi'_{\mathcal{H}'}}\}_{[(\Phi'_{\mathcal{H}'},\delta'_{\mathcal{H}'})]}$ *is a compatible choice of admissible smooth rational polyhedral cone decomposition data for* $\vec{\mathsf{M}}^{\text{ord}}_{\mathcal{H}'}$, *which is a* g_h*-refinement of* $\Sigma^{\text{ord}} = \{\Sigma_{\Phi_{\mathcal{H}}}\}_{[(\Phi_{\mathcal{H}},\delta_{\mathcal{H}})]}$ *as in Definition* 5.2.2.1. *Consider the sets* $\mathbf{K}^{\text{ord}}_{Q,\mathcal{H}'} \subset \mathbf{K}^{\text{ord},+}_{Q,\mathcal{H}'} \subset \mathbf{K}^{\text{ord},++}_{Q,\mathcal{H}'}$ *and* $\mathbf{K}^{\text{ord}}_{Q,\mathcal{H}',\Sigma^{\text{ord},\prime}} \subset \mathbf{K}^{\text{ord},+}_{Q,\mathcal{H}',\Sigma^{\text{ord},\prime}} \subset \mathbf{K}^{\text{ord},++}_{Q,\mathcal{H}',\Sigma^{\text{ord},\prime}}$ *as in Definitions* 7.1.1.11 *and* 7.1.1.19 *(for* $\mathcal{H}'$ *and* $\Sigma^{\text{ord},\prime}$*), with compatible directed partial orders, parameterizing generalized ordinary Kuga families and their compactifications with properties as in* (1), (2), *and* (3) *above. The sets* $\mathbf{K}^{\text{ord},++}_{Q,\mathcal{H}}$ *etc and* $\mathbf{K}^{\text{ord},++}_{Q,\mathcal{H}'}$ *etc (and the objects they parameterize) satisfy the compatibility with* $\widehat{g}$ *(and* g_h*) in the sense that the following are true:*

(a) *For each* $\kappa = (\widehat{\mathcal{H}}, \widehat{\Sigma}^{\mathrm{ord}}) \in \mathbf{K}_{Q,\mathcal{H}}^{\mathrm{ord},++}$ *(resp.* $\mathbf{K}_{Q,\mathcal{H}}^{\mathrm{ord},+}$*, resp.* $\mathbf{K}_{Q,\mathcal{H}}^{\mathrm{ord}}$*), and for each open compact subgroup* $\widehat{\mathcal{H}}' \subset \widehat{\mathrm{G}}(\hat{\mathbb{Z}})$ *such that* $\widehat{\mathcal{H}}' \subset \widehat{g}\widehat{\mathcal{H}}\widehat{g}^{-1}$ *(so that* $\mathcal{H}_\kappa = \widehat{\mathcal{H}}_{\mathrm{G}}$ *and* $\mathcal{H}_{\kappa'} = \widehat{\mathcal{H}}'_{\mathrm{G}}$ *satisfy* $\mathcal{H}_{\kappa'} \subset g_h \mathcal{H}_\kappa g_h^{-1}$*), there exists an element* $\kappa' = (\widehat{\mathcal{H}}', \widehat{\Sigma}^{\mathrm{ord},\prime}) \in \mathbf{K}_{Q,\mathcal{H}'}^{\mathrm{ord},++}$ *(resp.* $\mathbf{K}_{Q,\mathcal{H}'}^{\mathrm{ord},+}$*, resp.* $\mathbf{K}_{Q,\mathcal{H}'}^{\mathrm{ord}}$*) such that there exists a (necessarily unique) quasi-finite flat surjection*

$$[\vec{\widehat{g}}]^{\mathrm{ord}} : \vec{\mathsf{N}}_{\kappa'}^{\mathrm{ord}} \to \vec{\mathsf{N}}_{\kappa}^{\mathrm{ord}} \tag{7.1.4.10}$$

covering the compatible surjections $[\vec{g_h}]^{\mathrm{ord}} : \vec{\mathsf{M}}_{\mathcal{H}'}^{\mathrm{ord}} \to \vec{\mathsf{M}}_{\mathcal{H}}^{\mathrm{ord}}$, $[\vec{g_h}]_{r_{\kappa'},r_\kappa}^{\mathrm{ord}} : \vec{\mathsf{M}}_{\mathcal{H}',r_{\kappa'}}^{\mathrm{ord}} \to \vec{\mathsf{M}}_{\mathcal{H},r_\kappa}^{\mathrm{ord}}$*, and* $[\vec{g_h}]^{\mathrm{ord}} : \vec{\mathsf{M}}_{\mathcal{H}_{\kappa'}}^{\mathrm{ord}} \to \vec{\mathsf{M}}_{\mathcal{H}_\kappa}^{\mathrm{ord}}$ *given by Proposition* 3.4.4.1*, inducing a finite flat surjection* $\vec{\mathsf{N}}_{\kappa'}^{\mathrm{ord}} \to \vec{\mathsf{N}}_{\kappa}^{\mathrm{ord}} \underset{\vec{\mathsf{M}}_{\mathcal{H}_\kappa}^{\mathrm{ord}}}{\times} \vec{\mathsf{M}}_{\mathcal{H}_{\kappa'}}^{\mathrm{ord}}$ *of abelian scheme torsors equivariant with the isogeny (not just a* $\mathbb{Q}^\times$*-isogeny)*

$$\vec{\mathsf{N}}_{\kappa'}^{\mathrm{ord,grp}} \to \vec{\mathsf{N}}_{\kappa}^{\mathrm{ord,grp}} \underset{\vec{\mathsf{M}}_{\mathcal{H}_\kappa}^{\mathrm{ord}}}{\times} \vec{\mathsf{M}}_{\mathcal{H}_{\kappa'}}^{\mathrm{ord}}$$

induced by $(\kappa')^{\mathrm{isog}}$*,* κ^{isog}*, and the* $\mathbb{Q}^\times$*-isogeny* $G_{\vec{\mathsf{M}}_{\mathcal{H}_{\kappa'}}^{\mathrm{ord}}} \to G_{\vec{\mathsf{M}}_{\mathcal{H}_\kappa}^{\mathrm{ord}}} \underset{\vec{\mathsf{M}}_{\mathcal{H}_\kappa}^{\mathrm{ord}}}{\times} \vec{\mathsf{M}}_{\mathcal{H}_{\kappa'}}^{\mathrm{ord}}$ *realizing* $G_{\vec{\mathsf{M}}_{\mathcal{H}_\kappa}^{\mathrm{ord}}} \underset{\vec{\mathsf{M}}_{\mathcal{H}_\kappa}^{\mathrm{ord}}}{\times} \vec{\mathsf{M}}_{\mathcal{H}_{\kappa'}}^{\mathrm{ord}}$ *as an ordinary Hecke twist of* $G_{\vec{\mathsf{M}}_{\mathcal{H}_{\kappa'}}^{\mathrm{ord}}}$ *by* g_h *(which is the pullback of the* $\mathbb{Q}^\times$*-isogeny* $G_{\vec{\mathsf{M}}_{\mathcal{H}'}^{\mathrm{ord}}} \to G_{\vec{\mathsf{M}}_{\mathcal{H}}^{\mathrm{ord}}} \underset{\vec{\mathsf{M}}_{\mathcal{H}}^{\mathrm{ord}}}{\times} \vec{\mathsf{M}}_{\mathcal{H}'}^{\mathrm{ord}}$ *realizing* $G_{\vec{\mathsf{M}}_{\mathcal{H}}^{\mathrm{ord}}} \underset{\vec{\mathsf{M}}_{\mathcal{H}}^{\mathrm{ord}}}{\times} \vec{\mathsf{M}}_{\mathcal{H}'}^{\mathrm{ord}}$ *as an ordinary Hecke twist of* $G_{\vec{\mathsf{M}}_{\mathcal{H}'}^{\mathrm{ord}}}$ *by* g_h*). (Here all the base changes from* $\vec{\mathsf{M}}_{\mathcal{H}}^{\mathrm{ord}}$ *to* $\vec{\mathsf{M}}_{\mathcal{H}'}^{\mathrm{ord}}$ *and from* $\vec{\mathsf{M}}_{\mathcal{H}_\kappa}^{\mathrm{ord}}$ *to* $\vec{\mathsf{M}}_{\mathcal{H}_{\kappa'}}^{\mathrm{ord}}$ *use the surjections denoted by* $[\vec{g_h}]^{\mathrm{ord}}$*.) The characteristic zero pullback* $[\vec{\widehat{g}}]^{\mathrm{ord}} \underset{\mathbb{Z}}{\otimes} \mathbb{Q}$ *is étale.*

(b) *For each* $\kappa = (\widehat{\mathcal{H}}, \widehat{\Sigma}^{\mathrm{ord}})$ *and* $\widehat{\mathcal{H}}'$ *as in* (4a) *such that* $\kappa \in \mathbf{K}_{Q,\mathcal{H}}^{\mathrm{ord},++}$ *(resp.* $\mathbf{K}_{Q,\mathcal{H}}^{\mathrm{ord},+}$*, resp.* $\mathbf{K}_{Q,\mathcal{H}}^{\mathrm{ord}}$*), there is an element* $\kappa' = (\widehat{\mathcal{H}}', \widehat{\Sigma}^{\mathrm{ord},\prime}) \in \mathbf{K}_{Q,\mathcal{H}'}^{\mathrm{ord},++}$ *(resp.* $\mathbf{K}_{Q,\mathcal{H}'}^{\mathrm{ord},+}$*, resp.* $\mathbf{K}_{Q,\mathcal{H}'}^{\mathrm{ord}}$*) such that* $[\vec{\widehat{g}}]^{\mathrm{ord}}$ *is defined as in* (4a) *(see* (7.1.4.10)*), and such that* $\widehat{\Sigma}^{\mathrm{ord},\prime}$ *is a* $\widehat{g}$*-refinement of* $\widehat{\Sigma}^{\mathrm{ord}}$ *(cf. Lemma* 1.2.4.42 *and Definition* 5.2.2.1*), which extends to a (necessarily unique) surjection*

$$[\vec{\widehat{g}}]^{\mathrm{ord,tor}} : \vec{\mathsf{N}}_{\kappa'}^{\mathrm{ord,tor}} \to \vec{\mathsf{N}}_{\kappa}^{\mathrm{ord,tor}} \tag{7.1.4.11}$$

such that

$$R^i [\vec{\widehat{g}}]_*^{\mathrm{ord,tor}} \mathscr{O}_{\vec{\mathsf{N}}_{\kappa'}^{\mathrm{ord,tor}}} = 0 \tag{7.1.4.12}$$

for all $i > 0$*. If* $\widehat{\Sigma}^{\mathrm{ord},\prime}$ *is* $\widehat{g}$*-induced by* $\widehat{\Sigma}^{\mathrm{ord}}$ *(cf. Lemma* 1.2.4.42 *and Definition* 5.2.2.1*), then* $[\vec{\widehat{g}}]^{\mathrm{ord,tor}}$ *is quasi-finite. The characteristic zero pullback* $[\vec{\widehat{g}}]^{\mathrm{ord,tor}} \underset{\mathbb{Z}}{\otimes} \mathbb{Q}$ *is log étale.*

Under (7.1.4.11), *the* $[(\breve{\Phi}'_{\widehat{\mathcal{H}}'}, \breve{\delta}'_{\widehat{\mathcal{H}}'}, \widehat{\tau}')]$*-stratum* $\vec{\widehat{\mathsf{Z}}}^{\mathrm{ord}}_{[(\breve{\Phi}'_{\widehat{\mathcal{H}}'}, \breve{\delta}'_{\widehat{\mathcal{H}}'}, \widehat{\tau}')]}$ *of* $\vec{\mathsf{N}}^{\mathrm{ord,tor}}_{\kappa'}$ *is mapped to the* $[(\breve{\Phi}_{\widehat{\mathcal{H}}}, \breve{\delta}_{\widehat{\mathcal{H}}}, \widehat{\tau})]$*-stratum* $\vec{\widehat{\mathsf{Z}}}^{\mathrm{ord}}_{[(\breve{\Phi}_{\widehat{\mathcal{H}}}, \breve{\delta}_{\widehat{\mathcal{H}}}, \widehat{\tau})]}$ *of* $\vec{\mathsf{N}}^{\mathrm{ord,tor}}_{\kappa}$ *if and only if there are representatives* $(\breve{\Phi}_{\widehat{\mathcal{H}}}, \breve{\delta}_{\widehat{\mathcal{H}}}, \widehat{\tau})$ *and* $(\breve{\Phi}'_{\widehat{\mathcal{H}}'}, \breve{\delta}'_{\widehat{\mathcal{H}}'}, \widehat{\tau}')$ *of* $[(\breve{\Phi}_{\widehat{\mathcal{H}}}, \breve{\delta}_{\widehat{\mathcal{H}}}, \widehat{\tau})]$ *and* $[(\breve{\Phi}'_{\widehat{\mathcal{H}}'}, \breve{\delta}'_{\widehat{\mathcal{H}}'}, \widehat{\tau}')]$*, respectively, such that* $(\breve{\Phi}'_{\widehat{\mathcal{H}}'}, \breve{\delta}'_{\widehat{\mathcal{H}}'}, \widehat{\tau}')$ *is a* $\widehat{g}$*-refinement of* $(\breve{\Phi}_{\widehat{\mathcal{H}}}, \breve{\delta}_{\widehat{\mathcal{H}}}, \widehat{\tau})$ *(cf. Lemma* 1.2.4.42 *and [62, Def. 6.4.3.1]). In this case, the compatible morphisms* $\vec{\widehat{\mathfrak{X}}}^{\mathrm{ord}}_{\breve{\Phi}'_{\widehat{\mathcal{H}}'}, \breve{\delta}'_{\widehat{\mathcal{H}}'}, \widehat{\tau}'} \to \vec{\widehat{\mathfrak{X}}}^{\mathrm{ord}}_{\breve{\Phi}_{\widehat{\mathcal{H}}}, \breve{\delta}_{\widehat{\mathcal{H}}}, \widehat{\tau}}$ *and* $\vec{\widehat{\mathfrak{X}}}^{\mathrm{ord}}_{\breve{\Phi}'_{\widehat{\mathcal{H}}'}, \breve{\delta}'_{\widehat{\mathcal{H}}'}, \widehat{\Sigma}'_{\breve{\Phi}'_{\widehat{\mathcal{H}}'}}} \to \vec{\widehat{\mathfrak{X}}}^{\mathrm{ord}}_{\breve{\Phi}_{\widehat{\mathcal{H}}}, \breve{\delta}_{\widehat{\mathcal{H}}}, \widehat{\Sigma}_{\breve{\Phi}_{\widehat{\mathcal{H}}}}}$ *induced by* (7.1.4.11) *(and the canonical isomorphisms in* (1) *above) coincide with the canonical morphisms as in* (7.1.2.19) *and* (7.1.2.21)*.*

If $\kappa \in \mathbf{K}^{\mathrm{ord},++}_{Q,\mathcal{H},\Sigma^{\mathrm{ord}}}$ *(resp.* $\mathbf{K}^{\mathrm{ord},+}_{Q,\mathcal{H},\Sigma^{\mathrm{ord}}}$*, resp.* $\mathbf{K}^{\mathrm{ord}}_{Q,\mathcal{H},\Sigma^{\mathrm{ord}}}$*), we may assume in the above that* $\kappa' \in \mathbf{K}^{\mathrm{ord},++}_{Q,\mathcal{H}',\Sigma^{\mathrm{ord},\prime}}$ *(resp.* $\mathbf{K}^{\mathrm{ord},+}_{Q,\mathcal{H}',\Sigma^{\mathrm{ord},\prime}}$*, resp.* $\mathbf{K}^{\mathrm{ord}}_{Q,\mathcal{H}',\Sigma^{\mathrm{ord},\prime}}$*), so that* (7.1.4.11) *covers the surjection*

$$\vec{[g_h]}^{\mathrm{ord,tor}}_{r_{\kappa'},r_\kappa} : \vec{\mathsf{M}}^{\mathrm{ord,tor}}_{\mathcal{H}',\Sigma^{\mathrm{ord},\prime},r_{\kappa'}} \to \vec{\mathsf{M}}^{\mathrm{ord,tor}}_{\mathcal{H},\Sigma^{\mathrm{ord}},r_\kappa}$$

induced by the surjection $\vec{[g_h]}^{\mathrm{ord,tor}} : \vec{\mathsf{M}}^{\mathrm{ord,tor}}_{\mathcal{H}',\Sigma^{\mathrm{ord},\prime}} \to \vec{\mathsf{M}}^{\mathrm{ord,tor}}_{\mathcal{H},\Sigma^{\mathrm{ord}}}$ *given by Proposition* 5.2.2.2*.*

(c) *Suppose* $\vec{[\widehat{g}]}^{\mathrm{ord,tor}}$ *is defined as in* (4b) *for some* $\kappa \in \mathbf{K}^{\mathrm{ord},+}_{Q,\mathcal{H},\Sigma^{\mathrm{ord}}}$ *and* $\kappa' \in \mathbf{K}^{\mathrm{ord},+}_{Q,\mathcal{H}',\Sigma^{\mathrm{ord},\prime}}$ *(not just in* $\mathbf{K}^{\mathrm{ord},++}_{Q,\mathcal{H},\Sigma^{\mathrm{ord}}}$ *and* $\mathbf{K}^{\mathrm{ord},++}_{Q,\mathcal{H}',\Sigma^{\mathrm{ord},\prime}}$*). Then there is a canonical morphism*

$$\begin{aligned}(\vec{[\widehat{g}]}^{\mathrm{ord,tor}})^* : (\vec{[\widehat{g}]}^{\mathrm{ord,tor}})^* \underline{H}^{a+b}_{\text{log-dR}}(\vec{\mathsf{N}}^{\mathrm{ord,tor}}_{\kappa}/\vec{\mathsf{M}}^{\mathrm{ord,tor}}_{\mathcal{H},\Sigma^{\mathrm{ord}}}) \\ \to \underline{H}^{a+b}_{\text{log-dR}}(\vec{\mathsf{N}}^{\mathrm{ord,tor}}_{\kappa'}/\vec{\mathsf{M}}^{\mathrm{ord,tor}}_{\mathcal{H}',\Sigma^{\mathrm{ord},\prime}})\end{aligned}$$

respecting the Hodge filtrations and compatible with the canonical morphisms

$$(\vec{[\widehat{g}]}^{\mathrm{ord,tor}})^* : (\vec{[\widehat{g}]}^{\mathrm{ord,tor}})^* \overline{\Omega}^1_{\vec{\mathsf{N}}^{\mathrm{ord,tor}}_{\kappa}/\vec{\mathsf{M}}^{\mathrm{ord,tor}}_{\mathcal{H},\Sigma^{\mathrm{ord}}}} \to \overline{\Omega}^1_{\vec{\mathsf{N}}^{\mathrm{ord,tor}}_{\kappa'}/\vec{\mathsf{M}}^{\mathrm{ord,tor}}_{\mathcal{H}',\Sigma^{\mathrm{ord},\prime}}},$$

$$(\vec{[g_h]}^{\mathrm{ord,tor}}_{r_{\kappa'},r_\kappa})^* : (\vec{[g_h]}^{\mathrm{ord,tor}})^* \underline{\mathrm{Lie}}_{\vec{\mathsf{N}}^{\mathrm{ord,ext},\vee}_{\kappa}/\vec{\mathsf{M}}^{\mathrm{ord,tor}}_{\mathcal{H},\Sigma^{\mathrm{ord}}}} \to \underline{\mathrm{Lie}}_{\vec{\mathsf{N}}^{\mathrm{ord,ext},\vee}_{\kappa'}/\vec{\mathsf{M}}^{\mathrm{ord,tor}}_{\mathcal{H}',\Sigma^{\mathrm{ord},\prime}}},$$

$$(\vec{[g_h]}^{\mathrm{ord,tor}}_{r_{\kappa'},r_\kappa})^* : (\vec{[g_h]}^{\mathrm{ord,tor}})^* \underline{\mathrm{Lie}}^\vee_{\vec{\mathsf{N}}^{\mathrm{ord,ext}}_{\kappa}/\vec{\mathsf{M}}^{\mathrm{ord,tor}}_{\mathcal{H},\Sigma^{\mathrm{ord}}}} \to \underline{\mathrm{Lie}}^\vee_{\vec{\mathsf{N}}^{\mathrm{ord,ext}}_{\kappa'}/\vec{\mathsf{M}}^{\mathrm{ord,tor}}_{\mathcal{H}',\Sigma^{\mathrm{ord},\prime}}},$$

$$(\vec{[g_h]}^{\mathrm{ord,tor}}_{r_{\kappa'},r_\kappa})^* : (\vec{[g_h]}^{\mathrm{ord,tor}})^* \underline{\mathrm{Lie}}_{G^\vee/\vec{\mathsf{M}}^{\mathrm{ord,tor}}_{\mathcal{H},\Sigma^{\mathrm{ord}}}} \to \underline{\mathrm{Lie}}_{G^\vee/\vec{\mathsf{M}}^{\mathrm{ord,tor}}_{\mathcal{H}',\Sigma^{\mathrm{ord},\prime}}},$$

$$(\vec{[g_h]}^{\mathrm{ord,tor}}_{r_{\kappa'},r_\kappa})^* : (\vec{[g_h]}^{\mathrm{ord,tor}})^* \underline{\mathrm{Lie}}^\vee_{G/\vec{\mathsf{M}}^{\mathrm{ord,tor}}_{\mathcal{H},\Sigma^{\mathrm{ord}}}} \to \underline{\mathrm{Lie}}^\vee_{G/\vec{\mathsf{M}}^{\mathrm{ord,tor}}_{\mathcal{H}',\Sigma^{\mathrm{ord},\prime}}},$$

and the canonical isomorphisms in (3) *for* $\vec{\mathsf{N}}^{\mathrm{ord,tor}}_{\kappa}$ *and* $\vec{\mathsf{N}}^{\mathrm{ord,tor}}_{\kappa'}$*. The characteristic zero pullbacks of these canonical morphisms are isomorphisms.*

(d) *If the levels* $\mathcal{H}_{\kappa,p}$ *and* $\mathcal{H}_{\kappa',p}$ *at* p *are* **equally deep** *as in Definition* 3.2.2.9 (*by Remark* 7.1.1.3, *this is equivalent to the condition that* $\widehat{\mathcal{H}}$ *and* $\widehat{\mathcal{H}}'$ *are equally deep as in Definition* 7.1.1.2), *or if* $g_{h,p}$ *is of twisted* U_p *type as in Definition* 3.3.6.1 *and* $\mathrm{depth}_{\mathsf{D}}(\mathcal{H}_{\kappa',p}) - \mathrm{depth}_{\mathsf{D}}(g_{h,p}) = \mathrm{depth}_{\mathsf{D}}(\mathcal{H}_{\kappa,p}) > 0$, *then the surjection* (7.1.4.10) *is finite, and the surjection* (7.1.4.11) *is proper.*

(e) *If* $g_{h,p} \in \mathrm{P}^{\mathrm{ord}}_{\mathsf{D}}(\mathbb{Z}_p)$, *then the morphism*

$$[\vec{\widehat{g}}]^{\mathrm{ord}}_{r_{\kappa'}} : \vec{\mathsf{N}}^{\mathrm{ord}}_{\kappa'} \to \vec{\mathsf{N}}^{\mathrm{ord}}_{\kappa} \underset{\vec{\mathsf{S}}_{0,r_\kappa}}{\times} \vec{\mathsf{S}}_{0,r_{\kappa'}}$$

canonically induced by $[\vec{\widehat{g}}]^{\mathrm{ord}} : \vec{\mathsf{N}}^{\mathrm{ord}}_{\kappa'} \to \vec{\mathsf{N}}^{\mathrm{ord}}_{\kappa}$ *is étale, and the surjection*

$$[\vec{\widehat{g}}]^{\mathrm{ord,tor}}_{r_{\kappa'}} : \vec{\mathsf{N}}^{\mathrm{ord,tor}}_{\kappa'} \to \vec{\mathsf{N}}^{\mathrm{ord,tor}}_{\kappa} \underset{\vec{\mathsf{S}}_{0,r_\kappa}}{\times} \vec{\mathsf{S}}_{0,r_{\kappa'}}$$

canonically induced by $[\vec{\widehat{g}}]^{\mathrm{ord,tor}} : \vec{\mathsf{N}}^{\mathrm{ord,tor}}_{\kappa'} \to \vec{\mathsf{N}}^{\mathrm{ord,tor}}_{\kappa}$ *is log étale, and the canonical morphisms in* (4c) *are all isomorphisms. If* $\widehat{\Sigma}'$ *is* $\widehat{g}$*-induced by* $\widehat{\Sigma}$, *then* $[\vec{\widehat{g}}]^{\mathrm{ord,tor}}_{r_{\kappa'}}$ *is quasi-finite étale* (*not just log étale*).

(f) *If we have an element* $\widehat{g}' \in \widehat{\mathrm{G}}(\mathbb{A}^{\infty,p}) \times \widehat{\mathrm{P}}^{\mathrm{ord}}_{\mathsf{D}}(\mathbb{Q}_p)$ *with image* $g'_h \in \mathrm{G}(\mathbb{A}^{\infty,p}) \times \mathrm{P}^{\mathrm{ord}}_{\mathsf{D}}(\mathbb{Q}_p)$ *under the canonical homomorphism* $\widehat{\mathrm{G}}(\mathbb{A}^{\infty,p}) \times \widehat{\mathrm{P}}^{\mathrm{ord}}_{\mathsf{D}}(\mathbb{Q}_p) \to \mathrm{G}(\mathbb{A}^{\infty,p}) \times \mathrm{P}^{\mathrm{ord}}_{\mathsf{D}}(\mathbb{Q}_p)$, *with a similar setup such that* $[\vec{\widehat{g}'}]^{\mathrm{ord}} : \vec{\mathsf{N}}^{\mathrm{ord}}_{\kappa''} \to \vec{\mathsf{N}}^{\mathrm{ord}}_{\kappa'}$ *and* $[\vec{\widehat{g}'}]^{\mathrm{ord,tor}} : \vec{\mathsf{N}}^{\mathrm{ord,tor}}_{\kappa''} \to \vec{\mathsf{N}}^{\mathrm{ord,tor}}_{\kappa'}$ *are compatibly defined for some* $\kappa'' \in \mathbf{K}^{\mathrm{ord},++}_{Q,\mathcal{H}''}$, *then* $[\vec{\widehat{g}'\widehat{g}}]^{\mathrm{ord}} : \vec{\mathsf{N}}^{\mathrm{ord}}_{\kappa''} \to \vec{\mathsf{N}}^{\mathrm{ord}}_{\kappa}$ *and* $[\vec{\widehat{g}'\widehat{g}}]^{\mathrm{ord,tor}} : \vec{\mathsf{N}}^{\mathrm{ord,tor}}_{\kappa''} \to \vec{\mathsf{N}}^{\mathrm{ord,tor}}_{\kappa}$ *are also compatibly defined and satisfy the identities* $[\vec{\widehat{g}'\widehat{g}}]^{\mathrm{ord}} = [\vec{\widehat{g}}]^{\mathrm{ord}} \circ [\vec{\widehat{g}'}]^{\mathrm{ord}}$ *and* $[\vec{\widehat{g}'\widehat{g}}]^{\mathrm{ord,tor}} = [\vec{\widehat{g}}]^{\mathrm{ord,tor}} \circ [\vec{\widehat{g}'}]^{\mathrm{ord,tor}}$. *If* $\kappa \in \mathbf{K}^{\mathrm{ord},+}_{Q,\mathcal{H}}$, $\kappa' \in \mathbf{K}^{\mathrm{ord},+}_{Q,\mathcal{H}'}$, *and* $\kappa'' \in \mathbf{K}^{\mathrm{ord},+}_{Q,\mathcal{H}''}$, *we also have* $([\vec{\widehat{g}'\widehat{g}}]^{\mathrm{ord}})^* = ([\vec{\widehat{g}'}]^{\mathrm{ord}})^* \circ ([\vec{\widehat{g}}]^{\mathrm{ord}})^*$ *and* $([\vec{\widehat{g}'\widehat{g}}]^{\mathrm{ord,tor}})^* = ([\vec{\widehat{g}'}]^{\mathrm{ord,tor}})^* \circ ([\vec{\widehat{g}}]^{\mathrm{ord,tor}})^*$ (*in both applicable senses above*).

(g) *The morphism*

$$[\vec{\widehat{g}}]^{\mathrm{ord}} : \vec{\mathfrak{N}}^{\mathrm{ord}}_{\kappa'} \to \vec{\mathfrak{N}}^{\mathrm{ord}}_{\kappa} \tag{7.1.4.13}$$

(*cf. Definition* 3.4.4.2) *induced by* (7.1.4.10) *is finite. The morphism*

$$[\vec{\widehat{g}}]^{\mathrm{ord,tor}} : \vec{\mathfrak{N}}^{\mathrm{ord,tor}}_{\kappa'} \to \vec{\mathfrak{N}}^{\mathrm{ord,tor}}_{\kappa} \tag{7.1.4.14}$$

induced by (7.1.4.11) *is proper, and is finite flat if* $\widehat{\Sigma}'$ *is* $\widehat{g}$*-induced by* $\widehat{\Sigma}$. *If* $\widehat{g}_p \in \widehat{\mathrm{P}}^{\mathrm{ord}}_{\mathsf{D}}(\mathbb{Z}_p)$, *then the morphism*

$$[\vec{\widehat{g}}]^{\mathrm{ord,tor}}_{r_{\kappa'}} : \vec{\mathfrak{N}}^{\mathrm{ord,tor}}_{\kappa'} \to \vec{\mathfrak{N}}^{\mathrm{ord,tor}}_{\kappa} \underset{\vec{\mathfrak{S}}_{0,r_\kappa}}{\times} \vec{\mathfrak{S}}_{0,r_{\kappa'}} \tag{7.1.4.15}$$

induced by (7.1.4.14) *is proper log étale* (*because it is log étale by* (4e)). *If moreover* $\widehat{\Sigma}'$ *is* $\widehat{g}$*-induced by* $\widehat{\Sigma}$, *then* (7.1.4.15) *is finite étale* (*because it is quasi-finite étale by* (4e); *cf.* [35, IV-3, 8.11.1]).

(h) *If* $\widehat{g} = (\widehat{g}_0, \widehat{g}_p) \in \widehat{\mathrm{G}}(\hat{\mathbb{Z}}^p) \times \widehat{\mathrm{P}}^{\mathrm{ord}}_{\mathsf{D}}(\mathbb{Z}_p)$, *if* $\widehat{\mathcal{H}}'^{,p} = \widehat{g}_0 \widehat{\mathcal{H}}^p \widehat{g}_0^{-1}$, *if* $(\widehat{\mathcal{H}}'_p)^{\mathrm{ord}} = (\widehat{g}_p \widehat{\mathcal{H}}_p \widehat{g}_p^{-1})^{\mathrm{ord}}$ *(cf.* (7.1.1.24)*), and if* $\widehat{\Sigma}'$ *is* $\widehat{g}$*-induced by* $\widehat{\Sigma}$*, then* $(r_{\kappa'} = r_\kappa$ *and) the induced morphisms* (7.1.4.13) *and* (7.1.4.14) *are isomorphisms. (These conditions are true, in particular, when* $\widehat{g} = 1$ *and when* $\widehat{\mathcal{H}} = \widehat{\mathcal{H}}^p \widehat{\mathcal{H}}_p$ *and* $\widehat{\mathcal{H}}' = \widehat{\mathcal{H}}'^{,p} \widehat{\mathcal{H}}'_p$ *satisfy* $\widehat{\mathcal{H}}'^{,p} = \widehat{\mathcal{H}}^p$ *and* $(\widehat{\mathcal{H}}'_p)^{\mathrm{ord}} = \widehat{\mathcal{H}}_p^{\mathrm{ord}}$*; cf. see the remark at the end of Corollary* 3.4.4.4.*)*

(i) *(elements of* U_p *type.) Suppose* $\widehat{g}_0 = 1$ *and* $\widehat{g}_p$ *is the image of an element* $\widetilde{g}_p$ *of* U_p *type in* $\widetilde{\mathrm{P}}^{\mathrm{ord},\prime}_{\widetilde{\mathsf{Z}},\widetilde{\mathsf{D}}}(\mathbb{Q}_p)$ *under the canonical morphism* $\widetilde{\mathrm{P}}^{\mathrm{ord},\prime}_{\widetilde{\mathsf{Z}},\widetilde{\mathsf{D}}}(\mathbb{Q}_p) \to \widehat{\mathrm{P}}^{\mathrm{ord}}_{\mathsf{D}}(\mathbb{Q}_p)$ *(cf. Definition* 7.1.1.22*). Then* $g_{h,0} = 1$*,* $g_{h,p}$ *is an element of* U_p *type, and* $\widehat{\Sigma}^{\mathrm{ord},\prime}$ *is also a* 1*-refinement of* $\widehat{\Sigma}^{\mathrm{ord}}$*. The morphism*

$$[\vec{\widehat{g}}]^{\mathrm{ord}} : \vec{\mathsf{N}}^{\mathrm{ord}}_{\kappa'} \underset{\mathbb{Z}}{\otimes} \mathbb{F}_p \to \vec{\mathsf{N}}^{\mathrm{ord}}_{\kappa} \underset{\mathbb{Z}}{\otimes} \mathbb{F}_p \tag{7.1.4.16}$$

induced by (7.1.4.10) *is finite flat and coincides with the composition of the (finite flat) absolute Frobenius morphism*

$$\mathrm{F}_{\vec{\mathsf{N}}^{\mathrm{ord}}_{\kappa'} \underset{\mathbb{Z}}{\otimes} \mathbb{F}_p} : \vec{\mathsf{N}}^{\mathrm{ord}}_{\kappa'} \underset{\mathbb{Z}}{\otimes} \mathbb{F}_p \to \vec{\mathsf{N}}^{\mathrm{ord}}_{\kappa'} \underset{\mathbb{Z}}{\otimes} \mathbb{F}_p$$

with the canonical finite flat morphism

$$[\vec{1}]^{\mathrm{ord}} : \vec{\mathsf{N}}^{\mathrm{ord}}_{\kappa'} \underset{\mathbb{Z}}{\otimes} \mathbb{F}_p \to \vec{\mathsf{N}}^{\mathrm{ord}}_{\kappa} \underset{\mathbb{Z}}{\otimes} \mathbb{F}_p. \tag{7.1.4.17}$$

On the other hand, the morphism

$$[\vec{\widehat{g}}]^{\mathrm{ord,tor}} : \vec{\mathsf{N}}^{\mathrm{ord,tor}}_{\kappa'} \underset{\mathbb{Z}}{\otimes} \mathbb{F}_p \to \vec{\mathsf{N}}^{\mathrm{ord,tor}}_{\kappa} \underset{\mathbb{Z}}{\otimes} \mathbb{F}_p \tag{7.1.4.18}$$

induced by (7.1.4.11) *is proper and coincides with the composition of the (finite flat) absolute Frobenius morphism*

$$\mathrm{F}_{\vec{\mathsf{N}}^{\mathrm{ord,tor}}_{\kappa'} \underset{\mathbb{Z}}{\otimes} \mathbb{F}_p} : \vec{\mathsf{N}}^{\mathrm{ord,tor}}_{\kappa'} \underset{\mathbb{Z}}{\otimes} \mathbb{F}_p \to \vec{\mathsf{N}}^{\mathrm{ord,tor}}_{\kappa'} \underset{\mathbb{Z}}{\otimes} \mathbb{F}_p$$

with the canonical proper morphism

$$[\vec{1}]^{\mathrm{ord,tor}} : \vec{\mathsf{N}}^{\mathrm{ord,tor}}_{\kappa'} \underset{\mathbb{Z}}{\otimes} \mathbb{F}_p \to \vec{\mathsf{N}}^{\mathrm{ord,tor}}_{\kappa} \underset{\mathbb{Z}}{\otimes} \mathbb{F}_p. \tag{7.1.4.19}$$

Suppose moreover that $\widehat{\Sigma}^{\mathrm{ord},\prime}$ *is* $\widehat{g}$*-induced by* $\widehat{\Sigma}^{\mathrm{ord}}$*. Then* $\widehat{\Sigma}^{\mathrm{ord},\prime}$ *is also* 1*-induced by* $\widehat{\Sigma}^{\mathrm{ord}}$*, and the above morphisms* (7.1.4.18) *and* (7.1.4.19) *are finite flat.*

If $(\widehat{\mathcal{H}}'_p)^{\mathrm{ord}} = \widehat{\mathcal{H}}_p^{\mathrm{ord}}$ *as open compact subgroups of* $\widehat{\mathrm{M}}^{\mathrm{ord}}_{\mathsf{D}}(\mathbb{Z}_p)$ *(see* (7.1.1.24)*), then we can take* $\widehat{\Sigma}^{\mathrm{ord},\prime}$ *to be* g*-induced by* $\widehat{\Sigma}^{\mathrm{ord}}$*, so that* $(r_{\kappa'} = r_\kappa$ *and) the canonical morphism* (7.1.4.19) *is an isomorphism by* (4h)*, and so that the composition*

$$\vec{\mathsf{N}}^{\mathrm{ord,tor}}_{\kappa} \underset{\mathbb{Z}}{\otimes} \mathbb{F}_p \overset{([\vec{1}]^{\mathrm{ord,tor}})^{-1}}{\overset{\sim}{\to}} \vec{\mathsf{N}}^{\mathrm{ord,tor}}_{\kappa'} \underset{\mathbb{Z}}{\otimes} \mathbb{F}_p \overset{[\vec{\widehat{g}}]^{\mathrm{ord,tor}}}{\to} \vec{\mathsf{N}}^{\mathrm{ord,tor}}_{\kappa} \underset{\mathbb{Z}}{\otimes} \mathbb{F}_p$$

coincides with the (finite flat) absolute Frobenius morphism

$$\mathrm{F}_{\vec{\mathsf{N}}^{\mathrm{ord,tor}}_{\kappa} \underset{\mathbb{Z}}{\otimes} \mathbb{F}_p} : \vec{\mathsf{N}}^{\mathrm{ord,tor}}_{\kappa} \underset{\mathbb{Z}}{\otimes} \mathbb{F}_p \to \vec{\mathsf{N}}^{\mathrm{ord,tor}}_{\kappa} \underset{\mathbb{Z}}{\otimes} \mathbb{F}_p.$$

(j) *Suppose* $\widehat{g}_0 = 1$ *and* $\widehat{g}_p$ *is the image of an element* $\widetilde{g}_p$ *of* U_p *type in* $\widetilde{\mathrm{P}}^{\mathrm{ord},\prime}_{\widetilde{\mathsf{Z}},\widetilde{\mathsf{D}}}(\mathbb{Q}_p)$ *under the canonical morphism* $\widetilde{\mathrm{P}}^{\mathrm{ord},\prime}_{\widetilde{\mathsf{Z}},\widetilde{\mathsf{D}}}(\mathbb{Q}_p) \to \widehat{\mathrm{P}}^{\mathrm{ord}}_{\mathsf{D}}(\mathbb{Q}_p)$ *(cf. Definition* 7.1.1.22*), so that* $g_{h,0} = 1$ *and* $g_{h,p}$ *is an element of* $\mathrm{P}_{\mathsf{D}}(\mathbb{Q}_p)$ *of* U_p *type (cf. Definition* 3.3.6.1*).*
Suppose that $\kappa = (\widehat{\mathcal{H}}, \widehat{\Sigma}^{\mathrm{ord}}) \in \mathbf{K}^{\mathrm{ord},+}_{Q,\mathcal{H},\Sigma^{\mathrm{ord}}}$ *and* $\kappa' = (\widehat{\mathcal{H}}', \widehat{\Sigma}^{\mathrm{ord},\prime}) \in \mathbf{K}^{\mathrm{ord},+}_{Q,\mathcal{H}',\Sigma^{\mathrm{ord},\prime}}$*, that* $\mathrm{depth}_{\mathsf{D}}(\widehat{\mathcal{H}}') - 1 = \mathrm{depth}_{\mathsf{D}}(\widehat{\mathcal{H}}) > 0$ *(see Definition* 7.1.1.2*), that* $(\widehat{\mathcal{H}}'_p)^{\mathrm{ord}} = \widehat{\mathcal{H}}^{\mathrm{ord}}_p$ *as open compact subgroups of* $\widehat{\mathrm{M}}^{\mathrm{ord}}_{\mathsf{D}}(\mathbb{Z}_p)$ *(see* (7.1.1.24)*), that* $\widehat{\Sigma}^{\mathrm{ord},\prime}$ *is* $\widehat{g}$*-induced by* $\widehat{\Sigma}$*, that* $\mathrm{depth}_{\mathsf{D}}(\mathcal{H}') - 1 = \mathrm{depth}_{\mathsf{D}}(\mathcal{H}) > 0$ *(see Definition* 3.2.2.9*), that* $(\mathcal{H}_{\kappa',p})^{\mathrm{ord}} = (\mathcal{H}_{\kappa,p})^{\mathrm{ord}}$ *as open compact subgroups of* $\mathrm{M}^{\mathrm{ord}}_{\mathsf{D}}(\mathbb{Z}_p)$ *(see* (3.3.3.5)*), and that* $\Sigma^{\mathrm{ord},\prime}$ *is* g_h*-induced by* Σ^{ord} *as in Definition* 5.2.2.1.
Then ($r_{\kappa'} = r_\kappa = r_{\mathcal{H}'} = r_{\mathcal{H}}$ *and) the canonical morphism*

$$[\vec{g_h}]^{\mathrm{ord,tor}} : \vec{\mathsf{M}}^{\mathrm{ord,tor}}_{\mathcal{H}',\Sigma^{\mathrm{ord},\prime}} \to \vec{\mathsf{M}}^{\mathrm{ord,tor}}_{\mathcal{H},\Sigma^{\mathrm{ord}}}$$

and (7.1.4.11) *are finite flat surjections, which induce (as in Corollary* 5.2.2.5 *and statement* (4i) *above, by composition with inverses of canonical forgetful isomorphisms) the absolute Frobenius morphisms* $\mathrm{F}_{\vec{\mathsf{M}}^{\mathrm{ord,tor}}_{\mathcal{H},\Sigma^{\mathrm{ord}}} \underset{\mathbb{Z}}{\otimes} \mathbb{F}_p}$ *and* $\mathrm{F}_{\vec{\mathsf{N}}^{\mathrm{ord,tor}}_{\kappa} \underset{\mathbb{Z}}{\otimes} \mathbb{F}_p}$*, respectively; and* (7.1.4.11) *induces a finite surjection*

$$\vec{\mathsf{N}}^{\mathrm{ord,tor}}_{\kappa'} \to ([\vec{g_h}]^{\mathrm{ord,tor}})^* \vec{\mathsf{N}}^{\mathrm{ord,tor}}_{\kappa} := \vec{\mathsf{N}}^{\mathrm{ord,tor}}_{\kappa} \underset{\vec{\mathsf{M}}^{\mathrm{ord,tor}}_{\mathcal{H},\Sigma^{\mathrm{ord}}}}{\times} \vec{\mathsf{M}}^{\mathrm{ord,tor}}_{\mathcal{H}',\Sigma^{\mathrm{ord},\prime}}, \tag{7.1.4.20}$$

which induces the relative Frobenius morphism

$$\mathrm{F}_{(\vec{\mathsf{N}}^{\mathrm{ord,tor}}_{\kappa} \underset{\mathbb{Z}}{\otimes} \mathbb{F}_p)/(\vec{\mathsf{M}}^{\mathrm{ord,tor}}_{\mathcal{H},\Sigma^{\mathrm{ord}}} \underset{\mathbb{Z}}{\otimes} \mathbb{F}_p)} :$$
$$\vec{\mathsf{N}}^{\mathrm{ord,tor}}_{\kappa} \underset{\mathbb{Z}}{\otimes} \mathbb{F}_p \to \mathrm{F}^*_{\vec{\mathsf{M}}^{\mathrm{ord,tor}}_{\mathcal{H},\Sigma^{\mathrm{ord}}} \underset{\mathbb{Z}}{\otimes} \mathbb{F}_p}(\vec{\mathsf{N}}^{\mathrm{ord,tor}}_{\kappa} \underset{\mathbb{Z}}{\otimes} \mathbb{F}_p).$$

(5) *(*$\mathbb{Q}^\times$*-isogenies.) Let* $g_l = (g_{l,0}, g_{l,p})$ *be an element of*

$$\mathrm{GL}_{\mathcal{O} \underset{\mathbb{Z}}{\otimes} \mathbb{A}^{\infty,p}}(Q \underset{\mathbb{Z}}{\otimes} \mathbb{A}^{\infty,p}) \times \mathrm{GL}_{\mathcal{O} \underset{\mathbb{Z}}{\otimes} \mathbb{Z}_p}(Q \underset{\mathbb{Z}}{\otimes} \mathbb{Q}_p) = \mathrm{GL}_{\mathcal{O} \underset{\mathbb{Z}}{\otimes} \mathbb{A}^\infty}(Q \underset{\mathbb{Z}}{\otimes} \mathbb{A}^\infty).$$

Then the submodule $g_l(Q \underset{\mathbb{Z}}{\otimes} \hat{\mathbb{Z}})$ *in* $Q \underset{\mathbb{Z}}{\otimes} \mathbb{A}^\infty$ *determines a unique* $\mathcal{O}$*-lattice* Q' *(up to isomorphism), together with a unique choice of an isomorphism*

$$[g_l]_Q : Q \underset{\mathbb{Z}}{\otimes} \mathbb{Q} \xrightarrow{\sim} Q' \underset{\mathbb{Z}}{\otimes} \mathbb{Q},$$

inducing an isomorphism $Q \underset{\mathbb{Z}}{\otimes} \mathbb{A}^\infty \xrightarrow{\sim} Q' \underset{\mathbb{Z}}{\otimes} \mathbb{A}^\infty$ *matching* $g_l(Q \underset{\mathbb{Z}}{\otimes} \hat{\mathbb{Z}})$ *with* $Q' \underset{\mathbb{Z}}{\otimes} \hat{\mathbb{Z}}$ *(and in particular* $g_l(Q \underset{\mathbb{Z}}{\otimes} \mathbb{Z}_p) = Q \underset{\mathbb{Z}}{\otimes} \mathbb{Z}_p$ *with* $Q' \underset{\mathbb{Z}}{\otimes} \mathbb{Z}_p$ *if* $g_{l,p} \in \mathrm{GL}_{\mathcal{O} \underset{\mathbb{Z}}{\otimes} \mathbb{Z}_p}(Q \underset{\mathbb{Z}}{\otimes} \mathbb{Z}_p)$*), and inducing a canonical* $\mathbb{Q}^\times$*-isogeny*

$$[g_l]^*_Q : \underline{\mathrm{Hom}}_{\mathcal{O}}(Q', G_{\vec{\mathsf{M}}^{\mathrm{ord}}_{\mathcal{H}}})^\circ \to \underline{\mathrm{Hom}}_{\mathcal{O}}(Q, G_{\vec{\mathsf{M}}^{\mathrm{ord}}_{\mathcal{H}}})^\circ$$

defined by $[g_l]_Q$. *Consider the sets* $\mathbf{K}^{\mathrm{ord}}_{Q',\mathcal{H}} \subset \mathbf{K}^{\mathrm{ord},+}_{Q',\mathcal{H}} \subset \mathbf{K}^{\mathrm{ord},++}_{Q',\mathcal{H}}$ *and* $\mathbf{K}^{\mathrm{ord}}_{Q',\mathcal{H},\Sigma^{\mathrm{ord}}} \subset \mathbf{K}^{\mathrm{ord},+}_{Q',\mathcal{H},\Sigma^{\mathrm{ord}}} \subset \mathbf{K}^{\mathrm{ord},++}_{Q',\mathcal{H},\Sigma^{\mathrm{ord}}}$ *as in Definitions* 7.1.1.11 *and* 7.1.1.19 (*with* Q *replaced with* Q'), *with compatible directed partial orders, parameterizing generalized ordinary Kuga families and their compactifications with properties as in* (1), (2), *and* (3) *above. The sets* $\mathbf{K}^{\mathrm{ord},++}_{Q,\mathcal{H}}$ *etc and* $\mathbf{K}^{\mathrm{ord},++}_{Q',\mathcal{H}}$ *etc* (*and the objects they parameterize*) *satisfy the compatibility with* g_l *in the sense that the following are true:*

(a) *For each* $\kappa = (\widehat{\mathcal{H}}, \widehat{\Sigma}^{\mathrm{ord}}) \in \mathbf{K}^{\mathrm{ord},++}_{Q,\mathcal{H}}$ (*resp.* $\mathbf{K}^{\mathrm{ord},+}_{Q,\mathcal{H}}$, *resp.* $\mathbf{K}^{\mathrm{ord}}_{Q,\mathcal{H}}$), *there is an element* $\kappa' = (\widehat{\mathcal{H}}', \widehat{\Sigma}^{\mathrm{ord},\prime}) \in \mathbf{K}^{\mathrm{ord},++}_{Q',\mathcal{H},\Sigma^{\mathrm{ord}}}$ (*resp.* $\mathbf{K}^{\mathrm{ord},+}_{Q',\mathcal{H},\Sigma^{\mathrm{ord}}}$, *resp.* $\mathbf{K}^{\mathrm{ord}}_{Q',\mathcal{H},\Sigma^{\mathrm{ord}}}$) *such that* $\mathcal{H}_{\kappa'} = \widehat{\mathcal{H}}'_{\mathrm{G}} \subset \mathcal{H}_\kappa = \widehat{\mathcal{H}}_{\mathrm{G}}$, *such that the* $\mathbb{Q}^\times$*-isogeny*

$$[\vec{g_l}]^{*,\mathrm{ord},\mathrm{grp}}_{\kappa',\kappa} := \kappa^{\mathrm{isog}} \circ [g_l]^*_Q \circ ((\kappa')^{\mathrm{isog}})^{-1} : \vec{\mathsf{N}}^{\mathrm{ord},\mathrm{grp}}_{\kappa'} \to \vec{\mathsf{N}}^{\mathrm{ord},\mathrm{grp}}_{\kappa} \underset{\vec{\mathsf{M}}^{\mathrm{ord}}_{\mathcal{H}_\kappa}}{\times} \vec{\mathsf{M}}^{\mathrm{ord}}_{\mathcal{H}_{\kappa'}}$$

is an **isogeny** (*not just a quasi-isogeny*), *and such that there is a* (*necessarily unique*) *quasi-finite flat surjection*

$$[\vec{g_l}]^{*,\mathrm{ord}}_{\kappa',\kappa} : \vec{\mathsf{N}}^{\mathrm{ord}}_{\kappa'} \to \vec{\mathsf{N}}^{\mathrm{ord}}_{\kappa} \tag{7.1.4.21}$$

inducing a finite flat surjection $\vec{\mathsf{N}}^{\mathrm{ord}}_{\kappa'} \to \vec{\mathsf{N}}^{\mathrm{ord}}_{\kappa} \underset{\vec{\mathsf{M}}^{\mathrm{ord}}_{\mathcal{H}_\kappa}}{\times} \vec{\mathsf{M}}^{\mathrm{ord}}_{\mathcal{H}_{\kappa'}}$ *of abelian scheme torsors equivariant with the isogeny* $[\vec{g_l}]^{*,\mathrm{ord},\mathrm{grp}}_{\kappa',\kappa}$. *The characteristic zero pullback* $[\vec{g_l}]^{*,\mathrm{ord}}_{\kappa',\kappa} \underset{\mathbb{Z}}{\otimes} \mathbb{Q}$ *is finite étale.*

(b) *For each* $\kappa = (\widehat{\mathcal{H}}, \widehat{\Sigma}^{\mathrm{ord}})$ *as in* (5a), *there is an element* $\kappa' = (\widehat{\mathcal{H}}', \widehat{\Sigma}^{\mathrm{ord},\prime}) \in \mathbf{K}^{\mathrm{ord},++}_{Q',\mathcal{H}}$ (*resp.* $\mathbf{K}^{\mathrm{ord},+}_{Q',\mathcal{H}}$, *resp.* $\mathbf{K}^{\mathrm{ord}}_{Q',\mathcal{H}}$) *such that* $[\vec{g_l}]^{*,\mathrm{ord}}_{\kappa',\kappa}$ *is defined as in* (5a) (*see* (7.1.4.21)) *and extends to a* (*necessarily unique*) *surjection*

$$[\vec{g_l}]^{*,\mathrm{ord},\mathrm{tor}}_{\kappa',\kappa} : \vec{\mathsf{N}}^{\mathrm{ord},\mathrm{tor}}_{\kappa'} \to \vec{\mathsf{N}}^{\mathrm{ord},\mathrm{tor}}_{\kappa}, \tag{7.1.4.22}$$

such that

$$R^i([\vec{g_l}]^{*,\mathrm{ord},\mathrm{tor}}_{\kappa',\kappa})_* \mathscr{O}_{\vec{\mathsf{N}}^{\mathrm{ord},\mathrm{tor}}_{\kappa'}} = 0 \tag{7.1.4.23}$$

for all $i > 0$. *The characteristic zero pullback* $[\vec{g_l}]^{*,\mathrm{ord},\mathrm{tor}}_{\kappa',\kappa} \underset{\mathbb{Z}}{\otimes} \mathbb{Q}$ *is proper log étale.*

If $\kappa \in \mathbf{K}^{\mathrm{ord},++}_{Q,\mathcal{H},\Sigma^{\mathrm{ord}}}$ (*resp.* $\mathbf{K}^{\mathrm{ord},+}_{Q,\mathcal{H},\Sigma^{\mathrm{ord}}}$, *resp.* $\mathbf{K}^{\mathrm{ord}}_{Q,\mathcal{H},\Sigma^{\mathrm{ord}}}$), *then we may assume in the above that* $\kappa' \in \mathbf{K}^{\mathrm{ord},++}_{Q',\mathcal{H},\Sigma^{\mathrm{ord}}}$ (*resp.* $\mathbf{K}^{\mathrm{ord},+}_{Q',\mathcal{H},\Sigma^{\mathrm{ord}}}$, *resp.* $\mathbf{K}^{\mathrm{ord}}_{Q',\mathcal{H},\Sigma^{\mathrm{ord}}}$). *Then* (7.1.4.22) *is compatible with the canonical morphisms* $f^{\mathrm{tor}}_\kappa : \vec{\mathsf{N}}^{\mathrm{ord},\mathrm{tor}}_{\kappa} \to \vec{\mathsf{M}}^{\mathrm{ord},\mathrm{tor}}_{\mathcal{H},r_\kappa}$, $f^{\mathrm{tor}}_{\kappa'} : \vec{\mathsf{N}}^{\mathrm{ord},\mathrm{tor}}_{\kappa'} \to \vec{\mathsf{M}}^{\mathrm{ord},\mathrm{tor}}_{\mathcal{H},r_{\kappa'}}$, *and* $\vec{\mathsf{M}}^{\mathrm{ord},\mathrm{tor}}_{\mathcal{H},r_{\kappa'}} \to \vec{\mathsf{M}}^{\mathrm{ord},\mathrm{tor}}_{\mathcal{H},r_\kappa}$.

(c) *Suppose* $\vec{[g_l]}^{*,\mathrm{ord},\mathrm{tor}}_{\kappa',\kappa}$ *is defined as in* (5b) *for some* $\kappa \in \mathbf{K}^{\mathrm{ord},+}_{Q,\mathcal{H},\Sigma^{\mathrm{ord}}}$ *and* $\kappa' \in \mathbf{K}^{\mathrm{ord},+}_{Q',\mathcal{H},\Sigma^{\mathrm{ord}}}$ *(not just in* $\mathbf{K}^{\mathrm{ord},++}_{Q,\mathcal{H},\Sigma^{\mathrm{ord}}}$ *and* $\mathbf{K}^{\mathrm{ord},++}_{Q',\mathcal{H},\Sigma^{\mathrm{ord}}}$*). Then, for each integer* $i \geq 0$*, there is a canonical morphism*

$$(\vec{[g_l]}^{*,\mathrm{ord},\mathrm{tor}}_{\kappa',\kappa})^* : \underline{H}^i_{\text{log-dR}}(\vec{\mathsf{N}}^{\mathrm{ord},\mathrm{tor}}_{\kappa}/\vec{\mathsf{M}}^{\mathrm{ord},\mathrm{tor}}_{\mathcal{H},\Sigma^{\mathrm{ord}}}) \to \underline{H}^i_{\text{log-dR}}(\vec{\mathsf{N}}^{\mathrm{ord},\mathrm{tor}}_{\kappa'}/\vec{\mathsf{M}}^{\mathrm{ord},\mathrm{tor}}_{\mathcal{H},\Sigma^{\mathrm{ord}}})$$

extending the canonical morphism

$$([g_l]^*_{\kappa',\kappa})^* : \underline{H}^i_{\mathrm{dR}}(\vec{\mathsf{N}}^{\mathrm{ord}}_{\kappa}/\vec{\mathsf{M}}^{\mathrm{ord}}_{\mathcal{H}}) \to \underline{H}^i_{\mathrm{dR}}(\vec{\mathsf{N}}^{\mathrm{ord}}_{\kappa'}/\vec{\mathsf{M}}^{\mathrm{ord}}_{\mathcal{H}})$$

induced by $[g_l]_Q$*, respecting the Hodge filtrations and inducing canonical morphisms*

$$(\vec{[g_l]}^{*,\mathrm{ord},\mathrm{tor}}_{\kappa',\kappa})^* : R^b\vec{f}^{\mathrm{ord},\mathrm{tor}}_*(\overline{\Omega}^a_{\vec{\mathsf{N}}^{\mathrm{ord},\mathrm{tor}}_{\kappa}/\vec{\mathsf{M}}^{\mathrm{ord},\mathrm{tor}}_{\mathcal{H}}}) \to R^b\vec{f}^{\mathrm{ord},\mathrm{tor}}_*(\overline{\Omega}^a_{\vec{\mathsf{N}}^{\mathrm{ord},\mathrm{tor}}_{\kappa'}/\vec{\mathsf{M}}^{\mathrm{ord},\mathrm{tor}}_{\mathcal{H}}})$$

(for integers $a+b=i$*) compatible (under the canonical isomorphisms in* (3) *for* $\vec{\mathsf{N}}^{\mathrm{ord},\mathrm{tor}}_{\kappa}$ *and* $\vec{\mathsf{N}}^{\mathrm{ord},\mathrm{tor}}_{\kappa'}$*) with the canonical morphisms*

$$(\vec{[g_l]}^{*,\mathrm{ord},\mathrm{ext}}_{\kappa',\kappa})^* : \underline{\mathrm{Lie}}^{\vee}_{\vec{\mathsf{N}}^{\mathrm{ord},\mathrm{ext}}_{\kappa}/\vec{\mathsf{M}}^{\mathrm{ord},\mathrm{tor}}_{\mathcal{H},\Sigma^{\mathrm{ord}}}} \to \underline{\mathrm{Lie}}^{\vee}_{\vec{\mathsf{N}}^{\mathrm{ord},\mathrm{ext}}_{\kappa'}/\vec{\mathsf{M}}^{\mathrm{ord},\mathrm{tor}}_{\mathcal{H},\Sigma^{\mathrm{ord}}}}$$

and

$$d\vec{[g_l]}^{*,\mathrm{ord},\mathrm{ext},\vee}_{\kappa',\kappa} : \underline{\mathrm{Lie}}_{\vec{\mathsf{N}}^{\mathrm{ord},\mathrm{ext},\vee}_{\kappa}/\vec{\mathsf{M}}^{\mathrm{ord},\mathrm{tor}}_{\mathcal{H},\Sigma^{\mathrm{ord}}}} \to \underline{\mathrm{Lie}}_{\vec{\mathsf{N}}^{\mathrm{ord},\mathrm{ext},\vee}_{\kappa'}/\vec{\mathsf{M}}^{\mathrm{ord},\mathrm{tor}}_{\mathcal{H},\Sigma^{\mathrm{ord}}}}$$

induced by the morphisms

$$\vec{[g_l]}^{*,\mathrm{ord},\mathrm{ext}}_{\kappa',\kappa} : \vec{\mathsf{N}}^{\mathrm{ord},\mathrm{ext}}_{\kappa'} \to \vec{\mathsf{N}}^{\mathrm{ord},\mathrm{ext}}_{\kappa}$$

and

$$\vec{[g_l]}^{*,\mathrm{ord},\mathrm{ext},\vee}_{\kappa',\kappa} : \vec{\mathsf{N}}^{\mathrm{ord},\mathrm{ext},\vee}_{\kappa} \to \vec{\mathsf{N}}^{\mathrm{ord},\mathrm{ext},\vee}_{\kappa'}$$

respectively, induced by $\vec{[g_l]}^{*,\mathrm{ord},\mathrm{grp}}_{\kappa',\kappa} : \vec{\mathsf{N}}^{\mathrm{ord},\mathrm{grp}}_{\kappa'} \to \vec{\mathsf{N}}^{\mathrm{ord},\mathrm{grp}}_{\kappa}$ *and its dual* $\vec{[g_l]}^{*,\mathrm{ord},\mathrm{grp},\vee}_{\kappa',\kappa} : \vec{\mathsf{N}}^{\mathrm{ord},\mathrm{grp},\vee}_{\kappa'} \to \vec{\mathsf{N}}^{\mathrm{ord},\mathrm{grp},\vee}_{\kappa}$ *over* $\vec{\mathsf{M}}^{\mathrm{ord},\mathrm{tor}}_{\mathcal{H},\Sigma^{\mathrm{ord}}}$*. (In fact, all these morphisms are induced by* $[g_l]^*_Q$*.) The characteristic zero pullbacks of these canonical morphisms are isomorphisms.*

(d) *If the levels* $\mathcal{H}_{\kappa,p}$ *and* $\mathcal{H}_{\kappa',p}$ *at* p *are* **equally deep** *as in Definition* 3.2.2.9*, then the surjection* (7.1.4.21) *is finite, and the surjection* (7.1.4.22) *is proper.*

(e) *If* $g_{l,p} \in \mathrm{GL}_{\mathcal{O}\underset{\mathbb{Z}}{\otimes}\mathbb{Z}_p}(Q \underset{\mathbb{Z}}{\otimes} \mathbb{Z}_p)$*, then* $\vec{[g_l]}^{*,\mathrm{ord},\mathrm{grp}}_{\kappa',\kappa}$ *is a* $\mathbb{Z}^{\times}_{(p)}$*-isogeny, the surjection* (7.1.4.21) *is quasi-finite étale, the surjection* (7.1.4.22) *is log étale, and the canonical morphisms in* (5c) *are all isomorphisms.*

(f) *If we have an element* g'_l *of* $\mathrm{GL}_{\mathcal{O}\underset{\mathbb{Z}}{\otimes}\mathbb{A}^\infty}(Q\underset{\mathbb{Z}}{\otimes}\mathbb{A}^\infty)$ *with a similar setup such that* $\vec{[g'_l]}^{*,\mathrm{ord}}_{\kappa'',\kappa'}$ *and* $\vec{[g'_l]}^{*,\mathrm{ord,tor}}_{\kappa'',\kappa'}$ *are compatibly defined for some* $\kappa''\in\mathbf{K}^{\mathrm{ord},++}_{Q'',\mathcal{H},\Sigma^{\mathrm{ord}}}$, *then* $\vec{[g_lg'_l]}^{*,\mathrm{ord}}_{\kappa'',\kappa}$ *and* $\vec{[g_lg'_l]}^{*,\mathrm{ord,tor}}_{\kappa'',\kappa}$ *are also compatibly defined and satisfy the identities* $\vec{[g_lg'_l]}^{*,\mathrm{ord}}_{\kappa'',\kappa}=\vec{[g_l]}^{*,\mathrm{ord}}_{\kappa',\kappa}\circ\vec{[g'_l]}^{*,\mathrm{ord}}_{\kappa'',\kappa'}$ *and* $\vec{[g_lg'_l]}^{*,\mathrm{ord,tor}}_{\kappa'',\kappa}=\vec{[g_l]}^{*,\mathrm{ord,tor}}_{\kappa',\kappa}\circ\vec{[g'_l]}^{*,\mathrm{ord,tor}}_{\kappa'',\kappa'}$. *If* $\kappa\in\mathbf{K}^{\mathrm{ord},+}_{Q,\mathcal{H},\Sigma^{\mathrm{ord}}}$, $\kappa'\in\mathbf{K}^{\mathrm{ord},+}_{Q',\mathcal{H},\Sigma^{\mathrm{ord}}}$, *and* $\kappa''\in\mathbf{K}^{\mathrm{ord},+}_{Q'',\mathcal{H},\Sigma^{\mathrm{ord}}}$, *we also have* $(\vec{[g_lg'_l]}^{*,\mathrm{ord}}_{\kappa'',\kappa})^*=(\vec{[g'_l]}^{*,\mathrm{ord}}_{\kappa'',\kappa'})^*\circ(\vec{[g_l]}^{*,\mathrm{ord}}_{\kappa',\kappa})^*$ *and* $(\vec{[g_lg'_l]}^{*,\mathrm{ord,tor}}_{\kappa'',\kappa})^*=(\vec{[g'_l]}^{*,\mathrm{ord,tor}}_{\kappa'',\kappa'})^*\circ(\vec{[g_l]}^{*,\mathrm{ord,tor}}_{\kappa',\kappa})^*$.

(g) *The morphism*

$$\vec{[g_l]}^{*,\mathrm{ord}}_{\kappa',\kappa}:\vec{\mathfrak{N}}^{\mathrm{ord}}_{\kappa'}\to\vec{\mathfrak{N}}^{\mathrm{ord}}_{\kappa}\qquad(7.1.4.24)$$

(*cf. Definition* 3.4.4.2) *induced by* (7.1.4.21) *is finite. The morphism*

$$\vec{[g_l]}^{*,\mathrm{ord,tor}}_{\kappa',\kappa}:\vec{\mathfrak{N}}^{\mathrm{ord,tor}}_{\kappa'}\to\vec{\mathfrak{N}}^{\mathrm{ord,tor}}_{\kappa},\qquad(7.1.4.25)$$

induced by (7.1.4.22) *is proper, and there exist choices of* κ', *which can be assumed to satisfy* $\kappa'\succ\kappa''$ *for any given* κ'', *such that* (7.1.4.25) *is finite flat. If* $g_{l,p}\in\mathrm{GL}_{\mathcal{O}\underset{\mathbb{Z}}{\otimes}\mathbb{Z}_p}(Q\underset{\mathbb{Z}}{\otimes}\mathbb{Z}_p)$, *then the morphism*

$$\vec{[g_l]}^{*,\mathrm{ord,tor}}_{\kappa',\kappa,r_{\kappa'}}:\vec{\mathfrak{N}}^{\mathrm{ord,tor}}_{\kappa'}\to\vec{\mathfrak{N}}^{\mathrm{ord,tor}}_{\kappa}\underset{\vec{\mathfrak{S}}_{0,r_\kappa}}{\times}\vec{\mathfrak{S}}_{0,r_{\kappa'}}\qquad(7.1.4.26)$$

induced by (7.1.4.25) *is proper log étale* (*because it is log étale by* (5e)), *and there exists choices of* κ', *which can be assumed to satisfy* $\kappa'\succ\kappa''$ *for any given* κ'', *such that* (7.1.4.15) *is finite étale* (*by* (5e); *cf. [35, IV-3, 8.11.1]*).

(h) *If* $g_l\in\mathrm{GL}_{\mathcal{O}\underset{\mathbb{Z}}{\otimes}\hat{\mathbb{Z}}}(Q\underset{\mathbb{Z}}{\otimes}\hat{\mathbb{Z}})$, *if* $\widehat{\mathcal{H}}'^{,p}=\widehat{\mathcal{H}}^p$, *and if* $(\widehat{\mathcal{H}}'_p)^{\mathrm{ord}}=\widehat{\mathcal{H}}^{\mathrm{ord}}_p$ (*cf.* (7.1.1.24)), *then there exist choices of* κ', *which can be assumed to satisfy* $\kappa'\succ\kappa''$ *for any given* κ'', *such that* ($r_{\kappa'}=r_\kappa$ *and*) *the induced morphisms* (7.1.4.24) *and* (7.1.4.25) *are isomorphisms.* (*These conditions are true, in particular, when* $g_l=1$ *and when* $\widehat{\mathcal{H}}=\widehat{\mathcal{H}}^p\widehat{\mathcal{H}}_p$ *and* $\widehat{\mathcal{H}}'=\widehat{\mathcal{H}}'^{,p}\widehat{\mathcal{H}}'_p$ *satisfy* $\widehat{\mathcal{H}}'^{,p}=\widehat{\mathcal{H}}^p$ *and* $(\widehat{\mathcal{H}}'_p)^{\mathrm{ord}}=\widehat{\mathcal{H}}^{\mathrm{ord}}_p$; *cf. see the remark at the end of Corollary* 3.4.4.4.)

(6) *After pulled back to the characteristic zero fibers, the objects and morphisms in this theorem are canonically compatible with those in Theorem* 1.3.3.15 (*cf. Proposition* 7.1.1.21). (*In particular,* $f^{\mathrm{tor}}_{\kappa',\kappa}\underset{\mathbb{Z}}{\otimes}\mathbb{Q}$, $\vec{[\hat{g}]}^{\mathrm{ord,tor}}\underset{\mathbb{Z}}{\otimes}\mathbb{Q}$, *and* $\vec{[g_l]}^{*,\mathrm{ord,tor}}_{\kappa',\kappa}\underset{\mathbb{Z}}{\otimes}\mathbb{Q}$, *where* $f^{\mathrm{tor}}_{\kappa',\kappa}$ *is as in* (1), $\vec{[\hat{g}]}^{\mathrm{ord,tor}}$ *is as in* (7.1.4.11), *and* $\vec{[g_l]}^{*,\mathrm{ord,tor}}_{\kappa',\kappa}$ *is as in* (7.1.4.22), *are always proper, without the additional conditions on depths of levels at* p.)

With the ingredients we have already provided, the proof is similar to that of [61, Thm. 2.15; see also the errata]. Nevertheless, for the sake of certainly, we will spell out the details in the next few sections. The proof will clarify that, when $\kappa \in \mathbf{K}_{Q,\mathcal{H}}^{\mathrm{ord},+}$, the locally free sheaves such as $\underline{\mathrm{Lie}}_{\vec{\mathsf{N}}_{\kappa}^{\mathrm{ord}}/\vec{\mathsf{M}}_{\mathcal{H}}^{\mathrm{ord}}}$, $\underline{\mathrm{Lie}}^{\vee}_{\vec{\mathsf{N}}_{\kappa}^{\mathrm{ord}}/\vec{\mathsf{M}}_{\mathcal{H}}^{\mathrm{ord}}}$, $\underline{\mathrm{Lie}}_{\vec{\mathsf{N}}_{\kappa}^{\mathrm{ord},\vee}/\vec{\mathsf{M}}_{\mathcal{H}}^{\mathrm{ord}}}$, and $\underline{\mathrm{Lie}}^{\vee}_{\vec{\mathsf{N}}_{\kappa}^{\mathrm{ord},\vee}/\vec{\mathsf{M}}_{\mathcal{H}}^{\mathrm{ord}}}$ over $\vec{\mathsf{M}}_{\mathcal{H}}^{\mathrm{ord}}$, and their extensions $\underline{\mathrm{Lie}}_{\vec{\mathsf{N}}_{\kappa}^{\mathrm{ord,ext}}/\vec{\mathsf{M}}_{\mathcal{H}}^{\mathrm{ord,tor}}}$, $\underline{\mathrm{Lie}}^{\vee}_{\vec{\mathsf{N}}_{\kappa}^{\mathrm{ord,ext}}/\vec{\mathsf{M}}_{\mathcal{H}}^{\mathrm{ord,tor}}}$, $\underline{\mathrm{Lie}}_{\vec{\mathsf{N}}_{\kappa}^{\mathrm{ord,ext},\vee}/\vec{\mathsf{M}}_{\mathcal{H}}^{\mathrm{ord,tor}}}$, and $\underline{\mathrm{Lie}}^{\vee}_{\vec{\mathsf{N}}_{\kappa}^{\mathrm{ord,ext},\vee}/\vec{\mathsf{M}}_{\mathcal{H}}^{\mathrm{ord,tor}}}$ over $\vec{\mathsf{M}}_{\mathcal{H}}^{\mathrm{ord,tor}}$, respectively, are the correct ones to use in the statements.

Remark 7.1.4.27. (Compare with Remarks 1.1.2.1, 1.3.1.4, 1.3.3.33, 3.4.2.8, and 5.2.1.5.) By varying the choices of L and Q inducing the same $L \underset{\mathbb{Z}}{\otimes} \mathbb{Z}_{(p)}$ and $Q \underset{\mathbb{Z}}{\otimes} \mathbb{Z}_{(p)}$, respectively, and hence varying the choices of $\widetilde{L}$ inducing the same $\widetilde{L} \underset{\mathbb{Z}}{\otimes} \mathbb{Z}_{(p)}$, we can (in practice) allow the $\widehat{\mathcal{H}}$ in the parameter $\kappa = (\widehat{\mathcal{H}}, \widehat{\Sigma})$ to be any open compact subgroup of $\widehat{\mathrm{G}}(\mathbb{A}^{\infty})$ of the form $\widehat{\mathcal{H}} = \widehat{\mathcal{H}}^{p}\widehat{\mathcal{H}}_{p}$, where $\widehat{\mathcal{H}}^{p} \subset \widehat{\mathrm{G}}(\mathbb{A}^{\infty,p})$ and $\widehat{\mathcal{H}}_{p}$ is of standard form as in Definition 7.1.1.2. Nevertheless, this can be achieved by varying the lattice Q alone, and hence is already incorporated in (5) of Theorem 7.1.4.1.

7.2 Main Constructions of Compactifications and Morphisms

7.2.1 *Partial Toroidal Boundary Strata*

Let $(\widetilde{L}, \langle \cdot\,, \cdot \rangle^{\sim}, \widetilde{h}_0)$, $(\widetilde{\mathsf{Z}}, \widetilde{\Phi}, \widetilde{\delta})$, *etc* be chosen as in Section 1.2.4, and let $\widetilde{\mathsf{D}}$ be defined by D as in Section 7.1.1. (The choice of D is implicit in the construction of $\vec{\mathsf{M}}_{\mathcal{H}}^{\mathrm{ord}}$ and hence a prerequisite of Theorem 5.2.1.1 and its consequences.) Let $\widetilde{\kappa} = (\widetilde{\mathcal{H}}, \widetilde{\Sigma}^{\mathrm{ord}}, \widetilde{\sigma})$ be any element in the set $\widetilde{\mathbf{K}}_{Q,\mathcal{H}}^{\mathrm{ord},++}$ as in Definition 1.2.4.11, and let $\kappa = [\widetilde{\kappa}] \in \mathbf{K}_{Q,\mathcal{H}}^{\mathrm{ord},++}$ be as in Definition 1.2.4.44.

By Theorem 3.4.2.5 and Proposition 3.4.6.3, the data of $\mathcal{O}$, $(\widetilde{L}, \langle \cdot\,, \cdot \rangle^{\sim}, \widetilde{h}_0)$, $\widetilde{\mathsf{D}}$, and $\widetilde{\mathcal{H}} = \widetilde{\mathcal{H}}^{p}\widetilde{\mathcal{H}}_{p}$ (where $\widetilde{\mathcal{H}}^{p}$ is neat by assumption) define a scheme $\vec{\widetilde{\mathsf{M}}}_{\widetilde{\mathcal{H}}}^{\mathrm{ord}}$ quasi-projective over $\vec{\mathsf{S}}_{0,r_{\widetilde{\mathcal{H}}}} = \vec{\mathsf{S}}_{0,r_{\kappa}}$ (see Definition 2.2.3.3 and Condition 7.1.1.5). Since $\widetilde{\mathcal{H}}^{p}$ is neat and $\widetilde{\Sigma}^{\mathrm{ord}}$ is projective (and smooth), by Theorems 5.2.1.1 and 6.2.3.1, we have a partial toroidal compactification $\vec{\widetilde{\mathsf{M}}}_{\widetilde{\mathcal{H}}}^{\mathrm{ord,tor}} = \vec{\widetilde{\mathsf{M}}}_{\widetilde{\mathcal{H}},\widetilde{\Sigma}^{\mathrm{ord}}}^{\mathrm{ord,tor}}$ which is quasi-projective and smooth over $\mathsf{S}_{0,r_{\widetilde{\mathcal{H}}}}$. Since $\widetilde{\mathcal{H}}$ satisfies Conditions 1.2.4.7 and 7.1.1.5, by construction (see Propositions 4.2.1.29 and 4.2.1.30), we have

$$\vec{\widetilde{\mathsf{M}}}_{\widetilde{\mathcal{H}}}^{\mathrm{ord},\widetilde{\Phi}_{\widetilde{\mathcal{H}}}} \cong \vec{\widetilde{\mathsf{M}}}_{\widetilde{\mathcal{H}}}^{\mathrm{ord},\widetilde{\mathsf{Z}}_{\widetilde{\mathcal{H}}}} \cong \vec{\mathsf{M}}_{\mathcal{H}_{\kappa}}^{\mathrm{ord}}, \tag{7.2.1.1}$$

where $\mathcal{H}_{\kappa} = \widehat{\mathcal{H}}_{\mathrm{G}} = \mathrm{Gr}_{-1}^{\widetilde{\mathsf{Z}}}(\widetilde{\mathcal{H}}_{\widetilde{\mathrm{P}}'_{\widetilde{\mathsf{Z}}}}) = \mathrm{Gr}_{-1}^{\widetilde{\mathsf{Z}}}(\widetilde{\mathcal{H}}_{\widetilde{\mathrm{P}}_{\widetilde{\mathsf{Z}}}})$, and $\vec{\widetilde{C}}^{\mathrm{ord}}_{\widetilde{\Phi}_{\widetilde{\mathcal{H}}},\widetilde{\delta}_{\widetilde{\mathcal{H}}}} \to \vec{\widetilde{\mathsf{M}}}_{\widetilde{\mathcal{H}}}^{\mathrm{ord},\widetilde{\Phi}_{\widetilde{\mathcal{H}}}}$ is an abelian scheme torsor under an abelian scheme $\vec{\widetilde{C}}^{\mathrm{ord,grp}}_{\widetilde{\Phi}_{\widetilde{\mathcal{H}}},\widetilde{\delta}_{\widetilde{\mathcal{H}}}}$ canonically $\mathbb{Q}^{\times}$-isogenous to $\underline{\mathrm{Hom}}_{\mathcal{O}}(Q, G_{\vec{\mathsf{M}}_{\mathcal{H}_{\kappa}}^{\mathrm{ord}}})^{\circ}$ (which is the C in the proof of Proposition 4.2.1.30). If $\widehat{\mathcal{H}}$ satisfies Condition 1.2.4.8, then we have $\mathcal{H}_{\kappa} = \mathcal{H}$ and hence $\vec{\mathsf{M}}_{\mathcal{H}_{\kappa}}^{\mathrm{ord}} = \vec{\mathsf{M}}_{\mathcal{H}}^{\mathrm{ord}}$. If $\widehat{\mathcal{H}}$ also satisfies Condition 1.2.4.9, then $\vec{\widetilde{C}}^{\mathrm{ord}}_{\widetilde{\Phi}_{\widetilde{\mathcal{H}}},\widetilde{\delta}_{\widetilde{\mathcal{H}}}} = \vec{\widetilde{C}}^{\mathrm{ord,grp}}_{\widetilde{\Phi}_{\widetilde{\mathcal{H}}},\widetilde{\delta}_{\widetilde{\mathcal{H}}}} \to \vec{\mathsf{M}}^{\mathrm{ord}}_{\mathcal{H}_\kappa} = \vec{\mathsf{M}}^{\mathrm{ord}}_{\mathcal{H}}$ is an abelian scheme, not just a torsor.

Remark 7.2.1.2. (Compare with Remark 1.3.4.1 and Proposition 4.1.6.1.) The isomorphism (7.2.1.1) means we do not need to consider nontrivial twisted objects $(\widetilde{\varphi}^{\sim}_{-2,\widetilde{\mathcal{H}}}, \widetilde{\varphi}^{\sim}_{0,\widetilde{\mathcal{H}}})$ (resp. $(\widetilde{\varphi}^{\mathrm{ord}}_{-2,\widetilde{\mathcal{H}}}, \widetilde{\varphi}^{\mathrm{ord}}_{0,\widetilde{\mathcal{H}}})$) above $(\widetilde{\varphi}_{-2,\widetilde{\mathcal{H}}}, \widetilde{\varphi}_{0,\widetilde{\mathcal{H}}})$ and $\widetilde{\varphi}_{-1,\widetilde{\mathcal{H}}} = \alpha_{\mathcal{H}_\kappa}$ (resp. $\widetilde{\varphi}^{\mathrm{ord}}_{-1,\widetilde{\mathcal{H}}} = (\alpha_{\mathcal{H}^p_\kappa}, \alpha^{\mathrm{ord}}_{\mathcal{H}_{\kappa,p}})$).

Since $\widetilde{\sigma}$ is a *top-dimensional* nondegenerate rational polyhedral cone in the cone decomposition $\widetilde{\Sigma}_{\widetilde{\Phi}_{\widetilde{\mathcal{H}}}}$ in $\widetilde{\Sigma}^{\mathrm{ord}}$, by (2) of Theorem 5.2.1.1, the locally closed stratum $\vec{\widetilde{\mathsf{Z}}}^{\mathrm{ord}}_{[(\widetilde{\Phi}_{\widetilde{\mathcal{H}}},\widetilde{\delta}_{\widetilde{\mathcal{H}}},\widetilde{\sigma})]}$ (not its closure) is a zero-dimensional torus bundle over the abelian scheme torsor $\vec{\widetilde{C}}^{\mathrm{ord}}_{\widetilde{\Phi}_{\widetilde{\mathcal{H}}},\widetilde{\delta}_{\widetilde{\mathcal{H}}}}$ over $\vec{\mathsf{M}}^{\mathrm{ord}}_{\mathcal{H}_\kappa}$. In other words, $\vec{\widetilde{\mathsf{Z}}}^{\mathrm{ord}}_{[(\widetilde{\Phi}_{\widetilde{\mathcal{H}}},\widetilde{\delta}_{\widetilde{\mathcal{H}}},\widetilde{\sigma})]}$ is canonically *isomorphic* to $\vec{\widetilde{C}}^{\mathrm{ord}}_{\widetilde{\Phi}_{\widetilde{\mathcal{H}}},\widetilde{\delta}_{\widetilde{\mathcal{H}}}}$. Let us define $\vec{\mathsf{N}}^{\mathrm{ord}}_{\widetilde{\kappa}}$ to be this stratum $\vec{\widetilde{\mathsf{Z}}}^{\mathrm{ord}}_{[(\widetilde{\Phi}_{\widetilde{\mathcal{H}}},\widetilde{\delta}_{\widetilde{\mathcal{H}}},\widetilde{\sigma})]}$, and denote the canonical morphism $\vec{\mathsf{N}}^{\mathrm{ord}}_{\widetilde{\kappa}} \to \vec{\mathsf{M}}^{\mathrm{ord}}_{\mathcal{H},r_\kappa} = \vec{\mathsf{M}}^{\mathrm{ord}}_{\mathcal{H}} \underset{\vec{\mathsf{S}}_{0,r_{\mathcal{H}}}}{\times} \vec{\mathsf{S}}_{0,r_\kappa}$ by $f_{\widetilde{\kappa}}$ (which is the composition of the canonical morphisms $\vec{\mathsf{N}}^{\mathrm{ord}}_{\widetilde{\kappa}} \to \vec{\mathsf{M}}^{\mathrm{ord}}_{\mathcal{H}_\kappa}$ and $\vec{\mathsf{M}}^{\mathrm{ord}}_{\mathcal{H}_\kappa} \to \vec{\mathsf{M}}^{\mathrm{ord}}_{\mathcal{H},r_\kappa}$). Let us denote the canonical $\mathbb{Q}^\times$-isogeny $\underline{\mathrm{Hom}}_{\mathcal{O}}(Q, G_{\vec{\mathsf{M}}^{\mathrm{ord}}_{\mathcal{H}_\kappa}})^\circ \to \vec{\mathsf{N}}^{\mathrm{ord,grp}}_{\widetilde{\kappa}} := \vec{\widetilde{C}}^{\mathrm{ord,grp}}_{\widetilde{\Phi}_{\widetilde{\mathcal{H}}},\widetilde{\delta}_{\widetilde{\mathcal{H}}}}$ by $\widetilde{\kappa}^{\mathrm{isog}}$. Note that $\vec{\mathsf{N}}^{\mathrm{ord}}_{\widetilde{\kappa}} = \vec{\widetilde{\mathsf{Z}}}^{\mathrm{ord}}_{[(\widetilde{\Phi}_{\widetilde{\mathcal{H}}},\widetilde{\delta}_{\widetilde{\mathcal{H}}},\widetilde{\sigma})]}$ is canonically isomorphic to $\vec{\widetilde{C}}^{\mathrm{ord}}_{\widetilde{\Phi}_{\widetilde{\mathcal{H}}},\widetilde{\delta}_{\widetilde{\mathcal{H}}}}$ for every $\widetilde{\Sigma}^{\mathrm{ord}}$ and every top-dimensional cone $\widetilde{\sigma}$ in $\widetilde{\Sigma}_{\widetilde{\Phi}_{\widetilde{\mathcal{H}}}}$.

Lemma 7.2.1.3. (*Compare with Lemma* 1.3.4.2.) *The abelian scheme torsor* $\vec{\widetilde{C}}^{\mathrm{ord}}_{\widetilde{\Phi}_{\widetilde{\mathcal{H}}},\widetilde{\delta}_{\widetilde{\mathcal{H}}}}$ (*see* (5) *of Theorem* 6.2.1.1 *and Definition* 1.2.1.15) *and the canonical isogeny* $\underline{\mathrm{Hom}}_{\mathcal{O}}(Q, G_{\mathsf{M}_{\mathcal{H}}})^\circ \to \vec{\widetilde{C}}^{\mathrm{ord,grp}}_{\widetilde{\Phi}_{\widetilde{\mathcal{H}}},\widetilde{\delta}_{\widetilde{\mathcal{H}}}}$ *of abelian schemes over* $\vec{\widetilde{\mathsf{M}}}^{\mathrm{ord},\widetilde{\Phi}_{\widetilde{\mathcal{H}}}}_{\widetilde{\mathcal{H}}} \cong \vec{\mathsf{M}}^{\mathrm{ord}}_{\mathcal{H}_\kappa}$ *depend* (*up to canonical isomorphism*) *only on the open compact subgroup* $\widehat{\mathcal{H}} = \widetilde{\mathcal{H}}_{\widehat{\mathrm{G}}}$ *of* $\widehat{\mathrm{G}}(\hat{\mathbb{Z}})$ (*see Definitions* 1.2.4.3 *and* 1.2.4.4) *determined by* $\widetilde{\mathcal{H}}$. *Moreover, if* $\widetilde{\mathcal{H}}'$ *is any open compact subgroup of* $\widetilde{\mathrm{G}}(\hat{\mathbb{Z}})$ *still satisfying Conditions* 1.2.4.7 *and* 7.1.1.4 *such that* $\widetilde{\mathcal{H}}'_{\widehat{\mathrm{G}}} = \widehat{\mathcal{H}}_{\mathrm{G}} \ltimes \widehat{\mathcal{H}}_{\widehat{\mathrm{U}}}$ *under the isomorphism* $\widehat{\mathrm{G}}(\hat{\mathbb{Z}}) \cong \mathrm{G}(\hat{\mathbb{Z}}) \ltimes \widehat{\mathrm{U}}(\hat{\mathbb{Z}})$ *induced by the splitting* $\widetilde{\delta}$ (*cf. Condition* 1.2.4.9), *then we have* $\vec{\widetilde{C}}^{\mathrm{ord}}_{\widetilde{\Phi}_{\widetilde{\mathcal{H}}'},\widetilde{\delta}_{\widetilde{\mathcal{H}}'}} = \vec{\widetilde{C}}^{\mathrm{ord,grp}}_{\widetilde{\Phi}_{\widetilde{\mathcal{H}}'},\widetilde{\delta}_{\widetilde{\mathcal{H}}'}} \cong \vec{\widetilde{C}}^{\mathrm{ord,grp}}_{\widetilde{\Phi}_{\widetilde{\mathcal{H}}},\widetilde{\delta}_{\widetilde{\mathcal{H}}}}$.

Proof. This follows from the very construction of $\vec{\widetilde{C}}^{\mathrm{ord}}_{\widetilde{\Phi}_{\widetilde{\mathcal{H}}},\widetilde{\delta}_{\widetilde{\mathcal{H}}}}$ (see the proof of Proposition 4.2.1.30), which is (up to canonical isomorphism) insensitive to replacing $\widetilde{\mathcal{H}}$ with an open compact subgroup still satisfying Conditions 1.2.4.7 and 7.1.1.4 that defines the same $\widehat{\mathcal{H}} = \widetilde{\mathcal{H}}_{\widehat{\mathrm{G}}}$. □

Consequently, $\vec{\mathsf{N}}^{\mathrm{ord}}_{\widetilde{\kappa}}$ and $\widetilde{\kappa}^{\mathrm{isog}}$ depend (up to canonical isomorphism) only on the open compact subgroup $\widehat{\mathcal{H}}$ of $\widehat{\mathrm{G}}(\hat{\mathbb{Z}})$ determined by $\widetilde{\mathcal{H}}$ (see Definitions 1.2.4.3 and 1.2.4.4).

Let us take $\vec{\mathsf{N}}^{\mathrm{ord,tor}}_{\widetilde{\kappa}}$ to be the closure of $\vec{\widetilde{\mathsf{Z}}}^{\mathrm{ord}}_{[(\widetilde{\Phi}_{\widetilde{\mathcal{H}}},\widetilde{\delta}_{\widetilde{\mathcal{H}}},\widetilde{\sigma})]}$ in $\vec{\widetilde{\mathsf{M}}}^{\mathrm{ord,tor}}_{\widetilde{\mathcal{H}},\widetilde{\Sigma}^{\mathrm{ord}}}$. Then we obtain the canonical open fiberwise dense immersion $\widetilde{\kappa}^{\mathrm{tor}} : \vec{\mathsf{N}}^{\mathrm{ord}}_{\widetilde{\kappa}} \hookrightarrow \vec{\mathsf{N}}^{\mathrm{ord,tor}}_{\widetilde{\kappa}}$. Certainly, $\vec{\mathsf{N}}^{\mathrm{ord,tor}}_{\widetilde{\kappa}}$ depends not only on $\widehat{\mathcal{H}}$ but also on the choices of $\widetilde{\Sigma}_{\widetilde{\Phi}_{\widetilde{\mathcal{H}}}}$ and $\widetilde{\sigma}$.

Lemma 7.2.1.4. (*Compare with Lemma* 1.3.4.3.) *Under the assumption that $\widetilde{\mathcal{H}}^p$ is neat (and hence $\widetilde{\mathcal{H}} = \widetilde{\mathcal{H}}^p\widetilde{\mathcal{H}}_p$ is neat), the closure of every stratum in $\vec{\widetilde{\mathsf{M}}}^{\mathrm{ord,tor}}_{\widetilde{\mathcal{H}},\widetilde{\Sigma}^{\mathrm{ord}}}$ has no self-intersection.*

Proof. The same argument of the proof of Lemma 1.3.4.3 (or rather of [61, Lem. 3.1]) works here. □

Corollary 7.2.1.5. (*Compare with Corollary* 1.3.4.4.) *For each $\widetilde{\kappa} = (\widetilde{\mathcal{H}}, \widetilde{\Sigma}^{\mathrm{ord}}, \widetilde{\sigma}) \in \widetilde{\mathbf{K}}^{++}_{Q,\mathcal{H}}$, the closure $\vec{\mathsf{N}}^{\mathrm{ord,tor}}_{\widetilde{\kappa}}$ of $\vec{\mathsf{N}}^{\mathrm{ord}}_{\widetilde{\kappa}} = \vec{\widetilde{\mathsf{Z}}}^{\mathrm{ord}}_{[(\widetilde{\Phi}_{\widetilde{\mathcal{H}}},\widetilde{\delta}_{\widetilde{\mathcal{H}}},\widetilde{\sigma})]}$ in $\vec{\widetilde{\mathsf{M}}}^{\mathrm{ord,tor}}_{\widetilde{\mathcal{H}},\widetilde{\Sigma}^{\mathrm{ord}}}$ is quasi-projective and smooth over $\vec{\mathsf{S}}_{0,r_\kappa}$, and the complement of $\vec{\mathsf{N}}^{\mathrm{ord}}_{\widetilde{\kappa}}$ in $\vec{\mathsf{N}}^{\mathrm{ord,tor}}_{\widetilde{\kappa}}$ (with its reduced structure) is a relative Cartier divisor with simple normal crossings.*

Proof. Combine Lemma 7.2.1.4, (3) of Theorem 5.2.1.1, and Theorem 6.2.3.1. □

The stratification of $\vec{\widetilde{\mathsf{M}}}^{\mathrm{ord,tor}}_{\widetilde{\mathcal{H}},\widetilde{\Sigma}^{\mathrm{ord}}}$ induces a stratification of $\vec{\mathsf{N}}^{\mathrm{ord,tor}}_{\kappa}$. By (2) of Theorem 5.2.1.1, the strata of $\vec{\mathsf{N}}^{\mathrm{ord,tor}}_{\kappa}$ are parameterized by equivalence classes $[(\breve{\Phi}_{\widetilde{\mathcal{H}}}, \breve{\delta}_{\widetilde{\mathcal{H}}}, \breve{\tau})]$ having $[(\widetilde{\Phi}_{\widetilde{\mathcal{H}}}, \widetilde{\delta}_{\widetilde{\mathcal{H}}}, \widetilde{\sigma})]$ as a face (as in Definition 1.2.2.19), which have been described in Sections 1.2.4 (following Definition 1.2.4.11) and 7.1.1 (following Definition 7.1.1.7), such that $[(\breve{\mathsf{Z}}_{\widetilde{\mathcal{H}}}, \breve{\Phi}_{\widetilde{\mathcal{H}}}, \breve{\delta}_{\widetilde{\mathcal{H}}})]$ is an *ordinary cusp label* (as in Definition 3.2.3.8). With $(\breve{\mathsf{Z}}, \breve{\Phi}, \breve{\delta})$ *etc* chosen as in Section 1.2.4, and with $\breve{\sigma}$ being the image of $\widetilde{\sigma} \subset \mathbf{P}^+_{\widetilde{\Phi}_{\widetilde{\mathcal{H}}}}$ under the first morphism in (1.2.4.20), by Corollary 1.2.4.26, we may take $\breve{\tau} \in \widetilde{\Sigma}^+_{\breve{\Phi}_{\widetilde{\mathcal{H}}},\breve{\sigma}}$ (see Definition 1.2.4.21) having $\breve{\sigma}$ as a face, whose $\Gamma_{\breve{\Phi}_{\widetilde{\mathcal{H}}}}$-orbit is well defined, such that $(\breve{\Phi}_{\widetilde{\mathcal{H}}}, \breve{\delta}_{\widetilde{\mathcal{H}}}, \breve{\tau})$ is a representative of some $[(\breve{\Phi}_{\widetilde{\mathcal{H}}}, \breve{\delta}_{\widetilde{\mathcal{H}}}, \breve{\tau})]$ having $[(\widetilde{\Phi}_{\widetilde{\mathcal{H}}}, \widetilde{\delta}_{\widetilde{\mathcal{H}}}, \widetilde{\sigma})]$ as a face.

By construction (see Propositions 4.2.1.37 and 4.2.1.46, (4.2.2.1), (4.2.2.2), and (4.2.2.3)), the scheme

$$\vec{\widetilde{\Xi}}^{\mathrm{ord}}_{\breve{\Phi}_{\widetilde{\mathcal{H}}},\breve{\delta}_{\widetilde{\mathcal{H}}}} \cong \underline{\mathrm{Spec}}_{\mathscr{O}_{\vec{C}^{\mathrm{ord}}_{\breve{\Phi}_{\widetilde{\mathcal{H}}},\breve{\delta}_{\widetilde{\mathcal{H}}}}}} \Big(\bigoplus_{\breve{\ell} \in \mathbf{S}_{\breve{\Phi}_{\widetilde{\mathcal{H}}}}} \vec{\widetilde{\Psi}}^{\mathrm{ord}}_{\breve{\Phi}_{\widetilde{\mathcal{H}}},\breve{\delta}_{\widetilde{\mathcal{H}}}}(\breve{\ell}) \Big) \tag{7.2.1.6}$$

is a torsor over $\vec{C}^{\mathrm{ord}}_{\breve{\Phi}_{\widetilde{\mathcal{H}}},\breve{\delta}_{\widetilde{\mathcal{H}}}}$ under (the pullback of) the split torus $E_{\breve{\Phi}_{\widetilde{\mathcal{H}}}} = \underline{\mathrm{Hom}}_{\mathbb{Z}}(\mathbf{S}_{\breve{\Phi}_{\widetilde{\mathcal{H}}}}, \mathbf{G}_{\mathrm{m}})$, where $\vec{\widetilde{\Psi}}^{\mathrm{ord}}_{\breve{\Phi}_{\widetilde{\mathcal{H}}},\breve{\delta}_{\widetilde{\mathcal{H}}}}(\breve{\ell})$ is the subsheaf of $\mathscr{O}_{\vec{\widetilde{\Xi}}^{\mathrm{ord}}_{\breve{\Phi}_{\widetilde{\mathcal{H}}},\breve{\delta}_{\widetilde{\mathcal{H}}}}}$ (considered as an $\mathscr{O}_{\vec{C}^{\mathrm{ord}}_{\breve{\Phi}_{\widetilde{\mathcal{H}}},\breve{\delta}_{\widetilde{\mathcal{H}}}}}$-algebra, by abuse of language) on which $E_{\breve{\Phi}_{\widetilde{\mathcal{H}}}}$ acts by the character $\breve{\ell}$, with its $\breve{\tau}$-stratum

$$\vec{\widetilde{\Xi}}^{\mathrm{ord}}_{\breve{\Phi}_{\widetilde{\mathcal{H}}},\breve{\delta}_{\widetilde{\mathcal{H}}},\breve{\tau}} \cong \underline{\mathrm{Spec}}_{\mathscr{O}_{\vec{C}^{\mathrm{ord}}_{\breve{\Phi}_{\widetilde{\mathcal{H}}},\breve{\delta}_{\widetilde{\mathcal{H}}}}}} \Big(\bigoplus_{\breve{\ell} \in \breve{\tau}^{\perp}} \vec{\widetilde{\Psi}}^{\mathrm{ord}}_{\breve{\Phi}_{\widetilde{\mathcal{H}}},\breve{\delta}_{\widetilde{\mathcal{H}}}}(\breve{\ell}) \Big) \tag{7.2.1.7}$$

defined by the sheaf of ideals $\widetilde{\mathscr{I}}^{\mathrm{ord}}_{\breve{\tau}} = \underset{\breve{\ell}\in\breve{\tau}_0^\vee}{\oplus} \vec{\widetilde{\Psi}}^{\mathrm{ord}}_{\breve{\Phi}_{\widetilde{\mathcal{H}}},\breve{\delta}_{\widetilde{\mathcal{H}}}}(\breve{\ell})$, together with the affine toroidal embedding

$$\vec{\widetilde{\Xi}}^{\mathrm{ord}}_{\breve{\Phi}_{\widetilde{\mathcal{H}}},\breve{\delta}_{\widetilde{\mathcal{H}}}} \hookrightarrow \vec{\widetilde{\Xi}}^{\mathrm{ord}}_{\breve{\Phi}_{\widetilde{\mathcal{H}}},\breve{\delta}_{\widetilde{\mathcal{H}}}}(\breve{\tau}) \cong \underline{\mathrm{Spec}}_{\mathscr{O}_{\vec{\widetilde{C}}^{\mathrm{ord}}_{\breve{\Phi}_{\widetilde{\mathcal{H}}},\breve{\delta}_{\widetilde{\mathcal{H}}}}}} \Big(\underset{\breve{\ell}\in\breve{\tau}^\vee}{\oplus} \vec{\widetilde{\Psi}}^{\mathrm{ord}}_{\breve{\Phi}_{\widetilde{\mathcal{H}}},\breve{\delta}_{\widetilde{\mathcal{H}}}}(\breve{\ell})\Big) \tag{7.2.1.8}$$

along $\breve{\tau}$. (All the schemes are relatively affine over $\vec{\widetilde{C}}^{\mathrm{ord}}_{\breve{\Phi}_{\widetilde{\mathcal{H}}},\breve{\delta}_{\widetilde{\mathcal{H}}}}$, and all the morphisms are the canonical ones dual to the obvious morphisms between $\mathscr{O}_{\vec{\widetilde{C}}^{\mathrm{ord}}_{\breve{\Phi}_{\widetilde{\mathcal{H}}},\breve{\delta}_{\widetilde{\mathcal{H}}}}}$-algebras.) The closure $\vec{\widetilde{\Xi}}^{\mathrm{ord}}_{\breve{\Phi}_{\widetilde{\mathcal{H}}},\breve{\delta}_{\widetilde{\mathcal{H}}},\breve{\sigma}}(\breve{\tau})$ of the $\breve{\sigma}$-stratum on $\vec{\widetilde{\Xi}}^{\mathrm{ord}}_{\breve{\Phi}_{\widetilde{\mathcal{H}}},\breve{\delta}_{\widetilde{\mathcal{H}}}}(\breve{\tau})$ is defined by the sheaf of ideals $\underset{\breve{\ell}\in\breve{\sigma}_0^\vee\cap\breve{\tau}^\vee}{\oplus} \vec{\widetilde{\Psi}}^{\mathrm{ord}}_{\breve{\Phi}_{\widetilde{\mathcal{H}}},\breve{\delta}_{\widetilde{\mathcal{H}}}}(\breve{\ell})$, and hence we have a canonical isomorphism

$$\vec{\widetilde{\Xi}}^{\mathrm{ord}}_{\breve{\Phi}_{\widetilde{\mathcal{H}}},\breve{\delta}_{\widetilde{\mathcal{H}}},\breve{\sigma}}(\breve{\tau}) \cong \underline{\mathrm{Spec}}_{\mathscr{O}_{\vec{\widetilde{C}}^{\mathrm{ord}}_{\breve{\Phi}_{\widetilde{\mathcal{H}}},\breve{\delta}_{\widetilde{\mathcal{H}}}}}} \Big(\underset{\breve{\ell}\in\breve{\sigma}^\perp\cap\breve{\tau}^\vee}{\oplus} \vec{\widetilde{\Psi}}^{\mathrm{ord}}_{\breve{\Phi}_{\widetilde{\mathcal{H}}},\breve{\delta}_{\widetilde{\mathcal{H}}}}(\breve{\ell})\Big), \tag{7.2.1.9}$$

whose $\breve{\tau}$-stratum is canonically isomorphic to the scheme $\vec{\widetilde{\Xi}}^{\mathrm{ord}}_{\breve{\Phi}_{\widetilde{\mathcal{H}}},\breve{\delta}_{\widetilde{\mathcal{H}}},\breve{\tau}}$ above, which (as a closed subscheme of $\vec{\widetilde{\Xi}}^{\mathrm{ord}}_{\breve{\Phi}_{\widetilde{\mathcal{H}}},\breve{\delta}_{\widetilde{\mathcal{H}}},\breve{\sigma}}(\breve{\tau})$) is defined by the sheaf of ideals

$$\widetilde{\mathscr{I}}^{\mathrm{ord}}_{\breve{\sigma},\breve{\tau}} := \underset{\breve{\ell}\in\breve{\sigma}^\perp\cap\breve{\tau}_0^\vee}{\oplus} \vec{\widetilde{\Psi}}^{\mathrm{ord}}_{\breve{\Phi}_{\widetilde{\mathcal{H}}},\breve{\delta}_{\widetilde{\mathcal{H}}}}(\breve{\ell}). \tag{7.2.1.10}$$

Let $\vec{\widetilde{\mathfrak{X}}}^{\mathrm{ord}}_{\breve{\Phi}_{\widetilde{\mathcal{H}}},\breve{\delta}_{\widetilde{\mathcal{H}}},\breve{\sigma},\breve{\tau}}$ denote the formal completion of $\vec{\widetilde{\Xi}}^{\mathrm{ord}}_{\breve{\Phi}_{\widetilde{\mathcal{H}}},\breve{\delta}_{\widetilde{\mathcal{H}}},\breve{\sigma}}(\breve{\tau})$ along $\vec{\widetilde{\Xi}}^{\mathrm{ord}}_{\breve{\Phi}_{\widetilde{\mathcal{H}}},\breve{\delta}_{\widetilde{\mathcal{H}}},\breve{\tau}}$, which can be canonically identified as a closed formal subscheme of $\vec{\widetilde{\mathfrak{X}}}^{\mathrm{ord}}_{\breve{\Phi}_{\widetilde{\mathcal{H}}},\breve{\delta}_{\widetilde{\mathcal{H}}},\breve{\tau}}$, the formal completion of $\vec{\widetilde{\Xi}}^{\mathrm{ord}}_{\breve{\Phi}_{\widetilde{\mathcal{H}}},\breve{\delta}_{\widetilde{\mathcal{H}}}}(\breve{\tau})$ along its $\breve{\tau}$-stratum $\vec{\widetilde{\Xi}}^{\mathrm{ord}}_{\breve{\Phi}_{\widetilde{\mathcal{H}}},\breve{\delta}_{\widetilde{\mathcal{H}}},\breve{\tau}}$, inducing the closures of the $[(\widetilde{\Phi}_{\widetilde{\mathcal{H}}},\widetilde{\delta}_{\widetilde{\mathcal{H}}},\widetilde{\sigma})]$-strata on every ordinary good formal $(\breve{\Phi}_{\widetilde{\mathcal{H}}},\breve{\delta}_{\widetilde{\mathcal{H}}},\breve{\tau})$-model. (See Section 5.1.1.7 for the definition of good formal models, and see Definition 5.1.2.9 for the labeling of the strata by equivalence classes of triples of the form $[(\widetilde{\Phi}_{\widetilde{\mathcal{H}}},\widetilde{\delta}_{\widetilde{\mathcal{H}}},\widetilde{\sigma})]$.) By (5) of Theorem 5.2.1.1, the *strata-preserving* canonical isomorphism

$$(\vec{\widetilde{\mathsf{M}}}^{\mathrm{ord,tor}}_{\widetilde{\mathcal{H}},\widetilde{\Sigma}^{\mathrm{ord}}})^\wedge_{\vec{\widetilde{\mathsf{Z}}}^{\mathrm{ord}}_{[(\breve{\Phi}_{\widetilde{\mathcal{H}}},\breve{\delta}_{\widetilde{\mathcal{H}}},\breve{\tau})]}} \cong \vec{\widetilde{\mathfrak{X}}}^{\mathrm{ord}}_{\breve{\Phi}_{\widetilde{\mathcal{H}}},\breve{\delta}_{\widetilde{\mathcal{H}}},\breve{\tau}} = \vec{\widetilde{\mathfrak{X}}}^{\mathrm{ord}}_{\breve{\Phi}_{\widetilde{\mathcal{H}}},\breve{\delta}_{\widetilde{\mathcal{H}}},\breve{\tau}}/\Gamma_{\breve{\Phi}_{\widetilde{\mathcal{H}}},\breve{\tau}}$$

(where $\Gamma_{\breve{\Phi}_{\widetilde{\mathcal{H}}},\breve{\tau}}$ is trivial by [62, Lem. 6.2.5.27]) induces a canonical isomorphism

$$(\vec{\mathsf{N}}^{\mathrm{ord,tor}}_{\widetilde{\kappa}})^\wedge_{\vec{\widetilde{\mathsf{Z}}}^{\mathrm{ord}}_{[(\breve{\Phi}_{\widetilde{\mathcal{H}}},\breve{\delta}_{\widetilde{\mathcal{H}}},\breve{\tau})]}} \cong \vec{\widetilde{\mathfrak{X}}}^{\mathrm{ord}}_{\breve{\Phi}_{\widetilde{\mathcal{H}}},\breve{\delta}_{\widetilde{\mathcal{H}}},\breve{\sigma},\breve{\tau}}.$$

(Alternatively, one may refer directly to the gluing construction of $\vec{\widetilde{\mathsf{M}}}^{\mathrm{ord,tor}}_{\widetilde{\mathcal{H}}}$ in Section 5.1.3, based on the crucial Proposition 5.1.2.7; *cf.* [62, Sec. 6.3.3].)

7.2.2 *Justification for the Parameters*

So far we have parameterized the objects we constructed by $\widetilde{\kappa} \in \widetilde{\mathbf{K}}^{++}_{Q,\mathcal{H}}$. The goal of this subsection is to show that the equivalence classes $\kappa = [\widetilde{\kappa}] \in \mathbf{K}^{\mathrm{ord},++}_{Q,\mathcal{H}}$, with the natural directed partial $\succ$ order among them (see Definitions 7.1.1.11 and 7.2.2.20), form a more natural parameter set for the objects we have constructed. (See Proposition 7.2.2.19 below.)

Construction 7.2.2.1. (Compare with Construction 1.3.4.6.) For each $\widetilde{\kappa} = (\widetilde{\mathcal{H}}, \widetilde{\Sigma}^{\mathrm{ord}}, \widetilde{\sigma})$ in $\widetilde{\mathbf{K}}^{\mathrm{ord},++}_{Q,\mathcal{H}}$, consider the degenerating family

$$(\widetilde{G}, \widetilde{\lambda}, \widetilde{i}, \widetilde{\alpha}_{\widetilde{\mathcal{H}}^p}, \widetilde{\alpha}^{\mathrm{ord}}_{\widetilde{\mathcal{H}}_p}) \to \vec{\mathsf{M}}^{\mathrm{ord,tor}}_{\widetilde{\mathcal{H}},\widetilde{\Sigma}^{\mathrm{ord}}} \tag{7.2.2.2}$$

of type $\vec{\widetilde{\mathsf{M}}}^{\mathrm{ord}}_{\widetilde{\mathcal{H}}}$ as in Theorem 5.2.1.1. As in Construction 5.2.4.15, let

$$(\widehat{G}, \widehat{\lambda}, \widehat{i}) \to \vec{\mathsf{N}}^{\mathrm{ord,tor}}_{\widetilde{\kappa}} \tag{7.2.2.3}$$

denote the pullback of (7.2.2.2) to $\vec{\mathsf{N}}^{\mathrm{ord,tor}}_{\widetilde{\kappa}}$, the closure of $\vec{\mathsf{N}}^{\mathrm{ord}}_{\widetilde{\kappa}} = \vec{\widetilde{\mathsf{Z}}}^{\mathrm{ord}}_{[(\widetilde{\Phi}_{\widetilde{\mathcal{H}}}, \widetilde{\delta}_{\widetilde{\mathcal{H}}}, \widetilde{\sigma})]}$ in $\vec{\mathsf{M}}^{\mathrm{ord,tor}}_{\widetilde{\mathcal{H}},\widetilde{\Sigma}^{\mathrm{ord}}}$. Note that $\vec{\mathsf{N}}^{\mathrm{ord}}_{\widetilde{\kappa}}$ is canonically isomorphic to $\vec{\widetilde{C}}^{\mathrm{ord}}_{\widetilde{\Phi}_{\widetilde{\mathcal{H}}},\widetilde{\delta}_{\widetilde{\mathcal{H}}}}$ because $\widetilde{\sigma}$ is top-dimensional. Although $\widetilde{\alpha}_{\widetilde{\mathcal{H}}^p}$ is defined only over $\vec{\widetilde{\mathsf{M}}}^{\mathrm{ord}}_{\widetilde{\mathcal{H}}}$, by proceeding as in Construction 5.2.4.15, we can define a (partial) pullback

$$(\widehat{G}, \widehat{\lambda}, \widehat{i}, \widehat{\alpha}_{\widehat{\mathcal{H}}^p}, \widehat{\alpha}^{\mathrm{ord}}_{\widehat{\mathcal{H}}_p}) := (\widetilde{G}^{\natural}, \widetilde{\lambda}^{\natural}, \widetilde{i}^{\natural}, \widetilde{\beta}^{\natural}_{\widetilde{\mathcal{H}}^p}, \widetilde{\beta}^{\natural,\mathrm{ord}}_{\widetilde{\mathcal{H}}_p}) \to \vec{\mathsf{N}}^{\mathrm{ord,tor}}_{\widetilde{\kappa}} \tag{7.2.2.4}$$

of the degenerating family (7.2.2.2) to $\vec{\mathsf{N}}^{\mathrm{ord,tor}}_{\widetilde{\kappa}}$, with the convention that (as in the case of $(\widetilde{G}, \widetilde{\lambda}, \widetilde{i}, \widetilde{\alpha}_{\widetilde{\mathcal{H}}^p}, \widetilde{\alpha}^{\mathrm{ord}}_{\widetilde{\mathcal{H}}_p})$ itself) $\widehat{\alpha}_{\widehat{\mathcal{H}}^p}$ is defined only over $\mathsf{N}_{\widetilde{\kappa}}$, while $(\widetilde{G}, \widetilde{\lambda}, \widetilde{i})$ (resp. $\widetilde{\alpha}^{\mathrm{ord}}_{\widetilde{\mathcal{H}}_p}$) is defined (resp. extends) over all of $\vec{\mathsf{N}}^{\mathrm{ord,tor}}_{\widetilde{\kappa}}$ as in (7.2.2.3). By construction, the pullback

$$(\widehat{G}_{\vec{\mathsf{N}}^{\mathrm{ord}}_{\widetilde{\kappa}}}, \widehat{\lambda}_{\vec{\mathsf{N}}^{\mathrm{ord}}_{\widetilde{\kappa}}}, \widehat{i}_{\vec{\mathsf{N}}^{\mathrm{ord}}_{\widetilde{\kappa}}}, \widehat{\alpha}_{\widehat{\mathcal{H}}^p}, \widehat{\alpha}^{\mathrm{ord}}_{\widehat{\mathcal{H}}_p}) \to \vec{\mathsf{N}}^{\mathrm{ord}}_{\widetilde{\kappa}} \cong \vec{\widetilde{C}}^{\mathrm{ord}}_{\widetilde{\Phi}_{\widetilde{\mathcal{H}}},\widetilde{\delta}_{\widetilde{\mathcal{H}}}}, \tag{7.2.2.5}$$

of (7.2.2.4) to $\vec{\mathsf{N}}^{\mathrm{ord}}_{\widetilde{\kappa}}$ determines and is determined by (the prescribed $(\widetilde{\mathsf{Z}}_{\widetilde{\mathcal{H}}}, \widetilde{\Phi}_{\widetilde{\mathcal{H}}}, \widetilde{\delta}_{\widetilde{\mathcal{H}}})$ and) the tautological object

$$((A, \lambda, i, \alpha_{\mathcal{H}^p_\kappa}, \alpha^{\mathrm{ord}}_{\mathcal{H}_{\kappa,p}}), (\widetilde{c}^{\mathrm{ord}}_{\widetilde{\mathcal{H}}}, \widetilde{c}^{\vee,\mathrm{ord}}_{\widetilde{\mathcal{H}}})) \to \vec{\widetilde{C}}^{\mathrm{ord}}_{\widetilde{\Phi}_{\widetilde{\mathcal{H}}},\widetilde{\delta}_{\widetilde{\mathcal{H}}}} \tag{7.2.2.6}$$

(up to isomorphisms inducing automorphisms of $\widetilde{\Phi}_{\widetilde{\mathcal{H}}}$; *i.e.*, elements of $\Gamma_{\widetilde{\Phi}_{\widetilde{\mathcal{H}}}}$). Here $(A, \lambda, i, \alpha_{\mathcal{H}^p_\kappa}, \alpha^{\mathrm{ord}}_{\mathcal{H}_{\kappa,p}})$ is the tautological object over $\vec{\widetilde{\mathsf{M}}}^{\mathrm{ord},\widetilde{\mathsf{Z}}_{\widetilde{\mathcal{H}}}}_{\widetilde{\mathcal{H}}} \cong \vec{\mathsf{M}}^{\mathrm{ord}}_{\mathcal{H}_\kappa}$. As explained in the proof of Proposition 4.2.1.29, and also in Remark 1.3.4.1, we do not need to consider nontrivial twisted objects $(\widetilde{\varphi}^{\mathrm{ord}}_{-2,\widetilde{\mathcal{H}}}, \widetilde{\varphi}^{\mathrm{ord}}_{0,\widetilde{\mathcal{H}}})$ above $(\widetilde{\varphi}_{-2,\widetilde{\mathcal{H}}}, \widetilde{\varphi}_{0,\widetilde{\mathcal{H}}})$ and $\widetilde{\varphi}^{\mathrm{ord}}_{-1,\widetilde{\mathcal{H}}} = (\alpha_{\mathcal{H}^p_\kappa}, \alpha^{\mathrm{ord}}_{\mathcal{H}_{\kappa,p}})$. With the fixed choice of $(\widetilde{\mathsf{Z}}, \widetilde{\Phi}, \widetilde{\delta})$, the tautological object (7.2.2.6) depends only on $\widehat{\mathcal{H}}$, and hence so is the tuple (7.2.2.5). Thus, the notation

$(\widehat{\alpha}_{\widehat{\mathcal{H}}^p}, \widehat{\alpha}^{\mathrm{ord}}_{\widehat{\mathcal{H}}_p})$ is justified. As at the end of Construction 5.2.4.15, by considering the degenerating family

$$(\widetilde{G}, \widetilde{\lambda}, \widetilde{i}, \widetilde{\alpha}_{\widetilde{\mathcal{H}}}) \to \vec{\widetilde{\mathsf{M}}}^{\mathrm{ord,tor}}_{\widetilde{\mathcal{H}}, \widetilde{\Sigma}^{\mathrm{ord}}} \tag{7.2.2.7}$$

of type $\widetilde{\mathsf{M}}_{\widetilde{\mathcal{H}}}$, with the same $(\widetilde{G}, \widetilde{\lambda}, \widetilde{i})$ as in (7.2.2.2), where $\widetilde{\alpha}_{\widetilde{\mathcal{H}}}$ is defined only over $\vec{\widetilde{\mathsf{M}}}^{\mathrm{ord}}_{\widetilde{\mathcal{H}}} \underset{\mathbb{Z}}{\otimes} \mathbb{Q}$, such that the pair $(\widetilde{\alpha}_{\widetilde{\mathcal{H}}^p}, \widetilde{\alpha}^{\mathrm{ord}}_{\widetilde{\mathcal{H}}_p}) \underset{\mathbb{Z}}{\otimes} \mathbb{Q}$ is induced by $\widetilde{\alpha}_{\widetilde{\mathcal{H}}}$ as in Proposition 3.3.5.1, we obtain a (partial) pullback

$$(\widehat{G}, \widehat{\lambda}, \widehat{i}, \widehat{\alpha}_{\widehat{\mathcal{H}}}) \to \vec{\mathsf{N}}^{\mathrm{ord,tor}}_{\widetilde{\kappa}} \tag{7.2.2.8}$$

as in (1.3.4.9), where $\widehat{\alpha}_{\widehat{\mathcal{H}}}$ is defined only over $\vec{\mathsf{N}}^{\mathrm{ord}}_{\widetilde{\kappa}} \underset{\mathbb{Z}}{\otimes} \mathbb{Q}$, such that the pair $(\widehat{\alpha}_{\widehat{\mathcal{H}}^p}, \widehat{\alpha}^{\mathrm{ord}}_{\widehat{\mathcal{H}}_p}) \underset{\mathbb{Z}}{\otimes} \mathbb{Q}$ is induced by $\widehat{\alpha}_{\widehat{\mathcal{H}}}$ in an obvious analogue of Proposition 3.3.5.1.

Construction 7.2.2.9. (Compare with Construction 1.3.4.12.) Let $(\widehat{G}, \widehat{\lambda}, \widehat{i}, \widehat{\alpha}_{\widehat{\mathcal{H}}^p}, \widehat{\alpha}^{\mathrm{ord}}_{\widehat{\mathcal{H}}_p}) \to \vec{\mathsf{N}}^{\mathrm{ord,tor}}_{\widetilde{\kappa}}$ be as in (7.2.2.4), and let $(\widehat{G}, \widehat{\lambda}, \widehat{i}, \widehat{\alpha}_{\widehat{\mathcal{H}}}) \to \vec{\mathsf{N}}^{\mathrm{ord,tor}}_{\widetilde{\kappa}}$ be as in (7.2.2.8) in Construction 7.2.2.1. Consider any morphism $\xi : \mathrm{Spec}(V) \to \vec{\mathsf{N}}^{\mathrm{ord,tor}}_{\widetilde{\kappa}}$ centered at a geometric point $\bar{s}$ of $\vec{\mathsf{N}}^{\mathrm{ord,tor}}_{\widetilde{\kappa}}$ such that V is a complete discrete valuation ring with fraction field K, and such that $\eta := \mathrm{Spec}(K)$ is mapped to the generic point of the irreducible component containing the image of $\bar{s}$. Suppose the image of $\bar{s}$ lies on the $[(\breve{\Phi}_{\widetilde{\mathcal{H}}}, \breve{\delta}_{\widetilde{\mathcal{H}}}, \breve{\rho})]$-stratum $\vec{\widetilde{\mathsf{Z}}}^{\mathrm{ord}}_{[(\breve{\Phi}_{\widetilde{\mathcal{H}}}, \breve{\delta}_{\widetilde{\mathcal{H}}}, \breve{\rho})]}$ of $\vec{\widetilde{\mathsf{M}}}^{\mathrm{ord,tor}}_{\widetilde{\mathcal{H}}, \widetilde{\Sigma}^{\mathrm{ord}}}$, where $[(\breve{\Phi}_{\widetilde{\mathcal{H}}}, \breve{\delta}_{\widetilde{\mathcal{H}}}, \breve{\rho})]$ is represented by some $(\breve{\Phi}_{\widetilde{\mathcal{H}}}, \breve{\delta}_{\widetilde{\mathcal{H}}}, \breve{\rho})$ with $(\breve{\mathsf{Z}}_{\widetilde{\mathcal{H}}}, \breve{\Phi}_{\widetilde{\mathcal{H}}} = (\breve{X}, \breve{Y}, \breve{\phi}, \breve{\varphi}_{-2,\widetilde{\mathcal{H}}}, \breve{\varphi}_{0,\widetilde{\mathcal{H}}}), \breve{\delta}_{\widetilde{\mathcal{H}}})$ representing some ordinary cusp label as in Section 7.1.1. (As in Construction 1.3.4.12, we avoid using the more familiar notation $(\breve{\Phi}_{\widetilde{\mathcal{H}}}, \breve{\delta}_{\widetilde{\mathcal{H}}}, \breve{\tau})$ because the symbol τ will be used for another purpose below.) For simplicity, let us fix compatible choices of representatives $(\widetilde{\mathsf{Z}}, \widetilde{\Phi} = (\widetilde{X}, \widetilde{Y}, \widetilde{\phi}, \widetilde{\varphi}_{-2}, \widetilde{\varphi}_0), \widetilde{\delta})$ and $(\breve{\mathsf{Z}}, \breve{\Phi} = (\breve{X}, \breve{Y}, \breve{\phi}, \breve{\varphi}_{-2}, \breve{\varphi}_0), \breve{\delta})$, as in Sections 1.2.4 and 7.1.1, in their $\widetilde{\mathcal{H}}$-orbits.

Since $\vec{\widetilde{\mathfrak{X}}}^{\mathrm{ord}}_{\breve{\Phi}_{\widetilde{\mathcal{H}}}, \breve{\delta}_{\widetilde{\mathcal{H}}}, \breve{\rho}}$ is formally smooth over $\vec{\mathsf{S}}_{0,r_\kappa}$, there exists a complete regular local ring $\widetilde{V}$ and an ideal $\widetilde{I} \subset \widetilde{V}$ such that $\widetilde{V}/\widetilde{I} \cong V$ and such that the morphism $\mathrm{Spec}(V) \to \vec{\mathsf{N}}^{\mathrm{ord,tor}}_{\widetilde{\kappa}}$ extends to a morphism $\widetilde{\xi} : \mathrm{Spf}(\widetilde{V}, \widetilde{I}) \to \vec{\widetilde{\mathfrak{X}}}^{\mathrm{ord}}_{\breve{\Phi}_{\widetilde{\mathcal{H}}}, \breve{\delta}_{\widetilde{\mathcal{H}}}, \breve{\rho}}$, which induces a dominant morphism from $\mathrm{Spec}(\widetilde{V})$ to $\mathrm{Spec}(\widetilde{R})$, where $\widetilde{R}$ is the local ring of $\vec{\widetilde{\mathfrak{X}}}^{\mathrm{ord}}_{\breve{\Phi}_{\widetilde{\mathcal{H}}}, \breve{\delta}_{\widetilde{\mathcal{H}}}, \breve{\rho}}$ at the image of $\bar{s}$. Let

$$(\widetilde{G}^{\ddagger}, \widetilde{\lambda}^{\ddagger}, \widetilde{i}^{\ddagger}, \widetilde{\alpha}^{\ddagger}_{\widetilde{\mathcal{H}}^p}, \widetilde{\alpha}^{\mathrm{ord},\ddagger}_{\widetilde{\mathcal{H}}_p}) \to \mathrm{Spec}(\widetilde{V}) \tag{7.2.2.10}$$

denote the pullback of (7.2.2.2) under the composition of $\widetilde{\xi}$ with the canonical morphism $\vec{\widetilde{\mathfrak{X}}}^{\mathrm{ord}}_{\breve{\Phi}_{\widetilde{\mathcal{H}}}, \breve{\delta}_{\widetilde{\mathcal{H}}}, \breve{\rho}} \to \vec{\widetilde{\mathsf{M}}}^{\mathrm{ord,tor}}_{\widetilde{\mathcal{H}}, \widetilde{\Sigma}}$, and let

$$(\widehat{G}^{\ddagger}, \widehat{\lambda}^{\ddagger}, \widehat{i}^{\ddagger}, \widehat{\alpha}^{\ddagger}_{\widehat{\mathcal{H}}^p}, \widehat{\alpha}^{\mathrm{ord},\ddagger}_{\widehat{\mathcal{H}}_p}) \to \mathrm{Spec}(V) \tag{7.2.2.11}$$

denote the pullback of (7.2.2.4) under ξ. Similarly, let

$$(\widetilde{G}^{\ddagger}, \widetilde{\lambda}^{\ddagger}, \widetilde{i}^{\ddagger}, \widetilde{\alpha}^{\ddagger}_{\widetilde{\mathcal{H}}}) \to \mathrm{Spec}(\widetilde{V}) \tag{7.2.2.12}$$

denote the pullback of (7.2.2.7) under the composition of $\widetilde{\xi}$ with the canonical morphism $\vec{\widetilde{\mathfrak{X}}}^{\mathrm{ord}}_{\breve{\Phi}_{\widetilde{\mathcal{H}}},\breve{\delta}_{\widetilde{\mathcal{H}}},\breve{\rho}} \to \vec{\widetilde{\mathsf{M}}}^{\mathrm{ord,tor}}_{\widetilde{\mathcal{H}},\widetilde{\Sigma}}$, and let

$$(\widehat{G}^{\ddagger}, \widehat{\lambda}^{\ddagger}, \widehat{i}^{\ddagger}, \widehat{\alpha}^{\ddagger}_{\widehat{\mathcal{H}}}) \to \mathrm{Spec}(V) \tag{7.2.2.13}$$

denote the pullback of (7.2.2.4) under ξ. Then $\widetilde{\alpha}^{\ddagger}_{\widetilde{\mathcal{H}}}$ induces $(\widetilde{\alpha}^{\ddagger}_{\widetilde{\mathcal{H}}^p}, \widetilde{\alpha}^{\mathrm{ord},\ddagger}_{\widetilde{\mathcal{H}}_p})$ over η as in Proposition 3.3.5.1, and $\widehat{\alpha}^{\ddagger}_{\widehat{\mathcal{H}}}$ induces $(\widehat{\alpha}^{\ddagger}_{\widehat{\mathcal{H}}^p}, \widehat{\alpha}^{\mathrm{ord},\ddagger}_{\widehat{\mathcal{H}}_p})$ over η in an analogous way. (We omit the details for simplicity.)

As in (6) of Theorem 5.2.1.1, (7.2.2.10) defines an object in the essential image of $\mathrm{DEG}_{\mathrm{PEL},\widetilde{\mathsf{M}}^{\mathrm{ord}}_{\widetilde{\mathcal{H}}}}(\widetilde{V}) \to \mathrm{DEG}_{\mathrm{PEL},\ddot{\widetilde{\mathsf{M}}}^{\mathrm{ord}}_{\widetilde{\mathcal{H}}}}(\widetilde{V})$, which corresponds to a tuple

$$(\widetilde{B}^{\ddagger}, \lambda_{\widetilde{B}^{\ddagger}}, i_{\widetilde{B}^{\ddagger}}, \widetilde{X}^{\ddagger}, \widetilde{Y}^{\ddagger}, \widetilde{\phi}^{\ddagger}, \widetilde{c}^{\ddagger}, \widetilde{c}^{\vee,\ddagger}, \widetilde{\tau}^{\ddagger}, [\widetilde{\alpha}^{\natural,\mathrm{ord},\ddagger}_{\widetilde{\mathcal{H}}}])$$

in the essential image of $\mathrm{DD}_{\mathrm{PEL},\widetilde{\mathsf{M}}^{\mathrm{ord}}_{\widetilde{\mathcal{H}}}}(\widetilde{V}) \to \mathrm{DD}_{\mathrm{PEL},\ddot{\widetilde{\mathsf{M}}}^{\mathrm{ord}}_{\widetilde{\mathcal{H}}}}(\widetilde{V})$ under (4.1.6.4) in Theorem 4.1.6.2, where $[\widetilde{\alpha}^{\natural,\mathrm{ord},\ddagger}_{\widetilde{\mathcal{H}}}]$ is represented by some

$$\begin{aligned}\widetilde{\alpha}^{\natural,\mathrm{ord},\ddagger}_{\widetilde{\mathcal{H}}} = (\widetilde{\mathsf{Z}}^{\ddagger}_{\widetilde{\mathcal{H}}}, \widetilde{\varphi}^{\mathrm{ord},\ddagger}_{-2,\widetilde{\mathcal{H}}}, \widetilde{\varphi}^{\mathrm{ord},\ddagger}_{-1,\widetilde{\mathcal{H}}} = &(\widetilde{\varphi}^{\ddagger}_{-1,\widetilde{\mathcal{H}}^p}, \widetilde{\varphi}^{\mathrm{ord},\ddagger}_{-1,\widetilde{\mathcal{H}}_p}),\\ &\widetilde{\varphi}^{\mathrm{ord},\ddagger}_{0,\widetilde{\mathcal{H}}}, \widetilde{\delta}^{\mathrm{ord},\ddagger}_{\widetilde{\mathcal{H}}}, \widetilde{c}^{\mathrm{ord},\ddagger}_{\widetilde{\mathcal{H}}}, \widetilde{c}^{\vee,\mathrm{ord},\ddagger}_{\widetilde{\mathcal{H}}}, \widetilde{\tau}^{\mathrm{ord},\ddagger}_{\widetilde{\mathcal{H}}})\end{aligned}$$

induced by some

$$\widetilde{\alpha}^{\natural,\ddagger}_{\widetilde{\mathcal{H}}} = (\widetilde{\mathsf{Z}}^{\ddagger}_{\widetilde{\mathcal{H}}}, \widetilde{\varphi}^{\ddagger,\sim}_{-2,\widetilde{\mathcal{H}}}, \widetilde{\varphi}^{\ddagger}_{-1,\widetilde{\mathcal{H}}}, \widetilde{\varphi}^{\ddagger,\sim}_{0,\widetilde{\mathcal{H}}}, \widetilde{\delta}^{\ddagger}_{\widetilde{\mathcal{H}}}, \widetilde{c}^{\ddagger}_{\widetilde{\mathcal{H}}}, \widetilde{c}^{\vee,\ddagger}_{\widetilde{\mathcal{H}}}, \widetilde{\tau}^{\ddagger}_{\widetilde{\mathcal{H}}})$$

over η as in Section 4.1.6. Note that $(\widetilde{X}^{\ddagger}, \widetilde{Y}^{\ddagger}, \widetilde{\phi}^{\ddagger}, [\widetilde{\alpha}^{\natural,\ddagger}_{\widetilde{\mathcal{H}}}])$ determines some cusp label $[(\widetilde{\mathsf{Z}}^{\ddagger}_{\widetilde{\mathcal{H}}}, \widetilde{\Phi}^{\ddagger}_{\widetilde{\mathcal{H}}}, \widetilde{\delta}^{\ddagger}_{\widetilde{\mathcal{H}}})]$ equivalent to the cusp label $[(\breve{\mathsf{Z}}_{\widetilde{\mathcal{H}}}, \breve{\Phi}_{\widetilde{\mathcal{H}}}, \breve{\delta}_{\widetilde{\mathcal{H}}})]$ represented by the $\widetilde{\mathcal{H}}$-orbit of the $(\breve{\mathsf{Z}}, \breve{\Phi}, \breve{\delta})$ introduced above (where the $(\widetilde{\varphi}^{\ddagger}_{-2,\widetilde{\mathcal{H}}}, \widetilde{\varphi}^{\ddagger}_{0,\widetilde{\mathcal{H}}})$ in $\breve{\Phi}_{\widetilde{\mathcal{H}}}$ is induced by $(\widetilde{\varphi}^{\ddagger,\sim}_{-2,\widetilde{\mathcal{H}}}, \widetilde{\varphi}^{\ddagger,\sim}_{0,\widetilde{\mathcal{H}}})$ as in the corrected [62, Def. 5.4.2.8] in the errata). For simplicity, we shall use entries in this last representative to replace their isomorphic (or equivalent) objects, and say in this case that $(\widetilde{\varphi}^{\ddagger,\sim}_{-2,\widetilde{\mathcal{H}}}, \widetilde{\varphi}^{\ddagger,\sim}_{0,\widetilde{\mathcal{H}}})$ and $(\widetilde{\varphi}^{\mathrm{ord},\ddagger}_{-2,\widetilde{\mathcal{H}}}, \widetilde{\varphi}^{\mathrm{ord},\ddagger}_{0,\widetilde{\mathcal{H}}})$ induce $(\breve{\varphi}_{-2,\widetilde{\mathcal{H}}}, \breve{\varphi}_{0,\widetilde{\mathcal{H}}})$.

By definition, the pullback of $(\widetilde{B}^{\ddagger}, \lambda_{\widetilde{B}^{\ddagger}}, i_{\widetilde{B}^{\ddagger}}, \breve{X}, \breve{Y}, \breve{\phi}, \widetilde{c}^{\ddagger}, \widetilde{c}^{\vee,\ddagger})$ to the subscheme $\mathrm{Spec}(V)$ of $\mathrm{Spec}(\widetilde{V})$ depends only on $(\widehat{G}^{\ddagger}, \widehat{\lambda}^{\ddagger}, \widehat{i}^{\ddagger}) \to \mathrm{Spec}(V)$. Let us denote it by

$$(\widehat{B}^{\ddagger}, \lambda_{\widehat{B}^{\ddagger}}, i_{\widehat{B}^{\ddagger}}, \breve{X}, \breve{Y}, \breve{\phi}, \widehat{c}^{\ddagger}, \widehat{c}^{\vee,\ddagger}).$$

Note that the $\widetilde{\mathcal{H}}$-orbit $(\widetilde{\mathsf{Z}}_{\widetilde{\mathcal{H}}}, \widetilde{\Phi}_{\widetilde{\mathcal{H}}} = (\widetilde{X}, \widetilde{Y}, \widetilde{\phi}, \widetilde{\varphi}_{-2,\widetilde{\mathcal{H}}}, \widetilde{\varphi}_{0,\widetilde{\mathcal{H}}}), \widetilde{\delta}_{\widetilde{\mathcal{H}}})$ is part of the data of $\widetilde{\kappa}$. By Lemma 1.2.4.16, it makes sense to consider $\breve{\mathsf{Z}}_{\widehat{\mathcal{H}}}$, $(\breve{\varphi}_{-2,\widehat{\mathcal{H}}}, \breve{\varphi}_{0,\widehat{\mathcal{H}}})$, and $\breve{\delta}_{\widehat{\mathcal{H}}}$, which are the $\widehat{\mathcal{H}}$-orbits of $\breve{\mathsf{Z}}$, $(\breve{\varphi}_{-2} : \mathrm{Gr}^{\breve{\mathsf{Z}}}_{-2} \xrightarrow{\sim} \mathrm{Hom}_{\hat{\mathbb{Z}}}(\breve{X} \underset{\mathbb{Z}}{\otimes} \hat{\mathbb{Z}}, \hat{\mathbb{Z}}(1)), \breve{\varphi}_0 : \mathrm{Gr}^{\breve{\mathsf{Z}}}_0 \xrightarrow{\sim} \breve{Y} \underset{\mathbb{Z}}{\otimes} \hat{\mathbb{Z}})$, and $\breve{\delta}$, respectively. Then we have the orbit $\widehat{\mathcal{H}}$-orbit $\breve{\delta}_{\widehat{\mathcal{H}}}$ of $\breve{\delta}$, which induces the $\widehat{\mathcal{H}}$-orbit $\widehat{\delta}^{\mathrm{ord}}_{\widehat{\mathcal{H}}}$ of $\breve{\delta}^{\mathrm{ord}}$. Moreover, by extending restrictions to subgroups of $\widetilde{L}/n\widetilde{L}$ (with $\widetilde{\mathsf{Z}}_{-1,n}$ replaced with its subgroup $\breve{\mathsf{Z}}_{-1,n}$) as in Constructions 1.3.4.6 and 7.2.2.1, $(\widetilde{\alpha}^{\ddagger}_{\widetilde{\mathcal{H}}^p}, \widetilde{\alpha}^{\mathrm{ord},\ddagger}_{\widetilde{\mathcal{H}}_p})$ induces an ordinary level-$\mathcal{H}_\kappa$ structure $\widetilde{\varphi}^{\ddagger}_{-1,\widetilde{\mathcal{H}}} = (\widetilde{\varphi}^{\ddagger}_{-1,\widetilde{\mathcal{H}}^p}, \widetilde{\varphi}^{\mathrm{ord},\ddagger}_{-1,\widetilde{\mathcal{H}}_p})$ of $(\widehat{B}^{\ddagger}, \lambda_{\widehat{B}^{\ddagger}}, i_{\widehat{B}^{\ddagger}})$

depending only on $(\widehat{\alpha}^{\ddagger}_{\widehat{\mathcal{H}}^p}, \widehat{\alpha}^{\mathrm{ord},\ddagger}_{\widehat{\mathcal{H}}_p})$, which we denote by $\widehat{\varphi}^{\mathrm{ord},\ddagger}_{-1,\widehat{\mathcal{H}}} = (\widehat{\varphi}^{\ddagger}_{-1,\widehat{\mathcal{H}}^p}, \widehat{\varphi}^{\mathrm{ord},\ddagger}_{-1,\widehat{\mathcal{H}}_p})$, which is compatible with the level-$\mathcal{H}_\kappa$ structure $\widehat{\varphi}^{\ddagger}_{-1,\widehat{\mathcal{H}}}$ of $(\widehat{B}^{\ddagger}, \lambda_{\widehat{B}^{\ddagger}}, i_{\widehat{B}^{\ddagger}}) \underset{\mathbb{Z}}{\otimes} \mathbb{Q}$ induced by $\widetilde{\alpha}^{\ddagger}_{\widetilde{\mathcal{H}}}$ as in Proposition 3.3.5.1. Then it also makes sense to consider the $\widehat{\mathcal{H}}$-orbit $(\widetilde{\varphi}^{\mathrm{ord},\ddagger}_{-2,\widehat{\mathcal{H}}}, \widetilde{\varphi}^{\mathrm{ord},\ddagger}_{0,\widehat{\mathcal{H}}})$, which we denote by $(\widehat{\varphi}^{\mathrm{ord},\ddagger}_{-2,\widehat{\mathcal{H}}}, \widehat{\varphi}^{\mathrm{ord},\ddagger}_{0,\widehat{\mathcal{H}}})$, which is a subscheme of $(\breve{\varphi}_{-2,\widehat{\mathcal{H}}}, \breve{\varphi}_{0,\widehat{\mathcal{H}}}) \underset{\breve{\mathsf{Z}}_{\widehat{\mathcal{H}}}}{\times} \widehat{\varphi}^{\mathrm{ord},\ddagger}_{-1,\widehat{\mathcal{H}}}$ which can be identified with a system of $\widehat{\mathcal{H}}/\widehat{\mathcal{U}}(n)$-orbits, where $n \geq 1$ are integers such that $\widehat{\mathcal{U}}(n) := \widetilde{\mathcal{U}}(n)_{\widehat{\mathrm{G}}} \subset \widehat{\mathcal{H}}$, which surjects under the two projections to the orbits defining $(\breve{\varphi}_{-2,\widehat{\mathcal{H}}}, \breve{\varphi}_{0,\widehat{\mathcal{H}}})$ and $\widehat{\varphi}^{\mathrm{ord},\ddagger}_{-1,\widehat{\mathcal{H}}}$ and is compatible with the $\widehat{\mathcal{H}}$-orbit $(\widehat{\varphi}^{\ddagger,\sim}_{-2,\widehat{\mathcal{H}}}, \widehat{\varphi}^{\ddagger,\sim}_{0,\widehat{\mathcal{H}}})$ defined as a subscheme of $(\breve{\varphi}_{-2,\widehat{\mathcal{H}}}, \breve{\varphi}_{0,\widehat{\mathcal{H}}}) \underset{\breve{\mathsf{Z}}_{\widehat{\mathcal{H}}}}{\times} \widehat{\varphi}^{\ddagger}_{-1,\widehat{\mathcal{H}}}$. In this case, we say that $(\widehat{\varphi}^{\ddagger,\sim}_{-2,\widehat{\mathcal{H}}}, \widehat{\varphi}^{\ddagger,\sim}_{0,\widehat{\mathcal{H}}})$ and $(\widehat{\varphi}^{\mathrm{ord},\ddagger}_{-2,\widehat{\mathcal{H}}}, \widehat{\varphi}^{\mathrm{ord},\ddagger}_{0,\widehat{\mathcal{H}}})$ induce the $\widehat{\mathcal{H}}$-orbit $(\breve{\varphi}_{-2,\widehat{\mathcal{H}}}, \breve{\varphi}_{0,\widehat{\mathcal{H}}})$.

By proceeding as in Construction 1.3.4.12, we can define

$$(\widehat{\iota}^{\ddagger} : Y \to \widehat{G}^{\natural,\ddagger}_{\eta}, \widehat{\iota}^{\vee,\ddagger} : X \to \widehat{G}^{\vee,\natural,\ddagger}_{\eta})$$

and

$$(\widehat{\tau}^{\ddagger} : \mathbf{1}_{Y \times \breve{X},\eta} \xrightarrow{\sim} (\widehat{c}^{\vee,\ddagger}|_Y \times \widehat{c}^{\ddagger})^* \mathcal{P}^{\otimes -1}_{\widehat{B}^{\ddagger},\eta},$$
$$\widehat{\tau}^{\vee,\ddagger} : \mathbf{1}_{\breve{Y} \times X,\eta} \xrightarrow{\sim} (\widehat{c}^{\vee,\ddagger} \times \widehat{c}^{\ddagger}|_X)^* \mathcal{P}^{\otimes -1}_{\widehat{B}^{\ddagger},\eta})$$

(satisfying certain familiar compatibility conditions, which we omit for simplicity), and define the $\widehat{\mathcal{H}}^p$-orbits $(\widehat{\iota}^{\ddagger}_{\widehat{\mathcal{H}}^p}, \widehat{\iota}^{\vee,\ddagger}_{\widehat{\mathcal{H}}^p})$ and $(\widehat{c}^{\ddagger}_{\widehat{\mathcal{H}}^p}, \widehat{c}^{\vee,\ddagger}_{\widehat{\mathcal{H}}^p}, \widehat{\tau}^{\ddagger}_{\widehat{\mathcal{H}}^p}, \widehat{\tau}^{\vee,\ddagger}_{\widehat{\mathcal{H}}^p})$ determined by $\widehat{\alpha}^{\ddagger}_{\widehat{\mathcal{H}}^p}$, which is induced by the $\widehat{\mathcal{H}}$-orbits $(\widehat{\iota}^{\ddagger}_{\widehat{\mathcal{H}}}, \widehat{\iota}^{\vee,\ddagger}_{\widehat{\mathcal{H}}})$ and $(\widehat{c}^{\ddagger}_{\widehat{\mathcal{H}}}, \widehat{c}^{\vee,\ddagger}_{\widehat{\mathcal{H}}}, \widehat{\tau}^{\ddagger}_{\widehat{\mathcal{H}}}, \widehat{\tau}^{\vee,\ddagger}_{\widehat{\mathcal{H}}})$ determined by $\widehat{\alpha}^{\ddagger}_{\widehat{\mathcal{H}}}$. On the other hand, $\widetilde{\iota}^{\mathrm{ord},\ddagger}_{\widetilde{\mathcal{H}}_p}$ carries no more information than $(\widetilde{c}^{\mathrm{ord},\ddagger}_{\widetilde{\mathcal{H}}_p}, \widetilde{c}^{\vee,\mathrm{ord},\ddagger}_{\widetilde{\mathcal{H}}_p})$ (see Proposition 4.1.5.20 and Definitions 4.1.5.22 and 4.1.5.23), and $(\widetilde{c}^{\mathrm{ord},\ddagger}_{\widetilde{\mathcal{H}}}, \widetilde{c}^{\vee,\mathrm{ord},\ddagger}_{\widetilde{\mathcal{H}}})$ is the pullback of the tautological object $(\widetilde{c}^{\mathrm{ord}}_{\widetilde{\mathcal{H}}}, \widetilde{c}^{\vee,\mathrm{ord}}_{\widetilde{\mathcal{H}}})$ over $\vec{\breve{C}}^{\mathrm{ord}}_{\breve{\Phi}_{\widetilde{\mathcal{H}}},\breve{\delta}_{\widetilde{\mathcal{H}}}}$, which depends only on $\widehat{\mathcal{H}}$. Hence, it makes sense to define

$$(\widehat{\iota}^{\mathrm{ord},\ddagger}_{\widehat{\mathcal{H}}}, \widehat{\iota}^{\vee,\mathrm{ord},\ddagger}_{\widehat{\mathcal{H}}}) := (\widehat{\iota}^{\ddagger}_{\widehat{\mathcal{H}}^p}, \widehat{\iota}^{\vee,\ddagger}_{\widehat{\mathcal{H}}^p}),$$
$$(\widehat{\tau}^{\mathrm{ord},\ddagger}_{\widehat{\mathcal{H}}}, \widehat{\tau}^{\vee,\mathrm{ord},\ddagger}_{\widehat{\mathcal{H}}}) := (\widehat{\tau}^{\ddagger}_{\widehat{\mathcal{H}}^p}, \widehat{\tau}^{\vee,\ddagger}_{\widehat{\mathcal{H}}^p}),$$
$$(\widehat{c}^{\mathrm{ord},\ddagger}_{\widehat{\mathcal{H}}}, \widehat{c}^{\vee,\mathrm{ord},\ddagger}_{\widehat{\mathcal{H}}}) := (\widetilde{c}^{\mathrm{ord},\ddagger}_{\widetilde{\mathcal{H}}}, \widetilde{c}^{\vee,\mathrm{ord},\ddagger}_{\widetilde{\mathcal{H}}}).$$

In summary, given the family (1.3.4.9) in Construction 1.3.4.6, each morphism $\xi : \mathrm{Spec}(V) \to \mathsf{N}^{\mathrm{tor}}_{\widetilde{\kappa}}$ as above determines a tuple

$$(\widehat{B}^{\ddagger}, \lambda_{\widehat{B}^{\ddagger}}, i_{\widehat{B}^{\ddagger}}, \breve{X}, \breve{Y}, \breve{\phi}, \widehat{c}^{\ddagger}, \widehat{c}^{\vee,\ddagger}, \widehat{\tau}^{\ddagger}, \widehat{\tau}^{\vee,\ddagger}, [\widehat{\alpha}^{\natural,\mathrm{ord},\ddagger}_{\widehat{\mathcal{H}}}]), \tag{7.2.2.14}$$

where $[\widehat{\alpha}^{\natural,\mathrm{ord},\ddagger}_{\widehat{\mathcal{H}}}]$ is an equivalence class of

$$\widehat{\alpha}^{\natural,\mathrm{ord},\ddagger}_{\widehat{\mathcal{H}}} = (\breve{\mathsf{Z}}_{\widehat{\mathcal{H}}}, \widehat{\varphi}^{\mathrm{ord},\ddagger}_{-2,\widehat{\mathcal{H}}}, \widehat{\varphi}^{\mathrm{ord},\ddagger}_{-1,\widehat{\mathcal{H}}}, \widehat{\varphi}^{\mathrm{ord},\ddagger}_{0,\widehat{\mathcal{H}}}, \widehat{\delta}^{\mathrm{ord}}_{\widehat{\mathcal{H}}}, \widehat{c}^{\mathrm{ord},\ddagger}_{\widehat{\mathcal{H}}}, \widehat{c}^{\vee,\mathrm{ord},\ddagger}_{\widehat{\mathcal{H}}}, \widehat{\tau}^{\mathrm{ord},\ddagger}_{\widehat{\mathcal{H}}}, \widehat{\tau}^{\vee,\mathrm{ord},\ddagger}_{\widehat{\mathcal{H}}}) \tag{7.2.2.15}$$

(whose precise definitions we omit for simplicity), induced by some

$$\widehat{\alpha}^{\natural,\ddagger}_{\widehat{\mathcal{H}}} = (\breve{\mathsf{Z}}_{\widehat{\mathcal{H}}}, \widehat{\varphi}^{\ddagger,\sim}_{-2,\widehat{\mathcal{H}}}, \widehat{\varphi}^{\ddagger}_{-1,\widehat{\mathcal{H}}}, \widehat{\varphi}^{\ddagger,\sim}_{0,\widehat{\mathcal{H}}}, \breve{\delta}_{\widehat{\mathcal{H}}}, \widehat{c}^{\ddagger}_{\widehat{\mathcal{H}}}, \widehat{c}^{\vee,\ddagger}_{\widehat{\mathcal{H}}}, \widehat{\tau}^{\ddagger}_{\widehat{\mathcal{H}}}, \widehat{\tau}^{\vee,\ddagger}_{\widehat{\mathcal{H}}}) \tag{7.2.2.16}$$

(*cf.* (1.3.4.16)) over η, in a way analogous to that in Section 4.1.6.

Given a tuple as in (7.2.2.14), if we set

$$(B^{\ddagger}, \lambda_{B^{\ddagger}}, i_{B^{\ddagger}}, \varphi^{\mathrm{ord}}_{-1,\mathcal{H}_{\kappa}}) := (\widehat{B}^{\ddagger}, \lambda_{\widehat{B}^{\ddagger}}, i_{\widehat{B}^{\ddagger}}, \widehat{\varphi}^{\mathrm{ord}}_{-1,\widehat{\mathcal{H}}})$$

and

$$(c^{\ddagger}, c^{\vee,\ddagger}, \tau^{\ddagger}) := (\widehat{c}^{\ddagger}|_X, \widehat{c}^{\vee,\ddagger}|_Y, \widehat{\tau}^{\ddagger}|_{\mathbf{1}_{Y\times X,\eta}}),$$

and define $[\alpha^{\natural,\mathrm{ord},\ddagger}_{\mathcal{H}_{\kappa}}]$ using similar restrictions, then the tuple

$$(B^{\ddagger}, \lambda_{B^{\ddagger}}, i_{B^{\ddagger}}, X, Y, \phi, c^{\ddagger}, c^{\vee,\ddagger}, \tau^{\ddagger}, [\alpha^{\natural,\mathrm{ord},\ddagger}_{\mathcal{H}_{\kappa}}]) \tag{7.2.2.17}$$

defines an object in the essential image of $\mathrm{DD}_{\mathrm{PEL},\mathsf{M}^{\mathrm{ord}}_{\mathcal{H}_{\kappa}}}(V) \to \mathrm{DD}_{\mathrm{PEL},\ddot{\mathsf{M}}^{\mathrm{ord}}_{\mathcal{H}_{\kappa}}}(V)$ (because we can define $[\alpha^{\natural,\ddagger}_{\mathcal{H}_{\kappa}}]$ as in (1.3.4.17) in Construction 1.3.4.12). On the other hand, the pullback $(\widehat{G}^{\ddagger}, \widehat{\lambda}^{\ddagger}, \widehat{i}^{\ddagger}, \widehat{\alpha}^{\ddagger}_{\widehat{\mathcal{H}}^p}, \widehat{\alpha}^{\mathrm{ord},\ddagger}_{\widehat{\mathcal{H}}_p}) \to \mathrm{Spec}(V)$ is determined up to isomorphism by its generic fiber $(\widehat{G}^{\ddagger}_{\eta}, \widehat{\lambda}^{\ddagger}_{\eta}, \widehat{i}^{\ddagger}_{\eta}, \widehat{\alpha}^{\ddagger}_{\widehat{\mathcal{H}}^p,\eta}, \widehat{\alpha}^{\mathrm{ord},\ddagger}_{\widehat{\mathcal{H}}_p,\eta}) \to \mathrm{Spec}(K)$, which (up to isomorphism) determines and is determined by a tuple $((G^{\ddagger}_{\eta}, \lambda^{\ddagger}_{\eta}, i^{\ddagger}_{\eta}, \alpha^{\ddagger}_{\mathcal{H}^p_{\kappa},\eta}, \alpha^{\mathrm{ord},\ddagger}_{\mathcal{H}_{\kappa,p},\eta}), (\widetilde{c}^{\mathrm{ord},\ddagger}_{\widetilde{\mathcal{H}},\eta}, \widetilde{c}^{\vee,\mathrm{ord},\ddagger}_{\widetilde{\mathcal{H}},\eta})) \to \mathrm{Spec}(K)$ parameterized by $\vec{\widetilde{C}}^{\mathrm{ord}}_{\widetilde{\Phi}_{\widetilde{\mathcal{H}}},\widetilde{\delta}_{\widetilde{\mathcal{H}}}}$. The abelian part $(G^{\ddagger}_{\eta}, \lambda^{\ddagger}_{\eta}, i^{\ddagger}_{\eta}, \alpha^{\ddagger}_{\mathcal{H}^p_{\kappa},\eta}, \alpha^{\mathrm{ord},\ddagger}_{\mathcal{H}_{\kappa,p},\eta})$ extends to a degenerating family

$$(G^{\ddagger}, \lambda^{\ddagger}, i^{\ddagger}, \alpha^{\ddagger}_{\mathcal{H}^p_{\kappa}}, \alpha^{\mathrm{ord},\ddagger}_{\mathcal{H}_{\kappa,p}}) \tag{7.2.2.18}$$

of type $\vec{\mathsf{M}}^{\mathrm{ord}}_{\mathcal{H}_{\kappa}}$ over $\mathrm{Spec}(V)$ (with $\alpha^{\ddagger}_{\mathcal{H}^p_{\kappa}}$ still defined only over $\eta = \mathrm{Spec}(K)$) which defines an object in the essential image of $\mathrm{DEG}_{\mathrm{PEL},\mathsf{M}^{\mathrm{ord}}_{\mathcal{H}_{\kappa}}}(V) \to \mathrm{DEG}_{\mathrm{PEL},\ddot{\mathsf{M}}^{\mathrm{ord}}_{\mathcal{H}_{\kappa}}}(V)$ (because we can defined $\alpha^{\ddagger}_{\mathcal{H}_{\kappa}}$ as in (1.3.4.18) in Construction 1.3.4.12). By the theory of two-step degenerations (see [28, Ch. III, Thm. 10.2] and [62, Sec. 4.5.6]), and by analyzing endomorphism structures and level structures as in [62, Sec. 5.1–5.3] and Section 4.1, under (4.1.6.4) in Theorem 4.1.6.2, this last object (7.2.2.18) corresponds to the above object (7.2.2.17) in the essential image of $\mathrm{DD}_{\mathrm{PEL},\mathsf{M}^{\mathrm{ord}}_{\mathcal{H}_{\kappa}}}(V) \to \mathrm{DD}_{\mathrm{PEL},\ddot{\mathsf{M}}^{\mathrm{ord}}_{\mathcal{H}_{\kappa}}}(V)$.

As in Construction 1.3.4.12, by checking the values of their entries on $\bar{K}$-points, where $\bar{K}$ is any fixed algebraic closure of K, the tuple over $\mathrm{Spec}(V)$ as in (7.2.2.14) determines and is determined by the tuple (7.2.2.11) (up to isomorphism, over $\mathrm{Spec}(K)$).

As in (4.2.2.7), the pair $(\widehat{\tau}^{\ddagger}, \widehat{\tau}^{\vee,\ddagger})$ defines compatible morphisms $\upsilon_{\widehat{\tau}^{\ddagger}} : Y \times \breve{X} \to \mathbb{Z}$ and $\upsilon_{\widehat{\tau}^{\vee,\ddagger}} : \breve{Y} \times X \to \mathbb{Z}$ (using the discrete valuation $\upsilon : \mathrm{Inv}(V) \to \mathbb{Z}$ of V), which define the same element

$$\upsilon_{\widehat{\tau}^{\ddagger}} = \upsilon_{\widehat{\tau}^{\vee,\ddagger}} \in (\widehat{\mathbf{S}}_{\breve{\Phi}_{\widehat{\mathcal{H}}}})^{\vee}_{\mathbb{R}}$$

(see (1.2.4.29)). On the other hand, as in (6) of Theorem 5.2.1.1, $\widetilde{\tau}^{\ddagger}$ defines a morphism $\upsilon_{\widetilde{\tau}^{\ddagger}} : \breve{Y} \times \breve{X} \to \mathbb{Z}$, which defines an element

$$\upsilon_{\widetilde{\tau}^{\ddagger}} \in \breve{\rho} \subset \mathbf{P}^{+}_{\breve{\Phi}_{\widetilde{\mathcal{H}}}},$$

where $\breve{\rho}$ is as above. Since $(\widehat{\tau}^{\ddagger}, \widehat{\tau}^{\vee,\ddagger})$ is defined by extending restrictions of $\upsilon_{\widetilde{\tau}^{\ddagger}}$, we see that

$$\upsilon_{\widehat{\tau}^{\ddagger}} = \upsilon_{\widehat{\tau}^{\vee,\ddagger}} \in \widehat{\rho} = \mathrm{pr}_{(\widehat{\mathbf{S}}_{\breve{\Phi}_{\widehat{\mathcal{H}}}})^{\vee}_{\mathbb{R}}}(\breve{\rho}) \subset \widehat{\mathbf{P}}_{\breve{\Phi}_{\widehat{\mathcal{H}}}}$$

(see (1.2.4.41)). If $\breve{\rho}$ is replaced with another representative, then $\widehat{\rho}$ is replaced with a translation under the action of $\Gamma_{\breve{\Phi}_{\widehat{\mathcal{H}}}}$. (This finishes Construction 7.2.2.9.)

Proposition 7.2.2.19. (*Compare with Proposition* 1.3.4.19.) *Suppose* $\widetilde{\kappa} = (\widetilde{\mathcal{H}}, \widetilde{\Sigma}, \widetilde{\sigma})$ *and* $\widetilde{\kappa}' = (\widetilde{\mathcal{H}}', \widetilde{\Sigma}', \widetilde{\sigma}')$ *are elements in* $\widetilde{\mathbf{K}}^{\mathrm{ord},++}_{Q,\mathcal{H}}$ *such that* $\kappa' = [\widetilde{\kappa}'] \succ \kappa = [\widetilde{\kappa}]$ *in* $\mathbf{K}^{\mathrm{ord},++}_{Q,\mathcal{H}}$ *(see Definition* 7.1.1.11*). Let* $(\widehat{G}, \widehat{\lambda}, \widehat{i}, \widehat{\alpha}_{\widehat{\mathcal{H}}^p}, \widehat{\alpha}^{\mathrm{ord}}_{\widehat{\mathcal{H}}_p}) \to \vec{\mathsf{N}}^{\mathrm{ord},\mathrm{tor}}_{\widetilde{\kappa}}$ *(resp.* $(\widehat{G}', \widehat{\lambda}', \widehat{i}', \widehat{\alpha}'_{\widehat{\mathcal{H}}',p}, \widehat{\alpha}^{\mathrm{ord},\prime}_{\widehat{\mathcal{H}}'_p}) \to \vec{\mathsf{N}}^{\mathrm{ord},\mathrm{tor}}_{\widetilde{\kappa}'}$*) denote the pullback of the degenerating family* $(\widetilde{G}, \widetilde{\lambda}, \widetilde{i}, \widetilde{\alpha}_{\widetilde{\mathcal{H}}^p}, \widetilde{\alpha}^{\mathrm{ord}}_{\widetilde{\mathcal{H}}_p}) \to \vec{\widetilde{\mathsf{M}}}^{\mathrm{ord},\mathrm{tor}}_{\widetilde{\mathcal{H}},\widetilde{\Sigma}^{\mathrm{ord}}}$ *(resp.* $(\widetilde{G}', \widetilde{\lambda}', \widetilde{i}', \widetilde{\alpha}'_{\widetilde{\mathcal{H}}',p}, \widetilde{\alpha}^{\mathrm{ord},\prime}_{\widetilde{\mathcal{H}}_p}) \to \vec{\widetilde{\mathsf{M}}}^{\mathrm{ord},\mathrm{tor}}_{\widetilde{\mathcal{H}}',\widetilde{\Sigma}^{\mathrm{ord},\prime}}$*), and let* $(\widehat{G}, \widehat{\lambda}, \widehat{i}, \widehat{\alpha}_{\widehat{\mathcal{H}}}) \to \vec{\mathsf{N}}^{\mathrm{ord},\mathrm{tor}}_{\widetilde{\kappa}}$ *(resp.* $(\widehat{G}', \widehat{\lambda}', \widehat{i}', \widehat{\alpha}'_{\widehat{\mathcal{H}}'}) \to \vec{\mathsf{N}}^{\mathrm{ord},\mathrm{tor}}_{\widetilde{\kappa}'}$*) denote the pullback of the degenerating family* $(\widetilde{G}, \widetilde{\lambda}, \widetilde{i}, \widetilde{\alpha}_{\widetilde{\mathcal{H}}}) \to \vec{\widetilde{\mathsf{M}}}^{\mathrm{ord},\mathrm{tor}}_{\widetilde{\mathcal{H}},\widetilde{\Sigma}^{\mathrm{ord}}}$ *(resp.* $(\widetilde{G}', \widetilde{\lambda}', \widetilde{i}', \widetilde{\alpha}'_{\widetilde{\mathcal{H}}'}) \to \vec{\widetilde{\mathsf{M}}}^{\mathrm{ord},\mathrm{tor}}_{\widetilde{\mathcal{H}}',\widetilde{\Sigma}^{\mathrm{ord},\prime}}$*), as in Construction* 7.2.2.1*. Then there is a canonical surjection* $f^{\mathrm{tor}}_{\widetilde{\kappa}',\widetilde{\kappa}} : \mathsf{N}^{\mathrm{tor}}_{\widetilde{\kappa}'} \to \vec{\mathsf{N}}^{\mathrm{ord},\mathrm{tor}}_{\widetilde{\kappa}}$ *under which* $(\widehat{G}', \widehat{\lambda}', \widehat{i}', \widehat{\alpha}'_{\widehat{\mathcal{H}}',p}, \widehat{\alpha}^{\mathrm{ord},\prime}_{\widehat{\mathcal{H}}'_p}) \to \vec{\mathsf{N}}^{\mathrm{ord},\mathrm{tor}}_{\widetilde{\kappa}'}$ *is canonically isomorphic to the pullback of* $(\widehat{G}, \widehat{\lambda}, \widehat{i}, \widehat{\alpha}_{\widehat{\mathcal{H}}^p}, \widehat{\alpha}^{\mathrm{ord}}_{\widehat{\mathcal{H}}_p}) \to \vec{\mathsf{N}}^{\mathrm{ord},\mathrm{tor}}_{\widetilde{\kappa}}$*, and under which* $(\widehat{G}', \widehat{\lambda}', \widehat{i}', \widehat{\alpha}'_{\widehat{\mathcal{H}}'}) \to \vec{\mathsf{N}}^{\mathrm{ord},\mathrm{tor}}_{\widetilde{\kappa}'}$ *is canonically isomorphic to the pullback of* $(\widehat{G}, \widehat{\lambda}, \widehat{i}, \widehat{\alpha}_{\widehat{\mathcal{H}}}) \to \mathsf{N}^{\mathrm{tor}}_{\widetilde{\kappa}}$.

In particular, for each $\widetilde{\kappa} = (\widetilde{\mathcal{H}}, \widetilde{\Sigma}, \widetilde{\sigma}) \in \widetilde{\mathbf{K}}^{\mathrm{ord},++}_{Q,\mathcal{H}}$*, the closure* $\vec{\mathsf{N}}^{\mathrm{ord},\mathrm{tor}}_{\widetilde{\kappa}}$ *of* $\vec{\mathsf{N}}^{\mathrm{ord}}_{\widetilde{\kappa}} = \vec{\widetilde{\mathsf{Z}}}^{\mathrm{ord}}_{[(\widetilde{\Phi}_{\widetilde{\mathcal{H}}}, \widetilde{\delta}_{\widetilde{\mathcal{H}}}, \widetilde{\sigma})]}$ *in* $\vec{\widetilde{\mathsf{M}}}^{\mathrm{ord},\mathrm{tor}}_{\widetilde{\mathcal{H}},\widetilde{\Sigma}^{\mathrm{ord}}}$ *and the open (fiberwise dense) embedding* $\widetilde{\kappa}^{\mathrm{tor}} : \vec{\mathsf{N}}^{\mathrm{ord}}_{\widetilde{\kappa}} \hookrightarrow \vec{\mathsf{N}}^{\mathrm{ord},\mathrm{tor}}_{\widetilde{\kappa}}$ *depend (up to canonical isomorphism) only on the pair* $\kappa = [\widetilde{\kappa}] = (\widehat{\mathcal{H}}, \widehat{\Sigma}^{\mathrm{ord}})$ *in* $\mathbf{K}^{\mathrm{ord},++}_{Q,\mathcal{H}}$*. It satisfies the descriptions of stratifications and completions in the third, fourth, and fifth paragraphs of* (1) *of Theorem* 1.3.3.15.

The morphism $f^{\mathrm{tor}}_{\widetilde{\kappa}',\widetilde{\kappa}} : \vec{\mathsf{N}}^{\mathrm{ord},\mathrm{tor}}_{\widetilde{\kappa}'} \to \vec{\mathsf{N}}^{\mathrm{ord},\mathrm{tor}}_{\widetilde{\kappa}}$ *induces a morphism* $\vec{\mathsf{N}}^{\mathrm{ord},\mathrm{tor}}_{\widetilde{\kappa}'} \to \vec{\mathsf{N}}^{\mathrm{ord},\mathrm{tor}}_{\widetilde{\kappa}} \underset{\vec{\mathsf{S}}_{0,r_{\kappa}}}{\times} \vec{\mathsf{S}}_{0,r_{\kappa'}}$ *which is étale locally given by equivariant morphisms between toric schemes mapping strata to strata, and hence (this induced morphism) is log étale essentially by definition (see [45, Thm. 3.5]). Moreover, as in [28, Ch. V, Rem. 1.2(b)] and in the proof of [62, Lem. 7.1.1.4], we have* $R^i(f^{\mathrm{tor}}_{\widetilde{\kappa}',\widetilde{\kappa}})_* \mathscr{O}_{\vec{\mathsf{N}}^{\mathrm{ord},\mathrm{tor}}_{\widetilde{\kappa}'}} = 0$ *for* $i > 0$ *by [50, Ch. I, Sec. 3].*

Proof. Suppose $\widehat{\mathcal{H}}$ (resp. $\widehat{\mathcal{H}}'$) is determined by some $\widetilde{\mathcal{H}}$ (resp. $\widetilde{\mathcal{H}}'$) satisfying Conditions 1.2.4.7 and 7.1.1.5. By Lemma 7.2.1.3, we may replace $\widehat{\mathcal{H}}'$ (resp. $\widetilde{\mathcal{H}}'$) with $\widehat{\mathcal{H}}' \cap \widehat{\mathcal{H}}$ (resp. $\widetilde{\mathcal{H}}' \cap \widetilde{\mathcal{H}}$), in which case we have a canonical (forgetful) morphism $f_{\widetilde{\kappa}',\widetilde{\kappa}} : \vec{\mathsf{N}}^{\mathrm{ord},\mathrm{tor}}_{\widetilde{\kappa}'} \cong \vec{\widetilde{C}}^{\mathrm{ord}}_{\widetilde{\Phi}_{\widetilde{\mathcal{H}}'}, \widetilde{\delta}_{\widetilde{\mathcal{H}}'}} \to \vec{\widetilde{C}}^{\mathrm{ord}}_{\widetilde{\Phi}_{\widetilde{\mathcal{H}}}, \widetilde{\delta}_{\widetilde{\mathcal{H}}}} \cong \vec{\mathsf{N}}^{\mathrm{ord},\mathrm{tor}}_{\widetilde{\kappa}}$ (by construction). Suppose $((G', \lambda', i', \alpha'_{\mathcal{H}^p_{\kappa'}}, \alpha^{\mathrm{ord},\prime}_{\mathcal{H}_{\kappa',p}}), (\widetilde{c}^{\mathrm{ord},\prime}_{\widetilde{\mathcal{H}}'}, \widetilde{c}^{\vee,\mathrm{ord},\prime}_{\widetilde{\mathcal{H}}'}))$ (resp. $((G, \lambda, i, \alpha_{\mathcal{H}^p_{\kappa}}, \alpha^{\mathrm{ord}}_{\mathcal{H}_{\kappa,p}}), (\widetilde{c}^{\mathrm{ord}}_{\widetilde{\mathcal{H}}}, \widetilde{c}^{\vee,\mathrm{ord}}_{\widetilde{\mathcal{H}}}))$) is the tautological object over $\vec{\widetilde{C}}^{\mathrm{ord}}_{\widetilde{\Phi}_{\widetilde{\mathcal{H}}'}, \widetilde{\delta}_{\widetilde{\mathcal{H}}'}}$ (resp.

$\vec{\widetilde{C}}^{\mathrm{ord}}_{\widetilde{\Phi}_{\widetilde{\mathcal{H}}},\widetilde{\delta}_{\widetilde{\mathcal{H}}}})$, as in Construction 7.2.2.1, which determines and is determined by $(\widehat{G}'_{\vec{\mathsf{N}}^{\mathrm{ord}}_{\widetilde{\kappa}'}},\widehat{\lambda}'_{\vec{\mathsf{N}}^{\mathrm{ord}}_{\widetilde{\kappa}'}},\widehat{i}'_{\vec{\mathsf{N}}^{\mathrm{ord}}_{\widetilde{\kappa}'}},\widehat{\alpha}'_{\widehat{\mathcal{H}}',p},\widehat{\alpha}^{\mathrm{ord},\prime}_{\widehat{\mathcal{H}}'_p}) \to \vec{\mathsf{N}}^{\mathrm{ord}}_{\widetilde{\kappa}'}$ (resp. $(\widehat{G}_{\mathsf{N}_{\widetilde{\kappa}}},\widehat{\lambda}_{\mathsf{N}_{\widetilde{\kappa}}},\widehat{i}_{\mathsf{N}_{\widetilde{\kappa}}},\widehat{\alpha}_{\widehat{\mathcal{H}}^p},\widehat{\alpha}^{\mathrm{ord},p}_{\widehat{\mathcal{H}}_p}) \to \vec{\mathsf{N}}^{\mathrm{ord}}_{\widetilde{\kappa}}$), the pullback of $(\widehat{G}',\widehat{\lambda}',\widehat{i}',\widehat{\alpha}'_{\widehat{\mathcal{H}}',p},\widehat{\alpha}^{\mathrm{ord},\prime}_{\widehat{\mathcal{H}}'_p}) \to \vec{\mathsf{N}}^{\mathrm{ord,tor}}_{\widetilde{\kappa}'}$ (resp. $(\widehat{G},\widehat{\lambda},\widehat{i},\widehat{\alpha}_{\widehat{\mathcal{H}}^p},\widehat{\alpha}^{\mathrm{ord}}_{\widehat{\mathcal{H}}_p}) \to \vec{\mathsf{N}}^{\mathrm{ord,tor}}_{\widetilde{\kappa}}$) to $\vec{\mathsf{N}}^{\mathrm{ord}}_{\widetilde{\kappa}'}$ (resp. $\vec{\mathsf{N}}^{\mathrm{ord}}_{\widetilde{\kappa}}$). Then $f_{\widetilde{\kappa}',\widetilde{\kappa}}$ is also the canonical morphism determined by the universal property of $\vec{\widetilde{C}}^{\mathrm{ord}}_{\widetilde{\Phi}_{\widetilde{\mathcal{H}}},\widetilde{\delta}_{\widetilde{\mathcal{H}}}}$, under which the pullback of $((G,\lambda,i,\alpha_{\mathcal{H}^p_\kappa},\alpha^{\mathrm{ord}}_{\mathcal{H}_{\kappa,p}}),(\widetilde{c}^{\mathrm{ord}}_{\widetilde{\mathcal{H}}},\widetilde{c}^{\vee,\mathrm{ord}}_{\widetilde{\mathcal{H}}}))$ is canonically isomorphic to the $\widehat{\mathcal{H}}$-orbit $((G',\lambda',i',\alpha'_{\mathcal{H}^p_\kappa},\alpha^{\mathrm{ord},\prime}_{\mathcal{H}_{\kappa,p}}),(\widetilde{c}^{\mathrm{ord},\prime}_{\widetilde{\mathcal{H}}},\widetilde{c}^{\vee,\mathrm{ord},\prime}_{\widetilde{\mathcal{H}}}))$ of $((G',\lambda',i',\alpha'_{\mathcal{H}^p_{\kappa'}},\alpha^{\mathrm{ord},\prime}_{\mathcal{H}_{\kappa',p}}),(\widetilde{c}^{\mathrm{ord},\prime}_{\widetilde{\mathcal{H}}'},\widetilde{c}^{\vee,\mathrm{ord},\prime}_{\widetilde{\mathcal{H}}'}))$; or, rather, such that the pullback of $(\widehat{G}_{\vec{\mathsf{N}}^{\mathrm{ord}}_{\widetilde{\kappa}}},\widehat{\lambda}_{\vec{\mathsf{N}}^{\mathrm{ord}}_{\widetilde{\kappa}}},\widehat{i}_{\vec{\mathsf{N}}^{\mathrm{ord}}_{\widetilde{\kappa}}},\widehat{\alpha}_{\widehat{\mathcal{H}}^p},\widehat{\alpha}^{\mathrm{ord}}_{\widehat{\mathcal{H}}_p}) \to \vec{\mathsf{N}}^{\mathrm{ord}}_{\widetilde{\kappa}}$ under $f_{\widetilde{\kappa}',\widetilde{\kappa}}$ is canonically isomorphic to the $\widehat{\mathcal{H}}$-orbit $(\widehat{G}'_{\vec{\mathsf{N}}^{\mathrm{ord}}_{\widetilde{\kappa}'}},\widehat{\lambda}'_{\vec{\mathsf{N}}^{\mathrm{ord}}_{\widetilde{\kappa}'}},\widehat{i}'_{\vec{\mathsf{N}}^{\mathrm{ord}}_{\widetilde{\kappa}'}},\widehat{\alpha}'_{\widehat{\mathcal{H}}^p},\widehat{\alpha}^{\mathrm{ord},\prime}_{\widehat{\mathcal{H}}_p}) \to \vec{\mathsf{N}}^{\mathrm{ord}}_{\widetilde{\kappa}'}$ of $(\widehat{G}'_{\vec{\mathsf{N}}^{\mathrm{ord}}_{\widetilde{\kappa}'}},\widehat{\lambda}'_{\vec{\mathsf{N}}^{\mathrm{ord}}_{\widetilde{\kappa}'}},\widehat{i}'_{\vec{\mathsf{N}}^{\mathrm{ord}}_{\widetilde{\kappa}'}},\widehat{\alpha}'_{\widehat{\mathcal{H}}',p},\widehat{\alpha}^{\mathrm{ord},\prime}_{\widehat{\mathcal{H}}'_p}) \to \vec{\mathsf{N}}^{\mathrm{ord}}_{\widetilde{\kappa}'}$.

Since $\vec{\mathsf{N}}^{\mathrm{ord,tor}}_{\widetilde{\kappa}'}$ is noetherian normal, by [92, IX, 1.4], [28, Ch. I, Prop. 2.7], or [62, Prop. 3.3.1.5], since the family $(\widehat{G}'_{\vec{\mathsf{N}}^{\mathrm{ord}}_{\widetilde{\kappa}'}},\widehat{\lambda}'_{\vec{\mathsf{N}}^{\mathrm{ord}}_{\widetilde{\kappa}'}},\widehat{i}'_{\vec{\mathsf{N}}^{\mathrm{ord}}_{\widetilde{\kappa}'}},\widehat{\alpha}'_{\widehat{\mathcal{H}}',p},\widehat{\alpha}^{\mathrm{ord},\prime}_{\widehat{\mathcal{H}}'_p}) \to \vec{\mathsf{N}}^{\mathrm{ord}}_{\widetilde{\kappa}'}$ is canonically isomorphic to the pullback of $(\widehat{G}_{\vec{\mathsf{N}}^{\mathrm{ord}}_{\widetilde{\kappa}}},\widehat{\lambda}_{\vec{\mathsf{N}}^{\mathrm{ord}}_{\widetilde{\kappa}}},\widehat{i}_{\vec{\mathsf{N}}^{\mathrm{ord}}_{\widetilde{\kappa}}},\widehat{\alpha}_{\widehat{\mathcal{H}}^p},\widehat{\alpha}^{\mathrm{ord}}_{\widehat{\mathcal{H}}_p}) \to \vec{\mathsf{N}}^{\mathrm{ord}}_{\widetilde{\kappa}}$ under $f_{\widetilde{\kappa}',\widetilde{\kappa}}$, as soon as $f_{\widetilde{\kappa}',\widetilde{\kappa}} : \vec{\mathsf{N}}^{\mathrm{ord}}_{\widetilde{\kappa}'} \to \vec{\mathsf{N}}^{\mathrm{ord}}_{\widetilde{\kappa}}$ extends to a morphism $f^{\mathrm{tor}}_{\widetilde{\kappa}',\widetilde{\kappa}} : \vec{\mathsf{N}}^{\mathrm{ord,tor}}_{\widetilde{\kappa}'} \to \vec{\mathsf{N}}^{\mathrm{ord,tor}}_{\widetilde{\kappa}}$, we know that $(\widehat{G}',\widehat{\lambda}',\widehat{i}',\widehat{\alpha}'_{\widehat{\mathcal{H}}',p},\widehat{\alpha}^{\mathrm{ord},\prime}_{\widehat{\mathcal{H}}'_p}) \to \vec{\mathsf{N}}^{\mathrm{ord,tor}}_{\widetilde{\kappa}'}$ is canonically isomorphic to the pullback of $(\widehat{G},\widehat{\lambda},\widehat{i},\widehat{\alpha}_{\widehat{\mathcal{H}}^p},\widehat{\alpha}^{\mathrm{ord}}_{\widehat{\mathcal{H}}_p}) \to \vec{\mathsf{N}}^{\mathrm{ord,tor}}_{\widetilde{\kappa}}$ under $f^{\mathrm{tor}}_{\widetilde{\kappa}',\widetilde{\kappa}}$. Such an extension $f^{\mathrm{tor}}_{\widetilde{\kappa}',\widetilde{\kappa}}$ is necessarily unique, because $\vec{\mathsf{N}}^{\mathrm{ord}}_{\widetilde{\kappa}}$ (resp. $\vec{\mathsf{N}}^{\mathrm{ord}}_{\widetilde{\kappa}'}$) is dense in $\vec{\mathsf{N}}^{\mathrm{ord,tor}}_{\widetilde{\kappa}}$ (resp. $\vec{\mathsf{N}}^{\mathrm{ord,tor}}_{\widetilde{\kappa}'}$). Hence, it suffices to show that $f_{\widetilde{\kappa}',\widetilde{\kappa}} : \vec{\mathsf{N}}^{\mathrm{ord}}_{\widetilde{\kappa}'} \to \vec{\mathsf{N}}^{\mathrm{ord}}_{\widetilde{\kappa}}$ extends locally.

Let $\bar{s}$ be any geometric point of $\vec{\mathsf{N}}^{\mathrm{ord,tor}}_{\widetilde{\kappa}'}$ on the $\vec{\widetilde{\mathsf{Z}}}^{\mathrm{ord}}_{[(\breve{\Phi}_{\widetilde{\mathcal{H}}'},\breve{\delta}_{\widetilde{\mathcal{H}}'},\breve{\rho}')]}$-stratum of $\vec{\widetilde{\mathsf{M}}}^{\mathrm{ord,tor}}_{\widetilde{\mathcal{H}}',\widetilde{\Sigma}^{\mathrm{ord},\prime}}$, where $[(\breve{\Phi}_{\widetilde{\mathcal{H}}'},\breve{\delta}_{\widetilde{\mathcal{H}}'},\breve{\rho}')]$ is represented by some $(\breve{\Phi}_{\widetilde{\mathcal{H}}'},\breve{\delta}_{\widetilde{\mathcal{H}}'},\breve{\rho}')$ with $(\breve{\mathsf{Z}}_{\widetilde{\mathcal{H}}'},\breve{\Phi}_{\widetilde{\mathcal{H}}'} = (\breve{X},\breve{Y},\breve{\phi},\breve{\varphi}_{-2,\widetilde{\mathcal{H}}'},\breve{\varphi}_{0,\widetilde{\mathcal{H}}'}),\breve{\delta}_{\widetilde{\mathcal{H}}'})$ representing some cusp label as in Section 1.2.4. For simplicity, let us fix compatible choices of representatives $(\widetilde{\mathsf{Z}},\widetilde{\Phi} = (\widetilde{X},\widetilde{Y},\widetilde{\phi},\widetilde{\varphi}_{-2},\widetilde{\varphi}_0),\widetilde{\delta})$ and $(\breve{\mathsf{Z}},\breve{\Phi} = (\breve{X},\breve{Y},\breve{\phi},\breve{\varphi}_{-2},\breve{\varphi}_0),\breve{\delta})$, as in Sections 1.2.4 and 7.1.1, in their $\widetilde{\mathcal{H}}'$-orbits. As in Construction 7.2.2.9, each morphism $\xi' : \mathrm{Spec}(V) \to \vec{\mathsf{N}}^{\mathrm{ord,tor}}_{\widetilde{\kappa}'}$ centered at a geometric point $\bar{s}$ of $\vec{\mathsf{N}}^{\mathrm{ord,tor}}_{\widetilde{\kappa}'}$, where V is a complete discrete valuation ring with fraction field K, and where $\eta := \mathrm{Spec}(K)$ is mapped to the generic point of the irreducible component containing the image of $\bar{s}$, determines a tuple

$$(\widehat{B}^{\ddagger},\lambda_{\widehat{B}^{\ddagger}},i_{\widehat{B}^{\ddagger}},\breve{X},\breve{Y},\breve{\phi},\widehat{c}^{\ddagger},\widehat{c}^{\vee,\ddagger},\widehat{\tau}^{\ddagger},\widehat{\tau}^{\vee,\ddagger},[\widehat{\alpha}^{\natural,\mathrm{ord},\ddagger}_{\widehat{\mathcal{H}}'}])$$

as in (7.2.2.14), where $[\widehat{\alpha}^{\natural,\mathrm{ord},\ddagger}_{\widehat{\mathcal{H}}'}]$ is an equivalence class of

$$\widehat{\alpha}^{\natural,\mathrm{ord},\ddagger}_{\widehat{\mathcal{H}}'} = (\breve{\mathsf{Z}}_{\widehat{\mathcal{H}}'},\widehat{\varphi}^{\mathrm{ord},\ddagger}_{-2,\widehat{\mathcal{H}}'},\widehat{\varphi}^{\mathrm{ord},\ddagger}_{-1,\widehat{\mathcal{H}}'},\widehat{\varphi}^{\mathrm{ord},\ddagger}_{0,\widehat{\mathcal{H}}'},\widehat{\delta}^{\mathrm{ord}}_{\widehat{\mathcal{H}}'},\widehat{c}^{\mathrm{ord},\ddagger}_{\widehat{\mathcal{H}}'},\widehat{c}^{\vee,\mathrm{ord},\ddagger}_{\widehat{\mathcal{H}}'},\widehat{\tau}^{\mathrm{ord},\ddagger}_{\widehat{\mathcal{H}}'},\widehat{\tau}^{\vee,\mathrm{ord},\ddagger}_{\widehat{\mathcal{H}}'})$$

as in (7.2.2.15), induced by some

$$\widehat{\alpha}^{\natural,\ddagger}_{\widehat{\mathcal{H}}'} = (\breve{\mathsf{Z}}_{\widehat{\mathcal{H}}'},\widehat{\varphi}^{\ddagger,\sim}_{-2,\widehat{\mathcal{H}}'},\widehat{\varphi}^{\ddagger}_{-1,\widehat{\mathcal{H}}'},\widehat{\varphi}^{\ddagger,\sim}_{0,\widehat{\mathcal{H}}'},\breve{\delta}_{\widehat{\mathcal{H}}'},\widehat{c}^{\ddagger}_{\widehat{\mathcal{H}}'},\widehat{c}^{\vee,\ddagger}_{\widehat{\mathcal{H}}'},\widehat{\tau}^{\ddagger}_{\widehat{\mathcal{H}}'},\widehat{\tau}^{\vee,\ddagger}_{\widehat{\mathcal{H}}'})$$

as in (7.2.2.16) over η, in a way analogous to that in Section 4.1.6, and the pair $(\widehat{\tau}^{\ddagger}, \widehat{\tau}^{\vee,\ddagger})$ defines an element $\upsilon_{\widehat{\tau}^{\ddagger}} = \upsilon_{\widehat{\tau}^{\vee,\ddagger}}$ in $\widehat{\rho}'$ for some $\widehat{\rho}' \subset \widehat{\mathbf{P}}^+_{\breve{\Phi}_{\widehat{\mathcal{H}}'}}$ in $\widehat{\Sigma}_{\breve{\Phi}_{\widehat{\mathcal{H}}'}}$. (We should have denoted all these entries with some extra $'$ in their superscripts, because they are determined by the pullback of $(\widehat{G}', \widehat{\lambda}', \widehat{i}', \widehat{\alpha}'_{\widehat{\mathcal{H}}'^{,p}}, \widehat{\alpha}^{\mathrm{ord},\prime}_{\widehat{\mathcal{H}}'_p}) \to \vec{\mathsf{N}}^{\mathrm{ord,tor}}_{\widetilde{\kappa}'}$. But we omit them for the sake of simplicity.) By forming $\widehat{\mathcal{H}}$-orbits, we obtain a tuple

$$(\widehat{B}^{\ddagger}, \lambda_{\widehat{B}^{\ddagger}}, i_{\widehat{B}^{\ddagger}}, \breve{X}, \breve{Y}, \breve{\phi}, \widehat{c}^{\ddagger}, \widehat{c}^{\vee,\ddagger}, \widehat{\tau}^{\ddagger}, \widehat{\tau}^{\vee,\ddagger}, [\widehat{\alpha}^{\natural,\mathrm{ord},\ddagger}_{\widehat{\mathcal{H}}}]),$$

where $[\widehat{\alpha}^{\natural,\mathrm{ord},\ddagger}_{\widehat{\mathcal{H}}}]$ is an equivalence class of

$$\widehat{\alpha}^{\natural,\mathrm{ord},\ddagger}_{\widehat{\mathcal{H}}} = (\breve{\mathsf{Z}}_{\widehat{\mathcal{H}}}, \widehat{\varphi}^{\mathrm{ord},\ddagger}_{-2,\widehat{\mathcal{H}}}, \widehat{\varphi}^{\mathrm{ord},\ddagger}_{-1,\widehat{\mathcal{H}}}, \widehat{\varphi}^{\mathrm{ord},\ddagger}_{0,\widehat{\mathcal{H}}}, \widehat{\delta}^{\mathrm{ord}}_{\widehat{\mathcal{H}}}, \widehat{c}^{\mathrm{ord},\ddagger}_{\widehat{\mathcal{H}}}, \widehat{c}^{\vee,\mathrm{ord},\ddagger}_{\widehat{\mathcal{H}}}, \widehat{\tau}^{\mathrm{ord},\ddagger}_{\widehat{\mathcal{H}}}, \widehat{\tau}^{\vee,\mathrm{ord},\ddagger}_{\widehat{\mathcal{H}}})$$

induced by some

$$\widehat{\alpha}^{\natural,\ddagger}_{\widehat{\mathcal{H}}} = (\breve{\mathsf{Z}}_{\widehat{\mathcal{H}}}, \widehat{\varphi}^{\ddagger,\sim}_{-2,\widehat{\mathcal{H}}}, \widehat{\varphi}^{\ddagger}_{-1,\widehat{\mathcal{H}}}, \widehat{\varphi}^{\ddagger,\sim}_{0,\widehat{\mathcal{H}}}, \breve{\delta}_{\widehat{\mathcal{H}}}, \widehat{c}^{\ddagger}_{\widehat{\mathcal{H}}}, \widehat{c}^{\vee,\ddagger}_{\widehat{\mathcal{H}}}, \widehat{\tau}^{\ddagger}_{\widehat{\mathcal{H}}}, \widehat{\tau}^{\vee,\ddagger}_{\widehat{\mathcal{H}}})$$

over η, and the pair $(\widehat{\tau}^{\ddagger}, \widehat{\tau}^{\vee,\ddagger})$ defines the same element $\upsilon_{\widehat{\tau}^{\ddagger}} = \upsilon_{\widehat{\tau}^{\vee,\ddagger}}$ in $\widehat{\rho}'$. By assumption, $\widehat{\Sigma}'$ is a refinement of $\widehat{\Sigma}$. Hence, under the canonical isomorphism $\widehat{\mathbf{P}}_{\breve{\Phi}_{\widehat{\mathcal{H}}'}} \cong \widehat{\mathbf{P}}_{\breve{\Phi}_{\widehat{\mathcal{H}}}}$, we have $\widehat{\rho}' \subset \widehat{\rho}$ for some cone $\widehat{\rho} \subset \widehat{\mathbf{P}}^+_{\breve{\Phi}_{\widehat{\mathcal{H}}}}$ in $\widehat{\Sigma}_{\breve{\Phi}_{\widehat{\mathcal{H}}}}$, so that $\upsilon_{\widehat{\tau}^{\ddagger}} = \upsilon_{\widehat{\tau}^{\vee,\ddagger}}$ lies in $\widehat{\rho}$.

By the universal property of $\vec{\widetilde{\mathsf{M}}}^{\mathrm{ord},\breve{\Phi}_{\widetilde{\mathcal{H}}}}_{\widetilde{\mathcal{H}}}$ (as the normalization of $\ddot{\widetilde{\mathsf{M}}}^{\mathrm{ord},\breve{\Phi}_{\widetilde{\mathcal{H}}}}_{\widetilde{\mathcal{H}}}$ in $\widetilde{\mathsf{M}}^{\mathrm{ord},\breve{\Phi}_{\widetilde{\mathcal{H}}}}_{\widetilde{\mathcal{H}}}$, which depends only on $\widehat{\mathcal{H}}$; see Definition 1.2.1.15, (4.2.1.27), the definition preceding (4.2.1.28), and Proposition 4.2.1.29), the data $(\breve{\mathsf{Z}}_{\widehat{\mathcal{H}}}, \breve{\Phi}_{\widehat{\mathcal{H}}} = (\breve{X}, \breve{Y}, \breve{\phi}, \breve{\varphi}_{-2,\widehat{\mathcal{H}}}, \breve{\varphi}_{0,\widehat{\mathcal{H}}}), \breve{\delta}_{\widehat{\mathcal{H}}})$, $(\widehat{\varphi}^{\mathrm{ord},\ddagger}_{-2,\widehat{\mathcal{H}}}, \widehat{\varphi}^{\mathrm{ord},\ddagger}_{0,\widehat{\mathcal{H}}})$ and $(\widehat{B}^{\ddagger}, \lambda_{\widehat{B}^{\ddagger}}, i_{\widehat{B}^{\ddagger}}, \widehat{\varphi}^{\mathrm{ord},\ddagger}_{-1,\widehat{\mathcal{H}}})$ on the torus and abelian parts, which are induced by the corresponding data $(\widehat{\varphi}^{\ddagger,\sim}_{-2,\widehat{\mathcal{H}}}, \widehat{\varphi}^{\ddagger,\sim}_{0,\widehat{\mathcal{H}}})$ and $(\widehat{B}^{\ddagger}, \lambda_{\widehat{B}^{\ddagger}}, i_{\widehat{B}^{\ddagger}}, \widehat{\varphi}^{\ddagger}_{-1,\widehat{\mathcal{H}}})$ over η, define a canonical morphism $\xi_1 : \mathrm{Spec}(V) \to \vec{\widetilde{\mathsf{M}}}^{\mathrm{ord},\breve{\Phi}_{\widetilde{\mathcal{H}}}}_{\widetilde{\mathcal{H}}}$. By the universal property of $\vec{\widetilde{C}}^{\mathrm{ord}}_{\breve{\Phi}_{\widetilde{\mathcal{H}}},\breve{\delta}_{\widetilde{\mathcal{H}}}} \to \vec{\widetilde{\mathsf{M}}}^{\breve{\Phi}_{\widetilde{\mathcal{H}}}}_{\mathrm{ord},\widetilde{\mathcal{H}}}$ (as the normalization of $\ddot{\widetilde{C}}^{\mathrm{ord}}_{\breve{\Phi}_{\widetilde{\mathcal{H}}},n}$ in $C^{\mathrm{ord}}_{\Phi_{\mathcal{H}},\delta_{\mathcal{H}}}$ for some n; see (4.2.1.26), the definition preceding (4.2.1.28), and Proposition 4.2.1.30), the additional data $(\widehat{c}^{\mathrm{ord},\ddagger}_{\widehat{\mathcal{H}}}, \widehat{c}^{\vee,\mathrm{ord},\ddagger}_{\widehat{\mathcal{H}}})$ lifting $(\widehat{c}^{\mathrm{ord},\ddagger}, \widehat{c}^{\vee,\mathrm{ord},\ddagger})$, induced by $(\widehat{c}^{\ddagger}_{\widehat{\mathcal{H}}}, \widehat{c}^{\vee,\ddagger}_{\widehat{\mathcal{H}}})$ over η, define a canonical morphism $\xi_0 : \mathrm{Spec}(V) \to \vec{\widetilde{C}}^{\mathrm{ord}}_{\breve{\Phi}_{\widetilde{\mathcal{H}}},\breve{\delta}_{\widetilde{\mathcal{H}}}}$ lifting ξ_1. By the construction of

$$\vec{\widetilde{\Xi}}^{\mathrm{ord}}_{\breve{\Phi}_{\widetilde{\mathcal{H}}},\breve{\delta}_{\widetilde{\mathcal{H}}},\breve{\sigma}} \cong \underline{\mathrm{Spec}}_{\mathscr{O}_{\vec{\widetilde{C}}^{\mathrm{ord}}_{\breve{\Phi}_{\widetilde{\mathcal{H}}},\breve{\delta}_{\widetilde{\mathcal{H}}}}}} \left(\bigoplus_{\breve{\ell} \in \breve{\sigma}^{\perp}} \vec{\widetilde{\Psi}}^{\mathrm{ord}}_{\breve{\Phi}_{\widetilde{\mathcal{H}}},\breve{\delta}_{\widetilde{\mathcal{H}}}}(\breve{\ell}) \right)$$

over $\vec{\widetilde{C}}^{\mathrm{ord}}_{\breve{\Phi}_{\widetilde{\mathcal{H}}},\breve{\delta}_{\widetilde{\mathcal{H}}}}$, which we can canonically identify as

$$\vec{\widetilde{\Xi}}^{\mathrm{ord}}_{\breve{\Phi}_{\widehat{\mathcal{H}}},\breve{\delta}_{\widehat{\mathcal{H}}}} \cong \underline{\mathrm{Spec}}_{\mathscr{O}_{\vec{\widetilde{C}}^{\mathrm{ord}}_{\breve{\Phi}_{\widehat{\mathcal{H}}},\breve{\delta}_{\widehat{\mathcal{H}}}}}} \left(\bigoplus_{\breve{\ell} \in \widehat{\mathbf{S}}_{\breve{\Phi}_{\widehat{\mathcal{H}}}}} \vec{\widetilde{\Psi}}^{\mathrm{ord}}_{\breve{\Phi}_{\widehat{\mathcal{H}}},\breve{\delta}_{\widehat{\mathcal{H}}}}(\breve{\ell}) \right)$$

over $\vec{\widetilde{C}}^{\mathrm{ord}}_{\breve{\Phi}_{\widehat{\mathcal{H}}},\breve{\delta}_{\widehat{\mathcal{H}}}}$ (see Proposition 7.1.2.6), it enjoys the universal property (similar to that of $\vec{\widetilde{\Xi}}^{\mathrm{ord}}_{\breve{\Phi}_{\widetilde{\mathcal{H}}},\breve{\delta}_{\widetilde{\mathcal{H}}}} \to \vec{\widetilde{C}}^{\mathrm{ord}}_{\breve{\Phi}_{\widetilde{\mathcal{H}}},\breve{\delta}_{\widetilde{\mathcal{H}}}}$; see (4.2.1.25), the definition preceding (4.2.1.28), and Propositions 4.2.1.30 and 4.2.1.46) such that the final part of the data $(\widehat{\tau}^{\mathrm{ord},\ddagger}_{\widehat{\mathcal{H}}}, \widehat{\tau}^{\vee,\mathrm{ord},\ddagger}_{\widehat{\mathcal{H}}})$ lifting $(\widehat{\tau}^{\ddagger}, \widehat{\tau}^{\vee,\ddagger})$, induced by $(\widehat{\tau}^{\ddagger}_{\widehat{\mathcal{H}}}, \widehat{\tau}^{\vee,\ddagger}_{\widehat{\mathcal{H}}})$ over η, determines a canonical morphism $\tilde{\xi}_K : \mathrm{Spec}(K) \to \vec{\widetilde{\Xi}}^{\mathrm{ord}}_{\breve{\Phi}_{\widetilde{\mathcal{H}}},\breve{\delta}_{\widetilde{\mathcal{H}}},\breve{\sigma}}$ lifting ξ_0 under the canonical morphism $\vec{\widetilde{\Xi}}^{\mathrm{ord}}_{\breve{\Phi}_{\widetilde{\mathcal{H}}},\breve{\delta}_{\widetilde{\mathcal{H}}},\breve{\sigma}} \to \vec{\widetilde{C}}^{\mathrm{ord}}_{\breve{\Phi}_{\widetilde{\mathcal{H}}},\breve{\delta}_{\widetilde{\mathcal{H}}}}$. Since the element $\upsilon_{\widehat{\tau}^{\ddagger}} = \upsilon_{\widehat{\tau}^{\vee,\ddagger}}$ defined by $(\widehat{\tau}^{\ddagger}, \widehat{\tau}^{\vee,\ddagger})$ lies in $\widehat{\rho}' \subset \widehat{\rho}$, by the construction of

$$\vec{\widetilde{\Xi}}^{\mathrm{ord}}_{\breve{\Phi}_{\widetilde{\mathcal{H}}},\breve{\delta}_{\widetilde{\mathcal{H}}},\breve{\sigma}}(\breve{\rho}) \cong \underline{\mathrm{Spec}}_{\mathscr{O}_{\vec{\widetilde{C}}^{\mathrm{ord}}_{\breve{\Phi}_{\widetilde{\mathcal{H}}},\breve{\delta}_{\widetilde{\mathcal{H}}}}}} \left(\bigoplus_{\breve{\ell} \in \breve{\sigma}^{\perp} \cap \breve{\rho}^{\vee}} \vec{\widetilde{\Psi}}^{\mathrm{ord}}_{\breve{\Phi}_{\widetilde{\mathcal{H}}},\breve{\delta}_{\widetilde{\mathcal{H}}}}(\breve{\ell}) \right)$$

(see (7.2.1.9)), which we can canonically identify as

$$\vec{\widehat{\Xi}}^{\mathrm{ord}}_{\breve{\Phi}_{\widehat{\mathcal{H}}},\breve{\delta}_{\widehat{\mathcal{H}}}}(\widehat{\rho}) = \underline{\mathrm{Spec}}_{\mathscr{O}_{\vec{\widehat{C}}^{\mathrm{ord}}_{\breve{\Phi}_{\widehat{\mathcal{H}}},\breve{\delta}_{\widehat{\mathcal{H}}}}}} \left(\bigoplus_{\breve{\ell} \in \widehat{\rho}^{\vee}} \vec{\widehat{\Psi}}^{\mathrm{ord}}_{\breve{\Phi}_{\widehat{\mathcal{H}}},\breve{\delta}_{\widehat{\mathcal{H}}}}(\breve{\ell}) \right)$$

(see (7.1.2.10)), which depends only on $\widehat{\mathcal{H}}$ and on $\widehat{\rho}^{\vee} \cong \breve{\sigma}^{\perp} \cap \breve{\rho}^{\vee}$, and by the same argument as in the proof of Proposition 4.2.2.8, the morphism $\tilde{\xi}_K$ extends to a morphism $\tilde{\xi} : \mathrm{Spec}(V) \to \vec{\widetilde{\Xi}}^{\mathrm{ord}}_{\breve{\Phi}_{\widetilde{\mathcal{H}}},\breve{\delta}_{\widetilde{\mathcal{H}}},\breve{\sigma}}(\breve{\rho})$ lifting ξ_0 under the canonical morphism $\vec{\widetilde{\Xi}}^{\mathrm{ord}}_{\breve{\Phi}_{\widetilde{\mathcal{H}}},\breve{\delta}_{\widetilde{\mathcal{H}}},\breve{\sigma}}(\breve{\rho}) \to \vec{\widetilde{C}}^{\mathrm{ord}}_{\breve{\Phi}_{\widetilde{\mathcal{H}}},\breve{\delta}_{\widetilde{\mathcal{H}}}}$, which maps the special point of $\mathrm{Spec}(V)$ to the $\breve{\rho}$-stratum $\vec{\widetilde{\Xi}}^{\mathrm{ord}}_{\breve{\Phi}_{\widetilde{\mathcal{H}}},\breve{\delta}_{\widetilde{\mathcal{H}}},\breve{\rho}}$ of $\vec{\widetilde{\Xi}}^{\mathrm{ord}}_{\breve{\Phi}_{\widetilde{\mathcal{H}}},\breve{\delta}_{\widetilde{\mathcal{H}}},\breve{\sigma}}(\breve{\rho})$. (Alternatively, we can noncanonically lift $\upsilon_{\widehat{\tau}^{\ddagger}} = \upsilon_{\widehat{\tau}^{\vee,\ddagger}}$ to elements of $\breve{\rho} \subset \mathbf{P}^{+}_{\breve{\Phi}_{\widetilde{\mathcal{H}}}}$, work with $\vec{\widetilde{\Xi}}^{\mathrm{ord}}_{\breve{\Phi}_{\widetilde{\mathcal{H}}},\breve{\delta}_{\widetilde{\mathcal{H}}}}$ and $\vec{\widetilde{\Xi}}^{\mathrm{ord}}_{\breve{\Phi}_{\widetilde{\mathcal{H}}},\breve{\delta}_{\widetilde{\mathcal{H}}}}(\breve{\rho})$ directly, and invoke the original Proposition 4.2.2.8.) Since V is complete, $\tilde{\xi}$ induces a morphism $\hat{\xi}$ from $\mathrm{Spf}(V)$ to $\vec{\widetilde{\mathfrak{X}}}^{\mathrm{ord}}_{\breve{\Phi}_{\widetilde{\mathcal{H}}},\breve{\delta}_{\widetilde{\mathcal{H}}},\breve{\sigma},\breve{\rho}}$, the formal completion of $\vec{\widetilde{\Xi}}^{\mathrm{ord}}_{\breve{\Phi}_{\widetilde{\mathcal{H}}},\breve{\delta}_{\widetilde{\mathcal{H}}},\breve{\sigma}}(\breve{\rho})$ along its $\breve{\rho}$-stratum $\vec{\widetilde{\Xi}}^{\mathrm{ord}}_{\breve{\Phi}_{\widetilde{\mathcal{H}}},\breve{\delta}_{\widetilde{\mathcal{H}}},\breve{\rho}}$. Then the composition of $\hat{\xi}$ with the canonical morphism $\vec{\widetilde{\mathfrak{X}}}^{\mathrm{ord}}_{\breve{\Phi}_{\widetilde{\mathcal{H}}},\breve{\delta}_{\widetilde{\mathcal{H}}},\breve{\sigma},\breve{\rho}} \to \vec{\mathsf{N}}^{\mathrm{ord,tor}}_{\widetilde{\kappa}}$ gives a canonical morphism $\xi : \mathrm{Spec}(V) \to \vec{\mathsf{N}}^{\mathrm{ord,tor}}_{\widetilde{\kappa}}$.

As explained in Construction 7.2.2.9, $\xi_\eta := \xi|_\eta : \eta = \mathrm{Spec}(K) \to \vec{\mathsf{N}}^{\mathrm{ord}}_{\widetilde{\kappa}}$ is determined by the pullback $(\widehat{G}'_\eta, \widehat{\lambda}'_\eta, \widehat{i}'_\eta, \widehat{\alpha}'_{\widehat{\mathcal{H}}'^{,p},\eta}, \widehat{\alpha}^{\mathrm{ord},\prime}_{\widehat{\mathcal{H}}_p,\eta}) \to \mathrm{Spec}(K)$ of $(\widehat{G}', \widehat{\lambda}', \widehat{i}', \widehat{\alpha}'_{\widehat{\mathcal{H}}'^{,p}}, \widehat{\alpha}^{\mathrm{ord},\prime}_{\widehat{\mathcal{H}}'_p}) \to \vec{\mathsf{N}}^{\mathrm{ord,tor}}_{\widetilde{\kappa}'}$ under $\xi'_\eta := \xi'|_\eta : \mathrm{Spec}(K) \to \vec{\mathsf{N}}^{\mathrm{ord,tor}}_{\widetilde{\kappa}'}$, whose $\widehat{\mathcal{H}}$-orbit $(\widehat{G}'_\eta, \widehat{\lambda}'_\eta, \widehat{i}'_\eta, \widehat{\alpha}'_{\widehat{\mathcal{H}}^p,\eta}, \widehat{\alpha}^{\mathrm{ord},\prime}_{\widehat{\mathcal{H}}_p,\eta}) \to \mathrm{Spec}(K)$ is (as explained in the first paragraph of this proof) isomorphic to the pullback of $(\widehat{G}, \widehat{\lambda}, \widehat{i}, \widehat{\alpha}_{\widehat{\mathcal{H}}^p}, \widehat{\alpha}^{\mathrm{ord}}_{\widehat{\mathcal{H}}_p}) \to \vec{\mathsf{N}}^{\mathrm{ord,tor}}_{\widetilde{\kappa}}$ under the composition of $\xi'_\eta : \mathrm{Spec}(K) \to \vec{\mathsf{N}}^{\mathrm{ord}}_{\widetilde{\kappa}'}$ with $f_{\widetilde{\kappa}',\widetilde{\kappa}} : \vec{\mathsf{N}}^{\mathrm{ord,tor}}_{\widetilde{\kappa}'} \to \vec{\mathsf{N}}^{\mathrm{ord,tor}}_{\widetilde{\kappa}}$. Hence, $\xi_\eta = f_{\widetilde{\kappa}',\widetilde{\kappa}} \circ \xi'_\eta$ by the universal property of $\vec{\mathsf{N}}^{\mathrm{ord}}_{\widetilde{\kappa}}$, and $\xi : \mathrm{Spec}(V) \to \vec{\mathsf{N}}^{\mathrm{ord,tor}}_{\widetilde{\kappa}}$ can be interpreted as a (necessarily unique) extension of $f_{\widetilde{\kappa}',\widetilde{\kappa}} \circ \xi'_\eta : \mathrm{Spec}(K) \to \vec{\mathsf{N}}^{\mathrm{ord}}_{\widetilde{\kappa}}$.

Since $\xi' : \mathrm{Spec}(V) \to \vec{\mathsf{N}}^{\mathrm{ord,tor}}_{\widetilde{\kappa}'}$ and $\bar{s}$ (the prescribed center of ξ') are arbitrary, and since $\vec{\mathsf{N}}^{\mathrm{ord,tor}}_{\widetilde{\kappa}'}$ is noetherian normal, this shows that $f_{\widetilde{\kappa}',\widetilde{\kappa}}$ extends to $f^{\mathrm{tor}}_{\widetilde{\kappa}',\widetilde{\kappa}}$, as desired. The argument also shows that the restriction of $f^{\mathrm{tor}}_{\widetilde{\kappa}',\widetilde{\kappa}}$ to the $[(\breve{\Phi}_{\widetilde{\mathcal{H}}'}, \breve{\delta}_{\widetilde{\mathcal{H}}'}, \breve{\rho}')]$-stratum $\vec{\widetilde{\mathsf{Z}}}^{\mathrm{ord}}_{[(\breve{\Phi}_{\widetilde{\mathcal{H}}'},\breve{\delta}_{\widetilde{\mathcal{H}}'},\breve{\rho}')]} \cong (\vec{\widetilde{\Xi}}^{\mathrm{ord}}_{\breve{\Phi}_{\widetilde{\mathcal{H}}'},\breve{\delta}_{\widetilde{\mathcal{H}}'}})_{\breve{\rho}'}$ of $\vec{\widetilde{\mathsf{M}}}^{\mathrm{ord,tor}}_{\widetilde{\mathcal{H}}',\widetilde{\Sigma}^{\mathrm{ord},\prime}}$ coincides with the canonical morphism $(\vec{\widetilde{\Xi}}^{\mathrm{ord}}_{\breve{\Phi}_{\widetilde{\mathcal{H}}'},\breve{\delta}_{\widetilde{\mathcal{H}}'}})_{\breve{\rho}'} \to \vec{\widetilde{\Xi}}^{\mathrm{ord}}_{\breve{\Phi}_{\widetilde{\mathcal{H}}},\breve{\delta}_{\widetilde{\mathcal{H}}},\breve{\rho}}$ on geometric points. Since the images of such morphisms cover $\widetilde{\Xi}^{\rightarrow\mathrm{ord}}_{\breve{\Phi}_{\widetilde{\mathcal{H}}},\breve{\delta}_{\widetilde{\mathcal{H}}},\breve{\rho}}$, because $\widehat{\rho} = \mathrm{pr}_{(\widehat{\mathbf{S}}_{\breve{\Phi}_{\widehat{\mathcal{H}}}})^\vee_{\mathbb{R}}}(\breve{\rho})$ is covered by the cones $\widehat{\rho}' = \mathrm{pr}_{(\widehat{\mathbf{S}}_{\breve{\Phi}_{\widehat{\mathcal{H}}}})^\vee_{\mathbb{R}}}(\breve{\rho}')$, the morphism $f^{\mathrm{tor}}_{\widetilde{\kappa}',\widetilde{\kappa}}$ is surjective.

By considering ordinary good algebraic models in Section 5.1.2 and by arguing as in the paragraph preceding [61, Lem. 5.10], the morphism $f^{\mathrm{tor}}_{\widetilde{\kappa}',\widetilde{\kappa}} : \vec{\mathsf{N}}^{\mathrm{ord,tor}}_{\widetilde{\kappa}'} \to \vec{\mathsf{N}}^{\mathrm{ord,tor}}_{\widetilde{\kappa}}$ is étale locally given by the canonical morphism $\widetilde{\Xi}^{\rightarrow\mathrm{ord}}_{\breve{\Phi}_{\widetilde{\mathcal{H}}'},\breve{\delta}_{\widetilde{\mathcal{H}}'},\breve{\sigma}'}(\breve{\rho}') \to \widetilde{\Xi}^{\rightarrow\mathrm{ord}}_{\breve{\Phi}_{\widetilde{\mathcal{H}}},\breve{\delta}_{\widetilde{\mathcal{H}}},\breve{\sigma}}(\breve{\rho})$, because the tautological data (as in (7.2.2.14)) over $\widetilde{\Xi}^{\rightarrow\mathrm{ord}}_{\breve{\Phi}_{\widetilde{\mathcal{H}}'},\breve{\delta}_{\widetilde{\mathcal{H}}'},\breve{\sigma}'}(\breve{\rho}')$ is the pullback of the one over $\widetilde{\Xi}^{\rightarrow\mathrm{ord}}_{\breve{\Phi}_{\widetilde{\mathcal{H}}},\breve{\delta}_{\widetilde{\mathcal{H}}},\breve{\sigma}}(\breve{\rho})$. By construction, the induced morphism $\widetilde{\Xi}^{\rightarrow\mathrm{ord}}_{\breve{\Phi}_{\widetilde{\mathcal{H}}'},\breve{\delta}_{\widetilde{\mathcal{H}}'},\breve{\sigma}'}(\breve{\rho}') \to \widetilde{\Xi}^{\rightarrow\mathrm{ord}}_{\breve{\Phi}_{\widetilde{\mathcal{H}}},\breve{\delta}_{\widetilde{\mathcal{H}}},\breve{\sigma}}(\breve{\rho}) \underset{\vec{\mathsf{S}}_{0,r_\kappa}}{\times} \vec{\mathsf{S}}_{0,r_{\kappa'}}$ is log étale and equivariant with respect to the canonical homomorphism $\widetilde{E}_{\breve{\Phi}_{\widetilde{\mathcal{H}}'},\breve{\sigma}'} \to \widetilde{E}_{\breve{\Phi}_{\widetilde{\mathcal{H}}},\breve{\sigma}}$ between tori, which (by Proposition 1.3.2.56 again) can be canonically identified with the canonical log étale morphism $\widehat{\Xi}^{\rightarrow\mathrm{ord}}_{\breve{\Phi}_{\widehat{\mathcal{H}}'},\breve{\delta}_{\widehat{\mathcal{H}}'}}(\widehat{\rho}') \to \widehat{\Xi}^{\rightarrow\mathrm{ord}}_{\breve{\Phi}_{\widehat{\mathcal{H}}},\breve{\delta}_{\widehat{\mathcal{H}}}}(\widehat{\rho})$, equivariant with respect to the canonical homomorphism $\widehat{E}_{\breve{\Phi}_{\widehat{\mathcal{H}}'}} \to \widehat{E}_{\breve{\Phi}_{\widehat{\mathcal{H}}}}$ between tori (dual to the canonical homomorphism $\widehat{\mathbf{S}}_{\breve{\Phi}_{\widehat{\mathcal{H}}}} \to \widehat{\mathbf{S}}_{\breve{\Phi}_{\widehat{\mathcal{H}}'}}$ up to canonical identifications; see Definition 1.2.4.29).

We note that the above argument shows that $\vec{\mathsf{N}}^{\mathrm{ord,tor}}_{\widetilde{\kappa}}$ satisfies the descriptions of stratification and formal completions as in the third, fourth, and fifth paragraphs of (1) of Theorem 7.1.4.1, because they follow from (2) and (5) of Theorem 5.2.1.1 (for $\widetilde{\mathsf{M}}^{\mathrm{tor}}_{\mathcal{H},\widetilde{\Sigma}^{\mathrm{ord}}}$), from Lemma 5.2.4.38, and from the justifications provided in Section 5.2.4.

The remainder of the proposition then follows. □

Thanks to Lemma 7.2.1.3 and Proposition 7.2.2.19, we can make the following:

Definition 7.2.2.20. (*Compare with Definition* 1.3.4.20.) *For* $\widetilde{\kappa} \in \widetilde{\mathbf{K}}^{\mathrm{ord},++}_{Q,\mathcal{H}}$ *which defines* $\kappa = [\widetilde{\kappa}] \in \mathbf{K}^{\mathrm{ord},++}_{Q,\mathcal{H}}$ (*see Definition* 7.1.1.11), *we shall denote* $\widetilde{\kappa}^{\mathrm{isog}} : \underline{\mathrm{Hom}}_{\mathcal{O}}(Q, G_{\vec{\mathsf{M}}^{\mathrm{ord}}_{\mathcal{H}_\kappa}})^\circ \to \vec{\mathsf{N}}^{\mathrm{ord,grp}}_{\widetilde{\kappa}}$ *and* $\widetilde{\kappa}^{\mathrm{tor}} : \vec{\mathsf{N}}^{\mathrm{ord}}_{\widetilde{\kappa}} \hookrightarrow \vec{\mathsf{N}}^{\mathrm{ord,tor}}_{\widetilde{\kappa}}$ *by* $\kappa^{\mathrm{isog}} : \underline{\mathrm{Hom}}_{\mathcal{O}}(Q, G_{\vec{\mathsf{M}}^{\mathrm{ord}}_{\mathcal{H}_\kappa}})^\circ \to \vec{\mathsf{N}}^{\mathrm{ord,grp}}_{\kappa}$ *and* $\kappa^{\mathrm{tor}} : \vec{\mathsf{N}}^{\mathrm{ord}}_{\kappa} \hookrightarrow \vec{\mathsf{N}}^{\mathrm{ord,tor}}_{\kappa}$, *respectively. For* $\widetilde{\kappa}$ *and* $\widetilde{\kappa}'$ *in* $\widetilde{\mathbf{K}}^{\mathrm{ord},++}_{Q,\mathcal{H}}$ *such that* $\kappa' = [\widetilde{\kappa}'] \succ \kappa = [\widetilde{\kappa}]$ *in* $\mathbf{K}^{\mathrm{ord},++}_{Q,\mathcal{H}}$, *we shall denote the canonical morphisms* $f_{\widetilde{\kappa}',\widetilde{\kappa}} : \vec{\mathsf{N}}^{\mathrm{ord}}_{\widetilde{\kappa}'} \to \vec{\mathsf{N}}^{\mathrm{ord}}_{\widetilde{\kappa}}$, $f^{\mathrm{grp}}_{\widetilde{\kappa}',\widetilde{\kappa}} := \widetilde{\kappa}^{\mathrm{isog}} \circ ((\widetilde{\kappa}')^{\mathrm{isog}})^{-1} : \vec{\mathsf{N}}^{\mathrm{ord,grp}}_{\widetilde{\kappa}} \to \vec{\mathsf{N}}^{\mathrm{ord,grp}}_{\widetilde{\kappa}} \underset{\vec{\mathsf{M}}^{\mathrm{ord}}_{\mathcal{H}_\kappa}}{\times} \vec{\mathsf{M}}^{\mathrm{ord}}_{\mathcal{H}_{\kappa'}}$, *and* $f^{\mathrm{tor}}_{\widetilde{\kappa}',\widetilde{\kappa}} : \vec{\mathsf{N}}^{\mathrm{ord,tor}}_{\widetilde{\kappa}'} \to \vec{\mathsf{N}}^{\mathrm{ord,tor}}_{\widetilde{\kappa}}$ *by* $f_{\kappa',\kappa} : \vec{\mathsf{N}}^{\mathrm{ord}}_{\kappa'} \to \vec{\mathsf{N}}^{\mathrm{ord}}_{\kappa}$, $f^{\mathrm{grp}}_{\kappa',\kappa} := \kappa^{\mathrm{isog}} \circ ((\kappa')^{\mathrm{isog}})^{-1} : \vec{\mathsf{N}}^{\mathrm{ord,grp}}_{\kappa'} \to \vec{\mathsf{N}}^{\mathrm{ord,grp}}_{\kappa} \underset{\vec{\mathsf{M}}^{\mathrm{ord}}_{\mathcal{H}_\kappa}}{\times} \vec{\mathsf{M}}^{\mathrm{ord}}_{\mathcal{H}_{\kappa'}}$, *and* $f^{\mathrm{tor}}_{\kappa',\kappa} : \vec{\mathsf{N}}^{\mathrm{ord,tor}}_{\kappa'} \to \vec{\mathsf{N}}^{\mathrm{ord,tor}}_{\kappa}$, *respectively.* (*That is, we drop the tildes in all such notations.*) *We shall denote by* $\vec{\mathsf{Z}}^{\mathrm{ord}}_{[(\breve{\Phi}_{\widehat{\mathcal{H}}},\breve{\delta}_{\widehat{\mathcal{H}}},\widehat{\tau})]}$ *the* $[(\breve{\Phi}_{\widehat{\mathcal{H}}},\breve{\delta}_{\widehat{\mathcal{H}}},\widehat{\tau})]$*-stratum of* $\vec{\mathsf{N}}^{\mathrm{ord,tor}}_{\kappa}$, *which is the* $[(\breve{\Phi}_{\widetilde{\mathcal{H}}},\breve{\delta}_{\widetilde{\mathcal{H}}},\breve{\tau})]$*-stratum* $\vec{\mathsf{Z}}^{\mathrm{ord}}_{[(\breve{\Phi}_{\widetilde{\mathcal{H}}},\breve{\delta}_{\widetilde{\mathcal{H}}},\breve{\tau})]} \cong \widetilde{\Xi}^{\rightarrow\mathrm{ord}}_{\breve{\Phi}_{\widetilde{\mathcal{H}}},\breve{\delta}_{\widetilde{\mathcal{H}}},\breve{\tau}} \cong \widehat{\Xi}^{\rightarrow\mathrm{ord}}_{\breve{\Phi}_{\widehat{\mathcal{H}}},\breve{\delta}_{\widehat{\mathcal{H}}},\widehat{\tau}}$ *of* $\vec{\mathsf{N}}^{\mathrm{ord,tor}}_{\widetilde{\kappa}}$ *under the canonical identification* *between* $\vec{\mathsf{N}}_{\kappa}^{\mathrm{ord,tor}}$ *and* $\vec{\mathsf{N}}_{\widetilde{\kappa}}^{\mathrm{ord,tor}}$ *(up to canonical isomorphism) (when* $(\breve{\Phi}_{\widehat{\mathcal{H}}}, \breve{\delta}_{\widehat{\mathcal{H}}}, \widehat{\tau})$ *is determined by* $(\breve{\Phi}_{\widetilde{\mathcal{H}}}, \breve{\delta}_{\widetilde{\mathcal{H}}}, \breve{\tau})$ *as in Section* 1.2.4*).*

Corollary 7.2.2.21. *Suppose that* $\kappa = (\widehat{\mathcal{H}}, \widehat{\Sigma})$ *and* $\kappa' = (\widehat{\mathcal{H}}', \widehat{\Sigma}')$ *are elements of* $\mathbf{K}_{Q,\mathcal{H}}^{\mathrm{ord},++}$ *satisfying* $\kappa' \succ \kappa$, *and that* $\mathcal{H}_{\kappa}$ *and* $\mathcal{H}_{\kappa'}$ *are equally deep as in Definition* 3.2.2.9*. Then the canonical morphism* $f_{\kappa',\kappa}^{\mathrm{tor}} : \vec{\mathsf{N}}_{\kappa'}^{\mathrm{ord,tor}} \to \vec{\mathsf{N}}_{\kappa}^{\mathrm{ord,tor}}$ *(see Proposition* 7.2.2.19*) is proper.*

Proof. Suppose $\kappa = [\widetilde{\kappa}]$ for some $\widetilde{\kappa} = (\widetilde{\mathcal{H}}, \widetilde{\Sigma}, \widetilde{\sigma}) \in \widetilde{\mathbf{K}}_{Q,\mathcal{H}}^{\mathrm{ord},++}$. Take a third element $\kappa'' = (\widehat{\mathcal{H}}'', \widehat{\Sigma}'') \in \mathbf{K}_{Q,\mathcal{H}}^{\mathrm{ord},++}$ such that $\widehat{\mathcal{H}}'' = \widehat{\mathcal{H}}'$ (and so that $\mathcal{H}_{\kappa''} = \mathcal{H}_{\kappa'}$), such that $\kappa'' \succ \kappa'$, and such that $\kappa'' = [\widetilde{\kappa}'']$ for some $\widetilde{\kappa}'' = (\widetilde{\mathcal{H}}'', \widetilde{\Sigma}'', \widetilde{\sigma}'') \in \widetilde{\mathbf{K}}_{Q,\mathcal{H}}^{\mathrm{ord},++}$ such that $\widetilde{\mathcal{H}}'' \subset \widetilde{\mathcal{H}}$ are equally deep as in Definition 3.2.2.9, such that $\widetilde{\Sigma}''$ is a refinement of $\widetilde{\Sigma}$, and such that $\widetilde{\sigma}''$ is contained in $\widetilde{\sigma}$. Then we have canonical surjections $f_{\kappa'',\kappa'}^{\mathrm{tor}} : \vec{\mathsf{N}}_{\kappa''}^{\mathrm{ord,tor}} \to \vec{\mathsf{N}}_{\kappa'}^{\mathrm{ord,tor}}$ and $f_{\kappa'',\kappa}^{\mathrm{tor}} : \vec{\mathsf{N}}_{\kappa''}^{\mathrm{ord,tor}} \to \vec{\mathsf{N}}_{\kappa}^{\mathrm{ord,tor}}$ such that $f_{\kappa'',\kappa}^{\mathrm{tor}} = f_{\kappa',\kappa}^{\mathrm{tor}} \circ f_{\kappa'',\kappa'}^{\mathrm{tor}}$ (see Proposition 7.2.2.19), and $f_{\kappa'',\kappa}^{\mathrm{tor}}$ is proper because it is the restriction (to a closed subscheme) of the canonical surjection $\vec{\mathsf{M}}_{\widetilde{\mathcal{H}}'',\widetilde{\Sigma}''}^{\mathrm{ord,tor}} \to \vec{\mathsf{M}}_{\widetilde{\mathcal{H}},\widetilde{\Sigma}}^{\mathrm{ord,tor}}$, the latter being proper because $\widetilde{\mathcal{H}}$ and $\widetilde{\mathcal{H}}''$ are equally deep (see Proposition 5.2.2.2). Consequently, $f_{\kappa',\kappa}^{\mathrm{tor}}$ is also proper, because $f_{\kappa'',\kappa'}^{\mathrm{tor}}$ is surjective. □

7.2.3 *Extensibility of* f_{κ}

The goal of this subsection is to show that, when $\kappa \in \mathbf{K}_{Q,\mathcal{H},\Sigma^{\mathrm{ord}}}^{\mathrm{ord},++}$ (see Definitions 7.1.1.11 and 7.1.1.19), the structural morphism $f_{\kappa} : \vec{\mathsf{N}}_{\kappa}^{\mathrm{ord}} \to \vec{\mathsf{M}}_{\mathcal{H},r_{\kappa}}^{\mathrm{ord}}$ (see Section 7.2.1) extends (necessarily uniquely) to a morphism $f_{\kappa}^{\mathrm{tor}} : \vec{\mathsf{N}}_{\kappa}^{\mathrm{ord,tor}} \to \vec{\mathsf{M}}_{\mathcal{H},r_{\kappa}}^{\mathrm{ord,tor}} = \vec{\mathsf{M}}_{\mathcal{H},\Sigma^{\mathrm{ord}},r_{\kappa}}^{\mathrm{ord,tor}}$ between the compactifications. Recall that f_{κ} is the composition of the structural morphism $\vec{\mathsf{N}}_{\kappa}^{\mathrm{ord}} \to \vec{\mathsf{M}}_{\mathcal{H}_{\kappa}}^{\mathrm{ord}}$ with the canonical morphism $\vec{\mathsf{M}}_{\mathcal{H}_{\kappa}}^{\mathrm{ord}} \to \vec{\mathsf{M}}_{\mathcal{H},r_{\kappa}}^{\mathrm{ord}}$.

Let us begin with an arbitrary element $\kappa \in \mathbf{K}_{Q,\mathcal{H}}^{\mathrm{ord},++}$ (not necessarily in $\mathbf{K}_{Q,\mathcal{H},\Sigma^{\mathrm{ord}}}^{\mathrm{ord},++}$), which is of the form $\kappa = [\widetilde{\kappa}]$ for some $\widetilde{\kappa} = (\widetilde{\mathcal{H}}, \widetilde{\Sigma}^{\mathrm{ord}}, \widetilde{\sigma}) \in \widetilde{\mathbf{K}}_{Q,\mathcal{H}}^{\mathrm{ord},++}$ (not necessarily in $\widetilde{\mathbf{K}}_{Q,\mathcal{H},\Sigma^{\mathrm{ord}}}^{\mathrm{ord},++}$).

By Construction 7.2.2.1, we have the pullback (7.2.2.4)

$$(\widehat{G}, \widehat{\lambda} : \widehat{G} \to \widehat{G}^{\vee}, \widehat{i}, \widehat{\alpha}_{\widehat{\mathcal{H}}^p}, \widehat{\alpha}_{\widehat{\mathcal{H}}_p}^{\mathrm{ord}}) \to \vec{\mathsf{N}}_{\kappa}^{\mathrm{ord,tor}}$$

of (7.2.2.2) to $\vec{\mathsf{N}}_{\kappa}^{\mathrm{ord,tor}} = \vec{\mathsf{N}}_{\widetilde{\kappa}}^{\mathrm{ord,tor}}$ (see Definition 7.2.2.20), such that the semi-abelian scheme $\widehat{G}_{\vec{\mathsf{N}}_{\kappa}^{\mathrm{ord}}} = \widetilde{G}_{\vec{\mathsf{N}}_{\kappa}^{\mathrm{ord}}}$ (resp. $\widehat{G}_{\vec{\mathsf{N}}_{\kappa}^{\mathrm{ord}}}^{\vee} = \widetilde{G}_{\vec{\mathsf{N}}_{\kappa}^{\mathrm{ord}}}^{\vee}$) (*cf.* (7.2.2.5)) is the extension of $G_{\vec{\mathsf{N}}_{\kappa}^{\mathrm{ord}}}$ (resp. $G_{\vec{\mathsf{N}}_{\kappa}^{\mathrm{ord}}}^{\vee}$), the pullback under $f_{\kappa} : \vec{\mathsf{N}}_{\kappa}^{\mathrm{ord}} \to \vec{\mathsf{M}}_{\mathcal{H},r_{\kappa}}^{\mathrm{ord}}$ of the abelian scheme $G_{\vec{\mathsf{M}}_{\mathcal{H},r_{\kappa}}^{\mathrm{ord}}}$ (resp. $G_{\vec{\mathsf{M}}_{\mathcal{H},r_{\kappa}}^{\mathrm{ord}}}^{\vee}$) over $\vec{\mathsf{M}}_{\mathcal{H},r_{\kappa}}^{\mathrm{ord}}$, by the split torus $\widetilde{T}_{\vec{\mathsf{N}}_{\kappa}^{\mathrm{ord}}}$ (resp. $\widetilde{T}_{\vec{\mathsf{N}}_{\kappa}^{\mathrm{ord}}}^{\vee}$) over $\vec{\mathsf{N}}_{\kappa}^{\mathrm{ord}}$ with character group $\widetilde{X}$ (resp. $\widetilde{Y}$), parameterized by the tautological object $\widetilde{c} : \widetilde{X} \to G_{\vec{\mathsf{N}}_{\kappa}^{\mathrm{ord}}}^{\vee}$ (resp. $\widetilde{c}^{\vee} : \widetilde{Y} \to G_{\vec{\mathsf{N}}_{\kappa}^{\mathrm{ord}}}$) over $\vec{\mathsf{N}}_{\kappa}^{\mathrm{ord}} \cong \vec{C}_{\widetilde{\Phi}_{\widetilde{\mathcal{H}}},\widetilde{\delta}_{\widetilde{\mathcal{H}}}}^{\mathrm{ord}}$. By taking the abelian parts of $\widehat{\lambda}$, $\widehat{i}$, and $(\widehat{\alpha}_{\widehat{\mathcal{H}}^p}, \widehat{\alpha}_{\widehat{\mathcal{H}}_p}^{\mathrm{ord}})$, we obtain a polarization, an $\mathcal{O}$-endomorphism structure, and the ordinary level structure $(\overline{\alpha}_{\mathcal{H}^p}, \overline{\alpha}^{\mathrm{ord}}_{\mathcal{H}_p})$ on the abelian part of $\widehat{G}_{\vec{\mathsf{N}}^{\mathrm{ord}}_{\kappa}}$, which agree with the pullbacks of the data λ, i, and $(\alpha_{\mathcal{H}^p}, \alpha^{\mathrm{ord}}_{\mathcal{H}_p})$ over $\vec{\mathsf{M}}^{\mathrm{ord}}_{\mathcal{H},r_\kappa}$ to $\vec{\mathsf{N}}^{\mathrm{ord}}_{\kappa}$ by $f_\kappa : \vec{\mathsf{N}}^{\mathrm{ord}}_{\kappa} \to \vec{\mathsf{M}}^{\mathrm{ord}}_{\mathcal{H},r_\kappa}$. By noetherian normality of (the closure) $\vec{\mathsf{N}}^{\mathrm{ord,tor}}_{\kappa}$ (of $\vec{\mathsf{N}}^{\mathrm{ord}}_{\kappa}$ in $\vec{\widetilde{\mathsf{M}}}^{\mathrm{ord,tor}}_{\widetilde{\mathcal{H}}}$), and by [92, IX, 2.4], [28, Ch. I, Prop. 2.9], or [62, Prop. 3.3.1.7], the embedding

$$\widetilde{T}_{\vec{\mathsf{N}}^{\mathrm{ord}}_{\kappa}} \hookrightarrow \widehat{G}_{\vec{\mathsf{N}}^{\mathrm{ord}}_{\kappa}} = \widetilde{G}_{\vec{\mathsf{N}}^{\mathrm{ord}}_{\kappa}}$$

of group schemes over $\vec{\mathsf{N}}^{\mathrm{ord}}_{\kappa}$ extends (uniquely) to an embedding

$$\widetilde{T}_{\vec{\mathsf{N}}^{\mathrm{ord,tor}}_{\kappa}} \hookrightarrow \widehat{G} = \widetilde{G}_{\vec{\mathsf{N}}^{\mathrm{ord,tor}}_{\kappa}}$$

of group schemes over $\vec{\mathsf{N}}^{\mathrm{ord,tor}}_{\kappa}$, and the quotient

$$\overline{G} := \widehat{G}/\widetilde{T}_{\vec{\mathsf{N}}^{\mathrm{ord,tor}}_{\kappa}} = \widetilde{G}_{\vec{\mathsf{N}}^{\mathrm{ord,tor}}_{\kappa}}/\widetilde{T}_{\vec{\mathsf{N}}^{\mathrm{ord,tor}}_{\kappa}} \tag{7.2.3.1}$$

is a semi-abelian scheme over $\vec{\mathsf{N}}^{\mathrm{ord,tor}}_{\kappa}$ whose restriction $\overline{G}_{\vec{\mathsf{N}}^{\mathrm{ord}}_{\kappa}}$ to $\vec{\mathsf{N}}^{\mathrm{ord}}_{\kappa}$ can be identified with $G_{\vec{\mathsf{N}}^{\mathrm{ord}}_{\kappa}}$, the abelian part of $\widehat{G}_{\vec{\mathsf{N}}^{\mathrm{ord}}_{\kappa}}$. Similarly, we obtain

$$\overline{G}^{\vee} := \widehat{G}^{\vee}/\widetilde{T}^{\vee}_{\vec{\mathsf{N}}^{\mathrm{ord,tor}}_{\kappa}} = \widetilde{G}^{\vee}_{\vec{\mathsf{N}}^{\mathrm{ord,tor}}_{\kappa}}/\widetilde{T}^{\vee}_{\vec{\mathsf{N}}^{\mathrm{ord,tor}}_{\kappa}}, \tag{7.2.3.2}$$

a semi-abelian scheme over $\vec{\mathsf{N}}^{\mathrm{ord,tor}}_{\kappa}$ whose restriction $\overline{G}^{\vee}_{\vec{\mathsf{N}}^{\mathrm{ord}}_{\kappa}}$ to $\vec{\mathsf{N}}^{\mathrm{ord}}_{\kappa}$ can be identified with $G^{\vee}_{\vec{\mathsf{N}}^{\mathrm{ord}}_{\kappa}}$, the abelian part of $\widehat{G}^{\vee}_{\vec{\mathsf{N}}^{\mathrm{ord}}_{\kappa}}$. By [92, IX, 1.4], [28, Ch. I, Prop. 2.7], or [62, Prop. 3.3.1.5], the semi-abelian $\overline{G}$ carries (unique) additional structures $\overline{\lambda} : \overline{G} \to \overline{G}^{\vee}$ and $\overline{i}$, such that the restriction of $(\overline{G}, \overline{\lambda}, \overline{i})$ to $\vec{\mathsf{N}}^{\mathrm{ord}}_{\kappa}$ is the pullback of the tautological tuple $(G_{\vec{\mathsf{M}}^{\mathrm{ord}}_{\mathcal{H},r_\kappa}}, \lambda_{\vec{\mathsf{M}}^{\mathrm{ord}}_{\mathcal{H},r_\kappa}}, i_{\vec{\mathsf{M}}^{\mathrm{ord}}_{\mathcal{H},r_\kappa}})$ over $\vec{\mathsf{M}}^{\mathrm{ord}}_{\mathcal{H},r_\kappa}$ by $f_\kappa : \vec{\mathsf{N}}^{\mathrm{ord}}_{\kappa} \to \vec{\mathsf{M}}^{\mathrm{ord}}_{\mathcal{H},r_\kappa}$. Together with the ordinary level structure $(\overline{\alpha}_{\mathcal{H}^p}, \overline{\alpha}^{\mathrm{ord}}_{\mathcal{H}_p})$ already defined on the abelian part $\overline{G}_{\vec{\mathsf{N}}^{\mathrm{ord}}_{\kappa}}$ of $\widehat{G}_{\vec{\mathsf{N}}^{\mathrm{ord}}}$, we obtain a degenerating family

$$(\overline{G}, \overline{\lambda}, \overline{i}, \overline{\alpha}_{\mathcal{H}^p}, \overline{\alpha}^{\mathrm{ord}}_{\mathcal{H}_p}) \to \vec{\mathsf{N}}^{\mathrm{ord,tor}}_{\kappa} \tag{7.2.3.3}$$

of type $\vec{\mathsf{M}}^{\mathrm{ord}}_{\mathcal{H}}$. By applying the same construction to (7.2.2.8), we obtain a degenerating family

$$(\overline{G}, \overline{\lambda}, \overline{i}, \overline{\alpha}_{\mathcal{H}}) \to \vec{\mathsf{N}}^{\mathrm{ord,tor}}_{\kappa} \tag{7.2.3.4}$$

of type $\mathsf{M}_{\mathcal{H}}$, with the same $(\overline{G}, \overline{\lambda}, \overline{i})$, where $\overline{\alpha}_{\mathcal{H}}$ is defined only on $\vec{\mathsf{N}}^{\mathrm{ord}}_{\kappa} \underset{\mathbb{Z}}{\otimes} \mathbb{Q}$, such that the pair $(\overline{\alpha}_{\mathcal{H}^p}, \overline{\alpha}^{\mathrm{ord}}_{\mathcal{H}_p}) \underset{\mathbb{Z}}{\otimes} \mathbb{Q}$ is determined by $\overline{\alpha}_{\mathcal{H}}$ as in Proposition 3.3.5.1.

Proposition 7.2.3.5. *Suppose $\kappa \in \mathbf{K}^{\mathrm{ord},++}_{Q,\mathcal{H},\Sigma^{\mathrm{ord}}}$; i.e., κ is an element of $\mathbf{K}^{\mathrm{ord},++}_{Q,\mathcal{H}}$ satisfying Condition* 7.1.1.17 *(for the same Σ^{ord} in the definition of $\vec{\mathsf{M}}^{\mathrm{ord,tor}}_{\mathcal{H},r_\kappa} = \vec{\mathsf{M}}^{\mathrm{ord,tor}}_{\mathcal{H},\Sigma^{\mathrm{ord}},r_\kappa}$). (In this case, $\kappa = [\widetilde{\kappa}]$ for some $\widetilde{\kappa} \in \widetilde{\mathbf{K}}^{\mathrm{ord},++}_{Q,\mathcal{H},\Sigma^{\mathrm{ord}}}$; i.e., $\widetilde{\kappa}$ is an element of $\widetilde{\mathbf{K}}^{\mathrm{ord},++}_{Q,\mathcal{H}}$ satisfying Condition* 7.1.1.15.) *Then the structural morphism $f_\kappa : \vec{\mathsf{N}}^{\mathrm{ord}}_{\kappa} \to \vec{\mathsf{M}}^{\mathrm{ord}}_{\mathcal{H},r_\kappa}$* **extends** *(necessarily uniquely) to a morphism $f^{\mathrm{tor}}_{\kappa} : \vec{\mathsf{N}}^{\mathrm{ord,tor}}_{\kappa} \to \vec{\mathsf{M}}^{\mathrm{ord,tor}}_{\mathcal{H},\Sigma^{\mathrm{ord}},r_\kappa}$ between the compactifications, which satisfies the descriptions concerning stratifications and formal completions in the second paragraph of* (2) *of Theorem* 7.1.4.1.

Proof. Let $\xi : \mathrm{Spec}(V) \to \vec{\mathsf{N}}^{\mathrm{ord,tor}}_{\kappa} = \vec{\mathsf{N}}^{\mathrm{ord,tor}}_{\widetilde{\kappa}}$ be as in Construction 7.2.2.9. As explained there, the pullback $(G^{\ddagger}, \lambda^{\ddagger}, i^{\ddagger}, \alpha^{\ddagger}_{\mathcal{H}^p_{\kappa}}, \alpha^{\mathrm{ord},\ddagger}_{\mathcal{H}_{\kappa,p}}) \to \mathrm{Spec}(V)$ of (7.2.3.3) under ξ is a degenerating family as in (7.2.2.18) in the essential image of $\mathrm{DEG}_{\mathrm{PEL},\mathsf{M}^{\mathrm{ord}}_{\mathcal{H}_{\kappa}}}(V) \to \mathrm{DEG}_{\mathrm{PEL},\ddot{\mathsf{M}}^{\mathrm{ord}}_{\mathcal{H}_{\kappa}}}(V)$, which corresponds to an object $(B^{\ddagger}, \lambda_{B^{\ddagger}}, i_{B^{\ddagger}}, X, Y, \phi, c^{\ddagger}, c^{\vee,\ddagger}, \tau^{\ddagger}, [\alpha^{\natural,\mathrm{ord},\ddagger}_{\mathcal{H}_{\kappa}}])$ as in (7.2.2.17) in the essential image of $\mathrm{DD}_{\mathrm{PEL},\mathsf{M}^{\mathrm{ord}}_{\mathcal{H}_{\kappa}}}(V) \to \mathrm{DD}_{\mathrm{PEL},\ddot{\mathsf{M}}^{\mathrm{ord}}_{\mathcal{H}_{\kappa}}}(V)$, and the morphism $\upsilon_{\tau^{\ddagger}} : Y \times X \to \mathbb{Z}$ defined by $\tau^{\ddagger}$ as in (4.2.2.7) (using the discrete valuation $\upsilon : \mathrm{Inv}(V) \to \mathbb{Z}$ of V) defines an element $\upsilon_{\tau^{\ddagger}} \in \mathbf{P}^{+}_{\Phi_{\mathcal{H}}}$ in the image ρ of $\widehat{\rho} = \mathrm{pr}_{(\widehat{\mathbf{S}}_{\breve{\Phi}_{\widetilde{\mathcal{H}}}})^{\vee}_{\mathbb{R}}}(\breve{\rho})$ under (1.2.4.37). If κ satisfies Condition 7.1.1.17, which means any such ρ is contained in some cone in the cone decomposition $\Sigma_{\Phi_{\mathcal{H}}}$ (in Σ^{ord}), then the condition in (6) of Theorem 5.2.1.1 is satisfied for all such ξ, and hence it follows that f_{κ} extends to a (necessarily unique) $f^{\mathrm{tor}}_{\kappa}$, as desired. It satisfies the descriptions concerning stratifications and formal completions in the second paragraph of (2) of Theorem 7.1.4.1 because the universal property of $\mathsf{M}^{\mathrm{tor}}_{\mathcal{H},\Sigma}$ given by (6) of Theorem 1.3.1.3 is given in terms of the degeneration data, which determines the (approximations of) the invertible sheaves such as $\vec{\Psi}^{\mathrm{ord}}_{\Phi_{\mathcal{H}},\delta_{\mathcal{H}}}(\ell)$, and the same is true for the constructions of the canonical morphisms in Lemmas 5.2.4.38 and 7.1.2.29, and Proposition 7.1.2.17, using the various universal properties (all given in terms of degeneration data). □

Now let us resume the notation system at the end of Section 7.2.1 (using $\breve{\tau}$ instead of $\breve{\rho}$). By construction (and the proof of Proposition 7.2.3.5), for $\kappa = [\widetilde{\kappa}] = [(\widetilde{\mathcal{H}}, \widetilde{\Sigma}, \widetilde{\sigma})] \in \mathbf{K}^{\mathrm{ord},++}_{Q,\mathcal{H},\Sigma^{\mathrm{ord}}}$, we have a commutative diagram

$$\begin{array}{ccccc} \vec{\mathsf{N}}^{\mathrm{ord,tor}}_{\kappa} & \longleftarrow & \vec{\widetilde{\mathfrak{X}}}^{\mathrm{ord}}_{\breve{\Phi}_{\widetilde{\mathcal{H}}},\breve{\delta}_{\widetilde{\mathcal{H}}},\breve{\sigma},\breve{\tau}} & \longrightarrow & \vec{\widetilde{C}}^{\mathrm{ord}}_{\breve{\Phi}_{\widetilde{\mathcal{H}}},\breve{\delta}_{\widetilde{\mathcal{H}}}} \\ {\scriptstyle f^{\mathrm{tor}}}\downarrow & & \downarrow & & \downarrow \\ \vec{\mathsf{M}}^{\mathrm{ord,tor}}_{\mathcal{H},r_{\kappa}} & \longleftarrow & \vec{\mathfrak{X}}^{\mathrm{ord}}_{\Phi_{\mathcal{H}},\delta_{\mathcal{H}},\tau,r_{\kappa}} & \longrightarrow & \vec{C}^{\mathrm{ord}}_{\Phi_{\mathcal{H}},\delta_{\mathcal{H}},r_{\kappa}} \end{array} \tag{7.2.3.6}$$

of canonical morphisms (where all subscripts "r_{κ}" mean base changes from $\vec{\mathsf{S}}_{0,r_{\mathcal{H}}}$ to $\vec{\mathsf{S}}_{0,r_{\kappa}}$) whenever the image of $\breve{\tau}$ under the (canonical) second morphism in (1.2.4.20) is contained in τ.

It is worth recording the following observations:

Lemma 7.2.3.7. *Suppose* $\kappa = (\widehat{\mathcal{H}}, \widehat{\Sigma}^{\mathrm{ord}}) \in \mathbf{K}^{\mathrm{ord},++}_{Q,\mathcal{H}}$ *extends to some* $\kappa' = (\widehat{\mathcal{H}}, \widehat{\Sigma}) \in \mathbf{K}^{\mathrm{std},++}_{Q,\mathcal{H}}$ *(see Definition* 7.1.1.20*), which means in particular that* $\widehat{\Sigma}^{\mathrm{ord}}$ *extends to some* $\widehat{\Sigma}$ *as in Lemma* 1.2.4.42*. That is,* $\kappa = [\widetilde{\kappa}]$ *for some* $\widetilde{\kappa} = (\widetilde{\mathcal{H}}, \widetilde{\Sigma}^{\mathrm{ord}}, \widetilde{\sigma}) \in \widetilde{\mathbf{K}}^{\mathrm{ord},++}_{Q,\mathcal{H}}$*, and* $\widetilde{\kappa}$ *extends to some* $\widetilde{\kappa}' = (\widetilde{\mathcal{H}}, \widetilde{\Sigma}, \widetilde{\sigma}) \in \widetilde{\mathbf{K}}^{\mathrm{ord},++}_{Q,\mathcal{H}}$*, which means in particular that the projective smooth* $\widetilde{\Sigma}^{\mathrm{ord}}$ *for* $\vec{\widetilde{\mathsf{M}}}^{\mathrm{ord}}_{\widetilde{\mathcal{H}}}$ *extends to some projective smooth* $\widetilde{\Sigma}$ *for* $\widetilde{\mathsf{M}}_{\widetilde{\mathcal{H}}}$*, which can always be achieved by Proposition* 5.1.3.4*. Hence, we set* $\kappa' = [\widetilde{\kappa}'] \in \mathbf{K}^{\mathrm{ord},++}_{Q,\mathcal{H}}$*. Then there are canonical open immersions*

$$\vec{\mathsf{N}}^{\mathrm{ord,tor}}_{\kappa} \underset{\mathbb{Z}}{\otimes} \mathbb{Q} \hookrightarrow \mathsf{N}^{\mathrm{tor}}_{\kappa',r_{\kappa}} = \mathsf{N}^{\mathrm{tor}}_{\kappa'} \underset{\mathsf{S}_0}{\times} \mathsf{S}_{0,r_{\kappa}} \tag{7.2.3.8}$$

inducing an isomorphism

$$\widehat{\vec{\mathsf{Z}}}^{\mathrm{ord}}_{[(\breve{\Phi}_{\widehat{\mathcal{H}}},\breve{\delta}_{\widehat{\mathcal{H}}},\widehat{\tau})]} \underset{\mathbb{Z}}{\otimes} \mathbb{Q} \xrightarrow{\sim} \widehat{\mathsf{Z}}_{[(\breve{\Phi}_{\widehat{\mathcal{H}}},\breve{\delta}_{\widehat{\mathcal{H}}},\widehat{\tau})],r_\kappa} := \widehat{\mathsf{Z}}_{[(\breve{\Phi}_{\widehat{\mathcal{H}}},\breve{\delta}_{\widehat{\mathcal{H}}},\widehat{\tau})]} \underset{\mathsf{S}_0}{\times} \mathsf{S}_{0,r_\kappa} \tag{7.2.3.9}$$

(see Definition 7.2.2.20; cf. Definition 2.2.3.4) when the underlying $\breve{\mathsf{Z}}_{\widehat{\mathcal{H}}}$ of $[(\breve{\Phi}_{\widehat{\mathcal{H}}},\breve{\delta}_{\widehat{\mathcal{H}}},\widehat{\tau})]$ (suppressed in the notation by our convention; cf. Lemma 1.2.4.42) is compatible with D; otherwise, the pullback of $\widehat{\mathsf{Z}}_{[(\breve{\Phi}_{\widehat{\mathcal{H}}},\breve{\delta}_{\widehat{\mathcal{H}}},\widehat{\tau})],r_\kappa}$ under (7.2.3.8) is empty.

Proof. This follows from (2) and (7) of Theorem 5.2.1.1, because by construction $\vec{\mathsf{N}}^{\mathrm{ord,tor}}_{\kappa}$ is the closure of $\mathsf{N}_\kappa = \widetilde{\vec{\mathsf{Z}}}^{\mathrm{ord}}_{[(\widetilde{\Phi}_{\widetilde{\mathcal{H}}},\widetilde{\delta}_{\widetilde{\mathcal{H}}},\widetilde{\sigma})]}$ in $\widetilde{\vec{\mathsf{M}}}^{\mathrm{ord,tor}}_{\widetilde{\mathcal{H}},\widetilde{\Sigma}^{\mathrm{ord}}}$. □

Lemma 7.2.3.10. *With the same setting as in Lemma 7.2.3.7, suppose there exist Σ^{ord} and Σ such that Σ^{ord} extends to Σ as in Proposition 5.1.3.2, and such that $\kappa = (\widehat{\mathcal{H}}, \widehat{\Sigma}^{\mathrm{ord}}) \in \mathbf{K}^{\mathrm{ord},++}_{Q,\mathcal{H},\Sigma^{\mathrm{ord}}}$ and $\kappa' = (\widehat{\mathcal{H}}, \widehat{\Sigma}) \in \mathbf{K}^{\mathrm{std},++}_{Q,\mathcal{H},\Sigma}$. (This is always possible up to replacing $\widehat{\Sigma}$ with a refinement satisfying Condition 7.1.1.17 for this Σ.) Then the diagram*

$$\begin{array}{ccc} \vec{\mathsf{N}}^{\mathrm{ord,tor}}_{\kappa} \underset{\mathbb{Z}}{\otimes} \mathbb{Q} & \xrightarrow{(7.2.3.8)} & \mathsf{N}^{\mathrm{tor}}_{\kappa',r_\kappa} \\ {\scriptstyle f^{\mathrm{tor}}_{\kappa} \underset{\mathbb{Z}}{\otimes} \mathbb{Q}} \downarrow & & \downarrow {\scriptstyle f^{\mathrm{tor}}_{\kappa'}} \\ \vec{\mathsf{M}}^{\mathrm{ord,tor}}_{\mathcal{H},\Sigma^{\mathrm{ord}},r_\kappa} \underset{\mathbb{Z}}{\otimes} \mathbb{Q} & \xrightarrow[(5.2.1.2)]{} & \mathsf{M}^{\mathrm{tor}}_{\mathcal{H},\Sigma,r_\kappa} \end{array} \tag{7.2.3.11}$$

of canonical morphisms is commutative and Cartesian.

Proof. The diagram (7.2.3.11) is commutative because $f^{\mathrm{tor}}_{\kappa}$ and $f^{\mathrm{tor}}_{\kappa'}$ are induced by compatible universal properties of $\vec{\mathsf{M}}^{\mathrm{ord,tor}}_{\mathcal{H},\Sigma^{\mathrm{ord}}}$ and $\mathsf{M}^{\mathrm{tor}}_{\mathcal{H},\Sigma}$ (see (6) of Theorem 1.3.1.3 and (6) and (7) of Theorem 5.2.1.1).

By (2) and (7) of Theorem 5.2.1.1, the open image of (5.2.1.2) is the union of the strata $\mathsf{Z}_{[(\Phi_{\mathcal{H}},\delta_{\mathcal{H}},\tau)],r_{\mathcal{H}}}$ of $\mathsf{M}^{\mathrm{tor}}_{\mathcal{H},\Sigma,r_\kappa}$ (as in (2) of Theorem 1.3.1.3) whose underlying cusp labels $[(\Phi_{\mathcal{H}},\delta_{\mathcal{H}})]$ are ordinary. Similarly, by Lemma 7.2.3.7 (which is based on the same argument), the open image of (7.2.3.8) is the union of the strata $\widetilde{\mathsf{Z}}_{[(\breve{\Phi}_{\widetilde{\mathcal{H}}},\breve{\delta}_{\widetilde{\mathcal{H}}},\breve{\tau})],r_\kappa}$ of $\widetilde{\mathsf{M}}^{\mathrm{tor}}_{\widetilde{\mathcal{H}},\widetilde{\Sigma},r_\kappa}$ having $\mathsf{N}_{\kappa'} = \widetilde{\mathsf{Z}}_{[(\widetilde{\Phi}_{\widetilde{\mathcal{H}}},\widetilde{\delta}_{\widetilde{\mathcal{H}}},\widetilde{\sigma})],r_\kappa}$ as a face, whose underlying cusp labels $[(\breve{\Phi}_{\widetilde{\mathcal{H}}},\breve{\delta}_{\widetilde{\mathcal{H}}})]$ are ordinary. Hence, it follows from Lemma 7.1.1.8 that the diagram (7.2.3.11) is Cartesian, as desired. □

Corollary 7.2.3.12. *For every $\kappa \in \mathbf{K}^{\mathrm{ord},++}_{Q,\mathcal{H},\Sigma}$, the characteristic zero fiber $f^{\mathrm{tor}}_{\kappa} \underset{\mathbb{Z}}{\otimes} \mathbb{Q}$ of the morphism $f^{\mathrm{tor}}_{\kappa} : \vec{\mathsf{N}}^{\mathrm{ord,tor}}_{\kappa} \to \vec{\mathsf{M}}^{\mathrm{ord,tor}}_{\mathcal{H},\Sigma^{\mathrm{ord}},r_\kappa}$ is proper (although $f^{\mathrm{tor}}_{\kappa}$ might not be).*

Proof. Up to replacing $\widehat{\Sigma}$ with some refinement, we may assume that the assumptions of Lemma 7.2.3.10 are satisfied. Then, since the diagram (7.2.3.11) is Cartesian, the properness of $f^{\mathrm{tor}}_{\kappa} \underset{\mathbb{Z}}{\otimes} \mathbb{Q}$ follows from that of $f^{\mathrm{tor}}_{\kappa} : \mathsf{N}^{\mathrm{tor}}_{\kappa'} \to \mathsf{M}^{\mathrm{tor}}_{\mathcal{H},\Sigma}$ (see (2) of Theorem 1.3.3.15). □

7.2.4 *Properness of f^{tor}*

Proposition 7.2.4.1. *With the setting as in Proposition 7.2.3.5, suppose moreover that $\mathcal{H}_\kappa$ and $\mathcal{H}$ are* **equally deep** *as in Definition 3.2.2.9. Then the morphism $f_\kappa^{\mathrm{tor}} : \vec{\mathsf{N}}_\kappa^{\mathrm{ord,tor}} \to \vec{\mathsf{M}}^{\mathrm{ord,tor}}_{\mathcal{H},\Sigma^{\mathrm{ord}},r_\kappa}$ is proper.*

Proof. Starting with $\mathcal{H}$, Σ^{ord}, and $\kappa = (\widehat{\mathcal{H}}, \widehat{\Sigma}) \in \mathbf{K}^{\mathrm{ord},++}_{Q,\mathcal{H},\Sigma^{\mathrm{ord}}}$, under the assumption that $\mathcal{H}_\kappa$ and $\mathcal{H}$ are equally deep, by making compatible choices of $\widetilde{\kappa} = (\widetilde{\mathcal{H}}, \widetilde{\Sigma}^{\mathrm{ord}}, \widetilde{\sigma}) \in \widetilde{\mathbf{K}}^{\mathrm{ord},++}_{Q,\mathcal{H},\Sigma^{\mathrm{ord}}}$ (such that $\kappa = [\widetilde{\kappa}]$), $\mathcal{H}'$, $\Sigma^{\mathrm{ord},\prime}$, $\widetilde{\kappa}' = (\widetilde{\mathcal{H}}', \widetilde{\Sigma}^{\mathrm{ord},\prime}, \widetilde{\sigma}') \in \widetilde{\mathbf{K}}^{\mathrm{ord},+}_{Q,\mathcal{H}',\Sigma^{\mathrm{ord},\prime}}$, $\kappa' = (\widehat{\mathcal{H}}', \widehat{\Sigma}^{\mathrm{ord},\prime}) := [\widetilde{\kappa}] \in \mathbf{K}^{\mathrm{ord},+}_{Q,\mathcal{H}',\Sigma^{\mathrm{ord},\prime}}$, $\mathcal{H}_{\mathrm{aux}}$, $\Sigma^{\mathrm{ord}}_{\mathrm{aux}}$, $\widetilde{\kappa}_{\mathrm{aux}} = (\widetilde{\mathcal{H}}_{\mathrm{aux}}, \widetilde{\Sigma}^{\mathrm{ord}}_{\mathrm{aux}}, \widetilde{\sigma}_{\mathrm{aux}}) \in \widetilde{\mathbf{K}}^{\mathrm{ord},+}_{Q_{\mathrm{aux}},\mathcal{H}_{\mathrm{aux}},\Sigma^{\mathrm{ord}}_{\mathrm{aux}}}$, and $\kappa_{\mathrm{aux}} = (\widehat{\mathcal{H}}_{\mathrm{aux}}, \widehat{\Sigma}^{\mathrm{ord}}_{\mathrm{aux}}) := [\widetilde{\kappa}_{\mathrm{aux}}] \in \mathbf{K}^{\mathrm{ord},+}_{Q_{\mathrm{aux}},\mathcal{H}_{\mathrm{aux}},\Sigma^{\mathrm{ord}}_{\mathrm{aux}}}$, which are all *equally deep* of some depth r at p (in the obvious sense), which can always be achieved by replacing the collections of cone decompositions with refinements, we obtain by Proposition 5.2.2.2 (with $g = 1$), by (7.1.4.11) (with $\widehat{g} = 1$) and (4d) of Theorem 7.1.4.1, and by Proposition 6.1.1.6 a diagram

$$\begin{array}{ccccc}
\widetilde{\vec{\mathsf{M}}}^{\mathrm{ord,tor}}_{\widetilde{\mathcal{H}},\widetilde{\Sigma}^{\mathrm{ord}}} & \xleftarrow[\mathrm{can.}]{\mathrm{proper}} & \widetilde{\vec{\mathsf{M}}}^{\mathrm{ord,tor}}_{\widetilde{\mathcal{H}}',\widetilde{\Sigma}^{\mathrm{ord},\prime}} & \xrightarrow{\mathrm{proper}} & \widetilde{\vec{\mathsf{M}}}^{\mathrm{ord,tor}}_{\widetilde{\mathcal{H}}_{\mathrm{aux}},\widetilde{\Sigma}^{\mathrm{ord}}_{\mathrm{aux}}} \\
{\scriptstyle \mathrm{can.}}\big\uparrow & & {\scriptstyle \mathrm{can.}}\big\uparrow & & \big\uparrow{\scriptstyle \mathrm{can.}} \\
\vec{\mathsf{N}}^{\mathrm{ord,tor}}_{\kappa} & \xleftarrow[\mathrm{can.}]{\mathrm{proper}} & \vec{\mathsf{N}}^{\mathrm{ord,tor}}_{\kappa'} & \xrightarrow{\mathrm{proper}} & \vec{\mathsf{N}}^{\mathrm{ord,tor}}_{\kappa_{\mathrm{aux}}} \\
{\scriptstyle f^{\mathrm{tor}}_{\kappa}}\big\downarrow & & {\scriptstyle f^{\mathrm{tor}}_{\kappa'}}\big\downarrow & & \big\downarrow{\scriptstyle f^{\mathrm{tor}}_{\kappa_{\mathrm{aux}}}} \\
\vec{\mathsf{M}}^{\mathrm{ord,tor}}_{\mathcal{H},\Sigma^{\mathrm{ord}},r_\kappa} & \xleftarrow[\mathrm{can.}]{\mathrm{proper}} & \vec{\mathsf{M}}^{\mathrm{ord,tor}}_{\mathcal{H}',\Sigma^{\mathrm{ord},\prime},r_{\kappa'}} & \xrightarrow{\mathrm{proper}} & \vec{\mathsf{M}}^{\mathrm{ord,tor}}_{\mathcal{H}_{\mathrm{aux}},\Sigma^{\mathrm{ord}}_{\mathrm{aux}},r}
\end{array}$$

in which the upward arrows are canonical closed immersions, the leftward arrows are canonical proper surjections, the rightward arrows are proper morphisms, (where the middle ones are induced by the top ones,) and the downward arrows are surjections. Since all the morphism are induced by universal properties (using (6) of Theorem 1.3.1.3), the diagram is commutative. Therefore, to show that f_κ^{tor} is proper, it suffices to show that $f^{\mathrm{tor}}_{\kappa_{\mathrm{aux}}}$ is proper (and hence so is $f^{\mathrm{tor}}_{\kappa'}$). Therefore, for the purpose of proving this proposition, by replacing objects and morphism in the first column of the above diagram with those auxiliary ones in the third column, we may assume the following:

- p is a good prime for $(\mathcal{O}, \star, L, \langle\,\cdot\,,\,\cdot\,\rangle, h_0)$, and hence also for $(\mathcal{O}, \star, \widetilde{L}, \langle\,\cdot\,,\,\cdot\,\rangle^{\sim}, \widetilde{h}_0)$ (see Remark 1.2.4.2);
- $\widetilde{\mathcal{H}} = \widetilde{\mathcal{H}}^p \widetilde{\mathcal{H}}_p$, where $\widetilde{\mathcal{H}}^p$ is neat and where $\widetilde{\mathcal{H}}_p = \widetilde{\mathcal{U}}^{\mathrm{bal}}_{p,1}(p^r)$;
- $\mathcal{H}_\kappa = \mathcal{H} = \mathcal{H}^p \mathcal{H}_p$, where $\mathcal{H}^p$ is neat and where $\mathcal{H}_p = \mathcal{U}^{\mathrm{bal}}_{p,1}(p^r)$;

and we are free to replace $\mathcal{H}$ and $\widetilde{\mathcal{H}}$ with subgroups that are equally deep at p, and to replace Σ^{ord} and $\widetilde{\Sigma}^{\mathrm{ord}}$ with refinements. (Then $r_\kappa = r_{\mathcal{H}}$ and hence we may drop the subscripts "r_κ" from the notation for base changes from $\vec{\mathsf{S}}_{0,r_{\mathcal{H}}}$ to $\vec{\mathsf{S}}_{0,r_\kappa}$.)

Since Assumption 5.2.3.1 is satisfied by Lemma 5.2.3.2, by Lemma 5.2.3.9 and Proposition 5.2.3.18, up to replacing $\widetilde{\Sigma}^{\mathrm{ord}}$ with a refinement, we have compatible canonical morphisms

$$\vec{\widetilde{\mathsf{M}}}^{\mathrm{ord,tor}}_{\widetilde{\mathcal{H}},\widetilde{\Sigma}^{\mathrm{ord}}} \hookrightarrow \vec{\widetilde{\mathsf{M}}}^{\mathrm{tor}}_{\widetilde{\mathcal{H}},\widetilde{d}_0\widetilde{\mathsf{pol}},r_\kappa} \tag{7.2.4.2}$$

(as in (5.2.3.19)) and

$$\vec{\widetilde{\mathsf{M}}}^{\mathrm{tor}}_{\widetilde{\mathcal{H}},\widetilde{d}_0\widetilde{\mathsf{pol}},r_\kappa} \to \widetilde{\mathsf{M}}^{\mathrm{tor}}_{\widetilde{\mathcal{H}}^p,\widetilde{\Sigma}^p} \tag{7.2.4.3}$$

(as in (5.2.3.12)) for some $(\widetilde{\Sigma},\widetilde{\mathsf{pol}})$ extending $(\widetilde{\Sigma}^{\mathrm{ord}},\widetilde{\mathsf{pol}}^{\mathrm{ord}})$ as in Proposition 5.1.3.4, some integer $\widetilde{d}_0 \geq 1$, and some collection $\widetilde{\Sigma}^p$ for $\widetilde{\mathsf{M}}_{\widetilde{\mathcal{H}}^p}$ inducing $\widetilde{\Sigma}$. Consider the tautological degenerating family

$$(\widetilde{G},\widetilde{\lambda},\widetilde{i},\widetilde{\alpha}_{\widetilde{\mathcal{H}}^p},\widetilde{\alpha}^{\mathrm{ord}}_{p^r}) \to \vec{\widetilde{\mathsf{M}}}^{\mathrm{ord,tor}}_{\widetilde{\mathcal{H}},\widetilde{\Sigma}^{\mathrm{ord}}} \tag{7.2.4.4}$$

of type $\vec{\widetilde{\mathsf{M}}}^{\mathrm{ord}}_{\widetilde{\mathcal{H}}}$, where we denote $\widetilde{\alpha}^{\mathrm{ord}}_{\widetilde{\mathcal{H}}_p}$ by $\widetilde{\alpha}^{\mathrm{ord}}_{p^r}$ because $\widetilde{\mathcal{H}}_p = \widetilde{\mathcal{U}}^{\mathrm{bal}}_{p,1}(p^r)$. Then the subtuple $(\widetilde{G},\widetilde{\lambda},\widetilde{i},\widetilde{\alpha}_{\widetilde{\mathcal{H}}^p}) \to \vec{\widetilde{\mathsf{M}}}^{\mathrm{ord,tor}}_{\widetilde{\mathcal{H}},\widetilde{\Sigma}^{\mathrm{ord}}}$ extends to a degenerating family

$$(\widetilde{G},\widetilde{\lambda},\widetilde{i},\widetilde{\alpha}_{\widetilde{\mathcal{H}}^p}) \to \vec{\widetilde{\mathsf{M}}}^{\mathrm{tor}}_{\widetilde{\mathcal{H}},\widetilde{d}_0\widetilde{\mathsf{pol}},r_\kappa} \tag{7.2.4.5}$$

of type $\widetilde{\mathsf{M}}_{\widetilde{\mathcal{H}}^p}$, by taking the pullback under (7.2.4.3) of the tautological degenerating family of the same type over $\widetilde{\mathsf{M}}^{\mathrm{tor}}_{\widetilde{\mathcal{H}}^p,\widetilde{\Sigma}^p}$. Hence, by (6) of Theorem 1.3.1.3, the image of (7.2.4.2) can be characterized as the maximal open subscheme of $\vec{\widetilde{\mathsf{M}}}^{\mathrm{tor}}_{\widetilde{\mathcal{H}},\widetilde{d}_0\widetilde{\mathsf{pol}},r_\kappa}$ such that the pullback of (7.2.4.5) carries an ordinary level structure $\widetilde{\alpha}^{\mathrm{ord}}_{p^r}$.

Since $\vec{\mathsf{N}}^{\mathrm{ord,tor}}_\kappa$ is quasi-projective over $\vec{\mathsf{S}}_{0,r_\kappa}$ and $\vec{\mathsf{M}}^{\mathrm{ord,tor}}_{\mathcal{H},\Sigma^{\mathrm{ord}}}$ is separated over $\vec{\mathsf{S}}_{0,r_\kappa}$, the morphism f^{tor}_κ is separated and finite type. Moreover, the restriction of f^{tor}_κ to $\vec{\mathsf{N}}^{\mathrm{ord}}_\kappa$ is the proper morphism $f_\kappa : \vec{\mathsf{N}}^{\mathrm{ord}}_\kappa \to \vec{\mathsf{M}}^{\mathrm{ord}}_{\mathcal{H}}$ (underlying an abelian scheme torsor), and $\vec{\mathsf{N}}^{\mathrm{ord}}_\kappa \underset{\mathbb{Z}}{\otimes} \mathbb{Q}$ is open and dense in $\vec{\mathsf{N}}^{\mathrm{ord,tor}}_\kappa$ by construction. Also, $f^{\mathrm{tor}}_\kappa \underset{\mathbb{Z}}{\otimes} \mathbb{Q} : \vec{\mathsf{N}}^{\mathrm{ord,tor}}_\kappa \underset{\mathbb{Z}}{\otimes} \mathbb{Q} \to \vec{\mathsf{M}}^{\mathrm{ord,tor}}_{\mathcal{H},\Sigma^{\mathrm{ord}},r_\kappa} \underset{\mathbb{Z}}{\otimes} \mathbb{Q}$ is proper by Corollary 7.2.3.12. Therefore, in order to show that f^{tor}_κ is proper (by the valuative criterion), it suffices to show that, for each complete discrete valuation ring V with fractional field K of characteristic zero and algebraically closed residue field k of characteristic p, each morphism $\breve{\xi}_0 : \mathrm{Spec}(K) \to \vec{\mathsf{N}}^{\mathrm{ord}}_\kappa \underset{\mathbb{Z}}{\otimes} \mathbb{Q}$ such that $\xi_0 := f_\kappa \circ \breve{\xi}_0 : \mathrm{Spec}(K) \to \vec{\mathsf{M}}^{\mathrm{ord}}_{\mathcal{H},r_\kappa} \underset{\mathbb{Z}}{\otimes} \mathbb{Q}$ extends to a morphism $\xi : \mathrm{Spec}(V) \to \vec{\mathsf{M}}^{\mathrm{ord,tor}}_{\mathcal{H},\Sigma^{\mathrm{ord}}}$ also extends to a morphism $\breve{\xi} : \mathrm{Spec}(V) \to \vec{\mathsf{N}}^{\mathrm{ord,tor}}_\kappa$. Note that $\breve{\xi}_0$ induces a morphism $\overline{\xi}_0 : \mathrm{Spec}(K) \to \vec{\widetilde{\mathsf{M}}}^{\mathrm{ord,tor}}_{\widetilde{\mathcal{H}},\widetilde{\Sigma}^{\mathrm{ord}}} \underset{\mathbb{Z}}{\otimes} \mathbb{Q}$. Since $\vec{\widetilde{\mathsf{M}}}^{\mathrm{tor}}_{\widetilde{\mathcal{H}},\widetilde{d}_0\widetilde{\mathsf{pol}},r_\kappa}$ is projective (and hence proper) over $\vec{\mathsf{S}}_{0,r_\kappa}$, the morphism $\overline{\xi}_0$ extends to a morphism $\overline{\xi} : \mathrm{Spec}(V) \to \vec{\widetilde{\mathsf{M}}}^{\mathrm{tor}}_{\widetilde{\mathcal{H}},\widetilde{d}_0\widetilde{\mathsf{pol}},r_\kappa}$. We would like to verify the following:

Claim 7.2.4.6. *The image of $\overline{\xi}$ is contained in the image of* (7.2.4.2).

Since all objects involved are locally of finite presentation, we may assume that V is the completion of a localization of an algebra of finite type over $\mathbb{Z}$. (This might not be absolutely necessary, but will simplify the various constructions.) Then, by lifting (the finitely many) generators to local rings of $\vec{\widetilde{\mathsf{M}}}^{\mathrm{tor}}_{\widetilde{\mathcal{H}},\widetilde{d}_0\widetilde{\mathsf{pol}},r_\kappa}$ (together with the generators of the ideals to localize, and by taking suitable completions), there exists a complete noetherian normal domain $\widetilde{V}$, with fractional field $\widetilde{K}$ and prime ideals $\widetilde{I} \subset \widetilde{J} \subset \widetilde{V}$, such that $\widetilde{V}$ is complete with respect to $\widetilde{J}$, such that $\widetilde{V}/\widetilde{I} \cong V$, and such that $\widetilde{J}/\widetilde{I}$ defines the maximal ideal of the complete discrete valuation ring V, together with a morphism $\widetilde{\xi} : \mathrm{Spec}(\widetilde{V}) \to \vec{\widetilde{\mathsf{M}}}^{\mathrm{tor}}_{\widetilde{\mathcal{H}},\widetilde{d}_0\widetilde{\mathsf{pol}},r_\kappa}$ inducing the morphism $\bar{\xi} : \mathrm{Spec}(V) = \mathrm{Spec}(\widetilde{V}/\widetilde{I}) \to \vec{\widetilde{\mathsf{M}}}^{\mathrm{tor}}_{\widetilde{\mathcal{H}},\widetilde{d}_0\widetilde{\mathsf{pol}},r_\kappa}$ extending $\breve{\xi}_0 : \mathrm{Spec}(K) \to \vec{\widetilde{\mathsf{M}}}^{\mathrm{ord,tor}}_{\widetilde{\mathcal{H}},\widetilde{\Sigma}^{\mathrm{ord}}}$, and inducing a morphism $\widetilde{\xi}_0 : \mathrm{Spec}(\widetilde{K}) \to \widetilde{\mathsf{M}}_{\widetilde{\mathcal{H}},r_\kappa}$. Now Claim 7.2.4.6 follows from the following:

Claim 7.2.4.7. *The image of* $\widetilde{\xi}$ *is contained in the image of* (7.2.4.2).

Let us denote by

$$(\widetilde{G}_{\widetilde{V}}, \widetilde{\lambda}_{\widetilde{V}}, \widetilde{i}_{\widetilde{V}}, \widetilde{\alpha}_{\widetilde{\mathcal{H}}^p,\widetilde{V}}) \to \mathrm{Spec}(\widetilde{V}) \tag{7.2.4.8}$$

the pullback of (7.2.4.5), and by

$$(\widetilde{G}_{\widetilde{K}}, \widetilde{\lambda}_{\widetilde{K}}, \widetilde{i}_{\widetilde{K}}, \widetilde{\alpha}_{\widetilde{\mathcal{H}},\widetilde{K}}) \to \mathrm{Spec}(\widetilde{K}) \tag{7.2.4.9}$$

the pullback under $\widetilde{K}$ of the tautological tuple $(\widetilde{G}_{\widetilde{\mathsf{M}}_{\widetilde{\mathcal{H}}}}, \widetilde{\lambda}_{\widetilde{\mathsf{M}}_{\widetilde{\mathcal{H}}}}, \widetilde{i}_{\widetilde{\mathsf{M}}_{\widetilde{\mathcal{H}}}}, \widetilde{\alpha}_{\widetilde{\mathcal{H}}})$ over $\widetilde{\mathsf{M}}_{\widetilde{\mathcal{H}}}$. Let $(\widetilde{\alpha}_{\widetilde{\mathcal{H}}^p,\widetilde{K}}, \widetilde{\alpha}^{\mathrm{ord}}_{\widetilde{\mathcal{H}}_p,\widetilde{K}}) = (\widetilde{\alpha}_{\widetilde{\mathcal{H}}^p,\widetilde{K}}, \widetilde{\alpha}^{\mathrm{ord}}_{p^r,\widetilde{K}} = (\widetilde{\alpha}^{\mathrm{ord},0}_{p^r,\widetilde{K}}, \widetilde{\alpha}^{\mathrm{ord},\#,0}_{p^r,\widetilde{K}}, \widetilde{\nu}^{\mathrm{ord}}_{p^r,\widetilde{K}}))$ be determined by $\widetilde{\alpha}_{\widetilde{\mathcal{H}},\widetilde{K}}$ as in Section 4.1.6. Then the pullback of (7.2.4.8) to $\mathrm{Spec}(\widetilde{K})$ is canonically isomorphic to $(\widetilde{G}_{\widetilde{K}}, \widetilde{\lambda}_{\widetilde{K}}, \widetilde{i}_{\widetilde{K}}, \widetilde{\alpha}_{\widetilde{\mathcal{H}}^p,\widetilde{K}}) \to \mathrm{Spec}(\widetilde{K})$, with the first three entries as in (7.2.4.9). We shall denote with subscripts "V" (resp. K) the pullbacks of these objects under $\bar{\xi}$ (resp. $\bar{\xi}_0$).

Since p is a good prime for $(\mathcal{O}, \star, \widetilde{L}, \langle\,\cdot\,,\,\cdot\,\rangle^{\sim}, \widetilde{h}_0)$ (see Remark 1.2.4.2), $\widetilde{\lambda}_{\widetilde{K}}$ and hence $\widetilde{\lambda}_{\widetilde{V}}$ are both prime-to-p polarizations, which induce an isomorphism $\widetilde{\lambda}_{\widetilde{V}} : \widetilde{G}_{\widetilde{V}}[p^r] \xrightarrow{\sim} \widetilde{G}^\vee_{\widetilde{V}}[p^r]$. Since $\widetilde{\nu}^{\mathrm{ord}}_{p^r,\widetilde{K}} \in (\mathbb{Z}/p^r\mathbb{Z})^\times_{\mathrm{Spec}(\widetilde{K})}$ always uniquely extends to some $\widetilde{\nu}^{\mathrm{ord}}_{p^r,\widetilde{V}} \in (\mathbb{Z}/p^r\mathbb{Z})^\times_{\mathrm{Spec}(\widetilde{V})}$, we see that $\widetilde{\alpha}^{\mathrm{ord}}_{p^r,\widetilde{K}}$ extends to some ordinary level structure $\widetilde{\alpha}^{\mathrm{ord}}_{p^r,\widetilde{V}} = (\widetilde{\alpha}^{\mathrm{ord},0}_{p^r,\widetilde{V}}, \widetilde{\alpha}^{\mathrm{ord},\#,0}_{p^r,\widetilde{V}}, \widetilde{\nu}^{\mathrm{ord}}_{p^r,\widetilde{V}})$ over $\widetilde{V}$ if and only if $\widetilde{\alpha}^{\mathrm{ord},0}_{p^r,\widetilde{K}} : (\mathrm{Gr}^0_{\widetilde{\mathsf{D}},p^r})^{\mathrm{mult}}_{\mathrm{Spec}(\widetilde{K})} \hookrightarrow \widetilde{G}_{\widetilde{K}}[p^r]$ extends to some $\widetilde{\alpha}^{\mathrm{ord},0}_{p^r,\widetilde{V}} : (\mathrm{Gr}^0_{\widetilde{\mathsf{D}},p^r})^{\mathrm{mult}}_{\mathrm{Spec}(\widetilde{V})} \hookrightarrow \widetilde{G}_{\widetilde{V}}[p^r]$. Thus, to verify Claim 7.2.4.7 (which is equivalent to the extensibility of $\widetilde{\alpha}^{\mathrm{ord}}_{p^r,\widetilde{K}}$ to $\widetilde{\alpha}^{\mathrm{ord}}_{p^r,\widetilde{V}}$), it suffices to verify the following:

Claim 7.2.4.10. *Let* $\widetilde{H}$ *denote the finite flat subgroup scheme of multiplicative type of* $\widetilde{G}_{\widetilde{V}}[p^r]$ *uniquely lifting the maximal finite flat subgroup scheme* $\widetilde{H}_0$ *of multiplicative type of* $\widetilde{G}_0[p^r]$ *(by [26, IX, 3.6 bis] and [35, III-1, 5.1.4]; see also [34, IX, 6.1]), where* $\widetilde{G}_0$ *is the special fiber of* $\widetilde{G}$ *over* $\mathrm{Spec}(k)$*. Then the generic fiber* $\widetilde{H}_{\widetilde{K}}$ *of* $\widetilde{H}$ **coincides** *with* $\mathrm{image}(\widetilde{\alpha}^{\mathrm{ord},0}_{p^r,\widetilde{K}})$ *(see Definition* 3.3.3.1*).*

Let $\widetilde{V}^1$ denote the localization of $\widetilde{V}$ at the kernel of $\widetilde{V} \twoheadrightarrow V$, which has residue field K. Since the image of $\mathrm{Spec}(K)$ under $\widetilde{\xi}$ is contained in the open image of (7.2.4.2), so is the image of $\mathrm{Spec}(\widetilde{V}^1)$. Since $\widetilde{\alpha}^{\mathrm{ord}}_{p^r,\widetilde{K}}$ already extends to $\mathrm{Spec}(\widetilde{V}^1)$, it makes sense to write $\widetilde{\alpha}^{\mathrm{ord},0}_{p^r,K} : (\mathrm{Gr}^0_{\widetilde{\mathsf{D}},p^r})^{\mathrm{mult}}_{\mathrm{Spec}(K)} \hookrightarrow \widetilde{G}_K[p^r]$, and (by [26, IX, 3.6 bis] and [35, III-1, 5.1.4] again, applied to $\widetilde{V}^1$), Claim 7.2.4.10 follows from the following:

Claim 7.2.4.11. *Let $\overline{H}$ denote the finite flat subgroup scheme of multiplicative type of $\widetilde{G}_V[p^r]$ uniquely lifting the maximal finite flat subgroup scheme $\overline{H}_0$ of multiplicative type of $\widetilde{G}_0[p^r]$, then the generic fiber $\overline{H}_K$ of $\overline{H}$* **coincides** *with* $\mathrm{image}(\widetilde{\alpha}^{\mathrm{ord},0}_{p^r,K})$.

Let us denote by

$$(G_V, \lambda_V, i_V, \alpha_{\mathcal{H}^p,K}, \alpha^{\mathrm{ord}}_{\mathcal{H}_p,V}) \to \mathrm{Spec}(V) \tag{7.2.4.12}$$

the pullback under $\xi : \mathrm{Spec}(V) \to \vec{\mathsf{M}}^{\mathrm{ord,tor}}_{\mathcal{H},\Sigma^{\mathrm{ord}}}$ of the tautological degenerating family $(G, \lambda, i, \alpha_{\mathcal{H}^p}, \alpha^{\mathrm{ord}}_{p^r}) \to \vec{\mathsf{M}}^{\mathrm{ord,tor}}_{\mathcal{H},\Sigma^{\mathrm{ord}}}$, and denote its further pullback to $\mathrm{Spec}(K)$ by replacing the subscript "V" with "K". (Here we understand that $\alpha_{\mathcal{H}^p,K}$ is defined only over $\mathrm{Spec}(K)$.)

Since the image of $\mathrm{Spec}(K)$ under $\widetilde{\xi}$ is on $\vec{\mathsf{N}}^{\mathrm{ord}}_{\kappa} \subset \vec{\widetilde{\mathsf{M}}}^{\mathrm{ord,tor}}_{\widetilde{\mathcal{H}},\widetilde{\Sigma}^{\mathrm{ord}}}$, we know that $\widetilde{G}_K$ is an extension of G_K by the pullback $\widetilde{T}_K$ of a torus $\widetilde{T}$ with character group $\widetilde{X}$, and $\mathrm{image}(\widetilde{\alpha}^{\mathrm{ord},0}_{p^r,K})$ is an extension of $\mathrm{image}(\alpha^{\mathrm{ord},0}_{p^r,K})$ by $\widetilde{T}_K[p^r]$. As in Section 7.2.3, since V is normal (as a discrete valuation ring), by [92, IX, 2.4], [28, Ch. I, Prop. 2.9], or [62, Prop. 3.3.1.7], the embedding $\widetilde{T}_K \hookrightarrow \widetilde{G}_K$ extends to an embedding $\widetilde{T}_V \hookrightarrow \widetilde{G}_V$. Therefore, by [26, IX, 3.6 bis] and [35, III-1, 5.1.4], $\widetilde{T}_V[p^r]$ is automatically a closed subgroup scheme of any $\overline{H}$ in Claim 7.2.4.11. By [92, IX, 1.4], [28, Ch. I, Prop. 2.7], or [62, Prop. 3.3.1.5], the quotient semi-abelian scheme $\widetilde{G}_V/\widetilde{T}_V$ is canonically isomorphic to G_V. Hence, Claim 7.2.4.11 follows from the following:

Claim 7.2.4.13. *Let H denote the finite flat subgroup scheme of multiplicative type of $G_V[p^r]$ uniquely lifting the maximal finite flat subgroup scheme H_0 of multiplicative type of $G_0[p^r]$, where G_0 is the special fiber of G over* $\mathrm{Spec}(k)$, *then the generic fiber H_K of H_V* **coincides** *with* $\mathrm{image}(\alpha^{\mathrm{ord},0}_{p^r,K})$.

But this is trivially verified because $\alpha^{\mathrm{ord}}_{p^r,K}$ extends to the ordinary level structure $\alpha^{\mathrm{ord}}_{p^r,V}$ of (G_V, λ_V, i_V), in which case $H = \mathrm{image}(\alpha^{\mathrm{ord},0}_{p^r,V})$, by comparison over the characteristic p special fiber over $\mathrm{Spec}(k)$ (and by [26, IX, 3.6 bis] and [35, III-1, 5.1.4] once again). □

Proposition 7.2.4.14. *Suppose no longer that $\mathcal{H}_\kappa$ and $\mathcal{H}$ are equally deep as in Definition* 3.2.2.9. *For each $\kappa \in \mathbf{K}^{\mathrm{ord},++}_{Q,\mathcal{H}}$, there is a canonical surjection $\vec{\mathsf{N}}^{\mathrm{ord,tor}}_{\kappa} \to \vec{\mathsf{M}}^{\mathrm{ord,min}}_{\mathcal{H}}$ extending the canonical surjection $\vec{\mathsf{N}}^{\mathrm{ord}}_{\kappa} \to \vec{\mathsf{M}}^{\mathrm{ord}}_{\mathcal{H}}$ (which is the composition of $f_\kappa : \vec{\mathsf{N}}^{\mathrm{ord}}_{\kappa} \to \vec{\mathsf{M}}^{\mathrm{ord}}_{\mathcal{H},r_\kappa}$ with the canonical morphism $\vec{\mathsf{M}}^{\mathrm{ord}}_{\mathcal{H},r_\kappa} \to \vec{\mathsf{M}}^{\mathrm{ord}}_{\mathcal{H}}$), and the latter is the pullback of the former under the canonical morphism $\vec{\mathsf{M}}^{\mathrm{ord}}_{\mathcal{H}} \hookrightarrow \vec{\mathsf{M}}^{\mathrm{ord,min}}_{\mathcal{H}}$ on the target* (*see Theorem* 6.2.1.1). *More generally, such a morphism maps the*

$[(\breve{\Phi}_{\widehat{\mathcal{H}}}, \breve{\delta}_{\widehat{\mathcal{H}}}, \widehat{\tau})]$-stratum $\vec{\widehat{\mathsf{Z}}}^{\mathrm{ord}}_{[(\breve{\Phi}_{\widehat{\mathcal{H}}}, \breve{\delta}_{\widehat{\mathcal{H}}}, \widehat{\tau})]}$ of $\vec{\mathsf{N}}^{\mathrm{ord,tor}}_{\kappa}$ to the $[(\Phi_{\mathcal{H}}, \delta_{\mathcal{H}})]$-stratum $\vec{\mathsf{Z}}^{\mathrm{ord}}_{[(\Phi_{\mathcal{H}}, \delta_{\mathcal{H}})]}$ of $\vec{\mathsf{M}}^{\mathrm{ord,min}}_{\mathcal{H}}$ if and only if the cusp label $[(\Phi_{\mathcal{H}}, \delta_{\mathcal{H}})]$ is assigned to the cusp label $[(\breve{\Phi}_{\widehat{\mathcal{H}}}, \breve{\delta}_{\widehat{\mathcal{H}}})]$ as in Lemma 1.2.4.15. Such surjections are compatible with the canonical morphisms $f^{\mathrm{tor}}_{\kappa',\kappa} : \vec{\mathsf{N}}^{\mathrm{ord,tor}}_{\kappa'} \to \vec{\mathsf{N}}^{\mathrm{ord,tor}}_{\kappa}$ (defined by Proposition 7.2.4.1) when $\kappa' \succ \kappa$ in $\mathbf{K}^{\mathrm{ord},++}_{Q,\mathcal{H}}$. When $\mathcal{H}_\kappa$ and $\mathcal{H}$ are equally deep, both $\vec{\mathsf{N}}^{\mathrm{ord}}_{\kappa} \to \vec{\mathsf{M}}^{\mathrm{ord}}_{\mathcal{H}}$ and $\vec{\mathsf{N}}^{\mathrm{ord,tor}}_{\kappa} \to \vec{\mathsf{M}}^{\mathrm{ord,min}}_{\mathcal{H}}$ are proper.

Proof. We may assume that $\kappa = (\widehat{\mathcal{H}}, \widehat{\Sigma}) \in \mathbf{K}^{\mathrm{ord},+}_{Q,\mathcal{H}}$ (*i.e.*, $\mathcal{H}_\kappa = \mathcal{H}$) because, by Proposition 6.2.2.1, there is a canonical quasi-finite surjection $\vec{\mathsf{M}}^{\mathrm{ord,min}}_{\mathcal{H}_\kappa} \to \vec{\mathsf{M}}^{\mathrm{ord,min}}_{\mathcal{H}}$, and $\vec{\mathsf{M}}^{\mathrm{ord}}_{\mathcal{H}_\kappa}$ (as the unique open stratum) is the pullback of $\vec{\mathsf{M}}^{\mathrm{ord}}_{\mathcal{H}}$. Let Σ^{ord} be any compatible choice for $\vec{\mathsf{M}}^{\mathrm{ord}}_{\mathcal{H}}$ as in Definition 1.2.2.13. Take any element $\kappa'' = (\widehat{\mathcal{H}}'', \widehat{\Sigma}'') \in \mathbf{K}^{\mathrm{ord},+}_{Q,\mathcal{H},\Sigma^{\mathrm{ord}}}$ such that $\widehat{\mathcal{H}} = \widehat{\mathcal{H}}''$, and such that $\widehat{\Sigma}''$ is a refinement of $\widehat{\Sigma}$, so that $\kappa'' \succ \kappa$ in $\mathbf{K}^{\mathrm{ord},+}_{Q,\mathcal{H}}$. Then $\mathcal{H}_{\kappa''} = \mathcal{H}_\kappa = \mathcal{H}$, and hence we have canonical proper surjections $f^{\mathrm{tor}}_{\kappa'',\kappa} : \vec{\mathsf{N}}^{\mathrm{ord,tor}}_{\kappa''} \to \vec{\mathsf{N}}^{\mathrm{ord,tor}}_{\kappa}$ (by Proposition 7.2.2.19 and Corollary 7.2.2.21) and $f^{\mathrm{tor}}_{\kappa''} : \vec{\mathsf{N}}^{\mathrm{ord,tor}}_{\kappa''} \to \vec{\mathsf{M}}^{\mathrm{ord,tor}}_{\mathcal{H},\Sigma^{\mathrm{ord}}}$ (by Proposition 7.2.4.1).

These proper surjections are their own Stein factorizations (see [35, III-1, 4.3.3]), by the normality of the target schemes, by [62, Lem. 7.2.3.1], and by Zariski's main theorem (see [35, III-1, 4.4.3, 4.4.11]). That is, the canonical morphisms

$$\mathscr{O}_{\vec{\mathsf{N}}^{\mathrm{ord,tor}}_{\kappa}} \to (f^{\mathrm{tor}}_{\kappa'',\kappa})_* \mathscr{O}_{\vec{\mathsf{N}}^{\mathrm{ord,tor}}_{\kappa''}} \tag{7.2.4.15}$$

and

$$\mathscr{O}_{\vec{\mathsf{M}}^{\mathrm{ord,tor}}_{\mathcal{H},\Sigma^{\mathrm{ord}}}} \to (f^{\mathrm{tor}}_{\kappa''})_* \mathscr{O}_{\vec{\mathsf{N}}^{\mathrm{ord,tor}}_{\kappa''}} \tag{7.2.4.16}$$

are isomorphisms. (These are special cases of (7.1.4.4) with $a = b = 0$.)

Let $\omega_{\vec{\mathsf{N}}^{\mathrm{ord,tor}}_{\kappa}} := \wedge^{\mathrm{top}} \underline{\mathrm{Lie}}^{\vee}_{\widehat{G}/\vec{\mathsf{N}}^{\mathrm{ord,tor}}_{\kappa}} = \wedge^{\mathrm{top}} e^*_{\widehat{G}} \Omega^1_{\widehat{G}/\vec{\mathsf{N}}^{\mathrm{ord,tor}}_{\kappa}}$, where $\widehat{G} \to \vec{\mathsf{N}}^{\mathrm{ord,tor}}_{\kappa}$ is the tautological semi-abelian scheme as in (7.2.2.4), and $\omega_{\vec{\mathsf{N}}^{\mathrm{ord,tor}}_{\kappa''}}$ be similarly defined, so that

$$\omega_{\vec{\mathsf{N}}^{\mathrm{ord,tor}}_{\kappa''}} \cong (f^{\mathrm{tor}}_{\kappa'',\kappa})^* \, \omega_{\vec{\mathsf{N}}^{\mathrm{ord,tor}}_{\kappa}}. \tag{7.2.4.17}$$

On the other hand, by (7.2.3.1), we have

$$\omega_{\vec{\mathsf{N}}^{\mathrm{ord,tor}}_{\kappa''}} \cong (\wedge^{\mathrm{top}}_{\mathbb{Z}} \widetilde{X}) \underset{\mathbb{Z}}{\otimes} (f^{\mathrm{tor}}_{\kappa''})^* \omega_{\vec{\mathsf{M}}^{\mathrm{ord,tor}}_{\mathcal{H},\Sigma^{\mathrm{ord}}}}$$

(*cf.* [62, Lem. 7.1.2.1]). Hence, by choosing any isomorphism $(\wedge^{\mathrm{top}}_{\mathbb{Z}} \widetilde{X}) \xrightarrow{\sim} \mathbb{Z}$, we have

$$\omega_{\vec{\mathsf{N}}^{\mathrm{ord,tor}}_{\kappa''}} \cong (f^{\mathrm{tor}}_{\kappa''})^* \omega_{\vec{\mathsf{M}}^{\mathrm{ord,tor}}_{\mathcal{H},\Sigma^{\mathrm{ord}}}}. \tag{7.2.4.18}$$

By composing $f^{\mathrm{tor}}_{\kappa''}$ with the canonical proper surjection $\vec{\oint}^{\mathrm{ord}}_{\mathcal{H}} : \vec{\mathsf{M}}^{\mathrm{ord,tor}}_{\mathcal{H},\Sigma^{\mathrm{ord}}} \to \vec{\mathsf{M}}^{\mathrm{ord,min}}_{\mathcal{H}}$, we obtain a canonical proper surjection

$$\vec{\mathsf{N}}^{\mathrm{ord,tor}}_{\kappa''} \to \vec{\mathsf{M}}^{\mathrm{ord,min}}_{\mathcal{H}}, \tag{7.2.4.19}$$

which is canonically determined by the canonical isomorphisms

$$\begin{aligned}\vec{\mathsf{M}}_{\mathcal{H}}^{\mathrm{ord,min}} &\cong \mathrm{Proj}\Big(\underset{k\geq 0}{\oplus}\Gamma(\vec{\mathsf{M}}_{\mathcal{H}}^{\mathrm{ord,tor}}, \omega^{\otimes k}_{\vec{\mathsf{M}}_{\mathcal{H}}^{\mathrm{ord,tor}}})\Big)\\ &\cong \mathrm{Proj}\Big(\underset{k\geq 0}{\oplus}\Gamma(\vec{\mathsf{N}}_{\kappa''}^{\mathrm{ord,tor}}, \omega^{\otimes k}_{\vec{\mathsf{N}}_{\kappa''}^{\mathrm{ord,tor}}})\Big)\end{aligned}$$

(by (3) of Theorem 6.2.1.1, and (7.2.4.15), (7.2.4.16), and (7.2.4.18)). By (7.2.4.17), we also have a canonical isomorphism

$$\vec{\mathsf{M}}_{\mathcal{H}}^{\mathrm{ord,min}} \cong \mathrm{Proj}\Big(\underset{k\geq 0}{\oplus}\Gamma(\vec{\mathsf{N}}_{\kappa}^{\mathrm{ord,tor}}, \omega^{\otimes k}_{\vec{\mathsf{N}}_{\kappa}^{\mathrm{ord,tor}}})\Big),$$

which induces the desired canonical morphism

$$\vec{\mathsf{N}}_{\kappa}^{\mathrm{ord,tor}} \to \vec{\mathsf{M}}_{\mathcal{H}}^{\mathrm{ord,min}}, \tag{7.2.4.20}$$

whose composition with $f^{\mathrm{tor}}_{\kappa'',\kappa} : \vec{\mathsf{N}}_{\kappa''}^{\mathrm{ord,tor}} \to \vec{\mathsf{N}}_{\kappa}^{\mathrm{ord,tor}}$ is (7.2.4.19).

Note that (7.2.4.20) is proper because (7.2.4.19) is proper and $f^{\mathrm{tor}}_{\kappa'',\kappa}$ is surjective. (This uses the simplifying assumption that $\mathcal{H}_{\kappa} = \mathcal{H}$.)

The preimage of $\vec{\mathsf{M}}_{\mathcal{H}}^{\mathrm{ord}}$ under (7.2.4.19) is the preimage of $\vec{\mathsf{M}}_{\mathcal{H}}^{\mathrm{ord}}$ under $f^{\mathrm{tor}}_{\kappa''} : \vec{\mathsf{N}}_{\kappa''}^{\mathrm{ord,tor}} \to \vec{\mathsf{M}}_{\mathcal{H},\Sigma^{\mathrm{ord}}}^{\mathrm{ord,tor}}$, which is nothing but $\vec{\mathsf{N}}_{\kappa''}^{\mathrm{ord}}$ (which can be shown in many ways—*e.g.*, by comparing the ranks of torus parts, or by comparing the dimensions of abelian parts). Since $\widehat{\mathcal{H}} = \widehat{\mathcal{H}}'$, by Lemma 7.2.1.3, $\vec{\mathsf{N}}_{\kappa''}^{\mathrm{ord}}$ is mapped isomorphically to $\vec{\mathsf{N}}_{\kappa}^{\mathrm{ord}}$ under $f^{\mathrm{tor}}_{\kappa'',\kappa}$. Hence, the preimage of $\vec{\mathsf{M}}_{\mathcal{H}}^{\mathrm{ord}}$ under (7.2.4.19) is $\vec{\mathsf{N}}_{\kappa}^{\mathrm{ord}}$. This shows that the canonical surjection $\vec{\mathsf{N}}_{\kappa}^{\mathrm{ord}} \to \vec{\mathsf{M}}_{\mathcal{H}}^{\mathrm{ord}}$ is the pullback of (7.2.4.20) under the canonical morphism $\vec{\mathsf{M}}_{\mathcal{H}}^{\mathrm{ord}} \hookrightarrow \vec{\mathsf{M}}_{\mathcal{H}}^{\mathrm{ord,min}}$ on the target. More generally, by (4) of Theorem 6.2.1.1, the preimage of the $[(\Phi_{\mathcal{H}},\delta_{\mathcal{H}})]$-stratum $\vec{\mathsf{Z}}^{\mathrm{ord}}_{[(\Phi_{\mathcal{H}},\delta_{\mathcal{H}})]}$ of $\vec{\mathsf{M}}_{\mathcal{H}}^{\mathrm{ord,min}}$ under the canonical morphism $\vec{\oint}^{\mathrm{ord}}_{\mathcal{H}} : \vec{\mathsf{M}}_{\mathcal{H},\Sigma^{\mathrm{ord}}}^{\mathrm{ord,tor}} \to \vec{\mathsf{M}}_{\mathcal{H}}^{\mathrm{ord,min}}$ is the union of the strata $\vec{\mathsf{Z}}^{\mathrm{ord}}_{[(\Phi_{\mathcal{H}},\delta_{\mathcal{H}},\sigma)]}$ with the same underlying cusp label $[(\Phi_{\mathcal{H}},\delta_{\mathcal{H}})]$. Hence, its further preimage in $\vec{\mathsf{N}}_{\kappa''}^{\mathrm{ord,tor}}$, and the image of this preimage in $\vec{\mathsf{N}}_{\kappa}^{\mathrm{ord,tor}}$, are described as in the statement of the proposition, because of the corresponding strata-mapping properties of the canonical morphisms $f^{\mathrm{tor}}_{\kappa'',\kappa} : \vec{\mathsf{N}}_{\kappa''}^{\mathrm{ord,tor}} \to \vec{\mathsf{N}}_{\kappa}^{\mathrm{ord,tor}}$ and $f^{\mathrm{tor}}_{\kappa''} : \vec{\mathsf{N}}_{\kappa''}^{\mathrm{ord,tor}} \to \vec{\mathsf{M}}_{\mathcal{H},\Sigma^{\mathrm{ord}}}^{\mathrm{ord,tor}}$ (see Propositions 7.2.2.19 and 7.2.3.5). □

From now on, let us fix a choice of some $\widetilde{\kappa} = (\widetilde{\mathcal{H}},\widetilde{\Sigma},\widetilde{\sigma}) \in \mathbf{K}^{\mathrm{ord},++}_{Q,\mathcal{H},\Sigma^{\mathrm{ord}}}$, inducing some $\kappa = [\widetilde{\kappa}] \in \mathbf{K}^{\mathrm{ord},++}_{Q,\mathcal{H},\Sigma^{\mathrm{ord}}}$. For simplicity, we shall suppress κ, $\widetilde{\Sigma}$, and Σ from the notation when their choices are clear from the context.

7.2.5 *Log Smoothness of* f^{tor}

We would like to show that the morphism f^{tor} is log smooth (as in [45, 3.3] and [43, 1.6]) if we equip $\vec{\mathsf{N}}^{\mathrm{ord,tor}}$ and $\vec{\mathsf{M}}_{\mathcal{H},r_{\kappa}}^{\mathrm{ord,tor}}$ with the canonical fine log structures given respectively by the relative Cartier divisors with simple normal crossings given by the complements $\vec{\mathsf{N}}^{\mathrm{ord,tor}} - \vec{\mathsf{N}}^{\mathrm{ord}}$ and $\vec{\mathsf{M}}_{\mathcal{H},r_{\kappa}}^{\mathrm{ord,tor}} - \vec{\mathsf{M}}_{\mathcal{H},r_{\kappa}}^{\mathrm{ord}}$ with their reduced structures. (Note that the log structure of $\vec{\mathsf{M}}_{\mathcal{H},r_{\kappa}}^{\mathrm{ord,tor}}$ is the pullback of the one of $\vec{\mathsf{M}}_{\mathcal{H}}^{\mathrm{ord,tor}}$ defined by $\vec{\mathsf{M}}_{\mathcal{H}}^{\mathrm{ord,tor}} - \vec{\mathsf{M}}_{\mathcal{H}}^{\mathrm{ord}}$. In what follows, we will freely state related facts about $\vec{\mathsf{M}}^{\mathrm{ord,tor}}_{\mathcal{H},r_\kappa}$ that are already know for $\vec{\mathsf{M}}^{\mathrm{ord,tor}}_{\mathcal{H}}$.) Moreover, for each of the sheaves to be introduced below, we will denote with the subscript "free" its free quotients defined by the image under $(\,\cdot\,) \to (\,\cdot\,) \underset{\mathbb{Z}}{\otimes} \mathbb{Q}$ (as in Definition 3.4.3.1).

According to [45, 3.12] (*cf.* [61, Lem. 3.11]), we have the following:

Lemma 7.2.5.1. *To show that the morphism f^{tor} is log smooth, it suffices to show that the first morphism in the canonical exact sequence*

$$\begin{aligned}(f^{\mathrm{tor}})^*(\Omega^1_{\vec{\mathsf{M}}^{\mathrm{ord,tor}}_{\mathcal{H},r_\kappa}/\vec{\mathsf{S}}_{0,r_\kappa}}[d\log\infty]) &\to \Omega^1_{\vec{\mathsf{N}}^{\mathrm{ord,tor}}/\vec{\mathsf{S}}_{0,r_\kappa}}[d\log\infty] \\ &\to \overline{\Omega}^1_{\vec{\mathsf{N}}^{\mathrm{ord,tor}}/\vec{\mathsf{M}}^{\mathrm{ord,tor}}_{\mathcal{H},r_\kappa}} \to 0\end{aligned} \tag{7.2.5.2}$$

is **injective**, *and that $\overline{\Omega}^1_{\vec{\mathsf{N}}^{\mathrm{ord,tor}}/\vec{\mathsf{M}}^{\mathrm{ord,tor}}_{\mathcal{H},r_\kappa}}$ is* **locally free** *of finite rank.*

By (4) of Theorem 5.2.1.1, the extended Kodaira–Spencer morphism [62, Def. 4.6.3.44] for $G \to \vec{\mathsf{M}}^{\mathrm{ord,tor}}_{\mathcal{H},r_\kappa}$ induces an isomorphism

$$\mathrm{KS}_{G/\vec{\mathsf{M}}^{\mathrm{ord,tor}}_{\mathcal{H},r_\kappa}/\vec{\mathsf{S}}_{0,r_\kappa}} : \underline{\mathrm{KS}}_{(G,\lambda,i)/\vec{\mathsf{M}}^{\mathrm{ord,tor}}_{\mathcal{H},r_\kappa},\mathrm{free}} \xrightarrow{\sim} \Omega^1_{\vec{\mathsf{M}}^{\mathrm{ord,tor}}_{\mathcal{H},r_\kappa}/\vec{\mathsf{S}}_{0,r_\kappa}}[d\log\infty]$$

over $\vec{\mathsf{M}}^{\mathrm{ord,tor}}_{\mathcal{H},r_\kappa}$, while the extended Kodaira–Spencer morphism for $\widetilde{G} \to \widetilde{\vec{\mathsf{M}}}^{\mathrm{ord,tor}}_{\widetilde{\mathcal{H}}}$ induces an isomorphism

$$\mathrm{KS}_{\widetilde{G}/\widetilde{\vec{\mathsf{M}}}^{\mathrm{ord,tor}}_{\widetilde{\mathcal{H}}}/\vec{\mathsf{S}}_{0,r_\kappa}} : \underline{\mathrm{KS}}_{(\widetilde{G},\widetilde{\lambda},\widetilde{i})/\widetilde{\vec{\mathsf{M}}}^{\mathrm{ord,tor}}_{\widetilde{\mathcal{H}}},\mathrm{free}} \xrightarrow{\sim} \Omega^1_{\widetilde{\vec{\mathsf{M}}}^{\mathrm{ord,tor}}_{\widetilde{\mathcal{H}}}/\vec{\mathsf{S}}_{0,r_\kappa}}[d\log\infty]$$

over $\widetilde{\vec{\mathsf{M}}}^{\mathrm{ord,tor}}_{\widetilde{\mathcal{H}}}$. By (7.2.3.1) and (7.2.3.2), we have a commutative diagram

$$\begin{array}{ccccccccc} 0 \to & \mathrm{Lie}^{\vee}_{\overline{G}/\vec{\mathsf{N}}^{\mathrm{ord,tor}}} & \to & \mathrm{Lie}^{\vee}_{\widetilde{G}_{\vec{\mathsf{N}}^{\mathrm{ord,tor}}}/\vec{\mathsf{N}}^{\mathrm{ord,tor}}} & \to & \mathrm{Lie}^{\vee}_{\widetilde{T}_{\vec{\mathsf{N}}^{\mathrm{ord,tor}}}/\vec{\mathsf{N}}^{\mathrm{ord,tor}}} & \to 0 \\ & {\scriptstyle \overline{\lambda}^*}\downarrow & & {\scriptstyle \widetilde{\lambda}^*}\downarrow & & {\scriptstyle \lambda^*_{\widetilde{T}} = \widetilde{\phi}_*}\downarrow & \\ 0 \to & \mathrm{Lie}^{\vee}_{\overline{G}^{\vee}/\vec{\mathsf{N}}^{\mathrm{ord,tor}}} & \to & \mathrm{Lie}^{\vee}_{\widetilde{G}^{\vee}_{\vec{\mathsf{N}}^{\mathrm{ord,tor}}}/\vec{\mathsf{N}}^{\mathrm{ord,tor}}} & \to & \mathrm{Lie}^{\vee}_{\widetilde{T}^{\vee}_{\vec{\mathsf{N}}^{\mathrm{ord,tor}}}/\vec{\mathsf{N}}^{\mathrm{ord,tor}}} & \to 0 \end{array} \tag{7.2.5.3}$$

in which the horizontal rows are exact. Using the sheaves and vertical arrows in this diagram, we can define $\underline{\mathrm{KS}}_{(\overline{G},\overline{\lambda},\overline{i})/\vec{\mathsf{N}}^{\mathrm{ord,tor}}}$, $\underline{\mathrm{KS}}_{(\widetilde{G},\widetilde{\lambda},\widetilde{i})_{\vec{\mathsf{N}}^{\mathrm{ord,tor}}}/\vec{\mathsf{N}}^{\mathrm{ord,tor}}}$, and $\underline{\mathrm{KS}}_{(\widetilde{T},\lambda_{\widetilde{T}},i_{\widetilde{T}})_{\vec{\mathsf{N}}^{\mathrm{ord,tor}}}/\vec{\mathsf{N}}^{\mathrm{ord,tor}}}$ as in [62, Def. 6.3.1] and Definitions 1.3.1.2 and 3.4.3.1. Since $(\overline{G},\overline{\lambda},\overline{i},\overline{\alpha}_{\mathcal{H}^p},\overline{\alpha}^{\mathrm{ord}}_{\mathcal{H}_p}) \to \vec{\mathsf{N}}^{\mathrm{ord,tor}}$ is canonically isomorphic to the pullback of $(G,\lambda,i,\alpha_{\mathcal{H}^p},\alpha^{\mathrm{ord}}_{\mathcal{H}_p}) \to \vec{\mathsf{M}}^{\mathrm{ord,tor}}_{\mathcal{H},r_\kappa}$ under $f^{\mathrm{tor}} : \vec{\mathsf{N}}^{\mathrm{ord,tor}} \to \vec{\mathsf{M}}^{\mathrm{ord,tor}}_{\mathcal{H},r_\kappa}$, we have a canonical isomorphism

$$(f^{\mathrm{tor}})^*\underline{\mathrm{KS}}_{(G,\lambda,i)/\vec{\mathsf{M}}^{\mathrm{ord,tor}}_{\mathcal{H},r_\kappa},\mathrm{free}} \cong \underline{\mathrm{KS}}_{(\overline{G},\overline{\lambda},\overline{i})/\vec{\mathsf{N}}^{\mathrm{ord,tor}},\mathrm{free}}.$$

On the other hand, $\underline{\mathrm{KS}}_{(\widetilde{G},\widetilde{\lambda},\widetilde{i})_{\vec{\mathsf{N}}^{\mathrm{ord,tor}}}/\vec{\mathsf{N}}^{\mathrm{ord,tor}}}$ is canonically isomorphic to the pullback of $\underline{\mathrm{KS}}_{(\widetilde{G},\widetilde{\lambda},\widetilde{i})/\widetilde{\vec{\mathsf{M}}}^{\mathrm{ord,tor}}_{\widetilde{\mathcal{H}}}}$ to $\vec{\mathsf{N}}^{\mathrm{ord,tor}}$, and $\underline{\mathrm{KS}}_{(\widetilde{T},\lambda_{\widetilde{T}},i_{\widetilde{T}})_{\vec{\mathsf{N}}^{\mathrm{ord,tor}}}/\vec{\mathsf{N}}^{\mathrm{ord,tor}}}$ is canonically isomorphic to the pullback of the sheaf

$$\underline{\mathrm{KS}}_{(\widetilde{T},\lambda_{\widetilde{T}},i_{\widetilde{T}})_{\vec{\mathsf{S}}_{0,r_\kappa}}/\vec{\mathsf{S}}_{0,r_\kappa}} := (\underline{\mathrm{Lie}}^\vee_{\widetilde{T}_{\vec{\mathsf{S}}_{0,r_\kappa}}/\vec{\mathsf{S}}_{0,r_\kappa}} \underset{\mathscr{O}_{\vec{\mathsf{S}}_{0,r_\kappa}}}{\otimes} \underline{\mathrm{Lie}}^\vee_{\widetilde{T}^\vee_{\vec{\mathsf{S}}_{0,r_\kappa}}/\vec{\mathsf{S}}_{0,r_\kappa}}) \Big/ \begin{pmatrix} \lambda^*_{\widetilde{T}}(y)\otimes z - \lambda^*_{\widetilde{T}}(z)\otimes y \\ (b^\star x)\otimes y - x\otimes(by) \end{pmatrix}_{\substack{x\in\underline{\mathrm{Lie}}^\vee_{\widetilde{T}_{\vec{\mathsf{S}}_{0,r_\kappa}}/\vec{\mathsf{S}}_{0,r_\kappa}} \\ y,z\in\underline{\mathrm{Lie}}^\vee_{\widetilde{T}^\vee_{\vec{\mathsf{S}}_{0,r_\kappa}}/\vec{\mathsf{S}}_{0,r_\kappa}} \\ b\in\mathcal{O}}}$$

similarly defined by the split tori $\widetilde{T}$ and $\widetilde{T}^\vee$ over $\vec{\mathsf{S}}_{0,r_\kappa}$ with respective character groups $\widetilde{X}$ and $\widetilde{Y}$. By the commutativity in the above diagram, there is a canonical surjection

$$\underline{\mathrm{KS}}_{(\widetilde{G},\widetilde{\lambda},\widetilde{i})_{\vec{\mathsf{N}}^{\mathrm{ord,tor}}}/\vec{\mathsf{N}}^{\mathrm{ord,tor}}} \twoheadrightarrow \underline{\mathrm{KS}}_{(\widetilde{T},\lambda_{\widetilde{T}},i_{\widetilde{T}})_{\vec{\mathsf{N}}^{\mathrm{ord,tor}}}/\vec{\mathsf{N}}^{\mathrm{ord,tor}}}, \tag{7.2.5.4}$$

whose kernel

$$\underline{\mathrm{K}} := \ker(\underline{\mathrm{KS}}_{(\widetilde{G},\widetilde{\lambda},\widetilde{i})_{\vec{\mathsf{N}}^{\mathrm{ord,tor}}}/\vec{\mathsf{N}}^{\mathrm{ord,tor}}} \twoheadrightarrow \underline{\mathrm{KS}}_{(\widetilde{T},\lambda_{\widetilde{T}},i_{\widetilde{T}})_{\vec{\mathsf{N}}^{\mathrm{ord,tor}}}/\vec{\mathsf{N}}^{\mathrm{ord,tor}}}) \tag{7.2.5.5}$$

contains $\underline{\mathrm{KS}}_{(\overline{G},\overline{\lambda},\overline{i})/\vec{\mathsf{N}}^{\mathrm{ord,tor}}}$ as a natural subsheaf.

Because the étale local structure of $\vec{\widetilde{\mathsf{M}}}^{\mathrm{ord,tor}}_{\widetilde{\mathcal{H}}}$ along the $[(\breve{\Phi}_{\widetilde{\mathcal{H}}},\breve{\delta}_{\widetilde{\mathcal{H}}},\breve{\tau})]$-stratum is the same as that of $\vec{\Xi}^{\mathrm{ord}}_{\breve{\Phi}_{\widetilde{\mathcal{H}}},\breve{\delta}_{\widetilde{\mathcal{H}}}}(\breve{\tau})$, the calculation in the proof of Proposition 4.2.3.5 shows that the isomorphism $\mathrm{KS}_{\widetilde{G}/\vec{\widetilde{\mathsf{M}}}^{\mathrm{ord,tor}}_{\widetilde{\mathcal{H}}}/\vec{\mathsf{S}}_{0,r_\kappa}}$ induces by restriction (to the closure $\vec{\mathsf{N}}^{\mathrm{ord,tor}}$ of the $[(\widetilde{\Phi}_{\widetilde{\mathcal{H}}},\widetilde{\delta}_{\widetilde{\mathcal{H}}},\widetilde{\sigma})]$-stratum) an isomorphism

$$\underline{\mathrm{K}}_{\mathrm{free}} \xrightarrow{\sim} \Omega^1_{\vec{\mathsf{N}}^{\mathrm{ord,tor}}/\vec{\mathsf{S}}_{0,r_\kappa}}[d\log\infty] \tag{7.2.5.6}$$

making the diagram

$$\begin{array}{ccc} (f^{\mathrm{tor}})^*\underline{\mathrm{KS}}_{(G,\lambda,i)/\vec{\mathsf{M}}^{\mathrm{ord,tor}}_{\mathcal{H},r_\kappa},\mathrm{free}} & \hookrightarrow & \underline{\mathrm{K}}_{\mathrm{free}} \\ {\scriptstyle \mathrm{KS}_{G/\vec{\mathsf{M}}^{\mathrm{ord,tor}}_{\mathcal{H},r_\kappa}/\vec{\mathsf{S}}_{0,r_\kappa}}}\downarrow\wr & & \wr\downarrow{\scriptstyle (7.2.5.6)} \\ (f^{\mathrm{tor}})^*(\Omega^1_{\vec{\mathsf{M}}^{\mathrm{ord,tor}}_{\mathcal{H},r_\kappa}/\vec{\mathsf{S}}_{0,r_\kappa}}[d\log\infty]) & \longrightarrow & \Omega^1_{\vec{\mathsf{N}}^{\mathrm{ord,tor}}/\vec{\mathsf{S}}_{0,r_\kappa}}[d\log\infty] \end{array}$$

commutative. In particular, the bottom arrow (which is the first morphism in (7.2.5.2)) is *injective*, and the isomorphism (7.2.5.6) induces a canonical isomorphism

$$\underline{\mathrm{K}}_{\mathrm{free}}/\underline{\mathrm{KS}}_{(\overline{G},\overline{\lambda},\overline{i})/\vec{\mathsf{N}}^{\mathrm{ord,tor}},\mathrm{free}} \xrightarrow{\sim} \overline{\Omega}^1_{\vec{\mathsf{N}}^{\mathrm{ord,tor}}/\vec{\mathsf{M}}^{\mathrm{ord,tor}}_{\mathcal{H},r_\kappa}} \tag{7.2.5.7}$$

of coherent sheaves over $\vec{\mathsf{N}}^{\mathrm{ord,tor}}$. We would like to verify the following:

Claim 7.2.5.8. $\underline{\mathrm{K}}_{\mathrm{free}}/\underline{\mathrm{KS}}_{(\overline{G},\overline{\lambda},\overline{i})/\vec{\mathsf{N}}^{\mathrm{ord,tor}},\mathrm{free}}$ *is locally free of finite rank.*

Then it will follow from Lemma 7.2.5.1 and the above that f^{tor} is log smooth, and the proof of (2) of Theorem 7.1.4.1 will be complete.

Let us define

$$\underline{\mathrm{K}}' := \underline{\mathrm{K}}/\underline{\mathrm{KS}}_{(\overline{G},\overline{\lambda},\overline{i})/\vec{\mathsf{N}}^{\mathrm{ord,tor}}} \tag{7.2.5.9}$$

over $\vec{\mathsf{N}}^{\mathrm{ord,tor}}$.

Lemma 7.2.5.10. *The canonical morphism*

$$\underline{\mathrm{K}}_{\mathrm{free}}/\underline{\mathrm{KS}}_{(\overline{G},\overline{\lambda},\overline{i})/\vec{\mathsf{N}}^{\mathrm{ord,tor}},\mathrm{free}} \to \underline{\mathrm{K}}'_{\mathrm{free}} \tag{7.2.5.11}$$

is an isomorphism.

Proof. It suffices to verify this over the completions of the strict local rings of $\vec{\mathsf{N}}^{\mathrm{ord,tor}}$ at points of characteristic p, which are complete noetherian normal domains whose spectra we denote by S. Over each such S, we have compatible (noncanonical) ordinary level structures

$$\begin{aligned}\widetilde{\alpha}^{\mathrm{ord}} = (\widetilde{\alpha}^{\mathrm{ord},0} &: (\mathrm{Gr}^0_{\widetilde{\mathsf{D}}})^{\mathrm{mult}}_S \to G^{\natural}_S[p^\infty],\\ \widetilde{\alpha}^{\mathrm{ord},\#,0} &: (\mathrm{Gr}^0_{\widetilde{\mathsf{D}}\#})^{\mathrm{mult}}_S \to G^{\vee,\natural}_S[p^\infty], \widetilde{\nu}^{\mathrm{ord}})\end{aligned}$$

and

$$\begin{aligned}\overline{\alpha}^{\mathrm{ord}} = (\overline{\alpha}^{\mathrm{ord},0} &: (\mathrm{Gr}^0_{\mathsf{D}})^{\mathrm{mult}}_S \to \overline{G}_S[p^\infty],\\ \overline{\alpha}^{\mathrm{ord},\#,0} &: (\mathrm{Gr}^0_{\mathsf{D}\#})^{\mathrm{mult}} \to \overline{G}^{\vee}_S[p^\infty], \overline{\nu}^{\mathrm{ord}}_S)\end{aligned}$$

(where the subscripts "S" mean pullbacks to S) defining a commutative diagram

$$\begin{array}{ccccccccc} 0 \longrightarrow & \widetilde{T}_S[p^\infty] & \longrightarrow & \mathrm{image}(\widetilde{\alpha}^{\mathrm{ord},0}) & \longrightarrow & \mathrm{image}(\overline{\alpha}^{\mathrm{ord},0}_S) & \longrightarrow 0 \\ & \downarrow {\scriptstyle \lambda_{\widetilde{T}}=\widetilde{\phi}^*} & & \downarrow {\scriptstyle \widetilde{\lambda}_S} & & \downarrow {\scriptstyle \lambda_S} & \\ 0 \longrightarrow & \widetilde{T}^\vee_S[p^\infty] & \longrightarrow & \mathrm{image}(\widetilde{\alpha}^{\mathrm{ord},\#,0}) & \longrightarrow & \mathrm{image}(\overline{\alpha}^{\mathrm{ord},\#,0}) & \longrightarrow 0 \end{array} \tag{7.2.5.12}$$

canonically dual to the pullback of (7.2.5.3) to S, under canonical isomorphisms as in (5) and (6) of Proposition 3.2.1.1. The diagram (7.2.5.12) admits (noncanonical) splittings, namely splittings of the two exact rows compatible with $\widetilde{\lambda}_S$, because $\langle\,\cdot\,,\,\cdot\,\rangle^\sim$ is the direct sum of the pairings on $Q_{-2} \oplus Q_0$ and on L (and because the ordinary level structures match the diagram (7.2.5.12) with the corresponding diagram of constant objects). Such splittings induce (by duality) splittings of the pullback of (7.2.5.3) to S, namely splittings of the two exact rows compatible with $\widetilde{\lambda}_S$. Hence, we have a noncanonical isomorphism between $\underline{\mathrm{K}}_S$ and $\underline{\mathrm{KS}}_{(\overline{G},\overline{\lambda},\overline{i})/\vec{\mathsf{N}}^{\mathrm{ord,tor}},S} \oplus \underline{\mathrm{K}}'_S$ over S, and also a corresponding one between their free quotients. Hence, (7.2.5.11) is an isomorphism, as desired. □

By Proposition 3.1.3.4, $\underline{\mathrm{Hom}}_{\mathcal{O}}(\widetilde{X}, G)$ (resp. $\underline{\mathrm{Hom}}_{\mathcal{O}}(\widetilde{X}, G^\vee)$, resp. $\underline{\mathrm{Hom}}_{\mathcal{O}}(\widetilde{Y}, G)$, resp. $\underline{\mathrm{Hom}}_{\mathcal{O}}(\widetilde{Y}, G^\vee)$) is relatively representable by an extension of a quasi-finite flat group scheme of étale-multiplicative type by a semi-abelian scheme $\underline{\mathrm{Hom}}_{\mathcal{O}}(\widetilde{X}, G)^\circ$ (resp. $\underline{\mathrm{Hom}}_{\mathcal{O}}(\widetilde{X}, G^\vee)^\circ$, resp. $\underline{\mathrm{Hom}}_{\mathcal{O}}(\widetilde{Y}, G)^\circ$, resp. $\underline{\mathrm{Hom}}_{\mathcal{O}}(\widetilde{Y}, G^\vee)^\circ$) over $\vec{\mathsf{M}}^{\mathrm{ord,tor}}_{\mathcal{H},r_\kappa}$. Then the homomorphisms $\widetilde{\phi} : \widetilde{Y} \hookrightarrow \widetilde{X}$ and $\lambda : G \to G^\vee$

over $\vec{\mathsf{M}}^{\mathrm{ord,tor}}_{\mathcal{H},r_\kappa}$ induce homomorphisms $\underline{\mathrm{Hom}}_{\mathcal{O}}(\widetilde{X},G^\vee)^\circ \to \underline{\mathrm{Hom}}_{\mathcal{O}}(\widetilde{Y},G^\vee)^\circ$ and $\underline{\mathrm{Hom}}_{\mathcal{O}}(\widetilde{Y},G)^\circ \to \underline{\mathrm{Hom}}_{\mathcal{O}}(\widetilde{Y},G^\vee)^\circ$ with kernels that are quasi-finite group schemes of étale-multiplicative type over $\vec{\mathsf{M}}^{\mathrm{ord,tor}}_{\mathcal{H},r_\kappa}$, and hence the fiber product

$$\ddot{\widetilde{C}}^{\mathrm{ord,ext}} := \underline{\mathrm{Hom}}_{\mathcal{O}}(\widetilde{X},G^\vee) \underset{\underline{\mathrm{Hom}}_{\mathcal{O}}(\widetilde{Y},G^\vee)}{\times} \underline{\mathrm{Hom}}_{\mathcal{O}}(\widetilde{Y},G) \tag{7.2.5.13}$$

is also an extension of a quasi-finite group scheme $\underline{\pi}_0(\ddot{\widetilde{C}}^{\mathrm{ord,ext}}/\vec{\mathsf{M}}^{\mathrm{ord,tor}}_{\mathcal{H},r_\kappa})$ of étale-multiplicative type by a semi-abelian scheme $\ddot{\widetilde{C}}^{\mathrm{ord,ext},\circ}$ over $\vec{\mathsf{M}}^{\mathrm{ord,tor}}_{\mathcal{H},r_\kappa}$.

Lemma 7.2.5.14. *We have compatible canonical isomorphisms*

$$\underline{\mathrm{K}}' \cong (f^{\mathrm{tor}})^*\underline{\mathrm{Lie}}^\vee_{\ddot{\widetilde{C}}^{\mathrm{ord,ext}}/\vec{\mathsf{M}}^{\mathrm{ord,tor}}_{\mathcal{H},r_\kappa}} \tag{7.2.5.15}$$

and

$$\begin{aligned}\underline{\mathrm{K}}'_{\mathrm{free}} &\cong \big((f^{\mathrm{tor}})^*\underline{\mathrm{Lie}}^\vee_{\ddot{\widetilde{C}}^{\mathrm{ord,ext}}/\vec{\mathsf{M}}^{\mathrm{ord,tor}}_{\mathcal{H},r_\kappa}}\big)_{\mathrm{free}} \\ &\cong (f^{\mathrm{tor}})^*\underline{\mathrm{Lie}}^\vee_{\ddot{\widetilde{C}}^{\mathrm{ord,ext}}/\vec{\mathsf{M}}^{\mathrm{ord,tor}}_{\mathcal{H},r_\kappa},\mathrm{free}} \\ &\cong (f^{\mathrm{tor}})^*\underline{\mathrm{Lie}}^\vee_{\ddot{\widetilde{C}}^{\mathrm{ord,ext},\circ}/\vec{\mathsf{M}}^{\mathrm{ord,tor}}_{\mathcal{H},r_\kappa}}.\end{aligned} \tag{7.2.5.16}$$

Since $\ddot{\widetilde{C}}^{\mathrm{ord,ext},\circ}$ is a semi-abelian scheme over $\vec{\mathsf{M}}^{\mathrm{ord,tor}}_{\mathcal{H},r_\kappa}$, it follows that $\underline{\mathrm{Lie}}^\vee_{\ddot{\widetilde{C}}^{\mathrm{ord,ext},\circ}/\vec{\mathsf{M}}^{\mathrm{ord,tor}}_{\mathcal{H},r_\kappa}}$ is locally free of finite rank over $\vec{\mathsf{M}}^{\mathrm{ord,tor}}_{\mathcal{H},r_\kappa}$, and that $\underline{\mathrm{K}}'_{\mathrm{free}}$ is locally free of finite rank over $\vec{\mathsf{N}}^{\mathrm{ord,tor}}$.

Proof. By definition, we have canonical isomorphisms

$$\begin{aligned}&\underline{\mathrm{Lie}}^\vee_{\underline{\mathrm{Hom}}_{\mathbb{Z}}(\widetilde{X},G^\vee)/\vec{\mathsf{M}}^{\mathrm{ord,tor}}_{\mathcal{H},r_\kappa}} \cong \underline{\mathrm{Hom}}_{\mathbb{Z}}(\widetilde{X}^\vee,\underline{\mathrm{Lie}}^\vee_{G^\vee/\vec{\mathsf{M}}^{\mathrm{ord,tor}}_{\mathcal{H},r_\kappa}}) \\ &\cong \underline{\mathrm{Hom}}_{\mathscr{O}_{\vec{\mathsf{M}}^{\mathrm{ord,tor}}_{\mathcal{H},r_\kappa}}}(\underline{\mathrm{Hom}}_{\mathbb{Z}}(\widetilde{X},\mathscr{O}_{\vec{\mathsf{M}}^{\mathrm{ord,tor}}_{\mathcal{H},r_\kappa}}),\underline{\mathrm{Lie}}^\vee_{G^\vee/\vec{\mathsf{M}}^{\mathrm{ord,tor}}_{\mathcal{H},r_\kappa}}) \\ &\cong \underline{\mathrm{Hom}}_{\mathscr{O}_{\vec{\mathsf{M}}^{\mathrm{ord,tor}}_{\mathcal{H},r_\kappa}}}(\underline{\mathrm{Lie}}_{\widetilde{T}_{\vec{\mathsf{M}}^{\mathrm{ord,tor}}_{\mathcal{H},r_\kappa}}/\vec{\mathsf{M}}^{\mathrm{ord,tor}}_{\mathcal{H},r_\kappa}},\underline{\mathrm{Lie}}^\vee_{G^\vee/\vec{\mathsf{M}}^{\mathrm{ord,tor}}_{\mathcal{H},r_\kappa}}) \\ &\cong \underline{\mathrm{Lie}}^\vee_{\widetilde{T}_{\vec{\mathsf{M}}^{\mathrm{ord,tor}}_{\mathcal{H},r_\kappa}}/\vec{\mathsf{M}}^{\mathrm{ord,tor}}_{\mathcal{H},r_\kappa}} \underset{\mathscr{O}_{\vec{\mathsf{M}}^{\mathrm{ord,tor}}_{\mathcal{H},r_\kappa}}}{\otimes} \underline{\mathrm{Lie}}^\vee_{G^\vee/\vec{\mathsf{M}}^{\mathrm{ord,tor}}_{\mathcal{H},r_\kappa}}\end{aligned}$$

and

$$\begin{aligned}&\underline{\mathrm{Lie}}^\vee_{\underline{\mathrm{Hom}}_{\mathbb{Z}}(\widetilde{Y},G)/\vec{\mathsf{M}}^{\mathrm{ord,tor}}_{\mathcal{H},r_\kappa}} \cong \underline{\mathrm{Hom}}_{\mathbb{Z}}(\widetilde{Y}^\vee,\underline{\mathrm{Lie}}^\vee_{G/\vec{\mathsf{M}}^{\mathrm{ord,tor}}_{\mathcal{H},r_\kappa}}) \\ &\cong \underline{\mathrm{Hom}}_{\mathscr{O}_{\vec{\mathsf{M}}^{\mathrm{ord,tor}}_{\mathcal{H},r_\kappa}}}(\underline{\mathrm{Hom}}_{\mathbb{Z}}(\widetilde{Y},\mathscr{O}_{\vec{\mathsf{M}}^{\mathrm{ord,tor}}_{\mathcal{H},r_\kappa}}),\underline{\mathrm{Lie}}^\vee_{G/\vec{\mathsf{M}}^{\mathrm{ord,tor}}_{\mathcal{H},r_\kappa}}) \\ &\cong \underline{\mathrm{Hom}}_{\mathscr{O}_{\vec{\mathsf{M}}^{\mathrm{ord,tor}}_{\mathcal{H},r_\kappa}}}(\underline{\mathrm{Lie}}_{\widetilde{T}^\vee_{\vec{\mathsf{M}}^{\mathrm{ord,tor}}_{\mathcal{H},r_\kappa}}/\vec{\mathsf{M}}^{\mathrm{ord,tor}}_{\mathcal{H},r_\kappa}},\underline{\mathrm{Lie}}^\vee_{G/\vec{\mathsf{M}}^{\mathrm{ord,tor}}_{\mathcal{H},r_\kappa}}) \\ &\cong \underline{\mathrm{Lie}}^\vee_{G/\vec{\mathsf{M}}^{\mathrm{ord,tor}}_{\mathcal{H},r_\kappa}} \underset{\mathscr{O}_{\vec{\mathsf{M}}^{\mathrm{ord,tor}}_{\mathcal{H},r_\kappa}}}{\otimes} \underline{\mathrm{Lie}}^\vee_{\widetilde{T}^\vee_{\vec{\mathsf{M}}^{\mathrm{ord,tor}}_{\mathcal{H},r_\kappa}}/\vec{\mathsf{M}}^{\mathrm{ord,tor}}_{\mathcal{H},r_\kappa}}.\end{aligned}$$

Since $\ddot{\widetilde{C}}^{\mathrm{ord,ext}}$ is (by definition) the subgroup scheme of the ordinary semi-abelian scheme $\underline{\mathrm{Hom}}_{\mathbb{Z}}(\widetilde{Y}, G) \underset{\vec{\mathsf{M}}^{\mathrm{ord,tor}}_{\mathcal{H},r_\kappa}}{\times} \underline{\mathrm{Hom}}_{\mathbb{Z}}(\widetilde{X}, G^\vee)$ over which the tautological object $(\widetilde{c}, \widetilde{c}^\vee)$ is $\mathcal{O}$-equivariant and satisfies the compatibility $\widetilde{\widetilde{c}\phi} = \lambda \widetilde{c}^\vee$, and since $(\overline{G}, \overline{\lambda}, \overline{i}, \overline{\alpha}_{\mathcal{H}^p}, \overline{\alpha}^{\mathrm{ord}}_{\mathcal{H}_p}) \to \vec{\mathsf{N}}^{\mathrm{ord,tor}}$ is canonically isomorphic to the pullback of $(G, \lambda, i, \alpha_{\mathcal{H}^p}, \alpha^{\mathrm{ord}}_{\mathcal{H}_p}) \to \vec{\mathsf{M}}^{\mathrm{ord,tor}}_{\mathcal{H},r_\kappa}$ under $f^{\mathrm{tor}} : \vec{\mathsf{N}}^{\mathrm{ord,tor}} \to \vec{\mathsf{M}}^{\mathrm{ord,tor}}_{\mathcal{H},r_\kappa}$, the pullback $(f^{\mathrm{tor}})^* \underline{\mathrm{Lie}}^\vee_{\ddot{\widetilde{C}}^{\mathrm{ord,ext}}/\vec{\mathsf{M}}^{\mathrm{ord,tor}}_{\mathcal{H},r_\kappa}}$ is canonically isomorphic to the quotient of

$$\begin{aligned}&\Big(\underline{\mathrm{Lie}}^\vee_{\widetilde{T}_{\vec{\mathsf{N}}^{\mathrm{ord,tor}}}/\vec{\mathsf{N}}^{\mathrm{ord,tor}}} \underset{\mathscr{O}_{\vec{\mathsf{N}}^{\mathrm{ord,tor}}}}{\otimes} \underline{\mathrm{Lie}}^\vee_{\overline{G}^\vee/\vec{\mathsf{N}}^{\mathrm{ord,tor}}}\Big)\\ &\quad + \Big(\underline{\mathrm{Lie}}^\vee_{\overline{G}/\vec{\mathsf{N}}^{\mathrm{ord,tor}}} \underset{\mathscr{O}_{\vec{\mathsf{N}}^{\mathrm{ord,tor}}}}{\otimes} \underline{\mathrm{Lie}}^\vee_{\widetilde{T}^\vee_{\vec{\mathsf{N}}^{\mathrm{ord,tor}}}/\vec{\mathsf{N}}^{\mathrm{ord,tor}}}\Big),\end{aligned}$$

as a subsheaf of

$$\begin{aligned}&\Big(\mathrm{Lie}^\vee_{\widetilde{G}_{\vec{\mathsf{N}}^{\mathrm{ord,tor}}}/\vec{\mathsf{N}}^{\mathrm{ord,tor}}} \underset{\mathscr{O}_{\vec{\mathsf{N}}^{\mathrm{ord,tor}}}}{\otimes} \mathrm{Lie}^\vee_{\widetilde{G}^\vee_{\vec{\mathsf{N}}^{\mathrm{ord,tor}}}/\vec{\mathsf{N}}^{\mathrm{ord,tor}}}\Big)\\ &\quad / \Big(\underline{\mathrm{Lie}}^\vee_{\overline{G}/\vec{\mathsf{N}}^{\mathrm{ord,tor}}} \underset{\mathscr{O}_{\vec{\mathsf{N}}^{\mathrm{ord,tor}}}}{\otimes} \underline{\mathrm{Lie}}^\vee_{\overline{G}^\vee/\vec{\mathsf{N}}^{\mathrm{ord,tor}}}\Big),\end{aligned}$$

by relations as in Definition 4.2.3.4, which is by definition $\underline{\mathrm{K}}'$. Hence, we have the canonical isomorphism (7.2.5.15).

Since $\ddot{\widetilde{C}}^{\mathrm{ord,ext},\circ}/\vec{\mathsf{M}}^{\mathrm{ord,tor}}_{\mathcal{H},r_\kappa}$ is a semi-abelian scheme and $\underline{\pi}_0(\ddot{\widetilde{C}}^{\mathrm{ord,ext}}/\vec{\mathsf{M}}^{\mathrm{ord,tor}}_{\mathcal{H},r_\kappa})$ is quasi-finite flat of étale-multiplicative type, $\underline{\mathrm{Lie}}^\vee_{\ddot{\widetilde{C}}^{\mathrm{ord,ext},\circ}/\vec{\mathsf{M}}^{\mathrm{ord,tor}}_{\mathcal{H},r_\kappa}}$ is locally of finite rank over $\vec{\mathsf{M}}^{\mathrm{ord,tor}}_{\mathcal{H},r_\kappa}$, and the canonical morphism $\underline{\mathrm{Lie}}^\vee_{\ddot{\widetilde{C}}^{\mathrm{ord,ext}}/\vec{\mathsf{M}}^{\mathrm{ord,tor}}_{\mathcal{H},r_\kappa}} \to \underline{\mathrm{Lie}}^\vee_{\ddot{\widetilde{C}}^{\mathrm{ord,ext},\circ}/\vec{\mathsf{M}}^{\mathrm{ord,tor}}_{\mathcal{H},r_\kappa}}$ induces a canonical isomorphism

$$\underline{\mathrm{Lie}}^\vee_{\ddot{\widetilde{C}}^{\mathrm{ord,ext}}/\vec{\mathsf{M}}^{\mathrm{ord,tor}}_{\mathcal{H},r_\kappa},\mathrm{free}} \xrightarrow{\sim} \underline{\mathrm{Lie}}^\vee_{\ddot{\widetilde{C}}^{\mathrm{ord,ext},\circ}/\vec{\mathsf{M}}^{\mathrm{ord,tor}}_{\mathcal{H},r_\kappa}}. \tag{7.2.5.17}$$

Since the formation of free quotients is compatible with pulling back under f^{tor}, the canonical isomorphisms (7.2.5.15) and (7.2.5.17) induce the canonical isomorphisms in (7.2.5.16). The remaining statements of the lemma are self-explanatory. □

Thus, we have verified Claim 7.2.5.8 and proved (2) of Theorem 7.1.4.1.

Now suppose $\kappa \in \mathbf{K}^{\mathrm{ord},+}_{Q,\mathcal{H},\Sigma^{\mathrm{ord}}}$ (not just in $\mathbf{K}^{\mathrm{ord},++}_{Q,\mathcal{H},\Sigma^{\mathrm{ord}}}$, so that $\mathcal{H}_\kappa = \mathcal{H}$ and $r_\kappa = r_{\mathcal{H}}$). This is the setting in (3) of Theorem 7.1.4.1, where the semi-abelian scheme $\vec{\mathsf{N}}^{\mathrm{ord,ext}} = \vec{\mathsf{N}}^{\mathrm{ord,ext}}_\kappa \to \vec{\mathsf{M}}^{\mathrm{ord,tor}}_{\mathcal{H}}$ extending $\vec{\mathsf{N}}^{\mathrm{ord,grp}} = \vec{\mathsf{N}}^{\mathrm{ord,grp}}_\kappa \to \vec{\mathsf{M}}^{\mathrm{ord}}_{\mathcal{H}}$ is also defined (as in Definition 7.1.3.11).

Consider any $n = n_0 p^r$ such that $p \nmid n_0$, $\mathcal{U}^p(n_0) \subset \mathcal{H}^p$, and $\mathcal{U}^{\mathrm{bal}}_{p,1}(p^r) \subset \mathcal{H}_p \subset \mathcal{U}_{p,0}(p^r)$. By condition (5) of Definition 3.4.2.10, (by abuse of language) $\alpha^{\mathrm{ord}}_{\mathcal{H}_p}$ extends to an ordinary level structure of (G, λ, i) over $\vec{\mathsf{M}}^{\mathrm{ord,tor}}_{\mathcal{H}}$. Although $\alpha^{\mathrm{ord},0}_{p^r}$ (resp. $\alpha^{\mathrm{ord},\#,0}_{p^r}$) is only étale locally defined by $\alpha^{\mathrm{ord}}_{\mathcal{H}_p}$, its schematic image $\mathrm{image}(\alpha^{\mathrm{ord},0}_{p^r}) = \alpha^{\mathrm{ord},0}_{p^r}((\mathrm{Gr}^0_{\mathsf{D}_{p^r}})^{\mathrm{mult}}_{\vec{\mathsf{M}}^{\mathrm{ord,tor}}_{\mathcal{H}}})$ (resp. $\mathrm{image}(\alpha^{\mathrm{ord},\#,0}_{p^r}) = \alpha^{\mathrm{ord},\#,0}_{p^r}((\mathrm{Gr}^0_{\mathsf{D}^\#_{p^r}})^{\mathrm{mult}}_{\vec{\mathsf{M}}^{\mathrm{ord,tor}}_{\mathcal{H}}})$) depends only on $\alpha^{\mathrm{ord}}_{\mathcal{H}_p}$ and descends to a finite flat subgroup scheme of multiplicative type of G (resp. $G^\vee$) over $\vec{\mathsf{M}}^{\mathrm{ord,tor}}_{\mathcal{H}}$, which we (by abuse of notation) still denote by $\mathrm{image}(\alpha^{\mathrm{ord},0}_{p^r})$ (resp. $\mathrm{image}(\alpha^{\mathrm{ord},\#,0}_{p^r})$). As in Section 4.2.1, let us define

$$G \twoheadrightarrow G^{\mathrm{ord}}_{p^r} := G/\mathrm{image}(\alpha^{\mathrm{ord},0}_{p^r}) \tag{7.2.5.18}$$

and

$$G^\vee \twoheadrightarrow G^{\vee,\mathrm{ord}}_{p^r} := G^\vee/\mathrm{image}(\alpha^{\mathrm{ord},\#,0}_{p^r}), \tag{7.2.5.19}$$

respectively, together with morphisms

$$G^{\mathrm{ord}}_{p^r} \twoheadrightarrow G \tag{7.2.5.20}$$

and

$$G^{\vee,\mathrm{ord}}_{p^r} \twoheadrightarrow G^\vee, \tag{7.2.5.21}$$

respectively, such that the compositions (7.2.5.20) $\circ$ (7.2.5.18) and (7.2.5.21) $\circ$ (7.2.5.19) are multiplications by p^r. (See (4.1.4.31), (4.1.4.32), (4.1.4.34), (4.1.4.33), (4.2.1.1), (4.2.1.2), (4.2.1.3), and (4.2.1.4). The restrictions of the morphisms (7.2.5.20) and (7.2.5.21) to $\vec{\mathsf{M}}^{\mathrm{ord}}_{\mathcal{H}}$ exist as duals of the restrictions of (7.2.5.19) and (7.2.5.18), respectively, and extends to $\vec{\mathsf{M}}^{\mathrm{ord,tor}}_{\mathcal{H}}$ by [92, IX, 1.4], [28, Ch. I, Prop. 2.7], or [62, Prop. 3.3.1.5].) Note that the kernels of (7.2.5.20) and (7.2.5.21) are quasi-finite étale, because they are étale locally subgroups of the constant group schemes $(\mathrm{Gr}^{-1}_{\mathsf{D}_{p^r}})_{\vec{\mathsf{M}}^{\mathrm{ord,tor}}_{\mathcal{H}}}$ and $(\mathrm{Gr}^{-1}_{\mathsf{D}^{\#}_{p^r}})_{\vec{\mathsf{M}}^{\mathrm{ord,tor}}_{\mathcal{H}}}$, respectively.

By Proposition 3.1.3.4 (as above), $\underline{\mathrm{Hom}}_{\mathcal{O}}(\frac{1}{n}\widetilde{X}, G^{\vee,\mathrm{ord}}_{p^r})$ (resp. $\underline{\mathrm{Hom}}_{\mathcal{O}}(\frac{1}{n}\widetilde{Y}, G^{\mathrm{ord}}_{p^r})$) is representable by an extension of a quasi-finite flat group scheme of étale-multiplicative type by a semi-abelian scheme over $\vec{\mathsf{M}}^{\mathrm{ord,tor}}_{\mathcal{H}}$, and the fiber product

$$\ddot{\widetilde{C}}{}^{\mathrm{ord,ext}}_{n} := \underline{\mathrm{Hom}}_{\mathcal{O}}(\tfrac{1}{n}\widetilde{X}, G^{\vee,\mathrm{ord}}_{p^r}) \underset{\underline{\mathrm{Hom}}_{\mathcal{O}}(\widetilde{Y}, G^\vee)}{\times} \underline{\mathrm{Hom}}_{\mathcal{O}}(\tfrac{1}{n}\widetilde{Y}, G^{\mathrm{ord}}_{p^r}) \tag{7.2.5.22}$$

is also an extension of a quasi-finite group scheme $\underline{\pi}_0(\ddot{\widetilde{C}}{}^{\mathrm{ord,ext}}_{n}/\vec{\mathsf{M}}^{\mathrm{ord,tor}}_{\mathcal{H}})$ of étale-multiplicative type by a semi-abelian scheme $\ddot{\widetilde{C}}{}^{\mathrm{ord,ext},\circ}_{n}$ over $\vec{\mathsf{M}}^{\mathrm{ord,tor}}_{\mathcal{H}}$. Note that $\ddot{\widetilde{C}}{}^{\mathrm{ord,ext}}_{n}$ and $\ddot{\widetilde{C}}{}^{\mathrm{ord,ext},\circ}_{n}$ are closed subgroup schemes of the semi-abelian scheme $\underline{\mathrm{Hom}}_{\mathbb{Z}}(\frac{1}{n}\widetilde{X}, G^{\vee,\mathrm{ord}}_{p^r}) \underset{\vec{\mathsf{M}}^{\mathrm{ord,tor}}_{\mathcal{H}}}{\times} \underline{\mathrm{Hom}}_{\mathbb{Z}}(\frac{1}{n}\widetilde{Y}, G^{\mathrm{ord}}_{p^r})$ over $\vec{\mathsf{M}}^{\mathrm{ord,tor}}_{\mathcal{H}}$. The group $\widetilde{H}^{\mathrm{ord}}_{n,\widetilde{\mathrm{U}}^{\mathrm{ess}}_{\widetilde{\mathrm{Z}}_n}}$ acts naturally on the (abelian scheme) pullback of this semi-abelian scheme to $\vec{\mathsf{M}}^{\mathrm{ord}}_{\mathcal{H}}$, and this action extends canonically over $\vec{\mathsf{M}}^{\mathrm{ord,tor}}_{\mathcal{H}}$ by [92, IX, 1.4], [28, Ch. I, Prop. 2.7], or [62, Prop. 3.3.1.5]. Let

$$\ddot{\widetilde{C}}{}^{\mathrm{ord,ext}}_{\widetilde{\mathcal{H}},n} := \ddot{\widetilde{C}}{}^{\mathrm{ord,ext}}_{n}/\widetilde{H}^{\mathrm{ord}}_{n,\widetilde{\mathrm{U}}^{\mathrm{ess}}_{\widetilde{\mathrm{Z}}_n}} \tag{7.2.5.23}$$

(*cf.* (4.2.1.14)—we do not form the quotient by $\widetilde{H}^{\mathrm{ord}}_{n,\widetilde{\mathrm{G}}^{\mathrm{ess}}_{h,\widetilde{\mathrm{Z}}_n}}$ and other groups here, because we are already working over $\vec{\mathsf{M}}^{\mathrm{ord,tor}}_{\mathcal{H}}$), so that

$$\ddot{\widetilde{C}}{}^{\mathrm{ord,ext},\circ}_{\widetilde{\mathcal{H}},n} \cong \ddot{\widetilde{C}}{}^{\mathrm{ord,ext},\circ}_{n}/\widetilde{H}^{\mathrm{ord}}_{n,\widetilde{\mathrm{U}}^{\mathrm{ess}}_{\widetilde{\mathrm{Z}}_n}}. \tag{7.2.5.24}$$

By construction, we have the following commutative diagram

$$\begin{array}{ccc} \dddot{\widetilde{C}}^{\mathrm{ord,ext},\circ}_{\widetilde{\mathcal{H}},n} & \hookrightarrow & \Big(\underline{\mathrm{Hom}}_{\mathbb{Z}}(\tfrac{1}{n}\widetilde{X}, G^{\vee,\mathrm{ord}}_{p^r}) \underset{\vec{\mathsf{M}}^{\mathrm{ord,tor}}_{\mathcal{H}}}{\times} \underline{\mathrm{Hom}}_{\mathbb{Z}}(\tfrac{1}{n}\widetilde{Y}, G^{\mathrm{ord}}_{p^r})\Big)/\widetilde{H}^{\mathrm{ord}}_{n,\widetilde{\mathrm{U}}^{\mathrm{ess}}_{\widetilde{\mathsf{Z}}_n}} \\ \downarrow & & \downarrow \\ \dddot{\widetilde{C}}^{\mathrm{ord,ext},\circ} & \hookrightarrow & \underline{\mathrm{Hom}}_{\mathbb{Z}}(\widetilde{X}, G^{\vee}) \underset{\vec{\mathsf{M}}^{\mathrm{ord,tor}}_{\mathcal{H}}}{\times} \underline{\mathrm{Hom}}_{\mathbb{Z}}(\widetilde{Y}, G) \end{array} \tag{7.2.5.25}$$

of canonical morphisms of semi-abelian schemes over $\vec{\mathsf{M}}^{\mathrm{ord,tor}}_{\mathcal{H}}$, in which all horizontal arrows are closed immersions.

Lemma 7.2.5.26. *In* (7.2.5.25), *the vertical arrows are* (*quasi-finite and*) *unramified. Also, the first vertical arrow is an étale isogeny between semi-abelian schemes.*

Proof. The first statement is true because the vertical arrows are homomorphisms with quasi-finite étale kernels (because (7.2.5.20) and (7.2.5.21) are). Since the first vertical arrow is a homomorphism between semi-abelian schemes, it is surjective and automatically flat by [35, IV-3, 11.3.10 a)⇒b) and 15.4.2 e′)⇒b)] (*cf.* the proof of [62, Lem. 1.3.1.11]), and hence is an étale isogeny. □

By construction, under the canonical isomorphism (7.2.1.1), we have a canonical isomorphism

$$\dddot{\widetilde{C}}^{\mathrm{ord},\circ}_{\widetilde{\Phi}_{\widetilde{\mathcal{H}},n}} \underset{\dddot{\widetilde{\mathsf{M}}}^{\mathrm{ord},\widetilde{\mathsf{Z}}_{\mathcal{H}}}_{\mathcal{H}}}{\times} \vec{\mathsf{M}}^{\mathrm{ord}}_{\mathcal{H}} \cong \dddot{\widetilde{C}}^{\mathrm{ord,ext},\circ}_{\widetilde{\mathcal{H}},n} \underset{\vec{\mathsf{M}}^{\mathrm{ord,tor}}_{\mathcal{H}}}{\times} \vec{\mathsf{M}}^{\mathrm{ord}}_{\mathcal{H}} \tag{7.2.5.27}$$

of abelian schemes over $\vec{\mathsf{M}}^{\mathrm{ord}}_{\mathcal{H}}$, where $\dddot{\widetilde{C}}^{\mathrm{ord},\circ}_{\widetilde{\Phi}_{\widetilde{\mathcal{H}},n}}$ is defined over $\dddot{\widetilde{\mathsf{M}}}^{\mathrm{ord},\widetilde{\mathsf{Z}}_{\mathcal{H}}}_{\mathcal{H}}$ as in (4.2.1.15), in the toroidal boundary construction of $\vec{\widetilde{\mathsf{M}}}^{\mathrm{ord}}_{\widetilde{\mathcal{H}}}$.

Recall (see Section 7.2.1) that $\vec{\mathsf{N}}^{\mathrm{ord}}_{\kappa} = \vec{\widetilde{\mathsf{Z}}}^{\mathrm{ord}}_{[(\widetilde{\Phi}_{\widetilde{\mathcal{H}}},\widetilde{\delta}_{\widetilde{\mathcal{H}}},\widetilde{\sigma})]} \cong \vec{\widetilde{C}}^{\mathrm{ord}}_{\widetilde{\Phi}_{\widetilde{\mathcal{H}}},\widetilde{\delta}_{\widetilde{\mathcal{H}}}}$, which (by Proposition 4.2.1.30) is a torsor under an abelian scheme $\vec{\mathsf{N}}^{\mathrm{ord,grp}}_{\kappa} = \vec{\widetilde{C}}^{\mathrm{ord,grp}}_{\widetilde{\Phi}_{\widetilde{\mathcal{H}}},\widetilde{\delta}_{\widetilde{\mathcal{H}}}}$, which (as in the proof of Proposition 4.2.1.30) is canonically isomorphic to $\dddot{\widetilde{C}}^{\mathrm{ord},\circ}_{\widetilde{\Phi}_{\mathcal{H}},n} \underset{\dddot{\widetilde{\mathsf{M}}}^{\mathrm{ord},\widetilde{\mathsf{Z}}_{\mathcal{H}}}_{\mathcal{H}}}{\times} \vec{\mathsf{M}}^{\mathrm{ord}}_{\mathcal{H}}$. Combining this with (7.2.5.27), we obtain a canonical isomorphism

$$\vec{\mathsf{N}}^{\mathrm{ord,grp}}_{\kappa} \xrightarrow{\sim} \dddot{\widetilde{C}}^{\mathrm{ord,ext},\circ}_{\widetilde{\mathcal{H}},n} \underset{\vec{\mathsf{M}}^{\mathrm{ord,tor}}_{\mathcal{H}}}{\times} \vec{\mathsf{M}}^{\mathrm{ord}}_{\mathcal{H}} \tag{7.2.5.28}$$

of abelian schemes over $\vec{\mathsf{M}}^{\mathrm{ord}}_{\mathcal{H}}$. Since $\vec{\mathsf{M}}^{\mathrm{ord,tor}}_{\mathcal{H}}$ is noetherian normal, by [92, IX, 1.4], [28, Ch. I, Prop. 2.7], or [62, Prop. 3.3.1.5], we obtain a canonical isomorphism

$$\vec{\mathsf{N}}^{\mathrm{ord,ext}}_{\kappa} \xrightarrow{\sim} \dddot{\widetilde{C}}^{\mathrm{ord,ext},\circ}_{\widetilde{\mathcal{H}},n}. \tag{7.2.5.29}$$

of semi-abelian schemes over $\vec{\mathsf{M}}^{\mathrm{ord,tor}}_{\mathcal{H}}$. Combining this with the first vertical arrow in (7.2.5.25), we obtain (see Lemma 7.2.5.26) a canonical quasi-finite étale isogeny

$$\vec{\mathsf{N}}^{\mathrm{ord,ext}}_{\kappa} \to \ddot{\widetilde{C}}^{\mathrm{ord,ext},\circ} \qquad (7.2.5.30)$$

of semi-abelian schemes over $\vec{\mathsf{M}}^{\mathrm{ord,tor}}_{\mathcal{H}}$. Hence, we obtain the following:

Corollary 7.2.5.31. *The canonical isogeny* (7.2.5.30) *induces a canonical isomorphism*

$$\underline{\mathrm{Lie}}^{\vee}_{\ddot{\widetilde{C}}^{\mathrm{ord,ext},\circ}/\vec{\mathsf{M}}^{\mathrm{ord,tor}}_{\mathcal{H}}} \xrightarrow{\sim} \underline{\mathrm{Lie}}^{\vee}_{\vec{\mathsf{N}}^{\mathrm{ord,ext}}_{\kappa}/\vec{\mathsf{M}}^{\mathrm{ord,tor}}_{\mathcal{H}}}. \qquad (7.2.5.32)$$

Consequently, we obtain a canonical isomorphism

$$(f^{\mathrm{tor}})^{*}\underline{\mathrm{Lie}}^{\vee}_{\vec{\mathsf{N}}^{\mathrm{ord,ext}}/\vec{\mathsf{M}}^{\mathrm{ord,tor}}_{\mathcal{H}}} \cong \overline{\Omega}^{1}_{\vec{\mathsf{N}}^{\mathrm{ord,tor}}/\vec{\mathsf{M}}^{\mathrm{ord,tor}}_{\mathcal{H}}} \qquad (7.2.5.33)$$

by combining (7.2.5.7), *Lemmas* 7.2.5.10 *and* 7.2.5.14, *and* (7.2.5.32).

This isomorphism (7.2.5.33) gives the desired isomorphism (7.1.4.2). (The restriction of (7.2.5.33) to $\vec{\mathsf{N}}^{\mathrm{ord}}$ is compatible with the composition of isomorphisms (7.1.4.3) because of the same calculation in the proof of Proposition 4.2.3.5.) Thus, we have also proved (3a) of Theorem 7.1.4.1.

The same argument above (based on Proposition 4.2.3.5) also shows the following:

Lemma 7.2.5.34. (*Compare with Lemma* 1.3.2.79. *This is a continuation of Lemma* 7.1.2.29.) *Consider the morphisms*

$$\vec{\widehat{C}}^{\mathrm{ord}}_{\breve{\Phi}_{\widehat{\mathcal{H}}},\breve{\delta}_{\widehat{\mathcal{H}}}} = \vec{\widetilde{C}}^{\mathrm{ord}}_{\breve{\Phi}_{\widetilde{\mathcal{H}}},\breve{\delta}_{\widetilde{\mathcal{H}}}} \to \vec{C}^{\mathrm{ord}}_{\Phi_{\mathcal{H}},\delta_{\mathcal{H}},r_{\kappa}} := \vec{C}^{\mathrm{ord}}_{\Phi_{\mathcal{H}},\delta_{\mathcal{H}}} \underset{\vec{\mathsf{S}}_{0,r_{\mathcal{H}}}}{\times} \vec{\mathsf{S}}_{0,r_{\kappa}} \qquad (7.2.5.35)$$

and

$$\vec{\widehat{\Xi}}^{\mathrm{ord}}_{\breve{\Phi}_{\widehat{\mathcal{H}}},\breve{\delta}_{\widehat{\mathcal{H}}},\widehat{\Sigma}_{\breve{\Phi}_{\widehat{\mathcal{H}}}}} \to \vec{\Xi}^{\mathrm{ord}}_{\Phi_{\mathcal{H}},\delta_{\mathcal{H}},\Sigma_{\Phi_{\mathcal{H}}},r_{\kappa}} := \vec{\Xi}^{\mathrm{ord}}_{\Phi_{\mathcal{H}},\delta_{\mathcal{H}},\Sigma_{\Phi_{\mathcal{H}}}} \underset{\vec{\mathsf{S}}_{0,r_{\mathcal{H}}}}{\times} \vec{\mathsf{S}}_{0,r_{\kappa}} \qquad (7.2.5.36)$$

induced by (7.1.2.2) *and* (7.1.2.34), *respectively. Over* $\vec{C}^{\mathrm{ord}}_{\Phi_{\mathcal{H}},\delta_{\mathcal{H}}}$ (*where* $G^{\natural}$ *etc are tautologically defined*), *we have an extension*

$$0 \to \ddot{E}^{\mathrm{ord},\circ}_{\widehat{\mathcal{H}},n} \to \ddot{\widetilde{C}}^{\mathrm{ord},\natural,\circ}_{\widehat{\mathcal{H}},n} \to \ddot{C}^{\mathrm{ord},\natural,\circ}_{\mathcal{H},n} \to 0 \qquad (7.2.5.37)$$

of a semi-abelian scheme by a torus, where the definition of $\ddot{\widetilde{C}}^{\mathrm{ord},\natural,\circ}_{\widehat{\mathcal{H}},n}$ (*resp.* $\ddot{C}^{\mathrm{ord},\natural,\circ}_{\mathcal{H},n}$, *resp.* $\ddot{E}^{\mathrm{ord},\circ}_{\widehat{\mathcal{H}},n}$) *is similar to that of* $\ddot{\widetilde{C}}^{\mathrm{ord,ext},\circ}_{\widetilde{\mathcal{H}},n}$, *but with* G *etc replaced with* $G^{\natural}$ *etc* (*resp.* B *etc, resp.* T *etc*), *which can be identified up to compatible* $\mathbb{Q}^{\times}$*-isogenies with an extension*

$$0 \to \underline{\mathrm{Hom}}_{\mathcal{O}}(\widetilde{X},T)^{\circ} \to \underline{\mathrm{Hom}}_{\mathcal{O}}(\widetilde{X},G^{\natural})^{\circ} \to \underline{\mathrm{Hom}}_{\mathcal{O}}(\widetilde{X},B)^{\circ} \to 0. \qquad (7.2.5.38)$$

Then (7.2.5.35) *is smooth,* (7.2.5.36) *is log smooth, and we have canonical isomorphisms*

$$\Omega^1_{\vec{\widehat{C}}^{\mathrm{ord}}_{\breve{\Phi}_{\widehat{\mathcal{H}}},\breve{\delta}_{\widehat{\mathcal{H}}}}/\vec{C}^{\mathrm{ord}}_{\Phi_{\mathcal{H}},\delta_{\mathcal{H}},r_\kappa}} \cong (\vec{\widehat{C}}^{\mathrm{ord}}_{\breve{\Phi}_{\widehat{\mathcal{H}}},\breve{\delta}_{\widehat{\mathcal{H}}}} \to \vec{C}^{\mathrm{ord}}_{\Phi_{\mathcal{H}},\delta_{\mathcal{H}}})^* \underline{\mathrm{Lie}}^\vee_{\ddot{C}^{\mathrm{ord},\natural,\circ}_{\mathcal{H},n}/\vec{C}^{\mathrm{ord}}_{\Phi_{\mathcal{H}},\delta_{\mathcal{H}}}} \tag{7.2.5.39}$$

(*cf.* (1.3.2.54)) *and*

$$\overline{\Omega}^1_{\vec{\overline{\widehat{\Xi}}}^{\mathrm{ord}}_{\breve{\Phi}_{\widehat{\mathcal{H}}},\breve{\delta}_{\widehat{\mathcal{H}}},\widehat{\Sigma}_{\breve{\Phi}_{\widehat{\mathcal{H}}}}}/\vec{\overline{\Xi}}^{\mathrm{ord}}_{\Phi_{\mathcal{H}},\delta_{\mathcal{H}},\Sigma_{\Phi_{\mathcal{H}}},r_\kappa}} \cong (\vec{\overline{\widehat{\Xi}}}^{\mathrm{ord}}_{\breve{\Phi}_{\widehat{\mathcal{H}}},\breve{\delta}_{\widehat{\mathcal{H}}},\widehat{\Sigma}_{\breve{\Phi}_{\widehat{\mathcal{H}}}}} \to \vec{C}^{\mathrm{ord}}_{\Phi_{\mathcal{H}},\delta_{\mathcal{H}}})^* \underline{\mathrm{Lie}}^\vee_{\ddot{C}^{\mathrm{ord},\natural,\circ}_{\widehat{\mathcal{H}},n}/\vec{C}^{\mathrm{ord}}_{\Phi_{\mathcal{H}},\delta_{\mathcal{H}}}} \tag{7.2.5.40}$$

(*cf.* (1.3.2.86)), *where*

$$\begin{aligned}\overline{\Omega}^1_{\vec{\overline{\widehat{\Xi}}}^{\mathrm{ord}}_{\breve{\Phi}_{\widehat{\mathcal{H}}},\breve{\delta}_{\widehat{\mathcal{H}}},\widehat{\Sigma}_{\breve{\Phi}_{\widehat{\mathcal{H}}}}}/\vec{\overline{\Xi}}^{\mathrm{ord}}_{\Phi_{\mathcal{H}},\delta_{\mathcal{H}},\Sigma_{\Phi_{\mathcal{H}}},r_\kappa}} &:= (\Omega^1_{\vec{\overline{\widehat{\Xi}}}^{\mathrm{ord}}_{\breve{\Phi}_{\widehat{\mathcal{H}}},\breve{\delta}_{\widehat{\mathcal{H}}},\widehat{\Sigma}_{\breve{\Phi}_{\widehat{\mathcal{H}}}}}/\vec{C}^{\mathrm{ord}}_{\Phi_{\mathcal{H}},\delta_{\mathcal{H}}}}[d\log\infty])/ \\ &((\vec{\overline{\widehat{\Xi}}}^{\mathrm{ord}}_{\breve{\Phi}_{\widehat{\mathcal{H}}},\breve{\delta}_{\widehat{\mathcal{H}}},\widehat{\Sigma}_{\breve{\Phi}_{\widehat{\mathcal{H}}}}} \to \vec{\overline{\Xi}}^{\mathrm{ord}}_{\Phi_{\mathcal{H}},\delta_{\mathcal{H}},\Sigma_{\Phi_{\mathcal{H}}},r_\kappa})^* \Omega^1_{\vec{\overline{\Xi}}^{\mathrm{ord}}_{\Phi_{\mathcal{H}},\delta_{\mathcal{H}},\Sigma_{\Phi_{\mathcal{H}}},r_\kappa}/\vec{C}^{\mathrm{ord}}_{\Phi_{\mathcal{H}},\delta_{\mathcal{H}}}}[d\log\infty])\end{aligned}$$

is the sheaf of modules of relative log 1*-differentials. Moreover, the canonical morphism*

$$\begin{aligned}\vec{\overline{\widehat{\Xi}}}^{\mathrm{ord}}_{\breve{\Phi}_{\widehat{\mathcal{H}}},\breve{\delta}_{\widehat{\mathcal{H}}},\widehat{\Sigma}_{\breve{\Phi}_{\widehat{\mathcal{H}}}}} &\to \vec{\overline{\Xi}}^{\mathrm{ord}}_{\Phi_{\mathcal{H}},\delta_{\mathcal{H}},\Sigma_{\Phi_{\mathcal{H}}}} \underset{\vec{C}^{\mathrm{ord}}_{\Phi_{\mathcal{H}},\delta_{\mathcal{H}}}}{\times} \vec{\widehat{C}}^{\mathrm{ord}}_{\breve{\Phi}_{\widehat{\mathcal{H}}},\breve{\delta}_{\widehat{\mathcal{H}}}} \\ &\cong \vec{\overline{\Xi}}^{\mathrm{ord}}_{\Phi_{\mathcal{H}},\delta_{\mathcal{H}},\Sigma_{\Phi_{\mathcal{H}}},r_\kappa} \underset{\vec{C}^{\mathrm{ord}}_{\Phi_{\mathcal{H}},\delta_{\mathcal{H}},r_\kappa}}{\times} \vec{\widehat{C}}^{\mathrm{ord}}_{\breve{\Phi}_{\widehat{\mathcal{H}}},\breve{\delta}_{\widehat{\mathcal{H}}}}\end{aligned} \tag{7.2.5.41}$$

(*induced by* (7.1.2.2) *and* (7.1.2.34), *or by* (7.2.5.35) *and* (7.2.5.36); *cf.* (1.3.2.87)) *induces a canonical short exact sequence*

$$\begin{aligned}0 &\to (\vec{\overline{\widehat{\Xi}}}^{\mathrm{ord}}_{\breve{\Phi}_{\widehat{\mathcal{H}}},\breve{\delta}_{\widehat{\mathcal{H}}},\widehat{\Sigma}_{\breve{\Phi}_{\widehat{\mathcal{H}}}}} \to \vec{\widehat{C}}^{\mathrm{ord}}_{\breve{\Phi}_{\widehat{\mathcal{H}}},\breve{\delta}_{\widehat{\mathcal{H}}}})^* \Omega^1_{\vec{\widehat{C}}^{\mathrm{ord}}_{\breve{\Phi}_{\widehat{\mathcal{H}}},\breve{\delta}_{\widehat{\mathcal{H}}}}/\vec{C}^{\mathrm{ord}}_{\Phi_{\mathcal{H}},\delta_{\mathcal{H}},r_\kappa}} \\ &\to \overline{\Omega}^1_{\vec{\overline{\widehat{\Xi}}}^{\mathrm{ord}}_{\breve{\Phi}_{\widehat{\mathcal{H}}},\breve{\delta}_{\widehat{\mathcal{H}}},\widehat{\Sigma}_{\breve{\Phi}_{\widehat{\mathcal{H}}}}}/\vec{\overline{\Xi}}^{\mathrm{ord}}_{\Phi_{\mathcal{H}},\delta_{\mathcal{H}},\Sigma_{\Phi_{\mathcal{H}}},r_\kappa}} \\ &\to \overline{\Omega}^1_{\vec{\overline{\widehat{\Xi}}}^{\mathrm{ord}}_{\breve{\Phi}_{\widehat{\mathcal{H}}},\breve{\delta}_{\widehat{\mathcal{H}}},\widehat{\Sigma}_{\breve{\Phi}_{\widehat{\mathcal{H}}}}}/\vec{\overline{\Xi}}^{\mathrm{ord}}_{\Phi_{\mathcal{H}},\delta_{\mathcal{H}},\Sigma_{\Phi_{\mathcal{H}}}} \underset{\vec{C}^{\mathrm{ord}}_{\Phi_{\mathcal{H}},\delta_{\mathcal{H}}}}{\times} \vec{\widehat{C}}^{\mathrm{ord}}_{\breve{\Phi}_{\widehat{\mathcal{H}}},\breve{\delta}_{\widehat{\mathcal{H}}}}} \to 0\end{aligned} \tag{7.2.5.42}$$

(*cf.* (1.3.2.88)), *where*

$$\begin{aligned}\overline{\Omega}^1_{\vec{\overline{\widehat{\Xi}}}^{\mathrm{ord}}_{\breve{\Phi}_{\widehat{\mathcal{H}}},\breve{\delta}_{\widehat{\mathcal{H}}},\widehat{\Sigma}_{\breve{\Phi}_{\widehat{\mathcal{H}}}}}/\vec{\overline{\Xi}}^{\mathrm{ord}}_{\Phi_{\mathcal{H}},\delta_{\mathcal{H}},\Sigma_{\Phi_{\mathcal{H}}}} \underset{\vec{C}^{\mathrm{ord}}_{\Phi_{\mathcal{H}},\delta_{\mathcal{H}}}}{\times} \vec{\widehat{C}}^{\mathrm{ord}}_{\breve{\Phi}_{\widehat{\mathcal{H}}},\breve{\delta}_{\widehat{\mathcal{H}}}}} &:= (\Omega^1_{\vec{\overline{\widehat{\Xi}}}^{\mathrm{ord}}_{\breve{\Phi}_{\widehat{\mathcal{H}}},\breve{\delta}_{\widehat{\mathcal{H}}},\widehat{\Sigma}_{\breve{\Phi}_{\widehat{\mathcal{H}}}}}/\vec{\widehat{C}}^{\mathrm{ord}}_{\breve{\Phi}_{\widehat{\mathcal{H}}},\breve{\delta}_{\widehat{\mathcal{H}}}}}[d\log\infty])/ \\ &((\vec{\overline{\widehat{\Xi}}}^{\mathrm{ord}}_{\breve{\Phi}_{\widehat{\mathcal{H}}},\breve{\delta}_{\widehat{\mathcal{H}}},\widehat{\Sigma}_{\breve{\Phi}_{\widehat{\mathcal{H}}}}} \to \vec{\overline{\Xi}}^{\mathrm{ord}}_{\Phi_{\mathcal{H}},\delta_{\mathcal{H}},\Sigma_{\Phi_{\mathcal{H}}}})^* \Omega^1_{\vec{\overline{\Xi}}^{\mathrm{ord}}_{\Phi_{\mathcal{H}},\delta_{\mathcal{H}},\Sigma_{\Phi_{\mathcal{H}}}}/\vec{\widehat{C}}^{\mathrm{ord}}_{\breve{\Phi}_{\widehat{\mathcal{H}}},\breve{\delta}_{\widehat{\mathcal{H}}}}}[d\log\infty])\end{aligned}$$

is the sheaf of modules of relative log 1*-differentials, which is exact and has locally free terms, which can be canonically identified with the pullback under* $\vec{\overline{\Xi}}^{\mathrm{ord}}_{\breve{\Phi}_{\widehat{\mathcal{H}}},\breve{\delta}_{\widehat{\mathcal{H}}},\widehat{\Sigma}_{\breve{\Phi}_{\widehat{\mathcal{H}}}}} \to \vec{C}^{\mathrm{ord}}_{\Phi_{\mathcal{H}},\delta_{\mathcal{H}}}$ *of the canonical short exact sequence*

$$0 \to \underline{\mathrm{Lie}}^\vee_{\ddot{C}^{\mathrm{ord},\natural,\circ}_{\mathcal{H},n}/\vec{C}^{\mathrm{ord}}_{\Phi_{\mathcal{H}},\delta_{\mathcal{H}}}} \to \underline{\mathrm{Lie}}^\vee_{\ddot{C}^{\mathrm{ord},\natural,\circ}_{\widehat{\mathcal{H}},n}/\vec{C}^{\mathrm{ord}}_{\Phi_{\mathcal{H}},\delta_{\mathcal{H}}}} \to \underline{\mathrm{Lie}}^\vee_{\ddot{E}^{\mathrm{ord},\circ}_{\widehat{\mathcal{H}},n}/\vec{C}^{\mathrm{ord}}_{\Phi_{\mathcal{H}},\delta_{\mathcal{H}}}} \to 0 \tag{7.2.5.43}$$

of locally free sheaves. Hence, (7.2.5.41) *is also log smooth (by [45, 3.12]).*

If $p \nmid [L^\# : L]$ as in Definition 1.1.1.6 *and hence λ is prime-to-p, and if $\mathcal{O}$ is maximal at p, then we may assume in the above that* (7.2.5.37) *and* (7.2.5.38) *can be identified up to $\mathbb{Q}^\times$-isogenies which are separable up to $\mathbb{Z}_{(p)}^\times$-isogenies, and hence that* (7.2.5.42) *can be identified up with the pullback under $\vec{\widehat{\Xi}}^{\mathrm{ord}}_{\check{\Phi}_{\widehat{\mathcal{H}}},\check{\delta}_{\widehat{\mathcal{H}}},\widehat{\Sigma}_{\check{\Phi}_{\widehat{\mathcal{H}}}}} \to \vec{C}^{\mathrm{ord}}_{\Phi_{\mathcal{H}},\delta_{\mathcal{H}}}$ of the canonical short exact sequence*

$$\begin{aligned} 0 \to \underline{\mathrm{Hom}}_{\mathcal{O}}(\widetilde{X}, \underline{\mathrm{Lie}}^\vee_{B/\vec{C}^{\mathrm{ord}}_{\Phi_{\mathcal{H}},\delta_{\mathcal{H}}}}) &\to \underline{\mathrm{Hom}}_{\mathcal{O}}(\widetilde{X}, \underline{\mathrm{Lie}}^\vee_{G^\natural/\vec{C}^{\mathrm{ord}}_{\Phi_{\mathcal{H}},\delta_{\mathcal{H}}}}) \\ &\to \underline{\mathrm{Hom}}_{\mathcal{O}}(\widetilde{X}, \underline{\mathrm{Lie}}^\vee_{T/\vec{C}^{\mathrm{ord}}_{\Phi_{\mathcal{H}},\delta_{\mathcal{H}}}}) \to 0 \end{aligned} \tag{7.2.5.44}$$

(cf. (1.3.2.89)*).*

Proof. The statements are self-explanatory. (For the last paragraph, note that, under the assumption, $[\widetilde{X} : \widetilde{\phi}(\widetilde{Y})]$, $[\check{X} : \check{\phi}(\check{Y})]$, and $[X : \phi(Y)]$ are also prime-to-p, and the canonical homomorphisms $G^{\mathrm{ord}}_{p^r} \to G$ and $G^{\vee,\mathrm{ord}}_{p^r} \twoheadrightarrow G^\vee$, as in (7.2.5.20) and (7.2.5.21), are étale.) □

7.2.6 *Equidimensionality of* f^{tor}

Let us resume the context of the diagram (7.2.3.6) and take a closer look at it. (Then we no longer suppose that $\kappa \in \mathbf{K}^{\mathrm{ord},+}_{Q,\mathcal{H},\Sigma^{\mathrm{ord}}}$.) By the construction of f^{tor}, given any stratum $\vec{\mathsf{Z}}^{\mathrm{ord}}_{[(\Phi_{\mathcal{H}},\delta_{\mathcal{H}},\tau)],r_\kappa}$ of $\vec{\mathsf{M}}^{\mathrm{ord,tor}}_{\mathcal{H},r_\kappa}$, the preimage

$$\vec{\widetilde{\mathsf{Z}}}^{\mathrm{ord}}_{[(\Phi_{\mathcal{H}},\delta_{\mathcal{H}},\tau)]} := (f^{\mathrm{tor}})^{-1}(\vec{\mathsf{Z}}^{\mathrm{ord}}_{[(\Phi_{\mathcal{H}},\delta_{\mathcal{H}},\tau)],r_\kappa})$$

has a stratification formed by $\vec{\widetilde{\mathsf{Z}}}^{\mathrm{ord}}_{[(\check{\Phi}_{\widetilde{\mathcal{H}}},\check{\delta}_{\widetilde{\mathcal{H}}},\check{\tau})]}$, where $\check{\tau}$ runs through cones in $\widetilde{\Sigma}_{\check{\Phi}_{\widetilde{\mathcal{H}}}}$ satisfying the following conditions:

(1) $\check{\tau} \subset \mathbf{P}^+_{\check{\Phi}_{\widetilde{\mathcal{H}}}}$.
(2) $\check{\tau}$ has a face $\check{\sigma}$ that is a $\Gamma_{\check{\Phi}_{\widetilde{\mathcal{H}}}}$-translation of the image of $\widetilde{\sigma} \subset \mathbf{P}^+_{\widetilde{\Phi}_{\widetilde{\mathcal{H}}}}$ under the first morphism in (1.2.4.20).
(3) The image of $\check{\tau}$ under the (canonical) second morphism in (1.2.4.20) is contained in $\tau \subset \mathbf{P}^+_{\Phi_{\mathcal{H}}}$.

The formal completion $(\vec{\mathsf{N}}^{\mathrm{ord,tor}})^\wedge_{\vec{\widetilde{\mathsf{Z}}}^{\mathrm{ord}}_{[(\Phi_{\mathcal{H}},\delta_{\mathcal{H}},\tau)]}}$ admits a canonical morphism

$$(\vec{\mathsf{N}}^{\mathrm{ord,tor}})^\wedge_{\vec{\widetilde{\mathsf{Z}}}^{\mathrm{ord}}_{[(\Phi_{\mathcal{H}},\delta_{\mathcal{H}},\tau)]}} \to \vec{C}^{\mathrm{ord}}_{\Phi_{\mathcal{H}},\delta_{\mathcal{H}},r_\kappa} = \vec{C}^{\mathrm{ord}}_{\Phi_{\mathcal{H}},\delta_{\mathcal{H}}} \underset{\vec{\mathsf{S}}_{0,r_{\mathcal{H}}}}{\times} \vec{\mathsf{S}}_{0,r_\kappa},$$

whose pre-composition with the canonical morphism

$$(\vec{\mathsf{N}}^{\mathrm{ord,tor}})^\wedge_{\vec{\widetilde{\mathsf{Z}}}^{\mathrm{ord}}_{[(\check{\Phi}_{\widetilde{\mathcal{H}}},\check{\delta}_{\widetilde{\mathcal{H}}},\check{\tau})]}} \to (\vec{\mathsf{N}}^{\mathrm{ord,tor}})^\wedge_{\vec{\widetilde{\mathsf{Z}}}^{\mathrm{ord}}_{[(\Phi_{\mathcal{H}},\delta_{\mathcal{H}},\tau)]}},$$

for every stratum $\vec{\widetilde{\mathsf{Z}}}^{\mathrm{ord}}_{[(\breve{\Phi}_{\widetilde{\mathcal{H}}},\breve{\delta}_{\widetilde{\mathcal{H}}},\breve{\tau})]}$ of $\vec{\widetilde{\mathsf{Z}}}^{\mathrm{ord}}_{[(\Phi_{\mathcal{H}},\delta_{\mathcal{H}},\tau)]}$, coincides with the composition of canonical morphisms $\vec{\widetilde{\mathfrak{X}}}^{\mathrm{ord}}_{\breve{\Phi}_{\widetilde{\mathcal{H}}},\breve{\delta}_{\widetilde{\mathcal{H}}},\breve{\sigma},\breve{\tau}} \to \vec{\widetilde{C}}^{\mathrm{ord}}_{\breve{\Phi}_{\widetilde{\mathcal{H}}},\breve{\delta}_{\widetilde{\mathcal{H}}}} \to \vec{C}^{\mathrm{ord}}_{\Phi_{\mathcal{H}},\delta_{\mathcal{H}},r_\kappa}$ by its very construction.

Since f^{tor} is étale locally given by morphisms between toric schemes equivariant under (surjective) morphisms between tori, to determine whether f^{tor} is equidimensional (*cf.* [28, Ch. VI, Def. 1.3 and Rem. 1.4] and [61, Sec. 3D]), it suffices to determine whether the relative dimension of each of the induced (smooth) morphism $\vec{\widetilde{\mathsf{Z}}}^{\mathrm{ord}}_{[(\breve{\Phi}_{\widetilde{\mathcal{H}}},\breve{\delta}_{\widetilde{\mathcal{H}}},\breve{\tau})]} \to \vec{\mathsf{Z}}^{\mathrm{ord}}_{[(\Phi_{\mathcal{H}},\delta_{\mathcal{H}},\tau)]}$ between strata is at most $\dim_{\vec{\mathsf{M}}^{\mathrm{ord}}_{\mathcal{H}}}(\vec{\mathsf{N}}^{\mathrm{ord}})$, the relative dimension of $f : \vec{\mathsf{N}}^{\mathrm{ord}} \to \vec{\mathsf{M}}^{\mathrm{ord}}_{\mathcal{H},r_\kappa}$.

By abuse of language, we define the $\mathbb{R}$-dimension of a cone to be the $\mathbb{R}$-dimension of its $\mathbb{R}$-span. Then the codimension of $\vec{\mathsf{N}}^{\mathrm{ord}} = \vec{\widetilde{\mathsf{Z}}}^{\mathrm{ord}}_{[(\widetilde{\Phi}_{\widetilde{\mathcal{H}}},\widetilde{\delta}_{\widetilde{\mathcal{H}}},\widetilde{\sigma})]}$ in $\vec{\widetilde{\mathsf{M}}}^{\mathrm{ord,tor}}_{\widetilde{\mathcal{H}}}$ is $\dim_{\mathbb{R}}(\widetilde{\sigma}) = \dim_{\mathbb{R}}((\mathbf{S}_{\widetilde{\Phi}_{\widetilde{\mathcal{H}}}})^\vee_{\mathbb{R}})$ because $\widetilde{\sigma}$ is top-dimensional. The codimension of

$$\vec{\widetilde{\mathsf{Z}}}^{\mathrm{ord}}_{[(\breve{\Phi}_{\widetilde{\mathcal{H}}},\breve{\delta}_{\widetilde{\mathcal{H}}},\breve{\tau})]} \cong \vec{\widetilde{\Xi}}^{\mathrm{ord}}_{\breve{\Phi}_{\widetilde{\mathcal{H}}},\breve{\delta}_{\widetilde{\mathcal{H}}},\breve{\tau}}$$

in $\vec{\widetilde{\mathsf{M}}}^{\mathrm{ord,tor}}_{\widetilde{\mathcal{H}}}$ is equal to $\dim_{\mathbb{R}}(\breve{\tau})$. Therefore, the codimension of $\vec{\widetilde{\mathsf{Z}}}^{\mathrm{ord}}_{[(\breve{\Phi}_{\widetilde{\mathcal{H}}},\breve{\delta}_{\widetilde{\mathcal{H}}},\breve{\tau})]}$ in $\vec{\mathsf{N}}^{\mathrm{ord,tor}}$ is equal to $\dim_{\mathbb{R}}(\breve{\tau}) - \dim_{\mathbb{R}}(\widetilde{\sigma}) = \dim_{\mathbb{R}}(\breve{\tau}) - \dim_{\mathbb{R}}((\mathbf{S}_{\widetilde{\Phi}_{\widetilde{\mathcal{H}}}})^\vee_{\mathbb{R}})$. On the other hand, the codimension of $\vec{\mathsf{Z}}^{\mathrm{ord}}_{[(\Phi_{\mathcal{H}},\delta_{\mathcal{H}},\tau)]} \cong \vec{\Xi}^{\mathrm{ord}}_{\Phi_{\mathcal{H}},\delta_{\mathcal{H}},\tau}$ in $\vec{\mathsf{M}}^{\mathrm{ord,tor}}_{\mathcal{H}}$ is $\dim_{\mathbb{R}}(\tau)$, and so is the codimension of $\vec{\mathsf{Z}}^{\mathrm{ord}}_{[(\Phi_{\mathcal{H}},\delta_{\mathcal{H}},\tau)],r_\kappa} \cong \vec{\Xi}^{\mathrm{ord}}_{\Phi_{\mathcal{H}},\delta_{\mathcal{H}},\tau,r_\kappa} := \vec{\Xi}^{\mathrm{ord}}_{\Phi_{\mathcal{H}},\delta_{\mathcal{H}},\tau} \underset{\vec{\mathsf{S}}_{0,r_{\mathcal{H}}}}{\times} \vec{\mathsf{S}}_{0,r_\kappa}$ in $\vec{\mathsf{M}}^{\mathrm{ord,tor}}_{\mathcal{H},r_\kappa}$. Hence, we have (as in [61, (3.16)])

$$\begin{aligned} &\dim_{\vec{\mathsf{Z}}^{\mathrm{ord}}_{[(\Phi_{\mathcal{H}},\delta_{\mathcal{H}},\tau)],r_\kappa}}(\vec{\widetilde{\mathsf{Z}}}^{\mathrm{ord}}_{[(\breve{\Phi}_{\widetilde{\mathcal{H}}},\breve{\delta}_{\widetilde{\mathcal{H}}},\breve{\tau})]}) \\ &= \dim_{\vec{\mathsf{M}}^{\mathrm{ord}}_{\mathcal{H}}}(\mathsf{N}) - (\dim_{\mathbb{R}}(\breve{\tau}) - \dim_{\mathbb{R}}((\mathbf{S}_{\widetilde{\Phi}_{\widetilde{\mathcal{H}}}})^\vee_{\mathbb{R}})) + \dim_{\mathbb{R}}(\tau). \end{aligned} \tag{7.2.6.1}$$

Let τ' denote the image of $\breve{\tau}$ in $(\mathbf{S}_{\Phi_{\mathcal{H}}})^\vee_{\mathbb{R}}$. By assumption on $\breve{\tau}$, we have $\tau' \subset \tau$. If $\tau' = \tau$, then

$$\dim_{\mathbb{R}}(\tau) = \dim_{\mathbb{R}}(\tau') \leq \dim_{\mathbb{R}}(\breve{\tau}) - \dim_{\mathbb{R}}((\mathbf{S}_{\widetilde{\Phi}_{\widetilde{\mathcal{H}}}})^\vee_{\mathbb{R}}),$$

and hence (7.2.6.1) implies

$$\dim_{\vec{\mathsf{Z}}^{\mathrm{ord}}_{[(\Phi_{\mathcal{H}},\delta_{\mathcal{H}},\tau)],r_\kappa}}(\vec{\widetilde{\mathsf{Z}}}^{\mathrm{ord}}_{[(\breve{\Phi}_{\widetilde{\mathcal{H}}},\breve{\delta}_{\widetilde{\mathcal{H}}},\breve{\tau})]}) \leq \dim_{\vec{\mathsf{M}}^{\mathrm{ord}}_{\mathcal{H},r_\kappa}}(\vec{\mathsf{N}}^{\mathrm{ord}}).$$

(If this is true for all $\vec{\widetilde{\mathsf{Z}}}^{\mathrm{ord}}_{[(\breve{\Phi}_{\widetilde{\mathcal{H}}},\breve{\delta}_{\widetilde{\mathcal{H}}},\breve{\tau})]}$, then f^{tor} is equidimensional.) On the other hand, suppose $\tau' \subsetneq \tau$. Then there exists a face of τ'' of τ' such that $\tau'' \subset \tau$ and $\dim_{\mathbb{R}}(\tau'') < \dim_{\mathbb{R}}(\tau)$. Note that τ'' is the image of at least one face of $\breve{\tau}$ satisfying the three conditions in the first paragraph of this subsection. By dropping redundant basis vectors, we may assume moreover that this face $\breve{\tau}''$ of $\breve{\tau}$ satisfies $\dim_{\mathbb{R}}(\tau'') = \dim_{\mathbb{R}}(\breve{\tau}'') - \dim_{\mathbb{R}}((\mathbf{S}_{\widetilde{\Phi}_{\widetilde{\mathcal{H}}}})^\vee_{\mathbb{R}})$. Then we have

$$\dim_{\mathbb{R}}(\tau) > \dim_{\mathbb{R}}(\tau'') = \dim_{\mathbb{R}}(\breve{\tau}'') - \dim_{\mathbb{R}}((\mathbf{S}_{\widetilde{\Phi}_{\widetilde{\mathcal{H}}}})^\vee_{\mathbb{R}}),$$

and hence (7.2.6.1) implies

$$\dim_{\vec{\mathsf{Z}}^{\mathrm{ord}}_{[(\Phi_{\mathcal{H}},\delta_{\mathcal{H}},\tau)],r_\kappa}}(\vec{\widetilde{\mathsf{Z}}}^{\mathrm{ord}}_{[(\breve{\Phi}_{\widetilde{\mathcal{H}}},\breve{\delta}_{\widetilde{\mathcal{H}}},\breve{\tau}'')]}) > \dim_{\vec{\mathsf{M}}^{\mathrm{ord}}_{\mathcal{H},r_\kappa}}(\vec{\mathsf{N}}^{\mathrm{ord}}),$$

which means f^{tor} cannot be equidimensional.

This motivates the following strengthening of Condition 7.1.1.15 on an element $\widetilde{\kappa} = (\widetilde{\mathcal{H}}, \widetilde{\Sigma}^{\mathrm{ord}}, \widetilde{\sigma})$ in $\widetilde{\mathbf{K}}^{\mathrm{ord},++}_{Q,\mathcal{H}}$:

Condition 7.2.6.2. (*Compare with [61, Cond. 3.17].*) *For each* $(\breve{\Phi}_{\widetilde{\mathcal{H}}}, \breve{\delta}_{\widetilde{\mathcal{H}}}, \breve{\tau})$ *such that* $\vec{\mathsf{Z}}^{\mathrm{ord}}_{[(\breve{\Phi}_{\widetilde{\mathcal{H}}},\breve{\delta}_{\widetilde{\mathcal{H}}},\breve{\tau})]}$ *is a stratum in* $\vec{\mathsf{N}}^{\mathrm{ord,tor}}$, *the image of* $\breve{\tau} \subset \mathbf{P}^+_{\Phi_{\mathcal{H}}}$ *under the* (*canonical*) *second morphism in* (1.2.4.20) *is* **exactly** *some cone* $\tau \subset \mathbf{P}^+_{\Phi_{\mathcal{H}}}$ *in the cone decomposition* $\Sigma_{\Phi_{\mathcal{H}}}$ (*in* Σ^{ord}).

As in the case of Condition 7.1.1.17, if $\kappa = [\widetilde{\kappa}] \in \mathbf{K}^{\mathrm{ord},++}_{Q,\mathcal{H}}$ is the element determined by $\widetilde{\kappa}$, then Condition 7.2.6.2 for $\widetilde{\kappa}$ is equivalent to the following condition for κ:

Condition 7.2.6.3. (*Compare with [28, Ch. VI, Def. 1.3].*) *For each* $\widehat{\tau} \in \widehat{\Sigma}_{\breve{\Phi}_{\widehat{\mathcal{H}}}}$ (*where* $\widehat{\tau} = \mathrm{pr}_{(\widehat{\mathbf{S}}_{\breve{\Phi}_{\widehat{\mathcal{H}}}})^\vee_{\mathbb{R}}}(\breve{\tau})$ *for some* $(\breve{\Phi}_{\widetilde{\mathcal{H}}}, \breve{\delta}_{\widetilde{\mathcal{H}}}, \breve{\tau})$ *is in the cone decomposition* $\widehat{\Sigma}_{\breve{\Phi}_{\widehat{\mathcal{H}}}}$ *in* $\widehat{\Sigma}^{\mathrm{ord}}$), *the image of* $\widehat{\tau}$ *in* $\mathbf{P}^+_{\Phi_{\mathcal{H}}}$ *under* (1.2.4.37) *is* **exactly** *some cone* $\tau \subset \mathbf{P}^+_{\Phi_{\mathcal{H}}}$ *in the cone decomposition* $\Sigma_{\Phi_{\mathcal{H}}}$ (*in* Σ^{ord}).

Proposition 7.2.6.4. (*Compare with [61, Prop. 3.18].*) *The following are equivalent:*

(1) *Condition* 7.2.6.3 *is satisfied.*
(2) *The morphism* $f^{\mathrm{tor}} : \vec{\mathsf{N}}^{\mathrm{ord,tor}} \to \vec{\mathsf{M}}^{\mathrm{ord,tor}}_{\mathcal{H},r_\kappa}$ *is equidimensional* (*with relative dimension equal to the one of* $f : \vec{\mathsf{N}}^{\mathrm{ord}} \to \vec{\mathsf{M}}^{\mathrm{ord}}_{\mathcal{H},r_\kappa}$).
(3) *The morphism* f^{tor} *is flat.*
(4) *The morphism* f^{tor} *is log integral* (*see [45, Def. 4.3]*).

Proof. The equivalence between Condition 7.2.6.3 and equidimensionality has been explained above. Since both $\vec{\mathsf{N}}^{\mathrm{ord,tor}}$ and $\vec{\mathsf{M}}^{\mathrm{ord,tor}}_{\mathcal{H},r_\kappa}$ are regular (because they are smooth over $\vec{\mathsf{S}}_{0,r_\kappa} = \mathrm{Spec}(\mathcal{O}_{F_0,(p)}[\zeta_{p^{r_\kappa}}])$), the equidimensionality and flatness of f^{tor} are equivalent by [35, IV-3, 15.4.2 b)⇔e')]. By [45, Prop. 4.1(2)], the log integrality of f^{tor} is equivalent to the flatness of each of the canonical homomorphisms $\mathbb{Z}[\tau^\vee] \hookrightarrow \mathbb{Z}[\breve{\tau}^\vee]$ (defined when $\vec{\mathsf{Z}}^{\mathrm{ord}}_{[(\breve{\Phi}_{\widetilde{\mathcal{H}}},\breve{\delta}_{\widetilde{\mathcal{H}}},\breve{\tau})]}$ is mapped to $\vec{\mathsf{Z}}^{\mathrm{ord}}_{[(\Phi_{\mathcal{H}},\delta_{\mathcal{H}},\tau)],r_\kappa}$), which is equivalent to the equidimensionality of each such homomorphism (by the smoothness of $\mathbb{Z}[\tau^\vee]$ and $\mathbb{Z}[\breve{\tau}^\vee]$ over $\mathbb{Z}$, and by [35, IV-3, 15.4.2 b)⇔e')] again), which is equivalent to Condition 7.2.6.3 by the same (dimension comparison) argument. □

Proposition 7.2.6.5. (*Compare with [28, Ch. VI, Rem. 1.4] and [61, Prop. 3.18].*) *Condition* 7.2.6.3 *can be achieved by replacing both the cone decompositions* $\widetilde{\Sigma}^{\mathrm{ord}}$ *and* Σ^{ord} *with some refinements.*

Proof. Since this is a question only about cone decompositions, the same argument of the proof of [61, Prop. 3.18] works here. □

Remark 7.2.6.6. (Compare with [61, Rem. 3.20].) We will not need Propositions 7.2.6.4 and 7.2.6.5 in what follows. We supply them here because knowing equidimensionality, flatness, or log integrality of f^{tor} is useful in many applications.

7.2.7 *Hecke Actions*

The aim of this subsection is to explain the proof of statements (4) and (5) of Theorem 7.1.4.1, with (4c) and (5c) conditional on (3b) and (3c) of Theorem 7.1.4.1. These statements might seem elaborate, but they are self-explanatory and based on the following simple idea: Since N and $\mathsf{N}^{\mathrm{tor}}$ are constructed using the toroidal compactifications of $\widetilde{\mathsf{M}}_{\widetilde{\mathcal{H}}}$, we can use the Hecke actions on $\widetilde{\mathsf{M}}_{\widetilde{\mathcal{H}}}$ and their (compatible) extensions to toroidal compactifications provided by Proposition 5.2.2.2.

Let $\widehat{g}$, $\widehat{\mathcal{H}}'$, $\Sigma^{\mathrm{ord},\prime}$, g_l, and Q' be as in (4) and (5) of Theorem 7.1.4.1. (For proving (4) and (5) of Theorem 7.1.4.1, we may assume in what follows that either $\widehat{g} = 1$ or $g_l = 1$, although the theory works in a more general context.) Let $\widetilde{g} = (\widetilde{g}_0, \widetilde{g}_p)$ be any element in $\widetilde{\mathrm{P}}_{\widetilde{\mathsf{Z}}}(\mathbb{A}^{\infty,p}) \times \widetilde{\mathrm{P}}^{\mathrm{ord}}_{\widetilde{\mathsf{Z}},\widetilde{\mathsf{D}}}(\mathbb{Q}_p)$ such that, under the canonical isomorphism

$$\widetilde{\mathrm{P}}_{\widetilde{\mathsf{Z}}}(\mathbb{A}^{\infty,p}) \times \widetilde{\mathrm{P}}^{\mathrm{ord}}_{\widetilde{\mathsf{Z}},\widetilde{\mathsf{D}}}(\mathbb{Q}_p) \cong \widetilde{\mathrm{G}}_{l,\widetilde{\mathsf{Z}}}(\mathbb{A}^{\infty}) \ltimes (\widetilde{\mathrm{P}}'_{\widetilde{\mathsf{Z}}}(\mathbb{A}^{\infty,p}) \times \widetilde{\mathrm{P}}^{\mathrm{ord},\prime}_{\widetilde{\mathsf{Z}},\widetilde{\mathsf{D}}}(\mathbb{Q}_p))$$

induced by the splitting $\widetilde{\delta}$ (as in Definition 1.2.4.3), $\widetilde{g}$ is mapped to $(g_l^{-1}, \widetilde{g}')$ for some element $\widetilde{g}' \in \widetilde{\mathrm{P}}'_{\widetilde{\mathsf{Z}}}(\mathbb{A}^{\infty,p}) \times \widetilde{\mathrm{P}}^{\mathrm{ord},\prime}_{\widetilde{\mathsf{Z}},\widetilde{\mathsf{D}}}(\mathbb{Q}_p)$ that is mapped to $\widehat{g}$ under the canonical morphism

$$\widetilde{\mathrm{P}}'_{\widetilde{\mathsf{Z}}}(\mathbb{A}^{\infty,p}) \times \widetilde{\mathrm{P}}^{\mathrm{ord},\prime}_{\widetilde{\mathsf{Z}},\widetilde{\mathsf{D}}}(\mathbb{Q}_p) \to \widehat{\mathrm{G}}(\mathbb{A}^{\infty,p}) \times \widehat{\mathrm{P}}^{\mathrm{ord}}_{\mathsf{D}}(\mathbb{Q}_p).$$

Suppose $\kappa = [\widetilde{\kappa}]$ for some $\widetilde{\kappa} = (\widetilde{\mathcal{H}}, \widetilde{\Sigma}, \widetilde{\sigma})$. Let $\widetilde{\mathcal{H}}'$ be a (necessarily neat) subgroup of $\widetilde{\mathrm{G}}(\hat{\mathbb{Z}})$ such that we have the following:

- $\widetilde{\mathcal{H}}' = \widetilde{\mathcal{H}}'^{,p}\widetilde{\mathcal{H}}'_p$ is of standard form (where $\widetilde{\mathcal{H}}'^{,p} \subset \widetilde{\mathcal{H}}^p$ is necessarily neat) such that $r_{\widetilde{\mathcal{H}}'} \geq r_{\widetilde{\mathcal{H}}}$, and such that $\widetilde{g}_p$ satisfies the conditions analogous to those given in Section 3.3.4.
- $\widetilde{\mathcal{H}}' \subset \widetilde{g}\widetilde{\mathcal{H}}\widetilde{g}^{-1}$.
- $\widetilde{\mathcal{H}}'$ also satisfies Conditions 1.2.4.7 and 7.1.1.5.
- $\widetilde{\mathcal{H}}'_{\widehat{\mathrm{G}}} = \widehat{\mathcal{H}}'$ when $\widehat{\mathcal{H}}'$ is prescribed (as in (4) of Theorem 7.1.4.1).
- $\widetilde{\mathcal{H}}'$ satisfies Condition 1.2.4.8 or 1.2.4.9 when $\widetilde{\mathcal{H}}$ does.

(These are possible by Lemma 1.2.4.45.) By Proposition 5.2.2.2, there exists some choice of projective smooth $\widetilde{\Sigma}^{\mathrm{ord},\prime}$ such that the canonical morphism $[\widetilde{g}] : \vec{\widetilde{\mathsf{M}}}^{\mathrm{ord}}_{\widetilde{\mathcal{H}}'} \to \vec{\widetilde{\mathsf{M}}}^{\mathrm{ord}}_{\widetilde{\mathcal{H}}}$ extends canonically to $\vec{[\widetilde{g}]}^{\mathrm{ord,tor}} : \vec{\widetilde{\mathsf{M}}}^{\mathrm{ord,tor}}_{\widetilde{\mathcal{H}}',\widetilde{\Sigma}^{\mathrm{ord},\prime}} \to \vec{\widetilde{\mathsf{M}}}^{\mathrm{ord,tor}}_{\widetilde{\mathcal{H}},\widetilde{\Sigma}^{\mathrm{ord}}}$. By replacing $\widetilde{\Sigma}^{\mathrm{ord},\prime}$ with a refinement such that it satisfies Condition 7.1.1.15 (with $\Sigma^{\mathrm{ord},\prime}$ and) with some choice of $\widetilde{\sigma}'$, and such that the morphism $\vec{[\widetilde{g}]}^{\mathrm{ord,tor}}$ sends the stratum $\vec{\widetilde{\mathsf{Z}}}^{\mathrm{ord}}_{[(\check{\Phi}_{\widetilde{\mathcal{H}}'},\check{\delta}_{\widetilde{\mathcal{H}}'},\widetilde{\sigma}')]}$ to $\vec{\widetilde{\mathsf{Z}}}^{\mathrm{ord}}_{[(\check{\Phi}_{\widetilde{\mathcal{H}}},\check{\delta}_{\widetilde{\mathcal{H}}},\widetilde{\sigma})]}$, we see that the induced morphism from the closure of

$\vec{\widetilde{\mathsf{Z}}}^{\mathrm{ord}}_{[(\check{\Phi}_{\widetilde{\mathcal{H}}'},\check{\delta}_{\widetilde{\mathcal{H}}'},\widetilde{\sigma}')]}$ to the closure of $\vec{\widetilde{\mathsf{Z}}}^{\mathrm{ord}}_{[(\check{\Phi}_{\widetilde{\mathcal{H}}},\check{\delta}_{\widetilde{\mathcal{H}}},\widetilde{\sigma})]}$ gives the existences of the morphisms $[\widehat{g}]$ and $[\widehat{g}]^{\mathrm{tor}}$ when $g_l = 1$ (resp. $[g_l]^*_{\kappa',\kappa}$ and $([g_l]^*_{\kappa',\kappa})^{\mathrm{tor}}$ when $\widehat{g} = 1$) as in (4a) and (4b) (resp. (5a) and (5b)) of Theorem 7.1.4.1, where $\kappa' = (\widetilde{\mathcal{H}}', \widetilde{\Sigma}', \widetilde{\sigma}')$ lies in $\mathbf{K}_{Q,\mathcal{H}',\Sigma^{\mathrm{ord},\prime}}$ (resp. $\mathbf{K}_{Q',\mathcal{H},\Sigma}$), which satisfy (4d), (4e), and (4f) (resp. (5e) and (5f)) thanks to the corresponding statements of Proposition 5.2.2.2 for $[\vec{\widetilde{g}}]^{\mathrm{ord,tor}}$, except that (7.1.4.12) and (7.1.4.23) still have to be explained. (As explained at the end of the proof of Proposition 7.2.3.5, the description concerning stratifications and formal completions are true because such a canonical morphism $[\vec{\widetilde{g}}]^{\mathrm{ord,tor}}$ is constructed using (6) of Theorem 5.2.1.1, which is consistent with the constructions of the canonical morphisms in Lemmas 5.2.4.38 and 7.1.2.29, and Proposition 7.1.2.17, using the various universal properties given in terms of degeneration data.) Also, (4g) and (5g) follow from Corollaries 3.4.4.3 and 5.2.2.3; (4h) and (5h) follow from Corollaries 3.4.4.4 and 5.2.2.4; (4i) follows from Corollaries 3.4.4.6 and 5.2.2.5; and (4j) follows from Corollary 5.2.2.5 and (4i) (where the finite flatness of (7.1.4.20), which is a finite morphism between regular schemes, is automatic by [35, IV-3, 15.4.2 e′)⇒b)]; *cf.* [62, Lem. 6.3.1.11]).

As in the case of showing $R^i(f^{\mathrm{tor}}_{\kappa',\kappa})_*\mathscr{O}_{\vec{\mathsf{N}}^{\mathrm{ord,tor}}_{\kappa'}} = 0$ for $i > 0$ in Proposition 7.2.2.19, since the morphisms $[\widehat{g}]^{\mathrm{tor}}$ and $([g_l]^*_{\kappa',\kappa})^{\mathrm{tor}}$ are étale locally given by equivariant morphisms between toric schemes, we have $R^i[\widehat{g}]^{\mathrm{tor}}_*\mathscr{O}_{\vec{\mathsf{N}}^{\mathrm{ord,tor}}_{\kappa'}} = 0$ and $R^i([g_l]^*_{\kappa',\kappa})^{\mathrm{tor}}_*\mathscr{O}_{\vec{\mathsf{N}}^{\mathrm{ord,tor}}_{\kappa'}} = 0$ for $i > 0$ (by [50, Ch. I, Sec. 3]), which are (7.1.4.12) and (7.1.4.23) of Theorem 7.1.4.1.

The remaining statements in (4c) and (5c) of Theorem 7.1.4.1 now follow if we assume statements (3b) and (3c) of Theorem 7.1.4.1. (See Section 7.3.6 below.)

7.3 Calculation of Formal Cohomology

7.3.1 *Setting*

Throughout this section, unless otherwise specified, we assume that $\kappa = (\widehat{\mathcal{H}}, \widehat{\Sigma}^{\mathrm{ord}}) = [\widetilde{\kappa}] \in \mathbf{K}^{\mathrm{ord},+}_{Q,\mathcal{H},\Sigma^{\mathrm{ord}}}$ for some $\widetilde{\kappa} = (\widetilde{\mathcal{H}}, \widetilde{\Sigma}^{\mathrm{ord}}, \widetilde{\sigma}) \in \widetilde{\mathbf{K}}^{\mathrm{ord},+}_{Q,\mathcal{H},\Sigma^{\mathrm{ord}}}$, so that $\mathcal{H}_\kappa = \mathcal{H}$ and $r_\kappa = r_{\mathcal{H}}$, and so that $f : \vec{\mathsf{N}}^{\mathrm{ord}} \to \vec{\mathsf{M}}^{\mathrm{ord}}_{\mathcal{H}}$ is a torsor under an abelian scheme $\vec{\mathsf{N}}^{\mathrm{ord,grp}} \to \vec{\mathsf{M}}^{\mathrm{ord}}_{\mathcal{H}}$, which extends to a semi-abelian scheme $\vec{\mathsf{N}}^{\mathrm{ord,ext}} \to \vec{\mathsf{M}}^{\mathrm{ord,tor}}_{\mathcal{H}} = \vec{\mathsf{M}}^{\mathrm{ord,tor}}_{\mathcal{H},\Sigma^{\mathrm{ord}}}$ (where the subscripts "κ" are suppressed for the sake of simplicity); and we fix the choice of an arbitrary (locally closed) stratum $\vec{\mathsf{Z}}^{\mathrm{ord}}_{[(\Phi_{\mathcal{H}},\delta_{\mathcal{H}},\tau)]}$ of $\vec{\mathsf{M}}^{\mathrm{ord,tor}}_{\mathcal{H}}$. The aim of this section is to calculate the relative cohomology of the pullback of the structural morphism $f^{\mathrm{tor}} : \vec{\mathsf{N}}^{\mathrm{ord,tor}} \to \vec{\mathsf{M}}^{\mathrm{ord,tor}}_{\mathcal{H}}$ to the formal completion $(\vec{\mathsf{M}}^{\mathrm{ord,tor}}_{\mathcal{H}})^{\wedge}_{\vec{\mathsf{Z}}^{\mathrm{ord}}_{[(\Phi_{\mathcal{H}},\delta_{\mathcal{H}},\tau)]}}$. (See (5) of Theorem 5.2.1.1 for a description of this formal completion. See also the first paragraph of Section 7.2.6 for a description of the formal completion $(\mathsf{N}^{\mathrm{tor}})^{\wedge}_{\vec{\widetilde{\mathsf{Z}}}_{[(\Phi_{\mathcal{H}},\delta_{\mathcal{H}},\tau)]}}$ of $\mathsf{N}^{\mathrm{tor}}$ along $\vec{\widetilde{\mathsf{Z}}}^{\mathrm{ord}}_{[(\Phi_{\mathcal{H}},\delta_{\mathcal{H}},\tau)]} = (f^{\mathrm{tor}})^{-1}(\vec{\mathsf{Z}}^{\mathrm{ord}}_{[(\Phi_{\mathcal{H}},\delta_{\mathcal{H}},\tau)]})$.)

Let $\mathscr{I}_{\vec{\mathsf{D}}^{\mathrm{ord}}_{\infty,\mathcal{H}}}$ be the $\mathscr{O}_{\vec{\mathsf{M}}^{\mathrm{ord,tor}}_{\mathcal{H}}}$-ideal defining the relative Cartier divisor $\vec{\mathsf{D}}^{\mathrm{ord}}_{\infty,\mathcal{H}}$ (with its reduced structure) in (3) of Theorem 5.2.1.1, and let $\mathscr{I}_{\vec{\mathsf{E}}^{\mathrm{ord}}_{\infty}} = \mathscr{I}_{\vec{\mathsf{E}}^{\mathrm{ord}}_{\infty,\kappa}}$ be the $\mathscr{O}_{\vec{\mathsf{N}}^{\mathrm{ord,tor}}}$-ideal defining the relative Cartier divisor $\vec{\mathsf{E}}^{\mathrm{ord}}_{\infty} = \vec{\mathsf{E}}^{\mathrm{ord}}_{\infty,\kappa}$ (with its reduced structure) in (1) of Theorem 7.1.4.1. (The subscripts "κ" are suppressed for the sake of simplicity.) Note that we have a canonical inclusion

$$(f^{\mathrm{tor}})^*\mathscr{I}_{\vec{\mathsf{D}}^{\mathrm{ord}}_{\infty,\mathcal{H}}} \hookrightarrow \mathscr{I}_{\vec{\mathsf{E}}^{\mathrm{ord}}_{\infty}} \tag{7.3.1.1}$$

of $\mathscr{O}_{\vec{\mathsf{N}}^{\mathrm{ord}}}$-ideals, realizing $\mathscr{I}_{\vec{\mathsf{E}}^{\mathrm{ord}}_{\infty}}$ as the radical of $(f^{\mathrm{tor}})^*\mathscr{I}_{\vec{\mathsf{D}}^{\mathrm{ord}}_{\infty,\mathcal{H}}}$.

For $\mathscr{M}$ being one of the following quasi-coherent $\mathscr{O}_{\mathsf{N}^{\mathrm{tor}}}$-modules $\mathscr{O}_{\mathsf{N}^{\mathrm{tor}}}$, $\mathscr{I}_{\vec{\mathsf{E}}^{\mathrm{ord}}_{\infty}}$, and $(f^{\mathrm{tor}})^*\mathscr{I}_{\vec{\mathsf{D}}^{\mathrm{ord}}_{\infty,\mathcal{H}}}$, we will show that the relative cohomology $R^b f^{\mathrm{tor}}_*(\mathscr{M})$ is locally free and canonically isomorphic to the (putative) answers given in the statements of Theorem 7.1.4.1. As a byproduct of the method, we will also investigate the cases of $\mathscr{M}$ being $\mathscr{I}_{\vec{\vec{\mathsf{Z}}}^{\mathrm{ord}}_{[(\Phi_{\mathcal{H}},\delta_{\mathcal{H}},\tau)]}} := \ker(\mathscr{O}_{\mathsf{N}^{\mathrm{tor}}} \twoheadrightarrow \mathscr{O}_{\vec{\vec{\mathsf{Z}}}^{\mathrm{ord}}_{[(\Phi_{\mathcal{H}},\delta_{\mathcal{H}},\tau)]}})$ or $\mathscr{O}_{\vec{\vec{\mathsf{Z}}}^{\mathrm{ord}}_{[(\Phi_{\mathcal{H}},\delta_{\mathcal{H}},\tau)]}}$. Since f^{tor} is proper (and since the choice of $\vec{\mathsf{Z}}^{\mathrm{ord}}_{[(\Phi_{\mathcal{H}},\delta_{\mathcal{H}},\tau)]}$ is arbitrary), by Grothendieck's fundamental theorem [35, III-1, 4.1.5] (and by fpqc descent for the property of local freeness [33, VIII, 1.11]), it suffices to show these over the pullback f^{tor} : $(\mathsf{N}^{\mathrm{tor}})^{\wedge}_{\vec{\vec{\mathsf{Z}}}^{\mathrm{ord}}_{[(\Phi_{\mathcal{H}},\delta_{\mathcal{H}},\tau)]}} \to (\vec{\mathsf{M}}^{\mathrm{ord,tor}}_{\mathcal{H}})^{\wedge}_{\vec{\mathsf{Z}}^{\mathrm{ord}}_{[(\Phi_{\mathcal{H}},\delta_{\mathcal{H}},\tau)]}}$ of f^{tor} to $(\vec{\mathsf{M}}^{\mathrm{ord,tor}}_{\mathcal{H}})^{\wedge}_{\vec{\mathsf{Z}}^{\mathrm{ord}}_{[(\Phi_{\mathcal{H}},\delta_{\mathcal{H}},\tau)]}}$. These will be carried out in the remainder of this section.

For simplicity of notation, we will denote by $\mathscr{O}^{+}_{\mathfrak{X}}$ (resp. $\mathscr{O}^{++}_{\mathfrak{X}}$, resp. $\mathscr{O}^{0+}_{\mathfrak{X}}$) the pullback of $\mathscr{I}_{\vec{\mathsf{E}}^{\mathrm{ord}}_{\infty}}$ (resp. $(f^{\mathrm{tor}})^*\mathscr{I}_{\vec{\mathsf{D}}^{\mathrm{ord}}_{\infty,\mathcal{H}}}$, resp. $\mathscr{I}_{\vec{\vec{\mathsf{Z}}}^{\mathrm{ord}}_{[(\Phi_{\mathcal{H}},\delta_{\mathcal{H}},\tau)]}}$) under any morphism $\mathfrak{X} \to \mathsf{N}^{\mathrm{tor}}$ from a formal scheme. For example, the pullback of $\mathscr{I}_{\vec{\mathsf{E}}^{\mathrm{ord}}_{\infty}}$ under the canonical morphism $(\mathsf{N}^{\mathrm{tor}})^{\wedge}_{\vec{\vec{\mathsf{Z}}}^{\mathrm{ord}}_{[(\Phi_{\mathcal{H}},\delta_{\mathcal{H}},\tau)]}} \to \mathsf{N}^{\mathrm{tor}}$ will be denoted $\mathscr{O}^{+}_{(\mathsf{N}^{\mathrm{tor}})^{\wedge}_{\vec{\vec{\mathsf{Z}}}^{\mathrm{ord}}_{[(\Phi_{\mathcal{H}},\delta_{\mathcal{H}},\tau)]}}}$.

The definitions and arguments in this section will follow those in [61, Sec. 4] very closely, but we will take this opportunity to clarify or correct some flaws in the exposition there.

7.3.2 *Formal Fibers of f^{tor}*

The definitions and arguments in this subsection will follow those in [61, Sec. 4A] very closely.

Definition 7.3.2.1. $\Gamma_{\breve{\Phi}_{\widetilde{\mathcal{H}}},\tau}$ *is the subgroup of elements in* $\Gamma_{\breve{\Phi}_{\widetilde{\mathcal{H}}}}$ *stabilizing (both)* X *and* Y *and inducing an element in* $\Gamma_{\Phi_{\mathcal{H}},\tau}$ *(the subgroup of* $\Gamma_{\Phi_{\mathcal{H}}}$ *formed by elements mapping* τ *to itself).*

Since we have tacitly assumed that $\Gamma_{\Phi_{\mathcal{H}},\tau}$ is trivial by Conditions 1.2.2.9 and [62, Lem. 6.2.5.27], $\Gamma_{\breve{\Phi}_{\widetilde{\mathcal{H}}},\tau}$ is also the subgroup of elements in $\Gamma_{\breve{\Phi}_{\widetilde{\mathcal{H}}}}$ fixing (both) X and Y.

Let $\Gamma_{\breve{\Phi}_{\widetilde{\mathcal{H}}},\Phi_{\mathcal{H}}}$ be as in Definition 1.2.4.21. By Lemma 1.2.4.25, $\Gamma_{\breve{\Phi}_{\widetilde{\mathcal{H}}},\Phi_{\mathcal{H}}}$ maps $\breve{\sigma}$, the image of $\widetilde{\sigma}$ in $\mathbf{P}_{\breve{\Phi}_{\widetilde{\mathcal{H}}}}$, to itself. On the other hand, by Condition 1.2.2.9 (and

Lemma 1.3.4.3), if a cone $\breve{\tau} \subset \mathbf{P}^+_{\breve{\Phi}_{\widetilde{\mathcal{H}}}}$ in $\widetilde{\Sigma}_{\breve{\Phi}_{\widetilde{\mathcal{H}}}}$ has a face that is a $\Gamma_{\breve{\Phi}_{\widetilde{\mathcal{H}}},\tau}$-translation of $\breve{\sigma}$, then it cannot have a different face that is also a $\Gamma_{\breve{\Phi}_{\widetilde{\mathcal{H}}},\tau}$-translation of $\breve{\sigma}$.

Definition 7.3.2.2. *$\widetilde{\Sigma}_{\breve{\Phi}_{\widetilde{\mathcal{H}}},\breve{\sigma},\tau}$ is the subset of $\widetilde{\Sigma}_{\breve{\Phi}_{\widetilde{\mathcal{H}}}}$ consisting of cones $\breve{\tau}$ satisfying the following conditions* (*cf. similar conditions in the first paragraph of Section* 7.2.6)*:*

(1) *$\breve{\tau} \subset \mathbf{P}^+_{\breve{\Phi}_{\widetilde{\mathcal{H}}}}$.*

(2) *$\breve{\tau}$ has $\breve{\sigma}$ as a face.*

(3) *The image of $\breve{\tau}$ under the* (*canonical*) *second morphism in* (1.2.4.20) *is contained in $\tau \subset \mathbf{P}^+_{\Phi_{\mathcal{H}}}$.*

In other words, $\widetilde{\Sigma}_{\breve{\Phi}_{\widetilde{\mathcal{H}}},\breve{\sigma},\tau}$ is the subset of $\widetilde{\Sigma}_{\breve{\Phi}_{\widetilde{\mathcal{H}}},\breve{\sigma}} = \widetilde{\Sigma}_{\breve{\Phi}_{\widetilde{\mathcal{H}}},\breve{\sigma}}$ (*see* (1) *of Definition* 1.2.4.21) *consisting of cones $\breve{\tau}$ whose image under the* (*canonical*) *second morphism in* (1.2.4.20) *is contained in $\tau \subset \mathbf{P}^+_{\Phi_{\mathcal{H}}}$.*

Thus, to obtain a complete list of representatives of the equivalence classes $[(\breve{\Phi}_{\widetilde{\mathcal{H}}}, \breve{\delta}_{\widetilde{\mathcal{H}}}, \breve{\tau})]$ parameterizing the strata of $\vec{\widetilde{\mathsf{Z}}}^{\mathrm{ord}}_{[(\Phi_{\mathcal{H}},\delta_{\mathcal{H}},\tau)]}$, it suffices to take representatives of $\widetilde{\Sigma}_{\breve{\Phi}_{\widetilde{\mathcal{H}}},\breve{\sigma},\tau}$ modulo the action of $\Gamma_{\breve{\Phi}_{\widetilde{\mathcal{H}}},\Phi_{\mathcal{H}}}$. (That is, we do not have to consider $\Gamma_{\breve{\Phi}_{\widetilde{\mathcal{H}}},\Phi_{\mathcal{H}}}$-translations of $\breve{\sigma}$.)

Let $\vec{\widetilde{\Xi}}^{\mathrm{ord}}_{\breve{\Phi}_{\widetilde{\mathcal{H}}},\breve{\delta}_{\widetilde{\mathcal{H}}}}(\tau)$ denote the toroidal embedding of $\vec{\widetilde{\Xi}}^{\mathrm{ord}}_{\breve{\Phi}_{\widetilde{\mathcal{H}}},\breve{\delta}_{\widetilde{\mathcal{H}}}}$ formed by gluing the affine toroidal embeddings $\vec{\widetilde{\Xi}}^{\mathrm{ord}}_{\breve{\Phi}_{\widetilde{\mathcal{H}}},\breve{\delta}_{\widetilde{\mathcal{H}}}}(\breve{\tau})$ over $\vec{\widetilde{C}}^{\mathrm{ord}}_{\breve{\Phi}_{\widetilde{\mathcal{H}}},\breve{\delta}_{\widetilde{\mathcal{H}}}}$, where $\breve{\tau}$ runs through cones in $\Sigma_{\breve{\Phi}_{\widetilde{\mathcal{H}}},\breve{\sigma},\tau}$. To minimize confusion, we shall distinguish between $\vec{\widetilde{\Xi}}^{\mathrm{ord}}_{\breve{\Phi}_{\widetilde{\mathcal{H}}},\breve{\delta}_{\widetilde{\mathcal{H}}}}(\breve{\tau}_1)$ and $\vec{\widetilde{\Xi}}^{\mathrm{ord}}_{\breve{\Phi}_{\widetilde{\mathcal{H}}},\breve{\delta}_{\widetilde{\mathcal{H}}}}(\breve{\tau}_2)$ even when $[(\breve{\Phi}_{\widetilde{\mathcal{H}}}, \breve{\delta}_{\widetilde{\mathcal{H}}}, \breve{\tau}_1)] = [(\breve{\Phi}_{\widetilde{\mathcal{H}}}, \breve{\delta}_{\widetilde{\mathcal{H}}}, \breve{\tau}_2)]$. For each $\breve{\tau}$ as above (having $\breve{\sigma}$ as a face), recall that we have denoted the closure of the $\breve{\sigma}$-stratum of $\vec{\widetilde{\Xi}}^{\mathrm{ord}}_{\breve{\Phi}_{\widetilde{\mathcal{H}}},\breve{\delta}_{\widetilde{\mathcal{H}}}}(\breve{\tau})$ by $\vec{\widetilde{\Xi}}^{\mathrm{ord}}_{\breve{\Phi}_{\widetilde{\mathcal{H}}},\breve{\delta}_{\widetilde{\mathcal{H}}},\breve{\sigma}}(\breve{\tau})$. Let $\vec{\widetilde{\Xi}}^{\mathrm{ord}}_{\breve{\Phi}_{\widetilde{\mathcal{H}}},\breve{\delta}_{\widetilde{\mathcal{H}}},\breve{\sigma}}(\tau)$ denote the union of all such $\vec{\widetilde{\Xi}}^{\mathrm{ord}}_{\breve{\Phi}_{\widetilde{\mathcal{H}}},\breve{\delta}_{\widetilde{\mathcal{H}}},\breve{\sigma}}(\breve{\tau})$, let $\vec{\widetilde{\Xi}}^{\mathrm{ord}}_{\breve{\Phi}_{\widetilde{\mathcal{H}}},\breve{\delta}_{\widetilde{\mathcal{H}}},\tau}$ denote the union of all such $\vec{\widetilde{\Xi}}^{\mathrm{ord}}_{\breve{\Phi}_{\widetilde{\mathcal{H}}},\breve{\delta}_{\widetilde{\mathcal{H}}},\breve{\tau}}$, and let $\vec{\widetilde{\mathfrak{X}}}^{\mathrm{ord}}_{\breve{\Phi}_{\widetilde{\mathcal{H}}},\breve{\delta}_{\widetilde{\mathcal{H}}},\breve{\sigma},\tau}$ denote the formal completion of $\vec{\widetilde{\Xi}}^{\mathrm{ord}}_{\breve{\Phi}_{\widetilde{\mathcal{H}}},\breve{\delta}_{\widetilde{\mathcal{H}}},\breve{\sigma}}(\tau)$ along $\vec{\widetilde{\Xi}}^{\mathrm{ord}}_{\breve{\Phi}_{\widetilde{\mathcal{H}}},\breve{\delta}_{\widetilde{\mathcal{H}}},\tau}$.

For each $\breve{\tau} \in \widetilde{\Sigma}_{\breve{\Phi}_{\widetilde{\mathcal{H}}},\breve{\sigma},\tau}$, consider the open subscheme $U_{\breve{\tau}}$ of $\vec{\widetilde{\Xi}}^{\mathrm{ord}}_{\breve{\Phi}_{\widetilde{\mathcal{H}}},\breve{\delta}_{\widetilde{\mathcal{H}}},\tau}$ formed by the union of all (locally closed) strata of $\vec{\widetilde{\Xi}}^{\mathrm{ord}}_{\breve{\Phi}_{\widetilde{\mathcal{H}}},\breve{\delta}_{\widetilde{\mathcal{H}}},\tau}$ that contains the stratum $\vec{\widetilde{\Xi}}^{\mathrm{ord}}_{\breve{\Phi}_{\widetilde{\mathcal{H}}},\breve{\delta}_{\widetilde{\mathcal{H}}},\breve{\tau}}$ in its closure, and consider the open formal subscheme $\mathfrak{U}_{\breve{\tau}}$ of $\vec{\widetilde{\mathfrak{X}}}^{\mathrm{ord}}_{\breve{\Phi}_{\widetilde{\mathcal{H}}},\breve{\delta}_{\widetilde{\mathcal{H}}},\breve{\sigma},\tau}$ supported on $U_{\breve{\tau}}$. The subscheme $U_{\breve{\tau}}$ of $\vec{\widetilde{\Xi}}^{\mathrm{ord}}_{\breve{\Phi}_{\widetilde{\mathcal{H}}},\breve{\delta}_{\widetilde{\mathcal{H}}},\tau}$ is the closed subscheme of $\vec{\widetilde{\Xi}}^{\mathrm{ord}}_{\breve{\Phi}_{\widetilde{\mathcal{H}}},\breve{\delta}_{\widetilde{\mathcal{H}}}}(\breve{\tau})$ given by the intersection of $\vec{\widetilde{\Xi}}^{\mathrm{ord}}_{\breve{\Phi}_{\widetilde{\mathcal{H}}},\breve{\delta}_{\widetilde{\mathcal{H}}}}(\breve{\tau})$ and $\vec{\widetilde{\Xi}}^{\mathrm{ord}}_{\breve{\Phi}_{\widetilde{\mathcal{H}}},\breve{\delta}_{\widetilde{\mathcal{H}}},\tau}$ (as an open subscheme of $\vec{\widetilde{\Xi}}^{\mathrm{ord}}_{\breve{\Phi}_{\widetilde{\mathcal{H}}},\breve{\delta}_{\widetilde{\mathcal{H}}},\tau}$), and the formal subscheme $\mathfrak{U}_{\breve{\tau}}$ of $\vec{\widetilde{\mathfrak{X}}}^{\mathrm{ord}}_{\breve{\Phi}_{\widetilde{\mathcal{H}}},\breve{\delta}_{\widetilde{\mathcal{H}}},\breve{\sigma},\tau}$ is the formal completion of $\vec{\widetilde{\Xi}}^{\mathrm{ord}}_{\breve{\Phi}_{\widetilde{\mathcal{H}}},\breve{\delta}_{\widetilde{\mathcal{H}}},\breve{\sigma}}(\breve{\tau})$

along $U_{\breve{\tau}}$. The collection $\{U_{\breve{\tau}}\}_{\breve{\tau}\in\widetilde{\Sigma}_{\breve{\Phi}_{\widetilde{\mathcal{H}}},\breve{\sigma},\tau}}$ forms an open covering of $\vec{\widetilde{\Xi}}^{\mathrm{ord}}_{\breve{\Phi}_{\widetilde{\mathcal{H}}},\breve{\delta}_{\widetilde{\mathcal{H}}},\tau}$. We can interpret $\vec{\widetilde{\mathfrak{X}}}^{\mathrm{ord}}_{\breve{\Phi}_{\widetilde{\mathcal{H}}},\breve{\delta}_{\widetilde{\mathcal{H}}},\breve{\sigma},\tau}$ as constructed by gluing the collection $\{\mathfrak{U}_{\breve{\tau}}\}_{\breve{\tau}\in\widetilde{\Sigma}_{\breve{\Phi}_{\widetilde{\mathcal{H}}},\breve{\sigma},\tau}}$ of formal schemes along their intersections (of supports).

Definition 7.3.2.3.

(1) $\breve{\tau}^{\vee}_{\breve{\sigma}}$ *is the intersection of* $(\breve{\tau}')^{\vee}_0$ *(in* $\mathbf{S}_{\breve{\Phi}_{\widetilde{\mathcal{H}}}}$*) for* $\breve{\tau}'$ *running through faces of* $\breve{\tau}$ *in* $\widetilde{\Sigma}_{\breve{\Phi}_{\widetilde{\mathcal{H}}},\breve{\sigma},\tau}$ *(including* $\breve{\tau}$ *itself).*

(2) $\breve{\tau}^{\vee}_{\breve{\sigma},+}$ *is the intersection of* $(\breve{\tau}')^{\vee}_0$ *(in* $\mathbf{S}_{\breve{\Phi}_{\widetilde{\mathcal{H}}}}$*) for* $\breve{\tau}'$ *running through faces of* $\breve{\tau}$ *in* $\Sigma_{\breve{\Phi}_{\widetilde{\mathcal{H}}}}$ *(including* $\breve{\tau}$ *itself) that also has* $\breve{\sigma}$ *as a face.*

(3) $\tau^{\vee}_{+}$ *is the intersection of* $(\tau')^{\vee}_0$ *(in* $\mathbf{S}_{\Phi_{\mathcal{H}}}$*) for* τ' *running through faces of* τ *in* $\Sigma_{\Phi_{\mathcal{H}}}$ *(including* τ *itself).*

Then we have the canonical isomorphism

$$U_{\breve{\tau}} \cong \underline{\mathrm{Spec}}_{\mathscr{O}_{\vec{\widetilde{C}}^{\mathrm{ord}}_{\breve{\Phi}_{\widetilde{\mathcal{H}}},\breve{\delta}_{\widetilde{\mathcal{H}}}}}}\Big(\big(\bigoplus_{\breve{\ell}\in\breve{\tau}^{\vee}} \vec{\widetilde{\Psi}}^{\mathrm{ord}}_{\breve{\Phi}_{\widetilde{\mathcal{H}}},\breve{\delta}_{\widetilde{\mathcal{H}}}}(\breve{\ell})\big)/\big(\bigoplus_{\breve{\ell}\in\breve{\tau}^{\vee}_{\breve{\sigma}}} \vec{\widetilde{\Psi}}^{\mathrm{ord}}_{\breve{\Phi}_{\widetilde{\mathcal{H}}},\breve{\delta}_{\widetilde{\mathcal{H}}}}(\breve{\ell})\big)\Big)$$

of schemes affine over $\vec{\widetilde{C}}^{\mathrm{ord}}_{\breve{\Phi}_{\widetilde{\mathcal{H}}},\breve{\delta}_{\widetilde{\mathcal{H}}}}$. As $\mathscr{O}_{\vec{\widetilde{C}}^{\mathrm{ord}}_{\breve{\Phi}_{\widetilde{\mathcal{H}}},\breve{\delta}_{\widetilde{\mathcal{H}}}}}$-modules, we have a canonical isomorphism

$$\big(\bigoplus_{\breve{\ell}\in\breve{\tau}^{\vee}} \vec{\widetilde{\Psi}}^{\mathrm{ord}}_{\breve{\Phi}_{\widetilde{\mathcal{H}}},\breve{\delta}_{\widetilde{\mathcal{H}}}}(\breve{\ell})\big)/\big(\bigoplus_{\breve{\ell}\in\breve{\tau}^{\vee}_{\breve{\sigma}}} \vec{\widetilde{\Psi}}^{\mathrm{ord}}_{\breve{\Phi}_{\widetilde{\mathcal{H}}},\breve{\delta}_{\widetilde{\mathcal{H}}}}(\breve{\ell})\big) \cong \bigoplus_{\breve{\ell}\in\breve{\tau}^{\vee}-\breve{\tau}^{\vee}_{\breve{\sigma}}} \vec{\widetilde{\Psi}}^{\mathrm{ord}}_{\breve{\Phi}_{\widetilde{\mathcal{H}}},\breve{\delta}_{\widetilde{\mathcal{H}}}}(\breve{\ell}).$$

If we equip $\breve{\tau}^{\vee}-\breve{\tau}^{\vee}_{\breve{\sigma}}$ with the semigroup structure induced by the canonical bijection $(\breve{\tau}^{\vee}-\breve{\tau}^{\vee}_{\breve{\sigma}}) \to \breve{\tau}^{\vee}/\breve{\tau}^{\vee}_{\breve{\sigma}}$, then we may interpret $\bigoplus_{\breve{\ell}\in\breve{\tau}^{\vee}-\breve{\tau}^{\vee}_{\breve{\sigma}}} \vec{\widetilde{\Psi}}^{\mathrm{ord}}_{\breve{\Phi}_{\widetilde{\mathcal{H}}},\breve{\delta}_{\widetilde{\mathcal{H}}}}(\breve{\ell})$ as an $\mathscr{O}_{\vec{\widetilde{C}}^{\mathrm{ord}}_{\breve{\Phi}_{\widetilde{\mathcal{H}}},\breve{\delta}_{\widetilde{\mathcal{H}}}}}$-algebra, with algebra structure given by the isomorphisms

$$\vec{\widetilde{\Delta}}^{\mathrm{ord},*}_{\breve{\Phi}_{\widetilde{\mathcal{H}}},\breve{\delta}_{\widetilde{\mathcal{H}}},\breve{\ell},\breve{\ell}'} : \vec{\widetilde{\Psi}}^{\mathrm{ord}}_{\breve{\Phi}_{\widetilde{\mathcal{H}}},\breve{\delta}_{\widetilde{\mathcal{H}}}}(\breve{\ell}) \underset{\mathscr{O}_{\vec{\widetilde{C}}^{\mathrm{ord}}_{\breve{\Phi}_{\widetilde{\mathcal{H}}},\breve{\delta}_{\widetilde{\mathcal{H}}}}}}{\otimes} \vec{\widetilde{\Psi}}^{\mathrm{ord}}_{\breve{\Phi}_{\widetilde{\mathcal{H}}},\breve{\delta}_{\widetilde{\mathcal{H}}}}(\breve{\ell}') \xrightarrow{\sim} \vec{\widetilde{\Psi}}^{\mathrm{ord}}_{\breve{\Phi}_{\widetilde{\mathcal{H}}},\breve{\delta}_{\widetilde{\mathcal{H}}}}(\breve{\ell}+\breve{\ell}')$$

inherited from those of $\mathscr{O}_{\vec{\widetilde{\Xi}}^{\mathrm{ord}}_{\breve{\Phi}_{\widetilde{\mathcal{H}}},\breve{\delta}_{\widetilde{\mathcal{H}}}}} \cong \bigoplus_{\breve{\ell}\in\mathbf{S}_{\breve{\Phi}_{\widetilde{\mathcal{H}}}}} \vec{\widetilde{\Psi}}^{\mathrm{ord}}_{\breve{\Phi}_{\widetilde{\mathcal{H}}},\breve{\delta}_{\widetilde{\mathcal{H}}}}(\breve{\ell})$ if $\breve{\ell}+\breve{\ell}' \in \breve{\tau}^{\vee}-\breve{\tau}^{\vee}_{\breve{\sigma}}$ and by

$$\vec{\widetilde{\Psi}}^{\mathrm{ord}}_{\breve{\Phi}_{\widetilde{\mathcal{H}}},\breve{\delta}_{\widetilde{\mathcal{H}}}}(\breve{\ell}) \underset{\mathscr{O}_{\vec{\widetilde{C}}^{\mathrm{ord}}_{\breve{\Phi}_{\widetilde{\mathcal{H}}},\breve{\delta}_{\widetilde{\mathcal{H}}}}}}{\otimes} \vec{\widetilde{\Psi}}^{\mathrm{ord}}_{\breve{\Phi}_{\widetilde{\mathcal{H}}},\breve{\delta}_{\widetilde{\mathcal{H}}}}(\breve{\ell}') \to 0$$

otherwise. Then we have a canonical isomorphism

$$U_{\breve{\tau}} \cong \underline{\mathrm{Spec}}_{\mathscr{O}_{\vec{\widetilde{C}}^{\mathrm{ord}}_{\breve{\Phi}_{\widetilde{\mathcal{H}}},\breve{\delta}_{\widetilde{\mathcal{H}}}}}}\Big(\bigoplus_{\breve{\ell}\in\breve{\tau}^{\vee}-\breve{\tau}^{\vee}_{\breve{\sigma}}} \vec{\widetilde{\Psi}}^{\mathrm{ord}}_{\breve{\Phi}_{\widetilde{\mathcal{H}}},\breve{\delta}_{\widetilde{\mathcal{H}}}}(\breve{\ell})\Big).$$

By definition, we have

$$\breve{\tau}^\vee - \breve{\tau}^\vee_{\breve{\sigma}} = \Big(\bigcup_{\substack{\breve{\tau}' \text{ face of } \breve{\tau} \\ \text{in } \widetilde{\Sigma}_{\breve{\Phi}_{\widetilde{\mathcal{H}}},\breve{\sigma},\tau}}} ((\breve{\tau}')^\perp \cap \breve{\tau}^\vee) \Big) \subset \breve{\sigma}^\perp \cap \breve{\tau}^\vee.$$

The formal scheme $\mathfrak{U}_{\breve{\tau}}$, being the formal completion of

$$\vec{\widetilde{\Xi}}^{\mathrm{ord}}_{\breve{\Phi}_{\widetilde{\mathcal{H}}},\breve{\delta}_{\widetilde{\mathcal{H}}},\breve{\sigma}}(\breve{\tau}) \cong \underline{\mathrm{Spec}}_{\mathscr{O}_{\vec{\widetilde{C}}^{\mathrm{ord}}_{\breve{\Phi}_{\widetilde{\mathcal{H}}},\breve{\delta}_{\widetilde{\mathcal{H}}}}}} \Big(\bigoplus_{\breve{\ell}\in\breve{\sigma}^\perp\cap\breve{\tau}^\vee} \vec{\widetilde{\Psi}}^{\mathrm{ord}}_{\breve{\Phi}_{\widetilde{\mathcal{H}}},\breve{\delta}_{\widetilde{\mathcal{H}}}}(\breve{\ell}) \Big)$$

along $U_{\breve{\tau}}$, can be canonically identified with the relative formal spectrum of the $\mathscr{O}_{\widetilde{C}_{\breve{\Phi}_{\widetilde{\mathcal{H}}},\breve{\delta}_{\widetilde{\mathcal{H}}}}}$-algebra $\hat{\bigoplus}_{\breve{\ell}\in\breve{\sigma}^\perp\cap\breve{\tau}^\vee} \vec{\widetilde{\Psi}}^{\mathrm{ord}}_{\breve{\Phi}_{\widetilde{\mathcal{H}}},\breve{\delta}_{\widetilde{\mathcal{H}}}}(\breve{\ell})$ over $\vec{\widetilde{C}}^{\mathrm{ord}}_{\breve{\Phi}_{\widetilde{\mathcal{H}}},\breve{\delta}_{\widetilde{\mathcal{H}}}}$, where $\hat{\oplus}$ denotes the completion of the sum with respect to the $\mathscr{O}_{\vec{\widetilde{\Xi}}^{\mathrm{ord}}_{\breve{\Phi}_{\widetilde{\mathcal{H}}},\breve{\delta}_{\widetilde{\mathcal{H}}},\breve{\sigma}}(\breve{\tau})}$-ideal $\bigoplus_{\breve{\ell}\in\breve{\sigma}^\perp\cap\breve{\tau}^\vee_{\breve{\sigma}}} \vec{\widetilde{\Psi}}^{\mathrm{ord}}_{\breve{\Phi}_{\widetilde{\mathcal{H}}},\breve{\delta}_{\widetilde{\mathcal{H}}}}(\breve{\ell})$. Note that all the above canonical isomorphisms correspond to canonical isomorphisms of $\mathscr{O}_{\vec{\widetilde{C}}^{\mathrm{ord}}_{\breve{\Phi}_{\widetilde{\mathcal{H}}},\breve{\delta}_{\widetilde{\mathcal{H}}}}}$-algebras formed by sums of sheaves of the form $\vec{\widetilde{\Psi}}^{\mathrm{ord}}_{\breve{\Phi}_{\widetilde{\mathcal{H}}},\breve{\delta}_{\widetilde{\mathcal{H}}}}(\breve{\ell})$ (with $\mathscr{O}_{\vec{\widetilde{C}}^{\mathrm{ord}}_{\breve{\Phi}_{\widetilde{\mathcal{H}}},\breve{\delta}_{\widetilde{\mathcal{H}}}}}$-algebra structures inherited from that of $\mathscr{O}_{\vec{\widetilde{\Xi}}^{\mathrm{ord}}_{\breve{\Phi}_{\widetilde{\mathcal{H}}},\breve{\delta}_{\widetilde{\mathcal{H}}}}}$). By abuse of language, let us write

$$\mathscr{O}_{\mathfrak{U}_{\breve{\tau}}} \cong \hat{\bigoplus_{\breve{\ell}\in\breve{\sigma}^\perp\cap\breve{\tau}^\vee}} \vec{\widetilde{\Psi}}^{\mathrm{ord}}_{\breve{\Phi}_{\widetilde{\mathcal{H}}},\breve{\delta}_{\widetilde{\mathcal{H}}}}(\breve{\ell}),$$

$$\mathscr{O}^{0+}_{\mathfrak{U}_{\breve{\tau}}} \cong \hat{\bigoplus_{\breve{\ell}\in\breve{\sigma}^\perp\cap\breve{\tau}^\vee_{\breve{\sigma}}}} \vec{\widetilde{\Psi}}^{\mathrm{ord}}_{\breve{\Phi}_{\widetilde{\mathcal{H}}},\breve{\delta}_{\widetilde{\mathcal{H}}}}(\breve{\ell}),$$

$$\mathscr{O}_{U_{\breve{\tau}}} \cong \bigoplus_{\breve{\ell}\in\breve{\tau}^\vee-\breve{\tau}^\vee_{\breve{\sigma}}} \vec{\widetilde{\Psi}}^{\mathrm{ord}}_{\breve{\Phi}_{\widetilde{\mathcal{H}}},\breve{\delta}_{\widetilde{\mathcal{H}}}}(\breve{\ell}).$$

By Condition 1.2.2.9 (and Lemma 1.3.4.3), the action of $\Gamma_{\breve{\Phi}_{\widetilde{\mathcal{H}}},\Phi_{\mathcal{H}}}$ induces only the trivial action on each stratum it stabilizes. Therefore, the quotient morphism

$$\vec{\widetilde{\mathfrak{X}}}^{\mathrm{ord}}_{\breve{\Phi}_{\widetilde{\mathcal{H}}},\breve{\delta}_{\widetilde{\mathcal{H}}},\breve{\sigma},\tau} \to \vec{\widetilde{\mathfrak{X}}}^{\mathrm{ord}}_{\breve{\Phi}_{\widetilde{\mathcal{H}}},\breve{\delta}_{\widetilde{\mathcal{H}}},\breve{\sigma},\tau}/\Gamma_{\breve{\Phi}_{\widetilde{\mathcal{H}}},\Phi_{\mathcal{H}}} \tag{7.3.2.4}$$

of formal schemes over $\vec{\mathsf{S}}_{0,r_{\mathcal{H}}}$ is a *local isomorphism.* The morphism (7.3.2.4) is not defined over $\vec{\widetilde{C}}^{\mathrm{ord}}_{\breve{\Phi}_{\widetilde{\mathcal{H}}},\breve{\delta}_{\widetilde{\mathcal{H}}}}$ when the action of $\Gamma_{\breve{\Phi}_{\widetilde{\mathcal{H}}},\Phi_{\mathcal{H}}}$ on $\vec{\widetilde{C}}^{\mathrm{ord}}_{\breve{\Phi}_{\widetilde{\mathcal{H}}},\breve{\delta}_{\widetilde{\mathcal{H}}}}$ is nontrivial. Nevertheless, since $\Gamma_{\breve{\Phi}_{\widetilde{\mathcal{H}}},\Phi_{\mathcal{H}}}$ acts trivially on $\Phi_{\mathcal{H}}$, it acts trivially on $\vec{C}^{\mathrm{ord}}_{\Phi_{\mathcal{H}},\delta_{\mathcal{H}}}$, and hence (7.3.2.4) is defined over $\vec{C}^{\mathrm{ord}}_{\Phi_{\mathcal{H}},\delta_{\mathcal{H}}}$.

Proposition 7.3.2.5. (*Compare with [61, Prop. 4.3].*) *There is a canonical isomorphism*

$$(\vec{\mathsf{N}}^{\mathrm{ord,tor}})^\wedge_{\vec{\mathsf{Z}}^{\mathrm{ord}}_{[(\Phi_{\mathcal{H}},\delta_{\mathcal{H}},\tau)]}} \cong \vec{\widetilde{\mathfrak{X}}}^{\mathrm{ord}}_{\breve{\Phi}_{\widetilde{\mathcal{H}}},\breve{\delta}_{\widetilde{\mathcal{H}}},\breve{\sigma},\tau}/\Gamma_{\breve{\Phi}_{\widetilde{\mathcal{H}}},\Phi_{\mathcal{H}}} \tag{7.3.2.6}$$

of formal schemes over $\vec{C}^{\mathrm{ord}}_{\Phi_{\mathcal{H}},\delta_{\mathcal{H}}}$, characterized by the identifications

$$(\vec{\mathsf{N}}^{\mathrm{ord,tor}})^\wedge_{\vec{\mathsf{Z}}^{\mathrm{ord}}_{[(\breve{\Phi}_{\widetilde{\mathcal{H}}},\breve{\delta}_{\widetilde{\mathcal{H}}},\breve{\tau})]}} \cong \vec{\widetilde{\mathfrak{X}}}^{\mathrm{ord}}_{\breve{\Phi}_{\widetilde{\mathcal{H}}},\breve{\delta}_{\widetilde{\mathcal{H}}},\breve{\sigma},\breve{\tau}}$$

of formal schemes over $\vec{\widetilde{C}}^{\mathrm{ord}}_{\breve{\Phi}_{\widetilde{\mathcal{H}}},\breve{\delta}_{\widetilde{\mathcal{H}}}}$ *(compatible with the canonical morphisms*

$$(\vec{\mathsf{N}}^{\mathrm{ord,tor}})^{\wedge}_{\vec{\widetilde{\mathsf{Z}}}^{\mathrm{ord}}_{[(\breve{\Phi}_{\widetilde{\mathcal{H}}},\breve{\delta}_{\widetilde{\mathcal{H}}},\breve{\tau})]}} \to (\vec{\mathsf{N}}^{\mathrm{ord,tor}})^{\wedge}_{\vec{\widetilde{\mathsf{Z}}}^{\mathrm{ord}}_{[(\Phi_{\mathcal{H}},\delta_{\mathcal{H}},\tau)]}}$$

and $\vec{\widetilde{C}}^{\mathrm{ord}}_{\breve{\Phi}_{\widetilde{\mathcal{H}}},\breve{\delta}_{\widetilde{\mathcal{H}}}} \to \vec{C}^{\mathrm{ord}}_{\Phi_{\mathcal{H}},\delta_{\mathcal{H}}}$*). (The formation of the formal completion here is similar to the one in* (5) *of Theorem* 5.2.1.1.*)*

Proof. Let $\breve{\tau} \in \widetilde{\Sigma}_{\breve{\Phi}_{\widetilde{\mathcal{H}}},\breve{\sigma},\tau}$. Let $\widetilde{\mathfrak{U}}_{\breve{\tau}}$ denote the completion of $\vec{\widetilde{\Xi}}^{\mathrm{ord}}_{\breve{\Phi}_{\widetilde{\mathcal{H}}},\breve{\delta}_{\widetilde{\mathcal{H}}}}(\breve{\tau})$ along $U_{\breve{\tau}}$, which contains $\mathfrak{U}_{\breve{\tau}}$ as a closed formal subscheme (with the same support $U_{\breve{\tau}}$).

Since $U_{\breve{\tau}}$ is the union of $(\vec{\widetilde{\Xi}}^{\mathrm{ord}}_{\breve{\Phi}_{\widetilde{\mathcal{H}}},\breve{\delta}_{\widetilde{\mathcal{H}}}})_{\breve{\tau}'}$ with $\breve{\tau}'$ running through faces of $\breve{\tau}$ in $\widetilde{\Sigma}_{\breve{\Phi}_{\widetilde{\mathcal{H}}},\breve{\sigma},\tau}$, which are cones in $\mathbf{P}^+_{\breve{\Phi}_{\widetilde{\mathcal{H}}}}$, the tautological degeneration data over $\widetilde{\mathfrak{U}}_{\breve{\tau}}$ satisfies the positivity condition (with respect to the ideal defining $U_{\breve{\tau}}$), and we obtain by Mumford's construction as in Section 4.2.2 a degenerating family $({}^\heartsuit\widetilde{G}, {}^\heartsuit\widetilde{\lambda}, {}^\heartsuit\widetilde{i}, {}^\heartsuit\widetilde{\alpha}_{\widetilde{\mathcal{H}}^p}, {}^\heartsuit\widetilde{\alpha}^{\mathrm{ord}}_{\widetilde{\mathcal{H}}_p}) \to \widetilde{\mathfrak{U}}_{\breve{\tau}}$ of type $\vec{\widetilde{\mathsf{M}}}^{\mathrm{ord}}_{\widetilde{\mathcal{H}}}$ (*cf.* Definition 3.4.2.10), which we call a *Mumford family* (*cf.* Definition 4.2.2.21). Note that a Mumford family is defined in the sense of relative schemes, namely as a functorial assignment to each affine open formal subscheme $\mathrm{Spf}(R)$ of $\widetilde{\mathfrak{U}}_{\breve{\tau}}$ a degenerating family of type $\vec{\widetilde{\mathsf{M}}}^{\mathrm{ord}}_{\widetilde{\mathcal{H}}}$ over $\mathrm{Spec}(R)$. Therefore, (6) of Theorem 5.2.1.1 applies, and implies the existence of a canonical (strata-preserving) morphism $\widetilde{\mathfrak{U}}_{\breve{\tau}} \to \vec{\widetilde{\mathsf{M}}}^{\mathrm{ord,tor}}_{\widetilde{\mathcal{H}}}$ under which $({}^\heartsuit\widetilde{G}, {}^\heartsuit\widetilde{\lambda}, {}^\heartsuit\widetilde{i}, {}^\heartsuit\widetilde{\alpha}_{\widetilde{\mathcal{H}}^p}, {}^\heartsuit\widetilde{\alpha}^{\mathrm{ord}}_{\widetilde{\mathcal{H}}_p}) \to \widetilde{\mathfrak{U}}_{\breve{\tau}}$ is the pullback of $(\widetilde{G}, \widetilde{\lambda}, \widetilde{i}, \widetilde{\alpha}_{\widetilde{\mathcal{H}}^p}, \widetilde{\alpha}^{\mathrm{ord}}_{\widetilde{\mathcal{H}}_p}) \to \vec{\widetilde{\mathsf{M}}}^{\mathrm{ord,tor}}_{\widetilde{\mathcal{H}}}$. Moreover, if $\breve{\tau}' \in \widetilde{\Sigma}_{\breve{\Phi}_{\widetilde{\mathcal{H}}},\breve{\sigma},\tau}$, then the morphisms from $\widetilde{\mathfrak{U}}_{\breve{\tau}}$ and from $\widetilde{\mathfrak{U}}_{\breve{\tau}'}$ to $\vec{\widetilde{\mathsf{M}}}^{\mathrm{ord,tor}}_{\widetilde{\mathcal{H}}}$ agree over the intersection $\widetilde{\mathfrak{U}}_{\breve{\tau}} \cap \widetilde{\mathfrak{U}}_{\breve{\tau}'}$.

By taking the closures of the $[(\widetilde{\Phi}_{\widetilde{\mathcal{H}}}, \widetilde{\delta}_{\widetilde{\mathcal{H}}}, \widetilde{\sigma})]$-strata (not as closed subschemes of the supports, but as closed formal subschemes), and by arguing as in the proof of Proposition 7.2.2.19, we obtain canonical morphisms $\mathfrak{U}_{\breve{\tau}} \to \vec{\mathsf{N}}^{\mathrm{ord,tor}}$ for all $\breve{\tau}$ in $\widetilde{\Sigma}_{\breve{\Phi}_{\widetilde{\mathcal{H}}},\breve{\sigma},\tau}$, which patch together, cover all strata above $[(\Phi_{\mathcal{H}}, \delta_{\mathcal{H}}, \tau)]$, and induce the desired isomorphism (7.3.2.6). □

Let us also consider $\mathscr{O}^+_{\mathfrak{U}_{\breve{\tau}}}$ and $\mathscr{O}^{++}_{\mathfrak{U}_{\breve{\tau}}}$. By definition, the $\mathscr{O}_{\mathfrak{U}_{\breve{\tau}}}$-ideal $\mathscr{O}^+_{\mathfrak{U}_{\breve{\tau}}}$ is isomorphic to the pullback of the $\mathscr{O}_{\vec{\widetilde{\Xi}}^{\mathrm{ord}}_{\breve{\Phi}_{\widetilde{\mathcal{H}}},\breve{\delta}_{\widetilde{\mathcal{H}}},\breve{\sigma}}(\breve{\tau})}$-ideal defining the complement of $\vec{\widetilde{\Xi}}^{\mathrm{ord}}_{\breve{\Phi}_{\widetilde{\mathcal{H}}},\breve{\delta}_{\widetilde{\mathcal{H}}},\breve{\sigma}}$ in $\vec{\widetilde{\Xi}}^{\mathrm{ord}}_{\breve{\Phi}_{\widetilde{\mathcal{H}}},\breve{\delta}_{\widetilde{\mathcal{H}}},\breve{\sigma}}(\breve{\tau})$. (In general, this is different from the $\mathscr{O}_{\vec{\widetilde{\Xi}}^{\mathrm{ord}}_{\breve{\Phi}_{\widetilde{\mathcal{H}}},\breve{\delta}_{\widetilde{\mathcal{H}}},\breve{\sigma}}(\breve{\tau})}$-ideal defining the closed subscheme $\vec{\widetilde{\Xi}}^{\mathrm{ord}}_{\breve{\Phi}_{\widetilde{\mathcal{H}}},\breve{\delta}_{\widetilde{\mathcal{H}}},\breve{\tau}}$ of $\vec{\widetilde{\Xi}}^{\mathrm{ord}}_{\breve{\Phi}_{\widetilde{\mathcal{H}}},\breve{\delta}_{\widetilde{\mathcal{H}}},\breve{\sigma}}(\breve{\tau})$.) The above descriptions imply the following simple but important facts:

Lemma 7.3.2.7. (*Compare with [61, Lem. 4.1].*) *Suppose* $\breve{\tau}$ *and* $\breve{\tau}'$ *are two cones in* $\widetilde{\Sigma}_{\breve{\Phi}_{\widetilde{\mathcal{H}}},\breve{\sigma},\tau}$ *such that* $\breve{\tau}'$ *is a face of* $\breve{\tau}$*. Then:*

(1) *We have a canonical open immersion* $\mathfrak{U}_{\breve{\tau}'} \hookrightarrow \mathfrak{U}_{\breve{\tau}}$ *(resp.* $U_{\breve{\tau}'} \hookrightarrow U_{\breve{\tau}}$*) of formal subschemes of* $\widetilde{\vec{\mathfrak{X}}}^{\mathrm{ord}}_{\breve{\Phi}_{\widetilde{\mathcal{H}}},\breve{\delta}_{\widetilde{\mathcal{H}}},\breve{\sigma},\tau}$.

(2) *The compatible canonical restriction morphisms*

$$\begin{aligned} \mathscr{O}_{\mathfrak{U}_{\breve{\tau}}} &\to \mathscr{O}_{\mathfrak{U}_{\breve{\tau}'}}, \\ \mathscr{O}^{+}_{\mathfrak{U}_{\breve{\tau}}} &\to \mathscr{O}^{+}_{\mathfrak{U}_{\breve{\tau}'}}, \\ \mathscr{O}^{++}_{\mathfrak{U}_{\breve{\tau}}} &\to \mathscr{O}^{++}_{\mathfrak{U}_{\breve{\tau}'}}, \\ \mathscr{O}^{0+}_{\mathfrak{U}_{\breve{\tau}}} &\to \mathscr{O}^{0+}_{\mathfrak{U}_{\breve{\tau}'}} \end{aligned}$$

correspond to the compatible canonical morphisms

$$\begin{aligned} \widehat{\bigoplus_{\breve{\ell}\in\breve{\sigma}^{\perp}\cap\breve{\tau}^{\vee}}} \widetilde{\vec{\Psi}}^{\mathrm{ord}}_{\breve{\Phi}_{\widetilde{\mathcal{H}}},\breve{\delta}_{\widetilde{\mathcal{H}}}}(\breve{\ell}) &\to \widehat{\bigoplus_{\breve{\ell}\in\breve{\sigma}^{\perp}\cap(\breve{\tau}')^{\vee}}} \widetilde{\vec{\Psi}}^{\mathrm{ord}}_{\breve{\Phi}_{\widetilde{\mathcal{H}}},\breve{\delta}_{\widetilde{\mathcal{H}}}}(\breve{\ell}), \\ \widehat{\bigoplus_{\breve{\ell}\in\breve{\sigma}^{\perp}\cap\breve{\tau}^{\vee}_{\breve{\sigma},+}}} \widetilde{\vec{\Psi}}^{\mathrm{ord}}_{\breve{\Phi}_{\widetilde{\mathcal{H}}},\breve{\delta}_{\widetilde{\mathcal{H}}}}(\breve{\ell}) &\to \widehat{\bigoplus_{\breve{\ell}\in\breve{\sigma}^{\perp}\cap(\breve{\tau}')^{\vee}_{\breve{\sigma},+}}} \widetilde{\vec{\Psi}}^{\mathrm{ord}}_{\breve{\Phi}_{\widetilde{\mathcal{H}}},\breve{\delta}_{\widetilde{\mathcal{H}}}}(\breve{\ell}), \\ \widehat{\bigoplus_{\breve{\ell}\in\breve{\sigma}^{\perp}\cap(\breve{\tau}^{\vee}+\tau^{\vee}_{+})}} \widetilde{\vec{\Psi}}^{\mathrm{ord}}_{\breve{\Phi}_{\widetilde{\mathcal{H}}},\breve{\delta}_{\widetilde{\mathcal{H}}}}(\breve{\ell}) &\to \widehat{\bigoplus_{\breve{\ell}\in\breve{\sigma}^{\perp}\cap((\breve{\tau}')^{\vee}+\tau^{\vee}_{+})}} \widetilde{\vec{\Psi}}^{\mathrm{ord}}_{\breve{\Phi}_{\widetilde{\mathcal{H}}},\breve{\delta}_{\widetilde{\mathcal{H}}}}(\breve{\ell}), \\ \widehat{\bigoplus_{\breve{\ell}\in\breve{\sigma}^{\perp}\cap\breve{\tau}^{\vee}_{\breve{\sigma}}}} \widetilde{\vec{\Psi}}^{\mathrm{ord}}_{\breve{\Phi}_{\widetilde{\mathcal{H}}},\breve{\delta}_{\widetilde{\mathcal{H}}}}(\breve{\ell}) &\to \widehat{\bigoplus_{\breve{\ell}\in\breve{\sigma}^{\perp}\cap(\breve{\tau}')^{\vee}_{\breve{\sigma}}}} \widetilde{\vec{\Psi}}^{\mathrm{ord}}_{\breve{\Phi}_{\widetilde{\mathcal{H}}},\breve{\delta}_{\widetilde{\mathcal{H}}}}(\breve{\ell}) \end{aligned}$$

of $\mathscr{O}_{\vec{C}^{\mathrm{ord}}_{\breve{\Phi}_{\widetilde{\mathcal{H}}},\breve{\delta}_{\widetilde{\mathcal{H}}}}}$*-algebras, respectively, where the two instances of* $\hat{\oplus}$ *in each expression denote the completions of the sums with respect to the sheaves of ideals* $\bigoplus_{\breve{\ell}\in\breve{\sigma}^{\perp}\cap\breve{\tau}^{\vee}_{\breve{\sigma}}} \widetilde{\vec{\Psi}}^{\mathrm{ord}}_{\breve{\Phi}_{\widetilde{\mathcal{H}}},\breve{\delta}_{\widetilde{\mathcal{H}}}}(\breve{\ell})$ *and* $\bigoplus_{\breve{\ell}\in\breve{\sigma}^{\perp}\cap(\breve{\tau}')^{\vee}_{\breve{\sigma}}} \widetilde{\vec{\Psi}}^{\mathrm{ord}}_{\breve{\Phi}_{\widetilde{\mathcal{H}}},\breve{\delta}_{\widetilde{\mathcal{H}}}}(\breve{\ell})$*, respectively, and where* $\breve{\tau}^{\vee}+\tau^{\vee}_{+}$ *and* $(\breve{\tau}')^{\vee}+\tau^{\vee}_{+}$ *are defined by viewing* $\mathbf{S}_{\Phi_{\mathcal{H}}}$ *as a subsemigroup of* $\mathbf{S}_{\breve{\Phi}_{\widetilde{\mathcal{H}}}}$ *using the (canonical) first morphism* $\mathbf{S}_{\Phi_{\mathcal{H}}} \hookrightarrow \mathbf{S}_{\breve{\Phi}_{\widetilde{\mathcal{H}}}}$ *in* (1.2.4.18).

(3) *The canonical restriction morphism*

$$\mathscr{O}_{U_{\breve{\tau}}} \to \mathscr{O}_{U_{\breve{\tau}'}}$$

corresponds to the canonical morphism

$$\bigoplus_{\breve{\ell}\in\breve{\tau}^{\vee}-\breve{\tau}^{\vee}_{\breve{\sigma}}} \widetilde{\vec{\Psi}}^{\mathrm{ord}}_{\breve{\Phi}_{\widetilde{\mathcal{H}}},\breve{\delta}_{\widetilde{\mathcal{H}}}}(\breve{\ell}) \to \bigoplus_{\breve{\ell}\in(\breve{\tau}')^{\vee}-(\breve{\tau}')^{\vee}_{\breve{\sigma}}} \widetilde{\vec{\Psi}}^{\mathrm{ord}}_{\breve{\Phi}_{\widetilde{\mathcal{H}}},\breve{\delta}_{\widetilde{\mathcal{H}}}}(\breve{\ell})$$

of $\mathscr{O}_{\vec{C}^{\mathrm{ord}}_{\breve{\Phi}_{\widetilde{\mathcal{H}}},\breve{\delta}_{\widetilde{\mathcal{H}}}}}$*-algebras, which maps* $\widetilde{\vec{\Psi}}^{\mathrm{ord}}_{\breve{\Phi}_{\widetilde{\mathcal{H}}},\breve{\delta}_{\widetilde{\mathcal{H}}}}(\breve{\ell})$ *to* $\widetilde{\vec{\Psi}}^{\mathrm{ord}}_{\breve{\Phi}_{\widetilde{\mathcal{H}}},\breve{\delta}_{\widetilde{\mathcal{H}}}}(\breve{\ell})$ *when*

$$\breve{\ell} \in (\breve{\tau}^{\vee} - (\breve{\tau}')^{\vee}_{\breve{\sigma}}) = (\breve{\tau}^{\vee} - \breve{\tau}^{\vee}_{\breve{\sigma}}) \cap ((\breve{\tau}')^{\vee} - (\breve{\tau}')^{\vee}_{\breve{\sigma}}),$$

and to zero otherwise.

(4) *The correspondences in* (2) *and* (3) *above are canonically compatible with each other.*

By (5) of Theorem 5.2.1.1, we have a canonical isomorphism
$$(\vec{\mathsf{M}}_{\mathcal{H}}^{\mathrm{ord,tor}})^{\wedge}_{\vec{\mathsf{Z}}^{\mathrm{ord}}_{[(\Phi_{\mathcal{H}},\delta_{\mathcal{H}},\tau)]}} \cong \vec{\mathfrak{X}}^{\mathrm{ord}}_{\Phi_{\mathcal{H}},\delta_{\mathcal{H}},\tau}. \tag{7.3.2.8}$$
By Proposition 7.3.2.5 and by the very constructions, we may identify the pullback of f^{tor} to $(\vec{\mathsf{M}}_{\mathcal{H}}^{\mathrm{ord,tor}})^{\wedge}_{\vec{\mathsf{Z}}^{\mathrm{ord}}_{[(\Phi_{\mathcal{H}},\delta_{\mathcal{H}},\tau)]}}$ with the canonical morphism $\vec{\widetilde{\mathfrak{X}}}^{\mathrm{ord}}_{\breve{\Phi}_{\widetilde{\mathcal{H}}},\breve{\delta}_{\widetilde{\mathcal{H}}},\breve{\sigma},\tau}/\Gamma_{\breve{\Phi}_{\widetilde{\mathcal{H}}},\Phi_{\mathcal{H}}} \to \vec{\mathfrak{X}}^{\mathrm{ord}}_{\Phi_{\mathcal{H}},\delta_{\mathcal{H}},\tau}$. By abuse of notation, we shall also denote this pullback by
$$f^{\mathrm{tor}} : \vec{\widetilde{\mathfrak{X}}}^{\mathrm{ord}}_{\breve{\Phi}_{\widetilde{\mathcal{H}}},\breve{\delta}_{\widetilde{\mathcal{H}}},\breve{\sigma},\tau}/\Gamma_{\breve{\Phi}_{\widetilde{\mathcal{H}}},\Phi_{\mathcal{H}}} \to \vec{\mathfrak{X}}^{\mathrm{ord}}_{\Phi_{\mathcal{H}},\delta_{\mathcal{H}},\tau}.$$

For each $\breve{\tau} \in \widetilde{\Sigma}_{\breve{\Phi}_{\widetilde{\mathcal{H}}},\breve{\sigma},\tau}$, let $\mathfrak{U}_{[\breve{\tau}]}$ denote the image of $\mathfrak{U}_{\breve{\tau}}$ under (7.3.2.4), which is isomorphic to $\mathfrak{U}_{\breve{\tau}}$ as a formal scheme over $\vec{C}^{\mathrm{ord}}_{\Phi_{\mathcal{H}},\delta_{\mathcal{H}}}$. By admissibility of $\Sigma_{\breve{\Phi}_{\widetilde{\mathcal{H}}}}$, we know that the set $\widetilde{\Sigma}_{\breve{\Phi}_{\widetilde{\mathcal{H}}},\breve{\sigma},\tau}/\Gamma_{\breve{\Phi}_{\widetilde{\mathcal{H}}},\Phi_{\mathcal{H}}}$ is *finite*. Then $\vec{\widetilde{\mathfrak{X}}}^{\mathrm{ord}}_{\breve{\Phi}_{\widetilde{\mathcal{H}}},\breve{\delta}_{\widetilde{\mathcal{H}}},\breve{\sigma},\tau}/\Gamma_{\breve{\Phi}_{\widetilde{\mathcal{H}}},\Phi_{\mathcal{H}}}$ can be constructed by gluing the finite collection $\{\mathfrak{U}_{[\breve{\tau}]}\}_{[\breve{\tau}]\in\widetilde{\Sigma}_{\breve{\Phi}_{\widetilde{\mathcal{H}}},\breve{\sigma},\tau}/\Gamma_{\breve{\Phi}_{\widetilde{\mathcal{H}}},\Phi_{\mathcal{H}}}}$ of formal schemes over their intersections. Let us denote by
$$f^{\mathrm{tor}}_{[\breve{\tau}]} : \mathfrak{U}_{[\breve{\tau}]} \to \vec{\mathfrak{X}}^{\mathrm{ord}}_{\Phi_{\mathcal{H}},\delta_{\mathcal{H}},\tau}$$
the restriction of f^{tor} to $\mathfrak{U}_{[\breve{\tau}]}$. If we choose a representative $\breve{\tau}$ of $[\breve{\tau}]$, then we can identify $f^{\mathrm{tor}}_{[\breve{\tau}]} : \mathfrak{U}_{[\breve{\tau}]} \to \vec{\mathfrak{X}}^{\mathrm{ord}}_{\Phi_{\mathcal{H}},\delta_{\mathcal{H}},\tau}$ with the canonical morphism $f^{\mathrm{tor}}_{\breve{\tau}} : \mathfrak{U}_{\breve{\tau}} \to \vec{\mathfrak{X}}^{\mathrm{ord}}_{\Phi_{\mathcal{H}},\delta_{\mathcal{H}},\tau}$ induced by the canonical morphism $\vec{\widetilde{\mathfrak{X}}}^{\mathrm{ord}}_{\breve{\Phi}_{\widetilde{\mathcal{H}}},\breve{\delta}_{\widetilde{\mathcal{H}}},\breve{\sigma},\tau} \to \vec{\mathfrak{X}}^{\mathrm{ord}}_{\Phi_{\mathcal{H}},\delta_{\mathcal{H}},\tau}$. Let us denote by
$$g_{\breve{\tau}} : \mathfrak{U}_{\breve{\tau}} \to \vec{\mathfrak{X}}^{\mathrm{ord}}_{\Phi_{\mathcal{H}},\delta_{\mathcal{H}},\tau} \underset{\vec{C}^{\mathrm{ord}}_{\Phi_{\mathcal{H}},\delta_{\mathcal{H}}}}{\times} \vec{\widetilde{C}}^{\mathrm{ord}}_{\breve{\Phi}_{\widetilde{\mathcal{H}}},\breve{\delta}_{\widetilde{\mathcal{H}}}},$$
$$h : \vec{\widetilde{C}}^{\mathrm{ord}}_{\breve{\Phi}_{\widetilde{\mathcal{H}}},\breve{\delta}_{\widetilde{\mathcal{H}}}} \to \vec{C}^{\mathrm{ord}}_{\Phi_{\mathcal{H}},\delta_{\mathcal{H}}},$$
and
$$h_{\tau} : \vec{\mathfrak{X}}^{\mathrm{ord}}_{\Phi_{\mathcal{H}},\delta_{\mathcal{H}},\tau} \underset{\vec{C}^{\mathrm{ord}}_{\Phi_{\mathcal{H}},\delta_{\mathcal{H}}}}{\times} \vec{\widetilde{C}}^{\mathrm{ord}}_{\breve{\Phi}_{\widetilde{\mathcal{H}}},\breve{\delta}_{\widetilde{\mathcal{H}}}} \to \vec{\mathfrak{X}}^{\mathrm{ord}}_{\Phi_{\mathcal{H}},\delta_{\mathcal{H}},\tau}$$
the canonical morphisms. Then we have a canonical identification $f^{\mathrm{tor}}_{\breve{\tau}} = h_{\tau} \circ g_{\breve{\tau}}$. (Note that $g_{\breve{\tau}}$ is a morphism between affine formal schemes over $\vec{\widetilde{C}}^{\mathrm{ord}}_{\breve{\Phi}_{\widetilde{\mathcal{H}}},\breve{\delta}_{\widetilde{\mathcal{H}}}}$, and that h_{τ} is the pullback of h to the affine formal scheme $\vec{\mathfrak{X}}^{\mathrm{ord}}_{\Phi_{\mathcal{H}},\delta_{\mathcal{H}},\tau}$ over $\vec{C}^{\mathrm{ord}}_{\Phi_{\mathcal{H}},\delta_{\mathcal{H}}}$.)

For simplicity, let us view $\mathscr{O}_{\vec{\mathfrak{X}}^{\mathrm{ord}}_{\Phi_{\mathcal{H}},\delta_{\mathcal{H}},\tau}}$ and $\mathscr{O}_{\vec{\mathsf{Z}}^{\mathrm{ord}}_{[(\Phi_{\mathcal{H}},\delta_{\mathcal{H}},\tau)]}}$ as sheaves over $\vec{C}^{\mathrm{ord}}_{\Phi_{\mathcal{H}},\delta_{\mathcal{H}}}$, and suppress $(\vec{\mathfrak{X}}^{\mathrm{ord}}_{\Phi_{\mathcal{H}},\delta_{\mathcal{H}},\tau} \to \vec{C}^{\mathrm{ord}}_{\Phi_{\mathcal{H}},\delta_{\mathcal{H}}})_*$ and $(\vec{\mathsf{Z}}^{\mathrm{ord}}_{[(\Phi_{\mathcal{H}},\delta_{\mathcal{H}},\tau)]} \to \vec{C}^{\mathrm{ord}}_{\Phi_{\mathcal{H}},\delta_{\mathcal{H}}})_*$ from the notation. For push-forwards (to $\vec{C}^{\mathrm{ord}}_{\Phi_{\mathcal{H}},\delta_{\mathcal{H}}}$) of sheaves over $\vec{\mathfrak{X}}^{\mathrm{ord}}_{\Phi_{\mathcal{H}},\delta_{\mathcal{H}},\tau}$, we shall use the notation $\hat{\oplus}$ to denote the completion with respect to (the push-forward of) the ideal of definition of $\mathscr{O}_{\vec{\mathfrak{X}}^{\mathrm{ord}}_{\Phi_{\mathcal{H}},\delta_{\mathcal{H}},\tau}}$.

Based on Lemma 7.3.2.7, we have the following important facts:

Lemma 7.3.2.9. (*Compare with [61, Lem. 4.6].*)

(1) *For each* $\breve{\tau} \in \widetilde{\Sigma}_{\breve{\Phi}_{\widetilde{\mathcal{H}}},\breve{\sigma},\tau}$, *and each integer* $d \geq 0$, *we have the canonical isomorphisms*

$$R^d(f^{\mathrm{tor}}_{\breve{\tau}})_*\mathscr{O}_{\mathfrak{U}_{\breve{\tau}}} \cong \hat{\bigoplus_{\breve{\ell}\in\breve{\sigma}^\perp\cap\breve{\tau}^\vee}} R^d(h_\tau)_*(\vec{\widetilde{\Psi}}^{\mathrm{ord}}_{\breve{\Phi}_{\widetilde{\mathcal{H}}},\breve{\delta}_{\widetilde{\mathcal{H}}}}(\breve{\ell})),$$

$$R^d(f^{\mathrm{tor}}_{\breve{\tau}})_*\mathscr{O}^+_{\mathfrak{U}_{\breve{\tau}}} \cong \hat{\bigoplus_{\breve{\ell}\in\breve{\sigma}^\perp\cap\breve{\tau}^\vee_{\breve{\sigma},+}}} R^d(h_\tau)_*(\vec{\widetilde{\Psi}}^{\mathrm{ord}}_{\breve{\Phi}_{\widetilde{\mathcal{H}}},\breve{\delta}_{\widetilde{\mathcal{H}}}}(\breve{\ell})),$$

$$R^d(f^{\mathrm{tor}}_{\breve{\tau}})_*\mathscr{O}^{++}_{\mathfrak{U}_{\breve{\tau}}} \cong \hat{\bigoplus_{\breve{\ell}\in\breve{\sigma}^\perp\cap(\breve{\tau}^\vee+\tau^\vee_+)}} R^d(h_\tau)_*(\vec{\widetilde{\Psi}}^{\mathrm{ord}}_{\breve{\Phi}_{\widetilde{\mathcal{H}}},\breve{\delta}_{\widetilde{\mathcal{H}}}}(\breve{\ell})),$$

$$R^d(f^{\mathrm{tor}}_{\breve{\tau}})_*\mathscr{O}^{0+}_{\mathfrak{U}_{\breve{\tau}}} \cong \hat{\bigoplus_{\breve{\ell}\in\breve{\sigma}^\perp\cap\breve{\tau}^\vee_{\breve{\sigma}}}} R^d(h_\tau)_*(\vec{\widetilde{\Psi}}^{\mathrm{ord}}_{\breve{\Phi}_{\widetilde{\mathcal{H}}},\breve{\delta}_{\widetilde{\mathcal{H}}}}(\breve{\ell})),$$

$$R^d(f^{\mathrm{tor}}_{\breve{\tau}})_*\mathscr{O}_{U_{\breve{\tau}}} \cong \bigoplus_{\breve{\ell}\in\breve{\tau}^\vee-\breve{\tau}^\vee_{\breve{\sigma}}} R^d(h_\tau)_*(\vec{\widetilde{\Psi}}^{\mathrm{ord}}_{\breve{\Phi}_{\widetilde{\mathcal{H}}},\breve{\delta}_{\widetilde{\mathcal{H}}}}(\breve{\ell}))$$

of $\mathscr{O}_{\vec{\mathfrak{X}}^{\mathrm{ord}}_{\Phi_{\mathcal{H}},\delta_{\mathcal{H}},\tau}}$*-modules.*

(2) *For each* $\gamma \in \Gamma_{\breve{\Phi}_{\widetilde{\mathcal{H}}},\Phi_{\mathcal{H}}}$, *we have a commutative diagram*

$$\begin{array}{ccc}
\mathfrak{U}_{\breve{\tau}} & \xrightarrow{\gamma} & \mathfrak{U}_{\gamma\breve{\tau}} \\
\downarrow{\scriptstyle g_{\breve{\tau}}} & & \downarrow{\scriptstyle g_{\gamma\breve{\tau}}} \\
\vec{\mathfrak{X}}^{\mathrm{ord}}_{\Phi_{\mathcal{H}},\delta_{\mathcal{H}},\tau} \underset{\vec{C}^{\mathrm{ord}}_{\Phi_{\mathcal{H}},\delta_{\mathcal{H}}}}{\times} \vec{\widetilde{C}}^{\mathrm{ord}}_{\breve{\Phi}_{\widetilde{\mathcal{H}}},\breve{\delta}_{\widetilde{\mathcal{H}}}} & \xrightarrow{\gamma} & \vec{\mathfrak{X}}^{\mathrm{ord}}_{\Phi_{\mathcal{H}},\delta_{\mathcal{H}},\tau} \underset{\vec{C}^{\mathrm{ord}}_{\Phi_{\mathcal{H}},\delta_{\mathcal{H}}}}{\times} \vec{\widetilde{C}}^{\mathrm{ord}}_{\breve{\Phi}_{\widetilde{\mathcal{H}}},\breve{\delta}_{\widetilde{\mathcal{H}}}} \\
\downarrow{\scriptstyle h_\tau} & & \downarrow{\scriptstyle h_\tau} \\
\vec{\mathfrak{X}}^{\mathrm{ord}}_{\Phi_{\mathcal{H}},\delta_{\mathcal{H}},\tau} & = & \vec{\mathfrak{X}}^{\mathrm{ord}}_{\Phi_{\mathcal{H}},\delta_{\mathcal{H}},\tau}
\end{array}$$

of formal schemes, (naturally) compatible with the commutative diagram

$$\begin{array}{ccc}
U_{\breve{\tau}} & \xrightarrow{\gamma} & U_{\gamma\breve{\tau}} \\
\downarrow{\scriptstyle g_{\breve{\tau}}} & & \downarrow{\scriptstyle g_{\gamma\breve{\tau}}} \\
\vec{\Xi}^{\mathrm{ord}}_{\Phi_{\mathcal{H}},\delta_{\mathcal{H}},\tau} \underset{\vec{C}^{\mathrm{ord}}_{\Phi_{\mathcal{H}},\delta_{\mathcal{H}}}}{\times} \vec{\widetilde{C}}^{\mathrm{ord}}_{\breve{\Phi}_{\widetilde{\mathcal{H}}},\breve{\delta}_{\widetilde{\mathcal{H}}}} & \xrightarrow{\gamma} & \vec{\Xi}^{\mathrm{ord}}_{\Phi_{\mathcal{H}},\delta_{\mathcal{H}},\tau} \underset{\vec{C}^{\mathrm{ord}}_{\Phi_{\mathcal{H}},\delta_{\mathcal{H}}}}{\times} \vec{\widetilde{C}}^{\mathrm{ord}}_{\breve{\Phi}_{\widetilde{\mathcal{H}}},\breve{\delta}_{\widetilde{\mathcal{H}}}} \\
\downarrow{\scriptstyle h_\tau} & & \downarrow{\scriptstyle h_\tau} \\
\vec{\Xi}^{\mathrm{ord}}_{\Phi_{\mathcal{H}},\delta_{\mathcal{H}},\tau} & = & \vec{\Xi}^{\mathrm{ord}}_{\Phi_{\mathcal{H}},\delta_{\mathcal{H}},\tau}
\end{array}$$

of their supports. Then the canonical morphisms in (1) *are compatible with the canonical isomorphisms* $\gamma^*\mathscr{O}_{\mathfrak{U}_{\gamma\breve{\tau}}} \to \mathscr{O}_{\mathfrak{U}_{\breve{\tau}}}$, $\gamma^*\mathscr{O}^+_{\mathfrak{U}_{\gamma\breve{\tau}}} \to \mathscr{O}^+_{\mathfrak{U}_{\breve{\tau}}}$, $\gamma^*\mathscr{O}^{++}_{\mathfrak{U}_{\gamma\breve{\tau}}} \to \mathscr{O}^{++}_{\mathfrak{U}_{\breve{\tau}}}$, $\gamma^*\mathscr{O}^{0+}_{\mathfrak{U}_{\gamma\breve{\tau}}} \to \mathscr{O}^{0+}_{\mathfrak{U}_{\breve{\tau}}}$, *and* $\gamma^*\mathscr{O}_{U_{\gamma\breve{\tau}}} \to \mathscr{O}_{U_{\breve{\tau}}}$ *induced by the canonical isomorphisms* $\gamma^* : \gamma^*\vec{\widetilde{\Psi}}^{\mathrm{ord}}_{\breve{\Phi}_{\widetilde{\mathcal{H}}},\breve{\delta}_{\widetilde{\mathcal{H}}}}(\gamma\breve{\ell}) \xrightarrow{\sim} \vec{\widetilde{\Psi}}^{\mathrm{ord}}_{\breve{\Phi}_{\widetilde{\mathcal{H}}},\breve{\delta}_{\widetilde{\mathcal{H}}}}(\breve{\ell})$ *over* $\vec{\widetilde{C}}^{\mathrm{ord}}_{\breve{\Phi}_{\widetilde{\mathcal{H}}},\breve{\delta}_{\widetilde{\mathcal{H}}}}$, *respectively.*

(3) For each integer $d \geq 0$, if $\breve{\tau}'$ is a face of $\breve{\tau}$ in $\widetilde{\Sigma}_{\breve{\Phi}_{\widetilde{\mathcal{H}}},\breve{\sigma},\tau}$, then the canonical morphisms

$$\begin{aligned} R^d(f^{\mathrm{tor}}_{\breve{\tau}})_*\mathscr{O}_{\mathfrak{U}_{\breve{\tau}}} &\to R^d(f^{\mathrm{tor}}_{\breve{\tau}'})_*\mathscr{O}_{\mathfrak{U}_{\breve{\tau}'}}, \\ R^d(f^{\mathrm{tor}}_{\breve{\tau}})_*\mathscr{O}^{+}_{\mathfrak{U}_{\breve{\tau}}} &\to R^d(f^{\mathrm{tor}}_{\breve{\tau}'})_*\mathscr{O}^{+}_{\mathfrak{U}_{\breve{\tau}'}}, \\ R^d(f^{\mathrm{tor}}_{\breve{\tau}})_*\mathscr{O}^{++}_{\mathfrak{U}_{\breve{\tau}}} &\to R^d(f^{\mathrm{tor}}_{\breve{\tau}'})_*\mathscr{O}^{++}_{\mathfrak{U}_{\breve{\tau}'}}, \\ R^d(f^{\mathrm{tor}}_{\breve{\tau}})_*\mathscr{O}^{0+}_{\mathfrak{U}_{\breve{\tau}}} &\to R^d(f^{\mathrm{tor}}_{\breve{\tau}'})_*\mathscr{O}^{0+}_{\mathfrak{U}_{\breve{\tau}'}}, \\ R^d(f^{\mathrm{tor}}_{\breve{\tau}})_*\mathscr{O}_{U_{\breve{\tau}}} &\to R^d(f^{\mathrm{tor}}_{\breve{\tau}'})_*\mathscr{O}_{U_{\breve{\tau}'}} \end{aligned}$$

induced by restriction from $\mathfrak{U}_{\breve{\tau}}$ to $\mathfrak{U}_{\breve{\tau}'}$ correspond to the morphisms

$$\hat{\bigoplus}_{\breve{\ell}\in\breve{\sigma}^\perp\cap\breve{\tau}^\vee} R^d h_*(\vec{\widetilde{\Psi}}^{\mathrm{ord}}_{\breve{\Phi}_{\widetilde{\mathcal{H}}},\breve{\delta}_{\widetilde{\mathcal{H}}}}(\breve{\ell})) \to \hat{\bigoplus}_{\breve{\ell}\in\breve{\sigma}^\perp\cap(\breve{\tau}')^\vee} R^d h_*(\vec{\widetilde{\Psi}}^{\mathrm{ord}}_{\breve{\Phi}_{\widetilde{\mathcal{H}}},\breve{\delta}_{\widetilde{\mathcal{H}}}}(\breve{\ell})),$$

$$\hat{\bigoplus}_{\breve{\ell}\in\breve{\sigma}^\perp\cap\breve{\tau}^\vee_{\breve{\sigma},+}} R^d h_*(\vec{\widetilde{\Psi}}^{\mathrm{ord}}_{\breve{\Phi}_{\widetilde{\mathcal{H}}},\breve{\delta}_{\widetilde{\mathcal{H}}}}(\breve{\ell})) \to \hat{\bigoplus}_{\breve{\ell}\in\breve{\sigma}^\perp\cap(\breve{\tau}')^\vee_{\breve{\sigma},+}} R^d h_*(\vec{\widetilde{\Psi}}^{\mathrm{ord}}_{\breve{\Phi}_{\widetilde{\mathcal{H}}},\breve{\delta}_{\widetilde{\mathcal{H}}}}(\breve{\ell})),$$

$$\hat{\bigoplus}_{\breve{\ell}\in\breve{\sigma}^\perp\cap(\breve{\tau}^\vee+\tau^\vee_+)} R^d h_*(\vec{\widetilde{\Psi}}^{\mathrm{ord}}_{\breve{\Phi}_{\widetilde{\mathcal{H}}},\breve{\delta}_{\widetilde{\mathcal{H}}}}(\breve{\ell})) \to \hat{\bigoplus}_{\breve{\ell}\in\breve{\sigma}^\perp\cap((\breve{\tau}')^\vee+\tau^\vee_+)} R^d h_*(\vec{\widetilde{\Psi}}^{\mathrm{ord}}_{\breve{\Phi}_{\widetilde{\mathcal{H}}},\breve{\delta}_{\widetilde{\mathcal{H}}}}(\breve{\ell})),$$

$$\hat{\bigoplus}_{\breve{\ell}\in\breve{\sigma}^\perp\cap\breve{\tau}^\vee_{\breve{\sigma}}} R^d h_*(\vec{\widetilde{\Psi}}^{\mathrm{ord}}_{\breve{\Phi}_{\widetilde{\mathcal{H}}},\breve{\delta}_{\widetilde{\mathcal{H}}}}(\breve{\ell})) \to \hat{\bigoplus}_{\breve{\ell}\in\breve{\sigma}^\perp\cap(\breve{\tau}')^\vee_{\breve{\sigma}}} R^d h_*(\vec{\widetilde{\Psi}}^{\mathrm{ord}}_{\breve{\Phi}_{\widetilde{\mathcal{H}}},\breve{\delta}_{\widetilde{\mathcal{H}}}}(\breve{\ell})),$$

$$\bigoplus_{\breve{\ell}\in\breve{\tau}^\vee-\breve{\tau}^\vee_{\breve{\sigma}}} R^d h_*(\vec{\widetilde{\Psi}}^{\mathrm{ord}}_{\breve{\Phi}_{\widetilde{\mathcal{H}}},\breve{\delta}_{\widetilde{\mathcal{H}}}}(\breve{\ell})) \to \bigoplus_{\breve{\ell}\in(\breve{\tau}')^\vee-(\breve{\tau}')^\vee_{\breve{\sigma}}} R^d h_*(\vec{\widetilde{\Psi}}^{\mathrm{ord}}_{\breve{\Phi}_{\widetilde{\mathcal{H}}},\breve{\delta}_{\widetilde{\mathcal{H}}}}(\breve{\ell}))$$

of $\mathscr{O}_{\vec{\mathfrak{X}}^{\mathrm{ord}}_{\Phi_{\mathcal{H}},\delta_{\mathcal{H}},\tau}}$-modules, respectively. All of these morphisms send $R^d h_*(\vec{\widetilde{\Psi}}^{\mathrm{ord}}_{\breve{\Phi}_{\widetilde{\mathcal{H}}},\breve{\delta}_{\widetilde{\mathcal{H}}}}(\breve{\ell}))$ (identically) to $R^d h_*(\vec{\widetilde{\Psi}}^{\mathrm{ord}}_{\breve{\Phi}_{\widetilde{\mathcal{H}}},\breve{\delta}_{\widetilde{\mathcal{H}}}}(\breve{\ell}))$ when it is defined on both sides, and to zero otherwise.

7.3.3 *Relative Cohomology and Local Freeness*

The definitions and arguments in this subsection will follow those in [61, Sec. 4B] very closely.

By (7.3.2.8), we shall identify $(\vec{\mathsf{M}}^{\mathrm{ord,tor}}_{\mathcal{H}})^{\wedge}_{\vec{\mathsf{Z}}^{\mathrm{ord}}_{[(\Phi_{\mathcal{H}},\delta_{\mathcal{H}},\tau)]}}$ with $\vec{\mathfrak{X}}^{\mathrm{ord}}_{\Phi_{\mathcal{H}},\delta_{\mathcal{H}},\tau}$, and identify $\vec{\mathsf{Z}}^{\mathrm{ord}}_{[(\Phi_{\mathcal{H}},\delta_{\mathcal{H}},\tau)]}$ with $\vec{\Xi}^{\mathrm{ord}}_{\Phi_{\mathcal{H}},\delta_{\mathcal{H}},\tau}$. For simplicity of notation, we shall use $\vec{\mathfrak{X}}^{\mathrm{ord}}_{\Phi_{\mathcal{H}},\delta_{\mathcal{H}},\tau}$ and $\vec{\mathsf{Z}}^{\mathrm{ord}}_{[(\Phi_{\mathcal{H}},\delta_{\mathcal{H}},\tau)]}$ more often than their counterparts.

Recall that $\vec{C}^{\mathrm{ord}}_{\Phi_{\mathcal{H}},\delta_{\mathcal{H}}}$ (resp. $\vec{\widetilde{C}}^{\mathrm{ord}}_{\widetilde{\Phi}_{\widetilde{\mathcal{H}}},\widetilde{\delta}_{\widetilde{\mathcal{H}}}}$) is a torsor under an abelian scheme $\vec{C}^{\mathrm{ord,grp}}_{\Phi_{\mathcal{H}},\delta_{\mathcal{H}}}$ (resp. $\vec{\widetilde{C}}^{\mathrm{ord,grp}}_{\widetilde{\Phi}_{\widetilde{\mathcal{H}}},\widetilde{\delta}_{\widetilde{\mathcal{H}}}}$) over the finite étale cover $\vec{\mathsf{M}}^{\mathrm{ord},\Phi_{\mathcal{H}}}_{\mathcal{H}}$ (resp. $\vec{\widetilde{\mathsf{M}}}^{\mathrm{ord},\widetilde{\Phi}_{\widetilde{\mathcal{H}}}}_{\widetilde{\mathcal{H}}}$) of $\vec{\mathsf{M}}^{\mathrm{ord},\mathsf{Z}_{\mathcal{H}}}_{\mathcal{H}}$ (resp. $\vec{\widetilde{\mathsf{M}}}^{\mathrm{ord},\widetilde{\mathsf{Z}}_{\widetilde{\mathcal{H}}}}_{\widetilde{\mathcal{H}}}$) (see Section 4.2 and, in particular, Propositions 4.2.1.29 and 4.2.1.30). Since the pairing $\langle\,\cdot\,,\,\cdot\,\rangle^{\sim}$ is the direct sum of the pairings on $Q_{-2}\oplus Q_0$ and on L, we have $\vec{\widetilde{\mathsf{M}}}^{\mathrm{ord},\widetilde{\Phi}_{\widetilde{\mathcal{H}}}}_{\widetilde{\mathcal{H}}} \cong \vec{\mathsf{M}}^{\mathrm{ord},\Phi_{\mathcal{H}}}_{\mathcal{H}}$ and $\vec{\widetilde{\mathsf{M}}}^{\mathrm{ord},\widetilde{\mathsf{Z}}_{\widetilde{\mathcal{H}}}}_{\widetilde{\mathcal{H}}} \cong \vec{\mathsf{M}}^{\mathrm{ord},\mathsf{Z}_{\mathcal{H}}}_{\mathcal{H}}$ (*cf.* Lemmas 5.2.4.1 and

5.2.4.5). Let $(B, \lambda_B, i_B, \varphi_{-1,\mathcal{H}^p}, \varphi^{\mathrm{ord}}_{-1,\mathcal{H}_p})$ and $(\varphi^{\mathrm{ord}}_{-2,\mathcal{H}}, \varphi^{\mathrm{ord}}_{0,\mathcal{H}})$ (resp. $(\widetilde{\varphi}^{\mathrm{ord}}_{-2,\widetilde{\mathcal{H}}}, \widetilde{\varphi}^{\mathrm{ord}}_{0,\widetilde{\mathcal{H}}})$) be the tautological tuples over $\vec{\mathsf{M}}^{\mathrm{ord},\Phi_{\mathcal{H}}}_{\mathcal{H}}$ (resp. $\vec{\mathsf{M}}^{\mathrm{ord},\widetilde{\Phi}_{\widetilde{\mathcal{H}}}}_{\widetilde{\mathcal{H}}}$). Let T (resp. $T^\vee$) be the split torus with character group X (resp. Y). For simplicity of notation, we shall denote the pullbacks of B, $B^\vee$, T, and $T^\vee$, respectively, by the same symbols. The pullback of G (resp. $G^\vee$) to $\vec{\mathfrak{X}}^{\mathrm{ord}}_{\Phi_{\mathcal{H}},\mathsf{Z}_{\mathcal{H}},\tau}$ (as a formal scheme, rather than as a relative scheme as in the case of Mumford families) is an extension of B (resp. $B^\vee$) by T (resp. $T^\vee$), and this extension is a pullback of the tautological extension $G^\natural$ (resp. $G^{\vee,\natural}$) over $\vec{C}^{\mathrm{ord}}_{\Phi_{\mathcal{H}},\delta_{\mathcal{H}}}$. For simplicity, we shall also denote the pullbacks of $G^\natural$ and $G^{\vee,\natural}$, respectively, by the same symbols. By Lemma 7.1.2.1, the morphism $h : \vec{\widetilde{C}}^{\mathrm{ord}}_{\breve{\Phi}_{\widetilde{\mathcal{H}}},\breve{\delta}_{\widetilde{\mathcal{H}}}} \to \vec{C}^{\mathrm{ord}}_{\Phi_{\mathcal{H}},\delta_{\mathcal{H}}}$ is proper and smooth, and is a torsor under the pullback to $\vec{C}^{\mathrm{ord}}_{\Phi_{\mathcal{H}},\delta_{\mathcal{H}}}$ of an abelian scheme $\mathbb{Q}^\times$-isogenous to $\underline{\mathrm{Hom}}_{\mathcal{O}}(\widetilde{X}, B)^\circ \to \vec{\mathsf{M}}^{\mathrm{ord},\Phi_{\mathcal{H}}}_{\mathcal{H}}$.

Consider the union $\widetilde{\mathfrak{N}}_{\breve{\sigma},\tau}$ of the cones $\breve{\tau}$ in $\widetilde{\Sigma}_{\breve{\Phi}_{\widetilde{\mathcal{H}}},\breve{\sigma},\tau}$, which has a *closed covering* by the closures $\breve{\tau}^{\mathrm{cl}}$ (in $\widetilde{\mathfrak{N}}_{\breve{\sigma},\tau}$) of the cones $\breve{\tau}$ in $\widetilde{\Sigma}_{\breve{\Phi}_{\widetilde{\mathcal{H}}},\breve{\sigma},\tau}$ (with natural incidence relations inherited from those of the cones $\breve{\tau}$ as locally closed subsets of $(\mathbf{S}_{\breve{\Phi}_{\widetilde{\mathcal{H}}}})^\vee_{\mathbb{R}}$). By definition, the nerve of the open covering $\{\mathfrak{U}_{\breve{\tau}}\}_{\breve{\tau}\in\widetilde{\Sigma}_{\breve{\Phi}_{\widetilde{\mathcal{H}}},\breve{\sigma},\tau}}$ of $\vec{\widetilde{\mathfrak{X}}}^{\mathrm{ord}}_{\breve{\Phi}_{\widetilde{\mathcal{H}}},\breve{\delta}_{\widetilde{\mathcal{H}}},\breve{\sigma},\tau}$, or equivalently the open covering $\{U_{\breve{\tau}}\}_{\breve{\tau}\in\widetilde{\Sigma}_{\breve{\Phi}_{\widetilde{\mathcal{H}}},\breve{\sigma},\tau}}$ of $\vec{\widetilde{\Xi}}^{\mathrm{ord}}_{\breve{\Phi}_{\widetilde{\mathcal{H}}},\breve{\delta}_{\widetilde{\mathcal{H}}},\breve{\sigma}}(\tau)$ (by the supports of the formal schemes $\{\mathfrak{U}_{\breve{\tau}}\}_{\breve{\tau}\in\widetilde{\Sigma}_{\breve{\Phi}_{\widetilde{\mathcal{H}}},\breve{\sigma},\tau}}$), is naturally identified with the nerve of the closed covering $\{\breve{\tau}^{\mathrm{cl}}\}_{\breve{\tau}\in\widetilde{\Sigma}_{\breve{\Phi}_{\widetilde{\mathcal{H}}},\breve{\sigma},\tau}}$ of $\widetilde{\mathfrak{N}}_{\breve{\sigma},\tau}$.

Remark 7.3.3.1. Our description of $\widetilde{\mathfrak{N}}_{\breve{\sigma},\tau}$ here differs from that in [61, Sec. 4B]. (The description in [61, Sec. 4B] is misleading because it abusively identifies the homology of the nerve with the cohomology of the dual one realized by the unions of closures of cones. We take this opportunity to present the clarified and corrected exposition here.)

Accordingly, if we set

$$\mathfrak{N}_{\breve{\sigma},\tau} := \widetilde{\mathfrak{N}}_{\breve{\sigma},\tau}/\Gamma_{\breve{\Phi}_{\widetilde{\mathcal{H}}},\Phi_{\mathcal{H}}},$$

and let $[\breve{\tau}]^{\mathrm{cl}}$ denote the closure of $[\breve{\tau}]$ in $\mathfrak{N}_{\breve{\sigma},\tau}$, for each $[\breve{\tau}] \in \widetilde{\Sigma}_{\breve{\Phi}_{\widetilde{\mathcal{H}}},\breve{\sigma},\tau}/\Gamma_{\breve{\Phi}_{\widetilde{\mathcal{H}}},\Phi_{\mathcal{H}}}$. Then the nerve of the open covering

$$\{\mathfrak{U}_{[\breve{\tau}]}\}_{[\breve{\tau}]\in\widetilde{\Sigma}_{\breve{\Phi}_{\widetilde{\mathcal{H}}},\breve{\sigma},\tau}/\Gamma_{\breve{\Phi}_{\widetilde{\mathcal{H}}},\Phi_{\mathcal{H}}}} \tag{7.3.3.2}$$

of $(\vec{\mathsf{N}}^{\mathrm{ord,tor}})^\wedge_{\vec{\mathsf{Z}}^{\mathrm{ord}}_{[(\Phi_{\mathcal{H}},\delta_{\mathcal{H}},\tau)]}} \cong \vec{\widetilde{\mathfrak{X}}}^{\mathrm{ord}}_{\breve{\Phi}_{\widetilde{\mathcal{H}}},\breve{\delta}_{\widetilde{\mathcal{H}}},\breve{\sigma},\tau}/\Gamma_{\breve{\Phi}_{\widetilde{\mathcal{H}}},\Phi_{\mathcal{H}}}$, or equivalently the open covering

$$\{U_{\breve{\tau}}\}_{[\breve{\tau}]\in\widetilde{\Sigma}_{\breve{\Phi}_{\widetilde{\mathcal{H}}},\breve{\sigma},\tau}/\Gamma_{\breve{\Phi}_{\widetilde{\mathcal{H}}},\Phi_{\mathcal{H}}}}$$

of $\vec{\widetilde{\mathsf{Z}}}^{\mathrm{ord}}_{[(\Phi_{\mathcal{H}},\delta_{\mathcal{H}},\tau)]} \cong \vec{\widetilde{\Xi}}^{\mathrm{ord}}_{\breve{\Phi}_{\widetilde{\mathcal{H}}},\breve{\delta}_{\widetilde{\mathcal{H}}},\tau}/\Gamma_{\breve{\Phi}_{\widetilde{\mathcal{H}}},\Phi_{\mathcal{H}}}$ of the supports of formal schemes, is naturally identified with the nerve of the closed covering

$$\{[\breve{\tau}]^{\mathrm{cl}}\}_{[\breve{\tau}]\in\widetilde{\Sigma}_{\breve{\Phi}_{\widetilde{\mathcal{H}}},\breve{\sigma},\tau}/\Gamma_{\breve{\Phi}_{\widetilde{\mathcal{H}}},\Phi_{\mathcal{H}}}} \tag{7.3.3.3}$$

of $\mathfrak{N}_{\breve{\sigma},\tau}$.

Let us analyze the topological structure of $\widetilde{\mathfrak{N}}_{\breve{\sigma},\tau}$ in closer detail. By choosing some (noncanonical) splitting of $s_{\breve{X}} \underset{\mathbb{Z}}{\otimes} \mathbb{Q} : \breve{X} \underset{\mathbb{Z}}{\otimes} \mathbb{Q} \twoheadrightarrow \widetilde{X} \underset{\mathbb{Z}}{\otimes} \mathbb{Q}$ (over $\mathbb{Q}$), we obtain the decomposition (1.2.4.31) of $(\mathbf{S}_{\breve{\Phi}_{\widehat{\mathcal{H}}}})^\vee_{\mathbb{R}}$, inducing the projection $\mathrm{pr}_{(\widehat{\mathbf{S}}_{\breve{\Phi}_{\widehat{\mathcal{H}}}})^\vee_{\mathbb{R}}}$ in (1.2.4.32) (defined over $\mathbb{Q}$). Note that $\breve{\sigma}^\perp = \widehat{\mathbf{S}}_{\breve{\Phi}_{\widehat{\mathcal{H}}}}$ (see Definition 1.2.4.29) because $\widetilde{\sigma}$ is a top-dimensional cone in $\mathbf{P}^+_{\widetilde{\Phi}_{\widetilde{\mathcal{H}}}}$. Then Lemmas 1.2.4.38 and 1.2.4.39 also imply the following:

Corollary 7.3.3.4. (*Compare with Corollary* 1.2.4.40.) *The set*

$$\{\mathrm{pr}_{(\widehat{\mathbf{S}}_{\breve{\Phi}_{\widehat{\mathcal{H}}}})^\vee_{\mathbb{R}}}(\breve{\tau})\}_{\breve{\tau}\in\widetilde{\Sigma}_{\breve{\Phi}_{\widehat{\mathcal{H}}},\breve{\sigma},\tau}}$$

of rational polyhedral cones defines a $\Gamma_{\breve{\Phi}_{\widehat{\mathcal{H}}},\Phi_{\mathcal{H}}}$**-admissible rational polyhedral cone decomposition** (*cf. Definition* 1.2.2.4) *of*

$$\mathrm{pr}_{(\widehat{\mathbf{S}}_{\breve{\Phi}_{\widehat{\mathcal{H}}}})^\vee_{\mathbb{R}}}(\widetilde{\mathfrak{N}}_{\breve{\sigma},\tau}) = \bigcup_{\breve{\tau}\in\widetilde{\Sigma}_{\breve{\Phi}_{\widehat{\mathcal{H}}},\breve{\sigma},\tau}} \left(\mathrm{pr}_{(\widehat{\mathbf{S}}_{\breve{\Phi}_{\widehat{\mathcal{H}}}})^\vee_{\mathbb{R}}}(\breve{\tau})\right) \tag{7.3.3.5}$$

in the sense that we have the following:

(1) *Every* $\mathrm{pr}_{(\widehat{\mathbf{S}}_{\breve{\Phi}_{\widehat{\mathcal{H}}}})^\vee_{\mathbb{R}}}(\breve{\tau})$ *is a nondegenerate rational polyhedral cone.*
(2) *The union* (7.3.3.5) *is disjoint and defines a stratification of* $\mathrm{pr}_{(\widehat{\mathbf{S}}_{\breve{\Phi}_{\widehat{\mathcal{H}}}})^\vee_{\mathbb{R}}}(\widetilde{\mathfrak{N}}_{\breve{\sigma},\tau})$.
(3) $\{\mathrm{pr}_{(\widehat{\mathbf{S}}_{\breve{\Phi}_{\widehat{\mathcal{H}}}})^\vee_{\mathbb{R}}}(\breve{\tau})\}_{\breve{\tau}\in\widetilde{\Sigma}_{\breve{\Phi}_{\widehat{\mathcal{H}}},\breve{\sigma},\tau}}$ *is invariant under the action of* $\Gamma_{\breve{\Phi}_{\widehat{\mathcal{H}}},\Phi_{\mathcal{H}}}$ *in the sense that* $\Gamma_{\breve{\Phi}_{\widehat{\mathcal{H}}},\Phi_{\mathcal{H}}}$ *permutes the cones in it. Under this action, the set of* $\Gamma_{\breve{\Phi}_{\widehat{\mathcal{H}}},\Phi_{\mathcal{H}}}$*-orbits is finite.*

Proof. The same argument of the proof of Corollary 1.2.4.40 also works here (see Definition 7.3.2.2). □

Corollary 7.3.3.6. *The projection* $\mathrm{pr}_{(\widehat{\mathbf{S}}_{\breve{\Phi}_{\widehat{\mathcal{H}}}})^\vee_{\mathbb{R}}}$ *in* (1.2.4.32) *induces a homotopy equivalence from* $\widetilde{\mathfrak{N}}_{\breve{\sigma},\tau}$ *to* $\mathrm{pr}_{(\widehat{\mathbf{S}}_{\breve{\Phi}_{\widehat{\mathcal{H}}}})^\vee_{\mathbb{R}}}(\widetilde{\mathfrak{N}}_{\breve{\sigma},\tau})$.

Proof. Any continuous section $\tilde{x}_0$ as in Lemma 1.2.4.39 defines a continuous map $\mathrm{pr}_{(\widehat{\mathbf{S}}_{\breve{\Phi}_{\widehat{\mathcal{H}}}})^\vee_{\mathbb{R}}}(\widetilde{\mathfrak{N}}_{\breve{\sigma},\tau}) \to \widetilde{\mathfrak{N}}_{\breve{\sigma},\tau}$ whose pre- and post-compositions with $\mathrm{pr}_{(\widehat{\mathbf{S}}_{\breve{\Phi}_{\widehat{\mathcal{H}}}})^\vee_{\mathbb{R}}}$ are homotopic to the identity morphisms. Hence, the corollary follows. □

Lemma 7.3.3.7. *The preimage of* $\mathrm{pr}_{(\widehat{\mathbf{S}}_{\breve{\Phi}_{\widehat{\mathcal{H}}}})^\vee_{\mathbb{R}}}(\widetilde{\mathfrak{N}}_{\breve{\sigma},\tau})$ *under the identification* $(\Gamma_{\breve{\Phi}_{\widehat{\mathcal{H}}},\Phi_{\mathcal{H}}})^\vee_{\mathbb{R}} \oplus (\mathbf{S}_{\Phi_{\mathcal{H}}})^\vee_{\mathbb{R}} \cong (\widehat{\mathbf{S}}_{\breve{\Phi}_{\widehat{\mathcal{H}}}})^\vee_{\mathbb{R}}$ (*in* (1.2.4.32), *induced by* (1.2.4.31)) *is the subset* $(\Gamma_{\breve{\Phi}_{\widehat{\mathcal{H}}},\Phi_{\mathcal{H}}})^\vee_{\mathbb{R}} \times \tau$.

Proof. By Condition 7.1.1.17, we know that $\breve{\tau} \subset (\Gamma_{\breve{\Phi}_{\widehat{\mathcal{H}}},\Phi_{\mathcal{H}}})^\vee_{\mathbb{R}} \times \tau$ for every $\breve{\tau} \in \widetilde{\Sigma}_{\breve{\Phi}_{\widehat{\mathcal{H}}},\breve{\sigma},\tau}$. The question is whether every point of $(\Gamma_{\breve{\Phi}_{\widehat{\mathcal{H}}},\Phi_{\mathcal{H}}})^\vee_{\mathbb{R}} \times \tau$ is contained in $\breve{\tau}'$

for some $\check{\tau} \in \widetilde{\Sigma}_{\check{\Phi}_{\widetilde{\mathcal{H}}},\check{\sigma},\tau}$. By definition of $\widetilde{\mathfrak{N}}_{\check{\sigma},\tau}$ as the union of the cones $\check{\tau}$ in $\widetilde{\Sigma}_{\check{\Phi}_{\widetilde{\mathcal{H}}},\check{\sigma},\tau}$ (see Definition 7.3.2.2), the answer is in the affirmative, because of the following elementary fact: If $\begin{pmatrix} 0 & y \\ {}^t y & z \end{pmatrix}$ is a (partitioned) real symmetric matrix with entries in $\mathbb{R}$ such that z is positive definite, and if x_0 is a particular positive definite matrix with entries in $\mathbb{R}$, then there exists a sufficiently large real number t_0 such that $\begin{pmatrix} tx_0 & y \\ {}^t y & z \end{pmatrix}$ is positive definite for all $t \geq t_0$. □

Lemma 7.3.3.8. *The projection*

$$\mathrm{pr}_{(\Gamma_{\check{\Phi}_{\widetilde{\mathcal{H}}},\Phi_{\mathcal{H}}})^\vee_{\mathbb{R}}} : (\Gamma_{\check{\Phi}_{\widetilde{\mathcal{H}}},\Phi_{\mathcal{H}}})^\vee_{\mathbb{R}} \oplus (\mathbf{S}_{\Phi_{\mathcal{H}}})^\vee_{\mathbb{R}} \to (\Gamma_{\check{\Phi}_{\widetilde{\mathcal{H}}},\Phi_{\mathcal{H}}})^\vee_{\mathbb{R}} : (y,z) \mapsto y$$

induces a homotopy equivalence from $(\Gamma_{\check{\Phi}_{\widetilde{\mathcal{H}}},\Phi_{\mathcal{H}}})^\vee_{\mathbb{R}} \times \tau$ *to* $(\Gamma_{\check{\Phi}_{\widetilde{\mathcal{H}}},\Phi_{\mathcal{H}}})^\vee_{\mathbb{R}}$.

Proof. By choosing any point z_0 in τ, we obtain a continuous section

$$(\Gamma_{\check{\Phi}_{\widetilde{\mathcal{H}}},\Phi_{\mathcal{H}}})^\vee_{\mathbb{R}} \to (\Gamma_{\check{\Phi}_{\widetilde{\mathcal{H}}},\Phi_{\mathcal{H}}})^\vee_{\mathbb{R}} \times \tau : y \mapsto (y, z_0)$$

whose pre- and post-compositions with $\mathrm{pr}_{(\Gamma_{\check{\Phi}_{\widetilde{\mathcal{H}}},\Phi_{\mathcal{H}}})^\vee_{\mathbb{R}}}|_{((\Gamma_{\check{\Phi}_{\widetilde{\mathcal{H}}},\Phi_{\mathcal{H}}})^\vee_{\mathbb{R}} \times \tau)}$ are homotopic to the identity morphisms. Hence, the lemma follows, as desired. □

Lemma 7.3.3.9. (*Compare with [61, Lem. 4.21].*) *The topological space* $\mathfrak{N}_{\check{\sigma},\tau}$ *is homotopic to the real torus*

$$\mathbf{T}_{\widetilde{\Phi}_{\widetilde{\mathcal{H}}},\Phi_{\mathcal{H}}} := (\Gamma_{\check{\Phi}_{\widetilde{\mathcal{H}}},\Phi_{\mathcal{H}}})^\vee_{\mathbb{R}} / \Gamma_{\check{\Phi}_{\widetilde{\mathcal{H}}},\Phi_{\mathcal{H}}},$$

whose cohomology groups (by contractibility of $(\Gamma_{\check{\Phi}_{\widetilde{\mathcal{H}}},\Phi_{\mathcal{H}}})^\vee_{\mathbb{R}}$*) are*

$$H^j(\mathbf{T}_{\widetilde{\Phi}_{\widetilde{\mathcal{H}}},\Phi_{\mathcal{H}}}, \mathbb{Z}) \cong H^j(\Gamma_{\check{\Phi}_{\widetilde{\mathcal{H}}},\Phi_{\mathcal{H}}}, \mathbb{Z}) \cong \wedge^j(\mathrm{Hom}_{\mathbb{Z}}(\Gamma_{\check{\Phi}_{\widetilde{\mathcal{H}}},\Phi_{\mathcal{H}}}, \mathbb{Z}))$$

for each integer $j \geq 0$. *Over* $\vec{C}^{\mathrm{ord}}_{\Phi_{\mathcal{H}},\delta_{\mathcal{H}}} \underset{\mathbb{Z}}{\otimes} \mathbb{Q}$, *we have a canonical isomorphism*

$$H^j(\Gamma_{\check{\Phi}_{\widetilde{\mathcal{H}}},\Phi_{\mathcal{H}}}, \mathbb{Z}) \underset{\mathbb{Z}}{\otimes} \mathscr{O}_{\vec{C}^{\mathrm{ord}}_{\Phi_{\mathcal{H}},\delta_{\mathcal{H}}}} \underset{\mathbb{Z}}{\otimes} \mathbb{Q} \cong \wedge^j(\mathrm{Hom}_{\mathscr{O}}(Q^\vee, \underline{\mathrm{Lie}}_{T^\vee/\vec{C}^{\mathrm{ord}}_{\Phi_{\mathcal{H}},\delta_{\mathcal{H}}}})) \underset{\mathbb{Z}}{\otimes} \mathbb{Q}. \quad (7.3.3.10)$$

Proof. By Corollary 7.3.3.6 and Lemma 7.3.3.8, the projection

$$(\mathbf{S}_{\widetilde{\Phi}_{\widetilde{\mathcal{H}}}})^\vee_{\mathbb{R}} \oplus (\Gamma_{\check{\Phi}_{\widetilde{\mathcal{H}}},\Phi_{\mathcal{H}}})^\vee_{\mathbb{R}} \oplus (\mathbf{S}_{\Phi_{\mathcal{H}}})^\vee_{\mathbb{R}} \to (\Gamma_{\check{\Phi}_{\widetilde{\mathcal{H}}},\Phi_{\mathcal{H}}})^\vee_{\mathbb{R}} : (x,y,z) \mapsto y$$

defines a homotopy equivalence from $\widetilde{\mathfrak{N}}_{\check{\sigma},\tau}$ to $(\Gamma_{\check{\Phi}_{\widetilde{\mathcal{H}}},\Phi_{\mathcal{H}}})^\vee_{\mathbb{R}}$. This homotopy equivalence (defined by projection) is compatible with the action of $(\Gamma_{\check{\Phi}_{\widetilde{\mathcal{H}}},\Phi_{\mathcal{H}}})^\vee_{\mathbb{R}}$, because the action is defined by translations on the second factor $(\Gamma_{\check{\Phi}_{\widetilde{\mathcal{H}}},\Phi_{\mathcal{H}}})^\vee_{\mathbb{R}}$. Therefore, $\mathfrak{N}_{\check{\sigma},\tau} = \widetilde{\mathfrak{N}}_{\check{\sigma},\tau} / \Gamma_{\check{\Phi}_{\widetilde{\mathcal{H}}},\Phi_{\mathcal{H}}}$ is homotopic to the real torus $\mathbf{T}_{\widetilde{\Phi}_{\widetilde{\mathcal{H}}},\Phi_{\mathcal{H}}} = (\Gamma_{\check{\Phi}_{\widetilde{\mathcal{H}}},\Phi_{\mathcal{H}}})^\vee_{\mathbb{R}} / \Gamma_{\check{\Phi}_{\widetilde{\mathcal{H}}},\Phi_{\mathcal{H}}}$.

The canonical isomorphism (7.3.3.10) then follows from the composition of the following canonical isomorphisms

$$\begin{aligned}
&H^j(\Gamma_{\breve{\Phi}_{\widetilde{\mathcal{H}}},\Phi_{\mathcal{H}}}, \mathbb{Z}) \underset{\mathbb{Z}}{\otimes} \mathscr{O}_{\vec{C}^{\mathrm{ord}}_{\Phi_{\mathcal{H}},\delta_{\mathcal{H}}}} \underset{\mathbb{Z}}{\otimes} \mathbb{Q}\\
&\cong (\wedge^j(\mathrm{Hom}_{\mathbb{Z}}(\Gamma_{\breve{\Phi}_{\widetilde{\mathcal{H}}},\Phi_{\mathcal{H}}}, \mathbb{Z}))) \underset{\mathbb{Z}}{\otimes} \mathscr{O}_{\vec{C}^{\mathrm{ord}}_{\Phi_{\mathcal{H}},\delta_{\mathcal{H}}}} \underset{\mathbb{Z}}{\otimes} \mathbb{Q}\\
&\cong (\wedge^j(\mathrm{Hom}_{\mathbb{Z}}(\mathrm{Hom}_{\mathcal{O}}(\widetilde{X}, X), \mathbb{Z}))) \underset{\mathbb{Z}}{\otimes} \mathscr{O}_{\vec{C}^{\mathrm{ord}}_{\Phi_{\mathcal{H}},\delta_{\mathcal{H}}}} \underset{\mathbb{Z}}{\otimes} \mathbb{Q}\\
&\cong \wedge^j(\mathrm{Hom}_{\mathcal{O}}(Q^\vee, \mathrm{Hom}_{\mathbb{Z}}(Y, \mathscr{O}_{\vec{C}^{\mathrm{ord}}_{\Phi_{\mathcal{H}},\delta_{\mathcal{H}}}}))) \underset{\mathbb{Z}}{\otimes} \mathbb{Q}\\
&\cong \wedge^j(\mathrm{Hom}_{\mathcal{O}}(Q^\vee, \underline{\mathrm{Lie}}_{T^\vee/\vec{C}^{\mathrm{ord}}_{\Phi_{\mathcal{H}},\delta_{\mathcal{H}}}})) \underset{\mathbb{Z}}{\otimes} \mathbb{Q}
\end{aligned}$$

induced by the canonical isomorphisms

$$\Gamma_{\breve{\Phi}_{\widetilde{\mathcal{H}}},\Phi_{\mathcal{H}}} \underset{\mathbb{Z}}{\otimes} \mathbb{Q} \cong \mathrm{Hom}_{\mathcal{O}}(\widetilde{X}, X) \underset{\mathbb{Z}}{\otimes} \mathbb{Q} \cong \mathrm{Hom}_{\mathcal{O}}(Q, Y) \underset{\mathbb{Z}}{\otimes} \mathbb{Q},$$

as desired. □

Lemma 7.3.3.11. *Let $\breve{\tau} = \mathbb{R}_{>0}v_1 + \cdots + \mathbb{R}_{>0}v_n$ be a nonzero smooth nondegenerate rational polyhedral cone in $(\mathbf{S}_{\breve{\Phi}_{\widetilde{\mathcal{H}}}})^\vee_{\mathbb{R}}$, where $v_1, \ldots, v_n$ are nonzero vectors, and let* **K** *be a cone in $(\mathbf{S}_{\breve{\Phi}_{\widetilde{\mathcal{H}}}})^\vee_{\mathbb{R}}$ (i.e., a subset stable under the multiplicative action of $\mathbb{R}^\times_{>0}$) such that $0 \notin \mathbf{K}$ and $\overline{\breve{\tau}}_{\mathbf{K}} := \overline{\breve{\tau}} \cap \mathbf{K}$ is convex. (Here $\overline{\breve{\tau}}$ is the closure of $\breve{\tau}$ in $(\mathbf{S}_{\breve{\Phi}_{\widetilde{\mathcal{H}}}})^\vee_{\mathbb{R}}$.) Up to reordering $v_1, \ldots, v_n$ if necessary, suppose moreover that, for some* **nonzero** *$m \leq n$, we have*

$$(\mathbb{R}_{\geq 0}v_1 + \cdots + \mathbb{R}_{\geq 0}v_m) - \{0\} \subset \mathbf{K} \tag{7.3.3.12}$$

but

$$(\mathbb{R}_{\geq 0}v_{m+1} + \cdots + \mathbb{R}_{\geq 0}v_n) \cap \mathbf{K} = \emptyset. \tag{7.3.3.13}$$

In this case, $\breve{\tau}' := \mathbb{R}_{>0}v_1 + \cdots + \mathbb{R}_{>0}v_m$ is the largest face of $\breve{\tau}$ such that its closure $\overline{\breve{\tau}'}$ (in $(\mathbf{S}_{\breve{\Phi}_{\widetilde{\mathcal{H}}}})^\vee_{\mathbb{R}}$) satisfies $\overline{\breve{\tau}'} - \{0\} \subset \mathbf{K}$ (so that $\overline{\breve{\tau}'} - \{0\} \subset \overline{\breve{\tau}}_{\mathbf{K}}$). Consider the continuous map

$$F : [0, 1] \times \overline{\breve{\tau}}_{\mathbf{K}} \to \overline{\breve{\tau}}_{\mathbf{K}}$$

defined by sending

$$(t, x_1v_1 + \cdots + x_mv_m + x_{m+1}v_{m+1} + \cdots + x_nv_n)$$

to

$$x_1v_1 + \cdots + x_mv_m + (1 - t)x_{m+1}v_{m+1} + \cdots + (1 - t)x_nv_n.$$

Then F defines a deformation retract from $\overline{\breve{\tau}}_{\mathbf{K}}$ to its subset $\overline{\breve{\tau}'} - \{0\}$. The construction of F is compatible with restrictions to faces $\breve{\tau}''$ of $\breve{\tau}$ that still satisfy the condition of this lemma.

Proof. The statements are self-explanatory. (The condition (7.3.3.12) is needed for the compatibility with restrictions to faces. The condition (7.3.3.13) is needed for the deformation retract F to be defined at $t = 1$.) □

Definition 7.3.3.14. *For each* $\breve{\ell} \in \mathbf{S}_{\breve{\Phi}_{\widetilde{\mathcal{H}}}}$, *define the following subsets of* $\widetilde{\mathfrak{N}}_{\breve{\sigma},\tau}$:

(1) $\widetilde{\mathfrak{N}}^{\breve{\ell}}_{\breve{\sigma},\tau}$ *is the union of* $\breve{\tau} \in \widetilde{\Sigma}_{\breve{\Phi}_{\widetilde{\mathcal{H}}},\breve{\sigma},\tau}$ *such that* $\breve{\ell} \in \breve{\sigma}^{\perp} \cap \breve{\tau}^{\vee}$.
(2) $\widetilde{\mathfrak{N}}^{\breve{\ell},+}_{\breve{\sigma},\tau}$ *is the union of* $\breve{\tau} \in \widetilde{\Sigma}_{\breve{\Phi}_{\widetilde{\mathcal{H}}},\breve{\sigma},\tau}$ *such that* $\breve{\ell} \in \breve{\sigma}^{\perp} \cap \breve{\tau}^{\vee}_{\breve{\sigma},+}$.
(3) $\widetilde{\mathfrak{N}}^{\breve{\ell},0+}_{\breve{\sigma},\tau}$ *is the union of* $\breve{\tau} \in \widetilde{\Sigma}_{\breve{\Phi}_{\widetilde{\mathcal{H}}},\breve{\sigma},\tau}$ *such that* $\breve{\ell} \in \breve{\sigma}^{\perp} \cap \breve{\tau}^{\vee}_{\breve{\sigma}}$.

Lemma 7.3.3.15. *Suppose there exist nonzero rational vectors* $v_1, \ldots, v_n$ *in* $(\mathbf{S}_{\breve{\Phi}_{\widetilde{\mathcal{H}}}})^{\vee}_{\mathbb{R}}$ *such that*

$$\breve{\sigma} = \mathbb{R}_{>0}v_1 + \cdots + \mathbb{R}_{>0}v_m,$$

and such that

$$\breve{\tau} = \mathbb{R}_{>0}v_1 + \cdots + \mathbb{R}_{>0}v_m + \mathbb{R}_{>0}v_{m+1} + \cdots + \mathbb{R}_{>0}v_n$$

is a cone in $\widetilde{\Sigma}_{\breve{\Phi}_{\widetilde{\mathcal{H}}},\breve{\sigma},\tau}$. *Then we have the following criteria:*

(1) $\breve{\ell} \in \breve{\sigma}^{\perp} \cap \breve{\tau}^{\vee}$ *if and only if* $\langle \breve{\ell}, v_i \rangle \geq 0$ *for all* $m+1 \leq i \leq n$.
(2) $\breve{\ell} \in \breve{\sigma}^{\perp} \cap \breve{\tau}^{\vee}_{\breve{\sigma},+}$ *if and only if* $\langle \breve{\ell}, v_i \rangle > 0$ *for all* $m+1 \leq i \leq n$.
(3) $\breve{\ell} \in \breve{\sigma}^{\perp} \cap \breve{\tau}^{\vee}_{\breve{\sigma}}$ *if and only if* $\langle \breve{\ell}, v_i \rangle \geq 0$ *for all* $m+1 \leq i \leq n$ *and* $\langle \breve{\ell}, v_i \rangle > 0$ *for all* $m+1 \leq i \leq n$ *such that* $v_i \in \mathbf{P}^{+}_{\breve{\Phi}_{\widetilde{\mathcal{H}}}}$ *and such that the image of* v_i *under the (canonical) second morphism in* (1.2.4.20) *is contained in* $\tau \subset \mathbf{P}^{+}_{\Phi_{\mathcal{H}}}$.

Proof. These follow immediately from the definitions. (See Definitions 7.3.2.2, 7.3.2.3, and 7.3.3.14.) □

Proposition 7.3.3.16. *For each* $\breve{\ell} \in \mathbf{S}_{\breve{\Phi}_{\widetilde{\mathcal{H}}}}$, *the subsets* $\widetilde{\mathfrak{N}}^{\breve{\ell}}_{\breve{\sigma},\tau}$, $\widetilde{\mathfrak{N}}^{\breve{\ell},+}_{\breve{\sigma},\tau}$, *and* $\widetilde{\mathfrak{N}}^{\breve{\ell},0+}_{\breve{\sigma},\tau}$ *of* $\widetilde{\mathfrak{N}}_{\breve{\sigma},\tau}$ *(in Definition* 7.3.3.14*) all have contractible or empty complements in* $\widetilde{\mathfrak{N}}_{\breve{\sigma},\tau}$.

Proof. We may and we shall assume that $\breve{\ell} \in \breve{\sigma}^{\perp}$, because otherwise all subsets in question will be empty (and the lemma becomes trivial). Since $\breve{\ell} \in \breve{\sigma}^{\perp}$, the conditions for each cone $\breve{\tau} \in \widetilde{\Sigma}_{\breve{\Phi}_{\widetilde{\mathcal{H}}},\breve{\sigma},\tau}$ to be in each of the four subsets (see Definition 7.3.3.14) depend only on the image $\widehat{\tau} = \mathrm{pr}_{(\widehat{\mathbf{S}}_{\breve{\Phi}_{\widehat{\mathcal{H}}}})^{\vee}_{\mathbb{R}}}(\breve{\tau})$ under the projection $\mathrm{pr}_{(\widehat{\mathbf{S}}_{\breve{\Phi}_{\widehat{\mathcal{H}}}})^{\vee}_{\mathbb{R}}}$ in (1.2.4.32). By Lemma 1.2.4.39, as in Corollary 7.3.3.6 (and its proof), the projection $\mathrm{pr}_{(\widehat{\mathbf{S}}_{\breve{\Phi}_{\widehat{\mathcal{H}}}})^{\vee}_{\mathbb{R}}}$ induces homotopy equivalences from $\widetilde{\mathfrak{N}}_{\breve{\sigma},\tau}$, $\widetilde{\mathfrak{N}}^{\breve{\ell}}_{\breve{\sigma},\tau}$, $\widetilde{\mathfrak{N}}^{\breve{\ell},+}_{\breve{\sigma},\tau}$, $\widetilde{\mathfrak{N}}^{\breve{\ell},0+}_{\breve{\sigma},\tau}$, $\widetilde{\mathfrak{N}}_{\breve{\sigma},\tau} - \widetilde{\mathfrak{N}}^{\breve{\ell}}_{\breve{\sigma},\tau}$, $\widetilde{\mathfrak{N}}_{\breve{\sigma},\tau} - \widetilde{\mathfrak{N}}^{\breve{\ell},+}_{\breve{\sigma},\tau}$, and $\widetilde{\mathfrak{N}}_{\breve{\sigma},\tau} - \widetilde{\mathfrak{N}}^{\breve{\ell},0+}_{\breve{\sigma},\tau}$ to their images $\widehat{\widetilde{\mathfrak{N}}}_{\breve{\sigma},\tau}$, $\widehat{\widetilde{\mathfrak{N}}}^{\breve{\ell}}_{\breve{\sigma},\tau}$, $\widehat{\widetilde{\mathfrak{N}}}^{\breve{\ell},+}_{\breve{\sigma},\tau}$, $\widehat{\widetilde{\mathfrak{N}}}^{\breve{\ell},0+}_{\breve{\sigma},\tau}$, $\widehat{\widetilde{\mathfrak{N}}}_{\breve{\sigma},\tau} - \widehat{\widetilde{\mathfrak{N}}}^{\breve{\ell}}_{\breve{\sigma},\tau}$, $\widehat{\widetilde{\mathfrak{N}}}_{\breve{\sigma},\tau} - \widehat{\widetilde{\mathfrak{N}}}^{\breve{\ell},+}_{\breve{\sigma},\tau}$, and $\widehat{\widetilde{\mathfrak{N}}}_{\breve{\sigma},\tau} - \widehat{\widetilde{\mathfrak{N}}}^{\breve{\ell},0+}_{\breve{\sigma},\tau}$, respectively. By Corollary 7.3.3.4, each of such images has an induced cone decomposition (in the obvious sense) by subsets of $\{\mathrm{pr}_{(\widehat{\mathbf{S}}_{\breve{\Phi}_{\widehat{\mathcal{H}}}})^{\vee}_{\mathbb{R}}}(\breve{\tau})\}_{\breve{\tau} \in \widetilde{\Sigma}_{\breve{\Phi}_{\widetilde{\mathcal{H}}},\breve{\sigma},\tau}}$. Using Lemma 7.3.3.7 to identify $\widehat{\widetilde{\mathfrak{N}}}_{\breve{\sigma},\tau}$ with $(\Gamma_{\breve{\Phi}_{\widehat{\mathcal{H}}},\Phi_{\mathcal{H}}})^{\vee}_{\mathbb{R}} \times \tau$, we have

$$\big(\overline{\mathrm{pr}_{(\widehat{\mathbf{S}}_{\breve{\Phi}_{\widehat{\mathcal{H}}}})^{\vee}_{\mathbb{R}}}(\breve{\tau})} - \{0\}\big) \cap \big((\Gamma_{\breve{\Phi}_{\widehat{\mathcal{H}}},\Phi_{\mathcal{H}}})^{\vee}_{\mathbb{R}} \times \{0\}\big) = \emptyset$$

for all $\breve{\tau} \in \widetilde{\Sigma}_{\breve{\Phi}_{\widehat{\mathcal{H}}},\breve{\sigma},\tau}$, because a real symmetric matrix $\begin{pmatrix} x & y \\ {}^t y & 0 \end{pmatrix}$ can be positive semi-definite only when $y = 0$. (The proof is elementary.)

For simplicity, let us denote $\widehat{\mathbf{P}}_{\breve{\Phi}_{\widehat{\mathcal{H}}}} - \big((\Gamma_{\breve{\Phi}_{\widehat{\mathcal{H}}},\Phi_{\mathcal{H}}})^{\vee}_{\mathbb{R}} \times \{0\}\big)$ by $\widehat{\mathbf{P}}'$. Let

$$\widehat{\mathbf{P}}'_{\breve{\ell}<0} := \{y \in \widehat{\mathbf{P}}' : \langle \breve{\ell}, y \rangle < 0\}$$

and

$$\widehat{\mathbf{P}}'_{\breve{\ell}\leq 0} := \{y \in \widehat{\mathbf{P}}' : \langle \breve{\ell}, y \rangle \leq 0\}.$$

Let $\widehat{\mathbf{P}}'_{\breve{\ell}<0+}$ denote the subset of $\widehat{\mathbf{P}}'_{\breve{\ell}\leq 0}$ consisting of $y \in \widehat{\mathbf{P}}'_{\breve{\ell}\leq 0}$ such that $\langle \breve{\ell}, y \rangle < 0$ if the image of y under the (canonical) second morphism in (1.2.4.20) is not contained in $\tau \subset \mathbf{P}^{+}_{\Phi_{\mathcal{H}}}$. By Lemma 7.3.3.15, $\widehat{\widehat{\mathfrak{N}}}_{\breve{\sigma},\tau} - \widehat{\widehat{\mathfrak{N}}}^{\breve{\ell}}_{\breve{\sigma},\tau}$ (resp. $\widehat{\widehat{\mathfrak{N}}}_{\breve{\sigma},\tau} - \widehat{\widehat{\mathfrak{N}}}^{\breve{\ell},+}_{\breve{\sigma},\tau}$, resp. $\widehat{\widehat{\mathfrak{N}}}_{\breve{\sigma},\tau} - \widehat{\widehat{\mathfrak{N}}}^{\breve{\ell},0+}_{\breve{\sigma},\tau}$) is the union of all $\mathrm{pr}_{(\widehat{\mathbf{S}}_{\breve{\Phi}_{\widehat{\mathcal{H}}}})^{\vee}_{\mathbb{R}}}(\breve{\tau})$ such that $\overline{\mathrm{pr}_{(\widehat{\mathbf{S}}_{\breve{\Phi}_{\widehat{\mathcal{H}}}})^{\vee}_{\mathbb{R}}}(\breve{\tau})} - \{0\}$ has a nonempty intersection with $\widehat{\mathbf{P}}'_{\breve{\ell}<0}$ (resp. $\widehat{\mathbf{P}}'_{\breve{\ell}\leq 0}$, resp. $\widehat{\mathbf{P}}'_{\breve{\ell}<0+}$).

Consider the canonical embeddings

$$\widehat{\widehat{\mathfrak{N}}}_{\breve{\sigma},\tau} \cap \widehat{\mathbf{P}}'_{\breve{\ell}<0} \hookrightarrow \widehat{\widehat{\mathfrak{N}}}_{\breve{\sigma},\tau} - \widehat{\widehat{\mathfrak{N}}}^{\breve{\ell}}_{\breve{\sigma},\tau}, \tag{7.3.3.17}$$

$$\widehat{\widehat{\mathfrak{N}}}_{\breve{\sigma},\tau} \cap \widehat{\mathbf{P}}'_{\breve{\ell}\leq 0} \hookrightarrow \widehat{\widehat{\mathfrak{N}}}_{\breve{\sigma},\tau} - \widehat{\widehat{\mathfrak{N}}}^{\breve{\ell},+}_{\breve{\sigma},\tau}, \tag{7.3.3.18}$$

and

$$\widehat{\widehat{\mathfrak{N}}}_{\breve{\sigma},\tau} \cap \widehat{\mathbf{P}}'_{\breve{\ell}<0+} \hookrightarrow \widehat{\widehat{\mathfrak{N}}}_{\breve{\sigma},\tau} - \widehat{\widehat{\mathfrak{N}}}^{\breve{\ell},0+}_{\breve{\sigma},\tau}. \tag{7.3.3.19}$$

Consider any $\breve{\tau} \in \widetilde{\Sigma}_{\breve{\Phi}_{\widehat{\mathcal{H}}},\breve{\sigma},\tau}$ such that $\overline{\mathrm{pr}_{(\widehat{\mathbf{S}}_{\breve{\Phi}_{\widehat{\mathcal{H}}}})^{\vee}_{\mathbb{R}}}(\breve{\tau})} - \{0\}$ has a nonempty intersection with $\widehat{\widehat{\mathfrak{N}}}_{\breve{\sigma},\tau} - \widehat{\widehat{\mathfrak{N}}}^{\breve{\ell}}_{\breve{\sigma},\tau}$ (resp. $\widehat{\widehat{\mathfrak{N}}}_{\breve{\sigma},\tau} - \widehat{\widehat{\mathfrak{N}}}^{\breve{\ell},+}_{\breve{\sigma},\tau}$, resp. $\widehat{\widehat{\mathfrak{N}}}_{\breve{\sigma},\tau} - \widehat{\widehat{\mathfrak{N}}}^{\breve{\ell},0+}_{\breve{\sigma},\tau}$). Each face $\mathrm{pr}_{(\widehat{\mathbf{S}}_{\breve{\Phi}_{\widehat{\mathcal{H}}}})^{\vee}_{\mathbb{R}}}(\breve{\tau}')$ of $\overline{\mathrm{pr}_{(\widehat{\mathbf{S}}_{\breve{\Phi}_{\widehat{\mathcal{H}}}})^{\vee}_{\mathbb{R}}}(\breve{\tau})}$ is still the image of some face $\breve{\tau}'$ of $\breve{\tau}$ having $\breve{\sigma}$ has a face, although $\breve{\tau}'$ might not satisfy the conditions in Definition 7.3.2.2. Up to replacing the cone decomposition with smooth locally finite refinements without changing the two sides of (7.3.3.17) (resp. (7.3.3.18), resp. (7.3.3.19)), we may assume that, for each $\breve{\tau}$ as above, at least one $\overline{\mathrm{pr}_{(\widehat{\mathbf{S}}_{\breve{\Phi}_{\widehat{\mathcal{H}}}})^{\vee}_{\mathbb{R}}}(\breve{\tau})} - \{0\}$ as above is contained in $\widehat{\widehat{\mathfrak{N}}}_{\breve{\sigma},\tau} \cap \widehat{\mathbf{P}}'_{\breve{\ell}<0}$ (resp. $\widehat{\widehat{\mathfrak{N}}}_{\breve{\sigma},\tau} \cap \widehat{\mathbf{P}}'_{\breve{\ell}\leq 0}$, resp. $\widehat{\widehat{\mathfrak{N}}}_{\breve{\sigma},\tau} \cap \widehat{\mathbf{P}}'_{\breve{\ell}<0+}$).

Since $\widehat{\widehat{\mathfrak{N}}}_{\breve{\sigma},\tau} = (\Gamma_{\breve{\Phi}_{\widehat{\mathcal{H}}},\Phi_{\mathcal{H}}})^{\vee}_{\mathbb{R}} \times \tau$ and $\widehat{\mathbf{P}}'_{\breve{\ell}<0}$ (resp. $\widehat{\mathbf{P}}'_{\breve{\ell}\leq 0}$, resp. $\widehat{\mathbf{P}}'_{\breve{\ell}<0+}$) are convex subsets of $\widehat{\mathbf{P}}'$, both being stable under the multiplicative action of $\mathbb{R}^{\times}_{>0}$, by Lemma 7.3.3.11, there are deformation retracts, compatible with restrictions to faces, from both $\overline{\mathrm{pr}_{(\widehat{\mathbf{S}}_{\breve{\Phi}_{\widehat{\mathcal{H}}}})^{\vee}_{\mathbb{R}}}(\breve{\tau})} - \overline{\mathrm{pr}_{(\widehat{\mathbf{S}}_{\breve{\Phi}_{\widehat{\mathcal{H}}}})^{\vee}_{\mathbb{R}}}(\breve{\tau}'')}$ and $\big(\overline{\mathrm{pr}_{(\widehat{\mathbf{S}}_{\breve{\Phi}_{\widehat{\mathcal{H}}}})^{\vee}_{\mathbb{R}}}(\breve{\tau})} - \{0\}\big) \cap \widehat{\mathbf{P}}'_{\breve{\ell}<0}$ (resp. $\big(\overline{\mathrm{pr}_{(\widehat{\mathbf{S}}_{\breve{\Phi}_{\widehat{\mathcal{H}}}})^{\vee}_{\mathbb{R}}}(\breve{\tau})} - \{0\}\big) \cap \widehat{\mathbf{P}}'_{\breve{\ell}\leq 0}$, resp. $\big(\overline{\mathrm{pr}_{(\widehat{\mathbf{S}}_{\breve{\Phi}_{\widehat{\mathcal{H}}}})^{\vee}_{\mathbb{R}}}(\breve{\tau})} - \{0\}\big) \cap \widehat{\mathbf{P}}'_{\breve{\ell}<0+}$) to $\overline{\mathrm{pr}_{(\widehat{\mathbf{S}}_{\breve{\Phi}_{\widehat{\mathcal{H}}}})^{\vee}_{\mathbb{R}}}(\breve{\tau}')} -$

$\{0\}$, where $\breve{\tau}'$ is the largest face of $\breve{\tau}$ also having $\breve{\sigma}$ as a face such that $\overline{\mathrm{pr}_{(\widehat{\mathbf{S}}_{\breve{\Phi}_{\widetilde{\mathcal{H}}}})^\vee_{\mathbb{R}}}(\breve{\tau}')} -$ $\{0\}$ is contained in $\widehat{\mathbf{P}}'_{\breve{\ell}<0}$ (resp. $\widehat{\mathbf{P}}'_{\breve{\ell}\leq 0}$, resp. $\widehat{\mathbf{P}}'_{\breve{\ell}<0+}$), and where $\breve{\tau}''$ is the largest face of $\breve{\tau}$ such that

$$\overline{\mathrm{pr}_{(\widehat{\mathbf{S}}_{\breve{\Phi}_{\widetilde{\mathcal{H}}}})^\vee_{\mathbb{R}}}(\breve{\tau}')} - \{0\} \subset \overline{\mathrm{pr}_{(\widehat{\mathbf{S}}_{\breve{\Phi}_{\widetilde{\mathcal{H}}}})^\vee_{\mathbb{R}}}(\breve{\tau})} - \overline{\mathrm{pr}_{(\widehat{\mathbf{S}}_{\breve{\Phi}_{\widetilde{\mathcal{H}}}})^\vee_{\mathbb{R}}}(\breve{\tau}'')}.$$

Hence, we see that (7.3.3.17) (resp. (7.3.3.18), resp. (7.3.3.19)) is a homotopy equivalence.

Since the left-hand sides of (7.3.3.17), (7.3.3.18), and (7.3.3.19) are all convex subsets of $\widehat{\mathbf{P}}'$, which are therefore contractible or empty, the proposition follows. □

Suppose $\mathscr{M}$ is a quasi-coherent $\mathscr{O}_{\vec{\widetilde{\mathfrak{X}}}^{\mathrm{ord}}_{\breve{\Phi}_{\widetilde{\mathcal{H}}},\breve{\delta}_{\widetilde{\mathcal{H}}},\breve{\sigma},\tau}/\Gamma_{\breve{\Phi}_{\widetilde{\mathcal{H}}},\Phi_{\mathcal{H}}}}$-module (which can be canonically viewed as a sheaf over the support $\vec{\widetilde{\Xi}}^{\mathrm{ord}}_{\breve{\Phi}_{\widetilde{\mathcal{H}}},\breve{\delta}_{\widetilde{\mathcal{H}}},\tau}/\Gamma_{\breve{\Phi}_{\widetilde{\mathcal{H}}},\Phi_{\mathcal{H}}}$ of the formal scheme $\vec{\widetilde{\mathfrak{X}}}^{\mathrm{ord}}_{\breve{\Phi}_{\widetilde{\mathcal{H}}},\breve{\delta}_{\widetilde{\mathcal{H}}},\breve{\sigma},\tau}/\Gamma_{\breve{\Phi}_{\widetilde{\mathcal{H}}},\Phi_{\mathcal{H}}}$). Let us define for each integer $d \geq 0$ the constructible sheaf $\mathscr{H}^d(\mathscr{M})$ over $\mathfrak{N}_{\breve{\sigma},\tau}$ which has stalk $H^d(U_{[\breve{\tau}]}, \mathscr{M}|_{U_{[\breve{\tau}]}})$ at each point of $[\breve{\tau}]$, where $[\breve{\tau}] \in \widetilde{\Sigma}_{\breve{\Phi}_{\widetilde{\mathcal{H}}},\breve{\sigma},\tau}/\Gamma_{\breve{\Phi}_{\widetilde{\mathcal{H}}},\Phi_{\mathcal{H}}}$ is viewed as a locally closed stratum of $\mathfrak{N}_{\breve{\sigma},\tau}$. For each representative $\breve{\tau} \in \widetilde{\Sigma}_{\breve{\Phi}_{\widetilde{\mathcal{H}}},\breve{\sigma},\tau}$ of $[\breve{\tau}]$, we have canonical isomorphisms $\mathfrak{U}_{[\breve{\tau}]} \cong \mathfrak{U}_{\breve{\tau}}$ and $U_{[\breve{\tau}]} \cong U_{\breve{\tau}}$ identifying $f^{\mathrm{tor}}_{[\breve{\tau}]} : \mathfrak{U}_{[\breve{\tau}]} \to \vec{\mathfrak{X}}^{\mathrm{ord}}_{\Phi_{\mathcal{H}},\delta_{\mathcal{H}},\tau}$ with $f^{\mathrm{tor}}_{\breve{\tau}} = h_\tau \circ g_{\breve{\tau}} : \mathfrak{U}_{\breve{\tau}} \to \vec{\mathfrak{X}}^{\mathrm{ord}}_{\Phi_{\mathcal{H}},\delta_{\mathcal{H}},\tau} \underset{\vec{C}^{\mathrm{ord}}_{\Phi_{\mathcal{H}},\delta_{\mathcal{H}}}}{\times} \vec{\widetilde{C}}^{\mathrm{ord}}_{\breve{\Phi}_{\widetilde{\mathcal{H}}},\breve{\delta}_{\widetilde{\mathcal{H}}}} \to \vec{\mathfrak{X}}^{\mathrm{ord}}_{\Phi_{\mathcal{H}},\delta_{\mathcal{H}},\tau}$, and $\mathfrak{U}_{\breve{\tau}}$ and $U_{\breve{\tau}}$ are relatively affine over $\vec{\mathfrak{X}}^{\mathrm{ord}}_{\Phi_{\mathcal{H}},\delta_{\mathcal{H}},\tau} \underset{\vec{C}^{\mathrm{ord}}_{\Phi_{\mathcal{H}},\delta_{\mathcal{H}}}}{\times} \vec{\widetilde{C}}^{\mathrm{ord}}_{\breve{\Phi}_{\widetilde{\mathcal{H}}},\breve{\delta}_{\widetilde{\mathcal{H}}}}$ under the morphism $g_{\breve{\tau}}$.

Lemma 7.3.3.20. *For each* $[\breve{\tau}] \in \widetilde{\Sigma}_{\breve{\Phi}_{\widetilde{\mathcal{H}}},\breve{\sigma},\tau}/\Gamma_{\breve{\Phi}_{\widetilde{\mathcal{H}}},\Phi_{\mathcal{H}}}$, *we have*

$$H^0([\breve{\tau}]^{\mathrm{cl}}, \mathscr{H}^d(\mathscr{M})) = \mathscr{H}^d(\mathscr{M})([\breve{\tau}]^{\mathrm{cl}}) \cong H^d(U_{[\breve{\tau}]}, \mathscr{M}|_{U_{[\breve{\tau}]}})$$

and

$$H^j([\breve{\tau}]^{\mathrm{cl}}, \mathscr{H}^d(\mathscr{M})) = 0$$

for each integer $j > 0$.

Proof. By [30, II, 5.2.1], the sheaf $\mathscr{H}^d(\mathscr{M})|_{[\breve{\tau}]^{\mathrm{cl}}}$ has a resolution by the Čech complex defined by the (locally finite) closed covering of $[\breve{\tau}]$ by $[\breve{\tau}']$, where $[\breve{\tau}']$ runs through faces of $[\breve{\tau}]$ in $\widetilde{\Sigma}_{\breve{\Phi}_{\widetilde{\mathcal{H}}},\breve{\sigma},\tau}/\Gamma_{\breve{\Phi}_{\widetilde{\mathcal{H}}},\Phi_{\mathcal{H}}}$. By definition of $\mathscr{H}^d(\mathscr{M})$, this corresponds to the resolution of $R^d(f^{\mathrm{tor}}_{[\breve{\tau}]})_*(\mathscr{M}|_{U_{[\breve{\tau}]}})$ by the Čech complex defined by the open covering of $U_{[\breve{\tau}]}$ by $U_{[\breve{\tau}']}$, where $[\breve{\tau}']$ are as above. Hence, the lemma follows because $\mathfrak{U}_{[\breve{\tau}]}$ and hence each open formal subscheme $\mathfrak{U}_{[\breve{\tau}']}$ is relatively affine over the same formal scheme $\vec{\mathfrak{X}}^{\mathrm{ord}}_{\Phi_{\mathcal{H}},\delta_{\mathcal{H}},\tau} \underset{\vec{C}^{\mathrm{ord}}_{\Phi_{\mathcal{H}},\delta_{\mathcal{H}}}}{\times} \vec{\widetilde{C}}^{\mathrm{ord}}_{\breve{\Phi}_{\widetilde{\mathcal{H}}},\breve{\delta}_{\widetilde{\mathcal{H}}}}$ under the morphism $g_{\breve{\tau}}$. □

Consequently, by comparing the nerve spectral sequences as in [30, II, 5.2.4 and 5.4.1] defined by open covering (7.3.3.2) for $\mathscr{M}$ and by the closed covering (7.3.3.3) for $\mathscr{H}^d(\mathscr{M})$ (for various d), there is a spectral sequence

$$E_2^{c,d} := H^c(\mathfrak{N}_{\breve{\sigma},\tau}, \mathscr{H}^d(\mathscr{M})) \Rightarrow H^{c+d}(\widetilde{\vec{\Xi}}^{\mathrm{ord}}_{\breve{\Phi}_{\widetilde{\mathcal{H}}},\breve{\delta}_{\widetilde{\mathcal{H}}},\tau}/\Gamma_{\breve{\Phi}_{\widetilde{\mathcal{H}}},\Phi_{\mathcal{H}}}, \mathscr{M}). \qquad (7.3.3.21)$$

(Here we are computing the left-hand side using a Čech spectral sequence defined by the locally finite closed covering $\{[\breve{\tau}]^{\mathrm{cl}}\}_{[\breve{\tau}]\in\widetilde{\Sigma}_{\breve{\Phi}_{\widetilde{\mathcal{H}}},\breve{\sigma},\tau}/\Gamma_{\breve{\Phi}_{\widetilde{\mathcal{H}}},\Phi_{\mathcal{H}}}}$ of $\mathfrak{N}_{\breve{\sigma},\tau}$.) The construction of $\mathfrak{N}_{\breve{\sigma},\tau}$ depends only on the cone decomposition $\widetilde{\Sigma}_{\breve{\Phi}_{\widetilde{\mathcal{H}}},\breve{\sigma},\tau}$, while the constructions of both $\mathscr{H}^d(\mathscr{M})$ and the spectral sequence (7.3.3.21) are compatible with restrictions to affine open subschemes of $\vec{\mathsf{Z}}^{\mathrm{ord}}_{[(\Phi_{\mathcal{H}},\delta_{\mathcal{H}},\tau)]}$. Therefore, we can define the sheaves $\underline{\mathscr{H}^d}(\mathscr{M})$ (of constructible sheaves over $\mathfrak{N}_{\breve{\sigma},\tau}$) over $\vec{\mathsf{Z}}^{\mathrm{ord}}_{[(\Phi_{\mathcal{H}},\delta_{\mathcal{H}},\tau)]}$, and obtain a spectral sequence

$$E_2^{c,d} := H^c(\mathfrak{N}_{\breve{\sigma},\tau}, \underline{\mathscr{H}^d}(\mathscr{M})) \Rightarrow R^{c+d} f_*^{\mathrm{tor}}(\mathscr{M}). \qquad (7.3.3.22)$$

Here $H^c(\mathfrak{N}_{\breve{\sigma},\tau}, \underline{\mathscr{H}^d}(\mathscr{M}))$ is interpreted as a sheaf over $\vec{\mathsf{Z}}^{\mathrm{ord}}_{[(\Phi_{\mathcal{H}},\delta_{\mathcal{H}},\tau)]}$, and the formation of (7.3.3.22) is compatible with morphisms in $\mathscr{M}$.

In particular, we have compatible spectral sequences

$$\begin{aligned} E_2^{c,d} &:= H^c(\mathfrak{N}_{\breve{\sigma},\tau}, \underline{\mathscr{H}^d}(\mathscr{O}^?_{(\vec{\mathsf{N}}^{\mathrm{ord,tor}})^\wedge_{\vec{\mathsf{Z}}^{\mathrm{ord}}_{[(\Phi_{\mathcal{H}},\delta_{\mathcal{H}},\tau)]}}})) \\ &\Rightarrow R^{c+d} f_*^{\mathrm{tor}}(\mathscr{O}^?_{(\vec{\mathsf{N}}^{\mathrm{ord,tor}})^\wedge_{\vec{\mathsf{Z}}^{\mathrm{ord}}_{[(\Phi_{\mathcal{H}},\delta_{\mathcal{H}},\tau)]}}}) \end{aligned} \qquad (7.3.3.23)$$

(for ? = $\emptyset$, +, ++, and 0+) and

$$E_2^{c,d} := H^c(\mathfrak{N}_{\breve{\sigma},\tau}, \underline{\mathscr{H}^d}(\mathscr{O}_{\vec{\mathsf{Z}}^{\mathrm{ord}}_{[(\Phi_{\mathcal{H}},\delta_{\mathcal{H}},\tau)]}})) \Rightarrow R^{c+d} f_*^{\mathrm{tor}}(\mathscr{O}_{\vec{\mathsf{Z}}^{\mathrm{ord}}_{[(\Phi_{\mathcal{H}},\delta_{\mathcal{H}},\tau)]}}). \qquad (7.3.3.24)$$

To calculate the left-hand sides of (7.3.3.23) and (7.3.3.24), we define the sheaves $\underline{\mathscr{H}^d}(\mathscr{O}^?_{\vec{\mathfrak{X}}^{\mathrm{ord}}_{\breve{\Phi}_{\widetilde{\mathcal{H}}},\breve{\delta}_{\widetilde{\mathcal{H}}},\breve{\sigma},\tau}})$ (for ? = $\emptyset$, +, ++, and 0+) and $\underline{\mathscr{H}^d}(\mathscr{O}_{\vec{\Xi}^{\mathrm{ord}}_{\breve{\Phi}_{\widetilde{\mathcal{H}}},\breve{\delta}_{\widetilde{\mathcal{H}}},\tau}})$ of constructible sheaves over $\widetilde{\mathfrak{N}}_{\breve{\sigma},\tau}$ (in the obvious way, with an analogue of Lemma 7.3.3.20 over $\widetilde{\mathfrak{N}}_{\breve{\sigma},\tau}$), which, by Lemma 7.3.2.9, carry canonical equivariant actions of the group $\Gamma_{\breve{\Phi}_{\widetilde{\mathcal{H}}},\Phi_{\mathcal{H}}}$, and descend to the sheaves $\underline{\mathscr{H}^d}(\mathscr{O}^?_{(\vec{\mathsf{N}}^{\mathrm{ord,tor}})^\wedge_{\vec{\mathsf{Z}}^{\mathrm{ord}}_{[(\Phi_{\mathcal{H}},\delta_{\mathcal{H}},\tau)]}}})$ (for ? = $\emptyset$, +, ++, and 0+) and $\underline{\mathscr{H}^d}(\mathscr{O}_{\vec{\mathsf{Z}}^{\mathrm{ord}}_{[(\Phi_{\mathcal{H}},\delta_{\mathcal{H}},\tau)]}})$ on $\mathfrak{N}_{\breve{\sigma},\tau}$, respectively. Hence, we obtain compatible spectral sequences

$$\begin{aligned} E_2^{c-e,e} &:= H^{c-e}(\Gamma_{\breve{\Phi}_{\widetilde{\mathcal{H}}},\Phi_{\mathcal{H}}}, H^e(\widetilde{\mathfrak{N}}_{\breve{\sigma},\tau}, \underline{\mathscr{H}^d}(\mathscr{O}^?_{\vec{\mathfrak{X}}^{\mathrm{ord}}_{\breve{\Phi}_{\widetilde{\mathcal{H}}},\breve{\delta}_{\widetilde{\mathcal{H}}},\breve{\sigma},\tau}}))) \\ &\Rightarrow H^c(\mathfrak{N}_{\breve{\sigma},\tau}, \underline{\mathscr{H}^d}(\mathscr{O}^?_{(\vec{\mathsf{N}}^{\mathrm{ord,tor}})^\wedge_{\vec{\mathsf{Z}}^{\mathrm{ord}}_{[(\Phi_{\mathcal{H}},\delta_{\mathcal{H}},\tau)]}}})) \end{aligned} \qquad (7.3.3.25)$$

(for ? = $\emptyset$, +, ++, and 0+) and

$$\begin{aligned} E_2^{c-e,e} &:= H^{c-e}(\Gamma_{\breve{\Phi}_{\widetilde{\mathcal{H}}},\Phi_{\mathcal{H}}}, H^e(\widetilde{\mathfrak{N}}_{\breve{\sigma},\tau}, \underline{\mathscr{H}^d}(\mathscr{O}_{\vec{\Xi}^{\mathrm{ord}}_{\breve{\Phi}_{\widetilde{\mathcal{H}}},\breve{\delta}_{\widetilde{\mathcal{H}}},\tau}}))) \\ &\Rightarrow H^c(\mathfrak{N}_{\breve{\sigma},\tau}, \underline{\mathscr{H}^d}(\mathscr{O}_{\vec{\mathsf{Z}}^{\mathrm{ord}}_{[(\Phi_{\mathcal{H}},\delta_{\mathcal{H}},\tau)]}})). \end{aligned} \qquad (7.3.3.26)$$

Lemma 7.3.3.27. (*Compare with [61, Lem. 4.16].*) *For all integers $d \geq 0$ and $e > 0$, we have*

$$H^e(\widetilde{\mathfrak{N}}_{\breve{\sigma},\tau}, \underline{\mathscr{H}}^d(\mathscr{O}^?_{\vec{\mathfrak{X}}^{\mathrm{ord}}_{\breve{\Phi}_{\widetilde{\mathcal{H}}},\breve{\delta}_{\widetilde{\mathcal{H}}},\breve{\sigma},\tau}})) = 0 \tag{7.3.3.28}$$

(*for* $? = \emptyset$, $+$, $++$, *and* $0+$) *and*

$$H^e(\widetilde{\mathfrak{N}}_{\breve{\sigma},\tau}, \underline{\mathscr{H}}^d(\mathscr{O}_{\vec{\Xi}^{\mathrm{ord}}_{\breve{\Phi}_{\widetilde{\mathcal{H}}},\breve{\delta}_{\widetilde{\mathcal{H}}},\tau}})) = 0. \tag{7.3.3.29}$$

Proof. By (1) of Lemma 7.3.2.9, we have

$$\underline{\mathscr{H}}^d(\mathscr{O}_{\vec{\mathfrak{X}}^{\mathrm{ord}}_{\breve{\Phi}_{\widetilde{\mathcal{H}}},\breve{\delta}_{\widetilde{\mathcal{H}}},\breve{\sigma},\tau}})(\breve{\tau}^{\mathrm{cl}}) \cong R^d(f^{\mathrm{tor}}_{\breve{\tau}})_*(\mathscr{O}_{\mathfrak{U}_{\breve{\tau}}}) \cong \widehat{\bigoplus_{\breve{\ell}\in\breve{\sigma}^\perp\cap\breve{\tau}^\vee}} R^d h_*(\vec{\widetilde{\Psi}}^{\mathrm{ord}}_{\breve{\Phi}_{\widetilde{\mathcal{H}}},\breve{\delta}_{\widetilde{\mathcal{H}}}}(\breve{\ell})),$$

and for each face $\breve{\tau}'$ of $\breve{\tau}$ also in $\widetilde{\Sigma}_{\breve{\Phi}_{\widetilde{\mathcal{H}}},\breve{\sigma},\tau}$, the canonical morphism

$$\underline{\mathscr{H}}^d(\mathscr{O}_{\vec{\mathfrak{X}}^{\mathrm{ord}}_{\breve{\Phi}_{\widetilde{\mathcal{H}}},\breve{\delta}_{\widetilde{\mathcal{H}}},\breve{\sigma},\tau}})(\breve{\tau}^{\mathrm{cl}}) \to \underline{\mathscr{H}}^d(\mathscr{O}_{\vec{\mathfrak{X}}^{\mathrm{ord}}_{\breve{\Phi}_{\widetilde{\mathcal{H}}},\breve{\delta}_{\widetilde{\mathcal{H}}},\breve{\sigma},\tau}})((\breve{\tau}')^{\mathrm{cl}})$$

sends the subsheaf $R^d h_*(\vec{\widetilde{\Psi}}^{\mathrm{ord}}_{\breve{\Phi}_{\widetilde{\mathcal{H}}},\breve{\delta}_{\widetilde{\mathcal{H}}}}(\breve{\ell}))$ (identically) to $R^d h_*(\vec{\widetilde{\Psi}}^{\mathrm{ord}}_{\breve{\Phi}_{\widetilde{\mathcal{H}}},\breve{\delta}_{\widetilde{\mathcal{H}}}}(\breve{\ell}))$ when $\breve{\ell} \in \breve{\sigma}^\perp \cap \breve{\tau}^\vee$. By Definition 7.3.3.14, $\breve{\ell} \in \breve{\sigma}^\perp \cap \breve{\tau}^\vee$ exactly when $\breve{\tau}^{\mathrm{cl}} \subset \widetilde{\mathfrak{N}}^{\breve{\ell}}_{\breve{\sigma},\tau}$. Since $\widetilde{\mathfrak{N}}_{\breve{\sigma},\tau}$ and $\widetilde{\mathfrak{N}}_{\breve{\sigma},\tau} - \widetilde{\mathfrak{N}}^{\breve{\ell}}_{\breve{\sigma},\tau}$ are either contractible or empty for each given $\breve{\ell} \in \breve{\sigma}^\perp$, by Proposition 7.3.3.16, we have (7.3.3.28) for $e > 0$ and $? = \emptyset$ as usual (*cf.* the argument in [50, Ch. I, Sec. 3]). (Since the nerves involve infinitely many cones, let us briefly explain why we can work weight-by-weight as in [50, Ch. I, Sec. 3]. This is because, up to replacing the cone decompositions with locally finite refinements not necessarily carrying $\Gamma_{\breve{\Phi}_{\widetilde{\mathcal{H}}},\Phi_{\mathcal{H}}}$-actions, which is harmless for proving this lemma, we can compute the cohomology as a limit using unions of finite cone decompositions on expanding convex polyhedral subcones, by proving inductively that the cohomology of one degree lower has the desired properties, using [103, Thm. 3.5.8]. Then we can consider the associated graded pieces defined by the completions, and work weight-by-weight with subsheaves of $\underline{\mathscr{H}}^d(\mathscr{O}_{\vec{\mathfrak{X}}^{\mathrm{ord}}_{\breve{\Phi}_{\widetilde{\mathcal{H}}},\breve{\delta}_{\widetilde{\mathcal{H}}},\breve{\sigma},\tau}})(\breve{\tau}^{\mathrm{cl}})$ of the form $R^d h_*(\vec{\widetilde{\Psi}}^{\mathrm{ord}}_{\breve{\Phi}_{\widetilde{\mathcal{H}}},\breve{\delta}_{\widetilde{\mathcal{H}}}}(\breve{\ell}))$, because taking cohomology commutes with taking infinite direct sums for Čech complexes defined by finite coverings, as desired.)

The cases of (7.3.3.28) for $? = +$ and $0+$ are similar using Lemma 7.3.3.16. The case of (7.3.3.28) for $? = ++$ follows from the case for $? = \emptyset$ (by the projection formula [35, 0_{I}, 5.4.10.1]) because $\mathscr{O}^{++}_{\vec{\mathfrak{X}}^{\mathrm{ord}}_{\breve{\Phi}_{\widetilde{\mathcal{H}}},\breve{\delta}_{\widetilde{\mathcal{H}}},\breve{\sigma},\tau}}$ is (by definition) the pullback of $\mathscr{I}_{\vec{\mathsf{D}}^{\mathrm{ord}}_{\infty,\mathcal{H}}}$ over $\vec{\mathsf{M}}^{\mathrm{ord,tor}}_{\mathcal{H}}$. The case of (7.3.3.29) then follows from the cases of (7.3.3.28) for $? = 0+$ and $\emptyset$ by considering the long exact sequence attached to the pullback of the short exact sequence

$$0 \to \mathscr{O}^{0+}_{\vec{\mathsf{N}}^{\mathrm{ord}}} \to \mathscr{O}_{\vec{\mathsf{N}}^{\mathrm{ord}}} \to \mathscr{O}_{\vec{\mathsf{Z}}^{\mathrm{ord}}_{[(\Phi_{\mathcal{H}},\delta_{\mathcal{H}},\tau)]}} \to 0$$

(by definition of $\mathscr{O}^{0+}_{\vec{\mathsf{N}}^{\mathrm{ord}}}$). □

Lemma 7.3.3.30. (*Compare with [61, Lem. 4.16].*) *For every integer $d \geq 0$, the canonical morphisms*

$$R^d h_*(\mathscr{O}_{\vec{\mathfrak{X}}^{\mathrm{ord}}_{\Phi_{\mathcal{H}},\delta_{\mathcal{H}},\tau} \underset{\vec{C}^{\mathrm{ord}}_{\Phi_{\mathcal{H}},\delta_{\mathcal{H}}}}{\times} \vec{\vec{C}}^{\mathrm{ord}}_{\breve{\Phi}_{\widetilde{\mathcal{H}}},\breve{\delta}_{\widetilde{\mathcal{H}}}}}) \to H^0(\widetilde{\mathfrak{N}}_{\breve{\sigma},\tau}, \underline{\mathscr{H}}^d(\mathscr{O}_{\vec{\vec{\mathfrak{X}}}^{\mathrm{ord}}_{\breve{\Phi}_{\widetilde{\mathcal{H}}},\breve{\delta}_{\widetilde{\mathcal{H}}},\breve{\sigma},\tau}})), \tag{7.3.3.31}$$

$$\begin{aligned} & R^d h_*(\mathscr{O}_{\vec{\mathfrak{X}}^{\mathrm{ord}}_{\Phi_{\mathcal{H}},\delta_{\mathcal{H}},\tau} \underset{\vec{C}^{\mathrm{ord}}_{\Phi_{\mathcal{H}},\delta_{\mathcal{H}}}}{\times} \vec{\vec{C}}^{\mathrm{ord}}_{\breve{\Phi}_{\widetilde{\mathcal{H}}},\breve{\delta}_{\widetilde{\mathcal{H}}}}}) \underset{\mathscr{O}_{\vec{\mathsf{M}}^{\mathrm{ord,tor}}_{\mathcal{H}}}}{\otimes} \mathscr{I}_{\vec{\mathsf{D}}^{\mathrm{ord}}_{\infty,\mathcal{H}}} \\ & \cong H^0(\widetilde{\mathfrak{N}}_{\breve{\sigma},\tau}, \underline{\mathscr{H}}^d(\mathscr{O}^{++}_{\vec{\vec{\mathfrak{X}}}^{\mathrm{ord}}_{\breve{\Phi}_{\widetilde{\mathcal{H}}},\breve{\delta}_{\widetilde{\mathcal{H}}},\breve{\sigma},\tau}})) \to H^0(\widetilde{\mathfrak{N}}_{\breve{\sigma},\tau}, \underline{\mathscr{H}}^d(\mathscr{O}^{+}_{\vec{\vec{\mathfrak{X}}}^{\mathrm{ord}}_{\breve{\Phi}_{\widetilde{\mathcal{H}}},\breve{\delta}_{\widetilde{\mathcal{H}}},\breve{\sigma},\tau}})) \end{aligned} \tag{7.3.3.32}$$

(*induced by the projection formula [35, 0_I, 5.4.10.1] and* (7.3.1.1)), *and*

$$R^d h_*(\mathscr{O}_{\vec{\mathsf{Z}}^{\mathrm{ord}}_{[(\Phi_{\mathcal{H}},\delta_{\mathcal{H}},\tau)]} \underset{\vec{C}^{\mathrm{ord}}_{\Phi_{\mathcal{H}},\delta_{\mathcal{H}}}}{\times} \vec{\vec{C}}^{\mathrm{ord}}_{\breve{\Phi}_{\widetilde{\mathcal{H}}},\breve{\delta}_{\widetilde{\mathcal{H}}}}}) \to H^0(\widetilde{\mathfrak{N}}_{\breve{\sigma},\tau}, \underline{\mathscr{H}}^d(\mathscr{O}_{\vec{\vec{\Xi}}^{\mathrm{ord}}_{\breve{\Phi}_{\widetilde{\mathcal{H}}},\breve{\delta}_{\widetilde{\mathcal{H}}},\tau}})) \tag{7.3.3.33}$$

are isomorphisms compatible with each other.

Proof. Let us continue with the setting in the proof of Lemma 7.3.3.27. Since

$$\underset{\breve{\tau}\in\widetilde{\Sigma}_{\breve{\Phi}_{\widetilde{\mathcal{H}}},\breve{\sigma},\tau}}{\cap} (\breve{\sigma}^{\perp} \cap \breve{\tau}^{\vee}) = \tau^{\vee},$$

we see that (7.3.3.31) is an isomorphism. Since $\mathscr{O}^{++}_{\vec{\vec{\mathfrak{X}}}^{\mathrm{ord}}_{\breve{\Phi}_{\widetilde{\mathcal{H}}},\breve{\delta}_{\widetilde{\mathcal{H}}},\breve{\sigma},\tau}}$ is (by definition) the pullback of $\mathscr{I}_{\vec{\mathsf{D}}^{\mathrm{ord}}_{\infty,\mathcal{H}}}$ over $\vec{\mathsf{M}}^{\mathrm{ord,tor}}_{\mathcal{H}}$, the first morphism in (7.3.3.32) is an isomorphism (by the projection formula [35, 0_{I}, 5.4.10.1]). Since

$$\underset{\breve{\tau}\in\widetilde{\Sigma}_{\breve{\Phi}_{\widetilde{\mathcal{H}}},\breve{\sigma},\tau}}{\cap} (\breve{\sigma}^{\perp} \cap \breve{\tau}^{\vee}_{\breve{\sigma},+}) = \tau^{\vee}_{+},$$

the composition of the two morphisms in (7.3.3.32) is an isomorphism. In particular, the second morphism in (7.3.3.32) is also an isomorphism. Finally, since

$$\underset{\breve{\tau}\in\widetilde{\Sigma}_{\breve{\Phi}_{\widetilde{\mathcal{H}}},\breve{\sigma},\tau}}{\cap} (\breve{\tau}^{\vee} - \breve{\tau}^{\vee}_{\breve{\sigma}}) = \tau^{\perp},$$

we see that (7.3.3.33) is also an isomorphism. □

Lemma 7.3.3.34. (*Compare with [61, Lem. 4.23].*) *Let $c \geq 0$ be any integer. The $\mathscr{O}_{\vec{\mathfrak{X}}^{\mathrm{ord}}_{\Phi_{\mathcal{H}},\delta_{\mathcal{H}},\tau}}$-module $R^d h_*(\mathscr{O}_{\vec{\mathfrak{X}}^{\mathrm{ord}}_{\Phi_{\mathcal{H}},\delta_{\mathcal{H}},\tau} \underset{\vec{C}^{\mathrm{ord}}_{\Phi_{\mathcal{H}},\delta_{\mathcal{H}}}}{\times} \vec{\vec{C}}^{\mathrm{ord}}_{\breve{\Phi}_{\widetilde{\mathcal{H}}},\breve{\delta}_{\widetilde{\mathcal{H}}}}})$ is locally free of finite rank, and there is a canonical isomorphism*

$$\begin{aligned} & R^d h_*(\mathscr{O}_{\vec{\mathfrak{X}}^{\mathrm{ord}}_{\Phi_{\mathcal{H}},\delta_{\mathcal{H}},\tau} \underset{\vec{C}^{\mathrm{ord}}_{\Phi_{\mathcal{H}},\delta_{\mathcal{H}}}}{\times} \vec{\vec{C}}^{\mathrm{ord}}_{\breve{\Phi}_{\widetilde{\mathcal{H}}},\breve{\delta}_{\widetilde{\mathcal{H}}}}}) \underset{\mathbb{Z}}{\otimes} \mathbb{Q} \\ & \cong \wedge^d(\underline{\mathrm{Hom}}_{\mathcal{O}}(Q^{\vee}, \underline{\mathrm{Lie}}_{B^{\vee}/\vec{\mathfrak{X}}^{\mathrm{ord}}_{\Phi_{\mathcal{H}},\delta_{\mathcal{H}},\tau}}) \underset{\mathbb{Z}}{\otimes} \mathbb{Q}. \end{aligned} \tag{7.3.3.35}$$

Similarly, the $\mathscr{O}_{\vec{\mathsf{Z}}^{\mathrm{ord}}_{[(\Phi_{\mathcal{H}},\delta_{\mathcal{H}},\tau)]}}$*-module* $R^d h_*(\mathscr{O}_{\vec{\mathsf{Z}}^{\mathrm{ord}}_{[(\Phi_{\mathcal{H}},\delta_{\mathcal{H}},\tau)]} \underset{\vec{C}^{\mathrm{ord}}_{\Phi_{\mathcal{H}},\delta_{\mathcal{H}}}}{\times} \vec{\vec{C}}^{\mathrm{ord}}_{\breve{\Phi}_{\widetilde{\mathcal{H}}},\breve{\delta}_{\widetilde{\mathcal{H}}}}})$ *is locally free of finite rank, and there is a canonical isomorphism*

$$\begin{aligned} & R^d h_*(\mathscr{O}_{\vec{\mathsf{Z}}^{\mathrm{ord}}_{[(\Phi_{\mathcal{H}},\delta_{\mathcal{H}},\tau)]} \underset{\vec{C}^{\mathrm{ord}}_{\Phi_{\mathcal{H}},\delta_{\mathcal{H}}}}{\times} \vec{\vec{C}}^{\mathrm{ord}}_{\breve{\Phi}_{\widetilde{\mathcal{H}}},\breve{\delta}_{\widetilde{\mathcal{H}}}}}) \underset{\mathbb{Z}}{\otimes} \mathbb{Q} \\ & \cong \wedge^d(\underline{\mathrm{Hom}}_{\mathcal{O}}(Q^\vee, \underline{\mathrm{Lie}}_{B^\vee/\vec{\mathsf{Z}}^{\mathrm{ord}}_{[(\Phi_{\mathcal{H}},\delta_{\mathcal{H}},\tau)]}})) \underset{\mathbb{Z}}{\otimes} \mathbb{Q}. \end{aligned} \tag{7.3.3.36}$$

The isomorphisms (7.3.3.35) *and* (7.3.3.36) *are compatible with each other.*

Proof. By Lemma 7.1.2.1, the morphism $h : \vec{\vec{C}}^{\mathrm{ord}}_{\breve{\Phi}_{\widetilde{\mathcal{H}}},\breve{\delta}_{\widetilde{\mathcal{H}}}} \to \vec{C}^{\mathrm{ord}}_{\Phi_{\mathcal{H}},\delta_{\mathcal{H}}}$ is a torsor under (the pullback of) an abelian scheme $\mathbb{Q}^\times$-isogenous to $\underline{\mathrm{Hom}}_{\mathcal{O}}(Q, B)^\circ$ (and hence has a section étale locally). Since the cohomology of abelian schemes (with coefficients in the structural sheaves) are free and are compatible with arbitrary base changes (see [6, Prop. 2.5.2] and [81, Sec. 5]), we see that $R^d h_*(\mathscr{O}_{\vec{\mathfrak{X}}^{\mathrm{ord}}_{\Phi_{\mathcal{H}},\delta_{\mathcal{H}},\tau} \underset{\vec{C}^{\mathrm{ord}}_{\Phi_{\mathcal{H}},\delta_{\mathcal{H}}}}{\times} \vec{\vec{C}}^{\mathrm{ord}}_{\breve{\Phi}_{\widetilde{\mathcal{H}}},\breve{\delta}_{\widetilde{\mathcal{H}}}}})$ and $R^d h_*(\mathscr{O}_{\vec{\mathsf{Z}}^{\mathrm{ord}}_{[(\Phi_{\mathcal{H}},\delta_{\mathcal{H}},\tau)]} \underset{\vec{C}^{\mathrm{ord}}_{\Phi_{\mathcal{H}},\delta_{\mathcal{H}}}}{\times} \vec{\vec{C}}^{\mathrm{ord}}_{\breve{\Phi}_{\widetilde{\mathcal{H}}},\breve{\delta}_{\widetilde{\mathcal{H}}}}})$ are locally free and finite rank over $\vec{\mathfrak{X}}^{\mathrm{ord}}_{\Phi_{\mathcal{H}},\delta_{\mathcal{H}},\tau}$ and $\vec{\mathsf{Z}}^{\mathrm{ord}}_{[(\Phi_{\mathcal{H}},\delta_{\mathcal{H}},\tau)]}$, respectively. Moreover, we obtain compatible canonical isomorphisms

$$\begin{aligned} & R^d h_*(\mathscr{O}_{\vec{\mathfrak{X}}^{\mathrm{ord}}_{\Phi_{\mathcal{H}},\delta_{\mathcal{H}},\tau} \underset{\vec{C}^{\mathrm{ord}}_{\Phi_{\mathcal{H}},\delta_{\mathcal{H}}}}{\times} \vec{\vec{C}}^{\mathrm{ord}}_{\breve{\Phi}_{\widetilde{\mathcal{H}}},\breve{\delta}_{\widetilde{\mathcal{H}}}}}) \underset{\mathbb{Z}}{\otimes} \mathbb{Q} \\ & \cong \wedge^d(\underline{\mathrm{Lie}}_{(\underline{\mathrm{Hom}}_{\mathcal{O}}(Q,B)^\circ)^\vee/\vec{\mathfrak{X}}^{\mathrm{ord}}_{\Phi_{\mathcal{H}},\delta_{\mathcal{H}},\tau}}) \underset{\mathbb{Z}}{\otimes} \mathbb{Q} \\ & \cong \wedge^d(\underline{\mathrm{Hom}}_{\mathcal{O}}(Q^\vee, \underline{\mathrm{Lie}}_{B^\vee/\vec{\mathfrak{X}}^{\mathrm{ord}}_{\Phi_{\mathcal{H}},\delta_{\mathcal{H}},\tau}})) \underset{\mathbb{Z}}{\otimes} \mathbb{Q} \end{aligned}$$

and

$$\begin{aligned} & R^d h_*(\mathscr{O}_{\vec{\mathsf{Z}}^{\mathrm{ord}}_{[(\Phi_{\mathcal{H}},\delta_{\mathcal{H}},\tau)]} \underset{\vec{C}^{\mathrm{ord}}_{\Phi_{\mathcal{H}},\delta_{\mathcal{H}}}}{\times} \vec{\vec{C}}^{\mathrm{ord}}_{\breve{\Phi}_{\widetilde{\mathcal{H}}},\breve{\delta}_{\widetilde{\mathcal{H}}}}}) \underset{\mathbb{Z}}{\otimes} \mathbb{Q} \\ & \cong \wedge^d(\underline{\mathrm{Lie}}_{(\underline{\mathrm{Hom}}_{\mathcal{O}}(Q,B)^\circ)^\vee/\vec{\mathsf{Z}}^{\mathrm{ord}}_{[(\Phi_{\mathcal{H}},\delta_{\mathcal{H}},\tau)]}}) \underset{\mathbb{Z}}{\otimes} \mathbb{Q} \\ & \cong \wedge^d(\underline{\mathrm{Hom}}_{\mathcal{O}}(Q^\vee, \underline{\mathrm{Lie}}_{B^\vee/\vec{\mathsf{Z}}^{\mathrm{ord}}_{[(\Phi_{\mathcal{H}},\delta_{\mathcal{H}},\tau)]}})) \underset{\mathbb{Z}}{\otimes} \mathbb{Q} \end{aligned}$$

for all $d \geq 0$. □

Proposition 7.3.3.37. (*Compare with [61, Prop. 4.24].*) *Let* $c, d \geq 0$ *be any integers. The* $\mathscr{O}_{\vec{\mathfrak{X}}^{\mathrm{ord}}_{\Phi_{\mathcal{H}},\delta_{\mathcal{H}},\tau}}$*-modules* $H^c(\mathfrak{N}_{\breve{\sigma},\tau}, \underline{\mathscr{H}}^d(\mathscr{O}^?_{(\vec{\mathsf{N}}^{\mathrm{ord,tor}})^\wedge_{\vec{\vec{\mathsf{Z}}}^{\mathrm{ord}}_{[(\Phi_{\mathcal{H}},\delta_{\mathcal{H}},\tau)]}}}))$, *for* $? = \emptyset$, $+$, *and* $++$, *are locally free of finite rank, and there are canonical isomorphisms*

$$\begin{aligned} & H^c(\mathfrak{N}_{\breve{\sigma},\tau}, \underline{\mathscr{H}}^d(\mathscr{O}_{(\vec{\mathsf{N}}^{\mathrm{ord,tor}})^\wedge_{\vec{\vec{\mathsf{Z}}}^{\mathrm{ord}}_{[(\Phi_{\mathcal{H}},\delta_{\mathcal{H}},\tau)]}}})) \underset{\mathscr{O}_{\vec{\mathsf{M}}^{\mathrm{ord,tor}}_{\mathcal{H}}}}{\otimes} \mathscr{I}_{\vec{\mathsf{D}}^{\mathrm{ord}}_{\infty,\mathcal{H}}} \\ & \cong H^c(\mathfrak{N}_{\breve{\sigma},\tau}, \underline{\mathscr{H}}^d(\mathscr{O}^{++}_{(\vec{\mathsf{N}}^{\mathrm{ord,tor}})^\wedge_{\vec{\vec{\mathsf{Z}}}^{\mathrm{ord}}_{[(\Phi_{\mathcal{H}},\delta_{\mathcal{H}},\tau)]}}})) \\ & \overset{\sim}{\to} H^c(\mathfrak{N}_{\breve{\sigma},\tau}, \underline{\mathscr{H}}^d(\mathscr{O}^{+}_{(\vec{\mathsf{N}}^{\mathrm{ord,tor}})^\wedge_{\vec{\vec{\mathsf{Z}}}^{\mathrm{ord}}_{[(\Phi_{\mathcal{H}},\delta_{\mathcal{H}},\tau)]}}})) \end{aligned} \tag{7.3.3.38}$$

(induced by the projection formula [35, 0_I, 5.4.10.1] and (7.3.1.1)*), together with canonical isomorphisms*

$$\begin{aligned} & H^c(\mathfrak{N}_{\breve{\sigma},\tau}, \underline{\mathscr{H}}^d(\mathscr{O}_{(\vec{\mathsf{N}}^{\mathrm{ord,tor}})^\wedge_{\vec{\mathsf{Z}}^{\mathrm{ord}}_{[(\Phi_{\mathcal{H}},\delta_{\mathcal{H}},\tau)]}}})) \underset{\mathbb{Z}}{\otimes} \mathbb{Q} \\ & \cong (\wedge^c(\underline{\mathrm{Hom}}_{\mathcal{O}}(Q^\vee, \underline{\mathrm{Lie}}_{T^\vee/\vec{\mathfrak{X}}^{\mathrm{ord}}_{\Phi_{\mathcal{H}},\delta_{\mathcal{H}},\tau}}))) \\ & \underset{\mathscr{O}_{\vec{\mathfrak{X}}^{\mathrm{ord}}_{\Phi_{\mathcal{H}},\delta_{\mathcal{H}},\tau}}}{\otimes} (\wedge^d(\underline{\mathrm{Hom}}_{\mathcal{O}}(Q^\vee, \underline{\mathrm{Lie}}_{B^\vee/\vec{\mathfrak{X}}^{\mathrm{ord}}_{\Phi_{\mathcal{H}},\delta_{\mathcal{H}},\tau}}))) \underset{\mathbb{Z}}{\otimes} \mathbb{Q} \end{aligned} \tag{7.3.3.39}$$

and

$$\begin{aligned} & H^c(\mathfrak{N}_{\breve{\sigma},\tau}, \underline{\mathscr{H}}^d(\mathscr{O}^+_{(\vec{\mathsf{N}}^{\mathrm{ord,tor}})^\wedge_{\vec{\mathsf{Z}}^{\mathrm{ord}}_{[(\Phi_{\mathcal{H}},\delta_{\mathcal{H}},\tau)]}}})) \underset{\mathbb{Z}}{\otimes} \mathbb{Q} \\ & \cong H^c(\mathfrak{N}_{\breve{\sigma},\tau}, \underline{\mathscr{H}}^d(\mathscr{O}^{++}_{(\vec{\mathsf{N}}^{\mathrm{ord,tor}})^\wedge_{\vec{\mathsf{Z}}^{\mathrm{ord}}_{[(\Phi_{\mathcal{H}},\delta_{\mathcal{H}},\tau)]}}})) \underset{\mathbb{Z}}{\otimes} \mathbb{Q} \\ & \cong (\wedge^c(\underline{\mathrm{Hom}}_{\mathcal{O}}(Q^\vee, \underline{\mathrm{Lie}}_{T^\vee/\vec{\mathfrak{X}}^{\mathrm{ord}}_{\Phi_{\mathcal{H}},\delta_{\mathcal{H}},\tau}}))) \\ & \underset{\mathscr{O}_{\vec{\mathfrak{X}}^{\mathrm{ord}}_{\Phi_{\mathcal{H}},\delta_{\mathcal{H}},\tau}}}{\otimes} (\wedge^d(\underline{\mathrm{Hom}}_{\mathcal{O}}(Q^\vee, \underline{\mathrm{Lie}}_{B^\vee/\vec{\mathfrak{X}}^{\mathrm{ord}}_{\Phi_{\mathcal{H}},\delta_{\mathcal{H}},\tau}}))) \underset{\mathscr{O}_{\vec{\mathsf{M}}^{\mathrm{ord,tor}}_{\mathcal{H}}}}{\otimes} \mathscr{I}_{\vec{\mathsf{D}}^{\mathrm{ord}}_{\infty,\mathcal{H}}} \underset{\mathbb{Z}}{\otimes} \mathbb{Q}. \end{aligned} \tag{7.3.3.40}$$

On the other hand, the $\mathscr{O}_{\vec{\mathsf{Z}}^{\mathrm{ord}}_{[(\Phi_{\mathcal{H}},\delta_{\mathcal{H}},\tau)]}}$*-module* $H^c(\mathfrak{N}_{\breve{\sigma},\tau}, \underline{\mathscr{H}}^d(\mathscr{O}_{\vec{\mathsf{Z}}^{\mathrm{ord}}_{[(\Phi_{\mathcal{H}},\delta_{\mathcal{H}},\tau)]}}))$ *is locally free of finite rank, and there is a canonical isomorphism*

$$\begin{aligned} & H^c(\mathfrak{N}_{\breve{\sigma},\tau}, \underline{\mathscr{H}}^d(\mathscr{O}_{\vec{\mathsf{Z}}^{\mathrm{ord}}_{[(\Phi_{\mathcal{H}},\delta_{\mathcal{H}},\tau)]}})) \underset{\mathbb{Z}}{\otimes} \mathbb{Q} \\ & \cong (\wedge^c(\underline{\mathrm{Hom}}_{\mathcal{O}}(Q^\vee, \underline{\mathrm{Lie}}_{T^\vee/\vec{\mathsf{Z}}^{\mathrm{ord}}_{[(\Phi_{\mathcal{H}},\delta_{\mathcal{H}},\tau)]}}))) \\ & \underset{\mathscr{O}_{\mathsf{Z}_{[(\Phi_{\mathcal{H}},\delta_{\mathcal{H}},\tau)]}}}{\otimes} (\wedge^d(\underline{\mathrm{Hom}}_{\mathcal{O}}(Q^\vee, \underline{\mathrm{Lie}}_{B^\vee/\vec{\mathsf{Z}}^{\mathrm{ord}}_{[(\Phi_{\mathcal{H}},\delta_{\mathcal{H}},\tau)]}}))) \underset{\mathbb{Z}}{\otimes} \mathbb{Q}. \end{aligned} \tag{7.3.3.41}$$

The isomorphisms (7.3.3.38)*,* (7.3.3.39)*,* (7.3.3.40)*, and* (7.3.3.41) *are compatible with each other.*

Proof. By Lemmas 7.3.3.30 and 7.3.3.27, the spectral sequences (7.3.3.25) and (7.3.3.26) degenerate and show that for each pair of integers c and d we have compatible canonical isomorphisms

$$\begin{aligned} & H^c(\mathfrak{N}_{\breve{\sigma},\tau}, \underline{\mathscr{H}}^d(\mathscr{O}_{(\vec{\mathsf{N}}^{\mathrm{ord,tor}})^\wedge_{\vec{\mathsf{Z}}^{\mathrm{ord}}_{[(\Phi_{\mathcal{H}},\delta_{\mathcal{H}},\tau)]}}})) \\ & \cong H^c(\Gamma_{\breve{\Phi}_{\widetilde{\mathcal{H}}},\Phi_{\mathcal{H}}}, H^0(\widetilde{\mathfrak{N}}_{\breve{\sigma},\tau}, \underline{\mathscr{H}}^d(\mathscr{O}_{\vec{\mathfrak{X}}^{\mathrm{ord}}_{\breve{\Phi}_{\widetilde{\mathcal{H}}},\breve{\delta}_{\widetilde{\mathcal{H}}},\breve{\sigma},\tau}}))) \\ & \cong H^c(\Gamma_{\breve{\Phi}_{\widetilde{\mathcal{H}}},\Phi_{\mathcal{H}}}, \mathbb{Z}) \underset{\mathbb{Z}}{\otimes} R^d h_*(\mathscr{O}_{\vec{\mathfrak{X}}^{\mathrm{ord}}_{\Phi_{\mathcal{H}},\delta_{\mathcal{H}},\tau} \underset{\vec{C}^{\mathrm{ord}}_{\Phi_{\mathcal{H}},\delta_{\mathcal{H}}}}{\times} \vec{C}^{\mathrm{ord}}_{\breve{\Phi}_{\widetilde{\mathcal{H}}},\breve{\delta}_{\widetilde{\mathcal{H}}}}}), \end{aligned} \tag{7.3.3.42}$$

$$\begin{aligned} & H^c(\mathfrak{N}_{\breve{\sigma},\tau}, \underline{\mathscr{H}}^d(\mathscr{O}^?_{(\vec{\mathsf{N}}^{\mathrm{ord,tor}})^\wedge_{\vec{\mathsf{Z}}^{\mathrm{ord}}_{[(\Phi_{\mathcal{H}},\delta_{\mathcal{H}},\tau)]}}})) \\ & \cong H^c(\Gamma_{\breve{\Phi}_{\widetilde{\mathcal{H}}},\Phi_{\mathcal{H}}}, H^0(\widetilde{\mathfrak{N}}_{\breve{\sigma},\tau}, \underline{\mathscr{H}}^d(\mathscr{O}^?_{\vec{\mathfrak{X}}^{\mathrm{ord}}_{\breve{\Phi}_{\widetilde{\mathcal{H}}},\breve{\delta}_{\widetilde{\mathcal{H}}},\breve{\sigma},\tau}}))) \\ & \cong H^c(\Gamma_{\breve{\Phi}_{\widetilde{\mathcal{H}}},\Phi_{\mathcal{H}}}, \mathbb{Z}) \underset{\mathbb{Z}}{\otimes} R^d h_*(\mathscr{O}_{\vec{\mathfrak{X}}^{\mathrm{ord}}_{\Phi_{\mathcal{H}},\delta_{\mathcal{H}},\tau} \underset{\vec{C}^{\mathrm{ord}}_{\Phi_{\mathcal{H}},\delta_{\mathcal{H}}}}{\times} \vec{C}^{\mathrm{ord}}_{\breve{\Phi}_{\widetilde{\mathcal{H}}},\breve{\delta}_{\widetilde{\mathcal{H}}}}}) \underset{\mathscr{O}_{\vec{\mathsf{M}}^{\mathrm{ord,tor}}_{\mathcal{H}}}}{\otimes} \mathscr{I}_{\vec{\mathsf{D}}^{\mathrm{ord}}_{\infty,\mathcal{H}}}, \end{aligned} \tag{7.3.3.43}$$

for ? = + and ++, and

$$\begin{aligned} &H^c(\mathfrak{N}_{\breve{\sigma},\tau}, \underline{\mathscr{H}}^d(\mathscr{O}_{\vec{\mathsf{Z}}^{\mathrm{ord}}_{[(\Phi_{\mathcal{H}},\delta_{\mathcal{H}},\tau)]}})) \\ &\cong H^c(\Gamma_{\breve{\Phi}_{\widetilde{\mathcal{H}}},\Phi_{\mathcal{H}}}, H^0(\widetilde{\mathfrak{N}}_{\breve{\sigma},\tau}, \underline{\mathscr{H}}^d(\mathscr{O}_{\vec{\Xi}^{\mathrm{ord}}_{\breve{\Phi}_{\widetilde{\mathcal{H}}},\breve{\delta}_{\widetilde{\mathcal{H}}},\tau}}))) \\ &\cong H^c(\Gamma_{\breve{\Phi}_{\widetilde{\mathcal{H}}},\Phi_{\mathcal{H}}}, \mathbb{Z}) \underset{\mathbb{Z}}{\otimes} R^d h_*(\mathscr{O}_{\vec{\mathsf{Z}}^{\mathrm{ord}}_{[(\Phi_{\mathcal{H}},\delta_{\mathcal{H}},\tau)]} \underset{\vec{C}^{\mathrm{ord}}_{\Phi_{\mathcal{H}},\delta_{\mathcal{H}}}}{\times} \vec{C}^{\mathrm{ord}}_{\breve{\Phi}_{\widetilde{\mathcal{H}}},\breve{\delta}_{\widetilde{\mathcal{H}}}}}). \end{aligned} \tag{7.3.3.44}$$

Now the proposition follows from Lemmas 7.3.3.9 and 7.3.3.34, and the compatible canonical isomorphisms (7.3.3.42), (7.3.3.43), and (7.3.3.44). □

Proposition 7.3.3.45. (*Compare with [61, Lem. 4.29].*) *The spectral sequence* (7.3.3.23) *degenerates at* E_2 *terms for* $? = \emptyset$, $+$, *and* $++$. *Consequently, since the choice of the stratum* $\vec{\mathsf{Z}}^{\mathrm{ord}}_{[(\Phi_{\mathcal{H}},\delta_{\mathcal{H}},\tau)]}$ *is arbitrary, by Grothendieck's fundamental theorem [35, III-1, 4.1.5] (and by fpqc descent for the property of local freeness [33, VIII, 1.11]), for* $? = \emptyset$, $+$, *and* $++$, *the* $\mathscr{O}_{\vec{\mathsf{M}}^{\mathrm{ord,tor}}_{\mathcal{H}}}$*-module* $R^b f^{\mathrm{tor}}_*(\mathscr{O}^{?}_{\vec{\mathsf{N}}^{\mathrm{ord,tor}}})$ *is locally free of the same rank as* $\wedge^b(\underline{\mathrm{Hom}}_{\mathcal{O}}(Q^\vee, \underline{\mathrm{Lie}}_{G^\vee/\vec{\mathsf{M}}^{\mathrm{ord,tor}}_{\mathcal{H}}}))$, *and the composition of canonical morphisms*

$$R^b f^{\mathrm{tor}}_*(\mathscr{O}_{\vec{\mathsf{N}}^{\mathrm{ord,tor}}}) \underset{\mathscr{O}_{\vec{\mathsf{M}}^{\mathrm{ord,tor}}_{\mathcal{H}}}}{\otimes} \mathscr{I}_{\vec{\mathsf{D}}^{\mathrm{ord}}_{\infty,\mathcal{H}}} \cong R^b f^{\mathrm{tor}}_*((f^{\mathrm{tor}})^* \mathscr{I}_{\vec{\mathsf{D}}^{\mathrm{ord}}_{\infty,\mathcal{H}}}) \to R^b f^{\mathrm{tor}}_*(\mathscr{I}_{\vec{\mathsf{E}}^{\mathrm{ord}}_{\infty}}) \tag{7.3.3.46}$$

(*induced by* (7.3.1.1) *and the projection formula [35,* 0_I*, 5.4.10.1]) is an isomorphism. On the other hand, if we have*

$$\begin{aligned} &\dim_{k(s)}((R^b f^{\mathrm{tor}}_*(\mathscr{O}_{\vec{\mathsf{Z}}^{\mathrm{ord}}_{[(\Phi_{\mathcal{H}},\delta_{\mathcal{H}},\tau)]}})) \underset{\mathscr{O}_{\vec{\mathsf{Z}}^{\mathrm{ord}}_{[(\Phi_{\mathcal{H}},\delta_{\mathcal{H}},\tau)]}}}{\otimes} k(s)) \\ &\geq \dim_{k(s)}((R^b f^{\mathrm{tor}}_*(\mathscr{O}_{(\vec{\mathsf{N}}^{\mathrm{ord,tor}})^\wedge_{\vec{\mathsf{Z}}^{\mathrm{ord}}_{[(\Phi_{\mathcal{H}},\delta_{\mathcal{H}},\tau)]}}})) \underset{\mathscr{O}_{\vec{\mathfrak{X}}^{\mathrm{ord}}_{\Phi_{\mathcal{H}},\delta_{\mathcal{H}},\tau}}}{\otimes} k(s)) \end{aligned} \tag{7.3.3.47}$$

at every **maximal point** s *of* $\vec{\mathsf{Z}}^{\mathrm{ord}}_{[(\Phi_{\mathcal{H}},\delta_{\mathcal{H}},\tau)]}$ *(see [36, 0, 2.1.2]), then the spectral sequence* (7.3.3.24) *degenerates at* E_2 *terms as well, and the canonical morphism*

$$R^b f^{\mathrm{tor}}_*(\mathscr{O}_{\vec{\mathsf{N}}^{\mathrm{ord,tor}}}) \underset{\mathscr{O}_{\vec{\mathsf{M}}^{\mathrm{ord,tor}}_{\mathcal{H}}}}{\otimes} \mathscr{O}_{\vec{\mathsf{Z}}^{\mathrm{ord}}_{[(\Phi_{\mathcal{H}},\delta_{\mathcal{H}},\tau)]}} \to R^b f^{\mathrm{tor}}_*(\mathscr{O}_{\vec{\mathsf{Z}}^{\mathrm{ord}}_{[(\Phi_{\mathcal{H}},\delta_{\mathcal{H}},\tau)]}}) \tag{7.3.3.48}$$

of $\mathscr{O}_{\vec{\mathsf{Z}}^{\mathrm{ord}}_{[(\Phi_{\mathcal{H}},\delta_{\mathcal{H}},\tau)]}}$*-modules is an isomorphism.*

Proof. Let $\mathrm{Spf}(R, I)$ be any connected affine open formal subscheme of $\mathfrak{X}_{\Phi_{\mathcal{H}},\delta_{\mathcal{H}},\tau}$, with the ideal of definition I satisfying $\mathrm{rad}(I) = I$ for simplicity. Since $\vec{\mathsf{M}}^{\mathrm{ord,tor}}_{\mathcal{H}}$ is smooth and of finite type over $\vec{\mathsf{S}}_{0,r_{\mathcal{H}}} = \mathrm{Spec}(\mathcal{O}_{F_0,(p)}[\zeta_{p^{r_{\mathcal{H}}}}])$, the ring R is a noetherian domain. (See [76, 33.I and 34.A].) Since the subscheme $\vec{\mathsf{Z}}^{\mathrm{ord}}_{[(\Phi_{\mathcal{H}},\delta_{\mathcal{H}},\tau)]}$ of $\vec{\mathsf{M}}^{\mathrm{ord,tor}}_{\mathcal{H}}$ is also smooth (and of finite type) over $\vec{\mathsf{S}}_{0,r_{\mathcal{H}}}$, the quotient R/I is also a noetherian domain. Let $K := \mathrm{Frac}(R)$ and $k := \mathrm{Frac}(R/I)$ be the fraction fields. Note that they are both of characteristic zero. By abuse of notation, we shall denote with subscripts "K" (resp. "k") the pullbacks of schemes to $\mathrm{Spec}(K)$ (resp. $\mathrm{Spec}(k)$).

Since we have an exact sequence

$$0 \to \underline{\mathrm{Lie}}_{T^\vee/\vec{\mathfrak{X}}^{\mathrm{ord}}_{\Phi_{\mathcal{H}},\delta_{\mathcal{H}},\tau}} \to \underline{\mathrm{Lie}}_{G^{\vee,\natural}/\vec{\mathfrak{X}}^{\mathrm{ord}}_{\Phi_{\mathcal{H}},\delta_{\mathcal{H}},\tau}} \to \underline{\mathrm{Lie}}_{B^\vee/\vec{\mathfrak{X}}^{\mathrm{ord}}_{\Phi_{\mathcal{H}},\delta_{\mathcal{H}},\tau}} \to 0$$

of locally free sheaves, we have an equality

$$\begin{aligned}&\sum_{c+d=b} \dim_K(\wedge^c(\mathrm{Hom}_{\mathcal{O}}(Q^\vee, \mathrm{Lie}_{B_K^\vee}))) \underset{K}{\otimes} (\wedge^d(\mathrm{Hom}_{\mathcal{O}}(Q^\vee, \mathrm{Lie}_{T_K^\vee}))) \\ &= \dim_K(\wedge^b(\mathrm{Hom}_{\mathcal{O}}(Q^\vee, \mathrm{Lie}_{G_K^{\vee,\natural}}))) = \dim_K(\wedge^b(\mathrm{Hom}_{\mathcal{O}}(Q^\vee, \mathrm{Lie}_{G_K^\vee}))),\end{aligned} \tag{7.3.3.49}$$

and an analogous equality with K replaced with k.

By the construction of the spectral sequences (7.3.3.23) and (7.3.3.24), by the canonical isomorphisms (7.3.3.39) and (7.3.3.41), and by the equality (7.3.3.49), we have

$$\begin{aligned}&\sum_{c+d=b} \dim_K(H^c(\mathfrak{N}_{\breve{\sigma},\tau}, \underline{\mathscr{H}}^d(\mathscr{O}^?_{(\vec{\mathsf{N}}^{\mathrm{ord,tor}})^\wedge_{\vec{\mathsf{Z}}^{\mathrm{ord}}_{[(\Phi_{\mathcal{H}},\delta_{\mathcal{H}},\tau)]}}})) \underset{\mathscr{O}_{\vec{\mathfrak{X}}^{\mathrm{ord}}_{\Phi_{\mathcal{H}},\delta_{\mathcal{H}},\tau}}}{\otimes} K) \\ &= \dim_K(\wedge^b(\mathrm{Hom}_{\mathcal{O}}(Q^\vee, \mathrm{Lie}_{G_K^\vee}))) \\ &\geq \dim_K((R^b f^{\mathrm{tor}}_*(\mathscr{O}^?_{(\vec{\mathsf{N}}^{\mathrm{ord,tor}})^\wedge_{\vec{\mathsf{Z}}^{\mathrm{ord}}_{[(\Phi_{\mathcal{H}},\delta_{\mathcal{H}},\tau)]}}})) \underset{\mathscr{O}_{\vec{\mathfrak{X}}^{\mathrm{ord}}_{\Phi_{\mathcal{H}},\delta_{\mathcal{H}},\tau}}}{\otimes} K)\end{aligned} \tag{7.3.3.50}$$

for $? = \emptyset$, $+$, and $++$, and

$$\begin{aligned}&\sum_{c+d=b} \dim_k(H^c(\mathfrak{N}_{\breve{\sigma},\tau}, \underline{\mathscr{H}}^d(\mathscr{O}_{\vec{\mathsf{Z}}^{\mathrm{ord}}_{[(\Phi_{\mathcal{H}},\delta_{\mathcal{H}},\tau)]}})) \underset{\mathscr{O}_{\vec{\mathsf{Z}}^{\mathrm{ord}}_{[(\Phi_{\mathcal{H}},\delta_{\mathcal{H}},\tau)]}}}{\otimes} k) \\ &= \dim_k(\wedge^b(\mathrm{Hom}_{\mathcal{O}}(Q^\vee, \mathrm{Lie}_{G_k^\vee}))) \\ &\geq \dim_k(R^b f^{\mathrm{tor}}_*(\mathscr{O}_{\vec{\mathsf{Z}}^{\mathrm{ord}}_{[(\Phi_{\mathcal{H}},\delta_{\mathcal{H}},\tau)]}}) \underset{\mathscr{O}_{\vec{\mathsf{Z}}^{\mathrm{ord}}_{[(\Phi_{\mathcal{H}},\delta_{\mathcal{H}},\tau)]}}}{\otimes} k).\end{aligned} \tag{7.3.3.51}$$

Let $?$ be either $\emptyset$, $+$, or $++$. Since the pullback of f^{tor} to the open dense subscheme $\vec{\mathsf{M}}^{\mathrm{ord}}_{\mathcal{H}}$ of $\vec{\mathsf{M}}^{\mathrm{ord,tor}}_{\mathcal{H}}$ is simply the abelian scheme torsor $f : \vec{\mathsf{N}}^{\mathrm{ord}} \to \vec{\mathsf{M}}^{\mathrm{ord}}_{\mathcal{H}}$, and since the canonical morphism $\mathrm{Spec}(K) \to \vec{\mathsf{M}}^{\mathrm{ord,tor}}_{\mathcal{H}}$ factors through some maximal point of $\vec{\mathsf{M}}^{\mathrm{ord}}_{\mathcal{H}}$, we have

$$\begin{aligned}&(R^b f^{\mathrm{tor}}_*(\mathscr{O}^?_{\vec{\mathsf{N}}^{\mathrm{ord,tor}}})) \underset{\mathscr{O}_{\vec{\mathsf{M}}^{\mathrm{ord,tor}}_{\mathcal{H}}}}{\otimes} K \cong R^b f_*(\mathscr{O}^?_{\vec{\mathsf{N}}^{\mathrm{ord}}}) \underset{\mathscr{O}_{\vec{\mathsf{M}}^{\mathrm{ord}}_{\mathcal{H}}}}{\otimes} K \\ &\cong (\wedge^b \underline{\mathrm{Lie}}_{\mathsf{N}^\vee/\mathsf{M}_{\mathcal{H}}}) \underset{\mathscr{O}_{\vec{\mathsf{M}}^{\mathrm{ord}}_{\mathcal{H}}}}{\otimes} K \cong \wedge^b(\underline{\mathrm{Hom}}_{\mathcal{O}}(Q^\vee, \underline{\mathrm{Lie}}_{G_K^\vee})).\end{aligned}$$

This implies that the inequality in (7.3.3.50) is an equality, and hence that the spectral sequence (7.3.3.23) degenerates at E_2 terms (for $? = \emptyset$, $+$, and $++$) after pulled back to K. Since all E_2 terms of this spectral sequence are locally free sheaves, this shows that (7.3.3.23) degenerates at E_2 terms after pulled back to R. Since the choice of R is arbitrary, this shows that (7.3.3.23) degenerates over the whole $\vec{\mathfrak{X}}^{\mathrm{ord}}_{\Phi_{\mathcal{H}},\delta_{\mathcal{H}},\tau}$. Hence, as explained in the statement of the proposition, since the choice of $\vec{\mathsf{Z}}^{\mathrm{ord}}_{[(\Phi_{\mathcal{H}},\delta_{\mathcal{H}},\tau)]}$ is arbitrary, this shows that $R^b f^{\mathrm{tor}}_*(\mathscr{O}^?_{\vec{\mathsf{N}}^{\mathrm{ord,tor}}})$ is locally free of the same rank as $\wedge^b(\underline{\mathrm{Hom}}_{\mathcal{O}}(Q^\vee, \underline{\mathrm{Lie}}_{G^\vee/\vec{\mathsf{M}}^{\mathrm{ord,tor}}_{\mathcal{H}}}))$ over $\vec{\mathsf{M}}^{\mathrm{ord,tor}}_{\mathcal{H}}$. (Nevertheless, since

f^{tor} is not necessarily flat, this does not imply that the formation of $R^b f^{\mathrm{tor}}_*(\mathscr{O}^?_{\vec{\mathsf{N}}^{\mathrm{ord,tor}}})$ is compatible with arbitrary base change.)

By Proposition 7.3.3.37, the canonical inclusion $\mathscr{O}^{++}_{\vec{\mathsf{N}}^{\mathrm{ord,tor}}} \hookrightarrow \mathscr{O}^{+}_{\vec{\mathsf{N}}^{\mathrm{ord,tor}}}$ (which is nothing but (7.3.1.1)) induces isomorphisms between the E_2 terms of the spectral sequences (7.3.3.23) for $? = +$ and $++$. Hence, the composition of canonical morphisms in (7.3.3.46) is an isomorphism.

Since the canonical morphism $\mathrm{Spec}(k) \to \vec{\mathsf{Z}}^{\mathrm{ord}}_{[(\Phi_{\mathcal{H}},\delta_{\mathcal{H}},\tau)]}$ factors through some maximal point of $\vec{\mathsf{Z}}^{\mathrm{ord}}_{[(\Phi_{\mathcal{H}},\delta_{\mathcal{H}},\tau)]}$, the inequality (7.3.3.47) implies that

$$\begin{aligned} &\dim_k(R^b f^{\mathrm{tor}}_*(\mathscr{O}_{\vec{\vec{\mathsf{Z}}}^{\mathrm{ord}}_{[(\Phi_{\mathcal{H}},\delta_{\mathcal{H}},\tau)]}}) \underset{\mathscr{O}_{\vec{\mathsf{Z}}^{\mathrm{ord}}_{[(\Phi_{\mathcal{H}},\delta_{\mathcal{H}},\tau)]}}}{\otimes} k) \\ &\geq \dim_k((R^b f^{\mathrm{tor}}_*(\mathscr{O}_{(\vec{\mathsf{N}}^{\mathrm{ord,tor}})^{\wedge}_{\vec{\vec{\mathsf{Z}}}^{\mathrm{ord}}_{[(\Phi_{\mathcal{H}},\delta_{\mathcal{H}},\tau)]}}})) \underset{\mathscr{O}_{\vec{\mathfrak{X}}^{\mathrm{ord}}_{\Phi_{\mathcal{H}},\delta_{\mathcal{H}},\tau}}}{\otimes} k) \\ &= \dim_K((R^b f^{\mathrm{tor}}_*(\mathscr{O}_{(\vec{\mathsf{N}}^{\mathrm{ord,tor}})^{\wedge}_{\vec{\vec{\mathsf{Z}}}^{\mathrm{ord}}_{[(\Phi_{\mathcal{H}},\delta_{\mathcal{H}},\tau)]}}})) \underset{\mathscr{O}_{\vec{\mathfrak{X}}^{\mathrm{ord}}_{\Phi_{\mathcal{H}},\delta_{\mathcal{H}},\tau}}}{\otimes} K), \end{aligned}$$

and hence the inequality in (7.3.3.50) being an *equality* implies that the inequality in (7.3.3.51) is also an equality, because

$$\dim_k(\wedge^b(\mathrm{Hom}_{\mathcal{O}}(Q^\vee, \mathrm{Lie}_{G^\vee_k}))) = \dim_K(\wedge^b(\mathrm{Hom}_{\mathcal{O}}(Q^\vee, \mathrm{Lie}_{G^\vee_K}))).$$

Therefore, by the same argument as in the case of (7.3.3.23), the spectral sequence (7.3.3.24) also degenerates at E_2 terms. Since the spectral sequences (7.3.3.23) and (7.3.3.24) are compatible with each other (by their very construction), their degeneracy implies that the canonical morphism

$$R^b f^{\mathrm{tor}}_*(\mathscr{O}_{(\vec{\mathsf{N}}^{\mathrm{ord,tor}})^{\wedge}_{\vec{\vec{\mathsf{Z}}}^{\mathrm{ord}}_{[(\Phi_{\mathcal{H}},\delta_{\mathcal{H}},\tau)]}}}) \underset{\mathscr{O}_{\vec{\mathfrak{X}}^{\mathrm{ord}}_{\Phi_{\mathcal{H}},\delta_{\mathcal{H}},\tau}}}{\otimes} \mathscr{O}_{\vec{\mathsf{Z}}^{\mathrm{ord}}_{[(\Phi_{\mathcal{H}},\delta_{\mathcal{H}},\tau)]}} \to R^b f^{\mathrm{tor}}_*(\mathscr{O}_{\vec{\vec{\mathsf{Z}}}^{\mathrm{ord}}_{[(\Phi_{\mathcal{H}},\delta_{\mathcal{H}},\tau)]}})$$

is an isomorphism (by comparing graded pieces). Since the choice of the stratum $\vec{\vec{\mathsf{Z}}}^{\mathrm{ord}}_{[(\Phi_{\mathcal{H}},\delta_{\mathcal{H}},\tau)]}$ is arbitrary, this shows that the canonical morphism (7.3.3.48) is also an isomorphism. □

Remark 7.3.3.52. (Compare with [61, Rem. 4.35].) By upper semicontinuity for proper flat morphisms (see [81, Sec. 5, Cor. (a)]), the assumption (7.3.3.47) is satisfied when f^{tor} is flat, or equivalently when Condition 7.2.6.3 is satisfied (by Proposition 7.2.6.4), which can be achieved by refining both $\widetilde{\Sigma}^{\mathrm{ord}}$ and Σ^{ord} (by Proposition 7.2.6.5).

Corollary 7.3.3.53. (*Compare with [61, Cor. 4.36].*) *For each integer* $b \geq 0$, *the canonical morphism* $\wedge^b(R^1 f^{\mathrm{tor}}_*(\mathscr{O}_{\vec{\mathsf{N}}^{\mathrm{ord,tor}}})) \to R^b f^{\mathrm{tor}}_*(\mathscr{O}_{\vec{\mathsf{N}}^{\mathrm{ord,tor}}})$ *(defined by cup product) is an isomorphism.*

Proof. As in Proposition 7.3.3.45, by properness of f^{tor}, this is true if and only if it is true over the formal completion along each stratum $\vec{\mathsf{Z}}^{\mathrm{ord}}_{[(\Phi_{\mathcal{H}},\delta_{\mathcal{H}},\tau)]}$, which is the case because the canonical morphism induces isomorphisms on all graded pieces defined by spectral sequences such as (7.3.3.23), which are compatible with cup products by the very construction (see [30, II, Sec. 5–6]). □

7.3.4 *Degeneracy of the (Relative) Hodge Spectral Sequence*

Let

$$\underline{H}^i_{\text{log-dR}}(\vec{\mathsf{N}}^{\text{ord,tor}}/\vec{\mathsf{M}}^{\text{ord,tor}}_{\mathcal{H}}) := R^i f^{\text{tor}}_* \overline{\Omega}^{\bullet}_{\vec{\mathsf{N}}^{\text{ord,tor}}/\vec{\mathsf{M}}^{\text{ord,tor}}_{\mathcal{H}}}$$

be the (relative) log de Rham cohomology as in (3c) of Theorem 7.1.4.1. By the definition, the natural (Hodge) filtration on the complex $\overline{\Omega}^{\bullet}_{\vec{\mathsf{N}}^{\text{ord,tor}}/\vec{\mathsf{M}}^{\text{ord,tor}}_{\mathcal{H}}}$ defines the (relative) Hodge spectral sequence (7.1.4.7):

$$E_1^{a,b} := R^b f^{\text{tor}}_*(\overline{\Omega}^a_{\vec{\mathsf{N}}^{\text{ord,tor}}/\vec{\mathsf{M}}^{\text{ord,tor}}_{\mathcal{H}}}) \Rightarrow \underline{H}^{a+b}_{\text{log-dR}}(\vec{\mathsf{N}}^{\text{ord,tor}}/\vec{\mathsf{M}}^{\text{ord,tor}}_{\mathcal{H}}).$$

By (3a) of Theorem 7.1.4.1 (which we have proved in Section 7.2.5), there is a canonical isomorphism

$$\begin{aligned}\overline{\Omega}^a_{\vec{\mathsf{N}}^{\text{ord,tor}}/\vec{\mathsf{M}}^{\text{ord,tor}}_{\mathcal{H}}} &\cong \wedge^a\Big[(f^{\text{tor}})^*(\underline{\text{Lie}}^{\vee}_{\vec{\mathsf{N}}^{\text{ord,ext}}/\vec{\mathsf{M}}^{\text{ord,tor}}_{\mathcal{H}}})\Big] \\ &\cong (f^{\text{tor}})^*\Big[\wedge^a(\underline{\text{Lie}}^{\vee}_{\vec{\mathsf{N}}^{\text{ord,ext}}/\vec{\mathsf{M}}^{\text{ord,tor}}_{\mathcal{H}}})\Big]\end{aligned}$$

of locally free sheaves over $\vec{\mathsf{N}}^{\text{ord,tor}}$. Then (by the projection formula [35, 0_{I}, 5.4.10.1]) we have canonical isomorphisms

$$\begin{aligned}&R^b f^{\text{tor}}_*(\overline{\Omega}^a_{\vec{\mathsf{N}}^{\text{ord,tor}}/\vec{\mathsf{M}}^{\text{ord,tor}}_{\mathcal{H}}}) \\ &\cong (R^b f^{\text{tor}}_*(\mathscr{O}_{\vec{\mathsf{N}}^{\text{ord,tor}}})) \underset{\mathscr{O}_{\vec{\mathsf{M}}^{\text{ord,tor}}_{\mathcal{H}}}}{\otimes} (\wedge^a(\underline{\text{Lie}}^{\vee}_{\vec{\mathsf{N}}^{\text{ord,ext}}/\vec{\mathsf{M}}^{\text{ord,tor}}_{\mathcal{H}}})).\end{aligned} \tag{7.3.4.1}$$

Lemma 7.3.4.2. (*Compare with [61, Rem. 4.38].*) *If $R^b f^{\text{tor}}_*(\mathscr{O}_{\vec{\mathsf{N}}^{\text{ord,tor}}})$ is locally free for every integer $b \geq 0$, then the spectral sequence* (7.1.4.7) *degenerates at the E_1 terms.*

Proof. By (7.3.4.1), if $R^b f^{\text{tor}}_*(\mathscr{O}_{\vec{\mathsf{N}}^{\text{ord,tor}}})$ is locally free for every integer $b \geq 0$, then all the E_1 terms $R^b f^{\text{tor}}_*(\overline{\Omega}^a_{\vec{\mathsf{N}}^{\text{ord,tor}}/\vec{\mathsf{M}}^{\text{ord,tor}}_{\mathcal{H}}})$ of the spectral sequence (7.1.4.7) are locally free. Therefore, to show that (7.1.4.7) degenerates at E_1 terms, it suffices to show that it degenerates at E_1 terms over the open dense subscheme $\vec{\mathsf{M}}^{\text{ord}}_{\mathcal{H}}$ of $\vec{\mathsf{M}}^{\text{ord,tor}}_{\mathcal{H}}$, which is true because $f^{\text{tor}}|_{\vec{\mathsf{N}}^{\text{ord}}} = f : \vec{\mathsf{N}}^{\text{ord}} \to \vec{\mathsf{M}}^{\text{ord}}_{\mathcal{H}}$ is an abelian scheme torsor. (See, for example, [6, Prop. 2.5.2].) □

This proves (3c) of Theorem 7.1.4.1, because the local freeness of $R^b f^{\text{tor}}_*(\mathscr{O}_{\vec{\mathsf{N}}^{\text{ord,tor}}})$ has been established in Section 7.3.3 for every integer $b \geq 0$.

7.3.5 *Extended Gauss–Manin Connections*

In Section 7.2.5, we proved the log smoothness of $f^{\text{tor}} : \vec{\mathsf{N}}^{\text{ord,tor}} \to \vec{\mathsf{M}}^{\text{ord,tor}}_{\mathcal{H}}$ by verifying Lemma 7.2.5.1. For simplicity, let us set

$$\overline{\Omega}^1_{\vec{\mathsf{M}}^{\text{ord,tor}}_{\mathcal{H}}/\vec{\mathsf{S}}_{0,r_{\mathcal{H}}}} := \Omega^1_{\vec{\mathsf{M}}^{\text{ord,tor}}_{\mathcal{H}}/\vec{\mathsf{S}}_{0,r_{\mathcal{H}}}}[d\log\infty]$$

and

$$\overline{\Omega}^1_{\vec{\mathsf{N}}^{\text{ord,tor}}/\vec{\mathsf{S}}_{0,r_{\mathcal{H}}}} := \Omega^1_{\vec{\mathsf{N}}^{\text{ord,tor}}/\vec{\mathsf{S}}_{0,r_{\mathcal{H}}}}[d\log\infty].$$

Then (7.2.5.2) can be rewritten as the exact sequence

$$0 \to (f^{\mathrm{tor}})^*(\overline{\Omega}^1_{\vec{\mathsf{M}}^{\mathrm{ord,tor}}_{\mathcal{H}}/\vec{\mathsf{S}}_{0,r_{\mathcal{H}}}}) \to \overline{\Omega}^1_{\vec{\mathsf{N}}^{\mathrm{ord,tor}}/\vec{\mathsf{S}}_{0,r_{\mathcal{H}}}} \to \overline{\Omega}^1_{\vec{\mathsf{N}}^{\mathrm{ord,tor}}/\vec{\mathsf{M}}^{\mathrm{ord,tor}}_{\mathcal{H}}} \to 0, \quad (7.3.5.1)$$

which induces the *Koszul filtration* [48, 1.2, 1.3]

$$\begin{aligned} &\mathsf{K}^a(\overline{\Omega}^\bullet_{\vec{\mathsf{N}}^{\mathrm{ord,tor}}/\vec{\mathsf{S}}_{0,r_{\mathcal{H}}}}) \\ &:= \mathrm{image}\big(\overline{\Omega}^{\bullet-a}_{\vec{\mathsf{N}}^{\mathrm{ord,tor}}/\vec{\mathsf{S}}_{0,r_{\mathcal{H}}}} \underset{\mathscr{O}_{\vec{\mathsf{N}}^{\mathrm{ord,tor}}}}{\otimes} (f^{\mathrm{tor}})^*(\overline{\Omega}^a_{\vec{\mathsf{M}}^{\mathrm{ord,tor}}_{\mathcal{H}}/\vec{\mathsf{S}}_{0,r_{\mathcal{H}}}}) \to \overline{\Omega}^\bullet_{\vec{\mathsf{N}}^{\mathrm{ord,tor}}/\vec{\mathsf{S}}_{0,r_{\mathcal{H}}}}\big) \end{aligned}$$

on $\overline{\Omega}^\bullet_{\vec{\mathsf{N}}^{\mathrm{ord,tor}}/\vec{\mathsf{S}}_{0,r_{\mathcal{H}}}}$, with graded pieces

$$\mathrm{Gr}^a_{\mathsf{K}}(\overline{\Omega}^\bullet_{\vec{\mathsf{N}}^{\mathrm{ord,tor}}/\vec{\mathsf{S}}_{0,r_{\mathcal{H}}}}) \cong \overline{\Omega}^{\bullet-a}_{\vec{\mathsf{N}}^{\mathrm{ord,tor}}/\vec{\mathsf{M}}^{\mathrm{ord,tor}}_{\mathcal{H}}} \underset{\mathscr{O}_{\vec{\mathsf{N}}^{\mathrm{ord,tor}}}}{\otimes} (f^{\mathrm{tor}})^*(\overline{\Omega}^a_{\vec{\mathsf{M}}^{\mathrm{ord,tor}}_{\mathcal{H}}/\vec{\mathsf{S}}_{0,r_{\mathcal{H}}}}).$$

On the other hand, we have the Hodge filtration

$$\mathsf{F}^a(\overline{\Omega}^\bullet_{\vec{\mathsf{N}}^{\mathrm{ord,tor}}/\vec{\mathsf{S}}_{0,r_{\mathcal{H}}}}) := \overline{\Omega}^{\bullet \geq a}_{\vec{\mathsf{N}}^{\mathrm{ord,tor}}/\vec{\mathsf{S}}_{0,r_{\mathcal{H}}}}$$

on $\overline{\Omega}^\bullet_{\vec{\mathsf{N}}^{\mathrm{ord,tor}}/\vec{\mathsf{S}}_{0,r_{\mathcal{H}}}}$, giving the Hodge filtration

$$\begin{aligned} &\mathsf{F}^a(\underline{H}^i_{\text{log-dR}}(\vec{\mathsf{N}}^{\mathrm{ord,tor}}/\vec{\mathsf{M}}^{\mathrm{ord,tor}}_{\mathcal{H}})) \\ &:= \mathrm{image}\big(R^i f^{\mathrm{tor}}_*(\mathsf{F}^a(\overline{\Omega}^\bullet_{\vec{\mathsf{N}}^{\mathrm{ord,tor}}/\vec{\mathsf{S}}_{0,r_{\mathcal{H}}}})) \to R^i f^{\mathrm{tor}}_*(\overline{\Omega}^\bullet_{\vec{\mathsf{N}}^{\mathrm{ord,tor}}/\vec{\mathsf{S}}_{0,r_{\mathcal{H}}}})\big) \end{aligned}$$

on $\underline{H}^i_{\text{log-dR}}(\vec{\mathsf{N}}^{\mathrm{ord,tor}}/\vec{\mathsf{M}}^{\mathrm{ord,tor}}_{\mathcal{H}})$. By applying $R^\bullet f^{\mathrm{tor}}_*$ to the short exact sequence

$$\begin{aligned} 0 &\to \overline{\Omega}^{\bullet-1}_{\vec{\mathsf{N}}^{\mathrm{ord,tor}}/\vec{\mathsf{M}}^{\mathrm{ord,tor}}_{\mathcal{H}}} \underset{\mathscr{O}_{\vec{\mathsf{N}}^{\mathrm{ord,tor}}}}{\otimes} (f^{\mathrm{tor}})^*(\overline{\Omega}^1_{\vec{\mathsf{M}}^{\mathrm{ord,tor}}_{\mathcal{H}}/\vec{\mathsf{S}}_{0,r_{\mathcal{H}}}}) \\ &\to \mathsf{K}^0(\overline{\Omega}^\bullet_{\vec{\mathsf{N}}^{\mathrm{ord,tor}}/\vec{\mathsf{S}}_{0,r_{\mathcal{H}}}})/\mathsf{K}^2(\overline{\Omega}^\bullet_{\vec{\mathsf{N}}^{\mathrm{ord,tor}}/\vec{\mathsf{S}}_{0,r_{\mathcal{H}}}}) \to \overline{\Omega}^\bullet_{\vec{\mathsf{N}}^{\mathrm{ord,tor}}/\vec{\mathsf{M}}^{\mathrm{ord,tor}}_{\mathcal{H}}} \to 0, \end{aligned} \quad (7.3.5.2)$$

we obtain in the long exact sequence the connecting homomorphisms

$$\begin{aligned} &\underline{H}^i_{\text{log-dR}}(\vec{\mathsf{N}}^{\mathrm{ord,tor}}/\vec{\mathsf{M}}^{\mathrm{ord,tor}}_{\mathcal{H}}) = R^i f^{\mathrm{tor}}_*(\overline{\Omega}^\bullet_{\vec{\mathsf{N}}^{\mathrm{ord,tor}}/\vec{\mathsf{M}}^{\mathrm{ord,tor}}_{\mathcal{H}}}) \\ &\overset{\nabla}{\to} R^{i+1} f^{\mathrm{tor}}_*(\overline{\Omega}^{\bullet-1}_{\vec{\mathsf{N}}^{\mathrm{ord,tor}}/\vec{\mathsf{M}}^{\mathrm{ord,tor}}_{\mathcal{H}}} \underset{\mathscr{O}_{\vec{\mathsf{N}}^{\mathrm{ord,tor}}}}{\otimes} \overline{\Omega}^1_{\vec{\mathsf{M}}^{\mathrm{ord,tor}}_{\mathcal{H}}/\vec{\mathsf{S}}_{0,r_{\mathcal{H}}}}) \\ &\cong \underline{H}^i_{\text{log-dR}}(\vec{\mathsf{N}}^{\mathrm{ord,tor}}/\vec{\mathsf{M}}^{\mathrm{ord,tor}}_{\mathcal{H}}) \underset{\mathscr{O}_{\vec{\mathsf{M}}^{\mathrm{ord,tor}}_{\mathcal{H}}}}{\otimes} \overline{\Omega}^1_{\vec{\mathsf{M}}^{\mathrm{ord,tor}}_{\mathcal{H}}/\vec{\mathsf{S}}_{0,r_{\mathcal{H}}}}. \end{aligned} \quad (7.3.5.3)$$

As explained in [48, 1.4], the pullback of ∇ in (7.3.5.3) to $\vec{\mathsf{M}}^{\mathrm{ord}}_{\mathcal{H}}$ is nothing but the usual Gauss–Manin connection on $\underline{H}^i_{\mathrm{dR}}(\vec{\mathsf{N}}^{\mathrm{ord}}/\vec{\mathsf{M}}^{\mathrm{ord}}_{\mathcal{H}})$. Since the sheaves involved in (7.3.5.3) are all locally free,

$$\begin{aligned} \nabla : &\underline{H}^i_{\text{log-dR}}(\vec{\mathsf{N}}^{\mathrm{ord,tor}}/\vec{\mathsf{M}}^{\mathrm{ord,tor}}_{\mathcal{H}}) \\ &\to \underline{H}^i_{\text{log-dR}}(\vec{\mathsf{N}}^{\mathrm{ord,tor}}/\vec{\mathsf{M}}^{\mathrm{ord,tor}}_{\mathcal{H}}) \underset{\mathscr{O}_{\mathsf{M}^{\mathrm{ord,tor}}_{\mathcal{H}}}}{\otimes} \overline{\Omega}^1_{\vec{\mathsf{M}}^{\mathrm{ord,tor}}_{\mathcal{H}}/\vec{\mathsf{S}}_{0,r_{\mathcal{H}}}} \end{aligned}$$

satisfies the necessary conditions for being an integrable connection with log poles (because its restriction to the dense subscheme $\vec{\mathsf{M}}^{\mathrm{ord}}_{\mathcal{H}}$ does). We call this ∇

over $\vec{\mathsf{M}}_{\mathcal{H}}^{\mathrm{ord,tor}}$ the *extended Gauss–Manin connection.* If we take the F-filtration on (7.3.5.2), we obtain

$$0 \to (\mathsf{F}^{a-1}(\overline{\Omega}^{\bullet}_{\vec{\mathsf{N}}^{\mathrm{ord,tor}}/\vec{\mathsf{M}}_{\mathcal{H}}^{\mathrm{ord,tor}}) \underset{\mathscr{O}_{\vec{\mathsf{N}}^{\mathrm{ord,tor}}}}{\otimes} (f^{\mathrm{tor}})^*(\overline{\Omega}^1_{\vec{\mathsf{M}}_{\mathcal{H}}^{\mathrm{ord,tor}}/\vec{\mathsf{S}}_{0,r_{\mathcal{H}}}}))[-1]$$
$$\to \mathsf{F}^a(\mathsf{K}^2/\mathsf{K}^0) \to \mathsf{F}^a(\overline{\Omega}^{\bullet}_{\vec{\mathsf{N}}^{\mathrm{ord,tor}}/\vec{\mathsf{S}}_{0,r_{\mathcal{H}}}}) \to 0$$

and hence the *Griffiths transversality*

$$\nabla(\mathsf{F}^a(\underline{H}^i_{\log\text{-dR}}(\vec{\mathsf{N}}^{\mathrm{ord,tor}}/\vec{\mathsf{M}}_{\mathcal{H}}^{\mathrm{ord,tor}})))$$
$$\subset \mathsf{F}^{a-1}(\underline{H}^i_{\log\text{-dR}}(\vec{\mathsf{N}}^{\mathrm{ord,tor}}/\vec{\mathsf{M}}_{\mathcal{H}}^{\mathrm{ord,tor}})) \underset{\mathscr{O}_{\vec{\mathsf{M}}_{\mathcal{H}}^{\mathrm{ord,tor}}}}{\otimes} \overline{\Omega}^1_{\vec{\mathsf{M}}_{\mathcal{H}}^{\mathrm{ord,tor}}/\vec{\mathsf{S}}_{0,r_{\mathcal{H}}}}$$

(as in [48, Prop. 1.4.1.6]). This proves (3e) of Theorem 7.1.4.1.

Remark 7.3.5.4. By (3c) of Theorem 7.1.4.1, the (relative) Hodge spectral sequence

$$E_1^{a,i-a} := R^{i-a} f_*^{\mathrm{tor}}(\overline{\Omega}^a_{\vec{\mathsf{N}}^{\mathrm{ord,tor}}/\vec{\mathsf{M}}_{\mathcal{H}}^{\mathrm{ord,tor}}}) \Rightarrow \underline{H}^i_{\log\text{-dR}}(\vec{\mathsf{N}}^{\mathrm{ord,tor}}/\vec{\mathsf{M}}_{\mathcal{H}}^{\mathrm{ord,tor}})$$

degenerates. Then we have

$$\mathrm{Gr}^a_{\mathsf{F}}(\underline{H}^i_{\log\text{-dR}}(\vec{\mathsf{N}}^{\mathrm{ord,tor}}/\vec{\mathsf{M}}_{\mathcal{H}}^{\mathrm{ord,tor}})) \cong R^{i-a} f_*^{\mathrm{tor}}(\overline{\Omega}^a_{\vec{\mathsf{N}}^{\mathrm{ord,tor}}/\vec{\mathsf{M}}_{\mathcal{H}}^{\mathrm{ord,tor}}}),$$

and we can conclude (as in [48, Prop. 1.4.1.7]) that the induced morphism

$$\nabla : \mathrm{Gr}^a_{\mathsf{F}}\,\underline{H}^i_{\log\text{-dR}}(\vec{\mathsf{N}}^{\mathrm{ord,tor}}/\vec{\mathsf{M}}_{\mathcal{H}}^{\mathrm{ord,tor}})$$
$$\to \mathrm{Gr}^{a-1}_{\mathsf{F}}\,\underline{H}^i_{\log\text{-dR}}(\vec{\mathsf{N}}^{\mathrm{ord,tor}}/\vec{\mathsf{M}}_{\mathcal{H}}^{\mathrm{ord,tor}}) \underset{\mathscr{O}_{\vec{\mathsf{M}}_{\mathcal{H}}^{\mathrm{ord,tor}}}}{\otimes} \overline{\Omega}^1_{\vec{\mathsf{M}}_{\mathcal{H}}^{\mathrm{ord,tor}}/\vec{\mathsf{S}}_{0,r_{\mathcal{H}}}}$$

agrees with the morphism

$$R^{i-a} f_*^{\mathrm{tor}}(\overline{\Omega}^a_{\vec{\mathsf{N}}^{\mathrm{ord,tor}}/\vec{\mathsf{M}}_{\mathcal{H}}^{\mathrm{ord,tor}}})$$
$$\to R^{i-a+1} f_*^{\mathrm{tor}}(\overline{\Omega}^{a-1}_{\vec{\mathsf{N}}^{\mathrm{ord,tor}}/\vec{\mathsf{M}}_{\mathcal{H}}^{\mathrm{ord,tor}}}) \underset{\mathscr{O}_{\vec{\mathsf{M}}_{\mathcal{H}}^{\mathrm{ord,tor}}}}{\otimes} \overline{\Omega}^1_{\vec{\mathsf{M}}_{\mathcal{H}}^{\mathrm{ord,tor}}/\vec{\mathsf{S}}_{0,r_{\mathcal{H}}}}$$

defined by cup product with the so-called *extended Kodaira–Spencer class* defined by the extension class of (7.3.5.1). We will revisit a special case of this in Chapter 8 (see, in particular, Proposition 8.1.3.6 below).

7.3.6 *Identification of $R^b f_*^{\mathrm{tor}}(\mathscr{O}_{\vec{\mathsf{N}}^{\mathrm{ord,tor}}})$*

Let us define

$$\underline{\overline{\mathrm{Der}}}_{\vec{\mathsf{N}}^{\mathrm{ord,tor}}/\vec{\mathsf{M}}_{\mathcal{H}}^{\mathrm{ord,tor}}} := \underline{\mathrm{Hom}}_{\mathscr{O}_{\vec{\mathsf{N}}^{\mathrm{ord,tor}}}}(\overline{\Omega}^1_{\vec{\mathsf{N}}^{\mathrm{ord,tor}}/\vec{\mathsf{M}}_{\mathcal{H}}^{\mathrm{ord,tor}}}, \mathscr{O}_{\vec{\mathsf{N}}^{\mathrm{ord,tor}}}).$$

Its restriction to $\vec{\mathsf{M}}_{\mathcal{H}}^{\mathrm{ord}}$ can be canonically identified with

$$\underline{\mathrm{Der}}_{\vec{\mathsf{N}}^{\mathrm{ord}}/\vec{\mathsf{M}}_{\mathcal{H}}^{\mathrm{ord}}} := \underline{\mathrm{Hom}}_{\mathscr{O}_{\mathsf{N}}}(\Omega^1_{\vec{\mathsf{N}}^{\mathrm{ord}}/\vec{\mathsf{M}}_{\mathcal{H}}^{\mathrm{ord}}}, \mathscr{O}_{\vec{\mathsf{N}}^{\mathrm{ord}}}).$$

Then, by (7.1.4.2) and (7.1.4.3), and by the projection formula [35, 0_{I}, 5.4.10.1], we have a canonical isomorphism

$$f_*(\underline{\overline{\mathrm{Der}}}_{\vec{\mathsf{N}}^{\mathrm{ord,ext}}/\vec{\mathsf{M}}_{\mathcal{H}}^{\mathrm{ord,tor}}}) \cong \underline{\mathrm{Lie}}_{\vec{\mathsf{N}}^{\mathrm{ord,ext}}/\vec{\mathsf{M}}_{\mathcal{H}}^{\mathrm{ord,tor}}} \tag{7.3.6.1}$$

extending the canonical isomorphism

$$f_*(\underline{\mathrm{Der}}_{\vec{\mathsf{N}}^{\mathrm{ord}}/\vec{\mathsf{M}}_{\mathcal{H}}^{\mathrm{ord}}}) \cong \underline{\mathrm{Lie}}_{\vec{\mathsf{N}}^{\mathrm{ord}}/\vec{\mathsf{M}}_{\mathcal{H}}^{\mathrm{ord}}} \tag{7.3.6.2}$$

(*cf.* [62, Lem. 2.1.5.11]).

Over $\vec{\mathsf{M}}_{\mathcal{H}}^{\mathrm{ord}}$, by Lemma 7.1.3.10, we have a canonical isomorphism

$$R^1 f_*(\mathscr{O}_{\vec{\mathsf{N}}^{\mathrm{ord}}}) \cong \underline{\mathrm{Lie}}_{\vec{\mathsf{N}}^{\mathrm{ord},\vee}/\vec{\mathsf{M}}_{\mathcal{H}}^{\mathrm{ord}}}. \tag{7.3.6.3}$$

The goal is to show that (7.3.6.3) extends to a canonical isomorphism

$$R^1 f_*^{\mathrm{tor}}(\mathscr{O}_{\vec{\mathsf{N}}^{\mathrm{ord,tor}}}) \cong \underline{\mathrm{Lie}}_{\vec{\mathsf{N}}^{\mathrm{ord,ext},\vee}/\vec{\mathsf{M}}_{\mathcal{H}}^{\mathrm{ord,tor}}}. \tag{7.3.6.4}$$

However, since $\vec{\mathsf{N}}^{\mathrm{ord,tor}}$ and $\vec{\mathsf{N}}^{\mathrm{ord,ext},\vee}$ (or rather $\vec{\mathsf{N}}^{\mathrm{ord,ext}}$) are not directly related, there is no obvious morphism between $R^1 f_*^{\mathrm{tor}} \mathscr{O}_{\vec{\mathsf{N}}^{\mathrm{ord,tor}}}$ and $\underline{\mathrm{Lie}}_{\vec{\mathsf{N}}^{\mathrm{ord,ext},\vee}/\vec{\mathsf{M}}_{\mathcal{H}}^{\mathrm{ord}}}$ (in either direction). Nevertheless, since $\vec{\mathsf{M}}_{\mathcal{H}}^{\mathrm{ord}}$ is normal (and, in fact, regular, being smooth over $\vec{\mathsf{S}}_{0,r_{\mathcal{H}}}$), we have the following:

Lemma 7.3.6.5. *To show that there exists an isomorphism* (7.3.6.4) (*extending* (7.3.6.3)), *it suffices to show that there exists a canonical isomorphism*

$$R^1 f_*^{\mathrm{tor}}(\mathscr{O}_{\vec{\mathsf{N}}^{\mathrm{ord,tor}}}) \underset{\mathbb{Z}}{\otimes} \mathbb{Q} \cong \underline{\mathrm{Lie}}_{\vec{\mathsf{N}}^{\mathrm{ord,ext},\vee}/\vec{\mathsf{M}}_{\mathcal{H}}^{\mathrm{ord,tor}}} \underset{\mathbb{Z}}{\otimes} \mathbb{Q} \tag{7.3.6.6}$$

extending (7.3.6.3) $\underset{\mathbb{Z}}{\otimes} \mathbb{Q}$.

Proof. Let us denote by U the union of $\vec{\mathsf{M}}_{\mathcal{H}}^{\mathrm{ord}}$ and $\vec{\mathsf{M}}_{\mathcal{H}}^{\mathrm{ord,tor}} \underset{\mathbb{Z}}{\otimes} \mathbb{Q}$ in $\vec{\mathsf{M}}_{\mathcal{H}}^{\mathrm{ord,tor}}$. Then U is an open subscheme of $\vec{\mathsf{M}}_{\mathcal{H}}^{\mathrm{ord,tor}}$ such that its complement $\vec{\mathsf{M}}_{\mathcal{H}}^{\mathrm{ord,tor}} - U$ is of codimension at least two in $\vec{\mathsf{M}}_{\mathcal{H}}^{\mathrm{ord,tor}}$. Since the canonical isomorphisms (7.3.6.6) and (7.3.6.3) agree with (7.3.6.3) $\underset{\mathbb{Z}}{\otimes} \mathbb{Q}$ over $\vec{\mathsf{M}}_{\mathcal{H}}^{\mathrm{ord}} \underset{\mathbb{Z}}{\otimes} \mathbb{Q}$, they define a canonical isomorphism

$$R^1 f_*^{\mathrm{tor}}(\mathscr{O}_{\vec{\mathsf{N}}^{\mathrm{ord,tor}}}) \underset{\mathscr{O}_{\vec{\mathsf{M}}_{\mathcal{H}}^{\mathrm{ord,tor}}}}{\otimes} \mathscr{O}_U \cong \underline{\mathrm{Lie}}_{\vec{\mathsf{N}}^{\mathrm{ord,ext},\vee}/\vec{\mathsf{M}}_{\mathcal{H}}^{\mathrm{ord,tor}}} \underset{\mathscr{O}_{\vec{\mathsf{M}}_{\mathcal{H}}^{\mathrm{ord,tor}}}}{\otimes} \mathscr{O}_U$$

over U. Since $R^1 f_*^{\mathrm{tor}}(\mathscr{O}_{\vec{\mathsf{N}}^{\mathrm{ord,tor}}})$ and $\underline{\mathrm{Lie}}_{\vec{\mathsf{N}}^{\mathrm{ord,ext},\vee}/\vec{\mathsf{M}}_{\mathcal{H}}^{\mathrm{ord,tor}}}$ are both locally free $\mathscr{O}_{\vec{\mathsf{M}}_{\mathcal{H}}^{\mathrm{ord,tor}}}$-modules of finite rank, and since $\vec{\mathsf{M}}_{\mathcal{H}}^{\mathrm{ord,tor}}$ is noetherian and normal, this uniquely extends to the desired canonical isomorphism (7.3.6.4) over $\vec{\mathsf{M}}_{\mathcal{H}}^{\mathrm{ord,tor}}$. □

Suppose $\mathcal{L}$ is any invertible sheaf over $\vec{\mathsf{N}}^{\mathrm{ord}}$. Since $\vec{\mathsf{N}}^{\mathrm{ord}} \to \vec{\mathsf{M}}_{\mathcal{H}}^{\mathrm{ord}}$ is a torsor under the abelian scheme $\vec{\mathsf{N}}^{\mathrm{ord,grp}} \to \vec{\mathsf{M}}_{\mathcal{H}}^{\mathrm{ord}}$, by [92, XIII, Prop. 1.1], we have a canonical isomorphism of relative Néron–Severi groups

$$\begin{aligned} &\underline{\mathrm{NS}}(\vec{\mathsf{N}}^{\mathrm{ord}}/\vec{\mathsf{M}}_{\mathcal{H}}^{\mathrm{ord}}) := \underline{\mathrm{Pic}}(\vec{\mathsf{N}}^{\mathrm{ord}}/\vec{\mathsf{M}}_{\mathcal{H}}^{\mathrm{ord}})/\underline{\mathrm{Pic}}^0(\vec{\mathsf{N}}^{\mathrm{ord}}/\vec{\mathsf{M}}_{\mathcal{H}}^{\mathrm{ord}}) \\ &\cong \underline{\mathrm{NS}}(\vec{\mathsf{N}}^{\mathrm{ord,grp}}/\vec{\mathsf{M}}_{\mathcal{H}}^{\mathrm{ord}}) := \underline{\mathrm{Pic}}(\vec{\mathsf{N}}^{\mathrm{ord,grp}}/\vec{\mathsf{M}}_{\mathcal{H}}^{\mathrm{ord}})/\underline{\mathrm{Pic}}^0(\vec{\mathsf{N}}^{\mathrm{ord,grp}}/\vec{\mathsf{M}}_{\mathcal{H}}^{\mathrm{ord}}). \end{aligned} \tag{7.3.6.7}$$

Let $\mathcal{L}^{\mathrm{grp}}$ be any invertible sheaf over $\vec{\mathsf{N}}^{\mathrm{ord,grp}} \to \vec{\mathsf{M}}^{\mathrm{ord}}_{\mathcal{H}}$ such that the class $[\mathcal{L}]$ of $\mathcal{L}$ in $\underline{\mathrm{NS}}(\vec{\mathsf{N}}^{\mathrm{ord}}/\vec{\mathsf{M}}^{\mathrm{ord}}_{\mathcal{H}})$ corresponds under the canonical isomorphism (7.3.6.7) to the class $[\mathcal{L}^{\mathrm{grp}}]$ of $\mathcal{L}^{\mathrm{grp}}$ in $\underline{\mathrm{NS}}(\vec{\mathsf{N}}^{\mathrm{ord,grp}}/\vec{\mathsf{M}}^{\mathrm{ord}}_{\mathcal{H}})$. Then the homomorphism

$$\lambda_{\mathcal{L}^{\mathrm{grp}}} : \vec{\mathsf{N}}^{\mathrm{ord,grp}} \to \vec{\mathsf{N}}^{\mathrm{ord,grp},\vee} \tag{7.3.6.8}$$

induced by $\mathcal{L}^{\mathrm{grp}}$ (as in [62, Constr. 1.3.2.7]) is well defined and does not depend on the choice of $\mathcal{L}^{\mathrm{grp}}$ (with the same $[\mathcal{L}^{\mathrm{grp}}]$ corresponding to $[\mathcal{L}]$).

By abuse of notation, we shall also denote (7.3.6.8) by

$$\lambda_{\mathcal{L}} : \vec{\mathsf{N}}^{\mathrm{ord,grp}} \to \vec{\mathsf{N}}^{\mathrm{ord,grp},\vee}.$$

Then $\lambda_{\mathcal{L}}$ (or rather $\lambda_{\mathcal{L}^{\mathrm{grp}}}$) induces a morphism

$$d\lambda_{\mathcal{L}} : \underline{\mathrm{Lie}}_{\vec{\mathsf{N}}^{\mathrm{ord}}/\vec{\mathsf{M}}^{\mathrm{ord}}_{\mathcal{H}}} \to \underline{\mathrm{Lie}}_{\vec{\mathsf{N}}^{\mathrm{ord},\vee}/\vec{\mathsf{M}}^{\mathrm{ord}}_{\mathcal{H}}} \tag{7.3.6.9}$$

(see Definition 7.1.3.9). On the other hand, as an invertible sheaf, $\mathcal{L}$ defines a global section of $R^1 f_*(\mathscr{O}^{\times}_{\vec{\mathsf{N}}^{\mathrm{ord}}})$, and the morphism

$$d\log : \mathscr{O}^{\times}_{\vec{\mathsf{N}}^{\mathrm{ord}}} \to \Omega^1_{\vec{\mathsf{N}}^{\mathrm{ord}}/\vec{\mathsf{M}}^{\mathrm{ord}}_{\mathcal{H}}} : a \mapsto a^{-1}da$$

defines a global section $d\log(\mathcal{L})$ of $R^1 f_*(\Omega^1_{\vec{\mathsf{N}}^{\mathrm{ord}}/\vec{\mathsf{M}}^{\mathrm{ord}}_{\mathcal{H}}})$. By reducing to the case of $\vec{\mathsf{N}}^{\mathrm{ord,grp}}$ by étale descent, we see that the cup product with $d\log(\mathcal{L})$ induces a composition of morphisms

$$\begin{aligned} f_*(\underline{\mathrm{Der}}_{\vec{\mathsf{N}}^{\mathrm{ord}}/\vec{\mathsf{M}}^{\mathrm{ord}}_{\mathcal{H}}}) &\overset{\cup\, d\log(\mathcal{L})}{\to} R^1 f_*(\underline{\mathrm{Der}}_{\vec{\mathsf{N}}^{\mathrm{ord}}/\vec{\mathsf{M}}^{\mathrm{ord}}_{\mathcal{H}}} \underset{\mathscr{O}_{\vec{\mathsf{N}}^{\mathrm{ord}}}}{\otimes} \Omega^1_{\vec{\mathsf{N}}^{\mathrm{ord}}/\vec{\mathsf{M}}^{\mathrm{ord}}_{\mathcal{H}}}) \\ &\overset{\mathrm{can.}}{\to} R^1 f_*(\mathscr{O}_{\vec{\mathsf{N}}^{\mathrm{ord}}}), \end{aligned} \tag{7.3.6.10}$$

and the following diagram

$$\begin{CD} f_*(\underline{\mathrm{Der}}_{\vec{\mathsf{N}}^{\mathrm{ord}}/\vec{\mathsf{M}}^{\mathrm{ord}}_{\mathcal{H}}) @>{(7.3.6.2)}>{\sim}> \underline{\mathrm{Lie}}_{\vec{\mathsf{N}}^{\mathrm{ord}}/\vec{\mathsf{M}}^{\mathrm{ord}}_{\mathcal{H}}} \\ @V{(7.3.6.10)}VV @VV{(7.3.6.9)}V \\ R^1 f_*(\mathscr{O}_{\vec{\mathsf{N}}^{\mathrm{ord}}}) @>{\sim}>{(7.3.6.3)}> \underline{\mathrm{Lie}}_{\vec{\mathsf{N}}^{\mathrm{ord},\vee}/\vec{\mathsf{M}}^{\mathrm{ord}}_{\mathcal{H}}} \end{CD} \tag{7.3.6.11}$$

over $\vec{\mathsf{M}}^{\mathrm{ord}}_{\mathcal{H}}$ is commutative (*cf.* [62, Prop. 2.1.5.13]).

Suppose moreover that $\mathcal{L}$ is relatively ample over $\vec{\mathsf{M}}^{\mathrm{ord}}_{\mathcal{H}}$. Then any choice of $\mathcal{L}^{\mathrm{grp}}$ over $\vec{\mathsf{N}}^{\mathrm{ord,grp}}$ is also relatively ample over $\vec{\mathsf{M}}^{\mathrm{ord}}_{\mathcal{H}}$, and the homomorphism $\lambda_{\mathcal{L}} : \vec{\mathsf{N}}^{\mathrm{ord,grp}} \to \vec{\mathsf{N}}^{\mathrm{ord,grp},\vee}$ is a polarization. Hence, (7.3.6.11) induces a commutative diagram

$$\begin{CD} f_*(\underline{\mathrm{Der}}_{\vec{\mathsf{N}}^{\mathrm{ord}}/\vec{\mathsf{M}}^{\mathrm{ord}}_{\mathcal{H}}) \underset{\mathbb{Z}}{\otimes} \mathbb{Q} @>{(7.3.6.2) \underset{\mathbb{Z}}{\otimes} \mathbb{Q}}>{\sim}> \underline{\mathrm{Lie}}_{\vec{\mathsf{N}}^{\mathrm{ord}}/\vec{\mathsf{M}}^{\mathrm{ord}}_{\mathcal{H}}} \underset{\mathbb{Z}}{\otimes} \mathbb{Q} \\ @V{(7.3.6.10) \underset{\mathbb{Z}}{\otimes} \mathbb{Q}}V{\wr}V @V{\wr}V{(7.3.6.9) \underset{\mathbb{Z}}{\otimes} \mathbb{Q}}V \\ R^1 f_*(\mathscr{O}_{\vec{\mathsf{N}}^{\mathrm{ord}}}) \underset{\mathbb{Z}}{\otimes} \mathbb{Q} @>{\sim}>{(7.3.6.3) \underset{\mathbb{Z}}{\otimes} \mathbb{Q}}> \underline{\mathrm{Lie}}_{\vec{\mathsf{N}}^{\mathrm{ord},\vee}/\vec{\mathsf{M}}^{\mathrm{ord}}_{\mathcal{H}}} \underset{\mathbb{Z}}{\otimes} \mathbb{Q} \end{CD} \tag{7.3.6.12}$$

of isomorphisms over $\vec{\mathsf{M}}^{\mathrm{ord}}_{\mathcal{H}} \underset{\mathbb{Z}}{\otimes} \mathbb{Q}$. (If $\lambda_{\mathcal{L}} : \vec{\mathsf{N}}^{\mathrm{ord,grp}} \to \vec{\mathsf{N}}^{\mathrm{ord,grp},\vee}$ is a separable polarization, then the same holds for the original morphisms in (7.3.6.11).) Thus, (at least) in characteristic zero, the canonical isomorphism (7.3.6.3) is determined by the other three isomorphisms (7.3.6.2), (7.3.6.9), and (7.3.6.10).

Since $\mathsf{M}^{\mathrm{ord,tor}}_{\mathcal{H}}$ is noetherian normal, by [92, IX, 1.4], [28, Ch. I, Prop. 2.7], or [62, Prop. 3.3.1.5] (see also [80, IV, 7.1] or [62, Thm. 3.4.3.2]), $\lambda_{\mathcal{L}} : \vec{\mathsf{N}}^{\mathrm{ord,grp}} \to \vec{\mathsf{N}}^{\mathrm{ord,grp},\vee}$ uniquely extends to a homomorphism

$$\lambda^{\mathrm{ext}}_{\mathcal{L}} : \vec{\mathsf{N}}^{\mathrm{ord,ext}} \to \vec{\mathsf{N}}^{\mathrm{ord,ext},\vee},$$

which is an isogeny between semi-abelian schemes and induces a morphism

$$d\lambda^{\mathrm{ext}}_{\mathcal{L}} : \underline{\mathrm{Lie}}_{\vec{\mathsf{N}}^{\mathrm{ord,ext}}/\mathsf{M}^{\mathrm{ord,tor}}_{\mathcal{H}}} \to \underline{\mathrm{Lie}}_{\vec{\mathsf{N}}^{\mathrm{ord,ext},\vee}/\mathsf{M}^{\mathrm{ord,tor}}_{\mathcal{H}}}. \tag{7.3.6.13}$$

In characteristic zero, (7.3.6.13) $\underset{\mathbb{Z}}{\otimes} \mathbb{Q}$ is an isomorphism because the (quasi-finite flat) kernel of the isogeny $\lambda^{\mathrm{ext}}_{\mathcal{L}} \underset{\mathbb{Z}}{\otimes} \mathbb{Q}$ is necessarily étale over $\mathsf{M}^{\mathrm{ord,tor}}_{\mathcal{H}}$.

To proceed further we will need a more specific choice of $\mathcal{L}$. Let $j_Q : Q^\vee \hookrightarrow Q$ be as in (3d) of Theorem 7.1.4.1. By construction (see Sections 1.3.3 and 7.2.1), $\widetilde{X}^\vee(1) \cong \mathrm{Hom}_{\mathcal{O}}(\widetilde{X}, \mathrm{Diff}^{-1}_{\mathcal{O}'/\mathbb{Z}}(1))$ is the submodule Q_{-2} of $Q^\vee \underset{\mathbb{Z}}{\otimes} \mathbb{Q}(1)$, and $\widetilde{Y}$ is the submodule Q_0 of $Q \underset{\mathbb{Z}}{\otimes} \mathbb{Q}$. Therefore, the embedding $j_Q : Q^\vee \hookrightarrow Q$ corresponds to an element $\widetilde{\ell}_{j_Q}$ of $\mathbf{S}_{\widetilde{\Phi}_{\widetilde{\mathcal{H}}}} \underset{\mathbb{Z}}{\otimes} \mathbb{Q}$. The positive definiteness of the induced pairing $\langle j_Q^{-1}(\,\cdot\,), \,\cdot\,\rangle_Q$ then translates to the positivity condition that $\langle \widetilde{\ell}_{j_Q}, y\rangle > 0$ for every $y \in \mathbf{P}_{\widetilde{\Phi}_{\widetilde{\mathcal{H}}}} - \{0\}$. By replacing j_Q with a multiple by a positive integer, we may assume that $\widetilde{\ell}_{j_Q} \in \mathbf{S}_{\widetilde{\Phi}_{\widetilde{\mathcal{H}}}}$ (without altering the above positivity condition). Then we obtain an invertible sheaf $\vec{\widetilde{\Psi}}^{\mathrm{ord}}_{\widetilde{\Phi}_{\widetilde{\mathcal{H}}}, \widetilde{\delta}_{\widetilde{\mathcal{H}}}}(\widetilde{\ell}_{j_Q})$ over $\vec{\mathsf{N}}^{\mathrm{ord}}$ (see (4.2.1.49) in Proposition 4.2.1.46). Note that $\widetilde{\ell}_{j_Q} \in \widetilde{\sigma}_0^\vee$.

Lemma 7.3.6.14. (*Compare with [61, Lem. 5.5].*) *The invertible sheaf $\vec{\widetilde{\Psi}}^{\mathrm{ord}}_{\widetilde{\Phi}_{\widetilde{\mathcal{H}}}, \widetilde{\delta}_{\widetilde{\mathcal{H}}}}(\widetilde{\ell}_{j_Q})$ is relatively ample over $\vec{\mathsf{M}}^{\mathrm{ord}}_{\mathcal{H}}$, and the induced homomorphism*

$$\lambda_{\vec{\widetilde{\Psi}}^{\mathrm{ord}}_{\widetilde{\Phi}_{\widetilde{\mathcal{H}}}, \widetilde{\delta}_{\widetilde{\mathcal{H}}}}(\widetilde{\ell}_{j_Q})} : \vec{\mathsf{N}}^{\mathrm{ord,grp}} \to \vec{\mathsf{N}}^{\mathrm{ord,grp},\vee}$$

is a positive $\mathbb{Q}^\times$-multiple of the $\mathbb{Q}^\times$-polarization

$$\lambda_{\vec{\mathsf{N}}^{\mathrm{ord}}, j_Q} : \vec{\mathsf{N}}^{\mathrm{ord,grp}} \to \vec{\mathsf{N}}^{\mathrm{ord,grp},\vee}$$

induced by the $\mathbb{Q}^\times$-polarization

$$\lambda_{\vec{\mathsf{M}}^{\mathrm{ord}}_{\mathcal{H}}, j_Q} : \underline{\mathrm{Hom}}_{\mathcal{O}}(Q, G_{\vec{\mathsf{M}}^{\mathrm{ord}}_{\mathcal{H}}})^\circ \to (\underline{\mathrm{Hom}}_{\mathcal{O}}(Q, G_{\vec{\mathsf{M}}^{\mathrm{ord}}_{\mathcal{H}}})^\circ)^\vee$$

in (3d) *of Theorem* 7.1.4.1. *The induced morphism*

$$d\lambda_{\vec{\widetilde{\Psi}}^{\mathrm{ord}}_{\widetilde{\Phi}_{\widetilde{\mathcal{H}}}, \widetilde{\delta}_{\widetilde{\mathcal{H}}}}(\widetilde{\ell}_{j_Q})} \underset{\mathbb{Z}}{\otimes} \mathbb{Q} : \underline{\mathrm{Lie}}_{\vec{\mathsf{N}}^{\mathrm{ord}}/\vec{\mathsf{M}}^{\mathrm{ord}}_{\mathcal{H}}} \underset{\mathbb{Z}}{\otimes} \mathbb{Q} \to \underline{\mathrm{Lie}}_{\vec{\mathsf{N}}^{\mathrm{ord},\vee}/\vec{\mathsf{M}}^{\mathrm{ord}}_{\mathcal{H}}} \underset{\mathbb{Z}}{\otimes} \mathbb{Q}$$

is a positive $\mathbb{Q}^\times$*-multiple of the isomorphism*

$$d\lambda_{\vec{\mathsf{N}}^{\mathrm{ord}},j_Q} \underset{\mathbb{Z}}{\otimes} \mathbb{Q} : \underline{\mathrm{Lie}}_{\vec{\mathsf{N}}^{\mathrm{ord}}/\vec{\mathsf{M}}^{\mathrm{ord}}_{\mathcal{H}}} \underset{\mathbb{Z}}{\otimes} \mathbb{Q} \xrightarrow{\sim} \underline{\mathrm{Lie}}_{\vec{\mathsf{N}}^{\mathrm{ord},\vee}/\vec{\mathsf{M}}^{\mathrm{ord}}_{\mathcal{H}}} \underset{\mathbb{Z}}{\otimes} \mathbb{Q}$$

induced by the isomorphism

$$d\lambda_{\vec{\mathsf{M}}^{\mathrm{ord}}_{\mathcal{H}},j_Q} \underset{\mathbb{Z}}{\otimes} \mathbb{Q} : \underline{\mathrm{Hom}}_{\mathcal{O}}(Q, \underline{\mathrm{Lie}}_{G_{\vec{\mathsf{M}}^{\mathrm{ord}}_{\mathcal{H}}}/\vec{\mathsf{M}}^{\mathrm{ord}}_{\mathcal{H}}}) \underset{\mathbb{Z}}{\otimes} \mathbb{Q} \xrightarrow{\sim} \underline{\mathrm{Hom}}_{\mathcal{O}}(Q^\vee, \underline{\mathrm{Lie}}_{G^\vee_{\vec{\mathsf{M}}^{\mathrm{ord}}_{\mathcal{H}}}/\vec{\mathsf{M}}^{\mathrm{ord}}_{\mathcal{H}}}) \underset{\mathbb{Z}}{\otimes} \mathbb{Q}.$$

(These are abuses of notation because $d\lambda_{\vec{\mathsf{N}}^{\mathrm{ord}},j_Q}$ *and* $d\lambda_{\vec{\mathsf{M}}^{\mathrm{ord}}_{\mathcal{H}},j_Q}$ *are not necessarily defined, because* $\lambda_{\vec{\mathsf{N}}^{\mathrm{ord}},j_Q}$ *and* $\lambda_{\vec{\mathsf{M}}^{\mathrm{ord}}_{\mathcal{H}},j_Q}$ *are merely* $\mathbb{Q}^\times$*-polarizations.) In particular,* $d\lambda_{\vec{\tilde{\Psi}}^{\mathrm{ord}}_{\tilde{\Phi}_{\tilde{\mathcal{H}}},\tilde{\delta}_{\tilde{\mathcal{H}}}}(\tilde{\ell}_{jQ})} \underset{\mathbb{Z}}{\otimes} \mathbb{Q}$ *is an isomorphism.*

Proof. The proof of [61, Lem. 5.5], which is in turn based on the proof of [61, Lem. 2.9, Prop. 2.10, and Cor. 2.12], works almost verbatim here. □

The upshot is the following:

Proposition 7.3.6.15. *Let us take* $\mathcal{L}$ *to be* $\vec{\tilde{\Psi}}^{\mathrm{ord}}_{\tilde{\Phi}_{\tilde{\mathcal{H}}},\tilde{\delta}_{\tilde{\mathcal{H}}}}(\tilde{\ell}_{jQ})$*. Then the canonical isomorphism* (7.3.6.10) $\underset{\mathbb{Z}}{\otimes} \mathbb{Q}$ *over* $\vec{\mathsf{M}}^{\mathrm{ord}}_{\mathcal{H}} \underset{\mathbb{Z}}{\otimes} \mathbb{Q}$ *extends to a canonical isomorphism*

$$f^{\mathrm{tor}}_*(\underline{\mathrm{Der}}_{\vec{\mathsf{N}}^{\mathrm{ord,tor}}/\vec{\mathsf{M}}^{\mathrm{ord,tor}}_{\mathcal{H}}}) \underset{\mathbb{Z}}{\otimes} \mathbb{Q} \xrightarrow{\sim} R^1 f_*(\mathscr{O}_{\vec{\mathsf{N}}^{\mathrm{ord,tor}}}) \underset{\mathbb{Z}}{\otimes} \mathbb{Q} \quad (7.3.6.16)$$

over $\vec{\mathsf{M}}^{\mathrm{ord,tor}}_{\mathcal{H}} \underset{\mathbb{Z}}{\otimes} \mathbb{Q}$*, so that the diagram* (7.3.6.12) *extends to a commutative diagram*

$$\begin{array}{ccc} f^{\mathrm{tor}}_*(\underline{\mathrm{Der}}_{\vec{\mathsf{N}}^{\mathrm{ord,tor}}/\vec{\mathsf{M}}^{\mathrm{ord,tor}}_{\mathcal{H}}}) \underset{\mathbb{Z}}{\otimes} \mathbb{Q} & \xrightarrow[\sim]{(7.3.6.1) \underset{\mathbb{Z}}{\otimes} \mathbb{Q}} & \underline{\mathrm{Lie}}_{\vec{\mathsf{N}}^{\mathrm{ord,ext}}/\vec{\mathsf{M}}^{\mathrm{ord,tor}}_{\mathcal{H}}} \underset{\mathbb{Z}}{\otimes} \mathbb{Q} \\ {\scriptstyle (7.3.6.16)}\downarrow\wr & & \wr\downarrow{\scriptstyle (7.3.6.13) \underset{\mathbb{Z}}{\otimes} \mathbb{Q}} \\ R^1 f_*(\mathscr{O}_{\vec{\mathsf{N}}^{\mathrm{ord,tor}}}) \underset{\mathbb{Z}}{\otimes} \mathbb{Q} & \overset{\sim}{\dashrightarrow} & \underline{\mathrm{Lie}}_{\vec{\mathsf{N}}^{\mathrm{ord,ext},\vee}/\vec{\mathsf{M}}^{\mathrm{ord,tor}}_{\mathcal{H}}} \underset{\mathbb{Z}}{\otimes} \mathbb{Q} \end{array} \quad (7.3.6.17)$$

in which the dotted arrow is induced by (7.3.6.16) *and gives the desired isomorphism* (7.3.6.6)*. Thus, we also obtain the desired isomorphism* (7.3.6.4) *by Lemma* 7.3.6.5.

Proof. The arguments in [61, Sec. 5] leading to the proof of [61, Prop. 5.14], based on the idea of log invertible sheaves $\overline{\mathcal{L}}$ extending $\mathcal{L}$ and on the étale local description of the toroidal boundary charts, works almost verbatim here. (Since we only need the facts in characteristic zero, and since all construction steps and proofs are canonically compatible with those for Theorem 1.3.3.15 after pulled back to the characteristic zero fibers, thanks to (7) of Theorem 5.2.1.1, this is not even an imitation but rather a logical repetition of the arguments in [61, Sec. 5].) □

Combining Proposition 7.3.6.15 and Corollary 7.3.3.53:

Corollary 7.3.6.18. *For each integer $b \geq 0$, we have a canonical isomorphism*

$$R^b f_*^{\mathrm{tor}}(\mathscr{O}_{\vec{\mathsf{N}}^{\mathrm{ord,tor}}}) \cong \wedge^b(\underline{\mathrm{Lie}}_{\vec{\mathsf{N}}^{\mathrm{ord,ext},\vee}/\vec{\mathsf{M}}_{\mathcal{H}}^{\mathrm{ord,tor}}})$$

of locally free sheaves over $\vec{\mathsf{M}}_{\mathcal{H}}^{\mathrm{ord,tor}}$, compatible with cup products and exterior products, extending the canonical isomorphism (7.1.4.6) *over $\vec{\mathsf{M}}_{\mathcal{H}}^{\mathrm{ord}}$.*

Together with Proposition 7.3.3.45, this completes the proof of (3b) and (3d) of Theorem 7.1.4.1, using respectively (3a) and (3c) of Theorem 7.1.4.1. As explained in Section 7.2.7, this also makes (4c) and (5c) of Theorem 7.1.4.1 unconditional. Finally, (6) of Theorem 7.1.4.1 also follows, because all construction steps and proofs are canonically compatible with those for Theorem 1.3.3.15 after pulled back to the characteristic zero fibers, thanks to (7) of Theorem 5.2.1.1 and the corresponding statements in Propositions 5.2.2.2 and 6.2.2.1, as remarked in the proof of Proposition 7.3.6.15. The proof of Theorem 7.1.4.1 is now complete.

Chapter 8

Automorphic Bundles and Canonical Extensions

In this chapter, we explain the construction of automorphic bundles and their canonical extensions over the partial and total toroidal compactifications defined in Chapters 5 and 2, respectively. These generalize the constructions in characteristic zero in Section 1.4.

Our constructions are somewhat ad hoc and mainly designed to extend the definitions in characteristic zero with the least number of assumptions. It is likely that for applications requiring more refined properties along the characteristic p fibers our constructions will have to be substantially improved, which is possible in many special cases. (Of course, users of this theory can freely decide how they want to construct their automorphic bundles and canonical extensions in mixed characteristics, as long as they are compatible with the known constructions in characteristic zero.)

8.1 Constructions over the Ordinary Loci

8.1.1 *Technical Assumptions*

Suppose that there exists a discrete valuation ring R_0 of mixed characteristics $(0, p)$, which is faithfully flat over $\mathbb{Z}_{(p)}$ by assumption, together with the following data:

(1) Two $\mathcal{O} \underset{\mathbb{Z}}{\otimes} R_0$-modules $\mathrm{Gr}_{\mathtt{D},0}^{-1}$ and $\mathrm{Gr}_{\mathtt{D}^\#,0}^{-1}$ (with subscripts "0", by abuse of notation, meaning "R_0-models" of $\mathrm{Gr}_{\mathtt{D}}^{-1}$ and $\mathrm{Gr}_{\mathtt{D}^\#}^{-1}$; see below for the precise meaning).

(2) A p-adic complete flat R_0-algebra $\tilde{R}_0$, which is equipped with a faithfully flat homomorphism $\mathbb{Z}_p \to \tilde{R}_0$ by assumption, together with two isomorphisms

$$\mathrm{Gr}_{\mathtt{D},0}^{-1} \underset{R_0}{\otimes} \tilde{R}_0 \xrightarrow{\sim} \mathrm{Gr}_{\mathtt{D}}^{-1} \underset{\mathbb{Z}_p}{\otimes} \tilde{R}_0$$

and

$$\mathrm{Gr}_{\mathtt{D}^\#,0}^{-1} \underset{R_0}{\otimes} \tilde{R}_0 \xrightarrow{\sim} \mathrm{Gr}_{\mathtt{D}^\#}^{-1} \underset{\mathbb{Z}_p}{\otimes} \tilde{R}_0.$$

(3) An embedding of $\mathcal{O} \underset{\mathbb{Z}}{\otimes} R_0$-modules

$$\phi^{-1}_{\mathtt{D},0} : \mathrm{Gr}^{-1}_{\mathtt{D},0} \to \mathrm{Gr}^{-1}_{\mathtt{D}^\#,0}$$

making the following diagram

$$\begin{array}{ccc} \mathrm{Gr}^{-1}_{\mathtt{D},0} \underset{R_0}{\otimes} \tilde{R}_0 & \xrightarrow{\phi^{-1}_{\mathtt{D},0} \underset{R_0}{\otimes} \tilde{R}_0} & \mathrm{Gr}^{-1}_{\mathtt{D}^\#,0} \underset{R_0}{\otimes} \tilde{R}_0 \\ \wr \downarrow & & \downarrow \wr \\ \mathrm{Gr}^{-1}_{\mathtt{D}} \underset{\mathbb{Z}_p}{\otimes} \tilde{R}_0 & \xrightarrow[\phi^{-1}_{\mathtt{D}} \underset{\mathbb{Z}_p}{\otimes} \tilde{R}_0]{} & \mathrm{Gr}^{-1}_{\mathtt{D}^\#} \underset{\mathbb{Z}_p}{\otimes} \tilde{R}_0 \end{array}$$

commutative.

(4) There exists some maximal order $\mathcal{O}'$ in $\mathcal{O} \underset{\mathbb{Z}}{\otimes} \mathbb{Q}$ containing $\mathcal{O}$ and satisfying the requirement in Condition 1.2.1.1 such that the $\mathcal{O} \underset{\mathbb{Z}}{\otimes} R_0$-module structures of $\mathrm{Gr}^{-1}_{\mathtt{D},0}$ and $\mathrm{Gr}^{-1}_{\mathtt{D}^\#,0}$ (necessarily uniquely) extend to $\mathcal{O}' \underset{\mathbb{Z}}{\otimes} R_0$-module structures, which then (automatically) compatible with the above embedding $\phi^{-1}_{\mathtt{D},0} : \mathrm{Gr}^{-1}_{\mathtt{D},0} \to \mathrm{Gr}^{-1}_{\mathtt{D}^\#,0}$. Moreover, the above diagram is (automatically) compatible with the $\mathcal{O}' \underset{\mathbb{Z}}{\otimes} \tilde{R}_0$-module structures induced by the $\mathcal{O}' \underset{\mathbb{Z}}{\otimes} R_0$-module structures of $\mathrm{Gr}^{-1}_{\mathtt{D},0}$ and $\mathrm{Gr}^{-1}_{\mathtt{D}^\#,0}$ (in the last sentence) and by the $\mathcal{O}' \underset{\mathbb{Z}}{\otimes} \mathbb{Z}_p$-module structures of $\mathrm{Gr}^{-1}_{\mathtt{D}}$ and $\mathrm{Gr}^{-1}_{\mathtt{D}^\#}$ (as in the proof of Lemma 3.2.2.6). (These automatic compatibilities follow from the flatness of R_0 and $\tilde{R}_0$ over $\mathbb{Z}_{(p)}$, and from the identity $\mathcal{O} \underset{\mathbb{Z}}{\otimes} \mathbb{Q} = \mathcal{O}' \underset{\mathbb{Z}}{\otimes} \mathbb{Q}$.)

Let

$$\mathrm{Gr}^{0}_{\mathtt{D},0} := \mathrm{Hom}_{R_0}(\mathrm{Gr}^{-1}_{\mathtt{D}^\#,0}, R_0(1))$$

and

$$\mathrm{Gr}^{0}_{\mathtt{D}^\#,0} := \mathrm{Hom}_{R_0}(\mathrm{Gr}^{-1}_{\mathtt{D},0}, R_0(1)),$$

and let

$$\phi^{0}_{\mathtt{D},0} : \mathrm{Gr}^{0}_{\mathtt{D},0} \to \mathrm{Gr}^{0}_{\mathtt{D}^\#,0}$$

be the dual of $\phi^{-1}_{\mathtt{D},0}$, and let

$$\phi_{\mathtt{D},0} := \phi^{0}_{\mathtt{D},0} \oplus \phi^{-1}_{\mathtt{D},0}.$$

Let

$$\langle \cdot , \cdot \rangle_{\phi_{\mathtt{D},0}} : (\mathrm{Gr}^{0}_{\mathtt{D},0} \oplus \mathrm{Gr}^{-1}_{\mathtt{D},0}) \times (\mathrm{Gr}^{0}_{\mathtt{D},0} \oplus \mathrm{Gr}^{-1}_{\mathtt{D},0}) \to R_0(1)$$

be the alternating pairing defined by

$$\begin{aligned} \langle (x_1, y_1), (x_2, y_2) \rangle_{\phi_{\mathtt{D},0}} &:= x_1(\phi^{-1}_{\mathtt{D},0}(y_2)) - x_2(\phi^{-1}_{\mathtt{D},0}(y_1)) \\ &= y_2(\phi^{0}_{\mathtt{D},0}(x_1)) - y_1(\phi^{0}_{\mathtt{D},0}(x_2)). \end{aligned}$$

Definition 8.1.1.1. (*Compare with [61, Def. 6.2] and Definition* 1.4.1.1.) *For each R_0-algebra R, set*

$$\mathrm{G}^{\mathrm{ord}}_{\mathtt{D},0}(R) := \left\{ \begin{array}{c} (g,r) \in \mathrm{GL}_{\mathcal{O} \underset{\mathbb{Z}}{\otimes} R}((\mathrm{Gr}^0_{\mathtt{D},0} \oplus \mathrm{Gr}^{-1}_{\mathtt{D},0}) \underset{R_0}{\otimes} R) \times \mathbf{G}_{\mathrm{m}}(R) : \\ \langle gx, gy \rangle = r\langle x, y\rangle, \forall x, y \in (\mathrm{Gr}^0_{\mathtt{D},0} \oplus \mathrm{Gr}^{-1}_{\mathtt{D},0}) \underset{R_0}{\otimes} R \end{array} \right\},$$

$$\mathrm{P}^{\mathrm{ord}}_{\mathtt{D},0}(R) := \{(g,r) \in \mathrm{G}_0(R) : g(\mathrm{Gr}^0_{\mathtt{D},0} \underset{R_0}{\otimes} R) = \mathrm{Gr}^0_{\mathtt{D},0} \underset{R_0}{\otimes} R\},$$

$$\mathrm{M}^{\mathrm{ord}}_{\mathtt{D},0}(R) := \mathrm{GL}_{\mathcal{O} \underset{\mathbb{Z}}{\otimes} R}(\mathrm{Gr}^0_{\mathtt{D},0} \underset{R_0}{\otimes} R) \times \mathrm{GL}_{\mathcal{O} \underset{\mathbb{Z}}{\otimes} R}(\mathrm{Gr}^{-1}_{\mathtt{D},0} \underset{R_0}{\otimes} R) \times \mathbf{G}_{\mathrm{m}}(R),$$

with a canonical homomorphism

$$\mathrm{P}^{\mathrm{ord}}_{\mathtt{D},0}(R) \to \mathrm{M}^{\mathrm{ord}}_{\mathtt{D},0}(R) : (g,r) \mapsto (\mathrm{Gr}^0(g), \mathrm{Gr}^{-1}(g), r).$$

The assignments are functorial in R and define group functors $\mathrm{G}^{\mathrm{ord}}_{\mathtt{D},0}$, $\mathrm{P}^{\mathrm{ord}}_{\mathtt{D},0}$, and $\mathrm{M}^{\mathrm{ord}}_{\mathtt{D},0}$ over R_0, which are affine group schemes over $\mathrm{Spec}(R_0)$.

In the generality we are working, it is not always possible to construct automorphic bundles in the same way as in the smooth case. For constructions involving $\mathrm{P}^{\mathrm{ord}}_{\mathtt{D},0}$, we will need to assume that the following holds:

Condition 8.1.1.2. *Suppose R is a flat R_0-algebra, M is an $\mathcal{O} \underset{\mathbb{Z}}{\otimes} R$-module of finite presentation, and $\langle \cdot , \cdot \rangle_M : M \times M \to R(1)$ is an alternating pairing satisfying $\langle bx, y\rangle = \langle x, b^\star y\rangle$ for all $x, y \in M$.*

Suppose there exists a short exact sequence

$$0 \to \mathrm{Gr}^0_{\mathtt{D},0} \underset{R_0}{\otimes} R \to M \to \mathrm{Gr}^{-1}_{\mathtt{D},0} \underset{R_0}{\otimes} R \to 0$$

such that the image of $\mathrm{Gr}^0_{\mathtt{D},0} \underset{R_0}{\otimes} R$ in M is totally isotropic under the pairing $\langle \cdot , \cdot \rangle_M$, and such that the induced pairing

$$\mathrm{Gr}^0_{\mathtt{D},0} \underset{R_0}{\otimes} R \times \mathrm{Gr}^{-1}_{\mathtt{D},0} \underset{R_0}{\otimes} R \to R(1)$$

coincides with the pairing induced by $\phi_{\mathtt{D},0}$, which is given by

$$\mathrm{Gr}^0_{\mathtt{D},0} \underset{R_0}{\otimes} R \times \mathrm{Gr}^{-1}_{\mathtt{D},0} \underset{R_0}{\otimes} R \overset{\mathrm{Id} \times \phi_{\mathtt{D},0}}{\to} \mathrm{Gr}^0_{\mathtt{D},0} \underset{R_0}{\otimes} R \times \mathrm{Gr}^{-1}_{\mathtt{D}^\#,0} \underset{R_0}{\otimes} R \overset{\mathrm{can.}}{\to} R(1).$$

There exists a faithfully flat ring extension $R \to R_1$ such that there is an isomorphism

$$\begin{aligned} &(M \underset{R}{\otimes} R_1, \langle \cdot , \cdot \rangle_M, \mathrm{Gr}^0_{\mathtt{D},0} \underset{R_0}{\otimes} R_1) \\ &\cong ((\mathrm{Gr}^0_{\mathtt{D},0} \oplus \mathrm{Gr}^{-1}_{\mathtt{D},0}) \underset{R_0}{\otimes} R_1, \langle \cdot , \cdot \rangle_{\phi_{\mathtt{D},0}}, \mathrm{Gr}^0_{\mathtt{D},0} \underset{R_0}{\otimes} R_1). \end{aligned} \tag{8.1.1.3}$$

Even when Condition 8.1.1.2 holds, we do not claim that either of $\mathrm{G} \underset{\mathbb{Z}}{\otimes} R_1$, $\mathrm{G}^{\mathrm{ord}}_{\mathtt{D},0} \underset{R_0}{\otimes} R_1$, or $\mathrm{P}^{\mathrm{ord}}_{\mathtt{D},0} \underset{R_0}{\otimes} R_1$ is smooth over R_1. In practice, it is sufficiently interesting to consider the theory only for $\mathrm{M}^{\mathrm{ord}}_{\mathtt{D},0}$. In what follows, we will make it clear when we need Condition 8.1.1.2. (We will not need Condition 8.1.1.2 for constructions involving only $\mathrm{M}^{\mathrm{ord}}_{\mathtt{D},0}$.)

Lemma 8.1.1.4. *If p is a good prime as in Definition* 1.1.1.6, *then Condition* 8.1.1.2 *always holds. Moreover, we may assume that the faithfully flat ring extension $R \to R_1$ in Condition* 8.1.1.2 *is* **finite étale**.

Proof. This follows from [62, Cor. 1.2.3.10]. □

Lemma 8.1.1.5. (*Compare with Lemma* 1.4.1.2.) *When Condition* 8.1.1.2 *holds, there exists a faithfully flat ring extension $\tilde{R}_0 \to R_1$ of $\mathbb{Z}_p$-algebras such that there exists an isomorphism*

$$\begin{aligned}&(L \underset{\mathbb{Z}}{\otimes} R_1, \langle\,\cdot\,,\,\cdot\,\rangle, \mathrm{Gr}_{\mathtt{D}}^{-1} \underset{\mathbb{Z}_p}{\otimes} R_1)\\ &\cong ((\mathrm{Gr}^0_{\mathtt{D},0} \oplus \mathrm{Gr}^{-1}_{\mathtt{D},0}) \underset{R_0}{\otimes} R_1, \langle\,\cdot\,,\,\cdot\,\rangle_{\phi_{\mathtt{D},0}}, \mathrm{Gr}^{-1}_{\mathtt{D},0} \underset{R_0}{\otimes} R_1)\end{aligned} \tag{8.1.1.6}$$

over R_1, which induces an isomorphism $\mathrm{G} \underset{\mathbb{Z}}{\otimes} R_1 \cong \mathrm{G}^{\mathrm{ord}}_{\mathtt{D},0} \underset{R_0}{\otimes} R_1$. (Consequently, for every R_1-algebra R, the group $\mathrm{P}^{\mathrm{ord}}_{\mathtt{D},0}(R)$ can be identified with a subgroup of $\mathrm{G}(R)$.)

Proof. This follows by applying the assertion in Condition 8.1.1.2 to $(M, \langle\,\cdot\,,\,\cdot\,\rangle_M, \mathrm{Gr}^{-1}_{\mathtt{D},0} \underset{R_0}{\otimes} \tilde{R}_0) = (L \underset{\mathbb{Z}}{\otimes} \tilde{R}_0, \langle\,\cdot\,,\,\cdot\,\rangle, \mathrm{Gr}_{\mathtt{D}}^{-1} \underset{\mathbb{Z}_p}{\otimes} \tilde{R}_0)$, with $R = \tilde{R}_0$ there. □

Lemma 8.1.1.7. *Given any F_0' as in Section* 1.4, (*without assuming that Condition* 8.1.1.2 *holds*) *there exist ring extensions $R_0 \to K$ and $F_0' \to K$* (*which are automatically flat*) *such that there is an isomorphism*

$$\begin{aligned}&((\mathrm{Gr}^0_{\mathtt{D},0} \oplus \mathrm{Gr}^{-1}_{\mathtt{D},0}) \underset{R_0}{\otimes} K, \langle\,\cdot\,,\,\cdot\,\rangle_{\phi_{\mathtt{D},0}}, \mathrm{Gr}^0_{\mathtt{D},0} \underset{R_0}{\otimes} K)\\ &\cong ((L_0 \oplus L_0^\vee(1)) \underset{F_0'}{\otimes} K, \langle\,\cdot\,,\,\cdot\,\rangle_{\mathrm{can.}}, L_0^\vee(1) \underset{F_0'}{\otimes} K)\end{aligned} \tag{8.1.1.8}$$

which induces an isomorphism $\mathrm{G}^{\mathrm{ord}}_{\mathtt{D},0} \underset{R_0}{\otimes} K \cong \mathrm{G}_0 \underset{F_0'}{\otimes} K$ over K inducing compatible isomorphisms $\mathrm{P}^{\mathrm{ord}}_{\mathtt{D},0} \underset{R_0}{\otimes} K \cong \mathrm{P}_0 \underset{F_0'}{\otimes} K$ and $\mathrm{M}^{\mathrm{ord}}_{\mathtt{D},0} \underset{R_0}{\otimes} K \cong \mathrm{M}_0 \underset{F_0'}{\otimes} K$ (also over K). (Note the choices of maximal totally isotropic submodules.)

Proof. This follows from Lemma 1.4.1.2 and from Assumption 3.2.2.10. □

8.1.2 *Automorphic Bundles*

Let $\mathcal{H}$, $\mathcal{H}^p$, and $\mathcal{H}_p$ be as at the beginning of Section 3.4.1, let $r_{\mathtt{D}}$ and $r_{\mathcal{H}}$ be as in Definition 3.4.2.1, and let $(A, \lambda, i, \alpha_{\mathcal{H}^p}, \alpha^{\mathrm{ord}}_{\mathcal{H}_p})$ be the tautological tuple over $\vec{\mathsf{M}}^{\mathrm{ord}}_{\mathcal{H}}$ (see Convention 3.4.2.9).

For simplicity, we shall assume the following in the remainder of this chapter:

Assumption 8.1.2.1. *Up to replacing R_0 with a faithfully flat extension, we shall assume that R_0 is an $\mathcal{O}_{F_0,(p)}$-algebra. Moreover, by abuse of notation, we shall replace $\vec{\mathsf{M}}^{\mathrm{ord}}_{\mathcal{H}}$ etc with their base changes from $\mathrm{Spec}(\mathcal{O}_{F_0,(p)}[\zeta_{p^{r_{\mathcal{H}}}}])$ to $\mathrm{Spec}(R_0[\zeta_{p^{r_{\mathcal{H}}}}])$, and replace $\vec{\mathsf{S}}_{0,r_{\mathcal{H}}} = \mathrm{Spec}(\mathcal{O}_{F_0,(p)}[\zeta_{p^{r_{\mathcal{H}}}}])$ with $\mathrm{Spec}(R_0[\zeta_{p^{r_{\mathcal{H}}}}])$.*

Note that the base changes of $\vec{\mathsf{M}}^{\mathrm{ord}}_{\mathcal{H}}$ and its partial toroidal compactifications $\vec{\mathsf{M}}^{\mathrm{ord,tor}}_{\mathcal{H}}$ remain smooth over $\vec{\mathsf{S}}_{0,r_{\mathcal{H}}}$ and hence regular; and the base change of the partial minimal compactification $\vec{\mathsf{M}}^{\mathrm{ord,min}}_{\mathcal{H}}$ remains normal, by Proposition 6.2.1.6.

Definition 8.1.2.2. (*Compare with Definition* 1.4.1.3.) *The* **principal $\mathrm{G}^{\mathrm{ord}}_{\mathsf{D},0}$-bundle** *over $\vec{\mathsf{M}}^{\mathrm{ord}}_{\mathcal{H}}$ is the relative scheme*

$$\begin{aligned}\vec{\mathcal{E}}^{\mathrm{ord}}_{\mathrm{G}^{\mathrm{ord}}_{\mathsf{D},0}} := \underline{\mathrm{Isom}}_{\mathcal{O}\underset{\mathbb{Z}}{\otimes}\mathscr{O}_{\vec{\mathsf{M}}^{\mathrm{ord}}_{\mathcal{H}}}}&((\underline{H}^{\mathrm{dR}}_1(A/\vec{\mathsf{M}}^{\mathrm{ord}}_{\mathcal{H}}),\langle\,\cdot\,,\,\cdot\,\rangle_\lambda,\mathscr{O}_{\vec{\mathsf{M}}^{\mathrm{ord}}_{\mathcal{H}}}(1)),\\ &((\mathrm{Gr}^0_{\mathsf{D},0}\oplus\mathrm{Gr}^{-1}_{\mathsf{D},0})\underset{R_0}{\otimes}\mathscr{O}_{\vec{\mathsf{M}}^{\mathrm{ord}}_{\mathcal{H}}},\langle\,\cdot\,,\,\cdot\,\rangle_{\phi_{\mathsf{D},\mathbb{Z}}},\mathscr{O}_{\vec{\mathsf{M}}^{\mathrm{ord}}_{\mathcal{H}}}(1))),\end{aligned}$$

the sheaf of isomorphisms of $\mathscr{O}_{\vec{\mathsf{M}}^{\mathrm{ord}}_{\mathcal{H}}}$-sheaves of symplectic $\mathcal{O}$-modules, over $\vec{\mathsf{M}}^{\mathrm{ord}}_{\mathcal{H}}$. (By definition, the group $\mathrm{G}^{\mathrm{ord}}_{\mathsf{D},0}$ acts as automorphisms on $((\mathrm{Gr}^0_{\mathsf{D},0}\oplus\mathrm{Gr}^{-1}_{\mathsf{D},0})\underset{R_0}{\otimes}\mathscr{O}_{\vec{\mathsf{M}}^{\mathrm{ord}}_{\mathcal{H}}},\langle\,\cdot\,,\,\cdot\,\rangle_{\phi_{\mathsf{D},0}},\mathscr{O}_{\vec{\mathsf{M}}^{\mathrm{ord}}_{\mathcal{H}}}(1))$. The third entries in the tuples represent the values of the pairings.)

Definition 8.1.2.3. (*Compare with Definition* 1.4.1.4.) *The* **principal $\mathrm{P}^{\mathrm{ord}}_{\mathsf{D},0}$-bundle** *over $\vec{\mathsf{M}}^{\mathrm{ord}}_{\mathcal{H}}$ is the relative scheme*

$$\begin{aligned}\vec{\mathcal{E}}^{\mathrm{ord}}_{\mathrm{P}^{\mathrm{ord}}_{\mathsf{D},0}} := \underline{\mathrm{Isom}}_{\mathcal{O}\underset{\mathbb{Z}}{\otimes}\mathscr{O}_{\vec{\mathsf{M}}^{\mathrm{ord}}_{\mathcal{H}}}}&((\underline{H}^{\mathrm{dR}}_1(A/\vec{\mathsf{M}}^{\mathrm{ord}}_{\mathcal{H}}),\langle\,\cdot\,,\,\cdot\,\rangle_\lambda,\mathscr{O}_{\vec{\mathsf{M}}^{\mathrm{ord}}_{\mathcal{H}}}(1),\underline{\mathrm{Lie}}^{\vee}_{A^\vee/\vec{\mathsf{M}}^{\mathrm{ord}}_{\mathcal{H}}}(1)),\\ &((\mathrm{Gr}^0_{\mathsf{D},0}\oplus\mathrm{Gr}^{-1}_{\mathsf{D},0})\underset{R_0}{\otimes}\mathscr{O}_{\vec{\mathsf{M}}^{\mathrm{ord}}_{\mathcal{H}}},\langle\,\cdot\,,\,\cdot\,\rangle_{\phi_{\mathsf{D},0}},\mathscr{O}_{\vec{\mathsf{M}}^{\mathrm{ord}}_{\mathcal{H}}}(1),\mathrm{Gr}^0_{\mathsf{D},0}\underset{R_0}{\otimes}\mathscr{O}_{\vec{\mathsf{M}}^{\mathrm{ord}}_{\mathcal{H}}})),\end{aligned}$$

the sheaf of isomorphisms of $\mathscr{O}_{\vec{\mathsf{M}}^{\mathrm{ord}}_{\mathcal{H}}}$-sheaves of symplectic $\mathcal{O}$-modules with maximal totally isotropic $\mathcal{O}$-submodules, over $\vec{\mathsf{M}}^{\mathrm{ord}}_{\mathcal{H}}$. (By definition, the group $\mathrm{P}^{\mathrm{ord}}_{\mathsf{D},0}$ acts as automorphisms on $((\mathrm{Gr}^0_{\mathsf{D},0}\oplus\mathrm{Gr}^{-1}_{\mathsf{D},0})\underset{R_0}{\otimes}\mathscr{O}_{\vec{\mathsf{M}}^{\mathrm{ord}}_{\mathcal{H}}},\langle\,\cdot\,,\,\cdot\,\rangle_{\phi_{\mathsf{D},0}},\mathscr{O}_{\vec{\mathsf{M}}^{\mathrm{ord}}_{\mathcal{H}}}(1),\mathrm{Gr}^0_{\mathsf{D},0}\underset{R_0}{\otimes}\mathscr{O}_{\vec{\mathsf{M}}^{\mathrm{ord}}_{\mathcal{H}}})$. The third entries in the tuples represent the values of the pairings.)

Definition 8.1.2.4. (*Compare with Definition* 1.4.1.5.) *The* **principal $\mathrm{M}^{\mathrm{ord}}_{\mathsf{D},0}$-bundle** *over $\vec{\mathsf{M}}^{\mathrm{ord}}_{\mathcal{H}}$ is the relative scheme*

$$\begin{aligned}\vec{\mathcal{E}}^{\mathrm{ord}}_{\mathrm{M}^{\mathrm{ord}}_{\mathsf{D},0}} := \underline{\mathrm{Isom}}_{\mathcal{O}\underset{\mathbb{Z}}{\otimes}\mathscr{O}_{\vec{\mathsf{M}}^{\mathrm{ord}}_{\mathcal{H}}}}&((\underline{\mathrm{Lie}}^{\vee}_{A^\vee/\vec{\mathsf{M}}^{\mathrm{ord}}_{\mathcal{H}}}(1),\underline{\mathrm{Lie}}_{A/\vec{\mathsf{M}}^{\mathrm{ord}}_{\mathcal{H}}},\langle\,\cdot\,,\,\cdot\,\rangle_\lambda,\mathscr{O}_{\vec{\mathsf{M}}^{\mathrm{ord}}_{\mathcal{H}}}(1)),\\ &(\mathrm{Gr}^0_{\mathsf{D},0}\underset{R_0}{\otimes}\mathscr{O}_{\vec{\mathsf{M}}^{\mathrm{ord}}_{\mathcal{H}}},\mathrm{Gr}^{-1}_{\mathsf{D},0}\underset{R_0}{\otimes}\mathscr{O}_{\vec{\mathsf{M}}^{\mathrm{ord}}_{\mathcal{H}}},\langle\,\cdot\,,\,\cdot\,\rangle_{\phi_{\mathsf{D},0}},\mathscr{O}_{\vec{\mathsf{M}}^{\mathrm{ord}}_{\mathcal{H}}}(1))),\end{aligned}$$

the sheaf of isomorphisms of $\mathscr{O}_{\vec{\mathsf{M}}^{\mathrm{ord}}_{\mathcal{H}}}$-sheaves of $\mathcal{O}$-modules, over $\vec{\mathsf{M}}^{\mathrm{ord}}_{\mathcal{H}}$. (We view the second entries in the pairs as an additional structure, inherited from the corresponding objects for $\mathrm{P}^{\mathrm{ord}}_{\mathsf{D},0}$. By definition, the group $\mathrm{M}^{\mathrm{ord}}_{\mathsf{D},0}$ acts as automorphisms on $(\mathrm{Gr}^0_{\mathsf{D},0}\underset{R_0}{\otimes}\mathscr{O}_{\vec{\mathsf{M}}^{\mathrm{ord}}_{\mathcal{H}}},\mathrm{Gr}^{-1}_{\mathsf{D},0}\underset{R_0}{\otimes}\mathscr{O}_{\vec{\mathsf{M}}^{\mathrm{ord}}_{\mathcal{H}}},\langle\,\cdot\,,\,\cdot\,\rangle_{\phi_{\mathsf{D},0}},\mathscr{O}_{\vec{\mathsf{M}}^{\mathrm{ord}}_{\mathcal{H}}}(1))$.)

Remark 8.1.2.5. The induced pairing
$$\langle\,\cdot\,,\,\cdot\,\rangle_\lambda : \underline{\mathrm{Lie}}^\vee_{A^\vee/\vec{\mathsf{M}}^{\mathrm{ord}}_{\mathcal{H}}}(1) \times \underline{\mathrm{Lie}}_{A/\vec{\mathsf{M}}^{\mathrm{ord}}_{\mathcal{H}}} \to \mathscr{O}_{\vec{\mathsf{M}}^{\mathrm{ord}}_{\mathcal{H}}}(1)$$
in Definition 8.1.2.4 coincides with the composition
$$\underline{\mathrm{Lie}}^\vee_{A^\vee/\vec{\mathsf{M}}^{\mathrm{ord}}_{\mathcal{H}}}(1) \otimes \underline{\mathrm{Lie}}_{A/\vec{\mathsf{M}}^{\mathrm{ord}}_{\mathcal{H}}} \overset{\mathrm{Id}\otimes d\lambda}{\to} \underline{\mathrm{Lie}}^\vee_{A^\vee/\vec{\mathsf{M}}^{\mathrm{ord}}_{\mathcal{H}}}(1) \otimes \underline{\mathrm{Lie}}_{A^\vee/\vec{\mathsf{M}}^{\mathrm{ord}}_{\mathcal{H}}} \overset{\mathrm{can.}}{\to} \mathscr{O}_{\vec{\mathsf{M}}^{\mathrm{ord}}_{\mathcal{H}}}(1).$$

Lemma 8.1.2.6. (*Compare with Lemma* 1.4.1.7.) *The relative scheme* $\vec{\mathcal{E}}^{\mathrm{ord}}_{\mathrm{M}^{\mathrm{ord}}_{\mathtt{D},0}}$ *over* $\vec{\mathsf{M}}^{\mathrm{ord}}_{\mathcal{H}}$ *is an étale torsor under (the pullback of) the group scheme* $\mathrm{M}^{\mathrm{ord}}_{\mathtt{D},0}$. *When Condition* 8.1.1.2 *holds, the relative scheme* $\vec{\mathcal{E}}^{\mathrm{ord}}_{\mathrm{G}^{\mathrm{ord}}_{\mathtt{D},0}}$ *(resp.* $\vec{\mathcal{E}}^{\mathrm{ord}}_{\mathrm{P}^{\mathrm{ord}}_{\mathtt{D},0}}$*) over* $\vec{\mathsf{M}}^{\mathrm{ord}}_{\mathcal{H}}$ *is an fpqc torsor under (the pullback of) the group scheme* $\mathrm{G}^{\mathrm{ord}}_{\mathtt{D},0}$ *(resp.* $\mathrm{P}^{\mathrm{ord}}_{\mathtt{D},0}$*).*

Proof. In the case of $\vec{\mathcal{E}}^{\mathrm{ord}}_{\mathrm{M}^{\mathrm{ord}}_{\mathtt{D},0}}$, as in the proof of Lemma 1.4.1.7, this follows from the theory of infinitesimal deformations of ordinary abelian schemes with additional structures parameterized by $\vec{\mathsf{M}}^{\mathrm{ord}}_{\mathcal{H}}$ (see Theorems 3.4.1.9 and 3.4.2.5, and Convention 3.4.2.9), and from Artin's approximation theory (*cf.* [3, Thm. 1.10 and Cor. 2.5]). More precisely, in characteristic zero, this is essentially Lemma 1.4.1.7 by Lemma 8.1.1.7; and, in positive characteristics, instead of using the Lie algebra condition as in the case of $\mathsf{M}_{\mathcal{H}}$, we use the theory in [47, 3.4] and the ordinary level structure $\alpha^{\mathrm{ord}}_{\mathcal{H}_p}$ to determine the isomorphism classes of the pullbacks of $\underline{\mathrm{Lie}}^\vee_{A^\vee/\vec{\mathsf{M}}^{\mathrm{ord}}_{\mathcal{H}}}$ and $\underline{\mathrm{Lie}}_{A/\vec{\mathsf{M}}^{\mathrm{ord}}_{\mathcal{H}}}$ to the completions of strict local rings of $\vec{\mathsf{M}}^{\mathrm{ord}}_{\mathcal{H}}$.

In the cases of $\vec{\mathcal{E}}^{\mathrm{ord}}_{\mathrm{G}^{\mathrm{ord}}_{\mathtt{D},0}}$ and $\vec{\mathcal{E}}^{\mathrm{ord}}_{\mathrm{P}^{\mathrm{ord}}_{\mathtt{D},0}}$, in addition to the above argument over the completions of strict local rings, we use the assertion in Condition 8.1.1.2 to match the pairing defined by λ with the pairing $\langle\,\cdot\,,\,\cdot\,\rangle_{\phi_{\mathtt{D},0}}$. (We do not need Condition 8.1.1.2 at points of characteristic zero.) □

Definition 8.1.2.7. (*Compare with Definition* 1.4.1.8.) *For each* R_0*-algebra* R, *we denote by* $\mathrm{Rep}_R(\mathrm{G}^{\mathrm{ord}}_{\mathtt{D},0})$ *(resp.* $\mathrm{Rep}_R(\mathrm{P}^{\mathrm{ord}}_{\mathtt{D},0})$*, resp.* $\mathrm{Rep}_R(\mathrm{M}^{\mathrm{ord}}_{\mathtt{D},0})$*) the category of* R*-modules with algebraic actions of* $\mathrm{G}^{\mathrm{ord}}_{\mathtt{D},0} \underset{R_0}{\otimes} R$ *(resp.* $\mathrm{P}^{\mathrm{ord}}_{\mathtt{D},0} \underset{R_0}{\otimes} R$*, resp.* $\mathrm{M}^{\mathrm{ord}}_{\mathtt{D},0} \underset{R_0}{\otimes} R$*).*

Following Lemma 8.1.2.6, by fpqc descent of quasi-coherent sheaves (see [33, VIII, 1.3]), we can make the following:

Definition 8.1.2.8. (*Compare with Definition* 1.4.1.9.) *Let* R *be any* R_0*-algebra. For each* $W \in \mathrm{Rep}_R(\mathrm{M}^{\mathrm{ord}}_{\mathtt{D},0})$, *we define*
$$\vec{\mathcal{E}}^{\mathrm{ord}}_{\mathrm{M}^{\mathrm{ord}}_{\mathtt{D},0},R}(W) := (\vec{\mathcal{E}}^{\mathrm{ord}}_{\mathrm{M}^{\mathrm{ord}}_{\mathtt{D},0}} \underset{R_0}{\otimes} R) \overset{\mathrm{M}^{\mathrm{ord}}_{\mathtt{D},0} \underset{R_0}{\otimes} R}{\times} W,$$
called the **automorphic sheaf** *over* $\vec{\mathsf{M}}^{\mathrm{ord}}_{\mathcal{H}} \underset{R_0}{\otimes} R$ *associated with* W. *It is called an* **automorphic bundle** *if* W *is locally free of finite rank over* R, *in which case* $\vec{\mathcal{E}}^{\mathrm{ord}}_{\mathrm{M}^{\mathrm{ord}}_{\mathtt{D},0},R}(W)$ *is also locally free of finite rank over* $\vec{\mathsf{M}}^{\mathrm{ord}}_{\mathcal{H}} \underset{R_0}{\otimes} R$. *When Condition* 8.1.1.2 *holds, we define similarly* $\vec{\mathcal{E}}^{\mathrm{ord}}_{\mathrm{G}^{\mathrm{ord}}_{\mathtt{D},0},R}(W)$ *(resp.* $\vec{\mathcal{E}}^{\mathrm{ord}}_{\mathrm{P}^{\mathrm{ord}}_{\mathtt{D},0},R}(W)$*) for* $W \in \mathrm{Rep}_R(\mathrm{G}^{\mathrm{ord}}_{\mathtt{D},0})$ *(resp.* $W \in \mathrm{Rep}_R(\mathrm{P}^{\mathrm{ord}}_{\mathtt{D},0})$*) by replacing* $\mathrm{M}^{\mathrm{ord}}_{\mathtt{D},0}$ *with* $\mathrm{G}^{\mathrm{ord}}_{\mathtt{D},0}$ *(resp.* $\mathrm{P}^{\mathrm{ord}}_{\mathtt{D},0}$*) in the above expression.*

Lemma 8.1.2.9. (*Compare with Lemma* 1.4.1.10.) *Let R be any R_0-algebra. The assignment $\vec{\mathcal{E}}^{\mathrm{ord}}_{\mathrm{M}^{\mathrm{ord}}_{\mathtt{D},0},R}(\,\cdot\,)$ defines an* **exact functor** *from* $\mathrm{Rep}_R(\mathrm{M}^{\mathrm{ord}}_{\mathtt{D},0})$ *to the category of quasi-coherent sheaves over* $\vec{\mathsf{M}}^{\mathrm{ord}}_{\mathcal{H}}$.

Proof. By étale descent, the proof is similar to that of Lemma 1.4.1.10. □

Lemma 8.1.2.10. (*Compare with Lemma* 1.4.1.10.) *Let R be any R_0-algebra. Suppose that Condition* 8.1.1.2 *holds.*

(1) *The assignment $\vec{\mathcal{E}}^{\mathrm{ord}}_{\mathrm{G}^{\mathrm{ord}}_{\mathtt{D},0},R}(\,\cdot\,)$ (resp. $\vec{\mathcal{E}}^{\mathrm{ord}}_{\mathrm{P}^{\mathrm{ord}}_{\mathtt{D},0},R}(\,\cdot\,)$) defines an* **exact functor** *from* $\mathrm{Rep}_R(\mathrm{G}^{\mathrm{ord}}_{\mathtt{D},0})$ *(resp.* $\mathrm{Rep}_R(\mathrm{P}^{\mathrm{ord}}_{\mathtt{D},0})$*) to the category of quasi-coherent sheaves over* $\vec{\mathsf{M}}^{\mathrm{ord}}_{\mathcal{H}}$.

(2) *If we consider an object $W \in \mathrm{Rep}_R(\mathrm{G}^{\mathrm{ord}}_{\mathtt{D},0})$ as an object of $\mathrm{Rep}_R(\mathrm{P}^{\mathrm{ord}}_{\mathtt{D},0})$ by restriction to $\mathrm{P}^{\mathrm{ord}}_{\mathtt{D},0}$, then we have a canonical isomorphism $\vec{\mathcal{E}}^{\mathrm{ord}}_{\mathrm{G}^{\mathrm{ord}}_{\mathtt{D},0},R}(W) \cong \vec{\mathcal{E}}^{\mathrm{ord}}_{\mathrm{P}^{\mathrm{ord}}_{\mathtt{D},0},R}(W)$.*

(3) *If we view an object $W \in \mathrm{Rep}_R(\mathrm{M}^{\mathrm{ord}}_{\mathtt{D},0})$ as an object of $\mathrm{Rep}_R(\mathrm{P}^{\mathrm{ord}}_{\mathtt{D},0})$ via the canonical homomorphism $\mathrm{P}^{\mathrm{ord}}_{\mathtt{D},0} \to \mathrm{M}^{\mathrm{ord}}_{\mathtt{D},0}$, then we have a canonical isomorphism $\vec{\mathcal{E}}^{\mathrm{ord}}_{\mathrm{P}^{\mathrm{ord}}_{\mathtt{D},0},R}(W) \cong \vec{\mathcal{E}}^{\mathrm{ord}}_{\mathrm{M}^{\mathrm{ord}}_{\mathtt{D},0},R}(W)$.*

(4) *Suppose $W \in \mathrm{Rep}_R(\mathrm{P}^{\mathrm{ord}}_{\mathtt{D},0})$ has a decreasing filtration by subobjects $\mathtt{F}^a(W) \subset W$ in $\mathrm{Rep}_R(\mathrm{P}^{\mathrm{ord}}_{\mathtt{D},0})$ such that each graded piece $\mathrm{Gr}^a_{\mathtt{F}}(W) := \mathtt{F}^a(W)/\mathtt{F}^{a+1}(W)$ can be identified with an object of $\mathrm{Rep}_R(\mathrm{M}^{\mathrm{ord}}_{\mathtt{D},0})$. Then $\vec{\mathcal{E}}^{\mathrm{ord}}_{\mathrm{P}^{\mathrm{ord}}_{\mathtt{D},0},R}(W)$ has a filtration $\mathtt{F}^a(\vec{\mathcal{E}}^{\mathrm{ord}}_{\mathrm{P}^{\mathrm{ord}}_{\mathtt{D},0},R}(W)) := \vec{\mathcal{E}}^{\mathrm{ord}}_{\mathrm{P}^{\mathrm{ord}}_{\mathtt{D},0},R}(\mathtt{F}^a(W))$ with graded pieces $\vec{\mathcal{E}}^{\mathrm{ord}}_{\mathrm{M}^{\mathrm{ord}}_{\mathtt{D},0},R}(\mathrm{Gr}^a_{\mathtt{F}}(W))$.*

Proof. By fpqc descent, the proof is still similar to that of Lemma 1.4.1.10. □

Lemma 8.1.2.11. (*Compare with Lemma* 1.4.1.11.) *For any R_0-algebra R, the pullback of $\underline{\mathrm{Lie}}_{A/\vec{\mathsf{M}}^{\mathrm{ord}}_{\mathcal{H}}}$ (resp. $\underline{\mathrm{Lie}}^{\vee}_{A/\vec{\mathsf{M}}^{\mathrm{ord}}_{\mathcal{H}}}$, resp. $\omega_{\vec{\mathsf{M}}^{\mathrm{ord}}_{\mathcal{H}}} = \wedge^{\mathrm{top}}\,\underline{\mathrm{Lie}}^{\vee}_{A/\vec{\mathsf{M}}^{\mathrm{ord}}_{\mathcal{H}}}$) to $\vec{\mathsf{M}}^{\mathrm{ord}}_{\mathcal{H}} \underset{R_0}{\otimes} R$ is canonically isomorphic to $\vec{\mathcal{E}}^{\mathrm{ord}}_{\mathrm{M}^{\mathrm{ord}}_{\mathtt{D},0},R}(W)$ for $W = \mathrm{Gr}^{-1}_{\mathtt{D},0} \underset{R_0}{\otimes} R$ (resp. $(\mathrm{Gr}^{-1}_{\mathtt{D},0})^{\vee} \underset{R_0}{\otimes} R$, resp. $\wedge^{\mathrm{top}}\,(\mathrm{Gr}^{-1}_{\mathtt{D},0})^{\vee} \underset{R_0}{\otimes} R$).*

Proof. This follows from Definitions 8.1.2.4 and 8.1.2.8, and from Lemma 8.1.2.9. □

8.1.3 *Canonical Extensions*

Now let us explain the construction of canonical extensions (along the partial toroidal compactifications) using Theorem 7.1.4.1. Let $(G, \lambda, i, \alpha_{\mathcal{H}^p}, \alpha^{\mathrm{ord}}_{\mathcal{H}_p})$ be the tautological degenerating family of type $\vec{\mathsf{M}}^{\mathrm{ord}}_{\mathcal{H}}$ over $\vec{\mathsf{M}}^{\mathrm{ord,tor}}_{\mathcal{H}}$ (as in Theorem 5.2.1.1).

Then its restriction to $\vec{\mathsf{M}}^{\mathrm{ord}}_{\mathcal{H}}$ is canonically isomorphic to the pullback of the tautological tuple $(A, \lambda, i, \alpha_{\mathcal{H}^p}, \alpha^{\mathrm{ord}}_{\mathcal{H}_p})$ over $\dddot{\mathsf{M}}^{\mathrm{ord}}_{\mathcal{H}}$ (used in the above constructions in this subsection).

By taking $Q = \mathcal{O}$, so that $\underline{\mathrm{Hom}}_{\mathcal{O}}(Q, G_{\vec{\mathsf{M}}^{\mathrm{ord}}_{\mathcal{H}}})^\circ \cong G_{\vec{\mathsf{M}}^{\mathrm{ord}}_{\mathcal{H}}}$ and so that there exists some $\mathbb{Q}^\times$-isogeny $\kappa^{\mathrm{isog}} : G_{\vec{\mathsf{M}}^{\mathrm{ord}}_{\mathcal{H}}} \to \vec{\mathsf{N}}^{\mathrm{ord,grp}}_{\kappa} = \vec{\mathsf{N}}^{\mathrm{ord}}_{\kappa}$ over $\vec{\mathsf{M}}^{\mathrm{ord}}_{\mathcal{H}}$ for some $\kappa \in \mathbf{K}^{\mathrm{ord}}_{Q,\mathcal{H}}$ as in Theorem 7.1.4.1. Since $Q = \mathcal{O}$, we have $Q_{-2} = \mathrm{Hom}_{\mathcal{O}}(\mathcal{O}, \mathrm{Diff}_{\mathcal{O}'/\mathbb{Z}}(1)) \cong \mathrm{Diff}^{-1}_{\mathcal{O}'/\mathbb{Z}}(1)$ and $Q_0 = \mathcal{O}'$, in which case $\widetilde{X} = \mathrm{Hom}_{\mathcal{O}}(Q_{-2}, \mathrm{Diff}^{-1}(1)) \cong \mathcal{O}'$ and $\widetilde{Y} = \mathcal{O}'$, where $\mathcal{O}'$ is the maximal order as in the beginning of Section 1.2.4 (satisfying the requirement of Condition 1.2.1.1).

By using the level structures in characteristic zero, and by [62, Cor. 1.3.5.4], the action of $\mathcal{O}$ on $G_{\vec{\mathsf{M}}^{\mathrm{ord}}_{\mathcal{H}}} \underset{\mathbb{Z}}{\otimes} \mathbb{Q}$ extends (canonically) to an action of $\mathcal{O}'$ on $G_{\vec{\mathsf{M}}^{\mathrm{ord}}_{\mathcal{H}}} \underset{\mathbb{Z}}{\otimes} \mathbb{Q}$. Since $\vec{\mathsf{M}}^{\mathrm{ord}}_{\mathcal{H}}$ is noetherian normal, by [92, IX, 1.4], [28, Ch. I, Prop. 2.7], or [62, Prop. 3.3.1.5], the action extends (uniquely) to an action of $\mathcal{O}'$ on the whole $G_{\vec{\mathsf{M}}^{\mathrm{ord}}_{\mathcal{H}}}$. Thus, we can define a canonical morphism

$$G_{\vec{\mathsf{M}}^{\mathrm{ord}}_{\mathcal{H}}} \cong \underline{\mathrm{Hom}}_{\mathcal{O}}(\mathcal{O}, G_{\vec{\mathsf{M}}^{\mathrm{ord}}_{\mathcal{H}}}) \to \underline{\mathrm{Hom}}_{\mathcal{O}}(\mathcal{O}', G_{\vec{\mathsf{M}}^{\mathrm{ord}}_{\mathcal{H}}}) \tag{8.1.3.1}$$

Lemma 8.1.3.2. *The canonical morphism*

$$G_{\vec{\mathsf{M}}^{\mathrm{ord}}_{\mathcal{H}}} \cong \underline{\mathrm{Hom}}_{\mathcal{O}}(\mathcal{O}, G_{\vec{\mathsf{M}}^{\mathrm{ord}}_{\mathcal{H}}}) \to \underline{\mathrm{Hom}}_{\mathcal{O}}(\mathcal{O}', G_{\vec{\mathsf{M}}^{\mathrm{ord}}_{\mathcal{H}}})^\circ \tag{8.1.3.3}$$

induced by (8.1.3.1) *is an isomorphism.*

Proof. Since the composition of (8.1.3.3) with the canonical restriction morphism

$$\underline{\mathrm{Hom}}_{\mathcal{O}}(\mathcal{O}', G_{\vec{\mathsf{M}}^{\mathrm{ord}}_{\mathcal{H}}})^\circ \to \underline{\mathrm{Hom}}_{\mathcal{O}}(\mathcal{O}, G_{\vec{\mathsf{M}}^{\mathrm{ord}}_{\mathcal{H}}}) \tag{8.1.3.4}$$

is an isomorphism, the induced morphisms between fppf sheaves of groups must be injective. Hence, (8.1.3.3) and (8.1.3.4) must be group scheme isomorphisms, because they are at the same time isogenies of abelian schemes. (There are many ways to show this last fact. For example, they are flat by fiberwise criterion of flatness, while over each geometric fiber we have injective group homomorphisms between abelian varieties of the same dimension.) □

Lemma 8.1.3.5. *There exists* $\kappa \in \mathbf{K}_{Q,\mathcal{H},\Sigma^{\mathrm{ord}}}$ *such that* κ^{isog} *is an* $\mathbb{Z}^\times_{(p)}$*-isogeny.*

Proof. Suppose $\kappa = (\widetilde{\mathcal{H}}, \widetilde{\Sigma}^{\mathrm{ord}}, \widetilde{\sigma})$. By the construction of $\vec{\mathsf{N}}^{\mathrm{ord}}_{\kappa}$ in Section 7.2.1, $\vec{\mathsf{N}}^{\mathrm{ord}}_{\kappa} = \vec{\widetilde{\mathsf{Z}}}^{\mathrm{ord}}_{[(\widetilde{\Phi}_{\widetilde{\mathcal{H}}}, \widetilde{\delta}_{\widetilde{\mathcal{H}}}, \widetilde{\sigma})]}$ is canonically isomorphic to $\vec{\widetilde{C}}^{\mathrm{ord}}_{\widetilde{\Phi}_{\widetilde{\mathcal{H}}}, \widetilde{\delta}_{\widetilde{\mathcal{H}}}}$. By Proposition 4.2.1.34, it suffices to show that the canonical $\mathbb{Q}^\times$-isogeny

$$\begin{aligned} & G_{\vec{\mathsf{M}}^{\mathrm{ord}}_{\mathcal{H}}} \cong \underline{\mathrm{Hom}}_{\mathcal{O}}(\mathcal{O}, G_{\vec{\mathsf{M}}^{\mathrm{ord}}_{\mathcal{H}}}) \\ & \to \Big(\underline{\mathrm{Hom}}_{\mathcal{O}}(\mathcal{O}', G_{\vec{\mathsf{M}}^{\mathrm{ord}}_{\mathcal{H}}}) \underset{\underline{\mathrm{Hom}}_{\mathcal{O}}(\mathcal{O}', G^\vee_{\vec{\mathsf{M}}^{\mathrm{ord}}_{\mathcal{H}}})}{\times} \underline{\mathrm{Hom}}_{\mathcal{O}}(\mathcal{O}', G^\vee_{\vec{\mathsf{M}}^{\mathrm{ord}}_{\mathcal{H}}})\Big)^\circ \\ & \cong \underline{\mathrm{Hom}}_{\mathcal{O}}(\mathcal{O}', G_{\vec{\mathsf{M}}^{\mathrm{ord}}_{\mathcal{H}}})^\circ \end{aligned}$$

is an isomorphism, which follows from Lemma 8.1.3.2. □

By (3) of Theorem 7.1.4.1, the locally free sheaf $\underline{H}^1_{\mathrm{dR}}(\vec{\mathsf{N}}^{\mathrm{ord}}_{\kappa}/\vec{\mathsf{M}}^{\mathrm{ord}}_{\mathcal{H}})$ extends to the locally free sheaf $\underline{H}^1_{\log\text{-}\mathrm{dR}}(\vec{\mathsf{N}}^{\mathrm{ord,tor}}_{\kappa}/\vec{\mathsf{M}}^{\mathrm{ord,tor}}_{\mathcal{H}})$ over $\mathscr{O}_{\vec{\mathsf{M}}^{\mathrm{ord,tor}}_{\mathcal{H}}}$. Let

$$\begin{aligned}&\underline{H}_1^{\log\text{-}\mathrm{dR}}(\vec{\mathsf{N}}^{\mathrm{ord,tor}}_{\kappa}/\vec{\mathsf{M}}^{\mathrm{ord,tor}}_{\mathcal{H}})\\ &:= \underline{\mathrm{Hom}}_{\mathscr{O}_{\vec{\mathsf{M}}^{\mathrm{ord,tor}}_{\mathcal{H}}}}(\underline{H}^1_{\log\text{-}\mathrm{dR}}(\vec{\mathsf{N}}^{\mathrm{ord,tor}}_{\kappa}/\vec{\mathsf{M}}^{\mathrm{ord,tor}}_{\mathcal{H}}), \mathscr{O}_{\vec{\mathsf{M}}^{\mathrm{ord,tor}}_{\mathcal{H}}}).\end{aligned}$$

Then, for $\kappa \in \mathbf{K}_{Q,\mathcal{H},\Sigma^{\mathrm{ord}}}$ such that $\kappa^{\mathrm{isog}} : G_{\vec{\mathsf{M}}^{\mathrm{ord}}_{\mathcal{H}}} \to \vec{\mathsf{N}}^{\mathrm{ord}}_{\kappa}$ is an $\mathbb{Z}^{\times}_{(p)}$-isogeny, this $\underline{H}_1^{\log\text{-}\mathrm{dR}}(\vec{\mathsf{N}}^{\mathrm{ord,tor}}_{\kappa}/\vec{\mathsf{M}}^{\mathrm{ord,tor}}_{\mathcal{H}})$ qualifies as the $\underline{H}_1^{\mathrm{dR}}(G_{\vec{\mathsf{M}}^{\mathrm{ord}}_{\mathcal{H}}}/\vec{\mathsf{M}}^{\mathrm{ord}}_{\mathcal{H}})^{\mathrm{can}}$ in the following:

Proposition 8.1.3.6. (*Compare with Proposition* 1.4.2.1.) *There exists a* **unique** *locally free sheaf* $\underline{H}_1^{\mathrm{dR}}(G_{\vec{\mathsf{M}}^{\mathrm{ord}}_{\mathcal{H}}}/\vec{\mathsf{M}}^{\mathrm{ord}}_{\mathcal{H}})^{\mathrm{can}}$ *over* $\mathscr{O}_{\vec{\mathsf{M}}^{\mathrm{ord,tor}}_{\mathcal{H}}}$ *satisfying the following properties:*

(1) *The sheaf* $\underline{H}_1^{\mathrm{dR}}(G_{\vec{\mathsf{M}}^{\mathrm{ord}}_{\mathcal{H}}}/\vec{\mathsf{M}}^{\mathrm{ord}}_{\mathcal{H}})^{\mathrm{can}}$, *which can be canonically identified with a subsheaf of the quasi-coherent sheaf* $(\vec{\mathsf{M}}^{\mathrm{ord}}_{\mathcal{H}} \hookrightarrow \vec{\mathsf{M}}^{\mathrm{ord,tor}}_{\mathcal{H}})_*\underline{H}_1^{\mathrm{dR}}(G_{\vec{\mathsf{M}}^{\mathrm{ord}}_{\mathcal{H}}}/\vec{\mathsf{M}}^{\mathrm{ord}}_{\mathcal{H}})$, *admits a pairing* $\langle\cdot,\cdot\rangle^{\mathrm{can}}_{\lambda}$ *induced by* $(\vec{\mathsf{M}}^{\mathrm{ord}}_{\mathcal{H}} \hookrightarrow \vec{\mathsf{M}}^{\mathrm{ord,tor}}_{\mathcal{H}})_*\langle\cdot,\cdot\rangle_{\lambda}$. *This pairing* $\langle\cdot,\cdot\rangle^{\mathrm{can}}_{\lambda}$ *is perfect after pulled back to* $\mathsf{M}^{\mathrm{ord}}_{\mathcal{H}} \cong \vec{\mathsf{M}}^{\mathrm{ord}}_{\mathcal{H}} \underset{\mathbb{Z}}{\otimes} \mathbb{Q}$.

(2) $\underline{H}_1^{\mathrm{dR}}(G_{\vec{\mathsf{M}}^{\mathrm{ord}}_{\mathcal{H}}}/\vec{\mathsf{M}}^{\mathrm{ord}}_{\mathcal{H}})^{\mathrm{can}}$ *contains* $\underline{\mathrm{Lie}}^{\vee}_{G^{\vee}/\vec{\mathsf{M}}^{\mathrm{ord,tor}}_{\mathcal{H}}}(1)$ *as a subsheaf totally isotropic under the pairing* $\langle\cdot,\cdot\rangle^{\mathrm{can}}_{\lambda}$.

(3) *The quotient sheaf* $(\underline{H}_1^{\mathrm{dR}}(G_{\vec{\mathsf{M}}^{\mathrm{ord}}_{\mathcal{H}}}/\vec{\mathsf{M}}^{\mathrm{ord}}_{\mathcal{H}}))^{\mathrm{can}}/\underline{\mathrm{Lie}}^{\vee}_{G^{\vee}/\vec{\mathsf{M}}^{\mathrm{ord,tor}}_{\mathcal{H}}}$ *can be canonically identified with the subsheaf* $\underline{\mathrm{Lie}}_{G/\vec{\mathsf{M}}^{\mathrm{ord,tor}}_{\mathcal{H}}}$ *of* $(\vec{\mathsf{M}}^{\mathrm{ord}}_{\mathcal{H}} \hookrightarrow \vec{\mathsf{M}}^{\mathrm{ord,tor}}_{\mathcal{H}})_*\underline{\mathrm{Lie}}_{G_{\vec{\mathsf{M}}^{\mathrm{ord}}_{\mathcal{H}}}/\vec{\mathsf{M}}^{\mathrm{ord}}_{\mathcal{H}}}$.

(4) *The pairing* $\langle\cdot,\cdot\rangle^{\mathrm{can}}_{\lambda}$ *induces a morphism* $\underline{\mathrm{Lie}}_{G/\vec{\mathsf{M}}^{\mathrm{ord,tor}}_{\mathcal{H}}} \to \underline{\mathrm{Lie}}_{G^{\vee}/\vec{\mathsf{M}}^{\mathrm{ord,tor}}_{\mathcal{H}}}$ *which coincides with (the pullback of)* $d\lambda$. *This morphism is an isomorphism after pulled back to* $\vec{\mathsf{M}}^{\mathrm{ord,tor}}_{\mathcal{H}} \underset{\mathbb{Z}}{\otimes} \mathbb{Q}$. *If* $p \nmid [L^{\#} : L]$ *as in Definition* 1.1.1.6 *and hence* λ *is prime-to-p, then* $d\lambda$ *is an isomorphism, and consequently* $\underline{H}_1^{\mathrm{dR}}(G_{\vec{\mathsf{M}}^{\mathrm{ord}}_{\mathcal{H}}}/\vec{\mathsf{M}}^{\mathrm{ord}}_{\mathcal{H}})^{\mathrm{can}}$ *is self-dual under* $\langle\cdot,\cdot\rangle^{\mathrm{can}}_{\lambda}$.

(5) *Let*

$$\underline{H}^1_{\mathrm{dR}}(G_{\vec{\mathsf{M}}^{\mathrm{ord}}_{\mathcal{H}}}/\vec{\mathsf{M}}^{\mathrm{ord}}_{\mathcal{H}})^{\mathrm{can}} := \underline{\mathrm{Hom}}_{\mathscr{O}_{\vec{\mathsf{M}}^{\mathrm{ord,tor}}_{\mathcal{H}}}}(\underline{H}_1^{\mathrm{dR}}(G_{\vec{\mathsf{M}}^{\mathrm{ord}}_{\mathcal{H}}}/\vec{\mathsf{M}}^{\mathrm{ord}}_{\mathcal{H}})^{\mathrm{can}}, \mathscr{O}_{\vec{\mathsf{M}}^{\mathrm{ord,tor}}_{\mathcal{H}}}).$$

Then the Gauss–Manin connection

$$\nabla : \underline{H}^1_{\mathrm{dR}}(G_{\vec{\mathsf{M}}^{\mathrm{ord}}_{\mathcal{H}}}/\vec{\mathsf{M}}^{\mathrm{ord}}_{\mathcal{H}}) \to \underline{H}^1_{\mathrm{dR}}(G_{\vec{\mathsf{M}}^{\mathrm{ord}}_{\mathcal{H}}}/\vec{\mathsf{M}}^{\mathrm{ord}}_{\mathcal{H}}) \underset{\mathscr{O}_{\vec{\mathsf{M}}^{\mathrm{ord}}_{\mathcal{H}}}}{\otimes} \Omega^1_{\vec{\mathsf{M}}^{\mathrm{ord}}_{\mathcal{H}}/\vec{\mathsf{S}}_{0,r_{\mathcal{H}}}}$$

extends to an integrable connection

$$\begin{aligned}&\nabla : \underline{H}^1_{\mathrm{dR}}(G_{\vec{\mathsf{M}}^{\mathrm{ord}}_{\mathcal{H}}}/\vec{\mathsf{M}}^{\mathrm{ord}}_{\mathcal{H}})^{\mathrm{can}}\\ &\to \underline{H}^1_{\mathrm{dR}}(G_{\vec{\mathsf{M}}^{\mathrm{ord}}_{\mathcal{H}}}/\vec{\mathsf{M}}^{\mathrm{ord}}_{\mathcal{H}})^{\mathrm{can}} \underset{\mathscr{O}_{\vec{\mathsf{M}}^{\mathrm{ord,tor}}_{\mathcal{H}}}}{\otimes} \overline{\Omega}^1_{\vec{\mathsf{M}}^{\mathrm{ord,tor}}_{\mathcal{H}}/\vec{\mathsf{S}}_{0,r_{\mathcal{H}}}}\end{aligned} \tag{8.1.3.7}$$

with log poles along $\vec{\mathsf{D}}^{\mathrm{ord}}_{\infty,\mathcal{H}}$, *called the extended Gauss–Manin connection, such that the composition (ignoring Tate twists; cf. Remark* 1.1.2.3)

$$\begin{aligned}\underline{\mathrm{Lie}}^{\vee}_{G/\vec{\mathsf{M}}^{\mathrm{ord,tor}}_{\mathcal{H}}} &\hookrightarrow \underline{H}^1_{\mathrm{dR}}(G_{\vec{\mathsf{M}}^{\mathrm{ord}}_{\mathcal{H}}}/\vec{\mathsf{M}}^{\mathrm{ord}}_{\mathcal{H}})^{\mathrm{can}} \\ &\overset{\nabla}{\to} \underline{H}^1_{\mathrm{dR}}(G_{\vec{\mathsf{M}}^{\mathrm{ord}}_{\mathcal{H}}}/\vec{\mathsf{M}}^{\mathrm{ord}}_{\mathcal{H}})^{\mathrm{can}} \underset{\mathscr{O}_{\vec{\mathsf{M}}^{\mathrm{ord,tor}}_{\mathcal{H}}}}{\otimes} \overline{\Omega}^1_{\vec{\mathsf{M}}^{\mathrm{ord,tor}}_{\mathcal{H}}/\vec{\mathsf{S}}_{0,r_{\mathcal{H}}}} \\ &\twoheadrightarrow \underline{\mathrm{Lie}}_{G^{\vee}/\vec{\mathsf{M}}^{\mathrm{ord,tor}}_{\mathcal{H}}} \underset{\mathscr{O}_{\vec{\mathsf{M}}^{\mathrm{ord,tor}}_{\mathcal{H}}}}{\otimes} \overline{\Omega}^1_{\vec{\mathsf{M}}^{\mathrm{ord,tor}}_{\mathcal{H}}/\vec{\mathsf{S}}_{0,r_{\mathcal{H}}}}\end{aligned} \tag{8.1.3.8}$$

induces by duality the **extended Kodaira–Spencer morphism**

$$\underline{\mathrm{Lie}}^{\vee}_{G/\vec{\mathsf{M}}^{\mathrm{ord,tor}}_{\mathcal{H}}} \underset{\mathscr{O}_{\vec{\mathsf{M}}^{\mathrm{ord,tor}}_{\mathcal{H}}}}{\otimes} \underline{\mathrm{Lie}}^{\vee}_{G^{\vee}/\vec{\mathsf{M}}^{\mathrm{ord,tor}}_{\mathcal{H}}} \to \overline{\Omega}^1_{\vec{\mathsf{M}}^{\mathrm{ord,tor}}_{\mathcal{H}}/\vec{\mathsf{S}}_{0,r_{\mathcal{H}}}}$$

as in [62, Def. 4.6.3.44], which factors through the analogue of $\underline{\mathrm{KS}}_{\mathrm{free}}$ *(as in Definition* 3.4.3.1) *over* $\vec{\mathsf{M}}^{\mathrm{ord,tor}}_{\mathcal{H}}$ *and induces the pullback of the extended Kodaira–Spencer isomorphism* $\mathrm{KS}_{G/\vec{\mathsf{M}}^{\mathrm{ord,tor}}_{\mathcal{H}}/\vec{\mathsf{S}}_{0,r_{\mathcal{H}}}}$ *in* (4) *of Theorem* 5.2.1.1.

With these characterizing properties, we say $(\underline{H}^{\mathrm{dR}}_1(G_{\vec{\mathsf{M}}^{\mathrm{ord}}_{\mathcal{H}}}/\vec{\mathsf{M}}^{\mathrm{ord}}_{\mathcal{H}})^{\mathrm{can}},\nabla)$ *is the* **canonical extension** *of* $(\underline{H}^{\mathrm{dR}}_1(G_{\vec{\mathsf{M}}^{\mathrm{ord}}_{\mathcal{H}}}/\vec{\mathsf{M}}^{\mathrm{ord}}_{\mathcal{H}}),\nabla)$.

Proof. The uniqueness of $\underline{H}^{\mathrm{dR}}_1(G_{\vec{\mathsf{M}}^{\mathrm{ord}}_{\mathcal{H}}}/\vec{\mathsf{M}}^{\mathrm{ord}}_{\mathcal{H}})^{\mathrm{can}}$ is clear by the first four properties. To show the existence, let us take any $\kappa \in \mathbf{K}_{Q,\mathcal{H},\Sigma^{\mathrm{ord}}}$ such that $\kappa^{\mathrm{isog}} : G_{\vec{\mathsf{M}}^{\mathrm{ord}}_{\mathcal{H}}} \to \vec{\mathsf{N}}^{\mathrm{ord,grp}}_{\kappa} = \vec{\mathsf{N}}^{\mathrm{ord}}_{\kappa}$ is a $\mathbb{Z}^{\times}_{(p)}$-isogeny (for $Q = \mathcal{O}$, as mentioned before this proposition, by Lemma 8.1.3.5), and take $\underline{H}^{\mathrm{dR}}_1(G_{\vec{\mathsf{M}}^{\mathrm{ord}}_{\mathcal{H}}}/\vec{\mathsf{M}}^{\mathrm{ord}}_{\mathcal{H}})^{\mathrm{can}}$ to be the sheaf $\underline{H}^{\text{log-dR}}_1(\vec{\mathsf{N}}^{\mathrm{ord,tor}}_{\kappa}/\vec{\mathsf{M}}^{\mathrm{ord,tor}}_{\mathcal{H}})$. It is locally free with a Hodge filtration by (3c) of Theorem 7.1.4.1. Moreover, by taking some sufficiently divisible integer $N > 0$ such that $N\,\mathrm{Diff}^{-1} \subset \mathcal{O}$, we obtain by multiplication by N a morphism $j_Q : Q^{\vee} \cong \mathrm{Diff}^{-1} \hookrightarrow Q = \mathcal{O}$ as in Lemma 1.2.4.1 such that pulling back by κ^{isog} identifies

$$\langle\,\cdot\,,\,\cdot\,\rangle_{\lambda_{\vec{\mathsf{M}}^{\mathrm{ord}}_{\mathcal{H}}},j_Q} : \underline{H}^1_{\mathrm{dR}}(\vec{\mathsf{N}}^{\mathrm{ord}}_{\kappa}/\vec{\mathsf{M}}^{\mathrm{ord}}_{\mathcal{H}}) \times \underline{H}^1_{\mathrm{dR}}(\vec{\mathsf{N}}^{\mathrm{ord}}_{\kappa}/\vec{\mathsf{M}}^{\mathrm{ord}}_{\mathcal{H}}) \to \mathscr{O}_{\vec{\mathsf{M}}^{\mathrm{ord}}_{\mathcal{H}}}(1)$$

with a nonzero multiple of

$$\langle\,\cdot\,,\,\cdot\,\rangle_{\lambda_{\vec{\mathsf{M}}^{\mathrm{ord}}_{\mathcal{H}}}} : \underline{H}^1_{\mathrm{dR}}(G_{\vec{\mathsf{M}}^{\mathrm{ord}}_{\mathcal{H}}}/\vec{\mathsf{M}}^{\mathrm{ord}}_{\mathcal{H}}) \times \underline{H}^1_{\mathrm{dR}}(G_{\vec{\mathsf{M}}^{\mathrm{ord}}_{\mathcal{H}}}/\vec{\mathsf{M}}^{\mathrm{ord}}_{\mathcal{H}}) \to \mathscr{O}_{\vec{\mathsf{M}}^{\mathrm{ord}}_{\mathcal{H}}}(1).$$

(This nonzero multiple is harmless because we will be comparing locally free sheaves of finite rank over a noetherian normal scheme flat over $\mathrm{Spec}(\mathbb{Z})$ which we already know outside a closed subscheme of codimension at least two.) Then (1) and (2) follow from (3d) of Theorem 7.1.4.1, and (3) and (4) follows from Proposition 7.3.6.15 and (7.3.6.4) (which was used to prove (3b) of Theorem 7.1.4.1). (The assertion in (4) about self-duality of the pairing when $p \nmid [L^{\#} : L]$ is self-explanatory.) It remains to verify (5). By definition, $\underline{H}^1_{\mathrm{dR}}(G_{\vec{\mathsf{M}}^{\mathrm{ord}}_{\mathcal{H}}}/\vec{\mathsf{M}}^{\mathrm{ord}}_{\mathcal{H}})^{\mathrm{can}} \cong \underline{H}^1_{\text{log-dR}}(\vec{\mathsf{N}}^{\mathrm{ord,tor}}_{\kappa}/\vec{\mathsf{M}}^{\mathrm{ord,tor}}_{\mathcal{H}})$. The existence of ∇ in (8.1.3.7) follows from (3e) of Theorem 7.1.4.1. As explained in [61, Rem. 4.42], the pullback of (8.1.3.8) to $\vec{\mathsf{M}}^{\mathrm{ord}}_{\mathcal{H}}$ is induced by the usual Kodaira–Spencer morphism. Since the extended Kodaira–Spencer morphism in [62, Def. 4.6.3.44] is defined exactly as a morphism induced by the usual Kodaira–Spencer morphism (by normality of $\vec{\mathsf{M}}^{\mathrm{ord,tor}}_{\mathcal{H}}$ and local freeness of the sheaves involved), it is induced by duality by (8.1.3.8), as desired. □

Then the principal bundle $\vec{\mathcal{E}}^{\mathrm{ord}}_{\mathrm{G}^{\mathrm{ord}}_{\mathsf{D},0}}$ extends canonically to a principal bundle $\vec{\mathcal{E}}^{\mathrm{ord,can}}_{\mathrm{G}^{\mathrm{ord}}_{\mathsf{D},0}}$ over $\vec{\mathsf{M}}^{\mathrm{ord,tor}}_{\mathcal{H}}$ by setting

$$\begin{aligned}
&\vec{\mathcal{E}}^{\mathrm{ord,can}}_{\mathrm{G}^{\mathrm{ord}}_{\mathsf{D},0}}\\
&:= \underline{\mathrm{Isom}}_{\mathcal{O}\underset{\mathbb{Z}}{\otimes}\mathscr{O}_{\vec{\mathsf{M}}^{\mathrm{ord,tor}}_{\mathcal{H}}}}((\underline{H}^{\mathrm{dR}}_1(G_{\vec{\mathsf{M}}^{\mathrm{ord}}_{\mathcal{H}}}/\vec{\mathsf{M}}^{\mathrm{ord}}_{\mathcal{H}})^{\mathrm{can}}, \langle\,\cdot\,,\,\cdot\,\rangle^{\mathrm{can}}_{\lambda}, \mathscr{O}_{\vec{\mathsf{M}}^{\mathrm{ord,tor}}_{\mathcal{H}}}(1)),\\
&\quad((\mathrm{Gr}^0_{\mathsf{D},0}\oplus\mathrm{Gr}^{-1}_{\mathsf{D},0})\underset{R_0}{\otimes}\mathscr{O}_{\vec{\mathsf{M}}^{\mathrm{ord,tor}}_{\mathcal{H}}}, \langle\,\cdot\,,\,\cdot\,\rangle_{\phi_{\mathsf{D},0}}, \mathscr{O}_{\vec{\mathsf{M}}^{\mathrm{ord,tor}}_{\mathcal{H}}}(1))).
\end{aligned}\tag{8.1.3.9}$$

The principal bundle $\vec{\mathcal{E}}^{\mathrm{ord}}_{\mathrm{P}^{\mathrm{ord}}_{\mathsf{D},0}}$ extends canonically to a principal bundle $\vec{\mathcal{E}}^{\mathrm{ord,can}}_{\mathrm{P}^{\mathrm{ord}}_{\mathsf{D},0}}$ over $\vec{\mathsf{M}}^{\mathrm{ord,tor}}_{\mathcal{H}}$ by setting

$$\begin{aligned}
&\vec{\mathcal{E}}^{\mathrm{ord,can}}_{\mathrm{P}^{\mathrm{ord}}_{\mathsf{D},0}}\\
&:= \underline{\mathrm{Isom}}_{\mathcal{O}\underset{\mathbb{Z}}{\otimes}\mathscr{O}_{\vec{\mathsf{M}}^{\mathrm{ord,tor}}_{\mathcal{H}}}}((\underline{H}^{\mathrm{dR}}_1(G_{\vec{\mathsf{M}}^{\mathrm{ord}}_{\mathcal{H}}}/\vec{\mathsf{M}}^{\mathrm{ord}}_{\mathcal{H}})^{\mathrm{can}}, \langle\,\cdot\,,\,\cdot\,\rangle^{\mathrm{can}}_{\lambda}, \mathscr{O}_{\vec{\mathsf{M}}^{\mathrm{ord,tor}}_{\mathcal{H}}}(1), \underline{\mathrm{Lie}}^{\vee}_{G^{\vee}/\vec{\mathsf{M}}^{\mathrm{ord,tor}}_{\mathcal{H}}}(1)),\\
&\quad((\mathrm{Gr}^0_{\mathsf{D},0}\oplus\mathrm{Gr}^{-1}_{\mathsf{D},0})\underset{R_0}{\otimes}\mathscr{O}_{\vec{\mathsf{M}}^{\mathrm{ord,tor}}_{\mathcal{H}}}, \langle\,\cdot\,,\,\cdot\,\rangle_{\phi_{\mathsf{D},0}}, \mathscr{O}_{\vec{\mathsf{M}}^{\mathrm{ord,tor}}_{\mathcal{H}}}(1), \mathrm{Gr}^0_{\mathsf{D},0}\underset{R_0}{\otimes}\mathscr{O}_{\vec{\mathsf{M}}^{\mathrm{ord,tor}}_{\mathcal{H}}})).
\end{aligned}\tag{8.1.3.10}$$

The principal bundle $\vec{\mathcal{E}}^{\mathrm{ord}}_{\mathrm{M}^{\mathrm{ord}}_{\mathsf{D},0}}$ extends canonically to a principal bundle $\vec{\mathcal{E}}^{\mathrm{ord,can}}_{\mathrm{M}^{\mathrm{ord}}_{\mathsf{D},0}}$ over $\vec{\mathsf{M}}^{\mathrm{ord,tor}}_{\mathcal{H}}$ by setting

$$\begin{aligned}
&\vec{\mathcal{E}}^{\mathrm{ord,can}}_{\mathrm{M}^{\mathrm{ord}}_{\mathsf{D},0}}\\
&:= \underline{\mathrm{Isom}}_{\mathcal{O}\underset{\mathbb{Z}}{\otimes}\mathscr{O}_{\vec{\mathsf{M}}^{\mathrm{ord,tor}}_{\mathcal{H}}}}((\underline{\mathrm{Lie}}^{\vee}_{G^{\vee}/\vec{\mathsf{M}}^{\mathrm{ord,tor}}_{\mathcal{H}}}(1), \underline{\mathrm{Lie}}_{G/\vec{\mathsf{M}}^{\mathrm{ord,tor}}_{\mathcal{H}}}, \langle\,\cdot\,,\,\cdot\,\rangle^{\mathrm{can}}_{\lambda}, \mathscr{O}_{\vec{\mathsf{M}}^{\mathrm{ord,tor}}_{\mathcal{H}}}(1)),\\
&\quad(\mathrm{Gr}^0_{\mathsf{D},0}\underset{R_0}{\otimes}\mathscr{O}_{\vec{\mathsf{M}}^{\mathrm{ord,tor}}_{\mathcal{H}}}, \mathrm{Gr}^{-1}_{\mathsf{D},0}\underset{R_0}{\otimes}\mathscr{O}_{\vec{\mathsf{M}}^{\mathrm{ord,tor}}_{\mathcal{H}}}, \langle\,\cdot\,,\,\cdot\,\rangle_{\phi_{\mathsf{D},0}}, \mathscr{O}_{\vec{\mathsf{M}}^{\mathrm{ord,tor}}_{\mathcal{H}}}(1))).
\end{aligned}\tag{8.1.3.11}$$

Lemma 8.1.3.12. (*Compare with Lemmas* 1.4.2.8 *and* 8.1.2.6.) *The relative scheme* $\vec{\mathcal{E}}^{\mathrm{ord,can}}_{\mathrm{M}^{\mathrm{ord}}_{\mathsf{D},0}}$ *over* $\vec{\mathsf{M}}^{\mathrm{ord,tor}}_{\mathcal{H}}$ *is an étale torsor under the pullback of the group scheme* $\mathrm{M}^{\mathrm{ord}}_{\mathsf{D},0}$. *When Condition* 8.1.1.2 *holds, the relative scheme* $\vec{\mathcal{E}}^{\mathrm{ord,can}}_{\mathrm{G}^{\mathrm{ord}}_{\mathsf{D},0}}$ (*resp.* $\vec{\mathcal{E}}^{\mathrm{ord,can}}_{\mathrm{P}^{\mathrm{ord}}_{\mathsf{D},0}}$) *over* $\vec{\mathsf{M}}^{\mathrm{ord,tor}}_{\mathcal{H}}$ *is an fpqc torsor under the pullback of the group scheme* $\mathrm{G}^{\mathrm{ord}}_{\mathsf{D},0}$ (*resp.* $\mathrm{P}^{\mathrm{ord}}_{\mathsf{D},0}$, *resp.* $\mathrm{M}^{\mathrm{ord}}_{\mathsf{D},0}$).

Proof. At points of residue characteristic zero, this is essentially Lemma 1.4.2.8 by Lemma 8.1.1.7 (and we do not need Condition 8.1.1.2). At points of residue characteristic $p > 0$, we shall rigidify the isomorphism classes of $\underline{\mathrm{Lie}}^{\vee}_{G^{\vee}/\vec{\mathsf{M}}^{\mathrm{ord,tor}}_{\mathcal{H}}}$ and $\underline{\mathrm{Lie}}_{G/\vec{\mathsf{M}}^{\mathrm{ord,tor}}_{\mathcal{H}}}$ using the extensions of the ordinary level structure $\alpha^{\mathrm{ord}}_{\mathcal{H}_p}$ to all of $\vec{\mathsf{M}}^{\mathrm{ord,tor}}_{\mathcal{H}}$ (*cf.* condition (5) in Definition 3.4.2.10). Then the same argument as in the proof of Lemma 8.1.2.6 works here. □

Definition 8.1.3.13. (*Compare with Definition* 1.4.2.9.) *Let R be any R_0-algebra. For each $W \in \mathrm{Rep}_R(\mathrm{M}^{\mathrm{ord}}_{\mathsf{D},0})$, we define*

$$\vec{\mathcal{E}}^{\mathrm{ord,can}}_{\mathrm{M}^{\mathrm{ord}}_{\mathsf{D},0},R}(W) := (\vec{\mathcal{E}}^{\mathrm{ord,can}}_{\mathrm{M}^{\mathrm{ord}}_{\mathsf{D},0}} \underset{R_0}{\otimes} R) \overset{\mathrm{M}^{\mathrm{ord}}_{\mathsf{D},0} \underset{R_0}{\otimes} R}{\times} W,$$

called the **canonical extension** *of $\vec{\mathcal{E}}^{\mathrm{ord}}_{\mathrm{M}^{\mathrm{ord}}_{\mathsf{D},0},R}(W)$, and define*

$$\vec{\mathcal{E}}^{\mathrm{ord,sub}}_{\mathrm{M}^{\mathrm{ord}}_{\mathsf{D},0},R}(W) := \vec{\mathcal{E}}^{\mathrm{ord,can}}_{\mathrm{M}^{\mathrm{ord}}_{\mathsf{D},0},R}(W) \underset{\mathscr{O}_{\vec{\mathsf{M}}^{\mathrm{ord,tor}}_{\mathcal{H}}}}{\otimes} \mathscr{I}_{\vec{\mathsf{D}}^{\mathrm{ord}}_{\infty,\mathcal{H}}},$$

called the **subcanonical extension** *of $\vec{\mathcal{E}}^{\mathrm{ord}}_{\mathrm{M}^{\mathrm{ord}}_{\mathsf{D},0},R}(W)$, where $\mathscr{I}_{\vec{\mathsf{D}}^{\mathrm{ord}}_{\infty,\mathcal{H}}}$ is the $\mathscr{O}_{\vec{\mathsf{M}}^{\mathrm{ord,tor}}_{\mathcal{H}}}$-ideal defining the relative Cartier divisor $\vec{\mathsf{D}}^{\mathrm{ord}}_{\infty,\mathcal{H}}$ (with its reduced structure) in* (3) *of Theorem* 5.2.1.1. *When Condition* 8.1.1.2 *holds, we define similarly $\vec{\mathcal{E}}^{\mathrm{ord,can}}_{\mathrm{G}^{\mathrm{ord}}_{\mathsf{D},0},R}(W)$ and $\vec{\mathcal{E}}^{\mathrm{ord,sub}}_{\mathrm{G}^{\mathrm{ord}}_{\mathsf{D},0},R}(W)$ (resp. $\vec{\mathcal{E}}^{\mathrm{ord,can}}_{\mathrm{P}^{\mathrm{ord}}_{\mathsf{D},0},R}(W)$ and $\vec{\mathcal{E}}^{\mathrm{ord,sub}}_{\mathrm{P}^{\mathrm{ord}}_{\mathsf{D},0},R}(W)$) with $\mathrm{M}^{\mathrm{ord}}_{\mathsf{D},0}$ and its principal bundle replaced accordingly with $\mathrm{G}^{\mathrm{ord}}_{\mathsf{D},0}$ (resp. $\mathrm{P}^{\mathrm{ord}}_{\mathsf{D},0}$) and its principal bundle.*

Then we have:

Lemma 8.1.3.14. (*Compare with Lemma* 1.4.2.10.) *Lemmas* 8.1.2.9 *and* 8.1.2.10 *remain true if we replace the automorphic sheaves with their canonical or subcanonical extensions.*

As in the case of Lemma 1.4.2.10, the same argument of the proof of Lemma 1.4.1.10 (by étale or fpqc descent) works here.

Lemma 8.1.3.15. (*Compare with Lemmas* 1.4.2.11 *and* 8.1.2.11.) *For any R_0-algebra R, the pullback of $\underline{\mathrm{Lie}}_{G/\vec{\mathsf{M}}^{\mathrm{ord,tor}}_{\mathcal{H}}}$ (resp. $\underline{\mathrm{Lie}}^{\vee}_{G/\vec{\mathsf{M}}^{\mathrm{ord,tor}}_{\mathcal{H}}}$, resp. $\omega_{\vec{\mathsf{M}}^{\mathrm{ord,tor}}_{\mathcal{H}}} = \wedge^{\mathrm{top}}\, \underline{\mathrm{Lie}}^{\vee}_{G/\vec{\mathsf{M}}^{\mathrm{ord,tor}}_{\mathcal{H}}}$) to $\vec{\mathsf{M}}^{\mathrm{ord,tor}}_{\mathcal{H}} \underset{R_0}{\otimes} R$ is canonically isomorphic to $\vec{\mathcal{E}}^{\mathrm{ord,can}}_{\mathrm{M}^{\mathrm{ord}}_{\mathsf{D},0},R}(W)$ for $W = \mathrm{Gr}^{-1}_{\mathsf{D},0} \underset{R_0}{\otimes} R$ (resp. $(\mathrm{Gr}^{-1}_{\mathsf{D},0})^{\vee} \underset{R_0}{\otimes} R$, resp. $\wedge^{\mathrm{top}}\, (\mathrm{Gr}^{-1}_{\mathsf{D},0})^{\vee} \underset{R_0}{\otimes} R$).*

Proof. This follows from (8.1.3.11), Definition 8.1.3.13, and Lemma 8.1.3.14. □

8.1.4 *Hecke Actions*

Proposition 8.1.4.1. (*Compare with Propositions* 1.4.3.1, 3.4.4.1, *and* 5.2.2.2, *and* (4) *of Theorem* 7.1.4.1.) *Let R be any R_0-algebra, and let $W \in \mathrm{Rep}_R(\mathrm{M}^{\mathrm{ord}}_{\mathsf{D},0})$. Suppose we have an element $g = (g_0, g_p) \in \mathrm{G}(\mathbb{A}^{\infty,p}) \times \mathrm{P}^{\mathrm{ord}}_{\mathsf{D}}(\mathbb{Q}_p) \subset \mathrm{G}(\mathbb{A}^{\infty})$ (see Definition* 3.2.2.7), *and suppose we have two open compact subgroups $\mathcal{H}$ and $\mathcal{H}'$ of $\mathrm{G}(\hat{\mathbb{Z}})$ such that $\mathcal{H}' \subset g\mathcal{H}g^{-1}$, and such that $\mathcal{H}$ and $\mathcal{H}'$ are of standard form as in Definition* 3.2.2.9. *Suppose moreover that g_p satisfies the conditions given in Section* 3.3.4, *so that $\vec{[g]}^{\mathrm{ord}} : \vec{\mathsf{M}}^{\mathrm{ord}}_{\mathcal{H}'} \to \vec{\mathsf{M}}^{\mathrm{ord}}_{\mathcal{H}}$ is defined as in Proposition* 3.4.4.1;

and that $g_p \in \mathrm{P}^{\mathrm{ord}}_{\mathsf{D}}(\mathbb{Z}_p)$ *when* p *is not invertible in* R. *Then there is (by abuse of notation) a canonical isomorphism*

$$([\vec{g}]^{\mathrm{ord}})^* : ([\vec{g}]^{\mathrm{ord}})^* \vec{\mathcal{E}}^{\mathrm{ord}}_{\mathrm{M}^{\mathrm{ord}}_{\mathsf{D},0},R}(W) \xrightarrow{\sim} \vec{\mathcal{E}}^{\mathrm{ord}}_{\mathrm{M}^{\mathrm{ord}}_{\mathsf{D},0},R}(W) \tag{8.1.4.2}$$

of quasi-coherent sheaves over $\vec{\mathsf{M}}^{\mathrm{ord}}_{\mathcal{H}'}$, *where the first* $\vec{\mathcal{E}}^{\mathrm{ord}}_{\mathrm{M}^{\mathrm{ord}}_{\mathsf{D},0},R}(W)$ *is defined over* $\vec{\mathsf{M}}^{\mathrm{ord}}_{\mathcal{H}}$, *and where the second is defined over* $\vec{\mathsf{M}}^{\mathrm{ord}}_{\mathcal{H}'}$.

Suppose $\Sigma^{\mathrm{ord}} = \{\Sigma_{\Phi_{\mathcal{H}}}\}_{[(\Phi_{\mathcal{H}},\delta_{\mathcal{H}})]}$ *and* $\Sigma^{\mathrm{ord},\prime} = \{\Sigma'_{\Phi'_{\mathcal{H}'}}\}_{[(\Phi'_{\mathcal{H}'},\delta'_{\mathcal{H}'})]}$ *are compatible choices of admissible smooth rational polyhedral cone decomposition data for* $\vec{\mathsf{M}}^{\mathrm{ord}}_{\mathcal{H}}$ *and* $\vec{\mathsf{M}}^{\mathrm{ord}}_{\mathcal{H}'}$, *respectively, such that* $\Sigma^{\mathrm{ord},\prime}$ *is a* g*-refinement of* Σ^{ord} *as in Definition* 5.2.2.1, *so that* $[\vec{g}]^{\mathrm{ord,tor}} : \vec{\mathsf{M}}^{\mathrm{ord,tor}}_{\mathcal{H}',\Sigma^{\mathrm{ord},\prime}} \to \vec{\mathsf{M}}^{\mathrm{ord,tor}}_{\mathcal{H},\Sigma^{\mathrm{ord}}}$ *is defined as in Proposition* 5.2.2.2. *Then there is (by abuse of notation) a canonical isomorphism*

$$([\vec{g}]^{\mathrm{ord,tor}})^* : ([\vec{g}]^{\mathrm{ord,tor}})^* \vec{\mathcal{E}}^{\mathrm{ord,can}}_{\mathrm{M}^{\mathrm{ord}}_{\mathsf{D},0},R}(W) \xrightarrow{\sim} \vec{\mathcal{E}}^{\mathrm{ord,can}}_{\mathrm{M}^{\mathrm{ord}}_{\mathsf{D},0},R}(W) \tag{8.1.4.3}$$

of quasi-coherent sheaves over $\vec{\mathsf{M}}^{\mathrm{ord,tor}}_{\mathcal{H}',\Sigma^{\mathrm{ord},\prime}}$, *where the first* $\vec{\mathcal{E}}^{\mathrm{ord,can}}_{\mathrm{M}^{\mathrm{ord}}_{\mathsf{D},0},R}(W)$ *is defined over* $\vec{\mathsf{M}}^{\mathrm{ord,tor}}_{\mathcal{H},\Sigma^{\mathrm{ord}}}$, *and where the second is defined over* $\vec{\mathsf{M}}^{\mathrm{ord,tor}}_{\mathcal{H}',\Sigma^{\mathrm{ord},\prime}}$. *There is also (by abuse of notation) a canonical morphism*

$$([\vec{g}]^{\mathrm{ord,tor}})^* : ([\vec{g}]^{\mathrm{ord,tor}})^* \vec{\mathcal{E}}^{\mathrm{ord,sub}}_{\mathrm{M}^{\mathrm{ord}}_{\mathsf{D},0},R}(W) \to \vec{\mathcal{E}}^{\mathrm{ord,sub}}_{\mathrm{M}^{\mathrm{ord}}_{\mathsf{D},0},R}(W) \tag{8.1.4.4}$$

of quasi-coherent sheaves over $\vec{\mathsf{M}}^{\mathrm{ord,tor}}_{\mathcal{H}',\Sigma^{\mathrm{ord},\prime}}$. *The canonical morphisms* (8.1.4.2), (8.1.4.3), *and* (8.1.4.4) *are compatible with each other.*

When Condition 8.1.1.2 *holds, the analogous statements are true if we replace* $\mathrm{M}^{\mathrm{ord}}_{\mathsf{D},0}$ *with* $\mathrm{P}^{\mathrm{ord}}_{\mathsf{D},0}$.

If $g = g_1 g_2$, *where* $g_1 = (g_{1,0}, g_{1,p})$ *and* $g_2 = (g_{2,0}, g_{2,p})$ *are elements of* $\mathrm{G}(\mathbb{A}^{\infty,p}) \times \mathrm{P}^{\mathrm{ord}}_{\mathsf{D}}(\mathbb{Q}_p)$, *each having a setup similar to that of* g, *then we have* $([\vec{g}]^{\mathrm{ord}})^* = ([\vec{g_1}]^{\mathrm{ord}})^* \circ ([\vec{g_2}]^{\mathrm{ord}})^*$ *and* $([\vec{g}]^{\mathrm{ord,tor}})^* = ([\vec{g_1}]^{\mathrm{ord,tor}})^* \circ ([\vec{g_2}]^{\mathrm{ord,tor}})^*$ *whenever the involved isomorphisms are defined.*

Proof. Let us first compatibly define canonical isomorphisms

$$([\vec{g}]^{\mathrm{ord}})^* : ([\vec{g}]^{\mathrm{ord}})^* \vec{\mathcal{E}}^{\mathrm{ord}}_{\mathrm{M}^{\mathrm{ord}}_{\mathsf{D},0}} \underset{R_0}{\otimes} R \xrightarrow{\sim} \vec{\mathcal{E}}^{\mathrm{ord}}_{\mathrm{M}^{\mathrm{ord}}_{\mathsf{D},0}} \underset{R_0}{\otimes} R \tag{8.1.4.5}$$

and

$$([\vec{g}]^{\mathrm{ord,tor}})^* : ([\vec{g}]^{\mathrm{ord,tor}})^* \vec{\mathcal{E}}^{\mathrm{ord,can}}_{\mathrm{M}^{\mathrm{ord}}_{\mathsf{D},0}} \underset{R_0}{\otimes} R \xrightarrow{\sim} \vec{\mathcal{E}}^{\mathrm{ord,can}}_{\mathrm{M}^{\mathrm{ord}}_{\mathsf{D},0}} \underset{R_0}{\otimes} R \tag{8.1.4.6}$$

between the pullbacks to R of the corresponding principal bundles and their canonical extensions. Under $[\vec{g}]^{\mathrm{ord,tor}} : \vec{\mathsf{M}}^{\mathrm{ord,tor}}_{\mathcal{H}',\Sigma^{\mathrm{ord},\prime}} \twoheadrightarrow \vec{\mathsf{M}}^{\mathrm{ord,tor}}_{\mathcal{H},\Sigma^{\mathrm{ord}}}$, the ordinary Hecke twist of the tautological family $(G, \lambda, i, \alpha_{\mathcal{H}'^{,p}}, \alpha^{\mathrm{ord}}_{\mathcal{H}'_p}) \to \vec{\mathsf{M}}^{\mathrm{ord,tor}}_{\mathcal{H}',\Sigma^{\mathrm{ord},\prime}}$ by g (defined by Proposition 3.3.4.21 and Lemma 3.1.3.2) is the pullback $(G', \lambda', i', \alpha'_{\mathcal{H}^p}, \alpha^{\mathrm{ord},\prime}_{\mathcal{H}_p}) \to \vec{\mathsf{M}}^{\mathrm{ord,tor}}_{\mathcal{H}',\Sigma^{\mathrm{ord},\prime}}$ of the tautological family $(G, \lambda, i, \alpha_{\mathcal{H}^p}, \alpha^{\mathrm{ord}}_{\mathcal{H}_p}) \to \vec{\mathsf{M}}^{\mathrm{ord,tor}}_{\mathcal{H},\Sigma^{\mathrm{ord}}}$, equipped with a $\mathbb{Q}^\times$-isogeny $[g^{-1}]^{\mathrm{ord}} : G \to G'$. (See Propositions 3.4.4.1 and 5.2.2.2 and their proofs.) Since $g_p \in \mathrm{P}^{\mathrm{ord}}_{\mathsf{D}}(\mathbb{Z}_p)$ when p is not invertible in R, the $\mathbb{Q}^\times$-isogeny $[g^{-1}]^{\mathrm{ord}} : G \to G'$ induces isomorphisms

$$\begin{aligned}
&[g^{-1}]^{\mathrm{ord}}_* : \underline{\mathrm{Lie}}_{G/\vec{\mathsf{M}}^{\mathrm{ord,tor}}_{\mathcal{H}',\Sigma^{\mathrm{ord},\prime}}} \underset{R_0}{\otimes} R \\
&\quad \xrightarrow{\sim} \underline{\mathrm{Lie}}_{G'/\vec{\mathsf{M}}^{\mathrm{ord,tor}}_{\mathcal{H}',\Sigma^{\mathrm{ord},\prime}}} \underset{R_0}{\otimes} R \cong (\vec{[g]}^{\mathrm{ord,tor}})^* \underline{\mathrm{Lie}}_{G/\vec{\mathsf{M}}^{\mathrm{ord,tor}}_{\mathcal{H},\Sigma^{\mathrm{ord}}}} \underset{R_0}{\otimes} R
\end{aligned} \tag{8.1.4.7}$$

and

$$\begin{aligned}
&[g^{-1}]^{\mathrm{ord}}_* : \underline{\mathrm{Lie}}^\vee_{G^\vee/\vec{\mathsf{M}}^{\mathrm{ord,tor}}_{\mathcal{H}',\Sigma^{\mathrm{ord},\prime}}} \underset{R_0}{\otimes} R \\
&\quad \xrightarrow{\sim} \underline{\mathrm{Lie}}^\vee_{G',\vee/\vec{\mathsf{M}}^{\mathrm{ord,tor}}_{\mathcal{H}',\Sigma^{\mathrm{ord},\prime}}} \underset{R_0}{\otimes} R \cong (\vec{[g]}^{\mathrm{ord,tor}})^* \underline{\mathrm{Lie}}^\vee_{G^\vee/\vec{\mathsf{M}}^{\mathrm{ord,tor}}_{\mathcal{H},\Sigma^{\mathrm{ord}}}} \underset{R_0}{\otimes} R,
\end{aligned} \tag{8.1.4.8}$$

which induce the desired isomorphism (8.1.4.6) (*cf.* the definition of $\vec{\mathcal{E}}^{\mathrm{ord,can}}_{\mathrm{M}^{\mathrm{ord}}_{\mathsf{D},0}}$ in (8.1.3.11)), whose restrictions to $\vec{\mathsf{M}}^{\mathrm{ord}}_{\mathcal{H}'}$ induce the desired isomorphism (8.1.4.5) (*cf.* Definition 8.1.2.4). By definition, these isomorphisms (8.1.4.5) and (8.1.4.6) then induce the desired morphisms (8.1.4.2), (8.1.4.3), and (8.1.4.4).

The analogous statements for $\mathrm{P}^{\mathrm{ord}}_{\mathsf{D},0}$ are similar, using the relative de Rham cohomology $\underline{H}^{\mathrm{dR}}_1(G_{\vec{\mathsf{M}}^{\mathrm{ord}}_{\mathcal{H}}}/\vec{\mathsf{M}}^{\mathrm{ord}}_{\mathcal{H}})$ and its canonical extension $\underline{H}^{\mathrm{dR}}_1(G_{\vec{\mathsf{M}}^{\mathrm{ord}}_{\mathcal{H}}}/\vec{\mathsf{M}}^{\mathrm{ord}}_{\mathcal{H}})^{\mathrm{can}}$ (see Proposition 8.1.3.6, which is based on Theorem 7.1.4.1—the last statement, for $\mathrm{P}^{\mathrm{ord}}_{\mathsf{D},0}$, follows from (4f) of Theorem 7.1.4.1). □

8.2 Higher Direct Images to the Minimal Compactifications

8.2.1 *Some Vanishing Theorems*

Let us begin with the statement in characteristic zero:

Theorem 8.2.1.1. *Let R be any F_0'-algebra, and let W be any object of* $\mathrm{Rep}_R(\mathrm{M}_0)$ *(see Definition* 1.4.1.8*). Consider the canonical morphism $\oint_{\mathcal{H}} : \mathsf{M}^{\mathrm{tor}}_{\mathcal{H}} \to \mathsf{M}^{\min}_{\mathcal{H}}$ (where $\mathsf{M}^{\mathrm{tor}}_{\mathcal{H}} = \mathsf{M}^{\mathrm{tor}}_{\mathcal{H},\Sigma}$ can be as in Theorem* 1.3.1.3 *for any Σ). Suppose $\mathcal{H}$ is* **neat***. Then*

$$R^b \oint_{\mathcal{H},*}(\mathcal{E}^{\mathrm{sub}}_{\mathrm{M}_0,R}(W)) = 0$$

for all $b > 0$ (see Definition 1.4.2.9*).*

(See Remark 8.2.1.5 below for more information.)

When p is a good prime (for the integral PEL datum $(\mathcal{O}, \star, L, \langle \cdot , \cdot \rangle, h_0)$; see Definition 1.1.1.6), and when there is no level structure at p, we have the following analogue:

Theorem 8.2.1.2. *In this theorem, let R_0 and other related objects be chosen and defined as in [61, Sec. 6], except that we shall denote the level by $\mathcal{H}^p$ to emphasize that it is a subgroup of $\mathrm{G}(\hat{\mathbb{Z}}^p)$. Let R be any R_0-algebra, and let W be any object of* $\mathrm{Rep}_R(\mathrm{M}_0)$ *(see [61, Def. 6.5]). Consider the canonical morphism $\oint_{\mathcal{H}^p} : \mathsf{M}^{\mathrm{tor}}_{\mathcal{H}^p} \to \mathsf{M}^{\min}_{\mathcal{H}^p}$ (where $\mathsf{M}^{\mathrm{tor}}_{\mathcal{H}^p} = \mathsf{M}^{\mathrm{tor}}_{\mathcal{H}^p,\Sigma^p}$ can be as in [62, Thm. 6.4.1.1] for any Σ^p—here we* *use the superscript "p" to emphasize that Σ^p is a choice for $\mathsf{M}_{\mathcal{H}^p}$). Suppose $\mathcal{H}^p$ is* **neat**. *Then*

$$R^b \oint_{\mathcal{H}^p,*}(\mathcal{E}^{\mathrm{sub}}_{\mathrm{M}_0,R}(W)) = 0$$

for all $b > 0$ (see [61, Def. 6.13]).

More generally (regardless of whether p is a good prime or not), we have the following analogue over the ordinary loci: (Here we reinstate the choice of R_0 and the other running assumptions, including Assumption 8.1.2.1, in Section 8.1.)

Theorem 8.2.1.3. *Let R be any R_0-algebra, and let W be any object of* $\mathrm{Rep}_R(\mathrm{M}^{\mathrm{ord}}_{\mathtt{D},0})$ *(see Definition* 8.1.2.7*). Consider the canonical morphism* $\vec{\oint}^{\mathrm{ord}}_{\mathcal{H}} : \vec{\mathsf{M}}^{\mathrm{ord,tor}}_{\mathcal{H}} \to \vec{\mathsf{M}}^{\mathrm{ord,min}}_{\mathcal{H}}$. *Suppose $\mathcal{H}^p$ is* **neat**. *Then*

$$R^b \vec{\oint}^{\mathrm{ord}}_{\mathcal{H},*}(\vec{\mathcal{E}}^{\mathrm{ord,sub}}_{\mathrm{M}^{\mathrm{ord}}_{\mathtt{D},0},R}(W)) = 0 \tag{8.2.1.4}$$

for all $b > 0$ (see Definition 8.1.3.13*).*

Remark 8.2.1.5. Theorems 8.2.1.1, 8.2.1.2, and 8.2.1.3 might be considered surprising because we often have $R^b \vec{\oint}^{\mathrm{ord}}_{\mathcal{H},*}(\vec{\mathcal{E}}^{\mathrm{ord,can}}_{\mathrm{M}^{\mathrm{ord}}_{\mathtt{D},0},R}(W)) \neq 0$ for $b > 0$. One has to realize that there is a substantial difference between $R^b \vec{\oint}^{\mathrm{ord}}_{\mathcal{H},*}(\vec{\mathcal{E}}^{\mathrm{ord,can}}_{\mathrm{M}^{\mathrm{ord}}_{\mathtt{D},0},R}(W))$ and $R^b \vec{\oint}^{\mathrm{ord}}_{\mathcal{H},*}(\vec{\mathcal{E}}^{\mathrm{ord,sub}}_{\mathrm{M}^{\mathrm{ord}}_{\mathtt{D},0},R}(W))$. We learned this possibility from Taylor in our joint work with Harris and Thorne [39] (in some unitary case, initially only in characteristic zero—but the key idea based on Shapiro's lemma is independent of characteristics and naturally generalizes to our setting here).

Remark 8.2.1.6. In the recent article [1], Andreatta, Iovita, and Pilloni independently discovered a special case of such vanishing in the (principally polarized) Siegel modular case, for trivial W. (Their treatment of nontrivial W requires the base ring to be annihilated by a power of p that is no higher than the level at p, so that the nontrivial automorphic bundles admit filtrations with trivial graded pieces induced by the tautological level structures. Nevertheless, it should be possible to modify their strategy and make it also work for nontrivial locally free coefficients. They also treated the case of formal models of strict neighborhoods of the ordinary loci, which is possible because of the existence of good reduction models of toroidal compactifications at the base level.) For sufficiently large p (or in characteristic zero), still for trivial W, Stroh informed us (in personal communications) another approach based on an analogue of the Grauert–Riemenschneider vanishing theorem [31]. Later we realized that, when $W = W_\nu$ for some weight ν as in [71], we can deduce Theorem 8.2.1.1, and also Theorem 8.2.1.2 when p is larger than an explicitly computable bound defined by ν, from [71, Thm. 8.13]. See [69] for more details. (However, the approach does not work for Theorem 8.2.1.3 when p is ramified or when p is not larger than the above-mentioned bound. Also, since [71, Thm.

8.13] depends on nontrivial inputs such as [27] and [84], this shorter approach is arguably simpler but less elementary than the one we will present below.)

Remark 8.2.1.7. There has been some more recent progress since [1], [69], and [39] were available as preprints and then published. Firstly, in [66], we discovered that the relative vanishing as in Theorems 8.2.1.1 and 8.2.1.2 (or in any cases where we have proper smooth models of toroidal compactifications) can be used to prove a generalization—which we call a "*higher Koecher's principle*"—of the classical Koecher's principle to the case of higher cohomology (see [66, Sec. 2] for the precise statements). In [66], we also generalized the proof of Theorem 8.2.1.3 (to be given below) to all cases where smooth models of toroidal compactifications are available (see [66, Sec. 3 and 10]). Later, we found a sharper argument which also works for bad reduction cases with no restriction at all on the levels, ramifications, and polarization degrees at p, and obtained the generalizations in [68, Sec. 8] and [64, Sec. 4.4]. On the other hand, the rather different approach in [69] has also been generalized in [67] to incorporate all locally symmetric varieties (over $\mathbb{C}$). Thus, we now have two kinds of proofs for the analogue of Theorem 8.2.1.1 for Shimura varieties of exceptional type! However, given how different the two kinds of proofs are, the precise nature of such an analogue remains very mysterious.

We shall omit the proofs of Theorems 8.2.1.1 and 8.2.1.2 because they are similar to (and simpler than) that of Theorem 8.2.1.3. The proof of Theorem 8.2.1.3 will be carried out in the remainder of this section (and will be finished in Section 8.2.5). Apart from the key idea based on Shapiro's lemma, the details we will present below uses techniques similar to those in [61] and in Section 7.3, which we developed independently from (and earlier than) those in [39]. We are inspired by similar strategies in [50, Ch. I, Sec. 3] and [40, Lem. 1.6.5], although our use of nerve spectral sequences based on [30, II, 5.2.1, 5.2.4, and 5.4.1] is somewhat different.

In the remainder of Section 8.2, let us fix the choice of an arbitrary (locally closed) stratum $\vec{\mathsf{Z}}^{\mathrm{ord}}_{[(\Phi_{\mathcal{H}},\delta_{\mathcal{H}})]}$ of $\vec{\mathsf{M}}^{\mathrm{ord,min}}_{\mathcal{H}}$. We shall assume that $\mathcal{H}^p$ is *neat*, so that $\vec{\mathsf{Z}}^{\mathrm{ord}}_{[(\Phi_{\mathcal{H}},\delta_{\mathcal{H}})]} \cong [\vec{\mathsf{M}}^{\mathrm{ord},\mathsf{Z}_{\mathcal{H}}}_{\mathcal{H}}] \cong \vec{\mathsf{M}}^{\mathrm{ord},\mathsf{Z}_{\mathcal{H}}}_{\mathcal{H}}$ (see Corollary 6.1.2.21), without having to deal with coarse moduli spaces and the groups $\mathrm{Aut}(\bar{x})$ of automorphisms of geometric points $\bar{x} \to \vec{\mathsf{M}}^{\mathrm{ord},\mathsf{Z}_{\mathcal{H}}}_{\mathcal{H}}$ as in Proposition 6.1.2.19. (When $\mathrm{Aut}(\bar{x})$ is nontrivial, we will have to assume conditions analogous to (6.2.1.9), without which most results concerning direct images and higher direct images will no longer hold. Moreover, we will have to pullback to $([\vec{\mathsf{M}}^{\mathrm{ord},\mathsf{Z}_{\mathcal{H}}}_{\mathcal{H}}])^{\wedge}_{\bar{x}}$ for most of our results and proofs.)

8.2.2 *Formal Fibers of* $\vec{\oint}^{\mathrm{ord}}_{\mathcal{H}}$

The aim of this subsection is to describe the pullback of the structural morphism $\vec{\oint}^{\mathrm{ord}}_{\mathcal{H}} : \vec{\mathsf{M}}^{\mathrm{ord,tor}}_{\mathcal{H}} \to \vec{\mathsf{M}}^{\mathrm{ord,min}}_{\mathcal{H}}$ to the formal completion $(\vec{\mathsf{M}}^{\mathrm{ord,min}}_{\mathcal{H}})^{\wedge}_{\vec{\mathsf{Z}}^{\mathrm{ord}}_{[(\Phi_{\mathcal{H}},\delta_{\mathcal{H}})]}}$. By abuse of notation, we shall denote this pullback morphism as

$$\vec{\oint}^{\mathrm{ord}}_{\mathcal{H}} : (\vec{\mathsf{M}}^{\mathrm{ord,tor}}_{\mathcal{H}})^{\wedge}_{\tilde{\vec{\mathsf{Z}}}^{\mathrm{ord}}_{[(\Phi_{\mathcal{H}},\delta_{\mathcal{H}})]}} := (\vec{\mathsf{M}}^{\mathrm{ord,tor}}_{\mathcal{H}})^{\wedge}_{\tilde{\vec{\mathsf{Z}}}^{\mathrm{ord}}_{[(\Phi_{\mathcal{H}},\delta_{\mathcal{H}})]}} \to (\vec{\mathsf{M}}^{\mathrm{ord,min}}_{\mathcal{H}})^{\wedge}_{\vec{\mathsf{Z}}^{\mathrm{ord}}_{[(\Phi_{\mathcal{H}},\delta_{\mathcal{H}})]}}, \tag{8.2.2.1}$$

where $\tilde{\vec{\mathsf{Z}}}^{\mathrm{ord}}_{[(\Phi_{\mathcal{H}},\delta_{\mathcal{H}})]} = (\vec{\oint}^{\mathrm{ord}}_{\mathcal{H}})^{-1}(\vec{\mathsf{Z}}^{\mathrm{ord}}_{[(\Phi_{\mathcal{H}},\delta_{\mathcal{H}})]})$ as in Section 6.1.2. (The formation of the formal completions here are similar to the one in (5) of Theorem 5.2.1.1.)

By Proposition 6.1.2.13, as in the proof of Proposition 6.1.2.19, we have

$$\tilde{\vec{\mathsf{Z}}}^{\mathrm{ord}}_{[(\Phi_{\mathcal{H}},\delta_{\mathcal{H}})]} = \bigcup_{[(\Phi_{\mathcal{H}},\delta_{\mathcal{H}},\sigma)]} \vec{\mathsf{Z}}^{\mathrm{ord}}_{[(\Phi_{\mathcal{H}},\delta_{\mathcal{H}},\sigma)]}, \tag{8.2.2.2}$$

where $\vec{\mathsf{Z}}^{\mathrm{ord}}_{[(\Phi_{\mathcal{H}},\delta_{\mathcal{H}},\sigma)]}$ runs through the strata of $\vec{\mathsf{M}}^{\mathrm{ord,tor}}_{\mathcal{H}}$ over $\vec{\mathsf{Z}}^{\mathrm{ord}}_{[(\Phi_{\mathcal{H}},\delta_{\mathcal{H}})]}$. More precisely, for each fixed representative $(\Phi_{\mathcal{H}},\delta_{\mathcal{H}})$ of the cusp label $[(\Phi_{\mathcal{H}},\delta_{\mathcal{H}})]$, the indices $[(\Phi_{\mathcal{H}},\delta_{\mathcal{H}},\sigma)]$ above are parameterized by $\Gamma_{\Phi_{\mathcal{H}}}$-orbits of cones in

$$\Sigma^+_{\Phi_{\mathcal{H}}} := \{\sigma \in \Sigma_{\Phi_{\mathcal{H}}} : \sigma \subset \mathbf{P}^+_{\Phi_{\mathcal{H}}}\}.$$

For each such $\sigma \in \Sigma^+_{\Phi_{\mathcal{H}}}$, let $\mathfrak{U}_\sigma$ denote the formal completion of $\vec{\Xi}^{\mathrm{ord}}_{\Phi_{\mathcal{H}},\delta_{\mathcal{H}}}(\sigma)$ along the closed subscheme U_σ formed by the union of the (locally closed) subschemes $\vec{\Xi}^{\mathrm{ord}}_{\Phi_{\mathcal{H}},\delta_{\mathcal{H}},\tau}$ of $\vec{\Xi}^{\mathrm{ord}}_{\Phi_{\mathcal{H}},\delta_{\mathcal{H}}}(\sigma)$, where τ runs through the faces of σ (which automatically satisfies $\tau \in \Sigma_{\Phi_{\mathcal{H}}}$) such that $\tau \in \Sigma^+_{\Phi_{\mathcal{H}}}$. Then the formal schemes $\mathfrak{U}_\sigma$ (resp. the scheme U_σ) is relatively affine over $\vec{C}^{\mathrm{ord}}_{\Phi_{\mathcal{H}},\delta_{\mathcal{H}}}$.

Definition 8.2.2.3. (*Compare with Definition* 7.3.2.3.)

(1) $\sigma^\vee_{0+}$ *is the intersection of* $\tau^\vee_0$ *(in* $\mathbf{S}_{\Phi_{\mathcal{H}}}$*) for* τ *running through faces of* σ *in* $\Sigma_{\Phi_{\mathcal{H}}}$ *(including* σ *itself) such that* $\tau \subset \mathbf{P}^+_{\Phi_{\mathcal{H}}}$.

(2) $\sigma^\vee_{+}$ *is the intersection of* $\tau^\vee_0$ *(in* $\mathbf{S}_{\Phi_{\mathcal{H}}}$*) for* τ *running through faces of* σ *in* $\Sigma_{\Phi_{\mathcal{H}}}$ *(including* σ *itself).*

By abuse of notation, we have the $\mathscr{O}_{\vec{C}^{\mathrm{ord}}_{\Phi_{\mathcal{H}},\delta_{\mathcal{H}}}}$-algebra isomorphism

$$\mathscr{O}_{\mathfrak{U}_\sigma} \cong \hat{\bigoplus_{\ell\in\sigma^\vee}} \vec{\Psi}^{\mathrm{ord}}_{\Phi_{\mathcal{H}},\delta_{\mathcal{H}}}(\ell) \tag{8.2.2.4}$$

where $\hat{\bigoplus}_{\ell\in\sigma^\vee}$ denotes the completion of the sum with respect to the $\mathscr{O}_{\vec{\Xi}^{\mathrm{ord}}_{\Phi_{\mathcal{H}},\delta_{\mathcal{H}}}(\sigma)}$-ideal $\bigoplus_{\ell\in\sigma^\vee_{0+}} \vec{\Psi}^{\mathrm{ord}}_{\Phi_{\mathcal{H}},\delta_{\mathcal{H}}}(\ell)$.

Let $\vec{\overline{\Xi}}^{\mathrm{ord}}_{\Phi_{\mathcal{H}},\delta_{\mathcal{H}}} = \vec{\overline{\Xi}}^{\mathrm{ord}}_{\Phi_{\mathcal{H}},\delta_{\mathcal{H}},\Sigma_{\Phi_{\mathcal{H}}}}$ and $\vec{\mathfrak{X}}^{\mathrm{ord}}_{\Phi_{\mathcal{H}},\delta_{\mathcal{H}}} = \vec{\mathfrak{X}}^{\mathrm{ord}}_{\Phi_{\mathcal{H}},\delta_{\mathcal{H}},\Sigma_{\Phi_{\mathcal{H}}}}$ be as in Section 4.2.2, where the latter is the formal completion of the former along the union of the σ-strata $\vec{\Xi}^{\mathrm{ord}}_{\Phi_{\mathcal{H}},\delta_{\mathcal{H}},\sigma}$ for $\sigma \subset \mathbf{P}^+_{\Phi_{\mathcal{H}}}$. By Condition 1.2.2.9 and [62, Lem. 6.2.5.27], and by our running assumption that $\mathcal{H}^p$ and hence $\mathcal{H} = \mathcal{H}^p\mathcal{H}_p$ are neat, the action of $\Gamma_{\Phi_{\mathcal{H}}}$ induces only the trivial action on each stratum it stabilizes. Therefore, the quotient morphism

$$\vec{\mathfrak{X}}^{\mathrm{ord}}_{\Phi_{\mathcal{H}},\delta_{\mathcal{H}}} \to \vec{\mathfrak{X}}^{\mathrm{ord}}_{\Phi_{\mathcal{H}},\delta_{\mathcal{H}}}/\Gamma_{\Phi_{\mathcal{H}}} \tag{8.2.2.5}$$

of formal schemes over $\vec{\mathsf{M}}^{\mathrm{ord},\mathsf{Z}_{\mathcal{H}}}_{\mathcal{H}}$ is a *local isomorphism*.

Proposition 8.2.2.6. (*Compare with Lemmas* 1.3.2.41 *and* 5.2.4.38*; see also [61, Prop. 4.3] and Proposition* 7.3.2.5.) *There is a canonical isomorphism*

$$(\vec{\mathsf{M}}_{\mathcal{H}}^{\mathrm{ord,tor}})^{\wedge}_{\bar{\vec{\mathsf{Z}}}^{\mathrm{ord}}_{[(\Phi_{\mathcal{H}},\delta_{\mathcal{H}})]}} \cong \vec{\mathfrak{X}}^{\mathrm{ord}}_{\Phi_{\mathcal{H}},\delta_{\mathcal{H}}}/\Gamma_{\Phi_{\mathcal{H}}}, \tag{8.2.2.7}$$

characterized by the identifications

$$(\vec{\mathsf{M}}_{\mathcal{H}}^{\mathrm{ord,tor}})^{\wedge}_{\vec{\mathsf{Z}}^{\mathrm{ord}}_{[(\Phi_{\mathcal{H}},\delta_{\mathcal{H}},\sigma)]}} \cong \vec{\mathfrak{X}}^{\mathrm{ord}}_{\Phi_{\mathcal{H}},\delta_{\mathcal{H}},\sigma},$$

compatible with (8.2.2.2) *and with the canonical morphisms*

$$(\vec{\mathsf{M}}_{\mathcal{H}}^{\mathrm{ord,tor}})^{\wedge}_{\vec{\mathsf{Z}}^{\mathrm{ord}}_{[(\Phi_{\mathcal{H}},\delta_{\mathcal{H}},\sigma)]}} \to (\vec{\mathsf{M}}_{\mathcal{H}}^{\mathrm{ord,tor}})^{\wedge}_{\bar{\vec{\mathsf{Z}}}^{\mathrm{ord}}_{[(\Phi_{\mathcal{H}},\delta_{\mathcal{H}})]}}.$$

For each geometric point $\bar{x}$ of $\vec{\mathsf{Z}}^{\mathrm{ord}}_{[(\Phi_{\mathcal{H}},\delta_{\mathcal{H}})]} \cong \vec{\mathsf{M}}_{\mathcal{H}}^{\mathrm{ord},\mathsf{Z}_{\mathcal{H}}}$, the pullback of these canonical morphisms to $(\vec{\mathsf{M}}_{\mathcal{H}}^{\mathrm{ord,min}})^{\wedge}_{\bar{x}}$ are morphisms of formal schemes over $(\vec{\mathsf{M}}_{\mathcal{H}}^{\mathrm{ord},\mathsf{Z}_{\mathcal{H}}})^{\wedge}_{\bar{x}}$ (using the canonical morphism $(\vec{\mathsf{M}}_{\mathcal{H}}^{\mathrm{ord,min}})^{\wedge}_{\bar{x}} \to (\vec{\mathsf{M}}_{\mathcal{H}}^{\mathrm{ord},\mathsf{Z}_{\mathcal{H}}})^{\wedge}_{\bar{x}}$ as in Proposition 6.1.2.19 *and the canonical morphism $\vec{\mathfrak{X}}^{\mathrm{ord}}_{\Phi_{\mathcal{H}},\delta_{\mathcal{H}}} \to \vec{\mathsf{M}}_{\mathcal{H}}^{\mathrm{ord},\mathsf{Z}_{\mathcal{H}}}$ by the very construction of $\vec{\mathfrak{X}}^{\mathrm{ord}}_{\Phi_{\mathcal{H}},\delta_{\mathcal{H}}}$).*

Proof. The first statement is already established in Lemma 5.2.4.38. The last statement follows from the proof of Proposition 6.1.2.19. □

For simplicity of notation, we will denote by $\mathscr{O}^{+}_{\mathfrak{X}}$ the pullback of $\mathscr{I}_{\vec{\mathsf{D}}^{\mathrm{ord}}_{\infty,\mathcal{H}}}$ under any morphism $\mathfrak{X} \to \vec{\mathsf{M}}_{\mathcal{H}}^{\mathrm{ord,tor}}$ from a formal scheme. For example, the pullback of $\mathscr{I}_{\vec{\mathsf{D}}^{\mathrm{ord}}_{\infty,\mathcal{H}}}$ under

$$\vec{\mathfrak{X}}^{\mathrm{ord}}_{\Phi_{\mathcal{H}},\delta_{\mathcal{H}},\sigma} = \vec{\mathfrak{X}}^{\mathrm{ord}}_{\Phi_{\mathcal{H}},\delta_{\mathcal{H}},\sigma}/\Gamma_{\Phi_{\mathcal{H}},\sigma} \cong (\vec{\mathsf{M}}_{\mathcal{H}}^{\mathrm{ord,tor}})^{\wedge}_{\vec{\mathsf{Z}}^{\mathrm{ord}}_{[(\Phi_{\mathcal{H}},\delta_{\mathcal{H}},\sigma)]}} \to \vec{\mathsf{M}}_{\mathcal{H}}^{\mathrm{ord,tor}}$$

(*cf.* (5) of Theorem 5.2.1.1 and [62, Lem. 6.2.5.27]) will be denoted $\mathscr{O}^{+}_{\vec{\mathfrak{X}}^{\mathrm{ord}}_{\Phi_{\mathcal{H}},\delta_{\mathcal{H}},\sigma}}$. By Proposition 8.2.2.6, the canonical isomorphism (8.2.2.4) induces compatibly a canonical isomorphism

$$\mathscr{O}^{+}_{\mathfrak{U}_{\sigma}} \cong \mathop{\hat{\oplus}}_{\ell \in \sigma^{\vee}_{+}} \vec{\Psi}^{\mathrm{ord}}_{\Phi_{\mathcal{H}},\delta_{\mathcal{H}}}(\ell) \tag{8.2.2.8}$$

of $\mathscr{O}_{\vec{C}^{\mathrm{ord}}_{\Phi_{\mathcal{H}},\delta_{\mathcal{H}}}}$-submodules (viewed as ideals of the two sides of (8.2.2.4)). By construction, we have:

Lemma 8.2.2.9. (*Compare with [61, Lem. 4.1] and Lemma* 7.3.2.7.) *Suppose σ and τ are two cones in $\Sigma^{+}_{\Phi_{\mathcal{H}}}$ such that τ is a face of σ. Then:*

(1) We have a canonical open immersion $\mathfrak{U}_{\tau} \hookrightarrow \mathfrak{U}_{\sigma}$ (resp. $U_{\tau} \hookrightarrow U_{\sigma}$) of formal subschemes of $\vec{\mathfrak{X}}^{\mathrm{ord}}_{\Phi_{\mathcal{H}},\delta_{\mathcal{H}}}$.

(2) The compatible canonical restriction morphisms

$$\mathscr{O}_{\mathfrak{U}_{\sigma}} \to \mathscr{O}_{\mathfrak{U}_{\tau}}$$

and

$$\mathscr{O}^{+}_{\mathfrak{U}_{\sigma}} \to \mathscr{O}^{+}_{\mathfrak{U}_{\tau}}$$

correspond to the compatible canonical morphisms

$$\underset{\ell\in\sigma^\vee}{\hat{\oplus}} \vec{\Psi}^{\mathrm{ord}}_{\Phi_{\mathcal{H}},\delta_{\mathcal{H}}}(\ell) \to \underset{\ell\in\tau^\vee}{\hat{\oplus}} \vec{\Psi}^{\mathrm{ord}}_{\Phi_{\mathcal{H}},\delta_{\mathcal{H}}}(\ell)$$

and

$$\underset{\ell\in\sigma^\vee_+}{\hat{\oplus}} \vec{\Psi}^{\mathrm{ord}}_{\Phi_{\mathcal{H}},\delta_{\mathcal{H}}}(\ell) \to \underset{\ell\in\tau^\vee_+}{\hat{\oplus}} \vec{\Psi}^{\mathrm{ord}}_{\Phi_{\mathcal{H}},\delta_{\mathcal{H}}}(\ell)$$

of $\mathscr{O}_{\vec{C}^{\mathrm{ord}}_{\Phi_{\mathcal{H}},\delta_{\mathcal{H}}}}$*-algebras, respectively, where the two instances of* $\hat{\oplus}$ *in each expression denote the completions of the sums with respect to the sheaves of ideals* $\underset{\ell\in\sigma^\vee_{0+}}{\oplus} \vec{\Psi}^{\mathrm{ord}}_{\Phi_{\mathcal{H}},\delta_{\mathcal{H}}}(\ell)$ *and* $\underset{\ell\in\tau^\vee_{0+}}{\oplus} \vec{\Psi}^{\mathrm{ord}}_{\Phi_{\mathcal{H}},\delta_{\mathcal{H}}}(\ell)$*, respectively.*

For each $\sigma \in \Sigma^+_{\Phi_{\mathcal{H}}}$, let $\mathfrak{U}_{[\sigma]}$ denote the image of $\mathfrak{U}_\sigma$ under (8.2.2.5), which is isomorphic to $\mathfrak{U}_\sigma$ as a formal scheme over $\vec{\mathsf{M}}^{\mathrm{ord},\mathsf{Z}_{\mathcal{H}}}_{\mathcal{H}}$. By the admissibility of $\Sigma_{\Phi_{\mathcal{H}}}$, we know that the set $\Sigma_{\Phi_{\mathcal{H}}}/\Gamma_{\Phi_{\mathcal{H}}}$ is finite. Let us denote by

$$f_{[\sigma]} : \mathfrak{U}_{[\sigma]} \to \vec{\mathsf{M}}^{\mathrm{ord},\mathsf{Z}_{\mathcal{H}}}_{\mathcal{H}}$$

denote the composition of the restriction of (8.2.2.1) to $\mathfrak{U}_{[\sigma]}$ with the canonical morphisms in Proposition 8.2.2.6, which can be identified with the canonical morphism

$$f_\sigma : \mathfrak{U}_\sigma \to \vec{\mathsf{M}}^{\mathrm{ord},\mathsf{Z}_{\mathcal{H}}}_{\mathcal{H}}$$

induced by the canonical morphism (8.2.2.5). Let us denote by

$$g_\sigma : \mathfrak{U}_\sigma \to \vec{C}^{\mathrm{ord}}_{\Phi_{\mathcal{H}},\delta_{\mathcal{H}}}$$

and

$$h : \vec{C}^{\mathrm{ord}}_{\Phi_{\mathcal{H}},\delta_{\mathcal{H}}} \to \vec{\mathsf{M}}^{\mathrm{ord},\mathsf{Z}_{\mathcal{H}}}_{\mathcal{H}}$$

the canonical morphisms, so that we have a canonical identification

$$f_\sigma = h \circ g_\sigma.$$

Note that g_σ is relatively affine, and h is an abelian scheme torsor over a finite étale cover, which was denoted $\vec{\mathrm{p}}^{\mathrm{ord}}_{\Phi_{\mathcal{H}},\delta_{\mathcal{H}}}$ in Section 6.1.2.

Based on Lemma 8.2.2.9, we have the following important facts:

Lemma 8.2.2.10. (*Compare with [61, Lem. 4.6] and Lemma* 7.3.2.9.)

(1) *For each* $\sigma \in \Sigma^+_{\Phi_{\mathcal{H}}}$*, and each integer* $d \geq 0$*, we have the canonical isomorphisms*

$$R^d(f_\sigma)_*\mathscr{O}_{\mathfrak{U}_\sigma} \cong \underset{\ell\in\sigma^\vee}{\hat{\oplus}} R^d h_*(\vec{\Psi}^{\mathrm{ord}}_{\Phi_{\mathcal{H}},\delta_{\mathcal{H}}}(\ell))$$

and

$$R^d(f_\sigma)_*\mathscr{O}^+_{\mathfrak{U}_\sigma} \cong \underset{\ell\in\sigma^\vee_+}{\hat{\oplus}} R^d h_*(\vec{\Psi}^{\mathrm{ord}}_{\Phi_{\mathcal{H}},\delta_{\mathcal{H}}}(\ell))$$

of $\mathscr{O}_{\vec{\mathsf{M}}^{\mathrm{ord},\mathsf{Z}_{\mathcal{H}}}_{\mathcal{H}}}$*-modules.*

(2) *For each* $\gamma \in \Gamma_{\Phi_{\mathcal{H}}}$*, we have a commutative diagram*

$$\begin{array}{ccc}
\mathfrak{U}_\sigma & \xrightarrow[\sim]{\gamma} & \mathfrak{U}_{\gamma\sigma} \\
{\scriptstyle g_\sigma}\downarrow & & \downarrow{\scriptstyle g_{\gamma\sigma}} \\
\vec{C}^{\mathrm{ord}}_{\Phi_{\mathcal{H}},\delta_{\mathcal{H}}} & \xrightarrow[\sim]{\gamma} & \vec{C}^{\mathrm{ord}}_{\Phi_{\mathcal{H}},\delta_{\mathcal{H}}} \\
{\scriptstyle h}\downarrow & & \downarrow{\scriptstyle h} \\
\vec{\mathsf{M}}^{\mathrm{ord},\mathsf{Z}_{\mathcal{H}}}_{\mathcal{H}} & = & \vec{\mathsf{M}}^{\mathrm{ord},\mathsf{Z}_{\mathcal{H}}}_{\mathcal{H}}
\end{array}$$

of formal schemes. Then the canonical morphisms in (1) *are compatible with the canonical isomorphisms* $\gamma^*\mathscr{O}_{\mathfrak{U}_{\gamma\sigma}} \xrightarrow{\sim} \mathscr{O}_{\mathfrak{U}_\sigma}$ *and* $\gamma^*\mathscr{O}^+_{\mathfrak{U}_{\gamma\sigma}} \xrightarrow{\sim} \mathscr{O}^+_{\mathfrak{U}_\sigma}$ *induced by the canonical isomorphisms* $\gamma^* : \gamma^*\vec{\Psi}^{\mathrm{ord}}_{\Phi_{\mathcal{H}},\delta_{\mathcal{H}}}(\gamma\ell) \xrightarrow{\sim} \vec{\Psi}^{\mathrm{ord}}_{\Phi_{\mathcal{H}},\delta_{\mathcal{H}}}(\ell)$ *over* $\vec{C}^{\mathrm{ord}}_{\Phi_{\mathcal{H}},\delta_{\mathcal{H}}}$*, respectively.*

(3) *For each integer* $d \geq 0$*, if* τ *is a face of* σ *in* $\Sigma^+_{\Phi_{\mathcal{H}}}$*, then the canonical morphisms*

$$R^d(f_\sigma)_*\mathscr{O}_{\mathfrak{U}_\sigma} \to R^d(f_\tau)_*\mathscr{O}_{\mathfrak{U}_\tau}$$

and

$$R^d(f_\sigma)_*\mathscr{O}^+_{\mathfrak{U}_\sigma} \to R^d(f_\tau)_*\mathscr{O}^+_{\mathfrak{U}_\tau}$$

induced by restriction from $\mathfrak{U}_\sigma$ *to* $\mathfrak{U}_\tau$ *correspond to the morphisms*

$$\hat{\bigoplus_{\ell\in\sigma^\vee}} R^d h_*(\vec{\Psi}^{\mathrm{ord}}_{\Phi_{\mathcal{H}},\delta_{\mathcal{H}}}(\ell)) \to \hat{\bigoplus_{\ell\in\tau^\vee}} R^d h_*(\vec{\Psi}^{\mathrm{ord}}_{\Phi_{\mathcal{H}},\delta_{\mathcal{H}}}(\ell))$$

and

$$\hat{\bigoplus_{\ell\in\sigma^\vee_+}} R^d h_*(\vec{\Psi}^{\mathrm{ord}}_{\Phi_{\mathcal{H}},\delta_{\mathcal{H}}}(\ell)) \to \hat{\bigoplus_{\ell\in\tau^\vee_+}} R^d h_*(\vec{\Psi}^{\mathrm{ord}}_{\Phi_{\mathcal{H}},\delta_{\mathcal{H}}}(\ell))$$

of $\mathscr{O}_{\vec{\mathsf{M}}^{\mathrm{ord},\mathsf{Z}_{\mathcal{H}}}_{\mathcal{H}}}$*-modules, respectively. All of these morphisms send* $R^d h_*(\vec{\Psi}^{\mathrm{ord}}_{\Phi_{\mathcal{H}},\delta_{\mathcal{H}}}(\ell))$ *(identically) to* $R^d h_*(\vec{\Psi}^{\mathrm{ord}}_{\Phi_{\mathcal{H}},\delta_{\mathcal{H}}}(\ell))$ *when it is defined on both sides, and to zero otherwise.*

8.2.3 *Relative Cohomology of Formal Fibers of* $\vec{\mathfrak{f}}^{\mathrm{ord}}_{\mathcal{H}}$

Consider the union $\widetilde{\mathfrak{N}}$ of the cones σ in $\Sigma^+_{\Phi_{\mathcal{H}}}$, which admits a closed covering by the closures σ^{cl} (in $\widetilde{\mathfrak{N}}$) of the cones σ in $\Sigma^+_{\Phi_{\mathcal{H}}}$ (with natural incidence relations inherited from those of the cones σ as locally closed subsets of $(\mathbf{S}_{\Phi_{\mathcal{H}}})^\vee_{\mathbb{R}}$). Let

$$\mathfrak{N} := \widetilde{\mathfrak{N}}/\Gamma_{\Phi_{\mathcal{H}}}.$$

By definition, the nerve of the open covering $\{\mathfrak{U}_\sigma\}_{\sigma\in\Sigma^+_{\Phi_{\mathcal{H}}}}$ of $\vec{\mathfrak{X}}^{\mathrm{ord}}_{\Phi_{\mathcal{H}},\delta_{\mathcal{H}}}$ is naturally identified with the nerve of the (locally finite) closed covering $\{\sigma^{\mathrm{cl}}\}_{\sigma\in\Sigma^+_{\Phi_{\mathcal{H}}}}$ of $\widetilde{\mathfrak{N}}$. Accordingly, the nerve of the open covering $\{\mathfrak{U}_{[\sigma]}\}_{[\sigma]\in\Sigma^+_{\Phi_{\mathcal{H}}}/\Gamma_{\Phi_{\mathcal{H}}}}$ of $\vec{\mathfrak{X}}^{\mathrm{ord}}_{\Phi_{\mathcal{H}},\delta_{\mathcal{H}}}/\Gamma_{\Phi_{\mathcal{H}}} \cong$

$(\vec{\mathsf{M}}^{\mathrm{ord,tor}}_{\mathcal{H},\Sigma^{\mathrm{ord}}})^{\wedge}_{\vec{\mathsf{Z}}^{\mathrm{ord}}_{[(\Phi_{\mathcal{H}},\delta_{\mathcal{H}})]}}$ (see (8.2.2.7)) is naturally identified with the nerve of the (finite) closed covering $\{[\sigma]^{\mathrm{cl}}\}_{[\sigma]\in\Sigma^+_{\Phi_{\mathcal{H}}}/\Gamma_{\Phi_{\mathcal{H}}}}$ of $\mathfrak{N}$, where $[\sigma]^{\mathrm{cl}}$ denotes the closure of $[\sigma]$ in $\mathfrak{N}$.

Definition 8.2.3.1. (*Compare with Definition* 7.3.3.14.) *For each $\ell \in \mathbf{S}_{\Phi_{\mathcal{H}}}$, define the following subsets of $\widetilde{\mathfrak{N}}$:*

(1) $\widetilde{\mathfrak{N}}^{\ell}$ is the union of $\sigma \in \Sigma^+_{\Phi_{\mathcal{H}}}$ such that $\ell \in \sigma^{\vee}$.
(2) $\widetilde{\mathfrak{N}}^{\ell,+}$ is the union of $\sigma \in \Sigma^+_{\Phi_{\mathcal{H}}}$ such that $\ell \in \sigma^{\vee}_{+}$.

Lemma 8.2.3.2. (*Compare with Lemma* 7.3.3.15.) *Suppose*

$$\sigma = \mathbb{R}_{>0}v_1 + \cdots + \mathbb{R}_{>0}v_n$$

is a cone in $\Sigma^+_{\Phi_{\mathcal{H}}}$, where $v_1, \ldots, v_n$ are nonzero rational vectors in $\mathbf{S}_{\Phi_{\mathcal{H}}})^{\vee}_{\mathbb{R}}$. Then we have the following criteria:

(1) $\ell \in \sigma^{\vee}$ if and only if $\langle \ell, v_i\rangle \geq 0$ for all $1 \leq i \leq n$.
(2) $\ell \in \sigma^{\vee}_{+}$ if and only if $\langle \ell, v_i\rangle > 0$ for all $1 \leq i \leq n$.

Proof. These follow immediately from the definitions. (See Definitions 8.2.2.3 and 8.2.3.1.) □

Proposition 8.2.3.3. (*Compare with Proposition* 7.3.3.16.) *For each $\ell \in \mathbf{S}_{\Phi_{\mathcal{H}}}$, the subsets $\widetilde{\mathfrak{N}}^{\ell}$ and $\widetilde{\mathfrak{N}}^{\ell,+}$ of $\widetilde{\mathfrak{N}}$ (in Definition* 8.2.3.1) *both have contractible or empty complements in $\widetilde{\mathfrak{N}}$.*

Proof. For simplicity, let us denote $\mathbf{P}_{\Phi_{\mathcal{H}}} - \{0\}$ by $\mathbf{P}'$. Let

$$\mathbf{P}'_{\ell<0} := \{y \in \mathbf{P}' : \langle \ell, y\rangle < 0\}$$

and

$$\mathbf{P}'_{\ell\leq 0} := \{y \in \mathbf{P}' : \langle \ell, y\rangle \leq 0\}.$$

Consider the canonical embeddings

$$\widetilde{\mathfrak{N}} \cap \mathbf{P}'_{\ell<0} \hookrightarrow \widetilde{\mathfrak{N}} - \widetilde{\mathfrak{N}}^{\ell} \tag{8.2.3.4}$$

and

$$\widetilde{\mathfrak{N}} \cap \mathbf{P}'_{\ell\leq 0} \hookrightarrow \widetilde{\mathfrak{N}} - \widetilde{\mathfrak{N}}^{\ell,+}. \tag{8.2.3.5}$$

Consider any $\sigma \in \Sigma^+_{\Phi_{\mathcal{H}}}$ such that $\overline{\sigma} - \{0\}$ has a nonempty intersection with $\widetilde{\mathfrak{N}} - \widetilde{\mathfrak{N}}^{\ell}$ (resp. $\widetilde{\mathfrak{N}} - \widetilde{\mathfrak{N}}^{\ell,+}$). Up to replacing the cone decomposition with smooth locally finite refinements without changing the two sides of (8.2.3.4) (resp. (8.2.3.5)), we may assume that, for each σ as above, there exists at least one face τ of σ such that $\overline{\tau} - \{0\}$ is contained in $\widetilde{\mathfrak{N}} \cap \mathbf{P}'_{\ell<0}$ (resp. $\widetilde{\mathfrak{N}} \cap \mathbf{P}'_{\ell\leq 0}$).

Since $\widetilde{\mathfrak{N}} = \mathbf{P}^+_{\Phi_{\mathcal{H}}}$ and $\mathbf{P}'_{\ell<0}$ (resp. $\mathbf{P}'_{\ell\leq 0}$) are convex subsets of $\mathbf{P}'$, both being stable under the multiplicative action of $\mathbb{R}^{\times}_{>0}$, by Lemma 7.3.3.11, there are deformation retracts, compatible with restrictions to faces, from both $\overline{\sigma} - \overline{\tau'}$ and

$(\overline{\sigma}-\{0\})\cap\mathbf{P}'_{\ell<0}$ (resp. $(\overline{\sigma}-\{0\})\cap\mathbf{P}'_{\ell\leq 0}$) to $\overline{\tau}-\{0\}$, where τ is the largest face of σ such that $\overline{\tau}-\{0\}$ is contained in $\mathbf{P}'_{\ell<0}$ (resp. $\mathbf{P}'_{\ell\leq 0}$), and where τ' is the largest face of σ such that

$$\overline{\tau}-\{0\}\subset\overline{\sigma}-\overline{\tau'}.$$

Hence, we see that (8.2.3.4) (resp. (8.2.3.5)) is a homotopy equivalence.

Since the left-hand sides of (8.2.3.4) and (8.2.3.5) are convex subsets of $\mathbf{P}'$, which are therefore contractible or empty, the proposition follows. □

Suppose $\mathscr{M}$ is a quasi-coherent $\mathscr{O}_{\vec{\mathfrak{X}}^{\mathrm{ord}}_{\Phi_{\mathcal{H}},\delta_{\mathcal{H}}}/\Gamma_{\Phi_{\mathcal{H}}}}$-module. As in Section 7.3.3, let us define for each integer $d\geq 0$ (by abuse of language) the constructible sheaf $\underline{\mathscr{H}}^d(\mathscr{M})$ over $\mathfrak{N}$ which has stalk $R^d(f_{[\sigma]})_*(\mathscr{M}|_{\mathfrak{U}_{[\sigma]}})$ at each point of $[\sigma]$, where $[\sigma]\in\Sigma^+_{\Phi_{\mathcal{H}}}/\Gamma_{\Phi_{\mathcal{H}}}$ is viewed as a locally closed stratum of $\mathfrak{N}$. Then we have, as in Section 7.3.3, the following spectral sequence (based on [30, II, 5.2.1, 5.2.4, and 5.4.1]):

$$E_2^{c,d}:=H^c(\mathfrak{N},\underline{\mathscr{H}}^d(\mathscr{M}))\Rightarrow R^{c+d}\vec{\oint}^{\mathrm{ord}}_{\mathcal{H},*}(\mathscr{M}). \tag{8.2.3.6}$$

Let us also denote by $\mathscr{M}$ its pullback to $\vec{\mathfrak{X}}^{\mathrm{ord}}_{\Phi_{\mathcal{H}},\delta_{\mathcal{H}}}$. Then the E_2 terms of (8.2.3.6) can be computed by the spectral sequence

$$E_2^{c-e,e}:=H^{c-e}(\Gamma_{\Phi_{\mathcal{H}}},H^e(\widetilde{\mathfrak{N}},\underline{\mathscr{H}}^d(\mathscr{M})))\Rightarrow H^c(\mathfrak{N},\underline{\mathscr{H}}^d(\mathscr{M})). \tag{8.2.3.7}$$

Definition 8.2.3.8. *A quasi-coherent* $\mathscr{O}_{\vec{\mathfrak{X}}^{\mathrm{ord}}_{\Phi_{\mathcal{H}},\delta_{\mathcal{H}}}/\Gamma_{\Phi_{\mathcal{H}}}}$*-module* $\mathscr{M}$ *is* **formally canonical** *if there exists a quasi-coherent* $\mathscr{O}_{\vec{C}^{\mathrm{ord}}_{\Phi_{\mathcal{H}},\delta_{\mathcal{H}}}}$*-module* $\mathscr{N}$ *satisfying the following conditions:*

(1) For each $\sigma\in\Sigma^+_{\Phi_{\mathcal{H}}}$*, the pullback* $\mathscr{M}|_{\mathfrak{U}_\sigma}$ *of* $\mathscr{M}$ *to* $\mathfrak{U}_\sigma$ *admits an isomorphism*

$$i_\sigma:\mathscr{M}|_{\mathfrak{U}_\sigma}\xrightarrow{\sim}g_\sigma^*\mathscr{N}.$$

(2) For each $\gamma\in\Gamma_{\Phi_{\mathcal{H}}}$ *and each* $\sigma\in\Sigma^+_{\Phi_{\mathcal{H}}}$*, under the isomorphism* $\gamma:\mathfrak{U}_\sigma\xrightarrow{\sim}\mathfrak{U}_{\gamma\sigma}$ *in* (2) *of Lemma* 8.2.2.10*, the canonical isomorphism*

$$\gamma^*\mathscr{M}|_{\mathfrak{U}_{\gamma\sigma}}\xrightarrow{\sim}\mathscr{M}|_{\mathfrak{U}_\sigma}$$

induces by pre- and post-compositions with i_σ *and* $\gamma^*(i_{\gamma\sigma})^{-1}$ *an isomorphism*

$$\gamma^*:\gamma^*g_{\gamma\sigma}^*\mathscr{N}\xrightarrow{\sim}g_\sigma^*\mathscr{N}$$

of $\mathscr{O}_{\mathfrak{U}_\sigma}$*-modules. This last isomorphism induces by adjunction an isomorphism*

$$\gamma^*:\gamma^*\mathscr{N}\xrightarrow{\sim}\mathscr{N} \tag{8.2.3.9}$$

of $\mathscr{O}_{\vec{C}^{\mathrm{ord}}_{\Phi_{\mathcal{H}},\delta_{\mathcal{H}}}}$*-modules, which induces an isomorphism*

$$\begin{aligned}\gamma^*:\gamma^*R^dh_*(\vec{\Psi}^{\mathrm{ord}}_{\Phi_{\mathcal{H}},\delta_{\mathcal{H}}}(\gamma\ell)\underset{\mathscr{O}_{\vec{C}^{\mathrm{ord}}_{\Phi_{\mathcal{H}},\delta_{\mathcal{H}}}}}{\otimes}\mathscr{N})\\ \xrightarrow{\sim}R^dh_*(\vec{\Psi}^{\mathrm{ord}}_{\Phi_{\mathcal{H}},\delta_{\mathcal{H}}}(\ell)\underset{\mathscr{O}_{\vec{C}^{\mathrm{ord}}_{\Phi_{\mathcal{H}},\delta_{\mathcal{H}}}}}{\otimes}\mathscr{N})\end{aligned} \tag{8.2.3.10}$$

of $\mathscr{O}_{\vec{\mathsf{M}}_{\mathcal{H}}^{\mathrm{ord},\mathsf{z}_{\mathcal{H}}}}$-modules for each $d \geq 0$. For simplicity, for each $\ell \in \mathbf{S}_{\Phi_{\mathcal{H}}}$, let us set (as in (6.1.2.4)*)*

$$\underline{\mathrm{FJ}}^{\mathrm{ord},d,(\ell)}_{\Phi_{\mathcal{H}},\delta_{\mathcal{H}}}(\mathscr{N}) := R^d h_*(\vec{\Psi}^{\mathrm{ord}}_{\Phi_{\mathcal{H}},\delta_{\mathcal{H}}}(\ell) \underset{\mathscr{O}_{\vec{C}^{\mathrm{ord}}_{\Phi_{\mathcal{H}},\delta_{\mathcal{H}}}}}{\otimes} \mathscr{N}). \tag{8.2.3.11}$$

(3) The isomorphisms as in (8.2.3.9) *satisfy the compatibility $\gamma^* = \gamma_1^* \circ (\gamma_1^*(\gamma_2^*))$ when $\gamma = \gamma_2 \circ \gamma_1$ in $\Gamma_{\Phi_{\mathcal{H}}}$. Hence, by* (2) *of Lemma* 8.2.2.10*, the isomorphisms as in* (8.2.3.10) *also satisfy the compatibility $\gamma^* = \gamma_1^* \circ (\gamma_1^*(\gamma_2^*))$ when $\gamma = \gamma_2 \circ \gamma_1$ in $\Gamma_{\Phi_{\mathcal{H}}}$, and define an action of $\Gamma_{\Phi_{\mathcal{H}}}$ on $\prod\limits_{\ell \in \Gamma_{\Phi_{\mathcal{H}}} \cdot \ell_0} \underline{\mathrm{FJ}}^{\mathrm{ord},d,(\ell)}_{\Phi_{\mathcal{H}},\delta_{\mathcal{H}}}(\mathscr{N})$, for each $\ell_0 \in \mathbf{S}_{\Phi_{\mathcal{H}}}$.*

A quasi-coherent $\mathscr{O}_{\vec{\mathfrak{X}}^{\mathrm{ord}}_{\Phi_{\mathcal{H}},\delta_{\mathcal{H}}}/\Gamma_{\Phi_{\mathcal{H}}}}$-module $\mathscr{M}'$ is **formally subcanonical** *if*

$$\mathscr{M}' \cong \mathscr{M}^+ := \mathscr{M} \underset{\mathscr{O}_{\vec{\mathfrak{X}}^{\mathrm{ord}}_{\Phi_{\mathcal{H}},\delta_{\mathcal{H}}}/\Gamma_{\Phi_{\mathcal{H}}}}}{\otimes} \mathscr{O}^+_{\vec{\mathfrak{X}}^{\mathrm{ord}}_{\Phi_{\mathcal{H}},\delta_{\mathcal{H}}}/\Gamma_{\Phi_{\mathcal{H}}}}$$

for some formally canonical $\mathscr{O}_{\vec{\mathfrak{X}}^{\mathrm{ord}}_{\Phi_{\mathcal{H}},\delta_{\mathcal{H}}}/\Gamma_{\Phi_{\mathcal{H}}}}$-module $\mathscr{M}$. In this case, we have isomorphisms

$$i^+_\sigma : \mathscr{M}'|_{\mathfrak{U}_\sigma} \xrightarrow{\sim} (g^*_\sigma \mathscr{N}) \underset{\mathscr{O}_{\mathfrak{U}_\sigma}}{\otimes} \mathscr{O}^+_{\mathfrak{U}_\sigma}$$

with properties analogous to those of i_σ above.

Lemma 8.2.3.12. (*Compare with [61, Lem. 4.16] and Lemma* 7.3.3.27.) *Suppose $\mathscr{M}$ is the pullback of a formally canonical $\mathscr{O}_{\vec{\mathfrak{X}}^{\mathrm{ord}}_{\Phi_{\mathcal{H}},\delta_{\mathcal{H}}}/\Gamma_{\Phi_{\mathcal{H}}}}$-module to $\vec{\mathfrak{X}}^{\mathrm{ord}}_{\Phi_{\mathcal{H}},\delta_{\mathcal{H}}}$. Then, for all integers $d \geq 0$ and $e > 0$, we have*

$$H^e(\widetilde{\mathfrak{N}}, \underline{\mathscr{H}}^d(\mathscr{M}^?)) = 0 \tag{8.2.3.13}$$

for $? = \emptyset$ and $+$. (*See Definition* 8.2.3.8 *for the notation $\mathscr{M}^?$.*)

Proof. By assumption, there exists some quasi-coherent $\mathscr{O}_{\vec{C}^{\mathrm{ord}}_{\Phi_{\mathcal{H}},\delta_{\mathcal{H}}}}$-module $\mathscr{N}$, together with an isomorphism

$$i^?_\sigma : \mathscr{M}^?|_{\mathfrak{U}_\sigma} \xrightarrow{\sim} (g^*_\sigma \mathscr{N}) \underset{\mathscr{O}_{\mathfrak{U}_\sigma}}{\otimes} \mathscr{O}^?_{\mathfrak{U}_\sigma}.$$

for each $\sigma \in \Sigma^+_{\Phi_{\mathcal{H}}}$, as in Definition 8.2.3.8 (with the desired properties).

By (1) of Lemma 8.2.2.10, by using the above isomorphism $i^?_\sigma$, and by the projection formula [35, 0_{I}, 5.4.10.1], we have

$$\underline{\mathscr{H}}^d(\mathscr{M}^?)(\sigma^{\mathrm{cl}}) \cong \hat{\bigoplus_{\ell \in \sigma^\vee_?}} \underline{\mathrm{FJ}}^{\mathrm{ord},d,(\ell)}_{\Phi_{\mathcal{H}},\delta_{\mathcal{H}}}(\mathscr{N}),$$

and for every face τ of σ also in $\Sigma^+_{\Phi_{\mathcal{H}}}$, the canonical morphism

$$\underline{\mathscr{H}}^d(\mathscr{M}^?)(\sigma^{\mathrm{cl}}) \to \underline{\mathscr{H}}^d(\mathscr{M}^?)(\tau^{\mathrm{cl}})$$

sends the subsheaf $\underline{\mathrm{FJ}}^{\mathrm{ord},d,(\ell)}_{\Phi_{\mathcal{H}},\delta_{\mathcal{H}}}(\mathscr{N})$ (identically) to $\underline{\mathrm{FJ}}^{\mathrm{ord},d,(\ell)}_{\Phi_{\mathcal{H}},\delta_{\mathcal{H}}}(\mathscr{N})$ when $\ell \in \sigma^\vee_?$. By Definition 8.2.3.1, $\ell \in \tau^\vee_?$ exactly when $\tau^{\mathrm{cl}} \subset \widetilde{\mathfrak{N}}^{\ell,?}$. Since $\widetilde{\mathfrak{N}}$ and $\widetilde{\mathfrak{N}} - \widetilde{\mathfrak{N}}^{\ell,?}$ are either contractible or empty for each given $\ell \in \mathbf{S}_{\Phi_{\mathcal{H}}}$, by Proposition 8.2.3.3, we have (8.2.3.13) for $e > 0$ as usual (*cf.* the argument in [50, Ch. I, Sec. 3] and the proof of Lemma 7.3.3.27). (Since the nerves involve infinitely many cones, let us briefly review why we can still work weight-by-weight as in [50, Ch. I, Sec. 3]. This is because, up to replacing the cone decomposition $\Sigma_{\Phi_{\mathcal{H}}}$ in Σ^{ord} with locally finite refinements not necessarily carrying $\Gamma_{\Phi_{\mathcal{H}}}$-actions, which is harmless for proving this lemma, we can compute the cohomology as a limit using unions of finite cone decompositions on expanding convex polyhedral subcones, by proving inductively that the cohomology of one degree lower has the desired properties, using [103, Thm. 3.5.8]. Then we can consider the associated graded pieces defined by the completions, and work weight-by-weight with subsheaves of $\underline{\mathscr{H}}^d(\mathscr{N}^?)(\sigma^{\mathrm{cl}})$ of the form $\underline{\mathrm{FJ}}^{\mathrm{ord},d,(\ell)}_{\Phi_{\mathcal{H}},\delta_{\mathcal{H}}}(\mathscr{N})$, because taking cohomology commutes with taking infinite direct sums for Čech complexes defined by finite coverings, as desired.) □

Let us denote by $\mathbf{P}^{\vee,+}_{\Phi_{\mathcal{H}}}$ the subset of $\mathbf{P}^{\vee}_{\Phi_{\mathcal{H}}}$ consisting of elements $\ell \in \mathbf{S}_{\Phi_{\mathcal{H}}}$ that pairs positively with all nonzero elements in $\mathbf{P}_{\Phi_{\mathcal{H}}}$. By definition, for both $? = \emptyset$ and $+$, we have

$$\bigcap_{\sigma \in \Sigma^+_{\Phi_{\mathcal{H}}}} \sigma^{\vee}_{?} = \mathbf{P}^{\vee,?}_{\Phi_{\mathcal{H}}}. \tag{8.2.3.14}$$

Lemma 8.2.3.15. (*Compare with [61, Lem. 4.16] and Lemma* 7.3.3.30.) *Suppose $\mathscr{M}$ is the pullback of a formally canonical $\mathscr{O}_{\vec{\mathfrak{X}}^{\mathrm{ord}}_{\Phi_{\mathcal{H}},\delta_{\mathcal{H}}}/\Gamma_{\Phi_{\mathcal{H}}}}$-module to $\vec{\mathfrak{X}}^{\mathrm{ord}}_{\Phi_{\mathcal{H}},\delta_{\mathcal{H}}}$, which is the pullback of some quasi-coherent $\mathscr{O}_{\vec{C}^{\mathrm{ord}}_{\Phi_{\mathcal{H}},\delta_{\mathcal{H}}}}$-module $\mathscr{N}$ as in Definition* 8.2.3.8. *For every integer $d \geq 0$, and for $? = \emptyset$ or $+$, we have a canonical isomorphism*

$$H^0(\widetilde{\mathfrak{N}}, \underline{\mathscr{H}}^d(\mathscr{M}^?)) \cong \hat{\bigoplus_{\ell \in \mathbf{P}^{\vee,?}_{\Phi_{\mathcal{H}}}}} \underline{\mathrm{FJ}}^{\mathrm{ord},d,(\ell)}_{\Phi_{\mathcal{H}},\delta_{\mathcal{H}}}(\mathscr{N}), \tag{8.2.3.16}$$

which carries the action of $\Gamma_{\Phi_{\mathcal{H}}}$ induced by that on $\underline{\mathrm{FJ}}^{\mathrm{ord},d,(\ell)}_{\Phi_{\mathcal{H}},\delta_{\mathcal{H}}}(\mathscr{N})$ (see Definition 8.2.3.8).

Proof. Let us continue with the setting in the proof of Lemma 8.2.3.12. Then the lemma follows from (8.2.3.14). □

Lemma 8.2.3.17. *Let ℓ_0 be any element of $\mathbf{P}^{\vee,+}_{\Phi_{\mathcal{H}}}$. Under the running assumption that $\mathcal{H}$ and hence $\mathcal{H} = \mathcal{H}^p\mathcal{H}_p$ are neat, the stabilizer $\Gamma_{\Phi_{\mathcal{H}},\ell_0}$ of ℓ_0 in $\Gamma_{\Phi_{\mathcal{H}}}$ is trivial.*

Proof. Up to choosing a $\mathbb{Z}$-basis of Y, the element ℓ_0 can be represented as a positive definite matrix, which implies that $\Gamma_{\Phi_{\mathcal{H}},\ell_0}$ is finite because it is a discrete subset of a compact real orthogonal group. Consequently, the eigenvalues of all elements in $\Gamma_{\Phi_{\mathcal{H}},\ell_0}$ are roots of unity, which must be 1 because $\mathcal{H}$ is neat. Hence, $\Gamma_{\Phi_{\mathcal{H}},\ell_0}$ is trivial for all $\ell_0 \in \mathbf{P}^{\vee,+}_{\Phi_{\mathcal{H}}}$, and the lemma follows. □

Proposition 8.2.3.18. *Suppose $\mathscr{M}$ is the pullback of a formally canonical $\mathscr{O}_{\vec{\mathfrak{X}}^{\mathrm{ord}}_{\Phi_{\mathcal{H}},\delta_{\mathcal{H}}}/\Gamma_{\Phi_{\mathcal{H}}}}$-module to $(\vec{\mathfrak{X}}^{\mathrm{ord}}_{\Phi_{\mathcal{H}},\delta_{\mathcal{H}}})^{\wedge}_{\bar{x}}$; that is, there exists some quasi-coherent*

$\mathscr{O}_{\vec{C}^{\mathrm{ord}}_{\Phi_{\mathcal{H}},\delta_{\mathcal{H}}}}$*-module* $\mathscr{N}$*, together with an isomorphism*

$$i^?_\sigma : \mathscr{M}^?|_{\mathfrak{U}_\sigma} \xrightarrow{\sim} (g^*_\sigma \mathscr{N}) \underset{\mathscr{O}_{\mathfrak{U}_\sigma}}{\otimes} \mathscr{O}^?_{\mathfrak{U}_\sigma}$$

for each $\sigma \in \Sigma^+_{\Phi_{\mathcal{H}}}$*, as in Definition* 8.2.3.8 *(with the desired properties). Suppose moreover that there exists one such* $\mathscr{N}$ *such that*

$$\underline{\mathrm{FJ}}^{\mathrm{ord},d,(\ell)}_{\Phi_{\mathcal{H}},\delta_{\mathcal{H}}}(\mathscr{N}) = R^d h_*(\vec{\Psi}^{\mathrm{ord}}_{\Phi_{\mathcal{H}},\delta_{\mathcal{H}}}(\ell) \underset{\mathscr{O}_{\vec{C}^{\mathrm{ord}}_{\Phi_{\mathcal{H}},\delta_{\mathcal{H}}}}}{\otimes} \mathscr{N}) = 0$$

for all $d > 0$ *and* $\ell \in \mathbf{P}^{\vee,+}_{\Phi_{\mathcal{H}}}$*. Then*

$$R^b \vec{\oint}^{\mathrm{ord}}_{\mathcal{H},*}(\mathscr{M}^+) = 0$$

for all $b > 0$*.*

Proof. Using the spectral sequence (8.2.3.6), it suffices to show that $H^c(\mathfrak{N}, \underline{\mathscr{H}}^d(\mathscr{M}^+)) = 0$ when either $c > 0$ or $d > 0$. Using the spectral sequence (8.2.3.7), by Lemmas 8.2.3.12 and 8.2.3.15, we have

$$H^c(\mathfrak{N}, \underline{\mathscr{H}}^d(\mathscr{M}^+)) \cong H^c\Big(\Gamma_{\Phi_{\mathcal{H}}}, \underset{\ell \in \mathbf{P}^{\vee,+}_{\Phi_{\mathcal{H}}}}{\hat{\oplus}} \underline{\mathrm{FJ}}^{\mathrm{ord},d,(\ell)}_{\Phi_{\mathcal{H}},\delta_{\mathcal{H}}}(\mathscr{N})\Big),$$

which admits a filtration with graded pieces given by subquotients of

$$H^c\Big(\Gamma_{\Phi_{\mathcal{H}}}, \prod_{\ell \in \Gamma_{\Phi_{\mathcal{H}}} \cdot \ell_0} \underline{\mathrm{FJ}}^{\mathrm{ord},d,(\ell)}_{\Phi_{\mathcal{H}},\delta_{\mathcal{H}}}(\mathscr{N})\Big) \tag{8.2.3.19}$$

for ℓ_0 running through representatives of $\Gamma_{\Phi_{\mathcal{H}}}$-orbits in $\mathbf{P}^{\vee,+}_{\Phi_{\mathcal{H}}}$. By assumption, we may assume that $\mathscr{N}$ is chosen such that $\underline{\mathrm{FJ}}^{\mathrm{ord},d,(\ell)}_{\Phi_{\mathcal{H}},\delta_{\mathcal{H}}}(\mathscr{N}) = 0$ for all $d > 0$ and $\ell \in \mathbf{P}^{\vee,+}_{\Phi_{\mathcal{H}}}$. Hence, we have $H^c(\mathfrak{N}, \underline{\mathscr{H}}^d(\mathscr{M}^+)) = 0$ when $d > 0$. On the other hand, suppose $c > 0$. By (3) of Definition 8.2.3.8, and by Lemma 8.2.3.17, we have

$$\begin{aligned}\prod_{\ell \in \Gamma_{\Phi_{\mathcal{H}}} \cdot \ell_0} \underline{\mathrm{FJ}}^{\mathrm{ord},d,(\ell)}_{\Phi_{\mathcal{H}},\delta_{\mathcal{H}}}(\mathscr{N}) &\cong \mathrm{Coind}^{\Gamma_{\Phi_{\mathcal{H}}}}_{\Gamma_{\Phi_{\mathcal{H}},\ell_0}}(\underline{\mathrm{FJ}}^{\mathrm{ord},d,(\ell_0)}_{\Phi_{\mathcal{H}},\delta_{\mathcal{H}}}(\mathscr{N}))\\ &= \mathrm{Coind}^{\Gamma_{\Phi_{\mathcal{H}}}}_{\{\mathrm{Id}\}}(\underline{\mathrm{FJ}}^{\mathrm{ord},d,(\ell_0)}_{\Phi_{\mathcal{H}},\delta_{\mathcal{H}}}(\mathscr{N}))\end{aligned}$$

(see [11, Ch. III, Sec. 5]), and hence (8.2.3.19) is equal to zero for all $c > 0$, by Shapiro's lemma (see [11, Ch. III, (6.2)]). Thus, we also have $H^c(\mathfrak{N}, \underline{\mathscr{H}}^d(\mathscr{M}^+)) = 0$ when $c > 0$, as desired. □

8.2.4 *Formal Fibers of Canonical Extensions*

By the construction of $\vec{\mathfrak{X}}^{\mathrm{ord}}_{\Phi_{\mathcal{H}},\delta_{\mathcal{H}}}$, we have a commutative diagram of canonical morphisms

$$\begin{array}{ccc}(\vec{\mathfrak{X}}^{\mathrm{ord}}_{\Phi_{\mathcal{H}},\delta_{\mathcal{H}}} \to \vec{\mathsf{M}}^{\mathrm{ord,tor}}_{\mathcal{H}})^* \underline{\mathrm{Lie}}_{G/\vec{\mathsf{M}}^{\mathrm{ord,tor}}_{\mathcal{H}}} & \xrightarrow{\sim} & (\vec{\mathfrak{X}}^{\mathrm{ord}}_{\Phi_{\mathcal{H}},\delta_{\mathcal{H}}} \to \vec{C}^{\mathrm{ord}}_{\Phi_{\mathcal{H}},\delta_{\mathcal{H}}})^* \underline{\mathrm{Lie}}_{G^\natural/\vec{C}^{\mathrm{ord}}_{\Phi_{\mathcal{H}},\delta_{\mathcal{H}}}} \\ \downarrow{\scriptstyle\lambda} & & \downarrow{\scriptstyle\lambda^\natural} \\ (\vec{\mathfrak{X}}^{\mathrm{ord}}_{\Phi_{\mathcal{H}},\delta_{\mathcal{H}}} \to \vec{\mathsf{M}}^{\mathrm{ord,tor}}_{\mathcal{H}})^* \underline{\mathrm{Lie}}_{G^\vee/\vec{\mathsf{M}}^{\mathrm{ord,tor}}_{\mathcal{H}}} & \xrightarrow{\sim} & (\vec{\mathfrak{X}}^{\mathrm{ord}}_{\Phi_{\mathcal{H}},\delta_{\mathcal{H}}} \to \vec{C}^{\mathrm{ord}}_{\Phi_{\mathcal{H}},\delta_{\mathcal{H}}})^* \underline{\mathrm{Lie}}_{G^{\vee,\natural}/\vec{C}^{\mathrm{ord}}_{\Phi_{\mathcal{H}},\delta_{\mathcal{H}}}}\end{array} \tag{8.2.4.1}$$

(see Lemma 5.2.4.38). As in Remark 8.1.2.5, consider the pairing

$$\langle\,\cdot\,,\,\cdot\,\rangle_{\lambda^\natural} : \underline{\mathrm{Lie}}^\vee_{G^{\vee,\natural}/\vec{C}^{\mathrm{ord}}_{\Phi_{\mathcal{H}},\delta_{\mathcal{H}}}}(1) \times \underline{\mathrm{Lie}}_{G^\natural/\vec{C}^{\mathrm{ord}}_{\Phi_{\mathcal{H}},\delta_{\mathcal{H}}}} \to \mathscr{O}_{\vec{C}^{\mathrm{ord}}_{\Phi_{\mathcal{H}},\delta_{\mathcal{H}}}}(1)$$

defined by the composition

$$\begin{aligned}&\underline{\mathrm{Lie}}^\vee_{G^{\vee,\natural}/\vec{C}^{\mathrm{ord}}_{\Phi_{\mathcal{H}},\delta_{\mathcal{H}}}}(1) \otimes \underline{\mathrm{Lie}}_{G^\natural/\vec{C}^{\mathrm{ord}}_{\Phi_{\mathcal{H}},\delta_{\mathcal{H}}}}\\ &\stackrel{\mathrm{Id}\otimes d\lambda^\natural}{\to} \underline{\mathrm{Lie}}^\vee_{G^{\vee,\natural}/\vec{C}^{\mathrm{ord}}_{\Phi_{\mathcal{H}},\delta_{\mathcal{H}}}}(1) \otimes \underline{\mathrm{Lie}}_{G^{\vee,\natural}/\vec{C}^{\mathrm{ord}}_{\Phi_{\mathcal{H}},\delta_{\mathcal{H}}}} \stackrel{\mathrm{can.}}{\to} \mathscr{O}_{\vec{C}^{\mathrm{ord}}_{\Phi_{\mathcal{H}},\delta_{\mathcal{H}}}}(1).\end{aligned}$$

Let us define (*cf.* (8.1.3.11))

$$\begin{aligned}&\vec{\mathscr{E}}^{\mathrm{ord},\Phi_{\mathcal{H}},\delta_{\mathcal{H}}}_{\mathrm{M}^{\mathrm{ord}}_{\mathsf{D},0}}\\ &:= \underline{\mathrm{Isom}}_{\mathcal{O}\underset{\mathbb{Z}}{\otimes}\mathscr{O}_{\vec{C}^{\mathrm{ord}}_{\Phi_{\mathcal{H}},\delta_{\mathcal{H}}}}}((\underline{\mathrm{Lie}}^\vee_{G^{\vee,\natural}/\vec{C}^{\mathrm{ord}}_{\Phi_{\mathcal{H}},\delta_{\mathcal{H}}}}(1), \underline{\mathrm{Lie}}_{G^\natural/\vec{C}^{\mathrm{ord}}_{\Phi_{\mathcal{H}},\delta_{\mathcal{H}}}}, \langle\,\cdot\,,\,\cdot\,\rangle_{\lambda^\natural}, \mathscr{O}_{\vec{C}^{\mathrm{ord}}_{\Phi_{\mathcal{H}},\delta_{\mathcal{H}}}}(1)),\\ &(\mathrm{Gr}^0_{\mathsf{D},0}\underset{R_0}{\otimes}\mathscr{O}_{\vec{C}^{\mathrm{ord}}_{\Phi_{\mathcal{H}},\delta_{\mathcal{H}}}}, \mathrm{Gr}^{-1}_{\mathsf{D},0}\underset{R_0}{\otimes}\mathscr{O}_{\vec{C}^{\mathrm{ord}}_{\Phi_{\mathcal{H}},\delta_{\mathcal{H}}}}, \langle\,\cdot\,,\,\cdot\,\rangle_{\phi_{\mathsf{D},0}}, \mathscr{O}_{\vec{C}^{\mathrm{ord}}_{\Phi_{\mathcal{H}},\delta_{\mathcal{H}}}}(1))), \qquad (8.2.4.2)\end{aligned}$$

which is an $\mathrm{M}^{\mathrm{ord}}_{\mathsf{D},0}$-torsor with the group $\mathrm{M}^{\mathrm{ord}}_{\mathsf{D},0}$ acting as automorphisms on $(\mathrm{Gr}^0_{\mathsf{D},0}\underset{R_0}{\otimes}\mathscr{O}_{\vec{C}^{\mathrm{ord}}_{\Phi_{\mathcal{H}},\delta_{\mathcal{H}}}}, \mathrm{Gr}^{-1}_{\mathsf{D},0}\underset{R_0}{\otimes}\mathscr{O}_{\vec{C}^{\mathrm{ord}}_{\Phi_{\mathcal{H}},\delta_{\mathcal{H}}}}, \langle\,\cdot\,,\,\cdot\,\rangle_{\phi_{\mathsf{D},0}}, \mathscr{O}_{\vec{C}^{\mathrm{ord}}_{\Phi_{\mathcal{H}},\delta_{\mathcal{H}}}}(1))$. By construction, and by (4) of Proposition 8.1.3.6, the commutative diagram (8.2.4.1) induces a canonical isomorphism

$$(\vec{\mathfrak{X}}^{\mathrm{ord}}_{\Phi_{\mathcal{H}},\delta_{\mathcal{H}}} \to \vec{\mathsf{M}}^{\mathrm{ord,tor}}_{\mathcal{H}})^*\vec{\mathscr{E}}^{\mathrm{ord,can}}_{\mathrm{M}^{\mathrm{ord}}_{\mathsf{D},0}} \cong (\vec{\mathfrak{X}}^{\mathrm{ord}}_{\Phi_{\mathcal{H}},\delta_{\mathcal{H}}} \to \vec{C}^{\mathrm{ord}}_{\Phi_{\mathcal{H}},\delta_{\mathcal{H}}})^*\vec{\mathscr{E}}^{\mathrm{ord},\Phi_{\mathcal{H}},\delta_{\mathcal{H}}}_{\mathrm{M}^{\mathrm{ord}}_{\mathsf{D},0}} \qquad (8.2.4.3)$$

of $\mathrm{M}^{\mathrm{ord}}_{\mathsf{D},0}$-torsors.

Definition 8.2.4.4. (*Compare with Definition* 8.1.3.13.) *Let R be any R_0-algebra. For each $W \in \mathrm{Rep}_R(\mathrm{M}^{\mathrm{ord}}_{\mathsf{D},0})$, we define*

$$\vec{\mathscr{E}}^{\mathrm{ord},\Phi_{\mathcal{H}},\delta_{\mathcal{H}}}_{\mathrm{M}^{\mathrm{ord}}_{\mathsf{D},0},R}(W) := (\vec{\mathscr{E}}^{\mathrm{ord},\Phi_{\mathcal{H}},\delta_{\mathcal{H}}}_{\mathrm{M}^{\mathrm{ord}}_{\mathsf{D},0}}\underset{R_0}{\otimes}R) \overset{\mathrm{M}^{\mathrm{ord}}_{\mathsf{D},0}\underset{R_0}{\otimes}R}{\times} W.$$

Lemma 8.2.4.5. *Let R be any R_0-algebra.*

(1) The assignment $\vec{\mathscr{E}}^{\mathrm{ord},\Phi_{\mathcal{H}},\delta_{\mathcal{H}}}_{\mathrm{M}^{\mathrm{ord}}_{\mathsf{D},0},R}(\,\cdot\,)$ defines an exact functor from $\mathrm{Rep}_R(\mathrm{M}^{\mathrm{ord}}_{\mathsf{D},0})$ to the category of quasi-coherent $\mathscr{O}_{\vec{C}^{\mathrm{ord}}_{\Phi_{\mathcal{H}},\delta_{\mathcal{H}}}}$-modules.

(2) For each $W \in \mathrm{Rep}_R(\mathrm{M}^{\mathrm{ord}}_{\mathsf{D},0})$, we have a canonical isomorphism

$$(\vec{\mathfrak{X}}^{\mathrm{ord}}_{\Phi_{\mathcal{H}},\delta_{\mathcal{H}}} \to \vec{\mathsf{M}}^{\mathrm{ord,tor}}_{\mathcal{H}})^*\vec{\mathscr{E}}^{\mathrm{ord,can}}_{\mathrm{M}^{\mathrm{ord}}_{\mathsf{D},0},R}(W) \cong (\vec{\mathfrak{X}}^{\mathrm{ord}}_{\Phi_{\mathcal{H}},\delta_{\mathcal{H}}} \to \vec{C}^{\mathrm{ord}}_{\Phi_{\mathcal{H}},\delta_{\mathcal{H}}})^*\vec{\mathscr{E}}^{\mathrm{ord},\Phi_{\mathcal{H}},\delta_{\mathcal{H}}}_{\mathrm{M}^{\mathrm{ord}}_{\mathsf{D},0},R}(W), \qquad (8.2.4.6)$$

functorial in W, of quasi-coherent $\mathscr{O}_{\vec{\mathfrak{X}}^{\mathrm{ord}}_{\Phi_{\mathcal{H}},\delta_{\mathcal{H}}}}$-modules.

Proof. The proof of the first statement is the same as that of Lemmas 1.4.1.10 and 8.1.2.10. The second statement follows from the very constructions of both sides of (8.2.4.6), and from the canonical isomorphism (8.2.4.3). □

Let $(c : X \to B^\vee, c^\vee : Y \to B)$ be the tautological pair of morphisms over $\vec{C}^{\mathrm{ord}}_{\Phi_{\mathcal{H}},\delta_{\mathcal{H}}}$. Then $\gamma = (\gamma_X : X \xrightarrow{\sim} X, \gamma_Y : Y \xrightarrow{\sim} Y) \in \Gamma_{\Phi_{\mathcal{H}}}$ acts on $\vec{C}^{\mathrm{ord}}_{\Phi_{\mathcal{H}},\delta_{\mathcal{H}}}$ by sending $(c, c^\vee)$ to the pair of pre-compositions $(c\gamma_X, c^\vee\gamma_Y)$. Since $\lambda_B c^\vee = c\phi$ and $\phi\gamma_Y = \gamma_X\phi$, we still have the compatibility relation $\lambda_B(c^\vee\gamma_Y) = c\phi\gamma_Y = (c\gamma_X)\phi$. By [62, Lem. 3.4.4.2], we have a commutative diagram

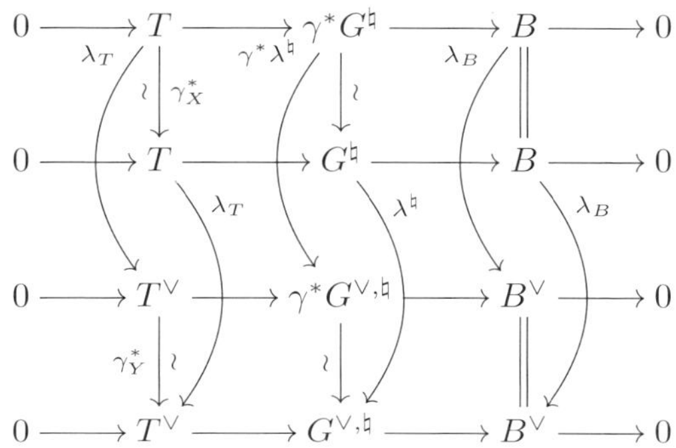

(8.2.4.7)

of canonically defined morphisms, in which the horizontal rows are exact, inducing the corresponding commutative diagrams between sheaves of relative Lie algebras and their duals.

Remark 8.2.4.8. Although $\gamma^*G^\vee \xrightarrow{\sim} G^\vee$ is an isomorphism of semi-abelian schemes that are extensions of B by T, and it induces the identity morphism on B, it does not induce the identity morphism on T. Hence, we cannot view $\gamma^*G^\vee$ as the tautological object over $\vec{C}^{\mathrm{ord}}_{\Phi_{\mathcal{H}},\delta_{\mathcal{H}}}$. Similarly, although $\gamma^*G^{\vee,\natural} \xrightarrow{\sim} G^{\vee,\natural}$ as extensions of $B^\vee$ by $T^\vee$, we cannot view $\gamma^*G^{\vee,\natural}$ as the tautological object over $\vec{C}^{\mathrm{ord}}_{\Phi_{\mathcal{H}},\delta_{\mathcal{H}}}$.

Lemma 8.2.4.9. *For each $\gamma \in \Gamma_{\Phi_{\mathcal{H}}}$, we have a canonically defined isomorphism*

$$\gamma^* : \gamma^*\vec{\mathcal{E}}^{\mathrm{ord},\Phi_{\mathcal{H}},\delta_{\mathcal{H}}}_{\mathrm{M}^{\mathrm{ord}}_{\mathsf{D},0}} \xrightarrow{\sim} \vec{\mathcal{E}}^{\mathrm{ord},\Phi_{\mathcal{H}},\delta_{\mathcal{H}}}_{\mathrm{M}^{\mathrm{ord}}_{\mathsf{D},0}}$$

of $\mathrm{M}^{\mathrm{ord}}_{\mathsf{D},0}$-torsors over $\vec{C}^{\mathrm{ord}}_{\Phi_{\mathcal{H}},\delta_{\mathcal{H}}}$. Such morphisms satisfy the compatibility $\gamma^* = \gamma_1^* \circ (\gamma_1^*(\gamma_2^*))$ when $\gamma = \gamma_2 \circ \gamma_1$ in $\Gamma_{\Phi_{\mathcal{H}}}$. Consequently, for each $W \in \mathrm{Rep}_R(\mathrm{M}^{\mathrm{ord}}_{\mathsf{D},0})$, we have a canonically defined isomorphism

$$\gamma^* : \gamma^*\vec{\mathcal{E}}^{\mathrm{ord},\Phi_{\mathcal{H}},\delta_{\mathcal{H}}}_{\mathrm{M}^{\mathrm{ord}}_{\mathsf{D},0},R}(W) \xrightarrow{\sim} \vec{\mathcal{E}}^{\mathrm{ord},\Phi_{\mathcal{H}},\delta_{\mathcal{H}}}_{\mathrm{M}^{\mathrm{ord}}_{\mathsf{D},0},R}(W),$$

also satisfying the compatibility $\gamma^* = \gamma_1^* \circ (\gamma_1^*(\gamma_2^*))$ when $\gamma = \gamma_2 \circ \gamma_1$ in $\Gamma_{\Phi_{\mathcal{H}}}$.

Proof. These follow from the very constructions of the objects and from the commutativity of (8.2.4.7). □

Proposition 8.2.4.10. *Let R be any R_0-algebra, and let W be any object of $\mathrm{Rep}_R(\mathrm{M}^{\mathrm{ord}}_{\mathsf{D},0})$. Then the pullback of the automorphic sheaf $\vec{\mathcal{E}}^{\mathrm{ord,can}}_{\mathrm{M}^{\mathrm{ord}}_{\mathsf{D},0},R}(W)$ (resp. $\vec{\mathcal{E}}^{\mathrm{ord,sub}}_{\mathrm{M}^{\mathrm{ord}}_{\mathsf{D},0},R}(W)$) over $\vec{\mathsf{M}}^{\mathrm{ord,tor}}_{\mathcal{H}}$ to $(\vec{\mathsf{M}}^{\mathrm{ord,tor}}_{\mathcal{H}})^\wedge_{\vec{\mathsf{Z}}^{\mathrm{ord}}_{[(\Phi_{\mathcal{H}},\delta_{\mathcal{H}})]}} \cong \vec{\mathfrak{X}}^{\mathrm{ord}}_{\Phi_{\mathcal{H}},\delta_{\mathcal{H}}}/\Gamma_{\Phi_{\mathcal{H}}}$ (see (8.2.2.7)) is formally canonical (resp. subcanonical) as in Definition 8.2.3.8.*

Proof. Let $\mathscr{M} := ((\vec{\mathfrak{X}}^{\mathrm{ord}}_{\Phi_{\mathcal{H}},\delta_{\mathcal{H}}}/\Gamma_{\Phi_{\mathcal{H}}}) \to \vec{\mathsf{M}}^{\mathrm{ord,tor}}_{\mathcal{H}})^* \vec{\mathcal{E}}^{\mathrm{ord,can}}_{\mathsf{M}^{\mathrm{ord}}_{\mathsf{D},0},R}(W)$ and $\mathscr{N} := \vec{\mathcal{E}}^{\mathrm{ord},\Phi_{\mathcal{H}},\delta_{\mathcal{H}}}_{\mathsf{M}^{\mathrm{ord}}_{\mathsf{D},0},R}(W)$. By construction, the isomorphism (8.2.4.6) induces an isomorphism $i_\sigma : \mathscr{M}|_{\mathfrak{U}_\sigma} \xrightarrow{\sim} g^*_\sigma \mathscr{N}$ for each $\sigma \in \Sigma^+_{\Phi_{\mathcal{H}}}$. Hence, (1) of Definition 8.2.3.8 is verified. Since the isomorphism (8.2.4.6) is based on (8.2.4.1), in which the horizontal isomorphisms (which is dependent on the rigidification of $\Phi_{\mathcal{H}}$ over $\mathfrak{U}_\sigma$) is twisted by γ (in the same way as the diagram of relative Lie algebras induced by (8.2.4.7)) under the isomorphism $\gamma : \mathfrak{U}_\sigma \xrightarrow{\sim} \mathfrak{U}_{\gamma\sigma}$ in (2) of Lemma 8.2.2.10, the canonical isomorphism $\gamma^*\mathscr{M}|_{\mathfrak{U}_{\gamma\sigma}} \xrightarrow{\sim} \mathscr{M}|_{\mathfrak{U}_\sigma}$ induces by pre- and post-compositions with i_σ and $\gamma^*(i_{\gamma\sigma})^{-1}$ the same isomorphism $\gamma^* : \gamma^* g^*_{\gamma\sigma}\mathscr{N} \xrightarrow{\sim} g^*_\sigma \mathscr{N}$ defined by adjunction by the isomorphism given by Lemma 8.2.4.9. Hence, (2) and (3) of Definition 8.2.3.8 are also verified, as desired. □

Lemma 8.2.4.11. *Let ℓ be any element of $\mathbf{P}^{\vee,+}_{\Phi_{\mathcal{H}}}$. Then the invertible sheaf $\vec{\Psi}^{\mathrm{ord}}_{\Phi_{\mathcal{H}},\delta_{\mathcal{H}}}(\ell)$ over $\vec{C}^{\mathrm{ord}}_{\Phi_{\mathcal{H}},\delta_{\mathcal{H}}} \to \vec{\mathsf{M}}^{\mathrm{ord},\mathsf{Z}_{\mathcal{H}}}_{\mathcal{H}}$ is relatively ample.*

Proof. Since ℓ pairs positively with all elements in $\mathbf{P}_{\Phi_{\mathcal{H}}}$, up to choosing a $\mathbb{Z}$-basis $y_1, \ldots, y_r$ of Y, and by completion of squares for quadratic forms, there exists some integer $N \geq 1$ such that $N \cdot \ell$ can be represented as a positive definite matrix of the form $ue\,{}^tu$, where e and u are matrices with integer coefficients, and where $e = \mathrm{diag}(e_1, \ldots, e_r)$ is diagonal with positive entries. Consider the *finite morphism* defined by the composition $\vec{C}^{\mathrm{ord}}_{\Phi_{\mathcal{H}},\delta_{\mathcal{H}}} \xrightarrow{\text{can.}} \underline{\mathrm{Hom}}_{\mathbb{Z}}(Y,B) \xrightarrow{u^*} \underline{\mathrm{Hom}}_{\mathbb{Z}}(Y,B)$ (see Section 4.2.1 for the first morphism), where B is the tautological abelian scheme over $\vec{\mathsf{M}}^{\mathrm{ord},\delta_{\mathcal{H}}}_{\mathcal{H}}$, under which a positive tensor power of $\vec{\Psi}^{\mathrm{ord}}_{\Phi_{\mathcal{H}},\delta_{\mathcal{H}}}(\ell)$ is isomorphic to a positive tensor power of the pullback of the ample line bundle $\underset{1\leq i\leq r}{\otimes} (\mathrm{pr}_i^*(\mathrm{Id}_B, \lambda_B)^*\mathcal{P}_B)^{\otimes e_i}$ over B. Since λ_B is a polarization (*cf.* [62, Def. 1.3.2.16]), and since all the e_i's are positive, we see that $\vec{\Psi}^{\mathrm{ord}}_{\Phi_{\mathcal{H}},\delta_{\mathcal{H}}}(\ell)$ is relatively ample, as desired. □

For each geometric point $\bar{x}$ of $\vec{\mathsf{M}}^{\mathrm{ord},\mathsf{Z}_{\mathcal{H}}}_{\mathcal{H}}$, we consider the completion $(\vec{\mathsf{M}}^{\mathrm{ord},\mathsf{Z}_{\mathcal{H}}}_{\mathcal{H}})^\wedge_{\bar{x}}$ of the strict localization of $\vec{\mathsf{M}}^{\mathrm{ord},\mathsf{Z}_{\mathcal{H}}}_{\mathcal{H}}$ at $\bar{x}$, and denote (abusively) by the same notation the pullback $h : (\vec{C}^{\mathrm{ord}}_{\Phi_{\mathcal{H}},\delta_{\mathcal{H}}})^\wedge_{\bar{x}} \to (\vec{\mathsf{M}}^{\mathrm{ord},\mathsf{Z}_{\mathcal{H}}}_{\mathcal{H}})^\wedge_{\bar{x}}$ of $h : \vec{C}^{\mathrm{ord}}_{\Phi_{\mathcal{H}},\delta_{\mathcal{H}}} \to \vec{\mathsf{M}}^{\mathrm{ord},\mathsf{Z}_{\mathcal{H}}}_{\mathcal{H}}$.

Corollary 8.2.4.12. *Suppose $\mathscr{N}$ is a coherent $\mathscr{O}_{\vec{C}^{\mathrm{ord}}_{\Phi_{\mathcal{H}},\delta_{\mathcal{H}}}}$-module such that, at each geometric point $\bar{x}$ of $\vec{\mathsf{M}}^{\mathrm{ord},\mathsf{Z}_{\mathcal{H}}}_{\mathcal{H}}$, the pullback $(\mathscr{N})^\wedge_{\bar{x}}$ of $\mathscr{N}$ from $\vec{C}^{\mathrm{ord}}_{\Phi_{\mathcal{H}},\delta_{\mathcal{H}}}$ to $(\vec{C}^{\mathrm{ord}}_{\Phi_{\mathcal{H}},\delta_{\mathcal{H}}})^\wedge_{\bar{x}}$ admits an exhaustive and separated filtration (depending on $\bar{x}$) by coherent $\mathscr{O}_{(\vec{C}^{\mathrm{ord}}_{\Phi_{\mathcal{H}},\delta_{\mathcal{H}}})^\wedge_{\bar{x}}}$-submodules $\{\mathscr{N}^a\}_{a\in\mathbb{Z}}$ such that each graded piece $\mathrm{Gr}^a(\mathscr{N}) := \mathscr{N}^a/\mathscr{N}^{a+1}$ is isomorphic to the pullback (from $\mathrm{Spec}(R_0)$ to $(\vec{C}^{\mathrm{ord}}_{\Phi_{\mathcal{H}},\delta_{\mathcal{H}}})^\wedge_{\bar{x}}$) of some (finitely generated) R_0-module $N^{(a)}$. Then*

$$\underline{\mathrm{FJ}}^{\mathrm{ord},d,(\ell)}_{\Phi_{\mathcal{H}},\delta_{\mathcal{H}}}(\mathscr{N}) = R^d h_*(\vec{\Psi}^{\mathrm{ord}}_{\Phi_{\mathcal{H}},\delta_{\mathcal{H}}}(\ell) \underset{\mathscr{O}_{\vec{C}^{\mathrm{ord}}_{\Phi_{\mathcal{H}},\delta_{\mathcal{H}}}}}{\otimes} \mathscr{N}) = 0$$

for all $d > 0$ and $\ell \in \mathbf{P}^{\vee,+}_{\Phi_{\mathcal{H}}}$.

Proof. Since $h : \vec{C}^{\mathrm{ord}}_{\Phi_{\mathcal{H}},\delta_{\mathcal{H}}} \to \vec{\mathsf{M}}^{\mathrm{ord},\mathsf{Z}_{\mathcal{H}}}_{\mathcal{H}}$ is proper, by [35, III-1, 4.1.5], for our purpose, we may fix the choice of an arbitrary geometric point $\bar{x}$ of $\vec{\mathsf{M}}^{\mathrm{ord},\mathsf{Z}_{\mathcal{H}}}_{\mathcal{H}}$, and replace all objects with their pullbacks from $\vec{\mathsf{M}}^{\mathrm{ord},\mathsf{Z}_{\mathcal{H}}}_{\mathcal{H}}$ to $(\vec{\mathsf{M}}^{\mathrm{ord},\mathsf{Z}_{\mathcal{H}}}_{\mathcal{H}})^{\wedge}_{\bar{x}}$. Since any exhaustive and separated filtration defines a filtration spectral sequence, we are reduced to the case that $\mathscr{N}$ is the pullback (from $\mathrm{Spec}(R_0)$ to $(\vec{C}^{\mathrm{ord}}_{\Phi_{\mathcal{H}},\delta_{\mathcal{H}}})^{\wedge}_{\bar{x}}$) of some finitely generated R_0-module N. Since R_0 is a Dedekind domain, by the same reduction argument as in the proof of [62, Lem. 7.1.1.4], we may assume that $N = R_0/\mathfrak{n}$ for some (possibly zero) ideal $\mathfrak{n}$ of R_0, and work over N after making the base change from R_0. Hence, we are reduced to showing (after base change from R_0 to $R_0/\mathfrak{n}$) that $R^d h_*((\vec{\Psi}^{\mathrm{ord}}_{\Phi_{\mathcal{H}},\delta_{\mathcal{H}}}(\ell))^{\wedge}_{\bar{x}}) = 0$ for all $d > 0$, which then follows from Lemma 8.2.4.11, [81, Sec. 16], and [35, III-1, 4.1.5], because h is an abelian scheme torsor over a finite étale cover. (Alternatively, without reducing to the case that $N = R_0/\mathfrak{n}$, but still using the fact that R_0 is a Dedekind domain, we may invoke the perfect complex construction as in [81, Sec. 5, Thm.] (see also [7, III, 3.7 and 3.7.1]) and the "universal coefficient theorem" as in the proof of [71, Thm. 8.2].) □

The question is how to verify the rather elaborate condition in Corollary 8.2.4.12. Let us define (*cf.* (8.2.4.2))

$$\begin{aligned}&\vec{\mathscr{E}}^{\mathrm{ord},\Phi_{\mathcal{H}},\delta_{\mathcal{H}}}_{\mathrm{M}^{\mathrm{ord}}_{\mathsf{D},0,\mathrm{fil}}}\\ &:= \underline{\mathrm{Isom}}_{\mathcal{O}\underset{\mathbb{Z}}{\otimes}\mathscr{O}_{\vec{C}^{\mathrm{ord}}_{\Phi_{\mathcal{H}},\delta_{\mathcal{H}}}}}((\underline{\mathrm{Lie}}^{\vee}_{G^{\vee,\natural}/\vec{C}^{\mathrm{ord}}_{\Phi_{\mathcal{H}},\delta_{\mathcal{H}}}}(1), \underline{\mathrm{Lie}}_{G^{\natural}/\vec{C}^{\mathrm{ord}}_{\Phi_{\mathcal{H}},\delta_{\mathcal{H}}}}, \langle\,\cdot\,,\,\cdot\,\rangle_{\lambda^{\natural}}, \mathscr{O}_{\vec{C}^{\mathrm{ord}}_{\Phi_{\mathcal{H}},\delta_{\mathcal{H}}}}(1), \text{filtrations}),\\ &\quad (\mathrm{Gr}^{0}_{\mathsf{D},0}\underset{R_0}{\otimes}\mathscr{O}_{\vec{C}^{\mathrm{ord}}_{\Phi_{\mathcal{H}},\delta_{\mathcal{H}}}}, \mathrm{Gr}^{-1}_{\mathsf{D},0}\underset{R_0}{\otimes}\mathscr{O}_{\vec{C}^{\mathrm{ord}}_{\Phi_{\mathcal{H}},\delta_{\mathcal{H}}}}, \langle\,\cdot\,,\,\cdot\,\rangle_{\phi_{\mathsf{D},0}}, \mathscr{O}_{\vec{C}^{\mathrm{ord}}_{\Phi_{\mathcal{H}},\delta_{\mathcal{H}}}}(1), \text{filtrations})), \quad (8.2.4.13)\end{aligned}$$

where the filtrations on $\underline{\mathrm{Lie}}^{\vee}_{G^{\vee,\natural}/\vec{C}^{\mathrm{ord}}_{\Phi_{\mathcal{H}},\delta_{\mathcal{H}}}}(1)$ and $\underline{\mathrm{Lie}}_{G^{\natural}/\vec{C}^{\mathrm{ord}}_{\Phi_{\mathcal{H}},\delta_{\mathcal{H}}}}$ are defined by the commutative diagram

$$\begin{array}{ccccccccc} 0 \longrightarrow & \underline{\mathrm{Lie}}_{T/\vec{C}^{\mathrm{ord}}_{\Phi_{\mathcal{H}},\delta_{\mathcal{H}}}} & \longrightarrow & \underline{\mathrm{Lie}}_{G^{\natural}/\vec{C}^{\mathrm{ord}}_{\Phi_{\mathcal{H}},\delta_{\mathcal{H}}}} & \longrightarrow & \underline{\mathrm{Lie}}_{B/\vec{C}^{\mathrm{ord}}_{\Phi_{\mathcal{H}},\delta_{\mathcal{H}}}} & \longrightarrow 0 \\ & \downarrow{\scriptstyle d\lambda_T} & & \downarrow{\scriptstyle d\lambda^{\natural}} & & \downarrow{\scriptstyle d\lambda_B} & \\ 0 \longrightarrow & \underline{\mathrm{Lie}}_{T^{\vee}/\vec{C}^{\mathrm{ord}}_{\Phi_{\mathcal{H}},\delta_{\mathcal{H}}}} & \longrightarrow & \underline{\mathrm{Lie}}_{G^{\vee,\natural}/\vec{C}^{\mathrm{ord}}_{\Phi_{\mathcal{H}},\delta_{\mathcal{H}}}} & \longrightarrow & \underline{\mathrm{Lie}}_{B^{\vee}/\vec{C}^{\mathrm{ord}}_{\Phi_{\mathcal{H}},\delta_{\mathcal{H}}}} & \longrightarrow 0 \end{array} \quad (8.2.4.14)$$

of canonically defined morphisms, in which the horizontal rows are exact; and where the filtrations on $\mathrm{Gr}^{0}_{\mathsf{D},0}(1)$ and $\mathrm{Gr}^{-1}_{\mathsf{D},0}$ can be any filtrations such that $\vec{\mathscr{E}}^{\mathrm{ord},\Phi_{\mathcal{H}},\delta_{\mathcal{H}}}_{\mathrm{M}^{\mathrm{ord}}_{\mathsf{D},0,\mathrm{fil}}}$ is indeed an étale torsor. (For example, they can be defined by any liftings of (4.1.4.14) and (4.1.4.15) to admissible filtrations over $\mathbb{Z}_p$ and the isomorphisms $\mathrm{Gr}^{-1}_{\mathsf{D},0}\underset{R_0}{\otimes}\tilde{R}_0 \overset{\sim}{\to} \mathrm{Gr}^{-1}_{\mathsf{D}}\underset{\mathbb{Z}_p}{\otimes}\tilde{R}_0$ and $\mathrm{Gr}^{-1}_{\mathsf{D}^{\#},0}\underset{R_0}{\otimes}\tilde{R}_0 \overset{\sim}{\to} \mathrm{Gr}^{-1}_{\mathsf{D}^{\#}}\underset{\mathbb{Z}_p}{\otimes}\tilde{R}_0$ in Section 8.1.1. Then we can deduce the existence of sections over geometric points in characteristic p by the deformation theory of ordinary abelian varieties, and in characteristic zero, or in any good characteristics, by classification of pairings. Afterwards, we can deduce the existence of étale locally defined sections as in the proof of Lemma 1.4.1.7.) We shall fix the choice of this filtrations on $\mathrm{Gr}^{0}_{\mathsf{D},0}(1)$ and $\mathrm{Gr}^{-1}_{\mathsf{D},0}$, and denote by $\mathrm{M}^{\mathrm{ord}}_{\mathsf{D},0,\mathrm{fil}}$

the subgroup scheme of $\mathrm{M}^{\mathrm{ord}}_{\mathtt{D},0}$ defined by elements stabilizing this chosen filtration. Then $\mathscr{E}^{\mathrm{ord},\Phi_{\mathcal{H}},\delta_{\mathcal{H}}}_{\mathrm{M}^{\mathrm{ord}}_{\mathtt{D},0,\mathrm{fil}}}$ is an $\mathrm{M}^{\mathrm{ord}}_{\mathtt{D},0,\mathrm{fil}}$-torsor over $\vec{C}^{\mathrm{ord}}_{\Phi_{\mathcal{H}},\delta_{\mathcal{H}}}$ as the subscript suggests.

Let us define the category $\mathrm{Rep}_{R_0}(\mathrm{M}^{\mathrm{ord}}_{\mathtt{D},0,\mathrm{fil}})$ as in Definition 8.1.2.7. Then, as in Lemmas 1.4.1.10 and 8.1.2.10 (whose proofs are essentially the same), we have an exact functor $\mathscr{E}^{\mathrm{ord},\Phi_{\mathcal{H}},\delta_{\mathcal{H}}}_{\mathrm{M}^{\mathrm{ord}}_{\mathtt{D},0,\mathrm{fil}},R_0}(\,\cdot\,)$ from $\mathrm{Rep}_{R_0}(\mathrm{M}^{\mathrm{ord}}_{\mathtt{D},0,\mathrm{fil}})$ to the category of quasi-coherent $\mathscr{O}_{\vec{C}^{\mathrm{ord}}_{\Phi_{\mathcal{H}},\delta_{\mathcal{H}}}}$-modules. Moreover, if we view an object W of $\mathrm{Rep}_{R_0}(\mathrm{M}^{\mathrm{ord}}_{\mathtt{D},0})$ as an object of $\mathrm{Rep}_{R_0}(\mathrm{M}^{\mathrm{ord}}_{\mathtt{D},0,\mathrm{fil}})$ by restriction to $\mathrm{M}^{\mathrm{ord}}_{\mathtt{D},0,\mathrm{fil}}$, then we have a canonical isomorphism

$$\mathscr{E}^{\mathrm{ord},\Phi_{\mathcal{H}},\delta_{\mathcal{H}}}_{\mathrm{M}^{\mathrm{ord}}_{\mathtt{D},0,\mathrm{fil}},R_0}(W) \cong \mathscr{E}^{\mathrm{ord},\Phi_{\mathcal{H}},\delta_{\mathcal{H}}}_{\mathrm{M}^{\mathrm{ord}}_{\mathtt{D},0},R_0}(W). \tag{8.2.4.15}$$

Let $\mathrm{M}^{\mathrm{ord}}_{\mathtt{D},0,\mathrm{uni}}$ be the normal subgroup scheme of $\mathrm{M}^{\mathrm{ord}}_{\mathtt{D},0,\mathrm{fil}}$ defined by sections of $\mathrm{M}^{\mathrm{ord}}_{\mathtt{D},0,\mathrm{fil}}$ inducing the trivial automorphisms on the graded pieces.

Lemma 8.2.4.16. *Suppose W is a finitely generated object of* $\mathrm{Rep}_{R_0}(\mathrm{M}^{\mathrm{ord}}_{\mathtt{D},0,\mathrm{fil}})$. *Then W admits an exhaustive and separated filtration $\{W^a\}_{a\in\mathbb{Z}}$ by (finitely generated) subobjects in* $\mathrm{Rep}_{R_0}(\mathrm{M}^{\mathrm{ord}}_{\mathtt{D},0,\mathrm{fil}})$, *such that* $\mathrm{M}^{\mathrm{ord}}_{\mathtt{D},0,\mathrm{uni}}$ *acts trivially on each graded pieces* $\mathrm{Gr}^a(W) := W^a/W^{a+1}$.

Proof. Let W' be the R_0-submodule of W spanned by $(u-\mathrm{Id})W$ for all sections u of $\mathrm{M}^{\mathrm{ord}}_{\mathtt{D},0,\mathrm{uni}}$. For any section p of $\mathrm{M}^{\mathrm{ord}}_{\mathtt{D},0,\mathrm{fil}}$, we have $p(u-1) = (pup^{-1}-1)p$, where pup^{-1} is also a section of the normal subgroup scheme $\mathrm{M}^{\mathrm{ord}}_{\mathtt{D},0,\mathrm{uni}}$ of $\mathrm{M}^{\mathrm{ord}}_{\mathtt{D},0,\mathrm{fil}}$. Therefore, W' and W/W' are subobject and quotient objects of W in $\mathrm{Rep}_{R_0}(\mathrm{M}^{\mathrm{ord}}_{\mathtt{D},0,\mathrm{fil}})$, respectively, and $\mathrm{M}^{\mathrm{ord}}_{\mathtt{D},0,\mathrm{uni}}$ acts trivially on the quotient object W/W'. Since W is finitely generated over R_0, the submodule W' of W is a proper (and possibly zero) submodule. (To show this, we can reduce modulo the maximal ideal of R_0, base extend to the algebraically closure, and apply the Lie–Kolchin theorem.) By replacing W with W', and by repeating this process (which terminates in finitely many steps because W is finitely generated), we obtain the desired exhaustive and separated filtration as in the statement of the lemma. □

Corollary 8.2.4.17. *Suppose W is a finitely generated object of* $\mathrm{Rep}_{R_0}(\mathrm{M}^{\mathrm{ord}}_{\mathtt{D},0,\mathrm{fil}})$. *Then $\mathscr{N} := \mathscr{E}^{\mathrm{ord},\Phi_{\mathcal{H}},\delta_{\mathcal{H}}}_{\mathrm{M}^{\mathrm{ord}}_{\mathtt{D},0,\mathrm{fil}},R_0}(W)$ satisfies the condition in Corollary* 8.2.4.12 *(and hence $\underline{\mathrm{FJ}}^{\mathrm{ord},d,(\ell)}_{\Phi_{\mathcal{H}},\delta_{\mathcal{H}}}(\mathscr{N}) = 0$ for all $d > 0$ and $\ell \in \mathbf{P}^{\vee,+}_{\Phi_{\mathcal{H}}}$).*

Proof. By Lemma 8.2.4.16 (and by the obvious analogue of Lemmas 1.4.1.10 and 8.1.2.10 for $\mathscr{E}^{\mathrm{ord},\Phi_{\mathcal{H}},\delta_{\mathcal{H}}}_{\mathrm{M}^{\mathrm{ord}}_{\mathtt{D},0,\mathrm{fil}},R_0}(\,\cdot\,)$), we may assume that W is a finitely generated object of $\mathrm{Rep}_{R_0}(\mathrm{M}^{\mathrm{ord}}_{\mathtt{D},0,\mathrm{fil}})$ on which $\mathrm{M}^{\mathrm{ord}}_{\mathtt{D},0,\mathrm{uni}}$ acts trivially.

Let $\bar{x}$ be an arbitrary geometric point of $\vec{\mathsf{M}}^{\mathrm{ord},\mathsf{Z}_{\mathcal{H}}}_{\mathcal{H}}$. By construction, over the formal completion $(\vec{C}^{\mathrm{ord}}_{\Phi_{\mathcal{H}},\delta_{\mathcal{H}}})^{\wedge}_{\bar{y}}$ of the strict localization of $\vec{C}^{\mathrm{ord}}_{\Phi_{\mathcal{H}},\delta_{\mathcal{H}}}$ at any geometric point $\bar{y}$ above $\bar{x}$, we have a section of $\mathscr{E}^{\mathrm{ord},\Phi_{\mathcal{H}},\delta_{\mathcal{H}}}_{\mathrm{M}^{\mathrm{ord}}_{\mathtt{D},0,\mathrm{fil}}}$ giving an isomorphism between the pullbacks to $(\vec{C}^{\mathrm{ord}}_{\Phi_{\mathcal{H}},\delta_{\mathcal{H}}})^{\wedge}_{\bar{y}}$ of

$$(\underline{\mathrm{Lie}}^{\vee}_{G^{\vee,\natural}/\vec{C}^{\mathrm{ord}}_{\Phi_{\mathcal{H}},\delta_{\mathcal{H}}}}(1), \underline{\mathrm{Lie}}_{G^{\natural}/\vec{C}^{\mathrm{ord}}_{\Phi_{\mathcal{H}},\delta_{\mathcal{H}}}}, \langle\,\cdot\,,\,\cdot\,\rangle_{\lambda^{\natural}}, \mathscr{O}_{\vec{C}^{\mathrm{ord}}_{\Phi_{\mathcal{H}},\delta_{\mathcal{H}}}}(1), \text{filtrations}) \tag{8.2.4.18}$$

and

$$(\mathrm{Gr}^0_{\mathtt{D},0} \underset{R_0}{\otimes} \mathscr{O}_{\vec{C}^{\mathrm{ord}}_{\Phi_{\mathcal{H}},\delta_{\mathcal{H}}}}, \mathrm{Gr}^{-1}_{\mathtt{D},0} \underset{R_0}{\otimes} \mathscr{O}_{\vec{C}^{\mathrm{ord}}_{\Phi_{\mathcal{H}},\delta_{\mathcal{H}}}}, \langle\,\cdot\,,\,\cdot\,\rangle_{\phi_{\mathtt{D},0}}, \mathscr{O}_{\vec{C}^{\mathrm{ord}}_{\Phi_{\mathcal{H}},\delta_{\mathcal{H}}}}(1), \text{filtrations}), \quad (8.2.4.19)$$

inducing isomorphisms between the graded pieces. Since the graded pieces of (8.2.4.18) are pullbacks of objects defined over $\vec{\mathsf{M}}^{\mathrm{ord},\mathsf{Z}_{\mathcal{H}}}_{\mathcal{H}}$, the section of $\vec{\mathscr{E}}^{\mathrm{ord},\Phi_{\mathcal{H}},\delta_{\mathcal{H}}}_{\mathrm{M}^{\mathrm{ord}}_{\mathtt{D},0,\mathrm{fil}}}$ over any such $(\vec{C}^{\mathrm{ord}}_{\Phi_{\mathcal{H}},\delta_{\mathcal{H}}})^{\wedge}_{\bar{y}}$ defines a section of

$$\begin{aligned} &\vec{\mathscr{E}}^{\mathrm{ord},\Phi_{\mathcal{H}},\delta_{\mathcal{H}}}_{\mathrm{M}^{\mathrm{ord}}_{\mathtt{D},0,\mathrm{gr}}} \\ &:= \underline{\mathrm{Isom}}_{\mathcal{O}\underset{\mathbb{Z}}{\otimes}\mathscr{O}_{\vec{C}^{\mathrm{ord}}_{\Phi_{\mathcal{H}},\delta_{\mathcal{H}}}}}(\mathrm{Gr}(\underline{\mathrm{Lie}}^{\vee}_{G^{\vee,\natural}/\vec{C}^{\mathrm{ord}}_{\Phi_{\mathcal{H}},\delta_{\mathcal{H}}}}(1), \underline{\mathrm{Lie}}_{G^{\natural}/\vec{C}^{\mathrm{ord}}_{\Phi_{\mathcal{H}},\delta_{\mathcal{H}}}}, \langle\,\cdot\,,\,\cdot\,\rangle_{\lambda^{\natural}}, \mathscr{O}_{\vec{C}^{\mathrm{ord}}_{\Phi_{\mathcal{H}},\delta_{\mathcal{H}}}}(1)), \\ &\quad \mathrm{Gr}(\mathrm{Gr}^0_{\mathtt{D},0} \underset{R_0}{\otimes} \mathscr{O}_{\vec{C}^{\mathrm{ord}}_{\Phi_{\mathcal{H}},\delta_{\mathcal{H}}}}, \mathrm{Gr}^{-1}_{\mathtt{D},0} \underset{R_0}{\otimes} \mathscr{O}_{\vec{C}^{\mathrm{ord}}_{\Phi_{\mathcal{H}},\delta_{\mathcal{H}}}}, \langle\,\cdot\,,\,\cdot\,\rangle_{\phi_{\mathtt{D},0}}, \mathscr{O}_{\vec{C}^{\mathrm{ord}}_{\Phi_{\mathcal{H}},\delta_{\mathcal{H}}}}(1)), \quad (8.2.4.20) \end{aligned}$$

in the image of the canonical morphism $\vec{\mathscr{E}}^{\mathrm{ord},\Phi_{\mathcal{H}},\delta_{\mathcal{H}}}_{\mathrm{M}^{\mathrm{ord}}_{\mathtt{D},0,\mathrm{fil}}} \to \vec{\mathscr{E}}^{\mathrm{ord},\Phi_{\mathcal{H}},\delta_{\mathcal{H}}}_{\mathrm{M}^{\mathrm{ord}}_{\mathtt{D},0,\mathrm{gr}}}$, over all of $(\vec{C}^{\mathrm{ord}}_{\Phi_{\mathcal{H}},\delta_{\mathcal{H}}})^{\wedge}_{\bar{x}}$ (not just over $(\vec{C}^{\mathrm{ord}}_{\Phi_{\mathcal{H}},\delta_{\mathcal{H}}})^{\wedge}_{\bar{y}}$).

Consequently, the construction of $(\mathscr{N})^{\wedge}_{\bar{x}} = (\vec{\mathscr{E}}^{\mathrm{ord},\Phi_{\mathcal{H}},\delta_{\mathcal{H}}}_{\mathrm{M}^{\mathrm{ord}}_{\mathtt{D},0,\mathrm{fil}},R_0}(W))^{\wedge}_{\bar{x}}$ depends only on the local choices of liftings of the above global section $\vec{\mathscr{E}}^{\mathrm{ord},\Phi_{\mathcal{H}},\delta_{\mathcal{H}}}_{\mathrm{M}^{\mathrm{ord}}_{\mathtt{D},0,\mathrm{gr}}}$ over $(\vec{C}^{\mathrm{ord}}_{\Phi_{\mathcal{H}},\delta_{\mathcal{H}}})^{\wedge}_{\bar{x}}$ to local sections of $\vec{\mathscr{E}}^{\mathrm{ord},\Phi_{\mathcal{H}},\delta_{\mathcal{H}}}_{\mathrm{M}^{\mathrm{ord}}_{\mathtt{D},0,\mathrm{fil}}}$ (under the canonical morphism $\vec{\mathscr{E}}^{\mathrm{ord},\Phi_{\mathcal{H}},\delta_{\mathcal{H}}}_{\mathrm{M}^{\mathrm{ord}}_{\mathtt{D},0,\mathrm{fil}}} \to \vec{\mathscr{E}}^{\mathrm{ord},\Phi_{\mathcal{H}},\delta_{\mathcal{H}}}_{\mathrm{M}^{\mathrm{ord}}_{\mathtt{D},0,\mathrm{gr}}}$), and such choices form a torsor under $\mathrm{M}^{\mathrm{ord}}_{\mathtt{D},0,\mathrm{uni}}$. Since $\mathrm{M}^{\mathrm{ord}}_{\mathtt{D},0,\mathrm{uni}}$ acts trivially on W, we see that $(\mathscr{N})^{\wedge}_{\bar{x}}$ is constant in the sense that it is isomorphic to the pullback (from $\mathrm{Spec}(R_0)$ to $(\vec{C}^{\mathrm{ord}}_{\Phi_{\mathcal{H}},\delta_{\mathcal{H}}})^{\wedge}_{\bar{x}}$) of the R_0-module W, as desired. □

Proposition 8.2.4.21. *Let R be any R_0-algebra, let W be any object of $\mathrm{Rep}_R(\mathrm{M}^{\mathrm{ord}}_{\mathtt{D},0})$, and let $\mathscr{N} := \vec{\mathscr{E}}^{\mathrm{ord},\Phi_{\mathcal{H}},\delta_{\mathcal{H}}}_{\mathrm{M}^{\mathrm{ord}}_{\mathtt{D},0},R}(W)$. Then $\underline{\mathrm{FJ}}^{\mathrm{ord},d,(\ell)}_{\Phi_{\mathcal{H}},\delta_{\mathcal{H}}}(\mathscr{N}) = 0$ for all $d > 0$ and $\ell \in \mathbf{P}^{\vee,+}_{\Phi_{\mathcal{H}}}$.*

Proof. Since this is a statement independent of the R-module structure of W, we may view W as an R_0-module. By construction, we have a canonical isomorphism $\vec{\mathscr{E}}^{\mathrm{ord},\Phi_{\mathcal{H}},\delta_{\mathcal{H}}}_{\mathrm{M}^{\mathrm{ord}}_{\mathtt{D},0},R}(W) \cong \vec{\mathscr{E}}^{\mathrm{ord},\Phi_{\mathcal{H}},\delta_{\mathcal{H}}}_{\mathrm{M}^{\mathrm{ord}}_{\mathtt{D},0},R_0}(W)$ of quasi-coherent $\mathscr{O}_{\vec{C}^{\mathrm{ord}}_{\Phi_{\mathcal{H}},\delta_{\mathcal{H}}}}$-modules. Since taking cohomology and tensor products commutes with direct limits, and since $\vec{\mathscr{E}}^{\mathrm{ord},\Phi_{\mathcal{H}},\delta_{\mathcal{H}}}_{\mathrm{M}^{\mathrm{ord}}_{\mathtt{D},0},R_0}(\,\cdot\,)$ is an exact functor (by Lemma 8.2.4.5), by writing W as a direct limit of finitely generated R_0-subobjects in $\mathrm{Rep}_R(\mathrm{M}^{\mathrm{ord}}_{\mathtt{D},0})$ (which is possible because $\mathrm{M}^{\mathrm{ord}}_{\mathtt{D},0}$ is of finite type over R_0), we may assume that W is finitely generated object of $\mathrm{Rep}_R(\mathrm{M}^{\mathrm{ord}}_{\mathtt{D},0})$. Then it follows from (8.2.4.15) and Corollary 8.2.4.17 that $\mathscr{N}$ satisfies the condition in Corollary 8.2.4.12, and hence the proposition follows. □

Remark 8.2.4.22. For general W, even when W is locally free of finite rank, it is not true that the pullback of the automorphic sheaf $\vec{\mathscr{E}}^{\mathrm{ord},\mathrm{can}}_{\mathrm{M}^{\mathrm{ord}}_{\mathtt{D},0},R}(W)$ to $(\vec{\mathsf{M}}^{\mathrm{ord},\mathrm{tor}}_{\mathcal{H}})^{\wedge}_{\vec{\mathsf{Z}}^{\mathrm{ord}}_{[(\Phi_{\mathcal{H}},\delta_{\mathcal{H}})]}} \cong \vec{\mathfrak{X}}^{\mathrm{ord}}_{\Phi_{\mathcal{H}},\delta_{\mathcal{H}}}/\Gamma_{\Phi_{\mathcal{H}}}$ (see (8.2.2.7)) admits an exhaustive and separated filtration such that each of the graded pieces is the pullback of some coherent sheaf from $\vec{\mathsf{M}}^{\mathrm{ord},\mathsf{Z}_{\mathcal{H}}}_{\mathcal{H}}$ (let alone from $\mathrm{Spec}(R_0)$). This is why we need a different trivialization over each $\mathfrak{U}_{\sigma}$ in (1) of Definition 8.2.3.8.

8.2.5 *End of the Proof*

Proof of Theorem 8.2.1.3. As in the proof of Proposition 8.2.4.21, we may reduce the question to the case that $R = R_0$ and W is a finitely generated object of $\mathrm{Rep}_R(\mathrm{M}^{\mathrm{ord}}_{\mathtt{D},0})$, so that $\vec{\mathscr{E}}^{\mathrm{ord,sub}}_{\mathrm{M}^{\mathrm{ord}}_{\mathtt{D},0},R}(W)$ is a coherent $\mathscr{O}_{\vec{\mathsf{M}}^{\mathrm{ord,tor}}_{\mathcal{H}}}$-module. Since $\vec{\oint}^{\mathrm{ord}}_{\mathcal{H}} : \vec{\mathsf{M}}^{\mathrm{ord,tor}}_{\mathcal{H}} \to \vec{\mathsf{M}}^{\mathrm{ord,min}}_{\mathcal{H}}$ is proper, by [35, III-1, 4.1.5], in order to show (8.2.1.4) for all $b > 0$, we may fix the choice of an arbitrary (locally closed) stratum $\vec{\mathsf{Z}}^{\mathrm{ord}}_{[(\Phi_{\mathcal{H}},\delta_{\mathcal{H}})]}$ of $\vec{\mathsf{M}}^{\mathrm{ord,min}}_{\mathcal{H}}$, and replace all objects with their pullbacks from $\vec{\mathsf{M}}^{\mathrm{ord,min}}_{\mathcal{H}}$ to $(\vec{\mathsf{M}}^{\mathrm{ord,min}}_{\mathcal{H}})^{\wedge}_{\vec{\mathsf{Z}}^{\mathrm{ord}}_{[(\Phi_{\mathcal{H}},\delta_{\mathcal{H}})]}}$. Then the theorem follows as a combination of Propositions 8.2.3.18, 8.2.4.10, and 8.2.4.21. □

8.3 Constructions over the Total Models

In this section, our goal is to extend the construction of $\vec{\mathscr{E}}^{\mathrm{ord}}_{\mathrm{M}^{\mathrm{ord}}_{\mathtt{D},0},R}(\,\cdot\,)$ over $\vec{\mathsf{M}}^{\mathrm{ord}}_{\mathcal{H}}$ to the construction of certain $\vec{\mathscr{E}}_{\mathrm{M}^{\mathrm{ord}}_{\mathtt{D},0},R}(\,\cdot\,)$ over $\vec{\mathsf{M}}_{\mathcal{H}}$ (see Proposition 2.2.1.1), and to extend the constructions of $\vec{\mathscr{E}}^{\mathrm{ord,can}}_{\mathrm{M}^{\mathrm{ord}}_{\mathtt{D},0},R}(\,\cdot\,)$ and $\vec{\mathscr{E}}^{\mathrm{ord,sub}}_{\mathrm{M}^{\mathrm{ord}}_{\mathtt{D},0},R}(\,\cdot\,)$ over $\vec{\mathsf{M}}^{\mathrm{ord,tor}}_{\mathcal{H},\Sigma^{\mathrm{ord}}}$ to the constructions of certain $\vec{\mathscr{E}}^{\mathrm{can}}_{\mathrm{M}^{\mathrm{ord}}_{\mathtt{D},0},R}(\,\cdot\,)$ and $\vec{\mathscr{E}}^{\mathrm{sub}}_{\mathrm{M}^{\mathrm{ord}}_{\mathtt{D},0},R}(\,\cdot\,)$ over $\vec{\mathsf{M}}^{\mathrm{tor}}_{\mathcal{H},d_0\mathsf{pol}}$, respectively, when $\mathcal{H}$ is neat and when Σ^{ord} extends to some projective Σ (so that $\vec{\mathsf{M}}^{\mathrm{tor}}_{\mathcal{H},d_0\mathsf{pol}}$ is defined as in Proposition 2.2.2.1 for some integer $d_0 \geq 1$). When Assumption 3.2.2.10 holds, we will also construct the analogues of these with $\mathrm{M}^{\mathrm{ord}}_{\mathtt{D},0}$ replaced with $\mathrm{P}^{\mathrm{ord}}_{\mathtt{D},0}$, such that they specialize to the constructions in the previous subsection. (Since we can canonically view objects of $\mathrm{Rep}_R(\mathrm{M}^{\mathrm{ord}}_{\mathtt{D},0})$ as objects of $\mathrm{Rep}_R(\mathrm{P}^{\mathrm{ord}}_{\mathtt{D},0})$, we would like $\vec{\mathscr{E}}^{\mathrm{can}}_{\mathrm{P}^{\mathrm{ord}}_{\mathtt{D},0},R}(\,\cdot\,)$ to be compatible with $\vec{\mathscr{E}}^{\mathrm{can}}_{\mathrm{M}^{\mathrm{ord}}_{\mathtt{D},0},R}(\,\cdot\,)$ as in Lemma 1.4.1.10, and as in Lemma 8.1.2.10 when Assumption 3.2.2.10 holds. On the other hand, since $\vec{\mathsf{M}}^{\mathrm{tor}}_{\mathcal{H},d_0\mathsf{pol}}$ is often far from smooth, we do not expect $\vec{\mathscr{E}}^{\mathrm{ord,can}}_{\mathrm{G}^{\mathrm{ord}}_{\mathtt{D},0},R}(\,\cdot\,)$ and the corresponding extended Gauss–Manin connections, induced by the one in Proposition 8.1.3.6, to further extend to the whole $\vec{\mathsf{M}}^{\mathrm{tor}}_{\mathcal{H},d_0\mathsf{pol}}$.)

Throughout this section, we shall retain Assumption 8.1.2.1; and we shall also replace $\vec{\mathsf{M}}_{\mathcal{H}}$ *etc* with the *normalizations* of their base changes from $\mathrm{Spec}(\mathcal{O}_{F_0,(p)})$ to $\mathrm{Spec}(R_0)$, and replace $\mathsf{S}_0 = \mathrm{Spec}(F_0)$ and $\vec{\mathsf{S}}_0 = \mathrm{Spec}(\mathcal{O}_{F_0,(p)})$ with $\mathrm{Spec}(R_0 \underset{\mathbb{Z}}{\otimes} \mathbb{Q})$ and $\mathrm{Spec}(R_0)$, respectively.

8.3.1 *Principal Bundles*

Let us begin with some preliminary constructions in characteristic zero.

Construction 8.3.1.1. Over S_0, we can define the principal $\mathrm{G}^{\mathrm{ord}}_{\mathtt{D},0}$-bundle (resp. $\mathrm{P}^{\mathrm{ord}}_{\mathtt{D},0}$-bundle, resp. $\mathrm{M}^{\mathrm{ord}}_{\mathtt{D},0}$-bundle) $\mathcal{E}_{\mathrm{G}^{\mathrm{ord}}_{\mathtt{D},0}}$ (resp. $\mathcal{E}_{\mathrm{P}^{\mathrm{ord}}_{\mathtt{D},0}}$, resp. $\mathcal{E}_{\mathrm{M}^{\mathrm{ord}}_{\mathtt{D},0}}$) over $\mathsf{M}_{\mathcal{H}}$ as in Definition 1.4.1.3 (resp. Definition 1.4.1.4, resp. Definition 1.4.1.5) by the tautological objects over $\mathsf{M}_{\mathcal{H}}$, which satisfy the analogue of Lemma 1.4.1.7. These are possible because of Lemma 8.1.1.7 (in characteristic *zero*). Thus, for each W in $\mathrm{Rep}_R(\mathrm{G}^{\mathrm{ord}}_{\mathtt{D},0})$ (resp. $\mathrm{Rep}_R(\mathrm{P}^{\mathrm{ord}}_{\mathtt{D},0})$, resp. $\mathrm{Rep}_R(\mathrm{M}^{\mathrm{ord}}_{\mathtt{D},0})$), if we set $R_{\mathbb{Q}} := R \underset{\mathbb{Z}}{\otimes} \mathbb{Q}$ and $W_{\mathbb{Q}} := W \underset{\mathbb{Z}}{\otimes} \mathbb{Q}$, then we can define $\mathcal{E}_{\mathrm{G}^{\mathrm{ord}}_{\mathtt{D},0},R_{\mathbb{Q}}}(W_{\mathbb{Q}})$ (resp. $\mathcal{E}_{\mathrm{P}^{\mathrm{ord}}_{\mathtt{D},0},R_{\mathbb{Q}}}(W_{\mathbb{Q}})$, resp. $\mathcal{E}_{\mathrm{M}^{\mathrm{ord}}_{\mathtt{D},0},R_{\mathbb{Q}}}(W_{\mathbb{Q}})$) over $\mathsf{M}_{\mathcal{H}} \underset{R_0}{\otimes} R_{\mathbb{Q}}$ as in Definition 1.4.1.9, which satisfy the analogue of Lemma 1.4.1.10. (We only define these sheaves for $W_{\mathbb{Q}}$, but not for W itself.) For each Σ of admissible smooth rational polyhedral cone decomposition data for $\mathsf{M}_{\mathcal{H}}$, we can define $\mathcal{E}^{\mathrm{can}}_{\mathrm{G}^{\mathrm{ord}}_{\mathtt{D},0}}$, $\mathcal{E}^{\mathrm{sub}}_{\mathrm{G}^{\mathrm{ord}}_{\mathtt{D},0}}$, $\mathcal{E}^{\mathrm{can}}_{\mathrm{P}^{\mathrm{ord}}_{\mathtt{D},0}}$, $\mathcal{E}^{\mathrm{sub}}_{\mathrm{P}^{\mathrm{ord}}_{\mathtt{D},0}}$, $\mathcal{E}^{\mathrm{can}}_{\mathrm{M}^{\mathrm{ord}}_{\mathtt{D},0}}$, $\mathcal{E}^{\mathrm{sub}}_{\mathrm{M}^{\mathrm{ord}}_{\mathtt{D},0}}$, $\mathcal{E}^{\mathrm{can}}_{\mathrm{G}^{\mathrm{ord}}_{\mathtt{D},0},R_{\mathbb{Q}}}(\,\cdot\,)$, $\mathcal{E}^{\mathrm{sub}}_{\mathrm{G}^{\mathrm{ord}}_{\mathtt{D},0},R_{\mathbb{Q}}}(\,\cdot\,)$, $\mathcal{E}^{\mathrm{can}}_{\mathrm{P}^{\mathrm{ord}}_{\mathtt{D},0},R_{\mathbb{Q}}}(\,\cdot\,)$, $\mathcal{E}^{\mathrm{sub}}_{\mathrm{P}^{\mathrm{ord}}_{\mathtt{D},0},R_{\mathbb{Q}}}(\,\cdot\,)$, $\mathcal{E}^{\mathrm{can}}_{\mathrm{M}^{\mathrm{ord}}_{\mathtt{D},0},R_{\mathbb{Q}}}(\,\cdot\,)$, and $\mathcal{E}^{\mathrm{sub}}_{\mathrm{M}^{\mathrm{ord}}_{\mathtt{D},0},R_{\mathbb{Q}}}(\,\cdot\,)$ as in Definition 1.4.2.9, which satisfy the analogue of Lemma 1.4.2.10. For each R_0-algebra K as in Lemma 8.1.1.7, set $\tilde{R}_{\mathbb{Q}} := R \underset{R_0}{\otimes} K$ and $\tilde{W}_{\mathbb{Q}} := W \underset{R_0}{\otimes} K$. Then $\mathcal{E}_{\mathrm{G}^{\mathrm{ord}}_{\mathtt{D},0},\tilde{R}_{\mathbb{Q}}}(\tilde{W}_{\mathbb{Q}})$ (resp. $\mathcal{E}_{\mathrm{P}^{\mathrm{ord}}_{\mathtt{D},0},\tilde{R}_{\mathbb{Q}}}(\tilde{W}_{\mathbb{Q}})$, resp. $\mathcal{E}_{\mathrm{M}^{\mathrm{ord}}_{\mathtt{D},0},\tilde{R}_{\mathbb{Q}}}(\tilde{W}_{\mathbb{Q}})$) is isomorphic to the usual $\mathcal{E}_{\mathrm{G}_0,\tilde{R}_{\mathbb{Q}}}(\tilde{W}_{\mathbb{Q}})$ (resp. $\mathcal{E}_{\mathrm{P}_0,\tilde{R}_{\mathbb{Q}}}(\tilde{W}_{\mathbb{Q}})$, resp. $\mathcal{E}_{\mathrm{M}_0,\tilde{R}_{\mathbb{Q}}}(\tilde{W}_{\mathbb{Q}})$) constructed as in [61, Sec. 6] and Section 1.4.1 over $\mathsf{M}_{\mathcal{H}} \underset{R_0}{\otimes} \tilde{R}_{\mathbb{Q}}$, and these isomorphisms extend to isomorphisms between the canonical (or subcanonical) extensions of the sources and the targets.

In mixed characteristics, we will only construct principal bundles over $\vec{\mathsf{M}}_{\mathcal{H}}$, over which we have the degenerating family $(\vec{A},\vec{\lambda},\vec{i},\vec{\alpha}_{\mathcal{H}})$ of type $\mathsf{M}_{\mathcal{H}}$ as in Proposition 2.2.1.1. Let us define the *principal* $\mathrm{M}^{\mathrm{ord}}_{\mathtt{D},0}$*-bundle*

$$\begin{aligned}\vec{\mathcal{E}}_{\mathrm{M}^{\mathrm{ord}}_{\mathtt{D},0}} := \underline{\mathrm{Isom}}_{\mathcal{O}\underset{\mathbb{Z}}{\otimes}\mathscr{O}_{\vec{\mathsf{M}}_{\mathcal{H}}}}&((\underline{\mathrm{Lie}}^{\vee}_{\vec{A}^{\vee}/\vec{\mathsf{M}}_{\mathcal{H}}}(1),\underline{\mathrm{Lie}}_{\vec{A}/\vec{\mathsf{M}}_{\mathcal{H}}},\langle\,\cdot\,,\,\cdot\,\rangle_{\vec{\lambda}},\mathscr{O}_{\vec{\mathsf{M}}_{\mathcal{H}}}(1)),\\ &(\mathrm{Gr}^{0}_{\mathtt{D},0}\underset{R_0}{\otimes}\mathscr{O}_{\vec{\mathsf{M}}_{\mathcal{H}}},\mathrm{Gr}^{-1}_{\mathtt{D},0}\underset{R_0}{\otimes}\mathscr{O}_{\vec{\mathsf{M}}_{\mathcal{H}}},\langle\,\cdot\,,\,\cdot\,\rangle_{\phi_{\mathtt{D},0}},\mathscr{O}_{\vec{\mathsf{M}}_{\mathcal{H}}}(1)))\end{aligned} \tag{8.3.1.2}$$

over $\vec{\mathsf{M}}_{\mathcal{H}}$ (*cf.* Definition 8.1.2.4), which extends the principal $\mathrm{M}^{\mathrm{ord}}_{\mathtt{D},0}$-bundle $\mathcal{E}_{\mathrm{M}^{\mathrm{ord}}_{\mathtt{D},0}}$ over $\mathsf{M}_{\mathcal{H}}$ in Construction 8.3.1.1. When Condition 8.1.1.2 holds, let us also define the *principal* $\mathrm{P}^{\mathrm{ord}}_{\mathtt{D},0}$*-bundle*

$$\begin{aligned}\vec{\mathcal{E}}_{\mathrm{P}^{\mathrm{ord}}_{\mathtt{D},0}} := \underline{\mathrm{Isom}}_{\mathcal{O}\underset{\mathbb{Z}}{\otimes}\mathscr{O}_{\vec{\mathsf{M}}_{\mathcal{H}}}}&((\underline{H}^{\mathrm{dR}}_1(\vec{A}/\vec{\mathsf{M}}_{\mathcal{H}}),\langle\,\cdot\,,\,\cdot\,\rangle_{\lambda},\mathscr{O}_{\vec{\mathsf{M}}_{\mathcal{H}}}(1),\underline{\mathrm{Lie}}^{\vee}_{\vec{A}^{\vee}/\vec{\mathsf{M}}_{\mathcal{H}}}(1)),\\ &((\mathrm{Gr}^{0}_{\mathtt{D},0}\oplus\mathrm{Gr}^{-1}_{\mathtt{D},0})\underset{R_0}{\otimes}\mathscr{O}_{\vec{\mathsf{M}}_{\mathcal{H}}},\langle\,\cdot\,,\,\cdot\,\rangle_{\phi_{\mathtt{D},0}},\mathscr{O}_{\vec{\mathsf{M}}_{\mathcal{H}}}(1),\mathrm{Gr}^{0}_{\mathtt{D},0}\underset{R_0}{\otimes}\mathscr{O}_{\vec{\mathsf{M}}_{\mathcal{H}}}))\end{aligned} \tag{8.3.1.3}$$

over $\vec{\mathsf{M}}_{\mathcal{H}}$ (*cf.* Definition 8.1.2.3), which extends the principal $\mathrm{P}^{\mathrm{ord}}_{\mathtt{D},0}$-bundle $\mathcal{E}_{\mathrm{P}^{\mathrm{ord}}_{\mathtt{D},0}}$ over $\mathsf{M}_{\mathcal{H}}$ in Construction 8.3.1.1. To justify these terminologies, we need the following:

Lemma 8.3.1.4. (*Compare with Lemmas* 1.4.1.7 *and* 8.1.2.6.) *The relative scheme* $\vec{\mathcal{E}}_{\mathrm{M}^{\mathrm{ord}}_{\mathtt{D},0}}$ *over* $\vec{\mathsf{M}}_{\mathcal{H}}$ *is an étale torsor under (the pullback of) the group scheme* $\mathrm{M}^{\mathrm{ord}}_{\mathtt{D},0}$. *When Condition* 8.1.1.2 *holds, the relative scheme* $\vec{\mathcal{E}}_{\mathrm{P}^{\mathrm{ord}}_{\mathtt{D},0}}$ *over* $\vec{\mathsf{M}}_{\mathcal{H}}$ *is an fpqc torsor under (the pullback of) the group scheme* $\mathrm{P}^{\mathrm{ord}}_{\mathtt{D},0}$.

Proof. Let $\mathcal{O}'$ be any maximal order in $\mathcal{O}\underset{\mathbb{Z}}{\otimes}\mathbb{Q}$ containing $\mathcal{O}$ as in the beginning of Section 8.1.1, which satisfies Condition 1.2.1.1, so that the $\mathcal{O}$-action on $L\underset{\mathbb{Z}}{\otimes}\hat{\mathbb{Z}}$

extends to an $\mathcal{O}'$-action on the same module. By using the level structures in characteristic zero, and by [62, Cor. 1.3.5.4], the $\mathcal{O}$-structure $i : \mathcal{O} \to \mathrm{End}_{\mathsf{M}_{\mathcal{H}}}(A)$ uniquely extends to an $\mathcal{O}'$-structure $i' : \mathcal{O}' \to \mathrm{End}_{\mathsf{M}_{\mathcal{H}}}(A)$, and $\underline{\mathrm{Lie}}_{A/\mathsf{M}_{\mathcal{H}}}$ with its $\mathcal{O} \underset{\mathbb{Z}}{\otimes} \mathbb{Q} = \mathcal{O}' \underset{\mathbb{Z}}{\otimes} \mathbb{Q}$-module structure given naturally by i' also satisfies the determinantal condition in [62, Def. 1.3.4.1] given by $(L \underset{\mathbb{Z}}{\otimes} \mathbb{R}, \langle\,\cdot\,,\,\cdot\,\rangle, h_0)$. Since $\vec{\mathsf{M}}_{\mathcal{H}}$ is noetherian and normal, by [92, IX, 1.4], [28, Ch. I, Prop. 2.7], or [62, Prop. 3.3.1.5], i' uniquely extends to an $\mathcal{O}'$-structure $\vec{i}' : \mathcal{O}' \to \mathrm{End}_{\vec{\mathsf{M}}_{\mathcal{H}}}(\vec{A})$. Since the determinantal condition is closed, $\underline{\mathrm{Lie}}_{\vec{A}/\vec{\mathsf{M}}_{\mathcal{H}}}$ with its $\mathcal{O}' \underset{\mathbb{Z}}{\otimes} \mathbb{Z}_{(p)}$-module structure given naturally by $\vec{i}'$ also satisfies the determinantal condition in [62, Def. 1.3.4.1] given by $(L \underset{\mathbb{Z}}{\otimes} \mathbb{R}, \langle\,\cdot\,,\,\cdot\,\rangle, h_0)$. Since the filtrations $\mathtt{D}$ and $\mathtt{D}^{\#}$ are $\mathcal{O}' \underset{\mathbb{Z}}{\otimes} \mathbb{Z}_p$-equivariant, as explained in the proof of Lemma 3.2.2.6, and by the assumptions made in the beginning of Section 8.1.1, $\mathcal{O}'$ acts equivariantly on the source and target of $\phi^0_{\mathtt{D},0}$. Over the formal completions of $\vec{\mathsf{M}}_{\mathcal{H}}$ at points of finite type over $\vec{\mathsf{S}}_0$, by Lemma 8.1.1.7 in characteristic zero, and by [91, 3.23 c) and d), and the reductions in Ch. 6] in characteristic p, the infinitesimal deformations of pullbacks of $(\vec{A}, \vec{\lambda}, \vec{i}')$ define sections of the pullback of the relative subscheme

$$\underline{\mathrm{Isom}}_{\mathcal{O}' \underset{\mathbb{Z}}{\otimes} \mathscr{O}_{\vec{\mathsf{M}}_{\mathcal{H}}}}((\underline{\mathrm{Lie}}^{\vee}_{\vec{A}^{\vee}/\vec{\mathsf{M}}_{\mathcal{H}}}(1), \underline{\mathrm{Lie}}_{\vec{A}/\vec{\mathsf{M}}_{\mathcal{H}}}, \langle\,\cdot\,,\,\cdot\,\rangle_{\vec{\lambda}}, \mathscr{O}_{\vec{\mathsf{M}}_{\mathcal{H}}}(1)),$$
$$(\mathrm{Gr}^0_{\mathtt{D},0} \underset{R_0}{\otimes} \mathscr{O}_{\vec{\mathsf{M}}_{\mathcal{H}}}, \mathrm{Gr}^{-1}_{\mathtt{D},0} \underset{R_0}{\otimes} \mathscr{O}_{\vec{\mathsf{M}}_{\mathcal{H}}}, \langle\,\cdot\,,\,\cdot\,\rangle_{\phi_{\mathtt{D},0}}, \mathscr{O}_{\vec{\mathsf{M}}_{\mathcal{H}}}(1)))$$

of $\vec{\mathcal{E}}_{\mathrm{M}^{\mathrm{ord}}_{\mathtt{D},0}}$ (note the difference between $\mathcal{O}$ and $\mathcal{O}'$), which induce sections of the pullback of $\vec{\mathcal{E}}_{\mathrm{M}^{\mathrm{ord}}_{\mathtt{D},0}}$. Hence, by Artin's approximation theory (*cf.* [3, Thm. 1.10 and Cor. 2.5]), as in the proof of Lemma 8.1.2.6, it follows that $\vec{\mathcal{E}}_{\mathrm{M}^{\mathrm{ord}}_{\mathtt{D},0}}$ is an étale $\mathrm{M}^{\mathrm{ord}}_{\mathtt{D},0}$-torsor, as desired. When Condition 8.1.1.2 holds, the analogous statements for $\mathrm{P}^{\mathrm{ord}}_{\mathtt{D},0}$ can be similarly proved. □

8.3.2 *Automorphic Bundles*

Following Lemma 8.3.1.4, by fpqc descent of quasi-coherent sheaves (see [33, VIII, 1.3]), we can make the following:

Definition 8.3.2.1. (*Compare with Definitions* 1.4.1.9 *and* 8.1.2.8.) *Let R be any R_0-algebra. For each W in* $\mathrm{Rep}_R(\mathrm{M}^{\mathrm{ord}}_{\mathtt{D},0})$, *we define*

$$\vec{\mathcal{E}}_{\mathrm{M}^{\mathrm{ord}}_{\mathtt{D},0},R}(W) := (\vec{\mathcal{E}}_{\mathrm{M}^{\mathrm{ord}}_{\mathtt{D},0}} \underset{R_0}{\otimes} R) \overset{\mathrm{M}^{\mathrm{ord}}_{\mathtt{D},0} \underset{R_0}{\otimes} R}{\times} W,$$

called the **automorphic sheaf** *over $\vec{\mathsf{M}}_{\mathcal{H}} \underset{R_0}{\otimes} R$ associated with W. It is called an* **automorphic bundle** *if W is locally free of finite rank over R, in which case $\vec{\mathcal{E}}_{\mathrm{M}^{\mathrm{ord}}_{\mathtt{D},0},R}(W)$ is also locally free of finite rank over $\vec{\mathsf{M}}_{\mathcal{H}} \underset{R_0}{\otimes} R$. When Condition* 8.1.1.2 *holds, we define similarly $\vec{\mathcal{E}}_{\mathrm{P}^{\mathrm{ord}}_{\mathtt{D},0},R}(W)$ for $W \in \mathrm{Rep}_R(\mathrm{P}^{\mathrm{ord}}_{\mathtt{D},0})$ by replacing $\mathrm{M}^{\mathrm{ord}}_{\mathtt{D},0}$ with $\mathrm{P}^{\mathrm{ord}}_{\mathtt{D},0}$ in the above expression.*

Lemma 8.3.2.2. (*Compare with Lemmas* 1.4.1.10 *and* 8.1.2.9.) *Let* R *be any* R_0*-algebra. The assignment* $\vec{\mathcal{E}}_{\mathrm{M}^{\mathrm{ord}}_{\mathtt{D},0},R}(\,\cdot\,)$ *defines an* **exact functor** *from* $\mathrm{Rep}_R(\mathrm{M}^{\mathrm{ord}}_{\mathtt{D},0})$ *to the category of quasi-coherent sheaves over* $\vec{\mathsf{M}}_{\mathcal{H}}$.

Proof. By étale descent, the proof is similar to that of Lemma 1.4.1.10. □

Lemma 8.3.2.3. (*Compare with Lemmas* 1.4.1.10 *and* 8.1.2.10.) *Let* R *be any* R_0*-algebra. Suppose that Condition* 8.1.1.2 *holds.*

(1) *The assignment* $\vec{\mathcal{E}}_{\mathrm{P}^{\mathrm{ord}}_{\mathtt{D},0},R}(\,\cdot\,)$ *defines an* **exact functor** *from* $\mathrm{Rep}_R(\mathrm{P}^{\mathrm{ord}}_{\mathtt{D},0})$ *to the category of quasi-coherent sheaves over* $\vec{\mathsf{M}}_{\mathcal{H}}$.

(2) *If we view an object of* $W \in \mathrm{Rep}_R(\mathrm{M}^{\mathrm{ord}}_{\mathtt{D},0})$ *as an object of* $\mathrm{Rep}_R(\mathrm{P}^{\mathrm{ord}}_{\mathtt{D},0})$ *via the canonical homomorphism* $\mathrm{P}^{\mathrm{ord}}_{\mathtt{D},0} \to \mathrm{M}^{\mathrm{ord}}_{\mathtt{D},0}$*, then we have a canonical isomorphism* $\vec{\mathcal{E}}_{\mathrm{P}^{\mathrm{ord}}_{\mathtt{D},0},R}(W) \cong \vec{\mathcal{E}}_{\mathrm{M}^{\mathrm{ord}}_{\mathtt{D},0},R}(W)$.

(3) *Suppose* $W \in \mathrm{Rep}_R(\mathrm{P}^{\mathrm{ord}}_{\mathtt{D},0})$ *has a decreasing filtration by subobjects* $\mathtt{F}^a(W) \subset W$ *in* $\mathrm{Rep}_R(\mathrm{P}^{\mathrm{ord}}_{\mathtt{D},0})$ *such that each graded piece* $\mathrm{Gr}^a_{\mathtt{F}}(W) := \mathtt{F}^a(W)/\mathtt{F}^{a+1}(W)$ *can be identified with an object of* $\mathrm{Rep}_R(\mathrm{M}^{\mathrm{ord}}_{\mathtt{D},0})$. *Then* $\vec{\mathcal{E}}_{\mathrm{P}^{\mathrm{ord}}_{\mathtt{D},0},R}(W)$ *has a filtration* $\mathtt{F}^a(\vec{\mathcal{E}}_{\mathrm{P}^{\mathrm{ord}}_{\mathtt{D},0},R}(W)) := \vec{\mathcal{E}}_{\mathrm{P}^{\mathrm{ord}}_{\mathtt{D},0},R}(\mathtt{F}^a(W))$ *with graded pieces* $\vec{\mathcal{E}}_{\mathrm{M}^{\mathrm{ord}}_{\mathtt{D},0},R}(\mathrm{Gr}^a_{\mathtt{F}}(W))$.

Proof. By fpqc descent, the proof is still similar to that of Lemma 1.4.1.10. □

Lemma 8.3.2.4. *With the setting as in Definition* 8.3.2.1, *we have a canonical isomorphism*

$$\vec{\mathcal{E}}_{\mathrm{M}^{\mathrm{ord}}_{\mathtt{D},0},R}(W) \underset{\mathbb{Z}}{\otimes} \mathbb{Q} \xrightarrow{\sim} \mathcal{E}_{\mathrm{M}^{\mathrm{ord}}_{\mathtt{D},0},R_{\mathbb{Q}}}(W_{\mathbb{Q}}) \tag{8.3.2.5}$$

over $\mathsf{M}_{\mathcal{H}} \underset{R_0}{\otimes} R_{\mathbb{Q}}$. *When Condition* 8.1.1.2 *holds, the analogous statement is true if we replace* $\mathrm{M}^{\mathrm{ord}}_{\mathtt{D},0}$ *with* $\mathrm{P}^{\mathrm{ord}}_{\mathtt{D},0}$.

Proof. We have (8.3.2.5) because we have $\vec{\mathcal{E}}_{\mathrm{M}^{\mathrm{ord}}_{\mathtt{D},0}} \underset{\mathbb{Z}}{\otimes} \mathbb{Q} \xrightarrow{\sim} \mathcal{E}_{\mathrm{M}^{\mathrm{ord}}_{\mathtt{D},0}}$ by definition (see Construction 8.3.1.1 and (8.3.1.2)). The analogous statement for $\mathrm{P}^{\mathrm{ord}}_{\mathtt{D},0}$ can be similarly proved. □

Lemma 8.3.2.6. (*Compare with Lemmas* 1.4.1.11 *and* 8.1.2.11.) *Let* $\vec{A}$ *and* $\omega_{\vec{\mathsf{M}}_{\mathcal{H}}}$ *be as in Proposition* 2.2.1.1. *For any* R_0*-algebra* R*, the pullback of* $\underline{\mathrm{Lie}}_{\vec{A}/\vec{\mathsf{M}}_{\mathcal{H}}}$ (*resp.* $\underline{\mathrm{Lie}}^{\vee}_{\vec{A}/\vec{\mathsf{M}}_{\mathcal{H}}}$*, resp.* $\omega_{\vec{\mathsf{M}}_{\mathcal{H}}}$) *to* $\vec{\mathsf{M}}_{\mathcal{H}} \underset{R_0}{\otimes} R$ *is canonically isomorphic to* $\vec{\mathcal{E}}_{\mathrm{M}^{\mathrm{ord}}_{\mathtt{D},0},R}(W)$ *for* $W = \mathrm{Gr}^{-1}_{\mathtt{D},0} \underset{R_0}{\otimes} R$ (*resp.* $(\mathrm{Gr}^{-1}_{\mathtt{D},0})^{\vee} \underset{R_0}{\otimes} R$*, resp.* $\wedge^{\mathrm{top}} (\mathrm{Gr}^{-1}_{\mathtt{D},0})^{\vee} \underset{R_0}{\otimes} R$).

Proof. This follows from the definition (see (8.3.1.2)), and from Lemma 8.3.2.2. □

8.3.3 *Canonical Extensions*

Definition 8.3.3.1. *With the setting as in Definition* 8.3.2.1*, suppose moreover that* $\mathcal{H}$ *is neat and that* Σ *is projective, with a collection* pol *of polarization functions, so that* $\vec{\mathsf{M}}^{\mathrm{tor}}_{\mathcal{H},d_0\mathsf{pol}}$ *over* $\vec{\mathsf{S}}_0 = \mathrm{Spec}(\mathcal{O}_{F_0,(p)})$ *is defined. Then, by Lemma* 8.3.2.4*, we can patch* $\vec{\mathcal{E}}_{\mathrm{M}^{\mathrm{ord}}_{\mathtt{D},0},R}(W)$ *(over* $\vec{\mathsf{M}}_{\mathcal{H}} \underset{R_0}{\otimes} R$*) and* $\mathcal{E}^{\mathrm{can}}_{\mathrm{M}^{\mathrm{ord}}_{\mathtt{D},0},R_{\mathbb{Q}}}(W_{\mathbb{Q}})$ *(over* $\mathsf{M}^{\mathrm{tor}}_{\mathcal{H},\Sigma} \underset{R_0}{\otimes} R_{\mathbb{Q}}$*) into a quasi-coherent sheaf*

$$\vec{\mathcal{E}}_{\mathrm{M}^{\mathrm{ord}}_{\mathtt{D},0},R}(W) \cup \mathcal{E}^{\mathrm{can}}_{\mathrm{M}^{\mathrm{ord}}_{\mathtt{D},0},R_{\mathbb{Q}}}(W_{\mathbb{Q}})$$

over the open subscheme

$$(\vec{\mathsf{M}}_{\mathcal{H}} \underset{R_0}{\otimes} R) \cup (\mathsf{M}^{\mathrm{tor}}_{\mathcal{H},\Sigma} \underset{R_0}{\otimes} R_{\mathbb{Q}})$$

of $\vec{\mathsf{M}}^{\mathrm{tor}}_{\mathcal{H},d_0\mathsf{pol}} \underset{R_0}{\otimes} R$ *(see Proposition* 2.2.2.1*), and define the quasi-coherent sheaf*

$$\vec{\mathcal{E}}^{\mathrm{can}}_{\mathrm{M}^{\mathrm{ord}}_{\mathtt{D},0},R}(W) := \Big(\big((\vec{\mathsf{M}}_{\mathcal{H}} \underset{R_0}{\otimes} R) \cup (\mathsf{M}^{\mathrm{tor}}_{\mathcal{H},\Sigma} \underset{R_0}{\otimes} R_{\mathbb{Q}})\big) \hookrightarrow \vec{\mathsf{M}}^{\mathrm{tor}}_{\mathcal{H},d_0\mathsf{pol}} \underset{R_0}{\otimes} R\Big)_* \Big(\vec{\mathcal{E}}_{\mathrm{M}^{\mathrm{ord}}_{\mathtt{D},0},R}(W) \cup \mathcal{E}^{\mathrm{can}}_{\mathrm{M}^{\mathrm{ord}}_{\mathtt{D},0},R_{\mathbb{Q}}}(W_{\mathbb{Q}})\Big)$$

over $\vec{\mathsf{M}}^{\mathrm{tor}}_{\mathcal{H},d_0\mathsf{pol}} \underset{R_0}{\otimes} R$*. Similarly, we define the quasi-coherent sheaf*

$$\vec{\mathcal{E}}^{\mathrm{sub}}_{\mathrm{M}^{\mathrm{ord}}_{\mathtt{D},0},R}(W)$$

over $\vec{\mathsf{M}}^{\mathrm{tor}}_{\mathcal{H},d_0\mathsf{pol}} \underset{R_0}{\otimes} R$ *by replacing* $\mathcal{E}^{\mathrm{can}}_{\mathrm{M}^{\mathrm{ord}}_{\mathtt{D},0},R}(W_{\mathbb{Q}})$ *with* $\mathcal{E}^{\mathrm{sub}}_{\mathrm{M}^{\mathrm{ord}}_{\mathtt{D},0},R}(W_{\mathbb{Q}})$ *in the above definition of* $\vec{\mathcal{E}}^{\mathrm{can}}_{\mathrm{M}^{\mathrm{ord}}_{\mathtt{D},0},R}(W)$*.*

When Condition 8.1.1.2 *holds, we also define similar assignments* $W \mapsto \vec{\mathcal{E}}^{\mathrm{can}}_{\mathrm{P}^{\mathrm{ord}}_{\mathtt{D},0},R}(W)$ *and* $W \mapsto \vec{\mathcal{E}}^{\mathrm{sub}}_{\mathrm{P}^{\mathrm{ord}}_{\mathtt{D},0},R}(W)$ *by replacing* $\mathrm{M}^{\mathrm{ord}}_{\mathtt{D},0}$ *with* $\mathrm{P}^{\mathrm{ord}}_{\mathtt{D},0}$ *in the above constructions.*

Lemma 8.3.3.2. *With the setting as in Lemma* 8.3.3.1*, we have compatible canonical isomorphisms*

$$\vec{\mathcal{E}}^{\mathrm{can}}_{\mathrm{M}^{\mathrm{ord}}_{\mathtt{D},0},R}(W) \underset{\mathbb{Z}}{\otimes} \mathbb{Q} \xrightarrow{\sim} \mathcal{E}^{\mathrm{can}}_{\mathrm{M}^{\mathrm{ord}}_{\mathtt{D},0},R_{\mathbb{Q}}}(W_{\mathbb{Q}})$$

and

$$\vec{\mathcal{E}}^{\mathrm{sub}}_{\mathrm{M}^{\mathrm{ord}}_{\mathtt{D},0},R}(W) \underset{\mathbb{Z}}{\otimes} \mathbb{Q} \xrightarrow{\sim} \mathcal{E}^{\mathrm{sub}}_{\mathrm{M}^{\mathrm{ord}}_{\mathtt{D},0},R_{\mathbb{Q}}}(W_{\mathbb{Q}})$$

over $\mathsf{M}^{\mathrm{tor}}_{\mathcal{H},\Sigma} \underset{R_0}{\otimes} R_{\mathbb{Q}}$*. When Condition* 8.1.1.2 *holds, the same is true if we replace* $\mathrm{M}^{\mathrm{ord}}_{\mathtt{D},0}$ *with* $\mathrm{P}^{\mathrm{ord}}_{\mathtt{D},0}$*.*

Proof. This follows immediately from the definitions. □

Lemma 8.3.3.3. *When Condition* 8.1.1.2 *holds, if we view an object* $W \in \mathrm{Rep}_R(\mathrm{M}^{\mathrm{ord}}_{\mathtt{D},0})$ *as an object of* $\mathrm{Rep}_R(\mathrm{P}^{\mathrm{ord}}_{\mathtt{D},0})$ *via the canonical homomorphism* $\mathrm{P}^{\mathrm{ord}}_{\mathtt{D},0} \to \mathrm{M}^{\mathrm{ord}}_{\mathtt{D},0}$*, then we have canonical isomorphisms* $\vec{\mathcal{E}}^{\mathrm{can}}_{\mathrm{P}^{\mathrm{ord}}_{\mathtt{D},0},R}(W) \cong \vec{\mathcal{E}}^{\mathrm{can}}_{\mathrm{M}^{\mathrm{ord}}_{\mathtt{D},0},R}(W)$ *and* $\vec{\mathcal{E}}^{\mathrm{sub}}_{\mathrm{P}^{\mathrm{ord}}_{\mathtt{D},0},R}(W) \cong \vec{\mathcal{E}}^{\mathrm{sub}}_{\mathrm{M}^{\mathrm{ord}}_{\mathtt{D},0},R}(W)$*, compatible with each other, and with the isomorphism in Lemma* 8.3.2.3*.*

Proof. This is because the analogues statements are true for $\mathcal{E}^{\mathrm{can}}_{\mathrm{P}^{\mathrm{ord}}_{\mathsf{D},0},R_{\mathbb{Q}}}(W_{\mathbb{Q}})$, $\mathcal{E}^{\mathrm{sub}}_{\mathrm{P}^{\mathrm{ord}}_{\mathsf{D},0},R_{\mathbb{Q}}}(W_{\mathbb{Q}})$, and $\vec{\mathcal{E}}^{\mathrm{sub}}_{\mathrm{P}^{\mathrm{ord}}_{\mathsf{D},0},R}(W)$. □

Lemma 8.3.3.4. *With the setting as in Definition* 8.3.3.1, *suppose moreover that R is an integral domain flat over R_0, and that W is locally free over R. Then $\vec{\mathcal{E}}^{\mathrm{can}}_{\mathrm{M}^{\mathrm{ord}}_{\mathsf{D},0},R}(W)$ and $\vec{\mathcal{E}}^{\mathrm{sub}}_{\mathrm{M}^{\mathrm{ord}}_{\mathsf{D},0},R}(W)$ are torsion-free over $\vec{\mathsf{M}}^{\mathrm{tor}}_{\mathcal{H},d_0\mathsf{pol}} \underset{R_0}{\otimes} R$, and their restrictions to the open subscheme $(\vec{\mathsf{M}}_{\mathcal{H}} \underset{R_0}{\otimes} R) \cup (\mathsf{M}^{\mathrm{tor}}_{\mathcal{H},\Sigma} \underset{R_0}{\otimes} R_{\mathbb{Q}})$ of $\vec{\mathsf{M}}^{\mathrm{tor}}_{\mathcal{H},d_0\mathsf{pol}} \underset{R_0}{\otimes} R$ (see Definition* 8.3.3.1) *are locally free.*

When Condition 8.1.1.2 *holds, the analogues statements are true if we replace* $\mathrm{M}^{\mathrm{ord}}_{\mathsf{D},0}$ *with* $\mathrm{P}^{\mathrm{ord}}_{\mathsf{D},0}$.

Proof. Since $\mathsf{M}_{\mathcal{H}}$ is smooth over S_0 and since $R_{\mathbb{Q}}$ is an integral domain, the local rings of $\mathsf{M}_{\mathcal{H}} \underset{R_0}{\otimes} R_{\mathbb{Q}}$ are integral domains. Since $\vec{\mathsf{M}}^{\mathrm{tor}}_{\mathcal{H},d_0\mathsf{pol}}$ is flat over R_0, since $\mathsf{M}_{\mathcal{H}}$ is open and dense in the noetherian normal scheme $\vec{\mathsf{M}}^{\mathrm{tor}}_{\mathcal{H},d_0\mathsf{pol}}$, and since R is flat over R_0, it follows that the local rings of $\vec{\mathsf{M}}^{\mathrm{tor}}_{\mathcal{H},d_0\mathsf{pol}}$ are also integral domains. By definition (see Construction 8.3.1.1 and Definition 8.3.2.1), when W is locally free over R, the sheaves $\vec{\mathcal{E}}_{\mathrm{M}^{\mathrm{ord}}_{\mathsf{D},0},R}(W) \cup \mathcal{E}^{\mathrm{can}}_{\mathrm{M}^{\mathrm{ord}}_{\mathsf{D},0},R_{\mathbb{Q}}}(W_{\mathbb{Q}})$ and $\vec{\mathcal{E}}_{\mathrm{M}^{\mathrm{ord}}_{\mathsf{D},0},R}(W) \cup \mathcal{E}^{\mathrm{sub}}_{\mathrm{M}^{\mathrm{ord}}_{\mathsf{D},0},R_{\mathbb{Q}}}(W_{\mathbb{Q}})$ are locally free over $(\vec{\mathsf{M}}_{\mathcal{H}} \underset{R_0}{\otimes} R) \cup (\mathsf{M}^{\mathrm{tor}}_{\mathcal{H},\Sigma} \underset{R_0}{\otimes} R_{\mathbb{Q}})$, and hence their direct images $\vec{\mathcal{E}}^{\mathrm{can}}_{\mathrm{M}^{\mathrm{ord}}_{\mathsf{D},0},R}(W)$ and $\vec{\mathcal{E}}^{\mathrm{sub}}_{\mathrm{M}^{\mathrm{ord}}_{\mathsf{D},0},R}(W)$ are also torsion-free over $\vec{\mathsf{M}}^{\mathrm{tor}}_{\mathcal{H},d_0\mathsf{pol}} \underset{R_0}{\otimes} R$, as desired. The analogous statement for $\mathrm{P}^{\mathrm{ord}}_{\mathsf{D},0}$ can be similarly proved. □

8.3.4 *Compatibility with the Constructions over the Ordinary Loci*

Proposition 8.3.4.1. *For each R flat over R_0, and for each W in* $\mathrm{Rep}_R(\mathrm{M}^{\mathrm{ord}}_{\mathsf{D},0})$ *that is flat over R_0, there is a canonical isomorphism*

$$\vec{\mathcal{E}}^{\mathrm{ord}}_{\mathrm{M}^{\mathrm{ord}}_{\mathsf{D},0},R}(W) \cong (\vec{\mathsf{M}}^{\mathrm{ord}}_{\mathcal{H}} \underset{R_0}{\otimes} R \to \vec{\mathsf{M}}_{\mathcal{H}} \underset{R_0}{\otimes} R)^* \vec{\mathcal{E}}_{\mathrm{M}^{\mathrm{ord}}_{\mathsf{D},0},R}(W), \tag{8.3.4.2}$$

(*cf.* (3.4.6.4)) *and there are compatible canonical isomorphisms*

$$\vec{\mathcal{E}}^{\mathrm{ord,can}}_{\mathrm{M}^{\mathrm{ord}}_{\mathsf{D},0},R}(W) \cong (\vec{\mathsf{M}}^{\mathrm{ord,tor}}_{\mathcal{H},\Sigma^{\mathrm{ord}}} \underset{R_0}{\otimes} R \to \vec{\mathsf{M}}^{\mathrm{tor}}_{\mathcal{H},d_0\mathsf{pol}} \underset{R_0}{\otimes} R)^* \vec{\mathcal{E}}^{\mathrm{can}}_{\mathrm{M}^{\mathrm{ord}}_{\mathsf{D},0},R}(W) \tag{8.3.4.3}$$

and

$$\vec{\mathcal{E}}^{\mathrm{ord,sub}}_{\mathrm{M}^{\mathrm{ord}}_{\mathsf{D},0},R}(W) \cong (\vec{\mathsf{M}}^{\mathrm{ord,tor}}_{\mathcal{H},\Sigma^{\mathrm{ord}}} \underset{R_0}{\otimes} R \to \vec{\mathsf{M}}^{\mathrm{tor}}_{\mathcal{H},d_0\mathsf{pol}} \underset{R_0}{\otimes} R)^* \vec{\mathcal{E}}^{\mathrm{sub}}_{\mathrm{M}^{\mathrm{ord}}_{\mathsf{D},0},R}(W) \tag{8.3.4.4}$$

(*cf.* (5.2.3.19)) *when $\mathcal{H}^p$ and hence $\mathcal{H} = \mathcal{H}^p\mathcal{H}_p$ are neat, and when Σ^{ord} extends to some projective Σ, with a collection* pol *of polarization functions (so that $\vec{\mathsf{M}}^{\mathrm{tor}}_{\mathcal{H},d_0\mathsf{pol}}$ is defined as in Proposition* 2.2.2.1). *These canonical isomorphisms are compatible with each other.*

When Condition 8.1.1.2 *holds, the analogous statements are true if we replace* $\mathrm{M}^{\mathrm{ord}}_{\mathsf{D},0}$ *with* $\mathrm{P}^{\mathrm{ord}}_{\mathsf{D},0}$.

Proof. We have the canonical isomorphism (8.3.4.2) because, by Proposition 3.4.6.3, the pullback of the tautological $(\vec{A}, \vec{\lambda}, \vec{i})$ (see Proposition 2.2.1.1) under (3.4.6.4) is canonically isomorphic to the tautological (A, λ, i) over $\vec{\mathsf{M}}_{\mathcal{H}}^{\mathrm{ord}}$.

As for (8.3.4.3) and (8.3.4.4), by construction, and thanks to the canonical isomorphism (8.3.4.2), we already have the isomorphisms (8.3.4.3) and (8.3.4.4) (compatible with (8.3.4.2) and with each other) in an open dense subscheme whose complement has codimension at least two. Now the question is whether the direct images of the open immersion from such an open dense subscheme to the whole scheme coincide with the canonical or subcanonical extensions; or, equivalently, whether the canonical or subcanonical extensions coincide with the direct images of their restrictions to such an open dense subscheme. (Note that we start with existing extensions here.) This is a local question, which we can verify after étale localization or rather just over the completions of strict local rings, without having to assume that the sheaves are automorphic. Since R and W are flat, by writing flat modules as direct limits of finitely generated free modules, and by using the fact that taking direct images (under quasi-compact and quasi-separated morphisms) commute with arbitrary direct limits, it suffices to treat the case that $R = R_0$ and $W = R$ is trivial, which is then known because the scheme $\vec{\mathsf{M}}_{\mathcal{H},\Sigma^{\mathrm{ord}}}^{\mathrm{ord,tor}}$ is regular.

The analogous statements for $\mathrm{P}_{\mathsf{D},0}^{\mathrm{ord}}$ can be similarly proved. □

8.3.5 *Pushforwards to the Total Minimal Compactifications*

Definition 8.3.5.1. *With the setting as in Definition* 8.3.2.1, *suppose moreover that* $\mathcal{H}$ *is neat. Let* Σ *be any collection for* $\mathsf{M}_{\mathcal{H}}$ *(which might not be projective), so that* $\mathsf{M}_{\mathcal{H},\Sigma}^{\mathrm{tor}}$ *is defined over* S_0. *Let the quasi-coherent sheaf*

$$\vec{\mathcal{E}}_{\mathrm{M}_{\mathsf{D},0}^{\mathrm{ord}},R}(W) \cup \mathcal{E}^{\mathrm{can}}_{\mathrm{M}_{\mathsf{D},0}^{\mathrm{ord}},R_{\mathbb{Q}}}(W_{\mathbb{Q}})$$

over the open subscheme

$$(\vec{\mathsf{M}}_{\mathcal{H}} \underset{R_0}{\otimes} R) \cup (\mathsf{M}_{\mathcal{H},\Sigma}^{\mathrm{tor}} \underset{R_0}{\otimes} R_{\mathbb{Q}})$$

of $\vec{\mathsf{M}}_{\mathcal{H},d_0\mathsf{pol}}^{\mathrm{tor}} \underset{R_0}{\otimes} R$ *be defined as in Definition* 8.3.3.1 *(by Lemma* 8.3.2.4*), and define the quasi-coherent sheaf*

$$\vec{\mathcal{E}}^{\mathrm{can,min}}_{\mathrm{M}_{\mathsf{D},0}^{\mathrm{ord}},R}(W) := \Big(((\vec{\mathsf{M}}_{\mathcal{H}} \underset{R_0}{\otimes} R) \cup (\mathsf{M}_{\mathcal{H},\Sigma}^{\mathrm{tor}} \underset{R_0}{\otimes} R_{\mathbb{Q}})) \to \vec{\mathsf{M}}_{\mathcal{H}}^{\min} \underset{R_0}{\otimes} R\Big)_* \Big(\vec{\mathcal{E}}_{\mathrm{M}_{\mathsf{D},0}^{\mathrm{ord}},R}(W) \cup \mathcal{E}^{\mathrm{can}}_{\mathrm{M}_{\mathsf{D},0}^{\mathrm{ord}},R_{\mathbb{Q}}}(W_{\mathbb{Q}})\Big)$$

over $\vec{\mathsf{M}}_{\mathcal{H}}^{\min} \underset{R_0}{\otimes} R$, *using Proposition* 2.2.1.2 *and the canonical morphisms* $\oint_{\mathcal{H}} : \mathsf{M}_{\mathcal{H}}^{\mathrm{tor}} \to \mathsf{M}_{\mathcal{H}}^{\min}$ *in* (3) *of Theorem* 1.3.1.5.

Similarly, we define the quasi-coherent sheaf

$$\vec{\mathcal{E}}^{\mathrm{sub,min}}_{\mathrm{M}_{\mathsf{D},0}^{\mathrm{ord}},R}(W)$$

over $\vec{\mathsf{M}}^{\min}_{\mathcal{H}} \underset{R_0}{\otimes} R$ *by replacing* $\mathcal{E}^{\mathrm{can}}_{\mathrm{M}^{\mathrm{ord}}_{\mathtt{D},0},R}(W_{\mathbb{Q}})$ *with* $\mathcal{E}^{\mathrm{sub}}_{\mathrm{M}^{\mathrm{ord}}_{\mathtt{D},0},R}(W_{\mathbb{Q}})$ *in the above definition of* $\vec{\mathcal{E}}^{\mathrm{can}}_{\mathrm{P}^{\mathrm{ord}}_{\mathtt{D},0},R}(W)$. *These are well defined (i.e., independent of the choice of* Σ*) because of the projection formula [35, 0_I, 5.4.10.1], because of the isomorphism* (1.4.3.3) *(with* $g = 1$ *and* $\mathcal{H}' = \mathcal{H}$ *there), and because, under any morphism* $\mathsf{M}^{\mathrm{tor}}_{\mathcal{H},\Sigma} \twoheadrightarrow \mathsf{M}^{\mathrm{tor}}_{\mathcal{H},\Sigma'}$ *as in [62, Prop. 6.4.2.3], the pushforward of the ideal sheaf defining the boundary of* $\mathsf{M}^{\mathrm{tor}}_{\mathcal{H},\Sigma'}$ *with reduced subscheme structure is canonically isomorphic to the ideal sheaf defining the boundary of* $\mathsf{M}^{\mathrm{tor}}_{\mathcal{H},\Sigma'}$ *with reduced subscheme structure (cf.* (1.3.3.19)*, with* $Q = \{0\}$ *there).*

When Condition 8.1.1.2 *holds, we also define similar assignments* $W \mapsto \vec{\mathcal{E}}^{\mathrm{can},\min}_{\mathrm{P}^{\mathrm{ord}}_{\mathtt{D},0},R}(W)$ *and* $W \mapsto \vec{\mathcal{E}}^{\mathrm{sub},\min}_{\mathrm{P}^{\mathrm{ord}}_{\mathtt{D},0},R}(W)$ *by replacing* $\mathrm{M}^{\mathrm{ord}}_{\mathtt{D},0}$ *with* $\mathrm{P}^{\mathrm{ord}}_{\mathtt{D},0}$ *in the above constructions.*

Lemma 8.3.5.2. *With the setting as in Lemma* 8.3.5.1*, we have compatible canonical isomorphisms*

$$\vec{\mathcal{E}}^{\mathrm{can},\min}_{\mathrm{M}^{\mathrm{ord}}_{\mathtt{D},0},R}(W) \underset{\mathbb{Z}}{\otimes} \mathbb{Q} \xrightarrow{\sim} \oint_{\mathcal{H},*} \mathcal{E}^{\mathrm{can}}_{\mathrm{M}^{\mathrm{ord}}_{\mathtt{D},0},R_{\mathbb{Q}}}(W_{\mathbb{Q}})$$

and

$$\vec{\mathcal{E}}^{\mathrm{sub},\min}_{\mathrm{M}^{\mathrm{ord}}_{\mathtt{D},0},R}(W) \underset{\mathbb{Z}}{\otimes} \mathbb{Q} \xrightarrow{\sim} \oint_{\mathcal{H},*} \mathcal{E}^{\mathrm{sub}}_{\mathrm{M}^{\mathrm{ord}}_{\mathtt{D},0},R_{\mathbb{Q}}}(W_{\mathbb{Q}})$$

over $\mathsf{M}^{\min}_{\mathcal{H}} \underset{R_0}{\otimes} R_{\mathbb{Q}}$, *where* $\oint_{\mathcal{H}} : \mathsf{M}^{\mathrm{tor}}_{\mathcal{H}} \to \mathsf{M}^{\min}_{\mathcal{H}}$ *is as in* (3) *of Theorem* 1.3.1.5*. When Condition* 8.1.1.2 *holds, the same is true if we replace* $\mathrm{M}^{\mathrm{ord}}_{\mathtt{D},0}$ *with* $\mathrm{P}^{\mathrm{ord}}_{\mathtt{D},0}$.

Proof. This follows immediately from the definitions. □

Lemma 8.3.5.3. *In Definition* 8.3.5.1*, if* Σ *is projective, with a collection* pol *of polarization functions, as in Definition* 8.3.3.1*, so that* $\vec{\mathsf{M}}^{\mathrm{tor}}_{\mathcal{H},d_0\mathsf{pol}}$ *over* $\vec{\mathsf{S}}_0$ *is defined, then we have compatible canonical isomorphisms*

$$\vec{\mathcal{E}}^{\mathrm{can},\min}_{\mathrm{M}^{\mathrm{ord}}_{\mathtt{D},0},R}(W) \cong \vec{\oint}_{\mathcal{H},*}\vec{\mathcal{E}}^{\mathrm{can}}_{\mathrm{M}^{\mathrm{ord}}_{\mathtt{D},0},R}(W)$$

and

$$\vec{\mathcal{E}}^{\mathrm{sub},\min}_{\mathrm{M}^{\mathrm{ord}}_{\mathtt{D},0},R}(W) \cong \vec{\oint}_{\mathcal{H},*}\vec{\mathcal{E}}^{\mathrm{sub}}_{\mathrm{M}^{\mathrm{ord}}_{\mathtt{D},0},R}(W)$$

over $\vec{\mathsf{M}}^{\min}_{\mathcal{H}}$, *where* $\vec{\oint}_{\mathcal{H}} : \vec{\mathsf{M}}^{\mathrm{tor}}_{\mathcal{H},d_0\mathsf{pol}} \to \vec{\mathsf{M}}^{\min}_{\mathcal{H}}$ *is as in Definition* 2.2.2.1*. Consequently, since* $(\vec{\mathsf{M}}_{\mathcal{H}} \underset{R_0}{\otimes} R) \cup (\mathsf{M}^{\mathrm{tor}}_{\mathcal{H},\Sigma} \underset{R_0}{\otimes} R_{\mathbb{Q}})$ *is the preimage of* $(\vec{\mathsf{M}}_{\mathcal{H}} \underset{R_0}{\otimes} R) \cup (\mathsf{M}^{\min}_{\mathcal{H}} \underset{R_0}{\otimes} R_{\mathbb{Q}})$ *under the canonical morphism* $\vec{\oint}_{\mathcal{H},*} : \vec{\mathsf{M}}^{\mathrm{tor}}_{\mathcal{H},d_0\mathsf{pol}} \underset{R_0}{\otimes} R \to \vec{\mathsf{M}}^{\min}_{\mathcal{H}} \underset{R_0}{\otimes} R$, *by Lemma* 8.3.5.2*,* $\vec{\mathcal{E}}^{\mathrm{can},\min}_{\mathrm{M}^{\mathrm{ord}}_{\mathtt{D},0},R}(W)$ *and* $\vec{\mathcal{E}}^{\mathrm{sub},\min}_{\mathrm{M}^{\mathrm{ord}}_{\mathtt{D},0},R}(W)$ *are canonically isomorphic to the pushforwards of the quasi-coherent sheaves*

$$\vec{\mathcal{E}}_{\mathrm{M}^{\mathrm{ord}}_{\mathtt{D},0},R}(W) \cup (\oint_{\mathcal{H},*} \mathcal{E}^{\mathrm{can}}_{\mathrm{M}^{\mathrm{ord}}_{\mathtt{D},0},R_{\mathbb{Q}}}(W_{\mathbb{Q}}))$$

and

$$\vec{\mathcal{E}}_{\mathrm{M}^{\mathrm{ord}}_{\mathtt{D},0},R}(W) \cup (\oint_{\mathcal{H},*} \mathcal{E}^{\mathrm{sub}}_{\mathrm{M}^{\mathrm{ord}}_{\mathtt{D},0},R_{\mathbb{Q}}}(W_{\mathbb{Q}})),$$

respectively, over the open subscheme

$$(\vec{\mathsf{M}}_{\mathcal{H}} \underset{R_0}{\otimes} R) \cup (\mathsf{M}_{\mathcal{H}}^{\min} \underset{R_0}{\otimes} R_{\mathbb{Q}})$$

of $\vec{\mathsf{M}}_{\mathcal{H}}^{\min} \underset{R_0}{\otimes} R$ *(cf. Definition* 8.3.3.1*).*

When Condition 8.1.1.2 *holds, the analogous statements are true if we replace* $\mathrm{M}_{\mathtt{D},0}^{\mathrm{ord}}$ *with* $\mathrm{P}_{\mathtt{D},0}^{\mathrm{ord}}$.

Proof. Since the sheaves $\vec{\mathcal{E}}^{\mathrm{can},\min}_{\mathrm{M}_{\mathtt{D},0}^{\mathrm{ord}},R}(W)$ and $\vec{\mathcal{E}}^{\mathrm{sub},\min}_{\mathrm{M}_{\mathtt{D},0}^{\mathrm{ord}},R}(W)$ in Definition 8.3.5.1 are (up to canonical isomorphisms) independent of the choice of Σ, we may assume (up to a refinement) that Σ is projective, with a collection pol of polarization functions. Then it suffices to note that the pushforward under a composition of morphisms is the composition of the pushforwards of the individual morphisms. The analogous statement for $\mathrm{P}_{\mathtt{D},0}^{\mathrm{ord}}$ can be similarly proved. □

Corollary 8.3.5.4. *For each* R *flat over* R_0*, and for each* W *in* $\mathrm{Rep}_R(\mathrm{M}_{\mathtt{D},0}^{\mathrm{ord}})$ *that is flat over* R_0*, there are compatible canonical isomorphisms*

$$\vec{\oint}^{\mathrm{ord}}_{\mathcal{H},*} \vec{\mathcal{E}}^{\mathrm{ord},\mathrm{can}}_{\mathrm{M}_{\mathtt{D},0}^{\mathrm{ord}},R}(W) \cong (\vec{\mathsf{M}}_{\mathcal{H}}^{\mathrm{ord},\min} \underset{R_0}{\otimes} R \to \vec{\mathsf{M}}_{\mathcal{H}}^{\min} \underset{R_0}{\otimes} R)^* \vec{\mathcal{E}}^{\mathrm{can},\min}_{\mathrm{M}_{\mathtt{D},0}^{\mathrm{ord}},R}(W) \tag{8.3.5.5}$$

and

$$\vec{\oint}^{\mathrm{ord}}_{\mathcal{H},*} \vec{\mathcal{E}}^{\mathrm{ord},\mathrm{sub}}_{\mathrm{M}_{\mathtt{D},0}^{\mathrm{ord}},R}(W) \cong (\vec{\mathsf{M}}_{\mathcal{H}}^{\mathrm{ord},\min} \underset{R_0}{\otimes} R \to \vec{\mathsf{M}}_{\mathcal{H}}^{\min} \underset{R_0}{\otimes} R)^* \vec{\mathcal{E}}^{\mathrm{sub},\min}_{\mathrm{M}_{\mathtt{D},0}^{\mathrm{ord}},R}(W) \tag{8.3.5.6}$$

(cf. (3) *and* (6) *of Theorem* 6.2.1.1*) when* $\mathcal{H}^p$ *and hence* $\mathcal{H} = \mathcal{H}^p\mathcal{H}_p$ *are neat, and when* Σ^{ord} *extends to some projective* Σ*, with a collection* pol *of polarization functions (so that* $\vec{\mathsf{M}}^{\mathrm{tor}}_{\mathcal{H},d_0\mathsf{pol}}$ *is defined as in Proposition* 2.2.2.1*).*

When Condition 8.1.1.2 *holds, the analogous statements are true if we replace* $\mathrm{M}_{\mathtt{D},0}^{\mathrm{ord}}$ *with* $\mathrm{P}_{\mathtt{D},0}^{\mathrm{ord}}$.

Proof. By choosing a collection Σ^{ord} for $\vec{\mathsf{M}}_{\mathcal{H}}^{\mathrm{ord}}$ that extends to some projective Σ, with a collection pol of polarization functions, these follow from Lemma 8.3.5.3, Corollary 6.2.3.2, and Proposition 8.3.4.1. □

Lemma 8.3.5.7. *With the setting as in Definition* 8.3.5.1*, suppose* R *is noetherian and flat over* R_0*, and suppose* W *is flat over* R_0 *and is finitely generated and* (S_1) *over* R *(see [35, IV-2, 5.7.2]). Then* $\vec{\mathcal{E}}^{\mathrm{can},\min}_{\mathrm{M}_{\mathtt{D},0}^{\mathrm{ord}},R}(W)$ *and* $\vec{\mathcal{E}}^{\mathrm{sub},\min}_{\mathrm{M}_{\mathtt{D},0}^{\mathrm{ord}},R}(W)$ *are coherent over* $\vec{\mathsf{M}}_{\mathcal{H}}^{\min} \underset{R_0}{\otimes} R$.

Proof. Let us explain the case of $\vec{\mathcal{E}}^{\mathrm{can},\min}_{\mathrm{M}_{\mathtt{D},0}^{\mathrm{ord}},R}(W)$, because the case of $\vec{\mathcal{E}}^{\mathrm{sub},\min}_{\mathrm{M}_{\mathtt{D},0}^{\mathrm{ord}},R}(W)$ is similar. Since $\vec{\mathsf{M}}_{\mathcal{H}}^{\min}$ is flat over R_0, and since $\vec{\mathsf{M}}_{\mathcal{H}}$ is fiberwise dense in $\vec{\mathsf{M}}_{\mathcal{H}}^{\min}$ by Proposition 2.2.1.7, the complement of $(\vec{\mathsf{M}}_{\mathcal{H}} \underset{R_0}{\otimes} R) \cup (\mathsf{M}_{\mathcal{H}}^{\min} \underset{R_0}{\otimes} R_{\mathbb{Q}})$ in $\vec{\mathsf{M}}_{\mathcal{H}}^{\min} \underset{R_0}{\otimes} R$ is of codimension at least two. Hence, by [35, IV-2, 5.11.4], it suffices to show that the restrictions $\vec{\mathcal{E}}_{\mathrm{M}_{\mathtt{D},0}^{\mathrm{ord}},R}(W)$ and $\oint_{\mathcal{H},*} \mathcal{E}^{\mathrm{can}}_{\mathrm{P}_{\mathtt{D},0}^{\mathrm{ord}},R_{\mathbb{Q}}}(W_{\mathbb{Q}})$ of $\vec{\mathcal{E}}^{\mathrm{can},\min}_{\mathrm{M}_{\mathtt{D},0}^{\mathrm{ord}},R}(W)$ to $\vec{\mathsf{M}}_{\mathcal{H}} \underset{R_0}{\otimes} R$

and $\mathsf{M}_{\mathcal{H}}^{\min} \underset{R_0}{\otimes} R_{\mathbb{Q}}$ (see Lemma 8.3.5.2), respectively, are coherent and (S_1) over some open subsets containing all points of codimension one. Since $\vec{\mathsf{M}}_{\mathcal{H}}$ is noetherian and normal, it is (S_2) by Serre's criterion [35, IV-2, 5.8.6]. Since $\vec{\mathsf{M}}_{\mathcal{H}}$ is flat over the Dedekind scheme $\vec{\mathsf{S}}_0$, by [35, IV-2, 5.7.6], the fibers of $\vec{\mathsf{M}}_{\mathcal{H}} \to \vec{\mathsf{S}}_0$ are (S_1). By Lemma 8.3.1.4 and Definition 8.3.2.1, the sheaf $\vec{\mathscr{E}}_{\mathrm{M}_{\mathsf{D},0}^{\mathrm{ord}},R}(W)$ over $\vec{\mathsf{M}}_{\mathcal{H}} \underset{R_0}{\otimes} R$, being étale locally isomorphic to the pullback of W, is coherent by étale descent, and is (S_1) by [35, IV-2, 6.4.2]. Similarly, since $\mathsf{M}_{\mathcal{H},\Sigma}^{\mathrm{tor}}$ is smooth over S_0, the sheaf $\mathscr{E}_{\mathrm{P}_{\mathsf{D},0}^{\mathrm{ord}},R_{\mathbb{Q}}}^{\mathrm{can}}(W_{\mathbb{Q}})$ over $\mathsf{M}_{\mathcal{H},\Sigma}^{\mathrm{tor}} \underset{R_0}{\otimes} R_{\mathbb{Q}}$ is also coherent and (S_1). By [62, Prop. 7.2.3.13], $\oint_{\mathcal{H}} : \mathsf{M}_{\mathcal{H},\Sigma}^{\mathrm{tor}} \to \mathsf{M}_{\mathcal{H}}^{\min}$ is an isomorphism over the open subscheme $\mathsf{M}_{\mathcal{H}}^1$ of $\mathsf{M}_{\mathcal{H}}^{\min}$ formed by the union of all strata (see (4) of Theorem 1.3.1.5) of codimension at most one. Thus, $\oint_{\mathcal{H},*} \mathscr{E}_{\mathrm{P}_{\mathsf{D},0}^{\mathrm{ord}},R_{\mathbb{Q}}}^{\mathrm{can}}(W_{\mathbb{Q}})$ is coherent and (S_1) over $\mathsf{M}_{\mathcal{H}}^1 \underset{R_0}{\otimes} R_{\mathbb{Q}}$, which contains all points of $\mathsf{M}_{\mathcal{H}}^{\min} \underset{R_0}{\otimes} R_{\mathbb{Q}}$ of codimension one, as desired. □

Corollary 8.3.5.8. (*Compare with Lemma* 8.3.3.4.) *With the setting as in Definition* 8.3.5.1, *suppose moreover that R is an integral domain flat over R_0, and that W is locally free over R. Then $\vec{\mathscr{E}}_{\mathrm{M}_{\mathsf{D},0}^{\mathrm{ord}},R}^{\mathrm{can},\min}(W)$ and $\vec{\mathscr{E}}_{\mathrm{M}_{\mathsf{D},0}^{\mathrm{ord}},R}^{\mathrm{sub},\min}(W)$ are torsion-free over $\vec{\mathsf{M}}_{\mathcal{H}}^{\min} \underset{R_0}{\otimes} R$.*

When Condition 8.1.1.2 *holds, the analogous statements are true if we replace* $\mathrm{M}_{\mathsf{D},0}^{\mathrm{ord}}$ *with* $\mathrm{P}_{\mathsf{D},0}^{\mathrm{ord}}$.

Proof. Since R is flat over R_0, the canonical morphism $\mathscr{O}_{\vec{\mathsf{M}}_{\mathcal{H}}^{\min} \underset{R_0}{\otimes} R} \to \vec{\oint}_{\mathcal{H},*} \mathscr{O}_{\vec{\mathsf{M}}_{\mathcal{H},d_0\mathrm{pol}}^{\mathrm{tor}} \underset{R_0}{\otimes} R}$ (*cf.* (2.2.2.2)) is an isomorphism, and it follows from Lemmas 8.3.5.3 and 8.3.3.4 that $\vec{\mathscr{E}}_{\mathrm{M}_{\mathsf{D},0}^{\mathrm{ord}},R}^{\mathrm{can},\min}(W)$ and $\vec{\mathscr{E}}_{\mathrm{M}_{\mathsf{D},0}^{\mathrm{ord}},R}^{\mathrm{sub},\min}(W)$ are torsion-free over $\vec{\mathsf{M}}_{\mathcal{H}}^{\min} \underset{R_0}{\otimes} R$ under the assumptions we made on R and W. □

Lemma 8.3.5.9. (*Compare with Lemmas* 1.4.2.11, 8.1.3.15, *and* 8.3.2.6.) *Let $a \geq 1$ be as in Lemma* 2.1.2.35, *and let $N_1 \geq 1$ be as in Proposition* 2.2.1.2, *so that $\omega_{\vec{\mathsf{M}}_{\mathcal{H}}^{\min}}^{\otimes aN_1}$ is defined. For any flat R_0-algebra R, and for each integer $k \geq 1$, the pullback of $\omega_{\vec{\mathsf{M}}_{\mathcal{H}}^{\min}}^{\otimes kaN_1}$ to $\vec{\mathsf{M}}_{\mathcal{H}}^{\min} \underset{R_0}{\otimes} R$ is canonically isomorphic to $\vec{\mathscr{E}}_{\mathrm{M}_{\mathsf{D},0}^{\mathrm{ord}},R}^{\mathrm{can},\min}(W_0^{\otimes kaN_1} \underset{R_0}{\otimes} R)$ for $W_0 := \wedge^{\mathrm{top}} (\mathrm{Gr}_{\mathsf{D},0}^{-1})^{\vee}$.*

Proof. By Definition 8.3.5.1, it suffices to verify this separately over $\vec{\mathsf{M}}_{\mathcal{H}} \underset{R_0}{\otimes} R$ and $\mathsf{M}_{\mathcal{H}}^{\min} \underset{R_0}{\otimes} R_{\mathbb{Q}}$. Over $\vec{\mathsf{M}}_{\mathcal{H}} \underset{R_0}{\otimes} R$, this follows from Lemma 8.3.2.6. Over $\mathsf{M}_{\mathcal{H}}^{\min} \underset{R_0}{\otimes} R_{\mathbb{Q}}$, this is a combination of Lemma 1.4.2.11 and the fact that $\omega_{\mathsf{M}_{\mathcal{H}}^{\min}}^{\otimes kaN_1} \cong \oint_{\mathcal{H},*} \omega_{\mathsf{M}_{\mathcal{H}}^{\mathrm{tor}}}^{\otimes kaN_1}$ (by (2) and (3) of Theorem 1.3.1.5, and by the projection formula [35, 0_{I}, 5.4.10.1]). □

Lemma 8.3.5.10. *Let $a \geq 1$ be as in Lemma 2.1.2.35, and let $\omega_{\vec{\mathsf{M}}_{\mathcal{H}}^{\min}}^{\otimes aN_1}$ and W_0 be as in Lemma 8.3.5.9. For each W in $\mathrm{Rep}_R(\mathrm{M}_{\mathtt{D},0}^{\mathrm{ord}})$, and for each integer $k \geq 1$, we have compatible canonical isomorphisms*

$$\vec{\mathcal{E}}_{\mathrm{M}_{\mathtt{D},0}^{\mathrm{ord}},R}^{\mathrm{can},\min}(W \underset{R_0}{\otimes} W_0^{\otimes ka_0N_1}) \cong \vec{\mathcal{E}}_{\mathrm{M}_{\mathtt{D},0}^{\mathrm{ord}},R}^{\mathrm{can},\min}(W) \underset{\mathscr{O}_{\vec{\mathsf{M}}_{\mathcal{H}}^{\min}}}{\otimes} \omega_{\vec{\mathsf{M}}_{\mathcal{H}}^{\min}}^{\otimes kaN_1}$$

and

$$\vec{\mathcal{E}}_{\mathrm{M}_{\mathtt{D},0}^{\mathrm{ord}},R}^{\mathrm{sub},\min}(W \underset{R_0}{\otimes} W_0^{\otimes k}) \cong \vec{\mathcal{E}}_{\mathrm{M}_{\mathtt{D},0}^{\mathrm{ord}},R}^{\mathrm{sub},\min}(W) \underset{\mathscr{O}_{\vec{\mathsf{M}}_{\mathcal{H}}^{\min}}}{\otimes} \omega_{\vec{\mathsf{M}}_{\mathcal{H}}^{\min}}^{\otimes kaN_1}.$$

Proof. As in the proof of Lemma 8.3.5.9, by Definition 8.3.5.1, it suffices to verify these separately over $\vec{\mathsf{M}}_{\mathcal{H}} \underset{R_0}{\otimes} R$ and $\mathsf{M}_{\mathcal{H}}^{\min} \underset{R_0}{\otimes} R_{\mathbb{Q}}$. Over $\vec{\mathsf{M}}_{\mathcal{H}} \underset{R_0}{\otimes} R$, this follows from the construction in Definition 8.3.2.1 and from Lemma 8.3.2.6 that

$$\mathcal{E}_{\mathrm{M}_{\mathtt{D},0}^{\mathrm{ord}},R}(W \underset{R_0}{\otimes} W_0^{\otimes k}) \cong \mathcal{E}_{\mathrm{M}_{\mathtt{D},0}^{\mathrm{ord}},R}(W) \underset{\mathscr{O}_{\vec{\mathsf{M}}_{\mathcal{H}}}}{\otimes} \omega_{\vec{\mathsf{M}}_{\mathcal{H}}}^{\otimes k}$$

for each integer k and each W in $\mathrm{Rep}_R(\mathrm{G}_{\mathtt{D},0}^{\mathrm{ord}})$. On the other hand, by Lemma 8.3.5.9, by (2) and (3) of Theorem 1.3.1.5, and by the projection formula [35, 0_{I}, 5.4.10.1], the verification of the lemma over $\mathsf{M}_{\mathcal{H}}^{\min} \underset{R_0}{\otimes} R_{\mathbb{Q}}$ can be reduced to the verification of the corresponding statements

$$\mathcal{E}_{\mathrm{M}_{\mathtt{D},0}^{\mathrm{ord}},R_{\mathbb{Q}}}^{\mathrm{can}}(W_{\mathbb{Q}} \underset{R_0}{\otimes} W_0^{\otimes kaN_1}) \cong \mathcal{E}_{\mathrm{M}_{\mathtt{D},0}^{\mathrm{ord}},R_{\mathbb{Q}}}^{\mathrm{can}}(W_{\mathbb{Q}}) \underset{\mathscr{O}_{\mathsf{M}_{\mathcal{H},\Sigma}^{\mathrm{tor}}}}{\otimes} \omega_{\mathsf{M}_{\mathcal{H}}^{\mathrm{tor}}}^{\otimes kaN_1}$$

and

$$\mathcal{E}_{\mathrm{M}_{\mathtt{D},0}^{\mathrm{ord}},R_{\mathbb{Q}}}^{\mathrm{sub}}(W_{\mathbb{Q}} \underset{R_0}{\otimes} W_0^{\otimes kaN_1}) \cong \mathcal{E}_{\mathrm{M}_{\mathtt{D},0}^{\mathrm{ord}},R_{\mathbb{Q}}}^{\mathrm{sub}}(W_{\mathbb{Q}}) \underset{\mathscr{O}_{\mathsf{M}_{\mathcal{H},\Sigma}^{\mathrm{tor}}}}{\otimes} \omega_{\mathsf{M}_{\mathcal{H}}^{\mathrm{tor}}}^{\otimes kaN_1}$$

over $\mathsf{M}_{\mathcal{H},\Sigma}^{\mathrm{tor}} \underset{R_0}{\otimes} R_{\mathbb{Q}}$, which are true by the construction of $\mathcal{E}_{\mathrm{M}_{\mathtt{D},0}^{\mathrm{ord}},R_{\mathbb{Q}}}^{\mathrm{can}}(\,\cdot\,)$ and $\mathcal{E}_{\mathrm{M}_{\mathtt{D},0}^{\mathrm{ord}},R_{\mathbb{Q}}}^{\mathrm{sub}}(\,\cdot\,)$ (*cf.* Construction 8.3.1.1), and by the analogue $\mathcal{E}_{\mathrm{M}_{\mathtt{D},0}^{\mathrm{ord}},R_{\mathbb{Q}}}^{\mathrm{can}}(W_0^{\otimes k} \underset{R_0}{\otimes} R_{\mathbb{Q}}) \cong \omega_{\mathsf{M}_{\mathcal{H}}^{\mathrm{tor}}}^{\otimes k} \underset{R_0}{\otimes} R_{\mathbb{Q}}$ of Lemma 1.4.2.11, for each integer k. □

8.3.6 *Hecke Actions*

Proposition 8.3.6.1. (*Compare with Propositions* 1.4.3.1, 1.3.1.14, *and* 2.2.3.1, (4) *of Theorem* 1.3.3.15, *and Proposition* 8.1.4.1.) *Suppose that we have an element $g = (g_0, g_p) \in \mathrm{G}(\mathbb{A}^{\infty,p}) \times \mathrm{G}(\mathbb{Z}_p) \subset \mathrm{G}(\mathbb{A}^\infty)$; that we have two open compact subgroups $\mathcal{H}$ and $\mathcal{H}'$ of $\mathrm{G}(\hat{\mathbb{Z}})$ such that $\mathcal{H}' \subset g\mathcal{H}g^{-1}$; and that the image $\mathcal{H}^p$ of $\mathcal{H}$ under the canonical homomorphism $\mathrm{G}(\hat{\mathbb{Z}}) \to \mathrm{G}(\hat{\mathbb{Z}}^p)$ is neat (and so that $\mathcal{H}$ is neat, and so that $[\vec{g}] : \vec{\mathsf{M}}_{\mathcal{H}'} \to \vec{\mathsf{M}}_{\mathcal{H}}$ is induced by the $[\vec{g}]^{\min} : \vec{\mathsf{M}}_{\mathcal{H}'}^{\min} \to \vec{\mathsf{M}}_{\mathcal{H}}^{\min}$ in Proposition 2.2.3.1). For each $W \in \mathrm{Rep}_R(\mathrm{M}_{\mathtt{D},0}^{\mathrm{ord}})$, there is (by abuse of notation) a canonical isomorphism*

$$[\vec{g}]^* : [\vec{g}]^* \vec{\mathcal{E}}_{\mathrm{M}_{\mathtt{D},0}^{\mathrm{ord}},R}(W) \xrightarrow{\sim} \vec{\mathcal{E}}_{\mathrm{M}_{\mathtt{D},0}^{\mathrm{ord}},R}(W) \tag{8.3.6.2}$$

of quasi-coherent sheaves over $\vec{\mathsf{M}}_{\mathcal{H}'}$*, where the first* $\vec{\mathcal{E}}_{\mathrm{M}^{\mathrm{ord}}_{\mathtt{D},0},R}(W)$ *is defined over* $\vec{\mathsf{M}}_{\mathcal{H}}$*, and where the second is defined over* $\vec{\mathsf{M}}_{\mathcal{H}'}$*.*

The canonical isomorphism (8.3.6.2) *is compatible with the canonical isomorphism*

$$[g]^* : [g]^*\mathcal{E}_{\mathrm{M}^{\mathrm{ord}}_{\mathtt{D},0},R_{\mathbb{Q}}}(W_{\mathbb{Q}}) \overset{\sim}{\to} \mathcal{E}_{\mathrm{M}^{\mathrm{ord}}_{\mathtt{D},0},R_{\mathbb{Q}}}(W_{\mathbb{Q}}) \tag{8.3.6.3}$$

over $\mathsf{M}_{\mathcal{H}}$ *(as in* (1.4.3.2)*). If* $g_p \in \mathrm{P}^{\mathrm{ord}}_{\mathtt{D}}(\mathbb{Z}_p)$ *(cf. Example* 3.3.4.5*), and if* $\mathcal{H} = \mathcal{H}^p\mathcal{H}_p$ *and* $\mathcal{H}' = \mathcal{H}'^{,p}\mathcal{H}'_p$ *are of standard form (with neat* $\mathcal{H}^p$*), then* (8.3.6.2) *is also compatible with the canonical isomorphism* (8.1.4.2) *(under the canonical isomorphism induced by* (8.3.4.2)*).*

If $g = g_1g_2$*, where* g_1 *and* g_2 *are elements of* $\mathrm{G}(\mathbb{A}^{\infty,p}) \times \mathrm{G}(\mathbb{Z}_p)$*, each having a setup similar to that of* g*, then we have* $\vec{[g]}^* = \vec{[g_1]}^* \circ \vec{[g_2]}^*$ *whenever the involved isomorphisms are defined.*

When Condition 8.1.1.2 *holds, the analogous statements are true if we replace* $\mathrm{M}^{\mathrm{ord}}_{\mathtt{D},0}$ *with* $\mathrm{P}^{\mathrm{ord}}_{\mathtt{D},0}$*.*

Proof. Since $\mathcal{H}^p$ is neat, $\mathcal{H}'^p \subset g_0\mathcal{H}^pg_0^{-1}$ is also neat (and so is $\mathcal{H}' = \mathcal{H}'^{,p}\mathcal{H}'_p$). By Proposition 2.2.3.1, the canonical surjection $[g] : \mathsf{M}_{\mathcal{H}'} \to \mathsf{M}_{\mathcal{H}}$ (over $\mathsf{S}_0 = \mathrm{Spec}(F_0)$) extends to a canonical finite surjection $\vec{[g]} : \vec{\mathsf{M}}_{\mathcal{H}'} \to \vec{\mathsf{M}}_{\mathcal{H}}$. By construction, the pullback of the tautological object $(A, \lambda, i, \alpha_{\mathcal{H}})$ over $\mathsf{M}_{\mathcal{H}}$ under $[g] : \mathsf{M}_{\mathcal{H}'} \to \mathsf{M}_{\mathcal{H}}$ is canonically isomorphic to the Hecke twist $(A', \lambda', i', \alpha'_{\mathcal{H}})$ of the tautological object $(A, \lambda, i, \alpha_m)$ over $\mathsf{M}_{\mathcal{H}'}$ by g, realized by a $\mathbb{Z}^\times_{(p)}$-isogeny $[g] : A \to A'$ over $\mathsf{M}_{\mathcal{H}'}$, because $g_p \in \mathrm{G}(\mathbb{Z}_p)$. (Here, for simplicity, we use the same notation (A, λ, i) for both the tautological objects over $\mathsf{M}_{\mathcal{H}}$ and over $\mathsf{M}_{\mathcal{H}'}$.) Since $\vec{\mathsf{M}}_{\mathcal{H}'}$ is noetherian and normal, and since $\mathsf{M}_{\mathcal{H}'}$ is dense in $\vec{\mathsf{M}}_{\mathcal{H}'}$, by [92, IX, 1.4], [28, Ch. I, Prop. 2.7], or [62, Prop. 3.3.1.5], the pullback of the $(\vec{A}, \vec{\lambda}, \vec{i})$ over $\vec{\mathsf{M}}_{\mathcal{H}}$ (see Proposition 2.2.1.1) under $\vec{[g]} : \vec{\mathsf{M}}_{\mathcal{H}'} \to \vec{\mathsf{M}}_{\mathcal{H}}$ is the unique extension of (A', λ', i') (up to canonical isomorphism) over $\vec{\mathsf{M}}_{\mathcal{H}'}$, which we denote by $(\vec{A}', \vec{\lambda}', \vec{i}')$, and the $\mathbb{Z}^\times_{(p)}$-isogeny $[g] : A \to A'$ uniquely extends to a $\mathbb{Z}^\times_{(p)}$-isogeny

$$\vec{[g]}^* : \vec{A} \to \vec{A}', \tag{8.3.6.4}$$

which induces isomorphisms $\vec{[g]}_* : \underline{\mathrm{Lie}}_{\vec{A}/\vec{\mathsf{M}}_{\mathcal{H}'}} \overset{\sim}{\to} \underline{\mathrm{Lie}}_{\vec{A}'/\vec{\mathsf{M}}_{\mathcal{H}'}} \cong \vec{[g]}^*\underline{\mathrm{Lie}}_{\vec{A}/\vec{\mathsf{M}}_{\mathcal{H}}}$ and $\vec{[g]}_* : \underline{\mathrm{Lie}}^\vee_{\vec{A}^\vee/\vec{\mathsf{M}}_{\mathcal{H}'}} \overset{\sim}{\to} \underline{\mathrm{Lie}}^\vee_{\vec{A}'^{,\vee}/\vec{\mathsf{M}}_{\mathcal{H}'}} \cong \vec{[g]}^*\underline{\mathrm{Lie}}_{\vec{A}^\vee/\vec{\mathsf{M}}_{\mathcal{H}}}$, which respect the pairings defined by $\vec{\lambda}$ and $\vec{\lambda}'$ up to the unique number r_0 in $\mathbb{Z}^\times_{(p),>0}$ such that $r_0\nu(g_0)\hat{\mathbb{Z}}^p = \hat{\mathbb{Z}}^p$. For each W in $\mathrm{Rep}_R(\mathrm{P}^{\mathrm{ord}}_{\mathtt{D},0})$, these two isomorphisms induced by (8.3.6.4) define the desired isomorphism (8.3.6.2) over $\vec{\mathsf{M}}_{\mathcal{H}'}$, which extends the canonical isomorphism (8.3.6.3).

The morphism (8.3.6.2) is compatible with (8.3.6.3) by construction. If $g_p \in \mathrm{P}^{\mathrm{ord}}_{\mathtt{D}}(\mathbb{Z}_p)$, then (8.3.6.2) is also compatible with the canonical morphism (8.1.4.2) (under the canonical isomorphism induced by (8.3.4.2)), because $\vec{[g]}^{\mathrm{ord}}$ and $\vec{[g]}$ are compatible (see (6.2.2.3)) and hence must induce the same $\mathbb{Z}^\times_{(p)}$-isogeny over $\vec{\mathsf{M}}^{\mathrm{ord}}_{\mathcal{H}'}$ extending the ordinary Hecke twist of the tautological object over $\mathsf{M}^{\mathrm{ord}}_{\mathcal{H}'}$ by $[g]$.

The second last paragraph of the proposition (for $\mathrm{M}^{\mathrm{ord}}_{\mathsf{D},0}$) is true because, by Proposition 2.2.3.1, we have the identity $[\vec{g}] = [\vec{g_2}] \circ [\vec{g_1}]$ between morphisms of schemes, which induces the identity $[\vec{g}] = [\vec{g_2}] \circ [\vec{g_1}]$ between $\mathbb{Z}^\times_{(p)}$-isogenies, and hence also the desired identity $[\vec{g}]^* = [\vec{g_1}]^* \circ [\vec{g_2}]^*$ between isomorphisms like (8.3.6.2).

When Condition 8.1.1.2 holds, the analogous statements for $\mathrm{P}^{\mathrm{ord}}_{\mathsf{D},0}$ can be similarly proved, by using the isomorphism

$$[\vec{g}]^* : \underline{H}_1^{\mathrm{dR}}(\vec{A}/\vec{\mathsf{M}}_{\mathcal{H}'}) \xrightarrow{\sim} \underline{H}_1^{\mathrm{dR}}(\vec{A}'/\vec{\mathsf{M}}_{\mathcal{H}'}) \cong [\vec{g}]^* \underline{H}_1^{\mathrm{dR}}(\vec{A}/\vec{\mathsf{M}}_{\mathcal{H}})$$

induced by (8.3.6.4). □

Corollary 8.3.6.5. *With the same setting as in Proposition* 8.3.6.1, *there are* (*by abuse of notation*) *a canonical isomorphism*

$$([\vec{g}]^{\min})^* : ([\vec{g}]^{\min})^* \vec{\mathcal{E}}^{\mathrm{can},\min}_{\mathrm{M}^{\mathrm{ord}}_{\mathsf{D},0},R}(W) \xrightarrow{\sim} \vec{\mathcal{E}}^{\mathrm{can},\min}_{\mathrm{M}^{\mathrm{ord}}_{\mathsf{D},0},R}(W) \tag{8.3.6.6}$$

and a canonical morphism

$$([\vec{g}]^{\min})^* : ([\vec{g}]^{\min})^* \vec{\mathcal{E}}^{\mathrm{sub},\min}_{\mathrm{M}^{\mathrm{ord}}_{\mathsf{D},0},R}(W) \to \vec{\mathcal{E}}^{\mathrm{sub},\min}_{\mathrm{M}^{\mathrm{ord}}_{\mathsf{D},0},R}(W) \tag{8.3.6.7}$$

of quasi-coherent sheaves over $\vec{\mathsf{M}}^{\min}_{\mathcal{H}'}$, *where the sheaves at the left-hand sides are defined over* $\vec{\mathsf{M}}^{\min}_{\mathcal{H}}$, *and where the sheaves at the right-hand sides are defined over* $\vec{\mathsf{M}}^{\min}_{\mathcal{H}'}$, *which are compatible with each other and with* (8.3.6.2).

The canonical morphisms (8.3.6.6) *and* (8.3.6.7) *are compatible with the canonical morphisms induced by* (1.4.3.3) *and* (1.4.3.4). *If* $g_p \in \mathrm{P}^{\mathrm{ord}}_{\mathsf{D}}(\mathbb{Z}_p)$ (*cf. Example* 3.3.4.5), *then they are also compatible with the canonical morphisms induced by* (8.1.4.3) *and* (8.1.4.4) *under the canonical isomorphisms induced by* (*the pushforwards from the partial toroidal compactifications to the partial minimal compactifications of*) (8.3.4.3) *and* (8.3.4.4).

If $g = g_1 g_2$, *where* g_1 *and* g_2 *are elements of* $\mathrm{G}(\mathbb{A}^{\infty,p}) \times \mathrm{G}(\mathbb{Z}_p)$, *each having a setup similar to that of* g, *then we have* $([\vec{g}]^{\min})^* = ([\vec{g_1}]^{\min})^* \circ ([\vec{g_2}]^{\min})^*$ *whenever the involved morphisms are defined.*

When Condition 8.1.1.2 *holds, the analogous statements are true if we replace* $\mathrm{M}^{\mathrm{ord}}_{\mathsf{D},0}$ *with* $\mathrm{P}^{\mathrm{ord}}_{\mathsf{D},0}$.

Proof. By the constructions of $\vec{\mathcal{E}}^{\mathrm{can},\min}_{\mathrm{M}^{\mathrm{ord}}_{\mathsf{D},0},R}(W)$ and $\vec{\mathcal{E}}^{\mathrm{sub},\min}_{\mathrm{M}^{\mathrm{ord}}_{\mathsf{D},0},R}(W)$ as pushforwards (see Definition 8.3.5.1), these follow from the corresponding statements of Proposition 8.3.6.1 over $\vec{\mathsf{M}}_{\mathcal{H}}$ and $\vec{\mathsf{M}}_{\mathcal{H}'}$, and of Proposition 1.4.3.1 over $\mathsf{M}^{\mathrm{tor}}_{\mathcal{H},\Sigma}$ and $\mathsf{M}^{\mathrm{tor}}_{\mathcal{H}',\Sigma'}$ (by Lemma 8.3.5.2; and from the compatibility stated in (6) of Theorem 7.1.4.1). □

Corollary 8.3.6.8. *With the same setting as in Proposition* 8.3.6.1, *suppose* $\mathcal{H}'$ *is a normal subgroup of* $\mathcal{H}$. *Then the canonical morphisms* (8.3.6.2), (8.3.6.6), *and* (8.3.6.7) *induce compatible actions of the finite group* $\mathcal{H}/\mathcal{H}'$ *on* $\vec{\mathcal{E}}^{\mathrm{can}}_{\mathrm{M}^{\mathrm{ord}}_{\mathsf{D},0},R}(W)$ *over* $\vec{\mathsf{M}}_{\mathcal{H}'}$ *and on* $\vec{\mathcal{E}}^{\mathrm{can},\min}_{\mathrm{M}^{\mathrm{ord}}_{\mathsf{D},0},R}(W)$ *and* $\vec{\mathcal{E}}^{\mathrm{sub},\min}_{\mathrm{M}^{\mathrm{ord}}_{\mathsf{D},0},R}(W)$ *over* $\vec{\mathsf{M}}^{\min}_{\mathcal{H}'}$, *covering those on* $\vec{\mathsf{M}}_{\mathcal{H}'}$ *and* $\vec{\mathsf{M}}^{\min}_{\mathcal{H}'}$ (*cf.* [62, *Cor.* 7.2.5.2] *and Corollary* 2.2.3.2).

Proof. The statements are self-explanatory, by taking $g \in \mathcal{H} \subset \mathrm{G}(\hat{\mathbb{Z}})$ in Proposition 8.3.6.1 and Corollary 8.3.6.5. $\square$

Proposition 8.3.6.9. *With the same setting as in Proposition* 8.3.6.1, *suppose moreover that* $\mathcal{H}'$ *is a normal subgroup of* $\mathcal{H}$, *so that the actions* $\mathcal{H}/\mathcal{H}'$ *as in Corollary* 8.3.6.8 *are defined; that* R *is flat over* R_0; *and that* W *is locally free over* R. *Then the canonical morphisms induced by* (8.3.6.2), (8.3.6.6), *and* (8.3.6.7) *by adjunction induce canonical isomorphisms*

$$\vec{\mathcal{E}}_{\mathrm{M}^{\mathrm{ord}}_{\mathsf{D},0},R}(W) \xrightarrow{\sim} \left([\vec{1}]_* \vec{\mathcal{E}}_{\mathrm{M}^{\mathrm{ord}}_{\mathsf{D},0},R}(W)\right)^{\mathcal{H}/\mathcal{H}'}, \tag{8.3.6.10}$$

$$\vec{\mathcal{E}}^{\mathrm{can},\min}_{\mathrm{M}^{\mathrm{ord}}_{\mathsf{D},0},R}(W) \xrightarrow{\sim} \left([\vec{1}]^{\min}_* \vec{\mathcal{E}}^{\mathrm{can},\min}_{\mathrm{M}^{\mathrm{ord}}_{\mathsf{D},0},R}(W)\right)^{\mathcal{H}/\mathcal{H}'}, \tag{8.3.6.11}$$

and

$$\vec{\mathcal{E}}^{\mathrm{sub},\min}_{\mathrm{M}^{\mathrm{ord}}_{\mathsf{D},0},R}(W) \xrightarrow{\sim} \left([\vec{1}]^{\min}_* \vec{\mathcal{E}}^{\mathrm{sub},\min}_{\mathrm{M}^{\mathrm{ord}}_{\mathsf{D},0},R}(W)\right)^{\mathcal{H}/\mathcal{H}'}, \tag{8.3.6.12}$$

where $[\vec{1}] : \vec{\mathsf{M}}_{\mathcal{H}'} \to \vec{\mathsf{M}}_{\mathcal{H}}$ *and* $[\vec{1}]^{\min} : \vec{\mathsf{M}}^{\min}_{\mathcal{H}'} \to \vec{\mathsf{M}}^{\min}_{\mathcal{H}}$ *are as in Proposition* 2.2.3.1 *and Corollary* 2.2.3.2, *and where the* $\vec{\mathcal{E}}_{\mathrm{M}^{\mathrm{ord}}_{\mathsf{D},0},R}(W)$, $\vec{\mathcal{E}}^{\mathrm{can},\min}_{\mathrm{M}^{\mathrm{ord}}_{\mathsf{D},0},R}(W)$, *and* $\vec{\mathcal{E}}^{\mathrm{sub},\min}_{\mathrm{M}^{\mathrm{ord}}_{\mathsf{D},0},R}(W)$ *at the left-hand sides (resp. right-hand sides) are defined over* $\vec{\mathsf{M}}_{\mathcal{H}}$ *and* $\vec{\mathsf{M}}^{\min}_{\mathcal{H}}$ *(resp.* $\vec{\mathsf{M}}_{\mathcal{H}'}$ *and* $\vec{\mathsf{M}}^{\min}_{\mathcal{H}'}$*).*

When Condition 8.1.1.2 *holds, the analogous statements are true if we replace* $\mathrm{M}^{\mathrm{ord}}_{\mathsf{D},0}$ *with* $\mathrm{P}^{\mathrm{ord}}_{\mathsf{D},0}$.

Proof. By Lemma 8.3.5.3, in order to justify the canonical isomorphisms (8.3.6.10), (8.3.6.11), and (8.3.6.12), it suffices to establish the canonical isomorphisms

$$\vec{\mathcal{E}}_{\mathrm{M}^{\mathrm{ord}}_{\mathsf{D},0},R}(W) \to \left([\vec{1}]_* \vec{\mathcal{E}}_{\mathrm{M}^{\mathrm{ord}}_{\mathsf{D},0},R}(W)\right)^{\mathcal{H}/\mathcal{H}'}, \tag{8.3.6.13}$$

$$\oint_{\mathcal{H},*} \mathcal{E}^{\mathrm{can}}_{\mathrm{M}^{\mathrm{ord}}_{\mathsf{D},0},R_{\mathbb{Q}}}(W_{\mathbb{Q}}) \to \left([1]^{\min}_* \oint_{\mathcal{H},*} \mathcal{E}^{\mathrm{can}}_{\mathrm{M}^{\mathrm{ord}}_{\mathsf{D},0},R_{\mathbb{Q}}}(W_{\mathbb{Q}})\right)^{\mathcal{H}/\mathcal{H}'}, \tag{8.3.6.14}$$

and

$$\oint_{\mathcal{H},*} \mathcal{E}^{\mathrm{sub}}_{\mathrm{M}^{\mathrm{ord}}_{\mathsf{D},0},R_{\mathbb{Q}}}(W_{\mathbb{Q}}) \to \left([1]^{\min}_* \oint_{\mathcal{H},*} \mathcal{E}^{\mathrm{sub}}_{\mathrm{M}^{\mathrm{ord}}_{\mathsf{D},0},R_{\mathbb{Q}}}(W_{\mathbb{Q}})\right)^{\mathcal{H}/\mathcal{H}'}, \tag{8.3.6.15}$$

where the $\mathcal{E}^{\mathrm{can}}_{\mathrm{M}^{\mathrm{ord}}_{\mathsf{D},0},R_{\mathbb{Q}}}(W_{\mathbb{Q}})$ and $\mathcal{E}^{\mathrm{sub}}_{\mathrm{M}^{\mathrm{ord}}_{\mathsf{D},0},R_{\mathbb{Q}}}(W_{\mathbb{Q}})$ at the left-hand sides (resp. right-hand sides) are defined over $\mathsf{M}^{\mathrm{tor}}_{\mathcal{H},\Sigma}$ (resp. $\mathsf{M}^{\mathrm{tor}}_{\mathcal{H}',\Sigma'}$) for some projective smooth collection Σ for $\mathsf{M}_{\mathcal{H}'}$ (resp. Σ' for $\mathsf{M}_{\mathcal{H}}$, which is a 1-refinement of Σ invariant under the action of $\mathcal{H}/\mathcal{H}'$), and where $[1]^{\min} : \mathsf{M}^{\min}_{\mathcal{H}'} \to \mathsf{M}^{\min}_{\mathcal{H}}$ is as in Proposition 1.3.1.14. Or, since $[1]^{\min} \circ \oint_{\mathcal{H}'} = \oint_{\mathcal{H}} \circ [1]^{\mathrm{tor}}$, where $[1]^{\mathrm{tor}} : \mathsf{M}^{\mathrm{tor}}_{\mathcal{H}',\Sigma'} \to \mathsf{M}^{\mathrm{tor}}_{\mathcal{H},\Sigma}$ is as in Proposition 1.3.1.15, we shall consider

$$\mathcal{E}^{\mathrm{can}}_{\mathrm{M}^{\mathrm{ord}}_{\mathsf{D},0},R_{\mathbb{Q}}}(W_{\mathbb{Q}}) \to \left([1]^{\mathrm{tor}}_* \mathcal{E}^{\mathrm{can}}_{\mathrm{M}^{\mathrm{ord}}_{\mathsf{D},0},R_{\mathbb{Q}}}(W_{\mathbb{Q}})\right)^{\mathcal{H}/\mathcal{H}'} \tag{8.3.6.16}$$

and

$$\mathcal{E}^{\mathrm{sub}}_{\mathrm{M}^{\mathrm{ord}}_{\mathsf{D},0},R_{\mathbb{Q}}}(W_{\mathbb{Q}}) \to \left([1]^{\mathrm{tor}}_* \mathcal{E}^{\mathrm{sub}}_{\mathrm{M}^{\mathrm{ord}}_{\mathsf{D},0},R_{\mathbb{Q}}}(W_{\mathbb{Q}})\right)^{\mathcal{H}/\mathcal{H}'}, \tag{8.3.6.17}$$

instead of (8.3.6.14) and (8.3.6.15).

Let us start with (8.3.6.13). By Corollary 2.2.3.2 (the argument of whose proof also works for the normalizations of the base changes from $\mathrm{Spec}(\mathcal{O}_{F_0,(p)})$ to $\mathrm{Spec}(R_0)$ of the schemes involved), we have a canonical isomorphism

$$\mathscr{O}_{\vec{\mathsf{M}}_{\mathcal{H}}} \xrightarrow{\sim} ([\vec{1}]_* \mathscr{O}_{\vec{\mathsf{M}}_{\mathcal{H}'}})^{\mathcal{H}/\mathcal{H}'}. \tag{8.3.6.18}$$

Since R is flat over R_0, we have an induced isomorphism

$$\mathscr{O}_{\vec{\mathsf{M}}_{\mathcal{H}} \underset{R_0}{\otimes} R} \xrightarrow{\sim} ([\vec{1}]_* \mathscr{O}_{\vec{\mathsf{M}}_{\mathcal{H}'} \underset{R_0}{\otimes} R})^{\mathcal{H}/\mathcal{H}'}. \tag{8.3.6.19}$$

Since W is locally free over R, by Lemma 8.3.1.4 and Definition 8.3.2.1, $\vec{\mathscr{E}}_{\mathsf{M}^{\mathrm{ord}}_{\mathsf{D},0},R}(W)$ is locally free over $\vec{\mathsf{M}}_{\mathcal{H}} \underset{R_0}{\otimes} R$. Therefore, by (8.3.6.2) (with $g = 1$) and by the projection formula [35, 0_{I}, 5.4.10.1], and by (8.3.6.19), the restriction of the canonical morphism induced by adjunction by (8.3.6.2) (with $g = 1$) induces the desired isomorphism (8.3.6.13).

Next let us consider (8.3.6.16) and (8.3.6.17). Since the morphism $[1]^{\mathrm{tor}}$: $\mathsf{M}^{\mathrm{tor}}_{\mathcal{H}',\Sigma'} \to \mathsf{M}^{\mathrm{tor}}_{\mathcal{H},\Sigma}$ is proper, its admits a Stein factorization (see [35, III-1, 4.3.3])

$$\mathsf{M}^{\mathrm{tor}}_{\mathcal{H}',\Sigma'} \to Z := \underline{\mathrm{Spec}}_{\mathscr{O}_{\mathsf{M}^{\mathrm{tor}}_{\mathcal{H},\Sigma}}} ([1]^{\mathrm{tor}}_* \mathscr{O}_{\mathsf{M}^{\mathrm{tor}}_{\mathcal{H}',\Sigma'}}) \to \mathsf{M}^{\mathrm{tor}}_{\mathcal{H},\Sigma},$$

where Z is a noetherian normal scheme finite over the projective smooth scheme $\mathsf{M}^{\mathrm{tor}}_{\mathcal{H},\Sigma}$. Hence, Z is also projective, and carries an action of $\mathcal{H}/\mathcal{H}'$ defined by the compatible isomorphisms $[g]^{\mathrm{tor}} : \mathsf{M}^{\mathrm{tor}}_{\mathcal{H}',\Sigma'} \to \mathsf{M}^{\mathrm{tor}}_{\mathcal{H}',\Sigma'}$ defined by Proposition 1.3.1.15 for each $g \in \mathcal{H}$. By Zariski's main theorem (see [35, III-1, 4.4.3, 4.4.11]), the induced morphism

$$Z/(\mathcal{H}/\mathcal{H}') \to \mathsf{M}^{\mathrm{tor}}_{\mathcal{H},\Sigma}$$

is an isomorphism, because it is generically the known isomorphism

$$\mathsf{M}_{\mathcal{H}'}/(\mathcal{H}/\mathcal{H}') \xrightarrow{\sim} \mathsf{M}_{\mathcal{H}}$$

(see [62, Cor. 7.2.5.2]). (Alternatively, one may define the $\mathcal{H}/\mathcal{H}'$-action on Z using the facts that the pullback of $Z \to \mathsf{M}^{\mathrm{tor}}_{\mathcal{H},\Sigma}$ under $\mathsf{M}_{\mathcal{H}} \hookrightarrow \mathsf{M}^{\mathrm{tor}}_{\mathcal{H},\Sigma}$ on the target is the canonical finite étale cover $\mathsf{M}_{\mathcal{H}'} \to \mathsf{M}_{\mathcal{H}}$ carrying a canonical action of $\mathcal{H}/\mathcal{H}'$, and that Z is noetherian normal and hence, by Zariski's main theorem, coincides with the normalization of $\mathsf{M}^{\mathrm{tor}}_{\mathcal{H},\Sigma}$ in $\mathsf{M}_{\mathcal{H}'}$ under the canonical morphism $\mathsf{M}_{\mathcal{H}'} \to \mathsf{M}_{\mathcal{H}} \to \mathsf{M}^{\mathrm{tor}}_{\mathcal{H},\Sigma}$. Therefore, it is not really necessary to assume that Σ' is invariant under the action of $\mathcal{H}/\mathcal{H}'$.) Thus, we obtain the canonical isomorphism

$$\mathscr{O}_{\mathsf{M}^{\mathrm{tor}}_{\mathcal{H},\Sigma}} \xrightarrow{\sim} ([1]^{\mathrm{tor}}_* \mathscr{O}_{\mathsf{M}^{\mathrm{tor}}_{\mathcal{H}',\Sigma'}})^{\mathcal{H}/\mathcal{H}'}, \tag{8.3.6.20}$$

which induces the canonical isomorphism

$$\mathscr{I}_{\mathsf{D}_{\infty,\mathcal{H}}} \xrightarrow{\sim} ([1]^{\mathrm{tor}}_* \mathscr{I}_{\mathsf{D}_{\infty,\mathcal{H}'}})^{\mathcal{H}/\mathcal{H}'}, \tag{8.3.6.21}$$

where $\mathscr{I}_{\mathsf{D}_{\infty,\mathcal{H}}}$ (resp. $\mathscr{I}_{\mathsf{D}_{\infty,\mathcal{H}'}}$) is defined over $\mathsf{M}^{\mathrm{tor}}_{\mathcal{H},\Sigma}$ (resp. $\mathsf{M}^{\mathrm{tor}}_{\mathcal{H}',\Sigma'}$) as in Definition 1.4.2.9, because $\mathsf{D}_{\infty,\mathcal{H}}$ and $\mathsf{D}_{\infty,\mathcal{H}'}$ are reduced subschemes defining normal crossings divisors, and $\mathsf{D}_{\infty,\mathcal{H}'} = \big(([1]^{\mathrm{tor}})^{-1}(\mathsf{D}_{\infty,\mathcal{H}})\big)_{\mathrm{red}}$. Since R is flat over R_0, we have induced isomorphisms

$$\mathscr{O}_{\mathsf{M}^{\mathrm{tor}}_{\mathcal{H},\Sigma}\underset{R_0}{\otimes} R_{\mathbb{Q}}} \xrightarrow{\sim} \big([1]^{\mathrm{tor}}_{*}\mathscr{O}_{\mathsf{M}^{\mathrm{tor}}_{\mathcal{H}',\Sigma'}\underset{R_0}{\otimes} R_{\mathbb{Q}}}\big)^{\mathcal{H}/\mathcal{H}'} \tag{8.3.6.22}$$

and

$$\mathscr{I}_{\mathsf{D}_{\infty,\mathcal{H}}\underset{R_0}{\otimes} R_{\mathbb{Q}}} \xrightarrow{\sim} \big([1]^{\mathrm{tor}}_{*}\mathscr{I}_{\mathsf{D}_{\infty,\mathcal{H}'}\underset{R_0}{\otimes} R_{\mathbb{Q}}}\big)^{\mathcal{H}/\mathcal{H}'}, \tag{8.3.6.23}$$

Since $\mathcal{E}^{\mathrm{can}}_{\mathrm{P}^{\mathrm{ord}}_{\mathtt{D},0},R_{\mathbb{Q}}}(W_{\mathbb{Q}})$ is locally free over $\mathsf{M}^{\mathrm{tor}}_{\mathcal{H},\Sigma}\underset{R_0}{\otimes} R_{\mathbb{Q}}$, the desired isomorphism (8.3.6.16) (resp. (8.3.6.17)) follows from (8.3.6.22) (resp. (8.3.6.23)), from (1.4.3.3) (with $g = 1$ and $\mathcal{H}' = \mathcal{H}$ there), and from the projection formula [35, 0_{I}, 5.4.10.1], as desired.

The analogous statements for $\mathrm{P}^{\mathrm{ord}}_{\mathtt{D},0}$ can be similarly proved. □

Bibliography

[1] F. Andreatta, A. Iovita and V. Pilloni, p-adic families of Siegel modular cuspforms, *Ann. Math. (2)* **181**, 2, pp. 623–697 (2015).

[2] M. Artin, Algebraic approximation of structures over complete local rings, *Publ. Math. Inst. Hautes Étud. Sci.* **36**, pp. 23–58 (1969a).

[3] M. Artin, Algebraization of formal moduli: I, in [96], pp. 21–71 (1969b).

[4] A. Ash, D. Mumford, M. Rapoport and Y. Tai, *Smooth Compactification of Locally Symmetric Varieties, Lie Groups: History Frontiers and Applications*, Vol. 4. Math Sci Press, Brookline, Massachusetts (1975).

[5] W. L. Baily, Jr. and A. Borel, Compactification of arithmetic quotients of bounded symmetric domains, *Ann. Math. (2)* **84**, 3, pp. 442–528 (1966).

[6] P. Berthelot, L. Breen and W. Messsing, *Théorie de Dieudonné Cristalline II, Lecture Notes in Mathematics*, Vol. 930. Springer-Verlag, Berlin, Heidelberg, New York (1982).

[7] P. Berthelot, A. Grothendieck and L. Illusie (eds.), *Théorie des intersections et théorème de Riemann–Roch (SGA 6), Lecture Notes in Mathematics*, Vol. 225. Springer-Verlag, Berlin, Heidelberg, New York (1971).

[8] A. Borel and W. Casselman (eds.), *Automorphic Forms, Representations and L-Functions, Proceedings of Symposia in Pure Mathematics*, Vol. 33, Part 2, held at Oregon State University, Corvallis, Oregon, July 11–August 5, 1977. American Mathematical Society, Providence, Rhode Island (1979).

[9] A. Borel and L. Ji, *Compactifications of symmetric and locally symmetric spaces*, Mathematics: Theory & Applications. Birkhäuser, Boston (2006).

[10] S. Bosch, W. Lütkebohmert and M. Raybaud, *Néron Models, Ergebnisse der Mathematik und ihrer Grenzgebiete, 3. Folge*, Vol. 21. Springer-Verlag, Berlin, Heidelberg, New York (1990).

[11] K. S. Brown, *Cohomology of Groups, Graduate Texts in Mathematics*, Vol. 87. Springer-Verlag, Berlin, Heidelberg, New York (1982).

[12] F. Bruhat and J. Tits, Groupes réductifs sur un corps local: I. Données radicielles valuées, *Publ. Math. Inst. Hautes Étud. Sci.* **41**, pp. 5–251 (1972).

[13] H. Carayol, Sur la mauvaise réduction des courbes de Shimura, *Compositio Math.* **59**, 2, pp. 151–230 (1986).

[14] W. Casselman, The unramified principal series of p-adic groups. I. The spherical function, *Compositio Math.* **40**, 3, pp. 387–406 (1980).

[15] C.-L. Chai, Arithmetic minimal compactification of the Hilbert–Blumenthal moduli spaces, *Ann. Math. (2)* **131**, pp. 541–554 (1990), appendix to "The Iwasawa conjecture for totally real fields" by A. Wiles.
[16] V. Cristante and W. Messing (eds.), *Barsottie Symposium in Algebraic Geometry, Perspectives in Mathematics*, Vol. 15. Academic Press Inc., Boston (1994).
[17] A. J. de Jong, The moduli spaces of polarized abelian varieties, *Math. Ann.* **295**, pp. 485–503 (1993).
[18] A. J. de Jong and F. Oort, On extending families of curves, *J. Algebraic Geom.* **6**, pp. 545–562 (1997).
[19] P. Deligne, Travaux de Shimura, in *Séminaire Bourbaki, exposé n° 389* (*février 1971*), *Lecture Notes in Mathematics*, Vol. 244. Springer-Verlag, Berlin, Heidelberg, New York, pp. 123–165 (1971).
[20] P. Deligne, Théorie de Hodge: III, *Publ. Math. Inst. Hautes Étud. Sci.* **44**, pp. 5–77 (1974).
[21] P. Deligne, Variétés de Shimura: Interprétation modulaire, et techniques de construction de modèles canoniques, in [8], pp. 247–290 (1979).
[22] P. Deligne and W. Kuyk (eds.), *Modular Functions of One Variable II*, *Lecture Notes in Mathematics*, Vol. 349. Springer-Verlag, Berlin, Heidelberg, New York (1973).
[23] P. Deligne and G. Pappas, Singularités des espaces de modules de Hilbert, en les caractéristiques divisant le discriminant, *Compositio Math.* **90**, pp. 59–79 (1994).
[24] P. Deligne and M. Rapoport, Les schémas de modules de courbes elliptiques, in [22], pp. 143–316 (1973).
[25] M. Demazure and A. Grothendieck (eds.), *Schémas en groupes* (*SGA 3*)*, I: Propriétés générales des schémas en groupes*, *Lecture Notes in Mathematics*, Vol. 151. Springer-Verlag, Berlin, Heidelberg, New York (1970a).
[26] M. Demazure and A. Grothendieck (eds.), *Schémas en groupes* (*SGA 3*)*, II: Groupes de type multiplicatif, et structure des schémas en groupes généraux*, *Lecture Notes in Mathematics*, Vol. 152. Springer-Verlag, Berlin, Heidelberg, New York (1970b).
[27] H. Esnault and E. Viehweg, *Lectures on Vanishing Theorems*, *DMV Seminar*, Vol. 20. Birkhäuser, Boston (1992).
[28] G. Faltings and C.-L. Chai, *Degeneration of Abelian Varieties*, *Ergebnisse der Mathematik und ihrer Grenzgebiete, 3. Folge*, Vol. 22. Springer-Verlag, Berlin, Heidelberg, New York (1990).
[29] J. Giraud, L. Illusie and M. Raynaud (eds.), *Surfaces Algébriques. Séminaire de Géométrie Algébrique d'Orsay 1976–78*, *Lecture Notes in Mathematics*, Vol. 868. Springer-Verlag, Berlin, Heidelberg, New York (1981).
[30] R. Godement, *Topologie algébrique et théorie des faisceaux*, *Actualités scientifiques et industrielles*, Vol. 1252. Hermann, Paris (1964).
[31] H. Grauert and O. Riemenschneider, Verschwindungssätze für analytische Kohomologiegruppen auf komplexen Räumen, *Invent. Math.* **11**, pp. 263–292 (1970).
[32] A. Grothendieck (ed.), *Cohomologie locale des faiseaux cohérents et théorèmes de Lefschetz locaux et globaux* (*SGA 2*). North-Holland Publishing Co., Amsterdam (1968).
[33] A. Grothendieck (ed.), *Revêtements étales et groupe fondamental* (*SGA 1*), *Lecture Notes in Mathematics*, Vol. 224. Springer-Verlag, Berlin, Heidelberg, New York (1971).

[34] A. Grothendieck (ed.), *Groupes de monodromie en géométrie algébrique (SGA 7 I)*, *Lecture Notes in Mathematics*, Vol. 288. Springer-Verlag, Berlin, Heidelberg, New York (1972).

[35] A. Grothendieck and J. Dieudonné, *Eléments de géométrie algébrique*, *Publications mathématiques de l'I.H.E.S.*, Vol. 4, 8, 11, 17, 20, 24, 28, 32. Institut des Hautes Etudes Scientifiques, Paris (1960, 1961, 1961, 1963, 1964, 1965, 1966, 1967).

[36] A. Grothendieck and J. Dieudonné, *Eléments de géométrie algébrique I: Le langage des schémas*, *Grundlehren der mathematischen Wissenschaften*, Vol. 166. Springer-Verlag, Berlin, Heidelberg, New York (1971).

[37] M. Hakim, *Topos annelés et schémas relatifs*, *Ergebnisse der Mathematik und ihrer Grenzgebiete*, Vol. 64. Springer-Verlag, Berlin, Heidelberg, New York (1972).

[38] M. Harris, Functorial properties of toroidal compactifications of locally symmetric varieties, *Proc. London Math. Soc. (3)* **59**, pp. 1–22 (1989).

[39] M. Harris, K.-W. Lan, R. Taylor and J. Thorne, On the rigid cohomology of certain Shimura varieties, *Res. Math. Sci.* **3**, 37, pp. 1–308 (2016).

[40] M. Harris and S. Zucker, Boundary cohomology of Shimura varieties I. — Coherent cohomology on toroidal compactifications, *Ann. Sci. Ecole Norm. Sup. (4)* **27**, pp. 249–344 (1994).

[41] H. Hida, *p-Adic Automorphic Forms on Shimura Varieties*, Springer Monographs in Mathematics. Springer-Verlag, Berlin, Heidelberg, New York (2004).

[42] J.-I. Igusa (ed.), *Algebraic Analysis, Geometry, And Number Theory*, Proceedings of the JAMI Inaugural Conference. The Johns Hopkins University Press, Baltimore (1989).

[43] L. Illusie, Logarithmic spaces (according to K. Kato), in [16], pp. 183–203 (1994).

[44] L. Ji, Y. S. Poon, L. Yang and S.-T. Yau (eds.), *Fifth International Congress of Chinese Mathematicians*, *AMS/IP Studies in Advanced Mathematics*, Vol. 51, part 1, Proceedings of the Fifth International Congress of Chinese Mathematicians held at Tsinghua University, Beijing, in December 2010. American Mathematical Society, Providence, Rhode Island, and International Press, Cambridge, MA (2012).

[45] K. Kato, Logarithmic structures of Fontaine–Illusie, in [42], pp. 191–224 (1989).

[46] N. Katz, p-adic properties of modular schemes and modular forms, in [56], pp. 70–189 (1973).

[47] N. Katz, Serre–Tate local moduli, in [29], pp. 138–202 (1981).

[48] N. M. Katz, Algebraic solutions of differential equations (p-curvature and the Hodge filtration), *Invent. Math.* **18**, pp. 1–118 (1972).

[49] N. M. Katz and B. Mazur, *Arithmetic Moduli of Elliptic Curves*, no. 108 in Annals of Mathematics Studies. Princeton University Press, Princeton (1985).

[50] G. Kempf, F. Knudsen, D. Mumford and B. Saint-Donat, *Toroidal Embeddings I*, *Lecture Notes in Mathematics*, Vol. 339. Springer-Verlag, Berlin, Heidelberg, New York (1973).

[51] M. Kisin, Integral models for Shimura varieties of abelian type, *J. Amer. Math. Soc.* **23**, 4, pp. 967–1012 (2010).

[52] D. Knutson, *Algebraic Spaces*, *Lecture Notes in Mathematics*, Vol. 203. Springer-Verlag, Berlin, Heidelberg, New York (1971).

[53] R. E. Kottwitz, Points on some Shimura varieties over finite fields, *J. Amer. Math. Soc.* **5**, 2, pp. 373–444 (1992).

[54] E. Kunz, Characterizations of regular local rings of characteristic p, *Amer. J. Math.* **91**, 3, pp. 773–784 (1969).
[55] E. Kunz, *Kähler Differentials*, Advanced Lectures in Mathematics. Friedr. Vieweg & Sohn, Braunschweig (1986).
[56] W. Kuyk and J.-P. Serre (eds.), *Modular Functions of One Variable III*, *Lecture Notes in Mathematics*, Vol. 350. Springer-Verlag, Berlin, Heidelberg, New York (1973).
[57] K.-W. Lan, Arithmetic compactification of PEL-type Shimura varieties, Ph. D. Thesis, Harvard University, Cambridge, Massachusetts (2008), errata and revision available online at the author's website.
[58] K.-W. Lan, Elevators for degenerations of PEL structures, *Math. Res. Lett.* **18**, 5, pp. 889–907 (2011), errata available online at the author's website.
[59] K.-W. Lan, Comparison between analytic and algebraic constructions of toroidal compactifications of PEL-type Shimura varieties, *J. Reine Angew. Math.* **664**, pp. 163–228 (2012a), errata available online at the author's website.
[60] K.-W. Lan, Geometric modular forms and the cohomology of torsion automorphic sheaves, in [44], pp. 183–208 (2012b).
[61] K.-W. Lan, Toroidal compactifications of PEL-type Kuga families, *Algebra Number Theory* **6**, 5, pp. 885–966 (2012c), errata available online at the author's website.
[62] K.-W. Lan, *Arithmetic compactification of PEL-type Shimura varieties*, *London Mathematical Society Monographs*, Vol. 36. Princeton University Press, Princeton (2013), errata available online at the author's website.
[63] K.-W. Lan, Boundary strata of connected components in positive characteristics, *Algebra Number Theory* **9**, 9, pp. 1955–2054 (2015a), an appendix to the article "Families of nearly ordinary Eisenstein series on unitary groups" by Xin Wan.
[64] K.-W. Lan, Compactifications of splitting models of PEL-type Shimura varieties, preprint (2015b), to appear in Trans. Amer. Math. Soc.
[65] K.-W. Lan, Compactifications of PEL-type Shimura varieties in ramified characteristics, *Forum Math. Sigma* **4**, e1, pp. 1–98 (2016a).
[66] K.-W. Lan, Higher Koecher's principle, *Math. Res. Lett.* **23**, 1, pp. 163–199 (2016b).
[67] K.-W. Lan, Vanishing theorems for coherent automorphic cohomology, *Res. Math. Sci.* **3**, 39, pp. 1–43 (2016c).
[68] K.-W. Lan, Integral models of toroidal compactifications with projective cone decompositions, *Int. Math. Res. Not. IMRN* **2017**, 11, pp. 3237–3280 (2017).
[69] K.-W. Lan and B. Stroh, Relative cohomology of cuspidal forms on PEL-type Shimura varieties, *Algebra Number Theory* **8**, 8, pp. 1787–1799 (2014).
[70] K.-W. Lan and J. Suh, Vanishing theorems for torsion automorphic sheaves on compact PEL-type Shimura varieties, *Duke Math. J.* **161**, 6, pp. 1113–1170 (2012).
[71] K.-W. Lan and J. Suh, Vanishing theorems for torsion automorphic sheaves on general PEL-type Shimura varieties, *Adv. Math.* **242**, pp. 228–286 (2013).
[72] R. P. Langlands and D. Ramakrishnan (eds.), *The Zeta Functions of Picard Modular Surfaces*, based on lectures delivered at a CRM Workshop in the spring of 1988. Les Publications CRM, Montréal (1992).
[73] J. Lubin, J.-P. Serre and J. Tate, Elliptic curves and formal groups, Summer Institute on Algebraic Geometry, Woods Hole (1964), available at `http://www.ma.utexas.edu/users/voloch/lst.html`.

[74] K. Madapusi Pera, Toroidal compactifications of integral models of Shimura varieties of Hodge type, preprint (2015).

[75] Y. I. Manin, The theory of commutative formal groups over fields of finite characteristic, *Russian Math. Surveys* **18**, pp. 1–83 (1963).

[76] H. Matsumura, *Commutative Algebra*, 2nd edn., Mathematics Lecture Note Series. The Benjamin/Cummings Publishing Company, Inc. (1980).

[77] H. Matsumura, *Commutative Ring Theory*, *Cambridge Studies in Advanced Mathematics*, Vol. 8. Cambridge University Press, Cambridge, New York (1986).

[78] J. S. Milne, The points on a Shimura variety modulo a prime of good reduction, in [72], pp. 151–253 (1992).

[79] B. Moonen, Models of Shimura varieties in mixed characteristic, in [94], pp. 267–350 (1998).

[80] L. Moret-Bailly, *Pinceaux de variétés abéliennes*, *Astérisque*, Vol. 129. Société Mathématique de France, Paris (1985).

[81] D. Mumford, *Abelian Varieties*, *Tata Institute of Fundamental Research Studies in Mathematics*, Vol. 5. Oxford University Press, Oxford (1970), with appendices by C. P. Ramanujam and Yuri Manin.

[82] D. Mumford, An analytic construction of degenerating abelian varieties over complete rings, *Compositio Math.* **24**, 3, pp. 239–272 (1972).

[83] D. Mumford, J. Fogarty and F. Kirwan, *Geometric Invariant Theory*, *Ergebnisse der Mathematik und ihrer Grenzgebiete*, Vol. 34, 3rd edn. Springer-Verlag, Berlin, Heidelberg, New York (1994).

[84] A. Ogus, *F-Crystals, Griffiths Transversality, and the Hodge Decomposition*, no. 221 in Astérisque. Société Mathématique de France, Paris (1994).

[85] G. Pappas and M. Rapoport, Local models in the ramified case, I. The EL-case, *J. Algebraic Geom.* **12**, 1, pp. 107–145 (2003).

[86] G. Pappas and M. Rapoport, Local models in the ramified case, II. Splitting models, *Duke Math. J.* **127**, 2, pp. 193–250 (2005).

[87] G. Pappas and M. Rapoport, Local models in the ramified case, III. Unitary groups, *J. Inst. Math. Jussieu* **8**, 3, pp. 507–564 (2009).

[88] G. Pappas and X. Zhu, Local models of Shimura varieties and a conjecture of Kottwitz, *Invent. Math.* **194**, pp. 147–254 (2013).

[89] R. Pink, *Arithmetic compactification of mixed Shimura varieties*, Ph.D. thesis, Rheinischen Friedrich-Wilhelms-Universität, Bonn (1989).

[90] M. Rapoport, Compactifications de l'espace de modules de Hilbert–Blumenthal, *Compositio Math.* **36**, 3, pp. 255–335 (1978).

[91] M. Rapoport and T. Zink, *Period Spaces for p-divisible Groups*, no. 141 in Annals of Mathematics Studies. Princeton University Press, Princeton (1996).

[92] M. Raynaud, *Faisceaux amples sur les schémas en groupes et les espaces homogènes*, *Lecture Notes in Mathematics*, Vol. 119. Springer-Verlag, Berlin, Heidelberg, New York (1970).

[93] I. Reiner, *Maximal Orders*, *London Mathematical Society Monographs. New Series*, Vol. 28. Clarendon Press, Oxford University Press, Oxford (1975).

[94] A. J. Scholl and R. L. Taylor, *Galois Representations in Arithmetic Algebraic Geometry*, *London Mathematical Society Lecture Note Series*, Vol. 254. Cambridge University Press, Cambridge, New York (1998).

[95] J.-P. Serre, Geométrie algébrique et géométrie analytique, *Ann. Inst. Fourier. Grenoble* **6**, pp. 1–42 (1955–1956).

[96] D. C. Spencer and S. Iyanaga (eds.), *Global Analysis. Papers in Honor of K. Kodaira*. Princeton University Press, Princeton (1969).

[97] B. Stroh, *Compactification de variétés de Siegel aux places de mauvaise réduction*, Ph.D. thesis, Université Henri Poincaré, Nancy, France (2008).

[98] B. Stroh, Compactification de variétés de Siegel aux places de mauvaise réduction, *Bull. Soc. Math. France* **138**, 2, pp. 259–315 (2010a).

[99] B. Stroh, Compactification minimale et mauvaise réduction, *Ann. Inst. Fourier. Grenoble* **60**, 3, pp. 1035–1055 (2010b).

[100] A. Vasiu, Integral canonical models of Shimura varieties of preabelian type, *Asian J. Mathematics* **3**, pp. 401–518 (1999).

[101] A. Vasiu and T. Zink, Purity results for p-divisible groups and abelian schemes over regular bases of mixed characteristic, *Documenta Math.* **15**, pp. 571–599 (2010).

[102] T. Wedhorn, Ordinariness in good reductions of Shimura varieties of PEL-type, *Ann. Sci. Ecole Norm. Sup. (4)* **32**, pp. 575–618 (1999).

[103] C. A. Weibel, *An Introduction to Homological Algebra*, *Cambridge Studies in Advanced Mathematics*, Vol. 38. Cambridge University Press, Cambridge, New York, Melbourne (1994).

[104] Y. G. Zarhin, Endomorphisms of abelian varieties and points of finite order in characteristic p, *Math. Notes* **21**, 6, pp. 415–419 (1977).

Index

$(\cdot)_S^{\mathrm{mult}}$, 156
$(\cdot)^{\mathrm{mult}}$, 156
$(\cdot,\cdot)_{\mathrm{aux}}$, 125
$(\cdot,\cdot)^*_{\mathrm{aux}}$, 126, 133
$(\!|\cdot,\cdot|\!)$, 34, 267
$[\cdot]$, 26
$^\circ$, 157
$\langle\cdot,\cdot\rangle$, 22, 34
$\langle\cdot,\cdot\rangle_{00,n}$, 78
$\langle\cdot,\cdot\rangle_{00,n_0}$, 232
$\langle\cdot,\cdot\rangle_{02}$, 219, 223
$\langle\cdot,\cdot\rangle_{10,n}$, 68
$\langle\cdot,\cdot\rangle_{10,n_0}$, 232
$\langle\cdot,\cdot\rangle^{\mathrm{ord}}_{10,p^r}$, 223, 232
$(\langle\cdot,\cdot\rangle^{\mathrm{ord}}_{10,p^r})^{\mathrm{mult}}_\eta$, 229, 232
$\langle\cdot,\cdot\rangle_{11}$, 30, 219
$\langle\cdot,\cdot\rangle_{11,\mathbb{R}}$, 30
$\langle\cdot,\cdot\rangle_{20}$, 28
$\langle\cdot,\cdot\rangle^{(g)}_{20}$, 40
$(\langle\cdot,\cdot\rangle_{20,n})^{\mathrm{mult}}_\eta$, 232
$\langle\cdot,\cdot\rangle_M$, 479
$\langle\cdot,\cdot\rangle_Q$, 43
$\langle\cdot,\cdot\rangle_{\mathrm{aux}}$, 126
$\langle\cdot,\cdot\rangle_{\mathrm{can.}}$, 118
$\langle\cdot,\cdot\rangle^{(g)}$, 39
$\langle\cdot,\cdot\rangle_{ij,n}$, 231
$\langle\cdot,\cdot\rangle_\lambda$, 24, 482
$\langle\cdot,\cdot\rangle_{\lambda_{\mathsf{M}_{\mathcal{H}},j_Q}}$, 101
$\langle\cdot,\cdot\rangle_{\lambda_{\vec{\mathsf{M}}^{\mathrm{ord}}_{\mathcal{H}}}}$, 486
$\langle\cdot,\cdot\rangle_{\lambda_{\vec{\mathsf{M}}^{\mathrm{ord}}_{\mathcal{H}},j_Q}}$, 397, 486
$\langle\cdot,\cdot\rangle_{\lambda_{\vec{\mathsf{M}}^{\mathrm{ord}}_{\mathcal{H}},j_Q}}\underset{\mathbb{Z}}{\otimes}\mathbb{Q}$, 397
$\langle\cdot,\cdot\rangle^{\mathrm{can}}_\lambda$, 121, 485
$\langle\cdot,\cdot\rangle_{\lambda^\natural}$, 502
$\langle\cdot,\cdot\rangle_\phi$, 28
$\langle\cdot,\cdot\rangle_{\phi_{\mathtt{D},0}}$, 478
$\langle\cdot,\cdot\rangle^\sim$, 44
$\sim_{\mathbb{Q}^\times\text{-isog.}}$, 72, 75
$\sim_{\mathbb{Z}^\times_{(p)}\text{-isog.}}$, 318, 323
$\sqrt{-1}$, 21
$\star$, 22
$\star_{\mathrm{aux}}$, 128
$\succ$, 53, 54, 99, 108, 113, 374, 411, 416
$[(0,0,\{0\})]$-stratum, 291
[0]-stratum, 291, 294
(1) (Tate twist), 21
$[\vec{1}]$, 521
$[\vec{1}]^{\mathrm{min}}$, 151, 521
$[\vec{1}]^{\mathrm{ord}}$, 211, 402
$[\vec{1}]^{\mathrm{ord,min}}$, 357, 359
$[\vec{1}]^{\mathrm{ord,tor}}$, 301, 304, 402
$[1]^{\mathrm{tor}}$, 3, 522

A, 23, 197
a, 139
a_0, 139
a_1, 125
a_2, 125
$\vec{A}$, 141
$\vec{A}_{\mathrm{aux}}$, 141
$A^\triangledown_{\mathrm{aux}}$, 126, 129, 199
$A^\vartriangle_{\mathrm{aux}}$, 126, 199
$A^{\vee,(p^i)}$, 180
$\mathcal{A}_{g_{\mathrm{aux}},d_{\mathrm{aux}}}$, 126, 142
$\mathcal{A}_{g_{\mathrm{aux}},d_{\mathrm{aux}},n}$, 127
$\mathcal{A}_{g,d}$, 126, 142
$\mathcal{A}_{g,d,n,n}$, 127
$A^{(p)}$, 190

$A^{(p^i)}$, 180
$A_s^\vee[p^\infty]$, 164
$A_s^\vee[p^\infty]^{\text{ét}}$, 164
$A_s^\vee[p^\infty]^{\text{mult}}$, 164
$A_s[p^\infty]$, 164
$A_s[p^\infty]^{\text{ét}}$, 164
$A_s[p^\infty]^{\text{mult}}$, 164
abelian scheme
  ordinary, *see* ordinary abelian scheme
absolute Frobenius morphism, 181, 190, 211, 304, 359, 362, 402
  flat, 212, 305
  not flat, 359
admissible boundary component, 35
admissible filtration, 27
admissible radical, 34
admissible rational polyhedral cone decomposition, 35, 52, 455
  projective, 35
  smooth, 35
    compatible choice, *see* compatible choice of admissible smooth rational polyhedral cone decomposition data
admissible submodule, 32, 34
admissible surjection, 32
affine toroidal embedding, 81, 89, 267, 326, 327, 383, 410
algebraic construction, 4, 8
$\vec{\alpha}_{\mathrm{G}_{\mathrm{aux}}(\hat{\mathbb{Z}}^p)}$, 141
$\vec{\alpha}_{\mathcal{H}}$, 141
$\overline{\alpha}_{\mathcal{H}}$, 422
$\overline{\alpha}_{\mathcal{H}^p}$, 422
$\overline{\alpha}^{\mathrm{ord}}_{\mathcal{H}_p}$, 422
$\alpha^{\mathrm{ord}}_{\mathrm{G}(\mathbb{Z}_p)}$, 212, 307
$\alpha_{\mathcal{H}}$, 23, 57, 58, 192
$[\alpha^\natural_{\mathcal{H}_{\mathrm{aux}}}]$, 138
$\alpha_{\mathcal{H}^p_{\mathrm{aux}}}$, 333
$\alpha^{\mathrm{ord}}_{\mathcal{H}_{\mathrm{aux},p}}$, 333
$\alpha^{\mathrm{ord},\triangledown}_{\mathcal{H}_{\mathrm{aux},p}}$, 200, 333
$\alpha^{\triangledown}_{\mathcal{H}^p_{\mathrm{aux}}}$, 130, 200, 333
$\alpha^{\triangledown}_{\mathcal{H}_{\mathrm{aux}}}$, 129, 130
$\alpha^\dagger_{\mathcal{H}}$, 59, 145
$\alpha^\ddagger_{\mathcal{H}}$, 59
${}^\diamond\alpha_{\mathcal{H}}$, 280
${}^\heartsuit\alpha_{\mathcal{H}}$, 59, 270, 271
$\alpha_{\mathcal{H}_\kappa}$, 107, 408
$\alpha^\ddagger_{\mathcal{H}_\kappa}$, 112
$[\alpha^{\natural,\ddagger}_{\mathcal{H}_\kappa}]$, 112
$[\alpha^{\natural,\mathrm{ord},\ddagger}_{\mathcal{H}_\kappa}]$, 415
$\alpha_{\mathcal{H}^p_\kappa}$, 408
$\alpha^\ddagger_{\mathcal{H}^p_\kappa}$, 415
$\alpha^{\mathrm{ord}}_{\mathcal{H}_{\kappa,p}}$, 408
$\alpha^{\mathrm{ord},\ddagger}_{\mathcal{H}_{\kappa,p}}$, 415
α_{H_n}, 192
$\alpha_{H_{n_0}}$, 173, 192, 237
$\alpha^{\natural,\mathrm{ord}}_{H_{n_0p^r}}$, 241
$\alpha^{\natural,\mathrm{ord}}_{H_n}$, 240
$[\alpha^\natural_{\mathcal{H}}]$, 138, 281
$\alpha^\natural_{\mathcal{H}}$, 243, 281
$[\alpha^{\natural,\ddagger}_{\mathcal{H}}]$, 59
$[\alpha^{\natural,\mathrm{ord}}_{\mathcal{H}}]$, 242, 281
$\alpha^{\natural,\mathrm{ord}}_{\mathcal{H}}$, 241, 281
$[\alpha^{\natural,\mathrm{ord},\ddagger}_{\mathcal{H}}]$, 292, 297
$\alpha_{\mathcal{H}^p}$, 174, 192, 197, 206, 237, 241, 295
  lifting, 174
$\alpha^\dagger_{\mathcal{H}^p}$, 296
$\alpha^\ddagger_{\mathcal{H}^p}$, 291, 296
${}^\diamond\alpha_{\mathcal{H}^p}$, 280
${}^\heartsuit\alpha_{\mathcal{H}^p}$, 270, 271, 296
$\alpha^{\mathrm{ord}}_{\mathcal{H}_p}$, 180, 193, 198, 206, 240, 241, 295
  lifting, 182
$\alpha^{\mathrm{ord},0}_{\mathcal{H}_p}$, 180
$\alpha^{\mathrm{ord},\dagger}_{\mathcal{H}_p}$, 296
$\alpha^{\mathrm{ord},\ddagger}_{\mathcal{H}_p}$, 291, 296
${}^\diamond\alpha^{\mathrm{ord}}_{\mathcal{H}_p}$, 280
${}^\heartsuit\alpha^{\mathrm{ord}}_{\mathcal{H}_p}$, 270, 271, 296
$\alpha^{\mathrm{ord},\#,0}_{\mathcal{H}_p}$, 180
$\alpha_{H_{p^r}}$, 192
$\alpha^{\mathrm{ord}}_{H^{\mathrm{ord}}_{p^r}}$, 179, 193, 240
$\alpha^{\mathrm{ord},0}_{H^{\mathrm{ord}}_{p^r}}$, 179
$\alpha^{\mathrm{ord},\#,0}_{H^{\mathrm{ord}}_{p^r}}$, 179
$\hat{\alpha}$, 165
$\widehat{\alpha}_{\widehat{\mathcal{H}}}$, 108, 412
$\widehat{\alpha}^\ddagger_{\widehat{\mathcal{H}}}$, 109, 413
$[\widehat{\alpha}^{\natural,\ddagger}_{\widehat{\mathcal{H}}}]$, 112
$\widehat{\alpha}^{\natural,\ddagger}_{\widehat{\mathcal{H}}}$, 112, 414
$[\widehat{\alpha}^{\natural,\mathrm{ord},\ddagger}_{\widehat{\mathcal{H}}}]$, 414
$\widehat{\alpha}^{\natural,\mathrm{ord},\ddagger}_{\widehat{\mathcal{H}}}$, 414
$\widehat{\alpha}_{\widehat{\mathcal{H}}^p}$, 411
$\widehat{\alpha}^\ddagger_{\widehat{\mathcal{H}}^p}$, 412
$\widehat{\alpha}^{\mathrm{ord}}_{\widehat{\mathcal{H}}_p}$, 411
$\widehat{\alpha}^{\mathrm{ord},\ddagger}_{\widehat{\mathcal{H}}_p}$, 412

$\hat{\alpha}_p$, 165
$\hat{\alpha}_{\bar{s}}$, 130, 193
$\hat{\alpha}_{\bar{s}}^{\mathrm{ord}}$, 182, 193, 199
$\hat{\alpha}_{\bar{s}}^{\mathrm{ord},0}$, 182, 193, 199
$\hat{\alpha}_{\bar{s}}^{\mathrm{ord},\#,0}$, 182, 193, 199
$\hat{\alpha}_{\bar{s},p}$, 193
$\hat{\alpha}_{\bar{s}}^{p}$, 174, 193, 199
$[\hat{\alpha}_{\bar{s}}^{p}]_{\mathcal{H}^p}$, 174
$\hat{\alpha}_{\bar{s}}^{\triangledown}$, 130
$\hat{\alpha}_{\bar{s}}^{\triangledown,p}$, 130, 200
$[\hat{\alpha}_{\bar{s}}^{\triangledown,p}]_{\mathcal{H}_{\mathrm{aux}}^p}$, 200
α_{n_0}, 173, 198, 204, 234
$(\alpha_{n_0}, \nu_{n_0})$, 172
$[\alpha_n^{\natural,\mathrm{ord}}]$, 234
$\alpha_n^{\natural,\mathrm{ord}}$, 231
  equivalence class, 234
α_{p^r}, 193
$\alpha_{p^r}^{\mathrm{ord}}$, 178, 179, 193, 198, 200, 204, 217, 229, 234
  compatible with degeneration, 217
$\alpha_{p^r}^{\mathrm{ord},0}$, 178, 179, 193, 206, 217, 229
$\alpha_{p^r}^{\mathrm{ord},0,\natural}$, 220, 230
$\alpha_{p^r}^{\mathrm{ord},0,\triangledown}$, 200
$\alpha_{p^r}^{\mathrm{ord},\#,0}$, 178, 179, 193, 206, 217, 229, 230
$\alpha_{p^r}^{\mathrm{ord},\#,0,\natural}$, 220, 230
$\alpha_{p^r}^{\mathrm{ord},\#,0,\triangledown}$, 200
$\alpha_{p^r}^{\mathrm{ord},\triangledown}$, 200
$\alpha_{p^r}^{\#,\mathrm{mult}}$, 193
$\alpha_{p^r,S}^{\mathrm{ord},0}$, 206, 220, 221
$\alpha_{p^r,S}^{\mathrm{ord},0,\natural}$, 220
$\alpha_{p^r,S}^{\mathrm{ord},\#,0}$, 220
$\alpha_{p^r}^{\mathrm{ord},\#,0}$, 206, 221
$\alpha_{p^r,S}^{\mathrm{ord},\#,0,\natural}$, 220
$\widetilde{\alpha}_{\widetilde{\mathcal{H}}}$, 108, 412
$\widetilde{\alpha}_{\widetilde{\mathcal{H}}}^{\ddagger}$, 109, 412
$[\widetilde{\alpha}_{\widetilde{\mathcal{H}}}^{\natural,\ddagger}]$, 110
$\widetilde{\alpha}_{\widetilde{\mathcal{H}}}^{\natural,\ddagger}$, 110, 413
$[\widetilde{\alpha}_{\widetilde{\mathcal{H}}}^{\natural,\mathrm{ord},\ddagger}]$, 413
$\widetilde{\alpha}_{\widetilde{\mathcal{H}}}^{\natural,\mathrm{ord},\ddagger}$, 413
$\widetilde{\alpha}_{\widetilde{\mathcal{H}}^p}$, 411
$\widetilde{\alpha}_{\widetilde{\mathcal{H}}^p}^{\ddagger}$, 412
$\widetilde{\alpha}_{\widetilde{\mathcal{H}}_p}^{\mathrm{ord}}$, 411
$\widetilde{\alpha}_{\widetilde{\mathcal{H}}_p}^{\mathrm{ord},\ddagger}$, 412
$\widetilde{\alpha}_{p^r}^{\mathrm{ord}}$, 426
analytic construction, 4
  comparison with, 124
Artin's approximation theory, 119, 122, 286, 482, 510
Artin's criterion, 198, 202
$\mathrm{Aut}(\bar{x})$, 343, 350
automorphic bundle, 4, 5, 8, 13, 120, 482, 510
  canonical extension, *see* canonical extension
  subcanonical extension, *see* subcanonical extension
automorphic sheaf, 120, 482, 510
auxiliary model, 9, 16, 17, 125–141, 198–202, 333–336, 425

$\underline{B}$, 268
B (abelian scheme), 66, 70, 138, 216, 248, 281, 313, 315, 454
B (homomorphism), 269
B_0, 216
$\vec{B}$, 339
B_{aux} (abelian scheme), 138
B_{aux} (homomorphism), 138
$B^{\ddagger}$ (abelian scheme), 59, 112, 292, 297, 415
$B^{\ddagger}$ (homomorphism), 60, 292, 297
$B^{\vee}$, 70, 216, 315
$B_{\eta,p^r}^{\vee,\mathrm{ord}}$, 224, 233, 245
$B_{\eta,p^r}^{\mathrm{ord}}$, 224, 233, 245
$\underline{B}(G)$, 279, 282
$\underline{B}(G^{\ddagger})$, 60, 292, 297
$\underline{B}({}^{\diamond}G)$, 281
$\widehat{B}^{\ddagger}$, 110, 413
$B_{p^r}^{\vee,\mathrm{ord}}$, 248, 379
$B_{p^r}^{\mathrm{ord}}$, 248, 379
$(B, \star, V, \langle \cdot\,, \cdot \rangle, h_0)$, 22
$\widetilde{B}^{\ddagger}$, 110, 413
bad prime, 27
balanced $\Gamma_1(p^r)$, 6, 13–15
$\beta_{\mathcal{H}}$, 80
β_{H_n}, 80
$\beta_{H_{n_0}}$, 325
$\beta_{H_{n_0}}^{\natural}$, 314
$\beta_{H_n}^{\natural}$, 69
$\beta_{\mathcal{H}}^{\natural}$, 69
$\beta_{\mathcal{H}^p}$, 325
$\beta_{\mathcal{H}^p}^{\natural}$, 314
$\beta_{\mathcal{H}_p}^{\natural,\mathrm{ord}}$, 314–316
$\beta_{\mathcal{H}_p}^{\mathrm{ord}}$, 325

$\beta^{\natural}_{\mathcal{H}^p,\vec{Z}^{\mathrm{ord}}}$, 321
$\beta^{\natural,\mathrm{ord}}_{\mathcal{H}_p,\vec{Z}^{\mathrm{ord}}}$, 321
$\beta^{\natural}_{\mathcal{H},\mathrm{Z}}$, 74
$[\hat{\beta}^{\natural}]_{\mathcal{H}_{\mathrm{G}_{1,\mathrm{Z}}}}$, 70
$[\hat{\beta}^{p,\natural}]_{\mathcal{H}^p_{\mathrm{G}_{1,\mathrm{Z}}}}$, 315, 316
$\hat{\beta}^{\natural}_{\bar{s}}$, 70, 316
$\hat{\beta}^{\natural,0}_{\bar{s}}$, 71
$[\hat{\beta}^{\natural}_{\bar{s}}]_{\mathcal{H}_{\mathrm{G}_{1,\mathrm{Z}}}}$, 70
$\hat{\beta}^{\natural,\#,0}_{\bar{s}}$, 71
$\hat{\beta}^{p,\natural,0}_{\bar{s}}$, 316
$[\hat{\beta}^{p,\natural}_{\bar{s}}]_{\mathcal{H}^p_{\mathrm{G}_{1,\mathrm{Z}}}}$, 316
$\hat{\beta}^{p,\natural,\#,0}_{\bar{s}}$, 316
β_n, 79
β^0_n, 78, 79
β_{n_0}, 324
$\beta^0_{n_0}$, 324
$\beta^{\natural}_{n_0}$, 313
$\beta^{\natural,0}_{n_0}$, 313
$\beta^{\natural,\#,0}_{n_0}$, 313
$\beta^{\#,0}_{n_0}$, 324
$\beta^{\natural}_n$, 69
$\beta^{\natural,0}_n$, 69
$\beta^{\natural,\#,0}_n$, 69
$\beta^{\#,0}_n$, 78, 79
$\beta^{\natural,\mathrm{ord}}_{p^r}$, 313, 314
$\beta^{\natural,\mathrm{ord},0}_{p^r}$, 314
$\beta^{\natural,\mathrm{ord},\#,0}_{p^r}$, 314
$\beta^{\mathrm{ord}}_{p^r}$, 324, 325
$\widetilde{\beta}^{\natural}_{\widetilde{\mathcal{H}}}$, 108
$\widetilde{\beta}^{\natural}_{\widetilde{\mathcal{H}}^p}$, 411
$\widetilde{\beta}^{\natural,\mathrm{ord}}_{\widetilde{\mathcal{H}}_p}$, 411
bottom level, 15, 17, 125
boundary moduli problem, 32
$\square$, 23

^{c} (complex conjugation), 22
c, 68, 70, 138, 216, 249, 281, 315
$\vec{C}^{\mathrm{ord}}_{\Phi_1,\delta_1}$, 262
$\vec{C}^{\mathrm{ord}}_{\Phi_{\mathcal{H}},\delta_{\mathcal{H}}}$, 260, 266, 295, 312, 314, 347, 379
$\vec{C}^{\mathrm{ord,grp}}_{\Phi_{\mathcal{H}},\delta_{\mathcal{H}}}$, 312, 379
$\vec{C}^{\mathrm{ord}}_{\Phi_{\mathcal{H}},\delta_{\mathcal{H}},r_\kappa}$, 423, 438
$\vec{C}^{\mathrm{ord}}_{\Phi_{\mathcal{H}},\delta_{\mathcal{H}},r_n}$, 312, 380
$\vec{C}^{\mathrm{ord}}_{\Phi_n,\delta_n}$, 312, 313
c_{aux}, 138
$c^{\vee}_{\mathrm{aux}}$, 138
$\breve{c}_{\widetilde{\mathcal{H}}}$, 85
$\breve{c}^{\vee}_{\widetilde{\mathcal{H}}}$, 85
$\breve{c}^{\vee,\mathrm{ord}}_{\widetilde{\mathcal{H}}}$, 379
$\breve{c}^{\mathrm{ord}}_{\widetilde{\mathcal{H}}}$, 379
$\breve{c}_n$, 85
$\breve{c}^{\vee}_n$, 85
$\breve{c}^{\vee,\mathrm{ord}}_n$, 379
$\breve{c}^{\mathrm{ord}}_n$, 379
$c^{\ddagger}$, 59, 112, 292, 297, 415
$\dddot{C}^{\mathrm{ord},\natural,\circ}_{\mathcal{H},n}$, 438
$\dddot{C}^{\mathrm{ord}}_{\Phi_{\mathcal{H}},n}$, 255
$\dddot{C}^{\mathrm{ord},\circ}_{\Phi_{\mathcal{H}},n}$, 256
$\dddot{C}^{\mathrm{ord,grp}}_{\Phi_{\mathcal{H}},n}$, 256
$\dddot{C}^{\mathrm{ord,grp},\circ}_{\Phi_{\mathcal{H}},n}$, 261
$\dddot{C}^{\mathrm{ord}}_{\Phi_n}$, 249
$\dddot{C}^{\mathrm{ord},\circ}_{\Phi_n}$, 250
$\dddot{C}^{\mathrm{ord},\circ\circ\circ}_{\Phi_n}$, 250
$c^{\vee}$, 68, 70, 138, 216, 249, 281, 315
$c^{\vee,\ddagger}$, 59, 112, 292, 297, 415
$c_{\mathcal{H}}$, 69, 243, 259, 266, 281
$c^{\vee}_{\mathcal{H}}$, 69, 243, 259, 266, 281
$c^{\vee,\mathrm{ord}}_{\mathcal{H}}$, 241, 255, 266, 270, 281, 314, 379
c_{H_n}, 243, 245
$c_{H_{n_0}}$, 236, 237
$c^{\vee}_{H_{n_0}}$, 236, 237
$c^{\vee}_{H_n}$, 243, 245
$c^{\vee,\mathrm{ord}}_{H_n}$, 240
$c^{\mathrm{ord}}_{H_n}$, 240
$c^{\mathrm{ord}}_{\mathcal{H}}$, 241, 255, 266, 270, 281, 314, 379
$c^{\vee,\mathrm{ord}}_{H^{\mathrm{ord}}_{p^r}}$, 239, 245
$c^{\mathrm{ord}}_{H^{\mathrm{ord}}_{p^r}}$, 239, 245
$\vec{\widehat{C}}^{\mathrm{ord}}_{\breve{\Phi}_{\widehat{\mathcal{H}}},\breve{\delta}_{\widehat{\mathcal{H}}}}$, 380, 394, 438
$\widehat{c}^{\ddagger}$, 110, 413
$\widehat{c}^{\vee,\ddagger}$, 110, 413
$\widehat{c}^{\ddagger}_{\widehat{\mathcal{H}}}$, 111, 414
$\widehat{c}^{\vee,\ddagger}_{\widehat{\mathcal{H}}}$, 111, 414
$\widehat{c}^{\vee,\mathrm{ord},\ddagger}_{\widehat{\mathcal{H}}}$, 414
$\widehat{c}^{\mathrm{ord},\ddagger}_{\widehat{\mathcal{H}}}$, 414
$\widehat{c}^{\ddagger}_{\widehat{\mathcal{H}}^p}$, 414
$\widehat{c}^{\vee,\ddagger}_{\widehat{\mathcal{H}}^p}$, 414
$\widetilde{c}^{\vee,\mathrm{ord}}_{\widetilde{\mathcal{H}}}$, 411
$\widetilde{c}^{\mathrm{ord}}_{\widetilde{\mathcal{H}}}$, 411
$\widehat{c}^{\ddagger}_n$, 111
$\widehat{c}^{\vee,\ddagger}_n$, 111

$\widehat{C}_{\check{\Phi}_{\widetilde{\mathcal{H}}},\check{\delta}_{\widetilde{\mathcal{H}}}}$, 86, 98
c_n, 68, 245
c_{n_0}, 233, 237, 249
$c_{n_0}^{\vee}$, 233, 237, 249
$c_n^{\ddagger}$, 111
$c_n^{\vee}$, 68, 245
$c_n^{\vee,\ddagger}$, 111
$c_n^{\vee,\mathrm{ord}}$, 233, 249, 313, 379
c_n^{ord}, 233, 249, 313, 379
$c_{p^r}^{\vee,\mathrm{ord}}$, 224, 229, 233, 239, 245, 249
$c_{p^r}^{\mathrm{ord}}$, 224, 229, 233, 239, 245, 249
$C_{\Phi_{\mathcal{H}},\delta_{\mathcal{H}}}$, 58, 62, 67, 69, 259
$C_{\Phi_{\mathcal{H}},\delta_{\mathcal{H}}}^{\mathrm{grp}}$, 67
$C_{\Phi_{\mathcal{H}},\delta_{\mathcal{H}}}^{\mathrm{ord}}$, 259
C_{Φ_n,δ_n}, 67, 68
$\overset{\rightarrow}{\widetilde{C}}{}^{\mathrm{ord}}_{\check{\Phi}_{\widetilde{\mathcal{H}}},\check{\delta}_{\widetilde{\mathcal{H}}}}$, 379, 409
$\overset{\rightarrow}{\widetilde{C}}{}^{\mathrm{ord,grp}}_{\check{\Phi}_{\widetilde{\mathcal{H}}},\check{\delta}_{\widetilde{\mathcal{H}}}}$, 379
$\overset{\rightarrow}{\widetilde{C}}{}^{\mathrm{ord}}_{\check{\Phi}_n,\check{\delta}_n}$, 380
$\overset{\rightarrow}{\widetilde{C}}{}^{\mathrm{ord}}_{\widetilde{\Phi}_{\widetilde{\mathcal{H}}},\widetilde{\delta}_{\widetilde{\mathcal{H}}}}$, 407
$\widetilde{c}^{\ddagger}$, 110, 413
$\overset{\because}{\widetilde{C}}{}^{\mathrm{ord},\natural,\circ}_{\widehat{\mathcal{H}},n}$, 438
$\overset{\because}{\widetilde{C}}{}^{\mathrm{ord,ext}}_{\widetilde{\mathcal{H}},n}$, 436
$\overset{\because}{\widetilde{C}}{}^{\mathrm{ord,ext},\circ}_{\widetilde{\mathcal{H}},n}$, 436
$\overset{\because}{\widetilde{C}}{}^{\mathrm{ord,ext}}_{n}$, 436
$\overset{\because}{\widetilde{C}}{}^{\mathrm{ord,ext},\circ}_{n}$, 436
$\overset{\because}{\widetilde{C}}{}^{\mathrm{ord,ext}}$, 434
$\overset{\because}{\widetilde{C}}{}^{\mathrm{ord,ext},\circ}$, 434
$\overset{\because}{\widetilde{C}}{}^{\mathrm{ord}}_{\check{\Phi}_{\widetilde{\mathcal{H}}},n}$, 418
$\overset{\because}{\widetilde{C}}{}^{\mathrm{ord},\circ}_{\widetilde{\Phi}_{\widetilde{\mathcal{H}},n}}$, 437
$\widetilde{c}^{\vee,\ddagger}$, 110, 413
$\widetilde{c}_{\widetilde{\mathcal{H}}}$, 109
$\widetilde{c}_{\widetilde{\mathcal{H}}}^{\ddagger}$, 110, 413
$\widetilde{c}_{\widetilde{\mathcal{H}}}^{\vee}$, 109
$\widetilde{c}_{\widetilde{\mathcal{H}}}^{\vee,\ddagger}$, 110, 413
$\widetilde{c}_{\widetilde{\mathcal{H}}}^{\vee,\mathrm{ord}}$, 414
$\widetilde{c}_{\widetilde{\mathcal{H}}}^{\vee,\mathrm{ord},\ddagger}$, 413
$\widetilde{c}_{\widetilde{\mathcal{H}}}^{\mathrm{ord}}$, 414
$\widetilde{c}_{\widetilde{\mathcal{H}}}^{\mathrm{ord},\ddagger}$, 413
$\widetilde{c}_n^{\ddagger,1}$, 111
$\widetilde{c}_n^{\vee,\ddagger,1}$, 111
$\widetilde{C}_{\check{\Phi}_{\widetilde{\mathcal{H}}},\check{\delta}_{\widetilde{\mathcal{H}}}}$, 85, 86
$\widetilde{C}^{\mathrm{grp}}_{\check{\Phi}_{\widetilde{\mathcal{H}}},\check{\delta}_{\widetilde{\mathcal{H}}}}$, 85
$\widetilde{C}_{\check{\Phi}_n,\check{\delta}_n}$, 86
$\widetilde{C}_{\widetilde{\Phi}_{\widetilde{\mathcal{H}}},\widetilde{\delta}_{\widetilde{\mathcal{H}}}}$, 106
$\widetilde{C}^{\mathrm{grp}}_{\widetilde{\Phi}_{\widetilde{\mathcal{H}}},\widetilde{\delta}_{\widetilde{\mathcal{H}}}}$, 106
canonical extension, 4, 5, 8, 13, 122, 123, 486, 488, 509, 512
  formal fiber, 501–507
canonical subgroup, 2, 13, 197
  partial, 197
Cartier dual, 162, 188, 193
coarse moduli space, 26, 61
compatible choice of admissible smooth rational polyhedral cone decomposition data, 3, 7, 37, 53, 135, 288, 373
  compatible, 135, 136
  existence, 37, 288, 289, 374, 376
  extension, 288, 289, 374, 376
  g_h-induced, 403
  g_h-refinement, 102, 398
  $\widehat{g}$-induced, 399
  $\widehat{g}$-refinement, 102, 399
  g-induced, 300
  g-refinement, 3, 65, 84, 123, 300
  induced, 136, 374
  projective, 37, 289
  refinement, 37, 288, 289, 373, 376
compatible collection of polarization functions, 4, 5, 8
compatible with D
  cusp label, 172
  Z, 172
cone, 34
  face, 34
  nondegenerate, 34
  rational polyhedral, 34
  smooth, 35
  supporting hyperplane, 34
cone decomposition, 35, 52, 455
${}^{\mathrm{conn}}$, 164
constant term, 340
convexity
  of polarization function, 35
cusp, 3, 32–33, 62
  ordinary, 7, 347
cusp label, 27, 29, 58
  assigment, 47, 85, 99
  compatible with D, 171, 172
  with cone, 36

assignment, 136
face, 38, 98, 393
$\widehat{g}$-refinement, 91, 103, 385, 400
g-refinement, 65, 84, 301, 329
induced, 136
with cone decomposition, 36
$\widehat{g}$-refinement, 91, 386
g-refinement, 84, 300, 329
refinement, 37
surjection, 36
g-assignment, 65, 84, 300, 322, 329, 354
$\widehat{g}$-assignment, 91, 385
$\widetilde{g}$-assignment, 91, 385
ordinary, 172, 373
surjection, 32, 33

$\mathtt{D}$, 6, 10, 165, 168, 177, 197, 311
^{D} (Cartier dual), 162
$\mathtt{D}^{-1}$, 165, 168
$\mathtt{D}_{-1}$, 171, 243
$\mathtt{D}_{-1}^{-1}$, 171
$\mathtt{D}_{-1}^{0}$, 171
$\mathtt{D}_{-1}^{1}$, 171
$\mathtt{D}_{-1,p^r}$, 171, 219–221, 231, 239, 243
$\mathtt{D}_{-1,p^r}^{-1}$, 171, 219
$\mathtt{D}_{-1,p^r}^{0}$, 171, 219
$\mathtt{D}_{-1,p^r}^{1}$, 171, 219
$\mathtt{D}_{-1,p^r}^{\#}$, 172, 219
$\mathtt{D}_{-1,p^r}^{\#,-1}$, 172, 219
$\mathtt{D}_{-1,p^r}^{\#,0}$, 172, 219
$\mathtt{D}_{-1,p^r}^{\#,1}$, 172, 219
$\mathtt{D}^{0}$, 165, 168, 243
d_0, 4, 5, 8, 63, 147, 359, 360
$d_{0,\mathrm{aux}}$, 334, 366
$\mathtt{d}_{00,n}$, 74, 78
$\mathtt{d}_{00,n_0}$, 233, 320
$\mathtt{D}^{1}$, 165, 168
$\mathtt{d}_{10,n}$, 68, 74, 78
$\mathtt{d}_{10,n_0}$, 233, 320
$\mathtt{d}_{10,p^r}^{\mathrm{ord}}$, 226, 229, 233
$\vec{\mathsf{D}}_{\infty,\mathcal{H}}^{\mathrm{ord}}$, 296
$\mathtt{D}_{\mathrm{aux}}$, 10, 198
$\mathtt{D}_{\mathrm{aux}}^{-1}$, 198
$\mathtt{D}_{\mathrm{aux}}^{0}$, 198
$\mathtt{D}_{\mathrm{aux}}^{1}$, 198
$\mathtt{D}_{\mathrm{aux}}^{\#}$, 198
$\mathtt{D}_{\mathrm{aux},\mathbb{Q}_p}^{0}$, 198
$[\vec{g_l}]_{\kappa',\kappa}^{*,\mathrm{ord},\mathrm{ext},\vee}$, 405
$\mathsf{D}_{\infty,\mathcal{H}}$, 58
$d\lambda_{\mathcal{L}}$, 473
$d\lambda_{\mathcal{L}}^{\mathrm{ext}}$, 474
$d\lambda_{\vec{\mathsf{M}}_{\mathcal{H}}^{\mathrm{ord}},j_Q}$, 475
$d\lambda_{\vec{\mathsf{M}}_{\mathcal{H}}^{\mathrm{ord}},j_Q} \underset{\mathbb{Z}}{\otimes} \mathbb{Q}$, 475
$d\lambda_{\vec{\mathsf{N}}^{\mathrm{ord}},j_Q}$, 475
$d\lambda_{\vec{\mathsf{N}}^{\mathrm{ord}},j_Q} \underset{\mathbb{Z}}{\otimes} \mathbb{Q}$, 475
$d\lambda_{\vec{\widetilde{\Psi}}^{\mathrm{ord}}_{\widetilde{\Phi}_{\widetilde{\mathcal{H}}},\widetilde{\delta}_{\widetilde{\mathcal{H}}}}(\widetilde{\ell}_{j_Q})} \underset{\mathbb{Z}}{\otimes} \mathbb{Q}$, 474
$d\log$, 473
$d\log(\mathcal{L})$, 473
$d\log(\vec{\Psi}^{\mathrm{ord}}_{\Phi_{\mathcal{H}},\delta_{\mathcal{H}}}(\ell))$, 277
$d\log q$, 281
$\mathtt{D}_{p^r}^{0}$, 169, 218
$\mathtt{D}_{p^{r_i}}^{i}$, 197
$\mathtt{D}_{p^r}^{\#,0}$, 169, 218
$d\mathsf{pol}$, 63
$\mathtt{D}^{\#}$, 165, 168
$\mathtt{D}^{\#,-1}$, 165, 168
$\mathtt{D}^{\#,0}$, 165, 168
$\mathtt{D}^{\#,1}$, 165, 168
$\mathtt{D}_{\mathbb{Q}_p}$, 168
$\mathtt{D}_{\mathbb{Q}_p}^{-1}$, 168
$\mathtt{D}_{\mathbb{Q}_p}^{0}$, 168
$\mathtt{D}_{\mathbb{Q}_p}^{1}$, 168
$\widetilde{\mathtt{D}}$, 371
$\widetilde{\mathtt{D}}^{-1}$, 371
$\widetilde{\mathtt{D}}^{0}$, 371
$\widetilde{d}_0$, 426
$\widetilde{\mathtt{D}}^{1}$, 371
$\mathtt{D}^{\mathtt{Z}}$, 311
$\mathtt{D}^{\mathtt{Z},0}$, 311
$\mathrm{DD}_{\mathrm{PE},\mathcal{O}}(R,I)$, 216
$\mathrm{DD}_{\mathrm{PEL},\dddot{\mathsf{M}}_{\mathcal{H}_\kappa}^{\mathrm{ord}}}(V)$, 415
$\mathrm{DD}_{\mathrm{PEL},\dddot{\mathsf{M}}_{\mathcal{H}}^{\mathrm{ord}}}(R,I)$, 242, 247, 281
$\mathrm{DD}_{\mathrm{PEL},\dddot{\mathsf{M}}_{\mathcal{H}}^{\mathrm{ord}}}(V)$, 297
$\mathrm{DD}_{\mathrm{PEL},\dddot{\mathsf{M}}_{n}^{\mathrm{ord}}}(R,I)$, 234
$\mathrm{DD}_{\mathrm{PEL},\mathsf{M}_{\mathcal{H}_{\mathrm{aux}}}}(V)$, 138
$\mathrm{DD}_{\mathrm{PEL},\mathsf{M}_{\mathcal{H}}}^{\mathrm{fil.\text{-}spl.}}(R,I)$, 243
$\mathrm{DD}_{\mathrm{PEL},\mathsf{M}_{\mathcal{H}_\kappa}^{\mathrm{ord}}}(V)$, 415
$\mathrm{DD}_{\mathrm{PEL},\mathsf{M}_{\mathcal{H}_\kappa}}(V)$, 112
$\mathrm{DD}_{\mathrm{PEL},\mathsf{M}_{\mathcal{H}}^{\mathrm{ord}}}(R,I)$, 247, 281
$\mathrm{DD}_{\mathrm{PEL},\mathsf{M}_{\mathcal{H}}^{\mathrm{ord}}}(V)$, 297
$\mathrm{DD}_{\mathrm{PEL},\mathsf{M}_{\mathcal{H}}}(R,I)$, 247, 281
$\mathrm{DD}_{\mathrm{PEL},\mathsf{M}_{\mathcal{H}}}(V)$, 59, 138
$\mathrm{DD}_{\mathrm{PEL},\dddot{\widetilde{\mathsf{M}}}_{\widetilde{\mathcal{H}}}^{\mathrm{ord}}}(\widetilde{V})$, 413
$\mathrm{DD}_{\mathrm{PEL},\widetilde{\mathsf{M}}_{\widetilde{\mathcal{H}}}^{\mathrm{ord}}}(\widetilde{V})$, 413

$\mathrm{DD}_{\mathrm{PEL},\widetilde{\mathsf{M}}_{\widetilde{\mathcal{H}}}}(\widetilde{V})$, 110
de Rham cohomology, 24
de Rham homology, 24
$\mathrm{DEG}_{\mathrm{PE},\mathcal{O}}(R,I)$, 216
$\mathrm{DEG}_{\mathrm{PEL},\ddot{\mathsf{M}}^{\mathrm{ord}}_{\mathcal{H}_\kappa}}(V)$, 415
$\mathrm{DEG}_{\mathrm{PEL},\ddot{\mathsf{M}}^{\mathrm{ord}}_{\mathcal{H}}}(R,I)$, 241, 247, 281
$\mathrm{DEG}_{\mathrm{PEL},\ddot{\mathsf{M}}^{\mathrm{ord}}_{\mathcal{H}}}(V)$, 297
$\mathrm{DEG}_{\mathrm{PEL},\ddot{\mathsf{M}}^{\mathrm{ord}}_{n}}(R,I)$, 234
$\mathrm{DD}_{\mathrm{PEL},\mathsf{M}_{\mathcal{H}_{\mathrm{aux}}}}(V)$, 138
$\mathrm{DEG}_{\mathrm{PEL},\mathsf{M}^{\mathrm{ord}}_{\mathcal{H}_\kappa}}(V)$, 415
$\mathrm{DEG}_{\mathrm{PEL},\mathsf{M}_{\mathcal{H}_\kappa}}(V)$, 112
$\mathrm{DEG}_{\mathrm{PEL},\mathsf{M}^{\mathrm{ord}}_{\mathcal{H}}}(R,I)$, 247, 281
$\mathrm{DEG}_{\mathrm{PEL},\mathsf{M}^{\mathrm{ord}}_{\mathcal{H}}}(V)$, 297
$\mathrm{DEG}_{\mathrm{PEL},\mathsf{M}_{\mathcal{H}}}(R,I)$, 247
$\mathrm{DEG}_{\mathrm{PEL},\mathsf{M}_{\mathcal{H}}}(V)$, 59, 138
$\mathrm{DEG}_{\mathrm{PEL},\ddot{\widetilde{\mathsf{M}}}^{\mathrm{ord}}_{\widetilde{\mathcal{H}}}}(\widetilde{V})$, 413
$\mathrm{DEG}_{\mathrm{PEL},\widetilde{\mathsf{M}}^{\mathrm{ord}}_{\widetilde{\mathcal{H}}}}(\widetilde{V})$, 413
$\mathrm{DEG}_{\mathrm{PEL},\widetilde{\mathsf{M}}_{\widetilde{\mathcal{H}}}}(\widetilde{V})$, 110
degenerating family, 8
  pullback to boundary stratum, 73–74, 319–321
  of type $\vec{\mathsf{M}}^{\mathrm{ord}}_{\mathcal{H}_{\mathrm{aux}}}$, 333
  of type $\vec{\mathsf{M}}^{\mathrm{ord}}_{\mathcal{H}}$, 205
  of type $\mathsf{M}_{\mathcal{H}}$, 57
  of type $\mathsf{M}_{\mathcal{H}_{\mathrm{aux}}}$, 137
  of type $\mathsf{M}_{\mathcal{H}^p_{\mathrm{aux}}}$, 137
  of type $(\mathrm{PE},\mathcal{O})$, 145, 215
degeneration datum, 27
  comparison with characteristic zero, 242–247
  of (G,λ,i), 217
  for general ordinary level structure, 235–242
  for principal ordinary level structure, 217–235
degeneration pattern, 8, 60
δ, 29, 47
$\vec{\Delta}^{\mathrm{ord},*}_{\Phi_{\mathcal{H}},\delta_{\mathcal{H}},\ell,\ell'}$, 266, 325
δ_{aux}, 134
$\breve{\delta}$, 47
$\breve{\delta}_{\widehat{\mathcal{H}}}$, 48, 110, 380, 413
$\breve{\delta}_{\widetilde{\mathcal{H}}}$, 47
$\delta^{(g)}$, 41
$\delta_{\mathcal{H}}$, 243, 248, 281
$\delta_{\mathcal{H}_{\mathrm{aux}}}$, 134
$\delta_{\mathcal{H}^p_{\mathrm{aux}}}$, 135
$\delta^{\ddagger}_{\mathcal{H}}$, 60, 292, 297
δ_{H_n}, 243
$\delta_{H_{n_0}}$, 236
$\delta^{\mathrm{ord}}_{H_n}$, 240
$\delta^{\mathrm{ord}}_{\mathcal{H}}$, 241, 255, 266, 270, 281
$\delta^{\mathrm{ord}}_{H^{\mathrm{ord}}_{p^r}}$, 239, 244
$\delta^{\mathrm{ord},0}_{H^{\mathrm{ord}}_{p^r}}$, 239, 244
$\delta^{\mathrm{ord},\#,0}_{H^{\mathrm{ord}}_{p^r}}$, 239, 244
$\hat{\delta}$, 29
$\widehat{\vec{\Delta}}^{\mathrm{ord},*}_{\breve{\Phi}_{\widehat{\mathcal{H}}},\breve{\delta}_{\widehat{\mathcal{H}}},\breve{\ell},\breve{\ell}'}$, 382
$\widehat{\delta}^{\mathrm{ord}}_{\widehat{\mathcal{H}}}$, 413
$\widehat{\Delta}^{*}_{\breve{\Phi}_{\widehat{\mathcal{H}}},\breve{\delta}_{\widehat{\mathcal{H}}},\breve{\ell},\breve{\ell}'}$, 88
$\Delta_{\vec{\mathsf{M}}^{\mathrm{ord},\mathrm{tor}}_{\mathcal{H}}}$, 299
δ_n, 29, 248, 313
δ_{n_0}, 232, 313
$\Delta^{\mathrm{ord},*}_{n,\ell,\ell'}$, 251
δ^{ord}_n, 232
δ_{p^r}, 244, 253
$\delta^{\mathrm{ord}}_{p^r}$, 232, 239, 244, 253, 314
$\delta^{\mathrm{ord},0}_{p^r}$, 222, 232, 239, 244
$(\delta^{\mathrm{ord},0}_{p^r})^{\mathrm{mult}}_{\eta}$, 230
$\delta^{\mathrm{ord},\#,0}_{p^r}$, 222, 232, 239, 244
$(\delta^{\mathrm{ord},\#,0}_{p^r})^{\mathrm{mult}}_{\eta}$, 230
$\Delta^{*}_{\Phi_{\mathcal{H}},\delta_{\mathcal{H}},\ell,\ell'}$, 80
$\widetilde{\delta}$, 45
$\widetilde{\vec{\Delta}}^{\mathrm{ord},*}_{\breve{\Phi}_{\widetilde{\mathcal{H}}},\breve{\delta}_{\widetilde{\mathcal{H}}},\breve{\ell},\breve{\ell}'}$, 447
$\widetilde{\delta}_{\widetilde{\mathcal{H}}}$, 46, 372
$\widetilde{\delta}^{\ddagger}_{\widetilde{\mathcal{H}}}$, 110, 413
$\widetilde{\delta}^{\mathrm{ord},\ddagger}_{\widetilde{\mathcal{H}}}$, 413
depth
  of elements of twisted U_p type, 195
  of subgroups of $\widehat{\mathrm{G}}(\hat{\mathbb{Z}})$, 372
  of subgroups of $\widehat{\mathrm{G}}(\mathbb{Z}_p)$, 371
  of subgroups of $\mathrm{G}(\hat{\mathbb{Z}})$, 170
  of subgroups of $\mathrm{G}(\mathbb{Z}_p)$, 170
$\mathrm{depth}_{\mathsf{D}}(g_p)$, 195
$\mathrm{depth}_{\mathsf{D}}(\mathcal{H})$, 170, 197
$\mathrm{depth}_{\mathsf{D}}(\widehat{\mathcal{H}})$, 372
$\mathrm{depth}_{\mathsf{D}}(\widehat{\mathcal{H}}_p)$, 371
$\mathrm{depth}_{\mathsf{D}}(\mathcal{H}_p)$, 170
$\mathrm{depth}_{\widetilde{\mathsf{D}}}(\widetilde{\mathcal{H}})$, 371
$\underline{\mathrm{Der}}_{\vec{\mathsf{N}}^{\mathrm{ord},\mathrm{tor}}/\vec{\mathsf{M}}^{\mathrm{ord},\mathrm{tor}}_{\mathcal{H}}}$, 471
$\underline{\mathrm{Der}}_{\vec{\mathsf{N}}^{\mathrm{ord}}/\vec{\mathsf{M}}^{\mathrm{ord}}_{\mathcal{H}}}$, 471
determinantal condition on Lie algebra, 6, 23, 131, 141, 164, 203, 510

Dieudonné–Manin classification, 156
Diff^{-1}, 43
$\mathrm{Diff}^{-1}_{\mathcal{O}/\mathbb{Z}}$, 43
Disc, 95, 158
$\mathrm{Disc}_{\mathcal{O}/\mathbb{Z}}$, 95, 158
dual lattice, 22
dual semi-abelian scheme, 57, 205, 215, 392

$\mathrm{e}_{00,n}$, 74
e_{00,n_0}, 320
$\mathrm{e}_{10,n}$, 74
e_{10,n_0}, 320
$\mathrm{e}^{\mathrm{ord}}_{10,p^r}$, 226
$\mathrm{e}_{A[p^r]}$, 178
$\mathrm{e}_{A_{\bar{s}}[p^\infty]}$, 182
$\mathrm{e}_{A_s[p^\infty]}$, 164
$\vec{\mathcal{E}}^{\mathrm{ord}}_{\mathrm{G}^{\mathrm{ord}}_{\mathtt{D},0}}$, 481
$\vec{\mathcal{E}}^{\mathrm{ord},\mathrm{can}}_{\mathrm{G}^{\mathrm{ord}}_{\mathtt{D},0}}$, 487
$\vec{\mathcal{E}}^{\mathrm{ord}}_{\mathrm{G}^{\mathrm{ord}}_{\mathtt{D},0},R}(\cdot)$, 482, 483
$\vec{\mathcal{E}}^{\mathrm{ord},\mathrm{can}}_{\mathrm{G}^{\mathrm{ord}}_{\mathtt{D},0},R}(\cdot)$, 488, 508
$\vec{\mathcal{E}}^{\mathrm{ord},\mathrm{sub}}_{\mathrm{G}^{\mathrm{ord}}_{\mathtt{D},0},R}(\cdot)$, 488
$\vec{\mathsf{E}}^{\mathrm{ord}}_{\infty,\kappa}$, 393
$\vec{\mathsf{E}}^{\mathrm{ord}}_{\infty}$, 397
$\vec{\mathcal{E}}^{\mathrm{ord},\Phi_{\mathcal{H}},\delta_{\mathcal{H}}}_{\mathrm{M}^{\mathrm{ord}}_{\mathtt{D},0,\mathrm{fil}}}$, 505
$\vec{\mathcal{E}}^{\mathrm{ord},\Phi_{\mathcal{H}},\delta_{\mathcal{H}}}_{\mathrm{M}^{\mathrm{ord}}_{\mathtt{D},0,\mathrm{fil}},R_0}(\cdot)$, 506
$\vec{\mathcal{E}}^{\mathrm{ord},\Phi_{\mathcal{H}},\delta_{\mathcal{H}}}_{\mathrm{M}^{\mathrm{ord}}_{\mathtt{D},0,\mathrm{gr}}}$, 507
$\vec{\mathcal{E}}_{\mathrm{M}^{\mathrm{ord}}_{\mathtt{D},0}}$, 509
$\vec{\mathcal{E}}^{\mathrm{ord}}_{\mathrm{M}^{\mathrm{ord}}_{\mathtt{D},0}}$, 481
$\vec{\mathcal{E}}^{\mathrm{ord},\mathrm{can}}_{\mathrm{M}^{\mathrm{ord}}_{\mathtt{D},0}}$, 487
$\vec{\mathcal{E}}^{\mathrm{ord},\Phi_{\mathcal{H}},\delta_{\mathcal{H}}}_{\mathrm{M}^{\mathrm{ord}}_{\mathtt{D},0}}$, 502
$\vec{\mathcal{E}}_{\mathrm{M}^{\mathrm{ord}}_{\mathtt{D},0},R}(\cdot)$, 508, 510–513, 518, 521
$\vec{\mathcal{E}}^{\mathrm{can}}_{\mathrm{M}^{\mathrm{ord}}_{\mathtt{D},0},R}(\cdot)$, 508, 512, 513, 515
$\vec{\mathcal{E}}^{\mathrm{can},\mathrm{min}}_{\mathrm{M}^{\mathrm{ord}}_{\mathtt{D},0},R}(\cdot)$, 514–517, 520, 521
$\vec{\mathcal{E}}^{\mathrm{ord}}_{\mathrm{M}^{\mathrm{ord}}_{\mathtt{D},0},R}(\cdot)$, 482, 483, 508, 513
$\vec{\mathcal{E}}^{\mathrm{ord},\mathrm{can}}_{\mathrm{M}^{\mathrm{ord}}_{\mathtt{D},0},R}(\cdot)$, 488, 508, 513, 516
$\vec{\mathcal{E}}^{\mathrm{ord},\Phi_{\mathcal{H}},\delta_{\mathcal{H}}}_{\mathrm{M}^{\mathrm{ord}}_{\mathtt{D},0},R}(\cdot)$, 502
$\vec{\mathcal{E}}^{\mathrm{ord},\mathrm{sub}}_{\mathrm{M}^{\mathrm{ord}}_{\mathtt{D},0},R}(\cdot)$, 488, 508, 513, 516
$\vec{\mathcal{E}}^{\mathrm{sub}}_{\mathrm{M}^{\mathrm{ord}}_{\mathtt{D},0},R}(\cdot)$, 508, 512, 513, 515
$\vec{\mathcal{E}}^{\mathrm{sub},\mathrm{min}}_{\mathrm{M}^{\mathrm{ord}}_{\mathtt{D},0},R}(\cdot)$, 514–517, 520, 521
$\vec{\mathcal{E}}_{\mathrm{M}^{\mathrm{ord}}_{\mathtt{D},0},R}(\cdot)\cup\mathcal{E}^{\mathrm{can}}_{\mathrm{M}^{\mathrm{ord}}_{\mathtt{D},0},R_{\mathbb{Q}}}((\cdot)_{\mathbb{Q}})$, 512, 514
$\vec{\mathcal{E}}_{\mathrm{M}^{\mathrm{ord}}_{\mathtt{D},0},R}(\cdot)\cup(\oint_{\mathcal{H},*}\mathcal{E}^{\mathrm{can}}_{\mathrm{M}^{\mathrm{ord}}_{\mathtt{D},0},R_{\mathbb{Q}}}((\cdot)_{\mathbb{Q}}))$, 515
$\vec{\mathcal{E}}_{\mathrm{M}^{\mathrm{ord}}_{\mathtt{D},0},R}(\cdot)\cup(\oint_{\mathcal{H},*}\mathcal{E}^{\mathrm{sub}}_{\mathrm{M}^{\mathrm{ord}}_{\mathtt{D},0},R_{\mathbb{Q}}}((\cdot)_{\mathbb{Q}}))$, 515
$\vec{\mathcal{E}}_{\mathrm{P}^{\mathrm{ord}}_{\mathtt{D},0}}$, 509
$\vec{\mathcal{E}}^{\mathrm{ord}}_{\mathrm{P}^{\mathrm{ord}}_{\mathtt{D},0}}$, 481
$\vec{\mathcal{E}}^{\mathrm{ord},\mathrm{can}}_{\mathrm{P}^{\mathrm{ord}}_{\mathtt{D},0}}$, 487
$\vec{\mathcal{E}}_{\mathrm{P}^{\mathrm{ord}}_{\mathtt{D},0},R}(\cdot)$, 508, 510, 511, 513, 521
$\vec{\mathcal{E}}^{\mathrm{can}}_{\mathrm{P}^{\mathrm{ord}}_{\mathtt{D},0},R}(\cdot)$, 508, 512, 513, 516
$\vec{\mathcal{E}}^{\mathrm{can},\mathrm{min}}_{\mathrm{P}^{\mathrm{ord}}_{\mathtt{D},0},R}(\cdot)$, 515–517, 521
$\vec{\mathcal{E}}^{\mathrm{ord}}_{\mathrm{P}^{\mathrm{ord}}_{\mathtt{D},0},R}(\cdot)$, 482, 483, 508, 513
$\vec{\mathcal{E}}^{\mathrm{ord},\mathrm{can}}_{\mathrm{P}^{\mathrm{ord}}_{\mathtt{D},0},R}(\cdot)$, 488, 508, 513, 516
$\vec{\mathcal{E}}^{\mathrm{ord},\mathrm{sub}}_{\mathrm{P}^{\mathrm{ord}}_{\mathtt{D},0},R}(\cdot)$, 488, 508, 513, 516
$\vec{\mathcal{E}}^{\mathrm{sub}}_{\mathrm{P}^{\mathrm{ord}}_{\mathtt{D},0},R}(\cdot)$, 508, 512, 513, 516
$\vec{\mathcal{E}}^{\mathrm{sub},\mathrm{min}}_{\mathrm{M}^{\mathrm{ord}}_{\mathtt{D},0},R}(\cdot)$, 515
$\vec{\mathcal{E}}^{\mathrm{sub},\mathrm{min}}_{\mathrm{P}^{\mathrm{ord}}_{\mathtt{D},0},R}(\cdot)$, 515–517, 521
$\mathrm{e}_{B[n]}$, 68
$\mathrm{e}_{B[p^r]_\eta}$, 228
$E_{\mathtt{D},\mathbb{Z}}$, 201
$E_{\mathtt{D},\mathbb{Z},\mathrm{free}}$, 204
$E_{\mathtt{D},\mathbb{Z},\mathrm{tor}}$, 204
$\dddot{E}^{\mathrm{ord},\circ}_{\widehat{\mathcal{H}},n}$, 438
$\ddot{E}_{\Phi_1}$, 264
$\ddot{E}^{\mathrm{ord}}_{\Phi_1}$, 264
$\dddot{E}^{\mathrm{ord}}_{\Phi_{\mathcal{H}},n,\mathrm{free}}$, 258
$\dddot{E}^{\mathrm{ord}}_{\Phi_{\mathcal{H}},n}$, 258
$\dddot{E}^{\mathrm{ord}}_{\Phi_{\mathcal{H}},n,\mathrm{tor}}$, 258
$\dddot{E}_{\Phi_n}$, 79
$\dddot{E}_{\Phi_{n_0}}$, 264
$\dddot{E}^{\mathrm{ord}}_{\Phi_n,\mathrm{free}}$, 251
$\dddot{E}^{\mathrm{ord}}_{\Phi_n}$, 251
$\dddot{E}^{\mathrm{ord}}_{\Phi_n,\mathrm{tor}}$, 251
$\mathcal{E}_{\mathrm{G}_0}$, 118
$\mathcal{E}^{\mathrm{can}}_{\mathrm{G}_0}$, 122
$\mathcal{E}_{\mathrm{G}_0,R}(\cdot)$, 120
$\mathcal{E}^{\mathrm{can}}_{\mathrm{G}_0,R}(\cdot)$, 122
$\mathcal{E}^{\mathrm{sub}}_{\mathrm{G}_0,R}(\cdot)$, 123
$\mathcal{E}_{\mathrm{G}_0,\tilde{R}_{\mathbb{Q}}}(\cdot)$, 509
$\mathcal{E}_{\mathrm{G}^{\mathrm{ord}}_{\mathtt{D},0}}$, 508
$\mathcal{E}^{\mathrm{can}}_{\mathrm{G}^{\mathrm{ord}}_{\mathtt{D},0}}$, 509

$\mathcal{E}_{\mathrm{G}^{\mathrm{ord}}_{\mathtt{D},0},R_{\mathbb{Q}}}(\,\cdot\,)$, 509
$\mathcal{E}^{\mathrm{can}}_{\mathrm{G}^{\mathrm{ord}}_{\mathtt{D},0},R_{\mathbb{Q}}}(\,\cdot\,)$, 509
$\mathcal{E}^{\mathrm{sub}}_{\mathrm{G}^{\mathrm{ord}}_{\mathtt{D},0},R_{\mathbb{Q}}}(\,\cdot\,)$, 509
$\mathcal{E}_{\mathrm{G}^{\mathrm{ord}}_{\mathtt{D},0},\tilde{R}_{\mathbb{Q}}}(\,\cdot\,)$, 509
$\mathcal{E}^{\mathrm{sub}}_{\mathrm{G}^{\mathrm{ord}}_{\mathtt{D},0}}$, 509
$\widehat{E}_{\check{\Phi}_{\widehat{\mathcal{H}}}}$, 88, 381
$\mathrm{e}^{\lambda_\eta}_{\mathrm{image}(\alpha^{\mathrm{ord},0}_{p^r})}$, 228
$\mathrm{e}_{\mathrm{image}(\varphi^{\mathrm{ord},0}_{-1,p^r})}$, 228
E_∞, 100
$\mathsf{E}_{\infty,\kappa}$, 98
e^{λ}, 173
$\mathrm{e}^{\lambda_\eta}$, 228
$\mathcal{E}_{\mathrm{M}_0}$, 119
$\mathcal{E}^{\mathrm{can}}_{\mathrm{M}_0}$, 122
$\mathcal{E}_{\mathrm{M}_0,R}(\,\cdot\,)$, 120
$\mathcal{E}^{\mathrm{can}}_{\mathrm{M}_0,R}(\,\cdot\,)$, 123
$\mathcal{E}^{\mathrm{sub}}_{\mathrm{M}_0,R}(\,\cdot\,)$, 123
$\mathcal{E}_{\mathrm{M}_0,\tilde{R}_{\mathbb{Q}}}(\,\cdot\,)$, 509
$\mathcal{E}_{\mathrm{M}^{\mathrm{ord}}_{\mathtt{D},0}}$, 508
$\mathcal{E}^{\mathrm{can}}_{\mathrm{M}^{\mathrm{ord}}_{\mathtt{D},0}}$, 509
$\mathcal{E}_{\mathrm{M}^{\mathrm{ord}}_{\mathtt{D},0},R_{\mathbb{Q}}}(\,\cdot\,)$, 509, 511, 519
$\mathcal{E}^{\mathrm{can}}_{\mathrm{M}^{\mathrm{ord}}_{\mathtt{D},0},R_{\mathbb{Q}}}(\,\cdot\,)$, 509, 512, 515, 521
$\mathcal{E}^{\mathrm{sub}}_{\mathrm{M}^{\mathrm{ord}}_{\mathtt{D},0},R_{\mathbb{Q}}}(\,\cdot\,)$, 509, 512, 515, 521
$\mathcal{E}_{\mathrm{M}^{\mathrm{ord}}_{\mathtt{D},0},\tilde{R}_{\mathbb{Q}}}(\,\cdot\,)$, 509
$\mathcal{E}^{\mathrm{sub}}_{\mathrm{M}^{\mathrm{ord}}_{\mathtt{D},0}}$, 509
e^{ϕ}_{m}, 231
e^{ϕ}_{n}, 232
$\mathcal{E}_{\mathrm{P}_0}$, 119
$\mathcal{E}^{\mathrm{can}}_{\mathrm{P}_0}$, 122
$\mathcal{E}_{\mathrm{P}_0,R}(\,\cdot\,)$, 120
$\mathcal{E}^{\mathrm{can}}_{\mathrm{P}_0,R}(\,\cdot\,)$, 123
$\mathcal{E}^{\mathrm{sub}}_{\mathrm{P}_0,R}(\,\cdot\,)$, 123
$\mathcal{E}_{\mathrm{P}_0,\tilde{R}_{\mathbb{Q}}}(\,\cdot\,)$, 509
$\mathcal{E}_{\mathrm{P}^{\mathrm{ord}}_{\mathtt{D},0}}$, 508
$\mathcal{E}^{\mathrm{can}}_{\mathrm{P}^{\mathrm{ord}}_{\mathtt{D},0}}$, 509
$\mathcal{E}_{\mathrm{P}^{\mathrm{ord}}_{\mathtt{D},0},R_{\mathbb{Q}}}(\,\cdot\,)$, 509, 511
$\mathcal{E}^{\mathrm{can}}_{\mathrm{P}^{\mathrm{ord}}_{\mathtt{D},0},R_{\mathbb{Q}}}(\,\cdot\,)$, 509, 512, 515
$\mathcal{E}^{\mathrm{sub}}_{\mathrm{P}^{\mathrm{ord}}_{\mathtt{D},0},R_{\mathbb{Q}}}(\,\cdot\,)$, 509, 515
$\mathcal{E}_{\mathrm{P}^{\mathrm{ord}}_{\mathtt{D},0},\tilde{R}_{\mathbb{Q}}}(\,\cdot\,)$, 509
$\mathcal{E}^{\mathrm{sub}}_{\mathrm{P}^{\mathrm{ord}}_{\mathtt{D},0}}$, 509
$E_{\check{\Phi}_{\widetilde{\mathcal{H}}}}$, 92, 387, 409
$E_{\Phi_{\mathcal{H}}}$, 62, 76, 80, 258, 324, 325
$E_{\Phi_{\mathcal{H}},\sigma}$, 62, 81, 326, 347
E_{Φ_n}, 76
$\widetilde{E}_{\check{\Phi}_{\widetilde{\mathcal{H}}},\check{\sigma}}$, 88, 381
easier unitary case, 18
elevator, 145
endomorphism structure, 23, 164, 172, 177, 191, 197
$\varepsilon^{(g)}$, 39, 41
equally deep
  subgroups of $\widehat{\mathrm{G}}(\hat{\mathbb{Z}})$, 372
  subgroups of $\widehat{\mathrm{G}}(\mathbb{Z}_p)$, 372
  subgroups of $\mathrm{G}(\hat{\mathbb{Z}})$, 170
  subgroups of $\mathrm{G}(\mathbb{Z}_p)$, 170
^{ét}, 164
extended ordinary Kuga family, *see* ordinary Kuga family, extended
extension of isogeny, 160–162

F, 22
f, 126, 199
^{f}, 161, 162
$\mathsf{F}^{(g)}_{-2}$, 39
$\mathsf{F}^{(g)}_{-2,h(\mathbb{C})}$, 42
$(\mathsf{F}^{(g)}_{-2,h(\mathbb{C})})^{\perp}$, 42
$\mathsf{F}^{(g)}_{-2,\mathbb{R}}$, 41
$\mathsf{F}^{(g)}_{-i}$, 39
F_0, 3, 23, 44
$F_{0,\mathrm{aux}}$, 9, 128
F'_0, 117
$\mathrm{F}^{(i)}_{A^\vee/S}$, 181
$\mathsf{F}^a(\vec{\mathcal{E}}^{\mathrm{ord}}_{\mathrm{P}^{\mathrm{ord}}_{\mathtt{D},0},R}(W))$, 483
$\mathsf{F}^a(\vec{\mathcal{E}}_{\mathrm{P}^{\mathrm{ord}}_{\mathtt{D},0},R}(W))$, 511
$\mathsf{F}^a(\mathcal{E}_{\mathrm{P}_0,R}(W))$, 120
$\mathsf{F}^a(\underline{H}^i_{\text{log-dR}}(\vec{\mathsf{N}}^{\mathrm{ord,tor}}/\vec{\mathsf{M}}^{\mathrm{ord,tor}}_{\mathcal{H}}))$, 470
$\mathrm{F}_{A/\vec{\mathsf{M}}^{\mathrm{ord}}_{\mathcal{H}'}\underset{\mathbb{Z}}{\otimes}\mathbb{F}_p}$, 211
$\mathsf{F}^a(\overline{\Omega}^{\bullet}_{\vec{\mathsf{N}}^{\mathrm{ord,tor}}/\vec{\mathsf{S}}_{0,r_{\mathcal{H}}}})$, 470
$\mathrm{F}_{A/S}$, 190
$\mathrm{F}^{(i)}_{A/S}$, 181
$\mathsf{F}^a(W)$, 120, 483, 511
$\mathrm{F}_{\mathrm{abs}}$, 362
$\mathrm{F}_{\mathrm{aux}}$, 362
$^{\mathrm{f,ét}}$, 162
$\mathsf{F}^{(g)}$, 39, 40
f_κ, 99, 395
  extensibility, 421–423
$f^{\mathrm{tor}}_{\kappa_{\mathrm{aux}}}$, 425
$f_{\kappa',\kappa}$, 99, 116, 394, 420
$f^{\mathrm{grp}}_{\kappa',\kappa}$, 99, 116, 394, 420

$f^{\mathrm{tor}}_{\kappa',\kappa}$, 99, 116, 394, 420
  properness, 421
$f_{\widetilde{\kappa}}$, 107, 408
$f_{\widetilde{\kappa}',\widetilde{\kappa}}$, 114, 416
$f^{\mathrm{grp}}_{\widetilde{\kappa}',\widetilde{\kappa}}$, 116, 420
$f^{\mathrm{tor}}_{\widetilde{\kappa}',\widetilde{\kappa}}$, 113, 416
$f^{\mathrm{tor}}_{\kappa}$, 99, 117, 395, 422
  equidimensionality, 441–443
  flatness, 442–443
  formal fiber, 445–453
  log integrality, 442–443
  log smoothness, 430–435
  properness, 424–428
$\mathrm{F}_{\vec{\mathsf{M}}^{\mathrm{ord,min}}_{\mathcal{H}} \underset{\mathbb{Z}}{\otimes} \mathbb{F}_p}$, 359
$\mathrm{F}_{\vec{\mathsf{M}}^{\mathrm{ord}}_{\mathcal{H}} \underset{\mathbb{Z}}{\otimes} \mathbb{F}_p}$, 211
$\mathrm{F}_{\vec{\mathsf{M}}^{\mathrm{ord,min}}_{\mathcal{H}'} \underset{\mathbb{Z}}{\otimes} \mathbb{F}_p}$, 359
$\mathrm{F}_{\vec{\mathsf{M}}^{\mathrm{ord}}_{\mathcal{H}'} \underset{\mathbb{Z}}{\otimes} \mathbb{F}_p}$, 211
$\mathrm{F}_{\vec{\mathsf{M}}^{\mathrm{ord,tor}}_{\mathcal{H}',\Sigma^{\mathrm{ord},\prime}} \underset{\mathbb{Z}}{\otimes} \mathbb{F}_p}$, 304
$\mathrm{F}_{\vec{\mathsf{M}}^{\mathrm{ord,tor}}_{\mathcal{H},\Sigma^{\mathrm{ord}}} \underset{\mathbb{Z}}{\otimes} \mathbb{F}_p}$, 304, 403
$\mathrm{F}_{\vec{\mathsf{N}}^{\mathrm{ord,tor}}_{\kappa} \underset{\mathbb{Z}}{\otimes} \mathbb{F}_p}$, 402
$\mathrm{F}_{(\vec{\mathsf{N}}^{\mathrm{ord,tor}}_{\kappa} \underset{\mathbb{Z}}{\otimes} \mathbb{F}_p)/(\vec{\mathsf{M}}^{\mathrm{ord,tor}}_{\mathcal{H},\Sigma^{\mathrm{ord}}} \underset{\mathbb{Z}}{\otimes} \mathbb{F}_p)}$, 403
$\mathrm{F}_{\vec{\mathsf{N}}^{\mathrm{ord,tor}}_{\kappa'} \underset{\mathbb{Z}}{\otimes} \mathbb{F}_p}$, 402
$\mathrm{F}_{\vec{\mathsf{N}}^{\mathrm{ord}}_{\kappa'} \underset{\mathbb{Z}}{\otimes} \mathbb{F}_p}$, 402
F_S, 181, 190
f_σ, 495
$f_{[\sigma]}$, 495
$f^{\mathrm{tor}}_{[\breve{\tau}]}$, 451
$f^{\mathrm{tor}}_{\breve{\tau}}$, 451
f^{tor}, 451, *see also* $f^{\mathrm{tor}}_{\kappa}$
face, 34
  of cusp label with cone, 38, 98, 393
failure of Hasse's principle, 26
fiber-wise geometric identity component, 157
  group scheme of, 158
filtration
  admissible, 27
  fully symplectic, 27
  fully symplectic-liftable, 27
  integrable, 27
  split, 27
  symplectic, 27
$\mathrm{FJ}^{\mathrm{ord}}_{\Phi_{\mathcal{H}},\delta_{\mathcal{H}}}$, 340
$\mathrm{FJ}^{\mathrm{ord},(0)}_{\Phi_{\mathcal{H}},\delta_{\mathcal{H}}}(f)$, 340
$(\underline{\mathrm{FJ}}^{\mathrm{ord},(0)}_{\Phi_{\mathcal{H}},\delta_{\mathcal{H}}})^{\wedge}_{\bar{x}}$, 344
$\underline{\mathrm{FJ}}^{\mathrm{ord},d,(\ell)}_{\Phi_{\mathcal{H}},\delta_{\mathcal{H}}}(\mathscr{N})$, 499
$\mathrm{FJ}^{\mathrm{ord}}_{\Phi_{\mathcal{H}},\delta_{\mathcal{H}}}(f)$, 340
$\underline{\mathrm{FJ}}^{\mathrm{ord},(\ell)}_{\Phi_{\mathcal{H}},\delta_{\mathcal{H}}}$, 339
$\mathrm{FJ}^{\mathrm{ord},(\ell)}_{\Phi_{\mathcal{H}},\delta_{\mathcal{H}}}(f)$, 340
$(\underline{\mathrm{FJ}}^{\mathrm{ord},(\ell)}_{\Phi_{\mathcal{H}},\delta_{\mathcal{H}}})^{\wedge}_{\bar{x}}$, 343
$\mathrm{FJ}^{\mathrm{ord}}_{\Phi_{\mathcal{H}},\delta_{\mathcal{H}},\sigma}$, 339
$\mathrm{FJ}^{\mathrm{ord}}_{\Phi_{\mathcal{H}},\delta_{\mathcal{H}},\sigma}(f)$, 339
$\mathrm{FJ}^{\mathrm{ord},(\ell)}_{\Phi_{\mathcal{H}},\delta_{\mathcal{H}},\sigma}(f)$, 339
$\mathrm{FJC}^{\mathrm{ord},(0)}_{\Phi_{\mathcal{H}},\delta_{\mathcal{H}}}(k)$, 340
$\mathrm{FJC}^{\mathrm{ord},(\ell)}_{\Phi_{\mathcal{H}},\delta_{\mathcal{H}}}(k)$, 340
formally canonical sheaf, 498
formally subcanonical sheaf, 499
Fourier–Jacobi
  coefficient
    along $(\Phi_{\mathcal{H}}, \delta_{\mathcal{H}})$, 340
    along $(\Phi_{\mathcal{H}}, \delta_{\mathcal{H}}, \sigma)$, 340
  expansion
    along $(\Phi_{\mathcal{H}}, \delta_{\mathcal{H}})$, 340
    along $(\Phi_{\mathcal{H}}, \delta_{\mathcal{H}}, \sigma)$, 340
    constant term, 340
  morphism
    along $(\Phi_{\mathcal{H}}, \delta_{\mathcal{H}})$, 340
    along $(\Phi_{\mathcal{H}}, \delta_{\mathcal{H}}, \sigma)$, 339
Fraktur, 210
Frobenius morphism
  absolute, 181, 190, 211, 304, 359, 362, 402
    flat, 212, 305
    not flat, 359
  relative, 181, 190, 211, 304, 362, 403
${}^{\mathrm{full\text{-}ord}}$, 365
full ordinary locus, 14, 365
fully symplectic filtration, 27
fully symplectic-liftable filtration, 27

G, 3, 22
G, 57, 58, 205, 215, 295
$[g]$, 3, 65, 83, 84, 322
$[[g]]$, 3, 65
$[g^{-1}]$, 194
$[g^{-1}]^{\mathrm{ord}}$, 191, 194
G_0, 118
$G^{[0]}$, 291
G_0, 216
$[g_0]$, 175
g_0, 149, 175, 191, 208, 518
$[g_0^{-1}]$, 175, 191

$\mathrm{G}_0(R)$, 118
$\mathrm{G}_{1,\mathsf{Z}}(R)$, 31, 56
$\mathrm{G}(\mathbb{A})$, 22
$\mathrm{G}(\mathbb{A}^\infty)$, 22
$\mathrm{G}(\mathbb{A}^{\infty,p})$, 22
$\mathrm{G}(\mathbb{A}^p)$, 22
$[[\vec{g}]]$, 5, 149
$[\vec{g}]$, 5, 149, 518–519
$[\vec{g}]^{\min}_{i',i}$, 152
$[\vec{g}]^{\min}$, 5, 149, 518
$([\vec{g}]^{\min})^*$, 520
$[[\vec{g}]^{\mathrm{ord}}]$, 7, 353
$[\vec{g}]^{\mathrm{ord}}$, 7, 209–211, 329, 519
$[\vec{g}]^{\mathrm{ord,min}}$, 7, 353, 357, 358
$([\vec{g}]^{\mathrm{ord,min}})^*$, 353
$([\vec{g}]^{\mathrm{ord}})^*$, 489
$[\vec{g}]^{\mathrm{ord,tor}}$, 7, 300, 303
$([\vec{g}]^{\mathrm{ord,tor}})^*$, 489
$[\vec{g}]^{\mathrm{ord}}_{r_{\mathcal{H}'}}$, 209, 210
$[\vec{g}]^{\mathrm{ord,tor}}_{r_{\mathcal{H}'}}$, 303
$[\vec{g}]^*$, 518, 519
$\mathrm{G}_{\mathrm{aux}}$, 9, 128
G_{aux}, 333, 361
$G^{\triangledown}_{\mathrm{aux}}$, 137, 333
$G^{\triangledown,\vee}_{\mathrm{aux}}$, 137
$G^{\vartriangle}_{\mathrm{aux}}$, 137
$G^{\vartriangle,\vee}_{\mathrm{aux}}$, 137
$\overline{G}$, 422
$\overline{G}^\vee$, 422
$\overline{G}_{\vec{\mathsf{N}}^{\mathrm{ord}}_\kappa}$, 422
$\overline{G}^\vee_{\vec{\mathsf{N}}^{\mathrm{ord}}_\kappa}$, 422
$([g]^{\mathrm{tor}})^*$, 123
$\mathrm{G}^{\mathrm{ord}}_{\mathsf{D},0}$, 479
$\mathrm{G}^{\mathrm{ord}}_{\mathsf{D},0}(R)$, 479
$G^\dagger$, 59, 145, 296
$G^\ddagger$, 59, 112, 291, 296, 415
${}^\diamondsuit G$, 280
$[\overset{\cdots}{g}]^{\mathrm{ord}}$, 208
$G^\vee$, 57, 205, 215
$G^{\vee,\natural}$, 68, 70, 313, 315, 454
$\mathrm{G}^{\mathrm{ess}}(R)$, 235
$\mathrm{G}^{\mathrm{ess}}(\mathbb{Z}/n\mathbb{Z})$, 235
$\mathrm{G}^{\mathrm{ess}}(\mathbb{Z}/p^r\mathbb{Z})$, 237
$[g_h]$, 102
g_h, 67, 102, 398
$g_{h,0}$, 398
$[\vec{g}_h]^{\mathrm{ord}}$, 399
$[\vec{g}_h]^{\mathrm{ord,tor}}$, 403
$[\vec{g}_h]^{\mathrm{ord}}_{r_{\kappa'},r_\kappa}$, 399
$[\vec{g}_h]^{\mathrm{ord,tor}}_{r_{\kappa'},r_\kappa}$, 400
$([\vec{g}_h]^{\mathrm{ord,tor}}_{r_{\kappa'},r_\kappa})^*$, 400
$g_{h,p}$, 398
$[g_h]^{\mathrm{tor}}$, 103
$\mathrm{G}_{h,\mathsf{Z}}$, 30
$\mathrm{G}^{\mathrm{ess}}_{h,\mathsf{Z}_n}$, 235
$\mathrm{G}^{\mathrm{ess,ord}}_{h,\mathsf{Z}_{p^r},\mathsf{D}_{p^r}}$, 238
$\mathrm{G}'_{h,\mathsf{Z}}(R)$, 31, 56
$\mathrm{G}_{h,\mathsf{Z}}(R)$, 30, 56, 378
$(\mathrm{G}_{h,\mathsf{Z}} \ltimes \mathrm{U}_{1,\mathsf{Z}})(R)$, 31
$\widehat{G}$, 108, 411
$[\widehat{g}]$, 91, 102
$\widehat{g}$, 102, 398
$\widehat{g}_0$, 398
$\widehat{\mathrm{G}}_{1,\breve{\mathsf{Z}}}(R)$, 56
$[\vec{\widehat{g}}]^{\mathrm{ord}}$, 385, 386, 399, 401, 402
$[\vec{\widehat{g}}]^{\mathrm{ord,tor}}$, 399, 401, 402
$([\vec{\widehat{g}}]^{\mathrm{ord,tor}})^*$, 400
$[\vec{\widehat{g}}]^{\mathrm{ord}}_{r_{\kappa'}}$, 401
$[\vec{\widehat{g}}]^{\mathrm{ord,tor}}_{r_{\kappa'}}$, 401
$\widehat{G}^\ddagger$, 109, 412
$\widehat{\mathrm{G}}'_{h,\breve{\mathsf{Z}}}(R)$, 56
$\widehat{\mathrm{G}}_{h,\breve{\mathsf{Z}}}(R)$, 56, 378
$\widehat{\mathrm{G}}'_{l,\breve{\mathsf{Z}}}(R)$, 56, 378
$\widehat{\mathrm{G}}_{l,\breve{\mathsf{Z}}}(R)$, 56
$\widehat{G}_{\vec{\mathsf{N}}^{\mathrm{ord}}_\kappa}$, 421
$\widehat{G}^\vee_{\vec{\mathsf{N}}^{\mathrm{ord}}_\kappa}$, 421
$\widehat{G}_{\vec{\mathsf{N}}^{\mathrm{ord}}_{\tilde\kappa}}$, 411
$\widehat{G}_{\mathsf{N}_{\tilde\kappa}}$, 109
$\widehat{g}_p$, 398
$\widehat{\mathrm{G}}(R)$, 45, 55
$[\widehat{g}]^{\mathrm{tor}}$, 102
$([\widehat{g}]^{\mathrm{tor}})^*$, 103
${}^\heartsuit G$, 59, 270, 271, 296
g_l, 103, 403
$g_{l,0}$, 403
$[\vec{g}_l]^{*,\mathrm{ord,tor}}_{\kappa',\kappa,r_{\kappa'}}$, 406
$[\vec{g}_l]^{*,\mathrm{ord}}_{\kappa',\kappa}$, 404, 406

$[\vec{g_l}]^{*,\mathrm{ord},\mathrm{ext}}_{\kappa',\kappa}$, 405
$[\vec{g_l}]^{*,\mathrm{ord},\mathrm{ext},\vee}_{\kappa',\kappa}$, 405
$([\vec{g_l}]^{*,\mathrm{ord},\mathrm{ext}}_{\kappa',\kappa})^*$, 405
$[\vec{g_l}]^{*,\mathrm{ord},\mathrm{grp}}_{\kappa',\kappa}$, 404, 405
$[\vec{g_l}]^{*,\mathrm{ord},\mathrm{grp},\vee}_{\kappa',\kappa}$, 405
$[\vec{g_l}]^{*,\mathrm{ord},\mathrm{tor}}_{\kappa',\kappa}$, 404, 406
$([\vec{g_l}]^{*,\mathrm{ord},\mathrm{tor}}_{\kappa',\kappa})^*$, 405
$[g_l]^*_{\kappa',\kappa}$, 104
$([g_l]^*_{\kappa',\kappa})^*$, 105, 405
$[g_l]^{*,\mathrm{tor}}_{\kappa',\kappa}$, 104
$([g_l]^{*,\mathrm{tor}}_{\kappa',\kappa})^*$, 104, 105
$g_{l,p}$, 403
$[g_l]_Q$, 103, 403
$[g_l]^*_Q$, 104, 403
$([g_l]^*_Q)^*$, 105
$\mathrm{G}^{\mathrm{ess}}_{l,\mathtt{Z}_n}$, 235
$\mathrm{G}^{\mathrm{ess}}_{l,\mathtt{Z}_{p^r},\mathtt{D}_{p^r}}$, 237
$\mathrm{G}^{\mathrm{ess},\mathrm{ord}}_{l,\mathtt{Z}_{p^r},\mathtt{D}_{p^r}}$, 238
$\mathrm{G}_{(L^{\mathtt{Z}},\langle\,\cdot\,,\,\cdot\,\rangle^{\mathtt{Z}})}$, 30
$\mathrm{G}'_{l,\mathtt{Z}}(R)$, 31, 56
$\mathrm{G}_{l,\mathtt{Z}}(R)$, 31, 56, 57
$G_{\vec{\mathsf{M}}^{\mathrm{ord}}_{\mathcal{H}}}$, 295
$G_{\vec{\mathsf{M}}^{\mathrm{ord}}_{\mathcal{H},r_\kappa}}$, 421
$G^\vee_{\vec{\mathsf{M}}^{\mathrm{ord}}_{\mathcal{H},r_\kappa}}$, 421
$G_{\mathsf{M}_{\mathcal{H}}}$, 58
$[g]^{\min}$, 3, 65, 358
$G_{\vec{\mathsf{N}}^{\mathrm{ord}}_\kappa}$, 421
$G^\vee_{\vec{\mathsf{N}}^{\mathrm{ord}}_\kappa}$, 421
$G^\natural$, 68, 70, 272, 313, 315, 454
$[g]^{\mathrm{ord}}$, 191, 209
g_p, 149, 183, 191, 208, 518
  of twisted U_p type, 195, 209
  of U_p type, 194, 211
$[g_p^{-1}]$, 211
$g_{p,-1}$, 183
$[g_p^{-1}]^{\mathrm{ord}}$, 184, 191
${}^t g_p^{-1}$, 194
${}^t g_{p,-1}$, 183
$g_{p,0}$, 183
$[g_p]^{\mathrm{ord}}$, 184
g_p^{ord}, 183
$g_{p,\#}$, 194
$g_{p,\#,0}$, 183
$\mathrm{G}(\mathbb{Q})$, 22
$\mathrm{G}(\mathbb{Q})_0$, 39
$\mathrm{G}(R)$, 22, 30, 56
$\mathrm{G}(\mathbb{R})$, 22
$\mathrm{G}(\mathbb{R})_0$, 39
$[g]^{\mathrm{ord}}_{r_{\mathcal{H}'}}$, 209
$[g]^{\mathrm{tor}}_{r_{\mathcal{H}'},r_{\mathcal{H}}}$, 301
g_σ, 495
$[g]^*$, 123, 519
$g_{\check{\tau}}$, 451
$\widetilde{\mathrm{G}}$, 45
$\widetilde{G}$, 108, 411
$\widetilde{\mathrm{G}}_{1,\widetilde{\mathtt{Z}}}(R)$, 45, 55
$\widetilde{G}^\ddagger$, 109, 412
$\widetilde{\mathrm{G}}'_{h,\widetilde{\mathtt{Z}}}(R)$, 45
$\widetilde{\mathrm{G}}_{l,\check{\mathtt{Z}},\widetilde{\mathtt{Z}}}(R)$, 56, 57, 378
$\widetilde{\mathrm{G}}'_{l,\widetilde{\mathtt{Z}}}(R)$, 45
$\widetilde{\mathrm{G}}_{l,\widetilde{\mathtt{Z}}}(R)$, 45
$\widetilde{G}_{\vec{\mathsf{N}}^{\mathrm{ord}}_\kappa}$, 421
$\widetilde{G}^\vee_{\vec{\mathsf{N}}^{\mathrm{ord}}_\kappa}$, 421
$\widetilde{G}_{\vec{\mathsf{N}}^{\mathrm{ord},\mathrm{tor}}_\kappa}$, 422
$\widetilde{G}^\vee_{\vec{\mathsf{N}}^{\mathrm{ord},\mathrm{tor}}_\kappa}$, 422
$\widetilde{G}^\natural$, 108, 411
$\widetilde{g}_p$, 402, 403
$(\mathrm{G}\ltimes\widehat{\mathrm{U}})(R)$, 45
$[g]^{\mathrm{tor}}$, 3, 65, 522
$\mathrm{G}(\mathbb{Z})$, 22
$\mathrm{G}^{\mathtt{Z}}$, 30
$G^\natural_{\overline{\mathtt{Z}}^{\mathrm{ord}}}$, 319
$G^\natural_{\overline{\mathtt{Z}}}$, 73
$\mathrm{G}(\hat{\mathbb{Z}})$, 22
$\mathrm{G}(\hat{\mathbb{Z}}^p)$, 22
$\mathrm{G}(\mathbb{Z}/n\mathbb{Z})$, 22
$\Gamma_0(p^r)$, 6
$\Gamma_1(p^r)$, 2, 6
  balanced, 6, 13–15
$\Gamma_{\check{\Phi}_{\widehat{\mathcal{H}}}}$, 48, 52, 98, 373, 394
$\Gamma_{\check{\Phi}_{\widehat{\mathcal{H}}},\Phi_{\mathcal{H}}}$, 48, 50, 373, 445
$(\Gamma_{\check{\Phi}_{\widehat{\mathcal{H}}},\Phi_{\mathcal{H}}})^\vee_{\mathbb{R}}$, 50, 455
$\Gamma_{\check{\Phi}_{\widetilde{\mathcal{H}}}}$, 48, 373
$\Gamma_{\check{\Phi}_{\widetilde{\mathcal{H}}},\Phi_{\mathcal{H}}}$, 48, 373
$\Gamma_{\check{\Phi}_{\widetilde{\mathcal{H}}},\Phi_{\mathcal{H}},\check{\sigma}}$, 48, 373
$\Gamma_{\check{\Phi}_{\widetilde{\mathcal{H}}},\tau}$, 445
$\Gamma_{\check{\Phi}_{\widetilde{\mathcal{H}}},\check{\tau}}$, 410
$\Gamma_{\Phi_{\mathcal{H}}}$, 34, 48, 66
$\Gamma_{\Phi_{\mathcal{H}},\ell_0}$, 500
$\Gamma_{\Phi_{\mathcal{H}},\sigma}$, 58, 62, 271, 290, 295, 347

$\Gamma_{\Phi_{\mathcal{H}},\tau}$, 445
$\Gamma_{\widetilde{\Phi}_{\widetilde{\mathcal{H}}}}$, 48
$\Gamma_{\widetilde{\Phi}_{\widetilde{\mathcal{H}}},\Phi_{\mathcal{H}}}$, 50
γ^*, 452, 496, 498, 503
Gauss–Manin connection, 25, 97, 101, 121, 391, 398, 470, 485
  extended, 101, 121, 398, 469–471, 486, 508
generalized Kuga family, 4, *see* Kuga family, generalized
  ordinary, *see* ordinary Kuga family, generalized
generalized ordinary Kuga family, *see* ordinary Kuga family, generalized
generalized U_p operator, 184
geometrically normal, 198, 350
geometrically unibranch, 198
good algebraic model
  ordinary, *see* ordinary good algebraic model
good formal model
  ordinary, *see* ordinary good formal model
good prime, 1, 9, 17, 23, 44, 128, 152–153, 198, 203, 212–213, 305–310, 349, 369, 425–428, 490–491
$(\mathrm{Gr}^{\mathsf{Z}}_{-1,n_0}, \langle\,\cdot\,,\,\cdot\,\rangle_{11,n_0})$, 231
$\mathrm{Gr}^{\mathsf{Z}}_{-1,p^r}$, 171
$\mathrm{Gr}^{\mathsf{Z}}_{-1,\mathbb{R}}$, 30
$\mathrm{Gr}^{\mathsf{Z}}_{-1}$, 30
$\mathrm{Gr}^{\mathsf{Z}}_{0}(g)$, 31
$\mathrm{Gr}_{-2}(\hat{\beta}^{\natural,0}_{\bar{s}})$, 71
$\mathrm{Gr}_{-2}(\hat{\beta}^{\natural,\#,0}_{\bar{s}})$, 71
$\mathrm{Gr}^{\mathsf{Z}}_{-2,p^r}$, 218, 221, 243
$\mathrm{Gr}^{\mathsf{Z}^\#}_{-2,p^r}$, 218, 221, 243
$(\mathrm{Gr}^{\mathsf{Z}^\#}_{-2,p^r})^{\mathrm{mult}}_S$, 221
$(\mathrm{Gr}^{\mathsf{Z}}_{-2,p^r})^{\mathrm{mult}}_S$, 221
$\mathrm{Gr}^{\mathsf{Z}}_{-2}(g)$, 31
$\mathrm{Gr}^{\mathsf{F}^{(g)}}_{-i}$, 39
$\mathrm{Gr}^{\mathsf{V}}_{-i}$, 39
$\mathrm{Gr}^{\mathsf{Z}}_{-i}$, 30
$\mathrm{Gr}^{\mathsf{Z}}_{-i}(g)$, 30
$\mathrm{Gr}^{\widetilde{\mathsf{Z}}}_{-i}$, 45
$\mathrm{Gr}^{a}(\mathscr{N})$, 504
Gr_{D}, 169
$\mathrm{Gr}^{-1}_{\mathsf{D}}$, 165, 168, 477
$(\mathrm{Gr}^{-1}_{\mathsf{D}_{-1,p^r}})_\eta$, 224
$\mathrm{Gr}^{0}_{\mathsf{D}_{-1,p^r}}$, 243
$(\mathrm{Gr}^{-1}_{\mathsf{D}^\#_{-1,p^r}})_\eta$, 224
$\mathrm{Gr}^{0}_{\mathsf{D}^\#_{-1,p^r}}$, 243
$\mathrm{Gr}^{0}_{\mathsf{D}}$, 165, 168, 314
$\mathrm{Gr}^{-1}_{\mathsf{D},0}$, 477
$\mathrm{Gr}^{0}_{\mathsf{D},0}$, 478
$(\mathrm{Gr}^{0}_{\mathsf{D}})^{\mathrm{mult}}_s$, 165
$\mathrm{Gr}^{i}_{\mathsf{D}}$, 170
$\mathrm{Gr}^{-1}_{\mathsf{D},p^r}$, 169, 177
$\mathrm{Gr}^{0}_{\mathsf{D},p^r}$, 169, 218
$(\mathrm{Gr}^{0}_{\mathsf{D}_{p^r}})^{\mathrm{mult}}_\eta$, 217
$(\mathrm{Gr}^{0}_{\mathsf{D},p^r})^{\mathrm{mult}}_S$, 177, 314
$(\mathrm{Gr}^{0}_{\mathsf{D}^\#_{p^r}})^{\mathrm{mult}}_\eta$, 217
$\mathrm{Gr}^{-1}_{\mathsf{D}^\#}$, 165, 168, 477
$\mathrm{Gr}^{0}_{\mathsf{D}^\#}$, 165, 169, 314
$\mathrm{Gr}^{-1}_{\mathsf{D}^\#,0}$, 477
$\mathrm{Gr}^{0}_{\mathsf{D}^\#,0}$, 478
$(\mathrm{Gr}^{0}_{\mathsf{D}^\#})^{\mathrm{mult}}_s$, 165
$\mathrm{Gr}^{-1}_{\mathsf{D}^\#,p^r}$, 169, 177
$\mathrm{Gr}^{0}_{\mathsf{D}^\#,p^r}$, 169, 218
$(\mathrm{Gr}^{0}_{\mathsf{D}^\#,p^r})^{\mathrm{mult}}_S$, 177, 314
$\mathrm{Gr}^{a}_{\mathsf{F}}(W)$, 120, 483, 511
$\mathrm{Gr}^{a}_{\mathsf{K}}(\overline{\Omega}^{\bullet}_{\vec{\mathsf{N}}^{\mathrm{ord,tor}}/\vec{\mathsf{S}}_{0,r_{\mathcal{H}}}})$, 470
$\mathrm{Gr}^{\mathsf{Z}}(g)$, 31
Gram–Schmidt process, 369
Grauert–Riemenschneider vanishing theorem, 491
Griffiths transversality, 101, 398, 471
group scheme
  étale, 155
  étale-multiplicative-type, 155
  of fiberwise geometric connected components, 158
  multiplicative-type, 155

$\mathcal{H}$, 3, 23, 197
h (morphism to $\vec{C}^{\mathrm{ord}}_{\Phi_{\mathcal{H}},\delta_{\mathcal{H}}}$), 451
h (morphism to $(\vec{\mathsf{M}}^{\mathrm{ord},\mathsf{Z}_{\mathcal{H}}}_{\mathcal{H}})^{\wedge}_{\bar{x}}$), 504
h (morphism to $\vec{\mathsf{M}}^{\mathrm{ord},\mathsf{Z}_{\mathcal{H}}}_{\mathcal{H}}$), 495, 504
h (polarization), 26
h_{-1}, 42
h_0, 22, 26
$(h_0)_{-1}$, 30, 42
$h_{0,\mathrm{aux}}$, 128
$\widetilde{h}_0$, 44
$\underline{H}^{\mathrm{dR}}_1(A/\mathsf{M}_{\mathcal{H}})$, 24
$\underline{H}^{\mathrm{dR}}_1(G_{\vec{\mathsf{M}}^{\mathrm{ord}}_{\mathcal{H}}}/\vec{\mathsf{M}}^{\mathrm{ord}}_{\mathcal{H}})$, 486, 490

$\underline{H}_1^{\mathrm{dR}}(G_{\vec{\mathsf{M}}_{\mathcal{H}}^{\mathrm{ord}}}/\vec{\mathsf{M}}_{\mathcal{H}}^{\mathrm{ord}})^{\mathrm{can}}$, 485, 490
$\underline{H}_1^{\mathrm{dR}}(G_{\mathsf{M}_{\mathcal{H}}}/\mathsf{M}_{\mathcal{H}})^{\mathrm{can}}$, 121
$\underline{H}_1^{\log\text{-dR}}(\vec{\mathsf{N}}^{\mathrm{ord,tor}}/\vec{\mathsf{M}}_{\mathcal{H}}^{\mathrm{ord,tor}})$, 485
$\underline{H}_1^{\log\text{-dR}}(\mathsf{N}^{\mathrm{tor}}/\mathsf{M}_{\mathcal{H}}^{\mathrm{tor}})$, 121
$\mathcal{H}_{\mathrm{aux}}$, 129, 199, 333
$\mathcal{H}_{\mathrm{aux}}^p$, 9, 129, 141, 199, 334
$\mathcal{H}_{\mathrm{aux},p}$, 199
$\underline{\mathscr{H}}^d(\mathscr{M})$ (over $\mathfrak{N}_{\check{\sigma},\tau}$), 461
$\mathscr{H}^d(\mathscr{M})$ (over $\mathfrak{N}_{\check{\sigma},\tau}$), 460
$\underline{\mathscr{H}}^d(\mathscr{M})$ (over $\mathfrak{N}$), 498
$\underline{\mathscr{H}}^d(\mathscr{O}_{(\vec{\mathsf{N}}^{\mathrm{ord,tor}})^\wedge_{\vec{\mathsf{Z}}^{\mathrm{ord}}_{[(\Phi_{\mathcal{H}},\delta_{\mathcal{H}},\tau)]}}})$, 461
$\underline{\mathscr{H}}^d(\mathscr{O}^{0+}_{(\vec{\mathsf{N}}^{\mathrm{ord,tor}})^\wedge_{\vec{\mathsf{Z}}^{\mathrm{ord}}_{[(\Phi_{\mathcal{H}},\delta_{\mathcal{H}},\tau)]}}})$, 461
$\underline{\mathscr{H}}^d(\mathscr{O}^{+}_{(\vec{\mathsf{N}}^{\mathrm{ord,tor}})^\wedge_{\vec{\mathsf{Z}}^{\mathrm{ord}}_{[(\Phi_{\mathcal{H}},\delta_{\mathcal{H}},\tau)]}}})$, 461
$\underline{\mathscr{H}}^d(\mathscr{O}^{++}_{(\vec{\mathsf{N}}^{\mathrm{ord,tor}})^\wedge_{\vec{\mathsf{Z}}^{\mathrm{ord}}_{[(\Phi_{\mathcal{H}},\delta_{\mathcal{H}},\tau)]}}})$, 461
$\underline{\mathscr{H}}^d(\mathscr{O}_{\vec{\mathfrak{X}}^{\mathrm{ord}}_{\check{\Phi}_{\widetilde{\mathcal{H}}},\check{\delta}_{\widetilde{\mathcal{H}}},\check{\sigma},\tau}})$, 461
$\underline{\mathscr{H}}^d(\mathscr{O}^{0+}_{\vec{\mathfrak{X}}^{\mathrm{ord}}_{\check{\Phi}_{\widetilde{\mathcal{H}}},\check{\delta}_{\widetilde{\mathcal{H}}},\check{\sigma},\tau}})$, 461
$\underline{\mathscr{H}}^d(\mathscr{O}^{+}_{\vec{\mathfrak{X}}^{\mathrm{ord}}_{\check{\Phi}_{\widetilde{\mathcal{H}}},\check{\delta}_{\widetilde{\mathcal{H}}},\check{\sigma},\tau}})$, 461
$\underline{\mathscr{H}}^d(\mathscr{O}^{++}_{\vec{\mathfrak{X}}^{\mathrm{ord}}_{\check{\Phi}_{\widetilde{\mathcal{H}}},\check{\delta}_{\widetilde{\mathcal{H}}},\check{\sigma},\tau}})$, 461
$\underline{\mathscr{H}}^d(\mathscr{O}_{\vec{\Xi}^{\mathrm{ord}}_{\check{\Phi}_{\widetilde{\mathcal{H}}},\check{\delta}_{\widetilde{\mathcal{H}}},\tau}})$, 461
$\underline{\mathscr{H}}^d(\mathscr{O}_{\vec{\mathsf{Z}}^{\mathrm{ord}}_{[(\Phi_{\mathcal{H}},\delta_{\mathcal{H}},\tau)]}})$, 461
$\underline{H}^1_{\mathrm{dR}}(A/\mathsf{M}_{\mathcal{H}})$, 24
$\underline{H}^1_{\mathrm{dR}}(G_{\vec{\mathsf{M}}_{\mathcal{H}}^{\mathrm{ord}}}/\vec{\mathsf{M}}_{\mathcal{H}}^{\mathrm{ord}})^{\mathrm{can}}$, 485
$\underline{H}^1_{\mathrm{dR}}(G_{\mathsf{M}_{\mathcal{H}}}/\mathsf{M}_{\mathcal{H}})^{\mathrm{can}}$, 121
$\underline{H}^i_{\mathrm{dR}}(\mathsf{N}^{\mathrm{grp}}/\vec{\mathsf{M}}_{\mathcal{H}}^{\mathrm{ord}})$, 391
$\underline{H}^i_{\mathrm{dR}}(\mathsf{N}^{\mathrm{grp}}/\mathsf{M}_{\mathcal{H}})$, 97
$\underline{H}^i_{\mathrm{dR}}(\mathsf{N}/\vec{\mathsf{M}}_{\mathcal{H}}^{\mathrm{ord}})$, 391
$\underline{H}^i_{\mathrm{dR}}(\mathsf{N}/\mathsf{M}_{\mathcal{H}})$, 97
$\mathcal{H}_{\mathrm{G}_{1,\mathrm{Z}}}$, 31, 56, 67, 72, 312
$\mathcal{H}_{\mathrm{G}_{h,\mathrm{Z}}}$, 31, 56, 66, 72, 311
$\mathcal{H}^{p,\prime}_{\mathrm{G}_{h,\mathrm{Z}}}$, 317
$\mathcal{H}_{\mathrm{G}_{h,\mathrm{Z}},\Phi}$, 66, 311
$\mathcal{H}_{\mathrm{G}'_{h,\mathrm{Z}}}$, 31, 56, 67, 70, 312
$\mathcal{H}_{\mathrm{G}_{l,\mathrm{Z}}}$, 31, 56
$\mathcal{H}_{\mathrm{G}'_{l,\mathrm{Z}}}$, 31, 56
$\mathcal{H}_h$, 32
$\mathcal{H}'_h$, 32
$\mathcal{H}''_h$, 32
$\widehat{\mathcal{H}}$, 45, 47, 53, 86, 97, 107, 371, 374, 393, 408
$\widehat{\mathcal{H}}_{\mathrm{G}}$, 45, 46, 56, 85, 97, 372, 393
$\widehat{\mathcal{H}}_{\widehat{\mathrm{G}}_{1,\check{\mathrm{Z}}}}$, 56
$\widehat{\mathcal{H}}_{\widehat{\mathrm{G}}_{h,\check{\mathrm{Z}}}}$, 56
$\widehat{\mathcal{H}}_{\widehat{\mathrm{G}}'_{h,\check{\mathrm{Z}}}}$, 56
$\widehat{\mathcal{H}}_{\widehat{\mathrm{G}}_{l,\check{\mathrm{Z}}}}$, 56
$\widehat{\mathcal{H}}_{\widehat{\mathrm{G}}'_{l,\check{\mathrm{Z}}}}$, 56
$\widehat{\mathcal{H}}_{\widehat{\mathrm{P}}_{\check{\mathrm{Z}}}}$, 56
$\widehat{\mathcal{H}}_{\widehat{\mathrm{P}}'_{\check{\mathrm{Z}}}}$, 56
$\widehat{\mathcal{H}}^{\mathrm{ord}}_{p,\mathrm{M}^{\mathrm{ord}}_{\mathrm{D}}}$, 377
$\widehat{\mathcal{H}}^{\mathrm{ord}}_{p}$, 377
$\widehat{H}_{p^r}$, 377
$\widehat{\mathcal{H}}^{\mathrm{ord}}_{p,\widehat{\mathrm{U}}^{\mathrm{ord}}_{\mathrm{D}}}$, 377
$(\widehat{\mathcal{H}}, \widehat{\Sigma})$, 53, 97
$(\widehat{\mathcal{H}}, \widehat{\Sigma}^{\mathrm{ord}})$, 374, 393
$\widehat{\mathcal{H}}_{\widehat{\mathrm{U}}}$, 45
$\widehat{\mathcal{H}}_{\widehat{\mathrm{U}}_{1,\check{\mathrm{Z}}}}$, 56
$\widehat{\mathcal{H}}_{\widehat{\mathrm{U}}_{2,\check{\mathrm{Z}}}}$, 56, 88, 381
$\widehat{\mathcal{H}}_{\widehat{\mathrm{U}}_{\check{\mathrm{Z}}}}$, 56
$\widehat{\mathcal{H}}_{\widehat{\mathrm{Z}}_{\check{\mathrm{Z}}}}$, 56
$\mathcal{H}_\kappa$, 97, 393
$\underline{H}^1_{\log\text{-dR}}(\vec{\mathsf{N}}^{\mathrm{ord,tor}}/\vec{\mathsf{M}}_{\mathcal{H}}^{\mathrm{ord,tor}})$, 397–398, 485
$\underline{H}^1_{\log\text{-dR}}(\mathsf{N}^{\mathrm{tor}}/\mathsf{M}_{\mathcal{H}}^{\mathrm{tor}})$, 101, 121
$\underline{H}^i_{\log\text{-dR}}(\vec{\mathsf{N}}^{\mathrm{ord,tor}}/\vec{\mathsf{M}}_{\mathcal{H}}^{\mathrm{ord,tor}})$, 397, 469
$\underline{H}^i_{\log\text{-dR}}(\mathsf{N}^{\mathrm{tor}}/\mathsf{M}_{\mathcal{H}}^{\mathrm{tor}})$, 100
H_n, 29, 66, 69, 80, 236
H_{n_0}, 172, 236, 314, 325
$H_{n_0,\mathrm{P}^{\mathrm{ess}}_{\mathrm{Z}_{n_0}}} \backslash H_{n_0}$, 236
$H_{n,\mathrm{G}^{\mathrm{ess}}_{h,\mathrm{Z}_n}}$, 62, 236
$H^{\mathrm{ord}}_{n,\mathrm{G}^{\mathrm{ess}}_{h,\mathrm{Z}_n}}$, 254
$H^{\mathrm{ord},\prime}_{n,\mathrm{G}^{\mathrm{ess}}_{h,\mathrm{Z}_n}}$, 254
$H'_{n,\mathrm{G}^{\mathrm{ess}}_{h,\mathrm{Z}_n}}$, 252
$H_{n,\mathrm{G}^{\mathrm{ess}}_{h,\mathrm{Z}_n} \ltimes \mathrm{U}^{\mathrm{ess}}_{1,\mathrm{Z}_n}}$, 62, 67, 252
$H^{\mathrm{ord}}_{n,\mathrm{G}^{\mathrm{ess}}_{h,\mathrm{Z}_n} \ltimes \mathrm{U}^{\mathrm{ess}}_{1,\mathrm{Z}_n}}$, 254
$H_{n,\mathrm{G}^{\mathrm{ess}}_{h,\mathrm{Z}_n} \ltimes \mathrm{U}^{\mathrm{ess}}_{\mathrm{Z}_n}}$, 76, 252
$H^{\mathrm{ord}}_{n,\mathrm{G}^{\mathrm{ess}}_{h,\mathrm{Z}_n} \ltimes \mathrm{U}^{\mathrm{ess}}_{\mathrm{Z}_n}}$, 254
$H_{n,\mathrm{G}^{\mathrm{ess}}_{l,\mathrm{Z}_n}}$, 236
$H^{\mathrm{ord}}_{n,\mathrm{G}^{\mathrm{ess}}_{l,\mathrm{Z}_n}}$, 254
$H^{\mathrm{ord},\prime}_{n,\mathrm{G}^{\mathrm{ess}}_{l,\mathrm{Z}_n}}$, 254
$H'_{n,\mathrm{G}^{\mathrm{ess}}_{l,\mathrm{Z}_n}}$, 252
H^{ord}_n, 254

$H_{n,\mathrm{P}^{\mathrm{ess}}_{\mathtt{Z}_n}}$, 236
$H^{\mathrm{ord}}_{n,\mathrm{P}^{\mathrm{ess}}_{\mathtt{Z}_n}}$, 254
$H_{n,*}$, 236
$H_{n,\mathrm{U}^{\mathrm{ess}}_{1,\mathtt{Z}_n}}$, 62, 67, 236
$H^{\mathrm{ord}}_{n,\mathrm{U}^{\mathrm{ess}}_{1,\mathtt{Z}_n}}$, 254
$H_{n,\mathrm{U}^{\mathrm{ess}}_{2,\mathtt{Z}_n}}$, 76, 236
$H^{\mathrm{ord}}_{n,\mathrm{U}^{\mathrm{ess}}_{2,\mathtt{Z}_n}}$, 254
$H_{n,\mathrm{U}^{\mathrm{ess}}_{\mathtt{Z}_n}}$, 236
$H^{\mathrm{ord}}_{n,\mathrm{U}^{\mathrm{ess}}_{\mathtt{Z}_n}}$, 254
$H_{n,\mathrm{Z}^{\mathrm{ess}}_{\mathtt{Z}_n}}$, 236
$H^{\mathrm{ord}}_{n,\mathrm{Z}^{\mathrm{ess}}_{\mathtt{Z}_n}}$, 254
$\mathcal{H}$-orbit, 241
$\mathcal{H}^p$, 6, 197, 490
$\mathcal{H}_p$, 6, 177, 197
$\mathcal{H}^{\mathrm{ord}}_p$, 179, 197
H_{p^r}, 236, 238
$H^{\mathrm{ord}}_{p^r,\mathrm{G}^{\mathrm{ess,ord}}_{h,\mathtt{Z}_{p^r},\mathtt{D}_{p^r}}}$, 238
$H^{\mathrm{ord},\prime}_{p^r,\mathrm{G}^{\mathrm{ess,ord}}_{h,\mathtt{Z}_{p^r},\mathtt{D}_{p^r}}}$, 253
$H^{\mathrm{ord}}_{p^r,\mathrm{G}^{\mathrm{ess,ord}}_{h,\mathtt{Z}_{p^r},\mathtt{D}_{p^r}}\ltimes\mathrm{U}^{\mathrm{ess,ord}}_{\mathtt{Z}_{p^r},\mathtt{D}_{p^r}}}$, 253
$H^{\mathrm{ord}}_{p^r,\mathrm{G}^{\mathrm{ess,ord}}_{h,\mathtt{Z}_{p^r},\mathtt{D}_{p^r}}\ltimes\mathrm{U}^{1,\mathrm{ess,ord}}_{\mathtt{Z}_{p^r},\mathtt{D}_{p^r}}}$, 253
$H^{\mathrm{ord}}_{p^r,\mathrm{G}^{\mathrm{ess,ord}}_{l,\mathtt{Z}_{p^r},\mathtt{D}_{p^r}}}$, 238
$H^{\mathrm{ord},\prime}_{p^r,\mathrm{G}^{\mathrm{ess,ord}}_{l,\mathtt{Z}_{p^r},\mathtt{D}_{p^r}}}$, 253
$H'_{p^r,\mathrm{G}^{\mathrm{ess}}_{l,\mathtt{Z}_{p^r}}}$, 253
$H^{\mathrm{ord}}_{p^r}$, 177, 197, 238, 315
$H^{\mathrm{ord}}_{p^r,\mathrm{P}^{\mathrm{ess,ord}}_{\mathtt{Z}_{p^r},\mathtt{D}_{p^r}}}$, 238
$H^{\mathrm{ord}}_{p^r,\mathrm{P}^{\mathrm{ess,ord}}_{\mathtt{Z}_{p^r},\mathtt{D}_{p^r}}}\backslash H^{\mathrm{ord}}_{p^r}$, 239
$H^{\mathrm{ord}}_{p^{r'}}$, 180
$H^{\mathrm{ord}}_{p^r,*}$, 238
$H^{\mathrm{ord}}_{p^r,\mathrm{U}^{\mathrm{ess,ord}}_{1,\mathtt{Z}_{p^r},\mathtt{D}_{p^r}}}$, 238
$H^{\mathrm{ord}}_{p^r,\mathrm{U}^{\mathrm{ess,ord}}_{2,\mathtt{Z}_{p^r},\mathtt{D}_{p^r}}}$, 238
$H^{\mathrm{ord}}_{p^r,\mathrm{U}^{\mathrm{ess,ord}}_{\mathtt{Z}_{p^r},\mathtt{D}_{p^r}}}$, 238
$H^{\mathrm{ord}}_{p^r,\mathrm{Z}^{\mathrm{ess,ord}}_{\mathtt{Z}_{p^r},\mathtt{D}_{p^r}}}$, 238
$\mathcal{H}_{\mathrm{P}_{\mathtt{Z}}}$, 31, 56, 66
$\mathcal{H}_{\mathrm{P}_{\mathtt{Z}}}/\mathcal{H}_{\mathrm{U}_{2,\mathtt{Z}}}$, 57
$\mathcal{H}_{\mathrm{P}_{\mathtt{Z}},\Phi}$, 66
$\mathcal{H}_{\mathrm{P}'_{\mathtt{Z}}}$, 31, 56, 76, 324
h_τ, 451
$\widetilde{\mathcal{H}}$, 45, 85, 371
$\widetilde{\mathcal{H}}_{\widehat{\mathrm{G}}}$, 45, 86, 107, 371, 372, 408
$\widetilde{\mathcal{H}}_{\widetilde{\mathrm{G}}_{1,\widetilde{\mathtt{Z}}}}$, 45
$\widetilde{\mathcal{H}}_{\widetilde{\mathrm{G}}_{h,\widetilde{\mathtt{Z}}}}$, 45
$\widetilde{\mathcal{H}}_{\widetilde{\mathrm{G}}'_{h,\widetilde{\mathtt{Z}}}}$, 45
$\widetilde{\mathcal{H}}_{\widetilde{\mathrm{G}}_{l,\breve{\mathtt{Z}},\widetilde{\mathtt{Z}}}}$, 56
$\widetilde{\mathcal{H}}_{\widetilde{\mathrm{G}}_{l,\widetilde{\mathtt{Z}}}}$, 45
$\widetilde{\mathcal{H}}_{\widetilde{\mathrm{G}}'_{l,\widetilde{\mathtt{Z}}}}$, 45
$\widetilde{\mathcal{H}}^p$, 371
$\widetilde{\mathcal{H}}_p$, 371
$\widetilde{\mathcal{U}}^{\mathrm{bal}}_{p,1}(p^r)_{\widehat{\mathrm{G}}}$, 371
$\widetilde{\mathcal{H}}_{\widetilde{\mathrm{P}}_{\breve{\mathtt{Z}},\widetilde{\mathtt{Z}}}}$, 56
$\widetilde{\mathcal{H}}_{\widetilde{\mathrm{P}}_{\breve{\mathtt{Z}},\widetilde{\mathtt{Z}}}}/\widetilde{\mathcal{H}}_{\widetilde{\mathrm{U}}_{2,\breve{\mathtt{Z}}}}$, 57
$\widetilde{\mathcal{H}}_{\widetilde{\mathrm{P}}_{\widetilde{\mathtt{Z}}}}$, 45
$\widetilde{\mathcal{H}}_{\widetilde{\mathrm{P}}'_{\widetilde{\mathtt{Z}}}}$, 45
$(\widetilde{\mathcal{H}},\widetilde{\Sigma}^{\mathrm{ord}},\widetilde{\sigma})$, 373
$(\widetilde{\mathcal{H}},\widetilde{\Sigma},\widetilde{\sigma})$, 46, 85
$\widetilde{\mathcal{H}}_{\widehat{\mathrm{U}}}$, 45
$\widetilde{\mathcal{H}}_{\widetilde{\mathrm{U}}_{1,\widetilde{\mathtt{Z}}}}$, 45
$\widetilde{\mathcal{H}}_{\widetilde{\mathrm{U}}_{2,\widetilde{\mathtt{Z}}}}$, 45
$\widetilde{\mathcal{H}}_{\widetilde{\mathrm{U}}_{\widetilde{\mathtt{Z}}}}$, 45
$\widetilde{\mathcal{H}}_{\widetilde{\mathrm{Z}}_{\widetilde{\mathtt{Z}}}}$, 45
$\mathcal{H}_{\mathrm{U}_{1,\mathtt{Z}}}$, 31, 56, 67, 312
$\mathcal{H}_{\mathrm{U}_{2,\mathtt{Z}}}$, 31, 56, 76, 80, 325
$\mathcal{H}_{\mathrm{U}_{\mathtt{Z}}}$, 31, 56
$\mathcal{H}_{\mathrm{Z}_{\mathtt{Z}}}$, 31, 56, 66
$\mathrm{Hasse}_{\mathrm{aux}}$, 362
$\mathrm{Hasse}^{a_0}_{\mathrm{aux}}$, 363
$\mathrm{Hasse}^{a_0 N_1}_{\mathrm{aux}}$, 364
$\mathrm{Hasse}^{a N_1}_{\mathcal{H}}$, 364
$\mathrm{Hasse}^{\mathrm{ord}}_{\mathcal{H}}$, 362
$(\mathrm{Hasse}^{\mathrm{ord}}_{\mathcal{H}})^a$, 363
Hasse invariant, 361–365
healthy regular, 162
Hecke action, 9, 13, 16, 38, 55
  of elements of U_p type, 211, 304, 358–359, 402–403
  of $\mathrm{G}_{1,\mathtt{Z}}(\mathbb{A}^\infty)$, 75, 83, 87
  of $\mathrm{G}_{1,\mathtt{Z}}(\mathbb{A}^{\infty,p})\times\mathrm{P}^{\mathrm{ord}}_{1,\mathtt{Z},\mathtt{D}}(\mathbb{Q}_p)$, 321, 322, 380
  of $\mathrm{G}(\mathbb{A}^\infty)$, 3, 65–66, 123–124
  of $\mathrm{G}(\mathbb{A}^{\infty,p})\times\mathrm{G}(\mathbb{Z}_p)$, 5, 149–150, 518–520
  of $\mathrm{G}(\mathbb{A}^{\infty,p})\times\mathrm{P}^{\mathrm{ord}}_{\mathtt{D}}(\mathbb{Q}_p)$, 7, 208–210, 300–303, 353–358, 488–490

of $\mathrm{G}_{h,\mathrm{Z}}(\mathbb{A}^\infty)$, 67, 75
of $\mathrm{G}_{h,\mathrm{Z}}(\mathbb{A}^{\infty,p}) \times \mathrm{P}^{\mathrm{ord}}_{h,\mathrm{Z},\mathrm{D}}(\mathbb{Q}_p)$, 311, 321
of $\widehat{\mathrm{G}}_{1,\breve{\mathrm{Z}}}(\mathbb{A}^\infty)$, 87, 90
of $\widehat{\mathrm{G}}_{1,\breve{\mathrm{Z}}}(\mathbb{A}^{\infty,p}) \times \widehat{\mathrm{P}}^{\mathrm{ord}}_{1,\breve{\mathrm{Z}},\mathrm{D}}(\mathbb{Q}_p)$, 380, 384
of $\widehat{\mathrm{G}}(\mathbb{A}^\infty)$, 102–103
of $\widehat{\mathrm{G}}(\mathbb{A}^{\infty,p}) \times \widehat{\mathrm{P}}^{\mathrm{ord}}_{\mathrm{D}}(\mathbb{Q}_p)$, 398–403, 443–444
of $\widehat{\mathrm{G}}'_{l,\breve{\mathrm{Z}}}(\mathbb{A}^\infty)$, 87, 90, 91, 381, 385
of $\mathrm{G}'_{l,\mathrm{Z}}(\mathbb{A}^\infty)$, 75, 83, 87, 322, 328, 381
of $\mathrm{G}'_{l,\mathrm{Z}}(\mathbb{A}^\infty) \times \mathrm{G}_{h,\mathrm{Z}}(\mathbb{A}^\infty)$, 75
of $\mathrm{G}'_{l,\mathrm{Z}}(\mathbb{A}^\infty) \times (\mathrm{G}'_{h,\mathrm{Z}}(\mathbb{A}^{\infty,p}) \times \mathrm{P}^{\mathrm{ord},\prime}_{h,\mathrm{Z},\mathrm{D}}(\mathbb{Q}_p))$, 322
of $\widetilde{\mathrm{G}}(\mathbb{A}^\infty)$, 117
of $\widetilde{\mathrm{G}}_{l,\breve{\mathrm{Z}},\widetilde{\mathrm{Z}}}(\mathbb{A}^\infty)$, 87, 381
of $\mathrm{G}^{\mathrm{Z}}(\mathbb{A}^\infty)$, 67
of $\mathrm{G}^{\mathrm{Z}}(\mathbb{A}^{\infty,p}) \times \mathrm{P}^{\mathrm{ord}}_{\mathrm{D}^{\mathrm{Z}}}(\mathbb{Q}_p)$, 311
of $\mathrm{G}(\hat{\mathbb{Z}}^p) \times \mathrm{P}^{\mathrm{ord}}_{\mathrm{D}}(\mathbb{Z}_p)$, 210–211, 303, 358
of $\mathrm{GL}_{\mathcal{O} \underset{\mathbb{Z}}{\otimes} \mathbb{A}^\infty}(Q \underset{\mathbb{Z}}{\otimes} \mathbb{A}^\infty)$, 103–105, 403–406, 443–444
of $\widehat{\mathrm{P}}_{\breve{\mathrm{Z}}}(\mathbb{A}^\infty)$, 90, 95
of $\widehat{\mathrm{P}}_{\breve{\mathrm{Z}}}(\mathbb{A}^{\infty,p}) \times \widehat{\mathrm{P}}^{\mathrm{ord}}_{\breve{\mathrm{Z}},\mathrm{D}}(\mathbb{Q}_p)$, 385, 389
of $(\widehat{\mathrm{P}}_{\breve{\mathrm{Z}}}(\mathbb{A}^{\infty,p}) \times \widehat{\mathrm{P}}^{\mathrm{ord}}_{\breve{\mathrm{Z}},\mathrm{D}}(\mathbb{Q}_p))/\widehat{\mathrm{U}}_{2,\breve{\mathrm{Z}}}(\mathbb{A}^\infty)$, 381, 385
of $\widehat{\mathrm{P}}_{\breve{\mathrm{Z}}}(\mathbb{A}^\infty)/\widehat{\mathrm{U}}_{2,\breve{\mathrm{Z}}}(\mathbb{A}^\infty)$, 87, 90
of $\widehat{\mathrm{P}}'_{\breve{\mathrm{Z}}}(\mathbb{A}^\infty)$, 90, 95
of $\widehat{\mathrm{P}}'_{\breve{\mathrm{Z}}}(\mathbb{A}^{\infty,p}) \times \widehat{\mathrm{P}}^{\mathrm{ord},\prime}_{\breve{\mathrm{Z}},\mathrm{D}}(\mathbb{Q}_p)$, 384, 389
of $\widetilde{\mathrm{P}}'_{\breve{\mathrm{Z}}}(\mathbb{A}^\infty)$, 94
of $\widetilde{\mathrm{P}}'_{\breve{\mathrm{Z}}}(\mathbb{A}^{\infty,p}) \times \widetilde{\mathrm{P}}^{\mathrm{ord},\prime}_{\breve{\mathrm{Z}},\widetilde{\mathrm{D}}}(\mathbb{Q}_p)$, 389
of $\widetilde{\mathrm{P}}_{\breve{\mathrm{Z}},\widetilde{\mathrm{Z}}}(\mathbb{A}^\infty)$, 94
of $\widetilde{\mathrm{P}}_{\breve{\mathrm{Z}},\widetilde{\mathrm{Z}}}(\mathbb{A}^{\infty,p}) \times \widetilde{\mathrm{P}}^{\mathrm{ord}}_{\breve{\mathrm{Z}},\widetilde{\mathrm{Z}},\widetilde{\mathrm{D}}}(\mathbb{Q}_p)$, 389
of $(\widetilde{\mathrm{P}}_{\breve{\mathrm{Z}},\widetilde{\mathrm{Z}}}(\mathbb{A}^{\infty,p}) \times \widetilde{\mathrm{P}}^{\mathrm{ord}}_{\breve{\mathrm{Z}},\widetilde{\mathrm{Z}},\widetilde{\mathrm{D}}}(\mathbb{Q}_p))/\widetilde{\mathrm{U}}_{2,\breve{\mathrm{Z}}}(\mathbb{A}^\infty)$, 380
of $\widetilde{\mathrm{P}}_{\breve{\mathrm{Z}},\widetilde{\mathrm{Z}}}(\mathbb{A}^\infty)/\widetilde{\mathrm{U}}_{2,\breve{\mathrm{Z}}}(\mathbb{A}^\infty)$, 87
of $\mathrm{P}_{\mathrm{Z}}(\mathbb{A}^\infty)$, 75, 83, 94, 95
of $\mathrm{P}_{\mathrm{Z}}(\mathbb{A}^{\infty,p}) \times \mathrm{P}^{\mathrm{ord}}_{\mathrm{Z},\mathrm{D}}(\mathbb{Q}_p)$, 322, 328, 389
of $(\mathrm{P}_{\mathrm{Z}}(\mathbb{A}^{\infty,p}) \times \mathrm{P}^{\mathrm{ord}}_{\mathrm{Z},\mathrm{D}}(\mathbb{Q}_p))/\mathrm{U}_{2,\mathrm{Z}}(\mathbb{A}^\infty)$, 322, 328, 381
of $\mathrm{P}_{\mathrm{Z}}(\mathbb{A}^\infty)/\mathrm{U}_{2,\mathrm{Z}}(\mathbb{A}^\infty)$, 75, 83, 87
of $\mathrm{P}'_{\mathrm{Z}}(\mathbb{A}^\infty)$, 75, 83, 94, 95
of $\mathrm{P}'_{\mathrm{Z}}(\mathbb{A}^{\infty,p}) \times \mathrm{P}^{\mathrm{ord},\prime}_{\mathrm{Z},\mathrm{D}}(\mathbb{Q}_p)$, 322, 328, 389
Hecke twist, 13
by $\mathrm{G}_{1,\mathrm{Z}}(\mathbb{A}^\infty)$, 75
by $\mathrm{G}(\mathbb{A}^\infty)$, 102
by $\mathrm{G}(\mathbb{A}^{\infty,p})$, 175–177
by $\mathrm{G}^{\mathrm{Z}}(\mathbb{A}^\infty) \cong \mathrm{G}_{h,\mathrm{Z}}(\mathbb{A}^\infty)$, 67
ordinary, *see* ordinary Hecke twist
by $\widehat{\mathrm{P}}_{\breve{\mathrm{Z}}}(\mathbb{A}^\infty)$, 90
by $\widehat{\mathrm{P}}'_{\breve{\mathrm{Z}}}(\mathbb{A}^\infty)$, 90
by $\widetilde{\mathrm{P}}'_{\breve{\mathrm{Z}}}(\mathbb{A}^\infty)$, 95
by $\widetilde{\mathrm{P}}_{\breve{\mathrm{Z}},\widetilde{\mathrm{Z}}}(\mathbb{A}^\infty)$, 95
by $\mathrm{P}_{\mathrm{Z}}(\mathbb{A}^\infty)$, 83
by $\mathrm{P}_{\mathrm{Z}}(\mathbb{A}^\infty)/\mathrm{U}_{2,\mathrm{Z}}(\mathbb{A}^\infty)$, 75
by $\mathrm{P}'_{\mathrm{Z}}(\mathbb{A}^\infty)$, 83
hereditary, 41, 169
higher Koecher's principle, 492
Hodge
filtration, 97, 101, 391, 397, 469, 470
invertible sheaf, 127, 331
spectral sequence, 100, 397, 469
Hodge invertible sheaf, 11
$\underline{\mathrm{Hom}}_{\mathcal{O}}(\frac{1}{n}X, B^{\vee,\mathrm{ord}}_{p^r})^\circ$, 250
$\underline{\mathrm{Hom}}_{\mathcal{O}}(\frac{1}{n}X, B^{\mathrm{ord}}_{p^r})^\circ$, 249
$\underline{\mathrm{Hom}}_{\mathcal{O}}(\frac{1}{n}Y, B^{\mathrm{ord}}_{p^r})^\circ$, 250
$\underline{\mathrm{Hom}}_{\mathcal{O}}(\underline{N}, Z)$, 157
$\underline{\mathrm{Hom}}_{\mathcal{O}}(N, \overline{Z})$, 163
$\underline{\mathrm{Hom}}_{\mathcal{O}}(N, \overline{Z})^\circ$, 163
$\underline{\mathrm{Hom}}_{\mathcal{O}}(N, Z)^\circ$, 158
$\underline{\mathrm{Hom}}_{\mathcal{O}}(Q^\vee, G^\vee_{\vec{\mathsf{M}}^{\mathrm{ord}}_{\mathcal{H}}})^\circ$, 390
$\underline{\mathrm{Hom}}_{\mathcal{O}}(Q^\vee, G^\vee_{\mathsf{M}_{\mathcal{H}}})^\circ$, 96
$\underline{\mathrm{Hom}}_{\mathcal{O}}(Q, G_{\vec{\mathsf{M}}^{\mathrm{ord}}_{\mathcal{H}}})^\circ$, 390
$\underline{\mathrm{Hom}}_{\mathcal{O}}(Q, G_{\mathsf{M}_{\mathcal{H}}})$, 95
$\underline{\mathrm{Hom}}_{\mathcal{O}}(Q, G_{\mathsf{M}_{\mathcal{H}}})^\circ$, 95
$\underline{\mathrm{Hom}}_{\mathcal{O}}(X, B)^\circ$, 67, 312
$\underline{\mathrm{Hom}}_{\mathcal{O}}(X, B^{\mathrm{ord}}_{p^r})^\circ$, 312
$\underline{\mathrm{Hom}}_{\mathcal{O}}(Y, B^\vee)^\circ$, 250
$\underline{\mathrm{Hom}}_{\mathbb{Z}}(Q^\vee, G^\vee_{\mathsf{M}_{\mathcal{H}}})$, 96
$\underline{\mathrm{Hom}}_{\mathbb{Z}}(Q^\vee, G^\vee_{\vec{\mathsf{M}}^{\mathrm{ord}}_{\mathcal{H}}})$, 390
$\underline{\mathrm{Hom}}_{\mathbb{Z}}(Q, G_{\vec{\mathsf{M}}^{\mathrm{ord}}_{\mathcal{H}}})$, 390
$\underline{\mathrm{Hom}}_{\mathbb{Z}}(Q, G_{\mathsf{M}_{\mathcal{H}}})$, 96

i, 57, 58, 197, 205, 215, 295
i_σ, 498
$i^{[0]}$, 291
$I_{0,0}$, 217
$\vec{i}$, 141
$\vec{i}_{\mathrm{aux}}$, 141
i_{aux}, 333
$i^{\triangledown}_{\mathrm{aux}}$, 129, 130, 137, 199, 333
$i^{\vartriangle}_{\mathrm{aux}}$, 130, 137, 199
i_B, 66, 70, 138, 216, 248, 281, 313, 316, 454
$i_{B_{\mathrm{aux}}}$, 138
$i_{B^\ddagger}$, 59, 112, 292, 297, 415
$i_{\widehat{B}^\ddagger}$, 110, 413
$i_{\widetilde{B}^\ddagger}$, 110

$i_{\widetilde{B}^\ddagger}$, 413
$\bar{i}$, 422
$\mathscr{I}_{\vec{\mathsf{D}}^{\mathrm{ord}}_{\infty,\mathcal{H}}}$, 397, 488
$\mathscr{I}_{\mathsf{D}_{\infty,\mathcal{H}}}$, 100, 123, 522
$i^\dagger$, 59, 145, 296
$i^\ddagger$, 59, 112, 291, 296, 415
$^\diamond i$, 280
$\mathscr{I}_{\vec{\mathsf{E}}^{\mathrm{ord}}_\infty}$, 397
$\mathscr{I}_{\mathsf{E}_\infty}$, 100
$\widehat{i}$, 108, 411
$\widehat{i}^\ddagger$, 109, 412
$\widehat{i}_{\vec{\mathsf{N}}^{\mathrm{ord}}_{\widetilde{\kappa}}}$, 411
$\widehat{i}_{\mathsf{N}_{\widetilde{\kappa}}}$, 109
$^\heartsuit i$, 59, 270, 271, 296
I_ℓ, 269
$i_{\vec{\mathsf{M}}^{\mathrm{ord}}_{\mathcal{H}}}$, 295
$i_{\mathsf{M}_\mathcal{H}}$, 58
$i^\natural$, 69, 70, 272, 313, 315, 316
$i^{\natural,0}$, 313
$i^{\natural,\#,0}$, 313
$i^{(p)}$, 190
$\mathscr{I}^{\mathrm{ord}}_{\Phi_\mathcal{H},\delta_\mathcal{H},\sigma}$, 269
i^+_σ, 499
$\widetilde{i}$, 108, 411
$\widetilde{i}^\ddagger$, 109, 412
$\widetilde{i}^\natural$, 108, 411
$\widetilde{\mathscr{I}}^{\mathrm{ord}}_{\breve{\sigma},\breve{\tau}}$, 410
$\widetilde{\mathscr{I}}^{\mathrm{ord}}_{\breve{\tau}}$, 410
$i_X(b)$, 216
$i_Y(b)$, 216
$I_{y,\chi}$, 217, 269
$I_{y,\phi(y)}$, 217
$i^\natural_{\vec{\overline{\mathsf{Z}}}^{\mathrm{ord}}}$, 319
$i^\natural_{\overline{\mathsf{Z}}}$, 73
$\mathscr{I}_{\vec{\overline{\mathsf{Z}}}^{\mathrm{ord}}_{[(\Phi_\mathcal{H},\delta_\mathcal{H},\tau)]}}$, 445
Igusa tower, v
image($\alpha^{\mathrm{ord},0}_{\mathcal{H}_p}$), 180
image($\alpha^{\mathrm{ord},\#,0}_{\mathcal{H}_p}$), 180
image($\alpha^{\mathrm{ord},0}_{H^{\mathrm{ord}}_{p^r}}$), 179
image($\alpha^{\mathrm{ord},\#,0}_{H^{\mathrm{ord}}_{p^r}}$), 179
image($\alpha^{\mathrm{ord},0}_{p^r}$), 178, 218
image($\alpha^{\mathrm{ord},\#,0}_{p^r}$), 178, 218
image($\beta^{\natural,\mathrm{ord},0}_{p^r}$), 314
image($\beta^{\natural,\mathrm{ord},\#,0}_{p^r}$), 314
image($\varphi^{\mathrm{ord},0}_{-1,p^r}$), 224, 248, 314
image($\varphi^{\mathrm{ord},\#,0}_{-1,p^r}$), 224, 248, 314
integrable filtration, 27
integral PEL datum, 2, 21
  auxiliary choice, 9
Inv(R), 269, 292
$\underline{\mathrm{Inv}}$(S), 279, 282
Inv(V), 60, 297
$\underline{\mathrm{Inv}}$($\vec{\overline{\Xi}}^{\mathrm{ord}}_{\Phi_\mathcal{H},\delta_\mathcal{H}}$), 268
inverse different, 43
ι, 77, 79, 227, 251, 272, 324
$\imath^{\mathrm{alg}}$, 286
$\iota^\vee$, 324
$\iota^{\mathrm{ord}}_\mathcal{H}$, 241
$\widehat{\iota}^\ddagger$, 111, 414
$\widehat{\iota}^{\vee,\ddagger}$, 111, 414
$\widehat{\iota}^\ddagger_{\widehat{\mathcal{H}}}$, 111, 414
$\widehat{\iota}^{\vee,\ddagger}_{\widehat{\mathcal{H}}}$, 111, 414
$\widehat{\iota}^{\vee,\mathrm{ord},\ddagger}_{\widehat{\mathcal{H}}}$, 414
$\widehat{\iota}^{\mathrm{ord},\ddagger}_{\widehat{\mathcal{H}}}$, 414
$\widehat{\iota}^\ddagger_{\widehat{\mathcal{H}}^p}$, 414
$\widehat{\iota}^{\vee,\ddagger}_{\widehat{\mathcal{H}}^p}$, 414
$\widehat{\iota}^\ddagger_n$, 111
$\widehat{\iota}^{\vee,\ddagger}_n$, 111
ι_n, 77
ι_{n_0}, 251
$\iota^\vee_n$, 78
ι^{ord}_n, 251
$\imath^{\mathrm{nat}}$, 286
ι_{p^r}, 228
$\widetilde{\iota}^\ddagger$, 110
$\widetilde{\iota}^{\ddagger,1}$, 111
$\widetilde{\iota}^{\vee,\ddagger}$, 110
$\widetilde{\iota}^{\vee,\ddagger,1}$, 111
$\widetilde{\iota}^{\mathrm{ord},\ddagger}_{\widetilde{\mathcal{H}}_p}$, 414
$\widetilde{\iota}^\ddagger_n$, 111
$\widetilde{\iota}^{\ddagger,1}_n$, 111
$\widetilde{\iota}^{\vee,\ddagger}_n$, 111
$\widetilde{\iota}^{\vee,\ddagger,1}_n$, 111
isotrivial torus, 216
Iwasawa decomposition, 368, 369

$\vec{\mathcal{J}}_{\mathcal{H},d_0\mathsf{pol}}$, 5, 147, 361
$\vec{\mathcal{J}}_{\mathcal{H},d_0\mathsf{pol}^{\mathrm{ord}}}$, 8, 360
$\vec{\jmath}_{\mathcal{H},d_0\mathsf{pol}^{\mathrm{ord}}}$, 360
$\vec{\mathcal{J}}_{\mathcal{H},d\mathsf{pol}}$, 147
$\vec{\mathcal{J}}_{\mathcal{H},d\mathsf{pol}^{\mathrm{ord}}}$, 359
$\vec{\jmath}_{\mathcal{H},d\mathsf{pol}^{\mathrm{ord}}}$, 359

$\vec{\mathcal{J}}_{\mathcal{H}',d_0\mathsf{pol}'}$, 148
$j^{\vee,\natural}$, 70, 315, 316
$\mathcal{J}_{\mathcal{H},d_0\mathsf{pol}}$, 4, 5, 63
$\jmath_{\mathcal{H},d_0\mathsf{pol}}$, 63
$\mathcal{J}_{\mathcal{H},d\mathsf{pol}}$, 63
$\jmath_{\mathcal{H},d\mathsf{pol}}$, 63
$\mathcal{J}_{\mathcal{H}^p,d\mathsf{pol}^p}$, 309
$\jmath_{\mathcal{H}^p,d\mathsf{pol}^p}$, 309
$\jmath_{\mathcal{H},\mathsf{pol}}$, 63
$\mathcal{J}^{(d)}_{\mathcal{H},\mathsf{pol}}$, 63
$\mathcal{J}_{\mathcal{H}',d\mathsf{pol}'}$, 64
$j^{\natural}$, 70, 315, 316
j_Q, 43, 101, 390, 397, 486

$\mathsf{K}^a(\overline{\Omega}^{\bullet}_{\vec{\mathsf{N}}^{\mathrm{ord,tor}}/\vec{\mathsf{S}}_{0,r_{\mathcal{H}}}})$, 470
$\mathbf{K}_{Q,\mathcal{H}}$, 53, 97
$\mathbf{K}^{\mathrm{ord}}_{Q,\mathcal{H}}$, 374, 392
$\mathbf{K}^{\mathrm{ord},+}_{Q,\mathcal{H}}$, 374, 392
$\mathbf{K}^{\mathrm{ord},++}_{Q,\mathcal{H}}$, 374, 392
$\mathbf{K}^{+}_{Q,\mathcal{H}}$, 53, 97
$\mathbf{K}^{++}_{Q,\mathcal{H}}$, 53, 97
$\mathbf{K}_{Q,\mathcal{H},\Sigma}$, 55, 97
$\mathbf{K}^{\mathrm{ord}}_{Q,\mathcal{H},\Sigma^{\mathrm{ord}}}$, 375, 392
$\mathbf{K}^{\mathrm{ord},+}_{Q,\mathcal{H},\Sigma^{\mathrm{ord}}}$, 375, 392
$\mathbf{K}^{\mathrm{ord},++}_{Q,\mathcal{H},\Sigma^{\mathrm{ord}}}$, 375, 392
$\mathbf{K}^{+}_{Q,\mathcal{H},\Sigma}$, 55, 97
$\mathbf{K}^{++}_{Q,\mathcal{H},\Sigma}$, 55, 97
$\widetilde{\mathbf{K}}_{Q,\mathcal{H}}$, 46
$\widetilde{\mathbf{K}}^{\mathrm{ord}}_{Q,\mathcal{H}}$, 373
$\widetilde{\mathbf{K}}^{\mathrm{ord},+}_{Q,\mathcal{H}}$, 373
$\widetilde{\mathbf{K}}^{\mathrm{ord},++}_{Q,\mathcal{H}}$, 373
$\widetilde{\mathbf{K}}^{+}_{Q,\mathcal{H}}$, 46
$\widetilde{\mathbf{K}}^{++}_{Q,\mathcal{H}}$, 46, 85
$\mathbf{K}^{\mathrm{std}}_{Q,\mathcal{H},\Sigma}$, 376
$\mathbf{K}^{\mathrm{std},+}_{Q,\mathcal{H},\Sigma}$, 376
$\mathbf{K}^{\mathrm{std},++}_{Q,\mathcal{H},\Sigma}$, 376
$\mathbf{K}^{\mathrm{std}}_{Q,\mathcal{H}}$, 376
$\widetilde{\mathbf{K}}^{\mathrm{std}}_{Q,\mathcal{H}}$, 376
$\mathbf{K}^{\mathrm{std},+}_{Q,\mathcal{H}}$, 376
$\widetilde{\mathbf{K}}^{\mathrm{std},+}_{Q,\mathcal{H}}$, 376
$\mathbf{K}^{\mathrm{std},++}_{Q,\mathcal{H}}$, 376
$\widetilde{\mathbf{K}}^{\mathrm{std},++}_{Q,\mathcal{H}}$, 376
κ, 53, 97, 108, 116, 374, 393, 420
κ_{aux}, 425
$\widetilde{\kappa}_{\mathrm{aux}}$, 425
κ^{isog}, 97, 116, 393, 420
$[\widetilde{\kappa}]$, 53, 108, 116, 374, 420
$\widetilde{\kappa}$, 46, 85, 373
$[\widetilde{\kappa}_{\mathrm{aux}}]$, 425
$\widetilde{\kappa}^{\mathrm{isog}}$, 107, 408
$\widetilde{\kappa}^{\mathrm{tor}}$, 107, 409
κ^{tor}, 98, 116, 393, 420
Kodaira–Spencer
class
extended, 471
isomorphism, 26, 207, 208
extended, 121, 273, 486
morphism, 25, 207, 208, 274–276, 487
extended, 59, 121, 272, 277, 281, 282, 284, 285, 287, 296, 431, 486, 487
Koecher's principle, 492
Koszul filtration, 470
$\underline{\mathrm{KS}}$, 57, 121, 206, 272
KS, 207
$\underline{\mathrm{KS}}_{(A,\lambda,i,\alpha_{\mathcal{H}})/\mathsf{M}_{\mathcal{H}}}$, 26
$\underline{\mathrm{KS}}_{(A,\lambda,i)/\mathsf{M}_{\mathcal{H}}}$, 26
$\mathrm{KS}_{A/\mathsf{M}_{\mathcal{H}}/\mathsf{S}_0}$, 25, 26
$\mathrm{KS}_{A/S/\vec{\mathsf{S}}_{0,r_{\mathcal{H}}}}$, 207
$\mathrm{KS}_{(B,c)/\vec{C}^{\mathrm{ord}}_{\Phi_{\mathcal{H}},\delta_{\mathcal{H}}}/\vec{\mathsf{S}}_{0,r_{\mathcal{H}}}}$, 274
$(\underline{\mathrm{KS}}_{(B,c,c^{\vee},\lambda^{\natural},i^{\natural})/\vec{C}^{\mathrm{ord}}_{\Phi_1,\delta_1}}/$
$\underline{\mathrm{KS}}_{(B,\lambda_B,i_B)/\vec{C}^{\mathrm{ord}}_{\Phi_1,\delta_1}})_{\mathrm{free}}$, 275
$\underline{\mathrm{KS}}_{(B,c,c^{\vee},\lambda^{\natural},i^{\natural})/\vec{C}^{\mathrm{ord}}_{\Phi_{\mathcal{H}},\delta_{\mathcal{H}}}}$, 274, 275
$\underline{\mathrm{KS}}_{(B,c,c^{\vee},\lambda^{\natural},i^{\natural})/\vec{C}^{\mathrm{ord}}_{\Phi_{\mathcal{H}},\delta_{\mathcal{H}}},\mathrm{free}}$, 274, 275
$(\underline{\mathrm{KS}}_{(B,c,c^{\vee},\lambda^{\natural},i^{\natural})/\vec{C}^{\mathrm{ord}}_{\Phi_{\mathcal{H}},\delta_{\mathcal{H}}}}/$
$\underline{\mathrm{KS}}_{(B,\lambda_B,i_B)/\vec{C}^{\mathrm{ord}}_{\Phi_{\mathcal{H}},\delta_{\mathcal{H}}}})_{\mathrm{free}}$, 275
$\mathrm{KS}_{(B^{\vee},c^{\vee})/\vec{C}^{\mathrm{ord}}_{\Phi_{\mathcal{H}},\delta_{\mathcal{H}}}/\vec{\mathsf{S}}_{0,r_{\mathcal{H}}}}$, 274
$\underline{\mathrm{KS}}_{(B,\lambda_B,i_B)/\vec{C}^{\mathrm{ord}}_{\Phi_{\mathcal{H}},\delta_{\mathcal{H}}}}$, 275
$\underline{\mathrm{KS}}_{(B,\lambda_B,i_B)/\vec{\mathsf{M}}^{\mathrm{ord},\Phi_{\mathcal{H}}}_{\mathcal{H}}}$, 275
$\underline{\mathrm{KS}}_{(B,\lambda_B,i_B)/\vec{\mathsf{M}}^{\mathrm{ord},\Phi_{\mathcal{H}}}_{\mathcal{H}},\mathrm{free}}$, 274
$\mathrm{KS}_{B/\vec{\mathsf{M}}^{\mathrm{ord},\Phi_{\mathcal{H}}}_{\mathcal{H}}/\vec{\mathsf{S}}_{0,r_{\mathcal{H}}}}$, 274
$\mathrm{KS}_{B/\vec{\Xi}^{\mathrm{ord}}_{\Phi_{\mathcal{H}},\delta_{\mathcal{H}}}/\vec{\mathsf{S}}_{0,r_{\mathcal{H}}}}$, 274
$\underline{\mathrm{KS}}_{\mathrm{free}}$, 206, 273, 486
$\underline{\mathrm{KS}}_{(\overline{G},\overline{\lambda},\overline{i})/\vec{\mathsf{N}}^{\mathrm{ord,tor}}}$, 431
$\underline{\mathrm{KS}}_{({}^{\diamond}G,{}^{\diamond}\lambda,{}^{\diamond}i)/S,\mathrm{free}}$, 281, 282, 284
$\mathrm{KS}_{{}^{\diamond}G/S/\vec{\mathsf{S}}_{0,r_{\mathcal{H}}}}$, 281, 282, 284
$\underline{\mathrm{KS}}_{(G,\lambda,i,\alpha_{\mathcal{H}^p},\alpha^{\mathrm{ord}}_{\mathcal{H}_p})/S}$, 206
$\underline{\mathrm{KS}}_{(G,\lambda,i,\alpha_{\mathcal{H}^p},\alpha^{\mathrm{ord}}_{\mathcal{H}_p})/S,\mathrm{free}}$, 206
$\underline{\mathrm{KS}}_{(G,\lambda,i,\alpha_{\mathcal{H}})/S}$, 57
$\underline{\mathrm{KS}}_{(G,\lambda,i)/\vec{\mathsf{M}}^{\mathrm{ord,tor}}_{\mathcal{H}},\mathrm{free}}$, 296

$\underline{\mathrm{KS}}_{(G,\lambda,i)/\vec{\mathsf{M}}^{\mathrm{ord,tor}}_{\mathcal{H},r_\kappa},\mathrm{free}}$, 431
$\underline{\mathrm{KS}}_{(G,\lambda,i)/\mathsf{M}^{\mathrm{tor}}_{\mathcal{H}}}$, 59
$\underline{\mathrm{KS}}_{(G,\lambda,i)/S}$, 57, 206
$\underline{\mathrm{KS}}_{(G,\lambda,i)/S,\mathrm{free}}$, 206, 286, 287
$\mathrm{KS}_{G/\vec{\mathsf{M}}^{\mathrm{ord,tor}}_{\mathcal{H}}/\vec{\mathsf{S}}_{0,r_{\mathcal{H}}}}$, 296, 486
$\mathrm{KS}_{G/\vec{\mathsf{M}}^{\mathrm{ord,tor}}_{\mathcal{H},r_\kappa}/\vec{\mathsf{S}}_{0,r_\kappa}}$, 431
$\mathrm{KS}_{G/\mathsf{M}^{\mathrm{tor}}_{\mathcal{H}}/\mathsf{S}_0}$, 59, 121
$\mathrm{KS}_{(G^\natural,\iota)/\vec{\vec{\Xi}}^{\mathrm{ord}}_{\Phi_{\mathcal{H}},\delta_{\mathcal{H}}}/\vec{\mathsf{S}}_{0,r_{\mathcal{H}}}}$, 272
$\mathrm{KS}_{(G^\natural,\iota)/\vec{\Xi}^{\mathrm{ord}}_{\Phi_{\mathcal{H}},\delta_{\mathcal{H}}}/\vec{\mathsf{S}}_{0,r_{\mathcal{H}}}}$, 276
$\underline{\mathrm{KS}}_{(G^\natural,\lambda^\natural,i^\natural)/\vec{\vec{\Xi}}^{\mathrm{ord}}_{\Phi_{\mathcal{H}},\delta_{\mathcal{H}}},\mathrm{free}}$, 273
$\underline{\mathrm{KS}}_{(G^\natural,\lambda^\natural,i^\natural)/\vec{\vec{\Xi}}^{\mathrm{ord}}_{\Phi_{\mathcal{H}},\delta_{\mathcal{H}}}}$, 272
$\mathrm{KS}_{G/S/\vec{\mathsf{S}}_{0,r_{\mathcal{H}}}}$, 286, 287
$\underline{\mathrm{KS}}_{(\widetilde{G},\widetilde{\lambda},\widetilde{i})/\widetilde{\vec{\mathsf{M}}}^{\mathrm{ord,tor}}_{\widetilde{\mathcal{H}}},\mathrm{free}}$, 431
$\underline{\mathrm{KS}}_{(\widetilde{G},\widetilde{\lambda},\widetilde{i})_{\vec{\mathsf{N}}^{\mathrm{ord,tor}}}/\vec{\mathsf{N}}^{\mathrm{ord,tor}}}$, 431
$\mathrm{KS}_{\widetilde{G}/\widetilde{\vec{\mathsf{M}}}^{\mathrm{ord,tor}}_{\widetilde{\mathcal{H}}}/\vec{\mathsf{S}}_{0,r_\kappa}}$, 431
$\underline{\mathrm{KS}}_{(T,\lambda_T,i_T)/\vec{\mathsf{S}}_{0,r_{\mathcal{H}}},\mathrm{free}}$, 276
$\underline{\mathrm{KS}}_{(T,\lambda_T,i_T)/\vec{\vec{\Xi}}^{\mathrm{ord}}_{\Phi_{\mathcal{H}},\delta_{\mathcal{H}}},\mathrm{free}}$, 276
$\underline{\mathrm{KS}}_{(T,\lambda_T,i_T)/\vec{\Xi}^{\mathrm{ord}}_{\Phi_1,\delta_1},\mathrm{free}}$, 276
$\underline{\mathrm{KS}}_{(T,\lambda_T,i_T)/\vec{\Xi}^{\mathrm{ord}}_{\Phi_{\mathcal{H}},\delta_{\mathcal{H}}},\mathrm{free}}$, 276
$\underline{\mathrm{KS}}_{(\widetilde{T},\lambda_{\widetilde{T}},i_{\widetilde{T}})_{\vec{\mathsf{N}}^{\mathrm{ord,tor}}}/\vec{\mathsf{N}}^{\mathrm{ord,tor}}}$, 431
$\underline{\mathrm{KS}}_{(\widetilde{T},\lambda_{\widetilde{T}},i_{\widetilde{T}})_{\vec{\mathsf{S}}_{0,r_\kappa}}/\vec{\mathsf{S}}_{0,r_\kappa}}$, 432
Kuga family, 4, 95
  generalized, 4, 96
  ordinary, *see* ordinary Kuga family
  toroidal compactification, 4, 97–106
Kunz's theorem, 212, 305, 359

$[\mathcal{L}]$, 473
L, 22
$\mathcal{L}$, 472
L_0, 117
L_{aux}, 125
$(L_{\mathrm{aux}}, \langle\cdot,\cdot\rangle_{\mathrm{aux}})$, 128, 198
$L^{\#}_{\mathrm{aux}}$, 126, 198
$\breve{\ell}$, 88
$L^{(g)}$, 39
$[\mathcal{L}^{\mathrm{grp}}]$, 473
$\mathcal{L}^{\mathrm{grp}}$, 473
$(L/n_0L, \langle\cdot,\cdot\rangle)$, 172, 173
$(L/p^rL, \langle\cdot,\cdot\rangle, \mathsf{D}_{p^r})$, 178, 179
$(L, \langle\cdot,\cdot\rangle)$, 22
$(L, \langle\cdot,\cdot\rangle, h_0)$, 22
$L^{\#}$, 22
$\widetilde{L}$, 43
$\widetilde{\ell}_{j_Q}$, 474
$(\widetilde{L}, \langle\cdot,\cdot\rangle^{\sim}, \widetilde{h}_0)$, 43, 85, 371
$(\widetilde{L}, \langle\cdot,\cdot\rangle^{\sim}, \widetilde{h}_0, \widetilde{\mathsf{Z}})$, 48
$(L \underset{\mathbb{Z}}{\otimes} \mathbb{A}^{\infty,p}, \langle\cdot,\cdot\rangle, \mathsf{Z} \underset{\hat{\mathbb{Z}}}{\otimes} \mathbb{A}^{\infty,p}, \Phi)$, 316
$(L \underset{\mathbb{Z}}{\otimes} \mathbb{A}^{\infty}, \langle\cdot,\cdot\rangle, \mathsf{Z} \underset{\hat{\mathbb{Z}}}{\otimes} \mathbb{A}^{\infty}, \Phi)$, 70, 76
$(L^{\mathsf{Z}}, \langle\cdot,\cdot\rangle^{\mathsf{Z}}, h_0^{\mathsf{Z}})$, 30, 66, 311
$(L^{\mathsf{Z}_{\mathcal{H}}}, \langle\cdot,\cdot\rangle^{\mathsf{Z}_{\mathcal{H}}}, h_0^{\mathsf{Z}_{\mathcal{H}}})$, 32
$(L^{\mathsf{Z}_n}, \langle\cdot,\cdot\rangle^{\mathsf{Z}_n}, h_0^{\mathsf{Z}_n})$, 30
$(L \underset{\mathbb{Z}}{\otimes} \hat{\mathbb{Z}}^p, \langle\cdot,\cdot\rangle)$, 173, 174
$(L \underset{\mathbb{Z}}{\otimes} \hat{\mathbb{Z}}^p, \langle\cdot,\cdot\rangle, \mathsf{Z} \underset{\hat{\mathbb{Z}}}{\otimes} \hat{\mathbb{Z}}^p)$, 313, 314, 324, 325
$(L \underset{\mathbb{Z}}{\otimes} \hat{\mathbb{Z}}, \langle\cdot,\cdot\rangle, \mathsf{Z})$, 69, 79, 80
$(L \underset{\mathbb{Z}}{\otimes} \mathbb{Z}_p, \langle\cdot,\cdot\rangle, \mathsf{D})$, 179, 180
$(L \underset{\mathbb{Z}}{\otimes} \mathbb{Z}_p, \langle\cdot,\cdot\rangle, \mathsf{Z} \underset{\hat{\mathbb{Z}}}{\otimes} \mathbb{Z}_p, \mathsf{D})$, 314, 315
Lagrange, 126
λ, 23, 57, 58, 79, 197, 205, 215, 295, 324
$\lambda^{[0]}$, 291
$\vec{\lambda}$, 141
$\vec{\lambda}_{\mathrm{aux}}$, 141
λ_{aux}, 333
$\lambda^{\triangledown}_{\mathrm{aux}}$, 126, 129, 137, 199, 333
$\lambda^{\vartriangle}_{\mathrm{aux}}$, 126, 137, 199
λ_B, 66, 70, 138, 216, 248, 281, 313, 315, 454
$\lambda_{B_{\mathrm{aux}}}$, 138
$\lambda_{B^\ddagger}$, 59, 112, 292, 297, 415
$\lambda^{\mathrm{ord}}_{B,\eta,p^r}$, 226
$\lambda_{\widehat{B}^\ddagger}$, 110, 413
$\underline{\lambda}_{\widetilde{B}^\ddagger}$, 110, 413
$\overline{\lambda}$, 422
$\lambda^\dagger$, 59, 145, 296
$\lambda^\ddagger$, 59, 112, 291, 296, 415
${}^\diamond\lambda$, 280
$\lambda_{\vec{\mathsf{M}}^{\mathrm{ord}}_{\mathcal{H}}}$, 295
$\widehat{\lambda}$, 108, 411
$\widehat{\lambda}^\ddagger$, 109, 412
$\widehat{\lambda}_{\vec{\mathsf{N}}^{\mathrm{ord}}_{\widetilde{\kappa}}}$, 411
$\widehat{\lambda}_{\mathsf{N}_{\widetilde{\kappa}}}$, 109
$\widehat{\lambda}^{\natural,\ddagger}$, 111
${}^\heartsuit\lambda$, 59, 270, 271, 296
$\lambda_{\mathcal{L}}$, 473
$\lambda^{\mathrm{ext}}_{\mathcal{L}}$, 474
$\lambda_{\mathcal{L}^{\mathrm{grp}}}$, 473
$\lambda_{\vec{\mathsf{M}}^{\mathrm{ord}}_{\mathcal{H}},j_Q}$, 390, 397, 474
$\lambda_{\vec{\mathsf{M}}^{\mathrm{ord}}_{\mathcal{H}},j_Q,\mathbb{Z}}$, 390
$\lambda_{\mathsf{M}_{\mathcal{H}}}$, 58
$\lambda_{\mathsf{M}_{\mathcal{H}},j_Q}$, 96, 101

$\lambda_{\mathsf{M}_{\mathcal{H}},j_Q,\mathbb{Z}}$, 96
$\lambda_{\vec{\mathsf{N}}^{\mathrm{ord}},j_Q}$, 397, 474
λ_{N,j_Q}, 101
$\lambda^{\natural}$, 68, 70, 272, 313, 315
$\lambda^{(p)}$, 190
$\lambda^{(p^i)}$, 181
$\lambda_{\vec{\widetilde{\Psi}}^{\mathrm{ord}}_{\widetilde{\Phi}_{\widetilde{\mathcal{H}}},\widetilde{\delta}_{\widetilde{\mathcal{H}}}}(\widetilde{\ell}_{j_Q})}$, 474
λ^*, 57
λ_T, 68, 70, 313, 315
λ_T^*, 70
$\widetilde{\lambda}$, 108, 411
$\widetilde{\lambda}^{\ddagger}$, 109, 412
$\widetilde{\lambda}^{\natural}$, 108, 411
$\widetilde{\lambda}^{\natural,\ddagger}$, 110
λ-Weil pairing, 173
$\lambda^{\natural}_{\vec{\overline{\mathsf{Z}}}^{\mathrm{ord}}}$, 319
$\lambda^{\natural}_{\overline{\mathsf{Z}}}$, 73
level structure
  comparison with characteristic zero, 192–194
  naive, *see* naive level structure
  ordinary, *see* ordinary level structure
  symplectic-liftability condition, 173
  of type $(L \underset{\mathbb{Z}}{\otimes} \mathbb{A}^{\infty,p}, \langle\cdot,\cdot\rangle, \mathsf{Z} \underset{\hat{\mathbb{Z}}}{\otimes} \mathbb{A}^{\infty,p}, \Phi)$, 316
    based at $\bar{s}$, 316
  of type $(L \underset{\mathbb{Z}}{\otimes} \mathbb{A}^{\infty}, \langle\cdot,\cdot\rangle, \mathsf{Z} \underset{\hat{\mathbb{Z}}}{\otimes} \mathbb{A}^{\infty}, \Phi)$, 70
    based at $\bar{s}$, 70
  of type $(L \underset{\mathbb{Z}}{\otimes} \hat{\mathbb{Z}}^p, \langle\cdot,\cdot\rangle)$, 174
    lifting, 174
    principal, 173
  of type $(L \underset{\mathbb{Z}}{\otimes} \hat{\mathbb{Z}}^p, \langle\cdot,\cdot\rangle, \mathsf{Z} \underset{\hat{\mathbb{Z}}}{\otimes} \hat{\mathbb{Z}}^p)$, 314, 325
    principal, 313, 324
  of type $(L \underset{\mathbb{Z}}{\otimes} \hat{\mathbb{Z}}, \langle\cdot,\cdot\rangle)$, 23
  of type $(L \underset{\mathbb{Z}}{\otimes} \hat{\mathbb{Z}}, \langle\cdot,\cdot\rangle, \mathsf{Z})$, 69, 80
    principal, 69, 79
level-$\mathcal{H}$ structure datum
  of type $(L \underset{\mathbb{Z}}{\otimes} \hat{\mathbb{Z}}, \langle\cdot,\cdot\rangle)$, 243
Lie–Kolchin theorem, 506
$\underline{\mathrm{Lie}}_{A^\vee/\mathsf{M}_{\mathcal{H}}}$, 24
$\underline{\mathrm{Lie}}^\vee_{A^\vee/\mathsf{M}_{\mathcal{H}}}$, 24
$\underline{\mathrm{Lie}}_{A/\mathsf{M}_{\mathcal{H}}}$, 24
$\underline{\mathrm{Lie}}^\vee_{A/\mathsf{M}_{\mathcal{H}}}$, 24
Lie algebra condition, 6, 23, 131, 141, 164, 203, 510
$\mathrm{Lie}^\vee(\mathrm{V}_{\mathrm{aux}})$, 362
$\underline{\mathrm{Lie}}_{G^\vee/S}$, 57
$\underline{\mathrm{Lie}}^\vee_{G^\vee/S}$, 57
$\underline{\mathrm{Lie}}_{G/S}$, 57
$\underline{\mathrm{Lie}}^\vee_{G/S}$, 57
$\underline{\mathrm{Lie}}_{\vec{\mathsf{N}}^{\mathrm{ord},\vee}_{\kappa}/\vec{\mathsf{M}}^{\mathrm{ord}}_{\mathcal{H}}}$, 396
$\underline{\mathrm{Lie}}^\vee_{\vec{\mathsf{N}}^{\mathrm{ord},\vee}_{\kappa}/\vec{\mathsf{M}}^{\mathrm{ord}}_{\mathcal{H}}}$, 396
$\underline{\mathrm{Lie}}_{\vec{\mathsf{N}}^{\mathrm{ord,ext},\vee}_{\kappa}/\vec{\mathsf{M}}^{\mathrm{ord,tor}}_{\mathcal{H}}}$, 396
$\underline{\mathrm{Lie}}^\vee_{\vec{\mathsf{N}}^{\mathrm{ord,ext},\vee}_{\kappa}/\vec{\mathsf{M}}^{\mathrm{ord,tor}}_{\mathcal{H}}}$, 396
$\underline{\mathrm{Lie}}_{\vec{\mathsf{N}}^{\mathrm{ord,ext}}_{\kappa}/\vec{\mathsf{M}}^{\mathrm{ord,tor}}_{\mathcal{H}}}$, 396
$\underline{\mathrm{Lie}}^\vee_{\vec{\mathsf{N}}^{\mathrm{ord,ext}}_{\kappa}/\vec{\mathsf{M}}^{\mathrm{ord,tor}}_{\mathcal{H}}}$, 396
$\underline{\mathrm{Lie}}_{\vec{\mathsf{N}}^{\mathrm{ord}}_{\kappa}/\vec{\mathsf{M}}^{\mathrm{ord}}_{\mathcal{H}}}$, 396
$\underline{\mathrm{Lie}}^\vee_{\vec{\mathsf{N}}^{\mathrm{ord}}_{\kappa}/\vec{\mathsf{M}}^{\mathrm{ord}}_{\mathcal{H}}}$, 396
$\underline{\mathrm{Lie}}_{\mathsf{N}^\vee/\vec{\mathsf{M}}^{\mathrm{ord}}_{\mathcal{H}}}$, 391
$\underline{\mathrm{Lie}}^\vee_{\mathsf{N}^\vee/\vec{\mathsf{M}}^{\mathrm{ord}}_{\mathcal{H}}}$, 391
$\underline{\mathrm{Lie}}_{\mathsf{N}^\vee/\mathsf{M}_{\mathcal{H}}}$, 96
$\underline{\mathrm{Lie}}^\vee_{\mathsf{N}^\vee/\mathsf{M}_{\mathcal{H}}}$, 96
$\underline{\mathrm{Lie}}_{\mathsf{N}^{\mathrm{grp},\vee}/\vec{\mathsf{M}}^{\mathrm{ord}}_{\mathcal{H}}}$, 391
$\underline{\mathrm{Lie}}^\vee_{\mathsf{N}^{\mathrm{grp},\vee}/\vec{\mathsf{M}}^{\mathrm{ord}}_{\mathcal{H}}}$, 391
$\underline{\mathrm{Lie}}_{\mathsf{N}^{\mathrm{grp},\vee}/\mathsf{M}_{\mathcal{H}}}$, 96
$\underline{\mathrm{Lie}}^\vee_{\mathsf{N}^{\mathrm{grp},\vee}/\mathsf{M}_{\mathcal{H}}}$, 96
$\underline{\mathrm{Lie}}_{\mathsf{N}^{\mathrm{grp}}/\vec{\mathsf{M}}^{\mathrm{ord}}_{\mathcal{H}}}$, 391
$\underline{\mathrm{Lie}}^\vee_{\mathsf{N}^{\mathrm{grp}}/\vec{\mathsf{M}}^{\mathrm{ord}}_{\mathcal{H}}}$, 391
$\underline{\mathrm{Lie}}_{\mathsf{N}^{\mathrm{grp}}/\mathsf{M}_{\mathcal{H}}}$, 96
$\underline{\mathrm{Lie}}^\vee_{\mathsf{N}^{\mathrm{grp}}/\mathsf{M}_{\mathcal{H}}}$, 96
$\underline{\mathrm{Lie}}_{\mathsf{N}/\vec{\mathsf{M}}^{\mathrm{ord}}_{\mathcal{H}}}$, 391
$\underline{\mathrm{Lie}}^\vee_{\mathsf{N}/\vec{\mathsf{M}}^{\mathrm{ord}}_{\mathcal{H}}}$, 391
$\underline{\mathrm{Lie}}_{\mathsf{N}/\mathsf{M}_{\mathcal{H}}}$, 96
$\underline{\mathrm{Lie}}^\vee_{\mathsf{N}/\mathsf{M}_{\mathcal{H}}}$, 96
liftability condition
  for β_n, 79
  for β_{n_0}, 324
  for $\beta^{\natural}_{n_0}$, 313
  for $\beta^{\natural}_n$, 69
  for $\beta^{\natural,\mathrm{ord}}_{p^r}$, 314
  for $(c_n, c_n^\vee)$, 68
  for τ_n, 78
local model, 15, 125
log de Rham cohomology, 100, 397, 469
log de Rham complex, 100, 397
log étale, 113, 416
log smooth, 99, 395

$\mathscr{M}$, 445, 460, 498, 504
M_0, 118
$\mathrm{M}_0(R)$, 118

$\vec{\mathsf{M}}_1^{\mathrm{ord},\Phi_1}$, 262
$\vec{\mathsf{M}}_1^{\mathrm{ord},\mathsf{Z}_1}$, 262
$\vec{\mathsf{M}}^{\mathrm{ord,tor}}_{\mathrm{G_{aux}}(\hat{\mathbb{Z}}),\Sigma^{\mathrm{ord}}_{\mathrm{aux}}}$, 335
$[\vec{\mathfrak{M}}_{\mathcal{H}}]$, 210
$[\vec{\mathsf{M}}_{\mathcal{H}}]$, 5, 141
$\vec{\mathfrak{M}}_{\mathcal{H}}$, 210
$\vec{\mathsf{M}}_{\mathcal{H}}$, 5, 9, 141, 336
$\vec{\mathsf{M}}_{\mathcal{H}_{\mathrm{aux}}}$, 336
$\vec{\mathsf{M}}^{\mathrm{tor}}_{\mathcal{H}_{\mathrm{aux}},d_{0,\mathrm{aux}}\mathsf{pol}_{\mathrm{aux}},r}$, 334
$\vec{\mathsf{M}}^{\mathrm{tor,full\text{-}ord}}_{\mathcal{H}_{\mathrm{aux}},d_{0,\mathrm{aux}}\mathsf{pol}_{\mathrm{aux}},r}$, 366
$\vec{\mathsf{M}}^{\mathrm{min}}_{\mathcal{H}_{\mathrm{aux}}}$, 336
$\vec{\mathsf{M}}^{\mathrm{ord}}_{\mathcal{H}_{\mathrm{aux}}}$, 213
$\vec{\mathsf{M}}^{\mathrm{ord,min}}_{\mathcal{H}_{\mathrm{aux}}}$, 334
$\vec{\mathsf{M}}^{\mathrm{min}}_{\mathcal{H}_{\mathrm{aux}},r}$, 334
$\vec{\mathsf{M}}^{\mathrm{ord,tor}}_{\mathcal{H}_{\mathrm{aux}},\Sigma^{\mathrm{ord}}_{\mathrm{aux}}}$, 333
$\vec{\mathsf{M}}^{\mathrm{tor}}_{\mathcal{H},d_0\mathsf{pol},i}$, 5, 151
$\vec{\mathfrak{M}}^{\mathrm{tor}}_{\mathcal{H},d_0\mathsf{pol},r_{\mathcal{H}}}$, 366
$\vec{\mathsf{M}}^{\mathrm{tor}}_{\mathcal{H},d_0\mathsf{pol},r_{\mathcal{H}}}$, 361
$\vec{\mathfrak{M}}^{\mathrm{tor,full\text{-}ord}}_{\mathcal{H},d_0\mathsf{pol},r_{\mathcal{H}}}$, 366
$\vec{\mathsf{M}}^{\mathrm{tor}}_{\mathcal{H},d_0\mathsf{pol}}$, 5, 9, 147
$\vec{\mathfrak{M}}^{\mathrm{tor,non\text{-}ord}}_{\mathcal{H},d_0\mathsf{pol}}$, 365
$(\vec{\mathsf{M}}^{\mathrm{tor}}_{\mathcal{H},d_0\mathsf{pol}} \underset{\mathbb{Z}}{\otimes} \mathbb{F}_p)^{\mathrm{non\text{-}ord}}$, 365
$\vec{\mathsf{M}}^{\mathrm{ord}}_{\mathcal{H}_{\mathrm{G}_{h,\mathsf{Z}}},r_{\mathcal{H}'}}$, 311
$\vec{\mathsf{M}}_{\mathcal{H},i}$, 5, 151
$\vec{\mathsf{M}}^{\mathrm{min}}_{\mathcal{H},i}$, 5, 151
$\vec{\mathfrak{M}}^{\mathrm{min,full\text{-}ord}}_{\mathcal{H},i}$, 367
$(\vec{\mathsf{M}}^{\mathrm{min}}_{\mathcal{H},i} \underset{\mathbb{Z}}{\otimes} \mathbb{F}_p)^{\mathrm{full\text{-}ord}}$, 367
$(\vec{\mathsf{M}}^{\mathrm{min}}_{\mathcal{H},i} \underset{\mathbb{Z}}{\otimes} \mathbb{F}_p)^{\mathrm{non\text{-}ord}}$, 367
$(\vec{\mathsf{M}}^{\mathrm{min}}_{\mathcal{H},i} \underset{\mathbb{Z}}{\otimes} (\mathbb{Z}/p^j\mathbb{Z}))^{\mathrm{full\text{-}ord}}$, 367
$\vec{\mathsf{M}}^{\mathrm{ord}}_{\mathcal{H}_{\kappa}}$, 407
$\vec{\mathsf{M}}^{\mathrm{min}}_{\mathcal{H}}$, 5, 143, 336
$\vec{\mathfrak{M}}^{\mathrm{min,non\text{-}ord}}_{\mathcal{H}}$, 365
$(\vec{\mathsf{M}}^{\mathrm{min}}_{\mathcal{H}} \underset{\mathbb{Z}}{\otimes} \mathbb{F}_p)^{\mathrm{non\text{-}ord}}$, 365
$[\vec{\mathfrak{M}}_{\mathcal{H}}]^{\mathrm{non\text{-}ord}}$, 365
$\vec{\mathfrak{M}}^{\mathrm{non\text{-}ord}}_{\mathcal{H}}$, 365
$[\vec{\mathsf{M}}^{\mathrm{ord}}_{\mathcal{H}}]$, 7, 10, 214, 338, 345
  quasi-projectivity, 10
$\vec{\mathsf{M}}^{\mathrm{ord}}_{\mathcal{H}}$, 7, 10, 203, 213
$\vec{\mathsf{M}}^{\mathrm{ord},1}_{\mathcal{H}}$, 343, 348
$\vec{\mathfrak{M}}^{\mathrm{ord,min}}_{\mathcal{H}}$, 357, 365, 367
$\vec{\mathsf{M}}^{\mathrm{ord,min}}_{\mathcal{H}}$, 5, 7, 11, 309, 332, 345
  flatness, 338
  normality, 338
$(\vec{\mathsf{M}}^{\mathrm{ord,min}}_{\mathcal{H}})^{\wedge}_{\bar{x}}$, 343, 494
$\vec{\mathsf{M}}^{\mathrm{ord},\Phi_{\mathcal{H}}}_{\mathcal{H}}$, 260, 265, 295, 311, 347, 379
$\vec{\mathsf{M}}^{\mathrm{ord,tor}}_{\mathcal{H}}$, 294, 295, *see also* $\vec{\mathsf{M}}^{\mathrm{ord,tor}}_{\mathcal{H},\Sigma^{\mathrm{ord}}}$
$(\vec{\mathsf{M}}^{\mathrm{ord,tor}}_{\mathcal{H}})^{\wedge}_{\bar{x}}$, 349
$[\vec{\mathsf{M}}^{\mathrm{ord},\mathsf{Z}_{\mathcal{H}}}_{\mathcal{H}}]$, 343, 346
$\vec{\mathsf{M}}^{\mathrm{ord},\mathsf{Z}_{\mathcal{H}}}_{\mathcal{H}}$, 259, 265, 295, 311, 346
$([\vec{\mathsf{M}}^{\mathrm{ord},\mathsf{Z}_{\mathcal{H}}}_{\mathcal{H}}])^{\wedge}_{\bar{x}}$, 343
$(\vec{\mathsf{M}}^{\mathrm{ord},\mathsf{Z}_{\mathcal{H}}}_{\mathcal{H}})^{\wedge}_{\bar{x}}$, 494, 504
$\vec{\mathsf{M}}^{\mathrm{tor}}_{\mathcal{H}',d_0\mathsf{pol}'}$, 148
$\vec{\mathsf{M}}_{\mathcal{H},r_{\mathcal{H}}}$, 214
$\vec{\mathfrak{M}}^{\mathrm{min}}_{\mathcal{H},r_{\mathcal{H}}}$, 365
$\vec{\mathsf{M}}^{\mathrm{min}}_{\mathcal{H},r_{\mathcal{H}}}$, 347
$\vec{\mathfrak{M}}^{\mathrm{min,full\text{-}ord}}_{\mathcal{H},r_{\mathcal{H}}} \underset{\mathbb{Z}}{\otimes} (\mathbb{Z}/p^j\mathbb{Z})$, 367
$\vec{\mathfrak{M}}^{\mathrm{min,full\text{-}ord}}_{\mathcal{H},r_{\mathcal{H}}}$, 365
$(\vec{\mathsf{M}}^{\mathrm{min}}_{\mathcal{H},r_{\mathcal{H}}} \underset{\mathbb{Z}}{\otimes} (\mathbb{Z}/p^j\mathbb{Z}))^{\mathrm{full\text{-}ord}}$, 367
$(\vec{\mathsf{M}}_{\mathcal{H},r_{\mathcal{H}}} \underset{\mathbb{Z}}{\otimes} \mathbb{F}_p)^{\mathrm{full\text{-}ord}}$, 368
$\vec{\mathsf{M}}^{\mathrm{ord}}_{\mathcal{H},r_{\kappa}}$, 393
$\vec{\mathsf{M}}^{\mathrm{ord,min}}_{\mathcal{H},S}$, 349
$[\vec{\mathsf{M}}^{\mathrm{ord,tor}}_{\mathcal{H},\Sigma^{\mathrm{ord}}}]$, 338
$\vec{\mathfrak{M}}^{\mathrm{ord,tor}}_{\mathcal{H},\Sigma^{\mathrm{ord}}}$, 303, 366
$\vec{\mathsf{M}}^{\mathrm{ord,tor}}_{\mathcal{H},\Sigma^{\mathrm{ord}}}$, 7, 11, 14, 289, 294–295, 309
$([\vec{\mathsf{M}}_{\mathcal{H}} \underset{\mathbb{Z}}{\otimes} \mathbb{F}_p])^{\mathrm{non\text{-}ord}}$, 365
$(\vec{\mathsf{M}}_{\mathcal{H}} \underset{\mathbb{Z}}{\otimes} \mathbb{F}_p)^{\mathrm{non\text{-}ord}}$, 365
$\vec{\mathsf{M}}^{\mathrm{min}}_{\mathcal{H}}$, 9
$\vec{\mathsf{M}}^{\mathrm{ord}}_{n_0p^r}$, 204
$\vec{\mathsf{M}}^{\mathrm{ord}}_{n}$, 204
$\vec{\mathsf{M}}^{\mathrm{ord},\Phi_n}_{n}$, 311, 313
$\vec{\mathsf{M}}^{\mathrm{ord},\mathsf{Z}_n}_{n}$, 311, 313, 380
$\mathrm{M}^{\mathrm{ord}}_{\mathtt{D},0,\mathrm{fil}}$, 505
$\mathrm{M}^{\mathrm{ord}}_{\mathtt{D},0}$, 479
$\mathrm{M}^{\mathrm{ord}}_{\mathtt{D},0}(R)$, 479
$\mathrm{M}^{\mathrm{ord}}_{\mathtt{D},0,\mathrm{uni}}$, 506
$\mathrm{M}^{\mathrm{ord}}_{\mathtt{D}}$, 170
$\mathrm{M}^{\mathrm{ord}}_{\mathtt{D}}(R)$, 169
$(\mathrm{M}^{\mathrm{ord}}_{\mathtt{D}} \ltimes \widehat{\mathrm{U}}^{\mathrm{ord}}_{\mathtt{D}})(R)$, 377
$\dddot{\mathsf{M}}^{\mathrm{ord}}_{\mathcal{H}_{\mathrm{aux}}}$, 199
$\dddot{\mathsf{M}}^{\mathrm{ord}}_{\mathcal{H}_h}$, 244, 255
$\dddot{\mathsf{M}}^{\mathrm{ord}}_{\mathcal{H}}$, 6, 10, 197
$\dddot{\mathsf{M}}^{\mathrm{ord},\Phi_{\mathcal{H}}}_{\mathcal{H}}$, 255
$\dddot{\mathsf{M}}^{\mathrm{ord}}_{\mathcal{H}}(S)$, 197
$\dddot{\mathsf{M}}^{\mathrm{ord},\mathsf{Z}_{\mathcal{H}}}_{\mathcal{H}}$, 255

$\dddot{\mathsf{M}}^{\mathrm{ord}}_{n_0p^r}$, 198
$\dddot{\mathsf{M}}^{\mathrm{ord}}_{n}$, 198
$\dddot{\mathsf{M}}^{\mathrm{ord},\mathsf{Z}_n}_{n}$, 248
$[\mathsf{M}_{\mathrm{G_{aux}}(\hat{\mathbb{Z}}^p)}]$, 141
$\mathsf{M}_{\mathrm{G_{aux}}(\hat{\mathbb{Z}}^p)}$, 129, 141, 335
$\mathsf{M}^{\mathrm{min}}_{\mathrm{G_{aux}}(\hat{\mathbb{Z}}^p)}$, 143
$(\mathsf{M}^{\mathrm{min}}_{\mathrm{G_{aux}}(\hat{\mathbb{Z}}^p)} \underset{\mathbb{Z}}{\otimes} \mathbb{F}_p)^{\mathrm{non\text{-}ord}}$, 365
$\mathsf{M}^{\mathrm{tor}}_{\mathrm{G_{aux}}(\hat{\mathbb{Z}}^p),\Sigma^p_{\mathrm{aux}}}$, 335
$\mathsf{M}^{\mathrm{tor}}_{\mathrm{G_{aux}}(\hat{\mathbb{Z}}^p)}$, 361
$(\mathsf{M}^{\mathrm{tor}}_{\mathrm{G_{aux}}(\hat{\mathbb{Z}}^p)} \underset{\mathbb{Z}}{\otimes} \mathbb{F}_p)^{\mathrm{non\text{-}ord}}$, 365
$[\mathsf{M}_{\mathcal{H}}]$, 3, 26, 61
$\mathsf{M}_{\mathcal{H}}$, 3, 8, 23, 57
$[\mathsf{M}_{\mathcal{H}_{\mathrm{aux}}}]$, 129
$\mathsf{M}_{\mathcal{H}_{\mathrm{aux}}}$, 129
$\mathsf{M}^{\mathrm{min}}_{\mathcal{H}_{\mathrm{aux}}}$, 139
$[\mathsf{M}_{\mathcal{H}^p_{\mathrm{aux}}}]$, 129
$\mathsf{M}_{\mathcal{H}^p_{\mathrm{aux}}}$, 9, 129
$\mathsf{M}_{\mathcal{H}^p_{\mathrm{aux}}\mathrm{G_{aux}}(\mathbb{Z}_p)}$, 129
$\mathsf{M}^{\mathrm{min}}_{\mathcal{H}^p_{\mathrm{aux}}}$, 9, 139
$\mathsf{M}^{\mathrm{tor}}_{\mathcal{H}^p_{\mathrm{aux}},\Sigma^p_{\mathrm{aux}}}$, 9, 136
$\mathsf{M}^{\mathrm{tor}}_{\mathcal{H}_{\mathrm{aux}},\Sigma_{\mathrm{aux}}}$, 136
$\mathsf{M}^{\mathrm{tor}}_{\mathcal{H},d_0\mathsf{pol}}$, 64
$\mathsf{M}_{\mathcal{H}_{\mathsf{G}_{h,\mathsf{Z}}}}$, 66
$\mathsf{M}_{\mathcal{H}_{\mathsf{G}_{h,\mathsf{Z}},\Phi}}$, 66
$\mathsf{M}_{\mathcal{H}_h}$, 32, 66, 244
$\vec{\mathsf{M}}^{\mathrm{ord}}_{\mathcal{H}_h}$, 274, 293, 311
$\mathsf{M}_{\mathcal{H}'_h}$, 32
$\mathsf{M}_{\mathcal{H}''_h}$, 32
$\vec{\mathsf{M}}^{\mathrm{ord}}_{\mathcal{H}_h,r_{\mathcal{H}}}$, 274, 293, 311
$\mathsf{M}_{\mathcal{H},i}$, 151
$\mathsf{M}^{\mathrm{min}}_{\mathcal{H},i}$, 151
$\mathsf{M}_{\mathcal{H}_\kappa}$, 97, 106
$\mathsf{M}^{\mathrm{min}}_{\mathcal{H}}$, 3, 8, 61
$\mathsf{M}^{\mathrm{ord}}_{\mathcal{H}}$, 6, 10, 203
$\mathsf{M}^{\mathrm{ord},\Phi_{\mathcal{H}}}_{\mathcal{H}}$, 259
$\mathsf{M}^{\mathrm{ord},\mathsf{Z}_{\mathcal{H}}}_{\mathcal{H}}$, 259
$[\mathsf{M}_{\mathcal{H}^p}]$, 152
$\mathsf{M}_{\mathcal{H}^p}$, 152, 212, 305
$\mathsf{M}^{\mathrm{min}}_{\mathcal{H}^p}$, 152, 305
$\mathsf{M}^{\mathrm{ord}}_{\mathcal{H}^p}$, 212
$\mathsf{M}^{\mathrm{ord,tor}}_{\mathcal{H}^p,\Sigma^p}$, 306
$\mathsf{M}^{\mathrm{tor}}_{\mathcal{H}^p,\Sigma^p}$, 305
$[\mathsf{M}^{\mathsf{Z}_{\mathcal{H}^p}}_{\mathcal{H}^p}]$, 308
$\mathsf{M}^{\Phi_{\mathcal{H}}}_{\mathcal{H}}$, 32, 58, 62, 66, 259
$\mathsf{M}^{\mathrm{tor}}_{\mathcal{H}',d_0\mathsf{pol}'}$, 64
$\mathsf{M}_{\mathcal{H},r_{\mathcal{H}}}$, 203
$\mathsf{M}^{\mathrm{min}}_{\mathcal{H},r_{\mathcal{H}}}$, 347
$\mathsf{M}^{\mathsf{Z}_{\mathcal{H}}}_{\mathcal{H},r_{\mathcal{H}}}$, 259
$\mathsf{M}_{\mathcal{H},r_\nu}$, 203
$\mathsf{M}^{\mathrm{tor}}_{\mathcal{H},\Sigma,i}$, 151
$\mathsf{M}^{\mathrm{tor}}_{\mathcal{H},\Sigma,r_{\mathcal{H}}}$, 297
$[\mathsf{M}^{\mathrm{tor}}_{\mathcal{H},\Sigma}]$, 124
$\mathsf{M}^{\mathrm{tor}}_{\mathcal{H},\Sigma}$, 3, 8, 58
$\mathsf{M}^{\mathrm{tor}}_{\mathcal{H}}$, 58
$[\mathsf{M}^{\mathsf{Z}_{\mathcal{H}}}_{\mathcal{H}}]$, 62
$\mathsf{M}^{\mathsf{Z}_{\mathcal{H}}}_{\mathcal{H}}$, 32, 58, 62, 66, 259
$\overset{\rightleftarrows}{\widehat{\mathsf{M}}}{}^{\mathrm{ord},\check{\Phi}_{\widehat{\mathcal{H}}}}_{\widehat{\mathcal{H}}}$, 380, 394
$\overset{\rightleftarrows}{\widehat{\mathsf{M}}}{}^{\mathrm{ord},\check{\mathsf{Z}}_{\widehat{\mathcal{H}}}}_{\widehat{\mathcal{H}}}$, 380, 394
$\widehat{\mathrm{M}}^{\mathrm{ord}}_{\mathsf{D}}(R)$, 377
$\widehat{\mathsf{M}}^{\check{\Phi}_{\widehat{\mathcal{H}}}}_{\widehat{\mathcal{H}}}$, 86, 98
$\widehat{\mathsf{M}}^{\check{\mathsf{Z}}_{\widehat{\mathcal{H}}}}_{\widehat{\mathcal{H}}}$, 86, 98
$\mathsf{M}^{\Phi_n}_{n}$, 68
$\mathsf{M}^{\mathsf{Z}_n}_{n}$, 30, 66, 68
$\mathrm{M}_{\mathrm{PE},\mathcal{O}}(R,I)$, 217
$\mathrm{M}_{\mathrm{PEL},\dddot{\mathsf{M}}^{\mathrm{ord}}_{\mathcal{H}}}(R,I)$, 242, 247
$\mathrm{M}_{\mathrm{PEL},\dddot{\mathsf{M}}^{\mathrm{ord}}_{n}}(R,I)$, 235
$\mathrm{M}_{\mathrm{PEL},\mathsf{M}^{\mathrm{ord}}_{\mathcal{H}}(R,I)}$, 247
$\mathrm{M}_{\mathrm{PEL},\mathsf{M}_{\mathcal{H}}(R,I)}$, 247
$\mathscr{M}^{+}$, 499
$\overset{\rightleftarrows}{\mathsf{M}}{}^{\mathrm{tor}}_{\widetilde{\mathcal{H}},\widetilde{d_0\mathsf{pol}},r_\kappa}$, 426
$\overset{\rightleftarrows}{\mathsf{M}}{}^{\mathrm{ord}}_{\widetilde{\mathcal{H}}}$, 372, 379, 407
$\overset{\rightleftarrows}{\mathsf{M}}{}^{\mathrm{ord},\check{\Phi}_{\widetilde{\mathcal{H}}}}_{\widetilde{\mathcal{H}}}$, 379
$\overset{\rightleftarrows}{\mathsf{M}}{}^{\mathrm{ord},\widetilde{\Phi}_{\widetilde{\mathcal{H}}}}_{\widetilde{\mathcal{H}}}$, 407
$\overset{\rightleftarrows}{\mathsf{M}}{}^{\mathrm{ord,tor}}_{\widetilde{\mathcal{H}}}$, 379, 407
$\overset{\rightleftarrows}{\mathsf{M}}{}^{\mathrm{ord},\check{\mathsf{Z}}_{\widetilde{\mathcal{H}}}}_{\widetilde{\mathcal{H}}}$, 380
$\overset{\rightleftarrows}{\mathsf{M}}{}^{\mathrm{ord},\widetilde{\mathsf{Z}}_{\widetilde{\mathcal{H}}}}_{\widetilde{\mathcal{H}}}$, 407
$\overset{\rightleftarrows}{\mathsf{M}}{}^{\mathrm{ord,tor}}_{\widetilde{\mathcal{H}},\widetilde{\Sigma}^{\mathrm{ord}}}$, 379, 407
$\overset{\rightleftarrows}{\mathsf{M}}{}^{\mathrm{ord},\check{\mathsf{Z}}_n}_{n}$, 380
$\widetilde{\mathrm{M}}^{\mathrm{ord}}_{\mathsf{D}}(R)$, 377
$\widetilde{\dddot{\mathsf{M}}}{}^{\mathrm{ord},\check{\Phi}_{\widetilde{\mathcal{H}}}}_{\widetilde{\mathcal{H}}}$, 418
$\widetilde{\mathsf{M}}^{(m)}_{\mathcal{H}}$, 25
$\widetilde{\mathsf{M}}_{\widetilde{\mathcal{H}}}$, 46, 85
$\widetilde{\mathsf{M}}_{\widetilde{\mathcal{H}}^p}$, 426
$\widetilde{\mathsf{M}}^{\mathrm{tor}}_{\widetilde{\mathcal{H}}^p,\widetilde{\Sigma}^p}$, 426
$\widetilde{\mathsf{M}}^{\check{\Phi}_{\widetilde{\mathcal{H}}}}_{\widetilde{\mathcal{H}}}$, 85, 86
$\widetilde{\mathsf{M}}^{\mathrm{tor}}_{\widetilde{\mathcal{H}},\widetilde{\Sigma}}$, 85, 106

$\widetilde{\mathsf{M}}^{\mathrm{tor}}_{\widetilde{\mathcal{H}}}$, 85, 106
$\widetilde{\mathsf{M}}^{\mathsf{Z}_n}_n$, 86
$\mathrm{M}^{\mathrm{ess,ord}}_{\mathsf{Z}_{p^r},\mathsf{D}_{p^r}}$, 238
maximal point, 266, 267, 466
maximal totally isotropic submodule, 168
minimal compactification
  of $\mathsf{M}_{\mathcal{H}}$, 3, 61–62
mixed Shimura varieties, 14, 16
module of universal finite differentials, 286
$\boldsymbol{\mu}_{p^\infty,S}$, 156
$\boldsymbol{\mu}_{p^\infty,s}$, 164
${}^{\mathrm{mult}}$, 156, 161, 162, 164
multi-rank, 28, 117
Mumford family, 59, 82, 272, 296, 449
Mumford's construction, 217

N, 96, 390
$\mathscr{N}$, 498, 504, 507
$\mathfrak{N}$, 496
N_0, 331, 345
n_0, 172
N_1, 139
$\mathscr{N}^a$, 504
$\vec{\mathsf{N}}^{\mathrm{ord,tor}}_{\kappa_{\mathrm{aux}}}$, 425
$\vec{\mathfrak{N}}^{\mathrm{ord}}_{\kappa}$, 401
$\vec{\mathsf{N}}^{\mathrm{ord}}_{\kappa}$, 393, 420
$\vec{\mathsf{N}}^{\mathrm{ord,ext}}_{\kappa}$, 396
$\vec{\mathsf{N}}^{\mathrm{ord,grp},\vee}_{\kappa}$, 396
$\vec{\mathsf{N}}^{\mathrm{ord,grp}}_{\kappa}$, 393, 420
$\vec{\mathfrak{N}}^{\mathrm{ord,tor}}_{\kappa}$, 401
$\vec{\mathsf{N}}^{\mathrm{ord,tor}}_{\kappa}$, 420
$\vec{\mathsf{N}}^{\mathrm{ord}}_{\widetilde{\kappa}}$, 408
$\vec{\mathsf{N}}^{\mathrm{ord,grp}}_{\widetilde{\kappa}}$, 408
$\vec{\mathsf{N}}^{\mathrm{ord,tor}}_{\widetilde{\kappa}}$, 409
$\mathsf{N}^{\vee}$, 96, 391
$\mathsf{N}^{\mathrm{ext}}$, 392
$\mathsf{N}^{\mathrm{grp}}$, 95, 96, 390
$\mathsf{N}^{\mathrm{grp},\vee}$, 96, 391
N_{κ}, 97, 116
$\mathsf{N}^{\mathrm{grp}}_{\kappa}$, 97, 116
$\mathsf{N}_{\widetilde{\kappa}}$, 107
$\mathsf{N}^{\mathrm{tor}}_{\widetilde{\kappa}}$, 107
$\mathsf{N}^{\mathrm{grp}}_{\widetilde{\kappa}}$, 107
$\mathsf{N}^{\mathrm{tor}}_{\kappa}$, 98, 116
$\mathfrak{N}_{\breve{\sigma},\tau}$, 454
$\widetilde{\mathfrak{N}}$, 496
$\widetilde{\widehat{\mathfrak{N}}}_{\breve{\sigma},\tau}$, 458
$\widetilde{\widehat{\mathfrak{N}}}^{\breve{\ell}}_{\breve{\sigma},\tau}$, 458
$\widetilde{\widehat{\mathfrak{N}}}^{\breve{\ell},0+}_{\breve{\sigma},\tau}$, 458
$\widetilde{\widehat{\mathfrak{N}}}^{\breve{\ell},+}_{\breve{\sigma},\tau}$, 458
$\widetilde{\mathfrak{N}}^{\ell}$, 497
$\widetilde{\mathfrak{N}}^{\ell,+}$, 497
$\widetilde{\mathfrak{N}}_{\breve{\sigma},\tau}$, 454
$\widetilde{\mathfrak{N}}^{\breve{\ell}}_{\breve{\sigma},\tau}$, 458
$\widetilde{\mathfrak{N}}^{\breve{\ell},0+}_{\breve{\sigma},\tau}$, 458
$\widetilde{\mathfrak{N}}^{\breve{\ell},+}_{\breve{\sigma},\tau}$, 458
$(\mathscr{N})^{\wedge}_{\bar{x}}$, 504, 507
∇, 25, 97, 101, 121, 391, 398, 470, 485
naive level structure
  of type $(\mathrm{Gr}^{\mathsf{Z}}_{-1,n_0}, \langle\,\cdot\,,\,\cdot\,\rangle_{11,n_0})$
    principal, 231
  of type $(L \underset{\mathbb{Z}}{\otimes} \hat{\mathbb{Z}}, \langle\,\cdot\,,\,\cdot\,\rangle)$, 173
    principal, 172
naive ordinary level structure
  of type $(L/p^rL, \langle\,\cdot\,,\,\cdot\,\rangle, \mathsf{D}_{p^r})$, 179
    principal, 178
  of type $\phi^0_{\mathsf{D}_{-1,p^r}} : \mathrm{Gr}^0_{\mathsf{D}_{-1,p^r}} \to \mathrm{Gr}^0_{\mathsf{D}^{\#}_{-1,p^r}}$
    principal, 221, 232
    symplectic condition, 221
  of type $\phi^0_{\mathsf{D},p^r} : \mathrm{Gr}^0_{\mathsf{D},p^r} \to \mathrm{Gr}^0_{\mathsf{D}^{\#},p^r}$
    principal, 178
naive ordinary level-n structure datum
  of type $(L/nL, \langle\,\cdot\,,\,\cdot\,\rangle, \mathsf{D}_{p^r})$, 233
    orbit of étale-locally-defined, 240
naive ordinary pre-level-n structure datum
  of type $(L/nL, \langle\,\cdot\,,\,\cdot\,\rangle, \mathsf{D}_{p^r})$, 231
    symplectic condition, 233
$\mathrm{NBl}_{\cdot}(\,\cdot\,)$, 64
$\mathrm{NBl}_{\vec{\mathcal{J}}_{\mathcal{H},d_0\mathsf{pol}}}(\vec{\mathsf{M}}^{\mathrm{min}}_{\mathcal{H}})$, 147
$\mathrm{NBl}_{\vec{\mathcal{J}}_{\mathcal{H},d_0\mathsf{pol}^{\mathrm{ord}}}}(\vec{\mathsf{M}}^{\mathrm{ord,min}}_{\mathcal{H}})$, 360
$\mathrm{NBl}_{\vec{\mathcal{J}}_{\mathcal{H},d_0\mathsf{pol}^{\mathrm{ord}}}}(\vec{\oint}^{\mathrm{ord}}_{\mathcal{H}})$, 360
$\mathrm{NBl}_{\vec{\mathcal{J}}_{\mathcal{H}',d_0\mathsf{pol}'}}(\vec{\mathsf{M}}^{\mathrm{min}}_{\mathcal{H}'})$, 148
$\mathrm{NBl}_{\mathcal{J}_{\mathcal{H},d_0\mathsf{pol}}}(\mathsf{M}^{\mathrm{min}}_{\mathcal{H}})$, 63
$\mathrm{NBl}_{\mathcal{J}_{\mathcal{H},d_0\mathsf{pol}}}(\oint_{\mathcal{H}})$, 63
$\mathrm{NBl}_{\mathcal{J}_{\mathcal{H}^p,d_0\mathsf{pol}^p}}(\mathsf{M}^{\mathrm{min}}_{\mathcal{H}^p})$, 309
$\mathrm{NBl}_{\mathcal{J}_{\mathcal{H}^p,d_0\mathsf{pol}^p}}(\oint_{\mathcal{H}^p})$, 309
$\mathrm{NBl}_{\mathcal{J}_{\mathcal{H}',d_0\mathsf{pol}'}}(\mathsf{M}^{\mathrm{min}}_{\mathcal{H}'})$, 64
neat, 3, 24
Néron model, 127, 145, 161, 202
Néron–Severi group, 472
nerve, 454, 496, 497
  spectral sequence, 461, 492

$^{\text{non-ord}}$, 365
nondegenerate cone, 34
nonemptiness of ordinary locus, 2, 13, 17, 18, 204, 369–370
nonordinary locus, 365
normal morphism, 350
$\underline{\mathrm{NS}}(\vec{\mathsf{N}}^{\mathrm{ord,grp}}/\vec{\mathsf{M}}_{\mathcal{H}}^{\mathrm{ord}})$, 472
$\underline{\mathrm{NS}}(\vec{\mathsf{N}}^{\mathrm{ord}}/\vec{\mathsf{M}}_{\mathcal{H}}^{\mathrm{ord}})$, 472
ν, 22, 170
ν_{-1,n_0}, 231
$\nu_{-1,p^r}^{\mathrm{ord}}$, 221, 232, 239
$\nu_{-1,p^r,S}^{\mathrm{ord}}$, 222
$\nu(\alpha_{H_{n_0}})$, 173
$\nu(\alpha_{\mathcal{H}_p}^{\mathrm{ord}})$, 180
$\nu(\alpha_{H_{p^r}^{\mathrm{ord}}}^{\mathrm{ord}})$, 179
$\nu(\alpha_{n_0})$, 173
$\nu(\alpha_{p^r}^{\mathrm{ord}})$, 178
$\nu(g)$, 22, 170
$\nu(H_{n_0})$, 173
$\nu_{\mathcal{H}_p}^{\mathrm{ord}}$, 180
$\nu_{H_{p^r}^{\mathrm{ord}}}^{\mathrm{ord}}$, 179
$\hat{\nu}$, 165
$\hat{\nu}_{-1,\bar{s}}$, 71
$\hat{\nu}_{-1,\bar{s}}^{p}$, 317
$\hat{\nu}_{\bar{s}}^{\natural}$, 71
$\hat{\nu}_{\bar{s}}^{\mathrm{ord}}$, 182, 193, 199
$\hat{\nu}_{\bar{s}}^{p}$, 174
$\hat{\nu}_{\bar{s}}^{p,\natural}$, 316
ν_n, 78, 79
ν_{n_0}, 173, 324
$\nu_{n_0}^{\natural}$, 313
ν_{p^r}, 193
$\nu_{p^r}^{\natural,\mathrm{ord}}$, 314
$\nu_{p^r}^{\mathrm{ord}}$, 178, 179, 193, 206, 217, 229, 230
$\nu_{p^r}^{\mathrm{ord},\triangledown}$, 200
$\nu_{p^r,S}^{\mathrm{ord}}$, 206, 221
$\widetilde{\nu}$, 371, 374

$\mathcal{O}$, 22
$\mathcal{O}(1)$, 61, 337, 345
$\mathcal{O}_{\mathrm{aux}}$, 128
$\mathcal{O}'_{\mathrm{aux}}$, 128
$(\mathcal{O}_{\mathrm{aux}}, \star_{\mathrm{aux}}, L_{\mathrm{aux}}, \langle\,\cdot\,,\,\cdot\,\rangle_{\mathrm{aux}}, h_{0,\mathrm{aux}})$, 9, 128, 198
$\mathcal{O}$-lattice, 22
$\mathcal{O}$-multi-rank, 28, 62, 346
$\mathcal{O}'$, 27, 43, 169, 371, 478, 484
$(\mathcal{O}, \star, L, \langle\,\cdot\,,\,\cdot\,\rangle, h_0)$, 2, 21, 22
$(\mathcal{O}, \star, \widetilde{L}, \langle\,\cdot\,,\,\cdot\,\rangle^{\sim}, \widetilde{h}_0)$, 44
$\mathscr{O}_{\mathfrak{U}_\sigma}$, 493–495
$\mathscr{O}_{\mathfrak{U}_\sigma}^{+}$, 494–495
$\mathscr{O}_{U_{\mp}}$, 448, 450
$\mathscr{O}_{\mathfrak{U}_{\mp}}$, 448, 450
$\mathscr{O}_{\mathfrak{U}_{\mp}}^{0+}$, 448, 450
$\mathscr{O}_{\mathfrak{U}_{\mp}}^{+}$, 449–450
$\mathscr{O}_{\mathfrak{U}_{\mp}}^{++}$, 449–450
$\mathscr{O}_{\mathfrak{X}}^{0+}$, 445
$\mathscr{O}_{\mathfrak{X}}^{+}$, 445
$\mathscr{O}_{\mathfrak{X}}^{+}$, 494
$\mathscr{O}_{\mathfrak{X}}^{++}$, 445
i, 23
$\omega_{\mathcal{A}_{g_{\mathrm{aux}},d_{\mathrm{aux}},n}}$, 127
$\omega_{\mathcal{A}_{g,d,n,n}}$, 127
$\omega_{\vec{B}}$, 339
$\overline{\Omega}^1_{\vec{\mathsf{M}}_{\mathcal{H}}^{\mathrm{ord,tor}}/\vec{\mathsf{S}}_{0,r_{\mathcal{H}}}}$, 469
$\overline{\Omega}^1_{\vec{\mathsf{N}}^{\mathrm{ord,tor}}/\vec{\mathsf{M}}_{\mathcal{H}}^{\mathrm{ord,tor}}}$, 396
$\overline{\Omega}^{\bullet}_{\vec{\mathsf{N}}^{\mathrm{ord,tor}}/\vec{\mathsf{M}}_{\mathcal{H}}^{\mathrm{ord,tor}}}$, 397
$\overline{\Omega}^1_{\vec{\mathsf{N}}^{\mathrm{ord,tor}}/\vec{\mathsf{S}}_{0,r_{\mathcal{H}}}}$, 469
$\overline{\Omega}^1_{\mathsf{N}^{\mathrm{tor}}/\mathsf{M}_{\mathcal{H}}^{\mathrm{tor}}}$, 100
$\overline{\Omega}^{\bullet}_{\mathsf{N}^{\mathrm{tor}}/\mathsf{M}_{\mathcal{H}}^{\mathrm{tor}}}$, 100
$\overline{\Omega}^1_{\vec{\overline{\widehat{\Xi}}}^{\mathrm{ord}}_{\check{\Phi}_{\widehat{\mathcal{H}}},\check{\delta}_{\widehat{\mathcal{H}}},\widehat{\Sigma}_{\check{\Phi}_{\widehat{\mathcal{H}}}}}/\vec{\overline{\Xi}}^{\mathrm{ord}}_{\Phi_{\mathcal{H}},\delta_{\mathcal{H}},\Sigma_{\Phi_{\mathcal{H}}},r_{\kappa}}}$, 439
$\overline{\Omega}^1_{\vec{\overline{\widehat{\Xi}}}^{\mathrm{ord}}_{\check{\Phi}_{\widehat{\mathcal{H}}},\check{\delta}_{\widehat{\mathcal{H}}},\widehat{\Sigma}_{\check{\Phi}_{\widehat{\mathcal{H}}}}}/\vec{\overline{\Xi}}^{\mathrm{ord}}_{\Phi_{\mathcal{H}},\delta_{\mathcal{H}},\Sigma_{\Phi_{\mathcal{H}}}}\underset{\vec{C}^{\mathrm{ord}}_{\Phi_{\mathcal{H}},\delta_{\mathcal{H}}}}{\times}\vec{\widehat{C}}^{\mathrm{ord}}_{\check{\Phi}_{\widehat{\mathcal{H}}},\check{\delta}_{\widehat{\mathcal{H}}}}}$, 439
$\overline{\Omega}^1_{\overline{\widehat{\Xi}}_{\check{\Phi}_{\widehat{\mathcal{H}}},\check{\delta}_{\widehat{\mathcal{H}}},\widehat{\Sigma}_{\check{\Phi}_{\widehat{\mathcal{H}}}}}/\overline{\Xi}_{\Phi_{\mathcal{H}},\delta_{\mathcal{H}},\Sigma_{\Phi_{\mathcal{H}}}}}$, 94
$\overline{\Omega}^1_{\overline{\widehat{\Xi}}_{\check{\Phi}_{\widehat{\mathcal{H}}},\check{\delta}_{\widehat{\mathcal{H}}},\widehat{\Sigma}_{\check{\Phi}_{\widehat{\mathcal{H}}}}}/(\overline{\Xi}_{\Phi_{\mathcal{H}},\delta_{\mathcal{H}},\Sigma_{\Phi_{\mathcal{H}}}}\underset{C_{\Phi_{\mathcal{H}},\delta_{\mathcal{H}}}}{\times}\widehat{C}_{\check{\Phi}_{\widehat{\mathcal{H}}},\check{\delta}_{\widehat{\mathcal{H}}}})}$, 94
$\widehat{\Omega}^1_{S/\vec{\mathsf{S}}_{0,r_{\mathcal{H}}}}$, 207, 281
$\widehat{\Omega}^1_{S/\vec{\mathsf{S}}_{0,r_{\mathcal{H}}}}[d\log\infty]$, 281
${}^{\heartsuit}\omega$, 339
$\omega_{\vec{\mathsf{M}}_{\mathcal{H}}}$, 141, 149, 511
$\omega_{\vec{\mathsf{M}}^{\mathrm{ord}}_{\mathcal{H}_{\mathrm{aux}}}}$, 213, 214
$\omega_{\vec{\mathsf{M}}^{\mathrm{ord,tor}}_{\mathcal{H}_{\mathrm{aux}},\Sigma^{\mathrm{ord}}_{\mathrm{aux}}}}$, 333, 336
$\omega_{\vec{\mathsf{M}}^{\mathrm{tor}}_{\mathcal{H},d_0\mathrm{pol}}}$, 147
$\omega_{\vec{\mathsf{M}}^{\mathrm{min}}_{\mathcal{H}}}$, 143, 149, 336, 337, 517
$\omega_{\vec{\mathsf{M}}^{\mathrm{ord}}_{\mathcal{H}}}$, 213, 214, 345, 483
$\omega_{\vec{\mathsf{M}}^{\mathrm{ord,min}}_{\mathcal{H}}}$, 337, 345, 347, 348
$\omega_{\vec{\mathsf{M}}^{\mathrm{ord,tor}}_{\mathcal{H}}}$, 346, *see also* $\omega_{\vec{\mathsf{M}}^{\mathrm{ord,tor}}_{\mathcal{H},\Sigma^{\mathrm{ord}}}}$
$\Omega^1_{\vec{\mathsf{M}}^{\mathrm{ord,tor}}_{\mathcal{H}}/\vec{\mathsf{S}}_{0,r_{\mathcal{H}}}}[d\log\infty]$, 296, 396

$\omega_{\vec{\mathsf{M}}^{\mathrm{min}}_{\mathcal{H},r_{\mathcal{H}}}}$, 347
$\omega_{\vec{\mathsf{M}}^{\mathrm{ord,tor}}_{\mathcal{H},\Sigma^{\mathrm{ord}}}}$, 11, 306, 331, 333, 336, 337, 429
$\omega_{\mathsf{M}^{\mathrm{tor}}_{\mathrm{G_{aux}}(\hat{\mathbb{Z}}^p),\Sigma^p_{\mathrm{aux}}}}$, 336
$\omega_{\mathsf{M}_{\mathcal{H}}}$, 61, 120, 139
$\omega_{\mathsf{M}_{\mathcal{H}_{\mathrm{aux}}}}$, 139
$\omega_{\mathsf{M}^{\mathrm{min}}_{\mathcal{H}_{\mathrm{aux}}}}$, 140
$\omega_{\mathsf{M}_{\mathcal{H}^p_{\mathrm{aux}}}}$, 139
$\omega_{\mathsf{M}^{\mathrm{min}}_{\mathcal{H}^p_{\mathrm{aux}}}}$, 139, 140, 363
$\omega_{\mathsf{M}^{\mathrm{tor}}_{\mathcal{H}^p_{\mathrm{aux}},\Sigma^p_{\mathrm{aux}}}}$, 139
$\omega_{\mathsf{M}^{\mathrm{tor}}_{\mathcal{H}^p_{\mathrm{aux}}}}$, 363, *see also* $\omega_{\mathsf{M}^{\mathrm{tor}}_{\mathcal{H}^p_{\mathrm{aux}},\Sigma^p_{\mathrm{aux}}}}$
$\omega_{\mathsf{M}^{\mathrm{tor}}_{\mathcal{H}_{\mathrm{aux}},\Sigma_{\mathrm{aux}}}}$, 138
$\omega_{\mathsf{M}^{\mathrm{min}}_{\mathcal{H}}}$, 61, 140
$\omega_{\mathsf{M}^{\mathrm{tor}}_{\mathcal{H}^p,\Sigma^p}}$, 306
$\omega_{\mathsf{M}^{\mathrm{min}}_{\mathcal{H},r_{\mathcal{H}}}}$, 347
$\omega_{\mathsf{M}^{\mathrm{tor}}_{\mathcal{H},\Sigma}}$, 138
$\omega'_{\mathsf{M}^{\mathrm{tor}}_{\mathcal{H},\Sigma}}$, 139
$\omega_{\mathsf{M}^{\mathrm{tor}}_{\mathcal{H}}}$, 61, 123, *see also* $\omega_{\mathsf{M}^{\mathrm{tor}}_{\mathcal{H},\Sigma}}$
$\Omega^1_{\mathsf{M}^{\mathrm{tor}}_{\mathcal{H}}/\mathsf{S}_0}[d\log\infty]$, 59, 100
$\omega_{\vec{\mathsf{N}}^{\mathrm{ord,tor}}_{\kappa}}$, 429
$\Omega^1_{\vec{\mathsf{N}}^{\mathrm{ord,tor}}/\vec{\mathsf{S}}_{0,r_{\mathcal{H}}}}[d\log\infty]$, 396
$\Omega^1_{\mathsf{N}^{\mathrm{tor}}/\mathsf{S}_0}[d\log\infty]$, 100
$\widetilde{\Omega}^1_{R/\mathcal{O}_{F_0,(p)}[\zeta_{p^{r_{\mathcal{H}}}}]}$, 286
$\widetilde{\Omega}^1_{S/\vec{\mathsf{S}}_{0,r_{\mathcal{H}}}}$, 286
$\widetilde{\Omega}^1_{S/\vec{\mathsf{S}}_{0,r_{\mathcal{H}}}}[d\log\infty]$, 286
openness of versality, 287–288
ord, 376
ordinary abelian scheme, 10, 155, 180, 198, 206, 267, 362
ordinary cusp, 7, 347
ordinary cusp label, 172, 373
ordinary good algebraic model, 286–287
ordinary good formal model, 283–284
ordinary Hecke twist, 183, 184
  by $\mathrm{G}_{1,\mathtt{Z}}(\mathbb{A}^{\infty,p}) \times \mathrm{P}^{\mathrm{ord}}_{1,\mathtt{Z},\mathtt{D}}(\mathbb{Q}_p)$, 321
  by $\mathrm{G}(\mathbb{A}^{\infty,p}) \times \mathrm{P}^{\mathrm{ord}}_{\mathtt{D}}(\mathbb{Q}_p)$, 190–192, 300, 301, 399
  by $\mathrm{G}(\mathbb{A}^{\infty,p}) \times \mathrm{P}^{\mathrm{ord}}_{\mathtt{D}}(\mathbb{Q}_p)$, 399
  by $\mathrm{G}_{h,\mathtt{Z}}(\mathbb{A}^{\infty,p}) \times \mathrm{P}^{\mathrm{ord}}_{h,\mathtt{Z},\mathtt{D}}(\mathbb{Q}_p)$, 311
  by $\mathrm{G}^{\mathtt{Z}}(\mathbb{A}^{\infty,p}) \times \mathrm{P}^{\mathrm{ord}}_{\mathtt{D}^{\mathtt{Z}}}(\mathbb{Q}_p)$, 311
  by $\mathrm{P}^{\mathrm{ord}}_{\mathtt{D}}(\mathbb{Q}_p)$, 183–190
  by $\mathrm{P}^{\mathrm{ord}}_{\mathtt{D}}(\mathbb{Z}_p)$, 183, 190
  by $\mathrm{P}_{\mathtt{Z}}(\mathbb{A}^{\infty,p}) \times \mathrm{P}^{\mathrm{ord}}_{\mathtt{Z},\mathtt{D}}(\mathbb{Q}_p)$, 328
  by $(\mathrm{P}_{\mathtt{Z}}(\mathbb{A}^{\infty,p}) \times \mathrm{P}^{\mathrm{ord}}_{\mathtt{Z},\mathtt{D}}(\mathbb{Q}_p))/\mathrm{U}_{2,\mathtt{Z}}(\mathbb{A}^{\infty})$, 322
  by $\mathrm{P}'_{\mathtt{Z}}(\mathbb{A}^{\infty,p}) \times \mathrm{P}^{\mathrm{ord},\prime}_{\mathtt{Z},\mathtt{D}}(\mathbb{Q}_p)$, 328
  by power of p, 184, 190
  by U_p operator, 184, 190, 304
    generalized, 184
ordinary Kuga family, 8, 13, 390
  extended, 392
  generalized, 8, 13, 390
  partial toroidal compactification, 8, 13, 14, 392–407
ordinary level structure, 6, 10, 177–183
  naive, *see* naive ordinary level structure
  symplectic-liftability condition, 179, 180
  of type $(L \underset{\mathbb{Z}}{\otimes} \mathbb{Z}_p, \langle\,\cdot\,,\,\cdot\,\rangle, \mathtt{D})$, 180
    lifting, 182
    principal, 179
  of type $(L \underset{\mathbb{Z}}{\otimes} \mathbb{Z}_p, \langle\,\cdot\,,\,\cdot\,\rangle, \mathtt{Z} \underset{\hat{\mathbb{Z}}}{\otimes} \mathbb{Z}_p, \mathtt{D})$, 315
    principal, 314
ordinary level-$\mathcal{H}$ structure datum
  of type $(L \underset{\mathbb{Z}}{\otimes} \hat{\mathbb{Z}}, \langle\,\cdot\,,\,\cdot\,\rangle, \mathtt{D})$, 241
ordinary level-n structure datum
  of type $(L/nL, \langle\,\cdot\,,\,\cdot\,\rangle, \mathtt{D}_{p^r})$
    naive, *see* naive ordinary level-n structure datum
  of type $(L \underset{\mathbb{Z}}{\otimes} \hat{\mathbb{Z}}, \langle\,\cdot\,,\,\cdot\,\rangle, \mathtt{D})$, 233
    orbit of étale-locally-defined, 240
ordinary locus, v, 2, 6–8, 10, 203–205
  full, 14, 365
  nonemptiness, 2, 13, 17, 18, 204, 369–370
ordinary pre-level-n structure datum
  of type $(L/nL, \langle\,\cdot\,,\,\cdot\,\rangle, \mathtt{D}_{p^r})$
    naive, *see* naive ordinary pre-level-n structure datum
  of type $(L \underset{\mathbb{Z}}{\otimes} \hat{\mathbb{Z}}, \langle\,\cdot\,,\,\cdot\,\rangle, \mathtt{D})$, 233
ordinary semi-abelian scheme, 155, 240, 362

$[p^{-r_0}g_{p,-1}]^{\mathrm{ord}}$, 184, 191
$[p^{-r_0}g_{p,0}]^{\mathrm{ord}}$, 185
$[p^{-r_0}g^{-1}_{p,\#,0}]^{\mathrm{ord}}$, 186
P_0, 118
$\mathrm{P}_0(R)$, 118
$\mathrm{P}^{\mathrm{ord}}_{1,\mathtt{Z},\mathtt{D}}(R)$, 172, 378
$\mathcal{P}_A$, 24, 164
$\vec{\mathsf{p}}^{\mathrm{ord}}_{\Phi_{\mathcal{H}},\delta_{\mathcal{H}}}$, 339, 495
$\mathcal{P}_B$, 77
$\mathrm{P}^{\mathrm{ord}}_{\mathtt{D},0}$, 479
$\mathrm{P}^{\mathrm{ord}}_{\mathtt{D},0}(R)$, 479

$\mathrm{P}^{\mathrm{ord}}_{\mathtt{D}}$, 6, 170
$\mathrm{P}^{\mathrm{ord}}_{\mathtt{D}}(R)$, 169
$\mathrm{P}^{\mathrm{ess}}_{\mathtt{D}_{p^r}}$, 237
$\mathrm{P}^{\mathrm{ord}}_{\mathtt{D}^{\mathtt{Z}}}$, 311
$\mathrm{P}^{\mathrm{ord}}_{\mathtt{D}^{\mathtt{Z}}}(R)$, 311
p-divisible group
attached to abelian variety, 164
connected part, 155, 164
Dieudonné–Manin classification, 156
étale, 156
étale part, 164
multiplicative-type, 156
multiplicative-type part, 164
$\mathrm{P}^{\mathrm{ord},\prime}_{h,\mathtt{Z},\mathtt{D}}(R)$, 172, 378
$\mathrm{P}^{\mathrm{ord}}_{h,\mathtt{Z},\mathtt{D}}(R)$, 172, 378
$(\mathrm{P}^{\mathrm{ord}}_{h,\mathtt{Z},\mathtt{D}} \ltimes \mathrm{U}_{1,\mathtt{Z}})(R)$, 172
$\widehat{\mathrm{P}}^{\mathrm{ord}}_{1,\check{\mathtt{Z}},\mathtt{D}}(R)$, 378
$\widehat{\mathrm{P}}^{\mathrm{ord}}_{\mathtt{D}}(R)$, 377
$\widehat{\mathrm{P}}^{\mathrm{ord},\prime}_{h,\check{\mathtt{Z}},\mathtt{D}}(R)$, 378
$\widehat{\mathrm{P}}^{\mathrm{ord}}_{h,\check{\mathtt{Z}},\mathtt{D}}(R)$, 378
$\widehat{\mathbf{P}}'_{\check{\ell}<0+}$, 459
$\widehat{\mathbf{P}}'_{\check{\ell}<0}$, 459
$\widehat{\mathbf{P}}'_{\check{\ell}\leq 0}$, 459
$\widehat{\mathbf{P}}_{\check{\Phi}_{\widehat{\mathcal{H}}}}$, 50, 373
$\widehat{\mathbf{P}}^{+}_{\check{\Phi}_{\widehat{\mathcal{H}}}}$, 50, 98, 373, 393
$\widehat{\mathbf{P}}'$, 459
$\widehat{\mathrm{P}}^{\mathrm{ord},\prime}_{\check{\mathtt{Z}},\mathtt{D}}(R)$, 378
$\widehat{\mathrm{P}}^{\mathrm{ord}}_{\check{\mathtt{Z}},\mathtt{D}}(R)$, 378
$\widehat{\mathrm{P}}'_{\check{\mathtt{Z}}}(R)$, 56
$\widehat{\mathrm{P}}_{\check{\mathtt{Z}}}(R)$, 55, 56, 378
$\mathbf{P}'_{\ell<0}$, 497
$\mathbf{P}'_{\ell\leq 0}$, 497
$\mathscr{P}^{m}_{\mathsf{M}_{\mathcal{H}}/\mathsf{S}_0}$, 25
$\mathbf{P}_{\check{\Phi}_{\widehat{\mathcal{H}}}}$, 48
$\mathbf{P}^{+}_{\check{\Phi}_{\widehat{\mathcal{H}}}}$, 48
$\mathbf{P}_{\check{\Phi}_{\widetilde{\mathcal{H}}}}$, 48
$\mathbf{P}^{+}_{\check{\Phi}_{\widetilde{\mathcal{H}}}}$, 48
$\mathbf{P}_{\Phi_{\mathcal{H}}}$, 34, 48, 267
$\mathbf{P}_{\Phi_{\mathcal{H}_{\mathrm{aux}}}}$, 134
$\mathbf{P}^{+}_{\Phi_{\mathcal{H}_{\mathrm{aux}}}}$, 134
$\mathbf{P}_{\Phi^{\ddagger}_{\mathcal{H}}}$, 60, 297
$\mathbf{P}^{\vee}_{\Phi_{\mathcal{H}}}$, 340
$\mathbf{P}^{\vee,+}_{\Phi_{\mathcal{H}}}$, 500
$\mathbf{P}^{+}_{\Phi_{\mathcal{H}}}$, 267
$\mathbf{P}_{\widetilde{\Phi}_{\widetilde{\mathcal{H}}}}$, 48
$\mathbf{P}^{+}_{\widetilde{\Phi}_{\widetilde{\mathcal{H}}}}$, 46, 48, 381
$\mathbf{P}'$, 497
$[p^{r_0}]$, 184, 191
$[p^{r_0} g^{-1}_{p,-1}]^{\mathrm{ord}}$, 185
$[p^{r_0} g^{-1}_{p,0}]^{\mathrm{ord}}$, 184, 191
$\widetilde{\mathrm{P}}^{\mathrm{ord}}_{1,\check{\mathtt{Z}},\widetilde{\mathtt{D}}}(R)$, 378
$\widetilde{\mathrm{P}}^{\mathrm{ord}}_{1,\widetilde{\mathtt{Z}},\widetilde{\mathtt{D}}}(R) = \widetilde{\mathrm{P}}^{\mathrm{ord},\prime}_{\widetilde{\mathtt{Z}},\widetilde{\mathtt{D}}}(R)$, 377
$\widetilde{\mathrm{P}}^{\mathrm{ord}}_{\widetilde{\mathtt{D}}}(R)$, 377
$\widetilde{\mathrm{P}}^{\mathrm{ord},\prime}_{\check{\mathtt{Z}},\widetilde{\mathtt{D}}}(R)$, 378
$\widetilde{\mathrm{P}}^{\mathrm{ord}}_{\check{\mathtt{Z}},\widetilde{\mathtt{D}}}(R)$, 378
$\widetilde{\mathrm{P}}_{\check{\mathtt{Z}}}(R)$, 378
$\widetilde{\mathrm{P}}^{\mathrm{ord}}_{\check{\mathtt{Z}},\widetilde{\mathtt{Z}},\widetilde{\mathtt{D}}}(R)$, 378
$\widetilde{\mathrm{P}}_{\check{\mathtt{Z}},\widetilde{\mathtt{Z}}}(R)$, 56, 378
$\widetilde{\mathrm{P}}_{\check{\mathtt{Z}},\widetilde{\mathtt{Z}}}(R)/\widetilde{\mathrm{U}}_{2,\check{\mathtt{Z}}}(R)$, 57
$\widetilde{\mathrm{P}}_{\widetilde{\mathtt{Z}}}$, 45
$\widetilde{\mathrm{P}}^{\mathrm{ord},\prime}_{\widetilde{\mathtt{Z}},\widetilde{\mathtt{D}}}(R)$, 377
$\widetilde{\mathrm{P}}^{\mathrm{ord}}_{\widetilde{\mathtt{Z}},\widetilde{\mathtt{D}}}(R)$, 377
$\widetilde{\mathrm{P}}'_{\widetilde{\mathtt{Z}}}(R)$, 45, 377
$\widetilde{\mathrm{P}}_{\widetilde{\mathtt{Z}}}(R)$, 45, 377
$\mathrm{P}^{\mathrm{ord},\prime}_{\mathtt{Z},\mathtt{D}}(R)$, 172, 378
$\mathrm{P}^{\mathrm{ord}}_{\mathtt{Z},\mathtt{D}}(R)$, 172, 378
$\mathrm{P}^{\mathrm{ess}}_{\mathtt{Z}_n}$, 235
$\mathrm{P}^{\mathrm{ess}}_{\mathtt{Z}_{p^r},\mathtt{D}_{p^r}}$, 237
$\mathrm{P}^{\mathrm{ess,ord}}_{\mathtt{Z}_{p^r},\mathtt{D}_{p^r}}$, 238
$\mathrm{P}'_{\mathtt{Z}}(R)$, 31, 56
$\mathrm{P}_{\mathtt{Z}}(R)$, 30, 56
$\mathrm{P}_{\mathtt{Z}}(R)/\mathrm{U}_{2,\mathtt{Z}}(R)$, 57
partial canonical subgroup, 197
partial minimal compactification, 2, 7
base change properties, 349–353
of $\vec{\mathsf{M}}^{\mathrm{ord}}_{\mathcal{H}_{\mathrm{aux}}}$, 334
of $\vec{\mathsf{M}}^{\mathrm{ord}}_{\mathcal{H}}$, 345–349
partial toroidal compactification, 2
of $\vec{\mathsf{M}}^{\mathrm{ord}}_{\mathcal{H}_{\mathrm{aux}}}$, 333
of $\vec{\mathsf{M}}^{\mathrm{ord}}_{\mathcal{H}}$, 7, 14, 295–299
of ordinary Kuga family, 8, 13, 14, 392–407
PEL datum, 22
integral, 2, 21
auxiliary choice, 9
PEL-type Shimura variety, 26
Φ, 28, 47
ϕ, 28, 47, 138, 216, 281
$\underline{\phi}$, 268

$\varphi_{-1,\mathcal{H}}$, 66, 72, 243, 259, 266, 281
$\varphi_{-1,\mathcal{H}_\kappa}$, 112
$\varphi^{\mathrm{ord}}_{-1,\mathcal{H}_\kappa}$, 415
φ_{-1,H_n}, 243
$\varphi_{-1,H_{n_0}}$, 236
$\varphi^{\mathrm{ord}}_{-1,H_n}$, 240
$\varphi^{\mathrm{ord}}_{-1,\mathcal{H}}$, 241, 281
$\varphi_{-1,\mathcal{H}^p}$, 255, 266, 270, 281, 454
$\varphi^{\ddagger}_{-1,\mathcal{H}^p}$, 293
$\varphi^{\mathrm{ord}}_{-1,\mathcal{H}_p}$, 255, 266, 270, 281, 315, 317, 454
$\varphi^{\mathrm{ord},\ddagger}_{-1,\mathcal{H}_p}$, 293
$\varphi_{-1,H_{p^r}}$, 244, 245
$\varphi^{\mathrm{ord}}_{-1,H^{\mathrm{ord}}_{p^r}}$, 239, 244, 245
φ_{-1,n_0}, 231, 236, 248, 313
$\varphi^{\mathrm{ord}}_{-1,n}$, 231, 313
φ_{-1,p^r}, 245
$\varphi^{\mathrm{ord}}_{-1,p^r}$, 220, 221, 232, 239, 248, 314
$\varphi^{\mathrm{ord},0}_{-1,p^r}$, 221, 232, 239
$\varphi^{\mathrm{ord},\#,0}_{-1,p^r}$, 221, 224, 232, 239
$\varphi^{\mathrm{ord},0}_{-1,p^r,S}$, 222
$\varphi^{\mathrm{ord},\#,0}_{-1,p^r,S}$, 222
φ_{-2}, 28, 47
$\varphi^{(g)}_{-2}$, 41
$\underline{\varphi}_{-2,\mathcal{H}}$, 268
$\varphi_{-2,\mathcal{H}}$, 66, 241, 243, 281
$\underline{\varphi}_{-2,\mathcal{H}}(G)$, 279
$(\varphi^{\sim}_{-2,H_{n_0}}, \varphi^{\sim}_{0,H_{n_0}})$, 236, 237
$\varphi^{\mathrm{ord}}_{-2,H_n}$, 240
$(\varphi^{\sim}_{-2,H_n}, \varphi^{\sim}_{0,H_n})$, 243
$\varphi^{\mathrm{ord}}_{-2,\mathcal{H}}$, 255, 266, 270, 281, 317, 454
$\varphi_{-2,H^{\mathrm{ord}}_{p^r}}$, 244
$\varphi^{\mathrm{ord}}_{-2,H^{\mathrm{ord}}_{p^r}}$, 239, 245
$(\varphi^{\mathrm{ord},\#}_{-2,H^{\mathrm{ord}}_{p^r}})^{\mathrm{mult}}_S$, 245
$(\varphi^{\mathrm{ord}}_{-2,H^{\mathrm{ord}}_{p^r}})^{\mathrm{mult}}_S$, 245
$(\varphi^{\#}_{-2,H^{\mathrm{ord}}_{p^r}})^{\mathrm{mult}}_S$, 244
$(\varphi_{-2,H^{\mathrm{ord}}_{p^r}})^{\mathrm{mult}}_S$, 244
$(\varphi^{\sim}_{-2,H_{p^r}}, \varphi^{\sim}_{0,H_{p^r}})$, 245
$\widetilde{\varphi}_{-2,\widetilde{\mathcal{H}}}$, 372
$(\varphi^{\sim}_{-2,\mathcal{H}}, \varphi^{\sim}_{0,\mathcal{H}})$, 66, 72, 243, 259, 266, 281
$\varphi_{-2,n}$, 29, 232, 313
φ_{-2,n_0}, 237, 313
$(\varphi_{-2,n})^{\mathrm{mult}}_\eta$, 232
φ_{-2,p^r}, 219–221, 239, 244, 245, 314
$(\varphi_{-2,p^r})^{\mathrm{mult}}_\eta$, 218
$\varphi^{\#}_{-2,p^r}$, 219, 221, 244
$(\varphi^{\#}_{-2,p^r})^{\mathrm{mult}}_\eta$, 218
$(\varphi^{\#}_{-2,p^r})^{\mathrm{mult}}_S$, 221, 232, 244
$(\varphi_{-2,p^r})^{\mathrm{mult}}_S$, 221, 232, 244
$\varphi^{\#}_{-2}$, 71, 317
φ_0, 28, 47
$\varphi^{(g)}_0$, 41
$\underline{\varphi}_{0,\mathcal{H}}$, 268
$\varphi_{0,\mathcal{H}}$, 66, 241, 243, 281
$\underline{\varphi}_{0,\mathcal{H}}(G)$, 279
$\varphi^{\mathrm{ord}}_{0,H_n}$, 240
$\varphi^{\mathrm{ord}}_{0,\mathcal{H}}$, 255, 266, 270, 281, 317, 454
$\varphi_{0,H^{\mathrm{ord}}_{p^r}}$, 244, 245
$\varphi^{\mathrm{ord}}_{0,H^{\mathrm{ord}}_{p^r}}$, 239, 245
$\varphi_{0,n}$, 29, 232, 313
φ_{0,n_0}, 237, 313
φ_{0,p^r}, 219–221, 239, 244, 245, 314
$\varphi_{0,p^r,S}$, 221
Φ_{aux}, 133
ϕ_{aux}, 133
$\varphi_{\mathrm{aux},-2}$, 133
$\varphi_{\mathrm{aux},0}$, 133
$\breve{\Phi}$, 47
$\breve{\phi}$, 47
$\breve{\varphi}_{-2}$, 47
$\breve{\varphi}_{-2,\widehat{\mathcal{H}}}$, 48, 110, 413
$\breve{\varphi}_0$, 47
$\breve{\varphi}_{0,\widehat{\mathcal{H}}}$, 48, 110, 413
$\breve{\Phi}_{\widehat{\mathcal{H}}}$, 48, 380
$[(\breve{\Phi}_{\widehat{\mathcal{H}}}, \breve{\delta}_{\widehat{\mathcal{H}}})]$, 52, 373
$(\breve{\Phi}_{\widehat{\mathcal{H}}}, \breve{\delta}_{\widehat{\mathcal{H}}}, \widehat{\tau})$, 98, 393
$[(\breve{\Phi}_{\widehat{\mathcal{H}}}, \breve{\delta}_{\widehat{\mathcal{H}}}, \widehat{\tau})]$, 53, 98, 393
$(\breve{\Phi}_{\widetilde{\mathcal{H}}}, \breve{\delta}_{\widetilde{\mathcal{H}}}, \breve{\tau})$, 46
$[(\breve{\Phi}_{\widetilde{\mathcal{H}}}, \breve{\delta}_{\widetilde{\mathcal{H}}}, \breve{\tau})]$, 46
$\phi^{-1}_{\mathtt{D}}$, 166, 169
$\phi^0_{\mathtt{D}_{-1,p^r}}$, 219, 221–222
$\phi^0_{\mathtt{D}}$, 169, 314
$\phi_{\mathtt{D},0}$, 478
$\phi^{-1}_{\mathtt{D},0}$, 478
$\phi^0_{\mathtt{D},0}$, 478
$(\phi^0_{\mathtt{D}})^{\mathrm{mult}}_s$, 166
$\phi^0_{\mathtt{D}_{\mathrm{aux}}}$, 198
$\phi^{-1}_{\mathtt{D},p^r}$, 169
$\phi^0_{\mathtt{D},p^r}$, 169, 177, 178, 218, 244
$\phi^\ddagger$, 59, 292, 297
$\Phi^{(g)}$, 40
$\underline{\phi}(G)$, 279
$\underline{\phi}^{(g)}$, 40

$\underline{\Phi}_{\mathcal{H}}$, 268, 279
$\Phi_{\mathcal{H}}$, 29, 243, 248, 281
$\Phi_{\mathcal{H}_{\mathrm{aux}}}$, 134
$\Phi_{\mathcal{H}^p_{\mathrm{aux}}}$, 135
$\Phi^{\ddagger}_{\mathcal{H}}$, 60, 292, 297
$(\Phi_{\mathcal{H}}, \delta_{\mathcal{H}})$, 29, 62
$[(\Phi_{\mathcal{H}}, \delta_{\mathcal{H}})]$, 29, 62
$[(\Phi_{\mathcal{H}}, \delta_{\mathcal{H}}, \sigma)]$, 38, 58
$\underline{\Phi}_{\mathcal{H}}(G^{\ddagger})$, 60, 292, 297
$\underline{\Phi}_{\mathcal{H}}({}^{\diamond}G)$, 280
$\Phi^{\sim}_{\mathcal{H}}$, 259
$\widehat{\varphi}^{\ddagger}_{-1,\widehat{\mathcal{H}}}$, 110, 414
$\widehat{\varphi}^{\mathrm{ord},\ddagger}_{-1,\widehat{\mathcal{H}}}$, 414
$\widehat{\varphi}^{\ddagger}_{-1,\widehat{\mathcal{H}}^p}$, 414
$\widehat{\varphi}^{\mathrm{ord},\ddagger}_{-1,\widehat{\mathcal{H}}_p}$, 414
$\hat{\varphi}_{-1,\bar{s}}$, 71
$\hat{\varphi}^{p}_{-1,\bar{s}}$, 317
$\hat{\varphi}^{p,\#}_{-1,\bar{s}}$, 317
$\hat{\varphi}^{\#}_{-1,\bar{s}}$, 71
$(\widehat{\varphi}^{\ddagger,\sim}_{-2,\widehat{\mathcal{H}}}, \widehat{\varphi}^{\ddagger,\sim}_{0,\widehat{\mathcal{H}}})$, 110, 414
$\widehat{\varphi}^{\mathrm{ord},\ddagger}_{-2,\widehat{\mathcal{H}}}$, 414
$\widehat{\varphi}^{\mathrm{ord},\ddagger}_{0,\widehat{\mathcal{H}}}$, 414
Φ_n, 28, 248, 313
ϕ_n, 249
$\underline{\phi}_{p^r}$, 226
$\underline{\Phi}$, 45
$\widetilde{\phi}$, 45, 372
$\widetilde{\varphi}_{-1,\widetilde{\mathcal{H}}}$, 107, 408
$\widetilde{\varphi}^{\ddagger}_{-1,\widetilde{\mathcal{H}}}$, 110, 413
$\widetilde{\varphi}^{\mathrm{ord}}_{-1,\widetilde{\mathcal{H}}}$, 408
$\widetilde{\varphi}^{\mathrm{ord},\ddagger}_{-1,\widetilde{\mathcal{H}}}$, 413
$\widetilde{\varphi}^{\ddagger}_{-1,\widetilde{\mathcal{H}}^p}$, 413
$\widetilde{\varphi}^{\mathrm{ord},\ddagger}_{-1,\widetilde{\mathcal{H}}_p}$, 413
$\widetilde{\varphi}^{\mathrm{ord},0}_{-1,p^r}$, 223
$\widetilde{\varphi}^{\mathrm{ord},\#,0}_{-1,p^r}$, 223
$\widetilde{\varphi}_{-2}$, 45
$(\widetilde{\varphi}^{\ddagger,\sim}_{-2,\widehat{\mathcal{H}}}, \widetilde{\varphi}^{\ddagger,\sim}_{0,\widehat{\mathcal{H}}})$, 110
$\widetilde{\varphi}_{-2,\widetilde{\mathcal{H}}}$, 46, 107, 408
$(\widetilde{\varphi}^{\ddagger,\sim}_{-2,\widetilde{\mathcal{H}}}, \widetilde{\varphi}^{\ddagger,\sim}_{0,\widetilde{\mathcal{H}}})$, 110, 413
$\widetilde{\varphi}^{\mathrm{ord}}_{-2,\widetilde{\mathcal{H}}}$, 408, 454
$\widetilde{\varphi}^{\mathrm{ord},\ddagger}_{-2,\widetilde{\mathcal{H}}}$, 413
$(\widetilde{\varphi}^{\sim}_{-2,\widetilde{\mathcal{H}}}, \widetilde{\varphi}^{\sim}_{0,\widetilde{\mathcal{H}}})$, 107, 408
$\widetilde{\varphi}_0$, 45
$\widetilde{\varphi}_{0,\widetilde{\mathcal{H}}}$, 46, 107, 372, 408
$\widetilde{\varphi}^{\mathrm{ord}}_{0,\widetilde{\mathcal{H}}}$, 408, 454
$\widetilde{\varphi}^{\mathrm{ord},\ddagger}_{0,\widetilde{\mathcal{H}}}$, 413
$\widetilde{\underline{\phi}}^{\ddagger}$, 110, 413
$\widetilde{\underline{\Phi}}_{\widetilde{\mathcal{H}}}$, 46, 372
physical Tate module, 165, 202
$\underline{\pi}_0(\dddot{C}^{\mathrm{ord,grp}}_{\Phi_{\mathcal{H}},n}/\dddot{\mathsf{M}}^{\mathrm{ord},\Phi_{\mathcal{H}}}_{\mathcal{H}})$, 256
$\underline{\pi}_0(\dddot{C}^{\mathrm{ord}}_{\Phi_{\mathcal{H}},n}/\dddot{\mathsf{M}}^{\mathrm{ord},\Phi_{\mathcal{H}}}_{\mathcal{H}})$, 256
$\underline{\pi}_0(\dddot{C}^{\mathrm{ord},\circ\circ\circ}_{\Phi_n}/\dddot{\mathsf{M}}^{\mathrm{ord}}_n)$, 250
$\underline{\pi}_0(\dddot{C}^{\mathrm{ord}}_{\Phi_n}/\dddot{\mathsf{M}}^{\mathrm{ord}}_n)$, 250
$\underline{\pi}_0(\widetilde{\dddot{C}}^{\mathrm{ord,ext}}_n/\vec{\mathsf{M}}^{\mathrm{ord,tor}}_{\mathcal{H}})$, 436
$\underline{\pi}_0(\widetilde{\dddot{C}}^{\mathrm{ord,ext}}/\vec{\mathsf{M}}^{\mathrm{ord,tor}}_{\mathcal{H},r_\kappa})$, 434
$\underline{\pi}_0(\underline{\mathrm{Hom}}_{\mathcal{O}}(N, \underline{Z})/\underline{U})$, 158
$\underline{\pi}_0(\underline{\mathrm{Hom}}_{\mathcal{O}}(N, \overline{Z})/\overline{U})$, 163
$\pi_0(W_{\bar{s}})$, 157
$\underline{\pi}_0(W/U)$, 158
$\underline{\mathrm{Pic}}^0(\mathsf{N}/\vec{\mathsf{M}}^{\mathrm{ord}}_{\mathcal{H}})$, 391
$\underline{\mathrm{Pic}}^0(\mathsf{N}/\mathsf{M}_{\mathcal{H}})$, 96
Poincaré
  biextension, 251
  invertible sheaf, 24, 77
pol, 4, 5, 37, 62
$\mathsf{pol}_{\mathrm{aux}}$, 334
$\mathsf{pol}^{\mathrm{ord}}_{\mathrm{aux}}$, 334
$\mathsf{pol}^{\mathrm{ord}}$, 8, 289, 359
$\mathsf{pol}^{\mathrm{ord},(\mathcal{H}')}$, 360
pol^{p}, 305, 308
$\mathsf{pol}_{\Phi_{\mathcal{H}}}$, 35, 37, 62, 289
polarization function, 4, 5, 8, 35
  convexity, 35
polarization type, 174
positivity condition, 217, 270, 292, 449, 474
$\mathrm{pr}_{(\widehat{\mathbf{S}}_{\check{\Phi}_{\widehat{\mathcal{H}}}})^\vee_{\mathbb{R}}}$, 50, 373
$\mathrm{pr}_{(\widehat{\mathbf{S}}_{\check{\Phi}_{\widehat{\mathcal{H}}}})^\vee_{\mathbb{R}}}(\check{\tau})$, 51
prime
  bad, *see* bad prime
  good, *see* good prime
principal bundle
  canonical extension, 122, 487
  for G_0, 118, 122
  for $\mathrm{G}^{\mathrm{ord}}_{\mathtt{D},0}$, 481, 487, 508
  for M_0, 119, 122
  for $\mathrm{M}^{\mathrm{ord}}_{\mathtt{D},0}$, 481, 487, 508, 509
  for P_0, 119, 122
  for $\mathrm{P}^{\mathrm{ord}}_{\mathtt{D},0}$, 481, 487, 508, 509
principal cusp label, 29

principal level structure
of type $(L \underset{\mathbb{Z}}{\otimes} \hat{\mathbb{Z}}^p, \langle \cdot, \cdot \rangle)$, 173
of type $(L \underset{\mathbb{Z}}{\otimes} \hat{\mathbb{Z}}^p, \langle \cdot, \cdot \rangle, \mathsf{Z} \underset{\hat{\mathbb{Z}}}{\otimes} \hat{\mathbb{Z}}^p)$, 313, 324
of type $(L \underset{\mathbb{Z}}{\otimes} \hat{\mathbb{Z}}, \langle \cdot, \cdot \rangle, \mathsf{Z})$, 69, 79
Proj, 61, 139, 140, 143, 147, 307, 308, 310, 331, 332, 334, 336, 346, 349–351, 355, 430
$\underline{\mathrm{Proj}}_S$, 349–350
projection formula, 148, 332, 338, 349, 462, 463, 465, 466, 469, 472, 499, 515, 517, 518, 522, 523
projective
admissible rational polyhedral cone decomposition, 35
projectivity
of $\mathsf{M}_{\mathcal{H}}^{\mathrm{tor}}$, 64
$\vec{\Psi}^{\mathrm{ord}}_{\Phi_{\mathcal{H}},\delta_{\mathcal{H}}}(\ell)$, 266, 325
relative ampleness, 504
$\vec{\Psi}^{\mathrm{ord}}_{\Phi_{\mathcal{H}},\delta_{\mathcal{H}}}(\ell)_K$, 269
$\vec{\Psi}^{\mathrm{ord}}_{\Phi_{\mathcal{H}},\delta_{\mathcal{H}}}(\ell)_R$, 269
$\widehat{\vec{\Psi}}^{\mathrm{ord}}_{\breve{\Phi}_{\widehat{\mathcal{H}}},\breve{\delta}_{\widehat{\mathcal{H}}}}(\breve{\ell})$, 382, 418
$\widehat{\Psi}_{\breve{\Phi}_{\widehat{\mathcal{H}}},\breve{\delta}_{\widehat{\mathcal{H}}}}(\breve{\ell})$, 88
$\Psi_n^{\mathrm{ord}}(\ell)$, 251
$\Psi_{\Phi_{\mathcal{H}},\delta_{\mathcal{H}}}(\ell)$, 80
$\Psi_{\Phi_n,\delta_n}(\ell)$, 79
$\widetilde{\vec{\Psi}}^{\mathrm{ord}}_{\breve{\Phi}_{\widetilde{\mathcal{H}}},\breve{\delta}_{\widetilde{\mathcal{H}}}}(\breve{\ell})$, 381, 409
$\widetilde{\vec{\Psi}}^{\mathrm{ord}}_{\widetilde{\Phi}_{\widetilde{\mathcal{H}}},\widetilde{\delta}_{\widetilde{\mathcal{H}}}}(\widetilde{\ell}_{j_Q})$, 474
$\widetilde{\Psi}_{\breve{\Phi}_{\widetilde{\mathcal{H}}},\breve{\delta}_{\widetilde{\mathcal{H}}}}(\breve{\ell})$, 88

Q, 43, 95, 371
Q_{-2}, 43, 371
Q_0, 43, 371
$Q^\vee$, 43
q-expansion, 33
$(\mathbb{Q}_p/\mathbb{Z}_p)_S$, 156
quasi-projectivity
of $\vec{C}^{\mathrm{ord}}_{\Phi_{\mathcal{H}},\delta_{\mathcal{H}}}$, 265
of $[\vec{\mathsf{M}}_{\mathcal{H}}]$, 141, 142
of $\vec{\mathsf{M}}^{\mathrm{ord}}_{\mathcal{H}}$, 214
of $\vec{\mathsf{M}}^{\mathrm{ord,min}}_{\mathcal{H}}$, 332, 334, 337
of $\vec{\mathsf{M}}^{\mathrm{ord},\Phi_{\mathcal{H}}}_{\mathcal{H}}$, 265
of $\vec{\mathsf{M}}^{\mathrm{ord,tor}}_{\mathcal{H}}$, 360
of $\vec{\mathsf{M}}^{\mathrm{ord},\mathsf{Z}_{\mathcal{H}}}_{\mathcal{H}}$, 265
of $[\mathsf{M}_{\mathcal{H}}]$, 24, 97
of $[\mathsf{M}_{\mathcal{H}_{\mathrm{aux}}}]$, 129
of $\vec{\mathsf{N}}^{\mathrm{ord,tor}}_{\kappa}$, 393
of $\vec{\mathsf{N}}^{\mathrm{ord,tor}}_{\widetilde{\kappa}}$, 409
of $\vec{\Xi}^{\mathrm{ord}}_{\Phi_{\mathcal{H}},\delta_{\mathcal{H}}}$, 265

R_0, 477, 490, 491
$\tilde{R}_0$, 477
R_{alg}, 286
$\vec{\mathsf{R}}^{\mathrm{ord}}_{\mathcal{H}}$, 289, 293
$\vec{\mathsf{R}}^{\mathrm{ord},[0]}_{\mathcal{H}}$, 291
r_{D}, 6, 203
$r_{\mathcal{H}}$, 6, 203
$r_{\mathcal{H}_{\mathrm{aux}}}$, 334
$r_{\widehat{\mathcal{H}}_{\mathrm{G}}}$, 372
$r_{\mathcal{H}_\kappa}$, 393
$r_{\widetilde{\mathcal{H}}}$, 371, 372
$r(j^\natural, j^{\vee,\natural})$, 70, 316
$r(j^\natural, j^{\vee,\natural})_{\bar{s}}$, 71, 316
r_κ, 393
r_n, 383
r_ν, 192, 203
$r_{\nu(\mathcal{H}_{\mathrm{aux}})}$, 334
$r_{\widetilde{\nu}}$, 371
$R_{\mathbb{Q}}$, 509
$\tilde{R}_{\mathbb{Q}}$, 509
radical
admissible, 34
rational boundary component, 27, 38, 40
rational level structure
of type $(L \underset{\mathbb{Z}}{\otimes} \mathbb{A}^{\infty,p}, \langle \cdot, \cdot \rangle, \mathsf{Z} \underset{\hat{\mathbb{Z}}}{\otimes} \mathbb{A}^{\infty,p}, \Phi)$, 316
based at $\bar{s}$, 316
of type $(L \underset{\mathbb{Z}}{\otimes} \mathbb{A}^{\infty}, \langle \cdot, \cdot \rangle, \mathsf{Z} \underset{\hat{\mathbb{Z}}}{\otimes} \mathbb{A}^{\infty}, \Phi)$, 70
based at $\bar{s}$, 70
rational polyhedral cone, 34
reflex field, 3, 23, 44, 128
regular singularity, 122
relative de Rham cohomology, 24, 97, 391, 490
relative de Rham homology, 24
relative Frobenius morphism, 181, 190, 211, 304, 362, 403
relative scheme, 57, 59, 205, 270, 296
relative Verschiebung morphism, 362
relatively representable functor, 213, 306
$\mathrm{Rep}_{R_0}(\mathrm{M}^{\mathrm{ord}}_{\mathsf{D},0,\mathrm{fil}})$, 506
$\mathrm{Rep}_R(\mathrm{G}_0)$, 120
$\mathrm{Rep}_R(\mathrm{G}^{\mathrm{ord}}_{\mathsf{D},0})$, 482
$\mathrm{Rep}_R(\mathrm{M}_0)$, 120
$\mathrm{Rep}_R(\mathrm{M}^{\mathrm{ord}}_{\mathsf{D},0})$, 482

$\mathrm{Rep}_R(\mathrm{P}_0)$, 120
$\mathrm{Rep}_R(\mathrm{P}^{\mathrm{ord}}_{\mathtt{D},0})$, 482
ρ, 81, 326
$\breve{\rho}$, 89, 383
$\rho^\vee$, 81, 326
$\widehat{\rho}$, 89, 383
$\widehat{\rho}^\perp$, 89, 384
$\widehat{\rho}$-stratum, 89
Rosati condition, 70, 164, 316

S_0, 216
S_0, 3, 8, 23
$\mathsf{S}_{0,\mathrm{aux}}$, 129
$\mathsf{S}_{0,i}$, 5, 151
$\mathsf{S}_{0,r_{\mathcal{H}}}$, 6, 10, 203
S_{0,r_ν}, 192, 203
(S_1), 516–517
(S_2), 517
$\vec{\mathsf{S}}_0$, 5, 141
$\vec{\mathsf{S}}_{0,\mathrm{aux}}$, 9
$\mathsf{S}_{0,\mathrm{aux}}$, 129
$\vec{\mathsf{S}}_{0,\mathrm{aux},r}$, 333
$\vec{\mathsf{S}}_{0,i}$, 5, 151
$\vec{\mathsf{S}}_{0,r_{\mathcal{H}}}$, 5, 7, 10, 203
$\vec{\mathsf{S}}_{0,r_{\widetilde{\mathcal{H}}}}$, 379, 407
$\vec{\mathsf{S}}_{0,r_\kappa}$, 393, 407
$\vec{\mathbf{S}}_{\mathtt{D}}$, 201
$\vec{\mathbf{S}}_{\mathtt{D},\mathrm{tor}}$, 201
$\vec{\mathbf{S}}_{\mathtt{D},\mathbb{Z}}$, 201, 203
$\vec{\mathbf{S}}_{\mathtt{D},\mathbb{Z},\mathrm{free}}$, 203
$\vec{\mathbf{S}}_{\mathtt{D},\mathbb{Z},\mathrm{tor}}$, 201, 204
$\vec{\oint}_{\mathcal{H}}$, 147, 515
$\vec{\oint}^{\mathrm{ord}}_{\mathcal{H}_{\mathrm{aux}}}$, 334
$\vec{\oint}_{\mathcal{H},i}$, 151
$[\vec{\oint}^{\mathrm{ord}}_{\mathcal{H}}]$, 338
$\vec{\oint}^{\mathrm{ord}}_{\mathcal{H}}$, 309, 332, 346, 492, 493
  formal fiber, 492–496
$\vec{\oint}^{\mathrm{ord}}_{\mathcal{H},\Sigma}$, 348
$\dddot{\mathbf{S}}_{\Phi_1}$, 264
$\dddot{\mathbf{S}}^{\mathrm{ord}}_{\Phi_1}$, 264
$\dddot{\mathbf{S}}^{\mathrm{ord}}_{\Phi_{\mathcal{H}},n,\mathrm{free}}$, 258
$\dddot{\mathbf{S}}^{\mathrm{ord}}_{\Phi_{\mathcal{H}},n}$, 258
$\dddot{\mathbf{S}}^{\mathrm{ord}}_{\Phi_{\mathcal{H}},n,\mathrm{tor}}$, 258
$\dddot{\mathbf{S}}_{\Phi_n}$, 33
$\dddot{\mathbf{S}}_{\Phi_{n_0}}$, 251, 264
$\dddot{\mathbf{S}}_{\Phi_n,\mathrm{free}}$, 33
$\dddot{\mathbf{S}}^{\mathrm{ord}}_{\Phi_n,\mathrm{free}}$, 251
$\dddot{\mathbf{S}}^{\mathrm{ord}}_{\Phi_n}$, 251
$\dddot{\mathbf{S}}^{\mathrm{ord}}_{\Phi_n,\mathrm{tor}}$, 251
S_{for}, 280
$\oint_{\mathcal{H}}$, 61, 515
$\oint_{\mathcal{H}^p}$, 307
$\widehat{\mathbf{S}}_{\breve{\Phi}_1}$, 88, 381
$\widehat{\mathbf{S}}^\vee_{\breve{\Phi}_1}$, 88, 381
$\widehat{\mathbf{S}}_{\breve{\Phi}_{\widehat{\mathcal{H}}}}$, 50, 87, 373, 381
$\widehat{\mathbf{S}}^\vee_{\breve{\Phi}_{\widehat{\mathcal{H}}}}$, 88, 381
$(\widehat{\mathbf{S}}_{\breve{\Phi}_{\widehat{\mathcal{H}}}})^\vee_{\mathbb{R}}$, 50, 373
$\mathbf{S}_{\breve{\Phi}_{\widetilde{\mathcal{H}}}}$, 48
$(\mathbf{S}_{\breve{\Phi}_{\widetilde{\mathcal{H}}}})^\vee_{\mathbb{R}}$, 48
$\mathbf{S}_{\Phi_{\mathcal{H}}}$, 33, 48, 80, 258, 324, 325
$\underline{\mathbf{S}}_{\underline{\Phi}_{\mathcal{H}}}$, 268
$\mathbf{S}_{\Phi_{\mathcal{H}_{\mathrm{aux}}}}$, 134
$\mathbf{S}^\vee_{\Phi_{\mathcal{H}_{\mathrm{aux}}}}$, 134
$\mathbf{S}_{\Phi_{\mathcal{H}^p_{\mathrm{aux}}}}$, 135
$\mathbf{S}^\vee_{\Phi_{\mathcal{H}^p_{\mathrm{aux}}}}$, 135
$(\mathbf{S}_{\Phi_{\mathcal{H}^p_{\mathrm{aux}}}})_{\mathbb{Q}}$, 135
$(\mathbf{S}_{\Phi_{\mathcal{H}^p_{\mathrm{aux}}}})^\vee_{\mathbb{Q}}$, 135
$(\mathbf{S}_{\Phi_{\mathcal{H}^p_{\mathrm{aux}}}})^\vee_{\mathbb{R}}$, 135
$(\mathbf{S}_{\Phi_{\mathcal{H}_{\mathrm{aux}}}})_{\mathbb{Q}}$, 134
$(\mathbf{S}_{\Phi_{\mathcal{H}_{\mathrm{aux}}}})^\vee_{\mathbb{Q}}$, 134
$(\mathbf{S}_{\Phi_{\mathcal{H}_{\mathrm{aux}}}})^\vee_{\mathbb{R}}$, 134
$\mathbf{S}_{\Phi^\ddagger_{\mathcal{H}}}$, 60, 292, 297
$\mathbf{S}^\vee_{\Phi_{\mathcal{H}}}$, 34, 267
$\mathbf{S}_{\underline{\Phi}_{\mathcal{H}}(G)}$, 279
$\underline{\mathbf{S}}_{\underline{\Phi}_{\mathcal{H}}(G^\ddagger)}$, 60, 292, 297
$\underline{\mathbf{S}}_{\underline{\Phi}_{\mathcal{H}}(\,^\diamond G)}$, 280
$(\mathbf{S}_{\Phi_{\mathcal{H}_{\mathrm{aux}}}})_{\mathbb{Q}}$, 134
$(\mathbf{S}_{\Phi_{\mathcal{H}}})^\vee_{\mathbb{R}}$, 34, 48, 267
$\mathbf{S}_{\Phi_{\mathcal{H}},\sigma}$, 62
$\mathbf{S}_{\Phi_n}$, 33
$\mathbf{S}_{\widetilde{\Phi}_{\widetilde{\mathcal{H}}}}$, 48
$(\mathbf{S}_{\widetilde{\Phi}_{\widetilde{\mathcal{H}}}})^\vee_{\mathbb{R}}$, 48
s_X, 32
$s_{\breve{X}}$, 47
s_Y, 32
$s_{\breve{Y}}$, 47
semi-abelian scheme
  dual, 57, 205, 215, 392
  ordinary, *see* ordinary semi-abelian scheme
  semistable reduction theorem, 145
Serre dual, 156, 185

Serre–Tate deformation theory, 10, 15, 202, 208
Serre's construction, 158–160
  semi-abelian extension, 163
Serre's criterion, 517
Serre's lemma, 53, 132
$\mathsf{Sh}_{\mathcal{H}}$, 26
$\mathsf{Sh}_{\mathcal{H}}^{\mathrm{min}}$, 124
$\mathsf{Sh}_{\mathcal{H},\Sigma}^{\mathrm{tor}}$, 124
Shapiro's lemma, 491, 492, 501
Shimura variety, 26
Siegel case, 17
Siegel parabolic subgroup, 18
Σ, 3, 37, 58, 63
$[\sigma]$, 496, 497
σ, 34, 81
$\sigma_0^\vee$, 34
$\sigma_{0+}^\vee$, 493
Σ_{aux}, 135, 334
σ_{aux}, 134
$\Sigma_{\mathrm{aux}}^{\mathrm{ord}}$, 12, 333
$\Sigma_{\mathrm{aux}}^{p}$, 12, 135, 335
$\sigma_{\mathrm{aux}}^{p}$, 135
$\overline{\sigma}$, 34
$\breve{\sigma}$, 48, 87, 373, 381, 445
$\breve{\sigma}^\perp$, 87, 381
$[\sigma]^{\mathrm{cl}}$, 497
σ^{cl}, 496
$\sigma^\ddagger$, 60, 292, 297
$\sigma^\vee$, 34
$\widehat{\Sigma}$, 52, 53, 97
$\widehat{\Sigma}^{(\widehat{\mathcal{H}}'')}$, 54
$\widehat{\Sigma}^{\mathrm{ord}}$, 373, 374, 393
$\widehat{\Sigma}_{\breve{\Phi}_{\widehat{\mathcal{H}}}}$, 52, 98, 373, 393
ς_n, 77
$\varsigma_n^\vee$, 78
Σ^{ord}, 7, 288, 289, 295, 359
$\Sigma^{\mathrm{ord},(\mathcal{H}')}$, 360
Σ^p, 305
ς_{p^r}, 228
$\varsigma_{p^r}^{\mathrm{ord},-1}$, 228
$\varsigma_{p^r}^{\mathrm{ord},0}$, 223
$\varsigma_{p^r}^{\mathrm{ord},\#,0}$, 223
$\sigma^\perp$, 34
$\Sigma_{\Phi_{\mathcal{H}}}$, 35
$\Sigma_{\Phi_{\mathcal{H}}^\ddagger}$, 60, 297
$\Sigma_{\Phi_{\mathcal{H}}}^+$, 493
$\sigma_+^\vee$, 493
σ-stratum, 62, 81, 269
$\widetilde{\Sigma}$, 46, 85
$\widetilde{\sigma}$, 46, 48, 87, 372, 381
$\widetilde{\Sigma}^{\mathrm{ord}}$, 372, 374
$\widetilde{\Sigma}^p$, 426
$\widetilde{\Sigma}_{\breve{\Phi}_{\widetilde{\mathcal{H}}},\breve{\sigma}}$, 48
$\widetilde{\Sigma}_{\breve{\Phi}_{\widetilde{\mathcal{H}}},\breve{\sigma}}^+$, 48, 49
$\widetilde{\Sigma}_{\breve{\Phi}_{\widetilde{\mathcal{H}}}}$, 48, 374
$\widetilde{\Sigma}_{\breve{\Phi}_{\widetilde{\mathcal{H}}},\breve{\sigma}}$, 48, 373
$\widetilde{\Sigma}_{\breve{\Phi}_{\widetilde{\mathcal{H}}},\breve{\sigma}}^+$, 48, 373
$\widetilde{\Sigma}_{\breve{\Phi}_{\widetilde{\mathcal{H}}},\breve{\sigma},\tau}$, 446
$\widetilde{\Sigma}_{\widetilde{\Phi}_{\widetilde{\mathcal{H}}}}$, 46
sgn(h), 42
similitude character, 22, 170
similitude isomorphism, 198, 235
slope, 156
smooth cone, 35
split filtration, 27
standard form
  subgroup of $\widehat{\mathrm{G}}(\hat{\mathbb{Z}})$, 372
  subgroup of $\widehat{\mathrm{G}}(\mathbb{Z}_p)$, 371
  subgroup of $\mathrm{G}(\hat{\mathbb{Z}})$, 170
  subgroup of $\mathrm{G}(\mathbb{Z}_p)$, 170
Stein factorization, 256, 307, 310, 522
strictly compatible stratifications, 283, 284
subcanonical extension, 4, 5, 8, 13, 123, 488, 509, 512
submodule
  admissible, 32, 34
supporting hyperplane, 34
surjection
  admissible, 32
  cusp label, 32, 33
    with cone decomposition, 36
symplectic condition
  for $(c_n, c_n^\vee)$, 68
  for τ_n, 78
symplectic filtration, 27
symplectic isomorphism
  similitude, 198, 235
symplectic-liftability condition
  for level structure, 173
  for ordinary level structure, 179, 180

T, 68, 70, 315, 454
T_0, 216
$\breve{T}$, 87

$\breve{T}^{\vee}$, 87
$T^{\vee}$, 68, 70, 315, 454
τ_n, 245
τ_n^{ord}, 245
$\mathrm{T}_p\, A$, 202
$\mathrm{T}_p\, A^{\vee}$, 202
$\mathrm{T}_p\, A_s$, 165
$\mathrm{T}_p\, A_s^{\vee}$, 165
$\mathbf{T}_{\widetilde{\Phi}_{\widetilde{\mathcal{H}}},\Phi_{\mathcal{H}}}$, 456
$\widetilde{T}_{\vec{\mathsf{N}}_{\kappa}^{\mathrm{ord}}}$, 421
$\widetilde{T}^{\vee}_{\vec{\mathsf{N}}_{\kappa}^{\mathrm{ord}}}$, 421
$\widetilde{T}_{\vec{\mathsf{N}}_{\kappa}^{\mathrm{ord,tor}}}$, 422
$\widetilde{T}^{\vee}_{\vec{\mathsf{N}}_{\kappa}^{\mathrm{ord,tor}}}$, 422
Tate module
  physical, 165, 202
Tate twist, 21
τ (cone), 34
τ (trivialization), 77, 79, 138, 216, 251, 281, 324
  compatibility with $\mathcal{O}$-actions, 216
  positivity, 216
τ_{aux}, 138
$\overline{\tau}$, 34
$[\breve{\tau}]$, 454, 460
$\breve{\tau}$, 48
$[\breve{\tau}]^{\mathrm{cl}}$, 454
$\breve{\tau}^{\mathrm{cl}}$, 454
$\breve{\tau}_{\widetilde{\mathcal{H}}}$, 92
$\breve{\tau}^{\mathrm{ord}}_{\widetilde{\mathcal{H}}}$, 386
$\breve{\tau}_n$, 89, 92
$\breve{\tau}_{n_0}$, 386
$\breve{\tau}_n^{\mathrm{ord}}$, 386
$\breve{\tau}^{\vee}_{\breve{\sigma}}$, 447
$\breve{\tau}^{\vee}_{\breve{\sigma},+}$, 447
$\tau^{\ddagger}$, 59, 112, 292, 297, 415
$\tau_{\mathcal{H}}$, 80, 243, 259, 266, 281
τ_{H_n}, 243
$\tau_{H_{n_0}}$, 236, 237
$\tau^{\mathrm{ord}}_{H_n}$, 240
$\tau^{\mathrm{ord}}_{\mathcal{H}}$, 241, 255, 266, 270, 281, 387
$\tau_{\mathcal{H}^p}$, 241
$\tau^{\mathrm{ord}}_{\mathcal{H}_p}$, 241
$\tau^{\mathrm{ord}}_{H^{\mathrm{ord}}_{p^r}}$, 239, 245
$\widehat{\tau}$ (cone), 52, 55, 375
$\widehat{\tau}^{\ddagger}$, 111, 414
$\widehat{\tau}^{\vee,\ddagger}$, 111, 414
$\widehat{\tau}_{\widehat{\mathcal{H}}}$, 88, 93
$\widehat{\tau}^{\ddagger}_{\widehat{\mathcal{H}}}$, 111, 414
$\widehat{\tau}^{\vee}_{\widehat{\mathcal{H}}}$, 88, 93
$\widehat{\tau}^{\vee,\ddagger}_{\widehat{\mathcal{H}}}$, 111, 414
$\widehat{\tau}^{\vee,\mathrm{ord}}_{\widehat{\mathcal{H}}}$, 382, 388
$\widehat{\tau}^{\vee,\mathrm{ord},\ddagger}_{\widehat{\mathcal{H}}}$, 414
$\widehat{\tau}^{\mathrm{ord}}_{\widehat{\mathcal{H}}}$, 382, 388
$\widehat{\tau}^{\mathrm{ord},\ddagger}_{\widehat{\mathcal{H}}}$, 414
$\widehat{\tau}^{\ddagger}_{\widehat{\mathcal{H}}^p}$, 414
$\widehat{\tau}^{\vee,\ddagger}_{\widehat{\mathcal{H}}^p}$, 414
$\widehat{\tau}_n$, 88, 382
$\widehat{\tau}_{n_0}$, 382
$\widehat{\tau}^{\vee}_{n_0}$, 382
$\widehat{\tau}^{\vee}_n$, 88, 382
τ_n, 76, 77
τ_{n_0}, 233, 237, 251, 387
$\tau_n^{\vee}$, 77
τ_n^{ord}, 233, 251, 324, 387
τ_{p^r}, 228
$\tau_{p^r}^{\mathrm{ord}}$, 230
$\tau_+^{\vee}$, 447
$\widetilde{\tau}^{\ddagger}$, 110, 413
$\widetilde{\tau}^{\ddagger}_{\widetilde{\mathcal{H}}}$, 110, 413
$\widetilde{\tau}^{\mathrm{ord},\ddagger}_{\widetilde{\mathcal{H}}}$, 413
$\widetilde{\tau}^{\ddagger}_n$, 111
theory of degeneration, 1, 4, 8, 11, 14, 215–247
toroidal compactification
  of Kuga family, 4, 97–106
  of $\mathsf{M}_{\mathcal{H}}$, 3, 58–61
    projectivity, 64
    universal property, 60
  partial, *see* partial toroidal compactification
toroidal embedding, 33, 81, 90, 267, 384, 493
  affine, *see* affine toroidal embedding
torus
  isotrivial, 216
torus argument, 28, 48
  at level $\mathcal{H}$, 29
  at level n, 28
    equivalence, 29
total model, 5–6, 12, 13, 141–153, 214, 308–310, 335–336, 347, 354, 361, 363–368, 508–523
$\mathrm{Tr}_{\mathcal{O}/\mathbb{Z}}$, 43
twisted U_p type element, 195, 209
two-step degeneration, 33, 112, 113, 415

type D, 23

$\mathcal{U}_1^{\mathrm{bal}}(n)$, 311
$\mathcal{U}_1^{\mathrm{bal}}(n)_{\mathrm{G}_{1,\mathsf{Z}}}$, 313
$\mathcal{U}_1^{\mathrm{bal}}(n)_{\mathrm{G}_{h,\mathsf{Z}}}$, 312
$\mathcal{U}_1^{\mathrm{bal}}(n)_{\mathrm{G}'_{h,\mathsf{Z}}}$, 312
$\mathcal{U}_1^{\mathrm{bal}}(n)_{\mathrm{P}'_{\mathsf{Z}}}$, 324
$\mathrm{U}^{\mathrm{ess}}_{1,\mathsf{Z}_n}$, 235
$\mathrm{U}^{\mathrm{ess}}_{1,\mathsf{Z}_{p^r},\mathsf{D}_{p^r}}$, 237
$\mathrm{U}^{\mathrm{ess,ord}}_{1,\mathsf{Z}_{p^r},\mathsf{D}_{p^r}}$, 238
$\mathrm{U}_{1,\mathsf{Z}}(R)$, 31, 56
$\mathrm{U}^{\mathrm{ess}}_{2,\mathsf{Z}_n}$, 235
$\mathrm{U}^{\mathrm{ess}}_{2,\mathsf{Z}_{p^r},\mathsf{D}_{p^r}}$, 237
$\mathrm{U}^{\mathrm{ess,ord}}_{2,\mathsf{Z}_{p^r},\mathsf{D}_{p^r}}$, 238
$\mathrm{U}_{2,\mathsf{Z}}(R)$, 31, 56, 378
$\vec{\mathsf{U}}^{\mathrm{ord}}_{\mathcal{H}}$, 289
construction, 290–291
stratification, 290
$\vec{\mathsf{U}}^{\mathrm{ord},[0]}_{\mathcal{H}}$, 291
$\vec{\mathsf{U}}^{\mathrm{ord}}_{\mathcal{H}}/\vec{\mathsf{R}}^{\mathrm{ord}}_{\mathcal{H}}$, 294
$\overline{\mathcal{U}}^{\mathrm{bal}}_{p,1}(p^r)$, 238
$\mathrm{U}^{\mathrm{ord}}_{\mathsf{D}}$, 170
$\mathrm{U}^{\mathrm{ord},i}_{\mathsf{D}}$, 170
$\mathrm{U}^{\mathrm{ord},i}_{\mathsf{D}}(R)$, 170
$\mathrm{U}^{\mathrm{ord}}_{\mathsf{D}}(R)$, 169
$\widehat{\mathcal{U}}_1^{\mathrm{bal}}(n)$, 382
$\widehat{\mathcal{U}}_1^{\mathrm{bal}}(n)_{\widehat{\mathrm{P}}_{\breve{\mathsf{Z}}}}$, 382
$\widehat{\mathrm{U}}_{1,\breve{\mathsf{Z}}}(R)$, 56, 378
$\widehat{\mathrm{U}}_{2,\breve{\mathsf{Z}}}(R)$, 56, 378
$\widehat{\mathrm{U}}^{\mathrm{ord}}_{\mathsf{D}}(R)$, 377
$\widehat{\mathcal{U}}(n)_{\widehat{\mathrm{P}}_{\breve{\mathsf{Z}}}}$, 88
$\widehat{\mathcal{U}}_{p,0}(p^r)$, 371
$\widehat{\mathcal{U}}^{\mathrm{bal}}_{p,1}(p^r)$, 371, 377
$\widehat{\mathcal{U}}^p(n_0)$, 382
$\widehat{\mathrm{U}}(R)$, 45
$\widehat{\mathrm{U}}_{\breve{\mathsf{Z}}}(R)$, 56, 378
$\mathcal{U}(n)$, 22
$\mathcal{U}(n)_{\mathrm{G}_{1,\mathsf{Z}}}$, 67
$\mathcal{U}(n)_{\mathrm{P}'_{\mathsf{Z}}}$, 76
$\mathcal{U}(n)_{\mathrm{U}_{1,\mathsf{Z}}}$, 67
$\mathcal{U}(n)_{\mathrm{U}_{2,\mathsf{Z}}}$, 76, 79
$\mathcal{U}^{\mathrm{ord}}(p^r)$, 170
$\mathcal{U}_{p,0}(p^r)$, 6, 170
$\mathcal{U}_{p,1,0}(p^{r_1},p^{r_0})$, 170, 211
$\mathcal{U}_{p,1,\mathrm{aux}}(p^r)$, 199
$\mathcal{U}^{\mathrm{bal}}_{p,1}(p^r)$, 6, 170
$\mathcal{U}^{\mathrm{bal}}_{p,1,0}(p^{r_1},p^{r_0})$, 170
$\mathcal{U}_{p,1}(p^r)$, 6, 170
$\mathcal{U}^p(n_0)$, 23, 172
U_p operator, 184, 190, 211, 304, 358–359, 402–403
generalized, 184
$\mathcal{U}_p(p^r)$, 170
U_p type element, 194, 211
twisted, 195, 209
U_σ, 493
$\mathfrak{U}_\sigma$, 82, 493
$\mathfrak{U}_{[\sigma]}$, 495
$U_{\check{\tau}}$, 446
$\mathfrak{U}_{[\check{\tau}]}$, 451
$\mathfrak{U}_{\check{\tau}}$, 446
$\widetilde{\mathrm{U}}_{1,\widetilde{\mathsf{Z}}}(R)$, 45
$\widetilde{\mathrm{U}}_{2,\widetilde{\mathsf{Z}}}(R)$, 45, 377, 378
$\widetilde{\mathrm{U}}^{\mathrm{ord}}_{\widetilde{\mathsf{D}}}(R)$, 377
$\widetilde{\mathcal{U}}(n)$, 53, 85
$\widetilde{\mathcal{U}}(n)_{\widehat{\mathrm{G}}}$, 53
$\widetilde{\mathcal{U}}_{p,0}(p^r)$, 371
$\widetilde{\mathcal{U}}_{p,0}(p^r)_{\widehat{\mathrm{G}}}$, 371
$\widetilde{\mathcal{U}}^{\mathrm{bal}}_{p,1}(p^r)$, 425
$\widetilde{\mathcal{U}}^p(n_0)$, 379
$\widetilde{\mathcal{U}}^{\mathrm{bal}}_{p,1}(p^r)$, 371
$\widetilde{\mathfrak{U}}_{\check{\tau}}$, 449
$\widetilde{\mathrm{U}}_{\breve{\mathsf{Z}}}(R)$, 378
$\widetilde{\mathrm{U}}_{\widetilde{\mathsf{Z}}}(R)$, 45, 377
$\mathrm{U}^{\mathrm{ess}}_{\mathsf{Z}_n}$, 235
$\mathrm{U}^{\mathrm{ess}}_{\mathsf{Z}_{p^r},\mathsf{D}_{p^r}}$, 237
$\mathrm{U}^{\mathrm{ess,ord}}_{\mathsf{Z}_{p^r},\mathsf{D}_{p^r}}$, 238
$\mathrm{U}_{\mathsf{Z}}(R)$, 31, 56
unitary case
easier, 18
universal property
of $\vec{\mathsf{M}}^{\mathrm{ord,tor}}_{\mathcal{H}}$, 296–297
of $\mathsf{M}^{\mathrm{tor}}_{\mathcal{H}}$, 59–60
of $\vec{\mathsf{M}}^{\mathrm{tor}}_{\widetilde{\mathcal{H}}}$, 105
of $\vec{\Xi}^{\mathrm{ord}}_{\Phi_{\mathcal{H}},\delta_{\mathcal{H}}}(\sigma)$, 268–269
of $\vec{\Xi}^{\mathrm{ord}}_{\Phi_{\mathcal{H}},\delta_{\mathcal{H}}}(\sigma)/\Gamma_{\Phi_{\mathcal{H}},\sigma}$, 281–282
unramified prime, 23

$\mathtt{V}$, 39
$\mathtt{V}_{-2}$, 39
$\mathtt{V}_{-i}$, 39
V_0, 23, 117
$V_{0,\mathrm{aux}}$, 128
V_0^c, 23

$\mathrm{V}_{\mathrm{aux}}$, 362
$\upsilon_{\tau^\ddagger}$, 423
$\upsilon_{\widehat{\tau}^\ddagger}$, 113, 415
$\upsilon_{\widehat{\tau}^{\vee,\ddagger}}$, 113, 415
$\upsilon_{\widetilde{\tau}^\ddagger}$, 113, 415
Verschiebung morphism
relative, 362

W, 120, 122, 123, 482, 488, 490, 491, 502, 503, 506–509, 512, 513, 516, 518
W_0, 517
$W_{\mathbb{Q}}$, 509
$\tilde{W}_{\mathbb{Q}}$, 509
Weil pairing, 173

X, 28, 47, 138, 281, 454
$\underline{X}$, 216, 268
X, 26, 39
X_0, 39
$\vec{\mathfrak{X}}^{\mathrm{ord}}_{\Phi_{\mathcal{H}},\delta_{\mathcal{H}},\Sigma_{\Phi_{\mathcal{H}}}}$, 270, 395, 493
$\vec{\mathfrak{X}}^{\mathrm{ord}}_{\Phi_{\mathcal{H}},\delta_{\mathcal{H}}}$, 270, 327, 395, 493
$\vec{\mathfrak{X}}^{\mathrm{ord}}_{\Phi_{\mathcal{H}},\delta_{\mathcal{H}},\rho}$, 387
$\vec{\mathfrak{X}}^{\mathrm{ord}}_{\Phi_{\mathcal{H}},\delta_{\mathcal{H}},\sigma}$, 270, 295, 327
$(\vec{\mathfrak{X}}^{\mathrm{ord}}_{\Phi_{\mathcal{H}},\delta_{\mathcal{H}},\sigma})^{\wedge}_{\bar{x}}$, 349
$\vec{\mathfrak{X}}^{\mathrm{ord}}_{\Phi_{\mathcal{H}},\delta_{\mathcal{H}},\Sigma_{\Phi_{\mathcal{H}}}}$, 327, 387
$\vec{\mathfrak{X}}^{\mathrm{ord}}_{\Phi_{\mathcal{H}},\delta_{\mathcal{H}},\sigma,\rho}$, 327
$\vec{\mathfrak{X}}^{\mathrm{ord}}_{\Phi_{\mathcal{H}},\delta_{\mathcal{H}},\tau}$, 395, 451, 453
$\vec{\mathfrak{X}}^{\mathrm{ord}}_{\Phi_{\mathcal{H}},\delta_{\mathcal{H}},\tau,r_\kappa}$, 423
X_{aux}, 133
$(X_{\mathrm{aux}}, Y_{\mathrm{aux}}, \phi_{\mathrm{aux}}, \varphi_{\mathrm{aux},-2}, \varphi_{\mathrm{aux},0})$, 133
$\breve{X}$, 47, 87
$(\breve{X}, \breve{Y}, \breve{\phi}, \breve{\varphi}_{-2}, \breve{\varphi}_0)$, 47
$(\breve{X}, \breve{Y}, \breve{\phi}, \breve{\varphi}_{-2,\widehat{\mathcal{H}}}, \breve{\varphi}_{0,\widehat{\mathcal{H}}})$, 48
$\underline{X}^\ddagger$, 59, 292, 297
$\underline{X}(G)$, 279
$X^{(g)}$, 40
$\underline{\mathbf{X}}({}^\diamond G)$, 280
$(X^{(g)}, Y^{(g)}, \phi^{(g)}, \varphi^{(g)}_{-2}, \varphi^{(g)}_0)$, 40
$\widehat{\vec{\mathfrak{X}}}^{\mathrm{ord}}_{\breve{\Phi}_{\widehat{\mathcal{H}}},\breve{\delta}_{\widehat{\mathcal{H}}},\widehat{\tau}}$, 393, 394
$\widehat{\vec{\mathfrak{X}}}^{\mathrm{ord}}_{\breve{\Phi}_{\widehat{\mathcal{H}}},\breve{\delta}_{\widehat{\mathcal{H}}}}$, 384, 394
$\widehat{\vec{\mathfrak{X}}}^{\mathrm{ord}}_{\breve{\Phi}_{\widehat{\mathcal{H}}},\breve{\delta}_{\widehat{\mathcal{H}}},\widehat{\rho}}$, 384
$\widehat{\vec{\mathfrak{X}}}^{\mathrm{ord}}_{\breve{\Phi}_{\widehat{\mathcal{H}}},\breve{\delta}_{\widehat{\mathcal{H}}},\widehat{\Sigma}_{\breve{\Phi}_{\widehat{\mathcal{H}}}}}$, 384
$\widehat{\mathfrak{X}}_{\breve{\Phi}_{\widehat{\mathcal{H}}},\breve{\delta}_{\widehat{\mathcal{H}}}}$, 90, 98
$\widehat{\mathfrak{X}}_{\breve{\Phi}_{\widehat{\mathcal{H}}},\breve{\delta}_{\widehat{\mathcal{H}}},\widehat{\rho}}$, 90
$\widehat{\mathfrak{X}}_{\breve{\Phi}_{\widehat{\mathcal{H}}},\breve{\delta}_{\widehat{\mathcal{H}}},\widehat{\Sigma}_{\breve{\Phi}_{\widehat{\mathcal{H}}}}}$, 90
$\widehat{\mathfrak{X}}_{\breve{\Phi}_{\widehat{\mathcal{H}}},\breve{\delta}_{\widehat{\mathcal{H}}},\widehat{\tau}}$, 98
$\mathfrak{X}_{\Phi_{\mathcal{H}},\delta_{\mathcal{H}}}$, 82
$\mathfrak{X}_{\Phi_{\mathcal{H}},\delta_{\mathcal{H}},\sigma}$, 58, 81
$\mathfrak{X}_{\Phi_{\mathcal{H}},\delta_{\mathcal{H}},\Sigma_{\Phi_{\mathcal{H}}}}$, 82
$\mathfrak{X}_{\Phi_{\mathcal{H}},\delta_{\mathcal{H}},\sigma,\rho}$, 82
$\mathbf{X}(T)$, 70, 315
$\mathbf{X}(T^\vee)$, 70, 315
$\widetilde{X}$, 45, 372
$\tilde{x}_0$, 51
$\widetilde{\vec{\mathfrak{X}}}^{\mathrm{ord}}_{\breve{\Phi}_{\widetilde{\mathcal{H}}},\breve{\delta}_{\widetilde{\mathcal{H}}},\breve{\rho}}$, 387
$\widetilde{\vec{\mathfrak{X}}}^{\mathrm{ord}}_{\breve{\Phi}_{\widetilde{\mathcal{H}}},\breve{\delta}_{\widetilde{\mathcal{H}}},\breve{\sigma},\breve{\rho}}$, 384
$\widetilde{\vec{\mathfrak{X}}}^{\mathrm{ord}}_{\breve{\Phi}_{\widetilde{\mathcal{H}}},\breve{\delta}_{\widetilde{\mathcal{H}}},\breve{\sigma},\breve{\tau}}$, 410, 423
$\widetilde{\vec{\mathfrak{X}}}^{\mathrm{ord}}_{\breve{\Phi}_{\widetilde{\mathcal{H}}},\breve{\delta}_{\widetilde{\mathcal{H}}},\breve{\sigma},\tau}$, 446
$\widetilde{\vec{\mathfrak{X}}}^{\mathrm{ord}}_{\breve{\Phi}_{\widetilde{\mathcal{H}}},\breve{\delta}_{\widetilde{\mathcal{H}}},\widetilde{\Sigma}_{\breve{\Phi}_{\widetilde{\mathcal{H}}}}}$, 387
$\widetilde{\vec{\mathfrak{X}}}^{\mathrm{ord}}_{\breve{\Phi}_{\widetilde{\mathcal{H}}},\breve{\delta}_{\widetilde{\mathcal{H}}},\breve{\tau}}$, 410
$\widetilde{X}^\ddagger$, 110, 413
$\widetilde{\mathfrak{X}}_{\breve{\Phi}_{\widetilde{\mathcal{H}}},\breve{\delta}_{\widetilde{\mathcal{H}}},\breve{\rho}}$, 92
$\widetilde{\mathfrak{X}}_{\breve{\Phi}_{\widetilde{\mathcal{H}}},\breve{\delta}_{\widetilde{\mathcal{H}}},\breve{\sigma},\breve{\rho}}$, 90
$\widetilde{\mathfrak{X}}_{\breve{\Phi}_{\widetilde{\mathcal{H}}},\breve{\delta}_{\widetilde{\mathcal{H}}},\widetilde{\Sigma}_{\breve{\Phi}_{\widetilde{\mathcal{H}}}}}$, 92
$(\widetilde{X}, \widetilde{Y}, \widetilde{\phi}, \widetilde{\varphi}_{-2,\widetilde{\mathcal{H}}}, \widetilde{\varphi}_{0,\widetilde{\mathcal{H}}})$, 46
$(\widetilde{X}, \widetilde{Y}, \widetilde{\phi}, \widetilde{\varphi}_{-2}, \widetilde{\varphi}_0)$, 45
$(X, Y, \phi, \varphi_{-2,\mathcal{H}}, \varphi_{0,\mathcal{H}})$, 29
$(X, Y, \phi, \varphi_{-2,n}, \varphi_{0,n})$, 28
$(X, Y, \phi, \varphi_{-2}, \varphi_0)$, 28, 47
$\overline{\vec{\Xi}}^{\mathrm{ord}}_{\Phi_{\mathcal{H}},\delta_{\mathcal{H}}}$, 267, 268, 326, 493
$\overline{\vec{\Xi}}^{\mathrm{ord}}_{\Phi_{\mathcal{H}},\delta_{\mathcal{H}},\Sigma_{\Phi_{\mathcal{H}}}}$, 267, 326, 387, 493
$\overline{\vec{\Xi}}^{\mathrm{ord}}_{\Phi_{\mathcal{H}},\delta_{\mathcal{H}},\Sigma_{\Phi_{\mathcal{H}}},r_\kappa}$, 438
$\vec{\Xi}^{\mathrm{ord}}_{\Phi_1,\delta_1}$, 264
$\vec{\Xi}^{\mathrm{ord}}_{\Phi_{\mathcal{H}},\delta_{\mathcal{H}}}$, 260, 266, 295
$\vec{\Xi}^{\mathrm{ord}}_{\Phi_{\mathcal{H}},\delta_{\mathcal{H}}}(\rho)$, 326
$\vec{\Xi}^{\mathrm{ord}}_{\Phi_{\mathcal{H}},\delta_{\mathcal{H}}}(\sigma)$, 267, 295, 326
$\vec{\Xi}^{\mathrm{ord}}_{\Phi_{\mathcal{H}},\delta_{\mathcal{H}},\rho}$, 326
$\vec{\Xi}^{\mathrm{ord}}_{\Phi_{\mathcal{H}},\delta_{\mathcal{H}},\sigma}$, 267, 269, 295, 326
$\vec{\Xi}^{\mathrm{ord}}_{\Phi_{\mathcal{H}},\delta_{\mathcal{H}},\sigma}(\rho)$, 326
$\vec{\Xi}^{\mathrm{ord}}_{\Phi_{\mathcal{H}},\delta_{\mathcal{H}},\tau,r_\kappa}$, 441
$\vec{\Xi}^{\mathrm{ord}}_{\Phi_n,\delta_n}$, 324
$\overline{\Xi}_{\Phi_{\mathcal{H}},\delta_{\mathcal{H}}}$, 81
$\overline{\Xi}_{\Phi_{\mathcal{H}},\delta_{\mathcal{H}},\Sigma_{\Phi_{\mathcal{H}}}}$, 81

$\dddot{\Xi}^{\mathrm{ord}}_{\Phi_{\mathcal{H}},n}$, 255
$\dddot{\Xi}^{\mathrm{ord}}_{\Phi_{\mathcal{H}},n,\mathrm{tor}}$, 258
$\dddot{\Xi}^{\mathrm{ord}}_{\Phi_n}$, 251
$\overline{\vec{\widehat{\Xi}}}^{\mathrm{ord}}_{\breve{\Phi}_{\widehat{\mathcal{H}}},\breve{\delta}_{\widehat{\mathcal{H}}},\widehat{\Sigma}_{\breve{\Phi}_{\widehat{\mathcal{H}}}}}$, 384, 438
$\overline{\vec{\widehat{\Xi}}}^{\mathrm{ord}}_{\breve{\Phi}_{\widehat{\mathcal{H}}},\breve{\delta}_{\widehat{\mathcal{H}}}}$, 384
$\vec{\widehat{\Xi}}^{\mathrm{ord}}_{\breve{\Phi}_{\widehat{\mathcal{H}}},\breve{\delta}_{\widehat{\mathcal{H}}}}$, 382, 394, 418
$\vec{\widehat{\Xi}}^{\mathrm{ord}}_{\breve{\Phi}_{\widehat{\mathcal{H}}},\breve{\delta}_{\widehat{\mathcal{H}}}}(\widehat{\rho})$, 383
$\vec{\widehat{\Xi}}^{\mathrm{ord}}_{\breve{\Phi}_{\widehat{\mathcal{H}}},\breve{\delta}_{\widehat{\mathcal{H}}}}(\widehat{\tau})$, 393
$\vec{\widehat{\Xi}}^{\mathrm{ord}}_{\breve{\Phi}_{\widehat{\mathcal{H}}},\breve{\delta}_{\widehat{\mathcal{H}}},r_n}$, 383
$\vec{\widehat{\Xi}}^{\mathrm{ord}}_{\breve{\Phi}_{\widehat{\mathcal{H}}},\breve{\delta}_{\widehat{\mathcal{H}}},\widehat{\rho}}$, 383
$\vec{\widehat{\Xi}}^{\mathrm{ord}}_{\breve{\Phi}_{\widehat{\mathcal{H}}},\breve{\delta}_{\widehat{\mathcal{H}}},\widehat{\tau}}$, 393
$\vec{\widehat{\Xi}}^{\mathrm{ord}}_{\breve{\Phi}_n,\breve{\delta}_n}$, 383, 386
$\overline{\widehat{\Xi}}_{\breve{\Phi}_{\widehat{\mathcal{H}}},\breve{\delta}_{\widehat{\mathcal{H}}}}$, 90
$\overline{\widehat{\Xi}}_{\breve{\Phi}_{\widehat{\mathcal{H}}},\breve{\delta}_{\widehat{\mathcal{H}}},\widehat{\Sigma}_{\breve{\Phi}_{\widehat{\mathcal{H}}}}}$, 90
$\widehat{\Xi}_{\breve{\Phi}_{\widehat{\mathcal{H}}},\breve{\delta}_{\widehat{\mathcal{H}}}}$, 88, 98
$\widehat{\Xi}_{\breve{\Phi}_{\widehat{\mathcal{H}}},\breve{\delta}_{\widehat{\mathcal{H}}},\widehat{\rho}}$, 89
$\widehat{\Xi}_{\breve{\Phi}_{\widehat{\mathcal{H}}},\breve{\delta}_{\widehat{\mathcal{H}}}}(\widehat{\rho})$, 89
$\widehat{\Xi}_{\breve{\Phi}_{\widehat{\mathcal{H}}},\breve{\delta}_{\widehat{\mathcal{H}}},\widehat{\tau}}$, 98
$\widehat{\Xi}_{\breve{\Phi}_{\widehat{\mathcal{H}}},\breve{\delta}_{\widehat{\mathcal{H}}}}(\widehat{\tau})$, 98
$\widehat{\Xi}_{\breve{\Phi}_n,\breve{\delta}_n}$, 89, 92
$\Xi_{\Phi_{\mathcal{H}},\delta_{\mathcal{H}}}$, 58, 76, 80, 259
$\Xi^{\mathrm{ord}}_{\Phi_{\mathcal{H}},\delta_{\mathcal{H}}}$, 259
$\Xi_{\Phi_{\mathcal{H}},\delta_{\mathcal{H}},\sigma}$, 58, 81
$\Xi_{\Phi_{\mathcal{H}},\delta_{\mathcal{H}}}(\sigma)$, 58, 81
$\Xi_{\Phi_{\mathcal{H}},\delta_{\mathcal{H}},\sigma}(\rho)$, 81
Ξ_{Φ_n,δ_n}, 76
$\overline{\vec{\widetilde{\Xi}}}^{\mathrm{ord}}_{\breve{\Phi}_{\widetilde{\mathcal{H}}},\breve{\delta}_{\widetilde{\mathcal{H}}},\widetilde{\Sigma}_{\breve{\Phi}_{\widetilde{\mathcal{H}}}}}$, 387
$\vec{\widetilde{\Xi}}^{\mathrm{ord}}_{\breve{\Phi}_{\widetilde{\mathcal{H}}},\breve{\delta}_{\widetilde{\mathcal{H}}}}$, 383, 409
$\vec{\widetilde{\Xi}}^{\mathrm{ord}}_{\breve{\Phi}_{\widetilde{\mathcal{H}}},\breve{\delta}_{\widetilde{\mathcal{H}}}}(\breve{\rho})$, 383
$\vec{\widetilde{\Xi}}^{\mathrm{ord}}_{\breve{\Phi}_{\widetilde{\mathcal{H}}},\breve{\delta}_{\widetilde{\mathcal{H}}}}(\tau)$, 446
$\vec{\widetilde{\Xi}}^{\mathrm{ord}}_{\breve{\Phi}_{\widetilde{\mathcal{H}}},\breve{\delta}_{\widetilde{\mathcal{H}}}}(\breve{\tau})$, 410
$\vec{\widetilde{\Xi}}^{\mathrm{ord}}_{\breve{\Phi}_{\widetilde{\mathcal{H}}},\breve{\delta}_{\widetilde{\mathcal{H}}},\breve{\rho}}$, 383
$\vec{\widetilde{\Xi}}^{\mathrm{ord}}_{\breve{\Phi}_{\widetilde{\mathcal{H}}},\breve{\delta}_{\widetilde{\mathcal{H}}},\breve{\sigma}}$, 381
$\vec{\widetilde{\Xi}}^{\mathrm{ord}}_{\breve{\Phi}_{\widetilde{\mathcal{H}}},\breve{\delta}_{\widetilde{\mathcal{H}}},\breve{\sigma}}(\breve{\rho})$, 384
$\vec{\widetilde{\Xi}}^{\mathrm{ord}}_{\breve{\Phi}_{\widetilde{\mathcal{H}}},\breve{\delta}_{\widetilde{\mathcal{H}}},\breve{\sigma}}(\tau)$, 446
$\vec{\widetilde{\Xi}}^{\mathrm{ord}}_{\breve{\Phi}_{\widetilde{\mathcal{H}}},\breve{\delta}_{\widetilde{\mathcal{H}}},\breve{\sigma}}(\breve{\tau})$, 410, 446
$\vec{\widetilde{\Xi}}^{\mathrm{ord}}_{\breve{\Phi}_{\widetilde{\mathcal{H}}},\breve{\delta}_{\widetilde{\mathcal{H}}},\breve{\tau}}$, 409
$\vec{\widetilde{\Xi}}^{\mathrm{ord}}_{\breve{\Phi}_{\widetilde{\mathcal{H}}},\breve{\delta}_{\widetilde{\mathcal{H}}},\tau}$, 446
$\vec{\widetilde{\Xi}}^{\mathrm{ord}}_{\breve{\Phi}_n,\breve{\delta}_n}$, 386
$\overline{\widetilde{\Xi}}_{\breve{\Phi}_{\widetilde{\mathcal{H}}},\breve{\delta}_{\widetilde{\mathcal{H}}},\widetilde{\Sigma}_{\breve{\Phi}_{\widetilde{\mathcal{H}}}}}$, 92
$\widetilde{\Xi}_{\breve{\Phi}_{\widetilde{\mathcal{H}}},\breve{\delta}_{\widetilde{\mathcal{H}}},\breve{\rho}}$, 89
$\widetilde{\Xi}_{\breve{\Phi}_{\widetilde{\mathcal{H}}},\breve{\delta}_{\widetilde{\mathcal{H}}}}(\breve{\rho})$, 89
$\widetilde{\Xi}_{\breve{\Phi}_{\widetilde{\mathcal{H}}},\breve{\delta}_{\widetilde{\mathcal{H}}},\breve{\sigma}}$, 88
$\widetilde{\Xi}_{\breve{\Phi}_{\widetilde{\mathcal{H}}},\breve{\delta}_{\widetilde{\mathcal{H}}},\breve{\sigma}}(\breve{\rho})$, 90
$\widetilde{\Xi}_{\breve{\Phi}_n,\breve{\delta}_n}$, 92

Y, 28, 47, 138, 281, 454
$\underline{Y}$, 216, 268
Y_{aux}, 133
$\breve{Y}$, 47, 87
$\underline{Y}^{\ddagger}$, 59, 292, 297
$\underline{Y}(G)$, 279
$Y^{(g)}$, 40
$\widetilde{Y}$, 45, 372
$\widetilde{Y}^{\ddagger}$, 110, 413

$\mathtt{Z}$, 27, 47, 310
  compatible with $\mathtt{D}$, 172
$\mathtt{Z}_{-1,n}$, 69
$\mathtt{Z}_{-1,n_0}$, 313
$\mathtt{Z}^{\#}_{-1,n_0}$, 313
$\mathtt{Z}^{\#}_{-1,n}$, 69
$\mathtt{Z}_{-1,p^r}$, 218
$\mathtt{Z}^{\#}_{-1,p^r}$, 218
$\mathtt{Z}_{-2,p^r}$, 218
$\mathtt{Z}^{\#}_{-2,p^r}$, 218
$\mathtt{Z}_{-3,p^r}$, 218
$\mathtt{Z}^{\#}_{-3,p^r}$, 218
$\mathtt{Z}_{-i}$, 47
$\mathtt{Z}_{-i,\mathbb{A}}$, 27
$\mathtt{Z}_{[(0,0)]}$, 62
$\mathtt{Z}_{0,p^r}$, 218
$\mathtt{Z}^{\#}_{0,p^r}$, 218
$\mathbb{Z}(1)$, 21
$\mathtt{Z}_{\mathbb{A}}$, 27
$\vec{\mathtt{Z}}^{\mathrm{ord}}_{[(0,0)]}$, 346
$\overline{\vec{\mathtt{Z}}}^{\mathrm{ord}}$, 319

$\vec{\overline{\mathsf{Z}}}^{\mathrm{ord}}_{[(\Phi_{\mathcal{H}},\delta_{\mathcal{H}},\sigma)]}$, 327
$\vec{\mathsf{Z}}^{\mathrm{ord}}_{[(\Phi_{\mathcal{H}_{\mathrm{aux}}},\delta_{\mathcal{H}_{\mathrm{aux}}},\sigma_{\mathrm{aux}})]}$, 333
$\vec{\mathsf{Z}}_{[(\Phi_{\mathcal{H}},\delta_{\mathcal{H}})]}$, 144
$\vec{\mathsf{Z}}_{[(\Phi_{\mathcal{H}},\delta_{\mathcal{H}})],i}$, 151
$\vec{\mathsf{Z}}^{\mathrm{ord}}_{[(\Phi_{\mathcal{H}},\delta_{\mathcal{H}})]}$, 341, 346, 394, 492
  $\mathcal{O}$-multi-rank, 346
$(\vec{\mathsf{Z}}^{\mathrm{ord}}_{[(\Phi_{\mathcal{H}},\delta_{\mathcal{H}})]})^{\wedge}_{\bar{x}}$, 344
$\vec{\mathsf{Z}}_{[(\Phi_{\mathcal{H}},\delta_{\mathcal{H}})],r_{\mathcal{H}}}$, 337, 341
$\vec{\mathsf{Z}}_{[(\Phi_{\mathcal{H}},\delta_{\mathcal{H}},\sigma)],d_0\mathsf{pol}}$, 148
$\vec{\mathsf{Z}}_{[(\Phi_{\mathcal{H}},\delta_{\mathcal{H}},\sigma)],d_0\mathsf{pol},i}$, 151
$\vec{\mathsf{Z}}^{\mathrm{ord}}_{[(\Phi_{\mathcal{H}},\delta_{\mathcal{H}},\sigma)]}$, 295, 333, 347
$\vec{\mathsf{Z}}^{\mathrm{ord}}_{[(\Phi_{\mathcal{H}},\delta_{\mathcal{H}},\tau)]}$, 395, 444, 453
$\vec{\mathsf{Z}}^{\mathrm{ord}}_{[(\Phi_{\mathcal{H}},\delta_{\mathcal{H}},\tau)],r_{\kappa}}$, 440
$\vec{\widetilde{\mathsf{Z}}}^{\mathrm{ord}}_{[(\Phi_{\mathcal{H}},\delta_{\mathcal{H}})]}$, 344, 493
$\mathsf{Z}_{\mathrm{aux}}$, 132
$\mathsf{Z}_{\mathrm{aux},-i}$, 132
$\mathsf{Z}_{\mathrm{aux},-i,\mathbb{A}}$, 133
$\mathsf{Z}_{\mathrm{aux},\mathbb{A}}$, 133
$\overline{\mathsf{Z}}$, 73
$\vec{\overline{\mathsf{Z}}}_{[(\Phi_{\mathcal{H}},\delta_{\mathcal{H}})]}$, 144
$\vec{\overline{\mathsf{Z}}}_{[(\Phi_{\mathcal{H}},\delta_{\mathcal{H}},\sigma)],d_0\mathsf{pol}}$, 148
$\overline{\mathsf{Z}}_{[(\Phi_{\mathcal{H}},\delta_{\mathcal{H}})]}$, 144
$\overline{\mathsf{Z}}_{[(\Phi_{\mathcal{H}},\delta_{\mathcal{H}},\sigma)]}$, 148
$\breve{\mathsf{Z}}$, 46, 85
$\breve{\mathsf{Z}}_{-i}$, 46
$\breve{\mathsf{Z}}_{\widehat{\mathcal{H}}}$, 48, 110, 380, 413
$(\breve{\mathsf{Z}}_{\widehat{\mathcal{H}}},\breve{\Phi}_{\widehat{\mathcal{H}}},\breve{\delta}_{\widehat{\mathcal{H}}})$, 48, 86, 380
$[(\breve{\mathsf{Z}}_{\widehat{\mathcal{H}}},\breve{\Phi}_{\widehat{\mathcal{H}}},\breve{\delta}_{\widehat{\mathcal{H}}})]$, 48
$(\breve{\mathsf{Z}}_{\widetilde{\mathcal{H}}},\breve{\Phi}_{\widetilde{\mathcal{H}}},\breve{\delta}_{\widetilde{\mathcal{H}}})$, 47
$[(\breve{\mathsf{Z}}_{\widetilde{\mathcal{H}}},\breve{\Phi}_{\widetilde{\mathcal{H}}},\breve{\delta}_{\widetilde{\mathcal{H}}})]$, 47, 373, 379
$(\breve{\mathsf{Z}},\breve{\Phi},\breve{\delta})$, 47
$\mathsf{Z}^{\ddagger}_{\mathcal{H}}$, 297
$\mathsf{Z}^{(g)}$, 40
$(\mathsf{Z}^{(g)},\Phi^{(g)},\delta^{(g)})$, 41
$[(\mathsf{Z}^{(g)},\Phi^{(g)},\delta^{(g)})]$, 41
$\mathsf{Z}_{\mathcal{H}}$, 32, 58, 241, 243, 248, 281
  compatible with D, 247
  ordinary, 247
$\mathsf{Z}_{\mathcal{H}_{\mathrm{aux}}}$, 134
$\mathsf{Z}_{\mathcal{H}^p_{\mathrm{aux}}}$, 135
$(\mathsf{Z}_{\mathcal{H}^p_{\mathrm{aux}}},\Phi_{\mathcal{H}^p_{\mathrm{aux}}},\delta_{\mathcal{H}^p_{\mathrm{aux}}})$, 135
$[(\mathsf{Z}_{\mathcal{H}^p_{\mathrm{aux}}},\Phi_{\mathcal{H}^p_{\mathrm{aux}}},\delta_{\mathcal{H}^p_{\mathrm{aux}}})]$, 135
$(\mathsf{Z}_{\mathcal{H}_{\mathrm{aux}}},\Phi_{\mathcal{H}_{\mathrm{aux}}},\delta_{\mathcal{H}_{\mathrm{aux}}})$, 134
$[(\mathsf{Z}_{\mathcal{H}_{\mathrm{aux}}},\Phi_{\mathcal{H}_{\mathrm{aux}}},\delta_{\mathcal{H}_{\mathrm{aux}}})]$, 134
$\mathsf{Z}^{\ddagger}_{\mathcal{H}}$, 60, 292
(Z,Φ,δ), 29
$[(\mathsf{Z},\Phi,\delta)]$, 29
Z_{H_n}, 32, 240, 243
$\mathsf{Z}_{H_{n_0}}$, 236
$(\mathsf{Z}_{\mathcal{H}^p},\Phi_{\mathcal{H}^p},\delta_{\mathcal{H}^p})$, 314, 325
$(\mathsf{Z}_{\mathcal{H}_p},\Phi_{\mathcal{H}_p},\delta_{\mathcal{H}_p})$, 315
$\mathsf{Z}_{H^{\mathrm{ord}}_{p^r}}$, 239, 243
$(\mathsf{Z}_{\mathcal{H}},\Phi_{\mathcal{H}},\delta_{\mathcal{H}})$, 29, 47, 58, 66, 248, 281
$[(\mathsf{Z}_{\mathcal{H}},\Phi_{\mathcal{H}},\delta_{\mathcal{H}})]$, 29, 47, 58, 373, 379
$\widehat{\mathsf{Z}}_{[(0,0,\{0\})]}$, 98
$\vec{\widehat{\mathsf{Z}}}^{\mathrm{ord}}_{[(0,0,\{0\})]}$, 393
$\vec{\widehat{\mathsf{Z}}}^{\mathrm{ord}}_{[(\breve{\Phi}_{\widehat{\mathcal{H}}},\breve{\delta}_{\widehat{\mathcal{H}}},\widehat{\tau})]}$, 393, 420
$\widehat{\mathsf{Z}}_{[(\breve{\Phi}_{\widehat{\mathcal{H}}},\breve{\delta}_{\widehat{\mathcal{H}}},\widehat{\tau})]}$, 98, 117
$\widehat{\mathsf{Z}}_{\breve{\mathsf{Z}}}(R)$, 55, 56, 378
Z_n, 27, 231, 248, 313
Z_{n_0}, 236
$(\mathsf{Z}_{n_0},\Phi_{n_0},\delta_{n_0})$, 314, 325
$(\mathsf{Z}_n,\Phi_n,\delta_n)$, 29, 66, 248
$[(\mathsf{Z}_n,\Phi_n,\delta_n)]$, 29
$\mathsf{Z}^{\#}_n$, 231
$(\mathbb{Z}_p(1))^{\mathrm{mult}}$, 156
Z_{p^r}, 218, 220, 221, 237, 239
$(\mathsf{Z}_{p^r},\Phi_{p^r},\delta_{p^r})$, 315
$\mathsf{Z}^{\#}_{p^r}$, 218
$\mathsf{Z}_{[(\Phi_{\mathcal{H}_{\mathrm{aux}}},\delta_{\mathcal{H}_{\mathrm{aux}}},\sigma_{\mathrm{aux}})]}$, 136
$\mathsf{Z}_{[(\Phi_{\mathcal{H}_{\mathrm{aux}}},\delta_{\mathcal{H}_{\mathrm{aux}}})]}$, 140
$\mathsf{Z}_{[(\Phi_{\mathcal{H}^p_{\mathrm{aux}}},\delta_{\mathcal{H}^p_{\mathrm{aux}}},\sigma^p_{\mathrm{aux}})]}$, 136
$\mathsf{Z}_{[(\Phi_{\mathcal{H}},\delta_{\mathcal{H}})]}$, 61, 98, 140
  $\mathcal{O}$-multi-rank, 62
$\mathsf{Z}_{[(\Phi_{\mathcal{H}},\delta_{\mathcal{H}})],i}$, 151
$\mathsf{Z}_{[(\Phi_{\mathcal{H}},\delta_{\mathcal{H}})],r_{\mathcal{H}}}$, 347
$\mathsf{Z}_{[(\Phi_{\mathcal{H}},\delta_{\mathcal{H}},\sigma)]}$, 58, 62
$\overline{\mathsf{Z}}_{[(\Phi_{\mathcal{H}},\delta_{\mathcal{H}},\sigma)]}$, 63, 81
$\mathsf{Z}_{[(\Phi_{\mathcal{H}},\delta_{\mathcal{H}},\sigma)],i}$, 151
$\mathsf{Z}_{[(\Phi_{\mathcal{H}},\delta_{\mathcal{H}},\sigma)],r_{\mathcal{H}}}$, 297
$\mathsf{Z}_{[(\Phi_{\mathcal{H}},\delta_{\mathcal{H}},\tau)]}$, 99
$\mathsf{Z}_{[(\Phi_{\mathcal{H}^p},\delta_{\mathcal{H}^p},\sigma)]}$, 307
$\mathsf{Z}^{\#}$, 69, 313
$\widetilde{\mathsf{Z}}_{-i}$, 44
$\vec{\widetilde{\mathsf{Z}}}^{\mathrm{ord}}_{[(\breve{\Phi}_{\widetilde{\mathcal{H}}},\breve{\delta}_{\widetilde{\mathcal{H}}},\breve{\tau})]}$, 410, 420
$\vec{\widetilde{\mathsf{Z}}}^{\mathrm{ord}}_{[(\Phi_{\mathcal{H}},\delta_{\mathcal{H}},\tau)]}$, 440, 444
$\vec{\widetilde{\mathsf{Z}}}^{\mathrm{ord}}_{[(\widetilde{\Phi}_{\widetilde{\mathcal{H}}},\widetilde{\delta}_{\widetilde{\mathcal{H}}},\widetilde{\sigma})]}$, 408
$\widetilde{\mathsf{Z}}_{\widetilde{\mathcal{H}}}$, 46, 372
$\widetilde{\mathsf{Z}}^{\ddagger}_{\widetilde{\mathcal{H}}}$, 110, 413
$(\widetilde{\mathsf{Z}}_{\widetilde{\mathcal{H}}},\widetilde{\Phi}_{\widetilde{\mathcal{H}}},\widetilde{\delta}_{\widetilde{\mathcal{H}}})$, 46, 372
$[(\widetilde{\mathsf{Z}}_{\widetilde{\mathcal{H}}},\widetilde{\Phi}_{\widetilde{\mathcal{H}}},\widetilde{\delta}_{\widetilde{\mathcal{H}}})]$, 46, 47, 372

$\widetilde{\mathsf{Z}}_{[(\breve{\Phi}_{\widetilde{\mathcal{H}}},\breve{\delta}_{\widetilde{\mathcal{H}}},\breve{\tau})]}$, 117
$(\widetilde{\mathsf{Z}}, \widetilde{\Phi}, \widetilde{\delta})$, 46, 47, 85, 371
$\widetilde{\mathsf{Z}}_{[(\widetilde{\Phi}_{\widetilde{\mathcal{H}}},\widetilde{\delta}_{\widetilde{\mathcal{H}}},\widetilde{\sigma})]}$, 107
$\widetilde{\mathrm{Z}}^{\mathrm{ord}}_{\breve{\mathtt{Z}},\widetilde{\mathtt{D}}}(R)$, 378
$\widetilde{\mathrm{Z}}_{\breve{\mathtt{Z}}}(R)$, 378
$\widetilde{\mathrm{Z}}_{\widetilde{\mathtt{Z}}}(R)$, 45
$\mathrm{Z}^{\mathrm{ess}}_{\mathtt{Z}_n}$, 235
$\mathrm{Z}^{\mathrm{ess}}_{\mathtt{Z}_{p^r},\mathtt{D}_{p^r}}$, 237
$\mathrm{Z}^{\mathrm{ess,ord}}_{\mathtt{Z}_{p^r},\mathtt{D}_{p^r}}$, 238
$\mathrm{Z}_{\mathtt{Z}}(R)$, 30, 56
Zarhin's trick, 17, 126
Zariski's main theorem, 64, 141, 142, 148, 150–152, 214, 303, 310, 337, 338, 343, 350, 356–358, 360, 368, 429, 522
$\zeta_{p^{r_\nu}}$, 192
$\zeta_{p^{r_\nu},S}$, 192